P9-CDX-133

DICTIONARY OF SCIENTIFIC BIOGRAPHY

CHARLES COULSTON GILLISPIE
Princeton University

EDITOR IN CHIEF

Volume 5

EMIL FISCHER—GOTTLIEB HABERLANDT

CHARLES SCRIBNER'S SONS · NEW YORK

First publication in an eight-volume edition 1981.

Library of Congress Cataloging in Publication Data

Main entry under title:

Dictionary of scientific biography.

"Published under the auspices of the American Council of Learned Societies."
Includes bibliographies and index.
1. Scientists—Biography. I. Gillispie, Charles Coulston. II. American Council of Learned Societies Devoted to Humanistic Studies.
Q141.D5 1981 509'.2'2 [B] 80-27830
ISBN 0-684-16962-2 (set)

ISBN 0-684-16963-0 Vols. 1 & 2
ISBN 0-684-16964-9 Vols. 3 & 4
ISBN 0-684-16965-7 Vols. 5 & 6
ISBN 0-684-16966-5 Vols. 7 & 8
ISBN 0-684-16967-3 Vols. 9 & 10
ISBN 0-684-16968-1 Vols. 11 & 12
ISBN 0-684-16969-X Vols. 13 & 14
ISBN 0-684-16970-3 Vols. 15 & 16

5 7 9 11 13 15 17 19 V/C 20 18 16 14 12 10 8 6 4

Printed in the United States of America

Panel of Consultants

Editorial Staff

MARSHALL DE BRUHL, *MANAGING EDITOR*

SARAH FERRELL, *Assistant Managing Editor*

LOUISE F. BILEBOF, *Administrative Editor*

DAVID L. GRAMBS, *Associate Editor*

JANET L. JACOBS, *Assistant Editor*

ROSE MOSELLE, *Editorial Assistant*

ELIZABETH I. WILSON, *Copy Editor*

JOEL HONIG, *Copy Editor*

DORIS ANNE SULLIVAN, *Proofreader*

MICHAEL KNIBBS, *Proofreader*

LELAND S. LOWTHER, *Associate Editor*

CLAIRE SOTNICK, *Proofreader*

Contributors to Volume 5

The following are the contributors to Volume 5. Each author's name is followed by the institutional affiliation at the time of publication and the names of articles written for this volume. The symbol † indicates that an author is deceased.

HANS AARSLEFF
Princeton University
FRANCK, S.; HAAK

KATHLEEN AHONEN
University of Michigan
GLAUBER

GARLAND E. ALLEN
Washington University
FOL

PETER AMACHER
University of California, Los Angeles
FREUD

G. C. AMSTUTZ
University of Heidelberg
GOLDSCHMIDT, V.; GRODDECK

DAVID L. ANDERSON
Oberlin College
GOLDSTEIN

RICHARD P. AULIE
Encyclopaedia Britannica
GAINES

WILLIAM H. AUSTIN
Rice University
GLANVILL

A. ALBERT BAKER, JR.
Grand Valley State College
FISCHER, H.; FRITZSCHE; GABRIEL; GRIGNARD

GEORGE BASALLA
University of Delaware
FITZROY

EDWIN A. BATTISON
Smithsonian Institution
GRAHAM, G.

LUIGI BELLONI
University of Milan
FONTANA

ALEX BERMAN
University of Cincinnati
FORDOS; GAULTIER DE CLAUBRY; GOBLEY; GUIGNARD

MICHAEL BERNKOPF
Pace College
FREDHOLM

ASIT K. BISWAS
Department of the Environment, Ottawa
FORTIN; GAMBEY

MARGARET BISWAS
McGill University
FORTIN; GAMBEY

L. J. BLACHER
Soviet Academy of Sciences
GURVICH

UNO BOKLUND
Royal Pharmaceutical Institute, Stockholm
GAHN

ALFRED M. BORK
University of California, Irvine
FITZGERALD

FRANCK BOURDIER
École Pratique des Hautes Études
GAUDRY; GEOFFROY SAINT-HILAIRE, É.; GEOFFROY SAINT-HILAIRE, I.

W. H. BROCK
University of Leicester
FRANKLAND, E.

JOAN BROMBERG
Niels Bohr Institute
FÖPPL

J. H. BROOKE
University of Lancaster
GERHARDT

THEODORE M. BROWN
City College, City University of New York
GALVANI

STEPHEN G. BRUSH
University of Maryland
FOWLER, R.

WERNER BURAU
University of Hamburg
GÖPEL

J. J. BURCKHARDT
University of Zurich
FUETER; GEISER; GRÄFFE; GROSSMANN, M.

JOHN G. BURKE
University of California, Los Angeles
FORBES, J.; FOUQUÉ

CONRAD BURRI
Swiss Federal Institute of Technology
GRUBENMANN

HAROLD BURSTYN
Carnegie-Mellon University
FOUCAULT

H. L. L. BUSARD
University of Leiden
FRENICLE; GULDIN

LUIGI CAMPEDELLI
University of Florence
FRISI; GHETALDI

ALBERT CAROZZI
University of Illinois
GUYOT

CARLO CASTELLANI
GAGLIARDI; GHISI

JOHN CHALLINOR
GEIKIE, A.; GEIKIE, J.

SEYMOUR L. CHAPIN
Los Angeles State College
FOUCHY; GODIN

JEAN CHÂTILLON
Catholic Institute of Paris
GILES OF ROME

GEORGES CHAUDRON
Laboratoire de Recherches Métallurgiques
GUILLAUME; GUILLET

MARSHALL CLAGETT
Institute for Advanced Study, Princeton
GERARD OF BRUSSELS

EDWIN CLARKE
University College London
FLECHSIG; FRITSCH

I. BERNARD COHEN
Harvard University
FRANKLIN, B.

ROBERT S. COHEN
Boston University
FRANK

WILLIAM COLEMAN
Northwestern University
GAIMARD; GRATIOLET

GEORGE W. CORNER
American Philosophical Society
FLEXNER

CARL W. CORRENS
FRANKENHEIM

ALBERT B. COSTA
Duquesne University
FOWNES; GELMO

PIERRE COSTABEL
École Pratique des Hautes Études
GALLOIS

RUTH SCHWARTZ COWAN
State University of New York at Stony Brook
GATES

CONTRIBUTORS TO VOLUME 5

A. C. CROMBIE
University of Oxford
GROSSETESTE

M. P. CROSLAND
University of Leeds
GAY-LUSSAC; GERHARDT

HALLOWELL DAVIS
Central Institute for the Deaf
FORBES, A.

GAVIN DE BEER
GOODRICH

ALLEN G. DEBUS
University of Chicago
FLUDD

RONALD K. DEFORD
University of Texas at Austin
GILBERT, G.

CLAUDE K. DEISCHER
University of Pennsylvania
GMELIN, L.

SUZANNE DELORME
Centre International de Synthèse
FONTENELLE

D. R. DICKS
University of London
GEMINUS

HERBERT DINGLE
University of London
FOWLER, A.

CLAUDE E. DOLMAN
University of British Columbia
FLEMING, A.

J. D. H. DONNAY
McGill University
FRIEDEL

J. G. DORFMAN
Soviet Academy of Sciences
FRENKEL; GOLITSYN

HAROLD DORN
Stevens Institute of Technology
GREGORY, O.

SIGALIA C. DOSTROVSKY
Barnard College
FOSTER, H.

STILLMAN DRAKE
University of Toronto
GALILEI, G.; GALILEI, V.

K. C. DUNHAM
Institute of Geological Sciences
FLETT; GODWIN-AUSTEN

A. HUNTER DUPREE
Brown University
GRAY, A.

BRUCE S. EASTWOOD
Kansas State University
GRIMALDI

FRANK N. EGERTON III
University of Wisconsin-Parkside
FORBES, E.; GRAUNT

GUNNAR ERIKSSON
University of Umeå
FORSSKÅL; FRIES, E.

JOSEPH EWAN
Tulane University
GAERTNER, J.

JOAN M. EYLES
FITTON; GRIFFITH, R.

V. A. EYLES
GREENOUGH

EDUARD FARBER†
FISCHER, E.; FRÉMY

KATHLEEN FARRAR
University of Manchester
GULLAND

FRANK FENNER
Australian National University
FLOREY

KONRADIN FERRARI D'OCCHIEPPO
University of Vienna
GRAFF

MARTIN FICHMAN
York University
GUIBERT

BERNARD S. FINN
Smithsonian Institution
GLAZEBROOK

WALTHER FISCHER
GROTH

C. S. FISHER
Brandeis University
GORDAN

HEINZ FLAMM
University of Vienna
GRUBER

MARCEL FLORKIN
University of Liège
FREDERICQ; GRAMME

ERIC G. FORBES
University of Edinburgh
FREUNDLICH

ROBERT FOX
University of Lancaster
GAUDIN

PIETRO FRANCESCHINI
GALEAZZI; GRASSI

H. C. FREIESLEBEN
GALLE

H. FREUDENTHAL
University of Utrecht
HAAR

JUNE Z. FULLMER
Ohio State University
GARNETT

GERALD L. GEISON
Princeton University
FLETCHER; FOSTER, M.; FRANKLAND, P.; GASKELL

WALTHER GERLACH
University of Munich
GEITEL

GEORGE E. GIFFORD, JR.
Harvard University
GOULD, A.

OWEN GINGERICH
Smithsonian Astrophysical Observatory
FLEMING, W.

EDWARD D. GOLDBERG
Scripps Institution of Oceanography
FORCHHAMMER; GOLDSCHMIDT, V. M.

MORRIS GORAN
Roosevelt University
HABER

J. B. GOUGH
Washington State University
FIZEAU; GOUY

I. GRATTAN-GUINNESS
Enfield College of Technology
FOURIER

FRANK GREENAWAY
Science Museum, London
GRIESS

JOSEPH T. GREGORY
University of California, Berkeley
GRANGER

NORMAN T. GRIDGEMAN
National Research Council of Canada
FISHER; GALTON

A. T. GRIGORIAN
Soviet Academy of Sciences
FRIEDMANN

M. D. GRMEK
Archives Internationales d'Histoire des Sciences
GERBEZIUS; GRISOGONO; GUIDI

V. GUTINA
Soviet Academy of Sciences
GAMALEYA

IAN HACKING
University of Cambridge
GOSSETT

A. RUPERT HALL
Imperial College of Science and Technology
'sGRAVESANDE

MARIE BOAS HALL
Imperial College of Science and Technology
FREIND

CONTRIBUTORS TO VOLUME 5

OWEN HANNAWAY
The Johns Hopkins University
GLASER, C.; GOHORY

BERT HANSEN
Fordham University
FÜCHSEL

R. S. HARTENBERG
Northwestern University
GRASHOF

JOHN L. HEILBRON
University of California, Berkeley
GRAY, S.

FRANZ HEIN
German Academy of Sciences
GUTBIER

DAVID HEPPELL
Royal Scottish Museum
GOODSIR

ARMIN HERMANN
University of Stuttgart
HAAS, A.

MAXIMILIAN HERZBERGER
Louisiana State University
GULLSTRAND

BROOKE HINDLE
New York University
GODFREY; GREENWOOD

ANN M. HIRSCH-KIRCHANSKI
University of California, Berkeley
GRIFFITH, W.

MICHAEL E. HOARE
Australian National University
FORSTER, G.; FORSTER, J.

J. E. HOFMANN
University of Tübingen
GUENTHER

GERALD HOLTON
Harvard University
FRANK

BRIGITTE HOPPE
Deutsches Museum
GAERTNER, K.

KARL HUFBAUER
University of California, Irvine
GREN

AARON J. IHDE
University of Wisconsin
FUNK; GOMBERG

JEAN ITARD
Lycèe Henri IV
GIRARD, A.

W. O. JAMES
Imperial College, London
GOEBEL

REESE V. JENKINS
Case Western Reserve University
FRAUNHOFER

DANIEL P. JONES
Oregon State University
GORE; GREGORY, W.

PHILLIP S. JONES
University of Michigan
GLAISHER, J. W.

HANS KANGRO
University of Hamburg
GEISSLER

GEORGE B. KAUFFMAN
Fresno State College
FRIEND; GENTH; GRAHAM, T.; GULDBERG

MARSHALL KAY
Columbia University
GRABAU

A. G. KELLER
University of Leicester
GHINI

SUZANNE KELLY, O.S.B.
Carroll College
GILBERT, W.

MARTHA B. KENDALL
Vassar College
FUHLROTT

H. C. KING
Royal Ontario Museum
GRUBB, H.; GRUBB, T.

STEFAN J. KIRCHANSKI
University of California, Berkeley
GRIFFITH, W.

GEORGE KISH
University of Michigan
GEMMA; GERMANUS

MARC KLEIN
University of Strasbourg
GRAAF

MARTIN J. KLEIN
Yale University
GIBBS, J.

OSKAR KLEIN
University of Stockholm
GORDON

DAVID M. KNIGHT
University of Durham
GIRTANNER

ZDENĚK KOPAL
University of Manchester
GOODRICKE

ELAINE KOPPELMAN
Goucher College
GREGORY, D. F.

SHELDON J. KOPPERL
Grand Valley State College
GADOLIN

HANS-GÜNTHER KÖRBER
Zentralbibliothek des Meteorologischen Dienstes, Potsdam
GUERTLER

FRITZ KRAFFT
University of Mainz
GUERICKE

EDNA E. KRAMER
Polytechnic Institute of Brooklyn
GERMAIN

CLAUDIA KREN
University of Missouri
GUNDISSALINUS

ABRAHAM D. KRIKORIAN
State University of New York at Stony Brook
GREGORY, F.

VLADISLAV KRUTA
Purkyně University
FLOURENS; GMELIN, J.; GRUBY

FRIDOLF KUDLIEN
University of Kiel
GALEN

H. G. KUHN
University of Oxford
FRANCK, J.

EMIL KUHN-SCHNYDER
University of Zurich
GAGNEBIN

P. G. KULIKOVSKY
Soviet Academy of Sciences
GERASIMOVICH

BENGT-OLOF LANDIN
University of Lund
GEER, C.; GYLLENHAAL

HENRY M. LEICESTER
University of the Pacific
FITTIG; FOLIN

JACQUES R. LÉVY
Paris Observatory
GAILLOT; GAUTIER, P.

DAVID C. LINDBERG
University of Wisconsin
FLEISCHER

G. A. LINDEBOOM
Free University, Amsterdam
GRIJNS

DAVID P. C. LLOYD
Rockefeller University
GASSER

MARIO LORIA
GIORGI

MICHAEL McVAUGH
University of North Carolina
FREDERIC II

MICHAEL S. MAHONEY
Princeton University
GOLDBACH

CONTRIBUTORS TO VOLUME 5

JEROME H. MANHEIM
Bradley University
FUCHS, I.

BRIAN G. MARSDEN
Smithsonian Institution Astrophysical Observatory
FROST; GOULD, B.

KIRTLEY F. MATHER
Harvard University
FOERSTE

KENNETH O. MAY
University of Toronto
GAUSS

A. MENIAILOV
Soviet Academy of Sciences
FYODOROV

CHARLES R. METCALFE
Royal Botanic Gardens, Kew
GREW

W. E. K. MIDDLETON
University of British Columbia
GLAISHER, J.

WYNDHAM DAVIES MILES
National Institutes of Health
GREEN

MICHAEL E. MITCHELL
University College, Galway
GLEICHEN-RUSSWORM

EDGAR W. MORSE
Sonoma State College
GASSIOT

ALPINOLO NATUCCI
University of Genoa
GRANDI

H. M. NOBIS
Deutsches Museum
FRIES, J.

LOWELL E. NOLAND
University of Wisconsin
GUYER

J. D. NORTH
Museum of the History of Science, Oxford
GELLIBRAND

LUBOŠ NOVÝ
Czechoslovak Academy of Sciences
GUDERMANN

MARY JO NYE
University of Oklahoma
GAUTIER, A.

ROBERT OLBY
University of Leeds
FLEMMING, W.; FRANKLIN, R.

C. D. O'MALLEY†
GEMINUS, T.; GUINTER

LEROY E. PAGE
Kansas State University
FLEMING, J.

JACQUES PAYEN
Conservatoire Nationale des Arts et Métiers
GRAMONT

JON V. PEPPER
Polytechnic of the South Bank
GUNTER

P. E. PILET
University of Lausanne
FOREL; GESNER, J.; GESSNER, K.; GLASER, J.

DAVID PINGREE
University of Chicago
GAṆEŚA

LUCIEN PLANTEFOL
University of Paris
GUILLIERMOND

J. A. PRINS
Technological University of Delft
HAAS, W.

HANS QUERNER
University of Heidelberg
GOETTE

SAMUEL X. RADBILL
College of Physicians of Philadelphia
GESELL

RHODA RAPPAPORT
Vassar College
GUETTARD

J. R. RAVETZ
University of Leeds
FOURIER

NATHAN REINGOLD
Smithsonian Institution
GIBBS, O.

SAMUEL REZNECK
Rensselaer Polytechnic Institute
FITCH

RUTH GIENAPP RINARD
Kirkland College
GRAEBE

GLORIA ROBINSON
Yale University
GAFFKY

BERNARD ROCHOT
École Pratique des Hautes Études
GASSENDI

JOEL M. RODNEY
Elmira College
FOLKES

GRETE RONGE
GEUTHER

B. VAN ROOTSELAAR
State Agricultural University, Wageningen
FRAENKEL; FREGE

HUNTER ROUSE
University of Iowa
FOURNEYRON; FROUDE; GIRARD, P.

CHARLES ROSENBERG
University of Pennsylvania
GOLDBERGER

K. E. ROTHSCHUH
University of Münster/Westphalia
FREY; GOLTZ

MORRIS H. SAFFRON
Rutgers University
GAY

A. P. M. SANDERS
State University of Utrecht
GARNOT; GARREAU

R. SATTLER
McGill University
HABERLANDT

EBERHARD SCHMAUDERER
FUCHS, J.

F. SCHMEIDLER
University of Munich
GROSSMANN, E.

CECIL J. SCHNEER
University of New Hampshire
GRESSLY

HANS HENNING SCHROTH
University of Bonn
GUDDEN

E. L. SCOTT
Stamford High School, Lincolnshire
GLADSTONE; GROVE

ROGER SERVAJEAN
Paris Observatory
FLAMMARION

ELIZABETH NOBLE SHOR
GABB; GILL, T.; GRINNELL; GUTENBERG

GEORGE G. SHOR, JR.
GUTENBERG

ROBERT H. SILLIMAN
Emory University
FRESNEL

DIANA M. SIMPKINS
Polytechnic of North London
FRAZER; FRERE; GOULD, J.; GWYNNE-VAUGHAN

W. A. SMEATON
University College London
FOURCROY; GEOFFROY, C.; GEOFFROY, É. F.; GUYTON DE MORVEAU

E. SNORRASON
Rigshospitalet, Copenhagen
GRAM

T. A. SOFIANO
Soviet Academy of Sciences
FRENZEL

CONTRIBUTORS TO VOLUME 5

PIERRE SPEZIALI
University of Geneva
FUBINI; GUCCIA

NILS SPJELDNAES
University of Aarhus
DE GEER

WILLIAM T. STEARN
British Museum (Natural History)
GERARD

FRANÇOIS STOCKMANS
Institut Royal des Sciences Naturelles de Belgique
GOSSELET; GRAND'EURY

R. H. STOY
Royal Observatory, Edinburgh
GILL, D.

J. P. STRADINS
Soviet Academy of Sciences
GROTTHUSS

D. J. STRUIK
Massachusetts Institute of Technology
FRENET; GERBERT; GERGONNE

ROGER H. STUEWER
University of Minnesota
GAMOW

CHARLES SÜSSKIND
University of California, Berkeley
FLEMING, J. A.

LOYD S. SWENSON, JR.
University of Houston
GODDARD; GUYE

F. SZABADVÁRY
Technical University, Budapest
FISCHER, N. W.; FRESENIUS; GÖRGEY

MANFRED E. SZABO
Sir George Williams University
GENTZEN

RENÉ TATON
École Pratique des Hautes Études
FONTAINE; FRANÇAIS, F.; FRANÇAIS, J. F.; GALOIS; GUA DE MALVES

SEVIM TEKELI
Ankara University
ḤABASH AL-ḤĀSIB

OWSEI TEMKIN
The Johns Hopkins University
GLISSON

ANDRÉE TÉTRY
École Pratique des Hautes Études
GIARD; GOLDSCHMIDT, R.

JEAN THÉODORIDÈS
Centre National de la Recherche Scientifique
GEOFFROY, É.-L.

VICTOR E. THOREN
Indiana University
FLAMSTEED; GASCOIGNE

V. V. TIKHOMIROV
Soviet Academy of Sciences
FRENZEL; GOEPPERT

HEINZ TOBIEN
University of Mainz
FREIESLEBEN

THADDEUS J. TRENN
University of Cambridge
GEIGER; GIESEL

HENRY S. TROPP
Smithsonian Institution
GOMPERTZ; GOURSAT

GEORGES UBAGHS
University of Liège
FRAIPONT

PETER W. VAN DER PAS
GOEDAERT

GRAZIELLA F. VESCOVINI
University of Turin
FRANCIS OF MARCHIA; FRANCIS OF MEYRONNES

A. I. VOLODARSKY
Soviet Academy of Sciences
FUSS; GRAVE

GERHARD WAGENITZ
Systematisch-Geobotanisches Institut, Berne
GRISEBACH

EARL WALKER
Johns Hopkins Hospital
FULTON

WILLIAM A. WALLACE, O.P.
Catholic University of America
GERARD OF SILTEO; GILES OF LESSINES

J. B. WATERHOUSE
University of Toronto
HAAST

GEORGE A. WELLS
University of London
GOETHE

D. T. WHITESIDE
Whipple Science Museum
GREGORY, D.; GREGORY, J.

GERALD J. WHITROW
Imperial College of Science and Technology
FORSYTH, A.

W. P. D. WIGHTMAN
King's College, Aberdeen
FRANCESCA

RONALD S. WILKINSON
GROTE

WESLEY C. WILLIAMS
Case Western Reserve University
FLOWER; GAYANT; GRAY, H.

MARY PICKARD WINSOR
University of Toronto
FORBES, S.

DENISE WROTNOWSKA
Institut Pasteur
FOURNEAU

H. WUSSING
University of Leipzig
FROBENIUS

A. P. YOUSCHKEVITCH
Soviet Academy of Sciences
GELFOND

ROBERT M. YOUNG
University of Cambridge
GALL

BRUNO ZANOBIO
University of Pavia
FRACASTORO; GOLGI

DICTIONARY OF SCIENTIFIC BIOGRAPHY

DICTIONARY OF SCIENTIFIC BIOGRAPHY

FISCHER—HABERLANDT

FISCHER, EMIL HERMANN (*b.* Euskirchen, near Bonn, Germany, 9 October 1852; *d.* Berlin, Germany, 15 July 1919), *chemistry.*

Fischer was the son of Laurenz Fischer, a successful merchant, and Julie Poensgen Fischer. His father hoped that he would become a businessman; but after a trial period in business ended in failure, the elder Fischer consented to his son's desire for a university education. Fischer entered the University of Bonn in 1871 and attended the lectures of Kekulé. He transferred to Strasbourg the following year, where he studied chemistry under Adolf von Baeyer and obtained the doctorate in 1874. He accompanied Baeyer to Munich in 1875, and qualified as *Privatdozent* in 1878 and as assistant professor in 1879. He became professor of chemistry at Erlangen in 1882, at Würzburg in 1885, and at Berlin in 1892.

Fischer married Agnes Gerlach in 1888; she died seven years later. Two of their three sons were killed in World War I, but the eldest, Hermann Fischer, became an outstanding organic chemist.

Emil Fischer received the Nobel Prize for chemistry in 1902 in recognition of his syntheses in the sugar and purine groups. During World War I he was active in organizing the German chemical resources and headed the commissions for chemical production and food supplies. After the war, he helped to reorganize the teaching of chemistry and to establish research facilities. His work in organic chemistry was primarily on the constitution and synthesis of substances present in organisms. He laid the chemical foundations for biochemistry by his study of sugars, enzymes, purines, and proteins.

Hydrazine Chemistry. Fischer's first publications (1875) dealt with the organic derivatives of hydrazine. He discovered this new group of compounds, considering them to be derivatives of the as yet unknown compound N_2H_4, which he named hydrazine to indicate its relation to nitrogen (azote). In 1866 Kekulé had formulated diazonium compounds as R—N=N—X, similar to the azo formula (R—N=N—R). Since azo compounds were very stable and diazonium compounds were not, chemists disputed Kekulé's formula. In 1871 Adolf Strecker obtained a salt from benzenediazonium nitrate and potassium hydrogen sulfite. Fischer repeated this work and proved that Strecker's salt was a reduction product—potassium phenylhydrazine sulfonate ($C_6H_5NHNH_2$—SO_3K)—of the diazonium compound. This work confirmed the Kekulé formula, and Strecker's compound was a salt of phenylhydrazine. Fischer then prepared phenylhydrazine itself and established its formula by 1878. He prepared many organic derivatives of hydrazine and explored their reactions. The reaction of hydrazines with carbon disulfide led to dyestuffs. Oxidation produced tetrazenes, compounds with chains containing four nitrogen atoms. Aryl hydrazines with ketones and keto acids condensed to form derivatives of indole (the Fischer indole synthesis, 1886).

In 1884 Fischer discovered that phenylhydrazine was a valuable reagent for aldehydes and ketones. It formed solid, crystalline compounds (phenylhydrazones) which had a definite melting point. He then found that it formed not only the hydrazone with carbohydrates but also attacked the hydroxyl group

$$\begin{matrix} \mathrm{H{-}C{=}O} \\ | \\ \mathrm{H{-}C{-}OH} \\ | \end{matrix} + C_6H_5NHNH_2 \longrightarrow \underset{\text{Hydrazone}}{\begin{matrix} \mathrm{H{-}C{=}NNHC_6H_5} \\ | \\ \mathrm{H{-}C{-}OH} \\ | \end{matrix}} \xrightarrow{C_6H_5NHNH_2} \underset{\text{Osazone}}{\begin{matrix} \mathrm{H{-}C{=}NNHC_6H_5} \\ | \\ \mathrm{C{=}NNHC_6H_5} \\ | \end{matrix}}$$

adjacent to the carbonyl group. He called these compounds osazones. The osazones were also crystalline solids and thus were useful in the identification of sugars. By 1888 he had established the structures of hydrazones and osazones. He was to utilize these reactions of phenylhydrazine in elucidating the chemistry and structure of the carbohydrates.

FISCHER

Aniline Dyestuffs. Fischer's doctoral thesis had been on the chemistry of colors and dyes. He extended this interest to the new synthetic dyestuffs. He and his cousin Otto Fischer examined the constitution of rosaniline, the basic dyestuff prepared by August von Hofmann in 1862 by the oxidation of toluidine and aniline. There were several conjectures on the constitution of this base but no satisfactory solution until the Fischers succeeded in showing that it was a triphenylmethane derivative. They reduced rosaniline to a colorless derivative, which they called leucaniline and converted, by removal of its nitrogen atoms, into a hydrocarbon of composition $C_{20}H_{18}$. They carried out similar reactions with pararosaniline (from *p*-toluidine and aniline), obtaining a hydrocarbon with the formula $C_{19}H_{16}$, which proved to be identical to triphenylmethane. In 1878 they proved that the rosaniline dyes were homologues and were triamine derivatives of triphenylmethane and its homologues, rosaniline being a derivative of metatolyldiphenylmethane and *p*-rosaniline of triphenylmethane.

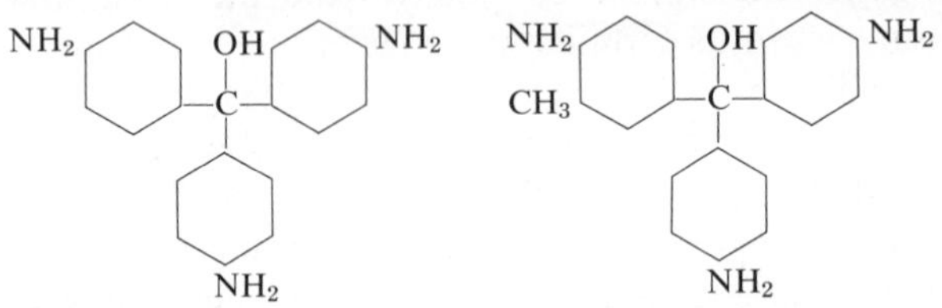

```
NH2-C6H4  OH  C6H4-NH2        NH2,CH3-C6H3  OH  C6H4-NH2
        \  |  /                          \  |  /
           C                                C
           |                                |
         C6H4                             C6H4
           |                                |
          NH2                              NH2
```

Pararosaniline   Rosaniline

Purines. Fischer began a study of uric acid and related substances in 1881 and continued his investigations until 1914, when he achieved the first synthesis of a nucleotide. These were biologically important substances. Xanthine, hypoxanthine, adenine, and guanine were present in the cell nucleus of animals. Theobromine, caffeine, and theophylline were stimulants in plants. Baeyer had studied these compounds in the 1860's and partially clarified their relations. The Würzburg chemist Ludwig Medicus proposed structural formulas for several of them in 1875. Fischer became the prime investigator in the field, and it is to him that almost all knowledge of the purines is due. He explored the whole series, established their structures, and synthesized about 130 derivatives by 1900.

Fischer studied the reactions and degradation products of the purines. In 1882 he ventured structural formulas for uric acid, caffeine, theobromine, xanthine, and guanine. He synthesized theophylline and caffeine (1895) and uric acid (1897); but further research convinced him that his structures were incorrect, since his reaction products were not reconcilable to his formulas. In 1897 he provided a new set of formulas. He had come to realize that uric acid and related compounds were oxides of a hypothetical base $C_5N_4H_4$, which he named purine:

$C_5N_4H_4O_3$	$C_5N_4H_4O_2$	$C_5N_4H_4O$
Uric Acid	Xanthine	Hypoxanthine

He proposed that purine was a heterocyclic compound and also proposed the notation system now used in purine chemistry:

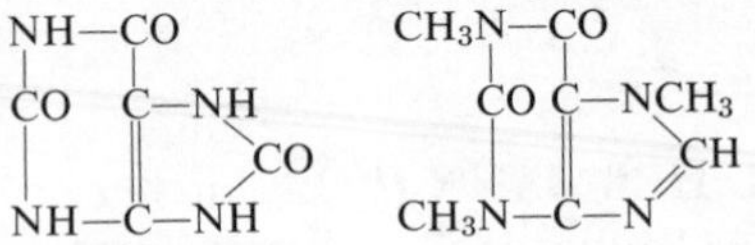

```
NH—CO                CH3N—CO
|   |                 |    |
CO  C—NH              CO   C—NCH3
|   ||   \            |    ||    \
|   ||    CO          |    ||     CH
|   ||   /            |    ||    //
NH—C—NH              CH3N—C—N
  Uric Acid              Caffeine

                                                  6
NH—CO              NH—CO              1N=CH
|   |               |   |              |    |     7
CO  C—NH            CH  C—NH          2HC 5C—NH  8
|   ||   \          ||  ||   \         ||   ||    \
|   ||    CH        ||  ||    CH       ||   ||     CH
|   ||   //         ||  ||   //        ||   ||    //
NH—C—N             N———C—N            3N—C—N
                                           4  9
  Xanthine          Hypoxanthine          Purine
```

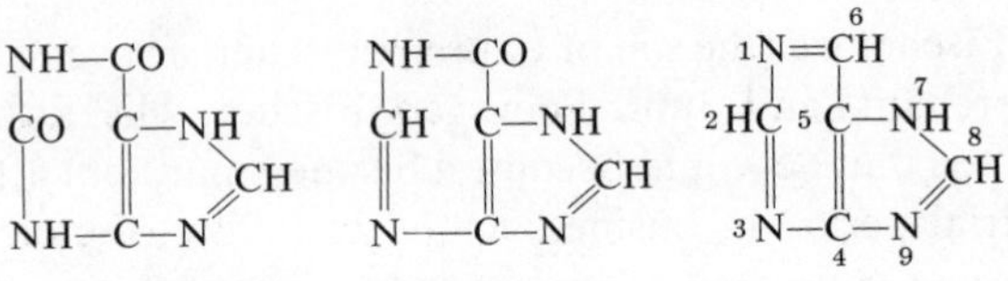

Subsequently, he synthesized hypoxanthine, xanthine, theobromine, adenine, and guanine. Finally, in 1898 he succeeded in reducing trichloropurine to purine, the parent substance of the class. These researches involved an immense series of preparations and very difficult reactions. He continued this work, combining it with his research on carbohydrates, and in 1914 prepared glucosides of theophylline, theobromine, adenine, hypoxanthine, and guanine. From theophylline-D-glucoside he prepared the first synthetic nucleotide, theophylline-D-glucoside phosphoric acid.

Fischer's purine research was of interest to the German drug industry. His laboratory methods became the basis for the industrial production of caffeine, theophylline, and theobromine. In 1903 he synthesized 5,5-diethyl-barbituric acid. Under various trade names—Barbital, Veronal, and Dorminal—this compound proved to be a valuable hypnotic. Another commercially valuable purine was phenyl, ethylbarbituric acid, prepared by Fischer in 1912 and known as Luminal or phenobarbital.

Carbohydrates. Fischer carried on his purine research simultaneously with his carbohydrate studies and became the prime investigator in both fields. When he began his carbohydrate studies in 1884, there were four known monosaccharides: two aldohexoses (glucose, galactose) and two ketohexoses (fructose, sorbose), all with the formula $C_6H_{12}O_6$. There were three known disaccharides (sucrose, maltose, lactose). The general structure of the simple sugars had been established. Glucose and galactose were straight-chain pentahydroxy aldehydes, and the ketohexoses were straight-chain pentahydroxy ketones. Fischer in an enormous effort elaborated the complex structures and chemistry of the carbohydrates, synthesized many of them, and established the configurations of the sixteen possible stereoisomers of glucose.

In 1885 Heinrich Kiliani developed the method of lengthening the carbon chain in sugars by means of the addition of hydrocyanic acid to the carbonyl group, followed by hydrolysis and reduction. Fischer utilized this method to convert pentoses into hexoses, the latter into heptoses, etc., synthesizing sugars with as many as nine carbon atoms:

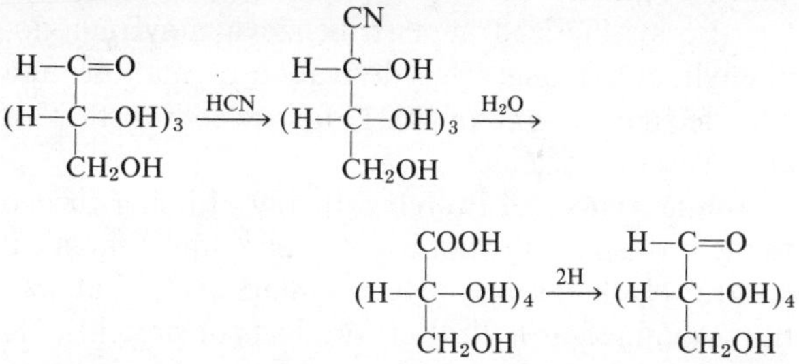

$$\begin{array}{c}H{-}C{=}O\\(H{-}C{-}OH)_3\\CH_2OH\end{array}\xrightarrow{HCN}\begin{array}{c}CN\\H{-}C{-}OH\\(H{-}C{-}OH)_3\\CH_2OH\end{array}\xrightarrow{H_2O}$$

$$\begin{array}{c}COOH\\(H{-}C{-}OH)_4\\CH_2OH\end{array}\xrightarrow{2H}\begin{array}{c}H{-}C{=}O\\(H{-}C{-}OH)_4\\CH_2OH\end{array}$$

Starting with glycerylaldehyde

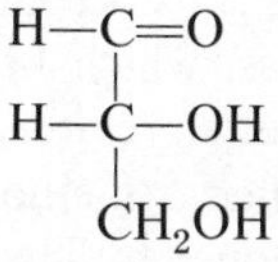

$$\begin{array}{c}H{-}C{=}O\\H{-}C{-}OH\\CH_2OH\end{array}$$

he built up the molecule step-by-step and synthesized several pentoses and hexoses, including glucose, fructose, and mannose, by the Kiliani method.

Fischer achieved his first synthesis of a sugar in 1887. He wanted to synthesize glyceraldehyde and use it as a starting point for building up the carbon chain in sugars. To prepare glyceraldehyde he combined acrolein dibromide and barium hydroxide:

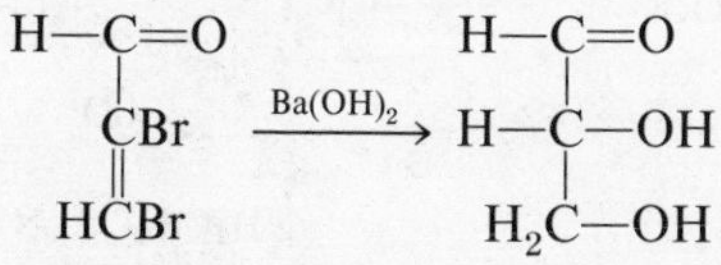

$$\begin{array}{c}H{-}C{=}O\\CBr\\\|\\HCBr\end{array}\xrightarrow{Ba(OH)_2}\begin{array}{c}H{-}C{=}O\\H{-}C{-}OH\\H_2C{-}OH\end{array}$$

Instead of glyceraldehyde he obtained a syrup which he named acrose:

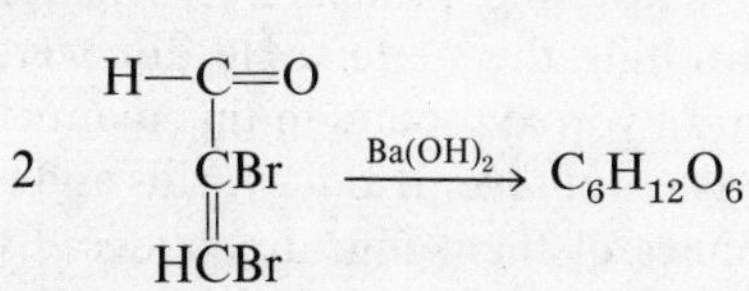

$$2\begin{array}{c}H{-}C{=}O\\CBr\\\|\\HCBr\end{array}\xrightarrow{Ba(OH)_2}C_6H_{12}O_6$$

With phenylhydrazine he obtained two different osazones and isolated from them two sugars. He proved that these were fructose and sorbose, the first naturally occurring sugars to be synthesized.

The reaction of sugars with phenylhydrazine yielded first the hydrazone and then the osazone. Fischer found that glucose, fructose, and mannose formed the same osazone. Therefore, the three sugars had the same configuration below the second carbon atom. Osazones on hydrolysis with hydrochloric acid eliminate phenylhydrazine and form osones, a new type of glucose derivative, possessing adjacent carbonyl groups. By reducing these, he obtained sugars, although an aldose is converted into a ketose:

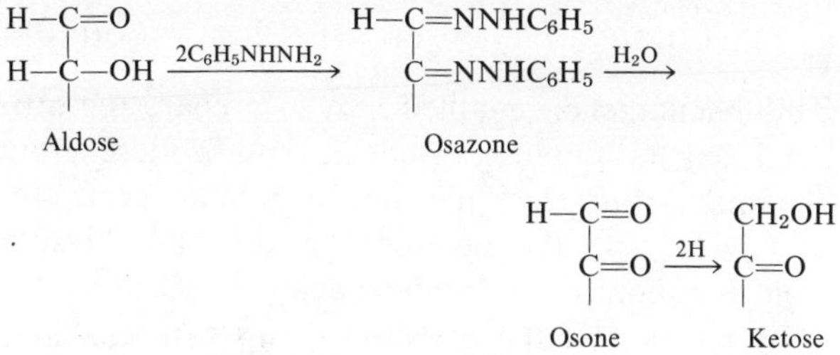

$$\underset{\text{Aldose}}{\begin{array}{c}H{-}C{=}O\\H{-}C{-}OH\\|\end{array}}\xrightarrow{2C_6H_5NHNH_2}\underset{\text{Osazone}}{\begin{array}{c}H{-}C{=}NNHC_6H_5\\C{=}NNHC_6H_5\\|\end{array}}\xrightarrow{H_2O}$$

$$\underset{\text{Osone}}{\begin{array}{c}H{-}C{=}O\\C{=}O\\|\end{array}}\xrightarrow{2H}\underset{\text{Ketose}}{\begin{array}{c}CH_2OH\\C{=}O\\|\end{array}}$$

At the other end of the carbon chain the primary alcohol could be reduced or oxidized. Oxidation of this group in glucose gave glucuronic acid; oxidation of the carbonyl group at the other end of the chain, a gluconic acid; and oxidation at both sites, a dicarboxylic acid. By differential reductions and oxidations Fischer could transfer the carbonyl group from one end of the chain to the other, and by testing the products for their properties and their optical rotation of the plane of polarized light, he could elucidate the structures of his compounds.

Using the van't Hoff theory of stereoisomers, Fischer realized that there were sixteen possible configurations for the aldohexoses. By his methods of oxidation, reduction, degradation, addition, etc., he identified the structures of these by 1891. He established the configurations for all members of the D-series of aldohexoses, i.e., those derived from D-glyceraldehyde, where D, according to Fischer's practice, refers to the hydroxyl group's being positioned to the right of the carbon atom next to the primary alcohol group:

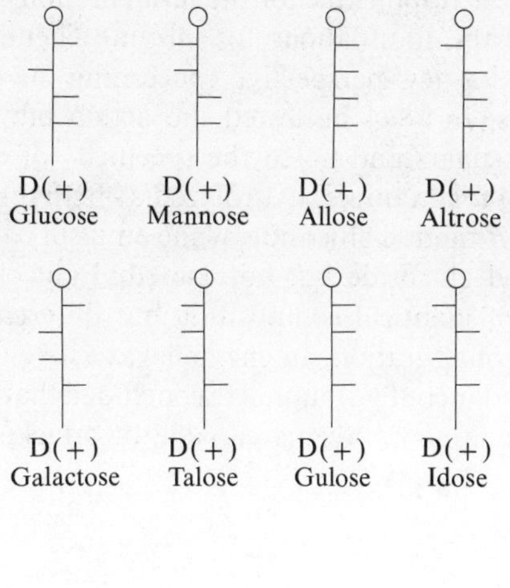

D(+) Glucose   D(+) Mannose   D(+) Allose   D(+) Altrose

D(+) Galactose   D(+) Talose   D(+) Gulose   D(+) Idose

Fischer prepared several artificial sugars. He had established the structures of the natural pentoses arabinose and xylose. He found that by extending the carbon chain of each of these pentoses he obtained two products. The Kiliani method introduced a new asymmetric atom and therefore two reaction products:

$$\begin{array}{c} H-C=O \\ (H-C-OH)_3 \\ CH_2OH \end{array} \xrightarrow{HCN} \begin{array}{c} CN \\ H-C-OH \\ (H-C-OH)_3 \\ CH_2OH \end{array} + \begin{array}{c} CN \\ HO-C-H \\ (H-C-OH)_3 \\ CH_2OH \end{array}$$

This phenomenon enabled him to prepare the artificial sugars *l*-gulose, *d*-talose, and *d*-idose from *l*-mannonic, *d*-galactonic, and *d*-gulonic acids, respectively; and the pentoses *l*-ribose and *d*-lyxose from *l*-arabonic and *l*-xylonic acids, respectively.

By reaction of the carbonyl group with alcohols, Fischer prepared α- and β-methyl glucoside, the first synthetic glycosides (1893). Since there were two methyl glucosides, he suggested that they must have a cyclic structure:

$$\begin{array}{c} CH_3O \quad H \\ C \\ \text{(ring with O)} \end{array} \qquad \begin{array}{c} H \quad OCH_3 \\ C \\ \text{(ring with O)} \end{array}$$

He never extended the ring structure to the sugars themselves, although in 1883 Tollens had suggested ring structures for glucose and fructose. Fischer thought that the disaccharides might have such rings and represented them as two hexoses united through an oxygen linkage: lactose was a glucose-β-galactoside, maltose a glucose-α-glucoside. He regarded the synthesis of glucosides as important because the polysaccharides were glucosides of the sugars themselves, and there was now the possibility of synthesizing polysaccharides.

Fischer examined the properties of enzymes, the substances responsible for the fermentation of sugars. He laid the foundations for enzyme chemistry and provided a new perspective concerning the action of enzymes. In 1894 he tested the action of yeasts on various sugars and noted the specificity of enzymes: maltase, for example, hydrolyzed α-methyl glucoside but not β-methyl glucoside, while emulsin hydrolyzed β-methyl glucoside but not α-methyl glucoside. For sugars of identical composition but different stereometric configuration, an enzyme was active only with a particular configuration. He concluded that enzymes were asymmetric agents capable of attacking molecules of only specific geometric configurations. The action of enzymes in hydrolyzing glucosides led him to use the analogy of a lock-and-key structural relationship between the enzyme and the sugar (1894). Molecular asymmetry gained new significance: the chemical transformations in the organism depended on asymmetry.

As an extension of his work in carbohydrates, Fischer from 1908 studied tannins, the gallic acid derivatives of sugars. In 1912 he showed that tannins were not glucosides but esters and synthesized a pentadigalloylglucose that had the properties of a tannin. In 1918 he established the composition of Chinese tannin as a penta(*meta*-digalloyl)glucose. He also synthesized hepta(tribenzoylgalloyl)-*p*-iodophenylmaltosazone. This derivative of maltose had a molecular weight of 4021, far exceeding that of any synthetic product.

Amino Acids and Proteins. In 1899 Fischer turned to the proteins in the hope of revealing their chemical nature. He knew of thirteen amino acids that had been obtained as hydrolysis products of proteins. He discovered additional amino acids, synthesized several of them, and resolved the *d-l* forms by fractional crystallization of the salts prepared from the benzoyl or formyl derivatives, which he combined with the optically active bases strychnine or brucine.

In 1901 he modified a method for the separation of amino acids developed by Theodor Curtius in 1883. A mixture of amino acids could be separated by esterifying the acids and distilling them at reduced pressure. Furthermore, Curtius showed that the ethyl ester of glycine eliminates alcohol to form a cyclic diketopiperazine, which on ring opening formed glycylglycine:

$$2NH_2CH_2COOC_2H_5 \xrightarrow{2C_2H_5OH} \begin{array}{c} O \\ \| \\ C-CH_2 \\ NH \qquad NH \\ CH_2-C \\ \| \\ O \end{array} \xrightarrow{H_2O} NH_2CH_2CO \cdot NHCH_2COOH$$

Fischer used Curtius' method to separate mixtures of amino acids from protein hydrolysates by fractionally distilling their esters. He discovered valine, proline, and hydroxyproline in this manner. He prepared the esters of several amino acids and condensed two molecules of them into dipeptides. By 1907 he was preparing polypeptides, the largest one consisting of fifteen glycyl and three leucyl residues and having a molecular weight of 1213: leucyl-triglycyl-leucyl-triglycyl-leucyl-octaglycylglycine. He suggested that the peptide linkage —CONH— was repeated in long chains in the polypeptide molecule. His synthetic methods involved either attacking the amino or the carbonyl group in the amino acid (e.g., using a halogen-containing acid to combine with the amino group and exchanging the halogen by another amino group):

$$CH_2Cl \cdot CO \cdot Cl + NH_2CH(CH_3) \cdot COOH \rightarrow CH_3Cl \cdot CO \cdot NHCH(CH_3) \cdot COOH$$

In this way he could introduce glycyl, leucyl, and other groups into a peptide.

Fischer recognized the complexity of proteins. Even his simple peptides would have numerous isomers, and it would be extremely difficult to establish the constitution and structure of any protein. By 1905 he had differentiated twenty-nine polypeptides and tested their behavior with various enzymes. He characterized proteins by the number, kind, and arrangement of amino acids. In 1916 he summarized his work on the synthesis of about 100 polypeptides and cautioned that these represented only a tiny fraction of the possible combinations that might be found in natural proteins.

Fischer used the Walden inversion in synthesizing amino acids. In 1895 Paul Walden had found that in some substitution reactions the optical antipode of the expected compound was obtained. Fischer examined such reversals of optical rotatory power with regard to amino acids. In 1906 he described the following reaction:

$$l\text{-alanine} \xrightarrow{NOBr} d\text{-}\alpha\text{-bromopropionic acid} \xrightarrow{NH_3} d\text{-alanine.}$$

Comparison with other inversion reactions failed to show at which step the inversion of optical rotation corresponded to a change in the atomic sequence on the asymmetric carbon atom. In 1911 Fischer developed a model to explain such rearrangements, his only excursion into theories of reaction mechanisms. He proposed that substitution is preceded by an addition step in which the entrant group need not take the place of the dislodged one, but that a relative distribution of substituents may take place. Thus, the configuration of the substituted compound may differ from the original.

Fischer's last work was on the esterification of glycerol by fatty acids. The aim of all his investigations was to apply the methods of organic chemistry to the synthesis and processes of substances in living matter.

BIBLIOGRAPHY

I. ORIGINAL WORKS. Fischer's publications were collected in eight large volumes: *Untersuchungen über Aminosäuren, Polypeptide und Proteine, 1899–1906* (Berlin, 1906); *Untersuchungen in der Puringruppe, 1882–1906* (Berlin, 1907); *Untersuchungen über Kohlenhydrate und Fermente, 1884–1908* (Berlin, 1909); *Untersuchungen über Depside und Gerbstoffe, 1908–1919* (Berlin, 1920); *Untersuchungen über Kohlenhydrate und Fermente. II, 1908–1918* (Berlin, 1922); *Untersuchungen über Aminosäuren, Polypeptide und Proteine. II, 1907–1919* (Berlin, 1923); *Untersuchungen über Triphenylmethanfarbstoffe, Hydrazine und Indole* (Berlin, 1924); and *Untersuchungen aus verschiedenen Gebieten, Vorträge und Abhandlungen allgemeinen Inhalts* (Berlin, 1924).

His autobiography is *Aus meinem Leben* (Berlin, 1922), and his Nobel lecture is in *Nobel Lectures. Chemistry 1901–1921* (Amsterdam–New York, 1966), pp. 21–35.

II. SECONDARY LITERATURE. Informative studies of Emil Fischer include Max Bergmann, in G. Bugge, ed., *Das Buch der grossen Chemiker,* II (Berlin, 1930), 408–420; Martin Onslow Forster, "Emil Fischer Memorial Lecture," in *Journal of the Chemical Society,* **117** (1920), 1157–1201; Burckhardt Helferich, in Eduard Farber, ed., *Great Chemists* (New York, 1961), pp. 981–995; and Kurt Hoesch, *Emil Fischer, sein Leben und sein Werk* (Berlin, 1921). "Gedächtnis—Feier für Emil Fischer," in *Berichte der Deutschen chemischen Gesellschaft,* **52A** (1919), 125–164, contains addresses by H. Wichelhaus, Ludwig Knorr, and Carl Duisberg.

EDUARD FARBER

FISCHER, HERMANN OTTO LAURENZ (*b.* Würzburg, Germany, 16 December 1888; *d.* Berkeley, California, 9 March 1960), *biochemistry.*

The son of Emil Fischer, professor of chemistry at the University of Würzburg and famous for his elucidation of the structures of pentoses and hexoses, Hermann was the oldest of three boys. The family moved to Berlin in 1892, when Emil Fischer became director of the Chemical Institute of Berlin University. Fischer attended a Gymnasium in Berlin and decided to become a chemist, while his two brothers chose medical careers. He studied at Cambridge in 1907, then fulfilled his year of military service at Lüneburg. He began his study of chemistry at Berlin and later transferred to Jena. Under Ludwig Knorr's direction he worked on tautomerism of diketones, preparing the enol form of acetyl acetone. After obtaining the doctorate at Jena, Fischer went to Berlin in 1912 to begin research under his father's guidance.

By 1912 Emil Fischer had ceased working on carbohydrates, purines, and proteins and turned his attention to depsides and tannins. He put his son to work on the synthesis of lecanoric acid, which occurs naturally in lichens. Hermann succeeded, showing that lecanoric acid is a didepside, *p*-diorsellinic acid. With the outbreak of World War I, Fischer went to the front in France; both of his brothers joined the military medical service and died in the line of duty. At the end of the war Fischer returned to his father's laboratory. He married in 1922 and was appointed assistant professor at the Chemical Institute of Berlin University.

With Gerda Dangschat, Fischer elucidated the structure of caffeic acid and chlorogenic acid, the depside of quinic acid. They showed that this depside carries the caffeine in coffee and determines the coffee's taste after roasting. Beginning an association with Erich Baer that was to continue for twenty-seven years, Fischer started work on the trioses, dihydroxyacetone, and glyceraldehyde. By 1932 Fischer and his co-workers had succeeded in synthesizing the calcium salt of D, L-glyceraldehyde-3-phosphoric acid which Warburg, Embden, and Meyerhof utilized in developing their schemes of alcoholic fermentation and glycolysis.

Fischer accepted a position at the University of Basel in 1932. He continued work on the trioses, with Baer preparing pure D- and L-glyceraldehydes and synthesizing D-fructose and L-sorbose. In Berlin, Gerda Dangschat continued the elucidation of the configuration of quinic acid and shikimic acid, which later proved to be an important intermediate in the formation of aromatic amino acids in bacterial metabolism.

Fischer accepted the invitation of Sir Frederick Banting, who with Charles Best had discovered insulin, to join the staff at the Banting Institute of the University of Toronto. He arrived there in the fall of 1937 with his wife, two sons and daughter, Erich Baer and other assistants, his private laboratory (including the 9,000 reference compounds representative of all of his father's work), and his father's library. Fischer continued work on trioses, glycerol derivatives, and glycerides. In his laboratory were Baer, J. M. Grosheintz, Leon Rubin, J. C. Sowden, and Henry Lardy, working on the synthesis of selachyl alcohol, the oxidation of glycols, the condensation of aldoses, the cyclization of glucose into myoinositol, and the synthesis of biologically interesting organic phosphates.

Fischer became a member of the new department of biochemistry at the University of California in 1948. On 9 October 1952 he dedicated the Emil Fischer Library, containing his father's books and reference compounds, to the university. Fischer became chairman of the biochemistry department in 1953, holding that position and lecturing on carbohydrates and lipids until his retirement in 1956. Among those who worked with him at Berkeley were D. L. MacDonald, C. E. Ballou, E. A. Kabat, and S. J. Angyal, who investigated sugar disulfones, sugar dialdehydes, inositol derivatives, tetrose phosphates, hexose dialdehydes, D-erythrose-4-phosphate, galactinol, and the confirmational analysis of sugars and inositols. On his retirement Fischer was given space in the Biochemistry and Virus Laboratory, where he conducted research on carbohydrates, assisted by Hans Helmut Baer, from the Max Planck Institute for Medical Research in Heidelberg. Starting from Fischer's observation that a sugar dialdehyde of the pentose series forms, with nitromethane, nitroinositols, Baer worked out the syntheses of 3-amino-3-deoxy-D-ribose and 3-amino-3-deoxy-D-mannose.

Fischer was a member of many professional societies, recipient of the Sugar Research Award of the American Chemical Society in 1949 and of the Adolf von Baeyer Memorial Gold Medal from the Society of German Chemists in 1955, and was named professor emeritus by the University of California in 1956.

BIBLIOGRAPHY

For a short autobiography with portrait, see Hermann O. L. Fischer, "Fifty Years 'Synthetiker' in the Service of Biochemistry," in *Annual Review of Biochemistry,* **29** (1960), 1–14. Early papers on carbomethoxy derivatives of acids written with Emil Fischer can be found in *Berichte der Deutschen chemischen Gesellschaft,* **46** (1913), 1138–1148, 2659–2664; and **47** (1914), 505–512, 768–780. Fischer and Dangschat's work on quinic acid, shikimic acid, and their derivatives is reported in papers found in *Naturwissenschaften,* **19** (1931), 310–311; *Berichte der Deutschen chemischen Gesellschaft,* **65B** (1932), 1009–1031, 1037–1040; and *Helvetica chimica acta,* **17** (1934), 1196–1200, 1200–1207; and **18** (1935), 1204–1206, 1206–1213. Some of the papers written in collaboration with Erich Baer and others on the trioses, dihydroxyacetone and glyceraldehyde, and their derivatives can be found in *Berichte der Deutschen chemischen Gesellschaft,* **60B** (1927), 479–485; **63B** (1930), 1732–1744, 1744–1748, 1749–1753; **65B** (1932), 337–345, 345–352; *Helvetica chimica acta,* **16** (1933), 534–547; **17** (1934), 622–632; **18** (1935), 514–521, 1079–1087; **19** (1936), 519–532; **20** (1937), 1213–1226, 1226–1236; *Naturwissenschaften,* **25** (1937), 588–589, 589; and *Chemical Reviews,* **29** (1941), 287–316.

Fischer published a review of his investigations of chemical and biological relationships between hexoses and inositols in *Harvey Lectures,* **40** (1944–1945), 156–178; and a review of his chemical synthesis of intermediate products of sugar metabolism in *Angewandte Chemie,* **69** (1957), 413–419.

A. ALBERT BAKER, JR.

FISCHER, NICOLAUS WOLFGANG (*b.* Gross-Meseritz, Bohemia [now Mezirici Velké, Czechoslovakia], 15 January 1782; *d.* Breslau, Germany [now Wrocław, Poland], 19 August 1850), *chemistry.*

Very little is known about Fischer's life. He studied medicine in Erfurt and practiced in Breslau, where he also gave lectures at the school of surgery. After the founding of the University of Breslau in 1811, he qualified there as *Privatdozent.* In 1815 the first chair in chemistry was established at the university, and Fischer was appointed to it. He held this position until his death.

In approximately sixty articles published in contemporary German journals, Fischer presented the results of his diversified chemical and medical research. Many of his investigations are of lasting importance.

At first he reported on the triple salt $K_3Co(NO_2)_6$, which entered inorganic chemistry as "Fischer's salt." He employed this salt in the analytical separation of nickel from cobalt.

He appears to have been one of the first to observe the semipermeability of animal bladders in solutions and pure water; moreover, he undertook osmotic investigations and hence observed endosmosis. He gave an account of this research in a heterogeneous article in the part entitled "Über die Eigenschaft der thierischen Blase, Flüssigkeiten durch sich hindurch zu lassen."

Fischer took a great interest in electrochemical phenomena. He investigated the electrochemical reducibility of metals in galvanic cells with the goal of finding the relationship between chemical affinity and galvanism. For this purpose he constructed various cells and observed the so-called galvanoplastic phenomenon (i.e., electroplating), which later, through other researchers, led to the elaboration of technical electroplating.

Fischer made a galvanic cell consisting of silver and zinc electrodes. The former was immersed in dilute sulfuric acid and the latter in moist silver chloride, which he contained in bladders. He thus obtained a voltaic pile which delivered a more or less constant electric current (1812). He therefore belongs among the pioneers in the construction of voltaic cells.

Fischer also concerned himself with, among other things, the investigation of the sensitivity to light of silver chloride, methods for the legal and medical detection of arsenic, and the chemical reactions of the then comparatively new elements tellurium and selenium.

BIBLIOGRAPHY

The article "Über die Wiederherstellung eines Metals durch ein anderes und über die Eigenschaft der thierischen Blase, Flüssigkeiten durch sich hindurch zu lassen und sie in einigen Fällen anzuheben" appeared in *Annalen der Physik,* **12** (1822), 289–307. Poggendorff gives a fairly complete listing of Fischer's articles; he wrote no books.

A biographical notice is J. Schiff, "N. W. Fischer, erster Chemie Professor der Universität Breslau," in *Archiv für die Geschichte der Naturwissenschaften und der Technik,* **8** (1917), 225, and **9** (1918), 29.

F. SZABADVÁRY

FISHER, RONALD AYLMER (*b.* London, England, 17 February 1890; *d.* Adelaide, Australia, 29 July 1962), *statistics, biometry, genetics.*

Fisher's father was a prominent auctioneer and the head of a large family; Fisher was a surviving twin. Chronic myopia probably helped channel his youthful mathematical gifts into the high order of conceptualization and intuitiveness that distinguished his mature work. At Cambridge, which he entered in 1909 and from which he graduated in 1912, Fisher studied mathematics and theoretical physics. His early postgraduate life was varied, including working for an investment house, doing farm chores in Canada, and teaching high school. Fisher soon became interested in the biometric problems of the day and in 1919 joined the staff of Rothamsted Experimental Station as a one-man statistics department charged, primarily, with sorting and reassessing a sixty-six-year accumulation of data on manurial field trials and weather records. In the following decade and a half his work there established him as the leading statistician of his era, and early in his tenure he published the epochal *Statistical Methods for Research Workers.*

Meantime, avocationally, Fisher was building a reputation as a top-ranking geneticist. He left Rothamsted in 1933 to become Galton professor of eugenics at University College, London. Ten years later he moved to Cambridge as Balfour professor of genetics. In 1959, ostensibly retired, Fisher emigrated to Australia and spent the last three years of his life working steadily and productively in the Division of Mathematical Statistics of the Commonwealth Scientific and Industrial Research Organization. Innumerable honors came to him, including election to fellowship of the Royal Society in 1929 and knighthood in 1952.

 FISHER

In 1917 Fisher married Ruth Eileen Guinness and, like his own parents, they had eight children—a circumstance that, according to a friend, was "a personal expression of his genetic and evolutionary convictions." Later in life he and Lady Fisher separated. Slight, bearded, eloquent, reactionary, and quirkish, Fisher made a strong impact on all who met him. The geniality and generosity with which he treated his disciples was complemented by the hostility he aimed at his dissenters. His mastery of the elegantly barbed phrase did not help dissolve feuds, and he left a legacy of unnecessary confusion in some areas of statistical theory. Nevertheless, Fisher was an authentic genius, with a splendid talent for intertwining theory and practice. He had a real feel for quantitative experimental data whose interpretation is unobvious, and throughout his career he was a happy and skillful performer on the desk calculator.

Fisher's debut in the world of mathematical statistics was occasioned by his discovery, as a young man, that all efforts to establish the exact sampling distribution of the well-known correlation coefficient had foundered. He tackled the problem in the context of points in an n-dimensional Euclidean space (n being the size of the sample), an original and, as it turned out, highly successful approach. In the following years he applied similar methods to obtain the distributions of many other functions, such as the regression coefficient, the partial and multiple correlation coefficients, the discriminant function, and a logarithmic function of the ratio of two comparable variances. Fisher also tidied up the mathematics and application of two important functions already in the literature: the Helmert-Pearson χ^2 (the sum of squares of a given number of independent standard normal variates, whose distribution is used to test the "goodness of fit" of numerical observations to expectations) and Gosset's z (the ratio of a normal sample mean, measured from a given point, in terms of its sample standard deviation). The latter was modulated by Fisher to the now familiar t, whose frequency distribution provides the simplest of all significance tests. He seized on Gosset's work to centralize the problem of making inferences from small samples, and he went on to erect a comprehensive theory of hypothesis testing.

The idea here was that many biological experiments are tests of a defined hypothesis, with the experimentalist wanting to know whether his results bolster or undermine his theorizing. If we assume, said Fisher in effect, that the scatter of results is a sample from a normal (Gaussian) distribution whose mean (suitably expressed or transformed) is the "null" hypothesis, we can, using the t distribution, compute the "tail" probability that the observed mean is a random normal sample—in much the same way that we can compute the probability of getting, say, seven or more heads in ten tosses of a fair coin. It might be thought that this probability was all an experimentalist would need, since he would subsequently make his own scientific judgment of its significance in the context of test and hypothesis. However, Fisher advocated the use of arbitrary "cutoff" points; specifically, he suggested that if the probability were $<1/20$ but $>1/100$, a weak judgment against the null hypothesis be made, and if the probability were $<1/100$ a strongly unfavorable judgment be made. This convention, helped by the publication of special tables, became popular, and it has often been used blindly. The real value of the discipline lay in its sound probabilistic structure rather than in its decision-making rules. It is noteworthy that other statistical theorists extended Fisher's ideas by introducing the power of a test, that is, its intrinsic ability, in terms of the parameters and distribution functions, to detect a given difference (between the null hypothesis and the observational estimate) at a prearranged probability level; but, for reasons never made plain, Fisher would not sanction this development.

Fisher devised his own extension of significance testing, the remarkable analysis of variance. This was bound in with his novel ideas on the wide subject of the theory of experimental design. He emphasized not merely the desirability but the logical necessity, in the design of an experiment whose results could not be freed from error, of maximizing efficiency (by devices such as blocks and confounding), and of introducing randomization in such a way as to furnish a valid estimate of the residual error. Whereas the time-honored practice in handling several factors had been to vary only one at a time during the experiment, Fisher pointed out that simultaneous variation was essential to the detection of possible interaction between factors, and that this change could be made without extra cost (in terms of experimental size and operational effort). Entrained with these innovations was the use of randomized blocks and Latin squares for the actual disposition of the test units.

Many would subscribe to the thesis that Fisher's contributions to experimental design were the most praiseworthy of all his accomplishments in statistics. It was to facilitate the interpretation of multifactor experiments carried out in the light of his ideas on design that Fisher introduced the analysis of variance (which he had originally devised to deal with hierarchical classification). In this scheme the variances due to different factors in the experiment are screened out and tested separately for statistical significance (that is, for incompatibility with the null hypothesis of "no effect"). The appeal of the analysis has been great, and here again misuses have not been uncommon among uncritical researchers—for example, some have contented themselves with, and even published, analyses of variance unaccompanied by tabulations of the group means to which the significance tests refer. This itself is a tribute, of a sort, to Fisher.

Arising out of his work on the analysis of variance was the analysis of covariance, a scheme in which the regression effect of concomitant nuisance factors could be screened out so as to purify further the significance tests. (For example, in a multifactor physiological experiment the weight of the animals might be a nuisance factor affecting response and calling for elimination or allowance by analysis of covariance.)

An early landmark in the Fisherian revolution was his long paper "On the Mathematical Foundations of Theoretical Statistics" (1922). Convinced that the subject had progressed too haphazardly and lacked a solid mathematical base, he set out to repair the situation. He drew attention to the shortcomings of the method of moments in curve fitting, stressed the importance of exact sampling distributions, and moved toward a realistic view of estimation. Discussing estimation of a parameter, he urged that a "satisfactory statistic" should be consistent (which means, roughly, unbiased), efficient (having the greatest precision), and sufficient (embracing all relevant information in the observations). His making "information" into a technical term, equatable to reciprocal variance, was a useful step (the reuse of the same word, much later, in C. E. Shannon's information theory is allied but not directly comparable).

To appreciate Fisher's handling of estimation theory, we must look upon it as a subdivision of the general problem of induction that has worried theoreticians since Hume's day. In its simplest and oldest form it is the question of how to arrive at a "best" value of a set of observations. Today we unthinkingly take the arithmetic mean, but in fact the conditions under which this procedure can be justified need careful definition. Broadly speaking, the arithmetic mean is usually the maximum likelihood estimate. This term expressed not a wholly new principle but one that Fisher transformed and named. He urged the need to recognize two kinds of uncertainty and proposed that the probability of an event, given a parameter, should be supplemented by the likelihood of a parameter, given an event. Likelihood has similarities to probability, but important differences exist (for instance, its curve cannot be integrated).

At this point, by way of illustration, we may bring in a modified form of one of Fisher's own examples. It concerns a discrete (discontinuous) distribution. We are given four counts, n_1 through n_4, of corn seedlings (each in a different descriptive category) that are hypothesized to arise from a parameter p, as shown below.

Descriptive category	(i)	1	2	3	4	Σ
Fractions expected	(f_i)	$(2+p)/4$	$(1-p)/4$	$(1-p)/4$	$p/4$	1
Numbers observed	(n_i)	1,997	906	904	32	3,839

The problem is to estimate p. Now, by definition, the likelihood, L, will be

$$(1) \qquad L = \prod_{i=1}^{4} f_i^{n_i}.$$

Instead of seeking the maximum of this, we shall find it more convenient to handle the logarithm, which is

$$(2) \quad L' = n_1 \log(2+p) + (n_2+n_3)\log(1-p) + n_4 \log p.$$

Differentiating this expression with respect to p, equating the result to zero, and replacing the n_i with the actual observations yields the quadratic

$$(3) \qquad 3839p^2 + 1655p - 64 = 0,$$

from which we find $\hat{p} = 0.03571$ (the caret is widely employed nowadays to denote "an estimate of"). Therefore, among all possible values of p this is the one most likely to have given rise to the particular tetrad of observed numbers. Fisher now went further and showed that the second derivative of (2) could be equated to the variance of $\hat{p}$. This gives us

$$(4) \quad \text{var}(\hat{p}) = \frac{2p(1-p)(2+p)}{(1+2p)(n_1+n_2+n_3+n_4)},$$

which, incidentally, is the minimum variance, indicating that the estimate of p is efficient in Fisher's special sense of that term. Substitution of $\hat{p}$ for p in (4) and insertion of the actual n_i, followed by extraction of the square root, gives us the best estimate of the standard error of $\hat{p}$. Thus we can submit $\hat{p} = 0.03571 \pm 0.00584$ as our final result.

Attachment of a standard error to an estimate, as above, is quite old in statistics, and this is yet another matter to which Fisher brought change. He was a pioneer in interval estimation, that is, the specification of numerical probability limits bracketing the point estimate. Fisher's approach is best described in the context of sampling the normal distribution. Imagine such a distribution, with mean μ and variance (second moment) σ^2, both unknown and from which we intend to draw a sample of ten items. Now, in advance of the sampling, various statements can be made about this (as yet) unknown mean, m. One such is

$$(5) \quad P\left(m < \mu - \frac{1.96}{\sqrt{10}}\,\sigma\right) = P\left(m > \mu + \frac{1.96}{\sqrt{10}}\,\sigma\right) = 0.025,$$

the factor 1.96 being taken from a table of the partial integrals of the normal function. The statement is formal, although without practical value. We draw the sample, finding, say, $m = 8.41$ with a standard deviation of $s = 6.325$; then, according to Fisher, a probability statement analogous to (5) can be cast into this form:

$$(6) \quad P\left(\mu < m - \frac{2.23}{\sqrt{10}}\,s\right) = P\left(\mu > m + \frac{2.23}{\sqrt{10}}\,s\right) = 0.025 \equiv P(\mu < 3.95) = P(\mu > 12.87) = 0.025,$$

the factor 2.23 being taken from the t table in the Gosset-Fisher theory of small samples.

Strictly speaking, this is dubious in that it involves a probable location of a parameter, which, by definition, is fixed—and unknown. But, said Fisher, the observed values of m and s have changed the "logical status" of μ, transforming it into a random variable with a "well-defined distribution." Another way of stating (6) is that the values 3.95–12.87 are the "fiducial limits" within which μ can be assigned with probability 0.95. There is no doubt that this is a credibility statement with which an experimentalist untroubled by niceties of mathematical logic should be satisfied. And indeed it might be thought that the notion of fiducialism could be rationalized in terms of a different definition of probability (leaning on credibility rather than orthodox limiting frequency). But Fisher always insisted that probability "obtained by way of the fiducial argument" was orthodox. The doctrine landed Fisher and the adherents of fiducialism in a logical morass, and the situation was worsened by the historical accident that an allied concept, the theory of confidence limits, was introduced by Jerzy Neyman about the same time (*ca.* 1930). Although it, too, had weaknesses, Neyman's theory, as the more mathematically rigorous and widely applicable, was espoused by many statisticians in preference to its rival. Battle lines were drawn, and the next few decades witnessed an extraordinarily acrimonious and indecisive fight between the schools. Fiducialism is still being explored by mathematical statisticians.

This by no means exhausts Fisher's contributions to statistics—topics such as multivariate analysis, bioassay, time series, contingency tables, and the logarithmic distribution are others that come to mind—and in fact it would be hard to do so, if only because he sometimes gave seminal ideas to colleagues to work out under their own names. We must end with some reference to another subject on which Fisher left a deep mark: genetics. In his young manhood natural selection and heredity were in a state of renewed ferment. The rediscovery in 1900 of Mendel's splendid work on particulate inheritance threw discredit not only on Karl Pearson's elaborate researches into blending inheritance but also on Darwinism itself, believed by some to be incompatible with Mendelism. Fisher thought otherwise, and in 1918 he published a paper, "The Correlation Between Relatives on the Supposition of Mendelian Inheritance," that brought powerful new mathematical tools to bear on the issue and that eventually swung informed opinion over to his views—which were, in brief, that blending inheritance is the cumulative effect of a large number of Mendelian factors that are individually insignificant. (This was to blossom into the modern discipline of biometric genetics, although Fisher himself never made any important contributions thereto.)

Fisher came to regard natural selection as a study in its own right, with evolution as but one of several sequelae. His work on the phenomenon of dominance was outstanding. He early pointed out that correlations between relatives could be made to furnish information on the dominance of the relevant genes. He demonstrated that the Mendelian selection process invariably favors the dominance of beneficial genes, and that the greater the benefit, the faster the process. Dominance, then, must play a major role in evolution by natural selection. It may here be added that Fisher's work in this area, as elsewhere, was a careful blending of theory and practice. He carried out breeding experiments with various animals (mice, poultry, and snails were some), often under trying circumstances (for example, much mouse breeding was done in his own home in Harpenden). One of his best experiments concerned the inheritance of whiteness of plumage in white leghorns: by "breeding back" into wild jungle fowl he showed that the responsible dominant gene is a result of artificial selection, and its dominance is quickly lessened when it is introduced into the ancestral wild species.

Fisher also enunciated a "fundamental theorem of natural selection" (for an idealized population) in this form: "The rate of increase in fitness of any organism at any time is equal to its genetic variance in fitness at that time." He was keenly interested in human genetics and, like most eugenicists, held alarmist views about the future of *Homo sapiens.* An important consequence of this special interest was his realization that a study of human blood groups could be instrumental in advancing both theory and practice, and in 1935 he set up a blood-grouping department in the Galton Laboratory. Many good things came out of this enterprise, including a clarification of the inheritance of Rhesus groups.

BIBLIOGRAPHY

I. ORIGINAL WORKS. Over a period of half a century Fisher turned out an average of one paper every two months. The best listing of these is M. J. R. Healy's, in *Journal of the Royal Statistical Society,* **126A** (1963), 170–178. The earliest item, in *Messenger of Mathematics,* **41** (1912), 155–160, is a brief advocacy of the method of maximum likelihood for curve fitting; and the last, in *Journal of Theoretical Biology,* **3** (1962), 509–513, concerns the use of double heterozygotes for seeking a statistically significant difference in recombination values. Key theoretical papers are "On the Mathematical Foundations of Theoretical Statistics," in *Philosophical Transactions of the Royal Society,* **222A** (1922), 309–368; "The Theory of Statistical Estimation," in *Proceedings of the Cambridge Philosophical Society,* **22** (1925), 700–725; and "The Statistical Theory of Estimation," in *Calcutta University Readership Lectures* (Calcutta, 1938). A publication of unusual interest is "Has Mendel's Work Been Rediscovered?" in *Annals of Science,* **1** (1936), 115–137, in which Fisher finds that, probabilistically, some of Mendel's celebrated pea results were a little too good to be true, but in which he also shows that Mendel's insight and experimental skill must have been outstandingly fine. Forty-three of Fisher's most important earlier papers are reproduced in facsimile, with the author's notes and corrections, in *Contributions to Mathematical Statistics* (New York, 1950).

Fisher published five holograph books: *Statistical Methods for Research Workers* (London, 1925; 13th ed., 1958), translated into five other languages; *The Genetical Theory of Natural Selection* (London, 1930; 2nd ed., New York, 1958); *The Design of Experiments* (London, 1935; 8th ed., 1966); *The Theory of Inbreeding* (London, 1949; 2nd ed., 1965); and *Statistical Methods and Scientific Inference* (London, 1956; 2nd ed., 1959). The statistical tables in his first book were subsequently expanded and published separately, with F. Yates, as *Statistical Tables for Biological, Agricultural, and Medical Research* (London, 1938; 6th ed., 1963).

II. SECONDARY LITERATURE. *Journal of the American Statistical Association,* **46** (1951), contains four informative papers by F. Yates, H. Hotelling, W. J. Youden, and K. Mather written on the occasion of the twenty-fifth anniversary of the publication of *Statistical Methods.* The principal commemorative articles published after Fisher's death will be found in *Biometrics,* **18** (Dec. 1962) and **20** (June 1964); *Biographical Memoirs of Fellows of the Royal Society of London,* **9** (1963), 92–129; *Journal of the Royal Statistical Society,* **126A** (1963), 159–170; and *Science,* **156** (1967), 1456–1462. The last contains "An Appreciation" by Jerzy Neyman, who differed with Fisher on several issues, and is therefore of special interest. The writings of Neyman and E. S. Pearson should be consulted for further information on controversial matters. Fisher's contributions to discussions at various meetings of the Royal Statistical Society are also illuminating in this regard (they are indexed in the Healy bibliography). A good philosophical study of statistical reasoning, with particular reference to Fisher's ideas, is Ian Hacking's *Logic of Statistical Inference* (London, 1965).

NORMAN T. GRIDGEMAN

FITCH, ASA (*b.* Salem, New York, 24 February 1809; *d.* Salem, 8 April 1879), *economic entomology, medicine.*

Born into a prominent old Connecticut family that had resettled in eastern New York, Fitch was named for his father, a distinguished physician and farmer. Despite a professional family background, young Asa had an erratic and disconnected education—due in part to the limited facilities that Salem afforded—and spent some time in search of a suitable career. In 1826, almost by chance, he came upon an announcement for the Rensselaer School in nearby Troy, headed by Amos Eaton. The school had an all-science curriculum and was the only one of its kind in the country.

Fitch was drawn to this new venture in scientific education and enrolled in its second class. Indeed, he came just in time to be admitted as a participant in another of Eaton's experiments in scientific education—a traveling school of science, set up on a barge on the newly opened Erie Canal, for the observation of geological formations and the collection of specimens. Fitch became the youngest member of a group of twenty men of all ages and conditions, including Joseph Henry and a son of Governor DeWitt Clinton. On this tour, which is recorded vividly in Fitch's diary, the young man already displayed an interest in and inclination toward the study of insects, which was to be his major occupation. There followed a year at the Rensselaer School, spent in the pursuit of a somewhat haphazard course that was at once all-embracing yet limited in scope and content.

Still in search of a career, Fitch next devoted himself to the study of medicine, attending lectures at medical schools in New York City, Albany, and Castleton, Vermont, and capping it with an apprenticeship to a practicing physician. He served briefly as an assistant professor of natural history at the Rensselaer School, then traveled to the Illinois frontier. Here, at Greenville, he spent an unhappy winter in 1830–1831, seeking to establish himself in the joint pursuit of medicine and science. Unsuccessful, he returned to his home state, where he remained for the rest of his life. Fitch's interest in medicine apparently was secondary to his zeal for natural history, acquired under Eaton's inspiration. He retired to the family farm, giving up medicine for agriculture. With it, however, he combined the assiduous collection and study of insects, especially in respect to their injurious or beneficial effects upon crops.

Fitch, who became known as the "Bug Catcher of Salem," began publishing reports about insects in 1845. Between 1854 and 1870 he received modest financial grants from New York State for his work, and thus he was, perhaps informally, the first entomologist in the service of a state. His numerous reports, published regularly in the *Transactions of the New York State Agricultural Society,* were widely circulated and acknowledged for their combination of sound scientific knowledge of insect life cycles with the conditions and problems of agriculture. From his obscure rural home in upstate New York, Fitch carried on a wide correspondence. His achievement, stemming from Eaton's zeal for applied science, laid the foundation of economic entomology as an American science. Entomology subsequently acquired a more professional character; but Fitch's role in it was perhaps the epitome of early American science, primitive but practical and dedicated.

BIBLIOGRAPHY

I. ORIGINAL WORKS. Asa Fitch's lifelong diary, begun at the age of twelve, is preserved in MS form in the Yale University Library. Many of his notebooks are in the United States National Museum. His writings on entomology began to appear in 1845 in the *American Quarterly Journal of Agriculture* and subsequently in the *Transactions of the New York State Agricultural Society.* Particularly to be mentioned is a series of fourteen "Reports on the Noxious, Beneficial and Other Insects of the State of New York," in *Transactions,* **14–30** (1855–1872). Fitch also prepared "An Historical, Topographical, and Agricultural Survey of Washington County," in *Transactions,* **8–9** (1849–1850). A full bibliography of his entomological work is in J. A. Lintner, *First Annual Report on the Injurious and Other Insects of the State of New York* (Albany, 1882), pp. 289–325.

II. SECONDARY LITERATURE. Aside from brief biographical sketches, especially in *Dictionary of American Biography,* III, 424, there is no full-length study of Fitch, and very little published about him. Fitch's diary is the basis of Samuel Rezneck, "A Traveling School of Science on the Erie Canal in 1826," in *New York History,* **40** (July 1959), 255 ff.; "Diary of a New York Doctor in Illinois, 1830–31," in *Journal of the Illinois Historical Society,* **54** (1961), 25 ff.; see also D. L. Collins, "The Bug Catcher of Salem," in *New York State Bulletin* (March 1954).

SAMUEL REZNECK

FITTIG, RUDOLPH (*b.* Hamburg, Germany, 6 December 1835; *d.* Strasbourg, Alsace [France], 19 November 1910), *chemistry.*

Fittig was the son of Johann Andreas Fittig, the director of a private school in Hamburg. He wished to become a teacher, and at the age of sixteen he taught in a private school. In April 1856 he entered the University of Göttingen, intending to become a teacher of natural science, with special attention to botany. The father of one of his school friends planned to open a dye factory and suggested to young Fittig that he might be employed there. Fittig was interested and therefore took up the study of chemistry, which was then taught at Göttingen by Heinrich Limpricht. He soon decided to make this his career and became assistant to Limpricht in the laboratory and to Friedrich Wöhler, who was still lecturing. In 1858 Fittig received his doctorate and in 1860 became a *Privatdozent* at the university. He married in 1864 and had three sons and three daughters. His wife died while the children were still young, and he raised them by himself.

In 1866 Fittig became extraordinary professor of chemistry and worked closely with Wöhler. At this time he established a friendship with Friedrich Beilstein. In 1865, with Beilstein and Hans Hübner, he took over the editorship of the *Zeitschrift für Chemie und Pharmacie,* which had been edited since 1859 by Emile Erlenmeyer but had lost nearly all its subscribers. The three new editors produced a more successful journal under the title *Zeitschrift für Chemie,* and the venture lasted until 1871. In 1870 Fittig became professor of chemistry at the University of Tübingen, where he remained until he replaced Adolf von Baeyer as professor of chemistry at the University of Strasbourg in 1876. Here he constructed a new chemical laboratory, begun in 1877 and completed in 1882. Fittig served as rector of the university in 1895–1896; he retired in 1902, but continued to publish the results of his researches until nearly the end of his life. During his last years he was made an honorary member of many chemical societies.

Fittig was a prolific author and editor. Besides his activities on the *Zeitschrift für Chemie* he served as an associate editor of the *Annalen der Chemie* from

1895 to 1910, wrote a massive textbook of chemistry which appeared in 1871 and went through a number of editions, and edited the tenth edition of Wöhler's textbook of organic chemistry in 1877. His bibliography lists 399 research papers. Fittig trained many chemists who subsequently became well-known, including a number of Englishmen and Americans. Among the best-known of these were William Ramsay, noted for his work on the inert gases, who received his degree in 1872; and Ira Remsen, whose doctorate was conferred in 1870, and who worked on saccharin and was later president of Johns Hopkins University.

Fittig was essentially an experimentalist, with little interest in theoretical chemistry. He was active at a time when the structural theory of organic chemistry was producing its most striking results, and his extensive studies on preparative organic chemistry contributed much to this development. For his doctoral dissertation he studied the action of sodium on anhydrous acetone, in the course of which work he discovered pinacol. This utilization of sodium in an organic reaction probably led him to extend the studies begun by Wurtz on the reaction of sodium with organic halogen compounds. The action of sodium on benzene halides led Fittig to the discovery of a number of homologous aromatic compounds, including biphenyl. This reaction is known to organic chemists as the Wurtz-Fittig reaction. Fittig was led by these studies to the investigation of other aromatic compounds, and he carried out work on mesitylene and its derivatives, naphthalene, and fluorene. He was an independent discoverer of phenanthrene in coal tar. In 1873 he proposed the quinoid structure for benzoquinone, a structure later used to explain the behavior of numerous organic dyestuffs. After 1873 Fittig worked chiefly on unsaturated acids and lactones. The extent and variety of his work helped greatly to advance the progress of organic chemistry during its period of very rapid development, and he is rightly considered one of the outstanding chemists of his day.

BIBLIOGRAPHY

I. ORIGINAL WORKS. A bibliography of the scientific papers of Fittig and his students is given in *Berichte der Deutschen chemischen Gesellschaft,* **44** (1911), 1383–1401. Among the more important papers are "Ueber das Monobrombenzol," in *Liebigs Annalen der Chemie,* **121** (1862), 361–365, in which the work on the Fittig synthesis is first described; "Ueber das Phenanthren, einen neuen Kohlenwasserstoff im Steinkohlenteer," in *Liebigs Annalen der Chemie,* **166** (1873), 361–382; and "Ueber Phenanthren und Anthracen," in *Berichte der Deutschen chemischen Gesellschaft,* **6** (1873) 167–169, in which the quinone formula is suggested.

II. SECONDARY LITERATURE. An extensive biography with many personal details is F. Fichter, *Berichte der Deutschen chemischen Gesellschaft,* **44** (1911), 1339–1383; and a shorter biography is "R. M.," in *Journal of the Chemical Society,* **99** (1911), 1651–1653.

HENRY M. LEICESTER

FITTON, WILLIAM HENRY (*b.* Dublin, Ireland, January 1780; *d.* London, England, 13 May 1861), *geology.*

Fitton was a son of Nicholas Fitton, a Dublin attorney. The family, long resident in Ireland, was descended from the Fittons of Gawsworth, Cheshire. Fitton entered Trinity College, Dublin, in 1794 and distinguished himself as a classical scholar, being awarded the senior scholarship in 1798 and graduating B.A. in 1799. In 1808 he went to Edinburgh University to study medicine and graduated M.D. in 1810. After a year or two in London he set up as a physician in the county town of Northampton in 1812 and remained there until 1820. During this period Fitton obtained other medical degrees, the M.B. and M.D. from Dublin in 1815 and the M.D. from Cambridge in 1816. In 1816 also he was admitted fellow of the Royal College of Physicians. In June 1820 he married Maria James (by which marriage he had five sons and three daughters) and, since she was a lady of some wealth, he gave up his practice and moved to London, thenceforth devoting himself to scientific, particularly geological, studies. He had been elected a fellow of the Royal Society in 1815 and had become a member of the Geological Society of London in 1816.

Fitton's interest in geology went back to his early years in Dublin, and the first volume of the *Transactions* of the Geological Society of London (1811) contained a paper about minerals found in the vicinity of Dublin, by Fitton and his friend Rev. Walter Stephens. In 1813, under the pseudonym "F," he published in Nicholson's *Journal of Natural Philosophy* two articles entitled "On the Geological System of Werner," a remarkably impartial review of Werner's theoretical views, their value, and their weaknesses. No doubt the fact that he had, while in Edinburgh, attended the lectures of Robert Jameson and T. C. Hope accounts for his being so well informed about the systems of both Werner and Hutton.

While in Northampton, Fitton began to write reviews of geological and medical books for the *Edinburgh Review.* Two published in 1817 and 1818 show that his interest in the early history of geology had been aroused. One of these in particular, his review of the publications of William Smith, directed the attention of geologists and others to the career and remarkable achievements of this still far from well-known man, who in 1815 had published a large-scale geological map of England and Wales but was not even a member of the Geological Society. Fitton made verbatim use of his reviews in some valuable articles, "Notes on the History of English Geology," which he contributed to the *Philosophical Magazine* (1832–1833). An article in the *Edinburgh Review* (1839) on Lyell's *Elements of Geology,* in which the Huttonian theory is discussed, and another on Murchison's *Silurian System* (1841), show that Fitton continued to be interested in the history of his favorite subject.

Fitton is perhaps better known for his contributions to stratigraphy and his elucidation of the succession of the Upper Jurassic and Lower Cretaceous strata of southern England. In his review of Smith's map he had endeavored to point out some of the errors, one of these being a confusion over the position in the stratigraphical succession of the beds of the Weald in southeast England. It was perhaps this which led Fitton, after his return to London from Northampton, to take a closer interest in this area, especially in Kent and Sussex. Then, in 1824, he examined sections in similar beds along the south coast of the Isle of Wight, already described by Thomas Webster in 1816.

In November 1824, Fitton published in Thomas Thomson's *Annals of Philosophy* a masterly account of the beds found below the Chalk in southeast England. Entitled "Inquiries Respecting the Geological Relations of the Beds Between the Chalk and the Purbeck Limestone in the South-east of England," it was accompanied by a colored geological map of the southern half of the Isle of Wight and geological sections showing the succession both there and on the neighboring coast of Dorset. In this article Fitton elucidated and corrected the errors which had arisen through the existence of different beds of ferruginous sands, with clays between.

Although it was some time before a final nomenclature was accepted, it was this paper which made it clear that there was an "Upper Greensand" just below the Chalk, with "Gault Clay" below it, as both William Smith and Thomas Webster had already recognized. Below this was the "Lower Greensand," and all three formations, Fitton pointed out, contained marine fossils. He also showed that the Weald Clay, containing many freshwater fossils, was present beneath the Lower Greensand not only in the Weald but also in the Isle of Wight and on the Dorset coast, and beneath these beds there were ferruginous sands, also with freshwater fossils (which clearly distinguished them from the Lower Greensand). These he named the Hastings Sands, after the town on the Sussex coast where they were well exposed. By providing a tabular "List of Strata From the Chalk to the Hastings Sands, With Synonimes [*sic*] of Different Geologists," he made it clear just what errors had arisen.

Fitton continued to study the Lower Cretaceous and Upper Jurassic strata, both in England and in northern France, and in 1836 published a monograph entitled "On Some of the Strata Between the Chalk and the Oxford Oolite (e.g., Corallian) in the South-east of England" (*Transactions of the Geological Society,* 2nd ser., **4**). This covered much the same field as his earlier paper, but was far longer and embodied many detailed observations, including a number on rocks exposed between Dorset and the Norfolk coast. Accompanied by maps and sections, it is regarded by geologists as a classic contribution to stratigraphy.

Later, between 1843 and 1847, he demonstrated the existence of a clay, the Atherfield Clay, at the base of the Lower Greensand but clearly different from the Weald Clay because it contained marine fossils and marked the readvance of the sea after a long nonmarine episode; he also provided detailed descriptions of an important section along the southwest coast of the Isle of Wight and discussed the equivalents of the Wealden beds in Europe.

Fitton's only separately published work was a small volume entitled *A Geological Sketch of the Vicinity of Hastings* (London, 1833), which gives a good account of the Wealden strata, with sections and a bibliography. He was always ready to instruct naturalist travelers in the principles of geology, and he took a particular interest in the geology of Australia. His last published paper, in the *Proceedings of the Geographical Society* in 1857, was entitled "On the Structure of North-west Australia."

Fitton was secretary of the Geological Society of London from 1822 to 1824 and president from 1827 to 1829. He was largely instrumental in establishing its printed *Proceedings,* which first appeared in 1827. In this publication the presidential addresses were printed, Fitton's being the first to appear, commencing that valuable series in which the progress of geology was annually reviewed. He served as vice-president of the society from 1831 to 1846 and in 1852 was awarded the Wollaston Medal, the society's premier award, in recognition of his important contribution to stratigraphy.

BIBLIOGRAPHY

I. Original Works. Except for his *Geological Sketch of the Vicinity of Hastings* (London, 1833), all of Fitton's contributions to geology appeared in scientific periodicals and are listed in Agassiz's *Bibliographia zoologiae et geologiae* (London, 1850) and the *Catalogue of Scientific Papers (1800–1863)* published by the Royal Society (London, 1868). Both should be consulted.

II. SECONDARY LITERATURE. An obituary notice in the *Quarterly Journal of the Geological Society of London,* **18** (1862), xxx–xxxiv, provides the fullest information about Fitton's life and work; other accounts, all brief, appear to be derived from it. W. Munk's *The Roll of the Royal College of Physicians,* III (London, 1878), 154, gives some additional facts. Biographies of his geological contemporaries do not throw much light on his activities, except for *The Journal of Gideon Mantell,* E. C. Curwen, ed. (London, 1940). J. Challinor, "Some Correspondence of Thomas Webster, Geologist (1773–1844)," in *Annals of Science,* **17–20** (1961–1964), includes twenty-seven letters from Fitton, and there are also interesting references to him in the other letters. A catalog of the sale of his library in 1856 is in the British Museum library.

JOAN M. EYLES

FITZGERALD, GEORGE FRANCIS (*b.* Dublin, Ireland, 3 August 1851; *d.* Dublin, 21 February 1901), *physics.*

FitzGerald was one of the initial group, which included Heaviside, Hertz, and Lorentz, that took Maxwell's electromagnetic theory seriously and began to explore its consequences. Very few others used Maxwell's theory to obtain results beyond those investigated by Maxwell himself. Among the first attempts to use the theory for such results was FitzGerald's paper, "Electromagnetic Theory of the Reflection and Refraction of Light," which Maxwell reviewed for the *Philosophical Transactions,* noting that it related to work carried out by H. A. Lorentz.

It is ironic that FitzGerald is best known for work that was probably of minor importance to him and was outside his work in electromagnetic theory. Together with Lorentz he is credited with being the first to explain the null results of the Michelson-Morley experiment as due to the contraction of an arm of the interferometer, which resulted from its motion through the ether. FitzGerald's ideas on the subject were published in *Science* (1889), and he also discussed the contraction hypothesis with Oliver Lodge. In a paper presented to the Physical Society in May 1892, Lodge commented, "Professor FitzGerald has suggested a way out of the difficulty by supposing the size of bodies to be a function of their velocity through ether."

In 1894 Lorentz wrote to FitzGerald about the hypothesis, and inquired whether he had indeed published on it. In his reply, FitzGerald mentioned his letter to *Science,* but at the same time admitted that he did not know if the letter had ever been printed and that he was "pretty sure" Lorentz had priority. Soon Lorentz began to refer to FitzGerald in his discussions.

Only after FitzGerald's death did English physicists begin to take any further notice. Thus in his Adam's Prize essay, published as *Aether and Matter,* Larmor discussed the Michelson-Morley experiment and the contraction effect in detail, but only Lorentz was mentioned in this connection. Two years later, Larmor, in his introduction to FitzGerald's papers, claimed priority for FitzGerald on the contraction effect. E. T. Whittaker, in the *History of the Theories of Aether and Electricity,* states that Lorentz obtained the hypothesis from FitzGerald; but it appears that Lorentz' concept was independent of FitzGerald's and that he was just giving due credit to FitzGerald.

A further piece of evidence gives additional weight to the argument that the contraction effect was not an important issue with FitzGerald. He carried on an extensive correspondence with Heaviside from 1888 to 1900, in which they discussed many major problems of the physics of the period. In all the surviving correspondence, the Michelson-Morley experiment is mentioned only once. The interest in the Michelson-Morley experiment from the time of the experiment until the development of the theory of relativity has perhaps been exaggerated.

We gather some insight into FitzGerald's view of his work from a letter to Heaviside dated 4 February 1889:

> I admire from a distance those who contain themselves till they worked to the bottom of their results but as I am not in the very least sensitive to having made mistakes I rush out with all sorts of crude notions in hope that they may set others thinking and lead to some advance.

The view is of a speculator, a scientist who generates ideas but does not necessarily develop them. Although FitzGerald's papers contain many examples of sound development, this description seems fair. In the same letter he was excited by Heaviside's work on the electromagnetic field caused by a moving charged sphere:

> I am very glad to hear that you have solved completely the problem. . . . I was anxious to find out how much energy is lost by the earth owing to its magnetisation rotating and going round the sun. . . . to what extent the energy of motion of molecules could be attributed to electrical charges on them and how this part of their energy would be radiated; this might lead to a theory of forces between molecules. . . . You ask what if the velocity be greater than that of light? I have often asked myself that but got no satisfactory answer. The most obvious thing to ask in reply is "Is it possible?"

In an 1893 letter to Heaviside, this speculation is extended supporting this view of his work:

> . . . have you considered what would be the extra mass of an atom owing to its atomic charge? A charge of electricity or magnetism acts like an added mass . . . this should interfere with Kepler's Laws. . . .

It is worth noting that FitzGerald took seriously his responsibilities as a teacher, both in Trinity College, Dublin, where he spent his academic life as student and teacher, and throughout Ireland through the boards he served on.

BIBLIOGRAPHY

See J. Larmor, ed., *The Scientific Writings of the Late George Francis FitzGerald* (Dublin, 1902).

FitzGerald's communication to *Science* concerning the Michelson-Morley experiment was published as "The Ether and the Earth's Atmosphere," **13** (1889), 390. Steven Brush, "Note on the History of the FitzGerald-Lorentz Contraction," in *Isis,* **58,** no. 2 (1967), 230–232, gives a full account and reproduces the original article and the correspondence between FitzGerald and Lorentz.

ALFRED M. BORK

FITZROY, ROBERT (*b.* Ampton Hall, Suffolk, England, 5 July 1805; *d.* Upper Norwood, London, England, 30 April 1865), *hydrography, meteorology.*

Fitzroy was the second son, by a second marriage, of Lord Charles Fitzroy; his paternal grandfather was Augustus Henry, third duke of Grafton, and his maternal grandfather was the first marquis of Londonderry. As a member of an aristocratic family noted for its association with seafaring, he entered the Royal Naval College at Portsmouth in 1819. He achieved the rank of lieutenant in 1824 and sailed on British naval vessels plying Mediterranean and South American waters.

Fitzroy received his first full command of a ship in 1828, when he was placed in charge of the *Beagle.* Under the command of Captain Philip Parker King, the *Beagle* and the *Adventure* had left England in 1826 with orders to survey the southern coasts of South America. Upon the death of the *Beagle*'s original commander, Pringle Stokes, Fitzroy was asked to complete the hydrographical tasks assigned to the ship. This first *Beagle* voyage to South America ended in the fall of 1830 with the expedition's return to England. During the summer of 1831, the *Beagle* was readied for her second surveying voyage to South America. Fitzroy, as the expedition's commander, chose the young Charles Darwin to accompany him on what was to be one of the most famous scientific expeditions in history. Fitzroy was promoted to captain in July 1835. Upon completion of the work in South America, the expedition returned to England in 1836 and Fitzroy married Mary Henrietta O'Brien.

After publishing the narrative of the two South American expeditions with which he was associated, Fitzroy sought a seat in Parliament and was elected a member for Durham in 1841. His short political career, marked by a violent public quarrel with his chief rival, was interrupted by Fitzroy's appointment to the governorship of New Zealand (1843). The New Zealand interlude proved disastrous for Fitzroy, even though he cannot be held personally accountable for all the troubles that arose during his term of office. By 1845 he had returned home, and in 1848 he was named superintendent of the dockyard at Woolwich. Before his retirement from active naval service in 1850, Fitzroy commanded the navy's first screw-driven steamship, the *Arrogant,* during her initial trial runs. Thus ended Fitzroy's active naval career, but he continued to rise in rank by reason of his seniority—becoming rear admiral in 1857 and vice admiral in 1863.

In the years of his retirement, Fitzroy turned his attention to the emerging science of meteorology and devoted all of his energies to the advancement of its practical aspects. His total absorption in his meteorological work—and his extreme sensitivity to criticisms of it—may have contributed to the mental illness that ended in his suicide.

Fitzroy deserves to be noted in the history of nineteenth-century science because of his association with Charles Darwin and because of his contributions to the fields of hydrography and meteorology.

Darwin, in his *Autobiography,* wrote: "The voyage of the *Beagle* has been by far the most important event in my life and has determined my whole career" (*The Autobiography of Charles Darwin,* Nora Barlow, ed. [New York, 1959], p. 76). Fitzroy, who was more concerned with science than were many naval officers of his day, made it possible for Darwin to visit tropical lands and study their flora, fauna, and geology. The two men shared the same cabin and Fitzroy was attentive to the scientific needs and interests of the young Darwin. Fitzroy's violent temper and his conservative opinions on religion and slavery were responsible for some disagreements between them, but Fitzroy and Darwin remained on friendly terms. So close was their relationship that on the journey home they sent a joint letter to a South African newspaper defending English missionary activity in the South Pacific.

Later in life Fitzroy became worried about the deleterious effect of scientific advances upon religious beliefs. As a result, he joined the opposition to Darwin's *Origin of Species.* At the famous Oxford meeting of the British Association for the Advancement of Science (1860), when Darwinism was the issue in a hot vocal debate, Fitzroy took a public stand against organic evolution. He told the audience he regretted Darwin's publication of the *Origin* and announced his refusal to accept the book as a logical arrangement of the facts of natural history. Darwin, who was not personally involved in the debate, preferred to remember his earlier, happier association with Fitzroy and always spoke kindly of his *Beagle* companion.

While Darwin made his observations in South America and collected his specimens, Fitzroy surveyed the southern coast of that continent. The years of the second *Beagle* voyage marked the beginning of a half-century of supremacy of British hydrography. In the period 1829–1855 Britain's Hydrographic Department was directed by her greatest hydrographer, Sir Francis Beaufort, who sent out some 170 major surveying expeditions. Through these expeditions the Admiralty was able to assemble a collection of accurate charts covering most of the earth's coastlines (with the exclusion of northeast Asia and Japan).

Beaufort ordered Fitzroy to continue the South American charting program begun by King in 1826. Fitzroy and his staff of surveyors furnished the Admiralty with eighty-two coastal sheets, eighty plans of harbors, and forty views covering portions of Patagonia, Tierra del Fuego, Chile, and Peru.

In addition to its surveying equipment, the *Beagle* was supplied with twenty-two chronometers. Utilizing these instruments, Fitzroy established a chain of meridian distances through the Pacific, Indian, and Atlantic oceans. That is, relying upon the accuracy of his chronometers and the celestial observations he made on the voyage, Fitzroy was able to make precise determinations of longitude at a series of positions around the globe.

The surveys he carried out in South American waters established Fitzroy as a first-rate hydrographer and won for him the gold medal of the Royal Geographical Society (1837). Because his marine surveys were accurate to such a high degree they are still used as the foundation for a number of charts of that area.

In 1851 Fitzroy was elected a fellow of the Royal Society of London on the strength of his contributions to hydrography and scientific navigation. These events marked his entrance into another field of scientific endeavor: meteorology.

Ever since his *Beagle* days Fitzroy had shown an interest in the study of the weather. Therefore, when the British government created (1855) the Meteorologic Office, instructed to gather weather information for shipping, it was not surprising that the Royal Society should ask that Fitzroy be placed in charge of it.

While a committee of the Royal Society deliberated about the exact nature of the work to be done by the Meteorologic Office, Fitzroy contacted the ship captains who would make meteorological observations for him. He was not satisfied merely to amass weather information; he wanted to warn sailors and others of approaching weather changes. He began by making available cheap barometers, with accompanying instructions, to coastal fishermen. His next move was to set up a series of stations which would telegraph weather data to the Meteorologic Office in London. Using this data, Fitzroy produced some of the first weather charts and began issuing weather forecasts, a term he helped to popularize. By the end of 1860 the *Times* was printing daily weather forecasts and Fitzroy was gaining recognition as the man who could predict coming weather conditions. At the height of his career as a pioneer forecaster, Fitzroy published a large, introductory *Weather Book* (1863) that summarized his meteorological work.

Fitzroy's position at the Meteorologic Office was clearly defined as "statist," that is, collector of weather information. Forecasting had carried him somewhat beyond his original instructions, and he soon encountered light criticism from the public when his forecasts failed—as they often did—and heavier criticism from scientists who argued that his whole approach to meteorology was marked by an excessive reliance upon the empirical at the expense of theoretical developments. These critics wanted the general theories of the science of meteorology established before weather predictions were made public. In the midst of this controversy Fitzroy took his own life. At his death the theorists won out, but public demand assured that at least his storm warnings would be continued. In retrospect it is clear that Fitzroy's difficulties arose more from the state of the new science of meteorology than from some fundamental failing of his own.

BIBLIOGRAPHY

I. Original Works. Fitzroy's account of the two *Beagle* voyages is to be found in the first two volumes of his *Narrative of the Surveying Voyages of His Majesty's Ships Adventure and Beagle, Between the Years 1826 and 1836,* 3 vols. (London, 1839). Vol. III of the *Narrative* was written by Charles Darwin and has frequently been reprinted as a separate volume. The results of Fitzroy's hydrographical work in South America were incorporated into a series of Admiralty charts issued by the Hydrographic Department. From the *Beagle* period there are also the Fitzroy-Darwin letter on missionary activity, "A Letter Containing Remarks on the Moral State of Tahiti, New Zealand, etc.," in *South African Christian Recorder,* **2,** no. 4 (Sept. 1836), 221–238; and Fitzroy's *Sailing Directions for South America* (London, 1848).

The meteorological contributions of Fitzroy can be studied in publications he wrote for the Board of Trade: *Annual Report of the Meteorologic Office* (London, 1856–1864) and *Barometer and Weather Guide* (London, 1858; final ed., 1877). Fitzroy's *The Weather Book: A Manual of Practical Meteorology* (London, 1863) was intended for a wider audience.

MSS relating to Fitzroy's career as meteorologist are located in the Meteorological Office Archives (London Road, Bracknell, Berkshire). They include correspondence, memoranda, and correspondence books. Other Fitzroy MSS can be found in the J. F. W. Herschel Collection of the Royal Society and at the National Maritime Museum, Greenwich.

II. Secondary Literature. H. E. L. Mellersh, *Fitzroy of the Beagle* (London, 1968), is a recent popular but reliable guide to Fitzroy's life. Darwin's remarks on Fitzroy and the *Beagle* voyage are found in Nora Barlow, ed., *The Autobiography of Charles Darwin* (New York, 1959), pp. 71–76. Other aspects of the Darwin-Fitzroy relationship can be pursued in the extensive published material dealing with Darwin's early years.

For Fitzroy as hydrographer see Sir Archibald Day, *The Admiralty Hydrographic Service, 1795–1919* (London, 1967), *passim;* G. S. Ritchie, *The Admiralty Chart* (London, 1967), pp. 183–190, 203–210, 215–220; and H. P. Douglas, "Fitzroy's Hydrographic Surveys," in *Nature,* **129,** no. 3249 (6 Feb. 1932), 200.

Fitzroy's meteorological contributions are discussed in Sir David Brunt, "The Centenary of the Meteorological Office: Retrospect and Prospect," in *Science Progress,* **44,** no. 174 (Apr. 1956), 193–195; Sir Napier Shaw, *Manual of Meteorology: Volume I. Meteorology in History* (Cambridge, 1932), pp. 149–153, 302–303, 311; and *Selected Meteorological Papers* (London, 1955), pp. 236–237; and Roger Prouty, *The Transformation of the Board of Trade* (London, 1957), pp. 52–54.

George Basalla

FIZEAU, ARMAND-HIPPOLYTE-LOUIS (*b.* Paris, France, 23 September 1819; *d.* Venteuil, near Jouarre, France, 18 September 1896), *experimental physics.*

Fizeau was the eldest son of a large and relatively wealthy family that had come to Paris from the Vendée. His father held the chair of internal pathology at the Paris Faculty of Medicine from 1823. Fizeau, aspiring to follow in his father's footsteps, began medical studies at the Collège Stanislas, but because of poor health he was obliged to interrupt his education in order to travel to a more agreeable climate. On returning to Paris, he gave up medicine and began an entirely new career in the physical sciences. At the Collège de France, he studied optics with H.-V. Regnault, and he followed the lectures given at the École Polytechnique through the notebooks compiled by one of his brothers. Fizeau's most fruitful educational experience, however, was the course of study he took at the Paris observatory under the tutelage of the famous astronomer François Arago. Arago recognized Fizeau's promise, encouraged his scientific endeavors, and brought his work to the attention of the Academy of Sciences.

Fizeau was made a member of the section of physics of the Academy of Sciences on 2 January 1860. He was vice-president of his section for 1877 and president for 1878. In 1878 he was elected to the Bureau of Longitudes. On 9 July 1856 the five academies constituting the Institut de France awarded him the Triennial Prize, a special award that had just been created by the emperor. From the Royal Society he received the Rumford Medal in 1866 and the title of foreign member in 1875.

On 19 August 1839 Arago made public a description of a new process of "light painting" or heliography that had been invented by L.-J.-M. Daguerre. The daguerrotype, as the result of this process soon came to be called, was a crude forerunner of the modern photograph. Fizeau's earliest work in science was an attempt to improve Daguerre's process and to make the heliograph an instrument of science. He showed that by covering the surface of the developed plate with a salt of gold, oxidation of the surface chemicals could be prevented and the contrasts between light and dark could be considerably heightened. He is often credited with the first use of bromine vapors to hasten the development of the photographic image, but this seems uncertain.[1] Fizeau also introduced a widely used but unpatented method for turning a photograph into a photoetching.

During this early period of his career, Fizeau often worked in collaboration with Léon Foucault, a young man who had also begun his education in medicine. Foucault was one of the most adept mechanicians of his age, and had he been able to tolerate the sight of blood, he might have become a great surgeon. Together the two young scientists worked on the improvement of photographic images, and in 1845 they opened a new and fruitful area of astronomy by taking what were probably the first clear photographs of the sun's surface.[2]

From 1844 Fizeau and Foucault undertook a series of precise and mechanically ingenious optical experiments that would ultimately have a profound effect on the course of physics. By the middle of the nineteenth century, most scientists had come to accept the wave theory of light, formulated near the beginning of the century by Thomas Young and Augustin Fresnel. There remained, however, several gaps in the investigation of the experimental consequences of the theory. For example, in the study of interference fringes produced by two rays of light issuing from the same source, only several dozen fringes on each side of the central band had been observed.

By analyzing the white light source into simpler constituents by means of a spectroscope, Fizeau and Foucault were able to observe fringes produced by interfering light rays with a difference of travel equal to more than 7,000 wavelengths, thus showing that light waves, like sound waves, remain geometrically constant over a large number of periods. But light waves, because of their transverse vibrations, are more complex than sound waves. Light can assume different forms or planes of vibration as well as different intensities. Using the same spectroscopic apparatus as in the preceding experiment, Fizeau and Foucault observed the interaction of two rays produced by passing a single polarized ray through a birefringent crystal. In this case, instead of obtaining alternating bands of light and dark, they obtained bands of light periodically polarized in different planes of vibration.

In 1800 William Herschel, the British astronomer, discovered a form of invisible radiation above the red end of the spectrum which produced a heating effect. The infrared rays (or calorific rays as they were usually called) were shown to follow most laws that had been established for visible light. By using extremely small and delicate thermometers, Fizeau and Foucault demonstrated that calorific rays could produce interference fringes like those produced by visible light, except that instead of appearing as alternating bands of light and dark, the fringes produced by infrared rays appeared as alternating bands of hot and cold.

One of the most important consequences of the wave theory of light that had not yet been demonstrated was that light traveled more slowly in dense than in rare mediums. In 1838 Arago had suggested using a rapidly rotating mirror for the purpose of this demonstration, a technique employed by Charles Wheatstone in 1834 in an unsuccessful attempt to measure the speed of electricity. In principle, Arago's idea was very simple. A narrow ray of light would be directed into a mirror rotating as rapidly as possible. The light from this mirror would be reflected back over the same path by a fixed mirror placed at a considerable distance. By the time the light had returned, the rotating mirror, having suffered a small angular displacement, would deflect the light off at an angle to the original path. If, in addition, the light returning from the fixed mirror were divided into two rays and one of them were sent through a tube of water, it would then be possible to establish directly, and without having to measure the absolute speeds of the two rays, whether the light had been slowed by its passage through the denser medium. Theoretically, the ray that had come through the tube filled with water would arrive at the rotating mirror a fraction of a second after the ray that had come through the air and would thus be deflected at a greater angle.

In practice, the essential problem was to arrange an optical system such that the narrow, intermittent rays of light were not dispersed in their passage between the two mirrors. After considerable trial, Fizeau and Foucault found a workable system, the essential element of which was a convergent reflector of convenient focal length. Unfortunately, once having solved this problem, the two experimenters broke up their partnership over a personal dispute. Each continued to work on the experiment. On 6 May 1850 two papers on the relative speeds of light in water and in air were submitted to the Academy of Sciences, one signed by Foucault, the other by Fizeau and Louis Bréguet. With almost identical apparatus, the two experiments had yielded substantially the same result. Light did, indeed, travel faster in air than in water.

Fizeau was not satisfied merely with determining the relative velocities of light. He wanted to measure with some precision the absolute velocity. In 1849 he had conceived an ingenious mechanism that would enable him to achieve his goal: a large toothed wheel was spun rapidly about its axis, and a beam of light sent through the spaces between the teeth was reflected back to its source by a fixed mirror. When the wheel was rotated rapidly enough, the intermittent light rays returning from the mirror intersected the path of the teeth and thus became invisible to the observer stationed behind the wheel. As the mechanism was turned faster and faster, the light reappeared and disappeared alternately. The time required for the light to travel through the carefully measured distance was a simple function of the angular displacement of the wheel.

In 1849 Fizeau made a trial of his new method between his father's house at Suresnes and Montmartre. The figure he obtained for the speed of light (about 315,000 kilometers per second) was not quite as accurate as the results of astronomical calculations, but the practicability of the method was established and became the basis of the more precise determinations made by Alfred Cornu in the 1870's.

By substituting for teeth alternating bands of conducting and nonconducting materials, Fizeau attempted with little success to adopt his mechanism to the measurement of the speed of electricity. (A galvanometer, of course, replaced the eye of the observer.) In 1849 Fizeau tried the experiment with the engineer E. Gounelle, but the results were indecisive because of the complex way in which electricity is propagated through a conductor.

In 1848 Fizeau published a paper that was to have a profound effect on the future of astrophysics. He showed that when a body emitting a continuous sound of unvarying frequency is moved, the sound waves do not dispose themselves symmetrically about the source. In front they come at shorter intervals, producing the effect of higher pitch; from behind the frequency appears lowered because of the larger interval between wave crests. (The sound of a passing railroad train is the classic example of this phenomenon.) Fizeau saw the implications of this principle for optics. A moving light source would undergo an analogous change in frequency. From behind, the light waves would be shifted toward the red end of the spectrum; from the front they would be shifted toward the violet.

Unknown to Fizeau, a physicist from Prague, Christian Doppler, had published a paper on exactly the same subject in 1842, six years before the appearance of Fizeau's work. Doppler, however, failed to understand correctly some of the consequences of his own idea. He supposed that light coming from a star moving relative to the earth would experience a change in color.[3] He was apparently unaware of the invisible radiations at the red and violet ends of the spectrum that would, by their shift into the visible range of the spectrum, compensate for the disappearance of any colored rays. The Doppler-Fizeau effect became useful in astronomy only after the work of Gustav Kirchhoff and Robert Bunsen showing that incandescent elements emit discrete frequencies of light. It then became possible, by measuring the shift in the spectra produced by the various elements in a given star, to ascertain the velocity of that star relative to the earth. The first such measurement was made in 1868 by the British astronomer William Huggins, and since then the technique has provided the science of astrophysics with one of its most important tools for measuring the size and structure of the universe.

Nearly all scientists in the nineteenth century believed that some sort of luminiferous ether filled the universe and provided the medium for the propagation of light. One of the many problems that arose with respect to the nature of the ether was whether it could participate in the motion of ponderous matter. In 1818 Fresnel discussed this question in a famous letter addressed to Arago. He assumed that bodies carried with them only as much ether as they contained in excess of that which was present in an equal volume of space void of all ponderous matter. By assuming in addition that the excess of ether contained in a given body of matter was proportional to its refractive index, Fresnel deduced that the percentage of a body's motion that could be communicated to light was equal to $1 - \frac{1}{n^2}$, where n represents the refractive index.

In 1851 Fizeau found a way of overcoming the seemingly impossible difficulty of measuring the small increment in the velocity of light that in theory would be produced by a body in motion. His method was simply to produce interference fringes from rays of light that had passed through two parallel tubes containing a fluid moving in opposite directions. Even a relatively small difference in the velocity of the two light rays would cause a perceptible displacement of the interference fringes. Using air as the test medium, Fizeau discerned no change in the speed of light, a result that he expected because the refractive index of air is equal almost to one. With water, however, the velocity of light was altered by an amount that accorded reasonably well with Fresnel's formula. In 1886 A. A. Michelson and E. W. Morley repeated the experiment on a larger scale and confirmed Fizeau's results.

In a world that was becoming increasingly professional, Fizeau was one of the last great amateurs of science. He was able to employ his personal wealth and virtually unlimited leisure in pursuit of his scientific researches. Except for the Doppler-Fizeau effect, he made no direct contributions to optical theory, but the ingenious experimental techniques that he invented were to supply an invaluable aid to the creation in this century of a new optics. That his reputation has not equaled his deeds is largely because once he had invented a new experimental method, he left it to his followers and collaborators to develop and perfect. Foucault went on to employ the rotating-mirror device in precise measurements of the speed of light; Cornu perfected the toothed-wheel mechanism for the same purpose; and much of the career of A. A. Michelson was built on Fizeau's unfinished business.

Fizeau married a daughter of the famous botanist Adrien de Jussieu. She died early in the marriage, after having given birth to two daughters and a son. After his wife's death, Fizeau retired to his home near Jouarre and came to Paris only rarely, to attend meetings of the Academy of Sciences and the Bureau of Longitudes. He died of cancer of the jaw just five days before his seventy-seventh birthday. His son and younger daughter survived him.

NOTES

1. Historians of photography sometimes give credit for this development to the Englishman J. F. Goddard, but since the photosensitive properties of silver bromide were widely known, the idea of trying bromine vapors in place of iodine was, perhaps, too obvious to be considered an important discovery.
2. A photograph of the moon had been taken as early as 1840 by J. W. Draper of New York.
3. Stars do change color, but not in the way that Doppler thought they did. The intensities of stars' electromagnetic emissions are not equitably distributed throughout the spectrum, and any shift in the maximum energy distribution of a star's visible light will appear as a color change; what Doppler failed to understand was the idea of frequency shift.

BIBLIOGRAPHY

Fizeau wrote no major publications. A list of his scientific articles is in Émile Picard, *Les théories de l'optique et l'oeuvre d'Hippolyte Fizeau* (Paris, 1924), pp. 57–64. There is also a list, presumably complete, in the Royal Society's *Catalogue of Scientific Papers.* Unfortunately there is no major biography of Fizeau. Virtually all the information we have about his life and work comes from Alfred Cornu, "Notice sur l'oeuvre scientifique d'Hippolyte Fizeau," in *Annuaire pour l'an 1898, publié par le Bureau des longitudes* (Paris, 1898[?]), notice C, pp. 1–40. The work by Picard, mentioned above, contains substantially the same information, but Picard has attempted to place his discussion in a broader historical context by relating Fizeau's work to the optical theories that came before and after him. A number of brief biographical notices appeared in various scientific journals just after Fizeau's death. A list of them is found in Picard, p. 64. The Academy of Sciences in Paris is reported to have some of Fizeau's MSS and a portrait.

J. B. GOUGH

FLAMMARION, CAMILLE (*b.* Montigny-le-Roi, France, 26 February 1842; *d.* Juvisy, France, 3 June 1925), *astronomy.*

At the time of Flammarion's birth his parents owned a small store, but his father had been a farmer and Flammarion often mentioned this with pride. He was the oldest of four children. His interest in astronomy dated from his early childhood, when on 9 October 1847 and 28 July 1851 he was able to observe solar eclipses. By the time he was eleven he was busily making astronomical and meteorological observations.

In 1856 his parents' disastrous financial condition led them to move to Paris. For the young Flammarion, this was a decisive event, since Paris offered him immensely greater opportunities for self-improvement. He found employment as apprentice engraver, attended evening courses at the Polytechnic Association, learned English, and pursued his studies of algebra and geometry.

A chance encounter in 1858 marked the start of his career; a physician who was treating Flammarion noticed a bulky, 500-page manuscript written by the young man and entitled *Cosmogonie universelle.* The doctor read it and was so impressed that he brought it to the attention of Le Verrier, who was director of the Paris Observatory. A few days later, Flammarion was hired by the observatory to work in the Bureau de Calcul as an apprentice astronomer.

In 1861 Flammarion wrote *La pluralité des mondes habités,* his first book to be published. In this he revealed the pleasant literary style that was to make him the most important popularizer of science at the turn of the twentieth century. In 1862 he was calculator to the Bureau des Longitudes. He wrote for the *Annuaire du cosmos* and published the *Annuaire astronomique et météorologique,* first in the *Magasin pittoresque,* then later in *L'astronomie.* During the same period he also wrote many popularizing articles for the newspapers and delivered very successful series of lectures in Paris, the provinces, and several other European capitals.

Flammarion became greatly interested in the problems of the atmosphere. Between 1867 and 1880 he made many balloon flights in order to study atmospheric phenomena. In 1871 he published *L'atmosphère. Les terres du ciel* appeared in 1877 and then, in 1880, the famous *Astronomie populaire,* his best-known work, a true best seller that was translated into many languages and which, more than any other book ever written, spread interest in astronomy. It was followed in 1882 by *Les étoiles et les curiosités du ciel.*

At this point in Flammarion's career, his scientific output was considerable and concerned with many subjects, including volcanology, atmospheric electricity, and climatology. Special mention must be made of his research concerning Mars. Scientific opinion of that time held that Mars was the only planet on which traces of life might be found. At the Juvisy Observatory, founded by him in 1883, Flammarion made numerous observations of the planet. As early as 1876 he had noticed the seasonal variations of the dark spots. In 1909 he completed *La planète Mars et ses conditions d'habitabilité,* a compilation of all known observations since 1636.

In 1887 Flammarion founded the French Astronomical Society, a model for all groups aiming to spread interest in science among the general public. For the first time relatively powerful astronomical instruments were put at public disposal, thus allowing numerous amateurs to indulge their taste for science. The activities of the French Astronomical Society created a reservoir of scientists from which emerged most of the outstanding French astronomers of this century.

It was inevitable that Flammarion, who possessed an extraordinary intellectual curiosity and imagination, take an interest in what today is called parapsychology. His taste for scientific precision and his intellectual honesty led him to unmask the inaccuracies, lies, and hoaxes that have always encumbered this field. He directed investigations of and performed experiments in psychic phenomena and gathered most of the results into several books, including *La mort et son mystère, L'inconnu et les problèmes psychiques,* and *Les maisons hantées.*

His love of life and his profound sensitivity led Flammarion to a literary as well as a scientific career. He published several novels, in which science serves as a backdrop.

BIBLIOGRAPHY

Flammarion's most important works in astronomy and geophysics, all published in Paris, are *Astronomie populaire* (1880); *Les étoiles et les curiosités du ciel* (1882); *Les terres du ciel* (1877); *Les merveilles célestes; La planète Mars et ses conditions d'habitabilité* (1892); *La planète Vénus; Les étoiles doubles; L'atmosphère* (1871); *Mes voyages aériens; Tremblements de terre et éruptions volcaniques; L'éruption du Krakatoa; Les caprices de la foudre; Les phénomènes de la foudre* (1905); and *Le monde avant la création de l'homme* (1885).

In philosophy, see *La pluralité des mondes habités; L'inconnu et les problèmes psychiques* (1900); *Les forces naturelles inconnues; La mort et son mystère,* 3 vols. (1920–1921); and *Les maisons hantées,* all published in Paris.

His autobiography is *Mémoires biographiques et philosophiques d'un astronome.* His novels include *Uranie* (1889) and *Stella.*

ROGER SERVAJEAN

FLAMSTEED, JOHN (*b.* Denby, England, 19 August 1646; *d.* Greenwich, England, 31 December 1719), *astronomy.*

The only son of Stephen Flamsteed, a prosperous businessman, and Mary Spateman, he was raised in comfortable circumstances in Derby, England. During a childhood marred by the deaths of his mother and stepmother, he attended the Derby free school and received the normal preparation for university study. Unfortunately, his educational plans were forestalled by a serious breakdown of his health when, at the age of fourteen, he was afflicted with a severe rheumatic condition and complications therefrom, which left him so debilitated as to render his health a subject of grave concern for the rest of his days. His later correspondence is filled with allusions to periodic incapacity and reports of (generally unsuccessful) medication for it. The most immediate consequence of his frailty was his father's refusal to send him on to university in 1662. Flamsteed was deeply disappointed, but his misfortune may well have been a disguised blessing, since, left to his own devices, he was able to follow his own interests to a degree that would otherwise have been impossible. With his introduction to Sacrobosco shortly after leaving school, the direction of those interests was established for life.

The period between 1662 and 1669 was for Flamsteed one of education in the details of astronomical science. The major impediment to his progress seems not to have been the lack of instruction, in which respect he would scarcely have been better off at Oxford or Cambridge, but the fact that his "studies were discountenanced by my father as much in the beginning as they have been since."[1] As late as 1673, Flamsteed was receiving mail through a friend, "that my father might not see all the letters that come to me."[2] The ostensible basis of the father's attitude was, of course, the son's weak constitution; but various of Flamsteed's remarks betray a suspicion that he had been kept away from university more because of the help that a capable only son could render in tending a motherless home and a flourishing business than for any other reason. In spite of the time lost to business and illness, however, Flamsteed persevered. By the end of 1669, he was ready to put himself up for professional consideration.

Flamsteed's debut was a cautious one. Rejecting certain of his efforts that might be "beyond the capacity of the vulgar,"[3] he chose to submit—anonymously—to the Royal Society a small ephemeris of lunar occultations for 1670. He was soon engaged in extensive correspondence with Henry Oldenburg and John Collins, the scientific "clearinghouses" of the day. Through them he was introduced to Sir Jonas Moore, whose interest and influence were to be decisive in the launching of his career. Already from their first meeting in 1670, he emerged with a micrometer and the promise of good telescope lenses, which equipment enabled him to inaugurate his serious observational work. By the winter of 1674–1675 the enthusiastic master of the royal ordinance was attempting to organize patronage for an observatory. In the midst of his labors there appeared at court

FLAMSTEED

a French dilettante (sponsored by the king's mistress) who claimed to have solved the problem of determining terrestrial longitudes, long recognized as the principal desideratum for safer navigation. When Moore called upon his protégé for judgment of the claim, Flamsteed replied convincingly that neither the positions of the stars nor the motions of the moon were well enough known to render the proposed method practicable. The result was the immediate realization of Moore's plans and more. The king founded an observatory and installed Flamsteed on 4 March 1675 as his "astronomical observator" at an annual stipend of £100.

With his appointment as astronomer royal and his removal to Greenwich, the pattern of Flamsteed's professional life was essentially established. At Easter of the same year he took orders, having the previous year taken the M.A. at Cambridge by letters-patent, after four years of nonresident enrollment. In 1684 he was granted the living of Burstow, in Surrey, not far from Greenwich; and in 1692, he was married to Margaret Cooke, the granddaughter of his predecessor at Burstow. At his death in 1719, he was succeeded at Greenwich by Edmond Halley and at Burstow by another astronomer, James Pound, an uncle of James Bradley.

The mandate of the newly created *mathematicus regius* was unequivocal: "Forthwith to apply himself with the most exact care and diligence to the rectifying the tables of the motions of the heavens, and the places of the fixed stars."[4] By no means a new idea, it was purely and simply the project conceived by Tycho Brahe a century earlier. The only thing at all remarkable about it was the extent to which it had been neglected during the intervening years. Incredible as it appears to later ages, the invention of the telescope had as yet had virtually no impact on fundamental astronomy. Two generations after Galileo's momentous discoveries, Tycho's star catalog remained the standard of excellence; and the one designed by Hevelius to replace it was likewise being constructed on the basis of naked-eye observations. With respect to the planets, the laws of Kepler were just winning general acceptance, while the observations from which they had been derived were being published (1666) because they still represented the most accurate information available. Flamsteed's assignment, then, was essentially that of dragging positional astronomy into the seventeenth century, of bringing it abreast of the new descriptive astronomy to which the telescope had thus far been almost exclusively applied. It was a project which coincided nicely with his own predilections and one to which he had already dedicated his efforts for some years before 1675. An indefatigable calculator and conscientious observer from the days of his youth, he had early learned that the existing tables and catalogs were unequal to the accuracy of even the most modest instruments. Unfortunately, however, it was one thing to recognize the scandalous state of astronomical science and quite another to do anything about it without professional apparatus. Prior to the fall of 1671, when he finally got his new micrometer-telescope fitted out, Flamsteed was able to do very little in the way of meaningful observation. What he did manage to do was to lay the foundation for his contributions to the solar and lunar theories.

Already several years before his post was created for the express purpose of "find[ing] out the so much-desired longitude of places for the perfecting the art of navigation,"[5] Flamsteed was "esteeming [him]self obliged" to publish predictions of the moon's occultations of stars. In the process of computing these "annual preadmonitions of the lunar appearances,"[6] he necessarily became very familiar with the various accounts of its motion. Since all were conspicuously inadequate, he took considerable interest in the news that the legendary Jeremiah Horrox had left writings pertaining to the lunar theory. Despite the fact that the material involved had been deemed so fragmentary as to be unworthy of publication in the *Opera posthuma* of Horrox—in press at the time—Flamsteed journeyed to Lancashire to look at it. What he found was two letters that together contained a sketch of the proposed model, some computing rules from which he was able to infer the mechanism, and some opinions as to the constants that would be appropriate. Intrigued by the scheme (which stemmed basically from Kepler's ruminations on the subject and featured an elliptical orbit with a librating line of apsides and a variable eccentricity), Flamsteed brought the constants up-to-date and constructed enough tables to put the theory on trial. When observational comparison convinced him that "Bullialdus's, Wing's, and Streete's theories were erroneous, and Horrox's near the truth,"[7] he naturally relayed the information to Collins, and ultimately Wallis, who was editing Horrox' works. The result was a last-minute inclusion of the lunar theory as reworked, tabulated, and elucidated by Flamsteed. Together with it was his defense of the equation of time, which restored it to its rightful status after three generations of confusion involving the annual equation of the lunar theory.

The lunar theory continued to occupy a special place in Flamsteed's work throughout his life. As a result of intermittent reconsiderations of the subject, he revised Horrox' original account at least three times: in 1680 for his *Doctrine of the Sphere,* in the late 1690's, and again about 1703 after the publication of Newton's second efforts on the theory. It was his model in terms of which Newton conceived the moon's motion, his observations by means of which Newton improved the theory, and his incorporation of Newton's revisions that rendered them subject to test, modification, and use. As late as 1746, Flamsteed's last version was published in Lemonnier's *Institutions astronomiques.* To his profound disappointment, however, neither it nor his similarly inspired work with the satellites of Jupiter ever achieved the accuracy requisite for longitude determinations.

Among the various features of Flamsteed's work on the motion of the sun, by no means the least interesting is the fact that he bothered to do it at all. Inherently simple and tractable by virtue of its small eccentricity and unique relation to the earth, the sun remained down to Flamsteed's day the least troublesome of the planets. Yet, for Flamsteed, relative virtue was not sufficient. Every aspect of astronomy had to be as perfect as his researches could make it; and since the solar theory figured in, and hence determined the upper limit of accuracy of, all the other planetary theories, it had a prior claim to the utmost precision attainable. For that reason, Flamsteed issued no fewer than three different sets of tables during his lifetime. The first, published with Horrox' lunar theory, amounted to little more than a computing exercise, since it was constructed before Flamsteed had any original observations on which to base his parameters. The second, computed to a new determination of the eccentricity which pegged it at almost exactly its true value of .01675, appeared in his *Doctrine of the Sphere* (1680). The third was printed in Whiston's *Praelectiones astronomicae* (1707). Common to all three versions was Flamsteed's unique denial of the reality of the generally accepted secular changes in the longitude of apogee and the obliquity of the ecliptic. Strictly a modern in such matters, Flamsteed held that neither the accuracy nor the coherence of the ancient determinations justified their being taken seriously.[8]

That the existing planetary theories were far from any respectable standard of accuracy was one of the early lessons in Flamsteed's astronomical education. Already with the acquisition of his micrometer-fitted telescope he began to look into the possibility of improving them; but neither those investigations nor any of his numerous subsequent attempts yielded the degree of satisfaction he demanded in his work. Aside from a tract on the angular diameters of the planets, composed in 1673 and given to Newton for his preparation of the *Principia,* all that resulted from his efforts was an occasional determination of an isolated orbital element. The hundreds of observations he published in his *Historia coelestis Britannica* were no doubt useful to succeeding generations, but even they have been overshadowed by the six inadvertent observations of Uranus later found in his star catalog. An interesting concomitant of his planetary work was his determination of solar parallax from observations of Mars's perihelion opposition of 1672. Using the rotation of the earth for a base line, he arrived at values of "certainly not 30″" for Mars, and accordingly, "not more than 10″" for the sun's parallax[9]—results essentially identical with those later released by the French and quite reasonably approximate to the true figures.

In 1676, two months after his installation at the Royal Observatory, Flamsteed inaugurated the observations that were to culminate in his celebrated 3,000-star "British Catalogue" (in volume III of his *Historia*). From the beginning, the task proved troublesome. King Charles's initial enthusiasm had sufficed only to the appropriation of funds for an observatory. Flamsteed was left to worry not only about such details as hiring computing assistants and obtaining instruments for the facility, but even whether he would receive his stipulated salary; by the close of 1676 some of it was already overdue, and as late as 1679 it was seventeen months in arrears and "in danger of a total retrenchment."[10] The prestige of the post ameliorated Flamsteed's pecuniary problems by drawing some 140 pupils to him for private mathematical instruction over the years. Sir Jonas Moore eased the want of instruments by giving him two clocks and a seven-foot sextant.

Between 1676 and 1689, Flamsteed made about 20,000 observations with the sextant. Accurate to about 10″, they constituted an improvement on Tycho's work by a factor of perhaps fifteen. Unfortunately, they consisted exclusively of relative distances, having no "anchor" in the celestial sphere. Refusing to refer his star places to Tycho's much less accurate bases, as he had criticized Halley for doing, Flamsteed resolved to underwrite a fixed-meridian instrument himself. A mural quadrant completed in 1683 proved too fragile; but a 140° mural arc made possible by the inheritance from his father's estate solved his problem. In the years following 1689, he did fundamental astronomy with an accuracy unsurpassed before Bradley. He achieved precise determinations of the latitude of Greenwich, the obliquity of the ecliptic, and the position of the equinox and then bypassed them all by devising an ingenious scheme for observing absolute right ascensions. Using matched occasions at which the sun had identical meridian altitudes near each equinox, he measured the time intervals between the passage of the sun and a bright star across the meridian. Halving the difference between the two time intervals then located the solstice and gave the star's right ascension. It is difficult to overstate the advance represented by this method. Not only did it do away with the errors formerly introduced by using an intermediary planet to measure the angle between sun and star, but it eliminated all uncertainties caused by parallax, refraction, and latitude. After using this method to obtain the positions of forty reference stars, Flamsteed computed the rest of his 3,000-star catalog from the intermutual readings already taken with the sextant.

Long before he had readied his observations for publication, Flamsteed had become resigned to the likelihood that he would have to underwrite the end of the project as he had every other aspect of it. Unfortunately, before he could carry the plan through to his own ideal of completion, his long-simmering relations with Newton and Halley boiled over. The wrangle began in 1704 and ended only with the unauthorized printing of Flamsteed's work in 1712. Because the perpetrators of this unorthodox operation were who they were, the incident has received more attention than it deserves. The essence of the situation seems to be that Newton and Halley, who lacked Flamsteed's passion for astronomical precision, felt that Flamsteed was being unnecessarily dilatory in the publication of his observations. Regarding the observations of the publicly supported astronomer royal as public property, they took steps, as officers of the Royal Society, to expedite the entrance of these observations into the public domain. Whatever their motives (and it is difficult to believe that they were purely objective), their action was quite questionable; and in any case, since Flamsteed had disbursed well over half of his life's salary for computing service, construction and repair of instruments, and other operating expenses, their basic premise was open to dispute. In 1714 a turn of the political wheel of fortune gave Flamsteed the satisfaction of burning all but ninety-seven pages of three-quarters of the spurious edition. By the time he died, he had pushed the work far enough that the three-volume *Historia coelestis Britannica* could be published in 1725. The companion *Atlas coelestis* appeared in 1729.

There can be little doubt that Flamsteed's reputation rests on his observational work. As Grant expressed it, "In carrying out views of practical utility, with a scrupulous attention to accuracy in the most minute details, in fortitude of resolution under adverse circumstances, and persevering adherence to continuity and regularity of observation throughout a long career, he had few rivals in any age or country." As is usually the fate of practical scientists, his other qualities and achievements have been either overlooked or denigrated. The fact is that no other astronomer royal until Airy manifested anything like the same concern for the reduction and manipulation of data. Far from bequeathing a mass of raw observations in the manner of Bradley, Flamsteed converted and applied his.

In addition to the numerous efforts already mentioned, one can cite tables of atmospheric refraction and tides. The first tabulation in England of the moon's elliptic inequality according to Kepler's second law was carried out under him.[11] Nor did Flamsteed lack his share of inventiveness in an age in which speculation was rife. He argued vigorously with Newton in behalf of the proposition that two comets of 1680–1681 were in fact appearances of the same comet. Prior to that, he had already noted that the comet of 1677 had appeared in the same place as those of 1653 and 1665. In publishing his observations of it, he posed the question of "what conformity there is betwixt the motions of this and them and whether it may probably be the same returned hither after two revolutions."[12] From reports of the coronal phenomena during a solar eclipse, he inferred the existence of a lunar atmosphere. He has been criticized for his "discovery" of stellar parallax in 1694: what he actually found was aberration of starlight, "the maximum value of which is deducible with an astonishing degree of accuracy" from his observations. The truth remains that whatever interpretive shortcomings might be attributed to him must also be laid at the door of Newton, Halley, Gregory, Wallis, and everyone else who could read the published account of his findings.[13]

One respect in which Flamsteed was conspicuously deficient was an aptitude for dealing with his fellowman. Possessed of an attitude that can only be described as uncompromising, he was an intemperate man even by the standards of an intemperate age. The particular and enduring subject of his passion was Edmond Halley. The last thirty years of Flamsteed's extensive correspondence is infused with vituperous remarks about the man who should have been his most natural and valuable ally. No single cause of such animosity has been convincingly advanced. Professional jealousy was, no doubt, an element: rare indeed was the occasion on which Flamsteed praised any third party, and even rarer was it that he passed up an opportunity to criticize. Basically, however, it was simply a personality clash. Halley's flamboyant nature, frivolous attitude toward religious matters, and hit-and-run approach to astronomy offended the dour Flamsteed, who took everything he did very seriously. Lurking behind it all was Flamsteed's perpetual ill health, which would surely have tried the patience of any man. That he managed to accomplish so much in spite of it and its effect on his relations with his contemporaries is a real tribute to his industry and ability.

NOTES

1. Baily, p. 10.
2. Rigaud, p. 130.
3. Bayle.
4. Baily, p. 111.
5. *Ibid.*
6. Rigaud, p. 120
7. Baily, p. 31.
8. Flamsteed MS, XXXVIII, 149.
9. *Philosophical Transactions,* **7,** 5118.
10. Baily, p. 118.
11. Baily, p. 704.
12. *Philosophical Transactions,* **12,** 873.
13. Wallis, III, 704.

BIBLIOGRAPHY

I. Original Works. Flamsteed's three major printed works were *The Doctrine of the Sphere* (London, 1680); *Historia coelestis Britannica* (London, 1725); and *Atlas coelestis* (London, 1729). Other lesser works appeared in Horrox' *Opera posthuma* (London, 1673); John Wallis' *Opera mathematica* (London, 1699); and the *Philosophical Transactions of the Royal Society.*

The bulk of the extant Flamsteed MSS is contained in his seventy-odd notebooks preserved at the Royal Observatory and recently published in microfilm, *Observations of the Royal Astronomers* (London, 1969). The most significant material in it is a serialized autobiographical statement, which was published in Francis Baily's *Account of the Revd. John Flamsteed* (London, 1835). Along with it was Baily's revision of the "British Catalogue" according to researches conducted by William and Caroline Herschel.

Flamsteed's extensive correspondence with various of his contemporaries provides a very interesting view of the scientific activity of his day. Many letters were published as early as 1738 in the English version of Bayle's *General Dictionary.* A large number were found in the nineteenth century. Baily printed about 200, while vol. II of Rigaud's *Correspondence of Scientific Men of the Seventeenth Century* (London, 1841) contains numerous others. See also Cudworth's *Life and Correspondence of Abraham Sharp* (London, 1889), and *The Correspondence of Isaac Newton* (London, 1959–). J. L. E. Dreyer found (*Observatory,* **45** [1922], 280–292) some seventy unpublished letters written to Richard Townley preserved in the rooms of the Royal Society at Burlington House. The Southampton Record Office contains Flamsteed's letters to Molyneux, while numerous other unpublished letters are preserved in the notebooks at the Royal Observatory.

II. Secondary Literature. The best secondary account of Flamsteed's work is Robert Grant, *History of Physical Astronomy* (London, 1852). Important additional material is presented in Agnes Clerke's article in the *Dictionary of National Biography;* and in E. F. MacPike, *Hevelius, Flamsteed, and Halley* (London, 1937).

Victor E. Thoren

FLECHSIG, PAUL EMIL (*b.* Zwickau, Germany, 29 June 1847; *d.* Leipzig, Germany, 22 July 1929), *neuroanatomy, psychiatry, neurology.*

His father, Emil Flechsig, a cultured man and a good friend of Robert Schumann, was a deacon of the Protestant church of St. Mary in Zwickau and much concerned with local social welfare. His mother, Ferdinande Richter, came from a wealthy family. Flechsig was educated in Zwickau, mainly at the Gymnasium, and at the medical school of the University of Leipzig from the spring of 1865 until June 1870, where he was taught in anatomy by E. H. Weber and E. F. W. Weber and in physiology by Karl Ludwig. Ludwig was especially impressed by Flechsig's histological work, and he encouraged and advised him in it.

Flechsig spent two years (1870–1871) in the army during the Franco-Prussian War and on 1 January 1872 became assistant to Ernst Wagner of the Institute of Pathology in the University of Leipzig. He also worked in the medical polyclinic. He was further able to develop his skills in histology and on 1 October 1873 was appointed head of the department of histology in the Institute of Physiology. Here he devoted all his time to research and benefited greatly from the facilities available and the contact with many outstanding German and foreign physiologists. By 1875 he was university lecturer and in 1877 became professor *extraordinarius* in the new chair of psychiatry; he soon was made *ordinarius.* After studying psychiatry at various European centers, on 2 May 1882 he returned to open the clinic of which he was director. Here he spent the rest of his working life and attracted many pupils and visitors, including Beevor, Bekhterev, Blumenau, Darkshevich, Donaldson, Held, Klimor, Martinotti, R. A. Pfeifer, Popov, Schütz, Tschirch, Oskar Vogt, and Yakovenko.

From 1894 to 1895 Flechsig was rector of the University of Leipzig and in 1901, along with the elder Wilhelm His, he helped to found the International Brain Commission, which planned to unify nomenclature, standardize methods, collect material, and encourage research in neuroanatomy. He was made an honorary member of the University of Dorpat in 1903, received the honorary D.Sc. of Oxford in 1904, and became honorary doctor of his alma mater in

1909. At the age of seventy-four he retired but continued working.

Flechsig was a true *Vogtländer,* with a thick neck, a large barrel-shaped trunk, and short legs. He wore a broad-brimmed hat and a velvet cloak with large glass buttons, resembling therefore, it was said by some, his psychiatric patients.

Flechsig had a cyclothymic personality, almost bordering on a true manic-depressive state. Years of intense activity—when he worked ceaselessly and poured out ideas, encouragement, and inspiration—alternated with years when he was irritable, arrogant, intolerant, tyrannical, and suffered from severe depression. Nevertheless his students and followers venerated him, and Pfeifer records that "his guidance was full of spirit and during discussions of various problems his whole youth was awakened." He was devoted to his work and had little time for anything else until late in life. He liked to mix with aristocrats, monarchs, and politicians, in part because of their interest in his work and in part because of his need for extramural research funds.

Flechsig married Auguste Hauff in 1870. After her death, in 1922 he married Irene Colditz, who was thirty years younger than he; she was able to interest him in social events during the closing years of his life.

Although Flechsig contributed to the clinical and pathological study of hysteria, epilepsy, neurosyphilis, and chorea, his fame is due mainly to his technique of myelogenesis for the examination of the brain and spinal cord. As Wagner's assistant, he was impressed by the work of Meynert on the brain and in 1872 began to investigate the myelination in the spinal cord and brain of premature, full-term, and early postnatal infants. He discovered that axones in different parts receive their myelin sheath at different stages of growth, and he could observe the chronological sequence of this process. He was therefore able to differentiate some of the innumerable pathways and concluded correctly that a tract functions adequately only when its axones become fully myelinated. His technique was thus the reverse of Ludwig Türck's and of those of other workers who used the process of secondary or Wallerian degeneration to trace tracts.

Flechsig first examined the spinal cord and, like Türck and others, identified several pathways. He reported this work in 1876 (*Leitungsbahnen im Gehirn und Rückenmark . . .*) and related it to that of Türck; a description of what is now called Flechsig's tract, the dorsal spinocerebellar, is included.

His monumental work on the pyramidal tract—in which for the first time he traced its origin to the cerebral cortex—appeared in parts in 1877 and 1878. It is the first clear account of the upper motor neurone, and the now familiar division of the internal capsule into knee and limbs is his.

In 1893 Flechsig began more intensive study of the cerebral hemispheres and supplemented his myelogenetic findings with clinical observations and data from degeneration experiments. He outlined the auditory radiation and could list twelve cortical areas that are myelinated (and therefore functional) before birth, as well as twenty-four in which myelinization occurs after birth; these he arranged chronologically according to the time of myelinization. He thus isolated primarily projection, or motor and sensory, areas, the fiber connections of which, both corticopetal and corticofugal, mostly mature prenatally; and association areas responsible, he claimed, for higher intellectual functions that develop after birth. From this he evolved a map of cortical function that appeared in a report of 1904 to the Central Committee for Brain Research. Flechsig's conclusions evoked considerable argument, especially from Leonardo Bianchi concerning frontal lobe function and Oskar Vogt on the techniques of myelogenesis. It is now clear that although Flechsig made many errors and ignored the work of others with which his results did not agree, he nevertheless stimulated much beneficial discussion and research.

Flechsig considered the parieto-temporo-occipital association zone to be essential for mental activity (which is at least partially correct), and this is discussed in his rector's oration of 1896, *Gehirn und Seele.* The book he published in 1920, *Anatomie des menschlichen Gehirn und Rückenmarks . . .*, contains most of his data on cortical localization.

BIBLIOGRAPHY

I. ORIGINAL WORKS. There is no bibliography or *opera omnia* of Flechsig, but Pfeifer (1930) has published a list of forty-nine of his most important writings from 1872 to 1920, all but one being on brain anatomy. Among these are *Die Leitungsbahnen im Gehirn und Rückenmark des Menschen auf Grund entwicklungsgeschichtlicher Untersuchungen* (Leipzig, 1876); "Über 'Systemerkrankungen' im Rückenmark," in *Archiv für der Heilkunde,* **18** (1877), 101–141, 289–343, 461–483, and *ibid.,* **19** (1878), 52–90, 441–447; "Zur Anatomie und Entwicklungsgeschichte der Leitungsbahnen im Grosshirn des Menschen," in *Archiv für Anatomie und Physiologie* (Anatomische Abteilung) (1881), pp. 12–75; *Plan des menschlichen Gehirns auf Grund einziger Untersuchungen* (Leipzig, 1883); *Gehirn und Seele* (Leipzig, 1896), his rectorial oration, with extensive notes; "Einige Bemerkungen über die Untersuchungs-methoden der Grosshirnrinde, insbesondere des Menschen. Dem Zentralkomitee für Hirnforschung vorgelegt von . . .," in

Berichte über die Verhandlungen der Sächsischen Akademie der Wissenschaften zu Leipzig, Math.-Phys. Klasse, **56** (1904), 50–104 (with 4 plates), 177–248; and *Anatomie des menschlichen Gehirns und Rückenmarks auf myelogenetischer Grundlage,* vol. I (Leipzig, 1920).

Useful English and French accounts of Flechsig's technique as applied to the cerebral cortex are "Developmental (Myelogenetic) Localisation of the Cerebral Cortex in the Human Subject," in *Lancet* (1901), **2,** 1027–1029; and "Les centres de projection et d'association du cerveau humain," in *Proceedings. XIIIe Congrès International de Médecine* (Section de Neurologie) (Paris, 1900), pp. 115–121. E. Clarke and C. D. O'Malley, *The Human Brain and Spinal Cord* (Berkeley–Los Angeles, 1968), contains English translations of some of Flechsig's writings: on spinal cord tracts, pp. 277–280, 287–290; on cerebral cortical localization, pp. 548–554; on cerebral white matter, pp. 611–619; and on the technique of myelogenesis, pp. 857–858; each contribution is discussed in relation to other work on the same subject.

II. Secondary Literature. There is no biography of Flechsig, and the various brief accounts of his life are based on his autobiography, *Meine myelogenetische Hirnlehre mit biographischer Einleitung* (Berlin, 1927): some of these are W. Haymaker, "Paul Emil Flechsig (1847–1929)," in his ed. of *The Founders of Neurology* (Springfield, Ill., 1953), pp. 31–35 (with a portrait); Henneberg, "Paul Flechsig †," in *Medizinische Klinik,* **25** (1929), 1490–1492; R. A. Pfeifer, "Paul Flechsig †. Sein Leben und sein Wirken," in *Schweizer Archiv für Neurologie und Psychiatrie,* **26** (1930), 258–264; F. Quensel, "Paul Flechsig zum 70. Geburtstag," in *Deutsche medizinische Wochenscrift,* **43** (1917), 818–819; and P. Schröder, "Paul Flechsig," in *Archiv für Psychiatrie and Nervenkrankheiten,* **91** (1930), 1–8.

Concerning Flechsig's work, see L. F. Barker, "The Phrenology of Gall and Flechsig's Doctrine of Association Centers in the Cerebrum," in *Bulletin of the Johns Hopkins Hospital,* **8** (1897), 7–14; and "The Sense-areas and Association-centers in the Brain as Described by Flechsig," in *Journal of Nervous and Mental Diseases,* **24** (1897), 325–356, 363–368 (discussion); W. W. Ireland, "Flechsig on the Localization of Mental Processes in the Brain," in *Journal of Mental Science,* **44** (1898), 1–17; M. P. Jacobi, "Considerations on Flechsig's 'Gehirn und Seele,'" in *Journal of Nervous and Mental Diseases,* **24** (1897), 747–768; J. M. Nielsen, "The Myelogenetic Studies of Paul Flechsig," in *Bulletin of the Los Angeles Neurological Society,* **28** (1963), 127–134; and F. R. Sabin, "On Flechsig's Investigations on the Brain," in *Bulletin of the Johns Hopkins Hospital,* **16** (1905), 45–49.

Flechsig's *Festschrift,* in *Monatsschrift für Psychiatrie und Neurologie,* **26** (1909), 1–416, with 19 plates, has an excellent photograph of him in his laboratory, but no biography or bibliography.

Edwin Clarke

FLEISCHER, JOHANNES (*b.* Breslau, Germany [now Wrocław, Poland], 29 March 1539; *d.* Breslau, 4 March 1593), *optics.*

Fleischer was born into a well-to-do family and received his education at the Goldberg Gymnasium near Breslau and later at the University of Wittenberg, where he matriculated in 1557. During his first years at Wittenberg, Fleischer studied under Philip Melancthon; he returned to Wittenberg after Melancthon's death to concentrate on Hebrew, astronomy, and theology, and in 1589 he was awarded the theological doctorate. Fleischer spent 1568–1569 teaching arts and languages at the Goldberg Gymnasium, and beginning in 1572 he was professor in the Gymnasium attached to St. Elizabeth's Church in Breslau. He also held a series of ecclesiastical posts in Breslau, in St. Maria Magdalena's Church as well as St. Elizabeth's.

Fleischer's only published scientific work is a treatise on the rainbow, which appeared in 1571. This book is remarkable, not for its correct solution to the problem of the rainbow, but for the precision and clarity of its argument, for its emphasis on what occurs in the individual drop of vapor, and for its insistence that both reflection and refraction of solar rays participate in the generation of the rainbow. The essential feature of Fleischer's explanation is that solar rays are refracted by individual drops as they enter the vapor cloud before being reflected back to the observer's eye by denser drops more interior to the cloud. The idea is drawn largely from Witelo, whom Fleischer cites repeatedly, although Fleischer has considerably clarified the mechanism of reflection and refraction described by Witelo.

BIBLIOGRAPHY

I. Original Works. See his *De iridibus doctrina Aristotelis et Vitellionis* (Wittenberg, 1571).

II. Secondary Literature. Fleischer's life is dealt with briefly in C. G. Jöcher, *et al., Allgemeines Gelehrten-Lexicon,* I (Leipzig, 1750), col. 636; and Gustav Bauch, *Valentin Trozendorf und die Goldberger Schule* (Berlin, 1921), pp. 235–237. On Fleischer's theory of the rainbow, see Carl B. Boyer, *The Rainbow: From Myth to Mathematics* (New York, 1959), pp. 163–166; and A. G. Kästner, *Geschichte der Mathematik,* II (Göttingen, 1797), 248–250.

David C. Lindberg

FLEMING, ALEXANDER (*b.* Lochfield, Ayrshire, Scotland, 6 August 1881; *d.* London, England, 11 March 1955), *bacteriology.*

Sir Alexander Fleming's professional career was devoted mainly to investigating the human body's defenses against bacterial infections. Late in life he achieved retrospective fame for discovering penicillin in 1928.

Descended from Lowland farmers, Alexander was the third of four children born to Grace Morton, second wife of Hugh Fleming. Four children of the first marriage survived. His father died when Alec was seven years old, leaving the widow to manage the farm with her eldest stepson. From age five to ten, the boy attended a tiny moorland school. Then he walked daily to school at Dorval, a small town four miles away. Two years later, he attended Kilmarnock Academy, twelve miles distant, which had exacting standards but meager resources. His basic education was thus hard-earned and rather primitive. He learned early to observe nature intimately, to enjoy simple pleasures, and to appreciate unaffectedness.

When just over thirteen, he followed a stepbrother, already a practicing physician, and his brother John to London. The youngest brother, Robert, soon joined them, and a sister kept house. After attending classes at the Regent Street Polytechnic for two years, Alec became a clerk in a shipping company. In 1900 he enlisted in the London Scottish Regiment, but the Boer War ended before he got overseas. He enjoyed life in the ranks and stayed attached to this regiment until 1914. Fleming was short but sturdy, blue-eyed, fair-haired, good at rifle shooting and water polo. At twenty, he inherited a small legacy and decided to study medicine.

Fleming excelled at competitive examinations. He won a scholarship to St. Mary's Hospital Medical School, Paddington, and in 1901 began his lifelong connection with that institution. Capturing numerous class prizes and trophies, he obtained the Conjoint Board Diploma in 1906, and two years later graduated M.B., B.S., with honors and a gold medal from the University of London. In 1909 he completed the fellowship examinations of the Royal College of Surgeons of England.

Upon qualifying in 1906, Fleming joined Sir Almroth Wright's disciples at St. Mary's Hospital. Although he was Wright's antithesis—cautious, unpretentious, and laconic—they were closely associated for forty years. Fleming became assistant director of the Inoculation Department in 1921. The department merged with the Institute of Pathology and Research in 1933; it was renamed the Wright-Fleming Institute in 1948. In that year, Fleming retired as professor of bacteriology, University of London, having held the chair since 1928. He retained the principalship of the institute (to which he succeeded when Wright retired in 1946) until January 1955. Two months later he died suddenly from coronary thrombosis in his Chelsea home. He is buried in St. Paul's Cathedral.

In 1915 Fleming married Sarah Marion McElroy, an Irish farmer's daughter, who had operated a private nursing home. Despite their dissimilar characters, the marriage was happy. They enjoyed gardening at their Suffolk country house, were very hospitable, and collected antiques. Their only child, Robert, born in 1924, became a physician. Fleming's wife died in 1949, and in 1953 he married Dr. Amalia Coutsouris-Voureka, a Greek bacteriologist working at the institute.

Fleming upheld and practiced Wright's doctrine of specific immunization against bacterial infection through vaccine therapy, without necessarily accepting every dictum. His earliest publications (1908) concerned "opsonic index" determinations—Wright's method of assaying the phagocytic power of a patient's blood for a particular microorganism. The procedure was of dubious utility and eventually fell out of favor. Fleming was among the first (1909) to treat syphilis with Salvarsan (606), samples of which Ehrlich had entrusted to Wright.

In World War I, Fleming joined the Royal Army Medical Corps as lieutenant, serving under Wright's colonelcy in a wound-research laboratory at Boulogne. By simple, ingenious techniques, he demonstrated the bactericidal power of pus and the inability of chemical antiseptics to sterilize tortuous wounds. He supported Wright in advocating hypertonic saline solution as a physiological irrigant for septic lacerations. Demobilized with captain's rank, Fleming resumed studying antibacterial mechanisms at St. Mary's in 1919.

His best work followed in the next decade. Using "slide cells," he showed that ordinary germicides damaged the leucocytes in artificially infected blood at dilutions harmless to the bacteria. He therefore condemned the intravenous administration of chemical antiseptics, asserting that ideal therapeutic antibacterial agents should arrest the growth of bacterial invaders without affecting host tissues. Fleming did not systematically search for such entities, but through sharp observation, pertinacious curiosity, and a prepared mind he discovered two outstandingly important antibacterial substances.

In 1921, while inspecting a contaminated culture plate, he observed nasal mucus dissolving a yellowish colony. The bacteriolytic agent was named "lysozyme," and the susceptible organism (at Wright's suggestion) *Micrococcus lysodeikticus.* With V. D. Allison's collaboration, Fleming detected lysozyme in human blood serum, tears, saliva, and milk; and in such diverse animal and plant substances as leucocytes, egg white, and turnip juice. Since inoffensive airborne bacteria were lyzed more readily than pathogenic species, chemical concentration of the active principle was attempted, without success. (Lysozymes were later crystallized in other laboratories; because of their specific disruptive action on the cell wall of certain gram-positive organisms, these enzymes have proved valuable in studies of bacterial cytology.) Fleming then became engrossed with another lytic agent.

In 1928 he noted a culture plate displaying *Penicillium* mold surrounded by lyzed colonies of staphylococci. Culture filtrates of the mold (later identified by C. Thom as *Penicillium notatum*) were antibacterial for many pathogenic species. He reported (1929) that "penicillin" did not interfere with leucocytic function, was nontoxic to laboratory animals, and "may be an efficient antiseptic for application to, or injection into, areas infected with penicillin-sensitive microbes." The intended clinical trials were abandoned because these crude preparations had unpredictable and fleeting potency, which Fleming's knowledge of chemistry was inadequate to overcome. Although he enlisted aid from junior colleagues and from the biochemist H. Raistrick, the problem remained unsolved. Thereafter, he used the "mold juice" mainly for isolating penicillin-insensitive bacteria from mixed cultures.

Following Domagk's work on prontosil (1935), Fleming studied the antibacterial properties of sulfonamides. But he never lost confidence that penicillin would be stabilized and purified, and rejoiced when Ernst Chain, Howard Florey, and their co-workers accomplished this at Oxford in 1940. Within two years, the antibiotic's remarkable powers were established. Mounting war casualties entailed securing the highest priority for its large-scale manufacture, and steps were taken to achieve this in the United States, Britain, and Canada.

As supplies of penicillin expanded and its efficacy became more widely known, tributes and honors showered upon Alexander Fleming. He was elected to fellowship of the Royal Society in 1943; knighted (K.B.) in 1944; received the Nobel Prize in medicine, jointly with Florey and Chain, in 1945; and was awarded numerous foreign decorations and medals, honorary memberships in medical and scientific societies, and doctorates from famous universities. He was elected president of the newly founded Society for General Microbiology (1945) and rector of Edinburgh University (1951) and was given the freedom of Dorval, Chelsea, and Paddington—where he had gone to school, lived, and worked. Between 1945 and 1953, despite his limited eloquence, he undertook a succession of speech-making triumphal tours of the United States and other countries.

Fleming at first accepted the honors and acclaim diffidently, but later came to enjoy them. His main characteristics persisted: the evening game of snooker at the Chelsea Arts Club (where he held a long-treasured honorary membership), the pawky humor, the taciturnity, the basic dedication to the health of mankind.

BIBLIOGRAPHY

I. Original Works. Fleming was sole or senior author of about 100 scientific papers, which appeared mainly in well-known British medical or scientific journals between 1908 and 1954. The majority were reports or reviews of original work on vaccine therapy, wound infection, antiseptics, lysozyme, penicillin, and other antibiotics. Occasionally he described novel techniques or wrote brief biographical memoirs. Philosophical or nonscientific topics were generally beyond his range. He contributed a few individual chapters or sections to composite works and committee reports and was author or coauthor of two books in his own special fields. Among his best papers were those prepared as addresses for endowed lectureships. Certain of these are included in the appended representative list.

No complete bibliography has been printed, but fairly comprehensive lists of his principal publications (containing several minor errors) were appended to Leonard Colebrook's memoir of Fleming and to the biography by André Maurois. A collection of his published works and unpublished manuscripts—including letters, diaries, laboratory notebooks, and many other documents relating to Fleming—is deposited with the Sir Alexander Fleming Museum in the Wright-Fleming Institute.

Fleming's works include "Some Observations on the Opsonic Index, With Special Reference to the Accuracy of the Method and to Some of the Sources of Error," in *Practitioner,* **80** (1908), 607–634; "On the Etiology of Acne Vulgaris and Its Treatment by Vaccines," in *Lancet* (1909), **1**, 1035–1038; "On the Use of Salvarsan in the Treatment of Syphilis," *ibid.* (1911), **1**, 1631–1634, written with L. Colebrook; "On the Bacteriology of Septic Wounds," *ibid.* (1915), **2**, 638–643; "The Action of Chemical and Physiological Antiseptics in a Septic Wound," in *British Journal of Surgery,* **7** (1919), 99–129, the Hunterian lecture; "On a Remarkable Bacteriolytic Element Found in Tissues and Secretions," in *Proceedings of the Royal Society,* **93B** (1922), 306–317; "Lysozyme—A Bacteriolytic Ferment Found Normally in Tissues and Secretions," in *Lancet* (1929), **1**, 217–220, the Arris and Gale lecture; "The Staphylococci," in *A System of Bacteriology in Relation to Medicine,* II (London, 1929), 11–28; "On the Antibacterial Action of Cultures of a Penicillium, With Special Reference to Their Use in the Isolation of *B. influenzae*," in *British Journal of Experimental Pathology,* **10** (1929), 226–236; "The Intravenous Use of Germicides," in *Proceedings of the Royal Society of Medicine,* **24** (1931), 46–58; *Recent Advances in Vaccine and Serum Therapy* (London, 1934), written with G. F. Petrie; "Serum and Vaccine Therapy in Combination With Sulphanilamide or M and B 693," in *Proceedings of the Royal Society of Medicine,* **32** (1939), 911–920; "The Effect of Antiseptics on Wounds," *ibid.,* **33** (1940), 487–502;

"Streptococcal Meningitis Treated With Penicillin," in *Lancet* (1943), **2,** 434–438; "Penicillin: Its Discovery, Development and Uses in the Field of Medicine and Surgery," in *Journal of the Royal Institute of Public Health and Hygiene,* **8** (1945), 36–49, 63–71, 93–105, the Harben lectures; "The Morphology and Motility of *Proteus vulgaris* and Other Organisms Cultured in the Presence of Penicillin," in *Journal of General Microbiology,* **4** (1950), 257–269, written with A. Voureka, I. R. H. Kramer, and W. H. Hughes; *Penicillin* (London, 1946); and "Twentieth Century Changes in the Treatment of Septic Infections," in *New England Journal of Medicine,* **248** (1953), 1037–1045, the Shattuck lecture.

II. SECONDARY LITERATURE. Obituaries include "Sir Alexander Fleming, M.D., D.Sc., F.R.C.P., F.R.C.S., F.R.S.," in *British Medical Journal* (1955), **1,** 732–735, unsigned; "Alexander Fleming, Kt., M.B., Lond., F.R.C.P., F.R.C.S., F.R.S.," in *Lancet* (1955), **1,** 624–626, unsigned; V. D. Allison, "Sir Alexander Fleming, 1881–1955," in *Journal of General Microbiology,* **14** (1956), 1–13; L. Colebrook, "Alexander Fleming, 1881–1955," in *Biographical Memoirs of Fellows of the Royal Society of London,* **2** (1956), 117–127; and R. Cruikshank, "Alexander Fleming, 6th August 1881–11th March 1955," in *Journal of Pathology and Bacteriology,* **72** (1956), 697–708.

A biography is André Maurois, *The Life of Sir Alexander Fleming,* Gerard Hopkins, trans. (London, 1959; Penguin ed., 1963). Lady Fleming persuaded Maurois to undertake this very readable biography, which shows occasional bias and contains some inaccuracies.

Various short appreciations and unusual photographs of Fleming were published in *St. Mary's Hospital Gazette,* **61,** no. 3 (1955), 58–74.

Special references are E. P. Abraham, E. Chain, C. M. Fletcher, H. W. Florey, A. D. Gardner, N. G. Heatley, and M. A. Jennings, "Further Observations on Penicillin," in *Lancet* (1941), **2,** 177–189; E. Chain, H. W. Florey, A. D. Gardner, N. G. Heatley, M. A. Jennings, J. Orr-Ewing, and A. G. Sanders, *ibid.* (1940), **2,** 226–228; R. D. Coghill, "Penicillin: Science's Cinderella," in *Chemical and Engineering News,* **22** (1944), 588–593; R. D. Coghill and R. S. Koch, "Penicillin: A Wartime Accomplishment," *ibid.,* **23** (1945), 2310–2316; L. Colebrook, *Almroth Wright: Provocative Doctor and Thinker* (London, 1954); R. Hare, *The Birth of Penicillin and the Disarming of Microbes* (London, 1970); and M. R. J. Salton, "The Properties of Lysozyme and Its Action on Microorganisms," in *Bacteriological Reviews,* **21** (1957), 82–99.

CLAUDE E. DOLMAN

FLEMING, JOHN (*b.* Kirkroads, near Bathgate, Linlithgowshire, Scotland, 10 January 1785; *d.* Edinburgh, Scotland, 18 November 1857), *zoology, geology.*

The son of Alexander Fleming, a tenant farmer of moderate means, and Catherine Nimmo Fleming, John Fleming completed his studies at the University of Edinburgh in 1805 and was licensed as a minister in the Church of Scotland in 1806. He served in the small parishes of Bressay, Shetland (1808–1810), and Flisk, Fifeshire (1810–1832), and in the larger parish of Clackmannan, near Edinburgh (1832–1834), before becoming professor of natural philosophy at King's College, Aberdeen, in 1834. When the Free Church broke away in 1843, Fleming joined it; and in 1845 he became professor of natural science at New College (Free Church), Edinburgh. He married Melville Christie in 1813, and they had two sons, of whom only one, Andrew, lived to adulthood.

A member of the Wernerian Natural History Society from its founding in 1808, Fleming was elected a fellow of the Royal Society of Edinburgh in 1814. He was later a member of the Royal Physical Society of Edinburgh and other scientific societies. After 1834 he began to suffer spells of ill health, which became more frequent until his death. Fleming's disposition tended to be grave and critical—what humor he had was often sarcastic—but his kindness and honesty were appreciated by those who knew him.

Fleming's scientific career may be divided into four periods: a Wernerian period (1808–1820), when he was a disciple of Robert Jameson; a period of productivity and controversy (1820–1832); a period of reduced productivity (1832–1845), due to the pressure of his duties at Clackmannan and Aberdeen; and a period of renewed activity at New College (1845–1857).

Regarded as Scotland's foremost zoologist as early as 1815, Fleming was concerned largely with the description and classification of freshwater and marine invertebrates. In his *Philosophy of Zoology* (1822), he advocated the binary or dichotomous system of classification, in which a twofold division is made at every level into those animals that possess a certain character and those that do not, a different character being used at each level. Despite his claim that this system was a practical approach to a true natural system, little notice was taken of it by other zoologists because of its artificiality. His *History of British Animals* (1828) was a detailed description and classification of the British fauna, although it omitted the insects, which were supposed to be covered in a succeeding volume that never appeared. The book was noteworthy for its inclusion of fossil species and for its application of the binary system throughout.

Although Fleming always had a low opinion of James Hutton's theory of the earth because of its plutonism and because he considered Hutton an incompetent mineralogist, he did agree with the Huttonians that knowledge of the present was the key to the past. Thus, in the *Philosophy of Zoology* one can find the following views that Fleming's friend,

Charles Lyell, would later develop as a part of his uniformitarian attack on catastrophism: (1) an emphasis on noncatastrophic causes of the extinction of species, particularly (in Fleming's case) man's activities; (2) a tendency to reject the evidence for a progressive increase in the complexity and perfection of life during geologic time; (3) the idea that world climate might have been affected by an increase in the amount of land during the course of geologic history; and (4) the rejection of the theory of an originally molten earth that has slowly cooled down.

From 1824 to 1826, with Jameson's secret encouragement, Fleming engaged in a controversy with William Buckland in which he showed that Buckland's theory of a violent deluge contradicted both the Bible and the scientific evidence. Fleming argued that the Bible represented the deluge as nonviolent, and he attributed the so-called "diluvial" phenomena to local river floods, the bursting of lakes, and uprisings of the sea. In a controversy with William Conybeare in 1829, Fleming disputed the fossil evidence for a warmer climate in the past, insisting that our knowledge of the habits of an existing species can tell us nothing about the behavior of a similar, but not identical, fossil species, since every species is fixed and unique in its behavior.

Fleming had argued in his controversy with Buckland that there must be free inquiry in science without regard for the Bible, yet in his *History of British Animals* he adopted a scheme of reconciliation between geology and Genesis that was originally proposed by a friend, the Reverend Thomas Chalmers, on the basis of Cuvier's idea of successive creations. This theory, which was elaborated by Fleming in his *Lithology of Edinburgh* (1859), assumed that the "pre-Adamic" life had been totally destroyed by some extraordinary cause (probably the darkening of the sun) accompanied by debacles of water rushing over the earth. The species of animals and plants of the present epoch had then been created during the six days (or periods) described in Genesis. Similar revolutions, he believed, had initiated at least five previous epochs in earth history. Thus it can be seen that Fleming rejected Lyell's uniformitarian views for earth history as a whole. The idea of evolution was, of course, anathema to him, and we find him in 1854 citing Scripture in opposition to it.

Fleming was by nature conservative, reluctant to alter his views once they had been firmly established. The discrediting of neptunism around 1820 only served to reinforce his already skeptical attitude toward geological theory. His contributions to geology were therefore essentially negative—the effective criticism of inadequate theories. His views, which were relatively enlightened in the 1820's, when they were opposed to the catastrophist excesses of Cuvier and Buckland, were definitely outmoded by the 1850's. Despite his extensive zoological and paleontological knowledge (he was, for example, the first to discover the remains of fish in the Old Red Sandstone), Fleming had been trained in the old mineralogical school of geology and apparently never fully accepted the new geology, which relied boldly upon paleontological criteria in correlating strata.

BIBLIOGRAPHY

I. ORIGINAL WORKS. Fleming's most important works were *The Philosophy of Zoology,* 2 vols. (Edinburgh, 1822); *A History of British Animals* (Edinburgh, 1828); and *The Lithology of Edinburgh,* ed., with a memoir, by Rev. John Duns (Edinburgh, 1859).

A list of Fleming's works by Alexander Bryson, *Transactions of the Royal Society of Edinburgh,* **22** (1861), 675–680, is fairly complete (129 items, including two, nos. 80 and 109, which are not by Fleming according to W. E. Houghton, ed., *The Wellesley Index to Victorian Periodicals, 1824–1900*).

The list in the Royal Society of London *Catalogue of Scientific Papers* (*1800–1863*), II, contains fifty-three items, of which six are not in Bryson's list. The Royal Society *Catalogue,* however, has fuller and sometimes more correct entries, and it usually gives all the journals in which each article appeared. These lists do not cite all the articles by Fleming in the *Edinburgh Encyclopaedia, Encyclopaedia Britannica,* supp., vol. III (1824), *Edinburgh Monthly Review,* and *New Edinburgh Review.*

There are a few of Fleming's letters still extant, notably several to Lyell at the American Philosophical Society, Philadelphia.

II. SECONDARY LITERATURE. The memoir by Duns (see above) contains extracts from Fleming's correspondence. A memoir by Bryson in *Transactions of the Royal Society of Edinburgh,* **22** (1861), 655–675, emphasizes his scientific work. See also "Scottish Natural Science," in *North British Review,* **28** (1858), 39–55, by Duns, which is largely devoted to Fleming.

L. E. PAGE

FLEMING, JOHN AMBROSE (*b.* Lancaster, England, 29 November 1849; *d.* Sidmouth, Devon, England, 18 April 1945), *electrical engineering.*

The son of a Congregational minister, Fleming moved to London in 1863 and attended University College School and later University College, graduating in 1870. After a period of alternate science teaching and additional study, he entered Cambridge University in 1877 to work under James Clerk Maxwell in the new Cavendish Laboratory. There he helped to repeat the century-old electrical experiments of Henry Cavendish, whose notes on them had recently come to light. Fleming was made demonstrator in 1880, and in the following year he became professor of physics and mathematics in the newly constituted University College at Nottingham.

He resigned after a year to become a consultant to the Edison Electric Light Co. in London. In 1885 he was appointed professor of electrical technology at University College, a post he held for forty-one years. He made many contributions to the design of transformers, to the understanding of the properties of materials at liquid-air temperatures, to photometry, and to electrical measurements in general. He was an outstanding teacher; the right-hand rule (a mnemonic aid relating direction of magnetic field, conductor motion, and the induced electromotive force) is attributed to him. He was also a highly successful popular scientific lecturer.

At University College he experimented widely with wireless telegraphy and gave special courses on the subject. He was aware of Edison's observation of the "Edison effect" (he had visited the United States and met Edison in 1884) of unilateral flow of particles from negative to positive electrode, and he repeated some of the experiments, with both direct and alternating currents, beginning in 1889. During the following years, he cooperated with Marconi in many of his experiments and helped to design the transmitter employed by Marconi in spanning the Atlantic in 1901. Thus it was not until 1904 that he returned to his experiments on the Edison effect, with a view to producing a rectifier that would replace the inadequate detectors then used in radiotelegraphy. He named the resulting device a "thermionic valve," for which he obtained a patent in 1904. This was the first electron tube, the diode, ancestor of the triode and the other multielectrode tubes which have played such an important role in both telecommunications and scientific instrumentation.

Fleming led an incredibly active scientific life. He read the first paper ever presented to the Physical Society on its foundation in 1874 and read his last paper to the same body sixty-five years later, in 1939. His career covered the time from Maxwell to the advent of electronic television. He published more than a hundred important papers. He was an active president of the Television Society from 1930 until his death and received many other honors, including the Hughes Medal in 1910 from the Royal Society (he was made F.R.S. in 1892); the Faraday Medal of the Institution of Electrical Engineers (1928); and the Gold Medal of the Institute of Radio Engineers (1933). He was knighted in 1929. He became professor emeritus after his retirement in 1926 but continued to be scientifically active nearly until his death at the age of ninety-five. In 1933, when he was eighty-four, Fleming, whose first wife (Clara Ripley Pratt) had died in 1917, was married to Olive May Franks, who survived him. He had no children.

BIBLIOGRAPHY

I. Original Works. Fleming wrote several important texts and handbooks on electrical engineering, beginning with the 2-vol. treatise *The Alternate Current Transformer* (London, 1889–1892) and leading up to the monumental *Principles of Electric Wave Telegraphy* (London, 1906), which went into several editions and remained the standard work for many years; it is a rich source of historical information as well. Another of his books to have long-term influence was *The Propagation of Electric Currents in Telephone and Telegraph Conductors* (London, 1911). A collection of his papers is in the library of University College, London, which also houses his own library and a number of the earliest diodes.

II. Secondary Literature. A short illustrated biography published on the fiftieth anniversary of the invention of the diode, J. T. MacGregor-Morris, *The Inventor of the Valve* (London, 1954), contains a list of Fleming's books. The same author prepared the entry in *Dictionary of National Biography, 1941–1950.* There is an obituary by W. H. Eccles in *Obituary Notices of Fellows of the Royal Society,* no. 14 (Nov. 1945), and a tribute in *Notes and Records of the Royal Society* (Mar. 1955).

Charles Süsskind

FLEMING, WILLIAMINA PATON (*b.* Dundee, Scotland, 15 May 1857; *d.* Boston, Massachusetts, 21 May 1911), *astronomy.*

Mrs. Fleming's father, Robert Stevens, a craftsman whose shop sold picture frames, died when she was seven; her mother was Mary Walker Stevens. After a public school education in Dundee, she married James Orr Fleming on 26 May 1877, and they immigrated to Boston at the end of 1878. Shortly thereafter the marriage fell apart, and Mrs. Fleming found it necessary to support herself and her infant son, Edward Pickering Fleming, who was born 6 October 1879. In 1881, after a period of domestic work for Edward C. Pickering, the new director of the Harvard College Observatory, she became a full-time copyist and computer at the observatory itself.

At that time Pickering had just embarked on an extensive program of celestial photography. Through her studies of the objective prism spectrum plates, usually in collaboration with Pickering, Mrs. Fleming became the leading woman astronomer of her day. Her suspicions aroused by the spectral peculiarities she observed, she discovered more than 200 variable stars and ten novae—the latter being a significant fraction of the twenty-eight novae recorded up to the time of her death.

Her most important contribution was the classification of 10,351 stars in the *Draper Catalogue of Stellar Spectra* (published as volume XXVII of the *Annals of Harvard College Observatory* [1890]). The spectra were organized into seventeen categories, lettered from *A* to *Q*, but 99.3 percent of the stars fell in the six classes *A, B, F, G, K,* and *M.* Although it was a great advance over the four types into which Angelo Secchi had visually classified about 4,000 stellar spectra, Mrs. Fleming's system was soon to be enormously refined at Harvard by Annie Jump Cannon.

Mrs. Fleming's keen eyesight, remarkable memory, and industrious nature enabled her to advance to a position of considerable authority at the observatory, so that ultimately she gave assignments to a corps of a dozen women computers. In 1899 she was appointed curator of astronomical photographs, and by 1910 she had examined nearly 200,000 plates. In 1906 she became the fifth woman member (honorary) of the Royal Astronomical Society. Dorrit Hoffleit has written, "Sparkling and friendly though she was, her reputation as a strict disciplinarian lived after her, and as late as the 1930's elderly ladies who had worked with her in their youth still regarded her with awe."

BIBLIOGRAPHY

Mrs. Fleming's principal work appeared in the *Annals of Harvard College Observatory,* including an important paper, "Stars Having Peculiar Spectra," published posthumously in vol. **56** (1912).

Most published biographical material derives from the obituaries by Annie Jump Cannon, in *Astrophysical Journal,* **34** (1911), 314–317; Edward C. Pickering, privately printed (1911), repr. in *Harvard Graduates Magazine,* **20** (1911), 49–51; or from the sketch by Grace A. Thompson, in *New England Magazine,* **48** (1912), 458–467.

The best biography, incorporating some new material, is Dorrit Hoffleit, in *Notable American Women* (Cambridge, Mass., 1971).

OWEN GINGERICH

FLEMMING, WALTHER (*b.* Sachsenberg, Mecklenburg, Germany, 21 April 1843; *d.* Kiel, Germany, 4 August 1905), *anatomy, cytology.*

Flemming's family was Flemish; his father, C. F. Flemming, had moved from Jüterbog to Mecklenburg to become director of a lunatic asylum there. In this region of Germany, Walther was born and brought up. He was educated at the Gymnasium Fridericianum, where he showed great aptitude for literature and philology, but for his university studies he chose medicine. After one semester at Göttingen he went to Tübingen. There followed a semester of intensive study at Berlin, then at Rostock, where, after a long illness from typhoid fever, he completed his medical studies with his thesis on the ciliary muscles of certain mammals (1868).

His subsequent work as assistant in the department of internal medicine under Thierfelder was short-lived, for in 1869 he was able to return to zoological research, first under his former teacher, Franz E. Schulze, and then as private assistant to Semper in Würzburg, studying the sensory epithelia of the mollusks. That autumn saw him in Amsterdam, where, under Willie Kühne, he completed a detailed study of the structure and physiology of fat cells. The Franco-Prussian War obliged him to interrupt his studies and join the medical corps reserve at Saarbrücken, in which he remained active until the autumn of 1871. Then he returned to Rostock and presented his *Habilitationsschrift,* "Über Bindessubstanzen und Gefässwandung bei Mollusken." Here as a *Privatdozent* he worked on connective tissue and gave a public lecture on the microscope, *Über die heutigen Aufgaben des Mikroskops . . .* (Rostock, 1872). That year he moved to Prague, where his work flourished and he formed a happy circle of friends among his colleagues. Only bad feeling among the students, two thirds of whom were Czech nationalists, caused him to leave in 1875 for Königsberg. Finally, in 1876, he settled at the small University of Kiel as professor of anatomy and director of the Anatomical Institute, a position he held until his retirement in 1901. Here he carried out the major part of his great work on cell division published in his classic book, *Zellsubstanz, Kern und Zelltheilung* (Leipzig, 1882). Flemming was a bachelor and for most of his time in Kiel lived with his sister Clara.

In the 1850's attempts to study the process of cell multiplication were vitiated by inadequate techniques of staining and the poor resolving power of lenses. By the 1860's it had become certain that before a cell divides the nucleus must first give rise to two daughter nuclei. This appeared to result either from direct fission or from a more indirect process in which dissolution of the mother nucleus was followed by coagulation of two daughter nuclei about two new centers; but the phases of this metamorphosis remained a mystery. It was seen that nucleoli came and went during indirect division, that their number was constant, and that the midpoint of the division process was marked by the appearance of nuclear granules. In the 1870's these granules were observed to elongate into threadlike structures which split up in some way, yielding the material for two daughter nuclei. It was Flemming's achievement to observe and interpret the stages correctly, to identify them in a wide variety of tissues, and to give indirect division the name by which it is still widely known—mitosis.

With his background in the study of connective tissues, epithelia, and fat cells, Flemming was well-prepared to attack the problem of cell division. His serious study of this subject would appear to date from 1874 while he was in Prague. His third paper in this field ("Beobachtungen über die Beschaffenheit des Zellkerns," in *Archiv für mikroskopische Anatomie,* **13** [1877], 693–717), on the constitution of the cell nucleus, described the use of Hermann's technique of overstaining followed by differentiation in alcohol. This led him to perceive the thread net (*Fadennetz*) with associated vacuoles and nucleoli as a constant feature of the living cell. That winter he still held to the direct-division theory of nuclear multiplication and lectured to that effect in Kiel in February 1878, but by the summer he had studied the same cells in living material as well as in fixed. Then he found what he called the *Äquatorial-platte* ("equatorial plate") stage (early anaphase), which is so easily missed when working only from fixed material, and realized that what he had called direct division was in fact the indirect process. His lecture and correction appeared in *Schriften des naturwissenschaftlichen Vereins für Schleswig-Holstein* (Kiel, February and August 1878).

By December 1879 he had investigated all the stages of mitotic division, the results of which appeared in his three famous papers: "Beiträge zur Kenntnis der Zelle und ihrer Lebenserscheinungen" (pts. 1, 2, and 3, 1879–1881). What had been taken for granules Flemming recognized as sections through threads; what hitherto had appeared as a vague and discontinuous process he saw as a continuous metamorphosis from the skein of the mother nucleus to thread loops (chromosomes) and back again to the skeins of the two daughter nuclei. In 1880 he coined the term "chromatin" for the stainable substance of the nucleus; the nonstaining portions he termed "achromatin." As the cell approaches division, the chromatin separates from the achromatin and becomes organized into a tangled skein (*Knäuel*) which proceeds to break up into figure-eight wreaths (*Kränze*), the loops of which open out to give stars (*Sterne*). At this stage the chromatin is organized with respect to a single center or pole (Flemming's "monocentric" phase), but this soon goes over to organization around two poles ("dicentric" phase). As the poles move apart, the star figures of chromatin pull apart; and because the arms of the stars have doubled along their length, four V-shaped loops (the chromatids) can be discerned, two traveling toward one pole and two toward the other. This is the equatorial plate stage, from which the barrel stage takes the chromatic loops back to the star, wreath, and skein stages of the nascent daughter nuclei.

Flemming was in error in believing that the chromatin loops arise by fragmentation of a continuous skein and that the loops fuse end-to-end at the close of mitosis to generate a skein once more. Hence it is natural that his terminology has been replaced: wreath by prophase, star by metaphase, equatorial plate by early anaphase, and nuclear barrel by telophase. Although he noted the same number of loops in successive divisions he did not perceive their continuity and individuality. This called for material with much lower chromosome numbers than he used. Flemming's contribution was to force the data from a wide range of apparently differing division processes into a single framework, to rule out so-called direct division of nuclei by simple fission, and to perceive longitudinal division of chromatin loops as the basis of nuclear multiplication, rather than transverse division as Strasburger believed.

In the summer of 1879 Flemming turned his attention to nuclear division in the testes and concluded that the spermatozoa are formed from cells whose nuclei have arisen by indirect division, the head of the spermatozoon being composed solely of chromatin. He also failed at this time to observe doubled threads in the closing nuclear figures. In 1882, in his book on cell division, he declared this to have been an error, for now he could detect doubled threads.

It was not until 1887 that he succeeded in clarifying the details of indirect division in the spermatozoa. Then he recognized two divisions, the first of which he termed heterotypic, the second homotypic. In the first, the threads formed curious rings and did not appear to double; in the second, doubling occurred and no ring structures were formed, the resulting nuclei having half the normal complement of chromosomes. In fact, doubling takes place in the first, not the second, division.

What Flemming called ring structures were the paired homologous chromosomes (bivalents) that Jansenns described accurately in 1909. The distinction between meiosis and mitosis is of course a deep and subtle one that Flemming, in his desire to unify the data, failed to appreciate. In 1880 he traveled to the zoological station at Naples to study cell division in the formation of the echinoderm egg and established there also indirect division as the mechanism of cell division.

Flemming's great merit as a theoretician lay in his attempt to find a single process to fit all forms of cell division. History has justified his vision.

In the 1880's attempts were frequently made to tie down the living substance to a particular class of compounds. Flemming would have none of this, for to him life resided in the whole body of the cell. Yet he took sides in the dispute between those who favored a network structure of protoplasm and those who preferred a filar structure. He belonged to the latter group. Such ideas were of course premature, and none has served as a foundation for later developments.

Among Flemming's other contributions may be mentioned his work on amitotic division, which he believed to characterize pathological and senescent tissues only; he improved Flesch's fixative in 1882 by addition of acetic acid to give Flemming's fluid; he introduced Flemming's stain (safranine-gentian violet-Orange G) in 1891. In his later years he returned to his earlier studies of connective tissue development and in 1902 wrote a valuable treatise, "Histogenese der Stützsubstanzen der Bindegewebesgruppe," for O. Hertwig's *Handbuch der vergleichenden und experimentellen Entwickelungsgeschichte der Wirbeltiere.*

BIBLIOGRAPHY

I. Original Works. Flemming published 80 papers and two books, all listed in Spee's obituary notice (below). The majority of his papers appeared in the *Archiv für mikroskopische Anatomie und Entwicklungsmechanik,* beginning with his doctoral dissertation of 1868 and ending with his paper on rod cells 30 years later. Most important are his "Beiträge zur Kenntnis der Zelle," pt. 1, **16** (1879), 302–436; pt. 2, **18** (1880), 152–159; and pt. 3, **20** (1881), 1–86. Pt. 2, setting out the details of mitosis, has been translated into English by Leonie Piternick and is in *Journal of Cell Biology,* **25** (1965), 3–69. Important also are his "Neue Beiträge zur Kenntnis der Zelle," pt. 1, **29** (1887), 389–463; pt. 2, **37** (1891), 685–751, the first analyzing meiosis and the second introducing the Flemming stain.

Flemming's first paper on cell division was "Über die ersten Entwickelungserscheinungen am Ei der Teichmuschel," **10** (1874), 257–292. The first chromatin figures of mitosis are "Studien zur Entwickelungsgeschichte der Najaden," in *Sitzungsberichte der mathematisch-naturwissenschaftlichen Classe der Kaiserlichen Akademie der Wissenschaften,* **71** (1875), pt. 3, 81–212. Major works are *Zellsubstanz, Kern und Zellteilung* (Leipzig, 1882), and "Histogenese der Stützsubstanzen der Bindesubstanzgruppe," in *Handbuch der vergleichenden und experimentellen Entwickelungsgeschichte der Wirbeltiere,* 3 vols. (Jena, 1902–1906).

II. Secondary Literature. No biographies and very few obituaries have been published. The present account is based largely on F. von Spee, "Walther Flemming," in *Anatomischer Anzeiger,* **28** (1906), 41–59, with portrait and complete bibliography, and on a short note by H. J. Conn, "Walther Flemming," in *Stain Technology,* **8** (1933), 48–49, with portrait.

Discussions of Flemming's work may be found in E. B. Wilson, *The Cell in Development and Inheritance* (New York, 1896); and A. Hughes, *A History of Cytology* (London–New York, 1959). Valuable also are W. Coleman, "The Nucleus as the Vehicle of Inheritance," in *Proceedings of the American Philosophical Society,* pt. 3, **109** (1965), 124–158; and A. Bolles Lee, *The Microtomist's Vade-Mecum,* 3rd ed. (London, 1893).

Robert Olby

FLETCHER, WALTER MORLEY (*b.* Liverpool, England, 21 July 1873; *d.* London, England, 7 June 1933), *physiology.*

Fletcher was the youngest of six sons in a family of ten children born to Alfred Evans Fletcher and his wife, Sarah Elizabeth Morley, cousin of the philanthropist Samuel Morley and of the future prime minister H. H. Asquith. Both parents were Congregationalists from Yorkshire. Fletcher's father, who in 1851 had won the gold medal in chemistry at University College, London, was inspector of alkali works for the local government board in Liverpool and later served as chief inspector in London.

Fletcher was educated at University College School in London, where he did not win great distinction, and in 1891 matriculated with a subsizarship at Trinity College, Cambridge. An elder brother, Herbert, had preceded him at Cambridge and, before going on to a medical career, had briefly pursued physiological research in the laboratory founded and directed by Michael Foster. Although Walter also intended to qualify for medical practice, physiology was from the first his chief aim.

Fletcher's intellectual powers matured rapidly in the Cambridge setting, and he took a first class in both parts of the natural sciences tripos in 1894 and 1895. After graduating B.A. in 1894, he was elected to the Coutts Trotter studentship in 1896 and to fellowship of Trinity College in 1897. In the same year he won the Wallingham Medal and an open scholarship at St. Bartholomew's Hospital, where by 1900 he had completed the clinical studies required for medical qualification. During this period Fletcher also taught at Cambridge. He received the M.A. from Cambridge in 1898, M.B. in 1900, M.D. in 1908, and Sc. D. in 1914. He was named senior demonstrator in physiology in 1903 and served as tutor at Trinity College from 1905 to 1914. He was elected a fellow of the Royal Society in 1915 and gave the

Croonian lecture in the same year. Of the students strongly influenced by Fletcher, the most notable is A. V. Hill, who won the Nobel Prize in medicine or physiology in 1922.

Fletcher left Cambridge and original research in 1914, when he was appointed first secretary of the Medical Research Committee, created in 1913 under terms of the National Insurance Act of 1911. He had already demonstrated his organizing powers in administrative work for Trinity College and the University of Cambridge. This previous experience was fortunate, for Fletcher was almost immediately required to mobilize the Medical Research Committee for the war effort. His success was recognized by the award of the K.B.E. in 1918. Of the work done for the committee during the war, Fletcher was particularly intrigued by that which established the cause of rickets, and he thereafter took a special interest in research that offered solutions to nutritional problems.

After the war Fletcher played an important role in securing a new charter for the committee, by which it was freed from the control of the Ministry of Health and established as the independent Medical Research Council under the aegis of the Privy Council (1920). He was criticized for a certain impatience with research that had no immediate practical benefit and for his sometimes naïve optimism about the clinical value of some basic research, but his contributions were on the whole very highly regarded. Until his death in 1933, he wrote an admirable introduction to each of the annual reports of the committee (and later the council).

Fletcher also served on a number of other important government bodies, including the Royal Commission on the Universities of Oxford and Cambridge (1919–1922) and the Indian Government Committee on the Organization of Medical Research (1928–1929), of which he was chairman. In the latter role he helped secure a private gift of £250,000 for research on leukemia. The trustees of the Sir William Dunn Foundation and the Rockefeller Foundation depended heavily on his advice in giving funds for biochemical laboratories at Oxford and Cambridge and for the London School of Hygiene and Tropical Medicine. Fletcher was appointed C.B. in 1929 and received honorary doctorates from the universities of Oxford, Edinburgh, Glasgow, Leeds, Birmingham, and Pennsylvania. In 1904 he married Mary Frances Cropper, who later attracted attention for her popular writing on religion, and by whom he had one son and one daughter.

Like his brother Herbert, Fletcher was a distinguished athlete in track and field, and this may have encouraged his interest in the problem of muscular metabolism and fatigue. All of his published research papers, with the exception of two early notes dealing with minor aspects of the involuntary nervous system, are devoted to this problem. His first major investigation, published in 1898, was the basis for his election to the Trinity fellowship. It presented a serious challenge to the prevailing conception of muscular metabolism, for which the German physiologists Ludimar Hermann and E. F. Pflüger were chiefly responsible. They supposed that there was in muscle a large and complex "inogen" molecule which stored up intramolecular oxygen and then exploded upon contraction to yield the recognized end products of muscular metabolism, chiefly carbon dioxide, water, and lactic acid. This basic conception had been extended by Max Verworn to other tissues, and the general phenomena of cellular metabolism were considered so complex as almost to defy experimental analysis.

The work that led to these ideas was open to the same general criticism that Claude Bernard had applied to the way some German physiologists approached the study of digestion. Their exclusive concern with the initial and end points of the process, said Bernard, was like trying to find out what went on in a house by observing what went in the door and what came out the chimney. Similarly, but without overtly discussing the theoretical implications of his work, Fletcher showed in 1898 that the prevailing theory of muscular metabolism was based on an inadequate understanding of the events taking place between the beginning and the end of the process. He was able to prove this point by adapting for his own uses a sensitive apparatus recently devised by the Cambridge plant physiologist F. F. Blackman for studying the gaseous exchange of leaves. With this apparatus, which made it possible to measure the discharge of even small quantities of carbon dioxide at frequent intervals, Fletcher showed that there was no sudden discharge of carbon dioxide upon contraction but, rather, that most of it was discharged gradually during the process of recovery from fatigue.

All of Fletcher's later research can be considered an extension of this paper. In 1902 he showed that an excised frog muscle would contract and relax in the absence of oxygen but that oxygen greatly facilitated its recovery from fatigue. The next important step was to measure simultaneously the discharges of carbon dioxide and of lactic acid during and after contraction and to study the relationship between them. For this work, which required the development of new methods, Fletcher secured the cooperation of the eminent Cambridge biochemist Frederick Gowland Hopkins.

In 1907 they showed that the confusing and often contradictory results obtained by earlier workers on lactic-acid fatigue could be traced largely to their treatment of the muscle before or during extraction. All of the previous methods—maceration, boiling, and alcohol solutions—tended to stimulate the artificial production of lactic acid and thus to obscure the entire process. They found that the muscle could be preserved in a truly resting and uninjured state by plunging it into ice-cold alcohol. In such a muscle they found very little lactic acid. As the muscle was stimulated to a series of contractions, they found a steady increase in the amount of lactic acid up to a certain maximum, at which irritability was lost. If the fatigued muscle was then placed in oxygen, the lactic acid was greatly reduced, irritability was restored, and carbon dioxide was evolved. The process could be repeated endlessly with the same results. Fletcher and Hopkins concluded that "excised but undamaged muscle when exposed to sufficient tension of oxygen has in itself the power of dealing in some way with the lactic acid which has accumulated during fatigue" ("Lactic Acid in Amphibian Muscle," p. 297).

In 1914, with G. M. Brown, Fletcher showed that none of the processes leading to the discharge of carbon dioxide was directly related to muscular contraction itself. In an extension of his earlier paper with Hopkins, he argued that the absorption of oxygen and the evolution of carbon dioxide were related instead to the removal of lactic acid and other fatigue products. With this established, Fletcher finally criticized the "inogen" theory directly. In a joint Croonian lecture to the Royal Society in 1915, Fletcher and Hopkins incorporated all of this and other work into a new synthesis. They emphasized above all that the main biochemical processes involved in muscular contraction were really quite simple and amenable to experimental analysis. No resort need be made to some complex and hypothetical "inogen" molecule. This point of view greatly stimulated further work on muscular metabolism and on cellular metabolism in general, and Fletcher's basic conception and approach have survived the alterations in detail which that further work inevitably produced.

BIBLIOGRAPHY

I. ORIGINAL WORKS. Fletcher's scientific papers are few enough that all may be cited here: "Preliminary Note on the Motor and Inhibitor Nerve Endings in Smooth Muscle," in *Journal of Physiology,* **22** (1898), xxxvii–xl; "The Vaso-Constrictor Fibres of the Great Auricular Nerve in the Rabbit," *ibid.,* 259–263; "The Survival Respiration of Muscle," *ibid.,* **23** (1898), 10–99; "The Influence of Oxygen Upon the Survival Respiration of Muscle," *ibid.,* **28** (1902), 354–359; "Preliminary Note on the Changes in the Osmotic Properties of Muscle Due to Fatigue," *ibid.,* xli–xlii; "The Relation of Oxygen to the Survival Metabolism of Muscle," *ibid.,* 474–498; "The Osmotic Properties of Muscle, and Their Modifications in Fatigue and Rigor," *ibid.,* **30** (1904), 414–438; "Lactic Acid in Amphibian Muscle," *ibid.,* **35** (1907), 247–309, written with F. G. Hopkins; "On the Alleged Formation of Lactic Acid in Muscle During Autolysis and in Post-survival Periods," *ibid.,* **43** (1911), 286–312; "Lactic Acid Formation, Survival Respiration and Rigor Mortis in Mammalian Muscle," *ibid.,* **47** (1913), 361–380; "The Carbon Dioxide Production of Heat Rigor in Muscle, and the Theory of Intra-molecular Oxygen," *ibid.,* **48** (1914), 177–204, written with G. M. Brown; and "Croonian Lecture (1915): The Respiratory Process in Muscle and the Nature of Muscular Motion," in *Proceedings of the Royal Society,* **89B** (1917), 444–467, written with F. G. Hopkins.

II. SECONDARY LITERATURE. Maisie [Mary Frances Cropper] Fletcher, *The Bright Countenance: A Personal Biography of Walter Fletcher* (London, 1957), a sometimes silly book by Fletcher's widow, at least bears an apt title. It also contains a complete bibliography of his writings, a list of his honors and of his obituary notices, and a valuable forty-page supplement by Sir Arthur MacNalty, which is particularly notable for the excellent survey it gives of the work done by the Medical Research Committee (Council) during Fletcher's period as secretary.

T. R. Elliot, who also worked in the Cambridge physiological laboratory at the turn of the century, has contributed three valuable sketches of Fletcher's life and work: *Dictionary of National Biography (1931–1940),* pp. 284–285; *Nature,* **132** (1933), 17–20; and *Obituary Notices of Fellows of the Royal Society of London,* **1,** no. 2 (1933), 153–163.

See also *The* [London] *Times* (8 June 1933), p. 14; *Cambridge Review* (13 Oct. 1933); *Lancet* (1933), **1,** 1319; and *British Medical Journal* (1933), **1,** 1085.

GERALD L. GEISON

FLETT, JOHN SMITH (*b.* Kirkwall, Orkney, Scotland, 26 June 1869; *d.* Ashdon, Essex, England, 26 January 1947), *geology, petrology.*

Educated at the Burgh School, Kirkwall, and George Watson's College, Edinburgh, Flett entered the University of Edinburgh at the age of seventeen. His academic career, of which he left an account (published posthumously), was one of remarkable promise. He read first for the arts degree, and the classics left him with a lifelong love of Horace, Catullus, and Lucretius; the other subjects in the course were mathematics, physics, metaphysics, moral philosophy, English, and political economy. Having graduated M.A. at nineteen, Flett proceeded to the B.Sc. in natural sciences and followed this with the medical degrees M.B. and C.M. in 1894.

After practicing medicine for a short time, Flett returned to the University of Edinburgh to become assistant to the professor of geology, James Geikie, and subsequently was promoted to a lectureship in petrology. His earliest researches were on the stratigraphy and petrology of the rocks of his homeland; this work was accepted for the D.Sc. in 1899. His five years as petrologist at Edinburgh gave Flett a sure grasp of the subject and a growing reputation, especially as a result of his studies of the monchiquite-camptonite suite of minor intrusions.

In 1901 Flett joined the Geological Survey of Great Britain, succeeding Sir Jethro Teall as petrographer and quickly attaining district geologist status. In 1911 he returned to Edinburgh as assistant director in charge of the Geological Survey in Scotland. He became director of the Geological Survey of Great Britain in 1921, occupying this position with great distinction until his retirement in 1935. He was elected a fellow of the Royal Society in 1913 and was created a Knight of the Order of the British Empire in 1925.

Flett's scientific achievements, in addition to his work on Orkney, include important contributions to the geology of Cornwall, especially his work on the sediments, metamorphic rocks, and igneous masses of the Lizard and adjacent areas. That some revision of the dating of the ancient sediments was needed as a result of the work of E. M. Hendriks does not detract from his basic work. Flett's joint study with C. E. Tilley of the unusual cordierite-anthophyllite rocks of the aureole of the Land's End granite is significant in metamorphic petrology, but he is remembered more generally for the formulation, with Henry Dewey, of the spilite suite, a worldwide type characteristic of early geosynclinal history. As Survey petrographer, Flett's work ranged widely over the richly varied rocks of the British Isles and is recorded in contributions to thirty-six issues of the Survey's *Memoirs,* covering areas as separated as Caithness and Meneage (1902–1947). Abroad, his best-known work arose from a Royal Society expedition to St. Vincent after the eruption of Soufrière in 1902; Tempest Anderson was his collaborator.

After 1911 Flett's energies turned to administration, and it was his powerful leadership that left its mark on science. Some of the Geological Survey's best work was done under him, such as the completion of the highly detailed mapping of the Tertiary volcanoes and the revision of the major coalfields. More than this, it was Flett's determination that brought to fruition the move of the Museum of Practical Geology (now called the Geological Museum) from its Victorian site behind Piccadilly to the present splendid building in South Kensington.

Flett became increasingly deaf, but there were those who believed that his success as an administrator was due to his hearing only those proposals according with his own wishes.

BIBLIOGRAPHY

I. Original Works. A full list of Flett's writings is given in the Royal Society obituary notice cited below. Among them are "The Old Red Sandstone of the Orkneys," in *Transactions of the Royal Society of Edinburgh,* **39** (1898), 383–424; "The Trap Dykes of the Orkneys," *ibid.,* **39** (1900), 865–918; "Report on the Eruption of Soufrière in St. Vincent in 1902, and on a Visit to Montagne Pélée, in Martinique," in *Philosophical Transactions of the Royal Society,* **200A** (1903), 353–553, written with T. Anderson; "On Some British Pillow-Lavas and the Rocks Associated With Them," in *Geological Magazine,* 5th ser., **8** (1911), 202–209, 241–248, written with H. Dewey; "Hornfelses from Kenidjack, Cornwall," in *Memoirs of the Geological Survey. Summary of Progress for 1929* (1930), 24–41, written with C. E. Tilley; "Geology of Lizard and Meneage (Explanation of Sheet 359)," in *Memoirs of the Geological Survey of Great Britain* (London, 1912, 2nd ed. [by Flett], 1946), written with J. B. Hill; *The First Hundred Years of the Geological Survey of Great Britain* (London, 1937); and "Memories of an Edinburgh Student, 1886–1894," in *University of Edinburgh Journal,* **15** (1950), 160–182.

II. Secondary Literature. See E. B. Bailey, *Geological Survey of Great Britain* (London, 1952), *passim;* and H. H. Read, "John Smith Flett, 1869–1947," in *Obituary Notices of Fellows of the Royal Society of London,* **5** (1948), 689–696.

K. C. Dunham

FLEXNER, SIMON (*b.* Louisville, Kentucky, 25 March 1863; *d.* New York, N.Y., 2 May 1946), *pathology, bacteriology.*

Flexner was the fourth child of Morris Flexner, member of an educated Jewish family in Czechoslovakia, who immigrated to Kentucky. Starting as a peddler, he became a successful wholesale merchant. Flexner's mother, Esther Abraham, was born in Alsace. Flexner attended public schools in Louisville and was apprenticed to a druggist who sent him to the Louisville College of Pharmacy, from which he was graduated in 1882. He then worked in his eldest brother's drugstore and studied medicine at the University of Louisville, receiving the M.D. degree in 1889. Although the medical school then provided little opportunity for laboratory study, Flexner acquired a microscope, with which he studied pathological tissues and made microscopic examinations for doctors who patronized the Flexner pharmacy.

In 1890, at the suggestion of another of his remarkable brothers, Abraham, Flexner went to Baltimore to study pathology and bacteriology at the Johns Hopkins Hospital with William H. Welch, who gave him a fellowship and in 1892, when the Johns Hopkins Medical School opened, made him his first assistant in the department of pathology. In 1893 Flexner visited Europe, working at Strasbourg with Friedrich von Recklinghausen and at Prague. On his return he became resident pathologist at the Johns Hopkins Hospital. By 1899 he reached full professorial rank. In that year, following the acquisition of the Philippine Islands by the United States, Flexner and two medical students spent several months in Manila studying health conditions. During this stay he isolated an organism that causes a prevalent form of dysentery. This *Bacillus* (now *Shigella*) *dysenteriae* is still commonly known as the "Flexner bacillus."

Soon after his return to Baltimore, Flexner was appointed professor of pathology at the University of Pennsylvania, where he organized an excellent staff, planned a new laboratory building, and carried out important researches on experimental dysentery, on experimental pancreatitis, and on immunological problems, especially with regard to hemolysis and hemagglutination. One of his associates was the brilliant young Japanese physician Hideyo Noguchi, who came from Japan inexperienced and penniless and found in Flexner a lifelong friend and guide.

When bubonic plague broke out in California in 1901, the federal government sent Flexner to San Francisco to study the epidemic. Within a month he and a few associates confirmed the presence of the plague bacillus and made a report to health authorities that aided them in eradicating the disease.

In 1901 John D. Rockefeller and his son John D. Rockefeller, Jr., were planning the creation of the Rockefeller Institute for Medical Research in New York City. Flexner, who by this time, at the age of thirty-eight, was beginning to be nationally known, was appointed to the institute's board of scientific directors, which was composed of seven eminent medical men and headed by his friend and mentor William H. Welch. In 1902 Flexner was chosen to lead a department of pathology and bacteriology in the institute, and soon he established himself as head of the whole enterprise. He brought together a strong group of investigators, including Hideyo Noguchi, S. J. Meltzer, P. A. T. Levene, Alexis Carrel, Jacques Loeb, Eugene Opie, Rufus I. Cole, and Peyton Rous. Flexner's colleagues found him a man of exceedingly keen intelligence, with a reserved manner that concealed a sympathetic heart. He directed his staff with great skill, giving free rein to those who showed independent competence while guiding with a wise hand those who needed advice. His financial acumen impressed the astute patron of the institute, who showed his confidence by successive additions to its funds.

To combat an epidemic of cerebrospinal meningitis in 1906, Flexner produced a serum that remained the best treatment until the sulfa drugs were introduced. When in 1910 poliomyelitis was epidemic in New York, he and his assistants were the first to transfer the virus from monkey to monkey. This success enabled the investigators to keep the virus alive in the laboratory and thus ultimately, after development by others of a less expensive method of perpetuating it (by cultivation in hens' eggs), led to the preparation, in the 1950's, of protective vaccines.

The Rockefeller Institute, quite early in its history, came under strong attack from organizations opposed to the use of animals in experiments on the causes of disease. Flexner's accomplishments in such work and his calm generalship made him a natural leader in the successful deterrence of these opponents.

In 1903 Flexner married Helen Whitall Thomas, member of a prominent Quaker family of Baltimore. She helped to expand his intellectual interests beyond the medical sciences, giving him an appreciation of literature and the arts. Of their two sons, William became a physicist and James Thomas a writer and historian of American culture.

Flexner's medical and biological accomplishments led to public service in various health fields, as chairman of the Public Health Commission of New York State, as medical consultant of the U.S. Army during World War I, and as member of the China Medical Board of the Rockefeller Foundation. When in 1902 William H. Welch wearied of the editorship of the *Journal of Experimental Medicine,* Flexner took it over and for about fifteen years was its chief editor, giving the task much time and attention.

His executive competence was recognized by trusteeships of the Rockefeller Foundation, the Carnegie Foundation for the Advancement of Teaching, and the Johns Hopkins University. A little-known but very important public service was his leadership in establishing fellowships of the National Research Council, provided by the Rockefeller Foundation, for promising young medical scientists. Oxford University called him in 1937–1938 to its Eastman professorship, at a time when his counsel was needed in the organization of Lord Nuffield's endowment of medical professorships. A book, *The Evolution and Organization of the University Clinic* (Oxford, 1939), resulted from this experience. Flexner was elected member of the American Philosophical Society in 1901, the National Academy of Sciences in 1908, and foreign member of the Royal Society in 1919.

During his long career Flexner published several hundred scientific papers, lectures, and essays. At the age of seventy-eight he published jointly with his son James a notable biography, *William H. Welch and the Heroic Age of American Medicine* (New York, 1941). He quietly resigned the directorship of the Rockefeller Institute in 1935.

BIBLIOGRAPHY

Flexner's papers are in the library of the American Philosophical Society, Philadelphia.

Secondary literature includes Stanhope Bayne-Jones, "Simon Flexner, 1863–1946," in *Year Book* [for] *1946. American Philosophical Society* (Philadelphia, 1947), pp. 284–297; George W. Corner, *History of the Rockefeller Institute* (New York, 1965), *passim; Memorial Meeting for Simon Flexner* (New York, 1946), a pamphlet issued by the Rockefeller Institute that contains personal characterizations by John D. Rockefeller, Jr., and others; and Peyton Rous, "Simon Flexner, 1863–1946," in *Obituary Notices of Fellows of the Royal Society of London,* **6** (1948–1949), 409–445, with portrait and complete list of publications.

GEORGE W. CORNER

FLOREY, HOWARD WALTER (*b.* Adelaide, Australia, 24 September 1898; *d.* Oxford, England, 21 February 1968), *pathology.*

Florey's scientific career was devoted to the experimental study of disease processes. His most notable contribution to science was the development of penicillin as a systemic antibacterial antibiotic suitable for use in man.

Florey was the third and last child, and the only son, of Joseph Florey and his second wife, Bertha Mary Wadham, a native of Australia. Joseph Florey owned a boot factory, and the family was in comfortable circumstances. The son attended St. Peter's Collegiate School, Adelaide, as a day boy. He had a brilliant scholastic career, obtaining scholarships at St. Peter's and subsequently for study at the University of Adelaide.

From an early age Florey had decided to study medicine and carry out medical research, rather than learn the management of the family business. He enrolled in the Faculty of Medicine at the University of Adelaide in 1917 and graduated M.B., B.S. five years later. Receipt of the Rhodes Scholarship for South Australia in 1922 enabled him to go to Oxford, where he studied in the Honours Physiology School. His exposure there to the great neurophysiologist Sir Charles Sherrington profoundly affected his outlook on pathology. He was so impressed by the value for aspiring young pathologists of the physiological and biochemical outlook conferred by the Honours Physiology School that when he later became professor of pathology at Oxford, he insisted that all candidates for the Ph.D. should study at the Honours School before beginning experimental work in the Sir William Dunn School. From Oxford, Florey went to Cambridge for a year and then spent a year in the United States as a Rockefeller Foundation traveling fellow. After a short period at the London Hospital he returned to Cambridge, where he took the Ph.D. degree. The most important formative influence at Cambridge came not from the department of pathology but from the biochemist Sir Frederic Gowland Hopkins, then at the height of his career.

In 1926 Florey married Mary Ethel Reed, who had been a fellow medical student at the University of Adelaide; they had a son, Charles, and a daughter, Paquita. Mrs. Florey died in 1966, and in 1967 he married the Hon. Margaret Jennings, a former colleague in his Oxford laboratories.

In 1931 Florey was appointed Joseph Hunter professor of pathology at the University of Sheffield. Four years later he moved to Oxford as professor in charge of the Sir William Dunn School of Pathology; Sir Edward Mellanby played an important part in this appointment. It was a milestone in the history of pathology in Britain, because for the first time a man trained in experimental physiology and viewing pathology with a physiologist's eye came into a position of influence in the teaching of the subject. Florey remained professor of pathology at Oxford from 1935 until 1962, when he resigned to become provost of Queen's College, Oxford.

Florey created a very lively and stimulating atmosphere at Oxford, which led to close contacts between department members, who had been selected to cover a wide range of scientific disciplines. Not only was this spirit of collaboration all-important in the early work on penicillin, but it also resulted in the Sir William Dunn School's becoming the leading center of experimental pathology in Europe, through which a succession of able and brilliant young men passed.

Although he remained in Britain after 1922, with a life centered at Oxford after 1935, Florey remained Australian in accent and outlook. In 1944 he was invited to Australia by the prime minister, John Curtin, to report on the Australian situation in medical research. His report led to the establishment of the Australian National University as a graduate university in 1946. Florey was very closely connected with this university for the next decade, as a senior adviser with a particular interest in the John Curtin School of Medical Research. He played the major role in establishing this school, and he visited the Australian National University for consultation almost every year until 1957. In 1965 he was appointed chancellor of the university and resumed his annual visits to Canberra.

The influence of both Sherrington and Hopkins can be traced throughout Florey's career. He was exceptional among contemporary pathologists in the United Kingdom in that he was interested in the study, by physiological and biochemical methods, of the functional changes of cells which lead to pathological changes, rather than merely in the morphological description of diseased tissues. He also had the pathologist's interest in structure, and in the latter part of his life he made extensive use of electron microscopy to study structure with the greater detail and precision made possible by that instrument. His basic tools were physiological operative techniques, in the use of which he displayed superb skill and ingenuity.

The idea of antibiosis, or microbial antagonism, was not new in Florey's time; Pasteur had made observations on the topic. Neither was penicillin new; Sir Alexander Fleming had discovered it in 1929, although he had looked on it only as a useful antiseptic for local application and had not realized that it was in fact a potent systemic antibacterial substance. Florey and his colleague E. B. Chain transformed what was a bacteriological curiosity into a clinical tool of immense value, and in so doing they opened up the new industry of antibiotic production. In the context of man's cultural evolution, it is interesting to reflect that the utilization of molds for the production of antibiotics represented the first important domestication of a species since prehistoric times.

Shortly after he arrived at Oxford, Florey sought Hopkins' advice on a suitable person to lead a biochemical unit in the Dunn School of Pathology. Hopkins recommended E. B. Chain, a young refugee from Nazi Germany then in Hopkins' laboratory, and Chain moved to Oxford in 1935. This was a critically important development as far as penicillin was concerned, for Chain had the biochemical insight that enabled him to purify penicillin without loss of its potency—a feat that had eluded Harold Raistrick, who had attempted this a decade earlier on behalf of Fleming.

Florey had long had an interest in natural antibacterial substances, and in 1930 he began a study of the antibacterial properties of lysozyme, an enzyme discovered by Fleming in 1921. This work was pursued until the enzyme was purified and the nature of its substrate determined.[1] It was against this background that Florey and Chain, in 1938–1939, initiated a systematic investigation of the biological and chemical properties of the antibacterial substances produced by bacteria and molds. As they recorded in the first publication on penicillin (1940) and in their major book on the subject, *Antibiotics* (1949), they were greatly encouraged by the success of René Dubos and his colleagues in isolating tyrothricin from the soil bacterium *Bacillus brevis* and in purifying its component antibiotic polypeptides, tyrocidin and gramicidin (only to find that although they were effective antibacterial agents, they were too toxic for systemic use).[2]

Florey emphasized that his research on antibacterial agents had been conceived as an academic study with possibilities of wide theoretical interest, not as "war work." By good fortune, and with excellent scientific judgment, Florey and Chain selected Fleming's penicillin as the first substance to be studied in detail. It proved so promising experimentally in mice with streptococci, staphylococci, and gas gangrene organisms (showing true systemic antibacterial potency combined with minimum toxicity) that all the resources of the Oxford laboratory were turned to its production on a scale that would allow clinical trials to be carried out. Many investigations, involving workers in several scientific disciplines, were necessary before penicillin could be used in human medicine. Because of the variety of skills possessed by the scientists Florey had gathered around him, developmental work proceeded rapidly. A simple and effective assay system was devised, the antibacterial spectrum of penicillin was determined, and pharmacological and toxicological studies were made in mice and later in man. In spite of efforts to increase the yield from the cultures of *Penicillium notatum,* it was necessary to process 2,000 liters of culture fluid to obtain enough penicillin to treat a single case of sepsis in man. In order to scale up production very unusual equipment was used, such as enameled bedpans for culture vessels; the "factory" on South Parks Road, Oxford, was far removed from the vast fermentation tanks and sophisticated chemical engineering of the modern antibiotics industry.

In 1941 penicillin was used in treating nine cases of human bacterial infection. All responded dramatically. The next stage called for other skills, at which Florey proved to be as adept as he had been in the laboratory. Industry in wartime Britain could not be expected to produce supplies of penicillin in the amounts so urgently needed, so Florey and his colleague Norman Heatley went to the United States (which had not yet entered World War II) to stimulate interest in its production there. The chairman of the Committee on Medical Research, Office of Scientific

Research and Development, was the pharmacologist A. N. Richards, with whom Florey had worked as a Rockefeller Foundation traveling fellow in 1929–1930. Florey's enterprise and Richards' perspicacity were responsible for the production of penicillin in sufficient quantities for the treatment of war casualties in 1944.[3] Florey journeyed widely to investigate the use of penicillin in the field, traveling to North Africa in 1943 and subsequently to the Soviet Union.

Having acquired such skill in antibiotic research, it was natural that Florey should continue it in the Oxford laboratories well after the end of World War II. The most successful outcome was the development of cephalosporin C. Florey played a part in the early work on the cephalosporins, but later developments were the work of his colleague E. P. Abraham.[4]

After about 1955 Florey returned to research in experimental pathology. His interests ranged widely but showed a continuing preoccupation with the structure and function of the smaller blood vessels and their relation to the movement of lymph and cells in the process of inflammation. In his studies on capillary function and cell migration he could fully indulge his interest in the fine structure of cells and tissues, his pleasure in skilled manipulation, and his enthusiasm for photography. These interests were manifested in his use of the rabbit-ear chamber and other techniques for *in vivo* microscopy, and later of the electron microscope.

The physiology of mucus secretion was another area in which Florey did valuable work, particularly in clarifying its protective function in the respiratory and intestinal tracts. His early recognition of the hormonal control of the secretion of Brunner's glands has recently been confirmed; while a third thread that runs through Florey's work was an interest in human reproduction, initially at the experimental level, where he was interested in the movement of spermatozoa in the female genital tract.

Florey remained an active laboratory investigator all his life. After penicillin, the main topics with which he was concerned were the relatively insoluble antibiotic micrococcin, the electron microscopy of blood vessels, and the nature of atherosclerosis; he published two major works on the structure and function of endothelial cells of blood vessels in 1967, the year before his death.

Florey's other great contribution to science was as president of the Royal Society (1960–1965), the highest office in British science. He was the first Australian and the first pathologist to hold that post. During his tenure, he established the Royal Society Population Study Group and acted as its chairman until the time of his death.

His work was recognized by numerous honors, both public and academic: many honorary degrees from British and Australian universities; the Nobel Prize for physiology or medicine (1945), which Florey shared with Fleming and Chain; the Lister Medal of the Royal College of Surgeons (1945); the Copley Medal of the Royal Society (1957); and the Lomonosov Medal of the Soviet Academy of Sciences (1965).

Florey was elected a fellow of the Royal College of Physicians in 1951, corresponding member of the Australian Academy of Science in 1958, fellow of the Postgraduate Medical Foundation of Australia in 1965, foreign member of the American Philosophical Society in 1963, foreign associate of the National Academy of Sciences of the United States (1963), and foreign honorary member of the American Academy of Arts and Sciences (1964). He was created knight in 1944 and commander of the Legion of Honor in 1946; in 1965 he received the O.M. and was created a life peer, Baron Florey of Adelaide and Marston.

In temperament Florey was reserved but sure of himself. His chief characteristics were his common sense and his intense sense of obligation toward, and responsibility for, his scientific colleagues. He had no liking for speculation: for Florey an idea was not worth having unless it could be used to help design an experiment which would, or could in principle, give a definitive result. Perhaps this is what made him one of the most effective medical scientists of his generation.

NOTES

1. L. A. Epstein and E. Chain, "Some Observations on the Preparation and Properties of the Substrate of Lysozyme," in *British Journal of Experimental Pathology,* **21** (1940), 339–355.
2. R. J. Dubos, "Studies on a Bactericidal Agent Extracted From a Soil Bacillus. I. Preparation of the Agent. Its Activity *in vitro,*" in *Journal of Experimental Medicine,* **70** (1939), 1–17; R. D. Hotchkiss and R. J. Dubos, "Fractionation of the Bactericidal Agent From Cultures of a Soil Bacillus," in *Journal of Biological Chemistry,* **132** (1940), 791–792; and "Chemical Properties of Bactericidal Substances Isolated From Cultures of a Soil Bacillus," *ibid.,* 793–794.
3. See "Alfred Newton Richards, Scientist and Man," in *Annals of Internal Medicine,* **71** (1969), supp. 8, 56; supp. 9, 63–64.
4. E. P. Abraham and G. G. F. Newton, "New Penicillins, Cephalosporin C, and Penicillinase," in *Endeavour,* **20** (1961), 92–100.

BIBLIOGRAPHY

Florey's works include "The Secretion of Mucus by the Colon," in *British Journal of Experimental Pathology,* **11** (1930), 348–361; "Some Properties of Mucus, With Special Reference to Its Antibacterial Functions," *ibid.,* 192–208, written with N. E. Goldsworthy; "Penicillin as a Chemotherapeutic Agent," in *Lancet* (1940), **2,** 226–228, written with E. Chain *et al.;* "Further Observations on Penicillin," *ibid.* (1941), **2,** 177–188, written with E. P. Abraham, E. Chain, *et al.; Antibiotics: A Survey of Penicillin, Streptomycin and Other Antimicrobial Substances From Fungi, Actinomycetes, Bacteria and Plants,* 2 vols. (London, 1949), written with E. Chain *et al.;* and his eds. of *General Pathology* (London, 1954, 1958, 1962, 1970).

A complete bibliography will be found in the memoir on Florey in *Biographical Memoirs of Fellows of the Royal Society* (1971).

FRANK FENNER

FLOURENS, MARIE-JEAN-PIERRE (*b.* Maureilhan, near Béziers, France, 13 April 1794; *d.* Montgeron, near Paris, France, 8 December 1867), *physiology, history of science.*

Pierre Flourens, as he signed his papers, was born into a humble family in a small town in southern France. He studied medicine at the University of Montpellier, graduating at the age of nineteen. The next year, with a letter of recommendation from the famous botanist Augustin de Candolle to Georges Cuvier, Flourens went to Paris, where he decided to abandon medicine and devote all his efforts and ingenuity to physiological research. The protégé of Cuvier, talented, unusually skillful in experimental work, industrious, persevering, and devoted to research and science, Flourens met with early success. In 1821 he was entrusted with lecturing on the physiological theory of sensations before the distinguished scientific Cercle Athénée, and this led him to deeper experimental study of nervous functions. The first results, presented to the Academy by Cuvier in 1822, earned Flourens notoriety and recognition among scientists.

In 1824 and 1825 Flourens received the Montyon Prize twice in succession, and in 1828 he became a member of the Academy of Sciences. That same year Cuvier made Flourens his deputy lecturer at the Collège de France, and he became professor in 1832. The next year, following a wish of Cuvier's expressed before his death, Flourens succeeded him as permanent secretary of the Academy of Sciences. One of his great achievements in this capacity was the founding, with Arago, of the *Comptes rendus,* reports of Academy meetings, which still constitute one of the most important scientific periodicals. In 1838 Flourens was elected deputy for Béziers, and two years later he won election to the French Academy against the celebrated poet Victor Hugo; the election was followed by bitter comments and criticism. In his remaining years Flourens devoted his activity mainly to scientific biographies and philosophical and popular writings. He died at his country house after a long illness.

Flourens' distinguished scientific career began in 1822 with Cuvier's presentation to the Academy of Sciences of the first in a series of his reports on the nervous system; they were collected into a volume in 1824, which was followed by a complementary volume in 1825, and were republished with supplementary material in 1842. These reports are a landmark in the history of the physiology of the nervous system. Flourens' idea was to break down the complicated facts—everything in the mechanisms of life is complex, phenomena as well as organs—into their simple components, to separate all diverse occurrences, to find all distinguishable parts. The art of separating simple facts was for Flourens the whole art of experimenting. In his studies of brain functions he used mainly the technique of ablation—surgical removal of different parts to study their functions—examining systematically one part after the other to differentiate their functions. His hand was sure and precise; his descriptions clear, trenchant, simple, and elegant.

Flourens distinguished three essentially distinct main faculties in the central nervous system: perception and volition (i.e., intelligence), reception and transmission of impressions (i.e., sensibility), and the excitation of muscular contractions. He distinguished excitability from contractility, which is the faculty of muscle to shorten when excited by an adequate stimulus. According to Flourens, the intellect and the faculty of perception reside in the brain proper (cerebral hemispheres), the faculty of immediate excitation of muscular contraction in the spinal cord; and the faculty of coordination of movements willed by the cerebral hemispheres resides in the cerebellum, lesions of which cause disturbances of coordination (i.e., disharmony of movement) and of equilibrium. The idea of coordination introduced by Flourens has played an important role in nervous physiology. For Flourens every part of the brain—"every organ"—had its specific function yet acted as a whole in respect to this function, just as the entire brain functioned as a whole. Thus he thought that there was no localization within each part: all perceptions could concurrently occupy the same seats in the forebrain. Flourens was strongly opposed to Gall's phrenology.

Another important advance was Flourens' discovery of compulsive movements of the head and disturbances of equilibrium after lesions of the semicircular canals of the inner ear (1824–1828). This was a puzzling phenomenon whose physiological background he could not elucidate. It was at that time extremely difficult to realize that the inner ear has not only the receptors of audition (in the cochlea), but also, in its vestibular part, another type of receptor reacting to gravity and accelerative forces. It was explained only fifty years later, in 1873–1874, by Ernst Mach, Josef Breuer, and Alexander Crum Brown simultaneously. Among Flourens' other important contributions to science were his classic localization of the respiratory center (*noeud vital*) in the medulla oblongata, the reunion of nerves (1827), the role of the periosteum in the formation and growth of bone (1842–1847), and the discovery of the anesthetic properties of chloroform on animals (1847).

Flourens had a great influence on the development of physiology, but sometimes it was not beneficial. He was often authoritarian, imposing his opinion without caution or comparison of his experimental results and interpretations with those of other scientists. He was usually right, as in his opposition to Gall's pseudoscience of phrenology, but sometimes wrong, as in his repudiation of every idea of localization in the brain. His most reproachable error was his criticism of Darwin's work (1864).

In his biographies of distinguished scientists Flourens tried to sum up their achievements, relating their work to what was done before and after along the same lines, in a clear, simple, elegant, and engaging style. Some biographies are accompanied by more general studies on the related problems of the history of science. They were very popular, and some are masterpieces which served as models for other biographies.

BIBLIOGRAPHY

I. Original Works. Flourens' writings are *Recherches expérimentales sur les propriétés et fonctions du système nerveux dans les animaux vertébrés* (Paris, 1824, 1842); *Expériences sur le système nerveux . . . faisant suite aux Recherches expérimentales . . .* (Paris, 1825); *Cours sur la génération, l'ovologie et l'embryologie* (Paris, 1836); *Examen de la phrénologie* (Paris, 1842); *Recherches sur le développement des os et des dents* (Paris, 1842); *Mémoires d'anatomie et de physiologie comparées* (Paris, 1843); *Histoire des travaux et des idées de Buffon* (Paris, 1844); *Histoire de la découverte de la circulation du sang* (Paris, 1854, 1857), also trans. into English (Cincinnati, 1859); *De la longévité humaine et de la quantité de vie sur le globe* (Paris, 1854), also trans. into English (London, 1855); *Cours de physiologie comparée* (Paris, 1856); *Recueil des éloges historiques*, 3 vols. (Paris, 1856–1862); *Éloge historique de François Magendie* (Paris, 1858); *De l'instinct et de l'intelligence des animaux. De la vie et de l'intelligence* (Paris, 1858); *Ontologie naturelle ou étude philosophique des êtres* (Paris, 1861); *Éloge historique de A. M. C. Duméril* (Paris, 1863); and *Examen du livre de M. Darwin sur l'origine des espèces* (Paris, 1864).

II. Secondary Literature. On Flourens or his work, see Claude Bernard, "Discours de réception," in *Recueil des discours, rapports et pièces diverses . . . de l'Académie française 1860–1869* (Paris, 1872), II, 319; E. G. Boring, *A History of Experimental Psychology,* 2nd ed. (New York, 1950), pp. 61–67, 69, 77–78; H. Buess, "Flourens, 1794–1867, l'un des créateurs de la neurophysiologie," in *Médecine et hygiène,* **25** (1967), 1377–1379; E. Clarke and C. D. O'Malley, *Human Brain and Spinal Cord. A Historical Study Illustrated by Writings from Antiquity to the Twentieth Century* (Berkeley–Los Angeles, 1968), pp. 483–488, 656–661; V. Kruta, "M. J. P. Flourens, J. E. Purkyně et les débuts de la physiologie de la posture et de l'équilibre," in *Conférences du Palais de la découverte,* no. D98 (1964); M. Neuburger, *Die historische Entwicklung der experimentellen Gehirn und Rückenmark's Physiologie vor Flourens* (Stuttgart, 1897); J. M. D. Olmsted, "Pierre Flourens," in *Science, Medicine and History. Essays in Honour of Charles Singer* (London–New York–Toronto, 1953), II, 290–302; M. Reynaud, *Des derniers ouvrages de M. Flourens et de l'origine des idées modernes sur la vie* (Paris, 1858); and A. Vulpian, *Éloge historique de M. Flourens . . .* (Paris, 1886).

Vladislav Kruta

FLOWER, WILLIAM HENRY (*b.* Stratford-on-Avon, England, 30 November 1831; *d.* London, England, 1 July 1899), *zoology.*

Flower first enrolled at University College, London, and then studied medicine and surgery at Middlesex Hospital, graduating in 1851 from London University. As a student he received medals in zoology and physiology. Soon after becoming a member of the Royal College of Surgeons in March 1854 (he became a fellow in 1857), he volunteered for the Royal Army Medical Service in the Crimea. Upon his return to London, Flower was appointed surgeon, lecturer in anatomy, and curator at the museum of Middlesex Hospital. He remained at Middlesex until late 1861 and during this period published most of his dozen and a half medical papers, including a chapter on injuries to the shoulder region in T. Holmes's *System of Surgery* (1860).

After the death of John Thomas Quekett in 1861 Flower became conservator of the Hunterian Museum at the Royal College of Surgeons. He held the conservatorship until 1884, when he succeeded Richard Owen as superintendent (later director) of the Natural History Departments of the British Museum, remaining in that position until 1898. He was an active member of numerous scientific organizations: he served several terms on the council and was vice-president of the Royal Society, president of the Zoological Society (1879–1899), president of the

Anthropological Institute (1883–1885), trustee of the Hunterian Collection (1885–1899), and president of the British Association for the Advancement of Science (1889). He was made K.C.B. in 1892.

Much of Flower's time and effort was spent on curatorial duties for anatomy and natural history museums, and he often spoke on the aims and organization of such museums. He strongly believed in their twofold function: the education of the public by a small number of expertly presented specimens which tell a story, and the provision of comprehensive collections and related library material for further education and research by knowledgeable experts. Under Flower's directorship the British Museum (Natural History) was developed so that it closely approached this goal.

Few of Flower's papers published before his appointment to the Hunterian conservatorship in 1861 were of a zoological nature. Of by far the greatest interest to Flower were the Mammalia, especially the Cetacea. A concern with problems of classification loomed large in many of his researches, and he made significant contributions to the clarification of the classification of the carnivores (1869), rhinoceroses (1875), and edentates (1882). He helped considerably to consolidate the class Mammalia through two papers on the Marsupialia and Monotremata, in which he demonstrated the essentially mammalian characteristics of the marsupial dentition and the cerebral commissures of both of these aberrant groups.

Beginning in the early 1860's the Cetacea became Flower's prime research interest, nearly a quarter of his published works being directed toward this order. Although hampered by a shortage of information and specimens, he became recognized as the British authority on the Cetacea through his assiduous collecting of all available material. He procured many specimens, first for the Royal College of Surgeons and later for the British Museum (Natural History), and was responsible for adding the whale room at the latter, with its skeletons and life-size models. Flower described and classified many species of whales and dolphins, which he obtained either from specimens washed up on the British coast or from correspondents throughout the world. Of particular interest in this field are his "On the Osteology of the Cachalot or Sperm-whale" (1867) and his long paper "On the Characters and Divisions of the Family Delphinidae" (1883). The latter is the most extensive of his papers on cetacean classification and was long the basis for the study of this family.

Although he was keenly interested in physical anthropology, Flower wrote only a few papers on nonhuman primates, and these were concerned chiefly with the controversy—sparked by Richard Owen in 1860 in opposition to Charles Darwin's *On the Origin of Species* and settled by T. H. Huxley—over certain characteristics of primate brains. Several papers by Flower provided strong evidence that Owen's distinctions between human and nonhuman primate brains were invalid. While at the Hunterian Museum, he added many anthropological specimens, particularly skulls, to the collection and produced a series of papers on the osteology of various primitive tribes. As a result of his researches he adopted the division of the human species into three races—Caucasian, Mongolian, and Ethiopian—which had been proposed by Cuvier sixty years earlier.

Flower was never a teacher in the sense of having students, but he did lecture extensively, particularly as the Hunterian professor of comparative anatomy and physiology (1870–1884). In the introductory lecture to his first Hunterian series he stressed the idea that they should be museum lectures and that he should be a mouthpiece for the specimens. This first series was published as *An Introduction to the Osteology of the Mammalia.* His 1883 series was on the horse, a topic of special interest to him; he published a volume treating the evolutionary development of the horse and related species. With Richard Lydekker, Flower published a comprehensive volume on living and extinct Mammalia (1891), which considers the evolution and classification of the mammals and their geological and geographical distribution.

BIBLIOGRAPHY

I. Original Works. Flower's principal separate publications are *An Introduction to the Osteology of the Mammalia* (London, 1870, 1876; 3rd ed., with Hans Gadow, 1885); *The Horse, a Study in Natural History* (London, 1891); *An Introduction to the Study of Mammals, Living and Recent* (London, 1891), written with Richard Lydekker; and *Essays on Museums and Other Subjects Connected With Natural History* (London, 1898), which contains a collection of essays on the organization and role of museums, on general biology (including whales), and on anthropology.

The following are perhaps the most significant of Flower's papers: "On the Commissures of the Cerebral Hemispheres of the Marsupialia and Monotremata, as Compared With Those of the Placental Mammals," in *Philosophical Transactions of the Royal Society,* **155** (1865), 633–651; "On the Development and Succession of the Teeth in the Marsupialia," *ibid.,* **157** (1867), 631–642; "On the Osteology of the Cachalot or Sperm-whale (*Physeter macrocephalus*)," in *Transactions of the Zoological Society of London,* **6** (1867), 309–372; "On Some Cranial and

Dental Characteristics of the Existing Species of Rhinoceroses," in *Proceedings of the Zoological Society of London* (1876), 443–457; "On the Mutual Affinities of the Animals Composing the Order Edentata," *ibid.* (1882), 358–367; and "On the Characters and Divisions of the Family Delphinidae," *ibid.* (1883), 466–513.

II. SECONDARY LITERATURE. See Charles J. Cornish, *Sir William Henry Flower. A Personal Memoir* (London, 1904), which includes a list of the topics of Flower's Hunterian lectures and a bibliography of Flower's writings compiled by Victor A. Flower; and Richard Lydekker, *Sir William Flower* (London, 1906).

WESLEY C. WILLIAMS

FLUDD, ROBERT (*b.* Milgate House, Bearsted, Kent, England, 1574; *d.* London, England, 8 September 1637), *alchemy, medicine.*

A son of Sir Thomas Fludd and Elizabeth Andros, Fludd came from a well-to-do family connected with the court. His father had been treasurer of war to Queen Elizabeth in France and the Low Countries, and Fludd himself was later to speak of James I as his patron. He attended St. John's College, Oxford, from which he graduated B.A. in 1596 and M.A. in 1598. The following six years he spent as a student of medicine, chemistry, and the occult sciences in France, Germany, Spain, and Italy. On his return to England, Fludd became a member of Christ Church, Oxford, where he received M.B. and M.D. degrees in 1605. After moving to London he sought admission as a fellow to the Royal College of Physicians. Largely because of his contempt for the Galenic system and his insolent manner he repeatedly failed the examination, but he was finally elected a fellow on 20 September 1609 and served as censor in 1618, 1627, 1633, and 1634. His London practice was highly successful, and he was wealthy enough to maintain his own apothecary and a secretary.

Although Fludd had already written a great deal, he had as yet published nothing when the appearance of the *Fama fraternitatis* (1614) initiated a Continental debate over the authenticity of the Rosicrucian texts. When the eminent iatrochemist Andreas Libavius attacked the Rosicrucians, Fludd rose to their defense in a short *Apologia* (1616), which reappeared in considerably expanded form as the *Tractatus apologeticus* (1617). Also in 1617 he began to publish his massive description of the macrocosm and the microcosm, the *Utriusque cosmi maioris scilicet et minoris, metaphysica, physica atque technica historia.* Here and in his other publications Fludd constantly attacked Aristotle, Galen, and the universities, which to him seemed dedicated to preserving the authority of the ancients. He sought instead a new understanding of nature based on Christian principles. His guides were primarily the Mosaic books of the Bible (especially the Creation account in Genesis, which he interpreted as a divine alchemical process) and the Hermetic and Neoplatonic works of late antiquity and the Renaissance, which seemed to mirror the Christian truths. Although Fludd was quite willing to use observational and experimental evidence, he thought that the eternal truths of Scripture and the mysteries of the ancient occultists carried far more weight than the evidence of the senses.

Fludd pictured the universe in terms of a double centrality, a central earth surrounded by the sun, moon, and planets (whose motions were explained by mechanical analogies) and a central sun situated midway between the center of the earth and God. Beyond the fixed stars were the heavens and the region of divinity. He suggested further that relative distances in the heavens might best be found through a study of the celestial monochord and the mathematical musical harmonies.

Fludd sought divine truths in the macrocosm-microcosm analogy and the doctrine of sympathy and antipathy. There was no question that man and divinity were linked through nature. Fludd placed the seat of the Holy Spirit in the sun, from which emanated light and the spirit of life. Life on earth was possible for man only through inspiration of this spirit from the atmosphere—a spirit which he identified as an aerial saltpeter. The source of this spirit affects the human body. Because of the circular motion of the sun, the spirit must have a circular motion impressed on it. Therefore the blood, which carries the spirit, must also circulate. This mystical description of the circulation of the blood was presented by Fludd in his *Anatomiae amphitheatrum* (1623). Yet Fludd was a trained anatomist and had watched Harvey carry out dissections at the Royal College of Physicians. In his later writings he referred to those dissections, and he was the first to support Harvey's *De motu cordis* in print, thinking that the views of his friend confirmed his own cosmological concept of the circulation of the blood (1629).

As a Hermeticist, Fludd had a special interest in the elements. In the first chapter of Genesis he found evidence only for darkness, light, and water as true elements. Therefore the four elements of Aristotle and the three principles of Paracelsus could at best be considered as secondary elements. Heat and cold corresponded to his elements of light and darkness, and he repeatedly employed a graduated thermoscope to show their effects. Here he seemed to have visual evidence of the doctrine of expansion and contraction. Similarly, Fludd entered into the contemporary dispute over the "weapon salve," which was an important test for the validity of sympathetic medicine. In the course of this debate he described William Gilbert's magnetic experiments in detail because they seemed to give valid examples of action at a distance. Here was support by analogy for the truth of the action of the weapon salve. And yet, although Fludd condemned the medicine of the Galenists in general, he accepted the humoral system of disease, which he described in relation to astral influences affecting the body.

The most detailed works on the macrocosm-microcosm universe in the early seventeenth century, Fludd's writings attracted a great deal of attention and controversy. Kepler attacked him after reading his views on the macrocosm and the mathematical harmony of the divine monochord. Mersenne wrote against him several times and was instrumental in having Gassendi write a detailed refutation of Fludd's philosophy. Fludd, in turn, found time to answer these opponents and others in detail. His own work was supported by a number of Continental authors, and in England his writings were proposed as a basis for a Christian understanding of the universe by John Webster in his plea for a reformation of the English universities in 1654.

BIBLIOGRAPHY

I. Original Works. The most complete list of Fludd's works is in J. J. Manget, *Bibliotheca scriptorum medicorum* (Geneva, 1731), I, pt. 2, 298. J. B. Craven's *Doctor Robert Fludd (Robertus de Fluctibus). The English Rosicrucian. Life and Writings* (Kirkwall, 1902; repr. New York, n.d.) also describes Fludd's complex bibliography.

Fludd's first publication, the *Apologia compendiaria Fraternitatem de Rosea Cruce suspicionis maculis aspersam veritatis quasi Fluctibus abluens et abstergens* (Leiden, 1616), was expanded from 23 to 196 pages in the 2nd ed., *Tractatus apologeticus integritatem Societatis de Rosea Cruce defendens* (Leiden, 1617). The work appeared in German translation by Ada Mah Booz (A[dam] M[elchior] B[irkholz]) as *Schutzschrift für die Aechtheit der Rosenkreutzergesellschaft* . . . (Leipzig, 1782). Also from this period are the *Tractatus theologo-philosophicus* (Oppenheim, 1617) and the first part of the *Utriusque cosmi maioris scilicet et minoris, metaphysica, physica atque technica historia,* subtitled *De macrocosmi historia* (Oppenheim, 1617). The second part appeared as the *De naturae simia seu technica macrocosmi historia* . . . (Oppenheim, 1618).

The reaction to these works was immediate, and Fludd defended both the Rosicrucians and his own writings to James I in his "Declaratio brevis" (*ca.* 1617). He returned to the same theme in "A Philosophicall Key . . . Wrighten as a Declaration Unto the Distrustfull and Suspicious, First to Manÿfest, That the Authour Flÿeth on his Owne Wings, and Then to Purifÿ the Adulterat Breath of Spurious Reports as Well of the Ignorant, as Envious Person." This work, probably composed between 1618 and 1620, was also dedicated to the king. Both MSS are currently being prepared for publication by Allen G. Debus.

Kepler attacked Fludd in an appendix to the *Harmonices mundi* (Linz, 1619), which was answered by Fludd in his *Veritatis proscenium . . . seu demonstratio quaedam analytica, in qua cuilibet comparationis particulae, in appendice quadam a J. Kepplero, nuper in fine harmoniae suae mundanae edita, facta inter harmoniam suam mundanam, et illam R. F., ipsissimis veritatis argumentis respondetur* (Frankfurt, 1621). Kepler replied in his *Prodromus dissertationum cosmographicum . . . Item ejusdem J. Kepleri pro suo opere Harmonices mundi, apologia adversus demonstrationem analyticam Roberti de Fluctibus* (Frankfurt, 1621–1622), which included a reprint of the *Mysterium cosmographicum* (1596); this, in turn, was answered by Fludd in the *Monochordum mundi replicatio R. F. . . . ad apologiam . . . J. Kepleri adversus demonstrationem suam analyticam nuperrime editam, in qua Robertus validoribus Joannis objectionibus harmoniae suae legi repugnantibus, comiter respondere aggreditur* (Frankfurt, 1622).

Fludd and the Hermeticists were attacked by Mersenne in the *Quaestiones celeberrimae in Genesim* . . . (Paris, 1623), to which he replied in the *Sophiae cum moria certamen* (Frankfurt, 1629). Gassendi's *Epistolica exercitatio, in qua principia philosophiae Roberti Fluddi, medici, reteguntur, et ad recentes illius libras adversus R. P. F. Marinum Mersennum . . . respondetur* (Paris, 1630) was answered at length by Fludd in the *Clavis philosophiae et alchymiae Fluddanae sive Roberti Fluddi armageri, et medicinae doctoris, ad epistolicam Petri Gassendi theologi exercitationem responsum* (Frankfurt, 1633).

Of lesser importance was Patrick Scot's *The Tillage of Light* (London, 1623), which was answered by Fludd in "Truth's Golden Harrow," first printed with commentary by C. H. Josten in *Ambix,* **3** (1948), 91–150. Fludd's defense of the weapon salve was criticized by a little-known pastor, William Foster, in the *Hoplocrisma-Spongus: Or a Sponge to Wipe Away the Weapon-Salve* (London, 1631); in reply to this there appeared *Doctor Fludds Answer Unto M. Foster. Or, the Squeesing of Parson Fosters Sponge, Ordained by Him for the Wiping Away of the Weapon-Salve* (London, 1631).

From Fludd's other major works three additional titles may be singled out: the *Anatomiae amphitheatrum effigie triplici, more et conditione varia disignatum* (Frankfurt, 1623), in which Fludd described both the scientific and the mystical anatomy of the body; the *Pulsus* (Frankfurt, n.d. [completed 1629]), which forms part of the *Medicina Catholica, seu mysticum artis medicandi sacrarium* and includes Fludd's first defense of Harvey; and the *Philosophia Moysaica* (Gouda, 1638), later trans. into English (London, 1659), which summarizes Fludd's cosmological views and then goes into the weapon-salve problem and magnetism at great length. A French translation of part of the work on the macrocosm exists: *Étude du macrocosme, annotée et traduite pour la première fois par Pierre Piobb. Traité d'astrologie générale* (*De astrologia*) (Paris, 1907).

II. Secondary Literature. The standard biography is that of J. B. Craven cited above. Additional material is given by Josten in the introduction to his "Truth's Golden Harrow" and in his "Robert Fludd's Theory of Geomancy and his Experiences at Avignon in the Winter of 1601 to 1602," in *Journal of the Warburg and Courtauld Institutes,* **27** (1964), 327–335. A general account of Fludd's work is in Allen G. Debus, *The English Paracelsians* (London, 1965; New York, 1966), pp. 105–127; and in "Renaissance Chemistry and the Work of Robert Fludd," in Allen G. Debus and Robert P. Multhauf, *Alchemy and Chemistry in the Seventeenth Century* (Los Angeles, 1966), pp. 1–29.

The Fludd-Kepler exchange is discussed by W. Pauli in "The Influence of Archetypal Ideas on the Scientific Theories of Kepler," in C. G. Jung and W. Pauli, *The Interpretation of Nature and the Psyche,* trans. by Priscilla Silz (New York, 1955), pp. 145–240. An older but still basic study is R. Lenoble, *Mersenne ou la naissance du mécanisme* (Paris, 1948), pp. 103–105, 367–370, and *passim.* Frances A. Yates discusses Fludd's controversies in her *Giordano Bruno and the Hermetic Tradition* (Chicago, 1964), pp. 432–455, and has also shown Fludd's connection with the Vitruvian revival in England and the work of John Dee in her *Theatre of the World* (London, 1969), pp. 42–79. The Gassendi controversy is discussed by L. Cafiero in "Robert Fludd e la polemica con Gassendi," in *Rivista di storia della filosofia,* **19** (1964), 367–410, and **20** (1965), 3–15.

Fludd's physiological concepts and his defense of Harvey are described in Walter Pagel's "Religious Motives in the Medical Biology of the 17th Century," in *Bulletin of the Institute of History of Medicine,* **3** (1935), 97–128, 213–231, 265–312 (see 265–297); and his *William Harvey's Biological Ideas* (Basel-New York, 1967), pp. 113–118; in Allen G. Debus, "Robert Fludd and the Circulation of the Blood," in *Journal of the History of Medicine and Allied Sciences,* **16** (1961), 374–393; and "Harvey and Fludd: The Irrational Factor in the Rational Science of the Seventeenth Century," in *Journal of the History of Biology,* **3** (1970), 81–105. Some aspects of Fludd's defense of the Rosicrucians are discussed by Debus in "Mathematics and Nature in the Chemical Texts of the Renaissance," in *Ambix,* **15** (1968), 1–28. Fludd's relation to the traditional art of memory is the subject of research by Frances A. Yates in *The Art of Memory* (Chicago, 1966), pp. 320–367; C. H. Josten discusses Fludd's alchemical experiment on wheat, taken from the "Philosophicall Key," in *Ambix,* **11** (1963), 1–23; his system of music is described by Peter J. Ammann in "The Musical Theory and Philosophy of Robert Fludd," in *Journal of the Warburg and Courtauld Institutes,* **30** (1967), 198–227. His relationship to Milton is the subject of research by Denis Saurat in *Milton et le matérialisme chrétien en Angleterre* (Paris, 1928); and the relationship of Fludd's thermoscope to the thermometer is described by F. Sherwood Taylor in "The Origin of the Thermometer," in *Annals of Science,* **5** (1942), 129–156 (see 142–150).

Robert Fludd's views on educational reform and the meaning of a new science are taken up in Allen G. Debus, *The Chemical Dream of the Renaissance* (Cambridge, 1968), pp. 20–23; and *Science and Education in the Seventeenth Century. The Webster-Ward Debate* (London–New York, 1970), pp. 23–26. Other aspects of Fludd's work are described by Debus in "The Paracelsian Aerial Niter," in *Isis,* **55** (1964), 43–61; "Robert Fludd and the Use of Gilbert's *De magnete* in the Weapon-salve Controversy," in *Journal of the History of Medicine and Allied Sciences,* **19** (1964), 389–417; and "The Sun in the Universe of Robert Fludd," in *Le soleil à la Renaissance—sciences et mythes, colloque international tenu en avril 1963 . . .* (Brussels, 1965), pp. 259–278.

Allen G. Debus

FOERSTE, AUGUST FREDERICK (*b.* Dayton, Ohio, 7 May 1862; *d.* Dayton, 23 April 1936), *invertebrate paleontology, stratigraphy.*

The son of John August and Louise Wilke Foerste, A. F. Foerste attended public schools in Dayton, graduating from the old Central High School in 1880. He then taught for three years in a small country school near Centerville, Ohio, before entering Denison University, where he received a B.A. degree in 1887. For graduate work he went to Harvard University, where he studied physical geography under W. M. Davis and petrography under J. E. Wolff, receiving the M.A. in 1888 and the Ph.D. in 1890. During this time he also served in the U.S. Geological Survey as part-time assistant to Nathaniel Shaler and Raphael Pumpelly.

Foerste's doctoral thesis in petrography led to advanced studies in that subject for the next two years at the University of Heidelberg and the Collège de France. He devoted vacations, as before, to work with the Geological Survey, apparently planning a career in that organization. In 1892, when its appropriation was drastically reduced, he had to seek employment elsewhere; for a year he worked as tutor to Pumpelly's children.

In 1893 Foerste returned to Dayton to teach science in the Steele High School, where he remained until his retirement in 1932, at the age of seventy. There were many invitations to teach in colleges and universities, but he felt that his position in Dayton, while providing him with a living as well as opportunities for service to burgeoning technological industries, interfered less with his research than would a more prestigious position elsewhere. During vacations he was employed at various times in the state geological surveys of Indiana, Ohio, and Kentucky, and by the Geological Survey of Canada. Beginning about 1920, his summers were spent at the U.S. National Museum, and after retiring he moved to Washington to continue research there as associate in paleontology. Foerste never married. His home in Dayton had been with his widowed sister and her three children, and it was while visiting them that he died of a heart attack.

Foerste was a fellow of the Geological Society of America and one of the founders of the Paleontological Society, which he served as president in 1928 and later as representative on the National Research Council. He was also a member of the American Association for the Advancement of Science, the Ohio Academy of Science, of which he was president in 1931, and the Washington Academy of Science. The Engineers Club of Dayton presented him with an honorary membership in 1926, and Denison University conferred an honorary Sc.D. on him in 1927.

Foerste's scientific interests were first directed toward flowering plants. Before graduating from high school he had accumulated a herbarium of over a thousand species, all collected within ten miles of Dayton. Attending a lecture by Edward Orton of Ohio State University while still in high school, he first learned the meaning of the word "fossil" and that fossils could be found in the nearby quarries. His interest in paleontology thus awakened, he collected fossils almost daily in the Soldier's Home quarry. Then, during the three years he taught near Centerville, he found many additional specimens in rocks of the same age in a large quarry in that locality. When he entered college he thus had a remarkably complete collection of fossils from the Silurian formation in the "Clinton group," which he later named the "Brassfield." The description of this fauna became the subject of his first papers and started him on his long-continuing specialization in early Paleozoic paleontology and stratigraphy.

At the beginning of Foerste's sophomore year at Denison, C. L. Herrick joined the faculty there as professor of natural history. Herrick was only four years older than Foerste, and the two spent much time together in geological fieldwork. Herrick shared with Foerste his plans for a new scientific publication, and the first issue of the *Bulletin* (now *Journal*) *of the Scientific Laboratories of Denison University* (1885) consists of one paper by Herrick and two by Foerste. In subsequent years about half of Foerste's scientific papers appeared in that publication, the latest during the month of his death. While he was responding to Herrick's influence, Foerste also became acquainted with E. O. Ulrich, with whom he had a lifelong friendship and close association. This had much to do with the devotion of his life to paleontology and stratigraphy rather than to petrography, which had attracted him while at Harvard (although even then he had close contact with Alphaeus Hyatt, curator of paleontological collections in the Boston Museum of Natural History).

Notable among Foerste's many contributions to paleontology was the restudy, redescription, and illustration of hundreds of species of invertebrate fossils inadequately described and figured by earlier writers. Of at least equal value was his systematic study of Ordovician and Silurian cephalopods, in which he emphasized the significance of internal structures rather than relying only on external forms.

BIBLIOGRAPHY

The biography by R. S. Bassler in *Proceedings of the Geological Society of America, 1936* (1937), pp. 143–156, includes a bibliography of 135 titles.

The following is a selected bibliography of Foerste's writings: "The Clinton Group of Ohio," in *Bulletin [Journal] of the Scientific Laboratories of Denison University,* **1** (1885), 63–120; **2** (1887), 89–110, 140–176; **3** (1888), 3–12; "Preliminary Description of North Attleboro Fossils," in *Bulletin of the Museum of Comparative Zoology at Harvard College,* no. 16 (1888), pp. 27–41, written with N. S. Shaler; "Notes on Clinton Group Fossils, With Special Reference to Collections From Indiana, Tennessee, and Georgia," in *Proceedings of the Boston Society of Natural History,* **24** (1889), 263–355; "Fossils of the Clinton Group in Ohio and Indiana," in *Report. Ohio Geological Survey,* **7** (1893), 516–601; and "A Report on the Geology of the Middle and Upper Silurian Rocks of Clark, Jefferson, Ripley, Jennings, and Southern Decatur Counties," in *Indiana Department of Geology and Natural Resources, 21st Annual Report* (1897), pp. 213–288.

See also "Geology of the Narragansett Basin," *Monographs of the U.S. Geological Survey,* no. 33 (1899), written with N. S. Shaler; "Silurian and Devonian Limestones of Western Tennessee," in *Journal of Geology,* **11** (1903), 554–583, 679–715; "The Silurian, Devonian, and Irvine Formations of East-Central Kentucky," *Bulletin of the Kentucky Geological Survey,* no. 7 (1906); "Strophomena and Other Fossils From Cincinnatian and Mohawkian Horizons," in *Bulletin of the Scientific Laboratories of Denison University,* **17** (1912), 17–172; and "The Phosphate Deposits in the Upper Trenton Limestones of Central Kentucky," in *Bulletin of the Kentucky Geological Survey,* 4th ser., **1** (1913), 387–439.

Of further interest are "Notes on the Lorraine Faunas of New York and the Province of Quebec," in *Bulletin of the Scientific Laboratories of Denison University,* **17** (1914), 247–339; "Upper Ordovician Formations in Ontario and Quebec," *Memoirs of the Geological Survey Branch, Department of Mines, Canada,* no. 83 (1916); "The Generic Relations of the American Ordovician Lichadidae," in *American Journal of Science,* 4th ser., **49** (1920), 26–50; "Notes on Arctic Ordovician and Silurian Cephalopods," in *Journal of the Scientific Laboratories of Denison University,* **19** (1921), 247–306; "Upper Ordovician Faunas of Ontario and Quebec," *Memoirs of the Geological Survey Branch, Department of Mines, Canada,* no. 138 (1924);

"Actinosiphonate, Trochoceroid, and Other Cephalopods," in *Journal of the Scientific Laboratories of Denison University,* **21** (1926), 285–383; "American Arctic and Related Cephalopods," *ibid.,* **23** (1928), 1–110; "Three Studies of Cephalopods," *ibid.,* **24** (1930), 265–381; "New Genera of Ozarkian and Canadian Cephalopods," *ibid.,* **30** (1935), 259–290, written with E. O. Ulrich; and "Silurian Cephalopods of the Port Daniel Area on Gaspé Peninsula in Eastern Canada," *ibid.,* **31** (1936), 21–92.

KIRTLEY F. MATHER

FOL, HERMANN (*b.* St. Mandé, France, 23 July 1845; *d.* at sea, March 1892), *biology.*

Although little is known of Fol's immediate family, he was descended from Gaspard Fol, a religious refugee who left Touraine to settle in Geneva in 1590. His father was rich, and one of his brothers became a prominent art scholar and curator of the Musée Fol, which he founded in Geneva; a cousin, Auguste Fol, was a watchmaker and director of Geneva's Caisse Hypothécaire from 1882 to 1891.

Fol himself always retained a strong attachment for Geneva. He received his early education at the Gymnasium there, where his interest was turned toward natural science by Edouard Claparède and F. J. Pictet. On Claparède's recommendation, he went on to study medicine and zoology at the University of Jena. Two of his teachers there were Gegenbauer and Haeckel, both of whom influenced him strongly. In the winter of 1866–1867 Fol joined Haeckel on an extended trip to the western and northern coasts of Africa and the Canary Islands; several collecting trips inland confirmed his enthusiasm for natural history.

In 1867 Fol began to study medicine at Heidelberg (and later at Zurich and Berlin as well); he took the M.D. in 1869, with a thesis on the anatomy and development of Ctenophora. On his return to Geneva in 1870 he continued his zoological researches in preference to practicing medicine and spent several winters collecting and studying marine invertebrates. He married a Mlle. Bourrit in the early 1870's.

In addition to his study on the Ctenophora, Fol worked on the embryology of Mollusca and made microscopic studies of fertilization, cell division, and early embryonic growth. In the 1870's little was actually known about the nature of fertilization and especially about the role in this process of such organelles as the nucleus and centrosome. It had been observed that the nucleus disappeared shortly before the onset of mitosis and that the centrosome and the mitotic spindle appeared by late prophase, but was the spindle system composed of the remains of the nucleus or was it produced by the centrosome? What was the function of the spindle, and what was the function of the chromatin bodies (chromosomes) within the nucleus? Was the sperm nucleus or the centrosome necessary to fertilize the egg, or were they both necessary to fertilization? And did the sperm contribute anything material to the fertilized egg, or was it simply an agent to development?

Among the many theories formulated to answer some of these questions were those of Oscar Hertwig, who noted that immediately after fertilization two distinct nuclei could be seen in the egg cell. From this he concluded that one nucleus had come from the sperm and one from the egg, and went on to theorize that the two nuclei fused and that all the nuclei in all the cells of the developing organism were therefore the descendants of this first fused pair. His inferences were open to dispute, however, and it was pointed out in criticism that he had not actually seen the sperm penetrate the egg. Moreover, some thought the germinal vesicle (the egg nucleus) to be a separate and autonomous cell within the larger cell.

Fol pursued the same line of investigation as Hertwig, apparently quite independently. In two papers, one presented in 1877 and the other in 1879, Fol showed that in sea urchins the egg nucleus is not a separate cell but rather an important structural and functional part of the ovum; that during maturation of the egg three daughter nuclei are cast off as polar bodies so that the mature ovum contains only one nucleus; and that sperm actually penetrates the egg. He made the last observation in 1877 and was thus able to provide more conclusive evidence than Hertwig's about the relationship between the two gametes at the moment of fertilization. Like Hertwig, Fol suggested that the daughter nuclei in all the cells of a growing embryo are descended from the original sperm-and-egg fusion pair.

At the time when Fol and Hertwig were carrying out their studies, the hereditary function of the nucleus was not suspected. Their emphasis on nuclear continuity during maturation of the gametes and during subsequent cleavage was, however, an important step in the understanding of this function. Fol himself did not go on to speculate that the nucleus was the vehicle of heredity; while he reaffirmed the idea of nuclear continuity, he also denied the purity of the fundamental nuclear components; that is, he did not believe that the nucleus had a special structure of its own whereby it could preserve hereditary information from one generation to the next. Fol did not characterize fertilization primarily as the fusion of the sperm-and-egg nuclei; rather, he considered the male nucleus to be a degeneration product of sperm components in contact with egg cytoplasm.

Fol's embryological work emphasized the structural and morphological aspects of fertilization by focusing attention on the material organelles (such as the nucleus); work done prior to his and Hertwig's was physiological in design. His investigations enabled later workers (including T. H. Boveri, August Weismann, and E. A. Strasburger) to clarify the actual hereditary function of the nucleus.

Fol conducted these researches while simultaneously engaged in an academic career. In 1876 he had declined the offer of a chair of comparative anatomy at the University of Naples; two years later he accepted a titular professorship (without pay) in comparative embryology and teratology at the University of Geneva. In addition to his work on sea urchins he applied himself to a wide variety of zoological problems and collected a great deal of material on the comparative embryology of invertebrates. He spent winters in Villefranche, near Nice, where in 1880 he established, at his own expense, a marine laboratory in an abandoned quarantine station; summers he gave courses in parasitology and general zoology at the university.

In 1882, when the International Congress on Hygiene was held in Geneva, Fol became greatly excited by the work of Louis Pasteur and Robert Koch. During the next year he made microscopic studies of various bacteria and, at the request of the city officials, made a detailed analysis of the Geneva water supply. At the same time he became an avid photographer and began to experiment with applying photographic techniques to his microscopic researches. He was founder of the Geneva Photographical Society and published articles in the *Revue suisse de photographie.* He also founded the *Recueil zoologique suisse.*

A somewhat unpredictable and eccentric person, Fol became involved in quarrels with some other members of the university and resigned in 1886 to retire to Villefranche. At the time of his teaching appointment he had given his laboratory there to the French government, which had appointed Jules Barrois to be its director. On his return, the government made Fol a codirector; and there he continued his studies on embryology, cytology, histology, and invertebrate zoology. He further collaborated with Eduard Serasin on a study of the penetration of light into seawater.

His last important microscopic research was a study of the centrosome. In 1889 Rádl had predicted, on strictly a priori grounds, that the centrosomes, like nuclei, would be found to unite at fertilization; by this theory, each gamete would contribute two centrosomes, the product of the fusion dividing to form the two poles of the mitotic spindle. In a paper of 1891, "Le quadrille des centres," Fol again used sea urchins in an experiment that confirmed Rádl's prediction. He showed that each gamete contributed either two centrosomes, or one centrosome that divided immediately after fertilization; the daughter centrosomes then united in such a way that one from the male parent always joined one from the female parent (the movements involved in this pairing reminded Fol of the distribution of cards in the eighteenth-century game of quadrille and thus gave him his paper's title). The paper became somewhat notorious and brought Fol under attack from both Boveri and Hertwig, who agreed with each other that centrosomes were not permanent cell organelles. At a slightly later date, Boveri and E. B. Wilson made specific studies of sea urchins and could observe nothing to substantiate the idea that the centrosomes actually congregate and fuse after fertilization; subsequent investigators also proved Fol wrong.

While working full-time at Villefranche, Fol also made several short collecting trips to the western Mediterranean. Ever since his trip with Haeckel in the 1860's, however, he had wished to make a longer expedition. In 1891 he was given a commission by the French government to lead an expedition to the coast of Tunisia to study the distribution of sponges; he set off with a two-man crew in a new yacht, the *Aster,* on 13 March 1892. Although the boat was rumored to have reached the port of Benodet a few days later, the expedition was never heard from again.

Fol received many honors in his lifetime, including the Legion of Honor for establishing the Villefranche laboratory. He belonged to a number of scientific societies, including those of Moscow and Belgium and the Leopoldina-Carolina.

BIBLIOGRAPHY

I. Original Works. An excellent bibliography is that by Maurice Bedot, cited below. A few of Fol's most crucial papers are "Le premier développement de l'oeuf chez les Géryonides," in *Archives des sciences physiques et naturelles,* 2nd ser., **48** (1873), 335–340; "Sur le commencement de l'hénogénie chez divers animaux," *ibid.,* **58** (1877), 439–472; "Recherches sur la fécondation et le commencement de l'hénogénie chez divers animaux," in *Mémoires de la Société de physique et d'histoire naturelle de Genève,* **26** (1878), 89–250; and "Le quadrille des centres, un épisode nouveau dans l'histoire de la fécondation," in *Archives des sciences physiques et naturelles,* 3rd ser., **25** (1891), 393–420.

II. Secondary Literature. The only available biographical sketch of Fol is Maurice Bedot, "Hermann Fol: sa vie et ses travaux," in *Archives des sciences physiques et naturelles,* **31** (1894), 1–22, which also includes a complete bibliography of Fol's writings from 1869 to 1891. A discussion of Fol's contributions may be found in Arthur Hughes, *A History of Cytology* (London, 1959), pp. 61–63, 70, 82–83; and William Coleman, "Cell, Nucleus, and Inheritance: An Historical Study," in *Proceedings of the American Philosophical Society,* **109** (1965), 124–158, esp. 138–139.

GARLAND E. ALLEN

FOLIN, OTTO (*b.* Asheda, Sweden, 4 April 1867; *d.* Boston, Massachusetts, 25 October 1934), *biochemistry.*

Folin was the son of Nils Magnus Folin, a tanner, and Eva Olson Folin, the village midwife. At the age of fifteen he immigrated to the United States to join his brother Axel, who was living in the lumbering town of Stillwater, Minnesota. He worked for a time on various farms in the region and in 1888 moved to Minneapolis to enter the University of Minnesota, from which he received his B.S. degree in 1892. He then studied with Stieglitz at the University of Chicago, completing a thesis on urethanes in 1896. Since he wished to study biochemistry, he spent two years in Europe, working with Albrecht Kossel at Marburg, Olof Hammarsten at Uppsala, and E. L. Salkowski in Berlin before returning to take his Ph.D. from Chicago in 1898. He was appointed assistant professor of analytical chemistry at the University of West Virginia in 1899, and in the same year he married Laura Grant. He was survived by a son and a daughter. In 1900 Folin took charge of the first laboratory for biochemical research to be established in a hospital, the McLean Hospital for the Insane at Waverley, Massachusetts. In 1907 he was appointed to the first chair of biochemistry in the Harvard Medical School, and there he remained for the rest of his life.

When he began work at the McLean Hospital, Folin decided to seek a method for detecting differences in metabolism between psychotic and normal individuals. He realized that urinary constituents reflect the metabolic state of the body and therefore began to study quantitative methods of urinalysis. He devoted particular attention to nitrogenous compounds and worked out colorimetric methods for their determination. Although his original hopes of obtaining results of psychiatric value were not realized, he grew more and more interested in developing analytical methods for biochemical research, and this field became his specialty. He soon recognized that analysis of blood constituents offered a better guide to metabolic reactions than did urinalysis, and most of his later work concerned blood analysis. He summed up much of this work in his classic paper on microchemical methods of blood analysis, which he published with Hsien Wu in 1919. His work with nitrogenous compounds led him to the concept of endogenous and exogenous metabolism which, although later greatly modified, was very fruitful at the time of its proposal. He established the fact that amino acids are absorbed from the intestine in free form rather than as proteins. His colorimetric methods made possible much of later biochemical analysis.

Folin was active in the organization and administration of various biochemical societies and in aiding the publication of papers on biochemical research. Most of his work appeared in the *Journal of Biological Chemistry,* of which he was a leading supporter. In the latter part of his life he received many honorary degrees, from both European and American universities. At the time of his death he was recognized as the leading authority on biochemical analysis.

BIBLIOGRAPHY

I. ORIGINAL WORKS. Most of Folin's early papers appeared in *Hoppe-Seyler's Zeitschrift für physiologische Chemie* and his later work in the *Journal of Biological Chemistry.* The paper "A System of Blood Analysis," written with Hsien Wu, appeared in *Journal of Biological Chemistry,* **38** (1919), 81–110.

II. SECONDARY LITERATURE. Biographical appreciations of Folin appear in *Science* (New York), **81** (1935), 35–38; and in *Medical Journal of Australia,* **1** (1935), 69.

HENRY M. LEICESTER

FOLKES, MARTIN (*b.* London, England, 29 October 1690; *d.* London, 29 June 1754), *antiquarianism.*

The eldest son of Martin Folkes, a solicitor, and his wife, Dorothy, he first attended the University of Saumur, in France. He entered Clare Hall (Clare College), Cambridge, in 1706, to study mathematics and matriculated in 1709. He was elected a fellow of the Royal Society in 1714, the year of his marriage to Lucretia Bradshaw. The university granted him an M.A. in 1717.

His interest in coins and artifacts led him to be chosen a fellow of the Society of Antiquaries in 1719, and he became vice-president of the Royal Society in 1722. After Newton's death in 1727, he was defeated by Sir Hans Sloan for the presidency but remained vice-president. He continued his numismatic and antiquarian pursuits, but his contributions to the *Philosophical Transactions* were minor.[1]

Upon Sloan's retirement from the presidency of the Royal Society in 1741, Folkes succeeded to the office. His "literary rather than scientific bent" was reflected in the society's meetings which, according to his friend William Stukeley, became "a most elegant and agreeable entertainment for a contemplative person."[2] Other comments on Folkes's leadership were less charitable: the *Philosophical Transactions* for the period of his presidency allegedly contained "a greater proportion of trifling and puerile papers than are anywhere else to be found," and the meetings merely allowed "personages acting the importants . . . to trifle away time in empty forms and grave grimaces."[3] John Hill, the society's severest critic, blamed Folkes for this state of affairs.[4]

Despite these criticisms, Folkes was elected to the Académie des Sciences in 1742, had his *Table of Silver Coins From the Conquest* published by the Society of Antiquaries in 1744, stood (unsuccessfully) for Parliament, received the D.C.L. from Oxford in 1746, and became president of the Society of Antiquaries in 1750. His health began to fail, and he resigned his office in the Royal Society in 1752.

Under Folkes the Royal Society lost much of its professional character. James Jaurin's epitaph best sums up his career: "He was Sir Isaac Newton's friend, and was often singled out . . . to fill his chair."[5]

NOTES

1. E.g., "Remarks on Standard Measures Preserved in the Capitol of Rome," cited in C. Weld, p. 478.
2. Stukeley, III, 472.
3. Weld, pp. 483, 487.
4. Hill, preface.
5. Cited in Weld, p. 479.

BIBLIOGRAPHY

Folkes's publications include *A Dissertation of the Weights and Values of Ancient Coins* (London, 1734); *A Table of Silver Coins From the Conquest* (London, 1744); and *A Table of English Gold Coins From the Eighteenth Year of King Edward II* (London, 1745).

No full-scale biography of Folkes exists. Sketches may be found in Charles R. Weld, *A History of the Royal Society: With Memoirs of the Presidents* (London, 1848); and Warwick Wroth, "Martin Folkes," in *Dictionary of National Biography,* V (London, 1884).

A less formal view of Folkes is presented in William Stukeley, *Family Memoirs,* 3 vols. (London, 1882–1884). The faults of his presidency are most clearly outlined in John Hill, *A Review of the Works of the Royal Society* (London, 1751).

JOEL M. RODNEY

FONTAINE (FONTAINE DES BERTINS), ALEXIS (*b.* Claveyson, Drôme, France, 13 August 1704; *d.* Cuiseaux, Saône-et-Loire, France, 21 August 1771), *mathematics.*

The son of Jacques, a royal notary, and of Madeleine Seytres, Fontaine studied at the Collège de Tournon before his introduction to mathematics at Paris under the guidance of Père Castel. About 1732 he acquired property in the vicinity of Paris and, having formed friendships with Clairaut and Maupertuis, presented several memoirs to the Académie des Sciences, which admitted him as *adjoint mécanicien* on 11 June 1733. Although he was promoted to geometer in 1739 and to pensionary geometer in 1742, Fontaine rarely participated in the work of the Academy and led a rather solitary existence. A difficult personality, he showed almost no interest in the work of others and incurred considerable enmity by claiming priority in certain discoveries. In 1765 he retired permanently to an estate in Burgundy, the purchase of which had almost ruined him. He broke his silence only in order to engage in an imprudent polemic with Lagrange, whom he had initially encouraged. He died before he was able to read Lagrange's reply, however.

Fontaine's work is of limited scope, often obscure, and willfully ignorant of the contributions of other mathematicians. Nevertheless, its inspiration is often original and it presents, amid confused developments, a number of ideas that proved fertile, especially in the fields of the calculus of variations, of differential equations, and of the theory of equations.

One of the first memoirs that Fontaine presented to the Academy in 1734 solved the problem of tautochrones in the case where the resistance of the medium is a second-degree function of the speed of the moving body; the method employed, more general than that of his predecessors (Huygens, Newton, Euler, Johann I Bernoulli, and others), heralds the procedures of the calculus of variations. It won deserved esteem for its author, but Fontaine erred in reconsidering the subject in 1767 and 1768 in order to criticize—unjustly—the method of variations presented by Lagrange in 1762.

On the subject of integral calculus, Fontaine was interested in the conditions of integrability of differential forms with several variables and in homogeneous functions; independently of Euler and Clairaut, he discovered the relation termed homogeneity. He gave particular attention to the problem of notation, utilizing both the Newtonian symbols of fluxions and fluents and the differential notation of Leibniz, which he usefully completed by introducing a coherent symbolism for partial differentials that was successful for a long time before being replaced by δ. One of the first to tackle the study of differential equations of the nth order, he failed in his ambitious plan to regroup all the types of equations that can be solved. He did, however, introduce several interesting ideas that foreshadowed in particular the theory of singular integrals.

In the theory of equations, Fontaine attempted to extend to higher degrees a method of studying equations based on their decomposition into linear factors that had shown its usefulness in the case of equations of the third and fourth degree. His memoir, complex and often unclear, was rapidly outclassed by the works of Lagrange and Vandermonde.

In his work of 1764 Fontaine included a study of dynamics dated 1739 and based on a principle closely analogous to the one that d'Alembert had made the foundation of his treatise of 1743. Although Fontaine did not raise any claim of priority, he attracted the hostility of a powerful rival who subsequently took pains to destroy the reputation of his work, which—without being of the first rank—still merits mention for its original inspiration and for certain fecund ideas that it contains.

BIBLIOGRAPHY

I. ORIGINAL WORKS. Fontaine's works are limited to several memoirs published in the *Histoire de l'Académie royale des sciences* for 1734, 1747, 1767, and 1768 and to a volume published as M. Fontaine, *Mémoires donnés à l'Académie royale des sciences non imprimés dans leur temps* (Paris, 1764), repr., without change, as *Traité de calcul différentiel et intégral* (Paris, 1770). Actually, the two successive titles are inexact, because this collection joins to the memoirs already published in 1734 and 1747 ten others, dealing essentially with infinitesimal geometry, integral calculus, mechanics, and astronomy. Only the three memoirs inserted in the volumes of the *Histoire de l'Académie royale des sciences* for 1767 and 1768 are not included in this work.

II. SECONDARY LITERATURE. On Fontaine and his work, see J. L. Boucharlat, in Michaud, ed., *Biographie universelle,* XV (Paris, 1816), 179–183, and new ed., XIV (Paris, 1856), 323–326; F. Cajori, *A History of Mathematical Notations,* II (Chicago, 1929), esp. 198–199, 206–207, 223–224; J. M. Caritat de Condorcet, "Éloge de M. Fontaine, prononcé le 13 novembre 1773," in *Histoire de l'Académie royale des sciences, 1771* (Paris, 1774), pp. 105–116; J. de Lalande, in *Bibliographie astronomique* (Paris, 1803), pp. 481, 486; M. Marie, *Histoire des sciences mathématiques et physiques,* VIII (Paris, 1886), 39–42; J. F. Montucla, in *Histoire des mathématiques,* new ed., III (Paris, 1802), 44, 177, 343, 627, 657; N. Nielsen, *Géomètres français du XVIIIe siècle* (Copenhagen–Paris, 1935), pp. 174–182; and Poggendorff, *Biographisch-literarisches Handwörterbuch,* I (Leipzig, 1863), col. 766.

RENÉ TATON

FONTANA, FELICE (*b.* Pomarolo, Italy, 15 [?] April 1730; *d.* Florence, Italy, 10 March 1805), *neurology, biology.*

Fontana was educated in Rovereto, where he was a student of G. Tartarotti, and at Verona, Parma, and the University of Padua. Toward the end of 1755 he went to Bologna, where he collaborated with L. M. A. Caldani in research on the irritability and sensitivity of the parts of the animal body, an advanced subject proposed to scholars in 1752 by Albrecht von Haller. In 1757 Fontana defended Haller's position in an epistolary dissertation which, published in the collection *Mémoires sur les parties sensibles et irritables du corps animal* (1760), marked the beginning of his fame. From Bologna, Fontana returned for a brief time to the Trentino and then moved to Rome; from there he went to Tuscany, which became his permanent residence until the time of his death.

In 1765 he was appointed to the chair of logic and, in 1766, to the chair of physics at the University of Pisa. Also in 1766, Leopold I, grand duke of Tuscany, who was very interested in natural sciences, summoned Fontana to Florence to organize and develop the court's physics laboratory, which was then located in the Pitti Palace. Fontana reorganized the surviving instruments of the Medici collection—including the relics of Galileo and of the Accademia del Cimento now in the Museum of the History of Science in Florence—and notably increased the collection through the acquisition of scientific instruments and natural objects, as well as by the production of wax models of the human anatomy, prepared under his supervision in an expressly established workshop. In 1775 the Museum of Physics and Natural History, with its important collection of wax models, was opened to the public; it is preserved in a building that was acquired in 1772 to house the growing material. The collection was greatly expanded under Fontana's supervision and during the course of the nineteenth century. In 1786 a duplicate of the collection was sent to Vienna to equip the Austrian medical-surgical military academy, which had been established the preceding year. After the museum had been inaugurated, Fontana was able to begin a long-planned trip in the autumn of 1775 to France and England, to observe, study, and make outstanding acquisitions. This trip, which lasted until January 1780, enabled him to make direct contact with the most significant scientists of the era.

Like his brother Gregorio, a celebrated mathematician, Felice Fontana was an abbot and, like his brother, he was sympathetic to the ideals of the French Revolution. In 1799 he was imprisoned—but only for a few days—by insurgents against the French. On 11 February 1805 he was stricken with apoplexy; he died the following month and was buried in the Church of Santa Croce.

The quality of Fontana's scientific accomplishment is evident from his first work, on irritability and sensitivity, a subject that he continued to pursue so intensely as to earn the praise of Haller in 1767: "Fontana leges irritabilitatis constituit, ingeniosus homo et accuratus." In 1767 there appeared the *De irritabilitatis legibus, nunc primum sancitis, et de spirituum animalium in movendis musculis inefficacia,* revised and translated into Italian as *Ricerche filosofiche sopra la fisica animale* (1775). According to the analysis made in 1955 by Marchand and Hoff,

> The first law concerned Haller's concept of contractility as a property of muscle fiber itself, and pointed out that a contraction follows only after some stimulus. The discussion displayed insight into the underlying nature of tetanic muscular contraction. The second principle was the refractory period discovered by Fontana in heart muscle and applied to better understanding of the function of other muscles. The original third principle was a disproof of the efficacy of a theoretical entity, the "animal spirits." It was a demonstration that the nervous system could excite, but not actually cause contraction of a muscle, and this proposition was illustrated by the classic spark and gunpowder analogy. Not actually a law, this principle was replaced by another in the later Italian version. The newer third law was a description of fatigue as a phenomenon occurring within the muscle fiber itself. In his fourth law, Fontana pointed out the loss of contractility which results from stretching or compressing a muscle, and certain medical applications of this principle. The fifth law was concerned with problems arising from atrophy of disuse. This chapter included a discussion of the behavior of muscles after a relatively brief rest and a progressive shortening related to the "treppe" effect observed on the kymograph record [*Journal of the History of Medicine and Allied Sciences,* **10** (1955), 202].

Initially unwilling to accept the idea of an identification of the nervous flux with electricity, Fontana gradually changed his mind in the course of his research so that finally "he thought in terms of an electric fluid and said the nerves would be the organs destined to conduct this electric fluid and perhaps even to excite it. This is, perhaps, the first suggestion that the production of electricity might be excited by the nerves rather than merely conducted by them as by wires" (Brazier, p. 110).

The research on the movements of the iris (1765) and on viper venom (1767, 1781) is strictly tied to irritability. Fontana observed that the reflex response to light of the pupil of one eye also occurs in the other eye, although it is not exposed to light; in a frightened or excited animal the pupil is dilated and remains so, even if the eye is struck by light; and the pupil of the animal eye is strongly contracted during sleep.

After a series of impressive and ingenious experiments, Fontana retraced the action of the bite of the viper to an alteration in the irritability of the fibers, which he maintained was mediated by the blood: in other words, the viper's poison directly alters the blood, coagulating it, and this in turn alters all parts of the organism—especially the nerve fibers—that the blood would normally nourish. Fontana extended his toxicological experiments to other substances, especially to curare.

Fontana also took advantage of microscopic investigations to complete the characterizations of the parts of the animal body which Haller had based upon irritability and sensitivity. The use of the microscope was at that time especially difficult, because of the illusory images abundantly produced by contemporary instruments. Although Fontana was unable to do away with these images—one can visualize his "tortuous primitive cylinders"—he nonetheless belongs, together with L. Spallanzani, among the major microscopists of the eighteenth century. In the nerve fibers (his "primitive nerve cylinders") he not only distinguished "the axone with myelin sheath and endoneural sheath" (Zanobio [1959], p. 307) but also recognized the fluidity of the axoplasm through accurate research of micromanipulation in which he took advantage of the use of the coverglass (Hoff [1959], p. 377). Thanks to the microscope, Fontana was able to demonstrate in 1778–1779 that the restoration of an interrupted nerve trunk may be traced to a real and actual regeneration of primitive nerve cylinders, or rather, of nerve fibers. Since 1776 Fontana had studied microscopically the "little red globes of the blood" and, discarding the illusory images observed by G. M. della Torre, he attributed to them a spheroidal configuration, modifiable with extreme ease under certain physiological conditions. In fact, the red corpuscles become noticeably elongated (by one third, by one half, or even to twice their diameter) when they cross a blood vessel: the corpuscles assume a cylindrical configuration and retain it as long as the canal remains narrow; but as soon as the canal increases in width, the corpuscles immediately contract and resume their original shape.

In 1766 Fontana demonstrated that the blight which had devastated the Tuscan countryside was caused by parasitic plants that feed on grain and that reproduce by means of spores. Again with the aid of the microscope, he studied the reproduction of cereal Anguillula and its anabiosis. In certain cellular elements (epithelial cells) he observed a nucleus equipped with nucleoli. His discovery of Fontana's canal in the ciliary body of the eye is also famous.

Fontana's biological works, which scholars of science have so far studied only fragmentarily, merit systematic investigation; and the exploration of Fontana's chemical works begun by Icilio Guareschi, also interesting in its applicative aspects, should be continued. Also noteworthy are Fontana's model of the eudiometer (for measuring the salubrity of air), an apparatus for oxygen therapy, and his studies on the absorbent powers of coal.

BIBLIOGRAPHY

I. ORIGINAL WORKS. Bibliographical essays on F. Fontana are contained in the works of Adami (1905), Guareschi, Marchand, and Hoff, cited below.

The principal works of Fontana are "Dissertation epistolaire de Mr. L'Abbé F. Fontana . . . au R. P. Urbain Tosetti . . .," in *Mémoires sur les parties sensibles et irritables du corps animal . . . ouvrage qui sert de suite aux Mémoires de Monsieur de Haller,* III (Lausanne, 1760), 157–243; *Dei moti dell'iride* (Lucca, 1765); *Nuove osservazioni sopra i globetti rossi del sangue* (Lucca, 1766); *De irritabilitatis legibus, nunc primum sancitis, et de spirituum animalium in movendis musculis inefficacia* (Lucca, 1767); *Ricerche fisiche sopra il veleno della vipera* (Lucca, 1767); *Osservazioni sopra la ruggine del grano* (Lucca, 1767); *Descrizione ed usi di alcuni stromenti per misurare la salubrità dell'aria* (Florence, 1774); *Ricerche fisiche sopra l'aria fissa* (Florence, 1775); *Saggio di osservazioni sopra il falso ergot, e tremella* (Florence, 1775); *Saggio del Real gabinetto di fisica e di storia naturale di Firenze* (Rome, 1775); *Traité sur le vénin de la vipère, sur les poisons américains, sur le laurier-cerise et sur quelques autres poisons végétaux . . .* (Florence, 1781); and *Opuscoli scientifici* (Florence, 1783).

II. SECONDARY LITERATURE. Works on Fontana are Casimiro Adami, *Di Felice e Gregorio Fontana scienziati pomarolesi del secolo XVIII* (Rovereto, 1905), and *Felice Fontana pomarolese narrato ai suoi conterranei* (Rovereto, 1930); Luigi Belloni, "Anatomia plastica: III. The Wax Models in Florence," in *Ciba Symposium,* **8** (1960), 129–132; Alberico Benedicenti, *Malati, medici e farmacisti,* II (Milan, 1951), 1180–1190; Guglielmo Bilancioni, "Felice Fontana trentino e gli studi sull'anatomia e sulla fisiologia dell'orecchio e di altri organi di senso nella seconda metà del secolo XVIII," in *Archeion,* **12** (1930), 296–362; Mary A. B. Brazier, "Felice Fontana," in Luigi Belloni, ed., *Essays on the History of Italian Neurology. Proceedings of the International Symposium on the History of Neurology, Varenna 1961* (Milan, 1963), pp. 107–116; Andrea Corsini, "La medicina alla corte di Pietro Leopoldo," in *Rivista Ciba,* **8** (1954), 1509–1540; Fielding H. Garrison, "Felice Fontana: A Forgotten Physiologist of the Trentino," in *Bulletin of the New York Academy of Medicine,* **11** (1935), 117–122; Icilio Guareschi, "Felice Fontana," in *Supplemento annuale 1908–1909 alla Enciclopedia di chimica scientifica e industriale* (Turin, 1909), pp. 411–448; Hebbel E. Hoff, "The History of the Refractory Period. A Neglected Contribution of Felice Fontana," in *Yale Journal of Biology and Medicine,* **14** (1942), 635–672; and "A Classic of Microscopy: An Early, If Not the First, Observation on the Fluidity of the Axoplasm, Micromanipulation, and the Use of the Cover-slip," in *Bulletin of the History of Medicine,* **33** (1959), 375–379; Giuseppe Mangili, *Elogio di Felice Fontana* (Milan, 1813); John Felix Marchand and Hebbel Edward Hoff, "Felice Fontana: The Laws of Irritability. A Literal Translation of the Memoir *De Irritabilitatis Legibus,* 1767; Added Material From *Ricerche Filosofiche sopra la Fisica Animale,* 1775; and Correlation of These Editions With the E. G. B. Hebenstreit German Translation, 1785," in *Journal of the History of Medicine and Allied Sciences,* **10** (1955), 197–206, 302–326, 399–420; Bruno Zanobio, "Le osservazioni microscopiche di Felice Fontana sulla struttura dei nervi," in *Physis,* **1** (1959), 307–320, "L'immagine filamentoso-reticolare nell'anatomia microscopica dal XVII al XIX secolo," *ibid.,* **2** (1960), 299–317, and "Ricerche di micrografia dell'eritrocita nel settecento," in *Actes du symposium international sur les sciences naturelles, la chimie et la pharmacie de 1630 à 1850, Florence–Vinci, 1960* (Florence, 1962), pp. 159–179.

LUIGI BELLONI

FONTENELLE, BERNARD LE BOUYER (or **BOVIER) DE** (*b.* Rouen, France, 11 February 1657; *d.* Paris, France, 9 January 1757), *dissemination of knowledge, mathematics, astronomy.*

Fontenelle's father, François Le Bouyer, *écuyer,* sieur de Fontenelle, was originally from Alençon; his mother, Marthe Corneille, sister of Pierre and Thomas Corneille, came from Rouen. The family was of modest means and lived in rented quarters in Rouen. His father, *sous-doyen des avocats* in the Parlement of Rouen, was "a man of quality but of mediocre fortune" and practiced his profession "with more honor than fame," according to Trublet. Fontenelle was said to resemble his mother, a woman of great intellect, who was also pious and exhorted her children to virtue. Two of them died at an early age, before Bernard was born; two more, Pierre and Joseph Alexis, were born after him—both were to become ecclesiastics. Bernard's two maternal uncles, especially his godfather Thomas Corneille, had a great influence on him; they often invited him to Paris, before he moved there permanently around 1687, and introduced him to the world of the French Academy, the theater, the salons of the *précieuses,* and the *Mercure galant,* which was directed by a friend of Thomas's, Donneau de Visé.

About 1664 the child was placed in the Jesuit *collège* in Rouen, where his uncles had studied. He was, according to his teachers, "a well-rounded child in all respects and foremost among the students." The logic and physics that he was taught seemed to him devoid of meaning: according to Trublet, "He did not find nature in them, but rather vague and abstract ideas which, so to speak, skirted the edge of things but did not really touch them at all." The Jesuits wished to make him one of their own, but Fontenelle did not have a vocation. In deference to his father he became a lawyer, but he pleaded only one case—which he lost—and quit the bar to devote himself to literature and philosophy, which were more to his taste.

Although his parents had dedicated him to St. Bernard and to the Virgin and had made him wear the habit of the Feuillants until the age of seven, Fontenelle never displayed any strong devotion. He maintained the appearance of a Catholic, however, especially toward the end of his life, and in 1684 won the Academy's prize for eloquence with a *Discours sur la patience* that would not have been out of place in a collection of sermons (but did he not take this as a joke?). Nevertheless, his scientific attitude led him to a certain skepticism toward religion. The spirit of tolerance animated him; he had, after all, Protestant paternal ancestors, and in Normandy, where Reformed churchgoers were numerous, he had friends such as the Basnages, to whom he remained faithful after the revocation of the Edict of Nantes.

Fontenelle was born with a very fragile constitution; in his childhood he spat blood and was forbidden to take any violent exercise. He was sparing and careful of himself all his life; this is undoubtedly why he was accused of egotism and of indifference toward others. Although self-centered and considering himself responsible only for his own actions, he was not at all insensitive to the needs of others; on the contrary, he was obliging toward his friends (Mlle. de Launy and Brunel, for example). He was even-tempered—perhaps that is the secret of his longevity. He loved the company of women but never married. Even as a nonagenarian he still frequented their salons, particularly that of Mme. Geoffrin, whom he made his general legatee. In his youth he had been received by Ninon de Lenclos and, from 1710 to 1733, by Mme. de Lambert, at whose home he met men of letters and scholars, such as Houdar de La Motte, Marivaux, Montesquieu, and Mairan. He also attended the duchesse du Maine at her court at Sceaux and was a frequent guest of Mme. de Tencin. He was affable and witty; his all-embracing curiosity made him an excellent listener. Above all, he prized his freedom of mind and independence in his relations with men of rank, like the regent, Philippe d'Orléans, who honored him with his friendship, lodged him in the Palais Royal (until 1730), and awarded him a pension.

Fontenelle received every possible academic honor, although he was refused four times before being accepted into the French Academy in 1691. On 9 January 1697 he entered the Académie des Sciences as *secrétaire perpétuel* and was confirmed in office on 28 January 1699. He was *sous-directeur* in 1706, 1707, 1719, and 1728; *directeur* in 1709, 1713, and 1723; and was made *pensionnaire vétéran* on 9 December 1740. He became a member of the Académie Royale des Inscriptions et Belles-Lettres in 1701 and requested veteran status in 1705. In 1733 he was elected a member of the Royal Society of London. In 1740 he contributed to the foundation of the Académie des Sciences, Belles-Lettres et Arts of Rouen, which received its charter in 1744 and of which he then became an honorary member. He became a member of the Berlin Academy on 4 December 1749, the Accademia dei Arcadi of Rome, and the Academy of Nancy. In 1702 he joined the society formed by the Abbé Bignon to direct the publication of the *Journal des sçavans.*

Commencing with his studies at the Jesuit *collège,* Fontenelle began to write poetry. In 1670 he competed for the prize of the Académie des Palinods of Rouen, writing in Latin on the Immaculate Conception, and his work was judged worthy of printing and was published that year in the *Revue des palinods.* In 1674 he translated an ode by his teacher P. Commire, addressed to the Grand Condé "on the fact that he is subsisting only on milk" (*Mercure galant,* July 1679). In 1677 the *Mercure galant* published his "L'amour noyé" with a very flattering introduction of the author as "nephew of the two Corneille poets." On several occasions Fontenelle competed for the poetry prize of the French Academy, but without great success. His operas, written under the name of Thomas Corneille and set to music by Lully, *Psyché* (1678) and *Bellérophon* (1679), were no more successful; even less so was his tragedy *Aspar* (1680), which was ridiculed by Racine. Under the name of Donneau de Visé he produced a comedy in 1681, *La comète,* inspired by the appearance of the comet of 1680 (the same referred to in Bayle's *Pensées sur la comète*). In it Fontenelle presents—obviously, in an amusing manner—various contemporary explanations of comets, including the most popular as well as the Cartesian theory; and the antiquated notions surrounding these celestial phenomena are held up to ridicule. In the work one can see the dawn of what was to make Fontenelle famous: his taste for the exposition of scientific ideas and his censorious and mocking attitude toward everything that seemed to him to be preconception or myth.

His "Lettre sur la Princesse de Clèves," which appeared in the *Mercure* of May 1678, revealed his talent as a literary critic sensitive to feelings, although he presented himself from this time on as a *géomètre,* with a "mind completely filled with measurements and proportions." Nevertheless, the first work of his period in Rouen was not a scientific one: it was, rather, the *Nouveaux dialogues des morts,* in two volumes, which he published anonymously in 1683. This was followed in 1684 by the *Jugement de Pluton* on the two parts of the first work. Fontenelle sometimes arranged the dialogues between the ancients, sometimes between the moderns, and sometimes between members of the two groups. From occasionally comical situations he draws subtle moral observations in a lively style. One can also find interesting considerations regarding the sciences, all of which have their chimera "which they run after without being able to seize . . . but on the way they trap other very useful knowledge" (dialogue between Artemis and Ramon Lull). He also comments on the role of instruments in the field of scientific knowledge (dialogue between Marcus Apicius and Galileo) and on the difficulty of discovering the truth (dialogue between the third Pseudo-Demetrius and Descartes).

At the same time as this work, which invites serious consideration despite its light touch, there appeared the *Lettres diverses de M. le Chevalier d'Her* * * * (or *Lettres galantes . . .,* depending on the edition), which were attributed to Fontenelle, who disavowed them. No one was deceived, for they clearly bear the mark of his style and his mind and reveal his ability to scrutinize a woman's soul.

In 1685 Fontenelle displayed his taste for mathematical reflection with the publication in the *Nouvelles de la république des lettres,* under the title of "Mémoire composé par M.D.F.D.R. [M. de Fontenelle de Rouen] contenant une question d'arithmétique," of a two-part article on the properties of the number nine. It was only a simple game that did not demonstrate the author's genius in these matters. Yet, if he did not solve the problem, he did pose the question for scholars; the *Nouvelles* published a reply by de Joullieu in February 1686 and a "Démonstration générale de la question . . . touchant les nombres multiples," by J. Sauveur, in October 1686.

This first, scarcely scientific essay was followed in 1686 by Fontenelle's most famous and most frequently published and translated work, *Entretiens sur la pluralité des mondes.* In five "Evenings" (*Soirs*), then six in the 1687 edition, Fontenelle undertook to set forth to a marquise who questioned him during evening promenades in a garden the different astronomical systems: those of Ptolemy, Copernicus, and Tycho Brahe. He spoke to her of the moon and the other worlds—Venus, Mercury, Mars, Jupiter, Saturn, the fixed stars—and discussed the possibility that they might be inhabited. He explained, in terms that could be understood by an intelligent but untrained mind, recent discoveries in the world of the stars, displaying a strong Cartesian bent in his account. In choosing this subject Fontenelle was undoubtedly inspired by a growing interest in the heavenly bodies, as well as by a work that appeared in Rouen in 1655, *Le monde de la lune* (a translation of the *Discovery of a New World* of John Wilkins), and by the *Discours nouveau prouvant la pluralité des mondes* of Pierre Borel (1657), not to mention two books by Cyrano de Bergerac, *L'autre monde: L'histoire comique des états et empires de la lune* (1657) and *L'histoire comique des états et empires du soleil* (1662).

Fontenelle was not an astronomer, and the earliest editions contained a number of errors which he continued to correct until 1742 in order to bring his text into agreement with the scientific data provided him by the members of the Academy of Sciences. The book offered him an opportunity to discuss problems that fascinated him: the relativity of knowledge and the desacralization of the earth—and hence man—attendant upon the recognition of a nongeocentric universe. Our world is not privileged: others might be inhabited, and our present knowledge is limited but grows unceasingly in the course of time. "The art of flying has only just been born; it will be brought to perfection, and someday we will go to the moon" ("Second Evening").

The work's success resulted from the author's having treated supposedly difficult subjects in a light style, playfully and with a touch of affectation that detracted nothing from the seriousness of the given explanations. All this was done in a slightly fictionalized form that permitted a certain lyricism on the enchantment of a summer evening and the immensity of the universe. It is the first example in French of a learned work placed within the reach of an educated but nonspecialized public. It is certainly to these aspects of his work that Fontenelle owed his later academic positions.

Meanwhile, he was active in other fields. He published "Éloge de Monsieur Corneille" in January 1685 in the *Nouvelles de la république des lettres.* (Revised as "Vie de Monsieur Corneille," it appeared in the 1742 edition of his *Oeuvres.*) This was followed in 1686 by *Doutes sur le système physique des causes occasionnelles,* on the theory that Malebranche had presented in the *Recherche de la vérité.* Also in 1686, the self-styled "author of the *Dialogue des morts,*" again under the veil of anonymity, published the *Histoire des oracles.* Actually, he had already set forth his reflections on history: he had sketched the treatise "Sur l'histoire," passages from which were to appear in *De l'origine des fables* (1724). Published along with them were several pages, "Sur le bonheur," also written much earlier and undoubtedly one of the best expressions of Fontenelle's practical philosophy, a human morality independent of religion.

In his reflections on history and on the origin of fables Fontenelle appears as one of the first to treat the history of religion comparatively. He espoused a critical history not only of human events but also of myths, legends, and religions. He studied their formation, showing the role of imagination and how "marvelous" phenomena can be explained by nonsupernatural causes. He found ideas similar to his in *De oraculis ethnicorum dissertationes duae,* a work published in 1683 by the Dutchman A. Van Dale, and he decided to translate it; in the end he preferred to rewrite it entirely in his own manner. This was again done under the cover of anonymity, of course, for it was dangerous to attack superstitions: it led to casting doubt on miracles—fundamental ideas of Christianity that do not agree with scientific truths discovered through reasoning and experiment. Thus, Fontenelle was later attacked by the Jesuits, in particular by Jean-François Baltus, in 1707 and 1708, following the fifth edition of the *Histoire des oracles;* in accordance with his temperamental dislike of dispute and perhaps counseled by his friends as well, he did not reply.

Fontenelle was not content, in 1686, to publish only this dangerous work. He had sent to his friend Basnage in Rotterdam (in order to forward it to Bayle, who published it in the *Nouvelles de la république des lettres* of January) a "Relation curieuse de l'Isle de Bornéo," a so-called extract from a "letter written from Batavia in the East Indies." Involved was a letter between two sisters, Mreo and Eenegu, who were, one quickly discovered, none other than Rome and Geneva. In other words, just after the revocation of the Edict of Nantes, Fontenelle stigmatized the struggle between Catholics and Protestants, besides making a clear allusion to an event of which he deeply disapproved. If he had not at this time had protectors as powerful as the lieutenant of police Marc René de Voyer de Paulmy, marquis d'Argenson, he would have received the *lettre de cachet* that Le Tellier, confessor to Louis XIV, attempted to obtain against him for his unorthodox writings.

Fontenelle settled in Paris around 1687 and resumed his literary activities, publishing in that year *Poésies pastorales de M.D.F., avec un traité sur la nature de l'églogue et une digression sur les anciens et les modernes.* Fontenelle belonged to the party of the moderns, the men of progress, together with his friend Houdar de La Motte, Charles Perrault, and the circle of the *Mercure galant,* in opposition to the party of the ancients, the men of tradition, among whom were Racine, Boileau, and La Bruyère. His relationship with the Corneille family obviously reinforced his hostility toward the partisans of Racine, but it is certain that the *Digression,* leaving aside the question of personalities, shows Fontenelle's reflections concerning science: it is owing to its progress that humanity is improved. Moreover, is not the notion of ancients and moderns really very relative?

Fontenelle wrote another libretto, for the opera *Enée et Lavinie* (1690), and a tragedy, *Brutus* (1691), under the pseudonym of Mlle. Bernard. Received into the French Academy two years before La Bruyère, who in the eighth edition of the *Caractères* (1694) was to mock him under the name of Cydias, Fontenelle published the *Recueil des plus belles pièces des poètes françois, depuis Villon jusqu'à Benserade, avec une préface et des petites vies des poètes* (1692) and the *Parallèle de Corneille et de Racine* (1693).

Thanks to his compatriot and friend Varignon, Fontenelle made the acquaintance of the Parisian scientific circle and became friendly with Nicolas de Malézieu and Guillaume de L'Hospital. For the latter's *Analyse des infiniment petits pour l'intelligence des lignes courbes* (1696), he composed a preface that might have been taken for the author's but which everyone was quite aware was by Fontenelle. In it are displayed his interest in the notion of infinity and his talent as a historian; in a few pages he retraces the history of the mathematical study of curved lines from Archimedes to Newton and Leibniz.

Fontenelle was a friend of the Abbé Bignon and of Pontchartrain, patrons of the Academy of Sciences; and his *Entretiens* was admired for its clear and elegant style, in contrast to the ponderous Latin of the Academy's secretary, Jean-Baptiste du Hamel. In 1697 Fontenelle was invited to replace the latter. The Academy's new statutes of January 1699, of which Fontenelle was in part the author, defined the role of the *secrétaire perpétuel:* he was required to publish each year the memoirs of the academicians drawn from the records, preceded by a sort of *histoire raisonnée* of the Academy's most remarkable accomplishments. He was also to deliver the *éloges* of those academicians who had died during the year and was to publish them in the *Histoire.*

Thus, under the facile pen of a writer who could simplify and clarify and who—without being a specialist—had sufficient knowledge in all areas of science to present its results without distortion, the works of the academicians could become accessible to a cultivated society that balked at Latin. From 1699 to 1740 Fontenelle devoted himself almost exclusively to his task of editing the *Histoire de l'Académie royale des sciences . . . avec les mémoires de mathématique et de physique pour la même année, tirés des registres de cette Académie.* The volume for the year 1699, which appeared in 1702, opens with an untitled preface usually called "Préface [sometimes "Discours préliminaire"] sur l'utilité des mathématiques et de la physique et sur les travaux de l'Académie," which contains essential material on the philosophy of science and is a sort of bridge between Descartes's *Discours de la méthode* and Claude Bernard's *Introduction à l'étude de la médecine expérimentale.* Here one finds the first literary expression of the idea of the interdependence of the sciences and of the constancy of the laws of nature. In 1733 there appeared the history of the early years of the Academy, under the title *Histoire de l'Académie royale des sciences. Tome Ier. Depuis son établissement en 1666, jusqu'à 1686.* Fontenelle covered only the years until 1679 but composed a preface that is an excellent history not only of the founding of the Academy but of the state of contemporary science as well.

Fontenelle eventually published forty-two volumes of the *Histoire de l'Académie,* containing sixty-nine *éloges.* He had already had some experience with this literary genre in the "Éloge de Monsieur Corneille" and especially in the "Éloge de Mons. Claude Perrault de l'Académie royale des sciences et docteur en médecine de la Faculté de Paris . . ." (*Journal des sçavans,* 28 February 1689). The first *éloges* read to the Academy were short, and one senses that Fontenelle had not yet attained complete mastery of the field in which he later proved to be without equal. His ability, evident as early as the *éloge* of Viviani (1703), was still apparent in the last one, that of Du Fay (1739).

No one before him had been able to evaluate so well the works of others nor to report on a life with such verve, nor to sprinkle his text with such subtle psychological and moral observations. The *éloges* were Fontenelle's greatest glory. They remain an astonishing—occasionally unique—source of biographical information on the scientists of the epoch. If one can sometimes reproach Fontenelle for being biased or too Cartesian at a time when science was already Newtonian, he was a good mirror of his times; and one finds in his writing what is undoubtedly the best approach in French to the works of Malebranche, Leibniz, Newton, Johann I Bernoulli, Jean-Dominique Cassini, Varignon, and Boerhaave, to cite only a few names.

The *éloges* enjoyed such success that Fontenelle saw the necessity, as early as 1708, of collecting them in a separate volume under the title *Histoire du renouvellement de l'Académie royale des sciences en M.DC.XCIX et les éloges historiques de tous les académiciens morts depuis ce renouvellement, avec un discours préliminaire sur l'utilité des mathématiques et de la physique.* In 1717 he brought out an edition with seventeen new *éloges,* in 1722 one with eleven more, and in 1733 the *Suite des éloges des académiciens . . . morts depuis l'an M.DCC.XXII.* Finally, in 1742, volumes V and VI of his *Oeuvres* contained the whole series of *éloges.*

As a member of the Academy of Sciences, Fontenelle also wished to do work of his own. In 1727, as a "Suite des mémoires de l'Académie royale des sciences," he published the *Élémens de la géométrie de l'infini.* Some doubted whether it was really the work of a mathematician, but the author believed it was and attached great value to it. He had worked on it for a long time, probably since the period of his preface to the *Analyse des infiniment petits.* The term *élémens* is to be understood in the sense of "first principles." According to Fontenelle, none of the geometers who had invented or employed the calculus of infinity had given a general theory of it; that is what he proposed to do. The work is divided into a preface relating the history of this branch of calculus and into two main parts: "Système général de l'infini" and "Différentes applications ou remarques." The author discusses "the infinite in series or in progressions of numbers" and then examines "the infinite in straight and curved lines," in the words of the Abbé Terrasson, who reviewed the work in the *Journal des sçavans* (July–October 1728).

There was a great deal of discussion in the scientific community about this work, in which mathematicians found numerous paradoxes. Johann I Bernoulli, for example, in his correspondence with Fontenelle allowed his criticisms to show through his praise: he did not understand what was meant by *finis indéterminables.* Fontenelle attempted to defend his theory and above all his distinction between metaphysical infinity and geometric infinity: one must ignore the metaphysical difficulties in order to further geometry, and the *finis indéterminables* ought to be considered "as a type of hypothesis necessary until now in order to explain several phenomena of the calculus" (letter to Johann I Bernoulli, 29 June 1729). "The orders of infinite and indeterminable quantities, like the magnitudes that they represent, are only purely relative entities, hypothetical and auxiliary. The subject matter of mathematics is only ideal," according to the terms of a "Projet de rapport" of Dortous de Mairan to the Academy on this work.

In 1731 the third edition of Thomas Corneille's *Dictionnaire des arts et des sciences* appeared, revised and augmented by Fontenelle with many scientific terms. When he retired from the Academy of Sciences, Fontenelle was feted at the French Academy on the fiftieth anniversary of his election to that body, and for this occasion he composed a "Discours sur la rime" (1741).

In 1743 a small, anonymous volume entitled *Nouvelles libertés de penser* appeared in Amsterdam; it included two articles believed to have been written by Fontenelle: "Les réflexions sur l'argument de M. Pascal et de M. Locke concernant les possibilités d'une vie à venir" and "Traité de la liberté," both of which are completely in accord with his way of thinking.

In 1752 Fontenelle published anonymously through his friend the physician Camille Falconet (who provided a preface) his *Théorie des tourbillons cartésiens avec des réflexions sur l'attraction.* Many were astonished to see the appearance at this time of a work conceived some years previously, and they tried to explain why Fontenelle had decided to present to the learned public a thoroughly outmoded scientific theory. Fontenelle agreed with Newton and the Newtonians to the degree that they did not attempt to give a meaning to "attraction" and contented themselves with calculations. Newton linked formulas with formulas; his method yielded results that corresponded to the facts, but he explained nothing in the sense that Fontenelle would wish, that is, through principles. Fontenelle wished to understand by going back to causes. It was all very well to take "attraction" as a simple word or a sign; one should not, however, endow it with content, and Newtonians who do this return to Scholastic notions and to "occult forces." If Fontenelle remained faithful to the Cartesianism of the *Entretiens,* it was certainly not owing to the stubbornness of age but to a profound conviction of the value of a mechanical explanation in Descartes's sense. This conviction, moreover, was supported by certain works that he analyzed at the Academy of Sciences, in particular those of Privat de Molières, who defended, with some modifications, the theory of vortices (*tourbillons*).

"One must always admire Descartes and on occasion follow him" (*Éloge d'Hartsoëker*): Fontenelle followed him in the matter of the vortices but not in such matters as his theory of animal machines. In his horror of systems that lull thought to sleep, he understood that the important thing is not the results acquired, which are always provisional, but the method of thinking, which consists in completely rejecting all "marvelous" facts, in questioning everything, and in believing only what reason supported by experiment clearly shows. This is the intellectual attitude inherited by the Encyclopedists that characterized the Enlightenment.

In most respects a man of the seventeenth century, Fontenelle was, in others, a man of the eighteenth—perhaps even of the twentieth—century in his unflagging intellectual curiosity and in his belief in the limitless progress of knowledge in a world in which everything must be open to rational explanation.

BIBLIOGRAPHY

I. Original Works. All of Fontenelle's works have been mentioned in the article. Cited here are only the principal eds. of selected and complete works, the most recent critical eds., and several selected texts.

Editions of selected and complete works include *Oeuvres diverses de M. de Fontenelle,* 3 vols. (Amsterdam, 1701, 1716); 2 vols. (London, 1707, 1713, 1714, 1716); 8 vols. (Paris, 1715); 3 vols. (Paris, 1724); 3 vols. (The Hague, 1728–1729), with an engraving by Bernard Picart, "le Romain"; and 5 vols. (The Hague, 1736). *Oeuvres de M. de Fontenelle* appeared in several eds.: 6 vols. (Paris, 1742); 8 vols. (Paris, 1751–1752); 6 vols. (Amsterdam, 1754); 10 vols. (Paris, 1758); 11 vols. (Paris, 1761, 1766); 12 vols. (Amsterdam, 1764); and 7 vols. (London, 1785). The oft-cited Bastien ed., *Oeuvres de Fontenelle,* 8 vols. (Paris, 1790–1792), is not always faithful to Fontenelle's text and cannot be recommended. Later eds. include G. B. Depping, ed., 3 vols. (Paris, 1818), with index; and J.-B. Champagnac, ed., 5 vols. (Paris, 1825).

Critical eds. include Louis Maigron, ed., *Histoire des oracles* (Paris, 1908; repr., 1934); *De l'origine des fables* (Paris, 1932), edited with intro., notes, and commentary by J.-R. Carré; Robert Shackleton, ed., *Entretiens sur la pluralité des mondes. Digression sur les anciens et les modernes* (Oxford, 1955); and *Entretiens sur la pluralité des mondes* (Paris, 1966), with intro. and notes by Alexandre Calame.

Among selected texts are (in chronological order) A. P. Le Guay de Prémontval, ed., *L'esprit de Fontenelle, ou Recueil des pensées tirées de ses ouvrages* (The Hague, 1753); *Oeuvres choisies de Fontenelle,* 2 vols. (Liège, 1779); J. Chass, ed., *Esprit, maximes et principes de Fontenelle* (Paris, 1788); Cousin d'Avallon, ed., *Fontenelliana, ou Recueil des bons mots . . . de Fontenelle* (Paris, 1801); *Oeuvres choisies de Fontenelle,* 2 vols. (Paris, 1883), with a preface by J.-F. Thénard; Henri Pothez, ed., *Pages choisies des grands écrivains, Fontenelle* (Paris, 1909); Émile Faguet, ed., *Textes choisis et commentés* (Paris, 1912); and *Textes choisis* (Paris, 1966), with intro. and notes by Maurice Roelens, the best current ed.

II. Secondary Literature. On Fontenelle's life, one can still profitably consult his first biographer, the Abbé Trublet, in "Mémoires pour servir à l'histoire de la vie et des ouvrages de M. de Fontenelle," a series of articles that first appeared in the *Mercure de France* (1756–1758). They were then collected in 2 vols. (Amsterdam, 1759) and were later added as vols. XI and XII of Fontenelle's *Oeuvres* (Amsterdam, 1764). They were summarized by Trublet in "Fontenelle," in Moreri's *Dictionnaire* . . . (1759).

The essential bibliography concerning Fontenelle and his work is S. Delorme, "Contribution à la bibliographie de Fontenelle," in *Revue d'histoire des sciences,* **10,** no. 4 (Oct.-Dec. 1957), 300–309.

Particularly noteworthy are Grandjeán de Fouchy, "Éloge de Fontenelle," in *Histoire de l'Académie royale des sciences, année 1757;* Louis Maigron, *Fontenelle, l'homme, l'oeuvre, l'influence* (Paris, 1906); J.-R. Carré, *La philosophie de Fontenelle, ou Le sourire de la raison* (Paris, 1932); M. Bouchard, *"L'histoire des Oracles" de Fontenelle* (Paris, 1947); John W. Cosentini, *Fontenelle's Art of Dialogue* (New York, 1952); S. Delorme, G. Martin, D. McKie, and A. Birembaut, in *Revue d'histoire des sciences,* **10,** no. 4 (1957); S. Delorme, A. Adam, A. Couder, J. Rostand, and A. Robinet, "Fontenelle, sa vie et son oeuvre, 1657–1757 (Journées Fontenelle)," in *Revue de synthèse,* 3rd ser., no. 21 (Jan.–Mar. 1961); François Grégoire, *Fontenelle, une philosophie désabusée* (Nancy, 1947); J. Rostand, *Hommes de vérité, Pasteur, Claude Bernard, Fontenelle* (Paris, 1942, 1955); and J. Vendryès, G. Canguilhem, A. Dupont-Sommer, R. Pintard, A. Adam, and A. Maurois, in *Annales de l'Université de Paris,* 27e année, no. 3 (July–Sept. 1957), p. 378 ff.

One should also consult the *Catalogue de l'Exposition Fontenelle à la Bibliothèque Nationale* (Paris, 1957), which is especially useful. See also the bibliographies in the latest critical eds. (Shackleton, Calame, and Roelens, *op. cit.*), and note, particularly from a philosophical point of view, Leonard M. Marsak, "Bernard de Fontenelle: The Idea of Science in the French Enlightenment," in *Transactions of the American Philosophical Society,* n.s. **49,** pt. 7 (1959), 1–64.

SUZANNE DELORME

FÖPPL, AUGUST (*b.* Grossumstadt, Germany, 25 January 1854; *d.* Ammerland, Germany, 12 August 1924), *engineering, physics.*

Föppl's choice of career was determined by the great program of German railway construction in the mid-nineteenth century. In the late 1860's, a line was begun through the Odenwald, near Grossumstadt. Föppl's father, a country physician, was appointed a director of a railway hospital, and both he and his family were thereby drawn into close association with the construction engineers. This contact was decisive for August, who was then just finishing at the Gymnasium. He started engineering studies in Darmstadt, later changing to Stuttgart, and finally graduating from the Polytechnic in Karlsruhe in 1874.

At the time, job opportunities were few for practicing engineers, and Föppl took a temporary, uncongenial position as a bridge engineer. Simultaneously, at the age of twenty-one, he published the results of his first independent research, a paper on bridge construction. The market for engineers was still poor in 1876, when Föppl completed his year of military service, and the young railway engineer hired himself out for a term as teacher in a building-trades school in Holzminden. The work proved unexpectedly to his liking. In the fall of 1877, Föppl accepted a permanent post at the Trades School in Leipzig.

Teaching at Leipzig was at a low level, and Föppl aspired to become a university professor. He decided to make himself known to the university world through publications. His first two books, *Theorie des Fachwerks* and *Theorie der Gewölbe* appeared in 1880 and 1881. In 1886 these two, together with a small textbook for trade school students, served to meet the requirements for a doctorate in his field at the University of Leipzig. The textbook was published in 1890 as *Leitfaden und Aufgabensammlung für den Unterricht in der angewandten Mechanik.*

Although these first books were well received, they did not bring Föppl the desired appointment as a professor of bridge engineering. He gave up the hope of being appointed in his specialty and resolved to master additional fields. In the self-study to which he now gave himself, the problems involved in writing textbooks suitable for independent reading first presented themselves to him. Meanwhile, throughout the early 1880's, electrical artifacts and machines, exhibited in fairs and commercial firms, brought the new field of electrical engineering to public attention. Föppl turned to this area and, to prepare himself, went in late 1893 to work in Gustav Wiedemann's laboratory at the University of Leipzig. He published his work in a series of papers in *Wiedemann's Annalen.* Among them were researches bearing on subjects as fundamental as the nature of electricity.

From Föppl's activity in electromagnetism, together with his interest in the independent student, grew an immensely successful text, the *Einführung in die Maxwellsche Theorie der Elektrizität* (Leipzig, 1894). It was one of the first German-language expositions of Maxwell's ideas. In addition, Föppl had been an early convert to the use of vector calculus in physics, and his *Einführung* was the first German text to incorporate this new mathematics.

During these fifteen years as a schoolteacher, Föppl also continued to work in engineering statics. He was hired by the city of Leipzig for a number of civil engineering assignments; the most noteworthy of these was the design of the iron framework of the Leipzig Markthalle. Föppl here achieved a new solution for roofing over an irregular polygonal space: it later became known as the *Föpplsche Flechtwerk-Hallendächer,* that is, "Föppl's wickerwork roof." He also published a succession of theoretical studies which were collected in 1892 and published as *Das Fachwerk im Raum* (Leipzig, 1892), one of his best books.

The summons to a university finally came in 1892.

The post was the newly created one of extraordinary professor of agricultural machinery and forestry at Leipzig. This was one field in which Föppl had had no experience. He threw himself into it with his customary wholeheartedness. He participated in the systematic program of tests on farm machinery organized by the German Agricultural Society and wrote an article on the theory of plows for the 1893 *Landwirtschaftliches Jahrbuch.* His labors in this new field were cut short, for in 1894 he was appointed to succeed Johannes Bauschinger at the Technische Hochschule in Munich, as ordinary professor of theoretical mechanics and director of the strength-of-materials laboratory.

In Munich, Föppl felt himself to have reached his life's goal and to have achieved a position at the top of his profession. He now turned almost exclusively to engineering mechanics and left it to Max Abraham to carry out the rewriting of subsequent editions of the *Einführung.* In his role as professor, Föppl published a number of texts, of which the most important was the six-volume *Vorlesungen über technischen Mechanik.* By 1927, the third of these volumes received its tenth edition; none of the others went through fewer than four. He also continued his researches, publishing works on gyroscopic phenomena and problems of relative and absolute motion, among other subjects.

In his role as laboratory director, Föppl succeeded in increasing the proportion of laboratory effort in basic science. Under Bauschinger the laboratory was mainly devoted to carrying out tests requested by industry and government. Under Föppl this part of the work was largely carried out by assistants. His own researches at the laboratory were published in a series of articles (making up vols. **24-33**) in the *Mittheilungen aus dem mechanisch-technischen Laboratorium der K. Technischen Hochschule München.* They appeared between 1896 and 1915 and were of diverse kinds: experimental studies of properties of materials, theoretical investigations of dynamical problems arising in engineering practice, and critical scrutinies of the foundations of engineering formulas and tests. In his evaluation of Föppl's Munich work, C. Prinz singled out the article on Laval waves, that is, disturbances in fast-moving waves, as having been of particular importance for both engineering and physics. Prinz also characterized the reorientation Föppl gave the laboratory as a change from an interest in tests on materials to an interest in the several components of a construction regarded in terms of their interrelations. This approach was unusual in Föppl's time and opened new paths.

Föppl gave up his professorship in 1921 but retained direction of the laboratory. In 1924 he had the satisfaction of seeing the appearance of a *Festschrift* on his seventieth birthday, written by former pupils. Among them were his two sons, Otto and Ludwig, as well as Theodor von Kármán, Prandtl, H. Thoma, and Timoschenko. Föppl died suddenly a few months later in his country house in Bavaria.

It was characteristic of Föppl to involve himself in problems as practical as bridge construction and as fundamental as the question of the existence of absolute motion. In both types of investigations, he concerned himself to an unusual degree with a critical examination of the underlying assumptions. His personal qualities were perhaps typical of his century and nation. He was a strict and conscientious father and husband. He was loyal to the cause of German nationalism and German greatness, and conscious of the significance of his field for the industrial and agricultural strength that lay behind Germany's power. Scrupulous and industrious in his work, he succeeded by his considerable effort in attaining the rather precisely defined goal he had set himself.

BIBLIOGRAPHY

In addition to the works mentioned in the text, see the lists in Poggendorff.

Of particular interest is Föppl's autobiography, *Lebenserinnerungen,* which appeared posthumously (Munich–Berlin, 1925).

The *Festschrift* is *Beiträge zur technischen Mechanik und technischen Physik. August Föppl zum siebsigsten Geburtstag* (Berlin, 1924); see esp. the articles "August Föppl," pp. v–viii (unsigned), and C. Prinz, "A. Föppl als Forscher und Lehrer," pp. 1–3.

JOAN BROMBERG

FORBES, ALEXANDER (*b.* Milton, Massachusetts, 14 May 1882; *d.* Milton, 27 March 1965), *physiology.*

Alexander Forbes was the youngest of seven children of a well-educated, well-to-do Unitarian Boston family. His father was William Hathaway Forbes, and his mother was Edith Emerson, the daughter of Ralph Waldo Emerson. Forbes graduated from Milton Academy in 1899 and then attended Harvard (B.A., 1904; M.A., 1905; M.D., 1910). He studied with Charles S. Sherrington at Liverpool (1911–1912) and with Keith Lucas at Cambridge (briefly in 1912). He returned to Harvard Medical School, where he remained until 1948. After 1948 he was professor emeritus and continued his research in the Harvard biological laboratories.

In 1910 Forbes married Charlotte Irving Grinell of New York. The couple had four children. Forbes lived his entire life in Milton, with summers at the Forbes family estate on Naushon Island (Massachusetts) or cruising the North Atlantic. His health was superb and he excelled at yachting, fancy skating, skiing, rock climbing, and canoeing; and for many years he piloted his private airplane. His only physical affliction was a moderate hearing loss, dating from about 1908.

Forbes's primary scientific interest was neurophysiology, but equally strong was his love of the outdoors, the woods, rivers, and hills of New England, particularly the coast of New England and Labrador. He made contributions to navigation and, at the suggestion of Sir Wilfred Grenfell, he mapped the coast of Labrador by aerial photography, using the technique of oblique photogrammetry. For this he was awarded (1938) the Charles P. Daly Gold Medal of the American Geographic Society. He served with the U.S. Navy in both World Wars; in the first he installed radio compasses, and in the second he mapped an aerial route across northern Greenland.

Forbes was a scientific amateur in the best sense of the word. As a man of independent means, he engaged in his laboratory experiments, outdoor sports, yachting, and explorations because he loved them. He did little formal teaching. Not only did he defray much of the expense of his own activities, but for many years he anonymously supported others in the department of physiology.

Forbes's greatest contributions to neurophysiology came early in his career. He was a technical innovator. About 1912 he installed what was probably the first string galvanometer in the New England area for the accurate measurement of the time relations of spinal-cord reflexes. Later, using his experience with radio compasses, he developed a capacity-coupled electronic amplifier for greater sensitivity, and in 1920 he became the first to report the use of electronic amplification in a physiological experiment.

The paper by Forbes (with Alan Gregg, 1915) on the flexion reflex of the decerebrate cat, timed by means of the string galvanometer, was a landmark in neurophysiology. His most influential paper was a review entitled "The Interpretation of Spinal Reflexes in Terms of Present Knowledge of Nerve Conduction" (1922). The properties of the spinal reflexes were chiefly those described by Sherrington and his pupils (including Forbes himself). The knowledge of nerve conduction was developed chiefly by Lucas and E. D. Adrian. Forbes visited Adrian in 1921.

Forbes's great contribution was to unite these two schools of thought and experimentation and thereby establish the form and direction of a major segment of American neurophysiology. In his paper he provided a coherent, consistent interpretation of the major features of reflex activity. Actually, many of the interpretations were erroneous, as Forbes readily admitted, but his ingenious suggestions inspired experiments and theorizing for at least twenty years. Through his influence on Norbert Wiener, Arturo Rosenblueth, and others, he contributed significantly to the development of the science of cybernetics.

Forbes himself considered his single most important scientific contribution to be the final establishment of the all-or-none law of nerve conduction. The experiments, planned by Forbes and carried out by his collaborators, showed that the strength of a nerve impulse is not diminished after it has passed through a local region of partial narcosis where the impulse was weakened but not extinguished. The impulse is a chain reaction, with local contribution of energy in amounts depending on the local condition of the nerve fiber, not on the previous history of the impulse elsewhere. The idea had been formulated previously by Adrian, but with inadequate experimental support. Forbes's experiment was performed independently and simultaneously, with the same result, by Genichi Kato and his associates in Japan.

Forbes's laboratory became a center for the training of both American and European neurophysiologists. Forbes participated in the early use of microelectrodes (large ones by modern standards) and fostered studies of the auditory system and the early development of electroencephalography in the United States. His studies of electrical responses of the brain under Nembutal narcosis paved the way for far-reaching later developments.

BIBLIOGRAPHY

Among the most important of Forbes's writings are the following: "Electrical Studies in Mammalian Reflexes. I. The Flexion Reflex," in *American Journal of Physiology,* **37** (1915), 118–176, written with A. Gregg; "Amplification of Action Currents With the Electron Tube in Recording With the String Galvanometer," *ibid.,* **52** (1920), 409–471, written with C. Thacher; "The All-or-Nothing Response of Sensory Nerve Fibres," in *Journal of Physiology,* **56** (1922), 301–330, written with E. D. Adrian; "The Interpretation of Spinal Reflexes in Terms of Present Knowledge of Nerve Conduction," in *Physiological Reviews,* **2** (1922), 361–414; "The Nature of the Delay in the Response to the Second of Two Stimuli in Nerve and in the Nerve-Muscle Preparation," in *American Journal of Physiology,* **66** (1923), 553–617, written with L. H. Ray and F. R. Griffith, Jr.; "Studies of the Nerve Impulse. II. The Question of Decrement," *ibid.,* **76** (1926), 448–471, written with H. Davis, D. Brunswick, and A. McH. Hopkins; "Tonus in Skeletal Muscle in Relation to Sympathetic Innervation," in *Archives of Neurology and Psychiatry,* **22** (1929), 247–264; "The Mechanism of Reaction," in C. Murchison, ed., *The Foundations of Experimental Psychology* (Worcester, Mass., 1929), pp. 128–168; "The Conflict Between Excitatory and Inhibitory Effects in a Spinal Center," in *American Journal of Physiology,* **95** (1930), 142–173, written with H. Davis and E. Lambert; "Chronaxie," in *Physiological Reviews,* **16** (1936), 407–441, written with H. Davis; "The Effects of Anesthetics on Action Potentials in the Cerebral Cortex of the Cat," in *American Journal of Physiology,* **116** (1936), 577–596, written with A. J. Derbyshire, B. Rempel, and E. F. Lambert; "Activity of Isocortex and Hippocampus: Electrical Studies With Micro-electrodes," in *Journal of Neurophysiology,* **3** (1940), 74–105, written with B. Renshaw and B. R. Morison; and "Electroretinogram of Fresh-water Turtle: Quantitative Responses to Color Shift," *ibid.,* **21** (1958), 247–262, written with H. W. Deane, M. Neyland, and M. S. Congaware.

HALLOWELL DAVIS

FORBES, EDWARD, JR. (*b.* Douglas, Isle of Man, England, 12 February 1815; *d.* Edinburgh, Scotland, 18 November 1854), *biogeography, invertebrate zoology, invertebrate paleontology.*

Both parents were natives of the Isle of Man. His father was involved in local fishery and lumber businesses, but he later switched to banking. His mother, Jane Teare Forbes, owned Manx property, some of which Edward was to inherit. He was the eldest of eight children surviving infancy. A brother, David, became a geologist.

As a child, Edward became deeply engrossed in the natural world which he found on the island and its shores. He also liked to draw, and in 1831 he went to London to study art. Since he did not show sufficient talent, in November of that year he went to Edinburgh to study medicine instead. There he was strongly influenced by Robert Graham in botany and Robert Jameson in geology, but he never channeled his enthusiasm for zoology and botany into an interest in their medical applications. After his mother's death in 1836 he abandoned his medical education. He went to Paris for the winter of 1836–1837 and attended the biological lectures of Henri M. D. de Blainville and Étienne Geoffroy Saint-Hilaire.

Forbes was remarkably friendly and enjoyed a wide popularity. At Edinburgh he organized a social club which, although rather puritanical compared with typical college fraternities, was a source of lifetime friendships. After 1842 he assumed the responsibility of annually persuading naturalists to attend meetings of the British Association for the Advancement of Science, and he organized younger members of the natural history section into a club called the Red Lions. Forbes enjoyed drawing whimsical animal cartoons and composing poems and songs, the best remembered being "The Song of the Dredge" (1839):

> Down in the deep, where the mermen sleep,
> Our gallant dredge is sinking;
> Each finny shape in a precious scrape
> Will find itself in a twinkling!
> They may twirl and twist, and writhe as they wist,
> And break themselves into sections,
> But up they all, at the dredge's call,
> Must come to fill collections.

He married Emily Marianne Ashworth on 31 August 1848 and remained devoted to her until his death. They had a son and a daughter. After his death, she married Major William Charles Yelverton in 1858.

The range of Forbes's interests was always broad, although from the start most strongly focused upon marine animals and the distribution of species. These interests early inspired him with a desire to travel, and in May 1833 he sailed with a friend to Norway, where he observed the distribution of vascular plants and mollusks. On that voyage and on later trips he concentrated on those groups because of the ease with which they could be collected. They could serve as indicators of the extent of biogeographical provinces. He published brief biogeographical papers in 1835, but more notable were his observations in 1837, "On the Comparative Elevation of Testacea in the Alps" (*Magazine of Zoology and Botany,* **1,** 257–259), which listed the species that appeared in four vegetation zones. This was a small step toward the concept of biotic communities.

He published five papers on mollusks and their distribution in 1839 and then a monograph on British starfish which appeared serially in 1839–1841. In the introduction he divided the British seas into eleven provinces and indicated in a table the distribution of each species within these provinces. Another table indicated the number of species in six families of starfish which are found in four depth zones. He also began studying the relationship between the distribution of fossil and living species, the importance of which subject Charles Lyell had capably expounded in volume II of his *Principles of Geology* (1832).

After leaving medical school Forbes lived on a yearly allowance of £150 from his father, supplemented by fees from occasional teaching and lectures. Gradually he became unhappy because of his inability to find a permanent position as a naturalist. In February 1841 he heard that Captain Thomas Graves of the Royal Navy, who was charged with surveying the coastal waters of Greece and Turkey, wanted a naturalist to accompany him. Like Charles Darwin before him, Forbes jumped at the opportunity for exploration—after asking his father. On this voyage, which began 1 April, Forbes carried out probably the most extensive and systematic dredging for marine animals ever conducted. Along coasts that naturalists had seldom visited since the time of Aristotle he discovered unknown species, both living and fossil. He also studied the rock formations, flora, and fauna of the islands and coastal regions. He planned extensive publications on his findings but was never able to produce them on the scale he desired.

While Forbes was in the Mediterranean, his father notified him that, because of financial reverses, his allowance must cease. His situation was worsened by a serious attack of malaria. Consequently, when his friend John Goodsir arranged for his appointment to the chair of botany at Kings College, London, Forbes reluctantly returned to London, arriving 28 October 1842.

During the next decade Forbes worked long and hard to earn a living as a naturalist. His professorship paid less than £100 per year, and therefore he also accepted the position of curator for the Geological Society of London. This job, which paid £150, involved managing the society's collections and library and editing its journal. He had scant time to prepare his Near Eastern researches for publication, and sickness, apparently the return of malaria, hampered him further. Nevertheless, in 1843 he managed to prepare a report for the British Association, "On the Molluscs and Radiata of the Aegean Sea, and on Their Distribution, Considered as Bearing on Geology." The generalizations in it were important for biogeography and paleontology and carried him closer to the concept of biotic communities. Extrapolating from his data, he postulated an "azoic zone" below 300 fathoms, but later research has revealed that animal life did exist below that depth. These generalizations were more explicitly stated in another paper the following year, "On the Light Thrown on Geology by Submarine Researches; Being the Substance of a Communication Made to the Royal Institution of Great Britain, Friday Evening, the 23d February 1844" (*Edinburgh New Philosophical Journal,* **36** [1844], 318–327).

In 1844 Forbes became a paleontologist for the Geological Survey, under Henry de la Beche, at £300 per year. He was elected a fellow of the Geological Society in 1844 and of the Royal Society in 1845. Although his new position often carried him afield, he still managed to conduct some classes, give lectures, and write articles and book reviews. He was coauthor, with Lieutenant Thomas A. B. Spratt of the Royal Navy, of the two-volume *Travels in Lycia* (London, 1846), but scientifically more important was Forbes's long paper of that year, "On the Connexion Between the Distribution of the Existing Fauna and Flora of the British Isles and the Geological Changes Which Have Affected Their Area" (*Memoirs of the Geological Survey of England and Wales,* **1,** 336–432). In this paper he emphasized hypothetical former land connections with the Continent and also glaciation to account for the discontinuous distribution of species. In 1848–1852 he published with Sylvanus Hanley a four-volume monograph, *The History of British Mollusca.*

An industrious worker who achieved broad experience and knowledge in both field and museum, Forbes was not always fortunate in the conclusions he drew from his studies. The nature of species was of great interest to him, but his thoughts on the subject were entangled by contemporary metaphysical ideas on typology and by his Anglican religion. He believed that his studies of living and fossil species had enabled him to discover God's plan for creating species, which he named "the principle of polarity." Forbes explained this principle in his presidential address to the Geological Society in February 1853 (*Quarterly Journal of the Geological Society of London,* **10** [1854], xxii–lxxxi, esp. pp. lxxvii–lxxxi). He believed that there existed, or had existed, a continuity of forms of species, but this continuity was supposedly the result of a creation plan rather than of evolution. He postulated two major periods of creation, which he labeled "palaeozoic" and "neozoic." He believed that there was a functional parallel between the species of these two epochs, that there had been a "replacement of one group by another, serving the same purpose in the world's economy." In reaction to this paper Alfred Russel Wallace published his paper of 1855 on the replacement of species as evidenced by the geological record. Forbes had earlier attacked the concept of evolution in two anonymous reviews, one on Robert Chambers' *Vestiges of Creation* (*Lancet* [1844], pp. 265–266), and the other on Adam Sedgwick's *Discourse* (*Literary Gazette* [4 January 1851], reprinted in Forbes's *Literary Papers* [London, 1855], pp. 10–24).

In April 1854 Professor Jameson's death resulted in the long-awaited vacancy in the Regius chair of natural history at Edinburgh, and Forbes was appointed to it. He now had the opportunity—it would prove short-lived—to teach and publish as he wished. He soon became sick with diarrhea and vomiting, and before fully recovering he was further weakened by exposure to a hard rain during a field trip. His death, reportedly from a kidney disease, soon followed.

In 1848 Forbes had begun a significant paleontological and stratigraphic study for the Geological

Survey and he was just starting to write the report at the time of his death. This report, *On the Tertiary Fluvio-Marine Formation of the Isle of Wight,* was completed by several of his former colleagues at the Survey, most notably Henry William Bristow, and was edited by Forbes's literary executor, Robert Godwin-Austen (London, 1856).

One of Forbes's most important works was the posthumous *The Natural History of European Seas* (London, 1859). He wrote the first five chapters (126 pages); it was completed by Robert Godwin-Austen. Published in the same year as Darwin's *Origin of Species,* it contained one of the last confident defenses of the idea of centers of creation. On the other hand, it was also a pioneering work, the first general study of oceanography. Although the chapters were organized mainly according to geographical regions, those by Forbes emphasized biogeography and those by Godwin-Austen emphasized physical aspects of oceanography, thereby providing a broad introduction to the science as a whole.

In spite of Forbes's antievolution commitment, it is a tribute to the value of his work that Darwin found numerous occasions to cite him with approval when writing the *Origin of Species.*

BIBLIOGRAPHY

Considering that he died at the age of thirty-nine, Forbes published a great many works. Some of those having particular theoretical interest are mentioned in the text. Almost all of his publications are listed in the short-title bibliography provided by George Wilson and Archibald Geikie, *Memoir of Edward Forbes, F.R.S., Late Regius Professor of Natural History in the University of Edinburgh* (Cambridge-London-Edinburgh, 1861), pp. 575-583, the only extensive study on Forbes.

Several shorter essays and necrologies provide additional information and assessments. Two that contain bibliographies are G. T. Bettany, "Forbes, Edward (1815-1854)," in *Dictionary of National Biography;* and Daniel Merriman, "Edward Forbes—Manxman," in *Progress in Oceanography,* **3** (1965), 191-206.

See also Nils von Hofsten, "Zur älteren Geschichte des Discontinuitätsproblems in der Biogeographie," in *Zoologischen Annalen,* **7** (1916), 197-353, esp. 301-306.

FRANK N. EGERTON III

FORBES, JAMES DAVID (*b.* Edinburgh, Scotland, 20 April 1809; *d.* Clifton, Scotland, 31 December 1868), *physics, geology.*

Forbes's discovery of the polarization of radiant heat strengthened the belief in the identity of thermal and luminous radiation and contributed to the development of the concept of a continuous radiation spectrum. His detailed studies of glaciers aided in the establishment of modern theories of their formation and movement.

Forbes was the youngest son of William Forbes, seventh baronet of Pitsligo, and Wilhelmina Belches, only child and heiress of John Belches Stuart of Fettercairn. Forbes was educated privately until the age of sixteen and developed an early interest in science. He entered the University of Edinburgh in 1825 to pursue legal studies in accordance with the wishes of his father. He distinguished himself in the natural philosophy courses offered by John Leslie and contributed several anonymous papers to David Brewster's *Philosophical Journal.* Upon revealing his authorship, he was encouraged in his scientific work by Brewster, who proposed his membership to the Royal Society of Edinburgh. Forbes was elected to this body upon reaching the age of twenty-one. He received a modest inheritance at his father's death in 1828 and abandoned law in favor of science. He rapidly became acquainted with prominent British scientists and was elected to the Royal Society of London in 1832. In the same year he participated actively in the formation of the British Association.

Forbes was appointed as Leslie's successor to the chair of natural philosophy at Edinburgh in 1833, despite his youth and relative scientific inexperience. This controversial election, in which Brewster was his principal opponent, caused a rift in their friendship which lasted several years. In 1830 Forbes had begun to investigate radiant heat phenomena without significant results. After learning of Melloni's successful detection of the refraction of thermal radiation, he visited Paris in 1833 to confirm Melloni's results, whereupon he requested Melloni to supervise the manufacture of a thermopile for use in his own investigations.

With this instrument Forbes, in November 1834, discovered the polarization of radiant heat by transmission through tourmaline and thin mica plates and by reflection through the latter. Using mica for dipolarization, he was successful in demonstrating the double refraction of thermal radiation, and in 1836 he found that heat could be circularly polarized by two reflections in a Fresnel rhomb of rock salt. He received the Rumford Medal of the Royal Society of London in 1838 for these discoveries and in 1845 was granted a royal pension for his scientific work.

After 1840 Forbes's central interest was geology, although he continued sporadically to study heat conduction in solids and various types of soils, as well as the effects of the atmosphere on solar radiation. He had become an experienced mountaineer during geological field trips in the Pyrenees, in southeastern France, and in the Pennine chain. He turned his attention to the glaciers of the Alps of Savoy when invited there by Agassiz in 1841 on an exploratory expedition. Forbes's early publications on glaciers, claiming priority in noting their veined structure and in demonstrating that the center of a glacier moves faster than its sides, occasioned controversy with Agassiz and others that did not terminate even at Forbes's death.

Forbes pursued detailed studies of the glaciers of the Alps and of Norway during many subsequent summers. He determined that the surface of a glacier moves faster than the ice vertically beneath it and that the velocity of a glacier increases directly with the steepness of its bed. He postulated that a glacier is a viscous body whose movement is due to the mutual pressure of its parts.

Forbes energetically supported the scientific institutions of his time. He served as secretary of the Royal Society of Edinburgh from 1840 to 1851, and he was obliged to decline its presidency in 1867 and that of the British Association in 1864 owing to ill health. He was a corresponding member of many European scientific societies, and he carried on a voluminous correspondence with the leading scientists of the British Isles and Europe. During his tenure at Edinburgh, Forbes sought successfully to reform the Scottish system of higher education by instituting examinations for degrees. He resigned from his chair at Edinburgh in 1860 after his election as principal of the United College of St. Andrews, a position he held until his death.

BIBLIOGRAPHY

I. ORIGINAL WORKS. Forbes's works include *Address to the British Association, 4th General Meeting at Edinburgh* (Edinburgh, 1834); *Travels Through the Alps of Savoy* (Edinburgh, 1843); *The Dangers of Superficial Knowledge* (London, 1849); *Norway and Its Glaciers* (Edinburgh, 1853); "A Review of the Progress of Mathematical and Physical Science in More Recent Times, and Particularly Between the Years 1775 and 1850," in *Encyclopaedia Britannica,* 8th ed. (London, 1853; repr. separately, Edinburgh, 1858); and *Occasional Papers on the Theory of Glaciers* (Edinburgh, 1859).

Forbes also published over 100 scientific papers. His extensive incoming and outgoing correspondence, several journals and notebooks, and his competent watercolors of Alpine landscapes are in the archives of the University of St. Andrews.

II. SECONDARY WORKS. See J. C. Shairp, P. G. Tait, and A. Adams-Reilly, *The Life and Letters of James David Forbes* (London, 1873); John Tyndall, *Principal Forbes and His Biographers* (London, 1873); and George E. Davie, *The Democratic Intellect: Scotland and Her Universities in the Nineteenth Century* (Edinburgh, 1961).

JOHN G. BURKE

FORBES, STEPHEN ALFRED (*b.* Silver Creek, Stephenson County, Illinois, 29 May 1844; *d.* Urbana, Illinois, 13 March 1930), *biology.*

The fifth of the six children of Isaac Forbes and Agnes Van Hoesen, Stephen A. Forbes was raised on a farm in relative poverty. When Stephen was ten, Isaac Forbes died and the elder son, Henry, supported the family. After a year at Beloit Academy, Stephen accompanied Henry into the cavalry on the Union side in the Civil War. Forbes later regarded these years of military service, from September 1861 to November 1865, including four months as a prisoner of war, as a valuable stimulus to his education.

After three years of studying medicine, Forbes decided to study natural history. He supported himself by teaching school in various Illinois towns between 1868 and 1872. Attending the Illinois State Normal University only briefly, Forbes pursued the study of natural history on his own. He married Clara Shaw Gaston in 1873 and had five children. The family was Unitarian, Forbes having moved from the orthodox religion of his childhood to a scientific agnosticism. Throughout his life Forbes was physically and mentally energetic and healthy, enjoying variety in exercise and in reading.

In 1872 he was made curator of the Museum of the State Natural History Society, Normal, Illinois, which he transformed into the Illinois State Laboratory of Natural History in 1877. The laboratory and museum were moved to Urbana when he became a professor at the University of Illinois. This was in 1884, the same year he received his Ph.D. from the University of Indiana. Forbes was appointed state entomologist in 1882. The Office of the Entomologist and the State Laboratory of Natural History were combined when the State Natural History Survey was created. Forbes was chief of the Survey until his death. Forbes's earliest love had been botany, but his first professional concern was how best to teach natural history in the public schools. He was active in the campaign, led by Louis Agassiz, for laboratory and field work. Forbes's first major research was a series of studies of the stomach contents of birds and fish that he collected in Illinois. Naturalists had, of course, made scattered notes of what a particular animal was observed to eat, but Forbes undertook a systematic program to determine directly, often by microscopic examination, what foods had been eaten and in what proportion. He became an authority not only on birds and fish but on the insects and crustacea that they ate. Throughout his career he maintained an interest in limnology, establishing a floating laboratory in the Illinois River in 1894. After his appointment as state entomologist in 1882, most of Forbes's work concerned insects harmful to agriculture. Highly respected in this field, he was president of the American Association of Economic Entomologists in 1893 and 1908, president of the Entomological Society of America in 1912, and elected to the National Academy of Sciences in 1918.

The scientific work of Stephen Alfred Forbes was dominated by his interest in ecology. From the outset he was concerned to investigate scientifically, and quantitatively where possible, the interrelations that make a group of individuals into a functioning system. His viewpoint was clearly and explicitly based upon that of Charles Darwin and Herbert Spencer. He studied the food of fish and birds because he saw the predator-prey relation as the most direct ecological link between species. He expected such studies to help explain not only the geographical distribution and abundance of a species but also its evolution. Forbes found in these studies considerably less specialization than expected, although there were characteristic differences in the proportion of various foodstuffs consumed by each species. The fact that each prey species had many enemies revealed the complexity of the system. Forbes decided that the oscillations in number inherent in a simple predator-prey relation, well described by Spencer (*Principles of Biology,* II, 399), would be damped when these relations were more complex, and he interpreted lack of specialization as an evolutionary adaptation to avoid harmful oscillations. Forbes also found that structural factors were correlated with feeding habits only in a very general sense. The physical adaptation of predator and prey seemed less exact than Darwinists assumed, while the invisible "psychological" factor of food preference, presumed heritable, responded to natural selection.

Spencer had described the stable balance generally found in nature as the result of a physiological law governing reproductive energies, for he thought natural selection unable to produce the necessary mutual adjustment of reproductive rates. But Forbes suggested that natural selection could act on the predator-prey pair as a unit, requiring them to achieve a profitable adjustment to each other or both lose out in the struggle for existence. These apparent enemies have, he said, a common interest. Forbes believed that in the isolated and constant environment of certain lakes, the assemblage of species was "sensible," that is, the entire system was sensitive to whatever affects one species; the assemblage is like an organism composed of interdependent organs. In the microcosm of a lake, the entire complex of species had a "community of interest," so that natural selection would tend to produce maximum productivity and stability of the whole.

Forbes repeated these ideas from his 1880 articles, in a somewhat popularized form, in a lecture read to the Scientific Association of Peoria, Illinois, on 25 February 1887. This lecture, "The Lake as a Microcosm," was later hailed as a minor classic by ecologists for its early statement of the concept of community. Karl Moebius' 1877 booklet on oysters (published in English in 1883) has been similarly hailed. The idea Moebius embodied in his definition of biocoenosis was that an area would hold a given sum of life, so that should the number of oysters be reduced, the number of mussels or other species would increase to maintain the sum.

In 1907 Forbes reviewed his data of thirty years of collecting fish, with the idea of analyzing statistically the geographic distribution of different species. The probability of cooccurrence of two species was predicted from their independent frequencies in his collection; comparing this with their actual cooccurrence yielded his so-called coefficient of association. He applied a similar analysis to data collected in a special survey of birds, giving mathematical expressions for the preference of different species for different types of fields. He had wondered in 1878 whether closely related species living together ever competed for the same food; in 1884 he had described how species of insects feeding on the strawberry plant avoid direct competition by separation in time; in 1907 he hoped to use his coefficient of association to uncover "evasions of competition, and the escape from its consequences, by those closely related and similarly endowed . . ." ("On the Local Distribution of Certain Illinois Fishes," in *Bulletin of the Illinois State Laboratory of Natural History,* **7** [1907], 275). Modern ecologists likewise find the competitive exclusion principle a fruitful basis for research.

While his understanding of the "sensibility" of an assemblage led him to warn of unwanted results if farmers and fishermen tried to alter the balance of nature with inadequate knowledge, his view of the flexibility of natural systems made him hopeful that man could learn how to alter the balance in his favor without disaster. For example, beginning in 1882 Forbes studied the natural diseases of insect pests in the hope of controlling them biologically rather than chemically. He interpreted the agricultural problems with which he dealt as state entomologist as a lack of adjustment of plant and insect, characteristic of the first, primitive stage of association, when evolution had not yet produced mutual adaptation.

In 1859 Darwin had plainly shown not only the possibility of explicitly analyzing the dependencies of organisms but indeed that such an analysis, along with the laws of variation and inheritance, was the key to the origin of species. Yet the existing momentum of academic biology made comparative anatomy, embryology, and paleontology the fields for evolutionary studies, while ecology, like genetics, became well established only in this century. To the modern ecologist Forbes therefore seems remarkably ahead of his time, for he worked on questions not widely appreciated until twenty or thirty years later. What is perhaps more remarkable is that the science of ecology, whose problems and methods had been so clearly defined in the 1880's, had such a long gestation period.

BIBLIOGRAPHY

I. ORIGINAL WORKS. Most of Forbes's work appeared in the various publications of the Office of the Illinois State Entomologist, the Illinois State Laboratory of Natural History, the Agricultural Experiment Station of the University of Illinois, and the Illinois State Natural History Survey. An extensive bibliography of his scientific articles, compiled by H. C. Oesterling, is given in *Biographical Memoirs. National Academy of Sciences,* **15** (1934), 26-54.

I have referred particularly to the following: "The Food of Birds," in *Transactions of the Illinois State Horticultural Society,* **10** (1877), 37-44; "The Food of Illinois Fishes," in *Bulletin of the Illinois State Laboratory of Natural History,* **1,** no. 2 (1878), 71-89; "Studies of the Food of Birds, Insects, and Fishes, Made at the Illinois State Laboratory of Natural History, at Normal, Illinois," the subtitle given to no. 3 (1880) and no. 6 (1883) of the same journal.

Among his articles on insects affecting the strawberry is "On the Life-histories and Immature Stages of Three Eumolpini," in *Psyche,* **4** (1884), 123-140. "The Lake as a Microcosm" was published in the first and only volume of the *Bulletin of the Scientific Association of Peoria, Illinois,* pp. 77-87, and reprinted in *Bulletin of the Illinois State Natural History Survey,* **15** (1925), 537-550. "Preliminary Report Upon the Invertebrate Animals Inhabiting Lakes Geneva and Mendota, Wisconsin, With an Account of the Fish Epidemic in Lake Mendota in 1884," is *Bulletin of the United States Fish Commission,* **8** (1890), 473-487.

See also "Summer Opening of the Biological Experiment Station of the University of Illinois," a pamphlet published by the University of Illinois (1896), 24 pp.; "On the Local Distribution of Certain Illinois Fishes: An Essay in Statistical Ecology," in *Bulletin of the Illinois Laboratory of Natural History,* **7** (1907), 273-303; "An Ornithological Cross-section of Illinois in Autumn," *ibid.,* 305-335; and "History of the Former State Natural History Societies of Illinois," in *Science,* **26** (1907), 892-898.

The third *Report on the Natural History Survey of the State of Illinois* (1909) is a monograph by S. A. Forbes and R. E. Richardson entitled *The Fishes of Illinois;* the 2nd ed. of this book was published in 1920.

Of additional interest are "Aspects of Progress in Economic Entomology," in *Journal of Economic Entomology,* **2** (1909), 25-35; "The General Entomological Ecology of the Indian Corn Plant," in *American Naturalist,* **43** (1909), 286-301; and "The Ecological Foundations of Applied Entomology," in *Annals of the Entomological Society of America,* **8** (1915), 1-19.

A posthumously published autobiographical note, written in 1923, is "Stephen Alfred Forbes," in *Scientific Monthly,* **30** (1930), 475-476.

II. SECONDARY LITERATURE. The fullest account of Forbes is Leland Ossian Howard, in *Biographical Memoirs. National Academy of Sciences,* **15** (1934), 3-54. This incorporates material from the autobiography cited above, from a memorial pamphlet published by the University of Illinois in 1930, and from an obituary by Henry Baldwin Ward, "Stephen Alfred Forbes—A Tribute," in *Science,* n.s. **71** (1930), 378-381.

MARY P. WINSOR

FORCHHAMMER, JOHAN GEORG (*b.* Husum, Denmark, 26 July 1794; *d.* Copenhagen, Denmark, 14 December 1865), *geology, oceanography, chemistry.*

Forchhammer was the son of Johan Ludolph Forchhammer, an educator, and Margrethe Elisabeth Wiggers. The elder Forchhammer was a teacher at the Citizens' School in Husum and later became rector of a similar school at Tønder and manager of the teachers' college.

Forchhammer's early education was at the schools in Husum and Tønder. Following the death of his father in 1810, he became an apprentice in a pharmacy at Husum, where he stayed for five years. In 1815 he enrolled at the University of Kiel, studying physics, chemistry, pharmacy, mathematics, and mineralogy. He went to Copenhagen in 1818 and became involved in an investigation of the coal and iron layers of Bornholm, a rocky island in the Baltic. The investigating commission included Hans Oersted, at that time lecturing in physics and chemistry at the University of Copenhagen, and Councillor of Justice L. Esmarch. In 1819 he enrolled at the University of Copenhagen, and upon completion of his thesis "De mangano," he received the doctorate in 1820.

Also in 1820 Forchhammer made a trip to England to further his understanding of geology, and there he became acquainted with such scientists as Prout, Davy, Dalton, Wollaston, Jameson, and Lyell. Together with Sir Walter C. Trevelyan he investigated the geology and coal formations of the Faeroe Islands; the resulting publication led to his becoming a member of the Royal Danish Academy of Sciences.

In 1821 Forchhammer became a lecturer in geology at the University of Copenhagen and also accepted employment at the Royal Copenhagen porcelain factory. Upon the opening of the Polytechnic Institute he became professor of chemistry and mineralogy and manager of one of its two chemical laboratories. He held this position until his death, and during the last fourteen years he was director of the institute. In 1831 Forchhammer was appointed professor of mineralogy and geology at the University of Copenhagen. From 1851 to his death he was secretary of the Royal Danish Academy of Sciences.

Forchhammer's fundamental researches on the composition of seawater brought him international acclaim. He began this work in 1843, more as a geologist than as a chemist, to explain the phenomena that give rise to the deposits on the sea floor. His immediate goals were the factors governing the marine precipitation of calcium carbonate and the influence of volcanic activity on the oceans. He carried out analyses of over 160 samples collected for him over a twenty-year period by the Danish and British navies. He measured chlorine, sulfur, magnesium, calcium, and potassium gravimetrically with 100-pound samples, obtaining sodium from the differences.

The principal consequence of this work was the proposition that although seawaters exhibit marked regional differences in total salt content, the ratios of the major dissolved constituents to each other are almost invariable. With but slight modifications this concept is valid today. Forchhammer also posed in an elegant form the "geochemical balance problem" arising from the major sedimentary cycle: "Thus the quantity of the different elements in sea water is not proportional to the quantity of elements which river water pours into the sea, but inversely proportional to the facility with which the elements are made insoluble by general or organo-chemical action in the sea." He reached this conclusion by noting that although calcium and silicon were often the principal constituents of river waters, they were less plentiful in the oceans. He correctly attributed the decrease in silicon to its incorporation into the skeletal material of the photosynthesizing diatoms but erroneously believed that calcium concentrations were regulated by carbonate-depositing animals.

Forchhammer's pioneering efforts in Danish geology earned him the appellation "the father of Danish geology." His fundamental work, *Danmarks geognostiske Forhold* (1835), was the first work on the structural geology of that country. This work and such other important investigations as those on the weathering of feldspars to clay minerals and the influences of biological materials upon the development of alums were always tinged with chemical insights. Forchhammer strengthened his arguments with chemical analyses. For example, he pointed out that trace quantities of heavy metals were present in almost all rocks as a result of their mobility in groundwaters circulating through fissures. His interest in soil chemistry extended to soil's effect on the growth of plants. As a member of several commissions of the Royal Danish Academy he investigated the origins of the "kitchen middens" along the Danish shores together with zoologists and archaeologists.

BIBLIOGRAPHY

I. ORIGINAL WORKS. Forchhammer published about 200 papers on geological and chemical subjects. *Om Sövandets Bestanddele* (Copenhagen, 1859) was translated into English as "On the Components of Sea Water" and published in *Philosophical Transactions of the Royal Society,* **155** (1865), 203-262. His treatise *Danmarks geognostiske Forhold* was published as a University of Copenhagen Report in 1835. He was also the author of several chemical textbooks.

II. SECONDARY LITERATURE. See S. A. Andersen, in *Dansk biografisk Leksikon,* VII (1935); Axel Garboe, "Fra myte til videnskab," in *Geologiens historie i Danmark,* 2 vols. (1959–1961); Hans Pauly, "G. Forchhammer, en af geokemiens pionerer," in *Naturens verden* (1967), pp. 24–32; I, 204–213, 224–246; and Johannes Steenstrup, "Forchhammer som menneske og personlighed," in *Meddelelser fra Dansk geologisk Forening,* **8** (1935), 438–476.

EDWARD D. GOLDBERG

FORDOS, MATHURIN-JOSEPH (*b.* Sérent, France, 3 November 1816; *d.* Paris, France, 1 July 1878), *chemistry.*

Fordos studied pharmacy in Paris, completed an internship in hospital pharmacy, and for the remainder of his professional career directed pharmacy services at three Paris municipal hospitals: Midi (1841–1842), Saint-Antoine (1842–1859), and Charité (1859–1878). A friendship with Amédée Gélis, a fellow pharmacy intern, led to many years of scientific collaboration. Fordos and Gélis were among a group of hospital pharmacy interns who in 1838 founded the Société d'Émulation pour les Sciences Pharmaceutiques, which for several decades provided an important scientific outlet for young pharmacy students serving internships in the Paris hospitals.

The bulk of Fordos's scientific work was carried on jointly with Gélis, who later made a career in industrial chemistry. The two were especially successful in their investigation of inorganic sulfur compounds. In 1842 they published their discovery of sodium tetrathionate, and in 1850 they elucidated the composition of sulfur nitride.

Of the several investigations which Fordos pursued alone, the most important was his chemical isolation of a blue crystalline pigment from purulent bandages, which he called pyocyanine and which Carle Gessard showed in 1882 to be produced by *Pseudomonas pyocyanea* (*Pseudomonas aeruginosa*). Fordos's procedure for obtaining pyocyanine and his description of its physical properties were published in 1860, and by 1863 he was able to perfect his method of extraction and obtain a much greater yield of this substance.

Noteworthy also was Fordos's contribution to public health aspects of lead toxicity. From 1873 to 1875 he conducted experiments showing that drinking water passing through lead pipes, as well as liquids stored in tin-lead alloy utensils, would absorb toxic amounts of lead. His interest in this problem led him to devise and publish (1875) a rapid industrial process for detecting lead in pots and vessels lined with tin.

BIBLIOGRAPHY

I. ORIGINAL WORKS. Among Fordos's most important publications dealing with inorganic compounds of sulfur, written jointly with Gélis, are the following: "Sur un nouvel oxacide du soufre," in *Comptes rendus hebdomadaires des séances de l'Académie des sciences,* **15** (1842), 920–923; and "Sur le sulfure d'azote," *ibid.,* **31** (1850), 702–705. For Fordos's isolation of pyocyanine, see "Recherches sur la matière colorante des suppurations bleues: pyocyanine," *ibid.,* **51** (1860), 215–217; and "Recherches sur les matières colorantes des suppurations bleues: pyocyanine et pyoxanthose," *ibid.,* **56** (1863), 1128–1131.

Representative of Fordos's papers on lead toxicity and the chemical detection of lead are "Action de l'eau aérée sur le plomb," *ibid.,* **77** (1873), 1099–1102; "Action de l'eau de Seine et de l'eau de l'Ourcq sur le plomb," *ibid.,* 1186–1188; "Du rôle des sels dans l'action des eaux potables sur le plomb," *ibid.,* **78** (1874), 1108–1111; "De l'action des liquides alimentaires ou médicamenteux sur les vases en étain contenant du plomb," *ibid.,* **79** (1874), 678–680; "De l'essai des étamages contenant du plomb; procédé d'essai rapide," *ibid.,* **80** (1875), 794–796. A comprehensive listing of Fordos's publications is given in Albert Goris *et al., Centenaire de l'internat en pharmacie des hôpitaux et hospices civils de Paris* (Paris, 1920), pp. 374–376.

II. SECONDARY LITERATURE. See Albert Goris, *op. cit.,* pp. 374, 805; and J. R. Partington, *A History of Chemistry,* IV (London–New York, 1964), 84, 391, 925.

ALEX BERMAN

FOREL, AUGUSTE-HENRI (*b.* near Morges, Switzerland, 1 September 1848; *d.* Yvorne, Switzerland, 27 July 1931), *medicine, neurology, entomology.*

Forel was fascinated very early by the life of insects and particularly by that of ants. Somewhat against his own desires, he studied medicine at the University of Zurich, registering in 1866. Attracted by the courses and clinical studies in psychiatry, he devoted himself to psychology but maintained a never-failing interest in the natural sciences. He became a friend of the famous botanist and paleontologist Oswald Heer, the specialist on Tertiary flora. Later, Forel went to Vienna to work on his thesis on the anatomy of the brain, under the guidance of Theodor Meynert. In 1872 he received his doctorate in medicine. Later that year Forel moved to Munich to work with B. A. von Güdden, who had been one of his teachers in Zurich and whose reputation as a brain specialist was international. In 1877 Forel returned to Zurich, where he was appointed *Privatdozent* at the university. He turned increasingly to psychiatry and in 1879 became professor of psychiatry at the University of Zurich Medical School. At the same time he took on the directorship of the important Burghölzli Clinic. Troubled by the effects of alcoholism, in 1889 he established the Asile d'Ellikon, which became one of the first institutions to treat alcoholics medically and give them the means of reestablishing themselves in society. In 1893 Forel retired prematurely and was thus able to devote himself without distraction to his first interest, the study of ants, as well as to problems concerning social reforms.

Although professionally Forel was one of the important psychiatrists of the last century, he is primarily known as an ant specialist. When very young he went on a study trip to southern Switzerland; the published results of his observations at once brought him high repute as an entomologist and earned him the Schläfli Foundation Prize. As an anatomist Forel studied the internal morphology of ants carefully and thus came to propose a new taxonomy of these members of the order Hymenoptera. In addition, having become engrossed in the psychology of these insects, he contributed greatly to the study of their social instincts. Forel was the first to describe the phenomena of parabiosis and lestobiosis in ants. Having gathered a considerable collection of hymenoptera he described the various species, finding more than 3,500 new ones. Thus he became a remarkable taxonomist.

Forel was also a great brain specialist. While in Munich, under the guidance of Güdden, he was the first to achieve histological preparations of human brain specimens. His specialized studies of particular brain regions made Forel a master of the development of the nervous system's microscopic anatomy. He made remarkable studies of the topography of the trigeminal, pneumogastric, and hypoglossal nerves and gave such a precise description of the hypothalamus that one of its regions was later named the campus Foreli in his honor.

Forel's teaching in Zurich, the direction of his clinic, and his interest in psychology led him to effect innumerable reforms that not only influenced psychiatry in Switzerland but brought about important changes in the penal code. His publications on alcoholism made him one of the pioneers in this field. He himself practiced complete abstinence, as an example, and fought in every way possible the effects of alcoholism on the working classes. Research on hypnotism also fascinated him, and he wrote many papers on that subject. A hygienist as well, Forel published the important book *La question sexuelle,* which was translated into nearly twenty languages.

BIBLIOGRAPHY

I. Original Works. Forel's writings include *La question sexuelle* (Paris, 1905; 5th ed., Paris, 1922); *L'activité psychique* (Geneva, 1919); *Les fourmis de la Suisse* (La Chaux-de-Fonds, 1920); and *Le testament d'A. Forel* (Lausanne, 1931).

II. Secondary Literature. On Forel and his work, see E. Bugnion, "A. Forel. Souvenirs myrmécologiques recueillis," in *Mitteilungen der Schweizerischen entomologischen Gesellschaft,* **15** (1931), 156; H. Kutter, "Verzeichnis des entomologischen Arbeiten von A. Forel," *ibid.,* **15** (1931), 180; A. Von Muralt, "A. Forel," in *Schweizerische medizinische Jahrbuch* (1929), p. 6; and E. Schwiedland, *Bibliographia Foreliana* (Vienna, 1908).

P. E. Pilet

FORSSKÅL (also **FORSSKÅHL** or **FORSKÅL**), **PETER** (*b.* Helsinki, Finland [then Sweden], 11 January 1732; *d.* Yarīm, Yemen, 11 July 1763), *botany.*

Forsskål was one of Linnaeus' most gifted pupils and had an unusually broad spectrum of interests. During his university years—1751-1753 in Uppsala, 1753-1756 in Göttingen, and 1756-1760 back in Uppsala—he did not confine himself to natural history; he also mastered economics and philosophy, theology, and the Oriental languages. The combination of knowledge of Arabic and botany made Forsskål unusually suited for the scientific expedition that led to both his fame and his death. In Denmark, under the sponsorship of King Frederick V, a major research voyage to Arabia was planned; its large scientific staff was to include a naturalist, an astronomer, a philologist, a physician, and an artist. Forsskål was accepted as a member of the expedition, received the title of professor, and moved to Copenhagen in 1760. In January 1761 the expedition departed; traveling via Marseilles, Malta, and Constantinople, it reached Egypt that autumn. In October 1762 the voyage continued toward southern Arabia, where Forsskål worked to complete his collections until his death from malaria in July 1763.

Forsskål's contribution to botany consists of a single work: the *Flora aegyptiaco-arabica,* which was saved for posterity by the only surviving member of the expedition, Carsten Niebuhr, and was published at Copenhagen in 1775. This work is of importance both for the greatly increased knowledge it provided about the vegetation in the areas visited (Forsskål proposed fifty new genera, half of which are still valid) and for the valuable and original morphological observations that are often found in the descriptions of the species. But today Forsskål's fame is based mainly upon the introduction to the *Flora,* in which he surveys the phytogeography of Egypt. By comparing the Scandinavian and the Egyptian flora he gave a precise characterization of Egyptian vegetation and clarified its relation to climate and soil. In this respect he can be seen as an often unfairly neglected precursor of Alexander von Humboldt.

BIBLIOGRAPHY

I. Original Works. Forsskål's most important publication in botany is *Flora aegyptiaco-arabica sive descriptiones plantarum quas per Aegyptum inferiorum et Arabiam felicem detexit, illustravit,* Carsten Niebuhr, ed. (Copenhagen, 1775). A MS of a more general character was published in Swedish as *Resa till lyckige Arabien* (Uppsala, 1950).

II. Secondary Literature. On Forsskål or his work, see C. Christensen, *Naturforskeren Per Forskål. Hans rejse til Aegypten og Arabien 1761-63 og hans botaniske arbejder og samlinger* (Copenhagen, 1918); B. Hildebrand and E. Matinolli, "Peter Forsskål," in *Svensk biografiskt lexikon,* XVI (Stockholm, 1965), 359-362; and E. Matinolli, *Petter Forsskål* (Turku, 1960), in Finnish, with a summary in German.

Gunnar Eriksson

FORSTER, (JOHANN) GEORG ADAM (*b.* Nassenhuben [or Nassenhof], near Danzig, Germany [now Gdansk, Poland], 27 November 1754; *d.* Paris, France, 10 January 1794), *natural philosophy, geography.*

Forster was the oldest son of Johann Reinhold Forster and Justina Elisabeth Forster. A precocious child, he was first educated by his father and acquired from him a lively and practical interest in natural history, as well as a thorough grounding in the numerous philological disciplines and languages which Johann Reinhold had mastered. In 1765 he accompanied his father on the survey of the German colonies on the Volga steppes and, for a short period while in Russia, attended the Petrisschule founded by the eminent geographer A. F. Büsching. In 1766 he went to England with his father and in 1767 published his first work, a translation of M. V. Lomonosov's history of Russia. By the age of thirteen he had a command of most of the major languages of Europe.

While his father was in Warrington, Lancashire, Forster was apprenticed to a merchant in London. In the autumn of 1767 he joined his father at the Dissenters' Academy, where he continued his own studies and assisted with the instruction. He also aided his father in the translation of Bougainville's *Voyage autour du monde.* When the elder Forster received the commission to sail on Cook's second voyage (1772–1775), he insisted that his son accompany him as assistant and artist. Afterward the younger Forster published his first major work, *A Voyage Round the World* (London, 1777). As a result of this work, issued without official sanction, Forster became engaged in a spirited polemic with William Wales, the astronomer on the voyage, over the ethics of publishing an independent narrative in defiance of the Admiralty. The *Voyage,* although deliberately lacking the systematic and scholarly presentation of geographic and scientific material found in his father's *Observations,* started a new genre of literary-scientific travel narratives, a genre ably developed later by Alexander von Humboldt, whom Forster influenced greatly by his work and ideas. In 1776 the Forsters issued *Characteres generum plantarum,* and in 1777 the younger Forster was elected a fellow of the Royal Society.

Although his preference was to continue his studies in England, Forster was forced by his family's circumstances to seek positions for himself and his father in Germany, and in 1779 he was appointed professor of natural history at the Collegium Carolinium in Kassel. He was soon in contact with the prominent men of science and letters in Germany, including J. F. Blumenbach, G. C. Lichtenberg, and S. T. Sömmering. Forster was particularly attracted by the intellectual climate of Göttingen. In 1784 he was appointed to the chair of natural history at Vilna, Poland, and the following year he married Therese Heyne, daughter of the eminent Göttingen philologist C. G. Heyne. Forster collaborated with Lichtenberg in editing and writing the *Göttingisches Magazin der Wissenschaften und Litteratur,* and he also published extensively in the *Göttingische Anzeigen von gelehrten Sachen.*

In Vilna, although isolated from the mainstream of European thought, Forster strove to correspond with men of science throughout Europe. In 1786 he published his M.D. dissertation (conferred by Halle), *De plantis esculentis insularum Oceani Australis commentatio botanica* (Berlin–Halle) and *Florulae insularum Australium prodromus* (Göttingen). The latter work was seen by Forster as the basis for a more comprehensive botanical work on the Pacific area, the "Icones plantarum in itinere ad insulas Maris Australis . . . ." He also intended to publish a major study of European exploration in the Pacific. In 1787 Forster published at Göttingen *Fasciculas plantarum Magellanicarum* and *Plantae Atlanticae.* J. D. Hooker, in his later work on the botany of the *Erebus* and *Terror* voyages, drew critically on the work of the Forsters, who in turn were indebted to Daniel Solander, Cook's *Endeavour* botanist. Apart from his botanical work Forster's main contributions to the natural history of Cook's second voyage were his drawings and, later, his philosophical and geographic essays. In 1786 he engaged in a polemic with Kant over his theory of the origins of man.

In 1787, prevented by war from taking up an appointment as naturalist to a Russian expedition, Forster returned to Göttingen; and in October 1788 he was appointed librarian at the University of Mainz. Between March and July 1790, accompanied by Humboldt, he traveled to England via the Rhineland and the Low Countries. His most important prose work, *Ansichten vom Niederrhein* (Berlin, 1791–1794), was a penetrating account of his journey with Humboldt. During the Mainz period his interest and writing turned more to social history and politics. He became absorbed in the French administration which governed Mainz from October 1792. In March 1793, Forster went as a Rhineland deputy to the National Convention in Paris, where he died of illness aggravated by scurvy contracted during the *Resolution* voyage.

Forster wrote of himself in 1789: "Natural science in its broadest sense and particularly anthropology have been my occupation hitherto. What I have written since my voyage is closely related to that." Cook's voyages opened up new areas of investigation to men of science in Europe. Forster, the universal scholar, was a remarkable apologist for the new era of scientific discovery. Fully alive to all the great movements of his day and in contact with the most eminent men in Germany and abroad, Forster, who had been well schooled by his father, did much to convey to the parochial world of German science and letters the significance of the great contemporary empirical advances in the geographic and biological sciences—in some of which disciplines German-speaking scientists were destined to have a profound influence in the ensuing century.

BIBLIOGRAPHY

I. ORIGINAL WORKS. The most complete collection of Forster's writings, edited by Gerhard Steiner, is *Georg Forsters Werke, Sämtliche Schriften, Tagebücher, Briefe,* which is being published by the Deutsche Akademie der Wissenschaften zu Berlin (1958-). To date only vols. I-III, VII, and IX have appeared. Two earlier, smaller collections are L. F. Huber, ed., *Kleine Schriften. Ein Beytrag zur Völker- und Länderkunde, Naturgeschichte und Philosophie des Lebens,* 6 vols. (Leipzig, 1789-1797); and G. G. Gervinus, ed., *Georg Forster's Sämmtliche Schriften,* 9 vols. (Leipzig, 1843).

Collected eds. of some of Forster's prolific correspondence are Therese Huber, ed., *Johann Georg Forster's Briefwechsel nebst einigen Nachrichten von seinem Leben* (Leipzig, 1829); and H. Hettner, ed., *Georg Forster's Briefwechsel mit S. Th. Sömmering* (Brunswick, 1877). Summaries of some of his English letters are in Warren R. Dawson, ed., *The Banks Letters. A Calendar of the Manuscript Correspondence of Sir Joseph Banks . . .* (London, 1958). MSS copies of his scientific papers, correspondence, etc. are extant in many collections throughout Europe, North America, and Australasia.

A good bibliography is in Johann Georg Meusel, *Lexikon der vom Jahr 1750 bis 1800 verstorbenen Teutschen Schriftsteller,* III (Leipzig, 1804), 419-430; some individual works are listed in Poggendorff, I, 776.

II. SECONDARY LITERATURE. Because of the universal nature of his work, Forster is cited by historians in many disciplines. He is also the subject of fictional writing. The fullest bibliography and assessment of his work are in Ludwig Uhlig, *Georg Forster. Einheit und Mannigfaltigkeit in seiner geistigen Welt* (Tübingen, 1965). Very little scholarly work on Forster is available in English: some appreciation of his science and writings can be found in E. D. Merrill, *The Botany of Cook's Voyages* (Waltham, Mass., 1954); and L. Bodi, in *Historical Studies, Australia and New Zealand,* **8** (1959), 345-363.

No satisfactory full-length biography of Forster exists. The standard note is still A. Dove, in *Allgemeine deutsche Biographie,* VII (1878), 173-181. Also useful is K. Karsten, *Der Weltumsegler* (Bern, 1957).

MICHAEL E. HOARE

FORSTER, JOHANN REINHOLD (*b.* Dirschau [now Tczew], Poland, 22 October 1729; *d.* Halle, Germany, 9 December 1798), *natural philosophy, geography.*

Forster was descended from a landed Yorkshire family that had emigrated to Germany about 1642. He was educated at Marienwerder and Berlin in 1743-1748 and then, against his wishes, as a Reformed clergyman at the University of Halle in 1748-1751. He became pastor at Nassenhuben (or Nassenhof), near Danzig, in 1753.

Forster began to maintain a wide scientific correspondence with such men as the elder Linnaeus, Thomas Pennant, and, later, Joseph Banks. In 1765 he was commissioned by the Russian government to undertake a survey of the Saratov-Tsaritsyn region of the lower Volga. In 1766, deprived of advancement in Russia and of his pastorate in Prussia, Forster went to England and soon succeeded Joseph Priestley as a tutor at the Dissenters' Academy in Warrington, Lancashire, teaching natural history and classical and modern languages.

Forster became known in Britain through his scientific writing. He was asked by the Royal Society of London to describe a collection presented by the Hudson's Bay Company. His papers on zoology, ornithology, and ichthyology were published in the *Philosophical Transactions of the Royal Society* (1772, 1773). Reprinted by the Willughby Society in 1882, they were recognized as having been written by "one of the earliest authorities on North American zoology." He published *An Introduction to Mineralogy* in 1768; two years later he issued *A Catalogue of British Insects,* and in 1771 he published works on American flora and on entomology. He also worked on the translation and editing of works on North America by the itinerant naturalists Peter Kalm and Nicholas Bossu and on Asia by Petrus Osbeck. In 1772 he published an edited translation of Bougainville's *Voyage autour du monde.* On 27 February 1772 Forster was elected fellow of the Royal Society.

In June 1772 Forster and his son Georg, "gentlemen skilled in Natural history and Botany but more especially the former," were appointed with ten days' notice to H.M.S. *Resolution,* bound, under Captain James Cook, to search for the hypothetical southern continent. Linnaeus the elder commended Forster as an outstanding man for such a charge. At Cape Town, Forster took on as an assistant Anders Sparrman, one of Linnaeus' pupils. The Forsters returned with Cook in the summer of 1775.

The voyage included a great part of the Pacific basin. In 1776 the Forsters published *Characteres generum plantarum,* a small, hurried, and preliminary account of the botany of the voyage. The remainder of the botanical specimens were dealt with later by Georg in Germany and by Daniel Solander privately. Forster's most significant publication was *Observations Made During a Voyage Round the World* (1778), the sum of his work "on physical geography, natural history and ethnic philosophy." Forster read deeply in the science of his day, including the work of Buffon and Torbern Bergman; he was able to test armchair hypotheses empirically against the facts of the field. The *Observations* is a remarkable systematic study of oceanographic, geographic, and ethnographic problems in the infancy of those sciences, a study characterized by perceptive observation, analogy, and experimentation. Forster predicted the scope and methods of Alexander von Humboldt's work and, in the same region, investigated those phenomena (e.g., coral reefs and volcanoes) which Charles Darwin examined sixty years later on the *Beagle* voyage. Forster also influenced the work of Blumenbach and the growing science of comparative anthropology.

After moving to Halle in 1780 as professor of natural history and mineralogy, Forster published extensively on zoological subjects, many of them connected with the *Resolution* voyage. He was associated with Pennant in the translation and later editions of his *Indian Zoology.* From 1790 he edited the *Magazin von merkwürdigen neuen Reisebeschreibungen,* which did a great deal to introduce the results of important scientific voyages to the German public. Forster published a history of northern maritime discovery in 1784. In the year of his death he published an essay containing his thoughts on a future theory of the history of the earth.

In 1844 H. Lichtenstein published *Descriptiones animalium . . .,* the zoological work of the *Resolution* voyage, which has since been attributed to Forster without reservation.

The whole of Forster's scientific career was affected by a tragic quarrel with the British Admiralty. He was a man of pride and quick temper. The considerable scientific data assembled on the *Resolution* voyage was, for this reason, scattered throughout the world, and the scientific publications of the Forsters, although competent, were fragmentary and obscure. The real merit of Forster's work has often been overlooked by authors more zealous to describe his weaknesses of character than to appraise his contribution to science.

BIBLIOGRAPHY

I. ORIGINAL WORKS. See bibliography in Johann Georg Meusel, *Lexikon der vom Jahr 1750 bis 1800 verstorbenen Teutschen Schriftsteller,* III (Leipzig, 1804), 430–439; and Poggendorff, I, 775–776.

No complete collection of Forster's works or papers exists. His unedited correspondence is scattered in Australasia, London, Germany, North America, and many European centers. Published works besides those cited in the text include *Florae Americae septentrionalis* (London, 1771); *Novae species insectorum centuria* (London, 1771); *Enchiridion historiae naturali inserviens . . .* (Halle, 1788); and numerous translations of accounts of voyages, with his own scientific appendixes and commentaries. He also contributed to a number of Continental scientific journals. His MS journal of his voyage (6 vols.) is in the Stiftung Preussischer Kulturbesitz, Berlin.

II. SECONDARY LITERATURE. There is no biography of Forster. The best biographical note is A. Dove, in *Allgemeine deutsche Biographie,* VII (1878), 167–172. For his scientific work, see J. C. Beaglehole, ed., *The Voyage of the Resolution and the Adventure 1772–1775* (Cambridge, 1961), pp. xlii–xlix; and M. E. Hoare, "Johann Reinhold Forster, the Neglected 'Philosopher' of Cook's Second Voyage (1772–1775)," in *Journal of Pacific History,* **2** (1967), 215–224.

MICHAEL E. HOARE

FORSYTH, ANDREW RUSSELL (*b.* Glasgow, Scotland, 18 June 1858; *d.* London, England, 2 June 1942), *mathematics.*

Forsyth was the son of John Forsyth, a marine engineer, and of Christina Glenn, of Paisley. The family moved to Liverpool, where Forsyth soon revealed his mathematical ability. He entered Trinity College, Cambridge, in 1877 and was senior wrangler in January 1881. He became a fellow of Trinity the same year with a remarkably powerful thesis on double theta functions. In 1882 he was appointed to the chair of mathematics at University College, Liverpool, but in 1884 he returned to Cambridge as a lecturer. He was elected a fellow of the Royal Society in 1886.

As a mathematician Forsyth belonged to the school of his Cambridge master, Cayley, and was outstanding in his ability to marshal complicated formulas. His importance in the history of British mathematics is due, however, to his being a great traveler and a good linguist; he was thus the first to realize the deficiencies of the Cambridge school, which was almost completely ignorant of Continental mathematics. Forsyth was determined to rectify this situation, and in 1893 he published his *Theory of Functions,* which, according to Sir Edmund Whittaker, "had a greater influence on British mathematics than any work since Newton's *Principia.*" As a result, for many years function theory dominated Cambridge mathematics.

In 1895 Forsyth succeeded Cayley as Sadlerian professor of pure mathematics but resigned in 1910 in order to marry Marion Amelia Boys, the former wife of the physicist C. V. Boys. After a short time in Calcutta, he was appointed chief professor of mathematics at Imperial College, London, in 1913. Although he retired in 1923, he continued to write mathematical treatises; but his point of view was antiquated, his work being based on manipulative skill rather than on logical processes.

Ironically, Forsyth's main achievement was having brought to Cambridge the modern style of mathematics that superseded his own, and as a result his reputation in his later years was less than it deserved to be.

BIBLIOGRAPHY

I. Original Works. Forsyth's most important books were *A Treatise on Differential Equations* (London, 1885; 6th ed., 1931), also trans. into German and Italian; *Theory of Differential Equations,* 6 vols. (Cambridge, 1890–1906); and *Theory of Functions of a Complex Variable* (Cambridge, 1893; 3rd ed., 1917). He also published *Lectures on the Differential Geometry of Curves and Surfaces* (Cambridge, 1912); *Lectures Introductory to the Theory of Functions of Two Complex Variables* (Cambridge, 1914); *Calculus of Variations* (Cambridge, 1927); *Geometry of Four Dimensions,* 2 vols. (Cambridge, 1930); and *Intrinsic Geometry of Ideal Space,* 2 vols. (London, 1935). He also contributed to British mathematical journals, *Proceedings of the Royal Society,* and other publications.

II. Secondary Literature. Biographical notices are E. H. Neville, in *Journal of the London Mathematical Society,* **17** (1942), 237–256; and E. T. Whittaker, in *Obituary Notices of Fellows of the Royal Society of London,* **4** (1942–1944), 209–227. The latter contains a complete bibliography of Forsyth's writings. See also the article on Forsyth by E. T. Whittaker, in *Dictionary of National Biography,* supp. VI (1941–1950), 267–268.

G. J. Whitrow

FORTIN, JEAN NICOLAS (*b.* Mouchy-la-Ville, Île-de-France, France, 9 August 1750; *d.* Paris, France, 1831), *scientific instruments.*

Fortin was one of the most skilled precision mechanics and scientific instrument makers of his time. Lavoisier realized his potential and asked him, at the beginning of his career, to make several new laboratory instruments for him. Of special note is a precision balance made in 1788, which had an arm one meter long and was sensitive to weights as slight as 1/400 ounce. For the Commission of Weights and Measures he made a similar precision balance with the necessary weights, and another instrument to compare the dimensions of the cylinders that constituted the standards for weights. In 1799 Fortin adjusted the platinum kilogram standard that was deposited in the National Archives of France. He also made instruments for the Paris observatory.

Around 1800 Fortin made a barometer whose distinctive feature was the combination of a leather bag containing mercury, an ivory pointer indicating the zero point of the barometric scale, and a glass cylinder. By this simple combination Fortin made it possible to adjust the level of the mercury surface in the cylinder to coincide with zero on the scale. The device made transportation of barometers easier, and later any barometer in which the mercury could be adjusted to touch a point was known as a Fortin barometer.

Fortin devised many instruments that were used by scientists and engineers in famous experiments: the study of the expansion of gases by Gay-Lussac; the verification of the Boyle-Mariotte law at high pressures by Arago and Dulong; the triangulations between Barcelona and Formentera by Biot and Arago; and so on. He was a member of the Bureau des Longitudes, and in 1776 he reduced Flamsteed's *Atlas céleste* to about a third of its former length.

BIBLIOGRAPHY

Fortin's only publication was his edition of J. Flamsteed's *Atlas céleste* (Paris, 1776).

On his work, see M. Daumas, *Les instruments scientifiques aux XVIIe et XVIIIe siècles* (Paris, 1953), *passim.*

Asit K. Biswas
Margaret R. Biswas

FOSTER, HENRY (*b.* Wood Plumpton, England, August 1797; *d.* Chagres River, Isthmus of Panama, 5 February 1831), *geophysics.*

Henry Foster was involved with geophysical observations throughout his career in the British navy. Foster joined the Royal Navy in 1812. Early projects included surveys and, on a trip to South America with Captain Basil Hall, determination of the acceleration of gravity. In 1824 Foster was made lieutenant and became a fellow of the Royal Society. He performed most of his investigations while on expeditions to the Arctic in 1824–1825 and to the South Seas in 1828–1831. He spent the winter of 1824–1825 at Port Bowen, north of the Arctic Circle, as astronomer of an expedition led by Sir William Edward Parry; he studied geomagnetism, the velocity of sound, atmospheric refraction, and the acceleration of gravity. The Board of Longitude printed a detailed account of his observations.[1] In 1827 Foster received the rank of commander and the Copley Medal of the Royal Society for these researches. In the spring of 1828 he sailed to the South Seas as commander of a sloop sent on a geophysical expedition, at the suggestion of the Royal Society, to study geomagnetism, gravity, meteorology, and oceanography.

Foster was on many occasions occupied with the indirect measurement of the acceleration of gravity. The project, generally referred to as a determination of the length of a seconds pendulum, was popular at the time and was sponsored officially. Foster used the method recently devised by Henry Kater[2] and observed coincidences between a pendulum of known length and the pendulum of a clock whose rate is determined by astronomical transit measurements. The final object of the observations at various latitudes was a determination of the ellipticity of the earth.

Foster was interested primarily in observations and performed them carefully.[3] He had a minor interest in theory—he speculated, for example, on the source of the diurnal variation in the earth's magnetic field. He was in some contact with other scientists: at Port Bowen, for example, he repeated some of Samuel Christie's experiments at the latter's request.[4]

NOTES

1. Published as *Philosophical Transactions of the Royal Society,* **26,** pt. 4 (1826).
2. Henry Kater, "An Account of Experiments for Determining the Length of the Pendulum Vibrating Seconds in the Latitude of London," *ibid.,* **18** (1818), 33–109; "An Account of Experiments for Determining the Variation in the Length of the Pendulum Vibrating Seconds," *ibid.,* **19** (1819), 336–508.
3. Such was the opinion of Gerard Moll, "On Captain Parry's and Lieutenant Foster's Experiments on the Velocity of Sound," *ibid.,* **28,** pt. 1 (1828), 97–104.
4. Henry Foster, "Account of the Repetition of Mr. Christie's Experiments on the Magnetic Properties Imparted to an Iron Plate by Rotation . . .," *ibid.,* **26,** pt. 4 (1826), 188–199.

BIBLIOGRAPHY

I. Original Works. Foster's writings include "Experiments With an Invariable Pendulum," in *Edinburgh Philosophical Journal,* **10** (1824), 91–95, written with Basil Hall; "Account of Experiments Made With an Invariable Pendulum . . .," in *Philosophical Transactions of the Royal Society,* **26,** pt. 4 (1826), 1–70; "Magnetical Observations at Port Bowen," *ibid.,* 73–117, written with W. E. Parry; and "Observations on the Diurnal Changes in the Position of the Horizontal Needle . . .," *ibid.,* 129–176. For a complete listing of Foster's publications, see the Royal Society of London, *Catalogue of Scientific Papers, 1800–1863,* II (London, 1868), 673–674.

II. Secondary Literature. See Francis Baily, "Report on the Pendulum Experiments Made by the Late Captain Henry Foster, R.N., in His Scientific Voyage in the Years 1828–1831, With a View to Determine the Figure of the Earth," in *Memoirs of the Royal Astronomical Society,* **7** (1834), 1–378. An account of Foster's expedition of 1828–1830, written by the surgeon of the sloop, is William H. B. Webster, *Narrative of a Voyage to the Southern Atlantic Ocean,* 2 vols. (London, 1834); the appendix contains measurements made on the expedition (II, 211–253). See also *Annual Biography and Obituary,* XVI (London, 1832), 436–437; and *Proceedings of the Royal Society,* **3A** (1830–1837), 82.

Sigalia Dostrovsky

FOSTER, MICHAEL (*b.* Huntingdon, England, 8 March 1836; *d.* London, England, 28 January 1907), *physiology.*

Foster was descended from a well-known family of religious Nonconformists who had farmed for many generations in Hertfordshire and Bedfordshire. His father, also named Michael, broke the yeoman tradition and became a medical practitioner. From 1831 to 1833 the elder Michael Foster studied medicine at the University of London (later University College), where he won many prizes. He practiced at Huntingdon from 1833 and was created fellow of the Royal College of Surgeons in 1852. A fervent Baptist, he was a prominent religious and civic leader in Huntingdon.

The younger Michael Foster, the eldest of ten children, was educated at the local grammar school in Huntingdon until 1849, when he was sent to University College School in London. In 1852 he entered University College, from which he graduated B.A. in 1854, placing first on the honors list in classics and receiving the college scholarship in that faculty. Foster might have pursued a career in classics at the University of Cambridge, but religious tests prevented his competing for a fellowship there. Instead, he immediately entered the medical school at University College, where in 1856 he won gold medals in anatomy and physiology and in chemistry. He graduated M.B. in 1858 and M.D. in 1859, and spent part of the next two years studying clinical medicine in the Paris hospital schools.

In the autumn of 1860, after impairment of his health led to fears of consumption, Foster signed on as ship's surgeon of H.M.S. *Union,* which was bound for the Red Sea to take part in the building of a lighthouse near Mt. Sinai. He signed on partly for the sake of his health and partly in the hope of studying the natural history of the area. After his attempts to make such studies were repeatedly discouraged or prevented, he returned to Huntingdon in March 1861. His health remained uncertain for many years and contributed to his later decision to leave London for Cambridge. From 1861 to 1866 he practiced medicine with his father in Huntingdon, but during all this time he longed for a career in science.

In January 1867, at the invitation of William Sharpey, Foster became instructor in practical physiology and histology at University College. In 1869 he was promoted to a professorship in the same subject. Meanwhile he had established a course of laboratory instruction, including elementary experimental physiology, that was the first of its kind in England. Also in 1869 he was appointed Fullerian professor of physiology at the Royal Institution.

In May 1870, Foster was appointed to a newly established prelectorship in physiology at Trinity College, Cambridge. In choosing both physiology as the subject of the prelectorship and Foster as the man to fill it, the Trinity seniority followed the recommendation of Thomas Henry Huxley. Foster remained prelector at Trinity College until 1883, when he was chosen to occupy the first chair of physiology in the university. Upon his resignation in 1903, the chair went to his former student John Newport Langley.

When Foster arrived at Cambridge in 1870, the biological sciences and the medical school were largely moribund. Neither Charles Babington, professor of botany, nor Alfred Newton, professor of zoology, was particularly receptive to the movement toward laboratory training in biology, and both were indifferent teachers. From the first Foster was determined to change this situation and to build a great school of biology and physiology at Cambridge, even though his original position made him responsible only to Trinity College. Before teaching his first class, Foster obtained the consent of the Trinity seniority to open his course to all students in the university. The university responded by giving him the use of one small room, which was furnished with the basic necessities by Trinity College. This accommodation very soon proved inadequate as Foster attracted ever larger numbers of students to his courses in physiology and elementary biology. New buildings were completed in 1879 and 1891, but even these were becoming overcrowded by the time Foster retired. From 1870 to 1883, the number of students attending his courses in physiology grew from about twenty to 130, while the number attending his course in elementary biology grew from about forty-five at his initial class in 1873 to more than eighty in 1883. In that year Foster turned the latter course over to two of his former students, and enrollment continued to grow dramatically. He then concentrated on the courses in physiology, which were drawing about 300 students when he retired in 1903.

Like other great teachers, Foster is probably best remembered for his students, many of whom remained at Cambridge to develop the principles and programs he had inaugurated. On leaving London for Cambridge in 1870, Foster had invited his two favorite students at University College to join him. Edward Sharpey-Schafer declined, on his father's advice, and remained at University College as assistant and later successor to John Burdon-Sanderson in the Jodrell chair of physiology. H. Newell Martin joined Foster at Cambridge and served as his right-hand man until 1876, when he was called to the United States as first occupant of the chair in biology at the newly established Johns Hopkins University. Martin thus carried to America the methods of teaching he had learned from Foster and Huxley.

While at Cambridge, Foster attracted a group of students as remarkable for the breadth of their interests as for their later eminence. Apart from physiologists, they include the embryologist Francis Maitland Balfour; the biologist G. J. Romanes; the anthropologist A. C. Haddon; the psychologist C. S. Myers; the neurologist Henry Head; the pathologist J. G. Adami; the botanists S. H. Vines, F. O. Bower, and H. Marshall Ward; and the morphologists A. Milnes Marshall, Adam Sedgwick, D'Arcy Wentworth Thompson, and A. E. Shipley. Of these, Balfour, Vines, Haddon, Ward, Sedgwick, and Shipley became leading members of the Cambridge faculty; and through them Foster remained a living influence on Cambridge biology long after his own direct role had come to an end.

But Foster was above all else the founder of the Cambridge School of Physiology, and the eminent physiologists trained while he was there are his chief contributions to science. Three of his earliest students were John Langley, Walter Holbrook Gaskell, and Arthur Sheridan Lea. Except for brief periods of study in Germany, all three remained at Cambridge throughout their careers, and all three were elevated to university lectureships in physiology when Foster was appointed to the professorship in 1883. Especially through the work of Langley on glandular secretion, of Gaskell on heart action, and of both on the involuntary nervous system, Cambridge soon became recognized as one of the world's leading centers for physiological research.

The list of physiologists trained at Cambridge later in Foster's career is one of almost staggering eminence. It includes Charles Scott Sherrington (matriculated at Cambridge in 1879, major work on reflexes and the integrative action of the nervous system); W. B. Hardy (1884, colloid chemistry); Walter Morley Fletcher (1891, muscle metabolism); Joseph Barcroft (1893, blood gases, respiration, and homeostasis); Henry H. Dale (1894, chemical transmission of nerve impulses); T. R. Elliott (1896, sympathomimetic drugs); and Keith Lucas (1898, conduction of nerve impulses). Sherrington and Dale went on to win the Nobel Prize in physiology or medicine in 1932 and 1936, respectively. Hardy, Fletcher, Barcroft, and Lucas joined the staff at Cambridge, with Barcroft succeeding Langley as professor of physiology in 1925.

Because Foster epitomizes the concept of a great teacher, it is interesting to consider his approach and the reasons for his success. He owed much to William Sharpey, who taught him at University College and first aroused his interest in physiology, and to Thomas Huxley, who very early perceived the remarkable kinship of mind and spirit between Foster and himself. Both taught Foster that physiology should be viewed broadly, as one of the biological sciences, and both encouraged his appreciation of the experimental approach, although neither made much use of that approach in his own work. Of the two, Huxley's influence was the more lasting and profound.

Foster first met Huxley in 1856, when the latter examined him in anatomy and physiology at University College. From then on, Huxley was his main guide and chief agent. The two rarely, if ever, disagreed on any issue of substance, and their educational philosophies are virtually indistinguishable. Both Huxley and Foster insisted that science must take equal place with mathematics and classics in the English educational system, and both urged students to undertake original laboratory research at an early stage. There are even two striking parallels in the course of their careers: Foster succeeded Huxley as Fullerian professor of physiology in 1869 and as biological secretary of the Royal Society in 1881.

In the summer of 1871, when Huxley introduced a laboratory course in elementary biology to a group of schoolmasters at South Kensington, he selected Foster as the first of his demonstrators. After a second summer as Huxley's demonstrator, Foster established his own one-term course in elementary biology at Cambridge in 1873. Both courses were taught on evolutionary principles, with a very few organisms being dissected and studied as representative "types." Huxley's course, which was the model, is considered the origin of the modern method of teaching introductory biology. Foster's very similar course seems to have been the first such course taught in a true university setting. It illustrates his broadly biological approach to physiology and helps to explain the breadth of his influence on Cambridge biology and physiology, for Foster designed this course in such a way that it became the standard means by which students were introduced to all of the biological sciences at Cambridge. In this way he was able to exert an influence on students who became botanists or morphologists as well as on those who became physiologists.

Another important factor in Foster's success was that the University of Cambridge was in a state of transition when he arrived. After decades of defending the virtual monopoly enjoyed by mathematics and classics at Oxford and Cambridge, many Cambridge dons were ready at last to give a more sympathetic hearing to the advocates of science and other "modern" studies. It was an excellent opportunity for Foster, who had a clear vision of how much could be done and how to go about doing it. With the support of several influential allies, especially G. M. Humphry, professor of anatomy, and Coutts Trotter, a fellow of Trinity College who was a leading force in university administration, he pressed the claims of physiology upon the university with remarkable success for a mere collegiate lecturer. That this success depended in part on a new attitude in the university at large is suggested by the comparable success of the famous Cambridge School of Physics, whose development paralleled almost exactly that of the School of Physiology.

Foster's achievement at Cambridge depended also on his own great clarity of aim and charm of manner. Even more important was his capacity for inspiring others to undertake original research. Many of his students have testified that they chose a career in research only because Foster promoted the enterprise with so much enthusiasm and with such a sense of adventure. Because Foster's own contributions to original research were few and largely ignored, he has acquired a reputation as a discoverer of men rather than of facts, as a great teacher of research who did not himself practice what he so effectively taught. But it is crucial to recognize that Foster did engage in original research. Throughout his brief career in research, the problem of the heartbeat held a special fascination for him.

The issue to be settled about the heartbeat was whether it depended ultimately on nerve discharges or rather on an inherent rhythmicity in active cardiac muscle—in other words, whether the heartbeat was neurogenic or myogenic. When Foster began his work, the issue had apparently been decided in favor of the neurogenic theory, especially because a number of German investigators had found that in the vertebrate heart, nerve ganglia were concentrated precisely in those regions of the cardiac tissue where excision or ligature disturbed the spontaneous rhythm of the heartbeat.

In 1859 Foster found that even very tiny pieces cut from the beating heart of a snail continued to beat rhythmically for some time. But no ganglia had been found anywhere in the snail's heart; and even if they were there, it seemed impossible to Foster that they could be so widely diffused as to appear in each and every part of the heart. He therefore concluded that, in the snail at least, the rhythmic beat must depend not on nerve ganglia but on the inherent properties of the general cardiac tissue.

In 1869 Foster published his first challenge to the neurogenic theory in vertebrate hearts. He showed that the lower two-thirds of a frog's ventricle, where no ganglia were known to exist, could nonetheless be induced to rhythmic pulsations by direct application of an interrupted current of appropriate strength. He therefore argued that the frog's cardiac musculature, like the snail's, must possess an inherent tendency to rhythmic pulsation.

Returning in 1871 to the snail's heart, Foster showed that when weak currents were applied directly to its muscular tissue, the result was an inhibition exactly like that produced in the vertebrate heart by stimulating the vagus nerve. Since in the ganglion-free snail's heart this inhibition could not be the result of any nervous mechanisms, Foster attributed it instead to the direct effects of weak electrical currents on the fundamental properties of active contractile tissue.

Between 1875 and 1877 Foster published four papers on the heartbeat, three of them in collaboration with his student A. G. Dew-Smith. Two of these were elaborate extensions of Foster's own earlier work on the snail's heart and on the frog's. The third, in 1877, was written in reply to a German scientist's claim that he had detected ganglia in the snail's heart. Foster and Dew-Smith argued that what the German had taken for ganglia were in fact pyriform connective-tissue cells.

The fourth paper, by Foster alone, concerned the effects of the poison upas antiar on the frog's heart. He emphasized the remarkable effect that vagus stimulation produced in a heart thus poisoned. The eventual result was not the usual inhibition but an opposite accelerating effect, which increased as the influence of the poison increased. Again Foster rejected the idea that nerve ganglia were somehow responsible. Antiar was known to be a muscular poison, and if it were admitted that vagus inhibition also resulted from direct action on the cardiac musculature, then the marked acceleration eventually observable under their combined influence could be explained simply as an exaggerated reaction by the muscle tissue to a previously exaggerated inhibition.

In all of this, the general trend of Foster's conclusions is clear. The causes both of inhibition and of the rhythmic heartbeat itself were to be sought not in nerve ganglia but in the basic properties of contractile tissue. Less obvious, but very definitely present in two of the papers with Dew-Smith, is the evolutionary basis for Foster's conviction. Impressed by the rhythmic capacity of undifferentiated protoplasm—in amoebas, in ciliates, and in the simple snail's heart—he saw no reason why differentiated nerve ganglia should be necessary for the same function in higher organisms. Apart from some ambiguity about whether the general cardiac tissue was purely muscular or rather neuromuscular in character, the chief difficulty in Foster's scheme was his concession that in the frog (and presumably other vertebrates) the ganglia might serve to coordinate the sequence of beats. In the undifferentiated and ganglion-free snail's heart, Foster attributed coordination to a muscular sense inherent in the general cardiac tissue, so by his own evolutionary criteria, differentiated ganglia should not have been required to perform the same function in higher organisms.

By 1900 the myogenic theory had largely replaced the neurogenic, but Foster's work was for a long time generally ignored and was not itself greatly influential in changing the direction of the debate. It is nonetheless important for at least three reasons. First, it reveals Foster as a competent research physiologist and suggests that this experience may have been a crucial factor in his success as a teacher. Without some such experience, it seems unlikely that he could have promoted research so convincingly or that he could have developed the critical acumen so essential to a director of research. Second, the problem of the heartbeat was the starting point or focus for much of the research carried out by the Cambridge School of Physiology during its crucial early years. Under Foster's direction Lea, Langley, Romanes, Francis Darwin, and Gaskell were all attracted to the problem during the 1870's. Like Foster himself, Romanes and Gaskell saw the problem of the heartbeat as part of the more general problem of rhythmic motion, and they pursued the issues with significant success. Although the others rather quickly found different research interests, their work on the heart was crucial in establishing a research tradition in physiology at Cambridge. Third, Foster's work on the heart provides a concrete illustration of his broadly biological and evolutionary approach to physiology and suggests again that this approach may have contributed greatly to his success and to that of the embryonic Cambridge School of Physiology. This seems even more likely because Gaskell, who did more than anyone else to resolve the problem in favor of the myogenic theory, depended very heavily on the evolutionary approach he had learned from Foster.

Foster's impact on physiology and science extended far beyond his contributions as original investigator and as founder and director of the Cambridge School of Physiology. His name was attached to several important textbooks, including the pioneering *Handbook of the Physiological Laboratory,* edited by John Burdon-Sanderson (1873), for which he wrote the section on nerve and muscle; *The Elements of Embryology,* which he published with his student F. M. Balfour in 1874; and *A Course of Elementary Practical Physiology and Histology,* which he published with Langley's assistance in 1876. The first edition of his famous *Text-Book of Physiology* appeared in 1877. Distinguished for its literary style, balanced judgment, and evolutionary perspective, this work was translated into Russian, Italian, and German. It went through six complete editions and part of a seventh.

Foster was also the leading figure in the professionalization of physiology in Victorian England. A conspicuous opponent of popular antivivisection sentiment, he was largely responsible for the founding in 1876 of the British Physiological Society. In 1878 he founded the *Journal of Physiology,* which he edited until 1894.

As he became increasingly occupied with these and other organizational activities, Foster abandoned original research in physiology and delegated to former students most of the responsibility for teaching the advanced classes at Cambridge and for directing the day-to-day activities of the research laboratory. By about 1880 the shift was complete. Foster concentrated thereafter on elementary teaching and on exercising his talent for organization. At Cambridge he led the fight for the establishment of a school of scientific agriculture in the 1890's. In 1898 he induced Frederick Gowland Hopkins to join his staff at Cambridge as teacher of biological chemistry. Hopkins went on to become founder of the Cambridge School of Biochemistry and to share the Nobel Prize in physiology or medicine in 1929.

Foster became a leader as well in the national and international organization of science. At the Royal Society, of which he was elected fellow in 1872, he served as biological secretary from 1881 to 1903. In this influential and burdensome office he supported a wide range of scientific expeditions and was a vigorously successful advocate of a closer partnership between the Society and the government. He was vice-president of the Society in 1903–1904. He was also active in the affairs of the Royal Horticultural Society and of the British Association for the Advancement of Science, of which he was president in 1899. In the founding of the International Physiological Congresses, the first of which was held in 1889, he played a major role, as he did in the establishment of the International Association of Academies and in the preliminary arrangements for the *International Catalogue of Scientific Literature.*

Foster served on national commissions dealing with vaccination, tropical disease, disposal of sewage, and the reorganization of the University of London. In 1901 he was designated chairman of the Royal Commission on Tuberculosis. From 1900 to 1906 he was Member of Parliament for the University of London. Because he opposed the Liberal bill for Irish home rule, he stood originally as a Conservative; later he joined the Liberal opposition because of his stand on the education bill. On seeking reelection in 1906, he lost by the narrow margin of twenty-four votes.

Despite all these administrative duties, Foster coedited *The Scientific Memoirs of T. H. Huxley* (1898–1902) and several times revised Huxley's *Lessons in Elementary Physiology.* In 1899 he published a biography of Claude Bernard and in 1901 *Lectures on the History of Physiology During the Sixteenth, Seventeenth and Eighteenth Centuries.* Even in gardening, which was his chief source of relaxation, Foster exhibited leadership. He hybridized several new varieties of iris and was for a long time the internationally acknowledged expert on the genus. In fact, he published more original papers in horticulture than in physiology.

A member or fellow of a vast number of scientific societies, both British and foreign, Foster received honorary doctorates from the universities of Glasgow, St. Andrews, McGill, and Dublin. At Cambridge he was made honorary M.A. in 1871, and the full degree was conferred upon him in 1884. He was created K.C.B. in 1899.

Foster was twice married to women from Huntingdon: in 1863 to Georgina Edmonds, who died in 1869, and in 1872 to Margaret Rust, who survived him. By his first wife he had a daughter, Mercy, and a son, Michael George.

BIBLIOGRAPHY

I. Original Works. Foster's papers on the heartbeat are the following: "On the Beat of the Snail's Heart," in

Report of the Twenty-ninth Meeting of the British Association for the Advancement of Science (London, 1860), transactions of the sections, p. 160; "Note on the Action of the Interrupted Current on the Ventricle of the Frog's Heart," in *Journal of Anatomy and Physiology,* **3** (1869), 400–401; "Ueber einen besonderen Fall von Hemmungswirkung," in *Pflüger's Archiv für die gesamte Physiologie des Menschen und der Tiere,* **5** (1872), 191–195; "On the Behaviour of the Hearts of Mollusks Under the Influence of Electric Currents," in *Proceedings of the Royal Society,* **23** (1875), 586–594, written with A. G. Dew-Smith; "The Effects of the Constant Current on the Heart," in *Journal of Anatomy and Physiology,* **10** (1876), 735–771, written with A. G. Dew-Smith; "Some Effects of Upas Antiar on the Frog's Heart," *ibid.,* 586–594; and "Die Muskeln und Nerven des Herzens bei einigen Mollusken," in *Archiv für mikroskopische Anatomie und Entwicklungsmechanik,* **14** (1877), 317–321, written with A. G. Dew-Smith.

The most complete bibliography may be found in the article by Henry Dale (see below). A slightly less complete bibliography can be obtained by combining the citations in the *Royal Society's Catalogue of Scientific Papers,* II, 674; VII, 692; IX, 906; XV, 69; and in the *British Museum General Catalogue of Printed Books,* LXXVI, cols. 241–243. Several publications are omitted from all of these sources. A series of popular articles on science (some unsigned), including one on the snail's heart, appeared in the *Christian Spectator* in 1863 and 1864. Foster expresses his views on education in the unsigned article "Science in the Schools," in *London Quarterly Review,* **123** (1867), 244–258; in "Vivisection," in *Macmillan's Magazine,* **29** (1874), 367–376, he defends physiologists against the charges of the antivivisectionists; and in "Reminiscences of a Physiologist," in *Colorado Medical Journal,* **6** (1900), 419–429, he gives a rambling account of his education and early career.

There is no central repository for Foster's private papers and correspondence. Of the Foster–Huxley correspondence, more than 200 letters are preserved in the Huxley Papers at Imperial College, London, and nearly as many, from Huxley to Foster only, at the Royal College of Physicians. Many letters from both collections are quoted in Leonard Huxley, *Life and Letters of Thomas Henry Huxley,* 2 vols. (London, 1900). Sir Robert Mordant Foster, grandson of Sir Michael, possesses more than twenty letters sent home by Foster during his brief career as ship's surgeon. The rest of Foster's extant letters are scattered among various collections in England, including the J. D. Hooker and W. T. Thistleton-Dyer letters at the Royal Botanical Gardens in Kew, and the E. A. Sharpey-Schafer Papers at the Wellcome Institute for the History of Medicine.

At the library of the Cambridge Physiological Laboratory is a bound MS volume in Foster's hand with the heading "Three Lectures on the 'Involuntary Movements of Animals' Delivered Before the Royal Institution of Great Britain, February 1869." This MS reveals the central thrust—as well as the difficulties and ambiguities—of Foster's later ideas on the rhythmic heartbeat.

II. Secondary Literature. This article is based on Gerald L. Geison, "Sir Michael Foster and the Rise of the Cambridge School of Physiology, 1870–1900," unpublished Ph.D. thesis (Yale, 1970). Of the many available sketches of Foster's work and career, the most valuable are J. N. Langley, in *Journal of Physiology* (London), **35** (1907), 233–246; and in *Dictionary of National Biography,* suppl. I (1901–1911), 44–46; W. H. Gaskell, in *Proceedings of the Royal Society,* **80B** (1908), lxxi–lxxxi; and Henry Dale, in *Notes and Records. Royal Society of London,* **19** (1964), 10–32.

For general background, see E. A. Sharpey-Schafer, *History of the Physiological Society During Its First Fifty Years, 1876–1926* (London, 1927); G. L. Geison, "The Stagnancy of English Physiology, 1850–1870," in *Bulletin of the History of Medicine* (in press); and Richard D. French, "Some Problems and Sources in the Modern Foundations of British Physiology," in *History of Science* (in press). French has also drawn attention to the evolutionary context of the cardiological research of Foster and Gaskell. See "Darwin and the Physiologists, or the Medusa and Modern Cardiology," in *Journal of the History of Biology,* **3** (1970), 253–274.

Gerald L. Geison

FOUCAULT, JEAN BERNARD LÉON (*b.* Paris, France, 19 September 1819; *d.* Paris, 11 February 1868), *experimental physics.*

The son of a bookseller-publisher, Foucault received his education at home because of his delicate health. An indifferent student, he passed the *baccalauréat* only after special coaching and began to study medicine, hoping to put to use as a surgeon the considerable dexterity he had demonstrated (from the age of thirteen) in making a number of scientific toys, including a steam engine. Revolted by the sight of blood and suffering and stimulated in new directions by the invention of daguerreotypy, Foucault abandoned his medical studies, although not before he had come to the attention of Alfred Donné, teacher of clinical microscopy at the École de Médecine. Donné made him assistant in the microscopy course, then coauthor of its textbook (published in 1844–1845). Foucault finally succeeded his master as science reporter for the newspaper *Journal des débats* (1845), thereafter writing, in a brilliant style at once lively and precise, a regular column in which he discussed for a general audience the latest from the world of science.

From 1844 to 1846 Foucault published geometry, arithmetic, and chemistry texts for the *baccalauréat.* Thereafter, except for his newspaper articles, he published only scientific papers. Foucault worked in a laboratory set up in his home until, following the award of the Cross of the Legion of Honor in 1851 (for his pendulum experiment) and the *docteur ès sciences physiques* in 1853 (for his thesis comparing the velocity of light in air and water), he was given a place as physicist at the Paris observatory by Napoleon III. Further honors followed: the Copley Medal of the Royal Society in 1855, officer of the Legion of Honor and member of the Bureau des Longitudes in 1862, and foreign member of the Royal Society (1864) and the academies of Berlin and St. Petersburg. Finally, after having failed to be elected in 1857, Foucault was chosen in 1865, following the death of Clapeyron, a member of the Académie des Sciences.

A nonobserving Catholic until his final illness returned him to the church, Foucault led a quiet life of total devotion to scientific research. Small and frail, he managed to preside gracefully over the group of scientific friends who gathered on Thursdays at his house in the rue d'Assas. He died of brain disease at the age of forty-eight after a seven-month illness.

Foucault is best-known for two of the most significant experiments of the mid-nineteenth century—the laboratory determination of the velocity of light (1850, 1862) and the mechanical demonstration of the earth's rotation (1851, 1852)—and for his advancement of the technology of the telescope. He also performed a number of other important experiments, chiefly in optics, and developed several devices which were widely used in both experimental science and technology.

In 1834 Charles Wheatstone developed a rotating-mirror apparatus to measure the velocity of electricity, and in 1838 Arago suggested that the same principle might be applied to determining the velocity of light terrestrially (earlier determinations were astronomical). A comparison of this velocity in air and in water would be a clear experimental test between the wave and particle theories of light, since the former required light to travel faster in air; the latter, in water. Arago's attempts to carry out the experiment were unsuccessful, and failing eyesight forced him to abandon them. Immediately Foucault and Hippolyte Fizeau, with whom Foucault had collaborated on optical researches between 1845 and 1847, began independently to attempt to overcome the obstacles that had defeated Arago.

Fizeau was the first to succeed; by replacing the rotating-mirror apparatus in the laboratory with a toothed wheel interrupting a ray of light traveling over a long terrestrial path, he obtained the first precision measurement of the velocity of light at the earth's surface in 1849. Fizeau returned to the rotating mirror to compare light's velocity in rare and dense media, but here he was beaten by Foucault, who announced on 30 April 1850 that "light travels faster in air than in water" (*Recueil,* p. 207). His apparatus is diagramed in Figure 1. A source of light at a is reflected by a mirror m, rotating at 800 revolutions per second, to a spherically concave stationary mirror M and back again to a'. (The glass plane g permits the observer at O to see both source and reflection.) By the use of both an air path (upper half of diagram, image a') and a water path (lower half of diagram,

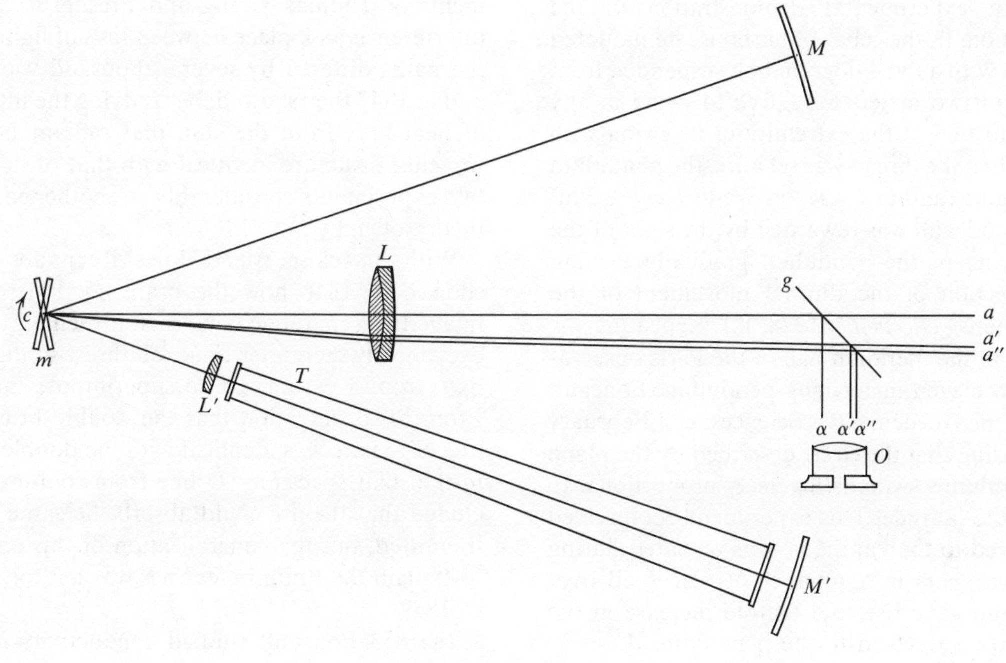

FIGURE 1

water-filled tube *T,* image *a''*), the velocity of light, which is a function of the displacement of the reflected image *a'* or *a''* from the source image *a,* can be compared in the two media. Since the water image *a''* is deflected more than the air image, light must travel faster in air than in water.

Foucault's first experiment, carried out in 1850 and written up in full in his doctoral thesis of 1853, was purely comparative; he announced no numerical values until 1862. Then, with an improved apparatus, he was able to measure precisely the velocity of light in air. This result, significantly smaller than Fizeau's of 1849, changed the accepted value of solar parallax and vindicated the higher value which Le Verrier had calculated from astronomical data. Foucault's turning-mirror apparatus was the basis for the later determinations of the velocity of light by A. A. Michelson and Simon Newcomb.

With Fizeau, Foucault had pioneered in astronomical photography by making the first daguerreotype of the sun in 1845. The long exposures necessary for photographing the stars required that the telescope remain continuously pointed at the heavenly object. To regulate the drive for such a telescope, Foucault in 1847 brought into practice Christian Huygens' abortive seventeenth-century project for a clock with a conical pendulum. Foucault's clock had a steel rod to support the bob of its pendulum, and he noticed that such a rod, set vibrating while clamped in the chuck of a lathe, tended to maintain its plane of vibration when the lathe was rotated by hand.

This unexpected behavior of the rod suggested to Foucault an experimental demonstration of the earth's rotation. In the cellar of his house he mounted a pendulum with a five-kilogram bob suspended from a steel thread two meters long, free to swing in any direction and tied at the extremity of its swing with a thread. When the thread was set afire, the pendulum began swinging, and at 2 A.M. on Wednesday, 8 January 1851, Foucault was rewarded by the sight of the plane of swing of the pendulum gradually turning "in the direction of the diurnal movement of the celestial sphere" (*Recueil,* p. 378, n.). Repeating the experiment in the meridian hall of the Paris observatory with an eleven-meter-long pendulum, Foucault reported to the Académie des Sciences on 3 February 1851 his finding that the circle described by the plane of the pendulum's swing is inversely proportional to the sine of the latitude. This experiment, soon scaled up and moved to the Panthéon, was repeated during the next two years in a number of places all over the world and gave rise to a tenfold increase in the scientific papers devoted to the pendulum.

As Foucault claimed in his report to the Academy, his finding illustrated Poisson's theoretical treatment of the deflecting force of the earth's rotation (*Journal de l'École polytechnique,* **16** [1838], 1–68), but Poisson had explicitly denied that the effect on the pendulum could be observed (p. 24).

Continuing to experiment on the mechanics of the earth's rotation, Foucault in 1852 invented the gyroscope, which, he showed, gave a clearer demonstration than the pendulum of the earth's rotation and had the property, similar to that of the magnetic needle, of maintaining a fixed direction. Foucault's pendulum and gyroscope had more than a popular significance (which continues to this day). First, they stimulated the development of theoretical mechanics, making relative motion and the theories of the pendulum and the gyroscope standard topics for study and investigation. Second, prior to Foucault's demonstrations the study of those motions on the earth's surface in which the deflecting force of rotation plays a prominent part (especially winds and ocean currents) was dominated by unphysical notions of how this force acted. Foucault's demonstrations and the theoretical treatments they inspired showed conclusively that this deflecting force acts in all horizontal directions, thus providing the sound physical insight on which Buys Ballot, Ferrel, Ulrich Vettin, and others could build.

Their daguerreotype of the sun was only one fruit of the collaboration between Foucault and Fizeau. Together, between 1844 and 1847, they carried out half a dozen researches. Two were of special importance: in 1845 and 1846 they extended the experiments of Thomas Young and Fresnel to show that interference took place between rays of light of which the paths differed by several thousand wavelengths, and in 1847 they showed, by studying the interference of heat rays from the sun, that radiant heat has a wavelike structure identical with that of light. These two experiments considerably strengthened the wave theory of light.

With his close friend Jules Regnault, Foucault showed in 1848 how the brain combines into one image two separate colors, each presented to a single eye. Shortly thereafter Foucault threw sunlight on the light from a carbon arc to superimpose the spectra. From his observation that the double bright-yellow line of the arc was identical with the double dark line in the solar spectrum (D line from sodium), he concluded that the arc could absorb the same light that it emitted, but the generalization of this observation to explain the Fraunhofer lines was left for Kirchhoff in 1859.

In 1853 Foucault studied conductivity in liquids, and in 1855 he demonstrated the conversion of mechanical work into heat by turning with a crank a copper disk placed between the poles of an electromagnet and measuring the heat produced in the disk.

No one in his time exceeded Foucault in technical inventiveness. From his first published papers on improvements in daguerreotypy (1841, 1843) to the completion of his siderostat shortly after his death, the devices designed by Foucault and executed, first by himself and later with the help of others, solved outstanding problems of practice in both science and technology. He developed a regulator for the arc lamp, which made it possible for gas to be supplanted by electricity in the supply of artificial light to the microscope (1843), and his improvement to this regulator (1849) brought the arc lamp into the theater. He designed a photometer (1855). His mercury interrupter (1856) improved the performance of Ruhmkorff induction coils, and his birefringent prism (1857), using air rather than balsam between the two pieces, made it possible to obtain plane polarized light into the ultraviolet. About 1860 he returned to the problem of making mechanical motion uniform, which had led him to the pendulum experiment, and he developed a whole series of mechanical regulators which went considerably beyond James Watt's governor in their effectiveness. These regulators were used first in machines which kept a telescope pointed continuously at the sun (heliostat) or a star (siderostat) and then in large steam engines, both in factories and at the Paris Exposition of 1867.

None of these inventions, however, was as significant for science as Foucault's introduction of the modern technique for silvering glass to make mirrors for reflecting telescopes (1857) and his simple but accurate methods for testing and correcting the figure of both mirrors and lenses (1858). Glass proved much superior to the speculum metal previously used in reflecting telescopes because it is much lighter in weight, easier to grind and figure, and easier to resurface if it becomes tarnished or damaged.

Foucault's extraordinary command of a precise language in both word and deed was not always taken at its true worth by his contemporaries among the masters of the French analytic tradition, for whom his sparing use of mathematics condemned him as merely a lucky tinkerer. His pungent newspaper articles, although never vicious, were also a source of hostility. Foucault's interest in astrophysics met the firm opposition of Le Verrier, director of the Paris observatory, a theoretical astronomer of the old school, and Foucault was therefore prevented from installing his siderostat in the observatory. Nevertheless, before he died, Foucault had acquired the respect of all as an outstanding experimentalist; and his reputation grew after his death as modern telescopic astronomy developed on the basis of the optical techniques he had inaugurated.

BIBLIOGRAPHY

I. Original Works. Foucault's papers, published mostly in the *Comptes rendus hebdomadaires des séances de l'Académie des sciences,* were collected and issued together with a number of unpublished papers in *Recueil des travaux scientifiques de Léon Foucault,* 2 vols. in one (Paris, 1878). Figure 1 in the text is taken from Plate 4 of the *Recueil,* which is in turn taken from Foucault's thesis, *Sur les vitesses relatives de la lumière dans l'air et dans l'eau* (Paris, 1853).

II. Secondary Literature. The two chief sources for Foucault's life and work are also in the *Recueil:* J. Bertrand, "Avertissement", I, i–iv, and "Des progrès de la mécanique," I, v–xxviii, the latter originally published in *Revue des deux mondes,* **51** (1 May 1864), 96–115, in order to help Foucault's candidacy for the Académie des Sciences; and J. A. Lissajous, "Notice historique sur la vie et les travaux de Léon Foucault," II, 1–18. Also useful is P. Gilbert, "Léon Foucault, sa vie et son oeuvre scientifique," in *Revue des questions scientifiques,* **5** (1879), 108–154, 516–563. Bertrand alludes in his article to the opposition Foucault faced in the Academy; the opposition of Le Verrier is mentioned in P. Larousse, *Grand dictionnaire universel du XIXe siècle,* VIII (Paris, 1872), 649.

Harold L. Burstyn

FOUCHY, JEAN-PAUL GRANDJEAN DE (*b.* Paris, France, 10 March 1707; *d.* Paris, 15 April 1788), *astronomy.*

He was the son of Marie-Madeleine Hynault and Philippe Grandjean de Fouchy, a Mâconnais noble who perfected the printing of deluxe editions under Louis XIV. Trained to succeed his father, Fouchy (who called himself alternately Grandjean, Grandjean de Fouchy, and de Fouchy) made some contributions in that area but soon found his art less appreciated and the demand for such work diminished. He became auditor of the Chambre des Comptes and secretary to the duc d'Orléans. More important, he undertook the study of science, devoting himself particularly to astronomy as a student of Joseph Nicolas Delisle.

In 1726 he became part of the newly formed Society of Arts in Paris. Among the several papers he presented to this group was one on the meridian of mean time, an innovation destined to be his most lasting contribution to astronomy. Named to the Academy of Sciences as a supernumerary assistant astronomer in 1731, he succeeded to regular membership in astronomy by the end of 1733. The first decade of his membership therein was his most productive scientific period.

Many of the memoirs he offered during that period were simply observational reports of specific phenomena such as eclipses, occultations, and the 1736 transit of Mercury. A few were more general: in 1731, a proposal for giving astronomical tables a more commodious form; in 1732, a memoir dealing with the reason for the disappearance of Jupiter's satellites from view before immersion and their reappearance only after a segment had emerged from Jupiter's shadow, and establishing rules to calculate the size of the segments involved based upon a new observational technique; in 1733, a method of employing bright spots on the moon for longitude determination; in 1737, his observation of Mercury's transit by a new means; in 1738, the proposal of a method to determine the eccentricity of the earth's orbit and that of the inner planets; in 1740, the extension of this method to any planet; and finally, also in 1740, a suggestion for improving Hadley's quadrant by substituting a telescope for open sights. Several other ideas for instrumental improvements, including a new level, a universal micrometer, and a device for moving a large quadrant, appeared in the collection *Machines et inventions approuvées par l'Académie.*

Unfortunately, these works were not as significant as they might seem. Few of them offered the advantages claimed by Fouchy, who, moreover, was much more prone to propose than to pursue. In the case of Jupiter's satellites, for example, it remained for Jean Sylvain Bailly to develop his idea and arrive at important results. Thus, Delambre's judgment that Fouchy was more an amateur than a true astronomer seems valid. This evaluation becomes even more appropriate after 1743, when Fouchy became the Academy's perpetual secretary. He served alone in that capacity for thirty years; but from 1773 until he resigned in 1776, he asked for and received the aid of Condorcet. During that period he wrote over sixty *éloges,* which, although lacking the style and philosophy of those of Fontenelle and Condorcet, were noteworthy for their information on, and analysis of, the scientific work of others.

During his secretariat his scientific contributions consisted mainly of meteorological observations but also included observations of eclipses and of the transits of Venus of 1761 and 1769. Some of these activities he continued thereafter, for he remained active nearly to the end of his life, even dispassionately describing a strange malady that afflicted him in 1784. His long career brought him membership both in the Academy of Sciences of Berlin and the Royal Society of London.

Fouchy was married twice, to Mlle. de Boistissandeau and to Mlle. Desportes-Pardeillan. The first union produced one daughter; the second, one daughter and two sons, both of whom pursued military rather than scientific careers.

BIBLIOGRAPHY

I. Original Works. The most important of Fouchy's contributions to the *Mémoires de l'Académie royale des sciences* were "Sur la forme la plus avantageuse qu'on puisse donner aux tables astronomiques" (1731), 433–442; "Sur la féconde inégalité des satellites de Jupiter" (1732), 419–427; "Observation du passage de Mercure sur le disque du Soleil, arrivé le 11 novembre 1736" (1737), 248–252; "Méthode pour déterminer par observation, l'excentricité de la Terre, et celle des planètes inférieures" (1738), 185–192; "Second mémoire sur l'excentricité des planètes" (1740), 235–242; and "Mémoire concernant la description et l'usage d'un nouvel instrument pour observer en mer les hauteurs et les distances des astres" (1740), 468–482. His longitude determination proposal was recorded, under the heading of "Sur une nouvelle méthode pour les longitudes," in the *Histoire de l'Académie royale des sciences* (1733), 76–79. His various instrument proposals may be read in *Machines et inventions approuvées par l'Académie royale des sciences depuis son établissement jusqu' à présent, avec leurs descriptions,* 7 vols. (Paris, 1735–1777), V, 91–92; VI, 45–47, 79–81, 113–114; and VII, 47–48. His *éloges* appeared in his annual histories of the Academy's work that preface the *Mémoires.* Those which appeared in the first sixteen years of his secretariat were also collected and published separately as *Éloges des académiciens de l'Académie royale des sciences, morts depuis l'an 1744* (Paris, 1761).

II. Secondary Literature. The archives of the Académie des Sciences contain an interesting MS of notes by one of Fouchy's sons for Condorcet's use in preparing the official *éloge.* Condorcet's product is in *Oeuvres complètes de Condorcet,* Marie Louise Sophie de Grouchy, Marquise de Condorcet, ed., 21 vols. (Paris, *an* XIII [1804]), IV, 3–26. A far more valuable estimation of his work is in J. B. J. Delambre, *Histoire de l'astronomie au dix-huitième siècle* (Paris, 1826), pp. 327–331. For a very brief treatment, see Niels Nielsen, *Géomètres français du dix-huitième siècle* (Paris, 1935), pp. 184–185.

Seymour L. Chapin

FOUQUÉ, FERDINAND ANDRÉ (*b.* Mortain, Manche, France, 21 June 1828; *d.* Paris, France, 7 March 1904), *geology, mineralogy.*

Fouqué made three important contributions to science. First, he added significantly to the knowledge of volcanic phenomena and volcanic products, in particular generalizing Henri Sainte-Claire Deville's explanation of the chemical composition of the emanations of fumaroles. Second, in collaboration with Auguste Michel-Lévy, he introduced into France the study of rocks by microscopical petrography. Third, again in collaboration with Michel-Lévy, he successfully synthesized a large number of igneous rocks in an attempt to determine the conditions necessary for the production of their mineralogical constituents.

Fouqué had some difficulty in settling upon a career. He attended Saint-Cyr (1847), the École d'Administration (1848), and the École Normale Supérieure (1849). He became a laboratory assistant at the latter and collaborated in 1853 with Sainte-Claire Deville in a memoir concerning the action of heat on topaz. After working briefly in the chemical industry, he commenced medical studies and received his doctorate in medicine in 1858. His lasting interest in volcanic phenomena was aroused when he accompanied Sainte-Claire Deville in 1861 to Vesuvius, then in eruption, to observe the fumaroles.

During the next twenty years, Fouqué traveled extensively to study volcanoes, both active and extinct. He was present at the eruption of Etna in 1865 and of Santorin (now Thíra, Greece) in 1866; and he investigated the volcanic chemistry of the Lipari Islands, Vesuvius, Solfatara, and the Cantal. His research resulted in several important publications: *Recherches sur les phénomènes chimiques qui se produisent dans les volcans* (1866), which was accepted as his thesis for the doctorate in physical sciences; *Les anciens volcans de la Grèce* (1867); and *Santorin et ses éruptions* (1879). His most important finding was that the chemical products of fumaroles are primarily a function of temperature, thus relating the product composition, the site of the fumarole with respect to the center of the eruption, and the elapsed time between the emergence of the vent and the beginning of the eruption.

Fouqué's studies of volcanoes led naturally to his other scientific activity, in which he collaborated closely with Michel-Lévy. Both had heard of Henry Clifton Sorby's work in the microscopical examination of thin sections of rocks, and they perfected this technique. Their two-volume work *Minéralogie micrographique* (1879) introduced this new petrographic method into France. Further, they laid the foundations of modern petrography by introducing a classificatory system based on the mineralogical composition, the structure, and the chemical composition of volcanic rocks.

From 1878 to 1882, Fouqué and Michel-Lévy worked continuously on the artificial synthesis of igneous rocks, primarily to determine the conditions surrounding their origins. They were successful in producing the majority of volcanic rocks with the identical mineralogical composition and structural peculiarities found in nature. Their work verified the importance of the rate of cooling on the extent of crystallization and the sizes of grain, and demonstrated that rocks of distinctly different mineralogical composition would be formed from the same magma, depending on the conditions of crystallization.

Fouqué received the Cuvier Prize in 1876, and in 1877 he became professor of natural history at the Collège de France. He was named to the French geological survey commission in 1880, and in this position he made contributions to the stratigraphic geology of the Haute-Auvergne region. He was elected to the Académie des Sciences in 1881 and presided over it in 1901. In 1884, following earthquakes in Andalusia, he directed a group sent there by the Institut de France to study these phenomena. This mission led to Fouqué's experiments on the speed of propagation of shock waves in a variety of soils. His last important work, completed in 1896, was a petrographic study of the plagioclase feldspars.

BIBLIOGRAPHY

I. Original Works. Fouqué's chief publications are *Recherches sur les phénomènes chimiques qui se produisent dans les volcans* (Paris, 1866); *Les anciens volcans de la Grèce* (Paris, 1867); *Santorin et ses éruptions* (Paris, 1879); *Minéralogie micrographique: Roches éruptives françaises,* 2 vols. (Paris, 1879), written with A. Michel-Lévy; *Synthèse des minéraux et des roches* (Paris, 1882), written with A. Michel-Lévy; and *Les tremblements de terre* (Paris, 1888). He published approximately 100 memoirs, some in collaboration with Michel-Lévy.

II. Secondary Literature. See A. Michel-Lévy, "Notice sur F. Fouqué," in *Bulletin de la Société française de minéralogie,* **28** (1905), 38–56; and Alfred Lacroix, *Notice historique sur Auguste Michel-Lévy* (Paris, 1914).

John G. Burke

FOURCROY, ANTOINE FRANÇOIS DE (*b.* Paris, France, 15 June 1755; *d.* Paris, 16 December 1809), *chemistry, medicine.*

A member of a noble family that had declined, Fourcroy was the son of Jean Michel de Fourcroy, an apothecary, and Jeanne Laugier. He left the Collège d'Harcourt in Paris at the age of fifteen, and after studying for a year under a writing master, became a clerk in the office of the chancellory. There he would have remained but for his good fortune in meeting F. Vicq d'Azyr, the anatomist, who persuaded his father to let Fourcroy study at the Paris Faculty of Medicine. Aided financially by members of the Société Royale de Médecine, of which Vicq d'Azyr was secretary, Fourcroy graduated as a doctor in 1780, but he did not practice medicine.

As a student he had shown great ability in chemistry and had lectured in the private laboratory of J. B. M. Bucquet, his teacher. Every winter from Bucquet's death in 1780 until 1791 or 1792 Fourcroy gave a course of seventy lectures in his own laboratory which was published as *Leçons élémentaires d'histoire naturelle et de chimie* (Paris, 1782), and from 1782 to 1784 he also gave a summer course in materia medica. In all his lectures Fourcroy emphasized the relations between chemistry and natural history and their application to medicine. Fieldwork in natural history led him to publish *Entomologia parisiensis* (Paris, 1785), a detailed account of the insects of the Paris region, and about this time he also did some research on the anatomy of muscles; but he soon decided to concentrate on chemistry.

In 1783 Fourcroy received his first public appointment as chemistry professor at the Ecole Royale Vétérinaire, at Alfort, near Paris, but this ended in 1787 when plans to expand the school were abandoned. His career did not suffer, for in 1784 he had succeeded P. J. Macquer in the important chair of chemistry at the Jardin du Roi. Here he lectured every summer to very large audiences and achieved fame by his brilliant exposition of a rapidly changing subject. From 1787 he added to his reputation by lecturing at the Lycée, a private educational institution on rue de Valois founded by J. F. Pilatre de Rozier.

The Société Royale de Médecine allowed Fourcroy to take part in its work while he was still a student and elected him to membership as soon as he graduated; he subsequently became one of its leading members, and his talent was further recognized in 1785 by his election to the Académie Royale des Sciences. Here he was in contact with A. L. Lavoisier, whose antiphlogistic theory he adopted in 1786, after several years of hesitation during which he had given his students a comparative account of the phlogistic and antiphlogistic theories. Most of the second edition of his *Leçons élémentaires,* retitled *Élémens d'histoire naturelle et de chimie* (Paris, 1786), was printed before 1786, and he announced his conversion in a specially written introduction. His *Principes de chimie* (Paris, 1787) was the first textbook written entirely according to the antiphlogistic theory.

In 1787 Fourcroy collaborated with Lavoisier, L. B. Guyton de Morveau, and C. L. Berthollet in the revision of chemical nomenclature, and he undertook the great task of completing the chemical section of the *Encyclopédie méthodique,* which Guyton had to abandon after completing volume I (Paris, 1789). Fourcroy was a teacher who always tried to arrange the fundamental principles of chemistry in a systematic order, and in his article "Axiomes" in the second volume of *Encyclopédie méthodique* (1792) he classified the chief facts of chemistry under twelve headings. When published separately as a little book entitled *Philosophie chimique* (Paris, 1792), this proved to be a very popular summary of antiphlogistic chemistry and was translated into eleven languages. Fourcroy also helped to advance the new chemistry as one of the editors of *Annales de chimie,* the journal founded in 1789 by Lavoisier and his colleagues; but he was more active as the editor of his own periodical, *La médecine éclairée par les sciences physiques,* which appeared fortnightly during 1791 and 1792 and was intended for medical practitioners wishing to keep up to date in all relevant branches of science.

While establishing his reputation as a professor and author, Fourcroy was also busy in the laboratory. One of the duties of the Société Royale de Médecine was to analyze mineral waters and assess their medicinal value, and in 1782 Fourcroy published a valuable account of the qualitative analysis of mineral waters by means of reagents, a method that was replacing the older analysis by evaporation to dryness. He considered that many of the reagents recommended in 1778 by T. O. Bergman were unnecessary and reduced the number from about twenty-five to eleven. Further, Bergman had not suggested any particular order for the reagents, but Fourcroy described a systematic analysis, using reagents that had the least effect on mineral waters before those that caused more complicated changes. A separate sample of the water was used for each test.

Fourcroy noticed that ammonia did not always completely precipitate magnesia from a mineral water containing it in solution. This led him in 1790 to investigate the reactions between salts of ammonia and of magnesia, and to the discovery of the crystalline double sulfates and double phosphates of the two bases. Bergman had recognized the existence of such salts in 1783, but few had been characterized.

Fourcroy's interest in the application of chemistry to medicine led him to study various solids and fluids of the human and animal body in health and sickness. In 1785 he found that both human and animal muscle fiber contained a substance chemically similar to the fibrous matter in coagulated blood. This fibrous matter must have been derived from the animal's food, which was inanimate; Fourcroy seems to have thought that it was converted into living muscle fiber by the agency of a vital force. C. W. Scheele and Berthollet had found nitrogen in animal matter, and in 1788 Fourcroy showed that there was a greater proportion of nitrogen in muscle fiber than in any other part of the body, and that the proportion of nitrogen contained in these fibers was the same for carnivorous and herbivorous animals. By 1789 he had found nitrogen in many vegetables, and it was therefore possible to account for its presence in herbivores without necessarily assuming that the animal absorbed it from the atmosphere.

Some parts of the body putrefied to form a white, waxy material resembling spermaceti, but in 1786 Fourcroy showed that it had a lower melting point than spermaceti and was more soluble in alcohol. Gallstones contained another similar substance (now known as cholesterol) which was only slightly soluble in alcohol and melted at a higher temperature than the others. This use of measurable physical properties to distinguish substances was very unusual before the nineteenth century.

Vegetable chemistry also interested Fourcroy, but to a lesser extent. His most important contribution was a detailed analysis of cinchona bark, which he extracted in a systematic manner with water, alcohol, alkalies, and acids. He did not isolate the active principle, but his analysis prepared the way for techniques that led to the extraction of cinchonine and quinine by P. J. Pelletier and J. B. Caventou in 1820.

About 1790 Fourcroy gave his first course in animal chemistry at the Lycée. He was assisted by N. L. Vauquelin, who collaborated in much of his research, but this joint work was done later, for Fourcroy's scientific activities were interrupted by his entry into politics.

Like most French scientists, Fourcroy held liberal opinions and supported the moves that led to the French Revolution. He was one of about 400 representatives of the Third Estate in Paris who met in April and May 1789 to elect twenty deputies to the Estates General, but he took no further part in politics until 1792. In the meantime he served on local committees of health, education, and public welfare and continued his scientific work.

The government called on scientists to assist in solving the country's economic difficulties. In 1790 Fourcroy successfully applied his chemical knowledge to the problem of extracting copper (needed for coinage and later for cannon manufacture and shipbuilding) from its alloy with tin, which, with the closing of many churches, was available in the form of bells. He heated the molten metal in air until the gain in weight showed that enough oxygen had been absorbed to oxidize the tin but not the copper. But at this stage some oxygen was combined with copper and some tin was uncombined, so, knowing that oxygen had a greater affinity for tin than for copper, he continued the heating in the absence of air. This caused all the oxygen to be transferred to the tin, leaving tin oxide and pure copper. The process gave a good yield of copper and was employed on a large scale for at least ten years.

The suspension of the monarchy and dissolution of the National Assembly on 10 August 1792 were followed by the election of a new National Convention and the declaration of the republic. Fourcroy became a member of the electoral assembly of Paris, the body that elected the deputies, but he had no ambition to enter politics actively and wished only to serve his country by continuing his scientific work. He stated this clearly on 10 September 1792, when he wrote to the government declining the position of *régisseur des poudres et salpêtres* (an administrative position concerned with gunpowder manufacture); but he failed in his attempt to remain a private citizen and on 21 September was elected fourth substitute deputy for Paris. He was called to take his seat in the Convention on 22 July 1793, after J. P. Marat's assassination. Fourcroy now showed some enthusiasm for politics. He joined the Jacobin Club and became a member of the Committee of Public Instruction of the Convention.

After 10 August 1792 Fourcroy supported the expulsion from the Société Royale de Médecine and the Académie Royale des Sciences of émigrés and counterrevolutionaries, but none of the active members still resident in France were affected. He also supported the decision of the Convention on 8 August 1793 to suppress these bodies and all other academies that had enjoyed privileges under the monarchy. This did not affect certain independent societies, such as the Lycée des Arts (not to be confused with the Lycée), which was particularly concerned with the applications of science and included among its members Fourcroy, Lavoisier, and other academicians.

The suppression of the academies was not intended to be an attack on their members, and as a member of the Committee of Public Instruction, Fourcroy became partly responsible for organizing commissions of scientists to continue the most important work in progress. But in fact, the political and military problems of the day made it possible to set up only the commission that developed the metric system of weights and measures. Lavoisier was a member of this commission until after his arrest, with the other farmers-general of taxes, in November 1793. His trial and execution on 8 May 1794 came as a great shock to his fellow scientists, and there is evidence that Fourcroy made an unsuccessful last-minute appeal on his behalf to Robespierre and the other members of the Committee of Public Safety.

Fourcroy served on the government committee that

founded the École Polytechnique in Paris (called the École Centrale des Travaux Publics when it opened in 1794) and new medical schools in Paris, Strasbourg, and Montpellier, which opened in 1795. These were urgently needed to train engineers and doctors for the army, but the École Polytechnique and the École de Médecine at Paris also became important research centers. Fourcroy was a professor at each until his death, and he also retained his chairs at the Lycée and the Jardin du Roi, which was reorganized in 1793 as the Muséum National d'Histoire Naturelle.

On 1 September 1794 Fourcroy was elected to the Committee of Public Safety, which had diminished powers after Robespierre's downfall, and for several months he was deeply involved in the organization of munitions manufacture. From July to October 1795 he was again on the Committee of Public Instruction and helped to prepare an ambitious plan for national education which was to include *écoles centrales* for boys aged eleven to eighteen. Much science was to be taught in them, but few were successful, largely because of the shortage of science teachers. Fourcroy was also one of the planners of the Institut National des Sciences et des Arts, which replaced the old learned societies, and he became a member soon after it opened in 1795.

Fourcroy was elected to the Conseil des Anciens, one of the two assemblies that succeeded the Convention in 1795, but he did not serve on any of its committees and was not reelected in 1797. His return to private life lasted only two years, for on 25 December 1799 Napoleon appointed him to the council of state. By this time he had resumed his scientific work, and he was able to continue it while a councillor. He published a new treatise on chemistry, *Système des connaissances chimiques* (Paris, 1801) and many research papers, but most of these were joint publications, generally with Vauquelin.

A useful contribution to inorganic chemistry was made by Fourcroy and Vauquelin in 1796, when they gave clear descriptions of the preparation and properties of sulfites and phosphites, including some new salts. In 1803, independently of H. V. Collet-Descotils, they examined the residue left when crude platinum dissolved in aqua regia and showed that it contained a new metal (which they named iridium), but they missed the second metal (osmium) that was discovered in 1804 by J. Smithson Tennant. In 1801 L. J. Thenard was Fourcroy's collaborator in a masterly study of the oxides and salts of mercury, which definitely established the existence of two series of compounds containing mercury in different degrees of oxidation.

The action of sulfuric acid on vegetable substances was studied by Fourcroy and Vauquelin in 1797, and they showed that it did not always act as an oxidizing agent, as was generally believed, but sometimes decomposed vegetable matter by removing water from it, even though the water was not originally present as such but only as its elements. The sulfuric acid was unaltered chemically, and the reaction ceased when it became too dilute. In the particular case of alcohol, which yielded ether when treated with sulfuric acid, they considered that hydrogen and oxygen were removed from the alcohol, forming water, but the liberation of carbon that they observed led them to believe that the reaction was more complicated. Later, chemists recognized that the carbon came from impurities in the alcohol, but Fourcroy and Vauquelin had made an important contribution to the development of the theory of etherification.

Animal chemistry was still of great interest to Fourcroy, and with Vauquelin he examined many solids and fluids, including brains, mucus, nasal humor, and bile, and tried to explain their formation and function in chemical terms and to find medicaments that would restore them to their original state when altered by disease. Like most animal chemists of the day, they did not characterize any organic constituents of these animal substances, which, unlike vegetables, rarely yield crystalline and easily purifiable compounds.

Fourcroy and Vauquelin achieved more when they examined the inorganic constituents of animal matter. They found, for example, that the phosphates of lime and magnesia were present in the same proportions in milk as in bones, and that phosphorus in the soft roe of a fish was combined in such a way that it did not give the usual reactions of a phosphate.

Hundreds of concretions from various parts of human and animal bodies were analyzed by Fourcroy and Vauquelin. Most were urinary calculi which, independently of W. H. Wollaston, they classified according to chemical composition from 1798 onward. They confirmed the frequent presence of uric acid and phosphate of lime (discovered in calculi by Scheele and George Pearson respectively) and also found urate of ammonia, the double phosphate of magnesia and ammonia, and occasionally other compounds. Fourcroy hoped that the analysis of urinary calculi would lead to the discovery of solvents suitable for dissolving them by injection into the bladder, but this was not achieved.

In an attempt to find why urinary calculi were formed, Fourcroy and Vauquelin investigated urine, and in 1799 they gave the first satisfactory account of urea, which they named. (H. Boerhaave and H. M. Rouelle had previously observed a crystalline substance in evaporated urine but had not examined its properties.) Fourcroy and Vauquelin isolated it by recrystallization from alcohol and, in 1808, achieved a purer state by adding alkali to the crystalline nitrate that they had discovered. They found that urea yielded carbonic and acetic acids and ammonia when its aqueous solution was boiled; these were also the products of putrefaction, and they thought that calculi containing ammonia might be formed by the partial fermentation of urea in the bladder. Such speculations provided a valuable stimulus to the next generation of animal chemists.

As councillor of state, Fourcroy played a large part in drafting a new educational system, from primary schools to advanced colleges, and in 1802 Napoleon appointed him director-general of public instruction, with the great task of implementing the proposals. He achieved considerable success, and it was a disappointment when, in March 1808, he was not made grand master of the Imperial University, the corporation that was to control the entire system. It is probable that Napoleon wanted a grand master who was completely acceptable to the Roman Catholic Church, and he knew that Fourcroy was a freethinker. The post was given to Louis de Fontanes, a man of letters with orthodox religious views.

The title of count of the empire was conferred on Fourcroy in 1808, and he remained a councillor of state. During 1809 he was occupied in drafting new mining legislation, and Napoleon may have intended to appoint him director-general of mines. But Fourcroy's health had begun to decline in 1808 and he died, aged fifty-four, before the mining law was passed.

Fourcroy was married to Anne Claude Bettinger in 1780; they had a son, an army officer who was killed in action in 1813, and a daughter. This marriage was dissolved in 1799, and in 1800 Fourcroy married Adelaide Flore Belleville, the widow of Charles de Wailly, a well-known architect. There were no children from the second marriage.

BIBLIOGRAPHY

I. Original Works. An extensive bibliography of Fourcroy's scientific writings is given by W. A. Smeaton (see below), pp. 211–252, with supplementary information in "Some Unrecorded Editions of Fourcroy's *Philosophie Chimique,*" in *Annals of Science,* **23** (1967), 295–298.

II. Secondary Literature. There are two comprehensive accounts of Fourcroy's life and work: W. A. Smeaton, *Fourcroy, Chemist and Revolutionary* (London, 1962); and Georges Kersaint, *Antoine François de Fourcroy, sa vie et son oeuvre* (Paris, 1966). Both books contain many references to manuscripts and printed sources.

W. A. Smeaton

FOURIER, JEAN BAPTISTE JOSEPH (*b.* Auxerre, France, 21 March 1768; *d.* Paris, France, 16 May 1830), *mathematics, mathematical physics.*

Fourier lost both his father (Joseph, a tailor in Auxerre) and his mother (Edmée) by his ninth year and was placed by the archbishop in the town's military school, where he discovered his passion for mathematics. He wanted to join either the artillery or the engineers, which were branches of the army then generally available to all classes of society; but for some reason he was turned down, and so he was sent to a Benedictine school at St. Benoît-sur-Loire in the hope that he could later pursue his special interests at its seminary in Paris. The French Revolution interfered with these plans, however, and without regret he returned in 1789 to Auxerre and a teaching position in his old school.

During the Revolution, Fourier was prominent in local affairs, and his courageous defense of the victims of the Terror led to his arrest in 1794. A personal appeal to Robespierre was unsuccessful; but he was released after Robespierre's execution on 28 July 1794 and went as a student to the ill-fated École Normale, which opened and closed within that year. He can have spent only a short time there, but nevertheless he made a strong impression; and when the École Polytechnique started in 1795 he was appointed *administrateur de police,* or assistant lecturer, to support the teaching of Lagrange and Monge. There he fell victim of the reaction to the previous regime and was, ironically, arrested as a supporter of Robespierre (who had declined his earlier appeal). But his colleagues at the École successfully sought his release, and in 1798 Monge selected him to join Napoleon's Egyptian campaign. He became secretary of the newly formed Institut d'Égypte, conducted negotiations between Napoleon and Sitty-Nefiçah (the wife of the chief bey, Murad), and held other diplomatic posts as well as pursuing research. He does not appear to have been appointed governor of southern Egypt, however, as has often been reported.

After his return to France in 1801, Fourier wished to resume his work at the École Polytechnique; but Napoleon had spotted his administrative genius and appointed him prefect of the department of Isère, centered at Grenoble and extending to what was then the Italian border. Here his many administrative achievements included the reconciliation of thirty-seven different communities to the drainage of a huge area of marshland near Bourgoin to make valuable farming land, and the planning and partial construction of a road from Grenoble to Turin (now route N91-strada 23), which was then the quickest route between Turin and Lyons. In 1808 Napoleon conferred a barony on him.

While in Egypt, Fourier suggested that a record be made of the work of the Institut d'Égypte, and on his return to France he was consulted on its organization and deputed to write the "Préface historique" on the ancient civilization and its glorious resurrection. This he completed in 1809, but some of its historical details caused controversy. Napoleon supported it, and it appeared in the *Description de l'Égypte.*

Fourier was still at Grenoble in 1814 when Napoleon fell. By geographical accident the town was directly on the route of the party escorting Napoleon from Paris to the south and thence to Elba; to avoid an embarrassing meeting with his former chief Fourier negotiated feverishly for a detour in the route of the cortege. But no such detour was conceivable on Napoleon's return and march on Paris in 1815, and so Fourier compromised, fulfilling his duties as prefect by ordering the preparation of the defenses—which he knew to be useless—and then leaving the town for Lyons by one gate as Napoleon entered by another. He did return, however, and the two friends met at Bourgoin. Fourier need have had no fears, for Napoleon made him a count and appointed him prefect of the neighboring department of the Rhône, centered at Lyons. But before the end of Napoleon's Hundred Days, Fourier had resigned his new title and prefecture in protest against the severity of the regime and had come to Paris to try to take up research full time (it had previously been only a spare-time activity).

This was the low point of Fourier's life—he had no job, only a small pension, and a low political reputation. But a former student at the École Polytechnique and companion in Egypt, Chabrol de Volvic, was now prefect of the department of the Seine and appointed him director of its Bureau of Statistics, a post without onerous duties but with a salary sufficient for his needs.

In 1816 Fourier was elected to the reconstituted Académie des Sciences, but Louis XVIII could not forgive his having accepted the prefecture of the Rhône from Napoleon, and the nomination was refused. Diplomatic negotiation eventually cleared up the situation, and his renomination in 1817 was not opposed. He also had some trouble with the second edition of the *Description de l'Égypte* (for now his references to Napoleon needed rethinking) but in general his reputation was rising fast. He was left in a position of strength after the decline of the Société d'Arcueil, led by Laplace in the physical sciences, and gained the favor and support of the aging Laplace himself in spite of the continued enmity of Poisson. In 1822 he was elected to the powerful position of *secrétaire perpétuel* of the Académie des Sciences, and in 1827—after further protests—to the Académie Française. He was also elected a foreign member of the Royal Society.

Throughout his career, Fourier won the loyalty of younger friends by his unselfish support and encouragement; and in his later years he helped many mathematicians and scientists, including Oersted, Dirichlet, Abel, and Sturm. An unfortunate incident occurred in 1830 when he lost the second paper on the resolution of equations sent to the Academy by Evariste Galois; but this would appear to be due more to the disorganization of his papers than—as Galois believed—to deliberate suppression.

The Egyptian period of his life had one final consequence in his last years. While there he had caught some illness, possibly myxedema, which necessitated his increasing confinement to his own heated quarters. He lived at 15 rue pavée St. André des Arts (now 15 rue Séguier) until 1829, then at 19 rue d'Enfer (now the site of 73 Boulevard St. Michel) until his death. On 4 May 1830 he was struck down while descending some stairs, and he allowed the symptoms to become worse until he died twelve days later. The funeral service took place at the Église St. Jacques de Haut Pas, and he was buried in the eighteenth division of the cemetery of Père Lachaise.

Various memorials have been made to Fourier. A bust by Pierre-Alphonse Fessard was subscribed in 1831 but was destroyed during World War II. A similar fate overtook the bronze statue by Faillot erected in Auxerre in 1849; the Nazis melted it down for armaments. During the night before its destruction, however, the mayor rescued two of its bas-reliefs, which were mounted on the walls of the town hall after the war. A medallion was founded in the town in 1952, and since 1950 the *Annales de l'Institut Fourier* have been published by the University of Grenoble. In 1968 the bicentenary of his birth was celebrated and the secondary school in Auxerre was renamed the Lycée Fourier.

Heat Diffusion and Partial Differential Equations. Fourier's achievements lie in the study of the diffusion of heat and in the mathematical techniques he introduced to further that study. His interest in the problem may have begun when he was in Egypt, but the substantial work was done at Grenoble. In 1807 he presented a long paper to the Academy on heat diffusion between disjoint masses and in special continuous bodies (rectangle, annulus, sphere, cylinder, prism), based on the diffusion equation

$$\frac{\partial^2 v}{\partial x^2} + \frac{\partial^2 v}{\partial y^2} + \frac{\partial^2 v}{\partial z^2} = k \frac{\partial v}{\partial t} \quad (1)$$

(in three variables). Of the examiners, Laplace, Monge, and Lacroix were in favor of accepting his work, but Lagrange was strongly opposed to it—due, to some extent, to the Fourier series

$$f(x) = \frac{1}{2\pi}\int_{-\pi}^{\pi} f(t)\,dt + \frac{1}{\pi}\sum_{r=1}^{\infty}\left[\cos rx \int_{-\pi}^{\pi} f(t)\cos rt\,dt + \sin rx \int_{-\pi}^{\pi} f(t)\sin rt\,dt\right] \quad (2)$$

required to express the initial temperature distribution in certain of these bodies, which contradicted Lagrange's own denigration of trigonometric series in his treatment of the vibrating string problem in the 1750's. The paper was therefore never published.

A prize problem on heat diffusion was proposed in 1810, however, and Fourier sent in the revised version of his 1807 paper, together with a new analysis on heat diffusion in infinite bodies. In these cases the periodicity of the Fourier series made it incapable of expressing the initial conditions, and Fourier substituted the Fourier integral theorem, which he wrote in forms such as

$$\pi f(x) = \int_{\infty}^{\infty} f(t)\,dt \int_{0}^{\infty} \cos q(x-t)\,dq. \quad (3)$$

The last sections of the paper dealt with more physical aspects of heat, such as the intensity of radiation, and these became more important in Fourier's thought during his later years. Fourier's paper won the competition, but the jury—probably at the insistence of Lagrange—made criticisms on grounds of "rigor and generality," which Fourier considered an unjustified reproach. He expanded the mathematical parts of the paper into his book *Théorie analytique de la chaleur.* An extended treatment of the physical aspects was first planned for further chapters of this book and then for a separate book, *Théorie physique de la chaleur,* but it was never achieved.

The history of Fourier's main work in mathematics and mathematical physics has long been confused by an exclusive concentration on only two results, Fourier series and Fourier integrals, and by the application of anachronistic standards of rigor in judgments on their derivation. Fourier's achievement is better understood if we see it as twofold: treating first the formulation of the physical problem as boundary-value problems in linear partial differential equations, which (together with his work on units and dimensions) achieved the extension of rational mechanics to fields outside those defined in Newton's *Principia;* and second, the powerful mathematical tools he invented for the solution of the equations, which yielded a long series of descendants and raised problems in mathematical analysis that motivated much of the leading work in that field for the rest of the century and beyond.

Fortunately for the historian, Fourier reproduced, nearly intact, almost all his successful results in the several versions of his basic work. A comparison of these, in correlation with unpublished sources, biographical information, and separate papers, enables a firm reconstruction of the sequence of his researches. Moreover, from this sequence and from his style of presentation, we can identify several crucial points at which he failed to solve problems as well as the reasons for his failure.

Fourier's first work in the rational mechanics of heat used a model of heat being transferred by a shuttle mechanism between discrete particles. The physical theory was a simple method of mixtures, and the mathematics was of the 1750's. Of the two problems he attacked, the second, with the n particles arranged in a ring, yielded a complete solution for the finite case. Fourier wished to extend this to the continuous case but could not, for as n increased, the time constants in the exponentials tended to zero, obliterating the time dependence of the solution. Only later did he understand how to modify his transfer model to avoid this anomaly; and by his concentration on the complete solution and its difficulties, he failed to notice that at $t = 0$ his solution gave an interpolation formula which would yield the Fourier series in the continuous case. (Lagrange's earlier failure to discover the Fourier series can be similarly explained; it had nothing to do with the scruples of rigor which are usually held to be the cause.)

Fourier's successful establishment of the equation of heat flow was probably indebted to early work of J. B. Biot on the steady temperatures in a metal bar, wherein Biot distinguished between internal conduction and external radiation. Biot's analysis was crippled by a faulty physical model for conduction which yielded an "inhomogeneous" equation $d^2v - kv\,dx = 0$; and Fourier was able to concoct a physical model which resolved the difficulty. The full time-dependent equations for one and two dimensions of the type (1) then came easily.

Fourier's masterstroke was in the choice of configuration for a problem in which to apply the equation. The semi-infinite strip, uniformly hot at one end and uniformly cold along the sides, combines the utmost simplicity with physical meaning, in the tradition of rational mechanics deriving from the Bernoullis and Euler. The steady-state case is simply Laplace's equation in Cartesian coordinates. Fourier probably tried complex-variable methods (a solution along these lines, probably retrospective, is in the *Théorie*) but then used separation of variables to yield a series solution and thus the boundary-condition equation

$$1 = \sum_{r=0}^{\infty} a_r \cos rx. \qquad (4)$$

The solution of the equation and its generalization for an arbitrary function $f(x)$ by infinite-matrix methods have been analyzed and criticized many times. It is well to remember that this work was done several decades before the Cauchy-Weierstrass orthodoxy was established. Fourier was not a naive formalist: he could handle problems of convergence quite competently, as in his discussion of the series for the sawtooth function. The leading technical ideas of several basic proofs, such as that of Dirichlet on the convergence of the Fourier series, can be found in his work. Moreover, he saw, long before anyone else, that term-by-term integration of a given trigonometric series, to evaluate the coefficients, is no guarantee of its correctness; the completeness of a series is not to be assumed. The great shock caused by his trigonometric expansions was due to his demonstration of a paradoxical property of equality over a finite interval between algebraic expressions of totally different form. Corresponding to any function in a very wide class, there could be constructed a trigonometric series whose values, on an assigned interval, are the same as those of the function. As he showed by example, the given function could even be a mixture of different algebraic expressions, each defined on disjoint subintervals of the basic one. Both trigonometric expansions and arbitrary functions had been used by others (including Poisson); but the former were restricted to problems involving periodic phenomena, and the latter, when they appeared in the solutions of partial differential equations, were assumed by their nature to be incapable of an algebraic expression.

The earliest records of this first successful investigation show its exploratory character and Fourier's excitement with his achievement. Also, in this work there are traces of the influence of Monge in the notation, in the representation of the solution as a surface, and in the separate expression of boundary values in determining the solution of a differential equation. Thenceforth, Fourier proceeded into new territory with assuredness. The three-dimension case caused some difficulties, which were resolved by splitting the original equation into two, one for interior conduction and the other relating radiation to the temperature gradient at the surface. Applied to the sphere, in spherical coordinates, this gave a nonharmonic trigonometric expansion, where the eigenvalues are the roots of a transcendental equation. Fourier used his knowledge of the theory of equations (see below) to argue for the reality of all roots, but the question caused him trouble for many years. The problem of heat conduction in a cylinder gave rise to a further generalization. Fourier's solution was in what are now called Bessel functions—derived several years before Friedrich Bessel—by techniques made fully general in the theory created by Fourier's later associates, J. Charles François Sturm and Joseph Liouville.

The study of diffusion along an infinite line, involving the development of the Fourier integral theorem, probably depended on the idea of Laplace of expressing the solution of the heat equation as an integral transform of an arbitrary function representing the initial temperature distribution. Fourier derived the cosine and sine transforms separately for configurations symmetrical and antisymmetrical about the origin, by the extension of the finite-interval expansion. Only gradually did he come to appreciate the generality of the odd-and-even decomposition of a given function.

Fourier's last burst of creative work in this field came in 1817 and 1818, when he achieved an effective insight into the relation between integral-transform solutions and operational calculus. There was at that time a three-cornered race with Poisson and Cauchy, who had started using such techniques by 1815. In a crushing counterblow to a criticism by Poisson, Fourier exhibited integral-transform solutions of several equations which had long defied analysis, and gave the lead to a systematic theory. This was later achieved by Cauchy, en route to the calculus of residues.

As a mathematician, Fourier had as much concern for practical problems of rigor as anyone in his day except Cauchy and N. H. Abel, but he could not conceive of the theory of limiting processes as a meaningful exercise in its own right. The famous referees' criticisms of the 1811 prize essay, concerning its defects of rigor and generality, have long been misinterpreted. Much of the motivation for them was political; Poisson and Biot, outclassed rivals in the theory of heat diffusion, tried for years to denigrate

Fourier's achievements. The criticism of rigor was probably based on Poisson's point that the eigenvalues in the sphere problem were not proven to be all real; and complex roots would yield a physically impossible solution. (Poisson himself solved the problem for Fourier years later.) The supposed lack of generality in Fourier's series solution (2) was probably by way of contrast to an integral solution already achieved by Laplace, in which the arbitrary function was neatly encased in the integrand.

Fourier's sensibility was that of rational mechanics. He had a superb mastery of analytical technique and notation ($\int_a^b$ is his invention, for example); and this power, guided by his physical intuition, brought him success. Before him, the equations used in the leading problems in rational mechanics were usually nonlinear, and they were solved by ad hoc approximation methods. Similarly, the field of differential equations was a jungle without pathways. Fourier created and explained a coherent method whereby the different components of an equation and its series solution were neatly identified, with the different aspects of the physical situation being analyzed. He also had a uniquely sure instinct for interpreting the asymptotic properties of the solutions of his equations for their physical meaning. So powerful was his approach that a full century passed before nonlinear differential equations regained prominence in mathematical physics.

For Fourier, every mathematical statement (although not all intermediate stages in a formal argument) had to have a physical meaning, both in exhibiting real motions and in being capable, in principle, of measurement. He always interpreted his solutions so as to obtain limiting cases which could be tested against experiment, and he performed such experiments at the earliest opportunity. He rejected the prevailing Laplacian orthodoxy of analyzing physical phenomena through the assumption of imperceptible molecules connected by local Newtonian forces; because of his approach to physical theory, together with his enmity to Poisson, he was adopted as philosophical patron by Auguste Comte in the development and popularization of *philosophie positive.*

Although the physical models of his earliest drafts were very sketchy, by the time of the 1807 paper he had fully incorporated physical constants into his theory of heat. The concern for physical meaning enabled him to see the potential in his formal technique for checking the coherence of the clumps of physical constants appearing in the exponentials of the Fourier integral solutions. From this came the full theory of units and dimensions (partly anticipated by Lazare Carnot), the first effective advance since Galileo in the theory of the mathematical representation of physical quantities. A comparison with the confused struggles of contemporaries such as Biot with the same problem illuminates Fourier's achievement.

Although Fourier studied the physical theory of heat for many years, his contributions, based primarily on the phenomena of radiation, did not long survive. His concern for applying his theory produced an analysis of the action of the thermometer, of the heating of rooms, and, most important, the first scientific estimate of a lower bound for the age of the earth. It is puzzling that in spite of his faith in the importance of heat as a primary agent in the universe, Fourier seems to have had no interest in the problem of the motive power of heat; and so, along with nearly all his contemporaries, he remained in ignorance of the essay on that topic by Lazare Carnot's son, Sadi.

On the side of real-variable analysis, the problems suggested especially by Fourier series lead directly through Dirichlet, Riemann, Stokes, and Heine to Cantor, Lebesgue, F. Riesz, and Ernst Fischer. Such deep results are not the chance products of algebraic doodling. None of Fourier's predecessors or contemporaries did—or could—exploit trigonometric expansions of arbitrary functions to their full effectiveness, nor could they recognize and accept their implications for the foundations of pure and applied analysis. Such achievements required a great master craftsman of mathematics, endowed with a lively imagination and holding a conscious philosophy of mathematics appropriate for his work. For Fourier, this was expressed in his aphorism, "Profound study of nature is the most fertile source of mathematical discoveries."

Theory of Equations. In contrast with his famous work on heat diffusion, Fourier's interest in the theory of equations is remarkably little known. Yet it has a much longer personal history, for it began in his sixteenth year when he discovered a new proof of Descartes's rule of signs and was just as much in progress at the time of his death. This rule may be stated as follows:

$$\text{Let } f(x) = x^m + a_1x^{m-1} + \cdots + a_{m-1}x + a_m.$$

Then there will be a sequence of signs to the coefficients of $f(x)$. If we call a pair of adjacent signs of the same type (i.e., $+ +$ or $- -$) a preservation and a pair of the opposite type a variation, then the number of positive (or negative) roots of $f(x)$ is at most the number of variations (or preservations) of sign in the sequence. Fourier's proof was based on multiplying $f(x)$ by $(x + p)$, thus creating a new polynomial which contained one more sign in its sequence and one more positive (or negative) root, according as p was less (or greater) than zero, and showing that the number of preservations (or variations) in the new sequence was not increased relative to the old sequence. Hence the number of variations (or preservations) is increased by at least one, and the theorem follows. The details of the proof may be seen in any textbook dealing with the rule, for Fourier's youthful achievement quickly became the standard proof, even if its authorship appears to be virtually unknown.

Fourier generalized Descartes's rule to estimate the number of real roots $f(x)$ within a given interval $[a,b]$, by taking the signs of the terms in the sequence

$$f^{(m)}(x), f^{(m-1)}(x), \cdots, f''(x), f'(x), f(x).$$

When $x = -\infty$ the series will be made up totally of the variations

$$+ - + - \cdots\cdots$$

while at $x = +\infty$ it is entirely preservations:

$$+ + + + \cdots\cdots.$$

Fourier showed that as x passes from $-\infty$ to $+\infty$ the variations are lost by the crossing of a real (possibly multiple) root, or the skirting of a pair of complex conjugate roots, and that the number of real roots within $[a,b]$ is at most the difference between the number of variations in the sequence when $x = a$ and the number when $x = b$. This theorem received an important extension in Fourier's own lifetime by Sturm, who showed in 1829 that the number of real roots is exactly the difference in the number of variations formulated above for the sequence of functions

$$f_m(x), f_{m-1}(x), \cdots, f_2(x), f'(x), f(x)$$

where $f_2(x), \cdots\cdot$ are defined algorithmically from $f(x)$ and $f'(x)$. This is the famous Sturm's theorem.

Fourier appears to have proved his own theorem while in his teens and he sent a paper to the Academy in 1789. However, it disappeared in the turmoil of the year in Paris, and the pressure of administrative and other scientific work delayed publication of the results until the late 1810's. Then he became involved in a priority row with Ferdinand Budan de Bois-Laurent, a part-time mathematician who had previously published similar but inferior results. At the time of his death, Fourier was trying to prepare these and many other results for a book to be called *Analyse des équations déterminées;* he had almost finished only the first two of its seven *livres.* His friend Navier edited it for publication in 1831, inserting an introduction to establish from attested documents (including the 1789 paper) Fourier's priority on results which had by then become famous. Perhaps Fourier was aware that he would not live to finish the work, for he wrote a synopsis of the complete book which also appeared in the edition. The synopsis indicated his wide interests in the subject, of which the most important not yet mentioned were various means of distinguishing between real and imaginary roots, refinements to the Newton-Raphson method of approximating to the root of an equation, extensions to Daniel Bernoulli's rule for the limiting value of the ratio of successive terms of a recurrent series, and the method of solution and applications of linear inequalities. Fourier's remarkable understanding of the last subject makes him the great anticipator of linear programming.

Fourier's other mathematical interests included a general search for problems in dynamics and mechanics, shown by a published paper on the principle of virtual work. In his later years his directorship of the Bureau of Statistics brought him in touch with the problems of probability and errors, and he wrote important papers on estimating the errors of measurement from a large number of observations, published in the Bureau's reports for 1826 and 1829.

BIBLIOGRAPHY

I. ORIGINAL WORKS. Fourier's most famous work is *Théorie analytique de la chaleur* (Paris, 1822; repr. Breslau, 1883). An English trans. was prepared by A. Freeman (Cambridge, 1878; repr. New York, 1955). A 2nd French ed. appeared in 1888 as vol. I of the *Oeuvres* of Fourier, Gaston Darboux, ed. Vol. II, containing the majority of the rest of Fourier's published works, appeared in 1890. The list of works given by Darboux in the intro. to this vol. shows that his principal omission was the *Analyse des équations déterminées* (Paris, 1831). Two German trans. of this book have been made: by C. H. Schnuse (Brunswick, 1836; notes added in 1846), and by A. Loewy, Ostwald's Klassiker, no. 127 (Leipzig, 1902). The other main omissions from Darboux's ed. were the first 79 articles of the 1811 prize paper on heat diffusion, in *Mémoires de l'Académie des sciences,* **4** (1819–1820), 185–555, which were largely in common with secs. of the book, and a joint paper with H. C. Oersted on thermoelectric effects, in *Annales de chimie et de physique,* **12** (1823), 375–389. Darboux's list also omitted the papers read by Fourier at the Institut d'Égypte, which are listed in Cousin's obit. of Fourier cited below and partially in Navier's introduction to the *Analyse.* As *secrétaire perpétuel* of the Académie des Sciences, Fourier wrote the *Analyse des travaux* (1823–1827), and *éloges* on Delambre, Herschel, Breguet, Charles, and Laplace. The references are in Darboux's list.

The main source of unpublished MSS is the twenty-nine vols. in the Bibliothèque Nationale (MSS fonds franç. 22501–22529), totaling about 5,200 sheets. Of these, 22501 is a set of miscellaneous studies and letters; 22502–22516 deal with the theory of equations, including topics only summarized in the *Analyse;* 22517–22522 cover extended work in mechanics, dynamics, and errors of measurement, etc.; and 22523–22529 are concerned with various aspects of heat diffusion and its associated mathematics and experimental work, including, in 22525, fols. 107–149, the "first draft" of his 1807 paper on heat diffusion. This paper itself was discovered by Darboux in the library of the École Nationale des Ponts et Chaussées (MS 267, now numbered 1851), which also contains several *cahiers* of lecture notes, totaling 386 pp., given by Fourier at the École Polytechnique in 1795–1796 (MS 668, and 1852). Another set of lecture notes, partly in common with this set, is in a vol. of 559 pp. in the Bibliothèque de l'Institut de France (MS 2044) where an early four-page article on Descartes's rule (MS 2038) may also be found. There is a scattering of letters in connection with his secretaryship of the Academy and his prefectures in various public MS collections in France and in the archives of various learned institutions.

With regard to his extrascientific writings, his notebook of the Egyptian campaign may be read in *Bibliothèque Égyptologique,* VI (Paris, 1904), 165–214. The various versions of the "Préface historique" are best compared on pp. 88–172 of the book by J. J. Champollion-Figeac cited below. Fourier wrote an article in the *Description* on the government of Egypt, and another on the astronomical monuments of the country (including a discussion of the Zodiacs). Details are given in the bibliographical account and collation of the *Description de l'Égypte* (London, 1838). He also contributed notes to a trans. by A. de Grandsagne of Pliny's *Natural History* and, although they are unsigned, it is clear that he annotated Pliny's discussion of Egypt, in *Histoire naturelle,* 20 vols. (Paris, 1829–1840), IV, 190–209. He also wrote the articles "Rallier des Ourmes," "Viète," and "Jean Wallis" in the *Biographie universelle,* 52 vols. (Paris, 1811–1829).

II. SECONDARY LITERATURE. The main biographical work is by the archaeologist Jacques Joseph Champollion-Figeac, who was encouraged by Fourier at Grenoble: *Fourier, Napoléon, l'Égypte et les cent jours* (Paris, 1844). The other primary sources of biography are Victor Cousin, *Notes biographiques sur M. Fourier* (Paris, 1831), with an addition; also in his *Fragments et souvenirs,* 3rd ed. (Paris, 1857), 283–392; François Arago, "Éloge historique de Joseph Fourier," in *Mémoires de l'Académie des Sciences,* **14** (1838), lxix–cxxxviii; also in his *Oeuvres* (Paris, 1854), V, 295–369, and in an English trans. in *Biographies of Distinguished Scientific Men* (London, 1857), 242–286; and in *Annual Reports of the Smithsonian Institution* (1871), 137–176. See also Aimé-Louis Champollion-Figeac, *Chroniques dauphinoises. Les savants du département de l'Isère . . .* (Vienne, 1880) and *Seconde période historique 1794–1810* (Vienne, 1881); Victor Parisot, "Fourier," in *Biographie universelle,* XIV (Paris, 1856), 525–534; and Georges Mauger, "Joseph Fourier," in *Annuaire statistique de l'Yonne* (1837), 270–276.

Some discussion of Fourier's work is to be found in Heinrich Burkhardt, "Entwicklungen nach oscillierenden Functionen . . .," in *Jahresbericht der Deutschen Mathematikervereinigung,* **10,** pt. 2 (1901–1908), esp. chs. 7 and 8; Gaston Bachelard, *Étude sur l'évolution d'un problème de physique . . .* (Paris, 1928), esp. chs. 2–4; and Ivor Grattan-Guinness in collaboration with Jerome Ravetz, *Joseph Fourier 1768–1830* (in press), which contains the full text of the 1807 paper on heat diffusion.

JEROME R. RAVETZ
I. GRATTAN-GUINNESS

FOURNEAU, ERNEST (*b.* Biarritz, France, 4 October 1872; *d.* Ascain, France, 5 August 1949), *chemistry.*

Of Spanish origin, Fourneau retained the appearance and elegant manner of that culture. His grandfather, the owner of a spinning mill, had established himself in France in the Basque country; his parents managed a large hotel in Biarritz. He received an excellent education; was fluent in English, German, and Spanish; and was interested in philosophy, literature, music, and painting, which he engaged in. He was also a brilliant conversationalist, owing to his education and to the cultivated circles that he frequented.

Following his secondary studies in Bayonne, he began studying pharmacy with Félix Moureu. In 1898 he obtained his diploma in Paris as a pharmacist; he then worked with the chemist Charles Moureu, brother of Félix. Having found his calling, he began to publish. In 1899 he commenced a three-year period of training in Germany under Theodor Curtius, Ludwig Gatterman, Emil Fischer (amino acids and barbiturate medications), and Richard Willstätter (chlorophyll). A witness to the birth of the German pharmaceutical industry, Fourneau returned to France and convinced the Poulenc brothers (Camille, Gaston, and Émile) of the necessity of creating a pharmaceutical chemistry laboratory. With the support of Camille Poulenc, he became the director of a laboratory in the factory at Ivry-sur-Seine. A born chemist and dextrous experimenter, there he was able to develop his talents. In 1904 he discovered an anesthetic that he named stovaine, a translation of the word *fourneau* ("stove").

In 1911, Émile Roux was director of the Institut Pasteur. Always alert to scientific progress, he welcomed Fourneau into the Grande Maison and named him chief of the new therapeutic chemistry service, which rapidly became world famous. Fourneau surrounded himself with remarkable researchers: chemists, microbiologists, physiologists, and physicians, notably Jacques and Thérèse Tréfouël (in 1921), and Daniel Bovet and Frédéric Nitti (in 1931). The service became a center for chemotherapeutic research.

From his marriage in 1906 to the daughter of the surgeon Paul Segond, he had three sons; one also became a chemist and director of a pharmaceutical laboratory. Fourneau reached retirement age in 1942 but continued to work at the Institut Pasteur until 1946. The Rhône-Poulenc chemical company then offered him a laboratory in Paris, where he continued his work. He was secretary-general of the Société Chimique de France, to which he gave great stimulus by his constant interest in the École de Pharmacie. He was an officer of the Légion d'Honneur (1903), and many orders of other societies were conferred upon him. He was a member of the Académie de Médecine (1919) and of a number of other French and foreign academies.

During World War I Fourneau was entrusted by the ministries of war and munitions with the study of various topics for the general dispensary (Pharmacie Générale) of military hospitals; in 1939 he was a member of the army's scientific commissions, and his laboratory was joined to the general staff.

Fourneau had lost his wife in 1942 at Ascain, to which he asked in 1949 to be transferred, being ill himself. Several days later, surrounded by his memorabilia, he passed away.

Fourneau published more than two hundred books, articles, and lectures in collaboration with other researchers on amino alcohols and ethylene oxides (stovaine). A master of this material, he was entrusted with the important chapter (XII) on it in the *Traité de chimie organique* of Victor Grignard (1941). As early as 1910 he had summarized his investigations with enumerations and descriptions of amino alcohols, oxaminated acids, *m*-acetylamino *p*-oxyphenylarsenic acid or stovarsol (Fourneau 190), and its isomer tryparsamide (Fourneau 270). As a natural continuation of his work he turned to the alkaloids (the ephedrines). He then studied corysanthine (the lysocythin of cobra venom) and glycerine esters and investigated the separation and the quantitative analysis of bismuth. Later he studied the stereochemistry of arsenic compounds, synthetic antipaludics (antimalarials), antihistamine derivatives and spasmolytics, and sulfur derivatives. He determined the formula of suramin sodium (Fourneau 309, Bayer 205) and its antibacterial action. Next he turned to the sulfamides (with the Tréfouëls, Bovet, and Nitti) and to sulfamidotherapy.

With his broad vision, Fourneau helped to establish the fundamental laws of chemotherapy that have saved so many human lives.

BIBLIOGRAPHY

I. Original Works. Fourneau's writings include "Stovaïne anesthésique local," in *Journal de pharmacie et de chimie,* 6th ser., **20** (1904), 108–109; "Ephédrines synthétiques," *ibid.,* **20** (1904), 481–499, and **25** (1907), 593–640; *Préparation des médicaments organiques* (Paris, 1921); "Anesthésiques locaux. Acides oxyaminés," in *Bulletin de la Société chimique de France,* 4th ser., **29** (1921), 413–416; "Recherches de chimiothérapie dans la série du 205 Bayer," in *Annales de l'Institut Pasteur,* **38** (1924), 81–114, written with J. Tréfouël and J. Vallée; "Sur une nouvelle méthode de sensibilité extrême pour la recherche, la séparation et le dosage du bismuth," in *Comptes rendus hebdomadaires des séances de l'Académie des sciences,* **181** (1925), 610–611, written with A. Girard; "Progrès récents dans le domaine des applications de la chimie à la thérapeutique," in *Bulletin des sciences pharmacologiques,* **35,** nos. 8 and 9 (1928), 499–516; "Préparation de dérivés en vue d'essais thérapeutiques: I, Amino alcools; II, Dérivés de l'atopian; III, Dérivés du carbostynyle; IV, Dérivés quinoléiniques et quiholéine arsinique," in *Annales de l'Institut Pasteur,* **44** (1930), 719–751, written with J. and T. Tréfouël and G. Benoit; "Sur une nouvelle classe d'hypnotique," in *Journal de pharmacie et de chimie,* **8** (1934), 49–54, written with J. R. Dilleter; "Chimiothérapie de l'infection pneumococcique par la di (p-acétylaminophényl) sulfone," in *Comptes rendus hebdomadaires des séances de l'Académie des sciences,* **205** (1937), 299–300, written with J. and T. Tréfouël, F. Nitti, and D. Bovet; "L'évolution de la chimiothérapie anti-bactérienne," in *Annales de l'Institut Pasteur,* **61** (1938), 799–811; "Quelques notions d'ordre chimique et rappel historique. L'antisepsie interne des maladies microbiennes par des dérivés du soufre," in *Gazette des hôpitaux civils et militaires,* **112,** no. 34 (1939), 577–581; "Aminoalcools," in Victor Grignard, *Traité de chimie organique,* XII (Paris, 1941), 393–635; and "La muscarine," in *Annales pharmaceutiques françaises,* **1,** no. 3 (1944), 120.

II. Secondary Literature. On Fourneau and his work, see M. Delépine, "Notice sur la vie et les travaux de Ernest Fourneau," in *Bulletin de la Société chimique de France,* 5th ser., **17** (1950), 953–982; P. Deloncle, "La science au service de l'action coloniale. Le laboratoire de chimiothérapie de l'Institut Pasteur," in *La dépêche coloniale* (27 Feb. 1927); T. A. Henry, "Ernest Fourneau," in *Journal of the Chemical Society,* **1** (1952), 261–266; R. Tiffeneau, "Ernest Fourneau," in *Paris-Médical,* **138** (1949), 470–471; J. Tréfouël, "Ernest Fourneau," in *Bulletin de la Société de pathologie exotique,* **42** (1949), 427–428; "Ernest Fourneau," in *Annales de l'Institut Pasteur,* **77** (1949), 644–647; and "Ernest Fourneau," in *Bulletin de l'Académie nationale de médecine,* **31-32** (1949), 589–595.

Denise Wrotnowska

FOURNEYRON, BENOÎT (*b.* Saint-Étienne, Loire, France, 31 October 1802; *d.* Paris, France, 8 July 1867), *hydraulic machinery.*

The son of a geometrician, Fourneyron prepared in mathematical sciences before entering at the age of fifteen a new school of mines at Saint-Étienne, from which he graduated at the top of the first class. His early activities were devoted to developing the mines at Le Creusot, prospecting for oil, laying out a railroad, and finally initiating the fabrication of tinplate —until then an English monopoly—at Pont-sur-l'Ognon, Haute-Saône. Involved in the latter process was a waterwheel of low efficiency, and Fourneyron became obsessed with the idea of producing a high-efficiency machine. At about the same time, one of his former professors at the school of mines, Claude Burdin, submitted to the Académie des Sciences a paper on hydraulic turbines; formally approved in 1824, it was noteworthy largely for its first use of the term. Both Burdin and Fourneyron then competed for a prize offered by the Société d'Encouragement pour l'Industrie Nationale for the first person to "succeed in applying at large scale, in a satisfactory manner, in mills and factories, the hydraulic turbines or wheels with curved blades of Bélidor." Burdin was a theoretician and was never able to produce a working model. Fourneyron, on the contrary, after four years of experimentation, had constructed by 1827 an operating unit of the outward-flow type, the power (6 h.p.) and efficiency (80 percent) of which he determined through the first practical application of the newly invented Prony brake. The Société Industrielle de Mulhouse that year awarded prizes to both inventors. Fourneyron patented the general design of his first three turbine installations in 1832. Although these were of the free-efflux type, he also foresaw the possibilities of efflux into a diffuser, and in 1855 he patented an outflow diffuser in the form of the present-day inflow scroll case. He was eventually to build more than one hundred hydraulic turbines of various forms for different parts of the world. His writings on water pressure, pipe design, and lock gates may be found in the *Mémoires* and *Comptes rendus* of the Académie des Sciences in the early 1840's.

BIBLIOGRAPHY

Relevant materials include B. Fourneyron, "Mémoire sur l'application en grand dans les mines et manufactures, des turbines hydrauliques ou roues à palettes courbes de Bélidor," in *Bulletin de la Société d'Encouragement pour l'Industrie Nationale,* **33** (1834), 3–17, 49–61, 85–96; M. Crozet-Fourneyron, *Invention de la turbine* (Paris, 1924); and H. Rouse and S. Ince, *History of Hydraulics* (New York, 1963), pp. 146–148.

HUNTER ROUSE

FOWLER, ALFRED (*b.* Wilsden, Yorkshire, England, 22 March 1868; *d.* Ealing, London, England, 24 June 1940), *astrophysics.*

Alfred Fowler was the eighth child and seventh son of Hiram Fowler and his wife, Eliza Hill Fowler. The family was of working-class origin, and Alfred attended elementary schools at Keighley, the largest neighboring town, to which the family moved about 1876. In 1880 he obtained a scholarship to the local trade and grammar school; in 1882, with the aid of a Devonshire exhibition, he proceeded to the Normal School of Science (later the Royal College of Science, now forming a constituent of the Imperial College of Science and Technology) at South Kensington. Here, after a successful career as a student of mechanics, he obtained an appointment as a teacher in training under Norman Lockyer, who had shortly before become director of the Solar Physics Observatory at South Kensington and lecturer in astronomy at the Normal School of Science. This began Fowler's close association with Lockyer and participation with him in the then new field of the application of the spectroscope to astronomy. The association lasted until Lockyer's retirement from the Royal College of Science in 1901, whereupon Fowler succeeded him as assistant professor, and later as professor, of astrophysics. He held this post until 1923, when he was appointed one of the first two Yarrow research professors of the Royal Society. He was thus enabled to continue research at the college but was relieved of teaching duties (other than the direction of research students). He remained there until his retirement in 1934.

In 1892, Fowler married Isabella Orr, who survived him, as did a daughter and a son. His life, apart from his astrophysical work, was uneventful, his interests being concentrated almost entirely on his scientific research and duties arising from his involvement in the organization of science. He successfully directed his students' activities, and many who later achieved distinction in spectroscopy owed much to their early research under his guidance. When the International Astronomical Union was formed in 1919, Fowler became its first general secretary, a position he retained until 1925. The original statutes of the union were drafted by Fowler and adopted almost without change.

During his association with Lockyer, much of Fowler's work was incorporated into that of the senior man, and it was not until he became an independent investigator that his own abilities began to be recognized. Consequently, honors came to him slowly at first, but later in good measure. Temperamentally he was a striking contrast to Lockyer; and their association, although not without occasional misunderstandings, was in many respects fortunate for both. Lockyer's impetuous development of an idea until it became a hypothesis too massive for its frail observational basis was tempered, and often supported, by Fowler's insistence on the primacy of facts and his great skill in acquiring them; while Fowler never forgot, or failed to acknowledge, the inspiration he received, especially in his early days, from Lockyer's enthusiasm. Fowler's life was marked by a quiet integrity and amiability that endeared him to all his associates.

His contributions to astrophysics were based on an exceptionally intimate knowledge of the characteristic spectra of the elements, acquired during his apprenticeship with Lockyer, and an almost uncanny skill in recognizing the identity of celestial spectra and those obtained under vastly different laboratory conditions. These abilities enabled him to assign the band spectra yielded by the cool stars (type M) to titanium oxide, to detect the presence of magnesium hydride in sunspots, to identify bands observed in comet tail spectra with those of low-pressure carbon monoxide, and, with R. J. Strutt (later Lord Rayleigh), to prove that the termination of solar and stellar spectra in the near ultraviolet was caused by ozone in the earth's atmosphere.

His outstanding achievements, however, followed the sudden enlargement of interest in spectroscopy created by Bohr's successful theory of the origin of spectra that appeared in 1913. This made possible a theoretical analysis of spectra that demanded a knowledge of their details, in which Fowler was unrivaled. His work as Lockyer's assistant had involved not only investigations of laboratory spectra from all available sources and of as great a range of celestial spectra as the atmospheric conditions at South Kensington allowed, but also participation in several expeditions to observe total eclipses of the sun. Moreover, under the influence of Lockyer's dissociation hypothesis Fowler had acquired a large amount of information on the variation of spectra with physical conditions, although the general disfavor extended to that hypothesis had discouraged its publication.

The Bohr theory showed the significance and value of such data, and Fowler accordingly took a leading part in the subsequent elucidation of the structure of the various atoms from the characteristics of their spectra. As a first reaction to the Bohr theory he pointed out a discrepancy between the wavelength of the so-called cosmic hydrogen line at λ4686 Å, which he had observed in the spectrum of the sun's chromosphere, and that calculated by Bohr as λ4688 Å and ascribed to ionized helium—an anomaly that resulted in the first refinement of the theory, in which account was taken of the finite ratio of the masses of the proton and the electron. There ensued a continuous interaction between theory and observation in which Fowler played a leading part.

It is fortunate that the period during which spectroscopy stood in the vanguard of physical advance coincided with that in which Fowler, with unique experience and possession of the necessary observational data, could place such data, with no preconceptions, at the service of theoretical investigators. His career offers one of the best examples we have of the variety of possible interactions of theory and observation in the advancement of science.

BIBLIOGRAPHY

Fowler wrote little beyond his original papers, mainly in *Proceedings of the Royal Society* and *Monthly Notices of the Royal Astronomical Society.* "I was too keenly interested in research," he wrote, "to give much thought to the writing of books." Nevertheless, he published a handbook entitled *Popular Telescopic Astronomy* (London, 1895), with the subtitle *How to Make a 2-inch Telescope and What to See With It,* and he contributed about 190 pages on "Geometrical Astronomy and Astronomical Instruments" to a volume entitled *The Concise Knowledge Astronomy* (London, 1896). His only work on his chief specialty was *Report on Series in Line Spectra* (London, 1922), which contains all the data then available on the regularities in spectra (much of which he had brought to light) together with a general account of the subject. He contributed a chapter to *Life and Work of Sir Norman Lockyer,* by T. Mary Lockyer and Winifred L. Lockyer (London, 1928), in which he gives some account of his relations with Sir Norman.

No biography of Fowler exists; the notice in *Obituary Notices of the Royal Society, 1940* is probably the fullest that has appeared.

HERBERT DINGLE

FOWLER, RALPH HOWARD (*b.* Roydon, Essex, England, 17 January 1889; *d.* Cambridge, England, 28 July 1944), *physics.*

Fowler was the oldest son of Howard Fowler, a London businessman, and Ena, daughter of George Dewhurst, a Manchester businessman. He was educated at Winchester and at Trinity College, Cambridge (B.A., 1911); as a student he showed considerable ability in golf and cricket as well as winning prizes in mathematics. After taking his degree he published some work on the theory of solutions of differential equations and as a result was elected to a fellowship at Trinity in October 1914.

By this time Fowler had obtained a commission in the Royal Marine Artillery. He was wounded in the Gallipoli campaign, and while convalescing in 1916 was persuaded to join a group of scientists, led by A. V. Hill, who were doing research on such military problems as tracking the flight of airplanes and computing trajectories of cannon shells. This early introduction to applied mathematics seems to have influenced Fowler's subsequent interest in physical problems, although he did not entirely abandon his earlier commitment to pure mathematics.

In 1921, two years after returning to Cambridge as a fellow, he married Eileen, the only daughter of Sir Ernest Rutherford. They had four children.

In 1922 Fowler and C. G. Darwin published a series of papers on statistical mechanics in which they developed methods for calculating the "partition functions" associated with the distribution of energy in quantum systems. (By using the theory of functions of a complex variable they were able to avoid some of the usual approximations.)

Fowler then extended these methods in statistical mechanics to deal with the equilibrium states of ionized gases at high temperatures. His results could be immediately applied to the interpretation of stellar spectra and provided a new method for estimating the temperatures and pressures of stellar interiors. Fowler also provided one of the earliest applications of the new "quantum statistics" of E. Fermi and P. A. M. Dirac when in 1926 he proposed that white dwarf stars consist of a "degenerate" gas of extremely high density. Thus Fowler was one of the founders of modern theoretical astrophysics. (His work on the solutions of Emden's equation was another contribution to this field.)

By the early 1920's Fowler was among the very few workers at Cambridge who maintained a continuing interest in the progress of the quantum theory; he kept in touch with recent developments through correspondence and visits to Copenhagen. Those students—such as Dirac—who turned their attention to the quantum theory had usually been introduced to it by Fowler, and it was he who gave Dirac the galley proofs of Heisenberg's "matrix article" of 1925, which led to Dirac's discovery of Poisson-Bracket relations (according to private communication from T. S. Kuhn, based on information in the Archive for History of Quantum Physics). Because of his connection with Rutherford, Fowler was particularly well placed to introduce problems from the quantum theory into the discussions of the more experimentally inclined physicists who gathered at the Cavendish Laboratory and in the Kapitza Club. Much of the early work at Cambridge on this aspect of physics was therefore stimulated by him.

Fowler was awarded the Adams Prize at Cambridge University in 1924 for an essay on statistical mechanics. In 1929 he published a revised version of this essay, including the application of quantum statistics and ionization theory to states of matter at high pressures and high temperatures. *Statistical Mechanics* became the standard reference work on the subject in English-speaking countries for the next decade; it was followed by a second edition (1936) and by *Statistical Thermodynamics* (1939), a book emphasizing applications to physical chemistry, written with E. A. Guggenheim.

Fowler was elected to the Plummer chair of theoretical physics at Cambridge in 1932 and continued to pioneer the applications of statistical mechanics and to explore other areas of theoretical physics. Together with such students and colleagues as E. A. Guggenheim and R. F. Peierls he developed the "Ising model" as a theory of phase transitions and cooperative phenomena in magnets, alloys, and solutions.

In 1938 he was appointed director of the National Physical Laboratory, to succeed Sir Lawrence Bragg, but had to decline the appointment almost immediately because of illness. In World War II he served as a consultant to the Ordnance Board and the Admiralty. In 1942 he was knighted for his services to the government, in particular for establishing scientific liaison between research efforts on military problems in England and in Canada, and for his accomplishments during an important mission to Canada and the United States.

BIBLIOGRAPHY

The best source of information about Fowler is the comprehensive memoir by his friend and colleague E. A. Milne, in the *Obituary Notices of Fellows of the Royal Society of London,* **5** (1945–1948), 61–78, which contains a portrait, bibliography, and many personal recollections.

STEPHEN G. BRUSH

FOWNES, GEORGE (*b.* London, England, 14 May 1815; *d.* Brompton, England, 31 January 1849), *chemistry.*

Fownes worked in his father's glove business until 1837, when he began to study science with Thomas Everitt, lecturer in chemistry at the Middlesex Hospital. In 1839 he earned the doctorate degree under

Liebig at Giessen. On his return to London he became Thomas Graham's assistant at University College. He held lectureships in chemistry at Charing Cross and Middlesex hospitals. In 1842 he became professor of chemistry to the Pharmaceutical Society and began a lecture series on organic chemistry at the Royal Institution. He was the first director of the newly established Birkbeck Laboratory at University College (1845). Pulmonary disease obliged him to resign his lectureships by 1846, and after three years of poor health he died of consumption.

Fownes accomplished the bulk of his work in only four years (1842–1846). His most notable achievement was the isolation of two new organic bases. In 1845 he prepared furfural by the action of sulfuric acid on bran. In the same year he isolated benzoline (hydrobenzamide) from the oil of bitter almonds. In 1839 he accurately determined the equivalent weight of carbon by means of the combustion of naphthalene; Fownes reported that the accepted value as determined by Berzelius and others was too high. He prepared potassium cyanide by passing nitrogen over potassium carbonate and charcoal at high temperature, a process that was used industrially for a time. In 1844 he discovered the presence of phosphate in igneous rocks and suggested that this was the original source of phosphate in clay and soil.

He published two widely read books. In 1843 he was awarded the Royal Institution's Acton Prize for his *Chemistry, As Exemplifying the Wisdom and Beneficence of God,* an argument for design in the universe based on the chemical constitution of the earth, sea, and atmosphere. His *Manual of Elementary Chemistry* (1844) was a very popular textbook for half a century.

BIBLIOGRAPHY

I. ORIGINAL WORKS. Fownes's *Manual of Elementary Chemistry, Theoretical and Practical* (London, 1844) appeared in many eds. under the editorship of H. B. Jones, A. W. Hofmann, H. Watts, and W. A. Tilden; the final ed. was published in 1889. His *Chemistry, As Exemplifying the Wisdom and Beneficence of God* had two eds. (London, 1844, 1849). Fownes also published *An Introduction to Qualitative Analysis* (London, 1846) and *Rudimentary Chemistry* (London, 1848). Significant papers include "On the Equivalent of Carbon," in *The Philosophical Magazine,* ser. 3, **15** (1839), 62–65; "On the Formation of Cyanogen From Its Elements," in *Pharmaceutical Journal and Transactions,* **1** (1842), 338–343; "On the Existence of Phosphoric Acid in Rocks of Igneous Origin," in *Philosophical Transactions of the Royal Society,* **134** (1844), 53–56; "An Account of the Artificial Formation of a Vegeto-Alkali," *ibid.,* **135** (1845), 253–262; and "On Benzoline, a New Organic Salt-Base From Bitter Almond Oil," *ibid.,* 263–268.

II. SECONDARY LITERATURE. The most detailed notice on Fownes is J. S. Rowe, "The Life and Work of George Fownes, F.R.S. (1815–49)," in *Annals of Science,* **6** (1948–1950), 422–435. Several brief obituary notices appeared at the time of his death: *Journal of the Chemical Society,* **2** (1849), 184–187; *Pharmaceutical Journal and Transactions,* **8** (1849), 449–450; and *Proceedings of the Royal Society,* **5** (1849), 882–883.

ALBERT B. COSTA

FRACASTORO, GIROLAMO (*b.* Verona, Italy, *ca.* 1478; *d.* Incaffi [now hamlet of Affi, Verona], 6 August 1553), *medicine, philosophy.*

Descendant of a patrician Veronese family and the sixth of seven brothers, Fracastoro received his first literary and philosophical instruction from his father; his mother, Camilla Mascarelli, seems to have died when he was still very young.

As an adolescent, he was sent to the Academy in Padua, where he was entrusted to a family friend, Girolamo Della Torre, a Veronese who taught and practiced medicine there. Fracastoro studied literature, mathematics, astronomy, philosophy (the latter under the guidance of Pietro Pomponazzi and Nicolò Leonico Tomeo), and medicine, in which he was instructed by Girolamo Della Torre and his son Marcantonio, and Alessandro Benedetti. Fracastoro was a fellow student of the brothers Giovan Battista and Raimondo Della Torre, of the future cardinals Ercole Gonzaga and Gaspare Contarini, and of Andrea Navagero; and he established relationships with Giovanni Battista Ramusio and Pietro Bembo that proved of primary importance. Immediately after receiving his degree (1502) he became an instructor in logic at the University of Padua, where he was also *conciliarius anatomicus.* His contacts with Copernicus, who had enrolled in medicine at Padua in 1501, date from this time. While he was still young, Fracastoro married (1500?) Elena de Clavis (or Schiavi), by whom he had five children: four sons—Giovanni Battista; Paolo and Giulio, who died at an early age and were lamented by their father in one of his odes; and Paolo Filippo, born in 1517 and the only son to survive his father—and a daughter, Isabella.

After the death of his father and the closing of the University of Padua and with the threat of war between Venice and the Emperor Maximilian I, Fracastoro left Padua and in 1508 followed Bartolomeo d'Alviano to Porto Naone (now Pordenone) in Friuli, where Alviano presented him at the Accademia Friulana. After a short stay, he followed Alviano to the border of the Veneto, apparently as a doctor. Alviano was taken prisoner at Giara d'Adda, after the Venetian defeat at the battle of Agnadello (1509). Fracastoro returned to Verona and established residence in the area of the church of Santa Eufemia.

He then dedicated himself to his studies, to reorganizing his estate, and, for a while, to medical practice, treating patients from all over Italy. He actively participated in the life of the local *collegio dei fisici,* where he had already matriculated in 1501 and of which he was four times prior and eight times councillor. Although interested in politics, he never held public office.

From 1511 he began to alternate his residence in Verona with long sojourns at his villa in Incaffi, on the slopes of Monte Baldo, where his learned friends gathered for philosophical and scientific meetings. Meanwhile he maintained relations with such leading figures as Gian Matteo Giberti, bishop of Verona, a man of great culture and patron of writers, scientists, and artists, whose guest Fracastoro was at Malcesine. He was also expanding his cultural interests, which touched not only on philosophy and medicine but on the liberal arts and natural sciences in general; and he attained noteworthy erudition and competence in each area, as his surviving writings testify.

Fracastoro's fame, esteem, and acquaintances in ecclesiastical circles contributed to his nomination by Pope Paul III in 1545 as *medicus conductus et stipendiatus* of the Council of Trent, to which he went upon request, a guest of Cardinal Madruzzo. His presentiment of a terrible epidemic seems to have influenced the transfer (1547), which the pope desired, of the Council from Trent to Bologna. Around 1546 Fracastoro was made canon of Verona, with special dispensations.

His mental faculties undimmed by age, Fracastoro suffered a stroke that killed him within the day, on 6 August 1553, almost certainly in his house at Incaffi. His body was transported to Verona and was buried in the church of Santa Eufemia, where it rested probably until 1740; the remains have since been lost. In 1555 a statue was erected to him in Verona, in the Piazza dei Signori, near the existing statues of Pliny and Catullus.

Fracastoro's scientific personality matured in the atmosphere of Padua, where he had ample opportunity to enter into the disputes of the Scholastics and the followers of Alexander of Aphrodisias and Ibn Rushd (Averroës). Philosophical considerations were thus always inherent in his more purely scientific work. His thought, although not always organic, is framed in those philosophies of nature which were developed by various writers of the Italian Renaissance and which are the result of two components, a diminished interest in theological subjects and metaphysics in general and an increased interest in the study of nature, in which man lives and which is held to be the only subject appropriate to his understanding, which requires certainty. (Significant in this regard is the beginning of the posthumously published *Turrius sive de intellectione dialogus.*) This interest in nature differs from that of the preceding era, that is, of the humanists: the contemplative aspect gives way to the operative one. That is, nature is considered as an autonomous reality, upheld by its own laws, in which a mixture of good and bad is inherent and before which any recourse to supernatural intervention is useless; to derive the most profit and happiness, man must rely only on himself and on his capacity for progressive understanding of the world's regulating principles. These ideas emerge in the narrative poem *Syphilis sive morbus Gallicus* (1530), which brought Fracastoro universal fame, as is attested by numerous editions and translations in various languages; in it the nature and cure of lues are illustrated.

Composed in 1521, the poem was initially divided into two books. In the final draft it was published in three books, despite advice to the contrary by Pietro Bembo (in his letter of 5 January 1526 to Fracastoro, in which he also firmly asserted that some passages be eliminated), to whom it is dedicated and by whom it was esteemed and praised, both when it was sent to him by Fracastoro for a preliminary reading in 1525 and subsequently.

The poem, drafted in Latin hexameter (about 1,300 verses) of exquisite beauty, occupies a prominent place in the literature of the times and represents a magnificent paradigm of formal sixteenth-century virtuosity in refined Latin of a didactic quality reminiscent of Vergil's *Georgics.* Through the work the name of the sickness became definitively established; the name was, in fact, considered to derive from that of the hero of Fracastoro's treatise, the unfortunate shepherd Sifilo. Others believe that the word *sifilo* derived from *sifilide,* a term already in use in the local dialect of the Veneto.

According to Fracastoro's mythological tale, the terrible disease originated as the punishment (an unclean ulcer on the body) inflicted by the sun god on the young shepherd Sifilo, who had become unfaithful to him. The misdeed was, however, forgiven, and the guaiacum, a great leafy tree, was born. Humanity learned to extract from it the medicament that cured the disease. Also effective against lues is mercury, which the nymph Lipare advised the shepherd Ilceo to use. The ample and exhaustive description of the various luetic manifestations demonstrates Fracastoro's lucid knowledge of the clinical events and the related course of the illness.

In *De morbo Gallico* the author lays the first foundations of his doctrine of infections, since he was already familiar with the *semina morbi* of Lucretius through Andrea Navagero's edition of *De rerum natura* (1515). The concepts of contagion indicated in *De morbo Gallico* were further developed in Fracastoro's prose treatise on syphilis, written in 1553 but not published until 1939, which served as preparation for the subsequent formulation of the Fracastorian doctrine of contagion. Some authors consider noteworthy in *De morbo Gallico* not so much the illustrations of the pathological phenomena as Fracastoro's manner, his feeling for human suffering, as exemplified in the episode of the death from syphilis of a young man from Brescia, and in the vivid description of the misfortunes that pervaded Europe, and especially Italy, in the first half of the sixteenth century. The work also provided the poet with an opportunity to celebrate the great geographical discoveries of the century.

The theory has also been advanced that the subject dealt with in *De morbo Gallico* is only a pretext for posing a problem of greater significance. The dominant theme of the first book is that of the mutations that take place in nature. Nature creates and destroys and gives misery and happiness, and it is useless to try to appease the gods. Science, whose power alone can give joy, dictates man's actions.

For the construction of a philosophy of nature that starts from the above premise, a fundamental question obviously is posed—that of method. For Fracastoro the only valid one is that of experience, as he does not hesitate to declare in the *Homocentrica sive de stellis* (1538), a work on astronomy in which the movements of the heavens and the celestial spheres with their orbits, the seasons, and various types of days (civil, solar, sidereal) are illustrated, and in which Fracastoro again reinstates in a place of honor the most ancient astronomical theory, the Eudoxian. Apart from the intrinsic value of the work, its attempts to solve certain problems in astronomical and terrestrial physics are interesting, as are the studies on refraction. In the course of the latter Fracastoro points out the apparent enlargement and approach of celestial objects (as well as the moon) observed through two superimposed lenses, analogous to the appearance of a body immersed in water, which varies exactly according to the quantity and density of the water itself.

The discourse on experience, begun in the *Homocentrica,* is developed in *De causis criticorum dierum libellus* (1538). Experience, in order to be fruitful, must be collected and examined by secure concepts; these keep it from degenerating into a dispersion of multiplicities or into fantasy and magic, which would constitute a renewed victory of the transcendentalist attitude toward nature. In *De causis* Fracastoro gives an example of badly interpreted experience: critical days really exist in the course of an illness, but it is an error to look for the explanation of this solely in astral influences or certain numerological relationships. The cause lies in the nature of the disease itself, that is, in the humoral modifications; the crisis is an expression of the organic actions and reactions determined by the qualitative and quantitative alteration of the humor or humors involved.

That which unifies experience is the Aristotelian concept of cause. The type of cause capable of unifying natural phenomena is not of the order of most general causes; Fracastoro considers these useless because they are *remotissimae a rebus.* It is, rather, that of the closest and most particular causes, that is, the middle causes; furthermore, one must strive to arrive at those causes which are *propinquissimae et propriae.* Thus the traditional position of philosophy is turned upside down: philosophy is such to the extent that it investigates not abstract but concrete nature. To proceed along the path of universals and principles of things is to condemn philosophy—so far as it concerns nature—and to leave *innumera intacta* and other things *non plana discussa.*

In *De sympathia ed antipathia rerum* (1546) Fracastoro recognizes that the principle immanent in nature and explaining it is *simpatia,* which Fracastoro conceives of in a sense different from that of the humanists. To Fracastoro sympathy is a principle of spiritual order; it is the *species spiritualis* that unifies the world. In particular, it is to be brought down to the plane of natural things that are to be studied naturalistically.

Bound to the cosmological principle of sympathy are Fracastoro's anthropological and esthetic concepts. It is sympathy, in fact, that gives nature an unbroken gradualism, so that there is no disruption in man's faculties; sense and intellect are both passive and both have a *species,* even though of a different unifying power. That which really belongs to the intellect is judgment, conceived as a synthesis of sensory data and of the universal. The pure sensations, on the other hand, attain a unity of their own in the forces of the *subnotiones,* which distinguish them from one another and place them in interrelation. Fracastoro expresses these ideas in the *Turrius sive de intellectione dialogus,* in which he conceived of knowledge as a progressive unification of multiplicities, rather as did Kant. Similarly, in the *Naugerius sive de poetica dialogus* (1549), the essence of poetry is rendered neither by the content with which the poet deals nor by the form, which can be various, but rather by the intuition of beauty, which is then the universal, present in all things and expressing itself in the sympathy that regulates them. This differentiates the poet from the historian, who deals with particulars.

Fracastoro's scientific thought culminates and concludes with *De contagione et contagiosis morbis et curatione* (1546), which assures him a lasting place in the history of epidemiology. In it he clearly describes numerous contagious diseases, with chapters of principal interest, such as that on phthisis, whose contagion and affinity for the lungs he affirms. In the work's most significant part Fracastoro illustrates the three means by which contagion can be spread: by simple contact (as in scabies and leprosy); by *fomites,* corresponding to carriers (clothing, sheets); and at a distance, without direct contact or carriers (as in plague, smallpox, and the like). Fracastoro imagines that in the last case the *seminaria* propagate either by choosing the humors for which they have the greatest affinity or by attraction, penetrating through the inspiration of the vessels. According to Fracastoro the seeds of contagion are in fact responsible for contagion; they are distinct imperceptible particles, composed of various elements. Spontaneously generated in the course of certain types of putrefaction, they present particular characteristics and faculties, such as that of increasing themselves, having their own motion, propagating quickly, enduring for a long time, even far from their focus of origin, exerting specific contagious activity, and dying.

In *De contagione* the epidemiological problems and the *principia contagionum* are delved into with great acuteness. Fracastoro's sheer prophetic intuition yielded hypotheses on causes and ways of infections that were verified in succeeding centuries.

In certain passages the Fracastorian *seminaria* seem to be like our microorganisms. Undoubtedly, the *seminaria* derive from Democritean atomism via the *semina* of Lucretius and the gnostic and Neoplatonic speculations renewed by St. Augustine and St. Bonaventura (*rationes seminales*); but the Fracastorian *seminaria* differ greatly from traditional *semina.* It is difficult—perhaps impossible—to establish incontrovertibly whether Fracastoro really foresaw, as some would like to believe, the existence of microbes. He seems to attribute certain vital faculties to his *seminaria* and to use suggestive terminology for them (such as generation, birth, and life), but in light of the state of knowledge at the time—the inability to distinguish clearly between the organic and inorganic and belief in spontaneous generation—Fracastoro could not assign to his *seminaria* all the typical characteristics of microorganisms.

Fracastoro left works on other subjects, including botany, geology, and medicine. Among those in which the philosophical and literary content merits mention are the *Fracastorius sive de anima dialogus,* in which he affirms the immortality of the soul; and the short poems *Alcon seu de cura canum venaticorum* and *Ioseph.*

BIBLIOGRAPHY

I. ORIGINAL WORKS. Fracastoro's works were published in one vol. as *Opera omnia* (Venice, 1555); there were various reprintings, of which the Cominiana ed. (Padua, 1739) is important. Included in the first ed. are *Homocentricorum sive de stellis, liber unus; De causis criticorum dierum libellus; De sympathia et antipathia rerum liber unus; De contagionibus et contagiosis morbis et eorum curatione libri tres; Naugerius sive de poetica dialogus; Turrius sive de intellectione dialogus; Fracastorius sive de anima dialogus; De vini temperatura sententia; Syphilis sive morbus Gallicus, libri tres; Ioseph libri duo;* and *Carminum liber unus.*

Some letters of Fracastoro to G. B. Ramusio, of interest because they refer to the scientific life of the era, were published in 1564 in Tommaso Poracchi, *Lettere di XIII huomini illustri.* A later ed. (Venice, 1632) also contains a letter from Fracastoro to Paolo Ramusio. See also the prose treatise on syphilis, Codice CCLXXV–I, Biblioteca Capitolare, Verona (Verona, 1939); and a vol. of *Scritti inediti* (Verona, 1955); other works are scattered in various publications. Many MSS, partially unpublished, are preserved in the Biblioteca Capitolare.

II. SECONDARY LITERATURE. The indicated works serve as bibliographical sources: L. Baumgartner and J. F. Fulton, *A Bibliography of the Poem Syphilis sive Morbus Gallicus by Girolamo Fracastoro of Verona* (New Haven, 1935); E. Cassirer, *Storia della filosofia moderna,* A. Pasquinelli, G. Colli, and E. Arnaud, trans., I (Turin, 1964), 258–264, *passim;* B. Croce, "Il dialogo di Fracastoro sulla poetica," in *Quaderni della critica,* no. 9 (Nov. 1947), 56–61; E. di Leo, *Scienze ed umanesimo in Girolamo Fracastoro,* 2nd ed. (Salerno, 1953); E. Garin, *Storia della filosofia italiana,* II. *Il Rinascimento,* 2nd ed. (Turin, 1966), 627–629, *passim;* F. Pellegrini, *Fracastoro* (Trieste, 1948); F. Pellegrini, G. Alberti, A. Spallicci, *et al., Studi e memorie nel IV centenario* (Verona, 1954); and G. Saitta, *Il pensiero italiano nell'umanesimo e nel Rinascimento,* II. *Il Rinascimento,* 2nd ed. (Florence, 1961), 177–212, *passim.*

BRUNO ZANOBIO

FRAENKEL, ADOLF ABRAHAM (*b.* Munich, Germany, 17 February 1891; *d.* Jerusalem, Israel, 15 October 1965), *mathematics.*

Fraenkel studied at the universities of Munich, Marburg, Berlin, and Breslau. From 1916 to 1921 he was a lecturer at the University of Marburg, where he became a professor in 1922. In 1928 he taught at the University of Kiel, and then from 1929 to 1959 he taught at the Hebrew University of Jerusalem. A fervent Zionist with a deep interest in Jewish culture, he engaged in many social activities. His interest in the history of mathematics appears in his papers "Zahlbegriff und Algebra bei Gauss" (1920), "Georg Cantor" (1930), and "Jewish Mathematics and Astronomy" (1960). As a mathematician he was interested in the axiomatic foundation of mathematical theories. His first works were on algebra, notably on the axiomatics of Hensel's p-adic numbers and on the theory of rings. He soon turned to the theory of sets, and in 1919 his remarkable *Einleitung in die Mengenlehre* appeared, which was reprinted several times. Engaged in a proof of the independence of the axiom system of Ernst Zermelo (1908), Fraenkel noticed that the system did not suffice for a foundation of set theory and required stronger axioms of infinity. At the same time he found a way to avoid Zermelo's imprecise notion of definite property.

Briefly stated, Zermelo's set theory is about a system B of objects closed under certain principles of set production (axioms). One of these axioms, the axiom of subsets, states that if a property E is definite in a set M, then there is a subset consisting precisely of those elements x of M for which $E(x)$ is true. Property E is definite for x if it can be decided systematically whether $E(x)$ is true or false. Another one is the famous axiom of choice, stating that the union of a set T of nonvoid disjoint sets contains a subset that has precisely one element in common with the sets of T.

Instead of Zermelo's notion of definite property Fraenkel used a notion of function, introduced by definition; and he replaced Zermelo's axiom of subsets by the following: If M is a set and ϕ and ψ are functions, then there are subsets M_E and $M_{E'}$ consisting of those elements x of M for which $\phi(x)$ is an element of $\psi(x)$, and $\phi(x)$ is not an element of $\psi(x)$ respectively. Using this axiom Fraenkel proved the independence of the axiom of choice, having recourse to an infinite set of objects that are not sets themselves. A proof avoiding such an extraneous assumption proved to be far more difficult and was given in 1963 by P. J. Cohen for a slightly revised system, ZFS, named after Zermelo, Fraenkel, and Thoralf Skolem. This system derives from a modification proposed by Skolem in 1922, consisting in the interpretation of definite property as property expressible in first-order logic.

In a series of papers Fraenkel developed ZF set theory to include theories of order and well-order. His encyclopedic knowledge of set theory is preserved in his works *Abstract Set Theory* (1953) and *Foundations of Set Theory* (1958). As early as 1923 he emphasized the importance of a thorough investigation of predicativism, based on ideas of H. Poincaré and undertaken much later by G. Kreisel, S. Feferman, and K. Schütte, among others.

BIBLIOGRAPHY

I. ORIGINAL WORKS. Fraenkel's writings include "Axiomatische Begründung von Hensels p-adischen Zahlen," in *Journal für die reine und angewandte Mathematik,* **141** (1912), 43–76; "Über die Teiler der Null und die Zerlegung von Ringen," *ibid.,* **145** (1915), 139–176; *Einleitung in die Mengenlehre* (Berlin, 1919); "Zahlbegriff und Algebra bei Gauss," in *Nachrichten von der Königlichen Gesellschaft der Wissenschaften zu Göttingen,* Math.-phys. Kl. (1920), pp. 1–49; "Über einfache Erweiterungen zerlegbarer Ringe," in *Journal für die reine und angewandte Mathematik,* **151** (1921), 121–167; "Über die Zermelosche Begründung der Mengenlehre," in *Jahresbericht der Deutschen Mathematikervereinigung,* **30** (1921), 97–98; "Zu den Grundlagen der Cantor-Zermeloschen Mengenlehre," in *Mathematische Annalen,* **86** (1922), 230–237; "Axiomatische Begründung der transfiniten Kardinal-zahlen. I," in *Mathematische Zeitschrift,* **13** (1922), 153–188; "Der Begriff 'definit' und die Unabhängigkeit des Auswahl-axioms," in *Sitzungsberichte der Preussischen Akademie der Wissenschaften,* Math.-phys. Kl. (1922), pp. 253–257; "Die neueren Ideen zur Grundlegung der Analysis und Mengenlehre," in *Jahresbericht der Deutschen Mathematikervereinigung,* **33** (1924), 97–103; "Untersuchungen über die Grundlagen der Mengenlehre," in *Mathematische Zeitschrift,* **22** (1925), 250–273; "Axiomatische Theorie der geordneten Mengen," in *Journal für die reine und angewandte Mathematik,* **55** (1926), 129–158; *Zehn Vorlesungen über die Grundlegung der Mengenlehre* (Leipzig–Berlin, 1927); "Georg Cantor," in *Jahresbericht der Deutschen Mathematikervereinigung,* **39** (1930), 189–226; "Das Leben Georg Cantors," in *Georg Cantor Gesammelte Abhandlungen,* E. Zermelo, ed. (Berlin, 1932; Hildesheim, 1966), pp. 452–483; "Axiomatische Theorie der Wohlordnung," in *Journal für die reine und angewandte Mathematik,* **167** (1932), 1–11; *Abstract Set Theory* (Amsterdam, 1953); *Axiomatic Set Theory* (Amsterdam, 1958), written with P. Bernays; *Foundations of Set Theory* (Amsterdam, 1958), written with Y. Bar-Hillel; "Jewish Mathematics and Astronomy," in *Scripta mathematica,* **25** (1960), 33–47; and *Lebenskreise, aus den Erinnerungen eines jüdischen Mathematikers* (Stuttgart, 1967).

II. SECONDARY LITERATURE. On Fraenkel and his work, see (in chronological order) T. Skolem, "Einige Bemerkungen zur axiomatischen Begründung der Mengenlehre," in *Wiss. Vorträge gehalten auf dem 5. Kongress der skandinav. Mathematiker in Helsingfors 1922* (1923), pp. 217–232; J. von Neumann, "Über die Definition durch transfinite Induktion und verwandte Fragen der allgemeinen Mengenlehre," in *Mathematische Annalen,* **99** (1928), 373–391; G.

Kreisel, "La prédicativité," in *Bulletin de la Société mathématique de France,* **88** (1960), 371–391; K. Schütte, *Beweistheorie* (Berlin, 1960); S. Feferman, "Systems of Predicative Analysis," in *Journal of Symbolic Logic,* **29** (1964), 1–30; P. J. Cohen, *Set Theory and the Continuum Hypothesis* (New York, 1966); and J. van Heijenoort, *From Frege to Gödel* (Cambridge, Mass., 1967).

B. VAN ROOTSELAAR

FRAIPONT, JULIEN (*b.* Liège, Belgium, 17 August 1857; *d.* Liège, 22 March 1910), *zoology, paleontology, anthropology.*

At the end of his intermediate studies, Fraipont entered the offices of the bank of which his father was director. However, attracted since childhood by the natural sciences, he attended at the same time the zoology courses given at the University of Liège by Edouard Van Beneden. He became one of Van Beneden's favorite students, and then abandoned the career for which he had seemed destined in order to pursue his scientific vocation. He was soon named student assistant and then became Van Beneden's *préparateur* (1878) and his assistant (1881). He was then hired to teach the following subjects at the University of Liège: animal paleontology (1884), animal geography (1885), and systematic zoology (1885). He was named professor in 1886 and, in 1909, a few months before his death, rector of the University of Liège. He was elected a foreign member of the Leopoldinisch-Karolinische Deutsche Akademie der Naturforscher in 1890, replacing L.-G. De Koninck, and a foreign member of the Imperial Society of Naturalists of Moscow in 1895. He became a member of the Académie Royale des Sciences de Belgique in 1895 and director of its science section in 1908.

Fraipont's zoological works (fifteen publications appeared between 1877 and 1908) deal with systematics, but above all with the morphology of Protozoa (*Acineta*), Hydrozoa (*Campanulariae*), Trematoda, Cestoda, and Archiannelida. They also include a monograph (1907) on the genus *Okapia* in which the author endeavors to demonstrate that this mammal, discovered in 1900 in the Belgian Congo, represents a form perfectly intermediate between the Cenozoic Giraffidae and present-day giraffes.

Fraipont's most important contribution to zoology probably consists of his studies on the Archiannelida (1884–1887), a group which had recently come to prominence through the investigations of B. Hatschek. Fraipont's studies began with the cephalic nephridia (1884) and the central and peripheral nervous systems of certain of these organisms (1884) and were completed by a monograph on the genus *Polygordius* (1887). As in all his preceding works, Fraipont was inspired by the example and the teaching of his mentor Van Beneden. He gave a minute description of the anatomy, histology, development, habits, and habitat of the genus, as well as of its geographic distribution and position in the class of the Annelida. In his conclusions he agreed, although not without some reservations, with Hatschek's opinion and accepted, like him, the group of the Archiannelida.

Fraipont was especially occupied with paleontological research during the period from 1883 to 1890. In particular he studied various fossils of the Upper Devonian and the Lower Carboniferous. In 1885 he published, in collaboration with De Koninck, the fifth part (devoted to the Lamellibranchia) of De Koninck's monumental work on the fauna of the Lower Carboniferous in Belgium.

During this period Fraipont also began to take a lively interest in prehistory, continuing the work done before him in Belgium by P.-C. Schmerling and E.-F. Dupont. Alone or with collaborators he explored several caves in the province of Liège, discovering numerous archaeological levels ranging from the Lower Mousterian to the Neolithic. He was also involved in study of the human fossils discovered at Spy, near Namur (Belgium), during the summer of 1886 by his friends the geologist Max Lohest and the prehistorian Marcel De Puydt.

This discovery played a considerable role in the history of human paleontology. The material found consisted of the remains of two human skeletons, associated with a great quantity of Quaternary mammalian bones and with lithic implements of the Mousterian type. This was the first discovery of relatively complete documents of Neanderthal man, exhumed in perfectly established stratigraphic conditions that fixed their age (known today to date from the Würm I stage) and guaranteed their authenticity. The principal observations and measurements of the two skeletons were carried out by Fraipont. They are remarkable for their precision, especially since Fraipont was not a professional anatomist. The results of these investigations were the subject of a memoir published jointly by Fraipont and Lohest in 1887.

This memorable discovery at Spy permitted the interpretation of fragmentary pieces previously brought to light, such as the jaw found at La Naulette (Belgium) in 1865 by E.-F. Dupont, and it completed and confirmed the knowledge of a type of human fossil whose special characteristics some in this period were still trying to explain by the action of pathological factors. Fraipont devoted several other articles to the Spy fossils between 1888 and 1893. He also published an interesting study on the tibia (1888) in its relation to the erect posture of man and the Pongidae, and in 1900 he presented a thorough study of certain Neolithic skeletons found in various Belgian caves.

Highly esteemed by everyone, Fraipont was a modest man, extremely kind and courteous, devoted to his teaching duties and to his students. His complete moral integrity is reflected in his work. One may reproach that work for too great a diversity in subject matter and for certain factual or interpretive errors; yet it preserves a fundamental unity of method and of thought, that of a zoologist devoted to the facts as he perceived them, rather than to constructing brilliant but hazardous speculative systems.

BIBLIOGRAPHY

I. Original Works. Fraipont published 46 works (15 in zoology, 11 in paleontology, and 20 in anthropology and prehistory) in addition to a great many reports and conference papers.

The principal works are "Faune du calcaire carbonifère, 5e partie, Lamellibranches," in *Annales du Musée royal d'histoire naturelle de Belgique,* **11** (1885), 1-33, written with L.-G. De Koninck; "Monographie du genre Polygordius," in *Fauna und Flora des Golfes von Neapel,* **14** (1887), 1-125; and "La race humaine de Neanderthal ou de Canstadt en Belgique. Recherches ethnographiques sur des ossements humains, découverts dans des dépôts quaternaires d'une grotte à Spy et détermination de leur âge géologique," in *Archives de biologie,* **7** (1887), 587-757, written with Max Lohest.

II. Secondary Literature. Of the obituaries published, the best, which includes a complete list of publications, is M. Lohest, C. Julin, and A. Rutot, "Notice sur Julien Fraipont," in *Annuaire de l'Académie royale de Belgique,* **91** (1925), 131-197. On the discoveries at Spy, see the interesting critical chapter in Aleš Hrdlička, *The Skeletal Remains of Early Man,* Smithsonian misc. coll., **83** (Washington, D.C., 1930), 178-212.

G. Ubaghs

FRANÇAIS, FRANÇOIS (JOSEPH) (*b.* Saverne, Bas-Rhin, France, 7 April 1768; *d.* Mainz, Germany, 30 October 1810); **FRANÇAIS, JACQUES FRÉDÉRIC** (*b.* Saverne, Bas-Rhin, France, 20 June 1775; *d.* Metz, France, 9 March 1833), *mathematics.*

The mathematical works of the Français brothers, François and Jacques Frédéric, are so poorly distinguished by most authors and their biographies so imprecise that it is necessary to devote a common article to the two of them. Sons of Jacques Frédéric Français, a grocer, and of Maria Barbara Steib, they were both born at Saverne, seven years apart. In 1789, François, the elder, became a seminarist. Named professor at the *collège* in Colmar in June 1791, he assumed the chair of mathematics at the *collège* in Strasbourg in September 1792. He participated actively in political life and was a secretary of the *société populaire* of Strasbourg. He took part in the Vendée campaign from May to October 1793 and, after a brief return to civilian life, went back to the army as an officer until October 1797, when he was named professor of mathematics at the École Centrale du Haut-Rhin in Colmar. He left this position in September 1803 to teach mathematics, first at the lycée in Mainz, then at the École d'Artillerie at La Fère (1804), and, finally, at the École d'Artillerie in Mainz. At his death he left four small children, whom his younger brother adopted soon afterward.

Jacques Frédéric Français, after having been an outstanding student at the *collège* of Strasbourg, enrolled as a volunteer in 1793 and was named assistant in the corps of engineers in September 1794. He was admitted to the École Polytechnique on 30 December 1797, and from there he went, in March 1800, to the École du Génie. A first lieutenant in January 1801, he participated in the expedition sent by Napoleon to attempt to save the French army in Egypt. On his return he was quartered at Toulon and was named captain of the sappers in December 1801; in November 1802, he became second in command of the staff headquarters of the corps of engineers. In this capacity he participated, with Admiral Villeneuve's squadron, in the expedition to the Antilles and in the naval battles of Cape Finisterre and Trafalgar (1805). Beginning at the end of that year, he was successively assigned to garrisons at Condé-sur-Escaut; then Kehl (1806); Strasbourg (1807), under the command of Malus; and Metz (from 1808), to the staff headquarters of the École d'Application. Promoted to first in command in July 1810, he was named, at the beginning of 1811, professor of military art at the École d'Application du Génie et de l'Artillerie in Metz. He held this last position until his death.

In July 1795 François Français presented a memoir on the integration of partial differential equations; a new version of this paper, addressed to the Académie des Sciences in 1797, is mentioned by S. F. Lacroix (*Traité du calcul différential et du calcul intégral,* III, 598) as an important contribution to the theory of these equations. According to the testimony of Biot (*Procès-verbaux de l'Académie des sciences,* III, 204-205), Français then assisted "his uncle" Louis-François Arbogast in the elaboration of the "calculus of derivations" and in the preparation of a treatise that he devoted to this subject (1800). Having inherited Arbogast's papers in April 1803, he continued to work on the development of the calculus of derivations and its applications; thus, in a memoir presented to the Academy in November 1804, he applied this calculus to the movement of projectiles in a resistant medium. Highly esteemed by the mathematicians of the Paris Academy—Lagrange, Legendre, Lacroix, and Biot—François Français pursued original mathematical investigations, but without publishing anything. Following his death his brother Jacques Frédéric included several brief extracts in the *Annales de mathématiques* from his unpublished papers (the application of the calculus of derivations; formulas concerning polygons and polyhedra; and a study of a special curve, the tractrix); these papers were, moreover, a precious source of inspiration to him.

After having presented a memoir, now lost, on the complete integral of first-order partial differential equations in 1800, Jacques Frédéric Français did not return to mathematics until 1807–1808. He then published, on the urging of Malus, two memoirs on analytic geometry; they treated the equation of the straight line and of the plane in oblique coordinates and the transformation of the systems of oblique coordinates. Applying his method to the famous problem of finding a sphere tangent to four given spheres, he gave a solution to it in 1808, which he corrected immediately before completing it in 1812.

The study of his brother's papers attracted him once more to infinitesimal calculus and especially to the development of the calculus of derivations, to which he in turn devoted two important memoirs: one in 1811 (published in 1813) on the separation of the scales of differentiation and the integration of those functions which they determine; and the other in 1815, on the principles of this calculus. In April 1811 he presented a memoir to the Academy, published in 1812 and 1813 in the *Annales de mathématiques,* in which he put forth an unusual example in the theory of the extrema of functions of several variables. In 1813 he published a rather fully developed study on the rotation of solid bodies, in which, in the words of Cauchy (*Procès-verbaux de l'Académie des sciences,* VIII, 523–525), "interesting research is mixed with several errors."

In September 1813 Français published in the *Annales de mathématiques* a resounding article in which he presents the principles of the geometric representation of complex numbers and draws from them several applications. However, in the final paragraph, he acknowledges having taken a portion of his ideas from a letter of 1806 in which Legendre gave his brother François information about a manuscript study on this same subject that had been entrusted to him by an anonymous young author, and he requested that this author reveal himself. In fact, in the following issue of the *Annales de mathématiques,* this author, Jean-Robert Argand, whose study, although printed, had remained practically unknown, replied with a summary of the main conceptions of his work (*Essai sur une manière de représenter les quantités imaginaires dans les constructions géométriques* [Paris, 1806]). A polemic then arose in the *Annales,* in 1813 and 1814, between Argand himself, Français, and François-Joseph Servois: the first two attempted to justify the principle itself of this geometric representation, while Servois was concerned above all else to preserve the rigor and purity of algebra. These publications had the great merit of widely diffusing an innovation whose essence, although presented by Caspar Wessel in Copenhagen in 1797 (and published in 1799), by the Abbé Adrien-Quentin Buée in London in 1805, and by Argand in Paris in 1806, had remained unnoticed by the leading mathematicians.

Although Jacques Frédéric Français's mathematical publications were interrupted rather suddenly at the end of 1815, it does not seem that his curiosity was extinguished, and Poncelet's long stay in Metz certainly contributed to maintaining it. While not of the first rank, the mathematical activity of the Français brothers merits mention for its originality and diversity.

J. F. Soleirol, in his *Éloge de Monsieur Français . . . prononcé sur sa tombe le 11 mars 1833 . . .* (Metz, 1833), points out that Français also composed a course on military art, a course on geodesy, and two memoirs, one on permanent fortifications and the other on the thrust of the earth.

A final point remains to be made. Upon the death of Arbogast in April 1803, his writings, his important mathematical library, and the rich collection of scientific manuscripts that he had gathered passed to his "nephew," François Français. When François died, he bequeathed this collection of manuscripts and books, augmented by his own writings, to his brother Jacques Frédéric, who announced in December 1823 that they were being placed on sale (see *Bulletin général et universel des annonces . . .* I, fasc. 3 [1823], 493–495). Upon the latter's death the essential portion of the collection, not yet sold, passed into the hands of a bookseller in Metz, at whose shop Count Libri was still able to find various valuable manuscripts in 1839. With the sale of Libri's library, these items were dispersed; some are now in the Biblioteca Medicea-Laurenziana in Florence (in particular, certain papers of François Français) and in the Bibliothèque Nationale de Paris, while others, which are very precious, have not yet been found. It is hoped that a thorough investigation will be undertaken in order to locate them.

BIBLIOGRAPHY

I. Original Works.

(1) François Français. His work amounts to four posthumous memoirs published by his brother in the *Annales de mathématiques pures et appliquées.* They deal with an aspect of the "Calcul des dérivations" ("Méthode de différentiation indépendante du développement des fonctions en séries," in *Annales,* **2** [May 1812], 325-331); with theorems concerning polyhedra and polygons (*ibid.,* **3** [Dec. 1812], 189-191, and **5** [May 1815], 341-350); and with the tractrix (**4** [Apr. 1814], 305-319). Two important memoirs presented to the Académie des Sciences remain unpublished but were known and utilized by different authors—the memoir on the integration of partial differential equations, presented in 1797 (see S. F. Lacroix, *Traité du calcul différentiel et du calcul intégral,* 2nd ed., III [Paris, 1819], 598); and the memoir on the movement of projectiles in resisting media, presented on 26 Nov. 1804, which, moreover, was the object of a flattering report by Biot on 22 April 1805 (*Procès-verbaux de l'Académie des sciences . . .*, III [Hendaye, 1913], 159, 204-205).

(2) Jacques Frédéric Français. His work includes an individual publication, *Mémoire sur le mouvement de rotation d'un corps solide libre autour de son centre de masse* (Paris, 1813); and a series of memoirs published in the *Correspondance sur l'École polytechnique* (*C.E.P.*), the *Journal de l'École polytechnique* (*J.E.P.*), and the *Annales de mathématiques pures et appliquées* of Gergonne (*Annales*). They are listed below by subject and in chronological order.

Analytic Geometry: letter to Hachette, in *C.E.P.,* **1,** no. 8 (May 1807), 320-321; on the straight line and the plane in oblique coordinates, in *C.E.P.,* **1,** no. 9 (Jan. 1808), 337-346; on a sphere tangent to four spheres in the following issues of *C.E.P.:* **1,** no. 9 (Jan. 1808), 346-349; **1,** no. 10 (Apr. 1808), 418-421; **2,** no. 2 (Jan. 1810), 63-66; **2,** no. 5 (Jan. 1813), 409-410; and in *Annales,* **3** (Nov. 1812), 158-161; on the transformation of oblique coordinates, in *J.E.P.,* **7,** *cahier* 14 (Apr. 1808), 182-190; and on various problems, in *C.E.P.,* **2,** no. 2 (Jan. 1810), 60-70.

Infinitesimal Calculus: on a singular case of the theory of the extrema of functions of several variables, in *Annales,* **3** (Oct. 1812), 132-137; and *ibid.,* **3** (June 1813), 197-206; on scales of differentiation and integration, *ibid.,* **3** (Feb. 1813), 244-272; on the calculus of derivations derived from its true principles, *ibid.,* **6** (Sept. 1815), 61-111.

Solid Mechanics: on rotation of solid bodies, in *Annales,* **3** (Jan. 1813), 197-206.

Geometric Representation of Imaginary Numbers: articles in *Annales,* **4** (Sept. 1813), 61-71; **4** (Jan. 1814), 222-227; **4** (June 1814), 364-366; and articles repr. in J. Hoüel, ed. (see below), pp. 63-74, 96-101, 109-110.

Other Topics: problems concerning the calendar, in *Annales,* **4** (Mar. 1814), 273-276, and *ibid.,* **4** (May 1814), 337-338; remarks on the tractrix, *ibid.,* 332-336; and a problem involving the pendulum and the flying bridge, *ibid.,* **6** (Oct. 1815), 126-129.

II. Secondary Literature. On the brothers Français, see M. Chasles, *Rapport sur les progrès de la géométrie* (Paris, 1870), p. 57 (on François), pp. 35, 61 (on Jacques Frédéric); S. F. Lacroix, *Traité du calcul différentiel et du calcul intégral* (Paris, 1819), II, pp. 656-658, 789, III, p. 598, 726, 752 (on François), pp. 631-632, 752 (on Jacques Frédéric); N. Nielsen, *Géomètres français sous la Révolution* (Copenhagen, 1929), pp. 96-97 (on François), pp. 97-103 (on Jacques Frédéric); and the Royal Society *Catalogue of Scientific Papers,* II (London, 1868), 694-695: nos. 4, 7, 14, and 16 are on François, the others on Jacques Frédéric. Baptism records may be found in the municipal archives of Saverne.

On François Français, see (in chronological order), *Almanach national* (later *Almanach impérial*) for *an* VII (1798-1799) to 1810; *Procès-verbaux du Comité d'instruction publique de la convention nationale,* VI (Paris, 1907), 452; *Procès-verbaux de l'Académie des sciences,* III (Hendaye, 1913), 59, 159, 204-205, 262, 504; and J. Joachim, *L'école centrale du Haut-Rhin* (Colmar, 1934), esp. pp. 151-155, where there is partial confusion with Louis François Français, war commissioner. See also the archives of the Département du Haut-Rhin.

On Jacques Frédéric Français, see R. Argand, *Essai sur une nouvelle manière de représenter les quantités imaginaires dans les constructions géométriques,* 2nd ed. (Paris, 1874), pp. v-xvi, 63-74, 96-101, 109-110; S. Bachelard, *La représentation géométrique des quantités imaginaires au début du XIXe siècle* (Paris, 1966), pp. 11-13, 30; C. B. Boyer, *History of Analytic Geometry* (New York, 1956), pp. 222-223; A. Fourcy, *Histoire de l'École polytechnique* (Paris, 1828), p. 403; G. Libri, in *Comptes rendus hebdomadaires des séances de l'Académie des sciences,* **9** (1839), 357-358, and "Fermat," in *Revue des deux mondes* (15 May 1845), pp. 679-707; G. Loria, "Origines, perfectionnement et développement de la notion de coordonnées," in *Osiris,* **8** (1948), esp. 220-223, where there is confusion with Frédéric Louis Lefrançois; and J. V. Poncelet, *Applications d'analyse et de géométrie,* II (Paris, 1864), 592-595; and J. F. Soleirol, *Éloge de Monsieur Français . . . prononcé sur sa tombe le 11 mars 1833 . . .* (Metz, 1833). Further material may be found in *Almanach national* (later *Almanach impérial,* then *Almanach royal*) for *an* VII (1798-1799) to 1833; Férussac, ed., *Bulletin général et universel des annonces et des nouvelles scientifiques,* **1,** fasc. 3 (1823), 493-495; and *Procès-verbaux de l'Académie des sciences,* III-V (Hendaye, 1913-1914), III, 265; IV, 475, 554; V, 152, 168, 524.

Part of the original documentation of this article comes from the archives of the École Polytechnique, the Service Historique de l'Armée, and the archives of the Legion of Honor.

René Taton

FRANCESCA, PIERO DELLA (or **Piero dei Franceschi**), also known as **Petrus Borgensis** (*b.* Borgo San Sepolcro [now Sansepolcro], Italy, between 1410 and 1420; *d.* Sansepolcro, 12 October 1492), *mathematics.*

Vasari's reason (in *Lives of the Artists*) for the adoption of the feminine form "Francesca" in Piero's name has been invalidated by the authentication of his father as "dei Franceschi" and his mother as Romana di Perino da Monterchi. Of Piero's early life—as the wide uncertainty of his birthdate reveals—nothing is known until 7 September 1439, when he was an associate of Domenico Veneziano at Florence. He is not named by Alberti in the famous dedication to his colleagues in *Della pittura* (1436), but Alberti's influence is revealed in the clearest manner in the architectural studies from Piero's workshop at Urbino.

Piero's value to science lay in his pioneering efforts to explore the nature of space and to construct it by his sophisticated study of linear perspective and masterly juxtaposition of color masses. Although his major work on the mathematics of painting, *De prospettiva pingendi,* was written only after his career as an artist was at an end, it can hardly be doubted that the diminution in the successive members of the black and white pavement in the *Flagellation* at Urbino must have been achieved by such complex calculations as are subsequently displayed in the *Prospettiva.* These represent a synthesis of the two operational diagrams that he probably learned from Alberti.

But a more strikingly original contribution of Piero's was his measuring of the distances between successive surfaces of a human head and transferring the plane sections thus obtained into a contoured plan. Luca Pacioli, whose influence in spreading the study of mathematics in the early *cinquecento* is well known, testified to the assistance of his fellow townsman but later paid him the dubious compliment of including (unacknowledged) in his *De divina proportione* a large part of Piero's last work, *De quinque corporibus regolaribus.*

BIBLIOGRAPHY

I. Original Works. The *De prospettiva pingendi* was written in the vernacular; a transcript exists in the Palatina at Parma and forms the basis of the definitive text edited by G. Fasola (Florence, 1942). There is a transcript of the (contemporary) Latin trans. in the Ambrosian Library at Milan. Piero's *De quinque corporibus regolaribus* exists in a transcript (with figs. by him) in the Vatican Library (Urbinas 632). Excerpts from the original works (in English) are available in E. G. Holt, ed., *A Documentary History of Art,* I (New York, 1957), 256–267.

II. Secondary Literature. A detailed and analytical study of Piero's life and work is Roberto Longhi, *Piero della Francesca* (London, 1930), Leonard Penlock, trans. The Introduction by Sir Kenneth Clark to the Phaidon review of his pictures, *Piero della Francesca* (London, 1951), is both readable and scholarly.

William P. D. Wightman

FRANCIS OF MARCHIA (*b.* Appignano, Italy; *fl.* first half of the fourteenth century), *theology, natural philosophy.*

Francis was a Friar Minor (Franciscan) whom Sbaralea identifies as a native of Pignano (Appignano), in the province of Ascoli Piceno, March of Ancona.[1] Many other names were incorrectly attributed to him: he was variously called di Apiniano (Esculo), D'Ascoli (Asculanus), and Rossi (Rubeus). He completed his studies at the University of Paris, where he received his degree as a teacher of theology. In all probability he commented on the *Sentences* during 1319 and 1320 in accordance with the theological program at Paris.[2] Later, around 1328, he was a lecturer at the Studio Generale of the Franciscans at Avignon.[3] In the fifteenth century he was given the honorary title of *doctor succinctus et praefulgens,* which can be seen in the inscriptions on one of the frescoes in the Franciscan convent at Bolzano.

Francis took an active part in the internal struggles regarding poverty that were then dividing the order. Together with Michael of Cesena, William of Ockham, and Bonagrazia of Bergamo, he supported a rule of absolute poverty for the successors of Christ and for the church. He rebelled against Pope John XXII, supporting his opponent, Emperor Louis IV the Bavarian.[4] He was excommunicated by the pope and joined Louis in Pisa in 1328 and once again rebelled to protest his excommunication (1329–1331).[5]

Francis was expelled from the order in 1329. He was persecuted by ecclesiastical authorities in Italy in 1341. In 1344 he made a formal recantation (which was to serve as an example to all later dissidents) and was reconciled with the church and with the order.[6] The date of his death is not known.

Francis' scientific thought is contained in his comments on Aristotle's *Physics* and in his theological writings. In these works he shows an original approach to certain problems of mechanics. He was the first of the medieval philosophers to employ the theory of impetus (anticipating the principle of inertia) to explain the movement of projectiles.[7] The term "impetus" was not used by him in the technical sense as it later was by Jean Buridan; rather he speaks of a force left or impressed (*vis derelicta*), which is the intrinsic cause—transmitted by the motor to the object moved—of the movement of the projectile. He was also the first to maintain that the movement of a projectile is not caused by something extrinsic (*ab alio*) to the object moved. It is not a movement transmitted by the motor to the projectile via the medium through which it moves. He supposes that the *proiciens* leaves in the projectile a part of its force that then causes subsequent motion. In other words, the moving force (*vis motrix*) in the launching of a projectile is not transmitted to the medium (air or water) and thence to the projectile, but rather directly to the body itself.[8]

Francis thus corrects Aristotle's doctrine. The cause of the movement of projectiles is not to be sought in the activity of a force derived through the vortical movement of air or in its heaviness or lightness; it does not depend on the form of the heavens and is not transmitted by the medium.[9] The cause is a *vis derelicta* impressed by the motor on the object itself.[10] The medium contributes to the movement of the projectile but is not the cause of it. The impressed or *derelicta* force is neither a permanent form (such as heat generated by fire) nor is it a *simpliciter fluens* form (that is, one which flows simply, as the heating of water) but rather an intermediate form—that is, one that has a form that lasts for only a limited period of time (*esse permanens ad determinatum tempus*).[11] The movement of the projectile diminishes in speed and exhausts itself not because of the destruction of the *subiectum*—the projectile—but because of the cessation of the motivating force, which occurs in two ways: a pure and simple slackening in the force of the motor or a slackening in the force of the motor in the projectile, whose movement lasts only a short time because of the existential imperfection of the movement.[12] In the latter case the movement of the projectile diminishes just as images impressed on the eye by a source of light are exhausted and disappear when the source of light is removed.[13]

Francis also invoked this principle to explain the movement of the celestial spheres. He suggested the idea that the divine intelligence impresses a driving power of this type on the celestial spheres—that is, an impetus implanted in the heavens themselves.[14] He thus gave a purely mechanical explanation for the movement of the celestial bodies.

Francis was also a proponent of the then new theory of actual infinity. He derived this concept from that of divine cause. There exists an infinity that is positively real, being the effect of divine causality or omnipotence. This is an actual infinite which exceeds any finite beyond any determined proportion, accepted or acceptable.[15] This actual infinite is so according to size, multiplicity, and extension. It is more a transfinite than a maximum.[16] Francis also admits as a variation of this actual infinity an actual infinity according to succession. Movement and time would be in this category and are therefore conceived of by Francis as an actual successive infinity. In other words, on admitting actual infinity according to succession, Francis came to conceive of a world of infinite space, a new concept in medieval cosmology.

NOTES

1. G. Sbaralea, *Supplementum et castigatio ad scriptores,* I (Rome, 1908), 257.
2. Cf. MS Naples, Biblioteca Nazionale, VII, C. 27: "Explicit fratris Francisci de Marchia super primum Sententiarum secundum reportationem factam sub eo tempore, quo legit Sententias Parisius anno Domini 1320."
3. Cf. Etienne Baluze and J. D. Mansi, *Miscellanea,* II (Lucca, 1761), 140.
4. Cf. M. D. Lambert, *Franciscan Poverty: The Doctrine of the Absolute Poverty of Christ and the Apostles* (London, 1961).
5. Cf. MS Florence, Biblioteca Laurenziana, Santa Croce, pluteo 31, sinistra 3, fols. 1–63.
6. Cf. Wadding, VII, 371–372.
7. Cf. Clagett, pp. 530–531.
8. *La teoria dell'impeto,* pp. 59 f. Cf. Maier, *Zwei Grundprobleme,* pp. 168 f., and Clagett, pp. 527 f.
9. *La teoria dell'impeto,* p. 10.
10. *Ibid.,* p. 9.
11. *Ibid.,* p. 11.
12. *Ibid.,* pp. 20–21.
13. *Ibid.,* p. 21.
14. *Ibid.,* pp. 18–19; cf. Clagett, p. 531.
15. *In Sententias,* d. 2, MS, Rome, Biblioteca Vaticana (Chigiano), B. VII 113, fols. 28v–33v; and Vat. lat. 4871, fols. 100r–101v.
16. Cf. A. Maier, *Ausgehendes Mittelalter,* I, 68 f.

BIBLIOGRAPHY

I. Original Works. Most of Francis' writings, not having been published, exist only in MS. The works fall into three categories, according to the subjects with which they are concerned: (1) politics, (2) theology, and (3) science and philosophy. A partial listing of MSS follows.

Political Works. MS Florence, Laurenziana, Santa Croce, pluteo 31, sinistra 3, fols. 1–63, contains his protest against the pope. His formal retraction was published in L. Wadding, *Annales minorum* (Florence, 1932), vol. VII.

Theological Works. Many of Francis' theological writings were in the form of commentaries on the *Sententiae* of Peter Lombard. Various MSS are cited in F. Stegmüller, *Repertorium commentariorum in Sententias* (Würzburg, 1947), pp. 237, 302; V. Doucet, "Commentaires sur les Sentences, Supplément au répertoire de F. Stegmüller," in *Archivum franciscanum historicum,* **47** (1954), 116–117; A. Maier, *Ausgehendes Mittelalter,* I (Rome, 1964), 68 ff.; and P. O. Kristeller, *Iter italicum,* II (London–Leiden, 1967), 445.

Scientific and Philosophical Works. Several of the works in this class took the form of *Quodlibeta* (MS Paris, Bibliothèque Nationale, lat. 16110, sec. XIV); and commentaries on Aristotle, among them *Quaestiones super primum et secundum librum Metaphysicorum* (MS Florence, Laurenziana, Fesulano, supp. 161, fols. 67–73) and *Expositio super Physicam* (Rome, Vaticana Ottoboniano, lat. 1816, fols. 30r–49r). Garcia y Garcia and Piana also attribute Bologna, Biblioteca del Real Collegio di Spagna, MS 104, fols. 48r*a*–102v*b*, to Francis, although P. Kuenzle credits it to Francis of Méyronnes. *La teoria dell'impeto,* G. Federici Vescovini, ed. (Turin, 1969), pp. 1–21, presents *In Sententias IV,* 1, MS Biblioteca Vaticana (Chigiano), B. VII, 113, fols. 175r*a*–177v*a,* as part of a collection of medieval Latin texts dealing with impetus.

II. SECONDARY LITERATURE. On Francis and his works, see M. Clagett, *The Science of Mechanics in the Middle Ages* (Madison, Wis., 1959), pp. 526–531, which includes an English trans. of the text on impetus; F. Ehrle, "Der Sentenzenkommentar Peters von Candia," in *Franziskanische Studien,* supp. IX (1925), pp. 253–259; A. Garcia y Garcia and C. Piana, "Los manuscritos filosofico, historico y cientificos del Real Colegio de Espagna de Bolonia," in *Salmanticensis,* **14** (1967), 81–169; P. Kuenzle, "Petrus Thomae oder Franciscus de Mayronis?," in *Archivum franciscanum historicum,* **61** (1968), 462–463; A. Maier, *Die Vorlaüfer Galileis im 14. Jahrhundert* (Rome, 1949), pp. 133 ff.; *Zwei Grundprobleme der scholastischen Naturphilosophie* (Rome, 1952), pp. 166–180, which contains a portion of the impetus text; *Ausgehendes Mittelalter,* I (Rome, 1964), 357, 461, and II (1967), 467, 478; M. Schmaus, "Der *Liber propugnatorius* des Thomas Anglicus," in *Beiträge zur Geschichte der Philosophie des Mittelalter,* **29** (1930), 34, n. 59; and A. Teetaert, "Pignano (François de)," in *Dictionnaire de théologie Catholique,* XII (1935), cols. 2104–2109.

GRAZIELLA FEDERICI VESCOVINI

FRANCIS OF MEYRONNES (*b.* Méyronnes, Provence, France, *ca.* 1285; *d.* Piacenza, Italy, *ca.* 1330), *theology, natural philosophy.*

The dates of Francis' birth and death are uncertain; he lived in the first half of the fourteenth century and was called de Mayronis, after his birthplace in the canton of St. Paul in the Basses-Alpes. He entered the Franciscan order of Provence, probably in the convent of Digne. Pope John XXII calls him Franciscus de Maironis de Digna in a letter of 23 May 1323.[1] He was a pupil of Duns Scotus and, according to the Benedictine scholar Johannes Trithemius, he taught in England; but it is not known if this was before or after he came to Paris (between 1302 and 1307).[2]

As bachelor of the faculty of theology at Paris, Francis lectured on Peter Lombard's *Sentences* and became a doctor of theology in 1323.[3] He is reported to have inaugurated, between 1315 and 1320, the *actus sorboniens.* This was the scholastic debate that took place every Friday during the summer season and lasted twelve hours consecutively; during these debates, speakers had to respond to and hold their own against all adversaries who appeared.[4] (The adoption of this scholastic practice, however, dates from before 1312 and is traceable to Robert de Sorbon.)[5] The debate between Francis himself and Pierre Roger (later Pope Clement VI) in 1321 was famous and is probably what is referred to as the *certamen Mayronicum.*[6]

In the spring of 1324 Francis was in Avignon. That year Pope John XXII sent him, together with the Dominican monk Domenico Grima, to Gascony to try to prevent a conflict between the armies of Charles IV of France and Edward III of England. Between 1323 and 1324 he was also minister provincial of Provence.[7] His death took place in the convent of Piacenza sometime between 1327 and 1333; according to Roth, he was still alive in 1328.[8]

Francis was a follower of the doctrines of Duns Scotus. His studies included science, metaphysics, and theology. As a theologian he commented on Aristotle's cosmology, correcting it in the light of the physics presented in the Scriptures, a methodological position that was later to be definitively examined and abandoned by Galileo. The facts of Aristotelian physics that Francis inherited from medieval science thus came to be integrated with those of the Bible.[9]

According to Francis, the universe was constituted of fourteen spheres: the empyrean, the crystalline, the firmament, Saturn, Jupiter, Mars, sun, Venus, Mercury, moon, fire, air, water, and earth; each of these has its own composition. At the center of these spheres, which are all in circular motion, is the immobile earth. Francis does not admit the possibility of the earth's movement or the immobility of the heavens, a possibility that was beginning to be argued in Paris—as he himself relates in connection with a teacher who would consider that the earth moves and that the skies are immobile.[10]

The world has been created by God, and Francis does not believe it to be *ab aeterno* or infinite. The reasoning that he uses to disavow the actual infinity of the world is of a philosophical and metaphysical nature. The world, because it was created, was begun, and what has a beginning must also have an end. But that which has an end or a conclusion is not infinite. Therefore, the physical world is not infinite. For this reason, Francis excludes the possibility that the world is actually infinite (*infinitum in actu*);[11] such infinity belongs only to divine omnipotence. Movement, time, discrete quantity, and number—which are determinations of the physical world and are successions—cannot be constituents of the infinite.

Francis does admit that God, as omnipotent infinity, can cause through his infinite power an infinite world, but with the limitation that it be according to continuous quantity and according to intensity or degree. That is, it can never be caused according to numerical succession, for the elements of a series can always be reenumerated. In other words, Francis admits that the physical world is potentially infinite, in the manner of the continuous quantity (infinitely divisible material) that constitutes it.[12]

In his doctrine on the movement of physical bodies Francis does not substantially modify the Aristotelian system of explaining the movement of projectiles. It is a movement that originates from without, and its cause is the medium, with the concurrence of four factors whereby the projectile is moved by whatever pushes it (*motus pellentis*), which divides the medium violently from behind; the medium then closes so that a void does not arise and this closing (*clausio*) pushes the moving body.[13]

More original, on the other hand, is Francis' philosophical explanation of motion, which he understands as a *fluxus formae* (flux of form) rather than as a *forma fluens* (flowing form). In his exposition he embellishes the doctrines of Aristotle with new content.[14]

Francis does not completely accept the fundamental rules of Aristotelian dynamics and kinematics (*Physics* VII 5, 250a, 1–20) as they had been formulated by the medieval scientific tradition. He asserts that the relationship (*comparatio*) established by Aristotle between the force of the mover, space, and time is not true; by means of this relationship it was argued that if a force can move an object in space for a certain time, the same force can move double the object through half the space in the same time. In fact, Francis argues, if, for example, Socrates can carry a quintal for a league, it does not follow that he can carry two quintals for half a league. According to Francis the inverse rule attributed to Aristotle is also not exact; for if a force can move an object in a given space for a certain time, it does not follow that half this force can move the entire object through half the space in the same time. For example, if thirty men can move a ship for thirty paces, it does not follow that fifteen men can move the same ship for fifteen paces in the same time.[15]

NOTES

1. Cf. P. W. Lampen, "Francis de Meyronnes," in *France franciscaine,* **9** (1926), 215–222; E. d'Alençon, "Francis de Meyronnes," in A. Vacant *et al.,* eds., *Dictionnaire de théologie catholique,* X (Paris, 1929), cols. 1634–1646.
2. J. Trithemius, *De scriptoribus ecclesiasticis* (Paris, 1494), fol. 123b.
3. H. Denifle and E. Chatelain, eds., *Chartularium universitatis Parisiensis,* II (Paris, 1891), 272, no. 823; P. Feret, *La faculté de théologie de Paris, Moyen-âge,* III (Paris, 1896), 323, no. 2.
4. Bartholomaeus de Rinonico Pisanus, *De conformitate vitae* IV (Florence, 1906), 339, 523, 540, 544; Gilberti Genebrardi, *Chronographiae libri,* IV (Cologne, 1518), 1014.
5. Cf. Vatican Library, MS Borghese 39; cf. A. Maier, *Ausgehendes Mittelalter,* I (Rome, 1964), 333; II (Rome, 1967), 257 ff.
6. Cf. P. Glorieux, "L'enseignement au moyen-âge, techniques et méthodes en usage à la Faculté de Théologie de Paris," in *Archives d'histoire doctrinale et littéraire du moyen-âge,* **35** (1968), 134.
7. Biblioteca Comunale, Assisi, MS 684. Cf. d'Alençon, col. 1646.
8. B. Roth, *Francis von Meyronnes, sein Leben, seine Werke, seine Lehre vom Formalunterschied in Gott* (Weil in Westfalen, 1936), p. 49.
9. *Commentum in secundum librum Sententiarum* (Venice, 1520), distinctio 14, quaestio V, fol. 150v, cols. a–b.
10. *Ibid.,* fol. 150v, col. b. Cf. P. Duhem, "Francis de Meyronnes et la question de la rotation de la terre," in *Archivum franciscanum historicum,* **6** (1913), 23–25.
11. *Commentum in primum librum Sententiarum,* dist. 43, qu. X, fol. 128v, col. b; dist. 43, qu. IX, fol. 127v, cols. a–b; dist. 44, qu. X, fol. 129v, col. a; *Expositio in Physicam,* bk. III (Ferrara, 1495), fol. Gv, cols. a–b; fol. Kr, cols. a–b.
12. *Ibid.,* fol. Kr, cols. a–b; fol. K^{ii}r, col. a ff., *Commentum in primum librum Sententiarum, loc. cit.*
13. *Ibid.,* dist. 14, qu. VII, fol. 151r, col. b.
14. *Ibid.,* dist. 16, qu. IV, fol. 68r, col. b; dist. 14, qu. IX, fol. 152r, col. b; *Expositio in physicam,* bk. V, fol. Mv, col. a.
15. *Ibid.,* bk. VII, fol. Ov, cols. a–b.

BIBLIOGRAPHY

I. ORIGINAL WORKS. Francis of Meyronnes was the author of numerous writings on various subjects, including theology, metaphysics, logic, physics, politics, and piety; almost all are in MS or in eds. published at the end of the fifteenth and beginning of the sixteenth centuries. There are no modern eds. of his scientific works. For a detailed indication of the MSS and first eds. of all his works, see B. Roth, pp. 50 ff. (see n. 8); on his political thought, see P. de Lapparent, "L'oeuvre politique de Francis de Méyronnes," in *Archives d'histoire doctrinale et littéraire du moyen âge,* **13** (1942), 57–74. On his comments to the *Sentences,* see F. Stegmüller, *Repertorium commentariorum in libros Sententiarum* (Würzburg, 1947), and V. Doucet, "Commentaires sur les Sentences, Supplément au répertoire de F. Stegmüller," in *Archivum franciscanum historicum,* **47** (1954), 114–116; and annals of the *Archivum:* XLVI, 164–166, 342; XLVII, 98, 114–116; 149, 153, 403; L, 203; LIV, 230; LV, 369, 531; LVI, 209; LVII, 363, 408, 573 (indications of MSS of his commentaries on Aristotle); LVIII, 187, 192, 264, 408; LIX, 86; LX, 263, 450, 466; LXI, 462–463. See also P. O. Kristeller, *Iter italicum,* I (London, 1963), 76, 312, 317, 420; II, 71, 216, 326, 390, 413, 465–466.

Francis' scientific thought is contained in his commentaries on Aristotle, especially in *Expositio in physicam* (Ferrara, 1495; Venice, 1517), in his comments to the *Sentences;* and in the *quaestiones quodlibetales.* For indications of MSS and rare eds., see especially Roth.

II. SECONDARY LITERATURE. Bibliographical indications are in Roth, which is the most nearly complete study to date. For more recent indications of Francis' political and theological thought, see the annals of the *Archivum franciscanum historicum, loc. cit.* On his scientific thought, in addition to the study by Duhem cited in note 10, see his *Système du monde,* vols. VI–X (Paris, 1956–1959), *passim,* and particularly VI, 451–474. See also A. Maier, *Zwischen Philosophie und Mechanik* (Rome, 1958), p. 96; *Ausgehendes Mittelalter,* I (Rome, 1964), 71, 247, 468; and *Zwei Grundprobleme der scholastischen Naturphilosophie* (Rome, 1951), pp. 51, 53, 56, 164, 197, 232, 238. See also B. Nardi, "La filosofia della natura nel Medioevo," in *Acts of the*

Third International Conference of Medieval Philosophy, La Mendola, 1964 (Milan, 1966), p. 23.

GRAZIELLA FEDERICI VESCOVINI

FRANCK, JAMES (*b.* Hamburg, Germany, 26 August 1882; *d.* Göttingen, Germany, 21 May 1964), *physics.*

Franck was the son of Jacob Franck, a banker, and Rebecca Franck. His scientific activity extended over about sixty years, from the beginning of the twentieth century, when the foundations of atomic physics and quantum theory were being laid, to a time when these disciplines had reached a high degree of sophistication. Although Franck was primarily a physicist, his work had a profound influence on chemistry and on the branch of biology concerned with the fundamental process by which the energy of sunlight is converted into the forms of energy that maintain life on earth. In all the varied phenomena that he studied one can recognize a unity of approach in his attempt to understand the processes of transfer of energy in atomic systems.

In the two semesters (1901–1902) of his studies at Heidelberg, Franck met Max Born and formed a friendship with him that lasted throughout his life. His serious study and research in physics began in 1902 when he moved to Berlin, at that time the center of physics in Germany. Rubens, Emil Warburg, and Planck (later Drude and Einstein) were professors in Berlin, and their joint colloquium was one of the great formative influences in Franck's life. He entered Warburg's laboratory and started work on corona discharges, a topic he soon abandoned in favor of the more fundamental study of ion mobilities. He found that collisions of electrons with noble gas atoms were mainly elastic, without loss of kinetic energy. His younger colleague, Gustav Hertz, joined him in a thorough study of elastic collisions, and this work led to the discovery of quantized transfer of energy in inelastic collisions between electrons and atoms. In their famous experiments, Franck and Hertz[1] showed that electrons could impart energy to a mercury atom only if they had a kinetic energy exceeding 4.9 ev., and that exactly this quantum of energy was taken up by the mercury atom, causing it to emit light of the resonance line Å 2537. It was the first direct proof of the quantized nature of the energy transfer and of the connection of the quantum ΔE of energy with the frequency $\nu = \Delta E/h$ of the light emitted as the result of the transfer. These experiments are rightly regarded as the first decisive proof of the reality of the quantized energy levels that had just been postulated by Niels Bohr. Misled by the observation of ions in their experiments and in those of other workers in the same conditions, Franck and Hertz initially believed ionization to occur simultaneously with emission of resonance radiation, so that $h\nu$ was to be regarded as the ionization energy; this was in accordance with current speculations by Stark and others but contradicted Bohr's theory.

The outbreak of World War I interrupted most scientific work and exchange of ideas. Franck served briefly in the German army but became seriously ill and was sent home to recover. It was probably due to this interruption of scientific activities and contacts in Europe that Franck and Hertz held to their views on ionization as late as 1916. The spurious origin of the ions was proved mainly by work in the United States and was recognized after the war by Franck and Hertz.[2] The fundamental importance of their experiments was acknowledged in 1926 by the award of the Nobel Prize to Franck and Hertz.

From 1917 to 1921 Franck was assistant professor and head of a section of the Kaiser Wilhelm Institut für Physikalische Chemie (later the Max Planck Institut), whose director was Fritz Haber. With a number of co-workers he extended the study of inelastic collisions of electrons with atoms and molecules and measured excitation and ionization potentials. With Knipping and Reiche he introduced the concept of metastable levels, excited states that can lose energy not by radiation but only by collisions. They play an important part in gas discharges and many other phenomena. The postwar years in Berlin marked the beginning of Franck's friendship with Niels Bohr, for whom he had a profound admiration as a scientist and a warm affection. The obituary of Bohr that Franck wrote not long before his own death[3] is a moving testimony to their friendship.

In 1921 Franck accepted the chair of experimental physics and directorship of the Zweite Physikalische Institut in Göttingen, where R. Pohl occupied the other chair as director of the Erste Physikalische Institut, located in the same building. Max Born had just accepted the chair of theoretical physics on the condition that a chair and department be established for Franck. For the next twelve years Franck and Born, linked by close ties of friendship and common interests, formed the nucleus of an active scientific community in Göttingen.

A central theme in the great variety of publications of that period may be described as the study of atoms in collision, and the formation and dissociation of molecules and their vibration and rotation. In two papers[4, 5] Born and Franck developed the use of the now familiar potential energy curves for treating two-atom systems, and they introduced the concept of quasi-molecules. Applying these ideas to the transfer of energy from electronic to vibrational motion in molecular spectra, Franck was led to the method of determining the energy of dissociation of molecules by extrapolation of vibrational levels and to the principle which, after its wave-mechanical formulation by Condon, became known as the Franck-Condon principle. It has since provided the key to the understanding of a wide range of phenomena in molecular physics, such as continuous molecular spectra, the intensity distribution in band spectra, predissociation, photodissociation, and pressure broadening of spectral lines.

Problems of energy transfer in collisions had occupied Franck since he started research, and in 1926 his only publication in book form[6] appeared; written with P. Jordan, it contains the basic ideas of most of his work to that date.

Political events in Germany in 1933, after Hitler came to power, brought most of the scientific work in Göttingen to an abrupt end. Franck, although Jewish, was initially allowed to continue in office, but new legislation would have forced him to dismiss co-workers and students who were either non-Aryan or politically committed. He refused to accept this, and in April 1933 he resigned his professorship and published a courageous statement of protest against the new laws. Within a few months not only Franck and Born but most of their co-workers had left Germany.

After spending over a year in Copenhagen, Franck immigrated to the United States in 1935 and accepted a professorship at Johns Hopkins University in Baltimore. In 1938 he was appointed professor of physical chemistry at the University of Chicago, where the Samuel Fels Foundation had established a laboratory for photosynthesis; he directed it until his retirement in 1949 and took an active part in it long afterward. At Göttingen and Baltimore, Franck and his colleagues had begun to extend the understanding of excitation and photodissociation from diatomic molecules to liquids and solids and finally to the process of photosynthesis in plants. This work was bound to involve Franck in all the complexities of biochemistry, but it attracted him by its fundamental importance. His contribution to the exciton theory and the photographic process, made jointly with Teller,[7] also belongs to this period, but it was to the problem of photosynthesis that most of his remaining work was devoted.

During World War II, he joined the metallurgical project in Chicago, which formed part of the atomic bomb project. After the surrender of Germany, he and many other scientists working on the project became seriously concerned about the consequences of using the new weapon. In a document later released and known as the Franck Report,[8] they urged the government to consider the use of the bomb a fateful political decision and not merely a matter of military tactics. After the end of the war, Franck resumed his research at Chicago. His wife Ingrid had died in 1942 after a long illness, and in 1946 he married Hertha Sponer, professor of physics at Duke University.

The work on photosynthesis involved Franck in much controversy. On the experimental side, he rejected the measurements of Warburg as being in conflict with basic thermodynamic principles and in disagreement with the work at other laboratories. On the theoretical side, Franck developed a model that assumed a two-step process in one single chlorophyll molecule and accounted for most of the experimental facts, although some details of his views are still contested. The award of the Rumford Medal of the American Academy of Arts and Sciences in 1955 for his work on photosynthesis showed the increasing recognition of his contribution to this field. It was one of the numerous honors he received in addition to the Nobel Prize and memberships in academies and learned societies, including the Royal Society of London. Honors also came to him after World War II from Germany: he received the Max Planck Medal of the German Physical Society and was made an honorary citizen of Göttingen, where he died while on a tour of Germany to visit old friends.

NOTES

1. J. Franck and G. Hertz, in *Verhandlungen der Physiologischen Gesellschaft zu Berlin,* **16** (1914), 512.
2. J. Franck and G. Hertz, in *Physikalische Zeitschrift,* **20** (1919), 132.
3. J. Franck, "Niels Bohr's Persönlichkeit," in *Naturwissenschaften,* **50** (1963), 341.
4. J. Franck and M. Born, in *Annalen der Physik,* **76** (1925), 225.
5. J. Franck and M. Born, in *Zeitschrift für Physik,* **31** (1925), 411.
6. J. Franck and P. Jordan, *Anregungen von Quantensprüngen durch Stösse* (Berlin, 1926).
7. J. Franck and E. Teller, in *Journal of Chemical Physics,* **6** (1938), 861.
8. Franck Report, *Bulletin of the Atomic Scientists* (*of Chicago*), **1,** no. 10 (1946), 1-5.

BIBLIOGRAPHY

A more detailed biography of James Franck is H. G. Kuhn, "James Franck 1882–1964," in *Biographical Memoirs of Fellows of the Royal Society,* **11** (1965), 53–74; it includes a complete bibliography by R. L. Platzman.

H. G. KUHN

FRANCK, SEBASTIAN (*b.* Donauwörth, Bavaria, Germany, 20 January 1499; *d.* Basel, Switzerland, 1542), *theology.*

We have no precise information about Franck's parents and early life. He may have attended the grammar school in Nördlingen before matriculating, in March 1515, in the Arts Faculty of the University of Ingolstadt, where he received a humanistic education that included Latin and Greek but not Hebrew. After graduation in December 1517, he went to Heidelberg in January 1518 to study theology at the Dominican college that was incorporated with the university. The theological faculty was then dominated by Aristotelian Scholasticism, but a few months after his arrival Franck heard the new Augustinian voice when he attended Luther's famous Heidelberg disputation. Among his fellow students were his later opponents Martin Frecht and the Strasbourg reformer Martin Bucer. We do not know when Franck left Heidelberg. He entered the Catholic priesthood, but by the end of 1527 he was a Protestant pastor in Gustenfelden, near Nuremberg. In 1528 he married Ottilie Beham, the sister of Albrecht Dürer's pupils Barthel and Hans Sebald Beham, both known for their Anabaptist leanings and for their association with Hans Denck. At this time, during the turbulent years that followed the Peasants' War, Franck adopted the spiritualist views that put him in strong opposition to Luther.

He left his pastorate before the end of 1528, and for the rest of his life he earned his living as a popular writer, printer, and, for a while, soapmaker, wandering from place to place with his family as he was banned from one town after another for his unorthodox writings. For a while he was in Nuremberg, where he published his first writings, but in 1529 he was in Strasbourg, then known for its liberal religious atmosphere and as a gathering place for radical reformers, who during these years included Michael Servetus, Hans Bünderlin, and Kaspar von Schwenkfeld. Here Franck published his great work *Chronica, Zeitbuch und Geschichtbibel* (1531), which immediately brought complaints from many sides, including one from Erasmus, whom Franck greatly admired. The book was confiscated, Franck was arrested, and at the end of the year he was expelled from Strasbourg.

In 1534 Franck became a citizen of Ulm, where he almost immediately faced new difficulties owing to his publications there, the *Paradoxa ducenta octoginta, das ist CCLXXX Wunderred und gleichsam Räterschaft, aus der Heiligen Schrift* (1534), his most characteristic theological work, and *Das theur und künstlich Büchlein Morie encomion* (1534), which Franck also called "die vier Kronbüchlein." It developed his spiritual doctrine by showing that all worldly piety and wisdom are folly before God. This work contained German versions of Erasmus' *In Praise of Folly* and of Agrippa von Nettesheim's *De incertitudine et vanitate omnium scientiarum et artium* as well as two pieces by Franck, *Ein Lob des thörichten Göttlichen Worts* and *Vom Baum des Wissens Gutes und Böses,* which alone among Franck's writings has been published in English. It was translated by the mystic John Everard, under the title *The Forbidden Fruit: Or a Treatise of the Tree of Knowledge* (1640).

Franck was now accused of rejecting the efficacy of preaching and the authority of the Bible but was allowed to remain in Ulm, provided he submitted to censorship. He published his next books in Tübingen, Augsburg, and Frankfurt. They included a book of geography (including the New World), folklore, and anthropology, *Weltbuch: Spiegel und Bildniss des ganzen Erdbodens* (1534); a history and description of Germany called *Germaniae chronicon* (1538); and *Die Güldin Arch* (1538), which is a sort of concordance to Scriptures designed to awaken the reader to the inward word. It was followed by *Das verbütschierte mit 7 Siegeln verschlossene Buch* (1539), a "discordance" in which Franck deliberately juxtaposed contradictory Scriptural passages. His presence in Ulm again became controversial and, with Schwenkfeld, he was banished from the city in January 1539. With his wife, his ten children, and his printing press he left for Basel, where he was allowed to live until his death in 1542. His first wife died in 1540, and the following year he married Margarete Beck, the daughter of the printer Reinhard Beck and the stepdaughter of Balthasar Beck, who had printed Franck's *Chronica.* Franck's last works, including a collection of German proverbs, were published at Basel.

Franck's thought presents a mixture of theology and philosophy. It was neither highly original nor entirely consistent, but what he borrowed he molded into a powerful and unusual statement of the spiritual freedom and self-sufficiency of the individual. It was guided by the principle that all men, regardless of time and place, are given equal capacity for moral, intellectual, and religious insight. From the beginning of history, God has in His creation revealed Himself unambiguously and uniformly to all mankind, and this revelation in nature is surer and more universal testimony to His power, wisdom, and goodness than the Scriptures are. The Scriptures are recorded in the dead letter of writing, full of contradictions and available only to part of mankind. Since God is impartial, faith and divine favor cannot depend on Scripture alone.

Hence the Incarnation and the historical Christ have no place in Franck's theology. Adam and Christ, the flesh and the spirit, the outward and the inward man, are qualities that lie in human nature. The common creation of the macrocosm and the microcosm ensures their conformity as well as the uniformity of human nature. Thus all men are born with the ability to gain divine insight. Reason, or the light of nature, combined with experience is the means man has been granted to gain this insight. Taken together with his individualism, this aspect of Franck's thought has a rationalist and naturalist quality that is absent in his German contemporaries but somewhat reminiscent of the philosophy of the late seventeenth century.

Franck is generally grouped with the spiritual reformers or mystics of his own century because, like them, he rebelled against the increasing dogmatism and growing institutional rigidity of the Protestant churches. But his rationalism and radical universalism set him apart from them. Unlike the true mystics—Valentin Weigel and Jacob Boehme, for instance—Franck never claimed authority and special insight by virtue of some unique personal revelation; the knowledge that was open to him was open to all. He was no enthusiast. Similarly, he had no fondness for esoteric and cabalistic lore, did not engage in fanciful verbal mysticism, and had no predilection for magic.

Although it may appear curious, Franck's outlook resembles that of John Locke (no influence is postulated). Both believed that the play of reason on experience was a God-given and certain avenue to all knowledge, both natural and moral; that this knowledge was equally open to all mankind; and that it agreed with the moral and religious precepts of the New Testament. Both had profound doubts about the Trinity and the divinity of Christ; both insisted on God's impartiality and therefore gave tolerance a prominent place in their concerns; and both took an interest in comparative anthropology. Their agreement on so many fundamental points is summed up in their mutual abhorrence of enthusiasm as an enemy of reason and tolerance. This position is not incompatible with the chiliasm which Franck shared with so many in his own and the following century, although it did not play a prominent role in his writings. Chiliasm gave a strong impulse to rationalism; being God-given and Godlike, only reason can discover the proper method for the speedy increase of knowledge.

Franck taught the most radical form of spiritualism. True belief depends on the illumination of the individual soul by the inward spirit, which is also called Christ, truth, and the inward word. Having created man in his own image, God has planted this spirit in man and has made it innate. All men are in this respect equal, whether they have heard the outward word or not. Just as many who have never heard of Adam live according to the flesh, so many who have not heard of Christ are filled with the spirit. God is impartial. He is wholly love, and this love is extended to all of creation, which gives testimony to his love and power. God is essentially without will. Self-will entered the world with the Fall, but the loving God will not use force against it. Union with God can occur only when man is rightly moved by the spirit, making himself altogether empty of will and thus becoming independent of outward things. This will-less state of the soul is called *Gelassenheit,* a term common among the spiritual reformers, who found it in the writings of late medieval mystics. It was also used by Luther and later regained importance among the Pietists. To ensure that man is indeed capable of this will-less spiritual state, Franck argued that man can actively exercise free will by prevenient grace alone, that is, grace before conversion and baptism. Predestination and election are contrary to the essential love and impartiality of God.

Within these terms, so very different from Luther's, Franck agreed with Luther that justification occurs by faith alone, but he firmly rejected Luther's scripturalism (the necessity of the outward word, even though insufficient, prior to the awakening of the inward spirit). God's word and truth cannot be written and read, spoken and taught. As reason will not submit to written rules, so the spirit cannot be contained in the dead letter. The only possible church is the invisible church of individual believers, each man gaining faith by his own private efforts. All men being endowed with the spirit and allowed full freedom of will, they are genuinely capable of making a responsible and free choice. The Old and New Testaments are written in the hearts of all men. When Franck revealed, in the *Paradoxa* and in other works, the contradictions contained in the Bible, he sought to weaken man's faith in the dead letter in order to guide him toward his own reliance on the spirit within. In this sense Franck can be said to have advocated an extreme and fundamental form of individualism. Still, the importance he attached to the Bible (when understood with spiritual guidance) is sufficiently strong to relieve him of mere pantheism.

Although Franck held that the hidden God can be understood only in the truth and faith that are the fruits of spiritual insight, he did not believe that this insight can be gained directly. Guided by the light of nature or reason, the experience of outward things, whether they were made by God or man, is the means by which man may learn to find the truth that lies hidden behind the mask of appearance. The world of man and his institutions has always been dominated by man's will, except among the apostles. It is the world of Antichrist, and it is an unending record of chaos, decay, violence, and intolerance. The events of history offer instruction when they are seen as the very opposite of truth, a view that forms the powerful theme of the *Chronica.* The world, Franck says in the preface, is God's carnival play; appearance is the reverse of truth. The most characteristic part of this work is the book devoted to the men who have been judged heretics by the Roman Church. Among them Franck included Luther, Erasmus, Zwingli, and the Anabaptists. As victims of mere human authority, will, and force, all heretics have become witnesses to truth by following their own consciences.

Tolerance and impartiality are duties man owes to man by virtue of being created in the image of God. Experience, truth, and faith will always be private and individual. For this reason Franck made no basic distinction between pagans and Christians. He accorded equal significance to citations from the Bible and from Plato, Seneca, Proclus, Plotinus, and Hermes Trismegistus, whom he knew from Ficino's Latin translation and commentary. They saw the good by means of experience and the inner light or reason that is common to all mankind. Although he cited them often, Franck admitted no special authority for the Church Fathers. He agreed with Erasmus that the wisdom of the learned is folly before God. In line with what may perhaps be called Franck's democratic spiritualism, he found wisdom in the proverbs of the common folk.

Franck's position in his own time was as independent as his theology. Although he had much sympathy for the Anabaptists, he was as little inclined to join them as any other sect. He was strongly influenced by Erasmus and the young Luther, as well as by late medieval German mysticism, especially by Johannes Tauler and the *Theologia Germanica.* He was indebted to a number of his contemporaries among the spiritual reformers, especially Hans Denck, Johann Bünderlin, and Michael Servetus. The distinctive quality of his thought was determined by his heavy debt to Renaissance humanism and the Neoplatonic tradition. It was his special accomplishment to make this tradition available to the public at large. As a writer of German prose, Franck was second only to Luther.

Franck and his teachings were condemned in the strongest terms by Luther, Zwingli, Calvin, and Melanchthon. He naturally formed no sect, but his writings, although often banned and burned, were reprinted with some frequency in Germany. Yet it was in the Netherlands that Franck gained his greatest following. He had a direct influence on David Joris and Dirck Coornhert, and some of his writings are preserved only in translations made for the Dutch spiritualists and published at Gouda. Valentin Weigel cites Franck with approval, but the true extent of his influence on later German spiritualists is not easily determined. The title of Gottfried Arnold's *Unparteiische Kirchen- und Ketzer-Historie* (1699–1700) is a reminder that Franck was not forgotten by the Pietists.

BIBLIOGRAPHY

I. Original Works. Although not complete, the best list of Franck's works is Karl Goedeke, *Grundrisz der Deutschen Dichtung aus den Quellen,* II, *Das Reformationszeitalter,* 2nd ed. (Dresden, 1886), 8–14. There is an unsatisfactory edition of the *Paradoxa* by Heinrich Ziegler (Jena, 1909). The original works are rare, but good selections will be found in the following: G. H. Williams, ed., *Spiritual and Anabaptist Writers* (Philadelphia, 1957), pp. 145–160 ("A Letter to John Campanus"); Heinold Fast, ed., *Der Linke Flügel der Reformation* (Bremen, 1962), pp. 217–248 ("Letter to Campanus" and preface to the book on the Roman heretics in *Chronica*); Kurt von Raumer, ed., *Ewiger Friede, Friedensrufe und Friedenspläne seit der Renaissance* (Freiburg–Munich), pp. 249–288 (extensive excerpts from *Kriegbüchlein des Friedes* [1539]). Good excerpts from various sources are also in Peter Meinhold, ed., *Geschichte der kirchlichen Historiographie,* I (Freiburg–Munich, 1967), 301–310; and in Ernst Staehelin, ed., *Die Verkündigung des Reiches Gottes in der Kirche Jesu Christi,* IV (Basel, 1957), 342–356.

II. Secondary Literature. The secondary literature is listed in Karl Schottenloher, *Bibliographie zur Deutschen Geschichte im Zeitalter der Glaubensspaltung 1517–1585,* I (Leipzig, 1933), 263–266; V (Leipzig, 1939), 92; VII (Stuttgart, 1962), 79–80. See also E. Teufel, "Die *Deutsche Theologie* und Sebastian Franck im Lichte der neueren Forschung," in *Theologische Rundschau,* **12** (1940), 99–129.

The standard work on Franck and still the best is Alfred Hegler, *Geist und Schrift bei Sebastian Franck, eine Studie zur Geschichte des Spiritualismus in der Reformationszeit* (Freiburg, 1892). This work should be supplemented by Hegler's "Sebastian Franck," in Albert Hauck, ed., *Realencyklopädie für protestantische Theologie und Kirche,* 3rd ed., VI (Leipzig, 1899), 142–150. The best biography is E. Teufel, *"Landräumig" Sebastian Franck, ein Wanderer an Donau, Rhein und Neckar* (Neustadt an der Aisch, 1954). Somewhat diffuse, with extensive quotations from the works, is Will-Erich Peuckert, *Sebastian Franck, ein Deutscher Sucher* (Munich, 1943). Wilhelm Dilthey devoted an influential section to Franck in *Weltanschauung und Analyse des Menschen seit Renaissance und Reformation, Gesammelte Schriften,* II (Leipzig–Berlin, 1914), 81–89.

General aspects are dealt with in Rudolf Stadelmann, *Vom Geist des ausgehenden Mittelalters, Studien zur*

Geschichte der Weltanschauung von Nikolaus Cusanus bis Sebastian Franck (Halle, 1929). A special issue of *Blätter für Deutsche Philosophie,* **2** (1928–1929), was devoted to Franck. See also Alexandre Koyré, "Sébastien Franck," in *Mystiques, Spirituels, Alchimistes du XVIe siècle allemand* (Paris, 1955), pp. 21–43; and Walter Nigg, *Das Buch der Ketzer* (Zurich, 1949), pp. 382–392. G. H. Williams, *The Radical Reformation* (Philadelphia, 1962), deals with Franck on pp. 264–268, 457–466, and 499–504. Good general introductions are offered in Rufus M. Jones, *Spiritual Reformers of the 16th and 17th Centuries* (New York, 1914), pp. 46–63; and Doris Rieber, "Sébastian Franck," in *Bibliothèque d'humanisme et renaissance,* **20** (1958), 218–228.

Special topics are dealt with in Kuno Räber, *Studien zur Geschichtsbibel Sebastian Francks* (Basel, 1952), which is vol. XLI in Basler Beiträge zur Geschichtswissenschaft. Joseph Lecler, *Histoire de la tolérance au siècle de la réforme,* I (Paris, 1955), 177–187, is excellent. Meinulf Barbers, *Toleranz bei Sebastian Franck* (Bonn, 1964), has a good bibliography (this is n.s. 4 in Untersuchungen zur allgemeinen Religionsgeschichte). See also Robert Stupperich, "Sebastian Franck und das münsterische Täufertum," in Rudolf Vierhaus and Manfred Botzenhart, eds., *Dauer und Wandel der Geschichte . . . Festgabe für Kurt von Raumer zum 15. Dezember 1965* (Münster, 1966), pp. 144–162. On Gottfried Arnold and Franck, see Erich Seeberg, *Gottfried Arnold, die Wissenschaft und die Mystik seiner Zeit* (Meerane, 1923; repr. Darmstadt, 1964), pp. 516–534. Franck's role in the study of comparative anthropology and folklore is demonstrated in Erich Schmidt, *Deutsche Volkskunde im Zeitalter des Humanismus und der Reformation,* Historische Studien, E. Eberling, ed., no. 47 (Berlin, 1904), pp. 108–131.

There is an excellent chapter on the general outlook of the Anabaptists in Claus-Peter Clasen, *Die Wiedertäufer im Herzogtum Württemberg und in benachbarten Herrschaften: Ausbreitung, Geisteswelt und Soziologie* (Stuttgart, 1965), pp. 69–117.

HANS AARSLEFF

FRANK, PHILIPP (*b.* Vienna, Austria, 20 March 1884; *d.* Cambridge, Massachusetts, 21 July 1966), *physics, mathematics, philosophy of science, education.*

Frank obtained his doctorate in physics in 1907 from the University of Vienna as a student under Ludwig Boltzmann. Frank later wrote of this period:

> . . . the domain of my most intensive interest was the philosophy of science. I used to associate with a group of students who assembled every Thursday night in one of the old Viennese coffee houses. . . . We returned again and again to our central problem: How can we avoid the traditional ambiguity and obscurity of philosophy? How can we bring about the closest possible *rapprochement* between philosophy and science?

As a physicist Frank was a creative contributor, working on fundamental problems of theoretical physics during an exciting period of its growth. Perhaps his most widely known publication of those years was the two-volume collection, edited with his lifelong friend Richard von Mises, *Die Differential- und Integralgleichungen der Mechanik und Physik.* Frank's own research was concerned with variational calculus, Fourier series, function spaces, Hamiltonian geometrical optics, Schrödinger's wave mechanics, and relativity theory. In an early paper with Hermann Rothe he derived the Lorentz transformation equations without assuming constancy of light velocity from the fact that the equations form a group.

But his first and most lasting love was the philosophy of science. From the beginning Frank was intrigued by Poincaré's neo-Kantian idea that many basic principles of science are purely conventional. In 1907 Frank took the bold step of using that idea to analyze the law of causality. This paper attracted Einstein's attention and started a lasting friendship. In 1912 Einstein recommended Frank as his successor as professor of theoretical physics at the German University of Prague, a position Frank held until 1938. Frank's original paper on causality—which Lenin criticized in his 1908 book on positivist philosophy and the sciences—was later expanded into his widely influential work *Das Kausalgesetz und seine Grenzen* (1932). In 1947 Frank published an authoritative biography, *Einstein: His Life and Times.*

Frank was a logical positivist, although a less doctrinaire one than many of those with whom he formed the Vienna circle in the 1920's. The breadth of interest which he exhibited in his work and fostered in his students made science a liberal discipline and reflected a style of life as well as of mind. As he once remarked, he sought always to achieve a balanced outlook on man and nature; and for him physics not only provided reliable answers to particular technical problems but also raised and illuminated important questions concerning the nature, scope, and validity of human knowledge. Indeed, Frank believed that a stable perspective on life can best be achieved through the critical, intellectual method of modern natural science.

He therefore saw it as a misfortune that science and philosophy are widely regarded as unrelated and incongruous. But it was also his conviction that this breach between a scientific and a humanist orientation toward life—a breach that he thought to be of relatively recent origin—could be diminished, if not overcome, by an adequate philosophy of science.

Holding that the meaning and validity of theoretical assumptions can be determined only if detailed consideration is given to the verifiable consequences which the assumptions entail, Frank called attention to certain misinterpretations of relativity theory and quantum mechanics and their fallacious use in support of questionable doctrines. The titles of some of his works indicate these concerns—"Das Ende der mechanistischen Physik" (1935), *Interpretations and Misinterpretations of Modern Physics* (1938), and *Philosophy of Science: The Link Between Science and Philosophy* (1957).

Frank was organizer or chief participant in the *International Encyclopedia of Unified Science,* the Philosophy of Science Association, *Synthèse,* the Institute for the Unity of Science, and the Boston Colloquium for the Philosophy of Science.

In 1938 Frank and his wife, Hania, came to the United States. After serving as a visiting lecturer, he remained as lecturer on physics and mathematics at Harvard, where his influential course on philosophy of science, his erudite mastery, and his warm and witty manner were remembered long after his retirement in 1954.

BIBLIOGRAPHY

I. ORIGINAL WORKS. Frank's books include *Die Differential- und Integralgleichungen der Mechanik und Physik,* 2 vols. (Brunswick, 1925; last rev. ed., 1935), trans. into Russian (Moscow, 1937), written with Richard von Mises; *Das Kausalgesetz und seine Grenzen* (Vienna, 1932), also trans. into French (Paris, 1937); the collection of papers in philosophy of science, *Between Physics and Philosophy* (Cambridge, Mass., 1941), later repr. and enl. as *Modern Science and Its Philosophy* (New York, 1949); *Einstein: His Life and Times* (New York, 1947; rev. 1953), published in German (Munich, 1949); *Relativity: A Richer Truth* (Boston, 1950); and *Philosophy of Science: The Link Between Science and Philosophy* (Englewood Cliffs, N.J., 1957).

Frank's papers in theoretical physics include "Das Relativitätsprinzip und die Darstellung der physikalischen Erscheinungen im vierdimensionalen Raum," in Ostwald's *Annalen der Naturphilosophie,* **10** (1911), 129–161; "Die statistische Betrachtungsweise in der Physik," in *Naturwissenschaften,* **7** (1919), 701–740; "Über die Eikonalgleichung in allgemein anisotropen Medien," in *Annalen der Physik,* 4th ser., **84** (1927), 891–898; "Relativitätsmechanik," in *Handbuch für physikalische und technische Mechanik,* II (Leipzig, 1928), 52 ff.; "Die Grundbegriffe der analytischen Mechanik als Grundlage der Quanten- und Wellenmechanik," in *Physikalische Zeitschrift,* **30** (1929), 209–228; "Statistische Mechanik Boltzmanns als Näherung der Wellenmechanik," in *Zeitschrift für Physik,* **61** (1930), 640–643, written with W. Glaser.

His epistemological writings include "Kausalgesetz und Erfahrung," in Ostwald's *Annalen der Naturphilosophie,* **6** (1908), 445–450; "Über die Anschaulichkeit physikalischer Theorien," in *Naturwissenschaften,* **16** (1928), 122–128; "Was bedeuten die gegenwärtigen physikalischen Theorien für die allgemeine Erkenntnislehre?," *ibid.,* **17** (1929), 971–977; "Das Ende der mechanistischen Physik," in *Einheitswissenschaft,* **5** (1935), 23–25; "The Mechanical Versus the Mathematical Conception of Nature," in *Philosophy of Science,* **4** (1937), 41–74; *Interpretations and Misinterpretations of Modern Physics* (Paris, 1938); "Physik und logischer Empirismus," in *Erkenntnis,* **7** (1938), 297–301; *Foundations of Physics,* I, no. 7 of the *International Encyclopedia of Unified Science* (Chicago, 1946); and "Metaphysical Interpretations of Science," in *British Journal for the Philosophy of Science,* **1** (1950), 60–91.

Frank's papers on sociological and cultural aspects of science include "Mechanismus oder Vitalismus? Versuch einer präzisen Formulierung der Fragestellung," in Ostwald's *Annalen der Naturphilosophie,* **7** (1908), 393–409; "Die Bedeutung der physikalischen Erkenntnistheorie Machs für das Geistesleben der Gegenwart," in *Naturwissenschaften,* **5** (1917), 65–72; "The Philosophical Meaning of the Copernican Revolution," in *Proceedings of the American Philosophical Society,* **87** (1944), 381–386; "Science Teaching and the Humanities," in *ETC: A Review of General Semantics,* **4** (1946), 3–24; "The Place of Logic and Metaphysics in the Advancement of Modern Science," in *Philosophy of Science,* **15** (1948), 275–286; "Einstein, Mach, and Logical Positivism," in *Albert Einstein: Philosopher-Scientist,* P. A. Schilpp, ed. (Chicago, 1949), pp. 271–286; "Einstein's Philosophy of Science," in *Review of Modern Physics,* **21** (1949), 349–355; "The Logical and Sociological Aspects of Science," in *Proceedings of the American Academy of Arts and Sciences,* **80** (1951), 16–30; "The Origin of the Separation Between Science and Philosophy," *ibid.* (1952), 115–139; "The Variety of Reasons for the Acceptance of Scientific Theories," in *The Validation of Scientific Theories,* Philipp Frank, ed. (Boston, 1956), pp. 3–17, first pub. in *Scientific Monthly,* **79,** no. 3 (1954), 139–145; and "The Pragmatic Component in Carnap's 'Elimination of Metaphysics,'" in *The Philosophy of Rudolf Carnap,* P. A. Schilpp, ed. (Chicago, 1963), pp. 159–164.

Frank edited a number of works, including *The Validation of Scientific Theories* (Boston, 1956) and *The International Encyclopedia of Unified Science* (Chicago, various dates). He also served on the editorial boards of the journals *Synthèse* (1946–1963) and *Philosophy of Science* (1941–1955).

II. SECONDARY LITERATURE. A *Festschrift* for Philipp Frank was published as vol. II of *Boston Studies in the Philosophy of Science,* R. S. Cohen and M. W. Wartofsky, eds. (Dordrecht–New York, 1965), with tributes by Peter G. Bergmann, Rudolf Carnap, R. Fürth, Gerald Holton, Edwin C. Kemble, Henry Margenau, Hilda von Mises, Ernest Nagel, Raymond J. Seeger, and Kurt Sitte, and essays in the philosophy of science.

A memorial booklet based on talks delivered by some of Frank's colleagues and friends at the memorial meeting of 25 October 1966 at Harvard University was distributed the following year, and an article "In Memory of Philipp Frank" appeared in *Philosophy of Science,* **35** (1968), 1–5.

GERALD HOLTON
ROBERT S. COHEN

FRANKENHEIM, MORITZ LUDWIG (*b.* Brunswick, Germany, 29 June 1801; *d.* Dresden, Germany, 14 January 1869), *crystallography.*

Frankenheim attended the Gymnasium in Wolfenbüttel and Brunswick. In 1820 he began his university studies in Berlin with philology but changed to mathematics and physics. In 1823 he received a doctorate for his dissertation, *De theoria gasorum et vaporum.* He qualified as a university lecturer in 1826, also in Berlin. In 1827 he was appointed assistant professor at Breslau, where he became professor of physics in 1850; he held this position until 1866.

Frankenheim's importance lies especially in the field of crystallography. In his work *Die Lehre von der Kohäsion . . .* (1835), he was the first to examine whether or not the geometrically possible types of crystal lattices agree in their symmetry relations with those actually observed in crystals. He showed that there could be only fifteen different "nodal," i.e., space lattice, type configurations. Bravais, in his "Mémoires sur les systèmes formés par des points distribués régulièrement sur un plan ou dans l'espace" (1848), acknowledged Frankenheim's achievement: "Frankenheim, in his beautiful researches in crystallography, arrived at the same classification." In 1856 he corrected himself: there could be only fourteen, because two of the proposed monoclinic subdivisions proved to be identical.

In his 1829 work *De crystallorum cohaesione* Frankenheim established that the hardness of crystals is always the same in the same crystallographic directions but varies with the direction through the crystal. He was also the first to investigate experimentally the influence of a crystal on oriented overgrowth from a crystal seed (epitaxy). In 1830 he investigated more exactly with the microscope the overgrowth of sodium carbonate on calcium carbonate that he had observed in his study of cohesion. In 1836 he grew potassium iodide on mica, a spectacular example of oriented overgrowth still used for demonstration. Frankenheim also introduced the concept of isodimorphism, which he derived from observations on sodium nitrate and potassium nitrate on the one hand and on calcite and aragonite on the other.

Frankenheim devised an experiment that even today is a suitable lecture demonstration. Out of a drop of warm supersaturated potassium nitrate solution, a rhombohedral unstable modification precipitates out onto the microscope slide. With further cooling needlelike orthorhombic crystals form outward from the edge, and the rhombohedrons in the vicinity of the orthorhombic needles dissolve. If one of the needles is touched by a rhombohedron, the latter is very quickly transformed into an aggregate of rhombic crystals.

Frankenheim repeatedly took a position on the question of amorphous minerals in a polemic with the Munich mineralogist J. N. von Fuchs. In 1851 he held that these bodies were aggregates of many crystals of imperceptible dimensions. Much later the introduction of X-ray investigation showed that this conception was correct in the case of a great many substances formerly considered amorphous.

As early as 1860, Frankenheim used a new kind of polarizing microscope; the specimen could be rotated so that the angles between two directions in the specimen, as well as its position relative to the orientation of the Nicol prisms, could be exactly determined.

BIBLIOGRAPHY

I. Original Works. Frankenheim's works include *De theoria gasorum et vaporum* (Berlin, 1823); *De crystallorum cohaesione* (Vratisl, 1829); *Die Lehre von der Kohäsion, umfassend die Elastizität der Gase, die Elastizität und Kohärenz der flüssigen und festen Körper und die Kristallkunde* (Breslau, 1835); "Über die Verbindung verschiedenartiger Krystalle," in *Annalen der Physik,* **37** (1836), 516–522; "System der Kristalle, ein Versuch," in *Nova acta Academiae Caesarae Leopoldina Carolinae germanicae naturae curiosorum,* Abt. II, **19** (1842), 471–660; "Krystallisation und Amorphie," in *Journal für praktische Chemie,* **54** (1851), 430–476; "Die Anordnung der Moleküle im Kristalle," in *Annalen der Physik,* **97** (1856), 337–382; "Entstehen und Wachsen der Kristalle, mikroskopische Beobachtungen," *ibid.,* **111** (1860), 37 ff.; *Zur Kristallkunde I. Charakteristik der Kristalle* (Leipzig, 1869), unfinished.

II. Secondary Literature. On Frankenheim and his work, see A. Bravais, "Mémoires sur les systèmes formés par des points distribués régulièrement sur un plan ou dans l'espace," in *Journal de l'École polytechnique,* **19** (1848), 1–128, presented to the Académie des Sciences on 11 Dec. 1848; it was translated by C. and E. Blasius as *Abhandlung über die Systeme von regelmässig auf einer Ebene oder im Raum verteilten Punkten* (Leipzig, 1897). See also P. Groth, *Entwicklungsgeschichte der mineralogischen Wissenschaften* (Berlin, 1926); and Poggendorff, I, 792, and III, 469.

Carl W. Correns

FRANKLAND, EDWARD (*b.* Catterall, near Churchtown, Lancashire, England, 18 January 1825; *d.* Golaa, Gudbrandsdalen, Norway, 9 August 1899), *chemistry.*

Frankland was the illegitimate son of Peggy Frankland, the daughter of a calico printer. After education in seven schools, including Lancaster Grammar School, Frankland was apprenticed by his stepfather, William Helm, to a Lancaster druggist, Stephen Ross. The drudgery of the years from 1840 to 1845, during which Ross taught him little, haunted Frankland's dreams for the remainder of his life. Through the efforts of two local doctors, Christopher and James Johnson, he was given facilities to perform chemical experiments in his spare time, and in 1845 they found him employment in Lyon Playfair's laboratory at the government's Museum of Economic Geology in London. There he met the brilliant German chemist A. W. H. Kolbe, who taught him Robert Bunsen's methods of gas analysis—a technique Frankland exploited in his later researches.

In 1846 Frankland became Playfair's assistant at the Civil Engineering College at Putney, London, and during the summer of 1847 he accompanied Kolbe to Marburg in order to study with Bunsen. From 1847 to 1848 he taught science with John Tyndall at the progressive Quaker school run by George Edmondson at Queenwood, Hampshire. Frankland completed his training with Bunsen at Marburg from 1848 to 1849, obtained his doctorate, and briefly studied with Justus Liebig at Giessen before returning to London to take Playfair's chair of chemistry at Putney from 1850 to 1851. He became professor of chemistry at Owens College, Manchester, in 1851, but this position proved unsatisfactory. In 1857 he returned to London, where, until 1864, he was lecturer in chemistry at St. Bartholomew's Hospital.

Frankland also indulged in the pluralism of a lectureship in science at Addiscombe Military College from 1859 to 1863, and from 1863 to 1869 he was professor of chemistry at the Royal Institution. Finally, in 1865, he succeeded A. W. Hofmann as professor of chemistry at the Royal College of Chemistry, a position he retained through the college's many transformations until his retirement in 1885. From 1865, Frankland made official monthly analyses of the water supplies of London, and from 1868 to 1874 he served on the important Royal Commission on Rivers Pollution. For these services he was knighted in 1897.

At the age of eighteen Frankland underwent an extreme form of evangelical conversion, but after 1848 he lapsed into skepticism. Together with Tyndall, T. H. Huxley, J. D. Hooker, and others he was an active member of an informal scientific pressure group which called itself the X Club. Yet the club was unable to gain for Frankland the presidency of the Royal Society or of the British Association for the Advancement of Science, owing to his modesty and poor ability in public debate. But he did serve as president of the Chemical Society from 1871 to 1873 and was the founder and first president of the Institute of Chemistry (the society for professional chemists) from 1877 to 1880.

Frankland is an outstanding example of a pure scientist who was deeply conscious of the significance and importance of applied science; but his public service in the improvement of water and gas supplies and his contributions to the development of British scientific education await proper assessment. Frankland possessed a voracious appetite for travel, which he combined with mountaineering, yachting, and fishing. He was also a keen gardener, music lover, and amateur astronomer. In 1874, following the death of his first wife, Sophie Fick, by whom he had three sons and two daughters, he married Ellen Grenside, by whom he had two daughters.

Frankland's extraordinary practical and manipulative ability, as well as his power, like Bunsen's, to combine physics with chemistry, was exemplified in all three of the broad categories of his research: organic, physical, and applied chemistry. In 1844 H. Fehling had obtained a new compound, benzonitrile, C_7H_5N (i.e., phenyl cyanide), by the dry distillation of ammonium benzoate. Following A. Schlieper's preparation of valeronitrile, C_5H_9N (i.e., butyl cyanide), in 1846, Kolbe and Frankland noted that both nitriles were easily hydrolyzed to their corresponding acids (i.e., benzoic and valeric acids). In their joint work of 1847 they pointed out that if these so-called nitriles were really cyanides, then their hydrolysis would agree with Berzelius' iconoclastic suggestion that acetic acid was a methyl radical conjugated with oxalic acid ($C_2H_3 \cdot C_2O_3 \cdot HO, C = 6, O = 8$). If both these assumptions were made, it followed that the homologues of acetic acid (e.g., propionic acid) arose from the conjugation of oxalic acid with ethyl (alkyl) radicals. Their production of propionic acid from ethyl cyanide in 1847[1] led Frankland and Kolbe to attempt separately the isolation of alkyl radicals from acids: Kolbe by the electrolysis of acids (1849) and Frankland by using a reaction between alkyl iodides and zinc based on analogy with Bunsen's celebrated isolation of cacodyl in 1837. But after much controversy and the reform of atomic weights, Frankland was forced to admit that the formulas of the radicals he prepared between 1848 and 1851 had to be doubled and that the radicals were in fact inert hydrocarbons of the paraffin series.

The work on radicals also led Frankland in 1849 to the isolation of a new reactive organometallic compound, zinc methyl; this, together with the alkyltin compounds which he prepared in 1850 by the action of sunlight on alkyl halides in the presence of tin, produced the following problem. If, as the conjugation theories of Berzelius and Liebig held, the different alkyl groups associated with oxalic acid (i.e., a carboxylic group) had little or no influence on the combining properties of the acid, why did alkyl-conjugated metals have combining powers different from those of the metals alone? For example, tin diethyl (stanethylium) formed only one oxide, whereas tin itself formed at least two oxides. Zinc methyl, on the other hand, seemed to possess the same singular combining power as zinc. Here was the seed of the concept of valence, which, with international agreement on atomic weight values, was to unite the rival theoretical schools of chemistry during the 1860's into the common aim of structural chemistry.

On 10 May 1852 Frankland read to the Royal Society a paper on organic metallic compounds in which he made the empirical observation that elements possessed fixed combining powers, or "only room, so to speak, for the attachment of a fixed and definite number of the atoms of other elements."[2] The expression "valence" or "valency" began to be used by other chemists only after 1865, whereas Frankland tended to use the misleading term "atomicity." Although the development of valence as an architectural concept for linking atoms together within a molecule owed more to the work of Kekulé in the 1850's and 1860's, Frankland's teaching position at the Royal College of Chemistry and his influence on the Department of Science and Art science examinations enabled him to spread the idea through the younger generation of British chemists. In 1866 he published an influential textbook, *Lecture Notes,* in which he adopted Crum Brown's graphic (structural) formulas and argued (against Kekulé) that elements could exhibit more than one valence below a fixed upper maximum. He also developed a special shorthand structural notation,[3] but it proved confusing and its use did not persist into the twentieth century.

Frankland was quick to see that the analytical techniques he had developed and the organometallic compounds he had prepared would be powerful aids to synthesis, by which he meant the chemist's ability to build up compounds "stone by stone" with a view to understanding their atomic configurations. From 1863 to 1870 he and Baldwin Duppa exploited zinc ethyl and other organic reagents, including ethyl acetate, in the synthesis of ethers, dicarboxylic acids, unsaturated monocarboxylic acids, and hydroxy acids. This meticulous work revealed clearly the structure and relationship of these compounds, and of course its methodology had great bearing on the growth of the chemical industry.

Intermittent work on combustion during the 1860's was initiated by a memorable ascent of, and night on, Mont Blanc with Tyndall in 1859. Frankland found that Humphry Davy's views on the nature of flame were unsound and that pressure variations produced striking changes in the illuminating power of flames.[4] He showed the relevance of this finding to the supply of domestic illuminating gas and, in 1868, to stellar spectroscopy. During the latter brief investigation in collaboration with the astronomer J. N. Lockyer, lines of helium were first observed in the sun; but Frankland did not agree with Lockyer's interpretation that helium was a new element.[5]

Frankland's wide interests included biology. In 1865, together with Adolf Fick and Johannes Wislicenus, he designed an experiment to test Liebig's theory that the source of muscular energy was the oxidation of nitrogenous muscular tissue. The two Germans performed this experiment by ascending Mt. Faulhorn in Switzerland while on a protein-free diet, then measuring the nitrogen output in their urine.[6] They confirmed their suspicion that muscular energy comes principally from the oxidation of non-nitrogenous materials. It remained for Frankland to confirm in the laboratory that the oxidation of carbohydrates and fats produces sufficient energy to account for the mechanical work of an organism.[7] His calorimetric experiments of 1866 on the energy values of common foodstuffs laid the foundation for quantitative dietetics.

In 1867, together with H. E. Armstrong, Frankland devised a method for analyzing water by combustion analysis of organic carbon and nitrogen *in vacuo.*[8] A rival method developed by J. A. Wanklyn in the same year,[9] which identified nitrogen content as ammonia, led to acrimonious disputes between the two men over the respective merits of their systems. Frankland's method, although extremely accurate, proved too cumbersome and difficult for the unskilled, so Wanklyn's simpler but less reliable technique was usually preferred by public analysts. Frankland's humanitarian and scientific interest in water analysis was continued by his son Percy.

NOTES

1. E. Frankland and H. Kolbe, "On the Chemical Constitution of Metacetonic Acid, and Some Other Bodies Related to It," in *Memoirs of the Chemical Society,* **3** (1845–1848), 386–391.
2. E. Frankland, "On a New Series of Organic Bodies Containing Metals," in *Philosophical Transactions of the Royal Society,* **142** (1852), 417–444, see p. 440. Publication of this paper was delayed by the oversight of the Society's secretary, G. Stokes (see Frankland's autobiography, 1902 ed., p. 187).
3. E. Frankland, "Contributions to the Notation of Organic and Inorganic Bodies," in *Journal of the Chemical Society,* **4** (1866), 372–395.
4. E. Frankland, "On the Influence of Atmospheric Pressure Upon Some of the Phenomena of Combustion," in *Philosophical Transactions of the Royal Society,* **151** (1861), 629–653.
5. Letter to Lockyer, 9 Sept. 1872, in the archives of the Sir J. N. Lockyer Observatory, Sidmouth, Devonshire.

6. A. Fick and J. Wislicenus, "On the Origin of Muscular Power," in *Philosophical Magazine,* 4th ser., **31** (1866), 485–503.
7. E. Frankland, "On the Origin of Muscular Power," *ibid.,* **32** (1866), 182–199.
8. E. Frankland and H. E. Armstrong, "On the Analysis of Potable Waters," in *Journal of the Chemical Society,* **6** (1868), 77–108.
9. J. A. Wanklyn, E. T. Chapman, and M. H. Smith, "Water Analysis: Determination of the Nitrogenous Organic Matter," *ibid.,* **5** (1867), 445–454.

BIBLIOGRAPHY

I. Original Works. Frankland published over 130 papers, of which the *Royal Society Catalogue of Scientific Papers* (London, 1867–1925) lists 107; see II, 699–700, VII, 700–701, IX, 918, and XV, 101; sixty-four were republished, some in a revised form, by Frankland in his 1877 book. To these should be added "A Course of Ten Lectures at the Royal Institution," in *Chemical News,* **3** (1861), 99–104, 118–122, 132–136, 166–170, 185–187, 201–203, 215–219, 291–299, 377–381, and **4** (1861), 51–54, 65–68, 93–97; "Chemical Research in England," in *Nature,* **3** (1870–1871), 445; an untitled paper on chemical apparatus read to the Kensington Science Conferences of 1876, in *Nature,* **14** (1876), 73–76—see also *South Kensington Museum. Conferences Held in Connection With the Special Loan Collection of Scientific Apparatus,* 3 unnumbered vols. (London, 1876), "Chemistry, Biology," pp. 1–13; the presidential address to the Institute of Chemistry, in *Chemical News,* **37** (1878), 57–59; reply to Lockyer's attack on the Institute of Chemistry, *ibid.,* **52** (1885), 305–306. Note also Frankland's important evidence to the Select Committee on Scientific Instruction for Industrial Classes, 1867–1868, in *Parliamentary Papers 1867–1868,* XV (432), pars. 8033–8177; and to the Devonshire Commission on Scientific Instruction and the Advancement of Science, 1871–1875, *ibid., 1872,* XXV (C.536); *1874,* XXII (C.1087); and *1875,* XXVIII (C.1298), pars. 40–47, 516–518, 758–835, 980–982, 2473–2488, 5667–5896, 11,053–11,108, and index. Finally, note Frankland's influence in George S. Newth, *Chemical Lecture Experiments. Non-metallic Elements* (London, 1892, 1896).

Frankland's books were *Ueber die Isolirung des Radicales Aethyl* (Marburg–Brunswick, 1849), his Ph.D. diss.; *Lecture Notes for Chemical Students* (*Embracing Mineral and Organic Chemistry*) (London, 1866), the 2nd ed., 2 vols., published as I, *Inorganic Chemistry* (1870), and II, *Organic Chemistry* (1872), and a 3rd ed. of II, rev. by F. R. Japp (1881)—see below for the 3rd ed. of I; *Reports of the Rivers Pollution Commission* (*1868*), 6 vols. (London, 1870–1874), also in *Parliamentary Papers, 1871,* XXV, XXVI; *1872,* XXXIV; and *1874,* XXXIII; *Experimental Researches in Pure, Applied, and Physical Chemistry* (London, 1877), Frankland's edited version of his papers, dedicated to Bunsen; *How to Teach Chemistry; Hints to Science Teachers and Students,* George Chaloner, ed. (London–Philadelphia, 1875); *Water Analysis for Sanitary Purposes* (London, 1880, 1890); *Inorganic Chemistry,* rev. by J. R. Japp, 3rd ed. of *Lecture Notes,* I; *Sketches From the Life of Edward Frankland* (London, 1901); and *Sketches From the Life of Sir Edward Frankland,* edited and completed by M. N. W. [West] and S. J. C. [Colenso] (Frankland's daughters).

For Kekulé's claim to priority in valence theory, see his unpublished MS "Zur Geschichte der Valenztheorie," in R. Anschütz, *August Kekulé,* I (Berlin, 1929), 555–569, repr. in facs. in R. Kuhn, ed., *Cassirte Kapitel aus der Abhandlung: Über die Carboxytartronsäure und die Constitution des Benzols* (Weinheim, 1965). Frankland's polemics with Wanklyn may be traced from *Chemical News,* **17** (1868), 45, 79, 97; **33** (1876), 85, 104–106; and **66** (1892), 103, 119. On the X Club, see Frankland's autobiography.

MS material is located in London in the Royal Institution (where the Tyndall papers may also be found), the Royal Society, and Imperial College archives. Other archives containing MS papers are those at Liverpool University (the Reade papers), the Lancaster Public Library (the Lancastrian Frankland Society), and the Sir J. N. Lockyer Observatory, Sidmouth. Unlisted papers held by the Frankland family are not yet available for study. Oddments of Frankland's apparatus are to be found at the Royal Institution and the City of Lancaster Museum.

II. Secondary Literature. The best obituaries are J. Wislicenus, in *Berichte der Deutschen chemischen Gesellschaft,* **33** (1900), 3847–3874, with photograph and list of papers; H. McLeod, in *Journal of the Chemical Society,* **87** (1905), 574–590; and [J. R. Japp], in *Minutes of Proceedings of the Institution of Civil Engineers,* **139** (1900), 343–349. A full version of H. E. Armstrong's Frankland memorial lecture to the Chemical Society was never published, but see his interesting "First Frankland Memorial Oration to the Lancastrian Frankland Society," in *Journal of the Society of Chemical Industry,* **53** (1934), 459–466. See also Sir W. Tilden, *Famous Chemists* (London, 1921), pp. 216–227; and J. R. Partington, *A History of Chemistry,* IV (London–New York, 1964), ch. 16. For an extremely thorough analysis of Frankland's contributions to valence theory, see C. A. Russell, *History of Valency* (Leicester, 1971). Frankland's period as a schoolteacher is sketched in D. Thompson, "Queenwood College, Hampshire," in *Annals of Science,* **11** (1955), 246–254; and his contribution to biochemistry in E. McCollum, *A History of Nutrition* (Boston, 1957), pp. 127–129. For Frankland's activities on behalf of professional chemists, see R. B. Pilcher, *The Institute of Chemistry of Great Britain and Ireland, History of the Institute, 1877–1914* (London, 1914), *passim.*

W H. Brock

FRANKLAND, PERCY FARADAY (*b.* London, England, 3 October 1858; *d.* House of Letterawe, on Loch Awe, Argyllshire, Scotland, 28 October 1946), *chemistry, bacteriology.*

Frankland was the second son of Edward Frankland, professor of chemistry at the Royal School of Mines in London. His middle name was given in honor of the eminent chemist Michael Faraday, who was his godfather. After studying at University College School in London from 1869 to 1874, Frankland entered the Royal School of Mines in 1875. His teachers there included his father and Thomas Henry Huxley. In 1877 he won a Brackenbury scholarship at St. Bartholomew's Hospital, but his father dissuaded him from a medical career and induced him to take up chemistry instead. From 1878 to 1880 he studied organic chemistry under Wislicenus at the University of Würzburg, taking his Ph.D. summa cum laude in the latter year. He was then appointed demonstrator under his father at South Kensington, where the Royal School of Mines had been transferred and its name changed to the Normal School of Science. He took his B.Sc. in 1881 from the University of London, which was then merely an examining and degree-granting body.

Frankland was professor of chemistry at University College, Dundee, from 1888 to 1894 and at Mason Science College (later the University of Birmingham) from 1894 to 1919. At the latter institution he also served as dean of the Faculty of Science from 1913 until his retirement. He was president of the Institute of Chemistry from 1906 to 1909 and of the Chemical Society in 1912 and 1913. During World War I, Frankland worked with the Chemical Warfare Committee on synthetic drugs, explosives, and mustard gas. These efforts led to his being named C.B.E. in 1920. Elected a fellow of the Royal Society in 1891, Frankland was awarded its Davy Medal in 1919. He was also awarded honorary doctorates by the universities of St. Andrews (1902), Dublin (1912), Birmingham (1924), and Sheffield (1926). Following his death a memorial lecture was established in his name at the Royal Institute of Chemistry.

Frankland's wife, Grace Coleridge Toynbee, whom he married in 1882, was the youngest daughter of Joseph Toynbee, the pioneer ear specialist. She was herself a research bacteriologist and frequently contributed to her husband's scientific work. Her death preceded his by a few weeks. They left a son, Edward.

Frankland's early research work seems to have been strongly influenced by his father. In the early 1880's he undertook a systematic study of the coal gas supplied to consumers in the larger British towns, thus following a path his father had trod thirty years before. Comparing his results with his father's, Frankland noted that the nitrogen content had increased because of a change in the methods of combustion. This study led to the publication of a series of five papers on the illuminating power of various hydrocarbons.

Frankland's interest in water analysis probably also derived originally from his father, who had concentrated on the purely chemical aspects of water analysis; Frankland was also attracted to its biological or bacteriological aspects. From 1885 to 1895 much of his research had as its goal the elucidation of the chemical reactions taking place in the presence of fermentative bacteria, and especially the development of effective methods for analyzing and preventing the bacterial contamination of water supplies. Largely as a result of his efforts, a monthly bacteriological examination of London's water supplies was inaugurated in 1885. He tested the efficacy of such materials as coke and greensand as agents for filtering bacteria from water and studied alterations in the viability and virulence of the anthrax and typhoid bacilli in drinking water. From 1892 to 1895 Frankland was coauthor, with Harry Marshall Ward, of four experimentally based reports to the Water Research Committee of the Royal Society. He also acted as private consultant to many of the largest water companies in Great Britain. His experience in original research added to the authority of his book, written with his wife, *Micro-organisms in Water: Their Significance, Identification and Removal* (London, 1894). Frankland also wrote a more popular book on bacteriology, *Our Secret Friends and Foes* (London, 1893), which went through four editions by 1899.

Most of the rest of Frankland's research concerned the stereochemistry of optically active substances. His interest in this topic was first aroused while he was working on his Ph.D. under Wislicenus, and it ultimately became his major preoccupation. By carrying out an exhaustive study of the rotatory effects of a large number of molecular groups, Frankland developed valuable methods for testing the quantitative relationship between molecular structure and degree of optical activity. Although he thought he had uncovered a few regularities, he admitted that his research had not produced any broad generalizations. His work showed mainly that the relationship between structure and optical activity was too complex to be explained by existing theories. No great original contributions resulted from his bacteriological work either.

In his scientific interests and approach Frankland recognized a kinship between himself and Louis Pasteur. With his wife he wrote an admirable biography bearing the simple title *Pasteur* (London, 1898), to which William Bulloch frequently referred in his *History of Bacteriology* (London, 1938). A leading advocate of original research by students, Frankland was considered an inspiring, if rather stern and demanding, teacher.

BIBLIOGRAPHY

I. ORIGINAL WORKS. Besides the books mentioned in the text, Frankland published well over 100 papers, several of which cover much the same ground and many of which were written in collaboration with his students and colleagues. Most of his early papers appeared in the *Journal of the Chemical Society* (London). A complete bibliography of his works published before 1900 may be found in the *Royal Society Catalogue of Scientific Papers,* IX (1891), 919; and XV (1916), 101–103. The most important of these and of his later papers are cited by Garner in the longer of his two biographical sketches of Frankland.

II. SECONDARY LITERATURE. See W. E. Garner, "Frankland, Percy Faraday," in *Dictionary of National Biography* (1941–1950), pp. 270–271; and "Percy Faraday Frankland," in *Obituary Notices of Fellows of the Royal Society of London,* **5,** no. 16 (1947), 697–715. The latter notice contains a bibliography of ninety-one "main publications" by Frankland between 1880 and 1927, as well as a detailed account of Frankland's research work, especially that on stereochemistry.

GERALD L. GEISON

FRANKLIN, BENJAMIN (*b.* Boston, Massachusetts, 17 January 1706; *d.* Philadelphia, Pennsylvania, 17 April 1790), *electricity, general physics, oceanography, meteorology, promotion and support of science and international scientific cooperation.*

Benjamin Franklin was the first American to win an international reputation in pure science and the first man of science to gain fame for work done wholly in electricity. His principal achievement was the formulation of a widely used theory of general electrical "action" (explaining or predicting the outcome of manipulations in electrostatics: charge production, charge transfer, charging by electrostatic induction). He advanced the concept of a single "fluid" of electricity, was responsible for the principle of conservation of charge, and analyzed the distribution of charges in the Leyden jar, a capacitor. He introduced into the language of scientific discourse relating to electricity such technical words as "plus" and "minus," "positive" and "negative," "charge" and "battery." By experiment he showed that the lightning discharge is an electrical phenomenon, and upon this demonstration (together with his experimental findings concerning the action of grounded and of pointed conductors) he based his invention of the lightning rod.

Franklin made contributions to knowledge of the Gulf Stream, of atmospheric convection currents, and of the direction of motion of storms. His observations on population were of service to Malthus. He was the principal founder of the American Philosophical Society, the New World's first permanent scientific organization.

Early Life and Career

Benjamin Franklin's father, Josiah, who was descended from a family of British artisans, immigrated to America, settling in Boston in October 1683. His mother, Josiah's second wife, was Abiah ("Jane") Folger, daughter of Peter Folger of Nantucket, a weaver, schoolmaster, miller, and writer of verses. On both sides of the family Franklin had forebears skilled in the use of their hands and with literary or intellectual gifts.

Franklin relates in his autobiography that he "was put to the Grammar School at eight years of Age," but remained "not quite one Year." His father then sent him "to a School for Writing and Arithmetic." Although Franklin by his own admission failed arithmetic, he later repaired this deficiency. In midlife, he took up "making magic Squares, or Circles," some of which were very complex and obviously required skill in computation. Published in England and in France from 1767 to 1773, they have attracted much attention and comment ever since.

At ten years of age, Franklin was taken home from school to assist his father, a tallow chandler and soap boiler. Since he was fond of reading and had in fact spent on books "all the little Money that came into . . . [his] Hands," it was decided that Benjamin should become a printer. He was, accordingly, at age twelve indentured to "Brother James." Within a few years he was able to break the indenture and secure his freedom. He left Boston to seek his fortune, first in New York (briefly and unsuccessfully) and then in Philadelphia.

Franklin had immediate success in Philadelphia. Before long he came to the attention of Governor Keith, who offered to subsidize him—although he was only eighteen—in the printing business. Franklin was sent to London to select types and presses and to make useful business contacts. Once at sea, Franklin discovered that the governor had sent him off without any letter of introduction and without funds for purchasing the printing equipment—indeed, that the governor had merely been "playing . . . pitiful Tricks . . . on a poor ignorant Boy!" On arrival, Franklin found work in Samuel Palmer's printing house, where he set type for William Wollaston's *The Religion of Nature Delineated.*

After two years away from Philadelphia (from November 1724 to October 1726) Franklin returned to his adopted city, skilled in the various aspects of the printing craft. He soon had his own shop and before long became a major figure in the town and, eventually, in the colony. With a partner, he published the *Pennsylvania Gazette;* when the partnership was dissolved in 1730, Franklin kept the newspaper and shortly began publication of *Poor Richard: An Almanack* (1733). He was Clerk of the Assembly, postmaster of Philadelphia (1737–1753), and publisher (1741) of the *General Magazine.* He was an organizer of the Library Company (1731), and the Union Fire Company (1736), and was a promoter of the Academy of Philadelphia (later the College and Academy of Philadelphia and now the University of Pennsylvania), of which he became president of the trustees (1749).

As he became more deeply concerned with civic affairs and public life, Franklin retired from active business (1748), setting up what would become an eighteen-year partnership with David Hall, his printing house foreman. He was elected a member of the Pennsylvania Assembly (1751) and alderman of Philadelphia and was appointed a deputy postmaster-general for the British colonies in North America (1753–1774). He was sent to England in 1757 and remained until 1762 as the Assembly's agent.

Preparation for Scientific Research

When, in 1757, Franklin sailed for England for the second time, he had already won a high place in world science. He had published articles in the world's leading scientific journal, the *Philosophical Transactions of the Royal Society,* and was a fellow of that society (elected 29 May 1756). For his research in electricity the Society had conferred upon him (on 30 November 1753) one of their highest awards—the Copley Medal. He had received honorary degrees from Harvard (1753), Yale (later in 1753), and William and Mary (1756). His book on electricity had already appeared in three editions in England and two in France, and one of his experiments—"proving the sameness of Lightning and Electricity"—was world-famous. Franklin was largely self-taught in science—as he was in other subjects—but this does not mean that he was uneducated. He had rigorously studied the science of his day in the writings of the best masters available.

In 1744 Franklin sponsored Adam Spencer's lectures on experimental science in Philadelphia and purchased his apparatus; he had previously attended Spencer's lectures in Boston. Also in 1744 Franklin published a pamphlet on the stove he had invented; in it he refers to, and quotes from, certain great masters of experimental science whose works he knew, including Boerhaave, Desaguliers, 'sGravesande, and Hales. He was also familiar with the writings of Robert Boyle and knew well the major treatise on experimental physics of the age, Newton's *Opticks.* He had also encountered expositions of the Newtonian natural philosophy in the published Boyle lectures, a series which included books by Samuel Clarke and William Derham. Having known Pemberton in London, he no doubt would have read Pemberton's *View of Sir Isaac Newton's Philosophy,* of which Peter Collinson had sent a copy to the Library Company in 1732. Thus, even though Franklin may have had no formal training, he was well educated in Newtonian experimental science.

Gadgets and Inventions

Benjamin Franklin's reputation in science was made by his experiments and the theories he conceived or modified to explain his results. The experimental scientist of Franklin's day had not only to be able to design but also to construct the devices he needed. Franklin the artisan had no aversion to manual labor and operations. A gifted gadgeteer and inventor, he was not only able to make the devices he conceived but he could also think in terms of the potential of gadgets and instruments in relation to the development of his ideas: a significant ability, since usually the conception of an experimental problem cannot be separated from the means of exploring or solving it.

Throughout his life Franklin found it (as he writes in his autobiography) a source of "Pleasure . . . to see good Workmen handle their Tools." He was aware of the great advantage to his research in being able "to construct little Machines for my Experiments while the Intention of making the Experiment was fresh and warm in my Mind." This aspect of Franklin's research was especially noted by William Watson in his review of Franklin's book on electricity (*Philosophical Transactions,* 1752); Franklin, the reviewer said, has both "a head to conceive" and "a hand to carry into execution" whatever he considers "may conduce to enlighten the subject-matter."

Among Franklin's notable inventions and gadgets are the rocking chair, bifocal glasses, and the Pennsylvania fireplace, or Franklin stove. He also conceived the idea of "summer time," or daylight saving time. His most important invention, the lightning conductor or lightning rod, is, however, in a different category altogether, an application to human needs (in the Baconian sense) of recent discoveries in pure science.

First Researches in Electricity

In the early 1740's Franklin encountered the new electrical experiments in at least two ways. He saw some experiments performed by Adam Spencer in Boston in 1743 and again in Philadephia in 1744. Then, in 1745 (or possibly 1746) the Library Company of Philadelphia received "from Mr. Peter Collinson, F.R.S., of London, a Present of a Glass Tube, with some Account of the Use of it in making such electrical Experiments." Franklin records that he "eagerly seized the Opportunity of repeating what I had seen in Boston, and by much Practice acquir'd great Readiness in performing those also which we had an account of from England, adding a Number of new Ones."

The first researches in electricity at Philadepia were made by a group of four experimenters: Franklin, Philip Syng, Thomas Hopkinson, and Ebenezer Kinnersley, who was Franklin's principal coexperimenter. One of Franklin's first recorded discoveries was the action of pointed bodies. A grounded pointed conductor, he found, could cause a charged, insulated conducting body to lose its charge when the point was six to eight inches away; but a blunt conductor would not produce such a discharge until it was an inch or so away, and then there would be an accompanying spark. A companion discovery was made by Hopkinson: a needle placed on top of a suspended iron rod would prevent it from becoming charged, the electrical fire "continually runing out silently at the point" as fast as it was accumulated; this discovery had been anticipated by William Watson.

Other discoveries led Franklin and his coexperimenters to the concept that "the electrical fire is a real element, or species of matter, not created by the friction, but *collected* only." Thus all kinds of electrification, or changes in electrification, were to be explained by the transfer of "electrical fire," which was "really an element diffused among, and attracted by other matter, particularly by water and metals." Each body has a "natural" quantity of "electrical fire"; if it loses some, Franklin would call it electrically negative, or minus; if it gains some and therefore has a "superabundance" of "electrical fire," it would be positive, or plus. "To electrise *plus* or *minus,*" Franklin wrote in a letter to Collinson of 25 May 1747, "no more needs to be known than this, that the parts of the tube or sphere that are rubbed, do, in the instant of the friction, attract the electrical fire, and therefore take it from the thing rubbing." In short, since one or more bodies must gain the "electrical fire" that a given body loses, plus and minus charges or states of electrification must occur in exactly equal amounts. This quantitative principle is known today as the law of conservation of charge. It is still fundamental to all science, from microphysics to the electrification of gross bodies.

The Analysis of the Leyden Jar

One of the earliest and most significant results of the new Franklinian theory was the successful analysis of the Leyden jar, a topic introduced in a letter to Collinson, sent sometime prior to 28 July 1747. The Leyden jar, a form of condenser, or capacitor, was discovered or invented in the 1740's and was named after one of the several claimants to the discovery, Musschenbroek of Leyden; Franklin knew the device as Musschenbroek's "wonderful bottle." Essentially the device was a nonconductor (glass) with a conductor on each side; before long it was used with the inside filled with water or metal shot, and the outside coated with metal. Electrical contact was made with the water or metal shot by means of a wire running through an insulating cork stuck into the neck of the bottle. When the outer coating was grounded, as by being held in the hands of an experimenter, and the wire was brought to a charged body, the jar seemed capable of "accumulating" and "holding" a vast amount of "electricity."

The first observation made by Franklin was that if the wire and water inside the bottle are "electrised *positively* or *plus,*" then the outer coating is simultaneously "electrised *negatively* or *minus* in exact proportion." The equilibrium could not be restored through the glass of the bottle unless a conducting material simultaneously made contact with the outer coating and with the wire connected to the water or inner conducting material. He was astonished at the "wonderful" way in which "these two states of Electricity, the *plus* and *minus*" are "combined and balanced in this miraculous bottle."

In a letter of 29 April 1748, containing "Farther Experiments and Observations in Electricity," Franklin described some new experiments showing that a charged Leyden jar always has charges of opposite signs on the two conductors and that the charges are of the same magnitude. Clearly, he concluded, the "terms of *charging* and *discharging*" a Leyden jar are misleading, since "there is really no more electrical fire in the phial after what is called its *charging,* than before, nor less after its *discharging*. . . ."

Franklin then announced the most astonishing discovery of all, that in the Leyden jar "the whole force of the bottle, and power of giving a shock, is in the GLASS ITSELF." He reached this conclusion by a series of ingenious experiments, which are known today as the Franklin experiments on "the dissectible condenser." A Leyden jar with a loosely fitting cork was charged in the usual way and then placed on a glass insulator. The cork was carefully removed, together with the wire that hung down into the water; it was then found that the jar could be discharged as before by an experimenter's putting one hand around the outside of the jar while bringing a finger of the other hand to the jar's mouth so as to reach the water. Thus, the "force" was not "in the wire." Next, a test was made to determine whether the force "resided in the water" and was "condensed in it." A jar was charged as before, set on glass, and the cork and wire removed. The water was then carefully decanted into an empty, uncharged jar resting on glass; this second jar showed no evidence whatever of being charged. Either the "force" must have been lost during the decanting, or it must have remained behind in the glass. The latter was shown to be the case by refilling the first bottle with "unelectrified water," whereupon it gave the shock as usual.

In the next stage Franklin looked into the question of whether this property of glass came from the nature of its substance, or whether it was related to shape—a relevant question, since Franklin had pioneered in studying the effect of shape in the action of pointed and blunt conductors. In this inquiry he constructed a parallel-plate condenser (or capacitor) consisting of two parallel lead plates separated by a flat pane of sash glass. This condenser produced the same electrical effects as a Leyden jar, thus demonstrating that the "force" is a property of the glass as glass and is not related to shape. Franklin ingeniously joined together a number of such parallel-plate condensers to make "what we called an *electrical-battery*" consisting of eleven panes of glass, each "armed" with lead plates pasted on both sides, hooked together in series by wire and chain; the battery could be discharged by a special contrivance.

Full Statement of the Mature Theory

On 29 July 1750, Franklin sent Collinson his "Opinions and Conjectures concerning the Properties and Effects of the electrical Matter, arising from Experiments and Observations, made at Philadelphia, 1749." This paper began with the proposition that the electrical matter consists of "extremely subtile" particles, since it can easily permeate all common matter, even metals, without "any perceptible resistance." Here Franklin used the term "electrical matter" for the first time. Although he indicated a cause for belief in its "subtility," he took its atomicity or particulate composition for granted. The difference between electrical matter and "common matter" lies in the mutual attraction of the particles of the latter and the mutual repulsion of the particles of the former (which causes "the appearing divergency in a stream of electrified effluvia"). In eighteenth-century terms, such electrical matter constitutes a particulate, subtle, elastic fluid. The particles of electrical matter, although mutually repellent, are attracted strongly by "all other matter." Therefore, if a quantity of electrical matter be applied to a mass of common matter, it will be "immediately and equally diffused through the whole." In other words, common matter is "a kind of spunge" to the electric fluid. Generally, in common matter there is as much electrical matter as it can contain; if more be added, it cannot enter the body but collects on its surface to form an "electrical atmosphere," in which case the body "is said to be electrified." All bodies, however, do not "attract and retain" electrical matter "with equal strength and force"; those called electrics per se (or non-conductors) "attract and retain it strongest, and contain the greatest quantity." That common matter always contains electrical fluid is demonstrated by the fact of experience that a rubbed globe or tube enables us to pump some out.

The "electrical atmospheres" said to surround charged bodies are a means for explaining the observed repulsion between them, but this explanation takes cognizance only of the repulsion between positively charged bodies (that is, those which have gained an excess of fluid over their normal quantity). It offers no aid whatever in understanding the repulsion between negatively charged bodies—a phenomenon that had been observed by Franklin and his colleagues and reported by him in an earlier paper.

The concept of "electrical atmospheres" was not wholly novel with Franklin. Franklin's original contribution lay in the particular use he gave to this concept in his theory of electrical action. For example, Franklin stated that it takes the "form . . . of the body it surrounds." A sphere will thus have a spherical atmosphere and a cylinder a cylindrical one. Others had supposed that both would have a sphere of effluvia.

Franklin's concept of "electrical atmospheres" was based on the idea that an uncharged body must have its "normal" quantity of electrical matter or fluid and that, therefore, any further electrical matter or fluid added to it will collect around the outside, like a cloud. If two such charged bodies came near one another, these two clouds would produce repulsion, since the particles of which they are made tend to repel one another. Similarly, a body that has lost some of its normal quantity of electrical matter or fluid will attract the particles in the electric atmosphere of a positively charged body, until the two draw together and make contact. Franklin applied the concept of "electrical atmospheres" to explain the unequal distribution of charge in bodies that were not completely symmetrical, such as those which might be pointed or pear-shaped. These explanations were qualitatively successful, but they do not always appear convincing and certainly constitute one of the weakest and least satisfactory parts of the theory. Even more important, the doctrine of "electric atmospheres" could not contribute to the solution of one outstanding unsolved problem in the Franklinian explanation of electrical phenomena: the "apparent" repulsion between negatively charged bodies. We shall see below that this major defect in the theory was remedied by the addition of a new and very radical postulate by Aepinus.

One of the major advantages of the Franklinian theory was that it enabled "electricians" to distinguish clearly between the concept of a "repelling force" which could act even through a sheet of glass, although the electric fluid itself does not penetrate through glass. This basic concept was used in the explanation of the action of the condenser, wherein Franklin explained clearly—for what was, so far as I know, the first time—the mechanism of induced charges, the phenomenon of a negative charge being induced on a grounded conductor when a positively charged conductor is brought near it, or when a nearby conductor acquires a positive charge.

In the Leyden jar, according to Franklin's doctrine, the application of a positive charge to the conductor on one side of the glass will not cause the jar to be charged until or unless the conductor on the other side can lose some of its normal electric fluid, that is, until or unless it is grounded. Then and only then will electric fluid move away from that grounded conductor, leaving it negatively charged. Franklin thus naturally predicted, and proved by experiment, that the jar could be charged through its outer coating when the wire leading into the water is grounded, just as easily as in the normal manner—when a positive charge is applied to the inner conductor (water and wire) and the outside is grounded.

Later, in a famous series of experiments and explanations based upon some earlier ones made by John Canton, Franklin developed more fully this explanation of what we call today induced charges, or the phenomenon of charging by (electrostatic) induction. There is no doubt that it was Franklin's clear understanding of this process that caused his theory to be so highly valued in the eighteenth century. The theory is still used, with slight modifications, in all laboratory circumstances when charged objects are moved in the neighborhood of conductors which may be grounded or insulated or which can undergo a change in their condition of grounding or insulation. Only Franklin, and those who accepted his doctrine, could easily explain such phenomena as this: A positively charged body is brought near a conducting metal object placed on an insulating base and temporarily grounded; then the grounding is interrupted before the charged body is removed; the effect will be to induce in that object a negative charge. Now let the second object be an insulated cylinder; it will plainly display an unequal charge distribution, the end near the first body becoming negative and the far end positive; when the first body is withdrawn, the cylinder returns to its normal state and no longer shows any indication of charge. In the eighteenth century many scientists adduced this feature of the Franklinian theory (its ability to predict exactly the outcome of such experiments) as its major asset. In our own time J. J. Thomson has explained that the service of the one-fluid theory "to the science of electricity, by suggesting and co-ordinating researches, can hardly be overestimated." We still use this theory in the laboratory, Thomson said: "If we move a piece of brass and want to know whether that will increase or decrease the effect we are observing, we do not fly to the higher mathematics, but use the simple conception of the electric fluid which would tell us as much as we wanted to know in a few seconds" (in *Recollections and Reflections* [London, 1936], p. 252).

Dissemination of Experiments and Theories

Franklin's experiments on pointed conductors, grounding, the Leyden jar, and the conservation of charge, together with the statement of his theory of electrical action, based on the principle of conservation of charge, were all assembled by Collinson into a ninety-page book issued by E. Cave of London in 1751, with an unsigned preface written by Dr. John Fothergill. Buffon, who had recently stated that in electrical phenomena there seemed to be no one law governing the outcome of experiments, and that indeed the subject was characterized more by "bizzareries" than by regularities, came upon the book and had it translated into French in the following year; the French version was done by the naturalist Dalibard.

Thus, within two years Franklin's concepts and experiments were available to "electricians" on both sides of the Channel and—but for a number of minor revisions and extensions to new phenomena—all the main elements of Franklin's contributions to electrical theory had appeared in print.

One of the most challenging parts of Franklin's book was his discussion of thunder, lightning, and the formation of clouds. In a letter addressed to John Mitchel in London, dated 29 April 1749, Franklin wrote out some "Observations and Suppositions" that had led him to the hypothesis that clouds tend to become electrified through the vaporization effect on water of "common fire" (or ordinary heat) and "electrical fire." Rain, dew, and flashes of lightning between land clouds and sea clouds formed part of Franklin's suppositions, but six years later he freely admitted that he was "still at a loss" about the actual process by which clouds "become charged with electricity; no hypothesis I have yet formed perfectly satisfying me." Nevertheless, before April 1749 Franklin had assumed that clouds are electrified and that the lightning discharge is a rapid release of electric fluid from clouds.

On 7 November 1749, Franklin drew up a list of twelve observable similarities between the lightning discharge and the ordinary spark discharges produced in the laboratory. Notably, he concluded that since the "electric fluid is attracted by points," we might find out "whether this property is in lightning . . . . Let the experiment be made." But even before this experiment could be performed, Franklin assumed a favorable outcome. Convinced that lightning must be an electrical phenomenon, he warned his readers that high hills, trees, towers, spires, masts, and chimneys will act "as so many prominencies and points" and so will "draw the electrical fire" as a "whole cloud discharges there." He therefore advised his readers never "to take shelter under a tree, during a thunder gust."

In the paper entitled "Opinions and Conjectures," sent to Collinson in July of 1750 (containing the full statement of his theory of electrical action), Franklin also discussed the possible electrification of clouds and the nature of the lightning discharge. Immediately following the presentation of the property of pointed bodies to "draw on" and "throw off" the electric fluid at great distances, Franklin indicated that this knowledge of the "power of points may possibly be of some use to mankind, though we should never be able to explain it." Just as a grounded needle with its point upright could discharge a charged body and prevent a "stroke" to another nearby body, so Franklin argued that sharpened upright rods of iron, gilded to prevent rusting, fixed "on the highest parts of . . . edifices" and run down the outside of a building into the ground, or down "one of the shrouds of a ship" into the water, would "probably draw the electrical fire silently out of a cloud before it came nigh enough to strike, and thereby secure us from that most sudden and terrible mischief." Later, when the experiments were made, Franklin found that another function of the lightning rod, apart from "disarming" a passing cloud, would be to conduct a lightning stroke safely into the ground.

The experiment that Franklin devised required a sentry box large enough to contain a man and "an electrical [insulating] stand." The sentry box was to be placed on a high building; a long, pointed rod was to rise out through the door, extending twenty or thirty feet in the air, terminating in a point. This rod was to be affixed to the middle of the insulated stand, which was to be kept clean and dry so as to remain an insulator. Then when clouds, possibly electrified, would pass low, the rod "might be electrified and afford sparks, the rod drawing fire to" the experimenter, "from a cloud." To avoid danger, Franklin advised the man to be well insulated and to hold in his hand a wax handle affixed to a "loop of a wire" attached to the ground; he could bring the loop to the rod so that "the sparks, if the rod is electrified, will strike from the rod to the wire, and not affect him." Some years later, when Richmann performed this experiment in St. Petersburg, he did not fully observe all of Franklin's warnings and was electrocuted.

The sentry-box experiment was first performed at Marly, France, in May 1752. After Franklin's book had appeared in a French translation in 1752, the experiments he described were performed for the king and court; Buffon, Dalibard, and De Lor were then inspired to test Franklin's conjectures "upon the analogy of thunder and electricity." On 13 May 1752 Dalibard reported to the Paris Academy of Sciences: "In following the path that Mr. Franklin has traced for us, I have obtained complete satisfaction."

The account of this experiment was printed in the second French edition of Franklin's book on electricity and was later included in the English editions. A letter addressed from France to Stephen Hales, describing both the presentation of the Philadelphia experiments to the king of France and the success of the sentry-box experiment, was published in the *Philosophical Transactions* and was also reprinted in Franklin's book. Soon the lightning experiments were repeated by others in France, Germany, and England; and Franklin had the satisfaction of achieving an immediate and widespread international renown.

Later, Franklin devised a second experiment to test the electrification of clouds, one which has become more popularly known: the lightning kite. Franklin reported this experiment to Collinson in a letter of 1 October 1752, written after Franklin had read "in the publick papers from Europe, of the success of the *Philadelphia-Experiment* for drawing the electrick fire from clouds by means of pointed rods of iron erected on high buildings . . . ." Actually, Franklin appears to have flown his electrical kite prior to having learned of Dalibard's successful execution of the sentry-box experiment. The kite letter, published in the *Philosophical Transactions,* referred to the erection of lightning rods on public buildings in Philadelphia.

The lightning experiments caused Franklin's name to become known throughout Europe to the public at large and not merely to men of science. Joseph Priestley, in his *History . . . of Electricity,* characterized the experimental discovery that the lightning discharge is an electrical phenomenon as "the greatest, perhaps, since the time of Sir Isaac Newton." Of course, one reason for satisfaction in this discovery was that it subjected one of the most mysterious and frightening natural phenomena to rational explanation. It also proved that Bacon had been right in asserting that a knowledge of how nature really works might lead to a better control of nature itself: that valuable practical innovations might be the fruit of pure disinterested scientific research.

No doubt the most important effect of the lightning experiments was to show that the laboratory phenomena in which rods or globes of glass were rubbed, to the accompaniment of sparks, and induced charges and electrical shocks, belong to a class of phenomena occurring naturally. Franklin's experiments thus proved that electrical effects do not result exclusively from man's artifice, from his intervention in phenomena, but are in fact part of the routine operations of nature. And every "electrician" learned that experiments performed with little toys in the laboratory could reveal new aspects of one of the most dramatic of nature's catastrophic forces. "The discoveries made in the summer of the year 1752 will make it memorable in the history of electricity," William Watson wrote in 1753. "These have opened a new field to philosophers, and have given them room to hope, that what they have learned before in their museums, they may apply, with more propriety than they hitherto could have done, in illustrating the nature and effects of thunder; a phaenomenon hitherto almost inaccessible to their inquiries."

Franklin's achievement of a highly successful career wholly in the field of electricity marked the coming of age of electrical science and the full acceptance of the new field of specialization. On 30 November 1753, awarding Franklin the Royal Society's Sir Godfrey Copley gold medal for his discoveries in electricity, the earl of Macclesfield emphasized this very point: "Electricity is a neglected subject," he said, "which not many years since was thought to be of little importance, and was at that time only applied to illustrate the nature of attraction and repulsion; nor was anything worth much notice expected to ensue from it." But now, thanks to the labors of Franklin, it "appears to have a most surprising share of power in nature."

Some Later Contributions to Electricity

Spurred on by the success of the sentry-box and kite experiments, Franklin continued to make investigations of the lightning discharge and the electrification of clouds. He erected a test rod on his house, so as to make experiments and observations on clouds passing overhead. One of the results was most interesting, because he discovered: *"That the clouds of a thunder-gust are most commonly in a negative state of electricity, but sometimes in a positive state."* This statement led him to the following astonishing conclusion: "So that, for the most part, in thunderstrokes, *it is the earth that strikes into the clouds, and not the clouds that strike into the earth.*" Of course, this discovery did not alter the theory or practice of lightning rods, which Franklin found perform two separate functions. One is to disarm a cloud and to prevent a stroke, while the other is to conduct a stroke safely to the ground. His theory of the direction of the stroke (from clouds to earth or from earth to clouds) depends upon the identification of vitreous electrification (glass rubbed with silk) with the positive state and of resinous electrification (amber rubbed with wool or fur) with the negative. Franklin was aware that he had no definitive evidence for this identification, and hoped that others might provide a crucial experimental test.

To this day one still talks of a "Franklinian" fictitious "positive" current in circuit theory, and also thinks physically of a flow of electrons in the opposite direction.

One question of great interest to Franklin was whether the gross dimensions or the mass of a body may be the determining factor in the amount of "electric fluid" it can acquire. He discovered that an "increase of surface" makes a given mass or quantity of matter "capable of receiving a greater amount of charge." The surface is what counts, not the mass. As usual, Franklin had a pretty experiment to support his conclusion. In this case he used a small silver can on an insulating wine glass; in the can there were three yards of brass chain, one end of which was attached to a long silk thread that went over a pulley in the ceiling so that the chain could be drawn partly or completely out, thereby increasing the "surface" and making the body (can and chain) capable of receiving an additional charge.

In a closely related experiment Franklin studied the distribution of charge on a metal can placed on an insulated base. He showed that the charge "resides" wholly on the outside of the can; that there is no charge inside. He did not know the reason at first, but he later concluded that the symmetry of the situation produced mutual repulsion that drove any charge from the inside surface of the can to the outer one. Joseph Priestley, arguing from the analogy of a cylinder to a sphere, showed that by the reasoning of Isaac Newton's *Principia,* it would be possible for one to conclude that the law of electrical force must, like gravitation, be a law of the inverse square of the distance.

A Major Defect Remedied by Aepinus

Franklin's theory failed to give a satisfactory explanation of the observed phenomenon of the mutual repulsion of two negatively charged bodies. This defect was remedied by Franz Aepinus. Perplexed by the difficulties in explaining repulsion, Kinnersley thought that perhaps one could get rid of the doctrine of repulsion altogether. Franklin disagreed, putting forth the argument that repulsion occurs "in other parts of nature."

Aepinus, who altered Franklin's system, was an ardent Franklinian and a teacher of and collaborator with J. C. Wilcke, who translated Franklin's book on electricity into German. Wilcke made the first major table of what we would call today a triboelectric series, thus accounting for the production of joint negative and positive charges in different combinations of two materials.

Aepinus aimed to establish a theory of magnetic phenomena based upon "principles extremely similar to those on which the Franklinian electric theory is built," that is, using the concept of a magnetic fluid, with laws of action much like those of Franklin's electric theory. To complete his analogy, however, Aepinus introduced the revolutionary idea that in solids, liquids, and gases the particles that Franklin called "common matter" would—in the pure state—repel one another just as the particles of the electric fluid did. Aepinus' revision introduced a complete duality, the particles of common matter and the particles of electric matter each having the property of repelling particles of their own kind while having the additional property of attracting particles of the other kind. Normally one does not encounter particles of pure matter repelling one another, because their natural repulsion is reduced to zero by the presence of the magnetic or the electric fluid in the normal state of bodies. Hence, the Newtonian universal gravitation remains unaffected by the new postulate. Repulsion exists only when we deprive bodies of a part of their normal complement of either electric fluid or magnetic fluid.

Furthermore, certain experiments devised by Aepinus and Wilcke, using condensers separated by air instead of glass, showed that the Franklin doctrine of "atmospheres" could not exist in a physical sense. This was a position that Franklin himself had eventually more or less adopted, coming to conceive that the concept of "electrical atmospheres" was no more than a way of describing collections or distributions of electric charge whose parts have repulsive forces acting at a distance.

In one set of experiments to test the effect of "electrical atmospheres," Aepinus blew a stream of dry air on a charged body and found, just as Franklin had, that the charge of the body was not diminished. Franklin had then assumed that such experiments indicated only that the "atmosphere" of a charged body is an integral part of it, and he even thought to make the atmosphere "visible" by dropping rosin on a hot piece of iron near a charged body. Aepinus carried the matter through to its logical conclusion, saying that by "electrical atmosphere" one intended only to denote the "sphere of action" of the electrical charge on a body. Franklin, in commenting on Aepinus' book, expressed admiration for the magnetic theory which Aepinus had constructed along lines analogous to his own electrical theory, and he himself began to write of a magnetic fluid in the terms introduced by Aepinus. We do not know whether Franklin read the book very thoroughly, since he never referred to the great revision of his theory which Aepinus introduced. Indeed, by the time Aepinus' book (1759) reached him, Franklin was no longer actively pursuing his researches into electricity.

Gulf Stream, Convection Currents, and Storms

From his boyhood days Franklin had a passion for the sea. In his eight crossings of the Atlantic, he was always fascinated by problems of seamanship, ship design, and the science of the seas; and he made careful observations of all sorts of marine phenomena. He made experiments to see if oil spread on the waters would still the waves, and he put on a spectacular exhibition of this phenomenon for a group of fellows of the Royal Society in Portsmouth harbor.

Franklin's name is associated with the Gulf Stream, of which he printed the first chart. His interest in this subject began about 1770, when the Board of Customs at Boston complained that it seemed to require two weeks more for mail packets to make the voyage to New England from England than the time of voyage for merchant ships. Franklin, then still postmaster general, discussed the matter with a Nantucket sea captain, who explained that the Nantucketers were "well acquainted with the Stream, because in our pursuit of whales, which keep to the sides of it but are not met within it, we run along the side and frequently cross it to change our side, and in crossing it have sometimes met and spoke with those packets who are in the middle of it and stemming it." Franklin asked the captain, Timothy Folger, to plot the course of the Gulf Stream; this was the basis of the chart he had engraved and printed by the General Post Office. As early as 1775 Franklin had conceived of using a thermometer as an instrument of navigation in relation to the Gulf Stream, and he made several series of surface temperature measurements during the Atlantic crossings. In 1785, on his last return voyage from France, Franklin devised a special instrument to attempt to measure temperatures below the surface to a depth of 100 feet.

Franklin's studies of cloud formation and the electrification of clouds constitute a major contribution to the science of meteorology. He appears to have been the earliest observer to report that northeast storms move toward the southwest. He is also the first to have observed the phenomenon of convection in air.

Heat and Light

Franklin rejected the currently accepted corpuscular theory of light because of a mechanical argument. If "particles of matter called light" be ever so small, he wrote, their momentum would nevertheless be enormous, "exceeding that of a twenty-four pounder, discharged from a cannon." And yet, despite such "amazing" momentum, these supposed particles "will not drive before them, or remove, the least and lightest dust they meet with." The sun does not give evidence of a copious discharge of mass, since its gravitational force on the planets is not constantly decreasing.

Franklin's arguments were long considered the primary statement of the mechanical inadequacy of the "emission" theory and were still cited in 1835 in Humphrey Lloyd's report on optical theories to the British Association. Bishop Horsley, editor of Newton's *Opera,* made the official Newtonian reply in the *Philosophical Transactions* in 1770, noting that: "Dr. Franklin's questions are of some importance, and deserve a strict discussion." And when Thomas Young revived the wave theory toward the beginning of the nineteenth century, he cited Franklin as one of those predecessors who had believed in the wave theory: "The opinion of Franklin adds perhaps little weight to a mathematical question, but it may tend to assist in lessening the repugnance which every true philosopher must feel, to the necessity of embracing a physical theory different from that of Newton."

Franklin was perhaps more successful in his doctrine of fire. Here he tried to apply the principle of conservation to heat, assuming that there is a constant amount of heat, which is simply distributed, redistributed, conducted, or nonconducted, according to the kind of material in question. Interested in problems of heat conductivity, he designed a famous experiment, still performed in most introductory courses, in which a number of rods of different metals are joined together at one end and fanned out at the other, with little wax rings placed on them at regular intervals. The ends that are joined together are placed in the flame, and the "conductivity" is indicated by the relative speeds with which the wax rings melt and fall off. Franklin (in France) never had the occasion to perform the experiment, although he did obtain the necessary materials for doing so, and he suggested that Ingenhousz and he might do the experiment together. Ingenhousz, however, did it on his own. Franklin's experiments on heat were not fully understood until Joseph Black introduced the concepts of specific heat and latent heat.

Franklin's only major contribution to the theory of heat is in the specific area of differential thermal conduction. The success of his fluid theory of electricity, and his writings on heat as a fluid, did, however influence the later development of the concept of "caloric." Lavoisier wrote in 1777 that if he were to be asked what he understood by "matter of fire," he would reply, "with Franklin, Boerhaave, and some of the olden philosophers, that the matter of fire or of light is a very subtle and very elastic fluid. . . ."

Medicine and Hospitals

Throughout his life Franklin had a passion for exercise (notably swimming), for which he was an active propagandist. He was always an advocate of fresh air and had many arguments in France with those who held the night air to be bad for health and who believed—then as now—in the evil effects of drafts. I have referred to his invention of bifocal glasses; he also designed a flexible catheter. He wrote on a variety of medical subjects: lead poisoning, gout, the heat of the blood, the physiology of sleep, deafness, nyctalopia, infection from dead bodies, infant mortality, and medical education.

Although Franklin at one time had opposed the practice of inoculation, he later regretted his action and lamented the death of his own son from smallpox—which he publicly admitted might have been prevented by inoculation. He gathered a set of impressive statistics in favor of the practice, which were published in a pamphlet (London, 1759) on the benefits of inoculation against smallpox, accompanying William Heberden's instructions on inoculation.

Like others of his day, Franklin gave electric shocks in the treatment of paralysis. He concluded from his experiences that "I never knew any advantage from electricity in palsies that was permanent." He would not "pretend to say" whether—or to what degree—

there might have been an "apparent temporary advantage" due to "the exercise in the patients' journey, and coming daily to my house" or even—we may note with special interest today—the "spirits given by the hope of success, enabling them to exert more strength in moving their limbs."

Franklin's opinion that the beneficial effects of electrotherapy might derive more from the patient's belief in the efficacy of the cure than from any true curative powers of electricity is very much like one of the conclusions of the royal commission appointed in 1784 to investigate mesmerism, of which he was a member. This Commission was composed of four prominent members of the faculty of medicine and five members of the Royal Academy of Sciences (Paris), including Franklin, Bailly, and Lavoisier. Its report gave the death blow to mesmerism, and Mesmer had to leave Paris. The commission, apparently, did not see the psychological significance of their finding that "The imagination does everything, the magnetism nothing."

Later Life and Career

In spite of his extraordinary scientific accomplishments, the public at large knows of Franklin primarily as a statesman and public figure, and as an inventor rather than as a scientist—possibly because he devoted only a small portion of his creative life to scientific research. One of the three authors (along with Thomas Jefferson and John Adams) of the Declaration of Independence, he was a member of the Second Continental Congress and drew up a plan of union for the colonies. Sent to Paris in 1776 as one of three commissioners to negotiate a treaty, his fame preceded him, both for his personification of many ideas cherished in the Age of Enlightenment and for his great reputation in electricity; in 1773 he had been elected one of the eight foreign associates of the Royal Academy of Sciences. To many Frenchmen, his simplicity of dress, his native wit and wisdom, and his gentle manners without affectation seemed to indicate the virtues of a "natural man." In September 1778 he was appointed sole plenipotentiary, and in 1781 he was one of three commissioners to negotiate the final peace with Great Britain.

In France, Franklin enjoyed contact with many scientists and made the acquaintance of Volta, a strong supporter of Franklin's one-fluid theory; Volta began the next stage of electrical science with his invention of the battery, which made possible the production of a continuous electric current. Franklin appears to have been the first international statesman of note whose international reputation was gained in scientific activity.

Franklin returned to America in 1785, served the state of Pennsylvania, and was a member of the Constitutional Convention. He died on 17 April 1790 and was buried in Christ Church burial ground, Philadelphia.

BIBLIOGRAPHY

I. Original Works. Franklin's scientific communications consist of pamphlets, reports, articles, and letters, published separately or in journals, especially *Gentleman's Magazine* and *Philosophical Transactions of the Royal Society.* His major scientific publication, *Experiments and Observations on Electricity, made at Philadelphia in America,* was assembled by his chief correspondent, Peter Collinson, and published with an unsigned preface by John Fothergill (London, 1751); supps. are *Supplemental Experiments and Observations . . .* (London, 1753) and *New Experiments and Observations . . .* (London, 1754), the latter with a paper by John Canton and a "Defence of Mr Franklin against the Abbe Nollet" by D. Colden. Subsequent eds. are described in *Benjamin Franklin's Experiments: A New Edition of Franklin's Experiments and Observations on Electricity,* ed., with a critical and historical intro., by I. Bernard Cohen (Cambridge, Mass., 1941). In addition to five eds. in English (1753–1774), translations appeared in French (1752, 1756, 1773), German (1758), and Italian (1774). See, further, Paul Leicester Ford, *Franklin Bibliography: A List of Books Written by, or Relating to Benjamin Franklin* (Brooklyn, N.Y., 1889), a work that is useful as a guide, although incomplete.

Franklin's complete writings and correspondence are in publication as *The Papers of Benjamin Franklin,* Leonard W. Labaree, inaugural ed. (New Haven, Conn., 1959–). Three earlier eds. of Franklin's works may be noted: Jared Sparks, *The Works of Benjamin Franklin . . .,* 10 vols. (Boston, 1836–1840); John Bigelow, *The Complete Works of Benjamin Franklin,* 10 vols. (New York–London, 1887–1888; a "Federal Edition" in 12 vols., 1904); and Albert Henry Smyth, *The Writings of Benjamin Franklin,* 10 vols. (New York, 1905–1907).

Information on Franklin MSS is available in Henry Stevens, *Benjamin Franklin's Life and Writings: A Bibliographical Essay on the Stevens' Collection of Books and Manuscripts Relating to Doctor Franklin* (London, 1881); Worthington C. Ford, *List of the Benjamin Franklin Papers in the Library of Congress* (Washington, D.C., 1905); and I. Minis Hays, *Calendar of the Papers of Benjamin Franklin in the Library of the American Philosophical Society* [and University of Pennsylvania], 5 vols. (Philadelphia, 1908). See also Francis S. Philbrick, "Notes on Early Editions and Editors of Franklin," in *Proceedings of the American Philosophical Society,* **97** (1953), 525–564.

Selections from Franklin's writings include Nathan G. Goodman, *The Ingenious Dr. Franklin, Selected Scientific Letters of Benjamin Franklin* (Philadelphia, 1931); Carl Van Doren, *Benjamin Franklin's Autobiographical Writings* (New York, 1945); and I. Bernard Cohen, *Benjamin Franklin: His Contribution to the American Tradition* (Indianapolis–New

York, 1953). See also *The Complete Poor Richard Almanacks Published by Benjamin Franklin, Reproduced in Facsimile,* intro. by Whitfield J. Bell, Jr., 2 vols. (Barre, Mass., 1970).

A parallel text ed. of Franklin's autobiographical writings, containing the text of the original MS, is Max Farrand, *Benjamin Franklin's Memoirs* (Berkeley-Los Angeles, 1949); the most recent and scholarly ed. based on MS sources is Leonard W. Labaree, Ralph L. Ketcham, Helen C. Boatfield, and Helene H. Fineman, eds., *The Autobiography of Benjamin Franklin* (New Haven-London, 1964).

I. Bernard Cohen has published, with an intro., a facs. ed. of Franklin's *Some Account of the Pennsylvania Hospital* (Baltimore, 1954).

II. SECONDARY LITERATURE. The standard biography is Carl Van Doren, *Benjamin Franklin* (New York, 1938), possibly the best biography of a scientist in English. An admirable shorter biography is Verner W. Crane, *Benjamin Franklin and a Rising People* (Boston, 1954). Paul Leicester Ford, *The Many-sided Franklin* (New York, 1899) is still useful, esp. ch. 9, "The Scientist"; Bernard Faÿ, *Franklin, the Apostle of Modern Times* (Boston, 1929), lacks the valuable "Bibliographie et étude sur les sources historiques relatives à sa vie" included in vol. III of the French ed. (Paris, 1929–1931).

On Franklin in Europe, see Alfred Owen Aldridge, *Franklin and his French Contemporaries* (New York, 1957); Edward E. Hale and Edward E. Hale, Jr., *Franklin in France,* 2 vols. (Boston, 1888); and Antonio Pace, *Benjamin Franklin and Italy* (Philadelphia, 1958).

On Franklin and medicine, see Theodore Diller, *Franklin's Contribution to Medicine* (Brooklyn, N.Y., 1912); and William Pepper, *The Medical Side of Benjamin Franklin* (Philadelphia, 1911).

On lightning rods, see I. B. Cohen, "Prejudice Against the Introduction of Lightning Rods," in *Journal of the Franklin Institute,* **253** (1952), 393–440; "Did Diviš Erect the First European Protective Lightning Rod, and Was His Invention Independent?," in *Isis,* **43** (1952), 358–364, written with Robert E. Scholfield; and "The Two Hundredth Anniversary of Benjamin Franklin's Two Lightning Experiments and the Introduction of the Lightning Rod," in *Proceedings of the American Philosophical Society,* **96** (1952), 331–366.

Some other specialized studies of value are Cleveland Abbe, "Benjamin Franklin as Meteorologist," in *Proceedings of the American Philosophical Society,* **45** (1906), 117–128; Lloyd A. Brown, "The River in the Ocean," in *Essays Honoring Lawrence C. Wroth* (Portland, Me., 1951), pp. 69–84; N. H. de V. Heathcote, "Franklin's Introduction to Electricity," in *Isis,* **46** (1955), 29–35; Edwin J. Houston, "Franklin as a Man of Science and an Inventor," in *Journal of the Franklin Institute,* **161** (1906), 241–316, 321–383; Henry Stommel, *The Gulf Stream* (Berkeley-Los Angeles, 1958), ch. 1, "Historical Introduction"; Francis Newton Thorpe, *Benjamin Franklin and the University of Pennsylvania* (Washington, D.C., 1893); and Conway Zirkle, "Benjamin Franklin, Thomas Malthus and the United States Census," in *Isis,* **48** (1957), 58–62.

A bibliography up to 1956 may be found in I. Bernard Cohen, *Franklin and Newton, an Inquiry Into Speculative Newtonian Experimental Science and Franklin's Work in Electricity as an Example Thereof* (Philadelphia, 1956; Cambridge, Mass., 1966; rev. repr. 1972).

I. BERNARD COHEN

FRANKLIN, ROSALIND ELSIE (*b.* London, England, 25 July 1920; *d.* London, 16 April 1958), *physical chemistry, molecular biology.*

The daughter of a banking and artistic family previously unconnected with science, Rosalind Franklin was a foundation scholar at St. Paul's Girls' School, London, from which she won an exhibition to Newnham College, Cambridge, in 1938. After graduating in 1941 she stayed on to investigate gas-phase chromatography under Ronald Norrish. In 1942 she joined the British Coal Utilisation Research Association where, under D. H. Bangham, she applied her expertise in physical chemistry to the problem of the physical structure of coals and carbonized coals. From 1947 to 1950 Franklin worked under Jacques Méring at the Laboratoire Central des Services Chimiques de l'État, Paris, where she developed her skill in X-ray diffraction techniques and applied them to a detailed and illuminating study of carbons and of the structural changes accompanying graphitization. In 1951 she joined Sir John Randall's Medical Research Council unit at King's College, London, to apply these techniques to the problems of the structure of DNA, and in 1953 she moved to Birkbeck College, London, to work similarly on the even more exacting problems of virus structure.

At the British Coal Utilisation Research Association, Franklin developed, with Bangham and other workers, a hypothesis of the micellar organization of coals which provided a satisfactory explanation of their absorptive behavior toward liquids and gases and their thermal expansion. From her study of the fine porosity of a range of coals, by measurements of true and apparent densities, Franklin concluded that their structure was best represented by a model with pore constrictions which gave coals the properties of molecular sieves. In Paris she turned her attention to the application of X-ray diffraction methods to the problems of carbon structure and developed a procedure for the detailed interpretation of the diffuse X-ray diagram of carbons. This allowed her to describe the structure in more precise quantitative terms than had been possible, and she made use of it to study in detail the structural changes that accompanied the formation of graphite when these carbons were heated to high temperatures.

In the course of this work Franklin developed a

relation between the apparent interlayer spacing of the partially graphitized carbons and the proportion of disoriented layers, which has proved of considerable value in the industrial study of carbons. In addition, by studying the changes in structure that chars of different origin underwent on heating, she established that there are two distinct classes of carbons—those which form graphite on heating to high temperatures (the graphitizing carbons) and those which do not (the nongraphitizing carbons)—and related these differences in behavior to structural differences in the parent chars. She showed, in particular, that the graphitizability increases with the fine-structure porosity and this, in turn, she believed to be related to the cross-linking between the crystallites.

Franklin's work on coals brought her into contact with Charles Coulson, through whom she was introduced to Randall, and with the award of a Turner Newall Fellowship she went to work in the King's College Medical Research Council Biophysics Unit. At that time (January 1951) Raymond Gosling, under M. H. F. Wilkins' direction, had obtained diffraction pictures of DNA showing a high degree of crystallinity; sharper pictures were obtained with higher ambient humidity.

Franklin and Gosling conducted a systematic study of the effect of humidity on the X-ray pattern produced. Using salt solutions to control humidity, they showed that there are two distinct intramolecular patterns, which they found to be producible from the same specimen: the crystalline "A" pattern at 75 percent relative humidity and a new "wet" paracrystalline pattern at 95 percent relative humidity. In a report which Franklin gave on this work in November 1951, she described this discovery and went on to show, as Wilkins had a year before in Cambridge, that the patterns were consistent with a helical conformation. She discussed how the $A \rightleftharpoons B$ transformation takes place and suggested, quite correctly, that the phosphates are on the outside of the helices and in the "A" form are held parallel to each other by electrostatic attraction between O^- and Na^+. When water is added, it penetrates between the helices, thus destroying the electrostatic attraction which holds them in parallel alignment. She said little about the forces operating inside the helices but mentioned hydrogen bonding between keto and amino groups of the bases.

Despite this promising beginning Franklin was too professional a crystallographer to proceed further in this way. Instead, she thought to solve the structure of DNA in an inductive manner by using Patterson functions and superposition. While publicly she heaped scorn on those who were convinced that DNA is helical, in her unpublished reports she stated that such a conformation is probable for the B form and not inconsistent with the A form. A spurious case of double orientation encountered in April 1952, which when indexed showed marked radial asymmetry (all left-hand reflections were indexed *hkl* and all right-hand ones $hk\bar{l}$), led her to seek nonhelical structures for the A form. Earlier ambiguities in the indexing of the A diagram had led Franklin to embark on a Patterson analysis. This helped her to obtain accurate parameters for the unit cell. Yet the cylindrical Patterson function obtained by Gosling in July 1952 strengthened her antihelical views, although the arrangement of peaks was consistent with a helix. She was misled, by what appeared to be clear evidence of a structural repeat at half the height of the unit cell, into ruling out helices for the A form, since no DNA chain could possibly be folded into a helix with a pitch equal to half the height of the unit cell (fourteen Å.).

At this time Franklin was thinking in terms of antiparallel rods in pairs back-to-back, forming a double sheet structure. Then she investigated diagonal rod structures such as would simulate the diffraction pattern of a helix, but by January 1953, when she started model building, she found such structures impossible to build. Still rejecting single- or multistrand helices, she investigated a figure-eight structure in which a single chain formed a long column of repeating eights. This, she believed, would account for the halving of the unit cell in the cylindrical Patterson function and clearly provided a form of tight packing which could be unfolded to give the dramatic increase in length (30 percent) when structure A changes to structure B. She knew that the helix in the extended B form is close-packed and was doubtful that the same type of structure could pack down even more densely.

At the end of February 1953 Franklin turned to the B pattern, and for two weeks she weighed the merits of single and multiple helices. In a paper dated 17 March which she wrote with Gosling, she ruled out triple-strand and equally spaced double-strand helices and stated that "if there are two non-equivalent, i.e., unequally spaced coaxial chains these are separated by 3/8th. of the fibre axis period." This is the conformation of the sugar-phosphate backbones as found in the Watson-Crick model. On the following day Franklin returned to the Patterson function of the A form, only to learn that Watson and Crick had solved the structure of the B form. She and Gosling quickly expanded and rearranged their draft paper of 17 March in the light of the Cambridge discovery so that it could appear in the 25 April issue of *Nature,* which contains Watson and Crick's paper on their model.

Franklin deserves credit for having discovered the A ⇌ B transformation and characterized the diffraction patterns of these forms of DNA; for providing Watson and Crick with vital data, in particular the parameters of the unit cell; for exposing the errors in their first unpublished model; and for marshaling the evidence in favor of the phosphates being on the outside of the helix. It was also she who, with the aid of the special tilting camera built by Gosling, discovered the meridional reflection on the eleventh layer line in the A pattern and was the first to show how the B form can pack down more tightly to give the A form with eleven residues in one turn of the helix. Although she had been misled by the cylindrical Patterson function, this did provide the most refined evidence in favor of the Watson-Crick model at the time of its discovery in 1953. Franklin and Gosling's rarely cited paper on this subject appeared in *Nature* on 25 July 1953.

For the last five years of her life Franklin worked in the Crystallography Laboratory of Birkbeck College, London, supported first by the Agricultural Research Council and later by the U.S. Department of Health. There she continued to publish on her earlier work on coals, completed the writing up of her DNA work, and took up the structure of tobacco mosaic virus (TMV). By 1956 she had greatly improved on J. D. Watson's X-ray pictures of 1954. With the aid of material supplied by Heinz Fraenkel-Conrat and by Gerhard Schramm, Franklin and her co-workers, A. Klug and K. C. Holmes, were able to reject the picture of TMV as a solid cylinder with the RNA in the center and the protein subunits, possibly of two types, on the outside. They showed that the particles are hollow, that the protein subunits are structurally of one type only, and that forty-nine such units are packed in helical array around the axis in the axial repeat period of sixty-nine Å. The greatest achievement of the Franklin team was the location of the RNA helix embedded within the protein fraction at a radial distance of forty Å. from the axis. From a study of the X-ray diagram of TMV, Franklin and Klug resolved the discrepancy between estimates of the maximum radius and the packing radius by postulating the morphology of the protein as "a helical array of knobs, one knob for each sub-unit." Shortly before her death from cancer, Franklin instituted work which was later to justify her conclusion that the RNA in TMV is present in the form of a single-strand helix.

Franklin was a deft experimentalist, keenly observant and with immense capacity for taking pains. As a result she was able with difficult material to achieve a remarkable standard of resolution in her X-ray diagrams. Although a bold experimentalist, she was critical of speculation, favoring an inductive approach which proved very successful in her work on coals and TMV but which allowed others to get ahead of her in her work on DNA. Where those with a more intuitive approach rejected antihelical data as spurious, Franklin felt obliged to invent other conformations which might yield a helical-type pattern. She would not trust to the principle of exclusion, nor was she confident of the "obvious" deductions dictated by physical intuition. Hence her work was not marked by great originality of thought. Her theory of graphitization, for instance, although the best of its day, was traditional in character and belongs to what is now regarded as the "classical" period. Her great strength lay in her technical innovations and her employment of precise techniques on difficult macromolecules. When she died at the age of thirty-seven, she had won international recognition both as an industrial chemist and as a molecular biologist.

BIBLIOGRAPHY

I. ORIGINAL WORKS. Franklin and her co-workers published about forty papers. A complete list of her publications on the structure of viruses is in her paper written with D. L. D. Casper and A. Klug, "The Structure of Viruses as Determined by X-Ray Diffraction," in C. S. Holton *et al.*, eds., *Plant Pathology: Problems and Progress, 1908–1958* (Madison, Wis., 1959), pp. 447–461. This paper provides a broad review of Franklin's work and contains a tribute to her by W. M. Stanley.

Of her other papers the following appeared in *Acta crystallographica:* "The Interpretation of Diffuse X-Ray Diagrams of Carbon," **3** (1950), 107–117; "A Rapid Approximate Method for Correcting Low-Angle Scattering Measurements for the Influence of the Finite Height of the X-Ray Beam," *ibid.,* 158–159; "The Structure of Graphitic Carbons," **4** (1951), 253–261; "The *a* Dimension of Graphite," *ibid.,* 561; and "The Structure of Sodium Thymonucleate Fibres. I, II, & III," **6** (1953), 673–677, 678–685; **8** (1955), 151–156.

Other important papers are "A Note on the True Density, Chemical Composition and Structure of Coals and Carbonized Coals," in *Fuel,* **27** (1948), 46–49; "A Study of the Fine-Structure of Carbonaceous Solids by Measurements of True and Apparent Densities. Part I. Coals. Part II. Carbonized Coals," in *Transactions of the Faraday Society,* **45** (1949), 274–286, 668–682; "Crystallite Growth in Graphitizing and Nongraphitizing Carbons," in *Proceedings of the Royal Society,* **209A** (1951), 196–218; "Molecular Configuration in Sodium Thymonucleate," in *Nature,* **171** (1953), 740–741; and "Evidence for 2-Chain Helix in

Crystalline Structure of Sodium Deoxyribonucleate," *ibid.*, **172** (1953), 156–157.

Her last paper, written with A. Klug, was published after her death: "Order-Disorder Transitions in Structures Containing Helical Molecules," in *Discussions of the Faraday Society*, **25** (1958), 104–110.

II. SECONDARY LITERATURE. Obituary notices appeared in *The Times* (19 Apr. 1958), p. 3; and *Nature*, **182** (1958), 154.

For two very different accounts of Franklin's work on DNA, see J. D. Watson, *The Double Helix* (New York, 1968), *passim;* and A. Klug, "Rosalind Franklin and the Discovery of the Structure of DNA," in *Nature*, **219** (1968), 808–810, 843–844; *corrigenda*, 879, 1192; correspondence, 880.

ROBERT OLBY

FRAUNHOFER, JOSEPH (*b.* Straubing, Germany, 6 March 1787; *d.* Munich, Germany, 7 June 1826), *optics, optical instrumentation.*

Fraunhofer represents the highest order of the union of the craftsman and the theoretician. His family and early acquaintances were closely associated with the skilled craft tradition and particularly concerned with the glass and optical trades. As he acquired mastery of lens grinding, lens design, and glassmaking—through apprenticeship and independent study of optical books—Fraunhofer sought not merely to produce lenses which surpassed the best on the market but also to design and produce lenses which approached the optical ideal. In this pursuit he turned to the theoretical study of optics and light, a study which, when combined with his practical experience and understanding, ultimately made him the master theoretical optician of Europe and, as a by-product, led him to make numerous significant contributions to science.

Fraunhofer was the eleventh and last child of a poor master glazier, Franz Xaver Fraunhofer, and Maria Anna Fröhlich. After receiving only a limited elementary education, he entered his father's workshop. In November 1798, following the death of his parents, Fraunhofer's guardian apprenticed him to Philipp A. Weichselberger, a dull, unintellectual Munich master mirror-maker and glass cutter. Fraunhofer found his apprenticeship degrading and miserable, as his master discouraged further schooling and isolated him from his peers. Nevertheless, he maintained his goal of becoming a spectacles maker.

The fortunes of the sickly boy took an ironic turn for the better when, on 21 July 1801, the workshop-house collapsed, pinning him under the wreckage for some time. Elector Maximilian Joseph heard of the accident and presented him with the handsome sum of eighteen ducats. With the money Fraunhofer purchased a glass-working machine, books on optics, and release from the last six months of his six-year apprenticeship.

After a short, abortive business venture (producing engraving plates for visiting cards), Fraunhofer returned in November 1804 to work as a journeyman for Weichselberger until May 1806, when he entered the optical shop of the Munich philosophical (scientific) instrument company founded in 1802 by Joseph von Utzschneider, Georg von Reichenbach, and Joseph Liebherr.

Influenced by Ulrich Schiegg, a trained astronomer, and by Josef Niggl, the optics master under whom he served as a journeyman, Fraunhofer developed expertise in practical optics and acquired an interest in and a knowledge of mathematics and optical science. In 1809, in accordance with a contract negotiated between Utzschneider and Pierre Louis Guinand, Utzschneider designated Fraunhofer, who had criticized the available optical glass, to receive from Guinand instruction in his closely kept secrets of glassmaking. Guinand had moved from Switzerland to Bavaria in 1805 at the initiative of Utzschneider in order to supply the optical firm with glass; his tutelage allowed Fraunhofer to combine his understanding of optics with the practical knowledge of glassmaking. This instruction led to a two-year collaboration between Fraunhofer and Guinand that resulted in substantial increases in the size of glass blanks for lenses. Fraunhofer's advance in the firm—from journeyman in 1806, to manager of the optical workshop in 1809, to business partner with Utzschneider and director of the glassmaking (over Guinand) in 1811—was a reflection of his quick grasp of and original contributions to optical science, practical optical work, and glassmaking.

From 1819, when the optical workshop was returned to Munich from Benediktbeuern, he participated actively in the affairs of the Bavarian Academy of Sciences in Munich. In 1823, while still maintaining his active business schedule, he accepted the post of director of the Physics Museum of the academy and received the honorary title royal Bavarian professor. Fraunhofer initiated lectures on physical and geometrical optics shortly thereafter but had to discontinue them because of his frail health. Although he enjoyed neighborhood walks, he had little time for relaxation. Late in 1825 the lifelong bachelor contracted tuberculosis, from which he never recovered. During the last few years of his life Fraunhofer was elected to several foreign societies, including the Society of Arts in England, and received state honors from Denmark and his native Bavaria. The University of Erlangen conferred upon him the title of doctor of philosophy.

Fraunhofer was a blend of mathematically inclined natural philosopher, optical technician, and glassmaker. Although he had little formal education, he sought to understand optical theory and apply it to the practical work of constructing aberration-minimizing lens combinations. At the time he entered Utzschneider's instrument shop, the optical trade of Europe centered in London, where the leading names were Dollond and Ramsden. Yet even in London the lack of large blanks of homogeneous, striae-free crown and flint glass and the comparatively crude determinations of the optical constants of the glass limited the size and quality of lenses and restricted opticians to trial-and-error methods of optical construction.

Early in the nineteenth century the Munich firm that Fraunhofer joined took the lead in Germany in the manufacture of precision optical instruments and gained a gradual advantage over the London opticians, initially by obtaining the services of Guinand, who improved the making of optical glass. Later, Guinand and Fraunhofer, working together from 1809 to 1813, further improved the homogeneity of optical glass and increased the size of the striae-free blanks, so that large-diameter lenses could be made. Fraunhofer also sought to determine, with significantly greater precision than before, the dispersion and refractive index for different kinds of optical glass, so that he could abandon the traditional trial-and-error methods and approach lensmaking according to optical theory and calculation.

In order to determine precisely the optical constants of glass, Fraunhofer in 1814 used the two bright-yellow lines in flame spectra as a source of monochromatic light. With improved values for optical constants, he hoped to design and construct lens combinations in which the spherical aberration and coma could be eliminated. While conducting these tests, he observed the effect of the refracting medium on light, comparing the effect of light from flames with light from the sun, and found that the solar spectrum was crossed with many fine dark lines, a few of which William Hyde Wollaston had observed and reported upon in 1802. Designating the more distinct lines with capital letters (*A, B, C, D, · · · , I*), he mapped many of the 574 lines that he observed between *B* on the red end and *H* on the violet end of the spectrum. Somewhat later he noted that some of these lines appeared to correspond to the bright doublet of lines in many flame spectra; yet he noted further that while the pattern observed for the sun and planets appeared identical, the patterns for the sun, Sirius, and other bright stars differed from one another. These observations stimulated considerable interest for the next half-century among natural philosophers, whose speculations culminated in the classical explanation of absorption and emission spectra made by Kirchhoff and Bunsen in 1859. For Fraunhofer, however, these observations were primarily of importance in his efforts to perfect the achromatic telescope.

In 1821 and 1823, shortly after Fresnel's studies of interference phenomena had received general attention, Fraunhofer published two papers in which he observed and analyzed certain diffraction phenomena and interpreted them in terms of a wave theory of light. In the 1821 paper he discussed his examination of the spectra resulting from light diffracted through a single narrow slit and quantitatively related the width of the slit to the angles of dispersion of the different orders of spectra. Extending his observations to diffraction resulting from a large number of slits, he constructed a grating with 260 parallel wires.

Although David Rittenhouse and Thomas Young had previously noted some effects of crude diffraction gratings, Fraunhofer made the first quantitative study of the phenomena. The presence of the solar dark lines enabled him to note that the dispersion of the spectra was greater with his grating than with his prism. Hence, he examined the relationship between dispersion and the separation of wires in the grating. Utilizing the dark lines as bench marks in the spectrum for his dispersion determinations, he concluded that the dispersion was inversely related to the distance between successive slits in the grating. From the same study Fraunhofer was able to determine the wavelengths of specific colors of light. Somewhat later he also constructed a grating by ruling lines on glass covered by gold foil and, even later, constructed a reflecting grating. The latter prompted him to consider the effects of light obliquely incident to the grating.

In the paper prepared in 1823, Fraunhofer revealed his continued investigation of diffraction gratings. Using a diamond point, he could rule up to 3,200 lines per Paris inch. He continued his study of the effect of oblique rays, developed formulations based on a wave conception, and calculated a revised set of wavelengths for the major spectral lines. Thus, his earlier observations of the dark lines in the solar spectrum enabled him to make the highly precise measurements of dispersion; then his use of the wave theory of light allowed him to derive, with suitable simplifications, the general formulation of the grating equation still in use today. His other papers focused principally upon the design and construction of new instruments, and one paper examined atmospheric light phenomena.

Fraunhofer's scientific studies were intimately related to his professional object: the design and production of the finest possible optical and mechanical instruments. Utilizing the lines in the solar spectrum as bench marks, he determined with unprecedented precision the optical constants of various kinds of glass. The combination of superior optical glass, the theoretical design and calculation of lens systems, the accurate determination of optical constants, and the use of Newton's rings for testing of lens surfaces enabled the Utzschneider-Fraunhofer shop in Munich to wrest leadership in the production of optical instruments from the London opticians during the first quarter of the century.

Although Fraunhofer openly published his observations of the spectral lines and his interpretation of diffraction spectra, he retained as trade secrets his knowledge of optical glassmaking and his methods of calculating and testing lenses. Among his most famous instruments were the nine-and-a-half-inch Dorpat refracting telescope and equatorial mounting used by Wilhelm Struve and the six-and-a-quarter-inch Königsberg heliometer with which Friedrich Bessel measured the parallax of 61 Cygni in 1838. Such a detection and measurement of parallax had been sought unsuccessfully since antiquity.

After Fraunhofer's death, Utzschneider and, later, Siegmund Merz continued the Munich business, actively participating in the movement initiated by Fraunhofer to replace the large reflecting telescopes with the large refractors. Although his Munich optical shop did not continue to lead in innovation during its remaining half-century of existence, Fraunhofer's approach of combining practical with theoretical knowledge and an understanding of both optics and glassmaking had not only made the German optical industry the leading one in the world but also continued to inspire generations of German optical scientists and industrialists. Fraunhofer's direct successors in the nineteenth century thus included Josef Max Petzval; Johann Friedrich Voigtländer, Peter Friedrich Voigtländer, and Friedrich von Voigtländer; Carl August Steinheil and Adolph Steinheil; Philipp L. Seidel; Carl Zeiss; Ernst Abbe; and Otto Schott.

BIBLIOGRAPHY

I. Original Works. Fraunhofer's published works were collected in Eugen C. J. Lommel, ed., *Joseph von Fraunhofer's gesammelte Schriften* (Munich, 1888). A review of an unpublished 1807 paper appears in *Forschungen zur Geschichte der Optik,* **1** (May 1929), 42–51.

II. Secondary Literature. Two important nineteenth-century studies are Joseph von Utzschneider, "Kurzer Umriss der Lebens-Geschichte des Herrn . . . Fraunhofer," in *Dinglers polytechnisches Journal,* **21** (1826), 161–181; and Siegmund Merz, *Fraunhofer's Leben und Wirken . . .* (Landshut, 1865). Two twentieth-century works have provided the bases for all subsequent studies: the most comprehensive biographical study, by the outstanding Zeiss historian of optics, Moritz von Rohr, *Joseph Fraunhofers Leben, Leistungen und Wirksamkeit* (Leipzig, 1929); and A. Seitz, *Josef Fraunhofer und sein optisches Institut* (Berlin, 1926). Upon the centennial of Fraunhofer's death, *Naturwissenschaften,* **14** (1926), 522–554, was devoted to an evaluation of his work. More recent studies include H. Jebsen-Marwedel, *Joseph von Fraunhofer und die Glashütte in Benediktbeuern* (1963); and W. Gerlach, "Joseph Fraunhofer und seine Stellung in der Geschichte der Optik," in *Optik,* **20** (1963), 279–292. Twentieth-century literature in English on Fraunhofer is very limited and mostly derived from the works of Rohr and Seitz: Moritz von Rohr, "Fraunhofer's Work and Its Present Day Significance," in *Transactions of the Optical Society,* **27** (1926), 277–294; W. H. S. Chance, "The Optical Glassworks at Benediktbeuern," in *Proceedings of the Physical Society,* **49** (1937), 433–443; and Henry C. King, *The History of the Telescope* (Cambridge, Mass., 1955), *passim.*

Reese V. Jenkins

FRAZER, JAMES GEORGE (*b.* Glasgow, Scotland, 1 January 1854; *d.* Cambridge, England, 7 May 1941), *anthropology.*

Frazer, the elder son of Daniel F. Frazer and his wife, Katherine Brown, grew up in Helensburgh, near Glasgow, in an educated Presbyterian household. He attended Springfield Academy, Larchfield Academy, and the University of Glasgow, graduating in 1874 after following a broadly based curriculum which included mathematics, physics, logic, moral philosophy, and English literature as well as the classics, which were his main interest. He then went to Trinity College, Cambridge, on a scholarship, took the classical tripos in 1878, and in 1879, following research on Plato's theory of Ideas, was elected to a fellowship at the college, which was renewed for the rest of his life. Frazer maintained his interest in the classics, and his only original research was some archaeological fieldwork undertaken for his translation of Pausanius' *Description of Greece,* published in 1898: his anthropological work was deeply rooted in his classical studies.

In 1883 William Robertson Smith came to Cambridge as professor of Arabic and aroused Frazer's interest in anthropology. He was also editor of the *Encyclopaedia Britannica* and asked Frazer to write a number of articles, first on classical subjects and later, after Frazer had read E. B. Tylor's *Primitive Culture* (1871), on "Totem" and "Taboo," both published in 1888; these were later expanded into the four-volume work *Totemism and Exogamy* (1910). As early as 1885 Frazer read to the Anthropological Institute a paper on burial customs which clearly showed the influence of Tylor, and in 1890 he published the first edition of *The Golden Bough.* This title was taken from book VI of the *Aeneid,* and the work started as an investigation of the rites surrounding the priest at the Grove of Diana in Aricia. But in the process of compilation, it was expanded into a detailed comparative study similar to *Primitive Culture* but was better documented and included more material on rites and practices in European countries which appeared similar to those of more primitive societies. This was extended further into a work of twelve volumes (1911–1915) and an appendix (1936); volume XII was an alphabetical bibliography of some 5,000 items in most European languages, of which less than 1 percent were asterisked to indicate that Frazer had not seen them himself.

Frazer was invited to Liverpool to occupy the first post of professor of social anthropology in any university, which he held for one year (1907–1908), giving an inaugural lecture entitled "The Scope of Social Anthropology." But he was not happy away from Cambridge and his library; and although he gave the Gifford lectures in 1924 and 1925, he did not like lecturing, any form of teaching, or even controversial discussion with professional anthropologists. Frazer was knighted in 1914, was elected a fellow of the Royal Society in 1920 under the statute governing special elections on grounds of "conspicuous service to the cause of science," was made a member of the Order of Merit in 1924, and received other honors, both British and foreign; in 1905 he was granted a Civil List pension. He married a Frenchwoman, Mrs. Lilly Grove, in 1896, and she energetically assisted him in his work and in ensuring its public recognition: she translated several of his works into French and worked on the abridgment of *The Golden Bough* published in 1922. Both died on 7 May 1941.

Frazer made no original observations in anthropology and is widely criticized for having little or no direct contact with the "savages" he wrote about. Lienhardt attributes some of his popular success to the simplifications encouraged by lack of personal experience. His method was to read the published literature, often for twelve to fifteen hours a day, and to make notes which he later classified, assembled in groups showing the relationships of practices from different parts of the world, and discussed. Leach has shown that his desire to present an elegant prose sometimes led him to distort the original report. Frazer also corresponded with fieldworkers, who valued his stimulating questions and comments, as shown by Bronislaw Malinowski and Sir Baldwin Spencer's *Scientific Correspondence.* He compiled a list of questions for fieldworkers, with advice on collecting data, clearly aimed at the amateur, which was published first in 1887 and in its final form in 1907. Many of his correspondents were missionaries and administrators, and although he collected a large number of letters, now at Trinity College, he rarely cited information from them.

The development of Frazer's work shows both the strengths and the weaknesses of the inductive method. No one before or since has brought together such a volume of data on customs and beliefs, classified and documented to stimulate other workers; yet most of his theories, arising from cogitation in a library on secondhand data, have been modified or superseded, particularly his conception of the evolutionary succession of magic, religion, and science in culture. His analysis of magic into sympathetic and contagious types still has some validity, and Frazer himself never maintained that his own theories were more than transitory and propounded to be superseded: "After all, what we call truth is only the hypothesis which is found to work best." He also believed that if his writings survived, it would be "less for the sake of the theories they propound than for the sake of the facts they record."

Although Frazer was unwilling to meet or discuss with anthropologists who did not agree with him, a generation of fieldworkers, of whom Malinowski is probably the best-known, were drawn to the subject by his inspiration, and it developed quickly. His other main importance in his time was simply the popularization of comparative anthropology and the beginning of acceptance of his thesis that mankind is one and that "when all is said and done our resemblances to the savage are still far more numerous than our differences from him." The prose style, consciously modeled on that of the eighteenth century, was attractive enough to be widely read, and his demonstration of numerous myths strikingly similar to the Christian stories undermined some conventional religious beliefs, a point picked up by the historian Arnold Toynbee.

Frazer's later influence was probably greatest on literature: T. S. Eliot acknowledged a debt to him, while Ezra Pound, D. H. Lawrence, and others probably were influenced. In his *Totem and Taboo* (1913) Freud quotes data from Frazer, often approving his conclusions. Although Frazer remarked that "the sexual instinct has moulded the religious consciousness of our race," he always refused to read Freud. He was, nevertheless, well aware of discrepancies between acts and the explanations given for them, and that motives are not always those men are conscious of; and he came near to an appreciation of the unconscious.

More recent reappraisals (beginning with the centenary of Frazer's birth in 1954) have been critical of his academic integrity as an anthropologist and have valued him predominantly as a literary figure: his work is, nevertheless, still read and referred to.

BIBLIOGRAPHY

I. ORIGINAL WORKS. The most comprehensive bibliography is T. Besterman, *A Bibliography of Sir James George Frazer* (London, 1934); it is arranged chronologically and contains 266 items, together with a note by Frazer on his own notebooks. The originals of the notebooks are now in the British Museum, and the contents of many were published as *Anthologia anthropologica,* R. A. Downie, ed., 4 vols. (London, 1938–1939): they comprise extracts from material already published but not previously quoted by Frazer and are arranged alphabetically with indexes.

Editions of *The Golden Bough* in Frazer's lifetime are detailed by Besterman. In 1957, Macmillan, the original publishers, issued a two-volume paperback edition in their St. Martin's Library. Subsequently T. H. Gaster edited *The New Golden Bough* (New York, 1959), which gives a new abridgment of Frazer's work, including selections from the appendix *Aftermath,* and attempts by excision and rearrangement to bring the text up to date; it is the only short edition to contain references, and it also includes the editor's additional notes, with references, and a sizable index.

Frazer's first publication on anthropology, already profusely documented, was "On Certain Burial Customs as Illustrative of the Primitive Theory of the Soul," followed by discussion, in *Journal of the Anthropological Institute,* **15** (1885), 64–104. The last edition of his questions was issued as *Questions on the Customs, Beliefs, and Languages of Savages* (Cambridge, 1907). His letters to Spencer are in Sir Baldwin Spencer, *Spencer's Scientific Correspondence With Sir J. G. Frazer and Others,* R. R. Marett and T. K. Penniman, eds. (Oxford, 1932).

II. SECONDARY LITERATURE. The only full-scale biography is by his assistant, R. A. Downie: *James George Frazer: The Portrait of a Scholar* (London, 1940); the actual biographical section is short and the evaluation appreciative, but the main body of the work is a useful summary of the contents of Frazer's various works. B. Malinowski, "Sir James George Frazer: A Biographical Appreciation," in his *A Scientific Theory of Culture and Other Essays* (Chapel Hill, N.C., 1944), pp. 177–221, is a professional appraisal. There is an article by E. O. James in the *Dictionary of National Biography, Supplement 1941–1950* (Oxford, 1959), pp. 272–278; and obituaries by H. J. Fleur in *Obituary Notices of Fellows of the Royal Society of London,* **3** (1939–1941), 897–914; by R. Marett in *Proceedings of the British Academy,* **27** (1941), 377–391; and three by Cambridge friends in *Cambridge Review,* **62** (1941), 439–440, 457.

More recent evaluations are R. G. Lienhardt, "James George Frazer," in *International Encyclopedia of the Social Sciences,* D. L. Sills, ed., V (New York, 1968), 550–553; J. B. Vickery, "The Golden Bough: Impact and Archetype," in *Virginia Quarterly Review,* **39** (1963), 37–57; T. H. Breen, "The Conflict in the Golden Bough: Frazer's Two Images of Man," in *South Atlantic Quarterly,* **66** (1967), 179–194; A. Kardiner and E. Preble, "James Frazer: Labor Disguised as Ease," in *They Studied Man* (London, 1962), pp. 78–197; A. Goldenweiser, "Sir James Frazer's Theories," in his *History, Psychology and Culture* (London, 1933), pt. II, ch. 2, pp. 167–176; and three articles in *The Listener,* **51** (1954), for the centenary: Gilbert Murray, pp. 13–14; V. White, pp. 137–139; and A. Macbeath, pp. 217–218.

Early criticism came from Sir William Ridgeway, "The Methods of Mannhardt and Sir J. G. Frazer . . .," in *Proceedings of the Cambridge Philological Society,* nos. 124–126 (1924), 6–19. Later criticism is F. Huxley, "Frazer With the Bloody Wood," in *New Statesman and Nation,* **59** (1960), 561–562.

Radical criticism by E. R. Leach began with his "Golden Bough or Gilded Twig?," in *Daedalus,* **90** (1959), 371–387; this was in their series "Reputations" and was followed by H. Weisinger, "The Branch That Grew Full Straight," *ibid.,* 388–399. Leach continued with a review of I. C. Jarvie, *The Revolution in Anthropology* (London, 1964), of which sec. VI, ch. 1, was entitled "Back to Frazer." There was an exchange between Leach and Jarvie in *Encounter,* **25** (1965), 24–36; **26,** (1966), 53–56, 92–93; it was conveniently reprinted as a discussion, with additional comments by other anthropologists and a full bibliography of relevant material, in *Current Anthropology,* **7** (1966), 560–576.

DIANA M. SIMPKINS

FREDERICK II OF HOHENSTAUFEN (*b.* Iesi, Italy, 26 December 1194; *d.* Castelfiorentino, Italy, 13 December 1250), *natural sciences.*

Frederick II was the son of Emperor Henry VI and Constance of Sicily and was thus the grandson of both Frederick Barbarossa and Roger II of Sicily. He was crowned king of Sicily at Palermo in May 1198, following his father's death in 1197. Upon his mother's death six months later, Pope Innocent III became Frederick's guardian and regent of Sicily, a situation not ended until 1208, when Frederick came of age. In 1210 Emperor Otto IV invaded Sicily; the young Frederick in turn challenged Otto's rule in Germany the next year. The victory of his ally Philip II of France at Bouvines (1214) strengthened his position, and at Otto's death in 1218 Frederick was left unchallenged in Germany and northern Italy; he was crowned Holy Roman emperor in 1220. Thereupon Frederick turned to the restoration of order in Sicily, a process finally completed with the promulgation of the Constitutions of Melfi (*Liber Augustalis*) in 1231. If taxation in the resulting nonfeudal, centralized state was high, the coinage was stable and justice relatively easy to obtain; with the creation of the University of Naples in 1224 the emperor sought to bring even professional education under his control.

A struggle between Frederick and the papacy was now inevitable: Rome was caught between the empire to the north and Sicily to the south and was threatened by their possible union. The conflict broke out first in 1227, when Gregory IX excommunicated Frederick as the latter was setting out on crusade; the pope was forced to release him in 1230, after he had returned from the East with possession of Jerusalem secured by negotiation. Frederick now momentarily favored the restoration of the German princes' privileges, hoping to use the princes against the increasingly powerful Lombard cities. By 1239 Gregory IX had allied with the Lombard League and excommunicated the emperor once again. This time the conflict did not die away; war became general between imperial and papal factions, Ghibelline and Guelf, in Germany and Italy. The imperial position was severely weakened by the emperor's defeat before Parma in 1248, but Frederick had begun to regain the advantage when he died of a sudden fever in 1250.

What immediately strikes anyone attempting to understand Frederick II is his intense curiosity about the particulars of nature, most unusual in an age that was forever seeking universals. His contemporaries were struck by this too, as the famous stories recounted by the monk Salimbene show—the story, for example, that Frederick once disemboweled two men after giving them a hearty meal in order to determine the relative effects of sleep and exercise (he had sent one hunting) upon digestion. True or not—Salimbene was a Guelf partisan—the tale shows how people expected the emperor to behave. Anecdotes like this were of a piece with others about the exotic menagerie (which at one time or another actually included monkeys, camels, a giraffe, and an elephant) that moved with him in Italy and Germany.

That a serious spirit of inquiry was behind all this show is easily seen in the "many-sided patronage of learning" so prominent at Frederick's court. His kingdom of Sicily, heir to both the Greek and the Islamic cultures, had been a center of science and translation in the twelfth century under his grandfather, Roger II. Frederick continued this tradition, drawing scholars of widely different backgrounds and interests to his court. Two seem to have been of particular importance as advisers: the famous Michael Scot (from *ca.* 1228 until his death, *ca.* 1236) and a Master Theodore (*ca.* 1235–1250). The mathematician Leonardo Fibonacci, although not attached to the court, was well known there; the revised version of his *Liber abaci* is dedicated to Michael Scot, and the *Liber quadratorum* to Frederick himself. In addition, Frederick was in communication with other scholars—Christian, Jewish, and Muslim—far from court.

The court philosophers were regularly called upon to satisfy Frederick's curiosity. Michael Scot has left us a questionnaire put to him by the emperor, containing a wide range of problems: Precisely where are heaven, hell, purgatory, and the several abysses in relation to the earth and to each other? Why are there both sweet and salt waters on the earth, and whence do they arise? What gives rise to volcanic fire and smoke? We know of still other questions—metaphysical, mathematical, and optical (why do objects partly immersed in water appear bent?)—sent to Muslim and Jewish philosophers in Spain and Egypt. No program lay behind these questions; it was simply that the accepted commonplaces of experience regularly stirred Frederick to inquiry.

Such questions may show the breadth of Frederick's interests but not the depth of his own knowledge; the latter is fully revealed only in the famous *De arte venandi cum avibus,* a composition developed by Frederick over some thirty years, that we possess in a draft completed *ca.* 1244–1248, in part later emended by his son Manfred. (A finished version has apparently been lost, probably at Parma.) The first of the extant six books, which was conceived by Frederick as a necessary preliminary to the technical substance of falconry, is a remarkable survey of general ornithology: it moves from the classification of birds to their feeding habits, migration, mating, nesting, anatomy and physiology, flight, and molting. As impressive as the collected material is the systematic personal observation on which it is obviously based; Frederick describes his experimental determination that vultures locate their food by sight rather than by smell (ch. 10) and tells of his successful efforts to duplicate in Apulia the artificial incubation of eggs by sunlight that he had observed in Egypt (ch. 23). The later books are more technical and specialized, treating the training and rearing of falcons (book II), the use of the lure (book III), and the techniques of hawking with various birds (books IV–VI). But still they rest upon the emperor's own experience. His passion for the sport is apparent even in what survives of his correspondence, which shows him turning all his administrative resources to the instruction and supervision of his falconers. He was in fact away hawking at the moment when his siege of Parma was so disastrously broken.

Frederick's attitude in *De arte venandi* toward Aristotle as zoologist is of some interest and might be likened to that of a field naturalist toward a research biologist: critical if sometimes grudgingly respectful of the other's specialized knowledge. He points out that experience shows that Aristotle's deductions cannot always be relied upon and notes regretfully that "he was ignorant of the practice of falconry," but he refers his readers for supplementary taxonomic and embryological detail to Aristotle's *Historia animalium* (which had been translated from Arabic by Michael Scot at least a decade before his arrival at Frederick's court, and included the *De animalibus, De partibus animalium,* and *De generatione animalium*). At one point (book I, ch. 27) he even draws upon the (pseudo-) Aristotelian *Mechanica* to argue that the primary wing feathers must have the greatest power to carry a bird forward in flight. It remains characteristic of the emperor that he would disdain no source of knowledge, as long as it could be controlled by his own experience and judgment.

BIBLIOGRAPHY

I. ORIGINAL WORKS. The *De arte venandi cum avibus* exists in MS in two traditions, one including only the first two books, the other including all six. The two earliest MSS of the former tradition (MS Vat. Pal. Lat. 1071, copied perhaps as early as 1260, and MS Paris BN Fr. 12400, probably copied from the Vatican MS *ca.* 1310) are illustrated with a series of remarkable miniatures that presumably reflect Frederick's archetype; Haskins considered that the emperor himself had given the directions for these illustrations, which in the Vatican MS are strikingly faithful to nature. This two-book tradition has been twice edited, by Johann Velser (Augsburg, 1596) and by Johann Gottlieb Schneider (Leipzig, 1788–1789); the six-book version has been published (although without editorial remarks of any sort) by Karl Arnold Willemsen, 2 vols. (Leipzig, 1942). Willemsen has also reproduced many of the illustrations accompanying the *De arte venandi* in the Paris MS under the title *Die Falkenjagd. Bilder aus dem Falkenbuch Kaiser Friedrichs II* (Leipzig, 1943). An English trans. of all six books of Frederick's work has been published by Casey A. Wood and F. Marjorie Fyfe (Stanford, 1943); on pp. lvii–lxxxvii the editors examine the known MSS of the *De arte venandi* and cite two German trans. (of 1756 and 1896) of the two-book tradition besides the Velser and Schneider eds. C. H. Haskins' article "The *De arte* . . .," referred to below, also discusses the MS tradition of the work and forms the basis of the Wood-Fyfe treatment.

II. SECONDARY LITERATURE. A full if somewhat overly romantic study of Frederick's life and thought is Ernst Kantorowicz, *Kaiser Friedrich der Zweite,* 2 vols. (Berlin, 1927; repr. Düsseldorf–Munich, 1963); an English trans. by E. O. Lorimer, *Frederick the Second, 1194–1250* (London, 1931; repr. New York, 1957), omits the original's bibliography and footnotes. Kantorowicz treats Frederick's attitude toward science and the natural world with considerable insight on pp. 308–327 of the German version (pp. 334–365 of the English version). His treatment owes a great debt to the still fundamental articles of Charles Homer Haskins: "Science at the Court of the Emperor Frederick II," "Michael Scot," and "The *De arte venandi cum avibus* of Frederick II," collected in his *Studies in the History of Mediaeval Science* (Cambridge, Mass., 1924; repr. New York, 1960), pp. 242–326. A number of more recent special studies have been brought together in Gunther Wolf, ed., *Stupor mundi. Zur Geschichte Friedrichs II. von Hohenstaufen,* vol. CI of *Wege der Forschung* (Darmstadt, 1966), which includes a considerable extract from Martin Grabmann's *Mittelalterliches Geistesleben,* "Kaiser Friedrich II. und sein Verhältnis zur aristotelischen und arabischen Philosophie" (pp. 134–177 of *Stupor mundi*). The question of Frederick's knowledge and use of contemporary medical and natural-philosophical doctrine has been further studied in two articles by Johannes Zahlten: "Medizinische Vorstellungen im Falkenbuch Kaiser Friedrichs II.," in *Sudhoffs Archiv für Geschichte der Medezin und der Naturwissenschaften,* **54** (1970), 49–103; and "Zur Abhängigkeit der naturwissenschaftlichen Vorstellungen Kaiser Friedrichs II. von der Medizinschule zu Salerno," *ibid.,* 173–210.

MICHAEL MCVAUGH

FREDERICQ, LÉON (*b.* Ghent, Belgium, 24 August 1851; *d.* Liège, Belgium, 2 September 1935), *physiology.*

After his secondary studies in Ghent, Fredericq entered the University of Ghent in October 1868. His vocation was clearly defined: he intended to study natural sciences. After obtaining his doctor's degree in 1871 he became *préparateur* for the physiology course at the Faculty of Medicine and at the same time followed the medical curriculum. After receiving his M.D. in 1875, Fredericq went to Paris, where he attended the lectures of Ranvier and G. Pouchet, and to Strasbourg, where he attended the lectures of Waldeyer, E. Tiegel, A. Kundt, and especially Hoppe-Seyler, professor of biochemistry, who allowed him to spend the afternoons in his laboratory to learn biochemical techniques. In the summer of 1876 he went to the marine biological laboratory at Roscoff, where, under Lacaze-Duthiers, he studied the nervous physiology of sea urchins.

Back in Ghent in October 1876, Fredericq began

his classic work on blood coagulation. In December he moved to the laboratory of Paul Bert in Paris, where he learned the techniques of blood-gas analysis. There he completed his first classic experiments comparing the distribution of carbon dioxide between blood cells and blood plasma. These experiments were continued in Ghent, to which he returned in March 1877. During the summer of the same year he went again to Hoppe-Seyler's laboratory in Strasbourg and began a study of digestion in invertebrates, which he continued in Ghent during that autumn and winter. In April 1878, in order to obtain the degree of *docteur spécial* in physiology, he presented an important paper on his work on blood coagulation and on blood gases. In this paper Fredericq defined the coagulable protein of the plasma, fibrinogen, after its isolation by heat coagulation at 56°C. In horse plasma he distinguished three protein entities: fibrinogen, paralbumin (now called serum globulin), and serum albumin. He showed that in the lungs, the exchanges of oxygen and carbon dioxide are controlled by simple diffusion. Regarding the transport of carbon dioxide from the tissues to the lungs, Fredericq compared the distribution of the gas between plasma and cells at different partial pressures of carbon dioxide, thus beginning the series of investigations leading to our present knowledge of the transport of carbon dioxide.

During the summer of 1878 Fredericq was again in Roscoff and within a short period completed a masterly study of the physiology of the octopus, in which he described and named the copper-containing oxygen carrier hemocyanin. In December 1878 he worked in Marey's laboratory in Paris on respiratory innervation. He subsequently returned to Ghent, where he and G. Vandevelde began a study on the speed of nerve impulse in lobster nerves which was completed at Roscoff during the summer of 1879.

In October 1879 Fredericq was appointed professor of physiology at the University of Liège, succeeding Theodor Schwann, originator of the cell theory. In order to prepare plans for a new physiological institute, Fredericq visited Emil du Bois-Reymond in Berlin, where he came under the lasting influence of the great physiologist. In the first days of 1880 Fredericq settled in Liège; there, in September 1881, he married Bertha Spring, a sister of the chemist Walthère Spring.

In 1882, while at the North Sea, Fredericq tasted the blood of a lobster and of other marine invertebrates and found it as salty as seawater. He then tasted the blood of a number of saltwater fishes but found them not as salty as seawater, or about as salty as the blood of freshwater fishes, which are in turn more salty than fresh water itself. Moreover, while cutting the legs off crabs in order to obtain the blood, Fredericq discovered the phenomenon he called autotomy, the reflex casting off of a part of the body when an animal is attacked, the mechanism of which he explained later. In his blood-tasting experiments, Fredericq discovered the equal salinity of the internal and external media of marine invertebrates, while the blood salinity of bony fishes was found to be independent of their external medium.

During the following years Fredericq continued experiments on osmoregulation and in 1901, at the marine station in Naples, he used cryoscopy to determine the lowering of the freezing point of the bloods and of the juices extracted from the tissues of several animals. He also determined the amount of ash in the two series of samples and accounted for the difference in molecular concentration resulting from the cryoscopy experiments and the weighing of the ash by postulating the existence of important amounts of small organic molecules in the tissues of marine invertebrates and in the blood and tissues of Elasmobranchs. Urea was identified later by E. Rodier (1900) in elasmobranchs, but the intracellular organic components of the marine invertebrates, which compensate for the lack of a high concentration of inorganic constituents in their blood, despite the fact that they must maintain osmotic equilibrium with seawater, were identified only recently as amino acids (M. C. Camien, H. Sarlet, G. Duchâteau, and M. Florkin, 1951).

In March 1882, Fredericq began a series of experiments on the regulation of temperature in mammals. For these studies, he devised a respiratory apparatus which allowed the estimation of the oxygen used by a simple volume determination, without a gas analysis. This apparatus was the ancestor of all devices used for the indirect measurement of metabolism in man. With the help of this apparatus Fredericq showed, among other things, that the curves of heat production and of oxygen consumption both show a minimum, the point of thermic neutrality. He also showed that homoiotherms resist an increase or a decrease in temperature by different mechanisms. The memorable experiments of Legallois, carried out in 1812, had led to the conclusion that the respiratory center must be located in the medulla oblongata. Several physiologists had been reluctant to accept this conclusion. By progressive cooling, Fredericq recorded a number of facts which strengthened the notion that the respiratory center is located in the medulla. This localization was also supported by his cross-circulation experiments, carried out from 1887, in which the blood from the carotid artery of one dog, A, goes to the head of another dog, B, and the head of the latter receives blood only from the first dog. If dog A inspires air poor in oxygen, it is dog B who shows the symptoms of dyspnea. The method of cross circulation has since been used in a great deal of important experimental research.

Research pursued in Pflüger's laboratory had been interpreted as showing that oxygen and carbon dioxide move from a higher-pressure region to a lower-pressure one, in accordance with the laws of diffusion. On the other hand, in 1888 and 1891, Christian Bohr had published a series of results which tended to show that the lung plays a secretory role in the absorption of oxygen and the elimination of carbon dioxide, both moving in a direction contrary to that which would agree with the laws of diffusion. Fredericq built an aerotonometer more exact than those of his predecessors and, in 1895, showed clearly that gas exchanges obey the laws of diffusion in the gills of aquatic animals as well as in the lungs of air-breathing animals.

The first topic studied by Fredericq in the field of circulation physiology concerned the oscillations of blood pressure. He defined three categories of these oscillations: small, numerous ones corresponding to cardiac beats; less frequent oscillations related to the respiratory movement (Traube–Hering curves); and vasomotor oscillations (Sigmund Mayer's curves).

Fredericq is one of the physiologists who has made the greatest contributions to the interpretation of the mechanisms of heart contraction. He recognized that a pulsation starts in the right auricle and spreads rapidly to both auricles, then travels slowly to the bundle of His to radiate through the muscle of both ventricles (1906). Fredericq also studied the phenomenon of fibrillation, described by Ludwig and M. Hoffa in 1850. He demonstrated that ventricular fibrillation has no effect on the auricles and does not cross the bundle of His, while auricular fibrillation, through irregular stimulations through the bundle of His, is the cause of ventricular arrhythmia.

Fredericq's work remains typical of the classical period of physiology. Centered on such topics as heat regulation, respiration, the heart, and circulation, it has played an important role in laying the bases of experimental medicine.

Fredericq was a man of many talents. An excellent draftsman and watercolorist, a great traveler, a botanist, and an entomologist, he devoted a large part of his activities after retirement to an extensive study of the subalpine region of the Hautes Fagnes in the Ardennes. There he established a scientific station, and a museum of the fauna, flora, and geology of the region bears his name.

Albert I, king of the Belgians, recognized Fredericq's outstanding merits by making him a baron in 1931.

BIBLIOGRAPHY

A complete list of Fredericq's publications is in his biography by Marcel Florkin, cited below (1943).

On Fredericq and his work, see M. Florkin, *Léon Fredericq et les débuts de la physiologie en Belgique* (Brussels, 1943); "Emil du Bois-Reymond et Léon Fredericq," in *Chronique de l'Université de Liège* (Liège, 1967), pp. 181–198; and "Léon Fredericq, 1851–1935," in *Florilège des sciences en Belgique* (Brussels, 1967), pp. 1015–1034; M. Florkin and Z. M. Bacq, *Un pionnier de la physiologie. Léon Fredericq* (Liège, 1953), a collection of extracts from Fredericq's works; and P. Nolf, "Léon Fredericq," in *Annuaire de l'Académie royale de Belgique* (1937), pp. 47–100.

MARCEL FLORKIN

FREDHOLM, (ERIK) IVAR (*b.* Stockholm, Sweden, 7 April 1866; *d.* Stockholm, 17 August 1927), *applied mathematics.*

Ivar Fredholm's small but pithy output was concentrated in the area of the equations of mathematical physics. Most significantly, he solved, under quite broad hypotheses, a very general class of integral equations that had been the subject of extensive research for almost a century. His work led indirectly to the development of Hilbert spaces and so to other more general function spaces.

Fredholm was born into an upper-middle-class family; his father was a wealthy merchant and his mother, née Stenberg, was from a cultured family. His early education was the best obtainable, and he soon showed his brilliance. After passing his baccalaureate examination in 1885, he studied for a year at the Polytechnic Institute in Stockholm. During this single year he developed an interest in the technical problems of practical mechanics that was to last all his life and that accounted for his continuing interest in applied mathematics. In 1886 Fredholm enrolled at the University of Uppsala—the only institution in Sweden granting doctorates at that time—from which he received the Bachelor of Science in 1888 and the Doctor of Science in 1898. Because of the superior instruction available at the University of Stockholm, Fredholm also studied there from 1888, becoming a student of the illustrious Mittag-Leffler. He remained at the University of Stockholm the rest of his life, receiving an appointment as lecturer in mathematical physics in 1898 and in 1906 becoming professor of rational mechanics and mathematical physics.

Fredholm's first major work was in partial differential equations. His doctoral thesis, written in

1898 and published in 1900, involved the study of the equation—written in Fredholm's own operator notation—

$$f\left(\frac{\partial}{\partial x_1}, \frac{\partial}{\partial x_2}, \frac{\partial}{\partial x_3}\right) u = 0,$$

where $f(\xi,\eta,\zeta)$ is a definite homogeneous form. This equation is significant because it occurs in the study of deformation of anisotropic media (such as crystals) subjected to interior or exterior forces. Initially, Fredholm solved only the particular equations associated with the physical problem; in 1908 he completed this work by finding the fundamental solution to the general elliptical partial differential equations with constant coefficients.

Fredholm's monument is the general solution to the integral equation that bears his name:

$$(1) \qquad \phi(x) + \int_0^1 f(x,y)\phi(y)\,dy = \psi(x).$$

In this equation the functions f (called the kernel) and ψ are supposed to be known continuous functions, and ϕ is the unknown function to be found. This type of equation has wide application in physics; for example, it can be shown that solving a particular case of (1) is equivalent to solving $\Delta u + \lambda u = 0$ for u, an equation that arises in the study of the vibrating membrane.

Equation (1) had long been under investigation, but only partial results had been obtained. In 1823, Niels Abel had solved a different form of (1) that also had a particular kernel. Carl Neuman had obtained, in 1884, a partial solution for (1) by use of an iteration scheme, but he had to impose certain convexity conditions to ensure convergence of his solution. By 1897 Vito Volterra had found a convergent iteration scheme in the case where f has the property that $f(x,y) = 0$ for $y > x$.

Fredholm began work on equation (1) during a trip to Paris in 1899, published a preliminary report in 1900, and presented the complete solution in 1903. His approach to this equation was ingenious and unique. Fredholm recognized the analogy between equation (1) and the linear matrix-vector equation of the form $(I + F)U = V$. He defined

$$D_f = 1 + \sum_{n=1}^{\infty} \frac{1}{n!} \int_0^1 \int_0^1 \cdots \int_0^1 [f(x_i,x_j)] dx_1\,dx_2 \cdots dx_n,$$

where $[f(x_i,y_j)]$ is the determinant of the $n \times n$ matrix whose ij^{th} component is $f(x_i,x_j)$. The quantity D_f, Fredholm showed, plays the same role in equation (1) that the determinant of $I + F$ plays in the matrix equation; that is, there is a unique solution ϕ of equation (1) for every continuous function ψ whenever D_f is not zero. Furthermore, he proved that the homogeneous equation associated with equation (1),

$$(2) \qquad \phi(x) + \int_0^1 f(x,y)\phi(x)\,dy = 0,$$

has a nontrivial solution (that is, one not identically zero) if and only if $D_f = 0$.

The analogy between the matrix and integral equations was further pointed up by Fredholm when he defined the nth order minor of f—denoted by $D_f\begin{pmatrix}\xi_1, \xi_2, \cdots, \xi_n \\ \eta_1, \eta_2, \cdots, \eta_n\end{pmatrix}$—by an expression similar to that for D_f. Then he showed that if D_f is not zero, an explicit representation for the solution of equation (1) similar to Cramer's rule was given by

$$\phi(x) = \psi(x) - \int_0^1 \frac{D_f\begin{pmatrix} x \\ y \end{pmatrix}}{D_f} \psi(y)\,dy.$$

Fredholm then went on to show that if D_f is equal to zero, the dimension of the null space (the vector space of the set of solutions) of equation (2) is finite dimensional. He did this by setting

$$\Phi_i(x) = \frac{D_f\begin{pmatrix}\xi_1, \xi_2, \cdots, \hat{\xi}_i, \cdots, \xi_n \\ \eta_1, \eta_2, \quad \cdots, \quad \eta_n\end{pmatrix}}{D_f\begin{pmatrix}\xi_1, \xi_2, \cdots, \xi_n \\ \eta_1, \eta_2, \cdots, \eta_n\end{pmatrix}}, \quad i = 1, 2, \cdots n,$$

where the denominator is a nonvanishing minor of least nth (finite) order (which he showed always exists) and where the same minor is denoted in the numerator, but with $\hat{\xi}_i$ replaced by the variable x. Then he proved the set $\{\Phi_i : i = 1,2,\cdots,n\}$ to be a basis for the null space of equation (2). Finally, to solve equation (1) in the case $D_f = 0$, Fredholm first showed that a solution will exist when and only when $\psi(x)$ is orthogonal to the null space of the transposed homogeneous equation—or, equivalently, when

$$\int_0^1 \psi(x)\Psi_i(x)\,dx = 0 \ (i = 1,2, \cdots, n),$$

where $\{\Psi_i : i = 1,2, \cdots, n\}$ is a basis for the null space of the equation

$$\phi(x) + \int_0^1 f(y,x)\phi(y)\,dy = 0.$$

In this case, the solutions are not unique but can be represented by

$$\phi(x) = \psi(x) + \int_0^1 g(x,y)\psi(y)\,dy + \sum_{i=1}^{n} a_i\Phi_i,$$

$$\text{where} \quad g(x,y) = \frac{-D_f\begin{pmatrix} x & \xi_1, & \xi_2, & \cdots, & \xi_n \\ y & \eta_1, & \eta_2, & \cdots, & \eta_n \end{pmatrix}}{D_f\begin{pmatrix} \xi_1, & \xi_2, & \cdots, & \xi_n \\ \eta_1, & \eta_2, & \cdots, & \eta_n \end{pmatrix}}$$

and $\{a_i : i = 1,2, \cdots, n\}$ is a set of arbitrary constants.

Thus, Fredholm proved that the analogy between the matrix equation $(I + F)U = V$ and equation (1) was complete and even included an alternative theorem for the integral equation. Yet he showed more. His result meant that the solution $\phi(x)$ for equation (1) could be developed in a power series in the complex variable λ

$$\phi(x) = \psi(x) + \sum_{p=1}^{\infty} \psi_p(x)\lambda^p$$

which is a meromorphic function of λ for every λ satisfying $D_{\lambda f} \neq 0$. (To see this, replace $f(x,y)$ with $\lambda f(x,y)$ in equation [1].) This result was so important that, unable to prove it, Henri Poincaré was forced to assume it in 1895–1896 in connection with his studies of the partial differential equation $\Delta u + \lambda u = h(x,y)$.

Fredholm's work did not represent a dead end. His colleague Erik Holmgren carried Fredholm's discovery to Göttingen in 1901. There David Hilbert was inspired to take up the study; he extended Fredholm's results to include a complete eigenvalue theory for equation (1). In the process he used techniques that led to the discovery of Hilbert spaces.

BIBLIOGRAPHY

The *Oeuvres complètes de Ivar Fredholm* (Malmö, 1955) includes an excellent obituary by Nils Zeilon.

Ernst Hellinger and Otto Toeplitz, "Integralgleichungen und Gleichungen mit unendlichvielen Unbekannten," in *Encyklopädie der mathematische Wissenschaften*, (Leipzig, 1923–1927), pt. 2, vol. III, art. 13, 1335–1602, was also published separately (Leipzig-Berlin, 1928). It presents an excellent historical perspective of Fredholm's work and the details of his technique; it also contains an excellent bibliography.

M. BERNKOPF

FREGE, FRIEDRICH LUDWIG GOTTLOB (*b.* Wismar, Germany, 8 November 1848; *d.* Bad Kleinen, Germany, 26 July 1925), *logic, foundations of mathematics.*

Gottlob Frege was a son of Alexander Frege, principal of a girl's high school, and of Auguste Bialloblotzky. He attended the Gymnasium in Wismar, and from 1869 to 1871 he was a student at Jena. He then went to Göttingen and took courses in mathematics, physics, chemistry, and philosophy for five semesters. In 1873 Frege received his doctorate in philosophy at Göttingen with the thesis, *Ueber eine geometrische Darstellung der imäginaren Gebilde in der Ebene.* The following year at Jena he obtained the *venia docendi* in the Faculty of Philosophy with a dissertation entitled "Rechungsmethoden, die sich auf eine Erweitung des Grössenbegriffes gründen," which concerns one-parameter groups of functions and was motivated by his intention to give such a definition of quantity as gives maximal extension to the applicability of the arithmetic based upon it. The idea presented in the dissertation of viewing the system of an operation *f* and its iterates as a system of quantities, which in the introduction to his *Grundlagen der Arithmetik* (1884) Frege essentially ascribes to Herbart, hints at the notion of *f*-sequence expounded in his *Begriffsschrift* (1879).

After the publication of the *Begriffsschrift,* Frege was appointed extraordinary professor at Jena in 1879 and honorary professor in 1896. His stubborn work toward his goal—the logical foundation of arithmetic—resulted in his two-volume *Grundgesetze der Arithmetik* (1893–1903). Shortly before publication of the second volume Bertrand Russell pointed out in 1902, in a letter to Frege, that his system involved a contradiction. This observation by Russell destroyed Frege's theory of arithmetic, and he saw no way out. Frege's scientific activity in the period after 1903 cannot be compared with that before 1903 and was mainly in reaction to the new developments in mathematics and its foundations, especially to Hilbert's axiomatics. In 1917 he retired. His *Logische Untersuchungen,* written in the period 1918–1923, is an extension of his earlier work.

In his attempt to give a satisfactory definition of number and a rigorous foundation to arithmetic, Frege found ordinary language insufficient. To overcome the difficulties involved, he devised his *Begriffsschrift* as a tool for analyzing and representing mathematical proofs completely and adequately. This tool has gradually developed into modern mathematical logic, of which Frege may justly be considered the creator.

The *Begriffsschrift* was intended to be a formula language for pure thought, written with specific symbols and modeled upon that of arithmetic (i.e., it develops according to definite rules). This is an essential difference between Frege's calculus and, for example, Boole's or Peano's, which do not formalize mathematical proofs but are more flexible in expressing the logical structure of concepts.

One of Frege's special symbols is the assertion sign $\vdash$—— (properly only the vertical stroke), which is interpreted if followed by a symbol with judgeable content. The interpretation of $\vdash\!\!\!-\!\!\!-\, A$ is "A is a fact." Another symbol is the conditional $\top\!\!\!\!\lfloor$, and $\top{A \atop \lfloor B}$ is to be read as "B implies A." The assertion $\vdash\!\top{A \atop \lfloor B}$ is justified in the following cases: (1) A and B are true; (2) A is true and B is false; (3) A and B are false.

Frege uses only one deduction rule, which consists in passing from $\vdash\!\!\!-\!\!\!-\, B$ and $\vdash\!\top{A \atop \lfloor B}$ to $\vdash\!\!\!-\!\!\!-\, A$. The assertion that A is not a fact is expressed by $\vdash\!\!\!\top\!\!\!-\, A$, i.e., the small vertical stroke is used for negation. Frege showed that the other propositional connectives, "and" and "or," are expressible by means of negation and implication, and in fact developed propositional logic on the basis of a few axioms, some of which have been preserved in modern presentations of logic. Yet he did not stop at propositional logic but also developed quantification theory, which was possible because of his general notion of function. If in an expression a symbol was considered to be replaceable, in all or in some of its occurrences, then Frege calls the invariant part of the expression a function and the replaceable part its argument. He chose the expression $\Phi(A)$ for a function and, for functions of more than one argument, $\Psi(A,B)$. Since the Φ in $\Phi(A)$ also may be considered to be the replaceable part, $\Phi(A)$ may be viewed as a function of the argument Φ. This stipulation proved to be the weak point in Frege's system, as Russell showed in 1902.

Generality was expressed by

$$\vdash\!\smile^{\mathfrak{a}}\!-\, \Phi(\mathfrak{a})$$

which means that $\Phi(a)$ is a fact, whatever may be chosen for the argument. Frege explains the notion of the scope of a quantifier and notes the allowable transition from $\vdash\!\!\!-\!\!\!-\, X(a)$ to $\vdash\!\smile^{\mathfrak{a}}\!-\, X(\mathfrak{a})$, where a occurs only as argument of $X(a)$, and from $\vdash\!\top{\Phi(a) \atop \lfloor A}$ to $\vdash\!\top\!\smile^{\mathfrak{a}}\!-{\Phi \atop \lfloor A}(\mathfrak{a})$, where a does not occur in A and in $\Phi(a)$ occurs only in the argument places.

Existence was expressed by $\vdash\!\top\!\smile^{\mathfrak{a}}\!\top\, \Lambda(\mathfrak{a})$. There was no explicitly stated rule of substitution.

It should be observed that Frege did not construct his system for expressing pure thought as a formal system and therefore did not raise questions of completeness or consistency. Frege applied his *Begriffsschrift* to a general theory of sequences, and in part III he defines the ancestral relation on which he founded mathematical induction. This relation was afterward introduced informally by Dedekind and formally by Whitehead and Russell in *Principia mathematica.*

The *Begriffsschrift* essentially underlies Frege's definition of number in *Grundlagen der Arithmetik* (1884), although it was not used explicitly. The greater part of this work is devoted to a severe and effective criticism of existing theories of number. Frege argues that number is something connected with an assertion concerning a concept; and essential for the notion of number is that of equality of number (i.e., he has to explain the sentence "The number which belongs to the concept F is the same as that which belongs to G."). He settled on the definition "The number which belongs to the concept F is the extension of the concept of being equal to the concept F," where equality of concepts is understood as the existence of a one-to-one correspondence between their extensions. The number zero is that belonging to a concept with void extension, and the number one is that which belongs to the concept equal to zero. Using the notion of f-sequence, natural numbers are defined, with ∞ the number belonging to the notion of being a natural number.[1]

Frege's theories, as well as his criticisms in the *Begriffsschrift* and the *Grundlagen,* were extended and refined in his *Grundgesetze,* in which he incorporated the essential improvements on his *Begriffsschrift* that had been expounded in the three important papers "Funktion und Begriff" (1891), "Über Sinn und Bedeutung" (1892), and "Über Begriff und Gegenstand" (1892). In particular, "Über Sinn und Bedeutung" is an essential complement to his *Begriffsschrift*. In addition, it has had a great influence on philosophical discussion, specifically on the development of Wittgenstein's philosophy. Nevertheless, the philosophical implications of the acceptance of Frege's doctrine have proved troublesome.

An analysis of the identity relation led Frege to the distinction between the sense of an expression and its denotation. If a and b are different names of the same object (refer to or denote the same object), we can legitimately express this by $a = b$, but $=$ cannot be considered to be a relation between the objects themselves.

Frege therefore distinguishes two aspects of an expression: its denotation, which is the object to which it refers, and its sense, which is roughly the thought expressed by it. Every expression expresses its sense. An unsaturated expression (a function) has no denotation.

These considerations led Frege to the conviction that a sentence denotes its truth-value; all true sentences denote the True and all false sentences denote the False—in other words, are names of the True and the False, respectively. The True and the False are to be treated as objects. The consequences of this distinction are further investigated in "Über Begriff und Gegenstand." There Frege admits that he has not given a definition of concept and doubts whether this can be done, but he emphasizes that concept has to be kept carefully apart from object. More interesting developments are contained in his "Funktion und Begriff." First, there is the general notion of function already briefly mentioned in the *Grundlagen,* and second, with every function there is associated an object, the so-called *Wertverlauf,* which he used essentially in his *Grundgesetze der Arithmetik.*

Since a function is expressed by an unsaturated expression $f(x)$, which denotes an object if x in it is replaced by an object, there arises the possibility of extending the notion of function because sentences denote objects (the True [T] and the False [F]), and one arrives at the conclusion that, e.g., $(x^2 = 4) = (x > 1)$ is a function. If one replaces x by 1, then, because $1^2 = 4$ denotes F, as does $1 > 1$, it follows that $(1^2 = 4) = (1 > 1)$ denotes T.

Frege distinguishes between first-level functions, with objects as argument, and second-level functions, with first-level functions as arguments, and notes that there are more possibilities. For Frege an object is anything which is not a function, but he admits that the notion of object cannot be logically defined. It is characteristic of Frege that he could not take the step of simply postulating a class of objects without entering into the question of their nature. This would have taken him in the direction of a formalistic attitude, to which he was fiercely opposed. In fact, at that time formalism was in a bad state and rather incoherently maintained. Besides, Frege was not creating objects but was concerned mainly with logical characterizations. This in a certain sense also holds true for Frege's introduction of the *Wertverlauf,* which he believed to be something already there and which had to be characterized logically.

In considering two functions, e.g., $x^2 - 4x$ and $x(x - 4)$, one may observe that they have the same value for the same argument. Therefore their graphs are the same. This situation is expressed by Frege true for Frege's introduction of the *Wertverlauf,* as $x^2 - 4x$." Without any further ado he goes on to speak of the *Wertverlauf* of a function as being something already there, and introduced a name for it. The *Wertverläufe* of the above mentioned functions $x(x - 4)$ and $x^2 - 4x$ are denoted by $\grave{\alpha}(\alpha - 4)$ and $\grave{\epsilon}(\epsilon^2 - 4\epsilon)$ respectively, and in general $\grave{\epsilon}f(\epsilon)$ is used to denote the *Wertverlauf* of function $f(\xi)$. This *Wertverlauf* is taken to be an object, and Frege assumes the basic logical law characterizing equality of *Wertverläufe:*

$$(\grave{\epsilon}f(\epsilon) = \grave{\alpha}\, g(\alpha)) = (\neg\!\overset{\mathfrak{a}}{\smile}\!\!-\, f(\mathfrak{a}) = g(\mathfrak{a})).$$

Frege extends this to logical functions (i.e., concepts), which are conceived of as functions whose values are truth-values, and thus extension of a concept may be identified with the *Wertverlauf* of a function assuming only truth-values. Therefore, e.g., $\grave{\epsilon}(\epsilon^2 = 1) = \grave{\alpha}(\alpha + 1)^2 = 2(\alpha + 1)$.

In the appendix to volume II of his *Grundgesetze,* Frege derives Russell's paradox in his system with the help of the above basic logical law. Russell later succeeded in eliminating his paradox by assuming the theory of types.

It is curious that the man who laid the most suitable foundation for formal logic was so strongly opposed to formalism. In volume II of the *Grundgesetze,* where he discusses formal arithmetic at length, Frege proves to have a far better insight than its exponents and justly emphasizes the necessity of a consistency proof to justify creative definitions. He is aware that because of the introduction of the *Wertverläufe* he may be accused of doing what he is criticizing. Nevertheless, he argues that he is not, because of his logical law concerning *Wertverläufe* (which proved untenable).

When Hilbert took the axiomatic method a decisive step further, Frege failed to grasp his point and attacked him for his imprecise terminology. Frege insisted on definitions in the classic sense and rejected Hilbert's "definition" of a betweenness relation and his use of the term "point." For Frege geometry was still the theory of space. But even before 1814 Bolzano had already reached the conclusion that for an abstract theory of space, one may be obliged to assume the term point as a primitive notion capable of various interpretations. Hilbert's answer to Frege's objections was quite satisfactory, although it did not convince Frege.

BIBLIOGRAPHY

I. ORIGINAL WORKS. Frege's writing includes *Ueber eine geometrische Darstellung der imaginären Gebilde in der Ebene,* his inaugural diss. to the Faculty of Philosophy at Göttingen (Jena, 1873); *Begriffsschrift, eine der arithmetischen nachgebildete Formelsprache des reinen Denkens* (Halle, 1879), 2nd ed., I. Angelelli, ed. (Hildesheim, 1964), English trans. in J. van Heijenoort, *From Frege to Godel* (Cambridge, 1967), pp. 1–82; *Die Grundlagen der Arithmetik* (Breslau, 1884), trans. with German text by J. L. Austin (Oxford, 1950; 2nd rev. ed. 1953), repr. as *The Foundations of Arithmetic* (Oxford, 1959); *Function und Begriff* (Jena, 1891), English trans. in P. Geach and M. Black, pp. 21–41 (see below); "Über Sinn und Bedeutung," in *Zeitschrift für Philosophie und philosophische Kritik,* n.s. **100** (1892), 25–50, English trans. in P. Geach and M. Black, pp. 56–78 (see below); "Über Begriff und Gegenstand," in *Vierteljahrschrift für wissenschaftliche Philosophie,* **16** (1892), 192–205, English trans. in P. Geach and M. Black, pp. 42–55 (see below); *Grundgesetze der Arithmetik,* 2 vols. (Jena, 1893–1903), repr. in 1 vol. (Hildesheim, 1962); and "Über die Grundlagen der Geometrie," in *Jahresbericht der Deutschen Mathematikervereinigung,* **12** (1903), 319–324, 368–375; **15** (1906), 293–309, 377–403, 423–430.

II. SECONDARY LITERATURE. On Frege and his work, see I. Angelelli, *Studies on Gottlob Frege and Traditional Philosophy* (Dordrecht, 1967); P. Geach and M. Black, *Translation of the Philosophical Writings of Gottlob Frege,* 2nd ed. (Oxford, 1960); H. Hermes, *et al., Gottlob Frege, Nachgelassene Schriften* (Hamburg, 1969); J. van Heijenoort, *From Frege to Godel* (Cambridge, 1967); P. E. B. Jourdain, "The Development of the Theories of Mathematical Logic and the Principles of Mathematics. Gottlob Frege," in *Quarterly Journal of Pure and Applied Mathematics,* **43** (1912), 237–269; W. Kneale and Martha Kneale, *The Development of Logic* (Oxford, 1962) pp. 435–512; J. Largeault, *Logique et philosophie chez Frege* (Paris–Louvain, 1970); C. Parsons, "Frege's Theory of Number," in M. Black, ed., *Philosophy in America* (London, 1965), pp. 180–203; G. Patzig, *Gottlob Frege, Funktion, Begriff, Bedeutung* (Göttingen, 1966); and *Gottlob Frege, Logische Untersuchungen* (Göttingen, 1966); B. Russell, *The Principles of Mathematics* (Cambridge, 1903), Appendix A, "The Logical and Arithmetical Doctrines of Frege"; M. Steck, "Ein unbekannter Brief von Gottlob Frege über Hilberts erste Vorlesung über die Grundlagen der Geometrie," in *Sitzungsberichte der Heidelberger Akademie der Wissenschaften,* Abhandlung 6 (1940); and "Unbekannte Briefe Frege's über die Grundlagen der Geometrie und Antwortbrief Hilbert's an Frege," *ibid.,* Abhandlung 2 (1941); H. G. Steiner, "Frege und die Grundlagen der Geometrie I, II," in *Mathematische-physikalische Semesterberichte,* n.s. **10** (1963), 175–186, and **11** (1964), 35–47; and J. D. B. Walker, *A Study of Frege* (Oxford, 1965).

B. VAN ROOTSELAAR

FREIESLEBEN, JOHANN KARL (*b.* Freiberg, Saxony, Germany, 14 June 1774; *d.* Nieder-Auerbach, Saxony, Germany, 20 March 1846), *geology, mineralogy, mining.*

Freiesleben came from an old Freiberg mining family. This circumstance, coupled with the active, centuries-old mining industry in the vicinity of Freiberg, determined his choice of mining science as his profession. While still a secondary school student he worked as a miner in pits and galleries. Freiesleben attended the Mining Academy in Freiberg from 1790 to 1792, and there he found a patron in Abraham Gottlob Werner, professor of geology and mineralogy. When Leopold von Buch and Alexander von Humboldt came to Freiberg in order to study under Werner, Freiesleben became friendly with both of them. He made his first scientific journey through Saxony and Thuringia with Buch, and with his friend E. F. von Schlotheim he explored the Thuringian Forest. With Humboldt he journeyed to Bohemia, and in 1795 they traveled in the Swiss Jura, the Alps, and Savoy. A lasting friendship developed between the two men. From 1792 to 1795 Freiesleben studied jurisprudence in Leipzig and often visited the Harz Mountains.

Upon returning from the journey to Switzerland, Freiesleben obtained a position as a mining official in Marienberg and Johanngeorgenstadt. In the latter city he married Marianne Caroline Beyer, the daughter of a clergyman, in 1800. In the same year he became director of the copper and silver mines in Eisleben. In this capacity he did much work in the technical aspects of mining and in science. He returned to Freiberg in 1808. There he joined the Bureau of Mines and was entrusted with the management of various governmental and corporate mines and metallurgical works. In 1838 he was placed in charge of all Saxon mining operations. He was pensioned in 1842. He died in 1846 after a short illness, while on an official tour.

Freiesleben was awarded a doctorate by the University of Marburg in 1817, and in 1828 he became a member of the Prussian Academy of Sciences in Berlin. He was a loyal friend and an adviser and benefactor to the lonely and needy; he was, moreover, closely bound to his family.

A product of Freiesleben's trips to the Harz Mountains was one of his first major works: *Bemerkungen über den Harz* (1795), in which mineralogical and technical mining observations and descriptions stand out. Freiesleben completed his most important work, *Beitrag zur Kenntniss des Kupferschiefergebirges* (1807–1815), during his stay in Eisleben. In this he presents a painstaking and detailed description of the Permian Kupferschiefer and of the accompanying formations of the Zechstein and the Lower Permian Rotliegend, as well as of the Triassic. For decades this work remained indispensable to science and technology. Only when he attempted to trace individual formations into neighboring regions did he make errors. Thus he equated the limestone in the Swiss Jura and the Swabian Alb (now known as Malm) with the similar-appearing dolomite of the Zechstein (Upper Permian) in Thuringia and Saxony because of its many caverns. He also compared the Cretaceous Alpine limestone with the Upper Permian limestone, and the Cretaceous Quadersandstein of the northern edge of the Harz with the Triassic Bunter sandstone of Saxony.

Beginning in 1820 Freiesleben published the *Magazin für die Oryktographie von Sachsen.* It consisted of individual volumes written by him and formed one of the most important sources of information about mineral occurrence in Saxony. Freiesleben himself composed twelve consecutive volumes and three special volumes; the remaining four volumes were published after his death. Of particular importance in this series is his work on the ore veins in Saxony (1843–1845). In this journal there are also descriptions of sediments and proof of the occurrence of fossil mammalian bones and teeth in Saxony (**7**, 1836).

Along with his scientific publications, Freiesleben was occupied with the administrative, technical, and mining matters connected with his official duties. He devised a series of technical improvements in mining and metallurgy, primarily in the mining of the Kupferschiefer in Eisleben and Mansfeld. He also found time to prepare for publication the extensive group of works on mineralogy, geology, and mining that he wrote while in office. Furthermore, during his years in Freiberg he maintained a constant interest in the administration and the social problems of his native city and was actively involved in them.

Freiesleben was one of the most gifted and most learned students of Werner, the leading geologist of the time in Europe. In his writings he helped to apply the theories of his teacher and to make them more widely known. Through his exact and reliable descriptions of the Permian and Triassic sedimentary formations in Saxony and Thuringia he pushed beyond Werner to a greater knowledge of the stratigraphic relationships of these periods in central Germany and helped to create the basis for stratigraphic comparisons with neighboring regions. His painstaking and accurate data on mining in Saxony and on the occurrence of minerals there are still valuable for historical purposes.

BIBLIOGRAPHY

I. Original Works. Freiesleben's writings include "Geognostisch-bergmännische Beobachtungen auf einer Reise durch Saalfeld, Camsdorf und einen Theil Thüringens," in *Lempe's Magazin für Bergbaukunde,* **10** (1793), 3–114; *Bemerkungen über den Harz,* 2 vols. (Leipzig, 1795); *Geognostischer Beitrag zur Kenntniss des Kupferschiefergebirges mit besonderer Hinsicht auf einen Theil der Grafschaft Mansfeld und Thüringens,* 4 vols. (Freiberg, 1807–1815); *Beyträge zur mineralogischen Kenntniss von Sachsen,* 2 vols. (Freiberg, 1817); and *Magazin für die Oryktographie von Sachsen,* **1–12** and 3 spec. vols. published by Freiesleben; **13–15** and spec. vol. 4 published after his death by C. H. Müller (Freiberg, 1820–1848).

II. Secondary Literature. On Freiesleben or his work, see H. Claus, "Beiträge zur Geschichte der geologischen Forschung in Thüringen," in *Beiträge zur Geologie von Thüringen,* I (Jena, 1927), 9–10; B. von Freyberg, *Die geologische Erforschung Thüringens in älterer Zeit* (Berlin, 1932), pp. 25, 29, 32, 36, 41, 51, 59, 71, 82, 83, 88, 89, 90, 94, 101, 102, 106, 109; C. W. von Gümbel, "Freiesleben, Johann Karl," in *Allgemeine deutsche Biographie,* VII (Leipzig, 1878), 339–340, with bibliography; B. F. Voigt, "Johann Karl Freiesleben," no. 52 in *Neuer Nekrolog der Deutschen,* XXIV, pt. 1 (Weimar, 1848), 191–196, with bibliography; and K. A. von Zittel, *Geschichte der Geologie und Paläontologie bis Ende des 19. Jahrhunderts* (Munich–Leipzig, 1899), pp. 120–121, also pp. 93, 268, 494, 500, 608, 609.

Heinz Tobien

FREIND, JOHN (*b.* Croughton, Northamptonshire, England, 1675; *d.* London, England, 26 July 1728), *chemistry, medicine.*

Freind's father, William, rector of Croughton, sent his three sons to Westminster School and Christ Church, Oxford, to follow in his footsteps. Freind's latinity attracted the favorable notice of Dean Henry Aldrich and led to his first publication of Latin translations. At Oxford he met Francis Atterbury, then a tutor and later bishop of Rochester, with whom he was to be associated politically. Freind was B.A. (1698), M.A. (1701), M.B. (1703), and, by diploma, M.D. (1707). In 1704 he gave, by invitation, nine lectures on chemistry at the Ashmolean Museum, later published as *Praelectiones chymicae.* These are notable for Freind's adoption of the principles of Newtonian attraction (which he derived from the lectures given by John Keill), in an attempt to make chemistry truly mechanical. He tried to estimate quantitatively the relative forces operating between particles in order to explain association, dissociation, calcination, distillation, fermentation, and all other chemical processes. The publication of a second edition at Amsterdam in 1710 provoked an unfavorable review in the *Acta eruditorum,* to which Freind replied in the *Philosophical Transactions of the Royal Society* for 1712. The Leipzig attack was part of the Leibniz–Newton polemic, and the criticism was based not upon Freind's chemistry but upon his Newtonianism, and was so regarded by Newton's friends (Arnold Thackray, "Matter in a Nutshell," in *Ambix,* **15** [1968], 35–36). Ironically, this may have assisted Freind's election as a fellow of the Royal Society in March 1712.

Freind had already left Oxford and begun his medical career, first as physician to the English forces in the 1705 campaign under the earl of Peterborough, whose defense he was soon to write, then in Italy and later in Flanders as physician to the duke of Ormonde. He married Anne Morice in 1709 and soon returned to London, where he practiced very successfully. In 1716 Freind became a fellow of the Royal College of Physicians, in whose affairs he was subsequently active, and he began to write on medical topics. *Emmenologia* (1717) displays a leaning toward mechanistic physiology, but most of his other medical works are concerned with therapeutics.

In 1722, having weathered the storm of controversy arising from his association with Peterborough, Freind became M.P. for Launceston and, having strong Jacobite leanings, became involved in Atterbury's plot and was for some months confined to the Tower on a charge of high treason. From the Tower he wrote to his friend Richard Mead—who was subsequently to secure his release—a letter on smallpox and also sent him his *History of Physick.* (Mead had sent him a copy of Daniel Leclerc's *Histoire de la médecine.*) This was long regarded as an authoritative work, especially on English medieval and Renaissance medicine, although the first volume is concerned entirely with post-Galenic Greek writers, and much of the second with Islamic physicians; it is especially strong on medical treatment. Soon after his release Freind was appointed physician to the royal children and in 1727 to Queen Caroline.

BIBLIOGRAPHY

I. Original Works. *Opera omnia medica,* John Wigan, ed. (London, 1733; Venice, 1733; Paris, 1735), contains all Freind's major scientific and medical writings: *Praelectiones chymicae* (London, 1709; Amsterdam, 1710), English trans. by "J. M." as *Chymical Lectures,* with app. (London, 1712); *Emmenologia, in qua fluxus mulieribus menstrui phaenomena . . . ad rationes mechanicas exiguntur* (Oxford, 1703; 2nd ed., London, 1717; Paris, 1727), English trans. by T. Dale (London, 1729), also a French trans. (Paris, 1730); Freind's revised version, with extensive commentary, of *Hippocratis de morbis popularibus liber primus et tertius* (London, 1717); *J. F. de purgantibus, in secunda variolarum confluentium febre adhibendis, epistola* (London, 1719), English trans. by T. Dale, with the commentaries on Hippocrates (London, 1730); *J. F. ad R. Mead de quibusdam variolarum generibus epistola* (London, 1723), English trans. by T. Dale (London, 1730); "Oratio anniversaria ex Harvaio instituto," delivered to the College of Physicians in 1720; and *Historia medicinae,* a Latin trans. by John Wigan of *The History of Physick; From the Time of Galen to the Beginning of the Sixteenth Century* (London, 1725, 1726; 2nd ed., London, 1727; 5th ed., 1758), French trans. by Stephen Coulot (Leiden, 1727). He also published three papers in the *Philosophical Transactions of the Royal Society:* "Concerning a Hydrocephalus," no. 256 (1699), 318-322; "A Case of an Extraordinary Cramp," no. 270 (1701), 799-804; and "A Vindication of His Chymical Lectures" [against the attack in *Acta eruditorum,* Leipzig, 1710], no. 331 (1712), 330-342, which is printed in English in the 1712 ed. of *Chymical Lectures.*

Freind's undergraduate eds. of Aeschines the Orator and of Ovid's *Metamorphoses* were published at Oxford in 1696. *An Account of the Earl of Peterborow's Conduct in Spain* (London, 1706, 1707, 1708) provoked considerable controversy and a number of replies.

There is a portrait of Freind in the Royal College of Physicians, an engraved portrait frontispiece to the *Opera omnia,* and a medal with his portrait executed by Saint Urbain. There is a contemporary monument in Westminster Abbey.

II. Secondary Literature. There is a short biography in the preface to the *Opera omnia* by John Wigan (in Latin). A much better account is in the abridgment of the *Philosophical Transactions* by Charles Hutton, George Shaw, and Richard Pearson (London, 1809), IV, 423. There is a fair account in William Munk, *Roll of the Royal College of Physicians* (London, 1851), II, 441-450. The *Dictionary of National Biography* has a very full account of his political activities and a fair summary of his medical interests.

Marie Boas Hall

FRÉMY, EDMOND (*b.* Versailles, France, 28 February 1814; *d.* Paris, France, 2 February 1894), *chemistry.*

Frémy began his career at the École Polytechnique as assistant to Pelouze and succeeded him as professor in 1846. He also became professor at the Muséum d'Histoire Naturelle when Gay-Lussac died in 1850 and was elected its director after Chevreul retired in 1879.

Frémy's first project was to continue Pelouze's studies of iron oxides, and he expanded them to include oxides of chromium, tin, and antimony that form salts with alkalies in the same way as manganese. In 1835 he published a memoir in the *Annales de chimie* on the splitting of fats by sulfuric acid, a process that was adopted by French industry. From then on, Frémy pursued scientific investigations as professor and industrial work as consultant (later as administrator of the Compagnie de Saint-Gobin). He proposed improvements in the chamber process for making sulfuric acid (low temperature and ample air and water), and he introduced the residue from burning pyrites as the raw material for iron production. From research on the setting of hydraulic cement, Frémy proceeded to the synthesis of rubies by heating alumina with potassium chromate and barium fluoride.

At the museum, from 1850 until 1879, he sought to prove the transformation of plant materials, especially "vasculose" (cellulose), into coal by way of lignite and ulmic acid.

Together with Pelouze, Frémy published a textbook that saw several editions until 1865; then he organized the collaboration of professors and industrialists on a chemical encyclopedia, which appeared in ninety-one parts between 1882 and 1901.

Paul Dehérain, his biographer and former student, said of Frémy: "He disliked theories, did not know them well, and thought them dangerous."

BIBLIOGRAPHY

Frémy's main work was *Encyclopédie chimique, publiée sous la direction de M. Frémy . . . par une réunion d'anciens élèves de l'École Polytechnique, de professeurs et d'industriels,* 10 vols. (Paris, 1882–1901). He wrote the "Discours préliminaire" (1882) and chs. in several vols., especially vol. V, *Applications de chimie inorganique;* see "Généralités sur quelques industries chimiques" (1883), in vol. V, sec. 1, pt. 2, in which he summarizes much of his own work and emphasizes its originality. He was coauthor, with T. J. Pelouze, of *Traité de chimie générale,* 6 vols. (Paris, 1854–1857).

A biography is P.-P. Dehérain, "Edmond Frémy," in *Revue générale des sciences* (18 Feb. 1894), 20–31.

EDUARD FARBER

FRENET, JEAN-FRÉDÉRIC (*b.* Périgueux, France, 7 February 1816; *d.* Périgueux, 12 June 1900), *mathematics.*

Frenet was the son of Pierre Frenet, a *perruquier.* In 1840 he entered the École Normale Supérieure and later studied at the University of Toulouse, where he received the doctorate for the thesis *Sur les fonctions qui servent à déterminer l'attraction des sphéroïdes quelconques. Programme d'une thèse sur quelques propriétés des courbes à double courbure* (1847). The latter part of the thesis was subsequently published in the *Journal de mathématiques pures et appliquées* (1852) and contains what are known in the theory of space curves as the Frenet-Serret formulas. Frenet, however, presents only six formulas explicitly, whereas Serret presents all nine. Frenet subsequently explained the use of his formulas in "Théorèmes sur les courbes gauches" (1853).

After a period as a professor in Toulouse, Frenet went to Lyons, where in 1848 he became professor of mathematics at the university. He was also director of the astronomical observatory, where he conducted meteorological observations. He retired in 1868 with the title of honorary professor and settled at Bayot, a family estate in his native Périgueux. Unmarried, he lived quietly with a sister until his death.

Frenet's constantly revised and augmented *Recueil d'exercises sur le calcul infinitésimal* (1856) was popular for more than half a century. It contains problems with full solutions and often with historical remarks.

Frenet was a man of wide erudition and a classical scholar who was respected in this community, but his mathematical production was limited.

BIBLIOGRAPHY

Frenet's best-known works are *Sur les fonctions qui servent à déterminer l'attraction des sphéroïdes quelconques. Programme d'une thèse sur quelques propriétés des courbes à double courbure* (Toulouse, 1847); "Sur quelques propriétés des courbes à double courbure," in *Journal de mathématiques pures et appliquées,* **17** (1852), 437–447; "Théorèmes sur les courbes gauches," in *Nouvelles annales de mathématiques,* **12** (1853), 365–372; and *Recueil d'exercises sur le calcul infinitésimal* (Paris, 1856; 7th ed., 1917).

Minor mathematical papers are "Note sur un théorème de Descartes," in *Nouvelles annales de mathématiques,* **13** (1854), 299–301; "Sur une formule de Gauss," in *Mémoires de la Société des sciences physiques et naturelles de Bordeaux,* **6** (1868), 385–392. Meteorological observations are in *Mémoires de l'Académie impériale de Lyon, Classe des sciences,* **3** (1853), 177–225; **6** (1856), 263–326; and **8** (1858), 73–121, continued afterward by A. Drian.

An obituary of Frenet is in *L'avenir de Dordogne* (17 June 1900).

D. J. STRUIK

FRENICLE DE BESSY, BERNARD (*b.* Paris, France, *ca.* 1605; *d.* Paris, 17 January 1675), *mathematics, physics, astronomy.*

Frenicle was an accomplished amateur mathematician and held an official position as counselor at the Cour des Monnaies in Paris. In 1666 he was appointed member of the Academy of Sciences by Louis XIV. He maintained correspondence with the most important mathematicians of his time—we find his letters in the correspondence of Descartes, Fermat, Huygens, and Mersenne. In these letters he dealt mainly with questions concerning the theory of numbers, but he was also interested in other topics. In a letter to Mersenne, written at Dover on 7 June 1634, Frenicle described an experiment determining the trajectory of bodies falling from the mast of a moving ship. By calculating the value of g from Frenicle's data we obtain a value of 22.5 ft./sec.^{2}, which is not far from Mersenne's 25.6 ft./sec.^{2}. In addition, Frenicle seems to have been the author, or one of the authors, of a series of remarks on Galileo's *Dialogue.*

On 3 January 1657 Fermat proposed to mathematicians of Europe and England two problems:

(1) Find a cube which, when increased by the sum of its aliquot parts, becomes a square; for example, $7^3 + (1 + 7 + 7^2) = 20^2$.

(2) Find a square which, when increased by the sum of its aliquot parts, becomes a cube.

In his letter of 1 August 1657 to Wallis, Digby says that Frenicle had immediately given to the conveyer of Fermat's problems four different solutions of the first problem and, the next day, six more. Frenicle gave solutions of both problems in his most important mathematical work, *Solutio duorum problematum circa numeros cubos et quadratos, quae tanquam insolubilia universis Europae mathematicis a clarissimo viro D. Fermat sunt proposita* (Paris, 1657), dedicated to Digby. Although it was assumed for a long time that the work was lost, four copies exist. In it Frenicle proposed four more problems: (3) Find a multiply perfect number x of multiplicity 5, provided that the sum of the aliquot parts (proper divisors) of $5x$ is $25x$. A multiply perfect number x of multiplicity 5 is one the sum of whose divisors, including x and 1, is $5x$. (4) Find a multiply perfect number x of multiplicity 7, provided that the sum of the aliquot divisors of $7x$ is $49x$. (5) Find a central hexagon equal to a cube. (6) Find r central hexagons, with consecutive sides, whose sum is a cube. By a central hexagon of n sides Frenicle meant the number

$$H_n = 1 + 6 + 2 \cdot 6 + 3 \cdot 6 + \cdots + (n-1) \cdot 6 = n^3 - (n-1)^3.$$

Probably in the middle of February 1657 Fermat proposed a new problem to Frenicle: Find a number x which will make $(ax^2 + 1)$ a square, where a is a (nonsquare) integer. We find equations of this kind for the first time in Greek mathematics, where the Pythagoreans were led to solutions of the equations $y^2 - 2x^2 = \pm 1$ in obtaining approximations to $\sqrt{2}$. Next the Hindus Brahmagupta and Bhaskara II gave the method for finding particular solutions of the equation $y^2 - ax^2 = 1$ for $a = 8$, 61, 67, and 92. Within a very short time Frenicle found solutions of the problem. In the second part of the *Solutio* (pp. 18–30) he cited his table of solutions for all values of a up to 150 and explained his method of solution. Fermat stated in his letter to Carcavi of August 1659 that he had proved the existence of an infinitude of solutions of the equation by the method of descent. He admitted that Frenicle and Wallis had given various special solutions, although not a proof and general construction. After noting in the first part of the *Solutio* (pp. 1–17) that he had made a fruitless attempt to prove that problem (1) is unsolvable for a prime x greater than 7, Frenicle investigated solutions of the problem for values of x that are either primes or powers of primes. At the end of this part he made some remarks about solutions of the equations $\sigma(x^3) = ky^2$ and $\sigma(x^2) = ky^3$, where $\sigma(x)$ is the sum of the divisors (including 1 and x) of x.

Also in 1657 Fermat proposed to Brouncker, Wallis, and Frenicle the problem: Given a number composed of two cubes, to divide it into two other cubes. For finding solutions of this problem Frenicle used the so-called secant transformation, which can be represented as

$$x_3 = x_1 + t(x_2 - x_1); \qquad y_3 = y_1 + t(y_2 - y_1).$$

Although Lagrange is usually considered the inventor of this transformation, it seems that Frenicle was first. Other works by Frenicle were published in the *Mémoires de l'Académie royale des sciences.* In the first of these, "Méthode pour trouver la solution des problèmes par les exclusions," Frenicle says that in his opinion, arithmetic has as its object the finding of solutions in integers of indeterminate problems. He applied his method of exclusion to problems concerning rational right triangles, e.g., he discussed right triangles, the difference or sum of whose legs is given. He proceeded to study these figures in his *Traité des triangles rectangles en nombres,* in which he established some important properties. He proved, e.g., the theorem proposed by Fermat to André Jumeau, prior of Sainte-Croix, in September 1636, to Frenicle in May (?) 1640, and to Wallis on 7 April 1658: If the integers a, b, c represent the sides of a right triangle, then its area, $bc/2$, cannot be a square number. He also proved that no right triangle has each leg a square, and hence the area of a right triangle is never the double of a square. Frenicle's "Abrégé des combinaisons" contained essentially no new things either as to the theoretical part or in the applications. The most important of these works by Frenicle is the treatise "Des quarrez ou tables magiques." These squares, which are of Chinese origin and to which the Arabs were so partial, reached the Occident not later than the fifteenth century. Frenicle pointed out that the number of magic squares increased enormously with the order by writing down 880 magic squares of the fourth order, and gave a process for writing down magic squares of even order. In his *Problèmes plaisants et délectables* (1612), Bachet de Méziriac had given a rule "des terrasses" for those of odd order.

BIBLIOGRAPHY

I. ORIGINAL WORKS. Copies of the *Solutio* are in the Bibliothèque Nationale, Paris: V 12134 and Vz 1136; in the library of Clermont-Ferrand: B.5568.R; and in the Preussische Staatsbibliothek, Berlin: Ob 4569. Pt. 1 of the *Traité des triangles rectangles en nombres* was printed at Paris in 1676 and reprinted with pt. 2 in 1677. Both pts. are in *Mémoires de l'Académie royale des sciences,* **5** (1729), 127–208; this vol. also contains "Méthode pour trouver la solution des problèmes par les exclusions," pp. 1–86; "Abrégé des combinaisons," pp. 87–126; "Des quarrez ou tables magiques," pp. 209–302; and "Table générale des quarrez magiques de quatres côtez," pp. 303–374, which were published by the Academy of Sciences in *Divers ouvrages de mathématique et de physique* (Paris, 1693).

II. Secondary Literature. There is no biography of Frenicle. Some information on his work may be found in A. G. Debus, "Pierre Gassendi and His 'Scientific Expedition' of 1640," in *Archives internationales d'histoire des sciences,* **63** (1963), 133–134; L. E. Dickson, *History of the Theory of Numbers* (Washington, D.C., 1919–1927), II, *passim;* C. Henry, "Recherches sur les manuscrits de Pierre de Fermat suivies de fragments inédits de Bachet et de Malebranche," in *Bullettino di bibliografia e di storia delle scienze matematiche e fisiche,* **12** (1870), 691–692; and J. E. Hofmann, "Neues über Fermats zahlentheoretische Herausforderungen von 1657," in *Abhandlungen der Preussischen Akademie der Wissenschaften,* Math.-naturwiss. Klasse, Jahrgang 1943, no. 9 (1944); and "Zur Frühgeschichte des Vierkubenproblems," in *Archives internationales d'histoire des sciences,* **54–55** (1961), 36–63.

H. L. L. Busard

FRENKEL, YAKOV ILYICH (*b.* Rostov, Russia, 10 February 1894; *d.* Leningrad, U.S.S.R., 23 January 1954), *physics.*

As a child Frenkel exhibited both interest and ability in music and painting; but later, in school, he was attracted to mathematics and physics. In 1911 he completed his first independent mathematical paper, in which he created a new type of calculus—but it proved to be already known under the name calculus of finite differences. In 1912 he independently developed a physical theory which he showed to A. F. Joffe, with whom he established a close relationship. In 1913 Frenkel entered the Physics and Mathematics Faculty of St. Petersburg University, from which he graduated with honors in 1916. In 1916–1917 he participated in a seminar led by Joffe at the Petrograd Polytechnic Institute, and in 1918 he taught at the newly created Tavrida University in Simferopol. Frenkel returned to Petrograd (Leningrad) in 1921 and worked at the Physico-Technical Institute, which was directed by Joffe, for the rest of his life; he also taught theoretical physics at Leningrad Polytechnic Institute. In 1929 Frenkel was elected an associate member of the Academy of Sciences of the U.S.S.R. He spent 1930–1931 in the United States, where he lectured at the University of Minnesota.

Frenkel published many scientific books and journal articles, and his research encompassed extremely varied fields of theoretical physics. He was one of the founders of the modern atomic theory of solids (metals, dielectrics, and semiconductors). In 1916 he conceived, on the basis of the Bohr model of the atom, the theory of the double electric layer on the surface of metals, which permitted the first evaluations of the surface tensions of metals and of the contact potential. In 1924, on the basis of virial theory, Frenkel demonstrated that during the condensation of a metal from vapor the valence electrons of the atoms must become itinerant, moving at a speed comparable to the rate of intra-atomic motion. This was a noteworthy contribution to the problem of the heat capacity of electrons in metals, which had been blocking progress of the theory.

In 1927 Frenkel became the first to attempt to construct a theory of metals based on the representations of quantum wave mechanics and was able to explain quantitatively the large mean free paths of electrons in metals. In 1928 he developed a simple, elegant deduction of the Pauli theory of the paramagnetism of electrons in metals, used in the majority of textbooks. He also offered in that year the first quantum mechanical explanation of the nature of ferromagnetism, which was independently developed somewhat later in Werner Heisenberg's theory. He simultaneously offered the theory of coercive force in metals.

Using the virial theorem, Frenkel established the connection between the electron theory of metals, the Thomas-Fermi atomic model, as well as the theory of the nucleus and high-density stars. The general fundamental questions first raised in these works have not lost their significance. In 1930 Frenkel and J. G. Dorfman offered the first theoretical substantiation of the breakup of a ferromagnetic substance into separate domains and predicted the existence of single-domain particles.

In 1930–1931 Frenkel made a detailed study of the absorption of light in solid dielectrics and semiconductors. He pointed out the possibility of the emergence of two different forms of excitation in a crystal. When light is absorbed, an excitation state without ionization may appear. Frenkel called this excitation state "exciton," since such a state has the properties of a quasi particle distributed inside the dielectric or semiconductor. The second type of excitation generated by light in solid bodies, according to Frenkel's theory, is associated with ionization, i.e., with formation of a free electron and a free hole.

When bound together the electron and hole form a unique neutral system that possesses a discrete energy spectrum; this system is called Frenkel's exciton.

Frenkel's work on the theory of electric breakdown in dielectrics and semiconductors (1938) has great significance. As early as 1926, in his work on thermal motion in solid and liquid bodies, Frenkel was the first to work out a model of a real crystal, in which a fraction of the molecules or ions oscillate around temporary equilibrium positions which are intermediate between lattice points and in which a fraction of the lattice points are correspondingly free; the vacancies thus formed (Frenkel's defects) migrate throughout the crystal.

In distinction to the generally held representation of the closeness of the liquid state to the gaseous, Frenkel put forward the new idea of an analogy between a liquid and a solid body. He considered a liquid to be a body possessing short-range but not long-range order. Frenkel's theory of diffusion and viscosity, which was built on this model, proved to be exceedingly fruitful. Frenkel systematically developed his thory of the liquid state in the monograph *Kineticheskaya teoria zhidkostey* ("The Kinetic Theory of Liquids," 1945), which earned him the first-degree State Prize in 1947.

Frenkel paid considerable attention to the theory of the mechanical properties of solid bodies. In papers published in conjunction with T. A. Kontorova (1937, 1938) it was first demonstrated theoretically that in distortion-free lattices a special form of particle motion is possible—a gradual, mutually concordant shift from certain equilibrium positions to others, which leads to a gradual, mutual displacement of the rows of atoms. This theory permitted the explanation of several specific particulars of the plastic deformation and twinning of crystals. The theory of the elasticity of rubbery substances, developed by Frenkel and S. E. Bresler in 1939, proved to be in good agreement with experimental data.

Frenkel's research had an essential influence on the development of electrodynamics and the theory of electrons, as well as the theory of atomic nuclei. His 1926 study served as the basis for the investigation of many questions concerning the dynamics of a spinning electron before the appearance, in 1928, of Dirac's theory of relativistic quantum mechanics. In *Elektrodinamika,* published by Frenkel in 1928, questions of classical electrodynamics were examined from a completely new point of view. In 1936 he was the first to attempt the construction of a statistical theory of heavy nuclei, considering the nucleus as a solid body and setting aside the individual motion of nucleons. In 1939, shortly after the discovery of the splitting of heavy nuclei by Otto Hahn and Fritz Strassman, Frenkel developed (independently of Bohr and J. A. Wheeler) a theory which explains the process of splitting as the result of the electrocapillary oscillation of electrically charged drops of nucleic liquid.

Frenkel also solved many problems in meteorology and geophysics. Between 1944 and 1949 he proposed the theory of atmospheric electrification in which the close connection between the electrification of clouds and the existence of fields in cloudless atmosphere was established. In 1945 he formulated a new theory of geomagnetism.

BIBLIOGRAPHY

I. Original Works. Frenkel's writings include *Lehrbuch der Elektrodynamik,* 2 vols. (Berlin, 1926–1928); *Kinetic Theory of Liquids* (Oxford, 1946); *Wave Mechanics. Elementary Theory* (New York, 1950); *Wave Mechanics. Advanced General Theory* (New York, 1950); *Sobranie izbrannykh trudov* ("Collection of Selected Works"), 3 vols. (Moscow–Leningrad, 1956–1958); *Prinzipien der Theorie der Atomkerne* (Berlin, 1957); and *Statistische Physik* (Berlin, 1957).

II. Secondary Literature. Articles and books on Frenkel and his work (in Russian) are A. I. Anselm, "Yakov Ilyich Frenkel," in *Uspekhi fizicheskikh nauk,* **47,** pt. 3 (1952), 470; J. G. Dorfman, "Yakov Ilyich Frenkel," in Frenkel's *Sobranie izbrannykh trudov,* II (Moscow–Leningrad, 1958), 3–15; V. Y. Frenkel, *Yakov Ilyich Frenkel* (Moscow–Leningrad, 1966); and I. E. Tamm, "Yakov Ilyich Frenkel," in *Uspekhi fizicheskikh nauk,* **76,** pt. 3 (1962), 327.

J. G. Dorfman

FRENZEL, FRIEDRICH AUGUST (*b.* Freiberg, Germany, 24 May 1842; *d.* Freiberg, 27 August 1902), *mineralogy.*

Frenzel was a member of a family of miners and from 1861 to 1865 studied at a mining school in Freiberg, graduating with the title of mine inspector. He then worked as timberman in a prospecting mine and simultaneously attended lectures and laboratory courses at the Mining Academy; he was permitted to do so because he had graduated from mining school with excellent marks. During this period Frenzel also gave lessons in mineralogy and chemistry to foreigners studying in Freiberg. In 1868 he became an employee of the state mines and in 1874 was promoted to mine chemist. He held this post for over a quarter of a century and, beginning in 1883, lectured also in mineralogy and geognosy at the Royal Mining School. Early in 1902 he became head of the laboratory of the main mining administration.

While studying at the Mining Academy, Frenzel attended lectures by prominent scientists: the chemist-mineralogist A. Breithaupt, the geologist Bernhard Cotta, the chemist Theodor Scheerer, and the mineralogist Theodor Richter, who influenced the formation of his interests. Decisive for all of Frenzel's subsequent scientific activity was Breithaupt's trusting him with technical work and, later, with scientific research as well.

In a short time Frenzel became a qualified specialist noted for the exceptional carefulness of his determinations. Beginning in 1870, he published articles on the description of individual minerals.

In 1871, while studying rocks from the Pucher mine in Saxony, Frenzel discovered a previously unknown bismuth vanadate and named the new mineral pucherite. A year later he described two other new minerals: miriquidite (arsenate and phosphate of lead and iron) and heterogenite (hydrous oxide of cobalt containing an admixture of copper).

In 1874 Frenzel's *Mineralogisches Lexikon für das Königreich Sachsen* was published at Leipzig. It contained descriptions of 723 minerals that gave their physical properties and chemical composition. This handbook became very popular among geologists and brought widespread recognition to its author.

During the next few years Frenzel discovered five more minerals: in 1881, lautite (CuAsS), named for the Lauta deposit in Saxony; in 1882, rezbanyite, a complex compound of copper, lead, and bismuth found in the ores of Rézbánya deposit in Hungary (now Băiţa, Rumania); in 1888, amarantite and hohmannite, iron sulfates from Ehrenfriedersdorf, that were of similar composition; and in 1893, cylindrite, a sulfostannate of lead and antimony found in silver-tin veins. He also described several other minerals which he thought were new species but which later proved to have been previously described. Frenzel published some fifty mineralogical papers, the majority of which are still valuable as references.

Frenzel was in constant contact with a number of prominent scientists of his time. He was especially friendly with the German mineralogists Gerhard vom Rath and Carl Hintze and the Russian geochemist-mineralogist Andreas Arzruni. He was a foreign member of the American Institute of Mining Engineers (1873) and a member of the German Geological Society (1875).

As a long-time officer of the Naturalists' Society of Freiberg he sought to arouse interest in research on natural history. A great help in this respect was his ability as a lecturer. A passionate collector, Frenzel assembled two large collections of minerals. He was also interested in ornithology; he wrote and edited numerous articles for the monthly publication of the German Society for the Protection of Birds.

BIBLIOGRAPHY

Frenzel's papers include "Pucherit," in *Journal für praktische Chemie,* **4** (1871), 227–231, 361–362; "Heterogenit," *ibid.,* **5** (1872), 401–408; "Miriquidit," in *Neues Jahrbuch für Mineralogie, Geologie und Paläontologie* (Stuttgart, 1874), 673–687; "Über Lautit und Trichlorit," in *Tschermaks mineralogische und petrographische Mitteilungen,* **4** (1882), 97; "Rezbanyit," *ibid.,* **5** (1883), 178–188; "Hohmannit und Amarantit," *ibid.,* **9** (1888), 397–400; and "Über den Kylindrit," in *Neues Jahrbuch für Mineralogie, Geologie und Paläontologie,* pt. 2 (1893), 125–128.

An obituary is R. Beck, "Friedrich August Frenzel. Nekrolog," in *Zentralblatt für Mineralogie, Geologie und Paläontologie* (1902), 641–646, which includes a bibliography of forty-six titles.

V. V. TIKHOMIROV
T. A. SOFIANO

FRERE, JOHN (*b.* Westhorpe, Suffolk, England, 10 August 1740; *d.* East Dereham, Norfolk, England, 12 July 1807), *archaeology.*

Frere was the son of a country gentleman, Sheppard Frere, and of Susanna Hatley. He was privately educated near his home before entering Gonville and Caius College, Cambridge, in 1758. He graduated with a B.A. in 1763 and was second wrangler; he took his M.A. in 1766 and was a junior fellow from 1766 to 1768. Frere's professional career was in law and politics: he was admitted to the Middle Temple in 1761, became high sheriff of Suffolk in 1766, and Member of Parliament for Norwich in 1799. In 1768 he married Jane Hookham and lived at the family seat, Roydon Hall. They had seven sons.

Frere's scientific work was mainly a hobby, and his election as a fellow of the Royal Society in 1771 was indicative of his general interest in science rather than of a distinction already achieved. He is said to have been active in the Royal Society, but there is little record of this. The only substantial publication is a two-page letter to the secretary of the Society of Antiquaries of London, that was read to the society on 22 June 1797 and published in *Archaeologia* in 1800. In this Frere records his discovery, at a brickyard near Hoxne, of shaped flints which were "evidently weapons of war, fabricated and used by a people who had not the use of metals." A careful examination of the strata showed that the gravel in which the flints were found had been covered for a very long period. He also heard of, but was not able to see, a very large jawbone which had been found in the same stratum, and concluded that the deposit was "of a very remote period indeed." Finally, Frere discovered from workmen on the site that they had already disposed of numerous such flints, and he presumed that this was "a place of their manufacture and not of their accidental deposit."

The discovery aroused little or no interest at the time, and it was not until 1840, when Boucher de Perthes made news by finding similar implements in the Somme, that Frere's perceptiveness was appreciated. These flints probably exhibited the best known workmanship of the lower Paleolithic period, and further excavations were made later in late Acheulean deposits at Hoxne. The weapons discovered by Frere are in the possession of the Society of Antiquaries of London and are deposited in the British Museum.

BIBLIOGRAPHY

Frere's "Account of Flint Weapons Discovered at Hoxne in Suffolk" was published in *Archaeologia,* **13** (1800), 204–205; there was a 2nd ed. of this volume in 1807. The paper was reprinted verbatim, with a discussion of Frere's work, by J. Reid Moir in his "A Pioneer in Palaeolithic Discovery," in *Notes and Records. Royal Society of London,* **2** (1939), 28–31, with portrait; and by Glyn Daniel, in *Origins and Growth of Archaeology* (London, 1967), pp. 57–58.

The main biographical sources for Frere are the article by Warwick Wroth in *Dictionary of National Biography,* XX (London, 1889), 267–268, which includes additional references; the memoir on the life of his son in John Hookham Frere, *Works,* I (London, 1872), xii–xv; and the entry in J. Venn, *Biographical History of Gonville and Caius College,* II, *1713–1897* (Cambridge, 1898), 75.

Frere's work was also discussed in J. Prestwich, "On the Accounts of Flint Implements Associated With the Remains of Animals of Extinct Species in Beds of a Late Geological Period, in France, at Amiens, and Abbeville, and in England at Hoxne," in *Philosophical Transactions of the Royal Society,* **150** (1860), 277–317; and in J. Reid Moir, *The Antiquity of Man in East Anglia* (Cambridge, 1927), p. 59.

DIANA M. SIMPKINS

FRESENIUS, CARL REMIGIUS (*b.* Frankfurt am Main, Germany, 20 December 1818; *d.* Wiesbaden, Germany, 11 June 1897), *analytical chemistry.*

Fresenius' father, Jakob Heinrich Fresenius, was a notary; his mother was Maria Veronika Finger. Until the Thirty Years' War most of his male ancestors had been Protestant ministers. After attending elementary and secondary school in Frankfurt and in Weinheim, Fresenius was apprenticed to an apothecary. He frequently attended the public science lectures at the Physical Society and other institutions, and in 1840 he entered the University of Bonn, where he studied science, history, and philosophy. In those days very few institutions of higher learning afforded any opportunity for practical experimentation. Bonn was not one of them, so Fresenius tested his newly acquired knowledge in the private laboratory of Ludwig Marquart. The principal aim of his experiments was identification of the different elements through qualitative analysis by the wet method.

When Fresenius began to conduct his own laboratory experiments with qualitative reactions, he was faced with a problem then doubtless confronting all young chemists: the lack of any guidelines for systematic qualitative analysis or any coherent sources from which the art could be learned. Older books on analytical chemistry dealt with the subject only from the standpoint of elements, which they listed in the order of their behavior in the presence of various reagents, but gave no systematic methods for identifying the constituents of a mixture of unknown substances. He therefore devised a method of his own for systematic identification and separation of the individual metals (cations) and nonmetals (anions), selecting from the great multitude of reactions those which struck him as most suitable. His system worked so well that, at Marquart's suggestion, he expanded it into a book, published in 1841 under the title *Anleitung zur qualitativen chemischen Analyse.*

By this time Fresenius was in Giessen to continue his chemical studies under the direction of Justus Liebig. In addition to Liebig's lectures, he attended those of Heinrich Buff and Hermann Kopp. The second edition of his *Anleitung,* with a preface by Liebig, appeared in 1842. It states that as early as 1841 inorganic analytical chemistry was being done according to the Fresenius text in Liebig's laboratory. The book gained Fresenius his doctorate on 23 July 1842, and Liebig made him his assistant. The book was an unprecedented success. A third edition was published in 1844 and a fourth in 1846. In a period of twenty years there were eleven German editions. By the time of Fresenius' death seventeen had appeared, each an improved and expanded version of the preceding one, incorporating the latest knowledge and results. The book was soon translated into English, French, Italian, Dutch, Russian, Spanish, Hungarian, and Chinese. The first English edition, translated by J. Lloyd Bullock, appeared in 1841 under the title *Elementary Instruction in Qualitative Analysis.* Eight English editions were published. This enormous success clearly shows the magnitude of the gap in scientific knowledge which the Fresenius system of qualitative analysis filled. The system was taught for a century in all colleges and universities, and while qualitative analysis by the wet method on a macroscopic scale has lost much of its practical importance in recent times, it is still regarded as a valuable instruction tool and continues to be taught in many places.

The Fresenius system is oviously based on Heinrich Rose's separation method. Fresenius divided the cations (or metal oxides, as they were then called under Berzelius' dualistic theory) into six groups. Classed in the same group are cations that behave in the same way (precipitate or, less frequently, dissolve) in the presence of a given reagent under specific experimental conditions. The basic reagent used was hydrogen sulfide; the determining property, the behavior of the various metallic sulfides in different situations. The breakdown was as follows:

Group 6—metals which form a precipitate with hydrogen sulfide in acid or alkaline solution: mercury (+1, +2), lead, bismuth, silver, copper, and cadmium. Silver, mercury (+1), and lead are precipitable with hydrochloric acid.

Group 5—gold, platinum, antimony, tin (+4, +2), arsenic (+3, +5), whose sulfides are soluble in ammonium sulfide.

Group 4—metals whose sulfides are precipitable only in alkaline or neutral solution: zinc, manganese, nickel, cobalt, and iron (+3, +2).

Group 3—metals whose sulfides are soluble and whose hydroxides are precipitable with ammonium sulfide: aluminum and chromium.

Group 2—metals whose sulfides are soluble and are precipitated with alkali carbonates and alkali phosphates: barium, strontium, calcium, and magnesium.

Group 1—both the sulfides and carbonates are soluble: sodium, potassium, and ammonium.

This classification matches exactly the one still in use today, except that Groups 4 and 3 were subsequently merged and the numeration was reversed, Fresenius' Group 6 becoming Class 1, etc. Fresenius did not content himself with the separation into groups but, through treatment with other suitable reagents, broke down each group into its individual member elements. He never used the blowpipe. The number of reagents that he employed was relatively small, which made his system simple and easy to learn.

Fresenius married his cousin Charlotte Rumpf in 1845. They had four daughters and three sons, two of whom, Heinrich and Wilhelm, became chemists and continued operation of the Fresenius Training and Research Institute and publication of the *Zeitschrift für analytische Chemie.* After more than twenty-five years of marriage Fresenius' wife died, and Fresenius married Auguste Fritze, a friend of his deceased wife.

In 1845 the Wiesbaden Agricultural College in the duchy of Nassau offered Fresenius a position as professor of chemistry, physics, and engineering. He accepted and moved to Wiesbaden. The college was very poorly equipped, and Fresenius had no laboratory for teaching or for his own experimental work. He decided to establish one. With a modest subsidy from the duchy, he bought a building and equipped it. This laboratory, which opened in 1848, served several purposes. It offered training in practical chemistry, especially analytic procedures. When it opened, five students started work there under the direction of an instructor, Emil Erlenmeyer, later professor of chemistry at the University of Munich. By 1854–1855 there were thirty-eight students and three instructors. A school of pharmacy was subsequently added. The duchy of Nassau allowed college credit for study at Fresenius' laboratory, but this was discontinued after Nassau was annexed to Prussia. The laboratory then switched to training food chemists and public health personnel. As the role of practical education began to increase at the universities, the laboratory turned more and more to the training of laboratory technicians. It also conducted analyses for industry, soon acquiring an international reputation in this field. Its arbitrational analyses settled many disputes in foreign countries. Fresenius ran the enterprise until his death, and his research institute still operates under the direction of the Fresenius family.

In 1845 Fresenius also published his *Anleitung zur quantitativen chemischen Analyse.* Although this book had had six printings by the time of his death, it is of less importance than his work on qualitative analysis. After describing the analytic operations, the book discusses the forms in which the individual elements can be determined. It then deals with the separations but fails to offer any particularly coherent system (which so far no one else has done). It is also noteworthy that the book does describe many examples of indirect analysis. In many places, too, Fresenius' book touches upon the thermal behavior of analytic precipitates, indicating their thermal stability and discussing the nature of thermal decomposition processes. He can therefore rightly be regarded as one of the pioneers of thermal analysis.

In 1862 Fresenius founded the journal *Zeitschrift für analytische Chemie.* The earliest chemical journals date from the last two decades of the eighteenth century. For almost a century, though, there was no differentiation within the general field of chemistry. The founding of the *Zeitschrift für analytische Chemie* marked the beginning of specialization.

In a special announcement, Fresenius explained his intention in publishing the journal:

> It is readily provable that all great advances in chemistry have been more or less directly related to new or improved analytic methods. The first usable procedures for analyzing the salts were followed by our discovery of the stoichiometric relationships; the progress in analysis of inorganic substances yielded ever more precise equivalent weights; the methods for exact determination of the elements in organic substances gave unexpected impetus to the development of organic chemistry. . . . In truth, therefore, our methods of analysis represent a great achievement in themselves, an important scientific treasure.

The nature of an independent science of analytical chemistry was thus proclaimed. The journal, consisting of original writings on all aspects of analytical chemistry and of systematically arranged reports, still appears regularly. Fresenius himself published many papers on the results of his experimental research. While all of them were scientifically precise, they dealt mostly with special cases and, as regards methodology, contained nothing remotely comparable in significance to his qualitative system. He reported his analyses of numerous mineral waters and explored in detail the possible analytic uses of potassium cyanide; and he was concerned with the detection and quantitative determination of arsenic in cases of poisoning and with the testing of potash, soda, acids, and pyrolusite. He reported also on the determination of nitric acid, lithium, a great many metal alloys, sulfuric acid, metal ores, and boric acid, and the separation of the salts of the alkaline earth metals. Most of those studies consisted of experimental and critical testing of existing methods and of selection of the most favorable operating conditions rather than of a quest for new methods and forms of analysis. To render the analytic process more precise and refine its methods, Fresenius determined the solubility of many analytic precipitates and, on the basis of those tests, recommended correction values for analytic calculations. He also engaged in food research, primarily in analysis of fruit and wine.

Fresenius was active in that period when analytical chemistry was serving not only to increase man's knowledge of the constituents of his environment but also was being increasingly used for day-to-day control of industrial products. Analytic laboratories became the natural and indispensable adjuncts of factories in which chemical analyses were an everyday routine. These laboratories required trained personnel, reliable and fast analytic techniques, and an expedient way to prepare information for the analytical literature. Fresenius recognized that chemical analysis had ceased to be a scholarly preoccupation of the few and had become the daily occupation of the many, and he made it his job to help satisfy those needs. This was the aim of his school, his analytic research institute, and his journal. Even most of his own scientific writings were oriented toward the practical and the industrial. This, in fact, was his second greatest accomplishment: he played a large role in shaping the science of chemical analysis to meet the requirements of an industrial age.

Fresenius' work gained him public recognition, and many honors were bestowed upon him. He was several times president of the Versammlung Deutscher Naturforscher und Ärzte and honorary member of the Gesellschaft Deutscher Chemiker. Contemporaries characterized him as a deeply religious man, with an excellent sense of humor, and an exemplary father. He loved hunting. In 1961 the Gesellschaft Deutscher Chemiker established a Fresenius Prize for outstanding achievement in the field of analytical chemistry.

BIBLIOGRAPHY

I. Original Works. Fresenius' writings include *Anleitung zur qualitativen chemischen Analyse* (Bonn, 1841; 2nd–17th eds., Brunswick, 1842–1896); *Anleitung zur quantitativen chemischen Analyse* (Brunswick, 1845; 2nd–6th eds., 1847–1887); many of his chief analytical works were published in *Justus Liebigs Annalen der Chemie* and in *Zeitschrift für analytische Chemie* and are listed in Poggendorff.

II. Secondary Literature. Works on Fresenius are E. Fischer, "Carl Remigius Fresenius," in *Zeitschrift für angewandte Chemie,* **10** (1897), 520; H. Fresenius, "Zur Erinnerung an Remigius Fresenius," in *Zeitschrift für analytische Chemie,* **36** (1897), 10; R. Fresenius, "Fresenius, Carl Remigius (1818–1897)," in *Nassauische Lebensbilder,* vol. I (Wiesbaden, 1940); W. Fresenius, "Remigius Fresenius," in *Zeitschrift für analytische Chemie,* **192** (1963), 3; A. J. Ihde, *The Development of Modern Chemistry* (New York–London, 1964), pp. 278–280; F. Szabadváry, *History of Analytical Chemistry* (New York–Oxford, 1966), pp. 161–172, 176–181; and *Geschichte der analytischen Chemie* (Brunswick, 1966), pp. 185–192, 196–200.

Ferenc Szabadváry

FRESNEL, AUGUSTIN JEAN (*b.* Broglie, France, 10 May 1788; *d.* Ville-d'Avray, France, 14 July 1827), *optics.*

Fresnel's father, Jacques, was a successful Norman architect and building contractor. In 1785, while directing improvements on the château of the maréchal de Broglie, he married Augustine Mérimée, the pious, well-educated daughter of the estate's overseer. Subsequently he was employed on the harbor construction project at Cherbourg; and when this work was interrupted by the Revolution in 1794, he retired with his family to Mathieu, north of Caen. Here Augustin spent the remainder of his childhood, deeply influenced by the home. In an atmosphere heavy with the values of a stern Jansenism his parents provided him with an elementary education. At twelve, undistinguished except for his practical ingenuity and mechanical talents, he entered the École Centrale in Caen. The school's progressive curriculum afforded Fresnel an introduction to science, and two of his masters made a lasting impression: F. J. Quesnot, the mathematics teacher, and P. F. T. Delarivière, the grammar instructor, whose course imparted the elements of *Idéologie.*

Intending a career in engineering, Fresnel was admitted to the École Polytechnique in Paris in 1804. For two years he benefited from the school's high-level scientific instruction. After an additional three years of technical courses and practical engineering experience at the École des Ponts et Chaussées, he completed his formal training and entered government service as a civil engineer. The Corps des Ponts et Chaussées first assigned him to Vendée, where he worked on the roads linking the department with its new *chef-lieu* at La Roche-sur-Yon. About 1812 he was sent to Nyon, France, to assist with the imperial highway which was to connect Spain with Italy through the Alpine pass at Col Montgenèvre. In moments snatched from his professional duties Fresnel diverted himself with a series of philosophical, technical, and scientific concerns.

By mid-1814 Fresnel had turned to optics and had begun to consider the claims of the wave or pulse hypothesis of light. This inquiry, barely begun, was suddenly thrust aside the following year. Seeing Napoleon's return from Elba as "an attack on civilization," Fresnel deserted his post and offered his services to the Royalist forces. With the reestablishment of the empire he found himself suspended from his duties and put under police surveillance. Returning home to Mathieu, he devoted his enforced leisure to optics, undertaking experiments on diffraction. These confirmed his belief in the wave nature of light and started him on a decade of research aimed at developing his hypothesis into a comprehensive mathematical theory. With the Second Restoration, Fresnel was reactivated by the Corps des Ponts et Chaussées and thereafter was forced to restrict his investigations to periods of leave. Through the intervention of such influential friends as François Arago, these were not infrequent. From the spring of 1818 his scientific work was made easier by assignments in Paris, and intensive research over the next few years produced important results. After 1824 his efforts slackened. Work with the Lighthouse Commission, including the development of his new "echelon" lenses, put severe demands on his time, and faltering health sapped his energy.

Tuberculosis, the cause of his early death, cast a shadow over Fresnel's entire career. Plagued continually by ill health, he sought consolation in a religious faith which offered belief in Divine Providence and the hope of an afterlife. But this was no theology of resignation. Summoning the will to struggle against bodily suffering and fatigue, Fresnel threw himself into difficult tasks. Behind his remarkable determination was a severe Puritan, middle-class outlook which saw the highest merit in personal achievement, performance of duty, and service to society. Serious, intent, haunted by thoughts of an early grave, Fresnel bound himself closely to these ideals, shunning pleasures and amusements and working to the point of exhaustion. Despite the urgency of everything he attempted, Fresnel was always attentive to detail, systematic, and thorough. In science no less than in politics he held tenaciously to his convictions and defended them with courage and vigor. As a functionary he voiced outrage when the behavior of others fell short of his own high ethical standards. At times this approached a rankling self-righteousness, but generally his contemporaries saw him as reserved, gentle, and charitable.

For his scientific achievements Fresnel received several important honors. In 1823 the Académie des Sciences elected him to membership by unanimous vote. He was a foreign associate of the Société de Physique et d'Histoire Naturelle of Geneva and a corresponding member of the Royal Society of London. In the last month of his life he received the Royal Society's Rumford Medal.

Confined almost exclusively to optics, Fresnel's scientific work shows an essential unity. Above all, his research found its motivation and direction in an attempt to demonstrate that light is undulatory and not corpuscular. Challenging the prevailing Newtonian view, he undertook a series of brilliant investigations which systematically elaborated the wave concept and established its conformity with experience. Fresnel brought to his research an ingenious mind, deft hands, and the discipline of an excellent scientific education. He was equally proficient in experiment and mathematics and effectively combined the two. Characteristically, he initiated his investigations with experiments and proceeded, via analysis, to theory. He set as his goal mathematical theories from which precise consequences could be deduced and tested by further experiments. For Fresnel a true theory was one that predicts experience and rests on a simple conceptual basis, free of all auxiliary hypotheses. The simplicity requirement, which served Fresnel as a constant guide in his theoretical formulations, was grounded on a deep-seated belief that nature aims at the production of the most numerous and varied effects by the fewest and most general causes. This is the meaning of the epigram placed at the head of his prize essay on diffraction: "Natura simplex et fecunda." The idea of the underlying unity of natural processes doubtless found a guarantee in Fresnel's Providentialism. A consideration of his close relationship with Ampère and others in the circle of Maine de Biran might also disclose certain philosophical influences contributing to this viewpoint.

It was Fresnel's belief in the essential unity and simplicity of nature that conditioned his preference for a wave conception of light. His earliest statement in favor of light as a form of motion (in a letter of 5 July 1814) envisioned the possibility of referring heat, light, and electricity to the modifications of a single, universal fluid. Apparently, then, he regarded the whole Newtonian scheme of imponderables with its multiple fluids as suspect from the very start. But within this general scheme the corpuscular theory of light had its own special burden of complexity. Ignorant of the elaborations of the theory undertaken to accommodate polarization, Fresnel was not yet aware of how complex corpuscular optics had become. His determination to overhaul optical theory was sparked by a dissatisfaction with the caloric view of heat and an appreciation of the analogies between heat and light. But after he learned of Biot's work, he regularly assailed the corpuscular theory for its lack of unity and simplicity.

Rejecting corpuscular optics, Fresnel was poorly acquainted with earlier theories that conceived light as waves or pulses. In France, as elsewhere, the views of Huygens and Euler had no following and were hardly discussed. Physics textbooks of the period took note of the "Cartesian" hypothesis but dismissed it in a few lines. Apparently it was only with the most general knowledge of the work of his predecessors that Fresnel began to construct his theory. If he knew of Huygens' principle when he undertook his first investigations, he did not reveal it. From the start, however, he possessed another important concept, which made his theory, unlike that of Huygens, a true wave theory. This was the idea that the pulses constituting light succeed one another at regular intervals. Fresnel may have taken the idea from Euler, but it is more probable that he hit upon it independently. Nor was he aided by the work of Thomas Young. He became familiar with Young's contributions to wave optics only after he was well into his own experiments.

Fresnel's first experimental investigation, a study of diffraction, gave him a firm foothold in undulatory optics and started him down a profitable path. By studying diffraction effects—the shadow and associated bands of color produced when a hair or other thin object is illuminated by a narrow beam—he hoped to counter objections to the wave hypothesis based on the apparent rectilinearity of the propagation of light and, if possible, to find positive support for the view of light as vibrations. The key to success was found in an application of the principle of interference, a concept drawn from acoustical theory. Attaching a slip of black paper to one edge of a diffracter, Fresnel observed that the bands of light within the shadow disappeared. By "a mere translation of the phenomenon" he concluded that the internal bands depend upon a crossing of rays inflected into the shadow from both edges of the diffracter. Since the bands outside the shadow on the side opposite the attached paper remained, the external bands appeared to arise from a crossing of rays proceeding directly from the light source and by reflection from one edge of the diffracter. Referred to the mechanical level, these effects seemed explicable only if light were undulatory. Bright bands would occur where the vibrations constituting light are in phase and reinforce one another. Intervening bands of darkness would correspond to places where the vibrations are out of phase by some odd number of half wavelengths and cancel one another.

To put the concept of constructive and destructive interference to the test, Fresnel worked out simple algebraic formulas correlating the positions of the bands with factors determining the occurrence of interference—the path differences of the intersecting rays and the wavelength of the light. Performing the experiment with monochromatic red light and gauging the positions of the bands for various intervals between the light source, the diffracter, and the receiving screen (or plane of observation, since it proved equally effective and more convenient to dispense with the screen and view the bands directly with a lens), he found a close correspondence between actual values and those predicted by his formulas.

Although a paper of October 1815, embodying these results, won Arago to the cause of wave optics and made a favorable impression on the Institut de France, Fresnel was still far from a complete theory of diffraction. His indiscriminate use of the terms "rays," "vibrations," "inflection," and "diffraction" bespoke a residue of corpuscular influences and was symptomatic of a lack of precision in his formulation. The mirror experiment, demonstrating interference in circumstances where the attractive forces of inflection could not be invoked to explain its effects, marked an important step forward. In front of two mirrors arranged end to end at an angle slightly less than 180° Fresnel set a minute light source. After the necessary adjustments to obtain precisely the right conditions, he saw bands of color produced as the rays reflected from one of the mirrors intersected and interfered with rays reflected from the other. The interpretation of the bands in terms of interference seemed all the more certain since band positions corresponded to theoretical values obtained by adapting the diffraction formulas.

Although inflection was thus effectively discredited, Fresnel saw the need for further refinements. His formulas, positing rectilinear "rays" and referring path measurements to the very edge of the diffracter, predicted band positions only if it were assumed that the rays turned aside at the diffracter lost half a wavelength. Otherwise there was an inexplicable reversal, the bright bands occurring where dark ones were predicted and vice versa. Spurred by a desire to eliminate the ad hoc hypothesis, Fresnel undertook to reconstruct his theory on a new basis, a step carrying him, for the first time, beyond Young. Boldly he conceived the idea of combining Huygens' principle with the principle of interference. Applying the idea to diffraction, he supposed that elementary waves arise at every point along the arc of the wave front passing the diffracter and mutually interfere. The problem was to determine the resultant vibration produced by all the wavelets reaching any point behind the diffracter. The mathematical difficulties were formidable, and a solution was to require many months of effort. In the first attempt, fashioned around the concept of "efficacious rays," Fresnel succeeded in reducing the discrepancy between theory and fact by half, and in a paper of 15 July 1816 that reported the investigation, he begged critics at the Institut de France to treat with indulgence "his essays in such a difficult theory."

Not until the spring of 1818 was Fresnel able to reach his goal. Restored to active service in the Corps des Ponts et Chaussées and assigned to Rennes, he bore heavily the yoke of his engineering duties and continually badgered his superiors for leave. Whenever possible he returned to Paris to pick up the thread of his research. Throughout 1817 he concerned himself with polarization, but the need to cope with the periodic effects of chromatic polarization immediately reintroduced the basic mathematical problem carried over from the study of diffraction: that of "calculating the influence of any number of systems of luminous waves on one another." Fresnel took a decisive step toward the solution when, aided by an analogy between the oscillations of an ether molecule and those of a pendulum, he derived a general expression for the velocity of ether molecules put into motion by a wave.

Considering next the combined effect of multiple waves, Fresnel worked through to an important result. Just as a force can be resolved into perpendicular components, so the amplitude of the oscillations imparted by any wave can be reduced to the amplitudes of two concurring waves following one another at an interval of a quarter wavelength. To find the net effect of multiple waves, then, it was sufficient to reduce each to its two components, add like components, and recombine the sums. Temporarily setting polarization aside, Fresnel hastened to apply this result to diffraction. Urgency was called for, because the Académie des Sciences had announced that diffraction would be the subject of its competition for 1819. Looking beyond the prize to the scientific "revolution" he hoped to effect, Fresnel was anxious to enter the contest. Not long before the closing date he put the final touches to his theory. Without any gratuitous hypotheses he could now calculate the light intensity at any point behind a diffracter.

In Figure 1 P is the point, AG the diffracter, C the light source, and AMI a partially intercepted wave

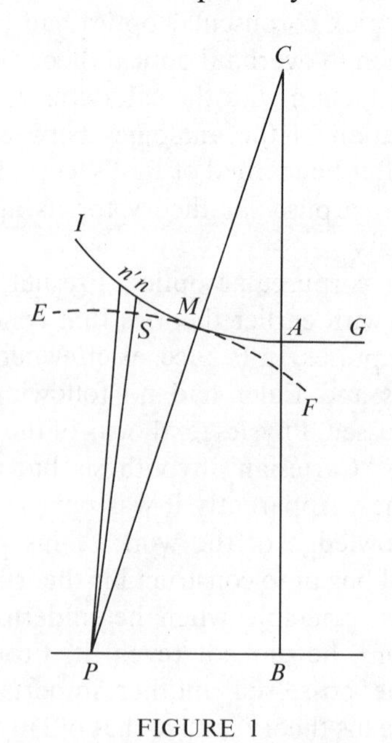

FIGURE 1

front. One of the infinitesimally small arcs into which AMI may be divided is shown as $n'n$. From each of these elements a train of wavelets is assumed to arise, and the problem is to determine the composite effect of all the wavelets at P. The procedure takes the form indicated above, the wavelets being related to a wave emanating from the point M on the line CP and to another wave differing in phase by a quarter undulation. Considering the effect at P produced by a wavelet proceeding from $n'n$, Fresnel represents this small portion of the original wave front as dz and specifies its distance to M as z. The interval nS between the wave AMI and the tangent arc drawn around P is determined to be $1/2\, z^2(a + b/ab)$, a and b representing the distances CA and AB. Substituting into the appropriate expression supplied by his general mathematical investigation, Fresnel wrote the component of the wavelet relative to the wave emanating from M as

$$dz \cos\left(\pi \frac{z^2(a + b)}{ab\lambda}\right),$$

λ representing the wavelength. The other component relative to a wave separated from the first by a quarter wavelength is

$$dz \sin\left(\pi \frac{z^2(a + b)}{ab\lambda}\right).$$

The sum of similar components of all the wavelets is then

$$\int dz \cos\left(\pi \frac{z^2(a + b)}{ab\lambda}\right) \text{ and } \int dz \sin\left(\pi \frac{z^2(a + b)}{ab\lambda}\right).$$

These expressions later passed into the textbooks as "Fresnel's integrals." The square root of the sum of their squares gives the amplitude of the resultant vibration at P, while the sum of their squares measures the observable light intensity at the point. Tested and strikingly corroborated by experiment, the new theory served as Fresnel's entry to the competition and was awarded the prize. During the judging it received a dramatic and unexpected confirmation. One of the commissioners, Poisson, had perceived in Fresnel's mathematics the seemingly improbable consequence that the center of the shadow of a small disk used as a diffracter would be brightly illuminated. An experiment performed to test the calculation confirmed it exactly.

That light under certain circumstances displays an asymmetric aspect remained the most serious challenge for the undulatory conception of light. Particles might have "sides," but longitudinal waves could not. Understandably, Young experienced "a descent from conviction to hesitation" when informed of Malus' discovery of polarization by reflection. With Fresnel, however, it was otherwise. Undaunted, he set out early in his investigations to find an accommodation between the asymmetry of light and the wave hypothesis. For clues about the nature of polarization his first tack was to pursue a comparative approach in which the effects of polarized light would be juxtaposed against the known characteristics of ordinary light. Specifically, he decided to substitute polarized light for ordinary light under conditions producing interference. Initial experiments carried out jointly with Arago early in 1816 afforded no new insights. Polarized light gave the same effects as ordinary light. Fresnel questioned the adequacy of these hasty tests, and several months later undertook new experiments, obtaining results that were quite different. Aided further by Arago, he showed convincingly that in circumstances where ordinary light would interfere, rays polarized in mutually perpendicular planes have no effect on one another.

The theoretical implications were puzzling. In a note to a preliminary draft of the paper reporting the investigation, Fresnel offered two hypotheses accounting for his findings, one his own and the other elicited from Ampère. The noninterference of rays polarized in mutually perpendicular planes suggested that the vibrations constituting light are either transverse or a combination of longitudinal and transverse motions. Neither hypothesis appeared tenable. Transverse waves, the prevailing theory of elasticity held, are possible only in a solid medium; but an all-pervading solid ether could not be reconciled with the free, unimpeded movements of the planets demonstrated by astronomy. When he submitted his paper to the Académie des Sciences, Fresnel deleted the note, and Arago's account of the investigation published in 1818 scrupulously avoided all theoretical considerations.

Lacking any alternative hypotheses, Fresnel continued his inquiry. On the basis of the recent experimental findings he next worked out a detailed explanation of chromatic polarization. But, as with diffraction, the calculation of precise theoretical values and full confirmation had to await the "general law of the reciprocal influence of luminous waves," available only at the beginning of 1818. Although it extended the sway of interference as an explanatory concept, the study of chromatic polarization disclosed nothing new about the basis of polarization.

In search of further clues, Fresnel turned his attention to the influence of reflection on polarized light. His first efforts, summarized in a paper of November 1817, resulted in the discovery of an unusual modification of light, later designated "circular polarization." The novel light appeared to be symmetric about an axis drawn in the direction of its motion, but in other respects it behaved like polarized light. Fresnel determined that these characteristics would follow if the light were supposed to consist of two components with mutually perpendicular planes of polarization and a phase difference of a quarter undulation. Yet for the moment he saw no way to translate this into a satisfactory mechanical hypothesis. Another important investigation, completed in March 1818, showed that the rotation of the plane of polarization associated with the passage of light through quartz and certain liquids depends upon a weak double refraction and the superposition of two circularly polarized rays.

As he weighed the implications of these studies, Fresnel was gradually brought to the realization that the vibrations constituting light could only be transverse. Consistently the characteristics of polarized light testified to forces acting at right angles to the rays, and finally the conclusion became inescapable. In an article appearing in the *Annales de chimie et de physique* in 1821, Fresnel publicly committed himself to the view that light waves are exclusively transverse. Attempting a mechanical rationale, he offered a brief account of a hypothetical ether that would lend itself to transverse vibrations and yet retain the essential properties of a fluid. He then proceeded to an interpretation of polarization. Ostensibly, polarized light had its basis in ether vibrations executed in a definite, fixed plane at right angles to the direction of the wave. As for ordinary light, it could be considered "the union, or more exactly, the rapid succession, of systems of waves polarized in all directions." For Fresnel the major support for this conception of the nature of light was that it gave meaning and order to his empirical findings. His essay into ether mechanics was weak and was intended only as a demonstration of the physical possibility of transverse waves. The proof came from all the indications of experience, and a rigorous mechanical justification, however desirable, seemed unnecessary. Few among Fresnel's contemporaries agreed. Even Arago, who gave his faithful support in everything else, deserted him here.

Fresnel found an effective answer to his critics in a successful application of the concept of transverse waves to double refraction. A start was made in the article of 1821, when he suggested that the two rays of double refraction correspond to perpendicular components of the vibrations of the ray incident on the doubling crystal. From this simple idea he rapidly traversed an arduous course to a full-blown mathematical theory. As usual, his approach was to work back from experience. Availing himself of the law of refraction and Huygens' law of extraordinary refraction, he proceeded to develop a series of unified constructions specifying the velocities of both the rays of double refraction. Initially he found that the velocities could be accurately represented by the semiaxes of the intersection of the wave surface with an ellipsoid of revolution. This sufficed for double refraction in uniaxial crystals, but a more general construction was needed to provide for the refracting characteristics of biaxial crystals. By substituting an ellipsoid with three unequal axes for the ellipsoid of revolution, Fresnel had the solution, or at least a partial solution.

In one crucial respect the new construction fell short of full generality. It proved valid for most doubling crystals, which show weak double refraction, but it was not adequate for those like Iceland spar, in which the separation of the rays is considerable. The final construction, in the form of an equation of the fourth degree, followed only a week after the paper of 19 November 1821, which recounted the previous results. Although he now had a general law of double refraction meeting every test of experience, Fresnel pressed on. As in the study of polarization, he was reluctant to end his investigation without showing the mechanical plausibility of his results. In two supplements to the November memoir he concerned himself with this problem. When later he prepared an account of the investigation for publication, the mechanical considerations were emphasized, and the law of double refraction was represented as a deduction from the general properties of an elastic fluid. That this obscured the actual route of discovery is unimportant. Fresnel's treatment of double refraction was an impressive synthesis, and while transverse wave motion may not have been rendered more acceptable, successful applications of the new theory built around it soon made it indispensable.

The study of double refraction was Fresnel's last major contribution to wave optics. Thereafter his responsibilities on the Lighthouse Commission absorbed the bulk of his energies. To the problems encountered here—the improvement of lenses and the design, construction, and location of the lighthouses—he brought the same inventiveness, concentration, and perseverance previously manifest in his scientific work. Yet science was not entirely forgotten in these later years. Fresnel found the time to carry out an investigation of partial reflection, to make a beginning toward a mathematical theory of dispersion, and to act as chief propagandist for the wave theory. In occasional notes and academic reports he addressed himself to topics outside of optics. Most noteworthy was his contribution to Ampère's electrodynamic molecular model. It is tempting to think that, given more time, he might have pursued further his youthful vision of restoring heat, light, and electricity to a common basis in the motions of a universal ether.

As it was, Fresnel succeeded fully in attaining his explicit goal, the establishment of the wave conception of light. Not long after his death scientific opinion definitely shifted in favor of waves and opened up the pathway leading to the deeper insights of Maxwell. In broad context Fresnel's work can be viewed as the first successful assault on the theory of imponderables and a major influence on the development of nineteenth-century energetics.

BIBLIOGRAPHY

I. Original Works. Fresnel's writings were collected in *Oeuvres complètes d'Augustin Fresnel,* Henri de Senaramont, Émile Verdet, and Léonor Fresnel, eds., 3 vols. (Paris, 1866–1870). Provided with a detailed analytical table of contents, this comprehensive edition lacks only the important notes on electrodynamics addressed to Ampère. For these, see *Mémoires sur l'électrodynamique,* Société Française de Physique, ed., 2 vols. (Paris, 1885–1887), I, 141–147.

II. Secondary Literature. The details of Fresnel's life are poorly known, and no full-length biography has been published. The most detailed account is François Arago's "Éloge historique," in Fresnel's *Oeuvres complètes,* III, 475–526. Other contemporary sources are A. J. C. Duleau, *Notice sur A. Fresnel* (Paris, 1827); an anonymous account in Vielh de Boisjoslin, ed., *Biographie universelle et portative des contemporains,* 4 vols. (Paris, 1830), II, 1770–1775; A. Marc, *Notice sur A. J. Fresnel* (Caen, 1845); and Léon Puiseux, "Fresnel," in *Notices sur Malherbe, La Place, Varignon, Rouelle, Vauquelin, Descotils, Fresnel et Dumont-d'Urville* (Caen, 1847).

Among recent accounts with biographical interest the most noteworthy are Charles Fabry, "La vie et l'oeuvre scientifique d'Augustin Fresnel," in *Revue internationale de l'enseignement,* **47** (1927), 321–345; G. A. Boutry, "Augustin Fresnel: His Time, Life and Work, 1788–1827," in *Science Progress,* **36** (1948), 587–604; and Pierre Speziali, "Augustin Fresnel et les savants genevois," in *Revue d'histoire des sciences et de leurs applications,* **10** (1957), 255–259.

The *Revue d'optique,* **6** (1927), is a centennial issue devoted entirely to Fresnel. The best starting point for an assessment of his scientific work is Verdet's introduction to the *Oeuvres complètes.* Brief but authoritative is the discussion in Vasco Ronchi, *Histoire de la lumière* (Paris, 1956), pp. 242–261. Fresnel's investigations leading to the recognition of transverse wave motion are analyzed in detail in André Chappert, "L'introduction de l'hypothèse des vibrations transversales dans l'oeuvre de Fresnel," an unpublished study done for the Diplôme d'Études Supérieures de Philosophie at the University of Paris. See also R. Silliman, "Augustin Fresnel (1788–1827) and the Establishment of the Wave Theory of Light" (thesis, Princeton University, 1968).

Robert H. Silliman

FREUD, SIGMUND (*b.* Freiberg, Moravia [now Příbor, Czechoslovakia], 6 May 1856; *d.* London, England, 23 September 1939), *psychology.*

Freud's father, Jakob, was a wool merchant in Freiberg. His mother, Amalie Nathanson, was Jakob's second wife and twenty years younger than he. Freud was the oldest child in the father's second family. An older half brother, about the age of Freud's mother and with a child of his own about Freud's age, lived nearby. Freud was to write that the confusion all this caused him as an infant sharpened his intellect and his curiosity. He also wrote of himself: "A man who has been the indisputable favorite of his mother keeps for life the feeling of a conqueror, that confidence of success that often induces real success."[1] The wool trade in Freiberg, which had made Jakob mildly prosperous, collapsed, and the family moved to Vienna in 1860. For the rest of his long life Jakob was often unemployed, and the family was at times on the brink of poverty. In this respect, Jakob provided an unheroic ideal for his son. The family was Jewish and kept to Jewish society and customs, but they were not strongly religious. The father was something of a freethinker, and the son had lost any religious beliefs by his adolescence. Freud attended Sperl Gymnasium in Vienna from the age of nine to the age of seventeen, graduating with distinction in 1873. The curriculum emphasized modern and classical languages and included mathematics. Freud was studious and was encouraged in this by his parents, who made considerable financial sacrifice for his education. They anticipated a distinguished career for their son, which anticipation he shared. Freud's unusual degree of ambition lasted well into his middle years.

In 1873 Freud entered the University of Vienna to study medicine. He chose medicine, not out of a desire to practice it, but with a vague intention of studying the human condition with scientific rigor. In choosing his career—and throughout his life—Freud placed a high ethical value on the physical sciences. He took Ernst von Brücke, professor of physiology at Vienna, as a model. Brücke was one of the founders of the Helmholtz school of German physiologists, who had accelerated the progress of that science with their own work and teaching. Freud spent three years more than was necessary in qualifying for his medical degree, which he finally received in 1881. This delay resulted from starting what he intended to be a career in biological research. He spent an increasing amount of time in Brücke's Physiological Institute from 1876 through 1882. His first studies were on the connections of a large nerve cell (Reissner's cell) that had been discovered in the spinal cord of a primitive genus of fish, and his observations made it possible to fit these cells into an evolutionary scheme. He also studied the structure of nerve fibers in living crayfish and devoted some time to the anatomy of the human brain. He had made a successful start on a research career when the poor economic prospects of his position forced him to change his plans. Brücke's two assistant professors were only ten years older than Freud, so the chance for moving up to a position with an adequate salary seemed remote. Freud met his future wife in 1882 and had to face the fact that he could not continue at the institute and support a wife. He decided to obtain the clinical experience that would gain him respectable status as a practitioner. He joined the resident staff of the Vienna General Hospital in July 1882 and remained there until August 1885, working in the various clinical departments of the hospital for short periods of time. He stayed fourteen months in the department of nervous diseases because he wished to specialize in neuropathology.

Freud did not limit his activities to training in clinical medicine while at the Vienna General Hospital. He found time to continue his anatomical research on the human brain, tracing the course of nerve tracts in the medulla oblongata. He also began a series of studies in clinical neurology. This work was in the tradition of patient investigation that he had learned at Brücke's institute, in contrast with his research on the therapeutic use of cocaine, which he began in 1884. He used the drug himself, finding it made him euphoric and able to work well. His letters to his fiancée show high hopes that the research would bring quick recognition and would enable him to afford marriage sooner. He published two articles on the use of cocaine as a stimulant, as an analgesic, and as an aid in withdrawal from morphine addiction. Within two years there were reports of cocaine addiction, and Freud's reputation was clouded. Freud spent four months studying with J. M. Charcot, the foremost French neurologist, after leaving the Vienna General Hospital.

Freud set up practice as a neuropathologist on his return from Paris and was soon married. His bride was Martha Bernays, the daughter of an intellectually distinguished German Jewish family. It had taken four years from his decision to give up a research career for marriage to become a financial possibility. His prolonged engagement had been full of near breakups and reconciliations. Freud had been extremely jealous of anyone for whom Martha showed any affection, including her mother. Since Martha was in Hamburg for much of their engagement, they left a vivid record of this tempestuous period in their letters. Freud was thirty at his marriage (1886); Martha was twenty-five. In the first decade of their marriage the couple had three sons and three daughters. Freud's professional offices were adjacent to the living quarters in their first and subsequent apartments. After he married, Freud's practice, home, family, and extensive writing occupied most of his time. In the early years of his practice he went several times a week to Kassowitz's Children's Clinic, where he headed the department of neurology. Throughout his career he was on the faculty (without chaired appointments) of the University of Vienna, where he lectured, first on neuropathology and then on psychoanalysis.

Freud's psychological life was not as smooth as the description of his everyday life might imply. Psychoanalysts have found a search for a father figure important in Freud's psychological history, his own father not having provided an adequate model. At least it is clear that he sought authoritative approbation of his career. Yet he was not swayed from a self-determined course by the successive candidates for father figure, most prominently Brücke, Josef Breuer, and Wilhelm Fliess. Brücke remained a loyal friend after Freud gave up his research career, but Freud's career was diverging too far from Brücke's for Freud to seek sanction from him. Breuer was fourteen years older than Freud, had made signal contributions to neurophysiology in his younger years, and was a distinguished physician. They culminated a decade of collaboration with their joint publication of *Studies in Hysteria* (1895), which described the clinical experience that was one of Freud's bases for psychoanalysis. Breuer was unwilling to join Freud in the radical innovations of psychoanalysis; he was dubious about the emphasis on sex that developed and, being established, did not share Freud's driving ambition. The publication of their joint work came near the end of their collaboration. Freud harshly criticized Breuer's personality in later years.

Fliess was a Berlin physician with whom Freud corresponded regularly from 1893 until 1900. During this period Freud was conducting his self-analysis and rapidly developing his psychoanalytic theories, the two being aspects of a single venture. His letters to Fliess rather fully disclose his thoughts during this crucial phase. Fliess also was developing radical theories on the periodicity of biological events (which have never appeared to have any validity). While there seems to have been little significant mutual influence, each provided the other an audience. Fliess took Breuer's place as someone who apparently could understand and who approved of Freud's ideas. Fliess broke with Freud in 1900 over a trivial matter. There is no equivalent of the Fliess correspondence, which exhibits Freud's bitterness toward Breuer, after their professional estrangement. Perhaps Freud was then intellectually more self-confident. In any event, he was soon to become the intellectual father figure of the psychoanalytic movement. The meetings of his followers began at Freud's house in 1902.

Freud was in robust health into his late sixties. He was bothered, but never disabled, by a number of afflictions, including some intestinal problems that he considered psychosomatic. World War I was difficult for the Viennese, with food and fuel in short supply. There was the stress of having two sons in combat. The shortage of cigars was also an affliction for Freud, who had been accustomed to smoking fifteen to twenty each day, preferably Schimmelpenninck cigars from Holland. It was a sign of Freud's vigor that he came through all this without a decline in health. In 1923 cancer of the jaw, the disease that led to his death, was detected. From then on, he had repeated operations, metal appliances were put in his jaw to replace the bone removed, and he was frequently in pain. He preferred to remain mentally alert rather than take pain-killing drugs. After 1923 he wrote three books and many articles, and continued his practice and his extensive correspondence.

Freud remained in Vienna even though he was well aware of the impending danger from the Nazis. He was offered foreign asylum in 1936 and 1937 but, partly out of his identification with less fortunate Jews, remained until June 1938, three months after the Nazis gained control in Austria. He was allowed to go to England after Ernest Jones managed some complicated diplomatic maneuvering and the payment of a ransom, and he died in London in September 1939.

Freud made a solid contribution to conventional neuropathology. His first book was *Aphasia,* published in 1891. It was a masterly review and critique of the literature on the subject and presented a synthetic view of the condition. Freud refuted the view, prevalent among German-speaking neurologists, that the losses of function in aphasia were due to lesions in anatomically circumscribed centers corresponding to the various functions involved in language. He demonstrated that the anatomical postulates would not fit with specific case studies and that it was necessary to assume that the cerebral areas involved in language were less circumscribed. Freud also incorporated into his synthesis the view that function could be reduced in an area, not simply canceled, by the disease. Here he relied on John Hughlings Jackson, the English neurologist. Hughlings Jackson used the term "disinvolution" to describe the lesser vulnerability to pathological weakening of cortical complexes acquired earlier in the life of the individual. Freud's book had little immediate impact, perhaps because it contained no new case material. Part of his motivation to write it must have been the desire to get at the neurological events underlying complex psychological processes. In this it foreshadowed his *Psychology for Neurologists,* which he wrote in 1895. Freud's three works on cerebral paralysis in children, published in 1891, 1893, and 1897, were immediately recognized as definitive works on the subject and have remained so valued. In them he presented his own cases from the Kassowitz clinic, as well as a review of the literature. His studies brought order out of a confusing array of paralyses.

The development of Freud's psychoanalytic thought can usefully be described as occurring in three phases. In the first phase he gradually developed his ideas during his experience in the therapy of hysteria. Breuer had some part—just how much is impossible to determine—in the formation of the ideas, and certainly held Freud back from making the large speculations that characterized the second phase. In the second phase, in the middle of the 1890's, Freud developed his ideas more rapidly and with less reference to clinical experience than he had before and would later. He formed a comprehensive theory of the determinants of human thought and behavior, which became his metapsychology. This was in large part based on previous theories. During this phase Freud had only Fliess for his critical audience. In the third phase, which lasted from the late 1890's until the end of his career, Freud elaborated greatly on the ideas developed during the first two phases. There was again much reference to his clinical experience, but it was often interpreted so that it fit his previous ideas.

The first phase of Freud's intellectual development occurred during his first years of treating hysteria. Although at the Kassowitz clinic he saw patients assumed to have definite physical damage to the nervous system, most of the patients who came to his office he considered hysterics. Hysteria was considered to have both physical and psychological causes. Most theories emphasized hereditary weakness of the nervous system as the physical cause. It was commonly considered that a psychologically traumatic event, with this background of a weak nervous system, brought on the condition. Physicians usually tried physical therapeutic approaches. In the first years of his practice Freud used bed rest and low-voltage stimulation to paralyzed limbs of hysteric patients. He had begun his practice with some knowledge of psychological therapy for hysteria; Breuer had told him in 1882 about what was to become known as the "case of Anna O." Breuer treated the patient, whom he considered a hysteric, from 1880 to 1882. Her symptoms included an intermittent paralysis of the limbs and severe speech and visual disturbances. Breuer found, rather by accident, that if the patient described in detail the manifestations of a symptom, it was relieved for a time. Breuer called this method "catharsis."

Breuer also used hypnotic suggestion with Anna O. and with other patients. He told the hypnotized patient that such-and-such a symptom would disappear and found that the symptom disappeared, at least temporarily. Hypnotic suggestion had been used for years by Liébeault and Bernheim at Nancy, and a few other physicians in Vienna were using it. When Freud left the Vienna General Hospital in 1885, he received a traveling grant to study in Paris with Charcot, who was then making hysteria the focus of his attention. He maintained that the manifestations of hysteria were regular and that the common medical opinion (prevalent in Vienna) that they were feigned by hysterics was therefore erroneous. Freud never accepted Charcot's elaborate systematization, but his reputation sanctioned Freud's taking the condition seriously, as deserving of scientific study, and his considering its psychological as well as physical aspects. Freud's growing doubt as to the efficacy of electric stimulation and other physical therapeutic techniques led him to employ Breuer's method of hypnotic suggestion. He considered the method useful and became an advocate of it in Vienna.

In 1889 Freud was using Breuer's carthartic method in conjunction with hypnosis. This gradually developed into the free association method. Instead of leading the patient to talk about the first occurrence of a symptom, he encouraged the patient to say whatever came into his mind, without exercising any conscious control over it. Freud believed that the patient exhibited parts of the network of associated ideas which had already been established during his life. The therapist could surmise those ideas which made up the neurosis from those which the patient disclosed in free association. Freud formed the essentials of his concepts of the unconscious, of repression, and of transference during the development of the free association method. Parts of the complex of associated ideas, unacceptable in the conscious thought of the patient, were repressed. They remained in the unconscious, influencing what came into consciousness, but never themselves came into consciousness. Freud made great progress in technique and concept with the case of Elizabeth von R., which he probably began in 1892. The patient could not accept her love for her brother-in-law, especially after the untimely death of her sister, repressed it, and developed hysteria. Using free association, Freud interpreted the unconscious ideas, related them to the patient, and gradually got her to accept the situation in her conscious thought. Freud found transference, the basically erotic feeling of the patient for the therapist, to be a regular development in an analysis and necessary for its success.

Freud wrote the crucial document of the second phase of the development of his psychoanalytic ideas in 1895. The "Project for a Scientific Psychology" was a comprehensive theory of the neurological events underlying human thought and behavior. The essentials of some of his central psychoanalytic concepts were in it, to be elaborated in his later writings. Freud did not publish it (it first appeared in 1954) but sent it to Fliess and discussed and revised it in their correspondence for several years. To attempt to explain human thought and behavior in terms of the structure and function of the nervous system was not unusual; an enthusiasm for physical science and a confidence in its methods for dealing with complex phenomena permeated German medicine. More specifically, two men with whom Freud was in close and frequent contact had made this attempt. Freud's professor of psychiatry at Vienna was Theodor Meynert, and he continued to work with Meynert while at the Vienna General Hospital. Meynert and Sigmund Exner, one of Brücke's two assistants at the Physiological Institute, both wrote large works correlating neurology with thought and behavior. These works resemble Freud's "Project" in their basic theories.

In the view of Meynert, Exner, and Freud, all nervous system function consisted of reflexes of a certain type. There was an analogy—sometimes implicit, sometimes explicit—between the flow of an electric current through a network of wires and the passage of nerve impulses through nervous pathways. A quantitative phenomenon (called excitation, nerve energy, or quantity in the "Project") flowed through the pathways with an unspecified force, tending to make it flow from the sensory periphery to the motor periphery of the system. The force was generated by the sense organs when they were stimulated, in proportion to the intensity and duration of the stimulation. The excitation was discharged at the motor periphery of the nervous system, primarily in triggering the contraction of muscles. Excitation was neither added nor lost during its passage through the system, so that the amount of motor activity was proportional to the amount of stimulation. There was ample evidence from neurological experiments that reflex contractions were, in general, proportional to the amount of stimulation. Since the middle of the eighteenth century, scientists had noted that the more painful the stimulation of a limb of an experimental animal, the greater the reflex flection of the limb. When Freud was at Brücke's institute, electrical recordings from nerve tracts were interpreted as supporting this view. An increase in voltage was recorded from sensory nerve tracts when the intensity of the stimulus increased. That this increase resulted from bringing into function more channels, or nerve cells, was not recognized. Experimentally based observations of spinal reflexes and the anatomical evidence that the gray matter of the brain was histologically similar to the gray matter of the spinal cord had led to the doctrine that the entire system functioned reflexly. It was assumed that food-seeking in response to somatic stimuli was a reflex response, with the central transfer of excitation taking place through brain pathways. Consciousness occurred when the excitation, on its way from sense organs to motor organs, passed through the cerebral cortex. In spinal reflexes, excitation passed through innately determined pathways. The reflex ended in a motor act, which brought about the end of the stimulation of the sense organ. The painfully stimulated limb was flexed and withdrawn. In the cortex, pathways were put into function during the life of the individual; this was the neurological event resulting in learning. The coming into function of new pathways took place when the innately determined pathways did not serve to end the excitation from the sense organs.

Freud followed Meynert in taking a baby's learning to nurse as a paradigm of the opening up of cortical pathways. Freud did so in the "Project" and in his two most important published works, *The Interpretation of Dreams* and *Three Essays on the Theory of Sexuality*. When excitation entered the infant's nervous system—the concomitant of hunger—it was first channeled through innately determined pathways. The baby kicked and screamed, which was to no avail in bringing about the ending of the influx of excitation from the somatic sense organs. Then the mother turned the baby's mouth to her nipple, the baby sucked, and the inflow from the sense organs ceased. The next time the baby was hungry, the excitation passed through the cortical pathways that had been opened up, those which served the sight of the nipple and the turning of the baby's head to the nipple. The nervous system did not transfer excitation through innate pathways or those serving learning until the sense organs were stimulated. There was no nervous function without stimulation. Meynert put it thus: "The brain . . . does not radiate its own heat; it obtains the energy underlying all cerebral phenomena from the world beyond it."[2] In the terminology of Freud's metapsychology, this became the pleasure principle, the tendency of the psychic apparatus to function so as to discharge the excitation that impinged upon it.

The outline of the distinction between the ego and the id is in the "Project." Freud defined the ego as that complex of cortical pathways that were put into function during the baby's learning to turn to the nipple and in other learning experiences. This was a term common in psychology. When the ego was again subject to the inflow of excitation, the correlate of hunger, the baby carried out the same motor acts that had previously ended the inflow. This reusing of pathways, without alteration of the pattern of transmission of excitation and without any change in the resulting behavior, Freud called the primary process in the ego. When the baby was hungry at a later time, part of the current stimuli to the sense organs was not the same as it had been when the pathways serving the primary process were put into function. For example, if the mother presented her other breast to the baby, the stimulation of the eyes would be different. To cover this situation, Freud postulated an inhibiting ego that did not allow discharge over the primary process pathways, which would result in an exact repetition of the first turning to the breast, but compared current perceptions with those making up the pathways serving the primary process. By a complex process, which Freud did not succeed in reducing to plausible mechanical terms, the necessary change in the motor act was determined by the inhibiting ego. In Freud's later formulation, the ego became roughly the equivalent of the inhibiting ego, while that part of the ego not under the control of the inhibiting ego became the id, the part of the psychic apparatus that mediated primary processes.

The whole construct depended to a large extent on certain ideas that were proper to psychology rather than neurology. Meynert and Freud accepted association psychology as an adequate statement of the determinants of human thought and behavior. For example, a law of association psychology—that simultaneous or temporally contiguous sense perceptions become associated, so that they tend to be recalled jointly—provided the background for their assumption that the sight of the nipple, the perception of hunger, etc. would have their nervous concomitants connected in the pathways of the primary process. There was of course no neurological evidence that the cortical pathways serving a visual perception and those serving the perception of a somatic state, when both had nervous excitation transmitted to them, would come into functional connection. The evidence was first behavioral, then incorporated into association psychology. German association psychology, which Freud had studied, also included the concept of affect as a quantitative entity attached to ideas. This agreed with the concept that the cortical pathway serving an idea has excitation passing through it. This association psychology was commonly described in mechanical terms and lent itself to theories of brain function.

Freud was led to his concept of the sexual origin of neuroses by this view of the function of the nervous system. In *Studies in Hysteria,* he and Breuer (Breuer is listed as the author of the theoretical section, but Freud took an equal—if not major—part in developing the ideas in it) wrote that the intense and long-lasting nervous and mental activity manifested in hysteria resulted from excitation impinging from the sex organs. The primary causes of the increase of excitation were the need for oxygen, food, and water. However, their patients were not deprived of oxygen, food, or water, so that the sex organs were "undoubtedly the most powerful source of persisting increases of excitation (and consequently of neuroses)."[3] Excitation was passing through the nervous pathways in quantity, the theory of nervous function required a sensory source for it, and the sex organs were the obvious choice. Certainly Freud's and Breuer's patients had disturbances in their sexual lives; it would be expected in lives so generally disordered. Yet the theory of nervous function led to selection of such disturbances as the essential etiology of neuroses.

Freud's idea that dreams are wish-fulfillment processes was a special instance of his view that all mental processes were wish-fufillment processes, which in turn followed from the theory of nervous function. He stated in the "Project" that by "wish" he meant the cortical pathway which had previously been opened up to discharge excitation impinging from the sense organs. The hungry baby wished for the mother's nipple because a cortical pathway representing the nipple had been part of the complex of channels put into function when the baby had first stopped the inflow of hunger excitation. In the simplest type of dream, a slightly hungry baby would dream of the nipple because there was a slight inflow of excitation from the sense organs serving hunger. If there was too much excitation flowing, the baby would wake up. Freud thought that excitation impingement of low quantity was a condition of sleep. He had not given much attention to the interpretation of dreams until after he had arrived at the theory of their wish-fulfillment nature. He made this momentous theoretical advance while he was writing the "Project" in the summer of 1895. The first dream he interpreted in detail was one he had during that summer and reported in the "Project." Thereafter he interpreted many of his own dreams as part of his regular self-analysis, using the wish-fulfillment theory as the essential interpretive tool.

Freud's observations and interpretations of his own mental states during the second phase of the development of his psychoanalytic thought were as important as all his case studies taken together. His wish-fulfillment theory was formed after the dream (which has become known as that of "Irma's injection") he had while writing the "Project." In July 1897, Freud began a regular analysis of himself, devoting some time to it each day, with the definite aim of unearthing the roots of his own character. In the fall of the same year, he reached his concept of infantile sexuality after interpreting his own dreams. The crucial phase of his self-analysis seems to have ended by 1900, but he continued it, on a regular basis, for the rest of his life.

Freud's theory of infantile sexuality was a momentous step forward in the understanding of human psychology. The theory was shaped by the "Project" and by Freud's powerfully working intuition. The influence of the "Project" can be traced, while only its results give evidence of his intuition. Freud's interpretation of his dreams was the central part of his accelerated self-analysis in 1897. Using the wish-fulfillment concept, he found evidence of intense mental activity, with an erotic content, in his own infancy. There is no detailed record of these dream interpretations, but the results of them are in his letters to Fliess. In October 1897, he wrote that when he was two, "*Libido* towards *matrem* was aroused; the occasion must have been the journey with her from Leipzig to Vienna, during which we spent the night together and I must have had the opportunity of seeing her *nudam*." Libido was intended in the sense he used it later, where he described it as the sexual counterpart of hunger. In his next letter Freud described his love of his mother and jealousy of his father, stated that this was a general occurrence in early childhood, and related it to the Oedipus legend. The next letter after that described infants' desire for "sexual experiences" they had already known.

Because he held the idea of the nervous reflex as a transfer of a quantity of excitation originating at the sensory periphery, Freud had to assume that the infant must have a source of considerable excitation in order to have aroused libido and "sexual experiences." By the next month, November 1897, Freud had taken this step. He wrote to Fliess: "We must suppose that in infancy sexual release is not so much localized as it becomes later, so that zones which are later abandoned (and possibly the whole surface of the body) stimulate to some extent the production of something that is analogous to the later release of sexuality."[4] The occurrence of sexual dreams implied the existence of previous sexual experiences. The theory of nervous function implied that excitation from the sense organs had been discharged over pathways opened up during these experiences; this in turn necessitated sense-organ sources of this excitation. Freud did not need to suppose anything about sources of excitation to make the generalization that infants have mental activity that becomes related to adult sexuality. But to save his comprehensive theory he needed to postulate infantile erogenous zones. His developed theory of psychosexual development, in *Three Essays on the Theory of Sexuality,* is a great elaboration of the ideas he presented to Fliess. His later ideas are elaborations of the 1897 concept that infantile sexual experiences are shaped by the character of inflows of excitation from the sense organs.

The third phase of the development of Freud's ideas is marked by the elaboration, in terms of a wealth of clinical experience, of the ideas he pioneered in the first two phases. He also applied his psychoanalytic understanding to social theory. The pace of development was slower, but his publications from this period (after 1900) fill nineteen volumes in his collected works. The work published previously and judged psychoanalytic by the editors of the *Standard Edition* of his writings fill four volumes. His writings in neuropathology, if similarly collected, would fill an additional three or four volumes.

The Interpretation of Dreams, published in 1901, has usually been considered Freud's most important work. In the famous seventh chapter he published for the first time much of the general theory he had formulated in the "Project" in 1895. He remarked that the book had been essentially finished in 1895, but not "written down" until the summer of 1899.[5] In the seventh chapter the brain becomes the "psychic apparatus," and most other neurological terms are replaced by psychological and psychoanalytic terms (but not all terms are replaced; the basic mode of function is still called the reflex). The volume contains many detailed accounts of dreams and many interpretations, primarily of his own dreams following his formulation of the wish-fulfillment theory. In carrying out these interpretations, Freud refined his understanding of the mode of operation of the unconscious. He discussed displacement, the transfer of hate or love from one person to another, when the transfer makes the resulting conscious emotion acceptable to the ego. For example, sexual desire for the mother might be transferred to another woman. He pointed to the extensive appearance in conscious thought of symbols for repressed thought. The consciously desired woman might be wearing shoes such as the unconsciously desired mother had worn. Regression, the tendency to think in a manner that was appropriate in the earlier life of the individual, was the neurotic equivalent of earlier thought patterns in dream consciousness.

Freud published *Three Essays on the Theory of Sexuality,* the second in importance of his books, in 1905. This and *The Interpretation of Dreams* were the only two of his books that he continually revised for succeeding editions. The same concern with the quantity of excitation, its origin, and the time of its impingement that led Freud to postulate infantile sources of sexual excitation in 1897 shaped his elaboration of infantile psychosexual development in this book. The paradigm of the infant learning to suck was, with additions, used in describing the oral stage of development. The lips were an erogenous zone which originated excitation (now libido). The baby did not know (there were no innate processes) how to end this inflow of excitation any more than he knew how to end the inflow of excitation caused by hunger. The inflow of excitation from the lips was ended by the act of sucking. Freud had to make all the erogenous zones consistent with the obvious model of the genitals, which cease to originate excitation after "some kind of manipulation that is analogous to sucking."[6] The child's dependence on the mother for stopping the inflow of excitation, or pleasure, resulted in a psychological makeup marked by receptive dependency.

In the anal phase of infantile development the source of excitation was the anus, and the inflow of excitation was brought to an end by the passage of feces. The child might retain feces so that there would be greater mechanical effect on the anus when the feces was passed or might manipulate the anus. Toilet training interfered with the child's management of excretion for his pleasure. The psychology of this phase is dominated by obstinacy and retentiveness, with toilet training imparting defensive prohibitions, such as disgust and cleanliness, against manipulating the anus. The phallic phase foreshadows adult sexuality, with the genitals the source of excitation and infantile masturbation the means of ending the inflow. The psychology of this phase is dominated by the Oedipus complex and its derivative competitiveness. Sexual maturity, or the genital phase, had of course the same source of excitation as the phallic phase, but with ego mastery of the drives that dominated the earlier phases.

Passage through the successive stages of psychosexual development influenced the behavior of the adult. In some individuals, the psychological patterns appropriate to infantile stages were dominant in their adult life. The traits associated with the successive phases were not as tidily stated as this summary implies. It was left to Karl Abraham, one of Freud's most capable followers, to work out the character types associated with the dominance of thought patterns appropriate to each stage of psychosexual development.

The Interpretation of Dreams and *Three Essays* made evident a novel feature of Freud's thought: his emphasis on the similarity between normal and abnormal thought and behavior. Dream consciousness was very much like neurotic thought processes. The difference between normal and neurotic sexual behavior could be only a matter of the relative strength of the processes established during the individual's passage through oral, anal, and phallic phases. That normal and abnormal phenomena were so similar was by no means a new idea, but the theorists of the dominant association psychology, psychiatrists, and theorists in neurology such as Meynert had made only trivial use of the idea.

Freud's last major contribution to psychoanalytic theory was *The Ego and the Id,* published in 1923, in which he elaborated on the concept of the superego. In the "Project" he had assumed that ego (then called inhibiting ego) processes were conscious and that those of the id (a term introduced later) were unconscious. The superego was a part of the ego that did not involve consciousness. In what Freud regarded as normal development, the infant first took his mother as the desired sexual object. Oedipal fear of the father (essentially fear of castration by the father) led the infant to give up this object. Fear preventing the mother from being the object of libido, she was replaced with the mental representation of the person himself. The discharge of excitation (now libido) could take place through acts that were in accord with parental commands and therefore did not produce fear. The superego, the result of parental criticism and prohibitions, was Freud's version of conscience. Two years later Freud wrote in *Inhibitions, Symptoms and Anxiety* that guilt feelings are the result of thoughts or acts not in accord with the superego, the internal representation of the parents. In the new scheme the aim of therapy remained the same: bringing unconscious processes under the control of the conscious ego. But it was now necessary to analyze not only the relatively simple unconscious processes in the id but also the more complex processes in the superego. Freud's beginning in ego analysis was advanced especially by his daughter Anna.

The focuses in Freud's writings fit with his view on the use of psychoanalysis. He regarded psychoanalysis as important primarily as a research tool and a theory of the determinants of human thought and behavior based on his research. The therapeutic usefulness of the method and theory he considered quite secondary. He did not believe psychoanalytic therapy was efficacious except in cases of hysteria and obsessional neuroses provided that the patient was relatively young and intelligent. In keeping with this view, he wrote only four detailed case studies after *Studies in Hysteria* (1895). He used fragments of cases in the rest of his publications to advance and buttress his theoretical expositions. The elaboration of his ideas in the third phase required that the therapist interpret clinical data differently, but there is little discussion in Freud's work on how to get at clinical data or how to impart interpretations to the patient so that the patient can bring the unconscious material under the control of his ego.

Totem and Taboo, published in 1913, was Freud's first and most important volume on social theory. He took cues from Darwin's theory that the first human society consisted of a horde of brothers led by a strong father and from Sir James Frazer's *Golden Bough,* which described a universal taboo against incest and against killing the totem animal. The sons, driven by their Oedipal urges, killed the father and became leaderless. Their need for a strong leader led them to deify the totem animal, thus establishing the forerunner of religious systems. Social development depended on overcoming, with institutions which forbade incest, the Oedipal desire for the mother. Freud emphasized the continuance of hostile impulses within developed societies in *Civilization and Its Discontents* (1927). Aggression against the father was repressed by the incorporated parental image, the superego. This repression was institutionalized in social justice. Discontent was an inevitable aspect of civilization because, even though Oedipal aggression had been repressed, the wish had not; and the wish had the same power to produce guilt that the act did.

The dispersion of Freud's thought in Europe centered in the psychoanalytic movement. The weekly meetings that began at Freud's house in 1902 developed into the International Psychoanalytic Association, established at Nuremberg in 1910. Swiss participants, including Carl Jung, and of course the

Viennese, including Alfred Adler and Sándor Ferenczi, were dominant. At this meeting Ferenczi, with Freud's encouragement, proposed an authoritarian structure, with an elite determining proper psychoanalytic doctrine. Freud's attitude was expressed in a statement he made about succession: "When the empire I founded is orphaned, no one but Jung must inherit the whole thing."[7] This aspect of the organization led to the resignation of Eugen Bleuler, professor of psychiatry in Switzerland and the only European member of the association with solid academic credentials in psychiatry. This was the first instance of a failure by the psychoanalytic movement to keep open the lines of communication with European academic psychiatry, and of course it decreased the European influences of Freud's ideas. Those who were not willing to subordinate their intellects to that of Freud left the association. Alfred Adler disagreed with Freud over the importance of the Oedipus complex and left in 1911. Jung departed in 1914 over differences on the importance of sexuality and also because of personal conflicts with Freud.

The response to the developing threat of Jung's schism was the formation of the Committee, a permanent elite to guarantee the maintenance of what had already become an orthodoxy. The members of the committee agreed that if any of them wished to depart from any of "the fundamental tenets of psychoanalytical theory—for example, the conception of repression, of the unconscious, of infantile sexuality—he would promise not to do so publicly before first discussing his views with the rest."[8] The founder and for many years chairman of the committee was Ernest Jones; the other members were Abraham, Hanns Sachs, Ferenczi, Otto Rank, and, later, Max Eitingon. Freud had passed from seeking to being an intellectual father figure: Breuer was fifteen years older than Freud, Fliess a year younger, and the members of the committee were between sixteen and twenty-nine years younger than Freud. All remained loyal except Rank, who broke with Freud in 1929 after several years of agonizing over the departure from his mentor. The committee had at least equal responsibility with Freud for the authoritarian nature of the movement. In the case of Rank, Freud first welcomed Rank's radical departure from his theory of psychosexual development, in which Rank made the birth trauma crucial, but was influenced by the committee not to accept this novelty. Freud became convinced that Rank's own neurotic psychology led to his revisionism, a reason too often used by Freud and his followers to explain objections to Freud's ideas.

The cultish aspect of the European psychoanalytic movement was one of the reasons for the relatively small influence of Freud's views there. (More important reasons will emerge in the discussion below of his far greater influence in America.) Most important was that their opportunity to influence was blocked when the Nazis came to power in the early 1930's. They regarded psychoanalysis as a Jewish doctrine, proscribed it, and forced many of its adherents, the majority of whom were Jewish, to flee. Often they went first to England, where psychoanalysis was more accepted than in any country except America. With appropriate employment limited in England, many came to America, where they greatly augmented the influence of Freud's ideas.

The United States has given Freud his stature in the history of thought. Long before he was accorded any equivalent honor in Europe, he was invited to give a series of lectures at Clark University in Worcester, Massachusetts, to mark its twentieth anniversary. G. Stanley Hall, a prominent psychologist and president of Clark, extended the invitation. Freud was enthusiastically received by James Putnam, professor of neurology at Harvard, who was to become president of the new American Psychoanalytic Society in 1910. Reputable American journals were open to Freud and his followers, while in Europe they published mostly in journals that they themselves had established. American physicians were the central element in the transmission of Freud's thought. From its beginnings, they dominated the American Psychoanalytic Association. (This led to the break of this association with the International Psychoanalytic Association over the question of whether psychoanalysts should be physicians in addition to having analytic training under the supervision of the psychoanalytic organization. In 1938 the American association, opting for medical qualification, separated from the international association. Freud had repeatedly stated that medical training was not of value to psychoanalysts.) The physicians' dominance of organized psychoanalysis, and the acceptance of parts of Freud's thought by physicians outside organized psychoanalysis, led American laymen, with their general respect for physicians, to respect psychoanalysis. Only the influence on American medicine will be outlined here, the rationale being that while Freud's thought affected all Americans who studied human thought and behavior, the influence on physicians was crucial for the general acceptance of Freud and the content of the influence tended to be the same for an anthropologist or a literary critic as it did for a physician.

By 1920 most American physicians interested in neurology and psychiatry had taken some account of Freud's theories. They prided themselves on their eclecticism, and many of them had accepted part of his thought. An important reason for this was that they were familiar with psychological therapies for illnesses assumed to have important, if not exclusively, psychological causes. Adolf Meyer had emphasized the patient's specific psychological history, as Freud did, and had used counseling as therapy. William A. White was ready to include a discussion of Freud's thought in a textbook of psychiatry he published in 1909, because his own dynamic psychiatry had prepared him to accept much of Freud's work as soon as he became familiar with it. Many American physicians had begun using psychotherapeutic techniques, two variants being called persuasion and reeducation, in the first decade of the twentieth century. These methods did not carry with them the theoretical luggage that Freud's free association method did. The Americans did or did not use such methods according to each physician's estimate of their efficacy; they approached psychoanalysis in the same pragmatic way. They thus reversed Freud's position that psychoanalysis was valuable chiefly as a theory and a research method to advance the theory and only secondarily as a means of therapy. There was an open-mindedness among American physicians not evident in Europe. A. A. Brill, the most energetic advocate of Freud in the United States, had patients referred to him by physicians who were opposed to Freud's thought. They apparently acted on the assumption that for these particular cases it might work, and they did not isolate Brill for his opinions. Psychoanalysts remained part of the medical establishment, albeit a small part.

Freud's name became the one most often associated with psychotherapy in general, and the distinction between the Freudian approach and any psychotherapy was blurred for many physicians. Nevertheless, there were influences more specific than Freud's furthering of psychotherapy in general. His thought brought about a greater emphasis on early childhood and on the sexual determinants of behavior. Psychotherapy used more exploratory techniques, in an attempt to reconstruct the etiology of a disease, and less exhortation to the patient to change himself. The few members of the American Psychoanalytic Association were of course more conversant with Freud's thought, and most of them used free association in relatively pure form.

The culmination of Freud's influence on American medicine came after World War II. From 1920 until the late 1930's the number of psychoanalysts in the United States did not greatly increase. The European refugees increased their numbers in the 1930's, but it was one small group joining another. In the late 1940's and 1950's there was a rapid increase in their number. Psychiatry shared in the great increase of federal funds available for medical research and education, and the disbursement of these funds was often controlled by people strongly inclined toward a Freudian approach. Federal funds after the war financed research and academic positions that were most often filled with psychoanalysts or with men who had indicated their acceptance of analysis by undergoing analysis themselves. Nearly all the chairmen of psychiatry departments since 1946 have been psychoanalyzed. Psychoanalysis became entrenched in the medical school curriculum, often being the core of the basic course in psychiatry. The general increase in prosperity in the United States was also essential to the increased number of psychoanalysts. Psychoanalysis required far more time from the physician than any other therapy, psychological or physical, and was therefore costly. Only an affluent people could afford psychoanalysis. The sponsorship of Freud's thought by the medical establishment was an important part of the context in which one implication of his thought became influential—an implication he never intended. He had emphasized the similarity of normal and abnormal behavior. Especially in his writing on social theory, he had found aspects of mental illness manifest in society at large. An extension of this emphasis was a radical redefinition of mental illness. Mental illness had been that which was marked by bizarre symptoms. Freud's patients, for example, had psychogenic paralyses, were unable to go out into the street, or repeatedly washed their hands. American psychoanalysts only rarely saw such cases. Their patients typically had an inability to form adequate personal relationships. Such inabilities were explicable in terms of Freud's thought, but he never proposed devoting most of the resources of psychiatry to curing them. The other side of this coin was a neglect of the mentally ill who did have bizarre symptoms, mainly those in institutions, in a period in which psychiatry had far more financial support than it had ever had before.

Psychoanalysis was no longer equated with Freud's thought, but his influence remained. The libidinous inflow of excitation was no longer taken as the sole motive power for the symptoms of mental illness but was one among several drives given equal theoretical status. Yet much of Freud's description of infantile psychosexual development as the primary determinant of adult psychology was incorporated into more complex theories. Above all, his general view of the unconscious remained influential. Most psychoanalytic theories included the unconscious as the sum of processes that, while not observable consciously, determine conscious thought and are organized so as to satisfy needs, although the needs are not necessarily consciously recognized. While the psychoanalytic hegemony over American psychiatry and the medical hegemony over psychotherapy began to break up in the 1960's, there was by 1970 no clear indication of how this would affect the influence of Freud's thought.

NOTES

1. Jones, I, 5.
2. Amacher, p. 24.
3. *Studies in Hysteria,* in *Standard Edition,* II, 199–200.
4. *The Origins of Psychoanalysis,* pp. 215–232.
5. Jones, I, 351.
6. *Three Essays,* in *Standard Edition,* VII, 184.
7. Ludwig Binswanger, *Sigmund Freud: Reminiscences of a Friendship* (New York, 1957), p. 31.
8. Jones, II, 152.

BIBLIOGRAPHY

I. ORIGINAL WORKS. The definitive ed. of Freud's work is in English: *Standard Edition of the Complete Psychological Works of Sigmund Freud,* James Strachey, ed., 23 vols. (London, 1953–1966), vol. XXIV in preparation. This is a magnificently edited work, with many annotations giving both details and interpretations. It is a variorum edition and includes with each article or book a bibliography of earlier eds. in German and English. Vol. XXIV will contain a full bibliography and an index to the entire work. Freud's neurological writings (and an incomplete bibliography of his works) are listed in *The Origins of Psychoanalysis: Letters to Wilhelm Fliess, Drafts and Notes: 1887–1902,* Marie Bonaparte, Anna Freud, and Ernst Kris, eds. (New York, 1954). This includes an annotated English trans. of the "Project for a Scientific Psychology" (also in vol. I of the *Standard Edition*). An incomplete listing of the various eds. of Freud's works and of secondary works on Freud is in Alexander Grinstein, ed., *The Index of Psychoanalytic Writings,* 9 vols. (New York, 1956–1966). These vols. list works published through 1960; vols. in preparation will list works published through 1968.

II. SECONDARY LITERATURE. The standard biography of Freud remains Ernest Jones, *The Life and Work of Sigmund Freud,* 3 vols. (New York, 1953–1957). This must be used with caution regarding two of its aspects. Jones brought together a great deal of material, but he was not consistently accurate in his use of it and was too much the disciple of Freud to make very critical interpretations. For Freud's life, a recent book by Paul Roazen is somewhat corrective to Jones and otherwise useful: *Brother Animal* (New York, 1969). It has valuable biographical information on Freud, references to material not available to Jones, and discussions of the tendentious editing for publication of Freud's letters by his followers.

The Freud Archive in the Library of Congress apparently contains most of the unpublished material relevant to Freud. The largest part of it was not to be used for fifty years, but this restriction has not been applied consistently. Serious scholars should consult the library as to the possibility of using the archive. The present study is in part based on the author's *Freud's Neurological Education and Its Influence on Psychoanalytic Theory,* Psychological Issues Monograph no. 16 (New York, 1965). The discussion of Freud's American influence relies heavily on John Chynoweth Burnham, *Psychoanalysis and American Medicine, 1894–1918: Medicine, Science, and Culture,* Psychological Issues Monograph no. 20 (New York, 1967). This work and David Shakow and David Rapaport, *The Influence of Freud on American Psychology,* Psychological Issues Monograph no. 13 (New York, 1964), are unusual in their careful analyses of Freud's influence.

PETER AMACHER

FREUNDLICH, ERWIN FINLAY (*b.* Biebrich, Germany, 29 May 1885; *d.* Wiesbaden, Germany, 24 July 1964), *astronomy.*

Freundlich was the son of a German businessman, E. Philip Freundlich, and his British wife, Ellen Elisabeth Finlayson. He had four brothers and two sisters. Like his brothers he received his primary schooling in Biebrich and completed a classical education at the Dilthey School in the neighboring and larger town of Wiesbaden. After leaving this school in 1903, Freundlich worked at the dockyard in Stettin before beginning a course in naval architecture at the Technical University of Charlottenburg. After a heart condition forced him to discontinue this course for about a year, he decided to begin anew and went to Göttingen to study mathematics, physics, and astronomy. With the exception of the winter semester 1905–1906, which he spent as a student in Leipzig, the rest of his higher education was confined to Göttingen University, from which he obtained his Ph.D. in 1910 with a thesis entitled "Analytische Funktionen mit beliebig vorgeschriebenem unendlich viel blättrigem Existenzbereiche."

At the suggestion of his tutor, Felix Klein, Freundlich applied for the post of assistant at the Royal Observatory in Berlin and was appointed on 1 July 1910. In the following year Albert Einstein, having heard that Freundlich was investigating the possibility of gravitational absorption, requested Freundlich's cooperation in observing the motion of the planet Mercury. Einstein himself had his own reasons for doubting that its position would coincide with that predicted on the basis of Newtonian mechanics. Freundlich's observations fully confirmed earlier evidence for such a discrepancy, and he insisted on publishing that discovery in 1913, against the wishes of the director of the Berlin observatory. That same year he married Käte Hirschberg in a civil ceremony in the Herder House in Weimar, and a small house was built for the couple at the Berlin observatory's new site in Neubabelsberg (Berlin). The German chemist Emil Fischer, impressed by Freundlich's insistence on the validity of his revolutionary conclusion, introduced him to his wealthy friend Gustav Krupp von Bohlen und Halbach, who financed Freundlich's first solar eclipse expedition to Feodosiya in the Crimea in 1914, which unfortunately had to be abandoned owing to the outbreak of World War I. Freundlich was interned for a short time before being allowed to return to Berlin.

The object of the expedition was to test the validity of a prediction of Einstein's still incomplete theory of general relativity relating to the deflection of a ray of light from a star by the sun's gravitational field, to which Freundlich drew attention in an article published in 1914. A further prediction, also mentioned in that article, which followed from the application of Einstein's principle of equivalence to light rays, was that the wavelength of light should be increased in the presence of a strong gravitational field such as the sun's; but Freundlich was obliged to conclude that the well-known phenomenon of the solar red shift could not be regarded as constituting a decisive verification of the theory.

Not having the necessary facilities at his disposal for making new reliable observations of the solar red shift, Freundlich turned his attention to the measurement of wavelength shifts in the spectra of stars in various star systems but obtained inconclusive results because so little was known about stellar masses. He later tried to reduce this deficiency by researches on the distribution of stars in globular star clusters. An increasing amount of attention was focused on Freundlich's work after Einstein's publication of 1916 had revealed the immense significance of his theory of general relativity for the future development of physics, and after Freundlich had discussed the means of testing it in his first book, *Die Grundlagen der Einsteinschen Gravitationstheorie* (Berlin, 1916). An article by him on this subject appeared in *Naturwissenschaften* (1917), and a second edition of his book was published at Leipzig in 1920. Freundlich resigned his post in 1918 to work full-time with Einstein, financed by the Kaiser Wilhelm Gesellschaft. He always modestly regarded himself as less of a collaborator with Einstein than as a butt for the latter's highly original ideas. His occasional inability to comprehend these ideas had the salutary effect of making Einstein seek to simplify their mathematical formulation, for if one of Felix Klein's pupils could not make sense of his equations who could? Through his intimate contact with Einstein, Freundlich was the first to become thoroughly acquainted with the fundamental principles of the new gravitational theory and, as Einstein himself remarks in the foreword of Freundlich's book, he was particularly well qualified as its exponent because he had been the first to attempt to put it to the test.

After Arthur Eddington confirmed that Einstein's theory accounted quantitatively for the discrepancy in the position of the planet Mercury, on whose reality Freundlich had boldly insisted in 1913, the validity of the principle of relativity itself could no longer be doubted and Freundlich's scientific reputation was fully vindicated. Thus, in 1920, the Prussian Ministry of Culture decided to support the creation of the Einstein Institute at the Astrophysical Observatory, Potsdam; Freundlich was appointed observer in 1921 and later chief observer and professor of astrophysics. This institute was designed specifically to strengthen the empirical foundations of Einstein's gravitational theory and was equipped with an astrophysical laboratory and a powerful tower telescope for solar work, the *Einsteinturm.* Here, from 1921 to 1933, Freundlich encouraged his co-workers to tackle problems that appeared at the time to be of particular significance in the solution of the solar red-shift problem, such as the origin of the ultraviolet cyanogen bands (1924) and the measurement of the center-to-limb variation in wavelength along different radii on the solar disk (1930). Simultaneously he was supervising the construction of specially designed equipment for observing light deflection; and he planned three further eclipse expeditions, two of which (in 1922 and 1926) were unsuccessful because of bad weather conditions. The third (to Sumatra in 1929) was a complete success—even though the final result proved to be significantly in excess of what Einstein's theory had predicted.

It was typical of Freundlich that he was less concerned about the negative role of this discrepancy in casting doubt upon Einstein's assumption that the laws of motion for matter are also valid for the energy contained in a light ray than about the positive and exciting possibility that its very existence might hold the key to the still unexplained rift between relativity theory and quantum mechanics, and thereby assist the unification of the macroscopic and microscopic patterns of nature. To say that Freundlich wished to disprove Einstein is to mistake entirely the motivation which caused him in later years continually to stress the excess of his observed value for the light-deflection over that predicted by general relativity theory—a fact, incidentally, which subsequent independent eclipse experiments have served only to confirm, and for which no satisfactory interpretation has as yet been forthcoming.

When Hitler came to power in 1933, Freundlich reluctantly resigned his post in Potsdam and emigrated to Turkey, a decision partially determined by the fact that his wife's sister had died earlier that year and he had assumed the guardianship of her two young children, whose lives he considered—with some justification—to be in danger.

From 1933 to 1937 Freundlich helped to reorganize the University of Istanbul and to create a modern observatory. He also wrote, for translation into Turkish, the first astronomical textbook published in that language (1937). He left Turkey for an appointment as professor of astronomy at Charles University in Prague, but Hitler's policy toward Czechoslovakia forced him to leave this post in January 1939. One month later, in Holland, he received an offer from Sir James Irvine, principal of St. Andrews University, of a lectureship there, on the understanding that he would be encouraged to build an observatory and create a new department of astronomy. The St. Andrews University Observatory was completed in 1940. In the following year Freundlich—or Finlay-Freundlich, as he preferred to call himself since he was now resident in Scotland—was elected a fellow of the Royal Society of Edinburgh. During the rest of World War II he lectured on astronomy to undergraduate students and taught celestial navigation to Air Ministry cadets who were taking special short courses at St. Andrews as part of their basic training.

Although useful for demonstration purposes, the instruments at Freundlich's observatory scarcely constituted the basis for observational research in a rapidly expanding and increasingly competitive field, so he had the inspiration to construct the first Schmidt-Cassegrain telescope, which reduces the spherical aberration and length of tube from those of the conventional Cassegrain type and gives a plane image conveniently located just outside the back of the main mirror. It is undoubtedly more than a mere coincidence that about this time Freundlich was occupied with further studies on the structure of globular star clusters, since the Schmidt-Cassegrain arrangement is eminently suited for astrographic work. An eighteen-inch-aperture pilot model was constructed under Freundlich's supervision in the workshops of the St. Andrews University Observatory and was mounted at the Mills Observatory, Dundee, in 1949. Theoretical calculations on the design of this new telescope were carried out by E. H. Linfoot of Cambridge University. The results of tests made with this pilot model in February 1950 proved to be so encouraging that work was begun on a larger-scale telescope thirty-seven inches in diameter. Experimental researches on the preparation of multilayer coatings for interferometer plates were simultaneously being carried out at the observatory by Alan Jarrett; these led to a long and fruitful program of research on the aurora borealis, resulting in the observation of certain emission lines of the solar corona with the aid of a Fabry-Pérot interferometer by Jarrett and H. von Klüber during the total solar eclipses of 1954, 1955, and 1958. Bad weather conditions during the eclipses in 1954 and 1955 foiled Freundlich's own attempts to repeat his light-deflection experiment with the instruments he had used in 1929.

On 1 January 1951, Freundlich became the first Napier professor of astronomy at St. Andrews University. In his inaugural lecture, delivered just over a year later, he characteristically stressed the broad cultural value of the study of astronomy as well as the more specialized aims of that science and again made reference to the importance of the light-deflection experiment as a means of verifying the fundamental principles of Einstein's theory. The related problem of interpreting the nonkinematic red shifts observed in solar, stellar, and galactic spectra was one that he began to reconsider seriously during his recovery from a heart attack in 1953; and his researches led him to jeopardize his scientific reputation once again by proposing the revolutionary hypothesis that the entire range of unexplained astrophysical data could be comprehended by an empirical formula relating the red shifts to the temperature of the radiation field and the distance through it which a ray of light would have to traverse before reaching a terrestrial observer.

Apart from the objection that this interpretation denied the existence of the gravitational red shift predicted from the general relativity theory, the majority of the criticisms which were immediately raised against Freundlich's red-shift formula centered on the unreliability of the observational data that he had cited in support of it. This led him to concentrate his attention on the particular problem of the solar red shift, for which reliable measurements could be made (at any rate, in principle); and he engaged a research assistant to examine whether or not the gravitational red shift was implicit in observations of this phenomenon. The initial aim of this research was inverted by the results of independent experiments in nuclear physics using gamma rays and the earth's gravitational field instead of light rays and the sun's gravitational field, which seemed to confirm the quantitative value of Einstein's prediction. Nevertheless it served to show that Freundlich's formula is merely an alternative empirical expression of the so-called relativity-radial current interpretation of the solar red shifts; moreover, it is one that is restricted to Fraunhofer lines of moderate intensity, for which the Doppler effects associated with the solar granulation are determinative in producing the observed shifts. This conclusion, derived from an analysis only of solar observations, would appear to imply that Freundlich's hypothesis is invalid as the basis of a general interpretation of the still unexplained red shifts of stars and galaxies; but this is an inference that Freundlich himself was not prepared to draw, as is evident from his last paper on the subject (1963).

Freundlich continued to act as director of the St. Andrews observatory until he resigned his chair in 1959. In the latter half of the 1950's he composed his book *Celestial Mechanics* (1958) and made plans for his retirement to Wiesbaden.

The closing years of Freundlich's life were marred by incidents arising out of the reluctance of his successor, D. W. N. Stibbs, to grant him open access to the St. Andrews observatory in order to witness the final stages of the work on the thirty-seven-inch Schmidt-Cassegrain telescope. The tensions that thus arose occasioned, *inter alia,* the resignation of his highly skilled technician, Robert L. Waland, before the optical components were satisfactorily completed and adjusted, and partly explain why that instrument has never yielded the results of which it might otherwise have been capable. At the time of his death Freundlich was an honorary professor at the University of Mainz.

BIBLIOGRAPHY

A complete list of Freundlich's publications, collated by Tadeusz B. Slebarski, is in an obituary notice by H. von Klüber in *Astronomische Nachrichten,* **288** (May 1964–December 1965), 281–286. A less detailed obituary by the same author is in *Quarterly Journal of the Royal Astronomical Society,* **6** (Mar. 1965), 82–84. *The Alumnus Chronicle* (University of St. Andrews), no. 36 (June 1951), 23–28, contains Freundlich's article "The Schmidt-Cassegrain Telescope"; and no. 40 (June 1953), 2–14, his inaugural address, "The Educational Value of the Study of Astronomy." A detailed account of researches on the solar red-shift problem, which gives due weight to Freundlich's contributions, is Eric G. Forbes, "A History of the Solar Red Shift Problem," in *Annals of Science,* **17** (1961), 129–164.

ERIC G. FORBES

FREY, MAXIMILIAN RUPPERT FRANZ VON (*b.* Salzburg, Austria, 16 November 1852; *d.* Würzburg, Germany, 25 January 1932), *physiology.*

Frey's father, Carl, was a well-to-do merchant in Salzburg. His mother, Anna Gugg, was the daughter of a high-ranking Austrian official. Frey began his studies in medicine at Vienna, where Ernst Brücke was then teaching physiology. From there he went to Leipzig and Freiburg. As early as 1876, in Carl Ludwig's physiology laboratory at Leipzig, he began examining the functioning of the vasodilating and vasoconstrictive nerves in the salivary glands. In 1877 he graduated from Leipzig. Frey returned in 1880 to Ludwig's laboratory and remained there until 1897. He became a lecturer in physiology at Leipzig in 1882 and in 1891 was made associate professor. In 1897 he accepted a professorship of physiology at Zurich and in 1899 at Würzburg, where he remained until his death.

Frey was charming but reserved and modest and of a critical temperament. He was musical (he played the flute) and loved the Alps. He possessed tremendous scientific imagination and great ingenuity in the techniques and variation of physiological experiments.

During the first period of his career (1880–1888) Frey was concerned primarily with muscle physiology; during the second period (1889–1892) with the mechanics of circulation, e.g., through analysis of pulse and blood pressure; and during the third period (1892–1932) he became a pioneer in the investigation of the "lower senses," i.e., the sensory organs of the skin and "deep sensibility."

In muscle physiology he worked on, among other things, a comparison of the extent of a single contraction and of tetanus, including when under a load. Together with Max Gruber, Frey discovered the increased oxygen consumption of muscle in the recovery phase (1880). He investigated the role of lactic acid in muscle metabolism, the influence of inorganic ions on muscle contraction and membrane permeability, and heat production in muscle. He built an interesting apparatus (1885) for perfusing a surviving, isolated muscle.

In circulatory physiology Frey developed distortion-free recording and measuring units which helped him to answer questions concerning the course of pulse curves, reflection phenomena, and the inertia of recording levers.

From 1894 Frey was particularly preoccupied with the physiology of the skin senses and found, identified, and localized the pressure points and sensory organs for heat and cold, using appropriate methods and working on a quantitative basis (irritating hair, prickling bristle). He examined the thresholds, the summation, the adequate and inadequate stimulation of the sensory receptors, the nature of itching, the sensation of vibration (1915), and tickling (1922).

Frey proved the existence of sensory muscle receptors for the development of strength and the changing of the muscle length, thus laying the foundations for the understanding of the so-called deep sensibility (from 1913).

His laboratory produced outstanding clinicians and researchers, such as L. Krehl, P. Hoffmann, H. Rein, H. Schriever, E. Wöhlisch, H. Strughold, and F. Schellong.

BIBLIOGRAPHY

I. Original Works. Among Frey's writings are *Die Untersuchung des Pulses und ihre Ergebnisse in gesunden und kranken Zuständen* (Berlin, 1892); "Physiologie der Haut," in E. Lesser, ed., *Enzyklopädie der Haut- und Geschlechtskrankheiten* (Leipzig, 1900), pp. 387–392; *Vorlesungen über Physiologie* (Berlin, 1904); "Allgemeine Physiologie der quergestreiften Muskeln," in *Nagels Handbuch der Physiologie,* IV (Brunswick, 1909), 427–543; "Allgemeine Muskelmechanik," in R. A. Tigerstedt, ed., *Handbuch der physiologischen Methodik,* II, pt. 3 (Leipzig, 1911), 87–119; "Physiologie der Sinnesorgane der menschlichen Haut," in *Ergebnisse der Physiologie,* **9** (1910), 351–368; "Die sensorischen Funktionen der Haut und der Bewegungsorgane," in R. A. Tigerstedt, ed., *Handbuch der physiologischen Methodik,* III, pt. 1 (Leipzig, 1914), 1–45; and "Physiologie der Haut," in *J. Jadassohns Handbuch der Haut- und Geschlechtskrankheiten,* I, pt. 2 (Berlin, 1929), 1–160, written with H. Rein.

II. Secondary Literature. See P. Hoffmann, "Die wissenschaftliche Persönlichkeit Max von Freys," in *Verhandlungen der Physikalisch-medizinischen Gesellschaft zu Würzburg,* n.s. **57** (1932), 56–66. Obituaries are in *Zeitschrift für Biologie,* **92** (1932), i–v; *Münchener medizinische Wochenschrift,* **79** (1932), 315–316; E. Wöhlisch, in *Verhandlungen der Physikalisch-medizinischen Gesellschaft zu Würzburg,* n.s. **57** (1932), 52–56; H. Rein, in *Klinische Wochenschrift,* **11** (1932), 439; and in *Ergebnisse der Physiologie,* **35** (1933), 1–9, with inaccurate bibliography.

Assessments are H. Schriever, in *Neue deutsche Biographie,* V (Berlin, 1961), 419–420; K. E. Rothschuh, in *Geschichte der Physiologie* (Gottingen–Berlin–Heidelberg, 1953), pp. 156–157, with portrait; and I. Fischer, *Biographisches Lexikon der hervorragenden Ärzte der letzten 50 Jahre,* I (Berlin–Vienna, 1932), 448.

K. E. Rothschuh

FRIBERGIUS, KALBIUS. *See* **Rülein von Calw, Ulrich.**

FRIEDEL, GEORGES (*b.* Mulhouse, France, 19 July 1865; *d.* Strasbourg, France, 11 December 1933), *crystallography.*

Georges Friedel was the son of the famous chemist Charles Friedel (1832–1899), who taught mineralogy and organic chemistry at the University of Paris and was, at the same time, the curator of the mineralogical collections of the School of Mines. Charles's father was a banker in Strasbourg; his maternal grandfather was Georges Duvernoy, a co-worker of Cuvier and his successor at the Collège de France. The Friedel family had left their Alsatian home before the Franco-Prussian War. Georges spent his childhood, until the age of fifteen, in Paris, where his parents' apartment was in the building of the School of Mines. This school was to exert a profound influence on his career. He entered the École Polytechnique in 1885, having placed first in the competitive entrance examination. Upon graduation he returned to the School of Mines for a three-year course (1887–1890). Mallard was his professor of mineralogy, and his father introduced him to research in mineral synthesis. He married Hélène Berger-Levrault (1888) while still in graduate school.

As a mining engineer Friedel received an appointment in the French civil service (1891) and was put in charge of the Moulins district. In 1893 he entered the School of Mines at Saint-Étienne, where he taught courses in assaying, ferrous metallurgy, physics, mineralogy, geology, and the applications of electricity to mining. From 1899 he lectured only on geology and mineralogy, and after he became director (1907), he limited himself to mineralogy. Friedel felt strong ties to the Saint-Étienne school and declined several calls from the School of Mines in Paris. But, after World War I, he did accept the chairmanship of the Institute of Geological Sciences at the newly reopened French University of Strasbourg, where his great-grandfather Duvernoy had been the dean of the Faculty of Sciences some eighty-five years before. The return to his liberated Alsace was one of the great joys of his life. For some time before his retirement in 1930, a painful illness prevented Friedel from giving his courses, and his son Edmond substituted for him. He was confined to his room, and one of his daughters, Marie, nursed him with great devotion. His wife had died in 1920. Racked with great physical suffering, he kept his intellectual curiosity and marvelous lucidity to the end.

The work of Friedel is remarkable for its diversity. It is essentially crystallographic and mineralogical, but it deals also with petrology, geology, and even engineering and pedagogy.

Jointly with his father, Friedel first published accounts of a number of syntheses produced in a steel tube lined with platinum, at about 500°C. and under high pressure. Synthetic minerals were prepared by letting group I hydroxides and silicates or salt solutions act on mica. Among nonminerals he obtained tricalcium aluminum hexahydroxytrichloride dihydrate (1897) and a calcium aluminate (1903), both known for their twinning, and lithium metasilicate (Li_2SiO_3), which syncrystallizes with beryllium orthosilicate (Be_2SiO_4). With François Grandjean, Friedel synthesized chlorites by attacking pyroxene with alkali solutions (1909). By preparing potassium nepheline (1912) he settled the question of "excess silica" in the nepheline formula.

Friedel's work (1896–1899) established the interstitial nature of zeolitic water, which can be replaced by many liquids and gases in the zeolitic "sponge." He found zeolitic water in compounds other than zeolites.

In 1893 Friedel developed a method for the accurate measurement of path-difference that is based on the restoration of elliptically polarized light to plane polarized light. This method was later applied to the study of stressed glass by R. W. Goranson and L. H. Adams.

In 1904 the law of Bravais, based as it was on speculative considerations, was far from being generally accepted. Friedel established its validity as a law of observation, regardless of theory. In this sense the law of Bravais is truly Friedel's: given any crystalline species with sufficient morphological development, it is always possible (and herein lies the law) to find a lattice such that the spacing $d(hkl)$ of a family of parallel nets is, to a first approximation, a measure of the frequency of occurrence of the corresponding form $\{hkl\}$. Friedel determined this unique morphological lattice for hundreds of substances, thereby removing the arbitrariness of the unit lengths chosen in accordance with the law of simple indices.

Another empirical law enunciated by Friedel, and shown by Alfred Liénard to be a consequence of the law of Bravais, is the law of mean indices (1908): the cell edges, a, b, c, are roughly proportional to the sums, $\Sigma h, \Sigma k, \Sigma l$, of the absolute values of the indices of the observed forms. After 1912, when the structural lattice, which expresses the periodicity of the crystal structure, could be determined by X-ray diffraction, Friedel noted that in many instances it did not coincide with his morphological lattice. This discrepancy, he pointed out, does not diminish but, rather, enhances the value of the morphological lattice, which remains the expression of duly observed facts. Final confirmation of the law of Bravais came with its generalization by J. D. H. Donnay and D. Harker in 1937, after Friedel's death, when it was found that the effective interplanar spacings depend not only on the lattice mode but also on the glide planes and screw axes in the space group.

In 1905 the Bravais lattice was, structurally speaking, only a hypothesis. Friedel proved its physical reality by noting that irrational threefold axes (compatible with the law of rationality but impossible in a lattice) had never been found in crystals. This fundamental observation is known as Friedel's law of rational symmetric intercepts.

Bravais and Mallard had begun the theory of twinning. In 1904 Friedel completed it and stated the general law that governs all twins: a lattice, the "twin lattice," extends through the whole crystalline edifice; it is the crystal lattice itself or one of its superlattices; its prolongation from one of the twinned crystals to another can be exact or approximate. Hence the four possibilities and the classification of twins into four classes. The theory accounted for all known twins but one: the Zinnwald twin in quartz. Friedel's very last paper (1933), correlating observations of J. Drugman and results of M. Schaskolsky and A. Schubnikow on alum, explains the exception: two pre-existing crystals unite during growth, one face of the smaller adhering to another face of the larger, in such an orientation as to have a lattice row in common. The theory was thus generalized: in addition to the four classes of triperiodic twins, monoperiodic twins, such as the Zinnwald twin, must be recognized; diperiodic ones, in which the twinned crystals would have a net in common (as in epitaxy), should also be possible.

Lehmann's so-called liquid crystals were thoroughly investigated by Friedel and his co-workers François Grandjean, Louis Royer, and his son Edmond from 1907 to 1931. Two new stases (structural types of matter), the nematic and the smectic, were found to exist between the amorphous and the crystalline. (Cholesteric substances belong to the nematic stasis.) The four stases are separated by discontinuous transformations, which justify the classification. A treatment in English is available in J. Alexander's *Colloid Chemistry* (1926); it summarizes the detailed review paper of 1922, which to this day remains the indispensable introduction to the field. Friedel's work on the mesomorphous stases is perhaps the most important of all his contributions: its many new observations and interpretations opened up most of the lines of research now pursued in this field, where it remains the basic reference.

Friedel took an immediate, although theoretical, interest in Laue's discovery of X-ray diffraction by crystals (1912). As early as 1913 he enumerated the eleven centrosymmetries that can be determined by X rays (Friedel's law, to X-ray diffractionists). Other papers deal with the role of the length of the X-ray wave train (1913), the calculation of intensities (1919), and diffraction by solid solutions (1926).

Friedel was responsible for a theory of crystal growth (1924–1927) that brings out the similarity of crystal corrosion by a slightly undersaturated solution and crystal growth in a slightly supersaturated one. The two phenomena are symmetrical with respect to the saturation point. The theory thus explains negative crystals. Curved faces are accounted for by convergent and divergent diffusion (angle effect, edge effect).

Friedel pointed out that a holoaxial hemihedry may be simulated by a holohedral crystal grown in an optically active medium. He also studied diamond, clarified its holohedry, discussed its inclusions, ascribed its birefringence to strain, and (with Ribaud, 1924) on rapid heating to 1885°C., found a new allotropic form, still unconfirmed but possibly the "white carbon" described by A. El Goresy and G. Donnay ("A New Allotropic Form of Carbon From the Ries Crater," in *Science,* **161** [1968], 363–364).

Friedel's chief geological contribution was the recognition of the first mylonite in France (with Pierre Termier, 1906). In 1907 he was awarded the Prix Joseph Labbé of the French Academy in recognition of the part he had played in the discovery of a new coalfield.

As a school administrator at Saint-Étienne, Friedel stressed laboratory work and introduced new courses in statistics, foreign languages, economics, and industrial hygiene; at Strasbourg, he planned the scientific training of geological engineers and was one of the founders of the Petroleum Institute. As a teacher he exerted an enormous influence, which is still felt: his admirable textbook *Leçons de cristallographie* (1926) was reprinted in 1964. Its often quoted preface, entitled "A Warning," is a sort of scientific testament, stressing the importance of meticulous observation and scrupulous acceptance of well-established facts.

BIBLIOGRAPHY

I. Original Works. For a complete list of Friedel's works, see Grandjean's article. Among his writings note especially "Les états mésomorphes de la matière," in *Annales de physique,* **18** (1922), 273–474; "The Mesomorphic States of Matter," in J. Alexander, *Colloid Chemistry,* I (New York, 1926), 102–136; and *Leçons de cristallographie professées à la Faculté des sciences de Strasbourg* (Paris, 1926; repr. 1964).

II. Secondary Literature. On Friedel or his work, see J. D. H. Donnay, "Memorial of Georges Friedel," in *American Mineralogist,* **19** (1934), 329–335, with condensed bibliography; F. Grandjean, "Georges Friedel 1865–1933," in *Bulletin de la Société française de minéralogie,* **57** (1934), 144–171, with bibliography, 172–183; *Memorial Volume. Georges Friedel 1865–1933* (Strasbourg–Paris, 1939), a limited ed. of 1,000 numbered copies; and A. F. Rogers, "Friedel's Law of Rational Symmetric Intercepts, With Bibliography of Irrational Three-Fold Axes of Symmetry," in *American Mineralogist,* **10** (1925), 181–187.

J. D. H. Donnay

FRIEDMANN, ALEKSANDR ALEKSANDROVICH (*b.* St. Petersburg, Russia, 29 June 1888; *d.* Leningrad, U. S. S. R., 16 September 1925), *mathematics, physics, mechanics.*

Friedmann was born into a musical family—his father, Aleksandr Friedmann, being a composer and his mother, Ludmila Vojáčka, the daughter of the Czech composer Hynek Vojáček.

In 1906 Friedmann graduated from the Gymnasium with the gold medal and immediately enrolled in the mathematics section of the department of physics and mathematics of St. Petersburg University. While still a student, he wrote a number of unpublished scientific papers, one of which, "Issledovanie neopredelennykh uravneny vtoroy stepeni" ("An Investigation of Second-degree Indeterminate Equations," 1909), was awarded a gold medal by the department. After graduation from the university in 1910, Friedmann was retained in the department to prepare for the teaching profession.

At the beginning of 1913, Friedmann began work at the aerological observatory located in Pavlovsk, near St. Petersburg. There he immersed himself in a study of the means of observing the atmosphere. In addition to synoptic and dynamic meteorology, he familiarized himself with the theory of the earth's magnetism and quickly became a prominent specialist in meteorology and related fields.

The year 1914 was marked for Friedmann by two important events: he passed the examinations for the degree of master of pure and applied mathematics at St. Petersburg University and he published in the *Geofizichesky sbornik* an important paper, "O raspredelenii temperatury vozdukha s vysotoyu" ("On the Relationship of Air Temperature to Altitude"). In this paper he examined theoretically the question of the existence of an upper temperature inversion point in the stratosphere.

In the fall of 1914, Friedmann volunteered for service in an aviation detachment, in which he worked, first on the northern front and later on other fronts, to organize aerologic and aeronavigational services. While at the front, Friedmann often participated in military flights as an aircraft observer. In the summer of 1917 he was appointed a section chief in Russia's first factory for the manufacture of measuring instruments used in aviation; he later became director of the factory. Friedmann had to relinquish this post because of the onset of heart disease. From 1918 until 1920, he was professor in the department of theoretical mechanics of Perm University.

In 1920 he returned to Petrograd and worked at the main physics observatory of the Academy of Sciences, first as head of the mathematical department and later, shortly before his death, as director of the observatory.

Friedmann's creative thought penetrated into every area of his knowledge and illuminated it with the brilliance of his disciplined mind and creative imagination. His scientific activity was concentrated in the areas of theoretical meteorology and hydromechanics. Here were manifested his mathematical talent and his unwavering striving for, and ability to attain, the concrete, practical application of solutions to theoretical problems.

Friedmann was one of the founders of dynamic meteorology. To him belong fundamental works in such areas as the theory of atmospheric vortices and vertical air fluxes. He also studied the problems of applying to aeronautics the theory of physical processes that occur in the atmosphere.

Friedmann's most important work in hydromechanics is *Opyt gidromekhaniki szhimaemoy zhidkosti* (1922). In this work he gave the fullest theory of vortical motion in a fluid and examined—and in a number of cases solved—the important problem of the possible motions of a compressible fluid under the influence of given forces.

Friedmann made a valuable contribution to Einstein's general theory of relativity. As always, his interest was not limited simply to familiarizing himself with this new field of science but led to his own remarkable investigations. Friedmann's work on the theory of relativity dealt with one of its most difficult questions, the cosmological problem. In his paper "Über die Krümmung des Raumes" (1922), he outlined the fundamental ideas of his cosmology: the supposition concerning the homogeneity of the distribution of matter in space and the consequent homogeneity and isotropy of space-time; that is, the existence of "world" time, for which, at any moment in time, the metrics of space will be identical at all points and in all directions. This theory is especially important because it leads to a sufficiently correct explanation of the fundamental phenomenon known as the "red shift." This solution of the Einstein field equations, obtained from the above propositions, is the model for any homogeneous and isotropic cosmological theory. It is interesting to note that Einstein thought that the cosmological solution to the equations of a field had to be static and had to lead to a closed model of the universe. Friedmann discarded both conditions and arrived at an independent solution. Einstein welcomed Friedmann's results because they showed the dispensability of the ad hoc cosmological term Einstein had been forced to introduce into the basic field equation of general relativity. Friedmann's interest in the theory of relativity was by no means a passing fancy. In the last years of his life, together with V. K. Frederiks, he began work on a multivolume text on modern physics. The first book, *The World as Space and Time,* is devoted to the theory of relativity, knowledge of which Friedmann considered one of the cornerstones of an education in physics.

In addition to his scientific work, Friedmann for several years taught courses in higher mathematics and theoretical mechanics at various colleges in Petrograd (the Polytechnical Institute, the Institute of Ways and Means of Communication, and the Military Naval Academy). He found time to create new and original courses, brilliant in their form and exceedingly varied in their content, which covered approximation and solution of numerical equations, differential geometry and tensor analysis, hydromechanics, applied aerodynamics, and theoretical mechanics. Friedmann's unique course in theoretical mechanics combined mathematical precision and logical continuity with original procedural and physical trends. He is rightfully considered a distinguished representative of a renowned pleiad of Russian students of mechanics to which also belonged such leading figures as Zhukovsky, Chaplygin, Krylov, and Kochin.

Friedmann died of typhoid fever at the age of thirty-seven. In 1931, he was posthumously awarded the Lenin Prize for his outstanding scientific work.

BIBLIOGRAPHY

I. Original Works. Friedmann's most important works include "Zur Theorie der Vertikaltemperaturverteilung," in *Meteorologische Zeitschrift,* **31,** no. 3 (1914), 154–156; "O raspredelenii temperatury vozdukha s vysotoyu" ("On the Relationship of Air Temperature to Altitude"), in *Geofizichesky sbornik,* **1,** pt. 1 (1914), 35–55; "Sur les tourbillons dans un liquide à température variable," in *Comptes* *rendus hebdomadaires des séances de l'Académie des sciences,* **163,** no. 9 (1916), 219–222; "O vikhryakh v zhidkosti s menyayushcheysya temperaturoy" ("On Whirlpools in a Liquid With Changing Temperature"), in *Soobshcheniya i protokoly Kharkovskago matematicheskago obshchestva,* 2nd ser., **15,** no. 4 (1917), 173–176; "Die Grössenordnung der meteorologischen Elemente und ihrer räumlichen und zeitlichen Abteilungen," in *Veröffentlichungen des Geophysikalischen Instituts der Universität Leipzig,* 2nd ser., no. 10 (1914), written with K. G. Hesselberg; "O vertikalnykh techeniakh v atmosfere" ("On Vertical Fluxes in the Atmosphere"), in *Zhurnal fiziko-matematicheskogo obshchestva pri Permskom gosudarstvennom universitete,* pt. 2 (1919), pp. 67–104; "O raspredelenii temperatury s vysotoy pri nalichnosti luchistogo teploobmena Zemli i Solntsa" ("On the Relation of Temperature to Altitude in the Presence of Radiation Heat Exchange Between the Earth and Sun"), in *Izvestiya Glavnoi geofizicheskoi observatorii,* no. 2 (1920), pp. 42–44; "O vertikalnykh i gorizontalnykh atmosfernykh vikhrvakh" ("On Vertical and Horizontal Atmospheric Whirlwinds"), *ibid.,* no. 3 (1921), pp. 3–4; *Opyt gidromekhaniki szhimaemoy zhidkosti* (Petrograd, 1922); "Über die Krümmung des Raumes," in *Zeitschrift für Physik,* **10,** no. 6 (1922), 377–386; *Mir kak prostranstvo i vremya* ("The World as Space and Time," Petrograd, 1923; 2nd ed., Moscow, 1965), written with V. K. Frederiks; "O dvizhenii szhimaemoy zhidkosti" ("On the Motion of a Compressible Liquid"), in *Izvestiya gidrologicheskogo instituta,* no. 7 (1923), pp. 21–28; "Über die Möglichkeit einer Welt mit konstanter negativer Krümmung des Raumes," in *Zeitschrift für Physik,* **21** (1924), 326–332; "O rasprostranenii preryvistosti v szhimaemoy zhidkosti" ("On the Extent of Discontinuity in a Compressible Liquid"), in *Zhurnal Russkago fiziko-khimicheskago obshchestva,* **56,** no. 1 (1924), 40–58; "O krivizne prostranstva" ("On the Curvature of Space"), *ibid.* (1924), 59–68; and "Théorie du mouvement d'un fluide compressible," in *Geographische Zeitschrift* (1924).

II. SECONDARY LITERATURE. On Friedmann and his work, see (in chronological order), A. F. Vangengeim, "A. A. Friedman," in *Klimat i pogoda,* nos. 2–3 (1925), pp. 5–7, an obituary; N. M. Gyunter, "Nauchnie trudy A. A. Freidmana" ("The Scientific Works of A. A. Friedmann"), in *Zhurnal Leningradskogo fiziko-matematicheskogo obshchestva,* **1,** no. 1 (1926), 5–11; and A. F. Gavrilov, "Pamyati A. A. Friedmanna" ("In Memory of A. S. Friedmann"), in *Uspekhi fizicheskikh nauk,* **6,** no. 1 (1926), 73–75.

See also *Geofizichesky sbornik* (Leningrad, 1927), V, pt. 1, which contains, in addition to an obituary by V. A. Steklov, E. P. Friedmann, "Pamyati A. A. Friedmanna" ("In Memory of A. A. Friedmann"), pp. 9–10; I. V. Meshchersky, "Trudy A. A. Friedmanna po gidromekhanike" ("A. A. Friedmann's Works on Hydromechanics"), pp. 57–60; and M. A. Loris-Melikov, "Raboty Friedmanna po teorii otnositelnosti" ("Friedmann's Works on the Theory of Relativity"), pp. 61–63.

Also useful is L. G. Loytsyansky and A. I. Lurie, "A. A. Friedmann," in *Trudy Leningradskogo politekhnicheskogo instituta imeni M. I. Kalinina,* no. 1 (1949), pp. 83–86.

A. T. GRIGORIAN

FRIEND, JOHN ALBERT NEWTON (*b.* Newton Abbot, Devonshire, England, 20 July 1881; *d.* Birmingham, England, 15 April 1966), *chemistry.*

The son of the Reverend Hilderic Friend, a Wesleyan missionary minister, Friend moved with his family every two or three years into a fresh circuit and consequently attended a number of schools. In 1890 the family moved to Idle-Eye Dale, near Bradford, Yorkshire. In 1896 the family moved to Ocker Hill, Tipton, Staffordshire, where he passed the examination admitting him to King Edward's High School in Birmingham. There he won a Foundation scholarship and passed the London matriculation examination in 1899 and, the following year, the London intermediate examination which qualified him to study for the B.Sc. The school awarded Friend a leaving exhibition which enabled him to proceed to Masons Science College, which had only recently become Birmingham University. There he studied chemistry under Percy Frankland and physics under J. H. Poynting. In 1902 he graduated with a B.Sc. degree with honors.

After receiving the Priestley scholarship, Friend spent a year carrying out research on Caro's permonosulfuric acid and was awarded the M.Sc. in 1903; he then became assistant science master at the Dual Watford Grammar Schools, where he taught chemistry and physics. Soon thereafter he began his researches on the corrosion of metals for which he was awarded the Carnegie Gold Medal by the Iron and Steel Institute in 1913. He demonstrated that the resistance of steel to neutral corroding media rises with increasing chromium content.

By living frugally and earning extra money from private pupils and evening classes, Friend was able to save enough to enroll in 1906 at the University of Würzburg. There under Wilhelm Manchot he carried out research on carbonyl derivatives of cuprous salts, for which he was awarded the Ph.D. in 1908.

In September 1908 Friend was appointed lecturer in chemistry at the Darlington Technical College. In October 1909 he began classes in the chemistry of paints, the first on the subject to be held outside London, and in 1910 he was awarded the D.Sc. by Birmingham University. Friend left Darlington for Worcester in 1912, to become head of the Victoria Institute Science and Technical Schools. In 1915 he became a fellow of the Institute of Chemistry, now the Royal Institute of Chemistry. The following year, in London, he was appointed lieutenant in the Anti-Gas Section of the Royal Army Medical Corps (later taken over by the Royal Engineers). In 1917 Friend was invited to act as scientific adviser to the Sea Action Committee of the Institution of Civil Engineers and carried out valuable researches on corrosion by seawater. He also conducted experiments at Southampton and Weston-super-Mare to compare the protective efficiencies of various paint coatings on steel plates exposed to similar corrosive influences.

At the end of World War I, Friend returned to his post at Worcester. In 1919 he was elected an honorary member of the Oil and Colour Chemists' Association and from 1922 to 1924 served as its honorary president. The following year (1920) he became head of the chemistry department at the Birmingham College of Advanced Technology, which in 1965 became the University of Aston in Birmingham. He was elected an honorary member of the British Association of Chemists in 1924 and, in 1958, an honorary life member of the Birmingham University Chemical Society.

After the outbreak of World War II in 1939, Friend joined the Local Defense Volunteers (later the Home Guard) and founded a chemical defense school, which was responsible for training some 50,000 Home Guard personnel in antigas measures. In 1946, having reached the age limit of sixty-five, he retired from Birmingham and was appointed lecturer in science in the service of the Central Advisory Council for Adult Education in His Majesty's Forces, in which capacity he lectured at camps in Europe, Africa, and the Near East. He returned to England in 1950 and began to lecture at various military camps in the Midlands for the Birmingham University Extra Mural Department, an activity which advancing age forced him to relinquish in 1958.

Friend's research activities constitute an unusual blend of pure and applied topics. Best known as a distinguished teacher and author, he vigorously pursued research on valence theory, persulfates, metallic corrosion, paints, linseed oil, rare earths, solubilities of salts, viscosities of organic liquids, analysis of ancient artifacts, and the history of science. Foremost among his twenty-four books is the *Textbook of Inorganic Chemistry,* of which he was editor and part author. The work, which began in 1914 and ultimately comprised twenty-two volumes, was not completed until 1930. Although most of his books deal with chemistry and related subjects, his volumes on iron in antiquity, mathematical puzzles, demonology, sympathetic magic, and witchcraft testify to the breadth of his interests.

BIBLIOGRAPHY

I. Original Works. Friend's books include *The Theory of Valency* (1909); *An Introduction to the Chemistry of Paints* (London, 1910); *Elementary Domestic Chemistry* (London, 1911); *The Corrosion of Iron and Steel* (London, 1911); *Textbook of Inorganic Chemistry,* 22 vols. (London, 1914–1930); *The Chemistry of Linseed Oil* (London, 1917); *The Chemistry of Combustion* (London, 1922); *Iron in Antiquity* (London, 1926); *A Textbook of Physical Chemistry,* 2 vols. (London, 1932–1935; rev. in 1 vol., 1948); *Man and the Chemical Elements* (London, 1951); *Numbers: Fun and Facts* (New York, 1954; 2nd ed., 1957); *Words: Tricks and Traditions* (New York, 1957); *Science Data* (London, 1960); *More Numbers: Fun and Facts* (New York, 1961); *Demonology, Sympathetic Magic and Witchcraft* (London, 1961); and *Still More Numbers: Fun and Facts* (New York, 1964).

II. Secondary Literature. G. B. Kauffman, ed., *Classics in Coordination Chemistry,* II: *Selected Papers (1798–1935)* (New York, in press), contains a short biography as well as annotated versions of two of Friend's papers on valency: "A Criticism of Werner's Theory and the Constitution of Complex Salts," in *Journal of the Chemical Society,* **93** (1908), 1006–1010, and "A Cyclic Theory of the Constitution of Metalammines and of Ferro- and Ferricyanides," *ibid.,* **109** (1916), 715–722.

George B. Kauffman

FRIES, ELIAS MAGNUS (*b.* Femsjö, Sweden 15 August 1794; *d.* Uppsala, Sweden, 8 February 1878), *botany.*

Fries was born in a parsonage in the southwestern part of the province of Småland, in southern Sweden. His father was interested in natural history and inspired the same interest in his son. The area is poor in phanerogams but extremely rich in fungi, which may have contributed to Fries's becoming, even in his teens, an advanced mycologist. He graduated from the Gymnasium in Växjö and in 1811 he enrolled at the University of Lund. There his botanical interest took on a more serious character and was guided principally by the professor of botany, C. A. Agardh. Under Agardh, Fries took his degree in philosophy (*filosofie magister*) in 1814, having defended the first part of his dissertation "Novitiae florae Suecicae." From then until his death Fries was totally devoted to botany. He became a docent at the university, which gave him the right to teach there without salary. In 1819 he advanced to adjunct, which brought him a small income, and in 1828 he became *botanices demonstrator,* a slightly more lucrative post. All this time he had to depend upon his father for his living.

During his first ten years at the university Fries devoted most of his studies to mycology, soon acquiring an international reputation. In 1815 he published the first volume of his first important work in mycological systematics, *Observationes mycologicae* (2 vols., Copenhagen, 1815–1818); and in 1821, when he was only twenty-seven he began publishing his *Systema mycologicum,* the great work which, more than anything else he wrote, brought him his fame (3 vols.: I, Lund, 1821; II, in 2 pts., Lund, 1822–1823; III, 2 pts., Greifswald, 1829–1832).

For the lichens, then regarded as a natural class or order, Fries worked out a general systematics in his *Lichenum dianome nova* (Lund, 1817), partly inspired by the great Swedish lichenologist Acharius. In the second half of his twenties he was intensely absorbed by the lichens, and with his friend Christian Stenhammar, a vicar and naturalist, he published *Lichenes Sueciae exsiccati* (Lund, 1824–1827), which consisted of dried specimens of lichens and an accompanying text. His lichenological studies were crowned by his authoritative *Lichenographia Europaea reformata* (Lund, 1831), containing a survey of all known lichens of Europe and information on their distribution.

From 1835 Fries was professor of botany at Uppsala. He soon rose to a central position in Swedish botany and was regarded as its unrivaled leader. Throughout his life he was an extremely prolific writer. He strengthened his international position in mycology through a number of works, such as *Epicrisis systematis mycologici,* 2 vols. (Uppsala, 1836–1838; new ed. with the title *Hymenomycetes Europaei,* Uppsala, 1874); *Monographia hymenomycetum Sueciae,* 2 vols. (Uppsala 1857–1863), which appeared both as a series of academic dissertations and as a separate work, *Sveriges ätliga och giftiga svampar,* 10 pts. (Stockholm, 1860–1866); and *Icones selectae hymenomycetum nondum delineatorum* (I, Stockholm, 1867–1875; II, Stockholm–Uppsala, 1877–1884), the second volume of which was published, following his wishes, by his sons Thore and Robert. Among the phanerogams such critical genera as *Hieracium* (*Epicrisis generis Hieraciorum,* Uppsala, 1862), *Salix,* and *Carex* were his favorites. His long floristic experience in Skåne was summarized in his *Flora Scanica* (Uppsala, 1835).

Fries's foremost accomplishment in botany was in systematics. Almost from the beginning he showed a passion for understanding the underlying principles of this discipline and for tackling the questions concerning the real and natural relationship between plants and plant groups—all problems of great difficulty in the age before Darwin. A factor that proved critical for all his work was his encounter with the romantic German *Naturphilosophie* in the years after 1810. During his first decade as a systematist he admired Lorenz Oken and his extremely speculative system of nature, as presented in Oken's *Lehrbuch der Naturphilosophie* (1809–1811). Oken conceived of the universe as built of spiritual principles, the four "elements," which in diverse combinations and refinements constitute animals and plants. At the same time the different classes of plants reflect in different degrees the main organs of an individual plant: the root, the stem, the flower, and the fruit. In *Systema mycologicum* Fries sought to apply Oken's principles to fungus systematics. His four main groups of fungi—Coniomycetes, Hyphomycetes, Gasteromycetes, and Hymenomycetes—were thus considered as expressing diverse "cosmic moments," very reminiscent of Oken's "elements." In this work Fries also took up ideas from Oken's friend C. G. Nees von Esenbeck, whose highly romantic *System der Pilze und Schwämme* had been published in 1817. Nees argued that the fungi represent a special aspect of the vegetable world, its negative pole and autumn side, manifested in their conspicuous fruit bodies (spore organs), which seemed to dominate their vegetable parts. In the same vein Fries spoke about a *nisus reproductivus* among the fungi, which separated this plant group from all other vegetables.

Already in *Systema orbis vegetabilis* (1825) Fries had to a great extent freed himself from the influence of Oken and Nees. Yet his views still reflected the same visionary and speculative romanticism, his conviction of nature's inner spirituality and unity. He expressed his belief in a close relationship between human logic and nature's way of separating the organisms into classes, orders, genera, and species. In every taxon he found four subdivisions, which in their turn could be divided in four units of lower degree. He called this consequent quartering a double dichotomy, reflecting the dichotomous method of ordinary logical division. Thus he divided the whole plant kingdom into the four groups Dicotyledoneae, Monocotyledoneae, Heteronemeae (ferns and mosses), and Homonemeae (algae and fungi).

Later, Fries gradually turned away from the extravagant speculation of his youth. More and more he became convinced that human reason could not grasp the great scheme according to which the Creator had distributed his creatures in systematic groups. Instead, he stated that one can trace the true relationships only through painstaking observation of the different species in their natural surroundings. In this context his vitalism became more and more apparent; life represents a secret and divine force, which determines the essence of each species, and the true characteristics of species can be determined only through empirical observation of the living specimens.

Yet Fries never completely abandoned his vision of a great natural system that would cover all of the vegetable kingdom in some detail. This was to a certain degree realized in 1835 in his *Flora scanica,* and even there one could retrace fundamental points of view which had their origin in German romantic philosophy.

Three characteristics of Fries's systematics are of special importance:

1. His idealistic conception of natural relationships. He thought that all organisms were related to certain types or ideas, which they expressed and resembled in a higher or lower degree, belonging to such a type being the basis for their true affinity. In his vision the taxa constituted "spheres"; the type was in the "center" and the other members of the taxon on the "radii" at different distances from the center and facing in different directions toward other groups.

2. The distinction he drew between affinity and analogy. By "analogy" he meant a kind of similarity in the outer appearance of two or more organisms which have no inner relationship or affinity. In his terminology this meant that the analogous forms are related to different types, belong to different spheres, but are situated at the same distance from the center of their respective spheres. These concepts found their way to British biologists through William Sharp MacLeay's article "Remarks on the Identity of Certain General Laws Which Have Been Lately Observed to Regulate the Natural Distribution of Insects and Fungi" (*Transactions of the Linnean Society,* **14** [1823]).

3. His ideas on evolution. Since the 1820's Fries had been convinced that evolutionary processes had taken place within the organic world and that through the ages the organisms had passed from more primitive to more perfect stages. But for a long while he could not accept any theory of descent. Thus, according to him, all species had existed from the beginning, rude and primitive but definitely different from each other. Through the ages they had separately and gradually reached their present forms. Later, Fries had to concede that not all species had an evolutionary history separate from all other species. He began to believe that all forms within a genus had only one common ancestor and that the different species now existing within it were *temporis filiae,* daughters of time. He considered that the driving force behind this evolution was mainly a tendency within the organisms to strive toward the perfect state of the respective types or ideas, a reflection of his basically romantic vision. When, in old age, Fries had read Darwin's *Origin of Species,* he could agree with the general theory of evolution, but he hesitated before the idea of the descent of nearly all organisms from one or a few original forms. And he absolutely could not accept the mechanism of "struggle for existence" and natural selection as the main force acting in evolution.

Today Fries's renown is based primarily upon his mycological work. His views on the large taxa, e.g., the "classes" within which he ordered the fungi, certainly are obsolete. He could never admit the importance of those microscopic characters which were detected with the new and better instruments available from the 1830's. Therefore he hesitated too long when confronted with the discovery of basidia and asci (reported to him in the 1830's by the British mycologist Miles Berkeley), fundamental for the modern grouping of the main categories of fungi. Yet his distinction of different spore colors still has important taxonomic value, and he is especially remembered for his ability to describe species. In fact, according to a decree of the Seventh International Botanical Congress (Stockholm, 1950), *Systema mycologicum* forms the basis for the nomenclature within all groups of fungi except Uredinales, Ustilaginales, and Gasteromycetes.

BIBLIOGRAPHY

I. Original Works. Fries's most important works are mentioned in the text. For a detailed bibliography, see T. O. B. N. Krok, *Bibliotheca botanica Suecana* (Uppsala–Stockholm, 1925), pp. 199–215. There are facsimiles of *Systema mycologicum* and *Elenchus fugorum* (New York, 1952; Weinheim–Bergstrasse, 1960); *Monographia hymenomycetum Sueciae* (Amsterdam, 1963); and *Hymenomycetes Europaei* (Leipzig, 1937; Amsterdam, 1963).

Unpublished correspondence is in the Uppsala University library and a number of other Swedish libraries (see *Svenskt biografiskt lexikon,* XVI [Stockholm, 1965], 526); the Botanisk Centralbibliotek, Copenhagen; the Muséum National d'Histoire Naturelle, Paris; the British Museum (Natural History), London; and the Arnold Arboretum and Gray Herbarium of Harvard University.

II. Secondary Literature. On his life and work, see J. Arrhenius, "Elias Magnus Fries," in *Levnadsteckningar över K. Svenska Vetenskapsakademiens ledamöter,* II (Stockholm, 1878–1885), 195–226; Gunnar Eriksson, *Elias Fries och den romantiska biologin* (Uppsala, 1952), with a summary in English, pp. 457–462; and *Svenski biografiskt lexikon,* XVI, 522–526; and R. Fries, "Elias Fries," in *Swedish Men of Science* (Stockholm, 1952), 178–185.

Gunnar Eriksson

FRIES, JAKOB FRIEDRICH (*b.* Barby, Germany, 23 August 1773; *d.* Jena, Germany, 10 August 1843), *philosophy, physics, mathematics.*

Fries was educated in Niesky at the Moravian

Academy of the United Brethren of Herrenhut. This upbringing had a lasting influence upon him, even though he early freed himself from the religious dogmas learned in his theological studies there. In 1795 Fries went to Leipzig to study philosophy and in 1797 he transferred to Jena to study with Fichte. By 1798 he had published five essays on the relation of metaphysics to psychology. Having completed his course at Jena, he spent a year as a private tutor in Switzerland. At the same time he worked to finish his thesis, "De intuitu intellectuali," with which he qualified as docent at Jena in 1801.

In 1805 Fries was called to become an assistant professor of philosophy and elementary mathematics at Heidelberg. In 1816 he returned to Jena as full professor of theoretical philosophy. He was suspended because of political pressures following his participation in the Wartburgfest, a demonstration by liberal students in October 1817. Fries did not obtain permission to teach again at Jena until 1824, when he received a professorship of physics and mathematics. In 1825 he became professor of philosophy, a post he held for the rest of his life.

Fries considered himself Kant's most loyal disciple. He believed that Kant had finished the philosopher's task for all time and that only individual elements of his doctrine were susceptible to correction. Despite this belief Fries himself decisively altered the Kantian formulation by psychologizing Kant's transcendental idealism in his major book, *Neue oder anthropologische Kritik der Vernunft* (Heidelberg, 1807). He was opposed to all contemporary speculative systems and considered the Romantic interpretation of nature good only for aesthetics. In opposition to Schelling and his school, he went back to the attitude of critical Kantianism that held *Naturphilosophie* to be the philosophy of exact sciences (see, for example, his *Die mathematische Naturphilosophie* [Heidelberg, 1822]).

Fries also spoke for contemporary positive scientific research, for which he gained authority from his works on physiological optics and the theory of probabilities, as well as from his textbook on experimental physics. With Kant he asserted the possibility of an a priori natural science, designed to show how the categories are applicable to experience as determinations of time through universal mathematical schema and how a system of universal laws of nature arises from this union. This system encompasses the theory of pure motion, which unites Newton's mathematical approach with Kantian philosophy. (Fries's interpretation of Newton was strongly influenced by the commentary of Le Seur and Jacquier to their four-volume 1739 edition of the *Principia,* which contains the development of those mathematical propositions that Fries presented in detail in his own *Die mathematische Naturphilosophie.*)

Fries's system almost yields the philosophy of applied mathematics, although there remains the field of derivable statements a priori. Among them Fries considers the Newtonian axioms to be foremost. On the other hand, attraction and its lawfulness cannot be deduced a priori. While Fries took over Kant's classification of natural knowledge into the four disciplines of phoronomy, dynamics, mechanics, and phenomenology, he added two new ones that he designated stochiology and morphology. Stochiology treats dynamics from a comprehensive viewpoint and attempts to establish a dynamic explanation of heat and light, as well as electrical and chemical phenomena.

Whereas Fries, like Kant, explained matter dynamically—and indeed he was much more sharply opposed to the atomists than was Kant—he deviated from him considerably with regard to dynamics itself. This is illustrated particularly by his rejection of apriorism in the concrete determination of the degree of force with which masses act upon each other. In further opposition to Kant, Fries posited four types of fundamental force in the constitution of matter: attractive and repulsive forces acting at a distance as well as attractive and repulsive contact forces. For the first two types of force the simplest mode of action is given by their proportionality to $1/v^2$, for the latter two, by the proportionality

$$k = \frac{\text{density}}{mv}.$$

Fries took a final important step beyond Kant in the introduction of the organic into his system of nature through the supposition of natural instincts. This doctrine of the instincts stated that along with the determination of the forms of interaction, the law of the counteraction of the fundamental forces (that is, the behavior of the moved mass in space) must also be considered. According to Fries, the succession of appearances is caused by such instincts, which display complete forms of reciprocal action. They alone, rather than particular materials and forces, provide the explanation of physical phenomena and processes.

Physical processes may be divided into four types: gravitational (heavy masses acting at a distance); chemical (heavy masses in admixtures and dissociations with contact action); phlogistic (in which the counteraction is determined by caloric); and morphotic (in which the counteraction is determined primarily by the rigidity of the moving forces). The instincts on which these processes are based may be further divided into two classes—instincts of mechanisms and instincts of organisms. In Fries's view, all processes that are freely determined by uniformly accelerating attractive forces must be attributed to organic instincts; and all processes that are determined by attraction of bodies in contact without acceleration, and are thus reaching a state of rest in reestablished equilibrium, must be attributed to mechanical instincts.

Planetary orbits and pendulum motion may, if air resistance and friction are ignored, serve as examples of the action of organic instincts. Organizing instincts here predominate over mechanical ones, since otherwise the world, beginning in chaos, would attain motionless equilibrium in finite time. The prevalence of organizing instincts is necessary to the closed system of periodic recurrence of events required for a world without beginning and without end. Lesser cycles of events are the only ones that can be altered from the outside alone. If one supposes these comprised in greater cycles of events in which, again, the organic force predominates, and then supposes these greater cycles included in still greater ones, and so on, a lack of exact periodicity in the whole is understandable.

The distinction between organic and mechanical instincts makes it possible to distinguish between organic and inorganic nature. An inorganic (dead) body, or mechanical formation, is any that is formed according to the law of equilibrium; a living body obeys the law of self-preservation of its motion—this principle of self-preservation is the corporeal soul. The corporeal soul itself exhibits only the form of the interaction of the corporeal parts. This definition reflects Kant's determination of the organism as the *causalitas mutua* of its parts.

According to Fries's speculative natural history, the cosmos is constructed out of three elements—earth, water, and air—according to weight. In the solid core of the earth the forces have long maintained themselves in equilibrium. Above the earth, in the water and air, the equilibrium is preserved by the action of sunlight, the daily and yearly motion of the earth, and the electrical reaction of the atmosphere to daily and annual heating and cooling cycles. In addition, the vigor of the life-movements and the circulation of life as the basic vital principle also depend upon thermal-electrical relationships. "All morphotic processes are dominated by the formative instincts [*Bildungstrieben*], and specifically the mineral and the two organic instincts," Fries wrote.

The laws of crystallization display the true morphotic principle. Without denying the differences between crystal and organism (including the difference in mode of growth, namely apposition as opposed to intussusception), Fries maintained that these laws are crucial to an understanding of organic form. Indeed, organic formation to the extent that it represents a higher, correspondingly altered process, is to be understood only on the basis of crystallization. Fries called for a theory of free crystallization to expedite a theory of organic formation; he thereby thought to explain a self-maintaining organic process by means of the circuit law of the voltaic cell. It is necessary to this hypothesis, however, that the electrical currents that arise in the play of chemical combinations and dissociations flow in such a way as to transform the oxides emerging from contact with the conductor of the second class, so that the material of this conductor be replaced. Fries could thus ask if a plant might be a self-maintaining open voltaic chain whose root acts as a negative conductor while the opposite pole puts forth leaves and flowers, or if an animal might be a self-maintaining closed voltaic chain that therefore also possesses its own inherent magnetism.

Fries devised a diagram

nitrogen
water ——|—— light
carbon

to represent the bases of all empirical theories of formation and dissolution in the three natural kingdoms. His representation combined the two polar principles of electrical and material opposition. In this system electrical opposition emerged as the principle of all animation through the activity of sunlight on water, while the differences between chemical materials, represented by the original opposition of carbon and nitrogen, is the principle of all production of form. The effect of these two original polarities on each other allowed the reconciliation of the three natural instincts posited by Fries. Specifically, the water-light polarity reconciled the mechanical-mineral instinct; the open circuit between water, carbon, and nitrogen, the vegetative one; and the closed circuit between nitrogen, light, carbon, and water (which encompasses the whole of nature), the animal instinct.

Fries's doctrine of nature thus can be understood only in connection with romantic *Naturphilosophie,* in which the idea of polarity and the importance of magnetism and galvanism were employed to account for far more than just physical events. Fries's doctrine of nature, however, employed the natural science available at the time in a serious attempt to overcome Kant's vitalism, which even a Newton of the grass-blades would not accept. Indeed, his speculations in natural philosophy were actually fruitful for the development of modern biology through the work of his student Matthias Schleiden, who sought to put into practice the idea of the reduction of the organic to crystallization processes and thereby became the founder of modern cytology.

BIBLIOGRAPHY

I. Original Works. Fries's principal work is *Neue oder anthropologische Kritik der Vernunft,* 3 vols. (Heidelberg, 1807; 2nd ed. 1828–1831). His other scientific works include *Entwurf eines Systems der theoretischen Physik* (1813); *Die mathematische Naturphilosophie nach philosophischer Methode bearbeitet* (Heidelberg, 1822); and *Versuch einer Kritik der Prinzipien der Wahrscheinlichkeitsrechnung* (Brunswick, 1842). A detailed bibliography of his books and essays is contained in the biography written by his son-in-law Henke, pp. 379 ff.

A 26-vol. ed. of Fries's collected works, ed. and with intro. and index by Gert Koenig and Lutz Geldsetzer, is in publication (Aalen, 1969–).

II. Secondary Literature. Works about Fries include T. Elsenhaus, *Fries und Kant. Ein Beitrag zur systematischen Grundlegung der Erkenntnistheorie,* 2 vols. (Giessen, 1906); Kuno Fischer, *Die beiden Kantischen Schulen in Jena,* Akademische Reden no. 2 (Stuttgart, 1862); M. Hasselblatt, *J. Fr. Fries, Seine Philosophie und seine Persoenlichkeit* (Munich, 1922); E. L. T. Henke, *J. Fr. Fries; Aus seinem handschriftlichen Nachlass dargestellt* (Leipzig, 1867); J. G. Meusel, in *Das gelehrte Teuschland oder Lexicon der jetzt lebenden teuschen Schriftsteller,* vol. II, and also in *Neuer Nekrolog der Deutschen Jgg.* (1823–1853), *Conversations-lexicon,* vol. X–A (Leipzig, 1851–1855); and C. Siegel, "Fries, Fortbildung der Kantischen Naturphilosophie," in *Geschichte der deutschen Naturphilosophie* (Leipzig, 1913), pp. 119–130.

H. M. Nobis

FRISI, PAOLO (*b.* Milan, Italy, 13 April 1728; *d.* Milan, 22 November 1784), *mathematics, physics, astronomy.*

Frisi was a member of the Barnabite order. In physics his research must be evaluated in relation to the concepts dominant in his time, which led him to justify and interpret certain phenomena of light and aspects of electricity, referring to the vibratory motion of ether and other properties attributed to it.

As an astronomer he concerned himself with the daily movement of the earth (in *De motu diurno terrae,* awarded a prize by the Berlin Academy), the obliquity of the ecliptic, the movement of the moon, the determination of the meridian circle, and matters concerning gravity in relation to Newton's general theories.

His mathematical activity included studies on kinematics (composition of rotatory movements, etc.) and, notably, on isoperimetry. He also did work in hydraulics and was called upon to plan works for the regulation of rivers and canals in various parts of northern Italy. He was responsible for laying out the canal built in 1819 between Milan and Pavia.

Frisi wrote critical notes in honor of Galileo, Cavalieri, Newton, and d'Alembert, illustrating the contributions each had made to science and the influence each had exerted. In Italy during his lifetime he was considered a scientific authority and was also well known abroad, so much so that his major works (which he wrote in Latin) were translated into French and English. *Algebra e geometria analitica* (1782), *Meccanica* (1783), and *Cosmografia* (1785), in which he brought together the best of his work, were, for that era, very up-to-date. Frisi was an editor of *Il caffè,* a newspaper that was influenced by the thought of the French Illuminati and that exerted a notable influence on the cultural, social, and political life of Milan in the second half of the eighteenth century.

BIBLIOGRAPHY

I. Original Works. Among Frisi's works are *Disquisitio mathematica in causam physicam figurae* (Milan, 1751); *De methodo fluxionum geometricarum* (Milan, 1753); *Nova electricitas theoria* (Milan, 1755); *De motu diurno terrae dissertatio* (Pisa, 1756); *Piano de' lavori da farsi per liberare* (Lucca, 1761); *Del modo di regolare i fiumi, e i torrenti* (Lucca, 1762); *De gravitate universali corporum* (Milan, 1768); *Danielis Melandri et Paulli Frisii alterius ad alterum de theoria lunae commentaria* (Parma, 1769); *Opuscoli filosofici* (Milan, 1781); *Opera,* 3 vols. (Milan, 1782–1785); and *Operette scelte* (Milan, 1825).

His many papers include "Dell'equilibrio delle cupole e delle volte," in *Atti della Società patriotica di Milano,* **1** (1773), 222 ff.; and "Dissertatio de quantitatibus maximis et minimis isoperimetricis," in *Atti dell'Accademia dei fisiocritici di Siena,* **6** (1781), 121 ff.

II. Secondary Literature. On Frisi and his work, see Girolamo Boccardo, in *Nuova enciclopedia italiana* (Turin, 1875–1888), IX, 1005–1006; Francesco Jacquier, *Elogio accademico* (Venice, 1786); Pietro Riccardi, *Biblioteca matematica italiana dall'origine della stampe . . .* (Modena, 1870; 7th ed., 1928); and Pietro Verri, *Memorie appartenenti alla vita ed agli studi del sig. D. Paolo Frisi* (Milan, 1784), which is reprinted in *Operette scelte.*

Luigi Campedelli

FRISIUS, GEMMA. See **Gemma Frisius, Reiner.**

FRITSCH, GUSTAV THEODOR (*b.* Cottbus, Germany, 5 March 1838; *d.* Berlin, Germany, 12 June

1927), *anatomy, physiology, zoology, anthropology, photography.*

Fritsch was the son of the royal Prussian inspector of buildings; his maternal grandfather was a well-known Silesian industrialist named Kramsta. After his Gymnasium education in Breslau he decided to work at Berlin under the physiologist Johannes Müller, whose early death prevented any personal association. After serving in a guards regiment, Fritsch began to study medicine and science at the University of Berlin and later attended the medical schools of Breslau and Heidelberg; at these schools he met Hermann von Helmholtz, Ludwig Traube, Friedrich von Frerichs, and Bernhard von Langenbeck. On 9 August 1862 he received the M.D. with a thesis on spinal cord structure. In the following year he received his medical license and, after a three-year visit to South Africa (1863–1866), during which he indulged in anthropological and geographical investigations, worked as an assistant to the Berlin anatomist Karl B. Reichert. Fritsch passed his habilitation in anatomy in 1869. In the meantime he had accompanied the Prussian solar eclipse expedition to Aden and had made a tour of Egypt. His photographic skill was valuable on these expeditions, and in Egypt he studied electric fishes.

After serving in the army during the Franco-Prussian War and winning the Iron Cross, Fritsch was appointed extraordinary professor of comparative anatomy under Reichert in 1874 but soon left to join the Prussian Venus expedition to Isfahan, Persia; he also visited Smyrna, where he made a comparative study of the fish brain, published in 1878. He now found working space in the Institute of Pathology, and Emil du Bois-Reymond created a position for him as chief of histology and photography in the department of physiology. Du Bois-Reymond revived Fritsch's interest in electric fishes, and these studies took him again to Africa.

After du Bois-Reymond retired, Fritsch worked under his successor, Theodor Wilhelm Engelmann; now, however, his interests changed to physical anthropology, in which field he again exploited his ability as a photographer. When Engelmann retired, the new director, Max Rubner, also had interests which did not coincide with Fritsch's. His friends Wilhelm Waldeyer and Adolf Fick gave him working accommodations in the Anatomy Institute. Fritsch retired in 1921, and before his death his vision became impaired by chemical trauma. In 1893 he had been made a member of the medical privy council and in 1899 an honorary ordinary professor.

Fritsch married the daughter of the University of Breslau's publisher, Ferdinand Hirt, in 1871. His financial independence, partly the result of this union, allowed him to indulge in activities which would otherwise have been impossible for him.

As Haller has pointed out, Fritsch belonged to the Kretschmer cyclothymic personality group. He demonstrated the many-sidedness of his character and his mental agility by working in several areas of science as well as having interests outside science. He had the characteristic tendency to collect data omnivorously and multisensorily but also a strong dislike of systematized and nonempirical methods. Philosophical and metaphysical subjects were avoided, and he preferred to deal with practical rather than theoretical issues. Fritsch's energy was immense, and he was happy only when active. This urge to create and to act, together with his independence of mind, made it difficult for him to accept the fact that his academic advancement had not gone beyond the level of extraordinary professor.

Fritsch's most important contribution to the medical sciences was his study of the electrophysiology of the brain. With Eduard Hitzig he published an epochal paper that established the existence of functional localization in the cerebral cortex of the dog (1870). Although the phrenologists had made similar claims at the beginning of the nineteenth century and such clinicians as Jean-Baptiste Bouillaud, Pierre-Paul Broca, and J. Hughlings Jackson had supported the concept of cerebral localization in the 1860's, Fritsch and Hitzig provided the first incontrovertible experimental evidence for it. They opened up a vast new field of cerebral physiology which is still being studied. Hitzig continued his interest in this subject but Fritsch did not, although his work on the electric fish contributed to the growing knowledge of electrophysiology.

Fritsch was a pioneer in photography, and throughout his life was a keen and able photographer. He applied his skill to photomicrography and stereoscopy in microscopic anatomy and also to other aspects of his medical and anthropological work; he was much interested in the artistic concept of the human figure. Fritsch was also a coeditor of the periodical *Internationale photographische Monatsschrift für Medizin und Naturwissenschaften,* which began publication at Leipzig in 1896.

Much of his work in anthropology and ethnology revolved about the concept of racial dominance. Fritsch believed that racial variations in visual acuity existed, and his 1908 book dealt with the "fovea" or "area centralis" (his terms). He also wrote a book on human anatomy for the anthropologist, as well as one on the physical features of the modern Egyptian. His last major publication, which appeared in 1912, dealt with the anthropological significance of scalp hair.

BIBLIOGRAPHY

I. Original Works. Fritsch's work in physiology includes "Ueber die elektrische Erregbarkeit des Grosshirns," in *Archiv für Anatomie und Physiologie,* **1** (1870), 300–332, written with Eduard Hitzig, English trans. in G. Von Bonin, *The Cerebral Cortex* (Springfield, Ill., 1960), pp. 72–96.

On anatomy and zoology he wrote *De medulla spinalis textura* (Berlin, 1862), his M.D. inaugural dissertation (not seen); "Zur vergleichenden Anatomie der Amphibienherzen," in *Archiv für Anatomie und Physiologie,* **1** (1869), 654–758, his *Habilitationsschrift; Ueber das stereoskopische Sehen im Mikrotypien auf photographischen Wege* (Breslau, 1872); and *Untersuchungen über den feinern Bau des Fischgehirns, mit besonderer Berücksichtigung der Homologien bei anderen Wirbelthierklassen* (Berlin, 1878) were on his work carried out in Smyrna. See Herrick (1892), below, for titles of other publications in several fields.

Electric fish are the subject of "Vorläufiger Bericht über die von Prof. Gustav Fritsch in Aegypten angestellten neuen Untersuchungen an elektrischen Fischen," in *Archiv für Anatomie und Physiologie,* Physiologische Abteilung, **2** (1882), 61–75, 307–413, based on letters from South Africa; *Die elektrische Fische im Lichte der Descendenslehre* (Berlin, 1884); *Sitzungsberichte der K. Preussischen Akademie der Wissenschaften zu Berlin* (1885), no. 1, 119–129, and (1891), no. 2, 941–970, reports from Africa on electric fish; *Die elektrische Fische: I. Malapterurus* (Leipzig, 1887); and *Die elektrische Fische. II. Torpede* (Leipzig, 1890).

Books on anthropology are *Drei Jahre in Südafrika* (Breslau, 1868), his first book on Africa, which included anthropology, zoology, and botany; *Die Eingeborenen Südafrika's ethnologisch und anatomisch beschrieben* (Breslau), an account of native races of South Africa with numerous portraits and color plates; *Süd-Afrika bis zum Zambesi* (Leipzig, 1885); *Die Gestalt des Menschen* (Stuttgart, 1893; 2nd ed., 1905), human anatomy for anthropologists; *Aegyptische Volkstypen der Jetztzeit* (Wiesbaden, 1904); *Ueber Bau und Bedeutung der Area centralis des Menschen* (Berlin, 1908); and *Die Haupthaar und seine Bildungstätte bei den Rassen des Menschen* (Berlin, 1912).

II. Secondary Literature. The best biography of Fritsch is Graf Haller, "Gustav Fritsch zum Gedächtnis," in *Anatomischer Anzeiger,* **64** (1927), 257–269. See also "Neurologists and Neurological Laboratories—No. 1. Professor Gustav Fritsch," in *Journal of Comparative Neurology,* **2** (1892), 84–88, probably by C. L. Herrick; C. Benda, "Gustav Fritsch zum 70. Geburtstage," in *Deutsche medizinische Wochenschrift,* **34** (1908), 605–606; and "Gustav Fritsch †," *ibid.,* **53** (1927), 1273, with portrait; "Death of Gustav Fritsch," in *Journal of the American Medical Association,* **89** (1927), 635; R. du Bois-Reymond, "Nachruf auf Gustav Fritsch," in *Medizinische Klinik,* **23** (1927), 1047–1048; and H. Grundfest, "The Different Careers of Gustav Fritsch (1838–1927)," in *Journal of the History of Medicine,* **18** (1963), 125–129 (with portrait), which contains several errors.

Edwin Clarke

FRITZSCHE, CARL JULIUS (*b.* Neustadt, Saxony, Germany, 29 October 1808; *d.* Dresden, Germany, 20 June 1871), *chemistry.*

Fritzsche was originally a pharmacist, served as assistant to the chemist Mitscherlich, and obtained the doctorate in botany at Berlin in 1833. He immigrated to Russia in 1834 and became manager of H. W. Struve's mineral-water works in St. Petersburg. He was elected a member of the Academy of Sciences in 1838, an associate in 1844, and full academician in 1852.

Fritzsche began a long series of researches on indigo in 1839 when he observed the action of nitric acid on indigo. In 1840 he distilled a mixture of indigo and potassium hydroxide; he correctly analyzed the base that he obtained and named it *Anilin* after the Spanish word for indigo, *añil,* derived in turn from the Arabic *an-nil.* Erdmann, editor of the *Journal für praktische Chemie,* recognized that Fritzsche's aniline was identical to *Krystallin,* which Unverdorben had prepared from indigo in 1826. Fritzsche began an association with Zinin of the University of Kazan—ultimately resulting in their sharing a laboratory at the St. Petersburg Academy—when he demonstrated in 1842 that *Benzidam,* which Zinin had obtained by reducing Mitscherlich's nitrobenzene with ammonium sulfide, was identical to aniline. In his studies of coal tar, Hofmann showed that the compounds of Unverdorben, Fritzsche, and Zinin were identical to *Kyanol,* which Runge had found in coal tar in 1834. Hofmann preferred the name, aniline, and he introduced a new method in 1845 for preparing it by using zinc and hydrochloric acid in reducing nitrobenzene.

Fritzsche continued his work on indigo, isolating and naming crysanilic and anthranilic acids. He found that when anthranilic acid is heated above its melting point it decomposes quantitatively into aniline and carbon dioxide. He also investigated uric and purpuric acids; osmium, iridium, vanadium, and their compounds; potassium bromate; and nitrogen oxides. He discovered *ortho-* and *para-*nitrophenols; compounds of hydrocarbons and picric acid; and gray tin. Fritzsche, who never became a classroom teacher, devoted his life to research and travel. Although stricken with paralysis, he finished his investigation of the dimorphism of tin shortly before he returned to Germany, where he died.

BIBLIOGRAPHY

I. ORIGINAL WORKS. Fritzsche's first publication on indigo was "Vorläufige Notiz über ein neues Zersetzungsproduct des Indigo durch Salpetersäure," in *Journal für praktische Chemie,* **16** (1839), 507–508, and was followed by his announcement of the discovery of aniline in "Ueber das Anilin, ein neues Zersetzungsproduct des Indigo," *ibid.,* **20** (1840), 453–457. His discovery of crysanilic and anthranilic acids was reported in "Ueber die Producte der Einwirkung des Kali auf das Indigblau," *ibid.,* **23** (1841), 67–83. For lists of other important papers see the Butlerov and Sheibley articles listed below.

II. SECONDARY LITERATURE. For biographical details and discussions of Fritzsche's contributions to chemistry, see Alexander M. Butlerov, "Carl Julius Fritzsche," in *Berichte der Deutschen chemischen Gesellschaft,* **5** (1872), 132–136, also printed in *Journal of the Chemical Society,* **25** (1872), 245–248; Henry M. Leicester, "N. N. Zinin, An Early Russian Chemist," in *Journal of Chemical Education,* **17** (1940), 303–306; and Fred E. Sheibley, "Carl Julius Fritzsche and the Discovery of Anthranilic Acid, 1841," *ibid.,* **20** (1943), 115–117.

A. ALBERT BAKER, JR.

FROBENIUS, GEORG FERDINAND (*b.* Berlin, Germany, 26 October 1849; *d.* Charlottenburg, Berlin, Germany, 3 August 1917), *mathematics.*

Frobenius was the son of Christian Ferdinand Frobenius, a parson, and Christiane Elisabeth Friedrich. He attended the Joachimthal Gymnasium in Berlin and then began his mathematical studies at Göttingen in 1867, completing them with a doctorate at Berlin in 1870. In the latter year he taught at the Joachimthal Gymnasium and moved to the Sophienrealschule the following year. In 1874, on the basis of his mathematical papers, Frobenius was appointed assistant professor at the University of Berlin. The next year he was made a full professor at the Eidgenössische Polytechnikum in Zurich. In 1876 he married Auguste Lehmann. Frobenius returned permanently to the University of Berlin in 1892, as professor of mathematics. Important publications led to his election to membership in the Prussian Academy of Sciences at Berlin in 1893.

Frobenius wrote many papers, a number of them of decisive importance. Several were done with other prominent researchers, particularly with Ludwig Stickelberger and Issai Schur.

Frobenius' major achievements were in group theory, which in the 1870's and 1880's, through the joining of its three historical roots—the theory of solutions of algebraic equations (Galois theory, permutation groups), geometry (finite and infinite transformation groups, Lie theory), and number theory (composition of quadratic forms, modules)—produced the concept of the abstract group, the first abstract mathematical structure in the modern sense.

Frobenius, who had become acquainted with the idea of abstract algebra in Berlin, through Leopold Kronecker and Ernst Kummer, made fundamental contributions to the concept of the abstract group in "Ueber Gruppen von vertauschbaren Elementen" (1879), written with Stickelberger, and in "Über endliche Gruppen" (1895). He exerted even greater influence on the development of group theory by means of the theory of finite groups of linear substitutions of *n* variables. This theory, which he and Schur completed in all its essential aspects, was conceived from the beginning as a representation theory of abstract groups. Its nucleus is the theory of group characters. Among the relevant works on this topic are "Über die Gruppencharaktere" (1896), "Über die Darstellung der endlichen Gruppen durch lineare Substitutionen" (1897, 1899), "Über die Komposition der Charaktere einer Gruppe" (1899), and "Über die reellen Darstellungen der endlichen Gruppen" (1906), written with Schur.

The representation theory of finite groups through linear substitutions was later to offer the possibility of surprising and important applications to difficult questions in the theory of finite groups, properly speaking, and, in the 1920's and 1930's, to group-theory questions in quantum mechanics.

BIBLIOGRAPHY

I. ORIGINAL WORKS. Frobenius' writings were brought together in *Gesammelte Abhandlungen,* J. P. Serre, ed., 3 vols. (Berlin–Heidelberg–New York, 1968). Among his works are "Ueber Gruppen von vertauschbaren Elementen," in *Journal für die reine und angewandte Mathematik,* **86** (1879), 217–262, written with L. Stickelberger; "Über endliche Gruppen," in *Sitzungsberichte der Preussischen Akademie der Wissenschaften zu Berlin* (1895), 81–112; "Über die Darstellung der endlichen Gruppen durch lineare Substitutionen," in *Monatsberichte der Preussischen Akademie der Wissenschaften zu Berlin* (1897), 994–1015, (1899), 482–500; and "Über die reellen Darstellungen der endlichen Gruppen," *ibid.* (1906), 186–208, written with I. Schur.

II. SECONDARY LITERATURE. Frobenius' work is discussed in H. Wussing, *Die Genesis des abstrakten Gruppenbegriffes* (Berlin, 1969), pp. 182–184. A short biography is N. Stuloff, "G. F. Frobenius," in *Neue Deutsche Biographie,* V (1961), 641. There are also biographies and information on works in *Deutsches biographisches Jahrbuch* (1917–1920; 2nd ed., Berlin–Leipzig, 1928), p. 654; Poggendorff, III, 481; IV, 463–464; V, 399, and VI, pt. 2, 824; and *Vierteljahrsschrift der Naturforschenden Gesellschaft in Zürich,* **62** (1917), 719.

H. WUSSING

FROST, EDWIN BRANT (*b.* Brattleboro, Vermont, 14 July 1866; *d.* Chicago, Illinois, 14 May 1935), *astronomy.*

The second son of Carlton Pennington Frost and Eliza Ann DuBois, Frost spent most of his youth at Hanover, New Hampshire, where his father held a professorship in medicine at Dartmouth College and was later dean of the medical school. He was much influenced by the astronomer C. A. Young. Frost's early education was at home, his first experience of a formal classroom being at the age of eleven. He graduated from Dartmouth in 1886 with honors in physics. After remaining at Dartmouth a few months to do graduate work in chemistry, Frost taught for a term in Hancock, New Hampshire. He spent the spring of 1887 learning practical astronomy under Young at Princeton and in the autumn returned to Dartmouth as instructor in physics and astronomy. In 1890 he left for two years' study in Europe, first for a semester at the University of Strasbourg and then with H. C. Vogel at the Potsdam Observatory. Frost returned to Dartmouth as assistant professor in astronomy, being promoted to full professor in 1895. In 1896 he married Mary Elizabeth Hazard. In 1898 he became professor of astrophysics at the new Yerkes Observatory, although he continued to spend part of his time teaching at Dartmouth until 1902. He succeeded Hale as director of the Yerkes Observatory in 1905, a position he held until 1932.

In 1915, while observing with the forty-inch telescope at Yerkes, his right retina became detached, and vision with this eye was completely lost within a year. A cataract developed in his left eye, and a hemorrhage occurred a few years later. Almost total blindness was only a minor inconvenience to him; if he had not been so afflicted, he would probably not have discovered that one can determine the temperature (in degrees Fahrenheit) by counting the number of chirps of the snowy tree cricket (*Oecanthus niveus*) in thirteen seconds and adding forty-two.

It was during his stay at Potsdam that Frost became involved in stellar spectroscopy, and the appearance of the nova T Aurigae (1891) prompted him to obtain photographs of the spectrum of this and other stars. On returning to Dartmouth he published under the title *A Treatise on Astronomical Spectroscopy* (1894) a translation and revision of Scheiner's *Die Spectralanalyse der Gestirne;* this was a standard text for many years. He made routine solar, cometary, and meteorological observations and participated in some of the first X-ray experiments outside Europe; he also made a qualitative study of the spectrum of Beta Lyrae (1895). At the 1900 eclipse he obtained photographs of the flash spectrum and also of the spectrum of the corona. He secured spectrograms of Comet Morehouse 1908 III and nova DI Lacertae 1910. He edited the extensive series of solar observations by C. H. F. Peters (1906) and subsequently Barnard's micrometric measurements of star clusters (1931).

Frost's principal research field, however, was stellar spectroscopy, specifically the determination of radial velocities of stars and especially stars of early spectral type. By 1895 radial velocities had been determined for only fifty stars. Soon after arriving at Yerkes, Frost designed for the forty-inch refractor the Bruce spectrograph; it is in no small measure due to the observations by Frost and his colleagues with this instrument that the number of stars whose radial velocities were known increased more than a hundredfold during the following forty years. Frost was the first to realize that there are systematic differences between the velocities of stars of different spectral types. He early recognized the need for calibrating the results obtained with different instruments and by a variety of methods. A natural outcome of radial velocity studies is the discovery of spectroscopic binaries; this is particularly true for stars of early spectral type, of which Frost found and determined the orbits of a considerable number.

Frost received honorary D.Sc. degrees from Dartmouth (1911) and Cambridge (1912). He was an associate of several foreign astronomical societies. He served as an assistant editor of the *Astrophysical Journal* from its inception in 1895 and as an editor from 1902 until his death.

BIBLIOGRAPHY

I. ORIGINAL WORKS. Frost's works include *A Treatise on Astronomical Spectroscopy* (Boston, 1894), a trans. and rev. of J. Scheiner's *Die Spectralanalyse der Gestirne;* "Spectroscopic Observations of Standard Velocity Stars," in *Astrophysical Journal,* **18** (1903), 237–277, written with W. S. Adams; "Radial Velocities of Twenty Stars Having Spectra of the Orion Type," in *Publications of the Yerkes Observatory,* **2** (1904), 145–250, written with W. S. Adams; "Radial Velocities of 368 Helium Stars," in *Astrophysical Journal,* **64** (1926), 1–77, written with S. B. Barrett and O. Struve; "Radial Velocities of 500 Stars of Spectral Class A," in *Publications of the Yerkes Observatory,* **7** (1929), 1–79, written with S. B. Barrett and O. Struve; and *An Astronomer's Life* (Boston-New York, 1933).

II. SECONDARY LITERATURE. An obituary notice by P. Fox appeared in *Astrophysical Journal,* **83** (1936), 1–9.

BRIAN G. MARSDEN

FROUDE, WILLIAM (*b.* Dartington, Devonshire, England, 1810; *d.* Simonstown, Cape of Good Hope [now Union of South Africa], 4 May 1879), *ship hydrodynamics.*

Froude (pronounced Frude) was the sixth son of Archdeacon Richard Hurrell Froude, rector at Dartington, and Margaret Spedding of Cumberland. He studied seven years at Oriel College , Oxford, where he was tutored in mathematics by his oldest brother, Robert Hurrell. (The latter was also a leader of the Oxford Movement, and it is noteworthy that William was the only member of the family who did not follow Newman into Roman Catholicism.) While subsequently occupied as a civil engineer, Froude came under the influence of I. K. Brunel, builder of both railways and oceangoing steamships, who stimulated his interest in naval architecture.

Froude retired from active civil engineering practice at the age of thirty-six, but he continued to give attention to various aspects of ship behavior, both recreational (he was an avid yachtsman) and technical. At Brunel's request he undertook in 1856 a resistance and rolling study of the *Great Eastern,* and his analytical and experimental work on the subject, at full as well as reduced scale, extended to many ships over many years. Of even greater importance than his control of rolling by use of bilge keels was his promotion of resistance studies on scale models. His efforts to secure the support of the Admiralty for the construction of a model towing tank at first aroused the opposition of John Scott Russell and other members of the Institution of Naval Architects, and it was not till 1870 that the sum of £2,000 was granted for this purpose. The original tank, 250 feet in length, was built on Froude's own land at Torquay only eight years before his death; he was ably assisted by his son, Robert Edmund Froude, who later built the Admiralty tank at Haslar.

William Froude's great manual skill was of inestimable value in the construction and operation of the tank, and many of his model and prototype processes and instruments continue to be employed: the use of paraffin and waterline cutting machines for models; resistance recorders; governors; roll indicators; and propeller-engine dynamometers. Use of the scale model for resistance studies was based upon his hypothesis that the total resistance could be considered the sum of wave formation and skin friction and that each could be scaled independently. He showed that the wave effects would be similar in model and prototype if the velocity were reduced in proportion to the square root of the length. This is known as Froude's law of similarity, even though it had been published by Ferdinand Reech, a professor in the school of naval architecture at Paris, in 1852 and purportedly introduced in his lectures as early as 1831. Froude formulated the law of skin-friction similarity after towing streamlined catamaran planks of various lengths and surface finishes through water over a wide range of speeds. The resistance of the smooth surfaces was found to vary with no more than the 1.85 power of the velocity, and only for the roughest did the power reach 2.0. His perceptive understanding of the effect of surface length was in close accord with present-day boundary-layer theory.

Froude was elected a fellow of the Royal Society of London in 1870. In 1876 he received both the honorary degree of LL.D. from the University of Glasgow and a Royal Medal of the Royal Society. His many writings are to be found in the *Transactions of the Institution of Naval Architects* and in reports to the British Association for the Advancement of Science. Froude's last paper, published a year before his death, was on the subject of screw propulsion, one of his early interests. While on a holiday trip to the Cape in 1879 he succumbed to dysentery just before his scheduled return to England.

BIBLIOGRAPHY

I. ORIGINAL WORKS. Froude's writings include "Experiments on the Surface-friction Experienced by a Plane Moving Through Water," in *British Association for the Advancement of Science Report,* 42nd Meeting, 1872; and "On Experiments with H.M.S. *Greyhound,*" in *Transactions of the Institution of Naval Architects,* **16** (1874), 36–73.

II. SECONDARY LITERATURE. On Froude and his work, see "Memoir of the Late William Froude, LL.D., F.R.S.," in *Transactions of the Institution of Naval Architects,* **20** (1879), 264–269; W. Abell, "William Froude," in *Transactions of the Institution of Naval Architects,* **76** (1934), which quotes his 1868 request to the Admiralty; and H. Rouse and S. Ince, *History of Hydraulics* (New York, 1963), pp. 243–256.

HUNTER ROUSE

FUBINI, GUIDO (*b.* Venice, Italy, 19 January 1879; *d.* New York, N.Y., 6 June 1943), *mathematics.*

Fubini was the son of Lazzaro Fubini, who taught mathematics at the Scuola Macchinisti in Venice, and Zoraide Torre. At the age of seventeen, after brilliantly completing secondary studies in his native city, he entered the Scuola Normale Superiore di Pisa, where Dini and Bianchi were among his teachers. In 1900 he defended a thesis on Clifford's parallelism in elliptical spaces, the results of which rapidly became classic because of their inclusion in the 1902 edition of Luigi Bianchi's treatise on differential geometry. Fubini remained in Pisa for another year to complete work on the diploma allowing him to teach at the university level. The important memoir that he wrote in this connection deals with the fundamental principles of the theory of harmonic functions in spaces of constant curvature, a subject quite different from that of his doctoral thesis.

Placed in charge of a course at the University of Catania toward the end of 1901, Fubini soon won the competition for nomination as full professor. From Catania he went to the University of Genoa and then, in 1908, to the Politecnico in Turin. There he taught mathematical analysis, and at the same time, at the University of Turin, higher analysis. In 1938 Fubini was forced to retire under the racial laws promulgated by the Fascist government. The following year, at the invitation of the Institute for Advanced Study at Princeton, he immigrated to the United States and was welcomed among the institute's members. His prudent decision to seek voluntary exile was in part dictated by his concern for the future of his two sons, both engineers. Already in poor health, he continued to teach at New York University until he died of a heart ailment at the age of sixty-four.

A man of great cultivation, fundamentally honorable and kind, Fubini possessed unequaled pedagogic talents. His witty banter and social charm made him delightful company; he was small in stature, and his voice was vigorous and pleasant. Deeply imbued with a sense of family, he wished toward the end of his life legally to add Ghiron—the maiden name of his wife, whom he had married in 1910—to his own. Those works on mathematical subjects designed to be of use to engineers resulted from his own interest in watching over his sons' studies. With regard to Luigi Bianchi, Fubini's gratitude was the equal of his respect and admiration for Bianchi as a model for both his life and his work. Upon Bianchi's death in 1928, Fubini succeeded him as coeditor of the *Annali di matematica pura ed applicata,* a position that he held until 1938. A member of several Italian scientific academies, Fubini received the royal prize of the Lincei in 1919.

Fubini was one of Italy's most fecund and eclectic mathematicians. His contributions opened new paths for research in several areas of analysis, geometry, and mathematical physics. Guided by an ever-alert geometric intuition and possessed of an absolute mastery of all the techniques of calculation, he was able to follow leads that had barely been glimpsed. His technical mastery often permitted him to discover simpler demonstrations of such theorems as those of Bernstein and Pringsheim on the development of Taylor series.

In analysis Fubini did work on linear differential equations, partial differential equations, analytic functions of several complex variables, and monotonic functions. He also studied, in the calculus of variations, the reduction of Weierstrass' integral to a Lebesgue integral; the possibility of expressing every surface integral by two simple integrations, and the converse; and the manner of deducing from the existence of $\delta^n f/\delta x^n$ and $\delta^n f/\delta y^n$ the existence of lower-order derivatives of the function $f(x,y)$. In addition, Fubini determined, with regard to the minimum-value principle, the limit of a series of functions that take on given values on the contour of a domain, by supposing that the corresponding Dirichlet integrals tend toward their lower limit; he also indicated how his procedure could be applied to the calculus of variations. Finally, he investigated nonlinear integral equations and those with asymmetric kernels.

In the field of discontinuous groups, Fubini studied linear groups and groups of movement on a Riemannian variety in order to establish their criteria of discontinuity, as well as to prove the existence of fundamental domains and to indicate the method of constructing them. He examined functions admitting of such groups, as well as the automorphic harmonic functions in a space of n dimensions, in this way generalizing certain theorems of Weierstrass. For continuous groups, he established the conditions required in order to be able to attribute a metric to them.

In the field of non-Euclidean spaces Fubini, in his thesis on Clifford's notion of parallelism, introduced sliding parameters, which made possible the transposition to elliptical geometry of certain results of ordinary differential geometry, such as Frenet's formulas and the determination of couples of applicable surfaces. His work on the theory of harmonic functions in spaces of constant curvature contains an extension of the Neumann method and of the Appell and Mittag-Leffler theorems.

The most extensive field that Fubini cultivated was that of differential projective geometry, for which he elaborated general procedures of systematic study that still bear his name. The difficulties to be surmounted in order to pass from classical to projective differential geometry arise mainly from mathematical techniques and their use. To succeed in this endeavor, Fubini utilized absolute differential calculus and certain contravariant differentials. First he defined the local application of two varieties with respect to a Lie group; then he introduced the "projective linear element" as the quotient of two covariant differential forms and demonstrated that the necessary and sufficient condition for a projective application is the equality of these elements. He envisaged homogeneous coordinates normalized from a variable point on the surface or hypersurface, and he defined the "projective normals," the "projective geodesics," and the more general geodesics. In a Euclidean space the transformation by affinity of a surface of constant curvature is characterized by its second projective normal's being extended to infinity. These fundamental investigations of metric, or affine, geometry, which were pursued by other researchers, are collected in *Geometria proiettiva differenziale* and *Introduction à la géométrie projective différentielle des surfaces,* both written in collaboration with Eduard Čech.

Fubini's contributions to mathematical physics are varied. They began during World War I with theoretical studies on the accuracy of artillery fire and then turned to such problems in acoustics and electricity as anomalies in the propagation of acoustic waves of large amplitude, the pressure of acoustic radiation, and electric circuits containing rectifiers. Fubini was also interested in the equations of membranes and vibrating diaphragms. The mathematical aspects of the engineering sciences likewise occupied his attention. A work on engineering mathematics and its applications appeared posthumously in 1954. Finally, one must note his textbooks—courses in analysis and collections of problems which have been used by many generations of students—to appreciate fully the many-faceted work of Fubini, one of the most luminous and original minds in mathematics during the first half of the twentieth century.

BIBLIOGRAPHY

I. Original Works. Fubini's writings were brought together in *Opere scelte,* 3 vols. (Rome, 1957–1962).

Among his articles are "Di un metodo per l'integrazione e lo studio delle equazioni alle derivate parziali," in *Rendiconti del Circolo matematico di Palermo,* **17** (1903), 222–235; "Nuove ricerche intorno ad alcune classi di gruppi discontinui," *ibid.,* **21** (1906), 177–187, in which, in a note, Fubini mentions seven of his articles on the same subject; "Sul principio di Dirichlet," *ibid.,* **22** (1906), 383–386; "Il principio di minimo e i teoremi di esistenza per i problemi al contorno relativi alle equazioni alle derivate parziali di ordine pari," *ibid.,* **23** (1907), 58–84, 300–301; "Applicabilità proiettiva di due superficie," *ibid.,* **41** (1916), 135–162; "Su una classe di congruenze W di carattere proiettivo," in *Atti della R. Accademia nazionale dei Lincei. Rendiconti,* **25** (1916), 144–148; "Invarianti proiettivo-differenziali delle curve tracciate su una superficie e definizione proiettivo-differenziale di una superficie," in *Annali di matematica pura ed applicata,* **25** (1916), 229–252; "Fondamenti di geometria proiettivo-differenziale," in *Rendiconti del Circolo matematico di Palermo,* **43** (1918–1919), 1–46; "Su alcune classi di congruenze di rette e sulle trasformazioni delle superficie R," in *Annali di matematica pura ed applicata,* **58** (1924), 241–257; and "Luigi Bianchi e la sua opera scientifica," *ibid.,* **62** (1929), 45–81.

Fubini's books are *Introduzione alla teoria di gruppi discontinui e delle funzioni automorfe* (Pisa, 1908); *Lezioni di analisi matematica* (Turin, 1913; 2nd ed., 1915); *Esercizi di analisi matematica (calcolo infinitesimale) con speciale riguardo alle applicazioni* (Turin, 1920), written with G. Vivanti; *Geometria proiettiva differenziale,* 2 vols. (Bologna, 1926–1927), written with E. Čech; *Introduction à la géométrie différentielle des surfaces* (Paris, 1931), written with E. Čech; *Anomalie nella propagazione di onde acustiche di grande ampiezza* (Milan, 1935); *Circuiti elettrici contenenti raddrizzatori* (Turin, 1936); *Acustica non lineare delle onde di ampiezza* (Milan, 1938); and *La matematica dell'ingegnere e le sue applicazioni* (Bologna, 1954), written with G. Albenga.

II. Secondary Literature. See the unsigned "Guido Fubini Ghiron," in *Annali di matematica pura ed applicata,* 4th ser., **25** (1946), ix–xii; and M. Picone, in *Bollettino dell'Unione matematica italiana,* 2nd ser., **1,** no. 1 (Dec., 1946), 56–58.

Pierre Speziali

FUCHS, JOHANN NEPOMUK VON (*b.* Mattenzell, Bavaria, Germany, 15 May 1774; *d.* Munich, Germany, 5 March 1856), *chemistry, mineralogy.*

Fuchs, the son of poor peasants, proved such an outstanding student in the convent school that he was sent on to the Gymnasium. Patrons enabled him to study medicine in Vienna. He graduated from the University of Heidelberg and passed the examinations for the physician's license in Munich. Under the influence of Jacquin, Fuchs became interested in chemistry and mineralogy. The Bavarian state, which was engaging chemists in the hope of promoting the chemical industry and of reviving interest in the University of Landshut, sent Fuchs on a fact-finding trip. He visited the Bergakademie in Freiburg and traveled to Berlin and Paris. He met with W. A. Lampadius, A. G. Werner, C. S. Weiss, Klaproth, V. Rose (the younger), Guyton de Morveau, Fourcroy, Berthollet, Vauquelin, and Haüy and widened his knowledge in the course of several mineralogical and geological field trips. Upon his return to Bavaria, he passed an examination before a commission of the Bavarian Academy of Sciences resulting in his appointment, in the fall of 1805, as lecturer at the University of Landshut; he received a full professorship in May 1807. In 1810 Fuchs married the daughter of a wealthy taverner named Fahrenbacher. In the fall of 1823 Fuchs went to Munich as a member of the Bavarian Academy of Sciences and curator of the state mineralogical collection. Following the transfer of the Ludwig-Maximilian University from Landshut to Munich in 1826, Fuchs rejoined its faculty as professor of mineralogy. From 1833 on, he served as chemist on the Obermedizinalausschuss and the Supreme School Board. In 1835 he was appointed Oberberg- und Salinenrat and served as president of the Polytechnischer Verein für das Königreich Bayern.

A devout Catholic, Fuchs was thrifty, kind, and helpful and abhorred scientific arguments. Although he had to regulate his living habits strictly after contracting a lung disease in 1805, he was able to carry on his scientific work until the age of eighty-two.

Fuchs was the first titular professor of chemistry at the Ludwig-Maximilian University, where chemistry was still considered a primitive empirical science even though its role as an auxiliary science in medicine, pharmacy, and economics was not to be gainsaid. Together with Stromeyer and Döbereiner, Fuchs was one of the first in Germany to introduce practical laboratory instruction at the university level. Under his guidance, advanced students were required to carry out mineral analyses.

Fuchs became, by royal appointment, a consultant to bleaching and dyeing enterprises, breweries, distillers, starch manufacturers, paper mills, and tobacco processors. He also introduced beet-sugar refining in Bavaria.

Fuchs's scientific work was directed mainly toward the practical and empirical. He improved the spirit lamp—important for the laboratory—and the soldering iron and developed a rapid beer-testing process, the hallymetric beer test. He was the first to produce water glass and tested its use in fire protection and as varnish. In the course of research on lime, mortar, and cement, he became the first to present a clear, scientific exposition of the process of hydraulic cement setting.

Fuchs continuously stressed the importance of chemistry in the study of mineralogy. By means of a great number of analyses he was able to determine the composition of many minerals and mineral waters used for medicinal purposes.

At a time when chemical formulas, quantitative analysis, and the theory of the atom were just beginning to become common knowledge, Fuchs endeavored to solve the mystery of the structure of minerals. He confirmed the stoichiometric laws, observed isomorphism—which he called *Varikierung* ("variation")—and the cation exchange of zeolites. Opal, quartz, diamond, and soot served as objects of study in his determination of the difference between crystalline and amorphous structures.

BIBLIOGRAPHY

I. Original Works. Fuchs's writings are collected in *Gesammelte Schriften des Johann Nep. von Fuchs,* edited and with an obituary by Cajetan Georg Kaiser (Munich, 1856). His *Naturgeschichte des Mineralreichs* was published as vol. III of *Handbuch der Naturgeschichte,* Johann Andreas Wagner, ed. (Kempten, Bavaria, 1842).

II. Secondary Literature. On Fuchs and his work, see Franz von Kobell, *Denkrede auf Johann Nepomuk von Fuchs vom 28. März 1856* (Munich, 1856); Wilhelm Prandtl, "Johann Nepomuk Fuchs," Ralph E. Oesper, trans., in *Journal of Chemical Education,* **28** (1951), 136–142; and Friedrich Quietmeyer, *Zur Geschichte der Erfindung des Portlandzementes* (Berlin, 1912), pp. 88–92.

Eberhard Schmauderer

FUCHS, IMMANUEL LAZARUS (*b.* Moschin, near Posen, Germany [now Poznan, Poland], 5 May 1833; *d.* Berlin, Germany, 26 April 1902), *mathematics.*

Fuchs was a gifted analyst whose works form a bridge between the fundamental researches of Cauchy, Riemann, Abel, and Gauss and the modern theory of differential equations discovered by Poincaré, Painlevé, and Picard.

By the time Fuchs was a student at the Friedrich Wilhelm Gymnasium, his unusual aptitude in mathematics had awakened a corresponding interest in the discipline. At the University of Berlin he studied with Ernst Eduard Kummer and Karl Weierstrass, and it was the latter who introduced him to function theory, an area that was to play an important role in his own researches. Fuchs received the doctorate from Berlin in 1858, then taught first at a Gymnasium and later at the Friedrich Werderschen Trade School. In 1865 he went to the University of Berlin as a *Privatdozent,* and there he began the study of regular singular points. In 1866 he became an extraordinary professor at the university. From 1867 to 1869 Fuchs was professor of mathematics at the Artillery and Engineering School and then went as ordinary professor to Greifswald. He remained there five years, spent one year at Göttingen and then, in 1875, went to Heidelberg. In 1882 Fuchs returned to Berlin, where he became professor of mathematics, associate director of the mathematics seminars, and a member of the Academy of Sciences. He remained at Berlin until his death. For the last ten years of his life he was editor of the *Journal für die reine und angewandte Mathematik.*

Except for a few early papers in higher geometry and number theory, all of Fuchs's efforts were devoted to differential equations.

In his monumental 1812 work, *Disquisitiones*

FUCHS

generales circa seriem infinitam, Gauss investigated the hypergeometric series and noted that, for appropriately chosen parameters, most known functions could find representation through this series. In 1857 Riemann conjectured that functions so expressed, which satisfy a homogeneous linear differential equation of the second order with rational coefficients, might be employed in the solution of any linear differential equation. This provided an alternate approach to the power series development which had been presented by Cauchy and extended by Briot and Bouquet. With Riemann's work as inspiration these methods were synthesized and extended by Fuchs in a series of papers that began to appear in 1865.

In the real domain the method of successive approximations was first applied by Liouville (1838) to homogeneous linear differential equations of the second order and later (1864) by M. J. Caqué to the nth-order case. A second method of proving the existence of solutions derives from a method suggested by Euler (1768), developed by Cauchy (1820–1830), and refined by Lipschitz in 1876.

For a first-order equation, $dw/dz = f(z,w)$, the Cauchy-Lipschitz method can be extended to the complex domain, as can the method of limits. It was Fuchs, however, who provided the proof of the existence of solutions, satisfying initial conditions for $z = z_0$, for the linear differential equation of order n. The general homogeneous linear differential equation of order n has the form

$$\frac{d^n w}{dz^n} + p_1(z)\frac{d^{n-1}w}{dz^{n-1}} + \cdots + p_n(z)w = 0, \tag{1}$$

and it is assumed that the $p_i(z)$ are analytic throughout a domain D in the Z plane. With z_0 and z in D, c_r are chosen so that the Taylor series,

$$w(z) = \Sigma\, c_r(z - z_0)^r,$$

formally satisfies the differential equation. The c_r are shown to be finite as long as the initial values are finite. Furthermore, if M_ν is the upper bound of $|p_\nu(z)|$ on $|z - z_0| = a$, then $p_\nu^{(r)} \leqslant M_\nu/a^r$, so that if

$$P_\nu(z) = \frac{M_\nu}{1 - \dfrac{z - z_0}{a}},$$

then

$$|p_\nu(z)| \leq |P_\nu(z)|$$

within $|z - z_0| = a$ and on the circumference.

If now w is replaced by W in equation (1), where, as above,

$$W(z) = \Sigma C_r(z - z_0)^r$$

where $C_i = |c_i|$, then

$$|w^{(r)}(z_0)| \leq W^{(r)}(z_0)$$

and, hence,

$$\Sigma|c_r(z - z_0)^r| \leq \Sigma C_r(z - z_0)^r.$$

But the circle of convergence of the dominant series can, as Fuchs showed, be readily found to be $|z - z_0| = a$. Consequently there is a solution to the differential equation, satisfying the given initial conditions, when $z = z_0$, which is expressible by a Taylor series that is absolutely and uniformly convergent within any circle, that has its center at z_0, in which the $p_i(z)$ are analytic.

The singularities of the solution are precisely the singularities of equation (1), so that, because of the linearity of this equation, there are neither movable singularities nor movable poles. To determine whether the point at infinity is a singular point, it is necessary only to effect the substitution $z = 1/Z$ and reduce the equation to the form given by (1). If the equation in Z has singularities at the origin, then the equation in z has a singular point at infinity. Thus, in all cases the singular points can be found simply by inspecting the equation itself.

Fuchs introduced the term "fundamental system" to describe n linearly independent solutions of the linear differential equation $L(u) = 0$. It is clear that for any nonsingular point such a fundamental set of solutions exists. The so-called Fuchsian theory is concerned with the same existence question in relation to an arbitrarily given singular point and, once the existence problem is solved, an investigation of the behavior of the solutions in the neighborhood of the singular point.

BIBLIOGRAPHY

Fuchs's works were published as *Gesammelte mathematische Werke,* Richard Fuchs and Ludwig Schlesinger, eds., 3 vols. (Berlin, 1904–1909). His speech as rector, given 3 Aug. 1900, is *Rede zur Gedächtnissfeier des Stifters der Berliner Universität* (Berlin, 1900).

Obituaries include E. J. Wilczynski, in *Bulletin of the American Mathematical Society,* **9** (1902), 46–49; *Atti dell'Accademia nazionale dei Lincei. Rendiconti,* 5th ser., **11** (1902), 397–398; *Bibliotheca mathematica,* 3rd ser., **3** (1902), 334; and *Enseignement mathématique,* **4** (1902), 293–294.

See also Otto Biermann, *Zur Theorie Fuchs'schen Functionen* (Vienna, 1885), reprinted from *Sitzungsberichte der Wien Akademie der Wissenschaften,* **92** (1885).

JEROME H. MANHEIM

FÜCHSEL, GEORG CHRISTIAN (*b.* Ilmenau, Germany, 14 February 1722; *d.* Rudolstadt, Germany, 20 June 1773), *geology.*

Füchsel studied medicine at the University of Jena and medicine, natural sciences, and theology at the University of Leipzig. After settling in Rudolstadt, he engaged in the "salon" science of natural history cabinets and mineral collections. He served as town physician and later court physician to the princes of Schwarzburg-Rudolstadt. His major work, "Historia terrae et maris, ex historia Thuringiae, per montium descriptionem, eruta," appeared in 1761 in the *Acta* of the Erfurt Academy of Sciences. Although written in German, this work was published in a very defective Latin translation made by a friend of Füchsel. Its frequent unintelligibility perhaps accounts in part for the relative neglect of the "Historia."

Most eighteenth-century writings on geology tended to be either submerged in extended and superficial accounts of general natural history and in cosmological speculation, or limited to the most detailed sort of descriptions of individual occurrences: hot springs, fossil finds, strange crystals, erratic boulders, mines, caves, and quarries. In contrast with such works, Füchsel's "Historia" is unusual for its purely geological orientation. This book-length work contains the enunciation and substantiation of general principles of historical geology, an extensive description of all the stratified rocks of the Thüringer Wald used to illustrate the general principles, an explanation of the causes of dynamic changes in the earth's crust, and an explanation of the origin of veins and their minerals. An appendix ("Usus historiae . . .") furnishes an extended discussion of the applications of such geological knowledge.

Most noteworthy among Füchsel's principles of geology are his assumptions of a rigorous actualism and the fruitful concept of a formation (which he calls a *series montana* or *ein Geburge*). After a thorough account of all the formations of the Thüringer Wald (including data on lithology, notable minerals, fossil content, inclusion of foreign rocks, spatial orientation, and geographical extent for each formation), Füchsel's work proceeds to construct, by induction, a historical account of the region's geology founded on the assumption of a uniformity of natural processes through time. "In truth we must take as the norm in our explanation [of the earth's history] the manner in which nature acts and produces solids at the present time; we know no other way."[1] He distinguished a formation as those strata formed at the same time, of the same material, and in the same manner.[2] In establishing the history Füchsel also used the principle of superposition and the method of correlation by index fossils to reconstruct a complete stratigraphic sequence for Thüringen and to assimilate separated occurrences of the same formations to their proper place in the sequence. To illustrate the stratigraphic data of this history, Füchsel produced the first published geological map. Unlike previous soil maps (such as Christopher Packe's *Chart of East Kent*), Füchsel's map indicated not only the distribution of rocks but also their arrangement in relation to each other and their relative ages.

The *Acta* of the Erfurt Academy had only a limited circulation, and in his other major work, *Entwurf zu der ältesten Erd- und Menschengeschichte . . .*, which appeared anonymously in 1773, Füchsel complained that only a few scholars seemed to have given the "Historia" much attention. Nonetheless, his ideas did exert an influence on the geological literature of the time, probably through a long notice by J. S. Schröter in the popular *Journal für Liebhaber des Steinreichs*[3] and through the writings of J. K. W. Voigt, J. E. I. Walch, and Goethe. A. G. Werner's system seems to owe much to Füchsel, but the precise connections have not yet been fully documented.

NOTES

1. "Historia," pp. 81–82, sec. 43. "Praecedentes observationes tam inter se quam cum aliis, combinare, atque eventus seu historiam exinde deducere liceat. Modus vero quo natura hodierno adhuc tempore agit, et corpora producit, in hac explanatione pro norma assumendus est; alium non novimus."
2. *Ibid.*, p. 48, sec. 4. "Montes eiusdem situs, ab eadem massa, eodemque modo constructos seriem montanam (ein Geburge) nominare liceat." A similar definition is found in p. 25, sec. 39, of the *Entwurf.*
3. **2** (1775), 54–63.

BIBLIOGRAPHY

I. Original Works. *Acta Academiae electoralis moguntinae scientiarum utilium, quae Erfordiae est,* **2** (1761), contains both the "Historia terrae et maris . . .," 44–208, and the "Usus Historiae terrae et maris," 209–254. Füchsel also published "Ansicht des Erfurthischen Gebietes als eines Theils von Thüringen" in *Neue oekonomische Nachrichten* (Leipzig), **3** (1766), 359–390. His last work appeared anonymously as *Entwurf zu der ältesten Erd- und Menschengeschichte, nebst einem Versuch, den Ursprung der Sprache zu finden* (Frankfurt–Leipzig, 1773). An English trans. of the "Historia" is in preparation.

II. Secondary Literature. The excellent and comprehensive biography by Rudolf Möller, "Mitteilungen zur Biographie Georg Christian Füchsels," in *Freiberger Forschungshefte,* **D43** (Leipzig, 1963), includes references to Füchsel's other published articles, to pertinent archival materials, and to relevant secondary literature. For further analysis of Füchsel's ideas see T. E. Gumprecht, "Einige Beiträge zur Geschichte der Geognosie," in *Archiv für Mineralogie, Geognosie, Bergbau und Hüttenkunde,* **23** (1850), 468–576.

BERT HANSEN

FUETER, KARL RUDOLF (*b.* Basel, Switzerland, 30 June 1880; *d.* Brunnen, Switzerland, 9 August 1950), *mathematics.*

Fueter was the son of Eduard Rudolf Fueter, an architect, and Adèle Gelzer. In 1908 he married Amélie von Heusinger.

After receiving his early education in Basel, Fueter began to study mathematics at Göttingen in 1899 and graduated in 1903. Under the supervision of David Hilbert he presented a work dealing with the theory of quadratic number fields. After further study in Paris, Vienna, and London and teaching in Marburg and Clausthal (now Clausthal-Zellerfeld), Fueter became a professor of mathematics in Basel in 1908; he accepted the same post at the Technische Hochschule in Karlsruhe in 1913 and at the University of Zurich in 1916. His field of interest was the theory of numbers as presented in Hilbert's work. He derived the class formula for the entire group of Abelian number fields over an imaginary quadratic base field. He gave a summary of these in his *Vorlesungen über die singulären Moduln und die komplexe Multiplikation der elliptischen Funktionen* (1924–1927). Later he founded his own school of thought on the theory of functions of a quaternion variable.

Fueter was cofounder and president of the Swiss Mathematical Society, rector of the University of Zurich, and president of the Euler Commission of the Swiss Society of Sciences (editors of the *Opera Omnia Leonhardi Euleri*). He held the rank of colonel in the artillery of the Swiss militia, and at the outbreak of World War II he served in the Department of Press and Radio. In *Spying for Peace* (London, 1961), Jon Kimche states, "Fueter restated the democratic rights of the press in almost classical form. . . . In his report of April 10, 1940 . . . Fueter developed his argument more fully. 'It is the duty of our press to reject the domestic and foreign policies of the National Socialists both clearly and forcefully.' " Fueter was, therefore, particularly noted for his opposition to Nazism and to the spread of its policies within Switzerland.

BIBLIOGRAPHY

Fueter's major works are *Synthetische Zahlentheorie* (Berlin, 1917; 3rd ed., 1950); *Vorlesungen über die singulären Moduln und die komplexe Multiplikation der elliptischen Funktionen,* 2 vols. (Leipzig–Berlin, 1924–1927); and *Das mathematische Werkzeug des Chemikers, Biologen und Statistikers* (Zurich, 1926; 3rd ed., 1947).

A biography is A. Speiser, in *Elemente der Mathematik,* **5** (1950), published with Fueter's autobiographical notes.

J. J. BURCKHARDT

FUHLROTT, JOHANN KARL (*b.* Leinefelde, Germany, 1 January 1804; *d.* Elberfeld [now Wuppertal], Germany, 17 October 1877), *natural history, human paleontology.*

Fuhlrott obtained his doctorate from the University of Bonn in 1830 and went from there to Elberfeld, where he became a science teacher and subsequently vice-director of the Realschule. He won modest recognition as a naturalist, publishing geological descriptions of the hills and caves in the Rhineland region between Düsseldorf and the Wupper River.

In August 1856, Fuhlrott received an assortment of fossilized bones found by two quarry workers in the Feldhofer cave of the Neander Valley. These men had uncovered what they thought to be the skeleton of a cave bear and were carelessly discarding it when the quarry owner persuaded them to save some of the remains for the Elberfeld teacher. Portions of the skull and pelvis, along with the larger limb bones, were delivered to Fuhlrott.

He studied the specimens and began to suspect that they were not bear bones but the remains of an ancient and primitive form of human being. Its physical build smaller than that of modern man, this creature with low, retracted forehead had plodded along on bowed legs, its head and chest hunched forward. Fuhlrott recognized the importance of this find and rushed to the grottoes in time to retrieve some ribs, the right radius, the left ulna, and part of the right scapula—all that remained of the probably perfect skeleton.

At Fuhlrott's request Hermann Schaaffhausen of Bonn examined the fragments and confirmed his diagnosis of their antiquity. Schaaffhausen presented a preliminary description of the fossils at the Lower Rhine Medical and Natural History Society on 4 February 1857; Fuhlrott was invited to discuss them fully before the Natural History Society of the Prussian Rhineland and Westphalia on 2 June of that year. Addressing this august body at Bonn, Fuhlrott was dismayed at the reaction to his find. Rudolf Virchow and Carter Blake dismissed the bones as the remains of an idiot ravaged by rickets in youth and arthritis in later life. They refused to credit any great age to them.

Schaaffhausen alone defended Fuhlrott's position, saying,

There is no reason whatever for regarding the unusual development of the frontal sinuses in the remarkable skull from the Neanderthal as an individual or pathological deformity: it is unquestionably a typical race characteristic and is physiologically connected with the uncommon thickness of the other bones of the skeleton, which exceeds by one-half the usual proportions [quoted in T. H. Huxley, *Collected Essays*, p. 176].

Reaction to his speech was hostile, but Fuhlrott and Schaaffhausen refused to quit their positions in the extensive controversy that ensued. They appealed to the public for support and managed to attract attention beyond the borders of Germany. They gained an important ally in Sir Charles Lyell, who journeyed from England in 1860 to investigate the discovery site of the disputed fossils. His visit to Fuhlrott convinced Lyell that the specimen was authentically human, *Homo neanderthalensis.* But it was not until after Fuhlrott's death and the discovery of fossil men at Spy, Belgium, and at Gibraltar that opposition to the notion of Neanderthal man was finally silenced.

BIBLIOGRAPHY

I. ORIGINAL WORKS. Fuhlrott's main publication is *Der fossile Mensch aus dem Neanderthal und sein Verhältniss zum Alter des Menschengeschlects* (Duisburg, 1865). Also important is his "Über die Kalksteinschichten im Neanderthale, Worin 1856 der Homo Neanderthalensis gefunden Wurde," in *Correspondenzblatt des Naturhistorischen Vereins für Rheinland und Westphalen,* **25** (1868), 62–70. In the *Jahresberichte des Naturwissenschaftlichen Vereins von Elberfeld und Barmen, Nebst wissenschaftlichen Beilagen* are "Felsenmeer im Odenwald," **3** (1858), 75–79; "Das Wisperthal und der Wisperwind," **4** (1863), 11–18; "Grundzüge der Quellenkunde," *ibid.,* 129–150; and "Die erloschenen Vulcane am Rhein und in der Eifel," **5** (1878), 3–25. In the *Verhandlungen des Naturhistorischen Vereins der Preussischenen Rheinlande und Westphalens* (Bonn) are "Paläontologisches" (1859), 125–126; and "Menschliche Ueberreste aus einer Felsengrotte des Düsselthals," *ibid.,* 131–153. There are also "Die Kalksteinschichten der Feldhofer Grotte im Neanderthale," in *Zeitschrift für gesammten Naturwissenschaften,* **33** (1869), 275–277; and "Ueber eine neu endeckte Höhle bei Barmen," in *Sitzungsberichte der Niederrheinischen Gesellschaft für Natur- und Heikunde zu Bonn* (1870), 208–209.

II. SECONDARY LITERATURE. See Aleš Hrdlička, *The Skeletal Remains of Early Man,* Smithsonian Miscellaneous Collections, **83** (Washington, D.C., 1930), 149–151. T. H. Huxley cites George Busk's trans. of Schaaffhausen's history of discovery and description of Neanderthal man in "Man's Place in Nature," in *Collected Essays,* **7** (New York, 1896; repr. 1968), 168–185. See also George Busk's trans. of Schaaffhausen's "On the Crania of the Most Ancient Races of Man," in *Natural History Review,* **1** (1861), 283.

MARTHA B. KENDALL

FULTON, JOHN FARQUHAR (*b.* St. Paul, Minnesota, 1 November 1899; *d.* New Haven, Connecticut, 28 May 1960), *physiology, history of medicine.*

The son of an ophthalmologist who assisted in the founding of the University of Minnesota, Fulton graduated from St. Paul High School at the age of sixteen. His studies at the University of Minnesota were interrupted by a period of military service in World War I. Upon discharge he transferred to Harvard University, where he received the Bachelor of Science degree, *magna cum laude,* in 1921. That year, as a Rhodes Scholar, Fulton went to Oxford and was admitted to Magdalen College. There he met C. S. Sherrington, and in 1923, when he received an appointment as Christopher Welsh Scholar, he had the privilege of working in Sherrington's laboratory. In the same year he married Lucia Pickering Wheatland. Fulton's return to Harvard for medical training brought him into contact with Harvey Cushing, then at the zenith of his neurosurgical career. He was so attracted by Cushing's clinical acumen and devotion that he spent a year working with him. This experience alerted the brilliant young physiologist to the possibility of using modern surgical techniques in the physiology laboratory in the analysis of the functions of the nervous system.

As a result, when he was appointed professor of physiology at Yale Medical School in 1929, Fulton organized the first primate laboratory for experimental physiology in America. His aim was to produce and analyze in higher primates such neurological syndromes as hemiplegia and ataxia. In operating rooms modeled after clinical neurosurgical theaters, he used his superb surgical skill on chimpanzees and orangutans, so that he might study the function of brains most closely resembling that of man. Stimulated by his enlightening analyses of spasticity and paralysis, would-be neurosurgeons, neurologists, and physiologists sought his laboratory for training and later became professors in universities in the United States and abroad. Perhaps Fulton's major contribution to this program was his enthusiasm, which stimulated his associates to explore—from the historical, bibliographical, or investigational viewpoint—the mysteries of the nervous system. Frequently this work was carried out more with his blessing than with his supervision, yet he was always eager to discuss and to integrate findings into the current neurological thinking, for his goal was to "aid those whose ultimate objective is the study of Clinical Medicine." In spite of these activities Fulton found time to write the textbook, *The Physiology of the Nervous System,* which was translated into French, German, Portuguese, Spanish, Japanese, and Russian; and he founded, with J. G. Dusser de Barenne, the *Journal of Neurophysiology,* which he nurtured through its early years.

Early in his career Fulton became interested in the motor system. His contacts with Sherrington stimulated him to study the mechanisms of neuromuscular transmission, which he reported in a monograph entitled *Muscular Contraction and the Reflex Control of Movement* (1926). His subsequent investigations concerned the cortical and subcortical influence on the spinal motor pathways. He used the techniques of stimulation and ablation to demonstrate the changing role of the cerebrum in the ascending phylogenetic scale. Fulton showed the differential influence on movement of the postcentral, central, and precentral cortex. He believed that lesions of the motor cortex produced essentially a flaccid paresis and that the addition of premotor lesions introduced a spastic element. He showed that cerebellar hemispheral and vermal lesions had varying effects upon tone and motor performance, and demonstrated that the cerebral motor cortex compensated for cerebellar lesions.

With the outbreak of World War II, the activities of the primate laboratory at New Haven became centered on physiology, particularly in its medical applications to aviation. Fulton had a decompression chamber built in his laboratory and devoted much time to important research in this field.

Throughout his physiological career Fulton maintained a deep interest in the history of medicine. Perhaps this stemmed from his contacts at Oxford with Sherrington and the posthumous influence of William Osler and may have been fostered further by his close friendship with Harvey Cushing. Fulton was particularly interested in the lives of the men who developed new concepts in medicine. There is evidence of this fascination in the many biographical and bibliographical sketches he wrote. The culmination of this activity was his life of Harvey Cushing, which presents a well-documented and, in view of Fulton's intimacy with Cushing, surprisingly well-balanced account.

With the aid of Cushing and Arnold Klebs, Fulton was able to establish a historical library for the Yale School of Medicine. It was to house the large collections of Cushing and Klebs and his own library. As his energies lessened because of impaired health, Fulton devoted more time to this hobby and less to the physiological laboratory. In 1951 he resigned as Sterling professor of physiology to become Sterling professor of the history of medicine and chairman of a newly created department of the history of medicine at Yale Medical School. He entered this new office with enthusiasm and developed graduate programs in the history of science and medicine.

The impact of Fulton's contributions was recognized at home and abroad. For his scientific assistance in the Allied war effort, he received honors from the governments of France, Belgium, Rumania, and Cuba. He was also awarded honorary degrees by nine American and foreign universities.

BIBLIOGRAPHY

I. ORIGINAL WORKS. Fulton's writings include *Muscular Contraction and the Reflex Control of Movement* (Baltimore, Md., 1926); *Selected Readings in the History of Physiology* (Springfield, Ill., 1930); *The Sign of Babinski. A Study of the Evolution of Cortical Dominance in Primates* (Springfield, Ill., 1932), written with A. D. Keller; *Physiology of the Nervous System* (New York, 1938); and *Harvey Cushing, A Biography* (Springfield, Ill., 1946).

II. SECONDARY LITERATURE. See H. E. Hoff, "The Laboratory of Physiology," in *Yale Journal of Biology and Medicine,* **28** (1955–1956), 165–167, with Fulton's bibliography appended, 168–190. Obituaries are in *Journal of Neurophysiology,* **23** (1960), 347–349; *Journal of Neurosurgery,* **17** (1960), 1119–1123; *Bulletin of the History of Medicine,* **35** (1961), 81–86; and *New England Journal of Medicine,* **262** (1960), 1340–1341.

A. EARL WALKER

FUNK, CASIMIR (*b.* Warsaw, Poland [then Russia], 23 February 1884; *d.* New York, N.Y., 20 November 1967), *biochemistry.*

Funk was the son of Jacques and Gustawa Zysan Funk. His father was a prominent dermatologist. In 1904 he received the Ph.D. in organic chemistry at the University of Bern, where he worked on the synthesis of stilbestrols. Following work at the Pasteur Institute, the Wiesbaden Municipal Hospital, and the University of Berlin, where he was an assistant to Emil Abderhalden, he took a post at the Lister Institute in London, where he was soon assigned to work on beriberi. In 1914 he married Denise Schneidesch of Brussels, by whom he had two children. In 1915 the Funks immigrated to New York, where he held several industrial and university positions. In 1920 he became a U.S. citizen. In 1923 the Rockefeller Foundation supported his return to Warsaw as chief of the biochemistry department in the State Institute of Hygiene, a post which he abandoned in 1927 because of political conditions in Poland. In Paris, from 1928 to 1939, Funk was consultant to a pharmaceutical firm and founder of the Casa Biochemica, a privately financed research institute. This was abandoned in the face of the German invasion, and Funk returned to New York as consultant to the U.S. Vitamin Corporation. From 1940 he was president of the Funk Foundation for Medical Research.

At the Lister Institute, Funk prepared a pyrimidine-related concentrate of rice polishings which was curative for beriberi in pigeons. In 1912 he proposed the term "vitamine" (for vital amine) for organic compounds responsible in trace amounts for the cure or prevention of beriberi, scurvy, rickets, and pellagra. His concentrates were primarily nicotinic acid (noneffective for beriberi but later shown by Elvehjem to be curative for pellagra) contaminated with the anti-beriberi vitamin. Besides his work on vitamins, Funk conducted extensive studies on animal hormones, particularly the male sex hormone, and on the biochemistry of cancer, ulcers, and diabetes. He theorized freely and saw close relationships between trace nutrients (vitamins and minerals), hormones, and enzymes. A number of substances developed in his laboratories were sold commercially in the pharmaceutical industry.

BIBLIOGRAPHY

I. ORIGINAL WORKS. A comprehensive but incomplete bibliography of Funk's publications appears in Harrow's biography (see below). His first paper on the nature of the anti-beriberi substance is "On the Chemical Nature of the Substance Which Cures Polyneuritis in Birds Induced by a Diet of Polished Rice," in *Journal of Physiology,* **43** (1911), 395–400. The paper introducing the term "vitamine" is "The Etiology of Deficiency Diseases," in *Journal of State Medicine,* **20** (1912), 341–368. *Die Aetiologie der Avitaminosen mit besonderer Berücksichtigung der physiologischen Bedeutung der Vitamine* (Wiesbaden, 1914) was printed in several German eds. and in an English trans. by H. E. Dubin as *The Vitamins* (Baltimore, 1922). See also his *L'histoire de la découverte des vitamines* (Paris, 1924).

II. SECONDARY LITERATURE. A popularized biography is Benjamin Harrow, *Casimir Funk, Pioneer in Vitamins and Hormones* (New York, 1955). An anonymous sketch of his life appeared in *Current Biography,* **6** (1945), 22–24; and an obituary notice appeared in the *New York Times* (21 Nov. 1967).

AARON J. IHDE

FUSS, NICOLAUS (or **Nikolai Ivanovich Fus**), (*b.* Basel, Switzerland, 30 January 1755; *d.* St. Petersburg, Russia, 4 January 1826), *mathematics, astronomy.*

Fuss was born into a Swiss family of modest means. His mathematical abilities, which manifested themselves quite early, attracted the attention of a number of prominent scholars, including Daniel Bernoulli, who in 1772 recommended him to Euler, then living in Russia, as a secretary. Fuss arrived in St. Petersburg at the age of seventeen and spent the rest of his life in Russia.

Fuss wrote his first papers, which had purely practical goals, under Euler's direct guidance. These were *Instruction détaillée pour porter les lunettes* . . . (1774) and *Éclaircissemens sur les établissemens publics en faveur tant des veuves* . . . (1776). The latter concerns problems of the insurance business.

In January 1776, Fuss was selected as a junior scientific assistant of the St. Petersburg Academy of Sciences; in February 1783, he became an academician in higher mathematics; and from September 1800 until his death he was the academy's permanent secretary.

The majority of Fuss's writings contain solutions to problems raised in Euler's works. They deal with several branches of mathematics (spherical geometry, trigonometry, the theory of series, the geometry of curves, the integration of differential equations) and with mechanics, astronomy, and geodesy. From 1774—the year of his first published paper—more than 100 of his articles appeared in the publications of the St. Petersburg Academy of Sciences.

Fuss's best papers deal with spherical geometry, the problems of which he worked out with the St. Petersburg academicians A. J. Lexell and F. T. Schubert. In his first paper on spherical geometry, which was published in *Nova acta Academiae scientiarum imperialis Petropolitanae* (1788), he gave solutions to three new problems concerning spherical triangles which are constructed on a given base, between two given great circles, and satisfy certain extremal conditions. In another article (1788) the characteristics of a spherical ellipse, i.e., of the geometrical locus of the vertexes of spherical triangles with a given base and a sum of two other sides, are studied in detail.

Fuss was also responsible for new solutions to a number of difficult problems in elementary geometry. These included Apollonius' problem of constructing a circle tangent to three given circles (1790) and Cramer's problem—which generalizes Pappus' problem—of inscribing a triangle inside a given circle, such that the sides of the triangle, or their extensions, pass through three given points (1783).

In differential geometry Fuss solved a number of problems concerning the determination of the properties of curves which are defined by certain relationships between the radius of curvature, the radius vector, and the length of an arc (1789). These papers partially bordered on so-called intrinsic geometry, which was developed into an independent mathematical discipline by Ernesto Cesàro and others at the end of the nineteenth century.

Fuss was an honorary member of the Berlin, Swedish, and Danish academies. In 1778 the Paris Academy of Sciences awarded him a prize for his astronomical paper "Recherche sur le dérangement d'une comète qui passe près d'une planète" (*Mémoires des savants étrangers,* **10** [1785]). In 1798 a prize was awarded to him by the Danish Society of Sciences for his paper *Versuch einer Theorie des Widerstandes zwei-und vierrädiger Wagen usw.* (Copenhagen, 1798).

Fuss also did much in the field of education. He taught for many years at the military and naval cadet academies. At the beginning of the nineteenth century he was active in the reform of the Russian national education system. He compiled a number of textbooks, including *Leçons de géométrie à l'usage du Corps impérial des cadets . . .* (St. Petersburg, 1798), *Nachalnye osnovania ploskoy trigonometrii, vysshey geometrii i differentsialnogo ischislenia* ("Foundations of Plane Trigonometry, Higher Geometry, and Differential and Integral Calculus," 3 vols., St. Petersburg, 1804), and *Nachalnye osnovania chistoy matematiki* ("Fundamentals of Pure Mathematics," 3 vols., St. Petersburg, 1810–1812). These textbooks show the influence of all of Euler's work, especially his *Vollständige Anleitung zur Algebra* (2 vols., St. Petersburg, 1770), which Fuss used as a model in compiling a handbook for the cadet corps and the first algebra textbook for Russian Gymnasiums.

BIBLIOGRAPHY

I. ORIGINAL WORKS. Fuss's writings include *Instruction détaillée pour porter les lunettes de toutes les différentes espèces au plus haut degré de perfection . . . tirée de la théorie dioptrique de M. Euler . . .* (St. Petersburg, 1774); *Éclaircissemens sur les établissemens publics en faveur tant des veuves que des morts, avec la déscription d'une nouvelle espece de tontine. . . . Calculés sous la direction de Mr. Léonard Euler par Mr. N[icolaus] F[uss]* (St. Petersburg, 1776); "Solutio problematis geometrici Pappi Alexandrini," in *Acta Academiae Scientiarum imperialis Petropolitanae,* **4,** pt. 1 (1783), 97–104; "Problematum quorundam sphaericorum solutio," in *Nova Acta Academiae Scientiarum imperialis Petropolitanae,* **2** (1788), 67–80; "De proprietatibus quibusdam ellipseos in superficie sphaerica descriptae," *ibid.,* **3** (1788), 90–99; "Solutio problematis ex methodo tangentium inversa," *ibid.,* **4** (1789), 104–128; "Recherches sur un problème de mécanique," *ibid.,* **6** (1790), 172–184.

A full bibliography of Fuss's mathematical works can be found in the following publications: *Matematika v izdaniakh Akademii Nauk (1728–1935). Bibliografichesky Ukazatel* ("Mathematics in the Publications of the Academy of Sciences [1728–1935]. Bibliographic Index"), compiled by O. V. Dinze and K. I. Shafranovsky (Moscow–Leningrad, 1936), see index; F. A. Brokgauz and I. A. Efron, eds., *Entsiklopedichesky slovar* ("Encyclopedic Dictionary"), XXXVI A (St. Petersburg, 1902), 913–914; Poggendorff, I, 822–823; and M. Cantor, *Vorlesungen über die Geschichte der Mathematik,* 3rd ed., IV (Leipzig, 1913), see index.

II. SECONDARY LITERATURE. On Fuss or his work, see V. V. Bobynin, "Fuss, Nikolai," in F. A. Brokgauz and I. A. Efron, eds., *Entsiklopedichesky slovar,* XXXVI A (St. Petersburg, 1902), 913–914; V. I. Lysenko, "O rabotakh peterburgskikh akademikov A. I. Lekselya, N. I. Fussa i F. I. Shuberta po sfericheskoy geometrii i sfericheskoy trigonometrii" ("On the Works of the Petersburg Academicians A. J. Lexell, N. I. Fuss, and F. T. Schubert in Spherical Geometry and Spherical Trigonometry"), in *Trudy Instituta istorii estestvoznaniya i tekhniki. Akademiya nauk SSSR,* **34** (1960), 384–414, which examines several of Fuss's unpublished compositions, which are preserved in the archives of the Soviet Academy of Sciences; and "Iz istorii pervoy peterburgskoy matematicheskoy shkoly" ("From the History of the First Petersburg School of Mathematics"), *ibid.,* **43** (1961), 182–205.

See also A. P. Youschkevitch, "Matematika" ("Mathematics"), in *Istoria estestvoznania v Rossii* ("History of Natural Science in Russia"), I, pt. 1 (Moscow, 1957), 215–272; and "Matematika i mekhanika" ("Mathematics and Mechanics"), in *Istoria Akademii nauk SSSR* ("History of the USSR Academy of Sciences"), I (Moscow–Leningrad, 1958), 350–352.

A. I. VOLODARSKY

FYODOROV (or **FEDOROV**), **EVGRAF STEPANOVICH** (*b.* Orenburg, Russia [now Chkalov, U.S.S.R.], 22 December 1853; *d.* Petrograd [now Leningrad], U.S.S.R., 21 May 1919), *crystallography, geometry, petrography, mineralogy, geology.*

His father, Stepan Ivanovich Fyodorov, came from a peasant family and was a major-general in the Engineer Corps. He was noted for the sharpness of his disposition, which he evidently passed on to his son.

Fyodorov's mother, Yulia Gerasimovna Botvinko, the daughter of a procurator in Vilna, was a progressive and cultured woman. She gave her son the elements of a musical education and, in particular, imparted a love of reading and accustomed him to steady work and discipline. Fyodorov later said that he was wholly indebted to his mother for his exceptional capacity for work. She made him knit large tablecloths with intricate figures, which probably developed his feeling for symmetry.

Fyodorov's mathematical abilities appeared very

early; when he was five, he had already mastered the rules of arithmetic. At the age of seven he studied with fascination and finished in two days a textbook of elementary geometry. In his own words, the content of the first pages of the text evoked "a resonance in my psyche, so that I was literally carried away." At the age of ten Fyodorov entered the second class of Annensky College. After the death of his father in 1866, the straitened circumstances of his family forced him to transfer to the military Gymnasium, which he could attend at state expense. Here he joined a small group of friends who were studying natural science and philosophy intensively. At the same time Fyodorov independently immersed himself in the mathematical disciplines, which for him were always invested with an aura of special beauty. In 1869 he transferred to the Petersburg Military Engineering School. There he became an active member of an illegal group devoted to self-education, in which, under the influence of the literary critic, Dmitry Pisarev, the works of the materialist natural scientists were studied.

In 1872 Fyodorov graduated and went to Kiev with the rank of second lieutenant in a combat engineering battalion; the following year he returned to St. Petersburg, and in 1874 he retired completely from military service. As a result of his enthusiasm for natural sciences he became a free auditor at the Military Medical and Surgical Academy. Having passed the necessary examinations, Fyodorov entered the second-year course of the Technological Institute, where he concentrated on physics and chemistry. All his thoughts and interests were already directed toward theoretical mathematics, particularly geometry; this made possible the completion of his important monograph, *Nachala uchenia o figurakh,* which he had begun at the age of sixteen. In it he touched on symmetry and the theory of crystal structure and set forth the principles of contemporary theoretical crystallography. His work on the theory of mathematical polyhedrons brought Fyodorov to questions relating to natural polyhedrons—the crystals of minerals—and to geometric mineralogy.

Because of his enthusiasm for crystallography Fyodorov chose the specialty closest to this science and in 1880, at the age of twenty-seven, he entered the third-year course at the Mining Institute, where a general course in crystallography and related mineralogy was taught. After graduating from the Mining Institute in 1883 (his name was placed at the top of the list carved in marble), Fyodorov joined a Mining Department expedition to investigate the northern Urals.

Fyodorov married Ludmila Vasilievna Panyutina, a vivacious and purposeful girl who had come from Kungur in the Urals to study medicine in St. Petersburg. She selflessly helped him in his scientific and revolutionary work. Fyodorov's son, Evgraf Evgrafovich, was a specialist in climatology and later a corresponding member of the Soviet Academy of Sciences.

In childhood Fyodorov had been sickly; in his middle years he was extremely robust. He always worked a great deal and was extraordinarily precise in carrying out the life plan he had set for himself. He went on expeditions to the Urals and other regions, distinguishing himself on them because of his great endurance. Fyodorov traveled under conditions of great privation and worked to the point of exhaustion; on expeditions he was sunburned and thin but cheerful and energetic. In his old age he was often ill.

In 1896 Fyodorov was elected a member of the Bavarian Academy of Sciences and in 1901 an adjunct of the Petersburg Academy of Sciences; but, not having received support for his demands for the creation of a mineralogical institute, he withdrew from the Petersburg academy in 1905. He was elected a member of the Soviet Academy of Sciences in 1919, when an institute in which mineralogy occupied an important place was organized.

Revolutionary activity gave meaning to Fyodorov's life. In 1876 he became a member of the populist Land and Freedom party; the following year he was commissioned to set up connections with revolutionary organizations in France, Belgium, and Germany, a task which he handled well. His apartment in Petersburg contained an underground revolutionary press.

Fyodorov's scientific and literary work is distinguished for its richness and extraordinary variety. Most numerous among his more than 500 scientific published works are those on crystallography, followed by geometry, mineralogy, petrography, geology, history of science, and philosophy.

Fyodorov devoted forty-three publications to mineralogy, but the only one to deal with the field as a whole is *Kritichesky peresmotr form mineralnogo tsarstva* (1903), in which the morphology of minerals is examined relative to crystal structure theory. Fyodorov exerted a very strong influence on the development of mineralogy and opened a new stage in it. He also contributed to the accumulation of significant new factual material and helped to change the methodological approach to the study of minerals. He used the analytical approach and attempted to understand natural processes and phenomena by starting from more general mathematical and physical-chemical principles. Through his work Fyodorov laid the foundations for the analytical period in the development of mineralogy.

The foundation of all of Fyodorov's scientific work was geometry. Among his first works was the monograph *Nachala uchenia o figurakh,* and his last article treated questions of the new geometry. Geometrical research led Fyodorov to the brilliant derivation of the 230 space groups—the symmetry groups governing the periodic distribution within crystalline matter. The derivation is the foundation of contemporary mineralogy and the basis of the atomic structure of minerals. His first work was an exposition of all those parts of the theory of figures which constitute the basis of contemporary crystallography.

The period of the first geometrical investigations and publications was crowned by the classic work *Simmetria pravilnykh sistem figur* (1890), which contained the first deduction of the 230 space groups. The publication in 1891 of Schoenflies' book with his derivation of the 230 space groups prompted Fyodorov to publish several articles in the *Zeitschrift für Kristallographie.* He compared his results with those of Schoenflies and made a series of essential corrections and notations. In these articles he gave a strict mathematical definition of thirty-two point groups for six crystallographic systems, which still retain their significance. At the basis of this classification he placed a number of single and symmetrically equivalent axes in crystals. This classification led to the working out of a new nomenclature of systems and point group symmetries, accepted throughout the world and known as Fyodorov-Groth nomenclature. For these accomplishments, and at the insistence of such eminent crystallographers as Groth and L. Sohncke, Fyodorov was elected a member of the Bavarian Academy of Sciences in 1896. Fyodorov himself considered these works as belonging to the theory of crystallography, specifically to geometric mineralogy. Fyodorov's group classifications are, with the aid of X-ray diffraction, at the base of modern mineralogical determinations of the atomic structure of minerals.

In the category of physical crystallography are Fyodorov's works that explain a universal method of optical research, which played an important role in mineralogy and petrography. In 1889, at a session of the Mineralogical Society, Fyodorov reported on his projected two-circle optical goniometer for the measurement of all angles in crystals with a single setting of the specimen. It differed from previous goniometers, which had only a single axis of rotation. (A two-circle goniometer had been proposed earlier by W. H. Miller.) This produced a revolution in the method of investigating minerals. Following Fyodorov and using his idea, Victor Goldschmidt, S. Czapski, Y. Flint, and others designed two-circle goniometers. Crystallographers and mineralogists throughout the world began to work exclusively with these instruments.

In 1891 Fyodorov proposed to the Geological Committee the construction of a universal stage for the petrographic microscope that would locate the specimen at the center of two glass hemispheres. In essence this method is crystallographic, but its application is to mineralogy and petrography. With what became known as the Fyodorov method the optical constants of minerals could be established and, without resorting to chemical analysis, the composition of the isomorphic lime-soda feldspars (plagioclases) was determined, as were those of other minerals.

Of all the instruments constructed and developed by Fyodorov, the universal stage (also called U-stage or Fyodorov table) has enjoyed the greatest popularity. Special courses in its use are given to students of mineralogy and petrography. Handbooks and special studies on the Fyodorov method have been published by Russian and foreign authors: S. N. Nikitin, M. A. Usov, V. S. Sobolev, L. Duparc, M. Reinhard, M. Berek, and others. The Fyodorov method has allowed researchers to carry out quickly the optical study of plagioclases, pyroxenes, and other minerals, using thin sections of rock. The present Fyodorov table is a refined instrument with five axes of rotation, convenient for the study of minerals in any cross section.

In describing Fyodorov's activity in petrography and mineralogy, it must be noted that he himself never drew a sharp boundary between these sciences. He considered that all the physical, mathematical, and natural sciences were used in them. He also believed that in the cycle of mineralogical and geological sciences one used the totality of knowledge of the earth—which, in his view, is located in infinite, starry space and interacts with it. These ideas are only now being worked out in detail by his followers in the fields of astrogeology and planetology.

Fyodorov gave much attention to the chemical composition of rocks and their graphic representation, and he introduced symbols of chemical composition, petrographic nomenclature, and rock classification. He conducted important research on the northern Urals, the Bogoslovsky (now Sverdlovsk) district, the coast of the White Sea, the Caucasus, and Kazakhstan. He was one of the first to show the great importance of the apatite resources lying in the depths of the Soviet north (1909).

In one of his first scientific works, the tract "Perfektsionizm" (1906), Fyodorov showed that he was a convinced materialist. Starting from the materialistic principle that natural conditions are in essence conditions of eternal change, he attacked authors who teach the theory of stability and equilibrium in nature. The essence of evolution, in his opinion, is not in the tendency toward a higher order, stability, and equilibrium of organisms but in their life movement. Criticizing the outlook of Herbert Spencer and other partisans of equilibrium in nature, Fyodorov showed that the main effect of such points of view when applied to the evolution of natural history was to give attention to the least changeable. But the creators of these systems of philosophy systematically failed to take into account the fact that equilibrium is attained only at the moment of death. As long as life is active, changing forms are developing. For this reason Fyodorov considered deeply erroneous the introduction into natural philosophy of the concept of the constant and the stable as the supreme mission of life. He asserted that life never finally achieves anything but eternally strives to achieve. It was in this that the true philosophy of nature lay for him.

Later, in published statements, Fyodorov defended science against positivism, which, in his view, reduces the significance of mind to simply a mold for the more convenient organization of material gathered by experimentation. Attacking the attempts of positivists and followers of Ernst Mach to disregard the atomic theory, Fyodorov wished to remove all metaphysical errors from contemporary science and to be led by the clear guide of atomic theory, which has produced so many valuable and stunning developments in contemporary theory.

Fyodorov's work harmoniously combined the achievements of varied fields of science. Mathematics was the basis of Fyodorov's theory of structure and symmetry of crystals. He was able to combine the methods drawn from mathematical analysis with the older empirical laws of crystallography and the methods of crystallographic research with descriptive mineralogy; he introduced principles of geometry into petrography; he combined chemistry with crystallography, thus creating crystal chemistry; and he introduced the principles of the new geometry into mine surveying. Fyodorov knew how to make generalizations and find simple solutions in the analysis of natural phenomena. His "simple" theodolite method in crystallography and petrography played a role in the history of these sciences no less than the most profound generalizations and theoretical achievements. Fyodorov presented all his scientific conclusions in mathematical form. He asserted that the crown of man's conscious activity, of man's intelligence, was the solution of the questions facing him by means of mathematical analysis.

The enormous *Tsarstvo kristallov* ("The Crystal Kingdom," 1920) was the fruit of forty years of work by Fyodorov and his colleagues. In it he noted that strongly developed sciences satisfy the spiritual needs of part of mankind and at the same time provide great power to direct the active forces of nature for man's use, thus forcing nature to serve man to a great degree. Lately it has become clear that special sciences can master nature in some respects and in certain areas of natural phenomena, directing them according to the wishes of man.

BIBLIOGRAPHY

I. ORIGINAL WORKS. Fyodorov's writings include "Teodolitny metod v mineralogii i petrografii" ("The Theodolitic Method in Mineralogy and Petrography"), in *Trudy Geologicheskogo komiteta,* **10,** no. 2 (1893), 1-191; "Iz itogov tridtsatipyatiletia" ("From the Results of Thirty-five Years"), in *Rech i otchet, chitannye v godichnom sobranii Moskovskogo selskokhozyaistvennago instituta, 26 sentyabrya 1904 g.* ("Speech and Report Given at the Annual Meeting of the Moscow Agricultural Institute, 26 September 1904," Moscow, 1904), pp. 1-15; "Perfektsionizm" ("Perfectionism"), in *Izvestiya S. Peterburgskoi biologicheskoi laboratorii,* **8,** pt. 1 (1906), 25-65; *ibid.,* pt. 2, 9-67; "Beloe more kak istochnik materiala dlya selskokhozyaystvennoy kultury" ("The White Sea as a Source of Material for Agriculture"), in *Izvestiya Moskovskago selskokhozyaistvennago instituta,* **1** (1908), 94-97; "Iz rezultatov poezdki v Bogoslovsky okrug letom 1911 g." ("From the Results of a Trip to the Bogoslov District in the Summer of 1911"), in *Zapiski Gornago instituta Imperatritsy Ekateriny II,* **3** (1912), 340-348; "Tsarstvo kristallov" ("The Crystal Kingdom"), in *Zapiski Rossiiskoi akademii nauk,* **36** (1920); and *Nachala uchenia o figurakh* ("Principles of the Theory of Figures," Leningrad, 1953).

II. SECONDARY LITERATURE. On Fyodorov or his work, see O. M. Ansheles, "100-letie so dnya rozhdenia velikogo russkogo uchenogo E. S. Fyodorova" ("Centenary of the Birth of the Great Russian Scientist E. S. Fyodorov"), in *Vestnik Leningradskogo gosudarstvennogo universiteta,* no. 1 (1954), 223-226; N. V. Belov, *Chetyrnadtsat reshetok Brave i 230 prostranstvennykh grupp simmetrii* ("The Fourteen Bravais Lattices and the 230 Three-Dimensional Groups of Symmetries," Moscow-Leningrad, 1962); N. V. Belov and I. I. Shafranovsky, "Rol E. S. Fyodorov v predistorii rentgenostrukturnoy kristallografii . . ." ("E. S. Fyodorov's Role in the Early Development of X-ray Structural Crystallography . . ."), in *Zapiski Vsesoyuznogo mineralogicheskogo obshchestva,* pt. 91, no. 4 (1962),

465–471; G. B. Boky, "O zakone raspolozhenia atomov v kristallakh" ("On the Law of Arrangement of Atoms in Crystals"), no. 5 in the collection *Kristallografiya* (Leningrad, 1956), pp. 25–36; F. Y. Levinson-Lessing, "Neskolko yubileynykh dat v petrografii (v tom chisle 'Sorokopyatiletie tak nazyvaemogo universalnogo, ili fyodorovskogo metoda v petrografii')" ("Several Jubilee Dates in Petrography [Including the 'Fortieth Anniversary of the So-called Universal or Fyodorov Method in Petrography']"), in *Priroda* (1938), no. 6, 137–144; I. I. Shafranovsky, *Evgraf Stepanovich Fyodorov* (Moscow-Leningrad, 1963); and N. M. Sokolov, "O mirovozzrenii E. S. Fyodorov" ("On the World View of E. S. Fyodorov"), in *Kristallografiya,* no. 5 (Leningrad, 1956), pp. 5–23.

A. MENIAILOV

GABB, WILLIAM MORE (*b.* Philadelphia, Pennsylvania, 20 January 1839; *d.* Philadelphia, 30 May 1878), *geology.*

A native of the leading American scientific city of the time, Gabb became acquainted early with the Academy of Natural Sciences of Philadelphia. His father, Joseph Gabb, was a salesman and may have died when the boy was fourteen, at which time his mother, known only as J. H. More Gabb, became a milliner. Gabb did well at Central High School in Philadelphia, where Martin Hans Boyé taught natural science. The boy was already collecting minerals and shells. He became a good friend of George H. Horn, later a paleontologist and entomologist, and George W. Tryon, Jr., later a conchologist.

In 1857 Gabb chose geology as his field and sought aid from the noted James Hall, whose pupil and assistant he became. In 1860 he spent some time at the Philadelphia Academy and then briefly joined the enthusiastic scientific group around Spencer F. Baird at the Smithsonian Institution. Recommended as a foremost authority on Cretaceous fossils, Gabb in 1862 joined the California State Geological Survey under Josiah Dwight Whitney. For six years he traveled throughout much of California and beyond, from Vancouver Island to the tip of Baja California. From 1869 to 1872 he conducted a topographical and geologic survey of Santo Domingo for its government and, from 1873 to 1876, a similar survey in Costa Rica. Tuberculosis ended his career.

Gabb was elected a member and curator of paleontology of the California Academy of Sciences in 1862 and was elected to the National Academy of Sciences in 1876.

A very competent field man, Gabb was a diligent worker and was early recognized as the authority on New World invertebrate Cretaceous fossils, of which he described many. As part of the California survey, he wrote monographs on the state's Upper Mesozoic and Tertiary rocks. Although his later paleontological work was unfinished, he contributed much on the geography of Baja California, Santo Domingo, and Costa Rica. On his extensive journeys he collected many geologic, ethnologic, entomological, and other natural history specimens for the Smithsonian Institution and other museums.

BIBLIOGRAPHY

I. ORIGINAL WORKS. Gabb's publications, listed in Dall's memoir (cited below), include a number of taxonomic papers and several valuable areal reports: "Description of the Triassic Fossils of California and the Adjacent Territories," in *Geological Survey of California, Paleontology,* I (Philadelphia, 1864), 17–35; "Description of the Cretaceous Fossils," *ibid.,* 55–243; "Cretaceous and Tertiary Fossils," *ibid.,* II (1869); and also "On the Topography and Geology of Santo Domingo," in *Transactions of the American Philosophical Society,* **15** (1872), 49–260. Several papers on the geology of Costa Rica appeared in *American Journal of Science* (1874, 1875). "Notes on the Geology of Lower California," in *Geological Survey of California,* II (Philadelphia, 1882), app. 137–148, was published after Gabb's death. George W. Tryon, Jr., completed several of Gabb's taxonomic papers after Gabb died.

II. SECONDARY LITERATURE. Gabb's early life is incompletely known. The available material and a summary of his professional life was presented by William H. Dall, in *Biographical Memoirs. National Academy of Sciences,* **6** (1909), 347–361. E. O. Essig's *History of Entomology* (New York, 1965), p. 638, refers briefly to Gabb's insect collecting.

ELIZABETH NOBLE SHOR

GABRIEL, SIEGMUND (*b.* Berlin, Germany, 7 November 1851; *d.* Berlin, 22 March 1924), *chemistry.*

The son of Aron and Golde Pollnow Gabriel, Siegmund Gabriel is noted for his method of preparing primary amines and his studies of the heterocyclic compounds of nitrogen. During his last year of secondary school he developed an interest in science, having been particularly impressed by *Kurzes Lehrbuch der Chemie,* Adolph F. L. Strecker's translation and revision of Henri Victor Regnault's popular textbook. Deciding on chemistry as a career, Gabriel enrolled at the University of Berlin and attended lectures in organic chemistry by August Wilhelm von Hofmann and in inorganic chemistry by E. A. Schneider for two semesters beginning in 1871.

At Heidelberg in the spring of 1872 Gabriel studied under Robert Wilhelm Bunsen. A year of military service interrupted his studies, but he returned to Heidelberg in the autumn of 1873 and passed his doctoral examinations *summa cum laude* in 1874. Gabriel received the degree without writing a dissertation, since at that time Bunsen did not customarily require it.

Returning to the University of Berlin, where he spent his entire career, Gabriel assisted Hofmann in the inorganic section of the university laboratory. Since his own interests coincided with those of Hofmann, Gabriel spent his years as assistant on problems in organic chemistry and subsequently devoted his life to that branch of chemistry.

In 1883 Gabriel married Anna Fraenkel; they had two sons, both of whom chose medical careers. In 1886 Gabriel was appointed professor of chemistry at Berlin. When Hofmann died in 1892, Emil Fischer, then at Würzburg, was chosen to succeed him. Fischer and Gabriel became great friends, and in the dedication of his book *Aus meinem Leben* Fischer praised Gabriel for his scientific capability, his technical knowledge, and his absolute reliability.

Gabriel's first research at Berlin, begun in 1876, was on the halogenation of aromatic azo compounds, and in the following year he began a collaboration with the American chemist Arthur Michael on the condensation products of phthalic anhydride. As a result of this work Gabriel elucidated the structure of phthalylacetic acid in 1884 and in 1886 first synthesized isoquinoline, which had been discovered in coal tar in 1885 by Sebastian Hoogewerff and W. A. van Dorp, who also established its constitution. From this time on, his research was concerned almost exclusively with the heterocyclic compounds of nitrogen.

One of Gabriel's most significant contributions to organic chemistry was made in 1887, when he announced the important general method for synthesizing pure primary amines, involving the reaction of potassium phthalimide with an alkyl halide, followed by hydrolysis. This reaction, called the Gabriel synthesis, was adapted by Gabriel in 1889 to a procedure for the preparation of amino acids.

In 1891 Gabriel synthesized pyrrolidine from 1-amino-4-chlorobutane, and in 1892, using the same procedure, he prepared piperidine from 1-amino-5-chloropentane. From these heterocyclic rings containing one nitrogen atom, Gabriel turned to a thorough investigation of the diazines; he was the first to prepare phthalazine in 1893 and, with his student James Colman, pyrimidine in 1899. In 1900 he devised a simpler method for obtaining pyrimidine using barbituric acid, and in 1903 he first prepared quinazoline. Gabriel also investigated oxazole, thiazole, and their derivatives. He retired from his professorship of organic chemistry at Berlin in 1921 and died in 1924 after a short illness.

BIBLIOGRAPHY

Some of Gabriel's most important contributions were first reported in the following papers: "Uber die Constitution der Phtalylessigsäure." in *Berichte der Deutschen chemischen Gesellschaft,* **17** (1884), 2521–2527; "Synthese des Isochinolins," *ibid.,* **19** (1886), 1653–1656; "Zur Kenntniss des Isochinolins und seiner Derivate," *ibid.,* 2354–2363; "Ueber eine Darstellungsweise primärer Amine aus den entsprechenden Halogenverbindungen," *ibid.,* **20** (1887), 2224–2236; "Ueber das Phtalazin," *ibid.,* **26** (1893), 2210–2216, written with Georg Pinkus; "Ueber das Pyrimidin," *ibid.,* **32** (1899), 1525–1538, 4083, and **33** (1900), 4189, written with James Colman; "Pyrimidin aus Barbitursäure," *ibid.,* **33** (1900), 3666–3668; and "Synthese des Chinazolin," *ibid.,* **36** (1903), 800–813.

For a short biography and review of Gabriel's research, see James Colman and August Albert, "Siegmund Gabriel," in *Berichte der Deutschen chemischen Gesellschaft,* **59A** (1926), 7–26.

A. ALBERT BAKER, JR.

GADOLIN, JOHAN (*b.* Åbo [now Turku], Finland, 5 June 1760; *d.* Wirmo, Finland, 15 August 1852), *chemistry, mineralogy.*

Gadolin's father, Jacob, was professor of physics and theology at the Finnish University at Åbo and later became bishop of Åbo. His maternal grandfather, Johan Brovallius, was also professor of physics at Åbo and a friend of Linnaeus. Gadolin studied chemistry under Pehr Adrian Gadd, the first professor of chemistry at Åbo. He then spent four years at Uppsala studying with Torbern Bergman, during which time he began his work on mineralogy and specific heats.

Upon Bergman's death in 1784 Gadolin became a candidate for the chair of chemistry at Uppsala, but Johann Afzelius, adjunct at Uppsala, was selected. After having become extraordinary professor at Åbo in 1785, Gadolin had sufficient time to travel in Europe and become well acquainted with Richard Kirwan, Adair Crawford, and Lorenz F. F. von Crell, to whose *Chemische Annalen* he later contributed frequently. When Gadd died in 1797, Gadolin became ordinary professor, a post which he held until his retirement in 1822. The great fire of 1827 destroyed his extensive mineral collection (and much of Åbo) and ended his scientific career. He retired to the country, where he died at the age of ninety-two.

As an educator Gadolin was significant for opening his chemical laboratory to students, preceding by many years Liebig's famous laboratory at Giessen. His *Inleding till chemien* (1798) was the first Swedish-language textbook written in the spirit of the new combustion theory.

Although he accepted the phlogiston theory early

in his career, Gadolin attempted to understand Lavoisier's ideas. In a paper published in 1788 he tried to define phlogiston and admitted that the French explanation of combustion was superior to some phlogiston theories, but for a long time he was not wholly converted. His lectures always made use of the new chemistry, and he eventually became the spokesman in Scandinavia for Lavoisier's nomenclature and combustion theory, often encountering Berzelius' opposition. Despite his willingness to accept these new ideas, he never made use of the work of Dalton, Davy, or Gay-Lussac.

Gadolin's chemical contributions cover a large area. By 1784 he had published two important papers on specific heat, and in 1791 he published one on the latent heat of steam. Having established the composition of Prussian blue, he made a significant contribution to analytical chemistry by suggesting the ferricyanide titration of ferrous iron (precipitating the ferrous ion quantitatively as ferro ferricyanide). This volumetric analysis preceded Gay-Lussac's classic work by forty years.

Best remembered for his studies in mineralogy, Gadolin in 1792–1793 analyzed a new black mineral (later named gadolinite) from Ytterby, Sweden, and discovered in it a new earth, yttria, later shown to contain several elements of the rare-earth series. In 1886 Jean Charles Marignac isolated a new rare-earth element and named it gadolinium, the first element named for a person.

Honored by his contemporaries with memberships in several European scientific societies, Gadolin declined a call to succeed J. F. Gmelin to a full professorship at Göttingen in 1804. Interested in politics, he was influential in bringing about the political separation of Finland from Sweden.

BIBLIOGRAPHY

I. ORIGINAL WORKS. Gadolin's most significant publication is his textbook *Inleding till chemien* (Åbo, 1798). A lengthy essay on chemical affinity is *Dissertatio academia historiam doctrinae de affinatibus chemicis exhibens* (Åbo, 1815). His classification of minerals was published as *Systema fossilium analysibus chemicis examinatorum secundum partium constitutivarum rationes ordinatorium* (Berlin, 1825). Reports of many of Gadolin's chemical investigations appeared in German in Crell's *Chemische Annalen für die Freunde der Naturlehre, Arzneygelahrheit, Haushaltungskeit und Manufacturen.* Among the more important publications in the *Chemische Annalen* are his theory of combustion (1788), **1,** 1–17; and his discovery of the new earth, yttria (1796), **1,** 313.

II. SECONDARY LITERATURE. Numerous short biographical sketches of Gadolin exist in the literature. Among the most useful and accessible are Vieno Ojala and Ernest R. Schierz, "Finnish Chemists," in *Journal of Chemical Education,* **14** (1937), 161–165; and T. E. T. [probably Thomas Edward Thorpe], "Johan Gadolin," in *Nature,* **86** (1911), 48–49. A book-length biography in Finnish is Robert A. A. Tigerstedt, *Johan Gadolin. Ett bidrag till de induktiva vetenskapernas historia in Finland* (Helsinki, 1877). A collection of German and Latin essays in honor of Gadolin is Edvard Hjelt and Robert Tigerstedt, eds., *Johan Gadolin 1760–1852, in memoriam*. . . (Leipzig, 1910).

SHELDON J. KOPPERL

GAERTNER, JOSEPH (*b.* Calw, Germany, 12 March 1732; *d.* Calw, 14 June 1791), *botany.*

While Gaertner is best known for his *De fructibus et seminibus plantarum* (1788–1792), which describes the fruits and seeds of 1,050 genera, Julius von Sachs considered his "valuable reflections" on sexuality in plants to be of great theoretical significance.

Gaertner was the son of a court physician. Orphaned at a young age, he was first destined for the church, then law, and finally medicine. In 1751 he entered the University of Tübingen, where he came under the influence of Haller. He received the M.D. degree from Tübingen in 1753 with his dissertation "De viis urinae ordinariis et extraordinariis," but he did not practice medicine. After visiting several cities in Italy, he arrived in Lyons, then spent six months each in Montpellier and Paris, and tarried in England for nearly a year (1755), pursuing mathematics, optics, and mechanics. During 1759 he attended with enthusiasm Adrian van Royen's botany lectures at Leiden and he soon commenced marine investigations. Peter Pallas published Gaertner's studies on zoophytes, and his "Account of the Urtica marina in a Letter to Mr. Peter Collinson" appeared in the *Philosophical Transactions of the Royal Society,* **52** (1761), 75–85.

Gaertner became in turn professor of anatomy at Tübingen, professor of botany at St. Petersburg, cataloger of the empress' cabinet of curiosities, and botanical traveler with Count Grigory Orlov in the Ukraine, where he discovered many undescribed plants before returning to Calw in 1770. Thenceforth he gave his attention to carpology.

Learning that Sir Joseph Banks, Daniel Solander, and the Forsters had brought back rich collections of plants and seeds from Cook's voyages around the world, Gaertner hastened to England in the spring of 1778, seeking to add these novelties to his survey of fruits and seeds. He found Banks openhanded in granting their study and in gifts of duplicates for use in *De fructibus.* At Rotterdam, Gaertner met Karl Thunberg, who, recently returned from South Africa and Japan, also generously assisted him. From these collections, as well as those found in botanical gardens at Leiden, Amsterdam, and Lyons, and pharmaceutical plots at Stuttgart, Gaertner proposed fifty new genera.

De fructibus was issued in five parts, the first late in 1788 (cf. Frans Stafleu, "Dates of Botanical Publications, 1788–1792" [1963], and *Taxonomic Literature* [1967]). The *Supplementum carpologicae,* issued in three parts (1805–1807), was published in Leipzig by Gaertner's son, Karl Friedrich. Volume I of *De fructibus* was fittingly dedicated to Banks and illustrated from Gaertner's sketches with 180 copperplate drawings by Gaertner and by Hermann Jakob Tyroff. The work, however, went almost unnoticed in Germany where only 200 copies were sold in three years, and this commercial failure even threatened its completion. Yet in France, it met with appreciation, coming as it did at the same time as Jussieu's *Genera plantarum* (Paris, 1789).

Gaertner demonstrated that the spores of Cryptogamia (which he called gemmae), being without an embryo yet capable of germination, were essentially different from seeds of Phanerogamia, which contain an embryo. He recognized that the early stages of an organ presented more significant information on origins and affinities of different forms than did a comparison of their mature condition. He established terminology, heretofore vague, for fruits and seeds. He distinguished between the pericarp, however dry and anomalous it may be in one-seeded fruits, and integuments. Further, he characterized endosperm (calling it albumen) as distinct from cotyledons, which he correctly interpreted as appendages of the embryo. His scheme of classifying fruits and seeds contributed importantly to Jussieu's emerging natural system of plant families.

The Achilles' heel of Gaertner's interpretation was his concept of what he called the vitellus, a term he used to embrace such diverse structures as the scutellum of grasses, cotyledons of *Zamia,* and the spore contents of various Cryptogamia. In the course of examining the reproductive structures of *Spirogyra,* Gaertner witnessed zygospore formation. This observation led Johann Hedwig to suggest, and Jean Vaucher subsequently to assert, that true sexuality occurs in algae.

BIBLIOGRAPHY

I. ORIGINAL WORKS. For bibliographic details of *De fructibus* see Frans Stafleu, "Dates of Botanical Publications, 1788–1792," in *Taxon,* **12** (1963), 60–62, and *Taxonomic Literature* (1967), 162–163. Gaertner's minor papers are listed in Jonas Dryander, *Catalogus bibliothecae historico-naturalis Josephi Banks* (London, 1798–1800).

II. SECONDARY LITERATURE. For interpretative commentary on Gaertner, see Julius von Sachs, *Geschichte der Botanik vom 16. Jahrhundert bis 1860* (Munich, 1875), 132–135, 222, 447; English trans. by H. E. F. Garnsey and Isaac Bayley Balfour (Oxford, 1906), 122–125, 207, 413; French trans. by Henry de Varigny (Paris, 1892), 128–132, 216, 428.

The fundamental biographical account is Joseph P. F. Deleuze, "Notice sur la vie et les ouvrages de Gaertner," in *Annales du muséum national d'histoire naturelle,* **1** (an XI [1802]), 207–233, to which a few details are added by François Chaumeton, in *Biographie universelle,* XV (1856), 342–343; and Paul Ascherson, in *Allgemeine deutsche Biographie,* VIII (1878), 377–380. A calendar of eight letters from Gaertner is given by Warren R. Dawson, *The Banks Letters* (London, 1958), 351–352. Stafleu (cited above) presents different evidence from George K. Brizicky, "Dates of Publication of Gaertner's *De Fructibus et Seminibus Plantarum,*" in *Rhodora,* **62** (1960), 81–84.

JOSEPH EWAN

GAERTNER, KARL FRIEDRICH VON (*b.* Göppingen, Germany, 1 May 1772; *d.* Calw, Württemberg, Germany, 1 September 1850), *botany.*

Gaertner was born out of wedlock to Joseph Gaertner and Maria Rebekka Mütschelin. His father, who never married, officially recognized him in 1773 and legally adopted him in 1787. Gaertner's paternal ancestors were apothecaries and physicians; his father also acquired a reputation as a botanist. Appointed professor of anatomy at Tübingen in 1761, Joseph Gaertner became professor of botany and natural history at St. Petersburg in 1768; until 1770 he was also in charge of the botanical garden and the natural history museum there. In addition, he undertook several scientific expeditions to various regions of Europe. His masterpiece was *De fructibus et seminibus plantarum* (1788–1792). The studies in comparative plant anatomy presented in this work contributed to the development of carpology and of the natural system of classification.

Gaertner spent his youth in his father's home. He attended the local Latin school and then, from October 1787, the lower convent school in Bebenhausen. In 1791, after a two-year apprenticeship at the royal pharmacy in Stuttgart, he began to study medicine at the Hohe Karlsschule. His interest in chemistry having been awakened by K. F. Kielmeyer, he went in 1794 to Johann Göttling's laboratory in Jena, where he also heard Christoph Hufeland's lectures. The following year he studied at Göttingen. In 1796, after earning his medical degree at Tübingen, he set

GAERTNER

up practice in Calw. He traveled to Paris, England, and Holland in 1802 and met the leading natural scientists of the period, many of whom had been friends of his father, including Georges Cuvier, A. L. de Jussieu, J. P. F. Deleuze, R. L. Desfontaines, Joseph Banks, and K. P. Thunberg.

In 1803 at Calw, Gaertner married Christine Sybille Wagner, the daughter of a wholesale merchant. His descendants include the political economist Gustav von Schmoller (a grandson) and the chemist Walter Hückel and the physicist Erich Hückel (great-grandsons). Gaertner was granted a title of personal nobility in 1846.

Like his father, Gaertner contracted an eye ailment in the course of his microscopical investigations. He therefore discontinued them and ceased his medical practice as well. Beginning about 1824 he devoted his energies entirely to research on plant hybridization. Gaertner, who conducted his studies as an independent scholar, became a member of the Leopoldina in 1826 and of the Royal Netherlands Academy of Sciences in 1849.

Gaertner's earliest research dealt with medical and physiological questions and with the chemical analysis of organic substances, including bone and urine. After 1800 he concentrated exclusively on botany. He prepared a supplementary fifth volume on the cryptogams to J. G. Gmelin's *Flora Sibirica* as well as a further supplement to that volume containing data on the plants his father had gathered in the Ukraine. Following his trip abroad in 1802, Gaertner worked until 1805 on editing another supplementary volume to the *Flora,* this one dealing with carpology and written by his father. He then began a series of systematic and morphological investigations of the grasses.

The writings of his father's friend J. G. Koelreuter had already drawn Gaertner's attention to the problems of the hybrid fertilization of plants. When F. J. Schelver and his student A. W. Henschel again brought the sexuality of plants into doubt on the basis of principles derived from *Naturphilosophie,* the question became the subject of much debate. In 1825, Gaertner began comprehensive research on this problem after planning and beginning a general treatise on plant physiology. He published his first results in 1826, and he made further contributions almost yearly. In 1837 he won a prize for his solution of a problem presented by the Netherlands Society of Sciences at Haarlem. The first part of his masterpiece, *Versuche und Beobachtungen über die Befruchtungsorgane der vollkommeneren Gewächse . . .,* appeared in 1844; the second and larger part in 1849, under the title *Versuche und Beobachtungen über die Bastarderzeugung im Pflanzenreich.*

The first part of this work treated the relationships and conditions of natural and artificial fertilization, as well as the functions and alterations of the individual parts of the flower during fertilization. The second part reported results of experiments carried on over decades; the experimental methods themselves were described in a supplement. In this treatise Gaertner set forth the different ways in which hybrid fertilization can occur and discussed the capacity for hybridization among the various systematic units (family, genus, species, variety). He also considered the question of the "regularities" in the behavior of several generations of hybrids, although he did not formulate any general rules. He referred to certain dominant characteristics as "decidirte Typen" ("definite types"). He observed alterations of individual characteristics but did not study them systematically. He denied the possibility of "the transformation of one species into another through hybridization" and believed in the stability of species, which he regarded as fixed types. Gaertner classified hybrids "according to their structure and origin" and collected data on the general "characteristics and properties of hybrids," as, for example, those pertaining to their fertility.

Although Gaertner's general theoretical conclusions were deficient in terms of contemporary biological knowledge, his writings were extremely rich in observations, presented new methodology, and, in sum, constituted the first comprehensive treatment of the problem of hybridization. In them the sexuality of plants was definitively established, and at the same time the attention of researchers was drawn to the biological problems connected with sexual reproduction. Charles Darwin read Gaertner closely when he was developing his theory of pangenesis. However, Gaertner's work achieved its greatest impact through its influence on Gregor Mendel, whose investigations were directly inspired by it.

BIBLIOGRAPHY

I. Original Works. Gaertner's major works are "Fortgesetzte Nachrichten über Bastardgewächse," in *Flora* (Regensburg), **10–21** (1827–1838); "Over de voortteling van bastaard-planten," in *Natuurkundige verhandelingen van de Hollandsche maatschappij der wetenschappen te Haarlem,* **24** (1844), 1–202; *Versuche und Beobachtungen über die Befruchtungsorgane der vollkommeneren Gewächse und über die naturliche und künstliche Befruchtung durch den eigenen Pollen* (Stuttgart, 1844); and *Versuche und Beobachtungen über die Bastarderzeugung im Pflanzenreich* (Stuttgart, 1849).

II. Secondary Literature. An obituary is in *Flora,* **34**

(1851), 135–143. See also F. Reinöhl, in *Schwäbische Lebensbilder,* III (Stuttgart, 1942), 190–198, with portrait, bibliography of his works, and list of the literature.

BRIGITTE HOPPE

GAFFKY, GEORG THEODOR AUGUST (*b.* Hannover, Germany, 17 February 1850; *d.* Hannover, 23 September 1918), *bacteriology, public health.*

Gaffky was the son of Georg Friedrich Wilhelm Gaffky, a shipping agent, and Emma Wilhelmine Mathilde Schumacher. After attending a Gymnasium in Hannover, he studied medicine at the University of Berlin. His studies were interrupted by the Franco-Prussian War, however, and he served as a hospital orderly, then returned to the university to take the M.D. in 1873, with a dissertation on the causal relation between chronic lead poisoning and kidney disease. He was an assistant at the Charité, and passed the *Staatsexamen* in 1875. He then spent several years as a military surgeon posted to various garrisons.

Gaffky's career took a new turn in 1880 when he and Friedrich Löffler were ordered to assist Robert Koch at the recently founded imperial health office in Berlin. They were the first two of the brilliant group of assistants that Koch was assembling there. Under Koch's tutelage, Gaffky participated in developing new bacteriological methods and in demonstrating the causes of infectious disease.

From the beginning of his work with Koch, Gaffky was drawn into a variety of researches at the public health laboratory. In 1881 he reported on experimentally produced septicemia in animals. Controverting Naegeli's view that pathogenic bacteria might eventually arise through the accommodation and indefinite variability of common, previously harmless forms, Gaffky maintained that these disease-producing bacteria were specific and derived only from forms like themselves. He took part with Löffler in Koch's work on steam disinfection, and participated in the investigations on anthrax, cholera, and tuberculosis.

Gaffky's most important contribution, however, was in the isolation and culture of the bacillus that is the causative agent of typhoid fever. In 1880 Karl Joseph Eberth had seen and described a bacillus which he believed to be the cause of this disease, while Koch had independently observed the organism and photographed masses of the bacilli. But Eberth and Koch had been able to discern the bacillus in no more than half the cases of typhoid fever in which they had made their examinations. It was Gaffky's hypothesis that this difficulty might be due in part to the culture methods employed, and for the next several years he sought ways to obtain the bacillus with high consistency in pure culture. To do so, it was necessary for him to differentiate between the causative agent of typhoid fever—or *Typhus abdominalis,* as he called it—and similar bacilli that might be present but represent secondary invaders of the diseased tissues.

The bacilli—typically short rods with rounded ends—could be found in the mesenteric glands, spleen, liver, and kidneys of typhoid victims, sometimes in isolation, but more frequently grouped in masses or threadlike arrangements. Using various techniques, Gaffky grew cultures of the bacillus in solid nutrient gelatin, on the surface of boiled potatoes (where a characteristic film formed), in solidified sheep blood serum (according to Koch's procedures), in fluid serum, and in bouillon. He identified the bacillus and believed that he had demonstrated its spores. The organism was clearly visible in stained sections, while in fluid media the living bacillus exhibited a distinctive spontaneous movement and there was a characteristic motion of the bacillar threads. Despite this success in culturing the bacillus in different media, Gaffky was never able to cultivate the bacillus in living animals to produce the disease, although he tried repeatedly to achieve this further proof. In vain he experimentally fed infective material to Java monkeys, mice, guinea pigs, and other animals in an attempt to induce typhoid fever in them.

Still, Gaffky could report that he had with the highest probability isolated the etiologic agent of typhoid fever, for his examinations had disclosed the presence of the bacillus in twenty-six of twenty-eight cases of the disease and he published these results of his investigations in the *Mittheilungen aus dem K. Gesundheitsamte* in 1884.

In 1883–1884 Gaffky was a member of the expedition, sponsored by the German state and led by Koch, that was sent to Egypt and India to investigate the outbreak of cholera there. On the commission's return, Koch reported the identification of the cholera bacillus and described the ways in which the infection was transmitted; Gaffky was responsible for preparing a detailed documentary report on the journey and its scientific results, published in the *Arbeiten aus dem K. Gesundheitsamte* in 1887. Its frontispiece showed a scene on the Hooghly River in which people were bathing, washing clothes, and carrying off drinking water, while boats lay nearby; the text was illustrated with photomicrographs of Koch's cultures of the vibrio and with maps and charts of epidemic statistics.

In 1885 Koch accepted the chair of hygiene at the University of Berlin and Gaffky succeeded him as director of the imperial health office. In 1888 Gaffky was appointed professor of hygiene at the University of Giessen and undertook the direction of the new Hygienic Institute there. While he was at Giessen in 1892 cholera broke out in Hamburg, and Gaffky interrupted his teaching to advise the government in combating the epidemic. He was rector of the university from 1894 to 1895.

Gaffky returned to India—this time as leader of the government commission—in 1897, when bubonic plague was rife in Bombay and other centers. Koch was at this time in South Africa, investigating rinderpest; he later assumed leadership of the plague commission, and Gaffky collaborated in its report.

Gaffky left Giessen in 1904 when Koch suggested that he succeed him as director of the Institut für Infektionskrankheiten. He was an able administrator and under his guidance the institute was enlarged by a division for tropical diseases, a rabies station, and a division for protozoology. He also served as coeditor of the *Zeitschrift für Hygiene und Infektionskrankheiten.*

In 1913 Gaffky left Berlin for the quieter surroundings of Hannover. He had intended to resume his own studies, but when World War I broke out he was again called to serve the government as adviser on hygiene and public health. He died just before the war ended.

In his career Gaffky followed in Koch's footsteps, but his own contributions to bacteriology and public health were significant. A genus of the family Micrococcaceae is named Gaffkya after him.

BIBLIOGRAPHY

I. ORIGINAL WORKS. Gaffky's writings include "Experimentell erzeugte Septicämie mit Rücksicht auf progressive Virulenz und accomodative Züchtung," in *Mittheilungen aus dem K. Gesundheitsamte,* **1** (1881), 80–133; "Ein Beitrag zum Verhalten der Tuberkelbacillen im Sputum," *ibid.,* 126–130; "Zur Aetiologie des Abdominaltyphus," *ibid.,* **2** (1884), 372–420, trans. by J. J. Pringle as "On the Etiology of Enteric Fever," in W. Watson Cheyne, ed., *Recent Essays by Various Authors on Bacteria in Relation to Disease* (London, 1886), pp. 203–257; "Versuche über Desinfection des Kiel- oder Bilgeraums von Schiffen," in *Arbeiten aus dem K. Gesundheitsamte,* **1** (1886), 199–221, written with R. Koch; "Bericht über die Thätigkeit der zur Erforschung der Cholera im Jahre 1883 nach Egypten und Indien entsandten Kommission," *ibid.,* **3** (1887), compiled with Koch; and "Die Verbreitung der orientalischen Beulenpest durch sogenannte 'Dauerausscheider' und 'Bazillenträger,'" in *Klinisches Jahrbuch,* **19** (1908), 491–496. In addition, he edited the *Gesammelte Werke von Robert Koch* (Leipzig, 1912) with E. Pfuhl. Olpp and Pagel (see below) provide further bibliographical data.

II. SECONDARY LITERATURE. See B. Harms, "Georg Gaffky zum Gedächtnis," in *Zentralblatt für Bakteriologie, Parasitenkunde, Infektionskrankheiten und Hygiene,* **156** (1950), 1–2; Wilhelm Katner, "Georg Theodor August Gaffky," in *Neue deutsche Biographie,* VI (Berlin, 1964), 28; Kirchner, "Georg Gaffky," in *Zeitschrift für ärztliche Fortbildung,* **15** (1918), 614–615; H. Kossel, "Georg Gaffky," in *Münchener medizinische Wochenschrift,* **65,** pt. 2 (1918), 1191–1192; E. Neufeld, "Georg Gaffky," in *Berliner klinische Wochenschrift,* **55,** pt. 2 (1918), 1062–1063; G. Olpp, *Hervorragende Tropenärzte in Wort und Bild* (Munich, 1932), pp. 139–141; "Georg Theodor August Gaffky," in J. Pagel, *Biographisches Lexikon hervorragender Ärzte des neunzehnten Jahrhunderts* (Berlin, 1901), pp. 577–578; and R. Pfeiffer, "Georg Gaffky," in *Deutsche medizinische Wochenschrift,* **44,** pt. 2 (1918), 1199.

GLORIA ROBINSON

GAGLIARDI, DOMENICO (*b.* Rome [?], Italy, 1660; *d.* Rome [?], *ca.* 1725), *anatomy.*

There is little or no information on Gagliardi's life. He was probably born in Rome or somewhere in the Papal States. The scanty information provided by biographers indicates that he was a professor of medicine at the University of Rome; but since his name does not appear in the *rotoli,* the lists of professors compiled each year for administrative purposes, such a statement cannot be considered reliably documented.

Gagliardi was also the *protomedico* of the Papal States, and of Rome in particular, a function in some measure similar to that of a chief provincial doctor, and he acquired great fame as a doctor and as an anatomist. His name is especially connected with anatomy, particularly the skeletal system, which he summarized in *Anatome ossium novis inventis illustrata* (1689). Gagliardi carried out morphological and microscopic investigations on human bones, using chemical reagents in order to bring out the fine structure; he also made comparative anatomical studies of the skull and vertebrae of man and calf.

His morphological and microscopic work was accompanied by anatomicopathologic research. His *Anatome ossium novis* contains the first description of a case of what is presumably tuberculosis of the bone. In order to emphasize the structure of bone lamellae, Gagliardi used solutions of various acid substances; on the basis of his results he proposed a theory that the "softening" of bones which he described was caused by the action of an "acid" present in the organism.

In 1720 Gagliardi had an opportunity to do a close study of the pneumonia epidemic raging in Rome;

the interest of this study lies in the fact that, before G. B. Morgagni, it was anatomicopathological in approach and based on carefully conducted autopsies. He was also interested in medical deontology and in what today would be called scientific popularization. He warned patients against the activity of charlatans, who in papal Rome appear to have had a large following. Two volumes, dedicated to moral and deontological topics and written expressly for the layman, make very clear the limits of the art of medicine. Written in a richly erudite and very ponderous style, according to the taste of the times, these works are full of precepts for reaching an advanced age by following certain rules of hygiene and sanitation.

Gagliardi left a third book of a more strictly deontological character, in which he gave young doctors advice on correct professional behavior toward patients and many rules of a more strictly medical-scientific nature.

BIBLIOGRAPHY

Gagliardi's writings are *Anatome ossium novis inventis illustrata* (Rome, 1689); *Idea del vero medico fisico e morale fermata secondo li documenti ed operazioni d'Ippocrate* (Rome, 1718); *L'infermo istruito nella scuola del disinganno* (Rome, 1719); and *De educatione filiorum* (Rome, 1723).

CARLO CASTELLANI

GAGNEBIN, ÉLIE (*b.* Liège, Belgium, 4 February 1891; *d.* Lausanne, Switzerland, 16 July 1949), *geology.*

Gagnebin was the eleventh of the twelve children of Henri Gagnebin, a minister, and the former Adolphine Heshuysen. In 1892 the family moved to Switzerland, where the father was pastor of the Free Church (Église Libre) in Biel and, from 1899, in Lausanne. Gagnebin attended the classical Gymnasium in Lausanne and took the *bachelier ès lettres* when he was eighteen; he then attended the University of Lausanne, where the program in natural sciences did not sufficiently engage his critical spirit and high intelligence. He became a member of the Société des Belles Lettres, a merry group of student revolutionaries.

In 1912 Gagnebin earned the *licence* in natural and physical sciences and became assistant to Maurice Lugeon, a specialist in the tectonics of Switzerland and one of the first advocates of the nappe theory of the Alps. He began geological researches toward his doctorate in 1913; they dealt with the Préalpes Bordières and the region of Châtel St. Denis, famous for its wealth of fossils. (The Préalpes, or fore-Alps, are divided into a northwest front, the Préalpes Bordières; a distinct central zone, the Préalpes Médianes; and a southeast Zone des Cols, or Préalpes Internes.) He received the doctorate at the University of Lausanne on 5 July 1920 with a dissertation (published in 1924) which he modestly entitled *Communication préliminaire.* In this *Communication,* Gagnebin set forth the main features of the stratigraphy and tectonics of the Préalpes Bordières, including the northward movement of the Ultrahelvetian nappes and klippes into the region. He continued to study this region during the next few years while furthering his professional knowledge by working under Wilfrid Kilian at Grenoble (1920) and Émile Haug at Paris (1921).

Gagnebin began his academic career in 1917, when he occasionally replaced Lugeon as lecturer in paleontology; he was officially assigned to this post in 1928. He was promoted to associate professor in 1933 and delivered an inaugural lecture, "La durée des temps géologiques," on 16 May of that year. In 1940 he succeeded Lugeon as full professor.

As a field geologist Gagnebin had made his mark as early as 1922 when he published his *Carte géologique des Préalpes entre Montreux et le Moléson et du Mt.-Pèlerin.* On the basis of this work the Swiss Geological Commission entrusted him with preparing the map of St. Maurice, a project that occupied Gagnebin's summers from 1925 to 1933 and was published in 1934. From 1931 he was also a member of the Service de la Carte Géologique de France.

In 1939 Gagnebin published his observations on the shredding of the Simme nappe in the Chablais Préalpes. In 1941, with Lugeon, he brought out the classic investigation "Observations et vues nouvelles sur la géologie des Préalpes romandes." In this study Gagnebin and Lugeon proposed continental drift as the primary process in the formation of the Alps; gravity then caused the resulting accumulation of faulting and nappes to slide into their present positions. (The gravity sliding hypothesis had earlier been advanced by Lugeon and Hans Schardt and had received new attention through the work of Daniel Schneegans in 1938.) Gagnebin extended this hypothesis to the Helvetic Alps of eastern Switzerland in 1945.

Gagnebin's other publications include a few short communications on chance paleontological finds and two popular works, *Le transformisme et l'origine de l'homme* (1943) and *Histoire de la terre et des êtres vivants* (1946). A study of the Quaternary led him to the concerns of the earlier book; in it he demonstrated man's descent from a branch of the anthropoids and stated the hope that man is still evolving and that "a different and more developed race will succeed ours." The success of this book encouraged him to write the *Histoire,* in the final chapter of which he formulated a finalistic ethics—"Good is that which goes in the direction of life . . ."—and called for a new Aquinas to create "a metaphysics and a religious doctrine capable of integrating the sum of truths of which man has become aware in the last three or four centuries. . . ."

Gagnebin was in addition an ardent champion of modern music. With his friend the Vaudois writer C. F. Ramuz, the composer Igor Stravinsky, and the conductor Ernest Ansermet, he created the drama *L'histoire du soldat* (1918), in which he took the role of the narrator.

BIBLIOGRAPHY

Gagnebin's most important works are *Carte géologique des Préalpes entre Montreux et le Moléson et du Mt.-Pèlerin,* carte spéciale no. 99, Matériaux pour la Carte de la Suisse (Bern, 1922); "Description géologique des Préalpes bordières entre Montreux et Semsales," in *Mémoires de la Société vaudoise des sciences naturelles,* **2,** no. 1 (1924), 1-70, diss.; "La finalité dans les sciences biologiques," in *Revue de théologie et de philosophie,* no. 78 (1931), 1-38; "St.-Maurice," map. no. 8 in *Atlas géologique de la Suisse au 1:25,000* (Bern, 1934), written with F. de Loys, M. Reinhard, M. Lugeon, N. Oulianoff, W. Hotz, and E. Poldini; "La durée des temps géologiques," in *Bulletin de la Société vaudoise des sciences naturelles,* **58** (1934), 125-146; "Notice explicative de la feuille St.-Maurice," in *Atlas géologique de la Suisse au 1:25,000* (1934), pp. 1-6; "Ossements de mammouth trouvés dans la moraine de Renens, près de Lausanne, et recensement des restes de mammouths connus dans la région lémanique," in *Bulletin de la Société vaudoise des sciences naturelles,* **58** (1935), 385-391; "Mécanisme ou vitalisme en biologie?" in *Revue de théologie et de philosophie,* no. 100 (1936), 1-7; "Découverte d'un lambeau de la nappe de la Simme dans les Préalpes du Chablais," in *Comptes rendus hebdomadaires des séances de l'Académie des sciences,* **208** (1939), 822; "Découverte d'une nouvelle défense de mammouth dans la terrasse du Boiron, près de Morges, et précisions sur quelques restes de mammouths de la région lémanique," in *Bulletin de la Société vaudoise des sciences naturelles,* **61** (1940), 291-296; "Observations et vues nouvelles sur la géologie des Préalpes romandes," in *Mémoires de la Société vaudoise des sciences naturelles,* **7,** no. 1 (1941), 1-90, written with M. Lugeon; *Le transformisme et l'origine de l'homme* (Lausanne, 1943; 2nd ed., 1947); "Quelques problèmes de la tectonique d'écoulement en Suisse orientale," in *Bulletin de la Société vaudoise des sciences naturelles,* **62** (1945), 476-494; *Histoire de la terre et des êtres vivants,* no. 10 in the series Gai Savoir (Lausanne, 1946); and "La notion d'espèce en biologie," in *Dialectica,* **1,** no. 3 (1947), 229-242.

An excellent appreciation is Maurice Lugeon, "E. Gagnebin, 1891-1949," in *Verhandlungen der Schweizerischen naturforschenden Gesellschaft* (1949), 382-399, with portrait.

EMIL KUHN-SCHNYDER

GAHN, JOHAN GOTTLIEB (*b.* Ovanåker, Sweden, 19 August 1745; *d.* Falun, Sweden, 8 December 1818), *mineralogy, chemistry.*

Gahn studied physics and chemistry at Uppsala from 1762 to 1770. When Torbern Bergman was appointed professor of chemistry there in 1767, Gahn became his laboratory assistant. After passing in 1770 the examination for mining engineer, he worked at the College of Mining, where he was assigned the task of applying new and more scientific methods to the copper smelting processes at the Falun mine in the Kopparberg district. For four years he worked exclusively with copper smelting, introducing important improvements and solving many technical problems. Above all, he modernized the methods for using the by-products of the smelting process, among them sulfur, iron sulfate, red pigment, copper mastic, and copper precipitate. Gahn performed his chemical research in a well-equipped laboratory that he installed at his own expense in his garden at Falun.

Although he seldom took the time to write down his observations and published almost nothing, rumors of Gahn's extensive chemical and technical abilities spread beyond Sweden; Falun became a mecca for scholars, factory owners, and industrialists seeking advice and guidance in technical problems. "Gahn is building in Sweden a real center for everything that happens in the technical field. The country still lacks a Polytechnic Institute where new ideas can be tried out and from which innovations and projects can emanate. Gahn supplies that" (Johann F. L. Hausman, *Reise durch Skandinavien in den Jahren 1806 und 1807* [Leipzig, 1811-1818]). Such merit did not remain unnoticed. In 1780 the College of Mining awarded Gahn its gold medal and two years later informed King Gustavus III of the improvements and growth that Gahn's work had brought to the refining of copper. On this basis the king conferred on him in 1782 the honorary title of superintendent of mines and in 1784 authorization as associate member at the College of Mining. In the same year he was elected member of the Academy of Science in Stockholm.

It was of great importance to contemporary Swedish chemistry that Scheele, who worked in the pharmacy Uplands Wapen in Uppsala from 1770 to 1775, was introduced by Gahn to Torbern Bergman. Gahn collaborated in the work of both of these men; and Bergman, who in many cases benefited from

Gahn's experimental ability, emphasized this both in his letters and published works. For instance, he mentions, concerning the mineral pyrolusite, that he himself had doubted that it contained any metal but that Gahn was the first to reduce the mineral and to discover, in 1774, the pure metal later named manganese.

Gahn shared a friendship and an exchange of ideas with Scheele that were fruitful for the work of both. Unfortunately, their correspondence provides no information about Gahn's contributions; although Gahn conscientiously preserved Scheele's letters during the 1770's, Scheele was so indifferent toward preserving his correspondence from Gahn that only a few writings saved at random still exist. Scheele's letters reveal that he often solicited and received valuable explanations for his experiments with pyrolusite and barium sulfate. It is interesting that Scheele thanks Gahn especially for the suggestion of an important study concerning what is now called solid-state reactivity.

Gahn was a capable chemical experimenter, but Scheele was his unchallenged superior in everything except blowpipe analysis, in which Gahn was unsurpassed. It is therefore not surprising that the possibility of conceptual cross-fertilization that existed here would materialize. A conversation with Scheele in the spring of 1770 concerning his research with inorganic substances in animal bones, the so-called animal earth, provided the incentive for Gahn to study this material more carefully; he was then able to show, with the aid of the blowpipe, the presence of phosphorus. This observation later led to Scheele's method of obtaining phosphorus from animal bones.

Preserved letters indicate that—at least in the first part of the 1770's—Gahn was the trusted friend for whose opinion Scheele first sent his scientific articles.

Gahn also worked with J. J. Berzelius. Among other things they were both financially and scientifically interested in a sulfuric acid factory near Gripsholm. Berzelius tried unsuccessfully to persuade Gahn to go to Stockholm, but ultimately he traveled to Falun to meet Gahn in the summers of 1813–1816. The two friends explored the area's rich mineral deposits and, as Berzelius wrote, "a number of entirely new minerals were discovered . . . and analysed at the time in Gahn's excellently equipped laboratory" (*Jöns Jacob Berzelius Autobiographical Notes,* trans. by Olof Larsell [Baltimore, 1934], p. 91).

BIBLIOGRAPHY

I. Original Works. Gahn's works include *Några anmärkningar i svenska bergs-lagfarenheten om författningar till befrämjande av god hushållning vid järnhyttor* (Uppsala, 1770), his doctoral diss.; "Yttrande över Kommerskollegii fråga om någon ljusare och gladare färg än rödfärg," in *Kongliga Vetenskaps Academiens nya Handlingar,* **25** (1804), 289–301; and *Underrättelse om upställningen och nyttjandet af herr assessor J. G. Gahns förbättrade appareil för vattens aererande med tabell* (Uppsala, 1804).

The principal part of Gahn's literary remains is kept in the library of the Royal Institute of Technology. Certain parts of his correspondence are preserved in the archives of the Nordic Museum and in the National Record Office. The important letters from Scheele and Bergman as well as Gahn's correspondence with Berzelius are to be found in the library of the Royal Academy of Science. Gahn's correspondence with Berzelius is in *Jac. Berzelius brev,* H. G. Söderbaum, ed., IX (Stockholm, 1922). The 38 surviving letters from Gahn to Bergman (1768–1778) are in the university library of Uppsala.

II. Secondary Literature. On Gahn and his work, see J. A. Almquist, *Bergskollegium och bergslagsstaterna* (Stockholm, 1909); J. G. Anrep, *Svenska slägtboken,* 3 vols. (Stockholm, 1871–1875); J. Berzelius, *Självbiografiska anteckningar* (Stockholm, 1901), and in the trans. by Olof Larsell, *Jöns Jacob Berzelius Autobiographical Notes* (Baltimore, 1934); B. Boethius, *Grycksbo 1382–1940* (Stockholm, 1942); U. Boklund, "När Gahn upptäckte Scheele på Lokks apotek," in *Lychnos* (1959), 217–222; Hans Järta, *Åminnelse-Tal öfver . . . Herr Joh. Gottl. Gahn . . . hållet inför Kongl. Vetenskaps-Academien den 8 October 1831* (Stockholm, 1832); AB Ferrolegeringar (publisher), *Av meteorernas ätt. En krönika om mangan . . .* (Stockholm, 1962); S. Lindroth, *Gruvbrytning och kopparhantering vid Stora Kopparberget . . .,* II (Stockholm, 1955); and C. Sahlin, "Johan Gottlieb Gahns laboratorium och samlingar," in *Blad för bergshanteringens vänner,* **16** (1919–1921). See also J. E. Jorpes, *Jac. Berzelius, His Life and Work* (Stockholm, 1966).

Uno Boklund

GAILLOT, AIMABLE JEAN-BAPTISTE (*b.* Saint-Jean-sur-Tourbe, Marne, France, 27 April 1834; *d.* Chartres, France, 4 June 1921), *astronomy, celestial mechanics.*

Gaillot spent his entire career at the Bureau of Computation of the Paris observatory, to which he was assigned at the time of his recruitment by Urbain Le Verrier in 1861, and was director of the bureau from 1873. He became astronomer in 1864 and chief astronomer in 1874, and was made assistant director of the Paris observatory in 1897.

In astronomy Gaillot concentrated on the calculations that would make his colleagues' observations most useful. He directed the publication of the *Catalogue de l'Observatoire de Paris,* which classified the 387,474 meridian observations made between 1837 and 1881. Despite poor health Gaillot saw this twenty-year task to completion before retiring in 1903.

He continued his research in celestial mechanics until he was eighty. His important contributions in this field brought him four awards from the Academy of Sciences, of which he became a corresponding member in 1908.

Gaillot was Le Verrier's only collaborator. The latter's widow wrote to him in 1877: "I especially want to express my deep gratitude to the devoted and intelligent collaborator who enabled my beloved husband to complete his colossal project. It pleased him to recognize that without you, this would have been impossible."

To complete Le Verrier's work, Gaillot amended the latter's analytical theories concerning Jupiter, Saturn, Uranus, and Neptune. By using a laborious but effective method of interpolation, he eliminated all discordances: his values for the mass of these planets turned out to be excellent and the positions he calculated were confirmed by observations to an accuracy of one or two seconds. Since their publication Gaillot's tables have served as a basis for the international ephemerides found in *Connaissance des temps.*

BIBLIOGRAPHY

In celestial mechanics three broad *mémoires* covering all of Gaillot's earlier research concern Le Verrier's theories on the motion of the planets; two of them furnish, for Saturn and Jupiter, additions to these theories and also corrected tables—see *Annales de l'Observatoire de Paris,* **24** (1904), 1-512, and **31** (1913), 1-317; the third, concerning Uranus and Neptune, includes Le Verrier's theory, which is considerably elaborated, and new tables: *ibid.,* **28** (1910), 1-649. See also "Contribution à la recherche des planètes ultraneptuniennes," in *Comptes rendus hebdomadaires des séances de l'Académie des sciences,* **148** (1909), 754-758; and "Le Verrier et son oeuvre," in *Bulletin des sciences mathématiques et astronomiques,* **2** (1878), 29-40.

In fundamental astronomy, see "Influence de l'attraction luni-solaire sur la verticale, la pesanteur et la marche des pendules," in *Bulletin astronomique,* **1** (1884), 113-118, 217-220; "Détermination géométrique des positions des circumpolaires," *ibid.,* 375-381, 577-583; "Sur la mesure du temps," *ibid.,* **3** (1886), 221-232; "Changements de la durée de l'année julienne . . .," in *Comptes rendus hebdomadaires des séances de l'Académie des sciences,* **97** (1883), 151-154; "Sur les mesures du temps," *ibid.,* 544-546; "Détermination de la constante de la réfraction . . .," *ibid.,* **102** (1886), 200-204, 247-250; "Sur les variations de la latitude . . .," *ibid.,* **111** (1890), 559-561; **112** (1890), 651-563; and "Sur les formules de l'aberration annuelle," *ibid.,* **116** (1893), 563-565. On the work on star catalogs, see "Discordances in d'Agelet's Observations," in *Astronomical Journal,* **16** (1896), 182-183; and, especially, the "Introduction" in *Catalogue de l'Observatoire de Paris,* I (Paris, 1887), (1)-(22).

For information about Gaillot, see B. Baillaud, "Notice nécrologique, A. Gaillot," in *Comptes rendus hebdomadaires des séances de l'Académie des sciences,* **172** (1921), 1393-1394.

JACQUES R. LÉVY

GAIMARD, JOSEPH PAUL (*b.* St. Zacherie, France, 31 January 1796; *d.* Paris [?], France, 10 December 1858), *natural history, scientific exploration, naval medicine.*

Born and educated in the French Midi, Paul Gaimard became one of the most widely traveled naturalists in the history of scientific exploration. Since his father had died in 1799 during uprisings in the Midi, Gaimard's early training was directed by relatives. He entered the naval medical school at Toulon in 1816 and, through success in academic competition, was named surgeon in the royal navy. His talents and background earned him a place as surgeon and naturalist aboard the *Uranie,* commanded by Louis Claude de Freycinet and charged with investigating the meteorology, oceanography, and natural history of vast areas of the South Pacific Ocean. Assisting him were Jean René Constant Quoy, surgeon and naturalist; Charles Gaudichaud-Beaupré, pharmacologist and botanist; and François Arago's youngest brother, Jacques, draftsman.

All were subject to regular naval discipline, Freycinet hoping to avoid the customary willfulness of scientific explorers. Upon their return from the circumnavigation of the globe (1817-1820) Gaimard and Quoy prepared a detailed account of their zoological discoveries. Gaimard thus early made his mark in one of the great periods of French maritime activity and earnest overseas scientific exploration.

Early in 1826 he toured Europe to inspect natural history collections and to prepare for his departure as first surgeon to the famed expedition of J. S. C. Dumont d'Urville. As captain of the *Astrolabe,* Dumont d'Urville's double task was to conduct a scientific survey of Oceania and to seek traces of the lost La Pérouse expedition. Between 1826 and 1829 Gaimard was again in the South Pacific, and once again he and Quoy prepared an account of their zoological collections and discoveries. While this work was in press, an outbreak of Asiatic cholera was reported from western Russia. The indefatigable and audacious Gaimard immediately set out to assess the epidemic. He spent several months observing the disease in eastern Europe and encountered it again upon his return to Paris in 1832. His report on the cholera, an affliction all the more terrifying for its utter novelty in western Europe, remains a classic account of the disease.

Gaimard soon set off on further exploratory voyages. He led a large scientific team aboard the *Recherche* to Iceland and Greenland (1835–1836), and a few years later, serving as director of the Scientific Commission for the North, he conducted extensive explorations in Lapland and on Spitsbergen and the Faeroes. With the latter journey (1838–1840) Gaimard's frenetic, albeit highly productive, wandering apparently came to an end. His later years remain a supreme mystery, but he evidently settled in Paris and was fully occupied with the preparation and publication of the official reports of the expeditions to Iceland and to northern Europe.

Of Gaimard's personality little is known save for effusive but perhaps accurate references to his uncommon benevolence and a readiness to serve France whatever be the task imposed. Details of his personal life also remain quite unknown. Clearly, Gaimard was devoted as much to the sheer pleasure of travel as to the joy of scientific discovery. His talents as a naturalist were indeed great, and he was assiduous and successful in seeing to completion the official reports of every expedition in which he participated. Those reports, for all—or the little—that we know of him, constitute the man himself.

BIBLIOGRAPHY

I. Original Works. Gaimard's principal publications are *Voyage autour du monde . . . exécuté sur les corvettes . . . Uranie et la Physicienne, pendant les années 1817, 1818, 1819, et 1820 . . . Zoologie par MM. Quoy et Gaimard,* 1 vol. and atlas (Paris, 1824), an assessment of which is given by Cuvier (see below); *Voyage de découvertes de l'Astrolabe . . . pendant les années 1826, 1827, 1828, 1829 . . . Zoologie par MM. Quoy et Gaimard,* 4 vols., and atlas, 2 vols. (Paris, 1830–1832); *Voyage en Islande et au Groënland exécuté pendant les années 1835 et 1836,* 8 vols., and atlas, 3 vols. (Paris, 1838–1852), of which only vol. I, *Histoire du Voyage,* was written by Gaimard himself, the other volumes being prepared under his editorship; *Voyages de la commission scientifique du Nord . . . pendant les années 1838, 1839 et 1840,* 17 vols., and atlas, 5 vols. (Paris, 1843–1855); and *Du choléra morbus en Russie, en Prusse et en Autriche, pendant les années 1831 et 1832* (Paris, 1832), written with Auguste Gérardin.

During the 1820's and 1830's Gaimard and Quoy published numerous reports on the zoology and ethnography of the Pacific area; these are listed in the Royal Society's *Catalogue of Scientific Papers,* II (London, 1868), 755–756. The Bibliothèque Centrale of the Muséum d'Histoire Naturelle, Paris, possesses a number of MSS by Gaimard; these relate principally to the zoological collections made on the *Astrolabe* (information courtesy of Yves Laissus).

II. Secondary Literature. Apparently no account of the full career of Gaimard's life exists. A sentimental and very brief report on his activities until about 1837 was published by the Société Montyon et Franklin: A. Jarry de Mancy, "Notice sur Paul Gaimard," in *Portraits et histoire des hommes utiles* (Paris, 1837), 192–196.

Notices in biographical dictionaries and encyclopedias are uniformly obscure. An exception is "Gaimard, Paul," in August Hirsch *et al.,* eds., *Biographisches Lexikon der hervorragenden Ärzte aller Zeiten und Völker,* 2nd ed., II (Berlin–Vienna, 1930), 656. On the background and motives of the naval expeditions in which Gaimard participated, see the excellent review by John Dunmore, *French Explorers of the Pacific,* II, *The Nineteenth Century* (Oxford, 1969), 63–108, 178–227. See also a review of Gaimard's work by Georges Cuvier, in *Annales des sciences naturelles,* **10** (1827), 239–243.

William Coleman

GAINES, WALTER LEE (*b.* Crete, Illinois, 17 March 1881; *d.* Urbana, Illinois, 20 November 1950), *dairy science.*

Gaines was raised on a farm near Crete, where he received his preliminary education and the direction for his lifework. The prospect of improving farm production guided his studies at the University of Illinois, from which he received the B.S. degree in 1908 and the M.S. in agriculture in 1910. He was able to focus the direction of his research career by graduate work at the University of Chicago, where he obtained the Ph.D. in 1915 with a contribution to the physiology of lactation.

After a brief interval at his home farm in Crete, Gaines returned to the University of Illinois to assist a federal program for increasing food production during World War I. He was professor of milk production at the University of Illinois from 1919 to 1949, during which time he wrote twenty-nine research reports, was co-author of seventeen others, and contributed frequently to farm journals, primarily on the problems of increasing milk production.

His studies on milk secretion included the production of lactose; the effects on lactation of pituitary hormones, blood transfusions, and pregnancy; the storage capacity of udders; the rates of secretion of milk constituents; and the relationship between milk yield and frequency of conception. By intravenous injections of pituitary extracts, Gaines demonstrated that milk is formed continuously by mammary alveoli. He studied the effect of weight changes on future milk production during early development and showed the efficacy of milk production per live weight as a measure of lactation in genetic studies. He also was able to distinguish between milk nutrients assimilated for maintenance and those transformed into milk. He devised a widely used "fat-corrected milk" (F.C.M.) formula for expressing the energy equivalent of milk by comparison with a base of 4 percent butterfat.

In 1949 the American Dairy Science Association gave Gaines its highest honor, the Borden Award, citing him as "the modest leader in the United States of the scientific approach to the problems of milk secretion."

BIBLIOGRAPHY

Gaines's works include "A Contribution to the Physiology of Lactation, in *American Journal of Physiology,* **38** (1915), 285-312; "Relative Rates of Secretion of Various Milk Constituents," in *Journal of Dairy Science,* **8** (1925), 486-496; "Milk Yield in Relation to the Recurrence of Conception," *ibid.,* **10** (1927), 117-125; "The Energy Basis of Measuring Milk Yield in Dairy Cows," in *Illinois Agricultural Experiment Station Bulletin* (1928), p. 308; "Size of Cow and Efficiency of Milk Production," in *Journal of Dairy Science,* **14** (1931), 14-21; and "Live Weight and Milk-Energy Yield in the Wisconsin Dairy Cow Competition," *ibid.,* **22** (1939), 49-53.

RICHARD P. AULIE

GALEAZZI, DOMENICO GUSMANO (*b.* Bologna, Italy, 4 August 1686; *d.* Bologna, 30 July 1775), *anatomy, biochemistry.*

Galeazzi (sometimes wrongly called Galeati), of whose background little is known, attended the Jesuit College in Bologna and studied medicine with the physician Matteo Bazzani, who is supposed to have discovered the coloring of bones in animals fed with madder root. He learned anatomy in the atmosphere created by Antonio Maria Valsalva, who in 1705 was named professor of anatomy at Bologna.

Galeazzi was graduated Doctor of Philosophy and Medicine in 1709 and took up science under the influence of Giacomo Bartolomeo Beccari, then professor of physics at Bologna. He was immediately appointed substitute lecturer in experimental physics; in 1734, when Beccari transferred to the chair of chemistry, Galeazzi succeeded him as professor of physics. His first published work in physics dealt with the construction of mercury thermometers.

In 1714 Galeazzi visited Paris, where he met Jacques Cassini, Louis Lémery, Malebranche, Réaumur, and other men of science and attended meetings of the Académie Royale des Sciences. During those meetings he became interested in the debate between Claude Geoffroy and Lémery concerning the significance of microscopic iron particles in living organisms: Geoffroy contended that these particles were produced by the organism, while Lémery asserted that they had been assimilated. On his return to Bologna, Galeazzi conducted experiments and demonstrated through chemical analysis that the iron particles were assimilated by the organism. He then made a systematic study to discover the connection between the iron in living organisms and iron salts in the soil. Later, in 1746, Galeazzi first ascertained the presence of iron in the human blood; his pupil Vincenzo Menghini detected whole hematic iron in erythrocytes.

In 1716 Galeazzi was appointed professor of philosophy at Bologna, a post he held for forty years. In 1719 he made geological observations in the Emilian Apennines. Later, in entomology, he discovered the endophagous generation of the fly and the oviparous generation of a cochineal, *Pulvinaria vitis* (L.). Galeazzi also had a successful medical practice and wrote on the use of Peruvian bark (cinchona), on jaundice, and on gallstones and kidney stones. He is remembered today primarily for his anatomical research.

Galeazzi began his anatomical work in 1711, when he observed corpora lutea in different stages of regression in pregnant women; but his most important anatomical discoveries were in the gastrointestinal system. He defined the positions of the three layers of muscle fibers in the stomach and described the peculiar arrangement of the superficial layer of longitudinal muscle fibers. He described two layers of muscle fibers in the small intestine: the interior circular and the external longitudinal; in the colon he considered only the circular layer important, because the external layer of longitudinal fibers forms only three longitudinal bands.

In the mucous coat of the intestines Galeazzi described the glands now called Lieberkühn's glands. Johann Nathanael Lieberkühn described them in 1745; Galeazzi made his observations in 1725 and published them in 1731. The existence of intestinal glands had already been claimed by Malpighi (1688), the villi of the small intestine had been described by Gaspare Aselli, and Thomas Bartholin had studied the villi in connection with the chyliferous vessels. In Galeazzi's time the question was whether the villi were hollow siphons, spongy perforated papillae, or unperforated papillae. Galeazzi clarified the structure of the villus: it can be long and cylindrical or short and squat, but he denied the existence of any free lymphatic opening on its surface.

Galeazzi discovered many minute pores not on the villi but between them, distributed over the entire intestinal surface. He concluded that a special sieve-like membrane exists on the interior surface of the intestines and that each of these numerous pores is an opening of a glandular structure in the intestinal wall. Galeazzi also wrote that these glandular structures discharge a secretion into the intestinal cavity.

In 1756 Galeazzi retired from teaching philosophy; although he was a distinguished anatomist, Galeazzi never held the chair of anatomy. He died at the age of eighty-eight and was buried in the Church of the Confraternity of St. Philip Neri in Bologna.

BIBLIOGRAPHY

I. ORIGINAL WORKS. Galeazzi's writings include "De muliebrium ovariorum vesiculis," in *Commentarii de Bononiensi Scientiarum et Artium Instituto atque Academia,* **1,** pt. 2 (1731), 127–130; "De calculis in cystifellea repertis," *ibid.,* 354–358; "De cribriformi intestinorum tunica," *ibid.,* 359–370; "De ferreis particulis quae in corporibus reperiuntur," *ibid.,* **2,** pt. 2 (1746), 20–38; "De thermometris Amontonianis conficiendis," *ibid.,* 201–209; "De carnea ventriculi et intestinorum tunica," *ibid.,* 238–243; "De insecto quodam in vite reperto," *ibid.,* 279–284; "De moscho," *ibid.,* **3** (1755), 177–193; "De morbis duobus," *ibid.,* **4** (1757), 26–43; "De renum morbis," *ibid.,* **5** (1757), 139–150, 249–260; "De cortice peruviano," *ibid.,* 216; and "De sudore quodam atque urina colore nigerrimo infectis," *ibid.,* **6** (1783), 1–12.

II. SECONDARY LITERATURE. On Galeazzi or his work, see A. Corti, "L'anatomico bolognese Domenico Gusmano Galeazzi e la sua esauriente descrizione delle ghiandole intestinali che molti dicono di Lieberkühn," in *Archivio italiano di anatomia e di embriologia,* **19** (1922), 407–434; and "Note storiche e biografiche su Bologna e il suo Studio," in *Rivista di storia delle scienze mediche e naturali,* **40** (1949), 19–51; J. N. Lieberkühn, *Dissertatio anatomico-physiologica de fabrica et actione villorum intestinorum tenuium hominis* (Leiden, 1745); M. Malpighi, "De structura glandularum conglobatarum," a letter to the Royal Society (1688), in *Marcelli Malpighii opera posthuma* (London, 1697), pp. 152–165; M. Medici, *Compendio storico della scuola anatomica di Bologna dal Rinascimento a tutto il secolo XVIII* (Bologna, 1857), 256–272; and V. Menghini, "De ferrearum particularum sede in sanguine," in *Commentarii de bononiensi scientiarum et artium instituto atque academia,* **2,** pt. 2 (1746), 244–266.

PIETRO FRANCESCHINI

GALEN (*b.* Pergamum, A.D. 129/130; *d.* 199/200), *medicine.*

The frequently cited forename "Claudius" is not documented in the ancient texts and seems to have been added in the Renaissance.[1]

In earlier research four years were accepted as possible birth dates: 128, 129, 130, and 131. After J. Ilberg, one of the foremost experts on Galen's biography, had committed himself to 129,[2] J. Walsh advocated the year 130, and the period around 22 September as the actual day of birth.[3] Ilberg defended his own date,[4] and Walsh again established his case with weighty and well-grounded arguments in a detailed reply to Ilberg,[5] whose sudden death prevented his responding. The writer is inclined, however, to accept Ilberg's chronology. Complete certainty cannot be achieved, but 128 and 131 are out of the question.

Parentage. Galen's father, Nikon, was an architect and geometer in Pergamum.[6,7] Galen's personal relationships with his parents were such that he spoke of his father with the greatest respect but compared his mother to Xanthippe.[8] Galen provided extensive data concerning his own life,[9] following the ancient tradition of autobiography.[10] The most important, composed in his later years, are *On the Arrangement of His Own Writings* and *On His Own Writings.*[11] A wealth of autobiographical statements are contained in many of Galen's other extant writings,[12] and additional details are in the Arabic translations.[13] The evocatively titled *On Slander, in Which Is Also Discussed His Own Life* is unfortunately lost.[14] Good introductions and surveys are Sarton[15] and Diepgen,[16] but Deichgräber[17] has justly remarked that these books are mainly biographies of Galen the writer.

Education. Continuing a family tradition, Galen's father had received an intensive education in mathematics and was generally a very cultivated man; similarly, he began to give his own son private lessons at an early age.[18] At fourteen, Galen received instruction in philosophy that encompassed the teachings of all the various schools.[19] Two years later he had to decide on an occupation; a dream allegedly caused him and his father to decide definitely that he should undertake medical studies.[20] (The medical literature of the time advised those interested in the profession to begin the study of medicine at the earliest possible age, usually at about fifteen.[21] Thus, Galen was not at all exceptional.) As to his motivation, the supposed dream is surely to be understood merely symbolically. It seems that at this time the young Galen was undergoing a kind of intellectual crisis as a result of his very extensive and eclectic philosophical education. According to his own testimony, he was rescued from this troubled state only through the help of mathematics—that is to say, geometry—with its certainty and its indubitable systematic foundation. Evidently, Galen was already seeking in the empirical realm the same certainty that he hoped to find in medicine, above all in a knowledge of the body.[22]

Galen's first medical teacher in Pergamum was Satyrus.[23] The latter, Galen reports, had been in that city for "four years already" along with Rufinus, the founder of the Asclepeum in Pergamum.[24] In all probability this Rufinus is Lucius Cuspius Rufinus, who had been consul in the year 142.[25] (The date Galen gives could refer to the year 146; this would fit with 130 as the year of his birth and at the same time agree with the report that Galen began his medical studies at the age of sixteen.) Hence it appears that even at this early date Galen was in contact with leading personalities of Pergamene society. It is not without interest in this connection that Satyrus treated the famous Sophist Aristides, and thus the young Galen very probably met the latter in Pergamum.[26] In these encounters, therefore, lie the origins of Galen's relation to the so-called Second Sophistic, of which he was later to become one of the leading medical representatives and which was so important in the intellectual life of the early Christian era.[27]

Galen composed several works while still a student in Pergamum. He himself names three of them: *On the Anatomy of the Uterus,* the *Diagnosis of Diseases of the Eye,* and *On Medical Experience.*[28] The second has not been preserved. The treatise on the uterus (written at the request of a midwife) confirms that Galen at first devoted his medical studies primarily to anatomy. The work on medical experience, on the other hand, shows that from the beginning Galen was also interested, both philosophically and practically, in the problem of certainty in the empirical sciences. This work was obviously written at the end of his first period in Pergamum—that is, about 150—since it reflects a two-day debate between the physicians Pelops and Philippus in Smyrna. This contact with Pelops induced Galen to go to Smyrna as his student.[29]

These three works, however, are clearly not the only ones Galen wrote during the first Pergamene period of study. In the treatise *On Medical Experience* he explicitly refers to an earlier work, *On the Best Sect,* which consequently must have been among his very earliest writings.[30] It is at the least very questionable whether it was identical with what has come down to us under Galen's name as *On the Best Sect, for Thrasybulos;* the latter work, as I. von Müller has shown, is very probably not genuine.[31] Here then, perhaps, is one of the not unusual cases in which a forgery has strayed into the gigantic *Corpus Galenicum.* This happened even in Galen's own lifetime, as he himself reports.[32]

Finally, in this group of earliest writings should be included, tentatively, the short treatise *On Pleuritis, for Patrophilus.*[33] If Patrophilus were identical with Calvisius Patrophilus, who held a high position in Egypt in the year 147–148,[34] this would provide further evidence of Galen's having had early links with influential people. Moreover, he later incorporated this short work verbatim into the larger *On the Constitution of the Medical Art* (also dedicated to Patrophilus),[35] a not unusual practice for Galen. The short treatise on pleuritis is constructed almost like a mathematical-geometric demonstration (it uses short, precise statements, and the phrase "from this it necessarily follows" recurs with striking frequency). This manner of construction seems, therefore, to be the first concrete application of the principle that he considered best suited to overcome his skeptical despair—namely, the employment of the form of the geometric proof in demonstrating the real facts of natural science and medicine.[36]

The problem of Galen's earliest writings has been discussed in detail as the best procedure for tracing what may be called the concretization of his original intellectual impulses. In what follows references to Galen's writings and their chronology will be made only occasionally and allusively.[37]

Galen's father died when his son was twenty and still living in Pergamum.[38] Not long afterward Galen went to Smyrna to study medicine with Pelops, whom he called his second medical teacher,[39] and Platonic philosophy with Albinus.[40] In philosophy Galen was most influenced by Platonism, just as later Hippocratism exercised the greatest influence on him in medicine;[41] indeed, he set forth a connection between the two in his great work *On The Doctrines of Hippocrates and Plato.*

From Smyrna, Galen went to Corinth, to continue his medical education with Numisianus,[42] and finally to Alexandria,[43] then the most famous center of research and training in medicine. It is generally assumed that Galen remained there for several years,[44] considerably longer than in Smyrna and Corinth. Yet, he himself says that after a stay in Smyrna, "I was in Corinth . . . in Alexandria and among several other peoples."[45] Doubtless Alexandria was very attractive to him. Moreover, it offered the only opportunity to examine human skeletons thoroughly (although not cadavers).[46] Nevertheless, Galen was highly critical of the research and pedagogical activity then being conducted at Alexandria, as is proved by numerous later sarcastic remarks about "Alexandrian prophets" and Alexandrian "scholasticism."[47]

Thus, Galen pursued his extended study of medicine purposively and intensively on the basis of definite intellectual presuppositions but was not, in fact, fully satisfied with them and at the conclusion of this study still had not completely decided on his own definitive approach. At the age of twenty-eight he returned to his native Pergamum as physician to the gladiators,[48] having studied medicine for about twelve years, much longer than was then customary. This does not mean, however, that he could not perhaps have practiced medicine on some occasions during this long period.[49]

Galen continued to treat the gladiators in Pergamum for several years. His experiences in this capacity brought him some chance discoveries (for example, the behavior of certain nerves and tendons), which were important for his later researches.[50] Yet his most important work was still before him.

Galen and Rome. In the year 161, at the beginning of the reign of the two Antonine emperors, Galen arrived at Rome where he quickly established a medical practice.[51,52] He succeeded in effecting several startling cures of influential patients.[53] Among them was the Peripatetic philosopher Eudemus, who introduced Galen to the high government official Flavius Boethus. The latter prompted Galen to compose his first major anatomical and physiological works. At this point a basic observation regarding Galen's literary activity should be made: as can be seen from the *Anatomical Procedures,* for example, Galen revised many of his works. The first version of the great anatomical work, written for Flavius Boethus during the first stay in Rome, contained only two books;[54] only later did it reach the dimensions in which it has been preserved. For certain reasons Galen later revised other of his works stylistically or substantively. On the whole, he had not at first planned on a "public edition" of some of the works, and such writings or versions of writings are fundamentally different from those destined for publication.[55]

Flavius Boethus also inspired Galen to hold public anatomical lectures and demonstrations. Such public medical lectures had come into fashion in the first century B.C.[56] in the context of an accelerated activity aimed at increasing the level of general knowledge, and they had been enthusiastically taken up again at the end of the first century A.D. by practitioners of the Second Sophistic. Among Galen's auditors were not only high Roman officials, including the consuls Lucius Sergius Paullus and Gnaeus Claudius Severus, but also famous Sophists and rhetoricians, such as Hadrian of Tyre and Demetrius of Alexandria. This closeness among highly placed Romans, Sophist rhetoricians, and scientific experts (particularly physicians) is typical of the Second Sophistic.

Galen himself vigorously insisted that it was not primarily *logoi sophistikoi* but rather medical successes that established his fame in Rome.[57] On the other hand, the Second Sophistic exercised a great influence on him, and he cannot be seen apart from this intellectual movement. His own teacher Satyrus was referred to as a physician and Sophist;[58] and the "iatrosophists" and "iatrophilosophers" were a typical phenomenon of the age. Among them were some extremely dubious individuals, and the whole school of thought, the physicians who subscribed to it in particular, occasionally ran the risk of slipping into nonsense or irrationality.[59] Nevertheless, this specific form of intellectual-social culture also contained some very positive features, which Bowersock has recently described. Galen was among the very best representatives in medicine of this tendency, but he too is not free of its negative traits.

Three such characteristics, which are frequently discussed and which can perhaps be judged most fairly by viewing Galen as a Sophist, are his prolixity, his vainglory, and his taste for dispute and polemic. Undoubtedly many of Galen's writings are fatiguingly diffuse—for which reason the great classical philologist Wilamowitz bestowed upon him the malicious epithet *Seichbeutel* ("windbag").[60] Comparison of these writings with several of his terse, mathematically precise earlier works will reveal a change in style and an increased boastfulness, surely attributable to the Sophistic influence, which became especially intense while he was practicing in Rome. The same is true of his proclivity for debate and polemic—Galen may here have inherited something from his mother. Yet, at least as important are the professional quarrels that, under the influence of Sophism, were completely typical of that period.[61] It is against this background that, for instance, Galen's polemic with his colleague Martialius during the first stay in Rome must be seen,[62] and no doubt his relations with other physicians were in general affected in the same way.[63]

At the time of his dispute with Martialius, Galen was thirty-four.[64] He stayed in Rome for three more years. Looking back on his public lectures and polemics of this period, he felt a certain degree of remorse and declared that he had decided not to act in this manner in public again—one of the good resolutions in which he did not persevere. Although he often repeated assurances that he wished to approach all things *sine ira et studio,*[65] he could free himself from neither his own temperament nor tradition.

Apparently, however, the many quarrels and polemical debates during his first Roman period upset him inwardly, so that after a while he longed to return to Pergamum.[66] He states that his Roman patrons wished to keep him in Rome and that this strengthened his resolve to go back to his native city.[67] In any case, Galen left Rome in the spring, shortly before the return of Lucius Verus to the capital. As he himself records: "When the great plague [an epidemic that accompanied the soldiers Lucius Verus was leading back from the Parthian War] broke out, I left the city and hastened home."[68] This departure has generally been explained as simply an escape in the face of the contagion—not exactly laudable behavior for a physician. On the other hand, Walsh has pointed out that even before this time Galen had become what might be called "Rome-weary." [69] Surely, many factors conjoined to produce his decision.[70]

Galen relates that following his return "he stuck to the customary things." [71] It seems that he also undertook other journeys for scientific purposes.[72] But this period did not last long: he received a letter from the two rulers Marcus Aurelius and Lucius Verus summoning him to Aquileia. He responded to this call and succeeded in escaping a renewed outbreak of the plague. He became personal physician to the young Commodus; he held this position for several years and was thereby able to pursue his medical research and literary activity in various Italian cities—although at first not in Rome. When Commodus became emperor in 180, Galen remained in close contact with him, just as he had been on friendly terms with Marcus Aurelius, and enjoyed his protection. He was likewise friendly with Septimius Severus, who became emperor in 193. Although Galen should not, on this account, automatically be considered "physician in ordinary" or "court physician" in a strict sense, beginning with his return to Italy he remained in contact with the imperial family as both a physician and a socially prominent personality; he was undoubtedly "a lion of society." [73] Moreover, he enjoyed the favor of such highly placed figures as the Sophist Aelius Antipater, secretary to Septimius Severus.[74]

Nor, in his second Roman period (which lasted several decades), was he spared public controversies. The most famous episode, concerning the originality of some of his anatomical conclusions, took place in the Temple of Peace in Rome.[75] A heavy personal blow for Galen was the loss of a large part of his library through a fire in the temple in the year 192. It is not known whether Galen spent the last years of his life in Rome or in his native city.

Standpoint and Position in Ancient Medicine. W. Pagel's general characterization of Galen is most concise: "a genius who is 'modern' and indispensable in so many ways and yet not easy to grasp in view of his limitations, obscurities, and apparent self-contradictions." [76] It is understandable then that there exists as yet no work that comprehensively and exhaustively treats Galen or even limited aspects of his personality and work. The following remarks are thus to be understood merely as hints or suggestions.

As a physician Galen accepted the "fourfold scheme" which brought the humors, the elementary qualities, the elements, the seasons, age, and other factors into common accord.[77] This fundamental theoretical system obviously satisfied Galen's striving for certainty, yet not infrequently he, who so often pleaded for a purely scientific basis and methodology in medicine, was forced to perform complicated intellectual maneuvers. This is especially evident in his making the mysterious black bile into a physiologically important humor, which subject he treated in *On the Black Bile.* Using the fourfold scheme, Galen attempted to restore medicine to its Hippocratic basis. After all the debates with many schools of medical thought, he still considered Hippocratism the most secure foundation and enunciated this belief most clearly in *On the Elements According to Hippocrates, On Mixtures,* and the commentary to the Hippocratic work *On the Nature of Man.* Galen constructed his own Hippocratism, however; Hippocrates himself (to the extent that we can grasp the genuine Hippocrates) was not acquainted with any fourfold system in Galen's sense. Nonetheless, Galen's suggestive construction was dominant for centuries.

Galen's anatomy suffers from a similar conflict. On the one hand, he was an energetic advocate of anatomy as the foundation of medicine, and his own accomplishments in anatomy, both as writer and as researcher and demonstrator, are without doubt considerable.[78] The great work *Anatomical Procedures* is proof of this, as is a series of his more specialized writings. On the other hand, his anatomy necessarily suffered from the lack of opportunity to examine human cadavers (human dissection was no longer possible for cultural reasons). Moreover, in attempting to call into question—in principle, at least—some of the anatomico-physiological achievements of Erasistratus (especially in regard to the role of the spleen) and in trying to place anatomy as well on a kind of Hippocratic basis, he necessarily introduced a speculative element into his anatomy.

The same is true of Galen's physiology. He had a clear conception of the importance of physiological experiment, and his knowledge of the physiology of the nerves was considerable and justly celebrated. Yet here again there is speculation, in substantial part teleological, as in the great physiologico-anatomical treatise *On the Usefulness of the Parts of the Body.*[79] As a teleologist in physiology Galen was a determined Aristotelian.[80] He consequently defended in almost hymnic praises the notion that the Demiurge has created everything for the best.

As a dietitian, Galen continued an illustrious and ancient tradition.[81] In this area conflict is less noticeable. Indeed, it is precisely in this field that Galen the author composed what even the layman would consider his most interesting and exciting works, especially the *Hygieina.*[82]

Among those writings that best illustrate Galen's Hippocratism are his commentaries on his predecessor. From them we learn that what Galen thought was genuinely Hippocratic was often subjectively established, as is obvious from his claims for the genuineness of the treatise *On the Nature of Man.* Many critics held that it was a work of Polybus; Galen, however, allegedly in agreement with most physicians, declared the work to be genuinely Hippocratic, because it provided the testimony he sought for his doctrine of the four humors.[83] He also composed a work now lost, *On Genuine and Ungenuine Hippocratic Writings.* His commentaries were not at all confined to technical medical explanation but also treated purely philological problems (see below). Originally these commentaries did not take into account the explanations of other authors.[84] Later, when he was planning a public edition of his writings, he considered such interpretations and recast his already existing commentary to integrate it with them. Here Galen can be seen at work as a polemicist who not only criticized content but also did not shrink from personal defamation.[85] Thus, for example, he reproached a Jewish colleague for being incapable of understanding the books of the ancients and, consequently, being unable to comment on Hippocrates.[86] In this enterprise he not infrequently abandoned the principle of *sine ira et studio,* for which he argued so strongly elsewhere.

It must be stressed in regard to Galen as a clinician that he was without doubt a brilliant diagnostician. He composed a series of important works on the division of diseases and on their various symptoms.[87] A splendid example of Galen's diagnostic art is his short treatise on malingerers;[88] surely the diagnostic accuracy of a physician is put to the test in this field.

If Galen had very great merits as a diagnostician (and naturally went far beyond his esteemed Hippocrates), the same cannot be said of his views on prognosis. In this field he remained far more of a Hippocratic: first, he developed, as a general principle, a self-reliance in prognosis that was unjustified. The resources he employed were the traditional ones—the famous doctrine of the "critical days," saturated with several prerational elements as it doubtless was, could nonetheless satisfy particularly well Galen's need for certainty and his predilection for mathematical regularity (apparent in his special studies *On Critical Days* and *On Crises*). As a Hippocratic he found in his model something that Lichtenthaeler has termed "logos mathématique."[89] To be sure, the theory of the critical days had an empirical root in the observation of fever (malaria) cycles. On the whole, however, it had been expanded into a speculative system under the influence of a "rationalization of an old arithmological foundation," as Joly has accurately expressed it.[90] As stated above, this speculative system of the critical days, penetrated by prerational elements, satisfied Galen's desire for certainty, just as it suited his taste for theorizing. The same is true of the two other traditional tools of prognosis, which he used willingly and extensively: the doctrines of the pulse and of the urine. The behavior of the pulse, which like all bodily "palpitations" was originally considered a mantic phenomenon, is analyzed in extreme detail in Galen's many works on sphygmology and is employed in both diagnosis and prognosis.[91] Yet in just this subtle division of the "qualities" of the pulse lay the danger of speculation. (Characteristically, Galen had little respect for the quantitative measurement of the pulse, despite the beginnings that Herophilus, among others, had made.)

In his knowledge of the formation of the urine, Galen went far beyond Hippocrates; but, as a clinician, he cherished the same speculative conception as Hippocrates, who had employed the consistency, the sediments, and the color of the urine not only in diagnosis but also in prognosis. The crucial thing is that Galen, like Hippocrates, did not apply the three criteria of consistency, sediments, and color in an exact fashion but rather in a vague, subjective way.[92]

Galen's knowledge of therapeutics is set forth mainly in the voluminous work *Therapeutic Method,* generally known in English as *On the Art of Healing.* Once again much interesting and correct material stands side by side with conjecture, above all for the medical preparations Galen employs and for his notions of their mode of action (on this, see *On the Mixture and Action of Simple Medicines, On the Composition of Medicines According to Locality,* and *On the Composition of Medicines According to Types*). In general, while Galen did in truth place the highest value on the empirical testing of medicines, his speculative conceptions of the way they worked constrained him occasionally to see a positive value in such "medicines" as excrement and amulets, in spite of his rejection of the magico-irrational medicine, which was then very popular.[93]

The impression that Galen possessed all clinical skills is only apparent. On closer examination he seems to have had no experience in operative gynecology[94] and obstetrics or in surgery in general, and it is obviously for this reason that he devoted none of his own writings to these fields.[95] Still more remarkable is his personal attitude toward surgery. On the whole, Galen was an inveterate internist and as such had a deep distaste for surgery, with the exception of surgery to repair injuries or suppurations, undertaken in treating gladiators. In those cases in which he did consider surgical questions, even regressive views can sometimes be detected. This prejudice is consistent with, among other things, his prolonged polemic against Erasistratus (and the so-called Erasistrateans), who had been one of the first to put operative surgery on a new basis. Galen, on the contrary, was, so to speak, a Hippocratic even as a surgeon; that is, he confined operative surgery—when he allowed it at all—to a relatively narrowed concept and area. Here the characteristic inconsistency and limitations of his otherwise universal mind show themselves with particular clarity.

In summary, then, as a physician, Galen was a Hippocratic and, as a scientist (anatomist and physiologist), an Aristotelian; and he adhered to these basic commitments even when he was ostensibly an eclectic. To this extent he was far from being primarily an eclectic, a designation he is not infrequently given. His inclination for philosophy went so far that he attempted to reconcile Hippocrates and Plato, and in a work whose title was chosen with this purpose in mind, too, he claimed "that the best physician is also a philosopher." On the other hand, he recognized and emphasized the boundary between medicine and philosophy.[96] All this, together with the contradictions in his behavior sketched above, should perhaps be viewed in the light of his constant striving for certainty. Thus, in the end, Galen became as much the "savior of a medicine which had become bankrupt" as the "executor of a faulty development."[97]

Philosophy and Philology. Aside from his medico-philosophical efforts, Galen not only interpreted the work of other philosophers (Plato, Aristotle, Theophrastus, Chrysippus, Epicurus) in some of his works[98] but also became known as a philosopher in his own right, above all in the field of logic, in such works as *On Scientific Proof* and *Introduction to Logic.* His many ethical writings have not been preserved, but this is not a significant loss, since as a philosopher Galen was essentially unoriginal.

His most outstanding appearance as a philologist and grammarian is in his commentaries on Hippocrates.[99] He also wrote a series of works dealing with lexicographical and stylistic problems, but they have been lost.[100] The work that has come down under his name as *Hippocrates Glossary* is quite possibly not genuine. Galen's philosophical and philological interests were, moreover, only part of the total activity of a man of truly universal education whose numerous writings contain an abundance of information.

Religion. This is a subject which perhaps merits separate examination. Here, too, a good deal of inconsistency seems to become evident. On the one hand, Galen has been called a typical representative of the "cultivated religion" of his time.[101] This means that as an enlightened man he quite possibly took a certain interest in religious phenomena, but one without genuine religious commitment behind it, and that, for the rest, he "officially" believed in the gods in the traditional manner. This view is supported by the lack of any real understanding in his utterances about Judaism and Christianity.[102] Furthermore, it is certain that he did not yield to his age's widespread passion for mysticism, often encountered even among the educated and particularly among physicians.[103] On the other hand, he has been termed "deeply religious"[104] because of the almost hymnic praise of the "Creator" (*demiourgos*) in his great physiological work *On the Usefulness of the Parts.* But, is teleology—the real subject of this treatise—identical with religiosity? Do we have here a secret religious yearning, or is it simply a question of the stiff, formal language of allegory? We would like to know the answer for several reasons. First, it would be significant to learn whether Galen considered religious commitment to be an unavoidable component of medical ethics. Second, Galen's revered predecessor Hippocrates was surely not "deeply religious" in the ordinary sense of the term.[105] In this connection one would wish to ask how Galen viewed the problem of Hippocratic religiosity. In addition, his relationship to Asclepius and to the "god-sent" dreams would have to be thoroughly analyzed once again.[106] Perhaps Galen's religiosity, too, must be considered in terms of that dichotomy characteristic of so much of his work.

Pseudo-Galenica. Those writings falsely attributed to Galen form a special chapter in the history of his influence. That such spurious writings existed during Galen's own lifetime has already been noted and is proof of the great authority and attractiveness of his name, an obvious incentive to forgers. In many instances the intellectual milieu out of which such falsifications arose is known (although more work should be done in this area). For example, the extant treatise *For Gauros, On the Question of How Embryos Are Ensouled* belongs to the circle of Porphyry, that is, to early Neoplatonism; and the lost work *On Medicine in Homer* appears to have come from the group around Sextus Julius Africanus, that is, from early Christianity and Neoplatonism.[107] In these cases, there was an attempt to legitimate certain views by placing them under a great, authoritative name.

Other spurious writings emerged when something written on a popular subject could more easily be sold under a distinguished name. For example, the forged works on urine, printed in volume XIX of Kühn's edition, must have originated in this way (witness the popularity of urine prognostication in late antiquity and the Middle Ages). Likewise, the treatise *On Sudden Death* was a contribution to the much discussed subject of the forecasting of death.[108]

Still other false Galenic works obviously had their origin in the medical teaching of late antiquity and the Middle Ages. Among these are the so-called *Summaria Alexandrina,* in which Galen's longer writings are presented in the form of a summary or abridged edition—and thereby often simplified and distorted.[109] Such compendia must have begun to be produced soon after Galen's death; an early example is the short tractate on *Galen's Hippocratic Principles.*[110] The great majority of these forgeries were probably written in Greek. The Greek originals of some of the works of this kind are extant, while others are available only in Latin or Arabic translations. The Arabs were well aware that such spurious works existed and to some extent made an effort to discover which ones they were.[111] Yet it is not out of the question that among the mass of Arabic writings in Galen's name one or another was actually composed by an Arab author and smuggled into the Galenic corpus. Works were still being forged in the Renaissance—e.g., the commentaries on the pseudo-Hippocratic *On Nourishment*[112] and *On the Humors.*[113]

The spurious Galenic writings will someday have to be studied as a whole. In view of the mass of Galen's works that are still unedited and in the absence of a critical, philologically sound complete edition of Galen, such a task would be arduous. In considering the extant Greek writings in the course of such an undertaking, one would have to employ, for example, stylistic comparison; but little preparatory work for a study of Galen's style has been done.

Galen's Influence. Far too little attention has been paid to the fact that, although he was very interested in medical pedagogy, Galen had no real students of his own[114] and, unlike many of his colleagues, founded no school. Even in his own lifetime he was a quite exceptional figure. His stature was explicitly acknowledged immediately after his death; it was the enormous range of his literary works, above all, that led to his being called divine.[115] In the medical schools of late antiquity and the Middle Ages, Galen's writings constituted the principal element of the curriculum (the *Summaria Alexandrina* has already been cited), and excerpts from Galen occupy considerable space in the great medical encyclopedias of Oribasius and Aetius of Amida. Byzantine physicians—Alexander of Tralles, for example—did not always accept Galen uncritically by any means; in general, however, they were all crucially dependent on him.

In the non-Greek world Galen's influence was based on innumerable translations of his works. Of those in Latin only a few need be cited: those by Cassius Felix, who in the fifth century translated Greek authors *logicae sectae,* including Galen, in his *De medicina liber;* and, in the medieval period, the important translations of Pietro d'Abano[116] and Nicola da Reggio.[117] In addition, Galen was early translated into Syriac;[118] but the Arabic translations had the greatest impact. In this regard the achievements of Ḥunayn ibn Isḥāq and his school are especially outstanding; moreover, he has also provided a survey of the Syriac and Arabic translations.[119]

The influence of Galen, transmitted equally by his own writings, in both the original Greek texts and translations, and by summaries, compendia, commentaries by other physicians, and even forgeries, created Galenism, which dominated the medicine of the Middle Ages. The real battle between the Galenists and the medical "revolutionaries" took place in the Renaissance. With the introduction of printing there occurred a revival of the genuine Galen in the form of text editions and commentaries.[120] The most important criticisms directed against him were in the fields of anatomy (where he was exposed as an "ape anatomist" and corrected), physiology (in which his dogma of the liver as the starting point of the blood was overthrown), and therapy (from which the bloodletting controversy linked primarily with the name of Brissot emerged). If Galen's authority was not destroyed in the Renaissance, it was seriously called into question.[121] Yet his influence was far from being eliminated thereby. For one thing, the results of the criticism of Galen (for example, Harvey's discovery of the circulation of the blood) occasionally required a long time to be definitively accepted among physicians. For another, the conception of, for example, humoral pathology in the form codified by Galen, encompassing such ideas as bad humors and blood purification, was so deeply rooted outside of the so-called school medicine that even around 1900 one could speak of a "Neogalenism."[122]

FRIDOLF KUDLIEN

GALEN: Anatomy and Physiology.

Galen's physiological system was, from the second century A.D. until the time of William Harvey, the basis for the explanation of the physiology of the body. His physiological theories are of particular interest because they included concepts of digestion, assimilation, blood formation, the maintenance of the tissues, nerve function, respiration, the heart beat, the arterial pulse, and the maintenance of vital warmth throughout the body—concepts which together formed a comprehensive and connected account of the functioning of the living animal body. His physiological system was based in large part on the work of such earlier anatomists as Aristotle, Praxagoras, Herophilus, and Erasistratus, but Galen made fundamental changes and additions to their theories and the resultant system was identifiably his own.

Aristotle had drawn attention to the role of the blood in forming the tissues of the developing embryo and a natural corollary of this role was that the blood should serve also to nourish and maintain the flesh of the adult body. Erasistratus had said that the food digested in the stomach and intestines was absorbed through the intestinal wall into the mesenteric veins as chyle, and the chyle was carried by these veins to the liver where it was transformed into blood. From the liver the blood was poured through the hepatic veins into the vena cava and thence distributed through the venous system to nourish the whole body. Galen adopted the same view. He considered the liver to be the chief organ governing the vegetative functions of the body, those functions which Aristotle said were governed by the vegetative soul. The liver attracted the blood-forming elements of the chyle and transformed them into blood; the gallbladder attracted those elements of the chyle unsuitable to form blood and discharged them as bile into the intestine. The bile was a by-product of blood formation.

Erasistratus had also thought that the blood which entered the right ventricle of the heart from the vena cava was prevented from returning by the tricuspid valve. The blood was then sent on via the pulmonary artery (or artery-like vein) to the lungs, which it served to nourish. Thus, the right ventricle and the pulmonary artery existed, according to Erasistratus, for nourishing the lungs, and Galen adopted the same view.

In order to study the distribution of the blood vessels Aristotle had advised that an animal intended for dissection should be killed by strangulation so as to retain the blood within the body. A result of this method of killing animals was that the left side of the heart and the arteries were left largely empty of blood. The arteries thus appeared as empty tubes running through the flesh, and Praxagoras of Cos had therefore distinguished them from the veins, or blood vessels, and had considered them to be air tubes. Erasistratus, working a generation or more after Praxagoras, in the early third century B.C., seems to have discovered and named the tricuspid and bicuspid valves guarding the entrances to the right and left ventricles of the heart respectively. He also understood that these structures functioned as valves, that is, as mechanical devices to allow the flow of materials in only one direction. Furthermore Erasistratus thought that the heart functioned like a bellows, that it distended itself actively in diastole, and the partial vacuum formed by the enlargement of the ventricles caused blood to flow into the right ventricle from the vena cava, and breath or *pneuma* to flow into the left ventricle from the lung through the pulmonary vein (vein-like artery). The contraction of the ventricles of the heart then forced blood from the right ventricle into the lungs and pneuma from the left ventricle into the arterial system. Thus, according to Erasistratus, the arterial pulse resulted from the filling of the arteries with pneuma by each contraction of the heart, and the arterial system served to convey pneuma to the whole body.

Erasistratus knew that when an artery was opened blood flowed from it, but he thought that the blood flowing from an opened artery was merely flowing through the artery from its principal reservoir in the veins. He considered that when an artery was opened the pneuma escaped from it, thereby creating a vacuum within its cavity and the blood then entered the artery from the veins through the *synanastomoses,* numerous minute, invisible passages connecting the arterial and venous systems throughout the body. He knew of the existence of the synanastomoses from the fact that when an animal was bled to death from an opened artery all of the blood in the veins was also drained away. Since there was no visible connection between the arterial and venous systems there must be a multitude of invisible connections. Erasistratus thought that these connections were normally closed, but that they opened in fevers when the presence of blood in the arteries was indicated by a flushed skin and a throbbing pulse. Similarly the opening of an artery in an animal created an abnormal and pathological condition which permitted blood to flow from the veins into the arteries.

The fundamental change which Galen made in the physiology of the heart, lungs, and vessels was to show that both the left ventricle of the heart and the arteries invariably contain blood and that this is their normal condition, not a sign of disease. By his demonstration of the normal presence of blood in the arteries, Galen destroyed Erasistratus' theory of how the pneuma was conveyed to the whole body.

Galen's proof of the normal presence of blood in the arteries is contained in his short work *Whether Blood Is Contained in the Arteries in Nature.* He observed that if Erasistratus were right that the pneuma escaped when an artery was opened the pneuma should be seen to escape first before the blood poured forth, but in fact the blood pours forth at once. Yet the pneuma, according to Erasistratus, was simply air taken into the body in breathing so that it could not be such a rarefied substance that all of the pneuma in the body could escape instantaneously through a mere pinprick in an artery. Galen also observed the presence of blood in the arteries of the transparent mesentery, and then he showed experimentally that when he isolated a portion of an artery in a living animal—he tied it off with ligatures so that no blood could flow into it from elsewhere and then opened it—he always found it full of blood. By opening the chest of a living animal he demonstrated that blood was present in the left ventricle of the heart.

Galen then had to devise new theories to account for the functions of the heart and arteries. He supposed that the prime function of respiration was to cool the excess heat of the heart. Since the lungs surround the heart in the chest cavity they might by that fact alone exert a cooling influence on the heart. In addition air might pass from the lungs along the pulmonary vein into the left ventricle, and there serve both to nourish and cool the innate heat of the heart and then return to the lungs accompanied by something like smoke. The pulse in the arteries was generated from the heart. In accounting for the arterial pulse Galen adopted an idea suggested originally by Herophilus that when the heart was in diastole a wave of dilatation passed along the walls of the arteries. The arteries thus dilated drew into themselves blood from the veins through the synanastomoses and pneuma from the surrounding air through pores in the skin. Thus the arterial pulse caused the whole body to breathe in and out and and served to nourish the innate heat throughout the body. If the pulse were cut off from a limb by a ligature, the limb became pale and cold, because, according to Galen, its innate heat was no longer nourished by the vital pneuma drawn into the arteries by the pulse.

Galen was obliged by his theory to suppose that the mitral valve, opening into the left ventricle of the heart, did not act as Erasistratus had seen that its structure would require it to act, as a device to allow the flow of materials in one direction only, into the heart. Since the mitral valve had only two flaps Galen argued that it would allow the return of air and smoky vapors from the heart to the lungs. In the right ventricle of the heart, however, Galen considered that the tricuspid valve with its three flaps was a tight valve. Blood entering the right ventricle from the vena cava could not return. Some of it passed through the pulmonary artery (artery-like vein) to the lungs, but since the pulmonary artery was smaller than the vena cava Galen thought that it could not remove all of the blood entering the right ventricle. Therefore, he said, some blood must pass through the interventricular septum into the left ventricle. Since there were synanastomoses between the venous and arterial systems throughout the body Galen thought they should also be present in the septum and in this way he explained how the left ventricle was supplied with blood. However, in his work *On the Usefulness of the Parts of the Body,* Galen considered that both ventricles contained both blood and pneuma, but that the left ventricle contained pneuma in larger proportion.

Although physiology remained for fourteen centuries after Galen's death basically Galenic it did not always coincide exactly with what Galen had taught. Galen had developed his theories in close relation to those of his predecessors, particularly Herophilus and Erasistratus, and he was not able to free them completely from inconsistencies. Furthermore, he frequently contradicted himself in different works. Galen's successors, in attempting to make his physiology simpler and more consistent, tended to revert to Erasistratus' view that the left ventricle of the heart and the arterial system contained pneuma rather than blood.

In his work *On the Doctrines of Hippocrates and Plato* Galen defended Plato's concept of a tripartite soul (that is, a nutritive soul, an animal soul, and a rational soul) against the Stoic doctrine of the soul as single and indivisible. Galen showed that in his physiological system the liver and the veins supplied the body with nutrition, the lungs, the left ventricle of the heart and the arteries maintained the pneuma and the innate heat throughout the body, while the brain and nerves controlled sensation and muscular movement through the medium of a special psychic pneuma. The later systematizers of Galen held that there were three kinds of pneuma or spirits corresponding to these three functional systems: the natural spirits formed in the liver, the vital spirits formed in the heart and arteries, and the animal spirits formed in the brain.

LEONARD G. WILSON

NOTES

1. Cf. W. Crönert, "Klaudios Galenos," in *Mitteilungen zur Geschichte der Medizin und der Naturwissenschaften und der Technik,* **1** (1902), 3 f.; K. Kalbfleisch, "Claudius Galenus," in *Berliner philologische Wochenschrift,* **22** (1902), 413.
2. J. Ilberg, "Aus Galens Praxis," in *Neue Jahrbücher für das klassische Altertum,* **15** (1905), 277, n. 1.
3. J. Walsh, "Date of Galen's Birth," in *Annals of Medical History,* n.s. **1** (1929), 378–382.
4. Ilberg, "Wann ist Galenos geboren?" in *Sudhoffs Archiv,* **23** (1930), 289–292.
5. Walsh, "Refutation of Ilberg as to the Date of Galen's Birth," in *Annals of Medical History,* n.s. **4** (1932), 126–146.
6. Crönert, p. 4.
7. H. Diller, "Nikon 18," in Pauly-Wissowa, *Real-Encyclopädie,* XVII, pt. 1 (1936), col. 507 f.
8. Galen V. 40 f.; here and following, references are, if not otherwise stated, to the Kühn ed.

GALEN

9. Cf. I. Veith, "Galen, the First Medical Autobiographer," in *Modern Medicine* (Minneapolis), **27** (1959), 232–245.
10. G. Misch, *Geschichte der Autobiographie,* 4 vols., 3rd ed., enl., I, pt. 1 (Bern, 1949).
11. Cf. *ibid.,* p. 344.
12. See the enumeration of these by J. C. G. Ackermann, in Galen I. xxi. n.*A.*
13. See M. Meyerhof, "Autobiographische Bruchstücke Galens aus arabischen Quellen," in *Sudhoffs Archiv,* **22** (1929), 72–86.
14. Cf. Galen XIX. 46.
15. G. Sarton, *Galen of Pergamon* (Lawrence, Kans., 1954).
16. P. Diepgen, *Geschichte der Medizin* (Berlin, 1949), I, 119 ff.
17. K. Deichgräber, *Galen als Erforscher des menschlichen Pulses* (Berlin, 1957), p. 32.
18. Diller, col. 507, 1. 20 ff.
19. Galen V. 41 f.; cf. X. 561, 609. A list of his teachers in philosophy is given in E. Groag and A. Stein, *Prosopographia imperii Romani,* 4 vols., 2nd ed., (Berlin, 1952), IV, art. G24
20. Galen X. 609.
21. See F. Kudlien, "Medical Education in Classical Antiquity," in C. D. O'Malley, ed., *History of Medical Education* (Berkeley–Los Angeles–London, 1970), p. 35, n. 83.
22. Cf. Sarton, p. 17; and Misch, p. 346.
23. Galen XIX. 57.
24. *Ibid.,* II. 224.
25. Cf. G. W. Bowersock, *Greek Sophists in the Roman Empire* (Oxford, 1969), pp. 60 f.
26. *Ibid.,* pp. 61 f.
27. *Ibid.,* ch. 5, *passim.*
28. Galen XIX. 16.
29. *Ibid.;* the treatise *On Medical Experience* is edited by R. Walzer (London–New York–Toronto, 1944).
30. Walzer, *Galen on Medical Experience,* p. 87; and p. viii, n. 5.
31. I. von Müller, "Über die dem Galen zugeschriebene Abhandlung Peri tes aristes haireseos," in *Sitzungsberichte der Bayerischen Akademie* (1898), pp. 53–162.
32. Galen XIX. 8 f.
33. See H. Diels, "Die Handschriften der antiken Ärzte," in *Abhandlungen der Preussischen Akademie* (1906), p. 129.
34. See A. Stein, "Calvisius 6," in Pauly-Wissowa, III, pt. 1 (1897), 1410.
35. Galen I. 274.2–279.5.
36. *Ibid.,* XIX. 40.
37. On the chronology of Galen's writings, see Ilberg, "Über die Schriftstellerei des Klaudios Galenos," in *Rheinisches Museum,* **44** (1889), 207–239; **47** (1892), 489–514; **51** (1896), 165–196; and **52** (1897), 591–623. See also K. Bardong, "Beiträge zur Hippokrates- und Galenforschung, Teil 2," in *Nachrichten. Akademie der Wissenschaften in Göttingen,* Phil.-hist. Kl., **7** (1942), pp. 603–640.
38. Galen VI. 756.
39. *Ibid.,* II. 217.
40. *Ibid.,* XIX. 16.
41. Cf. Deichgräber, pp. 33–36: "Zu Galen als Platoniker."
42. Galen II. 217.
43. On Galen's Alexandrian period, see Walsh, "Galen's Studies at the Alexandrian School," in *Annals of Medical History,* **9** (1927), 132–143.
44. Cf., for example, J. Mewaldt, "Galenos 2," in Pauly-Wissowa, VII, pt. 1 (1910), 579, lines 20 f.
45. Galen II. 217 f.
46. For this, cf. Kudlien, "Antike Anatomie und menschlicher Leichnam," in *Hermes,* **97** (1969), 79 f.
47. For this, cf. Kudlien, in O'Malley, pp. 23, 36, notes 97, 98.
48. Galen XIII. 599.
49. The early *Diagnosis of Diseases of the Eye* apparently had its source in practical experience; see Galen XIX. 16.
50. Cf. Galen XIII. 599 ff.
51. *Ibid.,* II. 215.
52. *Ibid.,* XVIIIA. 347.
53. On what follows, cf. Mewaldt, p. 579, lines 30 ff.; and Bowersock, pp. 62 ff., 82–84.
54. See Galen II. 216.
55. For this, cf., for example, Kudlien, *Die handschriftliche Überlieferung des Galenkommentars zu Hippokrates De articulis* (Berlin, 1960), pp. 19 f. In this special case, the earlier as well as the later, "official" version can be reconstructed from the manuscript tradition.
56. Cf. Kudlien, in O'Malley, pp. 20, 35, notes 84 and 85.
57. Galen VIII. 144.
58. Cf. Bowersock, p. 67.
59. Cf. Kudlien, "The Third Century A.D.—A Blank Spot in the History of Medicine?" in L. G. Stevenson and R. Multhauf, eds., *Medicine, Science and Culture: Historical Essays in Honor of Owsei Temkin* (Baltimore, 1968), pp. 32 f.
60. Wilamowitz later revised his judgment; see Deichgräber, p. 33.
61. For them, see Bowersock, ch. 7.
62. See Mewaldt, p. 579, lines 55 ff.
63. See J. Kollesch, "Galen und seine ärztlichen Kollegen," in *Das Altertum,* **11** (1965), 47–53.
64. For this and what follows, see Galen XIX. 15.
65. Cf. Deichgräber, p. 33.
66. Galen XIV. 622, 624.
67. *Ibid.,* pp. 647 f.
68. *Ibid.,* XIX. 15.
69. Walsh, "Refutation of the Charges of Cowardice Made Against Galen," in *Annals of Medical History,* n.s. **3** (1931), 195–208.
70. For a negative interpretation of his flight from Rome, see Ilberg, "Aus Galens Praxis," reprinted in H. Flashar, ed., *Antike Medizin* (Darmstadt, 1971), pp. 388 f.
71. On this and what follows, cf. Mewaldt, p. 580, lines 37 ff.
72. Cf. P. E. M. Berthelot, "Sur les voyages de Galien et de Zosime dans l'Archipel et en Asie et sur la matière médicale dans l'antiquité," in *Journal des savants* (1895), 382–387.
73. Bowersock, p. 66.
74. *Ibid.,* pp. 63 f. Some questions dealing with Galen's accomplishments as court physician are discussed in Kollesch, "Aus Galens Praxis am römischen Kaiserhof," in E. C. Welskopf, ed., *Neue Beiträge zur Geschichte der alten Welt,* II, 57–61.
75. On this and what follows, see Mewaldt, p. 580, lines 55 ff.
76. Cf. W. Pagel's review of May's trans. of *On the Usefulness of the Parts of the Body,* in *Medical History,* **14** (1970), 408.
77. Cf. W. Schöner, *Das Viererschema in der antiken Humoralpathologie* (Wiesbaden, 1964), pp. 86 ff., esp. 92.
78. For an evaluation of Galen's anatomy, see, for example, O. Temkin and W. L. Strauss, "Galen's Dissection of the Liver and of the Muscles Moving the Forearm," in *Bulletin of the History of Medicine,* **19** (1946), 167–176.
79. For an evaluation of Galen's physiology, see, for example, O. Temkin, "On Galen's Pneumatology," in *Gesnerus,* **8** (1961), 180–189; "A Galenic Model for Quantitative Physiological Reasoning?" in *Bulletin of the History of Medicine,* **35** (1961), 470–475; and "The Classical Roots of Glisson's Doctrine of Irritation," *ibid.,* **38** (1964), 297–328.
80. See Sarton, pp. 56 ff.
81. See L. Edelstein, "The Dietetics of Antiquity," in *Ancient Medicine. Selected Papers of Ludwig Edelstein,* O. and C. L. Temkin, eds. (Baltimore, 1967), pp. 303–316.
82. Mewaldt, p. 585, lines 53 ff.
83. Cf. Galen, in *Corpus medicorum Graecorum,* V, pt. 9, 1, 7 f.
84. Cf. Kudlien, *Die handschriftliche Überlieferung,* pp. 19 f.
85. Cf. L. Bröcker, "Die Methoden Galens in der literarischen Kritik," in *Rheinisches Museum,* **40** (1885), 415–438.
86. Cf. Galen, in *Corpus Medicorum Graecorum,* V, pt. 10, 2, 2, p. 413, lines 37 f.
87. Cf. Mewaldt, p. 586, lines 9 ff.
88. See Kudlien, "Wie erkannten die antiken Ärzte einen Simulanten?" in *Das Altertum,* **7** (1961), 226–233.
89. C. Lichtenthaeler, *Quatrième série d'études hippocratiques (VII–X)* (Geneva, 1963), pp. 109–135.
90. R. Joly, *Le niveau de la science hippocratique* (Paris, 1966), p. 234.

91. See Mewaldt, p. 585, lines 23 ff.
92. Cf. H. Koelbing, *Der Urin im medizinischen Denken,* pt. 5: "Die antiken Grundlagen der Harnschau," Documenta Geigy (Basel, 1967), *passim.*
93. Cf. Rothkopf, *Zum Problem des Irrationalen in der Medizin der römischen Kaiserzeit* (Kiel, 1969), 8 f., 16–19.
94. On other achievements of Galen in gynecology, see Ilberg, in Flashar, pp. 383 f., 408.
95. On what follows, see M. Michler, *Das Spezialisierungsproblem und die antike Chirurgie* (Bern–Stuttgart–Vienna, 1969), 50 ff.
96. Cf. O. Temkin, "Greek Medicine as Science and Craft," in *Isis,* **44** (1953), 224 f.
97. Michler, p. 62.
98. Cf. Mewaldt, p. 588, lines 42 ff.
99. See I. von Müller, "Galen als Philologe," in *Verhandlungen der 41. Versammlung deutscher Philologen und Schulmänner* (Munich, 1891), pp. 80–91.
100. See Mewaldt, p. 589, lines 9 ff.
101. G. Strohmaier, "Galen als Vertreter der Gebildetenreligion seiner Zeit," in Welskopf, II, 375–379.
102. Cf. R. Walzer, *Galen on Jews and Christians* (Oxford, 1949).
103. Cf. Deichgräber, p. 35.
104. See Sarton, p. 56; on Galen's religiosity, see pp. 82 ff.
105. See V. Schöllkopf, *Zum Problem der Religiosität älterer griechischer Ärzte* (Kiel, 1968).
106. Cf. Ilberg, in Flashar, p. 365.
107. Cf. Kudlien, "Zum Thema 'Homer und die Medizin,'" in *Rheinisches Museum,* **108** (1965), 299.
108. See M. Issa, *Die "galenische" Schrift "Über den plötzlichen Tod"* (Kiel, 1969).
109. It is to be hoped that O. Temkin will continue his "Studies on Late Alexandrian Medicine" (cf. *Bulletin of the History of Medicine,* **3** [1935], 405–430) and therein deal with the problem of the *Summaria Alexandrina.*
110. Cf. Kudlien, in Stevenson and Multhauf, pp. 29 f.
111. Cf. G. Bergsträsser, *Neue Materialien zu Hunain Ibn Ishaq's Galen-Bibliographie* (Nendeln, Liechtenstein, 1966), 95–98.
112. Cf. H. Diels, "Bericht über das Corpus Medicorum Graecorum," in *Sitzungsberichte der Preussischen Akademie der Wissenschaften zu Berlin* (1914), p. 128.
113. *Ibid.* (1913), p. 115; (1915), pp. 92 f.; and (1916), pp. 138 f.
114. Whether Epigenes—to whom Galen dedicated *On Prognosis*—was a doctor is not certain. Cf. Ilberg, in Flashar, p. 375.
115. On this and what follows, see Kudlien, in Stevenson and Multhauf, pp. 27 f.
116. See L. Thorndike, "Translations of the Works of Galen From the Greek by Peter of Abano," in *Isis,* **33** (1942), 649–653.
117. See Thorndike, "Translations of the Works of Galen by Nicola da Reggio," in *Byzantina Metabyzantina,* **1** (1946), 213–235; cf. I. Wille, "Überlieferung und Übersetzung. Zur Übersetzungstechnik des Nikolaus v. Rhegium in Galens Schrift De temporibus morborum," in *Helikon,* **3** (1963), 259–277.
118. Cf. M. Meyerhof, "Les versions syriaques et arabes des écrits Galeniques," in *Byzantion,* **3** (1926), 33–51.
119. Cf. Bergsträsser, *Hunain Ibn Ishaq: Über die syrischen und arabischen Galenübersetzungen,* repr. (Nendeln, 1966).
120. Cf. R. J. Durling, "A Chronological Census of Renaissance Editions and Translations of Galen," in *Journal of the Warburg and Courtauld Institutes,* **24** (1961), 230–305.
121. Cf. H. Heinrichs, *Die Überwindung der Autorität Galens durch die Denker der Renaissancezeit* (Bonn, 1914).
122. See Bachmann, "Neo-Galenismus," in *Janus,* **7** (1902), 455–459.

BIBLIOGRAPHY

The only complete ed. of Galen (Greek text with Latin trans.) is C. G. Kühn, *Claudii Galeni Opera Omnia,* 20 vols. (repr., Hildesheim, 1964–1965). The Greek text of this ed. is very defective. The *Corpus Medicorum Graecorum,* in its Abt. V and in supplementary vols., is publishing a series of philological-critical eds. of Galenic writings.

The Galenic MSS (for original texts and translations) are listed in H. Diels, "Die Handschriften der antiken Ärzte, Griechische Abteilung," in *Abhandlungen der königlich Preussischen Akademie der Wissenschaften* (Berlin, 1906), pp. 58–158; a supp. is *ibid.* (1907), 29–41. See also R. J. Durling, "Corrigenda et Addenda to Diel's Galenica," in *Traditio,* **23** (1967), 461–476.

There is no complete trans. of the works of Galen in a modern language. Among the English trans. of individual writings are *On Anatomical Procedures,* Charles Singer, trans. (Oxford, 1956), the surviving books; and W. L. H. Duckworth (Cambridge, 1962), the later books; *On the Natural Faculties,* A. J. Brock, trans., Loeb Classics (London–Cambridge, Mass., 1952); *Hygiene,* R. M. Green, trans. (Springfield, Ill., 1951); *On the Passions and Errors of the Soul,* P. W. Harkins and W. Riese, trans. (Columbus, Ohio, 1963); *On the Usefulness of the Parts of the Body,* M. T. May, trans., 2 vols. (Ithaca, N.Y., 1968); and *On Medical Experience,* R. Walzer, trans. (London–New York–Toronto, 1944), based on the 1st ed. of the Arabic version.

The most extensive modern bibliography is K. Schubring, in the Kühn ed., XX, xvii–lxii.

See also Jerome J. Bylebyl, *Cardiovascular Physiology in the Sixteenth and Early Seventeenth Centuries,* unpub. doctoral diss. (Yale University, 1969), pp. 10–137; Donald Fleming, "Galen on the Motions of the Blood in the Heart and Lungs," in *Isis,* **46** (1955), 14–21; and Leonard G. Wilson, "Erasistratus, Galen and the *Pneuma,*" in *Bulletin of the History of Medicine,* **33** (1959), 293–314.

GALERKIN, BORIS GRIGORIEVICH (*b.* Polotsk, Russia, 4 March 1871; *d.* Moscow, U.S.S.R., 12 June 1945), *mechanics, mathematics.*

For a detailed study of his life and work, see Supplement.

GALILEI, GALILEO (*b.* Pisa, Italy, 15 February 1564; *d.* Arcetri, Italy, 8 January 1642), *physics, astronomy.*

The name of Galileo is inextricably linked with the advent, early in the seventeenth century, of a marked change in the balance between speculative philosophy, mathematics, and experimental evidence in the study of natural phenomena. The period covered by his scientific publications began with the announcement of the first telescopic astronomical discoveries in 1610 and closed with the first systematic attempt to extend the mathematical treatment of physics from statics to kinematics and the strength of materials in 1638. The same period witnessed Kepler's mathematical transformation of planetary theory and Harvey's experimental attack on physiological dogma. Historians are divided in their assessment of this widespread scientific revolution with respect to its elements of continuity and innovation, both as to method and as to content. Of central importance to its understanding are the life and works of Galileo, whose personal conflict with religious authority dramatized the extent and profundity of the changing approach to nature.

Early Years. Galileo's father was Vincenzio Galilei, a musician and musical theorist and a descendant of a Florentine patrician family distinguished in medicine and public affairs. He was a member of the Florentine *Camerata,* a cultural group which included musicians whose devotion to the revival of Greek music and monody gave birth to opera. It was headed by Giovanni Bardi, who sponsored Vincenzio's musical studies under Gioseffo Zarlino at Venice around 1561. In 1562 he married Giulia Ammannati of Pescia, with whom he settled at Pisa. Galileo was the eldest of seven children. His brother Michelangelo became a professional musician and spent most of his life abroad. Two of his sisters, Virginia and Livia, married and settled in Florence. Of the other children no record survives beyond that of their births.

Galileo was first tutored at Pisa by one Jacopo Borghini. Early in the 1570's, Vincenzio returned to Florence, where he resettled the family about 1575. Galileo was then sent to school at the celebrated monastery of Santa Maria at Vallombrosa. In 1578 he entered the order as a novice, against the wishes of his father, who removed him again to Florence and applied unsuccessfully for a scholarship on his behalf at the University of Pisa. Galileo resumed his studies with the Vallombrosan monks in Florence until 1581, when he was enrolled at the University of Pisa as a medical student.

The chair of mathematics appears to have been vacant during most of Galileo's years as a student at Pisa. His formal education in astronomy was thus probably confined to lectures on the Aristotelian *De caelo* by the philosopher Francesco Buonamici. Physics was likewise taught by Aristotelian lectures, given by Buonamici and Girolamo Borro. As a medical student, Galileo may have received instruction from Andrea Cesalpino. His interest in medicine was not great; he was instead attracted to mathematics in 1583, receiving instruction from Ostilio Ricci outside the university. Ricci, a friend of Galileo's father and later a member of the Academy of Design at Florence, is said to have been a pupil of Niccolò Tartaglia. Galileo's studies of mathematics, opposed at first by his father, progressed rapidly; in 1585 he left the university without a degree and returned to Florence, where he pursued the study of Euclid and Archimedes privately.

From 1585 to 1589 Galileo gave private lessons in mathematics at Florence and private and public instruction at Siena. In 1586 he composed a short work, *La bilancetta,* in which he reconstructed the reasoning of Archimedes in the detection of the goldsmith's fraud in the matter of the crown of Hieron and described an improved hydrostatic balance. During the same period he became interested in problems of centers of gravity in solid bodies. During a visit to Rome in 1587, he made the acquaintance of the Jesuit mathematician Christoph Klau (Clavius). In 1588 he was invited by the Florentine Academy to lecture on the geography of Dante's *Inferno* treated mathematically. In the same year he applied for the chair of mathematics at the University of Bologna, seeking and obtaining from Guidobaldo del Monte an endorsement based on his theorems on the centers of gravity of paraboloids of revolution. The chair was awarded, however, to Giovanni Antonio Magini, probably on the basis of his superiority in astronomy, a subject in which Galileo appears to have shown little interest up to this time.

While Galileo was residing in Florence, his father was engaged in a controversy with Zarlino over musical theory. To destroy the old numerical theory of harmony, Vincenzio performed a series of experimental investigations of consonance and its relation to the lengths and tensions of musical strings. These he embodied in a published polemic of 1589, the *Discorso intorno all'opere di messer Gioseffo Zarlino da Chioggia,* and two unpublished treatises that survive among Galileo's papers. It is probable that Galileo's interest in the testing of mathematical rules by physical observations began with the musical experiments devised by his father during these years.

Professorship at Pisa. In 1589, on the recommendation of Guidobaldo, Galileo gained the chair of mathematics at the University of Pisa. The philosopher Jacopo Mazzoni, who came to Pisa at the same time, and Girolamo Mercuriale, professor of medicine, were close friends of the young mathematician. Luca Valerio, a Roman mathematician noted particularly for his later treatise on centers of gravity, met Galileo on a visit to Pisa and later corresponded with him. With other professors at Pisa, however, Galileo's relations were not so cordial, chiefly because of his campaign to discredit the prevailing Aristotelian physics to the advantage of his mathematical chair. His alleged demonstration at the Leaning Tower of Pisa that bodies of the same material but different weight fall with equal speed—if actually performed—was clearly not an experiment but a public challenge to the philosophers.

During Galileo's professorship at Pisa, he composed an untitled treatise on motion against the Aristotelian physics, now usually referred to as *De motu.* Its opening sections developed a theory of falling bodies derived from the buoyancy principle of Archimedes, an idea previously published by Giovanni Battista Benedetti in 1553–1554 and again in 1585. In the same treatise, Galileo derived the law governing equilibrium of weights on inclined planes and attempted to relate this law to speeds of descent. The result did not accord with experience—as Galileo noted—which may be the principal reason for his having withheld the treatise from publication. The discrepancy arose from his neglect of acceleration, a phenomenon that he then considered to be evanescent in free fall and that he accounted for by a Hipparchian theory of residual impressed force. In order to reconcile that theory with fall from rest, Galileo introduced a conception of static forces closely allied to Newton's third law of motion. Equality of action and reaction, together with the idea of virtual velocities, pervades much of Galileo's physics. From his earliest demonstrations of equilibrium on inclined planes, Galileo limited the action of tendencies to motion to infinitesimal distances, unlike his ancient and medieval predecessors. In so doing, he was able to relate vertical fall to descent along circular arcs and tangential inclined planes, an achievement that was to provide him with the key to many phenomena after he recognized the essential role of acceleration.

In his *De motu,* Galileo undertook to destroy the Aristotelian dichotomy of all motions into natural and forced motions. He did this by introducing imaginary rotations of massive spheres. Rotations of homogeneous spheres, or of any sphere having its geometric center or its center of gravity at the center of the universe, he declared to be "neutral" motions, neither natural nor forced. Motions on the horizontal plane, or on imaginary spheres concentric with the earth's center, were likewise neutral—a conception that led Galileo to his restricted concept of inertia in terrestrial physics. His discussion of spheres in *De motu* shows further that in 1590 Galileo had not yet abandoned the geocentric astronomy, but suggests that he saw no difficulty in the earth's rotation as assumed in the semi-Tychonic astronomy.

Vincenzio Galilei died in 1591, leaving Galileo, as eldest son, with heavy domestic and financial responsibilities. Galileo's position at Pisa was poorly paid; he was out of favor with the faculty of philosophy and he had offended Giovanni de' Medici by criticizing a scheme for the dredging of the harbor of Leghorn. His disrespectful attitude toward the university administration is reflected in a jocular poem he composed against the wearing of academic robes. Thus, at the end of his three-year contract, Galileo had no hope of strengthening his position at Pisa and little promise even of reappointment. Once more with the aid of Guidobaldo, he moved to the chair of mathematics at Padua. The rival candidate was again Magini, whose hostility toward Galileo after this defeat became extreme.

Professorship at Padua. The atmosphere at Padua was propitious in every way to Galileo's development. He quickly made the acquaintance of free and erudite spirits, in such men as G. V. Pinelli and Paolo Sarpi. Among his students were Gianfrancesco Sagredo and Benedetto Castelli. A conservative professor, Cesare Cremonini, became his personal friend while staunchly opposing his anti-Aristotelian views. Padua was a gathering point of the best scholars in Italy and drew students from all over Europe. Under the Venetian government, the university enjoyed virtually complete freedom from outside interference.

Galileo lectured publicly on the prescribed topics: Euclid, Sacrobosco, Ptolemy, and the pseudo-Aristotelian *Questions of Mechanics.* Privately he gave instruction also on fortification, military engineering, mechanics, and possibly also on astronomy, although we lack concrete evidence of his having become deeply interested in that subject much before 1604. He composed several treatises for the use of his students. One, usually known as *Le meccaniche,* survives in three successive forms, dating probably from 1593, 1594, and about 1600. In this treatise, besides developing further his treatment of inclined planes, he utilized as a bridge between statics and dynamics the remark that an infinitesimal force would serve to disturb equilibrium. This move, although itself not unobjectionable, removed serious existing obstacles (which had been raised on logical grounds by Guidobaldo and Simon Stevin) from the mathematical analysis of dynamic problems. Galileo's treatise, before it was first published in a French translation by Marin Mersenne in 1634, circulated widely in manuscript, and an English manuscript translation was made in 1626. Its authorship was not always known to readers even in Italy, because Galileo's treatises composed for his students were invariably supplied in copies bearing no title or signature.

In May 1597 Galileo wrote to his former colleague at Pisa, Jacopo Mazzoni, defending the Copernican system against a mistaken criticism. In August of the same year he received copies of the *Mysterium cosmographicum,* the first book by Johannes Kepler, to whom he wrote expressing his sympathies with Copernicanism. Kepler replied, urging him to support Copernicus openly, but Galileo allowed this correspondence to languish. His preference for Copernicus at this time seems to have had a mechanical rather than an astronomical basis; he wrote to Kepler that it afforded an explanation of physical effects not given by its rivals. This referred to a tidal theory of Galileo's in which the double motion of the earth was invoked to account for the periodic disturbance of its water. The first notation concerning this theory occurs in the notebooks of Sarpi in 1595. Galileo wrote a treatise on it early in 1616, and wished to make it the central theme of his Copernican *Dialogue* of 1632, considering the tides to offer a compelling argument for the double motion of the earth.

It was also in 1597 that Galileo began the production—for sale—of a mathematical instrument, the sector or proportional compass. The idea for this instrument probably came to him from Guidobaldo, whose knowledge of it may in turn have been derived from Michel Coignet. Galileo transformed it from a simple device of limited use to an elaborate calculating instrument of varied uses and of great practical utility by adding to it a number of supplementary scales. He employed a skilled artisan to produce it (and other mathematical instruments) in his own workshop and wrote a treatise on its use for engineers and military men.

During his residence at Padua, Galileo took a Venetian mistress named Marina Gamba, by whom he had two daughters and a son. The elder daughter, Virginia, who was born in 1600, later became Galileo's chief solace in life. The vivacity of her mind and the sensitivity of her spirit—as well as her many impositions on her father's good nature—are evident in the letters that Galileo received and treasured. Both she and her sister Livia were entered in a nunnery near Florence at an early age, Virginia taking the name Maria Celeste. Livia, who took the name Arcangela, was of a peevish disposition and frail health. The son, Vincenzio, was later legitimized. After periods of estrangement from his father, Vincenzio became reconciled with him in his last years but did not long survive him. Marina Gamba remained at Venice when Galileo returned to Florence, and shortly afterward she married.

Early Work on Free Fall. Toward the end of 1602, Galileo wrote to Guidobaldo concerning the motions of pendulums and the descent of bodies along the arcs and chords of circles. His deep interest in phenomena of acceleration appears to date from this time. The correct law of falling bodies, but with a false assumption behind it, is embodied in a letter to Sarpi in 1604. Associated with the letter is a fragment, separately preserved, containing an attempted proof of the correct law from the false assumption. No clue is given as to the source of Galileo's knowledge of the law that the ratios of spaces traversed from rest in free fall are as those of the squares of the elapsed times. The law is algebraically derivable from the medieval mean-degree theorem known as the Merton rule, but Galileo's false assumption in 1604 contradicts the specific association of speed and time that is always found in medieval derivations of that theorem. Moreover, Galileo's faulty demonstration invoked no single instantaneous velocity as a mean or representative value; instead, it proceeded by comparison of *ratios* between infinite sets of instantaneously varying velocities. It is probable either that he observed a rough 1, 3, 5, ... progression of spaces traversed along inclined planes in equal times and assumed this to be exact, or that he reasoned (as Christian Huygens later did) that only the odd-number rule of spaces would preserve the ratios unchanged for arbitrary changes of the unit time. From this fact, the times-squared law follows immediately. Galileo's derivation of it from the correct definition of uniform acceleration followed only at a considerably later date.

The appearance of a supernova in 1604 led to disputes about the Aristotelian idea of the incorruptibility of the heavens, in which Galileo took an active part. He delivered three lectures to overflow crowds at Padua and prepared to publish an astronomical work; he did not do so, however, and only a short fragment of the manuscript survives. Lodovico delle Colombe, who published a theory of new stars at Florence, suspected Galileo of having written a pseudonymous attack on him, and it is certain that Galileo's ideas are reflected in still another pseudonymous work, published in rustic dialect at Padua in 1605, which ridiculed the professors of philosophy. In 1606, however, Galileo's attention was diverted from this dispute by the plagiarism of his proportional compass by Simon Mayr (or Marius, in the Latinized form used for publication), a German then at Padua, and Mayr's pupil Baldassar Capra. Galileo had privately printed a small edition of his treatise on the use of the compass in that year; Mayr and Capra produced a Latin book on the construction and use of the same instrument, claiming that Galileo had stolen it from them. Mayr had returned to Germany, so Galileo brought his action against Capra. The book was suppressed and Capra was expelled from the university. In the following year Galileo published a full account of the case in his first publicly circulated printed work, the *Difesa . . . contro alle calunnie & imposture di Baldessar Capra.*

Early in 1609, Galileo began the composition of a systematic treatise on motion in which his studies of inclined planes and of pendulums were to be integrated under the law of acceleration, known to him at least since 1604. In the composition of his treatise, he became aware that there was something wrong with his attempted derivation of 1604, which had assumed proportionality of speed to space traversed. Accordingly, he introduced in its place two propositions drawn from mechanics, which he submitted for criticism to Valerio. Galileo received Valerio's reply in July 1609, just after his attention had again been diverted from mechanics, this time by news of the invention of the telescope.

The Telescope. A Dutch lens-grinder, Hans Lipperhey, had applied in October 1608 to Count Maurice of Nassau for a patent on a device to make distant objects appear closer. Sarpi, whose extensive correspondence (maintained for theological and political reasons) kept him currently informed, learned of this device within a month. Somewhat skeptical, he applied for further information to Jacques Badovere (Giacomo Badoer), a former pupil of Galileo's then at Paris. In due course the report was confirmed. Galileo heard discussions of the news during a visit to Venice in July 1609, learned from Sarpi that the device was real, and probably heard of the simultaneous arrival at Padua of a foreigner who had brought one to Italy. He hastened back to Padua, found that the foreigner had left for Venice, and at once attempted to construct such a device himself. In this he quickly succeeded, sent word of it to Sarpi, and applied himself to the improvement of the instrument. Sarpi, who had meanwhile been selected by the Venetian government to assess the value of the device offered for sale to them by the stranger, discouraged its purchase. Late in August, Galileo arrived at Venice with a nine-power telescope, three times as effective as the other. The practical value of this instrument to a maritime power obtained for him a lifetime appointment to the university, with an unprecedented salary for the chair of mathematics. The official document he received, however, did not conform to his understanding of the terms he had accepted. As a result, he pressed his application for a post at the Tuscan court, begun a year or two earlier.

Galileo's swift improvement of the telescope continued until, at the end of 1609, he had one of about thirty power. This was the practicable limit for a telescope of the Galilean type, with plano-convex objective and plano-concave eyepiece. He turned this new instrument to the skies early in January 1610, with startling results. Not only was the moon revealed to be mountainous and the Milky Way to be a congeries of separate stars, contrary to Aristotelian principles, but a host of new fixed stars and four satellites of Jupiter were promptly discovered. Working with great haste but impressive accuracy, Galileo recited these discoveries in the *Sidereus nuncius,* published at Venice early in March 1610.

His sudden fame assisted Galileo in his negotiations at Florence. Moreover, the new discoveries made him reluctant to continue teaching the old astronomy. In the summer of 1610, he resigned the chair at Padua and returned to Florence as mathematician and philosopher to the grand duke of Tuscany, and chief mathematician of the University of Pisa, without obligation to teach.

Galileo's book created excitement throughout Europe and a second edition was published in the same year at Frankfurt. Kepler endorsed it in two small books, the *Dissertatio cum Nuncio Sidereo,* published before he had personally observed the new phenomena, and the *Narratio de observatis a se quatuor Jovis satellitibus,* published a few months later. Other writers attacked the claimed discoveries as a fraud. Galileo did not enter the controversy but applied himself to further observations. He discovered, later in 1610, the oval appearance of Saturn and the phases of Venus. His telescope was inadequate to resolve Saturn's rings, which he took to be satellites very close to the planet. The phases of Venus removed a serious objection to the Copernican system, and he saw in the satellites of Jupiter a miniature planetary system in which, as in the Copernican astronomy, it could no longer be held that all moving heavenly bodies revolved exclusively about the earth.

Early in 1611 Galileo journeyed to Rome to exhibit his telescopic discoveries. The Jesuits of the Roman College, who had at first been dubious, confirmed them and honored Galileo. Federico Cesi feted Galileo and made him a member of the Lincean Academy, the first truly scientific academy, founded in 1603. The pope and several cardinals also showed their esteem for Galileo.

Controversies at Florence. Shortly after his return to Florence, Galileo became involved in a controversy over floating bodies. In that controversy an important role was played by Colombe, who became the leader of a group of dissident professors and intriguing courtiers that resented Galileo's position at court. Maffeo Barberini—then a cardinal but later to become pope—took Galileo's side in the dispute. Turning again to physics, Galileo composed and published a book on the behavior of bodies placed in water (*Discorso . . . intorno alle cose che stanno in su l'acqua, o in quella si muovono),* in support of Archimedes and against Aristotle, of which two editions appeared in 1612. Using the concept of moment and the principle of virtual velocities, Galileo extended the scope of the Archimedean work beyond purely hydrostatic considerations.

While this work was in progress, Galileo received from Marcus Welser of Augsburg a short treatise on sunspots that Welser had published pseudonymously for the Jesuit Christoph Scheiner, asking Galileo's opinion of it. Galileo replied in three long letters during 1612, demolishing Scheiner's conjecture that the spots were tiny planets. He asserted also that he had observed sunspots much earlier and had shown them to others at Rome early in 1611. This set the stage for a deep enmity of Scheiner toward Galileo, which, however, did not take active form at once.

Galileo's *Letters on Sunspots* was published at Rome in 1613 under the auspices of the Lincean Academy. In this book Galileo spoke out decisively for the Copernican system for the first time in print. In the same book he found a place for his first published mention of the concept of conservation of angular momentum and an associated inertial concept. During its composition he had taken pains to determine the theological status of the idea of incorruptibility of the heavens, finding that this was regarded by churchmen as an Aristotelian rather than a Catholic dogma. But attacks against Galileo and his followers soon appeared in ecclesiastical quarters. These came to a head with a denunciation from the pulpit in Florence late in 1614.

In December 1613 it had happened that theological objections to Copernicanism were raised, in Galileo's absence, at a court dinner, where Galileo's part was upheld by Benedetto Castelli. Learning of this, Galileo wrote a long letter to Castelli concerning the inadmissibility of theological interference in purely scientific questions. After the public denunciation in 1614, Castelli showed this letter to an influential Dominican priest, who made a copy of it and sent it to the Roman Inquisition for investigation. Galileo then promptly sent an authoritative text of the letter to Rome and began its expansion into the *Letter to Christina,* composed in 1615 and eventually published in 1636. Galileo argued that neither the Bible nor nature could speak falsely and that the investigation of nature was the province of the scientist, while the reconciliation of scientific facts with the language of the Bible was that of the theologian.

The book on bodies in water drew attacks from four Aristotelian professors at Florence and Pisa, while a book strongly supporting Galileo's position appeared at Rome. Galileo prepared answers to his critics, which he turned over to Castelli for publication in order to avoid personal involvement. Detailed replies to two of them (Colombe and Grazia), written principally by Galileo himself, appeared anonymously in 1615, with a prefatory note by Castelli implying that he was the author and that Galileo would have been more severe.

Late in 1615 Galileo went to Rome (against the advice of his friends and the Tuscan ambassador) to clear his own name and to prevent, if possible, the official suppression of the teaching of Copernicanism. In the first, he succeeded; no disciplinary action against him was taken on the basis of his letter to Castelli or his Copernican declaration in the book on sunspots. In the second objective, however, he failed. Pope Paul V, irritated by the agitation of questions of biblical interpretation—then a bone of contention with the Protestants—appointed a commission to determine the theological status of the earth's motion. The determination was adverse, and Galileo was instructed on 26 February 1616 to abandon the holding or defending of that view. No action was taken against him, nor were any of his books suspended. A book by the theologian Paolo Antonio Foscarini reconciling the earth's motion with the Bible was condemned, and the work of Copernicus and a commentary on Job by Diego de Zuñiga were suspended pending the correction of a few passages. One contemporary document, bound into the proceedings but of uncertain reliability, states that Galileo was also ordered never to discuss the forbidden doctrine again. If such an order was given, it was in contravention of certain specific instructions of the pope and had no legal force.

Returning to Florence, Galileo took up a practical and noncontroversial problem, the determination of longitudes at sea. He believed that this could be solved by the preparation of accurate tables of the eclipses of the satellites of Jupiter, which were of frequent occurrence and could be observed telescopically from any point on the earth. As a practical matter, the eclipses could neither be predicted with sufficient accuracy nor observed at sea with sufficient convenience to make the method useful.

It is probable that Galileo also returned during this period to his mechanical investigations, interrupted in 1609 by the advent of the telescope. A Latin treatise by Galileo, *De motu accelerato,* which correctly defines uniform acceleration and much resembles the definitive text reproduced in his final book, seems to date from this intermediate period, and copies of many of his propositions in kinematics exist in the handwriting of Mario Guiducci, who studied under Galileo at this time.

In 1618 three comets attracted the attention of Europe and became the subject of many pamphlets and books. One such book was printed anonymously by Orazio Grassi, the mathematician of the Jesuit Roman College. Galileo was bedridden at the time, but he discussed his views on comets with Guiducci, who then delivered lectures on them to the Florentine Academy and published them over his own name. In these lectures, which were largely dictated or corrected by Galileo, the anonymous Jesuit was subjected to criticism. The result was a direct attack on Galileo by Grassi, under the pseudonym of Lotario Sarsi, published in 1619.

Galileo replied, after much delay, with one of the most celebrated polemics in science, *Il saggiatore* (*The Assayer*). It was addressed to Virginio Cesarini, a young man who had heard Galileo debate at Rome in 1615–1616 and had written to him in 1619 to extol the method by which Galileo had opened to him a new road to truth. Since he could no longer defend Copernicus, Galileo avoided the question of the earth's motion; instead, he set forth a general scientific approach to the investigation of celestial phenomena. He gave no positive theory of comets, but developed the thesis that arguments from parallax could not be decisive concerning their location until it was first demonstrated that they were concrete moving objects rather than mere optical effects of solar reflection in seas of vapor. No such proof appeared to him to be available. In the course of his argument, Galileo distinguished physical properties of objects from their sensory effects, repudiated authority in any matter that was subject to direct investigation, and remarked that the book of nature, being written in mathematical characters, could be deciphered only by those who knew mathematics.

The *Saggiatore* was printed in 1623 under the auspices of the Lincean Academy. Just before it emerged from the press, Maffeo Barberini became pope as Urban VIII. The academicians dedicated the book to him at the last minute. Cesarini was appointed chamberlain by the new pope, who had long been Galileo's friend and was a patron of science and letters. Galileo journeyed to Rome in 1624 to pay his respects to Urban, and secured from him permission to discuss the Copernican system in a book, provided that the arguments for the Ptolemaic view were given an equal and impartial discussion. Urban refused to rescind the edict of 1616, although he remarked that had it been up to him, the edict would not have been adopted.

Dialogue on the World Systems. The *Dialogue Concerning the Two Chief World Systems* occupied Galileo for the next six years. It has the literary form of a discussion between a spokesman for Copernicus, one for Ptolemy and Aristotle, and an educated layman for whose support the other two strive. Galileo thus remains technically uncommitted except in a preface which ostensibly supports the anti-Copernican edict of 1616. The book will prove, he says, that the edict did not reflect any ignorance in Italy of the strength of pro-Copernican arguments. The contrary is the case; Galileo will add Copernican arguments of his own invention, and thus he will show that not ignorance of or antagonism to science, but concern for spiritual welfare alone, guided the Church in its decision.

The opening section of the *Dialogue* critically examines the Aristotelian cosmology. Only those things in it are rejected that would conflict with the motion of the earth and stability of the sun or that would sharply distinguish celestial from terrestrial material and motions. Thus the idea that the universe has a center, or that the earth is located in such a center, is rejected, as is the idea that the motion of heavy bodies is directed to the center of the universe rather than to that of the earth. On the other hand, the Aristotelian concept of celestial motions as naturally circular is not rejected; instead, Galileo argues that natural circular motions apply equally to terrestrial and celestial objects. This position appears to conflict with statements in later sections of the book concerning terrestrial physics. But uniform motion in precise circular orbits also conflicts with actual observations of planetary motions, whatever center is chosen for all orbits. Actual planetary motions had not been made literally homocentric by any influential astronomer since the time of Aristotle. Galileo is no exception; in a later section he remarked on the irregularities that still remained to be explained. Opinion today is divided; some hold that the opening arguments of the *Dialogue* should be taken as representative of Galileo's deepest physical and philosophical convictions, while others view them as mere stratagems to reduce orthodox Aristotelian opposition to the earth's motion.

Important in the *Dialogue* are the concepts of relativity of motion and conservation of motion, both angular and inertial, introduced to reconcile terrestrial physics with large motions of the earth, in answer to the standard arguments of Ptolemy and those added by Tycho Brahe. The law of falling bodies and the composition of motions are likewise utilized. Corrections concerning the visual sizes and the probable distances and positions of fixed stars are discussed. A program for the detection of parallactic displacements among fixed stars is outlined, and the phases of Venus are adduced to account for the failure of that planet to exhibit great differences in size to the naked eye at perigee and apogee. Kepler's modification of the circular Copernican orbits is not mentioned; indeed, the Copernican system is presented as more regular and simpler than Copernicus himself had made it. Technical astronomy is discussed with respect only to observational problems, not to planetary theory.

To the refutation of conventional physical objections against terrestrial motion, Galileo added two arguments in its favor. One concerned the annual variations in the paths of sunspots, which could not be dynamically reconciled with an absolutely stationary earth. Geometrically, all rotations and revolutions could be assigned to the sun, but their conservation would require very complicated forces. The Copernican distribution of one rotation to the sun and one rotation and one revolution to the earth fitted a very simple dynamics. The second new argument concerned the existence of ocean tides, which Galileo declared, quite correctly, to be incapable of any physical explanation without a motion of the earth. His own explanation happened to be incorrect; he argued that the earth's double motion of rotation and revolution caused a daily maximum and minimum velocity, and a continual change of speed, at every point on the earth. The continual variation of speed of sea basins imparted different speeds to their contained waters. The water, free to move within the basins, underwent periodic disturbances of level, greatest at their coasts; the period depended on sizes of basins, their east-west orientations, depths, and extraneous factors such as prevailing winds. In order to account for monthly and annual variations in the tides, Galileo invoked an uneven speed of the earth-moon system through the ecliptic during each month, caused by the moon's motion with respect to the earth-sun vector; for annual seasonal effects, he noted changes of the composition of rotational and revolutional components in the basic disturbing cause.

The *Dialogue* was completed early in 1630. Galileo took it to Rome, where it was intended to be published by the Lincean Academy. There he sought to secure a license for its printing. This was not immediately granted, and he returned to Florence without it. While the matter was still pending, Federico Cesi died, depriving the Academy of both effective leadership and funds. Castelli wrote to Galileo, intimating that for other reasons he would never get the Roman imprimatur and advising him to print the book at Florence without delay. Negotiations ensued for permission to print the book at Florence. Ultimately these were successful, and the *Dialogue* appeared at Florence in March 1632. A few copies were sent to Rome, and for a time no disturbance ensued. Then, quite suddenly, the printer was ordered to halt further sales, and Galileo was instructed to come to Rome and present himself to the Inquisition during the month of October.

The Trial of Galileo. The background of the action is fairly clear. Several ecclesiastical factions were hostile to the book but at first produced only shallow pretexts to suppress it. More serious charges were lodged against Galileo when Urban was persuaded that his own decisive argument against the literal truth of the earth's motion—that God could produce any effect desired by any means—had been put in the mouth of the simpleminded Aristotelian in the dialogue as a deliberate personal taunt by Galileo. Next, a search of the Inquisition files of 1616 disclosed the questionable document previously mentioned, which contained a specific threat of imprisonment for Galileo if he ever again discussed the Copernican doctrine in any way. Urban, having known nothing of any personal injunction at the time Galileo sought his permission to write the book, assumed that Galileo had deceitfully concealed it from him. The case was thereafter prosecuted with vindictive hostility. Galileo, who had either never received a personal injunction or had been told that it was without force, was unaware of any wrongdoing in this respect.

Confined to bed by serious illness, he at first refused to go to Rome. The grand duke and his Roman ambassador intervened stoutly in his behalf, but the pope was adamant. Despite medical certificates that travel in the winter might be fatal, Galileo was threatened with forcible removal in chains unless he capitulated. The grand duke, feeling that no more could be done, provided a litter for the journey, and Galileo was taken to Rome in February 1633.

The outcome of the trial, which began in April, was inevitable. Although Galileo was able to produce an affidavit of Cardinal Bellarmine to the effect that he had been instructed only according to the general edict that governed all Catholics, he was persuaded in an extrajudicial procedure to acknowledge that in the *Dialogue* he had gone too far in his arguments for Copernicus. On the basis of that admission, his *Dialogue* was put on the Index, and Galileo was sentenced to life imprisonment after abjuring the Copernican "heresy." The terms of imprisonment were immediately commuted to permanent house arrest under surveillance. He was at first sent to Siena, under the charge of its archbishop, Ascanio Piccolomini. Piccolomini, who is said to have been Galileo's former pupil, was very friendly to him. Within a few weeks he had revived Galileo's spirits—so crushed by the sentence that his life had been feared for—and induced him to take up once more his old work in mechanics and bring it to a conclusion. While at Siena, Galileo began the task of putting his lifelong achievements in physics into dialogue form, using the same interlocutors as in the *Dialogue*.

Piccolomini's treatment of Galileo as an honored guest, rather than as a prisoner of the Inquisition, was duly reported to Rome. To avoid further scandal, Galileo was transferred early in 1634 to his villa at

Arcetri, in the hills above Florence. It was probably on the occasion of his departure from Siena that he uttered the celebrated phrase "Eppur si muove," apocryphally said to have been muttered as he rose to his feet after abjuring on his knees before the Cardinals Inquisitors in Rome. The celebrated phrase, long considered legendary, was ultimately discovered on a fanciful portrait of Galileo in prison, executed about 1640 by Murillo or one of his pupils at Madrid, where the archbishop's brother was stationed as a military officer.

Galileo was particularly anxious to return to Florence to be near his elder daughter. But she died shortly after his return, in April 1634, following a brief illness. For a time, Galileo lost all interest in his work and in life itself. But the unfinished work on motion again absorbed his attention, and within a year it was virtually finished. Now another problem faced him: the printing of any of his books, old or new, had been forbidden by the Congregation of the Index. A manuscript copy was nevertheless smuggled out to France, and the Elzevirs at Leiden undertook to print it. By the time it was issued, in 1638, Galileo had become completely blind.

Two New Sciences. The title of his final work, *Discourses and Mathematical Demonstrations Concerning Two New Sciences* (generally known in English by the last three words), hardly conveys a clear idea of its organization and contents. The two sciences with which the book principally deals are the engineering science of strength of materials and the mathematical science of kinematics. The first, as Galileo presents it, is founded on the law of the lever; breaking strength is treated as a branch of statics. The second has its basis in the assumption of uniformity and simplicity in nature, complemented by certain dynamic assumptions. Galileo is clearly uncomfortable about the necessity of borrowing anything from mechanics in his mathematical treatment of motion. A supplementary justification for that procedure was dictated later by the blind Galileo for inclusion in future editions.

Of the four dialogues contained in the book, the last two are devoted to the treatment of uniform and accelerated motion and the discussion of parabolic trajectories. The first two deal with problems related to the constitution of matter; the nature of mathematics; the place of experiment and reason in science; the weight of air; the nature of sound; the speed of light; and other fragmentary comments on physics as a whole. Thus Galileo's *Two New Sciences* underlies modern physics not only because it contains the elements of the mathematical treatment of motion, but also because most of the problems that came rather quickly to be seen as problems amenable to physical experiment and mathematical analysis were gathered together in this book with suggestive discussions of their possible solution. Philosophical considerations as such were minimized.

The book opens with the observation that practical mechanics affords a vast field for investigation. Shipbuilders know that large frameworks must be strongly supported lest they break of their own weight, while small frameworks are in no such danger. But if mathematics underlies physics, why should geometrically similar figures behave differently by reason of size alone? In this way the subject of strength of materials is introduced. The virtual lever is made the basis of a theory of fracture, without consideration of compression or stress; we can see at once the inadequacy of the theory and its value as a starting point for correct analysis. Galileo's attention turns next to the problem of cohesion. It seems to him that matter consists of finite indivisible parts, *parti quante,* while at the same time the analysis of matter must, by its mathematical nature, involve infinitesimals, *parti non quante.* He does not conceal—but rather stresses—the resulting paradoxes. An inability to solve them (as he saw it) must not cause us to despair of understanding what we can. Galileo regards the concepts of "greater than," "less than," and "equal to" as simply not applicable to infinite multitudes; he illustrates this by putting the natural numbers and their squares in one-to-one correspondence.

Galileo had composed a treatise on continuous quantity (now lost) as early as 1609 and had devoted much further study to the subject. Bonaventura Cavalieri, who took his start from Galileo's analysis, importuned him to publish that work in order that Cavalieri might proceed with the publication of his own *Geometry by Indivisibles.* But Galileo's interest in pure mathematics was always overshadowed by his concern with physics, and all that is known of his analysis of the continuum is to be found among his digressions when discussing physical problems.

Galileo's *parti non quante* seem to account for his curious physical treatment of vacua. His attention had been directed to failure of suction pumps and siphons for columns of water beyond a fixed height. He accounted for this by treating water as a material having its own limited tensile strength, on the analogy of rope or copper wire, which will break of its own weight if sufficiently long. The cohesion of matter seemed to him best explained by the existence of minute vacua. Not only did he fail to suggest the weight of air as an explanation of the siphon phenomena, but he rejected that explanation when it was clearly offered to him in a letter by G. B. Baliani.

Yet Galileo was not only familiar with the weight of air; he had himself devised practicable methods for its determination, set forth in this same book, giving even the correction for the buoyancy of the air in which the weighing was conducted.

Phenomena of the pendulum occupy a considerable place in the *Two New Sciences.* The relation of period to length of pendulum was first given here, although it probably represents one of Galileo's earliest precise physical observations. Precise isochronism of the pendulum appears to have been the one result he most wished to derive deductively. In discussing resistance of the air to projectile motion, he invoked observations (grossly exaggerated) of the identity of period between two pendulums of equal length weighted by bobs of widely different specific gravity. He deduced the existence of terminal constant velocity for any body falling through air, or any other medium, but mistakenly believed increase of resistance to be proportional to velocity.

Like the pendulum, the inclined plane plays a large role in Galileo's ultimate discussion of motion. The logical structure of his kinematics, as presented in the *Two New Sciences,* is this: He first defines uniform motion as that in which proportional spaces are covered in proportional times, and he then develops its laws. Next he defines uniform acceleration as that in which equal increments of velocity are acquired in equal times and shows that the resulting relations conform to those found in free fall. Postulating that the path of descent from a given height does not affect the velocity acquired at the end of a given vertical drop, he describes an experimental apparatus capable of disclosing time and distance ratios along planes of differing tilts and lengths; finally, he asserts the agreement of experiment with his theory. The experiments have been repeated in modern times, precisely as described in the *Two New Sciences,* and they give the results asserted. Following these definitions, assumptions, and confirmation by experiment, Galileo proceeds to derive a great many theorems related to accelerated motion.

In the last section Galileo deduces the parabolic trajectory of projectiles from a composition of uniform horizontal motion and accelerated vertical motion. Here the concept of rectilinear inertia, previously illustrated in the *Dialogue* ("Second Day"), is mathematically applied but not expressly formulated. This is followed by additional theorems relating to trajectories and by tables of altitude and distance calculated for oblique initial paths. Because of air resistance at high velocities, the tables assumed low speeds and hence were of no practical importance in gunnery. But like Galileo's theory of fracture, they opened the way for rapid successive refinements at the hands of others.

Last Years. Galileo lived four years, totally blind, beyond the publication of his final book. During this time, he had the companionship of Vincenzio Viviani, who succeeded him (after Evangelista Torricelli) as mathematician to the grand duke and who inherited his papers. Viviani wrote a brief account of Galileo's life in 1654 at the request of Leopold de' Medici, which, despite some demonstrable errors, is still a principal source of biographical information, in conjunction with the voluminous correspondence of Galileo that has survived and with the autobiographical passages in his works. Near the end of his life, Galileo was also visited by Torricelli, a pupil of Castelli and the ablest physicist among Galileo's immediate disciples. Galileo's son, Vincenzio, also assisted in taking notes of his father's later reflections, in particular the design of a timekeeping device controlled by a pendulum.

Galileo died at Arcetri early in 1642, five weeks before his seventy-eighth birthday. The vindictiveness of Urban VIII, who had denied even Galileo's requests to attend mass on Easter and to consult doctors in nearby Florence when his sight was failing, continued after Galileo's death: The grand duke wished to erect a suitable tomb for Galileo but was warned to do nothing that might reflect unfavorably on the Holy Office. Galileo was buried at Santa Croce in Florence, but nearly a century elapsed before his remains were transferred, with a suitable monument and inscription, to their present place in the same church.

Sources of Galileo's Physics. The habitual association of Galileo's name with the rapid rise of scientific activity after 1600 makes the investigation of his sources a matter of particular interest to historians of science.

All agree that Archimedes was a prime source and model for Galileo, who himself avowed the fact. The work of Aristotle and the pseudo-Aristotelian *Questions of Mechanics* were likewise admitted inspirations to Galileo, although often only as targets of criticism and attack. The astronomy of Copernicus and the magnetic researches of William Gilbert were obvious and acknowledged sources of his work. Beyond these, there is little agreement.

Among sixteenth-century writers, Galileo probably drew chiefly on Niccolò Tartaglia, Girolamo Cardano, and Guidobaldo del Monte. Parallels between his early unpublished work and that of Benedetti are very striking, but the establishment of a direct connection is difficult. As with the case of Stevin, the parallels in thought may result from the Archimedean revival and a common outlook rather than from early and direct knowledge of Benedetti's work.

Similarly, a direct influence of medieval writers on Galileo, although widely accepted by most historians, is still largely conjectured on the basis of specific parallels. The statics of Jordanus de Nemore was widely known in Italy after 1546, when Tartaglia published in Italian and endorsed the "science of weights" as necessary to an understanding of the balance; yet all subsequent writers (at least in Italy) condemned it in favor of the Archimedean approach. Writings of the Merton school, published repeatedly in Italy up to about 1520, continued to be discussed thereafter at Paris and in Spain. Galileo's reasoning about acceleration, after his recognition of its importance around 1602, invariably proceeded by comparison of ratios, whereas medieval writers adopted a mean speed as representative of uniformly changing velocities. Medieval impetus theory, which Galileo adopted at first for the explanation of projectile motion, had no place in the concept of neutral motions that led him eventually to an inertial terrestrial physics. A connection of Galileo's own physical thought with medieval sources may yet be convincingly established, but at present this has not been done.

Experiment and Mathematics. The role of experiment in Galileo's physics was limited to the testing of preconceived mathematical rules and did not extend to the systematic search for such rules. It is probable that his use of experiment had its roots in the musical controversy conducted by his father rather than in philosophical considerations of method. Appeal to experiment in his published works was resorted to by Galileo chiefly as a means of confuting rival theories, as in the dispute over bodies in water and in his rejection of proportionality of speed to space traversed in free fall.

It is difficult to find older sources for Galileo's attitude toward mathematics, which was strikingly modern. He considered mathematics to enjoy a superior certainty over logic. Where a mathematical relation could be found in nature, Galileo accepted it as a valid description and discouraged further search for ulterior causes. He attributed discrepancies between mathematics and physical events to the investigator who did not yet know how to balance his books. Galileo did not adopt the traditional Platonist view that our world is a defective copy of the "real" world, and he derided philosophical speculation about a world on paper.

The Influence of Galileo. Except with respect to the acceptance of Copernican astronomy, Galileo's direct influence on science outside Italy was probably not very great. After 1610 he published his books in Italian and made little effort to persuade professional scholars either at home or abroad. His influence on educated laymen both in Italy and abroad was considerable; on university professors, except for a few who were his own pupils, it was negligible. Latin translations of his *Dialogue* appeared in Holland in 1635, in France in 1641, and in England in 1663; but the only Latin translation of the *Two New Sciences* was published in 1700, long after Newton's *Principia* had superseded it.

Between Galileo and Newton, science was Cartesian rather than Galilean. Indirectly, Galileo's science exerted some influence in France through Marin Mersenne, Pierre Gassendi, and Nicholas Fabri de Peiresc; in Germany through Kepler; and in England through John Wilkins and John Wallis. Descartes, who repudiated Galileo's approach to physics because of its neglect of the essence of motion and physical causation, did not mention him in any published work. Newton seems not to have read Galileo's *Two New Sciences,* at least not before 1700, but knew his *Dialogue* as early as 1666. Aware of his achievements in physics only indirectly, Newton, in the *Principia,* mistakenly credited Galileo with a derivation of the laws of falling bodies from the law of inertia and the force-acceleration relationship.

Within Italy, Galileo had a strong following both in scientific and nonscientific circles. His ablest pupil, Castelli, was the teacher of Torricelli and Cavalieri, both of whom also had personal acquaintance with Galileo. His last pupil, Viviani, did much to extend Galileo's influence in the succeeding generation, editing the first collection of his works in 1655–1656. But by that time physics and astronomy had both progressed well beyond the point where Galileo had left them.

Outside scientific circles, Galileo's influence was strongly felt in the battle for freedom of inquiry and against authority. English translations of his *Dialogue* and *Letter to Christina,* published in 1661, carried this influence outside academic circles. John Milton cited the fate of Galileo in his *Areopagitica.* French writers during the Enlightenment also made Galileo a symbol of religious persecution.

Personal Traits. Galileo was of average stature, squarely built, and of lively appearance and disposition. Viviani remarks that he was quick to anger and as quickly mollified. His unusual talents as a speaker and as a teacher are beyond question. Among those who knew him personally, even including adversaries, few seem to have disliked him. Many distinguished men became his devoted friends, and some sacrificed their own interests in his support at crucial periods. On the other hand, there were many contemporary rumors discreditable to Galileo, and demonstrable slanders occur in letters of Georg Fugger, Martin Horky, and others. Pugnacious rather than belligerent, he refrained from starting polemic battles but was ruthless in their prosecution when he answered an attack at all. His friends included artists and men of letters as well as mathematicians and scientists; cardinals as well as rulers; craftsmen as well as learned men. His enemies included conservative professors, several priests, most philosophers, and those scientists who had publicly challenged him and felt the bite of his sarcasm in return.

Caution and daring both had a place in Galileo's personality. His reluctance to speak out for the Copernican system until he had optical evidence against the rival theories is evidence of scientific prudence rather than of professorial timidity. Once convinced by his own eyes and mind, he would not be swayed even by the advice of well-informed friends who urged him to proceed with caution. In the writings he withheld from publication, as in his surviving notes, many errors and wrong conjectures are to be found; in his published works, very few. He was as respectful of authority in religion and politics as he was contemptuous of it in matters he could investigate for himself. It is noteworthy that before his Copernican stand was challenged by an official Church edict, he had composed and submitted to the authorities a carefully documented program, based on positions of Church fathers, that would have obviated official intervention against his science—a program that was in fact adopted by a pope nearly three centuries later as theologically sound.

BIBLIOGRAPHY

I. Original Works. All works by Galileo and virtually all known Galilean correspondence and manuscripts are contained in *Le opere di Galileo Galilei,* Antonio Favaro, ed., 20 vols. (Florence, 1890–1909); repr. with some additions (Florence, 1929–1939; 1965). English translations of Galileo's principal works are listed below. Following the translator's name are the English book title, the abbreviated original title of each work included, and date of first ed. or approximate date of composition.

T. Salusbury, *Mathematical Collections and Translations,* I (London, 1661; repr. 1967), *Lettera a Madama Cristina* (*ca.* 1615) and *Dialogo* (Florence, 1632); II (London, 1665; repr. 1967), *La bilancetta* (*ca.* 1586); *Le meccaniche* (*ca.* 1600); *Discorso . . . intorno alle cose che stanno in su l'acqua* (Florence, 1612); and *Discorsi* (Leiden, 1638).

T. Weston, *Mathematical Discourses Concerning Two New Sciences* (London, 1730; 2nd ed. 1734): *Discorsi* (Leiden, 1638).

E. Carlos, *The Sidereal Messenger* (London, 1880; repr. 1959): *Sidereus nuncius* (Venice, 1610).

H. Crew and A. De Salvio, *Dialogues Concerning Two New Sciences* (New York, 1914; repr. n.d.): *Discorsi* (Leiden, 1638).

G. de Santillana, ed., *Dialogue on the Great World Systems* (Chicago, 1953), the Salusbury trans.: *Dialogo* (Florence, 1632).

S. Drake, *Dialogue Concerning the Two Chief World Systems* (Berkeley, Cal., 1953; rev. 1967): *Dialogo* (Florence, 1632).

S. Drake, *Discoveries and Opinions of Galileo* (New York, 1957): *Sidereus nuncius* (Venice, 1610); *Lettere sulle macchie solari* (Rome, 1613); *Lettera a Madama Cristina* (*ca.* 1615); and *Il saggiatore* (Rome, 1623).

I. Drabkin and S. Drake, *Galileo on Motion and on Mechanics* (Madison, Wis., 1960): *De motu* (*ca.* 1590), and *Le meccaniche* (*ca.* 1600).

S. Drake and C. D. O'Malley, *The Controversy on the Comets of 1618* (Philadelphia, 1960): *Discorso sulle comete* (Florence, 1619), and *Il saggiatore* (Rome, 1623).

S. Drake, ed., *Galileo on Bodies in Water,* (Urbana, Ill., 1960), the Salusbury trans.: *Discorso* (Florence, 1612).

L. Fermi and G. Bernadini, *Galileo and the Scientific Revolution* (New York, 1961): C. S. Smith, trans., *La bilancetta* (*ca.* 1586).

S. Drake and I. Drabkin, *Mechanics in Sixteenth-Century Italy* (Madison, Wis., 1969): *Dialogus de motu* (*ca.* 1589).

II. Secondary Literature. Nearly 6,000 titles relating to Galileo are listed in the following bibliographies: *Bibliografia Galileiana, 1568–1895,* A. Carli and A. Favaro, eds. (Rome, 1896); *Bibliografia Galileiana, Primo Supplemento, 1896–1940,* G. Boffito, ed. (Rome, 1943); "Bibliografia Galileiana, 1940–1964," in *Galileo, Man of Science,* E. McMullin, ed. (New York, 1967); E. Gentili, *Bibliografia Galileiana fra i due centenari* (*1942–1964*) (Varese, 1966).

Selected biographies are A. Banfi, *Galileo Galilei* (Milan, 1948); J. Fahie, *Galileo: His Life and Works* (London, 1903); A. Favaro, *Galileo e lo studio di Padova* (Florence, 1883), and *Galileo Galilei e Suor Maria Celeste* (Florence, 1891); K. von Gebler, *Galileo Galilei and the Roman Curia* (London, 1879), English trans., Mrs. G. Sturge; L. Geymonat, *Galileo Galilei* (Milan, 1957), English trans., S. Drake (New York, 1965); T. Martin, *Galilée* (Paris, 1868); L. Olschki, *Galilei und seine Zeit* (Halle, 1927; repr., Vaduz, 1965); M. Allen-Olney, *The Private Life of Galileo* (London, 1870); P. Paschini, *Vita e opere di Galileo Galilei* (Rome, 1965); F. Reusch, *Der Process Galilei's und die Jesuiten* (Bonn, 1879); G. de Santillana, *The Crime of Galileo* (Chicago, 1955); F. Taylor, *Galileo and the Freedom of Thought* (London, 1938); E. Wohlwill, *Galilei und sein Kampf* (Hamburg-Leipzig, 1909, 1926).

Fundamental to the study of Galileo's scientific work are the publications of A. Favaro listed in G. Favaro, *Bibliografia Galileiana di A. Favaro* (Venice, 1942); A. Koyré, *Études Galiléennes* (Paris, 1939; repr., Paris, 1966); and M. Clavelin, *La philosophie naturelle de Galilée* (Paris, 1968).

Collections of modern Galilean studies include *Nel terzo*

centenario della morte di Galileo Galilei (Milan, 1942); M. Kaplon, ed., *Homage to Galileo* (Cambridge, Mass., 1965); *Nel quarto centenario della nascita di Galileo Galilei* (Milan, 1966); C. Golino, ed., *Galileo Reappraised* (Berkeley, Cal., 1966); *Atti del Symposium Internazionale . . . "Galileo nella storia e nella filosofia della scienza"* (Vinci, 1967); C. Maccagni, ed., *Saggi su Galileo Galilei* (Florence, 1967–); E. McMullin, ed., *Galileo: Man of Science* (New York, 1967); *Galilée, Aspects de sa vie et de son oeuvre,* preface by Suzanne Delorme (Paris, 1968); and S. Drake, *Galileo Studies* (Ann Arbor, Mich., 1970).

Separate articles and monographs are listed in the bibliographies cited above.

STILLMAN DRAKE

GALILEI, VINCENZIO (*b.* Santa Maria a Monte, Italy, *ca.* 1520; *d.* Florence, Italy, July [?] 1591), *music theory, acoustics.*

Vincenzio, father of Galileo Galilei, was of a Florentine patrician family originally surnamed Bonajuti, renamed in the fourteenth century. The son of Michelangelo Galilei and Maddalena di Bergo, he began the study of music at Florence about 1540. After establishing his reputation as a lutenist, he studied at Venice under Gioseffo Zarlino, the foremost music theorist of the time, probably about 1561–1562. On 5 July 1562 Galilei married Giulia Ammannati of Pescia and settled near Pisa. Galileo was the eldest of their seven children.

Through correspondence with Girolamo Mei at Rome during the 1570's, Galilei became interested in ancient Greek music and was encouraged to put to direct experimental test the teachings of Zarlino concerning intonation and tuning. The result was a bitter polemic with Zarlino, who in 1580–1581 appears to have used his influence to oppose the publication of Galilei's principal theoretical work at Venice and its sale there after it was printed at Florence.

Galilei's *Dialogo della musica antica e della moderna* (1581) attacked the prevailing basis of musical theory. This was rooted in the Pythagorean doctrine that the cause of consonance lay in the existence of the "sonorous numbers," two, three, and four, which in their ratios with one another and with unity were considered to produce the only true consonances. A modified tuning given by Ptolemy (the syntonic diatonic) was favored by Zarlino, who rationalized this tuning by extending the sonorous numbers to six. Galilei observed that musical practice did not conform to this (or any other) numerical system based on superparticular ratios (which, expressed as fractions, have numerators exceeding their denominators by unity). He declared that neither the authority of ancient writers nor speculative number theories could be valid against the evidence of the musician's ear. Although he recommended placing frets on lute and viol in the ratio 18:17, he recognized this as merely approximate in obtaining an equally tempered scale suitable for unrestricted modulation, in the direction of which musical practice was rapidly moving.

Renaissance physicists had already recognized the inadequacy of speculative mathematical acoustics. Giovanni Battista Benedetti had questioned the older tunings in letters to Cipriano da Rore, whom Zarlino succeeded as choirmaster at St. Mark's in Venice in 1565. Simon Stevin, in an unpublished treatise on music, advocated the outright abandonment of rational numbers and the division of the scale in true equal temperament based on the twelfth root of two. As mathematics matured, modern harmony replaced polyphony.

Zarlino defended his system based on the number six (the *senario*) in his *Sopplementi musicali,* published at Venice in 1588. His former pupil Galilei was a principal target of attack in this book, although Zarlino did not name him and although Bernardino Baldi, in a short biography of Zarlino, wrongly identified the adversary as Francisco de Salinas. Galilei replied with a spirited polemic, the *Discorso,* published in 1589. In this work he stated the law that a given musical interval between similar strings is produced either by different lengths, or by tensions inversely as the squares of those lengths. Thus the perfect fifth, which is produced by lengths related as 3:2, is also given when weights in the ratio of 4:9 are hung from strings of equal length. This is probably the first mathematical law of physics to have been derived by systematic experimentation, or at any rate the first to replace a universally accepted rival law, for a standard illustration in music books showed the Pythagorean sonorous numbers applying to weights on equal strings as well as to lengths of unequal strings (or air columns).

Galilei employed this experimental result to show that the traditional association of numbers with particular musical intervals was capricious. The musical qualities of intervals had to be determined by the ear, he argued, and mathematics had no authority where the senses were concerned.

Galilei's empirical attitude toward musical theory had an ancient counterpart in Aristoxenus, a prominent pupil of Aristotle's who shared his distrust of Pythagorean numerology. But in the sixteenth century music was again regarded as a branch of mathematics. Galilei's *Discorso* foreshadowed the subordination of mathematics to experience and the discovery of unexpected laws through close observation that was to distinguish science in the seventeenth century from its predecessors. Galilei was driven to experiment in order to refute erroneous entrenched musical theory, as his son Galileo later attacked ancient physical theory. Among the manuscripts inherited by Galileo is Vincenzio Galilei's untitled treatise beginning with the words "L'arte et la pratica del moderno contrapunto . . .," of which Claude Palisca has said: "For prophetic vision, originality, and integrity, it has few equals in the history of music theory."

Galileo gave two separate accounts of his introduction to mathematics, both indicating that his father opposed this introduction. Vincenzio's mathematical skills seem inconsistent with this attitude. His writings, however, reveal a deep hostility toward specious reasoning in practical matters induced by fascination with numerical relations and geometrical designs. It is understandable if he did not want his eldest son to be so beguiled.

BIBLIOGRAPHY

I. ORIGINAL WORKS. Galilei's writings, excluding musical compositions, are *Fronimo, Dialogo . . . del intavolare la musica nel liuto* (Venice, 1568; 2nd ed., 1584; facs. ed., Bologna, 1969); *Dialogo della musica antica e della moderna* (Florence, 1581; 2nd ed., 1602; facs. of 1st ed., Rome, 1934; abr. ed. of 1st ed., Milan, 1947); and *Discorso intorno all'opere di messer Gioseffo Zarlino da Chioggia* (Florence, 1589; facs. ed., Milan, 1933). MSS of unpublished works are preserved in the Biblioteca Nazionale, Florence, MSS Galileiani.

An excerpt from Galilei's *Dialogo* of 1581 is translated in O. Strunk, ed., *Source Readings in Music History* (New York, 1950; repr. New York, 1965), vol. II, *The Renaissance Era,* 112–134. The same vol. contains part of a "Discourse on Ancient Music and Good Singing" (pp. 100–111), published as the work of Giovanni Bardi but perhaps written for him by Galilei about 1578.

II. SECONDARY LITERATURE. The principal biography is Claude Palisca, "V. Galilei," in F. Blume, ed., *Die Musik in Geschichte und Gegenwart,* IV (Kassel–Basel, 1955), cols. 1903–1905, with bibliography to 1950. See also C. Palisca, "Vincenzio Galilei's Counterpoint Treatise: A Code for the *Seconda Pratica,*" in *Journal of the American Musicological Society,* **9** (1956), 81–96; "Scientific Empiricism in Musical Thought," in S. Toulmin and D. Bush, eds., *Seventeenth Century Science and the Arts* (Princeton, 1961), 91–137; "Vincenzio Galilei's Arrangements for Voice and Lute," in G. Reese and R. J. Snow, eds., *Essays in Musicology in Honor of Dragan Plamenac* (Pittsburgh, Pa., 1969), 207–232; and "Ideas of Music and Science," in P. P. Wiener and C. E. Pettie, eds., *Dictionary of the History of Ideas* (New York, in press); and S. Drake, "Renaissance Music and Experimental Science," in *Journal of the History of Ideas,* **31** (1970), 483–500; and "Vincenzio Galilei and Galileo," in *Galileo Studies* (Ann Arbor, Mich., 1970), 43–62.

STILLMAN DRAKE

GALITZIN, B. B. See **Golitsyn, B. B.**

GALL, FRANZ JOSEPH (*b.* Tiefenbronn, near Pforzheim, Germany, 9 March 1758; *d.* Paris, France, 22 August 1828), *neuroanatomy, psychology.*

Gall's father, Joseph Anthony Gall, was a modest merchant and sometime mayor of the village of Tiefenbronn. He was of Italian extraction (the original name was Gallo); and both he and his wife, Anna Maria Billingerin, were devout Roman Catholics. They intended Franz for the church; but although he remained nominally religious and even included an organ for religion in his theory of cerebral structure, it cannot be said that he was devout, that he led a morally conventional life, or that his work was well received by the church. His passions for science and gardening were complemented by strong appetites for money and women. He had many mistresses and once mentioned an illegitimate son. Gall's books were placed on the Index; and he was denied a religious burial, even though he claimed that the existence of the "organ of religion" was a new proof for the existence of God.

Gall married a young Alsatian girl surnamed Lieser, who had cared for him when he had typhus; the marriage was an unhappy one. They had no children, but his wife's niece and nephews lived with them at various times. After his wife died at Vienna in 1825, Gall married Marie Anne Barbe, with whom he had had a long-standing relationship. In 1826 signs of cerebral and coronary sclerosis appeared, and he died of an apoplectic stroke two years later.

Gall received his early education from his uncle, who was a priest, and in schools at Baden and Bruchsal. He began to study medicine at Strasbourg in 1777 and married while he was there. In 1781 he moved to Vienna, where he received the M.D. in 1785. In Vienna he carried on an active and successful medical practice which included many eminent patients. When he moved to Paris he was equally successful and numbered Stendhal, Saint-Simon, and Metternich, along with the staffs of twelve embassies, among his patients. On the other hand, he never held an academic post; and his relations with authority and orthodoxy were almost uniformly bad. His lectures at Vienna were proscribed by Emperor Francis I, and Napoleon took steps to restrict his influence in Paris. His doctrines were rejected by the Institut de France in 1808; and in 1821 he failed to gain admission to the Academy, although his candidacy was supported by Étienne Geoffroy Saint-Hilaire.

Gall had a flamboyant personality and was something of a showman. He gave numerous courses of public lectures in Vienna, Paris, and other cities throughout Europe. He was heavily criticized for charging admission to his scientific demonstrations; but he was generous in spending his considerable earnings from this source and from his practice on the pursuit and publication of his research, as well as on his full social life. Gall was as vehement and effective a controversialist as he was a devoted bon vivant. Indeed, his life-style was consistent with his major intellectual preoccupation: the integration of the scientific problems of mind and brain with those of life and society.

The first publication of the principles of his lifework was a treatise on the philosophy of medicine in 1791. Gall developed his views in public lectures and demonstrations in Vienna between 1796 and 1801, when the emperor, in a personal letter, forbade these activities, on the ground that his doctrines were conducive to materialism, immorality, and atheism. Repeated appeals and a long petition and remonstrance to the emperor failed to alter the position. In 1800 Gall had been joined by Johann C. Spurzheim, who served as research assistant and collaborator; and in 1805 they went on an extended and highly successful tour of the intellectual centers of Germany, Switzerland, Holland, and Denmark, visiting schools, hospitals, prisons, and insane asylums to gather evidence and demonstrate their doctrines. Gall also visited his parents during this period; and he and Spurzheim eventually found their way to Paris in November 1807. Gall remained there until his death, except for a brief trip to England in 1823. He became a French citizen in 1819.

Beginning in 1800, with the assistance of Spurzheim, Gall made a number of important neuroanatomical discoveries. Their full significance was not appreciated until the development of histological and neurophysiological findings was integrated with the influence of his theoretical and speculative conceptions many decades after his death. The unifying theme in his neuroanatomical work was the conception of the nervous system as a hierarchically ordered series of separate but interrelated ganglia designed on a unified plan. Higher structures developed from lower ones, receiving reinforcement from other nerve pathways along the way. The gray matter was the matrix of the nerves, and the fibrous white matter served a conducting function. The inclusion of the cerebral cortex in this scheme was an important development away from lingering glandular and humoral conceptions. The spinal cord was, Gall argued, arranged in the same way; and he noticed its segmental structure and successive swellings. He also discovered the origins of the first eight cranial nerves and traced the fibers of the medulla oblongata to the basal ganglia. In the cerebellum he described the systems of fibers now known as projection and commissural. In the cerebral cortex he finally established the contralateral decussation of the pyramids and drew attention to the detailed anatomy of the convolutions.

Gall and Spurzheim's investigations gave considerable impetus to the study of neuroanatomy, and both their findings and their general conceptions proved very important when they were later integrated with an evolutionary view of the nervous system and with the neuron theory. Gall vehemently opposed the contemporary practice of brain dissection by successive slicing and insisted on following the brain's own structural organization. In 1863, when his best-known theories were almost totally discredited, his most effective critic, Pierre Flourens, recalled that when he had first seen Gall dissect a brain, he felt as though he had never seen the organ before; and he called Gall "the author of the true anatomy of the brain."

Gall's conceptions were importantly influenced by the theories of J. G. von Herder and involved strong emphasis on comparative and developmental studies, along with more general themes from *Naturphilosophie,* such as the unity of plan and analogies drawn from botany. In addition to his specific discoveries, Gall's neuroanatomical work helped to alter the context of the study of the brain from the prevailing mechanical and humoral theories to an organic, biological perspective.

Yet it would be almost totally misleading to suggest that Gall's best-known and most influential theories grew inductively out of his neuroanatomical research. On the contrary, he had published the basic principles of his theory of the functions of the brain in 1791: the plurality and independence of the cerebral organs. His public lectures contained sufficiently detailed and provocative findings to lead to their suppression. In 1798 he spelled out the main argument of his major work in a letter to Baron von Retzer. This was two years before he undertook detailed dissections of the central nervous system, work which he did as a consequence of his general doctrines.

Nor can it be argued that Gall's neuroanatomical findings led to important elaborations or modifications of his general doctrines. It was pointed out by a commission of the Institut de France, including such eminent scientists as J. R. Tenon, Antoine Portal, R. B. Sabatier, Philippe Pinel, and Georges Cuvier, that the two aspects of his work were not inconsistent—but neither were they closely integrated. The commission sought to separate the two aspects of Gall's work for philosophical reasons; but it was, nevertheless, correct in its evaluation of the relationship between them. The conception of the nervous system as a series of relatively independent ganglia was common to both, but neither aspect was based on the other. Gall granted that any doctrine of the functions of the brain which was incompatible with its structure must necessarily be false. Nevertheless, the fundamental principle of his lifework was that it was essential that the issue be approached from the other side. He said that the knowledge of the functions had always preceded that of the parts and that all his physiological discoveries had been made without the anatomy of the brain; these discoveries might have existed for ages without their agreement with the organization of the brain having been detected. The commission was striking at the heart of Gall's doctrine by refusing, in principle, to consider the relationship between structure and function; but he and Spurzheim had presented no compelling evidence of that relationship.

Gall's theory of the functions of the brain and each of its parts calls for careful historical treatment, since the important features of his work and his influence are bounded on all sides by what are now seen as undoubted absurdities, although this was not at all clear in the context of contemporary science. With one notable exception, none of his localizations of cortical functions has been substantiated by subsequent research. The detailed methodology on which he based his physiological conclusions provides an excellent case study of the dangers of anecdotal and correlative methods, uncontrolled by statistical tests and attempts to seek out potentially falsifying evidence. Finally, the popular application of his theories in the form of phrenology soon came to be seen as a classic example of pseudoscience and its practice a form of quackery.

Yet those who would attempt sharply to demarcate science from pseudoscience and the internal history of science from external factors would stumble as badly over Gall's work as they have over Robert Chambers' and Herbert Spencer's. Embedded in his crude methodology and his detailed, although wholly incorrect, findings were a set of principles and a biological, adaptive, and functional approach to the study of mind and brain which have led to the recognition of his work as seminal in three spheres: (1) the origination of the modern doctrine of cerebral localization of functions, (2) the establishment of psychology as biological science, and (3), at a more general level, the use of his work and its popularizations as the vehicle for a naturalistic approach to the study of man which was very influential in the development of evolutionary theory, physical anthropology, and sociology. These points should be borne in mind when considering the curious amalgam which makes up Gall's systematic writings.

His psychophysiology had its origins in childhood experiences. As a schoolboy he noticed that those who were better than he at memorizing had "large flaring eyes." It was a popular contemporary doctrine that all aspects of character had external signs, and the initial theoretical context for his ideas was therefore a straightforward physiognomical correlation of the kind which J. C. Lavater had made popular—it had no detailed causal basis. When Gall later noticed the same correlation among his fellow medical students, he reflected on a possible physiological basis for it. Every physiological function had its own organ, as did each of the five external senses. Why should it not be the same with the talents and propensities of men? If this could be established, a doctrine of the nature of man could be founded on a doctrine of the functional organization of the brain. Localization of cerebral functions was also a long-held idea with contemporary exponents, but neither the functions nor their localizations were being actively studied in detail; Gall set out to till an existing, although fallow, field.

He immediately found himself faced with a number of conceptual boundary disputes; and in attempting to work out a consistent theory he had to address himself to fundamental problems in the borderlands of ontology, epistemology, psychology, physiology, and general biology. These issues lie unresolved at the heart of the assumptions of modern science and its philosophy of nature: mind and body, primary and secondary qualities. Gall pioneered the hope that the problems of a dualistic ontology and epistemology could be resolved by taking a biological point of view. The most fundamental result of his work stems, therefore, from the way he asked the question. He treated the problem of brain and mind analogously to that of any other organ and its function, thereby bringing the mind-body problem into the domain of dynamic physiology and biology. If one traces the concept of "function" as applied to psychological and social phenomena back from its late-nineteenth and twentieth-century uses, one finds its source in the writings of Gall and his followers.

This is not to say that Gall was original in arguing that the brain is the organ of the mind or that mental phenomena should be treated as analogous to physiological function. As he pointed out in his petition to Francis I, the conception of the brain as the organ of mind had been reiterated since the beginnings of anatomy and physiology. In Gall's own period the *idéologues* had treated mental phenomena in physiological terms, but it was Gall who united these conceptions and treated them in consistently biological terms. He argued that the sensationalism of Étienne Condillac and the *idéologues,* especially P. J. G. Cabanis and A. L. C. Destutt de Tracy, could not account for the observed differences between the talents and propensities of individuals and those between species. The origins of character and personality could not be adequately explained by experience alone. Gall claimed that the causes of the behavior of men and animals were innate, although modifiable by experience. Sensationalist psychology had been elaborated in opposition to idealist belief in innate ideas, but Gall was not treating the problem from a primarily epistemological point of view. He saw the talents and propensities as inherited instincts based on cerebral endowment.

Gall's organic conception was also opposed to the related doctrine of Charles Bonnet, that sensationalism could be related in mechanical terms to the fibrous connections in the nervous system. Thus, Gall rejected both sensationalism and its putative cerebral basis. Instead of synthesizing complex mental phenomena from simple ones by the mechanism of the association of ideas connected in the fibers of the nervous system, Gall argued for unitary faculties based on cerebral ganglia which served as centers for each determinate talent or propensity. Of course, his faculty psychology raised at least as many problems as it solved, but there seemed to be more hope of solving them in a biological context which was relatively free from attempts to interpret mental phenomena by analogy to the concepts of corpuscular physics.

Gall was interested not only in the intellectual functions but also in the passions, and it was not generally conceded that the latter had their seat in the brain. For example, the eminent physiologists Cabanis and Xavier Bichat still claimed that the passions had their seat in the thorax and abdomen. Gall claimed that the brain was the organ of all mental functions. As a result of his systematic investigations and his consistent reiteration of this claim, he succeeded in gaining final acceptance for the principle. Once again, it was Flourens who granted that although the proposition that the brain was the exclusive organ of the mind existed in science before Gall appeared, it was as a result of his work that it reigned in science by the middle of the nineteenth century.

Even more important was Gall's insistence that neither the faculties nor their localizations in the brain were known and that they had to be determined by empirical, naturalistic studies of men in society and of other species in nature. The prevailing categories of psychological analysis were derived from philosophical—especially epistemological—preoccupations: reason, memory, imagination, perception, and so on. These were concerned with the attributes of mind in general, not with a differential psychology which could account for individual and species differences. Gall called for faculties, the different distributions of which determine the behavior of different species of animals, and the different proportions of which explain individual human differences. The result of the application of this principle was the origination of the systematic empirical search for a natural classification of fundamental variables in animal and human nature and personality. When Alexander Bain turned to this issue again in 1861, he described phrenology as the only system of character hitherto elaborated; and subsequent research in personality theory, as well as comparative psychology, can be traced, in large part, to Gall's direct or indirect influence. Although the issues remain unresolved—and can be argued to have no unique scientific resolution—it was Gall who raised them in their modern form.

In his attempt to arrive at a list of determinate faculties, Gall sought out people who showed extremes of talents or other striking propensities, including manias. He related these to the behavior of animals; and although his analogies are often extremely farfetched, his approach extended the comparative method to psychology in a systematic way. His extensive case records from insane asylums, prisons, schools, and public life were supplemented by large collections of craniums and plaster casts, the last of which was bought by the French government and deposited in the Musée de l'Homme. Gall concluded that men and animals shared nineteen of the twenty-seven fundamental faculties.

The results of his work were published at his own expense in four quarto volumes and an atlas of 1,000 plates as *Anatomie et physiologie du système nerveux en général, et du cerveau en particulier, avec des observations sur la possibilité de reconnoitre plusieurs dispositions intellectuelles et morales de l'homme et des animaux, par la configuration de leurs têtes* between 1810 and 1819. (Spurzheim was coauthor of the first two volumes, but they parted company after that.) An inexpensive edition appeared between 1822 and 1825 as *Sur les fonctions du cerveau et sur celle de chacune de ses parties;* the atlas was omitted and a volume of replies to his critics was added. Gall summarized his theory in four fundamental suppositions: (1) that the moral and intellectual faculties are innate; (2) that their exercise or manifestation depends on organization; (3) that the brain is the organ of all the propensities, sentiments, and faculties; (4) that the brain is composed of as many particular organs as there are propensities, sentiments, and faculties which differ essentially from each other. The Achilles heel of the elaboration of these suppositions—the "special organology"—was a set of related beliefs: that the activity of a given organ varies with its size, that all of the cerebral organs impinge on bony structures of the cranium, and that in most cases the cranium faithfully reflects the conformation of the underlying cerebrum. The result of these subsidiary beliefs was the pseudoscience of cranioscopy, later popularized as phrenology—the reading of character from the conformation of the skull. During the nineteenth century this progressively became an object of ridicule; and although there are still practitioners of phrenological delineation, its critics have obscured Gall's more important contributions.

Gall was cautious—but not cautious enough—about making inferences to the brain from the study of the skull. In practice, most of his faculties were discovered by correlating a striking talent, propensity, or passion with prominences on the skulls of men and animals. From this evidence he went on to infer the existence of an innate faculty and a cortical organ. Having formulated a hypothesis about a given faculty, he collected a great deal of evidence to confirm his correlations. This was done uncritically; and although the theoretical aspects of his systematic edifice have passed into the foundations of the assumptions of modern psychology, neurophysiology, biology, and social theory, the structure itself has not stood. Gall made no claim to finality for his list of faculties or for precision in his cerebral localizations. He considered the detailed working out of the system to be a problem for the future. In any case, he was more interested in the nature of the functions than in their localizations. His opponents rejected the entire basis of his work, while his followers prematurely codified his system.

Neither Gall's detailed classification nor his faculty psychology have appealed to subsequent investigators. However, the questions which he asked have remained leading topics in neurology, psychology, and ethology. Thus, for example, his localization of sexual passion in the cerebellum has been totally discredited by experimental findings; but the study of the neurophysiological basis of sexual and other emotional functions plays a leading part in current research in physiological psychology. Similarly, although it was set aside for over a century, his insistence that the study of the organization of the brain should march side by side with that of its functions is once again the basic principle of biological psychology. Whatever the judgment of Gall's contemporaries, the organ-function paradigm which he elaborated for the study of man in society has become the predominant approach.

The reception of Gall's doctrines and his influence are as confusing and complex as would be expected from the intimate mixture of important principles, methodological crudity, and detailed nonsense which made up his work. There was vehement and sustained opposition from those who saw his theories leading to materialism, immorality, fatalism, and atheism. His division of the mind and its organ into separate compartments was anathema to those who followed Descartes in claiming that the mind is indivisible. At the other extreme, sensationalists opposed his faculty psychology and his belief in innate instincts. Physiologists who were beginning to find experimental support for the interpretation of the nervous system in sensory-motor terms could not, in principle, find any basis for Gall's conception of the fundamental faculties. These promising findings began to be made only in 1822; and Gall criticized both the methods and the generalizations of experimental neurophysiology, arguing that they would lead to the reduction of life and character to sensibility, motion, and association. He was quite prescient in arguing that those who took this approach would fail to address themselves to fundamental human problems.

Gall was preoccupied with the psychological question "What are the functions of the brain?" while the experimentalists were concerned with the narrower—but scientifically more fruitful—question of how the brain functions. Since 1822, the approaches of experimental neurophysiology and of personality psychology have diverged, and efforts to relate them have not been notably successful. As the experimental tradition developed, it lost sight of the significance of Gall's questions and concentrated on the localization of sensory, motor, and associative functions, with little thought about how they were to be related to the concepts of the layman, thereby reverting to the analytic categories which Gall had set aside at the beginning of his inquiries. The model of the elements of mind as analogous to corpuscular physics prevailed over the holistic characterological approach.

There were many devotees of Gall's special organology and his cranioscopic method, especially in France, Britain, and America. Societies with eminent medical and scientific members sprang up and were immensely popular in France and Britain until the 1840's and even later in America. Although their influence is not significant for the history of science in the narrow sense, failure to appreciate the importance of popular phrenology would blind one to the most important vehicle of scientific naturalism in the decades before evolutionary theory assumed this role. The list of eminent political, philosophical, and literary figures who took it seriously is astonishing and includes G. W. F. Hegel, Otto von Bismarck, Marx,

Balzac, the Brontës, George Eliot, President James Garfield, Walt Whitman, and Queen Victoria. Its leading popularizers were Spurzheim and George Combe; and it has been said that homes in Britain which contained only three books would have the Bible, Bunyan's *Pilgrim's Progress,* and Combe's *System of Phrenology.* The particular influence of phrenology can be traced in the writings of educationalists, advocates of public health, penal reform, and improvements in the care of the insane, as well as in the scientific writings of Auguste Comte, G. H. Lewes, Spencer, Chambers, and A. R. Wallace. The adaptive, biological view of man and mind was carried by phrenology into the formulation of theories of evolution and into the use of biological analogies in theories of society. This is most striking in psychology, where Gall was the main figure in altering the context of the study of mind from that of epistemology to that of general biology.

More straightforward scientific influences can be traced in physical anthropology—especially its preoccupation with skulls throughout the nineteenth century—and in somatist psychiatry's fundamental belief that all mental disease is brain disease. Of course, the most obvious influence of Gall's work lay in neurology and neurophysiology. Beginning in 1861 with Paul Broca's clinicopathologic localization of the lesion causing aphasia in the place where Gall had localized the faculty of "memory for words," localization of function has been the central conception in neurology, as it later became in neurosurgery. Once again, Gall's specific concepts were set aside, while his general principles were adopted. The same can be said of the localization of functions in experimental neurophysiology, which began in 1870 with the work of Gustav Fritsch and Eduard Hitzig and was carried on by workers in France, Germany, Italy, Britain, and America. This tradition was called, only half jokingly, "the new phrenology" by C. S. Sherrington, who claimed that all students of the correlations of brain, mind, and behavior are phrenologists of sorts.

In 1857, as Gall's reputation was waning, G. H. Lewes wrote a history of the development of thought on positivist lines. Although he was critical of Gall's detailed findings, he said that by placing man firmly in nature, Gall had rescued the problem of mental functions from metaphysics and made it one of biology. Gall's vision of psychology as a biological science may be said, he concluded, to have given the science its basis. Auguste Comte and Herbert Spencer, the founders of modern sociology, also acknowledged Gall's fundamental contribution to their views on man and society. His theory played an important role in the evolutionary theories of Robert Chambers, Spencer, and A. R. Wallace, the last of whom considered the neglect of phrenology as one of the greatest failures of nineteenth-century thought. But in rejecting the details of his work, modern science, and its twin brother scientism, embraced Gall's principles and his point of view, as a result of which he can be said to have made a central contribution to scientific naturalism in the biological and human sciences. Along with astrology, alchemy, Hermetism, mesmerism, and spiritualism, Gall's science and its manifold influence challenge any attempt to establish neat demarcations between the origins, the substance, the applications, and the validity of scientific ideas in their philosophical, theological, and social contexts.

BIBLIOGRAPHY

I. Original Works. Gall's writings include *Philosophisch-medicinische Untersuchungen über Natur und Kunst im kranken und gesunden Zustande des Menschen* (Vienna, 1791); *Recherches sur le système nerveux en général, et sur celui du cerveau en particulier; mémoire présenté à l'Institut de France, le 14 mars 1808; suivi d'observations sur le rapport qui en a été fait à cette compagnie par les commissaires* (Paris, 1809), written with Spurzheim; *Anatomie et physiologie du système nerveux* (see text for full title), 4 vols. and atlas (Paris, 1810–1819); *Sur les fonctions du cerveau et sur celles de chacune de ses parties,* 6 vols. (Paris, 1822–1825), English trans. by W. Lewis, Jr., with a biography including the letter to Baron von Retzer, 6 vols. (Boston, 1835); and Gall *et al., On the Functions of the Cerebellum by Drs Gall, Vimont and Broussais,* English trans. by G. Combe, including Gall's petition and remonstrance to Emperor Francis I (Edinburgh, 1838). Gall's letters have been published in M. Neuburger, "Briefe Galls an Andreas und Nanette Streicher," in *Archiv für Geschichte der Medizin,* **10** (1917), 3–70; and E. Ebstein, "Franz Joseph Gall im Kampf um seine Lehre," in C. Singer and H. E. Sigerist, eds., *Essays on the History of Medicine Presented to Karl Sudhoff* (London–Zurich, 1924), pp. 269–322.

II. Secondary Literature. The best single source is O. Temkin, "Gall and the Phrenological Movement," in *Bulletin of the History of Medicine,* **21** (1947), 275–321. See also the unsigned "Researches of Malcarne and Reil—Present State of Cerebral Anatomy," in *Edinburgh Medical and Surgical Journal,* **21** (1824), 98–141; and "Recent Discoveries on the Physiology of the Nervous System," *ibid.,* 141–159; A. Bain, *On the Study of Character, Including an Estimate of Phrenology* (London, 1861); M. Bentley, "The Psychological Antecedents of Phrenology," in *Psychological Monographs,* **21,** 4, no. 92 (1916), 102–115; C. Blondel, *La psycho-physiologie de Gall* (Paris, 1914); [R. Chevenix], "Gall and Spurzheim—Phrenology," in *Foreign Quarterly Review,* **2** (1828), 1–59; G. von Bonin, *Some Papers on the Cerebral Cortex* (Springfield,

Ill., 1960), which includes the papers of Broca and of Fritsch and Hitzig on cerebral localization; K. M. Dallenbach, "The History and Derivation of the Word 'Function' as a Systematic Term in Psychology," in *American Journal of Psychology,* **26** (1915), 473–484; P. Flourens, *Examen de la phrénologie* (Paris, 1842), English trans. by C. de L. Meigs (Philadelphia, 1846); and *De la phrénologie et des études vraies sur le cerveau* (Paris, 1863); [J. Gordon], "Functions of the Nervous System," in *Edinburgh Review,* **24** (1815), 439–452; H. Head, *Aphasia and Kindred Disorders of Speech,* 2 vols. (Cambridge, 1926); C. W. Hufeland, *Dr. Gall's New Theory of Physiognomy* (London, 1807); J. Hunt, "On the Localisation of Functions in the Brain, With Special Reference to the Faculty of Language," in *Anthropological Review,* **6** (1868), 329–345, and **7** (1869), 100–116, 201–214; T. Laycock, "Phrenology," in *Encyclopaedia Britannica,* 8th ed. (1859), XVII, 556–567; G. H. Lewes, "Phrenology in France," in *Blackwood's Edinburgh Magazine,* **82** (1857), 665–674; and *The History of Philosophy From Thales to Comte,* 3rd ed., 2 vols. (London, 1867–1871); A. Macalister, "Phrenology," in *Encyclopaedia Britannica,* 9th ed. (1885), XVIII, 842–849; C. S. Sherrington, "Sir David Ferrier, 1843–1928," in *Proceedings of the Royal Society,* **103B** (1928), viii–xvi; J. Soury, *Le système nerveux central* (Paris, 1899); H. Spencer, *Principles of Psychology* (London, 1855); J. R. Tenon *et al.,* "Report on a Memoir of Drs Gall and Spurzheim, Relative to the Anatomy of the Brain, Presented to and Adopted by the Class of Mathematical and Physical Sciences of the National Institute," in *Edinburgh Medical and Surgical Journal,* **5** (1809), 36–66; A. R. Wallace, *The Wonderful Century, Its Successes and Failures* (London, 1898); and S. Wilks, "Notes on the History of the Physiology of the Nervous System, Taken More Especially From Writers on Phrenology," in *Guy's Hospital Reports,* 3rd ser., **24** (1879), 57–94.

On the context of sensationalism and sensory-motor physiology, see O. Temkin, "The Philosophical Background of Magendie's Physiology," in *Bulletin of the History of Medicine,* **20** (1946), 10–35; on Gall's neuroanatomy, see Temkin's "Remarks on the Neurology of Gall and Spurzheim," in E. A. Underwood, ed., *Science, Medicine and History, Essays in Honor of Charles Singer,* 2 vols. (London, 1953), II, 282–289. See also E. H. Ackerknecht and H. V. Vallois, *Franz Joseph Gall, Inventor of Phrenology, and His Collection* (Madison, Wis., 1956); J. C. Greene, "Biology and Social Theory in the Nineteenth Century: Auguste Comte and Herbert Spencer," in M. Clagett, ed., *Critical Problems in the History of Science* (Madison, Wis., 1959); G. Jefferson, *Selected Papers* (London, 1960); E. Lesky, "Structure and Function in Gall," in *Bulletin of the History of Medicine,* **44** (1970), 297–314; W. Reise and E. C. Hoff, "A History of the Doctrine of Cerebral Localization," in *Journal of the History of Medicine,* **5** (1950), 51–71, and **6** (1951), 439–470; R. M. Young, "The Functions of the Brain: Gall to Ferrier (1808–1886)," in *Isis,* **59** (1968), 251–268; and *Mind, Brain and Adaptation in the Nineteenth Century: Cerebral Localization and Its Biological Context From Gall to Ferrier* (Oxford, 1970); and O. L. Zangwill, "The Cerebral Localization of Psychological Functions," in *Advancement of Science,* **20** (1963–1964), 335–344.

ROBERT M. YOUNG

GALLE, JOHANN GOTTFRIED (*b.* Pabsthaus, near Gräfenhainichen, Germany, 9 June 1812; *d.* Potsdam, Germany, 10 July 1910), *astronomy.*

Galle was the son of J. Gottfried Galle and Henriette Pannier. He was born in an isolated house on the Dübener Heide, a wooded heath between the Elbe and the Mulde, where his father was manager of a tar distillery. He attended school at Radis, his mother's birthplace. There the local clergyman prepared both one of his own sons and Galle for the secondary school at Wittenberg.

Galle was at Wittenberg from April 1825 until April 1830, when he went to study in Berlin. His teachers there included Hegel, Dirichlet, Dirksen, Dove, Ideler, and—most important—Encke, who was to be highly influential in his later career. In 1833 Galle was granted the *facultas docendi* to teach mathematics and physics at the Gymnasium level. He spent the required probationary year teaching at Guben and Berlin, where he was made assistant teacher at the Friedrich-Werder Gymnasium in March 1834. While he was teaching in secondary school, Galle kept in touch with Encke; and in 1835 Encke, who had become director of the Berlin Observatory (and had had it newly rebuilt to his own specifications), had Galle appointed to an assistantship that had been created especially for him.

Galle spent the next sixteen years at the observatory, where his duties concerned him largely with astrometry. He became in addition an avid observer of comets, including Halley's comet in its appearance of 1835 (he was to live to see it again in 1910); the comet newly discovered by Boguslavsky; and Encke's comet. In 1839 and 1840 Galle himself discovered, in quick succession, three new comets, and thus attracted the attention of experts in the field as well as royal recognition.

In 1836 Alexander von Humboldt invited Galle to participate in the computation of the astronomical material that he had collected during his journeys and thereby initiated a professional association that was to last fifteen years. During this same period, Galle again attended Encke's lectures in order to further his theoretical knowledge, and Encke entrusted him with further computational work involving the minor planets, especially Pallas, which he had previously observed. In about 1839 Galle began to compute the ephemerides of this planet for the *Berliner astronomisches Jahrbuch;* he continued these calculations for thirty years. He made other computations of the elements and ephemerides of comets, including two of those that he had discovered. In 1838 he observed the crepe ring of Saturn, although he did not publish this discovery.

Having continued his theoretical studies, Galle wished to obtain the doctorate. The government gave him financial aid and he received the degree on 1 March 1845. His thesis, *Olai Roemeri triduum observationum astronomicarum,* was based upon unanalyzed data from three days of exceptionally good meridian observations made by the Danish astronomer in 1706. (Except for these three days, the contents of Roemer's other valuable observations had been destroyed by fire.) Galle sent a copy of this thesis to Le Verrier, to whom he thought Roemer's observations would be of value.

Le Verrier did not immediately acknowledge the receipt of Galle's work, but when he did he also informed Galle of the presumed position of a planet beyond Uranus whose orbit he had computed from the perturbations of Uranus' motion. He encouraged Galle to look for this planet, since he thought the telescopes available to him at the Paris observatory inadequate to this purpose. Galle began to look for the planet the same evening that he received Le Verrier's letter; on 23 September 1846 he and d'Arrest, who was at that time studying in Berlin, searched the region cited, but without success. Galle had made no special preparations for his search, since the diameter given by Le Verrier seemed to be sufficient for recognizing the planet as such. Additional data were necessary, however; fortunately the *Berliner akademische Sternkarten* were being readied for publication and the chart covering the area of observation had just been printed. The chart was not yet available commercially, but Encke had a copy. Galle borrowed it, and described what happened next:

> Returning with the chart to the telescope I discovered a star of the eighth magnitude—not at first glance, to tell the truth, but after several comparisons. Its absence from the chart was so obvious that we had to try to observe it. Encke, who had been informed of all the details, took part in the observation on the same night. We observed the star until early morning; but, despite all duplications of effort, we did not succeed in discerning a definite motion, although a trace of change in the required sense seemed to occur. Full of excitement, we had to wait for the evening of 24 September, when our research was also favored by the weather and when the existence of the planet was proved [*Astronomische Nachrichten,* **89** (1877), 349–352].

The planet was at first called "Le Verrier's planet," but its name was shortly thereafter changed to Neptune. The location of the hypothetical planet had been computed simultaneously by Le Verrier and by John Couch Adams, working at Cambridge. The Cambridge astronomers were not able to find the planet and a long controversy arose concerning the priority of its computation. The possession of the new *Sternkarten* was of great advantage to the observers in Berlin; it was in all probability Arago, a close friend of Humboldt's, who knew of the chart and suggested to Le Verrier that help might be available. Both Le Verrier and Adams were aware that the time to search for the planet was ripe, despite the uncertainties inherent in their computations; possible errors in calculating both the mass and the distance of the undiscovered body disturbing Uranus would only be magnified at a later date. (It is interesting to note that Galle found Neptune less than one degree from where Le Verrier predicted it would be.)

Following the discovery of Neptune, Galle contributed observations and computations of a provisional circular orbit toward the further tracking of the planet. His modesty prevented him from capitalizing on his discovery, and it was Encke who reported in detail on it to the Berlin Academy and in the *Astronomische Nachrichten* (of which volume **23** contains several articles on Neptune, including some account of the theoretical work that preceded its discovery). Galle's achievement was nevertheless widely hailed.

Galle continued to work in Berlin as Encke's assistant—he was even referred to as his teacher's mirror image. (Encke's influence, indeed, is to be seen throughout Galle's lifework.) Among other projects, Galle made numerous distance measurements of double stars and, in 1847, published a supplement to the new edition of Olbers' *Abhandlungen . . . die Bahn eines Kometen zu berechnen,* a list of all comet orbits computed up to that time, with important emendations and references to the literature.

In June 1851 Boguslavsky died at Breslau and Galle was offered the post of director of the observatory and professor of the university there. It was not easy for him to decide to leave the well-equipped Berlin Observatory for a small, almost obsolete observatory situated in the very center of the provincial town, but he accepted the opportunity to do independent work. Galle stayed at Breslau for forty-six years; in 1874–1875 he performed the duties of rector of the university. He taught all aspects of astronometry and meteorology, but devoted much of his classroom activity (as well as his research) to studies of comets and planetoids. He was a vivid lecturer and attracted large audiences—as many as sixty auditors are recorded at one time.

The primitive equipment available at the Breslau

observatory did not permit Galle to do any pioneer work. He did, however, often participate in astronomical-geodetical tasks for the Europäische Gradmessung; as late as 1885 and 1888, he took part in the determinations of longitude between Berlin and Breslau. He also continued to observe comets, although he was mainly concerned with meteors, a continuation of his work in Berlin. He had already found that there is a relationship between the meteor showers recorded over the centuries and the appearance of comets.

Galle therefore tried to compute the orbit of the Lyrid meteor shower around the sun and to demonstrate its connection with comet 1861 I, discovered by Biela. He proved that meteors were to be expected to attend the descending node of the comet's orbit; his theoretical assumptions were confirmed by a great number of shooting stars as predicted on the night of 28 November 1872, establishing the relation between meteor showers and the decomposition of a parent comet. Galle continued these investigations, examining a variety of significant meteor appearances and computing the cosmic orbits of such meteors, which he classified as often hyperbolic.

It is known from a notice in a newspaper that at this time Galle was also considering the possible existence of a planet between Mercury and the sun, a hypothesis repeatedly put forth by Le Verrier. He seems to have dismissed its likelihood, however, reasoning that such a planet of any notable magnitude would be visible during total solar eclipses or on other occasions. He also, in 1864, issued a new edition of Olbers' *Abhandlungen,* this time including the orbits of 231 comets (in a supplement of 1885, he increased the number to 286).

Galle's interest in the minor planets led him to propose in 1872 that corresponding data on these bodies, observed at a close approach to the earth, be used to determine the solar parallax. The oppositions of Mars and lower conjunctions of Venus, particularly its passages in front of the sun's disk, had already been observed with this objective; but Galle, who was widely experienced in the observation of the larger planets, correctly stressed that observations of the planetoids should be free from systematic errors. Galle corresponded extensively with astronomers of leading observatories (particularly those in the southern hemisphere) on this proposal; his suggestions were adopted and a series of simultaneous observations of Flora were made. These showed close agreement with the values derived by Simon Newcomb from other measurements. Galle took active part in these observations; he also witnessed the great advance in the method he had designed that resulted from the discovery of the planetoid Eros, although he did not participate in the discovery itself.

As did many of his fellow astronomers, Galle made regular meteorological observations or had them made for him. As conditions for astronomical observations became worse at the Breslau observatory, he placed increasing emphasis on meteorological and even geomagnetic measurements. He conducted the latter from 1869 to 1897; these were considerably impaired, however, by the construction of a streetcar line near the university, started in 1893. Through these abortive observations Galle wished to examine the magnetism of the earth in relation to "northern lights and other terrestrial and even cosmic conditions." He also published a series of papers on climatology and weather forecasting; he was convinced that accurate scientific forecasting had not yet become feasible. Further works that he published late in his life touched upon several minor matters in a variety of fields.

In 1857 Galle married C. E. M. Regenbrecht, the daughter of a professor from Breslau. She died in 1887. They had two sons, one of whom, Andreas, was for many years an astronomer and geodesist at Potsdam. Throughout his long life Galle received numerous honors, especially memberships in scientific societies all over the world. He attained the age of ninety-eight in good physical and mental health and exerted a great influence on several generations of German astronomers. In his eulogy of Galle, W. Foerster accurately summarized his pedagogical career: "Without the men trained in theory and computation at the Breslau school . . ., it would not have been possible to cope with the enormous amount of computational work that resulted in the last fifty years from the discovery of more than half a thousand small planets between the orbits of Mars and Jupiter."

BIBLIOGRAPHY

I. Original Works. Poggendorff lists Galle's major works. Many of his most important papers were published in *Astronomische Nachrichten;* these include "Einige Messungen des Durchmessers des Saturns," in *Astronomische Nachrichten,* **32** (1851), 187–190; "Über den mutmasslichen Zusammenhang der periodischen Sternschnuppen des 24 April mit dem 1. Kometen des Jahres 1861," *ibid.,* **69** (1867), 33–36; "Sternschnuppenbeobachtungen in Breslau 27 Nov. 1872," *ibid.,* **80** (1873), 279–282; "Über die Berechnung der Bahnen heller und an vielen Orten beobachteter Meteore," *ibid.,* **83** (1874), 21–50; and "Nachtrag zu den in Band 23 der Astronomischen Nachrichten gegebenen Berichten über die erste Auffindung des Planeten Neptun," *ibid.,* **89** (1877), 349–352.

Individual works include *Grundzüge der schlesischen Klimatologie* (Breslau, 1857); *Über die Verbesserung der Planetenelemente* (Breslau, 1858); *Über eine Bestimmung der Sonnenparallaxe aus korrespondierenden Beobachtungen des Planeten Flora* (Breslau, 1875); and *Mitteilungen der Königlichen Universitäts-Sternwarte Breslau über hier bisher gewonnene Resultate für die geographischen und klimatologischen Ortsverhältnisse* (Breslau, 1879).

II. SECONDARY LITERATURE. Works on Galle and his work are W. Foerster, "J. G. Galle," in *Vierteljahrsschrift der Astronomischen Gesellschaft,* **46** (1911), 17–22; and D. Wattenberg, *J. G. Galle* (Leipzig, 1963).

H. C. FREIESLEBEN

GALLOIS, JEAN (*b.* Paris, France, 11 June 1632; *d.* Paris, 19 April 1707), *history of science.*

The son of a counsel to the Parlement of Paris, Gallois seems to have distinguished himself in that city around 1664 by the breadth of his learning, by his knowledge of Hebrew and of both living and classical languages, by his interest in the sciences, and by a genuine literary talent. Today his name is associated with the famous *Journal des sçavans.* He collaborated with its founder, Denys de Sallo, from January to April 1665, during the brief period in which the new publication provoked the violent polemics that led to its suspension; and it was to him that Colbert assigned its resumption. Gallois made the periodical a success, publishing forty-two issues as sole editor, beginning in 1666. The Académie Royale des Sciences, established that year, found a vehicle of expression in the *Journal;* yet, despite its support, the number of issues published decreased to sixteen in 1667 and to thirteen in 1668. Named a member of the Academy in 1667, Gallois temporarily assumed the duties of the perpetual secretary, Jean Baptiste Duhamel, who was on a diplomatic mission to England. The *Journal des sçavans* continued to appear under Gallois's editorship, but with steadily decreasing frequency.

Gallois entered the Académie Française in 1673 and in 1675 turned over the editorship of the *Journal* to the Abbé Jean-Paul de La Roque, although he became involved with the *Journal* again in 1684. Meanwhile, the death of his patron Colbert had led him to seek the position of custodian of the Royal Library. A few years later he was appointed professor of Greek at the Collège Royal. His name is mentioned in conjunction with various publications planned by the Académie Royale des Sciences, especially in 1692–1693. Starting in this period Gallois became an opponent of the introduction of infinitesimal methods in mathematics.

With the reorganization of the Académie Royale des Sciences in 1699 Gallois was made pensionary geometer with Michel Rolle and Pierre Varignon. He stated his intention of publishing a critical translation of Pappus, but nothing came of this project. Instead, he stimulated the quarrel between his two colleagues concerning differential calculus and impeded its settlement until 1706.

Despite this negative attitude, the consequences of which might have been disastrous, Gallois deserves recognition by historians of science for his activities as a publicist. Although he wrote somewhat fancifully and with little concern for coherence, he was of service in his time as a disseminator of ideas and his work is still valuable as an historical source.

BIBLIOGRAPHY

I. ORIGINAL WORKS. Gallois's translations and other works include *Traduction latine du traité de paix des Pyrénées* (Paris, 1659); *Breviarum Colbertinum* (Paris, 1679); "Extrait du livre intitulé: Observations physiques et mathématiques envoyées des Indes et de la Chine . . . par les P. P. Jésuites . . . à Paris . . . par l'abbé Galloys," in *Mémoires de mathématiques et de physique tirés des registres de l'Académie royale des sciences* (31 July 1692), pp. 113–120; "Extrait d'un écrit composé par Dom François Quesnet, religieux bénédictin, et envoyé à l'Académie royale des sciences, touchant les effets extraordinaires d'un écho," *ibid.* (30 Nov. 1692), pp. 158–160; "Extrait du livre intitulé: Divers ouvrages de mathématiques et de physique par messieurs de l'Académie royale des sciences," *ibid.* (30 Apr. 1693), pp. 49–64; and "Réponse à l'écrit de David Gregory touchant les lignes appelées Robervalliennes qui servent à transformer les figures," in *Mémoires de l'Académie royale des sciences pour l'année 1703* (Paris, 1705), pp. 70–77.

Many accounts of sessions are in the MS registers of *Procès-verbaux des séances de l'Académie royale des sciences* (1668–1699), *passim.* Correspondence with Leibniz is Hannover, LBr 295, 35 fol.

II. SECONDARY LITERATURE. See Denis-François Camusat, *Histoire critique des journaux* (Amsterdam, 1734), pp. 214–310; Bernard de Fontenelle, "Éloge de Mr l'abbé Gallois," in *Histoire et mémoires de l'Académie royale des sciences pour l'année 1707;* and "Bibliographie de Jean Galloys," in *Histoire de l'Académie royale des sciences depuis 1666 jusqu'à son renouvellement en 1699,* II (Paris, 1733), p. 360.

PIERRE COSTABEL

GALOIS, EVARISTE (*b.* Bourg-la-Reine, near Paris, France, 25 October 1811; *d.* Paris, 31 May 1832), *mathematics.*

There have been few mathematicians with personalities as engaging as that of Galois, who died at the age of twenty years and seven months from wounds received in a mysterious duel. He left a body of work—for the most part published posthumously—of less than 100 pages, the astonishing richness of which was revealed in the second half of the nineteenth century. Far from being a cloistered scholar, this extraordinarily precocious and exceptionally profound genius had an extremely tormented life. A militant republican, driven to revolt by the adversity that overwhelmed him and by the incomprehension and disdain with which the scientific world received his works, to most of his contemporaries he was only a political agitator. Yet in fact, continuing the work of Abel, he produced with the aid of group theory a definitive answer to the problem of the solvability of algebraic equations, a problem that had absorbed the attention of mathematicians since the eighteenth century; he thereby laid one of the foundations of modern algebra. The few sketches remaining of other works that he devoted to the theory of elliptic functions and that of Abelian integrals and his reflections on the philosophy and methodology of mathematics display an uncanny foreknowledge of modern mathematics.

Galois's father, Nicolas-Gabriel Galois, an amiable and witty liberal thinker, directed a school accommodating about sixty boarders. Elected mayor of Bourg-la-Reine during the Hundred Days, he retained this position under the second Restoration. Galois's mother, Adelaïde-Marie Demante, was from a family of jurists and had received a more traditional education. She had a headstrong personality and was eccentric, even somewhat odd. Having taken charge of her son's early education, she sought to inculcate in him, along with the elements of classical culture, the principles of an austere religion and respect for a Stoic morality. Affected by his father's imagination and liberalism, the varying severity of his mother's eccentricity, and the affection of his elder sister Nathalie-Théodore, Galois seems to have had an early youth that was both happy and studious.

Galois continued his studies at the Collège Louis-le-Grand in Paris, entering as a fourth-form boarder in October 1823. He found it difficult to submit to the harsh discipline imposed by the school during the Restoration at the orders of the political authorities and the Church, and although a brilliant student, he presented problems. In the early months of 1827 he attended the first-year preparatory mathematics courses given by H. J. Vernier, and this first contact with mathematics was a revelation for him. But he rapidly tired of the elementary character of this instruction and of the inadequacies of certain of the textbooks and soon turned to reading the original works themselves. After appreciating the rigor of Legendre's *Géométrie,* Galois acquired a solid grounding from the major works of Lagrange. During the next two years he followed the second-year preparatory mathematics courses taught by Vernier, then the more advanced ones of L.-P.-E. Richard, who was the first to recognize his indisputable superiority in mathematics. With this perceptive teacher Galois was an excellent student, even though he was already devoting much more of his time to his personal work than to his classwork. In 1828 he began to study certain recent works on the theory of equations, number theory, and the theory of elliptic functions. This was the period of his first memorandum, published in March 1829 in Gergonne's *Annales de mathématiques pures et appliquées;* making more explicit and demonstrating a result of Lagrange's concerning continuous fractions, it reveals a certain ingenuity but does not herald an exceptional talent.

By his own account, in the course of 1828 Galois wrongly believed—as Abel had eight years earlier—that he had solved the general fifth-degree equation. Rapidly undeceived, he resumed on a new basis the study of the theory of equations, which he pursued until he achieved the elucidation of the general problem with the help of group theory. The results he obtained in May 1829 were communicated to the Académie des Sciences by a particularly competent judge, Cauchy. But events were to frustrate these brilliant beginnings and to leave a deep mark on the personality of the young mathematician. First, at the beginning of July came the suicide of his father, who had been persecuted for his liberal opinions. Second, a month later he failed the entrance examination for the École Polytechnique, owing to his refusal to follow the method of exposition suggested by the examiner. Seeing his hopes vanish for entering the school which attracted him because of its scientific prestige and liberal tradition, he took the entrance examination for the École Normale Supérieure (then called the École Préparatoire), which trained future secondary school teachers. Admitted as the result of an excellent grade in mathematics, he entered this institution in November 1829; it was then housed in an annex of the Collège Louis-le-Grand, where he had spent the previous six years. At this time, through reading Férussac's *Bulletin des sciences mathématiques,* he learned of Abel's recent death and, at the same time, that Abel's last published memoir contained a good number of the results he himself had presented as original in his memoir to the Academy.

Cauchy, assigned to report on Galois's work, had to counsel him to revise his memoir, taking into account Abel's researches and the new results he had obtained. (It was for this reason that Cauchy did not present a report on his memoir.) Galois actually composed a new text that he submitted to the Academy at the end of February 1830, hoping to win the *grand prix* in mathematics. Unfortunately this memoir was lost upon the death of Fourier, who had been appointed to examine it. Brusquely eliminated from the competition, Galois believed himself to be the object of a new persecution by the representatives of official science and of society in general. His manuscripts have preserved a partial record of the elaboration of this memoir of February 1830, a brief analysis of which was published in Férussac's *Bulletin des sciences mathématiques* of April 1830. In June 1830 Galois published in the same journal a short note on the resolution of numerical equations and a much more important article, "Sur la théorie des nombres," in which he introduced the remarkable theory of "Galois imaginaries." That this same issue contains original works by Cauchy and Poisson is sufficient testimony to the reputation Galois had already acquired, despite the misfortune that plagued him. The July Revolution of 1830, however, was to mark a severe change in his career.

After several weeks of apparent calm the revolution provoked a renewal of political agitation in France and an intensification in republican propaganda, especially among intellectuals and students. It was then Galois became politicized. Before returning for a second year to the École Normale Supérieure in November 1830, he already had formed friendships with several republican leaders, particularly Blanqui and Raspail. He became less and less able to bear the strict discipline in his school, and he published a violent article against its director in an opposition journal, the *Gazette des écoles.* For this he was expelled on 8 December 1830, a measure approved by the Royal Council on 4 January 1831.

Left to himself, Galois devoted most of his time to political propaganda and participated in the demonstrations and riots then agitating Paris. He was arrested for the first time following a regicide toast that he had given at a republican banquet on 9 May 1831, but he was acquitted on 15 June by the assize court of the Seine. Meanwhile, to a certain extent he continued his mathematical research. His last two publications were a short note on analysis in Férussac's *Bulletin des sciences mathématiques* of December 1830 and "Lettre sur l'enseignement des sciences," which appeared on 2 January 1831 in the *Gazette des écoles.* On 13 January he began a public course on advanced algebra in which he planned to present his own discoveries; but this project seems not to have had much success. On 17 January 1831 Galois presented to the Academy a new version of his "Mémoire sur la résolution des équations algébriques," hastily written up at the request of Poisson. Unfortunately, in his report of 4 July 1831 on this, Galois's most important piece of work, Poisson hinted that a portion of the results could be found in several posthumous writings of Abel recently published and that the remainder was incomprehensible. Such a judgment, the profound injustice of which would become apparent in the future, could only stiffen Galois's rebellion.

Galois was arrested again during a republican demonstration on 14 July 1831 and placed in detention at the prison of Sainte-Pélagie, where in a troubled and often painful situation he pursued his mathematical investigations, revised his memoir on equations, and worked on the applications of his theory and on elliptic functions. On 16 March 1832, upon the announcement of a cholera epidemic, he was transferred to a nursing home, where he resumed his research, wrote several essays on the philosophy of science, and became involved in a love affair, of which the unhappy ending grieved him deeply.

Provoked to a duel in unclear circumstances following this breakup, Galois felt his death was near. On 29 May he wrote desperate letters to his republican friends, hastily sorted his papers, and addressed to his friend Auguste Chevalier—but really intended for Gauss and Jacobi—a testamentary letter, a tragic document in which he attempted to sketch the principal results he had achieved. On 30 May, mortally wounded by an unknown adversary, he was hospitalized; he died the following day. His funeral, on 2 June, was the occasion for a republican demonstration heralding the tragic riots that bloodied Paris in the days that followed.

Galois's work seems not to have been fully appreciated by any of his contemporaries. Cauchy, who would have been capable of grasping its importance, had left France in September 1830, having seen only its first outlines. Moreover, the few fragments published during Galois's lifetime did not give an overall view of his achievement and, in particular, did not afford a means of judging the exceptional interest of the results obtained in the theory of equations and rejected by Poisson. The publication in September 1832 of the famous testamentary letter does not appear to have attracted the attention it deserved. It was not until September 1843 that Liouville, who prepared Galois's manuscripts for publication, announced officially to the Academy that the young mathematician had effectively solved the problem, already considered by Abel, of deciding whether an irreducible first-degree equation is or is not "solvable with the aid of radicals." Although announced and prepared for the end of 1843, the publication of the celebrated 1831 memoir and of a fragment on the "primitive equations solvable by radicals" did not occur until the October–November 1846 issue of the *Journal de mathématiques pures et appliquées.*

It was, therefore, not until over fourteen years after Galois's death that the essential elements of his work became available to mathematicians. By this time the evolution of mathematical research had created a climate much more favorable to its reception: the dominance of mathematical physics in the French school had lessened, and pure research was receiving a new impetus. Furthermore, the recent publication of the two-volume *Oeuvres complètes de Niels-Henrik Abel* (1839), which contained fundamental work on the algebraic theory of elliptic functions and an important, unfinished memoir, "Sur la résolution algébrique des équations," had awakened interest in certain of the fields in which Galois has become famous. Lastly, in a series of publications appearing in 1844–1846, Cauchy, pursuing studies begun in 1815 but soon abandoned, had—implicitly—given group theory a new scope by the systematic construction of his famous theory of permutations.

Beginning with Liouville's edition, which was reproduced in book form in 1897 by J. Picard, Galois's work became progressively known to mathematicians and exerted a profound influence on the development of modern mathematics. Also important, although they came to light too late to contribute to the advance of mathematics, are the previously unpublished texts that appeared later. In 1906–1907 various manuscript fragments edited by J. Tannery revealed the great originality of the young mathematician's epistemological writings and provided new information about his research. Finally, in 1961 the exemplary critical edition of R. Bourgne and J. P. Azra united all of Galois's previously published writings and most of the remaining mathematical outlines and rough drafts. While this new documentary material provides no assistance to present-day mathematicians with their own problems, it does permit us to understand better certain aspects of Galois's research, and it will perhaps help in resolving a few remaining enigmas concerning the basic sources of his thought.

To comprehend Galois's work, it is important to consider the earlier writings that influenced its initial orientation and the contemporary investigations that contributed to guiding and diversifying it. It is equally necessary to insist on Galois's great originality: while assimilating the most vital currents of contemporary mathematical thought, he was able to transcend them thanks to a kind of prescience about the conceptual character of modern mathematics. The epistemological texts extracted from his rough drafts sketch, in a few sentences, the principal directions of present-day research; and the clarity, conciseness, and precision of the style add to the novelty and impact of the ideas. Galois was undoubtedly the beneficiary of his predecessors and of his rivals, but his multifaceted personality and his brilliant sense of the indispensable renewal of mathematical thinking made him an exceptional innovator whose influence was long felt in vast areas of mathematics.

Galois's first investigations, like Abel's, were inspired by the works of Lagrange and of Gauss on the conditions of solvability of certain types of algebraic equations and by Cauchy's memoirs on the theory of substitutions. Consequently their similarity is not surprising, nor is the particular fact that the principal results announced by Galois in May–June 1829 had previously been obtained by Abel. In the second half of 1829 Galois learned that Abel had published his findings in Crelle's *Journal für die reine und angewandte Mathematik* a few days before he himself died young. The interest that Galois took from that time in the work of Abel and of his other youthful rival, Jacobi, is evident from numerous reading notes. If, as a result of the progressive elaboration of group theory, Galois pursued the elucidation of the theory of algebraic equations far beyond the results published by Abel, beginning with the first months of 1830 he directed a large proportion of his research toward other new directions opened by both Abel and Jacobi, notably toward the theory of elliptic functions and of certain types of integrals.

The advances that Galois made in his first area of research, that of the theory of algebraic equations, are marked by two great synthetic studies. The first was written in February 1830 for the Academy's grand prize; the summary of it that Galois published in April 1830 in Férussac's *Bulletin des sciences mathématiques* establishes that he had made significant progress beyond Abel's recent memoir but that certain obstacles still stood in the way of an overall solution. The publication in Crelle's *Journal für die reine und angewandte Mathematik* of some posthumous fragments of Abel's work containing more advanced results (the unfinished posthumous memoir on this subject was not published until 1839) encouraged Galois to persevere in his efforts to overcome the remaining difficulties and to write a restatement of his studies. This was the purpose of the new version of the "Mémoire sur la résolution des équations algébriques" that he presented before the Academy.

Despite Poisson's criticisms Galois rightly persisted in thinking that he had furnished a definitive solution to the problem of the solvability of algebraic equations and, after having made a few corrections in it, he gave this memoir the first place in the list of his writings in his testamentary letter of 29 May 1837. This was the "definitive" version of his fundamental memoir, and in it Galois continued the studies of his predecessors but at the same time produced a thoroughly original work. True, he formulated in a more precise manner essential ideas that were already in the air, but he also introduced others that, once stated, played an important role in the genesis of modern algebra. Moreover, he daringly generalized certain classic methods in other fields and succeeded in providing a complete solution—and indeed a generalization—of the problem in question by systematically drawing upon group theory, a subject he had founded concurrently with his work on equations.

Lagrange had shown that the solvability of an algebraic equation depends on the possibility of finding a chain of intermediate equations of binomial type, known as resolvent equations. He had thus succeeded in finding the classic resolution formulas of the "general" equations of second, third, and fourth degree but had not been able to reach any definitive conclusion regarding the general fifth-degree equation. The impossibility of solving this last type of equation through the use of radicals was demonstrated by Paolo Ruffini and in a more satisfactory manner by Abel in 1824. Meanwhile, in 1801, Gauss had published an important study of binomial equations and the primitive roots of unity; and Cauchy in 1815 had made important contributions to the theory of permutations, a particular form of the future group theory.

In his study of the solvability of algebraic equations, Galois, developing an idea of Abel's, considered that with each intermediate resolvent equation there is associated a field of algebraic numbers that is intermediate between the field generated by the roots of the equation under study and the field determined by the coefficients of this equation. His leading idea, however, was to have successfully associated with the given equation, and with the different intermediate fields involved, a sequence of groups such that the group corresponding to a certain field of the sequence associated with the equation is a subgroup distinct from the one associated with the antecedent field. Such a method obviously presupposes the clarification of the concept of field already suspected (without use of the term) by Gauss and Abel, as well as a searching study of group theory, of which Galois can be considered the creator.

Galois thus showed that for an irreducible algebraic equation to be solvable by radicals, it is necessary and sufficient that its group be solvable, i.e., possess a series of composition formed of proper subgroups having certain precisely defined properties. Although this general rule did not in fact make the actual resolution of a determinate equation any simpler, it did provide the means for finding, as particular cases, all the known results concerning the solvability of the general equations of less than fifth degree as well as binomial equations and certain other particular types of equations; it also permitted almost immediate demonstration that the general equation of higher than fourth degree is not solvable by radicals, the associated group (permutation group of n objects) not being solvable. Galois was aware that his study went beyond the limited problem of the solvability of algebraic equations by means of radicals and that it allowed one to take up the much more general problem of the classification of the irrationals.

In his testamentary letter, Galois summarized a second memoir (of which several fragments are extant) that dealt with certain developments and applications of the theory of equations and of group theory. The article "Sur la théorie des nombres" is linked with it; it contained, notably, a daring generalization of the theory of congruences by means of new numbers that are today called Galois imaginaries and its application to research in those cases where a primitive equation is solvable by radicals. Beyond the precise definition of the decomposition of a group, this second memoir included applications of Galois's theory to elliptic functions; in treating the algebraic equations obtained through the division and transformation of these functions, it presents, without demonstration, the results concerning the modular equations upon which the division of the periods depends.

The third memoir that Galois mentions in his testamentary letter is known only through the information contained in this poignant document. This information very clearly demonstrates that, like Abel and Jacobi, Galois passed from the study of elliptic functions to consideration of the integrals of the most general algebraic differentials, today called Abelian integrals. It seems that his research in this area was already quite advanced, since the letter summarizes the results he had achieved, particularly the classification of these integrals into three categories, a result obtained by Riemann in 1857. This same letter alludes to recent meditations entitled "Sur l'application à l'analyse transcendante de la théorie de l'ambiguïté," but the allusion is too vague to be interpreted conclusively.

Galois often expressed prophetic reflections on the spirit of modern mathematics: "Jump with both feet on the calculus and group the operations, classifying them according to their difficulties and not according to their forms; such, in my view, is the task of future mathematicians" (*Écrits et mémoires,* p. 9).

He also reflected on the conditions of scientific creativity: "A mind that had the power to perceive at once the totality of mathematical truths—not just those known to us, but all the truths possible—would be able to deduce them regularly and, as it were, mechanically . . . but it does not happen like that" (*ibid.,* pp. 13–14). Or, again, "Science progresses by a series of combinations in which chance does not play the smallest role; its life is unreasoning and planless [*brute*] and resembles that of minerals that grow by juxtaposition" (*ibid.,* p. 15).

Yet we must also recall the ironic, mordant, and provocative tone of Galois's allusions to established scientists: "I do not say to anyone that I owe to his counsel or to his encouragement everything that is good in this work. I do not say it, for that would be to lie" (*ibid.,* p. 3). The contempt that he felt for these scientists was such that he hoped the extreme conciseness of his arguments would make them accessible only to the best among them.

Galois's terse style, combined with the great originality of his thought and the modernity of his conceptions, contributed as much as the delay in publication to the length of time that passed before Galois's work was understood, recognized at its true worth, and fully developed. Indeed, very few mathematicians of the mid-nineteenth century were ready to assimilate such a revolutionary work directly. Consequently the first publications that dealt with it, those of Enrico Betti (beginning in 1851), T. Schönemann, Leopold Kronecker, and Charles Hermite, are simply commentaries, explanations, or immediate and limited applications. It was only with the publication in 1866 of the third edition of Alfred Serret's *Cours d'algèbre supérieure* and, in 1870, of Camille Jordan's *Traité des substitutions* that group theory and the whole of Galois's *oeuvre* were truly integrated into the body of mathematics. From that time on, its development was very rapid and the field of application was extended to the most varied branches of the science; in fact, group theory and other more subtle elements included in Galois's writings played an important role in the birth of modern algebra.

BIBLIOGRAPHY

I. Original Works. Galois's scientific writings have appeared in the following versions: "Oeuvres mathématiques d'Evariste Galois," J. Liouville, ed., in *Journal de mathématiques pures et appliquées,* **11** (Oct.–Nov. 1846), 381–448; *Oeuvres mathématiques d'Evariste Galois,* J. Picard, ed. (Paris, 1897), also in facs. repro. (Paris, 1951) with a study by G. Verriest; "Manuscrits et papiers inédits de Galois," J. Tannery, ed., in *Bulletin des sciences mathématiques,* 2nd ser., **30** (Aug.–Sept. 1906), 246–248, 255–263; **31** (Nov. 1907), 275–308; *Manuscrits d'Evariste Galois,* J. Tannery, ed. (Paris, 1908); and *Écrits et mémoires mathématiques d'Evariste Galois,* R. Bourgne and J.-P. Azra, eds. (Paris, 1962), with pref. by J. Dieudonné. These eds. will be designated, respectively, as "Oeuvres," *Oeuvres,* "Manuscrits," *Manuscrits,* and *Écrits et mémoires.* Since the *Oeuvres* and *Manuscrits* are simply reeditions in book form of the "Oeuvres" and of the "Manuscrits," they are not analyzed below; the contents of the other three are specified according to date in the following list.

1. Scientific texts published during his lifetime.

Apr. 1829: "Démonstration d'un théorème sur les fractions continues périodiques," in Gergonne's *Annales de mathématiques pures et appliquées,* **19,** 294–301.

Apr. 1830: "Analyse d'un mémoire sur la résolution algébrique des équations," in Férussac's *Bulletin des sciences mathématiques,* **13,** 271–272.

June 1830: "Note sur la résolution des équations numériques," *ibid.,* 413–414.

June 1830: "Sur la théorie des nombres," *ibid.,* 428–436.

Dec. 1830: "Notes sur quelques points d'analyse," in Gergonne's *Annales de mathématiques pures et appliquées,* **21,** 182–184.

Jan. 1831: "Lettre sur l'enseignement des sciences," in *Gazette des écoles,* no. 110 (2 Jan. 1831).

2. Posthumous publications.

Sept. 1832: "Lettre à Auguste Chevalier," in *Revue encyclopédique,* **55,** 568–576.

Oct.–Nov. 1846: "Oeuvres," considered definitive until 1906; in addition to the memoirs published in Galois's lifetime (except for the last) and the letter to Auguste Chevalier, this ed. contains the following previously unpublished memoirs: "Mémoire sur les conditions de résolubilité des équations par radicaux," pp. 417–433; and "Des équations primitives qui sont solubles par radicaux," pp. 434–444.

Aug.–Sept. 1906: "Manuscrits," pt. 1, which contains, besides a description of Galois's MSS, the text of the following previously unpublished fragments (titles given are those in *Écrits et mémoires*): "Discours préliminaire"; "Projet de publication"; "Note sur Abel"; "Préface" (partial); "Discussions sur les progrès de l'analyse pure"; "Fragments"; "Science, hiérarchie, écoles"; and "Catalogue, note sur la théorie des équations."

Nov. 1907: "Manuscrits," pt. 2, containing "Recherches sur la théorie des permutations et des équations algébriques"; "Comment la théorie des équations dépend de celle des permutations"; "Note manuscrite"; "Addition au second mémoire"; "Mémoire sur la division des fonctions elliptiques de première espèce"; "Note sur l'intégration des équations linéaires"; "Recherches sur les équations du second degré."

Jan.–Mar. 1948; entire text of the "Préface" and of the "Projet de publication," R. Taton, ed., in *Revue d'histoire des sciences,* **1,** 123–128.

1956: "Lettre sur l'enseignement des sciences," repr. in

A. Dalmas, *Evariste Galois . . .* (Paris, 1956), pp. 105–108.

1962: *Écrits et mémoires mathématiques d'Evariste Galois,* R. Bourgne and J.-P. Azra, eds. (Paris, 1962). This remarkable ed. contains all of Galois's *oeuvre:* the articles published in his lifetime and a critical ed., with corrections and variants, of all his MSS, including his rough drafts. The majority of the many previously unpublished texts presented here are grouped in two categories: the "Essais," dating from the period when Galois was a student (pp. 403–453, 519–521) and the "Calculs et brouillons inédits" (pp. 187–361, 526–538), classed under five headings—"Intégrales eulériennes," "Calcul intégral," "Fonctions elliptiques," "Groupes de substitutions," and "Annexe." Galois's nine known letters are reproduced and described (pp. 459–471, 523–525). Galois's MSS, preserved at the Bibliothèque de l'Institut de France (MS 2108), are the subject of a detailed description that provides many complementary details (App. I, 478–521; App. II, 526–538).

II. SECONDARY LITERATURE. At the present time there is no major synthetic study of Galois's life and work. The principal biographical source remains P. Dupuy, "La vie d'Evariste Galois," in *Annales scientifiques de l'École normale supérieure,* 3rd ser., **13** (1896), 197–266, with documents and two portraits; reiss. as *Cahiers de la quinzaine,* 5th ser., no. 2 (Paris, 1903).

Among the few earlier articles the only ones of any documentary value are the two brief obituaries in *Revue encyclopédique,* **55** (Sept. 1832): the first (pp. 566–568), unsigned, is very general; the second ("Nécrologie," pp. 744–754), by Auguste Chevalier, Galois's best friend, is a source of valuable information. See also an anonymous notice, inspired by Evariste's younger brother, Alfred Galois, and by one of his former classmates, P.-P. Flaugergues, in *Magasin pittoresque,* **16** (1848), 227–228; and a note by O. Terquem in *Nouvelles annales de mathématiques,* **8** (1849), 452.

Of the later biographical studies a few present new information: J. Bertrand, "La vie d'Evariste Galois par P. Dupuy," in *Journal des savants* (July 1899), pp. 389–400, reiss. in *Éloges académiques,* n.s. (Paris, 1902), pp. 331–345; R. Taton, "Les relations scientifiques d'Evariste Galois avec les mathématiciens de son temps," in *Revue d'histoire des sciences,* **1** (1947), 114–130; A. Dalmas, *Evariste Galois, révolutionnaire et géomètre* (Paris, 1956); the ed. of *Écrits et mémoires mathématiques* by R. Bourgne and J.-P. Azra cited above; C. A. Infantozzi, "Sur la mort d'Evariste Galois," in *Revue d'histoire des sciences,* **21** (1968), 157–160; art. by J.-P. Azra and R. Bourgne in *Encyclopaedia universalis,* VII (Paris, 1970), 450–451; and R. Taton, "Sur les relations mathématiques d'Augustin Cauchy et d'Evariste Galois," in *Revue d'histoire des sciences,* **24** (1971), 123–148.

G. Sarton, "Evariste Galois," in *Scientific Monthly,* **13** (Oct. 1921), 363–375, repr. in *Osiris,* **3** (1937), 241–254; and E. T. Bell, *Men of Mathematics* (New York, 1937), pp. 362–377, were directly inspired by Dupuy. L. Infeld, *Whom the Gods Love. The Story of Evariste Galois* (New York, 1948); and A. Arnoux, *Algorithme* (Paris, 1948), mix facts with romantic elements.

Galois's scientific work has not yet received the thorough study it merits, although numerous articles attempt to bring out its main features. Among the older ones, beyond the "commentaries" of the first disciples, particularly Betti and Jordan, are the following: J. Liouville, "Avertissement" to the "Oeuvres," in *Journal de mathématiques pures et appliquées,* **11** (1846), 381–384; S. Lie, "Influence de Galois sur le développement des mathématiques," in *Le centenaire de l'École normale* (Paris, 1895), pp. 481–489; E. Picard, "Introduction" to *Oeuvres* (Paris, 1897), pp. v–x; J. Pierpont, "Early History of Galois's Theory of Equations," in *Bulletin of the American Mathematical Society,* **4** (Apr. 1898), 332–340; J. Tannery, "Introduction" to "Manuscrits" in *Bulletin des sciences mathématiques,* **30** (1906), 1–19, repr. in *Manuscrits,* pp. 1–19.

The most important recent studies are G. Verriest, *Evariste Galois et la théorie des équations algébriques* (Louvain–Paris, 1934; reiss. Paris, 1951); L. Kollros, *Evariste Galois* (Basel, 1949); J. Dieudonné, "Préface" (pp. v–vii), R. Bourgne, "Avertissement" (pp. ix–xvi), and J.-P. Azra, "Appendice" (pp. 475–538), in *Écrits et mémoires mathématiques* (cited above); N. Bourbaki, *Éléments d'histoire des mathématiques,* 2nd ed. (Paris, 1969), pp. 73-74, 104-109; and K. Wussing, *Die Genesis des abstrakten Gruppenbegriffes* (Berlin, 1969), esp. pp. 73–87, 206–211.

RENÉ TATON

GALTON, FRANCIS (*b.* Birmingham, England, 16 February 1822; *d.* Haslemere, Surrey, England, 17 January 1911), *statistics, anthropometry, experimental psychology, heredity.*

Galton's paternal ancestors were bankers and gunsmiths, of the Quaker faith, and long-lived. His mother was Erasmus Darwin's daughter, and thus he was Charles Darwin's cousin. Galton's intellectual precocity has become a textbook item, and Lewis Terman estimated his IQ to have been of the order of 200. His education, though, was desultory, its formal peaks being a few mathematics courses at Cambridge (he took a pass degree) and some unfinished medical studies in London. He quit the latter at the age of twenty-two when his father died, leaving him a fortune. He then traveled. Journeying through virtually unknown parts of southwestern Africa in 1850–1852, Galton acquired fame as an intrepid explorer. His immediate reward was a gold medal from the Geographical Society, and his later reports led to election as a fellow of the Royal Society in 1860. In 1853 he married, and in 1857 he settled into a quiet London home, where he remained, except for occasional European vacations, until his death over half a century later. Galton was knighted in 1909. He died childless.

Galton was perhaps the last of a now extinct breed—the gentleman scientist. He never held any academic or professional post, and most of his experiments were done at home or while traveling, or were farmed out to friends. He was not a great reader, and his small personal library was said to consist mainly of autographed copies of fellow scientists' books. He composed no *magnum opus,* but he kept up a rich flow of original ideas. An endless curiosity about the phenomena of nature and mankind was nicely coupled with mechanical ingenuity and inventiveness. Secure and contented in the employment of his wide-ranging talents, Galton was an unusually equable person. Anger and polemic were alien to him. In his later years he was fortunate in having the ebullient Karl Pearson as champion and extender of his ideas. Pearson subsequently became the first holder of the chair of eugenics at University College, London, that Galton had endowed in his will.

Galton's earliest notable researches were meteorologic, and it was he who first recognized and named the anticyclone.

Foremost in Galton's life was a belief that virtually anything is quantifiable. Some of his exercises in this direction are now merely amusing—a solemn assessment of womanly beauty on a pocket scale, a study of the body weights of three generations of British peers, and a statistical inquiry into the efficacy of prayer are examples—but there can be little doubt that his general attitude was salutary in its day. Moreover, against the trivia have to be set such good things as his developing Quetelet's observation that certain measurable human characteristics are distributed like the error function. Galton initiated an important reversal of outlook on biological and psychological variation, previously regarded as an uninteresting nuisance. In his own words: "The primary objects of the Gaussian Law of Errors were exactly opposed, in one sense, to those to which I applied them. They were to get rid of, or to provide a just allowance for, errors. But these errors or deviations were the very things I wanted to preserve and know about." In psychology Galton sowed the seeds of mental testing, of measuring sensory acuity, and of scaling and typing. In statistics he originated the concepts of regression and correlation.

Galton's best-known work was on the inheritance of talent—scholarly, artistic, and athletic—the raw data being the records of notable families. He found strong evidence of inheritance. Upholders of the rival nurture-not-nature theory attacked the work, on the ground that the children of gifted and successful parents are environmentally favored; but even when allowance was made for this truth, Galton's contention could not be wholly denied. One outcome of the investigation was a conviction in many people's minds—and particularly deeply in Galton's own mind—that a eugenic program to foster talent and healthiness and to suppress stupidity and sickliness was a *sine qua non* in any society that wished to maintain, let alone promote, its quality and status. (Galton coined the word "eugenics" in 1883.)

Galton's views on genetics are historically curious. Influenced by Darwin's belief that inheritance is conditioned by a blending mechanism, Galton propounded his law of ancestral heredity, which set the average contribution of each parent at 1/4, of each grandparent at 1/16, and so forth (the sum, over all ancestors of both parents, being asymptotic to unity). Karl Pearson and his colleagues pursued the notion in a series of sophisticated researches, but Galton's law received withering criticisms after the rediscovery, in 1900, of Mendel's work on particulate inheritance. Yet Galton had himself toyed with the notion of particulate inheritance, and in a remarkable correspondence with Darwin in 1875 he sketched the essence of the theory and even discussed something very like what we now know as genotypes and phenotypes under the names "latent" and "patent" characteristics. He did not press these views, perhaps because of the strong climate of opinion in favor of blending inheritance at that time.

Galton's establishment of fingerprinting as an easy and almost infallible means of human identification transformed a difficult subject, and his taxonomy of prints is basically that used today. He was disappointed, however, to find no familial, racial, moral, or intellectual subgroupings in the collections he examined.

BIBLIOGRAPHY

I. ORIGINAL WORKS. Galton wrote sixteen books and more than 200 papers. Of the books, recent printings are *Hereditary Genius* (London, 1869; 3rd ed., 1950); *Art of Travel* (5th ed., London, 1872; repr. Harrisburg, Pa., 1971); and *Finger Prints* (London, 1893; facs., New York, 1965). An unpublished utopian book, "The Eugenic College of Kantsaywhere," written toward the end of his life, is excerpted in Karl Pearson's biography (see below). His autobiography, *Memories of My Life* (London, 1908), is worth reading. The best listing of Galton's publications is appended to Blacker's book (see below).

II. SECONDARY LITERATURE. Immediately after Galton's death his friend Karl Pearson started a biography that was to become one of the most elaborate and comprehensive works of its kind in this century: *The Life, Letters and Labours of Francis Galton,* 4 vols. (London, 1914–1930). A treatment emphasizing the interests of his later years is C. P. Blacker, *Eugenics, Galton and After* (London, 1952). A good survey of his psychologic contributions is H. E. Garratt, *Great Experiments in Psychology* (New York, 1951), ch. 13. The 1965 repr. of *Finger Prints* (see above) contains a biographical intro. by Harold Cummins that places Galton's fingerprint work in historic context.

NORMAN T. GRIDGEMAN

GALVANI, LUIGI (*b.* Bologna, Italy, 9 September 1737; *d.* Bologna, 4 December 1798), *anatomy, physiology, physics.*

Galvani, who is most famous for his work relating to the discovery of current electricity, received his professional training in medicine. He studied at Bologna with several leading medical teachers of his time, including Jacopo Bartolomeo Beccari and Domenico Galeazzi. After receiving his degree in medicine and philosophy on 15 July 1759, Galvani divided the first years of his professional career between medical and surgical practice, anatomical research, and lecturing on medicine. After spending several years as an honorary lecturer, on 22 June 1768 he became a paid lecturer at the college he had attended, and on 12 December 1775 he became Galeazzi's adjunct in anatomy at the University of Bologna. The Senate of Bologna had installed Galvani as curator and demonstrator of the anatomical museum in March 1766, and on 26 February 1782 it elected him professor of obstetric arts at the Istituto delle Scienze. During the last years of his life Galvani suffered several personal misfortunes. In 1790 his beloved wife, Lucia Galeazzi, daughter of his anatomical preceptor, died; and a few years later he was deprived of his offices at the university and the Istituto delle Scienze because of his refusal to swear allegiance to Napoleon's Cisalpine Republic. He died in poverty and sorrow.

Galvani devoted most of his early scientific efforts to important but rather straightforward anatomical topics. His first publication, in 1762, was a dissertation on the structure, function, and pathology of bones. He described the chemical and anatomical elements from which bones are constructed, their pattern of growth, and various diseases to which they are subject. In 1767 he published an essay on the kidneys of birds, in which he described, among other things, the three-layered ureteral wall and its peristaltic and antiperistaltic movement upon irritation. Galvani also devoted several papers to the anatomy of the ear in birds, just before Antonio Scarpa published on this subject. He recounted with particular precision the comparative anatomy of the auditory canal in several species of birds, devoting some attention to the distribution of blood vessels, muscles, and nerves in the middle and inner ear.

Galvani addressed his most important and best-remembered investigations to problems of animal electricity. During the 1770's his research interests shifted to a considerable extent from largely anatomical to more strictly physiological studies, specifically on nerves and muscles. In 1772 Galvani read a paper on Hallerian irritability to the Istituto delle Scienze, and in 1773 he discussed the muscle movement of frogs before the same body. In 1774 he read a paper on the effect of opiates on frog nerves. These researches fused in his mind with slightly earlier eighteenth-century studies, several of them by Italians, on the electrical stimulation of nerves and muscles. Picking up where Beccaria, Leopoldo Caldani, Felice Fontana, and Tommaso Laghi had recently left off, Galvani began in late 1780 an extensive and meticulous series of investigations into the irritable responses elicited by static electricity in properly prepared frogs.

Galvani's frog preparations consisted of the spinal cords, crural nerves, and lower limbs dissected as a unit. Using these preparations, he at first touched the conductor of a static electrical machine directly to the spinal cord (kept on a pane of glass) and watched the convulsive contractions of the muscles in the lower limbs, which rested on a so-called "magic square," a flat plate condenser made by attaching a sheet of metal foil to both sides of a single pane of glass. Galvani was apparently trying to arrive at general laws relating the forcefulness of muscle contraction directly to the quantity of electric fluid applied and inversely to the distance of the nerve and muscle from the conductor. After much repetition and sometimes complex variation of this basic procedure, Galvani was faced with one quite unanticipated result: the lower limbs contracted even when the frog was completely insulated from the machine and removed some distance from it. As long as the crural nerves were touched by a grounded conductor, the muscles contracted whenever a spark was drawn from an electrical machine, even though the spark did not directly strike the frog preparation.

In the course of investigating this strange result, Galvani in the mid-1780's uncovered an even stranger one. He and his research associates had begun to explore the effects of atmospheric electricity on frog preparations, on the assumption that some analogy existed between convulsions induced by distant electrical machines and those sometimes induced by static discharge in the atmosphere. The expected analogous results were obtained. But then Galvani made the unanticipated observation that muscle contractions occurred even without discharge of atmospheric electricity. As he explained later in his *De viribus electricitatis in motu musculari commentarius* (1791), Galvani at one point fastened some prepared frogs by "brass hooks in their spinal cord to an iron railing which surrounded a certain hanging garden of my house." He noticed that these frogs went into contractions "not only when lightning flashed but even at times when the sky was quiet and serene," and he was able to intensify these effects by deliberately pressing the brass hooks in the spinal cord to the iron railing. He obtained similar results indoors by placing the frog on an iron plate and pushing the brass hook against it. Contractions resulted indoors only when metals, rather than glass or resin, were used; and these contractions seemed stronger with certain metals than with others. In a follow-up series of investigations, Galvani experimented with metallic arcs. He tried various bent metal conductors, touching one end to the hook in the spinal cord and the other to the muscles in the frog's leg. Contractions resulted, their strength depending on the metals used for the hook and the arc. Contractions did not result when a nonconductor replaced the metal in the arc.

Galvani had here hit upon the central phenomenon of galvanism: the production of electric current from the contact of two different metals in a moist environment. He did not, however, interpret his own discovery this way. Instead, Galvani thought that he had finally obtained confirmation for the suspicion, entertained from time to time during the eighteenth century, that animals possess in their nerves and muscles a subtle fluid quite analogous to ordinary electricity. He himself had occasionally flirted with this idea but had never previously made much of it. But his experiments with the metallic arcs seemed to provide clear and unmistakable proof of a special "animal electricity," and he spent considerable effort in specifying and elaborating his theory.

Galvani's fullest statement is in part IV of his *Commentarius.* He explains that the muscle can be compared to a small Leyden jar charged with a dual electrical charge, and the nerve to the jar's conductor. Animal electrical fluid is generated from the blood in the brain and passes via the nerves into the core of the muscles, which thus become positively charged while the outside becomes negative. Electrical equilibrium in the muscle, as in a Leyden jar, can be disrupted by applying an arc between conductor and core or by drawing a spark from an electrical machine. When the muscle discharges in either of these ways, its fibers are stimulated to violent, irritable contraction. Both the original anomaly of convulsive contraction upon distant sparking and the subsequent observation of contractions provoked by the metallic arc were thus explained in terms of "animal electricity" and its special discharge pathways.

Reaction to Galvani's published reflections was vigorous although somewhat confused. Alessandro Volta, the noted Italian electrician, was among the first to take up the new theory of animal electricity, but by 1792/1793 his original support turned to skeptical reserve. In papers published in the *Philosophical Transactions of the Royal Society,* Volta professed belief in Galvani's theory but simultaneously advanced the thesis that the "metals used in the experiments, being applied to the moist bodies of animals, can by themselves . . . excite and dislodge the electric fluid from its state of rest; so that the organs of the animal act only passively." By the end of 1793 Volta had discarded Galvani's animal electricity for his own theory of "contact," according to which conducting bodies of certain kinds, especially metals, can by their mere contact excite electrical fluid, which can in turn stimulate various irritable responses. Galvani was not prepared to concede defeat, and he and his nephew Giovanni Aldini mounted a campaign in the mid-1790's to establish beyond doubt the existence of a special animal electricity. In 1794 and 1797 he announced experiments employing only frog nerve-muscle preparations (without metals) and showed that convulsive contractions could be produced merely by touching nerves to muscles.

At the same time, Galvani extensively examined the electrical properties of marine torpedoes. He found that the strong electrical discharge is generated in these animals in structures analogous to ordinary nerves and muscles, and this seemed to supply additional support for the theory of animal electricity. Volta's counterattack led in 1799 to his invention of the pile, a stack of metal-metal-moist-conductor elements which was, in fact, the first primitive wet-cell battery. When Galvani died, prospects for the survival of his theory were very uncertain. Nevertheless, support for the concept of animal electricity survived into the nineteenth century and ultimately led in the 1840's to the basic work of Emil du Bois-Reymond.

BIBLIOGRAPHY

I. ORIGINAL WORKS. Galvani's most famous work is *De viribus electricitatis in motu musculari commentarius* (Bologna, 1791). It has been published several times since, reproduced in facsimile, and issued in several translations. A facsimile of the original Latin ed., together with an English trans., was issued by the Burndy Library (Norwalk, Conn., 1953). Fuller eds. of Galvani's writings include *Opere edite ed inedite* (Bologna, 1841), which contains several of his early anatomical papers and a report on then known MSS; *Memorie ed esperimenti inediti* (Bologna, 1937), which includes a transcription of Galvani's notes for his experiments in the early 1780's and a few draft papers on animal electricity from the same period; and a facsimile of *Taccuino* (Bologna, 1937), a notebook of Galvani's investigations into torpedoes in the mid-1790's.

II. SECONDARY LITERATURE. There is no full-length modern biography of Galvani, but several older *éloges*, e.g., by J. L. Alibert (Paris, 1806), are still useful and are supplemented by some extremely useful monographic work. Hebbel E. Hoff, "Galvani and the Pre-Galvanian Electrophysiologists," in *Annals of Science*, **1** (1936), 157–172, is a basic source, as is I. B. Cohen's "Introduction" to the Burndy Library ed. of the *Commentarius*. Also of fundamental importance are Giulio C. Pupilli's "Introduction" to the ed. of the *Commentarius* published by Richard Montraville Green (Cambridge, Mass., 1953); and John F. Fulton and Harvey Cushing, "A Bibliographic Study of the Galvani and Aldini Writings on Animal Electricity," in *Annals of Science*, **1** (1936), 239–268. Also worth consulting is Marc Sirol, *Galvani et le galvanisme* (Paris, 1939).

THEODORE M. BROWN

GAMALEYA, NIKOLAY FYODOROVICH (*b.* Odessa, Russia, 17 February 1859; *d.* Moscow, U.S.S.R., 29 March 1949), *microbiology*.

Gamaleya came from a Ukrainian family that had risen through service to the country since the seventeenth century. His father, Fyodor Mikhailovich Gamaleya, was a soldier; his mother, Karolina Vikentievna Gamaleya, was of Polish extraction.

Having graduated from the Gymnasium in 1876, Gamaleya enrolled in the Physics and Mathematics Faculty at Novorossysky University. While a student there he became fascinated with biology. One of his teachers was E. I. Mechnikov, and in Strasbourg, where Gamaleya went for vacations, he studied biochemistry under Hoppe-Seyler.

After graduation from the university in 1881, Gamaleya enrolled in the Military Medical Academy at St. Petersburg, then the center of medical education in Russia. His teachers included such prominent figures as S. P. Botkin, V. V. Pashutin, and V. A. Manassein. After graduation in 1883 with the title of physician, Gamaleya returned to Odessa. The young doctor became actively interested in bacteriology, a science then in its infancy, and conducted research in a bacteriological laboratory that he had set up in his apartment.

Pasteur's successful inoculation against rabies in 1885 definitively determined Gamaleya's scientific interests. In 1886 the Odessa Society of Physicians commissioned him to familiarize himself at Pasteur's laboratory with the technique of performing antirabies inoculations. His persistence and curiosity, medical knowledge, and microbiological training enabled him to master the method. The acquaintance with Pasteur was the beginning of creative collaboration and of a personal friendship that was strengthened by the struggle with opponents of Pasteur's method. At the time of especially sharp criticism of his method in England, Pasteur asked Gamaleya to defend it. Gamaleya was the first to inoculate himself with the antirabies vaccine, thereby proving its harmlessness to a healthy organism.

In 1886 the world's second bacteriological station—there was already one in Paris—was established in Odessa, with the participation of Mechnikov and Gamaleya. Here antirabies inoculations were successfully administered according to Pasteur's method, which undoubtedly was its best propaganda and defense. An ardent supporter of this method, Gamaleya used it widely and introduced important additions to its theoretical basis and valuable practical refinements.

In preparations containing the living virus, Gamaleya established that the effectiveness of antirabies vaccination depends on its quantitative content. On the basis of this principle he developed an intensive method of vaccination through the utilization of brain tissue less subject to drying. In addition, he discovered that inoculative antirabies immunity is physiologically limited and that vaccination is ineffective against manifest rabies as well as during the latent period of infection (about fourteen days).

In the 1880's, Gamaleya studied questions relating to the preparation of a vaccine against Siberian plague (anthrax). In 1887 he discovered a vibrio similar to that of cholera in the intestines of sick birds, which he named the Mechnikov bacillus. The study of this bacillus marked the beginning of many years of research in cholera.

In Pasteur's laboratory, as well as in those of Charles Bouchard and Joseph Strauss, Gamaleya studied the phenomena of inflammation and the processes whereby microbes are destroyed in an organism. He believed that microbes invading a living organism are subjected to the action of two closely related factors—humoral and cellular, that is, the action of soluble antibodies produced by the cells of the reticuloendothelial system. This research produced new data and concepts concerning these phenomena.

Returning to Russia in 1892 from France, where he had worked for a total of six years, Gamaleya initiated his study of cholera. In 1893 he defended his doctoral dissertation, *Etiologia kholery s tochki zrenia eksperimentalnoy patologii* ("The Etiology of Cholera From the Point of View of Experimental Pathology"). The study of cholera and the struggle against this disease occupied a conspicuous position in Gamaleya's scientific work and in his activities as a physician.

In 1899 Gamaleya published the textbook *Osnovy obshchey bakteriologii* ("Foundations of General Bacteriology"); its fruitful generalizations and original views on fundamental questions in bacteriology had great significance for the development of the new science. The hypothesis of a viral origin for cancer was first stated in this book, and in 1910 Mechnikov supported this hypothesis.

Until 1910 Gamaleya worked in Odessa at the Bacteriological-Physiological Institute, which he had founded, lectured on general bacteriology at the stomatology school, and published many works.

Gamaleya's importance in the history of bacteriology is as an outstanding researcher and fighter against bubonic plague. In 1902, in connection with a plague epidemic that had broken out in Odessa, Gamaleya began a theoretical investigation of its epidemiology. The system of practical measures he developed had a decisive significance in the liquidation and prevention of this dreaded disease.

In the period preceding the 1917 Revolution, Gamaleya actively concerned himself with prevention of epidemics. In 1908–1909 he conducted investigations of typhus; he was the initiator of a program of fumigation in Russia. From 1912 through 1928 he studied smallpox, which was endemic in Russia. As director of the Smallpox Inoculation Institute, he developed a new, refined means for obtaining smallpox detritus.

Exhaustive study of the theory and practical use of inoculations against rabies enabled Gamaleya to explain the causes of failures that had been observed in the application of the method and to propose the so-called intensive method, which was immediately accepted by Pasteur and introduced into wide use in critical cases of rabies. Gamaleya's work in paralytic rabies, then unstudied, was important. His research gained the high appreciation of Pasteur, who in 1887 conveyed his "keen appreciation for your rare services."

Gamaleya's proposals regarding the fight against cholera were exceptionally valuable in pre-Revolutionary Russia, where the low level of sanitation led to wide propagation of epidemic diseases. In contradistinction to the then accepted idea that cholera was spread exclusively by personal contact, Gamaleya contended that epidemics resulted from colossal multiplication of cholera bacilli in stagnant water. In this connection, he insisted on maximal observance of sanitation measures in densely populated areas. Moreover, Gamaleya proposed that cholera vaccinations be administered as prophylaxis. The success of this arrangement led to the complete elimination in the 1920's of cholera in the Soviet Union.

In 1883 Mechnikov had voiced his phagocyte theory of immunity. Gamaleya turned to a study of the mechanism of immunity against anthrax. Extensive and careful experiments in the preparation of vaccines and microscopic study of their action on anthrax bacilli in an organism enabled Gamaleya to establish the important regularity of the relationship between fever in the vaccinated organism and the manufacture of antibodies.

Study of the epidemiology of bubonic plague confirmed that it was transferred by the fleas on rodents. Having explained, in particular, the role of gray rats as carriers of the plague, Gamaleya launched a campaign during a plague epidemic for their complete extermination in cities. He also demonstrated that epidemic jaundice, mange, and typhus are also spread by rats. Following Gamaleya's suggestion, rats were annihilated not only by poison but also with the aid of microbes belonging to the paratyphoid group.

Gamaleya's many investigations of typhus were the result of much work on the surveillance of public sanitation. As early as 1874 the physician G. N. Minkh, having inoculated himself with the blood of a person suffering from relapsing fever, proved the contagiousness of this disease and put forth the hypothesis that it was carried by lice. In 1908 Gamaleya confirmed this hypothesis by epidemiological investigations. Studying methods for the annihilation of lice, he found that the only effective method was dry heat treatment (100°C.) of the infected insects, since their behavior is determined not by chemotaxis, as had been supposed, but solely by thermotaxis.

In studying tuberculosis, Gamaleya discovered various types of microbes that cause the disease. In 1910 he discovered a method for the cultivation of the tubercle bacillus in an artificial medium. He persistently worked on the creation of tuberculosis immunity and specific methods for treating the disease.

Gamaleya contributed greatly to the history of virology. He was the first to state, as early as 1886, that filterable viruses are pathogens of various illnesses. The subsequent development of virology has confirmed this brilliant vision.

Study of inflammation and the processes for destroying microbes led Gamaleya to the discovery in 1898 of certain bacteriolytic substances that destroy microbes. These previously unknown agents turned out to be bacteriophages, whose presence in nature was confirmed by d'Hérelle.

After 1917 Gamaleya successfully worked on problems of immunology, virology, and tuberculosis. Questions of sanitation, hygiene, and prophylactic medicine continued to remain the center of his attention. He was the scientific director of the Central

Institute of Microbiology and Epidemiology (1929–1931), which now bears his name. In 1931 he headed the organization of the Institute of Epidemiology and Microbiology in Yerevan. From 1938 Gamaleya headed the department of microbiology at the Second Moscow Institute of Medicine. He served as the organizer and permanent chairman of the All-Union Society of Microbiologists, Epidemiologists, and Infectionists.

Of Gamaleya's more than 350 works, over 100—primarily fundamental works and monographs—were written after 1917. Many have been published in translation.

BIBLIOGRAPHY

I. ORIGINAL WORKS. Gamaleya's collected works were published as *Sobranie sochineny* (Moscow, 1956). Among them are *Etiologia kholery s tochki zrenia eksperimentalnoy patologii* ("The Etiology of Cholera From the Point of View of Experimental Pathology," St. Petersburg, 1893), his diss.; *Bakterynye yady* ("Bacterial Poisons," Moscow, 1893); *Osnovy obshchey bakteriologii* ("Foundations of General Biology," Odessa, 1899); *Osnovy immunologii* ("Foundations of Immunology," Moscow-Leningrad, 1928); *Filtruyushchiesya virusy* ("Filterable Viruses," Moscow-Leningrad, 1930); *Ospoprivivanie* ("Smallpox Inoculation," Moscow-Leningrad, 1934); and *Uchebnik meditsinskoy mikrobiologii* ("Textbook of Medical Microbiology," Moscow, 1943).

II. SECONDARY LITERATURE. On Gamaleya or his work, see E. Finn, *Akademik Gamaleya. Ocherk zhizni i deyatelnosti* ("Academician Gamaleya. An Essay on His Life and Career," Moscow, 1963); N. P. Gracheva, *Bolshaya zhizn* ("A Great Life," Moscow, 1959); I. Gryaznov, *Nikolay Fyodorovich Gamaleya* (Moscow, 1949); Y. I. Milenushkin, *N. F. Gamaleya. Ocherk zhizni i deyatelnosti* ("N. F. Gamaleya. An Essay on His Life and Career," Moscow, 1954); and N. A. Semashko, "Pochetny akademik N. F. Gamaleya" ("Honorary Academician N. F. Gamaleya"), in *Nauka i zhizn,* no. 2 (1949), pp. 39–40.

V. GUTINA

GAMBEY, HENRI-PRUDENCE (*b.* Troyes, France, 8 October 1787; *d.* Paris, France, 28 January 1847), *precision instrumentation.*

Gambey was a workman and then supervisor at the École des Arts et Métiers in Compiègne. He then worked for a time in Châlons-sur-Marne; on the death of his father he returned to Paris, where he started a small shop in St. Denis. There he manufactured precision instruments for physicists and astronomers.

The high quality of Gambey's instruments soon brought him to the attention of French scientific circles. In 1819 he was asked by the director of the Paris Exhibition to display some of his work there (perhaps as an attempt to regain the international prestige of French instrumentation, lost to Ramsden in England and Fraunhofer and Georg von Reichenbach in Germany). Gambey had only two months in which to prepare his work for the exposition; nevertheless, his instruments were awarded the gold medal and the Royal Society of London characterized them as being unsurpassed in Europe for elegance and precision.

Shortly thereafter Gambey built a portable theodolite for the Bureau des Longitudes. He also made the first cathetometer, for Dulong and Petit; a heliostat for Fresnel; and a vastly improved compass for Coulomb. Most important, however, he constructed a number of major instruments for the Paris observatory, of which the mural circle that he finished just before he died is his masterpiece. (A gigantic new equatorial was built from his plans after his death.)

Gambey won further gold medals at the Paris exhibitions of 1824 and 1829. He was a member of the Bureau des Longitudes and was elected to the Académie des Sciences in 1837 to replace Mollart.

At one time Gambey planned to emigrate to America, but was persuaded to stay in France by François Arago. Arago later said that whenever French scientists needed new and delicate instruments they turned to Gambey, who invariably solved the problem to their satisfaction.

ASIT K. BISWAS
MARGARET R. BISWAS

GAMOW, GEORGE (*b.* Odessa, Russia, 4 March 1904; *d.* Boulder, Colorado, 20 August 1968), *physics.*

Gamow's father, Anton Gamow, taught Russian language and literature. Gamow was an outstanding student at the Odessa Normal School (1914–1920) but, owing to the turbulent political conditions of the time, his early education in general was rather sporadic. In 1922 he enrolled in the Physico-Mathematical Faculty of Novorossysky University, but within a year he transferred to the University of Petrograd (Leningrad). There, in 1925, he carried out experimental researches on optical glasses and briefly studied relativistic cosmology under A. A. Friedmann before his attention was drawn to the exciting and profound discoveries being made in quantum theory in Europe: his first publication (1926) involved an attempt to consider Erwin Schrödinger's wave function as the fifth dimension (the other four being the usual spatial and temporal dimensions).

In the summer of 1928, the year he received his Ph.D., Gamow traveled to Göttingen, where he made his first major contribution to physics: his theory of nuclear α decay. Ernest Rutherford had found (1927) that RaC α particles incident on uranium cannot penetrate the nucleus, although their energy is roughly double that of α particles emitted by uranium. Gamow immediately recognized that the apparent paradox vanished if the emitted α particles were "tunneling through" the nuclear potential barrier—a characteristic wave mechanical effect. Quantitative calculations proved that the empirically established relationship between the nuclear decay constant and the energy of the emitted α particles (the Geiger-Nuttall law) could be completely understood. This same conclusion was reached virtually simultaneously (see *Nature,* **122** [22 Sept. 1928]) by R. W. Gurney and E. U. Condon at Princeton University.

Niels Bohr, impressed by Gamow's achievement, offered him a Carlsberg fellowship to enable him to spend 1928–1929 at his Copenhagen Institute of Theoretical Physics, where Gamow continued to study problems in theoretical nuclear physics—for example, the parameters governing the yield of protons in α-bombardment reactions. In addition, through correspondence and personal contact with F. A. Houtermans and Robert Atkinson, he helped make pioneering contributions to the theory of thermonuclear reaction rates in stellar interiors. In the fall of 1929, after a visit to the Soviet Union, Gamow went to the Cavendish Laboratory at Cambridge on a Rockefeller fellowship. There he recognized that Heinz Pose's recent results on the α bombardment of aluminum indicated that the α particles were undergoing nuclear resonance. Later in the year Rutherford asked Gamow to estimate the energy required to split the nucleus by means of artificially accelerated protons and, encouraged by the result, set J. D. Cockcroft and Ernest Walton to work on the construction of the accelerator, with well-known results.

In 1930–1931 Gamow received further fellowship aid to return to Bohr's institute in Copenhagen, where a major part of his time was devoted to preparing a paper on the quantum theory of nuclear structure, which he had been invited to deliver at Rome in October 1931, to the first International Congress on Nuclear Physics. After returning to the Soviet Union to renew his visa in the spring of 1931 he was denied permission to attend the Rome conference. Gamow spent the next two years as professor of physics at the University of Leningrad; then he and his wife, Lyubov Vokhminzeva, whom he had married in 1931, were permitted to attend the Solvay Conference at Brussels—an opportunity they took to leave the Soviet Union for good. After the conference was over, they spent successive two-month periods in Paris at the Pierre Curie Institute, in Cambridge at the Cavendish Laboratory, and in Copenhagen at Bohr's institute, before going to the University of Michigan. In the fall of 1934 Gamow was appointed professor of physics at George Washington University in Washington, D.C. He remained at George Washington University until 1956, when he transferred to the University of Colorado. At the same time, after twenty-five years of marriage, he and his wife were divorced; two years later he married Barbara Perkins.

Soon after accepting his position at George Washington University, Gamow persuaded Edward Teller to join him. By mid-1936 they had jointly discovered what is now known as the Gamow-Teller selection rule for β decay—Gamow's last major contribution to "pure" nuclear theory. Subsequently he concerned himself largely with applying nuclear physics to astronomical phenomena. Early in 1938, for example, he used his knowledge of nuclear reactions to interpret stellar evolution, that is, the Hertzsprung-Russell diagram and the mass-luminosity relation. At about the same time he organized a conference on thermonuclear reactions, the discussions at which contributed significantly to Hans Bethe's discovery of the carbon cycle. In 1939 Gamow and Teller, both of whom were strong advocates of the expanding-universe theory, traced the origin of the great nebulae to the formation of ancient stellar condensations which subsequently began separating from each other; in addition, they investigated the energy production in red giants. In 1940–1941 Gamow and M. Schoenberg explicated the role of neutrino emission in the production of the rapid and tremendously large increase in luminosity associated with novae and supernovae (exploding stars).

Concurrently, Gamow was establishing his reputation among nonscientists as one of the most talented and creative popularizers of science of all time. His first book-length venture, the well-known *Mr. Tompkins in Wonderland,* grew out of a popular article on relativity entitled "A Toy Universe" which he wrote in 1937 but which was rejected by *Harper's Magazine* and several other magazines. Not until C. P. Snow, then editor of *Discovery,* read it, published it, and solicited more was Gamow's career launched. In all, Gamow wrote almost thirty books, most of which were of a popular nature and most of which he illustrated himself. In 1956 his popular writings brought him the UNESCO Kalinga Prize and a lecture tour to India and Japan.

During World War II, Gamow served as a consultant to the Division of High Explosives in the Bureau of Ordnance of the U.S. Navy Department, studying, for example, the propagation of shock and detonation waves in various conventional explosives. Immediately after the war he went as an observer to the Bikini atomic bomb test, contributed to the theory of war games for the U.S. Army, and (after gaining top security clearance in 1948) worked with Teller and Stanislaw Ulam on the hydrogen bomb project at Los Alamos.

Yet Gamow's thoughts were never far from relativity and cosmology. In 1948 he predicted that all matter in the universe is in a state of general rotation about some distant center; at the same time he began developing his ideas on the origin and frequency distribution of the chemical elements, postulating that before the "big bang" there existed a primordial state of matter ("ylem") consisting of neutrons and their decay products, protons and electrons, mixed together in a sea of high-energy radiation—the basic ingredients necessary for the formation of deuterons and heavier and heavier nuclei as the universe subsequently expanded. Most of the detailed theoretical calculations were carried out by R. Alpher (assisted by R. Herman), which resulted in the well-known Alpher-Bethe-Gamow letter in *Physical Review* of 1 April 1948—Bethe's name, in one of Gamow's more famous jokes, being added gratuitously to conform to the Greek alphabet. This work also led to the prediction of a residual blackbody radiation spectrum, the remnant from the primordial "big bang," corresponding to a few degrees Kelvin. This radiation was first detected in early 1965 by A. A. Penzias and R. W. Wilson; much more definite evidence was found the following year by P. G. Roll and D. T. Wilkinson (in experiments initiated by R. H. Dicke and P. J. E. Peebles) at Princeton University. Cosmological questions concerned Gamow to the end, one of his last investigations being on the possible inconstancy of the gravitational constant and the charge of the electron.

In early 1954, less than a year after J. D. Watson and Francis Crick discovered the double helical structure of DNA, Gamow recognized that the information contained in the four different kinds of nucleotides (adenine, thymine, guanine, cytosine) constituting the DNA chains could be translated into the sequence of twenty amino acids which form protein molecules by counting all possible triplets one can form from four different quantities. This remarkable way in which Gamow could rapidly enter a more or less unfamiliar field at the forefront of its activity and make a highly creative contribution to it, often far more by intuition than by calculation, led Ulam to characterize his work as "perhaps the last example of amateurism in scientific work on a grand scale." It earned him membership in a number of professional societies—American Physical Society, Washington Philosophical Society, International Astronomical Union, American Astronomical Society, U.S. National Academy of Sciences, Royal Danish Academy of Sciences and Letters—as well as an overseas fellowship in Churchill College, Cambridge.

Gamow was a tremendously prolific writer, having roughly 140 technical and popular articles, in addition to his many books, to his credit. (On the negative side, his historical writings, which like most of his books are of a basically "popular" character, are of marginal value.) He was tall, fair-haired, blue-eyed, and possessed a legendary sense of humor. He was very widely traveled, greatly enjoyed reading and memorizing poetry, spoke six languages (all dialects of "Gamowian"), and loved collecting photographs and other memorabilia.

BIBLIOGRAPHY

I. Original Works. A bibliography of Gamow's scientific and popular writings is included in his autobiography, *My World Line* (New York, 1970). The most important scientific papers consulted and referred to in text are the following: "Zur Wellentheorie der Materie," in *Zeitschrift für Physik,* **39** (1926), 865–868, written with D. D. Ivanenko; "Zur Quantentheorie des Atomkernes," *ibid.,* **51** (1928), 204–212; "Selection Rules for the β-Disintegration," in *Physical Review,* **49** (1936), 895–899, written with E. Teller; "Nuclear Energy Sources and Stellar Evolution," *ibid.,* **53** (1938), 595–604; "The Expanding Universe and the Origin of the Great Nebulae," in *Nature,* **143** (1939), 116–117, 375, written with E. Teller; "On the Origin of Great Nebulae," in *Physical Review,* **53** (1939), 654–657, written with E. Teller; "Energy Production in Red Giants," *ibid.,* 719, written with E. Teller; "The Possible Role of Neutrinos in Stellar Evolution," *ibid.,* **58** (1940), 117, written with M. Schoenberg; "Neutrino Theory of Stellar Collapse," *ibid.,* **59** (1941), 539–547, written with M. Schoenberg; "Rotating Universe?" in *Nature,* **158** (1946), 549; "The Origin of Chemical Elements," in *Physical Review,* **73** (1948), 803–804, written with R. A. Alpher and H. Bethe; "Possible Relation Between Deoxyribonucleic Acid and Protein Structures," in *Nature,* **173** (1954), 318; "Statistical Correlation of Protein and Ribonucleic Acid Composition," in *Proceedings of the National Academy of Sciences of the United States of America,* **41** (1955), 1011–1019, written with M. Yčas; and "History of the Universe," in *Science,* **158** (1967), 766–769.

II. Secondary Literature. See *American Men of Science; Current Biography, 1951; Physics To-day,* **21** (1968), 101–102; and *Nature,* **220** (1968), 723. See also P. G. Roll and D. T. Wilkinson, "Measurement of Cosmic Background Radiation at 3.2-cm. Wavelength," in *Annals of Physics,* **44** (1967), 289–321.

Roger H. Stuewer

GAṆEŚA (*b.* Nandod, Gujarat, India, 1507), *astronomy.*

Gaṇeśa was born into a Brâhmaṇa family of astronomers and astrologers. He was the son of Keśava of the Kauśikagotra and his wife Lakṣmî, and studied under his famous father, on many of whose works he eventually wrote commentaries. In his turn Gaṇeśa trained Nṛsiṃha (*b.* 1548), the son of his brother Râma, and Nṛsiṃha both commented on Gaṇeśa's *Grahalâghava* and wrote, in 1603, a set of astronomical tables entitled *Grahakaumudî* based on that work. Gaṇeśa also taught Divâkara of Golagrâma, many of whose descendants commented on various of his master's books. Gaṇeśa's last dated work, the *Vivâhadîpikâ,* was written in 1554; he must, however, have lived at least a decade longer in order to have been his nephew's teacher. So far as is known, he never left his native village.

Gaṇeśa wrote a number of works on *jyotiḥśâstra* (astronomy and astrology) and *dharmaśâstra* (Hindu law). These are listed by his nephew, Nṛsiṃha, in his commentary, *Harṣakaumudî,* on the *Grahalâghava:*

1. *Grahalâghava* (see essay in Supplement).
2. *Laghutithicintâmaṇi* (see essays in Supplement).
3. *Bṛhattithicintâmaṇi* (see essays in Supplement).
4. *Siddhântaśiromaṇivivṛti* (see essay in Supplement).
5. *Lîlâvatîvyâkṛti* (see essay in Supplement).
6. *Vṛndâvanaṭîkikâ.*
7. *Muhûrtatattvavivṛti.*
8. *Śrâddhâdivinirṇaya.*
9. *Chandorṇavavivṛti.*
10. *Sudhîrañjana.*
11. *Tarjanîyantraka.*
12. *Kṛṣṇâṣṭamînirṇaya.*
13. *Holikânirṇaya.*

To these the following can be added:

14. *Pâtasâraṇî.*
15. *Câbukayantra.*
16. *Pratodayantra.*
17. *Dhruvabhramaṇayantravyâkhyâ.*

The *Grahalâghava* or *Siddhântarahasya,* Gaṇeśa's main work on astronomy, was composed in 1520, when he was thirteen. It contains sixteen chapters:

1. On the mean longitudes of the planets.
2. On the true longitudes of the sun and moon.
3. On the true longitudes of the five "star-planets."
4. On the three problems involving diurnal motion.
5. On lunar eclipses.
6. On solar eclipses.
7. On calendrical problems.
8. On eclipses.
9. On heliacal risings and settings.
10. On the planets' altitudes.
11. On the altitudes of the fixed stars.
12. On the lunar crescent.
13. On planetary conjunctions.
14. On the *pâtas* of the sun and moon.
15. On calculating lunar eclipses with a calendar.
16. Conclusion.

The *Grahalâghava* has been the most popular Sanskrit astronomical treatise in northern and western India since the sixteenth century. Its popularity is reflected in the hundreds of manuscripts of it that are extant, in the several commentaries on it, and in the numerous sets of astronomical tables based on its parameters. The known commentaries are the following (for editions, see the list of editions of the *Grahalâghava* itself given below):

1. *Ṭîkâ* of Mallâri (*fl. ca.* 1600), the son of Divâkara of Golagrâma (published).
2. *Harṣakaumudî* of Nṛsiṃha (*b.* 1548), Gaṇeśa's nephew.
3. *Manoramâ* of Gaṅgâdhara (1586).
4. *Siddhântarahasyodâharaṇa* of Viśvanâtha (1612), the son of Divâkara of Golagrâma (published).
5. *Manoramâ* of Kamalâkara, the great-grandson of Divâkara of Golagrâma.
6. *Udâhṛti* of Nârâyaṇa (1635[?]).
7. *Sadvâsanâ* of Sudhâkara Dvivedin (1904, published).
8. *Sudhâmañjarîvâsanâ* of Sîtârâma Jhâ (1932, published).
9. *Mâdhurî* of Yugeśvara Jhâ (1946, published).
10. *Ṭîkâ* of Bâlagovinda.

The following astronomical tables are based on the *Grahalâghava*:

1. *Grahalâghavasâriṇî I* (the initial epoch is 1520).
2. *Grahakaumudî* of Nṛsiṃha (1603).
3. *Grahasâraṇî* of Gaṅgâdhara (1630).
4. *Grahalâghavasâriṇî* of Premamiśra (1656).
5. *Grahaprabodhasâriṇî* of Yâdava (1663).
6. *Grahalâghavasâriṇî II* (1754).
7. *Grahalâghavîyasâriṇî* of Gaṅgâdhara Varman (Bombay, 1907; 2nd ed., 1923).

Most of these sets of tables are described in D. Pingree, "Sanskrit Astronomical Tables in the United States," in *Transactions of the American Philosophical Society,* n.s. **58,** no. 3 (1968), *passim;* "On the Classification of Indian Planetary Tables," in *Journal of the History of Astronomy,* **1** (1970), 95–108; and "Sanskrit Astronomical Tables in England," in *Journal of Oriental Research* (to be published).

The *Grahalâghava* has often been published in India:

1. Edited with the *Ṭîkâ* of Mallâri by L. Wilkinson (Calcutta, 1848).
2. Edited with the *Ṭîkâ* of Mallâri and the

Udâharaṇa of Viśvanâtha by Bhâlacandra (Benares, 1864).

3. Edited with the *Udâharaṇa* of Viśvanâtha and a Marâṭhî translation by Kṛṣṇa Śâstrî Goḍabole and Vâmana Kṛṣṇa Jośî Gadre (2nd ed., Bombay, 1873; 5th ed., Poona, 1914; 6th ed., Poona, 1926).

4–7. Edited with the *Ṭîkâ* of Mallâri (Bombay, 1875; Benares, 1877; Delhi, 1877; Bombay, 1883).

8. Edited with the *Udâharaṇa* of Viśvanâtha and a Bengâlî translation by Rasikamohana Cattopâdhyâya (Calcutta, 1887).

9. Edited with the Hindî translation of Jiyârâma Śâstrî by Râmeśvara Bhaṭṭa (Kalyâna-Bombay, 1899).

10. Edited with the *Ṭîkâ* of Mallâri by Hariprasâda Śarman (Bombay, 1901).

11. Edited with the *Ṭîkâ* of Mallâri, the *Udâharaṇa* of Viśvanâtha, and his own *Sadvâsanâ* by Sudhâkara Dvivedin (Benares, 1904; repr. Bombay, 1925).

12. Edited with the *Ṭîkâ* of Mallâri and the *Ândhraṭîkâ* of Mañgipûḍi Vîrayya Siddhântigâr (Masulipatam, 1915).

13. Edited with his own *Sudhâmañjarîvâsanâ* and a Hindî *bhâṣâ* by Sîtârâma Jhâ (Benares, 1932; repr. Benares, 1941).

14. Edited with the *Udâharaṇa* of Viśvanâtha, the *Mâdhurî* of Yugeśvara Jhâ, and a Hindî *ṭîkâ* by Kapileśvara Śâstrî, Kâśî Sanskrit Series 142 (Benares, 1946).

The *Laghutithicintâmaṇi* consists of tables for determining *tithis, nakṣatras,* and *yogas* accompanied by a short introductory text; Gaṇeśa composed it in 1525. Of this work also there are hundreds of manuscripts as well as several commentaries:

1. *Ṭîkâ* of Nṛsiṃha (*b.* 1586), the grandson of Divâkara of Golagrâma and the nephew of Mallâri, the commentator on the *Grahalâghava.*

2. *Udâharaṇa* of Viśvanâtha (1634), the commentator on the *Grahalâghava.* Published.

3. *Ṭippaṇa* of Vyeñkaṭa, alias Bâpû.

4. *Ṭîkâ* of Yajñeśvara.

The *Laghutithicintâmaṇi* has been published twice:

1. Edited with his own Hindî commentary, *Vijayalakṣmî* (1924), by Mâtṛprasâda Pâṇḍeya, Haridas Sanskrit Series 76 (Benares, 1938).

2. Edited with the *Udâharaṇa* of Viśvanâtha by V. G. Âpṭe, Ânandâśrama Sanskrit Series 120 (Poona 1942), part 1.

The tables of the *Laghutithicintâmaṇi* are discussed in D. Pingree, "Sanskrit Astronomical Tables in the United States," pp. 47b–50b; and "Sanskrit Astronomical Tables in England."

The *Bṛhattithicintâmaṇi,* also consisting of tables for computing *tithis, nakṣatras,* and *yogas* and an introductory text, was written in 1552. It was much less popular than the *Laghutithicintâmaṇi;* there are only a dozen manuscripts, and the unique commentary is the *Sùbodhinî* composed by Viṣṇu, the son of Divâkara of Golagrâma and the brother of Mallâri. The text alone with Viṣṇu's *Subodhinî* is published by V. G. Âpṭe, Ânandâśrama Sanskrit Series 120 (Poona, 1942), part 2. The tables are described in D. Pingree, "Sanskrit Astronomical Tables in the United States," pp. 50b–51a; and "Sanskrit Astronomical Tables in England."

The *Lîlâvatîvyâkṛti* or *Buddhivilâsinî,* a commentary on the *Lîlâvatî* of Bhâskara II, was composed by Gaṇeśa in 1545. It was published in the edition of the *Lîlâvatî* produced by Dattâtreya Âpṭe, Ânandâśrama Sanskrit Series 107, 2 vols. (Poona, 1937).

The *Vṛndâvanaṭîkikâ* or *Vivâhadîpikâ,* a commentary on the *Vivâhavṛndâvana* of Keśavârka (a work on astrology applied to marriage), was written by Gaṇeśa in 1554. It was published at Benares in 1868.

The *Muhûrtatattvavivṛti* or *Muhûrtadîpikâ,* a commentary on the *Muhûrtatattva* of his father, Keśava (a work on catarchic astrology), was written by Gaṇeśa before the *Vivâhadîpikâ,* which refers to it. The *Muhûrtadîpikâ* has not yet been published.

The *Śrâddhâdivinirṇaya* is evidently a work on offerings to one's ancestors. No manuscripts are known.

The *Chandorṇavavivṛti* is a commentary on an unidentified work on metrics entitled *Chandorṇava.* No manuscripts are known.

The *Sudhîrañjana* is a work on the astronomical instrument of the same name. It has not yet been published.

The *Tarjanîyantraka* is presumably a work on another astronomical instrument called the *tarjanî.* No manuscripts are known.

The *Kṛṣṇâṣṭamînirṇaya* is a work on the festival of Kṛṣṇa's birthday, which falls on the eighth *tithi* of the *kṛṣṇapakṣa* of the month Śrâvaṇa. No manuscripts are known.

The *Holikânirṇaya* is a work on the festival called Holikâ which falls on the full moon of the month Phâlguna. No manuscripts are known.

The *Pâtasâriṇî* or *Pâtasâdhana* is a set of tables for computing the dates of *pâtas* of the sun and moon, accompanied by a brief explanatory text; Gaṇeśa wrote it in 1522. There are three commentaries:

1. *Vivṛti* of Divâkara (*b.* 1606), a great-grandson of Divâkara of Golagrâma.

2. *Vivṛti* of Viśvanâtha (1631), the son of Divâkara of Golagrâma.

3. *Vivṛti* of Dinakara (1839).

Neither the *Pâtasâriṇî* itself nor any of its commentaries has yet been published.

The *Câbukayantra* and *Pratodayantra* are works on the astronomical instruments called by these names. A manuscript of the latter is said to be dated 1516, when Gaṇeśa was only nine years old. Neither work has been published.

The *Dhruvabhramaṇayantravyâkhyâ* is a commentary on the second *adhikâra* of Padmanâbha's *Yantraratnâvalî* (*ca.* 1360). This *adhikâra* describes the *dhruvabhramaṇayantra,* which is an instrument for observing the north pole star. Its ascription to Gaṇeśa is uncertain. It has not yet been published.

BIBLIOGRAPHY

The editions of Gaṇeśa's works have already been mentioned. Very little else has been written of him or his astronomical system save my articles and books on astronomical tables, to which reference has been made. There are articles on him by Sudhâkara Dvivedin in his *Gaṇakataraṅgiṇî* (Benares, 1933; repr. from *Pandit,* n.s. **14** [1892], 58–63); by Ś. B. Dîkṣita, in *Bhâratîya Jyotiḥśâstra* (Poona, 1931; repr. of 1896 ed.), pp. 259–267; and by G. Thibaut, in *Astronomie, Astrologie und Mathematik* (Strasbourg, 1899), pp. 61–62. M. G. Inamdar, "An Interesting Proof of the Formula for the Area of a (Cyclic) Quadrilateral and a Triangle Given by the Sanskrit Commentator Ganesh in About 1545 A.D.," in *Nagpur University Journal,* **11** (1945), 36–42, deals with a passage in the *Buddhivilâsinî.*

David Pingree

GARNETT, THOMAS (*b.* Casterton, Westmorland, England, 21 April 1766; *d.* London, England, 28 June 1802), *medicine, natural philosophy.*

Garnett's importance derives from his influence on the aims, style, and method of operation of the Royal Institution in London, where he was the first professor of natural philosophy and chemistry. He was a famed lecture demonstrator who pleased intelligent public audiences. After indifferent schooling, he was voluntarily articled in 1781 to the mathematician and surgeon John Dawson.

In 1785 he matriculated at Edinburgh, where he was profoundly influenced by the chemical lectures of Joseph Black and the medical lectures of John Brown. He took the M.D. in 1788 and finished his medical education in London in 1789. Later that year he wrote the article "Optics" for the *Encyclopaedia Britannica.* He supplemented his medical practice, conducted in the north of England, with chemical analyses and lecture demonstrations, using equipment that he himself had designed and built. In 1795 he married Grace Cleveland. While waiting for passage to America, where he hoped to teach chemistry, he accepted the professorship of natural philosophy at Anderson's Institution in Glasgow. He resigned in 1799 to join the Royal Institution, then being organized. Count Rumford, who knew Garnett by reputation only, accepted his suggestions about necessary facilities and the design of the lectures. On 4 March 1800 Garnett opened the lectures; his first season was highly successful. Unfortunately, he became the victim of bouts of melancholy induced by the death of his wife in childbirth on 25 December 1798, and his second lecture season was not well received. Rumford's high-handed treatment of him only increased the tension growing between Garnett and the managers of the Royal Institution, leading to his resignation on 15 June 1801. Garnett subsequently set himself up in Great Marlborough Street as a lecturer, and he also edited the first volume of the *Annals of Philosophy, Natural History, Chemistry, Literature, Agriculture, and the Mechanical and Fine Arts.*

BIBLIOGRAPHY

I. Original Works. The library of the Royal Institution owns some Garnett letters; the minute books of the Institution for this early period are regrettably brief, but the information is helpful. J. R. Partington, *History of Chemistry,* IV (London, 1964), 32, lists Garnett's chief publications. Garnett's *Observations on a Tour Through the Highlands and Part of the Western Isles of Scotland, Particularly Staffa and Icolmkill,* 2 vols. (London, 1800), II, 193–205, contains a description of the aims, plans, and *modus operandi* for Anderson's Institution in Glasgow.

II. Secondary Literature. All accounts of Garnett derive from an anonymous introduction to his posthumously published *Popular Lectures on Zoonomia, or the Laws of Animal Life, in Health and Disease* (London, 1804), pp. [v]–xii. Garnett's portrait is the frontispiece. H. Bence-Jones, in *The Royal Institution: Its Founder and Its First Professors* (London, 1871), pp. 162–172, supplements this material with excerpts from Garnett's letters, including one of 23 December 1799, which outlines a plan for the operation of the Royal Institution. Richard Garnett wrote the biographical entry in *Dictionary of National Biography,* VII, 886–887. K. D. C. Vernon, "The Foundation and Early Years of the Royal Institution," in *Proceedings of the Royal Institution,* **39,** no. 179 (1963), 364–402, expands Jones's account.

June Z. Fullmer

GARNOT, PROSPER (*b.* Brest, France, 13 January 1794; *d.* Paris, France, 8 August 1838), *medicine, zoology, anthropology, ethnology.*

Garnot became an assistant surgeon in the French navy in 1811. After several voyages to Cayenne and Martinique (1817–1818) as a naturalist, he worked in the Antilles from 1819 to 1820. He received the M.D. in 1822 with the thesis "Essais sur le choléra morbus." In August of that year he joined Duperrey's world voyage on the French corvette *Coquille.* Garnot and the pharmacist R. P. Lesson were to serve as naturalists for the expedition.

After visiting the Falkland Islands and adding much to geographical knowledge of them, the expedition rounded Cape Horn and crossed the Pacific. In the autumn of 1823 Garnot visited the southern Moluccas, New Zealand, and the island of New Guinea, and in January 1824 the *Coquille* went to Port Jackson for repairs. Garnot fell ill and returned to Europe on a merchantman, taking a great part of the expedition's collected material with him. Nearly all of this material was lost by shipwreck in July 1824.

During the *Coquille*'s voyage, Garnot paid special attention to the vertebrate animals and to several human tribes in the South Pacific. He collected and measured a number of skulls from these tribes and described the Alfurs, a little-known people who inhabit the interior of New Guinea. In addition, he found a plant, which was named garnotia.

The results of the voyage were published by Duperrey as *Voyage autour du monde exécuté par ordre du roi sur la corvette La Coquille pendant les années 1822–1825* (Paris, 1828–1832). The first section of this work, dealing with zoology, was written by Garnot and Lesson.

Garnot became surgeon first-class in 1825, and from 1827 to 1828 he worked at hospitals in Brest. He then became second surgeon in Martinique, where he worked as an obstetrician after his retirement in 1833. He was a corresponding member of the Académie Royale des Médecins and a member of several scientific institutions.

BIBLIOGRAPHY

I. ORIGINAL WORKS. Garnot's works include *Remarques sur la zoologie des îles Malonines, faites pendant le voyage autour du monde de la corvette La Coquille exécuté en 1822–1825* (Paris, 1826); "Lettre sur les préparations anatomiques artificielles du docteur Auzoux," in *Annales maritimes et coloniales* (1827); *Leçons élémentaires sur l'art des accouchements destinées aux élèves sages-femmes dans les colonies françaises,* 2nd ed. (Paris, 1834); and *De l'homme considéré sous le rapport de ses caractères physiques* (Paris, 1836). Articles by Garnot are in *Bulletin de l'Académie ébroicienne d'Évreux, France maritime, Journal des voyages, découvertes et navigations modernes,* and *Bulletin des sciences médicales,* **8** (1826), 273–275.

II. SECONDARY LITERATURE. For information on Garnot, see *Almanac général de médecine* (1839), p. 370; Charles Berger and Henry Ray, "Répertoire bibliographique des travaux des médecins et des pharmaciens de la marine française, 1698–1873," app. to *Archives de médecine navale* (Paris, 1874); A. C. P. Callisen, *Medicinisches Schriftsteller Lexicon* (Copenhagen, 1831; repr. Nieukoop, 1963), VII, 57; XXVIII, 155–156; and A. Hirsch, *Biographisches Lexikon der hervorragenden Aerzte,* II (Munich-Berlin, 1962), 689.

A. P. M. SANDERS

GARREAU, LAZARE (*b.* Autun, France, 16 March 1812; *d.* Lille, France, 1892), *botany.*

After military service as assistant surgeon in Maubeuge and Strasbourg (1836–1838) and as surgeon and pharmacist in Algeria (1839–1844), Garreau became professor of natural history at the University of Lille in 1844. His earliest researches dealt with the relative values of both surfaces of a leaf as sites of gaseous exchange, especially the exhalation of water vapor. He measured the amount of water excreted by placing small glass domes, containing a water-absorbing substance, on opposite sides of a leaf and found no direct correlation between the amount of evaporation and the number of stomata. He concluded that the epidermis determines transpiration, the cuticular layer being very important. On the veins, where transpiration was the most intensive, he found almost no cuticle.

Garreau also confirmed that leaves are able to absorb water, as Bonnet had proposed in 1754, and determined the osmotic properties of the epidermis and cuticle. Here he observed a more direct correlation between the exhalation of carbon dioxide and the number of stomata. Garreau discovered that there was no relation between the cuticle and the cells of the epidermis (1850). The cuticle already exists before differentiation of the epidermal cells occurs. He thought that the cuticle was a living tissue and that the younger the organ producing the cuticle, the stronger its osmotic activity.

Garreau also worked on the theory of respiration and nutrition of green plants, proposed by Ingen-Housz in 1779. In 1851 he confirmed the results of the work of H. B. de Saussure, who had shown that the great mass of the vegetable body is derived from the carbon dioxide of the atmosphere and the constituents of water. Garreau observed that reduction of carbon dioxide, which he called the nutritive function, was dependent on light and was independent of respiration. Although not strictly separating the effects of assimilation and respiration, Garreau protested against distinguishing a "diurnal" and a "nocturnal" respiration in green plants.

Further, Garreau showed that there was a direct

relation between respiration and heat production. The idea of a vital force within the plant body was deprived of one of its chief supports when it was recognized that the natural heat of organisms is the result of chemical processes induced by respiration. Garreau explained the high intensity of this phenomenon in *Arum* inflorescences by showing that the surface area is large in relation to volume.

Garreau married in 1846; he had four children. He was an officer of the Académie des Sciences and the Legion of Honor. In 1862 he became a member of the Institut Impérial des Sciences.

BIBLIOGRAPHY

I. ORIGINAL WORKS. Garreau's works include "Sur la nature de la cuticule, ses relations avec l'ovule," in *Annales des sciences naturelles* (*Botanique*), 3rd ser., **13** (1850), 304–315; "Recherches sur l'absorption et l'exhalation des surfaces aériennes des plantes," *ibid.*, 321–346; "De la respiration chez les plantes," *ibid.*, **15** (1851), 5–36; "Mémoire sur les relations qui existent entre l'oxygène consommé par le spadice de l'Arum Italicum en état de paroxysme, et la chaleur qui se produit," *ibid.*, 250–256; "Nouvelles recherches sur la respiration des plantes," *ibid.*, **16** (1852), 271–292; "Mémoire sur la formation des stomates dans l'épiderme des feuilles de l'éphémère des jardins, et sur l'évolution des cellules qui les avoisinent," *ibid.*, 4th ser., **1** (1854), 213–219; "Recherches sur les formations cellulaires, l'accroissement et l'exfoliation des extrémités radiculaires et fibrillaires des plantes," *ibid.*, **10** (1858), 181–192, written with Brauwers; *Recherches expérimentales: 1° sur les causes qui concourrent à la distribution des matières minérales fixes dans les divers organes des plantes; 2° sur la matière vivante des plantes et la circulation intracellulaire* (Lille, 1859), a diss. presented to the Faculty of Sciences of Strasbourg; "Recherches sur la distribution des matières minérales fixes dans les divers organes des plantes," in *Annales des sciences naturelles,* **13** (1860), 145–218; and "Mémoire sur la composition élémentaire des faisceaux fibro-vasculaires des fougères," in *Comptes rendus hebdomadaires des séances de l'Académie des sciences,* **50** (1860), 854–855.

II. SECONDARY LITERATURE. Short descriptions of Garreau's work are to be found in M. Duchartre, *Rapport sur le progrès de la botanique physiologique* (Paris, 1868). An obituary notice appeared in *Journal de pharmacie et de chimie,* 5th ser., **28** (1893), 109.

A. P. M. SANDERS

GASCOIGNE, WILLIAM (*b.* Middleton, Yorkshire, England, *ca.* 1612;[1] *d.* Marston Moor, Yorkshire, 2 July 1644), *optics, astronomy.*

The eldest son of Henry Gascoigne by his first wife, Margaret Cartwright, Gascoigne appears to have spent most of his short life at the family home in Middleton, between Wakefield and Leeds. By his own testimony his formal education was slight, and there is no hint whatever as to the origin of his interest or competence in scientific matters.[2] One can only say that both were fully developed by 1640, when he entered into scholarly correspondence, and that his work was cut off not long thereafter by his participation in the English Civil War. He died in the royalist disaster at Marston Moor.

From the time of the appearance of the telescope on the scientific scene in 1610, its utility for purely descriptive purposes was taken for granted. Nearly two generations were to pass, however, before its use was extended into the traditional business of positional astronomy. This great advance depended on three quite distinct developments: the conversion of Galileo's terrestrial (concave eyepiece) telescope to obtain a real image, the introduction of cross hairs into the image (focal) plane to enable accurate pointing of the telescope (and the instrument to which it was attached), and the invention of a micrometer to measure small angular distances within the field of view. The first of these was suggested by Johannes Kepler in 1611 and implemented by Christoph Scheiner shortly thereafter. For practical purposes, the remaining two steps had to await the work of Adrien Auzout and Jean Picard in the late 1660's: in fact, however, they were both taken by Gascoigne in the late 1630's. By the beginning of 1641 he had not only a fully developed account of the optical ideas involved but also a working model of the instrument and a limited number of satisfactory observational results.[3] Unfortunately, Gascoigne's work essentially died with him. His micrometer survived in the hands of Richard Towneley but was used by him primarily to dispute the priority claims of the French.[4] A manuscript treatise on optics that Gascoigne is supposed to have left ready for the press had already become untraceable by 1667.[5]

NOTES

1. Until the mid-nineteenth century Gascoigne was believed to have been born sometime around 1620. In 1863 W. Wheater (*Gentlemen's Magazine,* **215,** 760–762) provided evidence suggesting that he was born no later than 1612. John Aubrey, who appears to have been responsible for the original tradition, also indirectly corroborates the newer one with his assertion (*Brief Lives*) that Gascoigne "gave [Sir Jonas Moore, *b.* 1617] good information in mathematicall knowledge."
2. Aubrey credits the Jesuits with Gascoigne's education. Gascoigne himself says only that he "entered upon these studies accidentally" after leaving "both Oxford and London [without knowing] what any proposition in geometry meant."
3. The primary information on Gascoigne's results is found in his letter of February 1641 to William Oughtred, printed by Stephen P. Rigaud in *Correspondence of Scientific Men of the 17th Cen-*

tury, I (Oxford, 1841), 33–59. Extracts from other letters of Gascoigne (to William Crabtree) in the Macclesfield Collection were given by William Derham (*Philosophical Transactions of the Royal Society,* **30** [1717], 603–610), but the letters have never been printed in full.

4. On behalf of the Royal Society, Robert Hooke provided a description, complete with plates, of Gascoigne's instruments (*Philosophical Transactions of the Royal Society,* **2** [1667], 541–544).
5. See Towneley's report in *Philosophical Transactions of the Royal Society,* **2,** 457–458. Various papers including Gascoigne's passed from the Towneley family to William Derham at the beginning of the eighteenth century. See *Philosophical Transactions,* **27** (1711), 270–290. A. Shapiro has kindly called my attention to (1) a statement by John Flamsteed (Francis Baily, *An Account of the Revd. John Flamsteed* [London, 1835], p. 31) attributing to Gascoigne some advanced ideas on geometrical optics and (2) an acknowledgment by William Molyneux ("Admonition to the Reader," *Dioptrica nova* [1692]) of his indebtedness to Gascoigne through Flamsteed. Flamsteed obtained his information from letters written by Gascoigne to Crabtree, which were in the possession of Derham (*op. cit.*) in 1717 but may already have been lost by 1753; at any rate, John Bevis, writing in that year (*Philosophical Transactions of the Royal Society,* **48,** 190–192), cited only the letter to Oughtred contained in the Macclesfield Collection.

VICTOR E. THOREN

GASKELL, WALTER HOLBROOK (*b.* Naples, Italy, 1 November 1847; *d.* Great Shelford, near Cambridge, England, 7 September 1914), *physiology, morphology.*

Gaskell was descended from a prominent Unitarian family in the north of England. He was the third child and younger twin son of John Dakin Gaskell, barrister of the Middle Temple, who practiced his profession only briefly before retiring to private life. His mother, Anne Gaskell, was his father's second cousin.

Gaskell attended the Highgate School, London, and in October 1865 matriculated at Trinity College, Cambridge, where he was elected to a scholarship in 1868. He graduated B.A. in 1869 as twenty-sixth wrangler in the Cambridge mathematical tripos. With the intention of making a career in medicine, he remained at Cambridge to study science and quickly fell under the influence of Michael Foster, who came to Cambridge in 1870 as Trinity College praelector in physiology. Although Gaskell completed clinical training at University College Hospital, London (1872–1874), and received an M.D. from Cambridge in 1878, he never practiced medicine. At Foster's urging, he devoted himself instead to physiological research, beginning in 1874, when he went to Leipzig to work under Carl Ludwig in the famous physiological institute there.

Soon after returning to England in the summer of 1875, Gaskell married Catherine Sharpe Parker, daughter of R. A. Parker, a solicitor, and settled near Cambridge, where he continued his research. His income apparently came chiefly from private sources. From 1883 until his death he was university lecturer in physiology. In 1889 he was elected to fellowship of Trinity Hall, Cambridge, where he also served as praelector in natural science. The Royal Society named Gaskell as Croonian lecturer in 1881, a fellow in 1882, gold medalist in 1889, and Baly medalist in 1895. He received honorary doctorates from the universities of Edinburgh and McGill and served on the Royal Commission on Vivisection (1906–1912). He was survived by two of his four daughters and by his son, John Foster Gaskell.

Gaskell's career in research can be conveniently divided into four periods, corresponding approximately to the following dates and dominant interests: (1) 1874–1879, vasomotor action; (2) 1879–1883, the problem of the heartbeat; (3) 1883–1887, the involuntary nervous system; and (4) 1888–1914, the origin of the vertebrates. Despite the apparent diversity of these interests, there is a remarkable internal unity to Gaskell's work, one investigation leading logically into the next. Much of his work and approach demonstrate clearly the powerful influence exerted upon him by Michael Foster.

In 1874, at the suggestion and with the help of Carl Ludwig, Gaskell followed up work done earlier in the Leipzig laboratory on circulation in skeletal muscle. He focused on the quadriceps extensor muscles of the dog and recorded with a kymograph the effects of nerve action on the rate of blood flow from a severed vein. Upon returning to Cambridge in 1875, Gaskell continued to work on the same general problem but chose to work on the mylohyoid muscle of the frog. In the simpler tissues of the frog, where the arterial diameters could be measured directly with a micrometer eyepiece, Gaskell was able to clarify greatly a number of issues left doubtful in his work on the dog. His most striking result was that stimulation of the mylohyoid nerve in the frog invariably produced a steady dilatation of the arteries in the mylohyoid muscle. According to John Langley, "this was the most decisive instance known at the time of [vasodilator] action in a purely muscular structure."[1]

In 1878, after Rudolf Heidenhain had disputed several of his results and conclusions,[2] Gaskell reinvestigated the effects of nerve action on circulation in the muscle arteries of the dog. He claimed that his new work supported the results he had reached with the frog. "In the dog as in the frog," he wrote, "the vasomotor system for the muscles consists essentially of vaso-dilator fibres. . . ."[3]

By about 1880 Gaskell had turned from vasomotor action to the problem of the heartbeat. The shift was not abrupt, however, and emerged fully only after a transitional study on the tonicity of the heart and arteries. Gaskell found that acidic and alkaline solutions produced the same effects on cardiac muscle as on the smooth muscle of arterial walls. Both in the heart and in the arteries, acidic solutions induced muscular relaxation, while alkaline solutions induced muscular contraction. Gaskell used this result to propose a new mechanism for vasodilatation. He wished to replace the then standard view that dilatation depended on the action of ganglionic nerve centers. He suggested instead that the determining factor was the chemical condition of the lymph fluid which surrounded the muscle walls of the arteries. When a muscle was inactive, this fluid was alkaline and would therefore contribute to vasoconstriction. But during muscular contraction, the surrounding lymph fluid became acidic, so that the muscle walls of the arteries would then relax and the end result would be vasodilatation. This hypothesis was supported by Gaskell's observation that dilatation occurred in a muscle artery whenever that muscle contracted.

But Gaskell was already concerned with a problem of much wider scope than vasodilatation alone. He presented his work on tonicity as a contribution to the general problem of rhythmical motion, two important examples of which were vasomotor action and the heartbeat. He was obviously skeptical toward the prevailing idea that all forms of physiological rhythmicity depended on the action of nerve cells or ganglia. In this he followed the example of his mentor Foster, who was convinced that vasomotor action and the heartbeat were analogous and that both depended not on ganglia but on the inherent properties of relatively undifferentiated muscle tissue. It was directly to the problem of the heartbeat that Gaskell next turned, and it was Foster who most decisively influenced his approach.

In the Croonian lecture for 1881, dealing with the frog heart, Gaskell presented an important new method for studying heart action (later named the "suspension method") and insisted that cardiac inhibition depended less on nerve or ganglionic mechanisms than on the inherent properties of the cardiac musculature. The role of the vagus nerve in inhibition was reduced to that of being the "trophic" (anabolic) nerve of the cardiac muscle. Yet in the same lecture Gaskell produced impressive evidence against Foster's myogenic theory of rhythmicity and advocated instead the neurogenic view that discontinuous ganglionic discharges are responsible for the rhythmicity of the normal heartbeat. The background to this defection was exceedingly complex, but it derived from an initial assumption (which Foster himself accepted) that ganglionic impulses—whatever their role in rhythmicity—are somehow involved in coordinating the normal sequence of the vertebrate heartbeat.

Within three years Gaskell had resolved the problem of the heartbeat in favor of the myogenic theory far more persuasively than Foster had ever thought possible. Experiments on the tortoise heart were the source of Gaskell's new conception of the heartbeat, presented at length in a classic monograph of 1883. He explained that he had turned from the frog heart to the tortoise out of conviction that "the study of the evolution of function . . . . is the true method by which the complex problems of the mammalian heart will receive their final solution." [4] What he had found was physiological and histologico-evolutionary evidence that both the rhythmicity and the sequential character of the heartbeat could be explained without reference to ganglionic action.

Gaskell insisted first that a small strip of muscle cut from the tortoise's ventricle—and therefore clearly isolated from nerve structures—could nonetheless develop rhythmic pulsations at a rate equal to that of the normal heartbeat. Since, moreover, such a strip could continue to beat rhythmically for at least thirty hours after all stimulation had been discontinued, Gaskell argued that rhythmicity could arise automatically in cardiac muscle, that it was in fact "due to some quality inherent in the muscle itself." [5] This conclusion was in keeping with the somewhat similar and earlier work of Foster and Wilhelm Theodor Engelmann.

Far more novel and important was Gaskell's evidence that the sequence, as well as the rhythm, of the heartbeat could be referred solely to the properties of cardiac muscle. As the heartbeat is followed in its course through all the cavities of the heart, distinct pauses are observed at the junctions between the separate cavities. And since it is precisely here, in these junctions, that ganglia are most abundant, it had been assumed even by Foster that ganglia must play some important role in producing the pauses and thus in regulating the sequences. Gaskell was now able to offer an alternative explanation. That he was able to do so depended crucially on his decision to study the tortoise heart instead of the frog heart.

In the tortoise the cardiac nerves and their accompanying ganglia lie outside the heart itself and are relatively easy to remove. Gaskell found that their removal in no way affected the sequence of the heartbeat. He then focused on the cardiac tissue itself and sliced through the auricle until it consisted of two parts (A_s and A_v) joined at an upper ligature by a narrow bridge of auricular tissue. When this bridge was made quite thin, Gaskell could see a wave of contraction pass up A_s and then, after a pause, down A_v to the junction between auricle and ventricle, where another brief pause preceded ventricular contraction. Gaskell concluded that "the ventricle contracts in due sequence with the auricle because a wave of contraction passes along the auricular muscle and induces a ventricular contraction when it reaches the auriculo-ventricular groove."[6] This conclusion was confirmed by continuing to narrow the tissue bridge until it seemed that another section would sever it completely. At that point, the waves of contraction passing up A_s were "blocked" at the bridge and were unable to pass down A_v. The sequence between auricular and ventricular beats was thereby destroyed.

Gaskell then showed that the three muscular cavities of the heart (sinus, auricle, and ventricle) were connected by two narrow rings of relatively undifferentiated muscle tissue through which the waves of contraction could be transmitted from one cavity to the next. To explain why there are normally pauses at the junctions between successive cavities, Gaskell began by positing an antagonism between the capacities for rhythmicity and for rapid conduction of contractile waves. He suggested that the capacity for rhythmicity was greatest in undeveloped muscle tissue, while the capacity for rapid conduction increased as muscle tissue underwent development and specialization. Since the least developed (least striated) muscle fibers are found in the sinus, it must possess the greatest capacity for rhythmicity, and so the heartbeat naturally begins there. The more highly developed auricular and ventricular fibers, on the other hand, are especially adapted to conduct the wave of contraction rapidly; but the muscle rings connecting the heart cavities consist of relatively embryonic tissue, and the contractile wave therefore passes more slowly through them. This, rather than ganglionic action, explained the pauses observed at the sinoauricular and auriculoventricular junctions.

Before the discovery of cardiac ganglia and vagus inhibition in the 1840's, the heartbeat had generally been viewed as a simple peristaltic wave of contraction passing from one end of the heart to the other. In a distinctly evolutionary context, Gaskell now advocated a return to this view and repeatedly insisted that in every really important respect—in its rhythmicity and in its sequence—the vertebrate heartbeat depends in the first place not on nerve influences but on the properties of the cardiac musculature.

Gaskell's work of 1883 did not immediately convince everyone, and the myogenic-neurogenic debate continued for some time, especially in Germany.[7] The task of extending the myogenic theory to the mammalian heart proved more difficult than Gaskell had perhaps expected it to be, and he did not himself contribute to this extension. By about 1910 the extension had been accomplished—chiefly through the work of A. F. S. Kent and Wilhelm His, Jr., on the atrioventricular bundle in mammalian hearts, and through the work of Arthur Keith and Martin Flack on the mammalian cardiac pacemaker. With the possible exception of His, all of these workers depended fundamentally on Gaskell's work. Before long, and especially through the British clinical cardiologists James Mackenzie, Thomas Lewis, and Arthur Cushny, Gaskell's myogenic theory and his concept of heart block became incorporated into the pathology, pharmacology, and therapeutics of the heart. His conclusions have formed the basis of concepts of heart action ever since.

Although Gaskell's interest in the heart did not end abruptly in 1883, it soon became bound up with and eventually submerged in a general study of the involuntary nervous system. The starting point for this work was Gaskell's discovery that cold-blooded animals possess augmentor, as well as inhibitory, cardiac nerves. That mammals possess augmentor or accelerator cardiac nerves had been known for some time, and their existence in cold-blooded animals had been supposed by some. Particularly to explain the bewildering range and variety of the effects produced by vagus stimulation in the frog, the hypothesis had been advanced that the vertebrate vagus was not a simple nerve, composed solely of inhibitory fibers, but a compound nerve containing augmentor fibers as well. Decisive evidence for this hypothesis was lacking, however, and Gaskell himself specifically rejected it both in his Croonian lecture of 1881 on the frog heart and in his monograph of 1883 on the tortoise heart.

But after the summer of 1884, when Gaskell succeeded in distinguishing both inhibitory and augmentor cardiac fibers in the crocodile, he considered it probable that both sets of fibers were also present in other cold-blooded animals. In an elegant paper of 1884 he confirmed this view in the all-important case of the frog. Tracing the frog's vagus from its origin in the medulla oblongata, he found it to consist of two branches which then joined in a large ganglion outside the cranial cavity to form a single nerve trunk which continued toward the heart. It was this trunk that was ordinarily used to examine the effects of vagus stimulation. Gaskell focused instead on the two preganglionic branches and found that stimulation of one branch resulted always in purely augmentor effects, while stimulation of the other resulted always in purely inhibitory effects. The so-called vagus, Gaskell concluded, was in fact the "vago-sympathetic," consisting of a mixture of purely inhibitory and purely augmentor fibers which could be clearly distinguished from one another prior to their merger in the large extracranial ganglion.

Gaskell seems to have been greatly impressed by this discovery. With the help of a Cambridge colleague, the morphologist Hans Gadow, he extended his investigation of the cardiac nerves to as many different species of cold-blooded animals as possible. They found that these nerves were distributed in basically similar ways in all the species they examined. In all cold-blooded vertebrates, as in mammals, there existed two sets of cardiac nerves performing separate, indeed opposing, functions. A flood of ideas now burst forth almost simultaneously from Gaskell, as functional and morphological considerations became intertwined and mutually reinforcing.

For one thing, Gaskell noticed while studying the cardiac nerves in a tortoise that the functional distinction between vagus and augmentor fibers was correlated with a striking morphological distinction: although both kinds of cardiac fibers originated from the spinal cord as medullated fibers, the accelerator fibers emerged from the sympathetic chain without medullas, while the vagus fibers retained their medullas throughout their course. By early 1885 Gaskell had confirmed this rule in a wide variety of vertebrate and mammalian species. Then, in 1886, he showed that vagus stimulation produced in the tortoise's heart an electrical variation opposite in sign to that produced by stimulating the accelerator nerves. In thus providing demonstrable evidence that the two kinds of cardiac nerves did indeed perform opposing functions, Gaskell contributed to the rapidly developing field of cardiac electrophysiology, a field from which the electrocardiogram was soon to emerge.

Already, though, Gaskell was occupied with ideas of far broader significance. For him, as for Foster, the heart was just one example of an involuntary muscle; and he was confident that his results on cardiac innervation could be extended to the smooth muscles of the arterial, alimentary, and glandular systems. Gaskell had long believed (again with Foster) that the inhibitory action of vagus fibers was a constructive, beneficial, or anabolic action. He therefore supposed that the action of the opposing augmentor fibers was destructive or catabolic, like that of a motor nerve, leading to exhaustion of muscle activity. When generalized to the involuntary system as a whole, this concept led to the notion that every involuntary muscle was innervated by two nerves of opposite action, one anabolic and the other catabolic. By further analogy with the cardiac nerves, Gaskell expected these anabolic and catabolic nerves to be histologically distinguishable from one another, particularly on the basis of their medullation after passing the sympathetic chain. It was under the inspiration of these leading themes that Gaskell undertook a full-scale, systematic investigation of the involuntary nervous system.

A classic paper of 1886 contains the major results of Gaskell's work on the involuntary system. He found that the visceral or involuntary nerves arise from the central nervous system in three distinct groups. There is a cervicocranial outflow, a thoracic outflow, and a sacral outflow. In all three groups the visceral fibers leave the central nervous system as peculiarly fine, white, medullated fibers. But the fibers issuing from the thoracic region lose their medullas in the sympathetic ganglia and pass to the viscera as nonmedullated fibers. The fibers issuing from the cervicocranial and from the sacral regions retain their medullas as they pass to the periphery. In action the fibers issuing from the thoracic region appeared to be antagonistic to both the cervicocranial and the sacral outflows. In broad outline, this plan is still accepted, although significant modifications in detail and in terminology were soon made, especially through the work of another of Gaskell's Cambridge colleagues, John Langley, and especially in light of the neuron theory.

From the point of view of basic physiological thought, perhaps the most important result of Gaskell's work on the involuntary system was his discovery that the connection between the central nervous system and the chain of sympathetic ganglia is unidirectional, with the peculiarly small white fibers (the "white rami") supplying the sole connection. Earlier in the century it had been thought that a system of gray rami returned from the sympathetic chain to the central nervous system, creating an interplay between two essentially independent nervous systems. Bichat had christened these two systems the "organic" (central) and the "vegetative" (sympathetic). Although this mode of thinking about the nervous system had since come under criticism, no broad generalization had taken its place until Gaskell clarified the relationship between the sympathetic chain and the central nervous system. He showed that the gray rami are in fact peripheral nerve fibers which supply the blood vessels of the spinal cord and its membranes and which issue not from the sympathetic chain but from the central nervous system, as do the white rami. There is, then, no real separation into "organic" and "vegetative" nervous systems,[8] and, wrote Gaskell in 1908, "no give and take between two independent nervous systems . . . as had been taught formerly, but only one nervous system, the cerebro-spinal."[9] So fundamentally did Gaskell alter the prevailing conceptions of the involuntary system that Walter Langdon-Brown could insist that "to read an account of this system before Gaskell is like reading an account of the circulation before Harvey."[10] After their elaboration and modification by Langley and others, Gaskell's conclusions found clinical application, not only in the interpretation of referred pain by James Mackenzie and Henry Head but, more generally, in the work of Walter B. Cannon.

After 1888 Gaskell devoted all of his research to the problem of the origin of the vertebrates. This interest may at first seem remote from his earlier work, but it evolved logically out of his work on the involuntary nervous system. For what had especially struck Gaskell then was that the involuntary nerves arise not only from three distinct regions of the spinal cord but also from clearly defined segments within these regions. Deeply impressed by the similarity between this vertebrate arrangement and the central nervous system of the segmented invertebrates, he gradually elaborated the extraordinary theory that the vertebrates are descended from an extinct arthropod stock of which the king crab is the nearest living representative. He agreed that earlier attempts to trace the vertebrates to the segmented invertebrates had failed, but only because they all began with the assumption that the transition required a reversal of dorsal and ventral surfaces. This supposition was thought necessary in order to explain how it happens that in vertebrates the nervous system is dorsal to the alimentary canal, while in invertebrates the arrangement is reversed.

To explain this fact, Gaskell proposed the revolutionary hypothesis that the vertebrates arose by the enclosure of the ancestral arthropod gut by the growing central nervous system, and the formation of a new alimentary canal ventral to the nervous system. According to this conception, the vertebrate infundibulum corresponds to the arthropod esophagus, the ventricles in the vertebrate brain to the arthropod stomach, and the vertebrate spinal canal to the arthropod alimentary canal. Perhaps the most controversial element in the theory was Gaskell's notion that in the transition from invertebrate to vertebrate, a new alimentary canal was formed by epidermal invagination. This notion was in direct violation of two settled morphological tenets: (1) that the alimentary canal is the one system which endures throughout evolutionary change, and (2) that in all cases the alimentary canal arises from the hypoblastic germ layer, and never from the epiblastic layer, as Gaskell proposed. Against the first of these tenets Gaskell argued that it was folly to insist upon the importance and durability of the alimentary canal in evolution when the central nervous system, especially the brain, was so obviously the engine of upward progress. In making this point, he coined the aphorism "The race is not to the swift, nor to the strong, but to the wise."[11] Against the second tenet Gaskell argued that morphologists applied the germ-layer theory in a circular manner, deducing the layer from which a structure arose merely from its ultimate morphological destination.

Gaskell developed his remarkable theory in a series of papers from 1888 to 1906, and then—convinced that his ideas were not being seriously considered—gathered the evidence together in a full-length book, *The Origin of Vertebrates* (1908). Despite some minor support and a few pleas for open-mindedness, it too met a chilly reception from most morphologists, with one opponent accusing Gaskell of "diabolical ingenuity."[12] The direction of research since has gone against Gaskell's brave attempt to trace the vertebrates to an arthropod ancestor. While they would probably acknowledge that Gaskell's work contains interesting and suggestive material—on the endocrine system, for example—most morphologists today consider the vertebrates of common origin with the echinoderms.

A large, generous man of open and genial disposition, Gaskell was both criticized and admired for his inclination to bold generalization. His final years were clouded by his wife's debilitating illness and by a feeling that his deeply loved theory of the origin of vertebrates was not receiving a fair hearing. Even at Cambridge, where Gaskell lectured on the topic until his death, his audience decreased over the years until, near the end, the poignant scene is drawn of Gaskell closing his course by shaking hands with a lone remaining auditor.[13]

NOTES

1. J. N. Langley, "Walter Holbrook Gaskell, 1847–1914," in *Proceedings of the Royal Society,* **88B** (1915), xxvii–xxxvi, see xxviii.
2. R. Heidenhain *et al.,* "Beiträge zur Kenntnisse der Gefässinnervation, I, II. Ueber die Innervation der Muskelgefässe," in *Pflügers Archiv für die gesammte Physiologie des Menschen und der Thiere,* **16** (1878), 1–46.
3. "Further Researches on Vasomotor Nerves," p. 281.
4. "On the Innervation of the Heart," p. 48.
5. *Ibid.,* p. 53.
6. *Ibid.,* p. 64.
7. See, e.g., E. Cyon, "Myogen oder Neurogen?," in *Pflügers*

Archiv für die gesammte Physiologie des Menschen und der Thiere, **88** (1902), 222–295.
8. See Donal Sheehan, "Discovery of the Autonomic Nervous System," in *Archives of Neurology and Psychiatry,* **35** (1936), 1081–1115.
9. *Origin of Vertebrates,* p. 2.
10. Walter Langdon-Brown, "W. H. Gaskell and the Cambridge Medical School," in *Proceedings of the Royal Society of Medicine,* **33** (1939), section of the history of medicine, 1–12, see 6.
11. *Origin of Vertebrates,* p. 19.
12. See the lively discussion following Gaskell's paper, "Origin of Vertebrates," in *Proceedings of the Linnean Society of London,* sess. 122 (1910), 9–15. The discussion (pp. 15–50) includes both supporters and opponents of Gaskell's approach and ideas. For an anonymous and largely unfavorable review of Gaskell's book, see *Nature,* **80** (1909), 301–303. Gaskell's response is *ibid.,* pp. 428–429. Even more critical of Gaskell's work was Bashford Dean, in *Science,* n.s. **29** (1909), 816–818.
13. This paragraph is based in part upon a private communication from Lord Edgar Douglas Adrian, O.M., Nobel laureate in physiology or medicine, who was working at Cambridge during Gaskell's final years.

BIBLIOGRAPHY

I. ORIGINAL WORKS. The Royal Society *Catalogue of Scientific Papers,* IX, 967; XII, 262; XV, 220–221, lists thirty-four papers by Gaskell up to 1900. The most important of these are "On the Innervation of the Heart, With Especial Reference to the Heart of the Tortoise," in *Journal of Physiology,* **4** (1883), 43–127; and "On the Structure, Distribution and Function of the Nerves Which Innervate the Visceral and Vascular Systems," *ibid.,* **7** (1886), 1–80.

Other papers discussed in the text are "On the Changes of the Blood-Stream in Muscles Through Stimulation of Their Nerves," in *Journal of Anatomy and Physiology,* **11** (1877), 360–402; "On the Vasomotor Nerves of Striated Muscles," *ibid.,* 720–753; "Further Researches on the Vaso-Motor Nerves of Ordinary Muscles," in *Journal of Physiology,* **1** (1878), 262–302; "On the Tonicity of the Heart and Arteries," in *Proceedings of the Royal Society,* **30** (1880), 225–227, and in *Journal of Physiology,* **3** (1882), 48–75; "The Croonian Lecture: On the Rhythm of the Heart of the Frog, and on the Nature of the Action of the Vagus Nerve [1881]," in *Philosophical Transactions of the Royal Society,* **173** (1882), 993–1033; "On the Action of the Sympathetic Nerves Upon the Heart of the Frog," in *Journal of Physiology,* **5** (1884), xiii–xv; "On the Augmentor (Accelerator) Nerves of the Heart of Cold-Blooded Animals," *ibid.,* 46–48; "On the Anatomy of the Cardiac Nerves in Certain Cold-Blooded Invertebrates," *ibid.,* 362–372, written with Hans Gadow; "On the Relationship Between the Structure and Function of the Nerves Which Innervate the Visceral and Vascular Systems," *ibid.,* **6** (1885), iv–x; and "On the Action of Muscarin Upon the Heart, and on the Electrical Changes in the Non-Beating Cardiac Muscle Brought About by Stimulation of the Inhibitory and Augmentory Nerves," *ibid.,* **8** (1887), 404–415. See also "On the Relations Between the Function, Structure, Origin, and Distribution of the Nerve-Fibres Which Compose the Spinal and Cranial Nerves," in *Transactions of the Medico-Chirurgical Society,* **71** (1888), 363–376.

Gaskell provides some historical background and an excellent account of his mature views on the heartbeat in "The Contraction of Cardiac Muscle," in E. A. Schafer, ed., *Textbook of Physiology,* II (Edinburgh [1900]), 169–227. Of uneven quality is his posthumous monograph, *The Involuntary Nervous System,* J. F. Gaskell, ed. (London, 1916).

The Origin of Vertebrates (London–New York, 1908), pp. 6–7, gives a complete bibliography of Gaskell's papers on that topic up to 1906. The only other paper known to the author, cited in n. 12 above, provides a clear and succinct account of Gaskell's theory.

There is apparently no central repository for Gaskell's letters and MSS, and few seem to have survived. The library of the Cambridge Physiological Laboratory possesses Gaskell's reprint collection, deposited in about 100 file boxes and fully indexed. A very few letters can be found in the Sharpey-Schafer Papers at the Wellcome Institute of the History of Medicine, London.

II. SECONDARY LITERATURE. This article is based chiefly on Gerald L. Geison, "Michael Foster and the Rise of the Cambridge School of Physiology, 1870–1900," unpub. Ph.D. diss. (Yale, 1970), pp. 382–475, 493–513, *passim.* For a clear analysis of Gaskell's major work on the heart, see also Richard D. French, "Darwin and the Physiologists, or the Medusa and Modern Cardiology," in *Journal of the History of Biology,* **3** (1970), 253–274, see 267–273.

Of the available accounts of Gaskell's life and work, the most valuable is that by John Langley (see n. 1 above). Also useful are Walter Langdon-Brown (see n. 10 above) and Henry Head, in *Dictionary of National Biography, 1912–1921,* pp. 207–209. A critical reading should be given to F. H. Garrison and F. H. Pike, in *Science,* n.s. **40** (1914), 802–807.

GERALD L. GEISON

GASSENDI (GASSEND), PIERRE (*b.* Champtercier, France, 22 January 1592; *d.* Paris, France, 24 October 1655), *philosophy, astronomy, scholarship.*

The Gassend family used the form Gassendi, according to the Italianism then in style, but Pierre always signed himself Gassend. When a very young man, he was already a principal professor at Digne. His family had him continue his studies, which he pursued at Aix.

In 1614 he was accepted into minor orders and obtained a doctorate at Avignon. Two years later he took holy orders at Aix, where, from 1617 to 1623, he was charged with the teaching of philosophy. He was then initiated into astronomy by Gaultier de la Valette and into humanism by Peiresc, who became his patron.

A partisan of new ideas, Gassendi had printed in Grenoble a first volume of *Exercitationes paradoxicae*

(1624) aimed against the Scholastics; he prudently withheld a second volume. His reputation—and the size of his correspondence—increased, and a canonry at Digne assured his independence (he became provost in 1634).

In Paris in 1624 and again in 1628, he met Mersenne, Mydorge, the du Puy brothers, and Luillier. In 1629–1630 he traveled with the latter in the Low Countries, where he met Isaac Beeckman.

On 7 November 1631 he observed the transit of Mercury, and in his *Mercurius in sole visus* (1632) he treated the event as a confirmation of Kepler's ideas. He returned to Digne at the end of 1632 and undertook an extensive study of Epicurus' thought, in the course of which he expressed his own. At some junctures he clearly departed from the ancient philosopher, but at others he placed statements inspired by materialism next to affirmations of orthodoxy with which they were difficult to reconcile. He was, however, in no hurry to publish and seems even to have interrupted his researches in 1637 when Peiresc died. He resumed them again under the protection of the new governor of Aix-en-Provence, Louis de Valois, at whose behest he returned to Paris after election to the Assembly of the Clergy, a position he was obliged to renounce in 1641. At the request of Mersenne, he immediately thereafter composed the *Cinquièmes objections* to the *Meditations* of Descartes. The *Instantiae* was published in 1644.

Gassendi's growing influence led Louis de Valois and Cardinal Alphonse de Richelieu, archbishop of Lyons, to appoint him professor of mathematics (i.e., astronomy) at the Collège Royal in Paris in 1645. He published a *Leçon inaugurale* and a *Cours*, in which he set forth the system of Copernicus, while prudently falling back on that of Tycho. He taught for only a short time, however. His health was uncertain, and in 1648 Louis de Valois called him back to Provence, where he spent several years. His *Animadversiones* of 1649 contains a portion of his works on Epicurus together with the Greek text and translation of book 10 of Diogenes Laertius.

In Paris once again in 1653, Gassendi produced a third version of his great work entitled *Syntagma philosophicum,* but he did not resume teaching. He died at the home of his host, Habert de Montmort, and was buried at St. Nicolas des Champs on 26 October 1655.

Gassendi's *Opera omnia* was published in six volumes by his friends in Lyons (1658), according to a plan he had established himself. The first two volumes contain the *Syntagma;* the third, a series of scientific works; the fourth, the astronomical lectures and observations; the fifth, the *Lives of Astronomers* and Epicurean works, as well as the *Life of Peiresc;* and the sixth, the Latin correspondence he had selected to preserve. The *Animadversiones* was not reprinted in its original form until 1675.

Although he excited the curiosity and attention of others, Gassendi did not seek to do so. He was not the leader of the "libertines" and the future "philosophes." Olivier Bloch, in his authoritative thesis, sees in Gassendi a belated humanist rather than an avant-garde thinker.[1] There is no reason to question the sincerity of his testimonies of allegiance to a church of which he was a respected dignitary, as were his best friends, Peiresc and Mersenne. His true intellectual master was Galileo. In the *Exercitationes* of 1624 Gassendi had demonstrated his philosophic independence, and as early as 12 July 1625 he wrote to Galileo that he shared his Copernican ideas. But he never had to suffer the anxieties of the great Florentine. His choice of Epicurean atomism as a framework for the exposition of his ideas appears to have been more a revolt against Scholasticism than the expression of any profound conviction. Moreover, his erudition embraced all doctrines, including those of the church fathers, whereas he rejected such important elements of Epicureanism as the vertical fall and swerving of atoms.

Gassendi's eclecticism was that of a skeptic assured that no one doctrine penetrates to the essence of things—indeed, this is a constant aspect of his thought. Yet he proceeded as would a historian for whom the human mind had exhausted all possibilities, in contrast to Descartes, who wrote as if unaware that anyone had ever done philosophy before him. Gassendi's first published letter (to Pibrac, 8 April 1621) reveals an extreme diversity in what he chose to adopt and a great deal of personal assurance; he rejected only dogmatism, even when Epicurean. Bound by no fixed viewpoint, he could more easily go along with the traditions of his peasant milieu. If his morality preached happiness, his method for attaining it was conformist. A worldly type like Saint-Évremond thought him timid. A fanatic like J.-B. Morin consigned him to the flames. Descartes accused him of nothing less than materialism—thereby contributing more than slightly to the suspicion in which he was held. Gassendi, in turn, treated Descartes as a dogmatist. Moreover, he disappointed the materialists. Gassendi wished, Karl Marx declared, to put a nun's habit on the body of Lais.[2] In reality, Gassendi, believing Aristotle's metaphysics to be pagan, attempted to establish a metaphysics that would be Christian, but in harmony with the fundamentally anti-Aristotelian contemporary science.

In this undertaking Gassendi may simply have become aware of his own ambiguities.[3] A thorough study of the philosophical manuscripts preserved at Carpentras, Tours, and the Laurentian Library, and also of the published works, which repeat and correct each other (*Disquisitio,* 1644; *Animadversiones,* 1649; and the posthumous *Syntagma,* 1658), reveals neither the duplicity nor the denial suspected by Pintard[4] but rather an effort to bring the Epicurean elements, accompanied by their materialist tendency, together with the traditional Christian elements. The two had previously been juxtaposed in Gassendi's writings without being mingled—but not without contradiction. This became evident after the beginning of the dispute with Descartes in 1641 and in the new drafts of the Epicurean works first undertaken in 1642. The factors that Gassendi emphasized to achieve a synthesis between Epicureanism and Christianity were nominalism, finality, and vitalistic or chemical analogies. A discussion of these factors is required before asking whether Gassendi felt that Descartes's reproaches really hit their target.

Nominalism had been born in a Christian atmosphere, where it remained a minority position, inspired by awareness of the limits of human understanding (*modulus intellectionis*). Feeble beings that they are, men (*homonciones*) cannot reach essential truth but only appearances, or phenomena, conditioned by laws that they did not make and cannot understand. God established these laws in order that things might endure and satisfy the needs of living creatures. Man establishes a system of signs, of names, which permits him to identify things perceived and to communicate with other men. But the concepts thus formed are conventions, not universal propositions. The universal does not exist ontologically. God has given man a mind capable only of conceiving the universal as the result of repeated contacts between the senses and well-ordered material realities. In animals imagination and memory record the facts to be retained. In man the rational spirit enables him to combine these representations with a view to action, guided by coherent predictions and based on reflections that take time and that are true inferences and not intuitions of some reality beyond the reach of sensation. But there is an evident providential finality in the Creation thus interpreted, and it is further illustrated by the wonders of the universe, of which man is the consummation and the goal. Hence, final causes are the "Royal Way." They demonstrate the existence of God. The view was opposed to that of Descartes; and Gassendi, incidentally, refuted the ontological argument on which Descartes relied in much the same way that Kant later did.

Gassendi held that the atoms were the first things created, not in infinite number, as Democritus had said, but in a number sufficient to create the finite universe we know. They are endowed with an unalterable (in French *inamissible*) movement propelling them without interference in all directions through the void. There is no swerving (no *clinamen*). The collisions that necessarily take place annul motion and result in the appearance of immobility. Collisions form molecules which are particles identifiable by several attributes. The homogeneous atomic particles for their part are endowed only with shape, resistance, minimum size, and a "weight" that is the effect of their elementary movement. Molecules combine in fewer ways than atoms to form sensible objects, possessing not powers, or internal qualities capable of activity, but mechanical forces. Various circumstances may liberate these forces in such a manner that impressions are made on other objects, notably the senses of living beings. At this level, other forces become effective—for example, chemical forces.[5]

The dynamism that is sometimes noticed in Gassendian physics, and that justifies the expression *semina rerum* (borrowed from Lucretius) to designate the atoms, was merely this accumulation of an energy potential, conceivable even in biology. For living bodies are subjected to the same laws as others. Life is composed of movements of the "flower of matter," the animal soul, which in a way resembles Descartes's animal spirits and subtle matter. Science is thus relative to our needs, a view in which there was both sensationalism and pragmatism. Thus, Gassendi was not only a belated humanist but also a precursor of Locke, Condillac, and the positivists and empiricists of the eighteenth and nineteenth centuries.

These ideas contained the entire arsenal upon which future materialists could draw. Yet Gassendi had no thought of being a materialist in the later sense of d'Holbach or Marx. The clash with Descartes had revealed to him the way in which his works, still unpublished, could scandalize certain readers; his role as a priest led him to take this danger into account. But until then he had been able to conjoin faith with Epicureanism with as little fear as Galileo had earlier felt in juxtaposing Copernicus and the Bible.

Galileo had pointed out in his letter to the grand duchess of Florence (see below) that the Bible had originally been addressed to the early Jews in terms that they could understand, while Copernicus, for his part, had offered his work to the pope, and it was not at first thought heretical. By the same token, in Gassendi's view, God had the power to make the world from atoms, as the Epicureans held, and was equally able to illuminate it by making the earth revolve around the sun on the Copernican hypothesis.

Galileo explained his theological position in relation to science in 1615 in his letter to the grand duchess of Florence, Christine of Lorraine. The argument was immediately and widely disseminated, and Gassendi undoubtedly saw it at Aix. It was published in Latin in Strasbourg as early as 1635,[6] although in response to the condemnation of 1633. Descartes's opposition also obliged Gassendi to take "precautions." The word is Mersenne's, who, by publishing the *Cinquièmes objections* had provoked the dispute with Descartes. He spoke of precautions in praising Gassendi's works in a letter to Rivet (8 February 1642).[7] That was precisely the date on which Gassendi undertook a new draft of his Epicurean works. Gassendi may probably have made these modifications in order to persevere in the same project, not to remove ambiguities or to modify it in some unexpected way. Mersenne gave his approbation to the earlier version, while expressing satisfaction with improvements in the new edition. Freethinkers were the only ones who judged differently and for their own reasons: they hoped that this physics would teach man to dispense with metaphysics.[8]

Was such a result what Gassendi wished? Not at all. In the seventeenth century it was possible to conceive of God's having created the universe in a single stroke, but after a model that permits the most convenient analysis. The "fable du monde," which Descartes imagined to be separate from dogma without contradicting it, played a finalist role despite its author's intentions. The atomic model could be employed in the same fashion. An admirer of Gassendi, the physician Deschamps, asked whether, without impiety, one could say that.[9]

Gassendi's influence on epistemology may now be stated more precisely. Koyré summarized it by saying that Gassendi contributed to the new science "the ontology that it needed."[10] In order to eliminate "powers" and "acts," "accidents" and "qualities," whether occult or not, it was necessary to suppose fixed and measurable data in a medium that in no way influences what is observed. Such are the atoms, endowed with shape, solidity, impenetrability, and a natural tendency to motion, which is weight. Such is the void in which bodies move without interference and without any change occurring in their nature through mere endurance. Time does not "eat away" at things; rather their mechanical and spatial relations change in the course of time. Contrary to the Scholastic view, space and time are neither substance nor accident. They exist when their content disappears and when nothing is happening. They establish the general frame of any knowledge of reality—with atoms redividing in a homogeneous void and moving in the unalterable course of time. Gassendi was one of the first to state this universal, categorial law of space and time.

Despite his influence on the ontology of classical physics, Gassendi's scientific successes were not of the first rank. He owed what he achieved to his fidelity to the Democritean schema. Thus his study of *Parhélies* (1630) suggests a corpuscular explanation of light. His patient and thorough method made him a pioneer of observational astronomy, in which field Galileo had already set the example in 1610.[11] But the observations, which almost fill the fourth volume of his *Oeuvres,* could serve only as a model for his contemporaries without leading him to any major discovery. For example, he corrected the geographical coordinates acknowledged for use in navigation in the Mediterranean, and he rejected the discovery of Jupiter's new satellites announced by de Rheita in 1643.

The observation of the transit of Mercury, in which he alone was successful and which confirmed Kepler and, indirectly, Copernicus, caused widespread discussion. Koyré, however, reproaches him for having disregarded the mathematical form that enabled Kepler to determine the elliptical orbits of the planets.[12] Numerous sketches of various aspects of Saturn did not suggest to him the ring hypothesis, which Huygens proposed in 1659 without access to information that was much superior. Gassendi remained a prisoner of what the senses, even when fortified, are able to show. The *Cours* of 1644 at the Collège Royal (published in 1647) prudently presented Tycho Brahe together with Copernicus, while leaning sufficiently toward the latter to shock J.-B. Morin. In the *De proportione qua gravia decidentia accelerantur* of 1645, as in the *De motu impresso,* Gassendi defended—against the criticism of Le Cazre—the law of freely falling bodies, in which velocity is proportional to the square of the time elapsed and not to the distance traversed. But he never understood the importance of its having been deduced either from simple observations of motion on an inclined plane or in any other way.

In 1654 Gassendi joined to his other lives of astronomers the *Life of Copernicus,* in which the trial of Galileo, although not omitted, is barely mentioned. He thus insisted on the hypothetical and mathematical character of Copernicus' work, whereas in 1647 the *Institutio astronomica* had explained the condemnation of Galileo by considerations relating to Galileo himself, but presenting no objections to Copernicus' theories.[13] It is further worth noting that Gassendi followed Galileo in the error of regarding the phenomenon of the tides as a proof of the motion of the earth. As was well known, the periodicity of the tides does not correspond to that of the diurnal movement, and Descartes did not make this mistake.[14]

On one point—and it is an important one—Gassendi was more successful than Galileo: he correctly stated the principle of inertia. The experiment of the *De motu impresso a motore translato,* performed in 1640 in Marseilles, overthrew the argument of Copernicus' opponents against the movement of the earth. Gassendi arranged to have a weight dropped from the top of a vertical mast on a moving ship in order to demonstrate that it fell at the foot of the mast and not behind it, thus sharing in its fall the forward motion of the ship. Galileo considered the experiment unnecessary; he foresaw the result by reasoning.[15] Others, notably Bruno, had already spoken of it. But Gassendi understood that the composition of motions is a universal phenomenon: Every movement impressed on a body in motion in any direction whatsoever persists in Democritean space, which has neither up nor down. Motion is, in itself, a physical state, a measurable quantity, not—as the Scholastics maintained—the change from one state to another. It changes only through the interposition of another movement or of an obstacle.

Furthermore, Gassendi also corrected the formulation given by Kepler, for whom inertia was a tendency to rest: in classical physics, inertia is indifference to both motion and rest. On this point, Gassendi was guided by Galileo's experiments on the pendulum, in which motion is maintained without any supplementary impetus. In addition, Kepler's idea of magnetic effluents or forces gave him an intimation of the existence of universal attraction or, rather, universal interaction—although he was no more successful than Descartes in conceiving its transmission otherwise than by contact.[16]

Gassendian atoms and Cartesian subtle matter belong, as has been seen, to a single period of thought. Moreover, the idea of inertia was common to Beeckman, Gassendi, and Descartes, who all knew each other, and we know that Newton read Gassendi, as did Boyle and Barrow.

In 1650, on a mountain near Toulon, another experiment repeated the famous one of the Puy-de-Dôme.[17] Gassendi fully appreciated the value of Pascal's work. But the latter, in the *Équilibre des liqueurs,*[18] speaks indiscriminately of "weight and pressure of the air," whereas, guided by the corpuscular picture and not by the hydrostatic scheme referred to in Pascal's title, Gassendi could differentiate weight (which is constant for a given mass of air) from pressure (which varies according to the state of agitation, dilation, or contraction of this same mass). It is variations in pressure that affect the barometer and that measure not only the approximate height of the "column of air" but also the changes of state of the atmosphere, which are capable of influencing subsequent weather conditions. Of course, the barometric vacuum proves that the natural vacuum is not impossible; but what happens in the tube depends only on what happens outside. Koyré rightly points out that in this regard Gassendi anticipated Boyle, who read him closely and regretted not having done so earlier.[19]

Gassendi applied his empirical and experimental sagacity to other fields, often in collaboration with Mersenne. Together they estimated the speed of sound as 1,038 feet per second, a passable approximation for the time.[20] Physiology and dissection also interested Gassendi, as did all of natural history. However, he never completely renounced a false observation made at Aix in his youth when Payen made him "see" a communication between the two parts of the heart; but at least he esteemed Harvey and Pecquet. Numismatics and music also occupied him on occasion.

It is evident that Gassendi's influence on science was more philosophical than technical and more critical than systematic. He rationalized physics by introducing quantity into it through the measurements he undertook but above all by introducing atoms, those mutually combinable units that are capable of joining together in molecules and of producing measurable bodies. It is regrettable that with excessive modesty he refrained from propounding general views of the sort that can direct and enrich experiment a priori and that he did not envisage the possibility of applying mathematics to concrete, physical cases.[21]

NOTES

1. In Gassendi one sees primarily a precursor of Locke and Condillac, mentioned later in this article, as well as Hume. See *Tricentenaire de Gassendi,* pp. 69, 227.
2. "Avant-propos" to "Mémoire sur Démocrite et Épicure," in *Oeuvres,* J. Molitor, trans., I (Paris, 1946), xxii.
3. This and the following three paragraphs have been freely inspired by the excellent thesis of M. Bloch (see below), who generously lent it to the author.
4. Cf. *Libertinage érudit* (Paris, 1943), p. 301, *passim.*
5. On this point, Bloch rehabilitates Étienne de Clave, a chemist condemned in 1624 by the Parlement of Paris.
6. Letter, in *Le opere di Galileo Galilei,* Favaro, ed. (Florence, 1890–1909), V, 309 ff. Gassendi does not approach the position of "double truth" to the extent that Bloch (see especially his ch. 11) thinks he does in his desire to reconcile Epicureanism and literal dogma. He thought he could juxtapose not two truths but facts equally real although differently expressed. Misunderstandings taught him what "precautions" (see fol-

lowing note) to take, precautions that Bloch sets forth with extreme precision; but these do not go as far as fideism.

7. *Correspondance du P. Mersenne,* XI, 38: "M. Gassendi réfute puissament, dans sa Philosophie Épicurienne, tout ce qui est contre le christianisme, et, comme vous avez fort bien remarqué, il y prend des précautions." Rivet did not necessarily see what Mersenne was talking about. Mersenne, however, knew the drafts that preceded the one begun on this date as well as the draft of the *Instantiae,* which was later joined to the *Cinquièmes objections* and Descartes's *Responsa* to form the *Disquisitio metaphysica* (1644).
8. The author's conclusions in this and the preceding paragraph are inspired by new material introduced by Gassendi in later editions that has been studied in depth by Bloch; the author's opinions differ, in accordance with his knowledge of the respective positions of Descartes, Galileo, Gassendi, and Mersenne in regard to each other.
9. Letter of 14 Aug. 1642, in *Correspondance du P. Mersenne,* XI, 229–231.
10. *Tricentenaire,* pp. 176, 186.
11. Galileo sent Gassendi a telescope through Diodati; see letter of 25 July 1634 from Galileo to Diodati.
12. *Tricentenaire,* p. 188, n. 9. However, the *Syntagma,* I, 639a–b, mentions the elliptical trajectories of Kepler.
13. *Opera omnia* (Lyons, 1658), V, 60b, end of book III, ch. 10.
14. *Principes,* IV, 49–52.
15. *Dialogo,* in *Le opere di Galileo Galilei,* VII, 171; and Koyré, *Études galiléennes,* pp. 215, 229, 249, 252; and in *Tricentenaire,* pp. 189 ff.
16. Despite everything that set them apart, Descartes and Gassendi were often bracketed by authors of the end of the seventeenth century. See also n. 5 and the corresponding text.
17. Gassendi had spoken of the Puy-de-Dôme experiment in a supp. to the *Animadversiones* (1649) and of his own in a letter (6 Aug. 1652) to Bernier, who had assisted him in that experiment. (Dating the letter "anno superiore," he called Bernier's memory into question: his own "diaire" testified that the experiment took place on 5 Feb. 1650.) All this is taken up again in the *Syntagma* (*Opera omnia,* I, 203–216). See Rochot's articles in *Aventure de l'esprit* (Mélanges Koyré) and in Koyré, *Tricentenaire,* pp. 184 ff.
18. Pléiade ed., pp. 383 ff.
19. *Tricentenaire,* pp. 184 ff.; see also Bloch, ch. 8, especially n. 190, opposing Koyré.
20. *Tricentenaire,* p. 180.
21. Did Gassendi read the *Saggiatore*? See *Le opere di Galileo Galilei,* VI, 232, as well as the letter to Liceti (Jan. 1641), *ibid.,* XVIII, 295: "The book of nature is written in mathematical language."

BIBLIOGRAPHY

I. ORIGINAL WORKS. The contents of the six vols. of the *Opera omnia* (Lyons, 1658), with a preface by Sorbière, are summarily described in the text. The work has been reprinted twice: N. Averrani, ed. (Florence, 1727); and in facs. (Stuttgart, 1964), with a pref. by T. Gregory.

Following is a list of Gassendi's principal individual works.

Scientific Works. Into this class fall *Mercurius in sole visus et Venus invisa* (Paris, 1632; 1658 ed., vol. IV); *De apparente magnitudine solis humilis et sublimis epistolae quatuor* (Paris, 1642; 1658 ed., vol. III); *De motu impresso a motore translato epistolae duae* (Paris, 1642; 1658 ed., vol. III), two letters to Dupuy, to which a third, to Gautier *contra* Morin and dated 1643, was added in the 1658 ed. (Gassendi's friends had published the Gautier letter earlier [Lyons, 1649] without his knowledge); *Oratio inauguralis habita in Regio Collegio, anno 1645, die Novembris XXIII, a P. Gassendo* (Paris, 1645; 1658 ed., vol. IV); *De proportione qua gravia decidentia accelerantur* (Paris, 1646; 1658 ed., vol. III); *Institutio astronomica juxta hypotheseis tam veterum quam Copernici et Tychonis. Dictata a Petro Gassendo. Ejusdem oratio inauguralis iterato edita* (Paris, 1647; 1658 ed., vol. IV); and *Tychonis Brahei . . . N. Copernici, G. Peurbachi et J. Regiomontani . . . vitae* (Paris, 1654; 1658 ed., vol. V).

Philosophical Works. This second class includes *Exercitationum paradoxicarum adversus Aristoteleos libri septem, in quibus praecipua totius Peripateticae doctrinae atque dialecticae excutiuntur; opiniones vero aut novae, aut ex vetustioribus obsoletae stabiliuntur, liber primus: In doctrinam Aristoteleorum universe,* issued independently (Grenoble, 1624); bk. 2, *In dialecticam Aristoteleorum,* did not appear until the 1658 ed. (vol. III) with the shortened title *Exercitationes paradoxicae adversus Aristoteleos, in quibus. . . .* It was separately published shortly afterward as *Exercitationum paradoxicarum liber alter in quo dialecticae Aristoteleae fundamenta excutiuntur* (The Hague, 1659); a text and French trans. appeared as *Dissertations en forme de paradoxes contre les aristotéliciens,* B. Rochot, ed. and trans. (Paris, 1959), in which bk. 2 is corrected according to the MS at the Laurentian Library (this MS was formerly at Tours but was stolen from there by Libri).

Epistolica exercitatio, in qua praecipua principia philosophiae R. Fluddi, medici, reteguntur, et ad recentes illius libros adversus R. P. F. Marinum Mersennum scriptos respondetur (Paris, 1630; 1658 ed., vol. III).

The *Disquisitio metaphysica seu dubitationes et instantiae adversus R. Cartesii metaphysicam, et responsa* (Amsterdam, 1644; 1658 ed., vol. III) consists of the *Objectiones quintae* of 1641 with the publisher Sorbière's addition of the *Instantiae* of 1642, after Descartes's *Responsa.* A text and French trans. of the *Disquisitio* was published as *Recherche de la métaphysique,* B. Rochot, ed. and trans. (Paris, 1962).

De vita et moribus Epicuri libri octo (Lyons, 1647; 1658 ed., vol. V).

Animadversiones in decimum librum Diogenis Laërtii, qui est de vita, moribus placitisque Epicuri, 3 vols. (Lyons, 1649; 2nd ed., 2 vols., 1675), was reproduced only in part in the 1658 ed. The Greek-Latin text of Diogenes, with philological notes, does appear in vol. V. The reworked doctrinal commentary was incorporated into the *Syntagma philosophicum* (see below). The *Philosophiae Epicuri syntagma, cum refutationibus dogmatum quae contra fidem christianam ab eo asserta sunt, oppositis per Petrum Gassendum* (1658 ed., vol. III), a sort of Epicurean breviary added as an appendix to vol. II of the *Animadversiones,* appeared separately (The Hague, 1659) with the preface that Sorbière had placed at the head of the 1658 ed.

His masterpiece, *Syntagma philosophicum* (*logica, physica, ethica*), was published posthumously (1658 ed., vols. I–II).

Correspondence. The *Lettres familières à Fr. Luillier* (*hiver 1632–33*), B. Rochot, ed. (Paris, 1944), is based on a MS that belonged to the heirs of the provost of Digne, now in the Bibliothèque Nationale (fonds latin 2643). The MS contains Gassendi's drafts of the Latin letters in vol. VI of the 1658 ed. Most of the letters addressed to him in the same vol. are in the Bibliothèque Nationale. The French correspondence with Peiresc is in *Lettres de Peiresc,* Tamizey de Larroque, ed., IV (Paris, 1893). Gassendi is frequently mentioned in correspondence of the period; see especially *Correspondance du P. Mersenne,* C. de Waard, Marie Tannery, and B. Rochot, eds. (Paris, 1932-). The bulk of his extensive correspondence in French and Latin is far from entirely known.

Miscellaneous Works. The biography *De Nicolai Claudii Fabricii de Peiresc, senatoris aquisextiensis, vita* (Paris, 1641; 1658 ed., vol. V) appeared in English as *The Mirrour of True Nobility and Gentility, Being the Life of. . . N. C. Fabricius, Lord of Peiresk,* W. Rand, trans. (London, 1657). It is especially useful as a source for the historian of early seventeenth-century science.

A curious, and anonymous, pamphlet of 1654 designed to calm widespread fears occasioned by an eclipse of the sun is reasonably attributed to Gassendi. It was reprinted by B. Rochot, ed., in *Bulletin de la Société d'étude du XVIIe siècle,* no. 27 (Apr. 1955), 161-177.

II. SECONDARY LITERATURE. The following items have been selected from the bibliography (343 items, including MSS, printed texts, biographical and doctrinal studies, and various articles) in the thesis of Olivier René Bloch, *La philosophie de Gassendi: Nominalisme, matérialisme et métaphysique* (Paris, 1971); F. Bernier, *Abrégé de la philosophie de Gassendi,* 2nd ed., 7 vols. (Lyons, 1684); Henri Berr, *Du scepticisme de Gassendi,* B. Rochot, trans. (Paris, 1960), a trans. of the 1898 thesis *An jure inter scepticos Gassendus numeratus fuerit;* [J. Bougerel], *Vie de Pierre Gassendi* (Paris, 1737), which should be examined carefully because the author had access to documents that are now lost; G. S. Brett, *Philosophy of Gassendi* (London, 1908); G. Cogniot, "Pierre Gassendi, restaurateur de l'épicurisme," in *La pensée,* no. 63 (Sept.-Oct. 1955); P. Damiron, *Histoire de la philosophie au XVIIe siècle* (Paris, 1846), I, 378-503; René Dugas, *La mécanique au XVIIe siècle* (Paris-Neuchâtel, 1954), ch. VI, pp. 103-116; Tullio Gregory, *Scetticismo ed empirismo. Studio su Gassendi* (Bari, 1961); Pierre Humbert, *L'oeuvre astronomique de Gassendi* (Paris, 1936), completed by *Philosophes et savants* (Paris, 1953), pp. 79-107; A. Koyré, *Études galiléennes* (Paris, 1939), pp. 237 ff., repr. (Paris, 1966), pp. 304 ff.; F. A. Lange, *Geschichte der Materialismus und Kritik seiner Bedeutung in der Gegenwart,* 2nd ed., 2 vols. (Iserlohn, 1873-1875), which appeared in French as *Histoire du matérialisme,* B. Pommerol, trans., 2 vols. (Paris, 1921), and in English as *The History of Materialism . . .,* E. C. Thomas, trans., 3rd ed. (London, 1957), contains a section on Gassendi; Kurd Lasswitz, *Geschichte der Atomistik vom Mittelalter bis Newton,* 2 vols. (Hamburg-Leipzig, 1890; 2nd ed., 1928), II, 126-188; L. Mabilleau, *Histoire de la philosophie atomistique* (Paris, 1895), pp. 400-422; P. Pendzig, *Pierre Gassendis Metaphysik . . .* (Bonn, 1908); René Pintard, *Libertinage érudit,* 2 vols. (Paris, 1943), which contains, in vol. I, numerous analyses in which Gassendi is portrayed as the leader of a libertine *tétrade* and, in vol. II, an important bibliography (see also the MSS examined in his *La Mothe Le Vayer, Gassendi, Guy Patin* [Paris, 1943]); B. Rochot, *Les travaux de Gassendi sur Épicure et l'atomisme* (Paris, 1944); G. Sortais, *La philosophie moderne depuis Bacon jusqu'à Leibniz,* II (Paris, 1922); J. S. Spink, *Free Thought From Gassendi to Voltaire* (London, 1960); and P. F. Thomas, *La philosophie de Gassendi* (Paris, 1889). More a summary than an interpretation, it does not take into account the evolution of Gassendi's thought as represented by the *Syntagma.* Two collections of studies are *Pierre Gassendi, sa vie et son oeuvre,* Centre International de Synthèse (Paris, 1955); and *Tricentenaire de Gassendi,* Actes du Congrès de Digne, 1955 (Paris-Digne, 1957).

The MSS enumerated by Bloch are in the Bibliothèque Nationale and in the libraries of Tours (706-710), Carpentras, and Florence (Laurentian). Biographical documents are at Aix-en-Provence, Digne, Grenoble, Marseilles, Munich, Oxford, Stuttgart, and Vienna; in the Archives du Ministère de la Guerre, Paris; and in the Bibliothèque Nationale (fonds français 12270 and fonds Dupuy).

Some texts have been translated into Polish by H. L. Kolakowski (Cracow, 1964) and into Russian by Sitkovsky (Moscow, 1966), with studies.

It should be noted that the important study by G. Gusdorf, *Révolution galiléenne,* vol. III of Les Sciences Humaines et la Pensée Occidentale, 2 vols. (Paris, 1969), was used in the preparation of this article.

BERNARD ROCHOT

GASSER, HERBERT SPENCER (*b.* Platteville, Wisconsin, 5 July 1888; *d.* New York, N.Y., 11 May 1963), *physiology.*

Gasser's father, Herman, was born in the Tyrol and emigrated as a boy to the United States, where he became a country doctor. His mother, Jane Elizabeth Griswold, came from an old Connecticut family. She trained as a teacher in the state Normal School of Platteville, which Gasser himself later attended. The controversies of the time concerning evolution, vitalism, and mechanism had led his father to acquire the works of Darwin, Huxley, and Herbert Spencer, which the younger Gasser read avidly. He entered the University of Wisconsin to major in zoology. Having completed quickly the requirements for a B.A. degree, he took courses in the newly organized medical school, where he first met Joseph Erlanger, with whom he was later to share the Nobel Prize. As the university was then only a half-school (two years) Gasser transferred to Johns Hopkins University, where the approach to medicine exactly suited his aims. Gasser's professional career was, in time, to involve him in teaching and administration as well as research. His other major interests were music, history, literature, and travel. His positions included instructor at the University of Wisconsin (1911-1916); physiologist at Washington University, St. Louis (1916-

1921); pharmacologist in the Chemical Warfare Service (1918); professor of pharmacology at Washington University, St. Louis (1921–1931); professor of physiology at Cornell University Medical College (1931–1935); and director of the Rockefeller Institute for Medical Research (1935–1953). On retirement he continued active research for nearly ten years. From 1923 to 1925 Gasser was in Europe working with A. V. Hill at University College, London; Sir Henry Dale at the National Institute for Medical Research; Walter Straub at Munich; and Louis Lapicque at the Sorbonne.

Gasser received academic honors from many universities both in the United States and Europe. He was a member of the National Academy of Sciences, the American Philosophical Society, and a number of professional societies. In 1944, Gasser and Erlanger were awarded the Nobel Prize for discoveries relating to nerve fibers, and in 1945 the Association of American Physicians awarded Gasser its Kober Medal.

Gasser's early work, dictated by the exigencies of World War I, was concerned largely with problems of traumatic shock and blood volume. Only after the war did his first work on nerves, written with H. S. Newcomer, appear. It concerned application of thermionic vacuum tubes to the study of nerve action currents. Then came the pioneering study, in association with Erlanger, on use of the cathode-ray oscilloscope as an inertialess instrument for recording action potentials of nerve and the initial analysis of their compound nature. Some problems arising at that time stayed with Gasser only to be resolved finally in his last paper, published in 1960. It is difficult in the days of near-universal television to imagine the early difficulties of oscillographic recording. Light intensity was so low that many repetitions of the nerve response were required to produce a photographic image, and tubes lasted but a few hours. Some consequences of these necessities were from one point of view essentially artifactual in nature. Typically, Gasser always was aware of, and concerned with, the possibilities of artifact. When it became possible to record single responses, it was apparent that they differed from those recorded during repetitive activity. A clue to the difference was found in subsequent study of the subnormal state of nerve, which, by responding repetitively, had influenced the early recordings.

Major problems that commanded Gasser's attention were the compound nature of the nerve action potential; the relation between nerve-fiber size and inpulse-conduction velocity; the excitability of nerve fibers in relation to after-potentials; potentials recordable from the spinal cord; the afferent fibers concerned with pain-producing impulses; and the morphology of unmyelinated fibers, with respect to their compound action potential and diameter-velocity relations.

Some of these studies were direct offshoots from prior work, while others represented an abrupt change in direction or a return to old problems still unsolved, for Gasser espoused the principle that there are two times for working on a problem—before anyone has thought of it and after everyone else has left it. As a result, Gasser was always the innovator or the finalist.

At the height of the controversy over which types of nerve fibers yield various compound action potentials, Gasser turned to the question of after-potentials and associated excitability changes. His studies showed the after-potentials and the excitability states associated with them to be closely correlated, but different in the several groups of nerve fibers. This proved crucial to the characterization of fiber groups, for by this means some somatic fibers and sympathetic preganglionic fibers overlapping in diameter and conduction velocity could be distinguished, as could be the somatic and sympathetic unmyelinated fibers.

Whenever possible, Gasser required that the results of two approaches to the same problem be congruent. A prime example is the convergent information from electron microscopy and oscilloscopic recording that he achieved virtually single-handedly with respect to unmyelinated nerve fibers. He, however, found minor incongruity between the division of somatic afferent myelinated fibers (A fibers) into the subgroups α, β, γ, δ recorded as elevations in the electrical response and the action potential reconstructed from anatomical fiber-size maps of the nerve made on the assumption that velocity varied in direct proportion to the diameter of the fiber. Unsatisfied, he finally identified the incongruity as a product of the method of leading from active nerve. Correcting this, he found that the potential of the skin nerve manifested but two elevations, α and δ. He wrote in his last published work, "Thus the action potential was brought into closer accord with the indications in the maps of fiber diameters." Characteristically, this remark was an understatement.

BIBLIOGRAPHY

A complete bibliography is included in Gasser's "An Autobiographical Memoir," with a preface by J. C. Hinsey, in *Experimental Neurology,* supp. 1 (1964), and in *Electrical Signs of Nervous Activity,* 2nd ed. (Philadelphia, 1969), written with J. Erlanger.

DAVID P. C. LLOYD

GASSICOURT. See **Cadet de Gassicourt.**

GASSIOT, JOHN PETER (*b.* London, England, 2 April 1797; *d.* Isle of Wight, 15 August 1877), *electricity.*

Gassiot, a wealthy wine merchant, was elected a fellow of the Royal Society in 1840 and was one of the founders of the Chemical Society in 1841. He also helped to endow the Kew observatory and for many years was the chairman of the Royal Society's Kew Observatory Committee. In 1863 Gassiot was awarded the Royal Society's Royal Medal in recognition of his work on voltaic electricity and on the discharge of electricity through gases at low pressure.[1]

In the late 1830's, when Gassiot began his investigations, the identity of static and voltaic electricity seemed likely. But if this were so, voltaic, like static, electricity ought to produce sparks before the circuit was completed. In 1839 Gassiot showed that even with a battery of 1,024 Daniell cells no sparks occurred. But if he used these cells either to charge a bank of nine Leyden jars or in conjunction with a circuit interrupter and transformer, he could produce sparks before contact. He also obtained sparks with a Zamboni dry pile of 10,000 cells and, later, of 1,000 cells.[2] But in 1843, using a massive battery of 3,520 zinc-copper rainwater cells, he produced sparks through 0.020 inch of air. Gassiot attributed his success to his great care in insulating the individual cells to prevent the loss of their electrical tension.[3]

At that time it had not been decided whether voltaic electricity is produced by contact between metals or by chemical reaction. In an attempt to decide this question Gassiot showed in the same paper that "the elements constituting the voltaic battery, when arranged in a series, assume polar tension before the circuit is completed. . . ." Yet a few months later he concluded that "to produce *static* effects in a voltaic battery, it is indispensible that the elements should be such as can combine by their chemical affinities. . . ." Furthermore, "in all the experiments I made, the higher the chemical affinities of the elements used, the greater was the evidence of tension." These discoveries gave further evidence to support the decision in favor of the chemical theory that had already been reached in 1839 by Gassiot's friend Michael Faraday.[4]

Faraday's discovery in 1838 of the negative dark space had revived interest in the glow discharge caused by conduction of electricity through gases at low pressure,[5] but Gassiot's interest in this discharge was directly stimulated by W. R. Grove's almost incidental report in 1852 that the discharge was "striated by transverse non-luminous bands. . . ."[6] In his initial investigations Gassiot showed that if enough care were exercised to achieve a sufficiently low pressure, striations could be produced in the Torricellian vacuum. Next he demonstrated that both a static electric machine and a Ruhmkorff coil with a Grove cell produced a striated discharge. This once again confirmed the identity of these two electricities.[7] He also noticed that a powerful electromagnet divided the striations into what appeared to be two distinct columns.[8] The paper in which Gassiot announced these discoveries was honored as the Royal Society's Bakerian lecture for 1858.

During the next two years Gassiot continued his efforts to obtain the striations, which he thought were caused by "pulsations or impulses of a force acting on highly attenuated but a resisting medium. . . ."[9] Although his theory here was not correct, his investigations produced much new information. First, because his 3,520-cell battery or a 400-cell Grove battery unassisted did produce striations, it was clear that the "induction coil is not necessary for the production of the striae. . . ."[10] Next Gassiot demonstrated experimentally that the striae exist only within a narrower range of pressure and temperature than the luminous discharge itself; that a sufficiently low pressure not only ends the discharge but also that this relative vacuum does not conduct electricity; that changes in the electrical resistance of the external circuit change the discharge; and that at least sometimes the luminous discharge is actually intermittent even though it appears to be continuous.[11]

In Gassiot's final group of experiments on the gaseous discharge he showed that in a series circuit containing two discharge tubes, if a magnet is used to interrupt one discharge, the electrical current in both is completely disrupted;[12] that excitation of a spiral "carbonic acid vacuum tube . . ." gives a brilliant white light;[13] that there is a mechanical disruption of the metal in the negative electrode;[14] and that changes in the external resistance in the electrical circuit also change the striae in the discharge.[15] Gassiot also perfected a rotating and vibrating mirror technique that he used to reconfirm his discovery that the discharge, under certain conditions, is intermittent. His last papers generally concerned improvements in spectroscopes. In particular he designed and had constructed a spectroscope with nine glass prisms and another with eleven prisms filled with carbon disulfide, which he presented to Kew observatory.[16]

NOTES

1. "Anniversary Meeting—President's Address," in *Proceedings of the Royal Society,* **13** (1864), 183-185.
2. "An Account of Experiments Made With the View of Ascer-

taining the Possibility of Obtaining a Spark Before the Circuit of the Voltaic Battery Is Completed," in *Philosophical Transactions of the Royal Society,* **130** (1840), 183–192.
3. "A Description of an Extensive Series of the Water Battery," *ibid.,* **134** (1844), 39–42.
4. *Ibid.;* see also Michael Faraday, *Experimental Researches in Electricity,* par. 2053.
5. *Ibid.,* pars. 1544–1560.
6. "On the Electro-Chemical Polarity of Gases," in *Philosophical Transactions of the Royal Society,* **142** (1852), 100.
7. "On the Stratification and Dark Band in Electrical Discharges as Observed in the Torricellian Vacua," *ibid.,* **148** (1858), 6.
8. *Ibid.,* p. 15.
9. *Ibid.,* p. 14.
10. "On the Electrical Discharge *in vacuo* With an Extended Series of the Voltaic Battery," in *Proceedings of the Royal Society,* **10** (1860), 36–37.
11. "On the Stratification in Electrical Discharges Observed in Torricellian and Other Vacua—Second Communication," in *Philosophical Transactions of the Royal Society,* **149** (1859), 137–160.
12. "On the Interruption of the Voltaic Discharge *in vacuo* by Magnetic Force," in *Proceedings of the Royal Society,* **10** (1860), 269–274.
13. *Ibid.,* p. 432.
14. *British Association Report* (London, 1861), section 2, 38–39.
15. "Experimental Investigations on the Stratified Appearance in Electrical Discharges," in *Proceedings of the Royal Society,* **12** (1863), 329–340.
16. *Philosophical Magazine,* **27** (1864), 143–144.

BIBLIOGRAPHY

I. ORIGINAL WORKS. Gassiot's papers are listed in Poggendorff, I, 849–850, and III, 495; and in Royal Society, *Catalogue of Scientific Papers,* II, 779–780; VII, 741–742. Michael Faraday assisted Gassiot with some of his experiments on gas discharge. Faraday's notes on these experiments are reprinted in *Faraday's Diary,* VII (London, 1936), 412–461. There is no collected edition of his works.

II. SECONDARY LITERATURE. There is no biography of Gassiot. Biographical information is based on the article on him in the *Dictionary of National Biography,* VII, 935–936, and on the references given there. The most useful survey of his scientific work is contained in the speech made by Edward Sabine, president of the Royal Society, in presenting Gassiot's Royal Medal. It is reprinted in *Proceedings of the Royal Society,* **13** (1864), 36–39. The debate about the "contact" and "chemical" theories of the voltaic battery is discussed in Edmund Whittaker, *History of the Theories of Aether and Electricity,* I (New York, 1960), 180–184. Whittaker's brief account of the work on conduction of electricity through rarefied gases, pp. 348–366, can be supplemented by J. J. Thomson, *Conduction of Electricity Through Gases,* 2 vols. (3rd ed., Cambridge, 1933).

EDGAR W. MORSE

GATES, REGINALD RUGGLES (*b.* Middleton, Nova Scotia, 1 May 1882; *d.* London, England, 12 August 1962), *genetics.*

Gates was one of several early geneticists who tried, unsuccessfully, to unravel the genetics of *Oenothera,* a particularly important botanical genus. The son of a farmer and fruit grower, he was educated in Canadian schools: B.A. (1903) and M.A. (1904) from Mount Allison University, Sackville, New Brunswick, and B.Sc. (1906) from McGill. During the summer of 1905 Gates was introduced to the complicated genetics of *Oenothera* (evening primrose) while studying at the Woods Hole Biological Laboratory; he pursued this problem in the research which later (1908) earned him the doctorate at the University of Chicago.

Oenothera had first come to the attention of biologists several years earlier, when Hugo de Vries discussed the genus in announcing the discovery of genetic mutations. De Vries had noticed several strikingly variant individuals in a field of wild evening primroses; these bred true when cultivated experimentally. In answer to a question which had long plagued biologists, de Vries had thus demonstrated the occurrence of new genetic types, appearing spontaneously and following the same inheritance patterns as older varieties.

De Vries' mutation theory was eventually validated on other species—particularly *Drosophila melanogaster*—but his original example did not stand the test of time; the mutant forms of *Oenothera* did not follow classic Mendelian patterns of inheritance. They were not, as biologists later learned, mutant forms in de Vries' original meaning of the phrase. At the time that Gates began his work, the genetics of *Oenothera* was regarded as extremely puzzling.

Gates studied the mutant forms of *Oenothera* cytologically rather than genetically. He discovered that one mutant species, *rubrinervis,* has chromosomes which form rings instead of aligning in pairs during meiosis; sometimes, as Gates demonstrated, this phenomenon can lead to an unequal division of chromosomes. Another species that Gates studied, *Oenothera lata,* had fifteen chromosomes instead of fourteen, the chromosome number of the parent species, *lamarckiana.* A third species, *gigas,* was tetraploid; it had twenty-eight chromosomes.

These cytological studies were widely applauded—Gates was awarded the Huxley Medal and Prize of Imperial College, London (1914), and the Mendel Medal (1911)—but they did little to explain the irregular breeding behavior of the primroses. Several decades passed before biologists were able to understand the genetics of *Oenothera.* Without a full understanding of the gene theory in general, as well as such particular phenomena as translocation, disjunction, and the balanced lethal system, the cytological facts that Gates discovered are difficult to interpret. He tried to interpret them in *The Mutation Factor in Evolution* (1915), but his analysis apparently did not win favor with the biological community.

After leaving Chicago, Gates spent two years at the Missouri Botanical Garden in St. Louis as an experimenter. In 1912 he crossed the Atlantic and began to teach, first as lecturer at St. Thomas Hospital, London, and then as reader (1919) and subsequently head (1921) of the botany department of King's College, University of London. As his administrative duties increased, his research activities decreased. He participated in the work of his graduate students; but until the early 1950's, by which time his interests had shifted from genetics to eugenics, Gates did not publish any additional significant work.

Gates had developed an interest in eugenics during the 1920's and published a textbook on the subject at that time, *Heredity and Eugenics* (1923). His subsequent researches, particularly into the pedigrees of Negro families, convinced him that mankind had had polyphyletic origins and that only a small number of chromosomes were needed to produce racial differences; he was also convinced that racial crossing was genetically harmful. Most anthropologists and geneticists did not agree with Gates on these points; this led him to found a new journal, *Mankind Quarterly,* in which he would be free to voice his opinions. His most significant eugenic discovery dealt with the gene for hairy ear rims, a characteristic of the Ainu of Japan; Gates located it on the Y chromosome.

Gates traveled widely, and late in life he often combined his travels with his eugenic investigations. From 1940 to 1950 he was in the United States, first on a lecture tour and then as honorary research fellow at Harvard. He was a fellow of the Royal Society and an officer of several other British scientific societies.

BIBLIOGRAPHY

A more complete biography of Gates and a complete list of his publications can be found in J. A. Fraser Roberts, "Reginald Ruggles Gates," in *Biographical Memoirs of Fellows of the Royal Society,* **10** (1964), 83–105.

Gates's studies of *Oenothera* mutants were published over several years: "Pollen Development in Hybrids of *Oenothera lata* x *O. lamarckiana,* and Its Relation to Mutation," in *Botanical Gazette,* **43** (1905), 81–115; "Hybridization and Germ Cells of *Oenothera* Mutants," *ibid.,* **44** (1907), 1–21; "A Study of Reduction in *Oenothera rubrinervis,*" *ibid.,* **46** (1908), 1–36; "Chromosomes of *Oenothera,*" in *Science,* n.s. **27** (1908), 193–195; and "Further Studies on the Chromosomes of *Oenothera,*" *ibid.,* 335. This work was summarized, and his theories of mutation advanced, in *The Mutation Factor in Evolution* (London, 1915); see also "The Cytology of *Oenothera,*" in *Bibliographia genetica,* **4** (1928), 401–492.

His eugenic studies and outlook can be sampled in *Heredity and Eugenics* (London, 1923); *Pedigrees of Negro Families* (Philadelphia, 1949); "The Inheritance of Hairy Ear Rims," in *Mankind Quarterly,* **1** (1962), written with P. N. Bhaduri; and his monumental textbook, *Human Genetics,* 2 vols. (London, 1946).

RUTH SCHWARTZ COWAN

GAUDIN, MARC ANTOINE AUGUSTIN (*b.* Saintes, France, 5 April 1804; *d.* Paris, France, 2 April 1880), *chemistry.*

Despite an unacademic family background (his father was a shopkeeper), Gaudin was active in science at an early age. In 1826, from Rochefort, he submitted his first paper to the Académie des Sciences in Paris, and in 1827 he gained the inspiration for much of his later work when he attended Ampère's lectures at the Collège de France. From 1835 to 1864 Gaudin was a calculator at the Bureau des Longitudes in Paris but, never holding a teaching or research post, he remained outside the Parisian scientific establishment. His one application for membership in the Académie des Sciences (1851) was unsuccessful, and the only formal recognition of his work by the Academy came in 1867, when he was awarded the Prix Trémont.

Gaudin's most important work was concerned with the arrangement of atoms within molecules and of molecules within crystals. The earliest statement of his views was contained in two short notes that he submitted to the Académie des Sciences in 1831 and 1832. The first of these notes, published in 1833, is remarkable since it contains a clear exposition of the gas hypothesis of Avogadro written some twenty-five years before the work of Cannizzaro and the Karlsruhe Congress made the hypothesis widely acceptable. Gaudin supported the hypothesis not in the form given it by Ampère but in its modern form. He correctly reconciled it with Gay-Lussac's law of combining volumes by supposing that the common elementary gases, such as hydrogen and oxygen, were diatomic, but that other gaseous substances, mercury vapor, for example, were monatomic, while others, notably many compound gases, were triatomic, and others again were of still greater complexity.

Of the other issues arising in the papers of 1831–1833 the one that was to prove most absorbing for Gaudin concerned the possible relationship between the physical and chemical properties of substances and the spatial arrangement of the atoms that composed them. It was not until 1847 that he returned to the problem, but from that year he wrote extensively on the subject until 1873, when he published his views in a definitive form in his *L'architecture du monde des atomes.* In the forty years that had elapsed since they were first expounded, his ideas had changed little, and they were largely obsolete by 1873. Not surprisingly, the book received little attention and stimulated no further work.

Despite his obvious debt to Ampère and to Haüy, Gaudin's treatment of molecular and crystalline structure showed a good deal of originality. He rejected Ampère's assumption that even the simplest molecules were polyhedral; and although he adopted a polyhedral form for the more complex molecules, he also abandoned Ampère's set of basic molecular shapes. Instead he chose his structures in accordance with a rigorously held belief that the atoms within a molecule were always arranged symmetrically. Hence any structure Gaudin proposed had not only to be consistent with the usual crystallographic data and the evidence of chemical composition but also to show symmetry, and this restriction caused his views on crystal structure to deviate considerably from those of Ampère and Haüy. His preoccupation with symmetry also affected his views on chemical combination. In particular it led him to reject the theory of radicals and the type theory, since he believed that symmetry would be destroyed by the simple replacement of certain atoms by others. According to Gaudin, it was only by a complete rearrangement of the atoms of the combining molecules that symmetry could be restored after a reaction.

In his long scientific career Gaudin was active in several other lines of research. He worked in microscopy, invented an ingenious pneumatic pump (1827), and showed a special interest in experimental work at high temperatures. It was Gaudin who prepared the fused quartz for Biot's work on optical activity in 1839, and he is noted for his method of preparing artificial rubies using an oxyhydrogen blowpipe. An important pioneer of photography, he wrote a comprehensive textbook on this subject in 1844. In his photographic work he was closely associated with his brother Alexis.

BIBLIOGRAPHY

I. Original Works. *L'architecture du monde des atomes dévoilant la structure des composés chimiques et leur cristallogénie* (Paris, 1873) is Gaudin's most important book, but he published much else. His *Traité pratique de photographie* (Paris, 1844) is an admirable practical handbook of photography, and "Recherches sur la structure intime des corps inorganiques définis," in *Annales de chimie et de physique,* 2nd ser., **52** (1833), 113-133, is of great historical interest. Most of his numerous communications to the Académie des Sciences are noted in *Comptes rendus hebdomadaires des séances de l'Académie des sciences.*

II. Secondary Literature. The best studies are M. Delépine, "Une étape de la notion d'atomes et de molécules," in *Bulletin de la Société chimique de France,* 5th ser., **2** (1935), 1-15, with supp. note by G. Urbain, 16-17; and S. H. Mauskopf, "The Atomic Structural Theories of Ampère and Gaudin: Molecular Speculation and Avogadro's Hypothesis," in *Isis,* **60** (1969), 61-74. Biographical information is scarce, but a useful supp. to the standard biographical dictionaries, such as the *Nouvelle biographie générale,* is in *Bulletin de la Société des archives historiques de la Saintonge et de l'Aunis,* **2** (1880), 163.

Robert Fox

GAUDRY, ALBERT JEAN (*b.* St.-Germain-en-Laye, France, 15 September 1827; *d.* Paris, France, 27 November 1908), *paleontology.*

Two events of Gaudry's youth strongly influenced his work. First, his mother died when he was quite young and, at the age of seventy, he still cried over her death; this emotional shock, from which he never recovered, may account for the tenderness mixed with mysticism that characterized both the man and his work. Second, his father, a renowned lawyer and historian, collected minerals and associated with geologists; one of these geologists, Alcide d'Orbigny, explorer of South America and founder of modern stratigraphic paleontology, married Gaudry's sister around 1845 and guided Gaudry's career.

Having completed his advanced studies, Gaudry entered the Muséum d'Histoire Naturelle at Paris in 1851, in the laboratory of the mineralogist P. L. A. Cordier. There he prepared a doctoral dissertation on the occurrence of flint in chalk strata, which he defended in 1852. He was then a timid young man, short and frail, with a shrill voice, a refined and gentle face, and blond hair. This fragile appearance concealed much courage, tenacity, and physical strength—qualities demonstrated in 1853, during a fatiguing but profitable geological mission in the countries of the eastern Mediterranean, including Cyprus.

While Gaudry was on this mission, the government, convinced of the scientific importance of fossil study, created a chair of paleontology at the museum for Alcide d'Orbigny, despite the opposition of that institution's professors, who believed that each of them should be entrusted with preserving the fossils relating to his own discipline: the professor of malacology, for example, would be curator of mollusks; the botany professor, of fossil plants; and so on. D'Orbigny made his young brother-in-law his assistant, and Gaudry henceforth devoted the major portion of his time to paleontology. In 1855 and 1860 he carried out two excavations in Attica, in the Tertiary mammal deposit at Pikermi. The fossil remains that he brought back enabled him to reconstruct several skeletons of new species. Some of them displayed characteristics intermediate to those of species already known; and in an article of February 1859 on the life and work of d'Orbigny, who had died in 1857, Gaudry explained that these intermediate species "restore the links which were missing in the great chain of beings." He repeated here almost verbatim the expression used in 1833 by the founder of evolutionary paleontology, Étienne Geoffroy Saint-Hilaire, regarding the fossil remains of Auvergne.

Published nine months before Darwin's *On the Origin of Species,* Gaudry's article drew no inspiration whatever from the ideas of the English naturalist, as Gaudry subsequently indicated. For him, biological evolution resulted from a continuous creation by God. He did not destroy His previous creations (as d'Orbigny believed); rather, He maintained the species through time, perfecting and transforming them until the sublime masterpiece—man—was finally attained. Each transformation reflected the infinite beauty of God, as Gaudry wrote (1862) in his great monograph on Attica, in which he established, following extensive research, remarkable genealogical trees of five large groups of mammals. Several years later he proposed dating stratigraphic terrains according to the degree of evolution of the fossils they contained, and he applied this new method successfully to the mammals of the Tertiary in *Animaux fossiles du Mont-Léberon* (1873).

From 1866 to 1892 Gaudry studied very small reptiles and batrachians, remarkable for their archaic anatomical type and great age, since they originated in the schists of Autun (Saône-et-Loire), which date from the Lower Permian. He also had a marked predilection for the Quaternary. His thorough excavations at St.-Acheul, near Amiens, in September 1859 removed the last doubts concerning the contemporaneity of man and the large extinct mammals. In 1894 Gaudry confirmed that the archaic chipped flints of Abbeville (Somme) were associated with the teeth of the advanced *Elephas meridionalis,* a finding which placed the appearance of man very far back in time. But the majority of geologists, even his favorite student, Marcellin Boule, refused—wrongly—to believe him, despite his great reputation.

Gaudry's friend and biographer, Gustave-Frédéric Dollfus, remarked: "There are some ideas so advanced that they triumph only after the disappearance of the generation which fought them; Gaudry lived long enough to witness the progressive spread of his doctrine." Nevertheless, during the major portion of his scientific career Gaudry encountered the animosity of the older naturalists, who reproached him for wanting to make paleontology an independent science capable of providing support for the theory of evolution. Fortunately, Gaudry's family and that of his wife were wealthy and had important connections. In 1868 Victor Duruy, the eminent minister of education under Napoleon III, placed Gaudry in charge of a course in paleontology at the Sorbonne and attended the inaugural lecture. In 1871, when Duruy was no longer minister, the course was canceled. In 1872 Gaudry was finally appointed professor of paleontology at the Muséum d'Histoire Naturelle. His colleagues removed from his laboratory the bulk of the paleontology collections, even those that he himself had assembled, and gave them to the professor of comparative anatomy; they were not returned to Gaudry until 1878. In 1885 he began to carry out, in a wooden shed, the project he had conceived in 1859: a museum of evolution where the public would follow, from room to room, the perfecting of the species.

For Gaudry the theory of evolution bore a spiritual message: Living beings form one family, one great unity, which has become more perfect through time; intelligence, created last, has not completed its development; God is the sole fixed point of this universe where everything is changing; if life is an immense progression, then "he who says progression says union, [and] he who says union says love. The great law which rules life is a law of love." Wishing to spread this message, so different from Darwin's, Gaudry devoted time to writing works of high-level popularization. They were presented in a limpid style that revealed his artistic sensibility and were illustrated by beautiful wood engravings. His most important book, published in three volumes under the title *Les enchaînements du monde animal* (1878–1890), was concluded by *Essai de paléontologie philosophique* (1896).

In 1902 paleontologists throughout the world celebrated Gaudry's jubilee. He was the first who ventured to reestablish the respectability of the evolutionary paleontology founded by Étienne Geoffroy Saint-Hilaire, and only paleontology could prove that biological evolution was a tangible reality.

BIBLIOGRAPHY

I. Original Works. A list of Gaudry's scientific writings, in the notice by Thévenin (see below), consists of 191 titles. The Royal Society's *Catalogue of Scientific Papers* (II, 784–785; VII, 744–745; IX, 972–973; XV, 228–229) lists 135 titles published in periodicals (excluding nonscientific journals) to 1900. Gaudry also published two *Notices sur les travaux scientifiques d'A. Gaudry* (Paris, 1878, 1881), in which he commented on his works.

His principal writings are "Alcide d'Orbigny, ses voyages,

et ses travaux," in *Revue des deux mondes,* **19** (1859), 816–847; "Contemporanéité de l'espèce humaine et des diverses espèces animales aujourd'hui éteintes," read to the Academy of Sciences on 3 Oct. 1859, in *L'Institut,* sec. 1, no. 1344 (5 Oct. 1859), pp. 317–318—despite its title this journal had no connection with the Académie des Sciences, which did not publish the note; *Animaux fossiles et géologie de l'Attique,* 2 vols. (Paris, 1862–1867); "La théorie de l'évolution et la détermination des terrains," in *Revue des cours scientifiques* (18 Dec. 1869); *Animaux fossiles du Mont-Léberon* (Paris, 1873), written with P. Fisher and R. Tournouër; "Les reptiles de l'époque permienne aux environs d'Autun," in *Bulletin de la Société géologique de France,* 3rd ser., **7** (1878), 62–77; *Les enchaînements du monde animal dans les temps géologiques,* 3 vols. (Paris, 1878–1890); *Mammifères tertiaires* (1878): *Fossiles primaires* (1883); *Fossiles secondaires* (1890)—the first vol. was translated into German as *Die Vorfahren der Säugetiere in Europa* (Leipzig, 1892); and *Matériaux pour l'histoire des temps quaternaires,* 4 fascs. (Paris, 1876–1892), the last fascicle in collaboration with Marcellin Boule.

II. SECONDARY LITERATURE. Armand Thévenin, "Albert Gaudry, notice nécrologique," in *Bulletin de la Société géologique de France,* 3rd ser., **10** (1910), 351–374, includes a portrait, a bibliography of his works, and a list of fourteen biographical notices. A notice by P. Glangeaud appears in English in *Report of the Board of Regents of the Smithsonian Institution* for 1919, publication no. 1969, pp. 417–429. See also the biography by Gustave-Frédéric Dollfus, *Albert Gaudry 1827–1908* (Paris, 1909), repr. from *Journal de conchyliologie,* **57** (1909), 274–278.

FRANCK BOURDIER

GAULTIER DE CLAUBRY, HENRI-FRANÇOIS (*b.* Paris, France, 21 July 1792; *d.* Paris, 4 July 1878), *chemistry, toxicology, public health.*

Gaultier de Claubry's father, Charles-Daniel, and his older brother, Charles-Emmanuel-Simon, were prominent surgeons and physicians. Following their example, Henri-François first embarked on a medical career but subsequently decided to devote his efforts to scientific pursuits. He served as a *répétiteur* at the École Polytechnique and in 1835 was named assistant professor of chemistry at the École de Pharmacie, a post he held until 1859, when he succeeded J.-B. Caventou as professor of toxicology. In 1825 he was appointed to the Council of Health of the Department of the Seine, and in 1848 he was elected to the Academy of Medicine (Paris).

A prolific author, Gaultier de Claubry is remembered largely for his discovery with J.-J. Colin of the blue color imparted to starch by free iodine (1814). In 1812 he translated into French William Henry's popular textbook, *Elements of Experimental Chemistry.* His investigation of the presence of iodine in seawater and in seaweed was published in 1815, as was his description of the properties of inulin. In succeeding years Gaultier de Claubry dealt with a great variety of subjects and worked in such fields as chemistry, toxicology, public health, medicine, and meteorology. This versatility, which extended to a multiplicity of largely unrelated projects, tended to make his work diffuse.

Nevertheless, Gaultier de Claubry was considered an able scientist by his contemporaries. Particularly noteworthy were his researches on the coloring matter in madder, carried out with J.-F. Persoz and published in 1831. A significant contribution to toxicology was his treatise on legal chemistry, which was incorporated in several editions of J. Briand and E. Chaudé's *Manuel complet de médecine légale,* one of the most authoritative textbooks on legal medicine in nineteenth-century France. The bulk of Gaultier de Claubry's writings on public health—dealing with food adulteration, environmental health, disinfection, industrial hygiene, and related topics—appeared in the *Annales d'hygiène publique et de médecine légale.*

BIBLIOGRAPHY

I. ORIGINAL WORKS. Some of Gaultier de Claubry's most important work is embodied in the following publications: "Mémoire sur les combinaisons de l'iode avec les substances végétales et animales . . .," in *Annales de chimie,* **90** (1814), 87–100; "Des recherches sur l'existence de l'iode dans l'eau de la mer et dans les plantes qui produisent la soude de varecks, et analyse de plusieurs plantes de la famille des algues," *ibid.,* **93** (1815), 75–110, 113–137; "Note sur une substance à laquelle on a donné le nom d'inuline," *ibid.,* **94** (1815), 200–208; and "Mémoire sur les matières colorantes de la garance," *ibid.,* 2nd ser., **48** (1831), 69–79.

A partial listing of Gaultier de Claubry's articles is in the Royal Society of London, *Catalogue of Scientific Papers (1800–1863),* II (London, 1867), 787–788; VII (London, 1877), 746; IX (London, 1891), 974. Most of his articles dealing with various aspects of public health that were published in the *Annales d'hygiène publique et de médecine légale* can be located in two separate index vols., covering the periods 1829–1855 and 1854–1878, respectively. His *Traité élémentaire de chimie légale* was included in J. Briand and E. Chaudé, *Manuel complet de médecine légale,* 4th ed. (Paris, 1846), and several later eds. For additional works, see *Catalogue général des livres imprimés de la Bibliothèque nationale,* LVIII (Paris, 1914), 42–46.

II. SECONDARY LITERATURE. See *Centenaire de l'École supérieure de Pharmacie de l'Université de Paris, 1803–1903* (Paris, 1904), pp. 330–331; Fritz Ferchl, *Chemisch-Pharmazeutisches Bio- und Bibliographikon* (Mittenwald, 1937), p. 173; and G. Vapereau, *Dictionnaire universel des contemporains,* 5th ed. (Paris, 1880), 785. See also J. C. F. Hoefer, ed., *Nouvelle biographie générale* (Paris, 1867),

679; and Pierre Larousse, ed., *Grand dictionnaire universel,* 15 vols. (Paris, 1866–1876), VIII, 1086.

ALEX BERMAN

GAUSS, CARL FRIEDRICH (*b.* Brunswick, Germany, 30 April 1777; *d.* Göttingen, Germany, 23 February 1855), *mathematical sciences.*

The life of Gauss was very simple in external form. During an austere childhood in a poor and unlettered family he showed extraordinary precocity. Beginning when he was fourteen, a stipend from the duke of Brunswick permitted him to concentrate on intellectual interests for sixteen years. Before the age of twenty-five he was famous as a mathematician and astronomer. At thirty he went to Göttingen as director of the observatory. There he worked for forty-seven years, seldom leaving the city except on scientific business, until his death at almost seventy-eight.

In marked contrast to this external simplicity, Gauss's personal life was complicated and tragic. He suffered from the political turmoil and financial insecurity associated with the French Revolution, the Napoleonic period, and the democratic revolutions in Germany. He found no mathematical collaborators and worked alone most of his life. An unsympathetic father, the early death of his first wife, the poor health of his second wife, and unsatisfactory relations with his sons denied him a family sanctuary until late in life.

In this difficult context Gauss maintained an amazingly rich scientific activity. An early passion for numbers and calculations extended first to the theory of numbers and then to algebra, analysis, geometry, probability, and the theory of errors. Concurrently he carried on intensive empirical and theoretical research in many branches of science, including observational astronomy, celestial mechanics, surveying, geodesy, capillarity, geomagnetism, electromagnetism, mechanics, optics, the design of scientific equipment, and actuarial science. His publications, voluminous correspondence, notes, and manuscripts show him to have been one of the greatest scientific virtuosos of all time.

Early Years. Gauss was born into a family of town workers striving on the hard road from peasant to lower middle-class status. His mother, a highly intelligent but only semiliterate daughter of a peasant stonemason, worked as a maid before becoming the second wife of Gauss's father, a gardener, laborer at various trades, foreman ("master of waterworks"), assistant to a merchant, and treasurer of a small insurance fund. The only relative known to have even modest intellectual gifts was the mother's brother, a master weaver. Gauss described his father as "worthy of esteem" but "domineering, uncouth, and unrefined." His mother kept her cheerful disposition in spite of an unhappy marriage, was always her only son's devoted support, and died at ninety-seven, after living in his house for twenty-two years.

Without the help or knowledge of others, Gauss learned to calculate before he could talk. At the age of three, according to a well-authenticated story, he corrected an error in his father's wage calculations. He taught himself to read and must have continued arithmetical experimentation intensively, because in his first arithmetic class at the age of eight he astonished his teacher by instantly solving a busy-work problem: to find the sum of the first hundred integers. Fortunately, his father did not see the possibility of commercially exploiting the calculating prodigy, and his teacher had the insight to supply the boy with books and to encourage his continued intellectual development.

During his eleventh year, Gauss studied with Martin Bartels, then an assistant in the school and later a teacher of Lobachevsky at Kazan. The father was persuaded to allow Carl Friedrich to enter the Gymnasium in 1788 and to study after school instead of spinning to help support the family. At the Gymnasium, Gauss made very rapid progress in all subjects, especially classics and mathematics, largely on his own. E. A. W. Zimmermann, then professor at the local Collegium Carolinum and later privy councillor to the duke of Brunswick, offered friendship, encouragement, and good offices at court. In 1792 Duke Carl Wilhelm Ferdinand began the stipend that made Gauss independent.

When Gauss entered the Brunswick Collegium Carolinum in 1792, he possessed a scientific and classical education far beyond that usual for his age at the time. He was familiar with elementary geometry, algebra, and analysis (often having discovered important theorems before reaching them in his studies), but in addition he possessed a wealth of arithmetical information and many number-theoretic insights. Extensive calculations and observation of the results, often recorded in tables, had led him to an intimate acquaintance with individual numbers and to generalizations that he used to extend his calculating ability. Already his lifelong heuristic pattern had been set: extensive empirical investigation leading to conjectures and new insights that guided further experiment and observation. By such means he had already independently discovered Bode's law of planetary distances, the binomial theorem for rational exponents, and the arithmetic-geometric mean.

During his three years at the Collegium, Gauss continued his empirical arithmetic, on one occasion finding a square root in two different ways to fifty decimal places by ingenious expansions and interpolations. He formulated the principle of least squares, apparently while adjusting unequal approximations and searching for regularity in the distribution of prime numbers. Before entering the University of Göttingen in 1795 he had rediscovered the law of quadratic reciprocity (conjectured by Lagrange in 1785), related the arithmetic-geometric mean to infinite series expansions, conjectured the prime number theorem (first proved by J. Hadamard in 1896), and found some results that would hold if "Euclidean geometry were not the true one."

In Brunswick, Gauss had read Newton's *Principia* and Bernoulli's *Ars conjectandi,* but most mathematical classics were unavailable. At Göttingen, he devoured masterworks and back files of journals, often finding that his own discoveries were not new. Attracted more by the brilliant classicist G. Heyne than by the mediocre mathematician A. G. Kästner, Gauss planned to be a philologist. But in 1796 came a dramatic discovery that marked him as a mathematician. As a by-product of a systematic investigation of the cyclotomic equation (whose solution has the geometric counterpart of dividing a circle into equal arcs), Gauss obtained conditions for the constructibility by ruler and compass of regular polygons and was able to announce that the regular 17-gon was constructible by ruler and compasses, the first advance in this matter in two millennia.

The logical component of Gauss's method matured at Göttingen. His heroes were Archimedes and Newton. But Gauss adopted the spirit of Greek rigor (insistence on precise definition, explicit assumption, and complete proof) without the classical geometric form. He thought numerically and algebraically, after the manner of Euler, and personified the extension of Euclidean rigor to analysis. By his twentieth year, Gauss was driving ahead with incredible speed according to the pattern he was to continue in many contexts—massive empirical investigations in close interaction with intensive meditation and rigorous theory construction.

During the five years from 1796 to 1800, mathematical ideas came so fast that Gauss could hardly write them down. In reviewing one of his seven proofs of the law of quadratic reciprocity in the *Göttingische gelehrte Anzeigen* for March 1817, he wrote autobiographically:

> It is characteristic of higher arithmetic that many of its most beautiful theorems can be discovered by induction with the greatest of ease but have proofs that lie anywhere but near at hand and are often found only after many fruitless investigations with the aid of deep analysis and lucky combinations. This significant phenomenon arises from the wonderful concatenation of different teachings of this branch of mathematics, and from this it often happens that many theorems, whose proof for years was sought in vain, are later proved in many different ways. As soon as a new result is discovered by induction, one must consider as the first requirement the finding of a proof by *any possible* means. But after such good fortune, one must not in higher arithmetic consider the investigation closed or view the search for other proofs as a superfluous luxury. For sometimes one does not at first come upon the most beautiful and simplest proof, and then it is just the insight into the wonderful concatenation of truth in higher arithmetic that is the chief attraction for study and often leads to the discovery of new truths. For these reasons the finding of new proofs for known truths is often at least as important as the discovery itself [*Werke,* II, 159–160].

The Triumphal Decade. In 1798 Gauss returned to Brunswick, where he lived alone and continued his intensive work. The next year, with the first of his four proofs of the fundamental theorem of algebra, he earned the doctorate from the University of Helmstedt under the rather nominal supervision of J. F. Pfaff. In 1801 the creativity of the previous years was reflected in two extraordinary achievements, the *Disquisitiones arithmeticae* and the calculation of the orbit of the newly discovered planet Ceres.

Number theory ("higher arithmetic") is a branch of mathematics that seems least amenable to generalities, although it was cultivated from the earliest times. In the late eighteenth century it consisted of a large collection of isolated results. In his *Disquisitiones* Gauss summarized previous work in a systematic way, solved some of the most difficult outstanding questions, and formulated concepts and questions that set the pattern of research for a century and still have significance today. He introduced congruence of integers with respect to a modulus ($a \equiv b \pmod{c}$ if c divides a-b), the first significant algebraic example of the now ubiquitous concept of equivalence relation. He proved the law of quadratic reciprocity, developed the theory of composition of quadratic forms, and completely analyzed the cyclotomic equation. The *Disquisitiones* almost instantly won Gauss recognition by mathematicians as their prince, but readership was small and the full understanding required for further development came only through the less austere exposition in Dirichlet's *Vorlesungen über Zahlentheorie* of 1863.

In January 1801 G. Piazzi had briefly observed and lost a new planet. During the rest of that year the astronomers vainly tried to relocate it. In September, as his *Disquisitiones* was coming off the press, Gauss decided to take up the challenge. To it he applied both a more accurate orbit theory (based on the ellipse rather than the usual circular approximation) and improved numerical methods (based on least squares). By December the task was done, and Ceres was soon found in the predicted position. This extraordinary feat of locating a tiny, distant heavenly body from seemingly insufficient information appeared to be almost superhuman, especially since Gauss did not reveal his methods. With the *Disquisitiones* it established his reputation as a mathematical and scientific genius of the first order.

The decade that began so auspiciously with the *Disquisitiones* and Ceres was decisive for Gauss. Scientifically it was mainly a period of exploiting the ideas piled up from the previous decade (see Figure 1). It ended with *Theoria motus corporum coelestium in sectionibus conicis solem ambientium* (1809), in which Gauss systematically developed his methods of orbit calculation, including the theory and use of least squares.

Professionally this was a decade of transition from mathematician to astronomer and physical scientist. Although Gauss continued to enjoy the patronage of the duke, who increased his stipend from time to time (especially when Gauss began to receive attractive offers from elsewhere), subsidized publication of the *Disquisitiones,* promised to build an observatory, and treated him like a tenured and highly valued civil servant, Gauss felt insecure and wanted to settle in a more established post. The most obvious course, to become a teacher of mathematics, repelled him because at this time it meant drilling ill-prepared and unmotivated students in the most elementary manipulations. Moreover, he felt that mathematics itself might not be sufficiently useful. When the duke raised

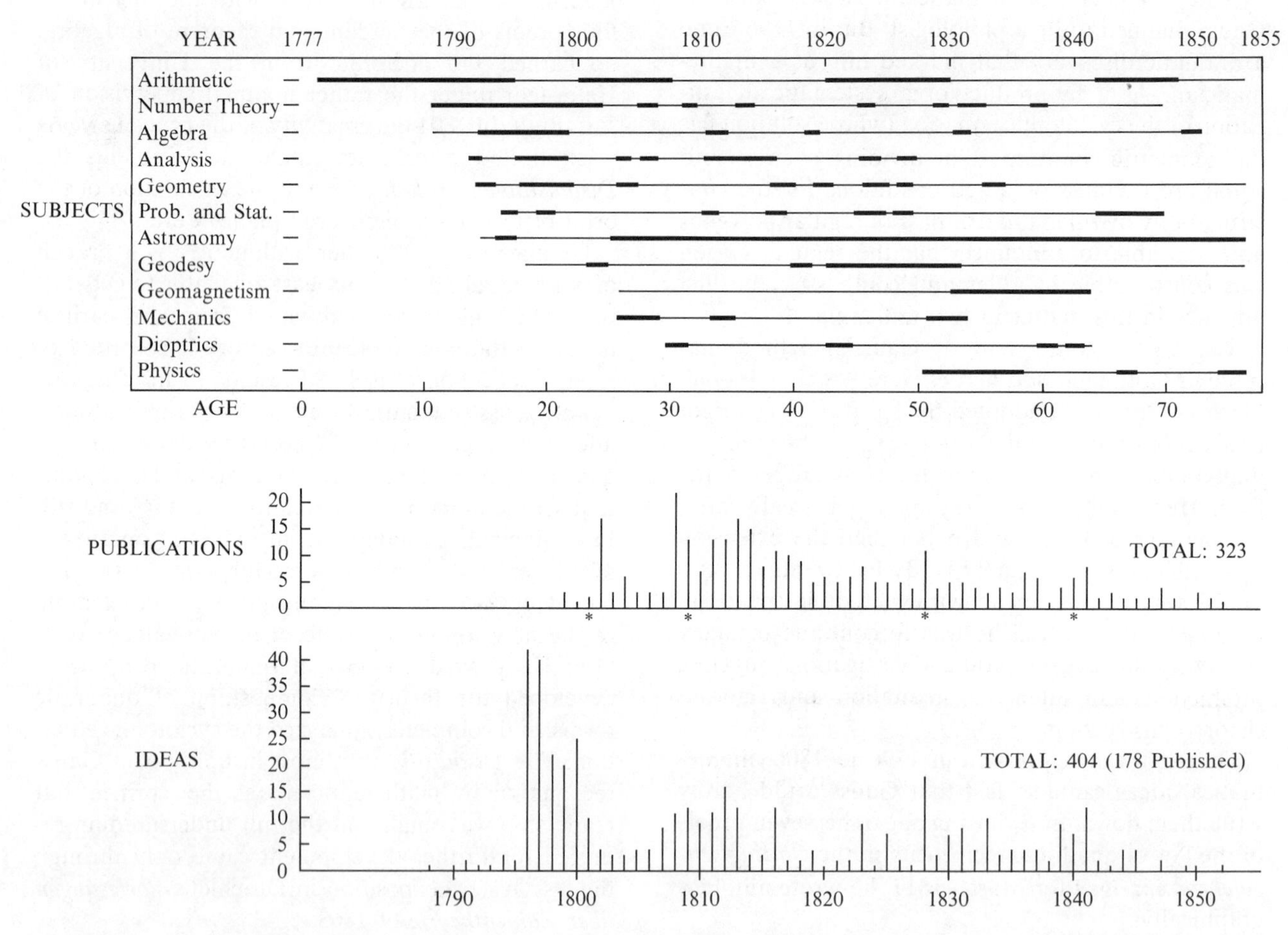

FIGURE 1. Interests, ideas, and publications. The horizontal lines show time spans of Gauss's interests in different subjects. Heavy lines indicate periods of intensive activity. The annual counts of recorded ideas include published and unpublished "results" (conjectures, theorems, proofs, concepts, hypotheses, theories), significant observations, experimental findings, and inventions. They are based on an examination of published materials, including correspondence and notebooks published after his death. Because of intrinsic ambiguities in dating, identification, and evaluation, this chart gives only an approximate picture of creative flux. The graph of publications shows the number of titles published in each year, including reviews. A count of pages would be similar except for surges (marked by *): 1801 (*Disquisitiones*), 1809 (*Theoria motus*), 1828 (least squares, surfaces, astronomy, biquadratic residues), and 1840 (geomagnetism).

his stipend in 1801, Gauss told Zimmermann: "But I have not earned it. I haven't yet done anything for the nation."

Astronomy offered an attractive alternative. A strong interest in celestial mechanics dated from reading Newton, and Gauss had begun observing while a student at Göttingen. The tour de force on Ceres demonstrated both his ability and the public interest, the latter being far greater than he could expect in mathematical achievements. Moreover, the professional astronomer had light teaching duties and, he hoped, more time for research. Gauss decided on a career in astronomy and began to groom himself for the directorship of the Göttingen observatory. A systematic program of theoretical and observational work, including calculation of the orbits of new planets as they were discovered, soon made him the most obvious candidate. When he accepted the position in 1807, he was already well established professionally, as evidenced by a job offer from St. Petersburg (1802) and by affiliations with the London Royal Society and the Russian and French academies.

During this decisive decade Gauss also established personal and professional ties that were to last his lifetime. As a student at Göttingen he had enjoyed a romantic friendship with Wolfgang Bolyai, and the two discussed the foundations of geometry. But Bolyai returned to Hungary to spend his life vainly trying to prove Euclid's parallel postulate. Their correspondence soon practically ceased, to be revived again briefly only when Bolyai sent Gauss his son's work on non-Euclidean geometry. Pfaff was the only German mathematician with whom Gauss could converse, and even then hardly on an equal basis. From 1804 to 1807 Gauss exchanged a few letters on a high mathematical level with Sophie Germain in Paris, and a handful of letters passed between him and the mathematical giants in Paris, but he never visited France or collaborated with them. Gauss remained as isolated in mathematics as he had been since boyhood. By the time mathematicians of stature appeared in Germany (e.g., Jacobi, Plücker, Dirichlet), the uncommunicative habit was too ingrained to change. Gauss inspired Dirichlet, Riemann, and others, but he never had a collaborator, correspondent, or student working closely with him in mathematics.

In other scientific and technical fields things were quite different. There he had students, collaborators, and friends. Over 7,000 letters to and from Gauss are known to be extant, and they undoubtedly represent only a fraction of the total. His most important astronomical collaborators, friends, and correspondents were F. W. Bessel, C. L. Gerling, M. Olbers, J. G. Repsold, H. C. Schumacher. His friendship and correspondence with A. von Humboldt and B. von Lindenau played an important part in his professional life and in the development of science in Germany. These relations were established during the period 1801–1810 and lasted until death. Always Gauss wrote fewer letters, gave more information, and was less cordial than his colleagues, although he often gave practical assistance to his friends and to deserving young scientists.

Also in this decade was established the pattern of working simultaneously on many problems in different fields. Although he never had a second burst of ideas equal to his first, Gauss always had more ideas than he had time to develop. His hopes for leisure were soon dashed by his responsibilities, and he acquired the habit of doing mathematics and other theoretical investigations in the odd hours (sometimes, happily, days) that could be spared. Hence his ideas matured rather slowly, in some cases merely later than they might have with increased leisure, in others more felicitously with increased knowledge and meditation.

This period also saw the fixation of his political and philosophical views. Napoleon seemed to Gauss the personification of the dangers of revolution. The duke of Brunswick, to whom Gauss owed his golden years of freedom, personified the merits of enlightened monarchy. When the duke was humiliated and killed while leading the Prussian armies against Napoleon in 1806, Gauss's conservative tendencies were reinforced. In the struggles for democracy and national unity in Germany, which continued throughout his lifetime, Gauss remained a staunch nationalist and royalist. (He published in Latin not from internationalist sentiments but at the demands of his publishers. He knew French but refused to publish in it and pretended ignorance when speaking to Frenchmen he did not know.) In seeming contradiction, his religious and philosophical views leaned toward those of his political opponents. He was an uncompromising believer in the priority of empiricism in science. He did not adhere to the views of Kant, Hegel, and other idealist philosophers of the day. He was not a churchman and kept his religious views to himself. Moral rectitude and the advancement of scientific knowledge were his avowed principles.

Finally, this decade provided Gauss his one period of personal happiness. In 1805 he married a young woman of similar family background, Johanna Osthoff, who bore him a son and daughter and created around him a cheerful family life. But in 1809 she died soon after bearing a third child, which did not long survive her. Gauss "closed the angel eyes in which for five years I have found a heaven" and was plunged into a loneliness from which he never fully recovered. Less than a year later he married Minna Waldeck, his deceased wife's best friend. She bore him two sons and a daughter, but she was seldom well or happy. Gauss dominated his daughters and quarreled with his younger sons, who immigrated to the United States. He did not achieve a peaceful home life until the younger daughter, Therese, took over the household after her mother's death (1831) and became the intimate companion of his last twenty-four years.

Early Göttingen Years. In his first years at Göttingen, Gauss experienced a second upsurge of ideas and publications in various fields of mathematics. Among the latter were several notable papers inspired by his work on the tiny planet Pallas, perturbed by Jupiter: *Disquisitiones generales circa seriem infinitam* (1813), an early rigorous treatment of series and the introduction of the hypergeometric functions, ancestors of the "special functions" of physics; *Methodus nova integralium valores per approximationem inveniendi* (1816), an important contribution to approximate integration; *Bestimmung der Genauigkeit der Beobachtungen* (1816), an early analysis of the efficiency of statistical estimators; and *Determinatio attractionis quam in punctum quodvis positionis datae exerceret planeta si eius massa per totam orbitam ratione temporis quo singulae partes describuntur uniformiter esset dispertita* (1818), which showed that the perturbation caused by a planet is the same as that of an equal mass distributed along its orbit in proportion to the time spent on an arc. At the same time Gauss continued thinking about unsolved mathematical problems. In 1813 on a single sheet appear notes relating to parallel lines, declinations of stars, number theory, imaginaries, the theory of colors, and prisms (*Werke,* VIII, 166).

Astronomical chores soon dominated Gauss's life. He began with the makeshift observatory in an abandoned tower of the old city walls. A vast amount of time and energy went into equipping the new observatory, which was completed in 1816 and not properly furnished until 1821. In 1816 Gauss, accompanied by his ten-year-old son and one of his students, took a five-week trip to Bavaria, where he met the optical instrument makers G. von Reichenbach, T. L. Ertel (owner of Reichenbach's firm), J. von Fraunhofer, and J. von Utzschneider (Fraunhofer's partner), from whom his best instruments were purchased. As Figure 1 shows, astronomy was the only field in which Gauss worked steadily for the rest of his life. He ended his theoretical astronomical work in 1817 but continued positional observing, calculating, and reporting his results until his final illness. Although assisted by students and colleagues, he observed regularly and was involved in every detail of instrumentation.

It was during these early Göttingen years that Gauss matured his conception of non-Euclidean geometry. He had experimented with the consequences of denying the parallel postulate more than twenty years before, and during his student days he saw the fallaciousness of the proofs of the parallel postulate that were the rage at Göttingen; but he came only very slowly and reluctantly to the idea of a different geometric theory that might be "true." He seems to have been pushed forward by his clear understanding of the weaknesses of previous efforts to prove the parallel postulate and by his successes in finding non-Euclidean results. He was slowed by his deep conservatism, the identification of Euclidean geometry with his beloved old order, and by his fully justified fear of the ridicule of the philistines. Over the years in his correspondence we find him cautiously, but more and more clearly, stating his growing belief that the fifth postulate was unprovable. He privately encouraged others thinking along similar lines but advised secrecy. Only once, in a book review of 1816 (*Werke,* IV, 364–368; VIII, 170–174), did he hint at his views publicly. His ideas were "besmirched with mud" by critics (as he wrote to Schumacher on 15 January 1827), and his caution was confirmed.

But Gauss continued to find results in the new geometry and was again considering writing them up, possibly to be published after his death, when in 1831 came news of the work of János Bolyai. Gauss wrote to Wolfgang Bolyai endorsing the discovery, but he also asserted his own priority, thereby causing the volatile János to suspect a conspiracy to steal his ideas. When Gauss became familiar with Lobachevsky's work a decade later, he acted more positively with a letter of praise and by arranging a corresponding membership in the Göttingen Academy. But he stubbornly refused the public support that would have made the new ideas mathematically respectable. Although the friendships of Gauss with Bartels and W. Bolyai suggest the contrary, careful study of the plentiful documentary evidence has established that Gauss did not inspire the two founders of non-Euclidean geometry. Indeed, he played at best a neutral, and on balance a negative, role, since his silence was considered as agreement with the public ridicule and neglect that continued for several decades and were only gradually overcome, partly by the revelation, beginning in the 1860's, that the prince of mathematicians had been an underground non-Euclidean.

Geodesist. By 1817 Gauss was ready to move toward geodesy, which was to be his preoccupation for the next eight years and a burden for the next thirty.

His interest was of long standing. As early as 1796 he worked on a surveying problem, and in 1799–1800 he advised Lt. K. L. E. von Lecoq, who was engaged in military mapping in Westphalia. Gauss's first publication was a letter on surveying in the *Allgemeine geographische Ephemeriden* of October 1799. In 1802 he participated in surveying with F. X. G. von Zach. From his arrival in Göttingen he was concerned with accurately locating the observatory, and in 1812 his interest in more general problems was stimulated by a discussion of sea levels during a visit to the Seeberg observatory. He began discussing with Schumacher the possibility of extending into Hannover the latter's survey of Denmark. Gauss had many motives for this project. It involved interesting mathematical problems, gave a new field for his calculating abilities, complemented his positional astronomy, competed with the French efforts to calculate the arc length of one degree on the meridian, offered an opportunity to do something useful for the kingdom, provided escape from petty annoyances of his job and family problems, and promised additional income. The last was a nontrivial matter, since Gauss had increasing family responsibilities to meet on a salary that remained fixed from 1807 to 1824.

The triangulation of Hannover was not officially approved until 1820, but already in 1818 Gauss began an arduous program of summer surveying in the field followed by data reduction during the winter. Plagued by poor transportation, uncomfortable living conditions, bad weather, uncooperative officials, accidents, poor health, and inadequate assistance and financial support, Gauss did the fieldwork himself with only minimal help for eight years. After 1825 he confined himself to supervision and calculation, which continued to completion of the triangulation of Hannover in 1847. By then he had handled more than a million numbers without assistance.

An early by-product of fieldwork was the invention of the heliotrope, an instrument for reflecting the sun's rays in a measured direction. It was motivated by dissatisfaction with the existing unsatisfactory methods of observing distant points by using lamps or powder flares at night. Meditating on the need for a beacon bright enough to be observed by day, Gauss hit on the idea of using reflected sunlight. After working out the optical theory, he designed the instrument and had the first model built in 1821. It proved to be very successful in practical work, having the brightness of a first-magnitude star at a distance of fifteen miles. Although heliostats had been described in the literature as early as 1742 (apparently unknown to Gauss), the heliotrope added greater precision by coupling mirrors with a small telescope. It became standard equipment for large-scale triangulation until superseded by improved models from 1840 and by aerial surveying in the twentieth century. Gauss remarked that for the first time there existed a practical method of communicating with the moon.

Almost from the beginning of his surveying work Gauss had misgivings, which proved to be well founded. A variety of practical difficulties made it impossible to achieve the accuracy he had expected, even with his improvements in instrumentation and the skillful use of least squares in data reduction. The hoped-for measurement of an arc of the meridian required linking his work with other surveys that were never made. Too hasty planning resulted in badly laid out base lines and an unsatisfactory network of triangles. He never ceased trying to overcome these faults, but his virtuosity as a mathematician and surveyor could not balance the factors beyond his control. His results were used in making rough geographic and military maps, but they were unsuitable for precise land surveys and for measurement of the earth. Within a generation, the markers were difficult to locate precisely or had disappeared altogether. As he was finishing his fieldwork in July 1825, Gauss wrote to Olbers that he wondered whether other activities might have been more fruitful. Not only did the results seem questionable but he felt during these years, even more than usual, that he was prevented from working out many ideas that still crowded his mind. As he wrote to Bessel on 28 June 1820, "I feel the difficulty of the life of a practical astronomer, without help; and the worst of it is that I can hardly do any connected significant theoretical work."

In spite of these failures and dissatisfactions, the period of preoccupation with geodesy was in fact one of the most scientifically creative of Gauss's long career. Already in 1813 geodesic problems had inspired his *Theoria attractionis corporum sphaeroidicorum ellipticorum homogeneorum methodus nova tractata,* a significant early work on potential theory. The difficulties of mapping the terrestrial ellipsoid on a sphere and plane led him in 1816 to formulate and solve in outline the general problem of mapping one surface on another so that the two are "similar in their smallest parts." In 1822 a prize offered by the Copenhagen Academy stimulated him to write up these ideas in a paper that won first place and was published in 1825 as the *Allgemeine Auflösung der Aufgabe die Theile einer gegebenen Fläche auf einer anderen gegebenen Fläche so auszubilden dass die Abbildung dem Abgebildeten in den kleinsten Theilen ähnlich wird.* This paper, his more detailed *Untersuchungen über Gegenstände der höhern Geodäsie* (1844–1847), and geodesic manuscripts later published in the *Werke* were further developed by

German geodesists and led to the Gauss-Krueger projection (1912), a generalization of the transverse Mercator projection, which attained a secure position as a basis for topographic grids taking into account the spheroidal shape of the earth.

Surveying problems also motivated Gauss to develop his ideas on least squares and more general problems of what is now called mathematical statistics. The result was the definitive exposition of his mature ideas in the *Theoria combinationis observationum erroribus minimis obnoxiae* (1823, with supplement in 1828). In the *Bestimmung des Breitenunterschiedes zwischen den Sternwarten von Göttingen und Altona durch Beobachtungen am Ramsdenschen Zenithsector* of 1828 he summed up his ideas on the figure of the earth, instrumental errors, and the calculus of observations. However, the crowning contribution of the period, and his last breakthrough in a major new direction of mathematical research, was *Disquisitiones generales circa superficies curvas* (1828), which grew out of his geodesic meditations of three decades and was the seed of more than a century of work on differential geometry. Of course, in these years as always, Gauss produced a stream of reviews, reports on observations, and solutions of old and new mathematical problems of varying importance that brought the number of his publications during the decade 1818–1828 to sixty-nine. (See Figure 1.)

Physicist. After the mid-1820's, there were increasing signs that Gauss wished to strike out in a new direction. Financial pressures had been eased by a substantial salary increase in 1824 and by a bonus for the surveying work in 1825. His other motivations for geodesic work were also weakened, and a new negative factor emerged—heart trouble. A fundamentally strong constitution and unbounded energy were essential to the unrelenting pace of work that Gauss maintained in his early years, but in the 1820's the strain began to show. In 1821, family letters show Gauss constantly worried, often very tired, and seriously considering a move to the leisure and financial security promised by Berlin. The hard physical work of surveying in the humid summers brought on symptoms that would now be diagnosed as asthma and heart disease. In the fall of 1825, Gauss took his ailing wife on a health trip to spas in southern Germany; but the travel and the hot weather had a very bad effect on his own health, and he was sick most of the winter. Distrusting doctors and never consulting one until the last few months of his life, he treated himself very sensibly by a very simple life, regular habits, and the avoidance of travel, for which he had never cared anyway. He resolved to drop direct participation in summer surveying and to spend the rest of his life "undisturbed in my study," as he had written Pfaff on 21 March 1825.

Apparently Gauss thought first of returning to a concentration on mathematics. He completed his work on least squares, geodesy, and curved surfaces as mentioned above, found new results on biquadratic reciprocity (1825), and began to pull together his long-standing ideas on elliptic functions and non-Euclidean geometry. But at forty-eight he found that satisfactory results came harder than before. In a letter to Olbers of 19 February 1826, he spoke of never having worked so hard with so little success and of being almost convinced that he should go into another field. Moreover, his most original ideas were being developed independently by men of a new generation. Gauss did not respond when Abel sent him his proof of the impossibility of solving the quintic equation in 1825, and the two never met, although Gauss praised him in private letters. When Dirichlet wrote Gauss in May 1826, enclosing his first work on number theory and asking for guidance, Gauss did not reply until 13 September and then only with general encouragement and advice to find a job that left time for research. As indicated in a letter to Encke of 8 July, Gauss was much impressed by Dirichlet's "eminent talent," but he did not seem inclined to become mathematically involved with him. When Crelle in 1828 asked Gauss for a paper on elliptic functions, he replied that Jacobi had covered his work "with so much sagacity, penetration and elegance, that I believe that I am relieved of publishing my own research." Harassed, overworked, distracted, and frustrated during these years, Gauss undoubtedly underestimated the value of his achievements, something he had never done before. But he was correct in sensing the need of a new source of inspiration. In turning toward intensive investigations in physics, he was following a pattern that had proved richly productive in the past.

In 1828 Alexander von Humboldt persuaded Gauss to attend the only scientific convention of his career, the Naturforscherversammlung in Berlin. Since first hearing of Gauss from the leading mathematicians in Paris in 1802, Humboldt had been trying to bring him to Berlin as the leading figure of a great academy he hoped to build there. At times negotiations had seemed near success, but bureaucratic inflexibilities in Berlin or personal factors in Göttingen always intervened. Humboldt still had not abandoned these hopes, but he had other motives as well. He wished to draw Gauss into the German scientific upsurge whose beginnings were reflected in the meeting; and especially he wished to involve Gauss in his own efforts, already extending over two decades, to organize worldwide geomagnetic observations. Humboldt had no success in luring Gauss from his Göttingen hermitage. He was repelled by the Berlin convention, which included a "little celebration" to which Humboldt invited 600 guests. Nevertheless, the visit was a turning point. Living quietly for three weeks in Humboldt's house with a private garden and his host's scientific equipment, Gauss had both leisure and stimulation for making a choice. When Humboldt later wrote of his satisfaction at having interested him in magnetism, Gauss replied tactlessly that he had been interested in it for nearly thirty years. Correspondence and manuscripts show this to be true; they indicate that Gauss delayed serious work on the subject partly because means of measurement were not available. Nevertheless, the Berlin visit was the occasion for the decision and also provided the means for implementing it, since in Berlin Gauss met Wilhelm Weber, a young and brilliant experimental physicist whose collaboration was essential.

In September 1829 Quetelet visited Göttingen and found Gauss very interested in terrestrial magnetism but with little experience in measuring it. The new field had evidently been selected, but systematic work awaited Weber's arrival in 1831. Meanwhile, Gauss extended his long-standing knowledge of the physical literature and began to work on problems in theoretical physics, and especially in mechanics, capillarity, acoustics, optics, and crystallography. The first fruit of this research was *Über ein neues allgemeines Grundgesetz der Mechanik* (1829). In it Gauss stated the law of least constraint: the motion of a system departs as little as possible from free motion, where departure, or constraint, is measured by the sum of products of the masses times the squares of their deviations from the path of free motion. He presented it merely as a new formulation equivalent to the well-known principle of d'Alembert. This work seems obviously related to the old meditations on least squares, but Gauss wrote to Olbers on 31 January 1829 that it was inspired by studies of capillarity and other physical problems. In 1830 appeared *Principia generalia theoriae figurae fluidorum in statu aequilibrii,* his one contribution to capillarity and an important paper in the calculus of variations, since it was the first solution of a variational problem involving double integrals, boundary conditions, and variable limits.

The years 1830–1831 were the most trying of Gauss's life. His wife was very ill, having suffered since 1818 from gradually worsening tuberculosis and hysterical neurosis. Her older son left in a huff and immigrated to the United States after quarreling with his father over youthful profligacies. The country was in a revolutionary turmoil of which Gauss thoroughly disapproved. Amid all these vexations, Gauss continued work on biquadratic residues, arduous geodesic calculations, and many other tasks. On 13 September 1831 his wife died. Two days later Weber arrived.

As Gauss and Weber began their close collaboration and intimate friendship, the younger man was just half the age of the older. Gauss took a fatherly attitude. Though he shared fully in experimental work, and though Weber showed high theoretical competence and originality during the collaboration and later, the older man led on the theoretical and the younger on the experimental side. Their joint efforts soon produced results. In 1832 Gauss presented to the Academy the *Intensitas vis magneticae terrestris ad mensuram absolutam revocata* (1833), in which appeared the first systematic use of absolute units (distance, mass, time) to measure a nonmechanical quantity. Here Gauss typically acknowledged the help of Weber but did not include him as joint author. Stimulated by Faraday's discovery of induced current in 1831, the pair energetically investigated electrical phenomena. They arrived at Kirchhoff's laws in 1833 and anticipated various discoveries in static, thermal, and frictional electricity but did not publish, presumably because their interest centered on terrestrial magnetism.

The thought that a magnetometer might also serve as a galvanometer almost immediately suggested its use to induce a current that might send a message. Working alone, Weber connected the astronomical observatory and the physics laboratory with a mile-long double wire that broke "uncountable" times as he strung it over houses and two towers. Early in 1833 the first words were sent, then whole sentences. This first operating electric telegraph was mentioned briefly by Gauss in a notice in the *Göttingische gelehrte Anzeigen* (9 August 1834; *Werke,* V, 424–425), but it seems to have been unknown to other inventors. Gauss soon realized the military and economic importance of the invention and tried unsuccessfully to promote its use by government and industry on a large scale. Over the years, the wire was replaced twice by one of better quality, and various improvements were made in the terminals. In 1845 a bolt of lightning fragmented the wire, but by this time it was no longer in use. Other inventors (Steinheil in Munich in 1837, Morse in the United States in 1838) had independently developed more efficient and exploitable methods, and the Gauss–Weber priority was forgotten.

The new magnetic observatory, free of all metal that might affect magnetic forces, was part of a network that Humboldt hoped would make coordinated measurements of geographical and temporal variations. In 1834 there were already twenty-three magnetic observatories in Europe, and the comparison of data from them showed the existence of magnetic storms. Gauss and Weber organized the Magnetische Verein, which united a worldwide network of observatories. Its *Resultate aus den Beobachtungen des magnetischen Vereins* appeared in six volumes (1836–1841) and included fifteen papers by Gauss, twenty-three by Weber, and the joint *Atlas des Erdmagnetismus* (1840). These and other publications elsewhere dealt with problems of instrumentation (including one of several inventions of the bifilar magnetometer), reported observations of the horizontal and vertical components of magnetic force, and attempted to explain the observations in mathematical terms.

The most important publication in the last category was the *Allgemeine Theorie des Erdmagnetismus* (1839). Here Gauss broke the tradition of armchair theorizing about the earth as a fairly neutral carrier of one or more magnets and based his mathematics on data. Using ideas first considered by him in 1806, well formulated by 1822, but lacking empirical foundation until 1838, Gauss expressed the magnetic potential at any point on the earth's surface by an infinite series of spherical functions and used the data collected by the world network to evaluate the first twenty-four coefficients. This was a superb interpolation, but Gauss hoped later to explain the results by a physical theory about the magnetic composition of the earth. Felix Klein has pointed out that this can indeed be done (*Vorlesungen über die Entwicklung der Mathematik im 19. Jahrhundert* [Berlin, 1926], pt. 1, p. 22), but that little is thereby added to the effective explanation offered by the Gaussian formulas. During these years Gauss found time to continue his geodesic data reduction, assist in revising the weights and measures of Hannover, make a number of electric discoveries jointly with Weber, and take an increasing part in university affairs.

This happy and productive collaboration was suddenly upset in 1837 by a disaster that soon effectively terminated Gauss's experimental work. In September, at the celebration of the 100th anniversary of the university (at which Gauss presented Humboldt with plans for his bifilar magnetometer), it was rumored that the new King Ernst August of Hannover might abrogate the hard-won constitution of 1833 and demand that all public servants swear a personal oath of allegiance to himself. When he did so in November, seven Göttingen professors, including Weber and the orientalist G. H. A. von Ewald, the husband of Gauss's older daughter, Minna, sent a private protest to the cabinet, asserting that they were bound by their previous oath to the constitution of 1833. The "Göttingen Seven" were unceremoniously fired, three to be banished and the rest (including Weber and Ewald) permitted to remain in the town. Some thought that Gauss might resign, but he took no public action; and his private efforts, like the public protest of six additional professors, were ignored. Why did Gauss not act more energetically? At age sixty he was too set in his ways, his mother was too old to move, and he hated anything politically radical and disapproved of the protest. The seven eventually found jobs elsewhere. Ewald moved to Tübingen, and Gauss was deprived of the company of his most beloved daughter, who had been ill for some years and died of consumption in 1840. Weber was supported by colleagues for a time, then drifted away and accepted a job at Leipzig. The collaboration petered out, and Gauss abandoned further physical research. In 1848, when Weber recovered his position at Göttingen, it was too late to renew collaboration and Weber continued his brilliant career alone.

As Gauss was ending his physical research, he published *Allgemeine Lehrsätze in Beziehung auf die im verkehrten Verhältnisse des Quadrats der Entfernung wirkenden Anziehungs- und Abstossungskräfte* (1840). Growing directly out of his magnetic work but linked also to his *Theoria attractionis* of 1813, it was the first systematic treatment of potential theory as a mathematical topic, recognized the necessity of existence theorems in that field, and reached a standard of rigor that remained unsurpassed for more than a century, even though the main theorem of the paper was false, according to C. J. de la Vallée Poussin (see *Revue des questions scientifiques,* **133** [1962], 314–330, esp. 324). In the same year he finished *Dioptrische Untersuchungen* (1841), in which he analyzed the path of light through a system of lenses and showed, among other things, that any system is equivalent to a properly chosen single lens. Although Gauss said that he had possessed the theory forty years before and considered it too elementary to publish, it has been labeled his greatest work by one of his scientific biographers (Clemens Schäfer, in *Werke,* XI, pt. 2, sec. 2, 189 ff.). In any case, it was his last significant scientific contribution.

Later Years. From the early 1840's the intensity of Gauss's activity gradually decreased. Further publications were either variations on old themes, reviews, reports, or solutions of minor problems. His reclusion is illustrated by his lack of response in 1845 to Kummer's invention of ideals (to restore unique factorization) and in 1846 to the discovery of Neptune by Adams, Le Verrier, and Galle. But the end of magnetic research and the decreased rate of publication did not mean that Gauss was inactive. He continued astronomical observing. He served several times as dean of the Göttingen faculty. He was busy during the 1840's in finishing many old projects, such as the last calculations on the Hannover survey. In 1847 he eloquently praised number theory and G. Eisenstein in the preface to the collected works of this ill-fated young man who had been one of the few to tell Gauss anything he did not already know. He spent several years putting the university widows' fund on a sound actuarial basis, calculating the necessary tables. He learned to read and speak Russian fluently, apparently first attracted by Lobachevsky but soon extending his reading as widely as permitted by the limited material available. His notebooks and correspondence show that he continued to work on a variety of mathematical problems. Teaching became less distasteful, perhaps because his students were better prepared and included some, such as Dedekind and Riemann, who were worthy of his efforts.

During the Revolution of 1848 Gauss stood guard with the royalists (whose defeat permitted the return of his son-in-law and Weber). He joined the Literary Museum, an organization whose library provided conservative literature for students and faculty, and made a daily visit there. He carefully followed political, economic, and technological events as reported in the press. The fiftieth anniversary celebration of his doctorate in 1849 brought him many messages and formal honors, but the world of mathematics was represented only by Jacobi and Dirichlet. The paper that Gauss delivered was his fourth proof of the fundamental theorem of algebra, appropriately a variation of the first in his thesis of 1799. After this celebration, Gauss continued his interests at a slower pace and became more than ever a legendary figure unapproachable by those outside his personal circle. Perhaps stimulated by his actuarial work, he fell into the habit of collecting all sorts of statistics from the newspapers, books, and daily observations. Undoubtedly some of these data helped him with financial speculations shrewd enough to create an estate equal to nearly 200 times his annual salary. The "star gazer," as his father called him, had, as an afterthought, achieved the financial status denied his more "practical" relatives.

Due to his careful regimen, no serious illnesses had troubled Gauss since his surveying days. Over the years he treated himself for insomnia, stomach discomfort, congestion, bronchitis, painful corns, shortness of breath, heart flutter, and the usual signs of aging without suffering any acute attacks. He had been less successful in resisting chronic hypochondria and melancholia which increasingly plagued him after the death of his first wife. In the midst of some undated scientific notes from his later years there suddenly appears the sentence "Death would be preferable to such a life," and at fifty-six he wrote Gerling (8 February 1834) that he felt like a stranger in the world.

After 1850, troubled by developing heart disease, Gauss gradually limited his activity further. He made his last astronomical observation in 1851, at the age of seventy-four, and later the same year approved Riemann's doctoral thesis on the foundations of complex analysis. The following year he was still working on minor mathematical problems and on an improved Foucault pendulum. During 1853–1854 Riemann wrote his great *Habilitationsschrift* on the foundations of geometry, a topic chosen by Gauss. In June 1854 Gauss, who had been under a doctor's care for several months, had the pleasure of hearing Riemann's probationary lecture, symbolic of the presence in Germany at last of talents capable of continuing his work. A few days later he left Göttingen for the last time to observe construction of the railway from Kassel. By autumn his illness was much worse. Although gradually more bedridden, he kept up his reading, correspondence, and trading in securities until he died in his sleep late in February 1855.

Mathematical Scientist. Gauss the man of genius stands in the way of evaluating the role of Gauss as a scientist. His mathematical abilities and exploits caused his contemporaries to dub him *princeps,* and biographers customarily place him on a par with Archimedes and Newton. This traditional judgment is as reasonable as any outcome of the ranking game, but an assessment of his impact is more problematic because of the wide gap between the quality of his personal accomplishments and their effectiveness as contributions to the scientific enterprise. Gauss published only about half his recorded innovative ideas (see Figure 1) and in a style so austere that his readers were few. The unpublished results appear in notes, correspondence, and reports to official bodies, which became accessible only many years later. Still other methods and discoveries are only hinted at in letters or incomplete notes. It is therefore necessary to reexamine Gauss as a participant in the scientific community and to look at his achievements in terms of their scientific consequences.

The personality traits that most markedly inhibited the effectiveness of Gauss as a participant in scientific activity were his intellectual isolation, personal ambition, deep conservatism and nationalism, and rather narrow cultural outlook. It is hard to appreciate fully the isolation to which Gauss was condemned in childhood by thoughts that he could share with no one. He must soon have learned that attempts to communicate led, at best, to no response; at worst, to the ridicule and estrangement that children find so hard to bear. But unlike most precocious children, who eventually find intellectual comrades, Gauss during his whole life found no one with whom to share his most valued thoughts. Kästner was not interested when Gauss told him of his first great discovery, the constructibility of the regular 17-gon. Bolyai, his most promising friend at Göttingen, could not appreciate his thinking. These and many other experiences must have convinced Gauss that there was little to be gained from trying to interchange theoretical ideas. He drew on the great mathematicians of the past and on contemporaries in France (whom he treated as from another world); but he remained outside the mathematical activity of his day, almost as if he were actually no longer living and his publications were being discovered in the archives. He found it easier and more useful to communicate with empirical scientists and technicians, because in those areas he was among peers; but even there he remained a solitary worker, with the exception of the collaboration with Weber.

Those who admired Gauss most and knew him best found him cold and uncommunicative. After the Berlin visit, Humboldt wrote Schumacher (18 October 1828) that Gauss was "glacially cold" to unknowns and unconcerned with things outside his immediate circle. To Bessel, Humboldt wrote (12 October 1837) of Gauss's "intentional isolation," his habit of suddenly taking possession of a small area of work, considering all previous results as part of it, and refusing to consider anything else. C. G. J. Jacobi complained in a letter to his brother (21 September 1849) that in twenty years Gauss had not cited any publication by him or by Dirichlet. Schumacher, the closest of Gauss's friends and one who gave him much personal counsel and support, wrote to Bessel (21 December 1842) that Gauss was "a queer sort of fellow" with whom it is better to stay "in the limits of conventional politeness, without trying to do anything uncalled for."

Like Newton, Gauss had an intense dislike of controversy. There is no record of a traumatic experience that might account for this, but none is required to explain a desire to avoid emotional involvements that interfered with contemplation. With equal rationality, Gauss avoided all noncompulsory ceremonies and formalities, making an exception only when royalty was to be present. In these matters, as in his defensive attitude toward possible wasters of his time, Gauss was acting rationally to maximize his scientific output; but the result was to prevent some interchanges that might have been as beneficial to him as to others.

Insatiable drive, a characteristic of persistent high achievers, could hardly in itself inhibit participation; but conditioned by other motivations it did so for Gauss. Having experienced bitter poverty, he worked toward a security that was for a long time denied him. But he had absorbed the habitual frugality of the striving poor and did not want or ever adopt luxuries of the parvenu. He had no confidence in the democratic state and looked to the ruling aristocracy for security. The drive for financial security was accompanied by a stronger ambition, toward great achievement and lasting fame in science. While still an adolescent Gauss realized that he might join the tiny superaristocracy of science that seldom has more than one member in a generation. He wished to be worthy of his heroes and to deserve the esteem of future peers. His sons reported that he discouraged them from going into science on the ground that he did not want any second-rate work associated with his name. He had little hope of being understood by his contemporaries; it was sufficient to impress and to avoid offending them. In the light of his ambitions for security and lasting fame, with success in each seemingly required for the other, his choice of career and his purposeful isolation were rational. He did achieve his twin ambitions. More effective communication and participation might have speeded the development of mathematics by several decades, but it would not have added to Gauss's reputation then or now. Gauss probably understood this well enough. He demonstrated in some of his writings, correspondence, lectures, and organizational activities that he could be an effective teacher, expositor, popularizer, diplomat, and promoter when he wished. He simply did not wish.

Gauss's conservatism has been described above, but it should be added here that it extended to all his thinking. He looked nostalgically back to the eighteenth century with its enlightened monarchs supporting scientific aristocrats in academies where they were relieved of teaching. He was anxious to find "new truths" that did not disturb established ideas. Nationalism was important for Gauss. As we have seen, it impelled him toward geodesy and other work that he considered useful to the state. But its most important effect was to deny him easy communication with the French. Only in Paris, during his most productive years, were men with whom he could have enjoyed a mutually stimulating mathematical collaboration.

It seems strange to call culturally narrow a man

with a solid classical education, wide knowledge, and voracious reading habits. Yet outside of science Gauss did not rise above petit bourgeois banality. Sir Walter Scott was his favorite British author, but he did not care for Byron or Shakespeare. Among German writers he liked Jean Paul, the best-selling humorist of the day, but disliked Goethe and disapproved of Schiller. In music he preferred light songs and in drama, comedies. In short, his genius stopped short at the boundaries of science and technology, outside of which he had little more taste or insight than his neighbors.

The contrast between knowledge and impact is now understandable. Gauss arrived at the two most revolutionary mathematical ideas of the nineteenth century: non-Euclidean geometry and noncommutative algebra. The first he disliked and suppressed. The second appears as quaternion calculations in a notebook of about 1819 (*Werke,* VIII, 357-362) without having stimulated any further activity. Neither the barycentric calculus of his own student Moebius (1827), nor Grassmann's *Ausdenunglehre* (1844), nor Hamilton's work on quaternions (beginning in 1843) interested him, although they sparked a fundamental shift in mathematical thought. He seemed unaware of the outburst of analytic and synthetic projective geometry, in which C. von Staudt, one of his former students, was a leading participant. Apparently Gauss was as hostile or indifferent to radical ideas in mathematics as in politics.

Hostility to new ideas, however, does not explain Gauss's failure to communicate many significant mathematical results that he did approve. Felix Klein (*Vorlesungen über die Entwicklung der Mathematik im 19. Jahrhundert,* pt. 1, 11-12) points to a combination of factors—personal worries, distractions, lack of encouragement, and overproduction of ideas. The last might alone have been decisive. Ideas came so quickly that each one inhibited the development of the preceding. Still another factor was the advantage that Gauss gained from withholding information, although he hotly denied this motive when Bessel suggested it. In fact, the Ceres calculation that won Gauss fame was based on methods unknown to others. By delaying publication of least squares and by never publishing his calculating methods, he maintained an advantage that materially contributed to his reputation. The same applies to the careful and conscious removal from his writings of all trace of his heuristic methods. The failure to publish was certainly not based on disdain for priority. Gauss cared a great deal for priority and frequently asserted it publicly and privately with scrupulous honesty. But to him this meant being first to discover, not first to publish; and he was satisfied to establish his dates by private records, correspondence, cryptic remarks in publications, and in one case by publishing a cipher. (See bibliography under "Miscellaneous.") Whether he intended it so or not, in this way he maintained the advantage of secrecy without losing his priority in the eyes of later generations. The common claim that Gauss failed to publish because of his high standards is not convincing. He did have high standards, but he had no trouble achieving excellence once the mathematical results were in hand; and he did publish all that was ready for publication by normal standards.

In the light of the above discussion one might expect the Gaussian impact to be far smaller than his reputation—and indeed this is the case. His inventions, including several not listed here for lack of space, redound to his fame but were minor improvements of temporary importance or, like the telegraph, uninfluential anticipations. In theoretical astronomy he perfected classical methods in orbit calculation but otherwise did only fairly routine observations. His personal involvement in calculating orbits saved others trouble and served to increase his fame but were of little long-run scientific importance. His work in geodesy was influential only in its mathematical by-products. From his collaboration with Weber arose only two achievements of significant impact. The use of absolute units set a pattern that became standard, and the Magnetische Verein established a precedent for international scientific cooperation. His work in dioptrics may have been of the highest quality, but it seems to have had little influence; and the same may be said of his other works in physics.

When we come to mathematics proper, the picture is different. Isolated as Gauss was, seemingly hardly aware of the work of other mathematicians and not caring to communicate with them, nevertheless his influence was powerful. His prestige was such that young mathematicians especially studied him. Jacobi and Abel testified that their work on elliptic functions was triggered by a hint in the *Disquisitiones arithmeticae.* Galois, on the eve of his death, asked that his rough notes be sent to Gauss. Thus, in mathematics, in spite of delays, Gauss did reach and inspire mathematicians. Although he was more of a systematizer and solver of old problems than an opener of new paths, the very completeness of his results laid the basis for new departures—especially in number theory, differential geometry, and statistics. Although his mathematical thinking was always concrete in the sense that he was dealing with structures based on the real numbers, his work contained the seeds of many highly abstract ideas that came later. Gauss, like Archimedes, pushed the methods of his time to the limit of their possibilities. But unlike his other ability peer, Newton, he did not initiate a profound new development, nor did he have the revolutionary impact of a number of his contemporaries of perhaps lesser ability but greater imagination and daring.

Gauss is best described as a mathematical scientist, or, in the terms common in his day, as a pure and applied mathematician. Ranging easily, competently, and productively over the whole of science and technology, he always did so as a mathematician, motivated by mathematics, utilizing every experience for mathematical inspiration. (Figure 2 shows some of the interrelations of his interests.) Clemens Schäfer, one of his scientific biographers, wrote in *Nature* (**128** [1931], 341): "He was not really a physicist in the sense of searching for new phenomena, but rather always a mathematician who attempted to formulate in exact mathematical terms the experimental results obtained by others." Leaving aside his personal failures, whose scientific importance was transitory, Gauss appears as the ideal mathematician, displaying in heroic proportions in one person the capabilities attributed collectively to the community of professional mathematicians.

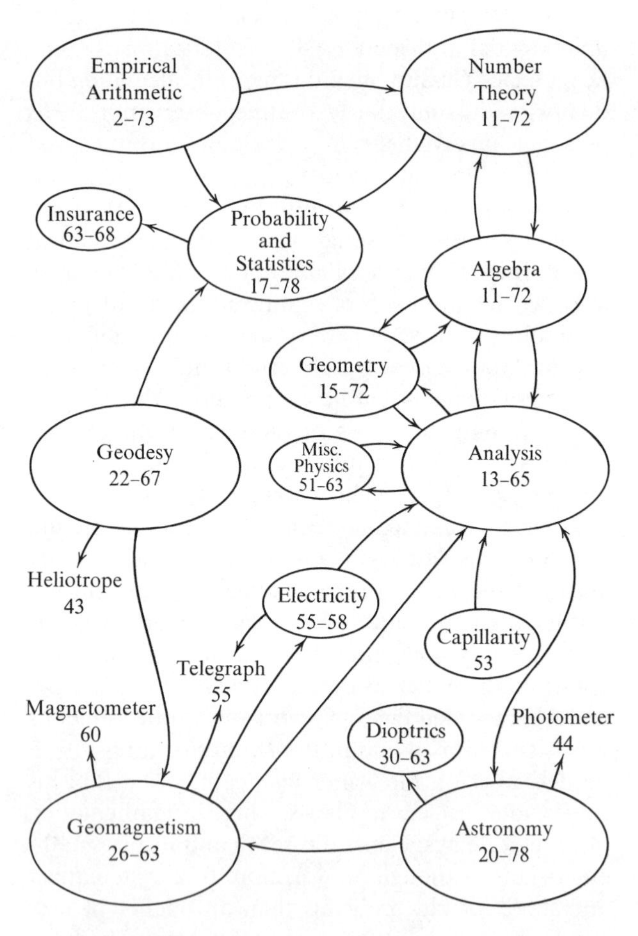

FIGURE 2. Main lines of development of Gauss's scientific ideas. Arrows suggest the most important directions of motivation and inspiration. Numerals indicate ages. His four most important inventions are given outside of any enclosing curves. The sizes of the ellipses suggest the weight of each field in his total effort, and the year span is indicative also of the number and variety of activities in each field. This figure should be compared with Figure 1.

BIBLIOGRAPHY

A complete Gauss bibliography would be far too large to include here, and the following is highly selective. Abbreviations used throughout are the following: *AMM: American Mathematical Monthly. AN: Astronomische Nachrichten. BA: Abhandlungen der (Königlichen) Bayerischen Akademie der Wissenschaften,* Mathematisch-naturwissenschaftliche Abteilung, II Klasse. *BAMS: Bulletin of the American Mathematical Society. BB: Bullettino (Bollettino) di bibliografia e di storia delle scienze matematiche (e fisiche)* (Boncompagni). *BSM: Bulletin des sciences mathématiques et astronomiques* (Darboux). *Crelle: Journal für die reine und angewandte Mathematik. DMV: Jahresbericht der Deutschen Mathematiker-vereinigung. FF: Forschungen und Fortschritte. GA: Abhandlungen der Akademie (K. Gesellschaft) der Wissenschaften zu Göttingen,* Mathematisch-naturwissenschaftliche Klasse. *GGM: Gauss-Gesellschaft Mitteilungen. GN: Nachrichten (Jahrbuch, Jahresbericht) der Gesellschaft der Wissenschaften zu Göttingen. HUB: Wissenschaftliche Zeitschrift der Humboldt-Universität Berlin,* Mathematisch-naturwissenschaftliche Reihe. *IINT: Trudy (Arkhiv) Instituta istorii nauki i tekhniki. IMI: Istoriko-matematicheskie issledovaniya. JMPA: Journal de mathématiques pures et appliquées* (Liouville). *LB: Berichte über die Verhandlungen der (Königlichen) Sächsischen Gesellschaft der Wissenschaften zu Leipzig. MA: Mathematische Annalen. MDA: Monatsberichte der Deutschen Akademie der Wissenschaften zu Berlin. NA: Nouvelles annales de mathématiques. NMM: National Mathematics Magazine.* OK: Ostwalds Klassiker der exacten Wissenschaften (Leipzig). *SM: Scripta mathematica. TSM: Scientific Memoirs, Selected from the Transactions of Foreign Academies and Learned Societies and From Foreign Journals* by Richard Taylor. *VIET: Voprosy istorii estestvoznaniya i tekhniki. Zach: Monatliche Correspondenz zur Beförderung der Erd- und Himmelskunde* (Zach). *ZV: Zeitschrift für Vermessungswesen.*

I. Original Works. All of Gauss's publications (including his fine reviews of his own papers) are reprinted in the *Werke,* published in 12 vols. by the Königliche Gesellschaft der Wissenschaften zu Göttingen (Leipzig–Berlin, 1863–1933). The *Werke* contains also a generous selection of his unpublished notes and papers, related correspondence, commentaries, and extensive analyses of his work in each field. The first 7 vols., edited by Ernst C. J. Schering, who came to Göttingen as a student in 1852 and taught mathematics there from 1858 until his death in 1897, contain Gauss's publications arranged by subject, as follows: I. *Disquisitiones arithmeticae* (1863; 2nd ed., with commentary, 1870). II. Number Theory (1863; 2nd ed., with the unpublished sec. 8 of the *Disquisitiones,* minor additions, and revisions, 1876). III. Analysis (1866; 2nd ed., with minor changes, 1876). IV. Probability, Geometry, and Geodesy (1873; 2nd ed., almost unchanged, 1880). V. Mathematical Physics (1867; unchanged 2nd ed., 1877). VI. Astronomy (1873). VII. *Theoria motus* (1871; 2nd ed., with new commentary by Martin Brendel and previously unpublished Gauss MSS, 1906).

After the death of Schering, work was continued under the aggressive leadership of Felix Klein, who organized a campaign to collect materials and enlisted experts in special fields to study them. From 1898 until 1922 he rallied support with fourteen reports, published under the title "Bericht über den Stand der Herausgabe von Gauss' Werken," in the *Nachrichten* of the Göttingen Academy and reprinted in *MA* and *BSM.* The fruits of this effort were a much enlarged Gauss Archive at Göttingen, many individual publications, and vols. VIII–XII of the *Werke,* as follows: VIII. Supp. to vols. I–IV (1900), papers and correspondence on mathematics (the paper on pp. 36–64 is spurious. See *Werke,* X, pt. 1, 137). IX. Geodesy (1903). Supp. to vol. IV, including some overlooked Gauss publications. X, pt. 1. Supp. on pure mathematics (1917), including the famous *Tagebuch* in which Gauss from 1796 to 1814 recorded mathematical results. Found in 1898 by P. Stäckel and first published by F. Klein in the *Festschrift zur Feier des hundertfünfzigjährigen Bestehens der Königlichen Gesellschaft der Wissenschaften zu Göttingen* (Berlin, 1901) and in *MA,* **57** (1903), 1–34, it was here reprinted with very extensive commentary and also in facsimile. A French trans. with commentary by P. Eymard and J. P. Lafon appeared in *Revue d'histoire des sciences et de leurs applications,* **9** (1956), 21–51. See also G. Herglotz, in *LB,* **73** (1921), 271–277. X, pt. 2. Biographical essays described below (1922–1933). XI, pt. 1. Supp. on Physics, Chronology, and Astronomy (1927). XI, pt. 2. Biographical essays described below (1924–1929). XII. Varia. *Atlas des Erdmagnetismus* (1929). A final volume, XIII, planned to contain further biographical material (especially on Gauss as professor), bibliography, and index, was nearly completed by H. Geppert and E. Bessel-Hagen but not published.

A. *Translations and Reprints.* The *Demonstratio nova* of 1799 together with the three subsequent proofs of the fundamental theorem (1815, 1816, 1849) were published in German with commentary by E. Netto under the title *Die vier Gauss'schen Beweise* . . . in OK, no. 14 (1890). The *Disquisitiones* (1801) is available in French (1807), German, with other works on number theory (1889; repr. New York, 1965), Russian (1959), and English (1966). Gauss's third published proof of the law of quadratic reciprocity (1808) is translated in D. E. Smith, *Source Book in Mathematics,* I (New York, 1929), 112–118. All his published proofs of this theorem are collected in *Sechs Beweise des Fundamentaltheorems über quadratische Reste,* E. Netto, ed., in OK, no. 122 (1901).

The *Theoria motus* (1809) was translated into English (1857), Russian (1861), French (1864), and German (1865). *Disquisitiones generales circa seriem* (1813) appeared in a German translation by H. Simon in 1888, and *Theoria attractionis* (1813) was translated in *Zach,* **28** (1813), 37–57, 125–234, and reprinted in OK, 19 (1890). The *Determinatio attractionis* (1818) was translated in OK, 225 (1927). The *Allgemeine Auflösung* (1825) was reprinted with related works of Lagrange in OK, 55 (1894). *Theoria combinationis* and supps. of 1823 appeared in French (by J. Bertrand, 1855), German (1887), and with other related work in *Abhandlungen zur Methode der kleinsten Quadrate,* translated by A. Börsch and P. Simon (Berlin, 1887), and in *Gauss's Work (1803–1826) on the Theory of Least Squares,* translated from French by H. F. Trotter (Princeton, N.J., 1957). The *Allgemeine Auflösung* of 1825 appeared in *Philosophical Magazine,* **4** (1828), 104–113, 206–215. *Disquisitiones generales circa superficies curvas* (1828) was translated into French in *NA,* **11** (1852), 195–252, and with notes by E. Roger (Grenoble, 1855); into German by O. Böklen in his *Analytische Geometrie des Raumes* (1884), and by Wangerin in OK, 5 (1889); into Russian (1895), Hungarian (1897); and English (1902). *Über ein neues allgemeines Grundgesetz* (1829) was translated in *NA,* **4** (1845), 477–479.

The *Intensitas vis magneticae* (1833) appears in the *Effemeridi astronomiche di Milano, 1839* (Milan, 1838); in OK, 53 (1894); and in W. F. Magie, *Source Book in Physics* (New York–London, 1935; repr., Cambridge, Mass., 1963), pp. 519–524. The *Allgemeine Theorie des Erdmagnetismus* of 1839 was promptly published in English in *TSM,* **2** (1841), 184–251, 313–316. The *Allgemeine Lehrsätze* (1840) was translated in *JMPA,* **7** (1842), 273–324, and reprinted in OK, 2 (1889). *Dioptrische Untersuchungen* (1841) appeared in English in *TSM,* **3** (1843), 490–498 (see also *Ferrari's Dioptric Instruments* [London, 1919]); and in French in *Annales de chimie,* **33** (1851), 259–294, and in *JMPA,* **1** (1856), 9–43. The *Untersuchungen über Gegenstände der höheren Geodäsie* (1844, 1847) was reprinted as OK, 177 (Leipzig, 1910).

Very little material from the *Nachlass* first printed in the *Werke* has been reprinted or translated. Parts of *Werke,* XI, pt. 1, on the arithmetic-geometric mean and modular functions appear in the OK, 255 (1927), translation of the *Determinatio attractionis* (1818). Some Gauss MSS and editor's commentary are translated from *Werke,* XII, by Dunnington in *Carl Friedrich Gauss, Inaugural Lecture on Astronomy and Papers on the Foundations of Mathematics* (Baton Rouge, La., 1937). Notes on Gauss's astronomy lectures by A. T. Kupffer are printed in A. N. Krylov, *Sobranie trudy* (Moscow–Leningrad, 1936), VI. The following selecta have appeared in Russian: *Geodezicheskie issledovania Gaussa . . .* (St. Petersburg, 1866); *Izbrannye trudy po zemnomu magnetizmu* (Leningrad, 1952); *Izbrannye geodezicheskie sochinenia* (Moscow, 1957).

B. *Correspondence.* Only the major collections are listed here. Many other letters have been published in journal articles and in bibliographies. G. F. J. A. von Auwers, *Briefwechsel zwischen Gauss und Bessel* (Leipzig, 1880). E. Schönberg and T. Gerardy, "Die Briefe des Herrn P. H.

L. von Bogulawski . . .," in *BA,* **110** (1963), 3-44. F. Schmidt and P. Stäckel, *Briefwechsel zwischen C. F. Gauss and W. Bolyai* (Leipzig, 1899). P. G. L. Dirichlet, *Werke,* II (Berlin, 1897), 373-387. C. Schäfer, *Briefwechsel zwischen Carl Friedrich Gauss und Christian Ludwig Gerling* (Berlin, 1927). T. Gerardy, *Christian Ludwig Gerling und Carl Friedrich Gauss. Sechzig bisher unveröffentlichte Briefe* (Göttingen, 1964). H. Stupuy, ed., *Oeuvres philosophiques de Sophie Germain* (Paris, 1879), pp. 298 ff.; and 2nd ed., pp. 254 ff. K. Bruhns, *Briefe zwischen A. v. Humboldt und Gauss* (Leipzig, 1877) (see also K.-R. Bierman, in *FF,* **36** [1962], 41-44, also in *GGM,* **4** [1967], 5-18). T. Gerardy, "Der Briefwechsel zwischen C. F. Gauss und C. L. Lecoq," in *GN* (1959), 37-63. W. Gresky, "Aus Bernard von Lindenaus Briefen an C. F. Gauss," in *GGM,* **5** (1968), 12-46. W. Valentiner, *Briefe von C. F. Gauss an B. Nicolai* (Karlsruhe, 1877). C. Schilling and I. Kramer, *Briefwechsel zwischen Olbers und Gauss,* 2 vols. (Berlin, 1900-1909). C. Pfaff, *Sammlung von Briefen, gewechselt zwischen Johann Friedrich Pfaff und . . . anderen* (Leipzig, 1853). P. Riebesell, "Briefwechsel zwischen C. F. Gauss und J. C. Repsold," in *Mitteilungen der mathematischen Gesellschaft in Hamburg,* **6** (1928), 398-431. C. A. Peters, *Briefwechsel zwischen C. F. Gauss und H. C. Schumacher,* 6 vols. (Altona, 1860-1865). T. Gerardy, *Nachtrage zum Briefwechsel zwischen Carl Friedrich Gauss und Heinrich Christian Schumacher* (Göttingen, 1969).

C. *Archives.* The MSS, letters, notebooks, and library of Gauss have been well preserved. The bulk of the scientific *Nachlass* is collected in the Gauss Archiv of the Handschriftenabteilung of the Niedersächsischen Staats- und Universitätsbibliothek, Göttingen, and fills 200 boxes. (See W. Meyer, *Die Handschriften in Göttingen* [Berlin, 1894], III, 101-113.) Theo Gerardy has for many years been working to arrange and catalog these materials. (See T. Gerardy, "Der Stand der Gaussforschung," in *GGM,* **1** [1964], 5-11.) Personal materials are concentrated in the municipal library of Brunswick. These include the contents of the Gauss Museum, removed from Gauss's birthplace before its destruction during World War II. (See H. Mack, "Das Gaussmuseum in Braunschweig," in *Museumskunde,* n.s. **1** [1930], 122-125.) Gauss's personal library forms a special collection in the Göttingen University Library. His scientific library was merged with the observatory library. There are also minor deposits of MSS, letters, and mementos scattered in the libraries of universities, observatories, and private collectors throughout the world. The best published sources on the Gauss archival material are Felix Klein's reports on the progress of the *Werke* mentioned above and in the yearly *Mitteilungen* of the Gauss-Gesellschaft (GGM), founded in Göttingen in 1962.

II. SECONDARY LITERATURE. There is no full-scale biography of the man and his work as a whole, although there are many personal biographies and excellent studies of his work in particular fields.

A. *Bibliography.* No complete Gauss bibliography has been published. The best ones are in Poggendorff, VII A, supp., Lieferung 2 (1970), 223-238; and in Dunnington's biography (see below).

B. *Biography.* The year after Gauss's death, Sartorius von Waltershausen, a close friend of his last years, published *Gauss zum Gedächtniss* (Leipzig, 1856). An English trans. by his great-granddaughter, Helen W. Gauss, was published as *Gauss, a Memorial* (Colorado Springs, Colo., 1966).

Other sources based on personal acquaintance and/or more or less reliable contemporary evidence are the following: L. Hänselmann, *K. F. Gauss. Zwölf Capital aus seinem Leben* (Leipzig, 1878); I. M. Simonov, *Zapiski i vospominaniya o puteshestvii po Anglii, Frantsii, Belgii i Germanii v 1842 godu* (Kazan, 1844); A. Quetelet, in *Correspondance mathématique et physique,* **6** (1830), 126-148, 161-178, 225-239, repr. in A. Quetelet, *Sciences mathématiques et physiques chez les Belges* (Brussels, 1866); Ernst C. J. Schering, *Carl Friedrich Gauss' Geburtstag nach hundertjähriger Wiederkehr, Festrede* (Göttingen, 1877); M. A. Stern, *Denkrede . . . zur Feier seines hundertjährigen Geburtstages* (Göttingen, 1877); F. A. T. Winnecke, *Gauss. Ein Umriss seines Lebens und Wirkens* (Brunswick, 1877); Theodor Wittstein, *Gedächtnissrede auf C. F. Gauss zur Feier des 30 April 1877* (Hannover, 1877); R. Dedekind, *Gauss in seiner Vorlesungen über die Methode der kleinsten Quadrate. Festschrift . . . Göttingen* (Berlin, 1901), repr. in Dedekind, *Gesammelte mathematische Werke,* II (1931), 293-306; Moritz Cantor lecture of 14 November 1899, in *Neue Heidelberger Jahrbucher,* **9** (1899), 234-255; and Rudolf Borch, "Ahnentafel des . . . Gauss," in *Ahnentafeln berühmter Deutscher,* I (Leipzig, 1929), 63-65.

Most of the personal biographical literature is derivative from the above sources and is of the "beatification forever" type, in which fact and tradition are freely mixed. Only a few works of special interest are mentioned here. Heinrich Mack, *Carl Friedrich Gauss und die Seinen* (Brunswick, 1927), contains substantial excerpts from family correspondence and a table of ancestors and descendants. F. Cajori published family letters in *Science,* n.s. **9** (19 May 1899), 697-704, and in *Popular Science Monthly,* **81** (1912), 105-114. Other studies based on documents are T. Gerardy, "C. F. Gauss und seine Söhne," in *GGM,* **3** (1966), 25-35; W. Lorey, in *Mathematisch-physikalische Semesterberichte* (Göttingen), **3** (1953), 179-192; and Hans Salié, in the collection edited by Reichardt described below. The most complete biography to date is G. W. Dunnington, *Carl Friedrich Gauss, Titan of Science* (New York, 1955), a useful derivative compendium of personal information and tradition, including translations from Sartorius, Hänselmann, and Mack, the largest bibliography yet published, and much useful data on genealogy, friends, students, honors, books borrowed at college, courses taught, etc.

During the Third Reich two rather feeble efforts—L. Bieberbach, *C. F. Gauss, ein deutsches Gelehrtenleben* (Berlin, 1938); and E. A. Roloff, *Carl Friedrich Gauss* (Osnabrück, 1942)—were made to claim Gauss as a hero, but it is clear that Gauss would have loathed the fascists as the final realization of his worst fears about bourgeois politics. Neither author mentions that Gauss's favorite mathematician, whom he praised extravagantly, was Gotthold Eisenstein.

Erich Worbs, *Carl Friedrich Gauss, Ein Lebensbild* (Leipzig, 1955), makes an effort to relate Gauss realistically to his times. W. L. Schaaf, *Carl Friedrich Gauss, Prince of Mathematicians* (New York, 1964), is a popularization addressed to juveniles.

C. *Scientific Work.* The literature analyzing Gauss's scientific work is expert and comprehensive, although its fragmentation by subject matter gives the impression of dealing with several different men. Beginning in 1911, F. Klein, M. Brendel, and L. Schlesinger edited a series of eight studies under the title *Materialien für eine wissenschaftliche Biographie von Gauss* (Leipzig, 1911–1920), most of which were later incorporated in the *Werke.* On the occasion of the hundredth anniversary of Gauss's death, there appeared *C. G. Gauss Gedenkband,* Hans Reichardt, ed. (Leipzig, 1957), republished as *C. F. Gauss, Leben und Werk* (Berlin, 1960); and I. M. Vinogradov, ed., *Karl Friedrich Gauss, 100 let so dnya smerti, sbornik statei* (Moscow, 1956). These collections will be abbreviated as Klein, Reichardt, and Vinogradov, respectively, when individual articles are listed below.

Brief anniversary evaluations by mathematicians are the following: R. Courant and R. W. Pohl, *Carl Friedrich Gauss, Zwei Vorträge* (Göttingen, 1955)—Courant's lecture also appeared in *Carl Friedrich Gauss . . . Gedenkfeier der Akademie der Wissenschaften . . . Göttingen anlässlich seines 100ten Todestages* (Göttingen, 1955) and was translated in T. L. Saaty and J. F. Weyl, eds., *The Spirit and the Uses of the Mathematical Sciences* (New York, 1969), pp. 141–155; J. Dieudonné, *L'oeuvre mathématique de C. F. Gauss* (Paris, 1962), a talk at the Palais de la Découverte, 2 December 1961; R. Oblath, "Megemlékezés halálának 100-ik évfordulóján," in *Matematikai lapok,* **6** (1955), 221–240; and K. A. Rybnikov, in *VIET,* **1** (1956), 44–53.

The following selected titles are arranged by topic.

Algebra. A. Fraenkel, "Zahlbegriff und Algebra bei Gauss," (Klein, VIII), in *GN,* supp. (1920); "Der Zusammenhang zwischen dem ersten und dem dritten Gauss'schen Beweis des Fundamentalsatzes der Algebra," in *DMV,* **31** (1922), 234–238; A. Ostrowski, "Über den ersten und vierten Gauss'schen Beweis des Fundamentalsatzes der Algebra," in *Werke,* X, pt. 2, sec. 3 (1933), 3–18 (an enlarged revision of Klein, VIII [1920], 50–58); R. Kochendörfer, in Reichardt, pp. 80–91; and M. Bocher, "Gauss's Third Proof of the Fundamental Theorem of Algebra," in *BAMS,* **1** (1895), 205–209.

Analysis. A. I. Markushevich, "Raboty Gaussa po matematicheskomu analizu," in Vinogradov, pp. 145–216, German trans. in Reichardt, pp. 151–182; K. Schröder, "C. F. Gauss und die reelle Analysis," in Reichardt, pp. 184–191; O. Bolza, "Gauss und die Variationsrechnung," in *Werke,* X, pt. 2, sec. 5 (1922), 3–93; L. Schlesinger, "Fragment zur Theorie des arithmetisch-geometrischen Mittels" (Klein, II), in *GN* (1912), 513–543; *Über Gauss' Arbeiten zur Funktionentheorie* (Berlin, 1933), also in *Werke,* X, pt. 2, sec. 2 (1933), 3–210—an enlarged revision of Klein II which appeared in *GN* (1912), 1–140; H. Geppert, "Wie Gauss zur elliptischen Modul-funktion kam," in *Deutsche Mathematik,* **5** (1940), 158–175; E. Göllnitz, "Über die Gauss'sche Darstellung der Funktionen sinlemn *x* und coslemn *x* als Quotienten unendlicher Produkte," in *Deutsche Mathematik,* **2** (1937), 417–420; P. Gunther, "Die Untersuchungen von Gauss in der Theorie der elliptischen Funktionen," in *GN* (1894), 92–105, and in trans. in *JMPA,* 5th ser., **3** (1897), 95–111; H. Hattendorff, *Die elliptischen Funktionen in dem Nachlasse von Gauss* (Berlin, 1869); A. Pringsheim, "Kritisch-historische Bemerkungen zur Funktionentheorie," in *BA* (1932), 193–200; (1933), 61–70; L. Schlesinger, "Über die Gauss'sche Theorie des arithmetisch-geometrischen Mittels . . .," in *Sitzungsberichte der Preussischen Akademie der Wissenschaften zu Berlin,* **28** (1898), 346–360; and "Über Gauss Jugendarbeiten zum arithmetisch-geometrischen Mittel," in *DMV,* **20** (1911), 396–403.

Astronomy. M. Brendel, "Über die astronomischen Arbeiten von Gauss," in *Werke,* XI, pt. 2, sec. 3 (1929), 3–254, enlarged revision of Klein, vol. VII, pt. 1 (Leipzig, 1919); M. F. Subbotin, "Astronomicheskie i geodesicheskie raboty Gaussa," in Vinogradov, pp. 241–310; and O. Volk, "Astronomie und Geodäsie bei C. F. Gauss," in Reichardt, pp. 206–229.

Geodesy and Surveying. A. Galle, "Über die geodätischen Arbeiten von Gauss," in *Werke,* XI, pt. 2, sec. 1 (1924), 3–161; W. Gronwald *et al., C. F. Gauss und die Landesvermessung in Niedersachsen* (Hannover, 1955); T. Gerardy, *Die Gauss'sche Triangulation des Königreichs Hannover (1821 bis 1844) und die Preussischen Grundsteuermessungen (1868 bis 1873)* (Hannover, 1952); G. V. Bagratuni, *K. F. Gauss, kratky ocherk geodezicheskikh issledovanii* (Moscow, 1955); M. F. Subbotin, in Vinogradov (see under Astronomy); W. Gäde, "Beiträge zur Kenntniss von Gauss' praktisch-geodätischen Arbeiten," in *ZV,* **14** (1885), 53–113; T. Gerardy, "Episoden aus der Gauss'schen Triangulation des Königreichs Hannover," in *ZV,* **80** (1955), 54–62; H. Michling, *Erläuterungsbericht zur Neuberechnung der Gauss-Kruegerischen Koordinaten der Dreiecks- und Polygonpunkte der Katasterurmessung* (Hannover, 1947); "Der Gauss'sche Vizeheliotrop," in *GGM,* **4** (1967), 27–30; K. Nickul, "Über die Herleitung der Abbildungsgleichung der Gauss'schen Konformen Abbildung des Erdellipsoids in der Ebene," in *ZV,* **55** (1926), 493–496; and O. Volk, in Reichardt (see under Astronomy).

Geomagnetism. Ernst Schering, "Carl Friedrich Gauss und die Erforschung des Erdmagnetismus," in *GA,* **34** (1887), 1–79; T. N. Roze and I. M. Simonov, in *K. F. Gauss, Izbrannye trudy po zemnomu magnitizmu* (Leningrad, 1952), pp. 237–336; H.-G. Körber, "Alexander von Humboldts und Carl Friedrich Gauss' organisatorisches Wirken auf geomagnetischen Gebiet," in *FF,* **32** (1958), 1–8; and K.-R. Biermann, "Aus der Vorgeschichte der Aufforderung A. v. Humboldts an der Präsidenten der Royal Society . . .," in *HUB,* **12** (1963), 209–227.

Geometry. P. Stäckel, "C. F. Gauss als Geometer," in *Werke,* X, pt. 2, sec. 4 (1923), 3–121, repr. with note by L. Schlesinger from Klein, V (1917), which appeared also in *GN,* **4** (1917), 25–140; A. P. Norden, "Geometricheskie raboty Gaussa," in Vinogradov, pp. 113–144; R. C. Archibald, "Gauss and the Regular Polygon of Seventeen Sides," in *AMM,* **27** (1920), 323–326; H. Carslaw, "Gauss and Non-Euclidean Geometry," in *Nature,* **84,** no. 2134 (1910), 362; G. B. Halsted, "Gauss and non-Euclidean Geometry," in *AMM,* **7** (1900), 247, and on the same subject, in *AMM,* **11** (1904), 85–86, and in *Science,* **9,** no. 232 (1904), 813–817; and E. Hoppe, "C. F. Gauss und der Euklidische Raum," in *Naturwissenschaften,* **13** (1925), 743–744, and in trans. by Dunnington in *Scripta mathematica,* **20** (1954), 108–109 (Hoppe objects to the story that Gauss measured a large geodesic triangle in order to test whether Euclidean geometry was the "true" one, apparently under the impression that this would have been contrary to Gauss's ideas. Actually, Gauss considered geometry to have an empirical base and to be testable by experience.); V. F. Kagan, "Stroenie neevklidovoi geometrii u Lobachevskogo, Gaussa i Boliai," in *Trudy Instituta istorii estestvoznaniya,* **2** (1948), 323–389, repr. in his *Lobachevskii i ego geometriya* (Moscow, 1955), pp. 193–294; N. D. Kazarinoff, "On Who First Proved the Impossibility of Constructing Certain Regular Polygons . . .," in *AMM,* **75** (1968), 647; P. Mansion, "Über eine Stelle bei Gauss, welche sich auf nichteuklidische Metrik bezieht," in *DMV,* **7** (1899), 156; A. P. Norden, "Gauss i Lobachevskii," in *IMI,* **9** (1956), 145–168; A. V. Pogorelov, "Raboty K. F. Gaussa po geometrii poverkhnostei," in *VIET,* **1** (1956), 61–63; and P. Stäckel and F. Engel, *Die Theorie der Parallelinien* (Leipzig, 1895); "Gauss, die beiden Bolyai und die nichteuklidische Geometrie," in *MA,* **49** (1897), 149–206, translated in *BSM,* 2nd ser., **21** (1897), 206–228.

Miscellaneous. K.-R. Biermann, "Einige Episoden aus den russischen Sprachstudien des Mathematikers C. F. Gauss," in *FF,* **38** (1964), 44–46; E. Göllnitz, "Einige Rechenfehler in Gauss' Werken," in *DMV,* **46** (1936), 19–21; and S. C. Van Veen, "Een conflict tusschen Gauss en een Hollandsch mathematicus," in *Wiskunstig Tijdschrift,* **15** (1918), 140–146. The following four papers deal with the ciphers in which Gauss recorded some discoveries: K.-R. Biermann, in *MDA,* **5** (1963), 241–244; **11** (1969), 526–530; T. L. MacDonald, in *AN,* **241** (1931), 31; P. Männchen, in *Unterrichtsblätter für Mathematik und Naturwissenschaften,* **40** (1934), 104–106; and A. Wietzke, in *AN,* **240** (1930), 403–406.

Number Theory. P. Bachmann, "Über Gauss' Zahlentheoretische Arbeiten" (Klein, I), in *GN* (1911), pp. 455–508, and in *Werke,* X, pt. 2, sec. 1 (1922), 3–69; B. N. Delone, "Raboty Gaussa po teorii chisel," in Vinogradov, pp. 11–112; G. J. Rieger, "Die Zahlentheorie bei C. F. Gauss," in Reichardt, pp. 37–77; E. T. Bell, "The Class Number Relations Implicit in the *Disquisitiones arithmeticae,*" in *BAMS,* **30** (1924), 236–238; "Certain Class Number Relations Implied in the *Nachlass* of Gauss," *ibid.,* **34** (1928), 490–494; "Gauss and the Early Development of Algebraic Numbers," in *NMM,* **18** (1944), 188–204, 219–223; L. E. Dickson, *History of the Theory of Numbers,* 3 vols. (Washington, D.C., 1919)—the indexes are a fairly complete guide to Gauss's extraordinary achievements in this field; J. Ginsburg, "Gauss' Arithmetization of the Problem of 8 Queens," in *SM,* **5** (1938), 63–66; F. Van der Blij, "Sommen van Gauss," in *Euclides* (Groningen), **30** (1954), 293–298; and B. A. Venkov, "Trudy K. F. Gaussa po teorii chisel i algebra," in *VIET,* **1** (1956), 54–60. The following papers concern an erroneous story, apparently started by W. W. R. Ball, that the Paris mathematicians rejected the *Disquisitiones arithmeticae:* R. C. Archibald, "Gauss's *Disquisitiones arithmeticae* and the French Academy of Sciences," in *SM,* **3** (1935), 193–196; H. Geppert and R. C. Archibald, "Gauss's *Disquisitiones Arithmeticae* and the French Academy of Sciences," *ibid.,* 285–286; G. W. Dunnington, "Gauss, His Disquisitiones Arithmeticae and His Contemporaries in the Institut de France," in *NMM,* **9** (1935), 187–192; A. Emch, "Gauss and the French Academy of Science," in *AMM,* **42** (1935), 382–383. See also G. Heglotz, "Zur letzten Eintragung im Gauss'schen Tagebuch," in *LB,* **73** (1921), 271–277.

Numerical Calculations. P. Männchen, "Die Wechselwirkung zwischen Zahlenrechnung und Zahlentheorie bei C. F. Gauss" (Klein, VI), in *GN,* supp. **7** (1918), 1–47, and in *Werke,* X, pt. 2, sec. 6 (1930), 3–75; and A. Galle, "C. F. Gauss als Zahlenrechner" (Klein, IV), in *GN,* supp. **4** (1917), 1–24.

Philosophy. A. Galle, "Gauss und Kant," in *Weltall,* **24** (1925), 194–200, 230, repr. in *GGM,* **6** (1969), 8–15; P. Mansion, "Gauss contre Kant sur la géométrie non-Euclidienne," in *Mathesis,* 3rd ser., **8,** supp. (Dec. 1908), 1–16, in *Revue néoscolastique,* **15** (1908), 441–453, and in *Proceedings of the Third (1908) International Congress of Philosophy in Heidelberg* (Leipzig, 1910), pp. 438–447; and H. E. Timerding, "Kant und Gauss," in *Kant-Studien,* **28** (1923), 16–40.

Physics. H. Falkenhagen, "Die wesentlichsten Beiträge von C. F. Gauss aus der Physik," in Reichardt, pp. 232–251; H. Geppert, "Über Gauss' Arbeiten zur Mechanik und Potentialtheorie," in *Werke,* X, pt. 2, sec. 7 (1933), 3–60; and C. Schäfer, "Gauss physikalische Arbeiten (Magnetismus, Elektrodynamik, Optik)," in *Werke,* XI, pt. 2, sec. 2 (1929), 2–211; "Gauss's Investigations on Electrodynamics," in *Nature,* **128** (1931), 339–341.

Probability and Statistics (Including Least Squares). B. V. Gnedenko, "O raboty Gaussa po teorii veroyatnostei," in Vinogradov, pp. 217–240; A. Galle, "Über die geodätischen Arbeiten von Gauss," in *Werke,* XI, pt. 2, sec. 6 (1924), 3–161; C. Eisenhart, "Gauss," in *International Encyclopedia of the Social Sciences,* VI (New York, 1968), 74–81; P. Männchen, "Über ein Interpolationsverfahren des jugendlichen Gauss," in *DMV,* **28** (1919), 80–84; H. L. Seal, "The Historical Development of the Gauss Linear Model," in *Biometrika,* **54** (1967), 1–24; T. Sofonea, "Gauss und die Versicherung," in *Verzekerings-Archive,* **32** (Aktuar Bijv, 1955), 57–69; and Helen M. Walker, *Studies in the History of Statistical Method* (Baltimore, 1931).

Telegraph. Ernst Feyerabend, *Der Telegraph von Gauss und Weber im Werden der elektrischen Telegraphie* (Berlin, 1933); and R. W. Pohl, "Jahrhundertfeier des elektromagnetischen Telegraphen von Gauss und Weber," in *GN* (1934), pp. 48–56, repr. in *Carl Friedrich Gauss, Zwei Vorträge* (Göttingen, 1955), pp. 5–12.

The author gratefully acknowledges many helpful suggestions and comments from Kurt-R. Biermann. Thanks are due also to the library staff at the University of Toronto for many services. The author claims undivided credit only for errors of fact and judgment.

KENNETH O. MAY

GAUTIER, ARMAND E.-J. (*b.* Narbonne, France, 23 September 1837; *d.* Cannes, France, 27 July 1920), *chemistry.*

Gautier, the son of Louis Gautier, a physician and landowner, studied chemistry at Montpellier under J. E. Bérard (a former assistant to Berthollet) and J. A. Béchamp. In contrast with his teachers, he favored the use of atomic representations, rather than equivalents, in chemical notation. He especially supported the ideas of Charles Gerhardt, who had taught at Montpellier until 1851. After receiving a medical degree in 1862, Gautier left Montpellier for the Paris laboratory of Adolphe Wurtz. His isolation in 1866 of the isonitriles (isomers of the nitriles), or carbylamines, as he called them, was an important contribution to the new chemical theories.

Gautier prepared his carbylamines in a double decomposition reaction between silver cyanide and a simple or compound ether. Despite the penetrating odor of the carbylamines, their formation had escaped earlier workers on cyanohydric ethers, since cyanogen compounds were believed to be completely analogous to halogen compounds; the production of isomers in a simple reaction of the cyanogen radical was unimaginable. Influenced, however, by Wurtz's researches on the "ammonia type," Gautier demonstrated that the isonitriles existed and were indeed amines, whereas ordinary nitriles were more like salts. He noted that the cyanogen carbon atom in the carbylamine permits polymerization into explosives as well as a direct union with sulfur or oxygen. Gautier's explanation illuminated analogous relations between the cyanic ethers and their isomers, and between the cyanates and fulminates. The use of silver cyanide with methyl or ethyl iodide to form the carbylamine suggested to Victor Meyer the reaction of these iodides with silver nitrate to produce nitrated aliphatics (nitromethane, nitroethane, etc.).

Gautier incorporated his researches on the carbylamines into his doctoral thesis ("Des nitriles des acides gras") in 1869 and so impressed Henri Sainte-Claire Deville that the young chemist was quickly appointed to the chemical laboratory of the École Pratique des Hautes-Études. In 1874 Gautier became director of the new laboratory of biological medicine at the Faculté de Médecine, and in 1884 he succeeded Wurtz in the chair of medical chemistry, a post which he held until 1912.

Gautier's researches in these years were prodigious. In 1873 he noticed the release of small quantities of volatile alkaloids during the bacterial fermentation of albuminous material. He demonstrated in 1882 that such alkaloids are constant products of the normal life of animal tissues and are eliminated from the healthy body in urine and saliva. He developed methods for the quantitative analysis of trace amounts of arsenic and demonstrated that such traces exist in healthy animals, especially in the skin. He further established the therapeutic value of arsenic compounds. Gautier analyzed iodine and free hydrogen in the air, iodine and fluorine in organic substances, the coloring matter of grapes, the composition of mineral waters, and chemical reactions related to volcanic phenomena.

He published numerous textbooks, the most significant of which was the three-volume *Cours de chimie minérale, organique et biologique* (1887–1892).

BIBLIOGRAPHY

I. ORIGINAL WORKS. In addition to the *Cours de chimie* (2nd ed., 1895–1897; Vol. II, 3rd ed., 1906), Gautier's texts include *Chimie appliquée à la physiologie, à la pathologie, et à l'hygiène,* 2 vols. (Paris, 1874); *La chimie de la cellule vivante,* 2 vols. (Paris, 1894–1898); and *L'alimentation et les régimes chez l'homme sain et chez les malades* (Paris, 1904; 3rd ed., 1908). An almost complete bibliography of Gautier's scientific and popular writings may be found in Lebon (see below).

II. SECONDARY LITERATURE. Ernest Lebon's biography, *Armand Gautier. Biographie, bibliographie analytique des écrits* (Paris, 1912), is the best single source of information about Gautier. An obituary notice in *Nature* (16 Sept. 1920), 85–86, is reprinted in *Journal of the Chemical Society* (London), **119,** pt. 1 (1921), 537–539. An *éloge* by Henri Deslandres, read at the Académie des Sciences (2 Aug. 1920), is reprinted in *Revue scientifique,* **58** (1920), 471–472.

MARY JO NYE

GAUTIER, PAUL FERDINAND (*b.* Paris, France, 12 October 1842; *d.* Paris, 7 December 1909), *astronomical instrumentation.*

Born into a family of modest means, Gautier was obliged to begin working at the age of thirteen. From the age of eighteen—when he was employed by M. L. F. Secrétan—until his death, he was occupied with construction of astronomical instruments.

Gautier's instruments were closely linked to the strides made by late nineteenth-century astronomy: many of the major refracting telescopes, astrographs, and transit instruments that he made from 1876 on are still used by the observatories, both in France and elsewhere, that commissioned them. His instruments performed perfectly; to the execution of precision screws, graduated circles, and telescopic mounts Gautier brought the unsurpassed competence that earned him recognition and membership in the Bureau des Longitudes in 1897.

Gautier supplied most of the double astrographs used in the international undertaking that produced the *Carte du ciel.* The mount for the first astrograph was built at his own expense, and the instrument's performance led to its adoption in 1887 as a prototype.

Among Gautier's accomplishments were all the equatorial telescopes of the *coudé* type. The classic *Atlas de la lune* of Loewy and Puiseux, compiled from 1896 to 1910, employed the instrument that he had built for the Paris observatory; the plates in this work compare favorably with the finest modern ones.

The career of this honest and unselfish man ended in undeserved failure. For the Paris Universal Exposition of 1900 Gautier constructed the largest refractor ever built. The lens, forty-nine inches in diameter, was mounted at the end of a horizontal tube more than 195 feet long and was joined to a large siderostat. The device operated at a huge financial loss and ruined Gautier. This instrument—which, with a few adjustments, might have played an important role in scientific research—was ultimately dismantled, and its components sold.

Gautier's most important work includes equatorial visual telescopes, double astrographs, reflectors, and *coudé* equatorial telescopes at leading observatories in France, Austria, Greece, the Netherlands, Vatican City, Spain, Algeria, Argentina, and Brazil.

BIBLIOGRAPHY

I. ORIGINAL WORKS. Gautier published two technical memoirs in *Comptes rendus hebdomadaires des séances de l'Académie des sciences:* "Sur un procédé de construction des vis de haute précision . . .," in **112** (1891), 991–992; and "Construction d'un miroir plan de 2 mètres par des procédés mécaniques," in **128** (1899), 1373–1375. Gautier left to others the task of describing his instruments but drafted "Note sur le sidérostat à lunette de 60 mètres de foyer," in *Annuaire publié par le Bureau des longitudes* (1899), C1–C26.

II. SECONDARY LITERATURE. On Gautier and his work, see H. Poincaré and B. Baillaud, "Funérailles de M. Paul Gautier," in *Annuaire publié par le Bureau des longitudes* (1911), D1–D11; and L. Vandevyver, "La grande lunette de 1900," in *Ciel et terre,* **19** (1899), 257–267.

JACQUES R. LÉVY

GAY, FREDERICK PARKER (*b.* Boston, Massachusetts, 22 July 1874; *d.* New Hartford, Connecticut, 14 July 1939), *bacteriology, pathology.*

Born into a prominent Boston family, Gay followed the tradition of attending the Boston Latin School and Harvard (B.A. 1897), where he developed a lifelong interest in art, music, and classical literature. At Johns Hopkins Medical School (1897–1901) the exceptional student attracted the attention of Simon Flexner, who made Gay his assistant and invited him to join the world tour of the Johns Hopkins Medical Commission for the study of bubonic plague and other diseases (1899). After one year in the Philippines investigating cholera and dysentery, Gay returned home via Paris, where he studied at the Pasteur Institute, was fascinated by the young and brilliant Jules Bordet, and became acquainted with the infant sciences of microbiology and immunology. After returning to America he was awarded the first fellowship of the recently established Rockefeller Institute (1901), serving from 1901 to 1903 as assistant demonstrator in pathology at the University of Pennsylvania, then headed by his mentor, Simon Flexner.

But pure pathology was not destined to retain Gay's exclusive attention. In 1903 he rejoined Bordet, now established in his own Pasteur Institute in Brussels, and for the next three years was deeply engrossed in the emerging problems of anaphylaxis, complement-fixation, and other aspects of immunology. In 1904 Gay married Catherine Mills Jones; they had three children. Returning to America in 1906, he served for one year as bacteriologist at the Danvers, Massachusetts, Insane Asylum, and from 1907 to 1909 was assistant and then instructor in pathology at Harvard Medical School. In 1909 Gay completed and published the first English translation of the classic *Studies in Immunology* by Bordet and his associates, an accomplishment that immediately brought him into national prominence.

In 1910 he accepted the position of professor of pathology at the University of California at Berkeley, a post he was to retain for thirteen years, with only a brief interruption for service in the army. At Berkeley he finally persuaded the authorities to establish a separate department of bacteriology, and it was as the department's first director that he spent his final two years in California. During World War I and later Gay served as a member of the medical section of the National Research Council, and in 1922 he was its chairman. He also served as chairman of the Council's Medical Fellowship Board from 1922 to 1926. In the latter year he traveled from one Belgian university to another as exchange professor.

In 1923 Gay had accepted his last academic position, as professor of bacteriology at the Columbia University College of Physicians and Surgeons. His monograph *Typhoid Fever* (New York, 1918) was already well-known, but at Columbia he produced his most famous work, *Agents of Disease and Host Resistance* (Springfield, 1935), the best exposition of the problems of bacteriology and immunology of the period. His humanistic interests also asserted themselves at this time; his last book, *The Open Mind* (Chicago, 1938), dedicated to the memory of his lifelong friend, the psychiatrist Elmer E. Southard, reveals the depth of his concern with current problems of psychology and sociology.

Gay's honors were many: Belgium accorded him the Order of the Crown for his work with the American Commission on Relief; George Washington University granted him an Sc.D. degree in 1932; and he was elected a member of the National Academy of Sciences a few months before his death. Membership in other learned societies included the Association of American Physicians, the American Association of Pathologists and Bacteriologists, the American Society for Experimental Pathology, the Society for Experimental Biology and Medicine, the Association of American Bacteriologists, and the American Association of Immunologists.

Any evaluation of Gay's wide and varied contributions to science and to society must deal with the unusual dichotomy of his interests. The influence of Bordet is evident in his studies on serum reactions (1905–1910) and on anaphylaxis (1905–1913). This led to a lasting concern with tissue immunity and especially with the roles played by the reticuloendothelial system and the clasmatocyte (histiocyte). As a bacteriologist he made substantial contributions to our knowledge of the carrier state in typhoid (1913–1919); hemolytic streptococcic infections (1919–1939); and viral diseases, especially the herpetic and encephalitic (1929–1939). Other specific entities with which he concerned himself included pneumonia, meningitis, influenza, and poliomyelitis. Always he explored the possibility of inducing antibody formation by the use of antigens, stressing the practical application of such reactions to the diagnostic problems of infectious disease. Among the multitude of related subjects on which Gay wrote are cowpox, tobacco mosaic, bacteriophage, protozoa, spirochetes, rickettsia, dental caries, Vincent's angina, bacterial mutation, chemotherapy, lysozyme, and the importance of hormones and vitamins in resistance to infection.

In all these areas his position is assured; yet he deserves special consideration as that relatively infrequent and unusual combination of scientist, social philosopher, and humanist. Gay was also a man of great sincerity and integrity. Possessed of a touch of compassion for the plight of humanity, he forged strong bonds of affection which made him a seminal influence in the lives of those who knew him well.

BIBLIOGRAPHY

On Gay or his work see J. M. Cattell and Jacques Cattell, eds., *American Men of Science* (New York, 1938), p. 508; A. R. Dochez, "Frederick Parker Gay 1874–1939," in *Biographical Memoirs. National Academy of Sciences,* **38** (1954), 99–116, with portrait and complete bibliography; and Claus W. Jungeblut, "Frederick Parker Gay," in *Science,* **20** (1939), 290–291.

MORRIS H. SAFFRON

GAY-LUSSAC, JOSEPH LOUIS (*b.* St. Léonard, France, 6 December 1778; *d.* Paris, France, 9 May 1850), *chemistry, physics.*

He was the eldest of five children of Antoine Gay, lawyer and *procureur royal* at St. Léonard, and Leonarde Bourigner. His father, to distinguish himself from others with the surname Gay in the Limoges region, had begun to call himself Gay-Lussac after the family property near St. Léonard. Joseph Louis, although baptized "Gay," adopted the same practice.

The comfortable social and economic position of the family was rudely disturbed by the Revolution. In September 1793, when Gay-Lussac was fourteen, his father was arrested as a suspect. The Abbé Bourdeix, who had been giving the son private lessons, fled the country. Joseph Louis was sent to a small private boarding school in Paris, where his lessons included mathematics and science. The opening of the École Polytechnique provided a splendid opportunity for an able boy without fortune. Gay-Lussac was successful in the competitive entrance examination and was admitted on 27 December 1797. Graduating on 22 November 1800, he followed the practice of many of the better students by entering the civil engineering school, the École Nationale des Ponts et Chaussées. In the winter of 1800–1801, the chemist Berthollet, impressed by the ability of the young man, took him to his country house at Arcueil as an assistant. Having already had an excellent mathematical education, Gay-Lussac received training in chemical research from Berthollet, who also played a key role in the professional advancement of his protégé. In 1808 Gay-Lussac married Geneviève Marie Josèphe Rojot; they had five children.

Gay-Lussac was successively *adjoint* (from 31 December 1802) and *répétiteur* (from 23 September 1804) at the École Polytechnique. On 31 March 1809 he was given the honorary title of professor of practical chemistry, but upon the death of Fourcroy he was appointed to succeed him as professor of chemistry (17 February 1810). On the creation of the Paris Faculty of Science in 1808, Gay-Lussac was appointed professor of physics; in 1832 he gave up this chair in favor of that of general chemistry at the Muséum National d'Histoire Naturelle. On 8 December 1806 Gay-Lussac obtained the coveted place of member of the first class of the Institute (physics section). He was already a member of the Société d'Arcueil and the Société Philomatique.

Nearly all of Gay-Lussac's life was devoted to pure and applied science, but he did have a brief political career. He was elected to the Chamber of Deputies in 1831, 1834, and 1837 but resigned on a matter of principle in 1838. On 7 March 1839, having earlier refused a title from Charles X, he was honored by Louis Philippe with nomination to the upper house.

Gay-Lussac's first major research was on the thermal expansion of gases.[1] It was carried out with the encouragement of Berthollet and Laplace in the winter of 1801–1802. There was conflicting evidence about the expansive properties of different gases when heated. Gay-Lussac improved on most earlier work by taking precautions to exclude water vapor from his apparatus and to use dry gases. After examining a variety of gases, including several soluble in water, and repeating each experiment several times, he concluded that equal volumes of all gases expand equally with the same increase of temperature. Over the range of temperature from 0°C. to 100°C. the expansion of gases was 1/266.66 of the volume at 0°C. for each degree rise in temperature. Similar research was carried out independently by Dalton at about the same time. Dalton's work, however, was considerably less accurate. About 1787 J. A. C. Charles had recognized the equal expansion of several gases but had never bothered to publish his findings. Although the quantitative law of thermal expansion is often called "Charles's law," Charles did not measure the coefficient of expansion; moreover, for soluble gases, he had found unequal expansion.

Gay-Lussac made an ascent in a hydrogen balloon with Biot on 24 August 1804. The primary objective of the ascent was to see whether the magnetic intensity at the earth's surface decreased with an increase in altitude.[2] They concluded that it was constant up to 4,000 meters. They also carried long wires to test the electricity of different parts of the atmosphere. Another objective was to collect a sample of air from a high altitude to compare its composition with that of air at ground level. Gay-Lussac made a second ascent, on 16 September 1804, but this time by himself, in order to lessen the weight of the balloon and thus reach a greater height.

He was able to repeat observations of pressure, temperature, and humidity and also make magnetic measurements. He had taken two evacuated flasks, which he opened to collect samples of air when he had attained an altitude of over 6,000 meters. His subsequent analysis of these samples showed that the proportion of oxygen was identical with that in ordinary air. Gay-Lussac reached a calculated height of 7,016 meters above sea level, a record not equaled for another half century.

One of Gay-Lussac's early collaborators was Alexander von Humboldt. Nearly ten years older than Gay-Lussac, Humboldt already had an international reputation as an explorer; yet he learned something about precision in scientific research from Gay-Lussac, who in turn had his horizons broadened by his German friend. They collaborated in an examination of various methods of estimating the proportion of oxygen in the air, particularly the use of Volta's eudiometer.[3] In this method the gas being tested (which was required to contain some oxygen) was sparked with hydrogen to form water vapor, which condensed.

The resulting contraction permitted an estimate of the proportion of oxygen in the sample. This method obviously presupposed a knowledge of the relative proportions by volume in which hydrogen and oxygen combine to form water; one of the principal objects of the work of Gay-Lussac and Humboldt was to determine the proportion with the greatest possible accuracy. They also determined the limiting proportions for an explosion to be possible. After carrying out a large number of experiments with an excess of first one gas and then the other, they calculated—making allowance for a slight impurity in the test oxygen—that 100 parts by volume of oxygen combined with 199.89 parts of hydrogen or, they said, in round numbers, 200 parts. Gay-Lussac clearly expressed his preference for volumes, pointing out that the presence of moisture, which would be difficult to estimate gravimetrically, did not alter the volumetric ratio. This memoir made a useful contribution to science not only for its accuracy but as a precursor of Gay-Lussac's famous research on the combining volumes of gases.

In March 1805 Gay-Lussac embarked on a year of European travel with Humboldt, going first to Rome and ending in Berlin. During this journey Gay-Lussac carried out various chemical analyses. Their principal object, however, was to record the magnetic elements at different points along their route.[4]

To obtain the magnetic intensity, the period of oscillation of a magnetized needle was determined. The magnetic intensity was then found to be proportional to the square of the number of oscillations made by the needle, displaced slightly from the magnetic meridian, in a given time. They did not think that magnetic intensity in any one place changed with time, since on taking readings at Milan on entering and leaving Italy at an interval of six months, they found no difference. A series of prolonged experiments to determine diurnal variation, both on Mount Cenis and in Rome, had not revealed any difference at different hours of the day and night. As regards the general accuracy of their readings, many of which were made under conditions that were far from ideal, they estimated that the greatest discrepancy between their angular readings could not have been more than ten minutes of arc. Their general conclusion was that the horizontal component of the earth's magnetic intensity increased from north (Berlin) to south (Naples) but that the total intensity decreased on approaching the equator.

In 1807 Gay-Lussac carried out a series of experiments designed principally to see whether there was a general relationship between the specific heats of gases and their densities.[5] He measured the change in the temperature of a gas (and thus heat capacity) as a function of density changes produced by the free expansion of the gas. From a modern viewpoint the importance of his work was his establishment of a basic principle of physics, since it follows from his experiments that (in modern terms) the internal energy of an ideal gas depends on the temperature only. He took two twelve-liter, double-neck flasks. To one neck a tap was fitted and to the other a sensitive alcohol thermometer. Each flask contained anhydrous calcium chloride to absorb all moisture. One of the flasks was then evacuated and the other filled with the gas under test. The flasks were then connected with a lead pipe, the taps opened, and the readings of the thermometers carefully noted. It was known that compression of gases was accompanied by evolution of heat and expansion by absorption of heat. Gay-Lussac, however, wished to find the relationship between heat absorbed and heat evolved in the two flasks, and from his experiments he drew the valuable conclusion that these were equal within the limits of experimental error. The change of temperature was, moreover, directly proportional to the change of pressure. This he found by connecting the flasks, equalizing the pressures by opening the tap (that is, reducing the pressure to half, since the volumes were equal), evacuating the second flask, and repeating this process until the temperature change was so slight as to make accurate measurement impossible.

Gay-Lussac's experiment with two connecting vessels was repeated nearly forty years later by Joule, who apparently knew nothing of the earlier work.

Probably Gay-Lussac's greatest single achievement is based on the law of combining volumes of gases, which he announced at a meeting of the Société Philomatique in Paris, on 31 December 1808. For Gay-Lussac himself, the law provided a vindication of his belief in regularities in the physical world, which it was the business of the scientist to discover. Gay-Lussac began his memoir by pointing out the unique character of the gaseous state.[6] For solids and liquids a particular increase in pressure would produce a change different in each case; it was only matter in the gaseous state that increased equally in volume for a given increase of pressure. His own statement was that "gases combine in very simple proportions . . . and . . . the apparent contraction in volume which they experience on combination has also a simple relation to the volume of the gases, or at least to one of them." He gives the following examples of the simple ratios of combining volumes of gases (modern symbols are used for brevity).

$$O:H = 1:2$$
$$HCl:NH_3 = 1:1$$
$$BF_3:NH_3 = 1:1 \text{ and } 1:2$$
$$CO_2:NH_3 = 1:1 \text{ and } 1:2$$
$$CO:O_2 = 2:1$$

Combinations

$$NH_3:N:H = 1:3$$
$$SO_3:SO_2:O_2 = 2:1$$
$$N_2O:N:O = 2:1$$
$$NO:N:O = 1:1$$
$$NO_2:N:O = 1:2$$

Analyses

These neat ratios do not, however, correspond exactly to his experimental results. He deduced his law from a few fairly clear cases (particularly the first few listed above) and glossed over discrepancies in some of the others. The simple reaction between hydrogen and chlorine, which is often used today as an elementary illustration of the law, was not discovered until 1809 and was included only as a footnote when this memoir was printed.

Gay-Lussac presented his law of combining volumes of gases as a natural consequence of his collaboration with Humboldt, with whom he had found that 100 parts by volume of oxygen combine with almost exactly 200 parts of hydrogen. That his work of January 1805 with Humboldt led naturally to the law of combining volumes may be logically true but historically the connection is less direct. One has to explain the interval of nearly four years between obtaining the first data and the announcement of the law. Probably something had happened earlier in the year 1808 that made Gay-Lussac turn his attention back to his earlier work and realize that the value he had obtained for the combining volumes of hydrogen and oxygen was more than a coincidence and was in fact only one example of a general phenomenon. In the autumn of 1808 Gay-Lussac and Thenard had discovered boron trifluoride. They were particularly impressed by one of the properties of this new gas, the dense fumes produced when it came into contact with the air; they compared these fumes with the fumes produced by the reaction of muriatic-acid gas and ammonia. It seems likely that Gay-Lussac, struck by the reaction of boron trifluoride with moist air, tried its reaction with other gases including ammonia. An obvious reaction for comparison would be that between hydrochloric-acid gas and ammonia. This reaction was given special prominence in the memoir on combining volumes of gases.

One of the points of strength of the memoir was that it took data from a wide variety of reputable sources. This was no suspect generalization based on the biased experimental work of its author. On the other hand, all the data had not, of course, been conveniently assembled for Gay-Lussac to publish. In many cases the analyses that had appeared in the chemical literature had given only the gravimetric composition, and Gay-Lussac, taking reliable data for the density, had to convert this to a volumetric ratio. A close examination of the provenance of the density data shows that much was derived from his associates in the Société d'Arcueil.

The influence of Berthollet is particularly prominent in Gay-Lussac's attempt to reconcile the opinions of Dalton, Thomson, and Wollaston on definite and multiple proportions with Berthollet's known conviction that compounds can always be formed in variable proportions except in special circumstances. It was possible to argue that the gaseous state provided such an exception, and Berthollet accepted Gay-Lussac's law. Considering the implications of the law for the atomic structure of matter, it would be reasonable to expect Dalton to have welcomed the law of combining volumes as additional evidence for his atomic theory. Dalton, however, refused to accept the accuracy of the results of the French chemist. The Italian physical chemist Avogadro, on the other hand, not only accepted Gay-Lussac's work but developed its implications for the relationship between the volumes of gases and the number of molecules they contain. His great debt to Gay-Lussac's memoir was explicit.

Although Berzelius and Gay-Lussac differed in the actual values given to "volume weights" (often by a factor of 2), Berzelius, especially in his earlier work, regarded Gay-Lussac's method of speaking of volumes as preferable to Dalton's atoms. It had the advantage of being based on more direct evidence, and there was no absurdity in dealing with half volumes, whereas there was a contradiction involved in speaking of "half atoms." The obvious disadvantage of translating "atom" as "volume" was that many compounds cannot exist in the gaseous state. Gay-Lussac himself was prepared to estimate the relative vapor density of mercury not by direct means but by calculation based on the weight combining with a given weight of oxygen in the solid state. Gay-Lussac later speculated about the proportion of "carbon vapor" in carbon compounds. His interest in volumes led him to devise an apparatus by which the vapor densities of liquids could be compared.[7] The vapor displaced mercury in a graduated glass tube immersed in a glass cylinder containing water which was heated. The apparatus was improved half a century later by A. W. Hofmann, and the method by which the volume of a given weight of a vaporized substance is found is now usually known as Hofmann's method.

The work of Volta inspired many chemists to investigate the chemical effects of the voltaic pile. Gay-Lussac and Thenard were among this number. They were influenced particularly by the news in the winter of 1807–1808 of Davy's isolation of potassium and sodium by the use of the giant voltaic pile at the Royal Institution. Napoleon ordered the construction of an even larger pile at the École Polytechnique and Gay-Lussac and Thenard were placed in charge of it. Their research, reported in part 1 of their *Recherches physico-chimiques* (1811), was basically a repetition of Davy's experiments. Although Davy seems to have exhausted the most obvious possibilities, Gay-Lussac and Thenard's report does contain the suggestion that the rate of decomposition of an electrolyte depends only on the strength of the current (and not, for example, on the size of the electrodes), and they used chemical decomposition as a measure of electric current thirty years before Faraday. The Institute's prize of 3,000 francs for work in the field of galvanism was awarded to Davy in December 1807 and to Gay-Lussac and Thenard in December 1809.

Gay-Lussac and Thenard's really important contribution stemming from Davy's work was their preparation (announced to the Institute on 7 March 1808) of potassium and sodium in reasonable quantities and by purely chemical means.[8] Davy's method of electrolysis, although spectacular, had produced only tiny amounts of the new metals. The two young Frenchmen, no doubt under the influence of Berthollet, had reasoned that the action of great heat should change the usual affinities. Thinking that the normal affinities of oxygen for iron and the alkali metals could thus be reversed, they fused the respective alkalies with iron filings subjected to a bright red heat in a bent iron gun barrel. The metal vapor distilled over into a receiver luted to the gun barrel. In this way they prepared samples of about twenty-five grams of each metal at a low cost.

They were then able to investigate the physical constants of potassium, finding its specific gravity to be 0.874 (modern value, 0.859 at 0°C.). Davy had been unable to produce a better result than 0.6. Gay-Lussac and Thenard also discovered the alloy of potassium and sodium that exists as a liquid at room temperature. They then began a program of research in which potassium was not the end product but a reagent used to make further discoveries. In particular, they investigated the reaction between potassium metal and various gases. They found that when potassium is strongly heated in hydrogen, it combines with it to form a gray solid, potassium hydride, which is decomposed by water. They proposed the use of heated potassium as a means of performing an accurate volumetric analysis of nitrous and nitric oxides. The data obtained in this way by Gay-Lussac about the composition of nitric oxide was used by him later as evidence for his law of combining volumes of gases. They found that heated potassium metal decomposed muriatic-acid gas, forming the muriate of potash and hydrogen. Unfortunately they were prevented from reaching the conclusion that the gas was a simple compound of hydrogen and the muriatic radical by the conviction that the reaction was really due to water vapor in the gas.

In a further memoir, Gay-Lussac and Thenard described an experiment in which potassium was heated in dry ammonia, forming a solid (KNH_2) and liberating hydrogen. Other related experiments seemed to indicate to them that potassium was not an element at all but a hydride, and they argued this at length with Davy. Despite their mistaken conclusions on this point, the French chemists deserve credit for their discovery of a new class of compounds, the amides of metals.

In their next memoir Gay-Lussac and Thenard made further use of potassium as a reagent, this time to decompose boric acid.[9] They were not, however, alone in this field, since in the early summer of 1808 Davy turned his attention to their method of using potassium as a reagent. In a memoir read to the Royal Society on 30 June 1808, Davy described in a footnote how he had ignited boric acid and heated the product with potassium in a gold tube; this process yielded a black substance, which he did not identify but which was later recognized to be boron. It was not until his fourth Bakerian lecture, read on 15 December 1808, that Davy made any claim to the discovery of a new substance. His experimental work was rather poor and hurried. He admitted that he had had a report that Thenard was investigating the decomposition of boric acid by potassium. In December 1808 Davy succeeded in decomposing boric acid. He doubted, however, whether the substance obtained was a "simple body."

Gay-Lussac and Thenard's discovery of boron was first announced in November 1808. On 20 June 1808 they had mentioned an olive-gray substance obtained by the action of potassium on fused boric acid, but it was not until 14 November that they claimed to have isolated a new element and discovered its properties. This is one case where the work of Gay-Lussac and Thenard is indubitably prior to that of Davy. Equal weights of potassium metal and fused boric acid were heated together in a copper tube, thus producing a mixture of potassium, potassium borate, and boron. The new element did not dissolve or react chemically with water; this property thus provided a method of separating it. Gay-Lussac and Thenard gave it the name *bore* ("boron") and noted the similarity of its properties to those of carbon, phosphorus, and sulfur. Boron was found to form borides similar to carbides.

Their success in decomposing boric acid and isolating its "radical" led Gay-Lussac and Thenard to apply their new reagent, potassium, to the isolation of other radicals. Although the natural limitations of their method prevented them from achieving their immediate objective, they made a number of interesting discoveries.[10] In the course of these experiments they tried to prepare pure "fluoric acid" by heating together a mixture of calcium fluoride and vitrified boric acid. Investigating the properties of their fluoric acid, they found it had no effect on glass (a well-known property of hydrofluoric acid), and they reasoned correctly that this must be because it was already combined with an element similar to the basis of silica, namely, the boron from the acid used in its preparation. They therefore named the gas fluoboric gas (boron trifluoride).

They were now ready to prepare true hydrofluoric

acid and attempt its decomposition. This difficult feat was not accomplished until 1886, but Gay-Lussac and Thenard managed to prepare nearly anhydrous acid by distilling calcium fluoride with concentrated sulfuric acid in a lead retort.

After an unsuccessful attempt to isolate the muriatic radical by the action of heated potassium on muriates (chlorides) of metals, they turned their attention to oxymuriatic acid (chlorine), hoping to decompose it by removing its supposed oxygen.[11] Potassium was useless, but they were even more surprised when they found that even strongly heated carbon would not decompose the gas. This was all the more unexpected since sunlight decomposed it so easily. This led them to carry out further experiments on the effect of light on chemical reactions. They prepared two mixtures of chlorine and hydrogen; one was placed in darkness and the other in feeble sunlight. The first mixture was still greenish-yellow in color after several days, but the second had reacted completely by the end of a quarter of an hour, judging by the disappearance of the color of the chlorine. The experiment was repeated with olefiant gas (ethylene) and oxymuriatic-acid gas, which were mixed and left for two days in total darkness. As soon as the mixture was exposed to bright sunlight, there was a violent explosion. This confirmed their hypothesis that the speed of the reaction was proportional to the intensity of the light.

Apart from their early contribution to photochemistry, Gay-Lussac and Thenard made a fundamental contribution to the realization that so-called oxymuriatic acid contained no oxygen and was an element. Their memoir, read at a meeting of the Institute on 27 February 1809, contains the following remark, the wording of which should be carefully noted:

> Oxygenated muriatic acid is not decomposed by charcoal, and it might be supposed from this fact and those which are communicated in this memoir, that this gas is a simple body. The phenomena which it presents can be explained well enough on this hypothesis; we shall not seek to defend it, however, as it appears to us that they are still better explained by regarding oxygenated muriatic acid as a compound body.

The explanation of this statement is provided by events at a meeting of the Société d'Arcueil. On the previous day, 26 February, Gay-Lussac and Thenard had read a first draft of their memoir. In the first reading the authors had suggested unequivocally that oxymuriatic gas was an element. Their patron, Berthollet, unfortunately persuaded them to alter their remarks to make this no more than a possibility—as in the above quotation. Because of the pressure he exerted on Gay-Lussac and Thenard, Davy is usually credited with the discovery of the elementary nature of chlorine, which he announced in 1810. He had been particularly impressed by the evidence of Gay-Lussac and Thenard that charcoal, even at white heat, could not affect the decomposition of oxymuriatic gas, a result which one would hardly expect in a gaseous oxide.

Another area in which the contributions of Gay-Lussac were eclipsed by Davy—at least, outside France—was in the understanding of the properties of iodine. We must avoid describing this as the "discovery" of iodine, since it was neither Gay-Lussac nor Davy but Courtois who was the first to study this substance. The details of the story would take considerable space to elaborate; here it is sufficient to do no more than describe the contributions of Gay-Lussac.

Courtois recognized iodine to be a distinctive substance from its purple vapor, but its compound with hydrogen was at first confused with hydrogen chloride. It was Gay-Lussac who gave it the name *iode* (from the Greek *ioeidēs,* "violet colored"). On 12 December 1813 an article appeared in *Le Moniteur* in which Gay-Lussac expressed his view that iodine was probably an element. But he allowed also for the possibility of its being a compound containing oxygen. He stressed the analogy of the properties of iodine with those of chlorine. He was able to prepare the related acid (HI) by the action of iodine on moist phosphorus. He had prepared successively potassium iodate and iodic acid by 20 December 1813; the former was also prepared independently by Davy. A large part of Davy's claim for the originality of his study of iodine depends on his complete honesty in claiming certain knowledge before that of Gay-Lussac and in particular in dating as 11 December a paper read to the Institute on 13 December (that is, the day following Gay-Lussac's publication).

Gay-Lussac's major publication on iodine was not ready to be read to the Institute until 1 August 1814,[12] by which time not only Davy but Vauquelin had explored the subject fairly extensively. Gay-Lussac, however, deserves full credit for his detailed study of hydrogen iodide, which he found to have a 50 percent hydrogen content by volume. He contrasted its thermal decomposition with the stability of hydrogen chloride. By the action of chlorine on iodine, he prepared, independently of Davy and at about the same time, iodine monochloride and trichloride. After further careful study of the properties of iodine, he prepared and examined a number of iodides and iodates. He prepared for the first time ethyl iodide by distilling together concentrated hydriodic acid with absolute alcohol.

The close analogy that he emphasized between chlorine and iodine led him to a further investigation of the former, and he discovered chloric acid by the action of sulfuric acid on a solution of barium chlorate. Later, in collaboration with Welter, he discovered dithionic acid ($H_2S_2O_6$).[13] He also showed that aqua regia contains chlorine (which attacks gold) and nitrosyl chloride (NOCl) and that both gases are evolved on heating.[14]

In 1809 Gay-Lussac established by purely empirical means the general principle that the weight of acid in salts is proportional to the oxygen in the corresponding oxide.[15] (Salts were then considered to be compounds of metallic oxides with "acids," that is, acid anhydrides.) Gay-Lussac used this principle (occasionally referred to as a law) to determine the composition of some soluble salts. The analysis of insoluble salts was comparatively straightforward, but there was little agreement about the composition of the majority of salts, those which were soluble and could not therefore be weighed as precipitates. For example, insoluble lead sulfate had been found by analysis to consist of lead, 100.00; oxygen, 7.29; and acid, 37.71. Knowing the proportions of oxygen in the corresponding oxides of lead and copper, soluble copper sulfate must therefore contain oxygen and acid in the same proportions as lead sulfate: copper, 100.00; oxygen, 24.57; acid, 127.09. This was in fairly close agreement with the value obtained experimentally by L. J. Proust.

An extension of Gay-Lussac's principle could be used to determine the composition of sulfites indirectly. Direct determination presented practical difficulties, because sulfites are easily oxidized by the atmosphere to sulfates. The principle could be used in reverse if the composition of the salt were known and if the weight of oxygen that would combine with a given weight of the metal were required. This was applicable to the newly isolated barium.

Among his other work in inorganic chemistry, Gay-Lussac investigated the thermal decomposition of sulfates.[16] Under suitable oxidizing conditions, he was able to convert the sulfides of zinc and iron to the sulfates. His thermal decomposition of sulfuric acid in a porcelain tube showed the volumetric composition of sulfur trioxide to be 100 parts sulfur dioxide and 47.79 parts oxygen. In 1808 Gay-Lussac was able to use this data to help establish his law of combining volumes of gases.

Gay-Lussac carried out research on sulfides. The most important part of this work for the subsequent history of qualitative analysis was his investigation of the precipitation of metal sulfides. It was generally considered that such metals as zinc, manganese, cobalt, and nickel could not be precipitated by passing hydrogen sulfide through solutions of their respective salts. Gay-Lussac successfully demonstrated that the sulfides of these metals could be precipitated if they were present as salts of acetic, tartaric, or oxalic acids (that is, weak acids) or, better, in the presence of an alkali, such as ammonia.[17]

In 1809 Gay-Lussac carried out a study of the combining volumes of nitric oxide and oxygen.[18] This was a more complex problem than he then realized, but he returned to it in 1816 after criticism of his earlier work by Dalton; this time his results were of permanent value.[19] He recognized five oxides of nitrogen, which he listed as follows:

	Vols. of Nitrogen	Vols. of Oxygen	Modern Formula
Oxide d'azote	100	50	N_2O
Gaz nitreux	100	100	NO
Acide pernitreux	100	150	N_2O_3
Acide nitreux	100	200	NO_2
Acide nitrique	100	250	N_2O_5

Lavoisier's oxygen theory of acids had been questioned by Berthollet, who had failed to find oxygen either in hydrogen sulfide or in prussic acid. Although both gave acid reactions, Berthollet was not fully satisfied that Lavoisier's theory was erroneous; it was therefore left to his pupil Gay-Lussac to demonstrate conclusively that there was a definite class of acids that, instead of containing oxygen, contained hydrogen. Gay-Lussac introduced the term "hydracid" to denote this class, which included hydrochloric acid, hydriodic acid, and hydrogen sulfide.[20] Gay-Lussac thus introduced the name hydrochloric acid (*acide hydrochlorique*) for what had been called muriatic acid. Yet, so firmly accepted was Lavoisier's idea of oxygen as an acidifying principle that Gay-Lussac was convinced that hydrogen had no connection with the acidic properties and was in fact a principle of alkalinity.[21] If, for example, a solution of hydrogen sulfide showed acidic properties, it would have to be attributed to the sulfur that it contained.

Gay-Lussac's important research on prussic acid began with his successful preparation in 1811 of the anhydrous acid by the action of hydrochloric acid on mercuric cyanide.[22] In 1815 he determined the physical constants of the acid, including its vapor density.[23] He expressed its composition as follows: one volume of carbon vapor, one-half volume hydrogen, one-half volume nitrogen. He also expressed this in gravimetric terms, his figures being reasonably accurate. When he heated mercuric cyanide, he found that it decomposed into mercury and an inflammable gas composed of carbon and nitrogen. He proposed for this gas the name *cyanogène* (from the Greek *kyanos*, "dark blue"). He carefully examined the new compound, establishing its composition and showing that it combined with alkalies to form salts (cyanides). He investigated the properties and composition of cyanogen chloride, previously discovered by Berthollet. He examined several compounds derived from cyanogen, including ferrocyanate, which he clearly recognized as a compound radical and which he would have written as $Fe(CN)_6$ with modern atomic weights.[24] He later analyzed Prussian blue and made suggestions about its composition.[25]

Gay-Lussac demonstrated the growth of crystals of ammonia alum over those of potash alum and suggested that "the molecules of the two alums have the same form."[26] This has sometimes been incorrectly interpreted as an anticipation of Mitscherlich's law of isomorphism.

Gay-Lussac's study of the solubility of salts is of considerable importance, since he was the first to construct a solubility curve showing the variation of solubility of various salts in water at different temperatures.[27] He recognized that the amount of solid has no influence on the ultimate solubility. He understood that the solubility of a salt in water at a given temperature is a constant in the presence of excess solute. He noticed the break in the solubility curve of hydrated sodium sulfate and that this occurs at the point of maximum solubility.

Gay-Lussac studied the effect of the material and form of different vessels on the constancy of boiling points of liquids.[28] He found that the vapor pressure of a solution is lower than that of the pure solvent, for example, a solution of sodium chloride with a specific gravity of 1.096 was 0.9 that of the vapor pressure of water.

Gay-Lussac, in considering the action of chlorine on alkalies, stated, "There is a general rule that in every case where the same elements can form compounds of different stability (but capable of existing simultaneously under the same given conditions), the first to be formed is the least stable."[29] This is a remarkably perceptive statement and is almost the "law of successive reactions" proposed by Ostwald in 1897.

As might be expected from a pupil of Berthollet, Gay-Lussac made several contributions to an understanding of chemical equilibrium and the realization of the relevance to a reaction of the mass of the reactants. He showed, for example, that the action of steam on heated iron is a reversible action.[30] Following Berthollet's ideas, Gay-Lussac considered that when sulfuric acid is added to a solution of borax, "the base is partitioned between the two acids . . . in proportion to the numbers of their atoms."[31] Toward the end of his career, Gay-Lussac wrote a long historical article on affinity. He considered the particles in a solution of different salts to be in state of random motion (*pêle-mêle*) at the moment of mixture. Eventually a situation of equilibrium was obtained, but any slight alteration of conditions could bring about a further exchange of the acid and basic parts of the salts in solution. He introduced the concept of permutation of the constituent part of salts (*équipollence*):

> At the moment of mixture of two neutral salts, two new salts are formed in certain ratios with the two original salts; and then according to whether one of the properties of insolubility, density, fusibility, volatility, etc. is greater for the new salts than for the original salts, there will be a disturbance of equilibrium and the separation of one or even several salts.[32]

Although Gay-Lussac is probably best known for his work in physical and inorganic chemistry, he also made a number of important contributions to organic chemistry. In January 1810 Gay-Lussac and Thenard developed the pioneer work of Lavoisier on the quantitative combustion analysis of organic compounds. Whereas Lavoisier had burned a few inflammable substances in oxygen gas, Gay-Lussac and Thenard greatly extended the generality of the method by the use of an oxidizing agent.[33] They at first proposed potassium chlorate, but as this was found to act too powerfully, Gay-Lussac suggested in 1815 that copper oxide was preferable and thereby established its use. They applied their method in 1810 to the analysis of twenty vegetable and animal substances. On the basis of these analyses, they divided vegetable substances into three classes according to the proportion of hydrogen and oxygen contained in them. One of these classes included compounds—such as starch, gum, and sugar—in which the proportion of oxygen to hydrogen was the same as in water. This classification was accepted by William Prout, who referred to this group as the saccharine class, later called carbohydrates.

Gay-Lussac's analysis of prussic acid in 1815 is particularly important because he drew attention to the existence of a radical (—CN) that is fully analogous to the chlorine in hydrochloric acid and the iodine in iodic acid, the essential difference being that "this radical is compound."[34] This was the first example of the analysis of a carbon-containing radical. If this had been generally considered as an organic radical, it might have anticipated the radical theory of organic chemistry of the 1830's.

In 1815 Gay-Lussac referred to prussic acid as a "true hydracid in which carbon and nitrogen *replace* [author's italics] chlorine in hydrochloric acid. . . ."[35] He later took the idea of replacement beyond the theoretical plane when he considered actual reactions in organic chemistry. When discussing the action of chlorine on oils and waxes, he observed that the chlorine removes part of the hydrogen from the oil-forming hydrochloric acid and also "part of the chlorine combines with the oil and takes the place of the hydrogen removed."[36] In 1834 Dumas began to develop the principle adumbrated by Gay-Lussac into a general theory of substitution in organic chemistry.

One of the consequences of Gay-Lussac's volumetric approach to chemistry was his conversion in 1815 of the gravimetric analysis of alcohol by Theodore de Saussure in terms of olefiant gas and water into a volumetric analysis.[37] This gave approximately equal volumes. Gay-Lussac confirmed that the sum of the vapor densities of olefiant gas and water is equal to the vapor density of absolute alcohol within the limits of experimental error. He therefore concluded that alcohol is composed of one volume of olefiant gas and one volume of water. By similar reasoning he concluded that "sulfuric ether" (diethyl ether) is composed of two volumes of olefiant gas and one volume of water. Gay-Lussac's work was the inspiration of the etherin theory of Dumas and Boullay in 1827.

Gay-Lussac also contributed to the early history of isomerism. In 1814 he remarked that acetic acid and *matière ligneuse* ("cellulose") have the same composition and concluded that it is the arrangement of the constituent particles of a compound which determines whether a substance has a neutral, acidic, or alkaline character.[38] In 1824, in his capacity as chemical editor of the *Annales de chimie et de physique,* he remarked that if Wöhler's analysis of silver cyanate was correct, it was identical with that of silver fulminate and that the explanation must lie in the different arrangement of the elements within the two compounds.[39] Later he showed experimentally that racemic acid (a name introduced by him) has the same composition as tartaric acid.[40]

We have left until now Gay-Lussac's vital work on volumetric analysis, since this belongs to the later part of his life, when he was largely occupied with applied science. French chemists made most of the basic contributions to the history of volumetric analysis. Some pioneer work was carried out by Henri Descroizilles, but the subsequent work of Gay-Lussac was even more influential. Although the French term *titre* was used about 1800, the concept of titration passed into general chemical practice from the method proposed by Gay-Lussac for estimating the purity of silver.

In 1824 Gay-Lussac extended earlier methods for the estimation of hypochlorite or chlorinated lime solution using indigo solution.[41] He later improved on the method by the use of standard solutions of certain reducing agents: arsenious oxide, mercurous nitrate, and potassium ferrocyanide.[42] His 1824 paper is important, as it contains the first use of the terms *pipette* and *burette* for the respective pieces of apparatus that have since become standard. Gay-Lussac's apparatus was an improvement on that used earlier by Descroizilles and Welter. Gay-Lussac's pipette had essentially the form of its modern counterpart, but his burette was more like a graduated cylinder with a connecting side arm. In a paper published in 1828 Gay-Lussac also used a one-liter volumetric flask.[43] In this paper he described *acide normale* ("normal acid") as a standard solution of 100 grams of sulfuric acid diluted to one liter. Gay-Lussac used litmus as an indicator and described accurately the color transition in different reactions. In 1829 he published a method for the determination of borax.[44] Sulfuric acid was added to a solution of the borax. The turning of the litmus present to the "color of onion skin" indicated the neutralization of the soda present in the borax. He even carried out an indicator correction, measuring the amount of acid required to change the color of the same amount of indicator as had been used in the titration.

Gay-Lussac made a major contribution to chemical analysis in 1832 when he introduced a volumetric method of estimating silver, which he justly claimed was much more accurate than the centuries-old method of cupellation.[45] He proposed two parallel procedures for this method, one gravimetric, which he said was the more accurate, and one volumetric, which had the advantage of simplicity. The principle of both methods was the precipitation of silver chloride. He prepared a standard solution of sodium chloride of such concentration that 100 milliliters precipitated rather less than one gram of silver. Another standard solution of sodium chloride one-tenth of the concentration of the first was also prepared. One gram of silver was accurately weighed and then dissolved in nitric acid; 100 milliliters of the concentrated sodium chloride solution was added, and the precipitate of silver chloride was allowed to settle. The dilute sodium chloride solution was then added in one-milliliter portions; after each addition, the flask was shaken and the precipitate allowed to settle. The procedure was continued until further addition caused no precipitation. This final excess of sodium chloride was found exactly by back-titrating with standard silver nitrate solution. It was characteristic of Gay-Lussac's standard solutions that they could be used only for specific analyses and for given weights of a sample, since the concentration of his solutions had no chemical basis related to equivalent weights. While, therefore, Gay-Lussac must be given credit for showing volumetric analysis to be convenient, rapid, and accurate, the establishment of a general system of volumetric analysis had to wait until the achievements of Fredrik Mohr in the next generation of chemists.

As early as 1807 Gay-Lussac had discussed the optimum temperature for the production of sulfuric acid in the lead chamber process.[46] His main contribution to this industry, however, was his suggestion of 1827 dealing with the spent gases discharged into the atmosphere at the end of the lead chamber process containing the expensive and noxious oxides of nitrogen. The latter were to be absorbed by passing them up a tower packed with coke over which trickled concentrated sulfuric acid. The adoption of this method had to wait until John Glover showed in 1859 that the oxides of nitrogen absorbed in the Gay-Lussac tower could be used again in the chamber if the acid, containing oxides of nitrogen, were passed through a second tower in which it could come into contact with water and sulfur dioxide. This released the oxides of nitrogen and simultaneously produced acid of the right concentration for use in the Gay-Lussac tower.

NOTES

1. "Sur la dilatation des gaz et des vapeurs," in *Annales de chimie,* **43** (1802), 137–175.
2. "Relation d'un voyage aérostatique," in *Journal de physique,* **59** (1804), 314–320, 454–461, written with J. B. Biot.
3. "Expériences sur les moyens eudiométriques et sur la proportion des principes constituants de l'atmosphère," *ibid.,* **60** (1805), 129–168.
4. "Observations sur l'intensité et l'inclinaison des forces magnétiques. . .," in *Mémoires de physique et de chimie de la Société d'Arcueil,* **1** (1807), 1–22, written with Humboldt.
5. "Premier essai pour déterminer les variations de température qu'éprouvent les gaz en changeant de densité. . .," *ibid.,* pp. 180–203.
6. "Mémoire sur la combinaison des substances gazeuses, les unes avec les autres," *ibid.,* **2** (1809), 207–234; translated in Alembic Club Reprint no. 4, pp. 8–24.
7. "Annonce d'un travail sur la densité des vapeurs de divers liquides," in *Annales de chimie,* **80** (1811), 218.
8. "Sur la décomposition de la potasse et de la soude," *ibid.,* **65** (1808), 325–326, written with Thenard; "Extrait de plusieurs notes sur les métaux de la potasse et de la soude," *ibid.,* **66** (1808), 205–217, written with Thenard.
9. "Sur la décomposition et la recomposition de l'acide boracique," *ibid.,* **68** (1808), 169–174, written with Thenard.
10. "Sur l'acide fluorique," *ibid.,* **69** (1809), 204–220, written with Thenard; "Des propriétés de l'acide fluorique et surtout de son action sur le métal de la potasse," in *Mémoires de physique et de chimie de la Société d'Arcueil,* **2** (1809), 317–331, written with Thenard.
11. "De la nature et des propriétés de l'acide muriatique et de l'acide muriatique oxigéné," *ibid.,* pp. 339–358, written with Thenard; translated in Alembic Club Reprint no. 13, pp. 34–48.
12. "Mémoire sur l'iode," in *Annales de chimie,* **91** (1814), 5–121.
13. "Sur un acide nouveau formé par le soufre et l'oxigène," in *Annales de chimie et de physique,* 2nd ser., **10** (1819), 312, written with Welter.
14. "Mémoire sur l'eau régale," *ibid.,* 3rd ser., **23** (1848), 203.
15. "Sur le rapport qui existe entre l'oxidation des métaux et leur capacité de saturation pour les acides," in *Mémoires de physique et de chimie de la Société d'Arcueil,* **2** (1809), 159–175.
16. "Sur la décomposition des sulfates par la chaleur," *ibid.,* **1** (1807), 215–251.
17. "Sur la précipitation des métaux par l'hydrogène sulfuré," in *Annales de chimie,* **80** (1811), 205–208.
18. "Sur la vapeur nitreuse, et sur le gaz nitreux considéré comme moyen eudiométrique," in *Mémoires de physique et de chimie de la Société d'Arcueil,* **2** (1809), 235–253.
19. "Sur les combinaisons de l'azote avec l'oxigène," in *Annales de chimie et de physique,* 2nd ser., **1** (1816), 394–410.
20. "Mémoire sur l'iode," *ibid.,* **91** (1814), 9, 148–149.
21. "Recherches sur l'acide prussique," *ibid.,* **95** (1815), 155.
22. "Note sur l'acide prussique," *ibid.,* **77** (1811), 128.
23. "Recherches sur l'acide prussique," *ibid.,* **95** (1815), 136–231.
24. "Sur l'acide des prussiates triples," *ibid.,* 2nd ser., **22** (1823), 320–323.
25. "Faits pour servir à l'histoire du bleu de Prusse," *ibid.,* **46** (1831), 73–80.
26. C. F. Bucholz and Meissner, "Expériences pour déterminer la quantité de strontiane contenue dans plusieurs espèces d'arragonite," *ibid.,* **2** (1816), 176.
27. "Premier mémoire sur la dissolubilité des sels dans l'eau," *ibid.,* **11** (1819), 296–315.
28. "Note sur la fixité du degré d'ébullition des liquides," *ibid.,* **7** (1817), 307–313.
29. "Sur les combinaisons du chlore avec les bases," *ibid.,* 3rd ser., **5** (1842), 302–303.
30. "Observations sur l'oxidation de quelques métaux," *ibid.,* 2nd ser., **1** (1816), 36–37.
31. "Sur la décomposition réciproque des corps," *ibid.,* **30** (1825), 291.
32. "Considérations sur les forces chimiques," *ibid.,* **70** (1839), 431.
33. "Sur l'analyse végétale et animale," in *Journal de physique,* **70** (1810), 257–266.
34. "Recherches sur l'acide prussique," in *Annales de chimie,* **95** (1815), 161.
35. *Ibid.,* 155 (my italics).
36. *Cours de chimie,* II, Leçon 28 (16 July 1828).
37. "Sur l'analyse de l'alcool et de l'éther sulfurique," in *Annales de chimie,* **95** (1815), 311–318.
38. "Mémoire sur l'iode," *ibid.,* **91** (1814), 149 n.
39. F. Wöhler, "Recherches analytiques sur l'acide cyanique," *ibid.,* 2nd ser., **27** (1824), 199–200 n.
40. *Cours de chimie,* II, Leçon 24, 1, 23.
41. "Instruction sur l'essai du chlorure de chaux," in *Annales de chimie et de physique,* 2nd ser., **26** (1824), 162–175.
42. "Nouvelle instruction sur la chlorométrie," *ibid.,* **60** (1835), 225–261.
43. "Essai des potasses du commerce," *ibid.,* **39** (1828), 337–368.
44. "Sur l'analyse du borax," *ibid.,* **40** (1829), 398.
45. *Instruction sur l'essai des matières d'argent par la voie humide* (Paris, 1832), *passim.*
46. "Sur la décomposition des sulfates par la chaleur," in *Mémoires de physique et de chimie de la Société d'Arcueil,* **1** (1807), 246.

BIBLIOGRAPHY

I. ORIGINAL WORKS. Gay-Lussac was the author of *Recherches physico-chimiques, faites sur la pile, sur la préparation chimique et les propriétés du potassium et du sodium . . .*, 2 vols. (Paris, 1811), written with Thenard; and *Instruction sur l'essai des matières d'argent par la voie humide . . . publiée par la Commission des monnaies et médailles* (Paris, 1832). Although he himself never compiled elementary textbooks, the following, based on his lectures, were published: *Cours de chimie par M. Gay-Lussac, comprenant l'histoire des sels, la chimie végétale et animale . . .*, 2 vols. (Paris, 1828); and *Leçons de physique de la Faculté des sciences de Paris, recueillies et rédigées par M. Grosselin,* 2 vols. (Paris, 1828).

Gay-Lussac's most important research papers include "Sur la dilatation des gaz et des vapeurs," in *Annales de chimie,* **43** (1802), 137–175; "Relation d'un voyage aérostatique," in *Journal de physique,* **59** (1804), 314–320, 454–461, written with J. B. Biot; "Expériences sur les moyens eudiométriques et sur la proportion des principes constituants de l'atmosphère," *ibid.*, **60** (1805), 129–168, written with Humboldt; "Observations sur l'intensité et l'inclinaison des forces magnétiques, faites en France, en Suisse, en Italie, etc.," in *Mémoires de physique et de chimie de la Société d'Arcueil,* **1** (1807), 1–22, written with Humboldt; "Premier essai pour déterminer les variations de température qu'éprouvent les gaz en changeant de densité, et considérations sur leur capacité pour le calorique," *ibid.*, pp. 180–203; "Extrait de plusieurs notes sur les métaux de la potasse et de la soude," in *Annales de chimie,* **66** (1808), 205–217, written with Thenard; "Sur la décomposition et la recomposition de l'acide boracique," *ibid.*, **68** (1808), 169–174, written with Thenard; "Mémoire sur la combinaison des substances gazeuses, les unes avec les autres," in *Mémoires de physique et de chimie de la Société d'Arcueil,* **2** (1809), 207–234; "Sur l'acide fluorique," in *Annales de chimie,* **69** (1809), 204–220, written with Thenard; "De la nature et des propriétés de l'acide muriatique et de l'acide muriatique oxigéné," in *Mémoires de physique et de chimie de la Société d'Arcueil,* **2** (1809), 339–358, written with Thenard; "Sur l'analyse végétale et animale," in *Annales de chimie,* **74** (1810), 47–64, written with Thenard; "Mémoire sur l'iode," *ibid.,* **91** (1814), 5–121; "Recherches sur l'acide prussique," *ibid.*, **95** (1815), 136–231; "Sur l'analyse de l'alcool et de l'éther sulfurique," *ibid.*, pp. 311–318; "Sur les combinaisons de l'azote avec l'oxigène," in *Annales de chimie et de physique,* 2nd ser., **1** (1816), 394–410; "Premier mémoire sur la dissolubilité des sels dans l'eau," *ibid.,* **11** (1819), 296–315; "Instruction sur l'essai du chlorure de chaux," *ibid.,* **26** (1824), 162–176; "Essai des potasses du commerce," *ibid.,* **39** (1828), 337–368; "Considerations sur les forces chimiques," *ibid.,* **70** (1839), 407–434; and "Mémoire sur l'eau régale," *ibid.,* 3rd ser., **23** (1848), 203–229.

II. SECONDARY LITERATURE. On Gay-Lussac and his work, see D. F. J. Arago, "Biographie lue en séance de l'Académie des sciences le 20 décembre 1852," in *Oeuvres de François Arago, Notices biographiques,* 2nd ed., III (Paris, 1865), 1–112; E. Blanc and L. Delhoume, *La vie émouvante et noble de Gay-Lussac* (Paris, 1950); M. P. Crosland, "The Origins of Gay-Lussac's Law of Combining Volumes of Gases," in *Annals of Science,* **17** (1961), 1–26; and *The Society of Arcueil. A View of French Science at the Time of Napoleon I* (Cambridge, Mass., 1967); J. R. Partington, *A History of Chemistry,* IV (London, 1964), pp. 77–90; and F. Szabadváry, *History of Analytical Chemistry* (Oxford, 1966).

M. P. CROSLAND

GAYANT, LOUIS (*b.* Beauvais, France; *d.* Maastricht, Netherlands, 19 October 1673), *comparative anatomy.*

Gayant was trained in medicine and was serving as a military physician at the time of his death. He was recognized as one of the most able anatomists of his time and was closely associated with the anatomists in the Académie Royale des Sciences. By 1667 he was collaborating with Jean Pecquet, Claude Perrault, and others in the anatomical work published later in the *Mémoires.* This work was part of the scheme of research that Perrault proposed to the Academy soon after its founding. The exact portion of the work done by each man in the group of anatomists in the Academy cannot generally be determined because of the policy of dissecting and writing up the results anonymously.

Gayant's name seems to be associated firmly with only three pieces of anatomical work—all dating from 1667—although scattered references indicate that he participated regularly in the collective work of the Parisian anatomists. In February 1667 he dissected a female cadaver to demonstrate a number of venous valves which were already known but not universally accepted. In his demonstration the pertinent parts of the venous system were both inflated with air and injected with milk. The following month he conducted similar demonstrations, again at the Academy, in collaboration with Perrault and Pecquet.

During 1667 the transfusion of blood as a potential remedy was much discussed both in London and in Paris. That year at least seven transfusion experiments on dogs were performed by the Parisian anatomists. One of these was related to the Royal Society of London by a correspondent who witnessed Gayant perform it. The writer claims that the vigor of the recipient dog immediately improved. The Parisians reported, however, that all seven recipients died or were enfeebled and that coagulated blood was ordinarily found in their heart or veins. The donors all carried on well. On the basis of these results the Parlement of Paris prohibited blood transfusions "as a useless and dangerous remedy."

The dissection subjects of the Parisian anatomists, including Gayant, covered a wide range of vertebrates, primarily mammals. These included a few domestic animals for comparative purposes, many wild European forms, and as many foreign species as they could obtain, the latter often from the royal menagerie at Versailles. Specifically on which parts of this extensive project Gayant worked cannot be determined. There is no doubt, however, that until his death he was an important member of the group working in comparative anatomy in the Academy and that he contributed substantially to their series of publications.

BIBLIOGRAPHY

I. ORIGINAL WORKS. The three demonstrations with which Gayant was associated in 1667 are described in *Histoire de l'Académie royale des sciences,* **1** (Paris, 1733), 36–39. The letter to the Royal Society is in *Philosophical Transactions,* **26** (3 June 1667), 479–480. The collected works published by the Parisian anatomists, to which Gayant certainly contributed, are *Extrait d'une lettre . . . sur un grand poisson . . .* (Paris, 1667); *Description anatomique d'un cameleon . . .* (Paris, 1669); and *Mémoires pour servir à l'histoire naturelle des animaux* (Paris, 1671–1676).

II. SECONDARY LITERATURE. See "Louis Gayant," in *Biographie médicale,* IV (Paris, 1820–1825), 366.

For a discussion of the Parisian anatomists as a group see F. J. Cole, *A History of Comparative Anatomy* (London, 1944), pp. 393–442; and Joseph Schiller, "Les laboratoires d'anatomie et de botanique à l'Académie des sciences au XVIIe siècle," in *Revue d'histoire des sciences et de leurs applications,* **17** (1964), 97–114.

WESLEY C. WILLIAMS

GEER, CHARLES DE (*b.* Finspång, Sweden, 10 February 1720; *d.* Leufsta, Sweden, 8 March 1778), *entomology.*

De Geer grew up in Holland and returned to Sweden in 1739, where, at the age of nineteen, he was invited to become a member of the Swedish Academy of Sciences. He had not yet published any scientific work but, since he was one of the country's wealthiest men, the academy expected him to make generous donations after his election. Nothing came of that hope, but De Geer became one of the academy's most outstanding and widely known scientists. He resided in Leufsta castle, which still belongs to the De Geer family and contains his large and extremely valuable library. His position brought him into close contact with the royal court, and he was made master of the royal household.

After Linnaeus, De Geer was Sweden's most important and internationally known biologist in the eighteenth century. In contrast with Linnaeus, he was not primarily interested in descriptive systematics; he considered the study of "dead insects" both meaningless and useless. That did not prevent him from describing and naming several species, but such activity was a necessity; he had to name the previously unobserved species which he studied and whose biology he described. Otherwise, as he acknowledged, the biological description would be of little value. It is, however, of interest that in his seven-volume *Mémoires* he did not employ the binary Linnaean nomenclature until the third volume (1773).

Although De Geer lacked a particular interest in classification and cataloging, his enthusiasm for studying the life and metamorphosis of insects was great. His position gave him plenty of time for all kinds of nature studies, the only work which could not be left to his large staff of servants. In the field De Geer made his brilliant observations; he placed insect larvae in cages and studied their metamorphosis; and with the microscope—the best and most expensive English make—he made morphological observations and drawings of insects and their structures. He was an exceptional draftsman, which is reflected in his illustrations.

De Geer's scientific models were Leeuwenhoek, Swammerdam, and Réaumur. From the first two he drew the technique of using the microscope, and from the latter the method of biological observation. It is no coincidence that his monumental *Mémoires pour servir à l'histoire des insectes* (1752–1778) carries exactly the same title as Réaumur's equally large work (1734–1742). When the first volume was published in 1752, De Geer was accused, doubtless on the basis of the title of the work, of compilation and lack of independence; it was only a rehash of Réaumur. The great injustice of these accusations strongly disturbed De Geer; he announced that he would not continue the work and had the remaining part of the edition of the first volume destroyed. Not until nineteen years later, in 1771, was it possible to persuade him to resume communication of his experience by publishing the other volumes. From then on, the volumes appeared in unbroken sequence and received the notice and appreciation they deserved. The seventh and last volume appeared posthumously, in the fall of 1778. As already mentioned, De Geer did not use Linnaeus' binary nomenclature consistently; but a commentary introducing that principle of naming was published by Retzius in 1783 on the basis of De Geer's work. The work was translated into German by the entomologist J. A. E. Goeze, who published it, together with an extensive and valuable commentary, in 1776–1783.

De Geer published about twenty entomological works, the majority in the *Handlingar* of the Royal Academy of Sciences. His first article, published in 1740, was a short and very valuable treatise on Collembola. His smaller articles were always of high quality; the majority were published in more or less revised form in the *Mémoires.* He also published excellent studies on protozoa, which testified to the author's abilities as well as to the quality of his microscope.

De Geer was in close contact with the greatest entomological authorities of his time, as the library at Leufsta shows. After his death his microscope and collections were turned over to the academy; the insect collections are now in the entomological department of the Swedish Museum of Natural History in Stockholm.

De Geer's importance for biological research and for the technique of observing nature is much greater than the quite limited literature about him would indicate. Torbern Bergman, himself one of the great Swedish scientists of the eighteenth century, gave fine testimony about De Geer's merits in the obituary that he read before the Academy of Sciences in 1778.

BIBLIOGRAPHY

I. Original Works. De Geer's writings include "Rön och Observation öfver små Insecter som kunna håppa i högden," in *Kungliga Svenska vetenskapsakademiens handlingar* (1740), pp. 265–281; "Beskrifning på en märkwärdig Fluga kallad Ichneumon ater, antennis ramosis," *ibid.*, pp. 458–463; "Beskrifning på en Insect af ett nytt slägte kallad Physapus," *ibid.* (1744), pp. 1–9; "Om maskar funne på snön om vintern," *ibid.* (1749), pp. 76–78; and *Mémoires pour servir à l'histoire des insectes,* 7 vols. (Stockholm, 1752–1778), translated into German by J. A. E. Goeze as *Des Herrn Baron Karl de Geer Abhandlungen zur Geschichte der Insecten,* 7 vols. (Leipzig–Nuremberg, 1776–1783).

II. Secondary Literature. See Torbern Bergman, *Åminnelse-Tal öfer Hof-Marschalken, Högvälborne Friherren Herr Carl De Geer, hållet 19 Decemb. 1778* (Stockholm, 1779); F. Bryk, in S. Lindroth, ed., *Swedish Men of Science 1650–1950* (Stockholm, 1952), pp. 113–121; and A. J. Retzius, *Caroli De Geer genera et species insectorum et generalissimi auctoris scriptis extraxit, digessit, latine quand. partem reddidit, et terminologiam insectorum Linneanam addidit* (Leipzig, 1783).

Bengt-Olof Landin

GEER, GERHARD JAKOB DE (*b.* Stockholm, Sweden, 2 October 1858; *d.* Saltsjöbaden, Sweden, 23 July 1943), *geology, geochronology.*

Geer belonged to one of the leading noble families in Sweden. Originally from Belgium, they settled in Sweden at the beginning of the seventeenth century, and many of them have since then been important in politics and economics. Both his father, Louis de Geer, and his older brother, Gerhard Louis de Geer, were prime ministers of Sweden (1858–1870, 1875–1880 and 1920–1921, respectively), and his father was the leading politician in Sweden in the last half of the nineteenth century. Geer himself was also involved in politics and was a member of the Swedish parliament from 1900 to 1905.

Geer grew up in a home which was a center for both the political and the cultural life of Stockholm. He received his master's degree in geology from Uppsala in 1879, having been appointed to the Swedish Geological Survey in 1878. After a few years of ordinary geological mapping, Geer turned his concentration to what was to be his lifetime interest, the study of Quaternary (Pleistocene) geology. The first main problem was that of raised beaches. Shorelines much above sea level were known both in Scandinavia and in other regions, but the complicated system by which they were uplifted isostatically was not understood. The rise of the land is highest where the ice was thickest and the depression largest. The uplift decreases in all directions from the center of glaciation. Superposed on this is the eustatic change in sea level, due partly to the melting of inland ice. This complicated system was discovered and elegantly described by Geer, who also coined the term "marine limit" (the highest shoreline of the sea at any particular locality). The summary of his work, published in 1896, was one of the classic works in Scandinavian geology.

In order to explain glacial phenomena Geer traveled to Spitsbergen, where glaciers somewhat similar to the Quaternary ones still exist. He took part in and led expeditions in 1882, 1896, 1899, 1901, and 1908, among them the Swedish-Russian meridian expedition, which he led. Geer introduced terrestrial photogrammetry as an aid in his studies, and it was later used by most other Arctic expeditions, in order to increase the precision of both geological and geodetical observations. In 1897 Geer became professor at the University of Stockholm and was its president from 1902 to 1910. At Stockholm he established the varve chronology. Varves are annual, cyclic sediments consisting of summer and winter bands of silt and clay deposited by glacial meltwater in fresh or brackish water. They vary in thickness, and Geer early had the idea that these variations could be used by correlating various sections through varve sequences. In 1904–1905 he had his students measure all varves in a 200-kilometer-long north-south section near Stockholm; they found that the last ice in the area had melted away over a period of 800 years and that they could pinpoint the position of the ice margin for every year. This system was soon extended from Scania to the mountains of central Scandinavia, where it covered a period of 15,000 years. Geer used the final stage of the melting of the inland ice as his zero year; later studies by his assistant Ragnar Lidén made in the Angerman River, where varves are formed today, showed that the zero year was 6739 B.C. Geer's varve method is cumbersome and restricted to the few areas where varves are found, but it gives an unprecedented accuracy in age determinations. Geer became world-famous when he presented his results at the International Geological Congress at Stockholm in 1910, but his final paper on the subject, "Geochronologica Suecica," did not appear until 1940.

In 1924 Geer retired and became head of the Institute of Geochronology at the University of Stockholm. After several years of travel he and a number of assistants tried to extend his system on a global scale. These "teleconnections" were not generally accepted, since it seemed unlikely that the variations in meltwater and sediment volume should be synchronous all over the globe. Geer worked intensely during his last years to refine his method and to prove his long-distance correlations. After his death studies by isotope methods (carbon 14) have shown that some of his correlations, especially those with North America, were remarkably precise.

BIBLIOGRAPHY

I. ORIGINAL WORKS. Complete bibliographies of Geer's more than 200 scientific publications are in Nilsson and in Post (see below). Geer's most important works were *Om Skandinaviens geografiska utveckling efter istiden* (Stockholm, 1896), written in unusually clear and lucid Swedish but not a popular book and not easily accessible to non-Scandinavians; and "Geochronologica Suecica, Principles," *Kungliga Svenska vetenskapsakademiens handlingar,* 3rd ser., **18,** no. 6 (1940).

II. SECONDARY LITERATURE. The best biographies of Geer are E. Nilsson, "Gerhard de Geer, geokronologen," in *Levnadsteckningar över Kungliga Svenska vetenskapsakademiens ledamoter,* **172** (1969), 213–250; and L. von Post, "H. J. de Geer," in *Svensk biografisk lexikon,* X (1931), 564–567.

NILS SPJELDNAES

GEGENBAUR, KARL (*b.* Würzburg, Germany, 21 August 1826; *d.* Heidelberg, Germany, 14 June 1903), *anatomy.*

For a detailed study of his life and work, see Supplement.

GEHLEN, ADOLF FERDINAND (*b.* Bütow, Germany, 15 September 1775; *d.* Munich, Germany, 15 July 1815), *chemistry.*

For a detailed study of his life and work, see Supplement.

GEIGER, HANS (JOHANNES) WILHELM (*b.* Neustadt an der Haardt [now Neustadt an der Weinstrasse], Rheinland-Pfalz, Germany, 30 September 1882; *d.* Potsdam, Germany, 24 September 1945), *physics.*

Geiger developed a variety of instruments and techniques for the detection and counting of individual charged particles. He was the eldest of the five children of Wilhelm Ludwig Geiger, professor of philology at Erlangen from 1891 to 1920. His only brother, Rudolf, became professor of meteorology at Munich. Geiger passed his *Abitur* at the Erlangen Gymnasium in 1901. After a brief period of military service he studied physics both at Munich and at Erlangen. He took his preliminary examination in 1904 and then began his research under Eilhard Wiedemann. In July 1906 Geiger defended his inaugural dissertation and received his doctorate from the University of Erlangen. He then took a position in England, followed by a series of appointments in Germany.

Geiger married Elisabeth Heffter in 1920; they had three sons. He completed his *Habilitationsschrift* at Berlin in 1924. Geiger was awarded the Hughes Medal by the Royal Society on 30 November 1929 "for his invention and development of methods of counting alpha and beta particles,"[1] and the Duddell Medal by the Physical Society (London) in 1938 for his contributions to scientific instrumentation. Geiger was elected to the Leopoldina in 1935 and to the Preussische Akademie der Wissenschaften in 1937.

Geiger was a perfectionist, always trying to obtain the most from both his students and his experiments. His enthusiastic and warmhearted nature inspired others to emulate his methods and share his goals. He was a talented lecturer, popular with both his colleagues and the public.

In 1906 Geiger was assistant to Arthur Schuster at Manchester. Ernest Rutherford, who succeeded Schuster in 1907, persuaded Geiger to remain at Manchester and continue research on radioactivity. In 1908, working out the probability variations and statistical error factor, Geiger extended the experimental confirmation of the 1905 theoretically predicted "Schweidler fluctuations" for the case of radioactive disintegrations.

In 1908 Geiger and Rutherford investigated the charge and nature of the α particle. They devised an electrical technique in order to count the individual α particles and compare results with those obtained by Erich Regener, who used the scintillation tech-

GEIGER

nique. Their ionization chamber was a cylinder at low pressure with a thin wire stretched coaxially (Fig. 1). The wall of the cylinder had an entrance window at one end and was at high negative potential to the wire. An α particle entering parallel to the wire generated a secondary avalanche, according to the 1900 collision principle of John S. Townsend, multiplying the primary ionization effect a thousandfold. The resulting voltage step on the central wire was measured by a sensitive electrometer. With this proto-"Geiger counter" they estimated that the total number of α particles emitted per second from the radium C (actually radium C′) present in one gram of radium in equilibrium was 3.4×10^{10}. From this value and a determination of the total charge in a beam of α particles, they established that the α particle was doubly charged. This led directly to the confirmation by Rutherford and T. Royds that α particles were doubly charged helium atoms.

The beam of α particles was observed to spread. Geiger investigated this scattering effect and was joined in 1909 by Ernest Marsden. Using a scintillation detector, they observed the number of particles scattered at various angles of incidence. They detected α particles reflected at angles sufficiently large to make inadequate a statistical interpretation based upon multiple scattering. On preliminary evidence Rutherford was led to propose in 1911 that this effect was due to single scattering from compact nuclei. He theoretically predicted the behavior of a set of scattering parameters based upon a nuclear model of the atom. Geiger and Marsden undertook a further series of experiments and verified the predicted behavior of these parameters by July 1912.[2]

In 1910 Geiger examined α particles from a variety of radioactive substances and noted a linear relationship between the maximum range and the third power of the velocity of expulsion. Following up the 1907 suggestion of Rutherford, Geiger and John Michael Nuttall empirically established in 1911–1912 what seemed an essentially linear relationship between the half-value period, on the one hand, and the maximum range or energy of disintegration, on the other, for α particles from various materials. Later, in 1921, Geiger revised this empirical Geiger-Nuttall rule, noting that actinium-X did not fit the

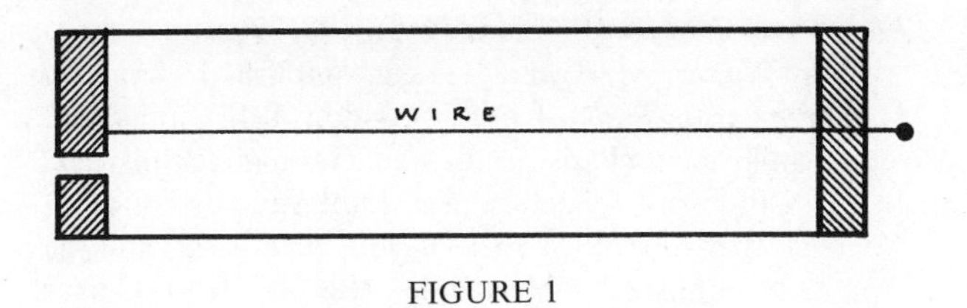

FIGURE 1

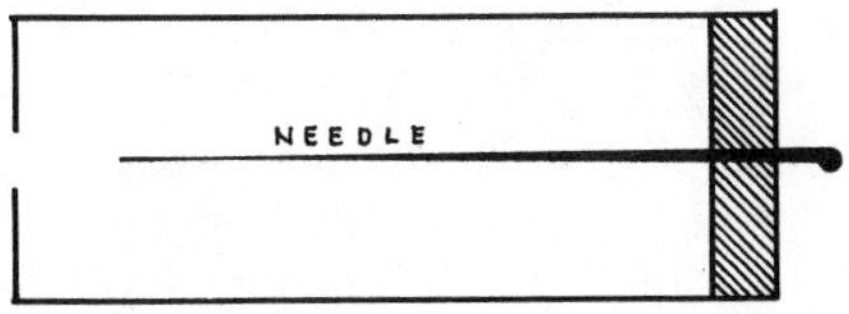

FIGURE 2

straight-line curve, and in 1928 George Gamow and others partially superseded this empirical rule altogether with a physical explanation of α disintegration in terms of the probability of α penetration of potential barriers.

In October 1912 Geiger, having turned down a call to Tübingen, took up the newly established position of director of the laboratory for radium research at the Physikalisch-Technische Reichsanstalt in Berlin. In addition to measuring radium samples, he continued experiments on the counting of α particles. Varying the form and dimensions of the central electrode, he found an arrangement which came to be known as the *Spitzenzähler* or "point counter," since "the whole working of the apparatus depends on the point of the needle" (Fig. 2).[3] The great advantage of this device was that in addition to α particles, it could, for the first time, count β particles as well as other types of radiation. Geiger noted that at ordinary pressure the size of the voltage step was proportional to the intensity of the incident radiation, whereas "the deflections *at low pressure* are all equal and independent of the intensity of the primary ionization."[4]

In November 1913 Geiger was joined by Walther Bothe, who investigated α scattering, and by James Chadwick, who counted β particles. In 1914 Chadwick was led by his investigations to note the presence of the continuous β spectrum in addition to the known line spectrum.

After the war Geiger returned to the Reichsanstalt. The 1924 statistical interpretation by Niels Bohr, H. A. Kramers, and J. C. Slater of the 1923 Compton effect stimulated Bothe and Geiger to devise, by that June, a technique to test its validity.[5] Using two *Spitzenzähler* in what was most likely the first coincidence experiment,[6] completed by April 1925, they noted that approximately every eleventh captured quantum was coincident to within 10^{-4} seconds of a recoiled electron.[7] This result, while unintelligible on the statistical interpretation, reconfirmed the validity of classic conservation principles for single atomic events.

In 1925 Geiger took his first teaching position as professor of physics at the University of Kiel. In addition to teaching and directing a large research team at the Institute of Physics, Geiger developed with

Walther Müller the counting device for which his name is best known. The Geiger-Müller counter was based upon the 1908 coaxial-wire principle (see Figure 1) and followed directly from the clarification by Geiger in July 1928 of an anomaly which Müller had encountered in his research.[8] The sensitivity of the counter was greatly improved so that it normally had to be shielded from background radiation. The recovery time was reduced through rapid quenching of the secondary avalanche. The construction of the central wire was improved so as to extend the operational lifetime for periods of months, and the signal could be amplified so as to trigger a mechanical register. In general, the device was designed to be compact, portable, and functional, and thus to meet a variety of laboratory requirements in particle detection and counting. The gradual production of these practical Geiger-Müller counters from 1928 marks the introduction of modern electrical devices into radiation research.

In October 1929 Geiger accepted a call to the University of Tübingen as professor of physics and director of research at the Institute of Physics. He continued to improve the counter, which by 1928 had already been extended in application to the study of canal rays. After noting that counters placed in separate rooms at the Institute periodically registered simultaneous bursts of radiation—the first detection of cosmic ray showers[9]—in 1931 Geiger refused a call to Lenard's chair at Heidelberg and began a series of investigations on cosmic radiation which he continued throughout the remainder of his career.

In October 1936 Geiger accepted the chair of physics at the Technische Hochschule in Berlin. While directing a large research team studying artificial radioactivity and the products of nuclear fission, he continued to make full use of his counting device, with which he had made possible the investigation of cosmic radiation. In 1937 Geiger and Otto Zeiller used nine such counting tubes in a circular arrangement to determine the angular distribution of a cosmic ray shower. His last lecture, given in April 1942, was on cosmic rays.

The war years brought a severe recurrence of a rheumatic condition, suffered during his front-line duty as an artillery officer in World War I, and contributed to his increasing absence from the Institute. However, it did not prevent him from going to the Institute even during 1944.[10] He also continued as editor of *Zeitschrift für Physik,* which he had taken over from Karl Scheel in 1936.[11]

In June 1945 Geiger was forced to abandon his home and possessions in Babelsberg and flee to nearby Potsdam, where he died shortly afterward.

NOTES

1. *Nature,* **124** (1929), 767.
2. Hans Geiger and Ernest Marsden, "Die Zerstreuungsgesetze der alpha Strahlen bei grossen Ablenkungswinkeln," in *Sitzungsberichte der Akademie der Wissenschaften zu Wien,* Math.-nat. Kl., Abt. II-a, **121** (1912), 2361–2390.
3. Geiger letter to Ernest Rutherford, 5 May 1913; Cambridge University Library Add. MSS 7653/G31.
4. *Ibid.*
5. Walther Bothe and Hans Geiger, "Ein Weg zur experimentellen Nachprüfung der Theorie von Bohr, Kramers und Slater," in *Zeitschrift für Physik,* **26** (1924), 44.
6. Professor Walther Gerlach informed me privately in Apr. 1971 that this was the first coincidence experiment.
7. Walther Bothe and Hans Geiger, "Über das Wesen des Comptoneffekts; ein experimenteller Beitrag zur Theorie der Strahlung," in *Zeitschrift für Physik,* **32** (1925), 639.
8. C. Schmidt-Schönbeck, *300 Jahre Kieler Universität,* p. 142.
9. Private confirmation by Walther Gerlach, Apr. 1971.
10. Private confirmation by Walther Gerlach, Apr. 1971.
11. Geiger was editor for eight years and issued the twenty vols. from **104** (1937) to **123** (1944); the partially complete last vol. was terminated in Oct. 1944.

BIBLIOGRAPHY

I. Original Works. A nearly complete bibliography of Geiger's writings is in M. von Laue, "Nachruf auf Hans Geiger," in *Jahrbuch der Deutschen Akademie der Wissenschaften zu Berlin: 1946–1949* (1950), pp. 150–158. In this bibliography, no. 46 should read "(1924)" instead of "(1927)" and the pagination of no. 61 is pp. 109–122. Besides the articles and scientific papers in Laue's bibliography, Geiger wrote "Erscheinungen bei sehr starken Strömen in Entladungsröhren," in *Comptes rendus du Premier Congrès International pour l'Étude de la Radiologie et de l'Ionisation, tenu a Liège, Septembre 1905* (Brussels, 1906), pp. 41–45, also published separately (Erlangen, 1905), an abstract of which is in *Physikalische Zeitschrift,* **6** (1905), 913–914; "Demonstrationsversuch zur Erläuterung der Temperaturverhältnisse in den Schichten des positiven Lichtes," in *Verhandlungen der Deutschen physikalischen Gesellschaft,* **8** (1906), 116–118, also published separately (Brunswick, 1906); "Die Kernstruktur der Atome und ihre experimentelle Begründung," in *Zeitschrift des Vereins deutscher Ingenieure,* **66** (1922), 221–225; "Der Einfluss der Atomphysik auf unser Weltbild," in *Deutschland in der Wende der Zeiten,* the open lectures for the summer semester of 1933 at the University of Tübingen (Stuttgart–Berlin, 1934), pp. 107–121; and "Memories of Rutherford in Manchester," in *Nature,* **141** (1938), 244, as well as "Some Reminiscences of Rutherford During His Time in Manchester," in J. Chadwick, ed., *The Collected Papers of Lord Rutherford,* II (London, 1963), 295–298.

With Walter Makower, Geiger published *Practical Measurements in Radio-Activity* (London, 1912), trans. and repub. in French (Paris, 1919) and German (Brunswick, 1920). With Karl Scheel he was joint general editor of *Handbuch der Physik,* 24 vols. (Berlin, 1926–1929) and the separately published subject index (Berlin, 1929). Geiger was also editor of vols. XXII–XXIV, dealing with the structure of matter and the nature of radiation. Geiger and Scheel extensively revised these three vols. in 1933, each of which appeared in two pts., and Geiger edited both pts. of XXII and XXIII. His own article "Durchgang von alpha Strahlen durch Materie," in XXIV (1927), 137–190, dealing with α radiation and detection principles, was revised, expanded, and reissued in XXII, pt. 2 (1933), 155–242.

Geiger contributed "Die Radioaktivität" to Leo Graetz, *Handbuch der Elektrizität und des Magnetismus,* III (Leipzig, 1914), 1–130. Written in 1913, the article contains an updated account of the relevant literature to the end of that year. Geiger was also instrumental in reworking the 13th and 14th eds. of Friedrich Kohlrausch, *Lehrbuch der praktischen Physik* (Leipzig-Berlin, 1921, 1923).

Extensive correspondence between Geiger and Rutherford from 1911 to 1937 is at Cambridge University Library, Add. MSS 7653/G.

II. SECONDARY LITERATURE. Biographical material is in M. von Laue, "Nachruf," which was included in his *Gesammelte Schriften und Vorträge,* III (Brunswick, 1961), 204–212; M. von Laue and R. W. Pohl, "Hans Geiger," in *Zeitschrift für Physik,* **124** (1947), 1; E. Stuhlinger, "Hans Geiger," in *Zeitschrift für Naturforschung,* **1** (1946), 50–52; and T. I. Williams, *A Biographical Dictionary of Scientists* (London, 1969), p. 211. Ewald Fünfer, who took his doctorate under Geiger at Tübingen, contributed the article in *Neue deutsche Biographie,* VI (Berlin, 1964), 141–142. Some early details are in the "Lebenslauf" of Geiger's inaugural dissertation (1906). See also Max Pollerman, in *Röntgen-Blätter,* **11** (1958), 33–35; and the brief account in *Reichshandbuch der Deutschen Gesellschaft,* I (Berlin, 1930), 528.

A brief account of Geiger's early research is in James Chadwick, "The Rutherford Memorial Lecture, 1953," in *Proceedings of the Royal Society,* **224A** (1954), 441–443. Valuable background information on Geiger is in A. S. Eve, *Rutherford* (Cambridge, 1939); and in L. Badash, ed., *Rutherford and Boltwood: Letters on Radioactivity* (New Haven, 1969), *passim.*

Physical descriptions of Geiger's work are given in Walter Dreblow, "Das Geiger-Müllersche Zählrohr," in *Kosmos* (Stuttgart), **47** (1951), 481–485; W. Christoph and W. Hanle, "Zum Mechanismus des Geiger-Müllerschen Zählrohrs," in *Physikalische Zeitschrift,* **34** (1933), 641–645; and Ewald Fünfer, *Zählrohre und Szintillationzähler,* 2nd ed. (Karlsruhe, 1959), pp. 1–5, which also contains a discussion of his work.

C. Schmidt-Schönbeck, *300 Jahre Physik und Astronomie an der Kieler Universität* (Kiel, 1965), pp. 136–146, focuses on Geiger's scientific work at Kiel.

Geiger's techniques and instrumentation are discussed in detail in W. Bothe, "Die Geigerschen Zählmethoden: Herrn Professor Hans Geiger zu seinem 60. Geburtstage am 30 September 1942," in *Naturwissenschaften,* **30** (1942), 593–599. The early work of Geiger is considered in A. T. Krebs, "Hans Geiger: Fiftieth Anniversary of the Publication of His Doctoral Thesis, 23 July 1906," in *Science,* **124** (1956), 166.

A full treatment of Geiger's work in its scientific context can be found in K. W. F. Kohlrausch, *Radioaktivität,* vol. XV of W. Wien and F. Harms, eds., *Handbuch der Experimentalphysik* (Leipzig, 1928), *passim.* The work for which he was awarded the Hughes Medal is considered in "Hughes Medal, Awarded to Professor Hans Geiger," in *Nature,* **124** (1929), 893.

An extensive bibliography containing over 300 references between 1913 and 1947 on Geiger and proportional counters was assembled by M. Healea, "Geiger and Proportional Counters," in *Nucleonics,* **1** (Dec. 1947), 68–75. Additional source material is listed in T. S. Kuhn, ed., *Sources for History of Quantum Physics* (Philadelphia, 1967), p. 164.

THADDEUS J. TRENN

GEIKIE, ARCHIBALD (*b.* Edinburgh, Scotland, 28 December 1835; *d.* Haslemere, Surrey, England, 10 November 1924), *geology.*

Geikie was the son of James Stuart Geikie, an Edinburgh businessman who was also a composer and music critic. His mother, Isabella Thom, was the daughter of a captain in the merchant marine. After three years at Black's School he went in 1845, at the age of ten, to Edinburgh High School, where he remained for four years. The education there was almost entirely classical, which was fortunate for Geikie, who showed a remarkable quality of mind and a great zest for study, and took full advantage of it to lay the foundation for his outstanding achievements in the literary exposition of science. Holiday pursuits, mostly of a scientific nature, suddenly became concentrated on geology when, with some of his schoolmates, Geikie found fossils in the limestone quarries of Burdiehouse, a few miles south of Edinburgh. His father introduced him to some of the professors and savants of the city; of these Robert Chambers, the publisher, and John Fleming, the distinguished naturalist, were particularly helpful and encouraging. Geikie himself eagerly read every book on geology he could find—Hugh Miller's *Old Red Sandstone* was his chief inspiration—and rambled among the strata and old volcanic rocks of Edinburgh and its environs.

In 1850 Geikie started a banking career, but he found the legal work that this involved "unspeakably dull." In 1851 he spent a holiday on the Isle of Arran and published an article, "Three Weeks in Arran by a Young Geologist," in an Edinburgh newspaper. This secured an introduction to Hugh Miller, who befriended him. His chief benefactor was the writer and lecturer on chemistry, George Wilson. Wilson introduced Geikie to Alexander Macmillan, head of the publishing firm, who became Geikie's intimate friend. Another eminent man to give Geikie encouragement was the geologist James David Forbes, a founder of the British Association.

In the summer of 1853 Andrew Ramsay, who held a responsible position in the Geological Survey of Great Britain, came to Scotland. Geikie was introduced to him and Ramsay held out hope that a place might sometime be found for him on the Survey. His banking career having been finally abandoned, it was decided that Geikie should meanwhile enter Edinburgh University. He matriculated in November 1854 as a student of classics and literature, and at the end of the session he had gained the reputation of being one of the best scholars of his year. He was then forced to leave the university owing to a financial crisis in the family. In 1855 he was recommended, by both Hugh Miller and Ramsay, to Sir Roderick Murchison, who had succeeded Sir Henry de la Beche as head of the Geological Survey. He was immediately appointed to that service as an assistant in the mapping of the Lothian counties. Between 1856 and 1859 Geikie was surveying on his own. In 1859 he visited London for the first time and was introduced into London learned circles, particularly by Leonard Horner and Charles Lyell.

During the next eight years Geikie vigorously carried out his official work in the field and also undertook highly significant researches in his holiday periods. In his spare time he wrote geological works and prepared papers for reading at the meetings of scientific societies. Geikie made several prolonged official excursions into various parts of Britain and traveled abroad to the Auvergne and to Norway. All the time he was thus widening and deepening his knowledge and experience. He was called on to deliver a course of lectures at the School of Mines in London and was also an examiner in the University of London. In 1861 he was elected to the Royal Society of Edinburgh, and he became a fellow of the Royal Society of London in 1865.

In 1867 Geikie was appointed director of the newly constituted Scottish branch of the Geological Survey; and for the next few years he was busy finishing his own personal mapping, superintending his staff, and making official and semiofficial visits abroad. He was taken seriously ill in the Lipari Islands in 1870, the same year Murchison founded a chair of geology at Edinburgh University and desired that Geikie be the first professor. After some difficulty in official circles, largely over the question of this post and that of the director of the Survey in Scotland being held by the same person, Geikie was appointed in 1871. In that year he married Alice Gabrielle Pignatel, whom he had met at Alexander Macmillan's house. She was a descendant of the Pignatelli family, southern Italians who had migrated into France some generations before and had settled in Lyons as merchants. The couple moved into a house that Geikie had already bought for himself on Castle Hill in Edinburgh.

Geikie was able to combine his professorial duties, which lasted from November to March, with the winter work of the Survey directed from Edinburgh. He threw himself enthusiastically into the conduct of his classes and excursions, drawing the necessary diagrams himself and introducing the newest techniques. In the evenings he was kept busy with his literary work, particularly with the large biography of Murchison which he had undertaken (Murchison had died in the autumn of 1871 and was succeeded by Ramsay), the preliminary work on his great textbook, and preparing a memoir (on the Old Red Sandstone) to be read before the Royal Society of Edinburgh. With the Geological Survey his most important charge, his life during the 1870's was fuller than ever. In the summer of 1879 he made a prolonged journey through the United States, meeting the geologists who had become famous for their explorations of the far west.

Ramsay, in failing health, retired as the head of the Geological Survey at the end of 1881, and Geikie was appointed to succeed him in this post, the highest in the profession. He and his family therefore moved to London. Reluctantly he had to relinquish his post as professor at Edinburgh, to which his brother James succeeded. Living in London enabled him to participate directly in the concerns of the learned scientific societies. In 1883 he was elected to the council of the Geological Society, which in 1881 had awarded him its Murchison Medal; and in 1885 he was elected to the council of the Royal Society, becoming its foreign secretary in 1889. In 1890 he was elected president of the Geological Society, and in 1891 he was knighted. Geikie was given the Geological Society's highest award, the Wollaston Medal, in 1895; and in 1896 he received the Royal Society's Royal Medal.

In 1901 Geikie, having reached the age of sixty-five, retired from the directorship of the Geological Survey and found relaxation and refreshment in turning to interests other than geological—largely English and classical literature. He was a member of the Classical Association from the time of its foundation in 1904 and in 1910 was elected its president—an extraordinary position to be held by a leading scientist. At the same time he continued to pursue his geological writing, translating, and editing. He was called upon to preside for the second time over the Geological

Society of London during the centenary celebrations in 1907 and was appointed its foreign secretary in 1908, an honor to which he was annually reelected until his death. In 1907 he was created Knight Commander of the Bath.

The most important and engrossing work that occupied Geikie during the years of his retirement was the conduct of the affairs of the Royal Society. In 1903 he became one of its two secretaries for five years and in 1908 was elected president, the only geologist ever so honored.

Geikie's natural genius, combined with his prodigious capacity for hard and sustained work, is clearly and abundantly revealed in his published writings, which, beginning in 1849, extend over a period of seventy-five years. His first was a schoolboy translation of a piece by Ovid into English verse (inserted in Stevens' *History of the High School of Edinburgh*), and the last was his autobiography, *A Long Life's Work*. His main publications fall under the following classifications:

1. The geological treatises setting forth the results of research, published in the journals of learned societies. Examples are those, mentioned below, concerning ancient volcanism, the Old Red Sandstone, and glaciation.

2. His contributions to the official Geological Survey memoirs. These are somewhat scattered and mixed with the contributions of his colleagues, some being published long after most of his own work concerning them had been finished. He was the main author of the first edition of the memoir on the Edinburgh neighborhood (1861) and of the memoirs on Fife (1900, 1902). He also was the originator and planner of the great series of stratigraphical monographs (on the Pliocene, Cretaceous, Jurassic, and Scottish Lower Paleozoic) published during his directorship.

3. Books on special themes. Two are outstanding and are considered below: *The Scenery of Scotland* (1865) and *Ancient Volcanoes* (1897).

4. Textbooks. While pursuing his researches in particular areas and on particular subjects, Geikie kept before him the whole panorama of geological science and produced a series of incomparable textbooks. The first was a largely rewritten edition (1872) of Joseph Jukes's *Student's Manual of Geology.* Then came (1873) two very small "primers," *Physical Geography* and *Geology.* They had an enormous success, immediate and continuing, and were translated into most European languages. *Field Geology* (1876), *Class-book of Physical Geography* (1877), and *Class-book of Geology* (1886) followed. All these works ran to many editions. In 1882 appeared the first edition of *Text-book of Geology,* which, in its successive editions of 1885, 1893, and 1903, served as a standard work which carried on the function of Lyell's famous *Principles of Geology.* It was based on a long article contributed to the *Encyclopaedia Britannica* in 1879. A storehouse of geological facts, a mine for the extraction of clear definitions and explanations, an authoritative exposition of established principles, and, with its copious notes and references, a worldwide guide to the records of geological discovery, the two-volume edition of 1903 is still indispensable, notwithstanding all the new information and techniques that have been made available during the years since its publication.

5. Biographies of geologists, annotated editions and translations, and histories of geology. Geikie wrote several well-known biographical and historical books. The memoirs of Murchison (1875) and Ramsay (1895), enlivened by contemporary letters and diaries, illuminate the progress of geology during the middle decades of the nineteenth century. He had previously written the greater part of the life of Edward Forbes (begun by George Wilson), the eminent naturalist and geologist (1861), and had communicated an obituary memoir of James David Forbes to the Edinburgh Geological Society (1869). His *Founders of Geology* was, as first published in 1897, the text of six lectures delivered at Johns Hopkins University, Baltimore. It was recast and greatly enlarged for the second edition of 1905, which treats the history of geology as revealed in the lives and work of the masters, both British and foreign. This is still the only work on its subject. The autobiography, written in his old age, shows an undiminished power to instruct, charm, and entertain.

Geikie's erudition was specially brought into play in the production of four works of editorship and commentary: (1) part of the third volume of Hutton's *Theory of the Earth* of 1795, the manuscript of which was found in the library of the Geological Society of London (1899); (2) a new translation of Faujas de Saint-Fond's *Voyage en Angleterre, en Écosse et aux Îles Hébrides* of 1797 (1907); (3) *Charles Darwin as Geologist,* a Cambridge University Rede lecture (1909); and (4) an appreciation of the work of John Michell, the eighteenth-century scientific savant who had been Woodwardian professor of geology at Cambridge (1918). These are all enriched with copious notes which illuminate the more obscure corners of the history of geology.

6. Belles-lettres. Geikie wrote innumerable essays and reviews—geological, geographical, biographical,

autobiographical, literary, historical, and educational. Fortunately, many of these were reprinted to form the two books *Geological Sketches at Home and Abroad* (1882) and *Landscape in History and Other Essays* (1905).

7. *Scottish Reminiscences* (1904). This book stands rather apart from his other publications. Here, more than anywhere else in his writing, Geikie's wit and humor and powers as a raconteur are allowed full play.

Geikie's fame as a geologist rests very largely on his work in the field of volcanic action in past geological times. He was born among the crags that mark the sites of some of these old volcanoes (of Carboniferous age) and was led in boyhood to interest himself in their structure and history. Little had been done before on these rocks beyond the recognition of their volcanic nature and the description of some details of local structure (chiefly by Charles Maclaren in 1839). As a result of his survey of the Lothians (1855–1859), Geikie was able to recognize a long series of eruptions through Devonian (Old Red Sandstone) and Carboniferous times. He summarized these results in a paper read before the British Association at Aberdeen (1859), which was expanded to form one read before the Royal Society of Edinburgh in 1861. In 1860 Geikie began his survey of the county of Fife. He tells us that he learned more here than he did anywhere else regarding the details of volcanic action as preserved in the earth's crust. The results were not published in official memoirs until much later: 1900 and 1902. His enthusiasm for this vast subject increased as his official work and holiday excursions took him to other parts of Scotland and the British Isles and to central France. He read his first paper on the Tertiary volcanic rocks of Scotland to the Royal Society of Edinburgh early in 1867. Later in 1867, in his presidential address to the Geological Section of the British Association at Dundee, on the history of volcanic action in the British Isles, Geikie made the first attempt to group in chronological order the sequence of eruptions in this westernmost part of Europe. His visits to the Eifel district of Germany in 1868 and to Naples and the Lipari Islands in 1870 extended his personal knowledge of recent, or comparatively recent, volcanic activity.

In 1879 the Royal Society of Edinburgh published a large memoir describing the volcanic history of the Carboniferous period in the basin of the Firth of Forth, containing in condensed form the results of Geikie's researches in that region during nearly a quarter of a century. In 1888 the same society published an even longer memoir, on the history of volcanic action during the Tertiary period in the British Isles. Geikie's journey in 1879 to western America, where Richthofen had shown the great volcanic plains to be due to massive fissure eruptions, caused him to advocate a similar origin for the volcanic rocks forming the plateaus of Scotland, Northern Ireland, the Faeroe Islands, and Iceland.

Geikie's presidential addresses to the Geological Society of London in 1891 and 1892 gave accounts of the history of volcanic action in the British Isles, and in 1896 this society published an extensive paper on the Tertiary basalt plateaus of northwestern Europe in which the results of the latest exploration were described.

Finally, in 1897 there appeared, in two large volumes, Geikie's masterpiece, *The Ancient Volcanoes of Great Britain.* Here we have a complete and detailed exposition of the essentials of igneous geology (other than the deeply plutonic). It is illustrated with a wealth of examples, described in word, diagram, and picture. As in all Geikie's writings, the subject is given depth by constant reference to the history of investigation. Exhaustive as this was of the knowledge of the time, Geikie himself was the foremost in realizing that its publication, far from closing the subject, opened up still wider horizons; for in 1895, two years before its appearance, he had launched a new campaign by the Geological Survey into the territory of the Tertiary volcanoes of Scotland by sending Alfred Harker to Skye. This campaign produced spectacular results during the next thirty years.

Apart from his work on the igneous rocks associated with the Old Red Sandstone strata, Geikie did important work on these strata themselves. He was the first to recognize a widespread unconformity throughout Scotland between an upper and a lower group (1860). His most important paper was that published by the Royal Society of Edinburgh on the Old Red Sandstone of Scotland, particularly on that of the northern region (1878). Further accounts were projected but did not materialize. In this paper he suggested that the Old Red Sandstone was deposited in separate lake basins, to which he gave names, the chief ones in Scotland being "Lake Orcadie" in the north and "Lake Caledonia" in the Central Valley. In each of these basins he considered the upper divisions to be equivalent in time; the same was true of the lower divisions. These large questions on which Geikie expressed his views are still not settled. It is considered that the general nature of the deposits certainly indicates a "Continental" origin, but probably not entirely a lacustrine one. The conception of separate basins of deposition is, however, taken to be valid. The general opinion now, based chiefly on the evidence of the fossil fish faunas, is that a tripartite time classification of the system is appropriate, at least in Scotland, a suggestion that had been put forward by Murchison in 1859. The lower division in "Orcadie" is mainly, if not entirely, Middle Old Red Sandstone, while the lower division in "Caledonia" is Lower Old Red Sandstone.

During his directorship of the Geological Survey, Geikie's most important operation was his direction of a vigorous and large-scale exploration of the extreme northwestern Highlands of Scotland. It had become realized that here lay a key to the unlocking of a knowledge of the structure and geological history of Scotland, but very different interpretations of what was to be seen on the ground had been put forward by some of the foremost geologists of the day. The official attack on the problem began in 1883 and lasted for about ten years. An interim report was given in 1888, and the complete description was finally published in the great Survey memoir of 1907. This memoir contains a detailed account of the history of research and ideas (with full documentation), particularly of the period before the Survey took over. Flett (1937) and Bailey (1952) provide commentaries, but each is rather disconnected.

Murchison, during the period of his directorship, had made geological excursions into the region, accompanied at one time (1859) by Ramsay and at another, during his most prolonged excursion (1860), by Geikie. Murchison had no doubt of the correctness of an interpretation which he expounded in full in 1861—essentially it was that there was an undisturbed upward succession of formations—and both Ramsay and Geikie agreed with this interpretation. Others, particularly James Nicol, Charles Callaway, and Charles Lapworth, had detected the great dislocations along what later came to be called the Moine thrust belt. Geikie sent his team onto the ground (1883) on the assumption that the Murchisonian interpretation would be found to be the correct one, and he proclaimed this assumption in his official report for that year. But the surveyors soon found that this view could not be sustained and that the interpretations given by Nicol, Callaway, and Lapworth were, on the main point at issue, correct. Geikie at once visited the ground, with every disposition to support Murchison's view, but he became entirely convinced that his surveyors were right. He immediately announced the important news in an article written by himself and his two chief surveyors, Benjamin Peach and John Horne, which was published in *Nature* in November 1884. His report for that year, published in 1885, contained the official announcement of the firm establishment as a known fact of what is perhaps the most striking geological feature in the structure of Britain, as well as a frank admission of his previous adherence to what had proved to be an erroneous assumption.

During the years 1861–1865, Geikie became, as he put it, "rather obsessed with glacial problems." The existence of a former "ice age" during which glacial conditions had been much more widespread than at the present time had been realized during the 1830's by geologists in the Alpine region; foremost among them was Louis Agassiz, whose chief publication appeared in 1840. It was shown that the Alpine glaciers had extended far beyond their present limits. At the same time the superficial deposits of stony clay, which in Britain had hitherto been generally called "diluvium" in the vague belief that they were relics of a universal flood, were being accounted for on the supposition that they had been transported and dropped by floating ice while the land was submerged beneath the sea. In the latter part of 1840, when Agassiz came to Scotland, he recognized the clear signs of the previous existence of land glaciers and succeeded at once in convincing such leading British geologists as William Buckland and Charles Lyell. All three gave their views to the Geological Society in November 1840. But the floating ice theory persisted and was the favored theory when three geologists—T. F. Jamieson, Andrew Ramsay, and Geikie—became independently interested in the matter; and each in his own way showed that it was untenable. Geikie investigated the deposits all over the southern counties of Scotland and gave a full description and discussion in a paper published by the Geological Society of Glasgow in 1863. At about this time it was decided that these superficial deposits (the "drifts") should be officially mapped. They had hitherto been regarded only as a nuisance obscuring the underlying rocks; now it was realized that they were not only of great interest in themselves but that the facts of their distribution were of great practical importance to agriculture. To carry out the mapping in Scotland a few new recruits were required, and among those appointed was Archibald's younger brother, James. Archibald Geikie had to initiate the new men into the mapping of the drifts, but after this had been done he abandoned his glacial work; it was taken up by his brother, who became one of the foremost authorities on this branch of geology. On his trip to arctic Norway in 1865 he, as he tells us, "caught the ice, as it were, in the very act of doing the work of which I had hitherto only seen the ancient results."

Geikie had an abiding interest in that earth science now called geomorphology, the study of landforms, but this interest was greatest during the early 1860's, when he was also especially interested in glacial geology. To some extent these two interests overlapped. In both fields there had been much doubt and argument during the preceding few decades, and in both it was Geikie who was among the foremost in settling the matter. In 1860 he made his tour with Murchison to the northwest Highlands of Scotland (primarily to study structure), and in 1861 he visited the Auvergne district of France (primarily to study volcanicity). On both these tours he was greatly impressed with the effects of erosion by weathering agents and the action of rivers (and, long ago, of ice in Scotland). The thesis he established was that the present relief of the land surface is the result of the subaerial erosion ("denudation") of a more or less even or gently curved surface which had been uplifted from beneath the sea. The sculpturing of the land is chiefly the result of the differential erosion of hard and soft rocks. This was in contrast with the thesis that valleys and hills and escarpments are due to the marine dissection of a land mass rising from beneath the sea. The subaerial erosion theory still left uncertain when the land surface was uplifted and what its height and form had been: these uncertainties remain today. Geikie's conclusions were given in his book *The Scenery of Scotland,* first published in 1865, which combines the virtues of a thoroughly scientific treatise with those of a popular account.

BIBLIOGRAPHY

I. ORIGINAL WORKS. Among Geikie's more important writings are *The Geology of the Neighbourhood of Edinburgh,* Memoirs of the Geological Survey of Great Britain (London, 1861); "On the Chronology of the Trap Rocks of Scotland," in *Transactions of the Royal Society of Edinburgh,* **22** (1861), 633–653; "On the Phenomena of the Glacial Drift of Scotland," in *Transactions of the Geological Society of Glasgow,* **1** (1863), 1–190; *The Scenery of Scotland* (London, 1865); *Geology,* in the series Science Primers (London, 1873); *Life of Sir Roderick Murchison* (London, 1875); *Outlines of Field Geology* (London, 1876); "On the Old Red Sandstone of Western Europe," in *Transactions of the Royal Society of Edinburgh,* **28** (1878), 345–452; "On the Carboniferous Volcanic Rocks of the Basin of the Firth of Forth," *ibid.,* **29** (1879), 437–518; *Geological Sketches at Home and Abroad* (London, 1882); *Text-book of Geology* (London, 1882; 4th ed., 1903); "The History of Volcanic Action During the Tertiary Period in the British Isles," in *Transactions of the Royal Society of Edinburgh,* **35** (1888), 23–184; "The History of Volcanic Action in the British Isles," printed as "Proceedings. Anniversary Address of the President," in *Quarterly Journal of the Geological Society of London,* **47** (1891), 63–122, and **48** (1892), 60–179; *Memoir of Sir Andrew Crombie Ramsay* (London, 1895); "The Tertiary Basalt-Plateaux of North-western Europe," in *Quarterly Journal of the Geological Society of London,* **52** (1896), 311–406; *The Ancient Volcanoes of Great Britain* (London, 1897); *The Geology of Fife,* Memoirs of the Geological Survey of Great Britain (London, 1902); *Scottish Reminiscences* (Glasgow, 1904); *The Founders of Geology* (London, 1905); and *A Long Life's Work: An Autobiography* (London, 1924).

II. SECONDARY LITERATURE. On Geikie or his work, see E. B. Bailey, *Geological Survey of Great Britain* (London, 1952), *passim;* E. B. Bailey and D. Tait, *Edinburgh's Place in Scientific Progress* (Edinburgh, 1921), pp. 89–91; R. J. Chorley *et al., The History of the Study of Landforms* (London, 1964), *passim;* "Eminent Living Geologists: Sir Archibald Geikie," in *Geological Magazine,* **27** (1890), 49–51; J. S. Flett, *The First Hundred Years of the Geological Survey of Great Britain* (London, 1937), *passim;* "A Century of Geology, 1807–1907: The Geological Society of London," in *Geological Magazine,* **44** (1907), 1–3; B. N. Peach and J. Horne, "The Scientific Career of Sir Archibald Geikie," in *Proceedings of the Royal Society of Edinburgh,* **45** (1926), 346–361; H. H. Thomas, in *Dictionary of National Biography, 1922–1930* (Oxford, 1937), pp. 332–334; and H. B. Woodward, *The History of the Geological Society of London* (London, 1908), *passim.*

Obituary notices include J. R. B. and J. H., in *Proceedings of the Royal Society,* **3A** (1926), xxiv–xxxix; and **99B** (1926), i–xvi; H. D., in *Proceedings of the Geologists' Association,* **36** (1925), 191–192; and A. S., in *Quarterly Journal of the Geological Society of London,* **81** (1925), "Proceedings," lii–lx.

JOHN CHALLINOR

GEIKIE, JAMES (*b.* Edinburgh, Scotland, 23 August 1839; *d.* Edinburgh, 1 March 1915), *geology.*

James Geikie was the son of James Stuart Geikie and Isabella Thom and the brother of Archibald Geikie. After leaving Edinburgh High School, for a few years he divided his time between working in a printer's office and attending the University of Edinburgh. Under the supervision of his brother, he joined the Geological Survey in 1861 as one of the men recruited to map the glacial deposits ("drifts") of parts of central Scotland. He devoted himself with sustained ardor to this work, with the result that the study of Pleistocene geology in general, and the glacial deposits in particular, became his lifework. In these fields he became the leading British authority. His official work in the mapping and investigation of the Paleozoic rocks and the superficial deposits overlying them in the areas allotted to him was of the highest quality but is hardly given full expression in the Survey publications.

It is on his unofficial papers and his books, *The Great Ice Age* (1874) in particular, that Geikie's reputation rests. This work was dedicated to Andrew Ramsay, the director general of the Geological Survey, who had already shown the effects of land ice and to whom he owed much as teacher and friend.

A second edition appeared in 1877, and in 1894 a third edition was published, extensively revised in accordance with new knowledge and with the advance in Geikie's own opinions. His book *Prehistoric Europe* (1880) supplemented it. In *The Great Ice Age* Geikie put forward the hypothesis that the glacial period as a whole had been interrupted by mild episodes or interglacial periods. This hypothesis was suggested to him by evidence in Scotland, where (as in the rest of Britain) the evidence is not very definite; thus he was at first hesitant. But support of a more conclusive nature was forthcoming from the Continent, and Geikie became a progressively stronger advocate for the existence of interglacial periods. His fundamental contention—now generally accepted—was strikingly supported by Albrecht Penck and Eduard Bruckner in their description of the glacial phenomena of the Alpine valleys published in 1909 in a volume dedicated to Geikie. He further taught that man lived in Europe throughout the glacial period, a theory that also has gained acceptance.

Geikie succeeded his brother Archibald as Murchison professor of geology at the University of Edinburgh in 1882. He had left the Geological Survey with great reluctance but enthusiastically took up the work of geological and geographical education. In 1884 he was one of the founders of the Royal Scottish Geographical Society and was its president from 1904 to 1910. For many years he was editor of its magazine. Geikie conducted his university teaching, both in the classroom and in the field, with great success and showed administrative ability within the university. For his students he wrote *Outlines of Geology* (1886; 4th ed., 1903) and *Structural and Field Geology* (1905). The latter was particularly successful and has been brought up to date in a sixth edition, published in 1953. He retired from his professorship in June 1914.

Geikie joined the Geological Society of London in 1873. He was awarded the Murchison Medal in 1889 and in the same year received the Brisbane Medal of the Royal Society of Edinburgh. He was elected a fellow of the Royal Society of London in 1875 and was president of the Royal Society of Edinburgh at the time of his death.

In 1875 Geikie married Mary Johnston, whose family lived in the region of the Cheviot Hills. A man of great activity and many interests, he published *Songs and Lyrics by Heinrich Heine,* translated from the German, in 1887.

BIBLIOGRAPHY

I. ORIGINAL WORKS. A full list of Geikie's writings is in Newbigin and Flett (see below). Among the more important are "On the Buried Forests and Peat Masses of Scotland," in *Transactions of the Royal Society of Edinburgh,* **24** (1867), 363–384; *The Great Ice Age* (London, 1874); *Prehistoric Europe* (London, 1880); "On the Geology of the Faroe Islands," in *Transactions of the Royal Society of Edinburgh,* **30** (1882), 217–269; *Outlines of Geology* (London, 1886); *Fragments of Earth Lore* (Edinburgh, 1893); *Earth Sculpture* (London, 1898); *Structural and Field Geology* (Edinburgh, 1905); *Mountains: Their Origin, Growth and Decay* (Edinburgh, 1913); and *Antiquity of Man in Europe* (Edinburgh, 1914).

II. SECONDARY LITERATURE. On Geikie or his work, see "Eminent Living Geologists: James Geikie," in *Geological Magazine,* **50** (1913), 241–248; J. H., in *Proceedings of the Royal Society,* **B91** (1920), "Obituary Notices," xxxiii–xxxv; J. Horne, "The Influence of James Geikie's Researches on the Development of Glacial Geology," in *Proceedings of the Royal Society of Edinburgh,* **36** (1917), 1–25; M. I. Newbigin and J. S. Flett, *James Geikie: The Man and the Geologist* (Edinburgh, 1917); and an obituary notice in *Quarterly Journal of the Geological Society of London,* **72** (1916), liii–lv.

JOHN CHALLINOR

GEISER, KARL FRIEDRICH (*b.* Langenthal, Bern, Switzerland, 26 February 1843; *d.* Küsnacht, Zurich, Switzerland, 7 May 1934), *mathematics.*

Geiser was the son of Friedrich Geiser, a butcher, and Elisabeth Geiser-Begert. Following graduation from the Polytechnikum in Zurich and the University of Berlin, where the influence of his great-uncle, Jakob Steiner, was of help to him, Geiser became *Dozent* in 1863. In 1873 he became professor at the Zurich Polytechnikum (later renamed the Eidgenössische Technische Hochschule), where he remained until his retirement.

Geiser made an outstanding contribution to the development of the Swiss system of higher education. Acquainted with many persons in the fields of politics and economics as well as with important mathematicians in the neighboring countries, and a close adviser of the chairman of school supervisors, Geiser worked effectively within the professoriate to attract first-rate teachers. There devolved upon him, above all, the instruction of candidates for the teaching of algebraic geometry, differential geometry, and invariant theory.

Geiser's scientific works are concerned especially with algebraic geometry. He explained the relation of the twenty-eight double tangents of the plane quadric to the twenty-seven straight lines of the cubic surface. An involution that he discovered bears his name. Minimal surfaces also engaged his attention: he investigated the intersection of an algebraic minimal surface with an infinite plane and determined all the algebraic minimal surfaces. In addition, Geiser edited Jakob Steiner's unpublished lectures and treatises. He was organizer and president of the first International Congress of Mathematicians, held at Zurich in 1897.

BIBLIOGRAPHY

I. ORIGINAL WORKS. Geiser's major works are *Einleitung in die synthetische Geometrie* (Leipzig, 1869); and *Zur Erinnerung an Jakob Steiner* (Schaffhausen, 1874). His papers are listed in the articles by Kollros and by Meissner and Scherrer (see below).

Die Theorie der Kegelschnitte, pt. 1 of Jakob Steiner's *Vorlesungen über Geometrie,* was compiled and edited by Geiser (Leipzig, 1867).

II. SECONDARY LITERATURE. See the following obituary notices: A. Emch, in *National Mathematics Magazine,* **12,** no. 6 (1938), 287–289, with portrait; L. Kollros, in *Verhandlungen der Schweizerischen naturforschenden Gesellschaft,* **115** (1934), 522–528, with list of publications and portrait; and E. Meissner and F. R. Scherrer, in *Vierteljahrsschrift der Naturforschenden Gesellschaft in Zürich,* **79** (1934), 371–376, with list of publications.

J. J. BURCKHARDT

GEISSLER, JOHANN HEINRICH WILHELM (*b.* Igelshieb, Thuringia, Germany, 26 May 1815; *d.* Bonn, Germany, 24 January 1879), *glassmaking, technology.*

Geissler's father, Johann Georg Jacob Geissler, was a maker of glass beads and a burgomaster. His mother, Johanne Rosine Eichhorn, was the daughter of a glassmaker. He was descended from other craftsmen active in the Thüringer Wald and in Böhmen, and two of his brothers worked as mechanicians, one in Berlin and the other in Amsterdam. Comparatively little is known of Geissler's life. He began working while very young and, having learned glassblowing in the duchy of Saxe-Meiningen, he practiced this trade at several universities, including Munich. He is said to have worked in the Netherlands for some eight years; he settled at The Hague in 1839 and is on census rolls of 1845 but not of 1849, although his workshop in Bonn is said to have been founded by him in 1841. In that year his brother, Friedrich Wilhelm Florenz, went to Amsterdam and made glass for the Gymnasium Illustre, especially for the physicist V. S. van der Willigen.

Geissler finally settled as a mechanic at the University of Bonn in 1852 (or earlier) and established a workshop for producing chemical and physical instruments. He provided many instruments for the mechanician W. H. Theodor Meyer and the physicist Julius Plücker, as well as for the mineralogist H.P.J. Vogelsang and the physiologist Eduard Pflüger. Later he was associated with Franz Müller, who succeeded him as owner of the workshop. On the fiftieth anniversary of the University of Bonn in 1868 he was awarded an honorary doctorate for his work.

Talented both in making instruments and in comprehending their physical bases, Geissler learned much from his association with the scientists at the university. He became an indispensable participant in the experimental work that was being conducted there.

The first account of Geissler's activity dates from 1852, when, with Julius Plücker, at Bonn, he constructed his famous standard thermometers. They differed from the thermometers then in use by their thin glass, by the application of capillarity, and by their high precision. For calibrating he used his new glass balance that had a sensitivity of 0.1 mg. of mercury. In 1863 Geissler constructed a maximum thermometer based on Casella's minimum thermometers. Also in 1852 Geissler, at the invitation of a Bonn industrialist, constructed an instrument for measuring the alcoholic strength of wine. This "vaporimeter" measured the pressure of vapors of alcohol and air against mercury. Plücker improved the apparatus by eliminating the air and was thus stimulated to investigate vapors scientifically. Geissler also used the instrument to measure the strength of liquid ammonia.

In 1858 it was stated that Plücker relied on Geissler's dexterity in using "Geissler's tubes," as Plücker called them—although, according to Plücker, Geissler was not the first to make such tubes. Geissler, in turn, stated in 1858 that he had made such tubes since 1857 and had sent many of them to Daniel Rühmkorff in Paris and to Bence Jones in London. There is a second invention connected with these tubes. The difficulty in obtaining a vacuum with the piston pumps then in use caused Geissler to construct a mercury air pump about 1855. Demonstrated to a wider public in 1858, it was an improvement on the idea of using Torricelli's vacuum for evacuation, which the theosopher Emanuel Swedenborg had described in 1722 and many others after him had tried to use. Geissler's pump, entirely of glass and thick rubber, was operated by manually moving a second tube up and down and therefore was slow, though effective.

Through use of this pump and his aptitude for glassblowing, Geissler was able to make rather small glass tubes with electrodes melted into the ends and filled with rarefied gases. By using these tubes Plücker was able to study discharges in very rarefied gases. The new and most interesting phenomenon, the stripes in the discharge light (*Schichtung im elek-*

trischen Licht), previously observed in the "electrical egg" by Rühmkorff and Jean Quet, could now, by means of "Geissler's tubes," be studied in detail. These tubes provided a much better vacuum and, in contrast with the "electrical egg," were not dependent on a pump. (John Gassiot had already tried to produce such a vacuum, and in 1856 V. S. van der Willigen had obtained the phenomenon using tubes constructed by Geissler's brother in Amsterdam.) At the thirty-ninth meeting of the Deutsche Naturforscher und Ärzte, held at Giessen in 1864, Geissler demonstrated his mercury pump; "Geissler's tubes" were shown there in experiments by J. C. Poggendorff to demonstrate induced currents. On the whole, the technology of "Geissler's tubes" helped to introduce a new branch of physics which led directly to the discovery of the cathode rays.

In addition, the thermometer tubes enabled Geissler to construct in 1852 an instrument for measuring the expansion of freezing water and ice. In 1858 Geissler suggested to Justus von Liebig that the chemical nature of gases could be identified by means of the discharge in "Geissler's tubes"; this was further implemented by Vogelsang and Geissler in 1868, when they described a vacuum tube in which liquids occluded in minerals could be identified chemically by means of the then new technique of spectral analysis. About 1860, in the inner of these tubes, Geissler converted white phosphorus into the red form "by electricity," as he put it. In 1873 he demonstrated the phenomenon before Anton von Schrötter at the world exhibition in Vienna, where he was awarded the Golden Cross of Merit in Art and Science.

BIBLIOGRAPHY

Descriptions of Geissler's work, by himself and his collaborators, are in *Annalen der Physik und Chemie,* **162** (1852), 233–279; **168** (1854), 199–200; **179** (1858), 88–106; **199** (1864), 657–658; **201** (1865), 153–158; **202** (1865), 222, 227; **211** (1868), 332–335; and **228** (1874), 171–173; and "Sitzungsberichte der Niederrheinischen Gesellschaft für Natur- und Heilkunde zu Bonn," in *Verhandlungen des naturhistorischen Vereines der preussischen Rheinlande und Westfalens,* **23** (Bonn, 1866), 12–14; W. H. T. Meyer, *Beobachtungen über das geschichtete electrische Licht* (Berlin, 1858).

There is an undated letter to an unknown correspondent in the Staatsbibliothek Preussischer Kulturbesitz, Berlin. Three letters to R. A. C. E. Erlenmeyer, dated 14 Jan. 1869, 9 June 1870, and 24 Mar. 1871, are in the Deutsches Museum, Munich. In the Bayerische Staatsbibliothek, Munich, are two letters to Justus von Liebig, one without date and one dated 2 Feb. 1858.

On Geissler's life and work, see also *Amtlicher Bericht über die 39. Versammlung Deutscher Naturforscher und Ärzte in Giessen im September 1864* (Giessen, 1865), pp. 81–86; *Amtlicher Bericht über die Wiener Weltausstellung im Jahre 1873,* 3 vols. (Brunswick, 1874–1875), II, 504 and III, 1, 221; *Bericht über das fünfzigjährige Jubilaeum der Rheinischen Friedrich-Wilhelms-Universität* (Bonn, 1868); *Dr. H. Geissler's Nachfolger, Gedenkblatt zur Erinnerung an Heinrich Geissler, Dr. phil., Glastechniker . . ., zur Feier des 50 jährigen Bestehens der Firma seinen Freunden mit beigeheftetem Bildnisse gewidmet* (Bonn, 1890); A. W. von Hofmann, in *Berichte der Deutschen chemischen Gesellschaft,* **12** (1879), 147–148; W. Huschke, *Forschungen über die Herkunft der Thüringischen Unternehmerschicht des 19. Jahrhunderts,* supp. 2 to *Tradition* (Baden-Baden, 1962), 34–35; H. Schimank, in *Lichttechnik* (Berlin), **6** (1954), 364; and in *Technikgeschichte in Einzeldarstellungen,* **19** (1971), 37–38; and A. Wissner, in *Neue deutsche Biographie,* VI (Berlin, 1964), 59.

HANS KANGRO

GEITEL, F. K. HANS (*b.* Brunswick, Germany, 16 July 1855; *d.* Wolfenbüttel, Germany, 15 August 1923), *experimental physics.*

Geitel studied from 1875 to 1877 at Heidelberg and from 1877 until 1879 at Berlin, where he took the examination for secondary-school teaching. From 1880 to 1920 he taught mathematics and physics in Wolfenbüttel. He published almost all his works in collaboration with Elster. Following the latter's death in 1920, Geitel published a paper on the photoelectric effect in very thin films of potassium, a topic which, like his earlier individual efforts, had been of interest to both (1922). Geitel wrote several comprehensive reports on atmospheric electricity, radioactivity, and photoelectric methods of measurement. Of particular note is "Die Radioaktivität der Erde und der Atmosphäre," in *Handbuch der Radiologie.*

In 1899 Geitel received a Ph.D. *honoris causa* from the University of Göttingen and in 1915 a doctorate in engineering *honoris causa* from the University of Brunswick. Following his retirement he became honorary professor at Brunswick. In 1910 Geitel represented Germany's physicists (and Otto Hahn its chemists) on the International Radium Standard Commission in Paris.

See also Elster article in Vol. 4.

BIBLIOGRAPHY

Geitel's writings include *Über die Anwendung der Lehre von den Gasionen auf die Erscheinungen der atmosphärischen Elektrizität* (Brunswick, 1901); *Bestätigung der Atomlehre durch die Radioaktivität* (Brunswick, 1913); "Die Radioaktivität der Erde und der Atmosphäre," in *Handbuch der Radiologie,* I (Leipzig, 1920), 399–457; "Photoelektrische Messmethoden," in *Handbuch der biologischen*

Methoden (Berlin–Vienna, 1921), pp. 1–38; and "Die Proportionalität von Photostrom und Beleuchtung an sehr dünnen Kaliumschichten," in *Annalen der Physik,* **67** (1922), 420–427.

There is an obituary notice by R. Pohl, in *Naturwissenschaften,* **12** (1924), 685–688.

WALTHER GERLACH

GELFOND, ALEXANDR OSIPOVICH (*b.* St. Petersburg [now Leningrad], Russia, 24 October 1906; *d.* Moscow, U.S.S.R., 7 November 1968), *mathematics.*

Gelfond was the son of Osip Isaacovich Gelfond, a physician who also did work in philosophy. From 1924 to 1927 he studied in the division of mathematics of the department of physics and mathematics at Moscow University; later he took a postgraduate course (1927–1930) under A. J. Khintchine and V. V. Stepanov. In 1929–1930 Gelfond taught mathematics at Moscow Technological College, and from 1931 until his death he was at Moscow University, where for a number of years he held the chair of analysis. He later held the chair of the theory of numbers, to which was subsequently added the history of mathematics. From 1933 he also worked in the Soviet Academy of Sciences Mathematical Institute. He became professor of mathematics in 1931 and doctor of mathematics and physics in 1935; he was elected corresponding member of the Academy of Sciences of the U.S.S.R. in 1939 and corresponding member of the International Academy of the History of Science in 1968.

Most important in Gelfond's scientific work were the analytical theory of numbers and the theory of interpolation and approximation of functions of a complex variable. Studies in both fields were closely related; he used and improved methods of the theory of functions in working on the problems of the theory of transcendental numbers.

In 1748 Euler had expressed the idea that logarithms of rational numbers with rational bases are either rational or transcendental. Generalizing that statement, among the famous twenty-three problems that Hilbert posed in 1900, was the hypothesis of the rationality or transcendence of logarithms of algebraic numbers with algebraic bases; i.e., he presumed the transcendence of a^b, where a is any algebraic number not 0 or 1 and b is any irrational algebraic number. For thirty years no approach to a solution of this, the seventh of Hilbert's problems, could be found. In 1929 Gelfond established profound connections between the growth and other properties of an entire analytic function and the arithmetic nature of its values when the values of argument belonged to a given algebraic field. This enabled him to find, proceeding from the expansion of the exponential function a^z, a being an algebraic number not 0 or 1, into the interpolating series of Newton,

$$a^z = \sum_{n=0}^{\infty} A_n (z - x_0)(z - x_1) \cdots (z - x_n),$$

where $x_0, x_1, x_2, \cdots$ are integers of an algebraic field $K(\sqrt{-D})$, $D > 0$, a solution of the problem in a particular case: the number a^b, where $b = i\sqrt{D}$ and D is a positive integer that is not a perfect square, is transcendental.

In 1930 R. O. Kuzmin extended Gelfond's method to real $b = \sqrt{D}$, and in 1932 C. L. Siegel applied it to the study of the transcendence of the periods of elliptic functions. Soon after, Gelfond consolidated his method with new ingenious ideas and, introducing linear forms of exponential function into consideration, confirmed in 1934 Hilbert's hypothesis in its entirety. His methods and results led to the most important contributions to the theory of transcendental numbers since Hermite's demonstration of the transcendence of e (1873) and K. L. F. Lindemann's of π (1882).

Applying his method to functions of p-adic variables, Gelfond made a number of new discoveries. Among them is the theorem that if α, β, γ are real algebraic numbers and at least one of them is not an algebraic unit, with γ being not equal to 2^n (n is a rational integer), the equation $\alpha^x + \beta^y = \gamma^z$ can possess only a finite number of solutions in rational integers x, y, z (1940). The further development of the method enabled Gelfond to solve a number of problems of mutual algebraic independence of numbers and to construct new classes of transcendental numbers. A considerable part of his discoveries in the theory of transcendental numbers is described in his monograph *Transtsendentnye i algebraicheskie chisla* (1952). Gelfond also wrote on other problems of the theory of numbers, including the diophantine approximations, the distribution of fractional parts of functions, and elementary methods of analytic theory.

In the theory of functions of a complex variable, Gelfond conducted numerous studies on problems of convergence of interpolation processes depending upon the density of a set of basic points of interpolation and upon the properties of the function to be approximated; on necessary and sufficient conditions for the determination of an entire analytical function on its given values or some other element; and on corresponding methods for the construction of functions. These studies were to a great extent summed up in *Ischislenie konechnykh raznostey* (1952).

Gelfond also promoted the history of mathematics, brilliantly characterizing Euler's work and his investigations in the theory of numbers. For many years he was the chairman of the scientific council that refereed theses on the history of physics and mathematics for the Soviet Academy of Sciences Institute of the History of Science and Technology.

Gelfond, creator of a large scientific school in the Soviet Union, profoundly influenced the advance of the theory of transcendentals and the theory of interpolation and approximation of functions of a complex variable.

BIBLIOGRAPHY

I. Original Works. Gelfond's writings include "Sur les nombres transcendants," in *Comptes rendus hebdomadaires des séances de l'Académie des sciences,* **189** (1929), 1224–1228; "Sur le septième problème de Hilbert," in *Izvestiya Akademii nauk SSSR,* **7** (1934), 623–630; "Sur la divisibilité de la différence des puissances de deux nombres entiers par une puissance d'un idéal premier," in *Matematicheskii sbornik,* **7 (49)** (1940), 7–26; *Ischislenie konechnykh raznostey* ("Calculus of Finite Differences"; Moscow–Leningrad, 1952; 3rd ed., 1967), translated into German as *Differenzenrechnung* (Berlin, 1958); *Transtsendentnye i algebraicheskie chisla* (Moscow, 1952), translated into English as *Transcendental and Algebraic Numbers* (New York, 1960); and *Elementarnye metody v teorii chisel* (Moscow, 1962), written with Y. V. Linnik.

Complete bibliographies of Gelfond's works are in Linnik and Marcushevich and Pjatetsky-Shapiro and Shidlovsky (see below).

II. Secondary Literature. On Gelfond or his work, see Y. V. Linnik and A. I. Marcushevich, "Alexandr Osipovich Gelfond," in *Uspekhi matematicheskikh nauk,* **11,** no. 5 (1956), 239–248; *Matematica v SSSR za sorok let* ("Forty Years of Mathematics in the U.S.S.R."), 2 vols. (Moscow, 1959), index; *Matematica v SSSR za tridtsat let* ("Thirty Years of Mathematics in the U.S.S.R."; Moscow–Leningrad, 1948), index; I. I. Pjatetsky-Shapiro and A. B. Shidlovsky, "Alexandr Osipovich Gelfond," in *Uspekhi matematicheskikh nauk,* **22,** no. 3 (1967), 247–256; and I. Z. Shtokalo, ed., *Istoria otechestvennoy matematiki* ("History of Native Mathematics"), III (Kiev, 1968), index.

A. P. Youschkevitch

GELLIBRAND, HENRY (*b.* London, England, 17 November 1597; *d.* London, 16 February 1636), *navigation, mathematics.*

Gellibrand was the son of a graduate of All Souls College, Oxford. He became a commoner of Trinity College, Oxford, in 1615, a few weeks after his father's death. After graduating in arts (B.A., 1619, M.A., 1623) he took holy orders, and before 1623 he held a curacy at Chiddingstone, Kent. Gellibrand was introduced to mathematics by one of Sir Henry Savile's lectures, and he had at least enough geometry to set up a sundial on the east side of his college quadrangle. When the professorship of astronomy at Gresham College, London, was vacated following the death of Edmund Gunter, Gellibrand was elected to the chair on 2 January 1627. He completed the second volume of his sponsor Henry Briggs's *Trigonometria Britannica* (left unfinished at his death in 1630) and saw it through the press in 1633.

By this time Gellibrand's Puritanism had brought him into conflict with William Laud, then bishop of London. Gellibrand and his servant were cited by Laud before the Court of High Commission in 1631 for the publication of an almanac in which the saints and martyrs from John Foxe's *Book of Martyrs* replaced those permitted by the Church of England. They were acquitted on the grounds that this was not the first almanac of its kind, and the case was later cited against Laud at his own trial in 1643.

Gellibrand's most widely appreciated scientific discovery, which he should share with John Marr, was that of the secular change in the magnetic variation (declination). It was announced, without much comment, in *A Discourse Mathematicall on the Variation of the Magneticall Needle, Together With Its Admirable Diminution Lately Discovered* (1635). His predecessor, Gunter, had noticed that the variation at Limehouse in 1622 differed from the value found by William Borough in 1580, but he ascribed the difference to an error on Borough's part. In 1633 some rough observations of his own and John Marr's convinced Gellibrand that the value was now even less, but not until 1634 was he sufficiently confident to make a categorical assertion of its secular change. As his main evidence he referred to an appendix to Edward Wright's *Certaine Errors in Navigation . . .* (1599, 1610). This contains a compendium of recorded values of variation at various places made by a number of physicists and navigators the world over. (Henry Bond, editor of *Tapp's Seaman's Kalendar,* spent many years elaborating upon Gellibrand's findings and argued that despite its change, variation could even now be used by sailors to determine terrestrial longitudes. This would have been easier, of course, granted constant variation.)

Gellibrand's position at Gresham College drew him into matters of mathematical navigation, and an example of his attempts at solving the problem of longitude is a three-page appendix to *The Strange and Dangerous Voyage of Captain Thomas James* (1633). James, a gentleman mariner who voyaged in 1631 to seek the Northwest Passage, had by prior arrangement observed at Charlton Island, James Bay, a lunar eclipse also observed by Gellibrand at Gresham.

James's position in longitude was thus calculated (79°30′ west of Gresham, being 15′ too low).

The essentially practical quality of Gellibrand's work, which is of very slight mathematical interest, may also be judged from four works: his "Treatise of Building of Ships," a manuscript mentioned by Anthony à Wood as belonging to Edward, Lord Conway; his textbook *An Institution Trigonometrical;* a longer Latin work translated by John Newton, *An Institution Trigonometrical . . . With the Application . . . to Questions of Astronomy and Navigation* (1652); and *An Epitome of Navigation* . . . (1674). This last and posthumous book (written after 1631 and before 1634) contains a number of logarithmic tables, including trigonometrical ones, and has an appendix on the use of the cross-staff, quadrant, and nocturnal in navigation. That it was found valuable is suggested by the appearance of later editions, in 1698 (by Euclid Speidell), 1706 and subsequently (by J. Atkinson), and 1759 (by William Mountaine under the title *A Short and Methodical Way to Become a Complete Navigator*).

Gellibrand's work was mainly derivative, leading influences on it having been Wright's *Certaine Errors* and Richard Norwood's *Trigonometrica* (1631). It can be said that he was a reasonably good calculator and a competent writer of textbooks which helped to raise English standards of navigation to new heights.

BIBLIOGRAPHY

I. ORIGINAL WORKS. Besides works mentioned in the text, Gellibrand wrote "Astronomia lunaris . . .," composed between 20 December 1634 and 22 January 1635. This belonged to Sir Hans Sloane but is not indexed in the catalog of the Sloane Collection in the British Museum. He also added a preface to and published *Sciographia, or the Art of Shadows* (London, 1635), written by J(ohn) W(ells), a Roman Catholic of Hampshire. A Latin oration, "In laudem Gassendi astronomiae," delivered in Christ Church Hall, Oxford, is now Brit. Mus. Add. MS 6193, f. 96.

II. SECONDARY LITERATURE. The two main sources for Gellibrand's life are Anthony à Wood, *Athenae Oxonienses,* rev. and enl. ed. (Oxford, 1721); and John Ward, *The Lives of the Professors of Gresham College* (London, 1740), pp. 81–85, 336. Gellibrand's work on navigation has received little attention from historians, but for some discussion of navigation in his time see D. W. Waters, *The Art of Navigation in England in Elizabethan and Early Stuart Times* (London, 1958), *passim;* and E. G. R. Taylor, *The Mathematical Practitioners of Tudor and Stuart England* (Cambridge, 1954), pp. 138, 164, 165, 175.

J. D. NORTH

GELMO, PAUL JOSEF JAKOB (*b.* Vienna, Austria, 17 December 1879; *d.* Vienna, 22 October 1961), *chemistry.*

Gelmo attended the Technische Hochschule in Vienna from 1898, obtaining an engineering diploma in 1903 and a doctorate in 1906. From 1904 to 1909 he was an assistant there to Wilhelm Suida. For twenty-eight years (1910–1938) he was chief chemist for the Austrian State Printing Office. In 1929 he became lecturer, and in 1954 professor, of the chemistry and technology of cellulose and paper at the Technische Hochschule.

Gelmo made his only noteworthy contribution to science while serving as assistant to Suida. Engaged in research with azo compounds and their usefulness as synthetic dyes, he prepared several new sulfonamides. One of his syntheses was sulfonanilamide (para-aminobenzenesulfonamide). After his work was published in 1908, I. G. Farbenindustrie used the new compound as a constituent in azosulfonamide dyes, but no one suspected that sulfonanilamide had curative powers. In 1932 Gerhard Domagk demonstrated the effectiveness of prontosil, a dye containing a sulfonamide group, in controlling streptococcal infections, which discovery led to the development of the sulfonamides as medicinals. Gelmo's sulfonanilamide came to be the most widely used of the sulfa drugs, and its commercial manufacture was accomplished essentially by his method.

BIBLIOGRAPHY

Gelmo's preparation of sulfonamides is presented in "Über Sulfamide der p-Amidobenzolsulfonsäure," in *Journal für praktische Chemie,* **77** (1908), 369–382.

Brief notices in Poggendorff, VIIa, pt. 2 (1957), 183; and *Who's Who in Austria* (1955), p. 136, constitute the secondary literature.

ALBERT B. COSTA

GEMINUS (*fl.* Rhodes [?], *ca.* 70 B.C.), *astronomy, mathematics.*

Geminus' name is Latin, but his works and manner are patently Greek—the forms Γεμῖνος and Γεμεῖνος, also found in the manuscripts, are probably false analogies based on true Greek forms such as Ἀλεξῖνος and Ἐργῖνος. He was author of an *Introduction* (*Isagoge*) *to Astronomy* (still extant) and of a work on mathematics (lost, except for quoted extracts from it). Nothing is known of the circumstances of his life, but his date and place of work may be inferred from internal evidence in the *Isagoge.* In chapter 8, sections 20–24 of the Manitius edition, Geminus corrects a widespread Greek view that the Egyptian festival of Isis coincided with the winter solstice; he explains that although this was so 120 years ago, in his own time there was a whole month's difference between the two dates, since every four years the Egyptian calendar (based on a "year" of twelve months of thirty days each plus five additional days)[1] became out of step with the solar year by one day. The festival of Isis took place from the seventeenth to the twentieth day of the Egyptian month Athyr, and a papyrus fragment tells us that in the calendar of Eudoxus of Cnidus the winter solstice occurred on 20 or 19 Athyr. By Julian reckoning the winter solstice in Eudoxus' time (*ca.* 350 B.C.) occurred on 25 December or 28 December, and the first years in which these Julian dates can coincide with 19 Athyr are 185 B.C. and 197 B.C.[2] Hence, the papyrus presumably was written at this time; and subtracting the 120 years mentioned by Geminus, we obtain a date for the *Isagoge* of 65 or 77 B.C., which agrees with other evidence indicating a date about 70 B.C.[3]

We have no definite evidence about where Geminus was born or worked, but the commonly held opinion that his place of work was Rhodes may be correct. It is the "clima" of Rhodes that he uses to illustrate his account of various astronomical phenomena (e.g., I, 10; I, 12; III, 15; V, 25); in XVII, 4, he refers to Mt. Atabyrius,[4] i.e., the modern Mt. Attaviros, in the center of Rhodes, without feeling it necessary to specify its location, whereas in the same passage he is careful to explain that Mt. Cyllene is in the Peloponnesus; and Rhodes had a reputation in the last two centuries B.C. as a center for those subjects (philosophy, astronomy, and mathematics) with which Geminus was concerned, Panaetius, Posidonius, and Hipparchus all having worked there.

On the other hand, the choice of Rhodes as a typical example may simply have been dictated by its position on the best-known and most central parallel of latitude, 36°N. (Geminus himself says [XVI, 12] that all globes were constructed for this "clima," which for practical purposes was regarded as the latitude of Greece as well),[5] or Geminus may merely have used the examples he found in his sources, in this case almost certainly Hipparchus (see below), just as Ptolemy in the *Almagest* (II, 3 and 4), some 200 years later, was to use the same example from the same source. Similarly, Geminus may have taken the examples of Mt. Cyllene and Mt. Atabyrius straight from Dicaearchus (who is cited for their heights in XVII, 5); and even though two of his chief authorities, Hipparchus and Posidonius, worked in Rhodes, it does not necessarily follow that Geminus did.

The *Isagoge* is an early example of an elementary astronomical handbook written to popularize the main ideas in the technical treatises of the scientists; it belongs to the same tradition as the *De motu circulari corporum caelestium* of Cleomedes, and there are many similarities in style and arrangement between the two works, although, as might be expected from its later date, Cleomedes' work is fuller and less elementary than its precursor. One interesting difference is that Geminus includes a chapter (*Isagoge,* II) on the astrological "aspects" of the zodiacal signs, i.e., their arrangement in pairs, triplets, quadruplets, etc., according to which the astrologers calculated the signs' influence on human affairs; Cleomedes has no such chapter and, in fact, does not mention astrological doctrines at all. Geminus gives a simplified description of basic astronomy as known in the time of Hipparchus, omitting most of the mathematics and giving (I, 23 f.) little information about the planets apart from their zodiacal periods (thirty years for Saturn, twelve for Jupiter, two and a half for Mars, one year each for Mercury, Venus, and the sun,[6] and $27\frac{1}{3}$ days for the moon); there is no mention of epicycles, but the eccentricity of the sun's path relative to the earth is carefully described (I, 31 ff.) and the Hipparchian values for the four astronomical seasons are given (I, 13 f.)—$94\frac{1}{2}$, $92\frac{1}{2}$, $88\frac{1}{8}$, and $90\frac{1}{8}$ days, respectively, starting from the vernal equinox.[7]

Chapter III describes the main constellations; chapters IV and V describe the chief circles of the celestial sphere; and chapter VI explains the variations in the lengths of day and night at different latitudes. Chapter VII (which is based largely on Aratus and, almost certainly, on Hipparchus' commentary on Aratus' astronomical poem entitled *Phaenomena*) deals with the rising of the zodiacal signs, and chapter VIII with the length of the lunar month, for which a figure of $29\frac{1}{2} + 1/33$ days is given (rounded off from the accurate Hipparchian value of 29 days, 12 hours, 44 minutes, 3 seconds);[8] this last chapter is especially valuable for its account of how the Greeks developed an astronomically based calendar with a scientific system of intercalation.[9] Chapter IX explains the phases of the moon, chapters X and XI solar and lunar eclipses, and chapter XII the general motions of the planets. Chapters XIII and XIV deal with the risings and settings and courses of the fixed stars, and chapters XV and XVI with the zones of the terrestrial globe, delineated according to the estimates of Eratosthenes, which Hipparchus also accepted.[10]

Chapter XVII discusses sensibly the principles on which the "parapegmata" (astronomical calendars containing weather prognostications connected with the risings and settings of certain stars and constellations)[11] were based. Chapter XVIII deals with the "exeligmus" (ἐξελιγμός), or shortest period containing a whole number of synodic months, of anomalistic months, and of days (cf. *Almagest,* IV, 2); this chapter, far more technical than the others and out of keeping with the elementary character of the rest of the book, may well be an unrelated fragment.[12] Finally, there is an astronomical calendar, or "parapegma," which begins with Cancer and is evidently based on figures for the astronomical seasons which are not Hipparchian;[13] it seems probable that this was not part of the original treatise but represents older material.

Geminus' exposition is usually sound and clear and includes some intelligent criticism, e.g., of the notion held "by many philosophers" that the planetary bodies do not in fact exhibit two opposite motions (the diurnal and the zodiacal) but only seem to do so because they revolve at different speeds (*Isagoge,* XII, 14 ff.); this idea, says Geminus, "is not in accordance with observed phenomena" (XII, 19, ἀσύμφωνός ἐστι τοῖς φαινομένοις). He also gives a shrewd appreciation of the limitations of parapegmata, points out that weather forecasting based on observation of the rising and setting of various stars is by no means an exact science but a rough codification of general trends, and asserts firmly that stellar risings do not cause changes in the weather but merely indicate them (οὐκ αὐταὶ παραίτιοί εἰσι τῶν περὶ τὸν ἀέρα μεταβολῶν, ἀλλὰ σημεῖα ἔκκεινται, XVII, 11; see the whole of this chapter, especially 6–41); this last assertion contrasts strangely with the belief underlying chapter II.

There are, of course, some errors and infelicities. The chapter which gives a rational discussion of parapegmata (XVII) also contains an absurdly exaggerated view of the restricted range of atmospheric phenomena, according to which no wind or rain is experienced at the top of a mountain under 10,000 feet high (XVII, 3–5). In chapters X and XI no mention is made that, for an eclipse to take place, the moon must be at a node; and in XVI, 13, it is wrongly stated that the magnitudes of eclipses are the same for people living on the same parallel of latitude. There are also minor errors in the description of a star's evening rising (XIII, 13) and morning setting (XIII, 16), which may be the result of scribal errors in the transmission of the text (cf. Manitius, Anmerkung 25, 26); but Manitius' theory (pp. 246–248) that our present text represents a Byzantine compilation from an epitome of Geminus' original work, although it cannot be disproved, rests on very slender evidence.[14]

A work on mathematics (known only from extracts quoted by later writers) is also attributed to Geminus; and although it has been thought that this is not the same man as the author of the *Isagoge,* there can, in fact, be little doubt of the identity of the two.[15] The title of the work was probably *Theory of Mathematics* (Θεωρία τῶν μαθημάτων),[16] and it must have extended to at least six books, since Eutocius quotes from the sixth. Our chief source for its contents is Proclus,[17] who quotes extensive extracts from it; parallel quotations appear in the scholia to book I of Euclid's *Elements*[18] and in the collection of writings attributed to Hero of Alexandria;[19] and there are isolated quotations elsewhere.[20] There is also an Arabic commentary on Euclid by al-Nayrīzī[21] which contains extracts from a commentary on the *Elements* by Simplicius, who quotes a certain "Aghanis" on the definition of parallels; from similarities with what we know from Proclus of Geminus' views on this subject, it has been suggested that Aghanis is actually Geminus, but there are grave objections to such an identification, which must therefore remain doubtful.[22] A Latin translation of this Arabic text by Gerard of Cremona is extant.[23]

The *Theory of Mathematics* apparently dealt with the logical subdivisions of the mathematical sciences, discussing the philosophical principles of their classification, distinguishing carefully between such terms as "hypothesis" and "theorem," "postulate" and "axiom," and paying particular attention to accurate definitions not only of the various branches of mathematics but also of concepts such as "line," "surface," "figure," and "angle." It seems to have been a more substantial work than the *Isagoge* and to have contained some very pertinent criticism of Euclid's postulates, particularly the fifth, the so-called parallel postulate, for which Geminus believed he had found a proof.[24]

We learn from Simplicius that Geminus wrote an exegesis on the *Meteorologica* of Posidonius;[25] Simplicius cites Alexander of Aphrodisias as transcribing a long extract from an "epitome" of Geminus' work (ἐκ τῆς ἐπιτομῆς τῶν Ποσειδωνίου Μετεωρολογικῶν ἐξηγήσεως), but an epitome of an exegesis sounds unlikely and it is tempting to excise the word ἐπιτομῆς as an otiose gloss that has crept into the text. The extract discusses the different aims of physics and astronomy and has obvious relevance to a work on meteorology, which has affinities with both these sciences. Simplicius ends his citation with the following: "In this manner, then, Geminus also, or rather Posidonius in Geminus, expounds the difference . . . ." (οὕτω μὲν οὖν καὶ ὁ Γεμῖνος ἤτοι ὁ παρὰ τῷ Γεμίνῳ Ποσειδώνιος τὴν διαφορὰν . . . παραδίδωσιν). From this it has been implied that Geminus did little else but reproduce the opinions of Posidonius.

Such an implication certainly overestimates Geminus' debt to Posidonius. There is no doubt that both writers subscribed to Stoic views of the universe, and both were concerned to combat the attacks on the validity of the mathematical sciences, which, as we learn from Proclus and Sextus Empiricus, were mounted by both Skeptic and Epicurean philosophers;[26] but Geminus shows his independence of Posidonius in several respects. In *Isagoge* XVI, it is Eratosthenes' estimate of 252,000 stades that is taken as the basis of the division of the earth into zones and not Posidonius' figure of 240,000 (Cleomedes, *De motu circulari,* I, 10) or 180,000 (Strabo, *Geography,* 95),[27] neither of which is mentioned (nor, in fact, is Posidonius named in the entire treatise). Chapters VIII and XVIII, on the calendar and the length of the lunar month, almost certainly owe nothing to Posidonius, as there is no evidence that he did any work on calendrical problems; and there are other indications that Geminus, although he must have been well acquainted with Posidonius' opinions and was probably Proclus' chief authority for the latter,[28] did not hesitate to differ from the latter in following other sources (particularly Eratosthenes and Hipparchus for geography and astronomy) or putting forward his own views, as in his criticism of Euclid's postulates.

NOTES

1. On this, see O. Neugebauer, *The Exact Sciences in Antiquity,* 2nd ed. (Providence, R.I., 1957), pp. 81, 94.
2. 25 December according to W. Kubitschek, *Grundriss der antiken Zeitrechnung* (Munich, 1928), p. 109; 28 December according to Böckh, cited by Manitius in his ed. of *Elementa,* p. 264. See also "Synchronistic Table," in E. J. Bickerman, *Chronology of the Ancient World* (London–Ithaca, N.Y., 1968), p. 150.
3. For details, see Manitius ed. of *Elementa,* Anmerkung 16, pp. 263–266.
4. The *Σαταβύριον* of the MSS must be corrected to *'Αταβύριον*; see D. R. Dicks, *Geographical Fragments of Hipparchus* (London, 1960), p. 30.
5. See V, 48; Dicks, *op. cit.,* pp. 123, 130, 176.
6. See D. R. Dicks, *Early Greek Astronomy to Aristotle* (London, 1970), notes 174, 345.
7. See Ptolemy, *Almagest,* III, 4.
8. *Ibid.,* IV, 2.
9. See Dicks, *Early Greek Astronomy,* pp. 86 f.
10. See Dicks, *Geographical Fragments,* p. 148.
11. See Dicks, *Early Greek Astronomy,* pp. 84 f.
12. See Manitius ed. of *Elementa,* p. 278.
13. *Ibid.,* Anmerkung 34.
14. See Tittel, "Geminos I," cols. 1031–1032.
15. *Ibid.,* cols. 1029–1030.
16. See Eutocius, *Commentaria in Conica I,* in *Apollonii Pergaei quae graece exstant,* J. L. Heiberg, ed., II (1893), 170, 25.
17. *Commentarii in primum Euclidis elementorum librum,* G. Friedlein, ed. (Leipzig, 1873).
18. *Euclidis opera omnia,* J. L. Heiberg and H. Menge, eds., V (Leipzig, 1888), 81, 4; 82, 28; 107, 20.
19. *Heronis Alexandri reliquiae,* F. Hultsch, ed. (Berlin, 1864), nos. 5–14, 80–86.
20. See Tittel, *op. cit.,* cols. 1039–1040.
21. *Codex Leidensis 399,1,* R. O. Besthorn, J. L. Heiberg, G. Junge, J. Raeder, and W. Thomson, eds., 3 pts. (Copenhagen, 1893–1932).
22. T. L. Heath, *History of Greek Mathematics,* II, 224, accepts it; but the same scholar, in *The Thirteen Books of Euclid's Elements,* 2nd ed. (Cambridge, 1925), pp. 27–28, rejects it. Compare A. I. Sabra, "Thabit Ibn Qurra on Euclid's Parallels Postulate," in *Journal of the Warburg and Courtauld Institutes,* **31** (1968), 13.
23. *Euclidis opera omnia. Supplementum: Anaritii in decem libros priores Elementorum Euclidis commentarii,* M. Curtze, ed. (Leipzig, 1899).
24. For details see Heath, *History of Greek Mathematics,* II, 223–231.
25. *Simplicii in Aristotelis Physicorum libros IV priores commentaria,* H. Diels, ed. (Berlin, 1882), pp. 291–292; this passage is also printed by Manitius in his ed. of the *Isagoge* as Fragmentum I, pp. 283–285.
26. Proclus, G. Friedlein, ed., pp. 199 f., 214 ff.; Sextus Empiricus, *Adversus mathematicos,* H. Mutschmann-J. Mau, eds., I, 1 ff.; compare the whole of III.
27. See E. H. Bunbury, *History of Ancient Geography,* II (London, 1879), 95–96; Dicks, *Geographical Fragments,* p. 150.
28. See Tittel, *op. cit.,* col. 1042.

BIBLIOGRAPHY

K. Manitius edited the *Gemini Elementa astronomiae* (Leipzig, 1898); chs. I, III–VI, VIII–XVI have been published by E. J. Dijksterhuis as *Gemini Elementorum astronomiae,* no. 22 in the series Textus Minores (Leiden, 1957).

See T. L. Heath, *History of Greek Mathematics,* II (Oxford, 1921; repr. 1960), 222–234; and Tittel, "Geminos I," in *Real-Encyclopädie,* Halbband XIII (1910), cols. 1026–1050.

D. R. DICKS

GEMINUS (also known as **Lambrit** or **Lambert**), **THOMAS** (*b.* Lixhe, Belgium, *ca.* 1510; *d.* London, England, June 1562), *medicine.*

Geminus—his assumed name may indicate he was the twin of his brother Jasper Lambrit—migrated about 1540 to England, where he practiced the arts of engraving, printing, and instrument making, the last possibly with the assistance of Humphrey Cole. The unique copper-engraved anatomical figure of a woman (*ca.* 1540), now in the Wellcome Historical Medical Library (*Catalogue of Printed Books,* I [London, 1962], no. 290), was probably the work of Geminus, as were the *Compendiosa totius anatomie delineatio* (1545) and the unique copy of *Morysse and Damashin Renewed and Encreased. Very Profitable for Goldsmiths and Embroiderars* (London, 1548), now in the Landesmuseum, Münster. Geminus both engraved and printed a map entitled *Nova descriptio Hispaniae* (London, 1555), on four sheets, and printed *Britanniae insulae. . . nova descriptio* (London, 1555); the unique copies of both are now in the Bibliothèque Nationale, Paris. He also printed Leonard Digges's

Prognostication of Right Good Effect in 1555 and a second edition in 1556, and he was possibly responsible for the diagram of the signs of the zodiac on the title page; he also brought out Digges's *Boke Named Tectonicon* (1562). In addition, he produced and engraved several astrolabes: one with the arms of the duke of Northumberland, Sir John Cheke, and Edward VI, dated 1552 (now in the Royal Belgian Observatory, Brussels), two about 1555 (National Maritime Museum, Greenwich, and Museo di Storia delle Scienze, Florence), and another for Queen Elizabeth (Museum of the History of Science, Oxford).

Geminus' most important work was the series of handsome, copper-engraved anatomical figures, most of them plagiarized from Vesalius' *Fabrica* and a few from his *Epitome.* Upon being presented with a set of these engravings in 1544, Henry VIII urged Geminus to issue them in the form of a book. This was done in the following year, under the title *Compendiosa totius anatomie delineatio,* published in London by J. Herford rather than by Geminus, who apparently had not yet established his own press. The illustrations were accompanied by the Latin text of Vesalius' *Epitome,* slightly defective in its concluding chapter as the result of verbal compression, and by the Latin text of Vesalius' descriptions (*indices*) of his illustrations. It appears likely that this first plagiarism of the major Vesalian texts and illustrations was intended as a treatise to accompany the anatomical lectures given at the recently (1540) organized United Company of Barber-Surgeons of London.

Although Geminus' book introduced Vesalian anatomy to England, it appears to have been too advanced for the surgeons, who, moreover, because of guild rather than university training, were unable to comprehend the Latin text. In consequence of what was probably a disappointing sale, the *Compendious Anatomy* was published again in 1553 ("Imprynted at London by Nycholas Hyll . . . for Thomas Geminus"), with an English "Treatyse" substituted for the Latin text of the *Epitome* and the Vesalian descriptions of the illustrations translated into English. The translations and the new text were the work of Nicholas Udall, the new text being basically the same as that employed by Thomas Vicary in his *Anatomie of Mans Bodie* (1548), i.e., a fourteenth-century compilation mostly representative of Henry of Mondeville's surgical anatomy, known today through the existence of a late fifteenth-century manuscript copy (Wellcome MS. no. 564). To this Udall added some sections from Ludovicus Vassaeus' Galenic *In anatomen corporis humani tabula quatuor* (1540) and a few fragments from Guy de Chauliac. The result therefore was essentially a medieval anatomical text factually denied by the accompanying Vesalian illustrations and their Vesalian descriptions; it could have been of little use for anatomical instruction despite the relatively low level of anatomical studies in England at that time. It was probably because of the spectacular Vesalian illustrations rather than the text that another edition of the *Compendious Anatomy* seems to have been called for in 1559, this time with some but very slight editorial assistance from Richard Eden ("Imprinted at London within the blacke fryars: by Thomas Gemini. Anno salutis. 1559").

Despite the questionable value of the text, Geminus' plagiarized illustrations were of considerable influence and were plagiarized in turn for Raynold's *Byrth of Mankynde* (London, 1545, and certain succeeding editions), William Bullein's *Government of Health* (London, 1558, 1559), Vicary's *Profitable Treatise of the Anatomie of Mans Body* (London, 1577), and Banester's *Historie of Man Sucked From the Sappe of the Most Approved Anathomistes* (London, 1579). In 1560 or shortly thereafter Geminus' plates were acquired by the Parisian printer André Wechsel and, with the editorial assistance of Jacques Grévin, were reproduced with Latin and French texts (1564, 1565, 1569) borrowed from Geminus' first edition of 1545. In turn these Parisian editions led to still others in France, Germany, and the Netherlands as late as the mid-seventeenth century.

Aside from these activities very little is known of Geminus' life. During the reign of Henry VIII he received wages as a royal surgeon, although there is no indication of his ever having had the necessary training, except perhaps for some mechanical skill and whatever he may have learned of anatomy from his plagiarism of the Vesalian illustrations and texts. Indeed, in 1555 he was penalized by the College of Physicians of London for practicing medicine without a license and undertook to print 200 copies of a proclamation against quackery for the college in place of his fine. No copy of this proclamation is known to exist today. Geminus' printing and engraving activities, located in Blackfriars, appear to have been financially successful, as is indicated by the distribution of his possessions according to his will, dated 22 May 1562 (O.S.).

BIBLIOGRAPHY

All of Geminus' works are today very rare. For bibliographical details of those pertaining to anatomy, see Harvey Cushing, *A Bio-bibliography of Andreas Vesalius,* 2nd ed.

(Hamden, Conn., 1962), pp. 121–130, 131–132, 135–136. The 1553 ed. of the *Compendiosa totius anatomie delineatio* has been published in a facs. ed. (London, 1959); the intro. to it by C. D. O'Malley contains the fullest account of Geminus and his activities. There is also a facs. ed. of the illustrations, curiously entitled *Andreas Vesalius* (Geneva, 1964). The accompanying explanatory text unfortunately contains many errors of fact.

C. D. O'MALLEY

GEMMA FRISIUS, REINER (*b.* Dokkum, Netherlands, 1508; *d.* Louvain, Belgium, 1555), *geography, mathematics.*

Gemma Frisius was a native of Friesland; hence his nickname Frisius. He received his medical degree at Louvain, practiced medicine there, and taught later at its medical faculty. Although he was a practicing physician, he is remembered for his contributions to geography and mathematics, his avocations.

At the age of twenty-one Gemma Frisius published *Cosmographicus liber Petri Apiani mathematici . . .* (Antwerp, 1529), an edition of Peter Apian's *Cosmography,* "carefully corrected and with all errors set to right." In 1530 he published at Antwerp his first original work, *Gemma Phrysius de principiis astronomiae & cosmographiae . . .*, which was translated into several languages and reprinted numerous times. The Spanish Netherlands was in close contact with court and business circles in Spain, and Brussels was an ideal place to gather current information on the discoveries. Gemma Frisius designed globes and astronomical instruments that were well known and much sought after throughout Europe. Several of them still survive and are of key importance in tracing the growth of knowledge of the newly discovered lands. Some of Gemma Frisius' globes were completed by Gerard Mercator, who had attended mathematical lectures that Gemma Frisius gave at his home.

Gemma Frisius made two significant contributions to the earth sciences. In a chapter added to the 1533 Antwerp edition of the *Cosmographicus,* entitled "Libellus de locorum describendorum ratione," he was first to propose—and illustrate—the principle of triangulation as a means of carefully locating places and accurately mapping areas. Twenty years later, in the 1553 Antwerp edition of *De principiis astronomiae,* he added a chapter entitled "De novo modo investigandi latitudinem regionis absq. meridiani vel loci solis cognitione," in which he was the first to suggest in explicit terms the use of portable timepieces to measure longitude by lapsed time. Although this important idea could not be put into practice until after the invention of optical instruments and accurate portable timepieces, the credit for first suggesting it rests with Gemma Frisius.

BIBLIOGRAPHY

The basic reference work, which includes the text of the few items of Gemma Frisius' surviving correspondence, is Fernand van Ostroy, "Biobibliographie de Gemma Frisius, fondateur de l'école belge de géographie . . .," in *Mémoires de l'Académie royale des sciences. . . de Belgique,* Classe des Lettres, 2nd ser., **11** (1920). The original text of the method of measuring longitude is reproduced in A. Pogo, "Gemma Frisius, His Method of Determining Longitude . . .," in *Isis,* **22** (1935), i–xix, 469–485. A brief biographical sketch and an appraisal of Gemma Frisius' work is George Kish, "Medicine • Mensura • Mathematica; The Life and Works of Gemma Frisius, 1508–1555," James Ford Bell lecture no. 4 (1967), published by the Associates of the James Ford Bell Collection, University of Minnesota Library.

GEORGE KISH

GENTH, FREDERICK AUGUSTUS (*b.* Wächtersbach, Hesse, Germany, 17 May 1820; *d.* Philadelphia, Pennsylvania, 2 February 1893), *chemistry, mineralogy.*

Genth exhibited a keen interest in natural history at an early age. After three years at the Hanau Gymnasium, in 1839 he entered the University of Heidelberg, where he studied chemistry, geology, and mineralogy under Leopold Gmelin, J. R. Blum, and K. C. Leonhard. From 1841 to 1843 he attended the University of Giessen, where he worked under Fresenius, Kopp, and Liebig. In 1844 he continued his chemical studies under Bunsen at the University of Marburg, receiving his doctorate there in 1845. He remained at Marburg as *Privatdozent* and Bunsen's assistant for three years. In 1848 he immigrated to the United States, making his home first in Baltimore and then in Philadelphia. After occupying several positions, establishing one of the first commercial analytical laboratories in America, and engaging in the instruction of special students, in 1872 he became professor of chemistry at the University of Pennsylvania, a post he held until 1888, when he returned to consulting work and research.

Genth's European background and education supplied him with technical skills possessed by few scientists in the United States during his lifetime, and he holds a place in the foremost rank of pioneer mineralogists in America. He was a chemist almost without peer, especially in the field of analysis. His best-known research involved the ammonia-cobalt bases (cobalt ammines), developed jointly with Oliver Wolcott Gibbs. His original memoir on this topic (1851) contained the first distinct recognition of the existence of perfectly defined and crystallized salts of the cobalt ammines. His joint monograph with

Gibbs (1856) described thirty-five salts of four bases—roseocobalt, purpureocobalt, luteocobalt, and xanthocobalt—and for the first time distinguished roseo salts from purpureo salts.

Genth served as chemist for the Second Geological Survey of Pennsylvania and chemist to the Board of Agriculture of Pennsylvania; his analyses of fertilizers did much to develop the state's agricultural industry. His chief chemical contributions to mineralogy are contained in fifty-four papers describing 215 mineral species; he himself discovered twenty-four new minerals. His contributions to chemistry and mineralogy total 102.

BIBLIOGRAPHY

I. Original Works. Genth's original memoir on cobalt ammines is in *Keller-Tiedemann's Nordamerikan Monatsbericht,* **2** (1851), 8–12; Gibbs and Genth's classic "Researches on the Ammonia-Cobalt Bases," described by Genth's student E. F. Smith as "among the finest chemical investigations ever made in this country," appears in *American Journal of Science,* 2nd ser., **23** (1856), 234, 319, and **24** (1856), 86, and as a separate publication in the series Smithsonian Contributions to Knowledge (Washington, D.C., 1856).

II. Secondary Literature. The constitution of the cobalt ammines was not understood until Alfred Werner proposed his coordination theory in "Beitrag zur Konstitution anorganischer Verbindungen," in *Zeitschrift für anorganische Chemie,* **3** (1893), 267–330, of which an English trans. is in G. B. Kauffman, *Classics in Coordination Chemistry,* I, *Selected Papers of Alfred Werner* (New York, 1968), pp. 9–88. On Genth's life see especially his student E. F. Smith, *Chemistry in America* (New York, 1918), pp. 261–263, and "Mineral Chemistry," in *Journal of the American Chemical Society,* **48,** no. 8A (1926), 71–75. Additional data are in George F. Barker, "Obituary Notice: Frederick Augustus Genth," in *Proceedings of the American Philosophical Society,* **40** (1901), x–xxii, and "Memoir of Frederick Augustus Genth 1820–1893," in *Biographical Memoirs. National Academy of Sciences,* **4** (1902), 201–231; W. M. Myers and S. Zerfoss, "Frederick Augustus Genth: 1820–1893, Chemist-Mineralogist-Collector," in *Journal of the Franklin Institute,* **241** (1946), 341–354; H. S. van Klooster, "Liebig and His American Pupils," in *Journal of Chemical Education,* **33** (1956), 493–497; and W. H. Wahl, H. F. Keller, and T. R. Wolf, "A Memoir of Frederick Augustus Genth," in *Journal of the Franklin Institute,* **135** (1893), 448–452.

George B. Kauffman

GENTZEN, GERHARD (*b.* Greifswald, Germany, 24 November 1909; *d.* Prague, Czechoslovakia, 4 August 1945), *logic, foundations of mathematics.*

Gentzen was a born mathematician. As a boy he declared his dedication to mathematics, and his short life constituted a realization of that promise. He benefited from the teaching of such renowned scholars as P. Bernays, C. Carathéodory, R. Courant, D. Hilbert, A. Kneser, E. Landau, and H. Weyl and, at the age of twenty-three, received his doctorate in mathematics. In 1934 he became one of Hilbert's assistants and held that position until 1943, with the exception of a two-year period of compulsory military service from 1939 to 1941, when he was requested to take up a teaching post at the University of Prague. He died at Prague of malnutrition three months after his internment by the liberating authorities in May 1945.

Gentzen combined in rare measure an exceptional inventiveness and the talent for coordinating diverse existing knowledge into a systematic conceptual framework. He invented "natural deduction" in order to create a predicate logic more akin to actual mathematical reasoning than the Frege-Russell-Hilbert systems, then used P. Hertz's "sentences" to transform his natural calculus into a "calculus of sequents." Thus he succeeded in making classical logic appear as a simple extension of intuitionist logic and in enunciating his *Hauptsatz* (chief theory), which he had discovered while studying more closely the specific properties of natural deduction. A formalization of elementary number theory based on sequents and his ingenious idea of using restricted transfinite induction as a metamathematical technique enabled Gentzen to carry out the first convincing consistency proof for elementary number theory, in spite of the limitations imposed on such proofs by Gödel's theorem. He eventually proved directly the nonderivability in elementary number theory of the required transfinite induction up to ϵ_0. In view of subsequent developments, Gentzen's consistency proof must be considered as the most outstanding single contribution to Hilbert's program.

The *Hauptsatz* says that in both intuitionist and classical predicate logic, a purely logical sequent can be proved without "cut" (*modus ponens,* in Hilbert-type systems). A corollary is the subformula property, to the effect that the derivation formulas in a cut-free proof are compounded in the sequent proved. This entails, for example, the consistency of classical and intuitionist predicate logic, the decidability of intuitionist propositional logic, and the nonderivability of the law of the excluded middle in intuitionist predicate logic. Gentzen succeeded in sharpening the *Hauptsatz* for classical logic to the midsequent theorem (Herbrand-Gentzen theorem) for sequents whose formulas are prenex, by proving that any such sequent has a cut-free proof consisting of two parts, the first part quantifier-free and the second consisting essentially of instances of quantification. The last quantifier-free sequent in the proof is the "midsequent" and corresponds closely to a "Herbrand tautology."

Among the consequences of the midsequent theorem are the consistency of arithmetic without induction, Craig's interpolation lemma, and Beth's definability theorem. Recent extensions of the *Hauptsatz* to stronger deductive systems, including infinitary logic, have yielded partial results at the second stage of Hilbert's program in the form of consistency proofs for subsystems of classical analysis. Only a few days before his death, Gentzen had in fact announced the feasibility of a consistency proof for classical analysis as a whole.

BIBLIOGRAPHY

Gentzen's writings include "Über die Existenz unabhängiger Axiomensysteme zu unendlichen Satzsystemen," in *Mathematische Annalen,* **107** (1932), 329-350; "Untersuchungen über das logische Schliessen," in *Mathematische Zeitschrift,* **39** (1935), 176–210, 405–431, his inaugural dissertation, submitted to the Faculty of Mathematics and Natural Science of the University of Göttingen in the summer of 1933; "Die Widerspruchsfreiheit der reinen Zahlentheorie," in *Mathematische Annalen,* **112** (1936), 493–565; "Die Widerspruchsfreiheit der Stufenlogik," in *Mathematische Zeitschrift,* **41,** no. 3 (1936), 357–366; "Der Unendlichkeitsbegriff in der Mathematik," in *Semester-Berichte, Münster in Westphalen,* 9th semester (Winter 1936/1937), pp. 65–80; "Unendlichkeitsbegriff und Widerspruchsfreiheit der Mathematik," in *Travaux du IXe Congrès International de Philosophie,* VI (Paris, 1937), 201-205; "Die gegenwärtige Lage in der mathematischen Grundlagenforschung," in *Forschungen zur Logik und zur Grundlegung der exakten Wissenschaften,* n.s. **4** (1938), 5–18, also in *Deutsche Mathematik,* **3** (1939), 255–268; "Neue Fassung des Widerspruchsfreiheitsbeweises für die reine Zahlentheorie," in *Forschungen zur Logik und zur Grundlegung der exakten Wissenschaften,* n.s. **4** (1938), 19–44; "Beweisbarkeit und Unbeweisbarkeit von Anfangsfällen der transfiniten Induktion in der reinen Zahlentheorie," in *Mathematische Annalen,* **119,** no. 1 (1943), 140–161, his Ph.D. *Habilitation* thesis, submitted to the Faculty of Mathematics and Natural Science of the University of Göttingen in the summer of 1942; and "Zusammenfassung von mehreren vollständigen Induktionen zu einer einzigen," in *Archiv für mathematische Logik und Grundlagenforschung,* **2,** no. 1 (1954), 1–3.

For a detailed account of Gentzen's life, extensive cross-references to his papers, and a critical appraisal of germane subsequent developments, see M. E. Szabo, *Collected Papers of Gerhard Gentzen,* in the series Studies in Logic (Amsterdam, 1969); of particular interest are the trans. of two articles Gentzen submitted to *Mathematische Annalen* but withdrew before publication (of which galley proofs are privately owned), "Über das Verhältnis zwischen intuitionistischer und klassischer Arithmetik" (1933) and sections 4 and 5 of "Die Widerspruchsfreiheit der reinen Zahlentheorie" (1935).

MANFRED E. SZABO

GEOFFROY, CLAUDE JOSEPH (*b.* Paris, France, 8 August 1685; *d.* Paris, 9 March 1752), *chemistry, botany.*

Geoffroy the Younger (le Cadet) was the second son of Matthieu François Geoffroy and Louise de Vaux, and the brother of Étienne François Geoffroy. He qualified as an apothecary in 1703 and took over the family pharmacy on his father's death in 1708. Highly esteemed in his profession, he was appointed inspector of the pharmacy at the Hôtel-Dieu (the Paris hospital) and in 1731 he served as a Paris alderman (*échevin*).

In 1707 Geoffroy, who had studied botany under J. P. de Tournefort and made a long field excursion in the south of France in 1704–1705, was elected to the Académie des Sciences as a botanist. He was already interested in chemistry and in his first research tried to find a chemical explanation of the colors of plants (1707). He discovered that the oil obtained by macerating and distilling thyme reacted with vinegar, potash, and other substances of vegetable origin to form colors similar to those in flowers and leaves, and this supported the belief that plants consisted of a limited number of principles combined in different ways and obtainable by simple processes such as distillation and fermentation.

After publishing a few botanical papers, Geoffroy transferred to the chemical section of the Academy in 1715. Much of his subsequent research arose from his pharmaceutical work, a good example being his discovery that sedative salt (boric acid), a medicament normally prepared in small quantities by subliming a mixture of borax and iron vitriol (ferrous sulfate), could be made on a larger scale by treating borax with diluted oil of vitriol (sulfuric acid) and crystallizing it (1732). This research was important in the development of theoretical as well as practical chemistry, for Geoffroy demonstrated the presence of a common constituent in borax, soda, and common salt, all of which yielded Glauber's salt (sodium sulfate) on treatment with sulfuric acid.

Geoffroy's son, Claude François Geoffroy (*ca.* 1728–1753; incorrectly called N. Geoffroy in *Nouvelle biographie générale*), succeeded him in his profession and was elected to the chemical section of the Academy in 1752.

BIBLIOGRAPHY

I. Original Works. About sixty papers by Geoffroy were published in *Histoire et mémoires de l'Académie royale des sciences.* A complete list is given by P. Dorveaux (see below).

II. Secondary Literature. A general account of Geoffroy's life and work is J. P. Grandjean de Fouchy, "Éloge de M. Geoffroy," in *Histoire de l'Académie royale des sciences* (1752, pub. 1756), 153–164. Further information is given by P. Dorveaux, "Claude-Joseph Geoffroy," in *Revue d'histoire de la pharmacie,* **3** (1932), 113–122, bib. on pp. 119–122. Dorveaux also gives an account of his son, "Claude François Geoffroy," *ibid.,* 122–126. There seems to be no connection between the Geoffroy family and that of Étienne and Isidore Geoffroy Saint-Hilaire, the naturalists, according to G. Planchon, "La dynastie des Geoffroy, apothicaires à Paris," in *Journal de pharmacie et de chimie,* 6th ser., **8** (1898), 289–293, 337–345 (esp. 344–345).

W. A. Smeaton

GEOFFROY, ÉTIENNE FRANÇOIS (*b.* Paris, France, 13 February 1672; *d.* Paris, 6 January 1731), *chemistry, medicine.*

Known as Geoffroy the Elder (l'Aîné) to distinguish him from his brother Claude Joseph, he was the son of Matthieu François Geoffroy, a wealthy pharmacist who had been a Paris alderman, and Louise de Vaux, daughter of a well-known surgeon. Such scientists as Wilhelm Homberg and J. D. Cassini visited their home, giving demonstrations and lectures that supplemented Geoffroy's education. His father, the fourth in a respected dynasty of pharmacists, hoping that Geoffroy would eventually take over the family business, sent him to Montpellier in 1692 for a year to learn pharmacy from a colleague, Pierre Sanche, whose son came to Paris. While in Montpellier, Geoffroy attended courses at the medical school; and although he qualified as a pharmacist in 1694 after his return to Paris, his real ambition was to become a physician. With his father's consent he therefore turned to the study of medicine.

While still a student Geoffroy was chosen as medical adviser by the comte de Talland, French ambassador extraordinary to England; and in 1698 he spent several months in London, where he became friendly with Hans Sloane. He was made a fellow of the Royal Society, of which Sloane was secretary; and in January 1699, after his return to Paris, he was elected to the Académie des Sciences as the student of Homberg; he became an *associé* later in 1699 and *pensionnaire* in 1715. When first elected, he offered to keep the Academy informed of scientific developments in England, and his correspondence with Sloane became a valuable medium for the transmission of scientific news between the two countries.

On his way home from London, Geoffroy had visited Holland; and in 1700 he made another long journey, to Italy. He eventually graduated M.D. at Paris in 1704 and began to practice a few years later. He acquired a considerable reputation and was often consulted by other physicians.

Geoffroy succeeded J. P. de Tournefort as professor of medicine at the Collège Royal (now the Collège de France) in 1709 and retained the chair until his death; but his lecturing career had begun in 1707, when he first deputized for G.-C. Fagan, professor of chemistry at the Jardin du Roi. Until 1710 he shared this duty with Louis Lemery and Claude Berger; then Berger acted alone until his death in 1712, when Fagan retired completely from the chair and Geoffroy was appointed. Geoffroy normally lectured on materia medica for two or three hours immediately after his two-hour lecture on chemistry, a feat that earned the praise of Bernard de Fontenelle; the comments of the students are not available.

In 1726 Geoffroy was elected dean of the Paris Faculty of Medicine, and after the customary two-year period he was reelected, serving until 1729, at a time when there was a serious dispute between the physicians and surgeons. The strains of this office, together with his chairs and his practice, weakened his health. He retired from the Jardin du Roi in 1730, and within a year he died of consumption at the age of fifty-eight. His son, Étienne Louis Geoffroy, became well known as a naturalist and was elected to the Institut de France in 1798.

A new Paris pharmacopoeia, *Codex medicamentarius seu pharmacopoeia parisiensis,* was published by the Faculty of Medicine in 1732 under the deanship of H. T. Baron. Largely the work of Geoffroy, it contained many chemical remedies in addition to the traditional galenicals, as did Geoffroy's unfinished book on materia medica, based on his lectures. The part that he had dictated, containing medicaments from the mineral kingdom and part of the vegetable kingdom, was published in Latin as *Tractatus de materia medica* in 1741 and was translated into French in 1743. Geoffroy's treatment of the vegetable kingdom was interesting. Exotic plants were not usually imported whole, so he classified them under such headings as roots, barks, and leaves, since these were the parts used in medicine; but he described the whole plant in the case of those native to France. Almost always he included the results of qualitative and quantitative analysis by distillation, recording the nature and amount of each oil, phlegm, salt, or earth obtained; and he endeavored to relate the medicinal properties to the products of analysis. This approach to the study of vegetable remedies, with the implication that the products of distillation were originally present as such in the plant, had been introduced by earlier chemists at the Academy; and many such analyses are recorded in its minute books. By the end of his life, though, Geoffroy had lost faith in the method, according to the editors of the supplement to his *Materia medica,* who excluded these analytical results.

Geoffroy's first publication, a study of heat effects observed when saline substances were mixed (1700), is of interest as an early attempt to use the thermometer in chemistry; but his results were inconclusive and it is clear that his air thermometer responded too slowly to temperature changes. He soon entered a field of experimental chemistry which he related to a theory of matter.

Homberg had "decomposed" common sulfur into an acid, a bituminous substance, and an earth. Geoffroy found in 1704 that it could be re-formed by heating oil of vitriol (sulfuric acid) and turpentine; after distillation iron, detected with a magnet, was present in the residue. In 1705 Geoffroy also found iron in the ashes of all vegetable matter, although it was not detectable in the original plant: he believed that the iron had been formed during these processes, a "sulfurous principle" being common to vegetable matter and metals and giving metals their properties of fusibility and ductility. Louis Lemery objected to this theory and argued that the iron was present all the time in the oil of vitriol or turpentine, and in the vegetable matter that had been burned (*Mémoires de l'Académie royale des sciences* for 1707 [1708], 5–11). He claimed that Geoffroy's reasoning was unsound, for in some circumstances—when dissolved in acid and then crystallized, for example—iron could not be detected with a magnet although it was certainly present (*Histoire de l'Académie royale des sciences* for 1708 [1709], 61–65). Yet Geoffroy persisted in his belief, and in 1720 he identified his "sulfurous principle" with Georg Stahl's phlogiston.

On fermentation, urine produced volatile alkali; and in 1717 Geoffroy considered that the acid in urine had been converted into alkali. He also believed that alkali could be formed from mineral acids; he explained these changes, as well as the formation of iron from vegetable matter, by a theory which was elaborated in his *Materia medica* but is also mentioned in some of his earlier publications.

Geoffroy recognized three "very simple substances": fire, which could exist only in combination with the others; water, composed of hard, smooth, oval or wedge-shaped particles that dissolved solids by forcing their particles apart; and earth, composed of irregular particles with many pores between them. These three combined to form two "principles," salt and sulfur, which were the usual products of analysis. Salt existed in two main forms, acid and alkali, which were interconvertible because they differed in the way their constituent fire, water, and earth were combined: the particles of acids were pointed at both ends; those of alkalies were spherical with projecting points that joined together, forming a porous globule into which acid particles could enter. Sulfur was formed by the combination of salt with more fire, water, and earth; this combination could take place in vegetables, or animals, or beneath the ground, where, according to the proportions of the constituents, the product was one of a variety of substances, such as petroleum, coal, common sulfur, or a metal.

It was well known before Geoffroy's time that certain substances could displace others from compounds; but in 1718 he advanced the first general proposition that if two substances in combination are approached by a third with which one of them has a greater relation (*rapport*), then that one will combine with the third, leaving the other free. He accompanied this with a sixteen-column table in which, using symbols, he showed the order of displacement of some common substances. The first column referred to compounds of mineral acids, with which fixed alkali (soda and potash were not yet distinguished) had the greatest relation, followed by volatile alkali (ammonia), absorbent earth (chalk and such), and metals. Since the order in which the metals displaced each other was not the same for all acids, the next three columns were devoted to the individual acids and their reactions with metals. There followed columns for absorbent earth, fixed alkali, volatile alkali, metals in combination with acids, and common sulfur; and then six columns for the "compounds" of the metals with each other and one for water in relation to salt and alcohol.

No details were given by Geoffroy of the experiments on which his table was based, and in 1720 he had to reply to some criticisms. For example, one critic challenged his statement that absorbent earth came below volatile alkali in the first column, for it was well known that lime expelled volatile alkali from sal ammoniac (ammonium chloride). Geoffroy replied that although it was formed by heating chalk, lime was not itself a pure absorbent earth; rather, it contained sharp and caustic particles of alkali, which might have been formed from acid in the wood that was burned when the lime was made, or perhaps from "aluminous vitriolic acid" present in the original chalk. This opinion was criticized by Louis Lemery, who, like his father, Nicolas Lemery, believed that lime owed its causticity to fire particles (*Histoire de l'Académie royale des sciences* for 1720 [1722], 33). Thus it is easy to understand why Geoffroy's table was not elaborated until after about 1750, when ideas on the nature of acids, alkalies, and earths were more clearly defined.

In 1718 and again in 1720 Geoffroy referred only to "rapport" and did not use the words "affinity" or "attraction." There have been suggestions that he was influenced by Newton's theory that chemical reactions were caused by attraction between particles, but his detailed account of pointed and porous particles that combine by interlocking seems to make it clear that his ideas were akin to those of Descartes, Pierre Gassendi, Nicolas Lemery, and Nicolaas Hartsoeker, and that he did not belong to the Newtonian school of chemists.

BIBLIOGRAPHY

I. ORIGINAL WORKS. According to Fontenelle, Geoffroy was the principal author of *Codex medicamentarius seu pharmacopoeia parisiensis* (Paris, 1732). The first version of his work on materia medica to be published was *A Treatise of the Fossil, Vegetable and Animal Substances, That Are Made Use of in Physick* (London, 1736), a partial English trans. by G. Douglas of a MS of Geoffroy's lectures, which were delivered in Latin at the Collège Royal; this contains a short account of the animal kingdom, which is not in the Latin and French eds., and an English version of Fontenelle's "Éloge." The Latin text was published (by Étienne Chardon de Courcelles, according to Dorveaux) as *Tractatus de materia medica,* 3 vols. (Paris, 1741); vol. I includes the Latin texts of five of Geoffroy's medical theses and the French texts of Fontenelle's "Éloge" and Geoffroy's papers written in 1713, 1718, and 1720. A French trans. (by A. Bergier, according to Dorveaux) appeared as *Traité de matière médicale,* 7 vols. (Paris, 1743; new ed. [unaltered], 1757). Vol. I contains Fontenelle's "Éloge," Geoffroy's papers written in 1713, 1718, and 1720, and a French trans. of one of his medical theses: "Question, si l'homme a commencé par être ver." The Latin and French eds. of *Materia medica* stop where Geoffroy's MS ended, at the word "Melilotus" in the alphabetical list of indigenous plants. The vegetable kingdom was completed in *Suite de la matière médicale de M. Geoffroy, par M***, docteur en médecine,* 3 vols. (Paris, 1750); the authors of this were apparently the men who published an account of the animal kingdom in *Suite de la matière médicale de M. Geoffroy, par Mrs Arnault de Nobleville & Salerne, médecins d'Orléans,* 7 vols. (Paris, 1756–1757). They claimed to follow Geoffroy's order but added new material and acknowledged their indebtedness to Bernard de Jussieu and others.

Seventeen papers read by Geoffroy to the Académie des Sciences are listed in Poggendorff, I, 873–874. The most important, to which reference has been made above, are "Observations sur les dissolutions et sur les fermentations que l'on peut appeller froides," in *Mémoires de l'Académie royale des sciences* for 1700 (1703), 110–121; "Manière de recomposer le souffre commun par la réunion de ses principes, et d'en composer de nouveau par le mêlange de semblables substances, avec quelques conjectures sur la composition des métaux," *ibid.,* for 1704 (1706), 278–286; "Problème de chimie: Trouver des cendres qui ne contiennent aucunes parcelles de fer," *ibid.,* for 1705 (1706), 362–363; "Éclaircissemens sur la production artificielle du fer, & sur la composition des autres métaux," *ibid.,* for 1707 (1708), 176–188; "Expériences sur les métaux, faites avec le verre ardent du Palais Royal," *ibid.,* for 1709 (1711), 162–176; "Observations sur le vitriol et le fer," *ibid.,* for 1713 (1716), 170–188; "Du changement des sels acides en sels alkalis volatiles urineux," *ibid.,* for 1717 (1719), 226–238; "Table des différens rapports observés en chymie entre différentes substances," *ibid.,* for 1718 (1719), 202–212; and "Éclaircissemens sur la table inserée dans les Mémoires de 1718, concernant les rapports observés entre différentes substances," *ibid.,* for 1720 (1722), 20–34. Some of his letters to Sloane were printed, in part, in *Philosophical Transactions of the Royal Society* and are listed in P. H. Maty, *A General Index to the Philosophical Transactions* (London, 1787) pp. 633–634.

II. SECONDARY LITERATURE. A general account of Geoffroy's life is in Bernard de Fontenelle, "Éloge de M. Geoffroy," in *Histoire de l'Académie royale des sciences* for 1731 (1733), 93–100. Additional information is given by P. Dorveaux, "Étienne-François Geoffroy," in *Revue d'histoire de la pharmacie,* **2** (1931), 118–126; J. P. Contant, *Enseignement de la chimie au Jardin Royal des Plantes de Paris* (Cahors, 1952), pp. 55–57; and J. Torlais, "Le Collège Royal," in R. Taton, ed., *Enseignement et diffusion des sciences en France au XVIIIe siècle* (Paris, 1964), pp. 261–286. Geoffroy's relations with British scientists (with extracts from his letters to Sloane which are now in the British Museum and the Royal Society) are discussed by I. B. Cohen, "Isaac Newton, Hans Sloane and the Académie Royale des Sciences," in I. B. Cohen and R. Taton, eds., *Mélanges Alexandre Koyré* (Paris, 1964), I, 61–116; Cohen's description of Geoffroy as "one of Newton's chemical disciples" is disputed by W. A. Smeaton, "E. F. Geoffroy Was not a Newtonian Chemist," in *Ambix,* **18** (1971), 212–214. A good account of tables of affinity is given by A. M. Duncan, "Some Theoretical Aspects of Eighteenth-Century Tables of Affinity," in *Annals of Science,* **18** (1962), 177–194, 217–232.

W. A. SMEATON

GEOFFROY, ÉTIENNE LOUIS (*b.* Paris, France, 2 October 1725; *d.* Chartreuve, near Soissons, France, 12 August 1810), *zoology, medicine.*

Geoffroy was the son of Étienne François Geoffroy, dean of the Faculty of Medicine of Paris and professor of medicine at the Collège de France, and the grandson of the apothecary Matthieu François

Geoffroy. His father died when the boy was five, and Étienne was brought up by his mother. After a brilliant career at the *collège* of Beauvais, he went to Paris to study medicine. There he attended the courses of Antoine Ferrein (anatomy), G. F. Rouelle (chemistry), Bernard de Jussieu (botany), and Jean Astruc (practical medicine). Geoffroy's thesis for the *licence,* which he defended in 1746, dealt with the manner in which the fetus is nourished. Received as a doctor in 1748, he henceforth applied himself to medical practice while pursuing research in zoology.

In 1762 Geoffroy published a work on the insects of the Paris area in which he used new criteria of classification: absence or presence, number, form, and texture of the wings; and distribution of the various orders according to the number of tarsomeres in the tarsi. The latter criterion attracted the attention of Linnaeus, who often quoted Geoffroy. In 1767 Geoffroy published a volume on the terrestrial and aquatic gastropods of the Paris area in which he used the characteristics of the animal and not, as was customary, those of the shell. His study on the auditory organ in man, the reptiles, and the fishes (1778) is an important work in comparative anatomy.

Geoffroy was a modest and unselfish scientist who did not seek honors; he declined Astruc's proposal that he succeed the latter in the chair of medicine at the Collège de France. He was elected an associate member of the Academy of Sciences on 24 April 1798.

Geoffroy was also the author of a Latin poem on hygiene (1771) and of a *Manuel de médecine pratique* (1800), which he wrote on his farm in the village of Chartreuve, near Soissons, where he had retired. He was also the mayor of this small town.

Geoffroy had several sons, of whom the best-known was René Claude Geoffroy, a physician and naturalist who traveled to Senegal and to Santo Domingo and became a member of the Academy of Medicine.

BIBLIOGRAPHY

I. Original Works. Geoffroy's principal writings are *Histoire abrégée des insectes qui se trouvent aux environs de Paris . . .,* 2 vols. (Paris, 1762); *Traité sommaire des coquilles tant fluviatiles que terrestres qui se trouvent aux environs de Paris* (Paris, 1767); *Hygieine, sive ars sanitatem conservandi* (Paris, 1771), also translated into French by P. Delaunay (Paris, 1774); *Dissertation sur l'organe de l'ouïe* 1) *de l'homme,* 2) *des reptiles,* 3) *des poissons* (Amsterdam–Paris, 1778); and *Manuel de médecine à l'usage des chirurgiens et des personnes charitables . . .* (Paris, 1800).

There are three letters from Geoffroy to Antoine-Laurent de Jussieu (1797–1798) in the archives of the Académie des Sciences, Paris.

II. Secondary Literature. See the unsigned "Notice sur M. Et. L. Geoffroy, docteur régent et professeur de l'ancienne Faculté de médecine de Paris . . .," in *Bibliothèque médicale* (1810); P. Delaunay, *Le monde médical parisien au dix-huitième siècle* (Paris, 1906), pp. 425–430; and E. Lamy, "Deux conchyliologistes français du XVIIIe siècle; les Geoffroy oncle et neveu," in *Journal de conchyliologie,* **73** (1929), 129–132.

Jean Théodoridès

GEOFFROY SAINT-HILAIRE, ÉTIENNE (*b.* Etampes, France, 15 April 1772; *d.* Paris, France, 19 June 1844), *zoology.*

Geoffroy's father, a procurator at the tribunal of Etampes, a small town near Paris, had little money and fourteen children. Étienne, the youngest, received as a child the surname Saint-Hilaire, which he later joined to his family name. His extraordinary career (professor at the Muséum at twenty-one) was furthered by priests who were captivated by his lively intelligence, his unusual imagination, and a sort of inner fire that added to his physical charm. Thanks to the cultivated *grand seigneur* the Abbé de Tressan, he became a canon at the age of fifteen with prospects for a splendid career in the church. He was introduced to natural history by the agronomist Abbé A. H. Tessier in Etampes and later by the ornithologist Brisson and Antoine de Jussieu at the Collège de Navarre in Paris, where Geoffroy was on a scholarship. With the outbreak of the Revolution, ecclesiastical careers were jeopardized; obeying his father's orders, he studied law, and then, pursuing his own desires, he began medical studies. In 1792 he was a *pensionnaire libre* at the Collège du Cardinal Lemoine in Paris. There he won the affection of that institution's most illustrious member, Abbé René Just Haüy, one of the founders of crystallography. In Haüy's simply furnished room he met all the famous scientists of the period. Consumed with enthusiasm for mineralogy, he became a student of the venerable Daubenton at the Collège de France.

While retaining a deep attachment to the priests who had supported his career, Geoffroy embraced revolutionary ideas. He frequented the clubs and committees and adopted a philosophical deism and a generous humanitarianism that he preserved for the rest of his life. At the beginning of the Terror (August 1792), Haüy was imprisoned because he was a priest; by actions as courageous as they were romantic, Geoffroy attempted to free him. Finally Haüy was liberated; in recognition of Geoffroy's efforts, Daubenton—a great friend of Haüy—had him named demonstrator in zoology at the Jardin des Plantes, replacing the count de Lacépède, who had been forced to flee (March 1793). In June 1793 the Jardin des Plantes became the Muséum d'Histoire Naturelle; and it was then, owing to his patrons, that Geoffroy, barely twenty-one years old, was named, in the absence of Lacépède, professor of quadrupeds, cetaceans, birds, reptiles, and fish. He became an intimate friend of Lamarck—his elder by almost thirty years—a botanist promoted to the study of insects, worms, and crustaceans. Each of them eagerly explored his new field.

At this period the Abbé Tessier, Geoffroy's former patron, recommended to him a poor young man living in Normandy who was making excellent drawings of careful dissections of fish and invertebrates; he was Georges Cuvier. Geoffroy was extremely friendly to him upon his arrival in Paris, and for one year (1795) they lived and worked together. In a note on tarsiers they stated that these animals constitute a link between the ape and the bat in the great chain of being. Then, following Buffon, they envisioned the possibility of deriving all species from a single type. Cuvier later renounced these audacious ideas, especially after the return to religious beliefs marked by Napoleon's coronation as emperor by the pope (1804). To the contrary, Geoffroy persisted on the "philosophical" road, which was also Lamarck's. In 1796 he began to study, perhaps with the latter, modifications of species due to the environment.

When Bonaparte organized the famous Egyptian campaign, he requested the assistance of numerous scientists. Cuvier avoided leaving, but Geoffroy accepted enthusiastically, and from 1798 to 1801, in the midst of adventures in which he often risked his life, he made many scientific observations. He proceeded up the Nile as far as Aswan. The English allies of the Egyptians were victorious but Geoffroy succeeded in rescuing from the English his important natural history collections. Cuvier, in his absence, had become in everyone's eyes France's leading naturalist.

Returning to Paris, Geoffroy devoted himself from 1802 to 1806 to descriptive zoology and classification, despite his predilection for great theoretical views. He pursued the research on the marsupials that he had begun in 1796. He also composed a large catalog of the mammal collections of the Muséum, the printing of which he stopped suddenly at page 272 in 1803, perhaps as a result of differences with Cuvier; the work remained unfinished. In 1807 he entered the Académie des Sciences, where Cuvier had dominated in natural history since 1796.

The study and publication of the rich material that he had gathered in Egypt proceeded very slowly in the sumptuous work *Description de l'Égypte par la Commission des sciences* (1808–1824). In the Egyptian tombs Geoffroy had found mummified animals more than three thousand years old. They were identical to existing species, and Cuvier saw in this fact the proof of the fixity of species. For Lamarck, on the other hand, this fixity was not demonstrable. In relation to geological ages, which Lamarck calculated in hundreds of millions of years in his *Hydrogéologie* (1802), three thousand years are but a moment; much too short a period for the evolution of creatures to be perceptible. Although Geoffroy does not seem to have accepted Lamarck's views on the duration of geological epochs, he continued to believe in the nonfixity of species. In the following years it was by comparative anatomy, teratology, and then paleontology that he strove to prove that species are transformed in the course of time, the simpler ones engendering the more complex.

Since Aristotle, the unity of plan (that is, the unity of anatomical structure) of vertebrates as a whole had been recognized. Buffon had proclaimed it as early as 1753. Today it seems quite obvious, but it was contested in Geoffroy's time, particularly by Cuvier. For the latter, the fish's fin bore no relation to the mammal's paw, each having been created separately by God. Fifteen years of work, beginning in 1806, enabled Geoffroy to establish two fundamental principles of comparative anatomy: (1) the principle of anatomical connections, which allows one to trace an organ from species to species despite its transformations; and (2) the principle of balance (that is, of the equilibration of the organs), which manifests itself in a reduction in the size of organs when a neighboring organ hypertrophies. The unity of the vertebrate plan that Geoffroy proposed implies a sort of kinship among the vertebrates, the more complex having issued from the simpler—in other words, the mammals having descended from the fish. In order to demonstrate this relationship, Geoffroy turned to comparative embryology.

Haüy greatly admired Daubenton's nephew, the anatomist Félix Vicq d'Azyr, with whose audacious ideas Geoffroy was undoubtedly acquainted. Vicq d'Azyr had observed in the human embryo the transitory appearance of the intermaxillary bone, which persists in the adult ape and is thus an indication of a kinship between man and ape. Furthermore, Cuvier had certainly presented to Geoffroy the ideas of his former professor and friend Karl Kielmeyer of the University of Stuttgart. Kielmeyer had shown that vertebrates, in the course of their embryonic life, go through phases recalling their supposed ancestors; hence, the human embryo possesses the preliminary forms of branchial fissures typical of the fish. Geoffroy, in turn, following J. H. F. Autenrieth, a disciple of Kielmeyer, demonstrated that the disposition of the centers of ossification of the fetal cranium conforms to the general plan of the vertebrate series, of which the prototype is found among the fish. Antoine Serres, a friend and disciple of Geoffroy, later developed this notion of the recapitulation by the embryo of certain characteristics of the supposed ancestors.

In the scientific language of the early nineteenth century, the word *evolution* designated the sum of the transformations undergone by an embryo. Girou de Buzareingnes, taking up Kielmeyer's and Serres's hypothesis of embryonic recapitulation, declared in 1828 that the embryological evolution of an individual follows the path of "the evolution of the animal kingdom itself." In 1831 Geoffroy adopted this new meaning of *evolution,* which implies a transformation of the species in the course of geologic time. The term was then promulgated by an admirer of Geoffroy, Frédéric Gérard, in the *Dictionnaire universel d'histoire naturelle* (see "Espèce," V [1844]; and "Géographie zoologique," VI [1845]), which was distributed throughout the world.

Geoffroy was confronted with the great problem of how the transition from the plan of the invertebrates to that of the vertebrates was accomplished. Carried away by his imagination, he supposed that the carapace of insects corresponds to the vertebra. This hypothesis, which contained only a small element of truth (the vertebrates and the insects are formed by more or less condensed metameres), rendered Geoffroy slightly ridiculous. In 1829 two young researchers, Laurencet and P. S. Meyranx, proposed an ingenious interpretation to account for a transition from cephalopods to fish. Geoffroy used it as an opportunity to attack Cuvier at the Academy. Thus there began the famous controversy of 1830 that so excited Goethe. (A full account of the controversy is in the article on Cuvier.)

It appears that Geoffroy, following the path of the botanists of the preceding century, saw in the study of monsters a means of explaining certain sudden transformations of species. From 1802 to 1840 he devoted more than fifty reports to descriptive teratology. Moreover, his *Essai de classification des monstres* (1821) marks the debut of scientific teratology. Between 1832 and 1837 his son, Isidore, regrouped and completed his father's publications, under the latter's direction, to form the first scientific treatise on teratology. Geoffroy went even further. According to Cuvier, it would seem that Geoffroy sought, by intervening in the development of the chicken embryo, to maintain the fish stage or to obtain the transition to the mammalian stage. He failed, but his attempts make him the founder of experimental embryology (1825–1826).

In 1824 a new way appeared for Geoffroy to demonstrate the transition from reptiles to mammals. In that year Cuvier introduced into the second edition of his *Recherches sur les ossements fossiles* a study, done a little too hastily, of the remains of a "crocodile" discovered near the city of Caen. Geoffroy, who had studied the living Egyptian crocodile extensively, announced that the Caen animal was in reality very different from the crocodile; he named it *Teleosaurus.* It presented, he stated, characteristics intermediate between those of saurians and mammals (1825). Although modern studies have somewhat modified Geoffroy's interpretation, it does nevertheless mark the starting point of evolutionary paleontology. Thereafter Geoffroy was obsessed by paleontological problems. In 1833 he studied the fossil mammals of the Perrier bed (old Pleistocene) in the Massif Central. He established that all the species of Perrier have disappeared from nature and that certain of them constitute "intermediate links" in the great "progressive" series of living beings. He also studied the fossil remains of the Oligocene deposit of St.-Gérand-le-Puy, near Vichy. The few lines that he wrote on each species demonstrate his profound knowledge of vertebrate osteology, and present-day science confirms the majority of his interpretations concerning St.-Gérand.

True to the hypotheses that had guided his researches in experimental embryology, he attributed the variations between species to physicochemical changes in the environment in the course of geologic ages, changes that, he presumed, had influenced not the adults, as Lamarck thought, but the embryos. It is regrettable that in paleontology Geoffroy never followed up his projects with detailed and illustrated memoirs in support of his theories, which after his death were challenged by the hypothesis of multiple creations put forward by Alcide Dessalines d'Orbigny. Multiple creation conflicted little with the Bible and contributed greatly to the progress of geology by justifying the creation of periods defined by their respective faunas. It was not until the beginning of 1859, seven months before the publication of Darwin's *Origin of Species,* that the young Albert Gaudry was able to demonstrate, through the study of the fossil mammals of Greece, that Geoffroy's "intermediate" species and "missing links" were tangible realities.

Starting around 1834, Geoffroy's writings became increasingly theoretical and vague, infused with a kind of grandiose cosmic poetry born of the concept of a unitary universe. Thereafter, the Academy published only the titles of his communications. In July 1840 he became blind as the result of a cataract, and by 1842 his mental powers had begun to fail. The news of his death was received with sorrow by a great many Parisians who had loved his enthusiasm and his liberal ideas.

In France, and perhaps even more in English-speaking countries, Geoffroy's work is often ranked below that of Cuvier, his implacable adversary. Geoffroy's awkward style and sometimes bizarre ideas harmed him considerably; yet Cuvier himself recognized that Geoffroy possessed a great talent for description and classification. Moreover, Geoffroy revitalized comparative anatomy in France and created scientific teratology, experimental embryology, and the concept of paleontological evolution. When Cuvier died in 1832, Geoffroy was only sixty years old; he could have tried for one last time to make his ideas prevail among younger naturalists. But already the fire of his genius was dimming. He and his friend Lamarck lived too early to be completely understood. Indeed, Geoffroy used to say that innovators, like Christ, wear a crown of thorns.

Nevertheless, in judging Geoffroy only on the basis of his publications, one risks underestimating his influence. His contemporaries assure us that he was bolder and clearer in his speech than in his writings. Much loved by his students, he put forth his ideas for forty-seven years in his courses at the Muséum and for thirty-two years in those at the Sorbonne. His home was frequented by the leading liberal thinkers of Paris. Balzac, who long esteemed Cuvier, ultimately concluded that Geoffroy was superior. Michelet admired Geoffroy and Lamarck equally: each had taught that the natural sciences and the human sciences ought to be closely joined, since man, by his origins, remains tied to the animal world.

Like Haüy and Lamarck, Geoffroy pursued the great, if somewhat mystical, idea of the fundamental unity of the universe, life, and human thought.

BIBLIOGRAPHY

I. Original Works. T. Cahn (see below) gives a bibliography of 259 titles. Isidore Geoffroy Saint-Hilaire (see below) classified 321 titles by subject matter and provided references to analyses and criticisms provoked by the works listed. Additional information may be found in the *Catalogue of Scientific Papers* of the Royal Society, which lists 247 titles.

Geoffroy's personal papers and MSS remained in the possession of his family until about 1950. Since then they have periodically appeared for sale in lots (on auction at the Hôtel Drouot and by Parisian autograph dealers); the library of the Muséum has been able to purchase a few of them.

Geoffroy's principal works are *Catalogue des mammifères du Muséum* (Paris, 1803), a very rare work, the printing of which was never completed and a copy of which is possessed by the library of the Muséum; "Considérations sur les pièces de la tête osseuse des vertébrés," in *Annales du Muséum d'histoire naturelle,* **10** (1807), 332–365; "Sur les déviations organiques provoquées et . . . les incubations artificielles," in *Mémoires du Muséum d'histoire naturelle,* **13** (1825), 289–296, in *Archives générales de médecine,* **13** (1826), 289, and in *Journal complémentaire des sciences médicales,* **24** (1826), 256; the article "Monstre," in *Dictionnaire classique d'histoire naturelle,* XI (Paris, 1827), 108–152, which bears the inexact signature "G.N."; *Principes de philosophie zoologique discutés en mars 1830 à l'Académie* (Paris, 1830), of which there is an analysis by Goethe (see below); and *Recherches sur les grands sauriens trouvés à l'état fossile* (Paris, 1831), repr. in *Mémoires de l'Académie des sciences,* **12** (1833), 1–138.

II. Secondary Literature. On his life and work, see I. E. Amlinsky, *Zhoffrua Sent-Iler i ego borba protiv Kyuvier* ("Geoffroy Saint-Hilaire and His Struggle Against Cuvier"; Moscow, 1955); F. Bourdier, "Geoffroy Saint-Hilaire Versus Cuvier: The Campaign for Paleontological Evolution (1825–1838)," in C. J. Schneer, ed., *Toward a History of Geology* (Cambridge, Mass., 1969), pp. 36–61; T. Cahn, *La vie et l'oeuvre d'Étienne Geoffroy Saint-Hilaire* (Paris, 1962); Isidore Geoffroy Saint-Hilaire, *Vie, travaux et doctrine scientifique d'Étienne Geoffroy Saint-Hilaire* (Paris, 1847); J. W. von Goethe, in *Jahrbuch für wissenschaftlichen Kritik,* nos. 52–53 (1830), and nos. 51–53 (1832); and Jean Rostand, "É. Geoffroy Saint-Hilaire et la tératogenèse expérimentale," in *Revue d'histoire des sciences et de leurs applications,* **17** (1964), 41–50.

Regarding Geoffroy's influence on historians and reformers, see R. van der Elst, *Michelet naturaliste* (Paris, 1914), and the works of Geoffroy's contemporaries, such as P. H. Buchez, Pierre Leroux, and Jean Raynaud.

Franck Bourdier

GEOFFROY SAINT-HILAIRE, ISIDORE (*b.* Paris, France, 16 December 1805; *d.* Paris, 10 November 1861), *zoology.*

The only son of Étienne Geoffroy Saint-Hilaire, Isidore wanted to become a mathematician; but his father saw in him the continuator of his work and engaged him in his laboratory as an *aide-naturaliste* in 1824, when he was only nineteen. In 1830 Isidore gave a course of lectures at the Athénée which attracted considerable attention. It dealt with a new subject: the interrelations of animal species and their relations with the environment. While undertaking extensive research on mammals and birds, he published in 1832 the first of three large volumes on monsters.

In 1833 Geoffroy Saint-Hilaire entered the Academy of Sciences, at the age of twenty-seven. For his candidacy, he published a notice on his works (1824–1833) which reveals their astonishing scope. Teaching, and especially administrative duties, subsequently reduced his scientific activity. He replaced his father as professor of comparative anatomy at the Faculté des Sciences in 1837, became inspector of the Académie de Paris in 1840, and replaced his father, who had become blind, as professor at the Muséum d'Histoire Naturelle in 1841. He assumed the considerable duties of inspector general of education in 1844 but gave them up in 1850, when he was named professor of zoology at the Sorbonne. His brilliant career was darkened by the death of his wife in 1855 and of the last of his sisters in 1860. The attacks of an undetermined illness grew increasingly severe, and he died in 1861, cared for by his aged mother, in the same room at the museum in which he had been born. He was only fifty-five.

Unlike his father, who possessed a vivid and intense imagination, was rash in both thought and action, and was given to making abstruse remarks, Isidore hid his feelings, was cold and reflective, and enjoyed precise reasoning and lucid exposition. He continued his father's work, which he strengthened and made more exact, although he sometimes dissembled its audacious aspects in the face of the all-powerful opposition of the partisans of Georges Cuvier. Was he a continuator without originality, as his biographers have implied? In truth, his work is too little known for this position to be upheld. His important views on the persistence of infantile characteristics among the primates and on "parallel" evolution appear to be original.

Although the idea of seeking laws governing the formation of monsters was his father's, Isidore Geoffroy Saint-Hilaire nonetheless grouped and brought into accord, judiciously and critically, a great number of scattered facts. In 1832 he coined the word *teratology,* to designate the science of monsters. His work on the description and classification of the mammals, especially of the apes, was original and successful. In 1832, taking up and refining the ideas of Buffon, the full significance of which had perhaps not been grasped, he showed that, in proportion to the entire body, the brain of young apes is larger than in adult apes and that young apes also possess relatively greater intelligence and adaptability. Geoffroy Saint-Hilaire believed that through the persistence of infantile characteristics certain apes possess a large brain and great possibility for adaptation throughout their lives. Long before L. Bolk enunciated his theory of neoteny (1921), Geoffroy Saint-Hilaire suggested that the adult human's large brain and potential for adaptation might likewise represent the persistence of an infantile form. In the same year, refining the notion of a genealogical tree of species, originated by Lamarck, he attempted to establish what he termed a "parallel" classification of beings, in which both the evolution of the phyla and their adaptive convergences are allowed for.

Primarily a theorist, Geoffroy Saint-Hilaire nevertheless took an interest in practical problems. For example, his duties as director of the menagerie of the Muséum d'Histoire Naturelle led him to experiments on hybridization among mammals and among birds. In 1856 he conducted an active campaign, with hippophagic banquets, to encourage the consumption of horsemeat, neglected until then because of traditional prejudice. Above all, however, he sought to develop the acclimatization of useful animals (1849) and founded two organizations which are still active: the Société d'Acclimatation and the Jardin d'Acclimatation in the Bois de Boulogne, Paris.

In almost all of his works Geoffroy Saint-Hilaire gave considerable space to the history of science. The large volume that he devoted to the life and work of his father remains a model of biography, although discretion sometimes made him tone down his father's conceptions concerning evolution. In 1859 he published *Résumé des vues sur l'espèce organique,* in which he quietly reminded Darwin of his predecessors in France: Buffon, Lamarck, and Étienne Geoffroy Saint-Hilaire.

BIBLIOGRAPHY

I. Original Works. A list of Geoffroy Saint-Hilaire's works was published in 1862, but it is both incomplete and inaccurate. It is preferable to refer to the printed catalog of the Bibliothèque Nationale, vol. LVI, cols. 130–136, for the individual publications and to Royal Society, *Catalogue of Scientific Papers,* II, 832–837, for the articles in scientific periodicals (125 titles). A large portion of his work is, however, to be found in dictionaries and nonscientific periodicals. The *Dictionnaire classique d'histoire naturelle,* 16 vols. (Paris, 1822–1830), contains a number of his articles on mammals beginning with vol. VII; see also X, 63–73, 199–206, 372–375; XI, 102–107; XII, 512–520; XV, 129–151; XVI, 141–149. A list of his many articles published in the *Gazette médicale,* the *Revue encyclopédique,* the *Encyclopédie moderne,* and the *Revue des deux-mondes,* among others, has never been compiled.

His chief works are *Histoire générale . . . des anomalies . . . ou Traité de tératologie,* 4 vols. (Paris, 1823–1837); *Description des mammifères . . . famille des singes* (Paris, 1839); *Essai de zoologie générale* (Paris, 1841); *Vie, travaux . . . d'Étienne Geoffroy Saint-Hilaire* (Paris, 1847); *Domestication et naturalisation* (Paris, 1854); and *Histoire naturelle générale des règnes organiques,* 3 vols. (Paris, 1854–1862), an undertaking of grandiose intentions which was never completed.

II. SECONDARY LITERATURE. See A. de Quatrefages de Bréau, "Éloge historique de M. Geoffroy Saint-Hilaire," in *Bulletin de la Société impériale d'acclimatation,* **9** (1862), 257–278; and "Éloge historique par J. B. Dumas," in *Mémoires de l' Académie des sciences,* 2nd ser., **38** (1873), 178–212, delivered at the Institut de France. Other, more brief, *éloges* are by Edouard Drouyn de Lhuys, Henri Milne-Edwards, Charles Delaunay, Émile Blanchard, and Nicolas Joly. The article in Hoeffer, ed., *Nouvelle biographie générale,* XX (Paris, 1857), 54–55, recounts the main facts of his administrative career, but no extended work has been devoted to him.

FRANCK BOURDIER

GERARD OF BRUSSELS (*fl.* first half of the thirteenth century), *geometry.*

Gerard played a minor but not unimportant role in the development of kinematics and the measure of geometrical figures. His career remains obscure except for his having written a treatise entitled *Liber de motu,* which remains in six manuscripts. Four of these date from the thirteenth century. Written sometime between 1187 and 1260, the *Liber de motu* quotes the translation of Archimedes' *De quadratura circuli* ("On the Measurement of the Circle") by Gerard of Cremona; the translation was completed before the translator died in 1187. On the other hand, the *Liber de motu* is mentioned in the *Biblionomia* of Richard of Fournival, who died in 1260.

Gerard seems to have known the *Liber philotegni de triangulis* of Jordanus de Nemore, whose exact dating is as difficult to determine as that of Gerard but who can be placed, with some confidence, in the early decades of the thirteenth century. The similarity in their names suggests that Gerard may be identified with the unknown mathematician Gernardus who wrote an arithmetical tract *Algorithmus demonstratus.* That Gerard is referred to as *magister* in the title of *Liber de motu* and his apparent knowledge of Jordanus suggest a university milieu for this work—perhaps the University of Paris.

The *Liber de motu* contains thirteen propositions, in three books. In these propositions the varying curvilinear velocities of the points and parts of geometrical figures in rotation are reduced to uniform rectilinear velocities of translation. The four propositions of the first book relate to lines in rotation, the five of the second to areas in rotation, and the four of the third to solids in rotation. Gerard's proofs are particularly noteworthy for their ingenious use of an Archimedean-type *reductio* demonstration, in which the comparison of figures is accomplished by the comparison of their line elements. In this latter technique Gerard assumed that if the ratio of the elements of two figures taken in pairs is the same, then the ratio of the totalities of the elements of the figures is the same. Such a technique resembles the procedure followed in Archimedes' *Method,* which Gerard could not have read. The proposition most influential on later authors was the first: "Any part as large as you wish of a radius describing a circle . . . is moved equally as its middle point. Hence the radius is moved equally as its middle point."[1] This is similar in a formal way to the rule for uniform acceleration, which appears to have originated with William of Heytesbury at Merton College, Oxford, in the 1330's.[2] This rule asserted that a body which is uniformly accelerated traverses the same space in the same time as a body which moves with a uniform velocity equal to the velocity that is the mean between the initial and final velocities of the accelerating body. Gerard's proposition concerns itself with movements that uniformly vary over some part or all of the linear magnitude rotating, while Heytesbury's rule concerns instantaneous velocities that uniformly vary through some period of time. The "middle velocity" is used in both rules to convert the movements to uniformity.

Gerard's influence on Thomas Bradwardine, the founder of the Merton school of kinematics, is evident, for Bradwardine knew and quoted Gerard's tract. Furthermore, Nicole Oresme's *De configurationibus qualitatum,* written in the 1350's, shows some possible dependency on the *De motu.*[3]

NOTES

1. *Liber de motu,* M. Clagett, ed., p. 112.
2. For a discussion of the Merton rule, see M. Clagett, *Science of Mechanics,* ch. 5.
3. M. Clagett, *Nicole Oresme and the Medieval Geometry of Qualities and Motions* (Madison, Wis., 1968), p. 466.

BIBLIOGRAPHY

The *Liber de motu* has been edited by M. Clagett, "The *Liber de motu* of Gerard of Brussels," in *Osiris,* **12** (1956), 73–175. See M. Clagett, *Science of Mechanics in the Middle Ages* (Madison, Wis., 1959; repr. 1961), ch. 3; G. Sarton, *Introduction to the History of Science,* II (Baltimore, 1931), 629; and V. Zubov, "Ob 'Arkhimedovsky traditsii' v srednie veka (Traktat Gerarda Bryusselskogo 'Odvizhenii')" ("The Archimedean Tradition in the Middle Ages [Gerhard of Brussels' Treatise on Motion]"), in *Istoriko-matematicheskie issledovaniya,* **16** (1965), 235–272.

MARSHALL CLAGETT

GERARD OF CREMONA (*b.* Cremona, Italy, 1114 [?]; *d.* Toledo, Spain, 1187), *transmission of knowledge.*

For a detailed study of his life and work, see Supplement.

GERARD OF SILTEO (SILETO) (*fl.* thirteenth century), *astronomy.*

Gerard was a Dominican friar about whom little is known except that he composed a *Summa de astris.* He is cited twice by Dominican chroniclers as from Silteo and once as from Sileto, but neither location is known. J. Quétif and J. Échard list him as belonging to a German province (*Scriptores ordinis,* II, 918), whereas M. Grabmann refers to him as an Italian (*Mittelalterliches Geistesleben,* II, 397). One manuscript of his work is preceded by a letter from the Dominican Gerard of Feltre, who is identified as the author of the work, to another Dominican called John, noted marginally (and in a different hand) as John of Vercelli, master general of the order from 1264 to 1283; perhaps Silteo or Sileto was a hamlet near Feltre, in the Italian Alps east of Trent and north of Padua. An early list gives the date 1291 for Gerard, which is possibly the year of his death.

The *Summa de astris* is divided into three parts, the first of which is concerned with astronomy, the second with astrology, and the third with a critical refutation of astrological excesses. Part 1 comprises twenty-three distinctions and deals with classical topics of medieval astronomy at approximately the level of the *Rudimenta* of al-Farghānī and the *Sphere* of Sacrobosco; the chief source cited is the Arab astrologer Abū Ma'shar. The final distinction, dealing with comets, has been edited and analyzed by Thorndike, who notes that Gerard saw and described the comet of 1264 (probably writing soon after that date) and that he otherwise borrowed extensively from Albertus Magnus. Generally, Gerard's discussion of the comet of 1264 is inferior to that of another Dominican, Giles (Aegidius) of Lessines; there seems to be no connection between the two. Gerard's criticisms of judicial astrology "may have had some influence on the subsequent attacks upon astrology by Nicolas Oresme, Henry of Hesse, and Pico della Mirandola" (Thorndike, p. 187).

BIBLIOGRAPHY

Lynn Thorndike, *Latin Treatises on Comets Between 1238 and 1368 A.D.* (Chicago, 1950), pp. 185–195, apart from the Latin text of the section of *Summa de astris* on comets, lists the distinction headings of the entire work. For biographical details, see Jacques Quétif and Jacques Échard, *Scriptores ordinis praedicatorum,* 2 vols. (Paris, 1719–1721; repr. New York, 1959), I, 725b; and Martin Grabmann, *Mittelalterliches Geistesleben,* 3 vols. (Munich, 1925–1956), II (1936), 397.

WILLIAM A. WALLACE, O.P.

GERARD, JOHN (*b.* at or near Nantwich, Cheshire, England, 1545; *d.* Holborn, London, England, February 1612), *botany, pharmacy, horticulture.*

Gerard belonged to a branch of the family of Gerard of Ince in Lancashire. He received a grammar school education at Willaston (Wistaston), Cheshire, and was apprenticed at the age of sixteen to a London barber-surgeon, Alexander Mason, for the customary seven years. Some time thereafter he traveled, presumably as a ship's surgeon, aboard a merchant ship of the Company of Merchant Adventurers in London trading in the Baltic, for he stated later that he had been in Denmark, Sweden (Swenia), Poland, and Russia (Muscovia). He then settled in London and probably carried on his profession of barber-surgeon while developing his horticultural interests.

By 1577 he had become superintendent of the gardens belonging to William Cecil, Lord Burleigh, at the Strand, London, and at Theobalds, Hertfordshire, a post he held for the next twenty-one years. He possessed a garden of his own at Holborn, London, so well stocked in 1597, according to George Baker, surgeon to Queen Elizabeth, with "all manner of strange trees, herbes, rootes, plants, flowers, and other such rare things, that it would make a man woonder, how one of his degree, not having the purse of another, could ever accomplish the same." He added, "Upon my conscience, I do not thinke for the knowledge of plants, that he is inferior to any." In 1596 Gerard issued a catalog of his plants, the first such in England, followed by a second edition in 1599. This period was one of horticultural expansion in England, many new plants being introduced from abroad through the powerful influence of Cecil and others. Gerard undoubtedly acquired a detailed first-hand knowledge of them, which is reflected in his *Herball.* His standing as a barber-surgeon, which necessitated a knowledge of medicinal plants, and as a practical gardener led to his appointment from 1586 to 1603 or 1604 as curator of a physic garden belonging to the College of Physicians of London. In 1597 he was elected junior warden of the Barber-Surgeon's Company and in 1608 master.

The Herball or Generall Historie of Plantes, first published in 1597, is the best-known and most often quoted herbal in the English language. Its lasting repute is due not so much to its originality and accuracy, which are ofttimes questionable, as to its entertaining Elizabethan descriptive style, its interspersed anecdotes and comments, its antique remedies, and its woodcuts.

It occupies 1,392 pages, plus introductory matter and index, and has nearly 2,200 woodcut illustrations, most of which had already been used by Tabernaemontanus (Bergzabern), whose woodblocks had been obtained from Frankfurt am Main. The work is divided into three books. The first (pp. 1–176) deals with monocotyledons, described as "Grasses, Rushes, Corne, Flags, Bulbose, or Onion-rooted plants"; the second (pp. 177–1076), with "all sorts of herbes for meate, medicine, or sweete smelling use"; and the third (pp. 1077–1392), with "trees, shrubs, bushes, fruit-bearing plants, Rosins, Gums, Roses, Heath, Mosses, Mushroms, Corall, and their severall kindes." These books are divided into numerous short chapters, each dealing with a small group of plants and setting forth "the kindes, description, place, time, names, nature, and vertues, agreeing with the best received opinions."

At the beginning Gerard genially states his intent: "Now with our friendly labors we will accompagnie thee, and lead thee through a grasse plot [i.e., an account of the *Gramineae*], little or nothing of many Herbarists heertofore touched. . . . Then little by little conduct thee through most pleasant gardens, and other delightfull places where any herbe or plant may be found, fit for meate or medicine." Such a vast work was necessarily compiled from other works, and much of it came from Dodoens' *Stirpium historiae pemptades sex* (1583). About 1584 a young London physician and Cambridge graduate, Robert Priest, was requested by the London printers and booksellers Bonham and John Norton (who later published Gerard's *Herball*) to translate Dodoens' Latin work into English, and they retained his services until 1590, but this translation had evidently not been completed when he died in 1596 or 1597.

The fate of Priest's manuscript is not known. Gerard in his preface "to the courteous and well-willing Readers" stated that "Doctor *Priest,* one of our London Colledge, hath (as I heard) translated the last edition of *Dodonaeus,* which meant to publish the same, but being prevented by death, his translation likewise perished." Stephen Bredwell, however, in his preface to the first edition of Gerard's *Herball* implied that it then still existed. Johnson in 1633 declared, presumably on the authority of Mathias de L'Obel, that "this translation became the groundworke whereupon Mr *Gerard* built up this work," thus directly accusing Gerard of dishonestly concealing his major source. The whole truth of this matter can never be known. It would seem probable, as indicated by Jeffers, that Gerard, with the help of L'Obel, was engaged in compiling a book about plants before Priest began his translation of Dodoens; that Gerard's book had not reached a state fit for publication when Priest relinquished his task; and that Norton then requested Gerard to produce a work of like character which became the *Herball.*

To what extent Gerard was indebted to Priest's work is quite uncertain. Although this book is probably not so gross an example of successful piracy and plagiarism as it is sometimes considered, Gerard's honesty has certainly been much questioned and little defended since the adverse comments and accusations rendered in the preface to the second edition (1633) by its editor and reviser Thomas Johnson (1604–1644): "His chiefe commendation is, that he, out of a propense good will to the publique advancement of this knowledge, endeavoured to performe therein more than he could well accomplish; which was partly through want of sufficient learning." The care bestowed by Johnson in correcting what Raven calls "the errors of Gerard's book, the misplaced pictures, the confused species, the blunders of fact" and in adding much new material made his edition (often cited as *Ger. emac.,* i.e., *Gerardus emaculatus*) a popular and standard work, which proved of especial value in promoting the study of the British flora well into the eighteenth century. Yet, the *Herball* as published contains so much that undoubtedly came from Gerard himself, and its production, even with the possible aid of Priest's translation, was so massive a task that it seems charitable to credit him with the whole. It remains a valuable source of information about the plants available in western European gardens at the end of the sixteenth century and about the Latin and vernacular names then applied to them.

BIBLIOGRAPHY

I. Original Works. The two eds. of *Catalogus arborum, fruticum ac plantarum tam indigenarum, quam exoticarum in horto Johannis Gerardi* (London, 1596; 2nd ed., 1599) are both reprinted in *A Catalogue of Plants Cultivated in the Garden of John Gerard, in the Years 1596–1599, Edited With Notes, References to Gerard's Herball, the Addition of Modern Names, and a Life of the Author,* B. D. Jackson, ed. (London, 1876). Gerard's major work is *The Herball or Generall Historie of Plantes Gathered by John Gerard, Master in Chirurgie* (London, 1597; repr. 1598); extracts from the "very much enlarged and amended" ed. by Thomas Johnson (London, 1633) are in Marcus Woodward, *Gerard's Herball, the Essence Thereof Distilled* (London, 1927; repr., 1964).

II. Secondary Literature. The best source of biographical information is in Jackson (see above), supplemented by R. H. Jeffers, *The Friends of John Gerard*

(1545–1612), Surgeon and Botanist (Falls Village, Conn., 1967). There is an excellent chapter on Gerard, particularly in relation to the British flora, in C. E. Raven, *English Naturalists From Neckam to Ray* (Cambridge, 1947), pp. 204–217.

WILLIAM T. STEARN

GERASIMOVICH, BORIS PETROVICH (*b.* Kremenchug, Russia, 19 March 1889; *d.* Moscow, U.S.S.R., October 1937), *astrophysics.*

Gerasimovich's father, a physician, was director of a district hospital. Gerasimovich attended the Poltava Gymnasium but did not graduate, having been expelled for participation in the revolutionary movement. Only in 1909, after passing the examination as an extern, did he receive his "certificate of maturity." In 1910 he studied at the Faculty of Physics and Mathematics of Kharkov University, where his teachers in astronomy were the director of the astronomical observatory, Ludwig Struve, and the astrometrist N. N. Yevdokimov.

Gerasimovich received a prize in 1911 for his student composition "Aberratsia sveta" ("Aberration of Light"). After graduating from the university in 1914, he remained to prepare for an academic career. In 1916 he was a probationer at Pulkovo Observatory, studying astrophysics under A. A. Belopolsky and Sergei Kostinsky. Having passed his master's degree examination, in 1917 Gerasimovich became a *Privatdozent* at Kharkov University. From 1920 to 1933 he was senior astronomer at the observatory, and from 1922 he was also professor of astronomy. At the same time Gerasimovich taught mathematics and mechanics at a number of higher educational institutions in Kharkov. In 1931 he was invited to head the astrophysics section of the Pulkovo Observatory, and in 1933 he was appointed director of the observatory.

In 1924 Gerasimovich made a scientific trip to England and France, and from 1926 to 1929 he was visiting professor at Harvard Observatory, where both independently and in collaboration with Harlow Shapley, Otto Struve, Willem Luyten, and Donald Menzel he conducted a number of scientific investigations. In 1926, 1932, and 1935 Gerasimovich participated in the Copenhagen Congress of the German Astronomical Society, and in Cambridge, Massachusetts, and Paris sessions of the General Assembly of the International Astronomical Union.

Gerasimovich's range of scientific interests was very broad, and there are about 170 publications in his bibliography. Among them are works on photometrical and spectral research on variable stars (about eighty works), planetary nebulas, emission stars, scientific works in the field of theoretical astrophysics, stellar astronomy, the structure of the Galaxy, stellar statistics, celestial mechanics, astrometry, solar physics, and other problems. His research on semiregular variable stars, which he conducted on materials from the rich collection of photographic plates of the Harvard Observatory, has retained its great value.

Gerasimovich was a pioneer in the study of planetary nebulas, which are expanded clouds thrown out at the time of the explosions of certain types of nonstationary stars. He studied the turbulent movement in gas nebulas and investigated the varied forms of the planetary nebulas as figures of equilibrium, which can accept masses of gas under the influence of the forces of gravity of the central star and the forces of light repulsion (light pressure). Gerasimovich also studied the processes of ionization of planetary nebulas, determined the luminosity of their nuclei (the central stars), and came to the important conclusion that the masses of their central stars are small. This hypothesis was confirmed by the later research of others.

Gerasimovich was one of the first to emphasize the important role of computation of interstellar light absorption (which weakens the visible brightness of stars and causes their reddening) in the study of the structure of the Galaxy. In particular he used an original method of studying the apparent change of the mean distance of variable stars of the Cepheid type from the galactic plane as a function of their distance from the sun for an estimate of the value of the interstellar absorption of light.

Gerasimovich was responsible for an important investigation of the dynamics of the stellar system as the site of the simultaneous action of regular and irregular forces. In this work the appearance of the Milky Way and its insufficient brilliance in the direction of the constellation Sagittarius (where the dynamic center of the Galaxy is found) again led him to the belief in the essential role of the interstellar substratum, which forms a substantial condensation in space and at the same time causes a weakening of the brightness of the Milky Way.

With the American astronomer W. Luyten, Gerasimovich determined in 1927 the altitude of the sun over the base plane of the Galaxy as thirty-four parsecs (contemporary determinations give fifteen–twenty).

The study of the sun and the organization of and participation in expeditions to observe total solar eclipses occupied a special place in Gerasimovich's scientific work. In the later years of his life, in addition to his other responsibilities, he was president of the

Commission for the Study of the Sun of the Astronomical Council of the Soviet Academy of Sciences and president of the special commission of the Soviet Academy of Sciences for the preparation of all expeditions to observe the total eclipse of 19 June 1936. This was the first eclipse of the sun for which, on Gerasimovich's initiative, the whole program's photographic observations were organized in a unified way, using six uniform, standard coronagraphs placed along the zone of visibility of the total phase of the eclipse. As a result, very valuable data on the movement of matter in the solar corona were obtained.

Gerasimovich's monograph *Solnechnaya fizika* ("Solar Physics," in Ukrainian, 1933; 2nd ed., in Russian, 1935) is an excellent description of the physics of the sun, summing up the achievements of Soviet and foreign scientists in the field. Another of his monographs, *Vselennaya pri svete teorii otnositelnosti* ("The Universe in Light of the Theory of Relativity," 1925), was of great importance in popularizing the theory of relativity and Einstein's cosmological ideas. An outstanding role in the preparation of Soviet astronomers of the following generations was played by *Kurs astrofiziki i zvezdnoy astronomii* ("Course in Astrophysics and Stellar Astronomy"), which contained, among other works, a number of his original articles and was published under his direction in 1934 and 1936.

Gerasimovich's scientific activity was recognized by awards in the Soviet Union (1924, 1926, 1936), the United States (1928), and France (1934). He was elected a member of the All-Union Astronomical-Geodesical Society, the American and German astronomical societies, the Royal Astronomical Society, the American Association of Observers of Variable Stars, and the International Astronomical Union.

BIBLIOGRAPHY

I. Original Works. Gerasimovich's writings include *Vselennaya pri svete teorii otnositelnosti* ("The Universe in Light of the Theory of Relativity"; Kharkov, 1925); "On the Radiative and Mechanical Equilibrium of Spherical Planetary Nebulae," in *Astronomische Nachrichten,* **225** (1925), 89; "Ionization in Nebular Matter," in *Harvard Reprints. Harvard College Observatory,* **38** (1927); "On the Distance of the Sun From the Galactic Plane," in *Proceedings of the National Academy of Sciences of the United States of America,* **13** (1927), 387-390, written with Willem Luyten; "On the Luminosities of the Nuclei of Planetary Nebulae," in *Publications of the Astronomical Society of the Pacific,* **39** (1927), 19; and "On the Spectroscopic Absolute Magnitude Effect," in *Circular. Astronomical Observatory of Harvard College,* **311** (1927), 1-11.

For the works on semiregular variable stars, see "V Vulpeculae," *ibid.,* **321** (1927), 1-11; "The Photographic Light Curves of R Scuti," *ibid.,* **323** (1928), 1-3; "Secular Changes in the Mean Period of R Scuti (1795-1927)," *ibid.,* **333** (1928), 1-5; "V Ursae Minoris and SW Persei," *ibid.,* **338** (1929), 1-9; "R Sagittae," *ibid.,* **340** (1929), 1-2, written with L. Hufnagel; "A General Study of RV Tauri Variables," *ibid.,* **341** (1929), 1-15; and "On Variables of the Intermediate Group," *ibid.,* **342** (1929), 1-9.

See also "Spectrophotometric Temperature of Early Stars," *ibid.,* **339** (1929), 1-27; "On the Photographic Absolute Magnitudes of the Nuclei of Planetary Nebulae," in *Bulletin. Harvard College Observatory,* **864** (1929), 9-14; "The Nuclei of Planetary Nebulae," in *Observatory,* **54** (1931), 108; "Probability Problems Connected With the Discovery of Variable Stars in a Photographic Way," in *Doklady Akademii nauk SSSR,* **4,** nos. 5-6 (1931), 43-45; "Rayleighsche Streung und anomale Sterntemperaturen," in *Zeitschrift für Astrophysik,* **4** (1932), 265-281; "Non-statistic Hydrogen Chromospheres and the Problem of B Stars," in *Monthly Notices of the Royal Astronomical Society,* **94** (1933), 737-765; *Kurs astrofiziki i zvezdnoy astronomii* ("Course in Astrophysics and Stellar Astronomy"), II (Moscow-Leningrad, 1936), chs. 3, 6-12; "The Zero Point of the Cepheids' Period-Luminosity Relation," in *Observatory,* **57** (1934), 22-23; "B Spectrum Variables," *ibid.,* **58** (1935), 115-124; *Solnechnaya fizika* ("Solar Physics"; 2nd ed., Kharkov, 1935); "Cosmic Absorption and the Galactic Concentration of Classical Cepheids," in *Tsirkular Glavnoi astronomicheskoi observatorii v Pulkove,* no. 10 (1936), 9-11; "On the Behavior of Absorption Lines in the Spectrum of P Cygni," *ibid.,* no. 17 (1936), 13-15; "The System of Polaris," *ibid.,* no. 19 (1936), 739; "A Unitary Model of the Galactic System," in *Nature,* **137** (1936), 739; and "On the Illumination of a Planet Covered With a Thick Atmosphere," in *Izvestiya Glavnoi astronomicheskoi observatorii v Pulkove,* no. 127 (1937), 1-32.

II. Secondary Literature. On Gerasimovich or his work, see *Astronomia v SSSR za 40 let 1917-1957* ("Astronomy in the U.S.S.R. for Forty Years"; Moscow, 1960), pp. 54, 120, 287, 294, 332; *Razvitie astronomii v SSSR* ("The Development of Astronomy in the U.S.S.R."; Moscow, 1967), pp. 206, 215-217, 219; Y. G. Perel, "B. P. Gerasimovich," in *Astronomichesky kalendar na 1964 g.* ("Astronomical Calendar for 1964"; Moscow, 1936), pp. 256-258; and O. Struve, "About a Russian Astronomer," in *Sky and Telescope,* **16,** no. 8 (1957), 379-381.

P. G. Kulikovsky

GERBERT, also known as **Gerbert d'Aurillac,** later **Pope Sylvester II** (*b.* Aquitaine, France, *ca.* 945; *d.* Rome, 12 May 1003), *mathematics.*

Gerbert received his early education at the Benedictine convent of Saint Géraud in Aurillac. He left it in 967 in the company of Borel, count of Barcelona. In Catalonia he continued his studies under Atto, bishop of Vich; he concentrated on mathematics, probably on the works of such authors as Boethius, Cassiodorus, and Martianus Capella. How much Moorish science Gerbert was able to study as far north as Vich is uncertain. In 970 he accompanied Borel and Atto to Rome, where he attracted the attention of Pope John XIII and, through him, of Otto I, Holy Roman Emperor, who was then residing in Rome.

This was the beginning of Gerbert's career, which was based not only on his intellectual gifts but also on his allegiance to the Saxon imperial house and its political dream—the restoration of the empire of Charlemagne, ruled in harmony by emperor and pope. Gerbert was assigned to Adalbero, the energetic and learned archbishop of Rheims; there he reorganized the cathedral school with such success that pupils began to flock to it from many parts of the empire. Equal attention was paid to Christian authors and to such pagan writers as Cicero and Horace; great pains were taken to enrich the library.

For many years experts have tried, without reaching full agreement, to date the many existing mathematical manuscripts dating from the tenth to the thirteenth centuries and ascribed to Boethius or Gerbert or their pupils. From what most authorities believe are transcripts of Boethius and Gerbert themselves, it seems that in arithmetic some Pythagorean number theory was taught in the spirit of Nicomachus, and in geometry, some statements (without proofs) of Euclid together with the mensuration rules of ancient Roman surveyors; the art of computing was taught with the aid of a special type of abacus. For his lessons in astronomy Gerbert constructed some armillary spheres.

Among Gerbert's pupils were Robert, son of Hugh Capet (later Robert II of France); Adalbold, later bishop of Utrecht; Richer of Saint-Remy, who wrote Gerbert's biography; and probably also a certain Bernelinus, the Paris author of a *Liber abaci* that mentions Gerbert as pope and may well represent his teachings. Gerbert's influence probably extended to other cathedral or monastic schools, especially in Lorraine. This school of training in the quadrivium (music, geometry, arithmetic, astronomy), which is also represented by the *Quadratura circuli* of Franco of Liège (*ca.* 1050), indicates the vivid interest in mathematics that was beginning to appear in western Europe. The only works available, however, were poor remains of Greek knowledge transmitted in the later Roman period. These prepared the way for Arabic science; a work entitled *De astrolabia,* which shows Arabic influence and of which the earliest manuscript is from the eleventh century, has occasionally been ascribed to Gerbert.

In 983 Otto II, a great admirer of Gerbert, appointed him abbot of Bobbio in the Apennines, but by 984 he was back in Rheims. From then on, he took an active part in the political schemes among the Saxons, Carolingians, and Capets; he and Adalbero were deeply involved on the side of the emperor. The Carolingian dynasty came to an end, and in 987 Gerbert assisted in the coronation of Hugh Capet as king of France. In 991 he became archbishop of Rheims, but in 997 he left his see amid controversy and intrigue. He followed the court of young Otto III through Germany to Italy. At Magdeburg between 994 and 995 he constructed an *oralogium,* either a sundial or an astrolabe, for which he took the altitude of the pole star. In 998 Gerbert became archbishop of Ravenna; and in 999, through Otto's influence, he was the first Frenchman to be elected pope. A lover of arts and sciences, Otto hoped that emperor and pope would revive the Carolingian Renaissance. Significantly, Gerbert assumed the name Sylvester II, Sylvester I having been pope at the time of Constantine and, thus, a participant in the first holy alliance of pope and emperor.

The great scheme, however, came to naught with the death of Otto in 1002 and of Gerbert the following year. Legendary ascriptions to Gerbert of supernatural and demonic powers (which even found their way into Victor Hugo's *Welf, castellan d'Osbor*) testify to the impression that his learning made on posterity.

Information about Gerbert's life is in the "Historiae" written from 996 to 998 by his pupil Richer, in 224 published letters, and in contemporary chronicles. Many theological and scientific writings testify to his work and influence. Among those that seem to be authentic are *Regulae de numerorum abaci rationibus* (also called *Libellus de numerorum divisione*), *De sphaera,* sections of a *Geometria,* a letter to Adalbold on the area of an isosceles triangle, and a *Libellus de rationali et ratione uti* (997 or 998) with considerations on the rational and the use of reason. A text that Olleris entitled *Regula de abaco computi* and ascribed to Gerbert is ascribed to Heriger by Bubnov. The authenticity of the *Geometria* has been the subject of much controversy, often in connection with that of Boethius' work of the same title.

The abacus connected with Gerbert was a board with as many as twenty-seven columns, combined in groups of three. From left to right they were headed by the letters *C* (*centum,* hundred), *D* (*decem,* ten), and *S* or *M* (*singularis* or *monad,* one); other columns (and other letters) were for higher decimal units. Numbers were expressed by counters (apices) which carried symbols equivalent to our 1, 2, $\cdots$ 9 (which may indicate some Arabic influence), so that 604, for instance, was expressed by an apex with 6 in the *C* column and an apex with 4 in the *S* column. There was no apex for zero. With such an apparatus Gerbert and his school were able to perform addition, subtraction, multiplication, and even division, something considered complicated. This art of computation probably remained confined to ecclesiastical schools, never replacing older forms of reckoning, and went out of fashion with the gradual introduction of Hindu-Arabic numerals. A letter from Adalbold to Gerbert "de ratione inveniendi crassitudinem sperae" mentions the equivalent of $\pi = 22/7$, which was probably considered as exact. For $\sqrt{3}$ Gerbert accepted 26/15 (at another place 12/7), and for $\sqrt{2}$, 17/12.

Low as the level of Gerbert's mathematical knowledge was, it surpassed that of his monastic contemporaries and their pupils. This is shown in eight letters exchanged between two monastic friends of Adalbold, edited by P. Tannery and Abbé Clerval (see *Mémoires scientifiques,* volume 5).

BIBLIOGRAPHY

I. Original Works. Gerbert's extant work can be found in *Oeuvres de Gerbert,* A. Olleris, ed. (Paris–Clermont-Ferrand, 1867); *Lettres de Gerbert,* J. Havet, ed. (Paris, 1889); and *Gerberti Opera mathematica,* N. Bubnov, ed. (Berlin, 1899). An added document is found in H. Omont, "Opuscules mathématiques de Gerbert et de Hériger de Lobbes," in *Notices et extraits des manuscrits de la Bibliothèque nationale et autres bibliothèques,* **39** (1909), 4–15.

II. Secondary Literature. Works on Gerbert include Moritz Cantor, *Vorlesungen über Geschichte der Mathematik,* 3rd ed., I (Leipzig, 1907), 848–878; O. G. Darlington, "Gerbert the Teacher," in *American Historical Review,* **52** (1947), 456–476, with titles of the articles published at the time of Gerbert's millennary commemoration in 1938; J. Leflon, *Gerbert, humanisme et chrétienté au Xe siècle* (Abbaye Saint Wandrille, 1946); F. Picavet, *Gerbert, un pape philosophe* (Paris, 1917); P. Tannery, ed., *Mémoires scientifiques,* V (Toulouse–Paris, 1922), arts. 5, 6, and 10; M. Uhlirz, *Untersuchungen über Inhalt und Datierung der Briefe Gerberts von Aurillac, Papst Sylvester II* (Göttingen, 1957); and J. M. Millás Vallicrosa, *Nueve estudios sobre historia de la ciencia española* (Barcelona, 1960).

Richer's "Historiae" were published in R. Latouche, *Richer, Histoire de France (888–995)* (Paris, 1937), with French trans., and in *Monumenta Germaniae historica, Scriptorum,* III, G. H. Pertz, ed. (Hannover, 1839). See also A. J. E. M. Smeur, "De verhandeling over de cirkelkwadratuur van Franco van Luik van omstreeks 1050," in *Mededelingen van de K. Vlaamsche academie voor wetenschappen, letteren en schoone kunsten van België,* Klasse der wetenschappen, **30,** no. 11 (1960).

D. J. Struik

GERBEZIUS, MARCUS, also known as **Marko Gerbec** (*b.* St. Vid, near Stična, Slovenia, 24 October 1658; *d.* Ljubljana, Slovenia, 9 March 1718), *medicine.*

Gerbezius was born into a family in modest circumstances, and it was only through a government scholarship that he was able to obtain a higher education. He first studied philosophy in Ljubljana, then medicine in Vienna, Padua, and Bologna, where he obtained the doctorate in philosophy and medicine in 1684. Gerbezius is one of the forty former students of the University of Padua whose portraits hang in the historic Sala dei Quaranta.

After returning to his native country Gerzebius was named chief physician of the province of Carniola and became the most sought-after practitioner in Ljubljana. Barely four years following receipt of his doctorate he was admitted to the Academia Leopoldina Naturae Curiosorum, to which he sent, from 1689 until his death, a great many medical, meteorological, and zoological observations. Gerbezius was also a founding member of the Academia Operosorum in Ljubljana (1701). In 1712–1713 he was president of this small learned society, which played quite a large role in the cultural development of Slovenia.

Although attracted to Bologna by the teaching of Marcello Malpighi, Gerbezius was not converted to the iatromechanist doctrine; nor was he fully aware of the proper importance of the orientation toward anatomy of contemporary Italian medicine. His own views are a mixture of Dutch and German iatrochemical ideas with English neo-Hippocratic ideas. Gerbezius stressed the importance of a minute clinical examination devoid of preconceptions. He published the first detailed observation of an auriculoventricular block (1692). It occurred in a ninety-year-old woman suffering from spells of dizziness and fainting. Gerbezius described the particulars of her pulse so well that it is possible to make a retrospective diagnosis of the Adams-Stokes syndrome. In 1717 Gerbezius observed two patients afflicted with the same illness—one with a complete and permanent block, the other with an intermittent block. George Cheyne described this syndrome in 1733, as did G. B. Morgagni in 1761. It must be emphasized that Morgagni cited Gerbezius' publications with praise and recognized their priority.

Gerbezius also observed hypertrophy of the myocardium, exanthematous typhus, malaria, removal of the spleen, mercury poisoning among miners, the harmful effect of wine on children, and the treatment of biliary disorders with mineral waters. In a series of publications he set forth, year by year, the relationships between meteorological and terrestrial factors, on the one hand, and clinical and epidemiological aspects of diseases current in Ljubljana, on the other. He confirmed the majority of Thomas Sydenham's conclusions and, emphasizing the importance of social factors, gave them even a broader meaning.

Gerbezius also carried out experimental investigations on the physical and chemical properties of the air and on fermentation. In his view, alcoholic and acetic fermentation were chemical processes induced by certain minuscule particles, "volatile bodies," which escape from organic matter and are found suspended in the atmosphere.

BIBLIOGRAPHY

I. ORIGINAL WORKS. Gerbezius' principal works are *Intricatum extricatum medicum, seu tractatus de morbis complicatis* (Ljubljana, 1692; 2nd ed., Frankfurt am Main, 1713); *Chronologia medica:* I (Ljubljana, 1699); II (Ljubljana, 1700); III (Ljubljana, 1702); IV (Augsburg, 1705); V (Frankfurt, 1710); 2nd ed., I–V (Frankfurt, 1713)—the following years were published in *Appendix ephemeridorum naturae curiosorum* (1712–1718); and *Vindiciae physico-medicae aurae Labacensis oder gründliche Verthädigung der laybacherischen Lufft* (Ljubljana, 1710). In addition Gerbezius published seventy-nine articles in *Miscellanea Academiae naturae curiosorum;* a list of these articles can be found in J. J. Manget, *Bibliotheca scriptorum medicorum veterum et recentiorum* (Geneva, 1731). The first description of the auriculoventricular block is in "Pulsus mira inconstantia," in *Miscellanea Academiae naturae curiosorum,* dec. 2, yr. 10 (1692), observatio 63, pp. 115–118. An English trans. of the cardiological fragments is in R. H. Major, *Classic Descriptions of Disease,* 3rd ed. (Springfield, Mass., 1945), pp. 326–327.

II. SECONDARY LITERATURE. A detailed study of Gerbezius' life and work is still lacking. All the previous publications on this subject are summarized in I. Pintar, "Dr. Marko Gerbec," in *Razprave. Slovenska akademija znanosti in umetnosti,* classis IV, pars medica, **3** (1963), 1–40. See also H. Tartalja, "Der slowenische Arzt Dr. Marko Gerbec als Vorgänger der Fermentationslehre," in *Vorträge der Hauptversammlung der Internationalen Gesellschaft für Geschichte der Pharmazie (Rotterdam, 1963)* (Stuttgart, 1965), pp. 173–180; and N. Flaxman, "The History of Heart-Block," in *Bulletin of the Institute of the History of Medicine,* **5** (1937), 115–130.

M. D. GRMEK

GERGONNE, JOSEPH DIAZ (*b.* Nancy, France, 19 June 1771; *d.* Montpellier, France, 4 May 1859), *geometry.*

Gergonne's father, a painter and architect, died when Joseph was twelve years old. Joseph studied at the religious *collège* of Nancy, did some private tutoring, and in 1791 became a captain in the National Guard. In 1792 he joined the volunteers to fight the Prussians. He saw action at Valmy, and later that year went to Paris as secretary to his uncle. After a year he was in the army again, this time as secretary to the general staff of the Moselle army. In 1794, after a month at the Châlons artillery school, Gergonne received a commission as lieutenant. Sent to the army in the east Pyrenees, he participated in the siege of Figueras in Catalonia. After the Treaty of Basel in 1795, Gergonne was sent with his regiment to Nîmes, where he obtained the chair of transcendental mathematics at the newly organized École Centrale. He then married, settled down, and began his mathematical career under the influence of Gaspard Monge, the guiding spirit of the École Polytechnique in Paris.

Not finding a regular outlet for mathematical papers in the existing journals, such as the *Mémoires* of the Academy of Sciences or the *Journal de l'École polytechnique,* Gergonne began to publish (1810) the *Annales de mathématiques pures et appliquées,* the first purely mathematical journal. It appeared regularly every month until 1832 and was known as the *Annales de Gergonne.* His colleague J. E. Thomas-Lavernède was coeditor of the first two volumes. Gergonne continued editing the journal after he had accepted, in 1816, the chair of astronomy at the University of Montpellier. In 1830 he became rector at Montpellier and discontinued publishing his *Annales* after twenty-one volumes and a section of the twenty-second had appeared. In these volumes alone he had published more than 200 papers and questions, dealing mainly with geometry but also with analysis, statics, astronomy, and optics.

By 1831 the *Annales* had ceased to be the only wholly mathematical journal. In 1825 there appeared at Brussels A. Quetelet's *Correspondance mathématique et physique* (1825–1839), and in 1826 at Berlin A. L. Crelle's *Journal für die reine und angewandte Mathematik.* In 1836 J. Liouville continued Gergonne's work in France through his *Journal de mathématiques pures et appliquées.* The latter two journals are still being published.

Although Gergonne had given up his journal, he continued to teach after 1830. It is said that during the July Revolution of that year, when rebellious students began to whistle in his class, he regained their sympathy by beginning to lecture on the acoustics of the whistle. He retired in 1844, and during the last years of his life suffered from the infirmities of advanced age.

Gergonne's *Annales* played an essential role in the creation of modern projective and algebraic geometry. It offered space for many contributors on these and other subjects. The journal contains papers by J. V. Poncelet, F. Servois, E. Bobillier, J. Steiner, J. Plücker,

M. Chasles, C. J. Brianchon, C. Dupin, and G. Lamé; in volume **19** (1828–1829) there is an article by E. Galois. The geometry papers stressed polarity and duality, first mainly in connection with conics, then also with structures of higher order. Here the terms "pole," "polar," "reciprocal polars," "duality," and "class" (of a curve) were first introduced. After Poncelet, in his monumental *Traité des propriétés projectives des figures* (1822), had given the first presentation of this new geometry in book form, a priority struggle developed between Gergonne and Poncelet. The result was that Poncelet switched to other journals, including Crelle's.

The discovery of the principle of duality in geometry can be said to have started with C. J. Brianchon, a pupil of Monge, who in 1806 derived by polar reciprocity, from Pascal's theorem, the theorem now named for him. This method of derivation was used by several contributors to the *Annales,* together with the polarity method typical of spherical trigonometry. In his *Traité,* Poncelet stressed polar reciprocity. Then, in three articles in the *Annales* (**15–17** [1824–1827]), Gergonne generalized this method into the general principle that every theorem in the plane, connecting points and lines, corresponds to another theorem in which points and lines are interchanged, provided no metrical relations are involved (*géométrie de la règle*).

In his "Considérations philosophiques sur les élémens de la science de l'étendue" (*Annales,* **16** [1825–1826], 209–232) Gergonne used the term "duality" for this principle and indicated the dual theorems by the now familiar device of double columns. He applied his principle first to polygons and polyhedrons, then to curves and surfaces; it is here that he made the now accepted distinction between curves of degree m and of class m (instead of "order" for both). These papers led to the controversy between him and Poncelet, which was partly based on the way Gergonne edited the papers for the *Annales.* In the meantime, however, A. F. Moebius had introduced duality for the plane in full generality in *Der barycentrische Calcul* (Leipzig, 1827).

One subject of contention between Gergonne and Poncelet was that Poncelet was the foremost representative of the synthetic (i.e., the purely geometric) method, while Gergonne believed in analytic methods. True, Gergonne said, the methods of analytic geometry were often clumsy, but this was only due to lack of *adresse.* He illustrated this point in "Recherche du cercle qui en touche trois autres sur un plan" (*Annales,* **7** [1816–1817], 289–303), in which he gave an elegant analytic solution of this, the "Apollonian," tangent problem. Then, in his third article on duality (*Annales,* **17** [1826–1827], 214–252), following ideas developed by Gabriel Lamé in a study published in 1818 (*Examen des différentes méthodes employées pour résoudre les problèmes de la géométrie*) of which Lamé already had had an abstract published in the *Annales* (**7** [1816–1817], 229–240), Gergonne showed the power of what we now call the "abbreviated" notation, in which, for instance, the pencil of circles in the plane is represented by $C_1 + \lambda C_2 = 0$. This method was fully developed by J. Plücker in his *Analytisch-geometrische Entwicklungen* (1828–1831).

In 1834 Plücker solved a problem which to both Gergonne and Poncelet had seemed something of a paradox. Poncelet (*Annales,* **8** [1817–1818], 215–217) had found that from a point outside a curve of degree m there can be drawn $m(m - 1)$ tangents to the curve. Gergonne missed this fact until (*Annales,* **18** [1827–1828], 151) he corrected himself on several points and introduced for the polar reciprocal of a curve of order m the term "curve of class m," which is therefore of order $m(m - 1)$. But the reciprocal of this curve is the original one, and this seems therefore of order $m(m - 1)\{m(m - 1) - 1\}$, which is greater than m except when $m = 2$. Poncelet had already stated that the answer was to be found in the fact that the polar curve was not fully general. Plücker gave the precise answer by means of his formulas on the number of singularities of a plane curve (*Journal für die reine und angewandte Mathematik,* **12** [1834], 105 ff.).

Among the many other theorems discovered by Gergonne is the following (*Annales,* **16** [1825–1826], 209–232): If two plane curves C_m of degree m intersect in such a way that mp points of intersection are on a C_p, then the $m(m - p)$ other points of intersection are on a C_{m-p}. This leads to a simple proof of Pascal's theorem by considering the six sides of the hexagon inscribed in a conic alternately as two C_3.

Gergonne liked to season his papers with "philosophic" remarks. In one such remark he said, "It is not possible to feel satisfied at having said the last word about some theory as long as it cannot be explained in a few words to any passerby encountered in the street" (M. Chasles, *Aperçu historique* . . . [Paris, 1889], p. 115).

BIBLIOGRAPHY

I. Original Works. Almost all of Gergonne's papers are in the *Annales,* and a bibliography is in Lafon (see below). The more important ones are mentioned in the text. Those on differential geometry include "Demonstration des principaux théorèmes de M. Dupin sur la courbure des surfaces," in *Annales,* **4** (1813–1814), 368–378; and "Théorie élémentaire de la courbure des lignes et des surfaces courbes," *ibid.,* **9** (1818–1819), 127–196. On statics, see "Démonstrations des deux théorèmes de géométrie," *ibid.,* **11** (1820–1821), 326–336. He returned to one of his old loves in an address at Lille: "Notes sur le principe de dualité en géométrie," in *Mémoires de l'Académie des sciences et lettres de Montpellier pour 1847,* Section des Sciences.

II. SECONDARY LITERATURE. A. Lafon, "Gergonne, ses travaux," in *Mémoires de l'Académie de Stanislas,* **1** (1860), xxv–lxxiv, includes a bibliography of Gergonne's works. For further information, see E. Kötter, "Die Entwicklung der synthetischen Geometrie von Monge bis auf Staudt (1847)," in *Jahresbericht der Deutschen Mathematiker-vereinigung,* **5,** no. 2 (1901), with details on the controversy between Poncelet and Gergonne (pp. 160–167). The documents in this case were reprinted by Poncelet in his *Traité des propriétés projectives des figures,* II (Paris, 1866), 351–396.

See also C. B. Boyer, *History of Analytic Geometry* (New York, 1956), ch. 9; and *A History of Mathematics* (New York, 1968), ch. 24; M. Chasles, *Aperçu historique sur l'origine et le développement des méthodes en géométrie* (Paris, 1837; 3rd ed., 1889); and *Rapport sur le progrès de la géométrie* (Paris, 1870), pp. 54–60; and H. de Vries, *Historische Studien,* II (Groningen, 1934), 114–142.

D. J. STRUIK

GERHARDT, CHARLES FRÉDÉRIC (*b.* Strasbourg, France, 21 August 1816; *d.* Paris, France, 19 August 1856), *chemistry.*

Gerhardt's father, Samuel Gerhardt, who came from a family of brewers, was born in Switzerland. As a young man he moved to Strasbourg, where he found employment in a bank and married an outstanding beauty of the town, Charlotte Henriette Weber. Samuel Gerhardt's position in the banking house of Turkheim gave his family the benefits of a prosperous and cultured home. Like many inhabitants of Strasbourg, the Gerhardts spoke and wrote French and German with almost equal facility. As was expected of the sons of the bourgeoisie, young Gerhardt attended the lycée, where he showed unusual ability. In 1825 Gerhardt *père* gave financial backing for the exploitation of a patent for white lead. The partner who was to supply the technical knowledge withdrew, and Gerhardt was left to run a factory, the scientific aspect of which was quite outside his competence. It was this situation which convinced Gerhardt to provide an education for his son Charles which would prepare him for the management of the factory. In 1831 young Gerhardt accordingly entered the newly founded Polytechnicum at Karlsruhe, where he followed intermediate courses in chemistry, physics, and mathematics. In 1833 his father decided to encourage the other aspect of his plan for Charles by sending him to a commercial college in Leipzig. Even there, however, he had lodgings in the house of the chemist Otto Erdmann, who encouraged his scientific interests.

A long paper on the revision of the formulas of natural silicates which Gerhardt wrote in 1834 was accepted for publication by the *Journal für praktische Chemie;* and in 1835, before his nineteenth birthday, his further research had won him the honor of election as corresponding member of the Société d'Histoire Naturelle de Strasbourg. Having seen the attraction of pure science, he came into open conflict with his father, who wanted him to return to the family chemical factory. After a violent quarrel the son left home to join a regiment of lancers. Here, too, he was unhappy and managed to borrow the 2,000 francs necessary to buy himself out. Realizing that his vocation was chemistry, Gerhardt went to Giessen, where Liebig was beginning to build up a reputation, and was enrolled in Liebig's course for 1836–1837.

To earn money for his studies, Gerhardt undertook the translation of one of Liebig's works into French and in so doing helped his own career. In April 1837 he left Liebig to make a final attempt to reconcile himself with his father. Yet he could not settle down to a life of commerce. To make his mark in chemistry he left for Paris. Arriving in October 1838, he enrolled in the course given by Dumas in the Faculty of Sciences. For three years he studied with Dumas and finally became his assistant.

Gerhardt's hard work and brilliant research won him the degrees of licentiate and doctor, both in April 1841 (his thesis was on helenin). Within a few days and on the recommendation of Dumas, Gerhardt was nominated to a vacancy in the Faculty of Sciences in Montpellier. In fact Gerhardt was only *chargé de cours* in 1841 and had to wait three years for nomination to the rank of titular professor. He soon found that the facilities for research were very limited and that it was difficult to settle down so far from Paris and Strasbourg. He went to Paris as often as possible and finally obtained leave of absence to work there. When in 1851 a request for further leave was refused, Gerhardt resigned. He did so the more easily as he was then, with the help of borrowed money, establishing his own private school for practical chemistry. This was not a financial success, but in 1854 he was able to secure a further university appointment. There were two vacancies for chairs of chemistry at Strasbourg, in the Faculty of Sciences and the School of Pharmacy, and both were offered to him. Although reluctant to leave Paris, Gerhardt was glad to return to his native city. Above all, his position in French chemistry was now recognized, and his financial problems were at an end. In 1844 Gerhardt had married a Scottish girl, Jane Sanders, whose brother was studying at the University of Montpellier. Gerhardt thus learned to speak and write English as well as French and German.

In 1844 there began a close friendship with Laurent, professor of chemistry at Bordeaux. They had much in common: zeal for a new approach to organic chemistry with poor facilities available for research, and above all, isolation from the center of power in Paris because of the remoteness of their provincial teaching posts. Wurtz justly commented: "In the history of science the great figure of Gerhardt cannot be separated from that of Laurent: their work was done in collaboration, their talent was complementary, their influence was reciprocal." Most commentators have agreed that Gerhardt had the merit of clarity. Yet he himself remarked how vital an impact was made on him by the profusion of ideas which poured from the pen and the lips of his friend.

Gerhardt was attracted several times in his career to scientific journalism. He was a collaborator in the first issues (1840) of Quesneville's *Revue scientifique et industrielle,* and it is typical of his literary zeal that he should produce a new journal from his association with Laurent. Their *Comptes rendus mensuels des travaux chimiques,* although a help in publicizing their own work, was not well received and lasted only from 1845 to 1851.

In national and academic politics Gerhardt was numbered among the radicals. He became bitterly opposed to Dumas, who had come to typify the scientific establishment. Gerhardt's position was not helped by his readiness to speculate, his lack of tact, and his dogmatism. He was involved in many disputes, even with his protector Liebig. Disputes which began in organic chemistry quickly descended to personalities, and it was characteristic of the organization of French science in the mid-nineteenth century that Gerhardt was not an active member of any important scientific society. On 21 April 1856 he was at last honored by election as a correspondent of the Institute, but within four months he was dead. Gerhardt's tragic death in his fortieth year, following a career of banishment to the provinces, did not prevent him from becoming one of the seminal figures in the history of nineteenth-century chemistry.

Gerhardt was sufficiently impressed by the substitution theories of Laurent and Dumas, which had invaded organic chemistry in the late 1830's, that he was prepared to regard each molecule as a unitary structure or "single edifice" within which substitution of different "residues" could occur. He did this in opposition to the more conservative chemists headed by Berzelius, who supported a dualistic theory, according to which all compounds were interpreted as consisting of an electropositive part and an electronegative part. In the case of many organic compounds this interpretation involved the arbitrary grouping of atoms without supporting evidence.

A major problem confronting chemists in the early 1840's was whether it was possible to ascertain the arrangement of atoms in a compound. Gerhardt reacted violently against the empiricist principle that the arrangement of the atoms could be inferred from a compound's mode of formation or from its reactions. He maintained that we cannot infer arrangements from reactions; all we can ever know are the reactions. So, whereas the view of Berzelius was that each compound had a rational formula and there remained only the practical difficulty of discovering what it was, Gerhardt's view became the exact opposite: We could never know the arrangement of atoms but only the reactions of compounds. Any formula must be based on a particular reaction. If an aldehyde behaves in some reactions like an oxide and in others like a hydride, it must be given two different formulas according to the occasion. Gerhardt took his operational definition of a formula to the logical conclusion and stated that a compound has "as many rational formulas as [it has different] reactions."[1] Consequently his chemistry has a positivist flavor. A chemical formula for Gerhardt was simply equivalent to a particular reaction of a compound and thus had little explanatory value.

A considerable part of Gerhardt's claim to our attention is connected with his use of certain formulas in chemistry. Like Lavoisier, he felt that chemistry could be successfully pursued only if it were based on a rational and systematic language. It was not merely that formulas are a reflection of our basic concepts but, more, that "our formulas *are* our ideas." It is not surprising, therefore, to find Gerhardt insisting on the importance of writing chemical equations: "The only approach which is at the same time rigorous and easy and which can be reconciled with the individual opinions of all chemists is that which consists in expressing reactions by equations from which all purely hypothetical entities are excluded."[2] Gerhardt was one of the first to make systematic use of equations in chemistry.

Gerhardt's most conspicuous contribution to the development of organic chemistry was his homologous series. His earliest publications were characterized by attempts to arrange organic compounds in series of increasing complexity: his "ladder of combustion," rising from water and carbon dioxide at the foot to albumin and fibrin at the summit, was an analogue of the biologists' ladder of nature. Another biological analogy was to underlie the application of his homologous series when they were refined in 1843:

Gerhardt presupposed a principle of plenitude in organic chemistry which dictated that hitherto undocumented members of any series must exist. In addition, the concept of homology itself was of biological origin, deriving from Cuvier. For Gerhardt, however, it did not carry that structural connotation which it had for Cuvier. On this subject Gerhardt simply asserted: "We call substances homologues when they exhibit the same chemical properties and when there are analogies in the relative proportions of their elements."[3]

Drawing on the work of Dumas—who, in 1842, had observed that a whole series of acids could be generated by the formula $(C^4H^4)_n\ O^4$ [$(CH_2)_n\ O_2$ in modern notation]—and also of Kopp, who had investigated "a great regularity in the physical properties of analogous organic compounds," Gerhardt generalized the concept of homologous series and introduced it into his *Précis de chimie organique* of 1844. He did so with particular reference to the four known alcohols, since he could demonstrate a significant numerical relation between their empirical formulas: They could be written in the forms $[(CH^2) + H^2O]$; $[(CH^2)^2 + H^2O]$; $[(CH^2)^5 + H^2O]$; and $[(CH^2)^{16} + H^2O]$. Furthermore, their respective oxidation, sulfonation, and halogenation products could all be denoted by formulas in which the characteristic (CH_2) unit recurred. The solution which Gerhardt found to the problem of prediction was then explicit in his enthusiastic conclusion: provided one knew how properties varied regularly along a series, "it would suffice to know the composition, the properties and the mode of formation of a single product obtained [from one of the above compounds] in order to be able to predict the composition, properties and mode of formation of all substances similar to this first product."[4]

Gerhardt systematically showed how one could argue from the reactions of one compound to those of another on the basis of a formal numerical analogy displayed by their respective empirical formulas. His enthusiasm for this system of classification led to his being accused of doing algebra rather than chemistry. Nevertheless, on this basis Gerhardt not only forecast the existence of many new compounds, such as those required to complete the alcohol series $[(CH^2)^n + H^2O]$, but also predicted the properties of many others, such as the boiling point (140°C.) of propionic acid. Within the space of twenty years organic compounds had been removed from what Wöhler had described as a primeval forest and had been transplanted in decorous straight lines. Within the next few years the carbon–carbon bonds of Kekulé were to explain the significance of the recurrent CH_2 unit; and in retrospect the only serious shortcoming of Gerhardt's understanding of homology was its failure to do justice to structural isomerism.

If the concept of homology was Gerhardt's most notable contribution, his most notorious was a reform of the presuppositions then underlying the determination of chemical equivalents. As a consequence of the prevalent disregard for Avogadro's hypothesis, current equivalents were not based on equal volumes of vapor. Accordingly, Berzelius saw nothing wrong in writing water as H^2O and acetic acid as $C^4H^8O^4$, in order that the latter might be considered the hydrate of an oxide $(C^4H^6O^3 + H^2O)$ analogous to an inorganic acid; compare sulfuric acid $(SO^3 + H^2O)$. Similarly, Gmelin's alternative system allowed an organic acid to be envisaged as an anhydride plus water $(C^4H^3O^3 + HO)$. But in both cases there was an inconsistency, and it was Gerhardt who faced the consequences of removing it. The inconsistency was this: if H^2O represented an equivalent of water corresponding to two volumes of vapor, how could the same equivalent of water be preformed within an organic acid if the overall equivalent of the acid corresponded to four volumes of vapor? To eliminate the inconsistency, it would be necessary to represent the water participating in organic reactions by H^4O^2 (Berzelius' system) or by H^2O^2 (Gmelin).

Consequently, in his famous paper read before the Paris Academy of Sciences in September 1842, Gerhardt enumerated a host of organic reactions in order to demonstrate that when water was produced or utilized in such reactions, it was in quantities which could be represented only by H^4O^2, on a four-volume system. From similar considerations with respect to the participation of carbon dioxide in organic reactions, Gerhardt argued that the only way to achieve consistency was either to double all two-volume inorganic formulas or to halve all four-volume organic formulas. But this was a particularly drastic measure—since it destroyed almost all the current constitutional analogies between inorganic and organic compounds—and such was the dogmatic and precocious tone of its advocate that Gerhardt's revision met constant hostility.

It was not until February 1845 that even the sympathetic Laurent finally agreed to the innovation, one corollary of which merits special attention. If all organic formulas corresponding to four volumes of vapor really had to be divided by two, then no four-volume formula could be tolerated if it included an odd number of atoms. Both Laurent and Gerhardt were therefore obliged to instigate a program of reanalysis in cases where four-volume formulas failed to conform to their scheme; and the implication that their contemporaries, particularly Liebig, had been incompetent did much more to increase their professional alienation during the 1840's. It has often been said that Gerhardt's standardization constituted a revival of Avogadro's hypothesis, but this is not strictly true. Nothing could be further from Avogadro's hypothesis than Gerhardt's conclusion that "atoms, equivalents, and volumes are synonymous. . . ." Certainly it was a corollary of Avogadro's hypothesis that molecular formulas should be standardized in the way Gerhardt had chosen, but Gerhardt himself was preoccupied with the corollary. Simplification and the elimination of the internal inconsistency had been ends in themselves, and it was left to Laurent and others to clarify further the relationship between atoms, equivalents, and molecules.

Another notable, although less well-known, contribution of Gerhardt's was his redefinition of acids. Since the standardization of inorganic and organic formulas, effected by Gerhardt, strictly precluded the preexistence of water in the molecule of a monobasic acid, some alternative convention for the formulation of acids was required. During the 1830's Liebig had become dissatisfied with the (anhydride + water) model but had not been prepared to introduce the alternative hydrogen theory in inorganic chemistry, even though he had commended it for the organic domain. Gerhardt's view of chemistry, however, was dominated by the attempt to introduce organic concepts into inorganic chemistry, and he felt no compunction in introducing a universal definition of acidity based on hydrogen. The idea that an acid could be defined with reference to its displaceable hydrogen appealed to Gerhardt for the additional reason that salt formation could then be construed as a displacement reaction rather than as an addition reaction.

In this way one of the main props of the prevalent electrochemical theory was removed. Hitherto, acid, base, and salt had often been defined in terms of each other, so it was an obvious merit of Gerhardt's definition that it broke the circle. The essential conceptual advance consisted in the obliteration of the artificial distinction between "oxacids" and "hydracids." In deference to their heritage from Lavoisier, chemists had been obliged to create a special class for acids, such as hydrochloric, which contained no oxygen. These were the "hydracids." Long after Lavoisier's erroneous explanation of acidic properties was obsolete, the classification of acids still paid lip service to it. A conceptual switch was necessary to recognize hydracids as the rule and to transpose the oxacids accordingly. Pursuing this transposition to its conclusion, Gerhardt was able to refine Liebig's criteria for the establishment of acid basicity: the diagnostic value of acid and double salts assisted in the determination of the number of replaceable hydrogen atoms in a given acid. The dibasicity of sulfuric acid was at last publicized; and when, in 1856, Alexander Williamson expressed the respective basicities of nitric, sulfuric, and phosphoric acids as NO_3H; SO_4H_2; PO_4H_3, he added that the "labors of Messrs. Laurent and Gerhardt greatly contributed to the establishment of these results which are uncontroverted."

During the 1830's the attempt to model the structure of organic compounds on the dualistic structure of inorganic compounds led to the postulation of a large number of hypothetical radicals, supposedly analogues of the inorganic elements. According to Liebig's famous definition, organic chemistry differed from inorganic chemistry in that it dealt with compound rather than simple radicals. Ethyl chloride, for example, could be envisaged as a salt in which the composite ethyl radical played a role analogous to the potassium of potassium chloride. As complex a compound as cacodyl chloride was regarded by Bunsen as $(C^4H^{12}As^2)Cl$, with the implication that the complex radical $(C^4H^{12}As^2)$ should be as capable of isolation as the inorganic elements. Bunsen was convinced that he had isolated the free cacodyl radical.

In the 1840's Kolbe and Frankland isolated what they thought were the methyl and ethyl radicals and thus appeared to corroborate the analogy between metal and radical, but there was a disconcerting feature of this corroboration. If one bowed before Gerhardt's insistence that molecular formulas should all refer to the same volume of vapor, it was impossible to equate what Bunsen had isolated with the hypothetical constituent of the cacodyl compounds, just as it was also impossible to equate Kolbe's species with the hypothetical methyl. The equivalent vapor densities of the isolated species corresponded to the dimers $(C^8H^{24}As^4)$ and dimethyl, respectively. How were these results to be interpreted? It was a clear case of circular verification: if one accepted the dualistic approach to organic chemistry, then Kolbe's work confirmed it; if one accepted Gerhardt's presuppositions, it did not. Eventually the inert character of the Kolbe-Frankland hydrocarbons, together with their respective boiling points, testified in Gerhardt's favor; and by 1851 Frankland himself was admitting that what he had taken to be methyl and ethyl were in reality stable dimers. Gerhardt's ideas were, therefore, instrumental in preventing a serious misinterpretation of hydrocarbon chemistry, as they were also in drawing attention to the legitimacy of postulating diatomic molecules X_2 (compare $(CH^3)_2$) in contradistinction to the tenets of electrochemical theories.

One of Gerhardt's principal claims to fame was his "type" theory. The concept of a chemical "type" has a long and intricate history. Chemists had long given similar compounds similar names (compare the "pyrites" of iron and copper), and when Berzelius had eventually represented the sulfates, selenates, and chromates by the formulas $MO.SO^3$, $MO.SeO^3$, and $MO.CrO^3$, he had recognized more than a superficial resemblance. The discovery of isomorphism by Mitscherlich and of chlorine substitution within organic compounds by Laurent and Dumas culminated in Dumas's "type theory" (1838–1840), which purported to explain the common properties of parent and chlorine derivative with reference to a common spatial arrangement of their constituents: acetic and trichloracetic acids, for example, belonged to the same "chemical" type. Impressive as Dumas's theory was, it had at least two outstanding defects. First, since there turned out to be no rigid distinction between substitutions with and without retention of properties, Dumas's division between "chemical" and "mechanical" types—which had been designed to cater to such a distinction—began to appear arbitrary. Second, during the 1840's, when the central problem became one of classification, his types proved to be too specific. They might illuminate the relation between any one acid and its chlorine derivatives, but there was no obvious way of correlating the types of different parent acids. It seemed that what was required was to expand the concept of "type" for an economic classification and for an understanding of all substitution reactions (irrespective of a change in properties), and yet escape the problems associated with the ever-increasing number of conflicting opinions about the precise arrangement of atoms within a given compound.

It was Gerhardt who provided a solution to this problem when, in his definitive exposition of a "new type theory" (1853–1856), he illustrated how one could envisage organic compounds as substitutionary derivatives of a minimal number of inorganic compounds: water, ammonia, hydrogen, and hydrogen chloride. Several chemists contributed the empirical foundation for this new theory. Williamson, for example, was able to prepare mixed ethers which admirably conformed to the water type—compare

$$\left.\begin{matrix}H\\H\end{matrix}\right\}O \quad \text{and} \quad \left.\begin{matrix}CH^3\\C^2H^5\end{matrix}\right\}O$$

—while Hofmann showed that the amines could be subsumed under ammonia as an all-embracing type—compare

$$\left.\begin{matrix}H\\H\\H\end{matrix}\right\}N \quad \text{and} \quad \left.\begin{matrix}R\\R'\\R''\end{matrix}\right\}N.$$

It was, in fact, in the context of establishing the utility of the water type that Gerhardt himself made what was his most prestigious contribution to practical chemistry. According to the water type theory, acetic acid could be written as $\left.\begin{matrix}C^2H^3O\\H\end{matrix}\right\}O$, and by a further substitution of an acyl group for a hydrogen atom there should result a compound $\left.\begin{matrix}C^2H^3O\\C^2H^3O\end{matrix}\right\}O$. Gerhardt's triumph was to prepare acetic anhydride by the reaction of acetyl chloride with sodium acetate. Furthermore, by producing mixed anhydrides, he established—against contemporary opinion—that two equivalents of a monobasic acid were involved in the process; compare

$$\left.\begin{matrix}H\\H\end{matrix}\right\}O,\quad \left.\begin{matrix}RCo\\RCo\end{matrix}\right\}O,\quad \left.\begin{matrix}RCo\\R'Co\end{matrix}\right\}O.$$

Gerhardt's type theory dominated the organic chemistry of the late 1850's, not simply because of its comprehensiveness but also because it could be extended in a number of profitable directions. The introduction of the methane type completed a series which was highly suggestive in the context of emerging ideas on valency:

$$HCl';\quad \left.\begin{matrix}H\\H\end{matrix}\right\}O'';\quad \left.\begin{matrix}H\\H\\H\end{matrix}\right\}N''';\quad \left.\begin{matrix}H\\H\\H\\H\end{matrix}\right\}C^{IV}.$$

Moreover, this theory was highly flexible, since types could be conjugated, condensed, or multiplied in order to accommodate more elusive species, such as ammonium hydroxide:

$$\begin{matrix}\left.\begin{matrix}H\\H\\H\end{matrix}\right\}N,\\ \left.\begin{matrix}H\\H\end{matrix}\right\{O\end{matrix}\quad \text{glycine}\quad C^2H^2\begin{matrix}\left.\begin{matrix}H\\H\end{matrix}\right\}N,\\ \left.\begin{matrix}O\\H\end{matrix}\right\}O\end{matrix}\quad \text{and glycol}\quad \left.\begin{matrix}C^2H^4\\H^2\end{matrix}\right\}O^2.$$

Although the purely formal nature of Gerhardt's types prevented their earning a permanent place in the body of chemical theory, at the time of their inception they exerted a powerful unifying influence on the development of chemistry. In two quite different senses Gerhardt had contributed to the unification of chemical theory. In the first place, by introducing the radicals, such as ethyl, into types such as water, Gerhardt was able to achieve some kind of rapprochement between radical and type concepts which had hitherto been in opposition. Second, by subsuming all organic compounds under four inorganic types, he was advocating analogies of unprecedented generality between organic and inorganic compounds.

Gerhardt, like Laurent, was convinced of the unity of chemical theory; and when Wurtz (1862) sought a proof of the artificiality of the inorganic-organic dichotomy, it was to Gerhardt's theory of types that he appealed.

Gerhardt will occupy a permanent position in the history of chemistry for his services to organic classification, for his concept of homology, for his preparation of acid anhydrides, and for his reform of equivalents. His unitary emphasis, while remaining indispensable for the comprehension of organic compounds, was eventually superseded in inorganic chemistry as ionic concepts marked the culmination of a return to a more mechanistic approach. Indeed, shortly after the death of Gerhardt chemistry was to change its sights. A new program, explicitly hostile to Gerhardt's positivist construction of chemistry, was promulgated by Kekulé and Scott-Couper. Reductionist in intent, it focused attention once more on the elements themselves and on the individual atoms, with the object of so elucidating the nature of chemical bonding that the properties of any compound could be demonstrated to follow from those of its constituent elements. Questions concerned with the real arrangement of atoms in a molecule and with what held them together could not be bypassed for long, and with the resuscitation of these questions certain features of Gerhardt's unitary chemistry were quietly forgotten: no longer was it reasonable to regard all reactions as double decompositions, and no longer could all the metals be regarded as diatomic M_2. Nevertheless, whenever chemical philosophers have gathered together, there have been those, like Benjamin Brodie, who have expressed their gratitude to those "great chemists, Laurent and Gerhardt who implanted in the science the germ of a more abstract philosophy, which it has ever since retained."

NOTES

1. Grimaux and Gerhardt, *Charles Gerhardt, sa vie . . .*, p. 490.
2. *Précis de chimie organique,* I, viii–ix.
3. *Annales de chimie et de physique,* 3rd ser., **8** (1843), 245.
4. *Revue scientifique et industrielle,* **14** (1843), 588.

BIBLIOGRAPHY

I. Original Works. Gerhardt's books are *Précis de chimie organique,* 2 vols. (Paris, 1844–1845); *Introduction à l'étude de la chimie par le système unitaire* (Paris, 1848); and *Traité de chimie organique,* 4 vols. (Paris, 1853–1856).

His major research papers are the following: "Sur la constitution des sels organiques à acides complexes, et leurs rapports avec les sels ammoniacaux," in *Annales de chimie,* 2nd ser., **72** (1839), 184–215; "Recherches chimiques sur les huiles essentielles," *ibid.,* 3rd ser., **1** (1841), 60–111; "On the Chemical Classification of Organic Substances," in *Revue scientifique et industrielle,* **7** (1841), 104; **8** (1842), 300; **10** (1842), 145; **12** (1843), 592; **14** (1843), 580; "Considérations sur les équivalents de quelques corps simples et composés," in *Annales de chimie,* **7** (1843), 129–143, and **8** (1843), 238–245; "Recherches sur la salicine," *ibid.,* **8** (1843), 215–229; "Recherches sur les alcalis organiques," *ibid.,* **7** (1843), 251–253; "Recherches chimiques sur l'essence de valériane et l'essence d'estragon," *ibid.,* pp. 275–295, also in *Annalen der Chemie,* **45** (1843), 29–41; "Action de l'acide sulfurique sur les matières," in *Comptes rendus hebdomadaires des séances de l'Académie des sciences,* **16** (1843), 458–460; "Sur les combinaisons de l'acide sulfurique avec les matières organiques," *ibid.,* **17** (1843), 312–317; "Recherches concernant les alcalis organiques," *ibid.,* **19** (1844), 1105–1107; "Sur la génération de l'éther," in *Revue scientifique,* 2nd ser., **3** (1844), 304–310; "Sur le point d'ébullition des hydrogènes carbonés," in *Annales de chimie,* **14** (1845), 107–114, also in *Journal für praktische Chemie,* **35** (1845), 300–305; "Sur une nouvelle classe de composés organiques," in *Annales de chimie,* **14** (1845), 117–125, and **15** (1845), 88–96; "Observations sur la formation des formules chimiques," in *Journal de pharmacie,* **8** (1845), i–viii; "Sur l'identité de l'essence d'estragon et de l'essence d'anis," in *Comptes rendus hebdomadaires des séances de l'Académie des sciences,* **20** (1845), 1440–1444; "Sur la loi de saturation des corps copulés," *ibid.,* pp. 1648–1657; "Sur les mellonures," *ibid.,* **21** (1845), 679–681; "Introduction à l'étude de la chimie," in *Journal de pharmacie,* **14** (1848), 63–67; "Recherches sur les anilides," in *Annales de chimie,* **24** (1848), 163–207; "Remarques sur les combinaisons des acides avec les alcalis organiques," in *Comptes rendus mensuels des travaux chimiques,* **5** (1849), 160–170; "Recherches sur les phénides, nouvelle classe de composés organiques," *ibid.,* pp. 429–437; "Sur la composition des mellonures et de leurs dérivés," *ibid.,* **6** (1850), 104–111, also in *Comptes rendus hebdomadaires des séances de l'Académie des sciences,* **30** (1850), 318–319; "Remarques sur un travail de M. Hofmann sur les radicaux, etc.," in *Comptes rendus mensuels des travaux chimiques,* **6** (1850), 233–236; "Remarques sur un travail de M. Williamson relatif aux éthers," *ibid.,* pp. 361–364; "Sur la constitution des composés organiques," *ibid.,* **7** (1851), 65–84; "Sur la basicité des acides," *ibid.,* pp. 129–156, also in *Journal für praktische Chemie,* **53** (1851), 460–488; "Recherches sur les acides organiques anhydres," in *Annales de chimie,* **37** (1853), 285; "Note sur la théorie des amides," in *Comptes rendus hebdomadaires des séances de l'Académie des sciences,* **37** (1853), 280–284; "Addition aux recherches sur les acides anhydres," *ibid.,* **36** (1853), 1050–1069; "Sur les amides," in *Annales de chimie,* **46** (1856), 129–172; and "Recherches sur les amides," *ibid.,* **53** (1858), 302–313.

Many of Gerhardt's most revealing articles were published in *Comptes rendus mensuels des travaux chimiques* between 1846 and 1852. For his correspondence, see M. Tiffeneau, ed., *Correspondance de C. Gerhardt* (Paris, 1918).

II. Secondary Literature. On Gerhardt or his work,

see E. Grimaux and C. Gerhardt, *Charles Gerhardt, sa vie, son oeuvre, sa correspondance* (Paris, 1900); J. Jacques, "Onze lettres inédites de Charles Gerhardt à J. B. Dumas," in *Bulletin de la Société chimique de France,* **156** (1856), 1315–1324; C. de Milt, "Auguste Laurent. Guide and Inspiration of Gerhardt," in *Journal of Chemical Education,* **28** (1951), 198–204; J. F. H. Papillon, *La vie et l'oeuvre de C. F. Gerhardt* (Paris, 1863); and J. R. Partington, *History of Chemistry,* IV (London, 1964), 405–424 and *passim.*

M. P. CROSLAND
J. H. BROOKE

GERMAIN, SOPHIE (*b.* Paris, France, 1 April 1776; *d.* Paris, 27 June 1831), *mathematics.*

Sophie Germain, France's greatest female mathematician prior to the present era, was the daughter of Ambroise-François Germain and Marie-Madeleine Gruguelu. Her father was for a time deputy to the States-General (later the Constituent Assembly). In his speeches he referred to himself as a merchant and ardently defended the rights of the Third Estate, which he represented. Somewhat later he became one of the directors of the Bank of France. His extensive library enabled his daughter to educate herself at home. Thus it was that, at age thirteen, Sophie read an account of the death of Archimedes at the hands of a Roman soldier. The great scientist of antiquity became her hero, and she conceived the idea that she too must become a mathematician. After teaching herself Latin and Greek, she read Newton and Euler despite her parents' opposition to a career in mathematics.

The Germain library sufficed until Sophie was eighteen. At that time she was able to obtain the lecture notes of courses at the recently organized École Polytechnique, in particular the *cahiers* of Lagrange's lectures on analysis. Students at the school were expected to prepare end-of-term reports. Pretending to be a student there and using the pseudonym Le Blanc, Sophie Germain wrote a paper on analysis and sent it to Lagrange. He was astounded at its originality, praised it publicly, sought out its author, and thus discovered that M. Le Blanc was Mlle. Germain. From then on, he became her sponsor and mathematical counselor.

Correspondence with great scholars became the means by which she obtained her higher education in mathematics, literature, biology, and philosophy. She wrote to Legendre about problems suggested by his 1798 *Théorie des nombres.* The subsequent Legendre-Germain correspondence was so voluminous that it was virtually a collaboration, and Legendre included some of her discoveries in a supplement to the second edition of the *Théorie.* In the interim she had read Gauss's *Disquisitiones arithmeticae* and, under the pseudonym of Le Blanc, engaged in correspondence with its author.

That Sophie Germain was no ivory-tower mathematician became evident in 1807, when French troops were occupying Hannover. Recalling Archimedes' fate and fearing for Gauss's safety, she addressed an inquiry to the French commander, General Pernety, who was a friend of the Germain family. As a result of this incident, Gauss learned her true identity and accorded even more praise to her number-theoretic proofs.

One of Sophie Germain's theorems is related to the baffling and still unsolved problem of obtaining a general proof for "Fermat's last theorem," which is the conjecture that $x^n + y^n = z^n$ has no positive integral solutions if n is an integer greater than 2. To prove the theorem, one need only establish its truth for $n = 4$ (accomplished by Fermat himself) and for all values of n that are odd primes. Euler proved it for $n = 3$ and Legendre for $n = 5$. Sophie Germain's contribution was to show the impossibility of positive integral solutions if x, y, z are prime to one another and to n, where n is any prime less than 100. In 1908 the American algebraist L. E. Dickson generalized her theorem to all primes less than 1,700, and more recently Barkley Rosser extended the upper limit to 41,000,000. In his history of the theory of numbers, Dickson describes her other discoveries in the higher arithmetic.

Parallel with and subsequent to her pure mathematical research, she also made contributions to the applied mathematics of acoustics and elasticity. This came about in the following manner. In 1808 the German physicist E. F. F. Chladni visited Paris, where he conducted experiments on vibrating plates. He exhibited the so-called Chladni figures, which can be produced when a metal or glass plate of any regular shape, the most common being the square or the circle, is placed in a horizontal position and fastened at its center to a supporting stand. Sand is scattered lightly over the plate, which is then set in vibration by drawing a violin bow rapidly up and down along the edge of the plate. The sand is thrown from the moving points to those which remain at rest (the nodes), forming the nodal lines or curves constituting the Chladni figures.

Chladni's results were picturesque, but their chief effect on French mathematicians was to emphasize that there was no pure mathematical model for such phenomena. Hence, in 1811 the Académie des Sciences offered a prize for the best answer to the following challenge: Formulate a mathematical theory of elastic surfaces and indicate just how it agrees with empirical evidence.

Most mathematicians did not attempt to solve the problem because Lagrange assured them that the mathematical methods available were inadequate for the task. Nevertheless, Sophie Germain submitted an anonymous memoir. No prize was awarded to anyone; but Lagrange, using her fundamental hypotheses, was able to deduce the correct partial differential equation for the vibrations of elastic plates. In 1813 the Academy reopened the contest, and Sophie Germain offered a revised paper which included the question of experimental verification. That memoir received an honorable mention. When, in 1816, the third and final contest was held, a paper bearing her own name and treating vibrations of general curved as well as plane elastic surfaces was awarded the grand prize—the high point in her scientific career.

After further enlargement and improvement of the prize memoir, it was published in 1821 under the title *Remarques sur la nature, les bornes et l'étendue de la question des surfaces élastiques et équation générale de ces surfaces.* In that work Sophie Germain stated that the law for the general vibrating elastic surface is given by the fourth-order partial differential equation

$$N^2\left[\frac{\partial^4\rho}{\partial s_1^4} + 2\frac{\partial^4\rho}{\partial s_1^2\partial s_2^2} + \frac{\partial^4\rho}{\partial s_2^4} - \frac{4}{S^2}\left(\frac{\partial^2\rho}{\partial s_1^2} + \frac{\partial^2\rho}{\partial s_2^2}\right)\right] + \frac{\partial^2\rho}{\partial t^2} = 0.$$

Here N is a physical constant if the "surface" is an elastic membrane of uniform thickness. The generality is achieved because S, the radius of mean curvature, varies from point to point of a general curved surface. The very concept of mean curvature $(1/S)$ was created by Sophie Germain.

The notion of the curvature of a surface generalizes the corresponding concept for a plane curve by considering the curvatures of all plane sections of the surface through the normal at a given point of the surface and then using only the largest and smallest of those curvatures. The extremes, called the principal curvatures, are multiplied to give the Gaussian total curvature at a point and are added to give the mean curvature. Sophie Germain, however, defined the mean curvature as half the sum, that is, the arithmetic mean, of the principal curvatures. Her definition seems more in accordance with the term "mean." Moreover, she indicated that her measure is a representative one, an average in the statistical sense, by demonstrating that if one passes planes through the normal at a point of a surface such that the angle between successive planes is $2\pi/n$ where n is very large (thus yielding sample sections in many different directions), the arithmetic mean of the curvatures of all the sections is the same as the mean of the two principal curvatures, a fact that remains true in the limit as n gets larger and larger. Also, while the Gaussian curvature completely characterizes the local metric geometry of a surface, the mean curvature is more suitable for applications in elasticity theory. A plane has zero mean curvature at all points. Hence $4/S^2 = 0$ in Germain's differential equation, and it reduces to the equation which she and Lagrange had derived for the vibration of flat plates. The same simplification holds for all surfaces of zero mean curvature, the so-called minimal surfaces (such as those formed by a soap film stretched from wire contours).

In later papers Sophie Germain enlarged on the physics of vibrating curved elastic surfaces and considered the effect of variable thickness (which emphasizes that one is, in fact, dealing with elastic solids).

She also wrote two philosophic works entitled *Pensées diverses* and *Considérations générales sur l'état des sciences et des lettres,* which were published posthumously in the *Oeuvres philosophiques.* The first of these, probably written in her youth, contains capsule summaries of scientific subjects, brief comments on the contributions of leading mathematicians and physicists throughout the ages, and personal opinions. The *État des sciences et des lettres,* which was praised by Auguste Comte, is an extremely scholarly development of the theme of the unity of thought, that is, the idea that there always has been and always will be no basic difference between the sciences and the humanities with respect to their motivation, their methodology, and their cultural importance.

BIBLIOGRAPHY

I. Original Works. Among Sophie Germain's scientific writings are *Remarques sur la nature, les bornes et l'étendue de la question des surfaces élastiques et équation générale de ces surfaces* (Paris, 1826); *Mémoire sur la courbure des surfaces* (Paris, 1830); *Oeuvres philosophiques de Sophie Germain* (Paris, 1879); and *Mémoire sur l'emploi de l'épaisseur dans la théorie des surfaces élastiques* (Paris, 1880).

II. Secondary Literature. On Sophie Germain or her work, see L. E. Dickson, *History of the Theory of Numbers* (New York, 1950), I, 382; II, 732–735, 757, 763, 769; M. L. Dubreil-Jacotin, "Figures de mathématiciennes," in F. Le Lionnais, *Les grands courants de la pensée mathématique* (Paris, 1962), pp. 258–268; and H. Stupuy, "Notice sur la vie et les oeuvres de Sophie Germain," in *Oeuvres philosophiques de Sophie Germain* (see above), pp. 1–92.

Edna E. Kramer

GERMANUS, HENRICUS MARTELLUS (*fl.* Florence, Italy, 1480–1496[?]), *geography.*

Henricus Martellus was probably Heinrich Hammer and assuredly was of German birth, since he referred to himself as "Germanus." He is a mysterious personality of whom no mention has ever been found except for the remarkable maps that he signed with his full name. It is established that he worked in Florence in the closing decades of the fifteenth century and was closely associated with the printer, engraver, and map publisher Francesco Rosselli.

Germanus' contribution is in the maps that he drew, remarkable not only for their high artistic quality but also for the new geographical concepts that they represented. His works include two sets of maps drawn to illustrate Ptolemy's *Geography,* five codices of an *Insularium,* or Book of Islands, and a large world map (43″ by 75″).

Roberto Almagià states that Germanus was the first mapmaker to append to the traditional set of maps illustrating the work of Ptolemy a set of maps showing the world of his time, the "tabulae modernae." Of these the most important are the world maps added to several of his *insularia* and the large world map now in the Yale University Library. These maps show the world on the eve of Columbus' voyages: the continent of Africa as reported by the Portuguese, an open Indian Ocean, a jumbled—but in places recognizable—eastern coast of Asia, and, on the Yale world map, Japan, identified as "Zipango."

There can be little doubt that Germanus' world maps served as the model for Martin Behaim's celebrated 1492 globe, the oldest surviving globe in the Western world: these two are the only "new" world maps prior to 1500 to be graduated in longitudes and latitudes. Further, it is highly probable that Germanus' map, or a copy thereof, inspired the geographic ideas of Columbus. According to Almagià, it can best explain the geographic premises of all of Columbus' voyages and would fully support his convictions, his conjectures, and his projects. The influence of Germanus' large world map, in all likelihood circulated in printed form throughout Europe, is evident in nearly all of the important maps of the Columbian age; and it is a landmark in the mapping of our world.

BIBLIOGRAPHY

On Germanus or his work, see Roberto Almagià, "I mappamondi di Enrico Martello e alcuni concetti geografici do Cristoforo Colombo," in *Bibliofilia,* **43,** nos. 8–10 (1940), 288–311, which has a list of Germanus' maps; and Marcel Destombes, ed., *Mappemondes, A.D. 1200–1500,* vol. I of Monumenta Cartographica Vetustioris Aevi (Amsterdam, 1964), pp. 229–233.

GEORGE KISH

GESELL, ARNOLD LUCIUS (*b.* Alma, Wisconsin, 21 June 1880; *d.* New Haven, Connecticut, 29 May 1961), *psychology.*

Trained in pedagogy, psychology, and medicine, Gesell developed a systematic program of research in human growth and development during many years of work with normal and problem children at Yale University's Clinic of Child Development; and through his associates, students, and a multitude of publications, he disseminated his conclusions for practical application and elaboration, initiating a new concept in child care.

Gesell was the oldest of the five children of Gerhard and Christine Giesen Gesell. His mother taught elementary school in the small town where he was born and raised. After graduation from high school in 1896, he went to the Stevens Point Normal School to train as a teacher, then taught for a short time at the Stevens Point High School.

Gesell next went to the University of Wisconsin, where he received the B.Ph. in 1903 with a thesis on higher education in Ohio and Wisconsin. He was principal of the high school in Chippewa Falls, Wisconsin, the following year; but having come under the influence of Joseph Jastrow, professor of psychology at Wisconsin, he enrolled in Clark University, Worcester, Massachusetts, where G. Stanley Hall, pioneer in child psychology, introduced him to this then-new field of investigation. There Gesell earned the Ph.D. in 1906 with a thesis on jealousy.

For the next two years Gesell taught at the California State Normal School in Los Angeles, where, on 18 February 1909, he married Beatrice Chandler. The following summer he went to study at the Pennsylvania Training School for Feeble-Minded Children, where Lightner Witmer (who founded this institution as well as the country's first psychological clinic at the University of Pennsylvania) was director. The Gesells also spent several weeks that summer at the Training School for Feeble-Minded Children in Vineland, New Jersey, where Henry Goddard, director of psychological research, was investigating the application of Binet tests to feebleminded children.

Gesell considered this visit to Vineland as the beginning of his professional interest in mentally defective children; and he later for several years conducted summer courses with Goddard at New York University, in which they trained special teachers for defective children. In 1910 he studied anatomy at the University of Wisconsin; and the following year, having joined the department of education at Yale University, he enrolled at the same time as a medical student there. The dean of the medical school, George Blumer, provided him with space in the New Haven Dispensary to study retarded children. Thus the Clinic of Child Development originated in 1911, while Gesell was still pursuing his medical degree. He received the M.D. in 1915 and was made professor in the Yale Graduate School of Medicine.

From 1915 to 1919 Gesell served as school psychologist for the Connecticut State Board of Education. This entailed identifying handicapped children and devising individualized programs for them. In 1918 he surveyed mental conditions in the elementary schools of New Haven, and from 1919 to 1921 he served on the Governor's Commission on Child Welfare. In 1924 he initiated the use of cinematography in psychological research and established, in 1925, the Photographic Library of Yale Research Films of Child Development, supervising for many years the production of still and motion pictures at Yale. With the cooperation of Louise Ames and Frances L. Ilg, Gesell issued a nationally syndicated column on child studies. Generous funds granted in 1926 by the Laura Spelman Rockefeller Memorial Foundation and other private foundations supported his research and enabled him to establish a nursery which provided for the observation and guidance of children as well as for aiding parents in caring for them. Here, in 1930, he devised a one-way observation dome for research and teaching of child behavior, a sort of "candid camera" technique, by means of which he could observe and record infant behavior without being seen.

From 1928 to 1948 Gesell was attending pediatrician at the New Haven Hospital. When he became professor emeritus of child hygiene upon his retirement in 1948, he directed the Yale Child Vision Research Project for two years; from 1948 to 1952 he also served as research associate on the Harvard Pediatric Study. In 1950 the Gesell Institute of Child Development was founded, superseding the Yale Clinic of Child Development, and here Gesell served as consultant and lecturer in the School for Social Research for the rest of his life. He was a member of the National Research Council (1937–1940), the American Psychological Association, the National Academy of Sciences, and the American Association for the Advancement of Science. The American Academy of Pediatrics decided in 1934 that a requirement for membership would be certification by the American Board of Pediatrics; an exception—and the only exception—was made for Gesell in 1948 because of his outstanding contributions in the field of child development. Clark University conferred an honorary D.Sc. in 1930, and the same degree was awarded him in 1953 by the University of Wisconsin.

Gesell's significance lies in his development of new methods in the study of children. He placed particular emphasis on the preschool years, a period that had previously received only cursory attention. His studies led him to conclusions about the psychological care of infants and the guidance of children that exerted an important influence on the attitudes and practices of nursery schools, kindergartens, and elementary schools. In addition, he inspired a group of active disciples who are continuing his work, and had a talent for popularizing and gaining support for his ideas. Some of his concepts related to predictive measurement of mental development in children have been adopted into current pediatrics.

BIBLIOGRAPHY

I. Original Works. Gesell's writings include *The First Five Years of Life* (New York, 1940); *Developmental Diagnosis* (New York, 1941; 2nd ed., rev. and enl., 1947), written with Catherine S. Armatruda, for medical students and practitioners, with translations into several other languages; *Infant and Child in the Culture of Today* (New York, 1943), written with Janet Learned and Louise B. Ames, which presents the significance of a democratic culture for the psychological welfare of infants and young children, and is based on studies of the relationship between the pressures of natural growth, which Gesell called "maturation," and the pressure of the social order, termed "acculturation"; and *The Child From Five to Ten* (New York, 1946), written with Frances Ilg, which supplements *The First Five Years of Life,* as both complement *Infant and Child in the Culture of Today.* An extensive bibliography is included by Miles, in his memoir listed below.

II. Secondary Literature. On Gesell's life and work, see Walter R. Miles, "Arnold Lucius Gesell," in *Biographical Memoirs. National Academy of Sciences,* **37** (1964), 55–96. See also *Who's Who in America,* **31** (1960–1961), 1066; and the obituary in *Journal of the American Medical Association,* **177** (1961), 75.

Samuel X. Radbill

GESNER, KONRAD (*b.* Zurich, Switzerland, 26 March 1516; *d.* Zurich, 13 March 1565), *natural sciences, medicine, philology.*

The godson and protégé of the Protestant reformer Ulrich Zwingli, Gesner was destined from early childhood to study theology. He attended the Carolinum in his native Zurich and later entered the Fraumünster seminary. After the death of his godfather and protector in 1531 at the Battle of Kappel, Gesner left Zurich; the following year he was at the Strasbourg Academy, where he attended the courses of Wolfgang Capito and soon became a Hebrew scholar. Theological studies no longer held much interest for him, however; and he turned to medicine, studying first at Bourges, then at Paris, and later at Basel, from which he received his doctorate in 1541. Meanwhile he was pursuing his studies of ancient languages. Such was his reputation that the government of Bern appointed him the first occupant of the chair of Greek, which he held from 1537 to 1540, at the newly founded Lausanne Academy.

On the advice of his close friend Christophe Clauser, then chief physician of Zurich, Gesner traveled to Montpellier to carry on work in botany. He settled permanently in Zurich, where he was named chief physician and was later elevated to *canonicus,* a position that substantially improved his financial status. Gesner was an ardent traveler: he explored the Alps and the Adriatic coast, bringing back from his many excursions documents that he used in preparing his treatises on botany and zoology. In 1555 Gesner climbed—under difficult conditions—Mont Pilate, overlooking Lake Lucerne, and brought back useful data on Alpine flora. He died during the plague epidemic that began in Brazil in 1560 and reached Zurich in 1565.

Two dominant poles of interest can be discerned in Gesner's work. At times they oriented him toward letters, but more often they directed him to the natural sciences.

In 1535 Gesner compiled a Greek-Latin dictionary, and from 1537 to 1540 he taught Greek at Lausanne. He published several treatises on philology, notably one in which he transcribed the Lord's Prayer in twenty-two languages. He devoted several years to the preparation of a treatise, the four-volume *Bibliotheca universalis* (1545–1555), an index to Greek, Latin, and Hebrew writers that earned him fame and marks him as the founder of bibliography and brought him into correspondence with all the scholars of his time.

But natural history remained his first interest. Fascinated by botany as a youth, Gesner continued his studies in that field at Lausanne and Montpellier. He never succeeded, however, in publishing the monumental treatise on which he had worked throughout his life. This *Opera botanica,* for which he himself drew nearly 1,500 plates, was published in two volumes from 1551 to 1571 through the efforts of C. Schmiedel. Gesner was virtually the only botanist of his time to grasp the importance of floral structures as a means of establishing a systematic key to the classification of vegetable life. He was also the first to stress the nature of seeds, which enabled him to establish the kinship of plants that seemed extremely dissimilar. Later, Linnaeus would frequently acknowledge his own debt to Gesner.

Zoology also attracted Gesner, who patterned his *Historia animalium* on a work published a few years earlier by Aelian. This massive work of more than 4,500 pages received immediate acclaim; even Georges Cuvier later delighted in recognizing its enduring interest. The attempt of the treatise to regroup animals recalls the principal themes that Gesner defended in proposing a classification of the vegetable kingdom according to flowering and nonflowering plants, vascular and nonvascular plants, and so on. Gesner was also drawn to animal physiology and pathology, and he is considered by some the founder of veterinary science.

Gesner's interests extended to the study of paleontology, in which he wrote several memoirs on vegetable forms no longer extant. He is also considered the first naturalist to have sketched fossils. In addition, he studied crystallography and was one of the first to include printed plates of crystals in his works.

BIBLIOGRAPHY

Works concerning Gesner are J. C. Bay, "Conrad Gesner: The Father of Bibliography," in *Papers of the Bibliographical Society of America,* **10,** no. 2 (1916), 53–86; H. Escher, "Die *Bibliotheca universalis* K. Gesners," in *Vierteljahrsschrift der Naturforschenden Gesellschaft in Zürich,* **79** (1934), 174–194; H. Günther, "K. Gesner als Tierarzt," thesis (Leipzig, 1933); W. Ley, *K. Gesner. Leben und Werk* (Munich, 1929); and K. Müller, *Der polyhistor Konrad Gesner als Freund und Förderer erdkundlicher Studien,* doctoral diss. (Munich, 1912).

P. E. Pilet

GESSNER (GESNER), JOHANNES (*b.* Zurich, Switzerland, 18 March 1709; *d.* Zurich, 6 May 1790), *botany, geology.*

Gessner was the son of Christophe Gessner, pastor of the Reformed Church of Zurich for many years, and Esther Maag, a member of a very distinguished Zurich family. For more than four centuries the Gessner family gave Switzerland theologians, physicians, and, above all, naturalists. One of Gessner's ancestors was Konrad Gesner, whose highly regarded *Historia animalium* earned its author an international reputation. Gessner's brother Johann Jakob (1704–1787) was a well-known theologian who taught Hebrew at the Carolinum and wrote extensively on numismatics.

After receiving a substantial education in Zurich, Gessner went to Leiden to study medicine. Among his teachers there was the famous Dutch physician and botanist Hermann Boerhaave. In Leiden, Gessner also met his fellow Swiss, Albrecht von Haller, and the two became friends. Boerhaave and Haller stimulated his interest in botany, which soon became his favorite area of research. Gessner next went to Paris and then to Basel, where he was initiated into higher mathematics by Johann I Bernoulli. He also became friendly with the latter's son Daniel, who later introduced probability theory into medicine. Gessner earned his medical degree in 1729, and the following year began to practice in his native city. In 1742 he was in Basel, teaching mathematics. He founded the Société de Physique in 1757 in Zurich, where he remained for the rest of his life.

Accomplished in medicine and physics as well as in botany and mathematics, Gessner was undoubtedly one of the last of the great humanists. He maintained an imposing correspondence with most of the scientists of his time, in particular with Linnaeus. One of the first to adopt the latter's system of classification, he did much to popularize it.

Gessner established his reputation as a naturalist primarily through his writings in systematic botany. His first important work was *Dissertationes de partium vegetationis et fructificationis structura* (1743). It was soon followed by *Dissertationes physicae de vegetabilibus* (1747) and the memoir *De Ranunculo bellidifloro et plantis degeneribus* (1753), in which he skillfully defended Linnaeus' conceptions. In *De petrificatorum variis originibus praecipuarum telluris mutationem testibus* (1756) Gessner showed some recognition of the real nature of fossil plants and in this regard can be considered one of the founders of paleobotany. Moreover, he devoted fourteen years (1759–1773) to the publication of an eleven-part treatise that summarized contemporary knowledge of plants, *Phytographia sacra generalis et specialis.*

Following Haller's example, Gessner specialized in the study of Alpine flora. From his very numerous excursions in the Alps and the Forealps he brought back many new species that now bear his name. Gessner was an innovator in his botanizing, since he was quite probably the first to describe plant habitats; for example, he provided altimetric data obtained with the aid of a barometer.

Toward the end of his life Gessner concentrated on geological observations. He was interested in the formation of mountain chains on the European continent and estimated the era and duration of their formation with great care. Gessner's research led him to attack the brief chronologies—based mainly on theological considerations—that were generally accepted at the time. A brilliant advocate of the principle of geochronological extrapolation, his work foreshadowed modern geophysics. Finally, his extremely precise observations of the mineral springs of the Alps were long authoritative.

The recipient of many honors, a member of countless scientific societies, and a correspondent of most of the European academies, Gessner was, in his own lifetime, considered one of the greatest naturalists of the age. Yet, shortly after his death he was somewhat forgotten, and new plants that he had been the first to describe were attributed, often wrongly, to his illustrious ancestor Konrad Gesner. He was also confused with Johann Matthias Gessner, a German pedagogue, humanist, and botanist, whose books were widely distributed. In addition, Gessner's memoirs on Alpine flora in particular could hardly withstand the inevitable comparison with those of Haller. It is only very recently, notably with the publication in 1949 of Gessner's correspondence with Linnaeus, that his originality and importance have finally gained recognition.

BIBLIOGRAPHY

I. Original Works. Gessner's works are *Dissertationes de partium vegetationis et fructificationis structura* (Zurich, 1743); *Dissertationes physicae de vegetabilibus* (Zurich, 1747); *De Ranunculo bellidifloro et plantis degeneribus* (Zurich, 1753); *De petrificatorum variis originibus praecipuarum telluris mutationem testibus* (Zurich, 1756); *Phytographia sacra generalis et specialis* (Zurich, 1759–1773); and *Tabulae phytographica analysin generum plantarum exhibentes,* with commentary and edited by C. S. Schinz (Zurich, 1795–1826).

II. Secondary Literature. On Gessner's life and work, see G. R. de Beer, "The Correspondence Between Linnaeus and Johannes Gessner," in *Proceedings of the Linnean Society of London,* **161** (1948–1949), 225; H. K. Hirzel, *Denkrede auf Johannes Gessner* (Zurich, 1790); and Rudolf Wolf, *Biographien zur Kulturgeschichte der Schweiz,* I (Zurich, 1858), 281–322.

P. E. Pilet

GEUTHER, ANTON (*b.* Neustadt, near Coburg, Germany, 23 April 1833; *d.* Jena, Germany, 23 August 1889), *chemistry.*

Geuther was the son of Christian Friedrich Geuther, a master weaver, brewer, and farmer who was also municipal treasurer of his native city. His mother was Anna Cordula Eichhorn. In accordance with his father's wishes, Geuther learned the weaver's trade and after his apprenticeship attended the Realschule in Coburg and, later, the one in Saalfeld. At Saalfeld he was more attracted by the scientific than by the commercial and technical subjects and thus, following the certificate examination, decided to study science. He entered the University of Jena, where among his teachers were H. W. F. Wackenroder (chemistry) and Matthias Schleiden (botany). He went to Göttingen in 1853 where, except for a semester in Berlin (the winter of 1853–1854), he remained for ten years. He attended lectures on mineralogy, physics, organic chemistry, and philosophy but, most important, he was for years an assistant to Friedrich Wöhler. He received his doctorate in 1855 and then advanced through various posts in Wöhler's institute, becoming successively lecture assistant, private assistant, head assistant, *Privatdozent* (1857), and associate professor (1862).

In 1863 Geuther went to the University of Jena as a full professor. In the same year he married Amalie Agnes Sindram, the daughter of Wilhelm Sindram, director of a hospital at Göttingen. They had one son and one daughter. Geuther remained in Jena for the rest of his life, achieving success both as a researcher and as a teacher. Great demands were made on Geuther's idealism, for although his summerhouse had been equipped for chemical research through the patronage of Grand Duchess Sophie of Saxe-Weimar, it was never adequate to his needs. A new structure had already been agreed to when, in 1889, he contracted typhus—and he did not live to see it built.

Under Wöhler's influence Geuther's years at Göttingen were devoted primarily to inorganic chemistry. Geuther was a strict adherent of J. J. Berzelius' dualistic, electrochemical conception; the substitution theory, which was rapidly gaining prominence, appeared to him insufficient "if it is a question of real insight and reduction to general principles." Geuther interpreted his extensive investigations of double compounds in the light of the dualistic theory and discussed the constitution of these compounds in a manner that suggests Alfred Werner's interpretation of complex compounds.

In Jena, where organic chemistry took precedence in his work, Geuther made his most important discovery, the synthesis of acetoacetic ester. The starting point for this achievement was his investigation of the constitution of alkyl compounds; among them was acetic acid, which, as the result of a then-common but false assumption concerning the atomic weights of carbon and oxygen, Geuther thought to be dibasic. Through the reaction of metallic sodium and acetic ester, he hoped to obtain the dibasic sodium salt of acetic acid; instead the result was acetoacetic ester. Geuther determined the composition of this previously unknown substance and ascertained its great reactivity. Moreover, from the color reaction with ferric chloride characteristic of phenoloid compounds and from the green color of the copper salts, he inferred the presence of an acidifying hydroxyl group in the molecule. A long controversy with Edward Frankland and Baldwin Duppa, who suspected that a ketonic form was involved, was not settled until 1911, when Geuther's successor Ludwig Knorr demonstrated that under normal conditions both substances exist in the compound. As the first example of a keto-enol tautomerism, acetoacetic ester was of great significance in the development of theoretical organic chemistry.

BIBLIOGRAPHY

I. Original Works. Lists of Geuther's publications may be found in Poggendorff, vols. III, IV, and VI. The most important are *Lehrbuch der Chemie* (Jena, 1869); "Untersuchungen über die einbasischen Säuren," in *Nachrichten von der Gesellschaft der Wissenschaften zu Göttingen* (1863), p. 281; "Untersuchungen über einbasische Kohlenstoffsäuren, über die Essigsäure," in *Jenaische Zeitschrift für Medizin und Naturwissenschaft,* **2** (1866), 388; and "Über die Constitution des Acetessigesters (Aethyldiacetsäure) und über die jenige des Benzols," in *Justus Liebigs Annalen der Chemie,* **219** (1883), 119–128.

II. Secondary Literature. Works by his pupils that show Geuther's immediate influence include C. Duisberg, "Beiträge zur Kenntniss des Acetessigesters," in *Justus Liebigs Annalen der Chemie,* **213** (1882), 133–181; and Wilhelm Wedel, "Über einige Abkömmlinge des Acetessigesters," *ibid.,* **219** (1883), 71–119.

On Geuther and his work see C. Duisberg and K. Hess, in *Berichte der Deutschen chemischen Gesellschaft,* **63A** (1930), 145–157, with portrait; C. Liebermann, *ibid.,* **22** (1889), 2388; and *Neue deutsche Biographie,* vol. VI.

Grete Ronge

GHETALDI (GHETTALDI), MARINO (*b.* Ragusa, Dalmatia [now Dubrovnik, Yugoslavia], 1566 [1568?]; *d.* Ragusa, 11 April 1626), *mathematics.*

Ghetaldi was born to a patrician family originally from Taranto, Italy. Ragusa was then an independent republic and very jealous of its Latinism. Ghetaldi spent only the latter part of his life (from 1603) there, holding various public and legal positions. As a young man, after his education in Ragusa, he had moved to Rome and then traveled extensively through Europe, returning to Rome briefly in 1603.

Ghetaldi lived the peripatetic life of a scholar participating in the intense scientific awareness of early seventeenth-century Italian culture, a last flowering of the Renaissance spirit. Galileo was its most notable example. Archimedes and Apollonius were its inspiration. In Rome Ghetaldi came under the influence of Christoph Clavius, famous as teacher and editor of Euclid. He then went to Antwerp to study with Michel Coignet. Thence he moved to Paris, where he associated with Viète, who entrusted him with an unpublished manuscript to revise and edit, although Ghetaldi had as yet published nothing of his own.

Ghetaldi's first publications appeared at Rome in 1603 and were part of the beginning of research on Archimedes. The first, *Promotus Archimedis,* dealt with the famous problem of the crown; it also included tables that Ghetaldi calculated from experiments on the specific weights of certain substances, with results that were, for the time, remarkably accurate. For this, and for research on burning glasses, a topic then of great interest, he also became known as a physicist. In his second work, *Nonnullae propositiones de parabola,* Ghetaldi treated parabolas which he had obtained as sections of a right circular cone of any proportions.

From the analysis of Apollonius' known work, Ghetaldi turned to the task of reconstructing the content of his lost works. He followed the example of his master Viète, who had attempted such reconstruction in his two books of 1600, *Tactionum* (Περὶ ἐπαφῶν), and was consequently nicknamed Apollonius Gallus. Ghetaldi took over and completed that work, although on less restricted problems, as *Supplementum Apollonii Galli* (1607). Ghetaldi later concentrated his attention on the last of Apollonius' books mentioned by Pappus, Περὶ νεύσεων (*Inclinationum libri duo*) and solved the four problems that were supposed to form the first book. The problems of insertion, as they were called, consisted of constructing certain segments with their extremes touching arcs of a circle or other given figure; the book was entitled *Apollonius redivivus seu restituta Apollonii Pergaei inclinationum geometria* (1607). Later, Ghetaldi reexamined the problem which, according to Pappus, made up the second book of *Inclinationum;* although it was rather complex, involving insertions between two semicircles, Ghetaldi declared he had needed only a few days to complete it. Published in 1613, it was entitled *Apollonius redivivus seu restitutae Apollonii Pergaei de inclinationibus geometriae, liber secundus.*

Meanwhile, Ghetaldi had produced a pamphlet with the solutions of forty-two geometrical problems, *Variorum problematum collectio* (1607). The method used in some of the solutions suggests that he was already applying methods of algebra to geometry, such as first-degree and second-degree problems, determinate or not, which he later treated specifically in a volume that appeared after his death, *De resolutione et de compositione mathematica, libri quinque* (1630). Because of this work, possibly his most significant, Ghetaldi has been considered the precursor of analytic geometry—a hypothesis difficult to support, especially in the light of the methods used by Descartes and Fermat.

Ghetaldi wrote in Latin, and his works were well and widely known—some for a long time. Pierre Herigone, for instance, included in the first volume of his *Cursus mathematicus* (1634) the first of Ghetaldi's two works devoted to the problems of insertion, translating it in a notation anticipating modern mathematical logic.

Ghetaldi was held in great esteem not only as a scientist, but also as a man. While still young he had been offered a chair at the University of Louvain, which he did not accept; and in 1621 his name was included in a list of scientists proposed for membership in the flourishing new Accademia dei Lincei. He was not nominated, however, because he returned to Ragusa without notice and the Academy did not know his whereabouts.

In letters of that time Ghetaldi was described as having "the morals of an angel" and Paolo Sarpi, his close friend, called him "a Ragusan gentleman of discernment." For his exceptional skill and intelligence Ghetaldi was good-naturedly called "a mathematical demon." Later, in Ragusa, he was even called a magician and gained the reputation of being a sorcerer because he made frequent astronomical observations and experimented with burning glasses; another explanation attributes the sobriquet to his using a nearby cave popularly called "the magician's den" for his research.

Ghetaldi had met Sarpi in Venice before 1600, during his frequent peregrinations between Rome, Padua, and Venice. In Rome his teacher Clavius had introduced him to another Jesuit scholar, Christopher Grinberg, author of a treatise on trigonometry and, later, famous as one of the four Jesuits whom Robert Cardinal Bellarmine consulted in April 1611 on the value of Galileo's *Sidereus nuncius.* Ghetaldi must certainly have met Grinberg again in Padua, where Grinberg lived from 1592 to 1610, and later the two maintained a correspondence, as shown by a letter of 1608 and another of 1614, which accompanied the second part of Ghetaldi's *Apollonius redivivus . . . liber secundus* and in which Ghetaldi declared he was sending the volume "as a sign of reverence and in memory of our old friendship."

In June 1606 the government of Ragusa charged Ghetaldi with a mission to the sultan of Constantinople. The task absorbed him considerably, and to it must be attributed the break in his scientific work that coincides with this period. The mission must have had its dangers, since rumors of his death began to circulate. So persistent were these rumors that even J. E. Montucla, in his *Histoire des mathématiques,* gave Ghetaldi's date of death as about 1609, "in the course of his mission to the [Sublime] Porte."

BIBLIOGRAPHY

Ghetaldi's writings were collected as *Opera omnia,* Žarko Dadić, ed. (Zagreb, 1968). Among his works are *Promotus Archimedes seu de variis corporum generibus gravitate et magnitudine comparatis* (Rome, 1603); *Nonnullae propositiones de parabola* (Rome, 1603); *Apollonius redivivus seu restituta Apollonii Pergaei inclinationum geometria* (Venice, 1607); *Supplementum Apollonii Galli seu exsuscitata Apollonii Pergaei tactionum geometriae pars reliqua* (Venice, 1607); *Variorum problematum collectio* (Venice, 1607); *Apollonius redivivus seu restitutae Apollonii Pergaei de inclinationibus geometriae, liber secundus* (Venice, 1613); and *De resolutione et compositione mathematica, libri quinque* (Rome, 1630).

See also A. Favaro, "Amici e corrispondenti de Galileo Galilei," in *Atti del Istituto veneto di scienze, lettere ed arti,* **69,** 303–324; E. Gelcich, in *Abhandlungen zur Geschichte der Mathematik,* **4,** 191–231; and H. Wieleitner, "Marino Ghetaldi," in *Bibliotheca mathematica,* **13,** 242–247.

LUIGI CAMPEDELLI

GHINI, LUCA (*b.* Croara d'Imola, Italy, *ca.* 1490; *d.* Bologna, Italy, 4 May 1556), *botany.*

A prominent pioneer in the creation of the first botanical gardens in sixteenth-century Italy and in the collection of the earliest herbaria, Ghini exerted his influence primarily through correspondence and teaching, for he wrote little and published nothing in his lifetime. His career reflects the growth and emancipation of botanical research within university medical schools.

His father, Ghino Ghini, was a notary in Imola; this profession may have been in the family, for Ghini's only son, Galeazzo, was also a notary. Ghini was sent to study medicine at Bologna, where he was appointed to read "medicina practica" in 1527. During the next decade the title of his post grew more specifically botanical: in 1535 it was noted "let him lecture on Simples" (i.e., botany as herbal therapeutics); these lectures led to a course on Galen's book *On Simples* (1537), then an associate chair "on Simples," and finally a professorial chair (1539). In 1544 he was invited to become professor of simples at Pisa, where he stayed until 1554. Nevertheless, he remained attached to Bologna, where he had married Gentile Sarti in 1528 and had been granted citizenship in 1535; he spent his vacations there and maintained a house, close to that of his wife's family, with a garden for his private work, where he could try to grow seeds sent him from abroad. Among his successes was "Medica" (*Medicago sativa*) from Spain.

While teaching at Bologna, Ghini introduced, probably for the first time, the herbarium or *hortus siccus,* the technique of pressing and drying plants which could then be attached to cards and filed as a source of reference more reliable than an illustration. His collection was also used for the dissemination of knowledge, for when inquiries were addressed to him, he could send a labeled card to exemplify his answer. For instance, in reply to an appeal by Pietro Mattioli, Ghini mentions enclosing specimens of two varieties of "lesser Horminum" (*Salvia sclarea*) treated in this way. The two oldest surviving herbaria were assembled by Gherardo Cibo da Roccacontrada and Michele Merini, who had been Ghini's pupils. Another pupil, John Falconer, compiled one much admired at the time.

Shortly after his arrival at Pisa, Ghini became actively involved in the creation of the botanical garden there, which disputes with Padua the title of the oldest in Europe set up as an aid to university teaching and research, as opposed to herb gardens intended purely for the immediate supply of medicinal plants. In a letter of July 1545, Ghini describes two expeditions to the mountains, in search of plants for the garden, the site of which was already being cleared. He also became its first prefect and therefore was responsible for the original collection. Ulisse Aldrovandi owned a catalog of the 610 plants that were in the garden in Ghini's time. Ghini had much to do also with the foundation in 1545 of another botanical garden, at Florence, capital of the duchy to which Pisa then belonged.

In 1554 Ghini returned to lecture at Bologna, where he died two years later. His influence was to be felt not only through the new institutions and techniques he helped to establish but also through his pupils, among them some of the outstanding naturalists of the next generation. While he was holidaying at Bologna in 1549, Aldrovandi studied with him, later claiming that his love for botany dated from that time. The affiliation was sustained by correspondence and the exchange of specimens; Aldrovandi also visited Ghini at Pisa, attended his lectures there, and went botanizing with him in the hills behind Lucca. Andrea Cesalpino studied under him at Pisa and succeeded him as prefect of the botanical garden in 1554. Luigi Anguillara, first prefect of the Padua botanical garden, was another pupil, as were two

English herbalists of the new school, William Turner and Falconer.

In 1551 Mattioli asked Ghini's help in the identification of a number of Dioscorides' plants on which he had no information. Ghini's reply, his "Placiti" on these plants, demonstrates his methods. The context obliged him to start with the accounts of Dioscorides or Pliny, then relate them to actual plants known to him; the leaves were his preferred indicator. To help him in these inquiries he procured specimens of the flora of Greece and the Levant from merchants and asked Greek soldiers or his Greek maid about the modern Greek names, in order to compare them with those in classical sources. Luckily, one of Ghini's brothers spent some years in Crete and could supply him with seeds or branches. Besides Crete, Ghini refers to receipt of material from Egypt, Syria, Spain, Sicily, and Calabria; and even "papyrus" leaves from the island of São Tomé, used for wrapping sugarloaves, were not unworthy of his investigation. Certainly he was far from being merely a bookish scholar: the "Placiti" often recall his collecting expeditions in the Apennines, along the Tuscan shore, and on a sail round Elba, where he admired the abundance of *Medicago marina.*

The "Placiti" is our only evidence on Ghini's researches. Although he is supposed to have begun compiling a pictorial herbal on plants of which there were as yet no illustrations, he abandoned it when Mattioli published his own commentaries on Dioscorides, to which Ghini had in effect contributed. Another manuscript, now lost, discussed the kindred topic of plants known to practicing apothecaries but not included in the written tradition of materia medica.

BIBLIOGRAPHY

Ghini's only published works—and those long after his death—were minor medical tracts. Much more important is the letter to Mattioli, published as "I placiti di Luca Ghini intorno a piante descritte nei commentarii al Dioscoride di P. A. Mattioli," G. B. de Toni, ed., in *Memorie del R. Istituto veneto di scienze, lettere ed arti,* **27,** no. 8 (1907). Toni's commentary has been used for the identifications in this article.

Toni is also the author of papers touching on aspects of Ghini's work, which is summarized in his "Luca Ghini," in A. Mieli, ed., *Gli scienziati italiani,* I (Rome, 1921), 1-4. See also A. Chiarugi, "Le date di fondazione dei primi orti botanici del mondo," in *Nuovo giornale botanico italiano,* **60** (1953); and "Nel quarto centenario della morte di Luca Ghini," in *Webbia,* **13** (1957), 1-14; and L. Sabbatani, "Alcuni documenti su la vita di Luca Ghini," in *Atti e memorie della R. Accademia di scienze, lettere ed arti* (Padua), n.s. **39** (1923), 243-248.

A. G. KELLER

GHISI, MARTINO (*b.* Soresina, Italy, 11 November 1715; *d.* Cremona, Italy, 11 May 1794), *medicine.*

Ghisi, who is unjustly forgotten by medical historians and on whom there is very little biographical information, was an obscure provincial doctor to whom we owe one of the first—if not the first—descriptions of diphtheria to be complete and valid both clinically and anatomicopathologically. Probably a handicap to recognition was that his ideas and observations were disseminated not from a university chair but in a small pamphlet with a limited circulation.

After completing his secondary education, Ghisi studied under Paolo Valcarenghi, a doctor of some renown who had founded a practical school of medicine in Cremona; he subsequently moved to Florence, where he graduated from the university. He returned to Cremona to practice and in 1747-1748 combated an epidemic which struck a large number of children and adolescents in the Cremona region. Ghisi made careful clinical and meteorological observations on the epidemic, publishing the results in a pamphlet entitled *Lettere mediche del Dottor M. Ghisi.* Of particular note is the section entitled "Istoria delle angine epidemiche," the first truly complete scientific description of diphtheria.

Ghisi began with a precise account of climatic and meteorological conditions prior to and during the epidemic—in accordance, apparently, with the Hippocratic tradition—but his attention was mainly centered on the clinical framework of the illness. He described in particular detail the diphtheric paralysis of the velum palatinum and the tumefaction of the submaxillary glands.

Ghisi was also interested in examining and describing the diphtheric membrane, which some patients would expel in coughing and which Ghisi compared to the fibrous crusts that formed in blood extracted by bleeding. He did not neglect the anatomicopathological aspect of the epidemic, even though he was able to conduct only one autopsy; this, thanks to his great knowledge of normal and pathological anatomy, was sufficient for him to compare accurately the diagnostic features and pathological alterations detected in the lungs and pleura of patients suffering from pneumonia and pleurisy with those found in diphtheria. He showed quite clearly that the bronchial and pulmonary edema caused by diphtheria resulted in a strain of the right side of the heart.

Ghisi limited himself to clinical and anatomico-pathological observations that could be made directly and objectively, and deliberately refrained from giving an opinion on the etiologic-pathogenic causes of diphtheria (which would in any case have been impossible to identify at that time). He confined himself to emphasizing the small amount of data that he had been able to collect, at the same time steadily refusing to classify it according to any of the several systems then in vogue.

BIBLIOGRAPHY

Ghisi's only publication is *Lettere mediche del Dottor M. Ghisi* (Cremona, 1749). The first letter deals with various illnesses cured by mercury; the second contains the history of epidemic angina in 1747 and 1748.

Secondary literature includes A. Caccia, *Elogio del celebre medico Martino Ghisi* (Cremona, 1794); and C. Castellani, "La 'Lettera medica' di Martino Ghisi relativa alla 'Istoria delle angine epidemiche,'" in *Rivista di storia della medicina,* **2** (1960), 163–188, which includes the text of the letter.

CARLO CASTELLANI

GIARD, ALFRED (*b.* Valenciennes, France, 8 August 1846; *d.* Paris, France, 8 August 1908), *botany, zoology, embryogeny, general biology.*

Giard was an extraordinarily gifted child who, by the age of fifteen, had already acquired, under the influence of his father, an extensive knowledge of insects and plants. Following secondary studies at the *lycée* in Valenciennes, he entered the École Normale Supérieure (1867), where he was named *préparateur* in 1871. He defended his doctoral thesis on the compound ascidians in 1872, then was successively professor at the Faculté des Sciences of Lille (1875), lecturer at the École Normale Supérieure (1887), and professor at the Faculté des Sciences of Paris, holding the chair of evolution of living organisms from 1888 until his death following a short illness. From 1882 to 1885 he sat in the Chamber of Deputies as the member from Valenciennes; but, failing to win reelection in 1885, he gave up politics. He was elected to the Académie des Sciences in 1900.

Giard was a morphologist, a phylogenist, an ethologist—a complete naturalist who was endowed with a remarkable memory and possessed prodigious factual knowledge. Moreover, he had the ability to rank facts and coordinate them to bring out the ideas of general biology.

In 1874, with his own funds, he founded the biological station at Wimereux in order to introduce his students to marine and terrestrial flora and fauna. He was a great laboratory director, and he rapidly formed a brilliant school of zoology in Lille.

Giard was the opposite of a specialist. He was a remarkable observer and the variety of his observations stimulated him to study diverse animals. He discovered the Orthonectida (parasites of the Ophiurida) in 1877, and a member of the Turbellaria (*Fecampia*), a parasite of the higher crustaceans; with J. Bonnier he carried out research on the crustaceans, notably on the Epicaridea (parasitic isopods) and on the Bopyridae.

Giard investigated several problems of general biology: regeneration (hypotypic regeneration), metamorphosis, sexuality, experimental parthenogenesis, merogenesis, hybridization, autotomy, convergence, mimetism, and anhydrobiosis. He defined poecilogony (in the same species animals will develop differently, according to their environment) and parasitic castration (all the morphological or physiological phenomena involved in the organization of a living being are a result of the presence of a parasite acting indirectly or directly on the genital function of the host).

Never an advocate of technique—be it injections or histological sections or preparations—Giard considered the examination of living creatures in their environment to be superior to that of materials that were preserved or cut in pieces. One ought always to begin with examination.

Giard was rapidly able to free himself from the tenor of instruction he had received, especially from J. H. Lacaze-Duthiers, who for a long time opposed his nomination to Paris. He became a convinced follower of transformism, a doctrine opposed by his contemporaries, and his courses were filled with new ideas already widespread abroad. The transformist hypotheses overturned the general classification of animals. Giard was among the first to bring together the mollusks and the Annelida, Brachiopoda, Bryozoa, and Gephyrea; in 1876 he united them under the name Gymnotoca—the name was forgotten, but the grouping was recognized by Emil Hatschek, who termed its members the Trochozoa.

In accepting transformism as fact and interpreting nature accordingly, Giard was under the influence of Ernst Haeckel. He thought that Lamarckism and Darwinism complemented each other. The primordial cause of variation resided in the actions of external agents, which constituted the primary factors of evolution (a Lamarckian idea); natural selection intervened only as a secondary factor of great power (a Darwinian idea). Yet Giard preferred Lamarckism to Darwinism and strove to glorify Lamarck. Heredity, sexual selection, and physiological selection were also secondary factors. He did not accept particulate theories of heredity, and he rejected all theories based on unverifiable internal tendencies, such as the orthogenesis conceived by Teodor Eimer.

Giard created new biological terms, some of which became classic. He was also greatly interested in scientific societies and attempted to guide their activity. In 1878 he became editor of the *Bulletin scientifique du Nord;* ten years later this local journal became the *Bulletin scientifique de la France et de la Belgique.*

BIBLIOGRAPHY

I. Original Works. Giard's writings include "Étude critique des travaux d'embryogénie relatifs à la parenté des Vertébrés et des Tuniciers," in *Archives de zoologie expérimentale,* **1** (1872), 233–288; "Deuxième étude critique des travaux d'embryogénie relatifs à la parenté des Vertébrés et des Tuniciers; recherches nouvelles du professeur Kuppfer," *ibid.,* 307–428; "Recherches sur les Ascidies composées ou Synascidies," *ibid.,* 501–704, his doctoral thesis; "Principes généraux de biologie," the introduction to T. H. Huxley's *Éléments d'anatomie comparée des Invertébrés* (1876); "Sur l'organisation et la classification des Orthonectidea," in *Bulletin scientifique du Nord,* **11** (1879), 338–341; "Nouvelles remarques sur les Entonisciens," in *Comptes rendus hebdomadaires des séances de l'Académie des sciences,* **102** (1886), 1173–1176, written with J. Bonnier; "Sur un Rhadocoele nouveau, parasite et nidulant (*Fecampia erythrocephala*)," *ibid.,* **103** (1886), 499–501; "La castration parasitaire et son influence sur les caractères extérieurs du sexe mâle chez les Crustacés décapodes," in *Bulletin scientifique du Nord,* **18** (1887), 1–28; "Contribution à l'étude des Bopyriens," in *Travaux de l'Institut de zoologie de Lille et du Laboratoire de zoologie maritime de Wimereux,* **5** (1887), written with J. Bonnier; "Leçon d'ouverture des cours d'évolution des êtres organisés," in *Bulletin scientifique,* **20** (1889), 1–26; "Sur la signification des globules polaires," *ibid.,* 95–103; and "Les facteurs de l'évolution, leçon d'ouverture du cours d'évolution des êtres organisés, 2ème année," in *Revue scientifique,* **44** (1889), 641–648.

Later works are "Prodrome d'une monographie des Epicarides du golfe de Naples," in *Bulletin scientifique du Nord,* **22** (1890), 367–391, written with J. Bonnier; "Le principe de Lamarck et l'hérédité des modifications somatiques, leçon d'ouverture du cours d'évolution des êtres organisés," in *Revue scientifique,* **46** (1890), 705–713; "L'anhydrobiose ou ralentissement des phénomènes vitaux sous l'influence de la déshydratation progressive," in *Comptes rendus de la Société de biologie,* **66** (1894), 497–500; "La direction des recherches biologiques en France et la conversion de M. Yves Delage," in *Bulletin scientifique du Nord,* **27** (1896), 432–458; *Titres et travaux scientifiques* (Paris, 1896); "Coup d'oeil sur la faune et note sur la flore du Boulonnais," in *Boulogne et le Boulonnais* (Paris, 1899); "Parthénogenèse de la macrogamète et de la microgamète des organismes pluricellulaires," in *Volume cinquantenaire de la Société biologique* (Paris, 1899), pp. 654–667; "Les faux hybrides de Millardet et leur interprétation," in *Comptes rendus de la Société de biologie,* **55** (1903), 779–782; "Les tendances actuelles de la morphologie et ses rapports avec les autres sciences," in *Bulletin scientifique du Nord,* **39** (1905), 455–486; "La poecilogonie," *ibid.,* 153–187; "L'évolution dans les sciences biologiques," *ibid.,* **41** (1907), 427–458; and "L'éducation du morphologiste," in *La méthode dans les sciences: Morphologie* (Paris, 1908), pp. 149–175.

II. Secondary Literature. There is a notice on Giard's life and works, with complete bibliography, by F. Le Dantec and M. Caullery in *Bulletin scientifique de la France et de la Belgique,* **42** (1909), i–lxxiii. Obituaries include those by H. Piéron, in *Scientia,* **5** (1909), 1–4; P. Pelseneer, in *Annales de la Société royale zoologique et malacologique de Belgique,* **43** (1908), 220–227; E. Rabaud, in *Bibliographie anatomique,* **18** (1909), 285–290; H. Fischer, in *Journal de conchyliologie,* **56** (1909), 294–301; and M. Caullery, in *Revue du mois,* **6** (1908), 385–399.

Andrée Tétry

GIBBS, JOSIAH WILLARD (*b.* New Haven, Connecticut, 11 February 1839; *d.* New Haven, 28 April 1903), *theoretical physics.*

Gibbs was the only son among the five children of Josiah Willard Gibbs and Mary Anna Van Cleve Gibbs. His father was a noted philologist, a graduate of Yale and professor of sacred literature there from 1826 until his death in 1861. The younger Gibbs grew up in New Haven and graduated from Yale College in 1858, having won a number of prizes in both Latin and mathematics. He continued at Yale as a student of engineering in the new graduate school, and in 1863 he received one of the first Ph.D. degrees granted in the United States. After serving as a tutor in Yale College for three years, giving elementary instruction in Latin and natural philosophy, Gibbs left New Haven for further study in Europe. By this time both his parents and two of his sisters were dead, and Gibbs traveled with his two surviving older sisters, Anna and Julia. He spent a year each at the universities of Paris, Berlin, and Heidelberg, attending lectures in mathematics and physics and reading widely in both fields. These European studies, rather than his earlier engineering education, provided the foundation for his subsequent career.

Gibbs returned to New Haven in June 1869. He never again left America and rarely left even New Haven except for his annual summer holidays in northern New England and a very occasional journey to lecture or attend a meeting. Gibbs never married and lived all his life in the house in which he had grown up, less than a block away from the college buildings, sharing it with Anna, Julia, and Julia's family. In July 1871, two years before he published his first scientific paper, Gibbs was appointed professor of mathematical physics at Yale. He held this position without salary for the first nine years, living on his inherited income. It was during this time that he wrote the memoirs on thermodynamics that constitute his greatest contribution to science. Gibbs had no problem about declining a paid appointment at Bowdoin College in 1873, but he was seriously tempted to leave Yale in 1880, when he was invited to join the faculty of the new Johns Hopkins University at Baltimore. Only then did Yale provide a salary for Gibbs, as tangible evidence of the esteem in which he was held by his colleagues and of his importance to the university, but this salary was still only two-thirds of what Johns Hopkins had offered him. Gibbs stayed on at Yale nevertheless and continued teaching there until his death, after a brief illness, in the spring of 1903.

Gibbs's first published paper did not appear until he was thirty-four years old, and it displays his unique mastery of thermodynamics. If there are even earlier signs of Gibbs's intellectual power, they must be sought in his previous engineering work. His doctoral thesis, "On the Form of the Teeth of Wheels in Spur Gearing," certainly shows Gibbs's unusually strong geometrical ability. He generally preferred "the niceties of geometrical reasoning" to analytical methods in his later work, and this is true in his thesis. The style of this early work also shows the same "austerity," the same "extreme economy (one might almost say parsimony) in the use of words,"[1] that made his later memoirs so difficult to read. Some of his engineering work, such as the design of an improved railway car brake, for which he received a patent in 1866, is hard to relate to the concerns of the future master of theoretical physics—but this is not true of all of it. After his return from Europe, Gibbs designed a new form of governor for steam engines by suitable mounting of a second pair of massive balls on the simple Watt governor. This new arrangement was planned to increase the responsiveness of the system to a change in the engine's running speed. Although Gibbs went no further with this invention than having a model of it built in the department workshop, it is interesting nevertheless: the problems of dynamic equilibrium and stability of this particular mechanical device foreshadow the related questions of equilibrium and stability that he would soon raise and answer for general thermodynamic systems.

When Gibbs first turned his attention to thermodynamics in the early 1870's, that science had already achieved a certain level of maturity. Rudolf Clausius had taken the essential step in 1850, when he argued that two laws and not just one are needed as the basis of a theory of heat. Only a year before that, William Thomson had been writing about the "very perplexing question," and the associated "innumerable" and "insuperable" difficulties, of choosing *the* correct axiom for the theory.[2] Should one hold fast to Carnot's postulate (that heat must pass from a hot body to a colder one when work is done in a cyclic process), even though Carnot's results seemed to depend on his use of the caloric theory of heat? Or should one accept the interconvertibility of heat and work, since James Joule's new experimental evidence clearly favored the mechanical theory of heat? Clausius showed that, despite the apparent need to choose one law or the other, both were necessary and both could be maintained without contradiction. One had only to drop Carnot's inessential requirement that heat itself be conserved. This change did have one major implication. The proof of Carnot's theorem (that the maximum motive power of heat depends solely on the temperatures between which the heat is transferred) now had to appeal to a new axiom: that heat "everywhere exhibits the tendency to annul temperature differences, and therefore to pass from a *warmer* body to a *colder* one."[3] Clausius' memoir demonstrated how one could develop a thermodynamics starting with both the equivalence of heat and work and his new axiom. These two laws of thermodynamics were restated in slightly different form a year later by Thomson, who proceeded to apply them to a variety of physical problems, including thermoelectricity.

Clausius tried as hard as he could to find the essence of the second law of thermodynamics, since he felt unable at first to "recognize . . ., with sufficient clearness, the real nature of the theorem."[4] This search led finally in 1865 to his most concise and ultimately most fruitful formulation of the two laws, the formulation Gibbs later used as the motto of his greatest work: "The energy of the universe is constant. The entropy of the universe tends to a maximum."[5] The two basic quantities, internal energy and entropy, were, in effect, defined by the two laws of thermodynamics. The internal energy U is that function of the state of the system whose differential is given by the equation expressing the first law,

$$dU = đQ + đW, \tag{1}$$

where $đQ$ and $đW$ are, respectively, the heat added to the system and the external work done on the system in an infinitesimal process. For a simple fluid the work $đW$ is given by the equation

$$đW = -PdV, \qquad (2)$$

where P is the pressure on the system and V is its volume. Neither the heat $đQ$ nor the work $đW$ is the differential of a function of state, and the inexactness or nonintegrability of these differentials is indicated by the symbol $đ$, whose use for this purpose goes back to Carl Neumann's lectures in the early 1870's.[6] The entropy S is that state function whose differential is given by the equation

$$dS = \frac{đQ}{T}, \qquad (3)$$

valid for reversible processes, where T is the absolute temperature. For irreversible processes equation (3) is replaced by the inequality

$$dS > \frac{đQ}{T}. \qquad (4)$$

The power and importance of the entropy concept were certainly not evident to Clausius' contemporaries upon the publication of his 1865 paper. Clausius himself considered it to be a summarizing concept and thought that the true physical significance of the second law was better expressed in terms of the disgregation, a concept he sought to interpret mechanically. Entropy plays no particular role in the thermodynamics texts by Georg Krebs (1874) and Carl Neumann (1875). The word was picked up by Peter Guthrie Tait in his *Sketch of Thermodynamics* (1868), but he completely changed its meaning, using it to denote available energy rather than the quantity Clausius had intended. Tait's misinterpretation was taken up and repeated by James Clerk Maxwell in his *Theory of Heat* (1871). The confusion and uncertainty about the thermodynamic significance of entropy were only aggravated by the bitter priority disputes between Clausius and Tait (on behalf of Thomson) that raged in the early 1870's. The basic structure of modern thermodynamics was implicitly present in the work of both Clausius and Thomson, but it was certainly not clearly visible to most writers on the subject.

It was in this context that Gibbs's first scientific paper, "Graphical Methods in the Thermodynamics of Fluids," appeared in 1873. His mastery and his quiet assurance in this paper are as remarkable as his scientific insight. Gibbs assumed from the outset that entropy is one of the essential concepts to be used in treating a thermodynamic system, along with energy, temperature, pressure, and volume. He immediately combined the first three equations given above to obtain the form

$$dU = TdS - PdV, \qquad (5)$$

a relation that contains only the state variables of the system, the process-dependent heat and work having been eliminated. As Gibbs pointed out, an equation expressing the internal energy of the system in terms of its entropy and volume could appropriately be called its fundamental equation; for equation (5) would then allow one to determine the two equations of state, expressing temperature and pressure as functions of the pair, volume and entropy. These remarks were the starting point for Gibbs's later work, but in this first paper he limited himself to a discussion of what could be done with geometrical representations of thermodynamic relationships in two dimensions.

James Watt's indicator diagram, in which pressure and volume are plotted on the two coordinate axes, had been in use for thermodynamic purposes since Émile Clapeyron's memoir of 1834. But Gibbs showed how other choices for the coordinate variables would produce representations even more useful for thermodynamic purposes; the temperature-entropy diagram, for example, had many advantages in the study of cyclic processes. The form of the basic equation (5), expressing both laws of thermodynamics, suggested that the volume-entropy diagram might be best suited for general thermodynamic considerations, and Gibbs discussed it in more detail. He also showed how some of the interrelations among the curves describing, respectively, states of equal pressure, equal temperature, equal energy, and equal entropy were independent of how the thermodynamic diagram was constructed and followed directly from the stability of equilibrium states.

In his second paper, which appeared later in 1873, Gibbs extended his geometrical discussion to three dimensions by analyzing the properties of the surface representing the fundamental thermodynamic equation of a pure substance. The thermodynamic relationships could be brought out most clearly by constructing the surface using entropy, energy, and volume as the three orthogonal coordinates. Gibbs pointed out that, as a consequence of equation (5), the temperature and pressure of the body in any state was determined by the plane tangent to the surface at the corresponding point, since one has the equations

$$T = \left(\frac{\partial U}{\partial S}\right)_V \qquad (6a)$$

and

$$-P = \left(\frac{\partial U}{\partial V}\right)_S. \qquad (6b)$$

This way of representing the thermodynamic properties of a body in thermodynamic equilibrium could be used just as well when different parts of the body were in different states (for example, a mixture of liquid and gas or of two different crystalline forms of the same pure substance). Gibbs showed how one could use the thermodynamic surface to discuss the coexistence of the various phases (liquid, solid, and gas) of a pure substance and the stability of these states under given conditions of temperature and pressure. One feature of particular interest was the critical point—the state at which liquid and gas become identical—a phenomenon discovered experimentally in carbon dioxide by Thomas Andrews only a few years earlier.

These early papers, as well as Gibbs's major memoir on thermodynamics that soon followed them, appeared in the *Transactions of the Connecticut Academy of Arts and Sciences,* a new and relatively obscure journal whose nonlocal circulation consisted largely of exchanges with other learned societies, including some 140 outside the United States. Gibbs did not count on finding his potential readers among those who checked the contents of the *Transactions;* he sent copies of his papers to an impressive list of scientists in many countries, a list that probably included all those he thought might really read and understand his work.[7] One of them was James Clerk Maxwell, who proved to be Gibbs's most enthusiastic and most influential reader. He immediately accepted Gibbs's clarification of what Clausius had intended by the term "entropy," correcting the error in his own *Theory of Heat* accordingly and informing Tait about his mistake.

Maxwell found Gibbs's use of geometric rather than algebraic arguments particularly attractive, since he too preferred geometric insight to calculations, even when others found algebraic procedures decidedly simpler. He was sufficiently impressed by Gibbs's paper on the thermodynamic surface to include a discussion of this subject in the fourth edition of *Theory of Heat* (1875) and actually to construct a model of the thermodynamic surface for water, which he sent to Gibbs. He talked about Gibbs's work to his colleagues at Cambridge and recommended it to his friends. "Read Prof. J. Willard Gibbs on the surface whose coordinates are Volume Entropy and Energy," he wrote to Tait, and then added for the benefit of his rather chauvinistic friend, "He has more sense than any German."[8] Maxwell even started to generalize Gibbs's thermodynamics of a pure substance to include the case of heterogeneous mixtures. This proved to be quite unnecessary and was dropped when he received a set of galley proofs of Gibbs's new memoir containing this generalization and a great deal more.

"On the Equilibrium of Heterogeneous Substances" contains Gibbs's major contributions to thermodynamics. In this single memoir of some 300 pages he vastly extended the domain covered by thermodynamics, including chemical, elastic, surface, electromagnetic, and electrochemical phenomena in a single system. The basic idea had been foreshadowed in his two earlier papers, in which Gibbs had directed his attention to the properties characterizing the equilibrium states of simple systems rather than to the heat and work exchanged in particular kinds of processes. In the abstract of his memoir that Gibbs published in the *American Journal of Science* in 1878, he began by stating the simple but profound idea underlying his work:

> It is an inference naturally suggested by the general increase of entropy which accompanies the changes occurring in any isolated material system that when the entropy of the system has reached a maximum, the system will be in a state of equilibrium. Although this principle has by no means escaped the attention of physicists, its importance does not appear to have been duly appreciated. Little has been done to develop the principle as a foundation for the general theory of thermodynamic equilibrium.[9]

Gibbs formulated the criterion for thermodynamic equilibrium in two alternative and equivalent ways. "For the equilibrium of any isolated system it is necessary and sufficient that in all possible variations of the state of the system which do not alter its energy [entropy], the variation of its entropy [energy] shall either vanish or be negative [positive]."[10] The bracketed form immediately indicates that thermodynamic equilibrium is a natural generalization of mechanical equilibrium, both being characterized by minimum energy under appropriate conditions. The consequences of this criterion could then be worked out as soon as the energy of the system was expressed in terms of the proper variables. Gibbs's first and probably most significant application of this approach was to the problem of chemical equilibrium. The result of his work was described by Wilhelm Ostwald as determining the form and content of chemistry for a century to come, and by Henri Le Chatelier as comparable in its importance for chemistry with that of Antoine Lavoisier.

The simplest case is that of a homogeneous phase—a liquid or gas, for example—containing n independent chemical species $S_1, \cdots, S_n$ whose masses $m_1, \cdots, m_n$ can be varied. Gibbs modified the basic equation (5) to include the change of internal energy due to a change in the mass of any of the chemical components by writing it in the form

$$dU = TdS - PdV + \sum_{i=1}^{n} \mu_i \, dm_i. \tag{7}$$

Here dm_i is the change in mass (conveniently expressed as the number of moles) and the new quantity μ_i is the (chemical) potential of the ith chemical species. The chemical potential is related to the energy by the equation

$$\mu_i = \left(\frac{\partial U}{\partial m_i}\right)_{S,V,m_j'}, \tag{8}$$

where the subscript m_j' means that μ_i represents the rate of change of energy with respect to the mass of the ith component in this phase, the masses of all other components being held constant along with entropy and volume.

In a heterogeneous system composed of several homogeneous phases the fundamental equilibrium condition leads to the requirement that temperature, pressure, and the chemical potential of each independent chemical component must have the same values throughout the system. From these general conditions Gibbs derived the phase rule, that cornerstone of physical chemistry, which specifies the number of independent variations δ in a system of r coexistent phases having n independent chemical components:

$$\delta = n + 2 - r. \tag{9}$$

Gibbs also showed how to obtain the specific conditions for equilibrium when chemical reactions could take place in the system. Instead of one's attention being restricted to a set of independent chemical components, all the relevant chemical species are considered. Suppose, for example, that a reaction of the type

$$\sum_j a_j A_j = 0 \tag{10}$$

can occur; here the a_j's are integers, the stoichiometric coefficients, and the A_j's stand for the chemical symbols of the reacting substances. (An illustration would be the reaction $H_2 + Cl_2 - 2\,HCl = 0$, where $a_1 = 1$, $a_2 = 1$, $a_3 = -2$ and the corresponding A_j are respectively H_2, Cl_2, and HCl.) The equilibrium condition that Gibbs derived for such a reaction has the simple form

$$\sum_j a_j \mu_j = 0, \tag{11}$$

obtained by replacing the chemical symbols A_j with the chemical potential μ_j of the corresponding substance in the reaction equation (10). Since the potentials could in principle be determined from experimental data, the equilibrium conditions were really established by equation (11).

The requirement that the energy have a minimum and not just a stationary value at equilibrium was used by Gibbs to explore the stability of equilibrium states. This stability depends ultimately on the second law of thermodynamics and manifests itself in the unique sign of certain properties of all substances; the heat capacity at constant volume, for example, must be positive, and the isothermal derivative of pressure with respect to volume must be negative for any substance. The most interesting aspect of Gibbs's investigation of stability was his theory of critical phases, those situations where the distinction between coexistent phases vanishes and the stability is of a lower order than that usually found.

Gibbs's memoir showed how the general theory of thermodynamic equilibrium could be applied to phenomena as varied as the dissolving of a crystal in a liquid, the temperature dependence of the electromotive force of an electrochemical cell, and the heat absorbed when the surface of discontinuity between two fluids is increased. But even more important than the particular results he obtained was his introduction of the general method and concepts with which all applications of thermodynamics could be handled.

Although Maxwell responded immediately to Gibbs's work and influenced a number of his English colleagues to apply Gibbs's results if not his methods, it took a little longer for Continental scientists to appreciate Gibbs. Such figures as Hermann von Helmholtz and Max Planck independently developed thermodynamic methods for chemical and electrochemical problems in the 1880's, quite unaware of Gibbs's prior work. This situation changed gradually—Ludwig Boltzmann referred to Gibbs in 1883—but it was Wilhelm Ostwald's German translation of Gibbs's papers in 1892 that made his ideas more readily available to German chemists.

Gibbs wrote no other major works on thermodynamics itself, restricting himself to a few small applications and developments of his extensive memoir. He rejected all suggestions that he write a treatise that would make his ideas easier to grasp. Even Lord Rayleigh thought the original paper "too condensed and too difficult for most, I might say all, readers."[11] Gibbs responded by saying that in his own view the memoir was instead "too *long*" and showed a lack of "sense of the value of time, of [his] own or others, when [he] wrote it."[12]

During the 1880's Gibbs seems to have concentrated on optics and particularly on Maxwell's electromagnetic theory of light. He was giving lectures on Maxwell's theory at Yale at least as early as 1885, and he published a series of papers in the *American Journal of Science* on double refraction and dispersion, that is, on the behavior of light as it passes through material media. Two aspects of Gibbs's optical work are of more than technical interest. He emphasized that a theory of dispersion requires one to treat the local irregularities of the electric displacement due to the atomic constitution of the medium. Since he was writing before H. A. Lorentz's theory of electrons (the first version of which did not appear until 1892), Gibbs had to make assumptions of a different kind from those of later theories, and he missed the essential contribution of the atomic structure of the medium—the frequency dependence of the dielectric constant. In the last two papers of this series, published in 1888 and 1889, Gibbs appeared as a defender of the electromagnetic theory of light against the latest versions of purely mechanical theories. These were based on special elastic ethers still being proposed by William Thomson. Gibbs showed that although such theories might account for the phenomena, they required rather artificial assumptions as to internal forces, while Maxwell's theory was "not obliged to invent hypotheses."[13]

Gibbs's reading of Maxwell's *Treatise on Electricity and Magnetism* led him to a study of quaternions, since Maxwell had used the quaternion notation to a limited extent in that work. Gibbs decided, however, that quaternions did not really provide the mathematical language appropriate for theoretical physics, and he worked out a simpler and more straightforward vector analysis. He wrote a pamphlet on this subject which he had printed in 1881 and 1884 for private distribution to his classes and to selected correspondents. No real publication of Gibbs's version of vector analysis took place until 1901, when his student Edwin B. Wilson prepared a textbook of the subject based on Gibbs's lectures. In 1891 Gibbs defended his use of vectors rather than quaternions against an attack by Tait, who was the great exponent of—and crusader for—William Rowan Hamilton's quaternions. Gibbs more than held his own in the ensuing controversy over quaternions, debating both Tait and his disciples in the pages of *Nature*, on minor matters of notation and on the "deeper question of notions underlying that of notations."[14]

During the 1890's the classic goal of the physicist—the explanation of natural phenomena in mechanical terms—was seriously questioned and sharply criticized. Among the most outspoken critics were such men as Ernst Mach, Pierre Duhem, and Wilhelm Ostwald, who rejected the concept of atomism along with that of mechanism. One of the spokesmen for the group who called themselves energeticists was Georg Helm, who in 1898 wrote a treatise on the historical development of energetics. In his book Helm claimed Gibbs as a fellow energeticist, pointing to Gibbs's writings on thermodynamics as a sign that he was free of any prejudice in favor of atomistic mechanical explanations. To Helm, Gibbs's thinking proceeded directly from the two laws of thermodynamics "without any hankering or yearning after mechanics"; his writings were "not decked out with molecular theories."[15] While it is true that Gibbs's papers on thermodynamics were based only on the two laws and required no assumptions about the molecular structure of matter, it is equally true that Gibbs believed in and constructed a (statistical) mechanical explanation of thermodynamics itself. He certainly did not share the scientific values of the energeticists.

Gibbs had carefully studied the writings of Maxwell and Boltzmann on the kinetic theory of gases. This is evident from his obituary notice of Clausius, written in 1889, in which he commented perceptively on a peculiar feature of Clausius' work in this field. Clausius never really accepted the statistical point of view. "In reading Clausius we seem to be reading mechanics; in reading Maxwell, and in much of Boltzmann's most valuable work, we seem rather to be reading in the theory of probabilities."[16] Maxwell had introduced statistical methods in his first paper on gases in 1860 and had emphasized the essentially statistical nature of the second law of thermodynamics with the help of the "Maxwell demon." Boltzmann then developed all the essential features of a theoretical explanation of the second law based on a combination of mechanics and the laws of probability applied to a large assemblage of molecules in a gas. Gibbs had already paid his own respects to this fundamental insight in his memoir on heterogeneous equilibrium, when he argued that "the impossibility of an uncompensated decrease of entropy seems to be reduced to improbability."[17]

During the academic year 1889–1890 Gibbs announced "A short course on the a priori Deduction of Thermodynamic Principles from the Theory of Probabilities,"[18] a subject on which he lectured repeatedly during the 1890's. He referred to this subject in a letter to Lord Rayleigh in 1892:

> Just now I am trying to get ready for publication something on thermodynamics from the *a priori* point of view, or rather on "Statistical Mechanics" of which the principal interest would be in its application to thermodynamics—in the line therefore of the work of Maxwell and Boltzmann. I do not know that I shall have anything particularly new in substance, but shall

be contented if I can so choose my standpoint (as seems to me possible) as to get a simpler view of the subject.[19]

In fact Gibbs did not publish anything more than a very brief abstract on this subject (1884) until 1902, when his book *Elementary Principles in Statistical Mechanics Developed With Special Reference to the Rational Foundation of Thermodynamics* appeared as one of the Yale Bicentennial series.

Gibbs thought of his book as offering a more general approach to statistical mechanics than that used by Boltzmann or Maxwell. Individual results previously obtained by others could now assume their proper places in the logical structure Gibbs gave to the subject. He considered an ensemble of systems, that is, a large number of replicas of the physical system of interest—which might be anything from a molecule to a sample of gas or a crystal. The replicas were all identical in structure but differed in the values of their coordinates and momenta. The ensemble was then characterized by its probability density ρ in phase space, where $\rho\, dq_1 \cdots dq_N dp_1 \cdots dp_N$ is the fractional number of systems in the ensemble whose coordinates $\{q_i\}$ and momenta $\{p_j\}$ lie in the intervals $\{q_j, q_j + dq_j\}$ and $\{p_j, p_j + dp_j\}$, respectively, at time t. The phase point representing any individual system moves with time, and so the probability density ρ at any point in the $2N$-dimensional phase space varies in a way determined by the mechanical equations of motion of the system.

If the average behavior of a system in the ensemble was to describe the behavior of the actual physical system, then a physical system in equilibrium had to be described by a stationary ensemble, one whose probability density ρ was constant in time. Gibbs analyzed several such stationary ensembles but found the one he called "canonical" to be most useful. In a canonical ensemble the probability density ρ is given by the equation

$$\rho(q_1 \cdots p_N) = \exp\{(\psi - E)/\Theta\}, \tag{12}$$

where $E(q_1 \cdots p_N)$ is the energy of the system having coordinates and momenta $q_1 \cdots p_N$, and ψ, Θ are constants in phase space. Gibbs showed that the energy of such an ensemble has a sharply peaked distribution, if the systems have many degrees of freedom; only a very small fraction of the systems in the canonical ensemble have energies appreciably different from the average.

The principal theme of Gibbs's book is the analogy, as he describes it, between the average behavior of a canonical ensemble of systems and the behavior of a physical system obeying the laws of thermodynamics. When this analogy is worked out, the modulus Θ of the canonical distribution is proportional to the absolute temperature, with a universal proportionality constant k,

$$\Theta = kT. \tag{13}$$

The average energy $\bar{E}$, where $\bar{E}$ satisfies the equation

$$\bar{E} = \frac{\int E\rho\, dq_1 \cdots dp_N}{\int \rho\, dq_1 \cdots dp_N}, \tag{14}$$

is identified with the thermodynamic internal energy U. The second parameter of the distribution, ψ, is fixed by the condition that ρ is a probability density, so one has the equation

$$\psi = -\Theta \ln \{\int \exp(-E/\Theta)\, dq_1 \cdots dp_N\} \tag{15}$$

and ψ is also related to the energy U (or $\bar{E}$) and the entropy S by the equation

$$\psi = U - TS, \tag{16}$$

a result Gibbs obtained by analyzing the meanings of heat and work in the ensemble.

Gibbs was very much aware of the gaps in his statistical mechanics. He had supplied a "rational foundation" for thermodynamics in statistical mechanics to the extent that thermodynamic systems could be treated as if they were conservative mechanical systems with a finite number of degrees of freedom. He could not incorporate the phenomena of radiation that were of so much interest at the turn of the century, nor could he surmount the long-standing difficulties associated with the theorem of the equipartition of energy. For these reasons he disclaimed any attempts "to explain the mysteries of nature" and put his work forward as the statistical "branch of rational mechanics."[20] He was also dissatisfied with the effort he had made in the twelfth chapter of his book to explain the irreversibility of nature embodied in the second law. His argument in this chapter was almost completely verbal rather than mathematical, and his statements were carefully qualified. Gibbs's manuscript notes suggest that he was still struggling with the problem of irreversibility and the nature of the entropy of systems not in equilibrium.

Despite these difficulties Gibbs's work in statistical mechanics constituted a major advance. His methods were more general and more readily applicable than Boltzmann's and eventually came to dominate the whole field. Gibbs did not live to see the real successes of statistical mechanics, for his fatal illness came within a year of the publication of his book.

NOTES

1. Everett O. Waters, "Commentary Upon the Gibbs Monograph 'On the Form of the Teeth of Wheels in Spur Gearing,'" in L. P. Wheeler, E. O. Waters, and S. W. Dudley, eds., *The Early Work of Willard Gibbs in Applied Mechanics* (New York, 1947), p. 43.
2. William Thomson, "An Account of Carnot's Theory of the Motive Power of Heat," in *Transactions of the Royal Society of Edinburgh,* **16** (1849), 541.
3. Rudolf Clausius, "On the Moving Force of Heat and the Laws of Heat Which May Be Deduced Therefrom," in *Annalen der Physik,* **79** (1850), 368, 500. Trans. into English by T. Archer Hirst as *The Mechanical Theory of Heat* (London, 1867), p. 45.
4. Clausius, "On a Modified Form of the Second Fundamental Theorem in the Mechanical Theory of Heat," in *Annalen der Physik,* **93** (1854), 481. Translation, *The Mechanical Theory of Heat,* p. 111.
5. Clausius, "On Several Convenient Forms of the Fundamental Equations of the Mechanical Theory of Heat," in *Annalen der Physik,* **125** (1865), 353. Translation, *The Mechanical Theory of Heat,* p. 365.
6. Carl Neumann, *Vorlesungen über die mechanische Theorie der Wärme* (Leipzig, 1875), p. ix.
7. The list is reprinted as app. IV in Lynde Phelps Wheeler, *Josiah Willard Gibbs* (New Haven, 1952).
8. James Clerk Maxwell to Peter Guthrie Tait, 13 Oct. 1874.
9. Gibbs, abstract of "Equilibrium of Heterogeneous Substances," in *American Journal of Science,* 3rd ser., **16** (1878), 441; *Scientific Papers,* I, 354.
10. *Ibid.*
11. Lord Rayleigh to J. W. Gibbs, 5 June 1892.
12. Gibbs to Lord Rayleigh, 27 June 1892.
13. Gibbs, "A Comparison of the Electric Theory of Light and Sir William Thomson's Theory of a Quasi-Labile Ether," in *American Journal of Science,* 3rd ser., **37** (1889), 144; *Scientific Papers,* II, 246.
14. Gibbs, "On the Role of Quaternions in the Algebra of Vectors," in *Nature,* **43** (1891), 511; *Scientific Papers,* II, 155.
15. Georg Helm, *Die Energetik* (Leipzig, 1898), p. 146.
16. Gibbs, "Rudolf Julius Emanuel Clausius," in *Proceedings of the American Academy of Arts and Sciences,* **16** (1889), 458; *Scientific Papers,* II, 265.
17. Gibbs, "On the Equilibrium of Heterogeneous Substances," in *Transactions of the Connecticut Academy of Arts and Sciences,* **3** (1875–1878), 108, 343; *Scientific Papers,* I, 167.
18. Henry A. Bumstead to Edwin B. Wilson, 22 Dec. 1915.
19. Gibbs to Lord Rayleigh, 27 June 1892.
20. Gibbs, *Elementary Principles in Statistical Mechanics* (New Haven, 1902), pp. ix–x.

BIBLIOGRAPHY

I. Original Works. Gibbs's scientific papers are repr. in *The Scientific Papers of J. Willard Gibbs,* H. A. Bumstead and R. G. Van Name, eds., 2 vols. (New York, 1906; repr. 1961). Vol. I contains the papers on thermodynamics; all others are in vol. II. There is also his *Elementary Principles in Statistical Mechanics* (New Haven, 1902; repr. New York, 1960). His work on vector analysis is in Edwin B. Wilson, *Vector Analysis Founded Upon the Lectures of J. Willard Gibbs* (New York, 1901). Gibbs's engineering work is reprinted in *The Early Work of Willard Gibbs in Applied Mechanics,* compiled by Lynde P. Wheeler, Everett O. Waters, and Samuel W. Dudley (New York, 1947). See also Edwin B. Wilson, "The Last Unpublished Notes of J. Willard Gibbs," in *Proceedings of the American Philosophical Society,* **105** (1961), 545–558.

Gibbs's notes, correspondence, and MSS are collected in the Beinecke Rare Book and Manuscript Library of Yale University.

II. Secondary Literature. There is extensive treatment of Gibbs's writings in F. G. Donnan and Arthur Haas, eds., *A Commentary on the Scientific Writings of J. Willard Gibbs,* 2 vols. (New Haven, 1936). The official biography is Lynde P. Wheeler, *Josiah Willard Gibbs. The History of a Great Mind* (New Haven, 1951; rev. ed., 1952). The appendixes include a bibliography of writings on Gibbs, a catalog of Gibbs's scientific correspondence, his mailing lists for reprints, the text of his first unpublished paper, and family information.

A more popularly written biography, valuable for period background and a poet's insight, is Muriel Rukeyser, *Willard Gibbs* (New York, 1942; repr. 1964).

Shorter accounts may be found in Henry A. Bumstead, "Biographical Sketch," repr. in Gibbs's *Scientific Papers,* I, xi–xxvi, and in J. G. Crowther, *Famous American Men of Science* (New York, 1937), pp. 229–297.

Two recent articles are Martin J. Klein, "Gibbs on Clausius," in Russell McCormmach, ed., *Historical Studies in the Physical Sciences,* I (Philadelphia, 1969), pp. 127–149; and Elizabeth W. Garber, "James Clerk Maxwell and Thermodynamics," in *American Journal of Physics,* **37** (1969), 146–155.

Martin J. Klein

GIBBS, (OLIVER) WOLCOTT (*b.* New York, N.Y., 21 February 1822; *d.* Newport, Rhode Island, 9 December 1908), *chemistry.*

Gibbs's career epitomizes much of the development of chemistry in America. He received his M.D. in 1845 from the College of Physicians and Surgeons with a dissertation on chemical classification. He then studied in Germany and France and, while in Berlin, was particularly influenced by Heinrich Rose. After fourteen years (1849–1863) teaching elementary students at what is now City College of the City University of New York, he became Rumford professor at Harvard in 1863. Until 1871, when the Lawrence Scientific School's laboratory was combined with the Harvard College laboratory, Gibbs trained a good number of professional chemists by the German practice of research as a method of teaching.

Gibbs had neither laboratory nor chemistry students from 1871 until his retirement in 1887. He did, however, lecture in the physics department on heat and spectroscopy. Publicly, this was ascribed to economy; privately, it was explained by some as President Charles Eliot's revenge for not getting the

Rumford professorship himself. Recent scholars explain the event as part of the policy of progressively attenuating the scientific schools at both Yale and Harvard in favor of a unified liberal arts undergraduate and graduate faculty.

Gibbs specialized in analytic and inorganic chemistry. His work on the platinum metals (1861–1864) and the development of the electrolytic method for the determination of copper are probably his principal contributions to the former. Gibbs was particularly interested in the structure of complex inorganic acids and their derivatives, especially those of tungsten, molybdenum, and vanadium.

BIBLIOGRAPHY

I. ORIGINAL WORKS. Gibbs's personal papers are in the Franklin Institute, Philadelphia. A good bibliography of his published works is F. W. Clarke, "Biographical Memoir of Wolcott Gibbs, 1822–1908," in *Biographical Memoirs. National Academy of Sciences,* **7** (1910), 1–22.

II. SECONDARY LITERATURE. Clarke's memoir remains the best biographical account. Gibbs, who was a member of the "Lazzaroni" around Alexander Dallas Bache, is mentioned in studies of such contemporaries as Agassiz, Asa Gray, Bache, and Joseph Henry—but hardly enough to illuminate his career. A recent work on the professionalization of chemistry is Edward A. Beardsley, *The Rise of the American Chemical Profession, 1850–1900,* University of Florida Monographs, Social Sciences, no. 23 (Gainesville, Fla., 1964).

NATHAN REINGOLD

GIESEL, FRIEDRICH OSKAR (*b.* Winzig, Silesia, Germany, 20 May 1852; *d.* Brunswick, Germany, 14 November 1927), *commercial chemistry.*

Giesel was an organic chemist who distinguished himself principally through his pioneering researches in radiochemistry. A physician's son who retained lifelong medical interests, Giesel pursued a career in chemistry. He studied from 1872 to 1874 at the Königliche Gewerbeakademie in Berlin. After a period of varied research Giesel received his doctorate from Göttingen in 1876. He then collaborated at the Gewerbeakademie until 1878 with Carl Liebermann, with whom he continued to publish jointly until 1897. Concentrating upon alkaloid research, they achieved a partial synthesis of cocaine and patented the technique in 1888. Working at Buchler & Co., a *Chininfabrik* in Brunswick, Giesel had introduced, by 1882, alkaloid extraction with benzol homologues.

In addition to this main line of research Giesel developed the use of radioactive luminous compounds and published over thirty papers on radioactivity between 1899 and 1909. A past master at the art of extracting and preparing pure substances in phytochemistry, after the Curies' 1898 discovery of polonium, he applied his craft to radiochemistry. By 1900 he had developed an improved method of fractional crystallization, producing a greater concentration of radium salts in a shorter time, by using bromide instead of chloride. One direct result of his highly influential efforts, by which pure radium bromide became commercially available for research, was the 1903 verification by William Ramsay and Frederick Soddy of the production of helium from radium. Giesel was the first to observe the decomposition of water by radium salts.

When his close friends and nearby colleagues Julius Elster and Hans Geitel obtained inconclusive results regarding magnetic influence upon Becquerel rays, Giesel provided a key to these rays' non-X-ray character by his decisive proof of their magnetic deflectability in October 1899. Three years later, using a zinc sulfide screen as a detector of alpha radiation, Giesel was able to isolate an emanating substance allied with lanthanum but free from thorium. He provisionally named his substance emanium (proven to be pure actinium by Otto Hahn and Otto Sackur in 1904) to distinguish it from André Debierne's 1899 thorium-contaminated actinium preparation. In 1903 Debierne also found the gaseous emanation (actinon) in his own actinium.

Giesel noted, as he had with radium in 1899, that the activity of his emanium increased with time. He thus proposed in 1904, by analogy with the 1902 analysis of Rutherford and Soddy, that an intermediate substance was the direct cause of the emanation. By 1905 this intermediate substance was named both by Giesel and by another, independent researcher, Tadeusz Godlewski. Giesel called it "emanium X," but Godlewski's terminology, actinium X, is the name that has survived.

Giesel was elected to the Leopoldina in 1903 and received the honorary title of professor (1903), as well as that of doctor of engineering (1916), all of which indicated the high esteem he enjoyed for his many contributions to science. Particular gratitude was expressed by Rutherford in a letter to Giesel dated 3 March 1904:

> I have followed with great interest your researches on radioactivity and I feel that I, as well as the scientific world, owe a debt of gratitude to you for your enterprise in preparing pure radium bromide for use of outside scientists. But for your aid, I feel confident most of us would have to be content with Barium chloride of Paris manufacture of activity about 20,000 [Rutherford Collection, Cambridge University Library, Add MSS 7653/G79].

The *Altmeister* of radium research in Germany, Giesel succumbed to radiation-induced lung cancer.

BIBLIOGRAPHY

I. ORIGINAL WORKS. A nearly complete list of Giesel's works is part of Otto Hahn, "Friedrich Giesel," in *Physikalische Zeitschrift,* **29** (1928), 353-357. Richard Lucas, *Bibliographie der radioaktiven Stoffe* (Leipzig, 1908), pp. 33-35, is a valuable supplement to Hahn's list.

Ferdinand Henrich, *Chemie und chemische Technologie radioaktiver Stoffe* (Berlin, 1918), contains a communication from Giesel (p. 344) concerning Buchler & Co.'s commercial provision of pure radium salts and a table (p. 283), beginning with 1902, indicating the rapidly increasing price per milligram of pure hydrated radium bromide. Some of Giesel's unpublished correspondence and other MS material is at Buchler & Co. in Brunswick, the Deutsches Museum in Munich, and the Darmstaedter Collection, Staatsbibliothek, Berlin.

II. SECONDARY LITERATURE. The following are partially derivative accounts: S. Loewenthal, "Leben und Werk des Professor Dr. Fritz Giesel," in *Braunschweigisches Magazin,* **36** (1930), 33-38; "Friedrich Giesel und die Radiumfabrikation," in Walther Buchler, *Dreihundert Jahre Buchler* (Brunswick, 1958), pp. 115-122.

Giesel's successful deflection experiment is considered by Lawrence Badash in "An Elster and Geitel Failure: Magnetic Deflection of Beta Rays," in *Centaurus,* **11** (1966), 236-240. Stefan Meyer and Egon von Schweidler, using similar equipment but different experimental technique, could not initially confirm that the effect they had observed was also due to deflection. Correspondence among Giesel, Elster and Geitel, and Meyer and von Schweidler dealing with this research during October and November 1899 was published by S. Meyer, "Zur Geschichte der Entdeckung der Natur der Becquerelstrahlen," in *Die Naturwissenschaften,* **36** (1949), 129-132.

Giesel's achievements concerning actinium, actinium X, and actinon are discussed in M. E. Weeks, *Discovery of the Elements,* 5th ed. (Easton, Pa., 1945), pp. 499-501; and H. W. Kirby, "The Discovery of Actinium," in *Isis,* **62** (1971), 290-308.

THADDEUS J. TRENN

GILBERT, GROVE KARL (*b.* Rochester, New York, 6 May 1843; *d.* Jackson, Michigan, 1 May 1918), *geology, geomorphology.*

Gilbert received the A.B. at the University of Rochester in 1862, having taken only one course in geology, which was taught by Henry A. Ward; he subsequently worked for the Ward Natural Science Establishment, which prepared and sold scientific materials to educational institutions, from 1863 to 1868. In 1869 he joined the second geological survey of Ohio as a volunteer assistant to J. S. Newberry. From 1871 to 1874 he was "geological assistant" on G. M. Wheeler's geographical and geological survey, conducted west of the 100th meridian.

On 2 December 1874 Gilbert joined John Wesley Powell's Rocky Mountain geographical and geological survey, which five years later, on Powell's recommendation, was combined with the other federal surveys to form the U.S. Geological Survey. Thus began the long and fruitful association of Powell, Gilbert, and Clarence Edward Dutton. Powell's fecundity in ideas found its effective foil in Gilbert's accurate observation and suspended judgment.

Powell's doctrine of subaerial erosion and baselevel was developed by Gilbert with emphasis on lateral planation in *Report on the Geology of the Henry Mountains* (1877). Importing these ideas into humid regions, W. M. Davis formulated the concept of geographic cycles. Thus were founded the fundamental principles of a new subscience, geomorphology, although that name, introduced by Powell, did not come into general use until the 1930's.

According to W. W. Rubey (p. 497), the essence of Gilbert's ideas was the concept of graded streams—the concept that, either by cutting down their beds or by building them up with sediment, streams tend always to make for themselves channels and slopes that, over a period of years, will transport exactly the load of sediment delivered into them from above. From 1907 to 1909, at Berkeley, California, Gilbert gratified his desire for quantitative data by conducting flume experiments on the transportation of debris by running water. The 1905-1908 investigation of hydraulic-mining debris and its sedimentary effects in the Sacramento River drainage system and in San Francisco Bay is a dispassionate account of the great power of man as a geologic agent.

Gilbert explained the structure of the Great Basin as the result of extension. The individual "basin ranges" are the eroded upper parts of tilted blocks, which were displaced along faults as "comparatively rigid bodies of strata." Gilbert analyzed Powell's diastrophism into orogeny, or mountain formation, and epeirogeny, or regional displacement. He described Lake Bonneville, the gigantic Pleistocene ancestor of Great Salt Lake, and related the displaced Bonneville shorelines and the displaced proglacial shorelines of the Great Lakes to epeirogenic isostatic rebound.

Gilbert studied the Henry Mountains in 1875 and 1876. He was the first to establish that an intrusive body may deform its host rock. He emphasized that the crust of the earth is "as plastic in *great masses* as wax is in small." But he exaggerated the fluidity of magma, and his laccoliths are now interpreted as "tongue-shaped masses . . . injected radially as satellites from stocks" (Hunt *et al.,* p. 142).

Gilbert was the very antithesis of a temperamental, erratic genius. He was scrupulous in giving credit due others, careless about receiving his own. In his presidential address to the Society of American Naturalists in December 1885, entitled "The Inculcation of Scientific Method," he stressed the importance of inventing and testing multiple hypotheses, maintaining: "The great investigator is primarily the man who is rich in hypotheses."

From January 1889 until August 1892 Gilbert served as chief geologist of the U.S. Geological Survey. His brilliant presidential address to the Philosophical Society of Washington in December 1893 gave cogent argument for the impact origin of craters on the moon. In 1899 he was a member of the Harriman expedition to Alaska.

BIBLIOGRAPHY

I. ORIGINAL WORKS. The B. D. Wood and G. B. Cottle bibliography that follows the Mendenhall memorial (see below), pp. 45-64, has 400 entries and is nearly complete. Gilbert's writings include *Report on the Geology of the Henry Mountains,* U.S. Geographical and Geological Survey of the Rocky Mountain Region (Washington, D.C., 1877; 2nd ed., 1880); *Lake Bonneville,* Monographs of the United States Geological Survey, no. 1 (Washington, D.C., 1890); *The Transportation of Débris by Running Water,* United States Geological Survey Professional Paper no. 86 (Washington, D.C., 1914); *Hydraulic-Mining Débris in the Sierra Nevada,* Professional Paper no. 105 (Washington, D.C., 1917); and *Studies of Basin-Range Structure,* Professional Paper no. 153 (Washington, D.C., 1928).

II. SECONDARY LITERATURE. W. C. Mendenhall, "Memorial to Grove Karl Gilbert," in *Bulletin of the Geological Society of America,* **31** (31 Mar. 1920), 26-45, includes a portrait and lists Gilbert's honors and biographical notices. W. M. Davis, "Biographical Memoir. Grove Karl Gilbert, 1843-1918," in *Biographical Memoirs. National Academy of Sciences,* **21** (1926), 5th memoir, presented 1922, includes pictures of Gilbert. See also Joseph Barrell, "Grove Karl Gilbert, an Appreciation," in *Sierra Club Bulletin,* **10,** no. 4 (Jan. 1919), 397-399; and, in the same number, p. 438, a letter from E. C. Andrews, Australia, dated 30 Apr. 1913. The references cited in text are W. W. Rubey, "Equilibrium-Conditions in Debris-Laden Streams," in *Transactions of the American Geophysical Union,* **13** (June 1933), 497-505; and C. B. Hunt, assisted by P. Averitt and R. L. Miller, *Geology and Geography of the Henry Mountains Region,* United States Geological Survey Professional Paper no. 228 (Washington, D.C., 1953).

RONALD K. DEFORD

GILBERT, J. H. For an account of his work, see the biography of **Lawes, J. B.**

GILBERT, WILLIAM (*b.* Colchester, Essex, England, 1544; *d.* London, England, 30 November 1603), *magnetism, electricity.*

Gilbert was born into a rising, middle-class family that had only recently acquired its well-to-do status. His great-grandfather, John Gilbert, had married Joan Tricklove, the only daughter of a wealthy merchant from Clare, Suffolk. Their son, William Gilbert of Clare, became a weaver and eventually sewer of the chamber to Henry VIII. This William married Margery Grey, and among their nine children was Jerome Gilbert. Jerome, who had some knowledge of law, moved from Clare to Colchester in the 1520's, became a free burgess and recorder there, and married Elizabeth Coggeshall. The oldest of their five children was William Gilbert of Colchester. After Elizabeth's death Jerome married Jane Wingfield, and after her death he married a third time. Little is known of this third marriage other than that the woman's name was Margery.

Nothing is known of Gilbert's early life and education. In May 1558 he was matriculated as a member of St. John's College, Cambridge, where he received his A.B. in 1561, his M.A. in 1564, and his M.D. in 1569. During this time he was appointed pensioner (1558), fellow of Mr. Symson's foundation (1561), mathematical examiner of St. John's College (1565 and 1566), and senior bursar (1569 and 1570). Seven months after receiving his M.D. he was elected a senior fellow of his college.

Frequently writers have stated that after receiving his doctor's degree, Gilbert went abroad to study. This is possible, but evidence for it is lacking; nothing is known of his life from the time he left Cambridge until he settled in London sometime in the mid-1570's. There he practiced medicine, obtained a grant of arms in 1577, and sometime before 1581 became a member of the Royal College of Physicians. By 1581 Gilbert was one of the prominent physicians in London, and for the rest of his life he was consulted by influential members of the English nobility. In 1600 he became physician to Elizabeth I and after her death was appointed physician to James I.

While in London, Gilbert was active in the Royal College and held several offices in that organization. In 1588 he was one of the four College physicians requested by the Privy Council to care for the health of the men in the Royal Navy. In 1589 he was assigned the topic "Philulae" for the College's *Pharmacopoeia,* and later in that year and again in

1594 he was mentioned among the examiners for this book on drugs. From 1582 on, Gilbert was an officer of the College and held the following positions: censor (1582, 1584–1587, 1589–1590); treasurer (1587–1594, 1597–1599); consiliarium (1597–1599); elect (1596–1597). In 1600 he was elected president.

Gilbert never married. He lived in London at Wingfield House (presumably a legacy from his step-mother), St. Peter's Hill. The house served as a laboratory but probably was not, as is sometimes stated, a center for the meetings of a scientific group. Little is known about Gilbert's life in London, for upon his death, presumably from the plague, he left his books, instruments, globes, and minerals to the Royal College of Physicians for their library. The Royal College, with its library, and Wingfield House were destroyed by the Great Fire in 1666.

Sometime during his life, presumably in the decade following his years at Cambridge, Gilbert studied magnetic phenomena. The results of these studies were published in 1600 under the title *De magnete, magneticisque corporibus, et de magno magnete tellure; physiologia nova, plurimis & argumentis, & experimentis demonstrata.* The book was an attempt to explain the nature of the lodestone and to account for the five movements connected with magnetic phenomena. It was well received both in England and on the Continent and was republished in 1628 and again in 1633.

Gilbert's other writings were not published until almost half a century after his death. His younger half brother, William Gilbert of Melford, collected and possibly edited his brother's papers and presented them, under the title *De mundo nostro sublunari philosophia nova,* to Prince Henry of England. Francis Bacon and Thomas Harriot were both acquainted with these writings, which were published at Amsterdam in 1651. The *De mundo* is, in part, an extension of the cosmological ideas Gilbert introduced in the last section of the *De magnete;* it is dependent on the latter work for much of its vocabulary and basic assumptions but lacks its finish and completeness. It is Gilbert's work in the *De magnete* that gives him a place in the history of science.

During the fifteenth century the widespread interest in navigation had focused much attention on the compass. Since at that time the orientation of the magnetic needle was explained by an alignment of the magnetic poles with the poles of the celestial sphere, the diverse areas of geography, astronomy, and phenomena concerning the lodestone overlapped and were often intermingled. Navigators had noted the variation from the meridian and the dip of the magnetic needle and had suggested ways of accounting for and using these as aids in navigation. The connection between magnetic studies and astronomy was less definite; but so long as the orientation of the compass was associated with the celestial poles, the two studies were interdependent to some extent. There were suggestions in Thomas Digges and Nicolas Rymers that perhaps the magnetic property was somehow innate in the earth, but these were slight hints or passing remarks. Gilbert provided the only fully developed theory dealing with all five of the then known magnetic movements and the first comprehensive discussion of magnetism since the thirteenth-century *Letter on the Magnet* of Peter Peregrinus.

Gilbert divided his *De magnete* into six books. The first deals with the history of magnetism from the earliest legends about the lodestone to the facts and theories known to Gilbert's contemporaries. The nature, properties, and behavior of the lodestone are discussed and ways of demonstrating them are suggested. Throughout the book Gilbert marked his own discoveries and experiments with asterisks; larger symbols were used for the more important discoveries and experiments, smaller asterisks for the less important ones. In the last chapter of book I, Gilbert introduced his new basic idea which was to explain all terrestrial magnetic phenomena: his postulate that the earth is a giant lodestone and thus has magnetic properties. The assertion is supported by a comparison of the earth and a lodestone (each has poles and an equator; each draws objects to itself), by an appeal to experience (lodestones are found in all parts of the earth; iron, the prime magnetic substance, lies deep within the earth), by a denial of the Aristotelian elements (elemental earth has never been found), and by Gilbert's repeated statements that this new idea is so.

The remaining five books of the *De magnete* are concerned with the five magnetic movements: coition, direction, variation, declination, and revolution. Before he began his discussion of coition, however, Gilbert carefully distinguished the attraction due to the amber effect from that caused by the lodestone. This section, chapter 2 of book II, established the study of the amber effect as a discipline separate from that of magnetic phenomena, introduced the vocabulary of electrics, and is the basis for Gilbert's place in the history of electricity.

Gilbert's distinction between magnetic phenomena and the amber effect was based upon the difference between their causes. He used a material cause to explain the amber effect and a formal one for magnetic attraction. That is, substances which had been formed from the fluid and humid matter in the earth would, after they had become solid, behave as amber does when it is rubbed. Gilbert's explanation was that the fluidity was never completely lost, and thus such substances emitted an effluvium which seized small particles and pulled them inward. Magnetic materials were those substances which shared in the specific, primary form of the earth. This form, implanted in the globe by the Creator, gave the earth its magnetic property. All parts of the earth which maintained a principal share in this form, i.e., lodestones and iron, were magnetic bodies. Unlike the electrics, magnetics did not depend on the emission of effluvia to draw bodies to themselves.

Having distinguished the magnetic and amber effects, Gilbert presented a list of many substances other than amber which, when rubbed, exhibit the same effect. These he called electrics. All other solids were nonelectrics. To determine whether a substance was an electric, Gilbert devised a testing instrument, the versorium. This was a small, metallic needle so balanced that it easily turned about a vertical axis. The rubbed substance was brought near the versorium. If the needle turned, the substance was an electric; if the needle did not turn, the substance was a nonelectric.

After disposing of the amber effect, Gilbert returned to his study of the magnetic phenomena. In discussing these, Gilbert relied for his explanations on several assumptions: (1) the earth is a giant lodestone and has the magnetic property; (2) the magnetic property is due to the form of the substance; (3) every magnet is surrounded by an invisible orb of virtue which extends in all directions from it; (4) pieces of iron or other magnetic materials within this orb of virtue will be affected by and will affect the magnet within the orb of virtue; and (5) a small, spherical magnet resembles the earth and what can be demonstrated with it is applicable to the earth. This small spherical magnet he called a terrella.

Gilbert was more negative than positive in his assertions about the form which accounts for the magnetic property. He tried to distinguish it from the commonly accepted forms of his time. Thus he denied that this form was the formal cause of Aristotle's four causes, or the specific cause alchemists associated with mixtures, or the secondary form of the philosophers, or the propagator of generative bodies. Yet when he attempted to describe or define this form, he could assert only that there is a primary, radical, and astral form that is unique to each body and that orders its own proper globe; for the earth this form was the anima of the earth and was associated with the magnetic property.

One of the effects of this primary form was the surrounding of the magnetic body with an orb of virtue. The orb extended in all directions from the body, and its extent and strength depended on the perfection and purity of the magnetic body. Magnetics within the orb would be attracted to the body; those outside would be unaffected. Thus Gilbert used this orb to eliminate the necessity for attraction at a distance in regard to the magnet just as he had used the effluvium in his explanation of the amber effect. Yet he was unable or unwilling to account for the magnetic effect solely in terms of substance, as he had done for the electrics. The primary form, once accepted, went beyond a material explanation and so, for Gilbert, provided another essential difference between the two phenomena. Once this was established as a fundamental part of his theory, he was free to discuss the five magnetic movements.

In discussing coition Gilbert was careful to distinguish magnetic coition from other attractions. For him magnetic coition was a mutual action between the attracting body and the attracted body. At the beginning of the *De magnete* he explained several terms that were necessary for understanding his work. One of these was "magnetic coition," which he said he "used rather than attraction because magnetic movements do not result from attraction of one body alone but from the coming together of two bodies harmoniously (not the drawing of one by the other)" (P. Fleury Mottelay, *William Gilbert of Colchester . . . on the Great Magnet of the Earth* [Ann Arbor, 1893], p. liv).

The coition occurred only if the bodies were within the orb of virtue of the magnet, and the action was dependent upon the size and purity of the magnet and the object. Larger and purer magnets were stronger than smaller and less pure ones. The removal of part of a lodestone would weaken it, while the addition of an iron cap would strengthen the orb of virtue.

Book III of the *De magnete* contains Gilbert's explanation of the orientation taken by a lodestone that is balanced and free to turn, that is, the behavior of the magnetic compass. Since the earth was a giant lodestone, it was surrounded by an orb of virtue; and magnetic substances within this orb behaved as they did in the orbs of small magnets. Thus the orientation of the compass was simply an alignment of the magnetic needle with the north and south poles of the earth. Gilbert gave numerous demonstrations of this with the terrella as well as directions for magnetizing iron.

By the end of the sixteenth century, navigators were well acquainted with variations from the meridian in the orientation of the compass. Thus, after discussing orientation, Gilbert turned in book IV to the variations in that orientation. Here he again used the comparison of the phenomena that can be demonstrated with the terrella and those that occur on the surface of the globe. Just as a very small magnetic needle will vary its orientation if the terrella on which it is placed is not a perfect sphere, so will the compass needle vary its orientation on the surface of the earth according to the proximity or remoteness of the masses of earth extending beyond the basic spherical core. Also, the purity of these masses (the amount of primary magnetic property retained by them) will affect the orientation of the compass just as stronger lodestones have greater attractive powers than weaker ones.

The next magnetic movement that Gilbert discussed was declination, the variation from the horizontal. This phenomenon had been described by Robert Norman in his book on magnetism, *The New Attractive* (1581). Although Norman had also given an effective means of constructing the compass needle so that it would not dip but would remain parallel to the horizontal, he had made no attempt to account for this strange behavior. As with the other magnetic effects of the compass, Gilbert explained declination in terms of the magnetic property of the earth and the experiments with the terrella. The small needle placed on the terrella maintained a horizontal position only when placed on the equator. When moved north or south of this position, the end of the needle closer to one pole of the terrella dipped toward that pole. The amount of dip increased as the needle was moved nearer the pole, until it assumed a perpendicular position when placed on the pole. A compass on the earth, according to Gilbert, behaved in a similar manner.

In discussing the variations from the meridian and the horizontal, Gilbert suggested practical applications of his theory. Navigators of the period were concerned with determining the longitude and latitude of their positions on the open seas. Since the deviation from the meridian was constant at a given point, Gilbert thought that if the seamen would record these variations at many points, an accurate table of variation for various positions could be compiled and the problem would be solved. He included detailed instructions for the construction of the instruments necessary for this task.

Gilbert thought that the variations from the horizontal could be obtained by means of experimentation with the terrella, since the dip depended on the position of the needle between the equator and the pole rather than on the configuration of the surface of the magnet. He does not seem to have had much data from navigators in this regard, as he did concerning the variation from the meridian, and was satisfied with his theoretical considerations.

The final book of the *De magnete,* book VI, deals with rotation and in this section Gilbert expounded his cosmological theories. Without discussing whether the universe is heliocentric or geocentric, Gilbert accepted and explained the diurnal rotation of the earth. From the time of Peter Peregrinus' *Letter on the Magnet,* written in the thirteenth century, rotation had been considered one of the magnetic movements. The assumption was that a truly spherical, perfectly balanced lodestone, perfectly aligned with the celestial poles, would rotate on its axis once in twenty-four hours. Since the earth was such a lodestone, it would turn upon its axis in that manner and thus the diurnal motion of the earth was explained. The theory was taken from Peter's *Letter;* the application to the earth was Gilbert's addition.

In advancing his theory Gilbert denied the existence of the solid celestial spheres, stated that the fixed stars were not equally distant from the earth, and accounted for the precession of the equinoxes and the tides in terms of the magnetic property of the earth. These statements were in general weakly, if at all, supported and were not well developed. Much of the criticism directed by Bacon and others against Gilbert's writing was based upon the sixth book of the *De magnete,* where Gilbert extended to the cosmos his magnetic theory and the results obtained from his experiments.

Throughout the *De magnete,* Gilbert discussed and usually dismissed previous theories concerning magnetic phenomena and offered observational data and experiments which would support his own theories. Most of the experiments are so well described that the reader can duplicate them if he wishes, and the examples of natural occurrences which support his theory are well identified. Where new instruments are introduced (for example, the versorium, to be used in identifying electrics), directions for their construction and use are included. The combination, a new theory supported by confirming evidence and demonstrations, is a pre-Baconian example of the new experimental philosophy which became popular in the seventeenth century.

Gilbert's other writings, those in the *De mundo,* do not follow this pattern. When the younger William Gilbert collected his half brother's papers for presentation to Prince Henry and eventual publication, he divided them into two sections. The first of these, "Physiologiae nova contra Aristotelem," is an expansion of the cosmology of the *De magnete;* the second,

"Nova meteorologia contra Aristotelem," follows the general pattern of Aristotle's *Meteorology.* There is nothing in the two works to indicate that William Gilbert of Colchester considered the two to be one work. Internal evidence indicates that the "Nova meteorologia" was written during the 1580's and left unfinished, while the "Physiologia nova" must have been written after the early 1590's. Also, since it is assumed in the "Physiologia nova" that the reader is familiar with the content of the *De magnete,* it appears that much of this work was written after the major portion of the *De magnete* was completed.

At the beginning of the "Physiologia nova" Gilbert denied the existence of the four terrestrial elements—earth, water, air, and fire—and replaced them with one element, earth. This earth was the one substance from which all terrestrial bodies were made; its primary attribute was its magnetic property; solids not exhibiting this property were degenerate forms of the element; and moist fluid substances on the earth were effluvia of this basic element. The surface of the globe consisted primarily of these degenerate forms and effluvia. The lodestones and iron were purer forms of this one element.

Gilbert also postulated a similar structure for all of the heavenly globes. Each consisted of a substance with a primary form and effluvia surrounding it. He then placed a void between the effluvia from one globe and that from the next. While the general scheme was the same for all, each globe or type of globe had its own distinguishing characteristics. Gilbert gave a more detailed description of the moon than of any other celestial body. This "Companion of the Earth," as he called it, was described as a miniature earth and possessed seas, continents, and islands. These Gilbert named and charted; the lighter parts of the moon were assumed to be bodies of water, the darker parts land masses. Since the moon was within the orb of virtue of the earth, there was a mutual attraction between the two bodies. Because the earth was larger, it held the moon in its power and thus the moon revolved around it. The lesser effect of the moon on the earth was seen in the tides.

The sun was designated as the center for the orbits of the five wandering stars and was the cause of motion for all the globes within its orb of virtue. Although the earth was located within this orb, Gilbert excluded it from those bodies affected by the sun. There are indications that he considered the planets to be earthlike bodies with continents and seas, but this was never definitively stated. The fixed stars were placed in the same category as the sun, the light-giving bodies, while moon, planets, and earth belonged to the group of light-reflecting bodies. The causes for the differences were not stated.

From the structure of the earth and the other globes, Gilbert moved to the structure of the universe. While he repeated his belief, expressed in the *De magnete,* that the earth had a diurnal rotation and that the fixed stars were not all equally distant from the earth, and while he discussed the motions of the earth according to Copernicus and Giordano Bruno, he did not affirm or deny the heliocentric system. At times he dismissed the system as not pertinent to the topic he was discussing; at other times he indicated that it would be taken up in another place. Gilbert dismissed the third motion described in the *De revolutionibus* as "no motion." This treatment of the motions of the earth is only one of the many indications that the *De mundo* was left in a fragmentary state.

Scattered throughout the latter part of the "Physiologia nova" are references to and statements about spices, twilight, putrefaction, the polarity of magnets, the buoyancy of a leaden vessel, the comparative densities of solid and liquid forms of the same substance, and light. All of these are mere statements or partial discussions. It seems as if the younger William Gilbert included everything he found, regardless of its relevance to the rest of the papers or its internal completeness.

The second part of the *De mundo,* the "Nova meteorologia," contains Gilbert's discussion of comets, the Milky Way, clouds, winds, the rainbow, the origin of springs and rivers, and the nature of the sea and tides. Most of these are summaries of other theories with Gilbert's ideas interspersed among them.

Comets, Gilbert considered, could be either above or below the moon. They were wandering bodies without polarity and with uncertain paths. He listed the various positions the vapor of a comet might take but said nothing about the centers of comets or the causes of their motions.

Several theories and mythological explanations for the Milky Way were reviewed and denied by Gilbert; then the hypothesis that the Milky Way is a collection of stars so numerous and so far from the earth as to appear to be a mist or cloud was given but was neither accepted nor denied. The section ends with a suggestion that the reader look at the Milky Way through a "specillis." The instrument is not described, and it is uncertain whether Gilbert meant a lens, a small mirror, or some other instrument.

Clouds, in Gilbert's meteorology, were exhalations and effluvia from the earth which rose to varying heights according to the density of their content. Some methods were given for estimating the heights of clouds, but there was little new information in this section.

Like the clouds, winds were part of the effluvia of the earth. They were described as expanding and swollen exhalations which escaped from the interior of the earth in search of more room in the region above the earth. The specific properties of any wind were determined by the location and circumstances of its origin and also by the positions of various stars at the time. Gilbert mentioned a "Table of Winds" he had composed, but this table was missing from the manuscript used for the printed edition. The editor inserted the table from Francis Bacon's "Historia ventorum," but since Bacon had four secondary winds and Gilbert mentioned five, it is unlikely that the two tables were similar.

In the section on the rainbow Gilbert included elaborate drawings showing the position and colors of both primary and secondary rainbows and the conditions necessary for the rainbow to be visible; he discussed the necessity for moisture, a dense object, and proper conditions in the air for the rainbow and digressed a little on the subject of mirror images. Again all the explanations are general and incomplete.

The last part of the "Meteorologia" concerns water phenomena—springs, rivers, the sea, and tides. Gilbert considered water to be a humor from the earth, and the motions of springs and rivers were explained in terms of water returning to its source; the tides were the results of the combined actions of the diurnal motion of the earth and the magnetic coition between the earth and the moon.

Throughout the "Nova meteorologia" Gilbert included numerous examples of specific instances of the phenomena he was describing. Some of these observations were his own, some he had received from others. There is an indication of an interest in astrology and, as in the *De magnete,* a concern for observational data to support his ideas. It appears that at one time Gilbert planned a detailed study of meteorology that would replace the existing theories, but he never completed the project.

The *De mundo* did not have the influence of the *De magnete,* since Gilbert's cosmology was less acceptable than either his magnetic theory or his electric theory. Gilbert's contemporaries generally praised the earlier work both for its content and for its methodology, and the idea of the earth's magnetism was incorporated into arguments for the support of the Copernican theory. Johann Kepler tried to use Gilbert's magnetic theory, with its orb of virtue, as a motive force for his astronomical theory but needed so many *ad hoc* postulates to do so that others found this use of magnetism unacceptable. Kepler also expressed interest in seeing Gilbert's theory of the void in the *De mundo,* but we do not know whether he did so.

Certainly the *De magnete* was the far more influential of Gilbert's books. The theory of the magnetic orb of virtue and the explanation of the amber effect in terms of an emitted effluvium provided mechanistic explanations for these phenomena and a starting point for the study of the two disciplines in the following centuries.

BIBLIOGRAPHY

I. ORIGINAL WORKS. Gilbert's writings are *De magnete, magneticisque corporibus, et de magno magnete tellure; physiologia nova, plurimis & argumentis, & experimentis demonstrata* (London, 1600), Eng. trans., P. Fleury Mottelay, *William Gilbert of Colchester . . . on the Great Magnet of the Earth* (Ann Arbor, 1893); and *De mundo nostro sublunari philosophia nova,* collected by his half brother, William Gilbert of Melford (Amsterdam, 1651).

II. SECONDARY LITERATURE. On Gilbert or his work, see Suzanne Kelly, *The De mundo of William Gilbert* (Amsterdam, 1965); and Duane H. D. Roller, *The De magnete of William Gilbert* (Amsterdam, 1959).

SUZANNE KELLY

GILES (AEGIDIUS) OF LESSINES (*b.* Lessines [now Hainaut, Belgium], *ca.* 1235; *d.* 1304 or later), *astronomy, natural philosophy.*

Giles entered the Dominican order, possibly at the priory of Valenciennes, and in all likelihood studied under Albertus Magnus at Cologne and under Thomas Aquinas at Paris during the latter's second professorship there (1269–1272). He was one of the first to develop, not merely expound, Thomistic doctrine, particularly on the unicity of substantial form (*De unitate formae,* completed July 1278). His *De usuris,* written between 1278 and 1284, is the most complete study of usury in the Middle Ages. He composed a letter to Albert asking his judgment on fifteen points of doctrine, thirteen of which were condemned by Étienne Tempier, bishop of Paris, on 10 December 1270; this elicited a reply from Albert, the important *De quindecim problematibus.* Giles wrote also a number of theological treatises and *De concordia temporum,* a concordance of historical chronology that ends with the year 1304—from which is conjectured the date of his death.

Works of scientific interest include a lost treatise, *De geometria;* a work on twilights, *De crepusculis;* and the classic *De essentia, motu et significatione cometarum* ("On the Nature, Movement, and Significance of Comets"), occasioned by a comet that was seen from the latter part of July to early October 1264. The work depends heavily on Aristotle's *Meteorologia,* which Giles knew in a translation made from the Greek in 1260 by his fellow Dominican, William of Moerbeke, and which shows Giles's awareness of problems of textual criticism. Giles cites classical authors as well as al-Bitrūjī, Abu Maʿshar, Robert Grosseteste, and Albertus Magnus, among others; evidently "he had access to a remarkably extensive library" [1] and used his sources intelligently. The treatise is divided into ten chapters: the first seven, concerned with the nature, causes, and properties of comets, are astronomical and meteorological in intent; and the last three, dealing with the significance of comets, are chiefly astrological. The final chapter includes a history of comets and their sequels; apart from those of antiquity, only the comets of 840, 1062 or 1066, 1222, 1239, and 1264 are mentioned. Incidental details of the comet of 1264, based on Giles's own observations, were used by him to falsify theories proposed by others; they were sufficiently precise to enable Richard Dunthorne, working from a manuscript of Giles's treatise in 1751, to compute the orbit of the comet.[2] Both Giles and Dunthorne were assailed by "an enlightened eighteenth-century *philosophe,*" Alexandre Guy Pingré, whose own work is criticized by Thorndike.[3]

NOTES

1. Thorndike, p. 95.
2. *Philosophical Transactions of the Royal Society,* **47** (1753), 282.
3. Pp. 97–99.

BIBLIOGRAPHY

The Latin text of Giles's treatise on comets, together with a critical intro., is in L. Thorndike, *Latin Treatises on Comets Between 1238 and 1368 A.D.* (Chicago, 1950), pp. 87–184. F. J. Roensch, *Early Thomistic School* (Dubuque, Iowa, 1964), lists all of Giles's works, summarizes his philosophical thought, and provides a bibliographical guide.

William A. Wallace, O.P.

GILES (AEGIDIUS) OF ROME (*b.* Rome, Italy, before 1247; *d.* Avignon, France, 22 December 1316), *physics, astronomy, medicine.*

Often called, probably mistakenly, Giles Colonna, he joined the Hermits of St. Augustine while very young. He pursued his studies in Paris where he was a disciple of Thomas Aquinas and became *baccalarius sententiarius* in 1276. In March 1277 Étienne Tempier, the bishop of Paris, delivered his famous condemnation of Aristotelianism and Averroism. His teaching and writings having thus been censured, Giles was obliged to leave the city. He did not return for several years. In 1285 he received the *licentia docendi* in Paris at the request of Pope Honorius IV, after having retracted several of his theses. From 1285 to 1291 he taught theology.

The Hermits of St. Augustine revealed exceptional confidence in deciding as early as 1287 that his opinions should be admitted and upheld throughout the order, and they chose him for their general on 6 January 1292. On 25 April 1295 Giles was named archbishop of Bourges by Pope Boniface VIII. He died during a stay at the papal court at Avignon.

Although mainly a philosopher and theologian, Giles frequently dealt with problems relating to natural philosophy, notably in his commentaries on Aristotle. Moreover, he did so in a style distinctive enough to place him in the first rank of those thinkers who have made a positive contribution to the scientific thought of their time (see Maier, *Die Vorläufer Galileis,* p. 2). One of the first theses that Giles defended was the unity of substantial form, which he presented—without, however, daring to apply it to man—in his early commentary on the *De anima,* written before 1275. He returned to it in his *Theoremata de corpore Christi* and then, in 1278, in his *Contra gradus formarum.* But it is chiefly in his commentary on the *Physics,* written around 1277, that he considered scientific problems.

Among Giles's theses that have attracted the attention of more recent historians of science are those relating to quantity, which led him to admit the existence of natural *minima* below which concrete material substance cannot exist and which thus imply an atomistic theory of matter. The study of movement induced him to investigate the nature of a vacuum, to which he attributed a kind of suction force, observable with the aid of the clepsydra, the cupping glass, or the siphon. He arrived at a curious theory according to which only the resistance of the material medium, and not the distance traversed, enables movement to occur in time: movement in a vacuum would be *non in tempore.* His observations on the accelerated motion of falling bodies have similarly been noted: he observed that the speed of a freely falling body depends not on the proximity to its destination but on the traversed distance from its point of departure.

Several of these theories reappear in his later works, mainly in the *Quodlibeta* (1286–1291) and the *Expo-*

sitio and *Quaestiones* on the *De generatione et corruptione.* The latter two became classics, and were often utilized by such fourteenth-century physicists as Buridan and Marsilius of Inghen, who considered Giles the *communis expositor* of the *De generatione.*

Giles was also interested in other questions, which he often dealt with in short treatises that are difficult to date. Especially noteworthy are *De materia coeli,* which takes the position—against Aristotle, Thomas Aquinas, and the majority of contemporary scholars—that celestial matter is identical to that of the sublunary world; *De intentionibus in medio,* on the nature of light and its propagation; and *De formatione corporis humani in utero,* an embryological treatise inspired by Ibn Rushd. Giles developed his cosmological views to their fullest at the end of his career, in his commentary on the second book of the *Sententiae* of Peter Lombard and in his *In hexaemeron,* both finished during his episcopacy. Undoubtedly influenced by the censure of 1277, he admitted the possibility of a plurality of worlds. Furthermore, he renounced Aristotle's theory of homocentric spheres in favor of that of eccentrics and epicycles, inherited from Ptolemy and Simplicius.

BIBLIOGRAPHY

I. Original Works. It is impossible to give a complete list here of the very numerous works that Giles has left us. For that, one should consult the bibliographies by Lajard, Boffito, Glorieux, and Bruni (see below). Early eds. of the majority of Giles's writings have recently been reproduced (18 vols., Frankfurt am Main, 1964–1968). A list of modern eds. whose introductions are of the most interest or which concern Giles's scientific thought follows.

De erroribus philosophorum: P. Mandonnet, ed., in *Siger de Brabant,* 2 vols. (Louvain, 1908), II, 3–25; and J. Koch, ed., with English trans. by J. O. Riedl (Milwaukee, 1944); Koch text repr. (Milan, 1965).

De ecclesiastica potestate: G. Boffito, ed., with intro. by G. M. Oxilia (Florence, 1908); R. Scholz, ed. (Weimar, 1929; repr., Aalen, 1961).

Theoremata de esse et essentia: E. Hocedez, ed. (Louvain, 1930); English trans. by M. V. Murray (Milwaukee, 1953).

De plurificatione intellectus possibilis: H. Bullotta Barracco, ed. (Rome, 1957).

Quaestio de natura universalis: G. Bruni, ed., in *Collezione di testi filosofici inediti,* II (Naples, 1935).

Other, previously unpublished "Quaestiones" were published by Bruni, in *Analecta augustiniana,* **17** (1939), no. 1, 22–66; no. 2, 125–157; no. 3, 197–207, 229–245; and by V. Cilento, in *Medio evo monastico e scolastico* (Milan, 1961), pp. 359–377.

II. Secondary Literature. For his biography and literary and doctrinal history, see G. Bruni, *Le opere di Egidio Romano* (Florence, 1936); "Saggio bibliografico sulle opere stampate di Egidio Romano," in *Analecta augustiniana,* **24** (1961), 331–355; "Rari e inediti egidiani," in *Giornale critico della filosofia italiana,* **40** (1961), 310–323; P. Glorieux, *Répertoire des maîtres en théologie de Paris au XIIIe siècle,* 2 vols. (Paris, 1933–1934), II, 293–308; E. Hocedez, "Henri de Gand et Gilles de Rome," in *Richard de Middleton* (Louvain, 1925), pp. 459–477; "Gilles de Rome et saint Thomas," in *Mélanges Mandonnet,* 2 vols. (Paris, 1930), I, 385–410; "La condamnation de Gilles de Rome," in *Recherches de théologie ancienne et médiévale,* **4** (1932), 34–58; F. Lajard, "Gilles de Rome, religieux augustin théologien," in *Histoire littéraire de la France,* XXX (Paris, 1888), 421–566; J. S. Makaay, *Der Traktat des Ägidius Romanus über die Einzigkeit der substantiellen Form* (Würzburg, 1924); P. Mandonnet, "La carrière scolaire de Gilles de Rome," in *Revue des sciences philosophiques et théologiques,* **4** (1910), 481–499; N. Mattioli, *Studio critico sopra Egidio Romano Colonna* (Rome, 1896); and Z. K. Siemiatkowska, "Avant l'exil de Gilles de Rome, au sujet d'une dispute sur les theoremata de esse et essentia de Gilles de Rome," in *Mediaevalia philosophica Polonorum,* **7** (1960), 3–48.

Further information on Giles's MSS has been provided by F. Pelster, in *Scholastik,* **32** (1957), 247–255; and for Polish MSS by W. Sénko, in *Mediaevalia philosophica Polonorum,* **7** (1960), 22–24, and **11** (1963), 146–151; in *Augustiniana,* **12** (1962), 443–450; and by Z. K. Siemiatkowska, in *Mediaevalia philosophica Polonorum,* **11** (1963), 5–22.

Giles's scientific thought is discussed in P. Duhem, *Études sur Léonard de Vinci,* II (Paris, 1909); *Le système du monde,* IV, VI–X (Paris, 1954–1959); A. Maier, *Die Vorläufer Galileis im 14. Jahrhundert* (Rome, 1949); *Zwei Grundprobleme der scholastischen Naturphilosophie,* 2nd ed. (Rome, 1951); *An der Grenze von Scholastik und Naturwissenschaft,* 2nd ed. (Rome, 1952); *Metaphysische Hintergründe der spätscholastischen Naturphilosophie* (Rome, 1955); and G. Sarton, *Introduction to the History of Science,* II, pt. 2 (Baltimore, 1931), 922–926.

Jean Châtillon

GILL, DAVID (*b.* Aberdeen, Scotland, 12 June 1843; *d.* London, England, 24 January 1914), *astronomy.*

Her Majesty's astronomer at the Cape of Good Hope and one of the foremost practical astronomers of his generation, Gill was the eldest surviving son of a well-established watch- and clockmaker in the city of Aberdeen. His father intended that David should follow him in the family business and educated him accordingly. After two years at Marischal College of the University of Aberdeen, where he attended classes conducted by James Clerk Maxwell, Gill was sent for another two years to learn the fundamentals of clockmaking in Switzerland, Coventry, and Clerkenwell. There is no doubt that this somewhat informal training was extremely valuable to him in his later career, both for the experience and knowledge of fine mechanisms that he acquired and for the mastery of the French language and of business methods.

Gill duly succeeded his father and ran the business for ten years. His days were given to clocks but his evenings and other free time to social duties, to rifle shooting—in which he was in the national-championship class—and, in ever increasing amount, to astronomy. His first astronomical project was undertaken soon after his return to Aberdeen; this was the provision of a reliable time service for Aberdeen based on astronomical observations made with a portable transit instrument set up in a small observatory at King's College.

Spurred on by the success of his time service and his delight in making precise observations, he soon acquired an instrument that made possible a wider range of astronomical work—a twelve-inch reflector. His main objective in its use was the measurement of the parallaxes, or distances, of stars, with the micrometric method that Struve had employed to find the distance of Vega. This was, at the time, one of the most interesting and difficult problems of practical astronomy, and it is intriguing to imagine what Gill would have made of it. He never completed it with this instrument, however, because at about this time he received an invitation to become private astronomer to Lord Lindsay, an invitation which he accepted with alacrity although it involved giving up his business and consequently entailed a heavy financial sacrifice. When Gill later returned to the measurement of stellar parallaxes, it was with a heliometer, a far more powerful instrument for the purpose.

Lord Lindsay was building an observatory at Dun Echt, about a dozen miles from Aberdeen, and it was Gill's job to help with the planning and to supervise the building. This was no small task as Lord Lindsay desired to have an observatory second to none and furnished with the best instruments available. The Dun Echt observatory later became the nucleus of the Royal Observatory on Blackford Hill, Edinburgh, to which it was removed in 1894. The work of fitting out the observatory gave Gill the opportunity of meeting and becoming friendly with most of the leading European astronomers and instrument makers and of acquiring much firsthand experience of practical details that was to be invaluable to him in his later work at the Cape observatory. It was also at Dun Echt that he first encountered and mastered a heliometer, which was to become his own special instrument. Heliometers, although potentially very accurate, require great dexterity of hand and eye to operate, skills that Gill had highly developed through astronomical observation, shooting, and watchmaking.

Gill remained at Dun Echt from 1872 to 1876, during which time he went on an expedition to Mauritius, where, with Lord Lindsay and others, he observed the 1874 transit of Venus. The object of the many transit of Venus expeditions of that year was to determine the distance of the sun and the associated constants, which must be accurate if the data computed in nautical almanacs are to be reliable. The method used was that originally proposed by Halley—combining the observed times of transit of Venus across the face of the sun as observed from a number of places as widely scattered across the earth as possible. As the last such transit had occurred in 1769, that of 1874 was eagerly anticipated, and several nations prepared elaborate expeditions to observe it. While on Mauritius, Gill used the Dun Echt heliometer to observe a near approach of the minor planet Juno and was able from relatively few observations to deduce a value of the solar parallax that was fully as reliable as that deduced from all the elaborate transit-of-Venus expeditions put together. The method he used was to measure the parallactic displacement of Juno resulting from the diurnal movement of the place of observation between the early evening and the late morning. Such observations are most effective when the place of observation is close to the equator and the object observed is close to opposition.

This method of determining the solar parallax and the related astronomical constants was clearly worth pursuing. Thus, in 1877 Gill and his wife went on a private expedition sponsored by the Royal Astronomical Society to Ascension Island, where he spent six months observing a near approach of Mars. The instrument he used was the same one that he had used in Mauritius—the four-inch heliometer lent to him by Lord Lindsay for the purpose. Gill was living in London, working up the results of this expedition, when in 1879 he was appointed Her Majesty's astronomer at the Cape of Good Hope. Before sailing for the Cape he toured the major European observatories, renewing his acquaintance with most of the leading astronomers and laying the foundations for future cooperative schemes, particularly those to determine the solar parallax.

The Royal Observatory at Cape Town had been founded in 1820 and was intended to make observations in the southern hemisphere strictly comparable with those made at Greenwich. For this purpose it had been supplied with similar instruments, but by 1879 they were in a very poor state of repair and largely obsolete. Moreover, many observations had not been fully reduced or published.

Gill's first work at the Cape was to clear away these arrears of reduction and publication and to recondition the various instruments. He paid particular attention to the Airy transit circle, the twin of that on longitude zero, which had been installed at the Cape in 1855 and with which the meridian observations—the main work of the observatory—were made. This instrument was not reversible and therefore, in Gill's opinion, was not really suitable for the determination of fundamental star positions. Nevertheless, Gill improved it as much as he could and kept it in active use until near the end of his term of office, when it was replaced by a fine reversible transit circle constructed by Troughton and Simms to Gill's own design. In 1900 this design was revolutionary, but it was so good that it has provided the pattern for most of the transit circles that have since been made.

Gill closely supervised observations with the transit circle but did most of his personal observing on the heliometers. His first heliometer was the four-inch instrument that he had bought from Lord Lindsay, the second, a seven-inch one made by Repsold of Hamburg and installed in 1887. This was probably the best—and certainly the most widely used—heliometer ever made. With it, and in cooperation with a number of northern observatories, he determined the solar parallax by systematic observations of three minor planets, Iris, Victoria, and Sappho. The value he deduced (8.80 arc seconds) was used in the computation of all almanacs until 1968, when it was replaced by 8.794 arc seconds derived by radar echo methods and by observations of Mariner space probes. Gill also used the heliometers to measure the distances of a score of the brighter and nearer southern stars and obtained results of which the accuracy was later confirmed by photographic observation.

Photographs of the bright comet of 1882 drew Gill's attention to the possibility of accurately charting and measuring star positions by means of photography. The immediate outcome was the *Cape Photographic Durchmusterung,* which gives the approximate positions and brightness of nearly half a million southern stars and which was the first major astronomical work to be carried out photographically.

While work on the *Durchmusterung* was still in progress, Gill became involved with the Paris astronomers, to whom he had sent copies of the 1882 comet photographs, in the initiation of the *Carte du ciel* astrographic project. A much larger and more ambitious undertaking, it aimed at preparing a photographic chart of the whole heavens showing stars to the fourteenth magnitude and a catalog giving precise positions for all stars to the eleventh magnitude—that is, for over two million stars. This vast project was divided among a dozen observatories but was too big for many of them; thus the catalog for the whole sky was not completed until 1961. The Cape observatory undertook a major section of this work, and a suitable telescope, the astrographic refractor, was acquired and made ready for use by 1892; all the necessary plates had been obtained by 1900.

The astrographic refractor proved to be an extremely useful all-round instrument. The British engineer and astronomical amateur Frank McClean used it with an objective prism to obtain the spectra of the brighter southern stars. A few years later he offered as a gift to the observatory a very fine, large modern telescope, fully equipped with a powerful spectrograph, which was installed in 1901. The first major program completed with it was a determination of the solar parallax from the observed radial velocities of a series of stars near the ecliptic. These radial velocities differ throughout the year because of the earth's motion around the sun; the amplitude of the variation in velocity is directly connected to the linear size of the earth's orbit.

Apart from his astronomical work, Gill acted as the organizer of geodetic and boundary surveys throughout southern Africa as well as of projects to determine the longitude and latitude of its various ports. His most ambitious project was for a triangulation of the thirtieth meridian of east longitude from South Africa to Norway, a total arc of 105°, the longest observable meridian in the world. Gill made himself responsible for the southern end of this arc, the survey of which as a whole was not completed until after the end of World War II. He was knighted on 24 May 1900.

On account of his own health and that of his wife, Gill retired from the Cape at age 63, two years before he need have done. On his arrival in 1879 he had found a small, rather run-down, dispirited institution; when he left in 1906 the Cape observatory was generally recognized as one of the best equipped in the world with a large, young, keen staff fully engaged in important astronomical projects. Throughout his directorate there was a constant stream of visiting astronomers and volunteer assistants, some of whom stayed for lengthy periods, quite often as house guests. Among them were Auwers, Kapteyn, De Sitter, Elkin, Newcomb, Franklin Adams, Bryan Cookson, McClean, and Agnes Clarke. On his retirement Gill and his wife went to live in London, where they were able to keep in close touch with astronomers and scientists from all over the world. Gill enjoyed good health until December 1913, when he caught pneumonia, which proved fatal.

No account of Gill's astronomical work would be

complete without some reference to the man himself and the very high esteem in which he was held by his contemporaries. Sir Arthur Eddington, who knew him well during his retirement, wrote:

> By his widespread activity, his close association with all the great enterprises of observational astronomy, and by the energy and enthusiasm of his character, he had come to hold an almost unique position in astronomical counsels. . . . By his individual achievements and by his leadership he has exerted an incalculable influence on the progress of all that pertains to precision of observation. . . . Those who came into contact with him felt the charm of his personality. In some indefinable way he could inspire others with his enthusiasm and determination. Enjoying a life crowded with activity, surrounded by an unusually wide circle of friends, he was ever eager to encourage the humblest beginner. It was no perfunctory interest that he displayed. He was quick to discern any signs of promise, and no less outspoken in his criticism; but whether he praised or condemned, few could leave him without the truest admiration and affection for his simple-hearted character (obituary, *Monthly Notices of the Royal Astronomical Society,* **75** [1915], 236).

BIBLIOGRAPHY

I. ORIGINAL WORKS. Gill's *History and Description of the Cape Observatory* (London, 1913), written during his retirement, contains a full account of his work. A complete list of his numerous scientific papers and of the very many honors bestowed upon him will be found in the full-length biography by George Forbes, *David Gill, Man and Astronomer* (London, 1916), which is more concerned with the personal details of his life than with his astronomical work. In a long introduction to his wife's book (see below) Gill outlined the history and general methods of obtaining the solar parallax.

II. SECONDARY LITERATURE. Gill was fortunate in those who wrote his obituary notices. Of these the most outstanding are Frank Dyson, in *Proceedings of the Royal Society of London,* **91A** (1915), xxvi–xlii; Arthur Eddington, in *Monthly Notices of the Royal Astronomical Society,* **75** (1915), 236–247; and J. C. Kapteyn, in *Astrophysical Journal,* **40** (1914), 161–172.

A charming account of the Ascension expedition is given by Mrs. Gill, *Six Months in Ascension: An Unscientific Account of a Scientific Expedition* (London, 1878).

R. H. STOY

GILL, THEODORE NICHOLAS (*b.* New York, N.Y., 21 March 1837; *d.* Washington, D.C., 25 September 1914), *ichthyology.*

His youthful interest in the Fulton Fish Market led Gill to a life's work in fishes and other animals despite the preference of his father, James Darrell Gill, for his son to become a minister. His mother, Elizabeth Vosburgh Gill, died when the boy was nine. Visits to the market and interest in natural history continued even after Gill had begun studying law, and a scholarship from the Wagner Free Institute of Science in Philadelphia enabled him to continue his preferred interests. Through William Stimpson he was introduced to Spencer F. Baird, who arranged for the Smithsonian Institution to publish Gill's report on the fishes of New York when he was only nineteen (1856). Almost his only fieldwork was an expedition to the West Indies in 1858, when he made collections especially of the freshwater fishes of Trinidad.

Gill became librarian of the Smithsonian Institution in 1862; and when the books were given to the Library of Congress in 1866, he went with them as assistant librarian until 1874. From 1860 he held various appointments at Columbian College (now George Washington University), including that of professor of zoology from 1884 to 1910. The college recognized his merit by awarding him the M.A. (1865), M.D. (1866), Ph.D. (1870), and LL.D. (1895). He was elected to the National Academy of Sciences, was a fellow and president (1897) of the American Association for the Advancement of Science, a member of many other scientific societies, and a founder of the Cosmos Club.

A bachelor, Gill lived and studied in cluttered offices in the Smithsonian throughout most of his scientific career. One of Baird's close-knit coterie in the U.S. Fish Commission, he was an outstanding taxonomist and synthesizer of scientific literature. His classifications of fishes, based primarily on skeletal structure, were especially valuable at the family and order levels and formed a major basis for the classification adopted and promulgated by David Starr Jordan. Many of Gill's papers were brief and succinct analyses of the genera of fishes, group by group. He was less keen in the recognition and description of fish species. His publications on the habits and life histories of fishes brought together the scattered observations of many workers. His taxonomic studies on birds and on mollusks have been generally superseded. Gill was unusually generous with advice and knowledge to colleagues and visitors at the Smithsonian Institution.

BIBLIOGRAPHY

I. ORIGINAL WORKS. Gill's publications consisted of a very large number of relatively short papers (with no single extensive monograph) which constituted a major contribution, primarily to ichthyology. Dall's biography of Gill (cited below) contains an almost complete bibliography.

II. SECONDARY LITERATURE. A full account of Gill's life and accomplishments is W. H. Dall, "Biographical Memoir of Theodore Nicholas Gill," in *Biographical Memoirs. National Academy of Sciences,* **8** (1916), 313–343. The same account, without the bibliography, appeared in *Smithsonian Report for 1916* (Washington, D.C., 1917), pp. 579–586. Brief references to Gill's contributions to ichthyology are found in C. L. Hubbs, "History of Ichthyology in the United States After 1850," in *Copeia,* no. 1 (1964), 46–48; and in David Starr Jordan, *The Days of a Man,* vol. I (Yonkers, N.Y., 1922).

ELIZABETH NOBLE SHOR

GIORGI, GIOVANNI (*b.* Lucca, Italy, 27 November 1871; *d.* Castiglioncello, Italy, 19 August 1950), *electrical theory, electrical engineering, mathematics.*

Giorgi's father was an eminent jurist, who served as president of the Council of State and senator of the kingdom. From him Giorgi inherited a respect for scholarship and an austere way of life. Giorgi's dedication to the doctrines of physics and their applications began early and lasted throughout his life; his more than 350 publications include works on engineering, pure physics, mathematical physics, electricity, magnetism, natural sciences, chemistry, and philosophy.

Giorgi took the degree in civil engineering from the Institute of Technology in Rome when he was twenty-two; his most important technological achievements include projects in steam-generated electrical traction, innovations in urban trolley systems, and pioneering concepts in hydroelectric installations (integral utilization of rivers) and distribution networks (as, for example, the secondary three-phase network with the fourth wire, used by him for the first time in Rome's municipal installation). The work in large part coincided with his tenure as director of the Technology Office of the city of Rome from 1906 to 1923.

Giorgi's teaching activities further reflect the scope of his interests. From 1913 to 1927 he taught courses in the Physics and Mathematics Faculty of the University of Rome, at the School of Aeronautics, and in the School of Engineering; he was later titular professor of mathematical physics and, by annual contract, head of the department of rational mechanics at the universities of Cagliari and Palermo. From 1934 he was professor of electrical communications at the University of Rome and in 1939 he became associate professor at the Royal Institute of Higher Mathematics. In addition to teaching and practical engineering, he did original scientific work (particularly in mathematics) and wrote popular treatments of scientific and technological subjects.

Giorgi's chief fame, however, arises from his concept of a new absolute system of measurement to be simultaneously applicable to all electrical, magnetic, and mechanical units. In a letter to the English periodical *Electrician,* dated 28 March 1895 and published in April 1896, Giorgi took issue with the French physicist Alfred Cornu about the rationality of retaining the c.g.s. system of Wilhelm Weber and the English physicists, standardized in 1898. Giorgi held that the system, whose basic energy unit was the erg—one gram cm./sec.^{2}, or one centimeterdyne —was ill-adapted to current physics, given the connection between electrical and magnetic phenomena long since revealed by the researches of Oersted and Ampère.

Giorgi then devoted considerable time to the systematization of electrical units, and on 13 October 1901 presented to a meeting of the Italian Electrical Engineering Association a report entitled "Unità razionali di elettromagnetismo"—the cornerstone of his subsequent work. In this paper he proposed a consistent measurement system based on the meter, the kilogram, and the mean solar second (and hence called the M.K.S. system, as well as the Giorgi International System). The Giorgi system, of which the basic energy unit is the joule (one kg. meter^2/sec.^{2}, or one meternewton), is adaptable to electrical, magnetic, and mechanical units; is entirely composed of the standard units of mechanics; and requires no conversion factors since it is applicable to both electrostatic and electromagnetic systems. It therefore offers fewer irrationalities and greater convenience than the c.g.s. system because of its establishment of a single basic unit of appropriate size for each application.

Giorgi's proposals were supported by Silvanus Thompson in England, Fritz Emde in Germany, and the U.S. Bureau of Standards, among others, but it was not until June 1935 that the plenary session of the International Electrical Engineering Commission, meeting in Scheveningen, Netherlands, and in Brussels, unanimously recommended the adoption of the new system of units to supersede the c.g.s. system. In October 1960 the General Conference of Weights and Measures confirmed the International System, based on the meter, the kilogram, and the second, as well as the ampere, kelvin, and candle. It is interesting to note that Giorgi himself had proposed the ohm or some other such unit as a fourth standard.

A half century thus elapsed between Giorgi's letter to *Electrician* and the final adoption of a system based upon his principles. The Giorgi system is the clear manifestation of his versatility and of his abilities as a synthesizer.

BIBLIOGRAPHY

I. Original Works. Giorgi reprinted several of his works that are most important to reforms in the study of electrical engineering and the system of units—together with biographical data and a bibliography to December 1948 that lists more than 300 publications—in *Verso l'elettrotecnica moderna* (Milan, 1949).

His textbooks include *Lezioni di costruzioni elettromeccaniche* (Rome, 1905); *Lezioni di meccanica generale* (*superiore*) (Rome, 1914); *Lezioni di fisica matematica* (*elettricità e magnetismo*) (Cagliari, 1926); *Lezioni di fisica matematica* (Rome, 1927); *Compendio delle lezioni di meccanica razionale* (Rome, 1928); *Lezioni di meccanica razionale,* 2 vols. (Rome, 1931–1934); *Lezioni del corso di communicazioni elettriche* (Rome, 1934–1939); *Meccanica razionale* (Rome, 1946); *Compendio di storia delle matematiche* (Turin, 1948); *Aritmetica per scuole medie* (Rome, 1948); and *Verso l'elettrotecnica moderna* (Milan, 1949).

His scientific popularizations include *Le ferrovie a trazione elettrica* (Bologna, 1905); *Che cos'è l'elettricità?,* no. 8 in Collezione Omnia (Rome, 1928); *Metrologia elettrotecnica antica e nuova* (Milan, 1937), a repr. of three arts. published in *Energia elettrica,* **14,** nos. 3–5 (Mar.-May, 1937); *L'etere e la luce* (*dall'etere cosmico alle moderne teorie della luce*), no. 32 in Collezione Omnia (Rome, 1939); and *La frantumazione dell'atomo* (Rome, 1946).

In addition, Giorgi wrote 37 papers on the new system of measurement, 32 on machinery and electrical installations, 14 on electrical traction, 45 on general electrical engineering, 93 on theories in mathematics and mathematical physics, 19 on the history of science, and 27 articles for the *Enciclopedia italiana Treccani.*

II. Secondary Literature. A very good summary of Giorgi's life and work was given by Basilio Focaccia at the commemoration ceremony in Rome on 26 April 1951 and was published in *Elettrotecnica,* **38** (1951).

Mario Loria

GIRARD, ALBERT (*b.* St. Mihiel, France, 1595; *d.* Leiden, Netherlands, 8 December 1632), *mathematics.*

Girard's birthplace is fixed only by the adjective *Samielois* that he often added to his name, an adjective the printers of St. Mihiel often applied to themselves in the seventeenth century. The city belonged at that time to the duchy of Lorraine. The exact date of Girard's birth is subject to dispute. That of his death is known from a note in the *Journal* of Constantijn Huygens for 9 December 1632. The place of death is only conjectured.

Girard was undoubtedly a member of the Reformed church, for in a polemic against Honorat du Meynier he accused the latter of injuring "those of the Reformed religion by calling them heretics." This explains why he settled—at an unknown date—in the Netherlands, the situation of Protestants being very precarious in Lorraine.

The respectful and laudatory tone in which he speaks of Willebrord Snell in his *Trigonometry* leads one to suppose that Girard studied at Leiden. According to Johann Friedrich Gronovius, in his *éloge* of Jacob Golius, in 1616 Girard engaged in scientific correspondence with Golius, then twenty years old.

When Golius succeeded Snell at Leiden in 1629, Constantijn Huygens wrote to him to praise the knowledge of Girard (*vir stupendus*), particularly in the study of refraction. On 21 July of the same year Pierre Gassendi wrote from Brussels to Nicholas de Peiresc that he had dined at the camp before Bois-le-Roi with ". . . Albert Girard, an engineer now at the camp." We thus know definitely that Girard was an engineer in the army of Frederick Henry of Nassau, prince of Orange; yet the only title that he gives himself in his works is that of mathematician.

The end of Girard's life was difficult. He complains, in his posthumously published edition of the works of Stevin, of living in a foreign country, without a patron, ruined, and burdened with a large family. His widow, in the dedication of this work, is more precise. She is poor, with eleven orphans to whom their father has left only his reputation of having faithfully served and having spent all his time on research on the most noble secrets of mathematics.

Girard's works include a translation from Flemish into French of Henry Hondius' treatise on fortifications (1625) and editions of the mathematical works of Samuel Marolois (1627–1630), of the *Arithmetic* of Simon Stevin (1625), and of Stevin's works (1634). He also prepared sine tables and a succinct treatise on trigonometry (1626; 1627; 2nd ed., 1629) and published a theoretical work, *Invention nouvelle en l'algèbre* (1629). Although in the preface to the trigonometric tables (1626) he promised that he would very soon present studies inspired by Pappus of Alexandria (plane and solid loci, inclinations, and determinations), no such work on these matters appeared. Likewise, his restoration of Euclid's porisms, which he stated he "hopes to present, having reinvented them," never appeared.

Contributions to the mathematical sciences are scattered throughout Girard's writings. It should be said at the outset that, always pressed for time and generally lacking space, he was very stingy with words and still more so with demonstrations; thus, he very often suggested more than he demonstrated. His notations were, in general, those of Stevin and François Viète, "who surpasses all his predecessors in algebra." He improved Stevin's writing of the radicals by proposing that the cube root be written not as $\sqrt{\textcircled{3}}$ but as $\sqrt[3]{\ }$ (*Invention nouvelle,* 1629) but, like Stevin, favored fractional exponents. He had his own symbols for $>$ and $<$, and in trigonometry he was one of the first to utilize incidentally—in several very clear tables—the abbreviations sin, tan, and sec for sine, tangent, and secant.

In spherical trigonometry, following Viète and like Willebrord Snell, but less clearly than Snell, Girard made use of the supplementary triangle. In geometry he generalized the concept of the plane polygon, distinguishing three types of quadrilaterals, eleven types of pentagons, and sixty-nine (there are seventy) types of hexagons (*Trigonométrie,* 1626). With the sides of a convex quadrilateral inscribed in a circle one can construct two other quadrilaterals inscribed in the same circle. Their six diagonals are equal in pairs. Girard declared that these quadrilaterals have an area equal to the product of the three distinct diagonals divided by twice the diameter of the circle.

Girard was the first to state publicly that the area of a spherical triangle is proportional to its spherical excess (*Invention nouvelle*). This theorem, stemming from the optical tradition of Witelo, was probably known by Regiomontanus and definitely known by Thomas Harriot—who, however, did not divulge it. Girard gave a proof of it that did not fully satisfy him and that he termed "a probable conclusion." It was Bonaventura Cavalieri who furnished, independently, a better-founded demonstration (1632).

In arithmetic Girard took up Nicolas Chuquet's expressions "million," "billion," "trillion," and so on. He "explains radicals extremely close to certain numbers, such that if one attempted the same things with other numbers, it would not be without greatly increasing the number of characters" (*Arithmétique de . . . Stevin,* 1625). He gave, among various examples, Fibonacci's series, the values 577/408 and 1393/985 for $\sqrt{2}$, and an approximation of $\sqrt{10}$. One should see in these an anticipation of continuous fractions. They are also similar to the approximation 355/113 obtained for π by Valentin Otho (1573) and by Adriaan Anthoniszoon (1586) and to the contemporary writings of Daniel Schwenter.

In the theory of numbers Girard translated books V and VI of Diophantus from Latin into French (*Arithmétique de . . . Stevin*). For this work he knew and utilized not only Guilielmus Xylander's edition, as Stevin had for the first four books, but also that of Claude Gaspar Bachet de Méziriac (1621), which he cited several times. He gave fourteen right triangles in whole numbers whose sides differ from unity. For the largest the sides are on the order of 3×10^{10} (*ibid.,* p. 629).

Girard stated the whole numbers that are sums of two squares and declared that certain numbers, such as seven, fifteen, and thirty-nine, are not decomposable into three squares; but he affirmed, as did Bachet, that all of them are decomposable into four squares (*ibid.,* p. 662). The first demonstration of this theorem was provided by Joseph Lagrange (1772). Girard also contributed to problems concerning sums of cubes by improving one of Viète's techniques (*ibid.,* p. 676).

In algebra, as in the theory of numbers, Girard showed himself to be a brilliant disciple of Viète, whose "specious logistic" he often employed but called "literal algebra." In his study of incommensurables Girard generally followed Stevin and the tradition of book X of Euclid, but he gave a very clear rule for the extraction of the cube root of binomials. It was an improvement on the method of Rafael Bombelli and was, in turn, surpassed by that which Descartes formulated in 1640 (*Invention nouvelle*).

Unlike Harriot and Descartes, Girard never wrote an equation in which the second member was zero. He particularly favored the "alternating order," in which the monomials, in order of decreasing degree, are alternately in the first member and the second member. That permitted him to express, without any difficulty with signs, the relations between the coefficients and the roots. In this regard he stated, after Peter Roth (1608) and before Descartes (1637), the fundamental algebraic theorem: "Every equation in algebra has as many solutions as the denominator of its largest quantity" (1629).

A restriction immediately follows this statement, but it is annulled soon after by the introduction of solutions which are "enveloped like those which have $\sqrt{-}$." From this point of view, Girard hardly surpassed Bombelli, his rare examples treating only equations of the third and fourth degrees. For him the introduction of imaginary roots was essentially for the generality and elegance of the formulas. In addition, Girard gave the expression for the sums of squares, cubes, and fourth powers of roots as a function of the coefficients (Newton's formulas).

Above all, Girard thoroughly studied cubic and biquadratic equations. He knew how to form the discriminant of the equations $x^3 = px + q$, $x^3 = px^2 + q$, and $x^4 = px^3 + q$. These are examples of the "determinations" that he had promised in 1626. The first equation is of the type solved by Niccolò Tartaglia and Girolamo Cardano, the second relates to book II of the *Sphere* of Archimedes, and the third to Plato's problem in the *Meno.* With the aid of trigonometric tables Girard solved equations of the third degree having three real roots. For those having only one root he indicated, beside Cardano's rules, an elegant method of numerical solution by means of trigonometric tables and iteration. He constructed equations of the first type geometrically by reducing them, as Viète did, to the trisection of an arc of a circle. This trisection was carried out by using a hyperbola, as Pappus had done. The figure then made evident the three roots of the equation.

Girard was the first to point out the geometric significance of the negative numbers: "The negative solution is explained in geometry by moving backward, and the minus sign moves back when the + advances." To illustrate this affirmation he took from Pappus a problem of intercalation that Descartes later treated in an entirely different spirit (1637). This problem led him to an equation of the fourth degree. The numerical case that he had chosen admitted two positive roots and two negative roots; he made the latter explicit and showed their significance.

BIBLIOGRAPHY

I. ORIGINAL WORKS. Girard's two books are *Tables des sinus, tangentes et sécantes selon le raid de 100,000 parties* . . . (The Hague, 1626; 1627; 2nd ed., 1629), which also appeared in Flemish but had the Latin title *Tabulae sinuum tangentium et secantium ad radium 100,000* (The Hague, 1626; 1629); and *Invention nouvelle en l'algèbre* (Amsterdam, 1629; repr. Leiden, 1884). The repr. of the latter, by D. Bierens de Haan, is a faithful facs., except for the notation of the exponents, in which parentheses are substituted for the circles used by Girard and Stevin. However, the parentheses had been used by Girard in the *Tables.*

Girard was also responsible for trans. and eds. of works by others: *Oeuvres de Henry Hondius* (The Hague, 1625), which he translated from the Flemish; Samuel Marolois's *Fortification ou architecture militaire* (Amsterdam, 1627), which he enlarged and revised, and also issued in Flemish as *Samuel Maroloys, Fortification* . . . (Amsterdam, 1627), and *Géométrie contenant la théorie et practique d'icelle, necessaire à la fortification* . . ., 2 vols. (Amsterdam, 1627–1628; 1629), which he revised and also issued in Flemish as *Opera mathematica ofte wis-konstige, Wercken . . . beschreven door Sam. Marolois* . . . (Amsterdam, 1630); and Simon Stevin's *L'arithmétique* (Leiden, 1625), which he revised and enlarged, and *Les oeuvres mathématiques de Simon Stevin* (Leiden, 1634), also revised and enlarged.

II. SECONDARY LITERATURE. On Girard or his work, see several articles by Henri Bosmans in *Mathesis,* **40** and **41** (1926); Antonio Favaro, "Notizie storiche sulle frazioni continue," in *Bullettino di bibliografia e di storia delle scienze matematiche e fisiche,* **7** (1874), 533–596, see 559–565; Gino Loria, *Storia delle matematiche,* 2nd ed. (Milan, 1950), pp. 439–444; Georges Maupin, "Étude sur les annotations jointes par Albert Girard Samielois aux oeuvres mathématiques de Simon Stevin de Bruges," in *Opinions et curiosités touchant le mathématique,* II (Paris, 1902), 159–325; Paul Tannery, "Albert Girard de Saint-Mihiel," in *Bulletin des sciences mathématiques et astronomiques,* 2nd ser., **7** (1883), 358–360, also in Tannery's *Mémoires scientifiques,* VI (Paris, 1926), 19–22; and G. A. Vosterman van Oijen, "Quelques arpenteurs hollandais de la fin du XVIème et du commencement du XVIIème siècle," in *Bullettino di bibliografia e di storia delle scienze matematiche e fisiche,* **3** (1870), 323–376, see 359–362.

See also *Nieuw Nederlandsch Woordenboek,* II (1912), cols. 477–481.

JEAN ITARD

GIRARD, PIERRE-SIMON (*b.* Caen, France, 4 November 1765; *d.* Paris, France, 30 November 1836), *hydraulic engineering.*

Educated at Caen, Girard was admitted to the École des Ponts et Chaussées at the age of twenty-one and in 1789 was appointed to the grade of engineer in the Corps des Ponts et Chaussées. His first subject of investigation, to which he later returned, was the strength of wood as a structural material, yet he won the 1790 competition of the Académie des Sciences on the theory and practice of canal and harbor lock construction. In 1798 he was among the scientific experts in many fields called to take part in Napoleon's expedition to Egypt, where he remained until 1803, after the last troops had left. At first assigned to the port of Alexandria, he soon undertook an extensive study of the surface elevation and bed characteristics of the Nile; this study eventually broadened to cover material on Egypt's agriculture, commerce, and industry, all to be included in the comprehensive report on the expedition, of which he was one of eight authors.

Upon Girard's return to France, Napoleon appointed him director of the Paris water supply, with the special task of connecting the Seine and Ourcq rivers with a ship canal to serve the capital. This led him to study the resistance of the flow of water through pipes and open channels, the most essential contribution of which was the attention that it called to an important analysis, buried since 1768 in the files of the Corps des Ponts et Chaussées, by Antoine de Chézy. The first barges reached Paris from the Ourcq in 1813, but the overthrow of Napoleon and the restoration of the monarchy delayed completion of the 100-kilometer canal until 1820. Girard's account of the project, including the causes of certain objectionable effects on the groundwater level in urban districts, is to be found in his major treatise, *Mémoire sur le canal de l'Ourcq* . . ., a two-volume work plus atlas (1831–1843).

Girard was elected to the first class of the Institut National des Sciences et des Arts in 1815—this class became the Académie des Sciences of the Institute the following year—and served as its president in 1830. Despite his having rallied to Napoleon during the Hundred Days, Girard retained his post as water commissioner until 1831, and in the latter period he was promoted to the grade of *officier* in the Legion of Honor. The day of his death, 30 November, is given incorrectly in several references.

BIBLIOGRAPHY

A complete list of Girard's writings is in *Nouvelle biographie générale* (see below), which refers to a projected collection of his works, but no trace of the latter can be found. His major work is *Mémoire sur le canal de l'Ourcq* . . ., 2 vols. plus atlas (Paris, 1831–1843).

Biographies are in *Nouvelle biographie générale,* XX (Paris, 1857), 661–668; and *Grand dictionnaire universel du XIXe siècle,* VIII (Paris, 1872), 1268. See also Charles Richet (Girard's great-grandson), "Pierre-Simon Girard . . .," in *Comptes rendus de l'Académie des sciences,* **197** (11 Dec. 1933), 1481–1486.

HUNTER ROUSE

GIRAUD-SOULAVIE, J. L. See **Soulavie, J. L. Giraud.**

GIRTANNER, CHRISTOPH (*b.* St. Gall, Switzerland, 7 December 1760; *d.* Göttingen, Germany, 17 May 1800), *medicine, chemistry.*

Girtanner's father, Hieronymus, was a banker; his mother, Barbara Felicitas, was the daughter of the burgomaster Christoph Wegelin. He studied first at Lausanne and then at Göttingen where he obtained his doctorate in 1782 with a thesis on chalk, quicklime, and the matter of fire. He studied pediatrics at St. Gall, and then visited Paris, Edinburgh, and London before returning to Göttingen in 1787, the year he met Georg Lichtenberg. After further travels in 1788–1789, he settled in Göttingen, and in 1793 became a privy councillor to the duke of Saxe-Coburg. In 1790 he had married Catherine Maria Erdmann; their two sons both became naturalists. Girtanner was of a contentious disposition and published antirevolutionary works.

Girtanner was attracted by the Brunonian theory and studied Lavoisier's work on oxygen which he believed might be the principle of irritability. In 1790 he suggested this possibility in Rozier's *Observations sur la physique* and was accused of plagiarizing John Brown. He then wrote critical expositions of the views of Brown and of Erasmus Darwin. Meanwhile he had also published a book on pediatrics and another on venereal disease, arguing forcibly for the American origin of syphilis.

Girtanner was an early convert to Lavoisier's doctrines, and in 1791 published the first German version of the new chemical nomenclature. But his term for nonacidic oxides, *Halbsäure,* proved unacceptable, and his scheme for distinguishing such acids as sulfuric and sulfurous was unsuccessful. In 1792 he published his *Anfangsgründe der antiphlogistischen Chemie,* a textbook modeled upon Lavoisier's, which saw three editions and was used by Berzelius. According to Lavoisier, muriatic acid (hydrogen chloride) must, like all acids, be an oxide. Girtanner thought he had proved it to be an oxide of hydrogen, and nitrogen another oxide which could be prepared from steam. But unlike contemporary Germans who believed that water was thus proved the basis of all gases, he refused to accept that his experiments entailed a return to the phlogiston principle.

BIBLIOGRAPHY

I. ORIGINAL WORKS. Girtanner's writings include *Dissertatio inauguralis chemica de Terra Calcarea cruda et calcinata* (Göttingen, 1782); *Abhandlung über die venerische Krankheit,* 3 vols. (Göttingen, 1788–1789), in which the first vol. is practical, the others bibliographical; "Mémoires sur l'irritabilité," in *Observations sur la physique,* **36** (1790), 422; **37** (1790), 139; *Neue chemische Nomenklatur für die Deutsche Sprache* (Berlin, 1791); *Anfangsgründe der antiphlogistischen Chemie* (Berlin, 1792; 2nd ed., 1795; 3rd ed., 1801); *Abhandlung über die Krankheiten der Kinder* . . . (Berlin, 1794), Italian trans., 2 vols. (Genoa, 1801); *Ueber das Kantische Prinzip für die Naturgeschichte* . . . (Göttingen, 1796); *Ausfürliche Darstellung des Brownischen Systemes der praktischen Heilkunde* . . ., 2 vols. (Göttingen, 1797–1798); Russian trans., *Iogona brovno sistema,* 3 vols. (St. Petersburg, 1806–1807); *Ausfürliche Darstellung des Darwinischen Systemes der praktischen Heilkunde* . . ., 2 vols. (Göttingen, 1799); "Sur l'analyse de l'azote," in *Annales de chimie,* **33** (1799), 229–231; **36** (1800), 3–40; a trans. is in *Philosophical Magazine,* **6** (1800), 152–153, 216–217, 335–354.

II. SECONDARY LITERATURE. For works about Girtanner, see M. P. Crosland, *Historical Studies in the Language of Chemistry* (London, 1962), pp. 207–210; G. W. A. Kahlbaum and A. Hoffman, *Die Einführung der Lavoisierischen Theorie im Besonderen in Deutschland* (Leipzig, 1897); *Neue deutsche Biographie,* VI (Berlin, 1964), 411–412; and J. R. Partington, *A History of Chemistry,* III (London, 1962), 589–590.

DAVID M. KNIGHT

GLADSTONE, JOHN HALL (*b.* London, England, 7 March 1827; *d.* London, 6 October 1902), *chemistry.*

Financially independent for the latter part of his life, Gladstone devoted much time to research as well as to philanthropic and religious work; in science he is best-known for his application of optical phenomena to chemical problems.

His father, John Gladstone, a junior partner in the firm of Cook and Gladstone, wholesale drapers, married a cousin, Alison Hall, whose father also owned a drapery business. John Hall was the eldest of their three sons. The boys were all educated at home under tutors and showed an early interest in natural science. At seventeen Gladstone wished to enter the Christian ministry but was dissuaded and entered University College, London, where he attended Thomas Graham's lectures and worked in his laboratory. He gained a gold medal for original research and in 1847 went to work under Justus Liebig at Giessen, from which he graduated Ph.D. He returned to London in 1848 and in 1850 became lecturer in chemistry at St. Thomas' Hospital, where he stayed for two years. In 1853 Gladstone was elected fellow of the Royal Society, and from 1874 to 1877 he was Fullerian professor of chemistry at the Royal Institution. He was a founder member of the Physical Society and its first president (1874–1876), and president of the Chemical Society from 1877 to 1879. Gladstone was married twice: in 1852 to May Tilt—she and their only son died in 1864—and in 1869 to Margaret King (niece of Lord Kelvin), who died in 1870, leaving a daughter.

In an early paper (1853) Gladstone arranged all the known elements in the order of their "atomic weights" (actually, equivalents), thus anticipating John Newlands and others in pointing out certain peculiarities and drawing attention to some surprising relationships existing between the atomic weights of related elements, including some relationships observed earlier by Johann Döbereiner. He did not mention Döbereiner, but acknowledged the work of Leopold Gmelin who, in 1843, had drawn attention to and enlarged upon Döbereiner's observations. (Gladstone's contribution to the evolution of the periodic table is assessed by J. W. van Spronsen in *The Periodic System of Chemical Elements* [Amsterdam–London–New York, 1969], pp. 76–78 and *passim.*)

In 1855 he carried out the first quantitative investigation of equilibria in homogeneous systems, particularly using solutions of various ferric salts and thiocyanates, choosing these reactions because of the red color of the ferric thiocyanate thus formed. (Gladstone's results were later examined mathematically by E. J. Mills, on the basis of the law of mass action—see *Philosophical Magazine,* **47** [1874], 241–247.) Since the reaction never went to completion in any one direction, the inadequacy of prevailing ideas on chemical affinity was demonstrated.

Gladstone's important pioneering work on refractivity, in collaboration with T. P. Dale, began in 1858 with the measurement of the decrease in the refractive indexes of a number of liquids with increase in temperature. Observations of this in connection with accompanying changes of density subsequently led them to the formulation of what they called the specific refractive energy (now called specific refractive index—$\frac{n-1}{d}$, where n is the refractive index and d is the density), which they found to be approximately constant for a given liquid. Hans Landolt termed the product of this and the atomic weight of an element the refraction equivalent, and Gladstone subsequently measured it for a number of elements, finding it to be additive in compounds. (Refractivity is both additive and constitutive and has subsequently been of importance in organic analysis, particularly to resolve structure.)

In a series of researches with Alfred Tribe the copper-zinc couple was introduced and used in a number of organic preparations. His work on essential oils in his refractivity experiments led Gladstone to analyze them, and he discovered a number of terpenes.

Gladstone was deeply involved in a number of religious movements (particularly the Y. M.C.A.) and in educational reform.

BIBLIOGRAPHY

I. ORIGINAL WORKS. About 200 papers by Gladstone (of which approximately one-third are collaborative) are listed in the Royal Society, *Catalogue of Scientific Papers,* II (London, 1868), 909–911; VII (London, 1877), 783–784; X (London, 1894), 2–3; XV (Cambridge, 1916), 327–328. The lists are not entirely reliable and contain a few duplications. The papers mentioned in the text are "On the Relations Between the Atomic Weights of Analogous Elements," in *Philosophical Magazine,* **5** (1853), 313–320 (not listed in the above); "On Circumstances Modifying the Action of Chemical Affinity," in *Philosophical Transactions of the Royal Society,* **145** (1855), 179–223—see also "Some Experiments Illustrative of the Reciprocal Decomposition of Salts," in *Journal of the Chemical Society,* **9** (1857), 144–156, and "Additional Notes on Reciprocal Decomposition Among Salts in Solution," *ibid.,* **15** (1862), 302–311; "On the Influence of Temperature on the Refraction of Light," in *Philosophical Transactions of the Royal Society,* **148** (1858), 887–894; "Researches on the Refraction, Dispersion and Sensitiveness of Liquids," *ibid.*, **153** (1863), 317–343; "Researches on Refraction-equivalents," in *Proceedings of the Royal Society*, **16** (1868), 439–444; "On the Refraction-equivalents of the Elements," in *Philosophical Transactions of the Royal Society*, **160** (1870), 9–32; "On Essential Oils," in *Journal of the Chemical Society*, **17** (1864), 1–21; **25** (1872), 1–12; **49** (1886), 609–623; and "Researches on the Action of the Copper-Zinc Couple on Organic Bodies," *ibid.*, **26** (1873), 445–452, 678–683, 961–970; **27** (1874), 208–212, 406–410, 410–415, 615–619; **28** (1875), 508–514; **35** (1879), 107–110; **47** (1885), 448–456, written with A. Tribe. Gladstone published a biography of Faraday, whom he knew well, *Michael Faraday* (London, 1872; 2nd ed., 1873); and, from articles which had appeared in *Nature*, *The Chemistry of the Secondary Batteries of Planté and Faure* (London, 1883), written with A. Tribe.

Gladstone also wrote a pamphlet on spelling reform, several pamphlets on religious matters, and some hymns.

II. SECONDARY LITERATURE. A biography is W. A. Tilden, "John Hall Gladstone," in *Journal of the Chemical Society*, **87** (1905), 591–597. Obituary notices are T. E. T., in *Proceedings of the Royal Society*, **75** (1905), 188–192; and W. C. R. A., in *Nature*, **66** (1902), 609–610.

E. L. SCOTT

GLAISHER, JAMES (*b.* Rotherhithe, England, 7 April 1809; *d.* Croydon, England, 7 February 1903), *meteorology*.

Glaisher seems to have been largely self-educated and to have acquired his interest in science on visits to Greenwich observatory. In 1833 he attracted the attention of George Airy, who appointed him assistant at Cambridge observatory; when Airy became astronomer royal in 1835, Glaisher soon followed him to Greenwich. In 1838 a magnetic and meteorological department was formed at Greenwich with Glaisher as superintendent, a post he held until his retirement at the statutory age in 1874. This appointment determined the course of Glaisher's life. He effectively organized meteorological observations and climatological statistics in the United Kingdom; and although more than 120 papers appeared under his name, his importance in the history of his chosen science lies chiefly in the great energy and persistence that he displayed in this work.

Glaisher's first extensive scientific paper was on the radiation of heat from the ground at night (1847), and in the same year he published his *Hygrometrical Tables Adapted to the Use of the Dry and Wet Bulb Thermometer*, which, although entirely empirical in construction, remained in use by British meteorologists for almost a century. His most spectacular activity, which brought him to the attention of the public, was a series of scientific balloon ascents with the aeronaut Henry Coxwell in 1862, under the auspices of the British Association for the Advancement of Science.

Glaisher was elected a fellow of the Royal Society in 1849 and took a leading part in the founding of the British (now the Royal) Meteorological Society in 1850. He was the first president of the Royal Microscopical Society (1865–1869), president of the Photographic Society for more than twenty years, and a member of the council of the Royal Aeronautical Society from its foundation in 1866 until his death.

BIBLIOGRAPHY

A bibliography compiled by W. Marriott is in *Quarterly Journal of the Royal Meteorological Society*, **30** (1904), 1–28, together with an account of Glaisher's scientific work. His writings include "Radiation of Heat at Night From the Earth . . .," in *Philosophical Transactions of the Royal Society*, **137** (1847), 119–216; *Tables Adapted to the Use of the Dry and Wet Bulb Thermometer* (London, 1847; 9th ed., 1902); and "Account of Meteorological and Physical Observations in Balloon Ascents," in *Report of the British Association for the Advancement of Science* (1862), 376–503.

W. E. K. MIDDLETON

GLAISHER, JAMES WHITBREAD LEE (*b.* Lewisham, Kent, England, 5 November 1848; *d.* Cambridge, England, 7 December 1928), *mathematics, astronomy*.

Glaisher was the eldest son of James Glaisher, an astronomer who was also interested in the calculation of numerical tables. His given names were derived from those of his father and his father's colleagues in the founding of the British Meteorological Society, S. C. Whitbread and John Lee.

Glaisher attended St. Paul's School, London (1858–1867), and Trinity College, Cambridge, where he graduated as second wrangler in 1871. Elected to a fellowship and appointed an assistant tutor at Trinity, he remained there the rest of his life. He never married.

A tall, slim, upright man who retained good health until his last few years, Glaisher enjoyed walking, bicycling, collecting, travel (often in the United States), and teaching as well as mathematical research and participation in the meetings of scientific societies. He became an authority on English pottery, writing parts of several books on the subject and leaving his fine collection to the Fitzwilliam Museum, Cambridge. Glaisher was active in the British Association for the Advancement of Science, as president in 1900 and as a member of several committees. He was the "reporter," as well as a member—along with A. Cayley, G. G. Stokes, W. Thomson, and H. J. S.

Smith—of the Committee on Mathematical Tables. Its 175-page *Report,* containing much historical and bibliographical data, appeared in 1873.

Glaisher's honors included memberships in the councils of the Royal Society (for three different periods), the London Mathematical Society, and the Royal Astronomical Society (from 1874 until his death), as well as the presidency of the last two societies. Cambridge University awarded him the new D.Sc. degree in 1887, and Trinity College of Dublin and Victoria University of Manchester awarded him honorary D.Sc. degrees. Glaisher was an honorary fellow of the Royal Society of Edinburgh, of the Manchester Literary and Philosophical Society, and of the National Academy of Sciences, Washington. He was awarded the De Morgan Medal of the London Mathematical Society in 1908 and the Sylvester Medal of the Royal Society in 1913.

Glaisher's first paper typified three of his continuing interests: special functions, tables, and the history of mathematics. It was written while he was an undergraduate and was communicated to the Royal Society by Arthur Cayley in 1870. It dealt with the integral sine, cosine, and exponential functions and included both tables which he had calculated and much historical matter. Glaisher's first astronomical paper also typified his interest: "The Law of the Facility of Errors of Observations and on the Method of Least Squares," published in the *Memoirs of the Royal Astronomical Society* for 1872. This paper was inspired by a historical note in an American journal giving Robert Adrain credit for the independent discovery of Gauss's law of errors. A. R. Forsyth labels it, along with a paper on Jacopo Riccati's differential equation and one on the history of plus and minus signs, as "classical."

Glaisher published nearly 400 articles and notes but never a book of his own. The nearest he came was the *Report* noted above, the *Collected Mathematical Papers of Henry John Stephen Smith,* which he edited, and volumes VIII and IX of the *Mathematical Tables* of the British Association for the Advancement of Science, published in 1940. The latter were revisions and extensions of number theoretical tables (divisors, Euler's ϕ function and its inverse, and others) which he had completed in 1884.

Glaisher served as editor of two journals, *Messenger of Mathematics* (1871–1928) and *Quarterly Journal of Mathematics* (from 1878 until his death). G. H. Hardy wrote, "A generation of well known English mathematicians began their careers as authors in the *Messenger,*" and stated that Glaisher was ". . . underestimated as a mathematician. He wrote a great deal of very uneven quality, and he was old-fashioned, but the best of his work is really good." He applied to number theory, especially to representations by sums of squares, the properties of special functions, especially elliptic modular functions.

Glaisher's interest in students and publications affected American mathematics. He befriended an American student at Cambridge, Thomas S. Fiske, and took him to meetings of the London Mathematical Society. When he returned to Columbia University, Fiske organized the New York Mathematical Society (later the American Mathematical Society) in 1888 and copied the format of the *Messenger* when the *Bulletin of the New York Mathematical Society* was initiated.

Forsyth's characterization of Glaisher as "a mathematical stimulus to others rather than a pioneer" seems sound.

BIBLIOGRAPHY

I. Original Works. For lists of papers see Poggendorff, III, 524–525; IV, 502; V, 427–428; VI, 900.

For Glaisher's contributions to number theory, see the author index in Leonard Eugene Dickson, *History of the Theory of Numbers* (New York, 1934). "On Riccati's Equation and its Transformations and on Some Definite Integrals Which Satisfy Them," is in *Philosophical Transactions of the Royal Society of London,* **172,** pt. 3 (1882), 759–828. His article on the history of plus and minus signs appeared in *Messenger of Mathematics,* vol. **51** (1922).

II. Secondary Literature. A. R. Forsyth published a biography in *Journal of the London Mathematical Society,* **4,** pt. 2, no. 14 (Apr. 1929), 101–112, repr. in *Proceedings of the Royal Society,* **126A,** no. A802 (22 Jan. 1929), i–xi, with a portrait facing p. i. Forsyth also wrote the biography in *Dictionary of National Biography. 1922–1930* (London, 1937), pp. 339–340, which records that there is a pencil drawing of Glaisher by Francis Dodd in Trinity College, Cambridge.

See also H. H. Turner, "James Whitbread Lee Glaisher," in *Monthly Notices of the Royal Astronomical Society,* **89** (Feb. 1929), 300–308; and G. H. Hardy, "Dr. Glaisher and the Messenger of Mathematics," in *Messenger of Mathematics,* **58** (1929), 159–160.

Phillip S. Jones

GLANVILL, JOSEPH (*b.* Plymouth, England, 1636; *d.* Bath, England, 4 November 1680), *theology, apologetics, history and philosophy of science.*

Little is known of Glanvill's early life. His father was a merchant; his family and early education, Puritan. He matriculated at Exeter College, Oxford, in 1652; obtained the B.A. in 1655; studied at Lincoln College both before and after taking the M.A. in 1658; and was ordained in 1660. The most important of Glanvill's ecclesiastical livings was the abbey church at Bath, which he held from 1666 until his death. He was elected fellow of the Royal Society on 14 December 1664 and was the first secretary of a Somerset affiliate established in 1669.

Neither experimenter nor theorist, Glanvill made a few minor contributions to natural history. In number 11 of the *Philosophical Transactions of the Royal Society,* Robert Boyle gave a set of general headings for the natural history of a country, and in number 19 he followed it up with a long list of inquiries concerning the procedures used in mines and the geographical characteristics of mining regions. Glanvill sent in two sets of concise replies for the Mendip lead mines near Bath, obtained through interviews with miners and local residents. Later, in response to queries from Henry Oldenburg, he provided a report on the medicinal springs at Bath, again confined to factual observations except for one modest theoretical suggestion (for which he quickly apologized).

Glanvill's principal contribution to the work of the Royal Society was to defend it against its critics. In a series of much-discussed works he argued that the new experimental philosophy was beneficial in practical terms, had already advanced knowledge beyond what antiquity could claim and would rapidly advance it still further, and was harmless—indeed, it was helpful—to the cause of religion. In the process he produced one of the earliest histories, and one of the earliest philosophies, of science.

While still at Oxford (where he was almost certainly familiar with John Wilkins' "Invisible College," although there is no record of his attending its meetings), Glanvill wrote an elaborate essay, published in 1661, called *The Vanity of Dogmatizing.* This was originally intended to be a preface, defending the use of reason in religion and attacking sectarians and enthusiasts, to a projected book on the immortality of the soul. The preface was to begin with a criticism of dogmatism in general and would lead into an attack on dogmatism and disputatiousness in religion. Throughout his career Glanvill was a prominent apologist for latitudinarian Anglicanism as well as for the new philosophy, and his two lines of apologetic endeavor were frequently intertwined. The fate of the intended preface suggests that defense of the new philosophy was a genuine and independent concern of Glanvill's and no mere adjunct or instrument of his theological apologetics, for the philosophical part grew to book length, while the sectaries were disposed of in two chapters at the end. There Glanvill argues that "confidence in opinions" has led to acrimonious disputes about obscure doctrines, while the essentials of religion—devotion to God and practical love of neighbor—are ignored. The philosophical study of nature, on the other hand, promotes piety both by teaching us to admire more justly the work of the Divine Architect and by elevating the mind above sensual concerns. (The virtues traditionally ascribed to quasi-mystical, world-scorning philosophers are here—with incongruously traditional rhetoric—credited to such men as "those illustrious Heroes, *Cartes, Gassendus, Galilaeo, Tycho, Harvey, More, Digby.*" But of course Glanvill had to counter the claim of the enthusiasts that "vain philosophy" turns men's thoughts from heaven to earth. In later works he chose to sidestep that charge.)

The book begins with a speculative discussion of Adam's knowledge before the Fall. Glanvill reasons that a being created in God's image would have senses that could perceive the hidden causes of things, as well as the wonders revealed by Galileo's telescope. The "circumference" of his senses must have been "the same with that of natures activity." But the Fall brought us very low; our ignorance is almost total. We cannot understand how the soul is united with and moves the body; we can give no account of sensation or memory; we cannot explain how plants and animals are formed, nor what holds the parts of material objects together. Not even the ingenious hypotheses of Descartes and Henry More—most admirable of philosophers—will do. The causes of our ignorance are many. Truth lies deeply hidden; our senses are weak; they deceive us or, rather, present misleading impressions, so that our precipitate judgment errs. Our attempts to imagine things which cannot be imagined, the mind's liability to fatigue in close reasoning, and such affectional factors as personal vanity and reverence for antiquity—all these lead to hasty judgment, the source of all error. The mention of antiquity occasions a denunciation of Aristotelianism: it delights in controversy, its terms are ambiguous, its occult qualities explain nothing, its astronomy is silly, it has led to no useful inventions or discoveries, it is inconsistent both with itself and with true divinity. In all these regards we can expect better things from the "*Neoterick* endeavours" now under way.

The influence of Descartes (whether direct or through Henry More) is plain, although it is also plain that Glanvill has picked and chosen among Cartesian doctrines. In particular he ignores Descartes's claims for certainty and follows that strand in his physical writings which claims only to offer useful hypotheses and calls for experiments.

The next section of *The Vanity of Dogmatizing,* where Glanvill tries to show by an analysis of causality that we can have no certain knowledge of nature, begins to seem more like a crude anticipation of Hume. To have "science" in the dogmatist's sense, we would need "knowledge of things in their true, immediate, necessary causes." We have no immediate perception of causal connections; at best we observe constant concomitances. To know that a causal connection is necessary, we would have to know that alternatives are impossible; but what is impossible on one set of hypotheses (for instance, those of Aristotle, or Descartes, or common sense) will often be possible on another, and there are phenomena (e.g., sympathetic cure of wounds or control of another's thoughts) which are impossible in all known thought systems yet are real. We do not know the true causes of things, the "first springs of natural motions"; and since even the proximate causes we know are often very dissimilar to their effects, the invisible "first springs" are probably quite unlike anything with which we are familiar. The argument is not logically tight. It is meant to be rhetorically persuasive, and much of it really aims only to show that no one is in a position to claim certain knowledge of nature. We can have certainty in mathematics, but it is about notions of our creation and not the world, and in the essentials of religion, which are "as demonstrable as Geometry" (although Glanvill does not tell how).

The Vanity of Dogmatizing was promptly attacked by Thomas White, a Catholic Aristotelian, as destructively skeptical and as, by its endorsement of the mechanical philosophy, an aid to Hobbesian atheism. Glanvill replied in 1664 by issuing a revised version, with additional essays directed at White and a long prefatory address to the Royal Society, the whole titled *Scepsis scientifica.* The way he had advocated is just that followed by the Royal Society: accumulating factual information (which only a theory-obsessed dogmatist could dispute) about natural phenomena while prudently refraining from premature theorizing. Moreover, we cannot guard against Hobbesian misuse of the "mechanical hypothesis" by appeal to a discredited Scholasticism; the only way is to show, as the virtuosos are showing, that when well worked out, the hypothesis provides overwhelming evidence for a wise and benevolent Designer.

The address won Glanvill election to the Royal Society and a reputation as a potentially valuable apologist. When Thomas Sprat's *History of the Royal Society* failed to silence the critics, Oldenburg encouraged Glanvill to write a supplementary defense, which appeared in 1668 under the title *Plus ultra.* Its main theme is that the experimental philosophy, and the Royal Society in particular, have accomplished more to advance useful knowledge in a few years "than all the *Philosophers* of the *Notional way,* since *Aristotle* opened his *Shop* in *Greece.*" The principal ways of advancing knowledge are by enlarging natural history and by improving communications. Under the former heading Glanvill includes "*investigation* of the *Springs* of *Natural Motions* as well as *fuller Accounts* of the . . . more *palpable Phaenomena.*" Recent additions to our knowledge of palpable phenomena are briefly cataloged; the emphasis is on the aids to "*deep Research*" provided by modern achievements in instrumentation and in chemistry, anatomy, and especially mathematics (under which he includes astronomy and optics). A rather detailed survey of these achievements is given, stressing experiment rather than theory. (While premature hypothesizing is continually deplored, Glanvill has no clear and consistent doctrine on the role of hypotheses in research.) As to communications, he mentions printing and the compass but dwells particularly on the Royal Society itself as a vehicle for cooperative efforts along the lines laid out by "the excellent Lord *Bacon.*"

Glanvill's answer to the challenge "What have they done?" is twofold. It is unreasonable to expect too much too soon; to clear away the rubble of Peripatetic philosophy and establish organized inquiry with a sound method is a creditable work for one generation. But of course he has pointed to many accomplishments, claiming credit for the Royal Society and its members and sympathizers where he can. The contributions of Boyle are reviewed as a particularly strong recommendation.

As for the charge of irreligion, Glanvill cites all the prelates and other pious men who are members and repeats the argument that the new philosophy provides the best proofs for the Deity. Investigations into the works of nature are, indeed, religious duties, for they enable us to discover the means God has provided for alleviating man's lot. They help to banish degrading superstitions, and they promote a religious temper of equanimity, charity, and modesty. To the charge that the new philosophy turns men from scriptural revelation to nature, he replies by (1) asserting that (considering the variety of revelations that are claimed) a solid natural theology must be established first and (2) simply affirming his reverence for the Scriptures.

Further controversies embroiled Glanvill, most notably that with Henry Stubbe, but the main points in his argument remained the same (and were repeatedly employed by other apologists). In his later years he wrote primarily on specifically religious and religiopolitical questions, and above all to combat disbelief in witches. As a pastor he observed that incredulity as to evil spirits usually led, if not to outright atheism, at least to religious indifference and scorn for the Bible. Further, he found that many unphilosophical worldlings, unimpressed by abstruse arguments for the existence of God, were shaken in unbelief by well-attested accounts of demonic activities. As early as 1666 he published a book on the subject, with his membership in the Royal Society advertised on the title page. Encouraged by Henry More (who edited and expanded the final version, *Saducismus triumphatus,* after the author's death), he published a succession of enlarged editions. In the 1668 version (only) he proposed investigation of the spirit world and its laws as a suitable project for the Royal Society.

Glanvill's argument has two steps. First, he tried to show that there is nothing absurd or inconceivable in the idea that there are evil spirits who act in the world. Those who say this is impossible can actually show only that we do not understand how it is done. But—and here Glanvill referred to the argument of *Scepsis scientifica*—there are many familiar phenomena whose causes we do not know. Matters of fact can be established only by sense or credible testimony; our inability to understand them is no ground for denying them. He buttressed this line of argument by offering conjectural natural accounts—in some cases vaguely mechanical—of how the spirits might work. He thought these explanations to be probably inadequate (much more likely the spirit world is governed by its own laws, of which we as yet have no inkling) but sufficient to show that the events in question are not impossible. The second step, of course, was to show that there are well-attested cases. To this end he offered a number of narratives, with detailed accounts of how he (or the source of his information) had observed the phenomena and tested for trickery and other purely natural sources of them. Bizarre as his occupation with witchcraft may now seem, it was consistent with his understanding of science and acceptable to most of his Royal Society colleagues.

BIBLIOGRAPHY

Glanvill's contributions to the *Philosophical Transactions of the Royal Society* are "Answers to Some of the Inquiries Formerly Published Concerning Mines," in **2** (1667), 525–527; "Additional Answers to the Queries of Mines," in **3** (1668), 767–771; and "Observations Concerning the Bath-Springs," in **4** (1669), 977–982.

Jackson I. Cope, *Joseph Glanvill: Anglican Apologist* (St. Louis, 1956; 2nd ed., 1958), is an excellent study and gives a complete list of Glanvill's works. See also Richard F. Jones, *Ancients and Moderns: A Study of the Rise of the Scientific Movement in Seventeenth-Century England* (St. Louis, 2nd ed., 1961), chs. 8 and 9; Henry G. van Leeuwen, *The Problem of Certainty in English Thought, 1630–1690* (The Hague, 1963), pp. 71–89; and Richard S. Westfall, *Science and Religion in Seventeenth-Century England* (New Haven, 1958), *passim.*

On special topics see Richard H. Popkin, "Joseph Glanvill: A Precursor of David Hume," in *Journal of the History of Ideas,* **14** (1953), 292–303; and "The Development of the Philosophical Reputation of Joseph Glanvill," *ibid.,* **15** (1954), 305–311; and Moody E. Prior, "Joseph Glanvill, Witchcraft, and Seventeenth-Century Science," in *Modern Philology,* **30** (1932), 167–193.

WILLIAM H. AUSTIN

GLASER, CHRISTOPHER (*b.* Basel, Switzerland, *ca.* 1615; *d.* Paris, France, 1672 [?]), *pharmacy, chemistry.*

Little is known about Glaser's early life, but he seems to have been trained as a pharmacist in his native city, and references in his published work indicate that he traveled in eastern Europe to observe mining practice. Sometime prior to 1662 he settled in Paris, where he opened an apothecary's shop in the Faubourg Saint-Germain. Here he prospered, becoming apothecary in ordinary to Louis XIV and to the king's brother, the duke of Orleans. He also enjoyed the patronage of Nicolas Fouquet, the ill-fated superintendent of finances. In 1662 he was appointed demonstrator in chemistry at the Jardin du Roi in Paris in succession to Nicolas Le Fèvre. The following year he published his only contribution to the literature of chemistry, *Traité de la chymie,* a textbook for his course. His most noted pupil in Paris was Nicolas Lemery.

The events of Glaser's later life are likewise elusive. In 1672 he was implicated in the famous Brinvilliers poison case when evidence came to light that the marquise de Brinvilliers and her accomplice Gaudin de Sainte-Croix had used a recipe of Glaser's to prepare the poison with which they disposed of the marquise's father (1666) and two brothers (1669–1670). At this point Glaser disappeared from public life in France. If the preface by the printer to the 1673 edition of Glaser's *Traité* is to be believed, Glaser died before completing the revisions for this new edition, which received the approbation of the Paris Faculty of Medicine on 15 October 1672. At her interrogation in 1676 the marquise de Brinvilliers alleged that Glaser had indeed prepared poison for Sainte-Croix but that he had been dead for a long time. One source maintains, however, that Glaser returned to Basel, where he died in 1678 (see C. de Milt, "Christopher Glaser").

In the series of French chemical manuals of the seventeenth century, that of Glaser appeared between the more famous works of Le Fèvre and Lemery. Whereas Le Fèvre's textbook drew on the Paracelsian-Helmontian tradition for its theoretical content and Lemery attempted a corpuscularian interpretation of the processes he described, Glaser largely eschewed theory and was content with a straightforward, concise recital of chemical operations and recipes. In spite of considerable competition, Glaser's textbook enjoyed some success. The fourteen editions recorded between 1666 and 1710 include one English and five German versions. The work is divided into two books: book I briefly describes the utility, definitions, principles, operations, and apparatus of chemistry; book II is devoted to a description of medicinal preparations drawn from the mineral, vegetable, and animal kingdoms. The section devoted to mineral remedies is by far the largest. Little is novel in these preparations, although Glaser displays individual refinements of technique. His recipe for a *sel antifebrile* (potassium sulfate made by heating saltpeter and sulfur and recrystallizing from water) became uniquely identified with him and was later known as *sel polychrestum Glaseri.* The naturally occurring mixed sulfate of sodium and potassium ($3K_2SO_4$: Na_2SO_4) was named glaserite in his honor.

Due to his influence on Lemery, Glaser's importance for the development of chemistry was greater than the contents of his book at first indicate. Although Fontenelle in his *éloge* of Lemery states that Lemery, finding Glaser obscure and secretive, abandoned studies with him after two months in 1666, the early editions of Lemery's highly successful *Cours de chymie* bear a remarkable resemblance both in organization and content to Glaser's textbook. There seems little doubt that Glaser's modest work served as a model for at least the practical part of the most popular chemical textbook of the late seventeenth century.

BIBLIOGRAPHY

I. ORIGINAL WORKS. For the various eds. and reprs. of Glaser's textbook see Partington, de Milt, and Neville in the works cited below. The first ed. has the title *Traité de la chymie, enseignant par une briève et facile méthode toutes ses plus nécessaires préparations* (Paris, 1663). The most important subsequent eds. (Paris, 1668, 1673) both contain the author's additions and corrections. An English trans., based on the 1668 ed., is *The Compleat Chymist, or A New Treatise of Chymistry* (London, 1677). Two separate German translations are *Chimischer Wegweiser* (Jena, 1677), and *Novum laboratorium medico-chymicum* (Nuremberg, 1677). The Jena vers. was reprinted several times, the last in 1710.

II. SECONDARY LITERATURE. Works on Glaser and his *Traité* are H. Lagarde, "Christopher Glaser, professeur de chimie au Jardin des Plantes, apothicaire du roi, fournisseur de la Brinvilliers," in *Mémoires de la Société d'émulation du Doubs,* 6th ser., **5** (1890), 407–421; H. Metzger, *Les doctrines chimiques en France du début du XVIIe à la fin du XVIIIe siècle,* repr. (Paris, 1969), pp. 82–86; C. de Milt, "Christopher Glaser," in *Journal of Chemical Education,* **19** (1942), 53–60; R. G. Neville, "Christopher Glaser and the *Traité de la Chymie,* 1663," in *Chymia,* **10** (1965), 25–52; J. R. Partington, *A History of Chemistry,* III (London, 1962), 24–26; and J. Read, *Humour and Humanism in Chemistry* (London, 1947), pp. 114–115.

Documents relating to Glaser and the Brinvilliers case are in F. Ravaisson, *Archives de la Bastille,* IV (Paris, 1870), 237, 244, 250; VII (Paris, 1874), 44–46. An extensive bibliography on Glaser and the Brinvilliers affair is given in J. Ferguson, *Bibliotheca Chemica,* I (Glasgow, 1906), 321.

OWEN HANNAWAY

GLASER, JOHANN HEINRICH (*b.* Basel, Switzerland, 6 October 1629; *d.* Basel, 5 February 1679), *anatomy, botany, surgery.*

Glaser's father had acquired a sound reputation as a painter and engraver in Basel, and it was there that his son began his studies. By 1645 he seemed committed to philosophy, but in 1648 he went to Geneva, where he studied medicine (he later practiced at Heidelberg). Afterward Glaser settled at Paris, where he became interested in the work of the botanists at the Muséum d'Histoire Naturelle. Following the resignation of his close friend, Felix Platter, he was considered for the chair of physics at the University of Basel. The chair went to someone else, however, and Glaser turned decisively to medicine.

His thesis, *De dolore colico,* submitted in July 1650, attracted the attention of the medical world, since in it Glaser considered anatomical and morphological data, as well as physiological. He returned to Basel and, in 1661, presented for his doctorate a classical dissertation, *Disputatio de rheumatismo.* In 1662 he began a medical practice that soon brought him international fame, and in 1665 he became full professor of Greek. Two years later, Glaser achieved his goal of being named professor of anatomy and botany at the Faculté de Médecine at Basel. In order to be prepared for this important post, he had published, in collaboration with J. J. Spörlin, the memoir *Positiones de respiratione ex Hippocrate et Galeno depromptae* (1661) and his *Theses opticae* (1664). In 1668 Glaser was named doctor-in-chief of the large municipal hospital at Basel, the crowning achievement of his career. He died of a fever contracted while caring for a patient.

Even more as a teacher than as a doctor, Glaser achieved a great reputation. The fame of the Basel hospital and of its doctors, many of whom had been his pupils, spread rapidly. Glaser particularly owed his fame to his clinical teaching and to the novelty of his methods. He introduced the application of theory through clinical teaching, spending hours with students at the bedsides of patients. Glaser was one of the first physicians to introduce hospital rounds, and many of his pupils later adopted this new conception of clinical teaching. He also was an innovator in surgery.

At Basel, Glaser held public dissections followed by surgical demonstrations. With his students he regularly examined the corpses of patients who had died in his hospital. He noted all his observations in memoranda that show how advanced he was in comparison with his colleagues. Glaser owes his lasting fame, however, to his work on the brain, the nerves, and the bones of the head. *Tractatus de cerebro* (1680) was a compilation of the numerous original observations that he had accumulated on the cerebral system. It is a classic that treats the anatomy, both normal and pathological, and the physiology of the central nervous system.

In memory of Glaser the fissure of the temporal bone through which the tympanic cords pass is known as the Glaserian petrotympanic fissure. Glaser was one of the first to carry out frequent dissections of animals, and he thereby discovered many anatomical details before they were found in man. Undoubtedly he was a great practitioner and certainly a master surgeon of his time.

BIBLIOGRAPHY

I. Original Works. Glaser's writings include *De summo hominis bono morale,* his diss. in philosophy (Basel, 1648); *De dolore colico,* his diss. in medicine (Paris, 1650); *Disputatio de rheumatismo,* his doctoral diss. (Basel, 1661); *Positiones de respiratione ex Hippocrate et Galeno depromptae,* written with J. J. Spörlin (Basel, 1661); *Theses opticae* (Basel, 1664); and *Tractatus de cerebro* (Basel, 1680).

II. Secondary Literature. Works on Glaser are H. Buess, *Recherches, découvertes et inventions de médecins suisses,* E. Kaech, trans. (Basel, 1946); A. Burckhardt, *Geschichte des medizinischen Fakultät zu Basel* (Basel, 1917); F. Husner, *Verzeichnis der Basler medizinischen Universitätsschriften von 1575–1829* (Basel, 1942); and Franciscus Pariz, *Sancta merx viri nobilissimi J. Henrici Glaseri* (Basel, 1675).

P. E. Pilet

GLAUBER, JOHANN RUDOLPH (*b.* Karlstadt, Germany, 1604; *d.* Amsterdam, Netherlands, March 1670), *chemistry, medicine, metallurgy.*

Glauber was the son of a barber, Rudolph Glauber von Hundsbach, and his second wife, Gertraut Gosenberger. Unlike most iatrochemists, he did not attend a university but instead set out in quest of spagyric wisdom, visiting laboratories in Paris, Basel, Salzburg, and Vienna. At one time he earned his living by casting metallic mirrors, and in 1635 he went to work as a court apothecary in Giessen. There he married Rebecca Jacobs but soon divorced her on grounds of infidelity; in 1641 he married Helene Cornelius, who bore him eight children.

The political uncertainties of the Thirty Years' War persuaded Glauber to leave Germany about 1639. Although his biographers often claim that he returned in the 1640's, Glauber himself tells of settling in Amsterdam and remaining there, except for brief stays in Utrecht and Arnhem, until 1650. During these years he invented his famous distillatory furnaces, which made it possible to obtain high temperatures and to heat substances under a variety of conditions. One of the furnaces had a chimney and may have been the first so equipped. Encouraged by these technical improvements, Glauber began to speak of himself as a chemical philosopher and, in a burst of creative activity, completed most of the practical work for which he is famous.

With the end of political strife Glauber happily returned to Germany to work in the wine industry in Wertheim and Kitzingen—at least this was his ostensible profession and the source of his livelihood. His main interest continued to be alchemy, but he found it prudent to conceal his activities because of the hostility to goldmakers. He also set aside one hour each day to administer free medicines, especially his *panacea antimonialis* (probably antimony pentasulfide). In 1655 Glauber again left Germany for Amsterdam, this time never to return. The move was undoubtedly related to a bitter dispute with Christopher Farner, who had stolen some of his processes and had slandered his work and character. Amsterdam was also more receptive to Glauber's religious beliefs; although born a Catholic, he argued that men would be judged by their deeds rather than by the idiosyncrasies of a particular sect.

In Amsterdam, Glauber outfitted what was surely the most impressive laboratory in Europe. Samuel Sorbière, a visitor to the laboratory in 1660, described it as "magnificent." There were workrooms both inside and outside the house, and the walls were covered with vessels and instruments of Glauber's own invention. Even the garden was utilized for agricultural experiments. After 1662 Glauber was plagued with ill health and was eventually bedridden for months at a time. He continued to write prolifically, but with more time for contemplation he began to emphasize the esoteric side of alchemy and to regret his years of toil in the laboratory. He believed that he had finally found the "secret fire of Artephius" and the material of the philosophers' stone. He died in 1670, poor, lonely, and embittered, and was buried in the Westerkerk, Amsterdam.

Glauber was an independent worker, boastful of his own achievements and suspicious of others. He had little to do with other chemists, and there is no evidence to suggest that he corresponded with important contemporaries. Since he refused to be associated with a patron, he was forced to live entirely on the sale of his products. His chief vehicle for luring customers was his writings, and he filled them with exaggerated claims—suggesting, indeed, that his inventions would usher in a kind of chemical apocalypse. Critics quickly charged Glauber both with revealing too much about alchemy and with peddling useless processes. As the criticism mounted, he withdrew even more into himself, taking comfort in the contrast of his own virtue with the pride and greed of other men.

Glauber was more influenced by the metallurgical tradition than other iatrochemists were, and it was undoubtedly from Agricola, Vannoccio Biringuccio, and Lazarus Ercker that he derived much of his practical good sense. On the other hand, alchemy, particularly that of Paracelsus, J. I. Hollandus, Michael Sendivogius, and J. B. van Helmont, determined his goals and his perception of his work. Glauber's own influence was widespread. The *Furni novi philosophici* (1646–1649) was quickly translated into Latin, English, and French and went through many editions; major compilations of his works appeared in Latin, English, and German. Robert Boyle, Nicholas Le Fèvre, and Johann Kunckel were all impressed by his labors, and Hermann Boerhaave spoke especially highly of him. In the eighteenth century his name continued to be associated with many processes, and even today hydrated sodium sulfate is familiarly known as Glauber's salt.

Glauber gave the best account of his practical work in the *Furni novi philosophici,* a book written with a clarity and an honesty almost unprecedented in early chemistry. With it he established his reputation as a master of laboratory skills. He carefully described the materials and dimensions for the construction of the furnaces and gave instructions for the necessary accessory equipment: vitrified earthen vessels to withstand the increased temperatures, a large quantity of cupels and crucibles, improved condensing apparatus, and jars with mercury seals or ground glass stoppers to store corrosive and volatile liquids.

The range of distillable substances was increased tremendously with these furnaces, and Glauber put into them almost anything he could lay hands on. From the mineral kingdom he prepared the mineral acids (hydrochloric, nitric, and sulfuric) in concentrated form and with them made chlorides, nitrates, and sulfates. He was probably the first to distill coal and obtain (with the help of hydrochloric acid) benzene and phenol. From the animal kingdom he distilled the "superfluities": hair, horns, feathers, silk, and urine. He extracted the aromatic oils of plants by first soaking their parts in salt water or hydrochloric acid, and he obtained acroleins by distilling burned clay balls presoaked in the fatty oils. The dry distillation of wood yielded wood vinegar, with which he produced metal acetates and acetone. The distillation of salt of tartar (potassium carbonate), effected by adding powdered flints, particularly intrigued him. Liquor of flints (potassium silicate) was obtained as a by-product, and the metallic trees that he grew by adding metal salts were a source of great delight to him.

Glauber's efficient production of such important chemical reactants as the mineral acids is particularly noteworthy since they are essential for other processes. He described several ways in which each can be prepared, realizing that the products are similar, although the methods by which they are prepared are different. He recognized, for example, that the spirit of sulfur produced by burning sulfur under a bell jar is of similar nature to the oil of vitriol distilled from green vitriol and that the vitriol used in his recommended recipe for hydrochloric acid functions as a catalyst. His preparation of nitric and hydrochloric acids by applying sulfuric acid to saltpeter and common salt, respectively, was long kept secret because of the purity obtainable.

The *Pharmacopoea spagyrica* (1654–1668) and *Dess Teutschlands-Wohlfahrt* (1656–1661) were more typical of Glauber's style than was the earlier *Furni.* Written intermittently over a period of several years, they lack both organization and a consistent point of view. Descriptions of processes are often squeezed between lengthy digressions or are obscured by metaphors and references to classical mythology.

The *Pharmacopoea spagyrica* is a collection of the medical preparations that Glauber found most reliable. Indeed, most of the products of Glauber's laboratory found eventual use in medicine. Like other iatrochemists, he complained about the sorry assortment of substances to be found in most apothecary shops while boasting of the high standards met by his own work. He believed that the most effective remedies were those prepared from the mineral kingdom, and he reported extensive work with chlorides, antimony and sulfur compounds, gold preparations, and a "magnesia of Saturn."

Although he preferred mineral remedies, Glauber nevertheless devoted considerable space in the *Pharmacopoea* to proving his skill in more traditional areas. He suggested a new way to prepare essences of herbs by separating and recombining their oils, spirits, and salts. By soaking plant substances in nitric or sulfuric acid and then adding potash, he precipitated fine powders that may well have been the alkaloids strychnine, brucine, and morphine. Frequently manifest in his choice of materials was the time-honored assumption that unpleasant substances yield the best medicines. He therefore praised the virtues of excrements and gave recipes using worms, beetles, and venomous toads.

The final sections of the *Pharmacopoea* show Glauber's immersion in esoteric alchemy in later life. His revelation of a "secret sal armoniack" (ammonium sulfate) was followed by the revelation of an even grander "most secret sal armoniack" (ammonium nitrate?). The latter was claimed to be the celebrated alchemical substance that Adam brought out of the Garden of Eden. Man thus carried within himself the means to transform the natural world—but in such a loathsome place that his pride kept him from finding it. In spite of the deliberate obscurity, it seems likely that Glauber prepared his "most secret" salt by combining ammonia and saltpeter, made from excrement and urine.

Glauber displayed a good sense of economic feasibility as well as his love for his homeland in *Dess Teutschlands-Wohlfahrt,* a work encouraging Germans to make better use of their natural resources and to become economically self-sufficient. He gave recipes for wine and beer concentrates that are both stable and easily exported and he mentioned a secret press for the efficient extraction of niter from wood. He proceeded to point out that niter can then be used in the extraction of metals, particularly gold and silver, and that these precious metals, in turn, could be directed into foreign trade. He dedicated a variety of other items to the fatherland: new medicines, a fertilizer of salt and lime, a seed preparation, and various techniques for processing metals. Finally, since all this was futile without adequate protection from the Turks, he disclosed a new weapon: a missile containing "fiery water" (a fuming acid, or perhaps essential oils to be ignited by nitric acid).

Glauber also devoted considerable attention to transmutation in *Dess Teutschlands-Wohlfahrt.* Although he sometimes used the word loosely, he usually meant to refer to two operations distinguished as "universal" and "particular" transmutation. The "universal" transmutation was effected through the philosophers' stone, which Glauber did not claim to have prepared in its final form. "Particular" transmutation, on the other hand, was effected through salts; and he claimed to be a master of these. In our terms, such operations usually involve the extraction of components present in the original material. Glauber believed that spiritual substances were transmuted into corporeal bodies, presumably because he could not detect components that vanished in the fire unless they were first fixed with salt.

Glauber's interest in the transmutation of metals and in industrial chemistry distinguished him from Paracelsus and other iatrochemists, who were more narrowly concerned with the preparation of chemical medicines. In the most general sense Glauber sought to perfect nature for the enhancement of human life—to render useless things useful through the release of their hidden virtues. Such changes were effected in his laboratory primarily through the "ripening" powers of salts.

Since art imitates nature, the role of salt in Glauber's laboratory corresponded to its role in the macrocosm. The sun was the fountain of all maturation and perfection in the natural world, and its fire was carried by an aerial salt to the earth below, where it was responsible for the growth of all things. This explained why gold and spices and sweet wine, the ripest of growths, came from southern lands, where sun and salt are most abundant. Sulfurous salts were operative in mines, and the active force of fertilizer was the salt it contained. Animals derived salt from the air they breathed and from their food. Glauber was by no means unusual among metallurgists in attributing growth to minerals; nor was he unusual, in the seventeenth century, in believing that the sun, operating through a universal salt, underlay the unity of all living things. His laboratory skill in preparing and handling salts, however, predisposed him to describe the role of the universal salt more specifically than did many of his contemporaries.

In the early parts of the *Miraculum mundi* (1653–1660) Glauber specified that the universal salt is niter. The claim was less preposterous than it sounds, since he was not referring to a single substance but to a family of substances: niter in its crude form (saltpeter); the "spirit" of niter (nitric acid), obtained through distillation; and "fixed niter" (potassium carbonate), the residue when niter is deflagrated with charcoal. Glauber's contention that niter was universally present in nature was grounded upon his ability to produce it from a variety of different sources and his realization that vegetable alkalies were similar to his "fixed nitre." (Even in the eighteenth century it was not always understood that these substances have the same chemical composition.) When Glauber called niter a universal dissolvent, he sometimes referred only to the fixed salt (alkahest: "Alkali est"), but usually he intended the term to apply to niter in its three forms. What one form did not dissolve, one of the others could. Even stones could be dissolved, since powdered flints fused with potassium carbonate yielded liquor of flints. He gave numerous uses for the ripening powers of niter, from the fertilization of grape vines to the removal of facial wrinkles and the coloring of shoe leather. As his critics were quick to note, Glauber had reduced the mysteries of alchemy to a rather mundane level.

The work with niter illustrates rather well that although Glauber generally had a better sense of chemical composition than his contemporaries, the animistic categories of alchemy still had a strong hold over his conception of matter. On the one hand, he displayed a good understanding of the acid and base constituents of salt: not only did he make potassium carbonate and nitric acid out of saltpeter, but he also took the further step of combining these two to yield saltpeter. (Boyle described a similar experiment in his *Essay on Nitre,* denying that it had been inspired by Glauber.) On the other hand, he regarded the constituents of saltpeter as spirit and body and described the three forms of niter as though they were preformed in one another and elaborated through sexual procreation.

In 1658 the *Tractatus de natura salium* appeared, and in 1660 a second part was added to the *Miraculum mundi.* Only then did Glauber recognize the significance of his "sal mirabile" (Glauber's salt) and begin to utilize it, not very successfully, in the central position that niter formerly held. It was produced in its most interesting form as a by-product of his secret process for hydrochloric acid: from common salt and sulfuric acid. Hence he considered it to be common salt brought to its highest degree of purity, and he argued plausibly that common salt is everywhere present in nature. Glauber drew on the analogy of microcosm and macrocosm and Harvey's discovery of the circulation of the blood to demonstrate the circulation of salt in the macrocosm. It was further argued that salts could be generated out of one another and that all were forms of the primordial common salt. Glauber therefore saw no incompatibility between his previous focus on niter and his new commitment to "sal mirabile." The applications suggested for the new salt in art, alchemy, agriculture, and medicine were as numerous as those for niter. One worth noting was the use of "sal mirabile" in its anhydrous form to remove excess water from oils, mineral acids, vinegar, and poor-quality wines.

Since the light (heat) of the sun was ultimately responsible for all changes on earth, Glauber was greatly intrigued by optical phenomena, particularly those associated with mirrors and colored glass. He had once made metallic mirrors for a living and claimed to own one of the finest in Europe. In the *Furni* he gave a careful description both of the casting procedure and of the speculum metal itself: an alloy of copper, arsenic, brass, and tin. He experimented on metals with these mirrors and was able to melt lead with a mirror two or three spans in diameter. The role of the sun intrigued Glauber, for he believed that its rays were concentrated by the mirror and subsequently materialized as the lead gained weight. He believed that a like process was responsible for the generation of metals in nature (an explicit account is given in *Operis mineralis,* pt. 2). Astral rays were concentrated into the center of the earth, where a fiery vacuum was produced, and were then turned back to be materialized in the bowels of the earth. Glauber also described hermetic medicines in optical terminology: the most efficacious substances were those whose circumferences (virtues) had been concentrated into their original centers.

Glauber's fascination with colored glass was closely tied to a rather attractive interpretation of the making of the philosophers' stone. The sulfur (tincture, soul) of a metal was its color, and this must be isolated and fixed in order to effect transmutations. He argued that distillates, however subtle, remain composed of the three Paracelsian principles: salt, sulfur, and mercury. For the isolation of sulfur alone, metals must be reduced to ashes in the fire and returned to their origin for rebirth. Since their origin was sand, and since glass was made from sand, the true colors of metals would be revealed when their ashes were added to glass. The colors were then to be extracted by a sulfurous menstruum to yield the universal medicine. Glauber conceded failure in this last step, but in the course of his labors he related much useful information on the coloring of glass and rediscovered the process for ruby glass, which had been lost for many years.

Since Glauber conceived of God as eternal light, the coloring of glass was weighty with symbolic implications for him. On the last day, he believed, our bodies will be reduced to ashes, from which we will arise with clarified bodies to stand before God, who is himself pure light, and will be revealed in our true colors.

Glauber has justly been called the best practical

chemist of his day and the first industrial chemist. His instructions for the improvement of laboratory technique were instrumental in preparing the way for the chemical revolution of the next century. In his own estimation, however, the final goal of his labors was the perfection of the material world, capped by the preparation of the philosophers' stone. Glauber attempted to renew the hermetic art by tying it to specific aspects of laboratory practice, but in so doing he interpreted the symbols of alchemy so concretely as to destroy their esoteric appeal. Later alchemists understandably found him too mundane, while chemists failed to appreciate his hermetic conception of the world.

BIBLIOGRAPHY

I. Original Works. The complete list of Glauber's works is available in J. R. Partington, *A History of Chemistry,* II (London, 1961), 343–347. His major works—*Furni novi philosophici oder Beschreibung einer newerfundener Destillirkunst* (Amsterdam, 1646–1649); *Miraculum mundi oder Ausführliche Beschreibung der wunderbaren Natur, Art, und Eigenschafft des grossmächtigen Subiecti* . . . (Amsterdam, 1653–1660); *Pharmacopoea spagyrica oder Gründlicher Beschreibung, wie man aus den Vegetabilien, Animalien, und Mineralien . . . gute, kräfftige und durchdringende Arztneyen zurichten und bereiten soll* (Amsterdam, 1654–1668); *Dess Teutschlands-Wohlfahrt* (Amsterdam, 1656–1661)—consist of several pts. and appendixes published separately and at different times. German compilations appeared in 1658–1659—*Opera chymica, Bücher und Schrifften* . . . and *Continuatio operum chymicorum* . . .—and in 1715—the much abbreviated *Glauberus concentratus, oder Kern der Glauberischen Schrifften*. . ., available in a facs. repr. (Ulm, 1961). Christopher Packe, ed., *The Works of the Highly Experienced and Famous Chymist, John Rudolph Glauber* . . . (London, 1689), includes almost all his writings, translated into English. The translation is generally reliable, even though often translated from the Latin collected ed. rather than from the original German. Much of the moralizing and polemics has been omitted.

II. Secondary Literature. P. Walden, "Glauber," in Günther Bugge, ed., *Das Buch der grossen Chemiker,* I (Berlin, 1929), 151–172, gives the best assessment of Glauber's practical work; a shortened English trans. of Walden's article can be found in Eduard Farber, ed., *Great Chemists* (New York, 1961), pp. 115–134. Glauber's biography was in some confusion until the twentieth century, even though it can be reconstructed fairly completely from his own apologetic writings. Kurt F. Gugel, *Johann Rudolph Glauber 1604–1670, Leben und Werk* (Würzburg, 1955), has compiled the recent literature to give the fullest account of Glauber's life; Gugel also gives an extensive bibliography, but his account of Glauber's practical work is wholly derivative from Walden. Erich Pietsch, *Johann Rudolph Glauber* (Munich, 1956), attempts to make Glauber's theory of salt and fire respectable by explicating it from the standpoint of modern energy concepts. A more historical appraisal of his theory and motivation is that of H. M. E. de Jong, "Glauber und die Weltanschauung der Rosenkreuzer," in *Janus,* **56** (1969), 278–304; however, there is little reason to believe that Glauber was a Rosicrucian.

Kathleen Ahonen

GLAZEBROOK, RICHARD TETLEY (*b.* West Derby, Liverpool, England, 18 September 1854; *d.* Limpsfield Common, England, 15 December 1935), *physics.*

The eldest son of Nicholas Smith Glazebrook, a surgeon, and Sarah Anne Tetley, Glazebrook was educated at Dulwich College until 1870, then Liverpool College, and in 1872 entered Trinity College, Cambridge. He received his B.A. as fifth wrangler in 1876 and his M.A. in 1879. After working under Maxwell at the Cavendish Laboratory from 1876 until Maxwell's death in 1879, he stayed on under Rayleigh and in 1880 was appointed demonstrator (with Napier Shaw). Glazebrook was a college lecturer in mathematics and physics (1881–1895) and university lecturer in mathematics (1884–1897).

Although he was disappointed in not being elected to succeed Rayleigh when the latter resigned in 1884, Glazebrook remained at the Cavendish Laboratory and was appointed assistant director in 1891. In 1895 he took on additional duties as senior bursar of Trinity College. He resigned these last positions to become principal of University College, Liverpool, in 1898, with the understanding that he might leave if offered the directorship of the National Physical Laboratory, then being established. This in fact occurred, and he left the college on the last day of 1899, taking up his new position on the first day of the new year. He remained in this post until his retirement in 1919. Glazebrook was elected to the Royal Society in 1882 and received numerous other honors, including presidencies of the Physical Society (1903–1905), the Optical Society (1904–1905, 1911–1912), the Institution of Electrical Engineers (1906), the Faraday Society (1911–1913), and the Institute of Physics (1919–1921).

Glazebrook's initial work under Maxwell was in optics, with considerable attention to electrical measurements. When Rayleigh became a member of the reconstituted British Association Committee on Electrical Standards in 1881, Glazebrook assisted him; and in 1883 he became secretary of the committee, a position he was to hold until 1913, when the work of the committee was taken over by the National Physical Laboratory. Glazebrook became increasingly interested in the precise measurement of electrical standards. This specialty, together with his talent as an administrator, made him an obvious candidate to head the National Physical Laboratory when it was formed.

As director of the new laboratory Glazebrook continued to press for the determination of the fundamental units for both scientific and industrial purposes. In 1909 work was begun in the field of aeronautics, leading to efforts that were greatly accelerated during the war.

Glazebrook was an accomplished experimentalist, and he wrote several physics textbooks which enjoyed widespread use. His true vocation, however, was scientific administration, as he demonstrated in leading the National Physical Laboratory through its first two decades.

BIBLIOGRAPHY

I. ORIGINAL WORKS. Glazebrook's textbooks, each of which went through several editions, include *Physical Optics* (London, 1883); *Laws and Properties of Matter* (London, 1893); *Heat* (Cambridge, 1894); and *Mechanics* (Cambridge, 1895). Two of his articles of particular interest are "Life and Works of James Clerk Maxwell," in *Cambridge Review,* **1** (1879), 70, 98–99, 118–120; and "The Aims of the National Physical Laboratory of Great Britain," in *Popular Science Monthly,* **60** (Dec. 1901), 124–144, reprinted in *Smithsonian Institution Annual Report* (1901), pp. 341–357.

A list of Glazebrook's pre-1900 papers is in the Royal Society, *Catalogue of Scientific Papers.* A bibliography of his books and papers is in Poggendorff, III, 526; IV, 504; V, 42; VI, 903–904. His correspondence is in the Public Record Office; the Royal Society; and the Forbes papers, University Library, St. Andrews, Fife.

II. SECONDARY LITERATURE. Biographical information appears in Rayleigh and F. J. Selby, *Obituary Notices of Fellows of the Royal Society of London,* **2** (1936–1938), 29–56, with portrait; *Proceedings of the Physical Society of London,* **48** (1936), 929–933; and W. C. D. Dampier, *Dictionary of National Biography, 1931–1940* (London, 1949), pp. 343–344.

BERNARD FINN

GLEICHEN-RUSSWORM, WILHELM FRIEDRICH VON (*b.* Bayreuth, Germany, 14 January 1717; *d.* Schloss Greifenstein, Bonnland, Hammelburg, Germany, 16 June 1783), *microscopy.*

Gleichen-Russworm was the elder son of Heinrich von Gleichen and Caroline von Russworm. He received little formal education and in 1734, after some years as a page at the court of Prince Thurn und Taxis in Frankfurt, he decided to make his career in the forces of the margrave of Bayreuth. He married Antoinette Heidloff in 1753 and they had seven children, of whom only two daughters survived to adulthood. Gleichen-Russworm remained in the army until 1756, when he resigned his commission in order to devote himself to the management of the Greifenstein estates, inherited from his mother in 1748.

His first published writings appeared after his departure from Bayreuth, in the periodical *Fränkische Sammlung aus der Naturlehre, Arzneigelahrtheit, Ökonomie und der damit verbundenen Wissenschaften;* they deal, *inter alia,* with natural history, physics, and chemistry but are, for the most part, quite fanciful. These articles involved Gleichen-Russworm in a certain amount of controversy, with the result that his subsequent writings were much less extravagant, although he did go on to publish, in 1782, a highly imaginative account of the origin and structure of the earth, which is now of interest only for its faint adumbration of evolutionary theory.

In the summer of 1760 Gleichen-Russworm made the acquaintance of Martin Ledermüller, who had already begun publication of his *Mikroskopische Gemüths- und Augenergötzungen* (1759–1762); it was this work which led Gleichen-Russworm to concentrate on microscopy. Ledermüller visited Schloss Greifenstein in 1762, and Gleichen-Russworm continued to benefit from his advice until the former took offense at certain criticisms of his work which appeared in *Geschichte der gemeinen Stubenfliege* (1764).

Gleichen-Russworm was particularly interested in the processes of fertilization in plants and animals, and in 1763 he published the first fascicle of *Das neueste aus dem Reiche der Pflanzen.* This work contains fifty-one colored plates illustrating numerous details of floral structure and various pollens; in addition, his interest in the construction of the microscope is reflected in the six plates devoted to the different modifications and accessories which he designed for the instrument. His account of the pollen of *Asclepias syriaca* L. in *Auserlesene mikroskopische Entdeckungen* (1777–1781) contains what appears to be the first observation of a pollen tube, although he remained unaware of its significance.

In 1778 Gleichen-Russworm made his most important contribution to science. In *Abhandlung über die Saamen- und Infusionsthierchen* he described the technique of phagocytic staining, which he had developed from earlier reports of the use of dyes as coloring agents for plant and animal tissues. In order to study the nutrition of a colony of ciliates, he added water colored with carmine and observed the subsequent staining of the food vacuoles, of which he provided an illustration. This technique did not become generally known until described by a number of nineteenth-century biologists, notably Christian Gottfried Ehrenberg, Theodor Hartig, and Joseph von Gerlach.

BIBLIOGRAPHY

I. ORIGINAL WORKS. Gleichen-Russworm's writings include *Das neueste aus dem Reiche der Pflanzen, oder mikroskopische Untersuchungen und Beobachtungen der geheimen Zeugungstheile der Pflanzen in ihren Blüthen, und der in denselben befindlichen Insekten,* 3 fascs. (Nuremberg, 1763–[?]1766); *Geschichte der gemeinen Stubenfliege* (Nuremberg, 1764); *Auserlesene mikroskopische Entdeckungen bey den Pflanzen, Blumen und Blüthen, Insekten und andere Merkwürdigkeiten,* 6 fascs. (Nuremberg, 1777–[?]1781); *Abhandlung über die Saamen- und Infusionsthierchen, und über die Erzeugung; nebst mikroskopischen Beobachtungen des Saamens der Thiere, und verschiedener Infusionen* (Nuremberg, 1778); and *Von Entstehung, Bildung, Umbildung und Bestimmung des Erdkörpers* (Dessau, 1782).

II. SECONDARY LITERATURE. Works on Gleichen-Russworm's life and career include (in chronological order): M. A. Weikard, *Biographie des Herrn Wilhelm Friedrich von Gleichen* (Frankfurt, 1783); Ascherson, "Wilhelm Friedrich von G. genannt Rusworm (Russworm)," in *Allgemeine deutsche Biographie,* IX (Leipzig, 1879), 226–228; Carl Willnau [Carl W. Naumann], *Ledermüller und v. Gleichen-Russworm. Zwei deutsche Mikroskopisten der Zopfzeit* (Leipzig, 1926); John R. Baker, "The Discovery of the Uses of Colouring Agents in Biological Microtechnique," in *Journal of the Quekett Microscopical Club,* 4th ser., **1**, no. 6 (1943), p. 12, rev. separate publication (London, 1945); Friedrich Klemm, "Wilhelm Friedrich Gleichen gen. v. Russwurm," in *Neue deutsche Biographie,* VI (Berlin, 1964), 447–448; and Frans A. Stafleu, *Taxonomic Literature* (Utrecht-Zug, 1967), p. 172.

MICHAEL E. MITCHELL

GLISSON, FRANCIS (*b.* England, 1597 [?]; *d.* London, England, 16 October 1677), *medicine, philosophy.*

According to tradition and evidence from the portraits in his books of 1672 and 1677, Francis Glisson was born in 1597. Upon matriculation in Gonville and Caius College, Cambridge, at Michaelmas 1617, Rampisham, Dorset, where he had been taught by Allot, was given as Glisson's home and eighteen as his age. The latter may be doubted in view of some inconsistencies in other matriculation entries. His father, William, was born in Bristol and was designated as "gentleman" in the son's Cambridge matriculation. His mother, Mary, was the daughter of John Hancock of Kingsweston, Somerset. Exactly when the family moved to Rampisham is uncertain, as is the place of Glisson's birth.

Glisson's academic career remained connected with Cambridge: B.A. in 1620–1621, M.A. and junior fellowship in 1624; Greek lecturer in 1625–1626; dean in 1629; and senior fellow 1629–1634. In 1634 he also completed his medical studies with the M.D. degree. Two years later, he was appointed regius professor of physic (John Wallis was among his students), a position he held until his death. There exists an isolated reference to Maria, daughter of Thomas Morgan, as the wife of Glisson (R. M. Walker, "Francis Glisson and his Capsule").

In the year of his medical graduation, Glisson was admitted as a candidate by the Royal College of Physicians of London. With the possible exception of practice in Colchester during the Civil War (contested by Walker), London remained the seat of his professional and scientific life. The College of Physicians made him a fellow in 1635, councilor from 1666 on, and president in 1667, 1668, and 1669. One of the scientific pillars of the College, he was also made a reader in anatomy and appointed to give the Gulstonian lecture in 1640. Thomas Wharton, in his *Adenographia,* mentions "his most faithful friend," Glisson, as his helper in dissecting. He belonged to the "Invisible College," and was an early member of the Royal Society, elected 4 March 1660/1661.

Around 1645 a group of the fellows of the College began to exchange notes on rickets, thought to have but recently spread in England, and Glisson, G. Bate, and A. Regemorter were assigned to publish a book on the subject. The investigation of the essential nature of the disease fell to Glisson, who impressed his co-workers so much that they entrusted him with drafting the whole book, into which their own observations and possibly those of authors like Daniel Whistler (cf. ch. 13) were incorporated. *De rachitide* appeared in 1650 with Glisson as the author, Bate and Regemorter as his associates, and with five additional contributors. It is hence hard to tell how much of the classic anatomical and clinical descriptions of the disease belongs to Glisson alone. He claimed originality specifically for chapters 3–14. These are concerned mainly with the nature of the disease, which he believed to be a cold and humid distemper in which the indwelling spirit (*spiritus insitus*) of the parts primarily affected (spinal cord and peripheral nerves) was deficient and torpid. This emphasis on an inner principle was to remain throughout Glisson's life.

In *De rachitide,* as well as his other publications, Glisson incorporated empirical findings into a scholastic framework of reasoning, trying to lay a broad basis for argumentation while discussing any problem encountered on the way. Thus this work dwells on such subjects as regulation of the circulation of the blood (which was assumed as a matter of course), mechanisms of nervous function, and the nature of hereditary disease. An English translation of the book appeared in 1651, testifying to the interest it aroused.

Until the Glisson papers in the British Museum, sporadically used by various biographers, are edited, Glisson's intellectual biography must rely mainly on his published works, the prefaces of which present a running commentary on their history and interconnection. Glisson's second work, the *Anatomia hepatis* (1654), rested largely on observations made in 1640, when he had lectured on the fine structure of the liver. The work begins with *Prolegomena quaedam ad rem anatomicam universe spectantia* ("Some Prolegomena Referring to Anatomy Generally"), where he tries to reconcile the Aristotelian doctrine of the elements with that of the chemists. In this work he advocates very advanced anatomical methods such as use of the microscope and injection of colored liquids.

In the *Anatomia hepatis* proper, a section of the book of the same title, Glisson denies the continuity of the branches of the portal vein into those of the hepatic veins. He contends that the branches cross, and that the blood carried in the portal vein is separated in the liver. Its bilious fraction is sucked up by the biliary vessels, because of an attraction which Glisson variously calls similar, magnetic, or natural, and which does not differ essentially from Galen's "attractive faculty." The remaining blood is attracted by the hepatic veins. The ramifications of the portal vein, together with the bile ducts, are encased in fibrous tissue, which Glisson calls *capsula communis,* now known as "Glisson's capsule."

The book ends with a chapter on the lymphatics. Stimulated by ideas of his friend George Ent, Glisson elaborated a theory which he revised in his last medical work, the *Tractatus de ventriculo et intestinis* (1677). The theory presented itself as follows: The nerves carry a nutritive juice (*succus nutritivus*) secreted by the brain between cortex and medulla from particles of the arterial blood. The psychic spirits are the "fixed spirits" of this juice, which serves nutrition rather than the function of body fibers. As a chemical substance, the psychic spirits cannot flow fast enough to assure simultaneity of events in the brain and the peripheral parts. Nerve action is transmitted by a vibration of the nerves (caused by localized contraction of the brain), and the muscle fibers then contract because of irritability, a property which they share with all fibers of the body.

In evidence of the independence of muscle contraction from any material influx, Glisson cited the experiment which Goddard had described and performed before the Royal Society in 1666. An arm was placed in a tube which was closed at one end and provided with a gauge. The tube was sealed around the arm and then filled with water. When the muscles were contracted, the gauge registered a fall of the water level rather than a rise.

Glisson had used the word "irritability" once before, in the *Anatomia hepatis,* where it connoted the ability of a part to become irritated, that is, to perceive an irritant and to try to rid itself of it. At that time he thought of irritation as being dependent on the presence of nerves. By 1677, however, natural irritability was a property attributed to almost all living parts of the body including the blood (an idea implicit in Harvey's theory), a property independent of the nerves. Irritability presupposed perception of the irritating object, appetite to attain it (if pleasant) or to flee it (if unpleasant), and motion to realize the appetite.

In sense organs connected with the brain, natural perception was elevated to sensitive, that is, conscious, perception, and it became psychic where the fibers followed commands coming from the brain. But these higher forms of perception, depending on organization, did not supersede natural perception, without which the fibers could not perceive messages from the brain. This metaphysical doctrine of natural perception and its interdependence with appetite and all motion (unless accidentally imparted) needed philosophical elaboration.

Glisson maintained that the first draft of the *Tractatus de ventriculo et intestinis* was written around 1662 but was set aside in favor of the *Tractatus de natura substantiae energetica* (1672), dedicated to Anthony Ashley Cooper, Lord Shaftesbury, whose family Glisson had long served as physician. The work attempts to prove there is life in all bodies. In so-called inanimate bodies it is specified by their forms, whereas in plants and animals life is modified to become the vegetative soul and the sensitive soul, respectively. In animals the implanted life (*vita insita*) is duplicated and triplicated by the influx of the blood (*vita influens*) and by the psychic regulations.

This philosophical work, even more than Glisson's medical books, has a strictly scholastic form of argumentation; large parts are a running debate with Francisco Suarez, whom Glisson held in the highest esteem. Among other modern authors, Glisson pays particular attention to Bacon, Scaliger, Harvey, and Descartes. He often refers to the *vis plastica,* which he identifies with van Helmont's *archeus.* Although the terminology is reminiscent of the Cambridge Platonists, it should not be overlooked that Glisson's metaphysics was fundamentally hylozoistic and thus hardly acceptable to Ralph Cudworth, who thought of "plastic nature" as incorporeal.

Glisson's doctrine of irritability acquired fame because in later years Haller traced the origin of the term back to Glisson. But in limiting irritability to muscle contractility, Haller defined it experimentally, depriving the concept of its broad biological significance.

The doctrine of irritability does not exhaust the content of the *Tractatus de ventriculo et intestinis,* which, apart from the treatise indicated by the title, also contains a treatise on skin, hair, nails, fat, abdominal muscles, peritoneum, and omentum. Together the *Anatomia hepatis* and the *Tractatus de ventriculo et intestinis* constitute a monumental work on general anatomy and on anatomy and physiology of the digestive organs. Moreover, in the latter treatise, Glisson goes far beyond the stomach and intestinal tract. Apart from discussing the theory of digestion (there is even an appendix on fermentation), Glisson manages to include theories of embryogenesis (in which the relationship to Harvey is particularly interesting). Aware of his discursiveness, Glisson in his apology referred to "the allurement and sweetness of speculation" (p. 333).

In the battle between the ancients and the moderns Glisson belongs to neither side. In a peculiar manner all his own, he adhered to the scholasticism of his formative years (possibly sustained by his professorship in Cambridge) combining it with Helmontian chemistry, Harvey's heritage, and the new science as represented by the Royal Society.

BIBLIOGRAPHY

I. Original Works. For the Glisson papers in the British Museum, see *Index to the Sloane Manuscripts in the British Museum* (London, 1904), pp. 217 ff. The following are the first eds. of his published works: *De rachitide, sive morbo puerili, qui vulgo the rickets dicitur, tractatus, opera primo ac potissimum Francisci Glissonii . . . adscitis in operis societatem Georgio Bate et Ahasvero Regemortero* (London, 1650); *Anatomia hepatis, cui praemittuntur quaedam ad rem anatomicam universe spectantia et ad calcem operis subjiciuntur nonnulla de lymphae ductibus nuper reperta* (London, 1654); *Tractatus de natura substantiae energetica, seu de vita naturae, ejusque tribus primis facultatibus . . .* (London, 1672); and *Tractatus de ventriculo et intestinis. Cui praemittitur alius, De partibus continentibus in genere; et in specie, de iis abdominis* (London–Amsterdam, 1677). The *Opera medico-anatomica, in unum corpus collecta . . .*, 3 vols. in a single pub. (Leiden, 1691), does not contain the *Tractatus de natura substantiae energetica.*

II. Secondary Literature. Although the literature on Glisson is considerable (he is discussed in almost all histories of medicine, biology, and science), there is no comprehensive monograph on his life and work. The main biographical sketches are John Aikin, *Biographical Memoirs of Medicine in Great Britain* (London, 1780), pp. 326–338; William Munk, *The Roll of the Royal College of Physicians of London,* I (London, 1878), 218–221; John Venn, *Biographical History of Gonville and Caius College 1349–1897,* I (Cambridge, 1897), 236 f.; and Norman Moore, *Dictionary of National Biography,* VII, 1316–1317.

R. Milnes Walker, "Francis Glisson and his Capsule," in *Annals of the Royal College of Surgeons of England,* **38,** no. 2 (1966), 71–91, adds many details to Glisson's biography and suggests that "he was born in 1598 or 1599, and probably in Bristol" (p. 77) and ascribes his alleged sojourn in Colchester (1640–1648) to a confusion with his younger brother, Henry, who was practicing medicine there (p. 78). Glisson's work on the liver is presented clearly and in detail by Nikolaus Mani, *Die historischen Grundlagen der Leberforschung,* II (Stuttgart, 1967), 104–120. For Glisson's relation to traditional medicine and his concept of irritability, see Walter Pagel, "The Reaction to Aristotle in Seventeenth-Century Biological Thought," in E. Ashworth Underwood, ed., *Science, Medicine, and History. Essays . . . in Honour of Charles Singer,* I (Oxford, 1953), 489–509; Owsei Temkin, "The Classical Roots of Glisson's Doctrine of Irritation," in *Bulletin of the History of Medicine,* **38,** no. 5 (1964), 297–323, where older literature on Glisson's philosophical and biological concepts is cited; and Walter Pagel, "Harvey and Glisson on Irritability," *ibid.,* **41,** no. 6 (1967), 497–514.

On Glisson's relationship to the London College of Physicians and the Royal Society, see Charles C. Gillispie, "Physick and Philosophy: A Study of the Influence of the College of Physicians of London Upon the Foundation of the Royal Society," in *Journal of Modern History,* **19** (1947), 210–225; and C. Webster, "The College of Physicians: 'Solomon's House' in Commonwealth England," in *Bulletin of the History of Medicine,* **41,** no. 5 (1967), 393–412. On the much debated question of the interrelationship of Glisson's *De rachitide* and Daniel Whistler's Leiden dissertation of 1645, *De morbo puerili anglorum,* see Edwin Clarke, "Whistler and Glisson on Rickets," *ibid.,* **36,** no. 1 (1962), 45–61, with ample literature.

Owsei Temkin

GMELIN, JOHANN GEORG (*b.* Tübingen, Germany, 10 August 1709; *d.* Tübingen, 20 May 1755), *botany, natural history, geography.*

Johann Georg Gmelin's father (also called Johann Georg), apothecary, chemist, and academician in Tübingen, was the founder of the older branch of the Gmelin family, which included several distinguished scholars and scientists. The younger Johann Georg was extremely gifted and was early encouraged in his scientific endeavors by his father, who had a natural history collection and a laboratory. In their travels, the elder Gmelin also introduced his son to the study of the Württemberg mineral springs. From the time Gmelin was fourteen he was able to follow university lectures. He held his first disputation when he was seventeen and a year later, in 1727, graduated in medicine. Among his teachers were the philosopher and mathematician Georg Bernhard Bilfinger and the botanist and anatomist Johann Georg Duvernoy; both went to St. Petersburg in 1725 and thus determined the destination of young Gmelin's first scientific voyage.

In St. Petersburg, with the help and guidance of his teachers, Gmelin was allowed to attend meetings of the Academy of Sciences. In 1728 he was offered a fellowship and in 1730 permitted to lecture at the Academy. He became professor of chemistry and natural history in 1731, and then academician. In 1733, when he had intended to return home, Gmelin took part instead in an imperial scientific expedition to eastern Siberia with the historian Gerhard Friedrich Müller and the astronomer Louis Delisle de la Croyère. Müller was to survey archives and records, Delisle de la Croyère to determine geographical coordinates, and Gmelin to study the natural history of the territories to be visited. They were supported by a party of six students, two painters, two hunters, two miners, four land surveyors, and twelve soldiers. They were expected to join, by land, the sea expedition to Kamchatka led by Captains Bering and Chirikov.

Gmelin's expedition left St. Petersburg on 8 July 1733 for Tobolsk, which they expected to reach early in 1734 and where they hoped to make a lengthy stay. They proceeded eastward with many side expeditions, exploring territories along the Irtysh, Ob, and Tom rivers, through Krasnoyarsk to Yeniseysk (January 1735) and then through Irkutsk to the Chinese (now Mongolian) frontier at Kyakhta. In 1735 they thoroughly explored the Transbaikal region proceeding through Selenginsk and Nerchinsk, then along the Lena River to the north. In September 1735 they reached Yakutsk (130° east) from which they undertook numerous expeditions.

In November 1736 a fire destroyed most of Gmelin's equipment, instruments, books, collections, and drawings. Facing additional difficulties, Gmelin and Müller realized they could not succeed in joining the Bering-Chirikov expedition and so received permission to continue explorations on their return journey. They left Yakutsk in May 1737 to explore the regions along the Angara and Tunguska rivers. At Yeniseysk they met Georg Wilhelm Steller, a bold and tough explorer who was sent from St. Petersburg to join them. Gmelin, however, sent him to the east with a small party. (Steller thus succeeded in joining Bering and distinguished himself by reaching the Alaskan coast; he was one of the few survivors of the disastrous winter of 1741–1742 on Bering Island, and went on to explore Kamchatka.)

Gmelin meanwhile traveled to the north along the Yenisey River to 66°N. latitude, then turned to the south and reached Krasnoyarsk in February 1740. Müller separated from Gmelin's party, which next explored the region between the Yenisey and Ob, the Baraba Steppe, and then advanced to the southwest to the Ishim and Wagai steppes and to the Caspian Sea. Eventually they explored the mines in the Ural Mountains. The party reached St. Petersburg on 28 February 1743 after nine and one-half years of travel.

Upon his return Gmelin resumed his academic functions at the Academy and worked on the scientific accounts of his journey. His four-volume *Flora sibirica* (1747–1769) contains descriptions of 1,178 species and illustrations of 294 of these. Although primarily a botanist, Gmelin had a good knowledge of other natural sciences of his time and with his travels contributed to the knowledge of the zoology, geography, geology, ethnography, and natural resources (e.g., location of coal, iron, salt, and mica) of the explored regions. He used the barometer to determine altitude and was the first to find (from the average of J. J. Lerche's eleven-month barometric pressure observation in Astrakhan) that the level of the Caspian Sea is below that of the Mediterranean and Black seas. Greatly astonishing the world's scientists, in January 1735 he recorded at Yeniseysk the lowest temperatures observed anywhere up to that time. In addition, he made another important finding in parts of eastern Siberia where a subsurface layer of soil, several feet thick, remained frozen even in summer. Gmelin attempted to measure its thickness.

In his preface to the *Flora sibirica,* Gmelin gave a remarkable overall picture of the nature of central Siberia, pointing out that western Siberia looks very much like eastern Europe, but that after crossing the Yenisey River he had the impression of being in another continent. Once on the other side, he saw rivers with clear water, new forms of plants and animals, a strange landscape with strange people—in short, a new world. Thus the Yenisey River seemed to him the natural frontier between Europe and Asia, an idea which had not occurred to any geographer before him. As a whole, the results of Gmelin's expedition represent the most important early contribution to the natural history and geography of the vast Siberian mainland.

In 1747 Gmelin was granted a year's leave from the Academy of Sciences and returned to Tübingen, where he married and remained until his death. In 1749 he became professor of medicine, botany, and chemistry at the University of Tübingen. In his inaugural lecture Gmelin reported that he had observed, in his St. Petersburg garden, the appearance of five or six new forms of the plant genus *Delphinium* from two original species brought from Siberia. He, with other leading scientists of the time, tried to reconcile, in the ensuing debate, such transmutations with belief in the original creation of all species and with the accepted Linnaean position on the fixity of species. He corresponded with Linnaeus, Haller, and Steller on this and other matters of scientific interest.

BIBLIOGRAPHY

I. ORIGINAL WORKS. Gmelin's principal scientific work, considered to be a masterpiece of scientific survey, is the *Flora sibirica sive historia plantarum Sibiriae,* 4 vols. (St. Petersburg, 1747-1769); the preface contains a short account of Gmelin's travels and results of his explorations. Vols. III and IV of this work were edited after his death by his nephew, Samuel Gottlieb Gmelin, who also worked as an explorer in Asia for the Russian Academy of Sciences. The most remarkable among Gmelin's shorter academic treatises is the *Sermo academicus De novarum vegetabilium post creationem divinam exortu* (Tübingen, 1749).

Gmelin's second major work, *Reise durch Sibirien von dem Jahr 1733 bis 1743,* 4 vols. (Göttingen, 1751-1752), is an adaptation of his travel notebooks for general publication. It contains a wealth of information, but makes rather dull reading. It was published in an abridged French version, *Voyage en Sibérie contenant la description des moeurs et usages des peuples de ce pays . . .,* 2 vols. (Paris, 1767), and in Dutch. Its publication in Russia was banned because of the work's severe criticism of the Russian bureaucracy for its inefficiency, incompetence, and even malevolence.

For Gmelin's correspondence with Linnaeus, Haller, Steller, and others see T. Plieninger, ed., *Johannis Georgii Gmelini reliquiae quae supersunt* (Stuttgart, 1861). Several vols. of MS notes from Gmelin's travels have been preserved in the archives of the Academy of Sciences in Leningrad. Their description is in D. J. Litvinov, *Bibliografia flory Sibiri* (St. Petersburg, 1909), pp. 53-64.

II. SECONDARY LITERATURE. Genealogical tables of the Gmelin family and other valuable information are in Moritz Gmelin, *Stammbaum der Familie Gmelin* (Karlsruhe, 1877); 2nd ed. by Edward Gmelin in 2 vols.: *Jungere Tübinger Linie* (Munich, 1922), and *Ältere Stuttgarter Linie und ältere Tübinger Linie* (Munich, 1929). M. Gmelin also published a short biography in *Allgemeine deutsche Biographie,* IX (1879), 269-270.

Another short biography is Eyries, *Biographie universelle,* J. F. Michaud, ed. (1856), pp. 644-646. Much information is in F. A. Golder, *Bering's Voyages, an Account of the Efforts of the Russians to Determine the Relations of Asia and America,* 2 vols. (New York, 1922-1925); R. Grandmann, *Johann Georg Gmelin, 1709-1755. Der Erforscher Sibiriens. Ein Gedenkbuch,* Otto Gmelin, ed. (Munich, 1911), which contains a German trans. of the preface to the *Flora sibirica,* a selection from the *Reise durch Sibirien,* and selections from Gmelin's letters; and L. Stejneger, *Georg Wilhelm Steller, the Pioneer of Alaskan Natural History* (Cambridge, Mass., 1936). See also P. Pekarsky, *Istoria imperatorskoy Akademii Nauk,* I (St. Petersburg, 1870), 431-457.

Gmelin's contributions to Siberian geology are reported by V. A. Obruchev in his *Istoria geologicheskovo issledovania Sibiri,* I (Leningrad, 1931).

VLADISLAV KRUTA

GMELIN, LEOPOLD (*b.* Göttingen, Germany, 2 August 1788; *d.* Heidelberg, Germany, 13 April 1853), *chemistry.*

Leopold Gmelin was the third and youngest son of Johann Friedrich Gmelin, professor variously of philosophy, medicine, chemistry, botany, and mineralogy at Tübingen and a distinguished historian of chemistry. (It was through the elder Gmelin's efforts that a student chemistry laboratory was built at the university in 1783.) The family had been physicians, ministers, teachers, scientists, and apothecaries from the beginning of the sixteenth century.

Gmelin's early education was at the hands of a private tutor, in addition to which he attended his father's lectures at the university. He then attended the Göttingen Gymnasium, from which he graduated in 1804. The following summer his father sent him to work in the family apothecary shop in Tübingen, in accordance with long-standing tradition. At the same time he attended lectures on materia medica and pharmacology given by Ferdinand Gmelin (his cousin) and on medicine by K. F. Kielmeyer (husband of his cousin Lotte Gmelin). Both lecturers were professors of medicine at the University of Tübingen. He also met a number of other medical students and professors—including Justinus Kerner, Ludwig Uhland, and J. H. F. Autenrieth—with whom he was to maintain professional contact.

Johann Friedrich Gmelin died in November 1804; on his return to Göttingen in that year Gmelin worked with F. X. Stromeyer, his father's successor at the university. He passed his examinations in 1809, then returned to Tübingen to study with his former teachers until Easter 1811. He simultaneously began his doctoral researches on the black pigmentation of cattle eyes, a study that he was to continue in the laboratory of Nicolas J. Jacquin in Vienna later in 1811. He was awarded the medical doctorate by Göttingen in 1812; he had also qualified himself as a chemist and had studied mathematics with Bernhard Thibaut.

Gmelin then decided to travel in Italy for a year to broaden his command of the natural sciences. He was particularly concerned with mineralogy and geology, and therefore concentrated his interest in the regions of Mt. Vesuvius and San Marco. He published his geological findings in 1815.

Upon his return to Göttingen he undertook the analysis of the mineral haüynite under the guidance of Stromeyer. At the same time he began his academic career; he was appointed docent at Heidelberg in fall of 1813, and soon thereafter, in 1814, became extraordinary professor. In a letter of that year, addressed to his mother, he stated that "medicine in Heidelberg is deplorably organized—but soon it will be better."

Gmelin's cousin Christian Gottlob Gmelin received his medical degree in the academic year 1814; together they then went to Paris to study and work in Vauquelin's laboratory. They stayed in Paris until spring 1815; in addition to their laboratory work they attended lectures by Gay-Lussac, Thenard, Vauquelin, and occasionally those of Haüy himself.

When Gmelin returned to Heidelberg he replaced F. K. Nägele on the faculty. F. Tiedemann became teacher of anatomy and physiology in the same year. M. J. von Chelius came to Heidelberg in 1817, and these men together set about to establish scientific method in the curriculum. Gmelin was appointed director of the Chemical Institute—which still, however, remained part of the physical institute within the medical faculty. In 1817 he was made full professor, having refused an offer to succeed Klaproth in Berlin.

It is possible that Gmelin was influenced in his decision to stay in Heidelberg by his wife, Luise Maurer, whom he had married on 1 October 1816. She was a singer and the daughter of the pastor of the nearby Kirchheim church. (Certainly Gmelin turned down attractive offers not only from Berlin but also, later, from Göttingen.) They had four children.

In 1818 the Chemical Institute was transferred to its own quarters in a former Dominican cloister, an installation that also included an apartment for Gmelin. This move made the institute virtually independent of the medical faculty, and thus fulfilled one of Gmelin's long-range plans. In a letter to his mother (28 February 1818) Gmelin described the auditorium of the new facility as being roomy and having elevated benches so that all the students could see the experiments, and added that the laboratory had running water and that there were four additional rooms, including a large one that could house the mineral collections. He further noted that he had thirty students in chemistry but only four in medicine.

Although Gmelin devoted a great deal of time in these early years to improving the teaching of chemistry, he also—and more importantly—made extensive laboratory studies that embraced physiology, organic chemistry, inorganic chemistry, and mineralogy, in addition to purely theoretical studies. He published papers on almost all these subjects, as well as teaching and publishing his great *Handbuch,* which first appeared in three volumes in 1817 and 1819. Tiedemann, Friedrich Wöhler, and Leonhard were his occasional collaborators. (It is interesting to note that Gmelin persuaded Wöhler, who was enrolled at Heidelberg as a medical student, to relinquish medicine and take up chemistry; and it was through Gmelin's efforts that upon his graduation in 1823 Wöhler went to work with Berzelius.)

The *Handbuch der theoretischen Chemie* was Gmelin's masterwork. The first edition bore that title; the fourth edition, of 1843–1852, had grown to five volumes (expanded to ten by 1870) and was entitled simply *Handbuch der Chemie.* Gmelin was solely responsible for the first three multivolume editions, and was sole author of the first four volumes of the five-volume fourth edition, the fifth being compiled by Karl List and Karl Kraut.

Little is known about how or when Gmelin decided to start work on the project that became the *Handbuch.* As early as 1808 Berzelius had begun work on a textbook of several volumes but had himself realized that it could not be an all-inclusive systematic presentation. Indeed, Gmelin's father had noted the difficulty involved in such a work as early as 1780. Gmelin sought the complete, objective presentation of the prevailing state of chemistry. His father had found the science in a state of flux, with each author altering his textbook to reflect his own ideas; it was Gmelin's task to unify it through his own knowledge and—more important—the existing literature. He planned, then, to adduce all pertinent facts, arrange them by element and compound, and give appropriate references. It was necessary for him to bring calm, scholarship, and a critical eye to the data at his disposal; it was likewise necessary that he avoid speculation, which he considered to be hazardous as well as demanding of an inappropriate amount of time and effort. He kept a card-file index, and it is said of him that whenever he found that a compound did not exist as a separate entity or that it was identical to another named substance, he would remark, "Thank God, that there is one less acid."

Gmelin was unable to objectify chemistry completely, however, and some confusion about atomic weights, equivalents, and molecular theories and compounds is evident in his book. He constantly sought means to simplify or resolve these conflicts, not always with success; some of the formulas he gave reflect such irresolution. Nor did he escape the charge of supporting certain theories above others, despite his announced intention to avoid personal advocacy.

The first edition of the *Handbuch* reported on only forty-eight elements; two volumes of this edition were devoted to inorganic chemistry and one to organic. By the fourth edition (1843), fifty-five elements were discussed and the work had grown to nine volumes, of which three were devoted to inorganic chemistry and six to organic—thus demonstrating the growth of interest in organic substances and the increase in their known number.

Gmelin was aware from the time of the first edition that the major problem that he must confront would be in the treatment of organic substances. He maintained that inorganic and organic compounds must be distinguished from each other and began working toward their definition. He first suggested that while simple inorganic compounds are composed of two elements, simple organic compounds require three, and accordingly considered methane, cyanogen, and other like compounds as inorganic. In addition, inorganic compounds could be created by the chemist out of their constituent elements, while organic compounds required a plant or animal for their synthesis, the chemist being able to produce only minor modifications in them.

In the first three editions of his book Gmelin used the terms stoichiometric number, combining weight, chemical equivalent, or mixing weight to obtain equivalents. He accepted Döbereiner's idea of triads, opposed Berthollet's theory of affinities, and accepted Laurent's nucleus theory as a basis for the systematization of organic compounds (a system that Beilstein was in turn to adopt in arranging organic compounds in his *Handbuch*). He devised a system in which compounds were assigned formulas on the basis of equivalents present, and suggested smaller values for them. By the fourth edition Gmelin had adopted the atomic hypothesis and had proposed that the chemical definition of an organic substance might be that it always includes in its composition carbon and hydrogen, with the frequent addition of oxygen or nitrogen or both.

Although the *Handbuch* may quite properly be considered Gmelin's masterwork, he did a considerable amount of original research throughout his career. With Tiedemann he did pioneering work in the chemistry of digestion, reported in their two-volume *Die Verdauung nach Versuchen* (Heidelberg–Leipzig, 1826); in this work they identified choline in bile cholesterol, hematin in blood, and taurine, which Gmelin had found in ox gall in 1824 (it was later synthesized by Kolbe)—to mention but a few of their discoveries. They also studied saliva and changes in the blood. By himself Gmelin prepared potassium ferricyanide (red prussiate of potash, or Gmelin's salt); cobalticyanides, platinocyanides, croconic acid (which resulted when potassium carbonate and coal were heated); rhodizonic acid; formic acid (by distilling alcohol with manganese dioxide and dilute sulfuric acid); uric acid; and selenium. In addition he developed a test for bile pigments.

Gmelin designed and described some chemical apparatus—a drying tube for gases, a straight tube condenser, and an inverted flask to contain water for washing precipitates—and introduced the terms "racemic acid," "ester," and "ketone" into the literature. He suggested that minerals should be classified by form and composition, and reported a number of experiments on galvanism.

Gmelin was highly regarded as a teacher. He was an engaging person with a friendly face surrounded by an aureole of snow-white hair—his friends compared him to a blossoming cherry tree. His stature as a scientist won him membership in many learned societies. He resigned from the Heidelberg faculty in 1851, because of failing health, and sought to obtain the appointment of Robert Bunsen as his successor. His efforts were rewarded when Bunsen became director of the Chemical Institute in 1852. Gmelin died the following year.

The fifth edition of the *Handbuch* was under way at the time of Gmelin's death. It appeared in three volumes and five parts (1871–1886), under the editorship of Karl Kraut. In this edition the organic section of the work was dropped and the remainder entitled *Handbuch der anorganische Chemie.* In 1922 the Deutsche Chemische Gesellschaft assumed the obligation to continue the monumental work; the eighth edition, now entitled *Gmelins Handbuch der anorganische Chemie,* began publication in 1924 and is still being published. The book maintains the same authoritative position that it has always had and is a fitting tribute to Gmelin's skill and scholarship.

BIBLIOGRAPHY

I. ORIGINAL WORKS. Poggendorff provides a list of Gmelin's individual writings, in addition to those cited in the text.

II. SECONDARY LITERATURE. For works about Gmelin and his life, see E. Beyer and E. H. E. Pietsch, "Leopold Gmelin—Der Mensch, sein Werk und seine Zeit," in *Berichte der Deutschen chemischen Gesselschaft,* **72** (1939), 5–33; Eduard Farber, ed., *Great Chemists* (New York, 1961), pp. 453–463; A. Ladenburg, *History of Chemistry* (Edinburgh, 1886); *Lectures on the History of the Development of Chemistry Since the Time of Lavoisier* (Edinburgh, 1886), nos. 347 and 682, both works trans. by L. Dobbin; M. Nikolas, "Das Werk von Friedrich Tiedemann und Leopold Gmelin—die Entwicklung der Ernährungslehre in der ersten Hälfte des 19. Jahrhunderts," in *Gesnerus,* **13** (1956), 190–214; and J. R. Partington, *A History of Chemistry,* vol. IV (London, 1964).

On the occasion of an anniversary celebration at the Gmelin Institute see E. H. E. Pietsch, *Die Familie Gmelin und die Naturwissenschaften; Ein Ruckblick auf drei Jahrunderte* (Frankfurt, 1964); and *Kinder und Jugenderinnerungen der Julie G. Mayer geb. Gmelin (1817–1896), der Tochter Leopold Gmelin* (Frankfurt, 1965). Also see P. Walden, "The Gmelin Dynasty," trans. by R. E. Oesper, in *Journal of Chemical Education,* **31** (1954), 534–541.

CLAUDE K. DEISCHER

GOBLEY, NICOLAS-THÉODORE (*b.* Paris, France, 11 May 1811; *d.* Bagnères-de-Luchon, France, 1 September 1876), *chemistry.*

As a youth Gobley was apprenticed to the eminent pharmacist and chemist Pierre Robiquet, whose son-in-law he later became. After studying pharmacy in Paris and completing an internship in hospital pharmacy, in 1837 he purchased a pharmacy on the rue du Bac, which he directed until 1861. Despite heavy professional obligations, Gobley found time for chemical pursuits and in due course achieved a reputation as a distinguished chemist. From 1842 to 1847 he served as *professeur agrégé* at the School of Pharmacy, and from 1850 until his death he was a member of the editorial board of the prestigious *Journal de pharmacie et de chimie.* In 1861 he was elected to the Academy of Medicine, and in 1868 he was named a member of the Council on Hygiene and Health of the Department of the Seine.

Gobley's most significant work concerned the chemistry of phosphatides. He investigated the fatty matter in egg yolk, milt and fish eggs, venous blood, bile, and brain tissue; and in 1845 he discovered a fatty substance containing phosphorus which in 1850 he named lecithin (from the Greek *lekithos,* egg yolk). Gobley was unable to elucidate the exact chemical composition of lecithin, which he obtained in impure form, but he noted that its hydrolysis yielded fatty acids as well as glycerophosphoric acid. In 1844 Gobley found phosphorus in oil from the ray's liver and recommended this oil as a more palatable substitute for cod liver oil.

Gobley collaborated with the physiologist J. L. M. Poiseuille in a study of blood levels of urea and its secretion from the kidneys, the results of which were published in 1859. He invented an instrument called the *élaïomètre* to test the purity of oils by determining their density and first described the device in 1843. He also carried out research on biliary calculi and vanillin.

BIBLIOGRAPHY

I. ORIGINAL WORKS. A listing of Gobley's scientific papers is in the Royal Society of London, *Catalogue of Scientific Papers (1800–1863),* II (London, 1868), 924–925; VII (London, 1877), 790; X (London, 1894), 11; and in A. Goris, *Centenaire de l'internat en pharmacie des hôpitaux et hospices civils de Paris* (Paris, 1920), pp. 404–405. Gobley's most important publications are "Note sur l'élaïomètre, nouvel instrument d'essai pour les huiles d'olives," in *Journal de pharmacie et de chimie,* 3rd ser., **4** (1843), 285–297; "Mémoire sur l'huile de foie de raie," *ibid.,* **5** (1844), 306–310; "De la présence du phosphore dans l'huile de foie de raie," *ibid.,* **6** (1844), 25–26; "Sur l'existence des acides oléique, margarique et phosphoglycérique dans le jaune d'oeuf," in *Comptes rendus hebdomadaires des séances de l'Académie des sciences,* **21** (1845), 766–769; "Recherches chimiques sur le jaune d'oeuf," in *Journal de pharmacie et de chimie,* 3rd ser., **9** (1846), 1–15, 81–91, 161–174; **11** (1847), 409–417; **18** (1850), 107–119; "Recherches chimiques sur la laitance de carpe," *ibid.,* **19** (1851), 406–421; "Recherches chimiques sur les matières grasses du sang veineux de l'homme," *ibid.,* **21** (1852), 241–254; "Recherches sur la nature chimique et les propriétés des matières grasses contenues dans la bile," *ibid.,* **30** (1856), 241–246; "Recherches sur le principe odorant de la vanille," *ibid.,* **34** (1858), 401–405; "Examen chimique d'un calcul biliaire, suivi de considérations sur les différentes phases de sa formation, et sur les meilleurs dissolvants des calculs biliaires," *ibid.,* **40** (1861), 84–91; "De l'action de l'ammoniaque sur la lécithine," *ibid.,* 4th ser., **12** (1870), 10–13; "Sur la lécithine et la cérébrine," *ibid.,* **19** (1874), 346–354; "Recherches chimiques sur le cerveau," *ibid.,* **20** (1874), 98–102, 161–166; and "Recherches sur l'urée,"' in *Comptes rendus hebdomadaires des séances de l'Académie des sciences,* **49** (1859), 164–167, written with J. L. M. Poiseuille.

II. SECONDARY LITERATURE. For additional information on Gobley's life and work, see "Discours prononcé par M. le Dr. Delpech, au nom de l'Académie de médecine," in *Journal de pharmacie et de chimie,* 4th ser., **24** (1876), 329–333; *Centenaire de l'École supérieure de pharmacie de l'Université de Paris, 1803–1903* (Paris, 1904), p. 348; E.

Bourquelot, *Le centenaire du Journal de pharmacie et de chimie, 1809–1909* (Paris, 1910), pp. 71–72; J. R. Partington, *A History of Chemistry,* IV (London–New York, 1964), 485; and D. L. Drabkin, *Thudicum, Chemist of the Brain* (Philadelphia, 1958), p. 173.

ALEX BERMAN

GODDARD, ROBERT HUTCHINGS (*b.* Worcester, Massachusetts, 5 October 1882; *d.* Baltimore, Maryland, 10 August 1945), *physics, rocket engineering.*

After the advent of ballistic missiles and space exploration, Goddard became posthumously world-famous as one of three scientific pioneers of rocketry. Like the Russian hero Konstantin Tsiolkovsky and the German pioneer Hermann Oberth, Goddard worked out the theory of rocket propulsion independently; and then almost alone he designed, built, tested, and flew the first liquid-fuel rocket on 16 March 1926 near Auburn, Massachusetts. Although Goddard seriously studied experimental physics throughout his life, whether teaching or doing applied research for the government, he began to dream of astronautics in 1899 and rocket engineering remained his prime preoccupation.

Raised by his old-line Yankee family in middle-class suburbs of Boston, Goddard was a studious child whose academic development was thwarted by ill health. He graduated from Worcester's South High School in 1904 and from Worcester Polytechnic Institute in 1908. Beginning graduate work in physics immediately at nearby Clark University, he obtained the M.A. and Ph.D. there in 1910 and 1911, respectively. Under the tutelage of A. G. Webster, Goddard studied radio devices, particularly the thermionic valve, electromagnetism in solids, and both solid and liquid propulsion for reaction engines. Following a year's research at Princeton (1912–1913), he returned to Clark to teach and rose to a full professorship by 1919.

Having explored the mathematical practicality of rocketry since 1906 and the experimental workability of reaction engines in laboratory vacuum tests since 1912, Goddard began to accumulate ideas for probing beyond the earth's stratosphere. His first two patents in 1914, for a liquid-fuel gun rocket and a multistage step rocket, led to some modest recognition and financial support from the Smithsonian Institution. During World War I, Goddard led research on tube-launched rockets that became the bazookas of World War II, and during the latter war he worked primarily on jet-assisted takeoff (jato) and variable-thrust rockets for aircraft, barely living to see evidence of the German V-2 rockets and to hear of Hiroshima.

The publication in 1919 of his seminal paper "A Method of Reaching Extreme Altitudes" gave Goddard distorted publicity because he had suggested that jet propulsion could be used to attain escape velocity and that this theory could be proved by crashing a flash-powder missile on the moon. Sensitive to criticism of his moon-rocket idea, he worked quietly and steadily toward the perfection of his rocket technology and techniques. With an eye toward patentability of demonstrated systems and with the aid of no more than a handful of technicians, Goddard achieved a series of workable liquid-fuel flights starting in 1926. Through the patronage of Charles A. Lindbergh, the Daniel and Florence Guggenheim Foundation, and the Carnegie and Smithsonian institutions, the Goddards and their small staff were able to move near Roswell, New Mexico. There, during most of the 1930's, Goddard demonstrated, despite many failures in his systematic static and flight tests, progressively more sophisticated experimental boosters and payloads, reaching speeds of 700 miles per hour and altitudes above 8,000 feet in several test flights. Among Goddard's successful innovations were fuel-injection systems, regenerative cooling of combustion chambers, gyroscopic stabilization and control, instrumented payloads and recovery systems, guidance vanes in the exhaust plume, gimbaled and clustered engines, and aluminum fuel and oxidizer pumps.

Although his list of firsts in rocketry was distinguished, Goddard was eventually surpassed by teams of rocket research and development experts elsewhere, particularly in Germany. By temperament and training Goddard was not a team worker, yet he laid the foundation from which team workers could launch men to the moon. Early in the 1960's the National Aeronautics and Space Administration named its first new physical facility at Greenbelt, Maryland, after Goddard; and the government awarded his estate one million dollars for all rights to the collection of over 200 Goddard patents.

BIBLIOGRAPHY

I. ORIGINAL WORKS. Goddard's writings include "A Method of Reaching Extreme Altitudes," in *Smithsonian Miscellaneous Collections,* **71,** no. 2 (1919), and "Liquid-Propellant Rocket Development," *ibid.,* **95,** no. 3 (1936), both reprinted in Goddard's *Rockets* (New York, 1946). See also *Rocket Development: Liquid-Fuel Rocket Research, 1929–1941,* Esther C. Goddard and G. Edward Pendray, eds. (New York, 1948); "An Autobiography," in *Astronautics,* **4** (Apr. 1959), 24–27, 106–109; and *The Papers of Robert H. Goddard,* Esther C. Goddard and G. Edward Pendray, eds., 3 vols. (New York, 1970), based on a voluminous MSS collection at Robert H. Goddard Memorial Library. Clark University, Worcester, Massachusetts. Microfilm and artifacts of Goddard's work are in the Goddard Wing of the Roswell, New Mexico, Museum and Art Center.

II. SECONDARY LITERATURE. On Goddard or his work, see Wernher von Braun and Frederick I. Ordway, III, *History of Rocketry and Space Travel,* rev. ed. (New York, 1969), pp. 40–59; Eugene M. Emme, ed., *The History of Rocket Technology* (Detroit, 1964), pp. 19–28; Bessie Z. Jones, *Lighthouses of the Skies: The Smithsonian Astrophysical Observatory, Background and History, 1846–1955* (Washington, D.C., 1965), pp. 241–276; Milton Lehman, *This High Man: The Life of Robert H. Goddard* (New York, 1963), the authorized biography; and Shirley Thomas, *Men of Space: Profiles of the Leaders in Space Research, Development, and Exploration,* I (Philadelphia, 1960), 23–46.

LOYD S. SWENSON, JR.

GODFREY, THOMAS (*b.* Bristol Township, Pennsylvania, 1704; *d.* Philadelphia, Pennsylvania, December 1749), *technology.*

Godfrey's major contribution was his invention of the double reflecting quadrant which became generally known as Hadley's quadrant and is, essentially, the navigational sextant used today. It quickly replaced other instruments for measuring elevation.

Godfrey produced his instrument in October 1730 and had it tested in Delaware Bay and on a voyage to Jamaica. James Logan, the most learned man in Pennsylvania, sent a description of the device to the astronomer royal, Edmond Halley, but received no reply and was soon surprised to find in the *Philosophical Transactions of the Royal Society* an account of an almost identical instrument invented by John Hadley, a fellow of the Royal Society. Fearing that Godfrey had unjustly lost credit for his invention, Logan collected affidavits on the chronology of the invention, obtained reports from Jamaica, and wrote again to Halley as well as to several of his friends in the Royal Society. Godfrey, too, fowarded a presentation of his case. The society heard these papers and published Logan's account in the next volume of *Philosophical Transactions.* Hadley clearly had the priority of publication, but Godfrey's invention was solely his own in a day when identical independent inventions were not easily accepted.

Godfrey was an important member of a small intellectual circle in Philadelphia. A glazier by trade, he developed an impressive command of mathematics and, with the help of Logan and Logan's extensive library, learned and used Latin. He was a founding member of Benjamin Franklin's 1727 Junto and a director of the Library Company of Philadelphia from its establishment in 1731. The 1743 American Philosophical Society included Godfrey as its "Mathematician."

Godfrey published almanacs from 1729 to 1736. He also contributed mathematical questions and answers, astronomical data, and general essays to the *Pennsylvania Gazette* and the *Pennsylvania Journal.* In 1740 he advertised instruction in navigation, astronomy, and mathematics. To fix the longitude of Philadelphia on his 1755 *General Map of the Middle British Colonies,* Lewis Evans used astronomical observations that he made with Godfrey.

BIBLIOGRAPHY

Godfrey issued *An Almanack for the Year 1730* (Philadelphia, 1729) and other annual almanacs until *The Pennsylvania Almanack for 1737* (Philadelphia, 1736). He made occasional contributions to the *Pennsylvania Gazette* and the *Pennsylvania Journal.*

On the quadrant, the key writings are Godfrey's letter to the Royal Society (9 Nov. 1732), in the Royal Society library (microfilm copy in American Philosophical Society Library), and James Logan, "An Account of Mr. T. Godfrey's Improvement of Davis's Quadrant," in *Philosophical Transactions of the Royal Society,* **38** (1733–1734), 441–450. Both publications and other Logan letters appear also in *American Magazine,* **1** (1757–1758), 475–480, 529–534. Other related Logan correspondence is in the Royal Society library and in the Logan Papers of the Historical Society of Pennsylvania.

Nathan Spencer, "Essay of a Memorial of Thomas Godfrey, September 8, 1809," MS., American Philosophical Society, is helpful. Frederick B. Tolles, *James Logan and the Culture of Colonial Pennsylvania* (Boston, 1957), pp. 202–204, is a brief but understanding account of Logan.

BROOKE HINDLE

GODIN, LOUIS (*b.* Paris, France, 28 February 1704; *d.* Cádiz, Spain, 11 September 1760), *astronomy.*

The son of François Godin, a lawyer in the Parlement, and Elisabeth Charron, Louis received his early training at the Collège de Beauvais. Although his courses in humanities were intended as a background for legal studies, he turned thereafter to philosophy and ultimately to astronomy, in which he received instruction under Joseph Delisle at the Collège Royal.

Having entered the Academy of Sciences in 1725, Godin presented his first memoir there the following year. Inspired by an appearance of a meteor which had frightened many people, Godin addressed himself to such transient phenomena, offering, for example, both a history and a physical explanation of displays of northern lights. It was partially the superiority of that historical analysis that led the Academy to involve Godin in its own historical project. For the next few years Godin concerned himself with editing the eleven volumes of the Academy's *Mémoires* from 1666 to 1699, the writing of its *Histoire* for nineteen of those years, and the preparation of a four-volume index of the materials included in this basic collection (*Histoire et les Mémoires de l'Académie Royale des Sciences*) from 1666 to 1730.

As time-consuming as these activities were, they did not prevent Godin from engaging in astronomical work. In addition to observing some eclipses at the royal observatory, he brought out, in 1727, an appendix to Philippe de La Hire's astronomical tables, a work which indicated his suitability for assuming, in 1730, the preparation of the Academy's annual ephemeris, the *Connaissance des temps;* he continued this task until 1735. Meanwhile, Godin had obtained his own observational site, where he viewed various eclipses of 1731, 1732, and 1733 and duly reported the observations to the Academy. The observation dealing with the lunar eclipse of 1732 was more than a simple report, since it offered a comparison with corresponding observations elsewhere and utilized these data to deduce longitudinal differences between observation sites. Moreover, his concern with the phenomenon in general led him to propose a method for the determination of lunar parallax by means of lunar eclipses.

Of Godin's other works of this period, about half were devoted to various instrumental and observational problems. These included a memoir and a later addendum on the construction, verification, and placing of a mural quadrant in the plane of the meridian, a description of a commodious observational tower, a means of determining the height of the pole independently of refraction, and a method for observing the variation of the magnetic needle at sea. The other half dealt with various aspects of planetary theory and positional changes of standard reference lines and points. Included here were memoirs concerned with the problem of the place of greatest reduction from the ecliptic to the equator and with a method for determining planetary nodes, with the apparent movements of the planets in epicycles, and with the diminution of the obliquity of the ecliptic and its amount. In none of these works was Godin responsible for any new insight or basic improvement.

Similar judgment would also apply to the one remaining memoir of this period, the 1733 paper on a means for tracing parallels of latitude. But because it contained reflections on the proportions of these circles in differing figures of the earth, this memoir led Godin soon thereafter to propose that the Academy send an expedition to the equator to resolve the issue between the "Cassinians" and the "Newtonians" with their respective views of the earth's prolateness or oblateness. Having accepted this plan, the Academy logically named Godin to undertake this task, along with Pierre Bouguer and Charles de La Condamine. He went first to England to consult with Edmond Halley and other astronomers; there he was received into the Royal Society and furnished with several instruments. As it turned out, however, Godin contributed little to the expedition.

Despite great and various difficulties, the members of the expedition did ultimately measure an arc of about three degrees in Ecuador, a province of the Spanish viceroyalty of Peru. Two slightly different figures for the length of a degree were arrived at by Bouguer and La Condamine, each of whom published an account of the voyage after returning to Paris. On the basis of his separate effort undertaken with Jorge Juan and Antonio de Ulloa, the two Spanish naval officers whose collaboration was one of the costs of Spain's cooperation, Godin produced still another figure. Because the later-dispatched but earlier-completed expedition to Lapland had already resolved the basic issue in favor of oblateness, these equatorial figures immediately served only as verifications, although subsequently they were employed in the calculations establishing the metric system.

Godin never published his account of the voyage, despite subsequent claims that he was working on it, but he did accomplish other works during this period. La Condamine related Godin's 1737 experiments on the speed of sound, and Godin himself reported to the Academy on the length of the seconds pendulum observed at Santo Domingo on the way to Ecuador and on a lunar eclipse viewed in Quito in 1737. Finally, in 1738, he submitted a memoir on a method for determination of solar parallax.

Bouguer and La Condamine left Peru in 1743; Godin stayed on, as professor of mathematics at the University of San Marcos, until 1751. After a year in Paris, during which he fruitlessly sought the return of his academic place and pension, he went to Spain and became the director of the Academy of Naval Guards at Cádiz. Although he returned to Paris briefly in 1756, reentering the Academy in "veteran" status and participating in a base-line verification, the preparation of a mathematics course for his Cádiz students was the principal occupation of his last years. Like his expedition account, a planned astronomical bibliography and a collection of astronomical observations remained unrealized.

Godin's 1728 marriage to Rose-Angélique le Moyne produced a son and a daughter, both of whom predeceased him. Godin died in 1760, following an attack of apoplexy.

BIBLIOGRAPHY

I. ORIGINAL WORKS. Godin's only publication independent of Academy sponsorship was his *Appendice aux tables astronomiques de Lahire* (Paris, 1727). Under the aegis of the Academy, he constructed five vols. of the *Connaissance des temps,* drew up and published the four-volume *Table alphabétique des matières contenues dans l'Histoire et les Mémoires de l'Académie royale des sciences* . . . (Paris, 1734), in the preparation of which collection he participated and presented many papers. His first contribution was "Sur le météore qui a paru le 19 octobre 1726," in the *Mémoires* for 1726, 287-302. His reports of eclipses may be seen in the *Mémoires* as follows: lunar (1729), 9-11, 346-349; (1731), 231-236; (1732), 484-494; (1733), 195-197; (1739), 389-392; solar (1726), 330-331; (1733), 149-150. His suggestions for lunar and solar parallax determinations appeared in the *Mémoires* as "Sur la parallaxe de la lune" (1732), 51-63; and "Méthode de déterminer la parallaxe du soleil par observation immédiate" (1738), 347-360.

Godin's pendulum observation, "La longueur du pendule simple, qui bat les secondes du temps moyen, observée à Paris et en petit Goave en l'île Saint-Domingue," was in *Mémoires* (1735), 505-521. Most of his other instrumental and observational offerings were also in the *Mémoires:* "Du quart de cercle astronomique fixe" (1731), 194-222; "Addition qu'il faut faire aux quarts-de-cercle fixes dans le méridien" (1733), 36-39; "Méthode nouvelle de trouver la hauteur de pôle" (1734), 409-416; "Méthode d'observer la variation de l'aiguille aimantée en mer" (1734), 590-593; his tower description, however, appeared in M. Gallon, ed., *Machines et inventions approuvées par l'Académie royale des sciences depuis son établissement jusqu'à présent, avec leurs descriptions,* 7 vols. (Paris, 1735-1777), VI, 49-52.

Godin's remaining contributions to the *Mémoires* were "Solution fort simple d'un problème astronomique d'où l'on tire une méthode nouvelle de déterminer les noeuds des planètes" (1730), 26-33; "Des apparences du mouvement des planètes dans un épicycle" (1733), 285-293; "Méthode pratique de tracer sur terre un parallèle par un degré de latitude donnée; et du rapport du même parallèle dans le sphéroïde oblong et dans le sphéroïde aplati" (1733), 223-232; and "Que l'obliquité de l'écliptique diminue, et de quelle manière; et que les noeuds des planètes sont immobiles" (1734), 491-502.

II. SECONDARY LITERATURE. The "official" *éloge* for the Academy was written by Grandjean de Fouchy, another Delisle student and Godin's friend, and appeared in the *Histoire de l'Académie* . . . (1760), 181-194. Although mentioning, and lauditorily analyzing, many of his memoirs and other works, this *éloge* does not provide explicit citations; better for the latter, although weak on analysis, are several subsequent biographical treatments: J. M. Quérard, *La France littéraire ou Dictionnaire bibliographique des savants* . . ., 10 vols. (Paris, 1827-1839), III, 391; J. F. Michaud, ed., *Biographie universelle,* 45 vols. (Paris, 1843-1858), XVII, 23; and Niels Nielsen, *Géomètres français du dix-huitième siècle* (Paris, 1935), pp. 192-195. The best general account of his astronomical work, with fair bibliographical information, is J. B. J. Delambre, *Histoire de l'astronomie au dix-huitième siècle* (Paris, 1827), pp. 331-336; he also provides a separate treatment of Godin's arc-measurement venture in G. Bigourdan, ed., *Grandeur et figure de la terre* (Paris, 1912), pp. 85-145.

Recent treatments of the problem of the shape of the earth are the brief but general account of Seymour L. Chapin, "The Size and Shape of the World," in *UCLA Library Occasional Papers,* no. 6 (1957), 1-7; and the large-scale account of the mid-1730's expeditions by Tom B. Jones, *The Figure of the Earth* (Lawrence, Kans., 1967), esp. ch. 6.

On Godin's observational site, see G. Bigourdan, *Histoire de l'astronomie d'observation et des observatoires en France,* II (Paris, 1930), 42-47.

SEYMOUR L. CHAPIN

GODWIN-AUSTEN, ROBERT ALFRED CLOYNE (*b.* Guildford, England, 17 March 1808; *d.* Guildford, 25 November 1884), *geology.*

Son of Sir Henry Edmund Austen of Shalford House, Guildford, and Anne Amelia Bate, Godwin-Austen was educated in France and subsequently at Oriel College, Oxford, taking his B.A. in 1830 and being elected to a fellowship of his college. He was also a student at Lincoln's Inn. His interest in the discipline to which he was to devote his life was kindled by William Buckland at Oxford, and among his early friends he numbered Charles Lyell, Leonard Horner, and Roderick Murchison, who sponsored, also in 1830, his election to the Geological Society of London. In 1833 Godwin-Austen married Maria Elizabeth Godwin, only daughter and heiress of General Sir Henry Thomas Godwin, who commanded the British army in Burma. Upon the death of his father-in-law in 1854, Austen added the name of Godwin to his own by royal license.

Godwin-Austen is remembered among geologists for his contributions to the stratigraphy of southern England, as one of the first European paleogeographers, and for his prediction that a coalfield would be discovered beneath the younger rocks of Kent. His first original work was devoted to the limestones and slaty rocks of southeast Devon, where he had settled after his marriage. Henry de la Beche, who in 1835 founded the Geological Survey of Great Britain, encouraged the young man by relying upon him for the geological lines on the map covering the district between Dartmouth and Chudleigh. Austen was, however, sufficiently independent of mind to resist the introduction, proposed by Adam Sedgwick, Murchison, and William Lonsdale, of the Devonian system.

After 1840, when he moved to Chilworth manor house, near Guildford, Godwin-Austen began to devote his considerable energies to the geology of Surrey. Here his interests included the fossil faunas of the Cretaceous rocks, the origin of the phosphatic deposits, and the succession in the Tertiary sands; his work on the structure of the Weald led him to conclude that the folding postdated the deposition of the lower Tertiaries. Now he was beginning to view the stratigraphical data in a wider context, to derive a picture of seas advancing and retreating over western Europe. Pursuing these conceptions, Godwin-Austen visited the coalfields of northern France and studied the structure of the Ardennes. In 1856 he produced what remains his best-known paper, suggesting a possible extension of the coal measures beneath southeast England. Maintaining, on theoretical grounds, that the coal-bearing strata of England, France, and Belgium once formed part of a continuous formation, he traced its breakup by folding and erosion, calling attention to the probability that in the east-west belt between the Ardennes and Bristol, coal measures basins other than those at the two extremities should exist. Godwin-Austen's views attracted interest, and in his last paper, published in 1879, he was still advocating them. They were not, however, vindicated until six years after his death, when a borehole drilled at the foot of Shakespeare Cliff near Dover proved coal measures beneath the chalk and led to the development of the Kent coalfield.

Godwin-Austen was also a pioneer in the elucidation of the history of the English Channel and among the first marine geologists. He was elected a fellow of the Royal Society in 1849. In awarding him the Wollaston Medal, premier award of the Geological Society of London, Murchison said in 1862 that he was "pre-eminently the physical geographer of bygone periods," a description amply justified by the essay on the European seas, begun by his friend Edward Forbes and completed by Godwin-Austen.

BIBLIOGRAPHY

I. ORIGINAL WORKS. A comprehensive list of Godwin-Austen's writings is given in the article by Woodward cited below. Among them are "On the Valley of the English Channel," in *Quarterly Journal of the Geological Society of London,* **6** (1850), 69–97; "On the Possible Extension of the Coal-Measures Beneath the South-Eastern Part of England," *ibid.,* **12** (1856), 38–73; *The Natural History of European Seas,* begun by E. Forbes, edited and completed by Godwin-Austen (London, 1859); and "On Some Further Evidence as to the Range of the Palaeozoic Rocks Beneath the South-East of England," in *Report of the British Association for the Advancement of Science for 1879* (1879), pp. 227–229.

II. SECONDARY LITERATURE. On Godwin-Austen or his work, see T. G. Bonney, "Anniversary Address of the President," in *Quarterly Journal of the Geological Society of London,* **41** (1885), 37–39; J. G. O. Smart, G. Bisson, and B. C. Worssam, "Geology of the Country Around Canterbury and Folkestone," in *Memoirs of the Geological Survey* (1966), pp. 16–30; and H. Woodward, "Robert Alfred Cloyne Godwin-Austen," in *Geological Magazine,* n.s. decade 3, **2** (1885), 1–10.

K. C. DUNHAM

GOEBEL, KARL (*b.* Billigheim, Baden, Germany, 8 March 1855; *d.* Munich, Germany, 9 October 1932), *botany.*

Although his full style was Karl Immanuel Eberhard Ritter von Goebel, he called himself simply Karl Goebel throughout his life. He was one of that group of independent German investigators who, in the latter half of the nineteenth century, revitalized botany and made it a wide-ranging experimental science. Goebel brought this activity well into the twentieth century; indeed, the *Organographie der Pflanzen,* his major work, was not fully published in its final form until 1933 (the year following his death).

Intended by his family for the church, Goebel was educated at the Evangelical College of Blaubeuren and thence, at the age of eighteen, went to study theology and philosophy at Tübingen. There he came under the influence of Wilhelm Hofmeister and realized his true inclination toward science. After some mental conflict—since he did not wish to disappoint his mother—he persuaded her of his interest and devoted himself to botany. At this time he also began his extensive botanical travels. Hofmeister fell ill before Goebel completed his training, and in 1876 he transferred to Strasbourg to study with Heinrich Anton de Bary. Here he took his doctorate.

Goebel then worked for a short time at the biological station at Naples. The following year Julius von Sachs appointed him his assistant at Würzburg, and he qualified as *Privatdozent.* He thus had the advantage of contact with three of the outstanding botanical figures of his time—he received inspiration from Hofmeister in the widest aspects of the newly arising morphology; from de Bary in plant anatomy and mycology; and from Sachs in physiology.

After four years as Sachs's assistant Goebel was appointed professor at Rostock and in 1887 at Marburg. In 1891 he moved to Munich, where he was professor of botany and later general director of the State Scientific Collections, retiring emeritus in 1931.

Goebel's principal administrative achievement in Munich was the removal of the botanical laboratories and gardens from their cramped quarters near the main railway station in the center of town to the edge of the Bavarian royal park at Nymphenburg. The gardens and greenhouses were his particular concern; through the years he enriched them with specimens collected on his many journeys (which took him to the Rockies, the Andes, the New Zealand Alps, and the Indian Ghats, among other places). Under his supervision the installation became second in Germany only to that in Berlin. An acute contemporary noted that Goebel had planned the greenhouses and gardens with such thought and consideration that, although their unity bore his unmistakable imprint, his assistants had not thought their ideas disregarded. Goebel also contributed several new editions of the guide to the collections. He wished to provide students and researchers in the institute itself with spacious, practical working and teaching laboratories, and there are indications that its vast, ornate entrance hall, with its marble and polychrome mosaics, was not of his planning or even desiring.

As a botanist, Goebel drew upon that great wealth of data accumulated in the late nineteenth century through use of the compound light microscope. One of the great problems of the time was to decide how this great assembly of facts could best be codified and studied, and it was Goebel's contribution to see and industriously apply a profitable method to the existing corpus of information and to his own collections. That he early established a reputation as a botanist of wide knowledge and great objectivity is testified to by Sachs's decision to ask Goebel to assist him in the preparation of a new edition of his textbook, entrusting him with the sections on systematics and special morphology.

Goebel preferred objective research to speculation. For him this meant observation and simple experimentation on the great variety of living plants. To a friend he expressed regret that he did not have the knowledge of the exact sciences necessary to carry his experimental studies to the biochemical level. He was, however, impatient of gadgetry and the niceties of preparation and rarely used a microtome for his sections.

An unfortunate by-product of Darwinism among botanists had been an excessive preoccupation with phylogenetic speculation. This Goebel despised. He once said, for example, that concern about the "natural system" of seed plants seemed about as hopeless as attempting to return to its original paper bags the confetti scattered during the Munich carnival. His attitude was best expressed in his introduction to the *Organographie:* "I take exactly the same view as Herbert Spencer. . . . He says 'Everywhere structures in great measure determine functions; and everywhere functions are incessantly modifying structures. In nature the two are inseparable co-operators; and science can give no true interpretation of nature without keeping their co-operation constantly in view.' "

The parts of plants might therefore be modified during their individual life by the effects of the surroundings upon their functions, as when a spiny plant, grown in a moist atmosphere under a bell jar, became leafy. Such changes were open to experimental measurement and proof. Nevertheless, Goebel clearly realized that organ primordia are inherited with properties that "belong to the capacity of the plant itself." The dependence of metamorphosis on both racial and individual characteristics and experiences would remain true, he taught, even if a general theory of descent were abandoned. Plants consist of operating and adjustable organs, and their study is an organography.

By the middle of the nineteenth century plant studies, which up until then had been mainly observational, had begun to harden into a rigid formalism. The significance of Goebel's organography for botanical science lay in the fact that it provided one of the main bridges from the achievements of observation to the fully fledged experimental science of the twentieth century. Earlier attempts had been premature; organography itself was transitional because methods were not yet in existence to enable it to be carried to its logical conclusion. Goebel, however, lived long enough to see the first developments of experimental plant physiology and biochemistry, although not perhaps their coordination with the subtler levels of structure, to which his organography had pointed.

Goebel's success as a teacher resulted from his clarity and impartiality, but he is said to have underrated his didactic powers—although adding, "If an angel from heaven came down to give the botany lectures, the medics would still not turn up." Yet the records show that many completed the full fifty hours prescribed. He gave freely of his time to his pupils and received many advanced students from abroad.

His aloofness in the lecture hall and laboratory appears to have been a pedagogic device which Goebel deemed useful. With his students in the field, or even during evening discussions, he was more relaxed; and those botanists who have left records of meeting him on his travels have all done so with affectionate admiration.

Beyond the vast knowledge of his special subject, Goebel had a cultivated and philosophic mind. He quoted freely from the Bible, although with age he became increasingly cool toward the church. His contemporaries regarded Goebel as the exponent of an extreme materialistic view of living things, yet he thought highly of Henri Bergson's concept of *élan vital,* could speak of the Logos in nature, and held Hegelian viewpoints to which he had been introduced as a student at Tübingen. The objectivity of his ideas and their freedom from the rigidity and speculation that were simultaneously besetting the older schools made his influence on later work very considerable. He died as the result of a fall suffered while botanizing in his native Swäbische Alb.

BIBLIOGRAPHY

I. Original Works. Goebel published more than 200 works, many of which appeared in *Flora,* which he edited until 1932. A full list of Goebel's publications is in the obituary by Karsten (see below). His principal work, *Organographie der Pflanzen,* had several eds.: 1st ed., 2 vols. (Jena, 1898–1901), English trans. (Oxford, 1900–1905); 2nd ed., 3 vols. (Jena, 1915–1923); 3rd ed., 3 vols. (Jena, 1928–1933).

II. Secondary Literature. An obituary notice is G. Karsten, in *Berichte der deutschen botanischen Gesellschaft,* **50** (1932), 131–162. Otto Renner, "Erinnerungen an K. Goebel," in *Flora,* **131** (1936), v–xi, is an excellent sketch of Goebel as man and scientist.

W. O. James

GOEDAERT, JOHANNES (*b.* Middelburg, Netherlands, *ca.* 19 March 1617; *d.* Middelburg, February 1668),[1] *entomology.*

Goedaert was the son of Pieter Goedaert and Judith Pottiers. (The family name is variously spelled Goedhart or Goedaerdt or latinized as Goedartius or Goedardus.) The occupation of the elder Goedaert and the religion of the family are unknown, but it is quite probable that they were members of the Dutch Reformed Church.[2] Little is known of Goedaert's life. He probably did not receive a secondary education. He apparently did not know Latin, and wrote his only book in Dutch; the Latin translation, which is the best known, was the work of others (see below). He certainly did not attend a university.

Goedaert is remembered as a painter, more particularly as a watercolorist, whose subjects were mainly birds and insects. According to his biographer C. de Waard, it appears that he had some knowledge of chemistry and pharmacy; it is reported that he knew how to make a remarkable extract of *Artemisia absinthium* and how to eliminate the tendency of antimony to cause vomiting. Goedaert lived all his life in Middelburg. He married Clara de Bock and by her had one daughter and one son, Johannes, who became a surgeon.

One of the earliest authors on entomology, Goedaert was the first to write on the insects of the Netherlands. More important, he was the first to base his discussions entirely on firsthand observation instead of making the traditional appeal to authority, citing and paraphrasing the work of predecessors. In his only work, *Metamorphosis naturalis,* he describes his observations of and experiments with insects made between 1635 and 1658.

Basically, Goedaert's technique was to catch "worms" (larvae) in the field and to rear them, feeding them with their natural nutrients and observing and recording their metamorphosis, until finally the mature animal could be observed and drawn. In this way, he studied the life cycles of a variegated collection of butterflies, bees, wasps, flies, and beetles. Goedaert's pioneering work is not without its faults. He made no attempt to devise a system of insect classification such as Swammerdam would do as early as 1669, and although aware that most of his "worms" originated from eggs, he believed some were produced by spontaneous generation. The year of his death saw the publication of Redi's *Esperienze intorno alla generatione degli insetti,* in which the possibility of spontaneous generation was denied for the first time. Goedaert's fieldwork was not sufficiently extensive to enable him to solve the problem of the ichneumon wasps. He reported that of two identical caterpillars, one yielded a beautiful butterfly and the other no less than eighty-two little flies. He also made anatomical errors, for example when he related the position of a caterpillar's legs to the position of the pupa in the cocoon.

Goedaert published two volumes of his book during his lifetime, in 1662 and in 1667. A third, posthumous volume was edited by Johannes de Mey from his papers at the request of his widow. De Mey also translated the first and third volumes into Latin, adding commentaries of his own on insects and comets, mostly the kind of material that Goedaert has been praised for omitting. The second volume was translated into Latin by P. Veezaerdt, who abstained from comment but added a chapter of philosophical speculations on insects. The zoologist M. Lister produced an English translation in which he tried to organize the contents of the book, and it is therefore difficult to compare it with the original. He also edited another Latin edition. Finally, a French translation was published in 1700. Several authors have offered identifications of the insects described by Goedaert.

The faults in Goedaert's work were recognized soon after his death. Swammerdam spent three folio pages criticizing him,[3] but he also added a few words of praise: "... but at the same time we own with satisfaction that this author alone observed and discovered, in the space of a few years, more singularities in the caterpillar kind, than had been done by all the learned men who treated the subject before him."

NOTES

1. Dates according to C. de Waard; the date given for birth is day of baptism. A. Schierbeek gives the birth year as 1620. P. J. Meertens states (p. 472, n. 280) that Goedaert was buried in Middelburg on 15 Jan. 1668.
2. It will never be possible to ascertain these points; the archives of Middelburg were destroyed, together with the city hall, in May 1940.
3. Jan Swammerdam, *The Book of Nature* (London, 1758), pp. 14–17, *passim.* This is Thomas Flloyd's trans. of the *Biblia naturae.* Goedaert is mentioned many times and sometimes criticized in this and in other of Swammerdam's works.

BIBLIOGRAPHY

I. Original Works. The bibliography of Goedart's only book is somewhat complicated; it is detailed in Kruseman's paper (see below). A summary account follows:

Metamorphosis naturalis, ofte historische beschrijvinge van den oirspronk, aerdt, eygenschappen ende vreemde veranderinghen der wormen, rupsen, maeden, vliegen, witjens, byen, motten en diergelijke dierkens meer; niet uit eenige boeken, maar aleenlyck door eygen ervarentheid uytgevonden, beschreven ende na de konst afgeteykent door Johannum Goedaerdt, 3 vols. (Middelburg, vol. I, 1662; vol. II, 1667; vol. III, 1669), carried the imprint of J. Fierens. This edition was reissued in 1700 (probably the same sheets of the 1662–1669 edition) but with a French title page as well as the original Dutch one. The French title page has the imprint of Adrian Moetiens of The Hague; there is an engraved frontispiece, which gives "Amsterdam, 1700."

Metamorphosis et historiae naturalis insectorum, autore Joanne Goedartio; cum commentariis D. Joannis de Mey... (Middelburg, 1662 [date of dedication]).

Metamorphoseos et historiae naturalis, pars secunda, De insectis, autore Joanne Goedartio, latine donata... a Paulo Veezaerdt (Middelburg, 1667 [date of preface]).

Metamorphoseos et historiae naturalis insectorum, pars tertia et ultima, autore Joanne Goedartio aucta observationibus et appendice D. Joannis de May (Middelburg [1669]).

Johannes Goedartius, Of Insects. Done Into English and Methodized, With the Addition of Notes by Martin Lister Esq. The Figures Etched Upon Copper by Mr. F. Pl. (York, 1682). The engraver was Francis Place (1647–1728). Only 150 copies were printed. See *Philosophical Transactions of the Royal Society,* **13** (1683), 22–23.

Johannes Goedartius, De insectis. In Methodum redactus cum notularum additione, opera M. Lister ... item ... (London, 1685). This translation is also "methodized." See *Philosophical Transactions,* **14** (20 Dec. 1684), 833–834.

Métamorphoses naturelles, ou histoire des insectes. Traduit en françois (Amsterdam–The Hague, 1700). It appears that undated copies exist. See *The History of the Works of the Learned,* III (1701), 597–602.

J. van Abcoude, in his *Naamregister van de ... Nederduitsche boeken* (Rotterdam, 1773), p. 153, has under the name of Goedaert *Historie van den oorsprong der wormen* (n.p., n.d.) and *Historie van de bloedelooze dieren,* 3 vols. (Haarlem, n.d.). I could not confirm these two titles. The title of the second book is the same as that of one of Swammerdam's books, which, however, was never printed in three volumes. Both works are probably ghosts.

II. Secondary Literature. On Goedaert and his work, see P. de la Ruë, *Geletterd Zeeland* (Middelburg, 1741), pp. 61–64; C. de Waard, "Johannes Goedaert," in *Nieuw Nederlandsch Biografisch Woordenboek,* I (Leiden, 1911), 944–945; P. J. Meertens, *Letterkundig leven in Zeeland in de zestiende en de eerste helft der zeventiende eeuw* (Amsterdam, 1943); F. Nagtglas, *Levensberichten van Zeeuwen* (Middelburg, 1890–1893), pp. 267–268; A. Schierbeek, *Schouwburg der dieren* (The Hague [1943]), pp. 122–127; and G. Kruseman, "The Editions of Goedaert's Metamorphosis naturalis," in *Entomologische Berichten,* **16** (1956), pp. 46–48.

Works especially concerned with the identification of Goedaert's illustrations are F. S. Bodenheimer, *Materialien zur Geschichte der Entomologie bis Linné,* 2 vols. (Berlin, 1928), II, 368–372; H. P. Snelleman Cz., "Johannes Goedaert," in *Album der Natuur,* **26** (1877), 203–212; S. C. Snellen van Vollenhoven, "Determinatie der platen in het werk van Johannes Goedaert," *ibid.,* pp. 307–318; and A. Werneburg, *Beiträge zur Schmetterlingskunde,* I (Erfurt, 1864), p. 24 ff.

Peter W. van der Pas

GOEPPERT, HEINRICH ROBERT (*b.* Sprottau, Lower Silesia, Germany [now Szprotawa, Poland], 25 July 1800; *d.* Breslau, Germany [now Wrocław, Poland], 18 May 1884), *paleobotany, botany.*

Goeppert, whose father owned a pharmacy, discovered his love of botany while still a schoolboy. In order to follow his inclination he left school early and worked for five years as a pharmacist. After finishing his education he entered the University of Breslau in 1821 to study medicine. In 1824 he went to the University of Berlin, where he earned his medical degree in 1825. He returned to Breslau in 1826 and established himself as a general practitioner, surgeon, and ophthalmologist. Goeppert soon realized however, that he would not be fully satisfied in this occupation. In 1827, therefore, he became *Privat-*

dozent at the Faculty of Medicine of the University of Breslau with a work on plant physiology. In the same year he became an assistant at the university's botanical garden, with which he was associated for more than fifty-six years. He was promoted to associate professor on the Faculty of Medicine in 1831 and to full professor in 1839. In 1852 Goeppert assumed the chair of botany and was appointed director of the botanical garden and museum. His lectures in these years covered many fields: pharmacology, toxicology, forensic chemistry, systematic botany, plant physiology, plant geography, and paleobotany. He was particularly interested in the cryptogams.

Besides his official duties and scientific studies Goeppert, who was extremely active in public life, participated in the promotion of the cultural and economic interests of the city of Breslau and of the province of Silesia. He was aided in this by his extraordinary organizational ability, as well as by his affability. He was especially concerned with the Schlesische Gesellschaft für Vaterländische Kultur, whose president he was from 1846 until his death. On the fiftieth anniversary of his doctorate, "old Goeppert," as he was called by the townspeople, was granted the honorary freedom of the city of Breslau. He was an honorary, corresponding, or regular member of more than a hundred learned societies and academies all over the world. Goeppert was married twice, first to the eldest daughter of his professor, Remer, and then—after her early death—to one of her younger sisters. He had one son and one daughter.

With the exception of a few medical topics, Goeppert's scientific publications were devoted to botany, especially paleobotany. His doctoral dissertation and *Habilitationsschrift* both dealt with plant physiology. He wrote other works in this field on the evolution of heat in living plants, especially during germination and blossoming, and on the influence of low temperatures on plants; in particular, he studied the problem of whether a plant exposed to cold dies at the moment of freezing or at thawing. He later resumed these investigations and collected them in a book (1883). In another series of works he considered the ecology, physiology, and pathology of forest trees and fruit trees, especially their reactions to mechanical interference and external injuries. For instance, he showed that in stands of spruces and silver firs the roots of all the trees grow together. Hence, if one trunk breaks off or is felled, the stump is nourished by the neighboring trees until the point of fracture or of cutting has grown over and healed. Goeppert also considered questions of plant anatomy and of descriptive botany. He published studies of the anatomical structure of the conifers, of several Casuarinaceae and Magnoliaceae and of tropical Balanophorales.

In 1833 Goeppert entered the field in which he was to accomplish his most distinguished work. Stimulated by Otto, an anatomist at the University of Breslau who had assembled a considerable collection of fossil animal remains found in Silesia, he began to examine this region's fossil plant remains. The two scientists issued a joint call to their fellow Silesians to assist this project by sending them fossil plants. Their appeal was very successful; Goeppert received rich and interesting materials from many areas. He studied both this material and his own collections with great industry and enthusiasm. The Carboniferous flora from the coal deposits of Upper and Lower Silesia provided his richest discoveries.

Goeppert's first paleontological work, "Die fossilen Farnkräuter," appeared in 1836. In it he discussed the Carboniferous ferns and compared them—following strict principles of comparative anatomy—with those of the modern period, thereby establishing his reputation as a paleobotanist. Five years later he began publication of *Gattungen der fossilen Pflanzen, vergliechen mit denen der Jetztwelt* (1841–1846). This large, illustrated work, with German and French texts, greatly advanced the knowledge of fossil plants. Among Goeppert's most important achievements was the demonstration that coal seams are formed from the same plants that are found in the clays and sandstones located above and below them. Furthermore, he showed that the coal seams had originated through the high pressures exerted by sedimentary coverings and through decomposition resulting from lack of air, and therefore were not structureless masses carbonized by fire. Goeppert's entry in the Haarlem Academy's prize competition concerning the question whether coal seams are autochthonous or allochthonous was awarded the double prize. His studies on the Silesian coal regions enabled Goeppert to give valuable advice on the seams that were worth mining. He generously provided this information to all who sought it.

Following his great success in the study of Carboniferous flora, Goeppert turned attention to the fossil plants of other stages of the earth's history and produced monographs on the fossil flora of almost all the geological periods. Among these are two masterpieces: *Die fossilen Coniferen* (1850) and *Die fossile Flora der Permischen Formation* (1863–1865). These works contain the results of his microscopic examination of various specimens, including chips and thin sections of siliceous trunks; this examination allowed him to provide the first detailed comparisons with the tissues of living woods.

Goeppert was especially attracted by the flora of the Tertiary. He described and reconstructed palm, yew, and plane forests and cypress stands from various fossil occurrences in Silesia. As in his work on the Paleozoic and Mesozoic, he also considered plant remains from other regions of Germany, as well as from the rest of Europe and from overseas. Thus he demonstrated that in the Tertiary deposits of central Europe the Japanese ginkgo, the Chilean *Libocedrus,* and the North American yew occur side by side, and that in the Tertiary the vegetation of Java had the same tropical character it has today. Goeppert took a special interest in the amber of east Prussia and throughout his life studied the plants in it. As early as 1837, for example, he realized that a certain species of conifer must have produced the resin of the east Prussian amber.

BIBLIOGRAPHY

I. ORIGINAL WORKS. A complete bibliography up to 1882 is in Goeppert's "Beiträge zur Pathologie und Morphologie fossiler Stämme," in *Palaeontographica,* **28** (1882), 141–145. A complete bibliography is in Conwentz (see below).

Goeppert's writings include "Die fossilen Farnkräuter," in *Nova acta Leopoldina,* **17** (1836), 1–258; *Die Gattungen der fossilen Pflanzen, vergliechen mit denen der Jetztwelt,* 6 pts. (Bonn, 1841–1846); *Die fossilen Coniferen, mit steter Berücksichtigung der lebenden* (Haarlem–Leiden, 1850); *Die fossile Flora der Permischen Formation* (Kassel, 1864–1865); and *Die Flora des Bernsteins und ihre Beziehungen zur Flora der Tertiärformation und der Gegenwart* (Danzig, 1883) and *Über das Gefrieren, Erfrieren der Pflanze und Schutzmittel dagegen* (Stuttgart, 1883), both written with Menge.

II. SECONDARY LITERATURE. See F. Cohn, "Heinrich Robert Göppert," in *Leopoldina,* **20** (1884), 196–199, 211–214; H. Conwentz, "Heinrich Robert Goeppert, sein Leben und Wirken," in *Schriften der naturforschenden Gesellschaft in Danzig,* n.s. **6** (1885), 253–285, with portrait, also in *Leopoldina,* **21** (1885), 135–139, 149–154; and K. Lambrecht and W. and A. Quenstedt, "Palaeontologi. Catalogus biobibliographicus," in *Fossilium catalogus,* **72** (The Hague, 1938), 166.

HEINZ TOBIEN

GOETHE, JOHANN WOLFGANG VON (*b.* Frankfurt am Main, Germany, 28 August 1749; *d.* Weimar, Germany, 22 March 1832), *zoology, botany, geology, optics.*

Born of middle-class parents—his father, Johann Kaspar Goethe, was a lawyer—Goethe obtained a degree in law at Strasbourg in 1771. He was summoned in 1775, on the basis of his literary fame, to the court of Weimar, where his duties soon included the supervision of mining in the duchy. He was raised to the nobility in 1782. After a sojourn in Italy (1786–1788) which constituted a decisive break with his turbulent youth, Goethe returned permanently to Weimar and established a lasting reputation as Germany's greatest poet. In religion he was never orthodox, although he did not deny God or immortality. Much of his theorizing in biology was based on belief in a Spinozistic God as Nature and on the conviction that his own mind could come to know the mind of this deity.

Goethe's first scientific paper (1784) claimed to demonstrate the presence of the intermaxillary (premaxillary) bone in man. It was published first in 1820, with a long postscript on the history of research on the problem and the controversy the manuscript had evoked. Long before Goethe it had been noted that, of the three sutures—external (facial), nasal, and palatal—which delimit the bone when it is present in the vertebrate upper jaw, the palatal is sometimes visible in human skulls, is more distinct in children than in adults, and can best be seen in embryos. Goethe was struck by the fact that in some mammals (for example, ruminants) the premaxilla is indisputably present even though the upper incisor teeth, which it normally supports, are absent. He inferred that if present even in such cases, it is unlikely to be absent in man, in whom upper incisors are well developed; and so he sought and found traces of the nasal and palatal sutures in human skulls.

J. C. Loder, the Jena anatomist, and later J. B. Spix accepted Goethe's inference that man has the bone, whereas Peter Camper, S. T. Sömmerring, and J. F. Blumenbach maintained that the inference would be justified only if the sutures were clearly visible. In fact the facial suture is never seen, and the two others are indistinct or absent. This rebuff led Goethe to regard the physicists' later rejection of his optical theories as yet another example of the impatience of the professional scientist with the amateur. His erroneous belief that Sömmerring and Blumenbach eventually accepted his findings arose because in his old age he no longer had clear memories of the controversy of the 1780's.

Goethe believed that to deny man the premaxilla would be to impugn the unity of nature. "Morphology" was his term for tracing out the unity underlying animal and plant diversity. He did not argue that similarities between genera are due to descent from common ancestors, for he understandably lacked the modern concept of specialization. Thus characters in apes and in sloths which are today attributed to a high degree of adaptation to arboreal conditions appeared to him as sheer lack of proportion. In botany Goethe found it difficult to divide some genera into distinct species with no transitional forms, since the classification was based on characters (particularly leaf structures) which were highly variable. This diversity suggested to him that species were in some way flexible, and he even allowed (following Georges Buffon) that differences in climate and food could lead to the evolution of one plant or animal species from another within the same genus. Thus he regarded an extinct species of bull, fossils of which were found near Stuttgart in 1820, as possibly the ancestor of the modern European and Indian bull. But to account for the unity of type pervading different genera he supposed that nature, regarded as a kind of creative artist, used a single archetype in constructing them; thus plants derive from "a supersensuous archetypal plant" (*Urpflanze*), which he thought of as an idea in the mind of nature, individual genera being modifications in one direction or another of this type.

Goethe thought that the biologist, by comparing a large number of plant and animal forms, can obtain a clear idea of the underlying archetypes. Having found at least traces of the premaxilla in cetaceans, amphibians, birds, and fishes, he inferred that a structure so widely distributed must be part of the vertebrate archetype and must therefore be represented in all vertebrates, including man.

Goethe also constructed his idea of the archetype from a study of function. A bone which is not only present in most vertebrates but also obviously serves an important feeding function (both when it supports upper incisor teeth which have a nipping action against the incisors of the lower jaw, and when it forms a toothless, hard pad against which the lower incisors bite) is likely, for both these reasons, to belong to the archetype. He stressed the stability of function and thought that a bone or organ which performs a function in one animal will be present to perform the same function in another—although he realized that in some few cases an organ functional in some animals may occur as a rudiment in others; and he emphasized that a functional organ may be drastically reduced if other structures are extended. This theory is in accordance with the principle of compensation that he derived from Aristotle, a *loi de balancement* (then being independently stated by Étienne Geoffroy Saint-Hilaire) which Goethe illustrated with the recession of the premaxilla in the walrus, whose canines are elongated into tusks.

The best-known of Goethe's examples illustrating the principle of compensation is the inverse development of horns and front teeth in the upper jaw. He said, for instance, that the lion, with upper incisors and canines, cannot have horns. Fossil evidence has since shown that there is no incompatibility, since some extinct horned ungulates have the full eutherian dentition. The connection between Goethe's principle of compensation and his idea of a vertebrate archetype appears in his criticism from his teleological standpoint of the grosser teleology of his day: that of the so-called physicotheologians, who supposed that all the organs of an animal were designed to be useful to it, for example, the horns of the ox for defense. Goethe countered by asking why Providence did not supply the sheep with horns, or, when they have horns, why they are curled round their ears so as to be useless. His view was that ruminants, with no upper incisors or canines, have horns because horns, or some alternative to them consistent with the principle of compensation, belong in the mammalian archetype.

Goethe extended his idea of unity of type to cover not only vertebrates but all animals. He pointed out, for instance, that insects, as well as vertebrates, have bodies consisting of three major divisions, each with its appropriate organs. He welcomed Geoffroy Saint-Hilaire's arguments that all animals are built upon a common plan, that all existing forms are modifications of a nonexistent *être abstrait.*

Goethe's views on the relationship between allegedly similar parts in different organisms are paralleled in his thought concerning the different parts of one and the same organism. Just as organisms consist of variations of a single type, so the type itself consists of a number of parts or segments, each of which is identical with the others. This, he said, is particularly clear in the case of plants: cotyledons, inflorescence, stamens, and pistils are all, he said (having observed transitional forms), variations of the foliage leaf. He did not mean that they develop from leaves during the growth of the plant or that they have evolved from leaves during the history of plants, but that an "ideal leaf" is the essential scheme which underlies them all. He attempted to give the metamorphosis a physical basis by arguing that forms more delicate than foliage leaves are produced by elaboration of the sap as it passes upward.

Goethe argued that the vertebral column preserves some indication of the underlying identity of the units which go to form the vertebrate archetype, and that the skull is really a series of bones which can be seen to be variations of vertebrae. Although Lorenz Oken was the first to publish such views, Goethe could prove that he had adumbrated them earlier in extant letters to friends. He summed up his services to morphology by saying that the recognition that man, too, possesses the premaxilla secured the admission that a single osteological type pervades all forms, and that the construction of the skull from vertebrae establishes the identity of all the segments of this osteological type.

The attribution of the premaxilla to man never attained the popularity of the vertebral theory of the skull (also closely connected with archetypal thinking), although Oken and Goethe were committed to both. After F. S. Leuckart's well-documented account of 1840 it was hardly possible to dispute that in man the premaxilla is eliminated during ontogeny. And so Richard Owen, who retained the premise of the vertebrate archetype—which he imagined as consisting of a number of modified vertebrae—did not find it necessary for the purpose to credit man with a premaxilla. The vertebral origin of the skull was finally refuted on an embryological basis by T. H. Huxley in 1858.

What Goethe sought in botany and zoology was nothing less than a theory that would explain all living forms. He had no interest in details for their own sake and undertook detailed study only because of his consciousness that he was working toward wide generalizations which far outran his observations. Although his theory posits "the original identity of all plant parts," he ignored the root and stem and studied only the lateral appendages of the annual herbs. This premature generalization was due neither to personal arrogance nor to an a priori method but to a conviction, religious in character, that he had penetrated to the mind of nature. This aspect of his work endeared Goethe to the *Naturphilosophen* of the early nineteenth century; and when they were discredited, his scientific reputation remained unaffected largely because Ernst Haeckel quite unjustifiably stamped him as one of the foremost precursors of Darwinism.

Goethe's concern with geology sprang from his superintending the reopening of the copper slate mines at Ilmenau in 1784. At that time most rocks were regarded as chemical precipitates from saline seas. Mountain chains such as the Harz, the Thuringian forest, and the Alps all have central cores of granite, which was therefore interpreted as an *Urgebirge,* a foundation against which all later deposits, precipitated from a universal ocean, rest: the granite is flanked by "transition rocks," believed to have been formed when the ocean had receded sufficiently to expose the highest granite. The steep inclination or dip of these transition strata was considered original, not the result of postdepositional tilting, and they were made partly of detritus (scree from the granite peaks) and partly of further chemical precipitate from the ocean; the chemical ingredients were believed sufficient to consolidate the gathering sediment with steep original dip on the submerged slopes. When the waters had retreated still further, the *Flöz,* or layered rocks, were deposited—steeply inclined where they rest against the mountain core but elsewhere mainly horizontal. The final retreat of the waters to their present level was accompanied by the deposition of recent gravels, often rich in mammalian remains.

Such a scheme underlies Goethe's geological thinking. In the Harz he saw granite in close contact with "transition" rock of an entirely different type (hornfels), and he envisaged the two as attracting each other as they crystallized. It is characteristic that whereas the modern geologist explains the facts by positing a long sequence of events (deposition of clay, intrusion of liquid granite, baking, cooling, and solidification), Goethe preferred to think in terms of events occurring more or less simultaneously.

It is an important part of his theory of rock origin that the joint planes which divide granite masses into blocks were original, not shrinkage cracks due to cooling or drying. Each block was, for Goethe, an original precipitated "crystal"; and the mountain mass was formed by piling them. Since an extra crystal would give an uneven top to this basement rock, later rocks would locally acquire a steep dip as they wrapped themselves around it. Goethe thus believed that the steep dip of the rocks leaning against the basement was original. If they had been originally horizontal, they could have become steeply inclined only as a result of considerable crustal dislocation; and he believed that nature produces her effects without violent disturbances, since there was no evidence in his day that cataclysms or catastrophes were then occurring. His geological thinking clearly lacked the crucial concepts of time and uplift.

Actual proof of relationships posited by the above theory seemed to be provided by the Ilmenau mines. We know that the commercial copper bed and parallel seams are there strongly upfolded by the contact with the Thuringian granite and porphyry. But to Goethe the steep inclination was original, the seam wrapping against an *Urgebirge* cliff. In explaining the vertical position of the strata without supposing any dislocation of the rocks, Goethe was, in 1785, in agreement with most professional geologists, although by the early nineteenth century they had revised this opinion and he had not.

If the steep limbs of *Flöz* beds were deposited in this steep position, then it was easier to regard the whole *Flöz* as a crystalline precipitate from water rather than as fragments carried and deposited (for particles in water tend to settle in horizontal layers). And so Goethe regarded nearly all *Flöz* horizons—limestones, sandstones, shales, and even conglomerates—as chemically deposited. Many quartz grains in the Thuringian Bunter sandstones are angular, which seemed to suggest that they had not been transported. These sandstones were also so thick that enormous periods of time would have been required to derive them by erosion of preexisting granites. And before the development of the polarizing microscope in the mid-nineteenth century, there was no decisive way of distinguishing the groundmass of a porphyry from the cement of a detrital rock. Field relationships encouraged Goethe to link the Thuringian porphyry with local red conglomerates, for as we now know, both are of Rotliegendes age.

Goethe's repugnance for theories involving terrestrial violence sometimes led him to pioneer a correct path. The most notable instance is his glacial interpretation of erratic blocks at a time when violent and catastrophic movement was being invoked to explain their remoteness from their parent rocks. He explained the Swiss erratics by arguing that the glaciers had extended to the lakes of Geneva and Lucerne at a time when the general sea level reached these lakes. The glaciers transported the boulders to the lakes, and they completed their journey on floating ice. He realized that the theory implied a cold climate, giving an ice sheet over most of northern Germany and floating ice when it melted. His failure to appreciate the time factor is well illustrated by this explanation of phenomena confined to the present-day land surface in terms of a change in sea level of 1,000 feet, which would point back to what he saw as the earliest stage of the earth's geological history. But his stress on the importance of glacier transport of the erratics was correct, and in 1841 Johann de Charpentier mentioned John Playfair and Goethe as pioneers of the idea.

Goethe's whole approach to rocks reflects the insistence on types which distinguishes his biological thinking. He could relate a conglomerate to a granite because he had in his mind an idea of a rock type of which the two were variations. His biological ideas proved useful because the likenesses he perceived could later be understood not in an archetypal way, but as genetic relationships, as due to common ancestry. But his rock analogies could not serve as a pointer to future development, since he linked rocks which are not genetically related at all.

Goethe's first publications on optics (1791) culminated in his *Zur Farbenlehre* (1810), his longest and, in his own view, best work, today known principally as a fierce and unsuccessful attack on Newton's demonstration that white light is composite. Goethe supposed that the pure sensation of white can be caused only by a simple, uncompounded substance. Not until 1826 did Johannes Müller establish that any nervous receptor, no matter how stimulated, can excite only a characteristic sensation peculiar to it.

Goethe propounded the ancient idea that colors arise from mixing light with darkness. He was aware that these normally mix to form gray but held that the intervention of a turbid medium produces color; that all bodies are to some extent turbid and hence may appear colored in daylight; and that even transparent refracting media are comparable with turbid ones. (This attempt to bring the color phenomena produced by prisms under the theory was further developed by Arthur Schopenhauer in 1816 but was designated "senseless" by Ernst Mach.) In 1827 H. W. Brandes pointed out that the color phenomena Goethe alleged are not shown by such turbid media as steam or water-saturated mist; and in 1852 Ernst Wilhelm von Brücke showed, in terms of the wave theory of light and on the basis of long-forgotten work by Thomas Young, that whether colors are produced depends on the size of the particles in the medium. But for Goethe the color effects of turbid media were an *Urphänomen,* an ultimate which cannot itself be explained—which is in fact not in need of explanation—but from which all that we observe can be made intelligible.

Goethe's chapter on physiological colors (those which depend more on the condition of the eye than on the illumination) is the most successful and also typifies his psychological approach to color. Color vision involves an exciting stimulus and a conscious sensation. Goethe was concerned with the latter, and he posited three primary sensations—yellow, blue, and purple (for him, the purest red). In his color circle these three primaries alternated with orange, violet, and green, each of which, he claimed, could be seen to be compounded of the two adjacent primaries. He admitted, however, that the eye can see no trace of another color in pure green, which he classed as mixed presumably because blue and yellow pigments together give green. The distinction between additive and subtractive mixing was not properly understood before Hermann von Helmholtz, and Goethe's color circle certainly confused a subjective or psychological classification with an objective one.

Violet was positioned in the circle above blue, as its intensified form, and orange likewise above yellow; purple was placed highest of all, as the fusion of these intensified forms of his two other primaries. By "purple" Goethe meant the color seen through a prism when spectral red and spectral violet are superposed by viewing a thin black strip on a white background. Although this purple was thus compounded, it was, Goethe insisted, as a sensation pure (facts which undermine his principal reason for rejecting Newton's view of white!); whereas spectral orange and violet—designated "pure" by the physicist in terms of wavelength—are impure sensations (the eye can see red and yellow in orange, and red and blue in violet). Later authorities have agreed with Goethe in taking purple in his sense as psychologically the purest red and in finding spectral red distinctly yellowish in comparison.

Goethe supposed that the eye, by virtue of its own vital activity, is impelled to change a given condition into its opposite. It cannot, he said (explaining the phenomena of simultaneous and successive contrast), remain for a moment in a specific state that has been evoked by an object presented to it; when offered one extreme, or one mean, it spontaneously posits the other. He extended this doctrine to all living substance and was convinced that reality must ultimately be explained in terms of polar opposites—a view which endeared him to F. W. J. Schelling and G. W. F. Hegel, and to the *Naturphilosophen* in general. He posited no neural mechanism to explain simultaneous and successive contrast and thought them sufficiently explained by reference to his principle of polarity. The phenomena are in fact to a large extent polar and have since been attributed by Ewald Hering to a neural mechanism which functions in a polar fashion.

The terms in which Goethe explained colors (light, darkness, and turbidity) can be readily visualized, and for him explanation was never adequate unless the explicans fulfilled this condition. He thus had no sympathy with mathematical physics. He insisted that concrete phenomena can be represented in numerical form only if some of their essential conditions are ignored, and that if we reason from such abstractions, we are bound to err. He argued that the student of nature must not transmute what he sees into concepts and these concepts into words, but must think only in terms of what he sees. A physicist would today find it quite impossible to implement such an injunction. Goethe's adherence to it explains why he was so much less successful with physical than with physiological optics.

BIBLIOGRAPHY

I. Original Works. Goethe's scientific writings are available in *Goethes Werke*, pt. 2, 13 vols. (Weimar, 1890–1904), edited by order of the Grand Duchess Sophie of Saxony. Another valuable (but as yet incomplete) ed. is *Goethe: Die Schriften zur Naturwissenschaft*, G. Schmid *et al.*, eds., pt. 1, text, 11 vols. (Weimar, 1947–1970); pt. 2, supplements and commentaries (Weimar, in progress). Details of other eds. are in *Goethes Werke*, Dorothea Kuhn *et al.*, eds., XIII (Hamburg, 1955), 598–600, 638–639.

II. Secondary Literature. The principal bibliographies are K. Goedeke, *Grundriss zur Geschichte der deutschen Dichtung aus den Quellen*, IV, pt. 5 (Berlin, 1960), 363–382, covering 1912–1950; H. Pyritz, *Goethe Bibliographie* (Heidelberg, 1965), pp. 483–528; M. Richter, *Das Schriftum über Goethes Farbenlehre* (Berlin, 1938); and G. Schmid, *Goethe und die Naturwissenschaften* (Halle, 1940).

Critical studies include A. Arber, "Goethe's Botany," in *Chronica botanica*, **10** (1946), 63–126; B. von Freyberg, *Die geologische Erforschung Thüringens in älterer Zeit* (Berlin, 1932); M. Gebhardt, *Goethe als Physiker* (Berlin, 1932); H. von Helmholtz, "Über Goethes naturwissenschaftliche Arbeiten," in *Populäre wissenschaftliche Vorträge* (Brunswick, 1876), I, 31–54; J. H. F. Kohlbrugge, "Historisch-kritische Studien über Goethe als Naturforscher," *Zoologische Annalen*, **5** (1913); R. Magnus, *Goethe als Naturforscher* (Leipzig, 1906); W. Ostwald, *Goethe, Schopenhauer und die Farbenlehre* (Leipzig, 1918); M. Semper, *Die geologischen Studien Goethes* (Leipzig, 1914); C. S. Sherrington, *Goethe on Nature and Science*, 2nd ed. (Cambridge, 1949); R. Trümpy, "Goethes geognostisches Weltbild," in *Eidgenössische technische Hochschule, kultur- und staatswissenschaftliche Schriften*, no. 127 (1968), 1–37; J. Walther, ed., *Goethe als Seher und Erforscher der Natur* (Halle, 1930); and G. A. Wells, "Goethe's Geological Studies," in *Publications of the English Goethe Society*, **35** (1965), 92–137; "Goethe's Scientific Method and Aims in the Light of His Studies in Physical Optics," *ibid.*, **38** (1968), 69–113; "Goethe and Evolution," in *Journal of the History of Ideas*, **28** (1967), 537–550; and "Goethe and the Intermaxillary Bone," in *British Journal for the History of Science*, **3** (1967), 348–361.

George A. Wells

GOETTE, ALEXANDER WILHELM (*b.* St. Petersburg, Russia, 31 December 1840; *d.* Heidelberg, Germany, 5 February 1922), *zoology.*

Goette was the son of Ernst Bernhard Goette, a physician and counselor of state in St. Petersburg, and the former Natalie Bagh. He studied medicine at the University of Dorpat from 1860 to 1865 and completed his training with an M.D. degree at the University of Tübingen under Franz Leydig. Goette then worked as an independent scholar and in 1872 qualified as a lecturer under the Strasbourg zoologist Oscar Schmidt, at the same time becoming an assistant in the Zoological Institute of the University of Strasbourg. Named an associate professor in 1877, he took the additional post in 1880 of director of the zoological collection of the Municipal Museum of Strasbourg. From 1882 to 1886 Goette was professor of zoology at the University of Rostock. He was called back to Strasbourg in 1886 as Schmidt's successor. In 1918 he left Strasbourg and spent his last years in Heidelberg.

Following several minor investigations in vertebrate embryology, Goette's principal work, *Die Entwicklungsgeschichte der Unke* (*Bombinator igneus*) *als Grundlage einer vergleichenden Morphologie der Wirbelthiere,* appeared in 1875. Along with a detailed description of the development of the organs of *Bombinator igneus,* he gave a detailed presentation of his ideas of the methods and problems of ontogenetic research. In opposition to the dominant phylogenetic interpretation of embryonic development, Goette emphasized the necessity of investigating purely ontological and physiological regularities as a basis for the understanding of morphological phenomena, thereby criticizing Ernst Haeckel's gastraea theory and the "biogenetic law," which said nothing about the causation involved in the events occurring during ontogenesis. Haeckel responded immediately with the polemic work *Ziele und Wege der heutigen Entwicklungsgeschichte* (Jena, 1875), in which he presented Goette as an opponent of Darwin's theory. Yet Goette had always acknowledged the theory of descent, although he was critical of the explanation of species transformation by means of the theory of selection, especially because of Darwin's (and Haeckel's) acceptance of the inheritance of acquired characteristics.

Goette sought to explain developmental processes through a particular "law of form" which ruled the world of living creatures. This conception superimposed purely mechanical operations upon a teleological and vitalistic interpretation of life itself. He viewed cell division as a purely chemicophysical process. With Wilhelm His and August Rauber, Goette was one of the pioneers in research in developmental physiology, which soon became a separate field of study in the "developmental mechanics" of Wilhelm Roux. Goette's own careful investigations on animal embryology remained purely descriptive. After 1875 they treated most invertebrate groups as well: sponges, coelenterates, worms, mollusks, and echinoderms.

Goette had a little-known dispute with August Weismann concerning the "duration of life" and the definition of death. Whereas Weismann accepted the concept of "natural death" only for multicell forms of life, Goette defended the position that death is the necessary concomitant of reproduction and that there is no absolute continuity of life.

Goette's reticence in the polemics over Darwinism, the cell theory, and conceptions of heredity soon caused his own views on these questions to recede into the background. Hence there are no assessments of his scientific work except for the short obituary by Karl Grobben. On the other hand, in accounts of zoology in the last decades of the nineteenth century his name is almost always mentioned. Besides his zoological writings, Goette published *Holbeins Totentanz und seine Vorbilder,* which shows his thorough knowledge of the history of art and culture.

BIBLIOGRAPHY

I. Original Works. Goette's writings include *Die Entwicklungsgeschichte der Unke* (*Bombinator igneus*) *als Grundlage einer vergleichenden Morphologie der Wirbelthiere* (Leipzig, 1875); *Über Entwicklung und Regeneration des Gliedmassenskeletts der Molche* (Leipzig, 1879); *Abhandlungen zur Entwicklungsgeschichte der Tiere,* 5 vols. (Leipzig, 1882–1890); *Über den Ursprung des Todes* (Leipzig, 1883); *Über Vererbung und Anpassung* (Strasbourg, 1898); *Lehrbuch der Zoologie* (Leipzig, 1902); and *Die Entwicklungsgeschichte der Tiere* (Berlin–Leipzig, 1921). A nonscientific work is *Holbeins Totentanz und seine Vorbilder* (Strasbourg, 1897).

II. Secondary Literature. See Karl Grobben, in *Almanach der Akademie der Wissenschaften in Wien,* **72** (1923), 171–173; Jürgen-Wilhelm Harms, in R. Dittler, *et al.*, eds., *Handwörterbuch der Naturwissenschaften,* V (Jena, 1934), 297; and Georg Uschmann, in *Neue deutsche Biographie,* VI (1964), 579.

Hans Querner

GOHORY, JACQUES (*b.* Paris, France, 1520; *d.* Paris, 15 March 1576), *natural history, alchemy, medicine.*

Jacques Gohory was the eldest of six children born to Pierre de Gohory, an advocate to the Parlement of Paris, and his wife Catherine de Rivière. The family had strong links with the Parlement and Court of Paris, and Jacques, like his father and two of his younger brothers, became an advocate to the Paris Parlement. As a young man Gohory served on various ambassadorial missions, including periods in Flanders, England (1546–1549), and Rome (1554–1556).

Finding himself unsuited to legal or courtly life, he decided on his return from Rome to devote himself to the study and pursuit of poetry, music, the occult arts including alchemy, natural history, and medical philosophy. He retained his title of advocate to the Parlement, however, until his death.

Gohory's wide-ranging interests place him in the mainstream of French Renaissance culture which surrounded the courts of the later Valois monarchy. Indeed, as references in his works make clear, he was a close friend of members of the Pléiade and of Jean Antoine de Baïf's circle. From 1572 Gohory maintained a private academy which he called the Lycium Philosophal San Marcellin, at his home in the Faubourg Saint-Marcel. This academy was a rival to Baïf's royally chartered Academy of Poetry and Music founded two years earlier. Both academies were devoted to the encyclopedic cultivation of the arts in the Italian Neoplatonic tradition, but whereas Baïf's emphasized poetry and music, Gohory's laid stress on alchemy, botany, and the magical arts. The Lycium had a botanical garden and a chemical laboratory; games and music were played in the alleys of the garden. The site of Gohory's Lycium was close to that of the later Jardin du Roi, founded in 1626, but there was no formal connection between the two institutions.

Gohory is important as an early disseminator of Paracelsian ideas in France. In his writings he refers to his discussions on Paracelsus' teachings with such distinguished medical figures as Jean Fernel, Ambroise Paré, Jean Chapelain, Honoré Chastellan, and Leonardo Botal. His Lycium became a center for the preparation of chemical medicines. His *Compendium* (1568) of the philosophy and medicine of Paracelsus contains a brief life of Paracelsus, a summary of his principal doctrines, a catalogue of his works, and a commentary on his *De vita longa.* Gohory was critical of contemporary commentators on Paracelsus, particularly Gerard Dorn, who published an immediate rebuttal in 1568. He linked Paracelsus with the medieval magical and alchemical tradition through Artephius, Roger Bacon (from whom he said Paracelsus borrowed much), Peter of Abano, Albert the Great, Arnald of Villanova, Raymon Lull, and John of Rupescissa. Although he recognized Paracelsus' debt to the later Neoplatonic magical tradition, in particular by pointing to the relationship of Paracelsus' *De vita longa* to Marsiglio Ficino's *De triplici vita* (1489), he was critical of both Ficino and Giovanni Pico della Mirandola for their religious scruples which, he alleged, prevented them from becoming truly great *magi.*

Gohory's short monograph on tobacco, *L'instruction sur l'herbe petum* (1572), is one of the earliest on the subject and contains recipes for chemical preparations derived from the plant. It is also notable for its information about the author and his Lycium.

Gohory's numerous literary works include translations of Machiavelli's *The Prince* (1571) and of an anonymous account of the conquest of Peru (1545). During the last three years of his life he was royal historiographer.

BIBLIOGRAPHY

I. Original Works. The most complete guide to Gohory's published work is contained in Hamy's art. cited below. Works of philosophical and medical interest are *De usu et mysteriis notarum liber* (Paris, 1550), a wide-ranging discussion of the occult; *Theophrasti Paracelsi philosophiae et medicinae . . . compendium* (Basel, 1568), published under the pseudonym Leo Suavius; *Livre de la fontaine perilleuse . . . contenant la steganographie des mystères secrets de la science minérale* (Paris, 1572), an ed. and commentary on a medieval poem which Gohory believed to be an alchemical allegory; *Discours responsif à celui d'Alexandre de la Tourette sur les secrets de l'art chymique et confection de l'or potable* (Paris, 1575), by L.S.S. [Leo Suavius Solitaire], a reply to Tourette's treatise on potable gold published in 1575.

Works of botanical interest are *Devis sur la vigne, vin et vendages* (Paris, 1550), published under the pseudonym Orl. de Suave; *L'instruction sur l'herbe petum . . .* (Paris, 1572), a monograph on the tobacco plant. His trans. of the account of Pizarro's conquest of Peru is *L'histoire de la Terre-Neuve du Péru* (Paris, 1545).

II. Secondary Literature. A very full account of Gohory's life and work is contained in E.-T. Hamy, "Un précurseur de Guy de la Brosse. Jacques Gohory et le Lycium Philosophal de Saint-Marceau-lès-Paris (1571–1576)," in *Nouvelles archives du Muséum d'histoire naturelle,* 4th ser., **1** (1899), 1–26. Gohory's commentary on Paracelsus is discussed in D. P. Walker, *Spiritual and Demonic Magic From Ficino to Campanella* (London, 1958), pp. 96–106. See also L. Thorndike, *A History of Magic and Experimental Science,* V (New York, 1941), 636–640; and J. R. Partington, *A History of Chemistry,* II (London, 1961), 162–163. The account of Peru and the treatise on tobacco are dealt with in W. H. Bowen, "L'histoire de la Terre-Neuve du Péru. A Translation by Jacques Gohory," in *Isis,* **28** (1938), 330–340; and "The Earliest Treatise on Tobacco: Jacques Gohory's 'Instruction sur l'herbe petum,'" *ibid.,* 347–363. In these arts. Bowen refers to his Harvard University diss. on Gohory (n.d.).

Owen Hannaway

GOLDBACH, CHRISTIAN (*b.* Königsberg, Prussia [now Kaliningrad, R.S.F.S.R.], 18 March 1690; *d.* Moscow, Russia, 20 November 1764), *mathematics.*

The son of a minister, Goldbach studied medicine and mathematics at the University of Königsberg before embarking, sometime around 1710, on a series of travels across Europe. Everywhere he went, he formed acquaintances with the leading scientists of his day, laying the basis for his later success as first corresponding secretary of the Imperial Academy of Sciences in St. Petersburg. Among others, he met Leibniz in Leipzig in 1711, Nikolaus I Bernoulli and Abraham de Moivre in London in 1712, and Nikolaus

II Bernoulli in Venice in 1721. At Nikolaus II's suggestion, in 1723 Goldbach initiated a correspondence with Daniel Bernoulli which continued until 1730. Back in Königsberg in 1724, Goldbach met Jakob Hermann and Georg Bilfinger on their way to participate in the formation of the Imperial Academy and decided to follow them. Writing from Riga in July 1725, he petitioned the president-designate of the new academy, L. L. Blumentrost, for a post in that body. Among his references he named General James Bruce, commander of the imperial forces, with whom he had exchanged ideas on a problem in ballistics around 1718. Although at first informed that no places were open, Goldbach soon received the position of professor of mathematics and historian of the academy at a yearly salary of 600 rubles. In the latter capacity he acted as recording secretary from the first meeting until January 1728, when he moved to Moscow.

That move resulted from Goldbach's new post as tutor to Tsarevich Peter II and his distant cousin Anna of Courland. Introduced into court circles by Blumentrost as early as 1726, Goldbach was in a position to benefit from the split between Peter II and Prince Menshikov by replacing the tutors appointed by the prince. Peter's sudden death in 1730 ended Goldbach's teaching career but not his connections with the imperial court. He continued to serve Peter's successor Anna and returned to St. Petersburg and the Imperial Academy only when she moved the court there in 1732. While in Moscow in 1729, Goldbach began the exchange of letters with Leonhard Euler that would continue regularly until 1763.

Returning to the Imperial Academy in 1732, Goldbach quickly rose to a commanding position. Under the presidency of Baron Johann-Albrecht Korf, he was first designated corresponding secretary (1732) and later named a *Kollegialrat* and, together with J. D. Schuhmacher, was charged with the administration of the Academy (1737). At the same time he rose steadily in court and government circles. The two roles began to conflict seriously in 1740, when Goldbach requested release from administrative duties at the Academy; and his promotion to *Staatsrat* in the Ministry of Foreign Affairs in 1742 ended his ties to the Imperial Academy. In 1744 his new position was confirmed with a raise in salary and (in 1746) a grant of land; in 1760 he attained the high rank of privy councilor at 3,000 rubles annually. That same year he set down guidelines for the education of the royal children that served as a model during the next century.

Coupled with a vast erudition that equally well addressed mathematics and science or philology and archaeology, and with a superb command of Latin style and equal fluency in German and French, Goldbach's polished manners and cosmopolitan circle of friends and acquaintances assured his success in an elite society struggling to emulate its western neighbors. But this very erudition and political success prevented Goldbach's obvious talent in mathematics from attaining its full promise. Unable or unwilling to concentrate his efforts, he dabbled in mathematics, achieving nothing of lasting value but stimulating others through his flashes of insight.

Goldbach's mathematical education set the pattern for his episodic career. Rather than engaging in systematic reading and study, he apparently learned his mathematics in bits and pieces from the various people he met, with the result that later he frequently repeated results already achieved or was unable to take full advantage of his insights. As he himself related in a letter to Daniel Bernoulli, he first encountered the subject of infinite series while talking to Nikolaus I Bernoulli at Oxford in 1712. Unable to understand a treatise by Jakob I Bernoulli on the subject, loaned to him by Nikolaus, he dropped the matter until 1717, when he read Leibniz's article on the quadrature of the circle in the *Acta eruditorum.* His reawakened interest led to his own article "Specimen methodi ad summas serierum," which appeared in the *Acta* in 1720. Only afterward did Goldbach discover that the substance of his article formed part of Jakob I's *Ars conjectandi,* published in 1713. In his article "De divisione curvarum . . ." Goldbach frankly admitted that Johann I Bernoulli had already solved the problem in question but that he could not remember the solution and so was deriving it again. Often Goldbach's mathematical knowledge showed surprising bare spots. Impressed by his solution of several cases of the Riccati equation (in "De casibus quibus integrari potest aequatio differentialis . . ." and "Methodus integrandi aequationis differentialis . . ." and in correspondence),[1] Daniel Bernoulli encouraged him to extend his results to exponential functions. Goldbach replied that he knew nothing about exponential functions and did not want to give the impression that he did.

Of Goldbach's other published articles, the two on infinite series—"De transformatione serierum" and "De terminis generalibus serierum"—and the one on the theory of equations, "Criteria quaedam aequationum . . .," show the greatest originality. "De transformatione serierum," read to the Imperial Academy in 1725, contains a technique for transforming one series, *A,* into another series, *B,* having the same sum, through term-by-term addition of A to, or subtraction of A from, a series, C, of which the sum is zero. Adjustment of the technique leads to a similar transformation through multiplication of the given series by a series of which the sum is one. In reply to objections that the multiplicative method may involve a divergent series as unit multiplier,[2] Goldbach defended the use of such a series provided that it leads to a convergent result. "De terminis generalibus serierum," read in 1728, continues the work begun in "Specimen methodi ad summas serierum" (1720) by addressing the problem of determining the "general term" of any sequence;[3] that is, it seeks a function (either explicit or finitely recursive) that yields the nth term of the sequence for a given n. Goldbach shows that the general term can always be expressed as an infinite series and that the problem therefore reduces to one of finding a general formula for the sum of that series. The general term of an infinite sequence proves useful, he argues, both for interpolation of missing terms and for the determination of terms for noninteger indices. Although Goldbach and Daniel Bernoulli corresponded on the specific problem of determining the general term of the sequence $\{n!\}$, neither could offer a solution (Euler later provided one).

In the article "Criteria quaedam aequationum quarum nulla radix rationalis est" Goldbach begins from results contained in "Excerpta a litteris C. G. ad * * * Regiomonte datis" and applies further some of the number-theoretical results worked out in correspondence with Euler to obtain a technique for testing quickly whether an algebraic equation has a rational root. For equations of the form $x^n = P(x)$, where P is an algebraic polynomial of degree $n - 1$ or less, the technique rests basically on determining all integers m for which P can be expressed in the form $mxR(x) + r$ and then ascertaining whether r is an nth-degree residue *modulo m*. If no such residue exists, the equation in question has no rational root.

If, in the realm of analysis, Goldbach's native talent could not substitute for thorough training in the subject, that talent did come into full play in his correspondence with Euler on number theory, a field then still at a rudimentary stage of development. Here Goldbach could be provocative on a fundamental level, as "Demonstratio theorematis Fermatiani . . ." and "Criteria quaedam aequationum quarum nulla radix rationalis est" show. Calling attention in his correspondence to Pierre de Fermat's assertion that all numbers of the form $2^{2^n} + 1$ are prime, he stimulated Euler's disproof for the case $n = 5$ (Euler's memoir in fact immediately follows Goldbach's "Criteria . . ."). Not all of his suggestions led to such positive results. In 1742 Goldbach conjectured that all even numbers may be expressed as the sum of two primes (taking 1 as a prime where necessary). Euler agreed with the assertion but could offer no proof, nor has any proof of "Goldbach's conjecture" yet been found. Goldbach also stated that every odd number may be expressed as the sum of three primes; in the form given it by Edward Waring (which excludes 1 as a prime) this assertion also remains an unproved conjecture. The above are only the outstanding results of the prolix correspondence with Euler on number theory.[4] That correspondence as a whole marks Goldbach as one of the few men of his day who understood the implications of Fermat's new approach to the subject.

NOTES

1. The solution includes the full conditions for the integrability of binomial differentials usually credited to Euler. See Youschkevich, *Istoria matematiki v Rossii,* p. 96.
2. I.e., the unit multiplier may not share the same domain of convergence as the resultant series.
3. Goldbach uses the same Latin term, *series,* to denote both series and sequences.
4. For details, consult Leonard E. Dickson, *History of the Theory of Numbers,* 2nd ed. (New York, 1952), I and II, *passim.*

BIBLIOGRAPHY

I. ORIGINAL WORKS. Goldbach's writings include "Temperamentum musicum universale," in *Acta eruditorum,* **36** (1717), 114–115; "Excerpta a litteris C[hristiani]. G[oldbachi]. ad * * * Regiomonte datis," in *Actorum eruditorum supplementum,* **6** (1718), 471–472; "Specimen methodi ad summas serierum," in *Acta eruditorum,* **39** (1720), 27–31; "Demonstratio theorematis Fermatiani, nullum numerum triangularem praeter l esse quadrato-quadratum," in *Actorum eruditorum supplementum,* **8** (1724), 483–484; "De casibus quibus integrari potest aequatio differentialis $ax^m dx + byx^p dx + cy^2 dx = dy$ observationes quaedam," in *Commentarii Academiae scientiarum imperialis Petropolitanae,* **1** (1728), 185–197; "Methodus integrandi aequationis differentialis $aydx + bx^n dx + cx^{n-1}dx + ex^{n-2}dx = dy$ ubi n sit numerus integer positivus," *ibid.,* 207–209; "De transformatione serierum," *ibid.,* **2** (1729), 30–34; "De divisione curvarum in partes quotcunque quarum subtensae sint in data progressione," *ibid.,* 174–179; "De terminis generalibus serierum," *ibid.,* **3** (1732), 164–173; and "Criteria quaedam aequationum quarum nulla radix rationalis est," *ibid,* **6** (1738), 98–102.

For Goldbach's correspondence with Nikolaus II and Daniel Bernoulli and Leonhard Euler, see Paul-Henri Fuss, *Correspondance mathématique et physique de quelques célèbres géomètres du XVIIIème siècle,* 2 vols. (St. Petersburg, 1843); for a more recent ed. of part of that correspondence, see *Leonhard Euler und Christian Goldbach, Briefwechsel, 1729–1764,* edited with introduction by A. P. Juškevič [Youschkevich] and E. Winter (Berlin, 1965).

II. SECONDARY LITERATURE. Piotr P. Pekarskii, *Istoria imperatorskoi akademii nauk v Peterburge,* I (St. Petersburg, 1870), 155–172, contains the most complete biography of Goldbach and quotes heavily from his nonmathematical correspondence and papers now in the State Archives, Moscow. A. P. Youschkevich includes a fairly complete account of Goldbach's mathematical work in his *Istoria matematiki v Rossii* (Moscow, 1968), pp. 92–97. See also the eds. cited above and the works cited in the notes.

MICHAEL S. MAHONEY

GOLDBERGER, JOSEPH (*b.* Girált, Hungary, 16 July 1874; *d.* Washington, D.C., 17 January 1929), *epidemiology.*

The son of poor Jewish immigrants, Goldberger was brought to the United States at the age of six by his parents, Samuel and Sarah Gutman Goldberger, who settled on New York's Lower East Side. He attended the city's public schools and entered the College of the City of New York in 1890 as an engineering student. In 1892 his career plans changed, and Goldberger became a student at the Bellevue Hospital Medical School, graduating second in his class three years later. After placing first on the highly competitive Bellevue internship examination, he spent eighteen months at the hospital as intern and house physician. Following two unhappy years of private practice in Wilkes-Barre, Pennsylvania (1897–1899), he took and passed the examination for an assistant surgeon's post in the U.S. Public Health Service. Appointed in 1899, he remained in the Public Health Service until his death.

Public health was then dominated by the infectious diseases; and during the next fifteen years Goldberger received intensive on-the-job training in classic epidemiology, beginning with a traditional apprenticeship as quarantine physician. When not on field assignments, he accumulated valuable experience in parasitology and bacteriology at the Public Health Service's Hygienic Laboratory. By 1910 Goldberger played an increasingly responsible role in field investigations of yellow fever, typhus, and dengue—as well as other, less dramatic, ills. During these journeyman years he became successively a victim of yellow fever, dengue, and typhus. In the course of his investigations he acquired a reputation in the Public Health Service as one of its most gifted epidemiologists. Goldberger also developed a familiarity with conditions in the southern United States and—in his typhus work—with Mexico as well.

During these years Goldberger made several important epidemiological contributions. Perhaps most significant was his demonstration, with J. F. Anderson, that measles is transmissible to monkeys by a filter-passing virus and that the virus is present in buccal and nasal secretions. In his typhus studies, also in collaboration with Anderson, Goldberger was able to show that head as well as body lice could act as vectors and that "Brill's disease," described in New York City, was actually typhus. In another, less significant but impressively elegant, field investigation, he demonstrated the role of a straw mite in causing a dermatological ailment.

While in the midst of directing a detailed study of diphtheria in Detroit in the winter of 1913–1914, Goldberger was requested by Surgeon-General Rupert Blue to undertake the direction of an expanded antipellagra program. Work on this disease was, with one or two brief diversions, to fill the rest of Goldberger's life.

Essentially unknown to American clinicians before 1900, pellagra had seemingly spread rapidly during the century's first decade. Its unpleasant symptoms, its novelty, and its rapid increase in an era proud of its public health accomplishments tended to focus both lay and medical attention on this new and terrifying disease. As early as 1909 the Public Health Service established a special committee on pellagra. Although its most dramatic incidence was in certain southern orphanages, insane asylums, and cotton-mill villages, few areas in the South were completely free of the disease. Southern senators and representatives were instrumental in passing a special appropriation to underwrite the extended pellagra study which Goldberger was chosen to direct.

Traditional explanations of the disease, long familiar to physicians in Italy and other Mediterranean countries, centered on the role of a diet based largely upon corn. This theory—in the form that spoiled corn somehow provided an appropriate substrate for the growth of a toxin-producing microorganism—dominated the conjectures of physicians in the generation before 1910. But by 1914 medical opinion had shifted toward a belief that the disease was infectious—that pellagra was caused by some as yet undiscovered microorganism (possibly a protozoon spread by an insect vector).

A few writers, most notably the biochemist Casimir Funk, had suggested that pellagra might be the consequence of an inadequate or unbalanced diet. The idea was hardly novel in itself. Clinicians had known empirically for many years of the role of diet in the etiology of scurvy, beri beri, and possibly rickets as well. F. G. Hopkins' and E. Willcocks' demonstration in 1906 of the pathological effects of specific amino acid deficiencies was well known to knowledgeable American workers; only a year or so before Goldberger began his pellagra work, two American laboratories had almost simultaneously discovered the presence of an accessory food substance in butterfat (vitamin A).

Goldberger decided, almost as soon as he had been put to work on the problem, that pellagra was a consequence of improper diet. (The well-attested immunity of staff and administrators at pellagra-ridden asylums and orphanages seems to have been the most significant factor in determining his conviction; it has been pointed out that such immunity would have been difficult for a survivor of typhus and yellow fever to have ignored.) Goldberger then proceeded with great care and ingenuity to prove his original intuition. In three major steps he succeeded by 1916 in marshaling extremely strong evidence for his position. By supplementing diets in particular institutional populations, Goldberger almost completely eliminated the disease. In a critical experiment, moreover, he was able to induce symptoms of pellagra in five of eleven Mississippi prison-farm volunteers by providing them with an abundant but protein-deficient diet. (The other prisoners served as a control group.) In a final and almost dismayingly heroic experiment Goldberger and co-workers were unable to produce symptoms of pellagra in themselves through ingestion and injection of excreta, vomitus, nasal secretions, and material from the skin lesions of pellagrins.

By 1917 Goldberger had convinced America's medical elite of the correctness of his views. Indeed, as early as November 1915 the Public Health Service had issued a press release reporting the Mississippi prison-farm experiment and urging that pellagra could be prevented by an appropriate diet; yet throughout the 1920's many practicing physicians, especially in the American South, were unwilling to accept diet as a more than predisposing cause of pellagra. Chronic resentment toward the East and the well-financed Public Health Service seems to have contributed to this incredulity.

In the decade after World War I, Goldberger turned his efforts toward the identification of the constituent or constituents lacking in a pellagra-producing diet; it seemed to him most likely that the substance he sought was some amino acid component of such protective foods as meat and yeast. Influenced by earlier work on protein chemistry, he experimented with the use of particular amino acids, including even tryptophan, in experimental therapeutic trials. It is significant that Goldberger's efforts were guided not only by the biochemist's desire to isolate a particular substance or substances but also by the pragmatic epidemiologist's desire to find an inexpensive and readily available food which might prove effective in preventing the disease.

The most striking aspect of Goldberger's antipellagra work was its flexibility and sensitivity to social and economic context. Goldberger and his co-workers, most prominently statistician and economist Edgar Sydenstricker, exhaustively studied conditions in a number of self-contained mill villages, in several of which the incidence of pellagra was atypically high. They explored every environmental factor which might shape the daily life of the villagers; diet, they assumed, was a function both of custom and of economics. For example, mill communities in diversified farming areas without urban markets or good transportation would naturally have a more varied food supply than villages in cotton-growing areas with ready access to railroads and roads, facilities which would tend to siphon off none-too-abundant truck crops and fresh meats to towns and cities. In the scale and complexity of their work, in their dependence on team techniques and interdisciplinary studies, Goldberger and his co-workers were forerunners of a new idiom in the social approach to disease, one appropriate to the problems and techniques of the twentieth century.

On 19 April 1906 Goldberger married Mary Humphreys Farrar, the daughter of a prominent New Orleans family; they had four children. With a salary never adequate for comfort and a father gone for long periods on field investigations, the Goldbergers' domestic life was often troubled. Goldberger died of cancer on 17 January 1929.

BIBLIOGRAPHY

A well-selected collection of Goldberger's most important papers has been reprinted with a brief intro.: *Goldberger on Pellagra,* edited, with intro., by Milton Terris (Baton Rouge, La., 1964). The most important source for Goldberger's life and work is his papers, deposited at the Southern Historical Collection, University of North Carolina Library, Chapel Hill. The collection contains many letters exchanged between Goldberger and his wife while he was on assignments in the field. The General Subject File of the U.S. Public Health Service, RG 90, boxes 150–155 in the National Archives, are devoted to the Service's pellagra work and provide a detailed record of Goldberger's place in their antipellagra campaign.

There is a full-length, popular biography: Robert P. Parsons, *Trail to Light. A Biography of Joseph Goldberger* (Indianapolis–New York, 1943); although largely uncritical, it does utilize the Goldberger papers extensively. See also Solomon R. Kagan, "Joseph Goldberger," in *Medical Life,* **40** (1933), 434–445; W. H. Sebrell, "Joseph Goldberger (July 16, 1874–January 17, 1929)," in *Journal of Nutrition,* **55** (1955), 3–12; James M. Phalen, "Joseph Goldberger," in *Dictionary of American Biography,* VII

(New York, 1931), 363–364. For Goldberger's pellagra work in perspective, see E. V. McCollum, *A History of Nutrition. The Sequence of Ideas in Nutrition Investigations* (Boston, 1957), pp. 296–317. For a clear presentation of the social assumptions which Goldberger held but never formally articulated, see Edgar Sydenstricker, *Health and Environment* (New York–London, 1933).

CHARLES ROSENBERG

GOLDSCHMIDT, RICHARD BENEDICT (*b.* Frankfurt am Main, Germany, 12 April 1878; *d.* Berkeley, California, 24 April 1958), *zoology, general biology.*

Goldschmidt belonged to a very old German-Jewish family that had included scientists, artists, bankers, and industrialists. His father managed a coffeehouse combined with a wine trade and a confectionery. The young Goldschmidt, who had a wide circle of friends, lived in a well-to-do milieu. He attended the Gymnasium in Frankfurt and planned from the first year of secondary school to study the natural sciences: his native city possessed the famous Senckenberg Museum, a powerful attraction; and his teacher F. C. Noll was a zoologist.

Goldschmidt entered the University of Heidelberg in 1896, and among his professors were the great Otto Bütschli and Karl Gegenbaur. In 1898 he continued his studies at the University of Munich, where Richard Hertwig was teaching. In 1902, at Heidelberg, he defended his thesis on the maturation, fertilization, and embryonic development of the worm *Polystomum integerrimum.* A year later Goldschmidt became Hertwig's assistant and in 1904, a *Privatdozent;* he remained at Munich until 1913. In that year Theodor Boveri and Carl Correns organized the Kaiser Wilhelm Institute for Biology in Berlin; Goldschmidt was appointed director of the genetics department, a post that he was to hold until 1935.

Having received a grant from the Club Autour du Monde Goldschmidt went to Japan in order to continue his research. When World War I broke out he was in Honolulu and went from there to San Francisco. He was detained in the United States, where he worked at various universities. In 1917 he was placed in an internment camp and finally was repatriated to Germany. In 1935, when conditions for Jewish scientists had become impossible under the Nazi regime, Goldschmidt decided to leave Germany. He received offers from England and Turkey, but he accepted a professorship at the University of California and left for America in July 1936. He began a new life in Berkeley, where he remained until his death. He rapidly organized a laboratory and formed a group of students and friends.

A zoologist, biologist, and geneticist of exceptional ability, an original thinker, a great traveler, and an indefatigable worker, Goldschmidt produced more than 250 memoirs and articles and about twenty books; and it is difficult to present a comprehensive view of his total output. Very broad general knowledge combined with great specialization enabled him to interpret new facts and to study individual problems thoroughly. Three orientations emerge quite clearly. He was interested at first in morphological problems and in the cytology, fertilization, meiosis, histology, comparative anatomy, and embryology of the trematodes, nematodes (*Ascaris*), and the Acrania. From the time of his appointment at Munich he was concerned with many students who were preparing dissertations. In 1906 he provided them with a vehicle for publication by founding a new journal, *Archiv für Zellforschung.*

During the same period Goldschmidt undertook a series of researches on moths of the genus *Lymantria.* The work lasted for twenty-five years. He was interested in a problem of microevolution: industrial melanism. Through recognizing that the melanic mutant possesses a selective advantage, he became one of the pioneers of population genetics. Goldschmidt experimented on the genetics of sex determination from 1911 to 1920 and pointed out the existence of intersexuality (a term coined in 1915), which he distinguished from gynandromorphism. He succeeded in obtaining at will all the degrees of intersexuality, up to a complete inversion of the genetic sex into the opposite sex. To account for these appearances, he constructed a coherent theory in which a quantitative balance intervenes between the male and female sex factors.

Since the *Lymantria* also presented enormous geographic variation, Goldschmidt undertook (1918–1933) an analysis of the genetics of geographic variation. He accepted the neo-Darwinian proposal that a geographic race represents a nascent species, but the numerous crossings carried out among geographic races from all parts of the world modified his opinion. While on a visit in Ithaca, New York (1933), Goldschmidt realized that geographic variation entails only a microevolution in the species and that it could not be the source of true evolution. He postulated the existence of macromutants, produced by alteration of the early embryonic processes; these he called "hopeful monsters." In 1940 he pursued his critique of neo-Darwinian conceptions and argued for the existence of macroevolution carried out by means of macromutations.

The *Lymantria* became the subject of a new series of investigations on the theory of the gene. As early as 1916 Goldschmidt had fashioned a physiological theory of heredity (one gene, one enzyme). He published it only in 1920 in a book which marks the beginning of physiological genetics. According to this theory it should be possible, by modifying the speeds of the chains of reactions, to produce insects among which the nonhereditary phenotype copies the phenotype of the mutations; these are the phenocopies (1935). With his students Goldschmidt was able to produce varied phenocopies. At this time he left the *Lymantria,* which had been widely studied, and selected the *Drosophila* for examination.

With this change in material the third major period of research began. Goldschmidt studied the physiological genetics of the *Drosophila* and established that in the vestigial series the genetically controlled clipping of the wings can be influenced by the introduction of dominance modifiers. He then proposed an unorthodox theory concerning the nature of the gene, rejecting its corpuscularity (1938). This position aroused violent reactions, but in 1951 Goldschmidt opened the symposium on the gene at Cold Spring Harbor, New York, with an exposition of his ideas. His last works concerned the podoptera effect; the homoeotic mutants of *Drosophila melanogaster,* Podoptera and Tetraltera, hold great morphological, genetic, and evolutionary interest.

An amateur of the history of science, Goldschmidt wrote biographies of biologists he had known throughout the world. In addition, he wrote popular articles and books on science. Goldschmidt had, in addition, refined and discriminating taste; he particularly loved music and oriental art.

BIBLIOGRAPHY

I. ORIGINAL WORKS. Goldschmidt's articles include the following: "Zur Entwicklungsgeschichte der Echinococcusköpfchen," in *Zoologische Jahrbücher,* Anatomie, **13** (1900), 467–494; "Untersuchungen über die Eireifung, Befruchtung und Zelltheilung bei *Polystomum integerrimum,*" in *Zeitschrift für wissenschaftliche Zoologie,* **71** (1902), 397–444; "Der Chromidialapparat lebhaftfunktionierender Gewebzellen," in *Zoologische Jahrbücher,* Anatomie, **21** (1904), 1–100; "Amphioxides," in *Wissenschaftliche Ergebnisse der Deutschen Tiefsee-Expedition auf dem Dampfer "Valdivia" 1898–1899,* **12** (1905), 1–92; "Lebensgeschichte der Mastigamöben *Mastigella vitrea* n. sp. und *Mastigina setosa* n. sp.," in *Archiv für Protistenkunde,* supp. **1** (1907), 83–168; "Das Nervensystem von *Ascaris lumbricoides* und *megalocephala,*" in *Zeitschrift für wissenschaftliche Zoologie,* **90** (1908), 73–136, and **92** (1909), 306–357, repr. in *Festschrift R. Hertwig,* II (Jena, 1910), 254–354; "Die cytologische Untersuchungen über Vererbung und Bestimmung des Geschlechtes," in *Die Vererbung und Bestimmung des Geschlechtes* (Berlin, 1913), pp. 73–149; "A Preliminary Report on Some Genetic Experiments Concerning Evolution," in *American Naturalist,* **52** (1918), 28–50; "Untersuchungen über Intersexualität," in *Zeitschrift für induktive Abstammungs- und Vererbungslehre,* **23** (1920), 1–199; **29** (1922), 145–185; **31** (1923), 100–133; **49** (1929), 168–242; **56** (1930), 275–301; **67** (1934), 1–40; "Untersuchungen zur Genetik der geographischen Variation," in *Archiv für Entwicklungsmechanik der Organismen,* **101** (1924), 92–337; **116** (1929), 136–201; **126** (1932), 277–324, 591–612, 674–768; **130** (1933), 266–339, 562–615; "*Lymantria,*" in *Bibliotheca genetica,* **11** (1934), 1–185; "The Time Law of Intersexuality," in *Genetica,* **20** (1938), 1–50; "The Structure of Podoptera, a Homoeotic Mutant of *Drosophila melanogaster,*" in *Journal of Morphology,* **77** (1945), 71–103; "Ecotype, Ecospecies and Macroevolution," in *Experientia,* **4** (1948), 465–472; "Fifty Years of Genetics," in *American Naturalist,* **84** (1950), 313–340; "The Maternal Effect in the Production of the Beaded-Minute Intersexes in *Drosophila melanogaster,*" in *Journal of Experimental Zoology,* **117** (1951), 75–110; "The Podoptera Effect in *Drosophila melanogaster,*" in *University of California Publications in Zoology,* **55** (1951), 67–294, written with A. Hannah and L. K. Piternick; "Homoeotic Mutants and Evolution," in *Acta biotheoretica,* **10** (1952), 87–104; "Heredity Within a Sex Controlled Structure of *Drosophila,*" in *Journal of Experimental Zoology,* **122** (1953), 53–96; "Materials for the Study of Dominant Personality Traits," in *Folia hereditaria et pathologica,* **2** (1953), 267–295; "The Genetic Background of Chemically Induced Phenocopies in *Drosophila,*" in *Journal of Experimental Zoology,* **135** (1957), 127–202, and **136** (1957), 201–228, written with L. K. Piternick.

His books include *Zoologisches Taschenbuch für Studierende* (Leipzig, 1907; 6th ed., 1912), written with E. Selenka; *Einführung in die Vererbungswissenschaft* (Berlin, 1911; 5th ed., 1928); *Die quantitativen Grundlagen von Vererbung und Artbildung* (Berlin, 1920); *Mechanismus und Physiologie der Geschlechtsbestimmung* (Berlin, 1920); *Der Mendelismus* (Berlin, 1920); *Physiologische Theorie der Vererbung* (Berlin, 1927); *Die Lehre von der Vererbung* (Berlin, 1927; 4th ed., 1953); *Les problèmes de la sexualité* (Paris, 1932); *Physiological Genetics* (New York, 1938); *The Material Basis of Evolution* (New Haven, Conn., 1940); *Understanding Heredity: An Introduction to Genetics* (New York, 1952); *Theoretical Genetics* (Berkeley, Calif., 1955); *Portraits From Memory: Recollections of a Zoologist* (Seattle, Wash., 1956); and *In and Out of the Ivory Tower. The Autobiography of Richard B. Goldschmidt* (Seattle, Wash., 1960), with twenty-seven plates and a complete bibliography.

II. SECONDARY LITERATURE. On Goldschmidt or his work, see A. Kühn, "Zum 70. Geburtstag Richard Goldschmidt am 12. April 1948," in *Experientia,* **4** (1948), 239–240; "R. B. Goldschmidt 1878–1958. Zoologo, geneticista, evolucionista," in *Revista de la Sociedad mexicana de historia natural,* **20** (1959), 185–193, with photo.; and a special vol. devoted to him of *Portugaliae acta biologica,* ser. A (1949–1951), published to honor his seventieth birthday: it contains a biography by

A. Quintanilha and twenty-seven memoirs contributed by European and American students and friends.

ANDRÉE TÉTRY

GOLDSCHMIDT, VICTOR (*b.* Mainz, Germany, 10 February 1853; *d.* Salzburg, Austria, 8 May 1933), *crystallography, harmonics.*

Born to a well-to-do family, Goldschmidt attended the Gymnasium in Mainz. He then entered the Freiberg Bergakademie and, after graduating, stayed on as an instructor in metallurgy, assaying, and blowpipe analysis under H. T. Richter (1875–1878). For graduate and research work he went to the universities of Munich, Prague, and Heidelberg; at the latter he obtained his Ph.D. under K. H. F. Rosenbusch in 1880, with a dissertation entitled "Ueber Verwendbarkeit einer Kaliumquecksilberjodidlösung bei mineralogischen und petrographischen Untersuchungen"; it concerned the determination of the specific gravity of minerals, a topic to which he returned in later papers.

From 1882 to 1887 Goldschmidt was at the University of Vienna; and these years, especially the work with Aristedes Brezina, appear to have determined his lifework. From Vienna he returned to Heidelberg, where in February 1888 he submitted "Ueber Projektion und graphische Kristallberechnung" as a *Habilitationsschrift* for a post in Rosenbusch's institute.

Also in 1888 Goldschmidt married Leontine von Portheim and settled in Heidelberg. His wife brought as a dowry a substantial part of the wealth which enabled him to work with little help from the university; she was also a very understanding companion who provided a homelike atmosphere for many of his co-workers and students. Except for a long journey to the Far East in 1894–1895 he spent almost all of his time in Heidelberg. In 1893 Goldschmidt was named associate professor, and somewhat later he became an honorary full professor. In 1916 he and his wife established the Eduard und Josefine von Portheim Stiftung, of which the Victor-Goldschmidt-Institut für Kristallforschung was a part.

Together with E. S. Fyodorov in St. Petersburg and Paul von Groth in Munich, Goldschmidt was the founder of modern crystallography. Until then that science's methods and mode of thought adhered rigidly to a purely geometric vision of crystals, with little or no interest in the physicochemical meaning of the wealth of geometric observations. The work of these three men opened the way and created the methods for Max von Laue's discovery of X-ray diffraction and for the discovery of the principles of crystal chemistry by Aleksandr Fersman, Victor Moritz, Goldschmidt, and Paul Niggli.

Goldschmidt's contribution centered mainly on a complete indexing and recording of mineral crystal forms, the final aim being to link these external variations of form to the physicochemical variations of composition and of physicochemical factors present during formation. He realized the vastness of the task and was pleased that he accomplished the mapping of most of the crystal forms available during his lifetime. The second *sine qua non* was his great skill in teaching and the enthusiasm he created in his students. Many students from various countries worked with him, including Fersman, William Nicol, Charles Palache, Friedrich Kolbeck, M. A. Peacock, and Ludwig Milch.

Goldschmidt's outstanding works form a sort of trilogy: *Index der Kristallformen der Mineralien* (1886–1891), *Kristallographische Winkeltabellen* (1897), and *Atlas der Kristallformen* (1913–1923). These works are a foundation of crystallography and are essential to much crystallographic work even today. They and the more than 100 papers on individual crystal forms or groups required the improvement of traditional methods, which were entirely inadequate, and the creation of new ones. Consequently, part of Goldschmidt's contribution to crystallography consisted of improvement of existing instruments and invention of new ones. The most important was the construction of the two-circle goniometer. His work required a great number of crystal models and much cutting and oriented polishing. In this work he was assisted by the skilled mechanic Stoe. Goldschmidt wrote many instruction booklets on the new methods and instruments. For about forty years—until he was past seventy-five—he taught courses in measurement and calculation of crystals, determinative mineralogy, and blowpipe analysis.

Other methods improved or initiated by Goldschmidt are the Goldschmidt symbols, the gnomonic projection, use of the position angles ϕ and ρ (borrowed from astronomy) to characterize crystal forms, the recognition of the importance of zones, and the mathematical periodicity of zone symbols from 0 to ∞ to a maximum number in the *Normalreihe* III:

$$N_3 = 0, 1/3, 1/2, 2/3, 1, 3/2, 2, 3, \infty.$$

Goldschmidt defined the task of crystallography in the first volume of the *Index* (1886) and made the definition the program for his own work in crystallography: "The main purpose of crystallography is to explore the molecular structures of solid substances and to determine the intensity of molecular forces and their manner of operation." Later in the *Index* he wrote: "Every surface is crystallographically possible, the perpendicular is the direction of molecular attraction." Thus, Goldschmidt hoped to gain insight into the atomic or molecular bond relations within the lattice, through the crystal form as the product of bond strength and bond direction. For his time this was most certainly an ingenious approach. As soon as X rays were applied to bond relations, the results confirmed most of Goldschmidt's findings.

But Goldschmidt was not satisfied with the geometric approach alone. He wanted to test the results by trying out the forces opposed to those active in crystal growth: he etched and dissolved crystals and compared the results. The statistical and geometric part of this work was excellent proof for his goniometrical work. But when attempting to interpret the resulting micromorphology dynamically, he made the mistake of applying analogies from erosion on the earth's surface. This revealed a weakness in his thinking: a rock is an aggregate and is subject to statistical laws different from those of the crystal lattice. Interest in intergrowth and related problems had not yet arisen. If it had—and if modern knowledge of phases and modern statistics had been available—an association of Goldschmidt and Rosenbusch would have been immensely fruitful. But such knowledge was not available; and, in addition, these men had very different personalities which would not allow a close friendship and cooperation.

Goldschmidt's work on crystal forms and on the forces active in the dissolution of crystals led him to extensive investigations of twinning, surface symmetries, accessories and vicinals, oblique surfaces, and multiple twinning; and the foundations were laid for the understanding of epitaxial overgrowth as a phenomenon related to unmixing or exsolution.

Among the monographs on individual crystal forms, the one on diamond written with Fersman is worthy of special mention. Fersman, who studied for almost two years with Goldschmidt, wrote in his obituary of the latter: "The three works, *Index, Winkeltabellen,* and *Atlas,* are henceforth basic materials for the study of crystals. Without them it is impossible to do crystallographic work. They translate the work of a complete century into a new language . . . the language of new, great ideas in . . . contemporary crystal chemistry. . . ."

Goldschmidt could not split himself into a scientific and a general personality; both parts of his life had to be a unit originating from the same source. Therefore his crystallographic thought—at least its pattern—extended, for him, without a break into the other domains of his personality: the aesthetic, the ethical, and perhaps the religious. Consequently he found excellent correspondence of the harmonic series of crystals with that in music, in fine arts, and with traits of human life in general. Attempts at such integrations were Goldschmidt's favorite philosophic themes and formed the subjects of some of his articles and books: *Ueber Harmonie und Complication* (1901), "Ueber harmonische Analyse von Musikstücken" (*Annalen der Naturphilosophie,* **3** [1904]), "Ueber Harmonie im Weltraum" (*ibid.,* **5**), "Beiträge zur Harmonielehre" (*ibid.,* **13** [1917]), and "Materialien zur Musiklehre" (*Heidelberger Akten der von-Portheim-Stiftung,* nos. 5, 8, 9, 11, 14, 15 [1923–1925]). These works showed that for Goldschmidt harmonic properties and symmetries occurred in all domains of human endeavor as a sort of pantheistic or mystic substratum. In the terms of his contemporary, C. G. Jung, these harmonic series and their ascending and descending differentiations ("complications") were archetypal properties of life.

BIBLIOGRAPHY

I. Original Works. Himmel's bibliography (see below) lists 180 articles or books, but this figure does not include numerous smaller articles and reviews. Goldschmidt's major works are *Index der Kristallformen der Mineralien,* 3 vols. (Berlin, 1886–1891); *Ueber Projektion und graphische Kristallberechnung* (Berlin, 1887); *Kristallographische Winkeltabellen* (Berlin, 1897); *Ueber Harmonie und Complication* (Berlin, 1901); *Der Diamant, eine Studie* (Heidelberg, 1911), written with A. Fersman; *Atlas der Kristallformen,* 9 vols. (Heidelberg, 1913–1923); and *Farben in der Kunst* (Heidelberg, 1919). In addition, Goldschmidt was founder of *Beiträge zur Krystallographie und Mineralogie,* of which he was editor from 1914 to 1926, and of the *Heidelberger Akten der von-Portheim-Stiftung* (1922).

II. Secondary Literature. See A. E. Fersman, "Victor Goldschmidt (10. February 1853 bis 8. Mai 1933)," in *Fortschritte der Mineralogie,* **37,** no. 2 (1959), 207–212, originally in *Reports of the Mineralogical Society of the USSR,* **87,** no. 6 (1958), 677; F. Herrmann, "Victor Goldschmidt," in *Neue deutsche Biographie,* vol. VI (1964); Hans Himmel, "Victor Goldschmidt zum Gedächtnis," in *Zentralblatt für Mineralogie, Geologie und Paläontologie,* sec. A (1933), 391–398; L. Milch, "Zum 75. Geburtstage von Victor Goldschmidt," in *Festschrift zum 75. Geburtstage von seinen Schülern und Freunden gewidmet* (Heidelberg, 1928); and P. Ramdohr, "Zum 100. Geburtstag von Victor Goldschmidt," in *Ruperto-Carola,* **5,** nos. 9–10 (1953), 160–161.

G. C. Amstutz

GOLDSCHMIDT, VICTOR MORITZ (*b.* Zurich, Switzerland, 27 January 1888; *d.* Oslo, Norway, 20 March 1947), *geochemistry, chemistry, mineralogy.*

Goldschmidt was the only son of the distinguished

physical chemist Heinrich Jacob Goldschmidt, who held professorships at Amsterdam, Heidelberg, and Oslo; his mother was Amelie Köhne. After secondary education at Heidelberg, Goldschmidt matriculated in 1905 at the University of Christiania (now Oslo) to study chemistry, mineralogy, and geology. During this year he obtained Norwegian citizenship. His university work was strongly influenced by W. C. Brøgger, the noted Norwegian petrologist and mineralogist, and by such earth scientists as Paul von Groth at Munich and Friedrich Becke at Vienna, in whose institutes he spent the winter terms of 1908 and 1911. Goldschmidt received the doctorate in 1911. Following two years as an instructor at the University of Christiania, in 1914 he was appointed full professor and director of its mineralogical institute.

Goldschmidt's doctoral thesis, "Die Kontaktmetamorphose im Kristianiagebiet," concerned the factors governing the mineral associations in contact-metamorphic rocks and was based upon samples collected in southern Norway. This investigation led to the mineralogical phase rule, which states that the maximum number of crystalline phases that can coexist in rocks in stable equilibrium is equal to the number of components. Goldschmidt continued these petrological studies on regional metamorphism as the first phase of his scientific career, until the middle of World War I. They culminated in the publication of five large reports with the common title *Geologisch-petrographische Studien im Hochgebirge des südlichen Norwegens,* published between 1912 and 1921.

In 1917 the Norwegian government called upon Goldschmidt to investigate the country's mineral resources, and he became chairman of the Government Commission for Raw Materials and director of the Raw Materials Laboratory. His dedication to these practical problems reflected his concern for the utilization of science for the benefit of society. These commitments involved finding local sources for previously imported chemicals, tasks which led Goldschmidt into the second phase of his scientific career—investigations seeking the factors governing the distribution of chemical species in nature.

The base for this geochemical work evolved from extensive crystallographic studies in the Oslo laboratory made by means of the newly developed X-ray techniques which utilized the discoveries of Max von Laue, W. H. Bragg, and W. L. Bragg. Goldschmidt and his associates worked out the crystal structures of 200 compounds of seventy-five elements to form the background for the elucidation of the laws of geochemical distribution. He was able to produce the first tables of atomic and ionic radii for many of the elements, and he investigated the substitution of one element for another in crystals and established patterns of elemental behavior in such processes. The complex formulas of such minerals as tourmaline and mica could be explained by the maintenance of charge neutrality for the positive and negative ions through substitutions based primarily on size. Goldschmidt related the hardness of crystals to their structures, ionic charges, and interatomic distances. This extensive work in geochemistry and mineralogy was published as the monographs *Geochemische Verteilungsgesetze der Elemente,* I-VIII.

In 1929 Goldschmidt became full professor in the Faculty of Natural Sciences at Göttingen and head of its mineralogical institute. Here he initiated geochemical investigations on germanium, gallium, scandium, beryllium, the noble metals, boron, the alkali metals, selenium, arsenic, chromium, nickel, and zinc. Analyses were performed on both terrestrial materials and extraterrestrial meteorites. A model of the earth was formulated in which elements were accumulated in various geological domains on the bases of their charges and sizes and the polarizabilities of their ions. The siderophilic elements, postulated to concentrate in the metallic liquid core of the earth, include iron, nickel, gold, and germanium. The lithophilic elements are enriched in the outer portions of the earth; silicon, magnesium, calcium, aluminum, and the alkalies are members of this class. A third group encompasses the chalcophilic elements, those which ally themselves to sulfur, such as lead and copper. The atmophilic elements have gaseous forms at the temperatures and pressures encountered in the earth's atmosphere and include the noble gases, nitrogen, oxygen, carbon, and hydrogen. Finally, there are the biophilic species, elements that are preferentially incorporated into organisms; carbon, hydrogen, oxygen, nitrogen, vanadium, calcium, and potassium fall within this group. A rather elegant verification of the first three categories is in Goldschmidt's study of the metallurgical products from the copper industry of the Mansfeld in Germany. Here the sulfide, pig iron, and silicate slags included elements that were predicted from his model of the earth.

Following a series of unpleasant confrontations with the emerging anti-Semitism of the Nazis, Goldschmidt abandoned his Göttingen chair in 1935 and returned to Oslo, where a similar position at the university was immediately offered to him. Here he collated his data on cosmic and terrestrial distributions of chemical elements in the ninth and final publication of the *Verteilungsgesetze* and entered into isotopic geology by considering the significances of the isotopic compositions of elements in minerals. While in Oslo, Goldschmidt reentered industrial work and developed techniques for utilizing Norwegian olivine rock in industrial refractories. The onset of World War II brought additional brushes with the Germans. He escaped concentration camps, although imprisoned several times, and, following periods of hiding, made his way to Sweden and then Great Britain. In the final phases of his scientific career at the Macaulay Institute for Soil Research and at Rothamsted, he applied his previously gained geochemical concepts to soil science. Goldschmidt's manuscripts for the definitive treatise on the science of geochemistry, which he had done so much to found, were edited after his death by A. Muir and were published in 1954.

The adversities and humiliations suffered by Goldschmidt at the hands of the Nazis were met with courage and wit. Under Nazi occupation in Norway and Germany, he carried a capsule of hydrocyanic acid for use as the final evasion of oppression. A university colleague in Oslo once asked Goldschmidt for a similar capsule. He replied, "This poison is for professors of chemistry only. You, as a professor of mechanics, will have to use the rope."

Goldschmidt stands as one of the pioneers in geochemistry who, utilizing the basic properties of matter, gave simple and beautiful explanations of the composition of our environment. He never married, but his students and associates provided him with warm personal friendships. His co-workers, such as Fritz Laves, T. F. W. Barth, and W. Zachariasen, became noted geochemists; and some of his students, including Theodor Ernst, H. Hauptmann, W. von Engelhardt, and C. Peters, became heads of university departments.

BIBLIOGRAPHY

I. Original Works. The complete list of Goldschmidt's some 200 papers may be found in *Norsk geologisk tidsskrift,* **27** (1949), 143-163. In addition, see his posthumously published *Geochemistry,* Alex Muir, ed. (London, 1954).

II. Secondary Literature. On Goldschmidt and his work see J. D. Bernal, "The Goldschmidt Memorial Lecture," in *Journal of the Chemical Society* (1949), pp. 2108-2114; Carl W. Correns, "Victor Moritz Goldschmidt," in *Naturwissenschaften,* **34** (1947), 129-131; and Ivar Oftedal, "Memorial to Victor Moritz Goldschmidt," in *Proceedings. Geological Society of America* (1948), pp. 149-154.

E. D. Goldberg

GOLDSTEIN, EUGEN (*b.* Gleiwitz, Upper Silesia [now Gliwice, Poland], 5 September 1850; *d.* Berlin, Germany, 25 December 1930), *physics.*

After attending Ratibor Gymnasium, Goldstein spent a year (1869-1870) at the University of Breslau. He then went on to the University of Berlin, where he worked with Helmholtz, taking his doctorate in 1881. He spent most of his exceptionally long professional career as a physicist at the Potsdam observatory. His first scientific paper was published in 1876, his last over fifty years later.

Almost all of Goldstein's published work was on topics which sprang naturally from his lifelong interest in electrical discharges in moderate to high vacuums. He is now known primarily as the discoverer, in 1886, of "Kanalstrahlen," as he called them—canal rays or positive rays, as they became known in English. He also made significant contributions to the study of cathode rays, which were discovered by Julius Plücker but named by Goldstein. Most of the rest of his work concerned various phenomena occurring in gaseous discharges.

In 1876 Goldstein showed that cathode rays could cast sharp shadows.[1] He was able to demonstrate that they were emitted perpendicularly to the cathode surface, a discovery that made it possible to design concave cathodes to produce concentrated or focused rays, which were useful in a wide range of experiments. But this same discovery cast some doubt on the idea then prevailing among German physicists that the rays consisted of some form of electromagnetic radiation. Further, Goldstein and others showed in 1880 that the rays could be bent by magnetic fields;[2] this discovery also gave aid and comfort to those physicists, predominantly British, who believed that the rays were streams of negative particles.

Sir William Crookes, for example, had suggested that the rays were charged "molecular torrents" rebounding from the cathode. To oppose this view, Goldstein conducted a series of experiments showing that cathode rays emitted light showing little if any Doppler shift and that they could traverse a distance some 150 times the mean free path for molecules at the pressures then being achieved in the discharge tubes.[3]

Over a span of many years Goldstein published several papers on other aspects of cathode rays. He showed (1895-1898) that they could make certain salts change color, that they could be "reflected" diffusely from anodes (1882), and that there was some evidence for electrostatic deflection of parallel beams. However, his "reflection" experiment may have been misleading: the "reflected" rays may well have been soft X rays produced in the anode by the impinging cathode rays (but of course X rays had not yet been discovered). An exceptionally clever experimentalist, Goldstein studied the effects of a wide range of cathode and anode configurations.

In 1886 Goldstein published his discovery of "Kanalstrahlen," rays which emerged from channels or holes in anodes in low-pressure discharge tubes.[4] His student Wilhelm Wien, who later became known primarily as a theoretical physicist, showed that the canal rays could be deflected by electric and magnetic fields, and that they had ratios of positive charge to mass approximately 10,000 times that of cathode rays.[5] Wien did not detect different ratios for different elements. The development of canal-ray apparatus into the important field of mass spectroscopy was, of course, carried out by others, notably J. J. Thomson and F. W. Aston.

Another of Goldstein's students, Johannes Stark, was able to show that light from canal-ray particles showed a Doppler shift.[6] This was the first clear-cut demonstration of an optical Doppler shift in a terrestrial source.

Goldstein continued to publish papers on various canal-ray topics, notably studies of the wavelengths of light emitted by various metals and oxides when they were struck by the rays. He found, for example, that the alkali metals, when hit by the rays, emitted their characteristic bright spectral lines, while they did not do so when hit by cathode rays. He also found that a constriction in a discharge tube could function as a source of positive rays.

In the last two decades of his life Goldstein devoted much attention to anode discharges and to the striations of the positive column in low-pressure discharge tubes. Such tubes present a wealth of beautiful and fascinating phenomena, and Goldstein's experimental virtuosity made it natural for him to pursue such topics. It is ironic that his work in these areas was of secondary importance and now is seldom mentioned in writings in the field, while his early work, and that of his students, was much more fundamental and lasting. But it is perhaps even more ironic that his last paper, published in 1928, reported detection of the synthesis of ammonia in discharge tubes containing various gases.[7] This virtually forgotten work foreshadowed an intriguing and interesting field of research that came to life over thirty years after Goldstein's death.

NOTES

1. *Monatsberichte der Königlichen Akademie der Wissenschaften zu Berlin* (1876), 284.
2. *Wiedemann's Annalen der Physik,* **11** (1880), 850.
3. *Philosophical Magazine,* **10** (1880), 234, originally in *Monatsberichte der Königlichen Akademie der Wissenschaften zu Berlin* (Jan. 1880).
4. "Über eine noch nicht untersuchte Strahlungsform an der Kathode inducirter Entladungen," in *Sitzungsberichte der Königlichen Akademie der Wissenschaften zu Berlin,* **39** (1886), 691.
5. "Deflection of Canal Rays," in *Berlin Physikalische Gesellschaft Verhandlungen,* **17** (1898), 10–12.
6. "Doppler Effect Exhibited by Canal Rays and the Spectrum of Positive Ions," in *Physikalische Zeitschrift,* **6** (1905), 892–897.
7. "Synthesis of Ammonia, Argon as Catalyst," in *Zeitschrift für Physik,* **47** (1928), 274.

BIBLIOGRAPHY

I. ORIGINAL WORKS. Most of Goldstein's work was published in such journals as *Wiedemann's Annalen der Physik* and *Zeitschrift für Physik.* Specific references can be found in *Science Abstracts.* A collection of papers was reprinted as no. 231 of Ostwald's Klassiker der Exacten Wissenschaften (Leipzig, 1930).

II. SECONDARY LITERATURE. As a tribute to Goldstein on his eightieth birthday, Rausch von Traubenberg wrote "Die Bedeutung der Kanalstrahlen für die Entwicklung der Physik," in *Naturwissenschaften,* **18** (5 Sept. 1930), 773–776. See also E. Rüchardt, "Zur Entdeckung der Kanalstrahlen vor fünfzig Jahren," and F. W. Aston, "Kanalstrahlen und Atomphysik," both in *Naturwissenschaften,* **24** (24 July 1936), 465–469. Goldstein's contributions to the understanding of cathode rays are briefly discussed in D. L. Anderson, *The Discovery of the Electron* (Princeton, 1964). A brief obituary note appeared in *Nature,* **127** (1931), 171.

DAVID L. ANDERSON

GOLGI, CAMILLO (Corteno [now Corteno Golgi], Brescia, Italy, 7 July 1843; Pavia, Italy, 21 January 1926), *histology, pathology.*

Golgi's family was from Pavia; his father Alessandro was a doctor. Golgi read medicine at the University of Pavia, where, together with Giulio Bizzozero and Enrico Sertoli, he studied under Eusebio Oehl, distinguished as the first in Pavia to develop systematically studies of microscopic anatomy and histology.

After obtaining a degree in medicine in 1865, he worked for a short time in the psychiatric clinic directed by Cesare Lombroso, but his main interest was in the histological research he was conducting in the laboratory of experimental pathology directed by Bizzozero.

Golgi's first publications, which appeared between 1868 and 1871, included some works on clinical topics but were mainly devoted to the anatomy and pathological anatomy of the nervous system. In his papers on neurology he described the morphological features of the glial cells and showed the relationships between their prolongations and blood vessels.

In 1871 he gave a private course on clinical microscopy, but in 1872 financial difficulties forced him to interrupt his scientific career temporarily and accept the modest post of principal doctor of the Pio Luogo degli Incurabili at Abbiategrasso. Even there he managed, although with difficulty, to continue his microscopic research on the structure of the nervous system; later he was able to publish the results in important papers.

Having gained a certain degree of fame, he became in 1875 a lecturer in histology at the University of Pavia. In 1879 he obtained the chair of anatomy at the University of Siena, but the following year he returned to Pavia, first as professor of histology and later of general pathology; he continued to teach histology as well until his obligatory retirement in 1918. Around him flourished a group of notable scholars and researchers.

In 1906 he shared the Nobel Prize for medicine or physiology with Santiago Ramón y Cajal, famous also for his studies on the fine anatomy of the nervous system.

Golgi became a senator in 1900 and took an active part in public and university life, especially in Pavia, where he was dean of the Faculty of Medicine and president of the university. He also concerned himself with problems of public health and university administration. A member of numerous scientific societies and academies, Italian and foreign, he had contacts with such personalities as the Swiss anatomist Albert von Koelliker, and the Norwegian explorer and scientist Fridtjof Nansen.

In the course of his long life Golgi penetrated various fields of biology and medicine with equal success, but the areas of research in which he earned the greatest distinction were neuroanatomy (for which work he won the Nobel Prize), cytology, and malariology.

In the second half of the nineteenth century considerable progress was made in the study of histology and microscopic anatomy; until the work of Golgi, however, little headway had been made in the study of the nervous system because of a lack of appropriate techniques; and theories on the function of nerve cells and their extensions were nebulous and conflicting. Golgi invented a completely original method based on the coloration of cells and nerve fibers by means of the prolonged immersion of samples, previously hardened with potassium bichromate or ammonium bichromate, in a 0.5 to 1 percent solution of silver nitrate. This technique brings out clearly the features of the nerve elements. Under controlled conditions, based on the length of the period of hardening in the bichromate, the "black reaction" permits the controlled staining of certain nerve elements (for example, either the nerve fibers with their fine branches, or only the nerve or connective cells) or even only certain parts of one (fibers) or the other (cells), thus allowing a better study of their interrelationships.

From 1873 Golgi published many articles on the results of his systematic observations, using his new technique, on the fine anatomy of the various organs of the nervous system (the gray matter of the brain, the cerebellum, the olfactory lobes, etc.). On the basis of his observations Golgi formulated a theory based on the following fundamental points:

(1) The function of the nerve extensions, or axons, is exclusively one of transmission of nerve impulses.

(2) The function of the protoplasmic extensions, or dendrites, is predominantly trophic, as can be deduced, for example, from their frequent relationships with the pia mater.

(3) There are two types of nerve cells, differing according to the characteristics of the nerve extension of each: nerve cells of the first type are those with an axon that, although serving a more or less large number of lateral fibrils, nevertheless preserves its individuality and continues directly into the cylindraxis of a medullary fiber. Nerve cells of the second type are those with an axon that within a relatively short distance of its origin subdivides within an indeterminate distance, and with no demonstrable spatial limit. The cells of the first type probably have a motive or psychomotive function; those of the second, hypothetically a sensorial or psychosensorial function.

(4) In the gray matter of the nerve centers there is a diffused nerve network of extreme fineness, continuous over the entire nerve substance and made of nerve fibrils finely and thickly interlaced. Golgi would not pronounce dogmatically on the question of whether it was a network in the true sense, made of anastomosed fibrils derived from various nerve elements, or simply interlaced, functionally independent filaments of different origin. Golgi considered the diffused nerve network to be the mediating organ that effectuated connections between various parts of the nervous system or between various functional activities related to that system.

Golgi's hypotheses superseded those formulated in 1872 by Josef von Gerlach but were soon challenged by the contributions made by Ramón y Cajal to the neuron theory.

Some of Golgi's contributions to neuroanatomy deserve mention. He discovered the existence along the length of nerve fibers of numerous special apparatuses that support the myelin (the corneal spires); the existence of special terminal bodies of a sensitive nature in muscle tendons that had never previously been described and the importance of which has been confirmed by recent physiological research; and the critical examination of the theory of brain localizations.

In the field of cytology Golgi was the first to describe, in 1898, the existence in the cytoplasm of the nerve cell of a special small organ, in the shape of a fine and elegant network of anastomosed and interlaced threads. Later Golgi and his co-workers demonstrated that this endocellular structure, called Golgi's internal reticular apparatus, was given special attention by classical cytologists, who recognized its undoubted individuality and suspected its importance in the cellular economy; it is now the object of particular studies, because it is considered of fundamental importance in cytometabolic processes.

The discovery in 1880 of the malaric parasite by C. L. Alphonse Laveran was not immediately accepted but was received at first skeptically and with diffidence. Among Italians the theory of its existence was accepted and developed by Ettore Marchiafava and Angelo Celli, whose research on it achieved (1885) results important to the understanding of the development cycle of the parasite.

Golgi followed for some time, in Rome, the research of Marchiafava and Celli; then in Pavia, between 1885 and 1893, he did important research on malaria, arriving at the verification of the following fundamental facts:

(1) the existence of the cycle of monogamic development of the tertian and quartan forms of malaria;

(2) the existence of specific differences between the parasites of the two forms;

(3) the correspondence of the cyclic development of the malaric parasites with the periodic succession of the fever fits;

(4) the constant relationship of the single fits with the development, maturing, and reproduction of one generation of parasites;

(5) the correspondence of various species or varieties of malaric parasites with the various fundamental classical types of intermittent fever.

From this knowledge Golgi immediately derived results capable of practical application: by the examination of the blood of malaria patients carried out with methods that he suggested, it was possible to diagnose the different forms of the disease and to establish the sequence in the appearance of the fever fits. Furthermore, since the plasmodes display different degrees of sensitivity to quinine according to the stage of their development (the young forms derived immediately from the segmentation or sporulation process are the most sensitive to quinine), the most efficient way to prevent the appearance of the fever fit and progressively extinguish the infection is to give quinine a few hours before the fit, so that it can act on the new generation of the parasite. Also basically important for diagnosis was the observation that the entire nosogenic process of malaric fevers sometimes occurs not in the circulating blood but in the internal organs, so that it is only later that the parasites spread in the blood.

BIBLIOGRAPHY

Golgi's works, which originally appeared in various journals, were collected in *Opera omnia,* 4 vols. (Milan, 1903–1929). A German trans. of his works is *Untersuchungen ueber den feineren Bau des centralen und peripherischen Nervensystems,* G. Fischer, ed. (Jena, 1894).

Golgi's MSS, partly published drawings (some completed by his students), and scientific and personal effects are in the Museo per la Storia dell'Università di Pavia. Material on Golgi may also be found at the Instituto di Patologia Generale dell' Università di Pavia.

The following works serve as bibliographical sources: Bruno Zanobio, "The Work of Camillo Golgi in Neurology," in *Essays on the History of Italian Neurology. Proceedings of the International Symposium on the History of Neurology. Varenna—30.VIII/1.IX.1961* (Milan, 1963), pp. 179–193; and "L'opera del biologo Camillo Golgi," in *Actes du III Symposium International d'Histoire des Sciences, Turin, 28–30 juillet 1961* (Florence, 1964), pp. 64–84; and Giorgio Pilleri, "Camillo Golgi" in Kurt Kolle, ed., *Grosse Nervenärzte,* 2nd ed., II (Stuttgart, 1970), 3–12.

BRUNO ZANOBIO

GOLITSYN, BORIS BORISOVICH (*b.* St. Petersburg [now Leningrad], Russia, 2 March 1862; *d.* Petrograd [now Leningrad], 16 May 1916), *physics, seismology.*

Golitsyn came from an old family of the nobility. In 1887 he graduated from the hydrographic section of the Maritime Academy; but since he did not wish to serve in the fleet, he went to study abroad, enrolling in the Physics and Mathematics Faculty of the University of Strasbourg. He graduated in 1890 and then returned to Russia. He began to teach at Moscow University in 1891. In 1893 Golitsyn presented to the Faculty of Mathematics and Physics his master's thesis, "Issledovania po matematicheskoy fizike" ("Investigations in Mathematical Physics"). The first part set forth the "general characteristics of dielectrics from the point of view of the mechanical theory of heat" (electrostriction, the dependence of dielectric constants on volume, pressure, and temperature). The second part discussed radiant energy (light pressure, the significance of absolute temperature, and the dependence of radiation on external factors).

What was significantly new in the second part was Golitsyn's departure from the electrodynamics of Michael Faraday and James Clerk Maxwell, by first considering the space occupied by radiation as a kind of medium to which the concept of temperature is applicable. He arrived at the following formulation: "Absolute temperature is conditioned by the sum total of all electrical displacements; and, in particular, the fourth power of the absolute temperature is directly proportional to the sum of the squares of all the electrical displacements which are carried away from the vacuum." In this work Golitsyn presented two adiabatic invariants of thermal radiation: $U\sqrt[3]{v}$ = constant and $T\sqrt[3]{v}$ = constant, where v is a "given volume of ether," U is the quantity of radiant energy in this volume, and T is the absolute temperature of the radiation.

After examining Golitsyn's dissertation, Aleksandr G. Stoletov and Aleksei P. Sokolov, both members of the Faculty of Mathematics and Physics, rendered a sharply negative review. They were not able to see Golitsyn's point of view on temperature radiation in its proper perspective. A heated discussion arose concerning the dissertation. Stoletov solicited opinions on the disputed questions from Hermann Helmholtz, Lord Kelvin, and Ludwig Boltzman—and they agreed generally with Stoletov and Sokolov. Golitsyn was not permitted to defend his dissertation. Court circles found this prohibition insulting to a member of the aristocracy, and consequently Stoletov was passed over as a candidate for membership in the Imperial Academy of Sciences, while Golitsyn was selected to be an adjunct member.

Before 1898 Golitsyn's scientific research was concerned mainly with molecular physics. He examined the problem of critical temperature in great detail. Against the theory of Thomas Andrews he advanced a new notion about conditions of quasi equilibrium based on extremely precise experimental data. Beginning in 1899, Golitsyn started to occupy himself mainly with seismology and seismometry. He laid the foundations of scientific seismometry and developed an improved type of seismograph, one with a galvanometric register. In such a seismograph the pendulum is equipped with coils and oscillates in a field of permanent magnets. Under these conditions an electric current is induced in the coils, the measurement of which permits the precise recording of seismic vibrations.

Through the efforts of Golitsyn, Russian seismometry occupied a leading place in world science at that time; and his seismographs were the prototypes for new apparatus for the study of earthquakes and mechanical vibrations, and for seismic prospecting for useful minerals. In 1908 Golitsyn was elected an academician. International recognition of his great services in seismology was expressed by his election as president of the International Seismic Association at its congress held at Manchester in 1911. In 1916 Golitsyn was made a foreign member of the Royal Society.

BIBLIOGRAPHY

Many of Golitsyn's writings were brought together as *Izbrannye trudy* ("Selected Works"), 2 vols. (Moscow, 1960). His works are listed (under "Galizin") in Poggendorff, IV, 474–475, and V, 408–409.

On Golitsyn's life or work, see G. P. Blok and N. V. Krutikova, "Rukopisi B. B. Golitsyna v arkhive Akademii nauk SSSR" ("Manuscripts of B. B. Golitsyn in the Archive of the Academy of Sciences of the U.S.S.R."), in *Trudy Arkhiva Akademii nauk SSSR,* pt. 10 (1952); A. N. Krylov, "Pamyati B. B. Golitsyna" ("In Memory of B. B. Golitsyn"), in *Priroda,* no. 2 (1918), 171–180; A. S. Predvoditelev, "O fizicheskikh rabotakh B. B. Golitsyna" ("On Golitsyn's Work in Physics"), in Golitsyn's *Izbrannye trudy,* I, 217–240; and A. S. Predvoditelev and N. V. Veshnyakov, "Zhizn i nauchnaya deyatelnost akademika B. B. Golitsyna" ("The Life and Scientific Activity of Academician B. B. Golitsyn"), in Golitsyn's *Izbrannye trudy,* I, 5–12.

See also *Materialy dlya biograficheskogo slovarya deystvitelnykh chlenov imperatorskoy Akademii nauk, 1889–1914* ("Materials for a Biographical Dictionary of Members of the Imperial Academy of Sciences, 1889–1914"), pt. 1 (Petrograd, 1915), 193–218, see entry for Golitsyn; and *Materialy dlya istorii akademicheskikh uchrezhdeny za 1889–1914 gg.* ("Materials for a History of Academic Institutions for the Years 1889–1914"), pt. 1 (Petrograd, 1917), 47–82, on Golitsyn's scientific activity.

J. G. DORFMAN

GOLTZ, FRIEDRICH LEOPOLD (*b.* Posen, Germany [now Poznan, Poland], 14 August 1834; *d.* Strasbourg, France, 4 May 1902), *physiology, encephalology.*

Goltz's father, Heinrich Goltz, a police inspector in Posen and Danzig, died when Friedrich was twelve. His mother, Leopoldine Friederike von Blumenberg, then moved the family to Thorn (now Torun, Poland), where an uncle, Bogumil Goltz, lived. A philosopher, natural scientist, and author, Bogumil Goltz came to have great influence on Friedrich.

Goltz started his medical studies in 1853 at Königsberg. There he attended Hermann Helmholtz' lectures on physiology and general pathology. He was not inclined toward Helmholtz' physical approach but leaned instead toward analysis of the morphological basis of physiological functions and the interpretation of simple observations and animal surgery to arrive at solutions of physiological questions. His doctoral dissertation (20 January 1858), *De spatii sensu cutis,* an investigation of the sense of touch, points in that direction. After two years of surgical training under Ernst Wagner, during which he acquired the techniques for his later brain operations, Goltz was appointed prosector in anatomy at Königsberg, under August Müller. He then became extraordinary professor of anatomy in 1865.

In 1868 Goltz married Agnes Simon, the daughter of Samuel Simon, city councillor in Königsberg. In 1870 he succeeded A. W. Volkmann as professor of physiology in Halle and two years later he was called to the new German university in Strasbourg, the former Faculté de Médecine. He was appointed rector of the university in 1888 and constantly concerned himself with maintaining good relations between the German and French populations of Alsace. He was spirited and witty, a popular teacher, and a great lover of animals. Among his colleagues were such gifted people as J. G. Gaule, Joseph von Mering, Jacques Loeb, Albrecht Bethe, A. Bickel, and J. Richard Ewald. Goltz died of progressive muscle paralysis, asthma, and sclerosis.

The principal subject of Goltz's research was the study of reflex phenomena, particularly in the spinal cord of the frog. Later he was concerned primarily with the analysis of localization phenomena in the brain. He originated a series of tests which for a long time were generally used in university lectures. The Goltz *Kochversuch* ("cooking test," 1860) called for the slow heating of a spinal (decerebrated) frog in a water bath. Because of the gradual increase in stimulation there was no reaction, and Goltz argued that its absence refuted E. F. W. Pflüger's concept of the *Rückenmarksseele* ("spinal cord soul"). His *Klopfversuch* ("tapping test," 1862) also became famous: When the abdominal wall of a frog is tapped, the heart stops momentarily because of the reflex vagus effect. Goltz next investigated the nerve mechanism of frogs during copulation (*Umklammerungsreflex,* or "embracing reflex," 1865). In the same year he demonstrated the famous "croaking reflex" (*Quakreflex*) by stroking the skin on the back of spinal frogs. He reported his findings in *Beiträge zur Lehre von den Funktionen des Nervensystems des Frosches.*

In the following years Goltz analyzed the functions of the labyrinth of the inner ear in frogs and pigeons, thereby succeeding in differentiating the functions of the *nervus octavus* into hearing and equilibrium. He was the first to recognize the importance of the semicircular canals for maintaining equilibrium. In 1874 Goltz proved the existence of reflex centers in dogs for erection, evacuation, and parturition.

Next Goltz concentrated on the study of the functions of the cerebrum (1876). On the basis of tests by Pierre Flourens, Gustav von Fritsch, Julius Hitzig, and Daniel Ferrier, he attempted to obtain, by means of careful surgery, information about the localization or local occurrence of cerebral functions in the center of the brain. At first he partially destroyed portions of the cerebrum, but in later experiments he was able to remove entire lobes, and even a hemisphere. Since he succeeded, with scrupulous care, in keeping the animals alive for several years, he was able to differentiate between the initial irritative symptoms produced by surgery and definite long-term results. He thus arrived at the concept of a reciprocal interchangeability, and even equivalence, of cerebral parts without completely rejecting the localization theory. A dog whose cerebrum had been removed was without intellect, memory, and intelligence.

Regarding animal research as unavoidable, and even essential, Goltz in 1883 protested the antivivisectionist movement. He wrote about his cerebral tests in *Über die Verrichtungen des Grosshirns* (1881) and produced a final report on the dog without a cerebrum in 1892. His research was carried on particularly by C. S. Sherrington and Harvey Cushing.

BIBLIOGRAPHY

I. Original Works. Goltz's books are *De spatii sensu cutis* (Königsberg, 1858); *Beiträge zur Lehre von den Funktionen des Nervensystems des Frosches* (Berlin, 1869); *Über die Verrichtungen des Grosshirns. Gesammelte Abhandlungen* (Bonn, 1881); and *Wider die Humanaster. Rechtfertigung eines Vivisektors* (Strasbourg, 1883).

His publications in journals include "Beiträge zur Lehre von den Funktionen des Rückenmarks der Frösche," in *Königsberger medizinische Jahrbuch,* **2** (1860), 189–226; "Über Reflexionen vom und zum Herzen," *ibid.,* **3** (1862), 271–274; "Über die physiologische Bedeutung der Bogengänge des Ohrlabyrinths," in *Pflügers Archiv für die gesamte Physiologie des Menschen und der Tiere,* **3** (1870), 172–192; "Über die Funktionen des Lendenmarks des Hundes," *ibid.,* **8** (1874), 460–498; "Über den Einfluss des Nervensystems auf die Vorgänge während der Schwangerschaft und des Gebäraktes," *ibid.,* **9** (1874), 552–565; and "Über die Verrichtungen des Grosshirns I–VII," *ibid.,* **13** (1876), 1–44; **14** (1876), 412–443; **20** (1879), 1–54; **26** (1881), 1–49; **34** (1884), 451–505; **42** (1888), 419–467; **51** (1892), 570–614.

II. Secondary Literature. Obituaries are A. Bickel, "Friedrich Goltz †," in *Deutsche medizinische Wochenschrift,* **28** (1902), 403; J. R. Ewald, "Friedrich Goltz," in *Pflüger's Archiv für die gesamte Physiologie des Menschen und die Tiere,* **94** (1903), 1–64, with portrait and complete bibliography; and Heinrich Kraft, "Friedrich Leopold Goltz," in *Münchener medizinische Wochenschrift,* **49** (1902), 965–970.

Shorter biographies are D. Trincker, in *Neue deutsche Biographie,* VI (Berlin, 1964), 636–637, with short bibliography; *Biographische Lexikon der hervorragenden Ärzte aller Zeiten und Völker,* 2nd ed., II (Berlin–Vienna, 1930), 792–793; and K. E. Rothschuh, *Geschichte der Physiologie* (Berlin–Göttingen–Heidelberg, 1953), esp. pp. 186–187, with portrait and list of students.

K. E. ROTHSCHUH

GOMBERG, MOSES (*b.* Elisavetgrad, Russia [now Kirovograd, U.S.S.R.], 8 February 1866; *d.* Ann Arbor, Michigan, 12 February 1947), *organic chemistry.*

Gomberg prepared the first stable free radical, triphenylmethyl, at the University of Michigan in 1900 and pioneered in the development of free-radical chemistry. He was the son of George and Marie Resnikoff Gomberg; his father was the owner of a modest estate in the Ukraine. Young Gomberg entered the Nicolau Gymnasium in Elisavetgrad in 1878. Six years later his father was accused of anticzarist activities. His estate was confiscated and he fled with his family to Chicago, where he and his son, neither of whom had any knowledge of English, worked at menial jobs in the stockyards and elsewhere in the city.

Moses quickly learned English and completed his high school education in Chicago. He then became a student at the University of Michigan, completing his B.S. in 1890. He continued at Michigan as an assistant in chemistry, receiving his M.S. in 1892 and his Ph.D. in 1894. His work was done under Albert B. Prescott, and his dissertation dealt with some reactions of caffeine. Appointed instructor in organic chemistry in 1893, he never severed his connection with the chemistry department at the University of Michigan. Promotion to assistant professor came in 1899, to junior professor in 1902, and to full professor in 1904. From 1927 until his retirement in 1936 he was chairman of the chemistry department.

In order to procure funds for European study, Gomberg carried out analyses of water, minerals, fatty oils, foods, drugs, and patent medicines and served as an expert witness in toxicology cases while he was still a graduate student. Taking a leave of absence in 1896–1897, he spent two terms in Adolf von Baeyer's laboratory in Munich and a term with Victor Meyer in Heidelberg. During this period he turned away from the concentration on analysis which characterized the work of Prescott and his students in order to concentrate on synthetic studies. At Munich he prepared some nitrogen derivatives of isobutyric acid; and at Heidelberg he set out to prepare tetraphenylmethane, a compound which a number of German chemists had unsuccessfully sought to synthesize and which Meyer believed probably could not be synthesized. Meyer suggested that Gomberg undertake a problem more likely to be successful. Gomberg persisted in his objective and, by oxidizing triphenylmethane hydrazobenzene, obtained the corresponding azo compound, which decomposed to tetraphenylmethane on heating at 110–120°C. Although Gomberg was successful in his endeavor, the yield of tetraphenylmethane was poor (2–5 percent).

Upon his return to Michigan, Gomberg sought to prepare hexaphenylethane, the next fully phenylated hydrocarbon of the series. He utilized the classical reaction of a metal on an appropriate halide:

$$2(C_6H_5)_3CX + \text{metal} \rightarrow (C_6H_5)_6C_2 + \text{metal halide.}$$

The use of either triphenylmethyl bromide or chloride with sodium failed to yield a product, but substitution of silver for sodium led to a reaction in which a white crystalline product began to separate after heating the reaction mixture for several hours at the boiling point of the benzene solvent. The crystalline product was assumed to be hexaphenylethane, but elementary analysis yielded 87.93 percent carbon and 6.04 percent hydrogen (calculated for hexaphenylethane, C = 93.83, H = 6.17). Carefully repeated syntheses yielded products giving similar analytical results, and Gomberg was forced to conclude that he was preparing an oxygenated compound (which proved to be the peroxide $[C_6H_5]_6C_2O_2$).

Gomberg repeated the reaction of triphenylmethyl chloride and silver in an atmosphere of carbon dioxide. Now he obtained no solid product at all, but the yellow color of his solution indicated that a reaction had occurred. Removal of the benzene solvent left a colorless solid of unexpectedly high reactivity toward oxygen and halogens. It had been expected that hexaphenylethane would be a colorless solid characterized by chemical inertness. In his first publication on the subject, Gomberg wrote, "The experimental evidence presented above forces me to the conclusion that we have to deal here with a free radical, triphenylmethyl, $(C_6H_5)_3C$. On this assumption alone do the results described above become intelligible and receive an adequate explanation" (*Journal of the American Chemical Society,* **22** [1900], 768).

The announcement of the preparation of a stable free radical was received with skepticism, for most contemporary chemists had become convinced that such chemical species could not exist. Gomberg set out to establish the soundness of his conclusion by carefully studying the properties of his substance and by preparing additional substances showing free-radical properties. He quickly developed an understanding of the experimental conditions necessary for successful synthesis of related compounds. His major research activities during the remainder of his career were aimed toward extending the understanding of free-radical chemistry.

Results of molecular weight determinations by L. H. Cone in Gomberg's laboratory proved to be variable in different solvents and fell closer to the theoretical value for the dimer hexaphenylethane than to that for the monomer triphenylmethyl. This was a matter of grave concern for Gomberg and caused him ultimately to accept the idea that the solution must contain an equilibrium mixture of both substances:

$$(C_6H_5)_3C - C(C_6H_5)_3 \rightleftarrows 2\ (C_6H_5)_3\ C\cdot$$

Other chemists believed the compound to be merely hexaphenylethane, assuming it to be unstable in the presence of oxygen, halogens, and other highly reactive substances. Gomberg argued that the presence of color supported the free-radical hypothesis. When critics sought to explain the color by use of quinoid structures, Gomberg showed that the addition of oxygen to the solution resulted in loss of color. On standing, the yellow color slowly returned, as if a new equilibrium were being established; but the color could be destroyed again by addition of more oxygen. Change of color with change of temperature also supported the concept of an equilibrium in solution. Jean F. Piccard of Munich showed in 1911 that the colored solution failed to obey Beer's law when diluted; dilution actually brought about intensification of the color, thereby lending support to the equilibrium concept.

Investigators in other laboratories, particularly Wilhelm Schlenk at Jena, brought forth evidence for free radicals in other hexaarylethane systems. By 1911 many organic chemists were willing to concede the existence of stable free radicals in solution, but only when dealing with compounds such as fully substituted ethanes carrying bulky substituents such as phenyl groups, or even better, substituted phenyl groups or naphthyl groups. Free radicals containing nitrogen, sulfur, and oxygen atoms carrying three aryl groups were also prepared in various laboratories, and in 1929 Fritz Paneth bolstered the free-radical concept by establishing evidence for the transient existence of free methyl radicals. Gradually free radicals came to take on great significance in reaction mechanisms.

Gomberg also carried out studies on organometallic compounds, and during World War I he worked on war gases. He was associated with the civilian chemists who originally worked in the Bureau of Mines and were later absorbed by the newly created Chemical Warfare Service. Although the idea of chemical warfare was abhorrent to him, Gomberg accepted the task of developing the commercial synthesis of ethylene chlorohydrin, required as an intermediate for the synthesis of mustard gas. Later, he was commissioned a major in the Ordnance Department and served as an adviser in the production of smokeless powder and high explosives.

During his forty-three years on the Michigan faculty Gomberg was a respected teacher and administrator. As an effective supervisor of graduate students he considered the development of the man as a scientist more important than the production of publications. Two of his students, C. S. Schoepfle and W. E. Bachmann, in their obituary memoir said of him: "Gifted with a remarkable memory, he presented his lectures with the full use of a wealth of historical material and so vividly that they left an indelible imprint on his students. A great teacher and scholar, he inspired his students by his methods and ideals, and his colleagues by the vigor and clarity of his mind. To this greatness, he added an innate kindliness and unassuming modesty that endeared him to all" (*ibid.*, **69** [1947], 2924).

Gomberg was a member of numerous professional societies and was honored by election to the National Academy of Sciences. In 1931 he served as president of the American Chemical Society and at various times was honored by awards administered by it: the Nichols Medal in 1914, the Willard Gibbs Medal in 1925, and the Chandler Medal in 1927.

Although he enjoyed the company of women in a somewhat courtly and reserved manner, Gomberg never married and firmly forbade his graduate students to marry before they finished their degrees. His younger sister Sonja served as his hostess and housekeeper at his cottage in Ann Arbor.

BIBLIOGRAPHY

I. Original Works. There is a bibliography of Gomberg's published papers in C. S. Schoepfle and W. E. Bachmann, "Moses Gomberg, 1866–1947," in *Journal of the American Chemical Society,* **69** (1947), 2924–2925. His original paper on triphenylmethyl appeared in *Berichte der Deutschen chemischen Gesellschaft,* **33** (1900), 3150–3163; and *Journal of the American Chemical Society,* **22** (1900), 757–771. Gomberg published review articles on free-radical chemistry in *Chemical Reviews,* **1** (1924), 91–141, and **2** (1925), 301–314; *Journal of Industrial and Engineering Chemistry,* **20** (1928), 159–164; *Journal of Chemical Educa-

tion, **9** (1932), 439–451; and *Science,* **74** (1931), 553–557.

II. SECONDARY LITERATURE. There is no lengthy biography of Gomberg. The best short sketch is C. S. Schoepfle and W. E. Bachmann, "Moses Gomberg, 1866–1947," in *Journal of the American Chemical Society,* **69** (1947), 2921–2925, repr. in E. Farber, ed., *Great Chemists* (New York, 1961), pp. 1209–1217. There is a short sketch by A. H. White, in *Industrial and Engineering Chemistry,* **23** (1931), 116–117. For an evaluation of Gomberg's role in the history of free-radical chemistry, see A. J. Ihde, in *Pure and Applied Chemistry,* **15** (1967), 1–13, repr. in International Union of Pure and Applied Chemistry, *Free Radicals in Solution* (London, 1967), pp. 1–13. These two publications carry the papers presented at the Centennial Symposium on Free Radicals in Solution, which was held under I.U.P.A.C. sponsorship at the University of Michigan in 1966. These papers reflect the consequences of Gomberg's work.

AARON J. IHDE

GOMPERTZ, BENJAMIN (*b.* London, England, 5 March 1779; *d.* London, 14 July 1865), *mathematics.*

One of three prominent sons of a distinguished mercantile family that emigrated from Holland in the eighteenth century, Gompertz appeared destined for a financial career. Denied matriculation at the universities because he was Jewish, he joined the Society of Mathematicians of Spitalfields in 1797 and educated himself by reading the masters, especially Newton, Colin Maclaurin, and William Emerson. He found in various learned societies the intellectual stimulation that led to many publications and a wide spectrum of accomplishments. Papers to the Royal Astronomical Society on the differential sextant and the aberration of light belie Gompertz's own statement that he was not a practicing astronomer. The Royal Society, of which he was elected a fellow in 1819; the London Mathematical Society, of which he was a charter member; the Society of Actuaries; and the Royal Statistical Society were only a few of the learned and philanthropic organizations to which he gave of his talent and energy.

In 1810 Gompertz married Sir Moses Montefiore's daughter Abigail and joined the stock exchange. In 1820, in a paper to the Royal Society, he applied the method of fluxions to the investigation of various life contingencies. In 1824 he was appointed actuary and head clerk of the newly founded Alliance Assurance Company. A year later he published what is now called Gompertz's law of mortality, which states ". . . the average exhaustion of man's power to avoid death to be such that at the end of equal infinitely small intervals of time he lost equal portions of his remaining power to oppose destruction which he had at the commencement of these intervals." His rigid adherence to Newton's fluxional notation prevented wide recognition of this accomplishment, but he must be rated as a pioneer in actuarial science and one of the great amateur scholars of his day. Augustus De Morgan called Gompertz "the link between the old and new" when he mourned "the passing of the last of the learned Newtonians."

BIBLIOGRAPHY

I. ORIGINAL WORKS. Gompertz's work on life contingencies appeared in the *Philosophical Transactions of the Royal Society:* "A Sketch of the Analysis and Notation Applicable to the Value of Life Contingencies," **110** (1820), 214–294; "On the Nature of the Function Expressive of the Law of Human Mortality, and on a New Mode of Determining the Value of Life Contingencies," **115** (1825), 513–585; and "A Supplement to the Two Papers of 1820 and 1825," **152** (1862), 511–559.

"The Application of a Method of Differences to the Species of Series Whose Sums Are Obtained by Mr. Landen by the Help of Impossible Quantities," *ibid.,* **96** (1806), 174–194, led to *The Principles and Applications of Imaginary Quantities,* 2 vols. (London, 1817–1818). The sequel to these two tracts is *Hints on Porisms* . . . (London, 1850).

A regular contributor to the *Gentleman's Mathematical Companion* from 1796, Gompertz was awarded their annual problem-solution prize every year from 1812 to 1822.

II. SECONDARY LITERATURE. P. F. Hooker, "Benjamin Gompertz," in *Journal of the Institute of Actuaries,* **91,** pt. 2, no. 389 (1965), 203–212, is a competent biography with a complete bibliography of Gompertz's works (twenty-two titles) and works about him (twenty-four titles). Augustus De Morgan, "The Old Mathematical Society," repr. in J. R. Newman, *The World of Mathematics,* IV, 2372–2376, contains a view by a close friend. Also worth reading is De Morgan's obituary in *The Atheneum* (22 July 1865), p. 117. Other informative obituaries are *Monthly Notices of the Royal Astronomical Society,* **26** (1865), 104–109; and M. N. Adler, "Memoirs of the Late Benjamin Gompertz," in *Journal of the Institute of Actuaries,* **13** (Apr. 1866), 1–20.

HENRY S. TROPP

GONSÁLEZ, DOMINGO. See **Gundissalinus, Dominicus.**

GOODRICH, EDWIN STEPHEN (*b.* Weston-super-Mare, England, 21 June 1868; *d.* Oxford, England, 6 January 1946), *comparative anatomy, embryology, paleontology, evolution.*

Goodrich was a son of Rev. Octavius Pitt Goodrich and Frances Lucinda Parker. Among his forebears was Thomas Goodrich, bishop of Ely and lord high chancellor of England, who helped to draw up the

Book of Common Prayer of the Church of England. Goodrich's branch of the family under John Goodrich came to New England in 1630, and settled at Nansewood, Virginia, in 1635. In 1775 the then John Goodrich returned to England, and with Goodrich's death this branch of the family became extinct. When Goodrich was two weeks old his father died, and his mother took him, another son, and a daughter to live with her mother at Pau, France, where he attended the local English school and a French lycée. In 1888 he entered the Slade School at University College, London, as an art student; and while there he became acquainted with E. Ray Lankester, who interested him in zoology. When Lankester became professor of comparative anatomy at Oxford, he made Goodrich his assistant in 1892; this marked the start of the researches which during half a century made Goodrich the greatest comparative anatomist of his day. In 1921 he was appointed Linacre professor of comparative anatomy, a post he held until 1945.

In 1913 Goodrich married Helen L. M. Pixell, a distinguished protozoologist, who helped greatly with his work. His artistic training always stood him in good stead in drawing diagrams of surpassing beauty and clarity while lecturing (students used to insist on photographing the blackboard before it was erased) and in illustrating his books and papers. He also held shows of his watercolor landscapes in London. Goodrich was elected fellow of the Royal Society in 1905 and received its Royal Medal in 1936. He was honorary member of the New York Academy of Science and of many other academies, and honorary doctor of many universities. In 1945 L. S. Berg of Leningrad sent him a message through Julian Huxley: "Please tell him [Goodrich] that though neither I nor my colleagues have ever met him, we all regard ourselves as his pupils." A dapper, tiny, thin man with a dry sense of humor, he always complained when traveling by air that he was not weighed together with his luggage, since his own weight was only half that of an average passenger.

From the start of his researches, most of which were devoted to marine organisms, Goodrich made himself acquainted at first hand with the marine fauna of Plymouth, Roscoff, Banyuls, Naples, Helgoland, Bermuda, Madeira, and the Canary Islands. He also traveled extensively in Europe, the United States, North Africa, India, Ceylon, Malaya, and Java. The most important area of his work involved unraveling the significance of the sets of tubes connecting the centers of the bodies of animals with the outside. There are nephridia, developed from the outer layer inward and serving the function of excretion. Quite different from them are coelomoducts, developed from the middle layer outward, serving to release the germ cells. These two sets of structures may acquire spurious visual similarity when each opens into the body cavity through a funnel surrounded by cilia which create a current of fluid. In some groups the nephridia may disappear (as in vertebrates, where the nephridia may have been converted into the thymus gland), and the coelomoducts then take on the additional function of excretion. This is why man has a genitourinary system. Before Goodrich's analysis, the whole subject was in chaos.

Goodrich established that a motor nerve remains "faithful" to its corresponding segmental muscle, however much it may have become displaced or obscured in development. He showed that organs can be homologous (traceable to a single representative in a common ancestor) without arising from the same segments of the body. Like a tune in music, they can be transposed up or down the scale, for example, the fins and limbs of vertebrates and the position of the occipital arch (the back of the skull), which varies in vertebrates from the fifth to the ninth segment. He distinguished between the different structures of the scales of fishes, living and fossil, by which they are classified and recognized, a fact of fundamental importance when boring into the earth's crust for mineral wealth because the different strata are identified by their fossils. Goodrich's attention was always focused on evolution, to which he made notable contributions, firmly adhering to Darwin's theory of natural selection.

BIBLIOGRAPHY

I. Original Works. A complete bibliography of Goodrich's writings is in the obituary by de Beer (see below). His books include *Cyclostomes and Fishes* (London, 1909); *Living Organisms: An Account of Their Origin and Evolution* (London, 1924); and *Studies on the Structure and Development of Vertebrates* (London, 1930).

II. Secondary Literature. On Goodrich and his work, see Gavin de Beer, "Edwin Stephen Goodrich," in *Obituary Notices of Fellows of the Royal Society of London,* **5** (1947), 477–490; and A. C. Hardy, "Edwin Stephen Goodrich," in *Quarterly Journal of Microscopical Science,* **87** (1947), 317–355.

Gavin de Beer

GOODRICKE, JOHN (*b.* Groningen, Netherlands, 17 September 1764; *d.* York, England, 20 April 1786), *astronomy.*

Goodricke, the British astronomical prodigy of the late eighteenth century whose discoveries laid the foundations of an important branch of stellar astronomy, died at not quite twenty-two years of age. But into this lamentably brief life—and despite the handicap of deafness and dumbness—he managed to compress enough accomplishment to earn a permanent place in the history of science. He was descended from an old family of English country squires, who, raised to baronetcy by the end of the fifteenth century, were occasionally called upon to perform minor diplomatic services. Thus Henry Goodricke, John's father, spent several years in consular service at Groningen, where in 1761 he married Levina Benjamina Sessler.

Scanty records reveal little of Goodricke's early childhood, beyond a suggestion that he became deaf and dumb as a result of a severe illness in early infancy. At the age of eight he was sent from the Netherlands to Edinburgh, to be educated at a school for deaf-mutes which Thomas Braidwood was conducting. Absence of school records conceals the early development of young Goodricke; but his progress must have been satisfactory, for in 1778 he was able to enter Warrington Academy—then a well-known educational institution in the north of England—which made no special provision for handicapped pupils. There, we are told by extant records, ". . . having in part conquered his disadvantage by the assistance of Mr. Braidwood, he attained a surprising proficiency becoming a very tolerable classicist and an excellent mathematician" (Turner, *Historical . . . Academy*). For the latter he had undoubtedly to thank William Enfield, an outstanding teacher and a mathematician of some renown. It was almost certainly he who awakened Goodricke's interest in astronomy and set him on his subsequent career.

Just when Goodricke left Warrington Academy we do not know; but certainly it was not later than 1781, for his *Journal of Astronomical Observations* contains a first entry dated 16 November of that year at York (to which his family had returned from Holland in 1776); it is to this source that we must turn for a description of Goodricke's discoveries.

On 12 November 1782, a few days before the first anniversary of the start of his diary, Goodricke recorded:

> This night I looked at β Persei, and was much amazed to find its brightness altered—it now appears to be of about 4th magnitude. I observed it diligently for about an hour—I hardly believed that it changed its brightness because I never heard of any star varying so quickly in its brightness. I thought it might perhaps be owing to an optical illusion, a defect in my eyes, or bad air; but the sequel will show that its change is true and that I was not mistaken. . . .

Goodricke was not the first to notice the variability of β Persei (or Algol, as it is more commonly known); the Italian astronomer Geminiano Montanari had done so more than a century before (1670) in Bologna. Goodricke was, however, the first to establish that these light changes were periodic. He continued his observations until the end of the season when Algol could be seen above the horizon at York; and it was not until 12 May 1783 that Goodricke communicated (through the good offices of Rev. Anthony Shepherd, then Plumian professor of astronomy at Cambridge) the results of his observations, in the form of a letter read before the Royal Society on 15 May.

This communication, which promptly appeared in print, created considerable interest in astronomical circles, and the Society's council awarded its youthful author one of the two Copley Medals for 1783. Goodricke did indeed deserve it, for not only did he discover the first known short-period variable star but also established a remarkably accurate estimate of its period. (Goodricke's original value for Algol's period was 2 days, 20 hours, 45 minutes—differing from its true period by only 4 minutes. A year later [1784] he revised this period to 2 days, 20 hours, 49 minutes, 9 seconds—a result on which all subsequent observations had little to improve.) At the end of his communication, we find the following sentence, which makes it truly prophetic: "If it were perhaps not too early to hazard even a conjecture on the cause of its variation, I should imagine it could hardly be accounted for otherwise than . . . by the interposition of a large body revolving around Algol . . ." (*Philosophical Transactions of the Royal Society*, **73** [1783], 474).

Nature had denied much to Goodricke but certainly not the gift of a splendid imagination; seldom in the annals of science has the first conjecture of a discoverer been more accurate. Within the remaining short life vouchsafed to him Goodricke discovered, besides that of Algol, the variability of two other naked-eye stars, β Lyrae (1785) and δ Cephei (1786), both of which became prototypes of other classes of variable stars. Unfortunately, Goodricke's bold suggestion that Algol (and β Lyrae) was an eclipsing variable, as they are now called, was made too early to gain speedy acceptance among contemporary astronomers. It was destined to remain a hypothesis until 1889, when the German astronomer Hermann Vogel discovered that Algol is also a spectroscopic binary, whose conjunctions coincide with the minima of light. This established beyond any doubt the binary nature of Algol and similar variables.

In the meantime, Goodricke's short life was fast running out. The last observation recorded in his diary is dated 24 February 1786. In April of that year the Royal Society elected him to fellowship, but he died at York only two weeks later, "in the consequence of a cold from exposure to night air in astronomical observations." The immediate cause of his death is unknown, for he died largely unnoticed. No stone over his tomb at Hunsingore (close to the former family seat of Ribston Hall, Yorkshire) commemorates his final resting place.

BIBLIOGRAPHY

Goodricke's papers published in the *Philosophical Transactions of the Royal Society* are "A Series of Observations On, and A Discovery of, the Period of the Variation of the Light of the Bright Star in the Head of Medusa, Called Algol," **73,** pt. 2 (1783), 474–482; "On the Periods of the Changes of Light in the Star Algol," **74** (1784), 287–292; "Observations of a New Variable Star," **75** (1785), 153–164; and "A Series of Observations on, and a Discovery of, the Period of the Variation of the Light of the Star Marked δ by Bayer, Near the Head of Cepheus," **76** (1786), 48–61.

Biographical information may be found in W. Turner, *Historical Account of Students Educated at the Warrington Academy* (Warrington, 1814).

ZDENĚK KOPAL

GOODSIR, JOHN (*b.* Anstruther, Fife, Scotland, 20 March 1814; *d.* Edinburgh, Scotland, 6 March 1867), *anatomy, marine zoology.*

Goodsir was the eldest of the five sons and one daughter of John Goodsir, a surgeon. Three of his brothers also studied medicine. After attending school in Anstruther until the age of twelve, Goodsir was sent to St. Andrews University, where he studied humanities and natural history, acquiring considerable proficiency in Latin and Greek. As was the custom at that time, he left without taking his degree. His natural predilections were for engineering and chemistry, but his father encouraged his interest in natural history while his mother taught him to draw.

Goodsir matriculated in Edinburgh University in 1830 and studied anatomy, surgery, and natural history under Robert Knox, James Syme, and Robert Jameson. From the beginning his anatomical work was characterized by a high degree of manual dexterity. He surprised his fellow students by making permanent plaster casts of his dissections and was meticulous in articulating skeletons and preserving pathological specimens, believing that "a piece of true dissection ought to turn out an object of wonder and beauty." His father apprenticed him to A. Nasmyth, a dental surgeon, but Goodsir grudged the time the dental work took from his anatomical investigations—which quickly became his absorbing hobby as well as his major study—and Nasmyth agreed to cancel his indentures before the legal term.

In 1835 Goodsir became a licentiate of the Royal College of Surgeons of Edinburgh, joined his father's practice in Anstruther, and began an investigation into the development of teeth. From a microscopic examination of developing jaws at different ages he demonstrated the independent origin of the deciduous and permanent dentitions. This study in the then new field of developmental anatomy was possibly Goodsir's finest piece of work. He communicated his conclusions "On the Origin and Development of the Pulps and Sacs of the Human Teeth" to the British Association in 1838. Although some of these observations on dental embryology had been anticipated by Friedrich Arnold in 1831, Goodsir believed most of his facts were new to science. This work was followed by a study "On the Follicular Stage of Dentition in the Ruminants; With Some Observations on That Process in the Other Orders of Mammals" (1840).

While a student Goodsir formed a close and important friendship with Edward Forbes. He joined local societies and read accounts of various natural history observations, from supposed new fossil fish to the structure of the cuttlefish eye. Goodsir spent a fortnight with Forbes dredging around the Orkney and Shetland Islands in 1839. Their results were jointly presented to the British Association and other joint publications on marine biology soon followed.

In 1840 Goodsir was appointed conservator of human and comparative anatomy in the university museum. He returned to Edinburgh and bought an apartment at 21 Lothian Street which he shared with Forbes. Their home became the headquarters of the Universal Brotherhood of Friends of Truth, a fellowship formed by Forbes some years before and in which Goodsir enjoyed the high rank of Triangle. (The site is now occupied by part of the Royal Scottish Museum.) Goodsir became a member of the Wernerian Society of Edinburgh, the Botanical Society of Edinburgh, the Anatomical and Physiological Society, and the Royal Physical Society, actively participating by communicating papers and holding office.

In 1841 Goodsir succeeded William Macgillivray as conservator of the museum of the Royal College of Surgeons, Edinburgh, and he initiated a series of lectures based on the collection. Some of these lectures were the basis of his *Anatomical and Pathological Observations* (1845) and include fundamental observations on cell structure, a subject for much debate at that time. Goodsir recognized the importance of cell division as the basis of growth and development. He differentiated between the embryonic growth centers of organs and the permanent "centers of nutrition" of the tissues, and established that cells are the active structures involved in glandular secretion.

He thus anticipated by a number of years the work of Rudolf Virchow, who dedicated the first edition of his *Cellularpathologie* (1859) to Goodsir "as one of the earliest and most acute observers of cell-life both physiological and pathological."

In May 1843 Syme offered the curatorship of the university anatomy and pathology museum to Goodsir, who willingly accepted. His brother Harry, a contributor to *Anatomical and Pathological Observations,* assumed his post at the College of Surgeons, but was later lost with John Franklin's polar expedition. Goodsir's fourth brother, Robert, twice voyaged to the Arctic in search of the expedition.

It was Goodsir's ambition to create a teaching museum second to none in Britain. By October 1845 he could report that "an individual studying the collection from the first to the last series may acquire a knowledge of the science from the structures themselves, instead of from books." The collection had been greatly supplemented by zoological specimens collected by Forbes, Harry Goodsir, and himself. Over a thousand carefully dissected and injected specimens were testimony to Goodsir's skill. He became demonstrator of anatomy in May 1844, and in December 1845 was appointed curator of the entire university museum. He communicated a paper "On the Supra-Renal, Thymus and Thyroid Bodies" to the Royal Society, London, in 1846 and was elected a fellow the same year.

With the retirement of Alexander Monro (Tertius) in the spring, a vacancy arose for the chair of anatomy, to which Goodsir had long aspired. Having provided satisfactory evidence of his fitness for anatomical teaching and—even more important in the Edinburgh of 1846—of his religious orthodoxy, Goodsir was elected. From that time he was less active as a scientific author, devoting himself instead to the reorganization of anatomical teaching. He emphasized the tutorial system and his methods came to be regarded as the best in any British university or medical school. He dearly wanted to illustrate his theoretical teaching with practical demonstrations in the surgical wards of a hospital and moved to a larger house in anticipation of becoming a consultant surgeon. When a vacancy did occur at the Edinburgh Royal Infirmary, Goodsir's application was turned down, leaving him a bitterly disappointed man.

In 1850 he started the *Annals of Anatomy and Physiology,* which contained original papers by his pupils and others. After three numbers had been issued the journal was discontinued in 1853, when Goodsir was obliged to withdraw for a year from active work. During the previous summer Goodsir had undertaken the natural history course in addition to his other duties. The extra imposition wore him out and at the end of the course he was "shrunk in features, worn in body, shattered in nerves, and almost a helpless invalid." His health had been deteriorating for a number of years and he now needed a complete rest. He spent a year in Germany and France, being treated for incipient paralysis. During this time he studied German and Italian language and literature. He returned to Germany in 1857 to study ichthyology and subsequently visited the Continent on several occasions to purchase physiological apparatus; he was the first to introduce these costly instruments to Scotland.

Disappointed in his hospital aims, Goodsir moved again, became careless in his domestic habits, and avoided visitors. The paralysis affecting his legs increased until he was able to walk only by concentrating intently on his feet. He keenly felt the loss of his friend Forbes, who had died in 1854. Restlessly he moved twice more before finally settling in the dingy cottage where Forbes had died. There, attended only by his sister, he became more and more of a hermit. But he did not rest from his work.

He began to concern himself with the mathematics of form, trying to perceive an underlying "crystal" arrangement of the fine structure of muscle, bone, and other tissues and organs. This led him to formulate a theory of the triangle as the universal image of nature—the mathematical figure from which both the organic and inorganic worlds are constructed. It is perhaps not without significance that the triangle was one of the outward signs by which Forbes's Universal Brethren were recognized. As his body became increasingly weak Goodsir's speculations became more metaphysical. He drew his strength from the triangle theory of formation which he hoped to complete as the greatest of his works.

Against all advice he commenced his usual course of lectures in November 1866, but before the end of the year he had to give up and confine himself mainly to bed. He died in March 1867, his triangle theory unwritten, and was buried alongside the grave of Forbes.

The significance of Goodsir's contribution to anatomical knowledge should be measured not only by his published writings, but also by his lucid practical demonstrations and inspirational teaching. Many valuable discoveries were incorporated in his lectures but never published. Before him (with the notable exception of John Hunter, whom Goodsir took for a model) anatomy had been regarded as a means to an end—medical practice; Goodsir had the perspective to make it a science in its own right.

BIBLIOGRAPHY

I. ORIGINAL WORKS. Goodsir's papers are listed in the Royal Society, *Catalogue of Scientific Papers,* II (1868). Most were republished posthumously in W. Turner, ed., *The Anatomical Memoirs of John Goodsir, F.R.S.,* 2 vols. (Edinburgh, 1868), which contains a number of unpublished lectures, including the ten lectures, "On the Dignity of the Human Body, Considered in a Comparison of Its Structural Relations With Those of the Higher Vertebrata," delivered to the class of anatomy in 1862; two unpublished papers on marine zoology; the full texts of some papers published in abstract only; and an appendix of selected observations from his notebooks on morphology and the action of muscles. The *Anatomical and Pathological Observations* (Edinburgh, 1845), which includes some work by Harry Goodsir, was also republished in the *Anatomical Memoirs.*

Other works for which Goodsir was responsible are *Annals of Anatomy and Physiology,* vol. I, nos. 1-3 (1850-1853); his ed. of Adolph Hannover, *On the Construction and Use of the Microscope* (Edinburgh, 1853); W. Turner, *Atlas of Human Anatomy . . . the Illustrations Selected and Arranged Under the Superintendence of John Goodsir* (Edinburgh, 1857); and his MS on the myology of the horse, which was incorporated in J. Wilson Johnston and T. J. Call, *Descriptive Anatomy of the Horse . . . Compiled from the Manuscripts of Thomas Strangeways . . . and the late Professor Goodsir* (Edinburgh, 1870).

II. SECONDARY LITERATURE. A comprehensive biographical memoir by H. Lonsdale is prefatory to vol. I of the *Anatomical Memoirs.* This memoir is the source for the quotations in the text and for the entry in *Dictionary of National Biography,* XXII, 137-139. A concise account of Goodsir's work is given by H. W. Y. Taylor in *Report of Proceedings. The Scottish Society of the History of Medicine* (1955-1956), pp. 13-19. Goodsir's reputation among his contemporaries is evidenced by the *Testimonials in Favour of John Goodsir . . . Candidate for the Chair of Anatomy in the University of Edinburgh* (Edinburgh, 1846), which also includes extracts from medical reviews of *Anatomical and Pathological Observations.* Obituaries are in *Edinburgh Medical Journal,* **12** (1867), 959-962; *Transactions and Proceedings of the Botanical Society of Edinburgh,* **9** (1868), 118-127; and *Proceedings of the Royal Society,* **16** (1868), xiv-xvi.

DAVID HEPPELL

GÖPEL, ADOLPH (*b.* Rostock, Germany, 29 September 1812; *d.* Berlin, Germany, 7 June 1847), *mathematics.*

The son of a music teacher, Göpel was able, thanks to an uncle, the British consul in Corsica, to spend several years of his childhood in Italy, where in 1825-1826 he attended lectures on mathematics and physics in Pisa. His real studies did not begin until 1829, at the University of Berlin. After earning his doctorate there in 1835, he taught at the Werder Gymnasium and at the Royal Realschule before becoming an official at the royal library in Berlin. Since he had little contact with his mathematical colleagues, all we know about him is what C. G. J. Jacobi and A. L. Crelle wrote in the brief accounts they contributed to Crelle's *Journal für die reine und angewandte Mathematik* shortly after his death. Of the two, only Crelle knew him personally, and for but a short time.

In his doctoral dissertation Göpel sought to derive from the periodic continued fractions of the roots of whole numbers the representation of those numbers by certain quadratic forms. Following an eight-year pause after his dissertation, he wrote several works for Grunert's *Archiv der Mathematik und Physik,* for which he was then working. In them he showed thorough familiarity with Jacob Steiner's style of synthetic geometry.

Göpel owes his fame to "Theoriae transcendentium Abelianarum primi ordinis adumbratio levis," published after his death in *Journal für die reine und angewandte Mathematik.* The investigations contained in this paper can be viewed as a continuation of the ideas of C. G. J. Jacobi. The latter had taught that elliptic functions of one variable should be considered as inverse functions of elliptic integrals, but later he also explained them in his lectures as quotients of theta functions of one variable. Moreover, Jacobi had formulated the inverse problem, named for him, for Abelian integrals of arbitrary genus p. From this arose the next task: to solve the problem for $p = 2$. This was done by Göpel and Johann Rosenhain in works published almost simultaneously. In "Theoriae transcendentium . . . ," Göpel started from sixteen theta functions in two variables (analogous to the four Jacobian theta functions in one variable) and showed that their quotients are quadruply periodic. Of the squares of these sixteen functions, four proved to be linearly independent. Göpel linked four more of these quadratics through a homogeneous fourth-degree relation, later named the "Göpel relation," which coincides with the equation of the Kummer surface. Göpel then presented differential equations satisfied by the sixteen theta functions and finally, after ingenious calculations, obtained the result that the quotients of two theta functions are solutions of the Jacobian inverse problem for $p = 2$.

BIBLIOGRAPHY

Göpel's major work is "Theoriae transcendentium Abelianarum primi ordinis adumbratio levis," in *Journal für die reine und angewandte Mathematik,* **35** (1847),

277–312, trans. into German as *Entwurf einer Theorie der Abelschen Transcendenten l. Ordnung,* Ostwalds Klassiker der Exacten Wissenschaften, no. 67 (Leipzig, 1895).

C. G. J. Jacobi and A. Crelle, "Notiz über A. Göpel," in *Journal für die reine und angewandte Mathematik,* **35** (1847), 313–318, was reprinted in the German version of "Theoriae. . . ."

WERNER BURAU

GORDAN, PAUL ALBERT (*b.* Breslau, Germany, 27 April 1837; *d.* Erlangen, Germany, 21 December 1912), *mathematics.*

The son of David Gordan, a merchant, Paul Albert attended Gymnasium and business school, then worked for several years in banks. His early interest in mathematics was encouraged by the private tutoring he received from N. H. Schellbach, a professor at the Friedrich Wilhelm Gymnasium. He attended Ernst Kummer's lectures in number theory at the University of Berlin in 1855, then studied at the universities of Breslau, Königsberg, and Berlin. At Königsberg he came under the influence of Karl Jacobi's school, and at Berlin his interest in algebraic equations was aroused. His dissertation (1862), which concerned geodesics on spheroids, received a prize offered by the philosophy faculty of the University of Breslau. The techniques that Gordan employed in it were those of Lagrange and Jacobi.

Gordan's interest in function theory led him to visit F. B. Riemann in Göttingen in 1862, but Riemann was ailing and their association was brief. The following year, Gordan was invited to Giessen by A. Clebsch, with whom he worked on the theory of Abelian functions. Together they wrote an exposition of the theory. In 1874 Gordan became a professor at Erlangen, where he remained until his retirement in 1910. He married Sophie Deuer, the daughter of a Giessen professor of Roman law, in 1869.

In 1868 Clebsch introduced Gordan to the theory of invariants, which originated in an observation of George Boole's in 1841 and was further developed by Arthur Cayley in 1846. Following the work of these two Englishmen, a German branch of the theory was developed by S. H. Aronhold and Clebsch, the latter elaborating the former's symbolic methods of characterizing algebraic forms and their invariants. Invariant theory was Gordan's main interest for the rest of his mathematical career; he became known as the greatest expert in the field, developing many techniques for representing and generating forms and their invariants. Correcting an error made by Cayley in 1856, Gordan in 1868 proved by constructive methods that the invariants of systems of binary forms possess a finite base. Known as the Gordan finite basis theorem, this instigated a twenty-year search for a proof in case of higher-order systems of forms. Making use of the Aronhold-Clebsch symbolic calculus and other elaborate computational techniques, Gordan spent much of his time seeking a general proof of finiteness. The solution to the problem came in 1888, when David Hilbert proved the existence of finite bases for the invariants of systems of forms of arbitrary order. Hilbert's proof, however, provided no method for actually finding the basis in a given case. Although Gordan was said to have objected to Hilbert's existential procedures, in 1892 he wrote a paper simplifying them. His version of Hilbert's theorem is the one presented in many textbooks.

Apparently unaware of James J. Sylvester's attempts in 1878, Gordan and a student, G. Alexejeff, applied the theory of invariants to the problems of chemical valences in 1900. Alexejeff went so far as to write a textbook on invariant theory that was intended for chemists. After some very hostile criticism from the mathematician Eduard Study and an indifferent reception by chemists, the project of introducing invariants into chemistry was dropped. Gordan made a few more contributions to invariant theory, but in the thirty years following Hilbert's work, interest in the subject declined among mathematicians.

The second major area of Gordan's contributions to mathematics is in solutions of algebraic equations and their associated groups of substitutions. Working jointly with Felix Klein in 1874–1875 on the relationship of icosahedral groups to fifth-degree equations, Gordan went on to consider seventh-degree equations with the group of order 168; and toward the end of his career, equations of the sixth degree with the group of order 360. His work was algebraic and computational, and utilized the techniques of invariant theory. Typical of Gordan's many contributions to these subjects are papers in 1882 and 1885 in which, following Klein's exposition of the general problem, he carries out the explicit reduction of the seventh-degree equation to the setting of the substitution group of order 168.

Gordan made other contributions to algebra and gave simplified proofs of the transcendence of e and π. The overall style of Gordan's mathematical work was algorithmic. He shied away from presenting his ideas in informal literary forms. He derived his results computationally, working directly toward the desired goal without offering explanations of the concepts that motivated his work.

Gordan's only doctoral student, Emmy Noether, was one of the first women to receive a doctorate in

Germany. She carried on his work in invariant theory for a while, but under the stimulus of Hilbert's school at Göttingen her interests shifted and she became one of the primary contributors to modern algebra.

BIBLIOGRAPHY

Further information on Gordan and his work may be found in Charles Fisher, "The Death of a Mathematical Theory," in *Archive for History of Exact Sciences,* **3,** no. 2 (1966), 137–159; and "The Last Invariant Theorists," in *European Journal of Sociology,* **8** (1967), 216–244. See also Felix Klein, *Lectures on the Icosahedron* (London, 1888), and *Lectures on Mathematics* (New York, 1911), lecture 9; Max Noether, "Paul Gordan," in *Mathematische Annalen,* **75** (1914), 1–41, which contains a complete bibliography of Gordan's works; and Hermann Weyl, "Emmy Noether," in *Scripta mathematica,* **3** (1935), 201–220.

C. S. FISHER

GORDON, WALTER (*b.* Apolda, Germany, 3 August 1893; *d.* Stockholm, Sweden, December 1940), *theoretical physics.*

After studying at the University of Berlin, Gordon obtained the Ph.D. there in 1921. He remained until 1929, when he became *Privatdozent*—and later associate professor—at the University of Hamburg. He lost his position there, like other professors of Jewish origin, in the spring of 1933. He became a member of the Institute of Mathematical Physics at the University of Stockholm in the fall of the same year. Through a grant from the Rockefeller Foundation and contributions from organizations for refugee aid and some private sources, he and his wife obtained a meager living. Poor conditions for science at the University of Stockholm, together with a general lack of understanding of existing German political conditions, prevented his obtaining a regular position.

Gordon's thorough mathematical foundation led him to rigorous solutions of important problems of quantum theory. He did not produce many writings, but his publications are of high quality. Some of his results were obtained by others about the same time, because of the intense development of quantum mechanics during the 1920's.

Soon after Erwin Schrödinger's publication of his first papers on wave mechanics in 1926, Gordon made several important contributions to the relativistic generalization of nonrelativistic quantum mechanics: the current-density vector of the scalar wave equation, and the quantitative formula for the Compton effect. That these results were not applicable to the electron—as was generally believed at that time—but to particles obeying Bose statistics, which, however, were discovered later, does not detract from the quality of this work. Soon after the appearance of Dirac's theory of the electron early in 1928, Gordon published two papers containing important contributions to this theory.

In the first paper he gave a rigorous treatment of both states of the Dirac equation in a Coulomb field: the bound states with the characteristic energy values given by the formula, derived earlier by Sommerfeld (before quantum mechanics and spin were known), and the continuous states, his treatment of which was of methodological importance. He returned to the continuous states, but in the nonrelativistic case, in a somewhat later paper containing a thorough study of the continuous wave functions in a Coulomb field, which was important for the problem of particle scattering.

In his next paper Gordon showed that the current-density vector, given by Dirac, can be split into two parts, one being formally equal to the one he himself had derived for the scalar equation—being, so to say, its kinematic part—while the other is connected with the spin of the electron. In his last paper, presented at the Congress of Scandinavian Mathematicians at Stockholm in 1934, he returned to a similar but more general problem, the possible states of a Schrödinger-type wave equation in a multidimensional space, applying it to the probability of a quantity given as a function of the momenta and the coordinates and ending the paper with establishment of the integral equation for the states in a Coulomb field as functions of the momenta.

During almost all of his stay in Sweden, Gordon participated eagerly in the seminars at the Institute of Mathematical Physics, to which his erudition, not merely in physics and mathematics, and his caustic but friendly humor gave a characteristic touch. He also gave lectures, among them a valuable course in group theory.

But Gordon's forced exile, taking him from the congenial and inspiring circle at the Hamburg Institute of Physics, and the uncertainty of his future brought an end to his creative powers. Early in 1937 his health declined, and inoperable stomach cancer was diagnosed. Good medical treatment and the care of his wife enabled him to live a reasonably normal life until the last months of 1940.

BIBLIOGRAPHY

Gordon's writings include "Der Comptoneffekt nach der Schrödingerschen Theorie," in *Zeitschrift für Physik,* **40** (1927), 117–133; "Die Energieniveaus des Wasserstoffatoms nach der Diracschen Quantentheorie des Elektrons," *ibid.,*

48 (1928), 11–14; "Über den Stoss zweier Punktladungen nach der Wellenmechanik," *ibid.,* 180–191; "Der Strom der Diracschen Elektronentheorie," *ibid.,* **50** (1928), 630–632; and "Eine Anwendung der Integralgleichungen in der Wellenmechanik," in *Comptes rendus du huitième Congrès des mathématiciens scandinaves tenu à Stockholm août 1934* (Lund, 1935), pp. 249–255.

On Gordon or his work, see Bertrand Russell, *Introduction to Mathematical Philosophy* (London–New York), trans. into German by E. J. Gumbel and W. Gordon as *Einführung in die mathematische Philosophie* (Munich, 1930), with a foreword by David Hilbert.

OSKAR KLEIN

GORE, GEORGE (*b.* Bristol, England, 22 January 1826; *d.* Birmingham, England, 20 December 1908), *electrochemistry.*

Gore was named for his father, who was a cooper. At the age of twelve he left school and went to work as an errand boy and later as apprentice to a cooper. All his life he was an avid reader and during his early years eagerly pursued an interest in science. In 1851 Gore moved to Birmingham, where he spent the rest of his life. There he was employed as a chemist by a local firm that manufactured phosphorus. Satisfactory phosphorus matches had been introduced into England only recently, and Birmingham had quickly become a center for their manufacture.

At this time Birmingham was also the center of a fast-growing electroplating industry, and Gore's interest in electricity led him to the investigation of plating techniques. From 1854 to 1863 he published many articles on electrodeposition of metals and acquired a reputation as a consultant for local manufacturers. Of particular interest was his study on the properties of electrodeposited antimony. In 1865 he was elected a fellow of the Royal Society for his work in the field of electrochemistry.

Gore published a study of the preparation and properties of anhydrous hydrofluoric acid, which he carried out from 1860 to 1870. In this work he repeated the electrolysis of hydrofluoric acid, using a variety of electrodes, in an attempt to isolate fluorine. In each case a fluoride compound was formed with the material of the anode, but with the use of a special carbon electrode Gore reported that he detected a faint odor resembling that of chlorine. A quantity of fluorine gas sufficient to permit its characterization was not isolated until 1886, by Henri Moissan, who used platinum-iridium electrodes. Gore also conducted an investigation of silver fluoride, which he published in 1870. He found that iodine combined with silver fluoride to produce IF_5.

From 1870 to 1880 Gore was a lecturer in physics and chemistry at King Edward's School in Birmingham. In 1880 he formed the Institute of Scientific Research and served as its director until his death at the age of eighty-two. During these years he was a consultant to industry and continued his research on electrolysis and voltaic cells.

BIBLIOGRAPHY

The 125 scientific papers of George Gore published to 1900 are listed in the Royal Society, *Catalogue of Scientific Papers,* vols. II, VII, X, XV. Among the most important are "On Hydrofluoric Acid," in *Philosophical Transactions of the Royal Society,* **159** (1869), 173–200; and "On Fluoride of Silver," *ibid.,* **160** (1870), 227–246. His books include *The Theory and Practice of Electro-Deposition* (London, 1856); *The Art of Electro-Metallurgy* (London, 1870); *The Scientific Basis of National Progress, Including That of Morality* (London, 1882); *The Art of Electrolytic Separation of Metals* (London, 1890); and *The Scientific Basis of Morality* (London, 1899).

Obituary notices appeared in *Nature,* **79** (1909), 290; *Electrician,* **62** (1909), 467; and *Proceedings of the Royal Society,* **84** (1911), xxi–xxii.

DANIEL P. JONES

GÖRGEY, ARTHUR (*b.* Toporc, Hungary [now Toporec, Czechoslovakia], 30 January 1818; *d.* Visegrád, Hungary, 21 May 1916), *chemistry.*

Görgey came from a very old but impoverished noble family. To please his father he entered the military engineering academy of the imperial army at Tulln, Austria. In 1837 he began active duty as a second lieutenant. Following his father's death in 1845 he left the army, in order to pursue his desire of studying chemistry. He entered the University of Prague, where, after completing his studies, he remained as an assistant to Joseph Redtenbacher. After the Hungarian Revolution of 1848 he returned to Hungary with the hope of obtaining a professorship. In the meantime, though, the Hungarian War of Independence had begun. Görgey immediately joined the newly created Hungarian national army and, as a trained officer, rapidly advanced through the ranks. He so distinguished himself in the battle of Ozora that he was promoted to general and was soon given command of an army. Following several defeats of the Hungarian army, Görgey was named commander-in-chief in 1849. In the so-called Winter Campaign he defeated the Austrian army in several battles and reconquered the capital city of Buda.

Unlike Louis Kossuth, the leader of the Hungarian Revolution, Görgey opposed a complete break with the Habsburg dynasty. When Russia began to give Austria military aid against Hungary and the situation became hopeless, Görgey forced Kossuth, who wanted to continue the fight, to resign. He then took command and surrendered to the Russians on 13 August 1849 near Világos. He was interned by the Austrians at Klagenfurt, where for a time he was a chemist in the gasworks. He returned to his native country following the Austro-Hungarian Agreement of 1867. He received a general's pension and spent the rest of his life justifying his actions.

Görgey remains one of the most disputed personalities in Hungarian history. His military and political activity is the subject of a vast number of books and even of plays; he is sometimes presented as the betrayer of the War of Independence and sometimes as a clever and realistic politician. It is almost completely forgotten in this literature that he is the same Görgey who is cited in organic chemistry textbooks as the discoverer of lauric acid. He made this discovery while carrying out an analysis of coconut oil during his stay in Prague. After saponifying and liberating the fatty acids, he separated them by means of distillation and found in the residue, through fractional crystallization, a component whose analysis indicated an unknown fatty acid consisting of twenty-four (today twelve) carbon atoms.

BIBLIOGRAPHY

Articles by Görgey on lauric acid are "Über die festen, flüchtigen, fetten Säuren des Cocosnussöls," in *Justus Liebigs Annalen der Chemie,* **66** (1848), 290; and *Sitzungsberichte der K. K. Akademie der Wissenschaften zu Wien,* Math.-naturwiss. Klasse (1848), 208. He also wrote an autobiography, *Mein Leben und Wirken in Ungarn,* 2 vols. (Leipzig, 1852).

On Görgey as a chemist, see F. Szabadváry, "Les recherches chimiques du général Görgey," in *Actes du XI Congrès international d'histoire des sciences,* IV (Warsaw, 1965), 78.

FERENC SZABADVÁRY

GOSSELET, JULES-AUGUSTE (*b.* Cambrai, France, 19 April 1832; *d.* Lille, France, 20 March 1916), *geology, paleontology.*

Son of Alexandre Gosselet, a pharmacist in Landrecies, Gosselet lived until the age of eleven in the open countryside, enjoying a freedom interrupted only by the lessons given him by his aunt, who lived in the same house. He then studied at the Institution Courboulis in Landrecies and finally at the *lycée* in Douai, where he received his *bachelier.*

Gosselet enrolled at the École de Pharmacie in Paris and remained in the French capital to take the *cours de candidature;* he then returned to Landrecies for his practical training, which at that time required three years. He occupied himself during this period with the education of his sisters; it proved so much to his taste that he abandoned the career that his father had planned for him in favor of one in education.

He entered the *collège* of Quesnoy, a small town near Landrecies, as assistant teacher of mathematics and remained there while preparing for his *licence,* for which he took the examination in Paris. Although he was unanimously rejected by the jury, his talents were recognized by his examiners; among them was Constant Prévost, who offered him the position of *préparateur* of the geology course at the Sorbonne. He held this post for seven years, identifying, preparing, and cataloging specimens. Of the two teachers who formed his views, Constant Prévost and Edmond Hébert, the former enjoyed trying out theoretical views, while the latter was a strict empiricist. As Charles Barrois aptly observed, Prévost taught his students to soar before they could walk, and Hébert taught them to walk but to remain earthbound.

While in Paris, Gosselet visited the nearby quarries and construction sites, thereby acquiring his views on the parallelism of marine and lacustrine facies.

In 1857 Gosselet presented a detailed section of the quarries at Etroeungt, situated near Landrecies and the Belgian frontier. Here he observed strata forming the transition from the Devonian to the Carboniferous. They were characterized by a mixture of fossils, a finding that opposed accepted ideas of clearly demarcated boundaries between the stages. This study was a prelude to his important researches on primary formations, which he submitted for the doctorate in natural sciences. In 1860 Gosselet became a teacher of physics and chemistry at the *lycée* in Bordeaux. He left secondary teaching in 1864 to become assistant professor of natural history in the Faculty of Science at Poitiers and several months later became professor of geology at Lille. He also assisted the Service de la Carte Géologique de France, for which he made some surveys, and worked for the Service de la Carte Géologique de Belgique.

Gosselet profited from his stay at Bordeaux by studying the shell marls of Saucats and Léognan, as well as the freshwater limestones of northeast Aquitaine and the limestone of Blaye; but it was his appointment to Lille which allowed him to devote himself to the study of the region bounded on the east by the Rhine, on the west by the English Channel, on the north by the Yser, and on the south by Paris. He began with the ancient massifs of the Ardennes.

With the presentation of "L'Ardenne" to the Société Géologique du Nord, Gosselet outlined the principles upon which his deductions were based: the intervention of the paleontological evidence; the notion of the stratigraphic basin, that is, of a depression similar to that of our present seas and surrounded by continents; the important role of faults; and the synchronism of facies. The application of these rules led him to several famous discoveries. He established that two basins had existed since the Devonian: the basin of Dinant and the basin of Namur, separated by an ancient shore, a narrow strip of Silurian. These two basins were gradually filled in, independently and under different bathymetric conditions. Gosselet did not propose an absolute synchronism of the same fauna. His principle is based on the variations of the sediments and the faunal succession under the geographic condition of the area. For Gosselet the paleogeographic reconstruction of a region furnished the essential data for the solution of local stratigraphic problems. He changed considerably the legends of the geological maps of Belgium in particular.

The study of numerous deep boreholes led Gosselet to recognize the continuity of the folding in the two distinct geographic massifs of the Boulonnais and the Ardennes. The paleontological horizons of the Boulonnais were repeated in the same order at the northern flank of the Namur basin. Thus, the coalfields of the Boulonnais were located in the Westphalian, not, as had been thought, in the Lower Carboniferous.

Gosselet determined that the Ardennes was the result of a fold. Since his work, inclined and even horizontal faults have been considered, as well as the vertical ones. He saw that the basin of Dinant—the entire Ardennes—was mobilized by tangential forces and its northern side overthrust the north of Belgium. These studies guided the German Geological Service in the interpretation of the Eifel plateau.

Gosselet was still interested in the cenozoic and mesozoic eras in the geology of the north of France. He mentioned successive seas which have invaded the region from the Triassic to the present. He showed that there was little uniformity in the accumulation conditions of the chalk deposits and no relationship to white marine chalk deposits. Moreover, he confirmed the existence of stepped faults of the Cretaceous (or epi-Cretaceous) strata that must be surmounted in approaching the Paris basin. The relief of the topographic surfaces of the north of France owes its fundamental features to a system of post-Cretaceous fractures. In Flanders these post-Cretaceous faults had, moreover, been preceded by post-Jurassic faults in Artois and post-Carboniferous ones in the north. "Gosselet thus revealed the evolution of a great tectonic line along which the earth's crust had contracted in a persistent and periodic manner since the beginning of time" (C. Barrois, p. 31).

Field studies and materials in historical archives led Gosselet to conclude that from the Tertiary to the present time, Flanders has been sinking at the foot of Artois in an uninterrupted but irregular movement. Gosselet was above all a field man whose observations gave rise to inspired hypotheses.

BIBLIOGRAPHY

I. ORIGINAL WORKS. Gosselet's most important writings include *Esquisse géologique du département du Nord et des contrées voisines,* 2 pts. (Lille, 1871–1876); *Esquisse géologique du Nord de la France et des contrées voisines,* 4 pts. (Lille, 1880–1903); *L'Ardenne* (Paris, 1888); and "L'Ardenne," in *Annales de la Société géologique du Nord,* **16** (1889), 64–104.

II. SECONDARY LITERATURE. On Gosselet or his work, see "Cinquantenaire scientifique de M. Jules Gosselet, 30 novembre 1902," in *Annales de la Société géologique du Nord,* **31** (1902), 157–296; Charles Barrois, "Jules Gosselet 1832–1916," *ibid.,* **44** (1919), 10–47, with portrait; and F. Stockmans, "Gosselet, Jules-Auguste, Alexandre," in *Biographie nationale publiée par l'Académie royale . . . de Belgique,* XXXIV (Brussels, 1967), cols. 425–429.

FRANÇOIS STOCKMANS

GOSSET, WILLIAM SEALY (also **"Student"**) (*b.* Canterbury, England, 13 June 1876; *d.* Beaconsfield, England, 16 October 1937), *statistical theory.*

The eldest son of Col. Frederic Gosset and Agnes Sealy, Gosset studied at Winchester College and New College, Oxford. He read mathematics and chemistry and took a first-class degree in natural sciences in 1899. In that year he joined Arthur Guinness and Sons, the brewers, in Dublin. Perceiving the need for more accurate statistical analysis of a variety of processes, from barley production to yeast fermentation, he urged the firm to seek mathematical advice. In 1906 he was therefore sent to work under Karl Pearson at University College, London. In the next few years Gosset made his most notable contributions to statistical theory, publishing under the pseudonym "Student." He remained with Guinness throughout his life, working mostly in Dublin, although he moved to London to take charge of a new brewery in 1935. He married Marjory Surtees in 1906; they had two children.

All of Gosset's theoretical work was prompted by practical problems arising at the brewery. The most famous example is his 1908 paper, "The Probable

Error of a Mean." He had to estimate the mean value of some characteristic in a population on the basis of very small samples. The theory for large samples had been worked out from the time of Gauss a century earlier, but when in practice large samples could not be obtained economically, there was no accurate theory of estimation. If an $n$fold sample gives values $x_1, x_2, \cdots x_n$, the sample mean

$$m = \frac{1}{n} \sum x_i$$

is used to estimate the true mean. How reliable is the estimate? Let it be supposed that the characteristic of interest is normally distributed with unknown mean μ and variance σ^2. The sample variance is

$$s^2 = \frac{1}{n} \sum (x_i - m)^2.$$

It was usual to take s as an estimate of σ; if it is assumed that $\sigma = s$, then for any error e, the probability that $|m - \mu| \leqslant e$ can be computed; and thus the reliability of the estimate of the mean can be assessed. But if n is small, s is an erratic estimator of σ; and hence the customary measure of accuracy is invalid for small samples.

Gosset analyzed the distribution of the statistic $z = (m - \mu)/s$. This is asymptotically normal as n increases but differs substantially from the normal for small samples. Experimental results m and s map possible values of z onto possible values of μ. Through this mapping a probability that $|x - \mu| \leqslant e$ is obtained. In particular, for any large probability, say 95 percent, Gosset could compute an error e such that it is 95 percent probable that $|x - \mu| \leqslant e$.

R. A. Fisher observed that the derived statistic $t = (n - 1)^{1/2}z$ can be computed for all n more readily than z can be. What came to be called Student's t-test of statistical hypotheses consists in rejecting a hypothesis if and only if the probability, derived from t, of erroneous rejection is small. In the theory of testing later advanced by Jerzy Neyman and Egon S. Pearson, Student's t-test is shown to be optimum. In the competing theory of fiducial probability advanced by R. A. Fisher, t is equally central.

Gosset was perhaps lucky that he hit on the statistic which has proved basic for the statistical analysis of the normal distribution. His real insight lies in his observation that the sampling distribution of such statistics is fundamental for inference. In particular, it paved the way for the analysis of variance, which was to occupy such an important place in the next generation of statistical workers.

BIBLIOGRAPHY

Gosset's *"Student's" Collected Papers* were edited by E. S. Pearson and John Wishart (Cambridge-London, 1942; 2nd ed., 1947).

For further biography, consult E. S. Pearson, "Student as Statistician," in *Biometrika*, **30** (1938), 210–250; and "Studies in the History of Probability and Statistics, XVII," *ibid.*, **54** (1967), 350–353; and ". . . XX," *ibid.*, **55** (1968), 445–457.

IAN HACKING

GOULD, AUGUSTUS ADDISON (*b.* New Ipswich, New Hampshire, 23 April 1805; *d.* Boston, Massachusetts, 15 September 1866), *conchology, medicine.*

Gould was born into an old Yankee family. He was the son of Nathaniel Duren Gould and Sally Andrews Prichard. At fifteen he took complete charge of the work on his father's farm while at the same time continuing his studies at the New Ipswich Appleton Academy. In 1821, when he was seventeen, he entered Harvard College where he worked diligently to support himself, and it was during these undergraduate years that his interest in natural history began to develop. Noted among his classmates for his industry and determination, he graduated with respectable grades.

After graduating with a B.A. from Harvard in 1825, he was employed as a private tutor by the McBlair family of Baltimore County, Maryland. Simultaneously, he began the study of medicine and from 1828 to 1829 he studied with James Jackson and Walter Channing at the Massachusetts General Hospital. He received the M.D. in 1830 from Harvard Medical School.

A quiet contemplative man, Gould married Harriet Cushing Sheafe on 25 November 1833. He was a religious man, and for more than thirty years was an active member of the Baptist Church.

Gould's interest in natural history remained keen throughout his life and led him to his first publication. "Lamarck's Genera of Shells." Following the publication of *Cicindelidae of Massachusetts* (Boston, 1833), Gould began his lifelong devotion to the study of mollusks. Six years later, in 1840, he described thirteen new species of shells from Massachusetts, the first of such descriptions that would number 1,100 at the time of his death. Also in 1840, Gould demonstrated his artistic skill by illustrating an article on pupa with thirty drawings of small land snails.

In 1837 the General Court of Massachusetts authorized a geological survey of the state which was to include reports on botany and zoology. Gould was assigned the Invertebrata, exclusive of insects. His preliminary findings were published in a paper entitled, "Results of an Examination of the Species of Shells of Massachusetts and Their Geographical Distribution" (*Boston Journal of Natural History,* **3** [1840], 483–494). This was an epoch-making work since the problem of geographical distribution had received very little attention in other countries and none in the United States. He noted that Cape Cod formed a barrier to some species; of 203 species, eighty were not found south of the Cape and thirty were not found north. Certain species, he noticed, appeared and disappeared suddenly in an area, and he stated that it is necessary to collect data over a period of years to be certain of the distribution.

His *Report on the Invertebrata of Massachusetts* (1841), an octavo volume of almost 400 pages, was the first monograph published in the United States that attempted to describe the entire molluskan fauna of a geographical region. The book was illustrated with more than 200 figures drawn by Gould himself, who said,

> Every species described, indeed almost every species mentioned, has passed under my own eye. The descriptions of species previously known, have been written anew; partly, that they may be more minute in particulars, and partly, with the hope of using language somewhat less technical than is ordinarily employed by scientific men.

The volume gave him an international reputation, and it remains the definitive text on New England mollusks.

In 1846 Gould began his major descriptive work on shells that had been collected by conchologist Joseph Pitty Couthouy during the United States Exploring Expedition, 1838–1842. This work, which forms volume XII of the *United States Exploring Expedition . . .* (1852) was also published under the title "Mollusca and Shells . . ." in *Proceedings of the Boston Society of Natural History* (1846–1850).

When Louis Agassiz came to the United States in 1846, he immediately became a close friend of Gould, with whom he had previously corresponded. Agassiz and Gould collaborated on *Principles of Zoology* (1848), which was published in Boston at the firm of Gould's brother. This work was revised in 1852 and had three additional printings: 1860, 1861, and 1872. A German edition came out in 1851 and a British edition, enlarged by Thomas Wright, was published in 1867.

One of the founders of the Boston Society of Natural History, Amos Binney, died in 1847, leaving an unfinished work, *The Terrestrial Air Breathing Mollusks of the United States,* and instructions in his will that someone be appointed to finish the work. Since Binney had been a man of wealth, no expense was spared. Gould completed the work while Joseph Leidy of Philadelphia did the anatomical drawings. The plates were engraved by Alexander Lawson and the result (published between 1851 and 1857) was one of the most artistic monographs on American Mollusca ever printed in the United States.

During the war with Mexico several collections of shells were made along the western coast of the United States and Mexico by army officers and Gould was selected to identify their collections and to describe new species. Gould was also selected to do the report on mollusks collected by the naturalist William Stimpson for the North Pacific Exploring Expedition of 1853–1855. This work appeared serially from 1859 to 1861 in the *Proceedings of the Boston Society of Natural History.*

The description of the mollusks of the North Pacific Exploring Expedition was Gould's last important work on new material. At the time of his death, he was working on a revision of the *Report on the Invertebrata of Massachusetts,* which was completed by Amos Binney's son, William. According to William H. Dall, Gould's 1841 publication of this report initiated a period in the study of natural history that "was characterized by the broader scope of investigation, the interest in geographical distribution, the anatomy of the soft parts, and the more precise definition and exact discrimination of specific forms" ("Some American Conchologists," p. 97). Fully convinced of Gould's influence, Dall termed this second epoch of American conchology, the "Gouldian Period."

Yet despite the fact that Gould's work in conchology was his greatest contribution to science, his profession always remained that of medicine. According to his daughter, Gould encouraged and advised W. T. G. Morton, the reputed discoverer of ether; helped arrange the first ether demonstration; suggested the use of a valve for the first ether apparatus; gave medical care to some of Morton's first patients; suggested "letheon" as a name for ether; and acted as a mediator between Charles T. Jackson (who claimed original discovery of ether) and Morton.

As if to demonstrate both aspects of his scientific endeavors, Gould was active in both the Boston Society of Natural History and the Massachusetts Medical Society. In the former he served as curator (1831–1838), corresponding secretary (1834–1850), and second vice-president (1860–1866); in the latter he served as president (1865).

BIBLIOGRAPHY

I. Original Works. An exhaustive bibliography of Gould's publications is in Jeffries Wyman, "Biographical

Memoir of Augustus Addison Gould," in *Biographical Memoirs. National Academy of Sciences,* **5** (1905), 91–113, with adds. by W. H. Dall. The Boston Museum of Science, which holds the material of the old Boston Society of Natural History, has 195 letters relating to Gould; his own annotated copy of *Otia Conchologica,* with notes and sketches laid in; his notebooks and drawings; the MSS for his natural history lectures at Harvard (1834–1836); and sixteen of his letters. Houghton Library at Harvard, the Rare Book Room at the Boston Public Library, the Academy of Natural Sciences of Philadelphia, and the Countway Library of Medicine all contain letters relating to Gould as well.

II. SECONDARY LITERATURE. On Gould or his work, see Harley H. Bartlett, "The Reports of the Wilkes Expedition; and the Work of the Specialists in Science: Gould's 'Mollusca and Shells,' Vol. 12," in *Proceedings of the American Philosophical Society,* **82** (1940), 650–655; Thomas T. Bouve, *Historical Sketch of the Boston Society of Natural History,* Anniversary Memoirs. Boston Society of Natural History (1880), with the life of Gould based on Jeffries Wyman's account (see below), pp. 112–116, with portrait; William H. Dall, "Some American Conchologists," in *Proceedings of the Biological Society of Washington.* **4** (1888), 120–122; and Daniel C. Haskell, *The United States Exploring Expedition, 1838–1842, and Its Publications, 1844–1874* (New York, 1942).

See also Jeffries Wyman, "An Account of the Life and Scientific Career of the Late Dr. A. A. Gould" in *Proceedings of the Boston Society of Natural History,* **11** (1867), 188–205; Richard I. Johnson, *The Recent Mollusca of Augustus Addison Gould,* Bulletin 239. United States National Museum (Washington, D.C., 1964); George E. Gifford, Jr., "The Forgotten Man in the Ether Controversy," in *Harvard Medical Alumni Bulletin,* **40,** no. 2 (1965), 14–19; *Dictionary of American Biography,* VII (New York, 1931), 446–447; and *Dictionary of American Medical Biography* (New York, 1928), pp. 483–484. For a list of societies and institutions of which Gould was a member see the "Biographical Memoir" cited above.

GEORGE E. GIFFORD, JR.

GOULD, BENJAMIN APTHORP (*b.* Boston, Massachusetts, 27 September 1824; *d.* Cambridge, Massachusetts, 26 November 1896), *astronomy.*

The eldest of the four children born to Benjamin Apthorp Gould and Lucretia Dana Goddard, Gould was educated at Boston Latin School (of which his father had been the principal) and Harvard College. Originally intending to take up classical languages, he came under the influence of Benjamin Peirce and his interests shifted to physics and mathematics. After graduation (1844) he taught for a while at Boston Latin School and then sailed to Europe for further study in astronomy; this included a year at Berlin and then work under Gauss at Göttingen, where he received his doctorate (1848).

On returning home, Gould was very depressed by the primitive level of scientific research in his country; but rather than accept a professorship at Göttingen, he vowed to dedicate his efforts to raising the reputation of American astronomy. He did much toward this by founding the *Astronomical Journal* in 1849; although at times forced to operate under extreme difficulties, financial and otherwise, Gould edited the *Journal* for a dozen years, until publication was suspended because of the Civil War.

From 1852 to 1867 Gould was head of the longitude department of the U.S. Coast Survey. He quickly appreciated the utility of the telegraph in determining longitudes and measured the longitude difference between Greenwich and Washington over the first transatlantic cable.

In 1852 Gould was approached concerning the directorship of the Dudley Observatory, recently established by the citizens of Albany, New York. Declining the directorship, he agreed in 1855 to serve, without compensation, as executive officer of the observatory's scientific council, the other members of which were Peirce, Joseph Henry, and A. D. Bache. These scientists, members of the Lazzaroni, were all aware of the poor state of American science and had long been trying to found national institutions for scientific research. Bache, as superintendent of the U.S. Coast Survey, provided the Dudley Observatory with instruments and observers; Gould in particular devoted much effort to converting the observatory into a worthy scientific institution and traveled to Europe to order equipment. The trustees of the observatory agreed to bear financial responsibility for the publication of the *Astronomical Journal,* and in 1857 the *Journal*'s headquarters were transferred there. Gould eventually accepted the directorship and moved to Albany early in 1858. The trustees felt that the observatory should serve the public and had all along been annoyed by delays and unforeseen expenses. Matters came to a head in a vicious newspaper campaign, in which Gould was charged with being incompetent, disloyal, and arrogant. The trustees resolved to remove him from the directorship and to dissolve the scientific council. But the director and the council had much invested in the observatory and refused to abandon it. Finally, on 3 January 1859, Gould was forcibly driven from his home by a band of toughs hired by the trustees, several of his papers being destroyed in the process. He then returned to Cambridge.

During Gould's early associations with the Dudley Observatory he attempted to determine the solar parallax from the Chilean observations of Mars and Venus made by James M. Gilliss, although this material was not entirely adequate for the purpose. After leaving Albany he prepared his "Standard Mean

Right Ascensions of Circumpolar and Time Stars" (1862), the first attempt to combine into one catalog stellar positions determined at a number of observatories. In 1861 he undertook the discussion of the observations made at the U.S. Naval Observatory during the preceding decade, and he subsequently reduced an important series of observations made by Joseph Dagelet at Paris between 1783 and 1785.

Meanwhile, from 1859 to 1864, Gould was greatly involved with his late father's mercantile business. He also became an actuary for the U.S. Sanitary Commission, in which capacity he accumulated extensive data on the vital statistics of military and naval personnel.

In 1861 Gould married Mary Apthorp Quincy, and his subsequent astronomical career owed much to her aid. She helped provide an observatory near Cambridge, and between 1864 and 1867 Gould made meridian observations of faint stars near the north celestial pole. In collaboration with Lewis Rutherfurd he investigated the application of photography to astrometry (1866), specifically to the stars in the Pleiades and Praesepe clusters.

About 1865 Gould resolved to travel to the southern hemisphere for the purpose of charting the southern stars with the detail achieved for the northern stars. With the cooperation of President Domingo Sarmiento he arrived in Córdoba, Argentina, in 1870 to found the Argentine National Observatory. Both the instruments and the accessory supplies had to be obtained from North America or Europe, and there was considerable delay before these materials reached Córdoba. Meanwhile, Gould and his four assistants were not idle; with nothing more than binoculars they determined the magnitudes and positions of all the naked-eye stars in the southern heavens. This was no easy task, and the results were published (1879) in the first volume of the *Resultados del Observatorio nacional argentino en Córdoba,* under the title "Uranometria argentina." This work clearly established "Gould's belt" of bright stars, spread in a broad band inclined at some 20 degrees to the galactic equator.

The observatory slowly took shape, and late in 1872 the first zone observations of the southern stars were made. Most of these observations were completed by 1877, but the onerous task of reduction took several years more; and the "Catálogo de las zonas estelares," comprising positions of 73,160 stars between 23 and 80 degrees south declination, was published as volumes 7 and 8 of the *Resultados* (1884). Parallel with this immense project the "Catálogo General" was prepared, giving more accurate positions, determined as the result of repeated measurements, of 32,448 stars (*Resultados,* **14** [1886]).

Gould returned to Massachusetts in 1885 with some 1,400 photographs of southern star clusters, which he spent much of his remaining years measuring and reducing. After a lapse of a quarter of a century he was also able to resume publication of the *Astronomical Journal,* which he continued to edit until his death.

BIBLIOGRAPHY

There is a complete bibliography of Gould's works following Comstock's memoir (see below), pp. 171–180.

On Gould or his work, see G. C. Comstock, "Biographical Memoir. Benjamin Apthorp Gould," in *Biographical Memoirs. National Academy of Sciences,* **17** (1924), 153–170; A. Hall, "Benjamin Apthorp Gould," in *Popular Astronomy,* **4** (1897), 337–340; and S. C. Chandler, "The Life and Work of Dr. Gould," *ibid.,* 341–347.

BRIAN G. MARSDEN

GOULD, JOHN (*b.* Lyme Regis, England, 14 September 1804; *d.* London, England, 3 February 1881), *ornithology.*

Gould was the son of a gardener and worked at first with his father at Windsor Castle; later he was a gardener in Yorkshire and had the opportunity to observe birds and teach himself taxidermy. The Zoological Society of London was formed in 1826, and after a competition Gould was appointed taxidermist under Nicholas Vigors. He remained with the society until his death. Elizabeth Coxen, whom he married in 1829, was skilled at drawing and took up lithography to help in her husband's publications. She also accompanied him on his travels. They had six children.

In 1830 Gould received a collection of bird skins from the Himalayas, and from them he produced a volume of colored illustrations with text by Vigors. The eighty plates, issued in twenty monthly groups, achieved a high level of accuracy in spite of the absence of living material. No publisher was willing to risk the volume, so Gould published it himself and continued as his own publisher with considerable financial success; his editions were limited to about 250 and were sold mainly on subscription. In all, he issued forty-one volumes in elephant folio containing some 3,000 plates, mostly of birds from all over the world. He also published numerous scientific papers, mainly on new species, which showed his ability in dealing with taxonomic details. The plates, all lithographed and hand-painted, are among the finest bird pictures ever produced: Gould experimented with new techniques and achieved an extraordinary effect conveying the sheen on feathers. The pictures show animals in their natural habitat, and some include fine illustrations of flowers as well; they are on the whole accurate, but Gould has sometimes been criticized for sacrificing correct detail to effect.

Gould's most significant work was *The Birds of Australia.* He issued two volumes of plates and then decided that he must visit Australia before continuing; he and his wife spent 1838–1840 there with an assistant, John Gilbert. They explored Australasia extensively and recorded their findings in notes, drawings, and letters. Issued between 1840 and 1869, the new series of plates, each with a page of description of the species, included notes on distribution and adaptation to the environment, an index of species, and a systematic table. Later he issued a series on Australian mammals, noting the parallels in form and function between marsupial and placental mammals. Gould is probably better remembered in Australia than in his home country; the Gould League of Bird Lovers was founded in Victoria in 1909.

He worked on birds collected by expeditions of the *Beagle* and the *Sulphur* and made plates for their reports. He also issued works on the birds of Europe, Asia, Britain, and New Guinea, and on special groups. In 1843 Gould was elected a fellow of the Royal Society; and during the exhibition of 1851 he displayed his collection of hummingbirds in the gardens of the Zoological Society. He later published a monograph on them. Volumes incomplete at the time of his death in 1881 were finished by R. Bowdler Sharpe, then at the British Museum (Natural History).

Gould, who was almost entirely self-taught, had a rare combination of qualities as naturalist, artist, and businessman which enabled him to leave an extremely valuable record of bird life.

BIBLIOGRAPHY

I. Original Works. His first publication was *A Century of Birds From the Himalaya Mountains* (London, 1831–1832), written with N. A. Vigors; the next was *The Birds of Europe,* 5 vols. (London, 1832–1837). The first attempt at a synopsis of the 4-part *The Birds of Australia and the Adjacent Islands,* 2 vols. (London, 1837–1838), is now very rare and was superseded by *The Birds of Australia,* 7 vols. (London, 1840–1848) and *Supplement* (London, 1851–1869). Of his works on special groups the most important is *A Monograph of the Trochilidae or Humming-Birds,* 5 vols. (London, 1849–1861), and a 5-part *Supplement* (London, 1880–1887). Modern reproductions of some of the plates were issued in a smaller format as *Plates of Birds of Europe, Reproduced,* 2 vols. (London, 1966), with text by A. Rutgers.

II. Secondary Literature. The most useful biography and bibliography of Gould are in R. Bowdler Sharpe, *An Analytical Index to the Works of the Late John Gould, F. R. S., With a Biographical Memoir and Portrait* (London, 1893); both are based on the obituary by Tommaso Salvadori in *Atti della R. Accademia delle scienze* (Turin), **16** (1881), 789–810. Two other good short accounts are G. T. Bettany, in *Dictionary of National Biography,* XXII (London, 1890), 287–288, which includes the bibliography of his separately published works and references to other useful obituaries; and A. H. Chisholm, in *Australian Dictionary of Biography,* I (Melbourne, 1966), 465–467. A popular account is C. L. Barrett, *The Bird Man: A Sketch of the Life of John Gould* (Melbourne–Sydney, 1938). The centenary of Gould's arrival in Australia was celebrated by a commemorative issue of *Emu,* **88,** pt. 2 (Oct. 1938), 89–244, which includes evaluations and information about the location of MSS by Gould, most of which went to Australian libraries.

Assessment of the artistic value of Gould's work can be found in the substantial review in *The Times,* no. 20,897 (3 Sept. 1851), 7; and in S. Sitwell *et al., Fine Bird Books* (London, 1953), pp. 25–40. The plates were dated by F. H. Waterhouse in *Dates of Publication of Some of the Zoological Works of the Late John Gould* (London, 1885); this includes a short biographical sketch but does not cover *The Birds of Europe,* for which Waterhouse did a MS volume of dates (1904), still in the library of the Zoological Society of London.

There is a portrait of Gould at the Linnean Society, of which he was a fellow. His collection of birds from Australia was sold to a collector in Philadephia, but his collection of hummingbirds was bought after his death by the British Museum (Natural History), which published a catalog by A. Günther, *A Guide to the Gould Collection of Humming Birds* (London, 1881; 2nd ed., 1883; 3rd ed., 1884).

Diana M. Simpkins

GOURSAT, ÉDOUARD JEAN-BAPTISTE (*b.* Lanzac, Lot, France, 21 May 1858; *d.* Paris, France, 25 November 1936), *mathematics.*

Goursat completed his elementary and secondary studies at the *collège* of Brive-la-Gaillarde and after only one preparatory year at the Lycée Henri IV in Paris was admitted in 1876 to the École Normale Supérieure. There he began a lifelong association with Émile Picard, whom he credited with being instrumental in his choice of a career. Claude Bouquet, Charles Briot, Jean Darboux, and Charles Hermite were among the faculty who provided inspiration and style to Goursat, who received his D.Sc. in 1881. "Hermite," Goursat said in 1935, "is the first who revealed to me the artistic side of mathematics."

Goursat was devoted throughout his academic career to research, teaching, and the training of future mathematics teachers. In 1879 he was appointed lecturer at the University of Paris, a post he held until 1881, when he was appointed to the Faculty of Sciences of Toulouse. He returned to the École Normale Supérieure in 1885 and remained there until 1897, when he was appointed professor of analysis at the University of Paris. He held this post until he reached the age of mandatory retirement, at which time he became an honorary professor. Simultaneously he was tutor in analysis at the École Polytechnique (1896–1930) and at the École Normale Supérieure, St.-Cloud (1900–1929).

Goursat received numerous honors, including the Grand Prix des Sciences Mathématiques (1886) for "Études des surfaces qui admettent tous les plans de symétrie d'un polyèdre régulier." He was awarded the Prix Poncelet in 1889 and the Prix Petit d'Ormoy in 1891. In 1919 he was elected to the Academy of Sciences. Goursat was also a chevalier of the Legion of Honor and president of the Mathematical Society of France. In 1936 an issue of the *Journal de mathématiques pures et appliquées* was dedicated to him on the occasion of the fiftieth anniversary of his becoming a teacher.

Goursat was a leading analyst of his day. At the University of Paris the "Goursat course" and "Goursat certificate" became synonyms for his course in analysis and its successful completion. One of his earliest works removed the redundant requirement of the continuity of the derivative in Augustin Cauchy's integral theorem. The theorem, now known as the Cauchy-Goursat theorem, states that if a function $f(z)$ is analytic inside and on a simple closed contour C, then

$$\int_c f(z)dz = 0.$$

Goursat's papers on the theory of linear differential equations and their rational transformations, as well as his studies on hypergeometric series, Kummer's equation, and the reduction of Abelian integrals, form, in the words of Picard, "a remarkable ensemble of works evolving naturally one from the other." Goursat introduced the notion of orthogonal kernels and semiorthogonals in connection with Erik Fredholm's work on integral equations. He made original contributions to almost every important area of analysis of his time. His *Cours d'analyse mathématique,* long a classic text in France, contained much material that was original at the time of publication.

Goursat brought warmth to his teaching and the same dedication that he applied to his research. He more than fulfilled the prediction of Darboux, who wrote in 1879: "Student [Goursat] whose development was extremely rapid, excellent mathematician, sure to become as superior a teacher as Appell and Picard." Former students and colleagues alike praised his clarity, precision, orderly teaching, and devotion to his students. His personal warmth and effectiveness are perhaps best summed up in the encomium of a former student and later collaborator, Gaston Julia: ". . . in the name of all those who received . . . not only the treasures of your science, but also the treasures of your heart, let me express . . . our faithful gratitude, . . . having received from you the nourishment of the soul, the bread of science and the example of virtue."

BIBLIOGRAPHY

I. Original Works. Goursat's doctoral thesis was "Sur l'équation différentielle linéaire qui admet pour intégrale la série hypergéometrique," in *Annales scientifiques de l'École normale superieure,* **10,** supp. (1881), 3–142. The Cauchy-Goursat theorem first appeared under the title "Démonstration du théorème de Cauchy," in *Acta mathematica,* **4** (1884), 197–200. This article is reproduced under the same title in *Bihang till K. Svenska vetenskapsakademiens handlingar,* **9,** no. 5 (1884), and essentially the same material appeared under the title "Sur la définition générale des fonctions analytiques, d'après Cauchy," in *Transactions of the American Mathematical Society,* **1** (1900), 14–16. "Études des surfaces qui admettent tous les plans de symétrie d'un polyèdre régulier" was published in *Annales scientifiques de l'École normale supérieure,* **4** (1887), 161–200, 241–312, 317–340.

Goursat's best-known work is *Cours d'analyse mathématique,* 2 vols. (Paris, 1902–1905; 2nd ed., 3 vols., 1910–1913); Earle Raymond Hedrick provided the English trans. of vol. I: *A Course in Mathematical Analysis* (Boston, 1904) and, with Otto Dunkel, of vol. II (Boston, 1917).

Other major works include *Leçons sur l'intégration des équations aux dérivées partielles du premier ordre,* Carlo Bourlet, ed. (Paris, 1891), trans. into German with a preface by Sophus Lie (1893), and a 2nd ed., rev. and enl. by J. Hermann (Paris, 1921); *Le problème de Backlund* (Paris, 1925); and *Leçons sur les séries hypergéométriques et sur quelques fonctions qui s'y rattachent* (Paris, 1936).

Notice sur les travaux scientifiques de M. Édouard Goursat (Paris, 1900), the best single source on Goursat's work up to that year, contains discussions by Goursat of his work in various branches of analysis, listed by topic. The bibliography (104 titles) is listed by journal of publication. The variety of topics and the level of their discussion clearly demonstrate Goursat's breadth and depth of accomplishments.

II. Secondary Literature. The notice on Goursat, in *Larousse mensuel,* no. 151 (Sept. 1919), p. 894, in honor of his election to the Academy of Sciences, is a good survey of Goursat's research contributions to that year.

Jubilé scientifique de M. Édouard Goursat (Paris, 1936) is a collection of speeches delivered to Goursat on 20 Nov. 1935 by former students, colleagues, and associates on the occasion of his fiftieth teaching anniversary. His responses to each address are included. This small vol. consists of encomiums relating to his mathematical and teaching accomplishments. The address of Gaston Julia, with its allusions to Kipling's *Jungle Book,* is particularly delightful. All of the quotations used in the body of this notice were translated from this source.

HENRY S. TROPP

GOUY, LOUIS-GEORGES (*b.* Vals-les-Bains, Ardèche, France, 19 February 1854; *d.* Vals-les-Bains, 27 January 1926), *general physics, optics.*

Almost nothing is known of Gouy's upbringing and education. He spent most of his productive life as a professor at the Faculty of Sciences at the University of Lyons. He was elected correspondent for the Section of General Physics of the Academy of Sciences on 25 November 1901. On 28 April 1913 he was made a nonresident member of the Academy.

Gouy was a prolific researcher, publishing dozens of articles in the major French scientific journals of his day. Most of his significant work was devoted to some of the more obscure problems of optics. By the 1870's, when Gouy began his researches, optical theory had been subjected to a highly developed and fairly rigorous mathematical analysis, and it was felt that most of the major optical problems had been solved. By applying his talents to some of the less obvious areas of optical theory, Gouy was able to make contributions of considerable significance and originality.

Gouy's first major optical paper, published in 1880, dealt with the velocity of light. He showed that in dispersive media it was necessary to distinguish between what is called the group velocity of light (the velocity of a series of light waves subject to direct measurement by J. B. Foucault's method) and the somewhat higher and less easily measured velocity of the individual waves. Lord Rayleigh (John William Strutt) in a later, independent demonstration of Gouy's theory labeled these two velocities the "group-speed" and the "wave-speed," respectively. Both Gouy and Rayleigh derived their results from mathematical theory. A. A. Michelson later confirmed them empirically, using carbon disulfide as a dispersive medium.

In another important paper concerning the propagation of light waves, Gouy demonstrated that spherical waves of weak emission advance more rapidly than plane waves emitted at the same time, by a value that rapidly approaches one-quarter of a wavelength. An analogous consideration led him in 1890 to show that when spherical light waves are sent through the focus of a concave mirror, they "advance" by one-half a wavelength (in other words, the sign of their amplitude is reversed). This he demonstrated from a simple calculation derived from Christian Huygen's analysis of point sources and also proved it experimentally. When two rays of white light are made to interfere with one another after being reflected from plane mirrors, the central fringe is always white. Gouy then showed that when one of these rays is reflected against a concave mirror, the central fringe is black because of the interference resulting from the reversal of the amplitude of the ray that passes through the focus of the concave mirror.

Another area in which Gouy carried on extensive and original experimentation concerned the diffraction produced by the passage of light across the edge of an opaque screen. By concentrating his light on the border of the screen by means of a convergent lens and by using very thin screens with sharply defined edges, he was able to observe deviations of light rays at very large angles. His experimental results showed—contrary to what had been the accepted theory—that the nature of the screen played a large role in determining the degree of diffraction and the manner in which the light waves were polarized. The thickness of the screen and the material of which it was composed were especially important considerations in all experiments of this nature.

Although Gouy's major achievements were in the realm of optics, he also devoted considerable attention to other areas of experimental and mathematical physics. He made important studies on the inductive powers of dielectrics, on electrocapillarity, and on the effects of a magnetic field on electrical discharge in rarefied gases. He was interested in spectroscopy. He investigated the emissive and absorptive powers of colored flames, and he invented a device to feed a constant supply of a salt to a flame in order to allow time to measure its spectroscopic emissions. The device made the air or gas take up the salt in the form of a fine spray before it reached the burner.

Gouy was also interested in phenomena produced by randomness of motion in nature. He was the first, perhaps, to point out that the nature of white light may best be understood as a complex disturbance resulting from a series of highly irregular impulses emanating from the source of light. The prism analyzes the irregular disturbance into constituents of definite wavelength, in the same way that a complex periodic function is analyzed mathematically into its simple harmonic components in a Fourier series. Gouy also made a detailed study of the randomness of Brownian movement in which he demonstrated that despite the irregularity of the movement, there is nevertheless a certain consistency independent of all adventitious circumstances.

Because of the number and variety of his researches, it is difficult to make a general evaluation of Gouy's career. His name is associated with no physical law or theory. He made no important breakthroughs; he opened no new areas of research. Great scientists work at the frontiers of knowledge; Gouy labored in the rear areas where most of the important work had already been accomplished. The significance of his researches was, in a sense, in tidying up the field of physical knowledge by extending already discovered theory into the obscurer areas that had been passed over in the first waves of discovery. His function was to integrate new phenomena into old theory, to extend and complete understanding rather than initiate it. If his work was not as significant as that of the great names in physics, it was nevertheless a vital and necessary part of the development of the science of his era.

BIBLIOGRAPHY

Gouy wrote no major works. A list of his many articles can be found in Poggendorff, IV, 520–521; V, 441–442; and VI, 933.

See the *éloge* of Gouy by Émile Picard in *Comptes rendus hebdomadaires des séances de l'Académie des Sciences,* **182** (1926), 293–295. Another notice by Picard was read to the Academy on 20 December 1937.

J. B. GOUGH

GRAAF, REGNIER DE (*b.* Schoonhoven, Netherlands, 30 July 1641; *d.* Delft, Netherlands, 21 August 1673), *medicine, anatomy, physiology.*

De Graaf began his medical studies in 1660 at Utrecht and continued them at Leiden, where he was a student of Franciscus Sylvius and Johannes van Horne. One of his fellow students was Jan Swammerdam; their friendship was later transformed into violent hostility as a result of priority disputes. As early as 1664 de Graaf published a work on the pancreatic juice; it was immediately translated into French and reprinted many times. After a period in France he received an M.D. degree at Angers in 1665. He established himself as a well-known practicing physician in Delft and privately did scientific research. In spite of his international reputation de Graaf held no university posts, presumably as a result of his being a Roman Catholic. According to a tradition reported by Antoine Portal, he was proposed as successor to Sylvius, but he refused the offer. During this period the Netherlands was involved in successive wars with England and France, a circumstance which did not prevent a remarkable artistic, philosophic, and scientific flowering. Living in Delft at the same time were Anton van Leeuwenhoek and the painter Vermeer. De Graaf was friendly with Leeuwenhoek, whom he introduced to the Royal Society of London in 1673 by a letter which is still preserved.

De Graaf is rightly considered one of the creators of experimental physiology. His reputation was great in his own lifetime, as is evident from the many editions and translations of his works which followed each other in rapid succession. In the eighteenth century Hermann Boerhaave praised him, and Portal devoted twenty pages to him in his *Histoire de l'anatomie et de la chirurgie.* In the nineteenth century Claude Bernard held him in very high esteem and dedicated his meditations on the role of the experimental physiologist to him.

De Graaf published works on very diverse subjects; he also devised the method of the pancreatic fistula. He is known, though, through the term "Graafian follicle," which commemorates his crucial role in the accurate and concrete description of the anatomy and physiology of the female mammalian reproductive organs. The problem of reproduction was vigorously debated around 1665. Many famous writers of the period devoted their works and their speculations to the problems of generation and claimed priority in bitter disputes. Only the name of de Graaf remains. In 1668 he had published a treatise on the male reproductive organs which was immediately reprinted several times; but its contents, showing little originality, are all but forgotten. On the other hand, his treatise on the female reproductive organs constitutes an important step in the history of biology.

For the female mammalian gonad de Graaf adopted the name "ovary," a term proposed at the time by such authors as van Horne and Swammerdam. During this period there appeared the completely new technique of injecting vessels with colored substances, an invention claimed by many, particularly by Swammerdam, who reproached de Graaf with having stolen it from him. The quarrel, often conducted with great bitterness, was sent in letters to the Royal Society for arbitration. According to a story spread by Leeuwenhoek twenty years later, de Graaf died from the exhaustion brought on by the polemic; but it is much more likely that he died of an epidemic illness.

It is easier to grasp the originality of de Graaf's discoveries from his illustrations than from his text. He examined and dissected the ovaries of numerous mammals, including the human, and he succeeded in isolating the ovarian vesicles with their envelopes. In the cow he described the ovary before and after mating and ascertained that its structure changed. He was thus the first to discover the morphological changes of the female gonad which accompany its physiological functions. De Graaf established that the vesicles disappeared to make room for a "glandulous substance projecting from the female testicle." Hence he was the first to recognize the glandular nature of the corpus luteum, a fundamental discovery that was not definitely established until around 1900 and that played an essential role in the development of modern sexual endocrinology. It is beyond doubt that de Graaf correctly depicted the stigma of the corpus luteum, which indicates rupture of the ovarian follicles. His only error was not recognizing the rupture of the follicle. Instead, he supposed that the follicle in its entirety constituted the egg expelled into the Fallopian tube.

The mammalian egg was not discovered until 1827, by Karl Ernst von Baer. The phenomenon of ovulation or follicular rupture was clarified in the course of extended research and debate throughout the nineteenth century but was not definitively established until the first years of the twentieth century. It is remarkable that de Graaf followed the progress of pregnancy in the rabbit from mating until birth. He left several plates illustrating it and was fully aware that the egg traveling in the tube was smaller than the ovarian follicle, but he was unable to explain this, not having observed the follicular rupture. It is very instructive to compare de Graaf's book on the female reproductive organs with the writings of his contemporaries and competitors. One can then see the fundamental differences between the precise details given by de Graaf and those of his colleagues, which are frequently inexact and allow free rein to an often unbridled imagination.

De Graaf's iconography deserves special mention. His likeness was engraved in 1666 by Gérard Edelinck, one of the most celebrated portraitists of the age, and is the frontispiece in a number of editions. A portrait of 1672 has likewise become well known. There is also a drawing of great artistic merit, attributed to Verkolje, which probably represents de Graaf in his laboratory, in the act of dissecting a cadaver. As a second frontispiece to his books there are some symbolic engravings of great iconographic interest. The one in the work on the pancreatic juice (1671) long held the attention of Claude Bernard, who considered it a symbol of experimental physiology in the service of medicine. The plates representing ovaries and internal reproductive organs have been reproduced, particularly in W. M. Bayliss' treatise, which, around 1920, constituted a veritable *summa* of general physiology. In 1943 the famous American endocrinologist and historian of medicine G. W. Corner published a translation of the most important chapters concerning the ovary along with the relevant plates. The present author has published a detailed commentary on these figures and believes that even today the physiology of the ovary can be illustrated by utilizing de Graaf's plates without modification.

BIBLIOGRAPHY

I. ORIGINAL WORKS. Complete bibliographies of de Graaf's writings are in the articles by Barge and Daniels. Collections of his works include *Opera omnia* (Lyons, 1678; Amsterdam, 1705, with a short biography). Individual writings mentioned in the text are *De succi pancreatici natura et usu exercitatio anatomico medica* (Leiden, 1664); *De virorum organis generationi inservientibus, de clysteris et de usu siphonis in anatomia* (Leiden–Rotterdam, 1668); *Tractatus anatomico-medicus de succi pancreatici natura et usu* (Leiden, 1671); and *De mulierum organis generationi inservientibus tractatus novus . . .* (Leiden, 1672).

II. SECONDARY LITERATURE. On de Graaf or his work, see J. A. J. Barge, "Reinier de Graaf, 1641–1941," in *Mededeelingen der Nederlandsche Akademie van Wetenschappen,* Afdeeling Letterkunde **5,** no. 5 (1942), 257–281; W. M. Bayliss, *Principles of General Physiology* (London, 1924), p. 882, cf. p. 253; A. M. Cetto, "Un portrait inconnu de Régnier de Graaf," in *Ciba-Symposium,* **5** (1958), 208–211; G. W. Corner, "On the Female Testes or Ovaries," in *Essays in Biology in Honor of H. M. Evans* (Los Angeles, 1943), p. 686, cf. pp. 121–137; and *The Hormones in Human Reproduction* rev. ed. (Princeton, 1947), p. 281; C. E. Daniels, "De Graaf," in *Biographisches Lexikon der hervorragenden Aerzte aller Zeiten und Völker,* II (Leipzig, 1885), 616; P. Delaunay, *La vie médicale aux XVIe, XVIIe, XVIIIe siècles* (Paris, 1935), p. 556; C. Dobell, *Antony van Leeuwenhoek and His Little Animals* (Amsterdam, 1932), p. 435; E. Gasking, *Investigations Into Generation* (Baltimore, 1967), p. 192; M. Klein, "Histoire et actualité de l'iconographie de l'ouvrage: *De mulierum organis generationi inservientibus* (1672) de R. de Graaf," in *Comptes rendus du 16e Congrès international d'histoire de la médecine, Montpellier, 1958,* I (Brussels, 1959), 316–320, also in *Yperman* (Brussels), **8,** fasc. 9 (1961), 3–7; and "Claude Bernard face au milieu scientifique de son époque," in *Philosophie et méthodologie scientifiques de Claude Bernard* (Paris, 1967), p. 170, cf. pp. 97, 98; J. Lévy-Valensi, *Les médecins et la médecine française au XVIIe siècle* (Paris, 1933), p. 668; J. Needham, *A History of Embryology,* 2nd ed., rev. (Cambridge, 1959), p. 304; E. Nordenskjöld, *Die Geschichte der Biologie* (Jena, 1926), p. 648; A. Portal, *Histoire de l'anatomie et de la chirurgie,* III (Paris, 1770), 214–235; R. C. Punnett, "Ovists and Animalculists," in *American Naturalist,* **62** (1928), 481–507; A. Rey, *De Sylvius à Régnier de Graaf* (Bordeaux, 1930), p. 86, an M.D. diss.; J. Roger, *Les sciences de la vie dans la pensée française du 18e siècle* (Paris, 1963), p. 842; and A. Schierbeek, *Jan Swammerdam 1637–1680, His Life and Works* (Amsterdam, 1967), p. 202.

MARC KLEIN

GRABAU, AMADEUS WILLIAM (*b.* Cedarburg, Wisconsin, 9 January 1870; *d.* Peking, China, 20 March 1946), *geology, paleontology.*

Grabau was a versatile scientist, a substantial contributor to systematic paleontology, an imaginative pioneer in stratigraphic geology, and a highly respected teacher and prolific writer. After spending the first half of his professional life in the United States, he went to China for the last twenty-five years. Grabau was the son of William Henry Grabau, a Lutheran pastor, and Maria von Rohr Grabau, who died when he was a small boy; he was the third of ten children. He was educated in parochial and public schools, becoming interested in natural history, first in botany and subsequently in paleontology and mineralogy. Correspondence with William O. Crosby at Massachusetts Institute of Technology led to his attending that institution; he received the B.S. in 1896. After a year as instructor he proceeded to graduate study at Harvard, gaining the M.S. and D.Sc. in 1898 and 1900.

Grabau soon became professor of paleontology at Columbia University after a short stay (1899–1901) at Rensselaer Polytechnic Institute in Troy, New York. In the succeeding twenty years he became a leading scientist in paleontology, stratigraphy, and sedimentary petrology, as well as a highly respected teacher. During the hysteria of World War I, Grabau's tenure at Columbia was embittered by accusations of pro-German sympathies and hints and rumors originating at the highest levels. Moving to China in 1920, he became professor of paleontology at the National University and chief paleontologist of the Geological Survey of China.

Grabau married Mary Antin, a Polish immigrant and a distinguished author and sociologist, in 1901. Her health was poor when he left for China, and she remained with a daughter in the United States. Grabau, rather a stocky man, suffered a deterioration of circulation that limited his capacity to work in the field during his later years in the United States; he became an invalid, requiring a wheelchair or requiring crutches after moving to China. Students, associates, and books increasingly became his sources of information.

The principal distinction of Grabau's work is his anticipation of several principles of stratigraphy and paleontology that were to become more generally recognized by later geologists. Of North American stratigraphers of the early years of the century, he seems to have been the best informed on the relationships in foreign lands. Yet he traveled to Europe only once, when he was over forty, and to Asia at the age of fifty. He was imaginative and philosophical. A pioneer in sedimentary petrology, he proposed a genetic classification of sedimentary rocks that strongly influenced advances in the field. Grabau early emphasized the importance of the environment of deposition in determining rock characters and organic assemblages: the field of paleoecology. He produced such theories as the polar control theory of climatic control through the movement of the crust over the interior of the earth, and the pulsation theory, which endeavored to attribute the changing distribution of lands and seas to fluctuations in sea level. Moreover, he made substantial contributions in paleontology, both in the systematic study of fossils of several classes and in the interpretation of their phylogeny and classification.

The relationship of marine bionomy to stratigraphy was the subject of a fifty-page paper published in 1899, a pioneering analysis of knowledge of the living conditions of modern organisms applied to the environment of ancient sedimentary rocks. The article is a masterly outline of the principles of what is now the science of paleoecology. Grabau early emphasized the impact of environment on the fauna and its relationship to the facies of the rocks, that the lateral changes in time-equivalent rocks might be analogous to their succession. In this respect he was a great admirer of Johannes Walther, who held similar views. Grabau was an antagonist of Edward Oscar Ulrich, the popular authority of the early twentieth century who held that faunas relate to marine invasions from several independent oceanic realms; and in general he concurred with Charles Schuchert in his emphasis on environment and lateral facies.

In paleontology Grabau was influenced in his interest in phylogeny and ontogeny by his association at Cambridge with the great paleobiologists Alpheus Hyatt and Robert T. Jackson. His own early studies were directed toward these ends, particularly his work on gastropods. Subsequently he prepared several monographs on such diverse subjects as Chinese Paleozoic corals, Devonian brachiopods, and Permian faunas; these were excellent systematic paleontologic treatises in the manner of the nineteenth-century classic monographs in North America, with attention to anatomic details that might relate to genetic relationships. He had the good fortune to be in Peking when the Peking man was discovered, and he advised on its study.

Grabau's "Classification of Sedimentary Rocks" (*American Geologist,* **33** [1904]) was a portent of the emphasis on the interpretation of origin, as well as texture and composition, in the classification of the deposited rocks. The original terminology was cumbersome in its having Latin-based names formed from terms for origin, texture, and composition; thus hydrosilicarenyte referred to a marine-laid quartz rock with the texture of sand. The use of a genetic term introduced a subjective element that deterred the direct application of the classification to the rock specimens. Adopted only reluctantly in the beginning, the textural-composition elements became widely used in the middle years of the century, only to be succeeded by other, more sophisticated classifications that further emphasize the aspects he recognized as most pertinent. Thus, Grabau had a great influence in directing the critical study of sedimentary petrology.

Early excursions in the Buffalo, New York, region brought Grabau into contact with the deposits of the continental glaciers. In the 1930's he developed the polar control theory of the distribution of climatic zones through the geologic record. He thought that the poles remained stable with respect to the earth's interior, retaining latitudinal climatic zones, but that the outer crust wandered from these poles. Thus the changing relations of continents to poles caused climatic changes, such as led to glaciation. In his day only a heretic could question the relative permanence of present relationships; half a century later, such a hypothesis came to be appreciably reasonable.

Grabau further believed that the continents once formed a single continental mass, Pangaea, that had been disrupted through relative movements among its dismembered parts. Thus, he was an early protagonist of a theory of continental drift, but one different from those devised by Frank B. Taylor and Alfred Wegener. He thought mountains were rising at the fore of the shifting continental plates and volcanism was at their rear. These theories were ingenious and nearly plausible, for few stratigraphic geologists had Grabau's broad grasp of world geology or his interest in its collation. With the great advances in geophysical science in the latter half of the twentieth century, many such conjectures became subject to more rigorous analysis. Although Grabau could not have anticipated some of this present knowledge, his concepts of the nature of continental movements and climatic zonations have much to commend them.

The pulsation theory attributed the distribution of the principal stratigraphic units to great rhythmic advances and regressions of the seas, which were in turn dependent on restriction and expansion of capacities of ocean basins: eustatic control. He gave distinctive names to his pulsation systems, such as Taconian, Cambrian, Cambrovician, Skiddavian, Ordovician, Silurian, and Siluronian. Fourteen of these cycles were placed in the Paleozoic era, five in the Mesozoic, and two in the Cenozoic, partially to replace the conventional systems. He thought each pulsation had had a duration of about 30 million years, the contraction of seas at the close of each period leading to marked changes in organisms.

Perhaps Grabau thought his pulsation theory was his greatest contribution, for he wrote many volumes endeavoring to relate the distribution of lands and seas to pulsing transgressions and retrogressions and saying that it was further controlled by provincial warping movements accentuating or reversing the effects of the eustatic movements. As has been the case with other endeavors to alter the general geologic classification, based appreciably on historical accidents rather than on clearly natural principles, authorities have never agreed on more ubiquitous natural spans and thus continue to use the established systems.

The greatest effect of Grabau's scientific work probably has been in his contributions to the principles of paleoecology and to the genetic aspects of sedimentary petrology. His stratigraphic work was influential in bringing about a three-dimensional attitude toward sedimentary rock distribution, rather than merely emphasizing the faunal correlation of exposed rock sections. His stratigraphy was dynamic, the source of understanding earth movements. He anticipated the attitudes that became prevalent when the petroleum industry added knowledge of subsurface sections to that of the surface outcrops. The concepts involved in his polar control theory, pulsation theory, and the separation of Pangaea encouraged imaginative syntheses of geologic evidence. This heritage, brought to his students and associates, has contributed far beyond the words and thoughts that he recorded.

The esteem in which Grabau was held was reflected in the honors and prizes that he received but was shown more fittingly in the commemorative volumes that were published by Chinese geologists on his sixtieth birthday and in his burial within the gates of the National University. Among his greatest contributions were the stimulus that he gave to scientific life in China and the instruction and enthusiasm that was productive in his many students. For example, in the first ten years of his residence in China, nineteen of the twenty-five monographs of Palaeontologia sinica were prepared by his students.

Grabau was a fellow of the Geological Society of America, the New York Academy of Science (vice-president, 1906–1907), and the Geological Society of China (vice-president, 1925); corresponding member of the Philadelphia Academy of Sciences and the Deutsche Akademie der Naturforscher; and an honorary member of the Peking Society of Natural History, the China Institute of Mining and Metallurgy, the Academia Sinica, and the Academia Peipinensis.

BIBLIOGRAPHY

I. Original Works. Grabau was the author of some 300 publications, of which more than a score were substantial monographs and other books. The full bibliography is listed in the publications in the secondary literature. The principal volumes are listed in three categories: paleontological studies, stratigraphic studies, and textbooks and collative works.

The paleontological treatises contain substantial systematic descriptions of fossil organisms. Because his associates were exploring in regions that had not been known, the greatest contributions are to faunas from China: *Phylogeny of Fusus and Its Allies* (Washington, D.C., 1904); *Ordovician Fossils From North China* (Peking, 1922); *Silurian Faunas of Eastern Yunnan* (Peking, 1926); *Paleozoic Corals of China* (Peking, 1928); *Devonian Brachiopods of China* (Peking, 1931); and *Early Permian Fossils of China,* 2 vols. (Peking, 1934–1936). *The Relations of Marine Bionomy to Stratigraphy* (Buffalo, 1899) is a substantial introduction to the field of paleoecology.

In stratigraphic geology, Grabau's first publications were descriptions of stratigraphic sequences in various localities: *Geology and Paleontology of Eighteen Mile Creek and the Lakeshore Sections of Erie County, New York* (Buffalo, 1898); *Guide to the Geology and Paleontology of Niagara Falls and Vicinity* (Albany, 1901); "Classification of Sedimentary Rocks," in *American Geologist,* **33** (1904), 228–247; and *Guide to the Geology and Paleontology of the Schoharie Valley in Eastern New York* (Albany, 1906). *The Monroe Formation of Southern Michigan and Adjacent Regions* (Lansing, Mich., 1910) was prepared with W. H. Sherzer. *The Permian of Mongolia* (New York, 1931) included description of faunas collected on the central Asia expeditions of the American Museum of Natural History. *The Stratigraphy of China* was essentially a summary of knowledge. Probably his best-known work is *The Principles of Stratigraphy* (New York, 1913), repr. with preface by Marshall Kay (New York, 1960), one of the most influential texts of the early twentieth century.

Grabau published *Textbook of Geology,* 2 vols. (New York, 1920–1921), which did not receive wide usage. His five-volume *Paleozoic Formations in the Light of the Pulsation Theory* (Peking, 1936–1938), intended to be the first encyclopedic summary of world stratigraphy, was developed to support his pulsation theory. *Rhythm of the Ages* (Peking, 1940) and the posthumously published, twenty-year-old MS of *The World We Live In* (Taipei, 1961), were popular summaries of his philosophy. *North American Index Fossils* (New York, 1909–1910), written with W. H. Shimer, with illustrations of more than 2,000 distinctive invertebrate fossils, was the standard reference work for more than thirty years. He also published *Principles of Salt Deposition* (New York, 1920).

II. Secondary Literature. The full bibliography of Grabau is contained in three biographic papers, H. W. Shimer, in *Proceedings of the Geological Society of America* for 1946 (1947), 161–166; V. K. Ting in *Grabau Anniversary,* the commemorative vol. presented to Grabau on his fiftieth birthday, *Bulletin of the Geological Society of China,* **10** (1931), ix–xviii; and the intro. to *The World We Live In* (Taipei, 1961), xii–xxv.

Marshall Kay

GRAEBE, KARL JAMES PETER (*b.* Frankfurt am Main, Germany, 24 February 1841; *d.* Frankfurt, 19 January 1927), *chemistry.*

Graebe's father, for whom he was named, was a soldier and a merchant; his mother, Emmeline Boeddinghaus, was a writer. He entered the Technische Hochschule at Karlsruhe in 1858 but, wishing to become a chemist, he studied with Robert Bunsen at Heidelberg from 1860 to 1862. He then studied at Marburg with Adolph Kolbe, learning a structural approach. After a semester he returned to Heidelberg as Bunsen's assistant.

In 1864 Graebe and his friend C. Diehl joined the firm of Farbwerk Meister, Lucius and Co., in Höchst. Here Graebe worked on the development of iodine dyes but soon developed vision problems and left the company for rest and travel. He then continued his studies with Emil Erlenmeyer at Heidelberg and worked with aromatic oxygen acids. From 1865 to 1869 he was Adolf von Baeyer's assistant at Berlin. Then, after spending a brief time at the Badische Anilin- und Sodafabrik in Mannheim, he went to Leipzig as a *Privatdozent.* In 1870 Graebe became professor at Königsberg, where he encountered difficulty because laboratory facilities were poor and both faculty and students showed little interest in chemistry. He suffered a breakdown and resigned. After his recovery he went to Zurich as a visiting professor. In 1878 he moved to Geneva and taught there until his retirement in 1906. He was a member of the Deutsche Chemische Gesellschaft and served as its president in 1907. He married Albertine Bergdorfer in 1896; they had no children.

Graebe took the work of F. A. Kekulé, published about the time he went to Berlin, as his starting point and studied compounds related to benzene, particularly the quinones. He collaborated with Carl Liebermann, who was then a student, on alizarin. Using Baeyer's zinc-dust reduction method, they showed that alizarin was reduced to anthraquinone, that is, was a derivative of anthracene, not of naphthalene. Within a short time they were also able to synthesize alizarin from anthraquinone and to give the formula for anthracene. The discovery of alizarin eliminated the use of natural madder and spurred the development of the synthetic dye industry. It also pointed out the lack of an adequate patent law.

Graebe continued to work on other organic dyes and quinone derivatives. He and Heinrich von Brunck worked on alizarin blue, found the correct formula, and recognized it as a quinone derivative. With Heinrich Caro, Graebe obtained acridine from anthracene. With Karl Glaser he discovered carbazole, and from this work he was able to analyze pyrene and chrysene. He showed martius yellow to be a derivative of naphthoquinone. Independently of Zdenko Skraup, he synthesized quinoline. He also worked with other organic dyestuffs, such as rosolic acid, euxanthon, and galloflavin.

Since his Berlin days, Graebe had been theoretically interested in the linkage of the oxygens in quinones. He introduced the terms "ortho," "meta," and "para" for disubstituted benzene compounds. He first thought that hydroquinone, which of the three dihydroxybenzenes forms quinone on oxidation, was an ortho compound. But he later decided that it was a para compound. Graebe was also interested in the relationship between color and constitution and proposed that colored compounds contain unsaturated valences or atoms more closely connected than is necessary.

To aid his teaching of chemistry, Graebe became interested in the history of his subject. He wrote articles on leading chemists and completed a history of organic chemistry, only the first volume of which was published.

BIBLIOGRAPHY

I. Original Works. Graebe's writings include "Ueber Methoxysalysäure," in *Annalen der Chemie und Pharmacie,* **136** (1865), 124–125; "Ueber eine neue Bildungsweise der Methylsalicylsäure," *ibid.,* **142** (1867), 327–330; "Ueber das Verhalten der aromatischen Säuren beim Durchgang durch den thierischen Organismus," *ibid.,* 345–350, written with O. Schultzen; "Untersuchungen ueber die Chinogruppe," *ibid.,* **146** (1868), 1–65; "Ueber Naphthalin," in *Berichte der Deutschen chemischen Gesellschaft,* **1** (1868), 36–38; "Ueber Alizarin und Anthracene," *ibid.,* 49–51, written with C. Liebermann; "Ueber den Zusammenhang zwischen Molecularconstitution und Farbe bei organischen Verbindungen," *ibid.* 106–108, written with C. Liebermann; "Ueber Synthese der Phenanthrens aus Toluol," *ibid.,* **7** (1874), 48–49, written with H. Caro; "Ueber Alizarinblau," in *Annalen der Chemie und Pharmacie,* **201** (1880), 333–354; "Ueber Acridin," in *Berichte der Deutschen chemischen Gesellschaft,* **16** (1883), 2828–2832; and *Geschichte der organischen Chemie* (Berlin, 1920).

Graebe's letters and notes can be found in the Deutsches Museum in Munich, in the Badische Anilin- und Sodafabrik archives at Ludwigshafen, and in the archives of the University of Geneva.

II. Secondary Literature. On Graebe or his work, see *Documents pour servir à l'histoire de l'Université de Genève,* III (Geneva, 1883), 36–38; IV (Geneva, 1896), 113–117; V (Geneva, 1909), 31–34; P. Duden and H. Decker, "Nachruf auf Carl Graebe," in *Berichte der Deutschen chemischen Gesellschaft,* **61A** (1928), 9–46; Frankfurt am Main Verein Deutscher Chemiker, "Carl Graebe," in *Zeitschrift für angewandte Chemie,* **40** (1927), 217–218; *Graebe-Feier, Cassel 20.9.1903* (Geneva, 1903); and W. Schlenk, "Carl Graebe," in *Berichte der Deutschen chemischen Gesellschaft,* **60A** (1927), 53.

Ruth Gienapp Rinard

GRAFF, KASIMIR ROMUALD (*b.* Prochnowo, Germany [now Próchnowo, Poland], 7 February 1878; *d.* Breitenfurt, near Vienna, Austria, 15 February 1950), *astronomy.*

After graduation from the secondary school in Poznan, Graff began his studies of astronomy and physics in 1897 at the University of Berlin, from which he obtained the Ph.D. degree in 1901. From 1898 he was employed at the Urania Observatory in Berlin. In 1902 Graff became assistant astronomer at the Hamburg observatory, and in 1909 when the latter was transferred to Bergedorf, he was appointed associate astronomer. In this position he was obliged to lecture on spherical astronomy. During World War I Graff served as an expert on geodetic surveys. In 1917 he received the honorary title of professor, and at the end of the war he continued photometric observations with instruments of his own design at Bergedorf.

In 1928 he was appointed full professor of practical astronomy at the University of Vienna. There he did his best to modernize the great university observatory. In order to avoid the difficulties in stellar photometry caused by the increasing electric illumination in Vienna, he spent several months every year on the islands of Mallorca and Šolta (Yugoslavia) equipped with instruments of moderate size. Forced to resign from the observatory after the German occupation of Austria in 1938, Graff reassumed the directorship in 1945 but retired three years later. He was a member of the Austrian Academy of Sciences and of the Academia Pontificia Vaticana.

Graff was one of the last of those pioneers in astrophysics who by visual observations promoted photometry and colorimetry, as well as planetary and lunar research. In his stellar photometers he used, instead of "nichols," wedges of gray glass (1914), even copying them photographically in circular form ("Kreiskeilphotometer" [1926]). A combination of assorted wedges of blue and yellow glass for the gradual adjustment of color of an artificial star to that of a natural star was the main part of his colorimeter (1928). Most of his papers contain long lists of stars classified by magnitude and color-type. He made many drawings of the surface features of Mars during its oppositions in 1898, 1901, 1909, and 1924. He occasionally made other observations which cannot be described here in detail. He published valuable textbooks on geographical position-finding and on astrophysics and, in collaboration with M. Beyer, a star atlas.

BIBLIOGRAPHY

I. ORIGINAL WORKS. Graff's writings include *Grundriss der geographischen Ortsbestimmung* (Berlin–Leipzig, 1914; 2nd ed., Berlin, 1941; 3rd ed., 1944); *Sternatlas,* 2 pts. (Hamburg, 1925–1927), in collaboration with M. Beyer; *Grundriss der Astrophysik* (Leipzig-Berlin, 1928); and "Physische Beschaffenheit des Planetensystems," in *Handbuch der Astrophysik,* IV (1929), 358–425, and VII (1934), 410–421.

Among his many papers are "Formeln und Hülfstafeln zur Reduktion von Mondphotographien und Mondbeobachtungen," in *Publikationen des Astronomischen Recheninstituts Berlin* (1901); and "Ortsverzeichnis von 580 veränderlichen Sternen," in *Astronomische Abhandlungen der Hamburger Sternwarte in Bergedorf,* **1** (1909), pt. 3. He published at least 170 other papers, the great majority of which are concerned with stellar photometry and colorimetry; the remainder deal with observations of the moon, planets, comets, and eclipses. Most of these papers appeared in the *Astronomische Abhandlungen der Hamburger Sternwarte in Bergedorf* and *Mitteilungen der Hamburger Sternwarte in Bergedorf* (1909–1926); and *Mitteilungen der Universitäts-Sternwarte, Wien,* **1** (1931–1938) and **4** (1947–1950).

II. SECONDARY LITERATURE. On Graff and his work, see Paul Ahnert, "Kasimir Graff," in *Sterne,* **26** (1950), 186–187, with portrait; Wilhelm Becker, "Kasimir Graff †," in *Astronomische Nachrichten,* **279** (1950), 141–142; and Victor Oberguggenberger, "Kasimir Romuald Graff," in *Almanach. Österreichische Akademie der Wissenschaften,* **100** (1950), 352–358, with portrait.

KONRADIN FERRARI D'OCCHIEPPO

GRÄFFE, KARL HEINRICH (*b.* Brunswick, Germany, 7 November 1799; *d.* Zurich, Switzerland, 2 December 1873), *mathematics.*

Gräffe was the son of Dietrich Heinrich Gräffe, a jeweler, and Johanna Frederike Gräffe-Moritz. Born in simple circumstances, he studied from 1813 until 1816 with a jeweler in Hannover. Then, almost ready to begin a career as a goldsmith, through unflagging industry he made up his educational deficiencies and in 1821 was accepted as a scholarship student at the Carolineum in Brunswick. In 1824 he entered the University of Göttingen, attended the classes of Bernhard Thibaut and C. F. Gauss, and concluded his studies with the prize-winning dissertation "Die Geschichte der Variationsrechnung vom Ursprung der Differential- und Integralrechnung bis auf die heutige Zeit" (1825).

In 1828 Gräffe became a teacher at the Technische Institut in Zurich, and in 1833 a professor at the Oberen Industrieschule there, working also as a *Privatdozent.* He was appointed extraordinary professor of mathematics at the University of Zurich in 1860. His name remains attached to a method for the numerical solution of algebraic equations, which he invented in response to a prize question posed by the Berlin Academy of Sciences. Let (1) $f(x) = x^n + ax^{n-1} + \cdots a_n = 0$, and let it then be supposed that all roots $\alpha_1, \cdots, \alpha_n$ are real and different from each other: (2) $|\alpha_1| > |\alpha_2| > \cdots > |\alpha_m|$. Let it further be possible to find an equation (3) $F(x) = x^n + A_1x^{n-1} + \cdots A_n = 0$, whose roots are the mth powers $\alpha_1^m, \cdots \alpha_n^m$ of (1). It follows from (2) that (4) $|\alpha_1^m| > \cdots > |\alpha_n^m|$. Since $\alpha_1^m + \cdots + \alpha_n^m = -A_1$, it follows from (4) that $|\alpha_1^m|$, for large m, is approximately equal to A_1, $|\alpha_1^m| \sim A_1$. Correspondingly, $|\alpha_2^m| \sim \left(\frac{A_2}{A_1}\right)$, and so on. One can find an equation (3) with $m = 2$ by constructing $g(x) = (-1)^n f(-x)f(x)$; proceeding in this manner one obtains, with $m = 2^k$, the equation $F(x) = 0$.

The method may be extended to equations with equal roots and to equations with complex roots. The method has found application in modern numerical mathematics.

BIBLIOGRAPHY

A bibliography of Gräffe's works may be found in *Historisch biographische Lexikon der Schweiz,* III (Neuenburg, 1926), 621 ff. His most important work is *Die Auflösung der höheren numerischen Gleichungen* (Zurich, 1837; with additions, 1839).

A biography is Rudolph Wolf, "Carl Heinrich Gräffe; Ein Lebensbild," in *Neue Zürcherzeitung,* nos. 30 and 31 (1874), also pub. separately (Zurich, 1874).

J. J. BURCKHARDT

GRAHAM, GEORGE (*b.* near Rigg, Cumberland, England, *ca.* 1674; *d.* London, England, 16 November 1751), *scientific instrumentation.*

Graham's father, also called George, died soon after Graham's birth; he was raised by a brother,

William, at nearby Sykeside. In 1688 he apprenticed himself to Henry Aske, a clockmaker in London. His apprenticeship lasted seven years and on 30 September 1695 he gained freedom of the Clockmaker's Company. He was soon employed by Thomas Tompion, the leading clock, watch, and instrument maker. He married Tompion's niece, Elizabeth Tompion, in 1704 and later became Tompion's partner, succeeding to the business in 1713 on Tompion's death. With Tompion, Graham made the original machine (later named an orrery) to demonstrate the motions of heavenly bodies by means of geared models.

In 1715 Graham began experiments to overcome the effects of temperature changes on pendulum length and rate of timekeeping. By 1722 he had devised his mercury-compensated pendulum, which he adjusted and tested carefully before making it public in a paper presented to the Royal Society in 1726. During this period he also invented the deadbeat escapement that certainly contributed to the overall success of his experiments. The combination of escapement and stable pendulum remained in precision clocks until the late nineteenth century. About 1725 Graham invented the cylinder escapement for watches, apparently an improvement inspired by the escapement (somewhat like a duplex in action) patented in 1695 by Tompion, Edward Booth (inventor of repeating clocks and watches), and William Houghton. Graham's design was also deadbeat and had the excellent quality of not being affected by changes in driving power. But his design never enjoyed the popularity in England that it eventually did in France and Switzerland, where it remained in production until the early part of the twentieth century.

In addition to clocks and watches, Graham also made scientific instruments, including barometers, and planetary models like the orrery described above. His precision instrument work began in 1725 with the construction of an eight-foot quadrant for Edmond Halley who had succeeded John Flamsteed as astronomer royal in 1720. This quadrant was used in Halley's work on the right ascension of the moon and adjacent stars. Both coordinates could be read from Graham's single instrument, and it afforded more accuracy and convenience than the separate instruments used by Flamsteed. In this quadrant Graham substituted the vernier for the diagonal scales previously used. The graduations, by his own hand, introduced a new level of accuracy. Besides the usual angular divisions, there was another division into ninety-six parts, produced by bisecting, and used for checking the angular scale. Curiously, although Graham was certainly aware of the different rates of expansion of iron and brass from his pendulum experiments, the graduations were on a brass scale attached to an iron frame. In 1750, to avoid distortions, this quadrant was replaced by one of the same design, but all brass, made by John Bird.

Graham's next important instrument, a twenty-four-and-one-quarter-foot zenith sector, was made for Samuel Molyneux's private observatory at Kew in 1727. With this instrument Graham introduced the great improvement of a micrometer screw to subdivide the vernier divisions. James Bradley used the sector for studies leading to the discovery of the aberration of light from fixed stars (1729). But most of Bradley's work was done with a twelve-and-a-half-foot zenith sector built by Graham particularly for this research. With provision to extend nearly six and one-half degrees on either side of the zenith, Graham's sector permitted observation of about 200 cataloged stars.

It was Graham to whom John Harrison was sent when he came to London to promote his concept of a precision marine clock for finding longitude at sea. Graham spent ten hours going over Harrison's work with him and advised him to build and test his timekeeper before submitting it to the Board of Longitude. Confirming his faith in Harrison's concepts, Graham generously encouraged him with an unsecured, interest-free loan.

Graham's work, particularly his astronomical instruments, gained international fame. His quadrant at Greenwich was copied by other instrument makers for French, Spanish, Italian, and West Indian astronomers. One of his precision clocks was used at Black River, Jamaica, and another at Uppsala, Sweden. When, in 1736, Louis XV sent Pierre Louis Moreau de Maupertius to Tornio, at the head of the Gulf of Bothnia, he took instruments by Graham, including an "instrument des passages," or transit, to help determine if the earth were flattened at the poles in accordance with the theories of Newton and Christian Huygens.

Collaboration between the Royal Society and the French Academy made it desirable, about 1741, that they be able to express weights and measures in mutually intelligible terms. Toward this end Graham employed Jonathan Sisson to prepare two substantial brass rods graduated in conformity with the English standard yard kept in the Tower of London. These were sent to Paris, where one was kept while the other was returned to England inscribed with a graduated half-toise (three Paris feet). Graham had Sisson subdivide both the yard and the half-toise into thirds or feet. Graham himself later devised a beam caliper with a micrometer screw of forty threads per inch and a graduated circular scale divided into twenty parts, read directly by an index pointer to 1/800 inch. He considered the accuracy of this instrument to be equal to half a graduation. This caliper was then used to compare the several standard yards in London at the Court of the Exchequer, Guildhall, Founders' Hall, and the Tower. The greatest variation found was about .040 inch.

Graham's most important contributions were kinematic designs, among them more accurate graduations and micrometer screws for precise subdivisions, including a micrometer eyepiece. His practical experience with scientific observations undoubtedly contributed to his designing and building skills. He trained such followers as Sisson and Bird. The esteem of his contemporaries is revealed both by his election to the council of the Royal Society (1722) and his burial in Westminster Abbey.

BIBLIOGRAPHY

I. ORIGINAL WORKS. Graham's works consist of papers presented to the Royal Society of London which appeared in the abridged *Philosophical Transactions of the Royal Society.* The most notable of these is "A Contrivance to Avoid the Irregularities in a Clock's Motion Occasion'd by the Action of Heat and Cold Upon the Rod of the Pendulum," in **34** (1726), no. 392, 40–44.

II. SECONDARY LITERATURE. For information on Graham, see James Bradley, "An Account of Some Observations Made in London, by Mr. George Graham; and at Black River in Jamaica, by Mr. Colin Campbell, Concerning the Going of a Clock; in Order to Determine the Difference Between the Lengths of Isochronal Pendulums in Those Places," in *Philosophical Transactions of the Royal Society,* **38** (1734), no. 432, 302–314; [Committee of the Royal Society] "An Account of a Comparison Lately Made by Some Gentlemen of the Royal Society, of the Standard of a Yard, and the Several Weights Lately made for Their Use . . .," *ibid.,* **42** (1742/1743), no. 470, 541–556; and "On the Proportions of the English and French Measures From the Standards of the Same Kept at the Royal Society," *ibid.,* pp. 604–606; Nicholas Goodison, *English Barometers, 1680–1860* (New York, 1968), 141–145; and C. Doris Hellman, "George Graham, Maker of Horological and Astronomical Instruments," in *Vassar Journal of Undergraduate Studies,* **5** (May 1931), 221–251.

Also see H. Alan Lloyd, "George Graham, Horologist and Astronomer," in *Horological Journal,* **93,** no. 1118 (Nov. 1951), 708–717; *Some Outstanding Clocks Over 700 Years* (London, 1958); Thomas Reid, *A Treatise on Clock and Watchmaking* (Edinburgh, 1826); and *Dictionary of National Biography* (London, 1891).

EDWIN A. BATTISON

GRAHAM, THOMAS (*b.* Glasgow, Scotland, 21 December 1805; *d.* London, England, 16 September 1869), *chemistry, physics.*

The son of a prosperous manufacturer, Graham entered the University of Glasgow in 1819, at the age of fourteen, and was convinced by the lectures of Thomas Thomson that his calling lay in the field of chemistry. His father, who wanted him to become a minister of the Church of Scotland, was opposed to this choice of vocation, but Graham received encouragement and help from his mother and sister. After receiving the M.A. at Glasgow in 1826, he worked for nearly two years in the laboratory of Thomas Charles Hope at the University of Edinburgh. He then returned to Glasgow, where he taught mathematics and chemistry in a private laboratory. In 1829 he became assistant at the Mechanics' Institution, and in 1830 he succeeded Alexander Ure as professor of chemistry at Anderson's College (later the Royal College of Science and Technology), where he produced his classic work on the phosphates and arsenates (1833).

In 1834 Graham became a fellow of the Royal Society. Three years later he succeeded Edward Turner as professor of chemistry at the University College, London (later the University of London). His time was then fully occupied in teaching, writing, advising on chemical manufactures, and investigating fiscal and other questions for the government. In 1841 he participated in the founding of the Chemical Society and became its first president. With the death of John Dalton in 1844, Graham was left as the acknowledged dean of English chemists, the successor of Joseph Black, Joseph Priestley, Henry Cavendish, William Wollaston, Humphry Davy, and John Dalton. He resigned his professorship in 1854 to succeed Sir John Herschel as master of the mint, a post which ceased to exist upon Graham's death. He died in 1869, an indefatigable but physically broken man.

As a lecturer Graham was well liked by his students, but he was somewhat nervous and hesitant. He was much in demand as a consultant. Most of his work lay in the field of inorganic and physical chemistry, and he is recognized as the real founder of colloid chemistry. His work, usually quantitatively accurate, was original in conception, simple in execution, and brilliant in the results to which it led. Much of his earlier experimental work, some of it not very accurate, is said to have been performed by students and assistants. He received the Royal Medal of the Royal Society twice (1837 and 1863), the Copley Medal of the Royal Society (1862), and the Prix Jecker of the Paris Academy of Sciences (1862). His original and admirable textbook *Elements of Chemis-*

try was widely used, not only in England but also on the Continent, in its much enlarged multivolume translation by Friedrich Julius Otto.

Graham's first original paper, which appeared during his twenty-first year, dealt with spontaneous gas movement, a subject that occupied him throughout his career. In fact, almost all his research is but a development, in different directions, of his early works on gaseous diffusion and water of hydration, as when he showed that Henry's law is not valid for very soluble gases. In another work he found that, like potash, ammonia forms a normal oxalate, binoxalate, and quadroxalate, but that soda forms only a normal oxalate and binoxalate. He also made interesting observations on the glow of phosphorus and the spontaneous flammability of phosphine.

In 1829 Graham published the first of his papers relating specifically to the subject of gaseous diffusion. Although this publication contains the essentials of Graham's law, known to every student of general chemistry, it was in a subsequent paper, for which he was awarded the Keith Prize of the Royal Society of Edinburgh, that he definitely established the principle:

> The diffusion or spontaneous intermixture of two gases in contact is effected by an interchange in position of indefinitely minute volumes of the gases, which volumes are not necessarily of equal magnitude, being, in the case of each gas, inversely proportional to the square root of the density of that gas . . . diffusion takes place between the ultimate particles of gases, and not between sensible masses ["On the Law of the Diffusion of Gases," in *Philosophical Magazine,* **2** (1833)].

Graham maintained that by means of this law the specific gravity of gases could be determined, through experiments on the principle of diffusion, with greater accuracy than by ordinary means. He also pointed out that mixtures of gases could be separated by diffusion, a process employed during World War II at Oak Ridge, Tennessee, to separate the fissionable isotope uranium 235 from the nonfissionable isotope uranium 238.

Graham also measured the effusion of gases through a small hole in a metal plate and found the velocities of flow to be inversely proportional to the square roots of the densities. Yet in his study of the rates of transpiration of gases through capillary tubes, he found that the rates became constant with a certain length of tube and were not simply related to the densities. Later in his career, in "On the Absorption and Dialytic Separation of Gases by Colloidal Septa," Graham began his studies of the penetration of hydrogen through heated metals, a phenomenon which he called "occlusion" and which he explained first by liquefaction of hydrogen and its dissolution in the metal. He later supposed hydrogen to be the vapor of a very volatile metal, hydrogenium, which forms an alloy with the metal. In 1863 he even suggested that the various chemical elements might "possess one and the same ultimate or atomic molecule existing in different conditions of movement."

Graham's major contribution to inorganic chemistry is his paper "Researches on the Arseniates, Phosphates, and Modifications of Phosphoric Acid," in which he elucidated the differences between the three phosphoric acids. This research and the style of the paper are reminiscent of Joseph Black's work on magnesia and the alkalies carried out in Glasgow eighty years earlier. Graham's discovery of the polybasicity of these acids provided Justus Liebig with the clue to the modern concept of polybasic acids. Of this classic work the eminent German chemist and historian of chemistry Albert Ladenburg has said, "so much has seldom been accomplished by a single investigation." Nevertheless, J. J. Berzelius insisted that the three phosphoric acids were isomers of P_2O_5, and as late as 1843 he wrote that Graham's "point of view lacks justification in several respects."

Before Graham's work the relationship between the various phosphates and phosphoric acids was a subject of the greatest confusion. Compounds of one and the same anhydrous acid with one and the same anhydrous base, in different proportions, had long been known, but Graham was the first to establish the concept of polybasic compounds, that is, a class of hydrated acids with more than one proportion of water replaceable by a basic metallic oxide so that several series of salts could be formed. Graham concluded that the individual properties of the phosphoric acids could not be expressed if they were regarded as anhydrides; they must contain chemically combined water essential to their composition. He therefore designated the three modifications of phosphoric acid as phosphoric acid, $\dot{H}^3\overset{\because\because}{P}$, that is, $3HO \cdot PO_5$ (modern, $3H_2O \cdot P_2O_5$ or H_3PO_4); pyrophosphoric acid, $\dot{H}^2\overset{\because\because}{P}$, i.e., $2HO \cdot PO_5$ (modern, $2H_2O \cdot P_2O_5$ or $H_4P_2O_7$); and metaphosphoric acid, $\dot{H}\overset{\because\because}{P}$, i.e., $HO \cdot PO_5$ (modern, $H_2O \cdot P_2O_5$ or HPO_3). In other words, he regarded them respectively as a triphosphate, a biphosphate, and a phosphate of water.

> When one of these compounds is treated with a strong base, the whole or a part of the water is supplanted, but the amount of base in combination with the acid

> remains unaltered. There are thus three sets of phosphates, in which the oxygen in the acid being five, the oxygen in the base is three, two, and one ["Researches on the Arseniates . . .," in *Philosophical Transactions of the Royal Society,* **123** (1833)].

Graham summarized the compositions of the three acids of phosphorus and of their sodium salts as shown below. Just as in his demonstration of the relationships to one another of phosphoric acid and the three sodium phosphates, Graham originated the concept of polybasic compounds; so, in his demonstration that the pyrophosphates and metaphosphates are compounds differing from the phosphates by loss of water or metallic base, he originated the concept of anhydro compounds.

Although some isolated investigations on colloids had been carried out before Graham, his publications in this field laid the foundations of colloid chemistry. In "On the Diffusion of Liquids," Graham applied to liquids the exact method of inquiry he had applied to gases twenty years before, and he succeeded in placing the subject of liquid diffusion on about the same footing as that to which he had raised the subject of gaseous diffusion prior to the discovery of his numerical law. He showed that the rate of diffusion was approximately proportional to the concentration of the original solution, increased with rise in temperature, and was almost constant for groups of chemically similar salts at equal absolute (not molecular) concentrations and different with different groups. He believed that liquid diffusion was similar to gaseous diffusion and vaporization with dilute solutions, but with concentrated solutions he noted a departure from the ideal relationship, similar to that in gases approaching liquefaction under pressure. Based on his work on osmosis, Graham developed what he called a "dialyzer," which he used to separate colloids, which dialyzed slowly, from crystalloids, which dialyzed rapidly. He prepared colloids of silicic acid, alumina, ferric oxide, and other hydrous metal oxides, and he distinguished between sols and gels.

Much of the terminology and fundamental concepts of this field are due to Graham:

> As gelatine appears to be its type, it is proposed to designate substances of the class as *colloids* [κόλλα, glue], and to speak of their particular form of aggregation as the *colloidal condition of matter.* Opposed to the colloidal is the crystalline condition. Substances affecting the latter form will be classed as *crystalloids.* . . . Fluid colloids appear to have always a pectous [πηκτός, curdled] modification; and they often pass under the slightest influences from the first into the second condition. . . . The colloidal is, in fact, a dynamical state of matter; the crystalloid being the statical condition.

Graham stated that crystals and crystalloids "appear like different worlds of matter," but he recognized that the essential difference is in the state and that the same substance can exist in the crystalloid or colloid state. He concluded that "in nature there are no abrupt transitions, and the distinctions of class are never absolute."

TABLE 1

		Oxygen in			Modern
		Soda	Water	Acid	Formulation
First Class	Phosphoric acid	0	3	5	H_3PO_4
	Biphosphate of soda	1	2	5	NaH_2PO_4
	Phosphate of soda	2	1	5	Na_2HPO_4
	Subphosphate of soda	3	0	5	Na_3PO_4
Second Class	Pyrophosphoric acid	0	2	5	$H_4P_2O_7$
	Bipyrophosphate of soda	1	1	5	$Na_2H_2P_2O_7$
	Pyrophosphate of soda	2	0	5	$Na_4P_2O_7$
Third Class	Metaphosphoric acid	0	1	5	HPO_3
	Metaphosphate of soda	1	0	5	$NaPO_3$

BIBLIOGRAPHY

I. ORIGINAL WORKS. Graham's writings include "On the Absorption of Gases by Liquids," in *Annals of Philosophy,* **12** (1826), 69; "A Short Account of Experimental Researches on the Diffusion of Gases Through Each Other, and Their Separation by Mechanical Means," in *Quarterly Journal of Science and the Arts* (Royal Institution), **27** (1829), 74; "On the Law of the Diffusion of Gases," in *Philosophical Magazine,* **2** (1833), 175, 269, 351; "Researches on the Arseniates, Phosphates, and Modifications of Phosphoric Acid," in *Philosophical Transactions of the Royal Society,* **123** (1833), 253, repr. as Alembic Club Reprint no. 10 (Edinburgh, 1961); *Elements of Chemistry, Including the Application of the Science in the Arts* (London,

1842), trans. into German and enlarged by F. J. Otto as *Ausführliches Lehrbuch der Chemie, physikalische, anorganische, organische* (Brunswick, 1854–1893); "On the Motion of Gases," in *Philosophical Transactions of the Royal Society,* **136** (1846), 573, and **139** (1849), 349; "On the Diffusion of Liquids," *ibid.,* **140** (1850), 1–46; "On Osmotic Force," *ibid.,* **144** (1854), 177–228; "Liquid Diffusion Applied to Analysis," *ibid.,* **151** (1861), 183–224; "Speculative Ideas Respecting the Constitution of Matter," in *Proceedings of the Royal Society,* **12** (1863), 620–623; "On the Absorption and Dialytic Separation of Gases by Colloidal Septa," in *Philosophical Transactions of the Royal Society,* **156** (1866), 399–439; and "On the Occlusion of Hydrogen Gas by Metals," in *Proceedings of the Royal Society,* **16** (1868), 422.

II. SECONDARY LITERATURE. Discussions of Graham's life and work are W. Odling, in *Report of the Board of Regents of the Smithsonian Institution* (1871), pp. 171–216, repr. in E. Farber, ed., *Great Chemists* (New York, 1961), pp. 553–571; J. R. Partington, *A History of Chemistry,* IV (New York, 1964), 265–270, 272–275, 729–732; and M. Speter, "Graham," in G. Bugge, ed., *Das Buch der grossen Chemiker,* II (Weinheim, 1965), 69–77. Discussions of Graham's law are E. A. Mason and R. B. Evans, "Graham's Law: Simple Demonstrations of Gases in Motion. Part I, Theory. Part II, Experiments," in *Journal of Chemical Education,* **46** (1969), 359–364, 423–427; E. A. Mason and B. Kronstadt, "Graham's Law of Diffusion and Effusion," *ibid.,* **44** (1967), 740–744; and A. Ruckstuhl, "Thomas Graham's Study of the Diffusion of Gases," *ibid.,* **28** (1951), 594–596.

GEORGE B. KAUFFMAN

GRAM, HANS CHRISTIAN JOACHIM (*b.* Copenhagen, Denmark, 13 September 1853; *d.* Copenhagen, 14 November 1938), *biology, medicine.*

Gram was the son of Frederik Terkel Julius Gram, a professor of jurisprudence, and Louise Christiane Roulund. He early took up studies in the natural sciences. After receiving a B.A. from the Copenhagen Metropolitan School (1871), he became an assistant in botany (1873–1874) to the zoologist Japetus Steenstrup. But he soon developed an interest in medicine, and in 1878 he obtained the M.D. from the University of Copenhagen. In the following years he was an assistant in various Copenhagen hospitals and in 1882 received the gold medal for a university essay concerning the number and size of human erythrocytes in chlorotics. The following year he defended at Copenhagen his doctoral thesis on the size of the human erythrocytes.

From 1883 to 1885 Gram traveled in Europe, studying pharmacology and bacteriology; in 1884, while working with Friedländer in Berlin, he published his famous microbiological staining method. Gram experimented with staining pneumococci bacteria by modifying Ehrlich's alkaline aniline solutions. Gram stained his preparations with aniline gentian violet, adding Lugol's solution for from one to three minutes. When he then removed the nonspecific attributed stain with absolute alcohol, certain bacteria (pneumococci, for example) retained the color (gram-positive), while other species bleached (gram-negative). Gram himself never used counterstaining for gram-negative microbes, as was later done by Weigert.

Gram spent the next few years as a hospital assistant. In 1891 he was appointed professor of pharmacology at the University of Copenhagen, a position he maintained with inspiring diligence until 1900, although he had also become chief physician in internal medicine at the Royal Frederiks Hospital in 1892. Gram took great interest in the clinical education of young students; he was appointed ordinary professor (1900) and from 1902 to 1909 he published his four-volume *Klinisk-therapeutiske Forelaesninger,* which shows his interest in rational pharmacotherapy in clinical science.

In addition to his university post, Gram had a large private practice in internal medicine; and as chairman of the Pharmacopoeia Commission (1901–1921) he cleared the field of many obsolete therapeutics. After his retirement in 1923 he resumed his former interest in the history of medicine.

Gram was made honorary member of Svenska Läkaresällskapet (1905), Verein für Innere Medizin (1907), and Dansk Selskab for Intern Medicin (1932). Kristiana University (now University of Oslo) awarded him the M.D. *honoris causa* in 1912; and the king awarded him the Dannebrog Commander's Cross, first-class (1912) and the Golden Medal of Merit (1924).

Gram married Louise I. C. Lohse in 1889; she died eleven years later.

BIBLIOGRAPHY

I. ORIGINAL WORKS. A full catalog of Gram's published writings is in O. Preisler, *Bibliotheca medica danica,* VII (Lyngby, 1919), 41; *Index medicus danicus 1913–1927,* II (Copenhagen, 1928), 370–371; and *ibid., . . . 1928–1947* (printed index cards). His more important works include *Blodet hos Klorotiske med Hensyn til Blodlegemernes Tal og Størrelse hos Mennesket* (Copenhagen, 1882); *Undersøgelser over de røde Blodlegemers Størrelse hos Mennesket* (Copenhagen, 1883); "Über die isolierte Färbung der Schizomyceten in Schnitt- und Trockenpräparaten," in *Fortschritte der Medizin,* **2** (1884), 185; *Laegemidlernes Egenskaber og Doser i Tabelform* (Copenhagen, 1897); and *Klinisk-therapeutiske Forelaesninger for de Studerende,* 4 vols. (Copenhagen, 1902–1909).

II. Secondary Literature. See P. Engelstoft, *Dansk biografisk Leksikon,* VIII (1936), 251–252; S. A. Gammeltoft, *Den farmakologiske Undervisnings Historie ved Københavns Universitet* (Copenhagen, 1952), pp. 80–94; H. Okkels, *Farvningsteckniken i den mikroskopiske Anatomi* (Copenhagen, 1947), pp. 48–49; C. Sonne, "Nekrolog," in *Acta medica scandinavica,* **98** (1939), 441–443; and H. R. Zeuthen, *Danske Farmakopeer indtil 1925* (Copenhagen, 1927), pp. 258–262.

E. Snorrason

GRAMME, ZÉNOBE-THÉOPHILE (*b.* Jehay-Bodegnée, Belgium, 4 April 1826; *d.* Bois-Colombes, near Paris, France, 20 January 1901), *technology.*

Gramme was born into an educated family of modest means; his father was a clerk in the tax department. Gramme showed no ability as a student but did not lack ingenuity and manual dexterity. He left school at an early age and became a joiner, practicing this trade in the small town of Hannut until he was twenty-two years old. He then moved with his family to Liège, where he remained until 1855. After visiting Brussels, Paris, Lyons, and Marseilles, he settled in Paris as a banister maker. He married Hortense Nysten, a dressmaker from Liège; they lived in Neuilly-sur-Seine, a suburb.

Shortly after he came to Paris, Gramme began to work as a model maker in a firm that specialized in the manufacture of electrical apparatus. This served as his apprenticeship in technology; by 1867 he had become interested in building an improved apparatus for producing alternating current. His success might be said to derive from his characteristic fastidiousness about his person, however. He was appalled by the dirt surrounding the batteries used to produce direct current, and by 1869 he had built a successful—and clean—direct-current dynamo, drawing on the work of Pacinotti (a version of whose machine he had improved) and other earlier physicists who had theorized autoexcitation in revolving machines. Gramme's dynamo, used in metallurgy as well as in the production of electric light, depended upon a ring winding to hold the conductors in place on the surface of the revolving armature. Gramme was the first to give final form to the collector that derives direct current from the revolving armature, and he rapidly saw the possibility of inverting the function of the dynamo to use it as an electrical engine.

Gramme's invention was presented to the Académie des Sciences by the physicist Jules Jamin at the meeting of 17 July 1871. It soon aroused the interest of scientific and industrial circles; and with the help of Marcel Deprez and Arsène d'Arsonval, Gramme was able to accomplish the long-distance transmission of direct-current electricity. Their results were announced to the Academy on 2 December 1872, 25 November 1874, and 11 June 1877. These four notes constitute the whole of Gramme's work published during his lifetime.

Gramme became associated with Hippolyte Fontaine in the further development of his machines; in 1871 they opened a factory—the Société des Machines Magnéto-Électriques Gramme—which manufactured the Gramme ring, Gramme armature, and Gramme dynamo, among other things. The factory grew to great size and the owners prospered. Gramme had a house in Bois-Colombes, complete with gardens and conservatories, built according to his specifications.

Gramme's wife died in 1890, and in 1891 he married Antonie Schentur, who was thirty-six years his junior. In 1901, following Gramme's death, she published a manuscript that he had written in the last two years of his life, containing a number of hypotheses about electricity and magnetism—hypotheses that, unfortunately, most eloquently illustrate Gramme's ignorance of contemporary science as well as his vivid imagination. Indeed, Gramme died semiliterate, without having advanced his mathematical training much beyond the four basic operations of elementary arithmetic.

Gramme was awarded the Volta Prize by Louis Napoleon in 1852. In 1898 he was made Commander of the Order of Leopold I of Belgium. His discoveries of the principles of the dynamo and the electrical engine were of the utmost importance to modern technology.

BIBLIOGRAPHY

I. Original Works. Gramme's writing are "Sur une machine magnéto-électrique produisant des courants continus," in *Comptes rendus hebdomadaires des séances de l'Académie des sciences,* **73** (1871), 175–178; "Sur les machines magnéto-électriques Gramme, appliquées à la galvanoplastie et à la production de lumière," *ibid.,* **75** (1872), 1497–1500; "Sur les nouveaux perfectionnements apportés aux machines magnéto-électriques," *ibid.,* **79** (1874), 1178–1182; "Recherches sur l'emploi des machines magnéto-électriques à courants continus," *ibid.,* **84** (1877), 1386–1389; and *Les hypothèses scientifiques émises par Zénobe Gramme en 1900* (Paris, 1902).

II. Secondary Literature. Biographies are the following (listed chronologically): O. Colson, *Zénobe Gramme, sa vie et ses oeuvres, d'après des documents inédits* (Liège, 1903; 5th ed., 1913); J. Pelseneer, *Zénobe Gramme* (Brussels, 1941); and L. Chauvois, *Histoire merveilleuse de Zénobe Gramme* (Paris, 1963).

Marcel Florkin

GRAMONT, ANTOINE ALFRED ARNAUD XAVIER LOUIS DE (*b.* Paris, France, 21 April 1861; *d.* Savennières, Maine-et-Loire, France, 31 October 1923), *physics, mineralogy.*

Gramont belonged to an aristocratic family and was able to devote himself to scientific research without regard for financial concerns. His first studies, in organic synthesis, were followed by the artificial production of several minerals, including boracite and datholite. He also investigated, in collaboration with Georges Friedel, the pyroelectricity of scolecite.

Beginning in 1894 Gramont specialized in spectroscopy, a field to which he soon contributed new methods. He found that the electric spark of a condenser that is constantly recharged by means of an induction coil (condensed spark) is brighter, shorter, and wider than the spark from a simple coil. In producing the spark on the surface of a compound, Gramont observed a complex spectrum in which each constituent element of the compound, upon being discharged, yielded its own spectrum independently. He termed this spectrum, resulting from the simple superposition of the line spectra of elements composing a body, the "dissociation spectrum."

By suppressing the condenser, Gramont eliminated the spectra of the metalloids and was left with the lines of the metals. At first he confined his studies to the visible portion of the spectrum, but it was obvious that such a technique could constitute a general method of investigation. Through the use of photography it could be extended to the portion of the ultraviolet that passes through the air.

For some twenty years Gramont perfected his method and broadened its field of application. About 1902, with Watteville and Hemsalech, he examined the effect of placing a self-induction coil in the discharge circuit; the result was a weakening of the high-temperature lines and a strengthening of the low-temperature lines. Enlarging the self-induction coil eliminated the lines of the air and then of the metalloids. In order to study the spark spectrum of liquids without interference from lines produced by the electrodes, Gramont generated the spark between the drops forming at the extremities of two capillary tubes (1907). He observed nonconducting substances in the form of solutions in fused salts. At about this same time he discovered the ultimate lines. If a substance is examined in increasingly smaller amounts, the lines likewise become steadily weaker, but their decrease is very irregular: the last visible lines are not the most intense ones of the ordinary spectrum. This discovery facilitated research on traces in general and also, to a degree, opened up the possibilities of quantitative analysis with the spectroscope. Gramont himself obtained some interesting results in this manner.

Gramont died very suddenly, shortly after he had finished correcting the proofs of a major work on spectroscopy written in collaboration with P. E. L. de Boisbaudran, whose career was somewhat similar to his own.

BIBLIOGRAPHY

I. ORIGINAL WORKS. From 1890 to 1921 Gramont published more than 100 communications to the Académie des Sciences in *Comptes rendus hebdomadaires des séances de l'Académie des sciences,* **110–173.** His last work was *Analyse spectrale appliquée à l'analyse chimique* (Paris, 1923), written with P. E. L. de Boisbaudran; Gramont was responsible for pt. 2. See also his *Notice sommaire sur les travaux scientifiques de M. A. de Gramont* (Paris, 1910).

II. SECONDARY LITERATURE. See Edouard Branly, *Rapport sur les travaux de M. A. de Gramont* (Paris, 1913); Charles Fabry, "Arnaud de Gramont (1861–1923)," in *Revue d'optique théorique et instrumentale,* **3** (1924), 153–156; and Albin Haller, "Notice biographique sur Arnaud de Gramont," in *Comptes rendus hebdomadaires des séances de l'Académie des sciences,* **180** (1925), 106–107.

J. PAYEN

GRAND'EURY, CYRILLE (*b.* Houdreville, Meurthe-et-Moselle, France, 9 March 1839; *d.* Malzéville, near Nancy, France, 22 July 1917), *paleobotany.*

Grand'Eury studied first at the École Loritz in Nancy, which later became the École Professionnelle de l'Est. He then attended the École des Mines at St.-Étienne, from which he graduated first in his class in 1859. Grand'Eury worked for several years as an engineer at the Roche-la-Molière mines, but being unable, for reasons of health, to spend prolonged periods in the mines, he then accepted a post as *répétiteur* at the École des Mines at St.-Étienne, where he later became professor of trigonometry. These were his only official positions (1863–1899). In 1885 the Institute named him a corresponding member.

Grand'Eury's first work, on the carboniferous flora of the department of the Loire (1877), touched upon all the topics with which he was to be especially concerned—paleobotanical stratigraphy, the reconstruction of Paleozoic plants, their ecology, and the conditions of formation of coal seams. It also contained a map of the subterranean topography of the basin of the upper Loire and listed new genera and species of plants.

La formation des couches de houille du terrain houiller (*géogénie*), published in 1887, was the result not only of Grand'Eury's researches in the Loire basin but also of his observations made during ten years of travel in northern Europe, Upper Silesia, and the Urals, and of his studies of the Gard basin. In this work Grand'Eury considered the coal strata, the coal's relationship to the encasing rocks, and the deposition and formation of the coal beds. In 1910 he published a new basic work, *La géologie et la paléontologie du bassin houiller du Gard,* which contained many geological cross sections and columns, together with illustrations of fossil plants and underclays.

Two sections of a planned larger work, *Recherches géobotaniques sur les forêts et sols fossiles et sur la végétation de la flore houillère,* were published before World War I. During the war Grand'Eury's only son and collaborator, Maurice, was killed in action, and the books were never completed.

The study of plant impressions led Grand'Eury to establish for the Massif Central a series of stages that he named for the most abundant representative plants: the cordaitean (lower portion of the St.-Étienne layer), the filicite (middle portion of the same layer), and the *Calamodendron.* He also used names of localities to designate other stages familiar to regional geologists.

His knowledge of the floral succession, applied to mine development, led Grand'Eury to advise the continuation of borings that had been stopped and, thus, to the discovery of new coal deposits. He also drew many scientific hypotheses from this work, notably some concerning the mutation of species.

The study of so many seeds excited Grand'Eury, and as early as 1875 he suspected that a great many of them that he found separately had come from Filicineen-type fronds. He did not describe *Pecopteris pluckeneti,* to which hundreds of tiny seeds are attached, until 1905—after the publications of Robert Kidston (1903) and Oliver and Dukinfield Scott (1904), which were also devoted to seeds in connection with fronds. He concluded that the vegetative organs were comparatively less variable than the reproductive organs, in other words, that a particular foliage will remain constant in the course of evolution but will have different seeds attributed to it.

Grand'Eury was convinced of the necessity of using special generic names for the various organs of a plant. He originated the terms *Cordaianthus* for the inflorescence of the Cordaites, *Cordaicladus* for the axis, *Cordaifloyos* for the bark, and *Cordaixylon* for certain woods. His reconstructions of *Lepidodendra,* of Cordaites, and of other plants from isolated remains of leaves, trunks, and seeds were scientific masterpieces that are still discussed in works on paleobotany.

The observation of mixtures of plant remains—particularly of various fossil rhizomes and roots *in situ*—and detailed surveys of the sections he investigated enabled Grand'Eury to reconstruct the vegetation, to identify both plants living in almost unmixed populations and social plants, and to establish certain conditions of growth.

Grand'Eury was active in the discussion of the deposition of coal. After having been convinced of allochthony, that is, origin by transport, he accepted the concept of deposition at the bottom of large marshy lakes. He stated that one always returned to the idea of a swamp, even if there is only a faint resemblance between coal and peat, in terms of the manner of accumulation of plant debris. The formation took place not on the spot but a short distance away. He also accepted, for certain basins, the existence of a cover of deep water surrounded by large, wooded marshes but held that a uniform conclusion for all basins was not possible. Grand'Eury clearly spoke of a subsidence of the ground necessary for the formation of a coal bed. In the marshy basins such as those he studied, he paid special attention to the schistification of coal, which is due to a water current's carrying sediments that mix with carbonaceous material.

BIBLIOGRAPHY

I. ORIGINAL WORKS. Grand'Eury's writings include "Mémoire sur la flore carbonifère du département de la Loire et du centre de la France étudiée aux trois points de vue botanique, stratigraphique et géognostique," in *Mémoires de l'Académie des sciences de l'Institut de France,* **24,** no. 1 (1877), 1–624; "La formation des couches de houille du terrain houiller (géogénie)," in *Mémoires de la Société géologique de France,* 3rd ser., **4** (1887), 109–196; *La géologie et la paléontologie du bassin houiller du Gard* (St.-Étienne, 1890); and *Recherches géobotaniques sur les forêts et sols fossiles et sur la végétation de la flore houillère,* 3 vols. (Paris–Liège, 1912–1914).

II. SECONDARY LITERATURE. See Paul Bertrand, "C. Grand'Eury. Notice nécrologique," in *Bulletin de la Société géologique de France,* 4th ser., **19** (1920), 148–162; H. Guyot, "Notes sur le paléobotaniste lorrain Cyrille Grand'Eury," in *Bulletin de la Société d'histoire naturelle de Metz,* 3rd ser., **10** (1935), 317–324; and Paul Vuillemin, "L'oeuvre de Cyrille Grand'Eury," in *Revue générale des sciences pures et appliquées,* **28** (1917), 601–604.

F. STOCKMANS

GRANDI, GUIDO (*b.* Cremona, Italy, 1 October 1671; *d.* Pisa, Italy, 4 July 1742), *mathematics.*

At the age of sixteen, Grandi entered the religious order of the Camaldolese and changed his baptismal name of Francesco Lodovico to Guido. His appointment in 1694 as teacher of mathematics in his order's monastery in Florence led him to study Newton's *Principia.* In order to understand it, he was obliged to increase his knowledge of geometry, and made such rapid progress that he was soon able to discover new properties of the cissoid and the conchoid and to determine the points of inflection of the latter curve. When in 1700 Grandi was called to Rome, Cosimo de' Medici encouraged him to stay in Tuscany, by making him professor of philosophy at Pisa. In 1707 he received the honorary post of mathematician to the grand duke, in 1709 he was made a member of the Royal Society of London, and in 1714 he became professor of mathematics at Pisa. Grandi's voluminous scientific correspondence preserved in the library of the University of Pisa testifies to the esteem he enjoyed among the mathematicians of his time.

Grandi also did successful work in theoretical and practical mechanics; his studies in hydraulics evoked considerable interest from the governments of central Italy (for example, the drainage of the Chiana Valley and the Pontine Marshes).

As a collaborator in the publication of the first Florentine edition of the works of Galileo, Grandi contributed to it a "Note on the Treatise of Galileo Concerning Natural Motion," in which he gave the first definition of a curve he called the *versiera* (from the Latin *sinus versus*): Given a circle with diameter AC, let BDM be a moving straight line perpendicular

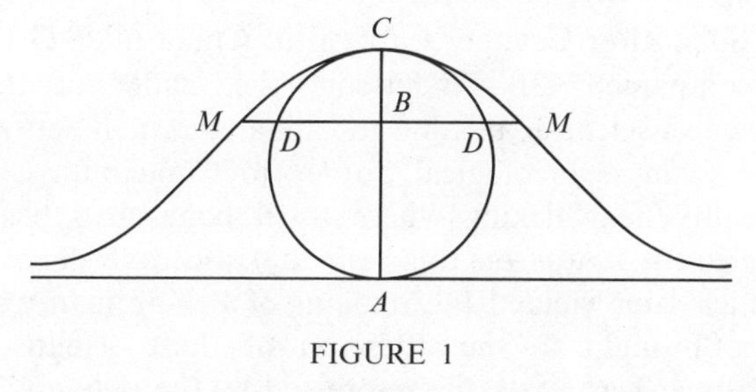

FIGURE 1

to AC at B, and intersecting the circumference of the circle at D. Let point M be determined by length BM satisfying the proportion $AB:BD = AC:BM$. The locus of all such points M is the *versiera*;[1] for a circle of diameter a, tangent to the x-axis at the origin, its Cartesian equation is $x^2y = a^2(a - y)$. The curve is more commonly known as the "witch of Agnesi" as the result of a mistranslation and a false attribution to Maria Gaetana Agnesi, who referred to it in her treatise *Istituzioni analitiche ad uso della gioventù italiana* (1748). Although Fermat had already investigated this particular equation,[2] Grandi's study extended to the more general family of curves of the form

$$y = \frac{a^{(m/n)+1}}{(a^2 + x^2)^{m/2n}},$$

where m and n are positive integers (1710). In an unpublished treatise Grandi also studied a curve known as the strophoid.

Grandi's reputation rests especially on the curves that he named "rodonea" and "clelia," after the Greek word for "rose" and the Countess Clelia Borromeo, respectively. He arrived at these curves in attempting to define geometrically the curves that have the shape of flowers, in particular the multileaved roses. They are represented in polar coordinates by equations of the form

$$\rho = R \sin a\vartheta,$$

in which R is a given line segment and a a positive integer. Grandi communicated the most significant properties of these curves to Leibniz in two letters dated December 1713 but did not make them generally known until ten years later in a memoir presented to the Royal Society of London. He later explained his complete theory in a special pamphlet (1728). Whereas analytic geometry now teaches the study of curves with given equations, Grandi here solved the inverse problem of determining the equations of curves having a preestablished form. The clelias are curves inscribed in a spherical zone, and their projection on a base plane of the zone yields the rodonea.

The *Acta eruditorum* of 4 April 1692 contained, under a pseudonym, the problem of constructing in a hemispheric cupola four equal-sized windows such that the remaining area of the cupola is quadrable. This is known as Viviani's problem, after Vincenzo Viviani who suggested it. It is an indeterminate problem and was solved shortly afterward by Leibniz and by Viviani himself;[3] Grandi also devoted a memoir to it (1699).

The curve

$$y = b \ln(x/a)$$

or

$$x = ae^{y/b},$$

called the logarithmic or logistic curve, was studied by Evangelista Torricelli as early as 1647; Huygens revealed its most important properties in a communication read before the Paris Academy in 1669. In 1701 Grandi demonstrated the theorems enunciated by Huygens.

On a more general level, Grandi's treatise on quadrature of 1703, in which he abandoned the Galilean methods of Cavalieri and Viviani in favor of those of Leibniz, marks the introduction of the

Leibnizian calculus into Italy. Grandi was also the author of several noteworthy and popular textbooks.

NOTES

1. Cf. also Grandi's *Quadratura circuli et hyperbolae.*
2. Cf. *Oeuvres de Fermat,* Charles Henry and Paul Tannery, eds. (Paris, 1891–1922), I, 279–280.
3. Vincenzo Viviani, *Formatione e misura di tutti i cieli* (Florence, 1692).

BIBLIOGRAPHY

I. ORIGINAL WORKS. Grandi's writings include *Geometrica divinatio Vivianeorum problematum* (Florence, 1699); *Geometrica demonstratio theorematum Hugenianorum circa logisticam seu logarithmicam* (Florence, 1701); *Quadratura circuli et hyperbolae* (Pisa, 1703, 1710); *De infinitis infinitorum et infinite parvorum* . . . (Pisa, 1710); "Florum geometricarum manipulus," in *Philosophical Transactions of the Royal Society* (1723); *Flores geometrici ex rhodonearum et cloeliarum curvarum descriptione resultantes* . . . (Florence, 1728); *Elementi geometrici piani e solidi di Euclide, posti brevemente in volgare* (Florence, 1731); *Istituzioni di aritmetica pratica* (Florence, 1740); and *Istituzioni geometriche* (Florence, 1741).

II. SECONDARY LITERATURE. Grandi's letters to Leibniz are in *Leibnizens mathematische Schriften, herausgegeben von C. J. Gerhardt,* 7 vols. and supp. (Berlin and Halle, 1848–1863), IV, 221, 224. There are three works of value by Gino Loria: *Curve sghembe speciali algebriche e trascendenti,* 2 vols. (Bologna, 1925), esp. II, 57; *Curve piane speciali algebriche e trascendenti,* 2 vols. (Milan, 1930), I, 94, 419 ff.; and *Storia delle matematiche,* 2nd ed. (Milan, 1950).

A. NATUCCI

GRANGER, WALTER WILLIS (*b.* Middletown Springs, Vermont, 7 November 1872; *d.* Lusk, Wyoming, 6 September 1941), *paleontology.*

The son of Charles H. Granger and Ada Byron Haynes, Granger acquired his keen interest in nature as a boy in rural Vermont and began his lifelong career at the American Museum of Natural History in 1890, after only two years of high school. He married Anna Dean of Brooklyn, New York, on 7 April 1904; they had no children. In 1932 he was awarded an honorary D.Sc. by Middlebury College, Vermont. Besides being a member of several scientific societies, he was particularly active in the Explorers Club, of which he was president in 1935–1937 and later an honorary member.

Granger's first years at the American Museum were divided between taxidermy and maintenance work. In 1894 he collected mammal and bird skins in the Rocky Mountains; his interest in fieldwork led to his transfer in 1896 to the department of vertebrate paleontology. In 1909 he was advanced to assistant curator, in 1911 to associate curator, and in 1927 to curator of fossil mammals.

Granger possessed the keen eye, steady hand, and infinite patience essential for finding and collecting delicate fossils. Between 1896 and 1918 he spent nineteen field seasons in the western United States; in 1907 he accompanied Henry F. Osborn to the Fayum in Egypt. From 1897 to 1901 he took part in the excavations for dinosaurs at Bone Cabin, Wyoming, and his first paleontological publication was on some of this material. His major efforts, however, were devoted to the early Tertiary of the Rocky Mountains. His extensive collections of well-preserved fossil mammals formed the basis for numerous systematic revisions and stratigraphic studies.

In 1921 Granger went to China as paleontologist and second in command of the American Museum's Central Asiatic Expeditions, led by Roy Chapman Andrews. During five summers in the Gobi Desert of Mongolia he found and collected a series of faunas ranging in age from Jurassic to Pleistocene. The excitement brought to the expedition members and to the scientific world by the discoveries of the small horned dinosaur *Protoceratops* and accompanying nests of eggs, of tiny Cretaceous mammal skulls, and of the giant *Baluchitherium,* largest of all land mammals, are well told in Andrews' report of the expedition.

Soon after Granger's arrival in China in 1923 he accompanied Dr. Andersson, the leader of the Swedish scientific mission to China, and members of the Chinese Geological Survey to Choukoutien, a locality near Peking where fossil bones had been collected. It was on this visit that the cave deposit which later yielded the remains of Peking man was first brought to the attention of these scientists. Granger was favorably impressed by the richness of the deposit and encouraged the Swedish and Chinese scientists to investigate it fully and advised them on suitable techniques for doing so.

During three winters between the Mongolian trips Granger visited a remote region of Szechwan and obtained an important series of Pleistocene mammals from Chinese collectors who dug out fossil bones and teeth for the Chinese drug markets.

After 1930 Granger's efforts were devoted largely to the necessary curatorial work on the Mongolian collections, to departmental administration, and to editing reports of the Asiatic expeditions. Each summer he found time to spend a few weeks in the field with his skilled preparator and long-time friend

Albert Thomson, and his life ended quietly in the middle of such a congenial excursion.

BIBLIOGRAPHY

I. ORIGINAL WORKS. The most extensive list of Granger's publications is included in the memorial by Simpson (1942). His technical papers are cited in the bibliographies of vertebrate paleontology by O. P. Hay, Carnegie Institution of Washington Publication 390, I (Washington, D.C., 1929), 198; and by C. L. Camp *et al.*, Geological Society of America Special Papers 27 (1940), p. 22; 42 (1942), p. 157; memoir 37 (1949), p. 80.

Granger published an important systematic study of primitive fossil horses in "A Revision of the American Eocene Horses," in *Bulletin of the American Museum of Natural History,* **24** (1908), 221-264; and of primitive ungulates known as condylarths in "A Revision of the Lower Eocene Wasatch and Wind River Faunas," *ibid.,* **34** (1915), 329-361; and, in the same bulletin, several papers on Tertiary geology of the Rocky Mountains. He more characteristically communicated his extensive knowledge of mammalian morphology and relationships orally to his co-workers. In this way he collaborated with William D. Matthew, William K. Gregory, George G. Simpson, *et al.* in twenty papers on the systematics of Eocene mammals; he also contributed importantly to Matthew's major monographs "The Carnivora and Insectivora of the Bridger Basin," in *Memoirs from the American Museum of Natural History,* **9,** pt. 6 (1909), 289-567; and "Paleocene Faunas of the San Juan Basin, New Mexico," in *Transactions of the American Philosophical Society,* n.s. **30** (1937). He was coauthor of 35 reports on vertebrate fossils from Mongolia and China.

Two chapters, "Paleontological Exploration in Eastern Szechuan . . ." and "A Reconnaissance in Yunnan, 1926-1927," were contributed by Granger to R. C. Andrews, *The New Conquest of Asia* (pp. 501-528 and 529-540). He also wrote popular accounts of the Gobi exploration and gracious memorials to W. D. Matthew, in *Journal of Mammalogy,* **12** (1931), 189-194; and to F. B. Loomis, in *Proceedings. Geological Society of America* for 1936 (1937), 173-178.

II. SECONDARY LITERATURE. On Granger or his work, see R. C. Andrews, *The New Conquest of Asia: Natural History of Central Asia,* Reports of Central Asiatic Expedition of the American Museum of Natural History, New York, I, pt. 1 (New York, 1932), 1-453; Donald R. Barton, "The Way of a Fossil Hunter," in *Natural History,* **47** (1941), 172-176; and G. G. Simpson, "Memorial to Walter Granger," in *Proceedings. Geological Society of America* for 1941 (1942), 159-172, with portrait and bibliography of 106 titles.

Brief obituary notices are C. F. Cooper, in *Nature,* **148** (1941), 654-655; J. J. Hickey and J. T. Nichols, in *Abstracts of Proceedings of the Linnaean Society of New York,* **52-53** (1941), 151-152; T. S. Palmer, in *Auk,* **59** (1942), 140; G. G. Simpson, in *Science,* **94** (1941), 338-339; and *News Bulletin. Society of Vertebrate Paleontology,* **4** (1941), 1-2.

JOSEPH T. GREGORY

GRASHOF, FRANZ (*b.* Düsseldorf, Germany, 11 July 1826; *d.* Karlsruhe, Germany, 26 October 1893), *applied mechanics, thermodynamics, machine design.*

Son of a teacher of the humanities, Grashof had a strong practical bent, interrupting his early schooling to work for a locksmith. He went to Berlin in 1844 and spent the following three years at the Gewerbe-Institut, studying mathematics, physics, and machine design as preparation for a career in metallurgy.

After a year as an army volunteer, a career as naval officer seemed attractive; Grashof shipped out as apprentice seaman on a sailing vessel, returning in 1851. Having found himself during this voyage, Grashof realized that he was not suited to a life of physical activity (in part because of myopia) and that his real inclination was to teach engineering sciences.

In 1852 Grashof resumed his studies at Berlin, where he was entrusted with lectures on applied mathematics. Elevated in 1854 to staff membership of the Gewerbe-Institut as teacher of mathematics and mechanics, he was also director of the Office of Weights and Measures.

On 12 May 1856, Grashof was among the twenty-three founders of the Verein Deutscher Ingenieure (VDI). Although no unified Germany then existed, this society was to include the engineers of the twenty-five German states and Grashof was to implement its organization; because of his scientific reputation he was made director of the society and editor of its *Zeitschrift.* The University of Rostock conferred an honorary doctorate in 1860.

Following Ferdinand Redtenbacher's death in 1863, Grashof was named his successor as professor of applied mechanics and the theory of machines at the Polytechnikum in Karlsruhe. He lectured on strength of materials, hydraulics, thermodynamics, and machine design with clarity and precision. He remained in Karlsruhe, rejecting an offer from Aachen and two from Munich.

Although he resigned the editorship of the *Zeitschrift,* Grashof remained director of the VDI and turned much of his attention to technical writing. He was the first to present the fundamental equations of the theory of elasticity, in a text on strength of materials in which he treated flexure, torsion, buckling, plates, and shells. His three-volume *Theoretische Maschinenlehre* (1871-1886) was characterized by sharp insight and critical observation with respect to the limits of accuracy and the admissibility of assumptions; no contemporary English or French work was its equal.

During his lifetime Grashof was recognized as an authority on mechanical engineering in its broadest sense. As teacher and engineer, and as founding member, editor, and long-time director of the VDI, he influenced a generation of engineers by bringing mathematical and scientific considerations to the burgeoning problems of the steam-engine age. Grashof used analysis, supporting and exploiting it with all available experimental work; he shunned the graphical approach of his contemporary Karl Culmann, who founded graphic statics.

Grashof's name is perpetuated in several ways. There is the dimensionless Grashof number of heat transfer in free-convection flow systems (a criterion similar to Reynolds' number of forced convection) for the transition from laminar to turbulent flow. Another Grashof criterion is used in kinematics for establishing whether one link of a four-bar chain can rotate completely. The VDI honored his memory by erecting the Grashof Monument in Karlsruhe and by establishing the annually awarded Grashof Medal, its highest honor for achievements in technology.

Grashof suffered a stroke in 1883 that restricted his activity. In that year he became a member of the Standards Commission; in 1887, he was named a trustee of the Bureau of Standards and received honorary membership in the VDI. A second stroke in 1891 disabled him, and he died two years later, survived by his wife and two children.

BIBLIOGRAPHY

I. ORIGINAL WORKS. Grashof published forty-two articles in *Zeitschrift des Vereins deutscher Ingenieure* (1857–1885). His books include *Angewandte Mechanik,* vol. V of Gustave Karsten's *Allgemeine Encyklopädie der Physik* (Leipzig, 1856–1862); *Festigkeitslehre mit Rücksicht auf den Maschinenbau* (Berlin, 1866), 2nd ed., under the title *Theorie der Elastizität und Festigkeit* (Berlin, 1878); *Resultate der mechanischen Wärmetheorie* (Heidelberg, 1870); and *Theoretische Maschinenlehre,* 3 vols. (Leipzig, 1871–1886). Other works are listed in Poggendorff, III, 543.

II. SECONDARY LITERATURE. On Grashof or his work, see H. Lorenz, "Die wissenschaftlichen Leistungen F. Grashofs," in *Beiträge zur Geschichte der Industrie und Technik,* **16** (1926), 1–12; C. Matschoss, *Männer der Technik* (Düsseldorf, 1925), p. 94; K. Nesselmann, in *Neue deutsche Biographie,* VI (Berlin, 1964), 746; R. Plank, "Franz Grashof als Lehrer und Forscher," in *Zeitschrift des Vereins deutscher Ingenieure,* **70** (1926), 28; and S. Timoshenko, *History of the Strength of Materials* (New York, 1953), p. 133.

Obituaries are in *Zeitschrift des Vereins deutscher Ingenieure,* **37** (1893), 48; and *Transactions of the American Society of Mechanical Engineers,* **15** (1894).

R. S. HARTENBERG

GRASSI, GIOVANNI BATTISTA (*b.* Rovellasca, Italy, 27 March 1854; *d.* Rome, Italy, 4 May 1925), *entomology, parasitology.*

The son of Luigi Grassi, a municipal official, and of Costanza Mazzuchelli, a peasant of unusual intelligence, Grassi was educated at Saronno. From 1872 he studied medicine at Pavia, graduating in 1878. He then went to Germany, where he worked at Heidelberg with the zoologist Otto Bütschli and the anatomist Carl Gegenbaur. There Grassi met his future wife, Maria Koenen. In 1883 he was appointed professor of zoology and comparative anatomy at Catania. In 1895 he became professor of comparative anatomy at Rome University, where he remained for the rest of his life. In 1908 he was appointed a senator of the kingdom.

As an anatomist Grassi studied the development of the vertebral column in bony fishes (1883), and as a physician he studied endemic goiter (1903–1917). Some of his more elegant studies were in pure entomology: on bees (1877–1884); on myriapods (1886–1889); and his monumental work on termites (1885–1893). For the latter work he was awarded a Darwin Gold Medal. He also studied the Chetognates (1881, 1883) and the reproduction of eels (1910–1919), and in 1885 he described a new species of spider (*Koenenia mirabilis*), naming it for his wife. Nevertheless, he is remembered today essentially for his studies in parasitology and in practical and applied entomology.

In 1876 in his native Rovellasca Grassi investigated the high mortality of cats and discovered in their bowels large numbers of *Dochmius balsami,* strongly hematophagous little worms very like *Anchylostoma.* In 1878, while still a student at the medical clinic of the University of Pavia and working under Francesco Orsi, he discovered anchylostomiasis in Italy. He made the diagnosis as a result of finding *Anchylostoma* eggs in feces.

Grassi developed a wide knowledge of helminthology, writing first on *Anguillula intestinalis* (or *Rhabdonema strongyloides*) in patients with marshy cachexia (1878–1887). He also studied *Filaria* (1887–1901), *Trichocephalus dispar* (1887), and *Bilharzia* (1888); but he was concerned particularly with the tapeworm. Grassi was the first to demonstrate that *Taenia nana* is able to go through its entire life cycle in one animal, without the need of an intermediate host, a fact that had long been denied. He was also the first to show that the flea *Pulex serraticeps* is the intermediate host of larvae of *Taenia cucumerina* or *elliptica.* Thus he wrote that the swallowing of infected fleas (for example, with milk) might be the reason for *Taenia* in children.

Grassi also made important studies on the parasitic and pathogenic protozoa (1879–1888). Of great practical importance were his studies on the fly (1879–1884): it could swallow, and expel still alive, the spores of *Botrytis* and *Oidium,* the eggs of *Taenia* or *Trichocephalus,* and even bacteria, particularly the cholera bacillus. Grassi then began a campaign to eradicate flies. Grassi made his first observations on malaria in 1890, when in collaboration with Raimondo Feletti he discovered *Plasmodium vivax.* (In 1889 Ettore Marchiafava and Angelo Celli had discovered *Plasmodium falciparum,* the deadliest form of the malarial parasite.) This confirmed the validity of Camillo Golgi's assertions (1885–1889) that the differences in the period (three or four days) and the severity of various malarial fevers arise because of different species of the malarial parasite. Also in collaboration with Feletti, Grassi worked on malaria in 1891 and 1892 and discovered the malaria parasite of birds (*Proteosoma praecox,* very like *Plasmodium vivax*). In 1891 he performed the first inoculation of malaria parasites from one bird into another. But all of Grassi's decisive investigations on the transmission of malaria in man were made between 15 July and 30 November 1898 and reported to the Accademia dei Lincei (see *Rendiconti,* meeting of 22 December 1898).

In 1894 Amico Bignami, in collaboration with Giuseppe Bastianelli, produced a typical malarial fever paroxysm in a healthy man by intradermically injecting a minute drop of blood from a malaria patient. But in 1896 all his attempts to produce malaria in man by the bite of mosquitoes failed. Nevertheless, human malaria from inoculation by mosquito bite—affirmed in 1896 by Bignami and by the English pathologist Patrick Manson in his lectures to the Royal College of Physicians of London—was accepted as most probable. Grassi was struck by the fact that there are always mosquitoes wherever malaria is found, which had already been observed in 1716 by the Italian physician Giovanni Maria Lancisi. Grassi, however, also noted that where mosquitoes abound, malaria is not necessarily present. After deducing that only a particular species of mosquito could transmit malaria to man, it became a question of identifying the species. In August 1898 Grassi discovered that the agent transmitting malaria to man is the female *Anopheles* mosquito, most frequently of the species *A. claviger.*

In November 1898, with the help of Bignami and Bastianelli, Grassi produced experimentally in a healthy man a typical malarial paroxysm of malignant tertian fever, resulting from bites of *Anopheles* mosquitoes. He then demonstrated that all anophelines are capable of becoming the hosts of human malaria parasites during sporogony. In consequence of these demonstrations, Grassi launched a great antimalaria campaign, emphasizing human protection through window screens, prophylaxis with quinine, and the destruction of *Anopheles* with *Gambusia,* which devours its larvae. Above all, Grassi recommended extensive distribution of quinine to all persons in malarial districts, because quinine kills the parasites of malaria and thereby prevents the infection of new anophelines. In 1899 he demonstrated that *Anopheles* is born uninfected and becomes able to transmit malaria only after biting an infected human. Grassi understood the importance of continuous suppressive treatment of malaria with quinine, so as to prevent the occurrence of the asexual blood stages in both uninfected and chronically infected persons.

That Grassi discovered the pathogenic activity of anophelines and, in consequence, is responsible for the victory over human malaria, is undeniable. But the English surgeon Ronald Ross claimed priority and the 1902 Nobel Prize for physiology or medicine was awarded to him; even today this decision is widely accepted. Indeed it was Grassi who first demonstrated the sporogonic cycle of the human malaria parasite (its schizogonic cycle had already been described in 1889 by Golgi). Although unappreciated at the time, Grassi also identified the true agent transmitting malaria in man. Today it is well accepted, in a zoological sense, that while man is the intermediate, the *Anopheles* mosquito is the definitive host of malaria parasites.

Ross made some very important observations on malaria in birds, working from 1896 to 1898 under Manson. In the summer of 1898 he achieved the transmission of experimental malaria in birds by the bite of mosquitoes. He also demonstrated the entire life cycle of *Proteosoma,* the malaria parasite of birds, which concludes as a sporogonic cycle, resulting in the formation of sporozoites that accumulate in the salivary glands of a mosquito. Thus, when the mosquito bites, it inoculates a bird with malaria parasites. Ross also stated that malaria is transmitted in man by the bite of mosquitoes but did not demonstrate this. In addition, in 1898 he did not know of the genus *Anopheles,* nor was he convinced of the mosquito's exclusive importance in transmitting malaria. In 1903 Grassi published his precisely documented vindication, but Ross never acknowledged priority.

Grassi turned to a new field of research—the study of the phylloxera of grapes—which he pursued for several years. On the strength of his first notes, *La questione fillosserica in Italia* (1904), the Italian Ministry of Agriculture requested him to do an exhaustive study of this subject. In 1912 he collected his own observations and those of his collaborators in a precise and monumental analysis of the morphology and biology of the Italian and other European genera of phylloxera. Thus it was possible to begin the fight against this agricultural pest.

Grassi was an extremely private person, and an affectionate husband and father. His forty-two years of teaching and research testify to his deep devotion to science. But his greatest source of pride was that he opened the way to the eradication of malaria. Thus he chose to be buried in the cemetery of Fiumicino, an area that his personal and persevering initiative had rid of the disease.

BIBLIOGRAPHY

Grassi's scientific production was enormous. Alone he wrote more than 250 papers, collaborating on another 100 with his students. For a full listing of his works, see A. Pazzini, "Giovanni Battista Grassi," in *Rivista di biologia,* **19** (1935), 1–46.

For information on Grassi or his work, see A. Corradetti, "L'opera protozoologica di Battista Grassi alla luce degli odierni sviluppi della scienza," in *Rivista di parassitologia,* **15** (1954), 190–199; A. Corti, "Battista Grassi e la trasmissione della malaria," in *Studia ghisleriana,* **1** (1961); C. Golgi, "Sul ciclo evolutivo dei parassiti malarici nella febbre terzana," in *Archivio per le scienze mediche,* **13** (1889), 173–196; and C. Jucci, *Nel centenario della nascita di Battista Grassi* (Milan, 1954).

See also S. Piccini, "Nel centenario della nascita di Battista Grassi," in *Atti del XIV congresso internazionale di Storia della medicina* (Rome–Salerno, Sept. 1954); F. Silvestri, "Commemorazione del Socio Nazionale Giovanni Battista Grassi," in *Atti dell'Accademia nazionale dei Lincei, memorie* (1926); and C. Tumiati, "Giovanni Battista Grassi," in *Vite singolari di grandi medici dell'800* (Florence, 1952), see esp. pp. 125–135 for the clearest assessment of Grassi's priority over Ross.

PIETRO FRANCESCHINI

GRASSMANN, HERMAN GÜNTHER (*b.* Stettin, Germany [now Szczecin, Poland], 15 April 1809; *d.* Stettin, 26 September 1877), *mathematics.*

For a detailed study of his life and work, see Supplement.

GRATIOLET, LOUIS PIERRE (*b.* Ste. Foy-la-Grande, Gironde, France, 6 July 1815; *d.* Paris, France, 16 February 1865), *anatomy, anthropology.*

Gratiolet was the son of a rural physician; his mother was of noble lineage. The father's royalist allegiance disturbed his practice and led to removal to Bordeaux. Here Gratiolet began his studies. Soon turning to Paris, he completed his secondary course at the Collège Stanislas and began, probably in 1834, formal preparation in medicine. While he quickly exhibited uncommon skill and interest in anatomy, he was also successful in the various academic competitions which spurred on the aspiring practical physician (he held internships at the Pitié and Salpetrière hospitals).

Adjacent to the Salpetrière was the Muséum d'Histoire Naturelle, an active center for anatomical studies. Gratiolet frequented the dissection halls of the museum and soon (1839) became a participant in Henri de Blainville's researches. By 1842, having been made Blainville's laboratory assistant, he had renounced practical medicine for a career in science. He nonetheless completed all requirements for a medical degree (1845).

Such auspicious beginnings merely introduced twenty years of acute professional frustration. Gratiolet lectured on anatomy at the museum as Blainville's deputy from 1844 until the latter's death in 1850. Gratiolet's candidacy for his teacher's chair was rejected. He did continue as laboratory assistant and in 1853 was placed in charge of anatomical studies at the museum. He lacked, however, the professorial chair which his scientific achievement and demonstrated instructional capacity deserved. Finally, in 1862, he was named deputy to the professor of zoology in the Faculty of Science, Paris, and received full rights to that chair at the close of 1863. Within sixteen months Gratiolet was dead of apoplexy. He had, reported Paul Broca, "lived only for science."

Gratiolet was an indefatigable investigator and adroit interpreter of animal structure and function. In the former capacity he excelled as a descriptive anatomist, dealing with some of the most difficult material which the organism can present: the vascular and nervous systems, with emphasis on the brain and cranium. Like Blainville, he studied mollusks and concentrated above all on man and the primates. He offered detailed descriptions of the vascular system of such disparate creatures as the hippopotamus and the physician's leech, molluscan organs of generation and generative products, and the osteology of mammals. Gratiolet was a descriptive and comparative anatomist and neither employed vivisection nor attended to pathological lesions and their putative functional correlates. So restricted an approach evoked criticism and should have imposed limits on the scope of the physiological inferences which he evidently conceived to be the primary impulse of his many inquiries. He was a pioneer in the use of embryological material for establishing general zoological affinities (and dissimilarities) and the assessment of the active roles of various structures.

Gratiolet's interpretation of life and particularly of the nature of man began with principles enunciated by Blainville. The organism was an integral whole ceaselessly coping with ever-shifting stimuli from the environment and with the physiological (and, in the case of man, the mental) needs of its own being. Organs of peculiar structure and functional capacity obviously were essential to these tasks. All parts of the body acted cooperatively to share in a given vital act. Body and mind were physiologically conjoined, their activity being most evident in movement guided by instinct or intelligence. Gratiolet was less concerned with animal taxonomy than Blainville had been and accorded slight attention to conspicuous, external parts suitable for classificatory purposes. He struck for the heart of the matter: the form and behavior of those neural and muscular mechanisms without which the higher forms of life are inconceivable.

Thus was produced Gratiolet's first major work: *Mémoire sur les plis cérébraux* (1854). Descriptions of the cerebral folds, or convolutions, of the human brain had long been available; Gratiolet added the careful comparative investigation of the brain form of a wide range of monkeys. On the basis of the distribution and degree of cerebral convolution he emphasized the distinctness of the primates as a group. Gratiolet's anatomical research culminated in the *Anatomie comparée du système nerveux* (1857), a major contribution to mammalian descriptive anatomy. But this comparative anatomy offered far more than description; it presented (part II) Gratiolet's statement on the nature and meaning of intelligence, the opposed roles of "sensation" (a consequence of external stimulus) and "sentiment" (evoked independently within the organism), and man's uniqueness founded on his capacity to reason. These themes were clarified, further developed, and augmented by reflections on the diversity of human races in a notable series of anthropological essays and discussions offered between 1860 and 1865.

Man alone, Gratiolet argued, is to be established as a rational being; only he can speak. Spoken language remained, as it had over the generations since Descartes, the most direct expression of intelligence and the essential criterion of humanity. "This innate and . . . ineffaceable potentiality [for speech] is certainly the most striking, the most noble character of man. . . . Only man can have an idea of an idea, and so on almost to infinity." Implicit in these conclusions was Gratiolet's conviction, and one which comparative anatomy seemed only to confirm and expand, that human intelligence was in some way a function of the cerebral convolutions. One must, he urged, focus on gross structure until the fortunate day arrived that "one might study directly the brain itself."

During the 1860's the question of a localization within the brain (cerebrum) of mental functions was much in dispute. Gratiolet adopted a conservative position: "Generally speaking, I agree with [M. J. P.] Flourens that intelligence is unitary, that the brain is one and that it acts above all as an integral organ [*organ d'ensemble*]." He took this position less on grounds of negative evidence than on the absence of indisputable data confirming any particular case of localization.

Gratiolet, laying groundwork for his anthropological views, accorded complex instinct and simple judgment to both man and animals. But complex judgments required intelligence and hence were, like imagination, man's alone. All men possessed intelligence but their shares, apparently, varied. Gratiolet introduced into the prospering multiple-origins conception of the human races (polygenic theory) the embryological criterion. Claiming that the frontal sutures of the developing cranium in whites closed later than those of other races, he found a splendid opportunity for curious "reflection." "Might not," he mused, "the long persistence of sutures in the white race have some relation to the almost indefinite perfectibility of intelligence among men of this race? . . . might not the brain, among these perfectible men, [thus] remain capable of a slow but continuous growth?" Regrettably, among "idiots and lower [*abruties*] races the cranium is closed upon the brain like a prison." Not only were the races of man different; it was part of their very nature to be so. Lower races were no degraded Caucasians. They were perfect beings but were placed lower on the scale of creation, a scale which, since Gratiolet dismissed transmutationist hypotheses, must be considered as temporally fixed.

Not wholly consistent with this viewpoint was Gratiolet's expression of another, and probably more cherished, notion of why some men are elevated and others depressed: The Caucasian manifests an "instinct for civilization." All whites, from fool to hero, recognized the horrors of that individualism which promotes egoism and scorn for other men and leads to a neglect of social obligation (*devoir*). All whites, Gratiolet's strange apology continued, understood the "usefulness of law [and] the necessity to submit to it." Each must and would sacrifice part of his liberty for the good of the social whole. This remarkable excursus by a cabinet anatomist was not original. It shared fully Blainville's emphasis on physiology as a model for society and thus that special interest, broadcast by Blainville's auditor, Auguste Comte, in the inviolable solidarity of the healthful social organism. It is conservative social doctrine entertained by an anatomist of royalist sentiment and tempered with the epoch's exaggerated interest in racial matters.

Gratiolet displayed patience and industry in the face of constant academic rebuffs. He was exceptionally well-read, particularly in the classics and the literature of philosophy. He commanded deepest friendship, evidencing firm loyalty to his doctrinal allies and candor to all. He was an able and frequent combatant in discussions at the Société d'Anthropologie of Paris, of which he was a founding member. Gratiolet apparently received few or none of the honors due a French scientist of his stature and was overlooked by the Academy of Sciences. Political considerations may have been operative here. Born just after Waterloo and raised in the royalist persuasion, he cast his intellectual foundations in the mold of an outspoken Christian royalist apologist, Blainville. He ably led troops of the National Guard against the republican insurgents of 1848 (but refused decoration for participation in what he called a "civil war") and was obviously neglected by the Bonapartist ministries of the 1850's. In good faith he argued a scientific brief for the autonomy of man, the primacy of the Caucasian race, and the necessary and desirable supremacy of traditional European social forms; and in support of his faith he brought great learning as a naturalist and equal facility as an anatomist.

BIBLIOGRAPHY

I. ORIGINAL WORKS. Gratiolet published little before 1850. His principal early contribution was a dissertation for the degree *docteur en médecine* at the Paris Faculty of Medicine: *Recherches sur l'organe de Jacobson* (Paris, 1845). His major anatomical works are two: *Mémoires sur les plis cérébraux de l'homme et des primates,* 1 vol. plus plates (Paris, 1854); *Anatomie comparée du système nerveux considerée dans ses rapports avec l'intelligence,* 2 vols. plus plates (Paris, 1839–1857)—this work was planned and begun by François Leuret but the research, organization, and interpretations of vol. II (1857) are due to Gratiolet.

Gratiolet published over fifty scientific papers, some with collaborators. They are listed in Royal Society, *Catalogue of Scientific Papers 1800–1863* (London, 1868), II, 989–991; and VII [*1864–1873*] (1877), 818. Among the more interesting of these papers are the following: "Mémoire sur les plis cérébraux de l'homme et des primates," in *Comptes rendus hebdomadaires des séances de l'Académie des sciences,* **31** (1850), 366–369, a valuable précis of the 1854 vol.; "Mémoire sur le développement de la forme du crâne de l'homme, et sur quelques variations qu'on observe dans la marche de l'ossification de ses sutures," *ibid.,* **43** (1856), 428–431; "Mémoire sur la microcéphalie considérée dans ses rapports avec la question des caractères du genre humain," in *Mémoires de la Société d'anthropologie de Paris,* **1** (1860), 61–67; "Sur la forme et la cavité cranienne d'un Totonaque, avec réflexions sur la signification du volume de l'encéphale," in *Bulletin de la Société d'anthropologie de Paris,* **2** (1861), 66–81; and "Recherches sur l'anatomie du *Troglydytes aubryi,* chimpanzé d'une nouvelle espèce," in *Nouvelles archives du Muséum d'histoire naturelle* (Paris), **2** (1866), 1–264, written with P. H. E. Alix.

On the evening of 20 January 1865 Gratiolet delivered a lecture on physiognomy at the Sorbonne. This remarkable lecture was soon published—"Considérations sur la physionomie en général et en particulier sur la théorie des mouvements d'expression," in *Annales des sciences naturelles.* Zoologie et paléontologie, 5th ser., **3** (1865), 143–179—as was a posthumous vol. on the subject: *De la physionomie et des mouvements d'expression, suivi d'une notice sur sa vie et ses travaux et de la nomenclature de ses ouvrages, par Louis Grandeau* (Paris, 1865; 4th ed., 1882). In the lecture and book Gratiolet explored the notion that, while spoken language was peculiar to man, animals shared with man another language: facial and bodily movement or "expression."

II. SECONDARY LITERATURE. Grandeau's essay on Gratiolet (cited above; also published separately [Paris, 1865]) is the most extensive account of the anatomist. Other notices, all quite personal, include Paul Bert, "Éloge de Pierre Gratiolet," in *Bulletin de la Société médicale de l'Yonne* (Auxerre) (1868), 17–37; Paul Broca, "Éloge funèbre de Pierre Gratiolet," in *Mémoires de la Société d'anthropologie de Paris,* **2** (1865), cxii–cxviii; and Edmond Alix, "Notice sur les travaux anthropologiques de Gratiolet," *ibid.,* **3** (1865), lxxi–ciii.

WILLIAM COLEMAN

GRAUNT, JOHN (*b.* London, England, 24 April 1620; *d.* London, 18 April 1674), *statistics, demography.*

Graunt, apparently the eldest of seven or eight children born to Henry and Mary Graunt, received some formal "English learning" and, after he was sixteen, was apprenticed in his father's profession of draper. He held various offices in the Freedom of the Drapers' Company and in the city government, and he prospered in his business. In February 1641 he married Mary Scott, who evidently bore him one son and three daughters. Graunt came to know prominent people in London, and before 1650 he had become a friend of William Petty.

After the publication of his only book in January 1662, Graunt was elected, at the request of Charles II, to membership in the Royal Society. He suffered serious losses from the great fire of 1666, and this crisis was worsened by legal harassments occurring after his conversion around that time to Catholicism (earlier he had converted from Puritanism to anti-Trinitarianism). In spite of assistance from Petty, Graunt remained in straitened circumstances until his death.

Graunt's *Natural and Political Observations . . . Upon the Bills of Mortality* was the foundation of both statistics and demography. He had never formally studied mathematics, and the computations in his book were not more complex than what a successful businessman of that time could be expected to know. There has been much speculation over how much assistance Graunt received from Petty in writing the book. Undoubtedly Petty encouraged the undertaking and most likely made some contributions to it, but Graunt seems to deserve the lion's share of credit. He got the idea for his investigation from "having (I know not by what accident) engaged my thoughts upon the *Bills of Mortality,*" which had been published for London since the end of the sixteenth century. These statistics were the primary basis for his study, although he supplemented them with parish christening records and data from a rural area, Romsey in Hampshire (Petty's birthplace).

Since Graunt's treatise was the starting point for two sciences, both his discoveries and the form of his presentation were important. He began by listing the kinds of knowledge that could be gained from analyzing vital statistics. Next, he discussed with impressive sophistication the various kinds of defects in his data—geographical inconsistencies, irregular intervals between recordings, lack of thoroughness, inaccurate age approximations, an ambiguous disease nomenclature, and a bias against honest reporting of certain causes of death, such as syphilis. He published tables of some of the data and some important statistical regularities which he discovered were evident from inspecting the data: a few more boys were born than girls; women tended to live longer than men; the sex ratio was about equal and was stable; the numbers of people dying from most causes except epidemic diseases were about the same from year to year; the mortality rate was high among infants; the frequency of death was higher in urban than in rural areas.

Graunt carried his analysis further by deducing various characteristics of populations from his data. These ingenious attempts indicate a good understanding of the kinds of questions that are significant for demography. Usually he explained his steps in solving problems, but he seldom included the actual calculations; and sometimes he omitted important information. Furthermore, his indirect approach sometimes went beyond the reliable use of his data, and the accuracy of some of his answers was difficult to evaluate. His calculations of the populations of England and Wales and of London are two examples.

Since he did realize the shortcomings of his data, on several occasions Graunt set an excellent example by seeking verification of his estimates by different indirect methods. He introduced the use of statistical samples but did not pursue this subject far enough to determine the sizes of samples or means of selection needed for insuring accuracy. He gave information on infant and old-age mortality which modern demographers have shown contained an implicit life table, but Graunt's method of computing it remains uncertain. He also realized that demographic procedures could be used to make projections concerning both past and future populations. In 1663 he furnished the Royal Society with a brief note on the rate of growth of salmon and the rate of increase of carp in a pond, which indicates that he also saw the value of studying animal populations.

BIBLIOGRAPHY

I. ORIGINAL WORKS. Graunt's only book is *Natural and Political Observations Mentioned in a Following Index, and Made Upon the Bills of Mortality* (London, 1662; 2nd ed., 1662; 3rd ed., 1665; 4th ed., Oxford, 1665; 5th ed., London, 1676). The 5th ed. was reprinted in *A Collection of the Yearly Bills of Mortality, From 1657 to 1758 Inclusive. Together With Several Other Bills of an Earlier Date . . .*, presumably edited by Thomas Birch (London, 1759). There is a German trans. by Gottfried Schultz (Leipzig, 1702). There is also a reprint of the 5th ed. in *The Economic Writings of Sir William Petty, Together With the Observations Upon the Bills of Mortality More Probably by Captain John Graunt,* Charles Henry Hull, ed., 2 vols. (Cambridge, 1899; repr. New York, 1963), II, 319–431. Hull also gives a full bibliography of earlier eds. in II, 658–660, 641. There are two reprs. of the 1st ed.: Walter F. Willcox, ed.

(Baltimore, 1939), and B. Benjamin, ed., in *Journal of the Institute of Actuaries,* **90** (1964), 1-61.

Graunt's notes on fish were first published by Thomas Birch in *The History of the Royal Society of London for Improving of Natural Knowledge, From Its First Rise,* 4 vols. (London, 1756-1757), I, 267, 294. Hull quoted the notes from p. 294 following his repr. of Graunt's book, II, 432.

II. SECONDARY LITERATURE. The most important contemporary accounts of Graunt are by John Aubrey and Anthony à Wood: *Aubrey's Brief Lives,* Oliver Lawson Dick, ed. (Ann Arbor, Mich., 1957), pp. 114-115; and Anthony à Wood, *Athenae Oxonienses,* 2nd ed., 2 vols. (London, 1721), I, col. 311. Wood's account has been quoted in full in James Bonar, *Theories of Population From Raleigh to Arthur Young* (London, 1931; facs. repr., 1966), pp. 69-71. There are two modern investigations of his life: C. H. Hull, in *Economic Writings of . . . Petty,* I, xxxiv-xxxviii; and D. V. Glass, "John Graunt and His *Natural and Political Observations,*" in *Notes and Records. Royal Society of London,* **19** (1964), 63-100, see 63-68, notes on 89-94.

The question of Petty's contribution to Graunt's book has been discussed in Glass, *op.cit.,* pp. 78-89, notes on pp. 97-100; Hull, *op. cit.,* I, xxxix-liv; Major Greenwood, *Medical Statistics From Graunt to Farr* (Cambridge, 1948), pp. 36-39; Walter F. Willcox, introduction to Graunt's *Natural and Political Observations* (Baltimore, 1939), pp. iii-xiii; and P. D. Groenewegen, "Authorship of the *Natural and Political Observations Upon the Bills of Mortality,*" in *Journal of the History of Ideas,* **28** (1967), 601-602.

Graunt's contributions to statistics and demography are surveyed and evaluated in B. Benjamin, "John Graunt," in *International Encyclopedia of the Social Sciences,* VI (1968), 253-255; Glass, *op. cit.,* pp. 69-78, notes on pp. 95-97; Hull, *op. cit.,* I, lxxv-lxxix; Greenwood, *op. cit.,* pp. 30-35; Harald Westergaard, *Contributions to the History of Statistics* (London, 1932), pp. 16-23; Ian Sutherland, "John Graunt: a Tercentenary Tribute," in *Journal of the Royal Statistical Society,* **126A** (1963), 537-556; and A. Wolf, F. Dannemann, A. Armitage, and Douglas McKie, *A History of Science, Technology and Philosophy in the 16th & 17th Centuries,* 2nd ed. (New York, 1950), pp. 588-598.

On the background situation for much of the bills of mortality used by Graunt, see Charles F. Mullett, *The Bubonic Plague and England. An Essay in the History of Preventive Medicine* (Lexington, Ky., 1956). Also relevant, and still useful, is the discussion by William Ogle, "An Inquiry Into the Trustworthiness of the Old Bills of Mortality," in *Journal of the Royal Statistical Society,* **55** (1892), 437-460. Norman G. Brett-James has written a very useful paper on the collection of the London data: "The London Bills of Mortality in the 17th Century," in *Transactions of the London & Middlesex Archaeological Society,* **6** (1933), 284-309.

The early reception of Graunt's book is discussed in Robert Kargon, "John Graunt, Francis Bacon, and the Royal Society: the Reception of Statistics," in *Journal of the History of Medicine and Allied Sciences,* **18** (1963), 337-348.

FRANK N. EGERTON, III

GRAVE, DMITRY ALEKSANDROVICH (*b.* Kirillov, Novgorod province, Russia, 6 September 1863; *d.* Kiev, U.S.S.R., 19 December 1939), *mathematics.*

In 1871, after the death of Grave's father, a petty official, the family moved to St. Petersburg. Grave entered the mathematics department of the Physics and Mathematics Faculty of St. Petersburg University in 1881 and studied under P. L. Chebyshev and his pupils A. N. Korkin, I. I. Zolotarev, and A. A. Markov. He began his research while still a student.

After graduating in 1885, Grave continued at St. Petersburg as a postgraduate, and in 1889 he defended his master's thesis. In the same year he started his teaching career at the university as a *Privatdozent.* In 1896 he defended his doctoral dissertation, and in 1899 he became a professor at the University of Kharkov. In 1902 Grave moved to the University of Kiev, where the rest of his work was done.

Grave's mathematical researches were originally connected with Chebyshev's school and were especially influenced by Korkin. In his master's thesis he developed methods originated by C. G. J. Jacobi and Korkin and, taking up a subject proposed by Korkin, contributed to the three-body problem. His doctoral dissertation, the subject of which also was proposed by Korkin, touched upon map projection researches by Euler, Lagrange, and Chebyshev. In it Grave presented a comprehensive study of equal-area plane projections of a sphere, with meridians and parallels being represented on the plane by straight lines and circumferences respectively.

At the beginning of his Kiev period Grave took up algebra and number theory. A brilliant speaker and organizer, he created a school which later became prominent. Among his pupils were Otto J. Schmidt, N. G. Chebotaryov, B. N. Delone, and A. M. Ostrovsky. In 1908-1914 Grave published several original and comprehensive works in algebra and number theory.

He continued his research and teaching activities well after the October Revolution, being elected to the Ukrainian Academy of Sciences (1920) and the Soviet Academy of Sciences (corresponding member from 1924, honorary member from 1929). In this period Grave's interest shifted to mechanics and applied mathematics, then returned to algebra in his last years. His last work on algebraic calculus was conceived as a comprehensive study, of which he was able to publish only two volumes.

BIBLIOGRAPHY

I. Original Works. Grave published a total of about 180 works; a comprehensive bibliography is in Dobrovolsky (see below). His main works are *Ob integrirovanii chastnykh differentsialnykh uravneny pervogo poryadka* ("On the Integration of Partial Differential Equations of the First Order"; St. Petersburg, 1889), his master's thesis; *Ob osnovnykh zadachakh matematicheskoy teorii postroenia geographicheskikh kart* ("On the Main Problems of the Mathematical Theory of Construction of Geographical Maps"; St. Petersburg, 1896), his doctoral dissertation; *Teoria konechnykh grupp* ("The Theory of Finite Groups"; Kiev, 1908); *Elementarny kurs teorii chisel* ("A Primer in Number Theory"; Kiev, 1909–1910; 2nd ed. 1913); *Arifmeticheskaya teoria algebraicheskikh velichin* ("Arithmetical Theory of Algebraic Quantities"), 2 vols. (Kiev, 1910–1912); *Entsiklopedia matematiki. Ocherk eyo sovremennogo polozhenia* ("Encyclopedia of Mathematics. An Essay on Its Current State"; Kiev, 1912); *Elementy vysshey algebry* ("Elements of Higher Algebra"; Kiev, 1914); and *Traktat po algebraicheskomu analizu* ("Treatise on Algebraic Calculus"), 2 vols. (Kiev, 1938–1939).

II. Secondary Literature. Biographies are N. G. Chebotaryov, "Akademik Dmitry Aleksandrovich Grave," in *Sbornik posvyashchenny pamyati akademika D. A. Grave* ("Collected Articles in Memory of Academician D. A. Grave"; Moscow–Leningrad, 1940), pp. 3–14; and V. A. Dobrovolsky, *Dmitry Aleksandrovich Grave* (Moscow, 1968).

The works of Grave are described in a number of general sources on the history of mathematics, such as *Istoria otechestvennoy matematiki* ("History of National Mathematics"), II (Kiev, 1967), 481–486; and A. P. Youschkevitch, *Istoria matematiki v Rossii do 1917 goda* ("History of Mathematics in Russia Until 1917"; Moscow, 1968), pp. 547–554.

A. I. Volodarsky

'sGRAVESANDE, WILLEM JACOB (*b.* 'sHertogenbosch, Netherlands, 26 September 1688; *d.* Leiden, Netherlands, 28 February 1742), *physics, mathematics, philosophy.*

'sGravesande was the earliest influential exponent of the Newtonian philosophy in continental Europe, his major work being widely read not only there but also in Britain. His family (originally known as Storm van 'sGravesande) was once important in Delft; like his brothers, he was educated at home by a tutor named Tourton, who was able to encourage his natural mathematical gifts. At Leiden University (1704–1707) he studied law, presenting a doctoral dissertation on the crime of suicide. Again like his brothers, 'sGravesande practiced law at The Hague, where he collaborated with Prosper Marchand and others in founding the *Journal littéraire de la Haye* (1713), a periodical of significance for twenty years in the history of science. He contributed several book reviews and some essays that were reprinted by J. N. S. Allamand in his *Oeuvres philosophiques et mathématiques de Mr. G. J. 'sGravesande* (Amsterdam, 1774). The most celebrated of these (in vol. **12** of the *Journal; Oeuvres,* I, 217–252) was his "Essai d'une nouvelle théorie du choc des corps fondée sur l'expérience" (1722), in which, departing from his customary attachment to the English school, 'sGravesande adopted the Huygens-Leibniz concept of *vis viva,* affirming (prop. X) that "La force d'un corps est proportionelle à sa masse multipliée par le quarré de sa vitesse." For this he was attacked by Samuel Clarke (1728), against whom he defended himself ably.

His association with the English Newtonian philosophers sprang from his appointment as secretary to the Dutch embassy (Wassenaer van Duyvenvoorde and Borsele van den Hooge) sent early in 1715 to congratulate George I on his accession to the English throne. This duty kept 'sGravesande in England for a year. His introduction to English learned society was facilitated by his acquaintance with the three sons of Gilbert Burnet, one of whom, William, proposed 'sGravesande as a fellow of the Royal Society in February 1715; he was elected on 9 June. On 24 March 1715 he was present (with other foreigners) at a demonstration of experiments by J. T. Desaguliers. There is no other mention of his name in the *Journal Book* until, on the brink of returning to The Hague in February 1716, 'sGravesande made a particular offer of his services to the Royal Society. Nevertheless, it is certain that he became acquainted with Newton and other fellows of the society, especially Desaguliers and John Keill, with whom he afterwards corresponded occasionally.

In June 1717, on the recommendation of Wassenaer van Duyvenvoorde, 'sGravesande was called to Leiden as professor of mathematics and astronomy. His inaugural lecture was on the usefulness of mathematics to all the sciences, physics above all (*Oeuvres,* II, 311–328). In 1734 he was additionally named professor of philosophy. By this time Hermann Boerhaave and 'sGravesande were established as the twin luminaries of Leiden, attracting hundreds of foreign students each year. From the outset of his teaching in both physics and astronomy 'sGravesande modeled his lectures on the example of Newton in the *Principia* and *Opticks,* although in later years they incorporated other influences, especially that of Boerhaave. Moreover, he adopted from Keill and Desaguliers the notion of demonstrating to his classes the experimental proof of scientific principles, accumulating an ever larger collection of apparatus, as may be seen from successive editions of his *Physices elementa mathematica, experimentis confirmata. Sive, introductio ad philosophiam Newtonianam* (Leiden, 1720, 1721). The scientific reputation of 'sGravesande is enshrined in this book, which he constantly corrected and amplified in later editions. An "official" English translation prepared by Desaguliers (to whom copies of the Latin original were sent in haste) was also issued in 1720 and 1721, and it passed through six editions. (The booksellers Mears and Woodward printed a rival version under the name of John Keill.) French translations appeared only in 1746 and 1747, but a critical review by L. B. Castel was published in the *Mémoires de Trévoux* in May and October 1721. The book was at once welcomed by British and a number of German scholars. 'sGravesande also published an abbreviated account for student use, *Philosophiae Newtonianae institutiones* (Leiden, 1723, 1728; and ed. Allamand 1744).

In 1721 and again in 1722 'sGravesande visited Kassel at the request of the landgrave to examine the secret perpetual-motion machine constructed by Orffyreus; he was unable to detect a fraud or (apparently) to convince himself that such a device is impossible.

In 1727 he published at Leiden, as a text for his mathematical teaching, *Matheseos universalis elementa. Quibus accedunt, specimen commentarii in Arithmeticam universalem Newtonii: ut et de determinanda forma seriei infinitae adsumtae regula nova* (*Oeuvres,* I, 89–214). This work, translated into Dutch (1728) and English (1752), is of didactic rather than original merit, but it was significant for its invitation to mathematicians to elucidate systematically Newton's *Universal Arithmetick,* which 'sGravesande exemplified by his own explanation of two passages from Newton's book. 'sGravesande found the light-hearted treatment of infinitesimals and the infinite in Bernard de Fontenelle's *Élémens de la géométrie de l'infini* (Paris, 1727) unacceptable, and he maintained his objections in the *Journal littéraire* against Fontenelle's rejoinder (1730).

After commencing the teaching of philosophy, 'sGravesande again published a textbook, *Introductio ad philosophiam, metaphysicam et logicam continens* (Leiden, 1736; repr. 1737, 1756, 1765; Venice, 1737, 1748; French ed., Leiden, 1748)—a work creating some odium for its author by its treatment of the question of necessity and free will. It was republished in *Oeuvres,* II, 1–215, together with some previously unprinted essays on metaphysics discovered by Allamand.

Apart from his own writings, 'sGravesande was active in promoting the publication at Leiden of the works of his greater countryman Christian Huygens, in *Opera varia* (1724) and *Opera reliqua* (1728), both of which he edited; in republishing the writings of his friend John Keill in 1725, as well as in editing Newton's *Arithmetica universalis* (1732); and in compiling the Dutch publication of the *Mémoires de l'Académie royale des sciences contenant les ouvrages adoptés* . . . (The Hague, 1731). Voltaire made a special journey to Leiden in 1736 to secure 'sGravesande's appraisal of his *Élémens de la philosophie de Newton* (London, 1738), writing afterward a warm appreciation of 'sGravesande's kindness and learning.

Although 'sGravesande was by no means the first semipopular exponent of Newtonian science and the experimental method (having been preceded in England by David Gregory, William Whiston, John Keill, and Desaguliers, among others), his *Mathematical Elements of Physics* was easily the most influential book of its kind, at least before 1750. It was a larger, better-argued, and more philosophical work than most of its predecessors; moreover, it leaned heavily on *Opticks* (including the queries) as well as on the *Principia.* One should therefore distinguish between 'sGravesande's roles as an exponent of Newtonian concepts (the rules of reasoning, the theory of gravitational attraction and its applications in celestial mechanics, theory of matter, theory of light, and so forth) and as an exponent of an empiricist methodology disdaining postulated hypotheses. Indeed, 'sGravesande contributed nothing to the progress of mathematical physics, for which one must look to the work of other contemporaries such as the Bernoullis, Pierre Varignon, and Alexis Clairaut. The strength of his exposition was in his perfection of the method of justifying scientific truths either by self-evidence or by appeal to experimental verification in the manner already begun by Keill and Desaguliers, perfected by him through the design of many new instruments constructed by the instrument maker Jan van Musschenbroek, brother of Pieter. (The extant instruments are preserved in the Rijksmuseum voor de Geschiedenis der Natuurwetenschappen at Leiden.) Yet, 'sGravesande's teaching and his *Elements* were by no means the sole vehicle for the introduction of British empiricism to the Continent, although probably they were the most important. He had been anticipated by Boerhaave (although Boerhaave did not employ didactic experiments) and was paralleled by Pieter van Musschenbroek at Utrecht (from 1730; he joined 'sGravesande at Leiden in 1739).

Unlike Newton, 'sGravesande commences his *Elements* with a prefatory discussion of metaphysics and epistemology directed against the Cartesians. The task of physics, he writes, is to determine the laws of nature laid down by the Creator and to unfold their regular operation throughout the universe. In thus examining the true works of God, fictitious hypotheses are to be set aside; but philosophers have differed in their methods of determining the laws of nature and the properties of bodies. "I have therefore thought fit," he continues, "to make good the Newtonian Method, which I have followed in this Work." Since the properties of bodies are not to be learned a priori, who can deny that there are in matter properties not known to us nor essential to matter, which flow from "the free Power of God"? How are the laws of nature to be sought and the three Newtonian laws of motion justified? 'sGravesande replies to these questions in a curious argument. First he asserts Newton's first rule of reasoning (Ockham's Razor). Next, distinguishing the truths of pure mathematics, which are verified by internal consistency, from those of physics ("mixed mathematics"), which depend on the senses, he argues that the latter are justified by analogy: "We must look up as true, whatever being denied would destroy civil Society, and deprive us of the means of living." This seemingly means that the consequences of induction must be true, for 'sGravesande goes on specifically to declare: "In Physics we are to discover the Laws of Nature by the Phenomena, then by induction prove them to be general Laws; all the rest is to be handled mathematically." The definitions of the scope of natural philosophy and of a law of nature (". . . the Rule and Law, according to which God resolved that certain Motions should always, that is, in all Cases, be performed") follow in chapter 1, which is concluded without further discussion by a statement of Newton's three rules of reasoning.

In volume I, 'sGravesande traverses the theory of matter (influences of the queries in *Opticks* are apparent but not marked), elementary mechanics, the five simple machines, Newton's laws of motion, gravity, central forces, hydrostatics and hydraulics, and pneumatics (including a treatment of sound and wave motion). His second volume opens with three chapters on fire, modeled on Boerhaave's ideas rather than Newton's, in whose manifestations he includes electrical phenomena. There follow two books on optics, one on the system of the world, and a final book entitled "The Physical Causes of the Celestial Motions," in which 'sGravesande can explain only that the cause of these motions is the operation of universal gravitation, whose cause is hidden "and cannot be deduced from Laws that are known." All this is treated with the aid of only trivial mathematics but is enriched with extremely numerous experimental illustrations and examples. Newton's ether does not appear, nor his "fits" of easy transmission and reflection, nor the extremely subtle physical speculation of the queries. No doubt the *Elements* owed almost as much of its success to its omissions and simplicity as to its clear and positive treatment of what it did contain. It was, obviously, very different from such later expositions as those of Henry Pemberton and Colin Maclaurin, and in many respects both more stimulating and more original.

BIBLIOGRAPHY

I. Original Works. Besides works mentioned in text, Allamand's *Oeuvres* (vol. I) include 'sGravesande's youthful *Essai de perspective* (Leiden, 1711; English trans., London, 1724) and other minor writings: *Usage de la chambre obscure; Remarques sur la construction des machines pneumatiques; Lettre à Mr. Newton sur une machine inventée par Orffyreus; Remarques touchant le mouvement perpétuel;* and *Lettres sur l'utilité des mathematiques.* The philosophical writings are in vol. II.

II. Secondary Literature. All biographies of 'sGravesande are based on the life by his friend Allamand, prefaced to the *Oeuvres.* See also Pierre Brunet, *Les physiciens hollandais et la méthode expérimentale en France au XVIIIe siècle* (Paris, 1926), *passim;* and *L'introduction des théories de Newton en France au XVIIIe siècle: avant 1738* (Paris, 1931), esp. pp. 97–107; I. Bernard Cohen, *Franklin and Newton* (Philadelphia, 1956), esp. pp. 234–243; C. A. Crommelin, *Descriptive Catalogue of the Physical Instruments of the 18th Century, Including the Collection 'sGravesande-Musschenbroek* (Leiden, 1951); P. C. Molhuysen, P. J. Blok, and K. H. Kossman, *Nieuw Nederlandsch biografisch woordenboek,* VI (Leiden, 1924), cols. 623–627; and A. Thackray, *Atoms and Powers* (Cambridge, Mass., 1970), pp. 101–104.

A. Rupert Hall

GRAY, ASA (*b.* Sauquoit, New York, 18 November 1810; *d.* Cambridge, Massachusetts, 30 January 1888), *botany.*

Gray was the son of Moses Gray and Roxana Howard Gray, who had migrated from New England to upstate New York after the American Revolution. He began his education in local schools at Sauquoit and for a time attended an academy at nearby Clinton, New York. In 1825 he entered Fairfield Academy and after a year began attending medical lectures at Fairfield's College of Physicians and Surgeons of the Western District of the State of New York. Here he came under the influence of a remarkable group of peripatetic medical teachers, including James Hadley, who introduced Gray to chemistry, mineralogy, and especially botany. He began to collect plants during his apprenticeship in Bridgewater, New York, where for a brief time after receiving his M.D. degree in 1831 he practiced medicine.

Already in touch with the leading botanist in the United States, John Torrey of New York, Gray abandoned the practice of medicine in 1832 and spent the next five years in a series of part-time teaching and library positions while increasingly concentrating on botany and making himself so useful to Torrey that he became a full collaborator on the *Flora of North America.* In 1836 he became a member of the scientific corps of the U.S. Exploring Expedition but, tired by the delays in its sailing, he resigned in 1838 to take a professorship at the newly organized University of Michigan. That position entailed Gray's spending a year in Europe, ostensibly to buy a general collection of books for the library but also to make the acquaintance of botanists and to study specimens of American plants in English and Continental herbaria. After his return in 1839, he worked on the *Flora of North America* while waiting in vain for a call from the nearly bankrupt state of Michigan. In 1842 Gray accepted the Fisher professorship of natural history at Harvard University, with the understanding that he could confine his activities to botany and the botanic garden. This milestone in the specialization of natural history made him the only adequately supported professional botanist in the United States and provided the home setting, both physically and institutionally, for the rest of his life.

Gray taught an elementary course in botany and offered a slight amount of advanced work to students of Harvard College, and later the Lawrence Scientific School, for thirty years. In 1848 he married Jane Lathrop Loring, daughter of a Boston lawyer and member of the Harvard Corporation. After 1873 he retired from teaching but continued to live in the house in the botanic garden and to develop the herbarium which ultimately became the property of Harvard. The only real breaks in the routine of his life after 1848 were a series of journeys—to Europe in 1850–1851, 1855, 1868–1869, 1880–1881, and 1887. He collected in the southern Appalachians as a young man and after the completion of the transcontinental railroad was fond of excursions to California, the trans-Mississippi West, and Mexico. Within this framework he lived an active and disciplined life until his last illness, which began in November 1887.

Gray became the leading botanical taxonomist in America in the nineteenth century, not because he was uninterested in physiology or unaware of the major advances made possible by the development of the achromatic microscope, but, rather, because the American setting demanded priority for a program of classification on a continental scale to match the programs of European nations in all parts of the globe not served by local collectors. Torrey and Gray's *Flora of North America* not only accomplished the shift from the Linnaean classification, still prevalent in America in the 1830's, to a natural system modeled on that of A. L. de Jussieu and A. P. de Candolle but also established the practice of thoroughly basing the taxonomy of American plants on the type specimens, until that time largely in the hands of European herbaria. By 1843 the *Flora* had proceeded in the Candollean system, beginning with the Ranunculaceae, through the Compositae, some seventy-six orders. At that point duties at Harvard and the flood of botanical returns from American expansion both westward and overseas in the era of Manifest Destiny made further progress by Torrey and Gray impossible.

Their response to the embarrassment of riches was twofold. Both Gray and Torrey devoted much of their time for the next thirty years to elaborating in reports the plants of collections sent in from the explorations. Some of these publications were in government documents, e.g., the *Reports* of Pacific Railroad Surveys (1855–1857) and the "everlasting" volumes of the botany of the U.S. Exploring Expedition, which were still incomplete at the time of Torrey's death in 1873. Other reports reflected Gray's sponsorship of individual collectors, who accompanied boundary surveys in the West and military expeditions, e.g., *Plantae Fendlerianae Novi-Mexicanae* (1849), *Plantae Wrightianae Texano-Neo-Mexicanae* (1852, 1853), and "Diagnostic Characters of New Species of Phaenogamous Plants, Collected in Japan by Charles Wright, Botanist of the U.S. North Pacific Exploring Expedition" (*Memoirs of the American Academy of Arts and Sciences,* **6** [1859], 377–452). In most cases Gray personally elaborated the orders through Compositae and called on others for other groups. Among the most regular and able collaborators besides Torrey were George Engelmann (especially Coniferae and Cactaceae), William Starling Sullivant (mosses), and Moses Ashley Curtis (fungi).

Gray's other strategy, forced on him by the incompleteness of the *Flora* and the competition of the textbook writer Alphonso Wood, was to modify his scholarly standards and to limit his range to the northeastern United States, thus producing for general use a manual which covered in one volume all the flowering plants and some of the lower plants as well. The *Manual of the Botany of the Northern United States* filled a need that the slower-moving *Flora* could not meet. It went through five editions in Gray's lifetime and has continued to be used in successive editions to the present time. After 1873 Gray minimized his writing of reports in order to return to the *Synoptical Flora of North America.* Much of the progress he made on it in his later years involved reworking in the light of accumulated scholarship the families treated in Torrey and Gray, so that it remained incomplete at his death. In matters of nomenclature and taxonomy Gray dominated American botany as no one before or after him; and if in the 1880's a younger generation was beginning to chafe at his authority his work still made an impressive contribution to the stream of science.

In 1851 Gray had lunch with Charles Darwin at Kew. By 1855 Gray's correspondence with Joseph Dalton Hooker on the geographical distribution of plants so impressed Darwin that he initiated an exchange of letters directly with Gray. Questions from Darwin led Gray to analyze the American flora on the basis of his *Manual* in an important paper, "Statistics of the Flora of the Northern United States" (*American Journal of Science,* **22** [1856], 204–232; **23** [1857], 62–84, 369–403). In 1857 Darwin let Gray in on the secret of the trend of his theory in a letter that became one of the bases of Darwin's priority for the idea of the origin of species by natural selection over Alfred Russel Wallace in the joint publication by Darwin and Wallace in the *Journal of the Linnean Society* in 1858.

Using Darwin's ideas and the collections of plants then coming to him from American expeditions to Japan, Gray explained species and genera of plants which appeared in eastern Asia and eastern North America not as separate creations but as descendants of a Tertiary circum-Boreal flora which had been pushed southward by the Pleistocene glaciation. This exercise in statistics led Gray to "admit that what are termed closely related species may in many cases be lineal descendants from a pristine stock, just as domestic races are." Thus he reached agreement with Darwin's main contention early in 1859, months before the publication of *Origin of Species.* Gray's announcement was the occasion for a full-dress debate with Louis Agassiz, his Harvard colleague who had imported an idealistic philosophy of natural history into the United States and gained an immense popular following.

After the publication of *Origin of Species* Gray was one of the leading reviewers on either side of the Atlantic, insisting on a fair hearing for Darwin in America and serving as agent to secure royalties on the American edition for the author. Until the Civil War distracted Gray's attention, he was through his letters a leading voice in the Darwin circle, urging with Charles Lyell an accommodating strategy in meeting religious objections. Darwin published Gray's commentary from the *Atlantic Monthly* at his own expense as a separate pamphlet under the motto "Natural Selection Not Inconsistent With Natural Theology." Eventually Darwin rejected the strategy suggested by Gray concerning theology and the assertion that natural selection had not damaged the argument from design. Yet Darwin's later years were largely spent on research that involved plants, and Gray figured prominently in his work on the coiling of tendrils and insectivorous plants. Darwin's *The Different Forms of Flowers on Plants of the Same Species* (1877) was dedicated to Gray "as a small tribute of respect and affection."

After the Civil War, Gray occasionally wrote anonymous articles attacking the religious opponents of Darwin, on the one hand, and those who followed T. H. Huxley into agnosticism, on the other. A clergyman, George Frederick Wright, eventually saw that these occasional pieces had a single author who was putting forward a consistent reconciliation of Darwinism and theism. Therefore he assisted Gray in collecting his occasional essays into a book, *Darwiniana* (1876), which firmly coupled Gray's name with the defense of Darwinism, Protestant Christianity, and the argument from design in nature. In 1881 Gray delivered a series of lectures at Yale Divinity School which were published as *Natural Science and Religion.* Their failure to cause a stir is a measure of the accommodation reached by that time between Darwinism and American Protestantism.

Without being a forceful lecturer or teacher, Gray nevertheless was a major force in scientific education in his day. His full line of textbooks shaped botanical education in the United States from the 1840's into the twentieth century. If few of his Harvard undergraduate students became professional botanists, he trained informally and assisted a whole generation of frontier collectors and part-time specialists who formed the rank and file of the botanical profession until German-trained Ph.D.'s began to appear in considerable numbers in the 1870's.

Gray was a more modest institution builder than his colleague Louis Agassiz, but in 1842 he had found Boston and Cambridge with few resources, and he left them a permanent center of botanical study in the Harvard Botanic Garden and the Gray Herbarium. In addition, when he retired from active teaching in 1873, Harvard engaged four men in his place, some of them—e.g., Charles Sprague Sargent—major institution builders themselves. While his relationship with Darwin marks the peak of Gray's career, his imprint on the pursuit of botany in the United States is also pervasive and enduring.

BIBLIOGRAPHY

I. Original Works. Gray's published writings, some 780 titles, are listed in [Sereno Watson and G. L. Goodale], "List of the Writings of Dr. Asa Gray, Chronologically Arranged, With an Index," in *American Journal of Science,* **36** (1888), app., 3–67. The major published collections of his works are *Scientific Papers of Asa Gray,* Charles S. Sargent, ed., 2 vols. (Boston, 1889); *The Letters of Asa Gray,* Jane Loring Gray, ed., 2 vols. (Boston, 1893); and *Darwiniana* (New York, 1876), A. Hunter Dupree, ed. (Cambridge, Mass., 1963).

The major MS collection is preserved in the Harvard University Herbarium, the library of which contains Gray's books, many of them annotated.

II. Secondary Literature. The recent full-scale biography is A. Hunter Dupree, *Asa Gray 1810–1888* (Cambridge, Mass., 1959), which includes notes and bibliography. Also relevant is A. D. Rodgers III, *American Botany, 1873–1892: Decades of Transition* (Princeton, 1944).

A. Hunter Dupree

GRAY, HENRY (*b.* London[?], England, 1825/1827; *d.* London, 8/13 June 1861), *anatomy, physiology.*

For a name as well-known as Gray's, extremely little is known about the man. He was one of four children of a private messenger to George IV and William IV; the family apparently had no financial problems. Essentially nothing is recorded of his preparatory education. On 6 May 1845 Gray entered St. George's Hospital as a perpetual student. He seems very early to have paid considerable attention to anatomical studies, and while still a medical student he won the Royal College of Surgeons' Triennial Prize for an essay entitled "The Origin, Connexion, and Distribution of the Nerves of the Human Eye and Its Appendages, Illustrated by Comparative Dissections of the Eye in the Other Vertebrate Animals." Part of this essay was incorporated into his later paper on the development of the retina.

Gray finished his medical studies and qualified as a member of the Royal College of Surgeons in 1849, and in June of the following year he was appointed house surgeon at St. George's Hospital for the customary twelve months.

Most of Gray's professional career was oriented around St. George's; in 1852 he was demonstrator of anatomy, and after 1853 he was lecturer in anatomy. He was also curator of the St. George's Hospital Museum. In 1852, after publication of his two major papers in the *Philosophical Transactions,* Gray was elected fellow of the Royal Society. In addition he became a fellow of the Royal College of Surgeons and was surgeon to St. James' Infirmary.

Besides his *Anatomy* Gray published several writings, the earliest of which was "On the Development of the Retina and Optic Nerve, and of the Membranous Labyrinth and Auditory Nerve," which incorporated some of the material in his prize essay. His observations were almost exclusively on the chick embryo. He clearly demonstrated that the retina develops from a protrusion of the brain, a point then still being debated. Gray also presented one of the earliest major accounts of the development of the layers of the retina. The labyrinth, he believed, develops in a mode analogous to that of the retina.

Gray's other anatomical paper was "On the Development of the Ductless Glands in the Chick," in which he dealt with the suprarenals, thyroids, and the spleen. On the basis of his observations Gray rejected much of the earlier writings on the embryological origin of each of these glands. From his work he considered it to be proved that these, with the thymus, should be classified in one group, the ductless glands—a classification by no means in general acceptance at that time, which has since developed into what are now known as the endocrine glands. He grouped these three glands on the basis of the similarity of their mode of origin, their structure in the first stages of development, and the manner in which their tissues develop throughout the fetal period. In both of these papers Gray showed a thorough familiarity with the pertinent literature and a high degree of competence in his microscopic observations.

With the support of a grant from the Royal Society, Gray continued his researches on the spleen. These studies culminated in a major treatise, *The Structure and Use of the Spleen,* which was awarded the Astley Cooper Prize in 1853 and was published in 1854. In a historical introduction Gray reviewed most of the previous writings on the spleen. His own observations included the origin of the spleen from the dorsal mesogastrium (often attributed to Johannes Müller) and early, if not initial, descriptions of the closed and open circulations, the lymphatics, and the nerves in the spleen. He also performed ligaturing and chemical experiments on the blood of the spleen.

The work for which Gray is justly most famous and which has become an institution in its own right is his *Anatomy, Descriptive and Surgical* (1858). The great success experienced by Gray's *Anatomy* was not due to lack of competition; there were successful works by Jones Quain, W. J. E. Wilson, Xavier Bichat, J. G. Cloquet, and others in use and readily available in England. Gray was described by a contemporary as a "lucid teacher of anatomy," a lucidity which carried over into his *Anatomy* not only in the logical arrangement of the material but also in the 363 new illustrations. The latter were done from drawings by Henry Vandyke Carter, who executed his drawings from fresh dissections that Gray and he performed. The arrangement of the material and the close relation between the text and the illustrations were Gray's work and show his clear understanding of the fundamentals of his subject. The literary style apparently was greatly polished by the assistance of Timothy Holmes, who also was editor of the third (1863) through ninth (1880) editions of the *Anatomy.* A major innovation, which greatly aided the success of the book was the introduction of remarks on surgical anatomy into an English textbook of anatomy. The reviews of the *Anatomy* often commented on Gray's ability to present, to students and practitioners alike, the practical information which they needed in an accessible form. This accessibility has been one of the great factors in the *Anatomy*'s success and has influenced other writers of anatomy textbooks.

Gray also wrote papers in pathology and is reputed to have made good progress on a major treatise on tumors at the time of his death. He died in June 1861 from smallpox contracted while tending a nephew.

BIBLIOGRAPHY

I. Original Works. Gray's two embryological papers are "On the Development of the Retina and Optic Nerve, and of the Membranous Labyrinth and Auditory Nerve," in *Philosophical Transactions of the Royal Society,* **140** (1850), 189–200; and "On the Development of the Ductless Glands in the Chick," *ibid.,* **142** (1852), 295–309. His books are *The Structure and Use of the Spleen* (London, 1854) and *Anatomy, Descriptive and Surgical* (London, 1858; Philadelphia, 1859). All the American eds. were closely based upon, but were not reprints of, the English eds. Goss (see below) lists both the English and the American series of eds., with their respective editors, through 1959.

II. Secondary Literature. There is a brief obituary in *Proceedings of the Royal Society,* **12** (1862–1863), xi. Two closely related articles which are the basis of much that has been written since are "Henry Gray," in *St George's Hospital Gazette,* **16** (1908), 49–54; and F. K. Boland, "Henry Gray, Anatomist: An Appreciation" in *American Journal of the Medical Sciences,* **136** (1908), 429–435. See also F. N. L. Poynter, "Gray's Anatomy, the First Hundred Years," in *British Medical Journal* (1958), **2,** 610–611; and Charles Mayo Goss, *A Brief Account of Henry Gray F. R. S. and His Anatomy, Descriptive and Surgical During a Century of Its Publication in America* (Philadelphia, 1959).

Wesley C. Williams

GRAY, STEPHEN (*b.* Canterbury, England, 1666; *d.* London, England, 15 February 1736), *electricity.*

The exact date of Gray's birth is uncertain, but records indicate that he was baptized on 26 December 1666. He came from a family of rapidly rising artisans; his grandfather was a carpenter, his father a dyer, his brothers a dyer, a carpenter, and a grocer, and his nephew a gentleman, a Cambridge graduate, and a doctor of medicine. The family understood the value of education. Although Gray followed his father's trade, he learned enough Latin to puzzle out Christoph Scheiner's interminable *Rosa ursina* (1630) when his omnivorous interests led him to sunspots. In science he was perhaps an autodidact, but his letters hint at a period of study in London or perhaps in Greenwich, under his "honoured Friend," the astronomer royal John Flamsteed. A stay in the metropolis would explain much: Gray's command of optics and astronomy; his loyalty to the much older Flamsteed, like himself the son of a tradesman; and his acquaintance with Henry Hunt of Gresham College, a minor functionary of the Royal Society of London. Hunt proved a valuable connection, supplying Gray with the *Philosophical Transactions* and transmitting to their editor the communications they called forth from Canterbury.

Gray's first published paper (1696) describes a microscope made of a water droplet inserted in a tiny hole in a brass plate. The globule "prodigiously magnified" animalcules swimming in it, a property both gratifying and perplexing, since, as Gray noticed, the standard optical theory required rays from bugs so positioned to diverge after refraction. His solution: the rays, if first reflected internally from the back wall of the globule, can be bent by its front surface into a parallel bundle. Although of little consequence itself, this first effort displays the characteristics which would bring Gray, when past sixty, to his capital discoveries in electricity; experiments "for the most part Naturall, being ushered in with very little assistance of Art"; alertness to effects unanticipated by theory; and cautious explanations of anomalies. Encouraged by the Royal Society's reception of his microscope, Gray communicated other ideas for instruments and reports of rarities like mock suns, magnetic sands, and the remains of antediluvian creatures (1699–1701). Then, from 1703 to about 1716, he devoted his scientific energies ("the far Greatest Part of my time that the avocations for a Subsistence would Permitt me") to accurate, quantitative observations of eclipses, sunspots, and (in the hope of improving navigation) the revolutions of Jupiter's satellites.

By his fortieth year Gray was widely known as a

careful and responsible observer. The scrupulous Flamsteed incorporated his results; William Derham solicited his help in experiments on the speed of sound: and the new Plumian professor of natural philosophy, Newton's protégé Roger Cotes, invited him to Cambridge to assist in establishing a new observatory. Against Flamsteed's advice Gray accepted and spent some months in 1707 and 1708 in Cambridge; but the observatory did not materialize, and Gray, who found his employers unexpectedly "mercenary" and their plan to redo Flamsteed's determinations of stellar positions both ignorant and insulting, returned to Canterbury. A bad back had made his trade too strenuous, however, and in 1711 he appealed to Hans Sloane, the secretary of the Royal Society, to intercede for him with the governors of Sutton's Hospital (the London Charterhouse). In June 1719, on the nomination of the prince of Wales, Gray became one of the Charterhouse's eighty "gentleman pensioners," having meanwhile, it is said, assisted in the public lectures of J. T. Desaguliers. The recent conjecture that Newton somehow delayed Gray's entry into the Charterhouse out of hatred for Flamsteed (and hence for his disciple) is baseless; the Newtonians tried to help, and even offered Hunt's old position to Gray, but "the poor man is so very bashful [wrote Sloane's successor, Brook Taylor, in 1713] that I can by no means prevail upon him to think of the business."

On 13 November 1706, Francis Hauksbee (the elder), demonstrator to the Royal Society, appeared before his "Philosophical Masters" armed with a tube of flint glass, with which to try the force of electricity. The tube appreciably outdid the customary generator, amber, and brought Hauksbee close to identifying the cause of electricity with that of the glow producible by chafing glass vessels. The "Strongness of the phenomena together with the facility of operation" of the portentous tube intrigued Gray, and during his stay in Cambridge he amused his nephew and others with experiments designed to map the course of the "Luminous and Electric Effluvium." He rediscovered Otto von Guericke's now famous demonstration that a feather, once drawn to the tube, might be made to hover above it; he found that light was "inherent in the Effluvia" of other electrics; and he conjectured that electrical motions and glows arose from a double stream of fine particles, one shot from the electric and the other an answer from the environment. This scheme, which effluvializes the incoming air current of Niccolo Cabeo's theory and anticipates the afflux of Nollet's system, was to guide Gray to the discovery of conduction.

Sloane did not print Gray's report of the Cantabrigian experiments, chiefly because Hauksbee, to whom it was referred, appropriated its novelties: the hovering feather, the luminous effluvia of wax and sulfur, and, in the form of a revived Cabean theory, the dual currents. Gray did not again bother the Royal Society with electricity until 1720, when he announced the discovery of a new class of nonrigid electrics, including hair, silk, feathers, and—if you please—gilded ox guts. Thereafter again silence, until February 1729, when, having conceded defeat in an attempt to electrify metals by friction, Gray thought to awaken their virtue by exposing them to effluvia from the tube: guided by his earlier theory, he imagined that just as the tube "communicated a Light to [bodies]," it might "at the same Time communicate an Electricity." He took a tube corked at both ends to keep out the dust, a precaution suggested by some old experiments of Hauksbee's. Thinking the corks might alter the tube's power, Gray brought a feather up to its far end, and in great amazement saw the fickle plume go to the cork, not the glass. "I then held the Feather over against the flat End of the Cork, which attracted and repelled many Times together; at which I was much surprized, and concluded that there was certainly an attractive Vertue communicated to the Cork by the excited Tube." It is a classic example of chance favoring the prepared mind.

Gray exploited his discovery by running sticks or threads from the cork to a "receiving Body," a teakettle, for example, or an ivory ball, which he supposed to emit its own effluvia, stimulated by those of the glass passed down the transmitting line. He extended the line to fifty-two feet, the greatest free drop available to him; he had also tried horizontal transmission, through thick threads hung up by pieces of the same material, but without the least success. At this point he visited his young friend Granville Wheler, F. R. S., a wealthy, able scientific amateur with a large country house admirably suited to the new experiments. Wheler suggested that they might send electricity down horizontal lines hung from silk threads narrow enough to prevent the loss of the "Vertue." They managed to transmit for some hundreds of feet before the silk parted under the strain. Quite naturally they replaced it with brass wire of similar bore, which most unexpectedly declined to behave like silk. And so, by attempting to increase the mere quantity of the effect, that is, the distance of transmission, they stumbled upon the fundamental qualitative distinction between insulators and conductors.

For thirty months Gray, Wheler, and another friend, a cousin of Flamsteed's named John Godfrey, enjoyed a monopoly in the study of communicated electricity. They followed two lines of inquiry: (1) the identification of substances which might serve as supporters (insulators) or as receivers and (2) the mapping of the course of transmitted electricity. These results, which conflated induction, conduction, and the mechanism of attraction and repulsion, required too much of the effluvia; serious contradictions, which Gray's group did not recognize, gradually came to light and, by adding their weight to the perplexities raised by the Leyden jar, assisted in forcing the rejection of the effluvial picture.

The publication of the results of Gray's group in 1732 awakened the interest of C. F. Dufay, whose extension and regularization of the phenomena attracted the attention of the learned to the study of electricity. Gray contributed a few further observations, particularly on the appearance of sparks drawn from bodies of different shapes and on the longevity of the electrification of objects encased in tight, dry boxes. These characteristic investigations gave way, in his last days, to a grand cosmic speculation based upon the discovery that a freely suspended conical pendulum would revolve about an electrified body precisely as the planets circle the sun. Cromwell Mortimer, secretary of the Royal Society (which had belatedly admitted Gray in 1732), got wind of the matter and hurried off to the Charterhouse. He found Gray dying. "He hoped [Mortimer reported], if God would spare his Life but a little longer, . . . to be able to astonish the World with a new Sort of *Planetarium* . . . [and] a certain Theory for accounting for the Motions of the Grand *Planetarium* of the Universe." Alas, the pendulum aped the planets only when supported from the hand, driven (as Dufay and Wheler showed) by motions largely involuntary. The Royal Society did Gray no favor by publishing, as his last paper, conjectures so out of keeping with his wonted care and sobriety.

BIBLIOGRAPHY

I. ORIGINAL WORKS. Gray's most important paper is "A Letter . . . Containing Several Experiments Concerning Electricity," in *Philosophical Transactions of the Royal Society,* **37** (1731–1732), 18–44. A bibliography of his published work, drawn up by R. A. Chipman, appears as an appendix to I. B. Cohen, "Neglected Sources for the Life of Stephen Gray," in *Isis,* **45** (1954), 41–50; to it should be added Gray's observations of the solar eclipse of 13 Sept. 1699, published by William Derham in *Philosophical Experiments and Observations of the Late Eminent Dr Robert Hooke . . . and Other Eminent Virtuosos* (London, 1726), p. 343.

A number of Gray's unpublished letters are preserved at the Royal Society of London, the British Museum, and the Royal Observatory (Herstmonceux). R. A. Chipman, "The Manuscript Letters of Stephen Gray," in *Isis,* **49** (1958), 414–433, provides a list which omits letters of 3 Feb. 1696 and 22 May 1696 to Henry Hunt (Royal Society, Guard Book G. 1, fols. 49–50) and an undated note on sunspots (British Museum, Sloane 4039, fol. 332). Chipman, "An Unpublished Letter of Stephen Gray on Electrical Experiments," in *Isis,* **45** (1954), 33–40, prints the letter appropriated by Hauksbee.

II. SECONDARY LITERATURE. Data concerning Gray's family and date of baptism were supplied by the Archivist, Canterbury Cathedral. For published biographical information see J. M. Cowper, *The Roll of the Freemen of the City of Canterbury* (Canterbury, 1903), p. 39; W. P. Courtney, "Stephen Gray, F. R. S.," in *Notes and Queries,* **6** (1906), 161–163, 354; F. Higenbottam, "The Apparition of Mrs. Veal to Mrs. Bargrave at Canterbury, 8th Sept., 1705," in *Archaeologia cantiana,* **78** (1959), 154–166; and the papers of Chipman and Cohen cited above. Estimates of Gray's work will be found in Chipman's papers; J. Priestley, *The History and Present State of Electricity,* 3rd ed. (London, 1775), I, 32–53; E. Hoppe, *Geschichte der Elektrizität* (Leipzig, 1884), pp. 8–11; I. B. Cohen, *Franklin and Newton* (Philadelphia, 1956), pp. 368–371; and, by indirection, in F. Baily, *An Account of the Revd John Flamsteed* (London, 1835), pp. 47, 310.

The important piece of Taylor's regarding Gray's succession to Hunt is in Royal Society Correspondence 82, fol. 5; Hauksbee's role in suppressing Gray's first paper on electricity appears from the Royal Society's *Journal Book,* X, fols. 175, 189–190, 192, and from his published work, for which see R. Home, "Francis Hauksbee's Theory of Electricity," in *Archives for History of Exact Science,* **4** (1967), 203–217.

JOHN L. HEILBRON

GREEN, GEORGE (*b.* Nottingham, England, 1793; *d.* Sneinton, Nottingham, 31 May 1841), *mathematics, natural philosophy.*

For a detailed study of his life and work, see Supplement.

GREEN, JACOB (*b.* Philadelphia, Pennsylvania, 26 July 1790; *d.* Philadelphia, 1 February 1841), *chemistry, biology, botany, dissemination of knowledge.*

Jacob Green was the son of Ashbel Green, prominent Presbyterian clergyman and eighth president of Princeton University. He attended the University of Pennsylvania (B.A., 1807), studied medicine briefly, sold books in Albany, New York, for a few years, took up law, and was admitted to the New York bar.

In 1816 Green moved to Princeton, New Jersey, to live with his father and study theology. He was sidetracked to science when the professor of natural philosophy, Henry Vethake, hired him as an assistant. In 1818 he was elected to a newly created professorship of chemistry, experimental philosophy, and natural history, and held this professorship until it was abolished in 1822.

Green then moved to Philadelphia, delivered a course of public lectures on chemistry, and joined several physicians in founding Jefferson Medical College. He was professor of chemistry at the medical school from 1825 until 1841. During several summers he traveled to Jefferson College (now Washington and Jefferson College) and to Lafayette College, Easton, Pennsylvania, to teach chemistry.

In 1828 Green visited Europe and later published his impressions in *Notes of a Traveller.* He met Michael Faraday, John Dalton, and other scientists but, to his disappointment, not Humphry Davy, who was abroad. In 1830 he published an edition of Davy's *Consolations in Travel.*

Green moved from one science to another, depending upon the circumstances of his life. At the age of nineteen he and his friend Erskine Hazard published *An Epitome of Electricity & Galvanism.* Learning botany largely by independent study, he wrote about plants of New York. Inspired by the skies during evening strolls, he wrote a popular book, *Astronomical Recreations.* As a teacher of chemistry he published three texts. In these activities Green was chiefly a disseminator of science. He advanced science through studies on shells, salamanders, and trilobites, contributing in a small way to the early knowledge of natural history of the United States.

Late in his life Green married and was the father of two children.

BIBLIOGRAPHY

I. Original Works. Green's works are *An Epitome of Electricity & Galvanism. By Two Gentlemen of Philadelphia* (Philadelphia, 1809), written with Erskine Hazard; *A Catalogue of the Plants Indigenous to the State of New York* (Albany, 1814); *Astronomical Recreations* (Philadelphia, 1824); *Electro-Magnetism* (Philadelphia, 1827); *Text-Book of Chemical Philosophy* (Philadelphia, 1829); *Notes of a Traveller, During a Tour Through England, France, and Switzerland, in 1828,* 3 vols. (New York, 1830); *Consolations in Travel, or the Last Days of a Philosopher. By Sir Humphry Davy . . . With a Sketch of the Author's Life, and Notes, by Jacob Green* (Philadelphia, 1830); *A Monograph on the Trilobites of North America* (Philadelphia, 1832; supp., Philadelphia, 1835); *Syllabus of a Course in Chemistry* (Philadelphia, 1835); and *Chemical Diagrams* (Philadelphia, 1837).

II. Secondary Literature. The primary biography is by his father, Ashbel Green; it may be found in James F. Gayley, ed., *History of the Jefferson Medical College of Philadelphia* (Philadelphia, 1858), pp. 31–34, with portrait. More recent accounts are Edgar F. Smith, *Jacob Green, 1790–1841, Chemist* (Philadelphia, 1923), with portrait, abridged in *Journal of Chemical Education,* **20** (1943), 418–427; and George W. Bennett, "Old Jakey Green at Canonsburg," in *Proceedings of the Pennsylvania Academy of Science,* **23** (1949), 218–221.

Wyndham Davies Miles

GREENOUGH, GEORGE BELLAS (*b.* London, England, 1778; *d.* Naples, Italy, 2 April 1855), *geology.*

Greenough's father, George Bellas, a lawyer, married Sarah, daughter of Thomas Greenough, a surgeon, in 1776. Both parents died when he was a child, and George was brought up by his maternal grandfather, whose surname he assumed in 1795. He entered Eton College in 1789 and Pembroke College, Cambridge, in 1794. He spent three years at Cambridge but did not graduate. Later he went to Göttingen University to study law, and to improve his German he attended the natural history lectures of J. F. Blumenbach. As a result he developed an enthusiasm for science, particularly geology, which lasted all his life. Considerable inherited wealth freed him from the necessity of following a profession.

After returning to England, the controversy between the Huttonians and the Wernerians aroused Greenough's interest. In 1805 he made a two-month tour of Scotland, closely examining the field evidence afforded by basalts and granites. Published extracts from his journal show that he did not fully accept the views of either neptunists or plutonists. A published statement that Greenough had studied under Werner at Freiberg is incorrect, but it is known that on some occasion he met Werner and recorded (in manuscript) an unfavorable impression of him.

In 1807 Greenough was elected both fellow of the Royal Society and a member of Parliament. Soon after, he joined with a dozen other enthusiasts in founding the Geological Society of London. As their first president he upheld the independence of the group and resisted the attempts of Sir Joseph Banks to bring it under the control of the Royal Society. He remained president until 1813 and was elected to further terms in 1818 and 1833. His presidential addresses for 1833 and 1834 were mainly summaries of recent advances in geological science.

Greenough was actively interested in the construction of geological maps, and in 1812 he presented to the Geological Society "nine maps of England with the principal strata sketched in." Its council then requested him to prepare a geological map of England and Wales on a larger scale, which task he undertook with the assistance of other members, although aware that William Smith was already preparing such a map for publication. The completed map was published on 1 May 1820 (see bibliography), five years after Smith's map. Greenough's map was on a slightly smaller scale than Smith's and had much more topographic detail, with hachuring to indicate valleys and escarpments. There was also more geological detail, and the Cretaceous and Upper Jurassic rocks in particular were more accurately delineated. More outcrops of granite and trap were shown, and an interesting feature was the attempt to show the distribution of diluvium, or drift. In the memoir that accompanied the map Greenough claimed that he had not seen Smith's map until it was published and had made very limited use of it. Nonetheless, it seems certain that Greenough did see and use manuscript maps based on Smith's work that were lent to him by John Farey. A revised edition of the map was issued by the Geological Society in 1840 and a third edition in 1865, ten years after Greenough's death. Only in the latter was belated acknowledgment made to William Smith.

In the memoir accompanying the maps, there is very little geology, but Greenough gave many of his notes to W. D. Conybeare and William Phillips, who made frequent use of them in their *Outlines of the Geology of England and Wales* (London, 1822), identifying them as "G. Notes" or simply "G." Greenough also spent many years compiling a geological map of India, which was published in 1854.

In 1819 he published *A Critical Examination of the First Principles of Geology; in a Series of Essays.* These essays were a challenge to those who saw uniformity and regularity in the strata, the skeptical Greenough quoting exceptions for every such assumption. "Before we yield or refuse assent to any proposition," he wrote, "we must sum up probabilities and improbabilities on both sides and strike a balance." But his balance was struck in such a manner that his own views or conclusions were seldom stated, except for his firm belief in the Deluge as the prime cause of valley excavation and a general agreement with the catastrophic theories of Cuvier. This publication, although not well received at the time, was based on wide reading and observation and can be used today as a source book.

When the Royal Geographical Society was founded in 1830, Greenough was elected to its council, and he served as president from 1839 to 1841. His continued interest in geology and geography led him to set off for the East at the age of seventy-six, but he was taken ill and died en route at Naples. In his will he left his extensive library to be divided between the Geological Society of London and the Royal Geographical Society. His collection of rocks and fossils went to University College, London, whose library also possesses a large collection of his manuscripts, including letters, journals, and memoranda.

BIBLIOGRAPHY

I. Original Works. Greenough's published works include *A Critical Examination of the First Principles of Geology; in a Series of Essays* (London, 1819); *Memoir of a Geological Map of England* (London, 1820), accompanying *A Geological Map of England & Wales* (London, 1819)—the map, although dated 1819, was not published until 1820; presidential addresses in *Proceedings of the Geological Society of London,* **2** (1838), 42–70, 145–175; and *General Sketch of the Physical and Geological Features of British India* (London, 1854), a map.

II. Secondary Literature. Obituary notices are in *Quarterly Journal of the Geological Society of London,* **12** (1856), xxvi–xxxiv; and *Journal of the Royal Geographical Society,* **25** (1855), lxxxviii–xc. Passing references are in H. B. Woodward, *The History of the Geological Society of London* (London, 1907); and H. R. Mill, *The Record of the Royal Geographical Society* (London, 1930). Two informative articles by M. J. S. Rudwick, based on Greenough's MSS, are "Hutton and Werner Compared: George Greenough's Geological Tour of Scotland in 1805," in *British Journal for the History of Science,* **1** (1963), 117–135; and "The Foundation of the Geological Society of London: Its Scheme for Co-operative Research and Its Struggle for Independence," *ibid.,* 325–355.

V. A. Eyles

GREENWOOD, ISAAC (*b.* Boston, Massachusetts, 11 May 1702; *d.* South Carolina, 12 October 1745), *natural philosophy, education.*

After graduating from Harvard College, studying science and perhaps medicine in England, and serving on occasional pulpits, Greenwood was installed at Harvard in 1727 as the first Hollis professor of mathematics and natural and experimental philosophy. His contribution to science in America lay in strengthening and modernizing the science program at Harvard College. He had, in 1726, offered the public an experimental course in mechanical philosophy, for which he published a prospectus. To his new professorship he brought this experience, the gift for teaching, a good knowledge of science, keen powers of observation, and an excellent collection of apparatus contributed by Thomas Hollis.

Greenwood's course made use of experiment and demonstration and probably rested heavily upon Newton's work, as did his public lectures. He took an important step to improve the level of Harvard preparation in mathematics in 1729, when he published *Arithmetick, Vulgar and Decimal,* a good textbook. He produced a manuscript text on algebra and seems to have taught Newtonian fluxions. His success was substantial but, after several attempts to reform him, the Harvard Corporation dismissed him in 1738 for excessive drinking. Thereafter, Greenwood sought to set up a private school of experimental philosophy in Boston, delivered lectures in Philadelphia, and went to sea as a tutor. At his death his career was in ruins, but his service to science was best measured by the tradition he developed at Harvard and by the students he inspired, including Professor John Winthrop.

During his professorship Greenwood published three papers in the *Philosophical Transactions of the Royal Society:* one urging the charting of winds, another a study of the effects of damps in wells, and the third a description of an aurora borealis. In response to an English request he made very careful drawings of the inscriptions on Dighton Rock, now understood to have been of Indian origin. Precision and care marked all of this work, but it was less significant than his teaching.

BIBLIOGRAPHY

I. ORIGINAL WORKS. Greenwood published a number of items related to his teaching: *An Experimental Course of Mechanical Philosophy* (Boston, 1726); *Course of Philosophical Lectures* (n.d.); *Arithmetick, Vulgar and Decimal: With the Application Thereof to a Variety of Cases in Trade and Commerce* (Boston, [1729]); and *Prospectus of Explanatory Lectures on the Orrery* (Boston, 1734). He also wrote *A Philosophical Discourse Concerning the Mutability and Changes of the Material World* (Boston, 1731) and three papers which appeared in the *Philosophical Transactions of the Royal Society:* "A New Method for Composing a Natural History of Meteors," **35** (1730–1731), 390–402; "A Brief Account of Some of the Effects and Properties of Damps," **36** (1731–1732), 184–191; and "Of an Aurora Borealis Seen in New-England, Oct. 22, 1730," **37** (1732–1733), 55–69. There are Greenwood MSS in the Harvard University archives and in the Massachusetts Historical Society.

II. SECONDARY LITERATURE. The only good account of Greenwood is Clifford K. Shipton, *Biographical Sketches of Those Who Attended Harvard College; Sibley's Harvard Graduates,* VI (Boston, 1942), 471–482, which includes Frederick G. Kilgour, "Isaac Greenwood and American Science," and the best bibliography.

BROOKE HINDLE

GREGORY, DAVID (*b.* Aberdeen, Scotland, 3 June 1659; *d.* Maidenhead, Berkshire, England, 10 October 1708), *mathematics, astronomy, optics.*

The eldest surviving son of the laird (also called David) of Kinnairdie in Banffshire, and nephew of James Gregory, David graduated from Marischal College, Aberdeen, and went on to Edinburgh University, where in October 1683—a month before taking his M.A.—he was elected to the chair of mathematics, vacant since his uncle's death in 1675, delivering an inaugural lecture "De analyseos geometricae progressu et incrementis." Staunchly supported by Archibald Pitcairne, an old friend from undergraduate days, he sought conscientiously in his professorial lectures (on elementary optics, astronomy, and mechanics) to impart to his students basic insights into the "new" science of Descartes, John Wallis and, after 1687 (if we are to believe William Whiston) Isaac Newton. Attempts by Gregory in 1684 and 1687 to start a correspondence with Newton failed, but an indirect link with Cambridge was formed in 1685 after a visit to Newton by a mutual acquaintance, John Craig(e).

Increasingly under attack by his fellow professors at Edinburgh for his radical views, Gregory jeopardized his position in 1690 by refusing to swear the required oath of loyalty to the English throne before a visiting parliamentary commission. The retirement of Edward Bernard from the Savilian professorship of astronomy at Oxford in 1691 offered an outlet. Backed by Newton's recommendation of him as "very well skilled in Analysis & Geometry both new & old. . . . understands Astronomy very well . . . & is respected the greatest Mathematician in Scotland," and with Flamsteed's support, Gregory was elected to the chair in face of strong opposition from Edmond Halley (later, after Wallis' death in 1703, to become his companion professor of geometry). In November 1692 he was elected fellow of the Royal Society, but he never took an active part in its affairs except for submitting several papers to its *Transactions.*

During his early years at Oxford, Gregory traveled widely to keep abreast of current developments in science, visiting Johann Hudde and Christian Huygens in Holland in May–June 1693 and Newton at Cambridge in May 1694 and on numerous later occasions in London. His extant Savilian lectures (from 1692) are for the most part a rehash of his Edinburgh *lectiones,* suitably updated; as he told Samuel Pepys, he was concerned to see that his students "should study some Euclid, trigonometry, mechanics, catoptrics and dioptrics, . . . the theory of planets and navigation." His appointment in 1699 as mathematical tutor to the young duke of Gloucester was thwarted by the latter's sudden death; his relations with Flamsteed, a competitor for the post, thereafter rapidly deteriorated, particularly after he joined Newton's committee set up to publish Flamsteed's *Historia coelestis.* Gregory's election in 1705 to the Royal College of Physicians at Edinburgh was purely honorary, but he took a more active role in the Act of Union between England and Scotland in 1707. He married in 1695 and was en route to London to visit his children, sick with smallpox, when he died.

No definitive assessment of Gregory's scientific achievement is possible until a detailed examination of his extant memoranda is made. Doubtless this will reinforce the impression gained from his printed work that a modicum of talent, effectively lacking originality, was stretched a long way. His earliest publication, *Exercitatio geometrica de dimensione figurarum* (1684), was a presentation of a number of manuscript *adversaria* bequeathed to him by his uncle James, interlarded with worked examples from René-François de Sluse's *Miscellanea,* Nicolaus Mercator's *Logarithmotechnia,* and James Gregory's *Geometriae pars universalis* and *Exercitationes* (all 1668) and a citation of Newton's series for the general circle zone communicated to John Collins in 1670 and passed forthwith to Scotland. Ignorant of the general binomial theorem which had been found independently by Newton and his uncle James, Gregory resorted to a brute-force development of the series expansion of the binomial square root by which he accomplished the "dimension" (quadrature and rectification) of various conics, conchoids, the cissoid, the Slusian pearl, and other algebraic curves, while the subtleties of his uncle's use of a Taylor expansion to invert Kepler's equation as an infinite series (first published here, but without any proof) clearly passed him by.

Gregory's *Treatise of Practical Geometry* and *Catoptricae et dioptricae sphaericae elementa* (1695) are printed versions of elementary lectures given at Edinburgh in the 1680's; the latter is often singled out for its appended remark (p. 98) suggesting, on the analogy of the crystalline and vitreous humours "in the Fabrick of the Eye," that an achromatic compound lens might be formed by combining simple lenses of different media, but this insight he might well have had from Newton. His thick folio text on foundations of astronomy, *Astronomiae . . . elementa* (1702), is a well-documented but unimaginative attempt to graft the gravitational synthesis propounded in the first book and especially the third book of Newton's *Principia* onto the findings of traditional astronomy. While respected as a source book it is now chiefly remembered for the remarks by Newton on the *prisca sapientia* of the ancients and their "knowledge" of the inverse-square law of universal gravitation and for the Latin version of Newton's short paper on lunar theory which it reproduces.

Gregory's first collected edition, following Bernard's wish, of *Euclidis quae supersunt omnia* (1703) is a competent gathering of the mathematical and physical writings attributed to Euclid of Alexandria (*Elements, Data, Introductio harmonica, Sectio canonis, Phaenomena, Optica, Catoptrica, Dioptrica, Divisions of figures, De levi et ponderoso*), but the one exciting passage in the preface (on the *Data,* especially 86) again stems from Newton. Of Gregory's articles in the *Philosophical Transactions of the Royal Society* that (1693) on Vincenzo Viviani's "testudo veliformis quadrabilis" is an elegant solution of a tricky but essentially elementary problem; that on the catenary (1697) erroneously derives the correct differential equation of the freely hanging uniform chain (he failed to see the necessity of compounding the tensions at both ends of the curve) and therefrom draws its logarithmic construction and main properties; that (1704) on the Cassini oval or cassinoid briefly sketches its main forms, determining, since it is not convex when its eccentricity is greater than $1/\sqrt{3}$, its inacceptability as a planetary orbit. The poverty of Gregory's astronomical observations merits Flamsteed's jibe of "closet astronomer."

In retrospect, Gregory's true role in the development of seventeenth-century science is not that of original innovator but that of custodian of certain precious papers and verbal communications passed to him by his uncle James and, as privileged information, by Newton.

BIBLIOGRAPHY

I. ORIGINAL WORKS. The brief "Index Chartarum," now in Edinburgh University Library, made by Gregory's son David after his father's death, outlines the content of some 400 MSS and memoranda on mathematical, physical, and astronomical topics gathered in four "M.S." (A–D), of which D is "plerumque Jacobi Gregorii." Those (the greater part) still extant are now scattered in the libraries of Edinburgh and St. Andrews universities and the Royal Society, London. Further memoranda are interleaved in "M.S." E (now Christ Church, Oxford, MS 346), essentially a journal of Gregory's scientific activities at Oxford between March 1696 and September 1708. No concordance to these papers is published, but I have in my possession a rough list of the location of the mathematical items made *ca.* 1950 by H. W. Turnbull. Selected extracts, only a small fraction of the total, are reproduced in W. G. Hiscock, *David Gregory, Isaac Newton and Their Circle* (Oxford, 1937) and in Turnbull's ed. of *The Correspondence of Isaac Newton,* III–IV (Cambridge, 1961–1967).

The MS (A57, Edinburgh) of Gregory's first published work, *Exercitatio geometrica de dimensione figurarum sive specimen methodi generalis dimetiendi quasvis figuras* (Edinburgh, 1684)—reviewed by Wallis in *Philosophical Transactions of the Royal Society,* **14,** no. 163 (20 Sept. 1684), 730–732—contains few variants. His "Lectiones opticae ad Acad. Edinburg. 1683" (B11, Edinburgh DC.1.75) remain unprinted, as does his "Geometria de motu: par[te]s [1-5] lect. ad Acad. Edinburg. [1684-1687]" (B12, B15, B16, Edinburgh DC.1.75: incomplete autographs are in the Royal Society and Christ Church; a complete contemporary copy is in Aberdeen University [MS 2171]) except for an Englished fragment "never printed till now" inserted by John Eames and John Martyn in their *Philosophical Transactions Abridged,* VI (London, 1734), 275-276.

Gregory's "Institutionum astronomicarum libri 1 et 2 in usum Academicorum Edinburgensium scripti 1685" (B7, Edinburgh) was later absorbed into his *Astronomia;* the parallel "Geometria practica . . . conscripta 1685" (B6, Edinburgh DC.1.75/DC.5.57; contemporary copy in Aberdeen MS 2171) was subsequently rendered into English (Aberdeen MS 672) by an unknown student and later published by Colin Maclaurin as *A Treatise of Practical Geometry . . . Translated from the Latin With Additions* (Edinburgh, 1745; 9th ed. 1780). His astronomical and medical lectures at Oxford during 1692 to 1697 are preserved in Aberdeen (MS 2206/8). His *Catoptricae et dioptricae sphaericae elementa* (B18) was published by him at Oxford in 1695 (2nd ed., Edinburgh, 1713); with addenda by William Brown it appeared in English as *Dr. Gregory's Elements of Catoptrics and Dioptrics* (London, 1715; enl. ed. by J. T. Desaguliers, London, 1735). The 1694 calculus compendium "Isaaci Newtoni methodus fluxionum ubi calculus differentialis Leibnitij et methodus tangentium Barrovij explicantur et exemplis plurimis omnis generis illustrantur"—variant autographs in St. Andrews (QA33G8D12) and Christ Church; contemporary copies by John Keill in the University Library, Cambridge, Lucasian Papers, and by William Jones, Shirburn 180.H.33—is unprinted.

Gregory's "Notae in Isaaci Newtoni *Principia philosophiae* . . . in anno 1693 conscripta"—original in the Royal Society, amanuensis copy in Christ Church; contemporary transcripts in Edinburgh and Aberdeen (MS GY)—was proposed for publication at Cambridge in 1714, but Nicholas Saunderson could find "nobody that can give me any account of it" (to Jones, February 1714); see S. P. Rigaud, *Correspondence of Scientific Men of the Seventeenth Century,* I (Oxford, 1841),* 264. His weighty *Astronomiae physicae & geometricae elementa* (Oxford, 1702; 2nd ed., Geneva, 1726) was "done into English" as *The Elements of Physical and Geometrical Astronomy* (London, 1715; 2nd ed., 1726); influential reviews appeared in *Philosophical Transactions of the Royal Society,* **23,** no. 283 (Jan.-Feb. 1703), 1312-1320; and *Acta eruditorum* (Oct. 1703), 452-462. Gregory's Latin (pp. 332-336) of Newton's "Theory of the Moon" (Cambridge, Add. 3966.10,82r-83v, published in *Correspondence,* IV [1967], 327-329; Gregory's copy [$C121_2$] is now in the Royal Society) appeared soon after in English as *A New and Most Accurate Theory of the Moon's Motion; Whereby All Her Irregularities May Be Solved* . . . (London, 1702). His supervised edition of ΕΥΚΛΕΙΔΟΥ ΤΑ ΣΩΖΟΜΕΝΑ. *Euclidis quae supersunt omnia. Ex recensione Davidis Gregorii* was published at Oxford in 1703. Gregory's abridgment of Newton's 1671 tract, his "Tractatus de seriebus infinitis et convergentibus" (A56, Edinburgh), is printed in *The Mathematical Papers of Isaac Newton,* III (Cambridge, 1969), 354-372.

In the *Philosophical Transactions of the Royal Society* Gregory published a solution of Viviani's Florentine problem (**18,** no. 207 [Jan. 1694], 25-29); two defenses of his uncle James against Jean Gallois's charges of plagiarism from Roberval (**18,** no. 214 [Nov.-Dec. 1694], 233-236; and **25,** no. 308 [autumn 1706], 2336-2341); a study of the "Catenaria" and a reply to Leibniz's "animadversion" (*Acta eruditorum* [Feb. 1699], 87-91) upon it (**19,** no. 231 [Aug. 1697], 637-652; and **21,** no. 259 [Dec. 1699], 419-426); his observations of the solar eclipse of 13 Sept. 1698 (**21,** no. 256 [Sept. 1699], 320-321); a remark on John Perk's quadrature of a circle lunule (**21,** no. 259 [Dec. 1699], 414-417); and a discourse "De orbita Cassiniana," refuting its claim to be a realistic planetary path (**24,** no. 293 [Sept. 1704], 1704-1706).

II. Secondary Literature. The documented assessment in *Biographia Britannica,* IV (London, 1757), 2365-2372, is still unreplaced. Some biographical complements are given in Agnes M. Stewart, *The Academic Gregories* (Edinburgh, 1901), 52-76. Gregory's Savilian "Oratio inauguralis" on 21 April 1692 is printed, with commentary, by P. D. Lawrence and A. G. Mollond in *Notes and Records. Royal Society of London,* **25** (1970), 143-178; see esp. 159-165; the only modern study in depth of any aspect of Gregory's mathematical and scientific output is C. Truesdell's examination of Gregory's spurious derivation of the catenary's differential equation: "The Rational Mechanics of Flexible or Elastic Bodies, 1638-1788," vol. II of *Euleri opera omnia,* 2nd ser. (Zurich, 1960), pt. 2; see esp. 85-86.

D. T. Whiteside

GREGORY, DUNCAN FARQUHARSON (*b.* Edinburgh, Scotland, 13 April 1813; *d.* Edinburgh, 23 February 1844), *mathematics.*

Duncan Gregory came from a family with a long tradition of interest in science. His great-grandfather, his grandfather, and his father, James, were each professor of medicine at the University of Edinburgh. His great-great-grandfather was the mathematician James Gregory. Gregory attended the Edinburgh Academy, studied for a year in Geneva, and attended the University of Edinburgh. He matriculated at Trinity College, Cambridge, in 1833 and ranked as fifth wrangler in 1837. He remained at Cambridge as lecturer and tutor, and in 1840 he became a fellow of Trinity. Gregory received his M.A. in 1841. At that time he was offered a position at the University of Toronto, but as he was in poor health, he declined it. In 1838 Gregory, together with Robert Ellis, founded the *Cambridge Mathematical Journal.* Gregory was the first editor, and in this role considerably aided George Boole, who submitted his earliest papers to that journal.

Gregory published two books, both designed for use at Cambridge: one on the calculus (1841) and one on applications of analysis to geometry (published posthumously in 1845). His major contribution to mathematics, however, was his theory of algebra. His earliest papers were on differential and difference equations, in which he used a method that came to be known as the calculus of operations. This method involved treating the symbols of operation

$$\frac{d}{dx} \quad \text{or} \quad \Delta$$

as if they were symbols of quantity. In his attempt to justify the validity of this method, Gregory examined the laws governing the combination of these symbols with constants and by iteration. As a result of these studies he came to a definition of algebra as the study of the combination of operations defined not by their specific nature but rather by the laws of combination to which they were subject. This is wholly modern in tone, and that Gregory's work is not more widely known is probably due to the fact that he did not live to create a large-scale abstract algebra to illustrate his view.

BIBLIOGRAPHY

I. ORIGINAL WORKS. Gregory's books are *Examples of the Processes of the Differential and Integral Calculus* (Cambridge, 1841) and *A Treatise on the Application of Analysis to Solid Geometry,* William Walton, ed. (Cambridge, 1845). See also *The Mathematical Writings by Duncan Farquharson Gregory,* William Walton, ed. (Cambridge, 1865), which contains almost all of Gregory's published papers. *The Royal Society Catalogue* lists Gregory's works but contains several errors, which are corrected in Clock's thesis (see below).

II. SECONDARY LITERATURE. Biographical material on Gregory includes a memoir by Robert Ellis, found in the *Mathematical Writings,* pp. xi–xxiv, and the article by H. R. Luard, "D. F. Gregory," in *Dictionary of National Biography.* The significance of his work is discussed in Daniel Arwin Clock, "A New British Concept of Algebra: 1825–1850," Ph.D. diss. (Univ. of Wis., 1964); Elaine Koppelman, "Calculus of Operations: French Influence on British Mathematics in the First Half of the Nineteenth Century," Ph.D. diss. (Johns Hopkins Univ., 1969); and Ernest Nagel, "'Impossible Numbers': A Chapter in the History of Logic," in Studies in the History of Ideas, III (New York, 1935), 429–475.

ELAINE KOPPELMAN

GREGORY, FREDERICK GUGENHEIM (*b.* London, England, 22 December 1893; *d.* London, 27 November 1961), *plant physiology.*

Gregory's unusual abilities in mathematics, physics, and chemistry enabled him to foresee the major role that biochemistry and physics would play in physiology and development. This ability, along with his voracious scientific curiosity, extended his work over an enormous range of topics; and it is difficult to select his chief contributions to botanical science. His development of new methods of growth analysis and introduction of the term "net assimilation rate" to denote average photosynthetic efficiency of leaves (i.e., the dry weight of the plant divided by the average area of leaf surface and number of hours of light) were the basis of his early reputation.

With O. N. Purvis he proved that the effect of controlled low (1°C.) temperatures which will convert a "winter" rye to a "spring" rye—the effect known as vernalization—is exerted upon the embryo itself. They showed that excised embryos can be vernalized in the presence of sugar and a minimal oxygen concentration and that the effect is specifically due to temperature. This work and other work with F. J. Richards on mineral nutrition, aimed at determining the physiological causes underlying crop growth, also attracted much attention because of their value to agriculture.

The outcome of Gregory's 1928 visit to the Gazira Research Station in the Anglo-Egyptian Sudan was an increased knowledge of the factors affecting cotton production and provided the basis for strengthening the economy of the Sudan. Although he never returned to Gazira, he advised from London and served on both the Scientific Advisory Committee of the Empire Cotton Growing Corporation and the London Advisory Committee in Agricultural Work in the Sudan. His invention of the resistance porometer and the diffusion porometer enabled sophisticated study of stomatal physiology and the factors controlling transpiration.

All of Gregory's scientific career was spent in association with the Imperial College of Science and Technology, a constituent college of the University of London. The vast number of students and visitors who were influenced by their contact with him represents one of his lasting achievements.

In 1915 Gregory graduated from the Royal College of Science with first-class honors in botany. Having been rejected by the army on physical grounds, he joined the recently founded Research Institute of Plant Physiology at Imperial College under Vernon Blackman. Gregory began his work at Cheshunt Experimental Station, moved to Rothamsted, and finally returned to Imperial College. In 1932 he was appointed assistant director of the Research Institute of Plant Physiology and later succeeded Blackman as professor of plant physiology and director. Elected a member of the Royal Society in 1940, he was awarded its coveted Royal Medal in 1957. In 1956 he was elected a foreign associate of the National Academy of Sciences of the United States.

BIBLIOGRAPHY

I. ORIGINAL WORKS. Gregory's works include "Physiological Conditions in Cucumber Houses," in *Report. Experimental and Research Station, Nursery and Market Garden Industries Development Society,* **3** (1917), 19–28; "Studies in the Energy Relation of Plants. I. The Increase in Area of Leaves and Leaf Surface of *Cucumis sativum,*" in *Annals of Botany,* **35** (1921), 93–123; "The Effect of Climatic Conditions on the Growth of Barley," *ibid.,* **40** (1926), 1–26; "Studies in Energy Relations of Plants. II. The Effect of Temperature on Increase in Area of Leaf Surface and in Dry Weight of *Cucumis sativum.* Part I. The Effect of Temperature on the Increase in Area of Leaf Surface," *ibid.,* **42** (1928), 469–507; and "Mineral Nutrition of Plants," in *Annual Review of Biochemistry,* **6** (1937), 557–578.

See also "Physiological Studies in Plant Nutrition. VI. The Relation of Respiration Rate to the Carbohydrate and Nitrogen Metabolism of the Barley Leaf as Determined by Nitrogen and Potassium Deficiency," in *Annals of Botany,* n.s. **1** (1937), 521–561, written with P. K. Sen; "Studies in Vernalization of Cereals. II. The Vernalization of Excised Mature Embryos and of Developing Ears," *ibid.,* n.s. **2** (1938), 237–251, written with O. N. Purvis; and "The Interrelation Between CO_2 Metabolism and Photoperiodism in *Kalanchoë,*" in *Plant Physiology,* **29** (1954), 220–229.

II. SECONDARY LITERATURE. On Gregory or his work see Helen K. Porter, "Prof. F. G. Gregory, F.R.S.," in *Nature,* **193** (1962), 118; Helen K. Porter and F. J. Richards, "Frederick Gugenheim Gregory 1893–1961," in *Biographical Memoirs of Fellows of the Royal Society,* **9** (1963), 131–153, with complete bibliography; F. C. Steward, "F. G. Gregory 1893–1961," in *Plant Physiology,* **37** (1962), 450; and *The Times* (London), an obituary (30 Nov. 1961), 15a; and a funeral notice (4 Dec. 1961), 12c.

A. D. KRIKORIAN

GREGORY (more correctly **GREGORIE**), **JAMES** (*b.* Drumoak, near Aberdeen, Scotland, November 1638; *d.* Edinburgh, Scotland, late October 1675), *mathematics, optics, astronomy.*

The youngest son of John Gregory, minister of the manse of Drumoak, James Gregory was descended through his father from the fiery Clan Macgregor and through his mother, Janet, from the more scholarly Anderson family, one of whom, Alexander, had been secretary to Viète. Somewhat sickly as a child, he received his early education (including an introduction to geometry) from his mother, but after his father's death in 1651 his elder brother David sent him to Aberdeen, first to grammar school and later to Marischal College. After graduating there and further encouraged by his brother, himself an enthusiastic amateur mathematician, James devoted himself to studies in mathematical optics and astronomy.

In 1662, aware of the lack of scientific opportunities in Scotland, he traveled to London, there publishing *Optica promota* (1663), in which he gathered his earliest researches, and making several influential friends, notably Robert Moray, interim president of the Royal Society in 1660. In April 1663 Moray sought to arrange Gregory's introduction to Christian Huygens in Paris, but this was thwarted by Huygens' absence. Subsequently, to improve his scientific knowledge Gregory went to Italy, studying geometry, mechanics, and astronomy under Evangelista Torricelli's pupil Stefano degli Angeli at Padua (1664–1667) and publishing *Vera circuli et hyperbolæ quadratura* (1667) and *Geometriæ pars universalis* (1668). About Easter 1668 he returned to London; there, backed by John Collins' glowing reviews of his two Italian treatises and much in demand for his fresh contact with recent developments in Italian science, he was elected to the Royal Society on 11 June. Soon after, he made Huygens' attack upon the originality and validity of his *Vera quadratura* and also the publication of Nicolaus Mercator's *Logarithmotechnia* an opportunity for publishing in riposte certain newly composed *Exercitationes geometricæ* of his own.

In late 1668, probably through Moray's intercession, he was nominated to the new chair of mathematics at St. Andrews in Scotland. In 1669, shortly after taking up the post, he married a young widow, Mary Burnet, who bore him two daughters and a son. Much of his time during the next five years was passed in teaching elementary mathematics and the principles of science to his students: "I am now much taken up," he wrote in May 1671, "& hath been so al this winter bypast, both with my publick lectures, which I have twice a week, & resolving doubts which som gentlemen & scholars proposeth to me, . . . al persons here being ignorant of these things to admiration." His London correspondent Collins, a good listener if incapable of appreciating Gregory's deeper insights, was his sole contact with mathematical and scientific developments in the outside world; through him he received extended transcripts of letters written by Isaac Barrow, René-François de Sluse, Huygens, and Newton on a variety of topics, and in return he made Collins privy to many of his researches into equations, infinite series, and number theory.

Early in 1671, when the Académie des Sciences made tentative plans to invite two "Englishmen" (one of them Mercator) to Paris as *pensionnaires,* Moray campaigned actively on Gregory's behalf, but the proposal was not implemented. In 1672 Gregory joined the St. Andrews University "clerk" William Sanders in drafting a scornful reply to a recently published book on hydrostatics by the Glasgow professor George Sinclair: to Sanders' *Great and New Art of Weighing Vanity,* whose title page named as its author "Patrick Mathers, Arch-Bedal to the University of St. Andrews," Gregory contributed a minute dynamical essay, "Tentamina quædam de motu penduli et projectorum." Backed in turn by Sanders, the next year he implemented a long-cherished desire in the face of considerable resistance from his fellow professors, founding at St. Andrews the first public observatory in Britain. Charged with the university's commission, he traveled to London in June 1673 to purchase telescopes and other instruments and to seek John Flamsteed's advice regarding its equipment. Whether or not he did, as he intended, break his return journey at Cambridge to see Newton is not known. His hopes for the new observatory were soon quashed. During his absence the students at St. Andrews had rebelled against their antiquated curriculum, publicly ridiculing the regents; and Gregory, with his radical ideas on introducing the "new" science, was made the scapegoat: "After this the servants of the Colleges got orders not to wait on me at my observations; my salary was also kept back from me; and scholars of most eminent rank were violently kept from me, . . . the masters persuading them that their brains were not able to endure [mathematics]." In 1674 Gregory was glad to accept the newly endowed professorship of mathematics at Edinburgh, "where my salary is double, and my encouragements much greater"; but within a year of his appointment a paralyzing stroke blinded him one evening as he showed Jupiter's satellites through a telescope to his students. A few days later he was dead.

Written in his twenty-fourth year, Gregory's *Optica promota* is—with the notable exception of its "Epilogus"—interesting more for its revelations of the inadequacies of his early scientific training than for its technical novelties. Deprived in Aberdeen of a comprehensive library and contact with any practicing scientist, Gregory nevertheless made good use of available books on optics (Friedrich Risner's 1572 edition of Ibn al-Haytham [Alhazen] and Witelo ["authores perobscuri et prolixi"], Kepler's *Paralipomena,* Kircher's *Ars magna lucis*) and astronomy (Galileo's *Nuncius sidereus,* Kepler's *Astronomia,* Seth Ward's *Astronomia geometrica*). Ignorant of Descartes's *Dioptrique* (1637) and of the sine law of refraction there first publicly announced, in his opening pages Gregory presents an analogical "proof" that all rays incident on a central conic parallel to its main axis are refracted to its further focus for a suitable value of its eccentricity (in fact, as Descartes had shown, when it is the inverse refractive ratio). Departing from the particular cases of infinite and zero refraction when the conic is a circle and a straight line respectively and the parabolic case of reflection (unit negative refraction) and relying on his intuition that the interface is a conic, he "interpolates" the general *mensura refractionis* and then gives his model—equivalent to the sine law he nowhere cites—an experimental basis by displaying its agreement with the refraction tables of Witelo and Kircher.

The following optical propositions (2–59) extend Gregory's Cartesian theorem to systems of conical lenses and also develop the allied properties of reflection in conical mirrors: a neusis construction of the generalized Ibn al-Haytham problem of finding the point(s) of reflection in a general surface (prop. 34) is attained by roughly determining the tangent members of the family of spheroids whose common foci are the object and image points. In his historically significant epilogue Gregory explains how the deficiencies of the conventional pure reflectors and refractors encouraged him to design a compound "catadioptrical" telescope in which their defects were minimized. As an example he sketches "unum hujus perfectissimi generis telescopium" in which a parabolic mirror reflects parallel incident rays to a primary focus, on whose further side they are reflected back through a hole in the center of the first mirror by a small concave elliptical one to a secondary focus and thence through a plano-convex lens to the eye.

In 1663 the London optician Richard Reive was commissioned by Gregory to construct a six-foot "tube" to this design but failed, according to Newton in 1672, to polish its conical mirrors correctly. Newton's own improved design (1668[?]) used a plane mirror to reflect the rays from a spherical main reflector to the side of the telescope tube, and in 1672 through Collins he and Gregory exchanged letters arguing the relative merits of the two mountings. (The 1672 Cassegrain design, in which rays converging on the primary focus are reflected to the secondary focus before they reach it, was dismissed by Gregory as "no great alteration.") The astronomical appendix to the *Optica* (props. 60–90) is of no importance, but it serves to reveal once again Gregory's limited awareness of current scientific research. Much influenced by Seth Ward's *Astronomia* (1656), he here describes at some length geometrical methods for computing solar, lunar, and (hopefully) stellar parallax. A remark (prop. 87, scholium) that the conjunctions of the sun and Earth with Venus or Mercury would have a "pulcherrimum usum" for this purpose ignores the practical difficulties earlier encountered by Jeremiah Horrocks in his observations of the Venus transit in 1639 (first published in 1672). His schemes of planetary computation embody either the Keplerian "hypothesis Ptolemaica" of motion in an excentric circle with equant at the bissextile point or the slightly better Boulliau-Ward hypothesis of elliptical motion round the sun at a focus with mean motion round the second focus.

A still unpublished addendum to the *Optica* (David Gregory, B29, Edinburgh), composed some time after Gregory's arrival in London in 1663, contains a revised discussion of reflections in mirrors and refractions in thin lenses according to the newly encountered sine law of refraction. One theorem, a "notion" of a "burning-glass" (concave leaded spherical glass mirror) was communicated without proof to Collins in March 1673 and published by William Brown in 1715.

By late 1667 a sheen of confidence gleams through Gregory's work. Having absorbed at Padua all that some of the finest intellects in Italian science (Angeli, Gabriele Manfredi, and others) could teach him, he at length emerges fully aware of his hitherto latent mathematical powers. Of the two treatises stemming from his Italian sojourn, the *Vera circuli et hyperbolae quadratura* is the more original. Generalizing a procedure used by Archimedes in his *Measurement of a Circle,* in the case of a general central conic Gregory recursively defines an unbounded double sequence i_n,I_n of inscribed/circumscribed *mixtilinea.* Given a conic arc bounded by its chord and the intersecting tangents at its end points, i_0,I_0 are the inscribed triangle and circumscribed quadrilateral bounding the central sector cut off by the arc and the lines joining its end points to the conic's center. By dividing the arc at the point where it is parallel to its chord, two half-arcs are formed, yielding i_1,I_1 as the total of the two triangles/quadrilaterals inscribed/circumscribed to the two component conic sectors; a similar bisection of the two half-arcs produces corresponding bounding *mixtilinea* i_2,I_2, and so on. Gregory proves (props. 1-5) that i_{n+1}, I_{n+1} are, respectively, the geometric mean of i_n, I_n and the harmonic mean of i_{n+1}, I_n; that is,

$$i_{n+1} = \sqrt{i_n \cdot I_n} \quad \text{and} \quad I_{n+1} = \frac{2i_{n+1} \cdot I_n}{i_{n+1} + I_n}.$$

In the terms of Gregory's "definitiones" 1-10 the general pair i_{n+1}, I_{n+1} forms a "series convergens" (monotonically increasing/decreasing double sequence) of terms "analyticè compositi" (recursively defined by addition, subtraction, multiplication, division, and root extraction) out of the preceding "termini" i_n, I_n; he also proves that as n increases indefinitely, the difference between i_n and I_n becomes arbitrarily small—whence (p. 19) the "ultimi termini convergentes" can be "imagined" to be equal and their common value ($I = \lim_{n\to\infty} i_n = \lim_{n\to\infty} I_n$) is defined to be the "terminatio" (limit) of the "series." As an example of the power of this new terminology and analytical structure he derives purely algebraically the generalized Snell-Huygens inequalities for the central conic: $\frac{1}{3}(i_0 = 2I_0) > I > \frac{1}{3}(4i_1 - i_0)$.

Most tellingly, Gregory reasons that if a "quantitas" (function) can be "compounded" in the "same way" from i_{n+1}, I_{n+1} as from i_n, I_n—say by $\phi(i_n,I_n) = \phi(i_{n+1},I_{n+1})$—then the "terminatio" I is defined by $\phi(i_0,I_0) = \phi(I,I)$. The function $\phi(i_n,I_n)$, i.e.,

$$I_n \sqrt{\frac{i_n}{I_n - i_n}} \cdot \cos^{-1} \sqrt{\frac{i_n}{I_n}}; \ \phi(I,I) = I,$$

appropriate to his particular double sequence he was unable to determine, but by considering the "imbalance" of the parametrization $i_n = a^2(a + b)$, $I_n = b^2(a + b)$, and so $i_{n+1} = ab(a + b)$, $I_{n+1} = 2ab^2$, he sought to "prove" not only that ϕ cannot be a rational function, which this makes plausible, but also that it cannot be algebraic, which does not follow at all; in that case I would not be analytically compounded from i_0, I_0 and hence the "true" (algebraic) quadrature of the general conic sector—and of the whole circle in particular—would be impossible. Gregory's ingenious if ultimately ineffective argument was somewhat impercipiently attacked by Huygens when he received a presentation copy of the *Quadratura*: his rebuff, still commonly allowed, that the limit sector I could conceivably be determined in a different, algebraic way from the initial *mixtilinea* i_0, I_0 is in fact invalid since Gregory's argument concerns the structure of the function ϕ, not the passage of i_n,I_n to the limit I (disposed of in the equality $\phi(I,I) = I$). The latter half of the *Quadratura* is of some computational interest: on setting $i_0 = 2$, $I_0 = 4$, then $I = \pi$; while if $i_0 = 99/20$, $I_0 = 18/11$, then $I = \log \text{nat } 10$. These and other circle/hyperbola areas are accurately calculated to fifteen places.

Gregory's second Italian treatise is more eclectic in spirit, being designedly a tool kit of contemporary geometrical analysis of tangent, quadrature, cubature, and rectification problems. In the preface to his *Geometriae pars universalis* he expresses his hope that by "transmuting" the essential defining property of a given curve, it may be changed into one of an already known kind; the "universal part of geometry" presaged in his title is that which comprehends such general methods of geometrical transformation. Under that manifesto Gregory produces a systematic exposition of elementary calculus techniques which he freely admits are largely reworkings and generalizations of approaches pioneered by others. Pierre de Fermat's assignment of linear bounds to a convex arc, itself an improvement of Christopher Wren's 1658 discussion of the cycloid, is developed into a general scheme, demonstrated by an extended Archimedean exhaustion proof, for rectifying an arbitrary "curva simplex et non sinuosa"; Grégoire de St. Vincent's use of a "ductus plani in planum" (the geometrical equivalent of a change of variables under a double integral) is applied to reduce the quadrature of a given plane curve to the "planification" of a "hoof" section of a cylindrical surface and thence to the quadrature of a second curve. Another method of quadrature, that by transform to the subtangential curve, stems from Roberval. A geometrical tangent method is borrowed, again by way of Fermat, from Wren's tract on the cycloid, while an analytical one making use of Jacques de Beaugrand's notation for the vanishing increment "nihil seu serum *o*" of the base variable is a revision of that expounded by Descartes in his 1638 *querelle* with Fermat and published by Claude Clerselier in 1667, illustrated by an example (a Slusian cubic) deriving from Michelangelo Ricci.

Above all, Wren's concept, earlier broached by Roberval and Torricelli in the instance of the Archimedean spiral and Apollonian parabola, of the arc length-preserving "convolution" of a spiral into an equivalent Cartesian curve reappears, much extended and given rigorous exhaustion proof, in Gregory's favorite transform of an "involute" into an "evolute," while vigorous use is made of the Pappus-Guldin theorems relating the quadrature and cubature of solids and surfaces of revolution to their cross-section and its center of gravity. On this basis Gregory was enabled to furnish simple proofs of results in the theory of higher curves and the "infinite" spirals beloved of his tutor Angeli, replacing their previous crude, disparate forms by a logically immaculate, standardized demonstration. But too modern an interpretation of Gregory's book should be avoided: what to us (in prop. 6) may seem a proof of the fundamental theorem of the calculus was for him merely a generalization of William Neil's method for rectifying the semicubical parabola, and its wider significance is not mentioned.

The *Geometria* also affords a glimpse of Gregory's scientific interests at the close of his Italian stay. A proposition on the Fermatian spiral allows him to discourse on its origin (in 1636) as a modified Galilean path of free fall to the earth's center and to comment on the current controversy between Angeli and Giovanni Riccioli on the motion of the earth (one which he reviewed for the Royal Society in June 1668, on his return to London). Again, certain appended nonmathematical passages deal briefly with the optical effect of the apparent twinkling of the stars and with the conjectured composition of cometary tails conceived of as a steamy "exhalation" lit up by the sun and, most important for future physical astronomy, offer the suggestion that the apparent brightnesses of stars of the same magnitude are inversely proportional to the squares of their distances with the corollary that Sirius—taken to be of the same magnitude and brightness as the sun—is 83,190 times its distance.

Mathematics retained its central place in Gregory's affection until his death. Back in London he published a compendium, *Exercitationes geometricæ*, containing primarily an "Appendicula" to his *Vera quadratura* which refuted Huygens' objections to its argument but also appending a number of miscellaneous theorems in geometrical calculus. The "Appendicula" itself is noteworthy for its concluding "theorema" (that if a_n, A_n; b_n, B_n are two convergent Gregorian sequences with respective terminations A, B and if for all r $\phi(a_n,A_n) > \phi(b_n,B_n)$, then $A > B$) and for its twenty-seven narrow upper and lower bounds to the sector of a central conic. Since an "approximatio" to the sector I is said to k-plicate the "true notes" of the *mixtilinea*

$$i_n = 2^{n-1} \sin (I/2^{n-1}), I_n = 2^n \tan (I/2^n)$$

when it compounds i_j, I_j, $j = 0, 1, 2, \ldots, n$ so as to equal $I + O(I^{2k+1})$, Gregory's method clearly made use of the series expansion of one or other of the elementary circle functions. To illustrate the power of the techniques elaborated in his *Geometria,* Gregory also, in ignorance of Harriot's prior resolution, reduced the theory of the plane chart (Mercator map) to "adding secants" (integrating sec x over a given interval, $0 \leq x \leq a$), effecting this elegantly if long-windedly by an involved appeal to a "ductus plani in planum." In addition, he gave analogous quadratures of the tangent, conchoid, and cissoid curves; and, further to expedite the "additio secantium naturalium" near the origin, he elaborated simple rules for integrating $y \approx ax^2 + bx$ and $y \approx ax^3 +$

bx^2, x small. The former of these is the first published instance of "Simpson's" rule. His rigorous geometrical deduction of the Mercator series for $\pm \log (1 \pm x)$ is of minor importance.

After his return to Scotland, Gregory made no further published contribution to pure mathematics, but his private papers reveal that the last half dozen years of his life were ones of intensive research. His executor William Sanders tells us, "His Elements of plain Geometry, with some few propositions of the solids; his Practicall Arithmetick, and Practicall Geometry taught at St. Andrews . . . are but of small moment, being contrived only for the use of such scholars as cannot be at pains to study the Elements." The lost *Tractatus trigonometricus,* in which he reduced "All Trigonometry rectilineal and spherical . . . unto five short canons," on the lines of Seth Ward's *Idea* (1654), was no doubt also intended for professorial lectures. But his real energy was reserved for deeper matters. At Collins' instigation Gregory spent much time on the theory of equations and the location of their roots: achieving success in the case of the reduced cubic and quartic by introducing an appropriate multiplying factor and equating all terms in the resolvent except those involving cube/fourth powers of the unknown to zero, he sought to solve the general quintic in a similar way by adjoining a factor of the fifteenth degree but failed to notice that the resulting equations to zero implied—ineluctably—a sextic eliminant. His papers on Fermatian equations, rational Heronian triangles, and other topics in Diophantine analysis are (much like Newton's contemporary studies, likewise inspired by the appearance of the Samuel Fermat-Jacques de Billy *Diophantus* in 1670) more workmanlike than profound: the "skailzy brods" (writing slates) found on his desk after his death contained his abortive calculations for Jacques Ozanam's unsolvable problem of cubes.

Gregory's letters to Collins are filled with a miscellany of calculus problems, among them his quadrature and rectification of the logarithmic spiral and "evolute" logarithmic curve, which had briefly made its introductory bow in the preface to his *Geometria,* and his construction of the tangent to the "spiralis arcuum rectificatrix" introduced by Collins for use in the "Mariners Plain Chart." His grasp of the subtleties of infinite series in particular quickly matured. His independent discovery of the general binomial expansion in November 1670—in disguised form as that of antilog $((a/c)(\log(b + d) - \log\ b))$—was matched a month later by his use of a "Newton-Gauss" interpolation formula to insert general means in a given sequence of sines. As a climax, in February 1671 Gregory communicated without proof a number of trigonometrical series, notably those for the natural and logarithmic tangent and secant, and in April 1672 a series solution of Kepler's problem (intended for publication in Collins' edition of Horrocks's *Opera posthuma,* but the bookseller took fright) regarding which he observed that "these infinite serieses have the same success in the roots of equations." Two examples—the series extraction of the root e of the conchoid's defining equation

$$L^2e^2 = (L + a)^2(L^2 - a^2)$$

and the inversion of the Kepler equation

$$a = \sqrt{2re - e^2} + (b - r) \sin^{-1} (\sqrt{2re - e^2}/r)$$

were later published by David Gregory, without direct acknowledgment, in his *Exercitatio geometrica* (1684). Until the printing in 1939 of Gregory's notes, jotted down on the back of a letter from the Edinburgh bookseller Gideon Shaw in January 1671, it seemed likely that these expansions were obtained by straightforward elementary methods, but we now know that he employed, twenty years before even Newton came upon the approach, a Taylor development of a function in terms of its nth-order derivatives.

Of Gregory's scientific pursuits during this last period of his life too little is known. His "Theory of the whole Hydrostaticks comprehended in a few definitions and five or sixe Theorems" (David Gregory, D18) is seemingly lost, although a short 1672 paper in which he proved Huygens' theorem relating atmospheric height logarithmically to barometric pressure still exists in several versions. In our present state of knowledge it seems impossible to determine how far William Sanders drew upon Gregory's hydrostatical ideas in his largely scurrilous *Great and New Art of Weighing Vanity,* but extant preliminary computations in Gregory's hand confirm contemporary report that the appended "Tentamina quædam de motu penduli et projectorum" is uniquely his. Of considerable historical importance as a bridging text between Galileo's *Discorsi* and Newton's *Principia,* these nine small duodecimo pages are a highly original contribution to dynamics. Independently deriving Huygens' Galilean generalization that the square of the instantaneous speed of a body falling freely under simple gravity in a smooth curve is proportional to the vertical distance fallen (and indeed anticipating an objection to Huygens' definition of the fall curve as the limit of a chain of line segments, which Newton put to Huygens in 1673), Gregory deduced the elliptical integral expressing the time of vibration in a circular pendulum and gave its infinite-series expansion for a small arc of swing. Subsequently, framing the supposition that the resistance is constant in magnitude and direction (opposite to that of initial motion), he determined that the resisted path of a projectile under simple gravity is a tilted parabola with main axis parallel to the resultant instantaneous force—a theorem, we now know, which had been found seventy years before by Harriot.

The "Fourty or thereabout of excellent Astronomical propositions invented . . . for the compleeting that art" found after his death doubtless originated in his correspondence with Colin Campbell during 1673–1674 on theoretical astronomy, during the course of which he solved—yet again in ignorance of a prior solution by Harriot—the Keplerian problem of constructing a planetary ellipse, given three focal radii in magnitude and position. Apart from his keen discussion with Newton in 1672 on the respective merits of their "catadioptrical" reflecting telescopes, little evidence has survived of Gregory's continuing interest in optics.

For all his talent and promise of future achievement, Gregory did not live long enough to make the major discovery which would have gained him popular fame. For his reluctance to publish his "several universal methods in Geometrie and analyticks" when he heard through Collins of Newton's own advances in calculus and infinite series, he posthumously paid a heavy price: the "Extracts from M[r] Gregories Letters" drawn up by Collins in 1676 for Leibniz' enlightenment were used by Newton in 1712 solely to further his claim to calculus priority and were thereafter forgotten. Gregory's published works had little contemporary impact; his *Vera quadratura* was successfully sabotaged by Huygens, his *Geometria* quickly overshadowed by Barrow's *Lectiones geometricæ*. We are only now beginning to realize the extent and depth of his influence, mathematically and scientifically, on Newton. A comprehensive edition of his work is sorely needed.

BIBLIOGRAPHY

I. ORIGINAL WORKS. Gregory's first published work, *Optica promota, seu abdita radiorum reflexorum & refractorum mysteria, geometricè enucleata; cui subnectitur appendix, subtilissimorum astronomiœ problematôn resolutionem exhibens* (London, 1663), is reprinted in C. Babbage and F. Maseres, *Scriptores optici* (London, 1823), 1–104. His *Vera circuli et hyperbolœ quadratura, in propria sua proportionis specie, inventa & demonstrata* (Padua, 1667; reviewed by John Collins in *Philosophical Transactions of the Royal Society,* **3,** no. 33 [16 Mar. 1668], 640–644) was reprinted by W. J. 'sGravesande in his ed. of *Christiani Hugenii opera varia,* I (Leiden, 1724), 405–482; it was reissued at Padua in 1668 together with Gregory's *Geometriœ pars universalis, inserviens quantitatum curvarum transmutationi & mensurœ* (reviewed by Collins in *Philosophical Transactions of the Royal Society,* **3,** no. 35 [18 May 1668], 685–688), which also contains (pp. 132–151) his discussion of "difficultates quædam physicomathematicæ ex principiis opticis geometricè enodatæ." The same year he published his *Exercitationes geometricœ* (London, 1668), comprising "Appendicula ad veram circuli & hyperbolæ quadraturam" (repr. in *Oeuvres complètes de Christiaan Huygens,* VI [The Hague, 1895], 313–321), sig. A2r–A4r/pp. 1–8; "N. Mercatoris quadratura hyperbolæ geometricè demonstrata/Analogia inter lineam meridianam planispherii nautici & tangentes artificiales geometricè demonstrata; seu, quod secantium naturalium additio efficiat tangentes artificiales. . . ." (repr. in F. Maseres, *Scriptores logarithmici,* II (London, 1791), 2–15), pp. 9–24; and "Methodus facilis & accurata componendi secantes & tangentes artificiales," pp. 25–27.

Gregory's "Tentamina quædam geometrica de motu penduli et projectorum" (repr. in Babbage and Maseres, *Scriptores optici,* pp. 372–376) first appeared as an anonymous appendix (pp. $_2$1–9) to "Patrick Mathers" [William Sanders], *The Great and New Art of Weighing Vanity* (Glasgow, 1672). His report on the moving earth dispute in Italy was published by Henry Oldenburg as "An Account of a Controversy betwixt Stephano de Angelis and John Baptista Riccioli," in *Philosophical Transactions of the Royal Society,* **3,** no. 36 (15 June 1668), 693–698 (repr. with commentary by A. Koyré in "A Documentary History of the Problem of Fall from Kepler to Newton," in *Transactions of the American Philosophical Society,* **45,** [1955], 329–395, esp. 354–358). His two "Answers" to the "Animadversions" of Huygens upon his *Quadratura* (*Journal des sçavans* [2 July and 12 Nov. 1668], repr. in Huygens' *Oeuvres,* VI, 228–230 and 272–276, and in Latin in *Opera varia,* I, 463–466 and 472–476) appeared in *Philosophical Transactions of the Royal Society,* **3,** no. 37 (13 July 1668), 732–735 and no. 44 (15 Feb. 1669), 882–886 (repr. in *Opera varia,* I, 466–471 and 476–481; also Huygens' *Oeuvres,* VI, 240–244 and 306–311).

A number of Gregory's minor mathematical and scientific papers are extant in the Royal Society, London, the University Library, Edinburgh, and also in private possession: these derive from John Collins and Gregory's nephew, David. An incomplete listing of those accessible to the public is given by H. W. Turnbull in *James Gregory Tercentenary Memorial Volume* (London, 1939), pp. 36–43. Extracts from Gregory's correspondence with Collins were published by Newton in *Commercium epistolicum D. Johannis Collins* (London, 1712), pp. 22–26, and by Jean Desaguliers in an appendix to *Dr.* [David] *Gregory's Elements of Catoptrics and Dioptrics,* 2nd ed. (London, 1735). His letters to Robert Bruce were printed by Leslie in *Scots Magazine,* **72** (Aug. 1810), 584–586; those to Colin Campbell by John Gregorson and Wallace in *Archaeologia Scotica,* **3,** Artic. 25 (Jan. 1831), 275–284; those to Collins (with Collins' draft replies) by S. P. Rigaud in his *Corre-*

spondence of Scientific Men of the Seventeenth Century, II (Oxford, 1841), 174–281; the originals of Collins' replies, invaluable for Gregory's mathematical notes upon them, were published by H. W. Turnbull in the *Gregory Volume,* pp. 45–343 (the notes themselves, with lavish commentary, follow on pp. 347–447). Gregory's earliest known letter (to Huygens, in Oct. 1667, accompanying a presentation copy of his *Vera quadratura*) is given in Huygens' *Oeuvres,* VI, 154.

II. SECONDARY LITERATURE. Thomas Birch's article on Gregory in the *Biographia Britannica,* IV (London, 1757), 2355–2365 remains unsuperseded, although it is now partially obsolete. For complements see Agnes M. Stewart, *The Academic Gregories* (Edinburgh, 1901), pp. 27–51; and *University of St Andrews James Gregory Tercentenary: Record of the Celebrations Held . . . July Fifth MCMXXXVIII* (St. Andrews, 1939), pp. 5–11 (H. W. Turnbull's commemoration address, repeated in expanded form in the *Gregory Volume,* pp. 1–15) and pp. 12–16 (G. H. Bushnell's notes on the St. Andrews' observatory). Section VII of the *Gregory Volume* contains summaries of Gregory's *Optica promota* and *Exercitationes* (by H. W. Turnbull, pp. 454–459 and 459–465), his *Quadratura* (by M. Dehn and E. Hellinger, pp. 468–478) and *Geometria* (by A. Prag, pp. 487–509), together with an account by E. J. Dijksterhuis of the Gregory–Huygens squabble (pp. 478–486). A short general survey of Gregory's researches in calculus is given by C. J. Scriba in *James Gregorys frühe Schriften zur Infinitesimalrechnung, Mitteilungen aus dem Mathem. Seminar Giessen,* no. 55 (Giessen, 1957). More specialist mathematical topics are explored by H. W. Turnbull in "James Gregory: A Study in the Early History of Interpolation," in *Proceedings of the Edinburgh Mathematical Society,* 2nd ser., **3** (1933), 151–172; and by J. E. Hofmann, in "Über Gregorys systematische Näherungen für den Sektor eines Mittelpunktkegelschnittes," in *Centaurus,* **1** (1950), 24–37. No study of any aspect of Gregory's scientific achievement exists.

D. T. WHITESIDE

GREGORY, OLINTHUS GILBERT (*b.* Yaxley, England, 29 January 1774; *d.* Woolwich, England, 2 February 1841), *applied mathematics, science education.*

Gregory was one of the band of self-taught or privately tutored mathematicians who swelled the ranks of British mathematics during the eighteenth and early nineteenth centuries. Despite his limited schooling he established a reputation as a writer on scientific subjects, and in 1803, through the patronage of Charles Hutton, he was appointed instructor of mathematics at the Royal Military Academy at Woolwich. In 1821 he succeeded to the professorship and held the post until his retirement in 1838.

Gregory's most important scientific publication, *A Treatise of Mechanics,* appeared in 1806 and went through at least four editions. Although it was a didactic compilation rather than a publication of original research, it was one of the most complete works on pure and applied mechanics that had appeared in English. In purpose and presentation it was an early example of what would now be described as "engineering mechanics." Its theoretical sections covered such topics as the analysis of the flexed beam and the theory of the loaded arch, while its descriptive sections dealt extensively with machine design. The book constituted a contribution to the tradition of applied mathematics and applied mechanics which was then being fostered by the Woolwich mathematicians.

In 1825 Gregory produced another book, *Mathematics for Practical Men,* devoted to "the principles and applications of the mechanical sciences for the use of the younger members of the Institution of Civil Engineers" (which had been founded in 1818 and of which Gregory later became an honorary member). Around this time he also did experimental research on the velocity of sound. From 1802 to 1819 he edited the *Gentleman's Diary* and from 1819 to 1840 the *Ladies' Diary.* On the strength of both his reputation in science and his status as a prominent Dissenter in religion he was included among the group that founded London University, the first nonsectarian university in England.

BIBLIOGRAPHY

Gregory's scientific publications include *A Treatise of Mechanics,* 3 vols. (London, 1806); the first volume of the *Treatise* was translated into German as *Darstellung der mechanischen Wissenschaften,* J. F. W. Dietlein, trans. (Halle, 1824); *Mathematics for Practical Men* (London, 1825); "An Account of Some Experiments Made in Order to Determine the Velocity With Which Sound Is Transmitted in the Atmosphere," in *Philosophical Magazine,* **63** (1824), 401–15.

Gregory also translated one of René Just Haüy's works, *An Elementary Treatise on Natural Philosophy,* 2 vols. (London, 1807).

For additional bibliography see *Dictionary of National Biography* and *British Museum Catalogue of Printed Books.*

HAROLD DORN

GREGORY, WILLIAM (*b.* Edinburgh, Scotland, 25 December 1803; *d.* Edinburgh, 24 April 1858), *chemistry, biology.*

William Gregory was the fourth son of James Gregory, professor of medicine at the University of Edinburgh. William was educated for the medical profession and graduated in 1828 from the University of Edinburgh. After graduation he chose to pursue his interest in chemistry rather than practice medicine. During the next few years he made extended visits to the Continent and worked as assistant to several chemists, most notably Justus Liebig at his Giessen laboratory in 1835. There he developed a primary interest in organic chemistry. Following his work with Liebig, Gregory returned to Edinburgh where he gave public lectures in chemistry.

In 1837 Gregory accepted a lectureship at Anderson College in Glasgow and the next year at a Dublin medical school. Gregory was appointed professor of chemistry at King's College, Aberdeen, in 1839 and remained there, except for an additional year of study with Liebig in 1841, until 1844. Gregory suffered from poor health most of his life and continued to make trips to the Continent to restore his strength. He returned to the University of Edinburgh in 1844 and held the chair of chemistry until his death.

William Gregory is important to the development of chemistry primarily because of his translations into English of the many works of Liebig on organic, agricultural, and physiological chemistry. His own research was devoted chiefly to organic chemistry, especially the separation and analysis of natural products. Gregory investigated the preparation of morphine and codeine from opium and was the first to describe the preparation of isoprene from crude rubber. He also wrote several successful chemical textbooks.

In 1846 Gregory abstracted for a British journal the studies which Karl von Reichenbach had performed in 1845 on animal magnetism. Gregory later translated and published, with a twenty-seven-page preface of his own, Reichenbach's *Researches on Magnetism, Electricity, Heat, Light, Crystallization, and Chemical Attraction, in Their Relations to the Vital Force* (1850). Criticized for both the abstract and the translation of Reichenbach's work, Gregory further incurred the disapproval of his colleagues at the University of Edinburgh with his publication of *Letters to a Candid Inquirer on Animal Magnetism* (1851). In this work he attempted to establish a scientific basis for phenomena such as clairvoyance, thought transference, and unusual sensitivity of subjects who were under the influence of hypnotism. Following Reichenbach, he attributed most of these cases to emanations of a physical fluid, called odyl. His work on animal magnetism went through four editions during the nineteenth century. In addition, he published many pamphlets and papers on this subject. During the last ten years of his life, Gregory also became interested in the study of diatoms, on which he wrote twelve papers.

BIBLIOGRAPHY

I. Original Works. A list of Gregory's published papers, including those on diatoms, can be found in the Royal Society's *Catalogue of Scientific Papers (1800–1863)*, **3** (1869), 8–10. This list does not include Gregory's papers on animal magnetism, many of which were published in *Zoist, A Journal of Cerebral Physiology and Mesmerism, and Their Applications to Human Welfare,* and the *Phrenological Journal* (Edinburgh). These articles include "On the True Scientific Spirit in Which the Claims of Phrenology and Mesmerism Ought to be Examined," in *Phrenological Journal* (1847), pp. 1–28; "On Animal Magnetism," in *Zoist,* **9** (1851), 423–424; and "On the Theory of Imagination as the Cause of Mesmeric Phenomena, and On Money Challenges in Mesmerism," *ibid.,* **10** (1852), 1–37.

Gregory's important books include *Outlines of Chemistry, for the Use of Students* (London, 1845); *Letters to a Candid Inquirer on Animal Magnetism* (Edinburgh, 1851); *Handbook of Organic Chemistry* (London, 1852); and *Elementary Treatise on Chemistry* (Edinburgh, 1855). He also edited, with Justus Liebig, the 1842 and 1847 rev. eds. of Edward Turner, *Elements of Chemistry.*

Gregory's translations of Liebig's works include *Instructions for Chemical Analysis of Organic Bodies* (Glasgow, 1839); *Animal Chemistry* (Cambridge, 1842); *Chemistry in its Applications to Agriculture and Physiology* (London, 1847); *Researches on the Chemistry of Food* (London, 1847); *Researches on the Motion of the Juices in the Animal Body* (London, 1848); *Familiar Letters on Chemistry* (London, 1851); and *Principles of Agricultural Chemistry* (London, 1855). See also his trans. of Karl von Reichenbach, *Researches on Magnetism, Electricity, Heat, Light, Crystallization, and Chemical Attraction, in Their Relations to the Vital Force* (London, 1850).

II. Secondary Literature. A brief sketch of Gregory's life may be found in the *Dictionary of National Biography,* VIII, 548. The following are obituaries: *Edinburgh New Philosophical Journal,* **8** (1858), 171–175; *Proceedings of the Royal Society of Edinburgh,* **4** (1857–1862), 121–122; *Transactions of the Botanical Society of Edinburgh,* **6** (1857–1860), 75–79; and *Journal of the Chemical Society,* **12** (1860), 172–175. An extensive review of Gregory's work on animal magnetism appeared in the *British and Foreign Medico-Chirurgical Review,* **8** (1851), 378–431.

Daniel P. Jones

GREN, FRIEDRICH ALBRECHT CARL (*b.* Bernburg, Germany, 1 May 1760; *d.* Halle, Germany, 26 November 1798), *chemistry, physics.*

The eldest son of a Swedish immigrant hatter, Gren was destined for the clergy. But the death of his father forced him to abandon his formal education and prepare for a pharmaceutical career. After an apprenticeship characterized by oppressive servitude and his own private study of botany and chemistry, Gren went to Offenbach am Main as a journeyman pharmacist in 1779. For health reasons, he proceeded to Erfurt the following year. There he administered the apothecary shop owned by Wilhelm B. Trommsdorff, professor of chemistry, botany, and materia medica, and father of the chemist Johann B. Trommsdorff.

Instructed and encouraged by his employer, Gren prepared a manuscript for a chemistry text and entered into correspondence with Lorenz von Crell, editor of Germany's leading chemical journal. Upon the elder Trommsdorff's death in 1782, Gren first attempted to establish a chemical factory in Bernburg, then entered Helmstedt University, in the duchy of Brunswick. There he assisted Crell (who had arranged a scholarship for him), studied medicine and science, and lectured on chemistry.

In 1783 Gren went on to Halle University, where he continued his studies, gave chemistry lectures, and served as research assistant to Wenceslaus Karsten, professor of mathematics and physics. He took an M.D. in 1786 and a Ph.D. in 1787 and then quickly rose to professor of physics and chemistry in Halle's medical faculty in 1788. He remained in this post until his death ten years later.

Gren made his mark on German scientific life as an author of texts, a journal editor, and a theorist in chemistry and physics. Both his sense of the inadequacy of existing works and his need for additional income led him to devote much time to writing textbooks. The books were well received, some continuing to appear long after his death. Chemistry was the subject of his first text, the *Systematisches Handbuch der gesammten Chemie* (1787–1790; 3rd ed., 1806). He subsequently published *Grundriss der Naturlehre* (1788; 6th ed., 1820), *Grundriss der Pharmakologie* (1790), *Handbuch der Pharmakologie* (1791–1792; 3rd ed., 1814–1815) and *Grundriss der Chemie* (1796; 4th ed., 1818; English translation, 1800).

Inspired by the example of his former teacher and patron Crell, Gren founded a periodical for the "mathematical and chemical branches of natural science." Under his editorship, the *Journal der Physik* (1790–1794), which was succeeded by the *Neues Journal der Physik* (1795–1797), soon became Germany's most exciting scientific journal. After Gren's death, it was continued by his colleague Ludwig Wilhelm Gilbert, and subsequent editors, as the *Annalen der Physik* (1799–present).

Gren first attracted attention as a theorist by proposing that phlogiston has negative weight in his *Dissertatio inauguralis physico-medica sistens observationes et experimenta circa genesin aëris fixi et phlogisticati* (1786). His pride in the phlogiston theory's German origins apparently led him to try to rescue it from the difficulties created by the new discoveries with gases. Although equally nationalistic, most German chemists rejected Gren's proposal as absurd, embracing instead Richard Kirwan's system, which identified hydrogen as phlogiston. Undaunted, Gren continued to campaign for the negative weight of phlogiston until 1790, when the physicist Johann Tobias Mayer persuaded him to abandon the view with arguments based on the motion of pendulums.

That same year Gren announced that the empirical cornerstone of Lavoisier's antiphlogistic theory lacked grounding—pure red calx of mercury (mercuric oxide) did not yield any gas when it was reduced. Two years later his claim was supported by Johann Friedrich Westrumb, a widely respected experimentalist. A bitter debate ensued between Lavoisier's German proponents (notably Sigismund Friedrich Hermbstädt and Martin Heinrich Klaproth in Berlin) and the German phlogistonists (notably Gren, Westrumb, and J. B. Trommsdorff). The turning point in the antiphlogistic revolution in Germany came by mid-1793, when Gren and his allies were discredited.

As a consequence of this defeat, Gren soon adopted the compromise phlogiston theory of Johann Gottfried Leonhardi and Jeremias Benjamin Richter. This theory differed but slightly from Lavoisier's, treating phlogiston (the basis of light) as a component of all substances which could be oxidized. Gren's support of this theory helped prepare the way for the ultimate acceptance of Lavoisier's theory. In the mid-1790's, Gren also helped prepare the way for the penetration of Kant's "dynamic system" into German chemistry and physics by giving it very favorable, if brief, attention in his publications.

BIBLIOGRAPHY

I. Original Works. A complete list of Gren's publications through 1795 appears in his autobiography in Johann Kaspar Philipp Elwert, *Nachrichten von dem Leben und den Schriften jeztlebender teutscher Aerzte, Wündärzte, Thierärzte, Apotheker und Naturforscher* (Hildesheim, 1799), pp. 171–185.

II. Secondary Literature. For information on Gren see Wolfram Kaiser and Karl-Heinz Krosch, "Zur Geschichte der Medizinischen Fakultät der Universität Halle," in *Wissenschaftliche Zeitschrift: Mathematisch-naturwissenschaftliche Reihe,* **13** (1964), 160–176; Dietrich Ludwig Gustav Karsten, "Kurze Nachrichten von dem Leben des Professors Gren zu Halle," in *Neue Schriften,* **2** (1799), 404–413; an article by Hans Schimank in *Neue deutsche Biographie,* VII (Berlin, 1966), 45–46 (Gren was a Lutheran, not a Calvinist as Schimank maintains); J. R. Partington, *A History of Chemistry,* III (London, 1962), 575–577, 620–625, 632–636; Alexander Nicolaus Scherer,

"Friedrich Albrecht Carl Gren," in *Allgemeines Journal der Chemie,* **2** (1799), 357–416, 615–618; and Johann Bartholomai Trommsdorff, "Kurze Biographie des verewigten Friedrich Albrecht Carl Gren," in *Journal der Pharmacie,* **6** (1799), 367–375.

For further information on Gren's role in the antiphlogistic revolution, see Karl Hufbauer, "The Formation of the German Chemical Community, 1700–1795," diss. (Univ. of Cal., Berkeley, 1970), chs. 6 and 7.

KARL HUFBAUER

GRESSLY, AMANZ (*b.* Bärschwyl, Switzerland, 17 July 1814; *d.* Bern, Switzerland, 13 April 1865), *geology, stratigraphy, paleontology.*

Born at La Verrerie, a glassworks of Bärschwyl established by his grandfather, Gressly was first educated at home, then at Laufon, Solothurn, Lucerne, and Fribourg. As a medical student at Strasbourg, he came under the influence of Phillipe Louis Voltz and Jules Thurmann of Porrentruy in the Swiss Jura. Their interest inspired him to become a geologist. This was during the period of subdivision of major units of the geological column. Some of the geological column (A. G. Werner's Secondary and Tertiary) had been subdivided earlier in relatively undisturbed sedimentary sequences, such as those of southern England (William Smith, Robert Bakewell) and the Paris basin (Georges Cuvier, Alexandre Brongniart).

To reach this stage of geological chronology, simple petrography and the principle of faunal succession had been sufficient. The extensions of the subdivisions to the disturbed sequences of mountainous regions required a new geometric perspective which the engineer Voltz could well appreciate and which in the hands of his protégé Thurmann (1832) were to begin the science of tectonics. But beyond the application of new mechanical and architectural considerations, the theoretical framework of geology required the abandonment of simplistic Wernerian doctrines of simultaneous worldwide depositions of lithologically similar formations—the "onionskin" view of stratigraphy. Gressly made a major contribution to this development, by his identification and definition of the concept of facies or "aspects de terrain," an accomplishment of his first significant work, "Observations géologiques sur le Jura Soleurois" (1838), written at the age of twenty-two. This work and his extensive collections of fossils (Rollier mentions 25,000 specimens [1911]) brought him to the attention of Louis Agassiz, who engaged him as assistant at Neuchâtel. Agassiz made free use of Gressly's abilities and fossils for his own monographs. When Agassiz departed for America in 1846, he carried with him a substantial part of Gressly's fossil collections. This departure aggravated Gressly's state of melancholia; already by 1845, he had spent time in a sanitarium. Nevertheless, Agassiz was always profuse in his published acknowledgments of Gressly's high abilities.

Gressly's close associations with Agassiz's other abandoned assistants, Eduard Desor and Carl Vogt, continued, although his employment after the departure of Agassiz was in what would now be described as engineering geology for the construction of the alpine railroads. In this capacity he described the geologically rewarding tunnels of Hauenstein, des Loges, and Mont-Sagne. In 1859 he experienced what Wegmann described as immense pleasure in finding, on the modern coast of Sète on the French Riviera, the ecological zones he had deduced from his studies of the Jura while a student. In 1861 Gressly accompanied Vogt on a six-month voyage to the high latitudes. An indefatigable field geologist and collector in his native Jura, Gressly was described perhaps romantically as something of a folk figure. The painter Auguste Bachelin was one of his few close friends and sketched him often at Combe-Varin, which he made headquarters with Desor. A rapid mental decline began in 1864, and he died within a year.

The chronostratigraphic rock unit of Thurmann's researches in the Bernese Jura had been the terrain, a term roughly equivalent to formation, which Gressly used (as one example) for the Portlandian series of the Upper Jurassic group of strata. Recognizing the striking variations within the horizontal extensions of the terrains of the Solothurn Jura, he characterized the facies as a distinguishable petrographic aspect always accompanied by the same faunal assemblage and rigorously excluding some of the genera and species common to other facies. Twenty-three years before he ever saw the sea, he described this law of dissociation as reflecting conditions at the time of sedimentation, with modern processes as a guide to the ancient environment. He proposed as a general law that every facies of a terrain presents quite distinct petrographic and geognostic or paleontologic characteristics in marked contrast with those of other facies, either on the same stratigraphic level or generally characteristic of the terrain. Further, petrographically and geognostically similar facies of distinct terrains are characterized by analogous faunal assemblages, even succeeding each other vertically through a series of superposed terrains.

Gressly derived a series of paleoecological rules from his observations, noting, for example, that the diversity of facies increases vertically with a rising series and decreases with a sinking series (corresponding to conditions of receding and advancing seas, respectively). This work of 1838 alone establishes Gressly as a pioneer in, if not the founder of paleogeography. Wegmann wrote of him, "In creating this vision of superposed paleogeographies, he added a fourth dimension to his perspective."

BIBLIOGRAPHY

I. Original Works. Gressly's own publications were very few. Among them are "Geognostische Bemerkungen über den Jura der nordwestlichen Schweiz, besonders des Kantons Solothurn und der Grenz-Partien der Kantone Bern, Aargau und Basel," in *Neues Jahrbuch für Mineralogie, Geognosie, und Petrefactenkunde* (1836), pp. 659-675, short version in *Bibliothèque universelle* (1837), pp. 194-197; "Résumé d'observations géologiques sur les modifications du Jura des cantons de Soleure et d'Argovie," in *L'Institut,* **4** (1836), 92-93; "Description géologique des montagnes du Jura Soleurois et Argovien," *ibid.,* 126-128: "Note sur les restes de mammiteres trouvés dans le portlandien de Soleure," *ibid.,* 165-166.

Also see "Observations géologiques sur les terrains des chaînes Jurassiques du canton de Soleure et des contrées limitrophes," in *Verhandlungen der Schweizerischen naturforschenden Gesellschaft* (1837), pp. 126-132; "Observations géologiques sur le Jura Soleurois," in *Neue Denkschriften der Schweizerischen naturforschenden Gesellschaft,* **2** (1838), 1-112; **4** (1840), 113-241; **5** (1841), 235-349; "Uebersicht der Geologie des nordwestlichen Aargau's," in *Neues Jahrbuch fur Mineralogie Geognosie, und Petrefactenkunde* (1844), pp. 153-163, short version in *Bulletin de la Société neuchâteloise des sciences naturelles,* **1** (1844-1846), 166-168; "Nouvelles données sur les faunes tertiaires d'Aljoi," in *Actes de la Société helvétique des sciences naturelles* (1853), pp. 251-261; "Ossements fossiles d'un saurien gigantesque de la famille des Dinosauriens," in *Bulletin de la Société neuchâteloise des sciences naturelles,* **4** (1856-1857), 13-16; "Études géologiques sur le Jura Neuchâtelois," in *Mémoires de la Société neuchâteloise des sciences naturelles,* **4** (1859), 1-159, written with E. Desor; *Briefe aus dem Norden. Der Bund* (Bern, 1861). pp. 246-251, 281-284; "Erinnerungen eines Naturforschers aus Südfrankreich," in *Album von Combe-Varin* (Zurich, 1861), pp. 201-296; "Differenzialheber (Wasserstandsmesser)," in *Mitteilungen der Naturforschenden Gesellschaft in Bern* (1866), 228-233; and "Uebersicht der geologischen Verhaltnisse der Umgebungen Oltens in Bezug auf den Hauenstein Tunnel (1853)," in *Mitteilungen der Naturforschenden Gesellschaft in Solothurn,* **8** (1928), 1-40.

II. Secondary Literature. For works on Gressly, see Kurt Meyer, "Amanz Gressly, ein Solothurner Geologe (1814-1865)," in *Mitteilungen der Naturforschenden Gesellschaft in Solothurn,* no. 22 (1966), 1-79; L. Rollier, "Lettres d'Amand Gressly, le géologue jurassien (1814-1865)," in appendices, *Actes de la Société Jurassienne d'émulation,* **16** (1911), **17** (1912), **18** (1913); and *Imprimerie du petit Jurassien* (Moutier, 1911), with biographical and bibliographical sketch and portrait.

Also see Emil Kuhn-Schnyder, in *Neue deutsche Biographie;* J. Thurmann. "Essai sur les soulèvements jurassiques du Porrentruy," 2 vols., I (Porrentruy, 1832), 1-84; II (Strasbourg, 1836), 1-51; and Eugene Wegmann, "L'exposé original de la notion de faciès par A. Gressly (1814-1865)," in *Sciences de la terre,* **9,** no. 1 (1962-1963), 83-119, with facsimile reproductions of critical parts of Gressly's 1838 definition of facies.

Cecil J. Schneer

GREW, NEHEMIAH (*b.* Mancetter, Warwickshire, England, 1641; *d.* London [?], England, 25 March 1712), *plant morphology, plant anatomy.*

Grew was the son of Obadiah Grew, a clergyman and schoolmaster, and Ellen Vicars. After early education at Coventry, he took his B.A. in 1661 at Cambridge, where he was a member of Pembroke Hall. Further study at Cambridge being impossible owing to his religious nonconformity, he qualified for the M.D. at the University of Leiden. Returning to England, Grew practiced first at Coventry and later in London. He relied almost entirely on medicine as a means of livelihood for the rest of his life. He first married Mary Huetson, who died in 1685, then Elizabeth Dodson, by whom he had at least one son and two daughters.

The Royal Society became the focal center of Grew's activities, for in 1672 he was persuaded by some of the fellows, notably John Wilkins, the bishop of Chester, to move from Coventry to London to take up more seriously the study of plant anatomy, in which he had become interested. Fifty pounds were raised by subscription among the fellows to induce him to make this change. Through the Royal Society, Grew also came into contact with Robert Hooke, whose diverse activities included pioneering studies in the field of microscopy. The compound microscope was just coming into use. Hooke was instructed to make the Society's microscope available to Grew. By 1677 both Hooke and Grew were secretaries of the Royal Society.

As a medical man Grew had been interested in the structure of animals before turning to plants. His philosophy and religious beliefs made him regard both plants and animals as "contrivances of the same Wisdom" and he therefore concluded that it would be just as rewarding to study the structure of plants as that of animals. Similar views had already been expressed by Francis Glisson, one of the founders of the Royal Society, in a published passage which Grew subsequently quoted in the preface to his first important book, *The Anatomy of Vegetables Begun.*

Grew was no narrow-minded specialist: besides plant anatomy he was interested in the occurrence of crystalline materials in plant tissues. He also described and illustrated the intestines and related organs of many different kinds of animals. In 1672 he discussed the nature of snow and noticed that it is composed of "icicles" of determined form. By 1675 he was interested in the taste of plants and attempted to classify them accordingly. In 1681 Grew's *Musaeum Regalis Societatis* appeared, a thick volume which not only listed but described in detail, sometimes with illustrations, the objects in the Society's museum at Gresham College. In 1701 Grew published his last great work, *Cosmologia sacra.* This religious and philosophical treatise reflects the beliefs in which he was brought up and which served as a background to his scientific work.

Grew's chief claim to scientific distinction rests on his outstanding contribution to plant anatomy. The suggestion made by some botanists that Grew merely copied Marcello Malpighi's results by referring to the manuscript of his *Anatome plantarum* between the time of its submission to the Royal Society for publication and its ultimate appearance cannot be taken seriously. A careful perusal of Thomas Birch's *History of the Royal Society,* as well as the writings of Agnes Arber and W. Carruthers, shows that there are no grounds for the view that his results were secondhand. Indeed, Grew went to some trouble to demonstrate to fellows of the Royal Society instances in which Malpighi was in advance of him. For example, Malpighi was the first to demonstrate spiral thickenings in vessels. Malpighi and Grew appear, in fact, to have held each other in high scientific regard. Yet communication must have been impeded, for Malpighi could not read English and was less able than Grew to express himself correctly in Latin.

Grew communicated his ideas on plant anatomy to the Royal Society in a series of "discourses" which were so well received that he was asked to publish them in book form. Grew's first three scientific books are much shorter than the fourth, which repeats and elaborates the contents of the first three. The publication of *The Anatomy of Plants* (1682) was therefore the highlight of Grew's career as a plant anatomist. An examination of the text and the profuse illustrations in this great work reveals the tremendous advance in knowledge which it represents. Grew was so successful partly because he started with naked-eye observations and then passed on to higher magnifications. He next elucidated the structures of stems and roots by the combined use of transverse, radial, and tangential longitudinal sections—still the practice today—and also studied obliquely cut surfaces.

Grew's primary aim was to discover the physiological functions of the various tissues. In this he was only partially successful, not surprisingly, for such mechanisms as the ascent of sap and the translocation of foodstuffs are still only partly understood. In Grew's time much energy was dissipated in trying to establish physiological similarities between plants and animals. For example, attempts were made to discover a circulatory system in plants comparable with that in animals. On 23 June 1672 Grew was "desired" by the Royal Society "to discover, whether, whilst plants are growing, there be a peristaltic motion in them." On 7 May 1673 he pointed out to the Society that roots have the power to overcome the resistance of the soil as they grow downward as well as to absorb nourishment. This suggested to some of those present that the downward movement of the roots was sustained by muscular action. In such circumstances it is not surprising that Grew became involved in a controversy with Martin Lister which was mainly an argument about the flow of fluids in the plant body. Grew was more successful in recognizing structural differences in plants with different taxonomic affinities, and in so doing he foreshadowed the modern study of systematic anatomy.

Grew confirmed the existence of cells, already seen by Robert Hooke, but he had no idea that they contain the living substance, protoplasm. But Grew went further than Hooke, for he noted the vessels in wood, the fibers in bark, and the parenchyma of the pith and cortex. Indeed, he was responsible for introducing the term "parenchyma." Grew found that the root consists of a skin, a "cortical body" commonly called the "barque," and a "ligneous body" or vascular core. He found that the cortical body pierces the ligneous body by "inserted pieces," which are evidently the structures now called medullary rays. In the ligneous body he described annual rings. He likened the vessels and fibers, together with the inserted pieces, to the warp and woof of a piece of cloth. He searched in vain to find valves in the vessels. The ascent of sap in the vessels was accounted for by capillarity, and he thought that the vessels were kept supplied with sap from neighboring parenchyma cells which served as cisterns. Grew believed that the sap rises through the wood only during the spring and that it moves through the bark at other times of the year. His concept of vessels was that their structure resembled a ribbon twisted spirally around an imaginary cylindrical object. He and many anatomists who followed him believed that the wood is in some way derived from the bark. Grew recognized the stomata as orifices or "passports" in the skin of leaves, but Malpighi seems to have understood their structure more completely.

Grew made important contributions to plant morphology as well as to anatomy. For example he studied flowers, fruits, and seeds, along with the vegetative organs. In the flower he termed the calyx the "emplacement," the corolla the "foliature," and the stamens and styles the "attire." On pollination he observed that the pollen "falls down upon the seed case or womb and touches it with a prolific virtue or subtle and vivific effluvia."

Unfortunately, Grew worked in circumstances that afforded no opportunity to teach students, and consequently, except for what he published, his knowledge died with him. Grew and Malpighi were more accurately informed about plant structure than their immediate successors, and it was not until the time of the German plant anatomist, Hugo von Mohl (1805–1872), that any really fundamental advances in the subject were made.

BIBLIOGRAPHY

I. Original Works. Grew's scientific writings are *The Anatomy of Vegetables Begun. With a General Account of Vegetation Founded Thereon* (London, 1672); *An Idea of a Phytological History Propounded. Together With a Continuation of the Anatomy of Vegetables, Particularly Prosecuted Upon Roots, and an Account of the Vegetation of Roots, Grounded Chiefly Thereupon* (London, 1673); *The Comparative Anatomy of Trunks, Together With an Account of Their Vegetation Grounded Thereupon* (London, 1675); and *The Anatomy of Plants With an Idea of a Philosophical History of Plants and Several Other Lectures Read Before the Royal Society* (London, 1682). He also wrote *Musaeum Regalis Societatis or a Catalogue and Description of the Natural and Artificial Rarities Belonging to the Royal Society and Preserved at Gresham College. Whereunto is Subjoyned the Comparative Anatomy of Stomachs and Guts* (London, 1681); and *Cosmologia sacra or a Discourse of the Universe as It Is the Creature and Kingdom of God. . . .* (London, 1701).

The following are among the MSS concerning Grew's activities to be found in the library of the Royal Society: "Letter Book," V (1672), 443–446, and VI (1673), 321, dealing with the controversy with Martin Lister; and "Register Book," III (4 Apr. 1672), dealing with the structure of snow; IV (25 Mar. 1675), the description and classification of tastes of plants; V (8 Feb. 1676), on animal anatomy, and (8 Mar.), on salts in plant tissues. There are others in both the *Letter Book* and the *Register Book*, but it is impossible to cite all of them here.

II. Secondary Literature. See Agnes Arber, "Tercentenary of Nehemiah Grew (1641–1712)," in *Nature*, **147** (1941), 630–632; "The Relation of Nehemiah Grew and Marcello Malpighi," in *Chronica botanica*, **6** (1941), 391–392; and "Nehemiah Grew and Marcello Malpighi," in *Proceedings of the Linnean Society of London* (1941), 218–238; Thomas Birch, *History of the Royal Society of London*, 4 vols. (London, 1660–1687); W. Carruthers, "On the Life and Work of Nehemiah Grew," in *Journal of the Royal Microscopical Society*, **129** (1902), 129–141; Robert Hooke, *Micrographia or Some Physiological Descriptions of Minute Bodies Made by Magnifying Glasses With Observations and Inquiries Thereupon* (London, 1665; facs. ed., New York, 1961); M. Malpighi, *Anatome plantarum* (London, 1675; 1679); C. R. Metcalfe, "A Vista in Plant Anatomy," in W. B. Turrill, ed., *Vistas in Botany* (London, 1959), pp. 76–98; and Julius von Sachs, *History of Botany (1530–1860)*, trans. by Henry E. F. Garnsey, rev. by Isaac Bayley Balfour (Oxford, 1906), p. 229–241.

Charles R. Metcalfe

GRIESS, JOHANN PETER (*b.* Kirchhosbach, Germany, 6 September 1829; *d.* Bournemouth, England, 30 August 1888), *chemistry.*

The son of a blacksmith, Griess began his advanced studies in Kassel at the Polytechnic, aimed at an agricultural career, then went on to Jena and Marburg. He was a rebellious and idle student, always in trouble with the authorities, but after a short period at the well-known tar distillery at Offenbach he became more subdued and returned to Marburg to work under A. W. H. Kolbe. His career was launched in 1858, when A. W. von Hofmann, who had been impressed by an early paper, invited him to London.

Griess struck everyone both by the eccentricity of his dress and by the excellence of his work. After three years a well-executed investigation for Allsopp and Sons, the brewers, brought him an appointment as chemist in their brewery at Burton-on-Trent, which he held until his death. Griess married Louisa Anna Mason in 1869; they had two sons and two daughters. He became a fellow of the Royal Society and was one of the founders of the Institute of Chemistry.

Griess's main contribution to chemistry had nothing to do with brewing but stemmed from the early discovery which Hofmann had noted, the formation of a new type of organic nitrogen compound by the action of nitrous acid on certain amines. Between 1860 and 1866 Griess developed the chemistry of this diazo reaction, which was mainly of theoretical interest at first, extending the work of Rafaelle Piria and Ernst Gerland, who had noted the action of nitrous fumes on anthranilic acid, yielding salicylic acid and nitrogen. Griess found that the reaction with picramic acid in alcohol yielded a new type of nitrogen compound for which he devised the name diazodinitrophenol, the first use of the term "diazo." Studies on aniline produced explosive compounds too unstable to have any application. Further studies with other reactions showed that the diazo reaction was a versatile route to new compounds, but it was not until 1864 that, by coupling diazotized aniline with naphthylamine, Griess opened up the general way to a new class of coloring substances. The azo dyes came under intense investigation and thousands were patented.

In 1884 Griess, simultaneously with Böttiger, discovered dyes capable of coloring cotton without a mordant. None of Griess's patents proved lucrative, although others made fortunes. He died content with his position as a practicing brewery chemist pursuing organic research as a hobby.

BIBLIOGRAPHY

A list of Griess's writings may be found in Poggendorff, III, 548–549; and IV, 533.

Several sketches of Griess's life derive from E. Fischer, in *Berichte der Deutschen Chemischen Gesellschaft,* **24** (1891), 1007–1078, with portrait. An appreciation of his chemical work in relation to the structure of dyes is in F. A. Mason, in *Journal of the Society of Dyers and Colourists,* **46** (Feb. 1930), 33–39. See also H. Grossmann, in G. Bugge, ed., *Das Buch der grossen Chemiker,* II (Berlin, 1930), 217–228.

FRANK GREENAWAY

GRIFFITH, RICHARD JOHN (*b.* Dublin, Ireland, 20 September 1784; *d.* Dublin, 22 September 1878), *geology.*

Griffith was the son of Richard Griffith, a wealthy merchant and a member of parliament. His father decided he should follow a career in engineering and mining, and in 1800 sent him to London to study chemistry and mineralogy under William Nicholson, chemist and editor of the *Journal of Natural Philosophy.* This was followed by visits to mining areas, first in England and Wales, and then in Scotland. In Edinburgh, Griffith attended the classes of Thomas Hope, professor of chemistry, and Robert Jameson, professor of natural history; in 1807 he was elected a fellow of the Royal Society of Edinburgh. In 1808 he was nominated as an honorary member of the recently formed Geological Society of London, a clear indication that he had already attracted some esteem.

Griffith returned to Ireland and in 1809 was appointed engineer to a commission inquiring into the nature and extent of the Irish bogs. His reports, published in 1810–1812, contained some geological details. When Richard Kirwan died in 1812, Griffith succeeded him as inspector general of the royal mines in Ireland; in the same year he was appointed mining engineer to the Royal Dublin Society. His duties for the Society were to investigate and report on the coalfields and mining areas, and he was also required to give public lectures on the geology of Ireland.

Griffith's friend G. B. Greenough, president of the Geological Society of London, suggested that he prepare a geological map of Ireland. This was the beginning of an undertaking that eventually resulted in the publication of the first geological map of Ireland. When Griffith was lecturing on geology in 1815, he exhibited a map he had prepared; but there is no record of its contents, and it was not published. Twenty years passed before the question of publication really arose.

The results of Griffith's first coalfield investigation, made for the Royal Dublin Society, were published in 1814 as *Geological and Mining Report on the Leinster Coal District.* This was accompanied by a colored geological map on a scale of .75 inch to the mile and several geological sections. A second report, on the Connaught coal district, also with geological map and sections, was published in 1818. In these reports Griffith stated that although he used Wernerian terms, he expressly dissociated himself from any theoretical implications. A third report, on the coal districts of the Ulster counties of Tyrone and Antrim, although prepared in 1818, did not appear in print until 1829. This was accompanied by sections but not a map.

There was still no suitable topographical map of Ireland to serve as a basis for a large-scale geological map, but in 1825 a trigonometrical survey under the Board of Ordnance was begun, principally to provide a basis for the equitable adjustment of local taxation. On Griffith's recommendation, the survey was to be executed on a scale of six inches to the mile (in England it was on a scale of two inches to the mile). The actual boundaries of the civil units had first to be determined; and since this was not the work of military surveyors, a special Boundary Department was set up with Griffith as director. This work was carried out in advance of the Ordnance Survey and was completed in 1844.

In 1829 Griffith was appointed commissioner for the general valuation of lands, an office which he held until 1868. Shortly afterward he resigned his post with the Royal Dublin Society but declared his intention of continuing his researches toward the completion of a geological map of all Ireland. With over 100 officials working under him, and traveling throughout the country, Griffith was well placed to obtain the information he required. The Board of Ordnance surveyors were also collecting geological information; but since they were following a plan that began in the north of Ireland, their notes were mostly confined to that area.

In 1835 the British Association for the Advancement of Science met in Dublin, and at the meeting Griffith exhibited a manuscript geological map of Ireland which he had prepared from his notes. Although this map does not seem to have been preserved, its main features were copied by John Phillips and shown in his *Index Geological Map of the British Isles,* issued in 1838, on a scale of about twenty-eight miles to the inch. Phillips acknowledged the use of Griffith's "valuable and yet unpublished map." On this map the southwest of Ireland is colored almost entirely as "clay slate and grauwacke slate," or "primary" rocks.

Soon after Phillips' map appeared, a colored geological map of Ireland by Griffith, on a scale of ten miles to the inch, was published in an atlas of maps accompanying the second report of the Irish Railway Commission, dated 13 July 1838. The report included a twenty-five-page "Outline of the Geology of Ireland," and this and the map were also issued separately. This map showed some major advances in Griffith's geological knowledge, such as the substitution of large areas of "old conglomerate" and "Old Red Sandstone" where Phillips' map had shown "clay slate."

The publication of a large-scale (four miles to the inch) map had been delayed by the fact that the topographical map was not yet complete, but in 1838 Griffith colored geologically an unfinished proof impression and exhibited it in August at the British Association meeting at Newcastle-on-Tyne. The engraved map, uncolored, was published in March 1839. Colored copies were supplied to order. It was remarkably detailed, with tablets for twenty-six different colors to indicate the various stratigraphic horizons, as well as different igneous and metamorphic rocks. A noteworthy feature was Griffith's division of the Carboniferous Limestone into five different groups. The main divisions of the strata were retained in later editions of the map, although the number of subdivisions was greatly increased. A reduction of the map (on a scale of about sixteen miles to the inch), dated 1853, had thirty-seven colored tablets; and the final revised edition of the large-scale map, issued in 1855, had over forty. The subdivisions were lithological and did not imply any relative age.

Griffith's particular interest in the Carboniferous Limestone rocks led him to amass a large collection of fossils from the formation; and he employed Frederick McCoy, a young paleontologist, to describe them in a well-illustrated and valuable publication, *A Synopsis of the Characters of the Carboniferous Limestone Fossils of Ireland,* published in 1844 at Griffith's own expense. This was followed by a similar work on the Silurian fossils of Ireland, collected by Griffith and described by McCoy.

In 1854 Griffith was awarded the Wollaston Medal of the Geological Society of London for his services to geological science and particularly for his geological map of Ireland. On the final revised edition, dated 1855, Griffith stated that he had taken some boundaries in the southeast from those on the recently published maps of the Irish Geological Survey. The official survey had begun in 1845, but publication had been delayed until 1855, when the one-inch topographical maps were ready. As the maps of the Survey and the accompanying memoirs were published, Griffith's map gradually became out-of-date.

Griffith was created a baronet in 1858. He continued to hold some public offices and was widely consulted. In 1869, when he was eighty-five, he testified before a select committee of inquiry into valuation; this evidence contains much of interest concerning Griffith's work.

BIBLIOGRAPHY

I. Original Works. A long, but not complete, list of Griffith's papers is in the unsigned obituary in *Geological Magazine* (see below). There is also a list in Royal Society, *Catalogue of Scientific Papers,* III (1866), 17–18.

Works by Griffith published separately are *First Report to the Commissioners on the Bogs of Ireland,* Parliamentary Report (1810) and subsequent reports (1811–1812); *Geological and Mining Report on the Leinster Coal District* (Dublin, 1814); *Geological and Mining Survey of the Connaught Coal District* (Dublin, 1818); *Geological and Mining Surveys of the Coal Districts of the Counties of Tyrone and Antrim* (Dublin, 1829); *Outline of the Geology of Ireland and Geological Map of Ireland,* accompanying *Second Report of the Railway Commissioners* (Dublin, 1838); *A General Map of Ireland to Accompany the Report of the Railway Commissioners Shewing the Principal Physical Features and Geological Structure of the Country* (Dublin, 1839, 1846, 1855); and *Geological Map of Ireland to Accompany the Instructions to Valuators* (Dublin, 1853). Griffith also wrote *Notice Respecting the Fossils of the Mountain Limestone of Ireland as Compared with Those of Great Britain, and Also With the Devonian System* (Dublin, 1842); and, jointly with F. McCoy, *A Synopsis of the Characters of the Carboniferous Limestone Fossils of Ireland* (Dublin, 1844) and *A Synopsis of the Silurian Fossils of Ireland* (Dublin, 1846).

II. Secondary Literature. A lengthy, unsigned obituary notice in *Geological Magazine,* **5** (11 Dec. 1878), 524–528, is the source of several other notices. Some additional biographical details are in H. F. Berry, *History of the Royal Dublin Society* (London, 1915), pp. 162 ff. A very detailed account of the progress of Griffith's map and its geological changes is given by Maxwell Close, "Anniversary Address to the Royal Geological Society of Ireland," in *Journal of the Royal Geological Society of Ireland,* n.s. **5** (1880), 132–148. Some further information is given in A. G. Davis, in "Notes on Griffith's Geological Maps of Ireland," in *Journal of the Society for the Bibliography of Natural History,* **2** (1950), 209–211. A valuable commentary is R. C. Simington and A. Farrington, "A Forgotten Pioneer, Patrick Ganly, Geologist, Surveyor, and Civil Engineer," in *Journal of the Department of Agriculture, Republic of Ireland,* **46** (1949), 2–16; in this paper the geological work of one of Griffith's assistants is described, with much background information.

JOAN M. EYLES

GRIFFITH, WILLIAM (*b.* Ham Common, Surrey, England, 4 March 1810; *d.* Malacca, India, 9 February 1845), *botany.*

Griffith, youngest son of Thomas Griffith, a London merchant, was the great-grandson of Jeremiah Meyer, historical painter to George II and a founder of the Royal Academy. He was educated for the medical profession and was apprenticed to a surgeon in the West End of London. In 1829 he began attending classes at the University of London, as one of John Lindley's students. Here Griffith became acquainted with Nathaniel Wallich, who had collected plants extensively in the Himalayas and Burma. Griffith studied in Paris under the anatomist Charles Mirbel, to whose famed dissertation on *Marchantia polymorpha* was appended Griffith's note on *Targionia hypophylla.* He also studied medical botany with William Anderson at Sir Hans Sloane's garden in Chelsea, where he became acquainted with Franz Bauer, the botanical artist at Kew Gardens, whom he admired for his accurate observations.

Griffith went to India in 1832 as an assistant-surgeon in the service of the East India Company and remained there until his death. In 1835, with Wallich and John MacClelland, a soil expert, he went to Assam as part of the delegation seeking to establish tea production in India. Afterward, Griffith traveled in India and neighboring countries, collecting plants; he was the first European to enter many of these areas. Griffith's goal was to write a flora of India. It was not to be an ordinary flora, since he planned to include information on the ecology, physiology, morphology, and anatomy of the native plants as well as a list of them. A fellow of the Linnean Society, Griffith regularly corresponded with J. D. Hooker, George Bentham, and Robert Wight. In 1842, at Hooker's recommendation, he became director of the Calcutta Botanical Gardens during Wallich's absence and served as professor of botany at Calcutta Medical College. In 1844, anticipating leave to England, Griffith married Miss Henderson, the sister of his brother's wife; but in January 1845 he contracted hepatitis (most likely a complication from repeated malaria attacks) and died the following month. His personal papers were willed to the East India Company, and a rough edition compiled by Griffith's nonbotanical friend MacClelland was published at Calcutta. The papers and a herbarium (estimated at 12,000 species) were shipped to England, where they are in the library of the herbarium at Kew Gardens.

Griffith observed the cryptogams as no earlier worker had, correctly describing the four-tiered antheridia of *Anthoceros* and the minute perispore elaters of *Equisetum.* He believed that all land plants reproduced by a sexual system similar to that of the angiosperms, involving pistils, anthers, and seeds. In attempting to make cryptogams and phanerogams conform, he divided the lower land plants into three classes: the "cryptogamic" plants, including ferns, anthocerotes, lycopods, and horsetails, believed to have no obvious sex organs; the "gymnospermous" plants, *Azolla, Salvinia,* and *Chara,* bearing naked "ovules" analogous to conifers; and the "pistilligerous" plants, the mosses and liverworts, fully equal to angiosperms because they possessed both pistils (archegonia) and anthers (antheridia).

His theories about seed plants were more accurate; Griffith was the first to observe pollen grains in the pollen chamber of a *Cycas* ovule. He attempted to explain the angiospermous ovule and established that the embryo sac exists prior to pollination. He also recognized the necessity of pollen-tube penetration into the nucellus for fertilization. His descriptions of ovules in the Loranthaceae and Santalaceae are noteworthy. Unfortunately, Griffith was unable to ascertain the ultimate fate of the pollen tube; he confused the suspensor of the embryo with the tip of the pollen tube, thus giving tentative approval to Matthias Schleiden's erroneous concept that the embryo comes from the tip of the pollen tube. Griffith's misunderstanding of fertilization in angiosperms undoubtedly contributed to his confusion about sexuality in cryptogams.

By 1850 Wilhelm Hofmeister had confirmed many of Griffith's observations, although his conclusions differed. Hofmeister discovered the function of the pollen tube in embryogeny, thus comprehending the true point of fertilization. He was able to extrapolate this knowledge to the cryptogams and thereby discovered alternation of generations in land plants. Griffith was an astute observer who possessed virtually all the data that Hofmeister later used. It is fair to suppose that had Griffith lived, he might have preceded Hofmeister in recognizing alternation of generations. Certainly he would have become a noted botanist of the nineteenth century.

BIBLIOGRAPHY

I. ORIGINAL WORKS. Griffith's shorter writings are listed in Royal Society, *Catalogue of Scientific Papers,* X, 18–19.

The following longer works were arranged by J. MacClelland and published as *Posthumous Papers: Journals of Travels in Assam, Burma, Bootan, and the Neighboring Countries* (Calcutta, 1847); *Icones plantarum asiaticarum,* 4 vols. (Calcutta, 1847–1854); *Notulae ad plantas asiaticas,* 4 vols. (Calcutta, 1847–1854); *Itinerary Notes of Plants Collected in the Khasyah and Bootan Mountains 1837–1838, in Afghanistan and Neighboring Countries 1839–1841* (Calcutta, 1848); and *Palms of British East India* (Calcutta, 1850).

II. SECONDARY LITERATURE. See the unsigned "Obituary of W. Griffith," in *Proceedings of the Linnean Society of London,* **1** (1838–1848), 239–244; "Obituary of W. Griffith," in *London Journal of Botany,* **4** (1845), 371–375; I. H. Burkill, *Chapters on the History of Botany in India* (Nasik, 1965), pp. 37–74; W. Hofmeister, "La formation de l'embryon des phanérogames," in *Annales des sciences naturelles,* 4th ser., **12** (1859), 1–71; W. J. Hooker, ed., "Works of the Late William Griffith, Esq., F. L. S.," in *London Journal of Botany,* **7** (1848), 446–449; J. M. Lamond, "The Afghanistan Collection of William Griffith," in *Notes from the Royal Botanic Garden, Edinburgh,* **30** (1970), 159–175; and W. H. Lang, "William Griffith, 1810–1845," in F. S. Oliver, ed., *Makers of British Botany* (Cambridge, 1913), 177–191.

ANN M. HIRSCH-KIRCHANSKI
STEFAN J. KIRCHANSKI

GRIGNARD, FRANÇOIS AUGUSTE VICTOR (*b.* Cherbourg, France, 6 May 1871; *d.* Lyons, France, 13 December 1935), *chemistry.*

Grignard developed the reaction that became one of the most fruitful methods of synthesis in organic chemistry. The son of Marie Hébert and Théophile Henri Grignard, foreman and sailmaker at the marine arsenal, Grignard attended the lycée at Cherbourg, the École Normale Spéciale at Cluny, and the University of Lyons. After fulfilling his military service from 1892 to 1893, he completed his studies in mathematics at Lyons and, influenced by a classmate from Cluny, overcame his dislike of chemistry and became an assistant in the chemistry department. He soon began a long association with Philippe Antoine Barbier, the head of the department, who in 1898 investigated the conversion of an unsaturated ketone into the corresponding tertiary alcohol by using methyl iodide and magnesium instead of zinc as called for by the Saytzeff method. When Grignard was looking for a doctoral thesis topic, Barbier recommended that he take up the study of this variation on the Saytzeff reaction.

A survey of the literature on organomagnesium compounds convinced Grignard that such an intermediate compound was formed in Barbier's reaction. He also learned of the difficulties other workers had experienced with organomagnesium compounds which ignite spontaneously in air or in carbon dioxide. He found, however, that E. Frankland in 1859 and J. Wanklyn in 1861 had solved a similar problem with zinc alkyls by keeping them in anhydrous ether. Adapting their method, Grignard treated magnesium turnings in anhydrous ether with methyl iodide at room temperature, preparing what came to be known as the Grignard reagent, which could be used for reaction with a ketone or an aldehyde without first being isolated. On hydrolyzing with dilute acid, the corresponding tertiary or secondary alcohol was produced in much better yield than Barbier had been able to obtain.

Grignard's discovery was reported in a short paper at a meeting of the Académie des Sciences in May 1900. Although he was frequently opposed, Grignard held to the view throughout his life that the organomagnesium compounds he prepared had the formula RMgX and that in anhydrous ether they existed as the etherate which most likely had the formula $(C_2H_5)_2O(R)MgX$.

Grignard submitted his thesis on organomagnesium compounds and their applications in synthesis and received the doctor of physical sciences degree at Lyons on 18 July 1901. The complete thesis was published by the university, and within the year a full abstract appeared in *Chemisches Zentralblatt.* Grignard's method of synthesis thus became widely known and firmly established. By 1908 more than 500 papers dealing with the Grignard reaction had been published. In his thesis Grignard described the preparation of carboxylic acids by the action of carbon dioxide on his reagent; secondary alcohols from aldehydes or formic esters; tertiary alcohols from ketones, esters, acid halides, or anhydrides; and unsaturated hydrocarbons in place of tertiary alcohols. He reported that alcohols react with organomagnesium compounds, as does water, to produce hydrocarbons and that aromatic bromides lead to products analogous to the aliphatic compounds.

Grignard was awarded the Cahours Prize of the Institut de France in 1901 and the Berthelot Medal in 1902. He became lecturer in chemistry at Besançon in 1905, returned to Lyons in 1906 and moved to Nancy in 1909 where he became professor of organic chemistry in 1910. That year he married Augustine Marie Boulant, and Roger, the first of two children, was born in 1911.

In November 1912 the Nobel Prize for chemistry was awarded jointly to Grignard, for his 1900 discovery of the role of organomagnesium compounds in synthesis, and Paul Sabatier, for his discoveries in catalytic hydrogenation made fifteen years earlier. Continuing his research on organomagnesium compounds, Grignard investigated their reactions with epoxides, glycols, and with cyanogen to produce nitriles. In army service from 1914 to 1919 he worked primarily on toluene production and war gases. He succeeded Barbier at Lyons in the fall of 1919, remaining there the rest of his life.

In addition to his organometallic researches, Grignard investigated terpenes; structure determination by ozonization; the condensation of carbonyls; and the cracking, hydrogenation, and dehydrogenation of hydrocarbons.

In recognition of his monumental contributions to chemistry, Grignard was a member or honorary member in the world's major chemical and scientific societies. Honorary doctorates were conferred on him by Louvain in 1927, Brussels in 1930, and an honorary professorship by Nancy in 1931.

BIBLIOGRAPHY

I. Original Works. Grignard announced the discovery of his method of synthesis in "Sur quelques nouvelles combinaisons organométalliques du magnésium et leur application à des synthèses d'alcools et d'hydrocarbures," in *Comptes rendus de l'Académie des sciences,* **126** (1898), 1322. His doctoral thesis, "Sur les combinaisons organomagnésiennes mixtes et leur application à des synthèses d'acides, d'alcools et d'hydrocarbures," in *Annales de l'Université de Lyon,* **6** (1901), 1–116, was also published in Paris later in 1901. Toward the end of his life Grignard began editing a handbook, *Traité de chimie organique,* 23 vols. (Paris, 1935–1954), and his son, Roger, in collaboration with Jean Cologne published his lectures as *Précis de chimie organique* (Paris, 1937).

Lists of Grignard's publications, positions, and honors appear in Charles Courtot, "Notice sur la vie et les travaux de Victor Grignard (1871–1935)," in *Bulletin. Société chimique de France,* **5**, no. 3 (1936), 1433–1472. A collection of addresses by and about Grignard was published by A. Rey, *Victor Grignard (In Memoriam)* (Lyons, 1936).

II. Secondary Literature. Biographical details and evaluations of Grignard's work are given by Charles Courtot in his paper listed above; Henry Gilman, "Victor Grignard," in *Journal of the American Chemical Society,* **59** (1937), 17–19; and Heinrich Rheinboldt, "Fifty Years of the Grignard Reaction," in *Journal of Chemical Education,* **27** (1950), 476–488. For a more recent treatment of the Grignard reaction see Rudolph M. Salinger, "The Structure of the Grignard Reagent and the Mechanism of Its Reactions," in Arthur F. Scott, ed., *Survey of Progress in Chemistry* (New York, 1963), pp. 301–324.

A. Albert Baker, Jr.

GRIJNS, GERRIT (*b.* Leerdam, Netherlands, 28 May 1865; *d.* Utrecht, Netherlands, 11 November 1944), *physiology.*

Grijns was the son of Cornelis Dirk Grijns, a merchant, and Janetta Christina Seret. He attended the Gymnasium at Delft and started his medical studies at the University of Utrecht in 1885. In 1901 he took his M.D. degree, offering a thesis entitled "Bijdrage tot de physiologie van den Nervus opticus." For this investigation he had to work at night, since the very sensitive galvanometer he used was disturbed by traffic during the day. In March 1893 Grijns passed the final examination that gave him the right to practice. A scholarship enabled him to study physiology for six months at Leipzig, under Carl Ludwig. After marrying Johanna Gesina de Wilde on 15 September 1893, Grijns left for the Netherlands East Indies as a medical officer. In his first years in the Far East he treated many patients suffering from beriberi, a disease then very common and of unknown origin.

The Dutch government had, in 1886, sent out the Pekelharing-Winkler commission with instructions to investigate the cause of beriberi. They concluded that most probably an infectious agent, a coccus, was the causative agent. Christiaan Eijkman, later winner of the Nobel Prize in physiology or medicine, was charged with the continuation of their studies. At Batavia, Java, he had a small laboratory for pathological anatomy and bacteriology at his disposal. Grijns became his co-worker but had to join the Atjeh expedition in Sumatra in 1895–1896. Here he observed patients with beriberi but had no time for thorough investigations.

On Eijkman's return to the Netherlands in 1896, Grijns was appointed to continue his investigations. The former had pointed to the close resemblance of human beriberi to polyneuritis gallinarum and had established that feeding only completely polished (overmilled) rice caused polyneuritis in fowls, but that incompletely polished (or husked by hand) rice prevented or even cured the disease. Eijkman firmly believed a bacterium or a poison to be the cause even several years after Grijns, in 1901, had advanced the idea that a deficiency of "protective substances" was the causative factor and that the absence in food of not only proteins, carbohydrates, fats, and minerals, but also of other (still unknown) substances, could result in disease. This idea of "partial hunger" became the starting point and the basis of the modern theory of vitamins. At the same time Grijns gave lessons in anatomy at the School for Native Doctors (S.T.O.V.I.A.) and later taught physiology and ophthalmology, on which he wrote a simple textbook. Except for an interruption during 1902–1904, when he was on leave in Europe because of ill health, he pursued his research until 1912, when he became director of the laboratory.

In 1917 Grijns returned to the Netherlands, and in 1921 he was appointed professor of animal physiology at the State Agricultural University, Wageningen. In the academic year 1929–1930 he served as vice-chancellor. In the year of his retirement (1935), on his seventieth birthday, a committee of honor presented him an English translation of his publications on nutrition (1900–1911) and of his thesis. Because of his brilliant and immensely fruitful idea on nutritional deficiency, he was awarded in 1940 the Swammerdam Medal, inscribed "Hodiernae Nutrimentorum Doctrinae Conditor atque Pater" ("Founder and Father of the Modern Doctrine of Nutrition").

BIBLIOGRAPHY

I. ORIGINAL WORKS. Grijns's classic publications appeared in *Geneeskundig tijdschrift voor Nederlandsch-Indië* (1901–1910). They appeared in English in his *Research on Vitamins 1900–1911* and *Physiology of the Nervus opticus,* his thesis, both translated and republished on the occasion of his seventieth birthday (Gorinchem, 1935). His inaugural address was *Nieuwere gezichtspunten in de voedingsleer* (Gorinchem, 1921).

II. SECONDARY LITERATURE. No full biography of Grijns is available. The best, although short, is inserted in *Research on Vitamins* (see above). Some short notes, all in Dutch, are the following (listed chronologically): E. Brouwer, "Prof. Dr. G. Grijns," in *Landbouwkundig tijdschrift,* **44,** no. 531 (Mar. 1932); N. H. Swellengrebel, "Toespraak tot Prof. Grijns," in *Nederlands tijdschrift voor geneeskunde,* **85** (1941), 120–123, with Grijns's answer; B. C. P. Jansen, "In Memoriam Prof. Dr. G. Grijns," *ibid.,* **90** (1946), 240–241, with portrait; S. Postmus, "Gerrit Grijns 1865–1944," in *Voeding,* **16** (1955), 3–4; and J. F. Reith, "Christiaan Eijkman en Gerrit Grijns," *ibid.,* **32** (1971), 180–195.

GERRIT A. LINDEBOOM

GRIMALDI, FRANCESCO MARIA (*b.* Bologna, Italy, 2 April 1618; *d.* Bologna, 28 December 1663), *astronomy, optics.*

His father, Paride Grimaldi, a silk merchant and member of a wealthy family of noble blood, settled in Bologna in 1589. Paride's first wife died childless, and, about 1614, he married Anna Cattani (or Cattanei). Of her six sons five survived; Francesco Maria was the fourth born, the third surviving. With his father deceased and his mother in possession of her grandfather's chemist's shop, Francesco Maria and his brother Vincenzo Maria, one year older, entered the Society of Jesus on 18 March 1632. Of Francesco's first three years in the novitiate, it is known that the third was spent at Novellara. Following this he went to Parma in 1635 to begin studying philosophy. Within a year that house was closed, and he was transferred to Bologna.

In 1636 he went to Ferrara for the second year of his three-year course in philosophy, while the third year, 1637–1638, was spent in Bologna again. From 1638 to 1642 Grimaldi taught rhetoric and humanities in the College of Santa Lucia at Bologna. From 1642 to 1645 he studied theology. Further study in philosophy brought him a doctorate in 1647, and he was then appointed to teach philosophy. Within a year, however, consumption undermined his health, making it necessary for him to transfer to a less time-consuming task, the teaching of mathematics. According to Riccioli, Grimaldi was well prepared to teach all branches of mathematics—geometry, optics, gnomonics, statics, geography, astronomy, and celestial mechanics. By 1651 he had determined to take the full vows for priesthood and did so on 1 May.

During the 1640's and especially in the 1650's Grimaldi was very active in astronomical and related studies. From 1655 to the end of his life his scientific efforts were devoted essentially to the preparation of *De lumine.* His death came shortly after finishing this work, at the end of an eight-day illness characterized by high fever and headaches.

The astronomical work of Grimaldi was closely tied to the career and interests of Giovanni Battista Riccioli (1592–1671), a Jesuit since 1614, who taught theology for a long time before gaining permission to pursue his love of astronomy. Riccioli was prefect of studies at Bologna and had been dispensed from all teaching in order to prepare his *Almagestum novum* when Grimaldi came under his influence. In 1640 Grimaldi conducted experiments on free fall for Riccioli, dropping weights from the Asinelli tower and using a pendulum as timer. He found that the square of the time is proportional to the distance of free fall from rest. Riccioli credited him as being absolutely essential to the completion, in 1651, of *Almagestum novum,* remarking especially on Grimaldi's ability to devise, build, and operate new observational instruments. Grimaldi's contributions included such measurements as the heights of lunar mountains and the height of clouds. He is responsible for the practice of naming lunar regions after astronomers and physicists.

An especially noteworthy contribution was his selenograph of the moon, a composite from telescopic observations of many phases, accurate and correct enough so that he must have used crossed hairs and a micrometer with his eyepiece. The use of a micrometer eyepiece seems also to have been made in the triangulation and leveling procedures carried out to establish the meridian line for Bologna. In this project, completed by 1655, Riccioli and Grimaldi collaborated with Montalbini and G. D. Cassini. The results were reported in Riccioli's *Geographiae et Hydrographiae Reformatae* (1661). Grimaldi appears to have been responsible for much of the tabular material in the second volume of Riccioli's *Astronomia Reformata* (1665), especially on the fixed stars.

Grimaldi's primary contribution to positive science was the discovery of optical diffraction. A comprehensive treatise on light, the complete descriptive title is *A physicomathematical thesis on light, colors, the rainbow and other related topics in two books, the first of which adduces new experiments and reasons deduced from them in favor of the substantiality of light. In the second, however, the arguments adduced in the first book are refuted and the Peripatetic teaching of the accidentality of light is upheld as probable.* The title page also states that he deals with "the previously unknown diffusion of light; the manner and causes of reflection, refraction, and diffraction; vision and the intentional species of visibles and audibles; the substantial effluvium of the magnet, which pervades all bodies; and in a special argument the atomists are attacked." A final descriptive element appears on the subtitle page preceding book II, where he notes that, in any case, "permanent colors are nothing other than light." If we take Grimaldi at his word, he is presenting two possible basic theses about the nature of light. It may be substance, or it may be accident, i.e., a quality of some other substance. His personal choice appears in his preface to the work, where he says he would be delighted by a student who would be persuaded that the experiments supporting the substantiality of light have no force and who could confirm better than he "the doctrine which we personally embrace and finally sustain in the present opuscule." At various places in book I he prescinds from a substantial theory of light in arguing a proposition, e.g., prop. 10, which deals with the nature of the propagation of light. His position in book I is thus not always in support of the substantial theory of light. As a philosopher he stands against the certainty of either hypothesis, each called an "opinion," on the nature of light. He says ultimately that the many experiments of book I, albeit persuasive, "do not in any way lead to the substantiality of light" (II, 2).

Grimaldi's position on the substance-accident question is better understood by a look at the whole book and what it deals with. Book I (sixty propositions, 472 pages) devotes the first twenty-seven propositions (229 pages) essentially to the four modes of light, the porous nature of bodies, and the propagation of light. Thereafter book I deals with colors and the rainbow (props. 28–60, 244 pages). The substance-accident question is not much debated after prop. 27, nor is it made a necessary basis for the treatment of colors and the rainbow. While books I and II present opposing views on the substantial nature of light, they agree on other major points. In both books Grimaldi opposes any corpuscular theory of light. In both books he is concerned to show color to be nothing more than a modification of light. Color is not the addition of something else to light. Both books agree on the fluid nature of light phenomena. Light may be a fluid substance or the accidents of some other fluid substance(s). Grimaldi expressly chooses the latter version of a fluid theory.

The discussion of diffraction (book I, prop. 1, pp. 1–11) is the basis for introducing a fluid, but not necessarily substantial, view of light. The experiments on diffraction are clear and well described by Grimaldi. He used bright sunlight introduced into a completely darkened room via a hole about 1/60 inch across. The cone of light thus produced was projected to a white screen at an angle so as to form an elliptical image of the sun on the screen. At a distance of ten to twenty feet from the slit he inserted a narrow opaque rod into the cone of light to cast a shadow on the screen. The border of this shadow, he noted, is not clear, and the size of the shadow is far beyond what rectilinear projection would predict. Having demonstrated this, he proceeded to his description of external diffraction bands. These bands are never more than three, and they increase in intensity and in width nearer to the shadow. The series of bands nearest the shadow has a wide central band of white with a narrow violet band nearer the shadow and a narrow red band away from the shadow. Grimaldi warned that the red and violet bands must be observed closely to avoid mistaking the series for alternating bands of light and dark. After describing these parallel bands, he turned to examine the effect of varying the shape of the opaque object. In place of the rod he used a step-shaped object to cast a shadow with two rectangular corners. Still describing external bands, he carefully described the curvature of the bands around the outer corner and continuing to follow the shadow border. When the series approaches the inner corner of the step-shaped shadow, it intersects perpendicularly another series approaching parallel to the other side of this corner. He noted that as they cross each other the colors "are either augmented intensively or are mixed." Nothing more about the appearance of these intersecting bands is found in the description.

In the diffraction experiments he now turned to a description of internal fringes. Here he omitted naming the colors or their order. His diagram shows two pairs of twin contiguous tracks following the border of an L-shaped shadow. These bands are said to appear only in pairs, while the number increases with the width of the obstacle and its distance from the screen. The bands bend around in a semicircle at the end of the L, remaining continuous. At the corner of the L he made a further observation. Here not only do the bands curve around to follow the shadow outline, but a shorter and brighter series of colors appears. He showed these as five feather-shaped fringes radiating from the inside corner of the L and perpendicularly crossing the previously described internal paired tracks of light. The nature of this phenomenon seems to have impressed him as being like the wash of a moving ship.

The final diffraction experiment allowed a cone of light to pass first through two parallel orifices, the first being 1/60 inch and the second being 1/10 inch in diameter. The distances between the holes and between the screen and second hole are equal, at least twelve feet each. The screen is parallel to the orifices. The screen holds a circle of direct illumination just over 1/5 inch across. The circle is significantly wider than rectilinear propagation allows and the border is colored red in part, blue in part. Neither the width nor order of these colors is given.

These diffraction experiments showed Grimaldi that a new mode of transmission of light had been discovered and that this mode contradicts the notion of an exclusively rectilinear passage of light. Diffraction thus gave prima facie evidence for a fluid nature of light. The name "diffraction" comes from the loss of uniformity observed in the flow of a stream of water as it "splits apart" around a slender obstacle placed in its path. He discussed other fluid phenomena analogously with light. To explain color and the varieties of color he decided that a "change in agitation" of the luminous flow is responsible. A light ray is conceived like a column of fluid in vibration, but not regular vibration. Lighter colors are said to result from a greater density of rays and darker colors from a lower density.

In performing his diffraction experiments, Grimaldi gives measurements only where they will show the nonrectilinear propagation of light. No quantities are given for the sizes or distances of the colored fringes in any of his experiments. No notion of periodicity occurred to him.

Knowledge of his work appears in the work of both Hooke and Newton. Hooke performed his first series of diffraction experiments later in 1672, after the notice of Grimaldi's book in the *Philosophical Transactions.* Hooke referred to it, however, as inflexion and may have encountered diffraction phenomena independently. Newton was aware of Grimaldi's work, but only at secondhand, crediting Honoré Fabri as the source of his knowledge on diffraction. At first (1675) Newton described and attempted to account for only the internal fringes. His description shows that he could not have performed the experiment. By 1686 he came to deny the existence of internal fringes on the basis of experiments. In the *Opticks* he described and tried to explain only the external fringes, which he never ceased to regard as a sort of refraction. The essence of Newton's contribution to the knowledge of diffraction is his set of careful measurements, which made clear the periodic nature of the phenomenon.

BIBLIOGRAPHY

I. Original Works. The sole work published under Grimaldi's name or written by him is the posthumous *Physico-mathesis de lumine, coloribus, et iride, aliisque adnexis libri duo, in quorum primo asseruntur nova experimenta, & rationes ab iis deductae pro substantialitate luminis. In secundo autem dissolvuntur argumenta in primo adducta, et probabiliter sustineri posse docetur sententia peripatetica de accidentalitate luminis. Qua occasione de hactenus incognita luminis diffusione, de reflexionis, refractionis, ac diffractionis modo et causis, de visione, deque speciebus intentionalibus visibilium et audibilium, ac de substantiali magnetis efluvio omnia corpora pervadente, non pauca scitu digna proferuntur, et speciale etiam argumento impugnantur atomistae* (Bologna, 1665).

II. Secondary Literature. The sources for Grimaldi's life and personality are minimal. A brief elogium by Giovanni Battista Riccioli is appended to the printed text of Grimaldi's book. Riccioli is also responsible for detailed information on Grimaldi's family in an "Epitome genealogiae Grimaldae gentis," in *Almagestum novum, astronomiani veterum novumque,* I, pt. 2 (Bologna, 1651). A useful biography appears in Angelo Fabrioni, *Vitae Italorum,* III (Pisa, 1779), 373–381.

Other sources with significant amounts of information are the brief (none as long as thirty pages) publications of Roberto Savelli, *Grimaldi e la rifrazione* (Bologna, 1951) and *Nel terzo centenario del "De lumine" di F. M. Grimaldi* (Ferrara, 1966); and esp. of Giorgio Tabarroni, *P. F. M. Grimaldi, bolognese iniziatore della ottica-fisica* (Bologna, 1964) and *Nel terzo centenario della morte de F. M. Grimaldi* (Bologna, 1964).

The best account of his astronomical work is the *Almagestum novum,* which indicates some forty items of which Grimaldi was the source. Jiří Marek, "Les notions de la théorie ondulatoire de la lumière chez Grimaldi et Huyghens," in *Acta historiae rerum naturalium necnon technicarum,* **1** (1965), 131–147, is too eager to attribute ideas to Grimaldi that are not his. The review of *De lumine* cited in the text is *Philosophical Transactions of the Royal Society of London,* **6,** no. 79 (22 Jan. 1672), 3068–3070.

By far the most useful discussion to date of Grimaldi

and his work is Francis A. McGrath, "Grimaldi's Fluid Theory of Light," M.Sc. diss. (University College, London, 1969).

For the importance of Grimaldi's work to Newton, see Roger H. Stuewer, "A Critical Analysis of Newton's Work on Diffraction," in *Isis*, **61** (1970), 188–205.

BRUCE S. EASTWOOD

GRINNELL, JOSEPH (*b.* Indian agency forty miles from Fort Sill [now Oklahoma], 27 February 1877; *d.* Berkeley, California, 29 May 1939), *zoology.*

Grinnell's Quaker father, Fordyce Grinnell, was a physician in the Indian service. He tried private practice in Tennessee but returned to the service in 1880 and went to Dakota Territory, where Joseph found welcome friends among the Indian children. Joseph's mother, Sarah Pratt Grinnell, also a Quaker, was, like her husband, descended from early New England stock. The family moved to Pasadena, California, in 1885, to Pennsylvania in 1888, and returned to Pasadena in 1891. Joseph attended Pasadena High School and earned his B.A. in 1897 at Throop Polytechnic Institute (now California Institute of Technology), spending all his free time making a collection of local birds.

In 1896 he seized an opportunity to spend the summer in Alaska, where he collected birds avidly, and he returned there in 1898 with a group of gold-seekers for eighteen months, during which time his success with birds was much greater than was the group's with gold. Grinnell began graduate work at Stanford University, but his studies were interrupted by typhoid fever. After his recovery (M.A., 1901) he taught at Throop Polytechnic from 1903 to 1908, first as instructor and then as professor.

A chance acquaintance with Annie M. Alexander, a generous benefactress of the new Museum of Vertebrate Zoology at the University of California, led the way for Grinnell to become director of the museum in 1908, a position he held until his death. Under his direction, the museum expanded from its original small building to become a large wing of the zoology building at Berkeley. After receiving his Ph.D. (Stanford, 1913), Grinnell also served at Berkeley, advancing from assistant to full professor of zoology.

Grinnell entered zoology at an exciting time, along with an enthusiastic circle at Stanford and vicinity which included, among others, Walter K. Fisher; Edmund Heller; Robert Evans Snodgrass; Grinnell's professor, Charles H. Gilbert; and the university's president, David Starr Jordan. The initial phase of exploration and classification in zoology had largely been completed and the role of the environment was coming under intense study.

Having informally agreed not to enter Stanford's preempted field of fishes, the Museum of Vertebrate Zoology elected to collect the terrestrial vertebrates of California and adjacent regions. Grinnell worked almost entirely within that ecologically diverse state, comparing animal species that were separated by natural barriers or that varied because of diversity of altitude or climate. He led a seven-year survey of the fauna of a cross section of the Sierra Nevada and another of the Mount Lassen area. He recognized from field studies that no two species can occupy the same ecologic niche and remain separate species, a concept usually attributed to G. F. Gause from later experimental studies.

A painstaking observer, a voluminous notetaker, and a precise writer, Grinnell contributed extensively to the knowledge of distribution and ecology of Californian vertebrates. With dismay he observed the deleterious effects of the state's growing population on the natural environment and became an active conservationist. A tree-surrounded meadow in the Northern California Coast Range Reserve, where he and his wife often camped, is dedicated to their memory.

BIBLIOGRAPHY

I. ORIGINAL WORKS. Grinnell's bibliography, listed in the memorial by his wife cited below, contains more than 550 titles. Among his most significant regional studies are "An Account of the Mammals and Birds of the Lower Colorado Valley With Especial Reference to the Distributional Problems Presented," in *University of California Publications in Zoology,* **12** (1914), 51–294; the valuable *Animal Life in the Yosemite: An Account of the Mammals, Birds, Reptiles, and Amphibians in a Cross Section of the Sierra Nevada* (Berkeley, 1924), written with Tracy I. Storer; and *Vertebrate Animals of Point Lobos Reserve, 1934–35*, Carnegie Institution Publication no. 481 (Washington, D. C., 1936).

Two definitive lists of special value are *Fur-Bearing Mammals of California: Their Natural History, Systematic Status, and Relations to Man,* 2 vols. (Berkeley, 1937), written with J. S. Dixon and J. M. Linsdale; and *Game Birds of California* (Berkeley, 1918), written with H. C. Bryant and T. I. Storer. Grinnell devoted his Sundays to compiling a bibliography on California birds, "Bibliography of California Ornithology," in 3 pts. in *Pacific Coast Avifauna,* no. 5 (15 May 1909); no. 16 (15 Sept. 1924); no. 26 (8 Dec. 1939).

II. SECONDARY LITERATURE. The memorial by his wife, Hilda Wood Grinnell, "Joseph Grinnell: 1877–1939," in *Condor,* **42,** no. 1 (1940), 3–34, is a remarkably straightforward detailed account of Grinnell's life. His Stanford days were touched on in Walter K. Fisher, "When

Joseph Grinnell and I Were Young," *ibid.,* pp. 35–38. His characteristics and impact on students were presented in Alden H. Miller, "Joseph Grinnell," in *Systematic Zoology,* **13,** no. 4 (1964), 235–242.

ELIZABETH NOBLE SHOR

GRISEBACH, AUGUST HEINRICH RUDOLF (*b.* Hannover, Germany, 17 April 1814; *d.* Göttingen, Germany, 9 May 1879), *botany, taxonomy.*

Grisebach was the son of the auditor general Rudolph Grisebach and Louise Meyer, his second wife. His uncle Georg Friedrich Wilhelm Meyer was a well-known botanist and the first to instruct the young Grisebach in botany. As a boy Grisebach began to collect plants and acquired a good knowledge of the native flora. He studied medicine and natural history at Göttingen (1832–1834) and at Berlin (1834–1836) and was *Privatdozent* at Berlin and, from 1837, at Göttingen. In 1839–1840 he traveled through the Balkan peninsula and northwestern Asia Minor. This most important journey of his life led him through regions that were for the greater part botanically unexplored. The two books he published about this journey established his reputation as a botanical taxonomist and phytogeographer. While a student he had explored the western Alps (1833), and later he traveled to Norway (1842), southern France and the Pyrenees (1850), and the Carpathian Mountains (1852). In 1841 he became associate professor, and in 1847 full professor, at the University of Göttingen. He declined various offers of professorships elsewhere.

His scientific career is marked by the close connection of traditional taxonomic investigations and phytogeographic studies. In taxonomy and floristic botany he began his work with a monograph on the genus *Gentiana.* He specialized in Malpighiaceae, Gramineae, and the genus *Hieracium,* and studied the flora of southeastern Europe, Central America and Argentina. His works on the flora of these regions are still well known and used, although of course outdated in detail. *Flora of the British West Indian Islands* has recently been reprinted, and there is a detailed commentary by Stearn.

Grisebach was far ahead of his time in proposing a work, "Flora Europaea," of which only a fragment appeared after his death. Grisebach was not one of the great taxonomists of the time. He was perhaps not primarily interested in the problems of systematics but rather in floristic botany as one of the cornerstones of the great structure of synthetic phytogeography that he envisioned.

In phytogeography, for which he coined the modern term "geobotany" (*Geobotanik*) in 1866, Grisebach was especially influenced by the ideas of Alexander von Humboldt about the effect of climate on the composition of flora, particularly on the so-called physiognomic plant types. Grisebach's main work, *Die Vegetation der Erde nach ihrer klimatischen Anordnung* (1872), drew on his floristic studies, various travels in Europe, his great herbarium, and an intensive study of the contemporary literature. The extent of his reading is apparent in his *Berichte über die Leistungen in der Pflanzengeographie* (1841–1853 and 1868–1876) forerunners of modern "progress reports." His herbarium was of use in such tasks as the calculation of the numbers of endemic species in different parts of the Mediterranean. In the *Vegetation der Erde* Grisebach gave a lively picture of the earth's plants emphasizing the effect of climate on the composition and distribution of the flora. It has been noted that he had an amazing ability to describe the vegetation of countries that he himself had never seen. Grisebach extended the system of physiognomic plant types (*Vegetationsformen*) founded by Humboldt to comprise fifty-four forms, an idea revived and refined in recent times. The limitations of his work are to be found in his relative disregard of historical factors and the imperfect knowledge of the physiological foundations of ecology of his time. Nevertheless, this book has been of great importance as one of the first comprehensive reviews of knowledge of the earth's vegetation.

BIBLIOGRAPHY

I. ORIGINAL WORKS. A complete bibliography of Grisebach's writings compiled by his son Eduard appeared in Grisebach's posthumous *Gesammelte Abhandlungen und kleinere Schriften zur Pflanzengeographie* (Leipzig, 1880) and was reprinted by Stearn, in *Journal of the Arnold Arboretum, Harvard University,* **46** (1965), 250. A short bibliography can be found in the necrology by Reinke (see below).

His most important works are *Genera et species Gentianearum* (Stuttgart–Tübingen, 1838); *Reise durch Rumelien und nach Brussa im Jahre 1839,* 2 vols. (Göttingen, 1841); *Spicilegium florae Rumelicae et Bithynicae,* 2 vols. (Brunswick, 1843–1844); *Flora of the British West Indian Islands* (London, 1859–1864; repr. 1963); *Catalogus plantarum Cubensium* (Leipzig, 1866); *Die Vegetation der Erde nach ihrer klimatischen Anordnung,* 2 vols. (Leipzig, 1872; 2nd ed., 1884–1885); *Plantae Lorentzianae* (Göttingen, 1874); and *Symbolae ad Floram argentinam* (Göttingen, 1879).

II. SECONDARY LITERATURE. The most detailed biographical notes have been published by J. Reinke in *Botanische Zeitung,* **37** (1879), 521–534. This is supplemented (especially as concerns his family) by E. Grisebach,

Geschichte der Familie Grisebach (Hamburg, 1936). The article by O. Drude, in *A. Petermanns Mitteilungen aus J. Perthes Geographischer Anstalt,* **25** (1879), 269–271, emphasizes the importance of his works for phytogeography. There are short biographies by E. Wunschmann, in *Allgemeine deutsche Biographie,* XLIX (Leipzig, 1904), and by H. Dolezal, in *Neue deutsche Biographie,* VII (Berlin, 1966), with an extensive bibliography of secondary literature.

GERHARD WAGENITZ

GRISOGONO, FEDERICO, also known as **Federicus De Chrysogonis** (*b.* Zadar, Dalmatia, Yugoslavia, 1472; *d.* Zadar, 2 January 1538), *cosmography, astrology.*

Grisogono, the son of Antonio de Grisogono and Catarina Giorgi, belonged to one of the most illustrious families of the town of Zadar (Zara). After military adventures in Italy and in France, he studied philosophy and medicine at Padua. He received a doctorate from the University of Padua (1506 or 1507) and then taught astrology and mathematics there. But the career of professor was hardly suitable for this rich aristocrat, and in 1508 he returned to Zadar. He spent the remainder of his life in his native city, administering his property, holding municipal offices, practicing medicine, and making astronomical observations. In 1512 he visited Venice and was prosecuted for his politico-astrological predictions.

In his medical publications Grisogono appears as an aggressive advocate of astrology. His chief contribution to science concerns the theory of the tides. He supposed that the tides result from the combined action of the sun and the moon and that each of these celestial bodies exerts an attraction on the waters lying not only below its zenith position but also, at the same time and with the same intensity, below its nadir. This hypothesis allowed Grisogono to construct a mathematical model which predicted high tide quite accurately, particularly its second appearance during the day.

BIBLIOGRAPHY

I. ORIGINAL WORKS. Only two books by Grisogono are known: *Speculum astronomicum terminans intellectum humanum in omni scientia* (Venice, 1507); and *De modo collegiandi, pronosticandi et curandi febres, nec non de humana felicitate ac denique de fluxu et refluxu maris* (Venice, 1528). The chapter on the tides from the latter was republished in J. P. Galluci, *Theatrum mundi et temporis* (Venice, 1588).

II. SECONDARY LITERATURE. Grisogono's life and medical work are described in M. D. Grmek, "Prinosi za poznavanje života i rada F. Grisogona," in *Radovi instituta Jugoslavenske akademije u Zadru,* **15** (1968), 61–91. An analysis of his hypothesis on the tides is given in Ž. Dadić, "Tumačenja pojave plime i oseke mora u djelima autora s područja Hrvatske," in *Rasprave i gradja za povijest nauka,* **2** (1966), 87–143. Remarks on Grisogono's astrological work can be found in L. Thorndike, *History of Magic and Experimental Science,* V (New York, 1941), 314; and in K. Sudhoff, *Iatromathematiker* (Breslau, 1902), pp. 47–48.

M. D. GRMEK

GRODDECK, ALBRECHT VON (*b.* Danzig, Germany [now Gdansk, Poland], 25 August 1837; *d.* Clausthal, Germany, 18 July 1887), *geology, mineralogy.*

An uncle was a well-known Prussian mine superintendent and through him Albrecht must have come in contact with the mining and metallurgical industry. Groddeck attended the Gymnasium in Danzig until 1856. In 1857 he decided to study metallurgy; but he first worked in the mining industry until 1860. Then he went to the universities of Berlin and Breslau for theoretical training. During vacations he visited the mines of Silesia, and subsequently also those of Mansfeld and the Oberharz, most of which are stratabound deposits. This fact may have left a lasting influence on his patterns of thought in ore genesis. Subsequently he spent two semesters at the mining school at Clausthal. Among his professors, F. D. A. Roemer had the strongest influence on him.

Groddeck worked for a short time as a chemist for mining companies, and in 1864 he was employed as an instructor in mining and ore dressing at Clausthal. In 1867 Roemer retired and Groddeck added mineralogy, geognosy, and paleontology to his teaching schedule and simultaneously became acting director of the school.

During the same year he had presented a doctoral thesis to the philosophy faculty of the University of Göttingen, published the year before under the title "Ueber die Erzgänge des nordwestlichen Oberharzes" (*Zeitschrift der Deutschen geologischen Gesellschaft,* **18** [1866], 693–776). On 1 January 1871 he moved up to the post of director of the School of Mines and on 16 June 1872 he obtained the title *königlicher Bergrath.* In 1880 he passed the technical subjects on to a younger professor, but he soon took over a course in ore geology, a subject he had introduced into the curriculum.

It is obvious that Groddeck's teaching and administrative responsibilities were extremely heavy. This may explain why he was not a prolific writer. Nevertheless, he cooperated in the detailed mapping of the Harz Mountains. He mainly concentrated on two research topics: Roemer had contributed to the paleontological knowledge of the Harz region and other areas; Groddeck continued this work by investigating the lithologic sequences, primarily in the Harz Mountains, as reflected in his booklet *Abriss der Geognosie des Harzes* (1871–1883). His second topic, which historically is probably much more significant and original, was the link between lithology and ore geology. Here he made a major step in a direction that was almost entirely lost for 80 years and which has been rediscovered only recently. It was an observational classification of ore deposits, taking into account the facts of congruence between the host rock and the deposits. In his textbook on economic ore deposits, he came very close to stating that conformable or congruent deposits were contemporaneous, and noncongruent deposits epigenetic. This idea is reflected in his book in the following classification of mineral deposits (p. 84):

I. Bedrock deposits (formed in situ)
- Formed with country rock:
  - A. Layered deposits
    - 1. massive ore strata
    - 2. coprecipitation ore strata (of disseminated ore matter)
    - 3. lenticular ore layers (or strings)
  - B. Massive (nonlayered) deposits
- Formed later than enclosing rock:
  - C. Cavity fillings
    - 1. fissure fillings or dikes
      - a. dikes in massive rocks
      - b. dikes in layered rocks
    - 2. fillings of caves
  - D. Metamorphic mineral deposits

II. Weathering deposits (detrital deposits)

Groddeck's fifty-seven types of ore deposits were classified first according to geometric criteria (layered, vein type, and so on) and second according to composition. In this morphologic trend he was closest of all his contemporaries to the general trend of objectivation, that is, of an introduction of observational as against interpretative criteria in scientific classifications. In botany, zoology, and crystallography, this observational pattern had been followed since the first half of the eighteenth century, whereas in geology, especially in ore geology, old mythologic theories of magic ore sources were still fashionable, and because of Pošepný, had again become accepted dogma in 1890. Consequently, Groddeck was clearly a forerunner of the modern approach, especially the modern French morphological school of thought.

Because of his teaching and administrative duties, his early death, and probably also his less active links with foreign researchers, Groddeck was not very influential in his field, and apparently not nearly as well known in Anglo-Saxon countries as Pošepný, who traveled in North America and whose book on ore deposits was translated into English as early as 1895. Pošepný, and in part also von Cotta, were strong proponents of an almost pan-epigenetic theory of ore genesis, whereas Groddeck showed an independent new approach, linking observations in the country rock with his genetic interpretations. For this independent observation and interpretation of ore features he was rediscovered after 1958; an English translation of his 1879 book on ore deposits is presently being prepared.

His keen scientific mind also led him to propose other new genetic solutions to old problems, thus far explained by complicated hypotheses based more on ideas (projections of ideas) than on observations. For example, he showed with both observation and a sound logic of relations, that the tectonic structure of the Oberharz diabase consists of a simple, compressed saddle-shaped fold. He also proved that the adinole schist of the Oberharz is a normal bed concordant with the siliceous Culm schist, again demonstrating that he was ahead of his time in regard to genetic understanding. He also pioneered observations and interpretations of wall rock alterations. Groddeck's work therefore deserves a more important place in the history of geology than it has up to now been accorded.

BIBLIOGRAPHY

I. ORIGINAL WORKS. A bibliography of Groddeck's works is in Poggendorff, III, 551–552; IV, 537. His major publications are *Abriss der Geognosie des Harzes,* 2 vols. (1871–1883); and his textbook, *Die Lehre von den Lagerstätten der Erze* (Leipzig, 1879).

II. SECONDARY LITERATURE. See A. K. Lossen, "Albrecht von Groddeck," in *Jahrbuch der Preussischen geologischen Landesanstalt u. Bergakademie zu Berlin, 1887* (1888), 109–132; and "Albrecht von Groddeck," in *Neues Jahrbuch für Mineralogie, Geologie und Paläontologie,* **1** (1888), 24.

W. Fischer, *Gesteins- und Lagerstättenbildung im Wandel der wissenschaftlichen Anschauung* (Stuttgart, 1961), refers often to Groddeck's work, but does not fully appreciate the role of his work as compared to that of Cotta and Pošepny.

G. C. AMSTUTZ

GROSSETESTE, ROBERT (*b.* Suffolk, England, *ca.* 1168; *d.* Buckden, Buckinghamshire, England, 9 October 1253), *natural philosophy, optics, calendar reform.*

Grosseteste was the central figure in England in the intellectual movement of the first half of the thirteenth century, yet the only evidence for his life before he became bishop of Lincoln in 1235 is to be found in fragmentary references by Matthew of Paris and other chroniclers, by Roger Bacon, and occasionally in charters, deeds and other records.[1] His birth has been variously dated between 1168 and 1175, but since he is described as "Magister Robertus Grosteste" (the first appearance of his name) in a charter of Hugh, bishop of Lincoln, of probably 1186–1190, the earlier date is the more likely. Tradition places his birth in Suffolk, of humble parentage. He may have been educated first at Lincoln, then at Oxford, and was in the household of William de Vere, bishop of Hereford, by 1198, when a reference by Gerald of Wales suggests that he may have had some knowledge of both law and medicine. After that it seems likely that he taught at Oxford in the arts school until the dispersion of masters and scholars during 1209–1214. He must have taken his mastership in theology, probably at Paris, during this period, some time before his appointment as chancellor of the University of Oxford, although with the title *magister scholarum,* probably about 1214–1221, when he must have lectured on theology.

Grosseteste was given a number of ecclesiastical preferments and sinecures, including the archdeaconry of Leicester in 1229; but in 1232 he resigned them all except for a prebend at Lincoln, writing to his sister, a nun: "If I am poorer by my own choice, I am made richer in virtues."[2] From 1229 or 1230 until 1235 he was first lecturer in theology to the Franciscans, who had come to Oxford in 1224. His influence there was profound and continued after he left Oxford in 1235 for the see of Lincoln, within the jurisdiction of which Oxford and its schools came. He contributed largely to directing the interests of the English Franciscans toward the study of the Bible, languages, and mathematics and natural science. Indispensable sources for this later period of his life are his own letters and those of his Franciscan friend Adam Marsh.

Grosseteste's career thus falls into two main parts, the first that of a university scholar and teacher and the second that of a bishop and ecclesiastical statesman. His writings fall roughly into the same periods: to the former belong his commentaries on Aristotle and on the Bible and the bulk of a number of independent treatises, and to the latter his translations from the Greek. Living at a time when the intellectual horizons of Latin Christendom were being greatly extended by the translations into that language of Greek and Arabic philosophical and scientific writings, he took a leading part in introducing this new learning into university teaching. His commentary on Aristotle's *Posterior Analytics* was one of the first and most influential of the medieval commentaries on this fundamental work. Other important writings belonging to the first period are his commentary on Aristotle's *Physics,* likewise one of the first; independent treatises on astronomy and cosmology, the calendar (with intelligent proposals for the reform of the inaccurate calendar then in use), sound, comets, heat, optics (including lenses and the rainbow), and other scientific subjects; and his scriptural commentaries, especially the *Moralitates in evangelica, De cessatione legalium, Hexaëmeron* and commentaries on the Pauline Epistles and the Psalms. Having begun to study Greek in 1230–1231, he used his learning fruitfully during the period of his episcopate by making Latin translations of Aristotle's *Nicomachean Ethics* and *De caelo* (with Simplicius' commentary), of the *De fide orthodoxe* of John of Damascus, of Pseudo-Dionysius and of other theological writings. For this work he brought to Lincoln assistants who knew Greek; he also arranged for a translation of the Psalms to be made from the Hebrew and seems to have learned something of this language.

Although in content a somewhat eclectic blend of Aristotelian and Neoplatonic ideas, Grosseteste's philosophical thinking shows a strong intellect curious about natural things and searching for a consistently rational scheme of things both natural and divine. His search for rational explanations was conducted within the framework of the Aristotelian distinction between "the fact" (*quia*) and "the reason for the fact" (*propter quid*). Essential for the latter in natural philosophy was mathematics, to which Grosseteste gave a role based specifically on his theory, expounded in *De luce seu de inchoatione formarum* and *De motu corporali et luce,* that the fundamental corporeal substance was light (*lux*). He held that light was the first form to be created in prime matter, propagating itself from an original point into a sphere and thus giving rise to spatial dimensions and all else according to immanent laws. Hence his conception of optics as the basis of natural science. *Lux* was the instrument by which God produced the macrocosm of the universe and also the instrument mediating the interaction between soul and body and the bodily senses in the microcosm of man.[3] Grosseteste's rational scheme included revelation as well as reason, and he was one of the first medieval thinkers to attempt to deal with the conflict between the Scriptures and the new Aristotle. Especially interesting are his discussions of the problems of the eternity or creation of the world, of the relation of will to intellect, of angelology, of divine knowledge of particulars, and of the use of allegorical interpretations of Scripture.

Grosseteste's public life as bishop of Lincoln was informed by both his outlook on the universe as a scholar and his conception of his duties as a prelate dedicated to the salvation of souls. Analogous to corporeal illumination was the divine illumination of the soul with truth. He extended the luminous analogy to illustrate the relationship between the persons of the Trinity, the operation of divine grace through free will like light shining through a colored glass,[4] and the relation of pope to prelates and of bishops to clergy: as a mirror reflects light into dark places, he said in asserting his episcopal rights against the cathedral chapter of Lincoln, so a bishop reflects power to the clergy.[5]

In practice Grosseteste was governed by three principles: a belief in the supreme importance of the cure of souls; a highly centralized and hierarchical conception of the church, in which the papacy, under God, was the center and source of spiritual life and energy; and a belief in the superiority of the church over the state because its function, the salvation of souls, was more vital. Such views were widely accepted, but Grosseteste was unique in the ruthlessness and thoroughness with which he applied them, for example, in opposing the widespread use of ecclesiastical benefices to endow officials in the service of the crown or the papacy. As a bishop he had attended the First Council of Lyons in 1245, and in a memorandum presented to the pope there in 1250 he expounded his views on the unsuitability of such appointments while accepting the papal right to dispose of all benefices. Likewise, his opposition to the obstruction of the disciplinary work of the church by any ecclesiastical corporation or secular authority brought him into conflict both with his own Lincoln chapter and with the crown over royal writs of prohibition when secular law clashed with church law and when churchmen were employed as judges or in other secular offices. Grosseteste was a close friend of Simon de Montfort and took charge of the education of his sons, but the degree to which he shared in or influenced Montfort's political ideals has probably been exaggerated. Above all he was a bishop with an ideal, an outstanding example of the new type of ecclesiastic trained in the universities.

Scientific Thought. Some of Grosseteste's scientific writings can be dated with reasonable certainty, and most of the others can be related to these in an order based on internal references and on the assumption that the more elaborated version of a common topic is the later.[6] From the evidence for his method of making notes on his reading and thoughts to be worked up into finished essays and commentaries,[7] and from these writings themselves, it may be assumed that many of them arose out of his teaching in the schools. Gerald of Wales's description of Grosseteste at Hereford as a young clerk with a manifold learning "built upon the sure foundation of the liberal arts and an abundant knowledge of literature"[8] is borne out by what is probably his earliest work, *De artibus liberalibus.* In this attractive introduction he described how the seven liberal arts at once acted as a *purgatio erroris* and gave direction to the gaze and inclination of the mind (*mentis aspectus et affectus*). Of particular interest is his treatment of music, of which his love became proverbial, and of astronomy. As for Boethius, music for him comprised the proportion and harmony not only of sounds produced by the human voice and by instruments but also of the movements and times of the celestial bodies and of the composition of bodies made of the four terrestrial elements—hence the power of music to mold human conduct and restore health by restoring the harmony between soul and body and between the bodily elements, and the related power of astronomy through its indication of the appropriate times for such operations and for the transmutation of metals. Related to this essay was his phonetical treatise *De generatione sonorum,* which he introduced with an account of sound as a vibratory motion propagated from the sounding body through the air to the ear, from the motion of which arose a sensation in the soul.

Grosseteste developed his mature natural philosophy through a logic of science based on Aristotle and through his fundamental theory of light. In their present form most of the works concerned were almost certainly written between about 1220 and 1235. *De luce* and *De motu corporali et luce,* with his cosmogony and cosmology of light, seem to date from early in this period. The structure of the universe generated by the original point of *lux* was determined, first, by the supposition that there was a constant proportion between the diffusion or "multiplication" of *lux,* corresponding to the infinite series of natural numbers, and the quantity of matter given cubic dimensions, corresponding to some finite part of that series. Second, the intensity of this activity of *lux* varied directly with distance from the primordial source. The result was a sphere denser and more opaque toward the center. Then from the outermost boundary of the sphere *lumen* emanated inward to produce another sphere inside it, then another, and so on, until all the celestial and elementary spheres of Aristotelian cosmology were complete. Another seemingly early work in this series, *De generatione stellarum,* shows Grosseteste dependent on Aristotle in many things but not in all, for he argued that the stars were composed of the four terrestrial elements. Later, in his commentary on the *Physics,* he contrasted the imprecise and arbitrary way man must measure spaces and times with God's absolute measures through aggregates of infinites.

In all these writings Grosseteste made it clear that by *lux* and *lumen* he meant not simply the visible light which was one of its manifestations, but a fundamental power (*virtus, species*) varying in its manifestation according to the source from which it was propagated or multiplied and in its effect according to its recipient. Thus he showed in *De impressionibus elementorum* how solar radiation effected the transformation of one of the four terrestrial elements into another and later, in *De natura locorum,* how it caused differences in climate. An explanation of the tides begun in *De accessione et recessione maris* or *De fluxu et refluxu maris* (if this work is by him)[9] was completed in *De natura locorum,* in which he argued that the rays of the rising moon released vapors from the depth of the sea which pushed up the tide until the moon's strength increased so much that it drew the vapors through the water, at which time the tide fell again. The second, smaller monthly tide was caused by the weaker lunar rays reflected back to the opposite side of the earth from the stellar sphere.

In *De cometis et causis ipsarum* Grosseteste gave a good example of his method of falsification in arguing that comets were "sublimated fire" separated from their terrestrial nature by celestial power descending from the stars or planets and drawing up the "fire" as a magnet drew iron. Later, in *De calore solis* (*ca.* 1230–1235), he produced perhaps his most elegant exercise in analysis by reduction to conclusions falsified either by observation or by disagreement with accepted theory, finally leaving a verified explanation. He concluded that all hot bodies generated heat by the scattering of their matter and that the sun generated heat on the earth in direct proportion to the amount of matter incorporated from the transparent medium (air) into its rays.

Grosseteste set out and exemplified the formal structure of his mature scientific method in his *Commentaria in libros posteriorum Aristotelis,* his *Commentarius in viii libros physicorum Aristotelis,*[10] and four related essays giving a geometrical analysis of the natural propagation of power and light. It seems likely that he began the commentary on the *Posterior Analytics* when he was still a master of arts, that is, before 1209, and completed it over a long period, finishing after 1220 and probably nearer the end of the decade. The commentary on the *Physics* was written later, likewise certainly over a period of years, probably around 1230. It has striking parallels with some of the scientific topics of the *Hexaëmeron* but shows less than even the limited knowledge of Greek found in this work, suggesting that it just precedes it.

For Grosseteste, as for Aristotle, a scientific inquiry began with an experienced fact (*quia*), usually a composite phenomenon. The aim of the inquiry was to discover the reason for the fact (*propter quid*), the proximate cause or natural agent from which the phenomenon could be demonstrated:

> Every thing that is to be produced is already described and formed in some way in the agent, whence nature as an agent has the natural things that are to be produced in some way described and formed within itself, so that this description and form itself, in the very nature of things to be produced before they are produced, is called knowledge of nature.[11]

His method of discovering the causal agent was to make first a *resolutio,* or analysis of the complex phenomenon into its principles, and then a *compositio,* or reconstruction and deduction of the phenomenon from hypotheses derived from the discovered principles. He verified or falsified these hypotheses by observation or by theory already verified by observation.

Besides this double method, Grosseteste used in the analysis of the causal agent as the starting-point of demonstration another Aristotelian procedure, that of the subordination of some sciences to others, for example, of astronomy and optics to geometry and of music to arithmetic, in the sense that "the superior science provides the *propter quid* for that thing of which the inferior science provides the *quia.*"[12] But mathematics provided only the formal cause; the material and efficient causes were provided by the physical sciences. Thus "the cause of the equality of the two angles made on a mirror by the incident ray and the reflected ray is not a middle term taken from geometry, but is the nature of the radiation generating itself in a straight path . . . ."[13] The echo belonged formally to the same genus as the reflection of light, but the material and efficient causes of the propagation of sound had to be sought in its fundamental substance: "the substance of sound is *lux* incorporated in the most subtle air . . . ."[14] This introduced a fundamental addition to the very similar discussion of the propagation of sound in *De artibus liberalibus* and *De generatione sonorum.*

Grosseteste developed his geometrical analysis of the powers propagated from natural agents in the four related essays written most probably in the period 1231–1235. He said in the first, *De lineis, angulis et figuris seu de fractionibus et reflexionibus radiorum:* "All causes of natural effects have to be expressed by means of lines, angles and figures, for otherwise it would be impossible to have knowledge *propter quid* concerning them."[15] The same power produced a physical effect in an inanimate body and a sensation in an animate one. He established rules for the operation of powers: for example, the power was greater the shorter and straighter the line, the smaller the incident angle, the shorter the three-dimensional pyramid or cone; every agent multiplied its power spherically. Grosseteste discussed the laws of reflection and refraction (evidently taken from Ptolemy) and their causes, and went on in *De natura locorum* to use Ptolemy's rules and construction with plane surfaces to explain refraction by a spherical burning glass. "Hence," he resumed, "these rules and principles and fundamentals having been given by the power of geometry, the careful observer of natural things can give the causes of all natural effects by this method." This was clear "first in natural action upon matter and later upon the senses . . . ."[16]

An example of the analysis of a power's producing sensation is provided by Grosseteste's *De colore.* The *resolutio* identified the constituent principles: color was light incorporated by a transparent medium; transparent mediums varied in degree of purity from earthy matter; light varied in brightness and in the multitude of its rays. In the *compositio* he asserted that the sixteen colors ranging from white (bright light, multitudinous rays, in a pure medium) to black were produced by the "intension and remission" of these three variable principles. "That the essence of color and a multitude of the same behaves in the said way," he concluded, "is manifest not only by reason but also by experiment, to those who know the principles of natural science and of optics deeply and inwardly. . . . They can show every kind of color they wish to visibly, by art [*per artificium*]."[17]

The last of these four essays, *De iride seu de iride et speculo,* is the most complete example of Grosseteste's method and his most important contribution to optics. The *resolutio* proceeds through a summary of the principle of subordination and its relation to demonstration *propter quid* into a discussion of the division of optics into the science of direct visual rays, of reflected rays, and of refracted rays, in order to decide to which part the study of the rainbow belonged. It was subordinate to the third part, "untouched and unknown among us until the present time";[18] and it is his treatment of refraction that has the greatest interest.

> This part of optics [*perspectiva*], when well understood, shows us how we may make things a very long distance off appear to be placed very close, and large near things appear very small, and how we may make small things placed at a distance appear as large as we want, so that it is possible for us to read the smallest letters at an incredible distance, or to count sand, or grain, or seeds, or any sort of minute objects.[19]

The reason, as he had learned from Euclid and Ptolemy, was "that the size, position and arrangement according to which a thing is seen depends on the size of the angle through which it is seen and the position and arrangement of the rays, and that a thing is made invisible not by great distance, except by accident, but by the smallness of the angle of vision." Hence "it is perfectly clear from geometrical reasons how, by means of a transparent medium of known size and shape placed at a known distance from the eye, a thing of known distance and known size and position will appear according to place, size and position."[20]

Grosseteste followed this account of magnification and diminution by refracting mediums with an apparently original law of refraction, according to which the refracted ray, on entering a denser medium, bisected the angle between the projection of the incident ray and the perpendicular to the interface. "That the size of the angle in the refraction of a ray may be determined in this way," he concluded, "is shown us by experiments similar to those by which we discovered that the reflection of a ray upon a mirror takes place at an angle equal to the angle of incidence."[21]

It was also evident from the principle that nature always acts in the best and shortest way. Grosseteste went on to use a construction of Ptolemy's to show how to locate the refracted image, claiming again that this "is made clear to us by the same experiment and similar reasonings"[22] as those used in a similar construction for locating the reflected image. The first of these references to experimental verification, since it would have been so inaccurate, may throw doubt on all such references by Grosseteste. As was true for the majority of medieval natural philosophers, most of these references came from books or from everyday experience. Clearly his interest was directed primarily toward theory. Yet he advocated and was guided by the principle of experiment and developed its logic.

Besides these works related to optics, Grosseteste wrote important treatises on astronomical subjects. In *De sphaera,* of uncertain date between perhaps 1215 and 1230, and *De motu supercaelestium,* possibly after 1230, he expounded elements of both Aristotelian and Ptolemaic theoretical astronomy. In a later work, *De impressionibus aëris seu de prognosticatione,* dating apparently from 1249, he discussed astrological influences and, again, his mature explanation of the tides.

More original were Grosseteste's four separate treatises on the calendar: *Canon in kalendarium* and *Compotus;* correcting these, *Compotus correctorius,* probably between 1215 and 1219; and *Compotus minor,* with further corrections, in 1244. He showed that with the system long in use, according to which nineteen solar years were considered equal to 235 lunar months, in every 304 years the moon would be one day, six minutes, and forty seconds older than the calendar indicated. He pointed out in the *Compotus correctorius* (cap. 10) that by his time the moon was never full when the calendar said it should be and that this was especially obvious during an eclipse. The error in the reckoning of Easter came from the inaccuracy both of the year of 365.25 days and of the nineteen-year lunar cycle.

Grosseteste's plan for reforming the calendar was threefold. First, he said that an accurate measure must be made of the length of the solar year. He knew of three estimates of this: that of Hipparchus and Ptolemy, accepted by the Latin computists; that of al-Battānī; and that of Thābit ibn Qurra. He discussed in detail the systems of adjustments that would have to be made in each case to make the solstice and equinox occur in the calendar at the times they were observed. Al-Battānī's estimate, he said in the *Compotus correctorius* (cap. 1), "agrees best with what we find by observation on the advance of the solstice in our time." The next stage of the reform was to calculate the relationship between this and the mean lunar month. For the new-moon tables of the *Kalendarium,* Grosseteste had used a multiple nineteen-year cycle of seventy-six years. In the *Compotus correctorius* he calculated the error this involved and proposed the novel idea of using a much more accurate cycle of thirty Arab lunar years, each of twelve equal months, the whole occupying 10,631 days. This was the shortest time in which the cycle of whole lunations came back to the start. Grosseteste gave a method of combining this Arab cycle with the Christian solar calendar and of calculating true lunations. The third stage of the reform was to use these results for an accurate reckoning of Easter. In the *Compotus correctorius* (cap. 10), he said that even without an accurate measure of the length of the solar year, the spring equinox, on which the date of Easter depended, could be discovered "by observation with instruments or from verified astronomical tables."[23]

As with Grosseteste's optics, it was Roger Bacon who first took up his work on the calendar; and Albertus Magnus first made serious use of his commentary on the *Posterior Analytics,* as did John Duns Scotus of that on the *Physics.* These attentions marked the beginning of a European reputation that continued into the early printing of his writings at Venice, the collecting of his scientific manuscripts by John Dee, and interest in them by Thomas Hobbes.[24]

NOTES

1. See D. A. Callus, ed., *Robert Grosseteste.*
2. *Epistolae,* H. R. Luard, ed., p. 44.
3. E.g., *Hexaëmeron,* British Museum MS Royal 6.E.V (14 cent.), fols. 147v–150v; L. Baur, "Das Licht in der Naturphilosophie des Robert Grosseteste," in *Abhandlungen aus dem Gebiete der Philosophie und ihrer Geschichte. Eine Festgabe zum 70. Geburtstag Georg Freiherrn von Hertling* (Freiburg im Breisgau, 1913), pp. 41–55.
4. *De libero arbitrio,* caps. 8 and 10, in L. Baur, *Die philosophischen Werke des Robert Grosseteste,* pp. 179, 202.
5. *Epistolae,* pp. 360, 364, 389.
6. For the basic work on this question, see Baur, *Die philosophischen Werke;* and S. H. Thomson, *The Writings of Robert Grosseteste*—with the revisions by Callus, "The Oxford Career of Robert Grossetest," *Robert Grosseteste;* A. C. Crombie, *Robert Grosseteste and the Origins of Experimental Science* (1953, 1971); and R. C. Dales, "Robert Grosseteste's Scientific Works," *Commentarius in viii libros.*
7. From William of Alnwick, as first noticed by A. Pelzer. See Callus, "The Oxford Career of Robert Grosseteste," pp. 45–47.
8. Giraldus Cambrensis, *Opera,* J. S. Brewer, ed., I (London, 1861), 249.
9. See R. C. Dales, "The Authorship of the *Questio de fluxu et refluxu maris* Attributed to Robert Grosseteste," in *Speculum,* **37** (1962), 582–588.
10. See the ed. by Dales. Grosseteste wrote probably about 1230 a summary of Aristotle's views in his *Summa super octo libros physicorum Aristotelis.*
11. *Commentarius in viii libros physicorum Aristotelis,* lib. I, Dales, ed., pp. 3–4.
12. *Commentaria in libros posteriorum Aristotelis,* I, 12 (1494), fols. 11r–12r.
13. *Ibid.,* I, 8, fol. 8r.
14. *Ibid.,* II, 4, fol. 29v.
15. *De lineis, angulis et figuris,* in Baur, *Die philosophischen Werke,* pp. 59–60.
16. *De natura locorum, ibid.,* pp. 65–66.
17. *De colore, ibid.,* pp. 78–79.
18. *De iride, ibid.,* p. 73. See L. Baur, *Die Philosophie des Robert Grosseteste,* pp. 117–118; Crombie, *Robert Grosseteste* (1971), pp. 117–124.
19. *De iride,* in Baur, *Die philosophischen Werke,* p. 74.
20. *Ibid.,* p. 75.
21. *Ibid.,* pp. 74–75.
22. *Ibid.,* p. 75.
23. *Compotus,* R. Steele, ed., pp. 215, 259.
24. See Crombie, *Robert Grosseteste* (1971); A. Pacchi, "Ruggero Bacone e Roberto Grossetesta in un inedito hobbesiano del 1634," in *Rivista critica di storia della filosofia,* **20** (1965), 499–502; and *Convenzione e ipotesi nella formazione della filosofia naturale di Thomas Hobbes* (Florence, 1965).

BIBLIOGRAPHY

I. Original Works. The earliest-dated printed ed. of a work by Grosseteste is *Commentaria in libros posteriorum Aristotelis* (Venice, 1494; 8th ed., 1552). It was followed by his *Summa super octo libros physicorum Aristotelis* (Venice, 1498; 9th ed., 1637); *Libellus de phisicis lineis angulis et figuris per quas omnes actiones naturales com-

plentur (Nuremburg, 1503); *De sphaera,* pub. as *Sphaerae compendium* (Venice, 1508; 5th ed., 1531); and *Compotus correctorius* (Venice, 1518). His *Opuscula* (Venice, 1514; London, 1690) includes *De artibus liberalibus, De generatione sonorum, De calore solis, De generatione stellarum, De colore, De impressionibus elementorum, De motu corporali, De finitate motus et temporis* (appearing first as the concluding section of his commentary on the *Physics*), *De lineis, angulis et figuris, De natura locorum, De luce, De motu supercaelestium,* and *De differentiis localibus.* All these essays, with *De sphaera* and the hitherto unprinted *De cometis, De impressionibus aëris* and *De iride,* were published by L. Baur in *Die philosophischen Werke des Robert Grosseteste* (see below). For further modern texts see *Canon in Kalendarium,* ed. by A. Lindhagen as "Die Neumondtafel des Robertus Lincolniensis," in *Archiv för matematik, astronomi och fysik* (Uppsala), **11,** no. 2 (1916); *Compotus, factus ad correctionem communis kalendarii nostri,* R. Steele, ed., in Roger Bacon, *Opera hactenus inedita,* VI (Oxford, 1926), 212 ff.; S. H. Thomson, "The Text of Grosseteste's *De cometis,*" in *Isis,* **19** (1933), 19–25; and "Grosseteste's *Questio de calore, de cometis* and *De operacionibus solis,*" in *Medievalia et humanistica,* **11** (1957), 34–43; *Commentarius in viii libros physicorum Aristotelis . . .,* R. C. Dales, ed. (Boulder, Colo., 1963); and R. C. Dales, "The Text of Robert Grosseteste's *Questio de fluxu et refluxu maris* with an English Translation," in *Isis,* **57** (1966), 455–474. See also *Roberti Grosseteste episcopi quondam Lincolniensis epistolae,* H. R. Luard, ed. (London, 1861).

II. SECONDARY LITERATURE. For the fundamental work of identifying and listing Grosseteste's writings see L. Baur, *Die philosophischen Werke des Robert Grosseteste, Bishop von Lincoln,* vol. IX of Beiträge zur Geschichte der Philosophie des Mittelalters (Münster, 1912); and S. H. Thomson, *The Writings of Robert Grosseteste Bishop of Lincoln 1235–1253* (Cambridge, 1940). For further discussions of his scientific writings with references to additional items, see D. A. Callus, "The Oxford Career of Robert Grosseteste," in *Oxoniensia,* **10** (1945), 42–72; D. A. Callus, ed., *Robert Grosseteste, Scholar and Bishop* (Oxford, 1955); A. C. Crombie, *Robert Grosseteste and the Origins of Experimental Science, 1100–1700* (Oxford, 1953; 3rd ed., 1971) and the comprehensive bibliography therein; and R. C. Dales, "Robert Grosseteste's Scientific Works," in *Isis,* **52** (1961), 381–402. The basic modern biography is still F. S. Stevenson, *Robert Grosseteste, Bishop of Lincoln* (London, 1899), while Callus, *Robert Grosseteste,* judiciously sums up more recent scholarship. The pioneering account of his scientific thought is L. Baur, *Die Philosophie des Robert Grosseteste, Bischofs von Lincoln,* XVIII, nos. 4–6 of Beiträge zur Geschichte der Philosophie des Mittelalters (Münster, 1917).

A. C. CROMBIE

GROSSMANN, ERNST A. F. W. (*b.* Rothenburg, near Bremen, Germany, 16 February 1863; *d.* Munich, Germany, 17 March 1933), *astronomy.*

Grossmann began to study astronomy in 1884 at Göttingen, where he took his doctorate under A. C. W. Schur and Leopold Ambronn in 1891. He was assistant at the Göttingen observatory from 1891 to 1896, at Moritz Kuffner's observatory in Vienna from 1896 to 1898, at the Leipzig observatory from 1898 to 1902, and at the Kiel observatory from 1902 to 1905. In 1905 he became observer at Munich, where he lived for the rest of his life. He retired in 1928.

Grossmann was an enthusiastic and important worker with meridian instruments. All of his work was devoted to questions concerning fundamental astrometric measurements. After examining systematic errors in measurements of double stars in his dissertation, he made careful observations with the meridian circles of all the observatories where he worked. One main area of his research was the theory of atmospheric refraction; his very important examination of existing observations resulted in a value for the constant of refraction of 60.15″, which is still used.

Grossmann's observations of fundamental right ascensions near the celestial pole indicated clearly that the values adopted at that time were affected by systematic errors. Although his attempts to measure stellar parallaxes by a meridian circle were unsuccessful, they convinced astronomers that the photographic method is better. In 1921 he showed that existing observations of the planet Mercury were not sufficiently accurate to permit determination of a reliable value of the relativistic motion of its perihelion. It was more than twenty years later that a new comprehensive discussion of all observations of Mercury made between 1765 and 1937, which was undertaken by the U.S. Naval Observatory at Washington, showed convincingly that the observed value of the motion of perihelion was in agreement with the theory of relativity.

BIBLIOGRAPHY

Grossmann's major works are "Untersuchungen über die astronomische Refraktion," in *Abhandlungen der K. Bayerischen Akademie der Wissenschaften,* Math.-phys. Kl., **28,** no. 9 (1917), 1–72; "Die Bewegung des Merkurperihels nach den Arbeiten Newcombs," in *Astronomische Nachrichten,* **214** (1921), 41–54; and "Parallaxenbestimmungen am Meridiankreise," in *Neue Annalen der K. Sternwarte in München,* **5,** no. 1 (1926), 1–173.

There is no secondary literature.

F. SCHMEIDLER

GROSSMANN, MARCEL (*b.* Budapest, Hungary, 9 April 1878; *d.* Zurich, Switzerland, 7 September 1936), *mathematics.*

Grossmann was the son of Jules Grossmann, a businessman, and Henriette Lichtenhahn. He took his final secondary school examination in 1896 in Basel, where his family had moved. He then studied mathematics at the Zurich Polytechnikum (later named the Eidgenössische Technische Hochschule) and in 1900 became an assistant to the geometer W. Fiedler. He earned his doctorate from the University of Zurich in 1912 with a work entitled *Über metrische Eigenschaften Kollinearer Gebilde* (Frauenfeld, 1902). He became a teacher at the cantonal school in Frauenfeld in 1901 and at the Oberrealschule in Basel in 1905. He was appointed professor of descriptive geometry at the Eidgenössische Technische Hochschule in 1907. In 1903 he married Anna Keller.

Grossmann was a classmate of Albert Einstein. When Einstein sought to formulate mathematically his ideas on general relativity theory, he turned to Grossmann for assistance. Grossmann discovered that the law of gravitation could be stated in terms of the absolute differential geometry first developed by E. Christoffel (1864), and later by M. M. G. Ricci together with T. Levi-Civita (1901). Grossmann and Einstein set forth their fundamental discoveries in the joint works cited in the bibliography.

Grossmann was a teacher of outstanding ability and he gave many mathematicians and engineers their training in geometry. His lectures were published in textbooks that enjoyed a large success.

BIBLIOGRAPHY

I. ORIGINAL WORKS. Grossmann's works include *Der mathematische Unterricht an der Eidgenössischen Technischen Hochschule,* Commission Internationale de l'Enseignement-mathématique, no. 7 (Basel–Geneva, 1911); "Mathematische Begriffsbildungen zur Gravitationstheorie," in *Vierteljahrsschrift der Naturforschenden Gesellschaft in Zurich,* **58** (1913), 291–297; and *Darstellende Geometrie* (Leipzig, 1915), with many other eds.

He collaborated with Einstein on "Entwurf einer verallgemeinerten Relativitätstheorie und einer Theorie der Gravitation," in *Zeitschrift für Mathematik und Physik,* **62** (1913), 1–38; the work is in 2 parts: I, "Physikalischer Teil" (pp. 1–22), is by Einstein, and II, "Mathematischer Teil" (pp. 23–38), is by Grossmann. The other collaboration is "Kovarianzeigenschaften der Feldgleichungen," *ibid.,* **63** (1914), 215–225.

II. SECONDARY LITERATURE. For information on Grossmann, see F. Bäschlin, "Marcel Grossmann," in *Schweizerische Zeitschrift für Vermessungswesen, Kulturtechnik und Photogrammetrie,* **34** (1936), 243 ff.; L. Kollros, "Prof. Dr. Marcel Grossmann," in *Verhandlungen der Schweizerischen naturforschenden Gesellschaft,* **118** (1937), 325–329, with portrait and bibliography; and W. Saxer, "Marcel Grossmann," in *Vierteljahrsschrift der Naturforschenden Gesellschaft in Zurich,* **81** (1936), 322–326, with bibliography.

JOHANN JAKOB BURCKHARDT

GROTE, AUGUSTUS RADCLIFFE (*b.* Aigburth, near Liverpool, England, 7 February 1841; *d.* Hildesheim, Germany, 12 September 1903), *entomology.*

Grote was the son of Friedrich Rudolf Grote, a German from Danzig, and Anna Radcliffe, daughter of a Welsh ironmaster. As a youth he immigrated to Staten Island, New York, where his parents had purchased a farm. Grote's formal education was interrupted by the panic of 1857, and although by his own account he continued his studies on the Continent, the only degree he is known to have taken was the honorary M.A. conferred in 1874 by Lafayette College in Pennsylvania.

Grote's first papers on the Lepidoptera were published in 1862, and he rapidly became an authority on the taxonomy of the order, especially that of the noctuid moths. Many of his early studies were written with Coleman T. Robinson. For almost two decades, except for a residence of several years in Demopolis, Alabama, Grote held various positions at the Buffalo (New York) Society of Natural Sciences. After the death of his father in 1880 he left his work at the society and the editorship of the *North American Entomologist* and returned to Staten Island.

Pressed by debts in the following year, Grote sold his valuable collection of Lepidoptera to the British Museum, and permanently left the United States in 1884. He took up residence in Bremen and later in Hildesheim, where he became an honorary curator of the Roemer-Museum. He died of endocarditis. Grote married twice. His first wife, Julia, died after the birth of their second child; his second wife, Gesa Maria, survived him.

One of the leading American entomologists of the nineteenth century, Grote was the first in the United States to study the Noctuidae in real depth, giving attention to the insufficient or confusing species descriptions of some European taxonomists. Although he investigated most areas of lepidopterology, his greatest contribution was the accurate description of a vast number of species. Almost 1,250 of Grote's names are included (many as synonyms) in current checklists, as are over 140 credited to Grote and Robinson.

Grote published over 600 papers in numerous journals. A composer and accomplished organist, he also found time to write poetry and popular articles on science. Although neither a Fiske nor a Huxley, he entered the controversy over science and religion, and several of his books suggested a logical conciliation on the basis of relative value.

BIBLIOGRAPHY

I. ORIGINAL WORKS. Citations for most of Grote's entomological papers and checklists are included in W. Horn and S. Schenkling, *Index Litteraturae Entomologicae . . . bis inklusive 1863,* II (Berlin, 1928), 465; and W. Derksen and U. Scheiding-Göllner, *Index Litteraturae Entomologicae 1864–1900,* II (Berlin, 1965), 212–221.

Among his full-length works are the semipopular *An Illustrated Essay on the Noctuidae of North America* (London, 1882) and the two books on science and religion, *Genesis I–II: An Essay on the Bible Narrative of Creation* (New York, 1880), and *The New Infidelity* (New York, 1881). Grote's poetry was collected as *Rip van Winkle: A Sun Myth and Other Poems* (London, 1882).

II. SECONDARY LITERATURE. The only extensive biographical summary is Ronald S. Wilkinson, intro. to the repr. ed. of Grote's *An Illustrated Essay on the Noctuidae of North America* (Hampton, Middlesex, 1971), in which the earlier lit. is cited. Grote's own autobiographical sketch was used by C. J. S. Bethune in his obituary, "Professor Augustus Radcliffe Grote," in *Report of the Entomological Society of Ontario . . . 1903* (1904), 109–112.

RONALD S. WILKINSON

GROTH, PAUL HEINRICH VON (*b.* Magdeburg, Germany, 23 June 1843; *d.* Munich, Germany, 2 December 1927), *mineralogy, crystallography.*

Following a trip to St. Petersburg in 1840 Groth's father, Philipp Heinrich August Groth, lived in Magdeburg and then, from 1845, worked in Dresden as a portrait painter. His mother, Marie Steffen, was a daughter of a businessman in Frankfurt an der Oder. After attending the Kreuzschule in Dresden from 1855 to 1862, Groth studied at the Freiberg Mining Academy and at the Dresden Polytechnical School. In 1865 he entered the University of Berlin to study physics and mineralogy. He received his doctorate in 1868 and until 1870 was an assistant to the physicist Gustav Magnus. He qualified as a lecturer at the University of Berlin in 1870 and from 1870 to 1872 taught mineralogy and geology at the mining academy in Berlin. In April 1872 he assumed the new professorship of mineralogy at the University of Strasbourg, where he established a mineral collection whose catalog (1878) was considered a model of the type. On 1 September 1883 he succeeded F. von Kobell as professor of mineralogy at the University of Munich and as director of the Bavarian State Collection, which he enlarged primarily in the areas of Alpine minerals and of mineral deposits. He retired on 1 April 1924 and devoted his time to the history of science.

Groth's first mineralogical work (1866) dealt with the titanite he discovered in the Plauenscher Grund, near Dresden, a substance that J. D. Dana named grothite in 1867. (A silicate probably related to harstigite was named grothine by F. Zambonini in 1913.) As a student of F. A. Breithaupt, Groth paid particular attention to paragenesis of minerals. In 1885, in Dauphiny, he accounted for the dependence of axinite-epidote occurrences in amphibole schists and of anatase-turnerite occurrences in gneiss by the leaching of the surrounding rock. His *Topographische Übersicht der Minerallagerstätten* (1917) was one of the best surveys in its time.

Groth's most important contribution to science was his explanation of the connections between chemical composition and crystal structure. Although he did not succeed in determining the optical properties of potassium permanganate through interpolation from isomorphic mixtures with potassium perchlorate, he did recognize the crystallographic peculiarities of mixed crystals. Comparison of the analogies between crystals of the same system with similar interfacial angles led him to a new definition of isomorphism (1874) as requiring the capacity to form homogeneous mixed crystals (isomorphic mixtures) as well as the growth of crystals of each end member in solution with the other.

Systematic measurements of the influences on the crystal form of benzene derivatives with the substitution of hydroxyl, nitro, and ammonia groups, or halides or alkali metals led Groth to call this influence "morphotropy" (1870) and to conceive of, for example, mononitrophenol, dinitrophenol, and trinitrophenol as a morphotropic series. In this regard he also spoke of the morphotropic force of an element or a group of atoms, asserting that the manifestation of such a force depended on the specific morphotropic force of the atom or group of atoms (or both) that is being substituted on (1) the chemical nature of the compound in which the substitution takes place, (2) the crystal system of the compound being altered, and (3) the position of the entering group relative to the other atoms in the molecule. Sometimes the elastic deforming force changes only an axial length, but it may also cause a predictable change in the crystal system. The deforming effect is necessarily greater in regular crystals than in other crystal systems, because in the former a change of angle is not possible without a change in the system.

In 1870 Groth began the lectures that he published in 1876 as the textbook *Physikalische Krystallographie.* With his students he systematically investigated the optical, thermal, elastic, magnetic, and electrical properties of crystals. In 1871 Groth improved the polariscope, the stauroscope, the axial-angle instrument, and the goniometer and combined them into a universal instrument. In 1890 he simplified the reflecting goniometer and modified Koch and Emil Warburg's device for determining the coefficients of elasticity in circular plates. His most important finding came in 1876, when he determined that crystallographically equivalent orientations are also always physically equivalent and hence that every geometric plane of symmetry of a crystal is also a physical plane of symmetry.

In 1895, in the third edition of *Physikalische Krystallographie,* Groth presented for the first time a derivation based on Leonard Sohncke's ideas of the crystal forms from the simplest to the highest symmetry and discussed the theory of the space lattice. In 1904 he provided this definition:

> A crystal consists of regular systems of points, placed within each other, each of which is formed of similar atoms; each of these systems of points belongs to a number of lattices, placed within each other, each of which is formed of similar atoms in parallel position; all the lattices of such a structure are congruent, that is, their elementary parallelepiped is the same [*Zeitschrift für Kristallographie,* **54** (1915), 67].

During the period in which other physicists and mineralogists showed scant interest in the space lattice theory, Groth "maintained, through his teaching in Munich, the Sohnckian tradition" (Max von Laue, *Geschichte der Physik* [1947], p. 119).

In accordance with this definition Groth treated structural change resulting from substitution of another atom or group of atoms (or both) as a homogeneous deformation and expressed the dimensions of the unit cell by the topical parameters ψ, χ, ω, of his co-worker W. Muthmann (1894), in the equations

$$\psi = \sqrt[3]{\frac{V}{ac \sin \beta \sin \gamma \sin A}}; \quad \chi = a\psi; \quad \omega = c\psi,$$

where in the triclinic case a, b, c are the axial lengths (expressed as a ratio $a:1:c$) and α, β, γ are the angles. V is molecular weight divided by density, or equivalent weight; and A is the angle opposite side α in the spherical triangle with sides α, β, γ (*Einleitung in die chemische Krystallographie,* p. 26).

In his *Chemische Krystallographie* (1906–1919) Groth compiled crystallographic and physical data on more than 7,000 substances, thereby facilitating their positive identification; but his data did not enable him to give a complete explanation of the relationships between chemical composition and crystal form—and above all he could not explain atomic structures. There is a certain element of tragedy in the fact that the first X-ray structural analyses of diamond, sphalerite, rock salt, fluorite, pyrite, and calcite misled Groth (1914) into thinking that molecules could no longer be mentioned in connection with crystals and were confined to gases, liquids, and colloids. In crystallization, molecules necessarily assumed a reciprocal orientation, a parallel or "twin" position.

> In the union of two or more molecules into a single crystal particle there emerge, in the place of the earlier, internal atomic bonds, bonds between the atoms of adjacent molecules. . . . That parts of the molecule's internal bonds enter into the crystal structure is shown by the . . . previously observed relationship between the structure of the chemical molecule and the crystal structure, that is, crystal form [*Berichte der Deutschen chemischen Gesellschaft,* **47** (1914), 2064].

Here Groth was thinking especially of the persistence of organic ring bonds. Earlier he had pointed out the limited tendency to crystallization of very large organic molecules and the preservation of enantiomorphic molecules in the crystal structure, but he was unwilling to accept a molecular lattice.

Groth's importance to modern structural research and to chemistry has been aptly expressed by E. H. Kraus:

> Many of his views on morphotropy and isomorphism, and on chemical crystallography in general have become firmly embodied in chemical literature. Furthermore, the remarkable advances in our knowledge of crystal structure as the result of the development of X-ray analysis, dating from 1912, are in large measure due to Groth's long and enthusiastic advocacy of the point system theory of crystal structure [*American Mineralogist,* **13** (1928), 96].

Groth was a member of the academies of science of Munich, St. Petersburg, and Vienna, of the National Academy of Sciences of the United States, the Accademia dei Lincei of Rome, the Royal Society, and the Geological Society of London, whose Wollaston Medal he received in 1908. He was also an honorary member of the Mineralogical Society of London, the Mineralogical Society of America, the French Society of Mineralogy and Geology, the Chemical Society of London, and the German Chemical Society, and received honorary doctorates from the universities of Cambridge, Geneva, and Prague. Groth founded the *Zeitschrift für Kristallographie und Mineralogie* in 1877 and from that year until 1920 edited its first fifty-five volumes.

BIBLIOGRAPHY

I. ORIGINAL WORKS. Groth's writings include *Tabellarische Übersicht der einfachen Mineralien* (Brunswick, 1874; 2nd ed., 1882; 3rd ed., 1889; 4th ed., 1898); *Physikalische Krystallographie* (Leipzig, 1876; 2nd ed., 1885; 3rd. ed., 1895; 4th ed., 1905); *Die Mineraliensammlung der Kaiser-Wilhelm-Universität Strassburg* (Strasbourg, 1878); *Grundriss der Edelsteinkunde* (Leipzig, 1887); *Führer durch die Mineraliensammlung des bayerischen Staates in München* (Munich, 1891); *Einleitung in die chemische Krystallographie* (Leipzig, 1904); *Chemische Krystallographie,* 5 vols. (Leipzig, 1906–1919), also in photocopy (University Park, Pa., 1959); *Elemente der physikalischen und chemischen Krystallographie* (Munich–Berlin, 1921); *Mineralogische Tabellen* (Munich, 1921), written with K. Mieleitner; *Entwicklungsgeschichte der mineralogischen Wissenschaften* (Berlin, 1926, repr. 1970); and "Vorgeschichte, Gründung und Entwicklung der *Zeitschrift für Kristallographie* in den ersten fünfzig Jahren," in *Zeitschrift für Kristallographie,* **66** (1928), 1–21.

There is a full bibliography by K. Mieleitner, "Verzeichnis der Arbeiten P. H. von Groth's" in *Zeitschrift für Kristallographie,* **58** (1923), 3–6, a special issue commemorating Groth's eightieth birthday; and in Poggendorff.

II. SECONDARY LITERATURE. On Groth or his work, see G. Menzer, in *Neue deutsche Biographie,* VII (1966), 167–168; and C. Schiffner, in *Aus dem Leben alter Freiberger Bergstudenten* (Freiberg, 1935), pp. 339–341. The numerous obituaries are listed by Menzer and Poggendorff.

WALTHER FISCHER

GROTTHUSS, THEODOR (CHRISTIAN JOHANN DIETRICH) VON (*b.* Leipzig, Germany, 20 January 1785; *d.* Geddutz, near Jelgava, Courland, Russia [now Lithuanian S.S.R.], 26 March 1822), *chemistry, physics.*

Grotthuss came from an old and distinguished family of Courland chancellery nobility. He was born while his parents were abroad. His father, an amateur composer and collector of natural science material, died while still young. Grotthuss lived on his mother's estate, Geddutz, and received a good education there. From 1803 to 1808 he completed his education in science in Leipzig, Paris, Naples, and Rome, and was an auditor at the École Polytechnique in Paris, studying with Antoine de Fourcroy, Claude Berthollet, Louis Vauquelin, and Domenico Morrichini, among others. In 1805, while in Italy, he presented an original explanation of the electrolysis of water, which postulated that molecules of water and salt are polarized and, under the influence of the electric poles, form in the solution electromolecular chains whose members at each end are discharged at the opposite poles of the current. The mechanism of electroconductivity according to Grotthuss was generally accepted until the appearance of the electrolytic dissociation theory and is now used to explain the anomalous high electroconductivity of hydrogen and hydroxyl ions.

After his return from France, Grotthuss spent the last part of his life on his mother's estate, not far from Jelgava, the capital of Courland, where in seclusion he conducted scientific experiments and constructed new theories; only in 1812, to save himself from Napoleon's invasion, did he go to St. Petersburg for six months. Grotthuss reported the results of his work to the Courland Society of Literature and Art, of which he was an active member; he had his articles published in the proceedings of this society, as well as in German journals of chemistry and physics (Johann Schweigger's *Journal für Chemie und Physik,* L. W. Gilbert's *Annalen der Physik,* Adolph Gehlen's *Neues allgemeines Journal der Chemie*) and in A. N. von Scherer's *Allgemeine nordische Annalen der Chemie,* published in St. Petersburg from 1819 to 1822. His articles and notes amount to more than seventy.

In the period from 1808 to 1822 Grotthuss discovered experimentally the basic laws of photochemistry (that a chemical reaction can be caused only by the light absorbed by a substance and that the chemical effect of light is proportional to the time of exposure [1818]), produced original theories on the nature of phosphorescence and color (1815), and attempted to develop a unified electromolecular conception of various chemical and physical phenomena (which anticipated certain elements of the modern kinetic-molecular theory). In studying the flames of gas mixtures Grotthuss came to the conclusion that components of the mixtures (i.e., individual gases such as H_2 and O_2) react among themselves only at a certain concentration (pressure), that a gas mixture in a narrow tube will not ignite, and that a spark or an open flame is necessary for an explosion (1811). Humphry Davy used these results in his construction of the miner's safety lamp (1815). In later years Grotthuss and Davy carried on a polemic concerning the explanation for the action of this lamp.

Grotthuss worked out detailed methods for obtaining, and studied the properties of, thiocyanic (sulfocyanic) acid and its salts, and discovered an analytical application of the reaction of trivalent iron and divalent cobalt with thiocyanides (1817–1818). At the same time as J. W. von Goethe, who was investigating sulfur sources at Bad Berka, Grotthuss suggested that sulfur sources in nature were formed as a result of the reduction of gypsum deposits by organic substances (1816). He worked on the analysis of meteorites and proposed original theories of their origins (1819–1821). He first observed the phenomenon of electrostenolysis in passing an electric current through very narrow cracks (1818).

Grotthuss was a very versatile chemist and physicist whose research received well-deserved recognition from his contemporaries, especially in Germany and Russia. He was elected a corresponding member of the Turin and Munich academies of science and an honorary member of the Société Galvanique in Paris. Many of Grotthuss' ideas contributed to the theoretical development of the kinetic theory, the theory of electrolytic dissociation, the electromagnetic theory of light, and contemporary theories of luminescence.

In the last years of his life Grotthuss' hereditary illness grew acute; as a result of his great suffering he committed suicide at the age of thirty-seven. He left his estate, archives, and library to found a chair of physics and chemistry at Jelgava, and in his will he freed his serfs from their taxes and obligations.

Grotthuss' scientific legacy in the field of electrochemistry was "discovered" by Wilhelm Ostwald, who did much to popularize it, at the end of the nineteenth century.

BIBLIOGRAPHY

I. Original Works. Grotthuss' writings include *Physisch-chemische Forschungen* (Nuremberg, 1820); and *Abhandlungen über Elektrizität und Licht,* R. Luther and A. von Oettingen, eds. (Leipzig, 1906), Ostwald's Klassiker der exacten Wissenschaften, no. 152.

II. Secondary Literature. On Grotthuss or his work, see W. Ostwald, *Elektrochemie, ihre Geschichte und Lehre* (Leipzig, 1896), pp. 309–316 and *passim;* J. Stradins, "The Work of Theodore Grotthuss and the Invention of the Davy Safety Lamp," in *Chymia,* **9** (1964), 125–145; and *Theodor Grotthuss, 1785–1822* (Moscow, 1966).

J. P. Stradins

GROVE, WILLIAM ROBERT (*b.* Swansea, Wales, 11 July 1811; *d.* London, England, 1 August 1896), *electrochemistry, physics.*

Grove was the only son of John Grove, magistrate and deputy lieutenant for Glamorganshire, and his wife, Anne Bevan. He was educated privately and at Brasenose College, Oxford, graduating B.A. in 1832 and M.A. in 1835. He became a barrister, but apparently because of ill health soon turned from law to science, toward which he had always had an inclination. He soon gained a reputation in the comparatively new but rapidly growing science of electrochemistry, particularly with his development of the Grove cell, an improved form of voltaic cell which became very popular. It was used, for example, by Faraday in his lecture demonstrations at the Royal Institution.

Grove was elected a fellow of the Royal Society in 1840 and from 1841 to 1846 was professor of experimental philosophy at the London Institution. In 1837 he had married Emma Maria Powles, who died in 1879; they had two sons and four daughters. In order to meet the financial needs imposed by a growing family, although without entirely abandoning scientific pursuits, Grove returned to the practice of law and became a Queen's Counsel in 1853. In 1856 he defended William Palmer, the "Rugeley poisoner," in a famous murder trial. He became a judge in 1871, and although it was thought that his special knowledge would be particularly valuable in trying cases involving infringement of patents, it was found that he became more interested in the subject of the patent, sometimes suggesting improvements, than in the bare legal aspects of the case.

One of the main defects of early zinc-copper cells was polarization, due to the accumulation of a film of hydrogen bubbles on the surface of the copper plate—this film not only had a high resistance, thus weakening the current, but produced a back emf. Polarization was overcome to some extent as early as 1829 by Antoine-César Becquerel, who used two liquids separated by a porous partition. In the first practical application of the two-liquid principle, devised by J. F. Daniell, the copper sulfate solution in contact with the copper plate was separated from the sulfuric acid containing the zinc plate by unglazed earthenware. This arrangement gave a reasonably constant emf of about 1.1 volts.[1]

After relating a number of experiments[2] using different metals and electrolytes as well as different containers, Grove described what was to become the standard form of his battery, consisting of zinc in dilute sulfuric acid and platinum in concentrated nitric acid (or a mixture of nitric and sulfuric acids), giving an emf of nearly two volts. In 1841 the platinum was replaced by carbon in Bunsen's adaptation of the cell.

It is important that the cell described above should not be confused with what Grove came to call his "gas battery," which was, in fact, the earliest fuel cell; its possibilities have only recently been exploited. In a postscript to the letter describing his first experiments on voltaic cells, Grove described how, when test tubes of hydrogen and oxygen were separately placed over two platinum strips, sealed into and projecting through the bottom of a glass vessel containing dilute sulfuric acid so that half of each strip was in contact with the acid and half exposed to the gas, a current flowed through a wire connecting the projecting ends.[3] In subsequent experiments Grove obtained a powerful current using hydrogen and chlorine, and appreciable currents with other pairs of gases. Grove realized that the electrical energy resulted from the chemical energy liberated when hydrogen and oxygen combined and that this electrical energy could be used to decompose water (he did in fact carry out the electrolysis of water with current from his gas battery). This realization stimulated thoughts which had been engaging him for some time: "This battery establishes that gases in combining and acquiring a liquid form evolve sufficient force to decompose a similar liquid and cause it to acquire a gaseous form. This is to my mind the most interesting effect of the battery; it exhibits such a beautiful instance of the correlation of natural forces."[4]

The concept underlying this observation was first briefly enunciated in a lecture given in January 1842 on the progress of physical science since the opening of the London Institution and was then developed in a series of lectures given during the following year. The substance of these lectures constituted the material for Grove's book, *On the Correlation of Physical Forces,* first published in 1846. New material was added to each of the five subsequent editions. The work was an early statement of the principle of the conservation of energy, one of several at about this time.[5]

Describing, in 1845, some experiments that he had carried out four or five years earlier on the possibility of using arc lighting in mines, Grove claimed that his lack of success led him to the idea of sealing a helix of platinum wire in a glass vessel and igniting it by an electrical current; the resulting device seems to have been the earliest form of the filament lamp.[6]

In 1846 Grove gave the first experimental proof of dissociation. He showed that steam in contact with a strongly heated platinum wire was dissociated into hydrogen and oxygen. He also showed that the reactions

$$CO_2 + H_2 = CO + H_2O$$
$$CO + H_2O = CO_2 + H_2$$

could take place under the same conditions. He expressed the view that the platinum wire merely rendered the chemical equilibrium unstable and that the gases restored themselves to a stable equilibrium according to the circumstances. Among other observations, he first drew attention to the striated appearance of rarefied gases in discharge tubes.

Grove was one of the original members of the Chemical Society, and at the jubilee meeting in 1891 he said, "For my part, I must say that science to me generally ceases to be interesting as it becomes useful."

There is therefore perhaps some irony in the fact that so much of his work led to important practical consequences, yet his contribution to the concept of energy conservation (for which, it is plain from the prefaces to the successive editions of his book, he felt he was insufficiently credited) was overshadowed by the work of others. A member of the Council of the Royal Society in 1846 and 1847 and one of its secretaries in the following two years, he played a leading part in the society's reform movement.[7] He was knighted in 1872.

NOTES

1. Grove denied that the ideas which led to the development of his cell owed anything to Daniell, a denial which led to a sharp exchange of letters between the two men; see *Philosophical Magazine,* **20** (1842), 294–304; **21** (1842), 333–335, 421–422; **22** (1843), 32–35.
2. The evolution of the cell is described in the papers listed in the bibliography. The fullest account of its refinements and mode of action is in *Philosophical Magazine,* **15** (1839), 287–293.
3. For an explanation in modern terms and the contemporary significance of this experiment, see K. R. Webb, "Sir William Robert Grove (1811–1896) and the Origins of the Fuel Cell," in *Journal of the Royal Institute of Chemistry,* **85** (1961), 291–293; and J. W. Gardner, *Electricity Without Dynamos* (Harmondsworth, 1963), pp. 42 and 49 ff.
4. *Philosophical Magazine,* **21** (1842), 420.
5. See T. S. Kuhn, "Energy Conservation as an Example of Simultaneous Discovery," in M. Clagett, ed., *Critical Problems in the History of Science* (Madison, Wis., 1959), pp. 321–356.
6. *Philosophical Magazine,* **27** (1845), 442–446.
7. See H. Lyons, *The Royal Society 1660–1940* (Cambridge, 1944), pp. 259 ff.

BIBLIOGRAPHY

I. Original Works. Grove's only book is *On the Correlation of Physical Forces* (London, 1846; 6th ed., with reprints of many of Grove's papers, 1874). His papers are listed in the Royal Society *Catalogue of Scientific Papers,* III (London, 1869), 31–33. The main papers on the Grove cell are "On Voltaic Series and the Combination of Gases by Platinum," in *Philosophical Magazine,* **14** (1839), 127–130 (see 129–130 for the postscript describing the first experiments on the "gas battery"); "On a New Voltaic Combination," *ibid.,* 388–390; and "On a Small Voltaic Battery of Great Energy; Some Observations on Voltaic Combinations and Forms of Arrangement; and on the Inactivity of a Copper Positive Electrode in Nitro-Sulphuric Acid," *ibid.,* **15** (1839), 287–293. See also *Report of the Ninth Meeting of the British Association for the Advancement of Science Held at Birmingham in August 1839* (London, 1840), pp. 36–38. Papers on the gas battery are "On a Gaseous Voltaic Battery," in *Philosophical Magazine,* **21** (1842), 417–420; and "On the Gas Voltaic Battery," in *Philosophical Transactions of the Royal Society,* **133** (1843), 91–112; **135,** (1845), 351–361.

Other papers referred to in the text are "On the Application of Voltaic Ignition to Lighting Mines," in *Philosophical Magazine,* **27** (1845), 442-446; "On Certain Phenomena of Voltaic Ignition, and the Decomposition of Water Into its Constituent Gases by Heat," in *Philosophical Transactions of the Royal Society,* **137** (1847), 1-21; and "On the Electro-Chemical Polarity of Gases," *ibid.,* **142** (1852), 87-101 (the first mention of his observation of "striae" appears at the end of this paper). See also "On the Striae Seen in the Electrical Discharge *in vacuo,*" in *Philosophical Magazine,* **16** (1858), 18-22; and "On the Electrical Discharge and Its Stratified Appearance in Rarefied Media," in *Proceedings of the Royal Institution of Great Britain,* **3** (1858-1862), 5-10.

II. SECONDARY LITERATURE. On Grove and his work, see the short obituary notice by A. Gray in *Nature,* **54** (1896), 393-394; K. R. Webb, "Sir William Robert Grove (1811-1896) and the Origins of the Fuel Cell," in *Journal of the Royal Institute of Chemistry,* **85** (1961), 291-293; and J. G. Crowther, "William Robert Grove," in *Statesmen of Science* (London, 1965), pp. 77-101, which is concerned mainly with Grove's contributions toward the reforms in the Royal Society.

E. L. SCOTT

GRUBB, HOWARD (*b.* Dublin, Ireland, February 1844; *d.* Monkstown, Ireland, 17 September 1931), *optical engineering.*

Grubb's father, Thomas, engineer to the Bank of Ireland, established a factory for the manufacture of machine tools and telescopes. Grubb studied civil engineering at Trinity College, Dublin, but in 1865 left his studies to assist his father in the construction of a Cassegrain reflecting telescope of forty-eight inches aperture for Melbourne, Australia. He took control of the factory in 1868 and moved into larger premises in Rathmines, Dublin. In 1871 he married Mary Hester Walker, the daughter of a physician from Louisiana.

At Rathmines, between 1890 and 1914, Grubb made upwards of ninety first-class telescope objectives from five inches to twenty-eight inches in diameter and most of the necessary tubes and mountings. The completion, in 1887, of a twenty-seven-inch equatorial refractor, together with a forty-five-foot dome and three smaller domes for the Royal Observatory, Vienna, established his reputation as a maker of large telescopes of improved design. One of his most important undertakings was the construction in the 1890's of seven identical photographic telescopes, each with a thirteen-inch objective and ten-inch guider. All seven were used in the *Carte du Ciel,* an international photographic survey of the entire heavens. He also worked on four larger photographic refractors from twenty-four inches to twenty-six and a half inches in aperture, and on several reflecting telescopes, among them a forty-inch for the Simeiz Observatory, Crimea.

Grubb patented (1900) a novel form of optical gunsight and perfected the submarine periscope, two instruments which he made in quantity during World War I. In 1914 the business was moved to St. Albans, England, and continued there until 1925, when a new company under the name of Sir Howard Grubb, Parsons and Company was formed, with headquarters at Newcastle-upon-Tyne. Grubb, then 81 years of age, returned to Dublin.

Grubb was elected a fellow of the Royal Society of London in 1883 and knighted in 1887. He received the Cunningham Gold Medal of the Royal Irish Academy in 1881 and the Boyle Medal of the Royal Dublin Society in 1912. He was appointed scientific advisor to the Commissioners of Irish Lights in 1913 in succession to Sir Robert Ball. He was an honorary member of the Royal Institute of Engineers of Ireland and held the honorary degree of master of engineering from the University of Dublin.

BIBLIOGRAPHY

I. ORIGINAL WORKS. Grubb's publications include "Telescopic Objectives and Mirrors: Their Preparation and Testing," in *Nature,* **34,** 85; "Polar Telescopes," in *Transactions of the Royal Dublin Society,* 2nd and 3rd series; "Automatic Spectroscope for Dr. Huggins' Sun Observations," in *Monthly Notices of the Royal Astronomical Society,* **31,** 36; "On the Choice of Instruments for Stellar Photography," *ibid.,* **47,** 309; "New Arrangement of Electric Control for the Driving Clock of Equatorials," *ibid.,* **48,** 352; "On a New Form of Ghost Micrometer," *ibid.,* **41,** 59, written with E. C. Burton; and "Telescopes of the Future," in *Observatory,* **1** (1877), 55.

II. SECONDARY LITERATURE. The main sketches of Grubb's life and work are obituary notices in *Proceedings of the Royal Society,* **135A** (1932), iv-ix; and *Monthly Notices of the Royal Astronomical Society,* **92,** no. 4 (1932), 253-255. Grubb's activities in telescope making are discussed in H. C. King, *The History of the Telescope* (London, 1955).

H. C. KING

GRUBB, THOMAS (*b.* 1800; *d.* Dublin, Ireland, 19 September 1878), *optical engineer.*

Grubb was a self-taught Irish mechanic, engaged by the Bank of Ireland to construct and develop machines for printing bank notes. He established a small private observatory near Charlemont Bridge, Dublin, and about 1830, on an adjoining site, he opened a factory for the manufacture of machine tools and reflecting telescopes.

In 1835 Grubb constructed a fifteen-inch equatorial newtonian-cassegrainian reflector for Armagh Observatory, in which, for the first time, a triangular system of balanced levers shared the weight of the primary speculum. He used a similar system in a twenty-inch reflector made for Glasgow Observatory, and with such success that it was adopted, with modifications, by William Parsons, earl of Rosse, for his thirty-six-inch and seventy-two-inch reflectors at Parsonstown.

Grubb's greatest achievement was his construction, at Charlemont Bridge Works, of a forty-eight-inch equatorial cassegrainian reflector for Melbourne, Australia. This telescope, completed in 1867, was hailed as a triumph of engineering and optical skill. Intended for photography, it had several novel features, among them counterpoises to reduce the pressure of the polar axis on the bearings. But once on site, the instrument required almost constant adjustment and the mirrors of speculum metal (through no fault of Grubb's) soon became tarnished. These defects and others temporarily destroyed confidence in this type of telescope and unfortunately delayed its development for some thirty years.

After 1865 Grubb was assisted by his son, Howard, who in 1868 took control of the factory and moved into larger premises in Rathmines, Dublin. A member of the Royal Irish Academy, Grubb became a fellow of the Royal Society in 1864 and a fellow of the Royal Astronomical Society in 1870.

BIBLIOGRAPHY

I. Original Works. A work by Grubb is "On Illuminating the Wires of Telescopes," in *Monthly Notices of the Royal Astronomical Society,* **3** (1836), 177–179.

II. Secondary Literature. Short biographies are given in *Dictionary of National Biography* and *Observatory,* **2** (1878), 203. Further references appear in "Obituary Notice of Sir Howard Grubb," in *Proceedings of the Royal Society,* **135A** (1932), iv–ix. Grubb's activities in telescope making are discussed in H. C. King, *The History of the Telescope* (London, 1955).

H. C. King

GRUBENMANN, JOHANN ULRICH (*b.* Trogen, Appenzell, Switzerland, 15 April 1850; *d.* Zurich, Switzerland, 16 March 1924), *mineralogy, petrography.*

Grubenmann was the only surviving child of Johann Kaspar Grubenmann and Katharina Eugster. Among Grubenmann's ancestors was the distinguished Johann Ulrich Grubenmann, who in 1775 erected the wooden bridge across the Rhine at Schaffhausen, a milestone in the history of wide-spanned wooden bridges. At the time of Grubenmann's birth the family was living in extremely reduced circumstances and the boy had to contribute to its means by hard work during his school years. Through scholarships and the help of friends he was able to complete his education and in 1874 he obtained the diploma of certified teacher in natural sciences from the Swiss Federal Institute of Technology. That year he was elected professor of chemistry, mineralogy, and geology at the cantonal school of Frauenfeld, which he also served as rector from 1880 to 1888.

In 1876 Grubenmann married Ida Caroline Baumer; she died in 1880, a month after the birth of their son Max Alfred. He married a second time in 1881 and, by his wife Lisette Augusta, had a son, Max Carl, and a daughter, Ida Clara.

In addition to his teaching responsibilities, Grubenmann carried out petrographic fieldwork in the volcanic area of Hegau, Germany, and in the Alps. He also visited the volcanic districts of Italy and spent some time studying in Munich (1875–1876) and Heidelberg (1886). He received the Ph.D. from the University of Zurich in 1886 with a thesis on the "basalts" (now called olivine melilithites) of Hegau. Two years later he qualified as *Privatdozent* at Zurich and on the death of G. A. Kenngott became professor of mineralogy and petrography and director of the Mineralogical and Petrographical Institute of the Institute of Technology and the University of Zurich. He also was dean of the Faculty of Philosophy (1896–1898) at Zurich and chairman of the division for natural sciences (1907–1909) and rector (1909–1911) at the Institute of Technology. He retired in 1920 but remained scientifically active. In 1921 Grubenmann founded *Schweizerische mineralogische und petrographische Mitteilungen,* which he edited until his death.

Grubenmann devoted himself to the study of metamorphic rocks at a time when petrographic research was mostly occupied with the seemingly genetically simpler rocks of magmatic origin. From the outset he recognized the importance of a physicochemical approach to his studies. In recognition of his achievements and those of his Viennese friends and colleagues F. Becke and F. Berwert, the Viennese Academy of Science entrusted the three scientists with the task of studying the crystalline schists of the eastern Alps. Their researches culminated in the publication of the classic treatise "Ueber Mineralbestand und Struktur der kristallinen Schiefer" (1903). A further result of these studies was Grubenmann's publication of *Die kristallinen Schiefer* (1904–1907), which went through two editions. Part 1 of the third edition, entitled *Die Gesteinsmetamorphose,* was published separately in 1924 in collaboration with Grubenmann's pupil and successor Paul Niggli.

Grubenmann's work was based entirely on observations in nature. He proceeded from the recognition that the mineral composition of metamorphic rocks of a given chemical composition must depend upon the conditions of pressure and temperature (P-T conditions) prevailing at the time of their formation. Accordingly, he created a rock classification in which for each of twelve chemically defined groups or orders the mineral compositions for varying P-T conditions were systematically studied.

The American scientist C. R. Van Hise had already distinguished between two main zones of rock formation within the earth's crust. One zone was characterized by the relatively low values of pressure and temperature normally found in the upper regions of the earth's crust. The second zone was dominated by relatively high values of pressure and temperature such as normally belong to the deeper regions. These two genetic zones were known as the epizone and katazone respectively. Studies carried out by Grubenmann on the rocks of the southern flank of the Gotthard massif indicated the necessity of recognizing a third, so-called mesozone, lying between the two other zones and having intermediate P-T conditions. This extension of the zonal principle proved most useful in later studies. For each of the twelve previously mentioned orders Grubenmann examined and defined the typomorphic mineral composition prevailing in each of the three zones. He proposed a nomenclature based on these relationships and his terminology is widely accepted.

Although rocks were Grubenmann's chief preoccupation, he was also interested in mineralogy. An important publication in this field was his monograph (1899) on the magnificent rutilated quartzes from Piz Aul, Graubünden, which he acquired for the Zurich collection. Grubenmann was also a successful organizer. He expanded the Minerological and Petrographical Institute by adding a chemical laboratory that produced numerous chemical analyses, mostly of metamorphic rocks, and where, perhaps for the first time in Europe, students were given the opportunity for systematic training in rock analyses. Grubenmann founded and for twenty-five years presided over the Swiss Geotechnical Commission whose task it was to locate natural raw materials in Switzerland. On his initiative a series of monographs dealing with building stones, roofing slates, clay deposits, and coal and peat occurrences were published.

BIBLIOGRAPHY

I. ORIGINAL WORKS. Grubenmann's works include "Die Basalte des Hegaus, eine petrographische Studie," inaug. diss. (Univ. of Zurich, 1886); "Prinzipien und Vorschläge zu einer Klassifikation der kristallinen Schiefer," in *Collected Works of the Tenth International Geological Congress, Mexico* (1896); "Ueber die Rutilnadeln einschliessenden Bergkristalle vom Piz Aul im Bündneroberland," in *Neujahrsblatt der Naturforschenden Gesellschaft in Zurich,* **101** (1899); *Die kristallinen Schiefer,* 2 vols. (Berlin, 1904–1907; 2nd ed. 1910); *Ueber einige schweizerische Glaukophangesteine* (Stuttgart, 1906); "Struktur und Textur der metamorphen Gesteine," in *Fortschritte der Mineralogie, Kristallographie und Petrographie,* **2** (1912); and *Die Gesteinsmetamorphose* (Berlin, 1923), written with P. Niggli.

The treatise "Ueber Mineralbestand und Struktur der kristallinen Schiefer" was published under Friedrich Becke's name in *Denkscriften der Akademie der Wissenschaften,* **75** (1903).

II. SECONDARY LITERATURE. For information on Grubenmann, see J. Jakob and A. H. Schinz, "Verzeichnis der Publikationen von Prof. Dr. U. Grubenmann, herausgegeben auf seinen 70. Geburtstag, den 15. April 1920," in *Vierteljahrsschrift der Naturforschenden Gesellschaft in Zurich,* **65** (1920), with complete bibliography; P. Niggli, "Prof. Dr. U. Grubenmann," in *Verhandlungen der Schweizerischen naturforschenden Gesellschaft* (Lucerne, 1924); and R. L. Parker, "Prof. Dr. U. Grubenmann," in *Zentralblatt für Mineralogie, Geologie, und Paläontologie* (1924).

The Swiss author Arnold Kübler (once a pupil of Grubenmann) gives an extremely lively portrait of the university professor in his novel *Oeppi der Student* (Zurich, 1947), in which Grubenmann appears under the pseudonym "Zwiesand."

CONRAD BURRI

GRUBER, MAX VON (*b.* Vienna, Austria, 6 July 1853; *d.* Berchtesgaden, Germany, 16 September 1927), *hygiene.*

Gruber, the youngest of five children, grew up in the center of the old section of Vienna. His father, Ignaz, a general practitioner and otologist who had published a two-volume textbook on chemistry, awakened Max's interest in that field; and from his mother, Gabriele Edle von Menninger, he inherited his love for nature. After graduating from the Schottengymnasium he entered the First Chemical Institute of the University of Vienna as demonstrator (5 April 1876) even before completing his medical studies and went on to become an assistant. From 1879 to 1883 Gruber improved his knowledge of chemistry and physiology under Max von Pettenkofer and Karl von Voit in Munich and Karl Ludwig in Leipzig.

Hans Buchner, who worked with Gruber under Pettenkofer, encouraged him to concentrate on bacteriology. He called Gruber's attention to the work of Carl Wilhelm von Naegeli, who concerned himself with bacteriology from the botanist's point of view. Whereas Naegeli and Theodor Billroth believed in the unlimited variability of bacteria and Ferdinand Cohn, followed by Robert Koch, supported a rigid constancy of bacterial characteristics, Gruber recognized that bacteria possess a variability within limits partially determined by the culture medium. The establishment of this theory was important for the differentiation of the categories of bacteria and gained significance for Gruber in his examinations of cholera vibrios, enabling him to distinguish them from other vibrios.

Gruber was made lecturer in Vienna at the age of twenty-nine; less than two years later he became associate professor and head of the newly established Institute for Hygiene at the University of Graz, Austria. He was particularly concerned with public health, and during this period he successfully combated the cholera epidemic which had broken out in southern Austria in 1885–1886.

After the death of Josef Nowak, Gruber became associate professor at Vienna on 23 March 1887; and on 10 December 1891 he was named full professor, the second to occupy the chair of hygiene established in 1875 at the University of Vienna. He was handicapped in his new post by the limited space in the makeshift quarters of the Institute for Hygiene and by the troublesome administrative duties and difficulties with the authorities. These obstacles weighed so heavily on Gruber that after the death of his first wife in 1888, and despite being a member of the Vienna Academy of Sciences, he attempted to resign his chair and find employment as head of a laboratory in Munich or at the Jenner Institute in London, under Joseph Lister.

In October 1902 Gruber succeeded Hans Buchner as director of the Institute for Hygiene in Munich. He held the post until his voluntary retirement in 1923, on the occasion of his seventieth birthday. From then until his death, he concentrated completely on his duties as president of the Bavarian Academy of Sciences.

While in Vienna, Gruber discovered the agglutination which gained him international fame. He and his English student Herbert Edward Durham found that the blood serum of animals inoculated with typhoid or cholera bacteria some time before agglutinated these bacteria. The significance of this phenomenon in nature is that even though the bacteria are not killed by the specific agglutinins, they become more susceptible to the attack of the unspecific alexins of the body. These results were announced by Durham on 3 January 1896, in *Proceedings of the Royal Society,* and by Gruber himself on 28 February 1896, to the Society of Physicians in Vienna. Shortly afterward these findings were published in *Münchener medizinische Wochenschrift* (3 March 1896), *Semaine médicale* (4 March 1896), and *Wiener klinische Wochenschrift* (12 March 1896). The priority of Gruber's discovery, contested at the time by Richard Pfeiffer, has long since been fully recognized.

The practical application of the agglutination reaction in the determination of unknown bacteria by means of artificially produced agglutinating animal sera was proved by Gruber in joint research with Albert Sidney Grünbaum, his other English student. The reverse problem of diagnosing typhoid fever by showing evidence of specific agglutinins in the serum of patients was correctly recognized and presented by Grünbaum on 9 April 1896 to the Fourteenth Congress of Internal Medicine in Wiesbaden. Gruber and Grünbaum could furnish proof of this with only two patients, however, because the occurrence of typhoid fever had greatly diminished since the introduction to Vienna of a central water supply from mountain springs in 1873. On 26 June 1896 Ferdinand Widal lectured at the Société Médicale des Hôpitaux de Paris on the serological diagnosis of typhoid fever, based on patients from Paris. This diagnostic method today is called the Gruber-Widal reaction.

The side-chain theory of antibodies, established by Paul Ehrlich and generally recognized around the turn of the century but nevertheless primarily hypothetical, was attacked by Gruber in 1901 in a paper published jointly with Clemens von Pirquet. Following a reply by Ehrlich, Gruber and Pirquet, whose views had been confirmed experimentally, voiced their opinion once more in 1903. They received recognition and support not only from the medical school in Vienna but also from far beyond the boundaries of their country.

BIBLIOGRAPHY

I. ORIGINAL WORKS. Gruber's writings include "Über die als 'Kommabacillen' bezeichneten Vibrionen von Koch und Finkler-Prior," in *Wiener medizinische Wochenschrift,* **35,** nos. 9–10 (1885), 261–264, 297–301; "Über active und passive Immunität gegen Cholera und Typhus, sowie über die bacteriologische Diagnose der Cholera und des Typhus," in *Wiener klinische Wochenschrift,* **9,** nos. 11–12 (1896), 183–186, 204–209; "Theorie der activen und passiven Immunität gegen Cholera, Typhus und verwandte Krankheitsprozesse," in *Münchener medizinische Wochen-*

schrift, **44,** no. 9 (1896), 206–207, written with H. Durham; "14. Congress für Innere Medizin, Wiesbaden 1896," in *Verhandlungen des Kongress für innere Medizin* (1896), pp. 207–227; "Neue Früchte der Ehrlich'schen Toxinlehre," in *Wiener klinische Wochenschrift,* **16** (1903), 791–793; "Wirkungsweise und Ursprung der aktiven Stoffe in den präventiven und antitoxischen Seris," *ibid.,* 1097–1105; "Geschichte der Entdeckung der spezifischen Agglutination," in R. Kraus and C. Levaditi, eds., *Handbuch der Immunitätsforschung und experimentellen Therapie,* I (Jena, 1914), 150–154; and "Dankrede anlässlich der Feier seines 70. Geburtstages," in *Münchener medizinische Wochenschrift,* **70** (1923), 1038–1039.

II. SECONDARY LITERATURE. On Gruber or his work, see N. W. Forst, "Max von Gruber," in *Geist und Gestalt,* II (Munich, 1959), 242–247; E. Glaser, "Max Gruber," in *Wiener medizinische Wochenschrift,* **74** (1927), 1330; R. Grassberger, "Max v. Gruber," in *Wiener klinische Wochenschrift,* **40** (1927), 1304–1306; K. B. Lehmann, "Max v. Gruber (6 Juli 1923)," in *Münchener medizinische Wochenschrift,* **70** (1923), 879–881; and "Zum Gedächtnis Max v. Gruber. 6 Juli 1853 bis 16 September 1927," *ibid.,* **74** (1927), 1838–1839; E. Lesky, *Die Wiener medizinische Schule im 19. Jahrhundert* (Graz–Cologne, 1965), 595–602 and *passim;* G. Rath, "Max(imilian) Franz Maria Ritter v. Gruber," in *Neue deutsche Biographie,* VII (Berlin, 1966), 177–178; and K. Süpfle, "Max v. Gruber zum Gedächtnis," in *Deutsche medizinische Wochenschrift,* **53** (1927), 1869–1870.

H. FLAMM

GRUBY, DAVID (*b.* Kis-Kér, Hungary [now Bačko Dobro Polje, Yugoslavia], 20 August 1810; *d.* Paris, France, 14 November 1898), *microbiology, medical mycology, parasitology.*

At the time of his death Gruby was known mainly as an eccentric physician famous for the extravagant cures prescribed for his distinguished patients, who had included Frédéric Chopin, Alexandre Dumas *père,* Heinrich Heine, Alphonse Lamartine, Alphonse Daudet, George Sand, Ambroise Thomas, and Franz Liszt. These prescriptions were actually clever applications of psychosomatic medicine. It was only slowly realized that in the short period of his scientific activity Gruby had made very original and important contributions to science—indeed, he founded an important branch of modern medicine, discovering the dermatomycoses, a group of skin diseases caused by parasitic lower plants.

Gruby was one of seven or eight children of a poor Jewish peasant in a village of Baczka, a fertile district of southern Hungary. Although in his birthplace he could have received only an elementary education, he very early showed great interest in reading and studying. He received his first instruction in secular knowledge from a medical student who worked as a substitute teacher in Kis-Kér and lodged in Gruby's father's home. Gruby left Kis-Kér about 1824 or 1825 to seek further education in Pest. His early years had been marked by great poverty, hardship, and prejudice, although some of the reported stories may be more fiction than truth. His great talent and determined pursuit of his aim enabled him to succeed, against all the odds, in completing his secondary studies at the Piarist Gymnasium in Pest. In 1828 Gruby went to Vienna to study medicine. According to the list of medical students of 1836 he was then in the fifth year of his studies. He passed the first examination on 13 February 1838, the second on 18 March 1839, and graduated on 5 August 1839.

The time he took for his studies seems unusually long for a gifted student, but in order to earn a living Gruby seems to have acquired some special technical knowledge: he is reported to have built himself an accurate clock and a microscope which was, for the time, an excellent instrument. In his microscopic studies Gruby was encouraged and guided by two young teachers who were beginning their distinguished careers at the Vienna medical school: Joseph Berres, from 1831 professor of anatomy, and Karl Rokitansky, from 1833 prosector and a year later associate professor of pathological anatomy. Some of Gruby's early microscopic observations on pathological morphology were included in his dissertation, which contains microscopic observations (with 103 illustrations) on the pathology of body fluids—mucus, sputum, pus, pseudomembranes, coagula, and saliva—and compares pathological with normal findings. His attempt at microscopic differentiation of pus from other pathological substances was a careful, original investigation in a new field of medicine. Gruby demonstrated, among other things, that every one of the studied body fluids contained living elements (leukocytes). He republished his dissertation as the first part of a larger treatise on microscopic pathology, but the other planned parts never appeared.

Gruby also made preparations to be sold to various institutions and gave courses in microscopy, which enabled him to meet many visiting foreign physicians, such as William Bowman and P. J. Roux. It was Roux who suggested to Gruby—who could not find a suitable position because he refused to give up his religion—that he move to Paris, the great center of medical learning. Microscopy was little practiced there, and thus Paris offered a promising field for an experienced young man.

Gruby settled in Paris in 1840. Assisted by friends, he worked in the Foundling Hospital under the distinguished pediatrician Jacques François Baron and, urged by some foreign students, he began to give courses in microscopic anatomy and pathology which were attended by Claude Bernard, François Magendie, Henri Milne-Edwards, Pierre Flourens, and many foreign scientists.

At this time Gruby began to announce his discoveries of various microscopic fungi that produce skin diseases. In 1841 he found a fungus in favus, a contagious skin disease marked by round yellow crusts resembling honeycomb, usually situated over hair follicles and accompanied by intense itching. The following year he described *Trichophyton ectothrix,* a microscopic cryptogam, found at the roots of a man's beard, which caused the disease *Sycosis barbae.* Shortly afterward he discovered *Oidium albicans* (*Monilia albicans*), the cause of thrush in infants. In 1843 Gruby described another fungus, which he called *Microsporum audouini,* in honor of Jean Victor Audouin. In man *Microsporum* causes a form of tinea (ringworm) that is also called microsporia or Gruby's disease. Another form of ringworm, caused by *Trichophyton tonsurans,* was discovered and described by Gruby in 1844. The cause of favus, *Achorion schoenleini,* was first described by J. L. Schönlein, who was not certain whether it was the cause or a manifestation of the disease. Gruby showed that the disease could be produced experimentally in man or animals by inoculating the specific mold. His clinical descriptions were generally inadequate, but his descriptions of the microscopic features were so excellent that when his findings were later rediscovered and confirmed, not much could be added to his original reports. The idea of a plant parasite as a cause of disease in man was something quite new in the era before Pasteur; and Gruby was responsible for the firm establishment of this conception by his findings and experiments, against the doubts, opposition, and ridicule of many contemporary physicians.

In 1843 Gruby discovered an animal parasite in the blood of the frog that he called, because of its corkscrew shape, *Trypanosoma;* the name has been used for this important genus ever since. In the same year he described, in association with his friend Onésime Delafond, professor at the veterinary school at Alfort, another parasitic hematozoon in the dog microfilaria. They also showed (1852) that the disease could be induced in a healthy dog by intravenous transfusion of defibrinated blood from an infected animal. In 1859 Gruby described and depicted a parasitic mite (*Acarus*) producing skin disease called *Erythema autumnale.*

In 1847–1848, in the early period of general anesthesia administered by inhalation, Gruby made experiments on animals with ether and chloroform, which contributed to the knowledge of their effects on several bodily functions. He also emphasized the higher toxicity and quicker action of chloroform as compared with ether.

After 1845 Gruby published only a few papers, eventually spending all his time and energy on his medical practice. The cause of his loss of interest in scientific endeavor remains unknown.

BIBLIOGRAPHY

I. ORIGINAL WORKS. Gruby's inaugural diss., *Observationes microscopicae ad morphologiam pathologicam* (Vienna, 1839), repub. in 1840 with the subtitle *Morphologia fluidorum pathologicorum, tomi primi, pars prima,* seems to be his only book on science. He published over 30 short papers, 21 of them in *Comptes rendus hebdomadaires des séances de l'Académie des sciences,* **13–34** (1841–1852). Six of them were written in collaboration with O. Delafond, and about 20 date from 1841–1845. Some of Gruby's papers were soon translated into German or English. Six of them, representing the foundation of medical mycology, were published "as a monument *aere perennius* of Gruby's scientific achievement" in an English trans. by S. J. Zakon and T. Benedek: "David Gruby and the Centenary of Medical Mycology 1841–1941," in *Bulletin of the History of Medicine,* **16** (1944), 155–168. A bibliography of Gruby's works was collected by R. Blanchard (1899) and is included (with some additions) in the biographies by L. Le Leu (1908), A. P. M. Salaun (1935), and B. Kisch (1954).

II. SECONDARY LITERATURE. A full biography is B. Kisch, "David Gruby (1810–1898)," pt. 2 of "Forgotten Leaders in Modern Medicine," in *Transactions of the American Philosophical Society,* n.s. **44** (1954), 193–226. Of the older appreciations the most important are R. Blanchard, "David Gruby (1810–1898)," in *Archives de parasitologie,* **2** (1899), 43–74; L. Le Leu, *Le Dr. Gruby. Notes et souvenirs* (Paris, 1908), a volume of notes and recollections by Gruby's last (from 1888) private secretary; and A. P. M. Salaun, *La vie et l'oeuvre de David Gruby* (Bordeaux, 1935). There are several short biographies, such as T. Rosenthal, "David Gruby (1810–1898)," in *Annals of Medical History,* n.s. **4** (1932), 339–346; and several appreciations of his discoveries in the field of dermatomycoses: E. Podolsky, "David Gruby (1810–1898) and the Fungus Growth," in *Medical Annals of the District of Columbia,* **26** (1957), 24–26, 60; J. H. Rille, "David Gruby," in *Dermatologische Wochenschrift,* **83** (1926), 512–526; and J. Théodoridès, "L'oeuvre scientifique du Docteur Gruby," in *Revue d'histoire de médecine hébraïque,* **27** (1954), 27–38, 138–143.

V. KRUTA

GUA DE MALVES, JEAN PAUL DE (*b.* near Carcassonne, France, *ca.* 1712; *d.* Paris, France, 2 June 1786), *mathematics, mineralogy, economics.*

Very little is known of de Gua's life, and even the

precise date and place of his birth are not established. According to Condorcet, he was struck by the contrast between the opulence of his first years and the privation that followed the ruin of his parents, Jean de Gua, baron of Malves, and Jeanne de Harrugue, in the wake of the bankruptcy of John Law in 1720. He planned an ecclesiastical career; while it seems that he never became a priest, this training nevertheless permitted him to obtain several benefices and pensions. After a stay in Italy he appears to have participated for a few years in the activities of the short-lived Société des Arts, a sort of scientific and technical academy founded in 1729 by Louis de Bourbon-Condé, prince of Clermont. In any case, he gradually acquired a thorough grounding in science.

De Gua's first publication (1740) was a work on analytic geometry inspired by both Descartes's *Géométrie* (1637) and Newton's *Enumeratio linearum tertii ordinis* (1704). Its principal aim was to develop a theory of algebraic plane curves of any degree (Descartes's "lignes géométriques") based essentially on algebra. Nevertheless, he drew on infinitesimal methods in order to simplify various calculations and recognized that their use is indispensable, particularly for everything involving the transcendental curves ("mécaniques"). De Gua was especially interested in tangents, asymptotes, and singularities: multiples, points, cusps, and points of inflection. In this area he skillfully used coordinate transformations and systematically made use of an "algebraic" or "analytic" triangle, obtained by a 45° rotation of Newton's parallelogram. The use of the latter had been popularized by the *Enumeratio* and by the commentaries of several of Newton's disciples, among them Brook Taylor, James Stirling, and s'Gravesande. The use of perspective allowed de Gua to associate the different types of points at a finite distance with various infinite branches of curves. Among his other contributions, he explicitly asserted that if a cubic admits three points of inflection, the latter are aligned. He also introduced two new types of cubics into the enumeration undertaken by Newton and Stirling, among others.

De Gua's treatise contributed to the rise of the theory of curves in the eighteenth century and partially inspired the subsequent works of Euler (1748), Gabriel Cramer (1750), A. P. Dionis du Séjour, and M. B. Goudin (1756). The fame of this work led to de Gua's election to the Royal Academy of Sciences as adjoint geometer on 18 March 1741, replacing P. C. Le Monnier. He presented several mathematical memoirs, two of which, published at the time, deal with the number of roots of an algebraic equation according to their nature and sign and with the famous rule of Descartes. However, on 3 June 1745, following a dispute de Gua renounced the pursuit of a normal academic career and requested that his modest position of associate be made honorary. Although he continued his scientific research, he seems to have moved away from the study of mathematics during this period. Not until the time of the reorganization of 23 April 1785 did he resume his place at the Academy, this time as a pensioner in the new class of natural history and mineralogy, which was closer to his new interests. Moreover, it seems that the contents of the several mathematical memoirs on spherical trigonometry and the geometry of polyhedra that he subsequently published in the *Histoire* of the Academy for 1783 date for the most part from the 1740's.

The career of de Gua was marked by several incidents which certainly resulted, at least in part, from difficulties inherent in his personality. His stay at the Collège Royal (Collège de France) was abnormally brief. Appointed on 30 June 1742 to the chair of Greek and Latin philosophy, which was vacant following the death of Joseph Privat de Molières, de Gua actually filled this post until 26 July 1748, when he resigned; he was replaced by P. C. Le Monnier. Like his predecessors and his successor, de Gua gave to his instruction an orientation having no connection with the official title of the chair; he dealt successively with Newtonian epistemology, differential and integral calculus, the principles of mathematics, arithmetic, and the philosophy of Locke—without, however, publishing anything based on his teaching.

Another incident took place during the same period. In 1745 the publisher A. F. Le Breton had joined with the Paris booksellers A. C. Briasson, M. A. David, and L. Durand for the purpose of publishing a much enlarged French version of the famous *Cyclopaedia* of Ephraim Chambers. Apparently appreciating de Gua's wide-ranging abilities, they made him responsible for the scientific material in the edition in a contract signed on 27 June 1746 in the presence of Diderot and d'Alembert, who acted both as witnesses and as consultants. The agreement was annulled on 3 August 1747. On 16 October 1747 de Gua, who had meanwhile mortgaged his other income to repay a portion of the advances he had received from the booksellers, was replaced by d'Alembert and Diderot as director of this project, which was to become the celebrated *Encyclopédie*.

De Gua next turned his attention to philosophy and political economy, translating works by George Berkeley and Matthew Decker, as well as a debate in the House of Commons that he introduced with a long "Avant-propos" on the problem of the interest rate on loans. At the same time he was actively interested in prospecting for gold in Languedoc and addressed several memoirs on this subject to the government. Having obtained, in 1764, an exploitation permit valid for twenty years, he undertook an unsuccessful venture that partially ruined him. In 1764 he published a work on mineral prospecting and composed the first six volumes of a series of *mémoires périodiques* on subjects in philosophy, science, economics, and so on; the series was never published—for lack, it seems, of official authorization. De Gua also was interested in lotteries, but beginning in the 1760's he specialized in mineralogy and conchology, which explains his change of sections at the Academy in 1785.

This disordered scientific activity and a taste for the unusual give to de Gua's work a special character. The interest of his first mathematical writings evokes regrets that he did not persevere in this direction.

BIBLIOGRAPHY

I. Original Works. Among de Gua's writings are *Usages de l'analyse de Descartes pour découvrir, sans le secours du calcul différentiel, les propriétés, ou affections principales des lignes géométriques de tous les ordres* (Paris, 1740); five mathematical memoirs in *Histoire de l'Académie royale des sciences:* "Démonstration de la règle de Descartes . . .," 72–96, and "Recherches des nombres des racines réelles ou imaginaires . . .," 435–494, in the volume for 1741 (Paris, 1744) and "Trigonométrique sphérique, déduite très brièvement . . .," 291–343, "Diverses mesures, en partie neuves, des aires sphériques et des angles solides . . .," 344–362, and "Propositions neuves . . . sur le tétraèdre . . .," 363–402, in the volume for 1783 (Paris, 1786); and *Projet d'ouverture et d'exploitation de minières et mines d'or et d'autres métaux aux environs du Cézé, du Gardon, de l'Eraut [sic] et d'autres rivières du Languedoc, du Comté de Foix, du Rouergue etc.* (Paris, 1764). He translated several works from English into French: George Berkeley, *Dialogues entre Hylas et Philonaüs contre les sceptiques et les athées* (Amsterdam, 1750); Matthew Decker, *Essai sur les causes du déclin du commerce étranger de la Grande Bretagne,* 2 vols. (n.p., 1757); and *Discours pour et contre la réduction de l'intérest naturel de l'argent, qui ayant été prononcés en 1737, dans la Chambre des communes du parlement de la Grande Bretagne, occasionnèrent en ce pays la réduction de 4 à 3%* . . . (Wesel–Paris, 1757)—"Avant-propos du traducteur," pp. i–clxviii, is by de Gua. With J. B. Romé de l'Isle he edited *Catalogue systématique et raisonné des curiosités de la nature et de l'art, qui composent le cabinet de M. Davila* . . . (Paris, 1767), for which he wrote I, pt. 2, 71–126: "Coquilles marines."

II. Secondary Literature. On de Gua or his work, see the following (listed chronologically): the *éloge* of M. J. A. N. Condorcet, read 15 Nov. 1786, in *Histoire de l'Académie royale des sciences pour l'année 1786* (Paris, 1788), pt. 1, 63–76; X. de Feller, ed., *Dictionnaire historique,* IV (Paris, 1808), 238; J. J. Weiss, in Michaud, ed., *Biographie universelle,* XVIII (Paris, 1817), 575–576, also in new ed., XVIII (Paris, 1857), 1–2; J. M. Quérard, *La France littéraire,* III (Paris, 1829), 494–495; Guyot de Fère, in F. Hoefer, ed., *Nouvelle biographie générale,* XXII (Paris, 1859), col. 278; Poggendorff, I, 967–968; *Intermédiare des mathématiciens,* VIII (1901), 158, and XI (1904), 148–149; P. Sauerbeck, "Einleitung in die analytische Geometrie der höheren algebraïschen Kurven nach der Methoden von Jean-Paul de Gua de Malves. Ein Beitrag zur Kurvendiskussion," in *Abhandlungen zur Geschichte der mathematischen Wissenschaften,* **15** (1902), 1–166; G. Loria, "Da Descartes e Fermat a Monge e Lagrange. Contributo alla storia della geometria analitica," in *Atti dell'Accademia nazionale dei Lincei. Memorie,* classe di scienze fisiche, matematiche e naturale, 5th ser., **14** (1923), 777–845; N. Nielsen, *Géomètres français du XVIIIe siècle* (Copenhagen–Paris, 1935), pp. 195–200; L. P. May, "Documents nouveaux sur l'*Encyclopédie,*" in *Revue de synthèse,* **15** (1938), 5–30; G. Loria, *Storia delle matematiche* (Milan, 1950), pp. 668–689, 739–740, 758, 851; and C. B. Boyer, *History of Analytic Geometry* (New York, 1956), pp. 174–175, 184, 194.

René Taton

GUCCIA, GIOVANNI BATTISTA (*b.* Palermo, Italy, 21 October 1855; *d.* Palermo, 29 October 1914), *mathematics.*

The son of Giuseppe Maria Guccia and Chiara Guccia-Cipponeri, Guccia belonged, through his father, to the noble Sicilian family of the marquis of Ganzaria. As a young man he was an ardent sportsman and was particularly interested in horsemanship. He studied first at Palermo, then at the University of Rome, where he was one of Luigi Cremona's best students. In 1880 he defended a thesis dealing with a class of surfaces representable, point by point, on a plane. Shortly before, he had presented a communication on certain rational surfaces dealt with in this work to the congress of the French Association for the Advancement of Science at Rheims and had been publicly congratulated by J. J. Sylvester. After returning to Palermo, Guccia pondered some grand schemes of theoretical research. In 1889 he was appointed to the newly created chair of higher geometry at the University of Palermo, a post he held for the rest of his life.

The path that Guccia followed throughout his career was that of the great Italian geometers of the nineteenth century. For them the synthetic method, aided by intuition, was the ideal instrument of discovery, more efficacious than the calculus, of which the artifices often conceal the logical structures and the relationships among the elements of a figure. To be sure, the role of algebra is not negligible, but it should be limited to what is linear, for the establishment of certain principles, and then give way to the intuitive method. Guccia's works concern primarily Cremona's plane transformations, the classification of linear systems of plane curves, the singularities of curves and of algebraic surfaces, and certain geometric loci which permit the projective properties of curves and surfaces to be deduced.

In studying the classification of linear systems of types 0 and 1, Guccia was inspired by the method used by Max Nöther to demonstrate that every Cremona transformation is the product of a finite number of quadratic transformations. Guccia's results were completed in 1888 by Corrado Segre, and the question was taken up in 1897 by Guido Castelnuovo in his memoir on linear systems of curves traced on an algebraic surface. In studying the singular points and singular curves of a surface, Guccia discovered theorems analogous to those for linear systems of curves. Although the majority of Guccia's publications are very short, they all contain original ideas and new relations profitably used by other geometers. This is particularly true of his researches on projective involutions, which laid the foundation for the generalizations of Federico Enriques and Francesco Severi. Occasionally, Guccia himself generalized from partial results, as in the case of the projective characteristics of plane algebraic curves and of their linear systems (where he introduced a projective definition of polars), which he extended to surfaces and to gauche curves.

In a period when knowledge of the geometry of algebraic surfaces was extremely limited, Guccia made a useful contribution. It was immediately exploited and absorbed by other mathematicians who, more attracted than he by analytical procedures, achieved greater fame. Compared with the work of his teacher Cremona, Guccia's is on a lower plane, if not in subtlety at least in extent and significance. Yet Guccia's chief merit lies elsewhere: his name remains associated with the foundation of the Circolo Matematico di Palermo.

In 1884, five years before his appointment to the university, Guccia had the idea of establishing a mathematical society in Palermo, for which he would furnish the meeting place, a library, and all necessary funds. His generous offer was favorably received, and on 2 March 1884 the society's provisional statutes were signed by twenty-seven members. The goal was to stimulate the study of higher mathematics by means of original communications presented by the members of the society on the different branches of analysis and geometry, as well as on rational mechanics, mathematical physics, geodesy, and astronomy. The group's activity was soon known abroad through the *Rendiconti del Circolo matematico di Palermo,* the first volume of which consisted of four sections appearing in July 1885, September 1886, December 1886, and September 1887. On 7 November 1887 Joseph Bertrand presented this volume to the Académie des Sciences of Paris, emphasizing its high scientific standard. On 26 February 1888 new statutes for the society authorized the election of foreign corresponding members, and the *Rendiconti* thereby became an international review. Guccia, who had placed his personal fortune at the disposal of the Circolo, established a mathematical publishing house in Palermo in 1893. To the *Rendiconti* he added *Supplemento ai Rendiconti del Circolo matematico di Palermo, Indici delle pubblicazioni del Circolo matematico di Palermo,* and *Annuario biografico del Circolo matematico di Palermo.* He also took personal charge of the editing of all these publications.

BIBLIOGRAPHY

I. Original Works. The list of mathematical works drawn up by de Franchis (see below) contains forty-four titles. Lectures given at the University of Palermo in 1889–1890 appeared as *Teoria generale delle curve e delle superficie algebriche* (Palermo, 1890). His longer arts. include "Teoremi sulle trasformazioni Cremoniane nel piano. Estensione di alcuni teoremi di Hirst sulle trasformazioni quadratiche," in *Rendiconti del Circolo matematico di Palermo,* **1** (1884–1887), 27, 56–57, 66, 119–132; "Generalizzazione di un teorema di Noether," *ibid.,* 139–156; "Sulla riduzione dei sistemi lineari di curve ellittiche e sopra un teorema generale delle curve algebriche di genere *p*," *ibid.,* 169–189; "Sui sistemi lineari di superficie algebriche dotati di singolarità base qualunque," *ibid.,* 338–349; "Sulle singolarità composte delle curve algebriche piane," *ibid.,* **3** (1889), 241–259; "Ricerche sui sistemi lineari di curve algebriche piane, dotati di singolarità ordinarie," *ibid.,* **7** (1893), 193–255, and **9** (1895), 1–64; and "Un théorème sur les courbes algébriques planes d'ordre *n*," in *Comptes rendus hebdomadaires des séances de l'Académie des sciences,* **142** (1906), 1256–1259.

II. Secondary Literature. See Michele de Franchis, "XXX anniversario della fondazione del Circolo matematico di Palermo . . .," in *Supplemento ai Rendiconti del Circolo matematico di Palermo,* **9** (1914), 1–68; and "G. B. Guccia, cenni biografici . . .," in *Rendiconti del Circolo matematico di Palermo,* **39** (1915), 1–14.

Pierre Speziali

GUDDEN, JOHANN BERNHARD ALOYS VON (*b.* Cleves, Germany, 7 June 1824; *d.* Lake Starnberg, near Schloss Berg, Germany, 13 June 1886), *psychiatry, neuroanatomy.*

Gudden was the third of seven sons of Johannes

Gudden, a landed proprietor in Lower Rhineland, and Bernhardine Fritzen. His feeling for exact observation and his aptitude for study became evident at an early age. After passing the final secondary school examination in the fall of 1843, he studied medicine in Bonn, Halle, and Berlin, where he passed the state medical examination with distinction in 1848. His dissertation, *Quaestiones de motu oculi humani* (1848), dealing with one of his fields of later research, revealed the originality of his investigations. It is possible that during his student years he was already in contact with the heads of psychiatric institutions, such as Maximilian Jacobi at Siegburg, near Bonn; August Damerow at Nietleben, near Halle; and Karl Wilhelm Ideler at the Charité in Berlin. Entry into the field of psychiatry was possible only through these institutions; the subject was rarely treated separately in the universities.

In 1849 Gudden obtained a position at the Siegburg asylum as an intern under Maximilian Jacobi, one of the leading German somatic psychiatrists. He married the latter's granddaughter Clarissa Voigt in 1855; they had nine children. From 1851 to 1855 Gudden worked with Wilhelm Roller at Illenau, a hospital known even outside Germany for its outstanding organization.

In 1855 Gudden was appointed director of the newly founded Werneck asylum near Würzburg. He supervised the transformation of the baroque palace, built by Balthasar Neumann, into Germany's most modern asylum; this achievement earned him a reputation as an excellent organizer. In the treatment of the mentally disturbed he rejected the methods of the older psychiatric schools at Siegburg and Illenau. Despite their humane conceptions these schools had continued the use of physical force and believed that "moral influence" and "educational" strictness were beneficial. An advocate of the principle of no restraint, Gudden championed, earlier than Wilhelm Griesinger and Ludwig Meyer, a liberal and humane orientation in the treatment of the mentally ill. Going beyond even John Conolly, he granted his patients an unprecedented measure of personal freedom. He insisted that proper treatment required communal social life for the patients, constant contact between physicians and patients, and a well-trained staff with a strong sense of duty.

In his articles "Ueber die Entstehung der Ohrblutgeschwulst" and "Ueber die Rippenbrüche bei Geisteskranken," Gudden demonstrated that reddening of the ears, rib fractures, and bedsores (which he considered an attendant symptom of mental illness, produced by injury to the "trophic nerve") were the consequences of mechanical therapy and insufficient care.

In 1869 Gudden became director of the recently constructed cantonal mental hospital in Burghölzli, near Zurich, and professor of psychiatry at the University of Zurich. Following the death of August Solbrig in 1872 he was named director of the district mental hospital in Munich; his practical talents were decisive in gaining him this appointment. Soon afterward he was also named professor of psychiatry at the University of Munich. The hospital, built by Solbrig in 1859, was enlarged and reorganized under Gudden's supervision, as was a second institution in Gabersee in 1883. Both were distinguished by the rational arrangement and distribution of their facilities. From 1870 Gudden was a coeditor of the *Archiv für Psychiatrie und Nervenkrankheiten.* He was ennobled in 1875.

In 1886, after he and other psychiatrists had examined records pertaining to the case, Gudden gave his opinion on the mental illness of Ludwig II of Bavaria. They diagnosed it as paranoia (what would now be called the paranoid form of schizophrenia). On the basis of this diagnosis, the king was relieved of all official duties. On 12 June 1886 he was taken by Gudden, who treated him with great consideration, to Schloss Berg, on Lake Starnberg, which was to serve as the king's residence. The following morning Gudden took a quiet walk with the king; they were accompanied by several attendants. That evening, at about 6:30, Gudden went for another walk with the king, this time without attendants. A few hours later both were found drowned in the lake, not far from the shore. (The evidence leaves little doubt that during the walk Ludwig suddenly ran into the shallow lake and that Gudden followed in order to restrain him; the powerful forty-year-old king then probably overpowered Gudden and drowned himself.)

In an early work, *Beiträge zur Lehre von den durch Parasiten bedingten Hautkrankheiten* (Stuttgart, 1855), Gudden conclusively verified through skillful clinical observations that scabies is a parasitic disease caused by mites. His major scientific work was, however, in three fields: care of the mentally ill, craniology, and cerebral anatomy. The last two were closely related, both by common experimental procedures and by the results obtained.

Through his practical work as well as through his publications on treatment of the mentally ill, most of which dealt with hospital administration, Gudden contributed significantly to liberating mental patients from treatment by physical force. In therapy his main concerns were that the hospital be rationally organized, that the personnel be properly trained, and that the curable—and even the incurable—patients be able to move about as freely as possible. Gudden's research did not deal with clinical psychopathology, an area which was investigated by his student Emil Kraepelin. Gudden was skeptical of systematic reflections that went beyond the individual case. In nosology he followed Griesinger's classification.

Gudden published his *Experimentaluntersuchungen über das Schädelwachstum* in 1874. In this work he showed that the growth of the cranium is essentially the result of interstitial processes; and he discovered that when sense organs and parts of the brain are extirpated, the cranial bones are also affected.

In cerebral anatomy, too, Gudden used extirpation in his research. Building on the observations of Ludwig Türck and Augustus Waller on secondary degeneration, he developed, through an ingenious combination of anatomical and experimental pathological investigations, the "Gudden method." By systematically destroying, on one side only, parts of the nervous system and of the brain in a newborn animal, he was able to induce atrophy of the conducting paths and centers; this made it possible, by means of a comparative examination of the two sides in the full-grown animal, to determine the functions of nerve fibers and nuclei. This method, which is still in use, allowed him to classify nerves that had previously been considered anatomically and physiologically similar into separate systems according to function and origin. Thus Gudden was the first to set forth many of the neuroanatomical facts generally accepted today concerning the paths, origins, and termini of the nerves, as well as many concerning the nuclei of the cranial nerves (the crossing of the optic nerve, the tractus opticus, the fornix, corpus mamillare, interpeduncular ganglion, and nuclei of the nerves of the eye muscle, among others). During his years in Munich he amassed, with his "Gudden microtome" (later improved by Auguste Forel), one of the world's largest collections of brain-tissue preparations.

Gudden was very cautious in drawing physiological conclusions from his research. He confined himself for the most part to recording morphological data, which he constantly reexamined under altered experimental conditions. His writings are characterized by their conciseness and by a wealth of carefully observed details. It took some time, however, for the scientific reliability of his findings to be recognized. He lacked the kind of intuitive inspiration possessed by, for instance, Theodor Meynert, whose research on cerebral anatomy he approved. Gudden's students Emil Kraepelin, Franz Nissl, Auguste Forel, and Sigbert Ganser all described him as having a commanding and magnetic personality. At the time of his death he was editing the results of his neuroanatomical research, the majority of which were still unpublished.

BIBLIOGRAPHY

I. Original Works. Gudden's writings include *Quaestiones de motu oculi humani* (Halle, 1848), his diss.; "Das Irrenwesen in Holland," in *Allgemeine Zeitschrift für Psychiatrie,* **10** (1853), 458–480; "Zur relativ verbundenen Irrenheil- und Pflegeanstalt," *ibid.,* **16** (1859), 627–632; "Über die Entstehung der Ohrblutgeschwulst," *ibid.,* **17** (1860), 121–138; **19** (1862), 190–220; **20** (1863), 423–430; *Beitrag zur Lehre von der Scabies,* 2nd ed. (Würzburg, 1863); *Der Tagesbericht der Kreisirrenanstalt Werneck* (Würzburg, 1869); "Ueber einen bisher nicht beschriebenen Nervenfaserstrang im Gehirn der Säugethiere und Menschen," in *Archiv für Psychiatrie und Nervenkrankheiten,* **2** (1870), 364–366; "Anomalien des menschlichen Schädels," *ibid.,* 367–373; "Ueber die Rippenbrüche bei Geisteskranken," *ibid.,* 682–692; "Experimentaluntersuchungen über das peripherische und centrale Nervensystem," *ibid.,* 693–723; *Experimentaluntersuchungen über das Schädelwachsthum* (Munich, 1874); "Ueber die Kreuzung der Fasern im Chiasma n. optici," in *Albrecht von Graefes Archiv für Ophthalmologie,* **20,** no. 2 (1874), 249–268; **21,** no. 3 (1875), 201–203; **25,** no. 1 (1879), 1–56; **25,** no. 4 (1879), 237–246; "Ueber ein neues Mikrotom," in *Archiv für Psychiatrie und Nervenkrankheiten,* **5** (1875), 229–244; "Ueber den Tractatus peduncularis transversus," *ibid.,* **11** (1881), 415–423; "Mittheilung über das Ganglion interpedunculare," *ibid.,* 424–427; "Ueber zwei verschiedene Fasersysteme im N. opticus," in *Tageblatt der Eisenacher Naturforscherversammlung* (1882), 307–310; "Ueber das Corpus mamillare und die sogenannten Schenkel des Fornix," in *Allgemeine Zeitschrift für Psychiatrie,* **41** (1884), 697–701; "Ueber die neuroparalytische Entzündung," *ibid.,* 714–715; *Jahresbericht der Kreisirrenanstalt München* (1885); "Ueber die Sehnerven, den Sehtractus, etc.," in *Allgemeine Zeitschrift für Psychiatrie,* **42** (1885), 347–348; "Ueber die Einrichtung von sogenannten Ueberwachungsstationen," *ibid.,* 454–456; and "Ueber die Frage der Lokalisation der Functionen der Grosshirnrinde," *ibid.,* 478–497. Other articles are in *Korrespondenzblatt für schweizer Ärzte* (1871), no. 5; (1872), no. 4; and in *Ärztliche Intelligenzblatt* (1884).

R. Grashey collected and edited Gudden's published and unpublished works on cerebral anatomy in the lavishly illustrated *B. v. Gudden. Gesammelte und hinterlassene Abhandlungen* (Wiesbaden, 1889).

II. Secondary Literature. On Gudden or his work, see Sigbert Ganser, in Theodor Kirchhoff, ed., *Deutsche Irrenärzte,* II (Berlin, 1924), 47–58; Hubert Grashey, "Nekrolog auf Bernhard von Gudden," in *Archiv für Psy-*

chiatrie und Nervenkrankheiten, **17** (1886), i-xxix; and "Nachtrag zum Nekrolog," *ibid.,* **18** (1887), 898; Ernst Grünthal, in Kurt Kolle, ed., *Grosse Nervenärzte,* I (Stuttgart, 1956), 128–134; Emil Kraepelin, "Bernhard von Gudden, ein Gedenkblatt," in *Münchener medizinische Wochenschrift,* **33** (1886), 577–580, 603–607; H. Laehr, in *Allgemeine Zeitschrift für Psychiatrie,* **43** (1887), 163–168; Theodor Meynert, in *Wiener medizinische Blätter,* **9** (1886), 24; and *Allgemeine Zeitschrift für Psychiatrie,* **43** (1887), 177–186; Franz Nissl, "Bernhard von Guddens hirnanatomische Experimentaluntersuchungen," in *Zeitschrift für Psychiatrie,* **51** (1895), 527; James W. Papez, in Webb Haymaker, ed., *The Founders of Neurology* (Springfield, Ill., 1953), pp. 45–48; E. Rehm, "König Ludwig II. und Professor Gudden," in *Psychologische neurologische Wochenschrift,* **38** (1936), 45; Hugo Spatz, "Bernhard von Gudden," in *Münchener medizinische Wochenschrift,* **103** (1961), 1277–1282; and Wallenberg, in *Archiv für Psychiatrie,* **76** (1925), 21–46.

HANS HENNING SCHROTH

GUDERMANN, CHRISTOPH (*b.* Vienenburg, near Hildesheim, Germany, 25 March 1798; *d.* Münster, Germany, 25 September 1852), *mathematics.*

Gudermann's father was a teacher. After graduating from secondary school Gudermann was to have studied to become a priest, but in Göttingen he studied, among other things, mathematics. From 1823 he was a teacher at the secondary school in Kleve; and from 1832 until his death he taught at the Theological and Philosophical Academy in Münster, first as associate professor, and from 1839 as full professor, of mathematics.

Gudermann's scientific work forms part of German mathematics in the second quarter of the nineteenth century. The characteristic feature of this period is that the ideas of transforming mathematics had been expressed or indicated, but understanding and realizing them in results or comprehensive theories was still beyond the capabilities of the mathematicians; this was achieved only in the second half of the century. Much preparatory work had to be carried out for this transformation. As soon as comprehensive, sufficiently accurate and general theories had been established, the preparatory work was forgotten. This was also the fate of Gudermann's work; he is known as the teacher of Karl Weierstrass rather than as an original thinker.

The depth of Gudermann's understanding of the contemporary trends in mathematics is substantiated by the topic which he discussed in his own work. C. F. Gauss's influence on Gudermann is still unclear, but the topic he chose is close to the intellectual environment of Gauss and his followers. Basically, Gudermann considered only two groups of problems: spherical geometry and the theory of special functions.

His book *Grundriss der analytischen Sphärik* (1830) deals with the former. He considered the study of spherical geometry important for several reasons. In the introduction he pointed out that a plane was a special case of a spherical surface, that is, a sphere with an infinite radius. For this reason and because of its constant curvature there exist many similarities between spherical geometry and plane geometry; yet at the same time Gudermann considered scientifically more interesting the study of cases in which this similarity no longer holds. As part of this program he sought to establish an analytical system for spherical surfaces akin to that formed by the coordinate system in planimetry. But he had to admit the existence of insurmountable difficulties if the required simplicity of the analytical means was to be preserved. At some points in the book Gudermann came close to problems which were important for non-Euclidean geometry but did not stress them, nor did he explicitly mention this aspect.

Gudermann devoted much more attention to the theory of special functions. After the earlier works of Leonhard Euler, John Landen, and A. M. Legendre (Gauss's results were still in manuscript), Niels Abel's studies on elliptical functions, published mostly in A. L. Crelle's *Journal für die reine und angewandte Mathematik,* represented an important divide in treating this area. In 1829 Carl Jacobi's book *Fundamenta nova theoriae functionum ellipticarum* was published. At the time Gudermann was one of the first mathematicians to expand on these results. Beginning with volume **6** (1830) of Crelle's *Journal,* he published a series of papers which he later summarized in two books: *Theorie der Potenzial- oder cyklisch-hyperbolischen Functionen* (1833) and *Theorie der Modular-Functionen und der Modular-Integrale* (1844), which were to have had a sequel which was never written.

In these books Gudermann went back to the origin of the theory of special functions—the problems of integral calculus—and stressed the genetic connection between simply periodic functions and elliptical functions. Since he did not neglect the requirements of integral calculus, he saw the necessity of arranging the theory to allow for numerical calculations. The key appeared to be in the development of the functions into infinite series and infinite products and in the use of suitable transformations. This made it possible to present extensive numerical tables in his first book and to work through to the nucleus of the theory of special functions in his second book, indicating the way which subsequently proved to be exceptionally fruitful. Thus he also came close to Gauss's intentions. Gudermann also introduced a notation for elliptical functions—*sn, cn,* and *dn*—which was adopted. He himself called elliptical functions "Modularfunctionen." It was pointed out later that Gudermann's work had an excess of special cases which in time lost interest.

Gudermann's work drew deserved attention in Germany in the 1830's. Since he was one of the few university professors to treat the problems of elliptical functions systematically, Karl Weierstrass came from the University of Bonn in 1839–1840 to attend Gudermann's lectures and presented "Über die Entwicklung der Modularfunctionen" as a *Habilitationsschrift* in 1841.

Gudermann was one of the first to realize Weierstrass' mathematical talent and scientific ability. Weierstrass, using Gudermann's idea of the development of functions into series and products, formed the principal, mighty, and accurate tool of the theory of functions.

BIBLIOGRAPHY

I. ORIGINAL WORKS. Gudermann's books include *Grundriss der analytischen Sphärik* (Cologne, 1830); *Theorie der Potenzial- oder cyklisch-hyperbolischen Functionen* (Berlin, 1833); *Theorie der Modular-Functionen und der Modular-Integrale* (Berlin, 1844); and *Über die wissenschaftliche Anwedung der Belagerungs-Geschütze* (Münster, 1850).

II. SECONDARY LITERATURE. See *Neue deutsche Biographie,* VII, 252–253; F. Klein, *Vorlesungen über die Entwicklung der Mathematik im 19. Jahrhundert,* I (Berlin, 1926), 278 f.; *Encyklopädie der mathematischen Wissenschaften,* II (Leipzig, 1913); and R. Sturm, "Gudermanns Urteil über die Prüfungsarbiet von Weierstrass (1841)," in *Jahresbericht der Deutschen Mathematiker-Vereinigung,* **19** (1910), 160.

LUBOŠ NOVÝ

GUENTHER, ADAM WILHELM SIEGMUND (*b.* Nuremberg, Germany, 6 February 1848; *d.* Munich, Germany, 3 February 1923), *mathematics, geography, meteorology, history of science.*

Guenther was the son of a Nuremberg businessman, Ludwig Leonhard Guenther, and Johanna Weiser. In 1872 he married Maria Weiser; they had one daughter and three sons, one of whom was the political economist and sociologist Gustav Adolf Guenther.

Guenther studied mathematics and physics from 1865 at Erlangen, Heidelberg, Leipzig, Berlin, and Göttingen. He received his doctorate from Erlangen with *Studien zur theoretischen Photometrie* (1872). He participated in the Franco-Prussian War and then took the teaching examination for mathematics and physics. In 1872 he became a teacher at Weissenburg, Bavaria, and immediately qualified for university lecturing at Erlangen with *Darstellung der Näherungswerte der Kettenbrüche in independenter Form* (Erlangen, 1872–1873). Guenther went to the Munich Polytechnicum as a *Privatdozent* in mathematics in 1874 and to Ansbach in 1876 as professor of mathematics and physics at the Gymnasium. From 1886 to 1920 he was professor of geography at the Munich Technische Hochschule, and from 1911 to 1914 he was rector of this school. A member of the Liberal party, he served in the German Reichstag from 1878 to 1884 and in the Bavarian Landtag from 1884 to 1899 and from 1907 to 1918. During World War I he headed the Bavarian flying weather service, beginning in 1917.

Guenther's numerous books and journal articles encompass both pure mathematics and its history and physics, geophysics, meteorology, geography, and astronomy. The individual works on the history of science, worth reading even today, bear witness to a thorough study of the sources, a remarkable knowledge of the relevant secondary literature, and a superior descriptive ability. Although it is true that his compendia contain a great many names and references, only hint at particulars, and are outdated today, they are nevertheless characteristic of their time.

BIBLIOGRAPHY

I. ORIGINAL WORKS. Guenther's principal writings are *Zur reinen Mathematik: Lehrbuch der Determinantentheorie* (Erlangen, 1875; 2nd ed., 1877); *Die Lehre von den gewöhnlichen verallgemeinerten Hyperbelfunktionen* (Halle, 1881); and *Parabolische Logarithmen und parabolische Trigonometrie* (Leipzig, 1882).

On physics, geography, and related fields, see *Einfluss der Himmelskörper auf Witterungsverhältnisse,* 2 vols. (Halle, 1877–1879); *Lehrbuch der Geophysik und physikalischen Geographie,* 2 vols. (Stuttgart, 1884–1885), 2nd ed. entitled *Handbuch der Geophysik* (1897–1899); *Handbuch der mathematischen Geographie* (Stuttgart, 1891); and *Didaktik und Methodik des Geographie-Unterrichtes* (Munich, 1895).

Works on the history of science include *Vermischte Untersuchungen zur Geschichte der mathematischen Wissenschaften* (Leipzig, 1876); *Ziele und Resultäte der neueren mathematisch-historischen Forschung* (Erlangen, 1876); *Antike Näherungsmethoden im Lichte moderner Mathematik* (Prague, 1878); "Geschichte des mathematischen Unterrichtes im deutschen Mittelalter bis zum Jahre 1525," in *Monumenta Germaniae paedagogica,* III (Berlin,

1887); "Abriss der Geschichte der Mathematik und Naturwissenschaften im Altertum," in *Handbuch der klassischen Altertumswissenschaften,* 2nd ed., V. supp. 1 (1894): *Geschichte der anorganischen Naturwissenschaften im 19. Jahrhundert* (Berlin, 1901); *Entdeckungsgeschichte und Fortschritte der Geographie im 19. Jahrhundert* (Berlin, 1902); *Geschichte der Erdkunde* (Vienna, 1904); and *Geschichte der Mathematik,* I, *Von den aeltesten Zeiten bis Cartesius* (Leipzig, 1908; repr., 1927).

Guenther was coeditor of *Zeitschrift für mathematischen und naturwissenschaftlichen Unterricht* (1876–1886); *Zeitschrift für das Ausland* (1892–1893); *Münchener geographische Studien* (from 1896); *Mitteilungen zur Geschichte der Medizin und der Naturwissenschaften* (from 1901); and *Forschungen zur bayerischen Landeskunde* (1920–1921).

II. SECONDARY LITERATURE. Obituaries include E. von Drygalski, in *Jahrbuch der bayerischen Akademie der Wissenschaften* (1920–1923), pp. 79–83; *Geographische Zeitschrift,* **29** (1923), 161–164; L. Günther, in *Lebensläufe aus Franken,* IV (1930), 204–219; W. Schüller, in *Zeitschrift für mathematischen und naturwissenschaftlichen Unterricht,* **56** (1925), 109–113; H. Wieleitner, in *Mitteilungen zur Geschichte der Medizin und der Naturwissenschaften,* **22** (1923), 1–2; and August Wilhelm, in *Neue deutsche Biographie,* VI (1966), 266–267.

J. E. HOFMANN

GUERICKE (GERICKE), OTTO VON (*b.* Magdeburg, Germany, 20 November 1602; *d.* Hamburg, Germany, 11 May 1686), *engineering, physics.*

Guericke was the son of Hans Gericke and Anna von Zweidorff. As the scion of a patrician family long established in Magdeburg he was destined to participate in political life. He was registered in the Faculty of Arts at the University of Leipzig from 1617 to 1620; attended the University of Helmstedt in 1620; and studied law at Jena in 1621 and 1622. Guericke then went to Leiden, where in addition to studying law he also attended lectures on mathematics and engineering, especially fortification.

Upon his return to Germany Guericke was elected an alderman of the city of Magdeburg in 1626; in the same year he married Margarethe Alemann, who died in 1645 (his second wife was Dorothea Lentke, whom he married in 1652). In 1630 he assumed the additional duties of city contractor. After the destruction of the city in 1631, Guericke worked in Brunswick and Erfurt as an engineer for the Swedish government; from 1635 he performed the same duties for the electorate of Saxony. This dual position allowed Guericke to serve Magdeburg throughout the Thirty Years' War, during which time he acted as envoy to the changing occupation powers. He further represented Magdeburg at the subsequent peace conferences and later at the Imperial Diet in Regensburg.

Diplomacy consumed much of Guericke's time from 1642 to 1666. He was also mayor of Magdeburg (1646–1676). He devoted his brief leisure to scientific experimentation, however, and his attendance at international congresses and princely courts allowed him to take part in the exchange of scientific ideas. Guericke presented some of his own experiments on several occasions at Regensburg in 1653–1654 and again in 1663 at the court of the Great Elector in Berlin. He also learned of new scientific developments in such circumstances; at Osnabrück in 1646 he first heard of Descartes's new physics and at Regensburg he was introduced to the experiments of Torricelli, who was working on the problem of the vacuum, as was Guericke himself, but from another point of view.

Indeed, Guericke had been preoccupied ever since his student days at Leiden with the question of the definition of space. A convinced Copernican, he was particularly concerned with three fundamental questions: (1) What is the nature of space? Can empty space exist, or is space always filled and empty space only a *spatium imaginarium,* a logical abstraction? (2) How can individual heavenly bodies affect each other across space, and how are they moved? (3) Is space, and therefore the heavenly bodies enclosed in it, bounded or unbounded?

Descartes's conception of space and matter as equivalent and his denial of a vacuum led Guericke to propose an experiment designed to resolve the old conflict between plenists and vacuists. Guericke posited that if the air were pumped out of a strong container and no other new material allowed to take its place the vessel would implode if Descartes's assertions were true. Soon after he returned from Osnabrück in 1647 Guericke made a suction pump using a cylinder and piston to which he added two flap valves; he then used this apparatus to pump water out of a well-caulked beer cask. Air entered the cask, however, as was evidenced by whistling noises. When Guericke repeated the experiment with the beer cask sealed within a second larger one that he had also filled with water, the water that he pumped out was replaced by water seeping in from the larger vessel.

In an attempt to solve the sealing problem Guericke ordered the construction of a hollow copper sphere with an outlet at the bottom. He pumped the air directly out of this apparatus which thereupon imploded. This result would seem to indicate that Descartes was right; but Guericke still thought otherwise on the basis of his earlier experiments. He had a new apparatus made, and with this his experiment succeeded. Guericke thus invented the air pump, or, rather, discovered the pumping capacity of air. He had thought that the air within the vessel would sink, as had the water in his previous devices, and that it would be evacuated from the bottom; later experiments, however, in which the outlet was placed at arbitrary points on the copper sphere proved that the air left in the container during the process of evacuation was distributed evenly throughout the interior space.

This discovery of the elasticity of the air represents perhaps the most important result of Guericke's experiments. From it he was led to investigate the decrease of the density of the air with height and to theorize concerning empty space beyond the atmosphere of heavenly bodies; to study variations of air pressure corresponding to changes in the weather (taking mean air pressure to correspond to a water column twenty Magdeburg ells high, he succeeded in 1660 in making barometric weather forecasts); to propose systematic weather reporting through a network of observation stations; to come to know the ponderability of air within air; and finally, to draw further conclusions about a variety of phenomena connected with vacuums, most of which he demonstrated experimentally, especially the work capacity of air, by which he refuted the theory of *horror vacui.*

The most famous of Guericke's public experiments is the one of the Magdeburg hemispheres, in which he placed together two copper hemispheres, milled so that the edges fit together snugly. He then evacuated the air from the resulting sphere and showed that a most heavy weight could not pull them apart. Contrary to legend, the demonstration was performed with a team of horses for the first time in Magdeburg in 1657 (not Regensburg in 1654) and repeated at court in Berlin in 1663. Guericke also made other, less dramatic, public demonstrations of the effectiveness of air pressure on several occasions in Regensburg; these Regensburg experiments were reported by Gaspar Schott in *Mechanica hydraulico-pneumatica* (1657) and *Technica curiosa* (1664), and were supplemented with additional information that Guericke communicated by letter.

Schott's books as well as other foreign publications of Guericke's experiments (for example, works of M. Cornaeus and S. Lubieniecky) stimulated Huygens and Boyle, among others, to repeat and extend the experiments and to set to work upon an improved air pump. Guericke himself was occupied with the same project; he improved his pump with hydraulic sealing and devised a stationary installation for it (it occupied two floors of his house). In 1663 Guericke developed a portable pump modeled on one of Boyle's and constructed one especially for his visit to Berlin in that year. (Three examples of this type of pump survive, one each in Munich, Lund, and Brunswick.)

Guericke's experimental work, however, represents only one facet of his attempt to reach a complete physical world view. He drew upon his Copernicanism to construct the foundations for such a system. Guericke's celestial physics were further based upon the notion that the heavenly bodies interacted with each other across empty space through magnetic force; here he turned to the earlier work of Gilbert and Kepler. Their magnetic hypotheses had been refuted by Athanasius Kircher (in *Magnes sive De arte magnetica,* 1641); joining the argument, Guericke sought to modify Gilbert's magnetism experiments by making use of materials mimicking the actual composition of the earth. To this end Guericke cast a sphere composed of a variety of minerals with a large proportion of sulfur—in later experiments he used pure sulfur—and showed that it possessed the *virtutes mundanae,* that is, such powers as attraction and the ability to move other bodies. By rubbing the sphere of sulfur, Guericke had actually produced static electricity; but since he did not recognize these electrical effects as special phenomena, but as demonstrations of the *virtutes* of a celestial body, he cannot properly be credited with the invention of the first electrical machine.

Having dealt with the problems of empty space and the movement of heavenly bodies, Guericke concerned himself further with the question of the boundedness of space and the number of worlds therein. He conceived of fixed stars as suns with planetary systems, each of which exerts a sphere of force (*orbis virtutis, sphaera activitatis*); these systems border on each other and do not interact—each heavenly body rather possesses a specific center of gravity for a specific *virtus conservativa,* which he interpreted as its source of cohesion. Thus, in opposition to Aristotelian cosmography, an immaterial boundary of space becomes inconceivable. Giordano Bruno had already speculated about an infinite universe containing an infinite number of worlds, but his ideas had been unacceptable because only God was considered infinite—all of God's creation must be finite. Guericke overcame this objection by redefining the notion of nothingness. By his reasoning, empty space as a mere receptacle for God's creations is nothingness and is not created. Empty space is therefore independent of God and the created universe—indeed, it precedes the latter. Therefore empty space cannot be bounded. Neither can the number of worlds be bounded, although such a number is not infinite (since there are no infinite numbers). Likewise it is not limited, since there is no greatest number and no end to the series of numbers. Infinite space is thus a conceptual possibility.

Such speculations about the heavenly bodies quite naturally led Guericke to the study of astronomy. He explained planetary orbits as exactly circular and concentric, effected by the rotating *orbis virtutis* of the sun, and interpreted the apparent eccentricities as a result of the different densities of the atmosphere.

In 1666 Guericke was made a noble and his family name became von Guericke. In 1681 he retired from his public offices and went to live in Hamburg with his son, a magistrate of Brandenburg. He spent the rest of his life there.

BIBLIOGRAPHY

I. ORIGINAL WORKS. Guericke's most important published work is *Experimenta nova, ut vocantur, de vacuo spatio* . . . (Amsterdam, 1672; repr. Aalen, 1962).

His letters to Schott, Lubieniecky, Leibniz, and others may be found in Hans Schimank, trans. and ed., with the collaboration of Hans Gossen, Gregor Maurach, and Fritz Krafft, *Otto von Guerickes Neue (sogenannte) Magdeburger Versuche über den leeren Raum, nebst Briefen, Urkunden und anderen Zeugnissen seiner Lebens- und Schaffensgeschichte* (Düsseldorf, 1968), which also contains a full bibliography by Krafft current through 1967.

II. SECONDARY LITERATURE. In addition to Schimank, above, see Alfons Kauffeld, *Otto von Guericke: Philosophisches über den leeren Raum* (Berlin, 1968); Fritz Krafft, "Experimenta nova. Untersuchungen zur Geschichte eines wissenschaftlichen Buches. I," in Eberhard Schmauderer, ed., *Buch und Wissenschaft* . . . (Düsseldorf, 1969), XVII, 103-129, in the series Technikgeschichte in Einzeldarstellungen; and "Sphaera activitatis—orbis virtutis," in *Sudhoffs Archiv: Zeitschrift für Wissenschaftsgeschichte,* **54** (1970), 113-140.

FRITZ KRAFFT

GUERTLER, WILLIAM MINOT (*b.* Hannover, Germany, 10 March 1880; *d.* Hannover, 21 March 1959), *metallography, metallurgy.*

Guertler was the son of Alexander Guertler, a physician. After graduating from the Gymnasium in Hameln in 1899 he studied variously until 1904 at the Technische Hochschule in Hannover, the University of Munich, and the University of Göttingen, from which he earned the doctorate in 1904 under Gustav Tammann. He worked as Tammann's assistant until 1906. From 1906 to 1908 Guertler was an assistant at the Technische Hochschule in Berlin; in 1908 he joined the faculty as a doctor of engineering.

In the academic year 1908-1909 Guertler was a research associate at the Massachusetts Institute of Technology; he worked there again in 1911. From 1909 to 1914 he worked, without pay, with A. Doeltz at the Metallurgical Institute of the Berlin Technische Hochschule. He was head assistant at the Institute in 1917, became professor there in the same year, and eventually served as deputy director (1921-1928) and director (1929). In 1930 he was appointed to teach at the Institute for Applied Metallurgy; he was promoted to professor and director of the institute in 1933. In 1936 he assumed the additional posts of professor and director of the Institute for Metallurgy and the Science of Materials at the Technische Hochschule in Dresden. He reached retirement age in 1945 and from then until 1956 was a guest professor at the Technical University in Istanbul and at M.I.T.

Guertler's lifework concerned pure and applied metallurgy, which he developed into an independent scientific discipline. He did not restrict himself to a purely academic approach to the subject but rather sought to consider all aspects of the study of metals—both theoretically and technologically—and to apply his results in the metal industry. At the beginning of his researches he systematically investigated the constitution of metals; he was especially interested in the conductivity of alloys. He discovered that the value of the conductivity of a compound composed of several metals is always less than the sum of the conductivities of the components. Through his work on nomenclature Guertler established many new metallurgical concepts, including segregation and peritectonics. He also developed new alloys, largely those of nonferrous metals and most notably those of silver and aluminum. In 1926 he was selected to give the Campbell Memorial Lecture; he also made lecture tours in India, Japan, and the United States.

Guertler was, moreover, a pioneer in the scientific organization of his field. Besides writing books—particularly the handbook *Metallographie* (begun in 1912)—he founded such periodicals as the *Internationale Zeitschrift für Metallographie* in 1911 (called *Zeitschrift für Metallkunde* after 1919). Further, he established the German Metallographic Society and the German Society for Technical Röntgenology.

BIBLIOGRAPHY

I. ORIGINAL WORKS. Guertler wrote more than 300 scientific papers and books and was awarded some 100 patents for his methods and devices. For bibliography, see Poggendorff, V, 462-463; VI, pt. 2, 973-974; and VIIa, pt. 2, 314-315.

His many papers and monographs include his diss. *Über wasserfreie Borate und über Entglasung* (Leipzig, 1904); "Die elektrische Leitfähigkeit der Legierungen," in *Zeit-*

schrift für anorganische Chemie, **51** (1906), 397–433; **54** (1907), 58–88; see also Guertler's paper on electric conductivity in *Physikalische Zeitschrift,* **9** (1908), 29–36, 404–405; **11** (1910), 476–479; "Stand der Forschung über die elektrische Leitfähigkeit der kristallisierten Metallegierungen," in *Jahrbuch der Radioaktivität und Elektronik,* **5** (1908), 17–81; **6** (1909), 127; "Beiträge zur Kenntnis der Elektrizitätsleitung in Metallen und Legierungen," *ibid.,* **17** (1920), 276–292, 298–299; "Vereinheitlichung der Benennung metallischer Produkte," in *Zeitschrift für Metallkunde,* **11** (1919), 200; *Richtlinien zur Gewinnung eines Überblicks über den Aufbau von Dreistoffsystemen,* no. 1 in the series Forschungsarbeiten zur Metallkunde, of which Guertler was editor (Berlin, 1923); *Sechs Vorlesungen zur Einführung in das Verständnis der modernen Spezialstähle,* Forschungsarbeiten zur Metallkunde, no. 8 (Berlin, 1928); "Colloidal Conditions in Metal Crystals," in J. Alexander, ed., *Colloid Chemistry,* III (New York, 1931), 439–448.

See also, under his editorship, *Metallographie. Ein ausführliches Lehr- und Handbuch der Konstitution und der physikalischen, chemischen und technischen Eigenschaften der Metalle und metallischen Legierungen,* I–III (Berlin, 1912–1935), of which Guertler is also author of numerous parts; *Metalltechnischer Kalender* (Berlin, 1922); *Vom Erz zum metallischen Werkstoff* (Leipzig, 1929), vol. I of *Der metallische Werkstoff,* ed. with W. Leitgebel; *Einführung in die Metallkunde,* 2 vols. (Leipzig, 1943), later published as *Metallkunde,* I (Berlin, 1954) and the new series of *Archiv für Metallkunde.*

II. SECONDARY LITERATURE. On Guertler and his work, see W. Claus, "William M. Guertler. Zum 60. Geburtstag am 10.3.1940," in *Metall: Wirtschaft, Wissenschaft, Technik,* **19** (1940), 175–176; F. Erdmann-Jesnitzer, "William Guertler †," in *Bergakademie,* **11** (1959), 334; B. Trautmann, "Guertler, William Minot," in *Neue deutsche Biographie,* VII (1966), 287–288; and "William Guertler †," in *Zeitschrift für Metallkunde,* **50** (1959), 239.

HANS-GÜNTHER KÖRBER

GUETTARD, JEAN-ÉTIENNE (*b.* Étampes, Seine-et-Oise, France, 22 September 1715; *d.* Paris, France, 6 January 1786), *geology, natural history, botany.*

A versatile scientist trained in medicine and chemistry, Guettard gradually acquired knowledge of the various branches of natural history. Although always concerned to some degree with all these fields, most of his career, especially after about 1746, was devoted to geology, and his reputation now rests upon two achievements: his discovery of the volcanic nature of Auvergne, and his attempt to construct a geological map of France.

Guettard's schooling in Étampes, nearby Montargis, and Paris was less important in shaping his career than was the influence of his maternal grandfather, François Descurain, a physician and apothecary in Étampes, as well as an amateur botanist and friend of Bernard de Jussieu. Guettard's early activities followed the pattern set by Descurain. While studying in Paris, he was introduced, probably by Jussieu, to naturalist-physicist René Antoine Ferchault de Réaumur and by 1741 had become curator of Réaumur's natural history collection. As Réaumur's assistant, he also conducted experiments on the regeneration of marine polyps. In 1742 he was admitted to the Faculté de Médecine de Paris, and on 3 July 1743 was elected to the Académie Royale des Sciences as an *adjoint botaniste.* At the same time he began to take field trips in the Loire Valley and in Normandy and to form his own natural history collection. His first major publication, *Observations sur les plantes* (1747), was a botanical study of the environs of Étampes, based upon a manuscript by his grandfather.

By 1747 Guettard had become *médecin botaniste* to Louis, duc d'Orléans, the two men having in common not only their religious views—both were devout Jansenists—but also their interest in the chemical analysis of minerals and rocks. After the duke's death in 1752, his son, Louis-Philippe, continued to support Guettard and he was given rooms in the Palais-Royal. With a laboratory at his disposal and an assured income, Guettard was free to devote himself entirely to scientific research. The remainder of his career is remarkable for his many long field trips, the development of a large scientific correspondence, and the publication of numerous articles often of the length and character of small monographs.

All of Guettard's scientific work bears the stamp of the Baconian naturalist who consciously avoided the formulation of theories. He attacked the ideas of the natural theologians, the cosmology and geology of Buffon, and the biology of Charles Bonnet, labeling all such systems premature, scientifically unsound, or philosophically dangerous and tending toward materialism. Although his own work is not free of preconceptions, Guettard tried to avoid drawing conclusions from his data. Thus, while his studies of sedimentation and erosion provided the kind of evidence other geologists were using to suggest an extension of the time scale beyond the biblical 6,000 years, Guettard's own writings contain no hint of such ideas. Similarly, his studies of the comparative anatomy of fossil and living forms led him only to deny repeatedly the likelihood that species can become extinct. It was therefore left to his contemporaries and successors to recognize the implications of Guettard's work.

Most of Guettard's field trips within France—he also traveled in the Low Countries, Italy, Switzerland, and Poland—were undertaken to supply data for a national geological survey and are thus of relatively small intrinsic interest apart from the larger project. A notable exception was his voyage to Auvergne in 1751, accompanied by his friend Chrétien-Guillaume de Lamoignon de Malesherbes. During this journey Guettard noticed that volcanic rocks were often used in the construction of local roads and dwellings; he examined the quarries and concluded that the whole region was volcanic. This discovery was announced in his "Mémoire sur quelques montagnes de la France qui ont été des volcans" (*Mémoires de l'Académie royale des sciences pour l'année 1752* [1756], pp. 27–59). He did not pursue the subject further, but this memoir induced several of his contemporaries to study the geological history of Auvergne, and the region very soon became a tourist attraction. Years later the priority of Guettard's discovery was challenged by Barthélemy Faujas de Saint-Fond in his *Recherches sur les volcans éteints du Vivarais et du Velay* (1778). Whether or not anyone had in fact anticipated Guettard, as Faujas claimed, it is certain that Guettard's discovery was made independently and was the first public announcement.

In the well-known controversy over the nature of columnar basalt, Guettard at first supported the view that these formations were not volcanic in origin. However, after visits to Italy in 1771 and 1772 and the vicinity of Montpellier in 1771, he began to have doubts, and these doubts were confirmed when in 1775 he explored the neighborhood of Montélimar in Dauphiné. His change of opinion about the origin of columnar basalt was announced in his *Mémoires sur la minéralogie du Dauphiné* (1779).

Guettard's work as a geological cartographer began before 1746, the year in which he presented what he called a preliminary "mineralogical map" of France to the Académie Royale des Sciences. His travels and reading had called his attention to "a certain regularity" in the distribution of minerals and rocks over the earth's surface, and he decided to plot his data on a map. The result was his "Mémoire et carte minéralogique sur la nature & la situation des terreins qui traversent la France & l'Angleterre" (*Mémoires de l'Académie royale des sciences pour l'année 1746* [1751], pp. 363–392), which was followed in later years by memoirs and maps dealing with the Middle East, part of North America, Poland, and Switzerland, as well as preliminary (unpublished) maps of Italy and Corsica. These maps show, by means of conventional chemical symbols, the location of rock formations and mineral deposits, and they are also marked off in regions labeled *Bandes.* Each *Bande* was characterized by the predominance of certain deposits, so that in France, for example, he could outline three such concentric zones: the Sandy Band (primarily sandstones and limestones) with its center near Paris, the Marly Band, and the Schistose or Metalliferous Band. Although superposition was sometimes noted, the scheme was basically not a stratigraphic one. Guettard hoped eventually to clarify the relationships between the systems of bands outlined for neighboring geographical regions, and he also expected that his maps would enable scientists to find or predict the locations of useful or valuable mineral deposits, building materials, and agricultural and industrial soils. As late as 1784, he was still perfecting his initial mineralogical map of France.

In 1766 Guettard and Lavoisier were commissioned by Henri Bertin, minister and secretary of state in charge of mining, to prepare a geological survey of France. The collaboration of Guettard and Lavoisier had actually begun before that date, and among the several field trips they took together was their famous geological tour of Alsace, Lorraine, and Franche-Comté in 1767. By 1777 they had completed sixteen quadrangles, while bringing an almost equal number to partial completion, out of a projected total of some 200 maps. The survey passed into the hands of Antoine Monnet in 1777, and he published thirty-one quadrangles in the *Atlas et Description minéralogiques de la France* (1780), which he later issued in a second edition containing forty-five quadrangles. The rest of the survey was never executed.

The maps of the *Atlas* feature the chemical symbols used by Guettard, but he intentionally omitted the system of bands so that the survey would not appear to be based on any one geological theory; in addition, Lavoisier used the margin of each quadrangle for a vertical section designed to show the stratigraphic arrangement of the earth's crust. Monnet's maps followed a somewhat similar pattern. Although employing chemical symbols on geological maps was popular for a time, the maps of the *Atlas* had no close imitators; contemporaries and later geologists, with the notable exception of Nicolas Desmarest, found these maps to be models of observational accuracy, but the cartographic techniques of Guettard and Lavoisier were superseded by those developed in subsequent decades by such men as A. G. Werner, William Smith, Georges Cuvier, and Alexandre Brongniart.

Among Guettard's many other achievements were his identification of trilobites in the slates of Anjou and his discovery in France of sources of kaolin and petuntse needed in the manufacture of good porcelain. As a botanist, he remained a defender of the Linnaean system against its many critics. When the Académie Royale des Sciences was reorganized in 1785, he became a pensionnaire in the division of botany and agriculture. Upon his death, his natural history collection and library of more than 3,500 titles were sold and their fate is uncertain. His unpublished papers include memoirs on subjects botanical and geological, the latter including studies of virtually every French province. Toward the end of his life, Guettard searched for funds to support the publication of these papers, but without success. The papers then passed into the hands of Lavoisier, Guettard's scientific executor, whose efforts to have some of them published also failed.

BIBLIOGRAPHY

I. ORIGINAL WORKS. Works by Guettard include *Observations sur les plantes,* 2 vols. (Paris, 1747); *Mémoires sur la minéralogie du Dauphiné,* 2 vols. (Paris, 1779); *Mémoires sur différentes parties de la physique, de l'histoire naturelle; des sciences et des arts, & c.,* 5 vols. in 6 (Paris, 1768–1786). Guettard contributed to *Atlas et Description minéralogiques de la France, Entrepris par ordre du Roi . . . Première Partie* (Paris, 1780), in which the maps are the work of Guettard, Lavoisier, and Monnet, and the text the work of Monnet; for an analysis, see the Lavoisier bibliographies cited below.

Guettard also contributed to J. B. de La Borde, E. Béguillet, *et al., Description générale et particulière de la France . . .,* 4 vols. (Paris, 1781–1784), continued as *Voyage pittoresque de la France,* 8 vols. (Paris, 1784–1800). He translated Pliny the Elder, *Histoire naturelle,* 12 vols. (Paris, 1771–1782), and published more than seventy articles in the *Mémoires de l'Académie royale des sciences,* with a few in *Observations sur la physique, sur l'histoire naturelle et sur les arts* and *Journal oeconomique.*

More than twenty-five cartons and volumes of Guettard's MSS are in the Central Library of the Muséum National d'Histoire Naturelle, Paris. Travel journals and other documents are in the archives of the Académie des Sciences, Paris, and the Olin Library, Cornell University, Ithaca, New York. One journal has been published, with minor modifications, by A. Vernière, "Note sur les environs de Vichy et sur la découverte des volcans éteints de l'Auvergne (d'après un manuscrit autographe de Guettard, 1751)," in *Revue scientifique du Bourbonnais et du centre de la France,* **14** (1901), 5–13. Additional letters are in the Municipal Library of Clermont-Ferrand, Puy-de-Dôme, and can be found among the papers of Pierre-Michel Hennin, Library of the Institut de France, Paris. Single letters of importance are published in Buffon, *Les époques de la nature,* Jacques Roger, ed., *Mémoires du Muséum National d'Histoire Naturelle,* ser. C, **10** (Paris, 1962), cxxxix, n. 7; and René Fric, "Une lettre de Guettard à Monnet au sujet des prismes basaltiques," in *Bulletin historique et scientifique de l'Auvergne,* **78** (1958), 91–96.

II. SECONDARY LITERATURE. For works on Guettard see Aimé de Soland, "Étude sur Guettard," in *Annales de la Société linnéenne de Maine-et-Loire,* **13, 14, 15** (1871–1873), 32–88; Gavin de Beer, "The Volcanoes of Auvergne," in *Annals of Science,* **18** (1962), 49–61; M. J. A. N. C. Condorcet, "Éloge de M. Guettard," in *Histoire de l'Académie royale des sciences pour l'année 1786* (1788), pp. 47–62; Denis I. Duveen and Herbert S. Klickstein, *A Bibliography of the Works of Antoine Laurent Lavoisier 1743–1794* (London, 1954); Denis I. Duveen, *Supplement to a Bibliography of the Works of Antoine Laurent Lavoisier 1743–1794* (London, 1965); Roland Lamontagne, "La participation canadienne à l'oeuvre minéralogique de Guettard," in *Revue d'histoire des sciences et de leur applications,* **18** (1965), 385–388; A.-L. Letacq, "Notice sur les travaux scientifiques de Guettard aux environs d'Alençon & de Laigle (Orne)," in *Bulletin de la Société linnéenne de Normandie,* **5** (1891), 67–85; R. Michel, "À propos de la découverte des Volcans éteints de l'Auvergne et du Vivarais: Notes sur deux géologues du XVIIIe siècle, Guettard (1715–1786) et Faujas de Saint-Fond (1741–1819)," in *Revue des sciences naturelles d'Auvergne,* **11** (1945), 37–53; and R. Rappaport, "The Geological Atlas of Guettard, Lavoisier, and Monnet: Conflicting Views of the Nature of Geology," in Cecil J. Schneer, ed., *Toward a History of Geology* (Cambridge, Mass., 1969).

RHODA RAPPAPORT

GUIBERT, NICOLAS (*b.* St. Nicolas-de-Port, Lorraine, *ca.* 1547; *d.* Vaucouleurs, France, *ca.* 1620), *chemistry.*

Little is known of Guibert's family and early life. A Catholic, he studied medicine at the University of Perugia and, after receiving his degree, traveled in Italy, France, Germany, and Spain. During this period, Guibert became well known as an alchemist, working for several important persons, including Francesco de' Medici, grand duke of Tuscany, and Cardinal Granvelle, viceroy of Naples and a leader of Philip II's Spanish faction at Rome. He also associated for a time with Giambattista della Porta in Naples. Settling in the Italian town of Casteldurante, Guibert established a successful medical practice and in 1578 was appointed chief medical authority of one of the papal states. He held this position until leaving Italy at the end of 1579 to work as alchemist for Otto Truchsess, archbishop of Augsburg, whom he advised to commission a translation of Paracelsus' complete works into Latin.

Guibert's growing frustration with alchemical pursuits, however, accentuated his dissatisfaction with the obscurity and pretensions of much of sixteenth-century alchemy, and he emerged finally as a vehement critic of the profession. His first published attack came in 1603. In *Alchymia ratione et experientia ita demum viriliter impugnata et expugnata,* Guibert attempted to refute the major alchemical literature by demonstrating that alchemy is false and that most important alchemical treatises are of no authority. He branded the *Tabula smaragdina* and other alchemical writings attributed to Ibn-Sīnā, Albertus Magnus, and Thomas Aquinas as spurious. Moreover, the works whose authorship he accepted as genuine, such as the writings of Arnald of Villanova, Roger Bacon, Agrippa, and Paracelsus (the "limb of Satan"), were condemned by him as quackery and heresy. Despite its often exaggerated tone and unsubstantiated claims concerning the literature, Guibert's *Alchymia* did serve to reinforce several significant, albeit not widely held, ideas. Most important was his demonstration that metals are distinct species and not transmutable; he rejected the common argument for the transmutation of metals based on analogy to the organic realm—such as the change from larva to butterfly—and contradicted the influential belief that iron can be changed into copper.

His attack on the fundamental tenets of alchemy elicited a vigorous response from Andreas Libavius, the famous German iatrochemist, whose *Defensio alchymiae transmutatoriae opposita Nicolai Guiberti* (Ursel, 1604) defends the apparent conversion of metals—as in iron-copper replacement reactions—as genuine transmutations. Libavius further asserted that the growth and change of plants and animals afforded a valid analogy for maintaining the reality of chemical transmutations. Guibert, in turn, attacked Libavius' position in detail in his second major work, *De interitu alchymiae* (1614). The controversy concerning alchemy was part of the broader debate, which persisted throughout the sixteenth and seventeenth centuries, on the relations between occult science, natural magic, and emerging modern science. Guibert's rejection of alchemy derived from his revulsion from the activities of those charlatans who styled themselves scientists; from an orthodox Catholic suspicion of heresy in Renaissance Neoplatonism and the Hermetic revival of the sixteenth century; and, finally, from a recognition of certain theoretical and experimental inconsistencies of alchemy.

BIBLIOGRAPHY

I. ORIGINAL WORKS. Guibert's scientific works include *Assertio de murrhinis, sive de iis quae murrhino nomine exprimuntur* (Frankfurt, 1597); *Alchymia ratione et experientia ita demum viriliter impugnata et expugnata* (Strasbourg, 1603); and *De interitu alchymiae metallorum transmutatoriae tractatus aliquot. Adiuncta est eiusdem apologia in sophistam Libavium, alchymiae refutatae furentem calumniatorem* (Toul, 1614).

II. SECONDARY LITERATURE. Information on Guibert's life and work may be found in Dom Calmet, *Contenant la bibliothèque Lorraine,* vol. IV of *Histoire de Lorraine* (Nancy, 1751), 454–455; and F. Hoefer, ed., *Nouvelle biographie générale,* XXII (Paris, 1858), 518. The most reliable assessment of Guibert's scientific work is Lynn Thorndike, *A History of Magic and Experimental Science* (New York, 1941), V, 648, and VI, 244–247, 451–452. See also James R. Partington, *A History of Chemistry* (London, 1961), II, 268.

MARTIN FICHMAN

GUIDI, GUIDO, also known as **Vidus Vidius** (*b.* Florence, Italy, 10 February 1508; *d.* Pisa, Italy, 26 May 1569), *anatomy, surgery.*

Guidi belonged through his father, Giuliano di Guido dei Guidi, to a family of physicians, and through his mother, Costanza, he was descended from the famous painter Domenico Ghirlandaio. A part of his success can be explained by his capacity to unite science and art harmoniously. After becoming a doctor of medicine Guidi practiced in Rome and Florence. In 1542 he went to Paris and Fontainebleau, bringing to Francis I two splendidly illustrated manuscripts containing the Greek transcription and Latin translation of several classic treatises on surgery. The illustrations in these manuscripts have long been attributed to Francesco Primaticcio, but they were very probably done by the painter Francesco Salviati and his pupils.

In Paris, Guidi was named royal physician and became the first professor of medicine at the Collège Royal. He lived in the private residence of Benvenuto Cellini, and it was there that he had his *Chirurgia* printed, one of the most beautiful scientific books of the Renaissance. Greatly envied by the Faculty of Medicine, Guidi had to leave Paris after the death of Francis I in 1547. His new patron, Cosimo I de' Medici, named him professor of philosophy and medicine at the University of Pisa in 1548. At Pisa, Guidi carried out important anatomical investigations, recorded in a manuscript (*Anatomia*) composed around 1560, which is preserved in Cosimo I de' Medici's library. Guidi became a priest and was given the high church office of provost of Pescia. He was also the consul of the Academy of Florence.

In his *Chirurgia* of 1544 Guidi presents himself above all as a humanist anxious for the faithful restoration of classical knowledge. On the other hand, the *Anatomia* is the work of a scientist fully conscious of the Vesalian revolution and seeking his inspiration from nature. Unfortunately, this treatise was printed, under the title *De anatome corporis humani,* in a posthumous edition with hideous illustrations and maladroit additions by Guido Guidi, Jr., Guidi's nephew. This explains the negative judgments of several historians of medicine and their claims that Guidi plagiarized Vesalius and Falloppio.

Guidi's true merits can be established only after study of the original version of his anatomical work (MS II, III 32, Biblioteca Nazionale, Florence), a study which has not yet been made. Guidi certainly described the vertebrae, the cartilaginous structures, and the bones of the cranium better than any of his predecessors. His name is still attached to the *canalis vidianus* of the sphenoid bone and to the nerve that traverses this canal. Moreover, he made important and original studies of the mechanism of the articulations in the human body resulting from its vertical position in relation to the mechanism of the quadruped articulations. It is interesting to note that his anatomical work concluded with a group of experiments on living animals (for example, ligature of the blood vessels). The first professor of medicine at the Collège de France thus inaugurated the method of vivisection that was to bring such fame to that chair.

In his writing on practical medicine Guidi remained within classical Galenism. Nevertheless, this conservatism did not prevent him from describing a new childhood disease (chicken pox) or from inventing an original method for tracheotomy.

BIBLIOGRAPHY

I. Original Works. The MSS of Guidi's works are preserved in the Bibliothèque Nationale, Paris (particularly MS. lat. 6866), and in the Biblioteca Nazionale and Biblioteca Riccardiana, Florence. The most important of his works to be printed in his lifetime is *Chirurgia e Graeco in Latinum conversa, Vido Vidio Florentino interprete, cum nonnulis ejusdem Vidij commentarijs* (Paris, 1554). The posthumous writings prepared by Guido Guidi, Jr., are *Universae artis medicinalis pars quae ad curationum morborum spectat* (Frankfurt, 1596); *Ars medicinalis,* 3 vols. (Venice, 1611), which contains the first printed edition of the treatise "De anatome corporis humani"; and *Opera omnia medica chirurgica et anatomica* (Frankfurt, 1668).

II. Secondary Literature. The older biographies are S. Salvini, "Guido Guidi consolo," in *Fasti consolari dell' Accademia Fiorentina* (Florence, 1717), pp. 115–123, and P. B., "Elogio di Monsig. Guidi," in *Elogi degli uomini illustri di Toscana,* III (Lucca, 1772), 250–256. More recent publications are H. Omont, *Collection des chirurgiens grecs avec dessins attribués au Primatice* (Paris, 1908); W. Brockbank, "The Man Who Was Vidius," in *Annals of the Royal College of Surgeons of England,* **19** (1956), 269–295; C. E. Kellett, "The School of Salviati and the Illustrations to the *Chirurgia* of Vidus Vidius," in *Medical History,* **2** (1958), 264–268; and M. D. Grmek, "La période parisienne dans la vie de Guido Guidi anatomiste de Florence et professeur au Collège de France," in *Atti della VI biennale dello Studio Firmano* (Fermo, 1965), pp. 191–200.

M. D. Grmek

GUIGNARD, JEAN-LOUIS-LÉON (*b.* Mont-sous-Vaudrey [Jura], France, 13 April 1852; *d.* Paris, France, 7 March 1928), *botany.*

After completing his preliminary education in Mont-sous-Vaudrey, where his parents earned a livelihood by farming, Guignard went on to receive his secondary school diploma (*baccalauréat ès lettres*) in Besançon. From 1871 to 1874 he worked as an apprentice in several pharmacies in Paris. Then in 1874 he enrolled in the Paris School of Pharmacy and concurrently pursued studies at the Faculty of Sciences. Guignard also competed successfully for an appointment to an internship in pharmacy in the Paris municipal hospitals where he served with distinction from 1876 to 1882.

From this formative period, Guignard emerged in 1882 with an advanced qualification in pharmacy (*diplôme supérieure*) and a doctorate in natural sciences from the Faculty of Sciences. Two outstanding theses crowned his scholastic achievement: a study of the embryo sac in angiosperms for his *diplôme supérieure* in pharmacy and a brilliant investigation of the embryogeny in leguminous plants for his *docteur ès sciences naturelles.* Both works immediately established him as a botanist of considerable ability.

His student years behind him, Guignard worked briefly as an assistant and aide-naturalist at the Museum of Natural History. In 1883 he left for Lyons where he became professor of botany at the Faculty of Sciences and director of the botanical garden of that city. Appointed professor of botany at the Paris School of Pharmacy in 1887, he served in that capacity until 1927 and was also director of the school from 1900 to 1910. Guignard was elected to many learned societies including the Botanical Society of France (president, 1894); the Paris Academy of Medicine; Academy of Sciences (president, 1919); Society of Biology (vice-president, 1894); and the National Society of Agriculture of France.

Guignard's publications dating from 1880 represented more than four decades of botanical investigation. His most important contributions were to embryology, cytology, fertilization, the morphology and development of the seed, and his study of reproductive organs in plants. Of considerable interest too was his research on the sites of specific plant principles, organs of secretion, and his work in bacteriology.

In 1882 Guignard demonstrated that the embryo sac in flowering plants always develops from one of the hypodermal cells of the nucellus and described the general character of the eight-nucleate embryo sac in thirty-six families of monocotyledons and dicotyledons. Shortly after, in 1883, Guignard observed the longitudinal division of chromosomes in karyokinesis, thus confirming W. Fleming's findings. But even more important was his confirmation in 1889 of meiosis in plants, a phenomenon discovered a year earlier by E. Strasburger.

Double fertilization in angiosperms was discovered by Guignard in 1899 independently of S. Nawaschin, who had announced his discovery of the same phenomenon at a scientific meeting in Kiev, Russia, in 1898. In the meantime, Guignard had expanded the scope of his work to include such studies as the morphology of bacteria; pollen formation; the role of the centrosome and related bodies; the development and structure of the male gamete in *Fucus,* liverworts, mosses and ferns; the localization of hydrocyanic acid, glycosides and enzymes in plants; the growth and development of the seed, especially the tegmen, and investigations of organs of secretion in *Laminaria* and *Copaifera.*

From 1900 to 1922, Guignard continued his research on fertilization, embryogeny, and pollen formation, while devoting a major portion of his time to studying the sites of production in plants of sulfurated and cyanogenetic glycosides and their enzymes. His findings, published in 1906, on the poisonous nature of the Java bean (*Phaseolus lunatus* L.) showed that the toxicity was due to its hydrocyanic content and led to a ban on the importation of the bean into France. Guignard also developed a sensitive test for determining hydrocyanic acid in plants by means of sodium picrate paper. Grafting experiments with cyanogenetic plants convinced Guignard that, except when two species of the same genus were grafted, there was no migration of cyanogenetic glycosides between graft and stock; each retained, in this respect, chemical autonomy.

After 1907 there was a marked decline in the number of Guignard's scientific publications. His most significant scientific accomplishments had been made by the turn of the century, and his work on the localization of plant principles had brought him to the threshold of a new field of inquiry, the study of plant biosynthesis.

BIBLIOGRAPHY

I. Original Works. For a comprehensive listing of Guignard's publications, see P. Guérin, "Léon Guignard, 1852–1928," in *Bulletin des sciences pharmacologiques,* **35** (1928), 374–380.

II. Secondary Literature. The fullest account of Guignard's life and work will be found in P. Guérin, cited above, pp. 354–380. See also R. Souèges, "Léon Guignard," in *Figures pharmaceutiques françaises* (Paris, 1953), pp. 203–208. A good discussion of Guignard's contributions to plant embryology is included in R. Souèges, *L'embryologie végétale, résumé historique,* 2 vols., II (Paris, 1934), 19–21 and *passim.*

For Guignard's work in cytology, see Maurice Hocquette, "Morphologie, anatomie, cytologie," in A. Davy de Virville, *et al., Histoire de la Botanique en France* (Paris, 1954), pp. 147–149, 151; and A. Hughes, *A History of Cytology* (London–New York, 1959), pp. 66, 70–72.

Alex Berman

GUILLANDINUS. See **Wieland, Melchior.**

GUILLAUME, CHARLES ÉDOUARD (*b.* Fleurier, Switzerland, 15 February 1861; *d.* Sèvres, France, 13 June 1938), *metallurgy, physics.*

Guillaume's father, Édouard Guillaume, returned to Switzerland, his family's original home, from London, where he had managed a clockmaking firm. His knowledge of science was considerable, and he was his son's first teacher. The latter was admitted at the age of seventeen to the Zurich Polytechnikum, where he studied not only the prescribed scientific subjects but also German and French literature. He used to say that François Arago's *Éloges académiques* had exerted a profound influence on him.

In 1883 Guillaume entered the International Bureau of Weights and Measures at Sèvres, near Paris. He remained there throughout his career and became its director. In 1911 he was elected a corresponding member of the physics section of the Académie des Sciences.

Guillaume's first works were devoted to the mercury thermometer; upon completion of these studies he published a treatise on thermometry which made available to physicists the methods perfected by the International Bureau of Weights and Measures. He next participated in the preparation of the national meters, a fundamental work which marked the origin of modern metrology and permitted the presentation in 1889, at the first Conference on Weights and Measures, of the complete collection of standardized meters destined for the different countries.

Since 1890, Guillaume was led to undertake investigations on metal alloys. Studies had been made at Sèvres on a ferronickel (an alloy of iron with 24 percent nickel and 2 percent chromium) that had just been created at the Imphy Works in Nièvre. This alloy was more expansible than the iron or the nickel composing it. While studying an alloy containing slightly more nickel, Guillaume observed that this small variation in composition resulted in an alloy less expansible than the constituent metals. He undertook a methodical study of ferronickels and showed that with 36 percent nickel, one obtained an alloy, which he called invar, that expanded ten times less than iron and that even possessed a zero coefficient of dilatation after appropriate tempering, drawing, and rolling.

This alloy immediately found numerous applications, particularly in clockmaking. Guillaume also helped to solve another problem—compensation in ordinary watches—through his discovery of elinvar, an alloy whose elasticity does not vary with temperature.

These successes gave Guillaume an important role in the International Physics Congress, held in Paris in 1900, and earned him the 1920 Nobel Prize in physics. Moreover, he had the pleasure of seeing that his work would be brilliantly continued at the International Bureau of Weights and Measures by the physicist Albert Pérard, who succeeded him as director. In addition, Albert Portevin and Pierre Chevenard obtained very satisfactory results in developing his researches on nickel alloys.

BIBLIOGRAPHY

Guillaume's works include "Sur la dilatation des aciers au nickel," in *Comptes rendus hebdomadaires des séances de l'Académie des sciences,* **124** (1897), 176; "Recherches sur les aciers au nickel. Propriétés métrologiques," *ibid.,* 752; "Recherches sur les aciers au nickel. Propriétés magnétiques et déformations permanentes," *ibid.,* 1515; "Recherches sur les aciers au nickel. Dilatations aux températures élevées; résistance électrique," *ibid.,* **125** (1897), 235, errata, p. 342; "Recherches sur les aciers au nickel. Variations de volumes des alliages irréversibles," *ibid.,* **126** (1898), 738; "Nouvelles recherches sur la dilatation des aciers au nickel," *ibid.,* **136** (1903), 303; "Changements passagers et permanents des aciers au nickel," *ibid.,* 357; "Variations du module d'élasticité des aciers au nickel," *ibid.,* 498; "Sur la théorie des aciers au nickel," *ibid.,* 1638; "L'anomalie de dilatation des aciers au nickel," *ibid.,* **152** (1911), 189; "Coefficient du terme quadratique dans la formule de dilatation des aciers au nickel," *ibid.,* 1450; "Modification de la dilatabilité de l'invar par des actions mécaniques ou thermiques," *ibid.,* **163** (1916), 655; "Écrouissage et dilatabilité de l'invar," *ibid.,* 741; "Homogénéité de dilatation de l'invar," *ibid.,* 966; "Recherches métrologiques sur les aciers au nickel," in *Travaux et mémoires du Bureau international des poids et mesures,* **17** (1927); "Les anomalies des aciers au nickel et leurs applications," in *Revue de métallurgie,* **25** (1928), 35.

GEORGES CHAUDRON

GUILLET, LÉON ALEXANDRE (*b.* Saint Nazaire, France, 11 July 1873; *d.* Paris, France, 9 May 1946), *metallurgy.*

Guillet entered the École Centrale des Arts et Manufactures near the top of his class and received his engineering degree in 1897. In 1902 he submitted his thesis for the doctorate in physical sciences to the Faculté des Sciences of Paris; it dealt with the alloys of aluminum. He was named *suppléant* professor of metallurgy at the Conservatoire National des Arts et Métiers in 1906, and in 1908 he became titular holder of that chair. In 1911 he was appointed to the chair of metallurgy at the École Centrale des Arts et Manufactures, and in 1923 he became that school's director. He was elected a member of the Académie des Sciences in 1925, in the division of applications of science to industry.

Throughout his career Guillet never separated science from its applications, for he was convinced that modern industry, and especially metallurgy, must no longer be content with the empiricism that had prevailed for so long in the factories; instead, factories should have research departments. In 1905 Guillet was named director of the laboratory of one of the largest automobile factories of de Dion and Bouton, in Puteaux, near Paris; and he transformed this laboratory into the first department of scientific research to be organized in an industrial plant.

Guillet's scientific work was related almost entirely to the theory of alloys. The research that contributed most to his reputation was that concerning special steels, that is, those made with nickel, manganese, chromium, and tungsten. From special steels he turned to the study of bronzes and brasses. He also made original contributions to experimental measurements, principally in his research on thermal treatments of alloys.

During World War I, Guillet was assigned to the naval yards of Penhoët, where he was concerned in particular with tempering projectiles. His results were used in France and in several allied countries.

Guillet, like Henry Le Chatelier, who was to a certain degree his patron, was a trainer of men. He had around him a group of disciples, some of whom became very well-known engineers and scientists. As director of the École Centrale des Arts et Manufactures, Guillet was responsible for the notable progress of this school. Speaking of his directorship, he declared: "I spent the happiest years of my life there, and that was because those were the years during which I was best able to work for others."

Guillet was an active member of the Académie des Sciences; he had a special gift for clearly mediating scientific discussions.

BIBLIOGRAPHY

Guillet's articles include "Contribution à l'étude des alliages d'aluminium," thesis (Paris, 1902); "Contribution à l'étude des alliages aluminium-fer et aluminium-manganèse," in *Comptes rendus hebdomadaires des séances de l'Académie des sciences,* **134** (1902), 236; "Sur la micrographie des aciers au nickel," *ibid.,* **136** (1903), 227; "Nouvelles recherches sur la cémentation des aciers au carbone et des aciers spéciaux," *ibid.,* **138** (1904), 1600; "Propriétés et constitution des aciers au chrome," *ibid.,* **139** (1904), 426; "Constitution et propriétés des aciers au tungstène," *ibid.,* 519; "Propriétés et constitution des aciers au molybdène," *ibid.,* 540; "Sur la trempe des bronzes," *ibid.,* **140** (1905), 307; "Comparaison des propriétés, essais de classification des aciers ternaires," *ibid.,* **141** (1906), 107; "Constitution des alliages cuivre-aluminium," *ibid.,* 464; "Sur les points de transformation et la structure des aciers nickel-chrome," *ibid.,* **156** (1913), 1774; "Sur les alliages de cuivre, de nickel et d'aluminium," *ibid.,* **158** (1914), 704; "Sur la trempe des laitons à l'étain," *ibid.,* **172** (1921), 1038; and "Influence de l'écrouissage sur la resistivité des métaux et des alliages," *ibid.,* **176** (1923), 1800.

His books include *Notice sur ses travaux scientifiques* (Paris, 1907, 1923); *Traitements thermiques des produits métallurgiques* (Paris, 1909); *L'enseignement technique supérieur à l'après guerre* (Paris, 1918); *Précis de métallographie microscopique et de macrographie* (Paris, 1918), written with A. Portevin; *Additif à la notice sur ses travaux scientifiques* (Paris, 1925); and *Les métaux légers et leurs alliages: Aluminium, magnésium, glucinium, métaux alcalins et alcalino-terreux* (Paris, 1936).

GEORGES CHAUDRON

GUILLIERMOND, MARIE ANTOINE ALEXANDRE (*b.* Lyons, France, 19 August 1876; *d.* Lyons, 1 April 1945), *botany.*

Born into a family of physicians in which science was held in esteem, Guilliermond experienced the premature death of his father and several years later of his mother, who had been remarried to a physician. Sensitive and shy, he wished to teach and conduct research. His teachers were Maurice Caullery, Eugène Bataillon, and, in cryptogamy, Louis Matruchot. His teaching career was brilliant: beginning as a lecturer in agricultural botany at the Faculté des Sciences in Lyons, he went to the University of Paris in 1913 as lecturer in botany at the Faculty of Sciences and in 1935 succeeded Pierre Augustin Dangeard to the chair of botany at the Sorbonne. Much tried by France's military misfortunes, he fell ill and retired in 1942. He died three years later. He had belonged to the Académie des Sciences since 1935.

Guilliermond's brilliant investigations were conducted in two profoundly different areas requiring different mental orientations. His first works dealt with lower organisms: blue-green algae (cyanophyceae), bacteria, and especially yeasts, to which he devoted his doctoral thesis, *Recherches cytologiques sur les levures et quelques moisissures à formes levures* (1902). Were yeasts, as Oscar Brefeld held, more highly evolved forms of fungi? It was believed that they possessed a primitive structure with diffuse chromatin, but Guilliermond's studies, which form the basis of our scientific knowledge of the yeasts, established as well the indisputable presence of a nucleus and its division at the time of budding. At first he thought the nucleus divided by amitosis. Later, progress in technique permitted him and his students to demonstrate that mitosis was in fact taking place. He also recognized the vacuoles and their content, which is precipitable in metachromatic corpuscles.

It was undoubtedly in the field of the sexuality of the yeasts that Guilliermond made the greatest progress: he established the occurrence of isogamous copulation before the formation of the ascus (*Schizosaccharomyces octosporus*) and of heterogamous copulation (*Zygosaccharomyces chevalieri*). He detected the copulation of the ascospores of various *Saccharomyces,* which allowed him to distinguish between haplobiontic and diplobiontic yeasts. Researches on the filamentous Endomycetaceae revealed analogous processes in this family and permitted the formulation of a classically accepted hypothesis regarding the phylogeny of the yeasts.

Guilliermond's studies on the formation of the ascus among the higher Ascomycetes yielded important results. For example, through the tiny Pezizaceae *Humaria rutilans,* which possess the largest chromosomes among the fungi, he was able to demonstrate the characteristics of three successive mitoses of the ascus: heterotypic, homeotypic, and typical. Along the same lines, his research on the cytology of bacteria and especially of the Cyanophyceae contributed to the establishment of the then classically accepted type of these cells. (The electron microscope has since revealed the inaccuracy of this type.)

Because the yeasts provided the greatest continuity to Guilliermond's first group of investigations, as a complete botanist he wished to become acquainted with the greatest number of forms. It was as a systematist that he published *Les levures* (1911) and later drew up the tables of yeasts in the *Tabulae biologicae.*

The second area of Guilliermond's work is completely different. In the period in which physicochemical biology was actively developing, he wished, as a plant cytologist, to establish a close contact with animal cytology. His researches therefore dealt with the morphological constituents of the cytoplasm.

It had been known since 1910 that the chondriosomes, or mitochondria, consisting of a lipoprotein complex much richer in lipids than is cytoplasm, must play an important role in secretory phenomena. Also known was the importance of the vacuole system for cellular physiological phenomena. Through various means, such as plasmolysis, vital stains, mitochondrial techniques, and the ultramicroscope, the attempt could be made to answer a series of questions: Are there organelles other than the vacuoles which can arise *de novo* in the cell? Is it possible for the amidon to be formed not in the plasts but in the cytoplasm and even in the chondriosomes? What is the significance of the Golgi apparatus, revealed by silver stains, which seems to be attached sometimes to the vacuome, sometimes to the chondriome, and sometimes to be an autonomous formation?

Since the electron microscope did not then exist, the ultramicroscope presented the cytoplasm as a homogeneous gel, and these problems were approached *in situ* in the cell (and often in the living state). It is then possible to understand how courageous was Guilliermond's research and how valuable the results obtained—for example, affirmation that only the vacuoles can be formed again. Furthermore, the attraction that this research held for many students provided Guilliermond with an excellent group with which to pursue this work.

BIBLIOGRAPHY

Works by Guilliermond include *Recherches cytologiques sur les levures et quelques moisissures à formes levures* (Lyons, 1902), his doctoral diss.; "Remarques sur la caryocinèse des Ascomycètes," in *Annales mycologici,* **3** (1905), 344; "Recherches cytologiques et taxinomiques sur les Endomycétacées," in *Revue générale de botanique,* **21** (1909), 353; "Nouvelles observations sur la sexualité des levures," in *Archiv für Protistenkunde,* **28** (1912), 52; *Traité de cytologie végétale* (Paris, 1933), written with G. Mangenot and L. Plantefol; *Précis de biologie végétale* (Paris, 1937), written with G. Mangenot; and *Introduction á l'étude de la cytologie,* 3 vols. (Paris, 1938).

LUCIEN PLANTEFOL

GUINTER, JOANNES (*b.* Andernach, Germany, *ca.* 1505; *d.* Strasbourg, France, 4 October 1574), *medicine.*

Nothing is known of Guinter's family, except that it was obscure and impoverished, or of his earliest education. He is said to have left Andernach at the age of twelve in quest of learning, studying successively at Utrecht, Deventer, and Marburg, in which last place he completed his humanistic and philosophical studies. Thereafter for a brief period he taught in a preparatory school at Goslar, Saxony, where he recouped his funds and was able to proceed to Louvain for further study and also some teaching of Greek, and then to Liège. At some undetermined earlier time Guinter seems to have begun the study of medicine at Leipzig, and about 1527 he proceeded from Liège to Paris to continue that study. He received the baccalaureate in medicine on 18 April 1528 after two witnesses had sworn to the fact of his previous studies at Leipzig. On 4 June 1530 he was promoted licentiate and on 29 October 1532 received the M.D. degree. He was accepted as a regent doctor by the Paris Faculty of Medicine on 6 February 1533, and on 7 November 1534 he was named one of the two professors of medicine at a salary of twenty-five livres.

As part of his academic duties Guinter was responsible for the annual winter course in human anatomy, and it was inevitable during the pre-Vesalian period that his approach would be Galenic. The procedure followed was in the medieval pattern, with Guinter lecturing to the class while a barber or surgeon performed the actual dissection in order merely to illustrate and confirm Galen's anatomy. However, Guinter himself appears occasionally to have dissected, although his technique left much to be desired. One of his pupils during the period 1533–1536, the later distinguished anatomist Andreas Vesalius, referred to Guinter's anatomical instruction in strongly condemnatory terms, even declaring: "I do not consider him an anatomist, and I should willingly suffer him to inflict as many cuts upon me as I have seen him attempt on man or any other animal—except at the dinner table." Nevertheless, it is to Guinter's credit that he did attempt to teach his students some comparative anatomy and was willing to allow them to gain some experience by participation in the actual dissection. It was in conjunction with his anatomical course that he published a dissection manual, *Institutiones anatomicae* (Paris, 1536), in four books, dealing first with the more corruptible internal organs and then with those less susceptible to putrefaction. Thus the work followed the form first made popular by Mondino da Luzzi (1316), that is, the medieval method of dissection based upon a limited amount of dissection material. Guinter acknowledged the assistance of his student Vesalius in preparation of the work, probably the dissection and preparation of anatomical specimens. Although Guinter's manual, preceded only by those of Mondino and Berengario da Carpi (1522), contained no genuine anatomical contributions, it did advocate that anatomy, hitherto considered as chiefly fit for study by surgeons, was fundamental to the education of the physician.

Guinter was one of the major Greek scholars of his day, a fact first disclosed by the publication of his *Syntaxis Graeca* (Paris, 1527). In particular he devoted this scholarship to translations of the classical writers on medicine, and in the *Commentaries* of the Faculty of Medicine of Paris he was recognized as having translated the larger part of Galen's writings and all those of Paul of Aegina. The considerable bulk of Guinter's translations is explained by his method, according to which, as he declared, he translated each day as much as his secretary could write out from dictation, after which Guinter edited the version for publication. Despite the speed with which he translated, his versions appear to have held up well before the criticism of later editors. Guinter's most important translations were his version of the first nine books of Galen's anatomical treatise, *De anatomicis administrationibus* (Paris, 1531), and *De Hippocratis et Platonis placitis* (Paris, 1534), the latter considered by Guinter to be his most significant contribution to knowledge of classical Greek medicine. He also translated the writings of Paul of Aegina, *Opus de re medica* (Paris, 1532); Caelius Aurelianus, *Liber celerum vel acutarum passionum* (Paris, 1533); and Oribasius, *Commentaria in aphorismos Hippocratis* (Paris, 1533). Guinter was responsible for the introduction and popularization of a number of Greek anatomical terms, such as dartos, pericranium, urachus, and colon, that were to replace the inexact and confusing medieval anatomical nomenclature and thus lead to greater precision in anatomical description.

Owing to the growing pressure of religious orthodoxy in France, Guinter, a Lutheran, left Paris in 1538 for Metz and after about two years went to Strasbourg, where he was provided with a chair of Greek studies at the Gymnasium. At the same time he developed a medical practice, but criticism of his double occupation compelled him to relinquish his academic position in 1556. Although he continued his studies of the classical Greek physicians, producing a translation of the writings of Alexander of Tralles in 1549 and a revised edition in 1556, most of his later publications reflected his interest as a practicing physician. His book of advice on how to avoid the plague, *De victus et medicinae ratione cum alio tum pestilentiae tempore observanda commentarius* (Strasbourg, 1542), was translated into French by Antoine Pierre in 1544 and by Guinter in 1547 as *Instruction très utile par laquelle un chacun se pourra maintenir en santé, tant au temps de peste, comme autre temps.* Further works on this subject were *Bericht, Regiment und Ordnung wie die Pestilenz und die pestilenzialische Fieber zu erkennen und zu kuriren* (Strasbourg, 1564) and *De pestilentia commentarius in quatuor dialogos distinctus* (Strasbourg, 1565). Guinter also produced *Commentarius de balneis et aquis medicatis* (Strasbourg, 1565); a general study of medicine containing some autobiographical material, *De medicina veteri et nova* (Basel, 1571); and a collection of writings on obstetrics published posthumously by Joannes Georg Schenck a Grafenberg, *Gynaeciorum commentarius, de gravidarum, parturientium, puerperarum et infantium cura* (Strasbourg, 1606). Guinter was entombed in the church of St. Gallus in Strasbourg.

BIBLIOGRAPHY

There are many references to Guinter, most of them unfortunately containing serious errors. In particular his name is usually given erroneously as Günther, Gonthier, Guinther, or even Winter, and an almost unshakable legend places his birth in 1487. These two points are given special attention in Edouard Turner, "Jean Guinter d'Andernach 1505–1574," in *Gazette hebdomadaire de médecine et de chirurgie,* 2nd ser., **18** (1881), 425, 441, 505. Turner's essay includes the best bibliography of Guinter's long list of publications, which is also given *in extenso* in J. J. Höveler, "Ioannes Guinterius Andernacus," in *Jahresbericht über das Progymnasium zu Andernach für das Schuljahr 1898–99* (Andernach, 1899), pp. 3–21. Höveler, however, accepts the legendary date for Guinter's birth. Some further information is to be found in *Commentaires de la Faculté de médecine de l'Université de Paris (1516–1560*), M.-L. Concasty, ed. (Paris, 1964). Guinter as an anatomist is treated in C. D. O'Malley, *Andreas Vesalius of Brussels 1514–1564* (Berkeley–Los Angeles, 1964); and there is a bibliography of eds. of Guinter's *Institutiones anatomicae,* including those revised by Vesalius, in Harvey Cushing, *Bio-bibliography of Andreas Vesalius,* 2nd ed. (Hamden, Conn., 1962). Some autobiographical information is to be found in the prefaces to Guinter's various translations and in his *De medicina veteri et nova* (Basel, 1571).

C. D. O'MALLEY

GULDBERG, CATO MAXIMILIAN (*b.* Christiania [now Oslo], Norway, 11 August 1836; *d.* Christiania, 14 January 1902), *chemistry, physics.*

Guldberg was the eldest of the nine children of Carl August Guldberg, a minister and owner of a bookshop and printing office, and Hanna Sophie Theresia Bull. When Guldberg was eleven years old, his father was appointed minister at Nannestad, about fifty miles north of Christiania. Although the foundation for Guldberg's later delight in outdoor life, hunting, and fishing was laid in his father's remote parish, he could not get a satisfactory education there, and at the age of thirteen he was sent to live with his maternal grandmother at Fredrikstad. There he entered secondary school, where he excelled in mathematics.

Because the school in Fredrikstad could not grant admission certificates for the university, in 1853 Guldberg went to Christiania, where he spent his last school year in a private Latin school; he then matriculated at the University of Christiania in 1854, the same year as his friend Peter Waage. At the university Guldberg majored in mathematics and studied physics and chemistry. While still a student, he worked independently on advanced mathematical problems, and his first published scientific paper, "On the Contact of Circles" (University of Christiania publication [Christiania, 1861]), won the crown prince's gold medal.

In 1859 he graduated in science and obtained a modest position as teacher at Nissen's secondary school in Christiania. In 1860 he was appointed teacher of mathematics at the Royal Military Academy. The next year, by means of a scholarship, he made a one-year study tour of France, Switzerland, and Germany. In 1862 he qualified for a position in applied mechanics at the Royal Military College, and in 1863 he was appointed a teacher of advanced mechanics at the same school. He held these two positions until his death. He was awarded a scholarship in 1867 at the University of Christiania, where he became professor of applied mathematics in 1869.

Guldberg and Waage, whose names are linked for their joint discovery of the law of mass action, were also closely related through marriage. Guldberg married his cousin Bodil Mathea Riddervold, the daughter of cabinet minister Hans Riddervold; the couple had three daughters. Waage married her sister. The collaboration between the two friends and brothers-in-law on the studies of chemical affinity that were to lead to the law of mass action began immediately after Guldberg's return from abroad in 1862. The first report of their results was presented by Waage on 14 March 1864 before the Norwegian Academy of Sciences and published the following year in the Academy's proceedings. But the report remained almost completely unknown to scientists, a fate also suffered by a more detailed description of their theory published in French in 1867. The theory did not become generally known until Wilhelm Ostwald, in a paper published in 1877, adopted the law of mass action and proved its validity by new experiments. Although the law had had several forerunners, the combined efforts of the theorist Guldberg and the empiricist Waage led to the first general mathematical and exact formulation of the role of the amounts of reactants in chemical equilibrium systems. In 1878 Jacobus Henricus van't Hoff, apparently without any knowledge of Guldberg and Waage's work, derived the law from reaction kinetics.

Although the law of mass action is Guldberg's greatest contribution to physical chemistry, it is not his only one. Some of his early work published in Norwegian did not get the publicity it deserved. He devoted much time to a search for a general equation of state for gases, liquids, and solids from a kinetic molecular approach. In 1867, nineteen years before van't Hoff, he introduced the ideal gas equation in the form $pV = 2T$. In 1869 he developed the concept of "corresponding temperatures" and deduced an equation of state valid for all liquids of certain types. In 1890 he formulated the rule that the reduced boiling temperatures of most liquids are close to 2/3, a relationship discovered independently by P. A. Guye. In addition, Guldberg made valuable contributions to the thermodynamics of solution and of dissociation, and he discovered and correctly explained cryohydrates. He wrote many articles on various practical problems and a number of textbooks on mathematics and mechanics. He was editor of the *Polyteknisk tidsskrift,* an active member and officer of scientific societies, and the recipient of many honors.

BIBLIOGRAPHY

I. Original Works. Guldberg and Waage's various papers on the law of mass action have been abridged and translated into German by Richard Abegg as *Untersuchungen über die chemischen Affinitäten,* in Wilhelm Ostwald, Klassiker der exakten Wissenschaften, CIV (Leipzig, 1899). Their first paper, "Studier over Affiniteten," published in Norwegian in *Forhandlinger i Videnskabsselskabet i Christiania,* (1865), 35–45, appears in facsimile, along with a number of articles on the law, in Haakon Haraldsen, ed., *The Law of Mass Action: A Centenary Volume 1864–1964* (Oslo, 1964).

II. Secondary Literature. A biography of Guldberg and a discussion of his work by Haakon Haraldsen appears on pp. 19–26 and 32–35 of the centenary volume cited above and on pp. 172–174 of Abegg's trans. Also see J. B. Halvorsen, "C. M. Guldberg," in *Norsk Forfatter-Lexikon,* II (Christiania, 1888), 447; Elling Holst, "C. M. Guldberg," in *Nordisk Universitetstidsskrift,* **2** (1902), 321; H. Goldschmidt, "C. M. Guldberg," in *Fordhandlinger Videnskabs-Selskabet Christiania,* no. 1 (1903), 1; Sophus Torup, "C. M. Guldberg," in *Norsk biografisk leksikon,* V (Oslo, 1931), 76; T. Hiortdahl, "Den Fysisk-Kemiske Forening, Tidsskr. Kemi, Farmaci og Terapi," in *Pharmacia,* **14** (1917), 240; Kåre Fasting, "Teknikk og Samfunn," in *Den Polytekniske Forening 1852–1952* (Oslo, 1952); and Gunnar Oxaal, *Teknisk Ukeblad* (1954), pp. 306, 308.

George B. Kauffman

GULDIN, PAUL (*b.* St. Gall, Switzerland, 12 June 1577; *d.* Graz, Austria, 3 November 1643), *mathematics.*

Guldin was of Jewish descent but was brought up as a Protestant. He began work as a goldsmith and as such was employed in several German towns. At the age of twenty he was converted to Catholicism and entered the Jesuit order, changing his first name, Habakkuk, to Paul. In 1609 he was sent to Rome for further education. Guldin taught mathematics at the Jesuit colleges in Rome and Graz. When a severe illness obliged him to suspend his lecturing, he was sent to Vienna, where he became professor of mathematics at the university. In 1637 he returned to Graz, where he died in 1643.

In 1582 the Gregorian calendar was introduced in western Europe, and it met with a great deal of opposition among both scientists and Protestants; one of the opponents was the famous chronologist Sethus Calvisius. To refute him and to defend Pope Gregory XIII and his fellow Jesuit Christoph Clavius, Guldin published his first work, *Refutatio elenchi calendarii Gregoriani a Setho Calvisio conscripti* (Mainz, 1618).

In 1622 Guldin published a physicomathematical dissertation on the motion of the earth caused by alteration of the center of gravity. In it he made the assumption that every unimpeded large body whose center of gravity does not coincide with the center of the universe is moved in such a way that it will coincide with the latter. In the fourteenth century the doctrine of centers of gravity had begun to play a role in the mechanics of large bodies. In his *Quaestiones super libros quattuor de caelo et mundo Aristotelis,* Jean Buridan argued that geological processes are always causing a redistribution of the earth's matter and therefore are continually changing its center of gravity. But the center of gravity always strives to be at the center of the universe, so the earth is constantly shifting about near the latter. Guldin accepted Buridan's hypothesis but was also well-informed about the objection which Nicole Oresme had formulated in his *Le livre du ciel et du monde.*

In 1627 a correspondence on religious subjects developed between Guldin and Johannes Kepler. On the occasion of his journey from Ulm to Prague, which he undertook to solicit funds from Emperor Rudolph II for the publication of the Rudolphine Tables (Ulm, 1627), Kepler wrote on his objections to the Catholic religion to Guldin. In his answer Guldin tried to refute them with theological arguments drawn up for him by a fellow Jesuit. Kepler's reply ended the correspondence.

Guldin's main work was *Centrobaryca seu de centro gravitatis trium specierum quantitatis continuae,* in four volumes (Vienna, 1635–1641). In the first volume Guldin determined the centers of gravity of plane rectilinear and curvilinear figures and of solids in the Archimedean manner. Against Niccolò Cabeo's attacks in *Philosophia magnetica* (1629) directed toward his theory concerning the motion of the earth, Guldin reproduced in volume I his dissertation of 1622 and a note in which he discussed Cabeo's arguments. The appendix to volume I contains tables of quadratic and cubic numbers and an exposition of the use of logarithms referring to Adriaan Vlacq's *Arithmetica logarithmica* (1628).

Volume II contains what is known as Guldin's theorem: "If any plane figure revolve about an external axis in its plane, the volume of the solid so generated is equal to the product of the area of the figure and the distance traveled by the center of gravity of the figure" (ch. 7, prop. 3, p. 147). This theorem has been much discussed in terms of possible plagiarism from the early part of book VII of Pappus' *Collectio* (*ca.* A.D. 300). However, the theorem cannot have been taken from the first published edition of the *Collectio,* the Latin translation of Federico Commandino (Venice, 1588), because that text shows obvious lacunae. Guldin attempted to prove his theorem by metaphysical reasoning, but Bonaventura Cavalieri pointed out the weakness of his demonstration and proved the theorem by the method of indivisibles. Volume II treats the properties of the Archimedean spiral and the conic sections, their lengths and surfaces, the determination of the center of gravity of a sector of a circle and of a segment of a circle and a parabola, the rise of solids of revolution, and the application of the Guldin theorem to them.

In volume III Guldin determined the surface and the volume of a cone, a cylinder, a sphere, and other solids of revolution and their mutual proportions. In his *Stereometria doliorum* (1615) Kepler determined the volumes of certain vessels and the areas of certain surfaces by means of infinitesimals, instead of the long and tedious method of exhaustions. In volume IV Guldin severely attacked Kepler for the lack of rigor in his use of infinitesimals. He also criticized Cavalieri's use of indivisibles in his *Geometria indivisibilibus* (1635), asserting not only that the method had been taken from Kepler but also that since the number of indivisibles was infinite, they could not be compared with one another. Furthermore, he pointed out a number of fallacies to which the method of indivisibles appeared to lead.

In 1647, after the death of Guldin, Cavalieri published *Exercitationes geometricae sex,* in which he defended himself against the first charge by pointing out that his method differed from that of Kepler in that it made use only of indivisibles, and against the second by observing that the two infinities of elements to be compared are of the same kind.

BIBLIOGRAPHY

Guldin's writings are listed in the text. A very good account of his works may be found in C. Sommervogel, *Bibliothèque de la Compagnie de Jésus,* II (Brussels–Paris, 1891), 1946–1947.

On Guldin or his work, see the following (listed chronologically): C. J. Gerhardt, *Geschichte der Mathematik in Deutschland* (Munich, 1877), pp. 129–130; L. Schuster, *Johann Kepler und die grossen kirchlichen Streitfragen seiner Zeit* (Graz, 1888), pp. 217–228, 233–243; M. Cantor, *Vorlesungen über Geschichte der Mathematik,* II (Leipzig, 1900), 840–844; H. G. Zeuthen, *Geschichte der Mathematik im 16. und 17. Jahrhundert* (Leipzig, 1903), pp. 240, 241, 293; G. A. Miller, "Was Paul Guldin a Plagiarist?," in *Science,* **64** (1926), 204–206; P. Ver Eecke, "Le théorème dit de Guldin considéré au point de vue historique," in *Mathésis,* **46** (1932), 395–397; R. C. Archibald, "Notes and Queries," in *Scripta mathematica,* **1** (1932), 267; P. Duhem, *Le système du monde,* IX (Paris, 1958), 318–321; C. B. Boyer, *The History of the Calculus and Its Conceptual Development* (New York, 1959), pp. 121, 122, 138, 139; and J. E. Hofmann, "Ueber die *Exercitatio geometrica* des M. A. Ricci," in *Centaurus,* **9** (1963), 151, 152.

H. L. L. BUSARD

GULLAND, JOHN MASSON (*b.* Edinburgh, Scotland, 14 October 1898; *d.* Goswick, England, 26 October 1947), *organic chemistry, biochemistry.*

Gulland's father was professor of medicine at Edinburgh University; his mother was the daughter of David Masson, professor of English literature at the same university. His own studies in chemistry at that university were interrupted by war service, and he graduated in 1921. At the universities of St. Andrews and Manchester he and Robert Robinson established the structures of an important group of alkaloids including morphine. From 1924 Gulland was at Oxford with W. H. Perkin, Jr., and worked on strychnine and brucine; but the routine of degradation and synthesis began to pall, and his interest turned to biochemical problems of wider significance. He is remembered mainly for his work on the chemistry of nucleic acids, for which he was elected a fellow of the Royal Society in 1945. He carried out this work at the Lister Institute of Preventive Medicine, London (1931–1936), and as professor of chemistry at University College, Nottingham (1936–1947).

Gulland was one of the first to use methods other than those of classical chemistry to study the structure of nucleic acids. In early work he showed spectroscopically that the pentose residue was attached to the 9 position rather than the 7 position in the purine nucleosides. Later, with Elisabeth Jackson (1938), he found that the enzymatic hydrolysis of ribonucleic acid (RNA) gave evidence that the RNA nucleosides, like those of deoxyribonucleic acid (DNA), were linked by phosphate ester groups through the 3′ and 5′ positions. Owing to the difficulty of preparing pure enzymes, such evidence was then regarded with suspicion, and there was subsequently much confusion about the nature of the internucleotide link in RNA. The role of cyclic phosphates in the hydrolysis of RNA was not understood until the 1950's, when Gulland's evidence for the 3′–5′ link in RNA was seen to be valid.

Electrometric titration of DNA, done with D. O. Jordan and H. F. W. Taylor (1947), proved the existence of the hydrogen bonding which was an essential feature of the famous "double helix" of J. D. Watson and Francis Crick (1953). An earlier type of helical structure put forward by Linus Pauling and E. J. Corey had envisaged the phosphate groups as closely packed inside a single helix, the bases projecting radially on the outside. The Watson-Crick structure, on the other hand, required that two helices be linked by hydrogen bonds between the base pairs adenine-thymine and guanine-cytosine.

Gulland and his co-workers found that the primary phosphoric acid groups of DNA were readily titratable and thus, as was later realized, were on the outside of the double helix. With D. O. Jordan and J. M. Creeth, Gulland showed that the amino and amido groups of the bases were titratable only after treatment at extreme acid or alkaline pH, that is, only after there had been a breakdown into smaller molecular units, as confirmed by a decrease in viscosity and disappearance of streaming birefringence (1947). Rosalind Franklin and Raymond Gosling used Gulland's titrations as the main evidence in favor of the double helix, to which they had been led by their own crystallographic studies (1953).

The full solution of most of the problems tackled by Gulland was usually a little beyond the reach of the techniques of the day; he did not think that "easy" research was worth doing. His striking appearance, personal charm, and skill with words made him a memorable teacher. He died in a railway accident and was survived by his wife and two daughters.

BIBLIOGRAPHY

I. ORIGINAL WORKS. There is a not quite complete bibliography of Gulland's works in the obituary notice of the Royal Society, cited below. Among his earliest publications, written with R. Robinson, are "The Morphine Group. Part I. A Discussion of the Constitutional Problem," in *Journal of the Chemical Society,* **123** (1923), 980; and "The Constitution of Codeine and Thebaine," in *Memoirs of the Manchester Literary and Philosophical Society,* **69** (1924–1925). For his work on purine nucleosides see "Spectral Absorption of Methylated Xanthines and Constitution of the Purine Nucleosides," in *Nature,* **132** (1933), 782, written with E. R. Holiday.

His papers on RNA nucleosides, written with Elisabeth M. Jackson, include "Phosphoesterases of Bone and Snake Venoms," in *Biochemical Journal,* **32** (1938), 590–596; "5-Nucleotidase," *ibid.,* 597–601; "The Constitution of Yeast Nucleic Acid," in *Journal of the Chemical Society* (1938), p. 1492; and "The Constitution of Yeast Ribonucleic Acid. Part III. The Nature of the Phosphatase-Resistant Group," *ibid.* (1939), p. 1842.

For works on Gulland's other researches, see "Some Aspects of the Chemistry of Nucleotides," *ibid.* (1944), p. 208; "Deoxypentose Nucleic Acids. Part II. Electrometric Titration of the Acidic and Basic Groups of the Deoxypentose Nucleic Acid of Calf Thymus," *ibid.* (1947), p. 1131, written with D. O. Jordan and H. F. W. Taylor; and "Deoxypentose Nucleic Acids. Part III. Viscosity and Streaming Birefringence of Solutions of the Sodium Salt of the Deoxypentose Nucleic Acid of Calf Thymus," *ibid.,* p. 1141, written with J. M. Creeth and D. O. Jordan.

II. SECONDARY LITERATURE. Obituaries by R. D. Haworth appeared in *Obituary Notices of Fellows of the Royal Society of London,* **6,** no. 17 (1948), 67–82, and in *Journal of the Chemical Society* (1948), pp. 1476–1482; the former has a portrait. See also J. W. Cook, in *Nature,* **160** (1947), 702–703.

For the significance of Gulland's work on nucleic acids, see D. M. Brown and A. R. Todd, "Evidence on the Nature of the Chemical Bonds in Nucleic Acids," in E. Chargaff and J. N. Davidson, eds., *The Nucleic Acids,* I (New York, 1955), 409–445; R. E. Franklin and R. G. Gosling, in *Nature,* **171** (1953), 740–741; D. O. Jordan, *The Chemistry of Nucleic Acids* (London, 1960), pp. 67–68, 140–153, 169–170, and *passim;* and J. D. Watson, *The Double Helix* (London, 1968), p. 183.

KATHLEEN R. FARRAR

GULLSTRAND, ALLVAR (*b.* Landskrona, Sweden, 5 June 1862; *d.* Uppsala, Sweden, 21 July 1930), *ophthalmology, geometrical optics.*

Allvar Gullstrand was the son of a prominent physician who was city physician of Landskrona. Under the influence of his father he began the study of medicine in 1880 and soon specialized in physiological optics. He studied in Uppsala, Vienna, and Stockholm. He finished his medical studies in 1884 and obtained the license to practice medicine in 1888. After receiving his doctorate in 1890, he was appointed in 1892 as lecturer at the Royal Caroline Institute. At the same time he worked as chief physician at an ophthalmological clinic and in 1892 was appointed head of the eye clinic in Stockholm.

In 1894 Gullstrand became professor of ophthalmology at the University of Uppsala, where he received an honorary degree in 1907. Six years later, the university created a special chair for him, without teaching obligations, in physiological and physical optics. Gullstrand also received an honorary degree from the University of Jena, and in 1911 from the University of Dublin. That same year, the Nobel Prize in physiology or medicine was awarded to him for his investigations of the dioptrics of the eye.

Allvar Gullstrand's greatest achievements lie in the field of ophthalmological optics, the study of the human eye as an optical system. This study engendered his interest in geometrical optics. He then drew the attention of the optical designers to several misconceptions and so made important contributions to this field as well.

Gullstrand started his work in ophthalmology with a paper on the astigmatism of the cornea. He became interested in the accommodation mechanism of the human eye and in an exact theory discussed the influence which the layers of the crystalline lens play. This was a difficult mathematical problem which had not been attacked in detail before. It led to the conception of a new and more accurate model of the human eye, a big step beyond Helmholtz. This is described in Gullstrand's masterly commentaries on the occasion of his reediting Helmholtz' *Handbook of Physiological Optics.* These commentaries contain by far the clearest and best description of all of Gullstrand's ideas on geometrical and physiological optics.

Gullstrand invented a slit lamp, which, in combination with a microscope, allowed him to locate exactly a foreign body in the eye with respect to all three dimensions. He designed aspheric lenses for aphakic eyes, that is, eyes from which the lens has been removed as a result of cataracts. He investigated the effect of the rotation of the eye around the fulcrum, and through his friendship with M. von Rohr many of his ideas led to the construction of optical instruments, particularly the great Gullstrand ophthalmoscope, which was manufactured by Zeiss.

In the field of geometrical optics Gullstrand wrote many extensive papers that went beyond the frontiers of optical knowledge for his time. He developed the theory of the fourth-order aberration of a general optical ray, independent of the axis of a rotational symmetry system. Especially, he made contributions to the knowledge of umbilic points, that is, points in which the two principal curvatures are the same. He then investigated how the characteristic quantities of general bundles change with refraction, thus obtaining what he called the system laws of optical systems. But Gullstrand did not restrict himself solely to the consideration of spherical surfaces; one of his longest papers deals with the construction and tracing through of aspheric surfaces.

Unacquainted with the work of H. R. Hamilton, he solved difficult mathematical problems simply by developing the necessary quantities in a series around the coordinates of the principal ray. He considered mathematical methods, such as the calculus of variation and vector methods, to be false ornaments. This prejudice makes his papers long and clumsy, but they contain a number of valuable and little-known results. H. Boegehold, C. W. Oseen, and the writer have endeavored to give simpler derivations of his beautiful results. However, there are limitations to his method. In the case of a branch point, for example, the series development does not work.

Gullstrand was a fighter, discovering several inaccuracies in the normal treatment of optical problems; he spent much of his time studying these inaccuracies, which were mostly a result of approximate pictures being applied to describe finite realities. For instance, the Sturm conoid described an astigmatic bundle as a bundle of rays going through two straight lines perpendicular to each other and to the principal ray. Gullstrand showed that such a manifold bundle of rays is not a normal system, that is, it cannot originate from an object point. Another fallacy was that the collinear image formation, which is the coordination of lines in object and image space such that the rays from any object point unite in a fixed image point, could not have been taken as an approximation to the real image formation, because the former cannot be obtained by optical means (with the trivial exception of the plane mirror). Unfortunately, books are still published disregarding these simple truths.

Gullstrand represents a scientist of very rigorous standards, and as such, he was highly respected by his peers for his intelligence and integrity. His advice was widely sought, even outside his special sphere of interest; among other honors, he was a member and later president of the Nobel Prize committee.

BIBLIOGRAPHY

I. ORIGINAL WORKS. Gullstrand's works include "Objektive Differential-Diagnostik und photographische Abbildung von Augenmuskellahmungen," in *Kungliga Svenska vetenskapsakademiens handlingar,* **18** (1892); "Allgemeine Theorie der monochromatischen Aberrationen und ihre nächsten Ergebnisse für die Ophthalmologie," in *Nova acta Regiae Societatis scientiarum upsaliensis* (1900); "Die Farbe der Macula centralis retinae," in *Archiv für Ophthalmologie,* **62** (1905), 1-72, 378; "Die reelle optische Abbildung," in *Kungliga Svenska vetenskapsakademiens handlingar,* **41** (1906), 1-119; "Tatsachen und Fiktionen in der Lehre von der optischen Abbildung," in *Archiv für Optik* (1907), 1-41, 81-97; and "Die optische Abbildung in heterogenen Medien und die Dioptrik der Kristallinse des Menschen," in *Kungliga Svenska vetenskapsakademiens handlingar,* **43** (1908), 1-58.

See also *Einführung in die Methoden der Dioptrik des Auges des Menschen* (Leipzig, 1911); "Die reflexlose Ophthalmoskopie," in *Archiv für Augenheilkunde,* **68** (1911), 101-144; "Das allgemeine optische Abbildungssystem," in *Kungliga Svenska vetenskapsakademiens handlingar,* **55** (1915), 1-139; "Ueber aspharische Flächen in optischen Instrumenten," *ibid.,* **60** (1919), 1-155; "Optische Systemsgesetze zweiter und dritter Ordnung," *ibid.,* **63** (1924), 1-175; and "Einiges über optische Bilder," in *Naturwissenschaften,* **14** (1926), 653-664.

II. SECONDARY LITERATURE. For information about Gullstrand and his work see H. Boegehold, "Ueber die Entwicklung der Theorie der optischen Instrumente seit Abbe," in *Ergebnisse der exakten Naturwissenschaften,* **8** (1929), 1-146; M. Herzberger, "Allvar Gullstrand," in *Optica acta,* **3** (1960), 237-241; J. W. Nordenson, "Allvar Gullstrand," in *Klinische Monatsblatter für Augenheilkunde,* pp. 560-566; C. W. Oseen, "Allvar Gullstrand," in *Kungliga Svenska vetenskapsakademiens årsbok,* (1937); "Une méthode nouvelle de l'optique géometrique," in *Kungliga Svenska vetenskapsakademiens handlingar,* **3** (1936), 1-41; and M. von Rohr, "Allvar Gullstrand," in *Zeitschrift für ophthalmologische Optik,* **18** (1930), 129-134.

MAXIMILIAN J. HERZBERGER

GUNDISSALINUS, DOMINICUS, also known as **Domingo Gundisalvo** or **Gonsález** (*fl.* Toledo, Spain, second half of the twelfth century), *science translation, philosophy of science.*

Gundissalinus' date of birth is unknown, although conjecture has offered 1110; there is some evidence that he was still alive in 1190. He was archdeacon of Segovia, but his intellectual activity was centered at Toledo, where a flourishing school of translators, under the patronage of such archbishops as Raymond of Toledo, introduced a considerable amount of Arabic and Judaic materials to the Latin West during the twelfth century.

Many of the translations were done with the collaboration of two scholars, one knowledgeable in Arabic, the other in Latin, with a vernacular serving as common ground. The translations attributed to Gundissalinus were probably done in this fashion, although only in the manuscripts of the translation of the *De anima* of Ibn Sīnā (Avicenna) is Gundissalinus' name specifically linked with that of a co-translator, Abraham ibn Daūd (Avendauth). In addition to the *De anima,* Gundissalinus' name has been connected with translations of Ibn Sīnā's *Sufficientia* and *Metaphysics,* as well as a portion of his *Posterior Analytics,* together with the *Logic* and *Metaphysics* of al-Ghazzālī; the *Fons vitae* of Ibn Gabirol; the *De intellectu,* the *Fontes questionum,* the *De scientiis,* the *Liber excitativus ad viam felicitatis,* and the *De ortu scientiarum* of al-Fārābī; the *De intellectu* of al-Kindī; and the *Liber de definitionibus* of Isaac Israeli.

Gundissalinus was the author of five philosophical works which drew heavily on the Arabic-Judaic materials of his translations as well as on Latin sources. He was the first to provide the Latin West with an introduction to Arabic-Judaic Neoplatonism and the first to blend this tradition with the Latin Christian Neoplatonism of Boethius and Augustine. His *De unitate* is such a syncretic work. It is rich in aphorisms which were quoted frequently during the Middle Ages, for example, "Quidquid est ideo est quia unum est." Gundissalinus' *De anima,* likewise a compilation from his translations, is essentially a presentation of Avicennian psychology and ideas from Ibn Gabirol, although it utilizes material from other sources, such as Augustine and the treatise *On the Difference Between Soul and Spirit* of Qusṭā ibn Lūqā.

Gundissalinus' *De processione mundi* is taken from numerous sources: Ibn Sīnā, Ibn Gabirol, al-Ghazzālī, al-Fārābī, Boethius, Porphyry, the *Epistola de anima* of Isaac de Stella, possibly the *De deo Socratis* of Apuleius, and his own *De unitate.* Its editor, Georg Bülow, considers it a late work. The *De processione* was used in the thirteenth century by both William of Auvergne (William of Paris) and Thomas Aquinas. Gundissalinus' *De immortalitate animae,* again dependent on Arabic materials, is a well-written treatise proving the indestructibility of the soul, using arguments based on the soul's own nature which were to become standard in the Middle Ages. The *De immortalitate* was reworked in the thirteenth century by William of Auvergne.

The *De divisione philosophiae* is a classification of the sciences which served as a source for later classification schemes. It incorporates al-Fārābī's work on the classification of the sciences (the *De ortu scientiarum*) and utilizes a wide variety of other sources: classical Latin, Arabic, and Aristotelian. Since it draws on Gerard of Cremona's translation of the Arabic mathematician al-Nayrīzī, the *De divisione* was likely written after 1140, since Gerard's translating activity probably did not begin before that year. The *De divisione* begins with a prologue followed by a section containing six definitions of philosophy taken from various sources.

The sciences are classified into three major groups: propaedeutic sciences, including grammar, poetics, and rhetoric; logic; and philosophical sciences. The latter are further divided into theoretical and practical sciences. The theoretical sciences are subdivided in turn into physics, mathematics, and theology. Physics contains eight subjects, and mathematics has seven. Following this discussion, Gundissalinus inserts a section from Ibn Sīnā's *Posterior Analytics.* Treatment of the practical sciences, which include politics, economics, and ethics, concludes the treatise.

Gundissalinus' classification transcends the conventional subject matter of the *trivium* and the *quadrivium.* He includes a section on medicine as a branch of physics, and the seven subjects subsumed under mathematics include discussions of *scientiae de aspectibus, de ponderibus,* and *de ingeniis,* in addition to the four subjects of the *quadrivium.* The *De divisione* was directly used by Robert Kilwardby in his own treatise on classification, and its influence is further revealed in the works of Michael Scot, Vincent of Beauvais, and Thierry of Chartres.

BIBLIOGRAPHY

I. Original Works. Editions of Gundissalinus' writings include M. Menéndez y Pelayo, *Historia de los heterodoxos españoles,* I (Madrid, 1880), 691–711, text of *De processione mundi;* Paul Correns, "Die dem Boethius fälschlich zugeschrieben Abhandlung des Dominicus Gundisalvi *De unitate,*" in *Beiträge zur Geschichte der Philosophie des Mittelalters,* **1,** no. 1 (1891), 1–11; Georg Bülow, "Des Dominicus Gundissalinus Schrift von der Unsterblichkeit der Seele," *ibid.,* **2,** no. 3 (1897), 1–38; Ludwig Baur, "Dominicus Gundissalinus *De divisione philosophiae,*" *ibid.,* **4,** nos. 2–3 (1903), 1–142; Georg Bülow, "Des Dominicus Gundissalinus Schrift von dem Hervorgange der Welt (*De processione mundi,*" *ibid.,* **24,** no. 3 (1925), 1–54; and J. T. Muckle, "The Treatise *De anima* of Dominicus Gundissalinus," in *Mediaeval Studies,* **2** (1940), 23–103.

II. Secondary Literature. On Gundissalinus or his work, see M. T. D'Alverny, "Avendauth?" in *Homenaje a Millás-Vallicrosa,* I (Barcelona, 1954), 19–43, esp. the arts. by P. Alonso listed in ftn. 14, pp. 24–25, including "Las fuentes literarias de Domingo Gundisalvo," in *Al-Andalus,* **11** (1947), 209–211; C. Bäumker, "Les écrits philosophiques de Dominicus Gundissalinus," in *Revue thomiste,* **5** (1897), 723–745; and "Dominicus Gundissalinus als philosophischer Schriftsteller," in *Beiträge zur Geschichte der Philosophie des Mittelalters,* **25,** nos. 1–2 (1927), 255–275; D. A. Callus, "Gundissalinus' *De anima* and the Problem of Substantial Form," in *New Scholasticism,* **13** (1939), 338–355; A. H. Chroust, "The Definition of Philosophy in the *De divisione philosophiae* of Dominicus Gundissalinus," *ibid.,* **25** (1951), 253–281; P. Duhem, *Le système du monde,* III (Paris, 1958), 177–181; E. Gilson,

History of Christian Philosophy in the Middle Ages (New York, 1955), pp. 235–239, 652–653; Nicholas M. Haring, "Thierry of Chartres and Dominicus Gundissalinus," in *Mediaeval Studies,* **26** (1964), 271–286; R. W. Hunt, "The Introductions to the *Artes* in the Twelfth Century," in *Studia Mediaevalia in Honor of R. J. Martin* (Bruges, 1948), pp. 85–112; L. Löwenthal, *Pseudo-Aristoteles über die Seele. Eine psychologische Schrift des 11. Jahrhunderts und ihre Beziehung zu Salomo ibn Gabirol* (*Avicebron*) (Berlin, 1891), pp. 77–113; J. Teicher, "Gundissalino e l'Agostonismo avicennizante," in *Rivista di filosofia neoscholastica* (May 1934), pp. 252–258; and L. Thorndike, *A History of Magic and Experimental Science,* II (New York, 1923), 78–82.

More general works are A. Jourdain, *Recherches critiques sur l'âge et l'origine des traductions d'Aristote* (Paris, 1819), pp. 107–119; Artur Schneider, "Die abenländische Spekulation des zwölften Jahrhunderts in ihrem Verhältnis zur aristotelischen und jüdisch-arabischen Philosophie," in *Beiträge zur Geschichte der Philosophie des Mittelalters,* **17,** pt. 4 (1915), 39–72; M. Steinschneider, *Die europäischen Übersetzungen aus dem arabischen bis Mitte des 17. Jahrhunderts* (Graz, 1956), pp. 40–50, 260–261; and R. de Vaux, *Notes et textes sur l'avicennisme latin aux confins des xiie et xiiie siècles* (Paris, 1934), pp. 141–142.

CLAUDIA KREN

GUNTER, EDMUND (*b.* Hertfordshire, England, 1581; *d.* London, England, 10 December 1626), *navigation, mathematics.*

Little is known of Gunter's origins or the details of his life. Of Welsh descent, he was educated at Westminster School and Christ Church, Oxford, graduating B.A. in 1603 and M.A. in 1605. He subsequently entered holy orders, became rector of St. George's, Southwark, in 1615, and received the B.D. degree later that year. In March 1619 he became professor of astronomy at Gresham College, London, retaining this post and his rectorship until his sudden death at the age of forty-five.

Gunter's contributions to science were essentially of a practical nature. A competent but unoriginal mathematician, he had a gift for devising instruments which simplified calculations in astronomy, navigation, and surveying; and he played an important part in the English tradition—begun in 1561 by Richard Eden's translation of Martín Cortes' *Arte de navegar* and furthered by William Borough, John Dee, Thomas Harriot, Thomas Hood, Robert Hues, Robert Norman, Edward Wright, and others—which put the theory of navigation into a form suitable for easy use at sea. Gunter's works, written in English, reflected the practical nature of his teaching and linked the more scholarly work of his time with everyday needs; the tools he provided were of immense value long afterward.

Gunter's first published mathematical work was the *Canon triangulorum* of 1620, a short table, the first of its kind, of common logarithms of sines and tangents. His account of his sector, in the *De sectore et radio* of 1623, had circulated in manuscript for sixteen years before its publication. The sector, a development from Hood's, included sine, tangent, logarithm, and meridional part scales; its uses included the solution of plane, spherical, and nautical triangles (the last formed from rhumb, meridian, and latitude lines). With improvements, the British navy used it for two centuries, and it was also a precursor of the slide rule. Gunter solved such problems as finding the sun's amplitude from its declination and the latitude of the observer by adding similar scales to the seaman's cross-staff. Comparison of the amplitude with the sun's direction, measured by a magnetic compass, was known to give the compass variation; but although Gunter's own observations in 1622 at Limehouse were about five degrees less than Borough's 1580 results there, a statement of the secular change of variation awaited the further decrease observed by Gunter's Gresham successor, Henry Gellibrand.

Gunter's other inventions may have included the so-called Dutchman's log for measuring a ship's way. Henry Briggs acknowledged his suggested use of arithmetical complements in logarithmic work and the terms cosine, contangent, and such are probably Gunter's own; his use of the decimal point and his decimal notation for degrees are to be noted. Gunter's chain, used in surveying, is sixty-six feet long and divided into 100 equal links, thus allowing decimal measurement of acreage. Largely following Willebrord Snell, Gunter took a degree of the meridian to be 352,000 feet; this decision gave English seamen a much improved result.

BIBLIOGRAPHY

I. ORIGINAL WORKS. Gunter's chief works went through six eds. by 1680 and were successively augmented by their editors. They are *Canon triangulorum, sive tabulae sinuum et tangentium artificialium ad radium 10000.0000. & ad scrupula prima quadrantis* (London, 1620)—the British Museum copy (C.54.e.10) is bound with Henry Briggs's rare *Logarithmorum chilias prima* (London, n.d. [probably 1617]) and contains copious MS additions; *De sectore et radio. The Description and Use of the Sector in Three Bookes. The Description and Use of the Crosse-Staffe in Other Three Bookes. . . .* (London, 1623), a work of great practical importance; and *The Description and Use of His Majesties Dials in White-Hall Garden* (London, 1624)—the British Museum copy (C.60.f.7) gives evidence of Gunter's friendship with Ben Jonson—describes the large complex of dials, which stood until about 1697. A copy of the enl. 2nd ed. of his works, entitled *The Description and Use of the Sector, Crosse-staffe, and Other Instruments: With a Canon of Artificiall Lines and Tangents, to a Radius of 100,000,000 Parts, and the Use Thereof in Astronomie, Navigation, Dialling and Fortification, etc.* . . . (London, 1636), was bought by Newton for five shillings in 1667 and may be seen, much thumbed, in the library of Trinity College, Cambridge (NQ.9.160); it includes the vexed method of "middle latitude," probably first put forth by Ralph Handson in his 1614 version of Bartolomäus Pitiscus' *Trigonometria* but not used by Gunter himself. The 1653 ed. of the works, amended by Samuel Foster and Henry Bond, contains an early printed statement of the logarithmic result for the integral of the secant function or meridional parts—Gunter's meridian scale, like Wright's earlier one, came from the simple addition of secants; and he was doubtless unaware of Harriot's unpublished calculation of them as (in effect) logarithmic tangents, completed in 1614: he was not, anyway, interested in such theoretical niceties.

II. Secondary Literature. There is little need to refer to the brief early biographical sketches by John Aubrey, Charles Hutton, and John Ward. Accounts of aspects of Gunter's scientific contributions and their contexts are given in James Henderson, *Bibliotheca tabularum mathematicarum Being a Descriptive Catalogue of Mathematical Tables. Part 1. Logarithmic Tables* (*A. Logarithms of Numbers*) (Cambridge, 1926); and, extensively, in David W. Waters, *The Art of Navigation in England in Elizabethan and Early Stuart Times* (London, 1958), which gives detailed references to the relevant work of his contemporaries, of whom Briggs, Harriot, and Wright are the most important in this context. Christopher Hill, *Intellectual Origins of the English Revolution* (Oxford, 1965), covers the wider background, with much detail on the Gresham College circles. E. G. R. Taylor, *The Mathematical Practitioners of Tudor and Stuart England* (London, 1954), is useful but often infuriating on documentation. A more recent survey of the mathematical and navigational references is in J. V. Pepper, "Harriot's Unpublished Papers," in *History of Science,* **6** (1968), 17–40. The scientific correspondence of the later seventeenth century contains references to Gunter but does not add much of substance.

Jon V. Pepper

GÜNTHER, JOHANN. See **Guinter, Joannes**

GURVICH, ALEKSANDR GAVRILOVICH (*b.* Poltava, Russia, 27 September 1874; *d.* Moscow, U.S.S.R., 27 July 1954), *biology.*

Gurvich was the son of a notary, G. K. Gurvich; his elder brother, L. G. Gurvich, was a prominent specialist in petroleum chemistry. He graduated from the Faculty of Medicine of the University of Munich in 1897 and from 1899 to 1901 was an assistant in the department of anatomy in the University of Strasbourg. From 1901 to 1905 he lived in Bern, where he did his early work on the histophysiology of kidney cells and studied mitoses in amphibian eggs that had been put through a centrifuge. In 1904 he published *Morphologie und Biologie der Zelle.*

From 1907 to 1917 Gurvich was professor of anatomy and histology at the Higher Courses for Women in St. Petersburg. *Atlas und Grundriss der Embryologie* appeared in German, Spanish, and Russian between 1907 and 1909; *Vorlesungen der allgemeinen Histologie* was published in 1913. At the same time, in 1912, he began the investigations into the processes of morphogenesis that were to lead him to the theory of the biological field.

Gurvich served as professor of histology at the University of Simferopol (Crimea) from 1918 to 1924; he held the same position from 1924 to 1929 at the University of Moscow. He was head of the department of experimental biology of the Institute of Experimental Medicine in Leningrad from 1930 to 1942. In 1942 he returned to Moscow to assume the same post at the All-Union Institute of Experimental Medicine, which became the Academy of Medical Sciences in 1944; his own department became the Institute of Experimental Biology, with him as its director.

From 1948 until the end of his life Gurvich continued his experimental work at his home laboratory. He had begun his scientific work in histology, cytology, and embryology, with a later concentration on the problem of mitosis—particularly on the causes of cell division. The latter led him to the discovery of the resolving factor of mitosis—that is, of weak shortwave ultraviolet radiation, which he called mitogenetic rays. His researches in this field paved the way for further developments in molecular biology and resulted in establishment of the chain of processes occurring in cells after mitogenetic irradiation and the applicability of spectral analysis of mitogenetic rays (various fermentative processes with various spectral characteristics being the source of radiation).

Gurvich's early researches on morphogenesis allowed him to establish that the arrangement of morphological structures—the regular movement of cells and change in their form in the process of development—is governed by the character of the vector field. This became known as the theory of the biological field. He published seventeen monographs and more than 120 special works. His ideas were developed in the works of his wife, L. D. Gurvich, his daughter, A. A. Gurvich, S. J. Salkind, G. M. Frank, M. A. Baron, L. J. Blacher, and L. V. Belousov, among others.

BIBLIOGRAPHY

Gurvich's publications include "Über Determination, Normierung und Zufall in der Ontogenese," in *Archiv für Entwicklungsmechanik der Organismen,* **30** (1910), 133–193;

"Über den Begriff des embryonalen Feldes," *ibid.,* **51** (1922), 383–415; "Die Natur des spezifischen Erregers der Zellteilung," in *Archiv für mikroskopische Anatomie und Entwicklungsmechanik,* **100** (1923), 11–40; *Das Problem der Zellteilung physiologisch betrachtet* (Berlin, 1926); "Sur les rayons mitogénétiques et leur identité avec les rayons ultraviolets," in *Comptes rendus hebdomadaires des séances de l'Académie des sciences,* **184** (1927), 903–904, written with G. M. Frank; *Die histologischen Grundlagen der Biologie* (Jena, 1930); *Die mitogenetische Strahlung* (Berlin, 1932; 2nd ed., Jena, 1959), written with L. D. Gurvich; and *Teoria biologicheskogo polya* ("The Theory of the Biological Field"; Moscow, 1944).

On Gurvich's life and work, see L. V. Belousov, A. A. Gurvich, S. J. Salkind, and N. N. Kanneguiser, *Aleksandr Gavrilovich Gurvich (1874–1954)* (Moscow, 1970).

L. J. BLACHER

GUTBIER, FELIX ALEXANDER (*b.* Leipzig, Germany, 21 March 1876; *d.* Jena, Germany, 4 October 1926), *chemistry.*

Gutbier was the son of Carl F. Gutbier, a factory owner, and Fanny Thilo. He studied chemistry at the Technische Hochschule in Dresden under Walter Hempel and Fritz Foerster, then at the University of Erlangen with Otto Fischer, and finally at the University of Zurich under Alfred Werner. He received his doctorate in 1899 under Otto Fischer with the dissertation "Beiträge zur Kenntnis der Rosinduline." After becoming Fischer's assistant, he qualified for university lecturing in 1902 with *Studien über das Tellur* and was appointed *Privatdozent* in chemistry at the University of Erlangen. He refused an offer in 1907 to move to the University of Montevideo, and in the same year he was named extraordinary professor at Erlangen. In 1912 he was called to the Technische Hochschule in Stuttgart as professor of electrochemistry and chemical technology. Later he became professor of inorganic chemistry there and director of the Institute for Inorganic Chemistry and Inorganic Chemical Technology. From 1920 to 1922 he was rector of the Technische Hochschule. He was called to the University of Jena as professor of inorganic chemistry and director of the chemistry laboratory in 1922. In Jena, from 1924 until 1926, he was dean of the Faculty of Mathematics and Natural Sciences, which was founded through his initiative and was independent of the Faculty of Philosophy. From Easter 1926 until his death he was rector of Jena. Gutbier's first wife was Olga Fischer, daughter of Otto Fischer; they had two sons. His second wife was Gertrud Gaugler.

Gutbier's scientific publications treat many branches of inorganic chemistry; only in the beginning of his career was he concerned with problems in organic chemistry with his teacher Otto Fischer. While in Erlangen he became involved in inorganic chemistry. His analytical, inorganic, and atomic weight investigations included a special interest in tellurium. Gutbier turned his attention to the chemistry of coordination complexes and colloid chemistry. Beginning with the colloids of tellurium, he went on to the description of the metallic colloids silver, gold, platinum and the platinum metals, and of other colloidal elements. He also examined protective colloids and their specific effectiveness. In addition, he obtained a wealth of results in the chemistry of coordination complexes. Particularly noteworthy are the findings on hexachloro and hexabromo salts and of many metallic acids.

In all these investigations very different reactions were studied: for example, those of hydrogen sulfide with selenious acid, of oxygen with ruthenium, of hydrogen peroxide with tellurium, the catalytic effect of platinum black on hydrazine, and the receptivity for hydrogen induced by the presence of palladium, platinum, rhodium, and iridium. In addition, Gutbier worked out quantitative determinations and methods of separation for tellurium, palladium, and selenium, and for tungstic acid by means of nitrone; he also formulated separation methods for palladium and tin by means of dimethylglyoxime and electrolysis, respectively.

Because of his ability in analytic chemistry and his work on palladium, tellurium, and bismuth, Gutbier also succeeded in the difficult field of atomic weight determination. His researches in physical chemistry include the electrolysis of bismuth salt solutions and the preparation of selenium colloids through electrolysis and of mercury colloids through sputtering.

Gutbier also had a great interest in the history of chemistry, as shown in his work on Henri Moissan (1908) and in his essay on Goethe, Grand Duke Karl August, and chemistry in Jena (1926).

His activity as a teacher found expression in Gutbier's *Lehrbuch der qualitativen Analyse,* his monograph *Chemiestudium und Chemieunterricht,* and in his collaborative efforts with L. Birckenbach: *Praktische Anleitung zur Massanalyse* and *Praktische Anleitung zur Gewichtsanalyse.* His technical aptitude is evident in his invention of the high-speed dialyzer in the course of his work on colloid chemistry.

Gutbier was above all an experimental chemist who was able to inspire numerous students and co-workers through his organizational skills. He published about 260 papers.

BIBLIOGRAPHY

Gutbier's books include *Zur Erinnerung an Henri Moissan* (Erlangen, 1908); *Praktische Anleitung zur Gewichts-*

analyse, 2nd ed. (Stuttgart, 1919), written with L. Birckenbach; *Lehrbuch der qualitativen Analyse* (Stuttgart, 1920); *Praktische Anleitung zur Mass analyse,* 4th ed. (Stuttgart, 1924), written with L. Birckenbach; and *Goethe, Karl August und die Chemie in Jena* (Jena, 1926).

Part of his doctoral diss. was published in the complete works of Otto Fischer, and part appeared as "Über Thio-N-methyl-Pyridon und -Chinolon," in *Berichte der Deutschen chemischen Gesellschaft,* **33** (1900), 3358–3359.

A systematic bibliography of his journal articles is in Poggendorff, VI, 983–984.

F. HEIN

GUTENBERG, BENO (*b.* Darmstadt, Germany, 4 June 1889; *d.* Pasadena, California, 25 January 1960), *seismology.*

Gutenberg was the son of Hermann Gutenberg, a soap manufacturer, and Pauline Hachenburger Gutenberg. He attended the Realgymnasium and Technische Hochschule in Darmstadt, taking intensive courses in mathematics, physics, and chemistry. He intended to specialize in mathematics and physics at the University of Göttingen, but an interest in weather forecasting and climatology led him to Emil Wiechert's course on instrumental observation of geophysical phenomena at the new geophysical institute there. He took all of Wiechert's courses until he was told that he had learned practically all that was known in seismology. For his Ph.D. (1911) he elected to study microseisms.

From 1911 until 1918 (with an interruption for army service) Gutenberg was assistant at the International Seismological Association in Strasbourg; in 1918 he became *Privatdozent* at the University of Frankfurt-am-Main, where he was appointed professor of geophysics in 1926. His father died that year, and Gutenberg also undertook the management of the family business.

In 1930 Gutenberg accepted a professorship at the California Institute of Technology. This post also provided him with research facilities at the Seismological Laboratory of the Carnegie Institution there, which had an extensive network of seismograph stations, together with good recording instruments; it became part of the California Institute of Technology in 1936 and Gutenberg was its director from 1947 to 1958.

Gutenberg began in seismology with the most complex and frustrating topic in the field: the origin of microseisms. In his thesis and a number of later papers he considered most of the presently known sources for microseismic disturbances. Soon thereafter he produced his most elegant piece of research: following earlier suggestions by Wiechert and by R. D. Oldham, he computed the travel times of waves that would be affected by a low-velocity core of the earth, searched seismograms for them, demonstrated the existence of the core and measured its depth (2,900 kilometers) to an accuracy that still stands.

Gutenberg's early interest in meteorology led him to studies of the structure of the upper atmosphere. Noting the curious ring zones of silence and signal around strong air blasts, he derived the general curves for temperature in the ionosphere.

With Charles F. Richter, he derived improved travel-time curves for earthquakes (and determinations of velocity within the earth) while similar work was being done by Harold Jeffreys and Keith Bullen. The difference in approach is well exemplified by the derivation of these curves. The Jeffreys-Bullen curves were derived by statistical methods from a large volume of data from many sources; those by Gutenberg and Richter were from fewer data, from seismograms individually examined. Gutenberg derived improved methods of epicenter and depth determinations (using advanced instruments developed by Hugo Benioff), extended Richter's magnitude scale to deep-focus shocks, and, with Richter, determined the quantitative relations between magnitude, energy, intensity, and acceleration.

Studies of amplitude variations of compressional waves gave Gutenberg initial evidence that low-velocity layers existed within the earth. Using precise determinations of focal depth, he determined the variation of travel times as a function of source depth and thus the fine-scale variations of velocity in the upper mantle, demonstrating the existence of a low-velocity channel at a depth of between 100 and 200 kilometers. This channel is essential to theories of crustal movements.

One widely quoted hypothesis of Gutenberg's has proved invalid. From studies of surface wave velocities, and of reflection of compressional waves beneath the oceans, he concluded that the Atlantic basin was nearly continental in average structure as contrasted with the "truly oceanic" structure of the Pacific. Later work at sea has shown that the anomaly he found was due primarily to the much larger proportion of the Atlantic occupied by the mid-ocean ridge.

With Richter, Gutenberg redetermined the locations of all major earthquakes, showing both the patterns of seismicity and the geometry of the deep-focus earthquakes. Both seismologists and geologists are indebted to Gutenberg. Among the former he will be remembered best for his studies of the core and for his travel times and thorough studies of the phases of earthquake arrivals in "On Seismic Waves"; the latter use *Internal Constitution of the Earth* (which he edited, and wrote large portions of) and *Seismicity of the Earth* as standard sources.

BIBLIOGRAPHY

I. ORIGINAL WORKS. Among Gutenberg's more outstanding works are "On Seismic Waves," in four pts., in *Beiträge zur Geophysik,* **43** (1934); **45** (1935); **47** (1936); and **54** (1939), written with C. F. Richter, which contains much of his work on earthquake travel times and studies in teleseisms; *Internal Constitution of the Earth* (New York, 1939; 2nd ed., 1951); and *Seismicity of the Earth* (New York, 1941), written with Richter, which gives the worldwide geography of earthquakes in a useful format.

A revision of the travel times is "Epicenter and Origin Time of the Main Shock on July 21 and Travel Time of Major Phases," in G. B. Oakeshott, ed., *Earthquakes in Kern County, California During 1952,* California Division of Mines Bulletin 171 (San Francisco, 1955), pp. 157–163. For Gutenberg's summary of seismologic research at three stages over a span of two decades, see *The Physics of the Earth's Interior,* International Geophysics Series, vol. I (New York, 1959). Richter's memorial, cited below, includes a bibliography.

II. SECONDARY LITERATURE. Gutenberg's life and contributions to seismology are summarized in H. Jeffreys, "Beno Gutenberg," in *Quarterly Journal of the Royal Astronomical Society,* **1** (1960), 239–242; and C. F. Richter, "Memorial to Beno Gutenberg, 1889–1960," in *Proceedings of the Geological Society of America for 1960* (1962), pp. 93–104. Sidelights on Gutenberg's early years appear in "Fifteenth Award of the William Bowie Medal," in *Transactions of the American Geophysical Union,* **34,** no. 3 (1953), 353–355.

GEORGE G. SHOR, JR.
ELIZABETH NOBLE SHOR

GUY DE CHAULIAC. See **Chauliac, Guy de.**

GUYE, CHARLES-EUGÈNE (*b.* St. Christophe, Switzerland, 15 October 1866; *d.* Geneva, Switzerland, 15 July 1942), *physics, electromagnetism, molecular physics.*

Guye was a member of a distinguished Swiss family. With his older brother, Philippe-Auguste (1862–1922), he pioneered in investigating phenomena on the borderline between physics and chemistry. Philippe-Auguste, primarily a chemist, was interested in electrochemical synthesis and is known for his precision studies of atomic weights. He was a founder of the *Journal de chimie physique.* Charles-Eugène became primarily a physicist. Interested in electromagnetism and molecular size determinations, he gained recognition for his precise measurements of variation of the mass of electrons as a function of their velocity, and as director of physical laboratories at the University of Geneva. Together and separately, the two Guyes achieved distinction in their different disciplines by devising experimental means for analyzing interface phenomena in physical chemistry and chemical physics.

Four years after Philippe-Auguste, Charles-Eugène left his native town in the canton of Vaud for Geneva, where he began his scientific career in the 1880's with experimental demonstrations of rotatory polarization in optically active crystals and liquids. Studying with Jacques-Louis Soret and Charles Soret, he obtained his doctorate with a thesis on this subject in 1889 at the University of Geneva. There both Guyes were active most of their lives. Significantly, in 1894 Charles-Eugène was called to a professorship at Zurich's Polytechnique. Remaining there for six years, he achieved intellectual independence and was aroused to new scientific interests through teaching electrical engineering. At Zurich Guye's research was in alternating currents, polyphasic generators, and hysteresis phenomena. Albert Einstein was one of his students.

In 1900 Guye returned to Geneva when offered a permanent professorship in experimental physics. He remained there through his retirement in 1930. Most active in the laboratory during the first two decades of the twentieth century, Guye studied electric arcs, their explosive potentials and spontaneous rotations. Bolometry, induction coefficients, and analysis of electrical measuring instruments were his specialties. By designing highly accurate instruments for such work, Guye found important new applications for his apparatus in determining the diameters of molecules and investigating the interior structures of solid-state materials.

With the advent of H. A. Lorentz' theory of the electron and Max Abraham's rival theory, Guye became interested in using his apparatus to test for evidences of the FitzGerald-Lorentz contraction hypothesis and transformation equations. It seemed to him that a crucial experiment ought to be possible to decide between Lorentz' idea of a deformable electron with a shape dependent on velocity and Abraham's notion of permanently spherical electrons. The opportunity seemed all the more inviting as Einstein's special theory of relativity, which was based upon Lorentz' work, began to stir controversy after 1905, while on the other hand, W. Kauffmann's experiments appeared to support Abraham and contradict the predictions of Lorentz and Einstein.

For fully a decade after 1907 Guye carried through a series of increasingly elaborate experiments with charged particles moving through electromagnetic fields. Collaborating with M. Ratnowsky and Charles Lavanchy, Guye was able to develop very precise techniques for measuring particle deflections within carefully controlled electric and magnetic fields. In 1916 and 1921 Guye published these methods and pronounced results in favor of the Lorentzian formulas and Einsteinian theory. Thereafter his reputation rose as a most able experimenter among the world's physicists, but the greater fame of his brother, whom he outlived by two decades, overshadowed Charles-Eugène.

Of French extraction, culture, and spirit, Guye served his profession and university long and well; first as dean of the Faculty of Sciences (1910–1914); then as consulting editor of *Helvetica physica acta* and editor of *Archives de Genève* (1919–1927). He served the Swiss government as a member of its Commission on Weights and Measures (1915–1931) and served as a member of the Solvay Institute of the University of Brussels (1925–1934). In 1927 he was honored to become a correspondent of the French Académie des Sciences, and many other French honors followed. The several small books that he published in later years on the evolution of statistical thermodynamics and on reductionism in physics and biology show the breadth of his interests and the vitality of his mind.

BIBLIOGRAPHY

I. ORIGINAL WORKS. Guye's works include *L'évolution physicochimique* (Paris, 1922), trans. by J. R. Clarke as *The Evolution of Physical Chemistry* (London, 1925); and *Les limites de la physique et de la biologie* (Geneva, 1936). There is a bio-bibliography in *Documents pour servir à l'histoire de l'Université de Genève,* VI (1938), 69–71; IX (1944), 32 f.

II. SECONDARY LITERATURE. An anonymous work about Guye is "Au Professor C. E. Guye à l'occasion de son soixante-dixième anniversaire," in *Helvetica physica acta,* **9** (1936), 511–514. See also Émile Briner, "Ch.-E. Guye (1866–1942)," in *Journal de chimie physique,* **40** (1943), 1–4; Louis de Broglie, "Notice sur la vie et les travaux de Charles-Eugène Guye," in *Comptes rendus de l'Académie des sciences,* **215** (1942), 209–211; *Historisch-biographisches Lexikon der Schweiz,* IV (Neuenburg, 1927), 25; and Poggendorff, IV, 986 f.; VIIa, 324.

LOYD S. SWENSON, JR.

GUYER, MICHAEL FREDERIC (*b.* Plattsburg, Missouri, 17 November 1874; *d.* New Braunfels, Texas, 1 April 1959), *zoology.*

Guyer was the son of Michael Guyer and Sarah J. Thomas. From 1890 to 1892 he attended the University of Missouri and then spent two years at the University of Chicago, where he received the B.S. degree in 1894. He was a teaching assistant in zoology at the University of Nebraska from 1895 to 1896 under Henry B. Ward. In 1897 he taught high school in Lincoln, Nebraska, and received the M.S. degree the same year. His election to Phi Beta Kappa and Phi Kappa Phi honorary societies attested to his scholastic ability. Guyer returned to the University of Chicago in 1897, holding a three-year fellowship; there he worked out his doctoral dissertation, on pigeon spermatogenesis, under Charles Otis Whitman. In 1899 he married Helen M. Stauffer; a son, Edwin Michael, was born on 25 November 1900, the year in which Guyer received his Ph.D.

Guyer became professor of zoology at the University of Cincinnati, also in 1900, and served as head of the department until 1911. At Cincinnati he continued his research on spermatogenesis of guinea fowl, chickens, and hybrids between them; wrote a text on *Animal Micrology* (1906); and, with W. O. Pauli, published a manual of physiology. He also advised both the Medical School (on premedical education) and the Cincinnati Zoological Garden. From 1908 to 1909 he studied at Paris and the Naples Biological Station.

In 1911 Guyer was brought to the University of Wisconsin by its president, Charles R. Van Hise, to be chairman of the department of zoology, a position he held until his retirement in 1945. At Wisconsin he taught animal biology, heredity and eugenics, and cytology. His book *Being Well Born* (1916) aroused widespread interest in human heredity and pointed out its significance, in certain cases, as a predisposing factor to crime, disease, and mental deficiency, as well as its possible role in the improvement of the human species. His *Animal Biology* (1931) quickly became a leading textbook of introductory zoology, going through four editions.

Guyer's continuing interest in medical education led to his appointment in the early 1920's to the National Commission on Medical Education and shortly thereafter to the Wisconsin Basic Science Board, an examining body for prospective Wisconsin physicians. In both of these bodies he exerted a strong influence, stressing the importance of basic sciences in the premedical and medical curricula.

As a cytologist he was one of the first to determine, with a margin of error of about 2 percent, the chromosome number in human spermatocytes. From 1917 to 1930 his main research effort was directed toward inducing hereditary eye defects by injecting antilens serum into pregnant rabbits as eyes were beginning to form in their unborn fetuses. This research held the intriguing possibility that the units of heredity in the germ cells might be altered by antibody action. His papers on these investigations stimulated research elsewhere, which, although it failed in the end to substantiate his main thesis, led to better knowledge about placental transmission of antibodies and other immunological problems. After 1930 he and his students turned to studies on the growth of cancer cells and their susceptibility to certain chemicals, such as palladium.

In his later years Guyer published biological reflections on his own species in *Speaking of Man* (1942). He retired from teaching in 1945 and subsequently spent much time in Arizona and Texas to conserve his health. During his years of teaching he supervised the doctoral research of over two dozen graduate students.

BIBLIOGRAPHY

Some of Guyer's more significant publications are *Animal Micrology* (Chicago, 1906; 5th ed., 1948); *Being Well Born* (Indianapolis, 1916; 2nd ed., 1927); "Transmission of Induced Eye Defects," in *Journal of Experimental Zoology,* **31** (1920), 171–223; "Soma and Germ," in *American Naturalist,* **59** (1925), 97–114; *Animal Biology* (New York, 1931; 4th ed., 1948); and *Speaking of Man* (New York, 1942).

LOWELL E. NOLAND

GUYONNEAU DE PAMBOUR, F. M. See **Pambour, F. M. Guyonneau de.**

GUYOT, ARNOLD HENRI (*b.* Boudevilliers, Switzerland, 28 September 1807; *d.* Princeton, New Jersey, 8 February 1884), *geography, glacial geology.*

At the University of Neuchâtel, Guyot's studies were classical while his early interest in nature was satisfied by collecting insects and plants. In 1825 he went to Germany to continue his education. He studied first at Karlsruhe, where he lived with the family of Alexander Braun, who was often visited by Louis Agassiz and Karl Schimper during their vacations. He later went to Berlin to prepare for the ministry. Guyot eventually abandoned theology for science, however, and terminated his education in 1835 with a doctoral dissertation on the natural classification of lakes.

Soon after, Guyot left Berlin for Paris, having accepted the responsibility of educating the sons of the Count de Pourtalès-Gorgier. While in this position he traveled extensively in Europe for four years. In the spring of 1838, Agassiz met Guyot in Paris, and finding him unconvinced about his new concept of a glacial age, urged him to visit the Alpine glaciers that summer. Guyot spent six weeks in the Alps making a series of fundamental observations on the moraines, the differential flow of glaciers, and the banded structure of the ice (blue bands). These results, although presented orally at the meeting of the Geological Society of France in Porrentruy, in September 1838, were not published because Agassiz and Guyot had decided to collaborate on a major work in which Agassiz would study the glaciers and Guyot the erratic boulders in the plains of Switzerland.

In the following years, Guyot saw with pleasure—and also with some bitterness—most of his conclusions confirmed by Agassiz, Edward Forbes, and others. But since Guyot's original work had not been published, he did not receive proper credit for these findings. In 1847 Agassiz published only one volume, *Système glaciaire,* of the joint work, which he alone had written. It was not until 1883, after Agassiz's death, that a short summary of Guyot's contributions to the project was published.

Guyot returned to Neuchâtel in 1839, and was appointed professor of history and physical geography at the academy there. His major study of the distribution of erratic boulders in Switzerland was undertaken between 1840 and 1847. By tracing the boulders to their original outcrops along the northern slope of the Alps, he recognized eight erratic basins demonstrating the former existence of gigantic Alpine glaciers, as postulated by Agassiz. But Guyot wrote very little on the subject, having planned to publish a complete account of his investigation in the second volume of Agassiz's work which never appeared.

The revolution of 1848 led to the suppression of the academy in June of that year, and Guyot followed Agassiz to America. At the Lowell Technological Institute in Boston he taught comparative physical geography. His lectures, *Earth and Man,* published in 1849, represent a far-reaching synthesis in which he visualized a divine law of progress common to Genesis, the evolution of the earth, and the history of humanity. In his later years, as shown by his work *Creation* (1884), he partially accepted the doctrine of evolution through natural causes.

In 1854 Guyot was appointed professor of physical geography and geology at Princeton. He spread the new concept of geographic education by means of field studies, and for that purpose prepared, between 1861 and 1875, a series of specially designed textbooks and wall maps which became very popular.

Under the auspices of the Smithsonian Institution, Guyot established the instrumental and geographic requirements for a national system of meteorological stations. In order to find the best location for these stations, he undertook a systematic topographical survey of the entire Appalachians from Vermont to North Carolina, a gigantic task which he completed in 1881 with the survey of the Catskills.

In his honor, the term "guyot" is applied to a seamount, generally deeper than 200 meters, whose top is a relatively smooth platform. Originally proposed by H. H. Hess in 1946 after extensive investigations in the Pacific, this term is now in common use throughout the world.

BIBLIOGRAPHY

I. ORIGINAL WORKS. Guyot's chief work is *Earth and Man, or Lectures on Comparative Physical Geography in Its Relation to the History of Mankind,* translated from the French by C. C. Felton (Boston, 1849). A collection of meteorological and physical tables, with other tables useful in practical meteorology, prepared for and published by the Smithsonian Institution are in Smithsonian Institution Publication no. 538 (Washington, D.C., 1852; other eds., 1859, 1884), p. 747.

Other works by Guyot include "On the Topography of the State of New York," in *American Journal of Science,* 2nd ser., **8** (1852), 272–276; "On the Appalachian Mountain System," *ibid.,* **31** (1861), 157–187; "On the Physical Structure and Hypsometry of the Catskill Mountain Region," *ibid.,* 3rd ser., **19** (1880), 429–451; "Observations sur les glaciers," in *Bulletin de la Société des sciences naturelles de Neuchâtel,* **13** (1883), 156–159, with a letter of introduction, pp. 151–156; and *Creation, or the Biblical Cosmogony in the Light of Modern Science* (New York, 1884).

II. SECONDARY LITERATURE. For works about Guyot, see J. D. Dana, "Memoir of Arnold Guyot (1807–1884)," in *Biographical Memoirs. National Academy of Sciences,* **2** (1886), 309–347, which has a complete list of Guyot's publications; and C. Faure, "Vie et travaux d'Arnold Guyot," in *Globe,* **23** (1884), 3–72.

ALBERT V. CAROZZI

GUYTON DE MORVEAU, LOUIS BERNARD (*b.* Dijon, France, 4 January 1737; *d.* Paris, France, 2 January 1816), *chemistry, aeronautics.*

The son of Antoine Guyton, a lawyer, and Marguerite Desaulle, Guyton was educated in Dijon at the Godran (Jesuit) College and the Faculty of Law, and from 1756 to 1762 he practiced there as an advocate. Dijon was the capital of the French province of Burgundy and the seat of one of the provincial *parlements,* or royal courts of law, which had both political and judicial functions; and in 1762 Guyton entered the *parlement* as *avocat-général du roi,* one of the public prosecutors. He then added "de Morveau" to his name, the designation being that of a family property, and until 1789 he was often called Monsieur de Morveau. During the French Revolution he became Guyton-Morveau, then Guyton, and finally Guyton-Morveau again.

The suppression of the Jesuits in France in 1763 resulted in the closing of many schools run by them. Various plans for educational reform were advanced, including Guyton's *Mémoire sur l'éducation publique* (1764), which contains detailed proposals for a large college in each province. He believed that a wide range of subjects should be taught, with less emphasis on classics than hitherto, and made the interesting suggestion that mathematics, physics, natural history, and chemistry should be included in the final two years. He quoted from many classical and modern authors and had obviously studied his subject with the thoroughness that was to characterize all his future work.

Guyton ably performed his heavy parliamentary duties until he retired in 1782 with a pension and the title of *avocat général honoraire.* Some of his speeches were published in *Discours publics et éloges* (3 vols., 1775–1782), one of the most important being his criticism, in 1767, of the local variations of the law in France—where there were, he said, one people, one legislator, and 285 legal codes. He was praised by Voltaire, and in 1771 he outlined a scheme for a new code applicable to the whole country; but the collapse of the old system, like that of education, was to come only with the Revolution, and its reform with Napoleon.

A long poem satirizing the Jesuits—*Le rat iconoclaste ou le Jésuite croqué*—was published anonymously at Dijon in 1763. It was known to be Guyton's work; and after hearing it read, the Académie des Sciences, Arts et Belles-Lettres of Dijon elected him as an *honoraire* on 20 January 1764. His early contributions to its meetings were literary but he became interested in chemistry, which was often discussed at the Academy, and in 1768 he installed a laboratory in his new house. He was entirely self-taught, studying initially the books of A. Baumé and P. J. Macquer.

Another member of the Dijon Academy, J. P. Chardenon, was engaged in research on combustion and calcination; and after his death in 1769 Guyton continued the work. His results were published in "Dissertation sur le phlogistique," the first essay in a volume entitled *Digressions académiques* (1772). Chardenon had speculated about the reason why metals gain weight on calcination, but Guyton pointed out that no one had, in fact, proved that every metal invariably gains weight. This he now did, in a careful and accurate piece of quantitative work that shows that he had developed into a competent chemist. He also proved that decreases in weight occasionally observed by earlier chemists were due to some effect other than calcination. In order to explain why a metal containing phlogiston weighs less than its calx, he modified Chardenon's theory, believing that phlogiston was specifically lighter than all other substances, however subtle, and therefore appeared to lighten anything containing it weighed in any medium whatsoever. Guyton was as unconvincing as Chardenon, who had considered only weighings in air, and his theory gained no support. The experimental part of Guyton's essay, however, was influential; for his proof of the gain in weight was one of the factors that led Lavoisier to investigate combustion and calcination.

Digressions académiques also contained a discussion of chemical affinity in which Guyton elaborated the theory, earlier suggested by Buffon, that ultimate particles of matter attracted each other by a force obeying Newton's inverse-square law—the relation was complicated in that at short distances their shapes had to be considered, for they could not be regarded as point masses. He hoped that the shapes of these particles might eventually be inferred from a study of crystals, but he gave no specific examples. In 1773 Guyton measured the forces of cohesion between mercury and other metals and thought that these could be related to the affinities supposed to be responsible for the formation of amalgams. He returned several times to this problem of measuring affinities, but with no more success than his contemporaries Richard Kirwan and C. F. Wenzel.

In 1772 Guyton became vice-chancellor of the Dijon Academy and was elected a correspondent of the Paris Académie des Sciences. During a visit to Paris in 1775 he was introduced to pneumatic chemistry by Lavoisier, and he soon became convinced that a portion of the air was absorbed during combustion and calcination, causing the gain in weight. He abandoned his former theory but still believed in phlogiston and thought, like Macquer, that it was released at the same time that air was absorbed.

This theory was taught in the public course of chemistry, published as *Élémens de chymie* (3 vols., 1777–1778), that Guyton gave in the Dijon Academy every year from 1776, assisted by Hugues Maret and J. F. Durande. The arrangement of the lectures and book was determined by Guyton's theory that the mutual attraction between the ultimate particles of different kinds of matter could cause one substance to dissolve in another, and that a chemical change was possible only as a result of such a solution. Every reaction therefore required a solvent, and a chapter was devoted to each of twenty solvents: fire, air, and water; nine acids; three alkalies; four oily substances; and mercury.

In order to keep his course up to date, Guyton read widely in several languages. He also translated a number of books and memoirs, his annotated edition of T. O. Bergman's *Opuscules physiques et chymiques* (2 vols., 1780–1785) being especially important. He also added notes to C. W. Scheele's *Mémoires de chymie* (1785), translated by his close friend Claudine Picardet, the wife of another Dijon academician.

A reformer by nature, Guyton became the leading critic of the current chemical nomenclature, in which the name of a substance was hardly ever related to its constitution but was derived from such unsystematic origins as the name of its discoverer, its place of occurrence, or its appearance. Macquer and Bergman proposed certain reforms; and Guyton had been influenced by both of them when, in 1780, he was commissioned by the publisher Panckoucke to write the chemical volumes of the *Encyclopédie méthodique.* All the articles had been arranged in one alphabetical sequence in the *Encyclopédie* of Diderot and D'Alembert and also in the supplementary volumes (1776–1777) to which Guyton contributed fourteen chemical articles, but in the new work each subject was to be treated in one or more separate volumes.

Guyton was about to write a comprehensive treatise on chemistry, and he now had a chance to reform the nomenclature completely. In 1782 he published his initial proposals. They were concerned mainly with acids, bases, and salts, but the principles he laid down were universally applicable. The most important was that the simplest substances should have the simplest names, and that names of compounds should recall their components. The old, unsystematic names were excluded. Thus, oil of vitriol (named from its oily appearance) and Epsom salt (named from its place of occurrence) became vitriolic acid and vitriol of magnesia, respectively. These reforms were welcomed by Macquer and Bergman and were adopted by chemists in France, England, and other countries.

From the beginning of his scientific career, Guyton was interested in metallurgy and mineralogy. In 1769 he investigated the use of coal instead of charcoal in blast furnaces, and in 1777 he described a special flux of powdered glass, borax, and charcoal for assaying iron ores. There were several different theories of the relation between iron and steel; and research led Guyton to discover in 1786, independently of G. Monge, C. A. Vandermonde, and C. L. Berthollet (*Observations sur la physique,* **29** [1786], 210–221), that cast iron, wrought iron, and steel differed only in carbon content. He was often consulted by directors of mines and foundries in Burgundy, and he devised a portable set of apparatus for analyzing minerals in the field. He wanted to reform the nomenclature of minerals on the same lines as chemical nomenclature, but in this he was less successful.

Guyton always attached importance to the applications of science, and he intended the laboratory of the Dijon Academy to be used for the public benefit. In 1782, for example, he examined several white pigments, hoping to find a substitute for the poisonous white lead. Zinc calx (oxide) proved satisfactory, and it was manufactured and sold at the Academy by the laboratory steward, J. B. Courtois, the father of Bernard Courtois. Several times Guyton was personally involved in industry. From about 1780, with three partners, he manufactured saltpeter at Dijon. The enterprise, which was taken over by the elder Courtois in 1788, led to the development of a new analytical method. Saltpeter (potassium nitrate) was made by mixing decayed animal manure (containing nitrates) with wood ashes (containing potash), leaching with water, and evaporating. If too much potash was added, the saltpeter was contaminated with potassium chloride, so generally some nitrate was wasted in the mother liquor. Guyton determined the amount of chloride in a sample of mother liquor by adding lead nitrate solution of known concentration until all the chloride was precipitated; this enabled him to calculate how much potash was needed to form the maximum quantity of saltpeter free from chloride. This was one of the earliest applications of volumetric analysis.

Soda manufacture was another of Guyton's interests. In 1783 he visited le Croisic, in Britanny, and set up a factory to prepare soda (sodium carbonate) by a method discovered by Scheele: the action of atmospheric carbon dioxide on a paste of slaked lime (calcium hydroxide) and concentrated brine (impure sodium chloride, prepared by solar evaporation of sea water). Some soda was made, but the enterprise lasted only a few years. Guyton's only profitable industrial venture was a glassworks, run in conjunction with a coal mine, which he opened in 1784 at St. Bérain sur Dheune, in Burgundy.

Despite his many activities in chemistry Guyton found time to contribute to a new and exciting application of science. In November and December 1783 the balloon flights of J. F. Pilatre de Rozier and J. A. C. Charles attracted widespread attention, and the Dijon Academy decided to make its own balloon, to be filled with "inflammable gas." Guyton tested various gases. The gas from zinc and sulfuric acid (hydrogen) was the lightest but expensive to prepare, so he rapidly developed a large-scale plant for generating a heavier but cheaper gas by the dry distillation of vegetable matter. The iron retorts leaked, however, and eventually hydrogen was used. Guyton made two flights, with Claude Bertrand, an astronomer, on 25 April 1784, and with C. A. H. Grossart de Virly, lawyer and amateur chemist, on 12 June 1784. During the second flight an attempt was made to steer the balloon with manually operated oars and a rudder, a method that was theoretically sound and seems to have been partly successful but required too much effort for sustained flight.

Full accounts of the preliminary calculations and experiments, as well as descriptions of the construction of the balloon and the large-scale production of gas, were published in *Description de l'aérostate* (1784), an important treatise that added to Guyton's international reputation. In Dijon, however, all was not well. For several years the Academy had been accumulating substantial debts, and some of the literary members believed that the expenses of the laboratory and the chemical course were responsible. A bitter dispute developed in 1786 when Maret, the secretary, died and Guyton accepted the office in addition to that of chancellor, which he had held since 1781. The atmosphere became so unpleasant that for over a year he stayed away from the Academy. When he returned at the end of 1787, he resumed his activities as chancellor, but not secretary, and gave the annual course, which was now an account of antiphlogistic chemistry.

After the publication in 1786 of volume I, part 1, of *Encyclopédie méthodique, chymie,* Guyton began to prepare the article "Air" for part 2. This was to include an account of combustion and descriptions of the gases in the atmosphere, and they would have to be named according to the theory that he accepted. He made a journey to Paris in February 1787 and stayed there for about seven months. Discussions with Lavoisier soon led him to adopt the antiphlogistic theory without reservation; and he collaborated with Lavoisier, Berthollet, and Fourcroy in writing *Méthode de nomenclature chimique* (1787), in which the nomenclature, more extensively revised than in 1782, was designed so that names of substances agreed with their constitutions according to the new theory. Vitriolic acid, for example, now considered to be a compound of sulfur and oxygen, was called sulfuric acid, and was distinguished from sulfurous acid, which contained less oxygen. Guyton also joined Lavoisier and his colleagues on the editorial board of *Annales de chimie,* the journal that was founded in 1789; but his scientific work, including the *Encyclopédie méthodique, chymie,* which he handed over to Fourcroy after the publication of part 2 in 1789, was now interrupted by the French Revolution.

In August 1789 Guyton became president of the Dijon Patriotic Club; and in 1790 he was elected *procureur général syndic* of the Côte d'Or, one of the new "departments" into which Burgundy was divided. He held this important administrative post until elected to the National Assembly in August 1791. This took him to Paris, where he remained for the rest of his life. In 1792 he became a deputy to the National Convention, which declared France a republic, and he was among the majority who voted for the execution of Louis XVI in January 1793. Guyton became secretary of the Committee of General Defense on 3 January 1793, and from 6 April to 11 July he was president of the first Committee of Public Safety, at a time when most of its nine members were men of moderate opinions trying to secure national unity while engaged in a desperate war. But in July the moderates, including Guyton, were removed; and under Robespierre the committee took steps that redeemed the military situation but led to the Terror.

During 1794 Guyton was concerned mainly with the applications of science to the war. He helped J. A. A. Carny to devise simplified methods of making saltpeter and gunpowder, and he was a lecturer at the intensive courses on gunpowder and cannon manufacture that were given at Paris in February and March to men from all parts of France. He was one of the organizers of the first military air force—the *Compagnie d'Aérostiers*—and on 26 June 1794 he witnessed the French victory over the Austrians at Fleurus, Belgium, when observers in a captive balloon threw out messages with reports on the Austrian positions. As political commissioner attached to the army he accompanied it to Brussels and returned to Paris on 31 July, four days after Robespierre's downfall. From 6 October 1794 to 3 February 1795 he again served on the Committee of Public Safety, which now had limited powers. Although elected to the Conseil des Cinq-Cents after the Convention was dissolved in 1795, he joined none of its committees and retired from politics in 1797.

One of the first members of the Institut de France when it was founded in 1795, Guyton was president of the class of mathematical and physical sciences in 1807. He was twice director of the École Polytechnique (1798–1799, 1800–1804), and as a professor from its founding in 1794 until 1811 he taught and did research there. In 1798 he liquefied ammonia by cooling the dry gas to $-44°C$. with a mixture of ice and calcium chloride. Under his direction C. B. Desormes and N. Clément proved in 1801 (independently of W. Cruickshank) that carbon formed two oxides, the lower one being the "heavy inflammable air" that had puzzled earlier chemists; in 1803 he devised a pyrometer consisting of a platinum rod which, as it expanded, caused a pointer to move over a circular scale. But, as in his Dijon days, Guyton's interests were too wide for him to be able to make great contributions to experimental chemistry. It was as a reformer of nomenclature, a teacher, and a systematizer that he made his name. And he was always concerned with the applications of chemistry.

Guyton did important research on the disinfection of air, a subject that first interested him in 1773, when he was consulted about the problem of putrid emanations from corpses in the crypt of a Dijon church. Believing the disease-carrying particles accompanied the volatile alkali (ammonia) given off by decaying flesh, he filled the church with marine (hydrochloric) acid fumes, which he hoped would precipitate the emanation with the ammonia. The treatment did, in fact, remove the odor, and it was later used successfully in prisons and hospitals. In England, Sir James Carmichael Smyth independently introduced the use of nitric acid fumes, and Guyton subsequently made the investigation described in his *Traité des moyens de désinfecter l'air* (1801). He found that his original theory was incorrect, for ammonia was not always evolved from decaying flesh; and he now thought that the disinfectant action was due to oxygen, which the antiphlogistic chemists assumed to be in all acids. Oxymuriatic acid (chlorine) was believed to contain a high proportion of oxygen, and Guyton found it to be an effective disinfectant—an interesting example of a satisfactory procedure based on a theory that was soon shown to be false. A simple apparatus for producing chlorine from common salt, sulfuric acid, and manganese dioxide was described in his book, which was translated into five languages. For this service to humanity he was admitted to the Legion of Honor in 1805, and in 1811 he became a baron of the empire.

In 1799, Napoleon appointed Guyton administrator of the mints, an important post, for there were nine mints in France and the number later increased. He left office at the Bourbon restoration in 1814 but resumed when Napoleon returned from Elba. He finally retired on 7 July 1815, three weeks after Waterloo.

Some of the men responsible for the execution of Louis XVI were exiled by Louis XVIII, but Guyton was left in peace. He died six months later and was survived by the former Mme. Picardet, whom he had married in 1798, after the death of her first husband. They had no children.

BIBLIOGRAPHY

I. Original Works. Details of the various eds. and trans. of Guyton's books, and references to his most important contributions to periodicals, are given in W. A.

Smeaton, "L. B. Guyton de Morveau: A Bibliographical Study," in *Ambix,* **6** (1957), 18–34.

II. SECONDARY LITERATURE. Georges Bouchard, *Guyton-Morveau, chimiste et conventionnel* (Paris, 1938) is a reliable biography but includes few details of Guyton's scientific work. Some aspects of this have been discussed in a series of articles by W. A. Smeaton: "The Contributions of P. J. Macquer, T. O. Bergman and L. B. Guyton de Morveau to the Reform of Chemical Nomenclature," in *Annals of Science,* **10** (1954), 87–106; "The Early History of Laboratory Instruction in Chemistry at the École Polytechnique, Paris, and Elsewhere," *ibid.,* 224–233; "Guyton de Morveau's Course of Chemistry in the Dijon Academy," in *Ambix,* **9** (1961), 53–69; "Guyton de Morveau and Chemical Affinity," *ibid.,* **11** (1963), 55–64; "Guyton de Morveau and the Phlogiston Theory," in I. B. Cohen and R. Taton, eds., *Mélanges Alexandre Koyré,* I (Paris, 1964), 522–540; "L. B. Guyton de Morveau: Early Platinum Apparatus," in *Platinum Metals Review,* **10** (1966), 24–28; "The Portable Chemical Laboratories of Guyton de Morveau, Cronstedt and Göttling," in *Ambix,* **13** (1966), 84–91; "Louis Bernard Guyton de Morveau and His Relations With British Scientists," in *Notes and Records. Royal Society of London,* **22** (1967), 113–130; and "Is Water Converted Into Air? Guyton de Morveau Acts as Arbiter Between Priestley and Kirwan," in *Ambix,* **15** (1968), 75–83.

There are numerous references to Guyton in Roger Tisserand, *Au temps de l'Encyclopédie: L'Académie de Dijon de 1740 à 1793* (Paris, 1936), a book based on a study of the archives of the Dijon Academy, which are now in the Archives Départementales de la Côte d'Or, Dijon. Guyton's early theory of calcination is discussed in J. R. Partington and D. McKie, "Historical Studies on the Phlogiston Theory. Part I," in *Annals of Science,* **2** (1937), 361–404; and, with more emphasis on the experimental work, in H. Guerlac, *Lavoisier—The Crucial Year* (Ithaca, N.Y., 1961), pp. 125–145. An account of Guyton's contributions to volumetric analysis is E. Rancke Madsen, *The Development of Titrimetric Analysis Till 1806* (Copenhagen, 1958), pp. 83–101. There is a discussion of his reform of chemical nomenclature in M. P. Crosland, *Historical Studies in the Language of Chemistry* (London, 1962), pp. 153–192. His work on disinfection is described in Lars Oberg, "De mineralsura rökningarna. En episod ur desinfektionsmedlens historia," in *Lychnos* (1965–1966), pp. 159–180, with English summary. There is an evaluation of his research on white pigments in R. D. Harley, *Artists' Pigments c. 1600–1835* (London, 1970), pp. 162–168. An annotated English trans. of his article "On the Nature of Steel and Its Proximate Principles" is in C. S. Smith, ed., *Sources for the History of the Science of Steel 1532–1786* (Cambridge, Mass.-London, 1968), pp. 257–274.

W. A. SMEATON

GWYNNE-VAUGHAN, DAVID THOMAS (*b.* Llandovery, Wales, 12 March 1871; *d.* Reading, England, 4 September 1915), *botany.*

Gwynne-Vaughan was the eldest child of Henry Thomas Gwynne-Vaughan and Elizabeth Thomas. He was educated at Monmouth Grammar School and Christ's College, Cambridge, where he held a scholarship and graduated with a first class in part I of the natural sciences tripos in 1893. He left Cambridge without taking part II and taught science for a year until, in 1894, he was invited to work at the Jodrell Laboratory, Kew, under D. H. Scott. There, using plants cultivated at Kew, he began to specialize in microscopic studies of plant anatomy, particularly the arrangement of vessels in the stem.

At the British Association for the Advancement of Science meeting in 1896 he read a paper on "The Arrangement of the Vascular Bundles in Certain Nymphaeaceae," and on the strength of this, Frederick Bower offered him a post in his laboratory in Glasgow. He held this post from 1897 to 1907, working mainly on the anatomy of Pteridophyta, while also lecturing and writing a book on practical botany with Bower.

Before settling in Glasgow, Gwynne-Vaughan was able to make two expeditions. During 1897 and 1898 he went up the Amazon and Purus rivers to report on rubber production for a commercial syndicate. Although he was fascinated by the wealth of plant life, this trip offered little opportunity for collection. After another short period at Kew he spent most of 1899 in the Malay Peninsula, collecting and observing with W. W. Skeat.

From 1907 to 1909 Gwynne-Vaughan was head of the department of botany at Birkbeck College, London, and then in 1909 moved, as professor of botany, to the Queen's University of Belfast where he stayed until 1914. He then assumed the chair of botany at University College, Reading, a position he held, in spite of illness, until his death in 1915. In 1911 he married Helen C. I. Fraser, a cytologist who had succeeded him at Birkbeck College and who continued to work there.

Gwynne-Vaughan was active in the British Association beginning with the formation of its botanical section in 1895, and was secretary and recorder of the section for some years. He was also a member of the Royal Irish Academy, a fellow of the Linnean Society of London, and a fellow of the Royal Society of Edinburgh.

His first research, published as "On a New Case of Polystely in Dicotyledons" (*Annals of Botany* [1896]), reported a complete series of transitions between polystely and astely within the Nymphaeaceae. This was expanded and illustrated in a paper to the Linnean Society in 1897 which gave details of the morphology of the leaf and showed how the ontogeny of the individual leaf repeats the successive forms of the seedling leaves. His work on polystely in *Primula* (1897) showed that in this genus the gametostelic condition is more primitive than the dialystelic condition, and the apical region is not simplified. These new examples and interpretations extended the basic work that had been done on polystely by P. E. Van Tieghem.

Next Gwynne-Vaughan moved to an extensive work, "Observations on the Anatomy of Solenostelic Ferns" (1901–1903), in which he showed that the apparent segmentation of the stele is late development rather than a primitive feature, and he named broken-up portions of the central stele "meristeles." Development of the dictyostele in both young plants and lateral shoots are illustrated to show that they are similar and that the structure of the dictyostele is due to overlapping leaf gaps.

He also worked briefly on the morphology of *Equisetum.* He studied some of the Marattiales (tree ferns) and Algae and in 1908 he published his view that the xylem vessels of ferns had openings in the vertical walls. But work by F. Halft (1910) and N. Bancroft (1911) did not support this.

Gwynne-Vaughan's interest in the ontogeny of fern vessels was now extended to the study of the phylogeny of the fossil Osmundaceae, undertaken in a fruitful collaboration with the paleontologist Robert Kidston of Stirling. Through their research the structure of specimens from all over the world was traced back to the Permian. The sequence was sufficiently complete to show how anatomic changes correlated with successive geological strata, with progression from the protostele to the solenostele. This work, "On the Fossil Osmundaceae" (1907–1914), was published in a series of five papers by the Royal Society of Edinburgh, which awarded the MacDougall-Brisbane Medal to Gwynne-Vaughan in 1910.

BIBLIOGRAPHY

I. Original Works. Gwynne-Vaughan's first paper, "On a New Case of Polystely in Dicotyledons," is in *Annals of Botany,* **10** (1896), 288–291; the expanded version, "On Some Points in the Morphology and Anatomy of the Nymphaeaceae," is in *Transactions of the Linnean Society of London,* 2nd ser., **5** (1897), 287–229. Other works include "On Polystely in the Genus *Primula,*" in *Annals of Botany,* **11** (1897), 307–325; and "Observations on the Anatomy of Solenostelic Ferns," *ibid.,* **14** (1901), 71–98; **17** (1903), 689–742.

Practical Botany for Beginners (London–New York, 1902), written with F. O. Bower, is Gwynne-Vaughan's only book. The classic series of papers "On the Fossil Osmundaceae," written with R. Kidston, is in *Transactions of the Royal Society of Edinburgh,* **45** (1907), 759–780; **46** (1908), 213–232; **46** (1909), 651–667; **47** (1910), 455–477; **50** (1914), 469–480.

II. Secondary Literature. The most useful obituaries are F. O. Bower in *Proceedings of the Royal Society of Edinburgh,* **36** (1916), 334–339; and D. H. Scott in *Annals of Botany,* **30** (1916), i–xxiv. Gwynne-Vaughan's collection of anatomical slides is held in the botany department of the University of Glasgow, and a typescript catalogue (1920) of the collection is in the library of the British Museum (Natural History).

Diana M. Simpkins

GYLLENHAAL, LEONHARD (*b.* Ribbingsberg, Sweden, 3 December 1752; *d.* Höberg, near Skara, Sweden, 13 May 1840), *entomology.*

Gyllenhaal was a major in the army, a landed proprietor, and an amateur scientist who became one of Sweden's foremost authorities on Coleoptera (a distinction that he shared with another amateur, C. J. Schönherr). His interest in nature was apparent early; as a child he was an avid collector of natural specimens, including plants, insects, and minerals. As he progressively concentrated on Coleoptera, Gyllenhaal had a special building constructed in the gardens of Höberg, his estate in Västergötland, for the housing and study of his collection. He generously shared his specimens with other collectors and institutions.

Gyllenhaal's chief importance rests in one work, *Insecta Suecica descripta* (1808–1827). The title of this four-volume work would indicate that the author had intended a survey of all of Sweden's insects, or at least more of them than the beetles. This wide design was not realized, and the book is concerned entirely with Coleoptera—conceived on the highest level and executed in the most minute detail, it became known and used throughout the world.

Gyllenhaal became a member of the Royal Swedish Academy of Science in 1809 but published only one paper in its *Handlingar.* He was never closely involved with the Academy or its work, in part because of his age (he was fifty-seven at the time of his election) and because he did not like to leave Höberg, preferring to work undisturbed in his research building.

Gyllenhaal's extensive collection of Coleoptera is now in the Zoological Museum of the University of Uppsala. Because of the number of its type specimens it is of great value to researchers even today.

BIBLIOGRAPHY

Gyllenhaal's works are "Instrumenta cibaria insectorum aliquot Sueciae descripta," in *Nova acta Regiae Societatis scientiarum upsaliensis,* **6** (1799), 117–132; *Insecta Suecica* *descripta. Classis I. Coleoptera sive Eleutherata,* 4 vols.: I–III (Skara, 1808–1813), IV (Leipzig, 1827); and "Ammärkningar rörande ett av Carl de Geer under namn af Attelabus glaber beskrifvet insect," in *Kungliga Svenska vetenskapsakademiens handlingar* (1817), pp. 137–141. In addition, he contributed many articles on beetles to Schönherr's monumental *Synonymia insectorum* (Paris, 1833–1845).

BENGT-OLOF LANDIN

HAAK, THEODORE (*b.* Neuhausen, near Worms, Germany, 25 July 1605; *d.* London, England, May 1690), *learned correspondence, translation.*

Haak's father, Theodor, attended the University of Heidelberg; but little is known about him before he married Maria Tossanus, the daughter of the Reformed theologian Daniel Tossanus (Toussaint), who after escaping the St. Bartholomew's Day massacre in France became court preacher at Heidelberg and later professor of theology and rector of the university. Through the marriages of his mother's sisters, Haak was related to a cousin, the Reformed theologian at Leiden, Friedrich Spanheim (1600–1649), and to J. F. Schloer (an uncle), who was counselor of state to Frederick IV, the elector palatine. A grandson of Schloer's settled in England, where he became a fellow of the Royal Society and of the College of Physicians; it was at his house that Haak died. Thus Haak was closely related to Continental families who were important in the Reformed church, in high state office, and in the learned world at Heidelberg and Leiden. As a conveyor of knowledge and information between England and the Continent, Haak's long and active life shows the importance of these family connections.

During 1625–1626 Haak studied at Oxford and Cambridge, then returned to Germany. In 1628 or 1629 he took up residence at Gloucester Hall (now Worcester College), then one of the Calvinist centers at Oxford, where he studied theology and mathematics. He left at the end of 1631, was ordained deacon in 1632, and settled in London. During the next few years he was active in the Palatine Collections for the benefit of the exiled ministers of the Palatinate, spending part of his time in Heidelberg. Little is known about his life during the late 1630's until he settled permanently in England toward the end of 1638, but by 1635 he was in touch with Samuel Hartlib and his circle, probably through his cousins in the Schloer family. In 1656 Haak married Elizabeth Genne, who died in 1669. He and his family were naturalized British citizens late in 1656.

Given his religious views, Haak naturally sided with Parliament during the 1640's. He undertook a largely unsuccessful diplomatic mission to the Continent and Denmark (1643–1644) but later turned down similar offers, preferring to remain in London, where he performed important work as a translator, for example, to the secretary of state, John Thurloe. His services gained him a pension and financial independence. He declined an offer to become secretary to the elector, Charles Louis, at Heidelberg, accepting instead a position as his unofficial agent in London. At this time John Wilkins was the elector's private chaplain. In 1645 the Westminster Assembly, whose secretary was John Wallis, engaged Haak to make a translation of the *Dutch Bible and Annotations,* a work that had been prepared at the request of the Synod of Dort. This immense task was completed in 1657, after many interruptions.

In 1647 Haak wrote to Marin Mersenne that the proper use and enjoyment of knowledge lie in its free communication, so that it may serve the general good, in accordance with the divine intent. Twenty years later, in a letter that accompanied a copy of Thomas Sprat's *History of the Royal Society,* Haak wrote to Governor John Winthrop of Connecticut about the need for mankind to improve "the treasures God hath communicated to them so abundantly throughout all the world," thus "more and more reconciling the estrangedness of the minds of mankind amongst themselves, that they may be willing to listen to more and more and still better Truths and Union." He hoped that the *History* would not only "revive and quicken" the governor himself to mind his "engagement and interest," but that he in turn would "excite and animate many others also to consort and cooperate for the advancement of so universal a Benefit as the scope of this Society holds forth, and their endeavours promise to all the world." In these words Haak stated the aspirations that were shared by Hartlib's Comenian group and by the members of the early Royal Society, two groups which show considerable overlap. Natural philosophy, i.e., the study of God's manifest revelation in creation, is an ennobling activity; it leads to piety, peace, order, and a sense of community among all mankind. The study of natural philosophy is the crucial prerequisite for the successful achievement of Comenian pansophy.

When Comenius arrived in London in September 1641, he was received by Hartlib, Joachim Hübner, John Pell, John Dury, and Haak. At that time Haak had already served Hartlib and his Comenian scheme. Late in 1639, through his command of French, Haak initiated a correspondence with Mersenne, then widely known as the chief figure in a group that was dedicated to the discussion and exchange of learned and scientific knowledge. The first letter was accompanied by copies of Comenius' *Pansophiae prodromus* (1639) and Pell's *Idea of Mathematics,* and both authors now began to correspond directly with Mersenne. Unfortunately we do not have Haak's letters; but the nineteen letters from Mersenne to Haak, written within little more than a year, show a lively exchange of books and information on the sorts of subjects that interested Hartlib and that later figured both in the early scientific meetings in England before 1660 and in the Royal Society after that date.

There is therefore good reason to accept John Wallis' account, although written many years later, that it was Haak who at London in 1645—"if not sooner"—"gave the first occasion, and first suggested those meetings" which have the best claim to be considered the beginnings of "the Royal Society of London for improving Natural Knowledge." In May 1647 Haak resumed his correspondence with Mersenne, who had been traveling extensively in Italy and France for much of the time between October 1644 and the early part of 1647; four more letters followed until July 1648, although it is not clear whether Mersenne answered. (He was ill much of that time and died on 1 September 1648.) Haak's letters appear unmistakably related to the scientific meetings then being held in London, for he refers to experiments being performed by a group of which he was a member.

During the following years Haak was busy with translation and was in touch with Pell, Hartlib, and Dury, among others. Within a year of the first meeting of the as yet unnamed Royal Society, Haak was proposed for membership by John Wilkins, thus becoming an original member of the Society. Less than a year later, in late August 1662, he received a letter from his old friend, the Leiden philologist J. F. Gronovius, "expressing his high sense of the usefulness of the design of the Society" (Birch, I, 108). It was Gronovius who in 1639 had taken Haak's letter with Pell's and Comenius' books to Mersenne in Paris. Until the end of his life nearly thirty years later, Haak regularly attended the meetings of the Royal Society, from time to time serving on various committees. He made a few communications but was chiefly active in the promotion and maintenance of correspondence with the learned world abroad, especially in Germany. He proposed a considerable number of visiting Germans for membership, he translated letters, and about 1680 he acted as intermediary in the correspondence between Hooke and Leibniz.

During the last two decades of his life, Haak was very close to Hooke; for periods of several years they saw each other at least once a week, sometimes eating or drinking tea together, holding conferences with learned friends, and playing chess. Like Hooke, Haak appears to have taken a special and continuing interest in that old Comenian project, a universal language. Haak and Hooke also talked much about books and book purchases. Haak had known Milton since the late 1640's, when both did translation work for the Council of State. He began but seems never to have finished a translation of *Paradise Lost* into German.

Haak's own letters and writings tell us very little about himself, and our knowledge of his private life is sparse; but all his contemporaries agree that he was an exceptionally kind-hearted and generous man. By virtue of his work, beliefs, and activities, he charted a well-nigh archetypal course during a long life that covered most of the seventeenth century.

BIBLIOGRAPHY

I. Original Works. Among Haak's translations are German versions of two pieces by the Puritan divine, Daniel Dyke the Elder: *The Mystery of Self-Deceiving* (London, 1615), in German under the title *Nosce teipsum* (Basel, 1638); and a *Treatise of Repentance* (London, 1631), in German as *Nützliche Betrachtung . . . der wahren Busse,* first published with *Nosce teipsum* (Frankfurt, 1643). Both were several times reissued together in German. The important Mersenne correspondence is in Cornelis de Waard *et al.,* eds., *Correspondance de Marin Mersenne* (Paris, 1932–), VIII (1963), IX (1965), X (1967), XI (1970), and in future volumes. Two of the letters from 1647 and 1648 are in Harcourt Brown, *Scientific Organizations in Seventeenth Century France* (Baltimore, 1934), pp. 268–272. The letters to John Winthrop are in "Correspondence of the Founders of the Royal Society With Governor Winthrop of Connecticut," in *Proceedings of the Massachusetts Historical Society 1878,* **16** (1879), 206–251. The Haak-Hooke-Leibniz letters have so far been only partially published: see C. I. Gerhardt, ed., *Die philosophischen Schriften von G. W. Leibniz,* 7 vols. (Berlin, 1875–1890), VII, 16–20. There is also material relevant to this correspondence and to Haak in general in the letters of Johann von Gloxin to Leibniz in Leibniz, *Sämtliche Schriften und Briefe,* 1st ser., *Allgemeiner und historischer Briefwechsel,* III (Leipzig, 1938). Haak's activity in the Royal Society can be traced in Thomas Birch, *The History of the Royal Society,* 4 vols. (London, 1756–1757).

II. Secondary Literature. There is an excellent monograph by Pamela R. Barnett, *Theodore Haak, F.R.S. (1605–1690)* (The Hague, 1962). The bibliography gives a full listing of the primary sources, both unprinted and printed, and of the secondary material; this book also presents the first printing of the translation of *Paradise Lost,* which covers bks. I–III and the opening lines of bk. IV. Miss Barnett's solid scholarship and good insight into the period transcend much recent but inferior writing on the events in which Haak was involved. Her "Theodore Haak and the Early Years of the Royal Society," in *Annals of Science,* **13** (Dec. 1957), 205–218, is also commendable.

Useful and suggestive information about the German intellectual milieus with which Haak was in touch will be found in Leopold Magon, "Die drei ersten deutschen Versuche einer Übersetzung von Miltons 'Paradise Lost,'" in Karl Bischoff, ed., *Gedenkschrift für Ferdinand Josef Schneider* (Weimar, 1956), pp. 39–82. Among much inferior writing on the subject, R. H. Syfret, "The Origins of the Royal Society," in *Notes and Records. Royal Society of London,* **5** (Apr. 1948), 75–137, still stands out as basically sound and reliable. Birch's *History* should be supplemented by Henry W. Robinson and Walter Adams, eds., *The Diary of Robert Hooke 1672–1680* (London, 1935); and R. T. Gunther, *Early Science in Oxford. The Life and Work of Robert Hooke,* X (Oxford, 1935), pp. 69–265, the diary from Nov. 1688 to Aug. 1693.

HANS AARSLEFF

HAAR, ALFRÉD (*b.* Budapest, Hungary, 11 October 1885; *d.* Szeged, Hungary, 16 March 1933), *mathematics.*

Alfréd was the son of Ignatz Haar and Emma Fuchs. While a student at the Gymnasium in Budapest, he was a collaborator on a mathematical journal for high schools and in 1903, his last year at the Gymnasium, he won first prize in the Eötvös contest in mathematics. He had started studying chemistry, but his success in the contest induced him to switch to mathematics. From 1904 he studied in Göttingen; in 1909 he took his Ph.D. degree as a student of D. Hilbert and that year became a *Privatdozent* at the University of Göttingen.

In 1912, after a short time at the Technical University of Zurich, he returned to Hungary and succeeded L. Fejér at Klausenburg University, first as extraordinary professor and then, from 1917, as ordinary professor. When Klausenburg became Rumanian he went to Budapest with his colleague F. Riesz. Together they continued their activity at Szeged University, where in 1920 they founded *Acta scientiarum mathematicarum,* a journal of great reputation. In 1931 Haar became a corresponding member of the Hungarian Academy of Sciences.

Haar did work in analysis. Although not formally abstract, it is so close to the abstract method that it still looks modern. His doctoral thesis had dealt with orthogonal systems of functions. Twenty years later he returned to the same subject. Haar first extended what was known on divergence, summation, and oscillation for the Fourier system to other orthogonal systems, in particular to solutions of Sturm-Liouville problems. He discovered a curious orthogonal system according to which every continuous function can be developed into an everywhere converging series; its elements are discontinuous functions admitting, at most, three values. Later he became interested in multiplicative relations of orthogonal systems and characterized their multiplication tables. This research led him to the character theory of commutative groups as a precursor of Pontryagin on duality.

In complex functions Haar did work on splitting lines of singularities and on asymptotics. As one of the first applications of Hilbert's integral methods in equations and of Dirichlet's principle, Haar in 1907 studied the partial differential equation $\triangle\triangle u = 0$; with T. von Kármán he put this method to use in elasticity theory. Haar also wrote a number of papers on Chebyshev approximations and linear inequalities.

Two of Haar's shorter papers (1927–1928) that greatly influenced problems and methods in partial differential equations in the 1930's concern the equation

$$F(x,y,z,p,q) = 0,$$

which is usually dealt with by the method of characteristics. Since this method presupposes the existence of the second instead of the first derivative of the solutions, one may ask whether there exist solutions which escape the methods of characteristics. Under rather broad conditions, Haar answered this question in the negative.

Haar's most important contribution to variational calculus (1917–1919) features an analogous principle, Haar's lemma, an extension of Paul du Bois-Reymond's to double integrals: If

$$\iint_B \left(u\frac{\partial f}{\partial x} + v\frac{\partial f}{\partial y} \right) dxdy = 0$$

for all continuously differentiable f which vanish on the boundary of B, then there is a w such that

$$\frac{\partial w}{\partial x} = -v, \frac{\partial w}{\partial y} = u.$$

Haar's lemma allows one to deal with variational problems like

$$\iint_B f(p,q,x,y,z)\, dxdy = \text{minimum},$$

without supplementary assumptions on the second derivative of the unknown function z. He applied his lemma to variational problems like Plateau's. A multitude of papers by others show the influence this lemma exerted on the whole area of variational calculus.

The notion to which Haar's name is most firmly attached is Haar's measure on groups. In 1932 Haar showed, by a bold direct approach, that every locally compact group possesses an invariant measure which assigns positive numbers to all open sets. An immediate consequence of this theorem was the analytic character of compact groups (J. Von Neumann). It was somewhat later applied to locally compact Abelian groups by Pontryagin. The theorem is now one of the cornerstones of those areas of mathematics where algebra and topology meet.

BIBLIOGRAPHY

See *Alfréd Haar: Gesammelte Arbeiten* (Budapest, 1959), B.S.-Nagy, ed.

H. FREUDENTHAL

HAAS, ARTHUR ERICH (*b.* Brünn, Moravia [now Brno, Czechoslovakia], 30 April 1884; *d.* Chicago, Illinois, 20 February 1941), *physics, history of physics.*

After studying physics at Vienna and Göttingen, Haas received his doctorate at Vienna in 1906 and then turned enthusiastically to the history of physics. In order to qualify as a lecturer he submitted to the Philosophy Faculty of the University of Vienna a dissertation on the history of the energy principle. His paper was the cause of considerable puzzlement to the physicists who were responsible for passing an initial judgment on it, and it was decided that he should prepare an additional work in pure physics.

In fulfilling the faculty's assignment, Haas followed the latest publications in physics and thereby came across the unsolved problem of black-body radiation toward the end of 1909. He studied J. J. Thomson's *Electricity and Matter,* the contents of which are reflections on atomic structure, while reading an essay by Wilhelm Wien in the *Encyclopädie der mathematischen Wissenschaften* in which the suggestion was put forth that the energy element "can be derived from a universal property of the atom." Seizing upon this idea, Haas became the first to apply a quantum formula to the clarification of atomic structure. In the process he substituted real atoms for the more formal than physical Planck oscillators in the radiation cavity.

Haas's quantum rule $E_{pot} = h\nu$ agrees, for the ground state, with the condition later stated by Niels Bohr; and thus Haas obtained the correct "Bohr" radius of the hydrogen atom. But, characteristically, he wrote down only the equation solved for the action quantum, i.e., $h = 2\pi e\sqrt{r \cdot m}$. Therefore, like Wien he considered the dimensions of the atom as fundamental, from which the action quantum can then be derived. Within a numerical factor of eight Haas also correctly derived the Rydberg constant from the action quantum h, the velocity of light c, and the fundamental magnitudes of the electron, e and m. He achieved this relation by a very formal second hypothesis, namely, that the frequency derived from his quantum rule corresponds with the constant of Balmer's equation.

Although Haas's theorem failed to take into account the excited states—and therefore the connection with spectroscopic data—it was nevertheless a remarkable forerunner of Bohr's atomic theory. Yet in February 1910 Haas's ideas were termed a "carnival joke" by Viennese physicists and only slowly found recognition.

In 1913, through the intervention of Karl Sudhoff, Haas became an associate professor of the history of science at the University of Leipzig. Sudhoff, then head of the German historians of science and medicine, had been favorably impressed by Haas's first address at Cologne in 1908 and also managed to get him the editorship of volume V of Poggendorff. At the end of World War I Haas returned to Vienna, where he gradually turned from the history of physics to physics. In 1920 he calculated—independently of F. Wheeler Loomis and Adolf Kratzer—the correct formulas for the isotope effect in rotational spectra. After several offers of guest lectureships he finally immigrated to the United States in 1935. From 1936 until his death he was professor of physics at the University of Notre Dame.

Haas's work in the history of physics was inspired by his interest in older as well as modern theories. (He was influenced by Mach's and Ostwald's interest in the history of science.) Haas possessed the "conviction that no other method is as suited as the historical for facilitating the understanding of physical principles and for clarifying and deepening the knowledge of their significance" (*Die Grundgleichungen der Mechanik . . .*, preface). His numerous books written from this point of view, often based on his lectures and addresses, are masterpieces of clear exposition which were widely disseminated and translated into many languages.

BIBLIOGRAPHY

I. ORIGINAL WORKS. Haas's writings include *Die Entwicklungsgeschichte des Satzes von der Erhaltung der Kraft* (Vienna, 1909), his *Habilitationsschrift; Die Grundgleichungen der Mechanik, dargestellt auf Grund der geschichtlichen Entwicklung* (Leipzig, 1914); *Einführung in die theoretische Physik,* 2 vols. (Leipzig, 1919–1921); and *Der erste Quantenansatz für das Atom* (Stuttgart, 1965), repr. of 1910 papers with extensive biography and bibliography by A. Hermann.

Among Haas's works in the history of physics are *Die Entwicklungsgeschichte des Satzes von der Erhaltung der Kraft* (Vienna, 1909); *Der Geist des Hellenentums in der modernen Physik* (Leipzig, 1914); and "Die ältesten Beobachtungen auf dem Gebiet der Dioptrik," in *Archiv für die Geschichte der Naturwissenschaften und der Technik,* **9** (1920–1922), 108–111.

II. SECONDARY LITERATURE. See A. Hermann, "Arthur Erich Haas und der erste Quantenansatz für das Atom," in *Sudhoffs Archiv für Geschichte der Medizin und der Naturwissenschaften,* **49** (1965), 255–268; and *Genesis of Quantum Theory* (Cambridge, Mass., 1971), ch. 5.

A. HERMANN

HAAS, WANDER JOHANNES DE (*b.* Lisse, Netherlands, 2 March 1878; *d.* Bilthoven, Netherlands, 26 April 1960), *physics.*

De Haas was first educated to be a notary, but then studied physics at Leiden University where he was assistant to H. Kamerlingh Onnes. In 1912 he wrote his doctoral thesis, "On the Compressibility of Hydrogen Gas at Low Temperatures." After working in Berlin and at the Physikalische Reichsanstalt in Potsdam, de Haas was assistant to H. A. Lorentz, his father-in-law and at that time director of the physics division in the Teyler Institute at Haarlem. De Haas first became a professor of physics at the Technische Hogeschool, Delft, in 1917. He left for Groningen University in 1922, and from 1924 to 1948 was professor at Leiden University.

Together with W. H. Keesom, de Haas was director of the Kamerlingh Onnes Laboratory of Experimental Physics, initially one of the few laboratories in the world where low-temperature work was systematically carried out. In 1922 he became a member of the Royal Netherlands Academy of Science and Letters at Amsterdam. His health was never very good, but with the help of his wife (a theoretical physicist) he was able to maintain his international scientific contacts and to execute the duties of his laboratory directorate. Although he specialized in magnetism, de Haas found time to dabble in many different branches of physics.

The general trend of de Haas's work is shown by his early work at Berlin. There, in 1915, he performed an experiment suggested by Einstein, known as the Einstein-de Haas effect: the sudden magnetization of a suspended iron cylinder in a vertical solenoid causes a momentary torque in the cylinder. The theoretical foundation for this effect is that the unidirectional aligning of the spinning electrons in the magnetic field also aligns their mechanical moments, resulting in a torque pulse.

The experimental results of later scientists indicated that the ratio of the magnetic to the mechanical moment differed by a factor of two from the original classic expectation. This is the fundamental "half-integer quantization" for the spinning electron, as compared to the integer quantization for the orbital moment, a difference which runs through the whole development of modern atomic physics.

Pioneering with simple apparatus was de Haas's favorite conception of experimental physics. But he was also aware of the need for organization and routine techniques that had been introduced in the Leiden laboratory by Kamerlingh Onnes. Together with E. C. Wiersma, de Haas was a leader in the production of extremely low temperatures by adiabatic demagnetization of precooled magnetized material. Other lines of research led to the so-called Van Alphen-de Haas effect on the anomalous behavior of the resistance of a metal crystal in a magnetic field, and to magneto-optic researches on crystals, mainly done and published by Jean Becquerel. During World War II de Haas succeeded in preventing a large amount of uranium ore from being taken to Germany. After the war this uranium was useful in starting the Netherlands-Norway joint establishment for nuclear energy at Kjeller (Norway).

BIBLIOGRAPHY

For information on the Einstein-de Haas effect see "Experimenteller Nachweis der Ampèreschen Molekularströme," in *Verhandlungen der Deutschen physikalischen Gessellschaft,* **17** (1915), 152–170, written with Einstein. Also see E. Beck, *Annalen der Physik,* **60** (1919), 109. Other of de Haas's publications are in *Proceedings. K. Nederlandse akademie van wetenschappen* and in *Physica* (The Hague). A biography of de Haas may be found in *Jaarboek van de K. Akademie van wetenschappen gevestigd te Amsterdam* (1959–1960), 300; on the World War II period see also S. A. Goudsmit, *ALSOS* (Schuman, N.Y., 1947), where de Haas is called "Professor X."

J. A. PRINS

HAAST, JOHANN FRANZ JULIUS VON (*b.* Bonn, Germany, 1 May 1822; *d.* Christchurch, New Zealand, 16 August 1887), *geology.*

Haast was the only son of Mathias Haast and Anna Ruth. His father, a merchant, was elected burgomaster at Bonn. Haast studied geology and mineralogy at the University of Bonn, where, tradition says, he rescued the prince consort from drowning in the Rhine. Although he did not graduate, he worked for a time with August Krantz, a mineral dealer—leading a somewhat undistinguished life gripped by wanderlust and rendered more unsettled by the early death of his first wife, Antonia Schmitt.

Haast's opportunity for a more rewarding career came in 1858 with his appointment to advise an English shipping firm on the prospects of encouraging German immigration to New Zealand. Exploration and scientific appraisal of resources was exactly the kind of work for which Haast longed. Immediately after fulfilling his contract to the shipping firm, he joined the explorer Ferdinand von Hochstetter in his explorations of New Zealand (1859), traveling extensively throughout both islands. Among the areas he visited was a volcanic wilderness closely controlled by the Maoris and, because of later outbursts of fighting, not revisited by Europeans for several decades.

Through the fame attendant on his journeys with Hochstetter, Haast was engaged to make a topographic and geologic survey of the west coast of South Island by the Nelson provincial government. He and his team discovered a new coalfield at Westport and reported on other coal and gold resources (1860).

A crisis arose at the South Island port of Lyttelton, situated in the heart of an ancient volcano which the authorities had hoped to pierce by a tunnel to the nearby city of Christchurch. The contractors had struck a hard lava flow and refused to go on. Hurrying to the scene, Haast made a rapid and competent survey of the geology to convince the workers that the amount of hard rock was limited. Work was resumed on the tunnel, which provided a vital link between city and port.

Haast was rewarded with appointment as geologist for Canterbury province (1861). He was naturalized as a British subject in the same year and made Christchurch his home. There he married Mary Dobson, daughter of the provincial engineer Edward Dobson and, like his first wife, musically gifted. Haast contributed enormously to the intellectual and cultural life of Christchurch. He founded the Philosophical Institute of Canterbury in 1862, the Canterbury Museum in 1870, the Canterbury Collegiate Union with Bishop Harper, later to become the University of Canterbury, and the Imperial Institute. When the provincial geological survey was concluded, he became director of the Canterbury Museum and professor of geology at the university. He also served as a member of the senate of the University of New Zealand.

A pioneer scientist in a largely unexplored land, Haast had interests that ranged far and wide. He collected new plants from the Southern Alps; he inquired into problems of early human settlement of New Zealand; and he gained great immediate attention through his discovery of bones of the gigantic moa and a giant extinct eagle which he named *Harpagornis.* Most of his research was devoted to geologic and topographic surveys of Canterbury province and the west coast. These activities were closely related, and he never perpetrated the modern error of divorcing current geological processes from the study of ancient rocks. Haast thus accurately recognized that most of the Canterbury plains were composed of rubble fans, formed as glacial outwash, rather than accepting more fanciful theories, then current, involving torrential deposition as drift.

A forceful, ebullient man who was well aware of having "made himself" and willing, according to C. A. Fleming, to prove and underline the point by festooning himself with honors, Haast naturally made both scientific and personal enemies. He scathingly condemned the brilliant government geologist Alexander McKay for "poaching" his moa bones, and he repeatedly resisted and opposed with his provincial experience geological theories advanced on a national scale by the small but excellent team of geologists in the New Zealand Geological Survey. As a result he found a natural ally in another great geologist, F. W. Hutton, in opposing the towering figure of nineteenth-century science in New Zealand, Sir James Hector. It would be difficult to allow that Haast matched either of these two contemporaries in scientific achievement, and politically and administratively he was much less significant than Hector. His ebullience, a certain lack of originality, and his geographically circumscribed field of work told against him. Yet some of his concepts live on, and to this day there is a recognizable Canterbury outlook on geology established by Haast.

BIBLIOGRAPHY

I. Original Works. Haast's writings include *Report of a Topographical and Geological Examination of the Western Districts of the Nelson Province* (Nelson, N.Z., 1861); "Notes on the Geology of the Province of Canterbury," in *Government Gazette, Province of Canterbury,* **9** (1862); "Report on the Geology of the Malvern Hills, Canterbury," in *Report. Geological Explorations,* **7** (1871–1872), 1–88; and *Geology of the Provinces of Canterbury and Westland, New Zealand* (Christchurch, 1879).

II. Secondary Literature. On Haast or his work, see P. Burton, *The New Zealand Geological Survey 1865–1965,* Information Series, Department of Scientific and Industrial Research no. 52 (1965); C. A. Fleming, "Haast, Sir Julius von, K.C.M.G., F.R.S. (1822–87)," in A. H. McLintock, ed., *An Encylopaedia of New Zealand,* I (Wellington, 1966), 892–893; H. F. von Haast, *The Life and Times of Sir Julius von Haast* (New Plymouth, N.Z., 1948); and J. B. Waterhouse, "A Historical Survey of the pre-Cretaceous Geology of New Zealand," in *New Zealand Journal of Geology and*

Geophysics, **8** (1965), pt. 6, 931–998; **10** (1967), pt. 4, 923–981.

J. B. WATERHOUSE

ḤABASH AL-ḤĀSIB, AḤMAD IBN ʿABDALLĀH AL-MARWAZĪ (*b.* Marw, Turkestan [now Mary, Turkmen S.S.R.]; *d.* 864–874), *trigonometry, astronomy.*

Little is known of Ḥabash's life and family. He worked at Baghdad as astronomer under the ʿAbbāsid caliphs al-Maʾmūn and al-Muʿtaṣim, but he may not have belonged to the small group that collaborated in the Mumtaḥan observations. He made observations from 825 to 835 in Baghdad. Abū Jaʿfar ibn Ḥabash, the son of Ḥabash, was also a distinguished astronomer and an instrument maker.

Works. The biographers Ibn al-Nadīm, Ibn al-Qifṭī, and Ḥājjī Khalīfa ascribe the following works to Ḥabash:

1. A reworking of the *Sindhind.*
2. The *Mumtaḥan Zīj,* the best known of his works, which relies on Ptolemy and is based on his own observations. Ibn Yūnus called it *al-Qānūn* ("The Canon").
3. The *Shāh Zīj,* the shortest of his *ziyajāt.*
4. The *Damascene Zīj.*
5. The *Maʾmūnī Zīj* (or *Arabic Zīj*). This and the *Damascene Zīj* are based on the Hijra calendar rather than on the Yazdigird or Seleucid eras.
6. On the *Rukhāmāt* and Measurements.
7. On the Celestial Spheres.
8. On Astrolabes.
9. On the Oblique and Perpendicular Planes.
10. On the Distances of the Stars.

Since not all of these works are extant, it is almost impossible to determine how many *ziyajāt* Ḥabash wrote and their titles. Two manuscripts on the tables of Ḥabash are preserved, one in Istanbul (Yeni Cami, no. 784) and the other in Berlin (no. 5750). These are not copies of his original works. There has been criticism of the Yeni Cami copy, suggesting that it is a revision of Ḥabash's *zīj* by Kūshyār ibn Labbān. In one way or another the introduction and the passages have come to us in their original forms and can be used, as can the Berlin manuscript, as the sources on Ḥabash.

Trigonometry. Ḥabash's trigonometric contributions are very important.

Sines. In the *Sūrya-Siddhānta* (A.D. 400) a table of half chords is given. A special name for the function which we call the sine is first found in the works of Āryabhaṭa I (A.D. 500). Besides half chord he also uses the term *jya* or *jiva.* In the Islamic world this word was transcribed as *jayb.* Al-Khwārizmī (*ca.* 825) was the first to prepare a table of sines. Ḥabash followed him by constructing such a table for

$$\theta = 0;0°, 0;15°, 0;30°, 0;45°, 1;0° \cdots 90;0°.$$

Versed sine. Among the trigonometric functions the versed sine (versine) also attracted attention. We know that it was mentioned in the *Sūrya-Siddhānta,* and a table for the versed sine is given in Āryabhaṭa. In Islam astronomers used special names to distinguish the versed sine, such as *jayb maʿkūs* (used by Ḥabash), *jayb mankūs* (used by al-Khwārizmī), and *sahm.* Ḥabash may be the first who clearly defined the sine and the versed sine as follows: "A perpendicular from the circumference to the diameter is the sine (*jayb mabsūṭ*) of the arc between the diameter and the perpendicular; the distance between the circumference and the perpendicular upon the diameter is the versed sine (*jayb maʿkūs*) of the abovementioned arc." He showed that if $A < 90°$, the versed sine $= 60^P - \cos A = 1 - \cos A$; and if $A > 90°$, the versed sine$= 60^P + \cos A = 1 + \cos A$. Also, if $A < 90°$, the versed sine $<$ sine; if $A > 90°$, the versed sine $>$ sine; and if $A = 90°$, the versed sine $=$ sine.

Tangent. The *Sūrya-Siddhānta* and other Hindu works mention the shadows, particularly in connection with astronomy. Ḥabash seems to have been the first to compile a table of tangents for

$$\theta = 0;0°, 0;30°, 1;0° \cdots 90;0°.$$

The function of *umbra extensa* (the length of shadow) is defined as

$$h = P\frac{\cos h}{\sin h},$$

$P =$ the length of gnomon. For the computation of the *umbra extensa* from the altitude of the sun, he gives the following steps (see Figure 1):

$$\frac{KR}{P} = \frac{RO}{S}$$

$$KR = \sin h$$
$$RO = \cos h$$
$$P = 12$$
$$S = \textit{umbra extensa}$$

$$\frac{\sin h}{12} = \frac{\cos h}{S}$$

$$S = \frac{\cos h}{\sin h}12.$$

In addition to finding the *umbra extensa* from the altitude of the sun, Ḥabash presents the following equations:

ḤABASH AL-ḤĀSIB

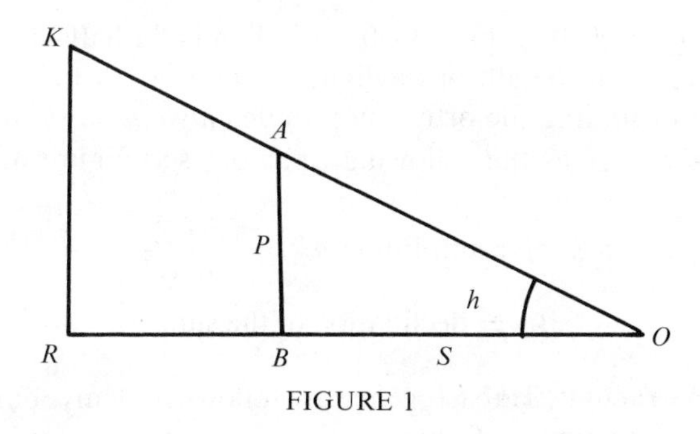

FIGURE 1

$$\text{hypotenuse} = \sqrt{S^2 + P^2}.$$

$$\sin h = \frac{P}{\sqrt{S^2 + P^2}} \cong.$$

Spherical Astronomy. For the solution of problems in spherical astronomy, transformations of coordinates, time measurements, and many other problems, Ḥabash gives astronomical tables of functions which are standard for all *ziyajāt*.

He gives the general rule for calculating the declination of the sun (the first declination, *al-mayl al-awwal*) (see Figure 2):

$$\sin \delta\odot = \sin \varepsilon \cdot \sin \lambda$$

$$\text{obliquity of ecliptic } \varepsilon = 23;35°.$$

The declination depends not only on λ but also on the value of the obliquity of the ecliptic.

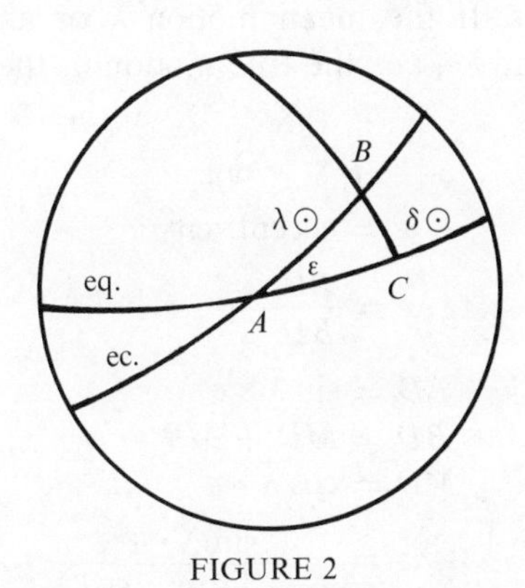

FIGURE 2

The culmination of the sun is defined. (See Figure 3.) If the declination of the sun is northern, $h = (90 - \phi) + \delta\odot$. If the declination of the sun is southern, $h = (90 - \phi) - \delta\odot$.

The time of day, measured from sunrise, is proportional to the altitude of the sun, i.e., the "arc of revolution" (*al-dā᾿ir min al-falak*). Islamic astronomers gave many trigonometric functions showing the relations between the time and the altitude of the sun. The first exact solution was given by Ḥabash and proved by Abu᾿l-Wafā᾿ and al-Bīrūnī. This function

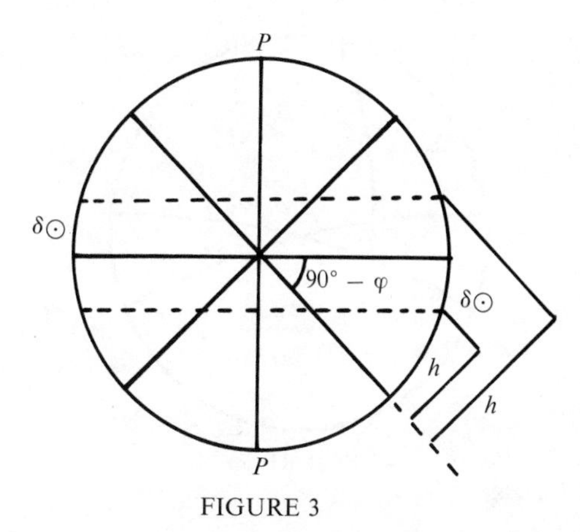

FIGURE 3

is equivalent to the function given by Brahmagupta in his *Khaṇḍakhādyaka:*

$$\text{vers } t = \text{vers } P - \frac{\sin h \cdot \text{vers } P}{\sin \text{alt} \cdot \text{merid}},$$

where P = half of the length of daylight
h = altitude of the sun
t = time
vers P = day sine (*jayb al-nahār;* Sanskrit, *antyā*).

Then he computes the altitude of the sun from the time:

$$\sin h = \frac{(\text{vers } P - \text{vers } t) \sin \text{alt} \cdot \text{merid.}}{\text{vers } P}.$$

Ḥabash calculates the length of daylight, i.e., equation of daylight (*taʿdīl al-nahār*), which al-Khwārizmī calls the ascensional difference. (See Figure 4.)

$$\frac{KD}{DG} = \frac{KL}{LM} \cdot \frac{ME}{EG}$$

$$\frac{\sin \phi}{\cos \phi} = \frac{\cos \delta\odot}{\sin \delta\odot} \cdot \frac{\sin ME}{R}$$

$$\sin ME = R \frac{\sin \phi \cdot \sin \delta\odot}{\cos \phi \cdot \cos \delta\odot}$$

$$\sin ME = R \, tg \, \phi \cdot tg \, \delta\odot.$$

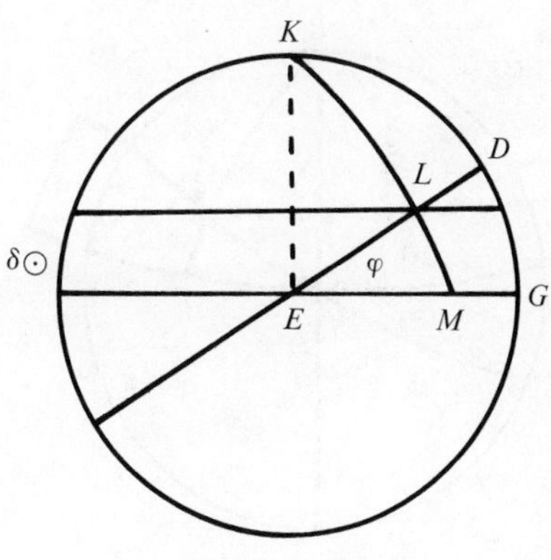

FIGURE 4

ḤABASH AL-ḤĀSIB

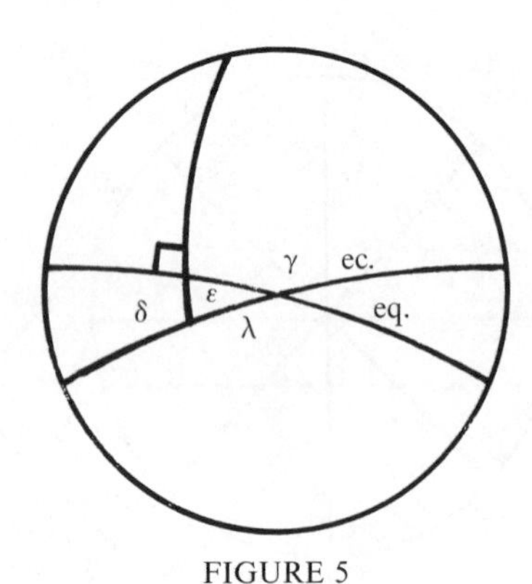

FIGURE 5

He shows that if the declination of the sun is northern, the length of daylight = the equation of daylight + 90°; and if the declination of the sun is southern, the length of daylight = 90° − the equation of daylight. The equation of daylight is tabulated for the planets, the sun, and the moon. With the aid of this table one can easily find the arc of daylight.

The ascensions (*maṭāli' al-burūj*) or rising times in the right sphere (*al-falak al-mustaqīm*), i.e., right ascension, is defined (see Figure 5) as

$$\frac{\sin \lambda \cdot \cos \varepsilon}{\cos \delta}.$$

Ḥabash prepared such tables because of the right ascension's astrological importance.

The ascension for a particular latitude is called the oblique ascension. Ḥabash showed that if the arbitrary point P on the ecliptic is between the vernal and autumnal equinoxes, the right ascension − 1/2 equation of daylight = oblique ascension; and if it is between the autumnal and vernal equinoxes, the right ascension + 1/2 equation of daylight = oblique ascension.

Ḥabash prepared tables for the seven climates. According to him the first climate (*iqlīm*) was that portion of the northern hemisphere in which

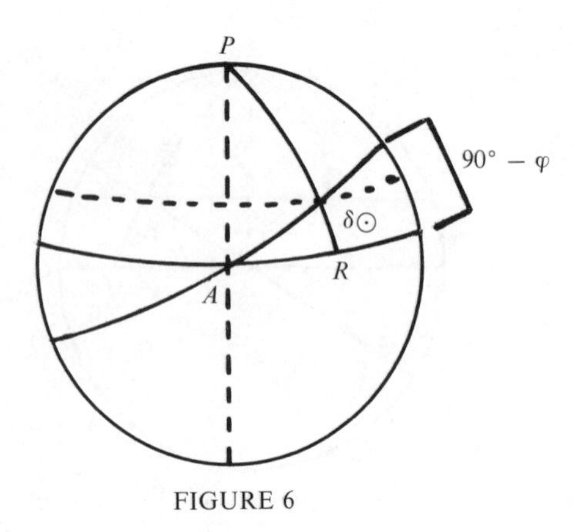

FIGURE 6

$13 \leq$ max. $D \leq 13 - 0.50$, i.e., a band of a half-hour advance in length of daylight.

For finding the ortive amplitude (*jayb al-mashriq*) Ḥabash gives the following function (see Figure 6):

$$\text{Ortive amplitude } AR = \frac{\sin \delta\odot}{\cos \phi}$$

$$\delta\odot = \text{declination of the sun.}$$

Astronomy. Ḥabash generally follows Ptolemy, but some sections of his work are distinctly non-Ptolemaic.

Theory of the Sun. Ḥabash compiled tables of mean motion of the sun for 1, 31, 61, 91, ··· 691 and 1, 2, 3, 4, 5, ··· 30 Hijra years; for 1, 2, 3, 4, ··· 29 days; for 1, 2, 3, 4, ··· 24 hours; and for 10, 20, 30, ··· 60 minutes. The mean motion of the sun is 384;55,14° per Hijra year and 0;59,8° per day (the value given in the *Almagest*). He computed the eccentricity of the sun 2^P5^1 (2; 1°).

Ḥabash divided half of the ecliptic into eighteen parts, each part called *kardaja.* The Arabic-Persian term *kardaja* (pl., *kardajāt*) is usually derived from the Sanskrit *kramajya.* It seems to have stood for a unit length of arc. He also prepared equation tables of the sun (*ta'dīl al-shams*) for each degree of anomaly.

Methods for the calculation of the equation of the sun were given by Ḥabash. This classical procedure was given in the *Almagest* and followed by the Islamic astronomers. If the mean motion $\bar{\lambda}$ or anomaly is given, to find λe, i.e., the true motion of the sun (see Figure 7):

$$\text{If } \bar{\lambda} < 90^\circ$$

$$e = \text{eccentricity}$$

$$\text{tg } \lambda e = \frac{ED}{BD}$$

$$ED = \sin \bar{\lambda} \cdot e$$

$$BD = MD + MB$$

$$MD = \cos \bar{\lambda} \cdot e$$

$$\text{tg } \lambda e = \frac{\sin \bar{\lambda} \cdot e}{\cos \bar{\lambda} \cdot e + 60^P}.$$

The converse of the problem, i.e., given λ, determine $\bar{\lambda}$, is set out. This equation gives the following approximate solution (see Figure 8):

$$\text{If } \lambda < 90^\circ$$

$$\text{tg } \lambda e = \frac{ED}{BD}$$

$$ED = \sin \lambda \cdot e.$$

Since the angle λe is small, Ḥabash supposed $BD = BC$, i.e., 60^P:

ḤABASH AL-ḤĀSIB

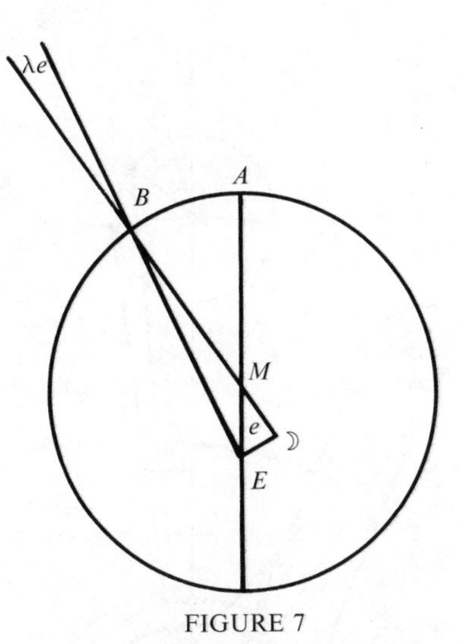

FIGURE 7

$$BE = 60^P - EC$$
$$EC = \cos\lambda \cdot e$$
$$BE = 60^P - \cos\lambda \cdot e$$
$$\operatorname{tg}\lambda e = \frac{e\sin\lambda}{60^P - (e\cdot\cos\lambda)}.$$

Ḥabash gave another rule to solve the above problem:

$$\sin\lambda e = \frac{ED}{60^P}$$
$$ED = \sin\lambda \cdot e$$
$$\sin\lambda e = \frac{\sin\lambda \cdot e}{60^P}.$$

This function is correct, but the independent variable is λ rather than $\bar{\lambda}$. If $\bar{\lambda}$ replaces λ, this will lead to the equation $\lambda = \lambda - e\sin\lambda$, which is known as Kepler's equation. The equivalent of this equation is found in Tamil astronomy in south India.

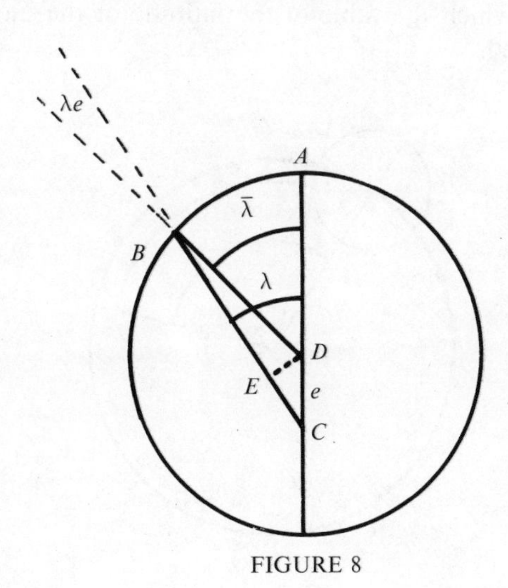

FIGURE 8

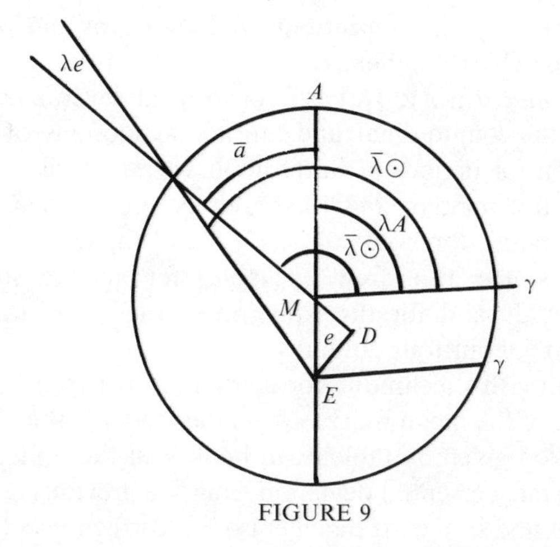

FIGURE 9

From the mean position Ḥabash found the true position of the sun (see Figure 9):

If $\quad \bar{\lambda}\odot > \lambda A$

and $\quad \bar{\lambda}\odot - \lambda A < 90°$,

anomaly $\bar{a} = \bar{\lambda}\odot - \lambda \mathrm{A}$,

where $\quad \bar{\lambda}\odot$ = mean longitude of the sun

and $\quad \lambda A$ = longitude of the apogee.

Thus, $\quad \operatorname{tg}\lambda e = \dfrac{\sin\bar{a}\cdot e}{\cos\bar{a}\cdot e + 60^P}$

and $\quad \bar{\lambda}\odot = \lambda e$ = true position of the sun.

He also computed from the true position the mean position of the sun (see Figure 10). The true position B, i.e., λB, is given:

$$\lambda B - \lambda A = \angle AEB = a$$
$$\sin\lambda e = \frac{\sin a\cdot e}{60^P}$$
$$\bar{\lambda}B = \lambda B + \lambda e.$$

By applying these methods Ḥabash calculated the

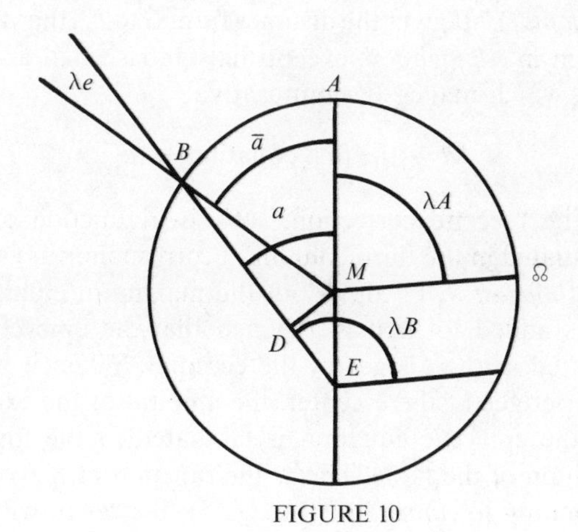

FIGURE 10

ḤABASH AL-ḤĀSIB

entrance of the sun into the zodiacal signs and compiled tables for them.

Lunar Theory. Ḥabash constructed several tables for the longitudinal and latitudinal motions of the moon for periods of thirty lunar years, for years, for months, for days (13;10,35°, the value given in the *Almagest*), for hours (0;32,27°), and for fractions of hours. He also drew up tables for general lunar anomaly and for the equation of the moon (*taʿdīl al-qamar*) in four columns.

Ḥabash's technique for computing the true longitude of the moon was based on the model of the lunar motion given by Ptolemy in book V of the *Almagest.* The most essential deviation from the previous tables consisted in the arrangement of all corrections of the mean positions so that they are never negative. As Neugebauer remarks, this constituted a great practical advantage over the Ptolemaic method. (See Figure 11.)

The true place of the moon is determined as follows:

$M =$ movable center of eccenter
$\delta_0 =$ apparent radius of epicyle when in apogee of eccenter
$G =$ "true apogee" of epicycle
$F =$ "mean apogee" of epicycle
$W_0 =$ maximum angular distance possible between F and G
$F_0 =$ point on the epicycle such that $F = W_0$
$\bar{a} =$ "mean anomaly" of moon counted from F_0
$W =$ angular distance between G and F_0
$a = \bar{a} + w_1$, "true anomaly" counted from G
$\lambda =$ "true longitude" of the moon counted from γ_0.

The "first correction," W_1:

$$\bar{\lambda}\odot - \bar{\lambda}☾ = K$$

The function of K is tabulated in the first column of the table, called "equation of the moon (*taʿdīl al-qamar*)." It gives the distance from G to F_0, the value given in *Almagest* V, except that Ḥabash had added W_0, which makes W nonnegative.

$$\bar{a} + \text{the first equation} = a.$$

The "second correction," W_2, is a function of a, tabulated in the third column. It corresponds in value to *Almagest* V, 8, col. 4, but the maximum equation δ_0 is added to it. It is assumed that the epicycle is located at the apogee of the eccenter. When it is in the perigee of the eccenter, the amount of the excess of the epicyclic equation is tabulated in the fourth column of the table. This is the function of K (corresponding to *Almagest* V, 8, col. 5); the result will be

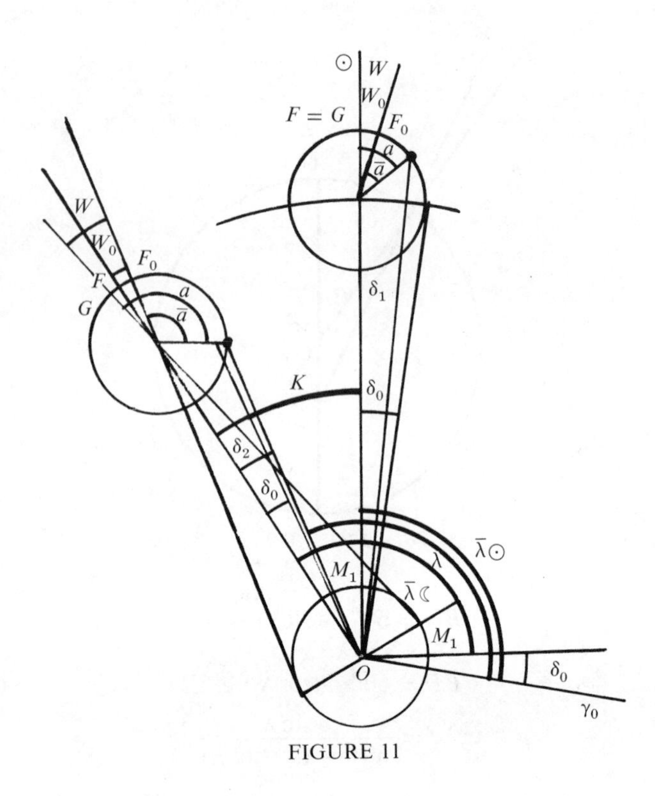

FIGURE 11

obtained by multiplying the value of the fourth column by the second: $\delta = W_2 + \mu\gamma$.

If $a < 180°$, the result will be subtracted from or added to the true center. Finally Ḥabash obtains

$$\lambda = \bar{\lambda} - \delta.$$

Latitude of the Moon. The latitude of the moon for a given moment is determined by means of a table prepared for one degree. The true place of the moon (λ☾) is added to the mean position of the ascending node (−λ☊). Because of the longitude of the ascending node, the distance of the node from γ_0 is counted in a negative direction. This total, A, is the argument with which the table of the latitude of the moon is entered.

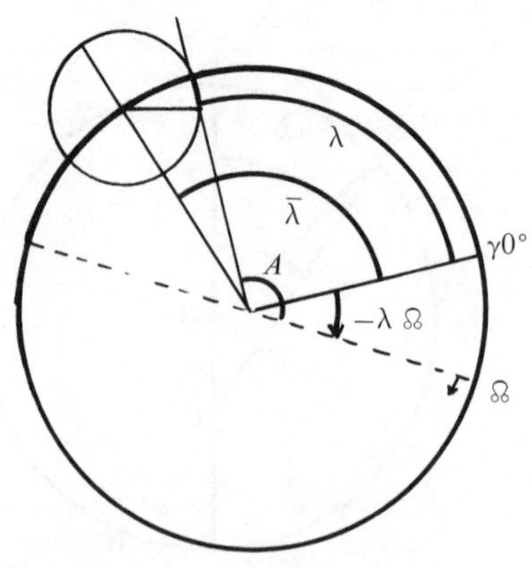

FIGURE 12

ḤABASH AL-ḤĀSIB

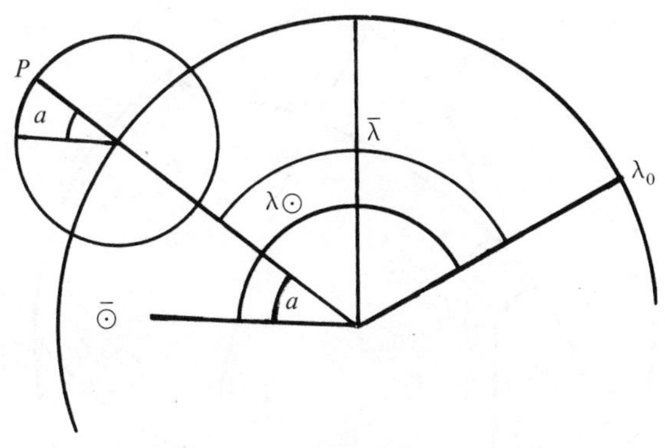

FIGURE 13

Theory of the Planets. For finding the longitudes of the planets Ḥabash prepared several tables for mean motions, in longitude and latitude, and the equations. His procedure for finding the true longitude of a planet for a given moment t is based on the Ptolemaic method (*Almagest* XI).

For outer planets (see Figure 13):

$$\bar{a} = \bar{\lambda}\odot - \bar{\lambda},$$

where $\lambda\odot$ = mean longitude of the sun
λ = mean longitude of the planet
$\bar{a}$ = anomaly or argumentum.

It implies that the radius of the planet on the epicycle is always parallel to the direction from 0 to the mean sun.

Inner Planets. For inner planets $\bar{\lambda} = \bar{\lambda}\odot$, and the anomaly can be found from the tables. (See Figure 14.)

For the outer planets he first found the mean longitude $\bar{\lambda}$, mean anomaly a, and the longitude of the apogee λA of the planet (see Figure 15): $\bar{\lambda} - \lambda A = \bar{K}$.

He found the distance of the center C of the epicycle from the apogee for the given moment. According to the Ptolemaic planetary theory, the planet makes its regular motion not around the point O but around E, i.e., the "equant." The center of the deferent falls between O and E. Then he found the epicyclic equation as seen from O. Thus the true anomaly, a, counted from true apogee, is found. Ḥabash tabulated this difference, $W_1 = a - \bar{a}$, as function $\bar{K}$, in the first column. This is called the "first correction": $a = W_1 + \bar{a}$.

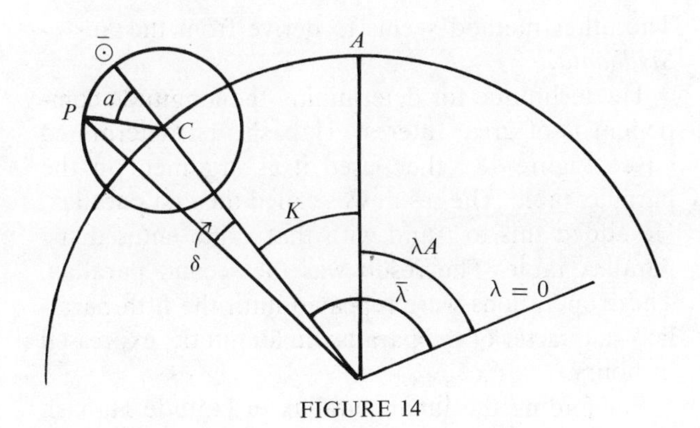

FIGURE 14

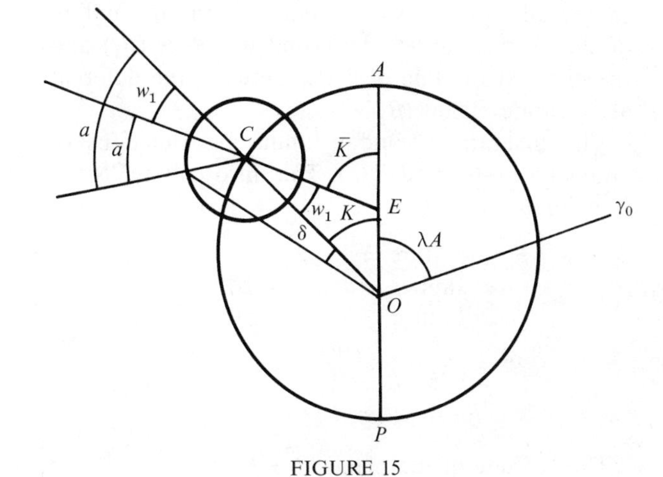

FIGURE 15

Then he computed the distance of C from A, i.e., K, as seen from A. This difference is also equal to the angle W_1:

$$K = \bar{K} - W_1.$$

If
$$\bar{K} < 180°$$
$$K = \bar{K} - W_1$$
$$a = \bar{a} + W_1.$$

Then comes the "second correction," W_2. This depends not only on the true anomaly a but also on the position of the epicycle. If it is exactly in the apogee, this amount of correction will be less than W_2 by the amount μA tabulated in column 4 as the function a:

$$\delta = W_2 - \mu A \cdot \gamma.$$

γ is found as a function of K in the second column.

All these procedures are correct only if the value found in the second column is negative. If it is positive, the second column is multiplied by the value found in the fifth column, then subtracted from the fourth. The true longitude of the planet is

$$\lambda = \lambda A + K + \delta.$$

Latitude of the Planets. The procedure for calculating the latitude of the superior planets is based on Ptolemaic method (*Almagest* XIII, 6). The table of latitudes was prepared in three columns. Ḥabash used the same numeric values as Ptolemy (*Almagest* XIII, table 5). According to him, the latitudes can be found by the addition of two components: the inclination of the epicycle about its second diameter (β_1) and the angle at the line of nodes between the deferent and ecliptic planes (β_2).

The first and second columns are functions of anomaly a:$b_1(a)$ and $b_2(a)$. The third column is the function θ.

$$\text{Mars} = \theta = \lambda$$
$$\text{Jupiter} = \theta = \lambda - 20^\circ$$
$$\text{Saturn} = \theta = \lambda + 50^\circ$$
$$c(\theta)$$
$$\beta_1 = b_1 \cdot c$$
$$\beta_2 = b_2 \cdot c$$

The latitude of the planet $\beta = \beta_1 + \beta_2$.

For the inferior planets (based on Ptolemaic method, *Almagest* XIII, 6), one enters the latitude table with the truly determined anomaly and records the corresponding numbers in the first and the second columns. These are the functions of a: $b_1(a)$ and $b_2(a)$.

One finds the determined true longitude of the planets. For Venus, $A = \lambda -$ the longitude of apogee. For Mercury, if the determined true anomaly is in the first fifteen rows,

$$A = \lambda - 10^\circ;$$

if it is in those that follow,

$$A = \lambda + 10^\circ.$$

Next,
$$A + 90^\circ = \theta \text{ for Venus}$$
$$A + 270^\circ = \theta \text{ for Mercury.}$$

One enters the table with that value and finds the corresponding number in the third column. This is the function $c(\theta)$.

Then, $b_1 \cdot c =$ the first latitude $= \beta_1$.

If θ is in the first fifteen rows, the planet is northern. If it is in those that follow, it is southern. If θ is after the first fifteen rows at the same time that a is in the first fifteen rows, the planet is northern.

Next, one enters the latitude table with

$$\theta \text{ for Venus}$$
$$\theta + 180^\circ \text{ for Mercury}$$

and finds the corresponding value in the third column. This is the function of θ or $\theta + 180^\circ$: $c(\theta)$ or $c(\theta + 180^\circ)$. Then $b_1c =$ the second latitude $= \beta_2$. If (θ) or $(\theta + 180^\circ)$ is in the first fifteen rows and $a < 180^\circ$, the planet has a northern latitude. If $a > 180^\circ$, it has a southern latitude. If (θ) or $(\theta + 180^\circ)$ is below the first fifteen rows and $a < 180^\circ$, the planet has a southern latitude; and if $a > 180^\circ$, the planet has a northern latitude. Then

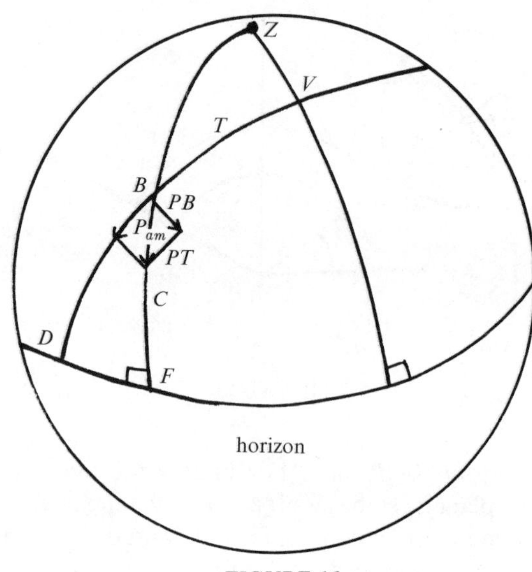

FIGURE 16

$c^2 + C^2\!/_6 = \beta_3$ for Venus. If the planet has northern latitude, $\beta = \beta_2 + \beta_3$ and $C^2 = 3C^2\!/_4 = \beta_3$ for Mercury. If the planet has southern latitude, $\beta = \beta_2 + \beta_3$.

Parallax (Ikhtilāf al-Manẓar) Theory. Ḥabash had two entirely different methods for determining the parallax, i.e., parallax in longitude P_λ and parallax in latitude P_β. One of them may seem a transition between that of Ptolemy and the later Islamic astronomers. This solution depends on the first sine (*al-jayb al-awwal*) that can be formulated (see Figure 16):

$$\frac{\sin BV}{\sin FB} \qquad \frac{\sin DB}{\cos DB}$$

equal cos B and the second sine, which is equal to sin B. Without proof he states that

$$\sin P_\lambda = (\text{first sine})\,(\sin P_{am}).$$

P_{am} is measured according to Ptolemy (see Figure 17):

$$\sin P_\beta = \frac{(\sin P_{am})\,(\text{first sine})}{\cos P_\beta}.$$

The other method seems to derive from the *Sūrya-Siddhanta*.

The technique for determining the longitude component is of great interest. Ḥabash first determined t (see Figure 18), then used it as argument in the parallax table. The result was called the first parallax. He added this to t and with that value entered the parallax table. The result was the second parallax. These operations were repeated until the fifth parallax—a quarter of the parallax in longitude, expressed in hours.

For finding the lunar parallax in latitude Ḥabash

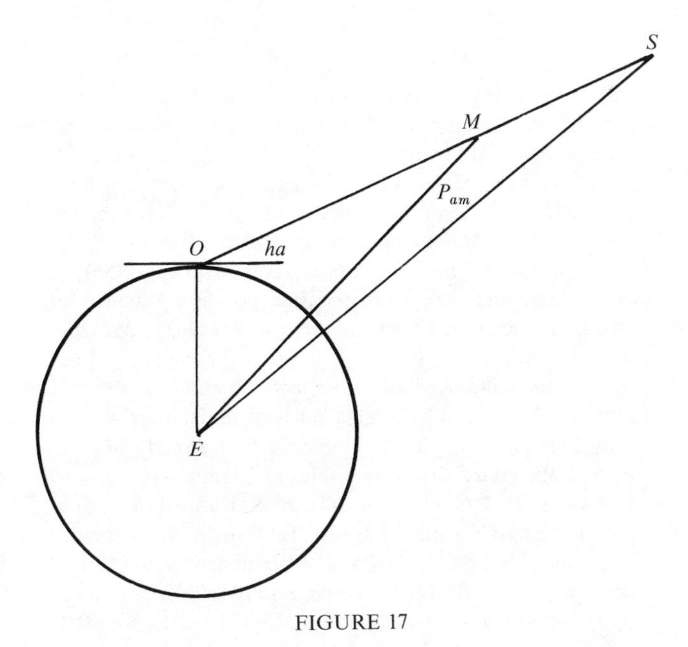

FIGURE 17

used A as an argument, and the corresponding value of the function was to be doubled (see Figure 18). This would be the lunar parallax in latitude.

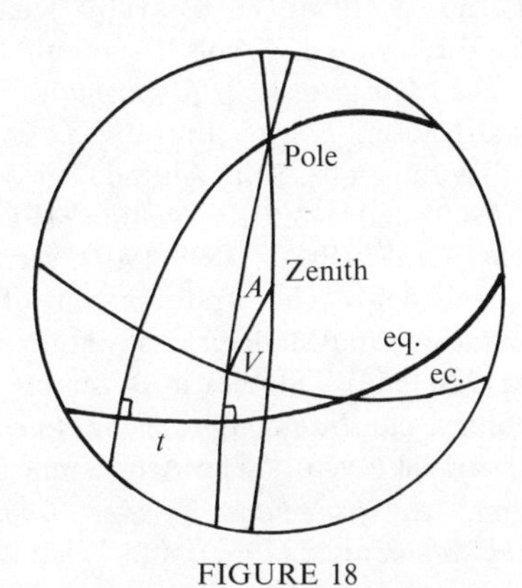

FIGURE 18

Visibility Theory (*Ru'yat al-Hilāl*). Habash may have been the first astronomer to engage in the computation of the new crescent. Like the ancient Babylonians and the Jews, the Muslims depended on visual observation of the new crescent for their religious and secular calendars. This led the Muslim astronomers to realize that the knowledge of the visibility of the new crescent is an essential task of astronomy. Ḥabash used the following method for the determination of the visibility of the new crescent. He added twenty or thirty minutes to the time of sunset, thus obtaining the mean position of the moon at the time when the new crescent becomes visible. Then the true position of the sun, the moon, and the head were needed for the above-mentioned time (see Figure 19). Thus, $\lambda - \lambda\odot = \lambda_1$, which Maimonides called the first elongation.

ḤABASH AL-ḤĀSIB

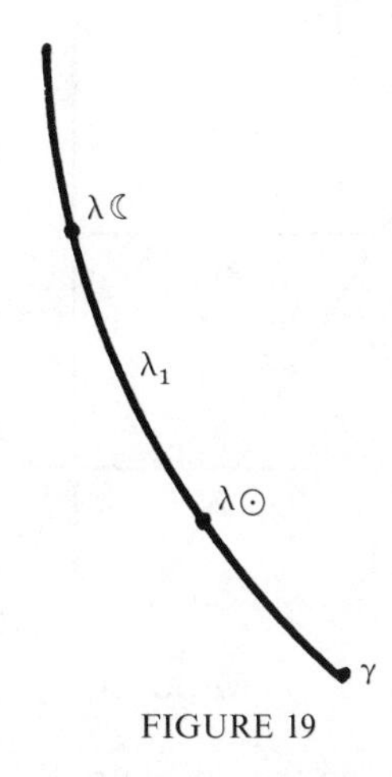

FIGURE 19

For an observer on the surface of the earth, the moon M would appear in a lower position M_1 because of the parallax.

$$\frac{\text{Diameter of the earth}}{\text{The distance of the moon to the center of the earth}} = \text{parallax of the moon } P_{am}.$$

Then the parallax in latitude P_β and longitude P_λ can be obtained (see Figure 20). Thus, $\lambda_1 - P\lambda = \lambda_2$, which Maimonides called the second elongation. The

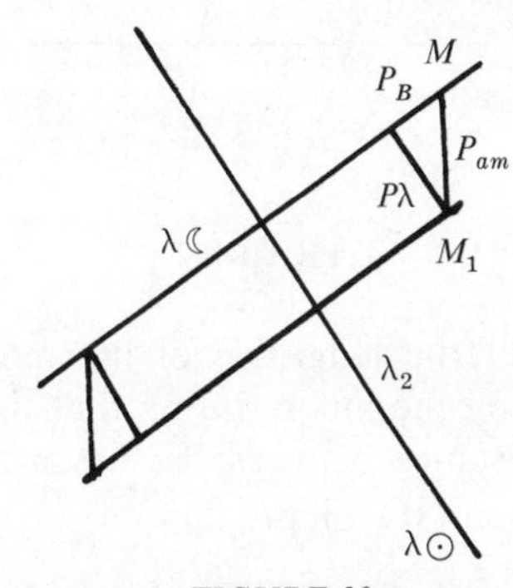

FIGURE 20

true latitude of the moon (Maimonides called it the first latitude) was subtracted from or added to the parallax in latitude, depending on the variable position of the moon:

$$M_\beta - P_\beta = M_{1\beta} \text{ (second latitude).}$$

From this second latitude one can derive half of the day arc of the moon, and from that the equation of the day of the moon is obtained. This equation of the moon is added to or subtracted from the longitude of the moon. Thus the point of the ecliptic O (see Figure 21), which sets simultaneously with the moon, is obtained: $\lambda_2 - C = \lambda_3 O$.

Then the arc of the equator QA, which sets simultaneously with the arc of the ecliptic, λ_3, is calculated

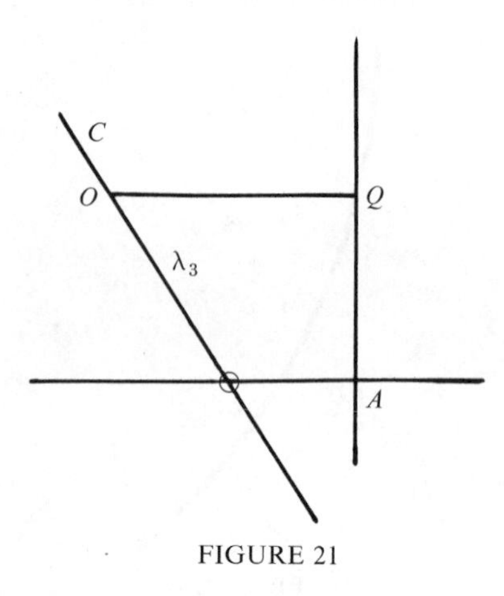

FIGURE 21

(see Figure 22). This is the difference between the rising times of the moon and the sun. This time difference is multiplied by the surplus of the moon in one hour and divided by fifteen. The result K is

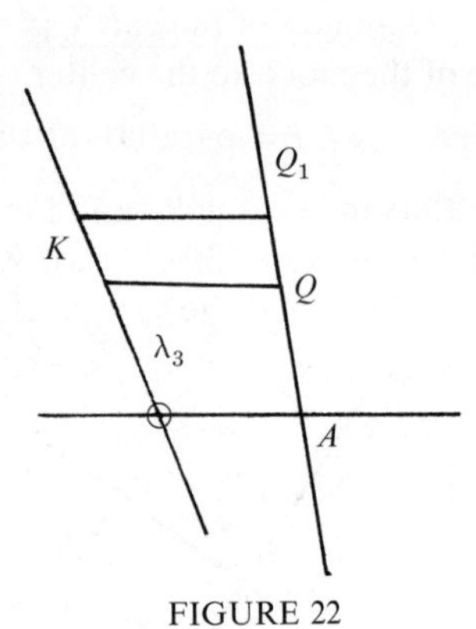

FIGURE 22

added to the true longitude of the moon, i.e., the distance cut by the moon during that time is added to get the distance between the moon and the sun at sunset. Then (see Figure 23)

$$\angle Q_1AR = 90^\circ - \Phi$$
$$\sin Q_1R = \cos \Phi \cdot \sin QA.$$

If $QR > 10^\circ$, the moon will be visible on that day. If $QR < 10^\circ$, the moon will not be visible.

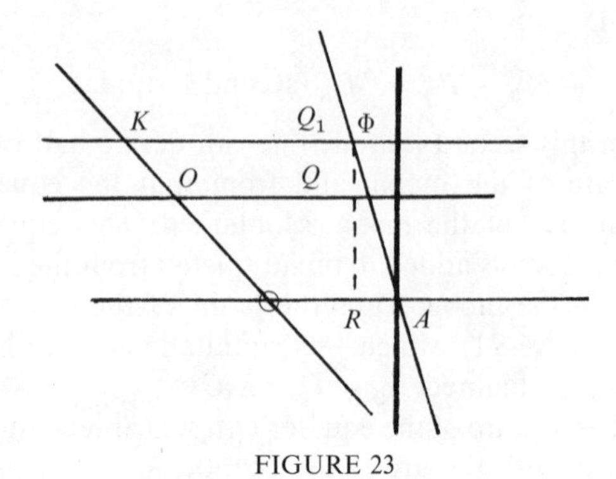

FIGURE 23

BIBLIOGRAPHY

The following works may be consulted for further information: A. Braunmühl, *Vorlesungen über Geschichte der Trigonometrie* (Leipzig, 1900); G. Caussin, *Le livre de la grande Hakémite,* vol. VII of Notices et Extraits des MSS (Paris, 1804); S. Gandz, J. Obemann, and O. Neugebauer, *The Code of Maimonides. Book Three. Treatise Eight; Sanctification of the New Moon* (New Haven, 1956); J. Hamadanizadeh, "A Medieval Interpolation Scheme for Oblique Ascensions," in *Centaurus,* **9** (1963), 257–265; E. S. Kennedy, "An Islamic Computer for Planetary Latitudes," in *Journal of the American Oriental Society,* **71** (1951), 12–21; and "Parallax Theory in Islamic Astronomy," in *Isis,* **47** (1956), 33–53; E. S. Kennedy and M. Agna, "Planetary Visibility Tables in Islamic Astronomy," in *Centaurus,* **7** (1960), 134–140; E. S. Kennedy and Janjanian, "The Crescent Visibility Table in Al-Khwārizmī's Zij," *ibid.,* **11** (1965), 73–78; E. S. Kennedy and Ahmad Muruwwa, "Bīrūnī on the Solar Equation," in *Journal of Near Eastern Studies,* **17** (1958), 112–121; E. S. Kennedy and Sharkas, "Two Medieval Methods for Determining the Obliquity of the Ecliptic," in *Mathematics Teacher,* **55** (1962), 286–290; E. S. Kennedy and W. R. Transue, "A Medieval Iterative Algorism," in *American Mathematical Monthly,* **63,** no. 2 (1956), 80–83; Ḥājjī Khalīfa, *Kashf al-Ẓunūn,* S. Yaltkaya, ed., 2 vols. (Istanbul, 1941–1943); E. Kramer, *The Main Stream of Mathematics* (New York, 1951); Ibn al-Nadīm, *Fihrist,* Flügel, ed., I (1871); N. Nadir, "Abūl-Wāfā' on the Solar Altitude," in *Mathematics Teacher,* **53** (1960), 460–463; C. A. Nallino, *Al-Battānī Opus astronomicum,* 3 vols. (Brera, 1899–1907), see vols. I and III; O. Neugebauer, "The Transmission of Planetary Theories in Ancient and Medieval Astronomy," in *Scripta mathematica,* **22** (1956); "Studies in Byzantine Astronomical Terminology," in *Transactions of the American Philosophical Society,* **50** (1960); "The Astronomical Tables of Al-Khwārizmī," in *Hist. Filos. Skrifter. Danske Videnskabernes Selskab,* **4,** no. 2 (1962); and "Thâbit ben Qurra 'On the Solar Year' and 'On the Motion of the Eighth Sphere,'" in *Transactions of the American Philosophical Society,* **106** (1962), 264–299; Ibn al-Qifṭī, *Ta'rīkh al ḥukamā',* Lippert, ed. (Berlin, 1903); G. Sarton, *Introduction to the History of Science,* I (Baltimore, 1927), 545, 550, 565, 667; A. Sayili, "Habeş el Hasib'in 'El Dimişki' adiyla Maruf Zîci'nin Mukaddemesi," in *Ankara üniversitesi dil ve tarih-coğrafya fakültesi dergisi,* **13** (1955), 133–151; C. Schoy, "Beiträge zur arabischen Trigonometrie," in *Isis,* **5** (1923), 364–399; D. E. Smith, *History of Mathematics,* II (London, 1925); and H. Suter, *Die Mathematiker und Astronomen der Araber und ihre Werke* (Leipzig, 1900).

S. TEKELI

HABER, FRITZ (*b.* Breslau, Germany [now Wrocław, Poland], 9 December 1868; *d.* Basel, Switzerland, 29 January 1934), *chemistry.*

Haber's mother died when he was born. His father, who sold pigments and dyestuffs, was one of

Germany's largest importers of natural indigo. Haber's early schooling was at the *Volksschule* and the St. Elisabeth Gymnasium. He attended the universities of Berlin and Heidelberg and the Charlottenburg Technische Hochschule; the latter school awarded him the Ph.D. in 1891. After little more than a year of employment at three different factories, Haber entered the Eidgenössische Technische Hochschule at Zurich, Switzerland, as a postdoctoral student in chemical technology and studied principally with Georg Lunge. Six months spent in his father's business proved to be unsatisfactory, and he became an assistant to Ludwig Knorr at the University of Jena and then to Hans Bunte at the Karlsruhe Technische Hochschule. He received *Privatdozent* status at the Baden school following publication of his first book, *Experimentelle Untersuchungen über Zertsetzgung und Verbrennung von Kohlenwasserstoffen* ("Experimental Studies on the Decomposition and Combustion of Hydrocarbons" [Munich, 1896]).

This record of his research exemplifies the work for which Haber became famous: theoretical studies, done with insight and thoroughness, in areas of growing practical importance. The thermal decomposition of hydrocarbons had been investigated extensively by Marcelin Berthelot twenty-five years earlier; Haber criticized Berthelot's conclusions as arbitrary. He found the carbon-to-carbon linkage in hydrocarbons to have a greater thermal stability than the carbon-to-hydrogen linkage in aromatic compounds; the reverse was true for aliphatic compounds. This rule has been shown to be subject to exceptions.

In 1901 Haber married Dr. Clara Immerwahr, also a chemist. A son, Hermann, was born in June 1902. During the autumn of 1917, two and one-half years after the death of his first wife, he married Charlotta Nathan. Two children, Eva and Ludwig, were born; the marriage ended in divorce in 1927.

After being named *Privatdozent,* Haber turned to problems of physical chemistry, although he had no formal education in this area. He had the help of his colleague Hans Luggin, a pupil of Svante Arrhenius, but Haber considered himself self-taught in the field. He first investigated the electrochemical reduction of nitrobenzene and showed the importance of electrode potential. He studied the nature and rate of the electrode process for the quinine-quinol system and, interested in the nature and rate of the electrode process, did not emphasize the application to the measurement of hydrogen ion concentration. Later he devised a glass electrode to measure hydrogen ion concentration through the electric potential across a piece of thin glass. Other electrochemical subjects investigated by Haber include fuel cells; measurement of the free energy of oxidation of hydrogen, carbon monoxide, and carbon; and the electrolysis of crystalline salts. At the end of his career he had an active interest in electrochemistry, studying autoxidation; application of Planck's quantum theory to chemistry was the basis of most of his later work. In 1898 he published his *Grundriss der technischen Elektrochemie auf theoretischer Grundlage* at Munich and was promoted to associate professor. As an indication of his growing reputation, in 1902 the Deutsche Bunsen-Gesellschaft sent him on a sixteen-week study tour of the United States. His report on chemical education and electrochemical industry in the country was acclaimed in Europe and America.

In 1905 Haber's *Thermodynamik technischer Gasreaktionen Vorlesungen* was published at Munich, and in 1906 he was given a full professorship. Gilbert Lewis and M. Randall's classic text, *Thermodynamics and the Free Energy of Chemical Substances*, published in 1923, described Haber's book as "a model of accuracy and critical insight."

Haber's outstanding accomplishment in chemistry, during the first decade of the twentieth century, involved a gas reaction. He was one of many scientists interested in nitrogen fixation. As in virtually all his work, the problem had both theoretical and practical significance, and he looked into several possible solutions. Walther Nernst, a leader in physical chemistry, obtained data at variance with Haber's for the combination of nitrogen and hydrogen to form ammonia. Nernst presented measurements from experiments done at high pressure and can be considered the first to accomplish the synthesis under these conditions. (Henry Le Chatelier had been the first to try the high-pressure synthesis, but an explosion induced him to forsake the venture.) Haber considered the difference in values a personal challenge. Working with Robert Le Rossignol, a student from the Isle of Jersey, and assisted by the mechanic Kirchenbauer, he performed high-pressure experiments and confirmed his earlier results, done at atmospheric pressure, which Nernst had questioned. A constant used by Nernst in his calculations—not his heat theorem—was later shown to be the cause of the erroneous values.

Haber went on to commercial exploitation of the synthesis of ammonia. His calculations showed that about 8 percent ammonia was available at pressures of 200 atmospheres and temperatures of 600°C. However, it was through the work of others that the process came to be the first successful high-pressure industrial chemical reaction. Such practical problems as a satisfactory, long-lasting container for the operation were solved under the direction of Carl Bosch and his associates at the Badische Anilin- und

Sodafabrik. Nonetheless, users and students of high-pressure techniques came to Haber's laboratory for instruction.

In 1912 Haber became director of the Kaiser Wilhelm Institute for Physical Chemistry and Electrochemistry at Dahlem, on the outskirts of Berlin. His friend Richard Willstätter was codirector, with Ernst Beckmann, of the first of these Kaiser Wilhelm research institutes, for chemistry.

At the start of World War I, Haber placed himself and his laboratory at the service of his country. Assigned problems involving the supply of war materials, he showed xylene and solvent naphtha to be good substitutes for toluene as an antifreeze in benzene motor fuel. The War Ministry consulted Walther Nernst about using irritants to drive the Allies out of their trenches so that open warfare might be resumed; Haber was given a share in solving this problem. Dianisidine chlorosulfonate, an irritant powder suggested by Nernst, and the lacrimator xylyl bromide proved to be ineffective. Haber's laboratory studied other irritants and the investigation came to a close in December 1914 with an explosion when a few drops of dichloromethylamine were added to a few cubic centimeters of impure cacodyl chloride. Otto Sackur, an outstanding physical chemist, was killed.

Haber developed the use of chlorine gas as a war weapon; by the end of January 1915 the preliminary laboratory research was completed. On 11 April 1915 about 5,000 cylinders of the gas were distributed, and the chemical was released over a 3.5-mile front near Ypres, Belgium. German military leaders later admitted that had massive attacks rather than the small test been done, German victory would have been assured. Instead, the Allies soon developed gases and the weapon on both sides was no longer intended to move men but to kill them.

In 1916 Haber became chief of the Chemical Warfare Service; and although he was only a captain, every detail of chemical offense, defense, supply, and research came under his supervision. By that time his process of nitrogen fixation was used in supplying Germany with nitrogen compounds, needed for fertilizers and for the explosives that provided staying power after the United States's entry into the war.

In November 1919 Haber was awarded the Nobel Prize in chemistry. The honor was denounced by some French, British, and American scientists, which dealt another blow to his spiritual and physical condition. Having put all his energies into the war, he was obliged to share personally in Germany's defeat. To be condemned as inhuman by fellow scientists also involved in war work deeply troubled him.

During the early postwar years Haber continued his patriotic efforts. He was the leading figure in appeals for the Notgemeinschaft, the Emergency Society for German Science. His most noteworthy contribution, although a failure, was to search the oceans for gold, in the hope of extracting enough to pay the war reparations demanded by the Allies. Nineteenth-century analyses had shown that some samples of seawater contained nearly twice as much as the lowest-grade land deposit that was profitable to operate. Unfortunately, Haber did not verify the published results, later established as much too high, and an extraction scheme was devised without ascertaining the exact amount of gold and its form in seawater. Several ocean trips were undertaken in vain, but a very accurate method for gold analysis was found.

Haber's other activities during the postwar years proved more fruitful. His institute became one of the great scientific research centers in the world. During his tenure as director the Kaiser Wilhelm Institute for Physical Chemistry and Electrochemistry was credited with more than 700 publications in scientific journals. As at Karlsruhe and as head of chemical warfare, Haber showed himself to be a talented leader. He displayed versatility in handling a variety of subject matters in both academic and military situations. He was able to conform to a military environment and provide at other times a liberal and independent atmosphere to his associates. Beginning as an organic chemist, he contributed to every branch of physical chemistry as well as to peripheral sciences. He was a pure and applied scientist able to bridge the gap between the purist and the engineer. Men rather than accomplishments were the products of his direction; many outstanding physical chemists of the first half of the twentieth century started their careers with him.

The Haber Colloquium, a research seminar at his institute, began in October 1919 and soon attracted scientists from all parts of Europe. Haber's contribution was clarity and the ability to abstract, spiced with satire and wit. His ability to think about and discuss material from the hydrogen atom to the flea, presented by expert lecturers, was greatly admired. As other commitments took him away from the meetings, they lost their verve and attendance dropped.

He was always engrossed in research projects and activities, and he had an extraordinary ability to concentrate, staying with problems both scientific and nonscientific. His speaking, reading, and writing habits were in this same mold. His best appreciated speeches were given at a commemoration for Justus Liebig, with whom he has been compared, and on the departure of his friend James Franck to join Göttingen University. He enjoyed mystery stories for recreational reading and composed verse for friends and relatives. He had an intellectual approach to these endeavors. Once he and Willstätter planned a vacation based on biblical quotations. As a youth he was interested in dramatics, and throughout his life he showed a flair for the theatrical.

Haber served Germany well in his relations with foreigners and in foreign countries. His laboratories at Karlsruhe and Dahlem always had foreign students. In 1929 half of the sixty members of his Institute were foreigners from a dozen different countries. After World War I, the number of foreign visitors increased and he traveled to other countries for vacations and scientific meetings. After a two-month visit to Japan, he helped establish the Japan Institute for promoting mutual understanding and cultural interests. He represented Germany on the Board of the Union Internationale de Chimie from 1929 to 1933.

The Kaiser Wilhelm Institute was affected soon after the Nazis came to power. The Ministry of Art, Science and Popular Education demanded the dismissal of the Jewish workers there. Haber formally resigned in a letter dated 30 April 1933, writing: "For more than forty years I have selected my collaborators on the basis of their intelligence and their character and not on the basis of their grandmothers, and I am not willing for the rest of my life to change this method which I have found so good."

Haber received an invitation to work at the Cambridge laboratory of William J. Pope, and he did so for four months. He also had an offer to head the physical chemistry section at the Daniel Sieff Research Institute in Israel and accepted, provided he found the climate and living conditions suitable. He died in Switzerland while on the way to the opening ceremonies for the Sieff Institute.

In 1935, on the first anniversary of his death, a number of learned societies in Germany did commemorate the occasion. Five hundred men and women gathered in Dahlem to pay tribute to him, despite the displeasure of the Nazis.

BIBLIOGRAPHY

I. ORIGINAL WORKS. A full list of Haber's works is in Morris Goran, *The Story of Fritz Haber* (Norman, 1967). His most important writings include "Bidioxymethylenindigo," in *Berichte der Deutschen chemischen Gesellschaft*, **23** (1890), 1566, written with C. Liebermann; "Über einige Derwate des Piperonals," *ibid.*, **24** (1891), 617; "Elektrolytische Darstellung von Phenyl-B-hydroxylamin," in *Zeitschrift für Elektrochemie,* **5** (1898), 77–78; "Über die elektrische Reduktion von Nichtelektrolyten," in *Zeitschrift für physikalische Chemie,* **32** (1900), 193–270; "Über den textilen Flachdruck," in *Zeitschrift für angewandte Chemie und Zentralblatt für technische Chemie,* **15** (1902), 1177–1183; and "Zur Theorie der Indigoreaktion," in *Zeitschrift für Elektrochemie,* **9** (1903), 607–608.

See also "Über das Ammoniakgleichgewicht," in *Berichte der Deutschen chemischen Gesellschaft,* **40** (1907), 2144–2154, written with R. Le Rossignol; "Zur Kenntnis des Hydroxylamins," in *Journal für praktische Chemie,* **79** (1909), 173–176; "Über die Darstellung des Ammoniaks aus Stickstoff und Wasserstoff," in *Zeitschrift für Elektrochemie,* **16** (1910), 244–246; "Die Schlagwetterpfeife," in *Naturwissenschaften,* **1** (1914), 1049; "Beitrag zur Kenntnis der Metalle," in *Sitzungsberichte der Preussischen Akademie der Wissenschaften zu Berlin* (1919), 506–518; "Über amorphe Niederschäge und Kristal-lisierte Sole," in *Berichte der Deutschen chemischen Gesellschaft,* **55** (1922), 1717–1733; and "Beitrag zur Kenntnis des Rheinwassers," in *Zeitschrift für anorganische und allgemeine Chemie,* **147** (1925), 156–170, written with J. Jaenicke.

II. SECONDARY LITERATURE. For works about Haber, see E. Berl, "Fritz Haber zum 60 Geburtstage," in *Zeitschrift für Elektrochemie,* **34** (1928), 797–803; J. E. Coates, "The Haber Memorial Lecture," in *Journal of the Chemical Society* (Nov. 1939), pp. 1642–1672; M. Goran, "Present Day Significance of Fritz Haber," in *American Scientist* (July 1947); and A. Stoll, ed., *The Memoirs of Richard Willstätter* (New York, 1965).

MORRIS GORAN

HABERLANDT, GOTTLIEB (*b.* Ungarisch-Altenburg, Hungary, 28 November 1854; *d.* Berlin, Germany, 30 January 1945), *botany.*

Haberlandt's father was professor of applied botany and introduced him to botany at an early age. Despite great interest and talent in music, painting, and German literature, he studied botany. Julius Wiesner in Vienna became his first teacher and supervised the work on his Ph.D., which he obtained in 1876. Julius von Sachs's textbook of botany influenced him greatly, but he decided to work under Simon Schwendener at Tübingen in 1877–1878. Schwendener's book *Das mechanische Prinzip im anatomischen Bau der Monokotylen* strengthened his belief that anatomy and physiology should be combined. After serving as *Privatdozent* at Vienna, Haberlandt moved to Graz, where he succeeded H. Leitgeb in 1888. In 1910 he replaced Schwendener at Berlin, where he established the Institute for Plant Physiology. After his retirement in 1923 he remained in Berlin, where he died in 1945.

Haberlandt's most influential book was *Physiologische Pflanzenanatomie* (1884), which he updated and enlarged through six editions. This work considers plant anatomy from a physiological point of view: structures are explained by their functions, not in a finalistic-vitalistic sense but teleologically, on the basis of selection theory. Using functional criteria, Haberlandt distinguishes twelve tissue systems and many more tissues in his anatomico-physiological classification. Although this specific classification was not adopted, many of his basic ideas have been incorporated into modern biology. At first Haberlandt's ideas met with great resistance and were ridiculed by many of his colleagues. H. A. de Bary called his book the "newest botanical novel," and it was said that some of his colleagues tried to hide it from their students so that they would not "go astray."

Another book of interest to the layman is Haberlandt's *Eine botanische Tropenreise* (1893), in which he describes his trip to Java and Ceylon in 1891–1892. Haberlandt wrote a number of other books and numerous papers. Many of these publications, especially the earlier ones, deal with functional investigations of plant structures such as stinging hairs, guard cells, oil glands, hydathodes, and all sorts of tissues. After a study on stimulus transmission in the sensitive plant (1890), Haberlandt concentrated on investigations of sense organs. He hypothesized—simultaneously with and independently of B. Nemeč—the statolith function of certain starch grains, thus exerting a lasting influence on plant physiology. His later work dealt mainly with experimental morphogenesis and physiology and established new experimental approaches. As early as 1902 Haberlandt reported on partially successful attempts at cell cultures. In 1921 he postulated the existence of wound hormones which induce mitoses in the cell cultures. He related this finding to fertilization, parthenogenesis, and formation of adventitious embryos and periderm. Several papers deal with the phenomenon of *Crataegomespilus,* graft hybrids between *Crataegus* and *Mespilus.* His last publication (1941) is on the nature of morphogenic substances.

BIBLIOGRAPHY

I. Original Works. Haberlandt's articles and books are listed in the paper by Guttenberg (see below). His books are *Die Entwicklungsgeschichte des mechanischen Gewebesystem der Pflanzen* (Leipzig, 1879); *Die physiologischen Leistungen der Pflanzengewebe,* vol. II of Heinrich Schenck, *Handbuch der Botanik* (Breslau, 1882); *Physiologische Pflanzenanatomie* (Leipzig, 1884; 6th ed., 1924), 4th ed. translated into English as *Physiological Plant Anatomy* (London, 1914); *Das reizleitende Gewebesystem der Sinnpflanze* (Leipzig, 1890); *Eine botanische Tropenreise. Indomalaische Vegetationsbilder und Reiseskizzen* (Leipzig, 1893; 2nd ed., 1910); *Sinnesorgane im Pflanzenreich* (Leipzig, 1901); *Die Lichtsinnesorgane der Laubblätter* (Leipzig, 1905); *Berliner Botaniker in der Geschichte der Pflanzenphysiologie* (Berlin, 1914); *Goethe und die Pflanzenphysiologie* (Leipzig, 1923); *Erinnerungen, Bekenntnisse und Betrachtungen* (Berlin, 1933); and *Botanisches Vademecum für Künstler* (Jena, 1936).

II. Secondary Literature. See H. von Guttenberg, "Gottlieb Haberlandt," in *Phyton,* **6** (1955), 1–88; A. D. Krikorian and D. L. Berquam, "Plant Cell and Tissue Cultures: The Role of Haberlandt," in *Botanical Review,* **35** (1969), 59–88, which includes an English trans. of Haberlandt's original paper (1902) reporting nonsterile culture of isolated plant cells (that did not divide) and pointing out the purposes and potentialities of culture techniques for physiological and morphological problems; A. C. Noé, "Gottlieb Haberlandt," in *Plant Physiology,* **9** (1934), 851–855; and O. Renner, in *Jahrbuch der bayerischen Akademie der Wissenschaften,* Math-nat. Kl., for 1944–1948 (1948), pp. 258–261.

R. Sattler

DICTIONARY
OF
SCIENTIFIC BIOGRAPHY

PUBLISHED UNDER THE AUSPICES OF
THE AMERICAN COUNCIL OF LEARNED SOCIETIES

The American Council of Learned Societies, organized in 1919 for the purpose of advancing the study of the humanities and of the humanistic aspects of the social sciences, is a nonprofit federation comprising thirty-five national scholarly groups. The Council represents the humanities in the United States in the International Union of Academies, provides fellowships and grants-in-aid, supports research-and-planning conferences and symposia, and sponsors special projects and scholarly publications.

MEMBER ORGANIZATIONS

AMERICAN PHILOSOPHICAL SOCIETY, 1743
AMERICAN ACADEMY OF ARTS AND SCIENCES, 1780
AMERICAN ANTIQUARIAN SOCIETY, 1812
AMERICAN ORIENTAL SOCIETY, 1842
AMERICAN NUMISMATIC SOCIETY, 1858
AMERICAN PHILOLOGICAL ASSOCIATION, 1869
ARCHAEOLOGICAL INSTITUTE OF AMERICA, 1879
SOCIETY OF BIBLICAL LITERATURE, 1880
MODERN LANGUAGE ASSOCIATION OF AMERICA, 1883
AMERICAN HISTORICAL ASSOCIATION, 1884
AMERICAN ECONOMIC ASSOCIATION, 1885
AMERICAN FOLKLORE SOCIETY, 1888
AMERICAN DIALECT SOCIETY, 1889
ASSOCIATION OF AMERICAN LAW SCHOOLS, 1900
AMERICAN PHILOSOPHICAL ASSOCIATION, 1901
AMERICAN ANTHROPOLOGICAL ASSOCIATION, 1902
AMERICAN POLITICAL SCIENCE ASSOCIATION, 1903
BIBLIOGRAPHICAL SOCIETY OF AMERICA, 1904
ASSOCIATION OF AMERICAN GEOGRAPHERS, 1904
AMERICAN SOCIOLOGICAL ASSOCIATION, 1905
AMERICAN SOCIETY OF INTERNATIONAL LAW, 1906
ORGANIZATION OF AMERICAN HISTORIANS, 1907
COLLEGE ART ASSOCIATION OF AMERICA, 1912
HISTORY OF SCIENCE SOCIETY, 1924
LINGUISTIC SOCIETY OF AMERICA, 1924
MEDIAEVAL ACADEMY OF AMERICA, 1925
AMERICAN MUSICOLOGICAL SOCIETY, 1934
SOCIETY OF ARCHITECTURAL HISTORIANS, 1940
ECONOMIC HISTORY ASSOCIATION, 1940
ASSOCIATION FOR ASIAN STUDIES, 1941
AMERICAN SOCIETY FOR AESTHETICS, 1942
METAPHYSICAL SOCIETY OF AMERICA, 1950
AMERICAN STUDIES ASSOCIATION, 1950
RENAISSANCE SOCIETY OF AMERICA, 1954
SOCIETY FOR ETHNOMUSICOLOGY, 1955

DICTIONARY
OF
SCIENTIFIC BIOGRAPHY

CHARLES COULSTON GILLISPIE
Princeton University
EDITOR IN CHIEF

Volume 6

JEAN HACHETTE—JOSEPH HYRTL

CHARLES SCRIBNER'S SONS · NEW YORK

Panel of Consultants

Contributors to Volume 6

The following are the contributors to Volume 6. Each author's name is followed by the institutional affiliation at the time of publication and the names of articles written for this volume. The symbol † indicates that an author is deceased.

GIORGIO ABETTI
Istituto Nazionale di Ottica
HORN D'ARTURO

H. B. ACTON
University of Edinburgh
R. B. HALDANE

MICHELE L. ALDRICH
Smith College
HAYDEN; HITCHCOCK

MAX ALFERT
University of California, Berkeley
M. HEIDENHAIN

MEA ALLAN
W. J. HOOKER

WILBUR APPLEBAUM
University of Illinois
HORROCKS

LAWRENCE BADASH
University of California, Santa Barbara
HAHN

ERNEST BALDWIN†
F. G. HOPKINS

MARGARET E. BARON
W. G. HORNER; C. HUTTON

EDWIN A. BATTISON
Smithsonian Institution
J. HARRISON

HEINRICH BECK
Pädagogische Hochschule Bamberg der Universität Würzburg
HERBART

ROBERT P. BECKINSALE
University of Oxford
W. HOPKINS

WHITFIELD BELL, JR.
American Philosophical Society Library
HARLAN

ENRIQUE BELTRÁN
Mexican Society of History of Science and Technology
HERRERA

JAMES D. BERGER
Indiana University
HENKING

MICHAEL BERNKOPF
Pace College
HALPHEN

KURT-R. BIERMANN
German Academy of Sciences
HUMBOLDT

ARTHUR BIREMBAUT
HASSENFRATZ

R. P. BOAS, JR.
Northwestern University
HUNTINGTON

UNO BOKLUND
Royal Pharmaceutical Institute, Stockholm
HIÄRNE

H. J. M. BOS
State University of Utrecht
HUYGENS

GERT H. BRIEGER
Duke University
HEKTOEN

BARUCH A. BRODY
Massachusetts Institute of Technology
W. HAMILTON

W. H. BROCK
University of Leicester
HOFMANN; T. S. HUNT

STEPHEN G. BRUSH
University of Maryland
J. HERAPATH

VERN L. BULLOUGH
San Fernando Valley State College
HENRY OF MONDEVILLE; HUNDT

IVOR BULMER-THOMAS
HIPPIAS OF ELIS; HIPPOCRATES OF CHIOS; HYPSICLES OF ALEXANDRIA

JOHN G. BURKE
University of California, Los Angeles
HESSEL

J. C. BURKILL
University of Cambridge
G. H. HARDY; HOBSON

H. L. L. BUSARD
State University of Leiden
C. HARDY; HENRY OF HESSE

JEROME J. BYLEBYL
University of Chicago
W. HARVEY

ALBERT V. CAROZZI
University of Illinois
W. HAMILTON; HAUG

JOHN CHALLINOR
HARKER

CARLETON B. CHAPMAN
Dartmouth Medical School
J. S. HALDANE

JEAN CHÂTILLON
Institut Catholique de Paris
HUGH OF ST. VICTOR

GEORGES CHAUDRON
Laboratoire de Recherches Métallurgiques
HÉROULT

ROBERT A. CHIPMAN
University of Toledo
HEFNER-ALTENECK

RICHARD J. CHORLEY
University of Cambridge
A. HEIM

J. G. VAN CITTERT-EYMERS
HARTING; HARTSOEKER

RONALD W. CLARK
J. B. S. HALDANE

EDWIN CLARKE
University College London
M. HALL; HITZIG; HORSLEY; J. J. HUBER; HUTCHINSON

ALBERT B. COSTA
Duquesne University
HANTZSCH; HILDITCH

PIERRE COSTABEL
École Pratique des Hautes Études
M.-G. HUMBERT; P. HUMBERT

MICHAEL J. CROWE
University of Notre Dame
H. HANKEL; HOÜEL

KARL H. DANNENFELDT
Arizona State University
HERMES TRISMEGISTUS

EDWARD E. DAUB
University of Wisconsin
HARCOURT

CLAUDE K. DEISCHER
University of Pennsylvania
W. B. HERAPATH

R. DESMOND
Royal Botanic Gardens Library, Kew
J. D. HOOKER

BERN DIBNER
Burndy Library
HOPKINSON

D. R. DICKS
University of London
HECATAEUS OF MILETUS; HICETAS OF SYRACUSE

CONTRIBUTORS TO VOLUME 6

SALLY H. DIEKE
The Johns Hopkins University
HOUGH

HERBERT DINGLE
University of London
HUGGINS

JESSIE DOBSON
Royal College of Surgeons Hunterian Museum
J. HUNTER; W. HUNTER

HAROLD DORN
Stevens Institute of Technology
HODGKINSON

A. G. DRACHMANN
HERO OF ALEXANDRIA

OLLIN J. DRENNAN
Western Michigan University
HITTORF; HORSTMANN

KINGSLEY DUNHAM
Institute of Geological Sciences, London
HOLMES

CAROLYN EISELE
Hunter College of the City University of New York
G. W. HILL

DAVID S. EVANS
University of Texas
J. F. W. HERSCHEL

P. P. EWALD
Polytechnic Institute of Brooklyn
C. H. HERMANN

JOSEPH EWAN
Tulane University
R. M. HARPER; HOSACK

V. A. EYLES
J. HALL; J. HUTTON

J. FABER
Hubrecht Laboratory, Utrecht
HUBRECHT

W. V. FARRAR
University of Manchester
HAMPSON

E. A. FELLMANN
Institut Platonaeum, Basel
J. HERMANN

KONRADIN FERRARI D'OCCHIEPPO
University of Vienna
HARTWIG; HELL

BERNARD FINN
Smithsonian Institution
E. H. HALL

WALTHER FISCHER
A. A. HEIM; HELMERT

DONALD W. FISHER
New York State Education Department
J. HALL, JR.

FREDERICK M. FOWKES
Lehigh University
HARKINS

H. C. FREIESLEBEN
HARDING; J. F. HARTMANN

HANS FREUDENTHAL
State University of Utrecht
HEINE; HERMITE; HILBERT; HOPF; HURWITZ

B. VON FREYBERG
University of Erlangen-Nuremberg
HOFF

JOSEPH S. FRUTON
Yale University
HOPPE-SEYLER

DAVID J. FURLEY
Princeton University
HERACLITUS OF EPHESUS; HERODOTUS OF HALICARNASSUS

GERALD L. GEISON
Princeton University
HENFREY

PATSY A. GERSTNER
Howard Dittrick Museum of Historical Medicine, Cleveland
J. HILL

OWEN GINGERICH
Smithsonian Astrophysical Observatory
A. HALL

J. ELISE GORDON
University of Oxford
HIGHMORE

STEPHEN JAY GOULD
Museum of Comparative Zoology, Harvard University
HYATT

RAGNAR GRANIT
National Institutes of Health
HOLMGREN

FRANK GREENAWAY
Science Museum, London
HADFIELD

HENRY GUERLAC
Cornell University
S. HALES; THE HAUKSBEES

ALICE A. GUIMOND
Holyoke Community College
HERBERT

KARLHEINZ HAAS
HESSE; HINDENBURG; HUDDE

K. HAJNIS
Charles University
HRDLIČKA

MARIE BOAS HALL
Imperial College of Science and Technology
HARTLIB; HELLOT; HOMBERG

THOMAS L. HANKINS
University of Washington
W. R. HAMILTON

R. S. HARTENBERG
Northwestern University
HIRN

HELMUT HASSE
Journal für die Reine und Angewandte Mathematik
HENSEL

MELVILLE H. HATCH
Thomas Burke Memorial State Museum, University of Washington
HORN; HOWARD

JEAN VAN HEIJENOORT
Brandeis University
HERBRAND

ERICH HINTZSCHE
University of Berne
HALLER; HENLE

TETU HIROSIGE
Nihon University
HONDA

ERNST HÖLDER
University of Mainz
HÖLDER

R. HOOYKAAS
State University of Utrecht
HAÜY

MICHAEL A. HOSKIN
University of Cambridge
C. L. HERSCHEL; W. HERSCHEL

KARL HUFBAUER
University of California, Irvine
HENCKEL

THOMAS PARKE HUGHES
Southern Methodist University
C. M. HALL

AARON J. IHDE
University of Wisconsin
HARDEN; HART

JEAN ITARD
Lycée Henri IV
HENRION

ROBERT JOLY
Free University of Brussels
HIPPOCRATES OF COS

DANIEL P. JONES
Oregon State University
HAHNEMANN

ROBERT H. KARGON
The Johns Hopkins University
HARRIS

M. KATĚTOV
HAUSDORFF

CONTRIBUTORS TO VOLUME 6

GEORGE B. KAUFFMAN
California State College, Fresno
HOWE

H. C. KING
Royal Ontario Museum
HADLEY

MANFRED KOCH
Bergbau Bücherei, Essen
C. F. A. HARTMANN; HECHT; HEYNITZ

ZDENĚK KOPAL
University of Manchester
P. A. HANSEN

SHELDON J. KOPPERL
Grand Valley State College
W. N. HAWORTH

HANS-GÜNTHER KÖRBER
Zentralbibliothek des Meteorologischen Dienstes der DDR, Potsdam
HALLWACHS; W. G. HANKEL; HOLBORN

EDNA E. KRAMER
Polytechnic Institute of Brooklyn
HYPATIA

CLAUDIA KREN
University of Missouri
HERMANN THE LAME

A. D. KRIKORIAN
State University of New York at Stony Brook
HOAGLAND

VLADISLAV KRUTA
Purkyně University
R. P. H. HEIDENHAIN; HERING

P. G. KULIKOVSKY
Academy of Sciences of the U.S.S.R.
HANSKY

GISELA KUTZBACH
University of Wisconsin
HANN

P. S. LAURIE
Royal Greenwich Observatory
HIND

WILLIAM LEFANU
Royal College of Surgeons of England
HAVERS; HEWSON: HOME

HENRY M. LEICESTER
University of the Pacific
G. H. HESS

JACQUES R. LÉVY
Paris Observatory
HAMY; P. M. HENRY; P. P. HENRY

G. A. LINDEBOOM
Free University, Amsterdam
HORNE

ROBERT BRUCE LINDSAY
Brown University
HARTREE

J. A. LOHNE
Municipal Gymnasium, Flekkefjord
HARRIOT

JAMES LONGRIGG
University of Newcastle Upon Tyne
HEROPHILUS

EDYTHE LUTZKER
HAFFKINE

RUSSELL MCCORMMACH
University of Pennsylvania
HERTZ

WILHELM MAGNUS
New York University
HELLINGER

MICHAEL S. MAHONEY
Princeton University
HERO OF ALEXANDRIA

S. MANDELBROJT
Collège de France
HADAMARD

M. V. MATHEW
Royal Botanic Garden Library, Edinburgh
HENSLOW

JOSEF MAYERHÖFER
HASENÖHRL; V. F. HESS

N. M. MERKOULOVA
Academy of Sciences of the U.S.S.R.
HUGONIOT

DANIEL MERRIMAN
Yale University
HJORT

WYNDHAM DAVIES MILES
National Institutes of Health
HARE

M. G. J. MINNAERT†
HOEK; HORTENSIUS

SAMUEL I. MINTZ
City College of the City University of New York
HOBBES

MICHAEL E. MITCHELL
University College, Galway
W. H. HARVEY

KR. PEDER MOESGAARD
University of Aarhus
C. HORREBOW

S. NAKAYAMA
University of Tokyo
HIRAYAMA

AXEL V. NIELSEN†
P. N. HORREBOW

ERIK NORIN
University of Uppsala
HEDIN

J. D. NORTH
Museum of the History of Science, Oxford
T. HENDERSON; HEVELIUS; HORNSBY

MARY JO NYE
University of Oklahoma
HAUTEFEUILLE

ROBERT OLBY
University of Leeds
W. A. O. HERTWIG

PETER D. OLCH
National Institutes of Health
W. S. HALSTED

EUGENIUSZ OLSZEWSKI
Polish Academy of Sciences
M. T. HUBER

JANE M. OPPENHEIMER
Bryn Mawr College
R. G. HARRISON;
K. W. T. R. VON HERTWIG

OYSTEIN ORE†
HOLMBOE

WALTER PAGEL†
HELMONT; HILDEGARD OF BINGEN

JOHN PARASCANDOLA
University of Wisconsin
L. J. HENDERSON; Y. HENDERSON

JOHN PASSMORE
Australian National University
HUME

KURT MØLLER PEDERSEN
University of Aarhus
HANSTEEN; HENRICHSEN

OLAF PEDERSEN
University of Aarhus
HARPESTRAENG

DAVID PINGREE
Brown University
HARIDATTA I; HARIDATTA II;
IBN HIBINTĀ

EMMANUEL POULLE
École Nationale des Chartes
HENRY BATE OF MALINES

JOHANNES PROSKAUER†
HOFMEISTER

HANS QUERNER
University of Heidelberg
HIS

ROY A. RAUSCHENBERG
Ohio University
A. HAWORTH; W. HUDSON

GERHARD REGNÉLL
University of Lund
HISINGER

CONTRIBUTORS TO VOLUME 6

NATHAN REINGOLD
Smithsonian Institution
HARKNESS; HASSLER; HAYFORD; J. HENRY

SAMUEL REZNECK
Rensselaer Polytechnic Institute
HORSFORD; HOUGHTON

P. W. RICHARDS
University College of North Wales
HEDWIG

ARNULF RIEBER
Pädagogische Hochschule Bamberg der Universität Würzburg
HERBART

GUENTER B. RISSE
University of Wisconsin
HOFFMANN

LUCILLE B. RITVO
Albertus Magnus College
G. HARTMANN

ANNE CLARK RODMAN
The Johns Hopkins University
HOWELL

COLIN A. RONAN
HALLEY

GRETE RONGE
HÖNIGSCHMID

PAUL G. ROOFE
University of Kansas
C. J. HERRICK; C. L. HERRICK

K. E. ROTHSCHUH
University of Münster/Westphalia
HENSEN

M. J. S. RUDWICK
University of Cambridge
L. HORNER

EUGENIUSZ RYBKA
Jagiellonian University
HUFNAGEL

A. I. SABRA
Harvard University
IBN AL-HAYTHAM

H. SCHADEWALDT
University of Düsseldorf
HELLRIEGEL; HIRSZFELD

F. SCHMEIDLER
University of Munich
HALM

RUDOLF SCHMITZ
University of Marburg
HARTIG; J. HARTMANN

BRUNO SCHOENEBERG
University of Hamburg
HECKE

E. L. SCOTT
Stamford High School, Lincolnshire
HATCHETT; T. HENRY; W. HENRY; HOPE

J. F. SCOTT†
HEATH

C. D. SHANE
University of California, Santa Cruz
HUSSEY

ELIZABETH NOBLE SHOR
G. S. HALL

DIANA M. SIMPKINS
Polytechnic of North London
R. A. HARPER; J. HUNT

PIETER SMIT
Catholic University Nijmegen
HOEVEN; HUSCHKE

CYRIL STANLEY SMITH
Massachusetts Institute of Technology
HUME-ROTHERY

I. SNAPPER
Veterans Administration Hospital, Brooklyn
HEURNE

H. A. M. SNELDERS
State University of Utrecht
HILDEBRANDT

E. SNORRASON
Rigshospitalet. Copenhagen
E. C. HANSEN

ERNEST G. SPITTLER
John Carroll University
HINSHELWOOD

NILS SPJELDNAES
University of Aarhus
HAMBERG

J. STEUDEL
Medizinhistorisches Institut, University of Bonn
HYRTL

K. AA. STRAND
U.S. Naval Observatory
HERTZSPRUNG

PER STRØMHOLM
HERIGONE; HYLLERAAS

D. J. STRUIK
Massachusetts Institute of Technology
HEURAET

CHARLES SÜSSKIND
University of California, Berkeley
W. W. HANSEN; HEAVISIDE; HULL

FERENC SZABADVÁRY
Technical University, Budapest
HEVESY

RENÉ TATON
École Pratique des Hautes Études
HACHETTE

MIKULÁŠ TEICH
University of Oxford
HEYROVSKÝ; HORBACZEWSKI

ARNOLD THACKRAY
University of Pennsylvania
B. HIGGINS; W. HIGGINS

JEAN THÉODORIDÈS
Centre National de la Recherche Scientifique
HALLIER; HÉRELLE

V. V. TIKHOMIROV
Academy of Sciences of the U.S.S.R.
HELMERSON

HEINZ TOBIEN
University of Mainz
HEER

CHRISTOPHER TOLL
University of Uppsala
AL-HAMDĀNĪ

HENRY S. TROPP
Smithsonian Institution
G. B. HALSTED

R. STEVEN TURNER
University of New Brunswick
HELMHOLTZ

GEORG USCHMANN
Deutsche Akademie der Naturforscher Leopoldina
HAECKEL; HATSCHEK

P. W. VAN DER PAS
HEISTER

ARAM VARTANIAN
New York University
D'HOLBACH

JUAN VERNET
University of Barcelona
IBN HAWQAL; HERNÁNDEZ

T. M. VOGELSANG
G. H. A. HANSEN

HELEN WALLIS
British Museum
HAKLUYT

RICHARD S. WESTFALL
Indiana University
HOOKE

FRANZ WEVER
Max-Planck-Institut für Eisenforschung, Düsseldorf
HEYN

JOYCE WEVERS
Vening Meinesz Laboratory, Utrecht
HAIDINGER

CHARLES A. WHITNEY
Smithsonian Astrophysical Observatory
HOLDEN

GERALD J. WHITROW
Imperial College of Science and Technology
HUBBLE

CONTRIBUTORS TO VOLUME 6

MARY E. WILLIAMS
Skidmore College
L. S. HILL

WESLEY C. WILLIAMS
Case Western Reserve University
HUXLEY

CURTIS A. WILSON
University of California, San Diego
HEYTESBURY

FRANK H. WINTER
Smithsonian Institution
W. HALE

MELVILLE L. WOLFROM†
C. S. HUDSON

HELEN WRIGHT
G. E. HALE

HATTEN S. YODER, JR.
Carnegie Institution of Washington Geophysical Laboratory
HAGUE

ROBERT M. YOUNG
University of Cambridge
HARTLEY

DICTIONARY
OF
SCIENTIFIC BIOGRAPHY

DICTIONARY OF SCIENTIFIC BIOGRAPHY

HACHETTE—HYRTL

HACHETTE, JEAN NICOLAS PIERRE (*b.* Mézières, Ardennes, France, 6 May 1769; *d.* Paris, France, 16 January 1834), *geometry, theory of machines, physics.*

The son of Jean Pierre Hachette, a bookseller, and Marie Adrienne Gilson, Hachette studied first at the *collège* of Charleville. He also attended the elementary technical courses organized at the École Royale du Génie of Mézières, where he favorably impressed Monge, Clouet, and C. Ferry. Beginning in 1788, after having completed his education at the University of Rheims (1785–1787), he was draftsman and technician at the École Royale du Génie of Mézières and assisted Ferry in teaching descriptive geometry, which Monge had introduced in this school. Following a competitive examination he was appointed professor of hydrography at Collioure and Port Vendres in 1792. The following year he returned to Mézières to teach mathematics, replacing Ferry, who had been elected a deputy to the Convention. A fervent revolutionary, Hachette was active in the political life of his native city, and at the École du Génie he sought rapid training of officers qualified for the revolutionary army and to remove teachers and students whose patriotism seemed doubtful to him.

Summoned to Paris by the Committee of Public Safety in 1794, Hachette carried out various technological and industrial assignments (military applications of balloons, manufacture of weapons, and so on) with Guyton de Morveau and Monge. He participated at the same time in the discussions concerning the reorganization of higher scientific and technical education. At the time of the creation of the École Polytechnique in November 1794—under the name École Centrale des Travaux Publics—Hachette took an active part in preparing the future instructors and then in the teaching of descriptive geometry, as assistant professor (1794) and later as full professor (1799) until April 1816.

Besides this course at the École Polytechnique, Hachette taught descriptive geometry at the short-lived École Normale de l'An III (from January to May 1795, as Monge's assistant) and then, from 1810, as assistant professor, at the Paris Faculty of Sciences and at the newly reestablished École Normale. In addition he taught at various schools that prepared students for the École Polytechnique and at the École des Pages, created by Napoleon in 1805. Through these various posts Hachette became one of the chief popularizers of the new methods that Monge had introduced in the various branches of geometry. An intimate friend and devoted collaborator of Monge—and editor of several of his works—Hachette shared his faith in the great value of science and technology as an element of social progress and in the importance of the École Polytechnique's role in this regard.

Having dedicated himself to the organization and launching of the École Polytechnique in 1794 and 1795, Hachette continued for more than twenty years to take an active interest both in its general orientation—he was several times a member of its Conseil de Perfectionnement—and in the life, work, and future of its students. Many of them were grateful to him for having guided their first research projects and for having kept in touch with them after graduation. Through his influence and contact with former students Hachette helped to raise the prestige of the École Polytechnique, inculcating in its best students a passion for scientific research, both pure and applied. In order to join this effort to the diffusion of his views, he was an editor of the *Journal de l'École polytechnique* and, in addition, created and directed an extremely valuable organ for the presentation of information and for the exchange of ideas, *Correspondance sur l'École polytechnique* (1804–1816), which contained the first works of some of the leading French scientists of the first half of the nineteenth century: Poisson, Fresnel, Cauchy, Malus, Brianchon, Chasles, and Lamé, among others.

In view of this great activity, Hachette was extremely pained when, in September 1816, the Restoration government excluded him from the École Polytechnique at the time of its reorganization. This political rancor manifested itself again in December

1823, when Louis XVIII refused to confirm his election to the mechanics section of the Académie des Sciences. (He was not elected to that body until October 1831, under the reign of Louis Philippe.) Yet, except at Mézières in 1793, Hachette does not seem to have played a notable role in politics, although he did remain faithful to the great ideas of the Revolution.

His exclusion from the École Polytechnique did not prevent Hachette from completing a series of pedagogical works for its students. In fact, it permitted him to concern himself more actively with the rise of new industrial and agricultural techniques. He was a member of the Société d'Encouragement à l'Industrie Nationale, of the Société Royale et Centrale d'Agriculture, and of the Comité Consultatif des Arts et Manufactures. This activity made concrete earlier preoccupations: his posts at the École Royale du Génie of Mézières and at the École Polytechnique, his work during 1794, and the influence of Monge had made him familiar with the problem of the relationships between technology and the mathematical and physical sciences.

By his marriage in 1810 to Jeanne Maugras, the daughter of a surgeon, Hachette had a son, Amédée Barthélémy, who became chief engineer of the Ministère des Ponts et Chaussées, and a daughter, who married the chemist J. J. Ebelmen, later director of the Sèvres porcelain factory.

Despite the internal unity of Hachette's scientific and technical work, the latter can be divided into three major parts: geometry, pure and applied mechanics (including the theory of machines), and physics (electricity, magnetism, optics, and the study of instruments).

Hachette collaborated with Monge in the writing of an exposition of three-dimensional analytic geometry that dealt especially with changes of coordinates and with the theory of second-degree surfaces (*Journal de l'École polytechnique,* **11** [1802], 143–169), a more complete version of which appeared in book form several years later as *Application de l'algèbre à la géométrie* (Paris, 1805). Hachette later drew from this work an analytic theory of second-degree surfaces (1813, 1817) enriched by the progress made in the meantime.

In pure and descriptive geometry Hachette disseminated and continued Monge's work, developing effective procedures for solving various problems and studying diverse properties of space curves and surfaces (tangents and tangent planes, elements of curvature, and so on) by the methods of synthetic geometry joined to perspective and projective geometry. The results he obtained heralded the development of projective geometry and modern geometry in the nineteenth century.

In physics Hachette was especially interested in optics, electricity, magnetism, and the theory of optical instruments. Several of his articles and his *Programme* of 1809 show the influence of Monge, Guyton de Morveau, and Oersted.

In the courses on the theory of machines that he gave at the École Polytechnique beginning in 1806, in the *Programme,* and in the *Traité* that he published in 1808 and 1811, Hachette developed Monge's ideas on the distinction between the motor, the mechanisms of transmission and their movements, and the classification of transmission mechanisms (or elementary machines) according to the nature of the transformations of movements that they produce. The *Traité,* which includes important advances in applied mechanics and detailed studies of many types of machines, exerted a great influence on the beginnings of the theory of machines. Hachette was also interested in applied hydrodynamics and in steam engines and their history.

Although not a scientist of the first rank, Hachette nevertheless contributed to the progress of French science at the beginning of the nineteenth century by his efforts to increase the prestige of the École Polytechnique and by making Monge's work widely known, especially in descriptive and analytic geometry and in the theory of machines.

BIBLIOGRAPHY

I. ORIGINAL WORKS. Besides about 100 memoirs, articles, and notes—a list of which is given in *Correspondance sur l'École polytechnique,* III, 421, and in Royal Society, *Catalogue of Scientific Papers,* III, 106–109—Hachette edited the three vols. of the *Correspondance sur l'École polytechnique* (Paris, 1804–1816) and published the following works: (1) *Application de l'algèbre à la géométrie* (Paris, 1805; reiss. 1807), written with Gaspard Monge, reiss. as *Traité des surfaces du second degré* (1813) and as *Éléments de géométrie à trois dimensions. Partie algébrique* (1817).

(2) *Programme d'un cours élémentaire sur les machines* (Paris, 1808), pub. with P. L. Lanz and A. de Bétancourt, *Essai sur la composition des machines,* and developed in *Traité élémentaire des machines . . .* (Paris, 1811; 4th ed., 1828).

(3) *Programme d'un cours de physique . . .* (Paris, 1809).

(4) *Supplément à la Géométrie descriptive* (Paris, 1811), *Cours de géométrie descriptive* (Paris, 1817), a collection of diagrams, and *Second supplément à la Géométrie descriptive* (Paris, 1818); the various elements of these works reappeared either in *Éléments de géométrie à trois dimensions, Partie synthétique* (Paris, 1817) or in *Traité de géométrie descriptive* (Paris, 1822; 2nd ed., 1828).

(5) *Histoire des machines à vapeur* . . . (Paris, 1830).

In addition, Hachette edited the first separately printed ed. of Monge's *Géométrie descriptive* (Paris, 1799), the reissues of that work, and the new ed. of 1811, as well as the fifth and succeeding eds. of Monge's *Traité élémentaire de statique* (Paris, 1809 ff.), and the third and fourth eds. of his *Application de l'analyse à la géométrie* (1807, 1809). In addition Hachette edited Auguste Comte's French trans. of John Leslie, *Elements of Geometry, Geometrical Analysis, and Trigonometry,* 2nd ed. (London, 1811), as *Analyse géométrique* (Paris, 1818) and published the French trans. of Thomas Young, *A Course of Lectures on Natural Philosophy and the Mechanical Arts,* 2 vols. (London, 1807), as *Précis de mécanique et de la science des machines* (Paris, 1829).

II. Secondary Literature. Hachette's life and works were the subject of the following accounts (in chronological order), some of which contain quite serious errors (e.g., date of birth, participation in the Egyptian expedition): F. Arago and S.-D. Poisson, *Funérailles de M. Hachette* (Paris, 1834); A. F. Silvestre, *Discours prononcé sur la tombe de M. Hachette* (Paris, 1834); C. Dupin and A. Quételet, in *Annuaire de l'Académie royale de Bruxelles* for 1836 (1836), 71–77; V. Parisot, in Michaud, ed., *Biographie universelle,* LXVI (supp.), 339–341, also in new ed., XVIII (Paris, 1857), 314–315; L. Louvet, in F. Hoefer, ed., *Nouvelle biographie générale,* XXIII (Paris, 1861), cols. 26–29; Poggendorff, I, cols. 985–986; A. Hannedouche, *Les illustrations ardennaises* . . . (Sedan, 1880), pp. 81–82; L. Sagnet, in *Grande encyclopédie,* XIX (Paris, n.d.), 698–699; and N. Nielsen, *Géomètres français sous la Révolution* (Copenhagen, 1929), pp. 121–125.

René Taton

HADAMARD, JACQUES (*b.* Versailles, France, 8 December 1865; *d.* Paris, France, 17 October 1963), *mathematics.*

Hadamard was the son of Amédée Hadamard, a Latin teacher in a noted Paris lycée; his mother, Claude-Marie Picard, was a distinguished piano teacher. After studying at the École Normale Supérieure from 1884 to 1888, he taught at the Lycée Buffon in Paris from 1890 to 1893 and received his *docteur ès sciences* degree in 1892. Hadamard was a lecturer at the Faculté des Sciences of Bordeaux from 1893 to 1897, lecturer at the Sorbonne from 1897 to 1909, then professor at the Collège de France from 1909 to 1937, at the École Polytechnique from 1912 to 1937, and at the École Centrale des Arts et Manufactures from 1920 to 1937.

Elected a member of the Académie des Sciences in 1912, Hadamard was also an associate member of several foreign academies, including the National Academy of Sciences of the United States, the Royal Society of London, the Accademia dei Lincei, and the Soviet Academy of Sciences. In addition he held honorary doctorates from many foreign universities.

Hadamard's interest in pedagogy led him to write articles about concepts in elementary mathematics that are introduced in the upper classes of the lycée. His *Leçons de géométrie élémentaire* (1898, 1901) still delight secondary school instructors and gifted pupils. The extent of his grasp on all domains of advanced research in France and abroad was evident in his famous seminar at the Collège de France; no branch of mathematics was neglected. The world's most famous mathematicians came there to present their own findings or those related to their specialty. But it was always Hadamard who had the last word and the surest judgment concerning the significance or the potential of the research presented.

Hadamard's first important works were concerned with analytic functions, notably with the analytic continuation of a Taylor series. Although Karl Weierstrass and Charles Méray were the first to define the meaning that must be attributed to the domain of existence of the analytic continuation of a Taylor series, their reflections amounted to a theorem of existence and uniqueness. Before Hadamard, little was known about the nature and distribution of the singularities of the series in terms of the nature of its coefficients, which define the function a priori. His thesis (1892), preceded by several notes in the *Comptes rendus* of the Academy, is one of his most beautiful works. For the first time an ensemble concept was introduced into function theory. In fact the upper limit—made more precise, explained, and applied to the ensemble composed of the coefficients—permits the determination of the radius of convergence (or, rather, its inverse) of the Taylor series. Conditions affecting the coefficients enable one to characterize the singular points on the circle of convergence. Hadamard's famous theorem on lacunary series (*"lacunes à la Hadamard"*) admitting the circle of convergence as a cut, the theorems on polar singularities, the introduction of the concept of *"écart fini"* and of the "order" of a singular point, and the theorem on the composition of singularities (1898), have remained fundamental in function theory. The results have inspired generations of highly talented mathematicians, especially those working at the beginning of the twentieth century and in the years between the two world wars. His *La série de Taylor et son prolongement analytique* (1901) was the "Bible" of all who were fascinated by the subject.

The year 1892 was one of the most fertile in the history of the theory of functions of a complex variable; it also marked the publication of a work by Hadamard that established the connection between the decrease of the modulus of the coefficients of the

Taylor series of an integral function and the genus of the function. This work (which received the Grand Prix of the Académie des Sciences) and the results of his thesis (especially those pertaining to polar singularities), applied to Riemann's ζ function, enabled Hadamard in 1896 to solve the ancient and famous problem concerning the distribution of the prime numbers. He demonstrated (in a less explicit form than is shown here but one easily reducible to it) that the function $\pi(x)$ designating the number of prime numbers less than x is asymptotically equal to $x/\log x$. This is certainly the most important result ever obtained in number theory. Charles de La Vallée-Poussin proved the theorem at the same time, but his demonstration is much less simple than Hadamard's. The total result seems to indicate—without, we believe, its ever having been mentioned in writing by Hadamard—that the research in his thesis and in his work on integral functions was implicitly directed toward the ultimate goal of indicating the properties of the function ζ, in order to derive from it the theorem on the prime numbers.

Returning to analytic functions, one should mention the 1896 theorem on the maximum modulus of an integral function (or of a holomorphic function in a disk). And, while remaining close to the essential principles of analytic functions but leaving aside those with a complex variable, one must emphasize Hadamard's introduction (1912, 1925) of the idea and of the problem of quasi analyticity, which consists in finding a relationship between the growth of the maxima of the moduli of the derivatives of a function on a segment and the fact of being determined in a unique way by the values that the function and its derivatives take at a point. It should be noted that it was Albert Holmgren's considerations relating to Augustin Cauchy's problem for the equation of heat that had led Hadamard to consider classes of infinitely differentiable functions, not necessarily analytic on a segment but nevertheless possessing the characteristic property of uniqueness on the segment. The idea of quasi analyticity plays a significant role in modern analysis.

It is important to emphasize a subject treated by Hadamard in which, avoiding analysis (under the circumstances, differential geometry) and replacing it with consideration of analysis situs or topology, he was able to display, in one of his most beautiful memoirs (1898), the philosophic character—referring to astronomical ideas—of the fundamental concept of "the problem correctly posed," although no concrete allusion to this term figured in it. This idea of the correctly posed problem played an essential part in Hadamard's later researches on equations with partial derivatives. The importance of analysis situs in the theory of differential equations was shown by Henri Poincaré, whom Hadamard admired greatly and to whose work he devoted several memoirs and monographs (1922, 1923, 1954).

The memoirs in question (1898) treat surfaces of negative curvature having a finite number of nappes extending to infinity. All analytic description is abandoned. On these surfaces the geodesics behave in three different ways: (1) they are closed or asymptotic to other such geodesics; (2) they extend to infinity on one of the nappes; (3) entire segments of these geodesics approach successively a series of closed geodesics, the length of these segments growing toward infinity. The striking thing is that the ensemble E of tangents to the geodesics passing through a point and remaining at a finite distance is perfect and never dense; and in each neighborhood of every geodesic whose tangent belongs to E (neighborhood of directions) there exists a geodesic which extends to infinity in an arbitrarily chosen nappe. In each of these neighborhoods there also exist geodesics of the third category. Hadamard states: "*Any change, however small, carried in the initial direction of a geodesic which remains at a finite distance is sufficient to produce any variation whatsoever in the final aspect of the curve,* the disturbed geodesic being able to take on any one of the forms enumerated above" (*Oeuvres,* p. 772).

But in a physics problem a slight modification in the circumstances at a certain moment ought to have little influence on the solution, since one never possesses conditions which are more than approximate. Hadamard concluded from this that the behavior of a trajectory might well depend on the arithmetic character of the constants of integration. One already sees here the genesis of the idea of the "problem correctly posed" which guided Hadamard in his researches on equations with partial derivatives. The problem of geodesics on the surfaces studied by Hadamard is not a correctly posed mechanics problem.

Hadamard fully set out the idea of the correctly posed problem for equations with partial derivatives in his excellent *Lectures on Cauchy's Problem in Linear Differential Equations* (1922; French ed., 1932). Thus, for Laplace's equation, Dirichlet's problem is a correctly posed problem; on the other hand, for an equation of the hyperbolic type, Cauchy's problem is the one which meets this criterion. These ideas have had a great influence on modern research because they have shown the necessity of introducing different types of neighborhoods and, in consequence, different species of continuity; these conceptions led to general topology and functional analysis. Also in the *Lectures* is the notion of the "elementary solution" which has so much in common with that of "distribution" (or "generalized function"). Also in connection with equations with partial derivatives, one should mention the concept of the "finite portion" of a divergent integral, which plays an essential role in the solution of Cauchy's problem.

Hadamard took a lively interest in Vito Volterra's functional calculus and suggested the term "functional" to replace Volterra's term "line function." Above all, in 1903 Hadamard was able to give a general expression for linear functionals defined for continuous functions on a segment. This was the ancestor of Friedrich Riesz's fundamental formula.

Few branches of mathematics were uninfluenced by the creative genius of Hadamard. He especially influenced hydrodynamics, mechanics, probability theory, and even logic.

BIBLIOGRAPHY

Hadamard's writings were collected as *Oeuvres de Jacques Hadamard,* 4 vols. (Paris, 1968). The years within parentheses in the text will enable the reader to find, in the bibliography of the *Oeuvres,* any *mémoire* that interests him.

On Hadamard or his work, see Mary L. Cartwright, "Jacques Hadamard," in *Biographical Memoirs of Fellows of the Royal Society,* **2** (Nov. 1965); P. Lévy, S. Mandelbrojt, B. Malgrange, and P. Malliavin, *La vie et l'oeuvre de Jacques Hadamard,* no. 16 in the series L'Enseignement Mathématique (Geneva, 1967); and S. Mandelbrojt and L. Schwartz, "Jacques Hadamard," in *Bulletin of the American Mathematical* Society, **71** (1965).

S. MANDELBROJT

HADFIELD, ROBERT ABBOTT (*b.* Attercliffe, Sheffield, England, 28 November 1858; *d.* Kingston, Surrey, England, 30 September 1940), *metallurgy.*

Hadfield was the only son of Robert Hadfield and Marianne Abbott. In 1872 the elder Hadfield initiated production of steel castings in England. The resultant breaking of the French monopoly in such articles as steel projectiles led to the development of a great arms industry. The younger Hadfield was educated at the Collegiate School in Sheffield, where he developed an interest in chemistry. On being employed in his father's works, he set up its first laboratory.

In 1882, seeking a solution to a local production problem, he began a systematic study of the alloys of iron with silicon and manganese. He prepared a steel with 12–14 percent manganese, possessing remarkable properties of resistance to crushing and abrasion—invaluable in such applications as railway points and grinding machinery. It also became grimly familiar as the material used for steel helmets in World War I.

Partly at the instigation of the British Association for the Advancement of Science, Hadfield worked on silicon steels, which were further investigated by William Barrett. These steels turned out to have exceptionally low magnetic hysteresis (1899) and, after a seven-year development period, made possible smaller, lighter, and more efficient electric transformers.

Hadfield also collaborated in other scientific research, for example, with Dewar and with the Leiden school on properties of metals at low temperatures. But after assuming the chairmanship of his father's company in 1888, his influence was chiefly felt in the systematic improvement of the production of steel and steel products, such as armor plate and armor-piercing shells.

Hadfield took a broad view of his subject and was a pioneer in the experimental investigation of historical metallurgical problems. He studied the Delhi Iron Pillar (fourth century A.D.) and the Faraday-Stodart alloys of 1818–1822. He was president of the Faraday Society (1914–1920) and was elected a fellow of the Royal Society of London in 1909. He was knighted in 1908 and created baronet in 1917.

BIBLIOGRAPHY

Hadfield's two books are *Metallurgy and Its Influence on Modern Progress* (London, 1925) and *Faraday and His Metallurgical Researches* (London, 1931). Information on his life and work is in *Obituary Notices of Fellows of the Royal Society of London,* no. 10 (1940).

FRANK GREENAWAY

HADLEY, JOHN (*b.* Hertfordshire, England, 16 April 1682; *d.* 14 February 1744), *optical instrumentation.*

Hadley was the son of George Hadley, a deputy lieutenant and, after 1691, high sheriff of Hertfordshire, England, and Katherine Fitzjames. Nothing is known of his early life or of the places of his education.

Hadley was the first to develop the form of reflecting telescope introduced by Newton in 1668. By 1719 he had produced paraboloidal mirrors of speculum metal superior to any made by the London master opticians. He then constructed two Newtonian reflectors with an aperture of 5 7/8 inches and a focus of 5 1/4 feet and, in 1726, a small Gregorian reflector.

He presented one of the Newtonian reflectors to the Royal Society of London, where it evoked great interest. James Bradley and James Pound compared it with an eight-inch object glass of 123-foot focus which Christiaan Huygens had made and presented to the Society. The reflector outperformed Huygens' refractor in both manageability and definition.

Hadley communicated his grinding and polishing methods to Bradley and Samuel Molyneux, who in turn instructed some of the London master opticians. He also befriended the Scottish optician James Short, then about to set up in London as a maker of Gregorian reflectors.

In 1731 Thomas Godfrey, a young American glazier, made a reflecting octant. In the same year, quite independently, Hadley produced a similar instrument. Both instruments, precursors of the modern nautical sextant, were based on a mirror arrangement proposed by Newton but not described in print until 1742. After Bradley had tested Hadley's octant at sea and obtained altitude readings down to one minute of arc, the instrument was universally adopted.

Hadley played an active part in the affairs of the Royal Society. Elected fellow in 1717, he was annually elected a member of council from 1726 until the year of his death and became vice-president in 1728. In 1726 he was one of the committee appointed by the Society to examine and report on the new instruments which Edmond Halley had obtained for the Royal Greenwich Observatory.

BIBLIOGRAPHY

I. ORIGINAL WORKS. Hadley's Newtonian reflector is described by him in "An Account of a Catadioptrick Telescope, Made by John Hadley, Esq; F.R.S. With the Description of a Machine Contriv'd by Him for Applying It to Use," in *Philosophical Transactions of the Royal Society,* **32** (1723), 303–312. Also see "A Letter from the Rev. Mr. James Pound, Rector of Wanstead, F.R.S., to Dr. Jurin, Secretary R.S. Concerning Observations Made With Mr. Hadley's Reflecting Telescope," *ibid.,* 382–384. The Society still possesses Hadley's mirror, five eyepieces, and the reflecting octant which is described in "The Description of a New Instrument for Taking Angles," *ibid.,* **37** (1731) 147–157. Hadley's own account of the grinding, polishing, and testing of concave specula comprises most of bk. 3, ch. 2, of R. Smith, *A Compleat System of Opticks* (Cambridge, 1738).

II. SECONDARY LITERATURE. A most complete sketch of Hadley's work is contained in *Biographical Account of John Hadley,* an unsigned and undated tract in the library of the Royal Astronomical Society, London. The library also contains a similar tract, *The Invention and History of Hadley's Quadrant.* The significance of Hadley's instruments in the history of applied optics and astronomy is discussed by H. C. King, *The History of the Telescope* (London, 1955), pp. 77–84.

H. C. KING

HAECKEL, ERNST HEINRICH PHILIPP AUGUST (*b.* Potsdam, Germany, 16 February 1834; *d.* Jena, Germany, 9 August 1919), *zoology.*

Haeckel's father, Carl Haeckel, was chief administrative advisor for religious and educational affairs in Merseburg. His mother, Charlotte Sethe, was the daughter of a privy councillor in Berlin.

Haeckel graduated from the Domgymnasium at Merseburg in 1852. After studying medicine at Berlin, Würzburg, and Vienna, he earned his medical degree at Berlin in 1857 and passed the state medical examination there in 1858. In 1861 he qualified as a lecturer in comparative anatomy at the Faculty of Medicine of the University of Jena. Appointed associate professor of zoology in the Faculty of Philosophy in 1862, he was promoted to full professor and director of the Zoological Institute in 1865. He retired in 1909.

Haeckel married his cousin Anna Sethe in 1862. She died in 1864 and in 1867 he married Agnes Huschke, the daughter of the anatomist Emil Huschke. Powerfully built, Haeckel enjoyed gymnastics and swimming, although for a time he suffered from rheumatoid arthritis. He worked quickly and intensively over long periods, offsetting this pace with long hikes and extended trips. Haeckel was also a member of more than ninety learned societies and scientific associations, including the Leopoldine Academy (1863), the Bavarian Academy of Sciences at Munich (corresponding member, 1870; foreign member, 1891), the Imperial Academy of Sciences at Vienna (corresponding member, 1872), the Royal Academy of Sciences at Turin (corresponding member, 1881; foreign member, 1898), the Royal Swedish Academy of Sciences at Stockholm (associate member, 1882), the Royal Lombard Institute of Sciences and Letters at Milan (corresponding member, 1884), the American Philosophical Society (1885), the Royal Society of Edinburgh (1888), and the Royal Academy of Sciences of the Institute at Bologna (1909). He was the recipient of many scientific honors.

During his school years Haeckel was an enthusiastic botanist and began an extensive herbarium that is still of scientific value.[1] After reading Matthias Schleiden's popular book *Die Pflanze und ihr Leben* (1848) and accounts of various expeditions, including those of Darwin, Humboldt, and Robert Schomburgk, Haeckel wanted to study botany under Schleiden at Jena and then to undertake scientific expeditions of his own. From a very early age drawing and painting were among his favorite pursuits. His aptitude for rapidly and accurately classifying plants, his love of collecting, and his pleasure in artistic activity marked all of his later work.

In 1852 Haeckel gave up his own plans to follow his parents' wish that he study medicine. After a period of resistance he realized that medical school offered him the most solid foundation for further scientific study. But this reconciliation to medicine did not extend to clinical medicine, since Haeckel never seriously intended to become a physician. While studying under Albert von Kölliker and Franz Leydig at Würzburg he became interested in comparative anatomy and embryology, as well as in microscopical investigations. At the same time, Haeckel was also influenced by the "mechanistic conception of the life processes" put forward by Rudolf Virchow, who at that time (1853) was in the midst of writing his *Cellularpathologie.*[2]

At Berlin in 1854–1855, however, Haeckel found in Johannes Müller "an authority recognized by all," and Müller became his "scientific ideal." Under his guidance Haeckel deepened his knowledge of comparative anatomy and was introduced to marine zoology, a field which, through Müller's studies of the lower marine animals on Helgoland and in the Mediterranean, was advancing the development of scientific zoology.

At Würzburg, moreover, Haeckel was confronted for the first time with materialistic conceptions of life. He considered such views—which he encountered in Virchow's lectures, in the writings of Carl Vogt, and in discussions with young scientists and physicians—to be the "opposite extreme" of the "caricature of the Christian religion" represented by dogmatic Catholicism. From the letters Haeckel wrote during these years it is evident that he was already leaning toward a compromise between Christianity and mechanistic materialism, a compromise he believed he had found by 1866 in his "monism."

Haeckel's first zoological work was his doctoral dissertation, *Über die Gewebe des Flusskrebses* (1857). He had intended, after finishing his medical studies in 1858, to complete his training in comparative anatomy and zoology at Berlin under Müller, but these plans were frustrated by Müller's death. At this time the anatomist Karl Gegenbaur offered Haeckel the attractive prospect of a future zoology professorship at Jena and encouraged him to undertake a zoological expedition in the Mediterranean. In the course of this trip (1859–1860) Haeckel discovered, following up Müller's last work on radiolarians at Messina, 144 new radiolarian species, thereby establishing the basis for the monograph *Die Radiolarien* (1862). This work contains Haeckel's first avowal of Darwinism, to which he was immediately converted upon reading the German translation of Darwin's *On the Origin of Species.*

Darwin's book provided a foundation and a direction for Haeckel's future work. His technical writings on zoology, some of them long monographs, treated the morphology, systematics, and embryology of the radiolarians, medusae, siphonophores, sponges, and echinoderms. The research for these works was carried out with the methods of the prephylogenetic period, and they clearly show Müller's influence. The new element consisted at first in the interpretation of the results in the light of Darwin's theory: the systems of the recent organisms were considered to be the reflected images of their phylogenetic development. But Haeckel was not satisfied with interpreting Darwin's theory of evolution and furnishing additional evidence for it. He thought his task lay in the further development of Darwinism. In his view this development ought to lead not only to a reform of the whole of biology; it should also provide the foundation for a science-based world view. Toward this goal he published his *Generelle Morphologie der Organismen* in 1866.

At the time Haeckel was writing this treatise, some eminent scientists had already publicly supported Darwin, while others were skeptical of his theory or rejected it completely. In addition the ideological consequences inherent in Darwinism had touched off vehement disputes. While Darwin himself did not take part in these debates, Haeckel deliberately refused to restrict himself to the field of biology.

The goal of the reformed morphology that Haeckel sought was not only to describe the forms of organisms but also to account for them in terms of the theory of evolution. This morphology was consequently divided into anatomy, or the science of developed forms (tectology and promorphology), and morphology, or the science of emerging forms (ontogeny and phylogeny). For Haeckel the correct method of research was "philosophical empiricism," the interaction of induction and deduction. The mechanical-causal approach was to take the place of any dogmatic or vitalistic-teleological way of viewing nature. Following the linguist August Schleicher,[3] Haeckel termed the philosophical system that corresponded to this approach "monism," the unity of mind and matter, in contrast with dualism, the separation of mind and matter.

Accordingly, in this system there are no absolute differences between organic and inorganic substances, only relative ones. Haeckel contended that the material basis of the true life phenomena, nourishment and reproduction, lay in the very intricate chemical composition of the carbon compounds and in the resultant unique physical properties (above all the capacity for imbibition). In contrast with Darwin, Haeckel asserted that the theory of evolution could be applied even to the emergence of the first primal organisms, which were formed spontaneously through abiogenesis. First complex molecules were formed, followed by a formless plasma clump, or *Moner.* From one or from several such *Moner* one could deduce the genealogical tree of the entire organic kingdom. In his classificatory scheme Haeckel inserted an intermediate kingdom of prostista between animals and plants. Each organic kingdom consisted of several *Stämmen,* or phyla. The *Stamm,* or phylum, was "the totality of all the organisms existing at present, or that are extinct, that are descended from one and the same common progenitor."

The "natural system" of the organisms is, according to Haeckel, "their natural family tree, the table of their genealogical relationships." He first published such genealogical tables or "family trees" in the *Generelle Morphologie* for organisms (plants, prostista, animals), plants, coelenterates, echinoderms, articulates (infusorians, worms, arthropods), mollusks, vertebrates, and mammals (including man).

Haeckel continually strove to produce a comprehensive theory of the process of evolution by demonstrating regularities, the majority of which were of a speculative nature. He set forth a series of laws of heredity that, unlike those of Mendel, were not based on experiment. Haeckel distinguished between conservative heredity (the inheritance of heritable characters) and progressive heredity (the inheritance of acquired characters). The interaction of progressive and conservative heredity, he held, makes possible the transmutation of species.

In Haeckel's view, the cell nucleus governs the inheritance of heritable characters and the plasma regulates the organism's adaptation to the environment. He repeatedly asserted that without the Darwinian theory "all the great and universal phenomena of organic nature" are incomprehensible and inexplicable. This insistence on the fundamental importance of Darwin's ideas is most apparent in his "ecology" and "chorology" of organisms. Haeckel defined ecology as "the comprehensive science of the relationships of the organism to the environment," encompassing all the conditions of existence of organic and of inorganic nature. Chorology he defined as "the entire science of the spatial distribution of organisms, that is, of their geographical and topographical extension over the earth's surface." Both concepts have won acceptance.

Haeckel considered the causal nexus of biontic and phyletic development to be an important law: "Ontogeny is the short and rapid recapitulation of phylogeny determined by the physiological functions of heredity (propagation) and adaptation (nourishment)." For this relationship, which had been formulated before him, he later (1872) coined the expression "fundamental biogenetic law." According to Haeckel this law was especially important because the embryological and systematic data available were much more complete than the paleontological. He divided phylogeny into three stages: *epacme* (blossoming), *acme* (peak flowering), and *paracme* (withering).

Drawing on the earlier writings of T. H. Huxley, Carl Vogt, Rolle, Filippo de Filippi, and Charles Lyell, Haeckel undertook a thorough study of the origin of man, setting the Tertiary as the time when man developed from the apes. He thought that the most important advance in that process was the "differentiation of the larynx, which resulted in the development of language and, consequently, of clearer communication and of historical tradition." In Haeckel's outline of a natural system, man is included among the tailless Catarrhinae; the term *Pithecanthropus,* which he coined, first appeared in this context.

For Haeckel anthropology was a part of zoology, since "man is separated from the other animals only by quantitative, not qualitative, differences," and since the methods of comparative anatomy can be applied to man.

In the concluding section of *Generelle Morphologie* Haeckel discussed "the unity of nature and the unity of science (system of monism)" as well as "God in nature (pantheism and monotheism)." Here he showed the significance of embryology for human knowledge generally, which in his view finds its most comprehensive expression in a "cosmology or nature philosophy." This philosophy was for him "identical with natural theology." Monism is conceived of here as the purest monotheism, in which God corresponds to the general causal law ("the unity of God in nature").

Generelle Morphologie contained all the essential aspects of Haeckel's later work. After 1866 he changed neither his methods nor his goal in any significant way. His zoological works included descriptions of approximately 4,000 new species of lower marine animals—mainly radiolarians, medusae, and sponges. For these groups of animals he established phylogenetically interpreted "natural systems."

Haeckel's concept of matter provides an especially clear illustration of the discrepancy between the scientific basis of his work and the pretentious theoretical structure he erected upon it. Even in his last publication, *Kristallseelen* (1917), he persisted in the defense of his thesis on the "ensoulledness" of inorganic nature.

In his monograph *Die Kalkschwämme* (1872) Haeckel distinguished the ascon, sycon, and leucon types of sponges, all of which, he held, were descended from a common primal form (olynthus). He derived this primal form from the gastrula stage by employing the "fundamental biogenetic law." In their ontogeny the sycon and leucon forms passed through the olynthus form.[4] He saw in the gastrula an image of the hypothetical primal form of all metazoans. This conception was the basis of his "gastraea theory" (1874–1877) of the homology of the two primary cotyledons. Although his demonstration of this theory rested on false assumptions, Haeckel had nevertheless taken up a problem that has since been the subject of an extensive literature.

Stimulated by Darwin's pangenesis hypothesis, Haeckel put forth his own hypothesis of the mechanism of heredity in 1876. According to it, heredity is "the transmission of the plastidial motion, [that is] the propagation of the individual molecular motion of the plastidial from the mother plastid to the daughter plastid." Haeckel thought that new adaptations could occur through alterations of the original plastidial motions resulting from the varying conditions of existence of the daughter cells. He persevered in his belief in "the inheritance of acquired characters" in Lamarck's sense and thereby became involved in a controversy with August Weismann, who about 1880 countered Lamarckism with his own "Neo-Darwinism." Although after 1900 Haeckel learned of Mendel's findings, he did not grasp the importance of experimental genetics.

Haeckel concluded his series of long zoological monographs in 1887–1889 with a treatment of the radiolarians, siphonophores, and deep-sea keratosa gathered on the *Challenger* expedition. Next, besides studies such as those on the plankton (1890), he wrote *Systematische Phylogenie* (1894–1896), subtitled *Entwurf eines natürlichen Systems der Organismen auf Grund ihrer Stammesgeschichte.* In this work Haeckel sought to present the advances made in phylogeny since the appearance of his *Generelle Morphologie.* Here again he contended that the foundation of the phylogenetic hypotheses lay in the direct empirical evidence of paleontology as well as in the indirect evidence of ontogeny and morphology.

In the chapter on the phylogeny of man Haeckel emphasized the incompleteness of the fossil record of the vertebrates. In his view, paleontological data was important primarily because it illustrated the sequence of descent among the individual vertebrate groups and their successive occurrences. At the same time he held that the unity of the vertebrates (including man) was already sufficiently demonstrated by comparative anatomy and ontogeny. "Laymen and one-sidedly trained specialists," he stated, place too great a value on the evidence of "fossil men" and on the "transition forms from ape to man."

It is therefore understandable that Haeckel did not discuss the prehistoric *Homo neanderthalensis* at all until 1900, whereas other scientists, including Huxley (1863) and Rolle (1866), early appreciated the importance of this discovery from the point of view of the theory of evolution. Only in his genealogical sketches of 1907 and 1908 did Haeckel upgrade Neanderthal man (*Homo primigenius*) to the intermediate stage (*Protanthropus*) "between *Pithecanthropus* and *Homo australis,* the lowest race of recent man" (1908).

Neither *Generelle Morphologie* nor *Systematische Phylogenie* had the success in scientific circles for which Haeckel had hoped. The evolutionary views he expressed in popular lectures, essays, and books had a far greater influence. In these writings, such as *Natürliche Schöpfungs-Geschichte* (1868) and *Anthropogenie* (1874), "monism" is always presented as a necessary consequence of the theory of evolution.

Convinced of the "truth of the monistic philosophy," Haeckel published a comprehensive statement of his beliefs in 1899 under the title *Die Welträthsel.* The book is divided into sections on anthropology (man), psychology (the soul), cosmology (the universe), and theology (God). The great success of this work, which was translated into many languages, was the result of the situation around 1900. Haeckel's attempt to establish a *Weltanschauung* in harmony with the advances of science answered a contemporary need. On the other hand, his harsh attack on church dogma and his often insecurely grounded generalizations led to heated controversies with scientists, theologians, and philosophers. On many occasions his rash statements were exploited to cast doubt on the validity of the theory of evolution. Typical in this regard was the controversy over Haeckel's far too schematized illustrations of various embryonic stages. Concerning these "forgeries," many distinguished anatomists and zoologists (including Theodor Boveri, Alexander Goette, Karl Grobben, Richard Hertwig, and Weismann) explained in 1909 that while they did not approve of Haeckel's methods, they nevertheless refused to attack him, since the concept of development "cannot suffer any damage through some incorrectly rendered embryological illustrations."

The nature of Haeckel's participation in the battle over concepts of development also affected his account of the prehistory of Darwinism. His historical sketches are characterized by a passionate defense of Lamarck, while Darwin severely criticized the latter's theory on many occasions.[5] According to Haeckel, Lamarck's *Philosophie zoologique* (1809) was "the first systematically founded presentation of the theory of the origin of species" and the first to "openly draw all its consequences"; it represented, moreover, "the beginning of a new period in the intellectual evolution of mankind." He thought that the reasons for Lamarck's failure lay above all in the authority of Georges Cuvier.

This view led Haeckel to make a series of misjudgments that, through his popular writings, were influential until recently. Among these was his judgment of the controversy between Étienne Geoffroy Saint-Hilaire and Cuvier at the French Academy of Sciences in 1830, which was "essentially over the theory of transformism." Haeckel interpreted Cuvier's catastrophist theory as a dogmatic theory of the absolute constancy and independent creation of species, although Cuvier himself did not take up these issues. Equally incorrect was Haeckel's interpretation of Goethe's writings on comparative anatomy and botany in terms of the theory of evolution.

Haeckel's enthusiastic defense of Lamarck can be understood from similarities in the two men's scientific careers, in their methods, and in their fixed goals. Both came to zoology from botanical systematics, and as zoologists both specialized in the systematics of the invertebrates. Both further employed the "natural system" of recent organisms to demonstrate phylogenetic relationships. Both were preoccupied by the religious and philosophical aspects of evolutionary theory. On the other hand, the differences between Haeckel and Darwin with regard to methods and argumentation are evident.

Haeckel's artistic endeavors were characteristic of him. During his many trips he produced numerous watercolors and vivid descriptions of his travels that are still charming. His *Kunstformen der Natur* (1899–1904) corresponded to his "monistic religion" with its three "cult ideals of the True, the Good, and the Beautiful."

Haeckel's historical importance consists principally in his suggestions that stimulated further work. His spirited advocacy of Darwin's ideas—not all of which he agreed with—contributed to the breakthrough of evolutionary thinking in the construction of biological theories. Moreover, concepts that Haeckel was the first to formulate, such as ontogeny, phylogeny, ecology, and chorology, have been adopted.

Haeckel—unlike Gegenbaur—did not form a school. Nevertheless, he did inspire many students (including Anton Dohrn, Richard and Oscar Hertwig, Arnold Lang, Hans Driesch, and W. Kükenthal) to take up zoology, especially research on marine animals. Yet these students early chose their own paths. In fact Haeckel scarcely participated in the development of modern experimental zoology that was then under way.

The characteristic elements of Haeckel's lifework were already evident in 1866 in his *Generelle Morphologie.* His striving for a scientifically based world view led him to statements on philosophical, political, and religious questions in which he advocated dubious conceptions drawn from social Darwinism. In this enterprise Haeckel was responding to the demands and needs of his time, which explains his work's success. Yet his lasting contribution lies not in the solutions he proposed but rather—and this is particularly true of his writings on the theory of evolution—in the questions he raised.

NOTES

1. Haeckel's herbaria are now in the Ernst Haeckel House and in the Haussknecht Herbarium of Friedrich Schiller University, Jena.
2. Haeckel wrote about this at length in his letters to his parents (1852–1856). In the summer semester of 1856 he was Virchow's assistant in Würzburg. They became friends at this time and remained on good terms until 1877, when Virchow publicly criticized certain of Haeckel's conceptions and suggestions (such as the teaching of Darwin's theory in the schools).
3. In 1863 August Schleicher had addressed an "Offenes Sendschreiben" to Haeckel entitled "Die Darwinsche Theorie und die Sprachwissenschaft."
4. Haeckel inferred the phylogeny of the olynthus from its ontogeny. The individual embryonic stages corresponded to the "original [phylogenetic] conditions." Thus he compared what he termed the "morula" stage with an amoeba colony (Synamoeba). According to Haeckel, the gastrula developed from the morula after passing through the "planula" stage.
5. Letter to J. D. Hooker of 11 Jan. 1844: "Heaven forfend me from Lamarck nonsense of a 'tendency of progression,' 'adaptions from the slow willing of animals,' etc.!" Revealing also are the letters to Lyell of 11 Oct. 1859 and 12 Mar. 1863, in Francis Darwin, ed., *Life and Letters of Charles Darwin* (London, 1887).

BIBLIOGRAPHY

I. ORIGINAL WORKS. Haeckel's most important writings are *Die Radiolarien* (*Rhizopoda radiaria*) (Berlin, 1862); *Generelle Morphologie der Organismen,* 2 vols. (Berlin, 1866); *Natürliche Schöpfungs-Geschichte* (Berlin, 1868); *Anthropogenie oder Entwickelungsgeschichte des Menschen. Keimes- und Stammes-Geschichte* (Leipzig, 1874); "Die Gastraea-Theorie, die phylogenetische Classification des Thierreichs und die Homologie der Keimblätter," in *Jenaische Zeitschrift für Naturwissenschaft,* **8** [n.s. **1**] (1874), 1–55; "Die Gastrula und die Eifurchung der Thiere," *ibid.,* **9** [n.s. **2**] (1875), 402–508; *Systematische Phylogenie, Entwurf* *eines natürlichen Systems der Organismen auf Grund ihrer Stammesgeschichte,* 3 vols. (Berlin, 1894–1896); *Die Welträthsel. Gemeinverständliche Studien über monistische Philosophie* (Bonn, 1899); *Fünfzig Jahre Stammesgeschichte. Historisch-kritische Studien über die Resultate der Phylogenie* (Jena, 1916); and *Kristallseelen. Studien über das anorganische Leben* (Leipzig, 1917).

Haeckel's extant MSS are in the Institute for the History of Medicine and Science, Ernst Haeckel House, Friedrich Schiller University, Jena.

II. SECONDARY LITERATURE. On Haeckel or his work, see Gerhard Heberer, ed., *Der gerechtfertigte Haeckel* (Stuttgart, 1968); Johannes Hemleben, *Ernst Haeckel* (Reinbek, 1964); Heinrich Schmidt, ed., *Was wir Ernst Haeckel verdanken* (Leipzig, 1914); and *Ernst Haeckel. Denkmal eines grossen Lebens* (Jean, 1934); Georg Uschmann, *Geschichte der Zoologie und zoologischen Anstalten in Jena 1779–1919* (Jena, 1959); and "Über das Verhältnis Haeckels zu Lamarck und Cuvier," in *Medizingeschichte unserer Zeit* (*Festschrift Heischkel-Artelt*) (Stuttgart, 1971), pp. 422–433; Georg Uschmann and Bernhard Hassenstein, "Der Briefwechsel zwischen Ernst Haeckel und August Weismann," in *Kleine Festgabe aus Anlass der hundertjährigen Wiederkehr der Gründung des Zoologischen Institutes der Friedrich-Schiller-Universität Jena* (Jena, 1965), pp. 6–68; and Georg Uschmann and Ilse Jahn, "Der Briefwechsel zwischen Thomas Henry Huxley und Ernst Haeckel," in *Wissenschaftliche Zeitschrift der Friedrich-Schiller-Universität Jena,* Math.-naturwiss. Reihe, **9** (1959–1960), 7–33.

GEORG USCHMANN

HAFFKINE, WALDEMAR MORDECAI WOLFE (*b.* Odessa, Russia, 15 March 1860; *d.* Lausanne, Switzerland, 25 October 1930), *bacteriology.*

Haffkine was the third of six children of Rosalie Landsberg and Aaron Khavkin (the Russian form of the name). The family was Jewish and of modest circumstances. Haffkine's mother died just before his seventh birthday and his father was frequently absent on business; his childhood was therefore lonely. He himself never married.

Haffkine attended the Gymnasium in Berdyansk, where he became interested in books, science, and physical fitness and received the highest grades. He attended the University of Odessa, supporting his studies with small sums he earned as a tutor and graduating doctor of science in 1884. Élie Metchnikoff was one of his teachers and influenced Haffkine toward devoting his life to science.

Haffkine was then offered a teaching position at the university on the condition that he convert to the Russian Orthodox Church, which he refused to do. Instead he accepted an appointment as assistant in the Odessa Museum of Zoology, which he held until 1888. While there he wrote two articles that were published in the *Annales des sciences naturelles* of Paris and became a member of the Society of Naturalists of Odessa. He left Odessa to teach physiology for a year under Moritz Schiff at the University of Geneva. In 1889 Metchnikoff, who was working at the Pasteur Institute, offered him the only position vacant there—that of librarian. Haffkine accepted eagerly and in 1890 became assistant to the director of the institute, Émile Roux. This event changed the entire course of Haffkine's life and brought him into the mainstream of research in preventive medicine.

The prevention of cholera had already occupied Metchnikoff and Robert Koch, and Haffkine took up cholera research during the 1888 epidemic. He conducted animal tests with a heat-killed culture of a highly virulent strain that he had created. By early July 1892 he was able to report success to the Biological Society of Paris; he then injected himself with a dose of four times the strength that was later used, recorded his reactions, and determined that his vaccine was safe for human use. His success brought him congratulations from Koch, Roux, and Pasteur.

Haffkine then sought to test the vaccine under epidemic conditions and decided to go to Siam. When Lord Dufferin, ambassador to France and formerly viceroy of India, learned of his project he persuaded Haffkine to go instead to India, where cholera was raging. Haffkine arrived in Calcutta in March 1893 and immediately set to work among a people totally strange to him and much divided among themselves. As cholera struck one village after another, Haffkine followed in its wake with two doctors, a few laboratory assistants, and two horse-drawn carriages with inoculation equipment. For two years, working without pay and in the face of hostility from the villagers, he inoculated volunteers. On one occasion stones thrown by the crowd broke glass instruments and a panic nearly ensued; Haffkine quickly pulled up his shirt and allowed another doctor to plunge a hypodermic into his side. The curiosity of the villagers was thus aroused and 116 of the 200 peasants assembled volunteered for inoculation (none were to die in the epidemic, although nine of those who refused inoculation did). Haffkine kept careful records of his subjects, including their sex, physique, age, race, religion, and caste. Within the two years 45,000 persons were inoculated, most of them twice, and the death rate from cholera was reduced by 70 percent. Haffkine was still unable, however, to evaluate the degree of immunity conferred or how long it lasted.

In 1895 Haffkine contracted malaria and left India for England to try to recover his health. He returned to Calcutta in March 1896; six months later he was reassigned to Bombay, where plague was epidemic—despite denials by the newspapers, the death toll was rising and many were fleeing the city. He improvised a laboratory in a corridor of Grant Medical College and, with a staff of one clerk and three servants, began experiments with his antiplague vaccine on laboratory animals.

By December, Haffkine was convinced of the efficacy of the vaccine on animals; on 10 January 1897 a doctor agreed to inoculate him in secret and the principal of the college agreed to be a witness. Once again the dosage was four times that which was later used. Haffkine developed high fever and pain at the site of the injection; nevertheless, he attended a meeting of the Indian Medical Service, no one present being aware of the injection he had undergone. The next morning, when he described his symptoms—admitting the pain—to the staff, faculty, and students of the college and asked for volunteers, hundreds responded.

Haffkine's proposals for the training of medical officers from epidemic areas in the preparation and administration of the vaccine were rejected by the Indian government. He further urged a program of public education and the inoculation of soldiers, prisoners, and coolies; again the government refused. Hearing of the high death rate from cholera and plague in Russia, he then offered his vaccine and his services in training doctors in his techniques free to the czarist government, and was again turned down. Nevertheless, scientists from many countries (including Russia) came to the Plague Research Laboratory in Bombay that Haffkine had founded and of which he was director in chief. There he taught them his methods of vaccine preparation and inoculation procedures. Requests for enormous quantities of vaccine arrived from all over the world. The laboratory, after much stress and agitation, was moved to the Old Government House, where it still functions. Haffkine himself was invited to lecture before the Royal Society of London and professional groups, and received many other honors. In 1897 Queen Victoria named him Companion of the Order of the Indian Empire, and in 1899 he applied for and was granted British citizenship.

In 1902 plague was epidemic in the Punjab and an all-out inoculation campaign was planned. Haffkine requested from England a dozen doctors and nurses and thirty soldiers—to be trained in his laboratory—to aid in this effort, but received only a handful. In the middle of this, at the end of October, nineteen people, of the tens of thousands inoculated, contracted tetanus and died. All had been inoculated from the same bottle of vaccine (brew No. 53N); no unusual results were traced to the five other bottles used the same day. The British medical officials publicly accused Haffkine and his laboratory of having sent contaminated vaccine to the Punjab, and Haffkine was suspended without pay before proper investigations were even begun. A Commission of Inquiry was appointed, headed by Sir Lawrence Jenkins, chief justice of Bombay. None of its members were bacteriologists. The commission lasted almost five years, including the preparation of its report. Haffkine was called before it several times, then returned to England in 1904. The report was withheld until 1907, when public and other pressures—most notably a letter to the *Times* (London), initiated by Sir Ronald Ross and signed by ten prominent bacteriologists, listing arguments why the charge against Haffkine must be disproved—forced its release. The letter concluded that "there is very strong evidence to show that the contamination took place when the bottle was opened at Mulkowal [the village where the deaths occurred], owing to the abolition by the Plague authorities of the technique prescribed by the Bombay laboratory and to the consequent failure to sterilize the forceps which were used in opening the bottle, and which during the process were dropped to the ground."

During this time medical journals and papers in India took positions on both sides, but Haffkine received tangible honors from other parts of the world. Haffkine was exonerated, and began negotiations that took him back to the work in India that he realized was unfinished. He was broken in morale but chose to return despite reduced status and at his original pay, contrary to the promises of local princes.

He was again in Calcutta in December 1907, in the laboratory of the Presidency General Hospital. He met with coolness from the British medical officers, as he had throughout his career, but received cooperation from the Indians. The Institut de France awarded him the Prix Briant, its highest honor, in 1909, and the Tata Institute of Science in Bangalore elected him to its Court of Visitors. In 1915 he reached compulsory retirement age and left India to spend some time in London and then Paris. For the rest of his life he occupied himself with Jewish affairs; in 1929 he created the Haffkine Foundation, which still exists, for fostering religious, scientific, and vocational education in Eastern European yeshivas—he bequeathed the Foundation his personal fortune of $500,000. In 1925 the Plague Research Institute that he founded in Bombay was renamed in his honor and still bears that name.

BIBLIOGRAPHY

I. Original Works. Some of Haffkine's articles published in professional journals are "Recherches biologiques sur l'*Astasia Ocellata,* n.s.," in *Annales des sciences naturelles* (1885); "Recherches biologiques sur l'*Euglena viridus,* Ehr.," *ibid.* (1886); "Vaccination Against Asiatic Cholera," a lecture given at Calcutta Medical College, 24 March 1893, in *Indian Medical Gazette,* vol. **28** (1893); "A Lecture on Vaccination Against Cholera," in *British Medical Journal* (21 Dec. 1895); "An Inoculation of Coolies," in *Indian Medical Gazette,* vol. **31,** no. 7 (1896); and "The Inoculation Accident in Manila in 1906: Contamination of Cholera Vaccine With Plague Virus," in *Journal of the American Medical Association,* vol. **52** (1909).

Separate publications include *Health of the Population After Plague Inoculation,* lecture given at Poona, 29 June 1901 (Bombay, 1901); and *On Prophylactic Inoculation Against Plague and Pneumonia* (Calcutta, 1914).

Some of Haffkine's articles written for popular journals include "Les nouvelles écoles techniques en Russie," in *Journal of the Norwegian Ministry of Public Instruction* (1889); "On the Primary Schools in Scandinavia," in Russian, in *Popular School* (1889); "Preventive Inoculation," in *Popular Science Monthly* (June–July, 1900); and "A Plea for Orthodoxy," in *Menorah Journal,* **2,** no. 2 (1916), 67–77.

Haffkine's official reports were submitted periodically to the government of India and dealt with his own research and the work of the Plague Research Laboratory. These reports are deposited in the National Archives of India, New Delhi and the Secretariat Record Office of the State of Maharashtra in the Elphinstone College Building, Bombay. Those written for the Bengal Government before 1900 are at Bhawani Dutta Lane, Calcutta, and those after 1900 are in Writers' Building, Dalhousie Square, Calcutta.

II. SECONDARY LITERATURE. A discussion of Haffkine's discourse on preventive inoculation, delivered at the Royal Society, London, 8 June 1899, chaired by Lord Lister, president, is in *British Medical Journal* (1 July 1899).

On Haffkine and his work see also Edythe Lutzker, "Waldemar M. Haffkine, His Contributions to Global Public Health," in *Actes du XIe Congrès International d'Histoire des Sciences* (Warsaw–Cracow, 1965), pp. 214–219, trans. into Italian by S. U. Nahon in *La rassegna mensile di Israel* (Rome–Milan, 1966), pp. 532–533; "Report on the Biography of Waldemar Haffkine (1860–1930); Life and Contributions to Global Public Health," in *Yearbook of the American Philosophical Society* (1967), pp. 577–580; "Some Missing Pages from the Histories of Nineteenth Century Medicine: Waldemar Mordecai Haffkine, C.I.E., D. Sc. (1860–1930)," in *Communicazione presentate al XXI Congresso Internazionale di Storia della Medicina* (Siena, 1968), pp. 22–28, repr. in *Medica judaica,* **1,** no. 1 (1970), 32–35, and in *Journal of the Indian Medical Profession,* **17,** no. 4 (1970), 7591–7595; "More on Waldemar Haffkine," *ibid.,* no. 7 (1970), 7711; "Waldemar M. Haffkine," in *Encyclopedia Judaica* (Jerusalem–New York, in press); and "In Honor of Waldemar Haffkine, C.I.E., on the 40th Anniversary of his Death," in *Actes du XXIIe Congrès International d'Histoire de Médecine* (Bucharest, in press); Mark Popovsky, *The Fate of Doctor Haffkine,* in Russian (Moscow, 1963); and *The Story of Dr. Haffkine,* trans. from the Russian by V. Vezey (Moscow, 1965); and Selman A. Waksman, *The Brilliant and Tragic Life of W. M. W. Haffkine, Bacteriologist* (New Brunswick, N.J., 1964).

EDYTHE LUTZKER

HAGUE, ARNOLD (*b.* Boston, Massachusetts, 3 December 1840; *d.* Washington, D.C., 14 May 1917), *geology.*

Hague was the son of the Reverend Dr. William Hague and Mary Bowditch Moriarty. His father urged him to pursue a business career, but he entered the Sheffield Scientific School of Yale after he failed to pass the physical examination for the army at the outbreak of the Civil War. His older brother, James, studied mining engineering at Lawrence Scientific School of Harvard and he may have influenced Hague's decision to pursue a career in geology. Hague's professors at Yale included James D. Dana, George J. Brush, and Samuel W. Johnson, and among his fellow students were J. Willard Gibbs, Ellsworth Daggett, Clarence King, and O. C. Marsh.

After graduation (Ph.B.), Hague, again rejected by the army, went to Germany—first to Göttingen and then to Heidelberg, where he studied in R. W. Bunsen's laboratory. He then attended the Bergakademie at Freiberg, Saxony. There he met S. F. Emmons and came under the personal guidance of Bernhardt von Cotta, author of a textbook on petrography.

In December 1866 Hague returned to Boston and shortly thereafter visited King, who invited him to join the proposed geological survey across the western cordilleras, if authorized by Congress. Hague immediately told Emmons of the planned survey and he too joined the expedition. Together these three men accomplished much for geology in their geological exploration of the fortieth parallel (1867–1872). Following preparation of the reports and atlases, Hague became government geologist for Guatemala in 1877, and in the following year went to northern China to study various mines for the Chinese government.

In 1879 the U.S. Geological Survey was established by Congress, and King was made its first director. Hague was appointed as government geologist with Joseph P. Iddings and, later, Charles D. Walcott and W. H. Weed were made assistants. From 1883 to 1889 Hague directed the survey of Yellowstone National Park and vicinity, returning again in 1893 with T. A. Jaggar, Jr., as an assistant. In subsequent visits Hague independently continued his observations on the hot springs and geysers of Yellowstone Park.

In collaboration with King and Emmons, Hague made a geological reconnaissance of a 100-mile-wide belt extending from the eastern California border to the Great Plains of Wyoming and Colorado, embracing the line of the first transcontinental railroad. This was the first of the extensive surveys which took note of the petrography of the extrusive rocks. Hague suggested that the name Laramie be used for a great series of sedimentary beds covering hundreds of square miles in the Rocky Mountains and Great Plains. The Laramie formation, which marks the end of the Mesozoic era, gave rise to one of the most prolonged controversies in the paleontological dating of rocks in the history of American geology. Hague also explored Mount Hood, Oregon, collecting volcanic rocks and studying the glacial phenomena. In addition, he mapped the famous silver-lead district of Eureka, Nevada.

Hague's Yellowstone survey covered more than 3,000 square miles. He was particularly interested in the volcanoes of the Absaroka Range, which poured out enormous volumes of rhyolitic material in single eruptions. His observations on the hot springs and geysers led to a theory on the origin of the thermal waters of the Yellowstone Park region. He was a strong advocate of the preservation of the region in its natural state and took an active part in advising the government on the development of the park for public enjoyment.

Hague was married late in life (1893) to Mary Bruce Howe of New York. He received honorary degrees from Columbia University (Sc.D., 1901) and the University of Aberdeen (LL.D., 1906). He was elected a member of the National Academy of Sciences (1885) and served as its home secretary from 1901 to 1913. He also served as president of the Geological Society of America (1910) and vice-president of the International Geological Congress on three occasions (1900, 1910, and 1913).

Hague was described as a gentleman, temperate in language and habits at all times—even with the pack mules. He had little interest in conveying his ideas to others or in influencing their opinions. Iddings, his assistant of many years, writes kindly of his liberal treatment in the matter of individual research and his interest in the work of the beginner. Like most men exploring the difficult wilderness, Hague found great beauty in nature, whether it was the Grand Canyon of the Yellowstone or the movements of the elk.

BIBLIOGRAPHY

I. Original Works. Works by Hague include "Descriptive Geology," *Report of the Exploration of the Fortieth Parallel,* vol. II, Professional Papers of the Engineer Department, U.S. Army, no. 18 (Washington, D.C., 1877), written with S. F. Emmons; "Notes on the Volcanoes of Northern California, Oregon, and Washington Territory," in *American Journal of Science,* 3rd ser., **26** (1883), 222–235, written with J. P. Iddings; "Notes on the Volcanic Rocks of the Great Basin," *ibid.,* **27** (1884), 453–463, written with J. P. Iddings; "Geological History of the Yellowstone National Park," in *Transactions of the American Institute of Mining Engineers,* **16** (1888), 783–803; *Geology of the Eureka District, Nevada,* U.S. Geological Survey Monograph no. 20 (Washington, D.C., 1892); and *Yellowstone National Park Folio, Wyoming; General Description,* Geological Atlas of the U.S., folio no. 30 (Washington, D.C., 1896).

See also "The Age of the Igneous Rocks of the Yellowstone National Park," in *American Journal of Science,* 4th ser., **1** (1896), 445–457; *Absaroka Folio, Wyoming,* Geological Atlas of the U.S., folio no. 52 (1899); "Early Tertiary Volcanoes of the Absaroka Range," in *Science,* n.s., **9** (1899), 425–442; "Descriptive Geology of Huckleberry Mountain and Big Game Ridge, Yellowstone Park," in *Geology of Yellowstone National Park,* U.S. Geological Survey monograph no. 32, pt. 2 (Washington, D.C., 1899), pp. 165–202; and "Origin of the Thermal Waters of the Yellowstone National Park," in *Bulletin of the Geological Society of America,* **22** (1911), 103–122.

H. S. Yoder, Jr.

HAHN, OTTO (*b.* Frankfurt am Main, Germany, 8 March 1879; *d.* Göttingen, Germany, 28 July 1968), *radiochemistry.*

Hahn was one of the first of the numerous great figures in Ernest Rutherford's circle, although his first fame dates from work performed even before their meeting. Early in the twentieth century he became a pioneer in radiochemistry, along with Frederick Soddy, Bertram Boltwood, and Kasimir Fajans. His long and distinguished career extended through the discovery of nuclear fission to the study of fission fragments and to the rebirth of German science following World War II.

His father, Heinrich, was descended from Rhenish peasant stock, but he was disinclined to follow the family tradition of farming. Instead, he pursued the family avocation and became a glazier, buying his own shop after settling in Frankfurt. His advance from artisan to businessman coincided with the building boom in his city which followed the Franco-Prussian War, and prosperity enabled the Hahn family to rise to middle-class respectability. Otto's mother, Charlotte Stutzmann, née Giese, had north German ancestry; most of her family were merchants, although a few were in the professions. In 1913 Otto married Edith Junghans, by whom he had one son.

Otto was a sickly youth, but after the age of fourteen he was quite healthy. At the local high school he was a good but not outstanding student. His interest in chemistry arose from some dabbling in the subject with a classmate and increased when he attended a series of lectures given to an adult audience. His father wished him to become an architect, but Otto prevailed and entered Marburg University in 1897. His autobiographical reminiscences suggest that he spent more time in the beer halls than in studying, and he expresses regret at his inattention to physics and mathematics. But he must have absorbed a respectable amount of chemistry; after receiving his doctorate in 1901 and following a year's infantry service, he returned to Marburg as assistant to his principal professor, Theodor Zincke.

This post was coveted, since one could obtain the professor's recommendation to any of the large chemical companies in Germany, which led the world in application of scientific talent to industry. For Hahn this was an important step since he had no thoughts of pursuing an academic career. Near the end of his two years with Zincke, Hahn was advised of a possible job which required command of a foreign language, since the firm might have need to send him abroad occasionally. At his own expense he went to England in September 1904, and Zincke, who did not want him to be idle, obtained a place for him in Sir William Ramsay's laboratory at University College, London.

Ramsay, famous for his discovery of several "inert" gases, developed an interest in radioactivity which was furthered by Soddy, who had first worked with Rutherford and then spent a year with Ramsay. The latter was without radiochemical help so, handing his young German visitor a dish containing about 100 grams of barium salt, he asked him to extract the few milligrams of radium in it according to Marie Curie's method. Hahn, an organic chemist whose dissertation had dealt with bromine derivatives of isoeugenol, was unfamiliar with this subject, but Ramsay observed that he would approach the work without preconceived ideas. Because the sample was small, Ramsay proposed that Hahn confirm Marie Curie's determination of the atomic weight of radium by preparing it in some organic compounds (thereby greatly increasing the total amount being examined) and calculating the atomic weight from the measured molecular weights.

Chance sometimes favors the unprepared mind, and Hahn, who familiarized himself with only the basics of radioactivity, followed the prescribed separations technique and found himself the discoverer of a new radioelement: radiothorium. The explanation was that the material given to him came from an ore which contained a large percentage of thorium in addition to the uranium. Thus, upon completion of the chemical procedure, not all the activity was confined in the radium-containing fraction; indeed, the new substance in the remainder was several hundred thousand times more active than thorium and ultimately yielded the characteristic one-minute half-life of thorium emanation. In this same year, with another young German, Otto Sackur, he examined A. L. Debierne's actinium and F. O. Giesel's emanium, showing them to be identical and resolving what was then a controversial issue.

Ramsay felt such research ability would be wasted in industry and urged his visitor to take a post which he secured for him in Emil Fischer's chemical institute at the University of Berlin. By this time Hahn's interest in organic chemistry had receded before the fascination of radioactivity, and he was amenable to the proposal. But first he wished to attain greater mastery over radioactivity by working under the leading figure in the field, Rutherford. Thus, in September 1905, he crossed the Atlantic to spend the next year at McGill University in Montreal. His reception was cordial but reserved, for Rutherford had a low opinion of Ramsay's competence in radioactivity and distrusted such work as came from his laboratory. Moreover, the New Zealander's good friend and prominent radiochemist at Yale, Boltwood, had characterized radiothorium as a "compound of Th-X and stupidity." Hahn, however, soon convinced the skeptics of the reality of his substance, established warm friendships with them, and again exhibited his talent for discovering radioelements by soon finding radioactinium. Such work was the means by which the constituents and their sequence in the radioactive decay series were determined.

Hahn arrived at Fischer's institute in the fall of 1906 and in order to continue these investigations he established a mutually profitable relationship with Knöfler and Company, producers of thorium preparations. While in Canada, he had measured a half-life for radiothorium of about two years; but Boltwood—who had tested a number of commercially prepared thorium salts, had found them deficient in radiothorium and had tried unsuccessfully to detect its growth—argued for a much longer half-life. From Knöfler, Hahn obtained samples prepared a number of years earlier and found that their activities decreased at first and then gradually increased. This was proof of his belief in a long-lived radioelement between thorium and radiothorium, which he separated in 1907 and named mesothorium. Because it was chemically inseparable from radium, which was difficult to obtain in Germany, and owing to the rising medical demand for radium, Knöfler successfully marketed high-activity mesothorium as "German radium."

Within a year of his return to his homeland, Hahn was appointed a *Privatdozent* in Fischer's institute, thereby joining the teaching faculty of the University of Berlin; he became a professor in 1910. He became friendly with physics professors Rubens, Nernst, and Warburg, and such younger colleagues as Max von Laue, Otto von Baeyer, James Franck, Gustav Hertz, Peter Pringsheim, and Erich Regener. But the most important physicist to enter his life was Lise Meitner, who came from Vienna in 1907 to do theoretical work under Max Planck and wished also to pursue some studies in experimental radioactivity. Thus began a fruitful collaboration that lasted thirty years. Since Hahn had an almost complete collection of known radioelements, they decided to examine all their beta radiations. This led to the proof that several elements, thought not to radiate as they decayed, actually were weak beta emitters. Further work on the magnetic deflection of the beta rays added much to the ultimate explanation of their continuous and line spectra. Hahn also pioneered the method of radioactive recoil in 1909 (done independently by Russ and Makower), with which he and Meitner found a few more radioelements.

When the new Kaiser Wilhelm Gesellschaft opened its research Institut für Chemie in Berlin-Dahlem, in late 1912, Hahn was made head of a small, independent department of radioactivity and invited Meitner to join him. Since this new laboratory was uncontaminated, he was able to study such weakly radioactive substances as rubidium and potassium and developed an enduring interest in the geological dating of rocks by means of these elements. This was also the time of the most profound theoretical advances in Hahn's own field of radiochemistry, but he seems to have taken little part in them. Fajans and Soddy independently in 1913 announced the group displacement laws, which placed the radioelements in appropriate boxes of the periodic table, and the concept of isotopy, which held that inseparable radioelements were not only similar but chemically identical. Like other radiochemists, Hahn had long been familiar with such facts as the inseparability of mesothorium and radium, and of radiothorium and thorium. But generalizations to explain these puzzles—and theoretical speculation in general—were not his style; Hahn was simply a superb experimentalist.

During World War I, Hahn served in the gas-warfare corps, under the scientific leadership of Fritz Haber. He was involved in research, development, testing, manufacturing, and using the new weapons. Even before the armistice, having had the opportunity to visit his laboratory in Berlin-Dahlem, Hahn and Meitner in 1917 discovered the most stable isotope of the element 91, which they named protactinium (the original discoverers of this element, Fajans and Göhring in 1913, had named their short-lived isotope brevium). This parent of actinium helped resolve the uncertain sequence in the actinium series, although recognition that it was entirely independent of the uranium series (descended from U^{238}) did not come until the discovery of actinouranium (U^{235}) (the existence of which was inferred from Aston's mass-spectrographic work in 1929), the ultimate source of this series. After the discovery of protactinium, Hahn believed that it descended, through uranium Y (Th^{231}), from primordial uranium in a branch parallel with the well-known uranium series. His subsequent examination of uranium and its products turned up in 1921 a small, but persistent and inexplicable, activity in the uranium series' protactinium isotope. Here was a case of branching, but not the one Hahn was looking for. He had found that the first example of nuclear isomerism, i.e., uranium Z, has the same parent and the same daughter product as uranium X_2; and both these protactinium isotopes are formed by, and decay by, beta emission. But their nuclei are at different energy levels and decay with different half-lives.

By the early 1920's almost all of the naturally occurring radioelements were known, and opportunities for basic research in radiochemistry were limited. Hahn turned toward applications of his specialty and developed the "emanation" method, by which changes in the surfaces and the formation of surfaces in finely divided precipitates could be studied. He also worked with tracer techniques, developed by Hevesy and Paneth, and extended the rules of Fajans and Paneth for the precipitation and adsorption of small quantities of matter.

Radiochemistry was resurrected and transformed into nuclear chemistry with the great events of the early 1930's: Chadwick's discovery of the neutron, the Joliot-Curies' discovery of artificial radioactivity, and Fermi's use of neutron bombardment to produce additional radioactive materials, including some thought to be new elements beyond uranium in the periodic table. There was much work now for nuclear chemists, and Hahn was deeply involved in identifying the many products and their decay patterns. The "transuranium" elements in particular excited his interest and, with Meitner and Fritz Strassmann, he endeavored to determine their chemical and physical properties. Along with these transuranium elements, the neutron bombardment of uranium seemed to produce several radioactive bodies which separated with barium and could only be, they thought, isotopes of radium. It was difficult enough to explain how uranium (element 92) changed to radium (88), especially as no alpha particles were observed, and virtually no thought was given to the possibility that these bodies were actually barium, an element in the middle of the periodic table. But when they next attempted to separate the "radium" from the barium carrier, the activity remained with the barium fraction.

At the end of 1938, writing as nuclear chemists, Hahn and Strassmann insisted upon the accuracy of their identification. As scientists familiar with nuclear physics, however, they could scarcely believe in a transmutation from uranium to barium. Hahn sent news of these findings to Meitner in Sweden, where she had fled to escape the Nazis. With her nephew, Otto Frisch, she correctly interpreted the phenomenon as a splitting of the uranium nucleus and named it "fission."

Hahn was little concerned with the energy released in fission and played no part in the German atomic bomb and reactor project during World War II. Instead, he devoted most of his efforts to the study of fission fragments. When the chemical institute, of which he had become director in 1928, was destroyed in an air raid, he moved his usable equipment to southern Germany and resumed work there. With several other nuclear physicists and chemists he was arrested in the spring of 1945 by Allied troops and interned for over half a year in England. There, to his profound dismay, he heard of the application of his discovery when nuclear weapons were detonated over Hiroshima and Nagasaki. He learned also of the award to him of the 1944 Nobel Prize in chemistry, and he received a request to become president of the Kaiser Wilhelm Gesellschaft.

On his release and return to Germany in early 1946, Hahn accepted leadership of this society, which was soon renamed the Max Planck Gesellschaft, at the instance of the occupation authorities. He played a major role in reestablishing not only the society's research institutes but German science as a whole. He also was responsible for the 1955 "Mainau Declaration" of Nobel laureates, warning of the danger in misuses of atomic energy, and was one of eighteen eminent German scientists who in 1957 protested publicly any German acquisition of nuclear arms.

BIBLIOGRAPHY

I. ORIGINAL WORKS. Hahn's only full-sized scientific text consists of his 1933 Baker Lectures at Cornell University, published as *Applied Radiochemistry* (Ithaca, 1936). He was, however, prolific in recording his reminiscences: "Einige persönliche Erinnerungen aus der Geschichte der natürlichen Radioaktivität," in *Die Naturwissenschaften,* **35** (1948), 67–74; *New Atoms, Progress and Some Memories,* W. Gaade, ed. (New York, 1950); a collection of papers; "Personal Reminiscences of a Radiochemist," in *Journal of the Chemical Society* (1956), 3997–4003, the Faraday Lecture; "The Discovery of Fission," in *Scientific American,* **198** (1958), 76–84; *Otto Hahn: A Scientific Autobiography* (New York, 1966), translated and edited by Willy Ley from the original *Vom Radiothor zur Uranspaltung* (Brunswick, 1962); *Otto Hahn: My Life* (London, 1970), which was translated by Ernst Kaiser and Eithne Wilkins from the original *Mein Leben* (Munich, 1968). An extensive bibliography of his scientific and other papers appears in the *Biographical Memoirs of Fellows of the Royal Society* reference below.

II. SECONDARY LITERATURE. Through Hahn's long and active life there appeared numerous articles about him, often on the occasion of a major birthday. Examples of this literature are Stefan Meyer, "Zur Erinnerung an die Jugendzeit der Radioaktivität," in *Die Naturwissenschaften,* **35** (1948), 161–163; Erich Regener, "Otto Hahn 70 Jahre," in *Zeitschrift für Elektrochemie,* **53** (1949), 51–53; O. R. Frisch, *et al.,* eds., *Trends in Atomic Physics; Essays Dedicated to Lise Meitner, Otto Hahn, Max von Laue on the Occasion of Their 80th Birthday* (New York, 1959).

The most extensive obituary notice in English is by R. Spence, in *Biographical Memoirs of Fellows of the Royal Society,* **16** (1970), 279–313. Concerning Hahn's greatest discovery, Hans G. Graetzer and David L. Anderson reprint numerous papers and furnish connecting narrative in *The Discovery of Nuclear Fission* (New York, 1971); an analysis is Esther B. Sparberg, "A Study of the Discovery of Fission," in *American Journal of Physics,* **32** (1964), 2–8.

LAWRENCE BADASH

HAHNEMANN, CHRISTIAN FRIEDRICH SAMUEL (*b.* Meissen, Germany, 10 April 1755; *d.* Paris, France, 2 July 1843), *medicine, chemistry.*

Hahnemann was the son of Christian Gottfried Hahnemann, a painter of porcelain. He received his early education at home, then at the local school in Meissen. He began his study of medicine at Leipzig in 1775, subsequently went to Vienna, and finally received his medical degree from the University of Erlangen in 1779. Hahnemann practiced medicine in several towns of Saxony before settling in Dresden. Here he temporarily abandoned medicine because of dissatisfaction with the treatments of the time, which were based upon the prescription of drugs the effects of which he claimed to be uncertain and often dangerous. In an important work of 1786, *Über Arsenikvergiftungen,* Hahnemann described the symptoms, remedies, and legal investigation of cases of arsenic poisoning. Following this his interests turned to chemistry, and from 1787 to 1792 he published eleven papers in *Chemische Annalen für die Freunde der Naturlehre.* Among these were descriptions of a test that used hydrogen sulfide for detecting the presence of lead in wine and the preparation of a mercury compound (mercurous oxide) soluble in acetic acid. In 1789 Hahnemann moved to Leipzig, where he published a work on the treatment of venereal diseases with mercurous oxide and other mercury preparations.

For the next twenty years Hahnemann practiced medicine, moving frequently from one town to another, until he returned to Leipzig, where he stayed from 1810 to 1821. From 1788 to 1796 he translated several English treatises on drugs, including William Cullen's *Materia medica,* and in 1796 published his first paper setting forth his own views, which later formed the basis of homeopathy. These ideas were more fully expressed in the first edition of the *Organon der rationellen Heilkunde* (1810) and the *Materia medica pura* (1811). Hahnemann's theory included the proving of drugs by administering them to healthy persons to ascertain their effects and to evaluate their essential action, the study of the symptoms of particular diseases, and treatment using the principle that a drug capable of evoking in the healthy body a response that is similar to the primary symptom of a disease is likely to produce a reaction in the body which will overcome the disease.

In 1812 Hahnemann was admitted to the faculty of the University of Leipzig, where he taught his theory of medicine. He believed in administering only one drug at a time—he himself prepared all his drugs—and thus incurred the anger of Leipzig's apothecaries. After the furor following the death of an Austrian prince who had placed himself under his care, Hahnemann was forbidden to dispense medicine and forced to resign from the university. In 1821 he moved to Köthen, where he remained until 1835. During this time his fame grew through his practice and the successive editions of his writings. In 1828 he published the first two volumes of *Die chronischen Krankheiten, ihre eigenthümliche Natur und homöopathische Heilung.* In this work he developed the doctrine of the psora, which maintained that the majority of chronic diseases are due to a morbid material present in the body, a material identical to that which produces a variety of scaly diseases on the surface of the skin.

In 1830 Hahnemann's wife of forty-two years died, and in 1835 he married a rich patient, Melanie d'Hervilly. They moved to Paris, where with the assistance of his wife Hahnemann managed a large medical practice until his death at the age of eighty-eight. He is considered the founder of homeopathy, and his followers increased throughout the nineteenth century.

BIBLIOGRAPHY

I. Original Works. Hahnemann's major writings include *Organon der rationellen Heilkunde* (Dresden, 1810); *Materia medica pura . . .* (Dresden, 1811); and *Die chronischen Krankheiten . . .* , 4 vols. (Dresden–Leipzig, 1828–1830), 2nd ed., 5 vols. (Dresden–Leipzig, 1835–1839). These went through many editions and translations. His lesser writings were collected and translated by Robert E. Dudgeon as *The Lesser Writings of Samuel Hahnemann* (London–New York, 1852); the preface of this work includes an extensive bibliography.

II. Secondary Literature. There are many biographies of Hahnemann and histories of homeopathy. Among the more useful are Linn J. Boyd, *A Study of the Simile in Medicine* (Philadelphia, 1936); R. E. Dudgeon, *Hahnemann, the Founder of Scientific Therapeutics* (London, 1882); Richard Haehl, *Samuel Hahnemann, sein Leben und Schaffen* (Leipzig, 1922); and Rudolf Tischner, *Geschichte der Homöopathie* (Leipzig, 1932).

Daniel P. Jones

HAIDINGER, WILHELM KARL (*b.* Vienna, Austria, 5 February 1795; *d.* Dornbach, near Vienna, 19 March 1871), *mineralogy, geology.*

After a basic education in Vienna, Haidinger in 1812 went to Graz, where he worked as an assistant, lodging in the home of Friedrich Mohs, professor of mineralogy at the recently established Landesmuseum Joanneum. They made frequent excursions in the adjacent mining regions, meeting A. G. Werner at Freiberg in 1816. In the company of Count Breunner (after whom he named the mineral breunnerite), Haidinger visited Georges Cuvier and J.-B. Biot at Paris and George Greenough, David Brewster, and Thomas Allan (a banker interested in mineralogy) at Edinburgh. In 1823 Haidinger moved to Edinburgh in order to arrange Allan's mineral collection. There he began publishing on the determination of mineral species, his first work being a translation, with many additions, of Mohs's famous textbook (1825), expounding for English readers what could be achieved in natural history through mineral determination and classification. The complications arising from isomorphism and polymorphism, then being clarified by Eilhard Mitscherlich and J. J. Berzelius, were still a stumbling block for Haidinger. Between 1827 and 1840 he worked in a china factory that had been founded by his two brothers in Elbogen. In 1840 he was appointed an inspector of mines (*Bergrath*) in Vienna.

Around 1827 Haidinger's chief interest turned to pseudomorphs, since they are one of the few unambiguous indicators of a mineral change that must have taken place in the past. Following Humphry Davy, he attributed the motion and replacement of particles to electrochemical forces, drawing a parallel between this interchange of constituents and the behavior of a solution in electrolytic dissociation. The calcite-dolomite transformation led Haidinger (1848) to the hypothesis that percolating saline solutions (*Gebirgsfeuchtigkeit*) containing $MgSO_4$ replace $2CaCO_3$ by $CaCO_3 + MgCO_3$ molecules. This process was thought to proceed at elevated pressure and temperature, with precipitation of gypsum. Most of the gypsum would be taken into solution and redeposited at greater depths. Under surface conditions the reverse reaction occurs, that is, dedolomitization, with lime replacing magnesium. This hypothesis was verified experimentally by the more geology-minded Charles von Morlet, who used a gun barrel equipped with inlet and outlet valves.

Haidinger also studied the absorption of light in crystals, most likely an interest acquired from Brewster during his Edinburgh period. He designed a charmingly simple and effective instrument (1848) that was later named Haidinger's dichroscope. With its aid Haidinger made many delicate observations of pleochroic minerals, first only with transmitted light, then also with reflected light. Observations on the connection between absorption and the direction of polarization of transmitted and reflected light led in 1848 to a well-founded theory bearing on Babinet's rule that a greater absorption (of the whole spectrum of visible light) corresponds to a higher index of refraction. Babinet's rule also tempted Haidinger to suggest a possible arrangement and "bonding" of iron particles in a crystal structure that was based on peculiar absorption phenomena (1855). Clearly aware that only the direction of vibration, and not the direction of propagation of the light, decides the degree to which light is absorbed by the crystal, Haidinger reached the fundamental conclusion (1852) that for linearly polarized light, the direction of vibration in the ether must be perpendicular to the direction of propagation—a conclusion leading to a coherent picture of the optical behavior of transparent crystals which is still followed in all treatises on this subject that are not sophisticated enough to identify this direction with the electric vector E.

At the Hof Mineralien Cabinet in Vienna, where Moritz Hoernes often received meteorites and information about them, Haidinger began to publish on these acquisitions in 1847. At first limited to mineralogical composition, this work later also considered the phenomena of light and sound emitted by meteorites when traversing the atmosphere and hitting the ground. His results, together with observations of the angles of incidence and depths of penetration, led to a summary and noteworthy explanation of these phenomena (1861). In the same publication, probably loath to lag behind his contemporaries, Haidinger ventured a theory on the origin of meteorites. He postulated an original *Weltkörper* made up of collected cold cosmic dust, which had been created out of nothing, and was subsequently heated through pressure and friction of the component particles. Differential internal tension caused an explosion of the body, flinging the resulting fragments apart.

Although not originally a geologist, Haidinger obstinately urged the need to create a geological survey, which in 1849 was established under the patronage of the royal and imperial Vienna Academy of Sciences. Haidinger, a member of the Academy since its founding, was appointed the first director of the geological *Reichsanstalt.* He was always respectful toward the authorities and grateful when he or his *Anstalt* was honored or praised, a situation that occurred frequently in later years. After a long illness Haidinger was granted a pension in 1866, upon which he moved to Dornbach, where he spent the last five years of his life.

BIBLIOGRAPHY

I. Original Works. Haidinger's writings include *Treatise on Mineralogy,* 3 vols. (Edinburgh, 1825), his trans. of Mohs's textbook; "On the Determination of the Species in Mineralogy," in *Transactions of the Royal Society of Edinburgh,* **10** (1826), 298–313; "On the Parasitic Formation of Mineral Species," *ibid.,* **11** (1831), 73–114; "Ueber die dichroskopische Loupe," in *Sitzungsberichte der k. Akademie der Wissenschaften in Wien,* Nat.-math. Kl., **1** (1848), 131–137; "Ueber den Zusammenhang des orientirten Flächenschillers mit der Lichtabsorption farbirger Krystalle," *ibid.,* 146–152; "Ueber Herrn von Morlot's Sendschreiben an Herrn Élie de Beaumont die Bildung des Dolomits betreffend," *ibid.,* 171–173; "Ueber die Richtung der Schwingungen des Lichtaethers im geradlinig polarisierten Lichte," in Poggendorffs *Annalen der Physik,* **86** (1852), 131–144; "Die grüne Farbe der oxalsauren Eisenoxyd-Alkalien und die weisse der Eisenoxyd-Alaune," *ibid.,* **94** (1855), 246–255; and "Ueber die Natur der Meteoriten in ihrer Zusammensetzung und Erscheinung," in *Sitzungsberichte der k. Akademie der Wissenschaften in Wien,* Nat.-math. Kl., 2nd ser. **43** (1861), 389–426.

II. Secondary Literature. On Haidinger or his work,

see "Die Haidinger Medaille," in *Jahrbuch der k. k. Geologischen Reichsanstalt,* **6** (1856), v–xix; F. von Hauer, "Zur Erinnerung an Wilhelm Haidinger," *ibid.,* **21** (1871), 31–41; and A. Johannsen, *A Descriptive Petrology of Igneous Rocks* (Chicago, 1938), II, 72, with portrait.

JOYCE WEVERS

HAKLUYT, RICHARD (*b.* London, England, *ca.* 1552; *d.* London, 23 November 1616), *geography, history, advocacy of English overseas expansion.*

Richard Hakluyt was the leading advocate and chronicler of English overseas expansion in the reigns of Elizabeth I and James I. His collections of voyages established in the English language a new kind of historical literature, which remained in vogue for over 200 years.

A member of an influential and long-established Herefordshire family—his father was a London merchant—Hakluyt was educated at Westminster School, London, and Christ Church, Oxford. His cousin Richard Hakluyt, a lawyer of the Middle Temple, introduced him as a schoolboy to maps and books on cosmography, thus firing his life-long interest in the new and rapidly developing subject of geography. At Oxford, where he held a studentship at Christ Church (1570–*ca.* 1588), his study of the humanities as undergraduate and bachelor of arts (1570–1577) was the prelude to a teaching career as master of arts (1577 to 1582 or 1583) in which geography was increasingly his concern. He claimed to be the first to give public lectures in "the olde imperfectly composed, and the new lately reformed Map, Globes, Spheares, and other instruments of this Art . . .," and he was among those consulted by the great Flemish geographer Abraham Ortelius when the latter visited London in 1577. By the time he was ordained in 1578 he was an accepted authority on maritime affairs.

Hakluyt's first pamphlet (MS, 1579–1580) was a memorandum recommending that England should colonize and fortify the Strait of Magellan and so command "the gate of entry into the tresure of both the East and the West Indies." With Spain and Portugal already in possession of rich empires in America and Asia, he saw the need for England to establish her own routes to the coveted regions of the Orient, and to acquire her own sphere of influence in lands not yet annexed. Hence his special interest in, and advocacy of, the colonization of North America and the search for the Northwest Passage to Asia. In 1580, when Sir Humphrey Gilbert was projecting a colony in North America, Hakluyt commissioned John Florio to translate the narrative of Cartier's voyages to Canada. This was the first of a series of foreign works for which Hakluyt sponsored publication in English, as propaganda for English enterprise and as intelligence about regions already discovered. His years in Paris as chaplain to the English ambassador (1583–1588) gave him valuable access to French and Spanish sources.

Hakluyt's major and most original contribution to knowledge and literature lay in his three great collections of voyages. In the first, *Divers Voyages Touching the Discoverie of America* (1582), published under the initials R.H., he sought to establish England's claim to North America on the basis of priority of discovery. This was followed in 1589 by a volume of much wider compass, *The Principall Navigations, Voiages and Discoveries of the English Nation. . . .* Based on such original sources as the journals of explorers, sailing directions, and reports by merchants and seamen, many received by Hakluyt in person, it was a handbook of Elizabethan exploration and discovery. In working design it owed its inspiration to Giovanni Battista Ramusio's great work *Delle navigationi et viaggi* (1550–1559). As in Ramusio's collection, the voyages were arranged regionally and systematically, and the text was a model of discreet but informative editing.

Finally came the great three-volume work, of similar title to its predecessor but much enlarged to bring the record up to date and to include foreign enterprises, *The Principal navigations . . . of the English Nation* (1598–1600), acclaimed by J. A. Froude as "the Prose Epic of the modern English nation." In his preface and three new dedicatory epistles, Hakluyt set out his own ideas for England's maritime destiny and affirmed his belief in geography as the "right eye" of history. He also made the practical proposal that a lecture in navigation be established in London, "for the banishing of our former grosse ignorance in Marine causes."

Hakluyt did not restrict his activities to the role of chronicler but participated actively in projects of overseas expansion. Twice in the early 1580's he had hopes of sailing on voyages to America, but others went instead. Hakluyt's 1584 manuscript treatise for the queen and Francis Walsingham, "The Discourse of Western Planting," urged the advantages of an American settlement as a national enterprise. From 1599 he acted as consultant to the East India Company. As patentee of the Virginia Company in 1606, he had plans to go to Jamestown, but these too did not materialize. When he died he had traveled no farther than Paris. Yet his work provided inspiration and a wealth of information for his own and future generations of British seamen and colonial entrepreneurs.

The Hakluyt Society, founded in London in 1846

for the publication of records of voyages and travel, carries on Hakluyt's work and commemorates his name. Hakluyt was buried in Westminster Abbey.

BIBLIOGRAPHY

I. ORIGINAL WORKS. Richard Hakluyt's major works are *Divers Voyages Touching the Discoverie of America* (London, 1582); *The Principall Navigations, Voiages and Discoveries of the English Nation* . . . (London, 1589); and *The Principal Navigations, Voiages, Traffiques and Discoveries of the English Nation* . . ., 3 vols. (London, 1598–1600).

Hakluyt's later collections, edited and augmented after his death by Samuel Purchas, are in *Hakluytus Posthumus, or Purchas His Pilgrimes* (London, 1625).

II. SECONDARY LITERATURE. Works on Hakluyt include G. B. Parks, *Richard Hakluyt and the English Voyages* (New York, 1928), revised in 1961 with a complete bibliography of Hakluyt's writings; E. G. R. Taylor, *The Original Writings and Correspondence of the Two Richard Hakluyts* (London, 1935), Hakluyt Society, 2nd ser., LXXVI–LXXVII; Edward Lynam, ed., *Richard Hakluyt & His Successors* (London, 1946); *The Principall Navigations* . . ., a photolithographic facs., with an intro. by D. B. Quinn and R. A. Skelton and a new index by Alison Quinn (Cambridge, 1965); and D. B. Quinn, *Richard Hakluyt, Editor. A Study Introductory to the Facsimile Edition of Richard Hakluyt's Divers Voyages (1582)* (Amsterdam, 1967).

HELEN WALLIS

HALDANE, JOHN BURDON SANDERSON (*b.* Oxford, England, 5 November 1892; *d.* Bhubaneswar, Orissa, India, 1 December 1964), *physiology, biochemistry, genetics.*

Haldane was the son of the Oxford physiologist John Scott Haldane, member of a Scottish family that traces its ancestry to the mid-thirteenth century. His mother was Louisa Kathleen Trotter, also a Scottish patrician. From both parents he inherited a self-confidence that enabled him to tackle the problems of science in the belief that to a Haldane nothing was impossible. An Eton education set him against established authority, and service in World War I confirmed an early tendency toward atheism. At Oxford he turned from mathematics and biology to "greats" and thus left the university without scientific qualification. Haldane was married twice: first to a journalist, Charlotte Franken; and then to Helen Spurway, a fellow biologist, who survived him.

A physiologist in the immediate postwar years, Haldane switched first to biochemistry under Frederick Hopkins at Cambridge and then to genetics at University College, London, where, in the precomputer age, his mathematical talents were fully employed. Throughout a varied working life he stressed the social responsibilities of science. He belonged naturally to the radical left and was for some years a member of the Communist party, to whose *Daily Worker* he contributed more than 300 articles on popular science. Despite his varied work in many fields Haldane is likely to be remembered mainly as a geneticist and as a popular expositor of the unity of science. His extensive writings, which continued until the year of his death, ranged from ten famous papers in which he made mathematical contributions to the theory of natural selection to many volumes of essays explaining science to the layman, an art of which he was one of the greatest practitioners since T. H. Huxley.

Haldane's first scientific training was provided by his father, whom he assisted from childhood in the latter's private laboratory and whom he accompanied on work as government investigator of mining accidents and as physiologist for the Admiralty. In 1901 his interest in genetics was aroused by a lecture on the recently rediscovered work of Gregor Mendel; it was increased in 1910, when he began to study the laws of inheritance as revealed by his sister's 300 guinea pigs. Reading an early paper by A. D. Darbishire, Haldane noted what appeared to be the first example of gene linkage in vertebrates; he later read an undergraduate paper on the subject but delayed publication until he had obtained his own data (1915).

Haldane saw service in World War I on the Western Front and in Mesopotamia, and was wounded in both campaigns. For a short while he worked with his father and C. G. Douglas, both hurriedly brought to France from England, on the improvisation of gas masks following the first German gas attacks. This gave him an interest in the physiological problems of respiration, an interest he retained for the rest of his life, and the material for a controversial book on gas warfare (1925).

On demobilization in 1919 Haldane took up a fellowship at New College, Oxford, and shortly afterward began teaching physiology. Respiration was the only part of the subject in which he was well versed, but a crash course provided by his father gave him, as he later wrote, "about six weeks' start on my future pupils." With Peter Davies, a young worker in the Oxford physiological laboratory, Haldane began to investigate how carbon dioxide in the human bloodstream enables the muscles to regulate breathing under different conditions. During the experiments both men consumed quantities of bicarbonate of soda and "smuggled" hydrochloric acid into their blood by drinking solutions of ammonium chloride. In further experiments, popularly described in "On Being One's Own Rabbit" (1927), he measured the changes in the sugar and phosphate content of his blood and urine which could be induced by various means.

In 1921 Haldane accepted a readership in biochemistry under Hopkins at Cambridge, where he concentrated on the study of enzymes. Using some elegant mathematics, he calculated the rates at which enzyme reactions take place (1931); with G. E. Briggs he produced the first proof that enzyme reactions obey the laws of thermodynamics (1925); and in *Enzymes* (1930), produced largely from his Cambridge lectures, he provided an overall picture of how enzymes work. In this, as in much other comparable work, his knowledge of physiology plus his mathematical expertise enabled him to bring a feeling of practical reality to what had previously been largely biochemical assumptions.

Meanwhile Haldane had been continuing his investigation of linkage, and as early as 1919 he had given a formula relating the extent of linkage to the interval on the chromosome. He investigated the variation of linkage with age (1925) and formulated Haldane's law (1922), covering the crossing of animal species to produce an offspring of which one sex is absent or sterile. Meanwhile Haldane was producing the first of his ten major papers on the mathematics of natural selection, later reprinted as an appendix to his classic *The Causes of Evolution* (1932). Both Ronald Fisher and Sewall Wright were working along similar lines; but both introduced novel ideas into their papers while Haldane tended, instead, to reinforce the conservative Darwinian theory that natural selection, rather than mutation, was the driving force behind evolution. The most famous example to which he applied his theory, the replacement of the light-colored moth *Biston betularia* by a dark mutant form (1924), was strikingly verified by field studies thirty years later.

In 1933 Haldane left Cambridge for University College, London, where he occupied first the chair of genetics and then that of biometry. Here he gave increasing time to human genetics, preparing in 1935 a provisional map of the X chromosome which showed the positions on it of the genes causing color blindness, severe light sensitivity of the skin, night blindness, a particular skin disease, and two varieties of eye peculiarity. In 1936 he and a colleague, Julia Bell, began an extensive investigation which showed the genetic linkage between hemophilia and color blindness. The same year he gave the first estimate of the mutation rate in man, and in 1937 he described the effect on a population of recurrent harmful mutations.

From 1927 until 1936 Haldane also held a part-time appointment at the John Innes Horticultural Institution, then at Merton, outside London, where he carried on the genetic research of the former director, William Bateson. Here he began the joint work by Hopkins' Cambridge laboratory and the Institute on variation in flower color; and with D. de Winton he contributed to linkage theory by developing the theory for polyploids to tetraploid *Primula sinensis* (1931, 1933). But his lack of botanical experience and of experimental dexterity combined with his personal aggressiveness to bring the appointment to an end.

Shortly before the outbreak of World War II, Haldane was retained by the Amalgamated Engineering Union to represent the interests of their members at the public inquiry into the loss of ninety-nine lives when the submarine *Thetis* sank while on trials. The physiological work, during which Haldane and four members of the International Brigade were sealed into a chamber in which conditions in the stricken submarine were simulated, led directly to Haldane's doing much wartime work for the Admiralty. This involved investigation of the physiological problems concerned in escape from submarines, the operations of midget submarines, and much other underwater work. For considerable periods Haldane and a band of personally recruited colleagues, including his future second wife, Helen Spurway, risked their lives regularly. Many of their results were described by Haldane and E. M. Case (1941).

This work, and later statistical investigations for the government, was carried on despite Haldane's chairmanship of the editorial board of the *Daily Worker*. He had joined the Communist party soon after the outbreak of the Spanish Civil War, during which he advised the Republican government on gas precautions; and his scientific experience was conscripted by the predominantly left-wing movement which before the outbreak of World War II demanded better air raid precautions in Britain. He was an early supporter of Trofim Lysenko, fighting a rearguard action in his defense—often, it is clear, against his better scientific judgment—until 1949, when his article "In Defence of Genetics" revealed far less than the unquestioning support that Communist orthodoxy demanded.

In 1957 Haldane emigrated to India, ostensibly in protest against the Anglo-French invasion of Suez but largely, in fact, because he was attracted by the country's facilities for research in genetics and biometry. He worked for the Indian Statistical Office in Calcutta under Prasanta Mahanalobis and, after a short and unsatisfactory spell with the Council for Scientific and Industrial Research, moved to Orissa, setting up a genetics and biometry laboratory in the state capital of Bhubaneswar, where he died of cancer in December 1964. It was typical that while apparently recovering from a cancer operation in London, Haldane should write for the *New Statesman* a short poem which he hoped would encourage people not to take the disease too seriously. Entitled "Cancer's a Funny Thing," and starting, "I wish I had the voice of Homer/To sing of rectal carcinoma," it was in many ways the apotheosis of Haldane. It brought him a large postbag of letters which, in almost equal numbers, complained of his lack of feeling and praised him for his courage.

The uniqueness of Haldane's contribution to science was that for much of his life he was able to bring to fresh fields the equipment and concepts he had acquired in other disciplines; for him "the cross-fertilisation of ideas" really worked. This was also true of his papers and books, which range from the highly technical to the popular and include one classic book on science for children (1937). It was typical that he should describe in "The Origin of Life" (1929) a mechanism for the synthesis of organic matter which Darwin had merely assumed, a speculation on the origin of life very comparable with Alexander Oparin's in Russia. It is no matter for surprise that his bibliography should occupy a dozen closely printed pages in *Biographical Memoirs of Fellows of the Royal Society*.

BIBLIOGRAPHY

I. ORIGINAL WORKS. A scientific bibliography as well as a list of Haldane's books of essays and a selection of his more important popular articles is in Pirie's biographical memoir (see below) and is reprinted in Clark's biography (see below).

Among his writings are "Reduplication in Mice," in *Journal of Genetics*, **5** (1915), 133–135, written with A. D. Sprunt and N. M. Haldane; "The Combination of Linkage Values, and the Calculation of Distances Between the Loci of Linked Factors," *ibid.*, **8** (1919), 299–309; "Sex Ratio and Unisexual Sterility in Hybrid Animals," *ibid.*, **12** (1922), 101–109; the 10-part "A Mathematical Theory of Natural and Artificial Selection": pt. 1. in *Transactions of the Cambridge Philosophical Society*, **23** (1924), 19–41; pts. 2–9 in *Proceedings of the Cambridge Philosophical Society*, **1**, *et seq.* (1924–1932); pt. 10 in *Genetics*, **19** (1934), 412–429; *Callinicus—A Defence of Chemical Warfare* (London, 1925): "Change of Linkage in Poultry With Age," in *Nature*, **115** (1925), 641, written with F. A. E. Crew; "A Note on the Kinetics of Enzyme Reaction," in *Biochemical Journal*, **29** (1925), 338–339, written with G. E. Briggs; *Possible Worlds, and Other Essays* (London, 1927); "The Origin of Life," in *Rationalist Annual* (1929), pp. 3–10; *Enzymes* (London, 1930); "The Molecular Statistics of an Enzyme Action," in *Proceedings of the Royal Society*, **108B** (1931), 559–567; "Linkage in the Tetraploid *Primula sinensis*," in *Journal of Genetics*, **24** (1931), 121–144, written with D. de Winton; *The Causes of Evolution* (London, 1932); "The Genetics of *Primula sinensis:* Segregation and Inter-action of Factors in the Diploid," in *Journal of Genetics*, **27** (1933), 1–44, written with D. de Winton; "A Provisional Map of a Human Chromosome," in *Nature*, **137** (1935), 397; "Natural Selection," *ibid.*, **138** (1936), 1053; *My Friend Mr. Leakey* (London, 1937); "The Effect of Variation on Fitness," in *American Naturalist*, **71** (1937), 337–349; "The Linkage Between the Genes for Colour-Blindness and Haemophilia in Man," in *Proceedings of the Royal Society*, **123B** (1937), 119–150, written with J. Bell; "Human Physiology Under High Pressure. 1. Effects of Nitrogen, Carbon Dioxide, and Cold," in *Journal of Hygiene*, **41** (1941), 225–249, written with E. M. Case; and "In Defence of Genetics," in *Modern Quarterly*, n.s. **4** (1949), 194.

Haldane's papers were with him in Calcutta and Bhubaneswar and, following his death, are believed to have been taken to Hyderabad by his widow.

II. SECONDARY LITERATURE. See Ronald W. Clark, *J. B. S.: The Life and Work of J. B. S. Haldane* (New York, 1969); K. R. Dronamraju, ed., *Haldane and Modern Biology* (Baltimore, 1968), edited by one of Haldane's former students; N. W. Pirie, in *Biographical Memoirs of Fellows of the Royal Society*, **12** (1966), 219–249; and *Science Reporter* (Delhi), **2** (1965), a special Haldane number containing articles on his life and work. The reminiscences of Haldane's mother are in L. K. Haldane, *Friends and Kindred* (London, 1961).

RONALD W. CLARK

HALDANE, JOHN SCOTT (*b.* Edinburgh, Scotland, 3 May 1860; *d.* Oxford, England, 15 March 1936), *physiology*.

In respiratory physiology, John Scott Haldane was the prime mover of modern times. A member of the Cloan branch of the centuries-old Haldane family of Gleneagles, he was the younger brother of Richard Burdon, Viscount Haldane of Cloan, and father of J. B. S. Haldane, the geneticist and philosopher. His own father, Robert, was a lawyer and writer to the signet in Edinburgh; his mother was Mary Burdon-Sanderson, sister of John Burdon-Sanderson, first Waynflete professor of physiology at Oxford.

Haldane was educated at Edinburgh Academy and Edinburgh University, from which he was graduated in medicine in 1884. He also spent short periods of time at Jena and, later, at Berlin. His first research work, on the composition of air in dwellings and schools, was done at Dundee; an account was published in 1887. Soon afterward he joined his uncle at Oxford as demonstrator in physiology, and Oxford was his base for the rest of his life.

One of Haldane's most impressive and constant characteristics, an impatience with artificial distinctions between theoretical and applied science, was soon manifest. Applying information gained from laboratory studies on the relation between carbon dioxide content of inspired air and respiratory volume, he began work on hazards to which coal miners were subjected. The result was a classic report on the causes of death in mine disasters which laid particular stress on the lethal effects of carbon monoxide. As a result of his inquiries into coal mine disasters, Haldane's curiosity took him back to the laboratory to establish the precise reasons for the toxicity of carbon monoxide. The result was a paper of enormous significance in which he showed that carbon monoxide binds hemoglobin, preventing it from serving as the body's oxygen carrier, and that the effect can be vitiated by placing the experimental subject (mice) in a hyperbaric environment. The full clinical implications of the work were not appreciated for over half a century. Seeing the need for better analytic methods, he devised, in principle, the well-known Haldane gas analysis apparatus in 1898; and a few years later, with Joseph Barcroft, he developed a method for determining blood gas content from relatively small amounts of blood. Both are still in use, although the Scholander apparatus has largely replaced the Haldane-Barcroft device for determination of blood gases.

Haldane's best-known paper, written with J. G. Priestley, set forth the view that pulmonary ventilation is controlled by the partial pressure of carbon dioxide in arterial blood reaching the respiratory center of the midbrain. It appeared in 1905 and, along with his analytic methods, immediately stepped up interest in respiratory physiology.

Haldane later modified and extended his concept of the chemical control of ventilation several times. Although now known to be much oversimplified, the work showed clearly that, except under extreme conditions, regulation of breathing depends far more on carbon dioxide content of inspired air than on oxygen content. It was probably Haldane's most influential work, and it is astonishing to note that it, like his work on carbon monoxide poisoning, received very little clinical application until after World War II.

Still influenced by his interest in the intact, integrated human organism under stressful conditions (as in mines and during deep-sea diving), Haldane proceeded to unravel the basic enigmas of heatstroke and caisson disease (bends). He worked out the method for stage decompression which is still in use in deep-sea diving operations and in underwater construction.

At the opposite extreme of the barometric pressure scale, Haldane and colleagues, including several American physiologists, undertook studies on the physiological effects of high altitude by making an expedition to the summit of Pikes Peak in 1911. The published work contains much of value but was marred by Haldane's stubborn conviction that passage of oxygen across the lining of the tiniest air sacs (alveoli) in the lung could not be due solely to passive diffusion along a gradient of partial pressure. Haldane believed that it had to be due in part, and under certain circumstances, to active secretion by the cells lining the sacs. The oxygen-secretion theory, to which Haldane clung for the rest of his life, never received substantiation and has, for all practical purposes, long since been abandoned by respiratory physiologists.

Other work was more solid. Studies on hemoglobin dissociation showed very clearly the manner in which the degree of oxygenation of hemoglobin affects the uptake of carbon dioxide in the tissues and its release in the lung. Others dealt with the reaction of the kidney to water content in the blood and with the physiology of sweating. The practical world of engineering turned often to Haldane for counsel which he, among basic scientists, was uniquely able to give because of his sympathetic and informed interest in its problems. He was enormously influential in planning safety measures for tunnel construction and mining and diving operations, and in solving ventilation problems in buildings, ships, and submarines. In devising methods for ventilating naval vessels, he was following the eighteenth-century precedent set by one of Britain's earliest and greatest physiologists, Stephen Hales.

Haldane summarized most of his work in the Silliman lectures at Yale (1916), which were published in book form in 1922. A new edition, prepared with Priestley in 1935, was for many years the standard textbook in respiratory physiology. Even a cursory look at the volume suffices to demonstrate the extent to which Haldane and his co-workers, especially Priestley and C. G. Douglas, laid the groundwork for respiratory physiology as it stands today. Two other books, dealing mainly with biophysical matters, took Haldane well out of his depth and received little acceptance. But nothing could diminish the importance to basic and applied scientists of the information contained in the Silliman lectures.

Like his brother, Viscount Haldane, J. S. Haldane had an abiding interest in philosophical topics and wrote extensively on the interface between science and philosophy. What he was grasping for is not easily understood, but a statement from the preface to the second edition of *Respiration* gives a few clues:

> Existing physical science can give no account of the characteristic features of life and conscious experience, or their assumed origin in the course of evolution. If we seriously endeavored to include the phenomena of life within the scope of physical science we should require to modify drastically the axioms on which existing physical science is based. . . . Physical science is certainly no more than a superficial aspect of ultimate philosophical truth [p. vii].

Haldane served on several royal commissions and was elected a fellow of the Royal Society in 1897. He was awarded the Royal Medal in 1916, received the Copley Medal in 1934, and was created Companion of Honour (for work in industrial hygiene) in 1928. He was a fellow of New College, Oxford, from 1901 to his death. He was also director of a research laboratory set up by the coal mining industry, first near Doncaster and later in Birmingham.

After 1921 Haldane spent a great deal of time at the Birmingham laboratory. But in the preceding twenty-five years he, probably more than anyone else, had brought the Oxford school of physiology into international prominence. He had looked searchingly at the process by which oxygen in the ambient air, in many different environments, enters the human body and arrives at the capillary. The reverse passage of carbon dioxide from capillary to exhaled air was of equal concern. But the mechanics of blood circulation seemed to interest him less than the total respiratory process, although he did devise a method for estimating cardiac output. Along the way he made many useful digressions and, by the ingenious application of many of his findings to industrial situations, he opened many doors and undoubtedly saved many lives.

An indefatigable worker and a thoroughly gifted scientist, Haldane showed by his example that the competent and committed scientific investigator must sometimes look beyond the laboratory and that the equilibrium between theoretical and applied science can be a very dynamic and constructive one.

BIBLIOGRAPHY

Haldane's writings include "The Relation of the Action of Carbonic Oxide to Oxygen Tension," in *Journal of Physiology*, **18** (1895), 201–217; *The Cause of Death in Colliery Explosions* (London, 1896); "The Influence of High Air Temperatures. No. 1," in *Journal of Hygiene*, **5** (Oct. 1905), 494–513; "The Regulation of the Lung-Ventilation," in *Journal of Physiology*, **32** (9 May 1905), 224–266, written with J. G. Priestley; "The Prevention of Compressed Air Disease," in *Journal of Hygiene*, **8** (June 1908), 342–443, written with A. E. Boycott and G. C. C. Damant; "Physiological Observations Made on Pike's Peak, Colorado, With Special Reference to Adaptation to Low Barometric Pressures," in *Philosophical Transactions of the Royal Society*, **203B** (Mar. 1913), 185–318, written with C. G. Douglas, Y. Henderson, and E. C. Schneider; "The Absorption and Dissociation of Carbon Dioxide by Human Blood," in *Journal of Physiology*, **48** (1914), 244–271, written with J. Christiansen and C. G. Douglas; and *Respiration*, new ed. (Oxford, 1935), written with J. G. Priestley, which includes the Silliman lectures.

A short biography is C. G. Douglas, "John Scott Haldane, 1860–1936," in *Obituary Notices of Fellows of the Royal Society of London*, **2**, no. 2 (Dec. 1936), 115–139.

CARLETON B. CHAPMAN

HALDANE, RICHARD BURDON (*b.* Edinburgh, Scotland, 30 July 1856; *d.* Cloan, Perthshire, Scotland, 19 August 1928), *philosophy.*

Haldane was the son of Robert Haldane, landowner and lawyer; his mother was Mary Elizabeth Burdon-Sanderson, a collateral descendant of Lord Eldon, a famous lawyer and member of Lord Liverpool's government. He studied philosophy at the University of Edinburgh and for some months at Göttingen under R. H. Lotze. Having commenced a legal career in London in 1877, Haldane became Liberal member of Parliament for East Lothian in 1885, supporting W. E. Gladstone and associating with Lord Rosebery as a "liberal imperialist."

Haldane's very successful career as a barrister led to his being given important administrative tasks, and he became secretary of state for war in the reforming Liberal government of 1905. Between 1905 and 1911 he transformed the organization of the British army, setting up the Expeditionary Force for service overseas, the locally recruited Territorial Army for home defense, the Officers' Training Corps (functioning in schools and universities), and the Imperial General Staff, to coordinate the policies of the Empire countries. He also reformed the army transport and medical services.

In 1912 Haldane became lord chancellor and went on a mission to Germany, seeking to persuade the German government to agree with Great Britain on common limitations on naval expansion. The mission failed; and when war broke out with Germany in 1914, a remark of Haldane's that Germany was his "spiritual home" was used by his political opponents to impugn his patriotism and force his resignation from the government in 1915. At the end of the war Earl Haig, the victorious British commander, wrote that Haldane was "the greatest Secretary of State for War that England has ever had." When the Labour party for the first time formed a government in 1924,

Haldane, who had long been friendly with Sidney Webb and other Labour party leaders, became lord chancellor for the nine months of its existence.

Another major concern of Haldane's life was the advancement of higher education. He helped to found the Imperial College of Science and Technology of the University of London and was active in the movement for enabling workingmen to obtain university-level education in evening classes. His chief work in the educational sphere was concerned with the formation of what he called civic universities in England and Wales so that university education could be available to more students than could be accommodated in the existing universities of Oxford, Cambridge, London, and Durham. He helped to obtain charters for universities at Liverpool, Manchester, and Leeds and advised the establishment of the University Grants Committee to allocate government funds to the universities without governmental control of their use.

Haldane's chief and abiding interest was in philosophy. In 1883 he joined with J. Kemp in a translation of Arthur Schopenhauer's *World as Will and Idea* and in the same year was joint editor with Andrew Seth of *Essays in Philosophical Criticism,* a collection of essays in memory of the idealist philosopher T. H. Green. This book was a sort of manifesto supporting the Hegelian movement in philosophy which was dominant in British philosophy for the next thirty years. Haldane contributed an article entitled "The Relation of Philosophy to Science" to this volume jointly with his younger brother, John Scott Haldane, who later became known for his work on the physiology of respiration and for his defense of vitalism in biology. In the article the Haldanes argued that for the immediate future "a new class of men" was required; they should be trained both in a scientific specialism and in "the critical investigations of Kant and Hegel." In 1903 Haldane published *The Pathway to Reality,* his Gifford lectures at the University of St. Andrews. In it he developed a Hegelian view of philosophy, arguing that philosophical criticism shows the inadequacies of the categories of common sense and the abstractness of the categories of the particular sciences.

Haldane's best-known philosophical book is *The Reign of Relativity* (1921), in which he sketches the mathematical context of the theories of general and special relativity in the work of C. F. Gauss, G. F. Riemann, and Hermann Minkowski, and maintains that Einstein's theory is only an illustration of "the principle of the relativity of knowledge" to a special subject. The exposition is general but lucid and critical. In the course of it Haldane discusses Moritz Schlick's view, expressed in *Raum und Zeit in der gegenwärtigen Physik* (1917), that physical space is "essentially dissimilar" from perceptual space although correlated with it and argues that such a view introduces "a splitting up of experience into sensations and conceptions which seems to have little warrant in the actual character of that experience" (p. 59). Noting, by reference to Arthur Eddington, that Einstein's equation for gravitation is "not so much a law as a definition," Haldane discusses the more metaphysical approach to relativity that A. N. Whitehead had taken in *The Concept of Nature* (1920), supporting Whitehead's rejection of "the bifurcation of nature."

Haldane believed that Einstein's theory supported the idealist thesis that the distinction between knowledge and what is known is a distinction within knowledge itself. He therefore mistakenly treated Einstein's "observer" as if it were akin to Kant's "transcendental unity of apperception." *The Reign of Relativity,* in consequence, became a compendium of idealist metaphysics, with discussions of the work of F. H. Bradley and Bernard Bosanquet, a defense of vitalism in biology continued in *The Philosophy of Humanism and Other Subjects* (1922), and even a vindication of the general will. When Einstein came to lecture at Kings' College, London, in 1921, he told Haldane he did not believe that his theory had metaphysical implications, and the archbishop of Canterbury that it had no religious implications. Haldane had wrongly supposed that "relative to an observer" entails "dependent on mind."

BIBLIOGRAPHY

I. Original Works. Haldane's writings include *Essays in Philosophical Criticism* (London, 1883), edited with Andrew Seth; *Life of Adam Smith* (London, 1887); *The Pathway to Reality,* 2 vols. (London, 1903); *Universities and National Life* (London, 1912); *Before the War* (London, 1920); *The Reign of Relativity* (London, 1921); *The Philosophy of Humanism and Other Subjects* (London, 1922), which contains discussions of vitalism; *Human Experience* (London, 1926); and *Richard Burdon Haldane. An Autobiography* (London, 1929).

II. Secondary Literature. See Stephen E. Koss, *Lord Haldane: Scapegoat for Liberalism* (New York–London, 1969); Sir Frederick Maurice, *Life of Lord Haldane of Cloan,* 2 vols. (London, 1937); and Dudley Sommer, *Haldane of Cloan. His Life and Times* (London, 1960).

H. B. Acton

HALE, GEORGE ELLERY (*b.* Chicago, Illinois, 29 June 1868; *d.* Pasadena, California, 21 February 1938), *astrophysics.*

George Hale was the eldest surviving son of

William Ellery Hale and Mary Scranton Browne. The family moved from Chicago to the suburb of Hyde Park before the great fire of 1871 destroyed his birthplace and a building erected by William Hale in the heart of Chicago. With the intense energy and engineering ability his son would inherit, William Hale turned to the manufacture of the hydraulic elevators that would make possible the tall buildings of the new Chicago. As his business expanded to other American cities, and even to London and Paris, he prospered.

William Hale's father had been a minister; Mary Hale, daughter of a Congregational minister who later became a doctor, was raised by her adopted grandfather, a stern Calvinist preacher. In his boyhood George Hale attended the Congregational church but, years later, when his wife asked him to go to church "for the sake of the children," he wrote: "Of course you must see that it is hard—really impossible—for me to reason one way through the week, and another way on Sunday. My creed is Truth, wherever it may lead, and I believe that no creed is finer than this."[1]

Although Hale failed to adopt the religious creed of his parents, he was grateful for the broad cultural outlook they gave him. To his mother, who had been educated at Catharine Beecher's famous Hartford Female Seminary, he was especially grateful for the love of literature and poetry that he considered vital to the development of his creative scientific imagination. His early reading ranged from *Grimm's Fairy Tales* to *Don Quixote,* from the *Iliad* and the *Odyssey* to the poetry of Keats and Shelley, from Cassell's *Book of Sports and Pastimes* to Jules Verne's *From the Earth to the Moon.*

These wide-ranging interests were consistently pursued in later years. Their diversity is reflected in the role Hale played not only in astronomical and other scientific institutions but also in those dealing with the humanities, in which he worked with equal fervor. The institutions range from the three great observatories he founded—Yerkes, Mt. Wilson, and Palomar (each in its time the greatest in the world)—to the California Institute of Technology and the Henry E. Huntington Library and Art Gallery; from the National Academy of Sciences, which he helped to reform; to the National Research Council, which he initiated; and the International Research Council, out of which evolved the International Council of Scientific Unions.

As a child Hale suffered from intestinal ailments and typhoid; and his mother, who was subject to migraine headaches, worried constantly about her high-strung son. In later life he had three serious breakdowns and was forced to give up work for long periods. At such times he suffered severe depression and acute pain at the back of his head which his doctors called brain congestion. They ascribed his troubles to overanxiety, overintensity, and an inability to relax.

Hale first attended the Oakland Public School, then the Allen Academy. He also took shopwork at the Chicago Manual Training School. Yet, as he wrote:

> I never enjoyed the confinements and the fixed duties of school life. Born a free lance, with a thirst for personal adventure, I preferred to work at tasks of my own selection. . . . As a boy, largely through the constant encouragement of my father, I became interested in tools and machinery at a very early age; I always had a small shop with tools, first in the house, and later in a building of my own construction in the yard. I also had a little laboratory where I performed simple chemical experiments, made batteries and induction coils, worked with a microscope etc. After construction of a small telescope for myself, my father bought me an excellent 4-inch Clark. I used this constantly, but my enthusiasm reached the highest pitch when I learned something about the spectroscope. My greatest ambition was to photograph a spectrum and this I soon succeeded in doing with a small one prism spectroscope purchased for me by my father. I think this was in 1884. Solar spectroscopic work appealed to me above all things and I read everything I could find on the subject. My father always bought for me any books that I needed, but in the case of instruments his policy always was to induce me to construct my own first and then to give me a good instrument if my early experiments were successful. In 1888 he built for me, after my designs, a spectroscopic laboratory, in which a Rowland concave grating of 10 feet focal length was erected. This was the nucleus of the Kenwood Observatory.[2]

This experimental approach to astronomy was in essence the basis for Hale's scientific goals. He entered astronomy at a time when the majority of astronomers, concerned with the positions, motions, and distances of the stars, evinced little interest in their physical nature. Pioneer work had been done by Angelo Secchi in Italy, William Huggins and J. N. Lockyer in England, P. J. C. Janssen in France, H. C. Vogel in Germany, and a few others. Hale's influence on the development of the embryonic science of astrophysics was so great that it is often said he contributed more than any other individual to the rise of modern astrophysics. His tools were those he had used in boyhood—the telescope, the spectroscope, the photographic plate; his working place was the observatory combined with laboratory and shop where astrophysical problems might be solved to discover the physical nature of the universe.

In 1886 Hale entered the Massachusetts Institute of Technology. He majored in physics but, in contrast with his own laboratory work, found most of the courses uninspiring. In his spare time he read and abstracted everything he could find on astronomy and spectroscopy at the Boston Public Library. He also persuaded E. C. Pickering, director of the Harvard College Observatory, to let him work there as a volunteer assistant. In August 1889, shortly after his twenty-first birthday, he was riding on a Chicago trolley car when the idea came to him "out of the blue" for an instrument that would solve the problem of photographing the solar prominences in full daylight and would provide a permanent record of these and other solar phenomena. He called it a spectroheliograph.

In 1868 Janssen and Lockyer had observed prominences visually outside of eclipse for the first time. C. A. Young, Károly Braun, and Wilhelm Lohse had tried to photograph the prominences spectroscopically in daylight but without practical success. Later Deslandres would claim priority but this was a full year after Hale had proved the success of his method. (Deslandres did develop successfully what he would call a *spectro-enrégisteur des vitesses,* or velocity recorder.) In the fall of 1889 Hale tried out his principle at the Harvard Observatory; and in his M.I.T. thesis, "Photography of the Solar Prominences," he described the results that proved the feasibility of the method. Out of these beginnings was born his lifelong interest in the "typical star"—our sun, the only star near enough to be studied in detail. His contributions to solar research that resulted were, as the Mt. Wilson astrophysicist Robert Howard noted, so vital that "Hale may be said to be the father of modern solar observational astronomy."[3]

The day after he graduated from M.I.T., Hale and Evelina Conklin of Brooklyn, New York, were married. On their honeymoon they visited the Lick Observatory, where Hale was inspired by the sight of James Keeler making his classic observations of radial velocities in planetary nebulae with a stellar spectroscope attached to the thirty-six-inch telescope, then the largest in the world.

Back in Chicago, Hale persuaded his father to provide the funds for a telescope with which he could continue his experiments with the spectroheliograph. In June 1891 his twelve-inch refractor with an object glass by John Brashear and mounting by W. R. Warner and Ambrose Swasey was dedicated in his small Kenwood Observatory, located behind their house at 4545 Drexel Boulevard. With it, in 1892, using the *H* and *K* lines of calcium in the ultraviolet, Hale photographed the bright calcium clouds (flocculi) and the prominences all around the sun's limb for the first time, thus proving the success of his instrument.

In 1892 Hale was appointed associate professor of astrophysics at the new University of Chicago. That summer he learned of the availability of two forty-inch lenses at the firm of Alvan Clark in Cambridgeport, Massachusetts, and persuaded the traction magnate Charles T. Yerkes to provide for a telescope that would surpass all others in focal length and light-gathering power. In 1893, while plans were being worked out for the new observatory, Hale went to the University of Berlin and worked there with Hermann von Helmholtz, Max Planck, A. A. Kundt, and Heinrich Rubens. At the Potsdam Observatory contacts with Vogel and Julius Scheiner increased his enthusiasm for astrophysical research. Yet before the year's end Hale abandoned his plans for a Ph.D. and never took time to earn one, although he received many honorary degrees, including one from Berlin. On his way home he attempted to photograph the solar corona outside of eclipse from Mt. Etna. This attempt, like every other he made, failed. It was not until 1930 that Bernard Lyot in France accomplished this difficult feat.

In 1897 the Yerkes Observatory at Williams Bay, Wisconsin, was dedicated. It was, as Hale noted, based on a revolutionary principle, "in reality a large physical laboratory as well as an astronomical establishment" where "all kinds of spectroscopic, bolometric, photographic and other optical work would be done in its laboratories."[4] Here he gathered a small but devoted staff that included future astronomical leaders. He also encouraged visits from foreign astronomers eager to use the superior facilities, while he himself continued the observation of sunspot spectra begun at Kenwood and designed the Rumford spectroheliograph to be attached to the forty-inch telescope. With this powerful tool, using the $H\beta$ hydrogen line, Hale found the dark hydrogen flocculi and investigated the calcium flocculi at different levels to gain knowledge of the complex circulatory processes in the sun. From the sun he turned to other stars, as he undertook (with Ferdinand Ellerman and J. A. Parkhurst) a study of the spectra of those late-type, low-temperature red stars, known as Secchi's fourth type, which showed certain marked similarities to sunspot spectra and had never before been photographed.

Meanwhile, Hale promoted the astrophysical cause in other ways. On a triumphal tour abroad in 1891 he had been welcomed by leaders in astronomy and physics. On his return, with their endorsement he and W. W. Payne, editor of the pioneer astronomical journal *Sidereal Messenger,* founded a journal called *Astronomy and Astro-Physics.* In 1895, with Keeler as joint editor, he founded the separate *Astrophysical Journal* with an international board of editors that included astronomers and physicists from England, France, Germany, Italy, and the United States. It is still the leading journal in its field.

In 1899 the first meeting of a new astronomical society was held at Yerkes. Fearful that astrophysics might be overlooked, Hale insisted to Simon Newcomb (who became the first president, with Charles Young and Hale as vice-presidents) that it should be called the American Astronomical and Astrophysical Society. In 1914, when astrophysics was more generally accepted, the name was changed to the American Astronomical Society.

In 1896 Hale persuaded his father to provide the disk for a sixty-inch reflecting telescope with which stellar spectra could be photographed "on so large a scale as to permit the study of their chemical composition, the temperature and pressure in their atmospheres, and their motions with that high degree of precision"[5] which could then be reached only in the case of the sun. William Hale offered this disk to the University of Chicago on condition that funds be found to mount it. This condition was never fulfilled.

In 1902 the Carnegie Institution of Washington was founded by Andrew Carnegie to "encourage investigation, research and discovery in the broadest and most liberal manner, and the application of knowledge to the improvement of mankind." On 20 December 1904, after overcoming many difficulties and gambling $30,000 on a successful outcome, Hale received $150,000 to found the Mt. Wilson Solar Observatory, under Carnegie auspices, on a peak above Pasadena. The story of the pioneer days on that mountain, when the astronomers lived under primitive conditions and all supplies had to be transported by burro and mule, has been dramatically told by Hale's colleague and successor as director of the Mt. Wilson Observatory, Walter Adams. He describes Hale's insight, courage, and enthusiasm and his unexpected reaction to the novel conditions:

> Apparently combined with a deep-seated love of nature in every form was the spirit of the pioneer, whose greatest joy is the adventure of starting with little and taking an active personal part in every phase of creation and growth. To both of these inborn characteristics of Hale, Mount Wilson in 1904 offered a rich field and scope for their full employment.[6]

The first instrument was the Snow telescope, brought on a temporary expeditionary basis from the Yerkes Observatory even before Mt. Wilson's founding. With this instrument, essentially a solar telescope fed by a coelostat, devised to accommodate larger, more powerful spectrographs than could be attached to the forty-inch, the first photograph of a sunspot spectrum was taken in 1905. By that time a small laboratory had been built on the mountain. Here spectroscopic results, obtained with the Snow telescope and other instruments, could be analyzed and compared with laboratory results obtained under controlled conditions. And here the significant observation was made by Hale, Adams, and Henry Gale that those lines which are strengthened in sunspots are exactly the lines that are strongest in low-temperature sources, such as the electric arc and furnace. Thus it became evident that sunspots are cooler than other regions of the solar disk, as Hale had long suspected.

In 1908, in the hope of overcoming the temperature problems that had plagued the low-lying Snow telescope, Hale designed and built a sixty-foot tower telescope with a thirty-foot spectrograph in an underground pit. With photographic plates sensitive to red light (developed by R. J. Wallace at Yerkes) he detected vortices in the hydrogen flocculi in the vicinity of sunspots. This observation led to the hypothesis that the widening of lines in sunspot spectra might be due to the presence of intense magnetic fields in sunspots. With the new sixty-foot tower telescope—which, with customary vision, he had planned in anticipation of the need—Hale was soon able to prove his hypothesis. Young and W. M. Mitchell at Princeton had observed double lines in sunspot spectra visually but had ascribed the effect to "reversal." Now Hale became convinced that the splitting was due to the Zeeman effect. In 1908 he compared his observations of the doubling of lines in sunspots with a similar doubling obtained with a powerful electromagnet in his Pasadena laboratory and showed for the first time the presence of magnetic fields in sunspots. This, his greatest discovery, was also the first discovery of an extraterrestrial magnetic field. The mathematical physicist R. S. Woodward, president of the Carnegie Institution of Washington, wrote: "This is surely the greatest advance that has been made since Galileo's discovery of those blemishes on the sun."[7]

This discovery was followed by Hale's recognition of the reversal of sunspot polarities with the sunspot cycle, and this in turn led to the formulation of his fundamental polarity law. In this law he stated the twenty-two- to twenty-three-year interval between successive appearances in high latitudes of spots of the same magnetic polarity.

Meanwhile, Hale had turned to the puzzling question of whether the sun itself is a magnet. In 1889 F. H. Bigelow, observing the corona during eclipse, had suggested that the sun might possess a magnetic field. In 1912 the 150-foot tower telescope with a seventy-five-foot vertical spectrograph, designed to obtain the spectral resolution needed to measure the sun's general field, was completed. Preliminary observations with this instrument indicated that the sun has a dipole field with a strength of about twenty gauss. In the 1930's observations by Hale, Theodore Dunham, Jr., John Strong, Joel Stebbins, and A. E. Whitford indicated a field of approximately four gauss. But these results were still inconclusive.

It was not until 1952 that H. D. and H. W. Babcock, using an electrooptic light modulator, developed the solar magnetograph in the Hale Solar Laboratory in Pasadena and obtained the first reliable method for measuring magnetic fields on the sun's surface. They found evidence of the existence of a polar field of the sun with a strength of about two gauss and a polarity opposite to that of the earth. At the next solar maximum the polarity was reversed. "It is clear to us now," Robert Howard said in 1969, "that magnetic fields hold the key to the phenomenon called solar activity, and it is a tribute to the genius of Hale that he recognized at such an early stage the great importance of these elusive magnetic fields."[8]

In 1908, twelve years after his father had given Hale the disk, the sixty-inch reflecting telescope, then the largest in the world, was set up on Mt. Wilson. At last, with its great light-gathering power, steps could be taken in the photographing of stellar spectra on a scale that might eventually approach the great dispersion available for the study of the solar spectrum. The way was prepared for an understanding of stellar evolution that would be realized only when knowledge of atomic processes gained in earthly laboratories could be applied to the interpretation of the nature of stars and nebulae and when, in turn, knowledge derived from studies of those "enormous crucibles," the stars, could be applied on the earth.

Even before this, in 1906, with the success of the sixty-inch still uncertain, Hale had described the possibilities of a 100-inch telescope to a Los Angeles businessman, John D. Hooker. It would, he said, give two and a half times as much light as the sixty-inch, seven times as much as any other telescope then in use for stellar astronomy. It would "enormously surpass all existing instruments in the photography of stars and nebulae, giving new information on their chemical composition and the temperature and pressure in their atmospheres."[9] Through his talent for convincing wealthy men of the urgent need for supporting his dreams he persuaded Hooker to provide for a 100-inch disk. The first observations with this telescope, built with Carnegie funds, were made in November 1917. Soon afterward, carrying on research begun with the sixty-inch, it was contributing to knowledge of the size and nature of the universe, solving problems that had previously seemed insoluble. With it, using Albert Michelson's interferometer, Francis Pease and J. A. Anderson measured the diameter of the giant red star Betelgeuse—an extremely difficult feat—and found it to be an astounding 300 million miles.

In 1920 the famous Heber Curtis-Harlow Shapley debate, "The Scale of the Universe," took place at the National Academy of Sciences. Curtis had made his observations with the thirty-six-inch Crossley at Lick; Shapley had made his with the sixty-inch at Mt. Wilson. Their observations led them to quite different conclusions. No definite answer could be given until the end of 1923, when Edwin Hubble, working with the 100-inch, identified a Cepheid variable in a spiral nebula and found the key to its distance. His results, as Allan Sandage points out, "proved beyond question that nebulae were external galaxies of dimensions comparable to our own. It opened the last frontier of astronomy, and gave, for the first time, the correct conceptual value of the universe. Galaxies are the units of matter that define the granular structure of the universe."[10] Without the 100-inch telescope this, like many other breakthroughs in our knowledge of the universe, would have been impossible.

For his building of large telescopes Hale has been called the "master builder," but he was also a builder of institutions. All his life he was interested first and foremost in research, yet early in life he realized that to achieve his goals in astronomy, in science, and in the humanities, he must divert some of his energies to the less appealing tasks of organization. In 1902 he was elected to the National Academy of Sciences. From the beginning he felt that this, the leading scientific academy in the United States, should accomplish much more than it was doing if it was ever to occupy its proper position in the scientific world and "acquire a commanding influence of a favorable character, favorable alike to the development of research and the public appreciation of science."[11] To change its hoary ways and increase its influence, Hale proposed an increase in the membership, with an emphasis on younger, more forward-looking scientists. To broaden its outlook, he urged that the membership be expanded to include such branches as engineering and archaeology. To enhance its international position, he urged programs of cooperation, especially in astronomy.

All his life Hale was an internationalist. In 1893 he helped to arrange an international astronomical congress in connection with the Columbian Exposition. In 1904 the Louisiana Purchase Exposition was to be held in St. Louis, and he proposed that a committee be formed to organize the International Union for Cooperation in Solar Research under Academy auspices. To the first meeting, held in St. Louis, came a number of European astronomers, including Henri Poincaré, who was made vice-president (Hale became president). The Union was formally organized at Oxford in 1905. At a large meeting on Mt. Wilson in 1910 its aims were expanded to include all branches of astronomy.

Soon after this, as World War I broke out in 1914, the outlook for international exchange dimmed. At the spring meeting of the Academy in 1915 Hale presented a resolution offering its services to President Woodrow Wilson in case of a diplomatic break with Germany. Out of this move the National Research Council was born in 1916, and Hale became its first chairman. The Council was, as the executive order stated, organized for the purpose "of stimulating research in mathematical, physical and biological sciences, and in the application of these sciences to engineering, agriculture, medicine and other useful arts, with the object of increasing knowledge, of strengthening the national defense, and of contributing in other ways to the public welfare."[12] Through it Hale saw the chance to develop cooperative research on an unprecedented scale, first for war and later as an instrument for peace. Representatives of scientific and technical agencies, medical and engineering bodies soon joined with the government's scientific bureaus and with the departments of the Army and Navy to solve the scientific problems posed by the war.

In 1918 Hale, foreseeing the possibilities of cooperation not only nationally but internationally, proposed the formation of the International Research Council, to which, as long as the war lasted, only the Allies would be admitted; later, all who wished could join. This Council would replace the moribund International Association of Academies. A preliminary meeting held at the Royal Society in London was followed by an organizational meeting at Paris in November 1918. The Council was formally inaugurated at Brussels in July 1919, at which time the International Astronomical Union and other unions were established under the Council. The Astronomical Union, evolved out of the earlier Solar Union, combined such international groups as the Carte du Ciel and the International Union for Determination of Time and Latitude. By 1931 forty countries had joined the Council, and eight unions had been established. In 1931 it was renamed the International Council of Scientific Unions, and in 1932 Hale became its president.

After the war Hale returned to his plans for the Academy itself. The most compelling of these was his dream for a permanent scientific headquarters in Washington and its official center in both a national and an international sense. Such a center, he had long felt, was fundamental to the Academy's future growth. In 1919 the Carnegie Corporation agreed to give $5,000,000 to the Academy and National Research Council, with a little over a quarter of it to be used for a building in Washington. This impressive building, designed by Bertram Goodhue, was dedicated in 1924. It stands today on Constitution Avenue across from the Lincoln Memorial.

One of the most significant programs in the development of the scientific life of the United States had been initiated on a limited scale during the war. In 1919 this program, in which Hale played a leading role, was expanded when the Rockefeller Foundation agreed to support the National Research Fellowships. Another scheme for which he had equally high hopes was the National Research Fund, launched in the early 1920's as a means of persuading industry to support basic research. Large sums were promised but the program failed, largely as a result of the depression in the 1930's as well as of the lack of vision of many industrial leaders. Nevertheless, all these developments, as well as others within the Academy itself (such as the publication of the *Proceedings* which Hale initiated in 1915), enhanced its position and increased its usefulness, so that it was no longer just the mutual admiration society that it had been in large part when Hale became a member in 1902.

The Mt. Wilson Observatory was founded in 1904. In 1906 Hale became a trustee of Throop Polytechnic Institute, a Pasadena school with meager resources where a range of courses was taught in an elementary school, a manual training division, an art school, and to a small number of college students. He proposed that its character be changed entirely so that it could become a scientific and technological institution of the first rank—like M.I.T. but broader in outlook. He wrote: "Fundamental science had been unduly subordinated to engineering in all American schools of technology and I therefore emphasized the importance of developing it on the highest plane."[13] To achieve this goal it was necessary to find the ablest scientists and teachers available and to persuade them to share his faith in the future of this unknown institute. It was also necessary to find the money to support it and raise its standards, as the 500-member student body was radically reduced to a small, select group of thirty students of college caliber.

It was not an easy task. Yet out of these small beginnings the California Institute of Technology, as Throop is known today, evolved. Hale's plan for

Throop was part of a larger dream for Pasadena as a center for scientific research, in which the new Throop would collaborate with Mt. Wilson in research on fundamental physical problems. It was part of a broader dream for Pasadena as a cultural center. In 1906 Hale learned from the transportation magnate Henry Huntington of his plan to give his collection of paintings and rare books to Los Angeles County. He urged that Huntington instead consider the possibilities of a center in the humanities to which scholars from the world over might come to do research. In 1919 Hale was made a trustee of what was to become the Henry E. Huntington Library and Art Gallery. Shortly before his death in 1927 Huntington provided the endowment for such a center as Hale had first proposed and had continued to urge with detailed plans over many years. For these contributions and for his work on a wide-ranging city plan for Pasadena, Hale received the city's highest award, the Noble Medal.

In 1923, as he continued to be plagued by ill health, Hale gave up the directorship of Mt. Wilson and built the Hale Solar Laboratory in Pasadena, where he could carry on his solar research. Here he invented and built the first spectrohelioscope, a special type of spectroscope, with an oscillating slit, for the visual study of solar phenomena. He also continued work on the sun's magnetic field with a spectroheliograph there.

Despite his "retirement" Hale launched the last great astronomical project of his career. As soon as the 100-inch had proved successful, he had begun thinking of a still larger telescope. In 1928 he wrote to Wickliffe Rose of the International Education Board of the Rockefeller Foundation. He emphasized once more the progress that had been attained through combining the spectroscope, telescope, and photographic plate with supplementary instruments and pointed out the importance of a 200-inch telescope to future advances in physics and astronomy. In a familiar vein he wrote: "In fact, the range of celestial temperatures, densities, masses and states of matter so enormously transcends that of the physical laboratory that many of the most fundamental advances in physics depend upon the utilization of these conditions."[14]

Forty years had passed since Hale first urged on a skeptical astronomical world the concept of an observatory as a physical laboratory. By 1928 this concept was no longer questioned. The International Education Board of the Rockefeller Foundation donated $6,000,000 to the California Institute of Technology for a 200-inch telescope, on condition that a cooperative plan be developed with the Mt. Wilson Observatory and its owner, the Carnegie Institution of Washington. This led to the formation of the Mt. Wilson and Palomar Observatories after the 200-inch instrument was set up on Palomar Mountain in southern California. Hale died before the telescope was finished. Since World War II intervened, ten years passed before the observatory was dedicated in 1948 and the Hale telescope was named after the man "whose vision and leadership made it a reality."[15] In December 1969 the Mt. Wilson and Palomar Observatories were renamed the Hale Observatories.

"It is perhaps symbolic of this man of great gifts and wide horizons," Walter Adams wrote, "that he who had devoted his life to the nearest star should find his last deepest interest in an instrument destined to meet the remotest objects of our physical universe."[16]

NOTES

1. Hale to E. C. Hale, 29 Apr. 1909, Hale Collection, Pasadena, Calif.
2. Hale to H. H. Turner, 17 Jan. 1903. Carbon copy in the Yerkes Observatory Records, Williams Bay, Wis. The Turner–Hale correspondence at Oxford was apparently destroyed.
3. Robert Howard, "Research on Solar Magnetic Fields from Hale to the Present," talk given at Hale Centennial Symposium at meetings of the American Association for the Advancement of Science at Dallas, Texas, Dec. 1968. The entire symposium will be published as *The Legacy of George Ellery Hale* by M.I.T. Press.
4. Hale to James Hall, 12 July 1895; copy of the letter is in the Yerkes Observatory Records, Williams Bay, Wis.
5. Letter to W. R. Harper, 9 May 1899, University of Chicago archives.
6. Walter Adams, "Early Days at Mount Wilson," in *Publications of the Astronomical Society of the Pacific,* **59** (1947), 213.
7. R. S. Woodward to G. E. Hale, 29 July 1908, Hale Collection, Pasadena, Calif.
8. Howard, *op. cit.*
9. Letter to John D. Hooker, 27 July 1906, Hale Collection, Pasadena, Calif.
10. Allan Sandage, *Hubble Atlas of Galaxies* (Washington, D.C., 1961), "Galaxies," p. 4.
11. Hale to C. D. Walcott, 25 Jan. 1908. The original letter is in the National Academy of Sciences Archives.
12. Executive order signed by Woodrow Wilson, 11 May 1918.
13. "Autobiographical Notes" (unpublished), Hale Collection, Pasadena, Calif.
14. Letter to Wickliffe Rose, 14 Feb. 1928, Hale Collection, Pasadena, Calif.
15. On the bronze plaque on the Hale bust in the foyer of the two-hundred-inch dome at Palomar these words are inscribed, "The two hundred inch telescope named in honor of George Ellery Hale 1868–1938 whose vision and leadership made it a reality."
16. *Astrophysical Journal,* **87** (1938), 388.

BIBLIOGRAPHY

I. Original Works. Hale's bibliography in Adams' biographical memoir for the National Academy of Sciences (see below) includes some 450 articles and books, in addition to his annual reports as director of Mt. Wilson. Among them are "Photography of the Solar Prominences," in *Technology Quarterly,* **3** (1890), 310–316, a condensed version of his thesis (which, in its original form, is in the M.I.T. archives); "The Astrophysical Journal," in *Astronomy and Astro-Physics,* **11** (1892), 17–22; "The Yerkes Observatory of the University of Chicago," *ibid.,* 741; "The Spectroheliograph," *ibid.,* **12** (1893), 241–257; "The Congress of Mathematics, Astronomy and Astrophysics—Section of Astronomy and Astrophysics," *ibid.,* 746–749; "On Some Attempts to Photograph the Solar Corona Without an Eclipse," *ibid.,* **13** (1894), 662–687; "The Astrophysical Journal," in *Astrophysical Journal,* **1** (1895), 80–84; "The Aim of the Yerkes Observatory," *ibid.,* **6** (1897), 310–321; and "The Dedication of the Yerkes Observatory," *ibid.,* 353–362.

In the first two decades of the twentieth century Hale wrote "The Spectra of Stars of Secchi's Fourth Type," in Decennial Publications, University of Chicago, 1st ser., VIII (Chicago, 1903), 251–385, written with F. Ellerman and J. A. Parkhurst; "The Rumford Spectroheliograph of the Yerkes Observatory," in *Publications of the Yerkes Observatory of the University of Chicago,* **3,** pt. 1 (1903), 1–26, written with F. Ellerman; "General Plan for Furthering Special Researches in Astronomy," in *Carnegie Institution Yearbook,* I (1902), 94–104; "Cooperation in Solar Research," in *Astrophysical Journal,* **20** (1904), 306–312; "The Solar Observatory of the Carnegie Institution of Washington," *ibid.,* **21** (1905), 151–172; "The Spectroscopic Laboratory of the Solar Observatory," *ibid.,* **24** (1906), 61–68; "A 100-Inch Mirror for the Solar Observatory," *ibid.,* 214–218; "A Vertical Coelostat Telescope," *ibid.,* **25** (1907), 68–74; "A Plea for the Imaginative Element in a Technical Education," in *The Technology Review,* **9,** no. 4 (1907), 467–481; *The Study of Stellar Evolution . . .,* Decennial Publications of the University of Chicago, 2nd ser., X (Chicago, 1908); "Solar Vortices," in *Astrophysical Journal,* **28** (1908), 100–116; "Solar Vortices and Magnetic Fields," in *Proceedings of the Royal Institution,* **19** (1909), 615–630; "Preliminary Results of an Attempt to Detect the Magnetic Field of the Sun," in *Astrophysical Journal,* **38** (1913), 27–98; "National Academies and the Progress of Research. I. Work of European Academies," in *Science,* n.s. **38** (1913), 681–698; "II. The First Half-Century of the National Academy of Sciences," *ibid.,* **39** (1914), 189–200; "III. The Future of the National Academy of Sciences," *ibid.,* **40** (1914), 907–919, and **41** (1915), 12–23; "IV. The Proceedings of the National Academy as a Medium of Publication," *ibid.,* 815–817; *Ten Years' Work of a Mountain Observatory,* Carnegie Institution of Washington Publication no. 235 (Washington, D.C., 1915); and "The National Value of Scientific Research," in *Technology Review,* **18** (1916), 801–817.

In the 1920's Hale produced "The International Organization of Scientific Research," in *International Conciliation,* no. 154 (1920), 431–441; "Introduction; Science and War" (ch. 1), "War Services of the National Research Council" (ch. 2), "The Possibilities of Cooperation in Research" (ch. 22), and "The International Organization of Research" (ch. 23), in Robert M. Yerkes, *The New World of Science* (New York, 1920); "Invisible Sun-spots," in *Monthly Notices of the Royal Astronomical Society,* **82** (1922), 168–169; "A Joint Investigation of the Constitution of Matter and the Nature of Radiation," in *Science,* n.s. **55** (1922), 332–334; "A National Focus of Science and Research," in *Scribner's Magazine* (Nov. 1922), 515–531; *The New Heavens* (New York, 1922); "The Possibilities of Instrumental Development," in *Report of the Board of Regents of the Smithsonian Institution* (1923), 187–193; "Sun-spots as Magnets and the Periodic Reversal of Their Polarity," in *Nature,* **113** (supp.) (1924), 105–112; *The Depths of the Universe* (New York, 1924); "Law of the Sunspot Polarity," in *Astrophysical Journal,* **62** (1925), 270–300, written with S. B. Nicholson; *Beyond the Milky Way* (New York, 1926); "The Huntington Library and Art Gallery: The New Plan of Research," in *Scribner's Magazine,* **82** (1927), 31–43; "Science and the Wealth of Nations," in *Harper's Magazine,* **156** (1928), 243–251; "The Possibilities of Large Telescopes," *ibid.,* 639–646; "The Spectrohelioscope and Its Work: I. History, Instruments, Adjustments, and Methods of Observation," in *Astrophysical Journal,* **70** (1929), 265–311; and "Building the 200-Inch Telescope," in *Harper's Magazine,* **159** (1929), 720–732.

The 1930's saw publication of "The Spectrohelioscope and Its Work: II. The Motions of the Hydrogen Flocculi Near Sunspots," in *Astrophysical Journal,* **71** (1930), 73–101; "III. Solar Eruptions and Their Apparent Terrestrial Effects," *ibid.,* **73** (1931), 379–412; "IV. Methods of Recording Observations," *ibid.,* **74** (1931), 214–222; *Signals From the Stars* (New York, 1931); "Solar Research for Amateurs," in *Amateur Telescope Making,* 1st ed. Albert G. Ingalls, ed. (New York, 1928), pp. 180–214; "The Astrophysical Observatory of the California Institute of Technology," in *Astrophysical Journal,* **82** (1935), 111–139; "Address of the President," International Council of Scientific Unions, Brussels, 1934, in *Reports of Proceedings of the International Council of Scientific Unions* (1935), 4–10; and *Magnetic Observations of Sunspots, 1917–1924,* Carnegie Institution of Washington, pub. no. 498, 2 vols. (Washington, D.C., 1938), written with S. B. Nicholson.

The bulk of the Hale MSS—original correspondence, unpublished papers, and other source materials—is divided between the Hale Observatories' offices in Pasadena and the archives in the Millikan Library at the California Institute of Technology. In addition, Hale's correspondence with H. M. Goodwin is in the Henry E. Huntington Library and Art Gallery in San Marino. All these collections have been microfilmed. There is also a large amount of correspondence at the Yerkes Observatory in Williams Bay, Wisconsin. This covers his years as director there and also includes correspondence dealing with the founding of the Mt. Wilson Observatory.

II. Secondary Literature. The only complete biography of Hale is Helen Wright, *Explorer of the Universe, A Biography of George Ellery Hale* (New York, 1966). Biographical articles on George Hale include Giorgio Abetti, "George Ellery Hale," in *Memorie della Società astronomica italiana,* **11** (1938), 3; Walter S. Adams, "Biographical

Memoir of George Ellery Hale," in *Biographical Memoirs. National Academy of Sciences,* **21** (1940), 181–241; "George Ellery Hale," in *Astrophysical Journal,* **87** (1938), 369–388; and in *Publications of the Astro. Soc. Pac.,* **50** (June 1938); Harold D. Carew, "A Man of Many Worlds, George Ellery Hale," in *Touring Topics* (Oct. 1928), 28–30, 48; Theodore Dunham, Jr., "Obituary Notice of George Ellery Hale," in *Monthly Notices of the Royal Astronomical Society* (Feb. 1939), 99, 322–328; Philip Fox, "George Ellery Hale," in *Popular Astronomy,* **46** (Oct. 1938), 423–430; F. R. Moulton, "Our Twelve Great Scientists VII," in *Technical World,* **22** (Nov. 1914), 342–347, 462–464; H. F. Newall, "Scientific Worthies, XLVII, George Ellery Hale," in *Nature,* **82** (1933), 1–5; F. H. Seares, "The Scientist Afield," in *Isis,* **30** (May 1939), 241–267; James Stokley, "A Tribute to George Ellery Hale, June 29, 1868–Feb. 21, 1938," in *The Sky,* **2** (July 1938), 10–11; "Obituary Notices of Dr. George Ellery Hale, Foreign Member of the Royal Society of London," in *Nature* (19 Mar. 1938), 501–503, which includes articles by F. W. Dyson, J. H. Jeans, H. F. Newall, and F. J. M. Stratton; "The Works of George Ellery Hale—A Survey of the Career of a Great Living Scientist" (in three parts), in *Telescope,* **3** (May–Dec., 1936), 64–71, 95–100, 117–120, 127; and H. H. Turner, "Address on George Ellery Hale Given at the Time of the Award of the Gold Medal of the Royal Astronomical Society," in *Monthly Notices of the Royal Astronomical Society,* **64** (1904), 388–401.

Additional works that contribute to a picture of the development of astronomy in particular, of science in general, and of Hale's role in that development include Charles G. Abbot, *Adventures in the World of Science* (Washington, D.C., 1958); Giorgio Abetti, "Solar Physics," *in Handbuch der Astrophysik,* IV (Berlin, 1929) and VII (Berlin, 1936); *The History of Astronomy,* Betty B. Abetti, trans. (New York, 1952); and *The Sun,* J. B. Sidgwick, trans. (New York, 1957); Walter Adams, "Some Reminiscences of the Yerkes Observatory," in *Science,* **106** (1947), 196–200; "Early Days at Mount Wilson," in *Publications of the Astronomical Society of the Pacific,* **59** (1947), 213–231, 285–304; "The History of the International Astronomical Union," *ibid.,* **61** (1949), 5–12; "The Founding of the Mount Wilson Observatory," *ibid.,* **66** (1954), 267–303; and "Early Solar Research at Mount Wilson," in Arthur Beer, ed., *Vistas in Astronomy* (London, 1955), pp. 619–623; Solon I. Bailey, *The History and Work of the Harvard Observatory* (New York, 1931); Robert Ball, *The Story of the Sun* (London, 1893); W. Valentine Ball, ed., *Reminiscences and Letters of Robert Ball* (Boston, 1915); Louis Bell, *The Telescope* (New York, 1922); Charles Breasted, *Pioneer to the Past* (New York, 1947); Agnes Clerke, *A Popular History of Astronomy During the 19th Century* (London, 1885); A. Hunter Dupree, *Science in the Federal Government* (Cambridge, Mass., 1957); Arthur Eddington, "Some Recent Results of Astronomical Research," in *Proceedings of the Royal Institution,* **19** (Mar. 1909), 561–576; Simon Flexner and James T. Flexner, *William Henry Welch and the Heroic Age of American Medicine* (New York, 1941); Raymond Fosdick, *Adventures in Giving. The Story of the General Education Board* (New York, 1962); George W. Gray, *The Advancing Front of Science* (New York, 1937); Wallace K. Harrison, "The Building of the National Academy and the National Research Council," in *Architecture,* **50,** no. 4 (1924), 328–334, with plates 145–152; Edwin P. Hubble, *The Realm of the Nebulae* (New Haven, 1936); Bernard Jaffe, *Outposts of Science* (New York, 1935); Gerard P. Kuiper, ed., *The Sun* (Chicago, 1953), esp. the introduction by Leo Goldberg; S. P. Langley, *The New Astronomy* (New York, 1884); J. Norman Lockyer, *Contributions to Solar Physics* (London, 1874); G. R. Miczaika and William M. Sinton, *Tools of the Astronomer* (New York, 1903); Robert A. Millikan, *Autobiography* (New York, 1950); Simon Newcomb, *Reminiscences of an Astronomer* (New York, 1903); H. W. Newton, *The Face of the Sun* (Harmondsworth, England, 1958); Alfred Noyes, *Watchers of the Sky* (New York, 1923); A. Pannekoek, *A History of Astronomy* (New York, 1961); G. Edward Pendray, *Men, Mirrors and Stars* (New York, 1935); Michael Pupin, *From Immigrant to Inventor* (New York, 1924); Angelo Secchi, *Le soleil,* 2 vols. (Paris, 1875–1877); Harlow Shapley, *Source Book in Astronomy, 1900–1950* (Cambridge, Mass., 1956); Allan Sandage, *The Hubble Atlas of Galaxies* (Washington, D.C., 1961); Otto Struve, "The Story of an Observatory," in *Popular Astronomy,* **55** (May 1947), 223–244; Otto Struve and Velta Zebergs, *Astronomy of the 20th Century* (New York, 1962); Carol Green Wilson, *California Yankee* (Claremont, Calif., 1946); Helen Wright, *Palomar, the World's Largest Telescope* (New York, 1952); Robert M. Yerkes, ed., *The New World of Science* (New York, 1920); and Charles A. Young, *The Sun* (New York, 1895).

In addition, M.I.T. Press is to publish a volume on the Hale centennial meeting in Dallas, Texas, *The Legacy of George Ellery Hale,* Helen Wright, Joan Warnow, and Charles Weiner, eds. It will include materials from the exhibit: photographs, letters, and other original documents; the republication of some of Hale's classic papers, including the first publication of his M.I.T. thesis on the spectroheliograph; and the symposium with papers presented by Donald Shane, Ira S. Bowen, Robert Howard, and Daniel Kevles.

HELEN WRIGHT

HALE, WILLIAM (*b.* Colchester, England, 21 October 1797; *d.* London, England, 30 March 1870), *rocketry.*

The son of a Colchester baker, William Hale appears to have been largely self-educated, but he probably received tutoring from his maternal grandfather, the scientific writer and schoolmaster William Cole. Hale's first scientific studies concerned hydrodynamics. In 1827 he patented a method of propelling vessels by the principle of the Archimedean screw: water was sucked in and expelled, driving the vessel forward by a crude form of jet propulsion. He read a paper on this principle to the Royal Society of London in 1832 and, constructing a clockwork model, successfully demonstrated the principle before the king and queen. At about the same time, he received the first-class gold medal from the Royal Society of Arts in Paris. Over thirty years later, in *Treatise on the Mechanical Means by Which Vessels Are Propelled by Steampower* (1868), Hale discussed, in greater detail, this hydrodynamic study of jet propulsion.

Hale's knowledge and application of Newton's third law of motion, as well as his research on the dynamics of propellers in fluids, may have led him to invent the rotating rocket, first patented in 1844. The Hale "rotary" or "stickless" rocket dispensed with the long guide stick of the Congreve variety by causing the exhaust gases to rotate the projectile on its own axis and thereby attain longitudinal stability through inertia and centrifugal force. Hale consequently wrote *Treatise on the Comparative Merits of a Rifle, Gun and Rotary Rocket* (London, 1863), one of the first works treating the exterior ballistics of spinning and nonspinning rockets. He also disproved the hypothesis that rockets move because the exhaust gases "push" against the air and correctly demonstrated rocket motion in terms of Newton's third law axiom. In addition, Hale developed the hydraulic method of loading rockets and investigated underwater rocket propulsion.

About 1828 Hale married Elizabeth Rouse, by whom he had two sons and three daughters; she died in 1846. He married Mary Wilson of Bath in 1867.

In 1970 the International Astronomical Union honored Hale's achievements by naming the Hale crater on the moon for him and George Ellery Hale.

BIBLIOGRAPHY

I. Original Works. The Royal Society of London has an unpublished paper, "An Account of a New Mode of Propelling Vessels" (1832). Hale's other works include *A Treatise on the Comparative Merits of a Rifle, Gun and Rotary Rocket* (London, 1863); *Hale's War Rockets.—Statement for the Referee, to Be Appointed by the Right Hon. Early de Gray and Ripon* (London, 1865), which is a statement of grievances against the government for noncompensation for the use of his rockets and contains a partial biography; and *Treatise on the Mechanical Means by Which Vessels Are Propelled by Steampower* (London, 1868).

II. Secondary Literature. On Hale and his work, see O. F. G. Hogg, *The Royal Arsenal,* II (London, 1965), 751, 767, 770, 824–827, 1377–1379. The Great Britain War Office, Treatise on Ammunition Series, deals with Hale's rockets in each treatise from 1870–1905; the 1870 ed. is typical, covering Hale's rockets in detail (pp. 179–186). See also *A Collection of Annual Reports . . . Vol. II—1845–1860,* U.S. Army, Ordnance Dept. (Washington, D.C., 1880), pp. 152–156, 190, 496.

Frank H. Winter

HALES, STEPHEN (*b.* Bekesbourne, Kent, England, 17 September 1677; *d.* Teddington, Middlesex, England, 4 January 1761), *physiology, public health.*

Stephen Hales, a clergyman without formal medical training, published his first discoveries in his fiftieth year, yet was soon recognized as the leading English scientist during the second third of the eighteenth century. As the acknowledged founder of plant physiology, he had no worthy successor until Julius von Sachs, a century later. In animal physiology he took " the most important step after Harvey and Malpighi in elucidating the physiology of the circulation."[1] His experiments concerning "fixed air"—and the apparatus he devised—laid the foundations of British pneumatic chemistry and stimulated the discoveries of Joseph Black, Henry Cavendish, and Joseph Priestley; Hales was a primary influence on the early researches of Lavoisier.

He was born of an old and distinguished Kentish family, but there is no record of his boyhood until, having been "properly instructed in grammar learning," he was sent to Cambridge, where he entered Benet College (now Corpus Christi) in 1696.[2] On receiving the B.A. degree, Hales became a fellow of his college in 1703 and was awarded the M.A. that same year. He was ordained deacon in 1709 and left Cambridge to become "perpetual curate," or minister, of Teddington, a village on the Thames between Twickenham and Hampton Court. He held this position for the rest of his life, and it was at Teddington that most of his scientific work was carried on.

An interest in science was awakened during his years at Cambridge, the university that boasted the great Isaac Newton (who had left for London the year that Hales entered the university) and the naturalist John Ray, whose earliest book was a catalog of the plants of Cambridgeshire. Something of a scientific renaissance took place during Hale's last years at the university. William Whiston, Newton's successor as Lucasian professor, was encouraged by Newton's old friend Richard Bentley, who became master of Trinity, the college of Newton and Ray, in 1700. Bentley helped secure the appointment of a gifted young fellow of the college, Roger Cotes, to the newly established Plumian professorship of astronomy and built for him an observatory over the Great Gate of Trinity. When John Francis Vigani became the first professor of chemistry at Cambridge, Bentley provided him with a laboratory "in the mediaeval chambers that look out on the Bowling Green."[3]

In 1703 William Stukeley, the future physician and antiquary, entered Benet College; intent on a medical career, he "began to make a diligent & near inquisition into Anatomy and Botany."[4] He became a close friend of Hales; with Stukeley and other students Hales went "simpling" in the surrounding countryside, Ray's catalog in hand. In a room that Stukeley's tutor had given him as a sort of laboratory, they performed chemical experiments and dissected frogs and other small animals. Together they devised a method of obtaining a lead cast of the lungs of a dog. It was at this time (about 1706) that Hales carried out his first blood-pressure experiments on dogs. He and Stukeley attended Vigani's chemical lectures and saw his demonstrations in the laboratory at Trinity. Hales, like Stukeley, must have seen the "many Philosophical Experiments in Pneumatic Hydrostatic Engines & Instruments performed at that time" by John Waller, rector of St. Benedict's Church, who later succeeded Vigani as professor of chemistry. Hales knew Waller, for about 1705 the two men "gathered subscriptions to make the cold bath about a mile & a half out of Town."[5] This introduction to pneumatic experiments was probably supplemented by the lectures in experimental physics given by Whiston and Cotes at the observatory in Trinity College. Cotes, in his share of the lectures, demonstrated the experiments of Torricelli, Pascal, Boyle, and Hooke.[6]

Newton's influence was strongly felt at Cambridge. In 1704 appeared his long-delayed *Opticks,* a work that, in its later editions, profoundly influenced Hales. We learn from Stukeley that students at Benet College read the Cartesian *Physics* of Jacques Rohault, but in the edition of Samuel Clarke,[7] who appended Newtonian footnotes to correct the text.

Hales, like Stukeley, doubtless witnessed Newton's arrival in Cambridge in April 1705, when he came to offer himself as the university's candidate for Parliament. On the sixteenth of that month Queen Anne visited Cambridge as the guest of the master of Trinity. As Stukeley recalled it: "The whole University lined both sides of the way from Emanuel college, where the Queen enter'd the Town, to the public Schools. Her Majesty dined at Trinity college where she knighted Sir Isaac, and afterward, went to Evening Service at King's college chapel."[8]

Although doubtless incapable of following the mathematical intricacies of Newton's *Principia,* Hales mastered the main features of the new system of the world. He showed both his mechanical ingenuity and some knowledge of celestial physics by devising a machine to show the motions of the planets. A drawing by Stukeley of Hales's orrery is preserved, along with Stukeley's diary, in the Bodleian Library, Oxford.[9]

At Teddington, Hales was first preoccupied with his parish duties; only several years later did he resume his scientific work; and it was later still before the scientific world heard from him. About 1712–1713 he took up again his experiments on animals, this time using as his victims two horses and a fallow doe. But he did not "pursue the Matter any further, being discouraged by the disagreeableness of anatomical Dissections." For several years his scientific endeavors lapsed, yet on 13 March 1717/18 (O.S.) he was elected a fellow of the Royal Society along with his old friend William Stukeley, who was now practicing medicine in London.[10] It was Stukeley, indeed, who brought Hales's name to the attention of the Royal Society.[11] Hales was soon to justify his election.

While conducting his experiments on animal blood pressure, Hales records that "I wished I could have made the like Experiments, to discover the force of the Sap in Vegetables" but "despaired of ever effecting it." Yet early in 1719 "by mere accident I hit upon it, while I was endeavouring by several ways to stop the bleeding of an old stem of a Vine, which was cut too near the bleeding season, which I feared might kill it." His account continues:

> Having, after other means proved ineffectual, tyed a piece of bladder over the transverse cut of the Stem, I found the force of the Sap did greatly extend the bladder; whence I concluded, that if a long glass Tube were fixed there in the same manner, as I had before done to the Arteries of several living Animals, I should thereby obtain the real ascending force of the Sap in that Stem.[12]

Hales was in no hurry to make an appearance at the Royal Society or to contribute to its proceedings; several months elapsed before he appeared in person to sign the required bond.[13] On 5 March 1718/19, perhaps at Stukeley's urging, Hales informed the president, Sir Isaac Newton, "that he had lately made a new Experiment upon the Effect wch ye Suns warmth has in raising ye sap in trees."[14]

Seven years of silence followed, during which, in the free time his parish duties allowed him, Hales followed out this original clue. The experiments on plants were virtually completed, and the *Vegetable Staticks* written out, by the middle of January 1724/25. He submitted his manuscript to the Royal Society, where it was read at successive meetings from January to March.[15] In this form it consisted of six chapters, not the seven of the published book; missing was the long chemical chapter "The Analysis of Air."

Although Hales was urged to publish, two years elapsed before the Royal Society heard from him again. During this time he performed the seventy chemical experiments of chapter 6 of the *Vegetable Staticks.* In three meetings during February 1726/27 this chemical chapter was read, and at the last of these meetings the book received the imprimatur of the Royal Society, signed by Sir Isaac Newton "Pr. Reg. Soc."[16] The book was already in press, for some of the early sections were read again to the Society in March, probably from advance sheets. On 13 April 1727 (O.S.) a copy of the *Vegetable Staticks,* dedicated to George, prince of Wales (the future George II), was presented to the Society; and the curator of experiments, J. T. Desaguliers, was asked to prepare an abridgment of it.[17]

Hales next turned to completing and publishing his experiments on animal circulation (mentioned only briefly in the *Vegetable Staticks*) under the title *Haemastaticks,* putting the two works together as the *Statical Essays.* Imbued with the empiricism of John Locke and the principles of Newton's "experimental philosophy," Hales was also influenced by the doctrines of the iatrophysicists—those physicians including Borelli, Baglivi, and the Scots doctors Archibald Pitcairne and James Keill who insisted, as a certain John Quincy put it, that the application of mechanical principles "to account for all that concerns the Animal Oeconomy" is the best means to "get clear of all suppositions and delusory Hypotheses" and "has appeared to be the only way by which we are fitted to arrive at any satisfactory Knowledge in the Works of Nature."[18]

An immediate stimulus came from James Keill, who, while Hales was still at Cambridge, published a book giving quantitative estimates of the amount of blood in the human body, the velocity of the blood as it left the heart, the amounts of various animal secretions, and so on.[19] Hales may have had Keill's book in mind when he wrote that

> . . . if we reflect upon the discoveries that have been made in the animal oeconomy, we shall find that the most considerable and rational accounts of it have been chiefly owing to the statical examination of their fluids, *viz.* by enquiring what quantity of fluids, and solids dissolved into fluids, the animal daily takes in . . . And with what force and different rapidities those fluids are carried about in their proper channels . . .[20]

What Hales called the "statical way of inquiry" he deemed the proper way to study living things. For his now obsolete use of the word "staticks" he had ample authority. The usage originated with Nicholas of Cusa; in his *De staticis experimentis*—a work several times reprinted in the fifteenth and sixteenth centuries and translated into English in 1650—Cusa outlined a series of "thought experiments" involving the use of the balance.[21] Statics, for Cusa, meant weighing. Santorio Santorio made the term familiar in medicine with his *De medicina statica aphorismi* (Venice, 1614), a work on "insensible perspiration" often reprinted and published in English translation in 1712 by John Quincy.[22] In 1718 James Keill published his *Tentamina medico-physica* (a Latin version of his earlier work) and appended to it some studies on perspiration called *Medicina statica britannica.*[23] Hales was familiar with this book, to which he referred several times.

The *Haemastaticks* describes the experiments on blood pressure begun at Cambridge, taken up again at Teddington, laid aside because of the "disagreeableness of the Work," but resumed again after the publication of the *Vegetable Staticks.* At first intended only as an addition to the earlier book, it grew "into the Size of another Volume, so fruitful are the Works of the great Author of Nature in rewarding, by farther Discoveries, the Researches of those *who have Pleasure therein.*"[24]

The first Cambridge experiments, Hales tells us, were stimulated by the confusion that existed as to the magnitude of the arterial blood pressure; some maintained that the pressure was enormous, and even that it might be the cause of muscle motion. The results of his series of investigations were of such importance that they have been described as "the most important step in knowledge of the circulation between Malpighi and Poiseuille."[25]

The *Haemastaticks* opens with an account of Hales's most dramatic experiment, a bold and bloody one. He tied a live mare on her back and, ligating one of her femoral arteries, inserted a brass cannula; to this he fixed a glass tube nine feet high; when he untied the ligature, the blood rose to the height of more than eight feet. Detaching the tube at intervals, he allowed a measured quantity of blood to flow out, noting how the pressure changed during exsanguination. He succeeded in inserting cannulas into the veins to record the venous pressure of a number of animals, including an ox, a sheep, a fallow doe, three horses, and several dogs.

His interest in the mechanics of the circulation now enhanced, Hales turned his attention to the chief factors that must maintain the blood pressure: the output of the heart per minute and the peripheral resistance in the small vessels. He made a rough estimate of cardiac output by multiplying the pulse rate of an animal by the internal volume of its left ventricle, of which he made a cast in wax after the animal had been killed. He noted that the pulse was faster in small animals than in large ones, and that the blood pressure was proportional to the size of the animal.

Hales next studied peripheral resistance with perfusion experiments. Injecting various chemical substances (brandy, decoction of Peruvian bark, various saline solutions), he compared the rate of flow of the perfusate and showed that certain substances had a pronounced effect on the rate at which the blood could flow through an isolated organ. He attributed this to changes in the diameter of the capillaries and so—although he did not observe the phenomenon directly—discovered vasodilatation and vasoconstriction.

Hales's experiments convinced him that the force of arterial blood in the capillaries "can be but very little" and wholly inadequate for "producing so great an Effect, as that of muscular Motion." This "hitherto inexplicable Mystery of Nature must therefore be owing to some more vigorous and active Energy, whose force is regulated by the Nerves." The recent experiments of Stephen Gray suggested to Hales that this energy, these "animal spirits," might be electrical, for Gray had shown that the electrical virtue from rubbed glass

> . . . will not only be conveyed along the Surface of Lines to very great Lengths, but will also be freely conveyed from the Foot to the extended Hand of a human Body suspended by Ropes in . . . the Air; and also from that Hand to a long Fishing Rod held in it, and thence to a String and a Ball suspended by it.[26]

Hales was therefore the first physiologist to suggest, with some evidence to support it, the role of electricity in neuromuscular phenomena.[27]

Despite these achievements, Hales's most original contribution was to apply to the study of plants the "statical method" which had brought such good results with animals. Like his contemporaries he was impressed by the analogies that he perceived between the animal and the vegetable worlds. Perhaps the most obvious was the fundamental similarity of the role of the sap in plants and of the blood in animals. Since the growth of plants "and the preservation of their vegetable life is promoted and maintained, as in animals, by the plentiful and regular motion of their fluids, which are the vehicles ordained by nature, to carry proper nutriment to every part," the same methods ought to be used which had illuminated the animal economy: "the statical examination of their fluids."[28] By an accident, as we saw, he was led to his first attempts to measure the force of the sap in vines and to determine the conditions under which it varied. Although he outstripped his predecessors, Hales was not the first to investigate the flow of sap.

The problem of sap flow had long interested the virtuosos of the Royal Society. In 1668 the *Philosophical Transactions* proposed to its readers a long series of "queries" concerning plants, "especially the Motion of the Juyces of Vegetables," asking, for example, whether the "Juyce ascends or descends" by the bark or the pith.[29] Among the responses was a letter from Francis Willughby describing some bleeding experiments on trees performed with John Ray which showed that the sap not only ascended but also seemed to descend and move laterally, and that the rise could not be attributed to capillarity, a common explanation.[30] When the letter was read to the Society, the two naturalists were requested to "try some experiments, to find, whether there be any circulation of the juice of vegetables as there is of the blood of animals."[31] That there might be such a circulation of sap, moving upward through the vessels of woody plants and downward by those between the wood and the bark, was a commonly held view in Hales's day; as late as 1720 Patrick Blair in his *Botanick Essays* tried to prove that this was the case.[32]

Nor was Hales the earliest to apply the "statical way of inquiry" to plants. J. B. van Helmont's famous willow tree experiment, which persuaded him that water was the sole principle or nutrient of plants, is a well-known example.[33] Closer to Hales's time were the quantitative experiments of John Woodward to determine whether water itself, or substances dissolved in it, accounted for the growth of plants. In these experiments Woodward discovered that the phenomenon of transpiration was of considerable magnitude. Growing mint in water, Woodward observed that the plant took up large quantities of water but gave off far more than it retained. He noted that solar heat played a part in the process, but he did not specify or prove that transpiration occurred through the leaves. Much the greatest part of the water imbibed by his plants, he wrote, "does not settle or abide there: but passes through the Pores of them, and exhales up into the Atmosphere."[34]

Hales's experiments on plants, begun in March 1719, were pursued with vigor during the years 1723–1725, using the resources of his own garden and plants and trees provided him from the nearby royal garden of Hampton Court.[35] In his early experiments, conducted during the bleeding season, he observed the rise of sap through long glass tubes fastened to the cut end of a branch of a grapevine. In one such experiment Hales joined glass tubes together to a height of thirty-eight feet. The sap was observed to rise in these tubes "according to the different vigor of the bleeding state of the Vine" from one foot up to twenty-five feet. He carefully observed how the sap flow varied with the weather and the time of day.

To measure the sap pressure Hales employed a "mercurial gage," a bent tube filled with mercury which he fixed to the cut branches of the vine, observing again the variations of the pressure at different times of day.

To determine the force with which trees imbibe moisture from the earth, Hales devised what he called "aqueo-mercurial" gauges. He laid bare the root of a small pear tree, cut it, and inserted it into a large glass tube, which in turn was fixed to a narrow tube eight inches long. When the tubes were filled with water and immersed in a vessel of mercury, the root, he reported, "imbibed the water with so much vigor" that in six minutes the mercury rose eight inches. These experiments were carried out in the summer months, when the trees and vines were in leaf. Hales noted that the more the sun shone on the plants, "the faster and higher the mercury rose"; it would subside toward evening and rise the next day. He observed that sometimes the mercury "rose most in the evening about 6 a clock, as the sun came on the Vine-branch." Such results may have suggested to him the role that transpiration—or, as he called it, "perspiration"—might play in causing the sap to rise.

Hales's experiments on transpiration—perhaps the most famous and brilliant of those he performed with plants—were carried out in the summer months of 1724. He grew a large sunflower in a garden pot covered tightly with a thin lead plate pierced by the plant, by a small glass tube to allow some communication with the air, and by another short, stoppered tube through which the plant could be watered. He weighed the pot and plant twice a day for fifteen days, then cut off the plant close to the lead plate, cemented the stump, and by weighing determined that the pot with its earth "perspired" two ounces every twelve hours. Subtracting this from his earlier weighings, he found that the plant perspired in that period an average of one pound, four ounces of water.

Hales then stripped off the leaves of the plant and divided them in groups according to their several sizes. Taking a sample leaf from each group, he measured their surface areas by placing over them a grid made of threads, composing quarter-inch squares. By multiplying the area of each sample leaf by the number of leaves in the group and adding his measurements together, he obtained the total surface area of the leaves. His figures for the loss of water from the leaves compare favorably with those obtained long after by Sachs. Hales also attempted to estimate the surface area of the roots to determine the rate of absorption per given area, but these figures are of no value because Hales "did not know how small a part of the roots is absorbent, nor how enormously the surface of that part is increased by the presence of root-hairs."[36] He was somewhat more successful in determining the rate of flow of the stem. Always hoping to find analogies between animals and plants, he estimated the total surface area of his sunflower and its weight so as to compare the quantity of water "perspired" by the plant in twenty-four hours with that of an average "well-sized man" over the same period, taking the latter figure from James Keill's *Medicina statica britannica.*[37]

Transpiration could not account for the powerful rise of sap in vines during the bleeding season. Hales devoted a chapter of his book to the experiments which led him to discover root pressure. He cut off a vine, leaving only a short stump with no lateral branches. To it, by means of a brass collar, he fixed a series of glass tubes reaching as high as twenty-five feet. The sap rose gradually nearly to the top of these tubes, both day and night, although much higher in daytime. From this experiment, Hales remarks, "we find a considerable energy in the root to push up sap in the bleeding season."[38]

Similar experiments using his mercurial gauge confirmed that the force of the rising sap was "owing to the energy of the root and stem." Comparing his results with his blood pressure experiments, Hales concluded that this force was "near five times greater" than that of the blood in the femoral artery of a horse and "seven times greater than the force of the blood in the like artery of a Dog."

Curious whether this force could be detected in vines when the bleeding season was over, Hales performed the same experiment in the month of July and found that the flow of sap ceased when the vine was cut from the stem, thus proving to his satisfaction that after the bleeding season the principal cause of the rise of sap was not root pressure but that which was "taken away, viz. the great perspiration of the leaves." This was evident, too, from a number of experiments which showed that branches stripped of their leaves did not imbibe water "for want of the plentiful perspiration of the leaves."[39]

A series of experiments to discover the direction of the flow of sap, and the portion of the stem through which it moved, were performed by cutting away the bark or slicing off a small section of it. These showed that while there must be some lateral communication, the sap moved upward between the bark and the wood, not downward "as many have thought," and that there is no circulation of the sap. Plants, Hales suggested, make up for the lack of a circulation by the much greater quantity of fluid that passes through them. Nature's "great aim in vegetables being only that the vegetable life be carried on and maintained, there was no occasion to give its sap the rapid motion, which was necessary for the blood of animals."[40]

Hales's explanation of the sap's motion invokes the Newtonian principle of attraction. The chief cause is "the strong attraction of the capillary sap vessels," greatly assisted "by the plentiful perspiration of the leaves, thereby making room for the fine capillary vessels to exert their vastly attracting power." This "perspiration" results from the sun's warmth acting on the leaves, which are fittingly broad and flat to serve this purpose of absorbing the sun's rays.

An experiment to show the "great force, with which vegetables imbibe moisture" was performed by filling an iron pot nearly to the top with peas and water. Over the peas Hales placed a cover of lead, and on the cover he placed a weight of 180 pounds, which—as the peas swelled with the imbibed water—was lifted up.

The role of attraction, "that universal principle which is so operative in all the very different works of nature, and is most eminently so in vegetables,"[41] was illustrated by Hales's modification of an experiment of Francis Hauksbee, described in query 31 of Newton's *Opticks,* showing the rise of water through a glass tube firmly packed with sifted wood ashes. Hales measured the imbibing force with his "aqueo-mercurial gage." He quotes Newton's words that "by the same principle, a sponge sucks in water, and the glands in the bodies of animals, according to their several natures and dispositions suck in various juices from the blood." Hales adds:

> And by the same principle it is, that . . . plants imbibe moisture so vigorously up their fine capillary vessels; which moisture, as it is carryed off in perspiration, (by the action of warmth,) thereby gives the sap vessels liberty to be almost continually attracting of fresh supplies.[42]

An influential experiment—it paved the way for some important researches by Sachs—demonstrated the unequal extent of growth in developing shoots and leaves. In the spring, using a comb-like device, Hales pricked, with homemade red paint, dots a quarter of an inch apart along a young vein shoot. Several months later, when the shoot was full-grown, he measured the distances between the dots. The shoots, he discovered, had grown chiefly by a longitudinal extension between the nodes; the oldest (basal) internode had grown the least and the youngest (apical) one, the most.[43]

Again, concerned with analogies between plants and animals, this experiment led Hales to see if a similar effect could be observed in the growth of the long bones in animals, with their tubelike cavities. He took a half-grown chick and pierced the thigh and shin bones with a sharp pointed iron, making small holes half an inch apart. After two months he killed the bird and found that although the bones had grown an inch in length, the marks remained the same distance apart. In contrast with what he had observed in his vineshoots, the growth had occurred not in the shaft but entirely at the junction of the shaft and its two ends, that is, at the symphyses.

In his experiments on plants Hales frequently noticed bubbles of air emerging from the cut stems of vines or rising through the sap, often in such quantity as to produce a froth. This, he remarked, "shews the great quantity of air which is drawn in thro' the roots and stem." The air, he thought for a time, was "perspired off" through the leaves; but an inconclusive experiment led him to suspect that "the leaves of plants do imbibe elastick air."[44] By 1725 he had performed a few experiments to prove that a considerable quantity of air is "inspired" by plants. The problem interested him so much that he deferred publication until he could make "a more particular enquiry into the nature of a Fluid," the air, "which is so absolutely necessary for the support of the life and growth of Animals and Vegetables."[45] These investigations, carried out between 1725 and 1727, were embodied in the long chapter, nearly half the final work, called "Analysis of Air." This chapter was to have momentous consequences for the later development of chemistry.

Since the investigations of Torricelli, Pascal, Otto von Guericke, and, of course, Robert Boyle, the physical properties of air had been pretty well understood: the law of its expansibility, its ability to refract light, its approximate density under standard conditions. But it was no longer thought by chemists to be an element.[46] Any apparent chemical activity, and its ability to sustain life and support combustion, could be explained by the properties of special substances dispersed through it, such as the nitro-aerial particles imagined by Hooke, John Mayow, and others.[47] Boyle's description of the atmosphere was widely accepted; it was composed, he wrote, of three kinds of particles: the permanently elastic particles making up the air properly speaking, a "thin, diaphanous, compressible and dilatable Body"; vapors and dry exhalations from the earth, water, vegetables, and animals; and, third, "magnetical steams of our terrestrial globe" and particles of light from the sun and stars.[48]

Yet Boyle, Hooke, and other fellows of the Royal Society had shown that "air" ("factitious air") could be produced from solid and liquid bodies in certain chemical reactions: the action of acids on oyster shells or coral, the reaction of dilute acids with iron nails, the explosion of gunpowder.[49] A particularly striking experiment was performed by Frederick Slare in 1694. He poured spirit of niter (nitric acid) over oil of caraway seeds, and the result was a violent explosion that blew up the glass container. Slare expressed amazement that so much air was produced from small amounts of these liquids.[50] This experiment made a profound impression; Slare's account was read, and the experiment perhaps repeated, by Roger Cotes in his lectures. It was described, too, by Newton in his *Opticks,* although without mentioning Slare by name. Hales was doubtless familiar with this experiment, although when he mentioned Slare in the *Vegetable Staticks* it was for a different experiment.[51] Newton's *Opticks,* to which Hales referred so often in his book, would have been sufficient authority for the existence of "factitious airs." At one point he quotes Newton's words: "Dense Bodies by Fermentation rarify into several sorts of Air, and this Air by Fermentation, and sometimes without it, returns into dense bodies."[52]

Of particular concern for Hales was evidence that air was thought to be of special importance to the plant economy. In France, Guy de La Brosse early in the seventeenth century had argued that plants cannot grow without the air from which they draw "la rosée & la manne."[53] Similar views were advanced by Robert Sharrock, a friend and collaborator of Robert Boyle.[54] This question was taken up in the early meetings of the Royal Society; John Beal suggested in 1663 that it should be determined "what effects would be produced on plants put into the pneumatic engine with the earth about their roots, and flourishing; whether they would not suddenly wither, if the air were totally taken from them."[55] Not long after, Robert Hooke showed that lettuce seed would not sprout and grow, and a thriving plant would wither and die, if kept in a vacuum.[56] In 1669 Beal felt able to conclude that a plant "feeds as well on the Air, as [on] the juice furnish'd through the root."[57] After the discoveries in plant anatomy by Malpighi and Nehemiah Grew, and their description of vessels in plants that appeared, like the trachea of insects, to be tubes for transmitting air, it was suggested that air contributed to the nutrition of plants, or—as John Ray put it—that plants have a kind of respiration.[58]

Except for Malpighi and Grew, whom he cites, Hales may have been unaware of these antecedents. But he was familiar with certain of Boyle's experiments admired by Roger Cotes. By these experiments, published in 1680–1682, Boyle showed, as Hales put it in the beginning of his "Analysis of Air," that

> . . . a good quantity of Air was producible from Vegetables, by putting Grapes, Plums, Gooseberries, Cheries, Pease, and several other sorts of fruits and grains into exhausted and unexhausted receivers, where they continued for several days emitting great quantities of Air.[59]

In this famous long chapter Hales describes a large number of experiments—some trivial, some confused, but some extremely interesting—performed to discover the amount of air "fixed" in different substances or given off or absorbed under various circumstances. Strictly speaking, Hales was not a chemist, although he had performed some chemical experiments during his Cambridge days, when he had read or consulted George Wilson's practical compendium, *A Compleat Course of Chemistry* (1699).[60] He knew Boyle's work and John Mayow's, and was familiar with Nicolas Lemery's popular textbook.[61] But his approach was more physical than chemical; and it is not surprising—since he thought of air as a unitary substance characterized by its physical property of elasticity—that he failed to note the different chemical properties of the airs he produced.[62]

Hales's true mentor was Newton, whose last query of the *Opticks* (1718) was in fact a monograph on the role of attractive and repulsive forces in chemical processes, and whose short "Thoughts About the Nature of Acids" Hales had also read.[63] He was familiar too with the *Chymical Lectures* in which John Freind attempted to explain chemical reactions in Newtonian terms.[64]

From Newton, Hales derived the fundamental principles by which he explained the effects he observed. Matter is particulate, and the particles are subject to very special laws of attraction and repulsion. In their free state the particles of air exert upon each other strong repulsive forces, which accounts for the air's "elasticity." Yet this elasticity is no immutable property, for Newton had remarked that "true permanent Air arises by fermentation or heat, from those bodies which the chymists call fixed, whose particles adhere by a strong attraction."[65] When air enters into "dense bodies" and becomes "fixed," its elasticity is lost because strong attractive forces overcome the forces of repulsion between its particles.

Hales's first experiments were distillations in which

different substances were strongly heated in a glass or iron retort. The retort was cemented and luted to a globular vessel with a long neck, called a bolthead.[66] This vessel, with a hole cut in the bottom, was immersed in a basin of water; and by means of a siphon the water level was raised in the neck to a point he carefully marked. The amount of air given off or absorbed was determined by allowing the vessel to cool and noting the change in the water level. With this apparatus Hales measured the air produced by weighed amounts of hog's blood, tallow, powdered oyster shell, amber, honey, and a variety of vegetable materials. Whereas he obtained little air from ordinary well water, a considerable quantity was yielded by Pyrmont water, leading Hales to comment that this air "contributes to the briskness of that and many other mineral waters." He distilled iron pyrites, known to be rich in sulfur, and from a cubic inch of this mineral obtained eighty-three cubic inches of air. When he heated minium or red lead (Pb_3O_4), he obtained a large quantity of air, remarking that this air might account for the increase in weight of lead when it is strongly heated to form minium. This air was doubtless what had "burst the hermetically sealed glasses of the excellent Mr. *Boyle,* when he heated the Minium contained in them by a burning glass."[67]

Two other contrivances were used by Hales to measure the air produced or absorbed in chemical reactions, or, as he put it, in "fermentations." One apparatus consisted of a bolthead placed in a basin of water; over its long neck he inverted a cylindrical vessel, using a siphon to draw up the water a given distance. As in the first apparatus, the amount of air given off or absorbed was determined by the change in the water level.[68] With this apparatus Hales measured the air produced by decomposing sheep's blood, by ale drawn from a fermenting vat, by the fermentation of raisins and apples, and by the action of vinegar on powdered oyster shells. Other experiments showed that salt of tartar (potassium carbonate) treated with acids yielded much air, a discovery that later put Joseph Black on the road to his major chemical discovery.[69] Hales also measured the large amount of air (hydrogen gas) produced from iron filings treated with dilute sulfuric acid. When the iron filings were dissolved in dilute nitric acid, he also obtained much air (in this case, nitric oxide). Of particular interest is Hales's measurement of air produced by the action of oil of vitriol on chalk and his further observation that lime (made from the same chalk) absorbed much air.[70]

His second contrivance has been called his pedestal apparatus.[71] A wooden pedestal is placed upright in a basin of water, and on its expanded top can be placed a candle, a weighed amount of some chemical substance to be ignited, or—in the larger form of this apparatus—a small animal. A •glass cylinder is suspended over the pedestal so that its mouth is a few inches under water. As in the other devices, air is withdrawn with a siphon or bellows to raise the water to a convenient level, and a change in the water level indicates the change in the volume of air in the cylinder. With his pedestal apparatus Hales discovered that when phosphorus and sulfur are burned, they absorb air. When he detonated niter (potassium nitrate) by means of a burning glass, he noted the large amount of air produced but observed that the volume steadily decreased, or as he put it, "the elasticity of this new air daily decreased."[72]

Repeating an experiment of John Mayow's, Hales placed a candle on the pedestal, ignited it with a burning glass, and noted the shrinkage in volume. When he used candles of equal size but in vessels of different capacities, he found that they burned longer in the larger ones and that "there is always more elastic air destroyed in the largest vessel." His burning glass, he found, could not light an extinguished candle "in this infected air."[73] Repeating another of Mayow's experiments, Hales placed a small animal on the pedestal and measured the air absorbed. Here, to be sure, two effects—both unknown to Hales—contributed to the rise of the water level: the intake of oxygen by the animal and its exhalation of carbon dioxide, much of which dissolved in the water. His results led to a series of rebreathing experiments carried out on himself which convinced him that animal respiration "vitiated" the air. His device, a bladder equipped with valves and breathing tube, enabled him to breathe repeatedly his own expired air. He found that he could continue in this fashion only about a minute. In a modification of his device, a series of diaphragms (flannel stretched over thin hoops) was placed in the bladder. When these were soaked with salt of tartar, especially when the salt was calcined (that is, causticized), he found that he could rebreathe for as long as eight and a half minutes. The salt, "a strong imbiber of sulphureous steams," in fact absorbs much carbon dioxide.

In his experiments, when he noticed a decrease in volume of air during certain reactions, Hales always spoke of a loss of elasticity and attributed this to the acid sulfurous fumes which "resorb and fix" the elastic particles of ordinary air.[74] Such fumes, he noted, were produced by burning sulfur, by a lighted candle—indeed, by all "flaming bodies"—and by the expired air of animals and man.

To obviate this effect, Hales devised his most famous apparatus: the first pneumatic trough. Substances were heated in an iron retort; to the long neck of the retort he fixed a bent lead tube which was immersed in a basin of water and projected upward into the open end of an "inverted chymical receiver" filled with water. The released air passing through the bent tube bubbled up through the water and was collected in the top of the glass vessel. Hales's purpose was not to measure the amount of air, as in his other experiments, but to wash the air by passing it through water, to intercept "a good part of the acid spirit and sulphureous fumes." By this means he could collect and store air and ascertain whether its elasticity could be preserved. By separating the generator from the collector, Hales invented the pneumatic trough, later used in modified form by Brownrigg, Cavendish, and Priestley.

With his trough Hales collected air from a variety of substances—horn, human bladder stones, pyrite, saltpeter, minium, salt of tartar, and various vegetable materials—and claimed that the greater part of the air remained for the most part "in a permanently elastick state" and so was true air, not a mere flatulent vapor. He did not explore the different chemical properties of the air produced from different substances—indeed, he had no great reason to believe they could be found. Yet he suspected that there were at least some physical differences. Newton had written of bodies rarefying into "several sorts of air," an opinion that Hales seems to have shared, for he suggested that since air arises from a great variety of "dense" bodies, it is probable that airs from different sources may differ in the size and density of their constituent particles and may have "very different degrees of elasticity." But his crude attempts to see if common air and the air produced by salt of tartar (carbon dioxide) differed in density and compressibility disclosed no difference.

Hales's explanation of combustion was a physical one.[75] He rejected the notion that fire is "a particular distinct kind of body inherent in sulphur," as the chemists Willem Homberg and Louis Lemery believed. Instead, he followed Newton in distinguishing between heat and fire: heat is the rapid intestine motion of particles; fire is merely "a Body heated so hot as to emit Light copiously," and flame is only a "Vapour, Fume or Exhalation heated red hot." Hales owed much also to the speculations of John Mayow, but he did not believe that combustion results from the activity of some nitro-aerial spirit. Candles and matches cease to burn not because they have rendered the air "effete, by having consumed its *vivifying spirit,*" but because of "acid fuliginous vapours" that destroy the air's elasticity. A continual supply of fresh elastic air is necessary to produce the rapid intestine motion of the fuel; this motion is the result of the "action and re-action" of acid sulfurous particles and the elastic particles of air. "Air cannot burn without sulphur, so neither can sulphur burn without air."

Despite the limitations of his achievement—he had prepared a number of gases without recognizing their differences—Hales passed on to the eighteenth century the conviction that there was such a thing as "fixed air" and that it abounds in all sorts of animal, vegetable, and mineral substances. Air is "very instrumental in the production and growth of animals and vegetables," serving in its fixed state as the bond of union "and firm connection of the several constituent parts" of bodies, that is, the chief elements or principles of which things are made: "their water, salt, sulphur and earth." He concluded that air should take the place of "mercury" or "spirit" as a fifth element:

> Since then air is found so manifestly to abound in almost all natural bodies; since we find it so operative and active a principle in every chymical operation . . . . may we not with good reason adopt this now fixt, now volatile *Proteus* among the chymical principles, and that a very active one, as well as acid sulphur; notwithstanding it has hitherto been overlooked and rejected by Chymists, as no way intitled to that denomination?[76]

For Hales, science was more than the avocation of a country minister: it was a natural extension of his religious life. If he was a devotee of the mechanistic world view and held that the living organism was a self-regulating machine, this was in no way incompatible with his faith. For him, as for many other "physical theologians," nature testified to the wisdom, power, and goodness of the all-wise Creator "in framing for us so beautiful and well regulated a world."[77]

But Hales never doubted what Robert Boyle called "the usefulness of experimental philosophy." Hales's study of plants would, he was confident, improve man's skill in "those innocent, delightful and beneficial arts" of agriculture and gardening. He was well aware, too, that his studies of the animal vascular system and respiration would prove of medical value. Like Benjamin Franklin, one of the many who read the *Statical Essays* and were influenced by them, he was constantly alert to the practical possibilities of his discoveries. In describing his perfusion experiments on animals, Hales took occasion to warn the heavy imbibers of spirituous liquors of the consequences of their vice. Indeed, he soon directed two pamphlets against this growing evil, "the Bane of the Nation," and, according to Gilbert White, was instrumental, under the patronage of Sir Joseph Jeckyll, in securing the passage of the Gin Act of 1736 "and stopping that profusion of spiritous liquors which threatened to ruin the morals and the constitution of the common people."

In the *Vegetable Staticks* he had described an ingenious mercury gauge used to determine the pressure exerted by peas expanding in water, and this led him to imagine its adaptation as a "sea gage" to measure the depths of the ocean. He applied his chemical knowledge to suggesting ways of keeping water sweet during long sea voyages and exploring the obstinate problem of distilling fresh from salt water.

With the publication of his *Haemastaticks,* Hales's career in pure science came to a close. From 1733 to the end of his life he devoted himself to applying scientific knowledge, technical skill, and his rich inventiveness to alleviating human problems, both medical and social. But even earlier he had turned his attention to a problem which had long challenged the resources of the medical profession: the painful affliction of kidney and bladder stones.

Early in 1727, while the *Vegetable Staticks* was in press, he obtained a specimen of such a human calculus from a friend, the famous surgeon John Ranby. On distilling this stone, Hales collected a much greater proportion of air than he had obtained from any other substance. Since various chemical agents were known to release this "strongly attracting, unelastic air," he thought it at last possible to find a solvent to dissolve the calculi and obviate the painful operation of being "cut for the stone." He carried out a number of experiments and published the results with his *Haemastaticks.* His attempts to find a useful solvent failed, and the paper is noteworthy chiefly for his success in perfusing a dog's bladder with one of his solutions and for his invention of a surgical forceps, which Ranby and other surgeons promptly used with success to remove stones from the human urethra. Ironically, it was for this largely useless work on human calculi—not for the remarkable experiments on plants and animals and on air published in the *Statical Essays*—that Hales was awarded the Royal Society's Copley Medal in 1739.

His newly acquired expertise entangled Hales in a rather notorious episode.[78] A Mrs. Joanna Stephens had for some years been treating victims of the stone with a secret proprietary remedy, supposedly with some success. Attempts to persuade her to divulge her secret led Parliament to vote a substantial reward and to set up a group of trustees to receive her disclosure and evaluate the effectiveness of her nostrum. Hales was one of the trustees, and he set to work to determine the effective ingredient in the odd mixture. Experiments convinced him that it was the lye used in soapmaking, and lime from eggshells used in her formula, that seemed to have the desired property of dissolving the stone. The result was Hales's suggestion—destined to be taken up by others—that limewater might prove an effective if somewhat corrosive remedy.

Hales's experiments on air and respiration were the stimulus for the invention that more than any other contributed to his contemporary fame: the ventilators he contrived to remove fetid air from prisons, hospitals, and slave ships. His experiments—especially the rebreathing experiments—had convinced him that "elastic" air, free from noxious fumes, was necessary for respiration, for there was great danger in respiring "vitiated air." These theories fitted well with the current belief that many diseases were attributable to bad air and "miasmas." After a victory over a rival inventor, Hales's ventilators were installed in His Majesty's ships, in merchant vessels, in slave ships, and in hospitals and prisons. The ventilators did not, of course, eliminate airborne bacterial or viral diseases, but they seem to have markedly reduced mortality rates. As one of the first to call attention to the importance of fresh air, Hales deserves his reputation as a pioneer in the field of public health.

These varied activities did not interfere with his parish duties. He preached regularly and presided with some severity over the morals of his village; he enlarged the churchyard and virtually rebuilt the old church. In 1754 Hales engineered a new water supply for the village and, as Francis Darwin remarks, "characteristically records, in the parish register, that the outflow was such as to fill a two-quart vessel in 3 swings of a pendulum, beating seconds, which pendulum was 39 + 2/10 inches long from the suspending nail to the middle of the plumbet or bob."[79]

Hales's later years were graced with honors. Oxford conferred on him the doctorate of divinity in 1733. He was one of the trustees of the Georgia colony; and John Ellis, the merchant-naturalist who was governor of the colony and a correspondent of Linnaeus, named after him a genus of American flowering shrubs (*Halesia*). Hales was one of the founders of what is now the Royal Society of Arts and became one of its vice-presidents in 1755. In 1753 he was chosen a foreign associate of the Paris Academy of Sciences, replacing Hans Sloane, who had died earlier that year. Hales's portrait was painted by Francis Cotes and by his neighbor at Twickenham, the popular Thomas Hudson.

He had many acquaintances in the neighborhood, among them Alexander Pope (he was one of the witnesses to Pope's will) and Horace Walpole (who called him "a poor, good, primitive creature"). He was patronized by Frederick Louis and Augusta, the prince and princess of Wales, who lived not far distant at Kew. The prince, it is said, enjoyed surprising Hales in his laboratory at Teddington.

Walpole's unflattering description bears out the opinion of contemporaries, who spoke of Hales's native innocence and simplicity of manner. Peter Collinson testified to "his constant serenity and cheerfulness of mind." He died after a brief illness and was buried under the tower of his beloved church. A monument in Westminster Abbey was erected to his memory by the princess of Wales, with a bas-relief of "the old philosopher" in profile. If there is anything in the church at Teddington recalling Hales to memory, the guidebooks make no mention of it. Instead, they single out a monument to Hales's most famous (and notorious) parishioner, the actress Peg Woffington.

NOTES

1. J. F. Fulton, *Selected Readings,* 2nd ed., p. 57.
2. *Gentleman's Magazine,* **34** (1764), 273. This article by Peter Collinson seems to have been based on information supplied by Stukeley. It is reproduced in *Annual Register* (1764), "Characters," pp. 42–49.
3. G. M. Trevelyan, *Trinity College* (Cambridge, 1946), p. 55. Cf. James Henry Monk, *Life of Richard Bentley,* 2nd ed., 2 vols. (London, 1833), I, 204.
4. *Family Memoirs of the Rev. William Stukeley, M.D.,* 3 vols. (London–Edinburgh, 1882–1887), I, 21. Henceforth referred to as *Family Memoirs.*
5. *Ibid.,* pp. 21–22.
6. Robert Smith, ed., *Hydrostatical and Pneumatical Lectures by Roger Cotes* (Cambridge, 1738; 2nd ed., 1747). These posthumously published lectures, as we have them, were delivered after 1706 (Cotes refers to Newton's Latin *Optice* of that year) and perhaps before 1710. Cotes died in 1716. Whiston's lectures on this subject were never published.
7. *Family Memoirs,* I, 21, where we read: "Mr. Danny read to us . . . Pardies Geometry, Tacquets Geometry by Whiston, Harris's use of the Globes, Rohaults Physics by Clark. He read to us Clarks 2 Volumes of Sermons at Boyles Lectures, Varenius Geography put out by Sr. Isaac Newton & many other occasional peices [*sic*] of Philosophy, & the Sciences subservient thereto."
8. A. Hastings White, ed., *Memoirs of Sir Isaac Newton's Life by William Stukeley* (London, 1936), p. 9. See also *Family Memoirs,* I, 23–24.
9. The sketch is reproduced in A. E. Clark-Kennedy, *Stephen Hales* (pl. IV), and in R. T. Gunther, *Early Science in Cambridge* (Oxford, 1937), p. 160. Stukeley says Hales "first projected, & gave the idea of horarys." *Family Memoirs,* I, 21. The name "orrery" was attached to such devices after the one later built by John Rowley for his patron, the fourth earl of Orrery.
10. After earning the degree of bachelor of medicine from Cambridge, Stukeley studied "the practical part of physick" under Richard Meade at St. Thomas' Hospital; early in 1717 he opened his own London practice.
11. On 6 Mar. 1717/18; see Royal Society *Journal Book,* V (1714–1720), 235. Stukeley, although formally elected the same day as Hales, had been nominated much earlier by Edmond Halley, and his nomination evidently approved by the Council.
12. *Vegetable Staticks* (1727), p. iii. Hales, probably writing his preface late in 1726, states that this accidental observation occurred "about seven years since."
13. "Mr. Hale [*sic*] having been formerly Elected, and lapsed the time of his admission, the same was dispensed with by the Society, and he Subscribed the Obligation and was admitted accordingly." *Journal Book,* V, entry of 20 Nov. 1718 (O.S.), pp. 250–251.
14. *Ibid.,* p. 289. A good summary of the experiment is here transcribed.
15. *Ibid.,* VI (1720–1726), 438–440 *et seq.*
16. *Ibid.,* VII (1726–1727), 44–45, 48–50.
17. *Philosophical Transactions of the Royal Society,* **35,** no. 398, 264–291; no. 399, 323–331. In Apr. and May 1727, Desaguliers repeated before the Royal Society certain of Hales's experiments. *Journal Book,* VII, 74, 83.
18. "Of Mechanical Knowledge, and the Grounds of Certainty in Physick," in his trans. of Santorio's *Medicina statica,* 2nd ed. (London, 1720), p. 1.
19. *An Account of Animal Secretion, the Quantity of Blood in the Human Body, and Muscular Motion* (London, 1708). James Keill was strongly influenced by his older brother, the mathematician and Newtonian disciple John Keill.
20. *Vegetable Staticks* (1727), pp. 2–3.
21. Cusa's *De staticis,* one of the *Idiota* dialogues, appeared in many eds., sometimes appended to eds. of the *De architectura* of Vitruvius. The English translation is *The Idiot in Four Books; the First and Second of Wisdome, the Third of the Minde, the Fourth of Statick Experiments, or Experiments of the Ballance. By the Famous and Learned C. Cusanus* (London, 1650).
22. *Medicina statica: Being the Aphorisms of Sanctorius, Translated Into English With Large Explanations* (London, 1712). The popular work was, of course, known in a number of Latin eds., some with commentaries by Giorgio Baglivi and Martin Lister. In a preface to the vol. of *Philosophical Transactions* for 1669 we read "The Ingenious Sanctorius hath not exhausted all the results of Statical indications." *Philosophical Transactions of the Royal Society.* **4,** no. 45, 897. The *Oxford English Dictionary* gives as the earliest example of the word in English Sir Thomas Browne's reference in the *Pseudodoxia epidemica* (1646) to "the statick aphorisms of Sanctorius." Quincy's was not the earliest English version; a trans. had been published by J. Davis (London, 1676).
23. James Keill, *Tentamina medico-physica quibus accessit medicina statica britannica* (London, 1718).
24. *Haemastaticks* (1733), preface.
25. Arturo Castiglioni, *History of Medicine,* E. B. Krumbaar, ed. and trans. (New York, 1941), p. 614. Malpighi was the first to observe the capillaries; J. L. M. Poiseuille studied blood viscosity and rate of flow and introduced the mercury manometer for the measurement of blood pressure.
26. *Haemastaticks* (1733), pp. 58–59.
27. Hales doubtless knew the passage in Francis Hauksbee's preface to his *Physico-Mechanical Experiments* (1709), in which Hauksbee wrote that electricity may possibly explain "the Production and Determination even of *Involuntary Motion* in the *Parts of Animals,*" for he quotes Hauksbee three times, but on other matters, in the "Analysis of Air." He surely also knew the concluding passage of the General Scholium of Newton's *Principia,* 2nd ed. (1713), in which Newton hints that "an electric and elastic spirit" may account for sensation and cause "the members of animal bodies [to] move at the command of the will, namely, by the vibrations of this spirit, mutually propagated along the solid filaments of the nerves, from the outward organs of sense to the brain, and from the brain to the muscles."
28. *Vegetable Staticks* (1727), pp. 2–3.
29. *Philosophical Transactions of the Royal Society,* **3,** no. 40 (1668), 787–801.

30. The letter was communicated 10 June 1669 (O.S.) and published in *Philosophical Transactions of the Royal Society,* **4**, no. 48, 963–965. See also Charles Raven, *John Ray* (Cambridge, 1950), pp. 187–188.
31. Thomas Birch, *History of the Royal Society of London,* 4 vols. (London, 1756–1757), II, 382.
32. The theory of sap circulation was advanced by Christopher Merret in 1664 and by Johann Daniel Major a year later. See Julius von Sachs, *History of Botany,* p. 456; and J. Reynolds Green, *History of Botany in the United Kingdom,* p. 76. This theory is clearly set forth by John Locke. See J. A. St. John, ed., *Philosophical Works of John Locke,* II (London, 1706), 487. Even later than Blair were the claims of a Mr. Fairchild in 1724 to have proved by experiments "a constant Circulation of the Sap in Trees and Plants." *Journal Book,* VI, 377.
33. For the background of this experiment, suggested by Cusa in his *De staticis,* see Herbert M. Howe, "A Root of van Helmont's Tree," in *Isis,* **56** (1965), 408–419. See also A. D. Krikorian and F. C. Steward, "Water and Solutes in Plant Nutrition," in *BioScience,* **18** (1968), 286–292.
34. "Some Thoughts and Experiments Concerning Vegetation," in *Philosophical Transactions of the Royal Society,* **21**, no. 253 (1699), 193–227. The quotation is from p. 208. Hales, when describing an experiment on the imbibition by a spearmint plant growing in water, wrote: "I pursued this Experiment no farther, Dr. *Woodward* having long since . . . given an account . . . of the plentiful perspirations of this plant." *Vegetable Staticks* (1727), p. 28.
35. Hales writes "by the favour of the eminent Mr. *Wise.*" *Vegetable Staticks* (1727), pp. 17–18. Hales owed something to his relations with "the skilful and ingenious Mr. Philip Miller" of the Chelsea Physic Garden and author of the popular *Gardener's Dictionary* (1724). On Miller see Green, *op. cit.,* pp. 156–157 and *passim.*
36. Francis Darwin, *Rustic Sounds,* p. 126.
37. *Vegetable Staticks* (1727), p. 10.
38. *Ibid.,* p. 103.
39. See, for example, Hales's experiments VII and XXVIII, *ibid.,* pp. 28–29, 90.
40. *Ibid.,* p. 136. See also pp. 13–14, where he writes that the sap has "probably only a progressive and not a circulating motion as in animals."
41. *Ibid.,* p. 96.
42. *Ibid.,* p. 100.
43. *Ibid.,* pp. 329–337. He was struck, as Sachs observes, by the fact that the longitudinal growth allows the capillary vessels to retain their hollowness, as when glass tubes are drawn out to fine threads.
44. *Ibid.,* pp. 102–103, 148. For the inconclusive experiment see experiment CXXII in the chapter "Of Vegetation." After the publication of the *Vegetable Staticks* Hales repeated the experiment and convinced himself that leaves imbibe air. He informed Desaguliers of these results by June 1727. See Desaguliers' postscript to his abstract of Hales's book in *Philosophical Transactions of the Royal Society,* **35**, no. 399, 331.
45. *Vegetable Staticks* (1727), pp. 155–156.
46. The prevailing view in the seventeenth century (of men like Jean Beguin, Lemery, and Homberg) was that there are five elements or principles: three active principles (variously described as spirit, oil, and salt or as mercury, sulfur, and salt) and two passive ones, water and earth. This was clearly a compromise between the Aristotelian theory of the four elements and the *tria prima* of the Paracelsians. For a clear statement of this view in Hales's day, see John Harris, *Lexicon technicum* (1704), article "Principle."
47. Henry Guerlac, "John Mayow and the Aerial Nitre," in *Actes du Septième Congrès d'histoire des sciences* (Jerusalem, 1953), pp. 332–349; and "The Poet's Nitre," in *Isis,* **45** (1954), 243–255.
48. Robert Boyle, *General History of Air* (London, 1692), p. 1. See also Harris, *Lexicon technicum* (1704), article "Air"; and Cotton Mather, *The Christian Philosopher* (London, 1721), p. 65.
49. When the experiment on powdered oyster shells was shown to the Society on 15 Mar. 1664/65, the air was collected in a deflated bladder. But when it was repeated a short time later, a large glass filled with water was inverted ("whelmed") over the reactants; and when the reaction was over, it was found that the "whelmed glass" was about a quarter full of an aerial substance. Birch, *op. cit.,* II, 22, 27. This early anticipation of the principle underlying the pneumatic trough seems to have escaped notice. It is not mentioned in John Parascandola and Aaron J. Ihde, "History of the Pneumatic Trough," in *Isis,* **60** (1969), 351–361.
50. *Philosophical Transactions of the Royal Society,* **18**, no. 212, 212–213.
51. Smith, ed., *Hydrostatical . . . Lectures by Roger Cotes,* 2nd ed., pp. 220–223; and Isaac Newton, *Optice* (1706), p. 325, and *Opticks* (1718), p. 353. Hales quotes the experiment in which Slare distilled or "calcined" an animal calculus and found that the greatest part of this stone "evaporated in the open fire." *Vegetable Staticks* (1727), pp. 188–189.
52. Query 30 of *Opticks* (1718), pp. 349–350; and *Vegetable Staticks* (1727), p. 312. Hales also quotes (*ibid.,* p. 165) from another long passage of the *Opticks* in which Newton speaks of airs formed from those bodies "which Chymists call fix'd, and being rarefied by Fermentation, become true permanent Air." Query 31, *Opticks* (1718), p. 372. Newton and Hales both use the word "fermentation" to mean chemical reactions that are accompanied by the production of heat and ebullition. The term originated with Thomas Willis in his *De fermentatione sive De motu intestino particularum in quovis corpore* (London, 1659).
53. *De la nature, vertu, et utilité des plantes* (Paris, 1628), p. 75; see also pp. 94–95.
54. *The History of the Propagation & Improvement of Vegetables* (Oxford, 1660), pp. 40–42, 84–85. Robert Sharrock, an Oxford graduate who became archdeacon of Winchester, supplied prefaces to three of Boyle's works. His book is dedicated to Boyle.
55. Birch, *op. cit.,* I, 304.
56. *Ibid.,* II, 54, 164; III, 418, 420–421.
57. *Philosophical Transactions of the Royal Society,* **3**, no. 42, 854.
58. *The Wisdom of God Manifested in the Works of the Creation,* 8th ed. (London, 1722), p. 72. Mather, *op. cit.,* p. 69, was clearly paraphrasing Ray when he wrote "Yea, *Malpighius* has discovered and demonstrated, that the *Plants* themselves have a kind of Respiration, being furnished with a Plenty of Vessels for the Derivation of *Air* to all their Parts."
59. *Vegetable Staticks* (1727), p. 156. Boyle's experiments, carried out with Denis Papin, using the latter's improved air pump, were published in Boyle's *A Continuation of New Experiments . . .,* which appeared in Latin in 1680 and in English in 1682.
60. For one such Cambridge experiment see *Vegetable Staticks* (1727), p. 195.
61. Hales also cited Hermann Boerhaave's *New Method of Chemistry.* An unauthorized version of Boerhaave's lectures had appeared in Latin in 1724; Hales seems to have used the English translation by Peter Shaw and E. Chambers, dated 1727. His references to it in the *Vegetable Staticks* were obviously added while his book was in press.
62. Although he records the combustibility of the gases produced by distilling peas, he failed to note the same property in coal gas. In describing the air produced by the action of dilute acid on iron filings (that is, hydrogen), he does not remark that it is inflammable.
63. *Vegetable Staticks* (1727), p. 291. Newton's paper was published by John Harris in the introduction to his *Lexicon technicum,* II (1710), where Hales consulted it.
64. "And Dr. Freind has from the same principles [as Newton] given a very ingenious Rationale of the chief operations in Chymistry." *Vegetable Staticks* (1727), preface, p. v.

65. *Ibid.*, p. 165. Cf. Newton, *Opticks* (1718), p. 372.
66. The bolthead was a chemist's globular flask with a long cylindrical neck, what Boyle called a "glass egg with a long neck." It was named for its resemblance to the head of a bolt or arrow.
67. *Vegetable Staticks* (1727), p. 287.
68. The method of collecting air by the displacement of water, used in all Hales's devices, was not original with him. It had been used as early as 1665, probably by Robert Hooke, in an experiment performed at the Royal Society. But it was doubtless from John Mayow that Hales learned of this method; Mayow used it extensively in his *Tractatus quinque medico-physici* (1674) and illustrated several modifications in an accompanying plate.
69. Henry Guerlac, "Joseph Black and Fixed Air," in *Isis,* **48** (1957), 435 and n. 141.
70. *Vegetable Staticks* (1727), p. 223.
71. Henry Guerlac, "Continental Reputation of Stephen Hales," *Archives internationales d'histoire des sciences,* **4,** no. 15 (1951), 396–397. See also Parascandola and Ihde, *op. cit.*, p. 355.
72. Hales (*Vegetable Staticks* [1727], p. 266) compared his results with the observations of Francis Hauksbee, who had noted the same effect. See Hauksbee's *Physico-Mechanical Experiments on Various Subjects* (1709), p. 83.
73. *Vegetable Staticks* (1727), p. 231.
74. *Ibid.*, p. 183.
75. *Ibid.*, pp. 272–275, 278–285.
76. *Ibid.*, pp. 315–316.
77. For Hales's "argument from design" to justify his scientific work, see in particular his eloquent preface to the *Haemastaticks* (1733).
78. For a detailed account of this episode, and its later influence on the work of Joseph Black, see Henry Guerlac, "Joseph Black and Fixed Air," 137–151.
79. "Hales, Stephen," in *Dictionary of National Biography.*

BIBLIOGRAPHY

I. Original Works. Hales's major writings in English are *Vegetable Staticks: Or, an Account of Some Statical Experiments on the Sap in Vegetables . . . Also, a Specimen of an Attempt to Analyze the Air . . .* (London, 1727), also repr. with a useful foreword by M. A. Hoskin (London, 1961); *Statical Essays: Containing Vegetable Staticks* (London, 1731), the 2nd ed., "with amendments"; *Statical Essays,* 2 vols. (London, 1733): vol. I is the 3rd ed. of *Vegetable Staticks,* and vol. II is the 1st ed. of *Haemastaticks; or an Account of Some Hydraulic and Hydrostatical Experiments Made on the Blood and Blood-Vessels of Animals,* with a separate preface, Hales's "Account of Some Experiments on Stones in the Kidnies and Bladder," an appendix with nine "Observations" relating to the motion of fluids in plants, seven additional experiments on air, and "Description of a Sea-gage, Wherewith to Measure Unfathomable Depths of the Sea"—vol. II repr. in facsimile as no. 22 in History of Medicine Series of the Library of the New York Academy of Medicine (New York, 1964), with a short introduction by Andre Cournand, M.D.

Statical Essays, 2 vols. (London, 1738–1740), vol. I is 3rd ed. of *Vegetable Staticks* and vol. II is 2nd ed., "corrected," of *Haemastaticks;* and *Statical Essays,* 2 vols. (London, 1769), vol. I is 4th ed. of *Vegetable Staticks,* and vol. II is 3rd ed. of *Haemastaticks.*

Translations of Hales's major works are *La statique des végétaux, et l'Analyse de l'air . . .*, G. L. L. Buffon, trans. (Paris, 1735), an influential French trans. which has the famous "Préface du traducteur," in which Buffon praises the experimental method, and includes Hales's appendix of 1733; *Haemastatique, ou la statique des animaux* (Geneva, 1744), the first French version of the *Haemastaticks,* translated by the physician and botanist François Boissier de Sauvages; *Statique des végétaux, et celle des animaux,* 2 pts. (Paris, 1779–1780); pt. I is Buffon's trans. of the *Vegetable Staticks* "revue par M. Sigaud de la Fond," and pt. II is Boissier de Sauvages's trans. of the *Haemastaticks; Statick der Gewächse* (Halle, 1747), translated, with a preface, by the philosopher Christian von Wolff; *Statick des Geblüts,* 2 pts. (Halle, 1748), pt. I is the *Haemastaticks,* and pt. II is Wolff's translation of the *Vegetable Staticks; Emastatica, ossia statica degli animali,* 2 vols. (Naples, 1750–1752), Italian trans. from the French of Boissier de Sauvages, vol. II has a trans. of Hales's work on bladder and kidney stones and two medical dissertations by Boissier de Sauvages; and *Statica de' vegetabili ed analisi dell' ari,* D. M. A. Ardinghelli, trans. (Naples, 1756, 1776), with commentary.

Hales's minor works were *A Sermon Preached Before the Trustees for Establishing the Colony of Georgia in America* (London, 1734); *A Friendly Admonition to the Drinkers of Brandy and Other Distilled Spirit* (London, 1734), anonymous, but attributed to Hales; *Distilled Spiritous Liquors the Bane of the Nation; Being Some Considerations Humbly Offered to the Hon. the House of Commons* (London, 1736); *Philosophical Experiments: Containing Useful and Necessary Instructions for Such as Undertake Long Voyages at Sea . . .* (London, 1739); *An Account of Some Experiments and Observations on Mrs. Stephens's Medicines for Dissolving the Stone* (London, 1740), also translated into French (Paris, 1742); *A Description of Ventilators . . .* (London, 1743), French trans. by P. Demours (Paris, 1744); *An Account of Some Experiments and Observations on Tar-Water* (London, 1745); *Some Considerations on the Causes of Earthquakes . . .* (London, 1750), French trans. by G. Mazeas (Paris, 1751), with the letter of the bishop of London, Thomas Sherlock, on the moral causes of the London earthquakes of 1750; *A Sermon Before Physicians, on the Wisdom and Goodness of God in the Formation of Man* (London, 1751), the annual Croonian sermon of the Royal College of Physicians, *not* the Croonian lecture of the Royal Society; *An Account of a Useful Discovery to Distill Double the Usual Quantity of Sea-water . . . and an Account of the Great Benefit of Ventilators . . .* (London, 1756); and *A Treatise on Ventilators . . .* (London, 1758).

II. Secondary Literature. General and biographical sources include the following (listed chronologically): "Some Account of the Life of the Late Excellent and Eminent Stephen Hales D.D., F.R.S. Chiefly From Materials Communicated by P. Collinson, F.R.S.," in *Gentleman's Magazine,* **34** (1764), 273–278, see also *Annual Register of World Events* (1764), pp. 42–49; Jean-Paul

Grandjean de Fouchy, "Éloge de M. Hales," in *Histoire de l'Académie royale des sciences* for 1762 (Paris, 1764), pp. 213–230; Robert Watt, *Bibliographia britannica,* 4 vols. (London, 1824), I, col. 457; F.D. [Francis Darwin], "Hales, Stephen," in *Dictionary of National Biography,* an excellent summary; Francis Darwin, *Rustic Sounds* (London, 1917), pp. 115–139, a useful essay by a distinguished botanist; G. E. Burget, "Stephen Hales," in *Annals of Medical History,* **7** (1925), 109–116; A. E. Clark-Kennedy, "Stephen Hales; Physiologist and Botanist," in *Nature,* **120** (1927), 228–231; George Sarton, "Stephen Hales's Library," in *Isis,* **14** (1930), 422–423; A. E. Clark-Kennedy, *Stephen Hales, D.D., F.R.S. An Eighteenth Century Biography* (Cambridge-NewYork, 1929; repr. Ridgewood, N.J., 1965), the only full-length biography; Jocelyn Thorpe, "Stephen Hales," in *Notes and Records. Royal Society of London,* **3** (1940), 53–63; and Lesley Hanks, *Buffon avant l'"Histoire naturelle"* (Paris, 1966), 73–101, which discusses Buffon's translation of the *Vegetable Staticks* and the Newtonianism of Hales and Buffon.

On his work in public health, see D. Fraser Harris, "Stephen Hales, the Pioneer in the Hygiene of Ventilation," in *Scientific Monthly,* **3** (1916), 440–454.

On animal physiology see the following (listed chronologically): John F. Fulton, *Selected Readings in the History of Physiology* (Springfield, Ill.-Baltimore, 1930), pp. 57–60, 75–79, 235, see also the greatly enl. ed., with material supplied by Leonard Wilson (Springfield, Ill., 1966); and *Physiology,* in the Clio Medica series (New York, 1931), pp. 35–36, 42–43; Thomas S. Hall, *A Source Book in Animal Biology* (New York, 1951), pp. 164–171, which reprints Hales's preface to the *Haemastaticks,* without the concluding acknowledgment, and experiment I; and Diana Long Hall, "From Mayow to Haller: A History of Respiratory Physiology in the Early Eighteenth Century" (Ph.D. thesis, Yale University, 1966), pp. 118–121.

Hales's work in plant physiology is discussed in the following (listed chronologically): Julius von Sachs, *Geschichte der Botanik* (Munich, 1875), pp. 514–521, 582–583, English trans. by Henry E. G. Garnsey, revised by I. B. Balfour, *History of Botany (1530–1860)* (Oxford, 1906), pp. 476–482, 539; J. Reynolds Green, *A History of Botany in the United Kingdom* (London, 1914), pp. 198–206 and *passim;* R. J. Harvey-Gibson, *Outlines of the History of Botany* (London, 1919), pp. 46–50 and *passim;* and Ellison Hawks and G. S. Boulger, *Pioneers of Plant Study* (London, 1928), pp. 228–230.

Hales's chemistry is discussed in the following (listed chronologically): Hermann Kopp, *Geschichte der Chemie,* 4 vols. (Brunswick, 1843–1847), III, 182–183 and *passim;* Ferdinand Hoefer, *Histoire de la chimie,* 2nd ed., 2 vols. (Paris, 1869), II, 338–342; Henry Guerlac, "The Continental Reputation of Stephen Hales," in *Archives internationales d'histoire des sciences,* **4,** no. 15 (1951), 393–404; Milton Kerker, "Hermann Boerhaave and the Development of Pneumatic Chemistry," in *Isis,* **46** (1955), 36–49; Henry Guerlac, "Joseph Black and Fixed Air, A Bicentenary Retrospective," in *Isis,* **48** (1957), 124–151, 433–456; Rhoda Rappaport, "G.-F. Rouelle: An Eighteenth-Century Chemist and Teacher," in *Chymia,* **6** (1960), 94; Henry Guerlac, *Lavoisier—The Crucial Year* (Ithaca, N.Y., 1961), *passim,* for Hales's influence upon Lavoisier; J. R. Partington, *History of Chemistry,* III (London, 1962), 112–123; and John Parascandola and Aaron J. Ihde, "History of the Pneumatic Trough," in *Isis,* **60** (1969), 351–361.

HENRY GUERLAC

HALL, ASAPH (*b.* Goshen, Connecticut, 15 October 1829; *d.* Goshen, 22 November 1907), *astronomy.*

Hall belonged to an old and once prosperous New England family. His father, also called Asaph Hall, a manufacturer of wooden clocks, died in 1842 on a clock-selling trip in Georgia, leaving the family in difficult circumstances. His mother, Hannah Palmer, attempted unsuccessfully for three years to pay the mortage on a family farm by operating a cheese factory. At age sixteen, Asaph became a carpenter's apprentice. Three years later, a strong, athletic lad, standing over six feet, he went to work as a journeyman carpenter, and in the six years that followed, employed skills that must have proved useful later in his career when he supervised the construction of observing shelters on several astronomical expeditions.

As a boy, Hall read the histories of Gibbon and Hume, which were in his father's varied library. He went at first to the district school, but later his formal education was less regular. During one winter of his apprenticeship he studied algebra and geometry at the Norfolk Academy, eight miles from his home, but soon found he was a better mathematician than his teacher. By 1854 he became impatient to continue his education and, hoping to become an architect, enrolled in Central College in McGrawville, New York. There, according to the *New York Tribune*, students could meet part of their expenses by manual labor. In McGrawville he found a motley crowd of adventurers and idealists who cared little for a classical education. Among the students, however, he met Chloe Angeline Stickney, a frail but determined suffragist, who taught mathematics while completing her senior year. Hall was among her pupils, and she soon became his fiancée. After their marriage in March 1856, they went to Ann Arbor, where for three months Hall studied under Franz Brünnow, director of the University of Michigan Observatory. Lacking money to continue his schooling, the couple took teaching posts for a year at the Shalersville Institute in Ohio.

Firmly determined to become an astronomer, Hall proceeded to Cambridge, Massachusetts, where, in spite of George Bond's admonition that he would starve, he took a low-paid job at the Harvard College Observatory. He quickly became both an observer and an expert computer of orbits. Having been taught German by his wife, he read Brünnow's *Astronomie* and by 1858 he was studying Gauss's *Theoria motus.* At this time he published the first of his numerous mathematical and astronomical articles in scientific journals. He supplemented his meager salary by computing almanacs and by observing moon culminations at a dollar per observation.

In 1862, enticed by an adequate income, Hall became an assistant astronomer at the U.S. Naval Observatory. But his first years in Washington were troubled by the Civil War; exhausted by the unwholesome climate as well as by exertions on behalf of wounded friends, Hall was so weakened by jaundice that it was two years before he fully recovered. A year after his arrival, a professorship in mathematics opened at the Naval Observatory. Hall, believing the office should seek the man, simply waited, but unknown to him, his wife proposed his name by letter to the superintendent of the observatory, and Hall was given the position.

From 1862 to 1866 he was assistant observer with the nine-and-one-half-inch equatorial, then considered a large instrument. For a year he took charge of the meridian circle and from 1868 to 1875 was again in charge of the nine-and-one-half-inch equatorial. His observations were primarily of asteroids and comets. In 1869 Hall traveled to the eastern coast of Siberia to observe a total solar eclipse, and the following year he went to Sicily for another eclipse. In 1874 he led a party to Vladivostok to observe the transit of Venus. Inclement weather and the lack of adequate photographic apparatus prevented these expeditions from fulfilling their expectations. Hall was more successful as the leader of expeditions to Colorado to observe the eclipse of 1878 and to Texas to observe the transit of Venus in 1882.

In 1875 Hall was placed in charge of the twenty-six-inch Clark equatorial at the Naval Observatory, then the largest refractor in the world. His first discovery with this telescope, in December 1876, was a white spot on the planet Saturn, which he measured through more than sixty rotations, thus finding the first reliable period of Saturn's rotation since Herschel's determination in 1794.

At the time of the unusually close approach of Mars in August 1877, Hall undertook a systematic search for possible satellites. In this search he was fortunately guided by theoretical considerations, which showed that any Martian satellite must revolve very close to the planet; otherwise the solar gravity would overpower the attraction of Mars. "The chance of finding a satellite appeared to be very slight," Hall wrote, "so that I might have abandoned the search had it not been for the encouragement of my wife." Angeline Hall was an enthusiast, and Angelo, the third of the four Hall sons, claimed that she "insisted upon her husband's discovering the satellites of Mars."

Hall first glimpsed the object that was eventually named Deimos on 11 August, and, after a few days of bad weather, by 17 August he convinced himself that it was indeed a satellite. He also found the second satellite, Phobos, on 17 August. He then disclosed his observations to Simon Newcomb, the scientific head of the observatory. Newcomb erroneously believed that Hall, in his modest conservatism, was reluctant to recognize the "Mars stars" as satellites, and hence took for himself an undeserved credit for this recognition in the wide press coverage that followed. For many years Hall quietly harbored a grudge against Newcomb, who eventually offered his apologies.

After these discoveries Hall gradually became known as the caretaker of the satellites not only of Mars, but of Saturn, Uranus, and Neptune. In 1884 he showed that the position of the elliptical orbit of Saturn's satellite Hyperion was retrograding by about twenty degrees per year. The celestial mechanician George William Hill called Hall's memoir on Iapetus (the outer satellite of Saturn) one of "the most admirable pieces of astronomical literature" and compared its clarity and precision to the work of Bessel.

Hall's principal papers on the satellites are listed in the bibliography. In addition to his work on planetary satellites, Hall was also an assiduous observer of double stars, with numerous investigations of binary star orbits. In 1892 he showed that the two components of 61 Cygni were physically related. He worked on determinations of stellar parallax and on the positions of faint stars in the Pleiades cluster.

For his discoveries Hall won the Gold Medal of the Royal Astronomical Society (1879), the Lalande Prize (1877) and Arago Medal (1893) of the French Academy of Sciences, and became a chevalier of the French Legion of Honor in 1896. Elected to the National Academy of Sciences in 1875, he served as its home secretary for twelve years and as vice-president for six. In 1902 he served as president of the American Association for the Advancement of Science. Beginning in 1888, he acted as a consulting astronomer to the Washburn Observatory in Madison, Wisconsin, and from 1897 to 1907 he was associate editor of the *Astronomical Journal.*

Following his mandatory retirement at age sixty-two from the Naval Observatory in 1891, Hall continued to work as a voluntary observer on the twenty-six-inch telescope. His wife died in 1892, and in 1894 Hall left Washington for Connecticut. Four years later he went to Harvard to teach celestial mechanics, becoming professor of mathematics. After five years of teaching, he married a longtime protégé of his brother's family, Mary Gauthier, and retired to his rural home in Connecticut, where he lived until his death.

BIBLIOGRAPHY

I. ORIGINAL WORKS. An extensive bibliography by William D. Horigan of 486 publications appears in the *Biographical Memoirs. National Academy of Sciences,* **6** (1908), 276–309. Hall's principal memoirs were published as separate appendices to the *Washington Observations* (more fully *Astronomical and Meteorological Observations Made During the Years . . . at the United States Naval Observatory*); these include "Catalogue of 151 Stars in Praesepe," appendix IV, 1867 (1870); "Observations and Orbits of the Satellites of Mars," in the vol. for 1875 (1878); "Observations of Double Stars," appendix VI, 1877 (1881); "The Orbits of Oberon and Titania," appendix I, 1881 (1885); "Orbit of the Satellite of Neptune," appendix II, 1881 (1885); "The Orbit of Iapetus," appendix I, 1882 (1885); "The Six Inner Satellites of Saturn," appendix I, 1883 (1886); "Observations for Stellar Parallax," appendix II, 1883 (1886); "Saturn and Its Ring. 1875–1889," appendix II, 1885 (1889); and "Observations of Double Stars," appendix I, 1888 (1892).

Other papers of note include "On the Determination of the Mass of Mars," in *Astronomische Nachrichten,* **86** (1875), cols. 327–334; "On the Rotation of Saturn," *ibid.,* **90** (1877), cols. 145–150; "The Motion of Hyperion," in *Monthly Notices of the Royal Astronomical Society,* **44** (1884), 361–365; "The Orbit of Iapetus," in *Astronomical Journal,* **11** (1891–1892), 97–102; and "Science of Astronomy," in *Proceedings of the American Association for the Advancement of Science,* **52** (1902–1903), 313–323, Hall's widely reprinted and translated address as retiring president of the Association.

The chief repository of Hall's MSS is the Library of Congress, with MSS under his own name and others in the Simon Newcomb Collection.

II. SECONDARY LITERATURE. For works about Hall, see Percival Hall, *Asaph Hall, Astronomer* (n.p., 1945), written and printed for private distribution by Hall's fourth son; George William Hill, "Biographical Memoir of Asaph Hall," in *Biographical Memoirs. National Academy of Sciences,* **6** (1908), 240–275; and H. S. Pritchett, "Asaph Hall," in *Science,* n.s. **26** (13 Dec. 1907), 805, repr. in *Popular Astronomy,* **16** (1908), 67–70. See also Owen Gingerich, "The Satellites of Mars: Prediction and Discovery," in *Journal for the History of Astronomy,* **1** (1970), 109–115; Angelo Hall, *An Astronomer's Wife* (Baltimore, 1908), a biography of his mother.

An extensive evaluation of Hall's scientific work up to 1879 is given in the Royal Astronomical Society's presidential address by Lord Lindsay at the presentation of the Gold Medal, in *Monthly Notices of the Royal Astronomical Society,* **39** (1879), 306–318.

OWEN GINGERICH

HALL, CHARLES MARTIN (*b.* Thompson, Ohio, 6 December 1863; *d.* Daytona, Florida, 27 December 1914), *commercial chemistry.*

Charles Martin Hall discovered and developed a commercial process for producing aluminum that brought about its widespread use. Because Paul T. L. Héroult discovered the same process independently and at about the same time in France, the discovery is known as the Hall-Héroult process.

Hall was the son of a Protestant clergyman, Heman Basset Hall, and Sophronia Brooks. Raised in Oberlin, Ohio, from the age of ten, he received his degree from Oberlin College and later became a trustee and benefactor of the school. Hall was determined to become an inventor and was interested in chemistry; he studied the latter subject at Oberlin under Frank F. Jewett, who predicted, in a lecture on aluminum, both financial and social rewards for the inventor of a cheap aluminum production process. Hall, then in his junior year, thereby envisaged how he might fulfill his financial and humanitarian aspirations. He devoted himself to the study of the metal.

At this time, aluminum was being produced expensively on a small scale by the process developed by Henri Sainte-Claire Deville, which used sodium as a chemical reducing agent. Hall read all he could about the element in such sources as were then available to aspiring young scientists and inventors, including the *Encyclopaedia Britannica* and the *Scientific American.* He also had access to information from German publications, provided by Jewett. In the family woodshed, he experimented tirelessly, and his life is said to have been divided between work and study—with the emphasis on work.

In 1859 Sainte-Claire Deville had described a means of plating aluminum on copper by electrolysis using fused cryolite (a double fluoride of aluminum and sodium) as an electrolyte. Almost thirty years later, Hall himself experimented with electrolysis using fused cryolite, but as a solvent for alumina, which he hoped to electrolyze. With a crucible of clay Hall's experiment failed, but after Hall ingeniously lined the clay with carbon, the alumina dissolved like sugar in water and globules of aluminum collected at the cathode.

This first success came on 23 February 1886. Two years later Hall founded a small company to produce aluminum commercially, now the Aluminum Company of America. He patented his production process in 1889. His major patent (No. 400,766, issued 2 April 1889) was challenged unsuccessfully on the grounds that Sainte-Claire Deville had anticipated him.

The not unusual problems of development and finance followed for the single-minded Hall. Among other improvements in his initial process was the abandonment of external heat for the fused cryolite. Largely because of his labors, the price of aluminum went from five dollars per pound in 1886 to seventy cents per pound in 1893. Seeking other major discoveries, Hall continued to experiment in chemistry, but commercial aluminum, for which he received the Perkin Medal in 1911, remains his greatest contribution.

BIBLIOGRAPHY

For works about Hall, see Junius Edwards, *The Immortal Woodshed: The Story of the Inventor Who Brought Aluminum to America* (New York, 1955), with Hall's letters; Alfred Cowles, *The True Story of Aluminum* (Chicago, 1958); *Decisions of the Commissioner of Patents . . . 1894* (Washington, D.C., 1895), pp. 573–594, 637–645, with the decision of Judge William Howard Taft, based on 1500 pages of testimony, in the patent case of "The Pittsburgh Reduction Co. [Hall's] v. The Cowles Electric Smelting and Aluminum Co."; *Addresses at the Memorial Service, Oberlin, Ohio, Jan. 22, 1915,* including contributions by Arthur V. Davis and F. F. Jewett, and Hall's Perkin address; and Harry N. Holmes, *Fifty Years of Industrial Aluminum,* Bulletin of Oberlin College, no. 346 (1937), which summarizes Hall's work authoritatively.

THOMAS PARKE HUGHES

HALL, EDWIN HERBERT (*b.* Great Falls [now North Gorham], Maine, 7 November 1855; *d.* Cambridge, Massachusetts, 20 November 1938), *physics.*

One of five children of Joshua Emery Hall and Lucy Ann Hilborn, only he and a younger brother survived to maturity. He prepared for college at Gorham Seminary for two years and graduated from Bowdoin in 1875. On the advice of John Trowbridge at Harvard, Hall entered the Johns Hopkins graduate school to study physics with Henry Rowland. He discovered the "Hall effect" in 1879 as a consequence of his dissertation research and received his Ph.D. in 1880. Hall remained at Johns Hopkins for another year and spent the summer of 1881 in Europe, visiting Hermann von Helmholtz' laboratory long enough to finish some measurements on the Hall effect. In the fall he went to Harvard as an instructor; he was appointed assistant professor in 1888, professor in 1895, and Rumford professor in 1914, becoming emeritus in 1921. He was elected to the National Academy of Sciences in 1911. Hall married Caroline Eliza Bottum in 1882; they had two children.

Hall is best known for the effect bearing his name; it formed the basis of his Ph.D. dissertation and was the subject of many of his later researches. He was stimulated by Rowland to question a statement in Maxwell's *Electricity and Magnetism* that the force acting on a conductor in a magnetic field acts on the conductor directly and not on the electric current. Hall's experimental persistence was rewarded when he found that a current through a gold conductor in a magnetic field produced an electric potential perpendicular to both the current and the field. The notion that this was due to simple interaction between the current and the field had to be abandoned, however, when other metals were found for which the effect was in a direction opposite to that predicted.

Hall's discovery was termed by Kelvin as comparable with the greatest ever made by Michael Faraday. It sparked interest in studies in this area, and three other transverse effects were soon discovered; they bear the names of Andreas von Ettingshausen, Walther Nernst, and Augusto Righi and Sylvestre Leduc.

Hall paid a great deal of attention to the possible methods of electron conduction in metals and was particularly interested in thermoelectric phenomena, especially in the years following 1914. He considered reactions between free electrons, bound electrons, and positive ions in the metallic structure; by setting up some rather arbitrary parameters and determining their values empirically, he managed to obtain reasonably consistent numbers for coefficients of the Thomson and Peltier effects. He was, however, unable to place his notions in a broader theoretical context.

Starting in 1911, Hall devoted considerable effort to very delicate experiments designed to determine precise values—measured on the same samples—for the four transverse effects. He was still at work on this problem shortly before his death.

Hall helped to stimulate the introduction of laboratory work into secondary schools and prepared a set of forty experiments that could be performed with simple apparatus. Published as *Harvard Descriptive List of Elementary Physical Experiments* in 1886 (later called *National Physics Course*), the list was designed to allow secondary schools to meet a new Harvard entrance requirement for laboratory work in physics. The effect on schools was considerable, as can be measured in part by instrument makers' catalogs which soon appeared, describing apparatus designed for Hall's experiments.

BIBLIOGRAPHY

I. ORIGINAL WORKS. A bibliography of Hall's scientific works is given in the Bridgman article mentioned below. The Hall effect was announced in "On a New Action of the Magnet on Electric Currents," in *American Journal of Mathematics,* **2** (1879), 287–292. His work on conduction theory is best represented by "On Electric Conduction and Thermoelectric Action in Metals," in *Proceedings of the American Academy of Arts and Sciences,* **50** (1914), 67–103; and by "Thermo-electric Action With Dual Conduction of Electricity," in *Proceedings of the National Academy of Sciences of the United States of America,* **4** (1918), 98–103; it is summarized in *A Dual Theory of Conduction in Metals* (Cambridge, Mass., 1938).

II. SECONDARY LITERATURE. A biographical sketch of Hall, written by his colleague at Harvard, P. W. Bridgman, appears in *Biographical Memoirs. National Academy of Sciences,* **21** (1939–1940), 73–94; it includes a portrait and an extensive bibliography. A short notice also appears in the *Dictionary of American Biography.*

BERNARD FINN

HALL, GRANVILLE STANLEY (*b.* Ashfield, Massachusetts, 1 February 1846; *d.* Worcester, Massachusetts, 24 April 1924), *psychology, education.*

Born on his grandfather's farm, Hall grew up in a conservative, rural environment. His parents, Granville Bascom Hall and Abigail Beals Hall, were descended from earliest New England colonists. Both had taught school but had moved onto a farm near Ashfield, Massachusetts, soon after their marriage. They urged education and instilled its value into their three children. Hall's father was a state legislator on the Know-Nothing ticket and active in civic affairs. His mother was especially pious.

For his early schooling Hall attended several one-room schools in and near Ashfield. He taught at such schools himself for a year but, determined to continue his education, attended Williston Academy in Easthampton, Massachusetts, for college preparation and in 1863 entered Williams College. He had overcome his shyness at his first entry into the wider world and at Williams was active in school events, especially in debates and in a literary club that he helped found. During his college years he developed a lifelong habit of omnivorous reading, chiefly in philosophy, literature, and all aspects of evolution. Professors at Williams, especially John Bascom, A. L. Perry, and Mark Hopkins, encouraged his wide-ranging selections. Hall received his B.A. in 1867 and M.A. in 1870.

Considering entering the ministry, Hall attended Union Theological Seminary in 1867 and took advantage of its location to explore thoroughly New York City. He became acquainted with Henry Ward Beecher, who, by arranging a loan, provided the young man with the chance to fulfill his keenest ambition, a trip to Europe. From 1868 to 1871 Hall studied at Bonn, Berlin, and Heidelberg. For the rest of his life he held European universities and teaching methods, especially German ones, in the highest regard.

On his return to the United States, uncertain of his future plans, Hall taught in boys' schools and tutored the family of banker Jesse Seligman. In 1872 he accepted an offer to teach English literature at Antioch College, and he later added modern languages and philosophy to his courses there. Very impressed by Wilhelm Max Wundt's *Grundzüge der physiologischen Psychologie* (1874), Hall resigned to study under Wundt but postponed the trip for a year to be an instructor of English at Harvard, where he began an acquaintance with William James. For the next two years he studied in Leipzig under Wundt, Carl Ludwig, and others, and in Berlin under Helmholtz and Hugo Kronecker; he also visited other European countries and centers of learning.

On his second return from Europe, Hall completed his work for the Ph.D. at Harvard, under Bowditch in physiology, then wrote and lectured until 1881, when an offer to give a semipublic lecture series at Johns Hopkins University led to his becoming professor of psychology and pedagogy there. In 1888 he accepted the presidency of Clark University, then being founded in Worcester, Massachusetts, by Jonas G. Clark, who proved to be rich but unpredictable. Hall spent the remainder of his career, until retirement in 1920, at Clark University, where he struggled against the early financial setbacks to establish a graduate and research institution of outstanding faculty. Much of his early effort was lost to the more securely financed fledgling University of Chicago.

Hall was a member of Phi Beta Kappa, received a B.D. degree from Union Theological Seminary in 1871, and was awarded the LL.D. by the University of Michigan (1888), by Williams College (1889), and by Johns Hopkins University (1902).

From his studies in philosophy, physiology, and psychology, Hall began his professional work "intensely impressed with the idea . . . of subjecting psychic processes to the control of scientific and experimental methods" (*Life and Confessions of a Psychologist,* p. 355). This work was directly influenced by G. T. Fechner's work in sensory stimuli, Helmholtz's measurements of visual and auditory responses, and especially by Wundt's experimental studies in physiological psychology. At Johns Hopkins, Hall established the first formal laboratory in psychology in the United States, one which drew to it such brilliant workers as John Dewey, Joseph Jastrow, and James McKeen Cattell. The studies were chiefly aimed at measuring psychic responses precisely. Hall encouraged others in these studies and participated in them somewhat, but his own early enthusiasm in laboratory work waned.

He turned his attention to what is considered his greatest contribution: studies of the mental development of children and adolescents. He adopted the questionnaire method of the German philosopher Moritz Lazarus to determine what children think and know but modified and enlarged his questionnaires to cover all aspects of a child's world, including such diverse subjects as toys, animals, and religion. The questionnaires were used especially in schools in Baltimore, Boston, and Worcester. Hall's pioneering work in this field, especially his early paper "The Contents of Children's Minds" (1883), gave a great impetus to many other studies on the development of children. Hall and his students, both at Johns Hopkins and at Clark, made great use of the 60,000 sheets of child-gathered information on traits of schoolchildren previously accumulated by E. H. Russell. *Adolescence, Its Psychology and Its Relations to Physiology, Anthropology, Sociology, Sex, Crime, Religion and Education* (1904) is considered Hall's most influential publication.

Hall's studies of children led him into pioneering work on educational methods. His approach to education and to teaching was historical, his interest chiefly in the development, or what he preferred to call the evolution, of education. His studies of children convinced him that education, which he considered the salvation of the world, must be adapted to the natures and needs of children, not the reverse. Hall participated in the development and extensive use of psychological and intelligence testing of students, an advance in educational psychology that considerably improved teaching methods. He was tolerant of, but not an advocate of, John Dewey's educational techniques.

A uniform philosophy running throughout all of Hall's publications was evolution. While a student he discovered Darwinism and the writings of many evolutionary philosophers, such as Thomas Henry Huxley, Herbert Spencer, John Tyndall, Ralph Waldo Emerson, and Ernst Heinrich Haeckel; and he developed a personal philosophy of evolution from the beginning of the cosmos to what he considered the ultimate product, man and his soul. His studies of children and his attitudes toward education, psychology, and religion were unified by his belief in a continual evolution. Children, he believed, recapitulated the development of the human race in their development. For many years Hall taught a broad course, which he called psychogenesis, in evolution, and another on the psychology of Christianity.

In psychology Hall was not a proponent of Freudianism—in fact, he was skeptical of its value, though he enjoyed taking psychological tests himself and was psychoanalyzed. His interest in psychology was concerned much more with the normal than the abnormal person.

Hall founded and for many years edited the first American journal in his profession, *American Journal of Psychology* (1887). In 1904 he began *Journal of Religious Psychology* but, to his regret, it survived only eleven years. At Clark University he helped found *Journal of Applied Psychology* (1917).

A pioneer in the application of psychology to education in the United States, Hall in his lifetime saw the profession advance from a scattered handful of workers to a multitude. His influence on education practice was extensive: through his many writings, through even more public speeches, through a great number of students at two universities, and through the emphasis he gave to educational research at Clark University.

BIBLIOGRAPHY

I. Original Works. Of more than 400 publications, Hall's most significant books were *Adolescence . . .* (New York, 1904); *Education Problems* (New York, 1911); *Founders of Modern Psychology* (New York, 1912); *Jesus the Christ in the Light of Psychology* (New York, 1917); *Senescence; the Last Half of Life* (New York, 1922); and two refreshing autobiographical accounts: *Recreations of a Psychologist* (New York, 1920) and *Life and Confessions of a Psychologist* (New York, 1923). His especially significant article "The Contents of Children's Minds," in *Princeton Review,* **11** (1883), 249-272, was reprinted in several other publications. A full bibliography is in Thorndike (see below).

II. Secondary Literature. *Life and Confessions of a Psychologist* (see above) provides a great deal of biographical material and an explanation of the early problems at Clark University. Other biographical material and an unusual assessment of Hall as a psychologist is in the memorial to him by Edward L. Thorndike in *Biographical Memoirs. National Academy of Sciences,* **12** (1925), 133-180.

Elizabeth Noble Shor

HALL, SIR JAMES (*b.* Dunglass, East Lothian, Scotland, 17 January 1761; *d.* Edinburgh, Scotland, 23 June 1832), *geology, chemistry.*

James Hall was the son and heir of Sir John Hall of Dunglass. He was educated in London at Elin's

Military Academy, Kensington. He succeeded to the baronetcy and a considerable fortune at the age of fifteen, while still at school. Thereafter his education was directed by his guardian and granduncle, Sir John Pringle, president of the Royal Society. He entered Christ's College, Cambridge, in 1777 but left after two years without graduating. He then spent some months in France and about a year in Geneva to further his education. In the autumn of 1781 he entered Edinburgh University and during the sessions 1781–1782 and 1782–1783 attended the lectures of John Robison, professor of natural philosophy, and Joseph Black the chemist, taking a particular interest in the latter. Later in 1783 he set out on a grand tour of Europe lasting three years. During this tour he met a number of European scientists and became actively interested in geology and chemistry. He investigated volcanic activity in Italy, the Lipari Islands, and Sicily. He then spent several months in Paris, studying the new chemical ideas of Antoine Lavoisier, to which he became a convert after meeting Lavoisier himself on a number of occasions.

Hall returned to Edinburgh in 1786 and on 9 November of that year married Lady Helen Douglas, second daughter of the earl of Selkirk. For the rest of his life he lived either at his country estate at Dunglass or in Edinburgh, where he took part in the social and scientific activities of the city. Hall was among the earliest British chemists to accept Lavoisier's new chemical views. In 1788 he addressed the Royal Society of Edinburgh on the subject and is said to have made several converts.

In 1791 Hall revisited Paris and renewed his friendship with Lavoisier. He was elected fellow of the Royal Society of Edinburgh in 1784 (and became president in 1812) and fellow of the Royal Society (London) in 1806. He served as member of parliament for a Cornish constituency, Michael Borough, from 1807 to 1812 but did not play a very active part in politics.

Hall is remembered chiefly for the experimental work he carried out to counter certain criticisms of James Hutton's *Theory of the Earth,* although he made other important contributions to geology. His first reaction to the *Theory* had been unfavorable, and it was only after numerous conversations with Hutton that he was persuaded to accept most of its fundamental principles. Once convinced, he provided strong support for Hutton, not only by his experiments but also by field observations.

Hall's first experiments were undertaken to refute the claim that if igneous rocks had originated as molten masses injected into overlying strata, they would be found to occur naturally as masses of glass, not as crystalline rocks. This claim was made because earlier experimenters had found that artificially melted basalt and similar rocks, when cooled, formed a glass. Hall had read accounts of René Réaumur's experiments on porcelain; and in a local glass foundry he had noticed that a mass of molten glass which had been allowed to cool slowly had congealed to a stony mass containing some crystals. He conceived the idea that igneous rocks, if they had cooled slowly, as seemed probable under natural conditions, would form crystalline rocks rather than a glass. Not long after Hutton's death in 1797 he carried out a series of experiments to prove this.

Hall melted specimens of intrusive and extrusive basalt ("whinstone and lava") from fifteen British and foreign localities and allowed the fused masses to cool slowly. The cooled melts he obtained were stony masses, sometimes containing obvious crystals; but none, so far as is known, resembled at all closely the rocks from which they had been derived. Some of his cooled melts have been preserved, and subsequent microscopic examination has shown that one of them contained small crystals of feldspar, augite, olivine, and iron ore, minerals characteristic of the rocks used in the experiments. Hall's results were sufficiently convincing to prove that fused basalt does not necessarily cool to a glass; but he had not taken into account the fact that, under natural conditions, igneous rocks take very much longer to cool than the time that he had allowed.

The second criticism dealt with by Hall was the obvious one that if the consolidation of limestones had been effected in the manner Hutton supposed—by the action of subterranean heat—they would have decomposed with loss of carbon dioxide. Hutton had in fact suggested that this would not happen if the limestones were heated under great pressure, such as that which would be exerted by an overlying mass of seawater. Hall proved this experimentally.

The task proved extremely difficult, but Hall showed great determination and remarkable experimental skill in bringing his investigation to a satisfactory conclusion. Between 1798 and 1805 he carried out more than 500 separate experiments. It was a classic case of proceeding by trial and error. No apparatus suitable for his purpose existed, and Hall had to design and construct his own. His method was to insert small weighed amounts of various types of limestone or carbonate of lime into a tubular container. Among many difficulties he encountered, the principal ones were the selection of suitable material for the container (which had to be nonporous and capable of withstanding both high temperature and high pressures) and the devising of an effective method of sealing the container after inserting the carbonate of lime.

Hall used Wedgwood pyrometers to regulate the temperature and related the Wedgwood scale to the melting point of silver. On this basis it seems probable that he attained temperatures in excess of 1000°C. To estimate the pressures reached, he adapted to his purpose a method devised by Count Rumford to measure the explosive power of gunpowder. He converted his results to a figure significant in relation to Hutton's theory, the highest pressure obtained being equivalent to a column of seawater 2,720 meters in height.

Hall certainly proved that limestone can be heated to high temperatures under high pressure without suffering decomposition. In the most successful of his experiments the loss in weight of the heated limestone was insignificant. It is probable that in some experiments he produced crystalline marble. He also claimed to have fused limestone; recent research suggests that possibly he may have done so, but this is uncertain. It was many years before Hall's experiments were repeated successfully, and his results aroused great interest in Europe. Some of the apparatus he used and the end products of his experiments on basalt and limestone are now in the Geological Museum in London and in the British Museum (Natural History).

Hall made other geochemical experiments of less importance, but his propensity for experiment found expression in another direction in his paper "On the Vertical Position and Convolutions of Certain Strata" (1815). He had recognized that the Lower Paleozoic rocks in southern Scotland occur in a series of closely packed folds with approximately vertical limbs, and he conceived the idea that these folds had been formed by lateral pressure. He constructed a machine in which layers of clay, representing strata, when subjected to lateral pressure from opposing directions, reproduced a series of folds closely comparable in appearance with those found in the rocks. Hall suggested that lateral pressures, exerted on partially consolidated sediments during the intrusion of large masses of granite, might have produced folding. In discussing this suggestion he described a number of detailed observations he had made on the junctions of granite and grauwacke in the south of Scotland.

These observations illustrate Hall's skill as a field geologist; but he was candid enough to admit that, so far as the south of Scotland was concerned, they did not support his proposed explanation (although they provided strong support for Hutton's conclusion that granite was an intrusive igneous rock). Hall's suggested explanation of this type of folding would not now be accepted, but it did contain the germ of later theories which accepted lateral pressure as an explanation of folding in strata; and it marked a stage in the advancement of geological thought.

Hall's next contribution to geology, "On the Revolutions of the Earth's Surface" (1815), again demonstrates his powers of observation in the field and records a further attempt to reproduce a geological process experimentally. In this paper Hall discussed certain surface features (later shown to have been caused by the ice sheets that covered Europe in the glacial period) and concluded that they resulted from the passage of enormous tidal waves (tsunamis).

Hall was familiar, partly from personal observation, with the distribution of erratic blocks in Switzerland and elsewhere in Europe. Rather surprisingly, because he once crossed much of Switzerland (including a glacier) on foot, he did not accept Hutton's suggestion that former glaciers might have distributed erratics. In fact, he attempted to disprove it, as far as Switzerland was concerned, although he did suggest that the transport of exceptionally large erratics by a tidal wave might have been assisted if they were embedded in ice.

Hall had also examined numerous manifestations of former glacial action in the region of Edinburgh, including deposits of boulder clay and fluvioglacial gravel, and features such as "crag and tail" and *roches moutonnées.* He had also noted the presence of glacial grooves and striae on the surface of the latter and had made careful records of their direction with a compass, showing they indicated movement in one well-defined direction. These observations led him to conclude that a tidal wave had crossed mid-Scotland in a particular direction, and he suggested that a similar explanation would account for the distribution of erratics in Europe.

Hall had studied the effects of tidal waves recorded in the literature, and his suggested explanation was based at least in part on observed facts. He supposed these tidal waves to have been caused by some deep-seated and powerful submarine manifestation of igneous activity. He attempted to produce a tidal wave by exploding gunpowder under water, but the experiment appears to have done no more than produce a violent upheaval of the water above the explosion.

In this paper Hall indicated certain points of disagreement with Hutton. He was inclined to believe that the elevation of the land had proceeded by occasional discontinuous upheavals, rather than slowly and continuously; and he rejected Hutton's view that the slow and continuous action of rivers accounted for valley systems.

In Hall's last paper, "On the Consolidation of the

Strata of the Earth" (1826), he attempted to prove experimentally that loose sand, heated in concentrated brine, would consolidate into a firm sandstone; but this was perhaps the least successful of his experiments, and the results were unconvincing.

One of Hall's interests was Gothic architecture, which came to his notice during his visits to France. He wrote a book on this subject, in which he expressed ingenious views as to its origin—but these are not now considered sound.

BIBLIOGRAPHY

I. ORIGINAL WORKS. The following papers were all published in the *Transactions of the Royal Society of Edinburgh:* "Sir James Hall on Granite," **3** (1794), pt. 1, History of the Society, 8–12 (a summary of two papers read 4 Jan. and 1 Mar. 1790, at Hall's request not published in full); "On the Origin and Principles of Gothic Architecture," **4** (1798), pt. 2, Papers of the Literary Class, 3–27, also published separately (London, 1813); "Experiments on Whinstone and Lava," **5** (1805), 43–75; "Account of a Series of Experiments, Shewing the Effects of Compression in Modifying the Action of Heat," **6** (1812), 71–185; "On the Vertical Position and Convolutions of Certain Strata, and Their Relations With Granite," **7** (1815), 79–108; "On the Revolutions of the Earth's Surface," *ibid.*, 139–211; and "On the Consolidation of the Strata of the Earth," **10** (1826), 314–329.

Hall also published his two most important papers in William Nicholson's *Journal of Natural Philosophy, Chemistry, and the Arts,* both in abstract and in full, no doubt to secure a wider circulation: "Curious Circumstances Upon Which the Vitreous or the Stony Character of Whinstone and Lava Respectively Depend; With Other Facts. In an Account of Experiments Made by Sir James Hall" (an abstract), **2** (1798), 285–288; "Experiments Upon Whinstone and Lava," **4** (1800), 8–18, 56–65; "Experiments on the Effects of Heat Modified by Compression" (an abstract), **9** (1804), 98–107; "Account of a Series of Experiments, Shewing the Effects of Compression in Modifying the Action of Heat," **13** (1806), 328–343, 381–405; **14** (1806), 13–22, 113–128, 196–212, 302–318—the last five pages are an appendix (not in the *Transactions* paper), a catalog of 31 specimens "shewing the results of Sir James Hall's experiments on the effect of heat modified by compression," deposited by him in the British Museum.

A French translation of the last paper by M. A. Pictet appeared as a separate book: *Description d'une suite d'expériences qui montrent comment la compression peut modifier l'action de la chaleur* (Geneva, 1807).

The Royal Society *Catalogue of Scientific Papers,* III (1866), 137, lists some translations and abstracts in foreign journals.

Hall's unpublished diaries and letters, the principal source of biographical information, are in the National Library of Scotland and the Scottish Record Office, Edinburgh.

II. SECONDARY LITERATURE. V. A. Eyles, "Sir James Hall, Bt. (1761–1832)," in *Endeavour,* **20** (1961), 210–216, illustrates some of Hall's original apparatus and melts of basalt, with microscopic sections of the latter; and his "The Evolution of a Chemist: Sir James Hall, Bt., F.R.S., P.R.S.E. . . . and His Relations With Joseph Black, Antoine Lavoisier, and Other Scientists of the Period," in *Annals of Science,* **19** (1963), 153–182, contains many biographical details of Hall's life. See also J. S. Flett, "Experimental Geology," presidential address to Section C, British Association for the Advancement of Science, Edinburgh, 1921, pp. 1–19, a general account of Hall's experimental work. J. A. Chaldecott, "Contributions of Fellows of the Royal Society to the Fabrication of Platinum Vessels," in *Notes and Records. Royal Society of London,* **22** (1967), 155–172, refers to Hall's use of platinum vessels in his experiments; and his "Scientific Activities in Paris in 1791," in *Annals of Science,* **24** (1968), 21–52, contains extracts from diaries kept by Hall during his visit to Paris in 1791.

V. A. EYLES

HALL, JAMES, JR. (*b.* Hingham, Massachusetts, 12 September 1811; *d.* Bethlehem, New Hampshire, 7 August 1898), *paleontology, geology.*

Hall's father emigrated from England to Boston in 1809, at age nineteen, to avoid the army (the traditional family career). En route he met Susan Dourdain, whom he married in 1810. Their first child, James Hall, Jr., attended Hingham schools but was a frequent absentee because he worked to assist a growing family of three sisters and one brother. His father died in 1836. His mother, of whom he was very fond and who accompanied him in the field during his early work, died in 1859. There is no intimation that he took any interest in the rocks near his boyhood home.

In assisting at Martin Gay's chemistry lectures by setting up apparatus and visual aids, Hall was brought into contact with a renowned coterie of organizers of the Boston Society of Natural History, including Amos Binney, George B. Emerson, Augustus Gould, and D. Humphreys Storer. His youthful veneration for these pioneer scientists strengthened his plans for a life of science. Fortunately, a new school with a novel approach to education had just opened at Troy, New York, under the patronage of Stephen van Rensselaer and with Amos Eaton as director. Students lectured while teachers listened; and fieldwork was an integral part of the curriculum. James Hall came into this progressive educational environment in 1830—after walking 220 miles. He sharpened his talents under the able, heterodox tutelage of Eaton and came into contact with Ebenezer Emmons, professor of mineralogy and chemistry. Hall became a bachelor of natural science with honors (1832) and master of arts with honors (1833).

In 1838 Hall married Sarah Aikin, daughter of a Troy lawyer. They had two daughters and two sons. His wife died in 1895 after many years of religious devotion and marital estrangement.

Persuaded by Eaton and prodded by DeWitt Clinton, the New York legislature in 1836 authorized Governor William Marcy to establish a geological survey in which the state was divided into four districts. Emmons, recently retired from Rensselaer, was given charge of the northern (second) district, with Hall as assistant. In 1837 young Hall was given charge of the western (fourth) district. As his assistants he chose fellow Rensselaer alumni: Eben Horsford, Ezra Carr, and George Boyd. A wealth of invertebrate fossils collected during a five-year survey of western New York afoot and on horseback touched off Hall's sometimes fiery but brilliant career and initiated the most influential and voluminous paleontologic work created in North America—the monumental thirteen-volume *Palaeontology of New York.*

The termination of the Geological Survey in 1843 found Emmons and Hall vying for the privilege of describing and illustrating the collection of ancient invertebrates; Timothy Conrad, the first state paleontologist, had become fatigued with the immensity of the project and readily left New York to resume his own studies of Tertiary fossils. Possibly owing to his youth, superior competence with fossils, or political favor, James Hall was appointed state paleontologist by Governor William C. Bouck in 1843, while Emmons became state agriculturist. Curiously, within the agriculture volume there appeared a lengthy description of the Taconic system—a concept that was an anathema to Hall. Thus began a controversy which endured in modified form, although with reduced animosity, into the mid-twentieth century.

As a collector Hall was unsurpassed. He knew no duplicates; no two specimens of a species seemed precisely alike. His first collection, the basis for volumes I and II of *Palaeontology of New York,* was sold to the American Museum of Natural History. In lieu of sufficient salary, he retained two-thirds of his collections. The sale of these and fossils gathered by his partner-collectors financed the continuation of *Palaeontology.*

In 1857 Hall constructed a roomy brick building (still standing in Lincoln Park, Albany) which served as laboratory until his death. His assistants, artists, and collectors included George B. Simpson (his nephew), Fielding Meek, Robert Parr Whitfield, Richard Rathbun, Orville Derby, Carl Rominger, Ebenezer Emmons, Jr. (one of Hall's most gifted artists), Ferdinand Hayden, Grove Gilbert, Charles Calloway, Charles E. Beecher, Charles D. Walcott, John M. Clarke, and Charles Schuchert. For a time, Hall's two sons James and Charles Edward (Ned) also assisted in the work. Hall criticized and reviewed all these men and taught not only the science but also the art of fossil collecting.

The problem of storing collections led concerned scientists to plead for the establishment of a state museum. Hall was appointed curator in 1865 and its first director in 1871. This was the precursor of the current New York State Museum.

The geological survey of 1836–1837 had established an orderly stratigraphic framework—the New-York system, a term never used outside New York although the component division names gained worldwide recognition. Hall's travels and correspondence with friends and contemporary scientists (including Louis Agassiz, Joachim Barrande, James D. Dana, Eduard Desor, Joseph Henry, Edward Hitchcock, William Logan, Charles Lyell, Jules Marcou, Ferdinand Roemer, Benjamin Silliman, and Eduard de Verneuil) expanded the knowledge of New York rocks and fossils and gained Hall an international reputation.

Hall was active in the establishment of the California, Iowa, Missouri, New Jersey, Ohio, and Wisconsin state geological surveys—in addition to being state geologist of New York (1837–1898). In a letter (1879) to President Rutherford B. Hayes, he urged the appointment of Clarence King as first director of the U.S. Geological Survey and later recommended another director, Hall's protégé, Walcott.

Hall received many honors and awards. He was president of the American Association for the Advancement of Science (1856); one of fifty charter members of the National Academy of Sciences (1863); organizing president of the International Geological Congress at Buffalo (1876); vice-president of the International Geological Congresses at Paris (1878), at Bologna (1881), and at Berlin (1885); first president of the Geological Society of America (1889); and honorary president of the International Geological Congress at St. Petersburg (1897). His awards included the Wollaston Medal of the Geological Society (1858); the Walker Prize of the Boston Society of Natural History (1884); and the Hayden Medal of the Philadelphia Academy of Natural Sciences (1890). In addition to numerous honorary memberships in scientific organizations, Hall was awarded many honorary degrees, including the LL.D. from Hamilton (1863), McGill (1884), and Harvard (1886).

Hall reported (1845) the first Mesozoic forms from the western United States, in collections of Col. John

C. Frémont. He also initiated two basic ideas in geology. In 1857 his presidential address, "Geological History of the North American Continent" (not published until 1883 because of its bizarre ideas), outlined a concept of crustal downfolds at the edges of continents that initially filled with sediments (later termed geosynclines by J. D. Dana), then evolved into mountain chains such as the Appalachians. The principle of isostasy was summarized as compensating responses within the continent to balance these downfoldings.

Picturesque throughout his life, Hall was self-reliant yet eager for enthusiastic aid, domineering yet attentive, irascible yet kind, complimentary of other's achievements yet parsimonious in his acknowledgment of assistance. His health was excellent. He stood erect and sported a thick, snow-white beard that grew high on his ruddy cheeks. With spectacles somewhat askew on a Moorish nose, stovepipe hat, cane, and knee-length coat buttoned at the neck, this round, pompous-looking figure was quite noticeable when he toured the Albany streets in a battered carriage drawn by an old horse.

Hall occupied his high position in paleontology primarily because he refused to be displaced. Repeatedly pestered and taunted by committees appointed not to investigate but to condemn, and maligned by scientific adversaries who were the targets of Hall's barbs, he weathered the onslaught to emerge as America's foremost invertebrate paleontologist. His biographer and successor as state paleontologist, John M. Clarke, said:

> And thus passed from life a very great man, not honoured in his family, not well understood in his own community, not always courteously entreated and appreciated by his scientific contemporaries; but on the other hand winning the admiration and acclaim of those great-minded enough to understand his inflexible purpose and the magnitude of his achievement [*James Hall of Albany,* p. 548].

BIBLIOGRAPHY

I. Original Works. Hall was author or coauthor of 302 scientific works, including the following: *Geology of New York,* pt. 4 (Albany, 1843), the survey of the fourth geological district; *Fremont's Exploring Expedition* (Washington, D.C., 1845), pp. 295–310; *New York State Natural History Survey: Palaeontology,* 8 vols. in 13 (Albany, 1847–1894); *Geological Survey of the State of Iowa,* I, pt. 2, *Paleontology of Iowa* (Albany, 1858), 473–724; "Descriptions of New Species of Crinoidea From the Carboniferous Rocks of the Mississippi Valley," in *Journal of the Boston Society of Natural History,* **7** (1861), 261–328; *Geological Survey of Canada, Figures and Descriptions of Canadian Organic Remains: Graptolites of the Quebec Group* (Montreal, 1865); *Geological Survey of the State of Wisconsin 1859–1863: Palaeontology,* pt. 3, *Organic Remains of the Niagara Group and Associated Limestones* (Albany, 1871); *Geological Survey of Ohio,* II, pt. 2, *Palaeontology, Description of Silurian Fossils* (Columbus, 1875), 65–161, written with R. P. Whitfield; "The Fauna of the Niagara Group in Central Indiana," in *Twenty-Eighth Annual Report. New York State Museum of Natural History* (Albany, 1879), 99–203; Clarence King, ed., *U.S. Geological Explorations of the Fortieth Parallel,* IV, pt. 2, *Palaeontology,* written with R. P. Whitfield, (Washington, 1877), 199–302; "Contributions to the Geological History of the American Continent," in *Proceedings of the American Association for the Advancement of Science,* **31** (1883), 29–71; and *A Memoir on the Paleozoic Reticulate Sponges Constituting the Family Dictyospongidae,* New York State Museum Memoir no. 2 (Albany, 1898).

II. Secondary Literature. On Hall or his work, see John M. Clarke, *James Hall of Albany* (Albany, 1921); George P. Merrill, *The First One Hundred Years of American Geology* (New Haven, 1924), esp. pp. 230–237; and John J. Stevenson, "Memoir of James Hall," in *Bulletin of the Geological Society of America,* **10** (1899), 425-451, with full bibliography.

Donald W. Fisher

HALL, MARSHALL (*b.* Basford, near Nottingham, England, 18 February 1790; *d.* Brighton, England, 11 August 1857), *physiology, clinical medicine.*

Hall's father, Robert, a successful Wesleyan cotton manufacturer, was the first to use chlorine gas on a large scale for bleaching cotton; nothing is known of his mother, other than that she was eighty-four when she died. After general education until the age of fourteen at the Nottingham Academy, Hall studied chemistry and anatomy at Newark. In October 1809 he entered the Edinburgh University Medical School, where he was graduated Doctor of Medicine with distinction three years later. His ability was further recognized by an appointment to the much-coveted post of resident medical officer at the Edinburgh Royal Infirmary, which he held for two years. The customary Continental tour (1814–1815) allowed him to visit the medical schools of Paris, Göttingen, and Berlin; he traveled alone and on foot from Paris to Göttingen.

From 1816 to 1826 Hall practiced medicine in Nottingham, where he built up a large practice and was elected an honorary physician to the Nottingham General Hospital on 12 October 1825. His reputation as a physician was established by means of his clinical acumen and ability, as well as by his 1817 book on diagnosis, then a new topic. His fame also rested on his advocacy of diminished bleeding, based on the revolutionary statistical analyses of the French physician P. C. A. Louis.

In 1826 Hall moved to London, where he stayed for the rest of his professional career. Although he occasionally lectured at medical schools, he was never on the staff of a hospital, as were other men of comparable clinical caliber. He conducted his large private practice from his home, where he also carried out his experimental work. Hall was elected a fellow of the Royal Society in 1832 but, although he served on its council, he received none of its honors. In 1841 he was made a fellow of the Royal College of Physicians of London and delivered there the Gulstonian and Croonian lectures. He retired from practice in 1853 and died four years later of an esophageal stricture; he was survived by his wife, whom he had married in 1829, and by his son, also named Marshall Hall, who became a famous barrister. The Marshall Hall Fund provided until 1911 a prize every five years for the best work done in the anatomy, physiology, and pathology of the nervous system; recipients include J. Hughlings Jackson, David Ferrier, and C. S. Sherrington.

Opinions differ concerning Hall's personality. Most seemed to find him insufferably conceited and overly aware of his brilliance and capacity for work. He was thus unable to make the usual personal contacts; and since he could not suffer injustice without protest, the resultant rancor and sense of persecution dominated his professional relationships. Consequently he had more detractors and opponents than friends and supporters. Yet the notorious Thomas Wakley, founder and editor of *Lancet,* was a firm friend and supported Hall's claim that the merits of his work equaled those of William Harvey's. Another acquaintance saw him as a courageous, extremely sensitive man who did not deserve the unjustifiable, bitter attacks leveled against his person and his work by those who envied his many talents. His wife's biography of him, as might be expected, is entirely laudatory.

A prolific writer, Hall published over 150 papers and nineteen books; his style was monotonously repetitive, for he was constantly claiming priority, defending, refuting, or attacking. Although aware of the work of others, he did not make adequate reference to it and instead usually emphasized the importance of his own.

Hall's importance lies in his studies on the physiology of reflex function. These began in 1832 and continued for twenty-five years; he claimed that he had spent 25,000 leisure hours on them.

The concept of the reflex has its origins in antiquity, but Hall's work was built upon the advances made in this field by Robert Whytt of Edinburgh, Albrecht von Haller of Göttingen, Georg Procháska of Prague, J. J. C. Legallois of Paris, and many others. By 1830 considerable knowledge existed of the isolated spinal cord and the reflex act, although virtually nothing was known of the underlying morphology. It was Hall's contribution to elaborate the reflex concept, from an isolated action of the cord as he found it, into an established and essential physiological function.

His numerous experiments were carried out on such animals as turtles, hedgehogs, frogs, toads, lizards, and eels, and from them Hall formulated what he considered to be an independent spinal cord system of nerves subserving only reflex function. This mechanism had nothing to do with the nerves of volition and sensation or with consciousness and psychic activity, functions which were mediated by the brain. Thus evolved what he termed "the excito-motory system"—afferent-efferent, in modern terminology. It existed in "the true spinal marrow," whereas the nerves connecting the brain with the body were mediated through "the spinal chord." Reflex activity took place through the spinal marrow, and thus Hall later termed his system "diastaltic"; this and many other terms he introduced have long since been forgotten.

Hall was, therefore, the first to provide a basis for the concept of the neural arc of the spinal cord. Admittedly Charles Bell had hinted at this in 1826 and Hall had made use of the earlier work of François Magendie and of Bell concerning the motor and sensory spinal roots of the cord, but his originality is unassailable. Unfortunately, he paid no attention to the new knowledge of the microscopic appearances of nervous tissue in the 1830's, and he totally ignored the possible influence of such cerebral mechanisms as psychic activity.

Opposition to Hall's concept was immediate and sustained. No doubt some of it was due to his unfortunate personality; but of equal importance was the fact that his reflex system excluded the soul, which many still held to be essential for all functions of the human body. He thus found himself embroiled in theological entanglements as well; the famous controversy centering on the "spinal cord soul" of E. F. W. Pflüger of Bonn lasted until the end of the century. The tenor of the ensuing polemics can be judged from the damning accusation leveled at Hall that he had plagiarized the work of Procháska. This episode has never been fully explained, but although the attack was probably unjustified, it illustrates the kind of reaction Hall engendered among contemporary scientists.

Support in Britain, other than from Wakley, who

crusaded tirelessly in Hall's favor, and from a few others, was small; yet abroad Hall found many powerful protagonists. The great physiologist Johannes Müller of Berlin, whose experiments, completed after those of Hall, were in accord with his, was an ardent supporter; his name is frequently associated with that of Hall in discussing this phase of growing knowledge of spinal-cord reflex activity. The more receptive atmosphere on the Continent may have been due partly to the fact that Hall's personality factors had little influence and partly to the materialism which was then beginning to pervade the physiological laboratories of France and Germany and which eventually brought an end to consideration of the soul in experimental medicine.

The work of Hall and Müller was faulty in detail, yet the basic principle it established was correct and of the greatest importance, even if Hall's contribution has usually been exaggerated. It led naturally to the more significant advances made at the end of the nineteenth century by Sherrington of Oxford, I. M. Sechenov of St. Petersburg, and F. L. Goltz of Strasbourg. Hall must also be credited with illustrating his physiological theories and observations with clinical examples and with making broad applications to diagnosis and treatment. He was often in error—as when, for example, he considered parturition to be a spinal reflex phenomenon—but he nevertheless pioneered an approach which was to increase in importance. Hall also extended the scope of the reflex to sneezing, coughing, and swallowing and described the grasp reflex, although its significance eluded him. Other aspects of reflex function that interested him were the effects on it of such drugs as strychnine and opium.

Hall's insistence that the cerebrospinal axis is a functional segmental series, although not original, was recognized by Sherrington as a significant contribution. Hall noted tonus in skeletal muscle as well as in sphincters, but his conclusion that it is maintained by the "diastaltic arc" was not warranted by his experimental evidence. Whytt had recognized spinal shock almost a century before Hall, who gave the first clear account of it and differentiated it from vascular collapse.

In clinical medicine the critical and scientific approach used by Hall in his experimental studies was often absent. His suggestion that epilepsy was due to irritation of the cervical spinal cord no doubt resulted from his misinterpretation of adversive seizures. His books on diseases of the nervous system were moderately successful but were displaced to some extent by the work of Moritz H. Romberg of Berlin. It is said that Hall coined the term "paralysis agitans."

Hall is remembered for a method of resuscitating the drowned, the Marshall Hall method, which was widely employed until other ways of restarting respiration were introduced by Henry R. Silvester and Edward Sharpey-Schafer. According to the Hall method, the subject was first placed in the prone position and pressed upon the back, causing an active expiration. He was then turned over on his side, with the shoulder raised, to bring about an active inspiration. Hall also perfected a biological test for strychnine.

Hall's versatility and diffuse interests are demonstrated by the wide range of topics, both medical and nonmedical, about which he wrote. In the latter area he published on geometry and Greek grammar and was always ready to use his pen and his tongue to attack social evils. He campaigned for the abolition of slavery in America and of flogging in the British army, as well as for the improvement of sewage disposal and for the safety of railway compartments.

BIBLIOGRAPHY

I. Original Works. There is an incomplete and inaccurate list of Hall's works on pp. 514–518 of his wife's biography (see below).

Hall's main reports on reflex function range from 1832 to 1850. The first is "A Brief Account of a Particular Function of the Nervous System," read at a meeting on 27 Nov. 1832 and reported in part in *Proceedings of the Zoological Society of London,* **2** (1832), 190–192. See also "On the Reflex Function of the Medulla Oblongata and Medulla Spinalis," in *Philosophical Transactions of the Royal Society,* **123** (1833), 635–665, also pub. separately as a repr. (1833); *Memoirs on the Nervous System* (London, 1837); *New Memoir on the Nervous System* (London, 1843); and *Synopsis of the Diastaltic Nervous System: Or the System of the Spinal Marrow, and Its Reflex Arcs; as the Nervous Agent in All the Functions of Ingestion and of Egestion in the Animal Economy* (London, 1850).

Only a few of Hall's clinical publications are worthy of mention: *On Diagnosis. In Four Parts* (London, 1817); *Researches Principally Relative to the Morbid and Curative Effects of Loss of Blood* (London, 1830); *On the Diseases and Derangements of the Nervous System, etc.* (London, 1841); and *Essays on the Theory of Convulsive Diseases, etc.,* Marshall Hall (his son), ed. (London, 1857).

Among several publications on nonclinical and nonmedical topics are *Work on the Thames and the Sewerage of London* (London, 1850); and *The Two-fold Slavery of the United States: With a Project of Self-Emancipation* (London, 1854).

II. Secondary Literature. The following biographical sketches (listed chronologically) are available, but an unbiased critical account of Hall has yet to be written: "Biographical Sketch of Marshall Hall, M.D., F.R.S.," in *Lancet* (1850), **2,** 120–128, probably by T. Wakley, and therefore effusively laudatory; obituary for Hall, *Lancet* (1857), **2,** 172–175, perhaps also by T. Wakley; Charlotte Hall, *Memoirs of Marshall Hall, M.D., F.R.S.* (London, 1861), a detailed work by his widow that tends to be a tedious eulogy of a misunderstood genius; J. F. Clarke, *Autobiographical Recollections of the Medical Profession* (London, 1874), pp. 327–330, also in praise of Hall (in the copy consulted, the owner had added the following revealing marginal comment: "Hall was the most pompous little man I ever met."); G. T. Bettany, in *Dictionary of National Biography,* VIII (1908), 964–967, mostly copied from *Lancet* material noted above; W. Hale-White, *Great Doctors of the Nineteenth Century* (London, 1935), pp. 85–105, a good account of Hall's clinical work; J. H. S. Green, "Marshall Hall (1790–1857): A Biographical Study," in *Medical History,* **2** (1958), 120–133, useful for biographical data but an inadequate account of the content of his writings; and T. J. Pettigrew, *Medical Portrait Gallery,* IV (London, n.d.), with a brief bibliography.

Hall's contribution to the physiology of the reflex is surveyed in "Reviews. A New Memoir on the Nervous System. By Marshall Hall, M.D., London: Baillière 1843," in *Lancet* (1846), **2,** 154–157, 187–189, 244–247, 250. Like the biographical material in *Lancet,* this article is uncritical and mostly effusively complimentary; the author is unknown but it may have been Wakley. The charge of plagiarism from Procháska and Hall's responses are in J. D. George, "Contributions to the History of the Nervous System," in *London Medical Gazette,* **2** (1837–1838), 40–47, 72–73, 93–96, 128, 160, 248–249, 252–254.

The briefest yet most accurate assessment of Hall's work is a passage by C. S. Sherrington, in W. Stirling, *Some Apostles of Physiology* (London, 1902), p. 86. See also D'Arcy Power, "Dr. Marshall Hall and the Decay of Blood-letting," in *Practitioner,* **82** (1909), 320–331; F. Fearing, *Reflex Action. A Study in the History of Physiological Psychology* (London, 1930), pp. 122–145; G. Jefferson, "Marshall Hall, the Grasp Reflex and the Diastaltic Spinal Cord," in E. A. Underwood, ed., *Science, Medicine and History . . . in Honour of Charles Singer,* II (London, 1953), 303–320; and E. G. T. Liddell, *The Discovery of the Reflexes* (Oxford, 1960), pp. 63–76.

EDWIN CLARKE

HALLER, (VICTOR) ALBRECHT VON (*b.* Bern, Switzerland, 16 October 1708; *d.* Bern, 12 December 1777), *anatomy, physiology, botany, bibliography.*

Haller's family had been established in Bern since 1550. He was the fifth and last child of Niklaus Emanuel Haller, a jurist, and Anna Maria Engel. His mother died when he was young, and he was raised by his stepmother, Salome Neuhaus. The family was neither rich nor well-connected and had little political influence. Many of its members were reputed to be nervous, secretive, and eccentric.

Haller received his earliest education from a former pastor. Later, after his father's death, he attended public school in Bern for a year and a half. A child whose health was delicate, he was precocious and gifted in languages. From 1722 to 1723 Haller lived in Biel in the house of his stepuncle Johann Rudolf Neuhaus, a physician who furthered his studies. Among other subjects, Neuhaus tried to instruct Haller in Cartesian philosophy, but Haller rejected it. At this time, however, he began to write poetry and decided to become a physician.

From January 1724 to April 1725 Haller studied medicine at Tübingen, where he learned the fundamentals of botany and anatomy from Johann Duvernoy. The fame of Hermann Boerhaave drew him to Leiden to continue his training; while there he also studied anatomy and surgery with Bernhard Siegfried Albinus. On 23 May 1727, at the age of eighteen, he graduated *doctor medicinae*—with a thesis proving that what had been called a salivary duct by Coschwiz was in reality a blood vessel.

In 1727–1728 Haller made an academic tour of London, Oxford, Paris, and Strasbourg that ended at Basel, where he studied advanced mathematics with Johann I Bernoulli in the spring and summer of 1728. During the following months he made an alpine journey to further his knowledge of botany. At the same time he began the botanical collection that was to form the basis for his massive work on the Swiss flora. His travels ended in Bern, which he left after only a few weeks in response to an invitation to lecture on anatomy at Basel during the winter of 1728–1729. He simultaneously began to conduct independent anatomical investigations.

Haller returned to Bern to practice medicine in 1729. He continued his anatomical studies, enlarged his herbarium, and gave private instruction, but he was unable to obtain a suitable appointment. Finally, in 1736, upon his own application, he was chosen professor of anatomy, surgery, and medicine at the new University of Göttingen.

In 1745, while still at Göttingen, Haller was elected a member of the cantonal council of Bern. Encouraged in his hope of having a political career, he visited Bern in the spring of 1753; he resigned his post at Göttingen and remained in Bern after having been selected for the office of *Rathausammann.* From 1758 to 1764 Haller lived at Roche as director of the Bern saltworks, then returned to Bern permanently. He combined his scientific and literary work with public service and was active both in political and administrative affairs, doing useful work for the school system, for orphans, and as a member of the sanitary council. In addition he codified the common law of the Aigle district and was later president of the Bern Oekonomische Gesellschaft.

Haller married three times; his wives were Marianne Wyss (*d.* 1736), Elisabeth Bucher (*d.* 1741), and Amalia Teichmeyer. Eight children survived to adulthood: two sons and a daughter from his first marriage, and two sons and three daughters from his third. A devoted Zwinglian, he was often tormented by doubts about the profundity of his own belief after the death of his first wife.

Until his thirtieth year Haller suffered constantly from headache; he was later to be plagued by gout, eye pain, dizziness, stomach distress, edema, inflammations of the bladder, a kidney pelvis, and despondency. He fought sleeplessness (probably caused by drinking inordinate amounts of tea) with opium, to which he became addicted—he later published a study of this illness. That he was in earlier times an active mountaineer may be seen from his alpine collecting trips; it was also during this period that he wrote the poetry (especially "Die Alpen," of 1732) that brought him youthful fame. In his old age his weight—238 pounds—hindered his taking even easy mountain strolls.

Haller's contemporaries found his character full of conflicting elements. He could be amiable and entertaining, but trivial matters caused him to lose his temper and become irritable and capricious. In both politics and religion Haller was intolerant and considered every expression of an opposing opinion a personal affront. Efficiency in his work was based on a sense of duty and personal ambition. Most of his co-workers found him petty and antisocial; he himself frequently bemoaned his shortcomings but was unsuccessful in overcoming them.

A list of Haller's honors and memberships in scientific societies may be found in the catalog of the Haller Exhibition, held at Bern in 1877.

As an anatomist Haller was especially influenced by Albinus and Jacques Benigne Winslow; Boerhaave introduced him to botanical and physiological studies. Haller himself considered anatomy and physiology a unit, calling physiology "anatomia animata." In his investigation of the nature of living substance, Haller drew upon Boerhaave's theory of fibers as the basic structural element of the body. He then considered membranes as aggregates of fibers and stated that the cell tissue—which he named the *tela cellulosa*—is constituted of either or both of these units, occurring separately or together. He used this term further to designate the cavities present within the tissue itself, holding that they were interconnected, as could be demonstrated by the injection into them of air or liquid. From Haller's account of the *tela cellulosa* it is apparent that it is in part identical with the loose connective tissue, although it also encompasses larger fiber bundles and muscles, carries the blood vessels, and acts as a sort of packing material to ease the mutual displacement of its parts. It is densified by the storage of fibers or membranes, which in turn shape the organs. For Haller the *tela cellulosa* is also the basis for the tendons, ligaments, cartilage, and bones—as well as the soft tissues and organs—and thereby permits arguments for a general theory of tissues. As will be shown later he went beyond Boerhaave in demonstrating specific functions of muscle and nerve fibers, and by so doing was the first to correlate defined functions with specific structures (Rothschuh, 1953).

Haller successfully employed injection techniques to investigate the distribution of blood vessels in the human body. In preparing his *Icones anatomicae* (1743) he used a decimalized system to number his observations in all cadavers examined. He obtained greater knowledge of the frequency of different variants and used the principle of greatest frequency as the anatomical norm.

Haller's investigations of monsters and deformities led him to observations from which he was later able to make significant generalizations. Although he studied many birth defects with great thoroughness, one example will serve to illustrate the nature of his concerns.

Haller chose to study a pair of premature twins joined at the chest and upper abdomen. The infants shared the organs of this region—the liver, spleen, diaphragm, and heart—but possessed all other organs individually. Most important, each had a separate nervous system and was therefore theoretically capable of expressing his own will; since the single heart received blood from both bodies and redistributed it, thoroughly mixed, to each, Haller considered this evidence that the anima did not reside in the blood, as had been thought previously.

Haller also drew upon the heart shared by both infants to redefine the entire concept of monstrosity. Since the heart was considered to be the first organ formed, it was apparent that the twins represented one body, unified from its beginning, rather than the conjunction of two formerly separated bodies. Therefore, this twinning was not a deformity but perhaps a new type of living creature and a proof of the manner in which divine wisdom can realize new human forms that are complete in their own ways. Haller wondered, then, if the same might be true of all fetuses classified as deformed—if indeed they might not simply be indications of the number and variety of existing forms. He was so certain that these compound forms were a unit that he did not observe that they were unable to survive because the single heart was insufficient for two bodies. Haller thought that malformations were not invariably caused by fortuity but sometimes might be the result of diseases of the fetus; he held that external factors might act upon the fetus only in rare instances. Haller extended his studies of malformations to the question of whether true hermaphrodites could exist among humans (concluding that most instances of disturbed development of the external genitalia were simply occurrences of split urethra in male individuals).

From the comparative anatomical studies of men and animals he adduced that the functions of the parts of the animal body are fully ascertainable only if a complete description of all visible details of a number of species is available. From comparison it may be demonstrated that structures common to a number of species have a common function; if a species shows a divergent detail in a structure analogous to those found in other species, however, that structure may be expected to perform a specific function.

Haller often combined his anatomical observations with studies of physiological function, particularly in his important work on the heart, in which he examined both structure and activity exhaustively. His researches enabled him to demonstrate, for example, that there are more veins that open from the walls of the atria and their auricles into these cavities than A. C. Thebesius and Raymond Vieussens had realized. He characterized the atrioventricular valves as folds in the endocardium continuous to an atrioventricular ring. He described exactly the alternate contraction of the atria and the ventricles, based on his observation of the changing shape and color of the parts of the heart during different phases of its activity, and recognized that the coronary vessels fill up during the systolic contraction of the ventricles (which he confirmed experimentally).

He rejected earlier notions of the self-regulation of the heart. In opposition to Boerhaave, he stated that the flaps of the aorta do not cover the exit openings of the coronary vessels and therefore have no effect on the regular sequence of heartbeats. Haller chose instead to explain the regular activity of the heart by the pronounced irritability of its muscles, stimulated by the filling of each section with blood. By his thesis, the muscles of the veins entering the atria of the heart drive blood into these cavities and cause them to contract. The blood then enters the ventricles; these in turn contract while the atria—which, unstimulated, have relaxed—begin to fill again. This recognition of the mechanical automatism of the heart has been celebrated as one of Haller's most important achievements (Rothschuh, 1953); with it, however, he restated rather than solved the problem, since he still could not account for the intensified irritability of the heart's muscles. (This irritability could be accounted for only later, when it was recognized that the heart possesses a system peculiar to itself whereby impulses are produced and conducted.)

Although Haller was aware of the innervation of the heart by the vagus nerve and the sympathetic trunk, he did not observe any effect of them in his experiments. His investigations of the heart here reached their limit. He was satisfied with his finding, especially when it was shown that the heart can continue to beat for some time after it has been removed from the body; this alone sufficed to refute Stahl, who held that the soul caused the action of the heart.

Haller's studies of the structure of the blood vessels are similarly pioneering and incomplete, and it is perhaps especially striking that he seldom mentioned capillaries as such. He wrote instead of very small arteries that enter very small veins, permitting the passage of only one corpuscle at a time. (Haller observed this, for example, in the loops of fine vessels in a limb richly supplied with veins and arteries.) He also mentioned nets of such small veins and arteries, and thereby acknowledged the theory put forth by Leeuwenhoek and Malpighi concerning a closed circulatory pathway of the blood. At the same time, however, Haller also accepted the idea (based in part on some unsuccessful injection experiments) that tributaries of the small arteries empty into lacunae in the cell tissue, or into the fat and glandular ducts, or end as "evaporation vessels" in the membranes or in the lung.

Haller rejected the notion that red corpuscles can break down into smaller units and denied the existence of vessels of appropriately small size. He thus rejected Boerhaave's conception of yellow globules and smaller entities, although he did not name Boerhaave in this context. Haller characterized lymph vessels as "backward flowing vessels filled with an almost pellucid liquid"; they originate in the cell tissue and are provided with a fine membrane and many small valves. He designated the lymph nodes as glands and considered them a tangle of lymphatic vessels, held together by loose cell tissue; their significance was not clear to him. (Peter Wobmann treats Haller's hemodynamics in detail in "Albrecht von Haller, der Begründer der modernen Hämodynamik" [1967].)

Haller turned to mechanics to define the role of the motion of the blood in the production of heat. He attributed heat to the friction produced by the blood corpuscles rubbing against each other and against the walls of the blood vessels. This friction seemed to Haller to be so strong that his notion of it led him to deny the lentiform blood cells reported—

correctly—by Leeuwenhoek; he thought blood cells subjected to such forces would be rounded off into spheres.

By careful observations Haller ascertained the effects of respiration on the motion of the blood in the veins. He recorded that during inhalation the blood is driven into the heart from the large veins in its near vicinity and in the lungs, thereby easing the flow of the blood (he did not, however, state that the blood is in reality sucked up, like air). During exhalation the blood wells up in the veins of the head, neck, chest, and abdomen, as may be seen most clearly in a dissected brain. Before Haller's investigations these movements of the blood in the brain were interpreted as evidence of contractions of the dura mater, which was thought to pump nervous fluid through the body, in analogy to the distribution of the blood.

Haller's concern with respiration did not end with his studies of its role in the motion of the blood. As early as 1729 he devoted his first independent anatomical researches to the structure of the human diaphragm. In 1733 he published his initial, imperfect work on this subject (imperfect because, for example, the tendons in the lumbar portion of the diaphragm are omitted in the illustration). By 1744, however, Haller had made repeated observations of the diaphragm and was able to supply the first accurate picture of it. He had also meanwhile been conducting experiments on animals to clarify its functions.

Haller was less successful in his interpretation of the role of the intercostal muscles in respiration, since he stated that both of these lifted the ribs and were therefore responsible for inspiration. On this point he became involved in a controversy with Georg Erhard Hamberger, whose functional analysis of the intercostal muscles was more nearly correct. Haller was no more fortunate in his explanation of the cause of the individual breath; unable to adduce anything demonstrable, he held the eliciting cause to be a sense of constriction resulting from the welling up and retention of the blood. Haller was right, however, in his assertion, which he demonstrated experimentally, that the pleural cavity contains no air—a discovery highly significant in understanding the process of breathing.

The results of Haller's investigations of the nervous system were still more important. He was able to demonstrate (against the assertion of Thomas Willis) that the cerebellum is not a primary regulatory mechanism for heart activity and respiration, and likewise refuted the theory that the corpus callosum is the seat of the soul.

The most important aspects of Haller's research were his findings on sensibility and irritability. Although the concept of irritability may be found earlier in the work of Francis Glisson and Giorgio Baglivi, Haller was responsible for its acceptance and wide dissemination. He came to study it through his work on the action of the heart; as early as 1740, in his notes to Boerhaave's lectures, Haller had assumed that the cause of cardiac activity—still unknown—must lie within the structure of the heart itself, and he gradually came to attribute such activity to muscle irritability. He next established that every animal muscle fiber contracts upon stimulation and explained that the continuous function of the vital organs requires a continuous stimulus, even when the animal organs are in a resting state. Relying upon his experimental data, he then designated as irritable all the parts of the human or animal body that contract on external contact. He classified his data on a scale graduated from highly irritable (parts that reacted to slight contact) to slightly irritable (parts that required a strong external impression).

What Haller understood as irritability is, then, identical with the contractibility of muscle fibers; it can be provoked by mechanical, thermal, chemical, or electrical stimuli. The nerves play no intermediate role in this process. Haller expressly stated that the ordinary stimulation of the muscle fibers indeed could not depend on an electrical process in the nerves. Despite such limitations, Haller's experiments and conclusions may well be taken as the basis for modern neurophysiology (Rudolph, 1967).

Haller's observations on the sensibility of parts of the body were derived, like his observations on irritability, from extensive experiments on animals. He viewed irritability and sensibility as independent phenomena and approached sensibility as a property of tissues imbued with nerves. His experimental method was simple; having determined which part of the animal he wished to examine, he stimulated that part in any of a number of ways (ranging from simply blowing on it to applying heat or chemicals or inflicting a mechanical injury, as by cutting or tearing). If the animal responded by showing signs of pain or discomfort, he classified the part in question as sensitive.

Although Haller conducted a great many experiments in an effort to ensure the reliability of his conclusions, some of his interpretations of the data were erroneous. For example, he held that tendons, ligaments, and meninges were insensible and that they therefore contained no nerves—although he later granted that tendons might be provided with very small nerves, from which a weak response might be expected. More important, he declared the periosteum, pleura, peritoneum, intestine, and cornea of the eye to be insensitive.

Despite his occasional misinterpretation of the facts in hand, Haller set an important example for his followers in drawing his conclusions from experiments rather than analogies. Moreover, in extended form, his ideas concerning irritability and sensibility became the basis of a medical system and a buttress of vitalism. Bichat drew upon Haller in his classifications of vital properties.

Haller applied his experimental methods to his studies of embryological development as well. He came to these studies through his investigations of the human gonads (he gave the first correct description of the rete testis, to which the designation *Halleri* is added in his honor). He then took up the chief generative problem of his time: the origin of the new individual. Controversies abounded on many sides—some investigators held the male parent to be the more important in creating the embryo, and others championed the female; the question of spontaneous generation was undecided, and ovists and animalculists argued for completely opposing views. A fundamental opposition also existed between evolutionists (preformationists), who based their theories on the development of an organism already formed in the egg or the sperm, and epigenicists, who asserted the new formation of all parts of the embryo body.

Haller began systematic investigations on hatching chicken eggs and, following the example of Harvey, on mammals. He repeatedly attempted the necessary microscopic observations but met with considerable difficulty, in part because of his poor eyesight and in part because the methods used to prepare embryos for study often resulted in the formation of artifacts. He had particular trouble with the eggs on which he based much of his embryological theory; he was unable to distinguish clearly between the yolk membrane and the yolk sac (Cole, 1930), and he failed to see the infolding process typical of epigenesis (Needham, 1959). He thus became an adherent of the theory of the predominance of the egg and the development of the preformed embryo on stimulation by the sperm—a position that he maintained until the end of his life, even after Kaspar Friedrich Wolff's demonstration (1768) that the intestinal tube of the developing chick emerges by folding out of an originally flat tissue area.

Of his conversion to evolutionism, Haller wrote, "It emerges sufficiently from my writings that I inclined toward the epigenetic theory, which seemed to me to agree better with the appearances. The latter, however, are so complicated and the evidence on both sides so disparate, that I hope for complete forbearance if I go over to the opposed opinion of the evolutionists" (*Operum anatomici argumenti minorum,* II [1767], 406–407). His authoritative adherence to this theory presented an obstacle to the further development of embryology for some time afterward.

Although he chose to pay allegiance to an older and already unsatisfactory general theory, Haller made important specific discoveries in embryology. He was thus able to correct an error of Malpighi, who thought that he had observed a passageway connecting the right and left ventricles in the embryonic chick heart. He was further able to refute Jean Mery by showing that the blood flows from right to left through the foramen ovale in the interatrial septum; and he perceived correctly that the branches of the umbilical vein leading into the fetal liver correspond to the branches of the portal vein in later development. Although he based his theory of ossification on the false assumption that the cartilage is directly transformed into bone, he described with great accuracy the vascular system that supplies the bones. Indeed, Haller's views on ossification were accepted until the advent of cell theory, although his finding that the periosteum—by which later workers (see especially Rita Schär) have established that he meant only its fibrous layer—has no bone-forming properties was frequently misunderstood.

Haller's most important finding in embryology again shows his statistical bias; he was able to devise a numerical method to demonstrate the rate of growth of the fetal body and its parts. By this quantitative determination he showed that fetal growth is relatively rapid in its earlier stages but that the tempo gradually decreases. These observations were entirely new and remain fundamentally correct (Needham, 1959). Their significance seems to have eluded Haller, however, since he does not mention them in a list of his own original anatomical and physiological discoveries.

Despite his considerable effort in anatomy and physiology, Haller did not neglect the botanical observations that had likewise occupied him from his early years in Basel. Characteristically, he set himself to create a complete, encyclopedic science of Swiss flora. He thus addressed himself to the most important botanical problem of the time—a comprehensive nomenclature. Unlike some of his contemporaries, Haller rejected the idea of the constancy of species; he would accept a plant species as such only after he had compared a large number of typical examples in order to establish the degree of their potential variability. To the same end he also studied cultivated plants and controlled stages of development. Nonetheless, he had difficulty in placing related species within families, and he remained entangled in verbose descriptive names.

Although they never worked together, Haller discussed with Linnaeus the problem of a natural system for botanical classification. Indeed, Haller's traditional orientation led him to reject the simplicity of Linnaeus' sexual system of classification and the resulting binary nomenclature. In his independent search for some other system, Haller did meet with some success, particularly with his work on the cryptogams, published in his *Enumeratio methodica stirpium Helvetiae indigenarum* (1742), which was recognized even by Linnaeus.

Nor did Haller neglect his herbarium, which was enriched by specimens sent by numerous correspondents. Preserved in the Muséum National d'Histoire Naturelle in Paris, this Swiss collection continues to aid science. Other of Haller's plant collections have been maintained in Göttingen, where they have been recently examined and rearranged.

After completion of his collection of plants at Roche, Haller published a new edition of his major treatise on Swiss flora, *Historia stirpium indigenarum Helvetiae inchoata* (1768), which long remained a model study of this thoroughly investigated area. Haller provided a geographical description of the country, together with a survey of the changing vegetation cover as influenced by climate, and illustrated the work with extraordinarily beautiful plates. The book brought him universal recognition, and as a result several plants were named for him.

A later botanical work, the two-volume *Bibliotheca botanica* (1771–1772), is still useful and serves to illustrate yet another of Haller's interests. Throughout his scientific career, Haller thoroughly studied everything that had been published on any given subject; it is therefore natural that he turned his systematizing instincts toward bibliography. His first such work was his annotation of his lecture transcripts of Boerhaave's *Institutiones medicae* (1739). Although used as a kind of textbook of physiology, it soon became outdated and Haller therefore wrote his *Primae lineae physiologiae* (1747), which enjoyed great popularity for many years. His completion of Boerhaave's *Methodus studii medici* (1751) contains many additions of a purely bibliographical nature—a 100-page listing of works on physics, a meager fifteen pages of works on chemistry, ninety-five pages of references on botany and pharmacy, and more than 300 pages of literature on anatomy and physiology. Haller here maintained the arrangement used by Boerhaave but was characteristically unsatisfied with it.

After returning to Bern, Haller began an eight-volume handbook of physiology, *Elementa physiologiae corporis humani* (1757). Certainly not the least important source of his immense knowledge of scientific literature had come from his numerous book reviews, especially for the *Göttingische Zeitungen von gelehrten Sachen,* a monthly journal which became well known under his direction (1747–1753) and to which he also contributed regularly in later years.

Haller planned a vast, comprehensive *Bibliotheca medica.* The completed parts list more than 50,000 titles; those marked with a small star were contained in Haller's own library, which is thus easily reconstituted. The works also comprise a number of brief biographical notes on the authors listed and cite historically interesting relationships between authors and between works. That Haller occasionally referred to the unified work as his *Historiae* indicates the point of view from which he planned to compose it.

In his old age Haller also turned to fiction and wrote three philosophical romances—*Usong* (1771), *Alfred* (1773), and *Fabius und Cato* (1774)—in which he drew upon his political experience and expounded his ideas of government. He also wrote on theology; in particular he defended Christianity and polemicized against atheism.

The many controversies which accompanied Haller's literary and scientific work throughout his life are intentionally omitted here. Although they may have been necessary to spur on the completion of his studies, they brought him much trouble and distress for he was unable to bear another's error in silence.

BIBLIOGRAPHY

I. ORIGINAL WORKS. A bibliography of Haller's works is Susanna Lundsgaard-Hansen-von Fischer, "Verzeichnis der gedruckten Schriften Albrecht von Hallers," *Berner Beiträge zur Geschichte der Medizin und der Naturwissenschaften,* no. 18 (1959).

His most important works are *Hermanni Boerhaave Praelectiones academicae in proprias institutiones rei medicae edidit . . .,* 7 vols. (Göttingen, 1739–1744); *Enumeratio methodica stirpium Helvetiae indigenarum* (Göttingen, 1742); *Icones anatomicae quibus praecipuae aliquae partes corporis humani delineatae proponuntur et arteriarum potissimum historia continuatur,* 4 fascs. (Göttingen, 1743–1754); *Primae lineae physiologiae in usum praelectionem academicarum* (Göttingen, 1747 and later eds., of which the 1786 ed. is repr. with a new intro. by Lester S. King, New York–London, 1966); *Opuscula botanica* (Göttingen, 1749); *Opuscula anatomica* (Göttingen, 1751); *Hermanni Boerhaave Methodus studii medici emaculata et accessionibus locupletata ab Alberto ab Haller* (Amsterdam, 1751); "De partibus corporis humani sensilibus et irritabilibus," in *Commentarii Societatis Regiae Scientiarum Gottingensis,* **2** (1753), 114–158, later separately pub., of which an English trans. of the 1755 ed. with intro. by Owsei Temkin is in *Bulletin of the Institute of History of Medicine,* **4** (1936), 651–699; *Elementa physiologiae corporis humani,* 8 vols. (Lausanne, 1757–1766); *Opera minora emendata, aucta, et renovata,* 3 vols. (Lausanne, 1763–1768); *Historia stirpium indigenarum Helvetiae inchoata,* 2 vols. (Bern, 1768); *Bibliotheca botanica,* 2 vols. (Zurich, 1771–1772); and *Bibliotheca anatomica qua scripta ad anatomen et physiologiam,* 2 vols. (Zurich, 1774–1777).

Haller's MS remains are listed in Letizia Pecorella Vergnano, *Il fondo Halleriano della Biblioteca Nazionale Braidense di Milano. Vicende storiche e catalogo dei manoscritti,* no. 8 in the series Studi e Testi, Istituto di Storia della Medicina (Milan, 1965), which includes references to MSS in the civic library of Bern and in the library of the University of Pavia.

Editions of Haller's correspondence are Erich Hintzsche, ed., *Albrecht von Haller-Giambattista Morgagni Briefwechsel 1745–1768* (Bern, 1964); *Albrecht von Haller-Ignazio Somis Briefwechsel 1754–1777* (Bern–Stuttgart, 1965); and *Albrecht von Haller-Marcantonio Caldani Briefwechsel 1756–1776* (Bern–Stuttgart, 1966); and Henry E. Sigerist, "Albrecht von Hallers Briefe an Johannes Gesner (1728–1777)," in *Abhandlungen der Königlichen Gesellschaft der Wissenschaften zu Göttingen,* Math.-phys. Kl., n.s. **11** (1923), viii, 576. Editions of the notebooks of Haller's student years are E. Hintzsche, ed., "Albrecht Hallers Tagebuch seiner Studienreise nach London, Paris, Strassburg und Basel, 1727–1728," in *Berner Beiträge zur Geschichte der Medizin und der Naturwissenschaften,* n.s. **2** (1968), and "Albrecht Hallers Tagebücher seiner Reisen nach Deutschland, Holland und England, 1723–1727" in *Berner Beiträge zur Geschichte der Medizin und der Naturwissenschaften,* n.s. **4** (1971).

II. SECONDARY LITERATURE. On Haller and his work see Heinrich Buess, "Zur Entwicklung der Irritabilitätslehre," in *Festschrift für Jacques Brodbeck-Sandreuter* (Basel, 1942), pp. 299–333; Michael Foster, *Lectures on the History of Physiology During the Sixteenth, Seventeenth and Eighteenth Centuries* (Cambridge, 1924); Eduard Frey, "Albrecht von Haller als Lichenologe," in *Mitteilungen der Naturforschenden Gesellschaft in Bern,* n.s. **21** (1964), 1–64; Baldur Gloor, "Die künstlerischen Mitarbeiter an den naturwissenschaftlichen und medizinischen Werken Albrecht von Hallers," in *Berner Beiträge zur Geschichte der Medizin und der Naturwissenschaften,* no. 15 (1958); Ernst Grünthal, *Albrecht von Haller, Johann Wolfgang von Goethe und ihre Nachkommen* (Bern–Munich, 1965); Kurt Guggisberg, "Albrecht von Haller als Persönlichkeit," in *Berner Zeitschrift für Geschichte und Heimatkunde* (1961), pp. 1–12; Erich Hintzsche, "Albrecht Hallers anatomische Arbeit in Basel und Bern 1728–1736," in *Zeitschrift für Anatomie und Entwicklungsgeschichte,* **111** (1941), 452–460; "Einige kritische Bemerkungen zur Bio- und Ergographie Albrecht von Hallers," in *Gesnerus,* **16** (1959), 1–15; "Neue Funde zum Thema: 'L'homme machine' und Albrecht Haller," *ibid.,* **25** (1968), 135–166; and "Boerhaaviana aus der Burgerbibliothek in Bern," in G. A. Lindeboom, ed., *Boerhaave and His Time* (Leiden, 1970), pp. 144–164; Erich Hintzsche and Jörn Henning Wolf, "Albrecht von Hallers Abhandlung über die Wirkung des Opiums auf den menschlichen Körper," in *Berner Beiträge zur Geschichte der Medizin und der Naturwissenschaften,* no. 19 (1962); Erna Lesky, "Albrecht von Haller und Anton de Haen im Streit um die Lehre von der Sensibilität," in *Gesnerus,* **16** (1959), 16–46; B. Milt, "Empirie und das statistisch fundierte biologisch-medizinische Denken in der Geschichte," *ibid.,* **13** (1956), 1–28; Joseph Needham, *A History of Embryology,* 2nd ed. (Cambridge, 1959), pp. 193–204; K. E. Rothschuh, *Geschichte der Physiologie* (Berlin–Göttingen–Heidelberg, 1953), pp. 76–80; G. Rudolph, "Hallers Lehre von der Irritabilität und Sensibilität," in K. E. Rothschuh, ed., *Von Boerhaave bis Berger* (Stuttgart, 1964), pp. 14–34; and "L'irritabilité Hallérienne point de départ de la neurophysiologie," in *Actualités neurophysiologiques,* 7th ser. (1967), 295–319; Rita Schär, "Albrecht von Hallers neue anatomisch-physiologische Befunde und ihre heutige Gültigkeit," in *Berner Beiträge zur Geschichte der Medizin und der Naturwissenschaften,* no. 16 (1958); Irmela Voss, *Das pathologisch-anatomische Werk Albrecht v. Hallers in Göttingen* (Göttingen, 1937); Peter Wobmann, "Albrecht von Haller, der Begründer der modernen Hämodynamik," in *Archiv für Kreislaufforschung,* **52** (1967), 96–128; Carlo Zanetti and Ursula Wimmer-Aeschlimann, "Eine Geschichte der Anatomie und Physiologie von Albrecht von Haller," in *Berner Beiträge zur Geschichte der Medizin und der Naturwissenschaften,* n.s. **1** (1968); and Heinrich Zoller, "A l'occasion du 250e anniversaire de Albrecht von Haller. Quelques remarques sur son oeuvre botanique et ses collections," in *Bulletin du Muséum National d'Histoire Naturelle,* 2nd ser., **30** (1958), 305–312; "Albrecht von Hallers Pflanzensammlungen in Göttingen, sein botanisches Werk und sein Verhältnis zu Carl von Linné," in *Nachrichten der Akademie der Wissenschaften in Göttingen,* Math.-phys. Kl. (1958), 217–252; and "Albrecht von Haller als Botaniker," in R. Blaser and H. Buess, eds., *Aktuelle Probleme aus der Geschichte der Medizin* (Basel–New York, 1966), pp. 461–463.

ERICH HINTZSCHE

HALLEY, EDMOND (*b.* London, England, 29 October 1656[?]; *d.* Greenwich, England, 14 January 1743), *astronomy, geophysics.*

Halley was the eldest son of Edmond Halley, a prosperous landowner, salter, and soapmaker of the City of London. There is doubt about when he was born, and the date given is that accepted by Halley himself. Although his father suffered some loss of property in the Great Fire of London in 1666, he remained a rich man and spent liberally on his son's education, arranging for him to be tutored at home before sending him to St. Paul's School and then, at the age of seventeen, to Queen's College, Oxford. Young Halley showed an early interest in astronomy and took to Oxford a valuable collection of astronomical instruments purchased by his father. Halley's mother died in 1672, the year before he went to Oxford; and after his father's disastrous second marriage ten years later, financial support became rather more restricted. Nevertheless, everything points to Halley's having private means, for although he married Mary Tooke, daughter of an auditor of the Exchequer, in 1682 and thus accepted wider financial liabilities, he was able to pay for the publication of Newton's *Principia* four years later. Halley and his wife had three children: Katherine and Margaret, born probably in 1688, and a son, Edmond, born in 1698. The daughters survived their father but young Edmond, a naval surgeon, predeceased his father by one year; Halley's wife died five years earlier, in 1736. Halley seems to have enjoyed life and to have possessed a lively sense of humor; religiously he was a freethinker and did not consider that the Bible should be taken literally throughout. Indeed, when he was thirty-five, he was considered for the Savilian professorship of astronomy at Oxford, but the appointment went to David Gregory.

A man of great natural diplomacy, at twenty-two Halley dedicated a planisphere of the southern hemisphere stars to Charles II and obtained a royal mandamus for his M.A. degree at Oxford, although he had not resided there for the statutory period. A year later, with the blessing of the Royal Society, of which he had been elected a fellow in 1678, Halley visited Johannes Hevelius at Danzig and, in spite of a forty-five-year difference in age, was able to pacify the older astronomer, who had received severe criticisms about his use of open instead of telescopic sights for the measurement of celestial positions. Again, when Newton was writing the *Principia,* it was Halley who contributed important editorial aid and persuaded him to continue, despite an argument with Robert Hooke about priority. In 1698, when Peter the Great visited Deptford to study British shipbuilding, Halley was his frequent guest, discussing with him all manner of scientific questions; perhaps it was this kind of success that led Queen Anne, in 1702 and 1703, to send him on diplomatic missions to Europe to advise on the fortification of seaports, a subject on which he had already shown himself adept by providing intelligence reports on French port fortifications while surveying the English channel in 1701.

Halley's interests were wide, even for a seventeenth-century savant. He showed a lively concern with archaeology, publishing in 1691 a paper on the date and place of Julius Caesar's first landing in Britain, using evidence from an eclipse of the moon and critically analyzing other accounts; in 1695 he published one on the ancient Syrian city of Palmyra, the ruins of which had been described by English merchants a few years previously. The latter paper aroused considerable interest and stimulated British antiquaries in the eighteenth century to make an exhaustive study. When he was elected to assist the honorary secretaries of the Royal Society in 1685—a paid post that obliged him to resign his fellowship—he was able to broaden his interests further by an extensive correspondence. Halley held this post for fourteen years, during which time he discussed microscope observations by letter with Anton van Leeuwenhoek and, with others, matters that ranged from medical abnormalities and general biology to questions of geology, geography, physics, and engineering, as well as his own more familiar subjects of astronomy and mathematics.

When he became deputy controller of the mint at Chester in 1696, during the country's recoinage, Halley retained his Royal Society office and reported everything of archaeological and scientific interest in the area. From 1685 to 1693 he also edited the *Philosophical Transactions of the Royal Society* with outstanding competence at a formative time in the journal's development. Halley was also fortunate in possessing great practical sense as well as intellectual ability, and he carried out many experiments in diving, designing a diving bell and a diver's helmet that were much in advance of anything available. Reports on the colors of sunlight that he observed at various depths were sent to Newton, who incorporated them in his *Opticks.* Halley also formed a public company for exploiting the bell and helmet by using them for salvaging wrecks; its shares were quoted between 1692 and 1696.

Halley's best-known scientific achievement was a scheme for computing the motion of comets and establishing their periodicity in elliptical orbits. Although he took a particular interest in the bright naked-eye comet of 1680, it was only in 1695, after the publication of Newton's *Principia,* that he was able to begin an intensive study of the movements of comets. The difficulty in determining cometary paths arose because a comet could be seen for only a short time and, in consequence, it was possible to fit a series of curves through the observed positions. A straight line had been favored for a long time, but by the mid-seventeenth century it was generally accepted that the path must be an ellipse, a parabola, or a hyperbola. Newton preferred the parabola, but Halley decided to consider in detail the possibility of an ellipse.

Utilizing this hypothesis that cometary paths are nearly parabolic, he made a host of computations that led him to consider that the bright comets of 1531, 1607, and 1682 were the same object, making a periodic appearance approximately every seventy-five years. Later he also identified this object with the bright comets of 1305, 1380, and 1456. Halley next set about calculating its return and, allowing for perturbations by the planet Jupiter, announced that it should reappear in December 1758. The comet was in fact observed on 25 December 1758, arriving some days later than Halley's calculations had indicated, but in that part of the sky he had predicted. He also believed that the bright comet of 1680 was periodic, taking 575 years to complete an orbit, but in this he was mistaken. Halley's cometary views were published in 1705 in the *Philosophical Transactions,* and separately at Oxford in the same year in Latin and at London in English with the title *A Synopsis of the Astronomy of Comets.* Although this work aroused the interest of astronomers, it was not until the 1682 comet reappeared as predicted in 1758 that the whole intellectual world of western Europe took notice. By then Halley had been dead fifteen years; but his hope that posterity would acknowledge that this return "was first discovered by an Englishman" was not misplaced, and the object was named "Halley's comet." This successful prediction acted as a strong independent confirmation of Newtonian gravitation, and it is often said, but without direct evidence, to have helped dissipate the superstitious dread attached to cometary appearances.

Halley's astronomical contributions were not confined to comets, and he made notable advances in the determination of the distance of the sun, in positional and navigational astronomy, and in general stellar astronomy. Determination of the distance of the sun from the earth was crucial, since a correct evaluation was necessary before the size of the planetary system or the distances of the stars could be determined as direct values. Halley proposed evaluating the distance by observing the transit of Venus across the sun, an idea first sketched by James Gregory in 1663. Halley first assessed the practicability of the idea when he observed and timed a transit of Mercury in 1677. By recording the local time at which Mercury appeared to enter the sun's disk and the time at which it left, and then comparing his results with those made at an observing station in a different latitude, the distance of Mercury was obtained. Using Johann Kepler's third law of planetary motion, the distance from the earth to the sun could be found.

Halley appreciated that greater precision could be obtained by observing a transit of Venus, since it lies nearly twice as close to the earth as Mercury and thus the same percentage of error in timing would result in smaller errors in distance determination. Transits of Venus are rare, and the next were to occur in 1761 and 1769, by which time he would doubtless be dead. Nevertheless, Halley worked out methods of observation and subsequent calculation in considerable detail, publishing his results in the *Philosophical Transactions* for 1691, 1694, and, most fully, 1716. Joseph Delisle, who planned to organize expeditions to observe the 1761 transit, came to London in 1724 and discussed the subject with him; and it was Delisle's arrangements for European observations that at last stimulated British astronomers to take action in June 1760, twelve months before the transit. Delisle had devised a method that was a slight modification of what Halley had proposed and, in June 1761, a total of sixty-two observing stations were in operation. For the 1769 transit a total of sixty-three stations sent in observations and a value of 95 million miles was obtained for the sun's distance, a figure that further analysis subsequently reduced to 93 million. This compares favorably with the present figure of 92.87 million miles, but even 95 million represented a great achievement in the mid-eighteenth century.

Halley began positional astronomy assisting Flamsteed in 1675. He broke this connection when he continued on his own, leaving Oxford in 1676 for the island of St. Helena, off the west coast of Africa at a latitude of sixteen degrees south. Here he cataloged the stars of the southern hemisphere and, incidentally, discovered a star cluster in Centaurus (ω Centauri). He compiled his results in *Catalogus stellarum Australium . . .,* which was published late in 1678 at London; a French translation by Augustin Royer appeared at Paris early in 1679. In addition Halley drew up a planisphere, a copy of which was presented in 1678 to the king. The Royal Society received both catalog and planisphere, and it was primarily on the strength of these that he was elected a fellow.

Halley's other positional work was carried out at Greenwich after he was appointed astronomer royal in 1720, succeeding John Flamsteed. Here he found no instruments, since those used by Flamsteed had been removed, but he immediately obtained financial aid from the government. He established the first transit instrument to be put to regular use and ordered a large mural quadrant that was set up in 1724. He then observed the planets and, in particular, studied the motion of the moon. Halley's observing program for the latter was as bold as it was ambitious, for although he was aged sixty-four when appointed astronomer royal, he set about planning observations to cover a complete saros of eighteen years, after which the relative positions of the sun and moon would be repeated with respect to the nodes of the lunar orbit. He adopted this program because he was convinced, correctly, that once the moon's orbit was really known precisely, the problem of determining longitude at sea would be solved.

Flamsteed had made excellent measurements of star positions and some of the moon, so Halley concentrated on completing a set of lunar observations and, surprisingly enough, was able to finish his self-imposed task. By 1731 he was already in a position to publish a method of using lunar observations for determining longitude at sea that gave an error of no more than sixty-nine miles at the equator, a result that showed a real improvement over previous methods and augured well for even greater precision. Halley's observations were later criticized for their lack of precision; but even if they were not all they might have been, he certainly established the viability of the "method of lunars" as a solution of the longitude problem. It is worth noting, too, that while Halley was astronomer royal he was visited by John Harrison, who explained his ideas for an accurate timepiece. On Halley's personal recommendation, the instrument maker George Graham lent Harrison money to enable him to make a clock for submission to the Board of Longitude and thus develop what ultimately was to prove another successful solution.

Halley's achievements in stellar astronomy were of considerable significance, although they were not as fully appreciated in his day as might have been expected. In 1715 he published a paper on novae, listing those previously observed, making comments, and drawing parallels with long-period variables such as *o* Ceti (Mira), which is sometimes visible to the naked eye and sometimes invisible. In the same year Halley also made known his thoughts on nebulae. A few had been detected with the naked eye but the number had increased after the telescope came into use astronomically. Without a telescope they often looked like stars; with a telescope they were clearly seen to be something different. Halley boldly suggested that they were composed of material spread over vast expanses of space, "perhaps not less than our whole Solar System," and were visible because each shone with its own light, which was due not to any central star but to the "lucid Medium's" behavior. In this explanation Halley anticipated some aspects of the later work of William Herschel and William Huggins.

Halley also studied the question of the size of the universe and the number of stars it contained. The problem was much discussed just then, even by Newton, although he had also stated that the universe was infinite—otherwise gravity would attract all matter to the center. Halley's approach was an observational one, and in 1720 he concluded that since every increase in telescopic power had shown the existence of stars fainter than any hitherto observed, it seemed likely that the universe was to be taken as "actually infinite." There was a physical argument, too, for Halley considered the effects of gravitation on material spread out in a finite part of an infinite space and came to a conclusion similar to Newton's.

One contemporary criticism (revived a few years later by Jean de Chésaux and again in 1823 by H. W. M. Olbers) stated that if the number of stars were infinite, the sky should be bright, not dark, at night: Halley believed that he had resolved this paradox. He calculated that if all the stars were as distant from each other as the nearest (to earth) was from the sun, then, in spite of an increase in numbers, they would occupy ever smaller areas of the sky, so that, at very large distances, their diminished brightness would render them too dim to observe. As a corollary, he pointed out that even when observed with the largest telescopes some stars were so dim that it was to be expected that there were others whose light did not reach us.

There was a fallacy in Halley's argument, for he seems to have confused linear and angular dimensions: star disks do become smaller with greater distance, but the solid angle subtended by the heavens does not. Nevertheless, it was a carefully reasoned attempt to analyze an important problem that was to exercise astronomers for many generations. In a subsequent paper Halley discussed the number of stars to be expected in a given volume of space, assuming a given separation between them, and the way in which their brightness would diminish with distance. In this he anticipated what John Herschel was to discover and express precisely a century later: that stars of magnitude six were 100 times dimmer than those of magnitude one. Again Halley worked out figures that led him to conclude that the most distant stars would still be too dim to be detectable; but whatever the faults in all this work, his methods of attack were new and paved the way for later investigators.

Halley's most notable achievement in stellar astronomy was his discovery of stellar motion. From earliest times the stars had been regarded as fixed, and there seemed no reason to question this assumption. In 1710 Halley, who took a great interest in early astronomy, settled down to examine Ptolemy's writings and paid particular attention to his star catalog. It soon became evident that there were discrepancies, even allowing for precession and observational errors; and Halley rightly decided that the differences between Ptolemy's catalog and those compiled some 1,500 years later were so gross that the only rational explanation was to assume that the stars possessed individual motions. Halley was able to detect such proper motion only in the case of three bright stars—Arcturus, Procyon, and Sirius—but he correctly deduced that others which were dimmer, and could therefore be expected to be further away, possessed motions too small to be detected. It was not until a century and a half later that the study of proper motions could really be extended, but this was due to insufficient instrumental accuracy and not to disregard of Halley's opinion. The limitations of precise measurement in Halley's time also prevented the successful determination of even one stellar distance. Claims to have achieved this were made nonetheless, notably in 1714 by Jacques Cassini, who believed he had obtained an annual parallax for Sirius. In 1720 Halley analyzed this claim, showed that it could not be upheld, and made suggestions for observations which he thought might be successful.

Halley's interest in early astronomy was coupled with an equally great interest in early mathematics; and when he was appointed Savilian professor of geometry at Oxford in 1704, Henry Aldrich, dean of Christ Church, suggested to him that he prepare a translation of the *Conics* of Apollonius. Aldrich made a similar proposal to David Gregory, who held the Savilian chair of astronomy; Halley and Gregory worked on the subject together until the latter's death in 1708, after which Halley carried on alone. Two Latin editions of books V–VII (from Arabic) existed, but since these lacked book VIII Halley used Greek lemmas by Pappus to aid him in his reconstruction of the whole work. The *Conics* had attracted other mathematicians, but Halley aimed at and prepared a definitive edition. He also translated Apollonius' *Sectio rationis* (and restored his *Sectio spatii*) and tracts by Serenus of Antinoeia, publishing these in 1706 and 1710. Oxford University recognized the scholarly achievement by conferring a Doctor of Civil Laws degree, and it is worth noting that his *Conics,* although partially supplanted by J. L. Heiberg's translation of books I–IV (Leipzig, 1891–1893), is still used for the remaining books (V–VII). Halley followed up this work on early mathematics by translating the *Sphaerica* of Menelaus of Alexandria, an elegant translation that has won praise even today; it was published posthumously in 1758.

Halley's mathematical interests were not purely historical: between 1687 and 1720 he published seven papers on pure mathematics, ranging from higher geometry and construction and delimitation of the roots of equations to the computation of logarithms and trigonometric functions. He also published papers in which he applied mathematics to the calculation of trajectories in gunnery and the computation of the focal length of thick lenses. Halley was also one of the pioneers of social statistics, demonstrating in 1693 how mortality tables could be used as a basis for the calculation of annuities, a suggestion that was later pursued by Abraham de Moivre.

Halley was not only an astronomer and mathematician; he was also the founder of scientific geophysics. His first major essay in this field was an important paper on trade winds and monsoons (1686) in which he specified solar heating as their cause, although he was aware that this was not a complete explanation and urged others to pursue the matter. To aid them he produced a meteorological chart of the winds, the first provision of data in such a form, in which he depicted the winds by short broken lines, each dash having a thick front and a pointed tail to indicate direction. He also studied tidal phenomena, in 1684 analyzing information received at the Royal Society about tides at Tonkin; his work on tides culminated in his survey of the English Channel in 1701.

Halley's most significant geophysical contribution was his theory of terrestrial magnetism, on which he published two important papers (1683, 1692); in both he developed his own theory, the second paper providing a physical basis for the proposals made in the first. Halley's suggestion was that the earth possessed four magnetic poles, one pair situated at the ends of the axis of an outer magnetic shell and the other at the extremities of the axis of an inner magnetic core. The shell and core had slightly different periods of diurnal rotation to account for observed variations. He also postulated that the space between core and shell was filled with an effluvium—a favorite theoretical device of the seventeenth century—and in 1716 used it as a basis for his suggestion that the aurora was a luminous effluvium that escaped from the earth and that its motion was governed by the terrestrial magnetic field.

Between 1698 and 1700 Halley was commissioned as a naval captain and, in spite of a mutiny on board, took the small ship *Paramore* across the Atlantic, reaching as far as fifty-two degrees south latitude and the same latitude north. He charted magnetic variation in the hope of using it as a means of determining longitude at sea; but although it proved unsatisfactory for this purpose, his chart, published in different editions in 1701, 1702, and 1703, was significant because it was the first to adopt isogonic lines (called "Halleyan lines" by contemporaries) to connect points of equal magnetic variation.

Halley's scientific attitude toward terrestrial physics led him to take an independent and novel approach to the question of the age of the earth. From investigations he made in 1693 on the rate of evaporation of water, he concluded that the salinity of lakes and oceans must gradually be increasing and suggested that if the rate of increase could be determined, it should be possible to obtain factual evidence about the earth's age. From approximate results Halley suggested that the figure derived from biblical genealogies was too low and that an alternative view, that the earth was eternal, was also incorrect. He further suggested a physical explanation for the Flood, postulating a very close approach of a comet to the earth. Although not now accepted, this was an interesting scientific explanation for a biblical event. These views did not commend him to some powerful ecclesiastics of his day.

Throughout much of his life Halley had to suffer the active disapproval of John Flamsteed, the first astronomer royal, who first encouraged and then turned against him. In 1712, at Newton's request, Halley prepared an edition of Flamsteed's observations using materials deposited at the Royal Society. Their publication as *Historia coelestis* . . . infuriated Flamsteed.

Halley was also involved in the Newton-Leibniz controversy to the extent of lending his name to the report of the supposed committee of the Royal Society which in effect sanctioned Newton's own version of the affair.

Recognition came to Halley early in life, with his M.A. and election to the Royal Society; but after that there was a long pause due, to a great extent, to Flamsteed. Nevertheless, he obtained the Savilian chair of geometry at Oxford in 1704, was appointed astronomer royal in 1720, and was elected a foreign member of the Académie des Sciences at Paris in 1729. At his death in 1743 Halley seems to have been widely mourned, for he was a friendly as well as a famous man and always ready to offer support to young astronomers.

BIBLIOGRAPHY

I. Original Works. For a complete list of Halley's publications, see E. F. MacPike, *Correspondence and Papers of Edmond Halley* (Oxford, 1932), pp. 272–278 (but note that the second item under the year 1700 should be dated 1710). Halley's most important publications in astronomy were "A Direct and Geometrical Method of Finding the Aphelia and Eccentricities of the Planets," in *Philosophical Transactions,* **11** (1676), 683–686; *Catalogus stellarum Australium* . . . (London, 1678); *Astronomiae cometicae synopsis* (Oxford, 1705), also in *Philosophical Transactions of the Royal Society,* **24** (1704–1705), 1882–1889; "An Account of Several Nebulae . . .," *ibid.,* **29** (1714–1716), 354–356; "Considerations of the Change of the Latitudes of Some of the Principal Fixt Stars," *ibid.,* 454–464; "Methodus singularis qua Solis Parallaxis . . . Veneris intra Solem conspiciendæ . . .," *ibid.,* **30** (1717–1719), 736–738; "Of the Infinity of the Sphere of Fix'd Stars," *ibid.,* **31** (1720–1721), 22–24; "Of the Number, Order and Light of the Fix'd Stars," *ibid.,* 24–26; "A Proposal . . . for Finding the Longitude at Sea Within a Degree . . .," *ibid.,* **37** (1731–1732), 185–195; and *Edmundi Halleii astronomi dum viveret regii tabulae astronomicae* . . ., John Bevis, ed. (London, 1749).

His main geophysical writings were "A Theory of the Variation of the Magnetical Compass," in *Philosophical Transactions of the Royal Society,* **13** (1683), 208–221; "An Historical Account of the Trade Winds, and Monsoons . . .," *ibid.,* **16** (1686–1687), 153–168; "An Account of the Cause of the Change in the Variation of the Magnetical Needle; With an Hypothesis of the Structure of the Internal Parts of the Earth," *ibid.,* **17** (1691–1693), 563–578; and "A Short Account of the . . . Saltness of the Ocean . . . With a Proposal . . . to Discover the Age of the World," *ibid.,* **29** (1714–1716), 296–300.

On mathematics and vital statistics they are "An Estimate of the Degrees of the Mortality of Mankind . . .," in *Philosophical Transactions of the Royal Society,* **17** (1693), 596–610, 654–656; "Methodus . . . inveniendi radices aequationum . . .," *ibid.,* **18** (1694), 136–148; *Apollonii Pergaei de sectione rationis* . . . (Oxford, 1706); *Apollonii conicorum libri III, posteriores* . . . (Oxford, 1710); *Apollonii Pergaei conicorum libri octo et Sereni Antissensis de sectione cylindri & coni* . . . (Oxford, 1710); and *Menelai sphaericorum* . . . (Oxford, 1758). Other important papers on mathematics which appeared in the *Philosophical Transactions* are in **16,** 335–343, 387–402, 556–558; and **19,** 58–67, 125–128, 202–214.

His main physics writing is "An Instance of . . . Modern Algebra, in . . . Finding the Foci of Optick Glasses Universally," in *Philosophical Transactions of the Royal Society,* **17** (1691–1693), 960–969. His archaeological paper on Palmyra is in *Philosophical Transactions,* **19** (1695–1697), 160–175.

II. Secondary Literature. Besides MacPike's book mentioned above, there are two biographies: A. Armitage, *Edmond Halley* (London, 1966); and C. A. Ronan, *Edmond Halley—Genius in Eclipse* (New York, 1969; London, 1970).

See also the following articles: E. Bullard, "Edmond Halley (1656–1741)," in *Endeavour* (Oct. 1956), pp. 189–199; and G. L. Huxley, "The Mathematical Work of Edmond Halley," in *Scripta Mathematica,* **24** (1959), 265–273.

Colin A. Ronan

HALLIER, ERNST HANS (*b.* Hamburg, Germany, 15 November 1831; *d.* Dachau, Germany, 21 December 1904), *botany, parasitology.*

Hallier began his studies in 1848 at the botanical garden in Jena, which he left in 1851 for Erfurt, Charlottenburg, and Berlin. In 1855 he commenced botanical and philosophical studies at the universities of Berlin, Jena, and Göttingen. He received his doctorate at Jena in 1858 and became an assistant to Matthias Schleiden. Following the completion of his dissertation, *De geometricis plantarum rationibus* (1860), he was appointed assistant professor. He published numerous works on the relationships between plant parasites and human health.

According to Hallier, fungi were the causative agents of cholera, exanthematous typhus, typhoid, measles, smallpox, gonorrhea, syphilis, and other diseases. These fungi supposedly took various forms: *Leptothrix, Mycothrix, Micrococcus, Cryptococcus, Arthrococcus,* and so on. For example, the fungus responsible for syphilis, *Coniothecium syphiliticum,* could be found in the shape of *Cladosporium, Mucor, Penicillium,* and *Micrococcus.* The cocci of cowpox would yield the fungus *Eurotium herbariorum;* furthermore, the micrococci of enteric fever (typhoid) supposedly constituted a stage of *Rhizopus nigricans* and those of gonorrhea a stage of *Coniothecium.*

Hallier isolated these microorganisms from human pathological fluids by means of his isolation device ("Isolir-apparat") and placed them in culture media ("Cultur-apparat"). Yet he did not take sufficient precautions, and all his cultures were in reality infested with the spores of common molds (*Penicillium, Aspergillus*) from the air. Oskar Brefeld wittily summarized Hallier's research with the statement "From it emerge only nonsense and *Penicillium glaucum.*"

Hallier's assertions were quickly criticized by such contemporary scientists as the mycologist Anton de Bary (1868), with whom he conducted a polemic, and the bacteriologist Ferdinand Cohn (1872), who showed that Hallier's culture experiments were without the slightest scientific value. His works were quickly forgotten, and in the last years of his life he devoted himself to the study of aesthetics.

Hallier's work is only of historical interest today. His sole merit is having been one of the first to maintain that infectious diseases are due to pathogenic microorganisms—which he did not succeed in isolating. In 1869 he founded an important journal, *Zeitschrift für Parasitenkunde,* which is still published.

BIBLIOGRAPHY

I. ORIGINAL WORKS. Hallier's writings include *De cycadeis quibusdam fossilibus in regione Apoldensi repertis* (Jena, 1858), his botanical thesis; *De geometricis plantarum rationibus* (Jena, 1860), his philosophical thesis; "Über einen pflanzlichen Parasiten auf dem Epithelium bei Diphteritis," in *Botanische Zeitung,* **23** (1865), 144–146; "Über *Leptothrix buccalis,*" *ibid.,* 181–183; *Die pflanzlichen Parasiten des menschlichen Körpers* (Leipzig, 1866); *Das Cholera-Contagium. Botanische Untersuchungen Aerzten und Naturforschern mitgetheilt* (Leipzig, 1867); *Gährungserscheinungen. Untersuchung über Gährung, Fäulniss und Verwesung mit Berücksichtigung der Miasmen und Contagien sowie der Desinfection* (Leipzig, 1867); *Parasitologische Untersuchungen bezüglich auf die pflanzlichen Organismen bei Masern, Hungertyphus, Darmtyphus, Blattern, Kuhpocken, Schafpocken, Cholera nostras . . .* (Leipzig, 1868); *Phytopathologie. Die Krankheiten der Culturgewächse* (Leipzig, 1868); "Researches Into the Nature of Vegetable Parasitic Organisms," in *Medical Times and Gazette,* **2** (1868), 222–223; "Über die Parasiten der Ruhr," in *Zeitschrift für Parasitenkunde,* **1** (1869), 71–75; "Die Parasiten der Infectionskrankheiten," *ibid.,* 117, 191; **2** (1870), 67, 113; **3** (1872), 7, 157; **4** (1873), 56; "Beweis dass der *Micrococcus* der Infectionskrankheiten keimfähig und von höheren Pilzformen abhängig ist und Widerlegung der leichtsinnigen Angriff des Herrn Collegen Bary zu Halle," *ibid.,* **2** (1870), 1–20; "Beweis dass der *Cryptococcus* keimfähig und von höheren Pilzformen abhängig ist und Widerlegung der Ansichten der Bary'schen Schule über die Bierhefe," *ibid.,* **3** (1872), 217–244; and *Die Parasiten der Infectionskrankheiten bei Menscher, Thieren und Pflanzen.* I, *Die Plastiden der niederen Pflanzen* (Leipzig, 1878).

II. SECONDARY LITERATURE. The only biographical material consists of a few short accounts in various biographical dictionaries. Hallier's work is discussed in W. Bulloch, *The History of Bacteriology* (London, 1938), pp. 178, 188–192, 195, 198, 219, 291, 321–322, 371; and C. J. Clemedson, "*Penicillium syphiliticum* och några andra teorier om syfilis orsak och uppkomst," in *Medicinhistorik årsbok* (1968), pp. 158–170, which discusses Hallier's ideas on syphilis.

JEAN THÉODORIDÈS

HALLWACHS, WILHELM LUDWIG FRANZ (*b.* Darmstadt, Germany, 9 July 1859; *d.* Dresden, Germany, 20 June 1922), *physics.*

Hallwachs was one of the pioneers of modern physics. An experimental physicist, he laid the foundations for research on photoelectric processes. He received his education at the universities of Strasbourg and Berlin, earning his doctorate in 1883 under A. A. Kundt at Strasbourg. He was then an assistant to Friedrich Kohlrausch at Würzburg from 1884 to 1886 and, from 1886 to 1888, to G. H. Wiedemann at Leipzig, where he qualified as lecturer in 1886. In 1888 he again became an assistant to Kohlrausch, this time at Strasbourg, and married the latter's daughter Marie in 1890. He became professor of electrical engineering in 1893 and of physics in 1900 at the Technische Hochschule in Dresden. During his teaching career he introduced and fostered the study of engineering physics.

Hallwachs constructed electrical measuring devices and built, among other things, a quadrant electrometer and a double refractometer of high precision. At Leipzig in 1888 he investigated, following the model of Heinrich Hertz's studies, photoelectric activity, establishing that through absorption of ultraviolet light, negatively charged metal plates discharge and uncharged metal plates become positively charged. This process, which is called the photoelectric effect or Hallwachs effect, forms the basis for the physics of the photoelectric cell and was theoretically interpreted in 1905 in Einstein's work on light quanta.

Hallwachs interrupted his photoelectric investigations and, beginning in 1890, worked with Kohlrausch on electrolytic questions. After 1904, when he was able to return to photoelectricity, he determined the value of the photoelectric work function, photoelectric fatigue, and related phenomena. He became the leading expert in this field, summarizing its development to 1914 in his treatise "Die Lichtelektrizität."

BIBLIOGRAPHY

I. ORIGINAL WORKS. A bibliography is in Poggendorff, IV, 572, and V, 489, and in Wiener's obituary (see below). Hallwachs wrote about 50 scientific papers. The most important are "Ueber den Einfluss des Lichtes auf electrostatisch geladene Körper," in *Annalen der Physik und Chemie,* n.s. **33** (1888), 301–312; "Ueber die Electrisierung von Metallplatten durch Bestrahlung mit electrischem Licht," *ibid.,* n.s. **34** (1888), 731–734; "Ueber den Zusammenhang des Electricitätsverlustes durch Beleuchtung mit der Lichtabsorption," *ibid.,* n.s. **37** (1889), 666–675; and "Die Lichtelektrizität," in Erich Marx, ed., *Handbuch der Radiologie,* III (Leipzig, 1916), 245–563. Hallwachs was the coeditor, with A. Heydweiller, K. Strecker, and O. Wiener, of F. Kohlrausch, *Gesammelte Abhandlungen,* 2 vols. (Leipzig, 1910–1911).

II. SECONDARY LITERATURE. The main biography is O. Wiener, "Wilhelm Hallwachs," in *Berichte über die Verhandlungen der Sächsischen Akademie der Wissenschaften zu Leipzig,* Math.-phys. Kl., **74** (1922), 293–316, with bibliography on 313–316; abridged version in *Physikalische Zeitschrift,* **23** (1922), 457–462. See also A. Hermann, in *Neue deutsche Biographie,* VII (1966), 565–566. For information on Hallwachs' work, see Erich Marx, "Anhang: Entwicklung der Lichtelektrizität von Januar 1914 bis Oktober 1915," in Marx's work cited above.

HANS-GÜNTHER KÖRBER

HALM, JACOB KARL ERNST (*b.* Bingen, Germany, 30 November 1866; *d.* Stellenbosch, Union [now Republic] of South Africa, 17 July 1944), *astronomy.*

Halm attended the Gymnasium at Bingen, then studied mathematics at the universities of Giessen (1884–1886), Berlin (1886–1887), and Kiel (1887–1889), where he took his doctorate with a dissertation on homogeneous linear differential equations. From 1889 he was assistant at the astronomical observatory in Strasbourg. In 1895 Halm was invited to become assistant astronomer of the Royal Observatory at Edinburgh; in 1907 he became chief assistant at the Cape Observatory, Cape Town, from which post he resigned in 1926. He was married in 1894 to Hanna Bader; they had one son and two daughters.

Halm's method of attacking astronomical questions was to make observations and then try to find a convincing explanation for them. A good example of this was his work at Edinburgh on the rotation of the sun. He used spectroscopic measurements of the radial velocity of the solar limb to determine the velocity of rotation in various heliographic latitudes. His results agreed fairly well with observations made previously at Uppsala—and with those made later at Mt. Wilson in California. But a careful analysis of his results led Halm to the hypothesis that the velocity of the solar rotation is slightly variable in the course of the eleven-year cycle of the variation of sunspots. The minuteness of the postulated variation makes it impossible even now to decide whether his hypothesis is correct.

At the Cape Observatory, Halm conducted extensive research on stellar statistics. He examined a great number of radial velocities of fixed stars and concluded that there was a systematic streaming of those stars which astronomers usually call B-type stars. By studying the distribution of the fixed stars in space he found indications that there was absorbing matter in the galactic system. The relation between a star's mass and luminosity that he suggested was later well established by A. S. Eddington. Halm also did very intensive work to create good standard sequences to be used for photographic photometry in the southern sky.

Halm did much to make astronomy popular in South Africa, assisting amateur astronomers in their work and publishing *A Universal Sundial* (1924), a little book on the construction of sundials. He was also an enthusiastic musician.

BIBLIOGRAPHY

I. ORIGINAL WORKS. Halm's writings include "Further Considerations Relating to the Systematic Motions of the Stars," in *Monthly Notices of the Royal Astronomical Society,* **71** (1911), 610–639; "On the Rotation of the Sun's Reversing Layer," *ibid.,* **82** (1922), 479–483; and *Magnitudes of Stars Contained in the Cape Zone Catalogue of 20,843 Stars* (London, 1927).

II. SECONDARY LITERATURE. See E. von der Pahlen, *Lehrbuch der Stellarstatistik* (Leipzig, 1937), pp. 261–267, 728–729; and M. Waldmeier, *Ergebnisse und Probleme der Sonnenforschung* (Leipzig, 1941), pp. 43–49, 112, 123.

F. SCHMEIDLER

HALPHEN, GEORGES-HENRI (*b.* Rouen, France, 30 October 1844; *d.* Versailles [?], France, 23 May 1889), *mathematics.*

Halphen's mathematical reputation rests primarily on his work in analytic geometry. Specifically, his principal interests were the study of singular points of algebraic plane curves, the study of characteristics of systems of conics and second-order surfaces, the enumeration and classification of algebraic space curves, the theory of differential invariants and their applications, and the theory of elliptic functions and their applications. His papers are marked by brilliance combined with dogged perseverance.

Halphen was raised in Paris, where his mother moved shortly after she was widowed in 1848. His early schooling was at the Lycée Saint-Louis, and he was admitted to the École Polytechnique in 1862. He served with great distinction in the Franco-Prussian War, and in 1872 he married the daughter of Henri Aron; she eventually bore him four sons and three daughters. Also in 1872 Halphen returned to the École Polytechnique, where he was appointed *répétiteur* and rose to *examinateur* in 1884. His doctorate in mathematics was awarded in 1878 upon the presentation of his thesis, *Sur les invariants différentiels.* In 1880 Halphen won the Ormoy Prize (Grand Prix des Sciences Mathématiques) of the Academy of Sciences in Paris for advances he had made in the theory of linear differential equations, and in 1882 he received the Steiner Prize from the Royal Academy of Sciences in Berlin for his work on algebraic space curves. He was elected to membership in the French Academy in 1886, an honor which he enjoyed for only three years before he died of what was called "overwork."

Halphen first came to the attention of the mathematical community in 1873, when he resolved Michel Chasles's conjecture: Given a family of conics depending on a parameter, how many of them will satisfy a given side condition? Chasles had found a formula for this, but his proof was faulty. Halphen showed that Chasles was essentially correct, but that restrictions on the kinds of singularities were necessary. Halphen's solution was ingenious: he transformed the given system of conics into one algebraic plane curve, and the side condition into another; his results were then obtained from the study of the two curves.

After solving Chasles's problem, Halphen went on to make significant contributions in the theory of algebraic plane curves, especially in the study of their singular points. He was the first to classify singular points and extended earlier work of Bernhard Riemann by giving a general formula for the genus of an algebraic plane curve. Then, considering curves in the same genus, he extended a theorem of Max Noether which proved that in any class there always exist curves with only ordinary singularities.

This work led Halphen to the subject of differential invariants. He had noticed in his earlier work that under projective (i.e., linear and one-to-one) transformations certain differential equations remained unchanged. He was able to characterize all such equations and presented the results in 1878 as his thesis. Henri Poincaré was so impressed that he said: ". . . the theory of differential invariants is to the theory of curvature as projective geometry is to elementary geometry" ("Notice sur Halphen," p. 154; also *Oeuvres,* I, xxxv). Later Halphen applied these results to the integration of linear differential equations, greatly extending the classes of these equations which could be solved. For the latter work he was awarded its prize for 1880 by the French Academy of Sciences.

Halphen's most significant original work was the paper which won the Steiner Prize. In it he made a complete classification of all algebraic space curves up to the twentieth degree. This problem is much more difficult than the corresponding one for algebraic plane curves. A plane curve of degree k can be considered to be a special case of the most general curve of degree k; thus the class and genus of the curve are known if the degree is known, perhaps modified by singularities. But for space curves there is no such thing as a most general curve of degree k (a space curve requires at least two equations) and so, in Halphen's words, ". . . one never knows any geometric entity which includes, as special cases, all space curves of given degree. One cannot, therefore, assert *a priori* for any property of space curves, no matter how general, that it will depend only on the degree" ("Sur quelques propriétés des courbes gauches algébriques," p. 69; also *Oeuvres,* I, 203). For example, the genus has no algebraic relation to the degree but instead satisfies certain inequalities.

Halphen's last work was a monumental treatise on elliptic functions. He intended that it consist of three volumes, but he died before he could finish the last. The aim of the work was to simplify the theory of elliptic functions to the point where they could be put to use by the nonspecialist without losing any of the essential points. In the first volume he realized this aim, proving everything he needed without recourse to more general function theory. In the process Halphen not only simplified the theory but also eliminated much of the very cumbersome notation then in use. The second volume is concerned principally with applications from mechanics, geometry, and differential equations. The problems solved are all difficult and are either new or show new insights. The third volume was to contain material on the theory of transformation and applications to number theory.

The amount and quality of Halphen's work is impressive, especially considering that his mathematically creative life covered only seventeen years. Why, then, is his name so little known? The answer lies partly in the fact that some of his work, the theory of differential invariants, is now only a special case of the more general Lie group theory and thus has lost its identity. But part of the answer is related to a larger question: Why is so much mathematics of even the recent past lost? In Halphen's case, he worked in analytic and differential geometry, a subject so unfashionable today as to be almost extinct. Perhaps with its inevitable revival, analytic geometry will restore Halphen to the eminence he earned.

BIBLIOGRAPHY

Halphen's writings are in *Oeuvres de Georges-Henri Halphen,* 4 vols. (Paris, 1916–1924), compiled for publication by C. Jordan, H. Poincaré, and É. Picard. Among them is "Sur quelques propriétés des courbes gauches algébriques," in *Bulletin de la Société mathématique de France,* **2** (1873–1874), 69–72. See also *Traité des fonctions elliptiques et de leurs applications,* 3 vols. (Paris, 1886–1891); the last vol. consists of fragments only.

Biographical material is in Henri Poincaré, "Notice sur Halphen," in *Journal de l'École polytechnique,* cahier 60 (1890), 137–161, repr. in *Oeuvres*, I, xvii–xliii.

MICHAEL BERNKOPF

HALSTED, GEORGE BRUCE (*b.* Newark, New Jersey, 23 November 1853; *d.* New York, N.Y., 16 March 1922), *mathematics, education.*

Halsted's father, Oliver Spencer Halsted, Jr., was a distinguished lawyer; his mother, Adela Meeker, was the only daughter of a wealthy Charleston, South Carolina, family. Halsted was the fourth generation of his family to attend Princeton, where he received his A.B. in 1875 and his A.M. in 1878. During this period he also attended the Columbia School of Mines.

He received his Ph.D. from Johns Hopkins University in 1879, where he was the first student of J. J. Sylvester. He also studied in Berlin, where he arrived with a flattering letter from Sylvester introducing him to the distinguished Carl Borchardt, then editor of *Crelle's Journal.* From 1879 to 1881 Halsted was tutor at Princeton, and from 1881 to 1894 he was instructor there in postgraduate mathematics. His most productive period occurred from 1894 to 1903, when he held the chair of pure and applied mathematics at the University of Texas. His academic career continued at St. John's College, Annapolis, Maryland (1903); Kenyon College, Gambier, Ohio (1903–1906); and Colorado State College of Education, Greeley (1906–1914). Halsted was married to Margaret Swearingen; they had three sons.

In a period when American mathematics had few distinguished names, the eccentric and sometimes spectacular Halsted established himself as an internationally known scholar, creative teacher, and promoter and popularizer of mathematics. He was a member of and active participant in the major mathematical societies of the United States, England, Italy, Spain, France, Germany, and Russia. His activities penetrated deeply in three main fields: translations and commentaries on the works of Nikolai Lobachevski, János Bolyai, Girolamo Saccheri, and Henri Poincaré; studies in the foundations of geometry; and criticisms of the slipshod presentations of the mathematical textbooks of his day.

Upon his retirement in Greeley, Halsted wrote somewhat bitterly: "I am working as an electrician, as there is nothing [for me] in cultivating vacant lots" ("Princeton University Biographical Questionnaire"). His withdrawal was not complete, however, for his annotated translation of Saccheri's *Euclides vindicatus* was published in 1920; and at the time of his death he was working on a translation of Saccheri's *Logica demonstrativa* from what he believed to be the only extant copy.

BIBLIOGRAPHY

I. ORIGINAL WORKS. No complete bibliography of Halsted's publications has been published. The most extensive appears to be in Poggendorff, III, 578; IV, 573–574; and V, 490. In *American Mathematical Monthly* alone he published over fifty articles, of which twenty were biographical sketches. His main works are the following:

"Bibliography of Hyper-Space and Non-Euclidean Geometry," in *American Journal of Mathematics,* **1** (1878), 261–276, 384–385, and **2** (1879), 65–70; *Basis for a Dual Logic* (Baltimore, 1879), his doctoral diss.; *Mensuration. Metrical Geometry* (Boston, 1881), the unacknowledged source for W. Thomson's article "Mensuration" in the 9th ed. of *Encyclopaedia Britannica* and also the work in which his "prismoidal formula" first appeared (4th ed., 1889, p.

130)—Halsted was unduly proud of this contribution to mensuration; also presented in "Two-Term Prismoidal Formula," in *Scientiae baccalaureus,* **1** (1891), 169–178; and *Rational Geometry* (New York, 1904), an attempt to write an elementary geometry text based on David Hilbert's axioms which, after much criticism, was revised (1907) and later translated into French, German, and Japanese.

Halsted's translations include J. Bolyai, *The Science Absolute of Space,* trans. from Latin (Austin, Texas, 1896); H. Poincaré, *The Foundations of Science,* with a special pref. by Poincaré and an intro. by Josiah Royce (New York, 1913); N. Lobachevski, *The Theory of Parallels* (La Salle, Ill., 1914); and Girolamo Saccheri's *Euclides vindicatus,* also ed. by Halsted (Chicago, 1920), portions of which also appeared in *American Mathematical Monthly,* **1-5** (June 1894–Dec. 1898).

II. SECONDARY LITERATURE. L. E. Dickson, "Biography. Dr. George Bruce Halsted," in *American Mathematical Monthly,* **1** (1894), 337–340, contains a good deal of the family history and some personal observations. See also A. M. Humphreys, "George Bruce Halsted," in *Science,* **56** (1921), 160–161; F. Cajori, "George Bruce Halsted," in *American Mathematical Monthly,* **29** (1922), 338–340; and H. Y. Benedict, "George Bruce Halsted," in *Alcalde,* **10** (1922), 1357–1359, a notice that is mainly anecdotal. "Princeton University Biographical Questionnaire," which was filled out by Halsted personally, contains details not available elsewhere.

HENRY S. TROPP

HALSTED, WILLIAM STEWART (*b.* New York, N.Y., 23 September 1852; *d.* Baltimore, Maryland, 7 September 1922), *surgery.*

Halsted was the son of William Mills Halsted, Jr., and Mary Louisa Haines. His grandfather and father were successful merchants in New York City, and the family occupied a prominent position financially and philanthropically. His early education included a private school in Monson, Massachusetts, and Phillips Academy, Andover, Massachusetts, prior to his entering Yale College in 1870. Halsted was a mediocre student but an exceptional athlete who first became interested in medicine in his senior year. He entered the College of Physicians and Surgeons of New York in 1874, when it was essentially a proprietary school allied to Columbia College in name only. His preceptor was Henry Burton Sands. Halsted absorbed much of the philosophy of John Call Dalton, professor of physiology, with whom he worked as a student assistant. He graduated among the top ten members of his class in 1877 and in April 1878 completed an eighteen-month period of training in the fourth surgical division of Bellevue Hospital, under the guidance of Frank Hastings Hamilton. He then served briefly as house physician at New York Hospital.

In the fall of 1878 Halsted went to Europe for two years of further study in Austria and Germany, chiefly in the basic sciences and particularly in anatomy under Emil Zuckerkandl and Moriz Holl. He attended many clinical lectures and first became acquainted with the German method of graduate surgical education which was to have a profound effect on his future. In 1880 he returned to New York City. Shortly thereafter he joined the faculty of the College of Physicians and Surgeons as a demonstrator in anatomy. He became associated with Sands at Roosevelt Hospital, where he initiated the outpatient department, and held visiting or attending positions at four other hospitals. He also established a private practice limited to surgery and a quiz session which was academically sound.

In 1884, while experimenting with cocaine hydrochlorate as a surgical anesthetic, Halsted and several of his colleagues and students became addicted. In an attempt to overcome the addiction, he was hospitalized in Butler Hospital, Providence, Rhode Island, for six months in 1886 and for nine months in 1887. This illness ended his professional career in New York City, and he moved to Baltimore, Maryland, to work in the laboratory of William H. Welch, professor of pathology at the Johns Hopkins University, in December 1886. When he had apparently regained his health and the authorities of the Johns Hopkins Hospital (and later the Johns Hopkins Medical School) were convinced of his capabilities and reliability, he was appointed surgeon in chief to the hospital in 1890 and professor of surgery in 1892. The question of Halsted's drug addiction and his apparent cure have been discussed for years. William Osler's "The Inner History of the Johns Hopkins Hospital" confirms that Halsted was treated for morphine addiction as late as 1898.

In New York City before his illness Halsted was an aggressive and extraordinarily active surgeon who was rapidly rising in the ranks of the gifted surgical specialists. His career in Baltimore was that of a thoughtful, painstaking operator who returned to the laboratory to study a succession of basic problems in surgery. In a sense he left the path of Sands and Hamilton to follow a career more akin to that of his former teacher John C. Dalton.

Halsted's important contributions included the development of neuroregional anesthesia through his cocaine experiments, a technique he used with some hesitation in future years because of his personal experience and for which he received no credit until shortly before his death; a radical operation for carcinoma of the breast, which incorporated certain modifications and improvements on the radical procedures developed by others; a radical operation for the treatment of inguinal hernia; physiologic studies of the thyroid and parathyroid glands and a technique for thyroidectomy; and the surgical treatment of vascular aneurysm. More important was the methodical manner in which he approached any surgical problem. Whether in the laboratory studying basic problems of the care and handling of wounds or at the operating table or bedside, his scholarly and painstaking approach was a model for many, although an annoyance to those surgeons who felt dexterity and rapidity were the hallmarks of greatness.

Halsted was an excellent teacher of the exceptional student and resident but devoted little time to others. Those selected few residents who trained under him for seven years or more were given complete patient responsibility, a significant alteration of the German system enthusiastically adopted by Halsted. This system of residency training is a major contribution of the Johns Hopkins Hospital to American medicine. Halsted felt that the leading surgeons in Germany, Austria, and Switzerland were the world's finest, and he made frequent trips to their clinics. Although often considered a classic example of the salaried, full-time clinical professor, he was in fact a public supporter of the geographic full-time system (salaried position supplemented by private fees) and had a modest but lucrative private practice prior to the institution of full-time clinical chairs at Johns Hopkins in 1914.

Halsted's meticulous nature and search for perfection in surgery were mirrored in his personal life, particularly in matters of dress and cuisine. To the majority of his colleagues he was cold and reserved, avoiding social intercourse whenever possible. To a few intimate friends he was warm and exceedingly hospitable, and displayed a rich sense of humor. He rebelled against his strict Presbyterian upbringing and was an agnostic in his adult life. In 1890 he married Caroline Hampton, a niece of Wade Hampton III of South Carolina. She was formerly the chief nurse in his operating room. They had no children. Following his marriage he retired to his estate, High Hampton, in Cashiers Valley, North Carolina, for a portion of each summer. In 1919 he underwent cholecystectomy, but in 1922 he had another attack of jaundice and pain that required an operation. He died the day after he had undergone surgery.

BIBLIOGRAPHY

I. ORIGINAL WORKS. Halsted's major publications appear in *Surgical Papers by William Stewart Halsted,* W. C. Burket, ed., 2 vols. (Baltimore, 1924; repr. 1952).

Halsted's papers are preserved at the Welch Medical Library, Johns Hopkins University.

II. SECONDARY LITERATURE. The most complete biography available is William G. MacCallum, *William Stewart Halsted, Surgeon* (Baltimore, 1930). An interesting view of Halsted is also found in George W. Heuer, "Dr. Halsted," in *Bulletin of the Johns Hopkins Hospital,* **90,** supp. (1952), 2. A detailed review of his career in New York City is found in Peter D. Olch, "William S. Halsted's New York Period, 1874–1886," in *Bulletin of the History of Medicine,* **40** (1966), 495–510. Osler's candid comments about Halsted are found in William Osler, "The Inner History of the Johns Hopkins Hospital," in *Johns Hopkins Medical Journal,* **125** (1969), 184–194. Another biographical sketch that includes biographical notes about his colleagues is Samuel J. Crowe, *Halsted of Johns Hopkins: The Man and His Men* (Springfield, Ill., 1957).

PETER D. OLCH

HAMBERG, AXEL (*b.* Stockholm, Sweden, 17 January 1863; *d.* Djursholm, Sweden, 28 June 1933), *geography, geology.*

Hamberg's father, N. P. Hamberg, was a chemist and pharmacologist who became professor in forensic chemistry at the University of Stockholm. The young Hamberg went to high school in Stockholm, entering the university in 1881. Hamberg took his licenciate degree in 1893, his doctorate in 1901, and in the same year became professor of physical geography and historical geology at the university. In 1907 he became professor in geography at the University of Uppsala, where he remained until his retirement in 1928.

At Stockholm, Hamberg had broad training in chemistry, physics, geology, and geography, and was soon attracted to hydrology and glaciology. He took part in Nils Adolf Nordenskjöld's expedition to Greenland (1883) as hydrologist, and in Alfred Nathorst's expedition to Spitsbergen (1898), where he gained extensive experience of arctic conditions. He also made an elaborate hydrological survey in Swedish Lapland (1884–1886), especially of the Sarek Mountains, the geology and physical geography of which became the subject not only of Hamberg's doctoral thesis but of a whole series of scientific papers, written with collaborators. He became a member of the Royal Swedish Academy of Science and was the president of the International Glaciological Commission from 1914 to 1916 and from 1927 to 1930.

Hamberg was a meticulous and careful organizer of expeditions, an enthusiastic and inspiring leader, and a popular teacher. Many aspects of modern glaciology, especially the study of the water-ice budget of glaciers, can be traced back to his research and the research of his pupils. His most valuable scientific work was his introduction of new, exact methods of measurement and new and improved instruments. These innovations, coupled with his hydrological surveys, have had great importance in the development of the hydroelectric power plants in northern Sweden and other boreal regions.

BIBLIOGRAPHY

Hamberg's most important scientific works are *Geologiska och fysisk geografiska undersökningar i Sarekfjällen* ("Geological and Physical-Geographical Studies in the Sarek Mountains," Stockholm, 1901); and *Naturwissenschaftliche Untersuchungen des Sarekgebietes im Schwedich-Lappland,* the series of monographs written with collaborators. For a complete bibliography, see H. Köhler, "Axel Hamberg," in *Svensk Geografisk årsbok* (1933), 175–184, and G. Aminoff, "Axel Hamberg," in *Kungliga Svenska vetenskapsakademiens årsbok* (1934), 265–272.

NILS SPJELDNAES

AL-HAMDĀNĪ, ABŪ MUḤAMMAD AL-ḤASAN IBN AḤMAD IBN YAʿQŪB, also known as **Ibn al-Ḥāʾik, Ibn Dhi ʾl-Dumayna,** or **Ibn Abī ʾl-Dumayna** (*b.* Ṣanʿāʾ, Yemen, 893 [?]; *d.* after 951 [?]), *geography, natural science.*

Al-Hamdānī belonged to a well-known South Arabian tribe, Hamdān, and his family had for four generations lived in Ṣanʿāʾ.[1] He traveled extensively, visiting Iraq and spending considerable time in Mecca. He corresponded with the intellectuals of his time, such as the Kūfa philologists Ibn al-Anbārī and Ghulām Thaʿlab and their student Ibn Khālawayh. Later he lived in the South Arabian cities of Rayda and Ṣaʿda. Involvement in political struggles led to his being jailed twice.

Al-Hamdānī passionately supported his kinsmen's side in the incessant antagonism between the North and South Arabian tribes. He expressed this most clearly in his poem *al-Dāmigha* ("The Crusher"). Other of his poems also have a political content. His national pride may have been the source of his decision to create the two monuments to his country and to his people: the historical work *al-Iklīl* ("The Crown"), written in 943, and the geographical work *Ṣifat Jazīrat al-ʿArab* ("Description of the Arabian Peninsula").

Only four of the ten books of *al-Iklīl* have been preserved. Books I, II, and X contain genealogies of South Arabian tribes, and book VIII describes the old castles erected by the Ḥimyarites in Yemen. Of the lost books, book III is said to have dealt with the merits of the South Arabian tribes, and books IV–VI with the history of South Arabia before Islam; book VII is said to have contained a criticism of false traditions, and book IX Ḥimyaritic inscriptions. It is said that scattered through the work were pieces on astronomy and physics as well as ancient conceptions of the world as being eternal or created.[2]

Ṣifat Jazīrat al-ʿArab is based primarily on al-Hamdānī's own observations. In a few cases he uses information from other geographers, such as al-Jarmī, Abu'l-Hasan al-Khuzāʿī, Aḥmad ibn al-Ḥasan al-ʿĀdī al-Falajī, and Muḥammad ibn ʿAbdallāh ibn Ismāʿīl al-Saksakī. In the introduction he cites Ptolemy's *Geography,* Hermes Trismegistus, and Dioscorides.[3] He also cites the Indian astronomical work *Sindhind* and its Arabian translator al-Fazārī, as well as Ṣanʿāʾ's own astronomers.[4] Aside from purely geographical information this work contains observations on fruits and vegetables, precious stones and metals, and linguistic matters. The work is often cited in the geographical lexicons of Yāqūt and al-Bakrī, the latter also containing many citations from *al-Iklīl.*

Before his geographical work al-Hamdānī wrote an astronomical one, *Saraʾir al-ḥikma fī ʿilm an-nujūm* ("The Secrets of Wisdom Concerning Astronomy"), of which only book X has been preserved. In it he quotes Dorotheus of Sidon and Ptolemy. Al-Hamdānī is also said to have compiled astronomical tables, but these have not survived. His medical work *al-Quwā* ("Powers"), also not extant, apparently was connected with his astronomical writings, for in it he demonstrated how the air temperature is influenced by the planets.[5]

From a trilogy concerned with property and consisting of *al-Ḥarth wa'l-ḥīla* ("Farming and Its Method"), *al-Ibil* ("The Camels"), and *Kitāb al-Jawharatayn al-ʿatīqatayn* ("The Two Precious Metals [gold and silver]"), only the last, written later than all of the other works mentioned, has been preserved. In it he is concerned with gold and silver in all possible aspects, including religious, literary, and linguistic. But chiefly the work is the first and most extensive Arabian account of the treatment of the metals: extraction, purification, the determination of the standard of fineness, gilding, soldering, and coinage—all built on al-Hamdānī's own observations in the mints of Yemen and on information obtained from craftsmen who worked there. In the theoretical part on the origin of the metals, their use in medicine, and such, he cites Aristotle, Dorotheus of Sidon, Dioscorides, and Hippocrates. Some technical terms for weights and coins are of Greek origin.

Greek and Persian influences combined in this work. South Arabia had been a Persian satrapy until 628, and Persian immigration had continued into the following centuries. The Persian influence is especially noticeable in the terminology for chemical substances and tools. Al-Hamdānī's work demonstrates a connected world picture, typical for his time, in which the influence of the heavenly bodies on the elements and qualities is decisive for the generation and characteristics of metals and other substances, for geography, for the conditions of mankind—and, consequently, is also a foundation for medicine.

Al-Hamdānī built first on his own observations of what is possible in fact and useful in practice. He did not use the elixir of the alchemists to transmute lower elements into gold or silver; according to him, gold was derived from gold ore and silver from silver ore, never from any other kind of metal. The metals were purified by a carefully described chemical-technical process, without magic or ritual procedures. Al-Hamdānī was very precise in the details; some instruments can be completely reconstructed by following his description. He did not accept uncritically the theories of predecessors and he would disagree with Aristotle or Ptolemy.[6] Contrary opinions on the same problem are compared: on linguistic questions the opinions of the philologists and those of laymen; the opinions of Greek philosophers and practical mining experts on problems concerning the origin of gold and silver; the opinions of Greek, Indian, and Chinese scholars on the extent of the inhabited world.[7] Al-Hamdānī was thus a good representative of the union of Greek, Persian, and Arabian culture.

NOTES

1. An account of al-Hamdānī's descent is given by himself in *al-Iklīl,* bk. X, 198 and preceding pages, and by al-Andalusī, 58, tr. 114 f.
2. Al-Andalusī, 58 f., tr. 115; al-Qiftī, I, 281.
3. *Ṣifat Jazīrat al-ʿArab,* index.
4. *Ibid.,* 27.
5. *Kitāb al-Jawharatayn,* 72a.
6. *Ibid.,* 15b; *Ṣifat Jazīrat al-ʿArab,* 29.
7. *Kitāb al-Jawharatayn,* 9b ff., 21a; *Ṣifat Jazīrat al-ʿArab,* 27.

BIBLIOGRAPHY

I. Original Works. Of al-Hamdānī's surviving works, the following are the principal editions and commentaries:

Al-Iklīl: bks. I–II, facs. ed. (Berlin, 1943); M. b. ʿAlī al-Akwaʿ al-Ḥiwālī, ed., 2 vols. (Cairo, 1963–1966). Bk. I: O. Löfgren, ed., 2 vols., vol. LVIII, no. 1 in Bibliotheca Ekmaniana (Uppsala, 1954–1965). An extract from bk. II is in *Südarabisches Muštabih,* O. Löfgren, ed., vol. LVII in Bibliotheca Ekmaniana (Uppsala, 1953). Bk. VIII: D. H. Müller, "Die Burgen und Schlösser Südarabiens nach dem Iklīl des Hamdānī," in *Sitzungsberichte der Akademie der Wissenschaften in Wien,* **94** (1879), 335–423; *ibid.,* **97** (1881), 955–1050; "Auszüge aus dem VIII. Buche des Iklīl," in *Südarabische Alterthümer im Kunsthistorischen Hofmuseum* (Vienna, 1899), 80–95; A. M. al-Karmalī, ed. (Baghdad, 1931); translated by N. A. Faris as *The Antiquities of South Arabia,* Princeton Oriental Texts no. 3 (Princeton, 1938); and N. A. Faris, ed., Princeton Oriental Texts no. 7 (Princeton, 1940). Bk. X: M. al-D. al-Khaṭīb, ed. (Cairo, 1949).

Ṣifat Jazīrat al-ʿArab: D. H. Müller, ed., *Al-Hamdānī's Geographie der arabischen Halbinsel,* 2 vols. (Leiden, 1884–1891; repr., Amsterdam, 1968); L. Forrer, *Südarabien nach al-Hamdānī's Beschreibung der arabischen Halbinsel,* vol. XXVII, no. 3 in Abhandlungen für die Kunde des Morgenlandes (Leipzig, 1942); C. Rabin, *Ancient West-Arabia* (London, 1951), pp. 43 ff. for a trans. of al-Hamdānī's observations on the linguistic state of affairs of the Arabian peninsula (pp. 134–136 of Müller's ed.); and M. b. ʿA. b. B. an-Najdī, ed. (Cairo, 1953).

Kitāb al-Jawharatayn al-ʿatīqatayn: edited and translated into German by Christopher Toll as *Die beiden Edelmetalle Gold und Silber,* Studia semitica upsaliensia no. 1 (Uppsala, 1968).

II. Secondary Literature. On al-Hamdānī and his work, see Carl Brockelmann, *Geschichte der arabischen Literatur,* I (Weimar, 1898), 229; 2nd ed., I (Leiden, 1943), 263 ff., and suppl. I (Leiden, 1937), 409; Oscar Löfgren, *Ein Hamdānī-Fund,* no. 7 in Uppsala Universitets Årsskrift för 1935 (Uppsala, 1935); and "al-Hamdānī," in *Encyclopedia of Islam,* 2nd ed., III (Leiden–London, in press), 124 ff.

See also Ṣāʿid al-Andalusī, *Ṭabaqāt al-umam,* L. Cheikho, ed. (Beirut, 1912), 58, also translated by R. Blachère as *Kitâb Ṭabaḳât al-umam* (Paris, 1935), 114; al-Bakrī, *Muʿjam mā 'staʿjam,* F. Wüstenfeld, ed., 2 vols. (Göttingen–Paris, 1876–1877), and M. al-Saqqā, ed., 4 vols. (Cairo, 1945–1951); al-Qifṭī, *Inbāh al-ruwāt ʿalā anbāh al-nuhāt,* M. A. Ibrāhīm, ed., I (Cairo, 1950), 279–284; *Taʾrīkh al-ḥukamā,* J. Lippert, ed. (Leipzig, 1903), 163; and Yāqūt, *Muʿjam al-buldān,* F. Wüstenfeld, ed., 6 vols. (Leipzig, 1866–1870).

Christopher Toll

ḤĀMID IBN KHIḌR AL-KHUJANDI. See **al-Khujandī.**

HAMILTON, WILLIAM (*b.* Glasgow, Scotland, 8 March 1788; *d.* Edinburgh, Scotland, 6 May 1856), *philosophy, logic.*

Hamilton's father, William Hamilton, professor of astronomy at the University of Glasgow, died in 1790, leaving William to be raised by his mother, Elizabeth Hamilton. After receiving a degree from the University of Edinburgh in 1807, Hamilton went to Balliol College, Oxford, with a Snell Exhibition. He quickly acquired the reputation of being the most learned authority in Oxford on Aristotle, and the list of books that he submitted for his final examination in 1810 was unprecedented. He did not, however, receive a fellowship, primarily because of the unpopularity of Scots at Oxford. He returned therefore to Edinburgh to study and there became an advocate in 1813. Because he had little interest in his career in the law, he applied in 1821 for the chair of moral philosophy at Edinburgh vacated by the death of Thomas Brown. Hamilton was a Whig, and the Tory town council therefore chose his opponent, John Wilson. When in 1829 Macvey Napier became the editor of the *Edinburgh Review,* he persuaded Hamilton to write a series of articles for that journal. The articles, which appeared between 1829 and 1836, were the basis of his international reputation, a reputation that forced the town council to elect him in 1836 to the chair in logic and metaphysics, which he held until his death.

Hamilton's three most important articles for the *Edinburgh Review* were those on Cousin (1829), on perception (1830), and on logic (1833). In the first two, he revealed his unique philosophical position, a combination of the Kantian view that there is a limitation on all knowledge and the Scottish view that man has, in perception, a direct acquaintance with the external world.

The first paper deals with the possibility of human knowledge of the absolute. In it Hamilton argued against Cousin's view that man has immediate knowledge of the absolute and against Schelling's view that man can know the absolute by becoming identical with it. Hamilton tried to show that neither of these views is coherent and that there is something incoherent about the very notion of thought about the absolute. Hamilton's own position was close to Kant's, but he wanted to go further than Kant and say that the mind cannot use the absolute even as a regulative idea.

The second article is a defense of Reid's view, that the direct object of perception is external, against the attacks of Brown. Hamilton had little trouble in showing that Brown neither understood Reid's position nor could offer arguments that disproved either Reid's position or the position mistakenly attributed to him by Brown. Hamilton did, however, agree with Brown's claim that Reid was mistaken when he identified the direct object of acts of memory with some previously existing external object.

These metaphysical positions were developed further during the twenty years that Hamilton was professor at Edinburgh. Many of his mature opinions, as expressed in the appendixes to his edition of Reid's major works and to his own published lectures, modify what he wrote in these two articles; but he never really gave up these basic positions, which were extremely influential during his lifetime and still have some interest today. They are, however, far less important for the history of thought in general and for the history of science in particular than his work in logic.

Hamilton was one of the first in that series of British logicians—a series that included George Boole, Augustus De Morgan, and John Venn—who radically transformed logic and created the algebra of logic and mathematical logic. To be sure, Hamilton only helped begin this development, and given his dislike of mathematics, he probably would not have been very happy with its conclusion. Nevertheless, his place in it must be recognized.

The traditional, Aristotelian analysis of reasoning allowed for only four types of simple categorical propositions:

(A) All A are B.
(E) No A are B.
(I) Some A are B.
(O) Some A are not B.

Hamilton's first important insight was that logic would be more comprehensive and much simpler if it allowed for additional types of simple categorical propositions. In particular, Hamilton suggested that one treat the signs of quantity ("all," "some," "no") in the traditional propositions as modifiers of the subject term A and that one introduce additional signs of quantity as modifiers of the predicate term B. Hamilton called this innovation the quantification of the predicate. Other logicians before Hamilton had made the same suggestion, but Hamilton was the first to explore the implications of quantifying the predicate, of admitting eight simple categorical propositions:

(1) All A are all B.
(2) All A are some B (traditional A).
(3) Some A are all B.
(4) Some A are some B (traditional I).
(5) Any A is not any B (traditional E).
(6) Any A is not some B.
(7) Some A are not any B (traditional O).
(8) Some A are not some B.

The first important inference that Hamilton drew from this modification had to do with the analysis of simple categorical propositions. There were, according to the traditional, Aristotelian logic, two ways of analyzing a simple categorical proposition such as "All A are B": extensively, that is, as asserting that the extension of the term A is contained within the extension of the term B; or comprehensively, that is, as asserting that the comprehension of the term B is contained within the comprehension of the term A. In either case, the proposition expresses a whole-part relation. But the new Hamiltonian modification, because it distinguished (1) from (2), (3) from (4), (5) from (6), and (7) from (8), enables one to adopt a different analysis of these propositions. According to this new analysis, each of these propositions asserts or denies the existence of an identity-relation between the two classes denoted by the quantified terms. Thus, "All A are all B" asserts that the classes A and B are identical, while "Some A are not some B" asserts that there is a subset of the class A which is not identical with any subset of the class B. One result, therefore, of the quantification of the predicate is that simple categorical propositions become identity claims about classes. This is just the analysis of simple categorical propositions that Boole needed and used in formulating the algebra of logic.

Hamilton's work facilitated a considerably simplified analysis of the validity of reasoning. The traditional, Aristotelian analysis of mediate reasoning, for example, involved many concepts (such as the figure of a syllogism, major and minor terms) that were based on the distinction between the subject of a proposition and its predicate. This subject-predicate distinction had some significance when simple categorical propositions were understood as expressing asymmetrical whole-part relations. But given the new analysis, where these propositions are understood as expressing symmetrical identity relations, there is little point to a distinction between the subject and the predicate of a proposition. Further, if the subject-predicate distinction is dropped, then all of the traditional cumbersome machinery based on it should also be dropped. As a result, the complicated traditional rules for the validity of syllogistic reasoning disappear. One is then left, as Hamilton pointed out in his theory of the unfigured syllogism, with two simple rules for valid syllogisms: If $A = B$ and $B = C$, then $A = C$; and If $A = B$ and $B \neq C$, then $A \neq C$. Similarly, the traditional Aristotelian analysis of immediate reasoning, based upon the complicated distinctions between simple conversion, conversion *per accidens,* and contraposition, is replaceable by the simple rule that all eight propositions are simply convertible. This rule follows directly from the fact that all eight propositions are concerned with symmetrical identity relations.

New advances in a given science, besides simplifying the treatment of previously solved problems, usually enable one to solve problems that one could not previously handle. Hamilton's quantification of the predicate is no exception to this rule. The logician could now explain the validity of many inferences that resisted traditional analysis. The simplest example of this is the inference to the identity of classes A and C from premises asserting that they are both identical with some class B. Traditional analysis did not even recognize the existence of propositions asserting that two classes are identical; it could not, therefore, explain the validity of such an inference. Hamiltonian analysis, however, could do so by referring to the first of the two rules for the validity of all mediate reasoning.

Some of Hamilton's new ideas, such as his class-identity analysis of propositions, were incorporated into Boole's far more sophisticated system. This contribution to mathematical logic would in itself be sufficient to earn for Hamilton an important place in intellectual history, but his claim to recognition is strengthened by the significance of his innovations to the history of logic. As is well known, Kant and most other important eighteenth-century philosophers thought that nothing of importance had been done in formal logic since the time of Aristotle, primarily because of the completeness and perfection of the Aristotelian system. The only people who saw a future for logic were those who wanted to change logic from a formal analysis of the validity of reasoning to an epistemological and psychological analysis of the conditions for, and limits on, human knowledge. Hamilton, by showing that the Aristotelian analysis could be greatly improved and supplemented, changed the minds of many philosophers, logicians, and mathematicians and helped produce the interest in formal logic that was so necessary for the great advances of the nineteenth century.

Despite its great historical significance, Hamilton's quantification of the predicate has had little direct influence in more recent times. This is partly due to the fact that both it and the Boolean algebra of logic, which it so greatly influenced, have been superseded by Frege's far more powerful quantificational analysis—an analysis so different that Hamilton's theory has no relevance to it. It is, however, also due to a certain internal weakness in Hamilton's initial quantification of the predicate, which was pointed out by Hamilton's great adversary, Augustus De Morgan, during their long and acrimonious quarrels.

There really were two quarrels between Hamilton and De Morgan. The first had to do with Hamilton's charge that De Morgan had plagiarized some of Hamilton's basic ideas. In 1846 De Morgan sent a draft of one of his most important papers on logic to William Whewell, who was supposed to transmit it to the Cambridge Philosophical Society. De Morgan then received from Hamilton, in the form of a list of requirements for a prize essay set for Hamilton's students, a brief account of Hamilton's quantification of the predicate. At about this time, De Morgan asked

Whewell to return the draft of his paper and then made some changes in it. Hamilton charged that the alterations were based on his communication to De Morgan. In his reply De Morgan claimed that he made the changes before he received the communication from Hamilton. Although it is not clear as to who was right about the date of the changes, it is clear that De Morgan did not plagiarize Hamilton's ideas. Even if De Morgan's ideas were suggested by Hamilton's communication, they are so different from Hamilton's that no one could consider them to be a plagiarism.

The second, far more important, quarrel was about the relative merits of their innovations in logic. This arose out of the first, since De Morgan was not content with pointing out the differences between the two systems. He also argued that Hamilton's innovations, unlike his own, were based on a defective list of basic propositions.

De Morgan first made this claim in an appendix to his book *Formal Logic* (1847). Although he offered several criticisms of Hamilton's list of eight basic categorical propositions, there was only one that was really serious—that Hamilton's first proposition is not a simple categorical proposition because it is equivalent to the joint assertion of the second and third propositions. Thus "All A are some B" and "Some A are all B" can both be true if, and only if, "All A are all B" is also true.

Hamilton was slow in responding to this argument, primarily because he had suffered an attack of paralysis in 1844 that made it very difficult for him to do any work. When, however, in 1852, he published a collection of his articles from the *Edinburgh Review,* he included in the book an appendix in which he argued that De Morgan's criticisms were based on a misunderstanding of the eight propositional forms. De Morgan thought that "some" meant "some, possibly all." If it did, then he would certainly be right in his claim that the first proposition is equivalent to the joint assertion of the second and third propositions. But Hamilton said that he had meant in his forms "some, and not all." Consequently, the conjunction of "All A are some B" and "Some A are all B" is inconsistent, and neither of these propositions can be true if "All A are all B" is true.

The controversy rested at this point until some years after Hamilton's death. Then, De Morgan renewed it in a series of letters in *Athenaeum* (1861–1862) and in his last article on the syllogism, in *Transactions of the Cambridge Philosophical Society* (1863). De Morgan began his new attack by casting doubt on the claim that Hamilton had meant "some, but not all." He did this by showing that much of what Hamilton had to say about the validity of particular inferences made sense only if we suppose that he meant "some, perhaps all." There is little doubt that De Morgan was right about this point. Yet De Morgan now had an even more crushing criticism of Hamilton's list of simple categorical propositions: Even if one grants, he said, that Hamilton meant "some, but not all," there is still something wrong with his list. After all, there are five, and only five, relations of the type discussed in categorical propositions that can hold between two classes: (1) the two classes are coextensive, (2) the first class is a proper subset of the second class, (3) the second class is a proper subset of the first class, (4) the two classes have some members in common but each has members that are not members of the other, and (5) the two classes have no members in common. Thus Hamilton's propositions 6–8 seem to be superfluous.

De Morgan's final critique clearly showed that Hamilton had not exercised sufficient care in laying the foundations for his new analysis of the validity of reasoning. This was quickly recognized by most logicians; Charles Sanders Peirce, the great American logician, described De Morgan's 1863 paper as unanswerable. While there is no doubt that De Morgan's critique helped lessen the eventual influence of Hamilton's work, it should not prevent the recognition of both the intrinsic merit of Hamilton's work and its role in the development of mathematical logic in Great Britain during the nineteenth century.

BIBLIOGRAPHY

I. Original Works. Hamilton's main writings are important essays supplementary to *The Works of Thomas Reid* (Edinburgh, 1846); *Discussions on Philosophy and Literature and Education and University Reform* (Edinburgh, 1852); *Lectures on Metaphysics,* H. Mansel and J. Veitch, eds. (Edinburgh, 1859); and *Lectures on Logic,* H. Mansel and J. Veitch, eds. (Edinburgh, 1861).

II. Secondary Literature. On Hamilton and his work, see T. S. Baynes, *An Essay on the New Analytic of Logical Forms* (Edinburgh, 1850); S. A. Grave, *The Scottish Philosophy of Common Sense* (Oxford, 1960); L. Liard, *Les logiciens anglais contemporains* (Paris, 1890); J. S. Mill, *An Examination of Sir William Hamilton's Philosophy* (London, 1889); S. V. Radmussen, *The Philosophy of Sir William Hamilton* (Copenhagen, 1925); and J. Veitch, *Sir William Hamilton* (Edinburgh, 1869).

Baruch A. Brody

HAMILTON, WILLIAM (*b.* Scotland, 13 December 1730; *d.* London, England, 6 April 1803), *archaeology, geology.*

William Hamilton is best known for his diplomatic career and perhaps also for his wife Emma's notorious affair with Horatio Nelson. His scientific reputation rests on his hobbies—the study of volcanism and the collection of antiquities. After extended military duties, he became the envoy of Great Britain at the court of Naples.

In his first four years in Naples, he climbed Vesuvius at least twenty-two times, during which he or Pietro Fabris, an artist he had trained, did numerous sketches of all the stages of the eruptions which they were able to observe. In order to maintain a complete record of volcanic activity around Naples when he could not observe himself, Hamilton employed, about 1791, a Dominican friar named Resina who kept a daily account of the phenomena.

Hamilton's scientific activity was not limited to the observation of volcanic eruptions but also included the collection of various types of lavas, ashes, and minerals produced by volcanism in the Naples vicinity. He donated this collection to the British Museum in 1767. Hamilton then extended his investigations to Etna and the Lipari Islands and in February 1783 traveled to Calabria to survey the effects of the earthquakes which had just devastated that area.

Elected a fellow of the Royal Society in 1766, Hamilton published the results of his studies on volcanoes in the form of numerous letters in the *Philosophical Transactions* (1767–1795). In addition, he published several beautifully illustrated volumes, unique by their association of a French and English text.

Hamilton was a famous collector of antiquities, especially of Greek vases. One collection was sold to the British Museum and the other to Thomas Hope. Hamilton said that the study of antiquities had taught him "the perpetual fluctuation of everything and that the present hour was the sweetest in life." This virtually Epicurean attitude appears in his complaisance toward his wife and Nelson. In 1800 he was recalled to England and although he hinted many times at a separation, the affair continued and both were at his bedside when he died.

Hamilton would have been considered only a collector of antiquities and an educated observer of volcanic eruptions had he not played an important role, probably unknown to him, in the basalt controversy. In 1769 Rudolf Erich Raspe, curator of the Fridericianum Museum and professor at the Collegium Carolinum in Kassel, wrote his paper entitled "Nachricht von einigen niederhessischen Basalten . . ." (published in 1771), which Goethe considered epochal because it introduced to Germany the volcanic origin of basalt as demonstrated by Desmarest's discoveries in Auvergne. Raspe, although convinced of the volcanic origin of basalt, had observed in the Habichtswald, near Kassel, prismatic basalt overlain by massive basalt and had interpreted this situation as representing the products of submarine eruptions overlain by those of subaerial eruptions. On the other hand, Desmarest, in Auvergne, had followed the subaerial lava flows from their terminal accumulations of prismatic basalt to the craters of the volcanoes from which they originated. Still Raspe remained unconvinced, considering that the occurrence in Auvergne could be accidental. He therefore wrote to Hamilton, whom he considered the most qualified naturalist on the question of recent prismatic basalt. In his letter, dated 31 October 1769, Raspe wrote in particular: "I would like to know from your Excellency, if you have found among the new lavas of the Vesuvius, anything which confirms the interpretation of Mr. Desmarest; in other words, do any of the cooled lava flows of the Vesuvius display toward their end something similar to prismatic basalt?" Hamilton replied (12 December 1769) that he had never found any kind of columnar or polygonal rock like basalt among the numerous types of lavas of Vesuvius, Sicily, and Ischia.

This statement was understood by Raspe as support for his hypothesis that prismatic basalt was characteristic of submarine eruptions. Subsequently, he further developed his ideas in *Beytrag zur allerältesten und natürlichen Historie von Hessen* (1774), translated into English during his self-imposed exile in Great Britain (1776), where he had fled to avoid prosecution for embezzlement.

The proof that Raspe thought he had obtained from Hamilton was actually unwarranted. Although Hamilton might have observed among Vesuvian and Sicilian lavas some true basaltic types comparable to those described by Raspe, the phenomenon of prismatic jointing is complex—not as yet fully understood—and certainly cannot be used as a criterion to distinguish between subaerial and submarine lavas.

BIBLIOGRAPHY

I. Original Works. A comprehensive and well-annotated bibliography of Hamilton's writings is in Doria (see below). Among them are *Antiquités étrusques, grecques et romaines, tirées du cabinet de M. Hamilton . . . Collection of Etruscan, Greek and Roman Antiquities . . .*, 4 vols. (Naples, 1766–1767), with intro. by P. F. Hugues d'Hancarville; "Antwort-Schreiben von Herrn W. Hamilton an R. E. Raspe, aus dem französischen," in *Deutsche Schriften der K. Gesellschaft der Wissenschaften zu Göttingen,* **1** (1771), 89–93; *Observations on Mount Vesuvius, Mount Etna, and Other Volcanos: In a Series of Letters Addressed *to the Royal Society* . . . (London, 1772; repr., 1773, 1774); "Voyage au mont Ethna et observations par M. Hamilton," in J. H. von Riedesel, *Voyage en Sicile et dans la Grande Grèce* (Lausanne, 1773), translated from English by Villebois; *Campi phlegraei. Observations on the Volcanos of the Two Sicilies . . . Observations sur les volcans des Deux-Siciles* (Naples, 1776); *Account of the Discoveries at Pompei* (London, 1777); *Supplement to the Campi phlegraei, Being an Account of the Great Eruption of Mount Vesuvius, in the Month of August 1779 . . . Supplément au Campi phlegraei* . . . (Naples, 1779); *Oeuvres complètes de M. le chevalier Hamilton* (Paris, 1781), a French trans., with comments by the Abbé Giraud-Soulavie; "An Account of the Earthquakes Which Happened in Italy From February to May 1783," in *Philosophical Transactions of the Royal Society*, **63** (1783), 169–208, repr. as *An Account of the Earthquakes in Calabria, Sicily, &c.* (Colchester, 1783) and translated into French by Lefebvre de Villebrune as *Détails historiques des tremblemens de terre arrivés en Italie* . . . (Paris, 1783); *Neuere Beobachtungen über die Vulkane Italiens und am Rhein* . . . (Frankfurt–Leipzig, 1784), a partial German trans. of the French ed. of his complete works; *Antiquités étrusques, grecques et romaines*, 5 vols. (Paris, 1785–1788), with text by Hancarville; and *Collection of Engravings From Ancient Vases, Mostly of Pure Greek Workmanship, Discovered . . . During the Course of the Years 1789 and 1790*. . ., 5 vols. (Naples, 1791–1795), which appeared in French trans. as *Recueil de gravures d'après des vases antiques* . . ., 2 vols. (Paris, 1803–1806).

II. SECONDARY LITERATURE. On Hamilton and his work, see Gino Doria, ed., *Campi phlegraei osservazioni sui volcani delle Due Sicilie* . . . (Milan, 1962), with bibliography by Uberto Limentani; B. Fothergill, *Sir William Hamilton: Envoy Extraordinary* (London, 1969); R. E. Raspe, "Anhang eines Schreibens an den königlichen Grossbrittanischen Gesandten Herrn William Hamilton, zu Neapolis," in *Deutsche Schriften der K. Gesellschaft der Wissenschaften zu Göttingen*, **1** (1771), 84–89; "A Letter Containing a Short Account of some Basalt Hills in Hassia," in *Philosophical Transactions of the Royal Society*, **41** (1771), read to the society on 8 Feb. 1770; and *An Account of Some German Volcanoes and Their Productions* . . . (London, 1776); and D. D. Stacton, *Sir William, or a Lesson in Love* (New York, 1963).

ALBERT V. CAROZZI

HAMILTON, WILLIAM ROWAN (*b.* Dublin, Ireland, 4 August 1805; *d.* Dunsink Observatory [near Dublin], 2 September 1865), *mathematics, optics, mechanics.*

His father was Archibald Rowan Hamilton, a Dublin solicitor, whose most important client was the famous Irish patriot Archibald Hamilton Rowan. Both Hamilton and his father took one of their Christian names from Rowan, and the matter is further complicated by the fact that Rowan's real name was Hamilton as well. But there is no evidence of kinship.

The fourth of nine children, Hamilton was raised and educated from the age of three by his uncle James Hamilton, curate of Trim, who quickly recognized his fabulous precocity. By his fifth year he was proficient in Latin, Greek, and Hebrew; and during his ninth year his father boasted of his more recent mastery of Persian, Arabic, Sanskrit, Chaldee, Syriac, Hindustani, Malay, Marathi, Bengali, "and others."

Mathematics also interested Hamilton from an early age, but it was the more dramatic skill of rapid calculation that first attracted attention. In 1818 he competed unsuccessfully against Zerah Colburn, the American "calculating boy"; he met him again in 1820. At about this time he also began to read Newton's *Principia* and developed a strong interest in astronomy, spending much time observing through his own telescope. In 1822 he noticed an error in Laplace's *Mécanique céleste.* His criticism was shown by a friend to the astronomer royal of Ireland, the Reverend John Brinkley, who took an interest in Hamilton's progress and was later instrumental in getting Hamilton appointed as his successor at Dunsink Observatory.

Hamilton's enthusiasm for mathematics caught fire in 1822, and he began studying furiously. The result was a series of researches on properties of curves and surfaces that he sent to Brinkley. Among them was "Systems of Right Lines in a Plane," which contained the earliest hints of ideas that later were developed into his famous "Theory of Systems of Rays." On 31 May 1823 Hamilton announced to his cousin Arthur Hamilton that he had made a "very curious discovery" in optics,[1] and on 13 December 1824 he presented a paper on caustics at a meeting of the Royal Irish Academy with Brinkley presiding. The paper was referred to a committee which reported six months later that it was "of a nature so very abstract, and the formulae so general, as to require that the reasoning by which some of the conclusions have been obtained should be more fully developed. . . ."[2] Anyone who has struggled with Hamilton's papers can sympathize with the committee, but to Hamilton it was a discouraging outcome. He returned to his labors and expanded his paper on caustics into the "Theory of Systems of Rays," which he presented to the Academy on 23 April 1827, while still an undergraduate at Trinity College. Hamilton considered his "Systems of Rays" to be merely an expansion of his paper on caustics. Actually the papers were quite different. The characteristic function appeared only in the "Theory of Systems of Rays," while "On Caustics" investigated the properties of a general rectilinear congruence.

Hamilton had taken the entrance examination for Trinity on 7 July 1823 and, to no one's surprise, came out first in a field of 100 candidates. He continued this auspicious beginning by consistently winning extraordinary honors in classics and science throughout his college career. Trinity College, Dublin, offered an excellent curriculum in mathematics during Hamilton's student years, owing in large part to the work of Bartholomew Lloyd, who became professor at the college in 1812 and instituted a revolution in the teaching of mathematics. He introduced French textbooks and caused others to be written in order to bring the students up to date on Continental methods. These reforms were essentially completed when Hamilton arrived at Trinity.

On 10 June 1827 Hamilton was appointed astronomer royal at Dunsink Observatory and Andrews professor of astronomy at Trinity College. He still had not taken a degree, but he was chosen over several well-qualified competitors, including George Biddell Airy.

As a practical astronomer Hamilton was a failure. He and his assistant Thompson maintained the instruments and kept the observations with the somewhat reluctant help of three of Hamilton's sisters who lived at the observatory. After his first few years Hamilton did little observing and devoted himself entirely to theoretical studies. On one occasion in 1843 he was called to task for not having maintained a satisfactory program of observations, but this protest did not seriously disturb his more congenial mathematical researches.

Life at the observatory gave Hamilton time for his mathematical and literary pursuits, but it kept him somewhat isolated. His reputation in the nineteenth century was enormous; yet no school of mathematicians grew up around him, as might have been expected if he had resided at Trinity College. In the scientific academies Hamilton was more active. He joined the Royal Irish Academy in 1832 and served as its president from 1837 to 1845. A prominent early member of the British Association for the Advancement of Science, he was responsible for bringing the annual meeting of the association to Dublin in 1835. On that occasion he was knighted by the lord lieutenant. In 1836 the Royal Society awarded him the Royal Medal for his work in optics at the same time that Faraday received the medal for chemistry. A more signal honor was conferred in 1863, when he was placed at the head of fourteen foreign associates of the new American National Academy of Sciences.

In 1825 Hamilton fell in love with Catherine Disney, the sister of one of his college friends. When she refused him, he became ill and despondent and was close to suicide. The pain of this disappointment stayed with Hamilton throughout his life and was almost obsessive. In 1831 he was rejected by Ellen De Vere, sister of his good friend the poet Aubrey De Vere, and in 1833 he married Helen Bayly. It was an unfortunate choice. Helen suffered from continual ill health and an almost morbid timidity. She was unable to run the household, which eventually consisted of two sons and a daughter, and absented herself for long periods of time.

When Catherine Disney was dying in 1853, Hamilton visited her twice. He was desperate to get mementos from her brother—locks of hair, poetry, a miniature that he secretly had copied in Dublin—and relieved his distress by writing his confessions to close friends, often daily, sometimes twice a day. Harassed by guilt over his improper feelings and the fear that his secret would become known, Hamilton sought further release in alcohol, and for the rest of his life he struggled against alcoholism.

It would be a mistake to picture Hamilton's life as constant tragedy, however. He was robust and energetic, with a good sense of humor. He possessed considerable eloquence; but his poetry, which he greatly prized, was surprisingly bad. Hamilton had many acquaintances in the Anglo-Irish literary community. He was a frequent visitor at the home of the novelist Maria Edgeworth, and in 1831 he began his long correspondence with De Vere, with whom he shared his metaphysical and poetical ideas and impressions. But his most important literary connection was with William Wordsworth, who took to Hamilton and seemed to feel an obligation to turn him from his poetic ambitions to his natural calling as a mathematician. Hamilton insisted, however, that the "spirit of poetry" would always be essential to his intellectual perfection.

A more important philosophical influence was Samuel Taylor Coleridge. Hamilton was greatly impressed by *The Friend* and *Aids to Reflection,* and it was through Coleridge that he became interested in the philosophy of Immanuel Kant. Hamilton's first serious venture into idealism came in 1830, when he began a careful reading of the collected works of George Berkeley, borrowed from Hamilton's friend and pupil Lord Adare. A letter written in July of the same year mentions Berkeley together with Rudjer Bošković. By 1831 he was struggling with Coleridge's distinction between Reason and Understanding as it appeared in the *Aids to Reflection.* A draft of a letter to Coleridge in 1832, which remained unsent, proclaims his adherence to Bošković's theory of point atoms in space. Coleridge had been very critical of atomism in the *Aids to Reflection,* and Hamilton inquired whether the mathematical atomism, which he believed was required for the undulatory theory of light, was acceptable to Coleridge. He had obtained a copy of Kant's *Critique of Pure Reason* in October 1831 and set about reading it with enthusiasm.

By 1834 Hamilton's idealism was complete. In the introduction to his famous paper "On a General Method in Dynamics" (1834), he declared his support for Bošković and argued for a more abstract and general understanding of "force or of power acting by law in space and time" than that provided by the atomic theory. In a letter also of the same year he wrote: "Power, acting by law in Space and Time, is the ideal base of an ideal world, into which it is the problem of physical science to refine the phenomenal world, that so we may behold as one, and under the forms of our own understanding, what had seemed to be manifold and foreign."[3]

While Hamilton was strongly attracted to the ideas of Kant and Bošković, it is difficult to see how they had any direct effect on his system of dynamics. The "General Method in Dynamics" of 1834 was based directly on the characteristic function in optics, which he had worked out well before he studied Kant or Bošković. The most significant contribution of his philosophical studies was to confirm him in his search for the most general application of mathematics to the physical world. It was this high degree of generality and abstraction that permitted him to include wave optics, particle optics, and dynamics in the same mathematical theory. In reading Kant, Hamilton claimed that his greatest pleasure was in finding his own opinions confirmed in Kant's works. It was more "recognizing" than "discovering."[4] He had the same reaction in talking to Faraday. Faraday, the eminently experimental chemist, had arrived at a view as antimaterialistic as his own, although his own view came completely from theoretical studies.[5]

Scientific Work. Hamilton's major contributions were in the algebra of quaternions, optics, and dynamics. He spent more time on quaternions than on any other subject. Next in importance was optics. His dynamics, for which he is best known, was a distant third. His manuscript notebooks and papers contain many optical studies and drafts of published papers, while there is relatively little on his dynamical theories. One is forced to conclude that the papers on dynamics were merely extensions of fundamental ideas developed in his optics.

The published papers are very difficult to read. Hamilton gave few examples to illustrate his methods, and his exposition is completely analytical with no diagrams. His unpublished papers are quite different. In working out his ideas he tested them on practical problems, often working through lengthy computations. A good example is his application of the theory of systems of rays to the symmetrical optical system—a very valuable investigation that remained in manuscript until his optical papers were collected and published in 1931.

In the "Theory of Systems of Rays" (1827) Hamilton continued the work of his paper "On Caustics" (1824), but he applied the analysis explicitly to geometrical optics and introduced the characteristic function. Only the first of three parts planned for the essay was actually published, but Hamilton continued his analysis in three published supplements between 1830 and 1832. In the "Theory of Systems of Rays," Hamilton considered the rays of light emanating from a point source and being reflected by a curved mirror. In this first study the medium is homogeneous and isotropic, that is, the velocity of light is the same at every point and for every direction in the medium. Under these conditions the rays filling space are such that they can be cut orthogonally by a family of surfaces, and Hamilton proved that this condition continues to hold after any number of reflections and refractions.

Malus had proved the case only for a single reflection or refraction. There had been more general proofs given subsequently by Dupin, Quetelet, and Gergonne, but Hamilton was apparently unaware of their work.[6] He proved the theorem by a modification of the principle of least action which he later called the principle of varying action. The principle of least or stationary action (which was identical to Fermat's principle in the optical case) determined the path of the ray between any fixed end points. By varying the initial point on a surface perpendicular to the ray, Hamilton was able to demonstrate that the final points of the original and varied rays fall on a surface perpendicular to both of them. Therefore, at any time after several reflections the end points of the rays determine a surface perpendicular to the rays.

Hamilton called these surfaces "surfaces of constant action," a term that made sense if light was considered as particles, since all the particles emanating together from the source reached the surface at the same time. But Hamilton continually insisted that this "remarkable analogy" between the principle of least action and geometrical optics did not require the assumption of any hypothesis about the nature of light, because the stationary integral to be found by the calculus of variations was of the same form whether light was considered as particles (in which case the integral is the action $\int v\, ds$) or as waves (in which case the integral is the optical length $\int \mu\, ds$).

From the property that all rays are cut perpendicularly by a family of surfaces, Hamilton showed that the differential form

$$\alpha\, dx + \beta\, dy + \gamma\, dz$$

has to be derived, where α, β, γ are direction cosines of the ray and are taken as functions of the coordinates (x,y,z). The direction cosines must then equal the partial differential coefficients of a function V of x, y, z, so that

$$\alpha = \frac{\partial V}{\partial x}, \beta = \frac{\partial V}{\partial y}, \gamma = \frac{\partial V}{\partial z}.$$

From the relation between the direction cosines $\alpha^2 + \beta^2 + \gamma^2 = 1$ Hamilton obtained the expression

$$\left(\frac{\partial V}{\partial x}\right)^2 + \left(\frac{\partial V}{\partial y}\right)^2 + \left(\frac{\partial V}{\partial z}\right)^2 = 1$$

by substitution. He noticed that a solution of this equation is obtained by making V the length of the ray. It is the function V that Hamilton called the "characteristic function," and he declared that it was "the most complete and simple definition that could be given of the application of analysis to optics." The characteristic function contains "the whole of mathematical optics." [7]

The long third supplement of 1832 was Hamilton's most general treatment of the characteristic function in optics and was essentially a separate treatise. Where the initial point was previously fixed so that V was a function only of the coordinates of the final point, the initial coordinates were now added as variables so that V became a function of both initial and final coordinates. The characteristic function now completely described the optical system, since it held for any set of incident rays rather than only for those from a given initial point. Hamilton further generalized his investigation by allowing for heterogeneous and anisotropic media, and this greater generality allowed him to introduce an auxiliary function T, which, in the case of homogeneous initial and final media, is a function of the directions of the initial and final rays.

The importance of the characteristic function came from the fact that it described the system as a function of variables describing the initial and final rays. The principle of least action determined the optical path between fixed points. The characteristic function made the optical length a function of variable initial and final points.

At the end of the third supplement, Hamilton applied his characteristic function to the study of Fresnel's wave surface and discovered that for the case of biaxial crystals there exist four conoidal cusps on the wave surface. From this discovery he predicted that a single ray incident in the correct direction on a biaxial crystal should be refracted into a cone in the crystal and emerge as a hollow cylinder. He also predicted that if light were focused into a cone incident on the crystal, it would pass through the crystal as a single ray and emerge as a hollow cone. Hamilton described his discovery to the Royal Irish Academy on 22 October 1832 and asked Humphrey Lloyd, professor of natural philosophy at Trinity College, to attempt an experimental verification. Lloyd had some difficulty obtaining a satisfactory crystal, but two months later he wrote to Hamilton that he had found the cone.

Hamilton's theoretical prediction of conical refraction and Lloyd's verification caused a sensation. It was one of those rare events where theory predicted a completely unexpected physical phenomenon. Unfortunately it also involved Hamilton in an unpleasant controversy over priority with his colleague James MacCullagh, who had come very close to the discovery in 1830. MacCullagh was persuaded not to push his claim, but after this incident Hamilton was very sensitive about questions of priority.

Hamilton's theory of the characteristic function had little impact on the matter of greatest moment in optics at the time—the controversy between the wave and particle theories of light. Since his theory applied equally well to both explanations of light, his work in a sense stood above the controversy. He chose, however, to support the wave theory; and his prediction of conical refraction from Fresnel's wave surface was taken as another bit of evidence for waves. He entered into the debates at the meetings of the British Association and took part in an especially sharp exchange at the Manchester meeting of 1842, where he defended the wave theory against attacks by Sir David Brewster.

Shortly after completion of his third supplement to the "Theory of Systems of Rays," Hamilton undertook to apply his characteristic function to mechanics as well as to light. The analogy was obvious from his first use of the principle of least action. As astronomer royal of Ireland he appropriately applied his theory first to celestial mechanics in a paper entitled "On a General Method of Expressing the Paths of Light and of the Planets by the Coefficients of a Characteristic Function" (1833). He subsequently bolstered this rather general account with a more detailed study of the problem of three bodies using the characteristic function. The latter treatise was not published, however, and Hamilton's first general statement of the characteristic function applied to dynamics was his famous paper "On a General Method in Dynamics" (1834), which was followed the next year by a second essay on the same subject.

These papers are difficult to read. Hamilton presented his arguments with great economy, as usual, and his approach was entirely different from that now commonly presented in textbooks describing the method. In the two essays on dynamics Hamilton first applied the characteristic function V to dynamics just as he had in optics, the characteristic function being the action of the system in moving from its initial to its final point in configuration space. By his law of varying action he made the initial and final coordinates the independent variables of the characteristic function. For conservative systems, the total energy H was constant along any real path but varied if the initial and final points were varied, and so the characteristic function in dynamics became a function of the $6n$ coordinates of initial and final position (for n particles) and the Hamiltonian H.

The function V could be found only by integrating the equations of motion—a formidable task, as Hamilton realized. His great achievement, as he saw it, was to have reduced the problem of solving $3n$ ordinary differential equations of the second order (as given by Lagrange) to that of solving two partial differential equations of the first order and second degree. It was not clear that the problem was made any easier, but as Hamilton said: "Even if it should be thought that no practical facility is gained, yet an intellectual pleasure may result from the reduction of the most complex . . . of all researches respecting the forces and motions of body, to the study of one characteristic function, the unfolding of one central relation."[8]

The major part of the first essay was devoted to methods of approximating the characteristic function in order to apply it to the perturbations of planets and comets. It was only in the last section that he introduced a new auxiliary function called the principal function (S) by the transformation $V = tH + S$, thereby adding the time t as a variable in place of the Hamiltonian H.

The principal function could be found in a way analogous to the characteristic function; that is, it had to satisfy the following two partial differential equations of the first order and second degree:

$$\frac{\partial S}{\partial t} + \sum \frac{1}{2m}\left[\left(\frac{\partial S}{\partial x}\right)^2 + \left(\frac{\partial S}{\partial y}\right)^2 + \left(\frac{\partial S}{\partial z}\right)^2\right] = U$$

$$\frac{\partial S}{\partial t} + \sum \frac{1}{2m}\left[\left(\frac{\partial S}{\partial a}\right)^2 + \left(\frac{\partial S}{\partial b}\right)^2 + \left(\frac{\partial S}{\partial c}\right)^2\right] = U_0.$$

The variables x, y, z of the first equation represent the position of the particles at some time t; and the variables a, b, c of the second equation are the initial coordinates. U is the negative of the potential energy.

In the second essay, Hamilton deduced from the principal function the now familiar canonical equations of motion and immediately below showed that the same function S was equal to the time integral of the Lagrangian between fixed points

$$S = \int_0^t (T + U)\,dt = \int_0^t L\,dt.$$

The statement that the variation of this integral must be equal to zero is now referred to as Hamilton's principle.

A solution to Hamilton's principal function was very difficult to obtain in most actual cases, and it was K. G. J. Jacobi who found a much more useful form of the same equation.[9] In Jacobi's theory the S function is a generating function which completely characterizes a canonical transformation even when the Hamiltonian depends explicitly on the time. Since the canonical transformation depends on a single function, Jacobi was able to drop the second of Hamilton's two equations and the problem was reduced to the solution of the single partial differential equation

$$\frac{\partial S}{\partial t} + H\left(q_1, \cdots, q_n; \frac{\partial S}{\partial q_1}, \cdots, \frac{\partial S}{\partial q_n}; t\right) = 0,$$

which is usually referred to as the Hamilton-Jacobi equation. Hamilton had shown that the principal function S, defined as the time integral of the Lagrangian L, was a special solution of a partial differential equation; but it was Jacobi who demonstrated the converse, that by the theory of canonical transformations any complete solution of the Hamilton-Jacobi equation could be used to describe the motion of the mechanical system.[10]

The difficulty of solving the Hamilton-Jacobi equation gave little advantage to the Hamiltonian method over that of Lagrange in the nineteenth century. The method had admirable elegance but little practical advantage. With the rise of quantum mechanics, however, Hamilton's method suddenly regained importance because it was the one form of classical mechanics that carried over directly into the quantum interpretation. A great advantage of the Hamiltonian method was the close analogy between mechanics and optics that it contained; and this analogy was exploited by Louis de Broglie and Schrödinger in their formulations of wave mechanics.

Hamilton's tendency to pursue his studies in their greatest generality led to other important contributions. He extended his general method in dynamics to create a "calculus of principal relations," which permitted the solution of certain total differential equations by the calculus of variations. Another important contribution was the hodograph, the curve defined by the velocity vectors of a point in orbital motion taken as drawn from the origin rather than from the moving point.

Hamilton also attempted to apply his dynamics to the propagation of light in a crystalline medium. Previous authors, said Hamilton, had written much on the preservation of light vibrations in different media, but no one had attempted to investigate the propagation of a wave front into an undisturbed medium; or as he explained it to John Herschel, "Much had been done, perhaps, in the dynamics of *light;* little, I thought, in the dynamics of *darkness.*"[11] This new science of the dynamics of darkness he named "skotodynamics." He was actually hampered by his enthusiasm for Bošković's theory because it led him to study the medium as a series of attracting points rather than as a continuum; but his research, as usual, led to important new ideas. One was the distinction between group velocity and phase velocity. Another was his valuable study of "fluctuating functions," an extension of Fourier's theorem, which in turn led him to give the first complete asymptotic expansion for Bessel functions.

All of Hamilton's work in optics and dynamics depended on a single central idea, that of the characteristic function. It was the first of his two great "discoveries." The second was the quaternions, which he discovered on 16 October 1843 and to which he devoted most of his efforts during the remaining twenty-two years of his life. In October 1828 Hamilton complained to his friend John T. Graves about the shaky foundations of algebra. Such notions as negative and imaginary numbers, which appeared to be essential for algebra, had no real meaning for him; and he argued that a radical rewriting of the logical foundations of algebra was badly needed.[12] In the same year John Warren published *A Treatise on the Geometrical Representation of the Square Roots of Negative Quantities.* Hamilton read it in 1829 at Graves's instigation. Warren's book described the so-called Argand diagram by which the complex number is represented as a point on a plane with one rectangular axis representing the real part of the number and the other axis representing the imaginary part. This geometrical representation of complex numbers raised two new questions in Hamilton's mind: (1) Is there any other algebraic representation of complex numbers that will reveal all valid operations on them? (2) Is it possible to find a hypercomplex number that is related to three-dimensional space just as a regular complex number is related to two-dimensional space? If such a hypercomplex number could be found, it would be a "natural" algebraic representation of space, as opposed to the artificial and somewhat arbitrary representation by coordinates.

On 4 November 1833 Hamilton read a paper on algebraic couples to the Royal Irish Academy in which he presented his answer to the first question. His algebraic couples consisted of all ordered pairs of real numbers, for which Hamilton defined rules of addition and multiplication. He then demonstrated that these couples constituted a commutative associative division algebra, and that they satisfied the rules for operations with complex numbers. For some mathematicians the theory of number couples was a more significant contribution to mathematics than the discovery of quaternions.[13] On 1 June 1835 Hamilton presented a second paper on number couples entitled *Preliminary and Elementary Essay on Algebra as the Science of Pure Time,* in which he identified the number couples with steps in time. He combined this paper with his earlier paper of 1833, added some *General Introductory Remarks,* and published them in the *Proceedings of the Royal Irish Academy* of 1837.

This was a time of intense intellectual activity for Hamilton. He was deeply involved in the study of dynamics as well as the algebra of number couples. This was also the time when he was most involved in the study of Kant and was forming his own idealistic philosophy. Manuscript notes from 1830 and 1831 (before he read Kant) already contained Hamilton's conviction that the foundation for algebra was to be found in the ordinal character of numbers, and that this ordering had an intuitive basis in time. Kant's philosophy must certainly have strengthened this conviction.[14] It was through the concept of time that Hamilton hoped to correct the weaknesses in the logical foundations of algebra. He recognized three different schools of algebra. The first was the practical school, which considered algebra as an instrument for the solution of problems; therefore it sought rules of application. The second, or philological school, considered algebra as a language consisting of formulas composed of symbols which could be arranged only in certain specified ways. The third was the theoretical school, which considered algebra as a group of theorems upon which one might meditate. Hamilton identified himself with the last school and insisted that in algebra it was necessary to go beyond the signs of the formalist to the things signified. Only by relating numbers to some real intuition could algebra be truly called a "science."

In the *Critique of Pure Reason,* Kant argued that the ordering of phenomena in space was an operation of the mind and that this ordering had to be part of the mode of perceiving things. The science that studied this aspect of perception in its purest form was geometry; therefore geometry could well be called the "science of pure space." According to Hamilton the intuition of order in time was even more deep-seated in the human mind than the intuition of order in space. We have an intuitive concept of pure or mathematical time more fundamental than all actual chronology or ordering of particular events. This intuition of mathematical time is the real referent of algebraic symbolism. It is "co-extensive and identical with Algebra, so far as Algebra itself is a Science."[15] Hamilton presented this idea to his fellow mathematicians Graves and De Morgan with some hesitation and received an unenthusiastic response, which he probably anticipated. Although the idea had been with him for some time Hamilton mentioned it casually to Graves for the first time in a letter of 11 July 1835 where he referred to it as this "crochet of mine."[16] But in spite of the adverse reaction Hamilton never wavered in his conviction that the intuition of time was the foundation of algebra.

Hamilton had less success in answering the second question posed above, whether it would be possible to write three-dimensional complex numbers or, as he called them, "triplets." Addition of triplets was obvious, but he could find no operation that would follow the rules of multiplication. Thirteen years after Hamilton's death G. Frobenius proved that there is no such algebra and that the only possible associative division algebras over the real numbers are the real numbers themselves, complex numbers, and real quaternions. In searching for the elusive triplets, Hamilton sought some way of making his triplets satisfy the law of the moduli, since any algebra obeying this law is a division algebra. The modulus of a complex number is that number multiplied by its complex conjugate, and the law of the moduli states that the product of the moduli of two complex numbers equals the modulus of the product. By analogy to complex numbers, Hamilton wrote the triplet as $x + iy + jz$ with $i^2 = j^2 = -1$ and took as its modulus $x^2 + y^2 + z^2$. The product of two such moduli can be expressed as the sum of squares; but it is the sum of four squares not the sum of three squares, as would be the case if it were the modulus of a triplet.

The fact that he obtained the sum of four squares for the modulus of the product must have indicated to Hamilton that possibly ordered sets of four numbers, or "quaternions," might work where the triplets failed. Thus he tested hypercomplex numbers of the form $(a + ib + jc + kd)$ to see if they satisfied the law of the moduli. They worked, but only by sacrificing the commutative law. Hamilton had to make the product $ij = -ji$.[17] Hamilton's great insight came in realizing that he could sacrifice commutativity and still have a meaningful and consistent algebra. The laws for multiplication of quaternions then followed immediately:

$$\begin{aligned} ij &= k = -ji, \\ jk &= i = -kj, \\ ki &= j = -ik, \\ i^2 &= j^2 = k^2 = ijk = -1. \end{aligned}$$

The quaternions came to Hamilton in one of those flashes of understanding that occasionally occur after long deliberation on a problem. He was walking into Dublin on 16 October 1843 along the Royal Canal to preside at a meeting of the Royal Irish Academy, when the discovery came to him. As he described it, "An electric circuit seemed to close."[18] He immediately scratched the formula for quaternion multiplication on the stone of a bridge over the canal. His reaction must have been in part a desire to commemorate a discovery of capital importance, but it was also a reflection of his working habits. Hamilton was an inveterate scribbler. His manuscripts are full of jottings made on walks and in carriages. He carried books, pencils, and paper everywhere he went. According to his son he would scribble on his fingernails and even on his hard-boiled egg at breakfast if there was no paper handy.

Hamilton was convinced that in the quaternions he had found a natural algebra of three-dimensional space. The quaternion seemed to him to be more fundamental than any coordinate representation of space, because operations with quaternions were independent of any given coordinate system. The scalar part of the quaternion caused difficulty in any geometrical representation and Hamilton tried without notable success to interpret it as an extraspatial unit. The geometrical significance of the quaternion became clearer when Hamilton and A. Cayley independently showed that the quaternion operator rotated a vector about a given axis.[19]

The quaternions did not turn out to be the magic key that Hamilton hoped they would be, but they were significant in the later development of vector analysis. Hamilton himself divided the quaternion into a real part and a complex part which he called a vector. The multiplication of two such vectors according to the rules for quaternions gave a product consisting again of a scalar part and a complex part.

$$\alpha = xi + yj + zk$$
$$\alpha' = x'i + y'j + z'k$$

$$\alpha\alpha' = -(xx' + yy' + zz') + i(yz' - zy') + j(zx' - xz') + k(xy' - yx')$$

The scalar part, which he wrote as $S.\ \alpha\alpha'$, is recognizable as the negative of the scalar or dot product of vector analysis, and the vector part, which he wrote as $V.\ \alpha\alpha'$, is recognizable as the vector or cross product. Hamilton frequently used these symbols as well as a new operator which he introduced,

$$\lhd = i\frac{d}{dx} + j\frac{d}{dy} + k\frac{d}{dz}$$

and

$$-\lhd^2 = \left(\frac{d}{dx}\right)^2 + \left(\frac{d}{dy}\right)^2 + \left(\frac{d}{dz}\right)^2,$$

and called attention to the fact that the applications of this new operator in physics "must be extensive to a high degree."[20] Gibbs suggested the name "del" for the same operator in vector analysis and this is the term now generally used.

Hamilton was not the only person working on vectorial systems in the mid-nineteenth century.[21] Hermann Günther Grassmann working independently of Hamilton published his *Ausdehnungslehre* in 1844 in which he treated n-dimensional geometry and hypercomplex systems in a much more general way than Hamilton; but Grassmann's book was extremely difficult and radical in its conception and so had very few readers. Hamilton's books on quaternions were also too long and too difficult to attract much of an audience. His *Lectures on Quaternions* (1853) ran to 736 pages with a sixty-four-page preface. Any reader can sympathize with John Herschel's request that Hamilton make his principles "clear and familiar down to the level of ordinary unmetaphysical apprehension" and to "introduce the new phrases as strong meat gradually given to babes."[22] His advice was ignored and the *Lectures* bristles with complicated new terms such as *vector, vehend, vection, vectum, revector, revehend, revection, revectum, provector, transvector,* etc.[23] Herschel replied with a cry of distress, but it did no good and the *Elements of Quaternions,* which began as a simple manual, was published only after Hamilton's death and was even longer than the *Lectures.*

The first readable book on quaternions was P. G. Tait's *Elementary Treatise on Quaternions* (1867). Tait and Hamilton had been in correspondence since 1858, and Tait had held up the publication of his book at Hamilton's request until after the *Elements* appeared. Tait was Hamilton's most prominent disciple, and during the 1890's entered into a heated controversy with Gibbs and Heaviside over the relative advantages of quaternions and vectors. One can sympathize with Tait's commitment to quaternions and his dissatisfaction with vector analysis. It was difficult enough to give up the commutative property in quaternion multiplication, but vector analysis required much greater sacrifices. It accepted *two* kinds of multiplication, the dot product and the cross product. The dot product was not a real product at all, since it did not preserve closure; that is, the product was not of the same nature as the multiplier and the multiplicand. Both products failed to satisfy the law of the moduli, and both failed to give an unambiguous method of division. Moreover the cross product (in which closure was preserved) was neither associative nor commutative.[24] No wonder a devout quaternionist like Tait looked upon vector analysis as a "hermaphrodite monster."[25] Nevertheless vector analysis proved to be the more useful tool, especially in applied mathematics. The controversy did not entirely die, however, and as late as 1940 E. T. Whittaker argued that quaternions "may even yet prove to be the most natural expression of the new physics [quantum mechanics]."[26]

The quaternions were not the only contribution that Hamilton made to mathematics. In 1837 he corrected Abel's proof of the impossibility of solving the general quintic equation and defended the proof against G. B. Jerrard, who claimed to have found such a solution. He also became interested in the study of polyhedra and developed in 1856 what he called the "Icosian Calculus," a study of the properties of the icosahedron and the dodecahedron. This study resulted in an "Icosian Game" to be played on the plane projection of a dodecahedron. He sold the copyright to a Mr. Jacques of Piccadilly for twenty-five pounds. The game fascinated a mathematician like Hamilton, but it is unlikely that Mr. Jacques ever recovered his investment.

In spite of Hamilton's great fame in the nineteenth century one is left with the impression that his discoveries had none of the revolutionary impact on science that he had hoped for. His characteristic function in optics did not hit at the controversy then current over the physical nature of light, and it became important for geometrical optics only sixty years later when Bruns rediscovered the characteristic function and called it the method of the eikonal.[27] His dynamics was saved from oblivion by the important additions of Jacobi, but even then the Hamiltonian method gained a real advantage over other methods only with the advent of quantum mechanics. The quaternions, too, which were supposed to open the doors to so many new fields of science turned out to be a disappointment. Yet quaternions were the seed from which other noncommutative algebras grew. Matrices and even vector analysis have a parent in quaternions. Over the long run the success of Hamilton's work has justified his efforts. The high degree of abstraction and generality that made his papers so difficult to read has also made them stand the test of time, while more specialized researches with greater immediate utility have been superseded.

NOTES

1. Graves, *Life of Sir William Rowan Hamilton,* I, 141.
2. *Ibid.,* 186.
3. Hamilton to H. F. C. Logan, 27 June 1834, Graves, II, 87–88.
4. Hamilton to Lord Adare, 19 July 1834, Graves, II, 96; and to Wordsworth, 20 July 1834, Graves, II, 98.
5. Graves, II, 95–96.
6. *Mathematical Papers,* I, 463, editor's note.
7. *Ibid.,* 17, 168.
8. *Ibid.,* II, 105.
9. *Crelle's Journal,* **17** (1837), 97–162.
10. The differences between Hamilton's and Jacobi's formulations are described in detail in the *Mathematical Papers,* II, 613–621, editor's app. 2; and in Lanczos, *Variational Principles,* 229–230, 254–262.
11. *Mathematical Papers,* II, 599.
12. Graves, I, 303–304.
13. C. C. MacDuffee, "Algebra's Debt to Hamilton," in *Scripta mathematica,* **10** (1944), 25.
14. MS notebook no. 25, fol. 1, and notebook no. 24.5, fol. 49. See also Graves, I, 229, where Hamilton in 1827 referred to "the sciences of Space and Time (to adopt here a view of Algebra which I have elsewhere ventured to propose)."
15. *Mathematical Papers,* III, 5.
16. Graves, II, 143.
17. E. T. Whittaker, "The Sequence of Ideas in the Discovery of Quaternions," in *Proceedings of the Royal Irish Academy,* **50A** (1945), 93–98.
18. Graves, II, 435.
19. *Mathematical Papers,* III, 361–362.
20. *Ibid.,* 262–263.
21. Crowe, *History of Vector Analysis,* pp. 47–101.
22. Graves, II, 633.
23. Crowe, p. 36.
24. *Ibid.,* pp. 28–29.
25. *Ibid.,* p. 185.
26. E. T. Whittaker, "The Hamiltonian Revival," in *Mathematical Gazette,* **24** (1940), 158.
27. J. L. Synge, "Hamilton's Method in Geometrical Optics," in *Journal of the Optical Society of America,* **27** (1937), 75–82.

BIBLIOGRAPHY

I. ORIGINAL WORKS. Hamilton's mathematical papers have been collected in three volumes, *The Mathematical Papers of Sir William Rowan Hamilton* (Cambridge, 1931–1967). These volumes are carefully edited with short introductions and very valuable explanatory appendices and notes. The collection is not complete, but the editors have selected the most important papers, including many that were previously unpublished. A complete bibliography of Hamilton's published works appears at the end of vol. III of Robert P. Graves, *Life of Sir William Rowan Hamilton,* 3 vols. (Dublin, 1882–1889). Graves collected Hamilton's papers and letters for his biography shortly after Hamilton's death. The bulk of these manuscripts is now at the library of Trinity College, Dublin, with a smaller collection at the National Library of Ireland, Dublin. The manuscript collection at Trinity is very large, containing approximately 250 notebooks and a large number of letters and loose papers.

II. SECONDARY LITERATURE. R. P. Graves's biography is composed largely of letters which have been edited to remove much of the mathematical content. Graves also suppressed some correspondence that he considered too personal.

Most of the secondary literature on Hamilton has been written by mathematicians interested in the technical aspects of his work. An exception is Robert Kargon, "William Rowan Hamilton and Boscovichean Atomism," in *Journal of the History of Ideas,* **26** (1965), 137–140. The best introduction to Hamilton's optics is John L. Synge, *Geometrical Optics; an Introduction to Hamilton's Method* (Cambridge, 1937). Also valuable are his "Hamilton's Method in Geometrical Optics," in *Journal of the Optical Society of America,* **27** (1937), 75–82; G. C. Steward, "On the Optical Writings of Sir William Rowan Hamilton," in *Mathematical Gazette,* **16** (1932), 179–191; and George Sarton, "Discovery of Conical Refraction by Sir William Rowan Hamilton and Humphrey Lloyd (1833)," in *Isis,* **17** (1932), 154–170.

Hamilton's work in dynamics is described in René Dugas, "Sur la pensée dynamique d'Hamilton: origines optiques et prolongements modernes," in *Revue scientifique,* **79** (1941), 15–23; and A. Cayley, "Report on the Recent Progress of Theoretical Dynamics," in *British Association Reports* (1857), pp. 1–42. Another valuable exposition of the method is in Cornelius Lanczos, *The Variational Principles of Mechanics,* 3rd ed. (Toronto, 1966).

The centenary of Hamilton's discovery of quaternions was the occasion for two very important collections of articles, in *Proceedings of the Royal Irish Academy,* **50A,** no. 6 (Feb. 1945), 69–121; and in *Scripta mathematica,* **10** (1944), 9–63. These collections cover not only the quaternions, but also contain biographical notices, an article on the mathematical school at Trinity College, Dublin, and articles on Hamilton's dynamics, his optics, and his other contributions to algebra. The relationship between quaternions and vector analysis is described in great detail in Michael Crowe, *A History of Vector Analysis; the Evolution of the Idea of a Vectorial System* (Notre Dame, Ind., 1967); and in Reginald J. Stephenson, "Development of Vector Analysis From Quaternions," in *American Journal of Physics,* **34** (1966), 194–201; and Alfred M. Bork, "'Vectors Versus Quaternions'—the Letters in Nature," in *American Journal of Physics,* **34** (1966), 202–211.

THOMAS L. HANKINS

HAMPSON, WILLIAM (*b.* Bebington, Cheshire, England, *ca.* 1854; *d.* London, England, 1 January 1926), *chemical engineering.*

Hampson was educated at Manchester Grammar School and Trinity College, Oxford, graduating M.A. in 1881. He went to the Inner Temple, evidently with the intention of becoming a barrister; but he does not appear in any *Law List,* and his activities are unknown until 1895, when he patented a machine for making liquid air. Independently of and slightly earlier than Carl von Linde and Georges Claude, Hampson applied the "cascade" principle: air cooled by the Joule-Thomson effect was used to precool incoming air before its expansion. This simple device transformed liquid air, and liquid gases in general, from laboratory curiosities to articles of commerce. The invention was taken up by Brin's Oxygen Company of Westminster (later the British Oxygen Company), with Hampson acting as consultant. He worked closely with William Ramsay and his colleagues at University College, London, who were then engaged in their classic work on the inert gases; the ample supplies of liquid air provided by Hampson proved invaluable and, indeed, led directly to the discovery of neon. He had the misfortune, however, to cross the path of the ungenerous James Dewar regarding priority over the liquefaction of hydrogen, and a pointless and unedifying controversy arose between them.

After taking out a few more patents modifying his invention, Hampson again disappeared into obscurity, except as the author of two books on popular science, *Paradoxes of Science* (1904) and *The Explanation of Radium* (1906), and an unnoticed political tract, *Modern Thraldom* (1907), in which he ascribed all the ills of the age to the institution of credit. He later qualified as a medical practitioner, worked in various London hospitals on the medical applications of electricity and X rays, and invented some devices of no lasting importance.

BIBLIOGRAPHY

Hampson described his machine for liquefying air in a lecture, "Self-Intensive Refrigeration of Gases: Liquid Air and Oxygen," reprinted in *Journal of the Society of Chemical Industry,* **17** (1898), 411—the lecture ended in an angry argument with Dewar. Their controversy was carried on in *Nature,* **55** (1897), 485, and **58** (1898), 77, 174, 246, 292. Of his medicoelectrical contributions probably the most important is "A Method of Reducing Excessive Frequency of the Heart Beat by Means of Rhythmical Muscle-Contractions Electrically Provoked," in *Proceedings of the Royal Society of Medicine,* Electrotherapy Sec., **5** (1912), 119.

No biography of Hampson has previously been written; the only sources are reference books and his own publications. His part in the discovery of the inert gases is described in M. W. Travers, *The Discovery of the Rare Gases* (London, 1928), pp. 89, 94, 98, 115; and *A Life of Sir William Ramsay* (London, 1956), pp. 172–176, 180.

W. V. FARRAR

HAMY, MAURICE THÉODORE ADOLPHE (*b.* Boulogne-sur-Mer, France, 31 October 1861; *d.* Paris, France, 9 April 1936), *celestial mechanics, astronomy, optics.*

Born into a Picard family, Hamy received his secondary education in the various cities where his father, an official in the postal service, was assigned. He came to Paris to prepare for his *licence* in science, which he obtained in 1884. Next he was a student astronomer at the Paris observatory, where he remained throughout his career, becoming astronomer in 1893 and chief astronomer in 1904. He retired in 1929.

A mathematician and physicist, Hamy did research in various areas of astronomy and in related fields. In celestial mechanics he studied the forms of heavenly bodies and demonstrated that their equipotential surfaces cannot be strictly ellipsoidal unless the ellipsoids are homofocal, a condition which cannot be met in the case of planets. The problem of planetary perturbations led him to consider the asymptotic value of coefficients of high degree. This in turn permitted Hamy to calculate certain long-term inequalities, such as those of the motion of Juno. Through this accomplishment he advanced the research on approximate values of functions of large numbers.

In instrumental astronomy Hamy devised several procedures for improving the determination of the constants of the meridian instrument. He also improved the technique of measuring radial velocities with the objective prism. In the course of his work in spectroscopy he made several important determinations on monochromatic radiations, notably on those of cadmium.

Hamy's investigations probably of most interest today are those concerning the study of stellar and planetary diameters through interferometry. The use of wide slits, necessary for collecting sufficient light, posed difficult mathematical problems that he overcame. Thus he was able to use the method successfully on the satellites of Jupiter and on Vesta.

Although Hamy suffered until his death from an intestinal disease contracted in 1905 in Spain while observing an eclipse, his activity was considerable. He left more than 100 scientific publications. In 1908 he was elected to the Academy of Sciences and in 1916 to the Bureau of Longitudes, of which he became chairman in 1921.

BIBLIOGRAPHY

Almost all of Hamy's works appeared in either *Comptes rendus hebdomadaires des séances de l'Académie des sciences* or *Bulletin astronomique* between 1887 and 1928. The most important in mathematics and interferometry are "Étude sur la figure des corps célestes," in *Annales de l'Observatoire de Paris,* **19** (1889), F1–F54, his doctoral dissertation; "Théorie générale de la figure des planètes," in *Journal de mathématiques pures et appliquées,* 4th ser., **6** (1890), 69–143; "Mesures interférentielles des faibles diamètres," in *Bulletin astronomique,* **10** (1893), 489–504, and **16** (1899), 257–274, also in *Comptes rendus hebdomadaires des séances de l'Académie des sciences,* **127** (1898), 851, 982 (errata); **128** (1899), 583; **174** (1922), 342, 904; **175** (1922), 1123; **176** (1923), 1849; "Développement approché de la fonction perturbatrice . . .," in *Journal de mathématiques pures et appliquées,* 4th ser., **10** (1894), 391–472, and 5th ser., **2** (1896), 381–439; and "L'approximation des fonctions de grands nombres," *ibid.,* n.s. **4** (1908), 203–281.

There are two biographical notices by E. Picard: "Notice nécrologique," in *Comptes rendus hebdomadaires des séances de l'Académie des sciences,* **202** (1936), 1317; and "La vie et l'oeuvre de Maurice Hamy," in *Annuaire publié par le Bureau des longitudes pour l'an 1943,* A1–A15.

JACQUES R. LÉVY

HANKEL, HERMANN (*b.* Halle, Germany, 14 February 1839; *d.* Schramberg, near Tübingen, Germany, 29 August 1873), *mathematics, history of mathematics.*

Hankel's father, the physicist Wilhelm Gottlieb Hankel, was associate professor at Halle from 1847 and full professor at Leipzig from 1849. Hankel studied at the Nicolai Gymnasium in Leipzig, where he improved his Greek by reading the ancient mathematicians in the original. Entering Leipzig University in 1857, he studied with Moritz Drobisch, A. F. Moebius, Wilhelm Scheibner, and his father. In 1860 Hankel proceeded to Göttingen, where from Georg Riemann he acquired his special interest in the theory of functions. At this time he published his prize-winning *Zur allgemeinen Theorie der Bewegung der Flüssigkeiten* (Göttingen). The following year he studied in Berlin with Karl Weierstrass and Leopold Kronecker, and in 1862 he received his doctorate at Leipzig for *Ueber eine besondere Classe der symmetrischen Determinanten* (Göttingen, 1861). He qualified for teaching in 1863 and in the spring of 1867 was named associate professor at Leipzig. In the fall of that year he became full professor at Erlangen, where he married Marie Dippe. Called to Tübingen in 1869, he spent the last four years of his life there.

Hankel's contributions to mathematics were concentrated in three areas: the study of complex and higher complex numbers, the theory of functions, and the history of mathematics. His most important contribution in the first area was *Theorie der complexen Zahlensysteme* (Leipzig, 1867), to which he had hoped to add a treatise on the functions of a complex variable. This work constitutes a lengthy presentation of much of what was then known of the real, complex, and hypercomplex number systems. In it Hankel presented algebra as a deductive science treating entities which are intellectual constructs. Beginning with a revised statement of George Peacock's principle of the permanence of formal laws, he developed complex numbers as well as such higher algebraic systems as Moebius' barycentric calculus, some of Hermann Grassmann's algebras, and W. R. Hamilton's quaternions. Hankel was the first to recognize the significance of Grassmann's long-neglected writings and was strongly influenced by them. The high point of the book lies in the section (pp. 106–108) in which he proved that no hypercomplex number system can satisfy all the laws of ordinary arithmetic.

In the theory of functions Hankel's major contributions were *Untersuchungen über die unendlich oft oscillirenden und unstetigen Functionen* (Tübingen, 1870) and his 1871 article "Grenze" for the Ersch-Gruber *Encyklopädie.* In the former, he reformulated Riemann's criterion for integrability, placing the emphasis upon measure-theoretic properties of sets of points. After making explicit that functions do not possess general properties, he attempted a fourfold classification of functions, discussed the integrability of each type, and presented a method, based on his principle of the condensation of singularities, for constructing functions with singularities at every rational point. Although he confounded the notions of sets of zero content and nowhere-dense sets, his work marked an important advance toward modern integration theory. In "Grenze" he pointed out for the first time the importance of Bernard Bolzano's work on infinite series and published an example of a continuous function that was nondifferentiable at an infinite number of points. In a series of papers in *Mathematische Annalen,* Hankel showed the significance of what are now known as "Hankel functions" or "Bessel functions of the third kind."

Among Hankel's historical writings the best-known are his short *Entwicklung der Mathematik in den letzten Jahrhunderten* (Tübingen, 1869) and his long *Zur Geschichte der Mathematik in Alterthum und Mittelalter* (Leipzig, 1875). Although Moritz Cantor pointed out many errors in the latter book he, G. J. Allman, Florian Cajori, T. L. Heath, and J. T. Merz have recognized the brilliance of Hankel's historical insight.

BIBLIOGRAPHY

I. ORIGINAL WORKS. A list of all Hankel's publications through 1875 will be found in *Bullettino di bibliographia e di storia delle scienze matematiche e fisiche,* **9** (1876), 297–308. This is completed by the following additions: *Untersuchungen über die unendlich oft oscillirenden und unstetigen Functionen,* republished in *Mathematische Annalen,* **20** (1882), 63–112, and as Ostwalds Klassiker der Exacten Wissenschaften, no. 153 (Leipzig, 1905), with comments by P. E. B. Jourdain. Also republished were *Entwicklung der Mathematik* (Tübingen, 1884) and, recently, *Zur Geschichte der Mathematik* (Hildesheim, 1965), with a foreword by J. E. Hofmann.

II. SECONDARY LITERATURE. Hankel's life was discussed by W. von Zahn in *Mathematische Annalen,* **7** (1874), 583–590; and by M. Cantor in *Allgemeinen deutsche Biographie,* X (Leipzig, 1879), 516–519. For his work on complex numbers, see M. J. Crowe, *A History of Vector Analysis* (Notre Dame, Ind., 1967). His contributions to analysis are discussed in P. E. B. Jourdain, "The Development of the Theory of Transfinite Numbers," in *Archiv der Mathematik und Physik,* 3rd ser., **10** (1906), 254–281; and in Thomas Hawkins, *Lebesgue's Theory of Integration: Its Origins and Development* (Madison, Wis., 1970). J. E. Hofmann's foreword to his republication of Hankel's *Zur Geschichte der Mathematik in Altertum und Mittelalter* (Hildesheim, 1965) contains a brief discussion of the quality of Hankel's historical writing as well as a portrait.

MICHAEL J. CROWE

HANKEL, WILHELM GOTTLIEB (*b.* Ermsleben, Harz, Germany, 17 May 1814; *d.* Leipzig, Germany, 17 February 1899), *physics, chemistry.*

Hankel belongs among the older nineteenth-century physicists who typically represented the classical scientist. The son of a choirmaster and teacher, he was greatly interested in practical questions even as a child. After graduating from the Gymnasium at Quedlinburg, he studied at the University of Halle under Johann Schweigger. In 1835 he was an assistant in the physics laboratory and, in 1836, a teacher at the newly founded Realschule of the Frankische Stiftung in Halle. In 1838 he married the daughter of a farmer from near Halberstadt; in 1839 they had a son, Hermann, who became famous as a mathematician. Also in 1839 Hankel earned his doctorate with a dissertation on the electricity of crystals, and in 1840 he qualified to lecture in chemistry at the University of Halle.

In 1842–1843 a severe case of pleurisy forced Hankel to give up his work in the chemistry laboratory, and he turned his attention completely to physics, a decision that he had not made previously out of respect for his teacher Schweigger. In 1847 he obtained a professorship in physics at the University of Halle and, in 1849, a similar position at the University of Leipzig, which he held until 1887.

As an experimenter Hankel investigated primarily piezoelectric and thermoelectric phenomena in crystals and became a pioneer in this specialized field. His thorough observations and measurements were based on the use of new, more reliable measuring instruments which he himself constructed or improved. In 1850 he developed a new electrometer of high sensitivity and low self-capacity, which was utilized in conjunction with a microscope. In his researches Hankel discovered the relationship in crystals between pyroelectric properties and the rotation of the plane of polarization of light. He drew attention to crystal structure and to crystals with and without inversion centers, thereby clarifying the peculiarities of their electrical properties. Moreover, he investigated the thermoelectric currents between metals and minerals as well as the photoelectricity of fluorite and the actinoelectricity of quartz. In addition, Hankel carried out more precise determinations of the galvanic electromotive series. He also studied electricity in flames and gas formation. He reduced his observations on atmospheric electricity, through the use of a torsion balance, to values of the system of absolute measurement by an experimental method (comparison of a known electrostatic field with the atmospheric electrical field) which was complicated but quite exact for the period (1858). In 1856 he wrote a thorough critique of the instruments used until then in studying atmospheric electricity.

Hankel proposed a new theory of electricity which postulated, instead of action at a distance, the existence of variously oriented rotational motions in a single fluid: ether. The theory excited little enthusiasm when it was announced and now merits only historical interest as one in a series of ultimately fruitless efforts to reduce electrodynamics to mechanics.

BIBLIOGRAPHY

I. ORIGINAL WORKS. Hankel wrote sixty-two scientific papers, most of them published in *Berichte über die Verhandlungen der K. Sächsischen Gesellschaft der Wissenschaften zu Leipzig* and in *Annalen der Physik und Chemie.* About twenty-five dealt with the pyroelectricity of crystals. A bibliography may be found in Poggendorff, I, 1011; III, 581–582; and IV, 580.

His scientific papers include "De thermo-electricitate crystallorum" (Halle, 1839), his doctoral dissertation, extracts from which were published as "Ueber die Thermo-Elektricität der Krystalle," in *Annalen der Physik und Chemie,* **49** (1840), 493–504, and **50** (1840), 237–250; and as "Nachtrag zu der Thermo-Elektricität des Topases," *ibid.,* **56** (1842), 37–58; "Ueber die Construction eines Elektrometers," *ibid.,* **84** (1851), 28–36; "Ueber die

Messung der atmosphärischen Elektricität nach absolutem Maasse," *ibid.,* **103** (1858), 209–240; "Maassbestimmungen der elektromotorischen Kräfte," *ibid.,* **115** (1862), 57–62, and **126** (1865), 286–298; "Neue Theorie der elektrischen Erscheinungen," *ibid.,* **126** (1865), 440–466, and **131** (1867), 607–621; "Ueber die thermoelektrischen Eigenschaften des Bergkrystalles," *ibid.,* **131** (1867), 621–631; "Ueber einen Apparat zur Messung sehr kleiner Zeiträume," *ibid.,* **132** (1867), 134–165; "Ueber die actino- und piezoelectrischen Eigenschaften des Bergkrystalles und ihre Beziehungen zu den thermoelectrischen," *ibid.,* n.s. **17** (1882), 163–175; "Neue Beobachtungen über die Thermo- und Actinoelektricität des Bergkrystalles . . .," *ibid.,* **19** (1883), 818–844; "Endgültige Feststellung der auf den Bergkrystallen an den Enden der Nebenaxen bei steigenden und sinkenden Temperaturen auftretenden Polaritäten," *ibid.,* **32** (1887), 91–108; "Das elektrodynamische Gesetz ein Punktgesetz," *ibid.,* **36** (1889), 73–93; and "Die galvanische Kette," *ibid.,* **39** (1890), 369–389.

Hankel's books include *Grundriss der Physik* (Halle, 1848). He also translated into German D. F. J. Arago's *Notices biographiques,* 3 vols. (Leipzig, 1856), and his *Astronomie populaire,* 2nd ed., 4 vols. (Leipzig, 1865). In addition, he edited Arago's *Sämmtliche Werke,* 16 vols. (Leipzig, 1854–1860).

II. SECONDARY LITERATURE. See C. Neumann, "Worte zum Gedächtnis an Wilhelm Hankel," in *Berichte über die Verhandlungen der K. Sächsischen Gesellschaft der Wissenschaften zu Leipzig,* Math.-phys. Kl., **51** (1899), lxii–lxvi; and P. Drude, "Wilhelm Gottlieb Hankel," *ibid.,* lxvii–lxxvi.

HANS-GÜNTHER KÖRBER

HANN, JULIUS FERDINAND VON (*b.* Mühlkreis, near Linz, Austria, 23 March 1839; *d.* Vienna, Austria, 1 October 1921), *meteorology, climatology.*

Hann's father, Joseph, was curator of the manor house in Mühlkreis; his early death in 1852 left the large family in a difficult economic situation. Hann's mother, Anna, subsequently opened a pension for pupils of the Gymnasium in Kremsmünster, where Hann's education began in 1853. After passing his *Abitur* (1860) he enrolled at the University of Vienna and studied mathematics, chemistry, and physics, then geology and paleontology under Eduard Suess and physical geography under Friedrich Simony. He passed his teaching examination in 1863 and taught at high schools in Vienna and Linz.

Hann had pursued meteorological investigations since his boyhood. In 1865 he was invited by Karl Jelinek, director of the Zentralanstalt für Meteorologie und Erdmagnetismus in Vienna, to participate in the editorship of the newly founded *Zeitschrift der Oesterreichischen Gesellschaft für Meteorologie* (which merged in 1885 with the German *Meteorologische Zeitschrift*). In 1867 he was appointed to the staff of the Zentralanstalt. Hann received his Ph.D. at Vienna in 1868, and in the following year he delivered his *Habilitationsschrift.* Appointed associate professor of physical geography at the University of Vienna in 1874, he taught meteorology, climatology, and oceanography. In 1877 he succeeded Jelinek as the director of the Zentralanstalt and was appointed professor of physics. In 1897, at the age of fifty-eight, he resigned his directorship and became professor of meteorology at Graz. Partly because of poor library facilities in Graz, Hann returned in 1900 to Vienna, where he held a professorship of cosmic physics until 1910. He edited the *Meteorologische Zeitschrift* until 1920 and carried on his scientific work until his death.

In 1878 Hann married Luise Weismayr, daughter of a district court president. He was fully absorbed in his work, and only summer journeys to the Alps and several meteorological congresses took him from it. Hann was one of the secretaries of the first international assembly of meteorologists at Leipzig in 1872 and was a member of the International Meteorological Committee (1878–1898). He was also a member of the Vienna Academy and honorary or foreign member of academies and societies throughout the world. He was knighted in 1910.

Hann was one of the most prominent meteorologists of his generation. His importance rested less on the creation of new theoretical concepts than on his efforts to coordinate empirical and theoretical results into a coherent structure. Hann was a very competent editor, and the *Meteorologische Zeitschrift* became the leading meteorological journal under his guidance. He exerted great influence through his capacity to stimulate thought and through clarifying debates by commentaries and critical reviews. He contributed to almost all branches of meteorology and attracted scientists of highest ability to Vienna, including Max Margules, J. M. Perntner, Wilhelm Trabert, F. M. Exner, and A. Merz.

Hann first became widely known when he entered the heated debate between H. W. Dove and the Swiss meteorologists on the origin of the warm Föhn winds occurring in the Alps (1866). He dismissed explanations based on mechanical displacement of warm air from tropical regions and demonstrated that the Föhn was produced locally when air passed over mountain ranges: moisture was removed on the windward side as a result of precipitation and the air was compressed during its descent on the lee. This thermodynamic theory included all katabatic wind phenomena in mountains. Unnoticed by meteorologists, the same explanation had briefly been offered by Helmholtz in 1865. Hann's article a year later and his subsequent papers on adiabatic changes of state in vertically displaced dry and moist air gave new directions to research, particularly in German-speaking countries.

Convinced that progress in science was possible only through extensive use of observational material, Hann made it his lifelong duty to bring rapidly accumulating meteorological data into precise and consistent forms. His work therefore formed the basis for many theories. Of far-reaching consequence in this respect were his investigations of cyclones and anticyclones. An early theory, based principally on surface observations, attributed the initial fall of pressure and the driving force of cyclones to rising columns of warm air and the release of latent heat during condensation of water vapor. Anticyclones, being relatively cool, were regarded as regions of descending cold air. Hann criticized this so-called convective theory, supported by William Ferrel, C. M. Guldberg, Henrik Mohn, Helmholtz, and others, as early as 1874. His most substantial objections were based on observations made on the mountains of Europe and the United States (1876). He demonstrated that up to the height of the mountains anticyclones were, on the average, warmer than cyclones except for a shallow surface layer. These temperature conditions contradicted the thermally induced independent circulation between cyclones and anticyclones supposed in the convective theory. Hann's observations received wide attention after 1890, when he corroborated them with new European data. Investigators in the United States, however, obtained different results. This discrepancy resulted from the statistical nature of the studies of Hann and others, which did not discriminate between the relatively cold mature cyclones and slow-moving warm anticyclones common in Europe and the initially warm-core young cyclones and migratory cold anticyclones frequently experienced in the United States. These differences were largely clarified by S. Hanszlik's studies on anticyclones in 1909 and the model of the thermal structure of cyclones developed by the Norwegian school of meteorologists after 1918.

From the beginning Hann had regarded cyclones and anticyclones as essential components of the general circulation of the atmosphere. His observational results convinced him that warm, deep anticyclones were produced by large-scale subsidence of the upper midlatitude westerlies. He suggested that extratropical cyclones originated as eddies in the westerlies, their kinetic energy deriving ultimately from the basic north-south temperature gradient (1879, 1890). Influenced by Helmholtz's vortex theory, Hann proposed that cyclones tended to develop in preexisting low-pressure areas by conversion of potential energy available in the horizontal pressure distribution into kinetic energy (1877, 1880). Latent heat of condensation and local temperature differences appeared to him of only secondary importance for the development and maintenance of cyclones.

Hann was a driving force in the establishment of mountain observatories because he realized the importance of aerological data. He studied upper-air data from many different aspects. One of his lifelong interests was the daily variation of meteorological elements, requiring harmonic analysis of high-altitude and low-altitude observations. He computed the harmonic coefficients of the diurnal oscillations (and their seasonal variations) of pressure, temperature, and wind for more than 100 stations throughout the world. In 1886 Hann produced conclusive evidence of the dominant and universal character of the twelve-hour solar pressure oscillation first suggested by Johann von Lamont in 1862. For many years Hann's publications, together with the results of C. A. Angot (1889), formed the basic material for theoretical studies on the cause of this oscillation. Hann presented details of a terdiurnal solar oscillation in 1917.

Hann also studied diurnal pressure and temperature variations in relation to the formation of mountain and valley winds. Two wind systems were required in his theory (1879), a thermal slope wind and a larger circulation between plain and mountain valley resulting from thermally induced horizontal pressure gradients. The development of this theory culminated in the work of A. Wagner in 1932.

The collection of long, homogeneous data series enabled Hann to make careful statistical studies of correlations between pressure, temperature, and precipitation anomaly patterns of distant regions and their causal relations (1904). He recognized future possibilities of such investigations for long-range forecasting, yet in his own time he considered practical weather forecasting a goal not scientific enough to merit interest.

Hann regarded his investigations in meteorology as inseparable from his climatological work. He viewed the statistical treatment of data as an indispensable basis for the knowledge of atmospheric motions and their explanation by physical laws, as well as for the knowledge of climatic conditions in the geographical sense. He developed valuable tabulation and statistical techniques in climatology and was instrumental in making accessible data series from remote places. He produced the first comprehensive climatologies of the tropics and the polar regions. Hann's classification of climates was based on geographical features and the variation of meteorological elements, primarily temperature and humidity. Still in use, this geographical-statistical approach has been supplemented by considerations of atmospheric dynamics.

One of Hann's greatest services to meteorology was the publication of two comprehensive treatises, *Handbuch der Klimatologie* (1883), the standard work for half a century, and *Lehrbuch der Meteorologie* (1901). Hann was never very active in organizational and administrative matters but, significantly, he insisted on the establishment of chairs of meteorology at all Austrian universities.

BIBLIOGRAPHY

I. ORIGINAL WORKS. Hann's comprehensive treatises on climatology and meteorology, containing extensive historical references, form a good guide to his contributions to science: *Die Erde als Ganzes, ihre Atmosphäre und Hydrosphäre* (Prague, 1872; rev. eds., 1875, 1881, 1886, 1897); *Handbuch der Klimatologie,* 3 vols., in the series Bibliothek geographischer Handbücher, F. Ratzel, ed. (Stuttgart, 1883; rev. eds., 1897, 1908, 1932), trans. into English by C. de Waard (New York, 1903); *Atlas der Meteorologie* (Gotha, 1887); *Lehrbuch der Meteorologie* (Leipzig, 1901; rev. eds., 1906, 1915, 1926, 1937–1951), eds. from 1915 with assistance of R. Süring.

Most of Hann's more than 1,000 other contributions, including numerous climatological notices, are scattered throughout *Meteorologische Zeitschrift* and the publications of the Academy of Sciences in Vienna. Over 300 are listed in the Royal Society's *Catalogue of Scientific Papers:* III, 162; VII, 902–903; X, 131–134; XV, 619–623. A large selection of Hann's publications is also listed in Poggendorff, III, 582–583; IV, 580–581; V, 493–494. Some of Hann's more important papers (most of them mentioned in the text) published in the *Zeitschrift der Oesterreichischen Gesellschaft für Meteorologie* are "Zur Frage über den Ursprung des Föhn," **1** (1866), 257–263; "Die Gesetze der Temperatur-Änderung in aufsteigenden Luftströmen und einiger wichtigsten meteorologischen Folgerungen aus denselben," **9** (1874), 321–329, 337–349, English trans. by C. Abbe in "The Mechanics of the Earth's Atmosphere," in *Report of the Board of Regents of the Smithsonian Institution* (1877), 397–419; "Ueber das Luftdruck-Maximum vom 23. Jänner bis 3. Februar 1876 nebst Bemerkungen über die Luftdruck-Maxima im Allgemeinen," **11** (1876), 129–135; "Bemerkungen über die Entstehung der Zyklonen," **12** (1877), 308–313, and **15** (1880), 313–321; "Einige Bemerkungen zur Lehre von den allgemeinen atmosphärischen Strömungen," **14** (1879), 33–41; and "Zur Theorie der Berg- und Talwinde," *ibid.,* 444–448.

Some important publications in the *Denkschriften der Wiener Akademie der Wissenschaften,* Math.-naturwiss. Kl., are "Untersuchungen über die tägliche Oscillation des Barometers," **55** (1889), 49–121; "Das Luftdruck-Maximum von November 1889 in Mitteleuropa," **57** (1890), 401–424; "Weitere Untersuchungen über die tägliche Oscillation des Barometers," **59** (1892), 297–356; "Der tägliche Gang der Temperatur in der inneren Tropenzone," **78** (1905), 249–366; "Der tägliche Gang der Temperatur in der äusseren Tropenzone," **80** (1907), 317–404, and **81** (1907), 21–113; and "Untersuchungen über die tägliche Oscillation des Barometers. Die dritteltägige (achtstündige) Luftdruckschwankung," **95** (1917), 1–64.

Some of Hann's papers in the *Sitzungsberichte der Wiener Akademie der Wissenschaften,* Math.-naturwiss. Kl., Abt. IIa (after 1888), are "Die Wärmeabnahme mit der Höhe an der Erdoberfläche und ihre jährliche Periode," **61** (1870), 65–81; "Bemerkungen zur täglichen Oscillation des Barometers," **93** (1886), 981–994; and "Die Anomalien der Witterung auf Island in dem Zeitraume 1851 bis 1900 und deren Beziehungen zu den gleichzeitigen Witterungsanomalien in Nord-Westeuropa," **113** (1904), 183–269.

See also "Die Verteilung des Luftdruckes über Mittel- und Südeuropa," in *Geographische Abhandlungen,* **2,** no. 2 (1887).

II. SECONDARY LITERATURE. Obituaries are in *Meteorologische Zeitschrift,* **38** (1921), 321–327; *Deutsches biographisches Jahrbuch,* III (Stuttgart, 1921), 118–122; *Bolletino della R. Società geografica italiana,* 5th ser., **10–11** (1921), 5–11; *Wetter,* **38** (1921), 161–168; *Nature,* **108** (1921), 249–251; *Mitteilungen der Geographischen Gesellschaft in Wien,* **64** (1922), 121–131; and *Naturwissenschaften,* **10** (1922), 49–52.

GISELA KUTZBACH

HANSEN, EMIL CHRISTIAN (*b.* Ribe, Denmark, 8 May 1842; *d.* Hornbaek, Denmark, 27 August 1909), *botany, physiology.*

Hansen's father, Joseph Christian Hansen, was a house painter who settled in the small provincial town of Ribe, Jutland, where he married Ane Dyhre. Their large family and poor circumstances often made it necessary for Emil Hansen to help his father. In 1850 he entered school and showed himself to be a diligent pupil and an avid reader. He wished to become an actor, but his father would not allow it. In 1860 he became a journeyman house painter; he also sought to become an artist, but the Academy of Fine Arts refused his application for admission.

Hansen became a private tutor in 1862 at the estate of Holsteinborg, where he prepared to become a teacher. During his stay at Holsteinborg the botanist Peder Nielsen, at that time schoolmaster in Ørslev, aroused Hansen's interest in botany and gave him emotional and financial support. In spite of illness Hansen completed a three-year teaching course at Copenhagen Polytechnical High School in 1869, earning money by publishing novels. The following two years he tutored in natural science, and in 1871 he became private assistant to the zoologist Japetus Steenstrup.

In 1873 Hansen discovered beech leaves in the deeper peat stratum of a moor in Femsølyng and concluded that the beech had existed in Denmark longer than Steenstrup believed. Hansen boasted of his discovery, then resigned from his job in the belief that he had made an enemy of Steenstrup. In the following years he prepared for the M.Sc. but did not achieve it. Nor did he make the further investigations of the moor regions for which he had obtained government support.

In 1876, however, Hansen received a gold medal from Copenhagen University for his essay on fungi growing on mammal dung, the subject of the 1874 competition. Hansen gave a detailed morphological and anatomical description of the fungi he had found and mentioned several new species (for example, *Peziza ripensis*) as a result of his culture experiments. His prize-winning work was published as *De danske Gjødningssvampe* (1876), and he spent the following years studying the biology and variation of species of fungi. He cultured *Coprinus stercorarius, Coprinus niveus,* and *Coprinus restrupianus* and demonstrated the phototropism of *Coprinus stercorarius:* the stalk turns toward the light while the spores are thrown away from it. He also described a new family of Ascomycetes, *Anixiopsis,* from the species *Eurotium stercorarium* (1878). These studies were published in French and German (both in 1880).

Inspired by the physiologist P. L. Panum, Hansen next studied fermentation in the zoophysiological laboratory of Copenhagen University. Despite some opposition from Steenstrup he was allowed to prepare and publicly defend a thesis for a Ph.D. On 1 July 1878 he obtained a job in the laboratory of the Carlsberg breweries—through Steenstrup's recommendation—and here Hansen began his microscopic studies of beer. From 1879 until his death Hansen was superintendent of the laboratories. In his thesis, "Om Organismer i Øl og Ølurt" (1879), he tried to demonstrate which organisms (yeast fungi, molds, bacteria) could be found free in nature and how they could be cultured on sterile nutrient liquids. By chemical reactions, when morphology failed, he could prove which organisms occurred in beer, which occurred in its foam, and which occurred in other organic liquids when they were exposed to air. He was able to demonstrate that there are two forms of Pasteur's bacterium: *Mycoderma aceti* and *Mycoderma pasteurianum.* These discoveries led him to assume that the common types of beer yeast are physiologically different.

Between 1881 and 1908 Hansen published thirteen papers under the general title "Undersøgelser over Alkoholgjaersvampenes Fysiologi og Morfologi," which provided many essential contributions to the knowledge of saccharomycetes. He chose the easily recognizable *Saccharomyces apiculatus* and demonstrated its life cycle in nature, its relation to sugar, its hibernation in the earth, and its presence on juicy fruits in summer. He demonstrated that it could neither invert saccharoses nor produce alcoholic fermentation (1881).

In order to follow the development of the microorganisms under the microscope Hansen constructed a special "moist chamber" (1881). Through further experiments he subsequently succeeded in proving the life cycle of other saccharomycete species. (Their morphologies and their spore formation were being investigated at that time not only in Copenhagen but also in Germany and Italy [1902].) In addition Hansen demonstrated that *Torula, Mucor,* and other bacteria react like yeasts. Finding that beer often acquired a bad taste from "wild" yeast types, which involved the breweries in heavy economic losses, Hansen was inspired to follow up his studies on various yeast types. A common airborne saccharomycete, which he named *Saccaromyces pastorianus Riess,* produced a beer with bitter taste and heavy sedimentation.

Hansen then took up the methods for pure cultivation developed by Pasteur, Robert Koch, and the Danish bacteriologist C. J. Salomonsen and succeeded in developing cultures from one cell. These cultures could be kept alive for years in a glass flask that he had made. He also proved that there were several different varieties and races of saccharomycete species, and through fermentation experiments he found that their effect on beer was very different. The common yeast of the old Carlsberg brewery, "Carlsberg bottom yeast I," which J. C. Jacobsen had obtained at Munich in 1845, produced excellent beer, but the admixture of "wild" yeast types had spoiled it. Although skeptical, Jacobsen allowed Hansen to use his cultured pure strains of yeast to brew experimentally on a large scale—and 12 November 1883 became a red-letter day in the history of brewing. The new beer was excellent, as Hansen had expected, and Jacobsen continued to use the cultivated yeast. Following the policy he had proclaimed for the Carlsberg laboratory, Jacobsen refused to have the method patented, and in a short time it was used all over the world. Hansen published his investigations and results in seven papers with the title "Undersøgelser fra Gjaeringsindustriens Praxis" (1888-1892), which were translated into French (1888), German (1888, 1890, 1893), and English (1896).

During the following years Hansen took up problems concerning acetic-acid bacteria—their film formation, variation, and life-span—and continued investigations on the life cycle, spore formation, variation, genetics, and systematics of saccharomycetes (1904). He had seen that pure yeast cultured through several generations in beer wort at a certain temperature lost its ability to form spores and never regained it even if the most favorable temperatures for spore formation were obtained. Hansen acknowledged here an environmentally produced genetic characteristic, a conclusion that caused a great stir—but his discovery was probably the result of a mutation (1899).

From 1900 Hansen studied the relationship between top yeast and bottom yeast and found that the changes in the nature of yeast were caused by mutation (1905, 1907). For years he had considered most of the yeast types differentiated by him as physiological races, but now he considered them as species and in 1904 published "Grundlinien zur Systematik der Saccharomyceten," which also appeared in Danish, French, and English journals.

Hansen was made a member of the Royal Danish Society for Sciences in 1890 and obtained large donations from the Carlsberg Foundation and the Brewers' Association. He left a large fortune for a foundation to bear his name, the income to be used for prizes for biological papers; the prizes were to be awarded by an international committee. He also left a fine library on art and the history of the natural sciences. Hansen was an honorary member of learned societies all over the world and held honorary doctorates from the universities of Uppsala (1907) and Geneva (1909) and the Technische Hochschule of Vienna (1908).

In 1879 Hansen married Mathilde Melchior. His difficult childhood had made him somewhat harsh and devoid of humor; he compelled strict respect from his co-workers but in later years was kind to the poor.

BIBLIOGRAPHY

I. Original Works. A full catalog of Hansen's works is in Carl Christensen, *Den danske Botaniks Historie,* II (Copenhagen, 1926), 441–457. Many of his scientific papers were reprinted in Albert Kløcker, ed., *Gesammelte theoretische Abhandlungen über Gärungsorganismen von Emil Chr. Hansen* (Jena, 1911).

II. Secondary Literature. See Carl Christensen, *Den danske Botaniks Historie,* I (Copenhagen, 1924), 718–731; E. Gotfredsen, *Medicinens Historie* (Copenhagen, 1964), p. 447; A. Kløcker, in *Meddelelser fra Carlsberg Laboratoriet,* **2** (1911), i–xxxvi; C. Nyrop, *J. C. Jacobsen* (Copenhagen, 1911), pp. 55–58; and Johannes Pedersen, *The Carlsberg Foundation,* XII (Copenhagen, 1956), 46–48.

E. Snorrason

HANSEN, GERHARD HENRIK ARMAUER (*b.* Bergen, Norway, 29 July 1841; *d.* Florø, Norway, 12 February 1912), *bacteriology.*

Hansen was the eighth of fifteen children. His mother was Elisabeth Concordia Schram, who was a member of a family of master joiners long established in Bergen. His father, Claus Hansen, was a wholesale merchant until the severe contraction of credit of 1848–1851 drove him into bankruptcy; he then worked as a cashier in a bank.

In 1859 Hansen began his medical studies at the University of Christiania (now Oslo). It was necessary for him to earn his own living while he was a student. He first taught at a girls' school and later spent a year as substitute for the prosector of anatomy. He then began his own tuition courses in anatomy. In later years he said that during this period he had known neither physical nor mental fatigue and had found that he did his best work between six and eight in the morning. He passed his degree with honors in 1866 and completed his internship at the National Hospital in Christiania. He then served as doctor to a community of some 6,000 fishermen at Lofoten, a group of islands off northern Norway.

In 1868 Hansen entered the service of the leprosy hospitals in Bergen. His new chief, Daniel Cornelius Danielssen, had, with C. W. Boeck, published the major work *Om Spedalsked* ("On Leprosy," 1847) and had helped to establish Bergen as the European center for leprosy research. Danielssen, like other investigators of the time, regarded the affliction as hereditary (and was to continue to do so throughout his life). Hansen, however, quickly concluded on the basis of epidemiological studies that leprosy was a specific disease which must have a specific cause.

Bacteriology was then in its infancy. In 1870 Hansen received a grant that allowed him to travel to improve his knowledge of histopathology. He went to Bonn and later to Vienna. Returning to Norway, using primitive staining methods and working with biopsy specimens from patients with leprosy, Hansen in 1873 discovered the rod-shaped bodies—*Mycobacterium leprae,* sometimes called Hansen's bacillus. By 1879 he was able, through the use of improved staining methods, to show great numbers of the rod-shaped bodies typically aggregated in parallel cells. He believed the bacillus to be the causative agent of leprosy and thereby became the first investigator to suggest that a chronic disease might be caused by microorganisms. (The tuberculosis bacillus, for example, was not discovered until 1882.)

Hansen conducted many experiments to prove that his bacillus was indeed the cause of leprosy. He attempted to find a method of cultivating the

Mycobacterium leprae on artificial media, without success. (The bacillus has not yet been cultivated *in vitro.*) He further tried to transmit leprosy to animals and humans; he failed in experiments on rabbits, and experimental reproduction of the disease has not so far been accomplished. In a less well-advised effort, Hansen inoculated the eye of a woman suffering from the neural form of the disease with material drawn from a leprous nodule of a patient suffering from the cutaneous form. There were no clinical consequences of the inoculation, but since Hansen had not asked permission to perform the experiment, he met with legal difficulties as a result of which he was removed from his post as resident physician of the Bergen leprosy hospitals in May 1880.

Hansen's sentence was less severe than it might seem, however, since he was allowed to retain his position as leprosy medical officer for the entire country of Norway—an appointment conferred on him in 1875 and one that he held until his death. He was thus able to implement changes in the methods of control of leprosy in Norway—changes that had been in part made necessary by his own hypotheses concerning the etiology of the disease. The Norwegian Leprosy Act of 1877 and the amended act of 1885 were the fruits of his untiring work. Under these laws health authorities could order lepers to live in precautionary isolation away from their families (subsequent studies have shown leprosy to be a familial affliction); enforcement of the laws led to a quick and steady decline of the disease in Norway. There were 1,752 known cases of leprosy in Norway in 1875; by the beginning of the twentieth century there were 577 (that there are only four known today may well reflect generally improved economic and hygienic conditions). The word "hansenarium" was suggested to replace the still more standard "leprosarium."

Hansen received many honors for his leprotic studies. He was elected honorary chairman of the first Conférence Internationale de la Lèpre, held in Berlin in 1897, and was president of the second such conference, held in Bergen in 1909. He was honorary chairman of the International Leprosy Committee, corresponding or honorary member of numerous scientific societies, and was decorated several times. In 1900 contributions toward a portrait bust of Hansen were solicited internationally; the bust was unveiled with great ceremony the following year.

While Hansen was chiefly known for his work on leprosy, he also played a role in the dissemination of Darwin's ideas. He learned of Darwin's doctrines during his trip to Vienna, early in his career, and he then set about to study Darwin's books. He sought to emulate Darwin's methods, which he considered a model of dispassionate observation; in 1886 he published a book on Darwinism in Norwegian. In his apostolic zeal for Darwin's work, Hansen also gave numerous lectures and published articles in the popular press. These evoked a great sensation, especially from the clergy and religious organizations—who reacted violently against the "blasphemer" in their midst. Hansen was, however, devoid of philosophical speculations and had no aptitude for martyrdom. He did not acknowledge the attacks made against him; he continued to sit at his microscope, smoking his pipe, and do his work.

Hansen married Danielssen's daughter, Stephanie Marie, on 7 January 1873; she died of pulmonary tuberculosis on 25 October of the same year. On 27 August 1875 he married Johanne Margrethe Tidemand, a widow related to almost the entire Bergen commercial patriciate. Their only child, a son, became a physician specializing in tuberculosis and in 1929 was appointed chief of the tuberculosis hospital in Bergen.

Hansen suffered the first symptoms of heart disease as early as 1900; in following years he had several severe heart attacks that confined him to bed for long periods of time. In the intervals of his illness, however, he continued to travel around the country on official inspection tours. In February 1912 he made such a trip to the fishing areas north of Bergen; in Florø, a little town on the western coast, he was invited to stay in the home of a friend, and it was there that he died. He was given a funeral at state expense; he had been president of the Bergen Museum and the ceremony took place from its hall.

BIBLIOGRAPHY

I. Original Works. Hansen's most important works concerning the bacillus include "Undersøgelser angaaende Spedalskhedens Aarsager" ("Investigations Concerning the Etiology of Leprosy"), in *Norsk magazin for laegevidenskaben,* 3rd ser., **4**, no. 9 (1874), supp. 1–88, case reports I–LIII; "On the Etiology of Leprosy," in *British and Foreign Medical Magazine,* **55** (1875), 459–489; "Bacillus leprae," in *Virchow's Arkiv für pathologische Anatomie und Physiologie und für klinische Medizin,* **79** (1880), 32–42; "Studien über Bacillus leprae," *ibid.,* **90** (1882), 542–548; *Leprosy. In Its Clinical and Pathological Aspects* (Bristol, 1895), written with C. Looft; and "Lepra," in Wilhelm Kolle and A. Wassermann, eds., *Handbuch der pathogenen Mikroorganismen,* II (Jena, 1903), 178–203.

II. Secondary Literature. On Hansen and his work, see W. H. Feldman, "Gerhard Henrik Armauer Hansen. What Did He See and When?," in *International Journal of Leprosy,* **33** (1965), 412–416; B. Helland-Hansen, "G. Armauer Hansen in Memoriam," in *Medical Review* (Bergen), **29** (1912), 125–128; O. Lasser, "Gerhard Armauer Hansen," *ibid.,* **18** (1901), 193–198; C. Looft, "G. Armauer Hansen," *ibid.,* **29** (1912), 164–166; P. Pallamary, "Translation of Gerhard Armauer Hansen: Spedalskhedens Aarsager (Cause of Leprosy)," in *International Journal of Leprosy,* **23** (1955), 307–309.

The following works about Hansen were written by T. M. Vogelsang: *Armauer Hansen og Spedalskhetens historie i Norge* ("Armauer Hansen and the History of Leprosy in Norway"), Universitetet i Bergeny Småskrifter no. 12 (1962); "Hansen's First Observation and Publication Concerning the Bacillus of Leprosy," in *International Journal of Leprosy,* **32** (1964), 330–331; and *Gerhard Henrik Armauer Hansen. 1841–1912* (Oslo, 1968), with a bibliography (51 references) and two appendixes, all of which have incomplete listings of Hansen's work.

T. M. VOGELSANG

HANSEN, PETER ANDREAS (*b.* Tondern, Schleswig, Germany, 8 December 1795; *d.* Seeberg, Germany, 28 March 1874), *astronomy.*

Hansen, a leading German theoretical astronomer of the mid-nineteenth century, was the son of a goldsmith. The straitened circumstances in which the family found itself after the Napoleonic Wars prevented him from embarking on a higher scholastic career in his youth and, like his older contemporary Friedrich Bessel, he arrived at it by a detour through trade. He was apprenticed to a clockmaker at Flensburg and eventually qualified as a master craftsman of this art, for a time becoming a clockmaker in his native town. During his spare time he privately studied French and Latin as well as mathematics, in which he had shown particular proficiency since early childhood.

Hansen's chance to turn his intellectual gifts to better account came in 1820, when Heinrich Christian Schumacher, then a leading astronomer in Denmark, was temporarily in need of a computing assistant. Hansen applied for the position; and even though unsuccessful, he volunteered to accompany Schumacher in his measurements of an arc of the meridian in Holstein. At the beginning of 1821 he worked in Tondern on the calculations connected with these measurements; later Schumacher summoned him to Copenhagen under an appointment sanctioned by the Danish government.

Between 1821 and 1825 Hansen continued to serve as assistant to Schumacher—mainly at Altona, where a small observatory was set up for him by the Danish king. In 1823 Schumacher started publishing the *Astronomische Nachrichten,* the oldest astronomical journal still appearing today, and Hansen became its editorial assistant and frequent contributor. In the summer of 1824 Schumacher and Hansen traveled to Helgoland to determine, together with a parallel expedition sent out by the British Admiralty, the accurate geographical position of that island.

Hansen's connection with Schumacher ended in 1825, when he was invited to succeed Johann Franz Encke as director of the private observatory of the duke of Mecklenburg at Seeberg, near Gotha; he remained its head until his death almost half a century later.

During this time Hansen's contributions to astronomy were so numerous and enriched so many branches of that field that he was considered among the foremost astronomers of his time. He was above all a theoretician, concerned with the representation of the motion of the moon and the planets in the sky in terms of Newtonian celestial mechanics. His first major work was an extensive study of the mutual perturbations of Jupiter and Saturn (1831), for which he received a prize from the Royal Academy of Sciences in Berlin in 1831 and a gold medal from the Royal Astronomical Society in 1842. Several of his important papers were devoted to the theory of the motion of comets or minor planets (1859), but his main work was concerned with the motion of the moon (1838).

The latter work became the basis of extensive tables of lunar motion, published in 1857 at the expense of the British government. These tables proved to be so accurate that, in the words of George Biddell Airy, then astronomer royal, "Probably in no recorded instance has practical science ever advanced so far by a single stride." The theoretical investigations on which these tables were founded were published later in two parts (1862–1864). In recognition of this work Hansen received a prize of £1,000 from the British Admiralty in 1860, and in the same year the Royal Astronomical Society awarded him their gold medal for the second time.

Hansen's work in celestial mechanics constitutes the main part of, but does not exhaust, his contributions to astronomy. Another lifelong interest was the theory of astronomical instruments, to which he contributed a number of studies concerning the theory and use of heliometers, of the astronomical equatorial, and of the transit instrument. Moreover, his early work with Schumacher led Hansen to many refinements of theoretical geodesy, and he also made several contributions to the calculus of probabilities.

The high regard in which Hansen was held by his contemporaries as a result of these contributions is reflected in the opinion recorded by Simon Newcomb, a leading American astronomer of his time, in his *Reminiscences of an Astronomer* (New York, 1903):

Modest as was the public position that Hansen held, he may now fairly be considered the greatest master of celestial mechanics since Laplace. In what order Leverrier, Delaunay, Adams, and Hill should follow him, it is not necessary to decide. To many readers it will seem singular to place any name ahead of that of the master who pointed out the position of Neptune before a human eye had ever recognized it. But this achievement, great as it was, was more remarkable for its boldness and brilliancy than for its inherent difficulty. If the work had to be done over again today, there are a number of young men who would be as successful as Leverrier; but there are none who would attempt to reinvent the methods of Hansen, or even to improve radically upon them [p. 315].

BIBLIOGRAPHY

Hansen's writings include *Untersuchung über die Gegenseitigen Störungen des Jupiters und Saturns* (Berlin, 1831); *Fundamenta nova investigationis orbitae verae quam luna perlustrat* (Gotha, 1838); *Tables de la lune construites d'après le principe newtonien de la gravitation universelle* (London, 1857); "Auseinandersetzung einer zweckmässigen Methode zur Berechnung der absoluten Störungen der kleinen Planeten," in *Abhandlungen der K. Sächsischen Gesellschaft der Wissenschaften,* **5** (1859), 1–148; and "Darlegung der theoretischen Berechnung der in den Mondtafeln angewandeten Störungen," *ibid.,* **6** (1862), 91–498, and **7** (1864), 1–399.

ZDENĚK KOPAL

HANSEN, WILLIAM WEBSTER (*b.* Fresno, California, 27 May 1909; *d.* Palo Alto, California, 23 May 1949), *physics, microwave electronics.*

Encouraged by his father, a hardware store owner of Danish ancestry, Hansen showed great precocity in mathematics and electricity as a child. He entered Stanford University at sixteen, where he first studied electrical engineering and later physics, in which he received the doctorate with a dissertation on X-ray excitation (1933). After a year and a half as a National Research fellow at the Massachusetts Institute of Technology, he returned to Stanford in 1934 as an assistant professor. He turned his attention to the problem of accelerating electrons for experiments in X-ray physics, eschewing large static voltages in favor of arrangements utilizing rapidly varying fields, to avoid the difficult insulation problems of the former.

At this time, northern California was becoming a great center of nuclear research: Ernest O. Lawrence had recently invented the cyclotron at Berkeley and his co-worker David H. Sloan had proposed an accelerator, the voltage of which was produced by a resonating coil. Hansen saw that, with a resonant cavity, power losses would be reduced for a given accelerating voltage, and he resolved to employ a high-quality ("high-Q") cavity resonator in place of the conventional coils and condensers: the inside of a hollow, closed-off conductor made of copper or some other highly conducting material. Although the resulting configuration came to underlie the design of all subsequent linear electron accelerators, it did not see immediate realization, for another important development intervened.

In 1937 two other Californians, the brothers Russel H. and Sigurd F. Varian, came to Stanford to develop a new source of ultra-high-frequency oscillations, which they foresaw would be useful in air defense. Hansen had employed electron tubes to excite his accelerator cavities. A new principle was now needed to avoid the pitfall of progressively smaller structures as frequency increased. Russel Varian solved the problem by the use of velocity modulation, in which the electrons traverse a cavity at the entrance to a drift tube, where they are concentrated into periodic bunches as a result of the varying accelerations imparted to them by the alternating cavity voltage, and then pass into a second cavity where the bunched electrons in turn induce oscillations (which may be fed back to the first cavity in a self-reinforcing manner). Hansen's resonant cavity (christened the rhumbatron after a popular dance of the period) became an essential part of the new tube, which was designated as the klystron.

During the years 1937–1940, Hansen helped elaborate the theory and practice of the new field he had founded, microwave electronics. He pioneered novel configurations, measurement techniques, and solutions of radiation problems generally. Among his contributions dating from this period is the classic paper he coauthored with his pupil John R. Woodyard in which the total effectiveness ("gain") of an antenna array was shown to be capable of being substantially increased when the elements do not radiate exactly in phase in the principal direction. (See "A New Principle in Directional Antenna Design," in *Proceedings of the Institute of Radio Engineers,* **26** [1938], 333–345.)

In 1941 Hansen and his collaborators moved to the Sperry Gyroscope Company's laboratory in Garden City, New York, where they remained until the end of World War II. There they worked on the klystron and on other electronic devices and their applications, including Doppler radar and blind landing systems for aircraft. Owing to his great versatility, Hansen was also able to contribute toward such diverse problems as the design of aircraft superchargers and the exploitation of atomic energy. The physicist Felix Bloch said of this period in Hansen's career:

> Equally versed in the methods and terminology of both, he was one of the first and most important links in the close connection between engineering and physics, which was responsible for the rapid development of radar. With his previous experience and clear insight in the principles of microwave technique he was asked to deliver a series of lectures at M.I.T. and for a considerable period of time he willingly submitted to the strain of commuting between Garden City and Cambridge [Mass.]. In these lectures he touched upon almost all of the central problems, restricting himself to those topics which he knew to be of basic significance. Many of the leading scientists, engaged or about to engage in radar research, were among the audience and gratefully acknowledge the important stimulus received from his masterly exposition [*Biographical Memoirs. National Academy of Sciences,* **27** (1952), 128].

Bloch also acknowledged Hansen's influence on the investigation of nuclear magnetic resonance, work for which Bloch shared the Nobel Prize in 1952.

Hansen returned to Stanford as a full professor in 1945 and laid the groundwork for the series of enormous linear electron accelerators (the accelerations of which are measured in billions of electron volts), powered by giant klystrons, that were subsequently constructed there. But he saw only the beginning of their realization. Hansen's constitution had been weakened by the hard work of the war years, and he died a few days before his fortieth birthday, just after he had been elected to the National Academy of Sciences. Among his other honors were the Liebmann Prize of the Institute of Radio Engineers (1945) and the Presidential Certificate of Merit (1948). He was survived by his wife, Betsy, the younger daughter of Stanford physicist P. A. Ross, but only briefly; she died (by her own hand) a few months later. The W. W. Hansen Laboratories of High Energy Physics at Stanford University are named in his honor.

BIBLIOGRAPHY

For information on Hansen, see Felix Bloch, *Biographical Memoirs. National Academy of Sciences,* **27** (1952), 121–137, which contains a complete bibliography of Hansen's twenty-eight papers and sixteen laboratory reports. Obituaries appear in *New York Times* (24 May 1949), and in *Proceedings of the Institute of Radio Engineers,* **37** (1949), 910.

CHARLES SÜSSKIND

HANSKY, ALEKSEY PAVLOVICH (*b.* Odessa, Russia, 20 July 1870; *d.* the Crimea, Russia, 11 August 1908), *astronomy.*

After completing his studies at the Gymnasium, Hansky entered the Faculty of Physics and Mathematics of Novorossisk University in Odessa (now Odessa University), from which he graduated in 1894. He was then retained at the university to prepare for a career in science. In 1896 he went to St. Petersburg as a probationer at Pulkovo observatory, which was then completing preparations for an expedition to Novaya Zemlya to observe the total solar eclipse of 8 August. Hansky participated actively in this expedition and obtained excellent photographs of the solar corona, which provided him with a beginning for his later research on the form of the solar corona in relation to the phases of solar activity.

To continue his research on the sun, Hansky in 1897 visited Pierre J. Janssen's observatory at Mont Blanc and the Meudon astrophysical observatory, near Paris. From 1897 to 1905 he made more than ten ascents of Mont Blanc and spent a total of a month and a half at the observatory. He was distinguished by great personal courage: in 1898–1900, in order to observe the Leonids, he made three balloon flights, in Paris and St. Petersburg. In 1901 Hansky participated in an expedition to Spitsbergen, where he made gravimetric measurements under very adverse conditions.

In 1905 Hansky became an astronomer at Pulkovo and a member of the scientific center organized by A. A. Belopolsky at the Academy of Sciences for study of the sun. Also in 1905 he traveled to Spain to observe the solar corona during total solar eclipse. His later expeditions to the Crimea and Central Asia to study zodiacal light stimulated his desire to create in southern Russia, and in the most favorable climatic conditions, an astrophysical section of the Pulkovo observatory. By chance Hansky learned of the existence in Simeiz, on the southern coast of the Crimea, of a modest amateur observatory belonging to N. S. Maltsov and persuaded him to donate it to the Pulkovo observatory as a southern astrophysical branch. Before the organization of the Simeiz section could be completed, Hansky, who had been named its director, drowned. The Simeiz observatory later became famous through the work of Grigory Neuymin (comets and asteroids) and Grigory Shayn (spectra of stars and gas nebulae).

During his short scientific career Hansky studied gravimetry, measuring gravitational force atop Mont Blanc and in the depths of a coal mine on Spitsbergen, conducted research on the zodiacal light, and made observations of Jupiter and the meteors. But his chief service to science was his solar research.

Having compared his photographs of the solar corona taken during the 1896 eclipse with data on the various phenomena of solar activity over a fifty-year period, Hansky postulated a relation between the form of the solar corona and the number of sunspots, that is, with the phase of solar activity. It appeared that when there is a minimum of spots, the corona is stretched along the plane of the solar equator but is scarcely observable at the poles, and its total luminosity is only slightly greater than the luminosity of the full moon. But during the period of maximum sunspots the corona is ten times brighter than the full moon and is rather evenly distributed on all sides of the solar disk. During later eclipses Hansky's predictions of the form of the corona were fully confirmed.

While on an ascent of Mont Blanc, Hansky attempted to determine the so-called solar constant (the quantity of ray energy crossing one square centimeter of surface set perpendicular to the solar rays outside the earth's atmosphere, that is, at a distance of one astronomical unit from the sun). He obtained a somewhat excessive value, but this was the first determination in scientific history. The problem was not successfully resolved until C. G. Abbot's determination of the solar constant ten years later. Hansky persistently sought a method of photographing the corona without waiting for an eclipse. Thus, in 1898 he fitted to the thirty-centimeter refractor at Janssen's observatory a special instrument to reduce the brightness of the sky by means of which he covered the solar disk with a metal circle and used a light filter that allowed only red rays to pass through. A fully satisfactory out-of-eclipse coronagraph was not obtained until 1931, by Bernard Lyot.

At Pulkovo, Hansky had striking success in photographing sunspots and details of solar granulation; only in the 1960's, with the launching of telescopes on special balloons, have better photographs been obtained. According to Hansky's research, the granules appeared to be very short-term phenomena, sometimes changing beyond recognition within a few seconds. He determined their size (diameters about one second of arc, i.e., up to thousands of kilometers). The granules provided information on the instability of the photosphere and the origins of the flow of hotter substances of the lower photosphere. Hansky noted the connection between the coronal rays and the protuberances and determined the velocity of movement of the substance in the coronal rays (about thirty kilometers per second) and in the protuberances (about 200 kilometers per second).

BIBLIOGRAPHY

I. Original Works. Hansky's principal works include: "Die totale Sonnenfinsternis am 8 August 1896. Über die Corona und den Zusammenhang zwischen ihrer Gestaltung und anderen Erscheinungsformen der Sonnentätigkeit," in *Izvestiya Imperatorskoi akademii nauk,* **6,** no. 3 (1897), 251-270; "Sur la détermination de la pesanteur au sommet du Mont-Blanc à Chamonix et à Meudon," in *Comptes rendus de l'Académie des sciences,* **127** (1898), 942-945; "Issledovanie 30-dyuymovogo obektiva Pulkovskoy observatorii po sposobu Gartmana" ("The Examination of the Pulkovo Observatory 30-Inch Objective by Hartmann's Method"), *ibid.,* **20,** no. 2 (1904), 77-92; "Sur la grande période de l'activité solaire," *ibid.,* **20,** no. 4 (1904), 145-148; *Intensité de la pesanteur. Missions scientifiques au Spitzberg, 1 Géodésie, 5-ème Section* (St. Petersburg, 1905); "Observations de l'éclipse totale du Soleil du 30 août 1905," in *Mitteilungen der Nikolai-Hauptsternwarte zu Pulkowo,* **1,** no. 10 (1906), 121-136; "Études des photographies de la couronne solaire," *ibid.,* **2,** no. 19 (1907), 107-118; "Bemerkungen über das Zodiakallicht," *ibid.,* 99; "Mouvements des granules sur la surface du soleil," *ibid.,* **3,** no. 25 (1908), 1-20; and "O dvizhenii veshchestva v korone solntsa" ("On the Movements of Matter in the Solar Corona"), in *Izvestiya Russkago astronomicheskago obshchestva,* pt. 13, no. 9 (1908), 295-304.

II. Secondary Literature. Papers by O. A. Baklund, G. A. Tikhov, and V. V. Akhmatov on the life and scientific activity of Hansky with a list of his scientific papers are in *Izvestiya Russkago astronomicheskago obshchestva,* pt. 14, no. 7 (1908), 232-249. See also Y. G. Perel, *Vydayushchiesya russkie astronomy* ("Outstanding Russian Astronomers," Moscow, 1951), 194-211; G. A. Tikhov, "A. Hansky (Necrologe)," in *Bulletin de la Société astronomique de France,* **22** (1908), 421, 461, with portrait; the article on Hansky in *Bolshaya sovetskaya entsiklopedia* ("Great Soviet Encyclopedia"), X, 2nd ed. (Moscow, 1952), 211; and an obituary in *American Journal of Science,* **26** (1908), 404.

P. G. Kulikovsky

HANSTEEN, CHRISTOPHER (*b.* Christiania [now Oslo], Norway, 26 September 1784; *d.* Christiania, 15 April 1873), *physics, astronomy.*

Hansteen's father, Johannes Mathias Hansteen, was a customs officer; his mother was Ane Cathrine Treschow, a niece of the philosopher Niels Treschow. In 1814 he married Johanne Cathrine Andrea Borch, a daughter of the head of a famous Danish public school in Sorø. In 1802 he began studying law at Copenhagen but, after having been introduced to the fashionable circle around H. C. Oersted, he dropped his law studies in 1806 and devoted the rest of his life to astronomy and physics, particularly geomagnetism. Hansteen became a schoolmaster at Elsinore until 1814, when he was appointed lecturer in applied mathematics and astronomy at the University of Christiania. He became professor in 1816 and retired from this position in 1861.

Geomagnetism had been studied quantitatively since about 1600. Edmond Halley was the first to publish, in 1701, a magnetic chart of the world distribution of the magnetic declination. It was not until near the beginning of the nineteenth century that the intensity of geomagnetism was measured, on the initiative of Jean-Charles Borda and Alexander von Humboldt. Hansteen's main contribution to science consisted in measurements and theories of geomagnetism. In 1819 he published a magnetic atlas, but at that time observations from large parts of the world were still missing. It was for this reason that Hansteen visited Paris and London in 1819 and Bergen in 1821 and in 1825 traveled round the Gulf of Bothnia to Finland to measure magnetic elements, of which he published the first reliable chart in 1826. In 1828–1830 he led an expedition to Siberia, where he carried out more than 400 measurements.

An announcement in 1811 of a prize to be awarded by the Royal Danish Academy of Sciences for a theory of geomagnetism directed Hansteen to original research in this field. Like Halley in 1683, Hansteen tried in his prize essay to explain the direction and intensity of the magnetic force at any point of the earth by a hypothesis of two magnets of unequal size and strength. He experienced difficulties in carrying out the mathematical consequences of his hypothesis, and it was not until 1839 that an adequate mathematical theory was given by Gauss. It dealt the death blow to Hansteen's hypothesis, but Gauss admitted having been inspired by Hansteen. In 1815 Hansteen established the first astronomical observatory in Norway, and in 1841 he founded a magnetic observatory. His most important contribution to astronomy was a method for time measurements with simple instruments by observing a star in the vertical plane of the polestar.

In 1817 Hansteen was appointed one of the presidents of the Geodetic Institute, and he played a leading role in the survey of Norway. He had several other public offices, and the elaboration of a new system of standards in Norway was mainly due to his indefatigable work as a member of a government commission in 1819–1824.

Hansteen corresponded with many of the leading scientists of his day; his correspondence with Oersted was particularly extensive and marked by their close friendship and zeal in exchanging scientific results.

BIBLIOGRAPHY

I. ORIGINAL WORKS. Hansteen's most important writings are *Untersuchungen über den Magnetismus der Erde I* (Christiania, 1819), a rev. version of his prize essay of 1812 (pt. II, never written, was to contain a theory of the aurora borealis and its influence upon geomagnetism); *Magnetischer Atlas gehörig zum Magnetismus der Erde* (Christiania, 1819); "Isodynamische Linien für der ganze Magnetkraft der Erde," in Poggendorff's *Annalen der Physik,* **9** (1827), 49–66, 229–244; and *Resultate magnetischer, astronomischer und meteorologischer Beobachtungen auf einer Reise nach dem östlichen Sibirien in den Jahren 1828–1830* (Christiania, 1863).

Hansteen kept a diary of his expedition to Siberia: *Reise-Erinnerungen aus Sibirien* (Leipzig, 1854); an enl. Norwegian version, *Reise-Erindringer* (Christiania, 1859), contains a short autobiography.

Hansteen's correspondence with H. C. Oersted is in M. C. Harding, ed., *Correspondance de H. C. Örsted avec divers savants,* I (Copenhagen, 1920), 77–251. An interesting letter from Hansteen to Faraday is in Henry Bence Jones, *Life and Letters of Faraday,* II (London, 1870), 131.

II. SECONDARY LITERATURE. A commentary on the letter from Hansteen to Faraday is in R. C. Stauffer, "Persistent Errors Regarding Oersted's Discovery of Electromagnetism," in *Isis,* **44** (1953), 307–310. Biographical information is in *Norsk biografisk leksikon,* V (Oslo, 1931), 432–448.

KURT MØLLER PEDERSEN

HANTZSCH, ARTHUR RUDOLF (*b.* Dresden, Germany, 7 March 1857; *d.* Dresden, 14 March 1935), *chemistry.*

The son of Rudolf Georg Hantzsch, a Dresden wine merchant, Hantzsch attended the Dresden Polytechnic (now Technische Hochschule) from 1875 to 1879. He studied chemistry under Rudolf Schmitt, a pupil of Hermann Kolbe, and did his doctoral work under Schmitt's direction. Dresden was not entitled to grant the doctor's degree, however, and Hantzsch therefore obtained the degree after attending the University of Würzburg for one semester. He was an assistant at the Institute for Physical Chemistry in Leipzig before becoming a professor at the Zurich Polytechnic in 1885. He succeeded Emil Fischer at Würzburg in 1893 and ten years later succeeded Johannes Wislicenus at Leipzig. In 1883 Hantzsch married Katherine Schilling, by whom he had three children. She died in 1904 and in 1911 he married Hedwig Steiner. Hantzsch was a member of the academies of Göttingen, Halle, Leipzig, Vienna, and Zurich, and of the German and London chemical societies.

Hantzsch's earliest work was in organic synthesis. In 1882 he announced a general method of synthesis for pyridine compounds from α-keto esters and aldehyde ammonia compounds. He then turned to the synthesis of other heterocyclic types. After Victor Meyer had noted the mimicry between benzene and thiophene, Hantzsch proposed a similar relationship between pyridine and thiazole.

Benzene 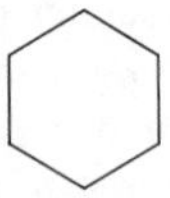 is to thiophene  as pyridine  is to thiazole.

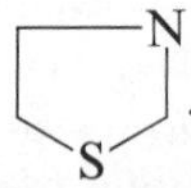

He synthesized thiazole in 1887 and suggested that other aromatics may exist, arguing that imidazole, oxazole, and selenazole were analogues of thiazole. By new synthetic methods he prepared all of these aromatic types.

In 1890 Hantzsch and Alfred Werner, his student at Zurich, launched the stereochemistry of nitrogen compounds. They explained the existence of two monoximes and three dioximes of benzil by proposing that the valences of trivalent nitrogen were nonplanar and disposed along three sides of a tetrahedron. Hence, compounds with carbon-nitrogen double bonds should exhibit geometrical isomerism. There should be two benzilmonoximes, which they named syn and anti forms, and three dioximes:

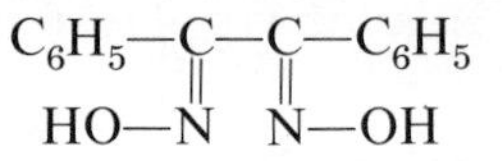

$C_6H_5—C(=N—OH)—C(=N—OH)—C_6H_5$

Anti

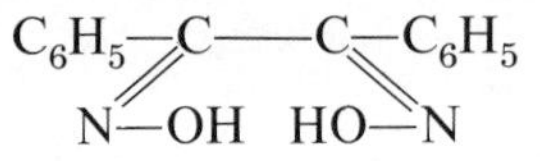

Syn

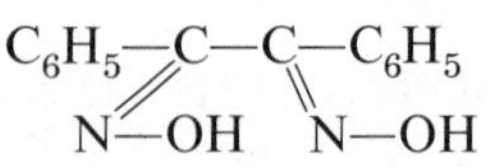

Amphi

The paper presented a new theory, although with very little experimental support. In one year of work Hantzsch provided overwhelming evidence for the stereochemistry of nitrogen. He also considered the assignment of configuration. He used the Beckmann rearrangement and various elimination reactions to determine oxime configurations, based on the assumption that *cis* groups rearrange and eliminate. All of Hantzsch's syn and anti configurations subsequently had to be reversed with the realization that trans groups were involved in these changes.

Hantzsch extended his theory to nitrogen-nitrogen double bonds in 1894 with his first paper on diazo compounds, which began a long, acrimonious controversy with Ludwig Bamberger. Three distinct diazo families existed: diazonium salts, normal diazotates, and isodiazotates. Bamberger argued that the isodiazo compounds were nitrosamines (Ar—NH—NO) and the isomeric normal diazotates were true diazo compounds (Ar—N=N—OH). Hantzsch established that they were not structural isomers but syn and anti forms.

The controversy with Bamberger, in which they exchanged many papers, was decisive in the development of organic chemistry. Hantzsch used physicochemical data from cryoscopic, conductivity, and absorption spectra studies. Bamberger used only reactions and syntheses for evidence of structure. He distrusted the physicochemical methods and arguments of Hantzsch, boasting that he used only pure organic chemical methods. Working with unstable compounds which changed into tautomeric forms in solution, he was at a disadvantage, whereas Hantzsch's methods enabled him to elucidate the complex interrelations of diazo compounds.

His work on stable and labile tautomeric diazo forms led to the discovery in 1896 that phenylnitromethane forms a salt not in the neutral nitro form but in the tautomeric aci form:

$$C_6H_5—CH_2—NO_2 \rightleftharpoons C_6H_5—CH{=}NO—OH.$$

He called the more stable nitro form a "pseudo acid."

From 1899 he developed a general theory of pseudo acids and bases as neutral compounds which can undergo reversible isomeric change into acids and bases respectively. Hantzsch proved that the true acids or bases corresponding to the pseudo forms were often much stronger than the common organic acids and bases. Methylquinolinium hydroxide or the di- and triphenylmethane bases were colored, ionized bases which underwent isomerization into the colorless, nonionized pseudo bases. Hantzsch proved the existence of such isomeric forms in many cases. In the aromatic series the colored form always possessed a quinonoid structure. He contributed to the theory of indicators by proposing that indicator action was an intramolecular change of quinonoid and nonquinonoid forms.

From 1906 Hantzsch was largely concerned with the study of the absorption spectra of organic compounds and the relation of color and constitution. He found evidence of a new constitutional type, the conjugated aci form, utilizing Alfred Werner's idea of partial valences and intramolecular complexes. Hantzsch's views on conjugation changed several times. By 1919 he recognized the limitations of symbolism and structural formulas and acknowledged that the difference between quinonoid and nonquinonoid structures in a conjugated complex did not really exist, and his formulas were accordingly simplified. Hantzsch's intent in eliminating the quinonoid character from his formulas was to show that all formulas were incomplete and that conjugation could not be expressed by any static formula or even as a dynamic equilibrium between different forms. His views came very near to those of the resonance theory, where forms are not expressible by ordinary structural formulas.

Hantzsch's last important investigations began in 1917 with a paper on the absorption spectra of carboxylic acids. He detected the presence of two forms of acid, depending on the solvent used, and proposed two carboxylic acid structures:

$$\mathrm{R{-}C}\left\langle\begin{matrix}\mathrm{O}\\\mathrm{O}\end{matrix}\right\}\mathrm{H} \qquad \mathrm{R{-}C}\begin{matrix}\mathrm{{=}O}\\\mathrm{{-}OH}\end{matrix}$$

These represented the true and pseudo acid forms respectively. The former possessed an ionizable hydrogen atom, whereas the pseudo acid in equilibrium with it did not.

Hantzsch thought that similar true and pseudo forms contributed to the nature of all acids, including the mineral acids. Furthermore, he obtained evidence that in aqueous solution the true acids were present as hydronium salts ($[H_3O]X$); in every case the acidic function attributed to the hydrogen ion in the Arrhenius-Ostwald theory was that of the hydronium ion.

From 1917 to 1927 he investigated the whole range of organic and inorganic acids by means of their absorption spectra, molecular refractivity, electrical conductivity, and relative stability of their compounds. In 1927 he summarized his results and stated his final views. He abandoned his earlier hypothesis that acids exist in true and pseudo forms. He thus considered all acids to be pseudo acids only, ionizing by forming a hydronium salt with a suitable solvent, the degree of ionization being dependent on the extent to which these salts are formed. There was no single favorable observation for the existence of free true acids.

Regarding all acids as pseudo acids, Hantzsch adopted as the criterion of acidity the tendency to form salts and measured the relative strength of acids by several different methods. He proved that the "strong" acids differed greatly, perchloric acid being the strongest, and established the now accepted sequence of mineral acid strength. Hantzsch's investigations broadened the conception of acids, showing that their properties depended on reaction with a solvent.

BIBLIOGRAPHY

I. ORIGINAL WORKS. Hantzsch wrote monographs on his major areas of investigation. His *Grundriss der Stereochemie* (Leipzig, 1893; 2nd ed., 1904) appeared in French and English eds. as *Précis de stéréochemie,* Guye and Gautier, trans. (Paris, 1896) and *The Elements of Stereochemistry,* C. G. L. Wolf, trans. (Easton, Pa., 1901). His *Die Diazoverbindungen* was published at Stuttgart in 1902; the 2nd ed. (Berlin, 1921) was written with G. Reddelien. His theory of true and pseudo forms is found in *Die Theorie der ionogenen Bindung als Grundlage der Ionentheorie* (Leipzig, 1923).

Among his important papers are "Über die synthese pyridinartiger Verbindungen aus Acetessigäther and Aldehydammoniak," in *Annalen der Chemie,* **215** (1882), 1–82; "Über Verbindungen des Thiazols," in *Berichte der Deutschen chemischen Gesellschaft,* **20** (1887), 3118–3132, written with J. H. Weber; "Über räumliche Anordnung der Atome in stickstoffhaltigen Molekülen," *ibid.,* **23** (1890), 11–30, written with Alfred Werner; "Die Bestimmung der räumlichen Configuration stereoisomerer Oxime," *ibid.,* **24** (1891), 13–31; "Über Stereoisomerie bei Diazoverbindungen und die Natur der Isodiazokörper," *ibid.,* **27** (1894), 1702–1725; "Über Isomerie beim Phenylnitromethan," *ibid.,* **29** (1896), 699–703, written with Otto Schultze; "Zur Constitutionsbestimmung von Körpern mit labilen Atomgruppen," *ibid.,* **32** (1899), 579–600; "Optische Untersuchungen von Diazo- und Azo-Verbindungen," *ibid.,* **45** (1912), 3011–3036, written with J. Lifschitz; "Die optischen und chemischen Veränderungen der organischen Nitroderivate und die stereochemische Erklärung ihrer Isomerien," in *Annalen der Chemie,* **492** (1931), 65–104; "Über die Konstitution der Carbonsäuren sowie über die optischen und chemischen Vorgänge bei der Bildung von Estern, Salzen, und Ionen," in *Berichte der Deutschen chemischen Gesellschaft,* **50** (1917), 1422–1457; and "Reaktionskinetische Untersuchungen an starken Säuren," in *Zeitschrift für physikalische Chemie,* **125** (1927), 251–263, written with A. Weissberger.

II. SECONDARY LITERATURE. The best accounts of Hantzsch's work were written by his former students: Arnold Weissberger, in Eduard Farber, ed., *Great Chemists* (New York, 1961), pp. 1065–1083; T. S. Moore, "The Hantzsch Memorial Lecture," in *Journal of the Chemical Society* (1936), 1051–1066; and Franz Hein, "Arthur Hantzsch 1857–1935," in *Berichte der Deutschen chemischen Gesellschaft,* **74** (1941), 147–163. See also A. Burawoy, "Arthur Hantzsch," in *Berichte der Deutschen chemischen Gesellschaft,* **68** (1935), 65–68; F. Hein, "A. Hantzsch" in *Zeitschrift für Elektrochemie und angewandte physikalische Chemie,* **42** (1936), 1–4; B. Helferich, "Nachruf auf Arthur Hantzsch," in *Berichte über die Verhandlungen der Sächsischen Akademie der Wissenschaften zu Leipzig,* **87** (1935), 213–222; and C. Paal, "Arthur Hantzsch zum 70. Geburtstage," in *Zeitschrift für angewandte Chemie,* **40** (1927), 301–303.

ALBERT B. COSTA

HARCOURT, A. G. VERNON (*b.* London, England, 24 December 1834; *d.* Hyde, Isle of Wight, 23 August 1919), *chemistry.*

After a typical classical preparation at Cheam and Harrow, Harcourt entered the newly developing chemistry program at Oxford. Initially a student assistant to Benjamin Brodie in the basement crypts at Balliol, he progressed to the Lee Laboratory of Christ

Church College and a chemistry professorship. Always active in the British Association for the Advancement of Science (his uncle William Vernon Harcourt having been the principal founder), he served as president of the chemical section in 1875 and later as general secretary for fourteen years. Elected to the Royal Society in 1863, Harcourt gave the Bakerian lecture in 1895, the same year that he was elected president of the Chemical Society.

Harcourt was no academic recluse, and his interest in the technical applications of science led to his appointment in 1872 to the board which prescribed tests and purity standards for London gas. His major contribution in the field of gas testing was to introduce the pentane lamp in place of the less reliable spermaceti candles hitherto used in brightness measurements. At the turn of the century he furthered anesthesia research by devising a method for determining the chloroform concentration in air and served as a consultant in the British Medical Association study of the anesthetic properties of chloroform.

In the realm of pure chemistry Harcourt in 1866, aided by his mathematician colleague William Esson, discovered independently of Guldberg and Waage the law of mass action in its simplest form: "The velocity of chemical change is directly proportional to the quantity of substance undergoing change." Attracted to reaction rate studies in 1864 by Friedrich Kessler's suggestion that manganous sulfate accelerates the reduction of potassium permanganate, Harcourt designed experiments which showed that every reactant gives a similarly shaped curve for the effect of concentration on the permanganate reduced during a fixed time period. When the two men examined the time rate of change, they obtained results which Esson identified as an exponential relation between reaction velocity and time.

In 1867 Harcourt and Esson confirmed these observations with data from the somewhat simpler reaction of hydrogen peroxide with potassium iodide but also found that the iodide gave a disproportionate increase at lower acidities. For some unknown reason, their further experimental studies of this acidity effect were not published until Harcourt's 1895 Bakerian lecture, in which they also reported Esson's empirical correlation between reaction rate k and absolute temperature T, that is, $k/k_0 = (T/T_0)^m$—which suggested that at absolute zero all chemical activity would cease. Feeling that due recognition had not been paid, they returned to this question in 1912, comparing their rate-temperature correlation with those more generally used in physical chemistry by van't Hoff and others. Their purely empirical arguments, however, gave no theoretical support for their contentions.

BIBLIOGRAPHY

Harcourt's Bakerian lecture is strangely omitted from Poggendorff: IV, 1560; V, 1306; VI, 2751. It may be found as "On the Laws of Connexion Between the Conditions of a Chemical Change and Its Amount," in *Philosophical Transactions of the Royal Society,* **186A** (1895), 817–895.

Harold Dixon, whom Harcourt led from apparent failure in classics to success in chemistry, wrote a highly appreciative obituary in *Proceedings of the Royal Society,* **97** (1920), vii–xi.

See also J. R. Partington, *A History of Chemistry,* IV (London–New York, 1964), 585–587.

EDWARD E. DAUB

HARDEN, ARTHUR (*b.* Manchester, England, 12 October 1865; *d.* Bourne End, Buckinghamshire, England, 17 June 1940), *biochemistry.*

Harden was the third child and only son of Albert Tyas Harden, a Manchester businessman, and Eliza MacAlister of Paisley; there were eight daughters. The family was nonconformist and their values were austere, a characteristic which Harden maintained throughout his life. He studied under Henry Roscoe at Owens College, University of Manchester, from which he graduated in 1885 with first-class honors in chemistry.

Having been awarded the Dalton scholarship in 1886, Harden undertook graduate study under Otto Fischer at the University of Erlangen. He received the Ph.D. in 1888 with a dissertation on the preparation and properties of β-nitrosonaphthylamine. He then became junior lecturer at the University of Manchester, where he soon advanced to senior lecturer and demonstrator. He was heavily involved in teaching and writing and published a paper (1897) on the composition of some bronze and iron tools discovered by Flinders Petrie.

In 1897 Harden became head of the chemistry department at the British Institute of Preventive Medicine (renamed the Jenner Institute in 1898 and the Lister Institute in 1903) and began research in microbiological chemistry. After 1905 the biochemistry department merged with Harden's department under his leadership.

In 1898 Harden began studies on the fermentation of sugars by coliform bacteria, hoping to discover a chemical means to distinguish varieties of *Escherichia coli* (then termed *Bacterium coli*). He discovered several compounds formed in the bacterial decomposition of sugars and developed a scheme for the breakdown process. He showed that acetylmethylcarbinol produced by *Bacterium coli aerogenes* was responsible for the Voges-Proskauer color reaction which was used empirically by bacteriologists for diagnostic purposes. Although he continued his work in bacterial chemistry until 1912, his major attention after 1900 was given to alcoholic fermentation by yeasts.

In 1897 Eduard Buchner had discovered that alcoholic fermentation could be carried out by a cell-free juice extracted from yeast, and he named the active enzyme present in the extract zymase. The first crude zymase preparations produced carbon dioxide and alcohol—even without the presence of added sugar. Harden investigated this reaction and found that glycogen was expressed from yeast cells when zymase was prepared, thus furnishing a source of sugar. Harden also investigated the fact that the yeast juice quickly lost its power to ferment sugar and discovered that a proteinase expressed from yeast cells destroyed the zymase in a short time.

Continuing his studies on alcoholic fermentation with his student William John Young (later professor of biochemistry at the University of Melbourne), Harden learned in 1904 that the capacity of yeast juice to ferment glucose was stimulated by the addition of boiled yeast juice. He and Young showed that by dialysis or filtration through a Martin gelatin filter it was possible to separate yeast juice into two fractions, neither of which had the capacity to ferment glucose. Combination of the fractions led to normal fermentation. The dialyzable, filterable portion was a low molecular weight substance, stable to boiling, and easily precipitated by 75 percent alcohol. Harden called this substance a coferment although the names coenzyme I, cozymase, and (after its constitution was fully established by Euler-Chelpin, H. Schlenk, and co-workers) diphosphopyridine nucleotide (DPN) came to be used. Harden showed that the coferment contained phosphate but failed to make further progress in clarifying its chemical nature.

Harden and Young also observed that phosphate salts stimulated yeast juice to produce carbon dioxide. They found that phosphate combined with glucose, fructose, or mannose to form a hexose diphosphate which Young later isolated and identified. They showed this compound to be hydrolyzed by a phosphatase present in the juice. In 1914 Robert Robison, working in Harden's laboratory, discovered hexose monophosphate as an intermediate in the fermentation process. Although Harden continued his studies on fermentation, he failed to make further major contributions toward understanding the nature of the process. However, his recognition of the presence of phosphate esters in fermentation liquors was important in directing the attention of other workers to phosphorus compounds as intermediates in fermentation and muscular respiration.

During World War I Harden abandoned fermentation studies and directed his attention toward vitamin problems. With S. S. Zilva he studied problems connected with the substances which prevented beriberi and scurvy. He established the synthesis of the antiberiberi factor by yeast and disproved the claims for the reported activity of α-hydroxypyridine and adenine. By removal of sugars, organic acids, and proteins from lemon juice he prepared a concentrate with enhanced antiscorbutic activity which was useful in treating infant scurvy.

Harden's postwar researches were concentrated on the nature of the enzymes in yeast and their mode of action. He confirmed Carl Neuberg's discovery of carboxylase in yeast and studied peroxidase and invertase. He also studied the role of inorganic salts in fermentation.

In 1929 Harden shared the Nobel Prize for chemistry with Euler-Chelpin for their studies of alcoholic fermentation. He was elected to the Royal Society in 1909, received the society's Davy Medal in 1935, and was knighted the next year.

Throughout his career, Harden wrote and edited many works. In addition to several early chemistry textbooks, he collaborated with Roscoe in a study of Dalton's notebooks, which led to the publication of *A New View of the Origin of Dalton's Atomic Theory* (1896). From 1913 to 1937 he was largely responsible for editing the *Biochemical Journal.*

Harden was married to Georgina Bridge of Christchurch, New Zealand, in 1890; they had no children. His wife died in 1928, two years before Harden retired from his professorship at the Lister Institute. A progressive nervous disease was responsible for his death.

BIBLIOGRAPHY

I. Original Works. There is no collected bibliography of Harden's publications. Most of his research papers were published in the *Proceedings of the Royal Society, Journal of the Chemical Society,* and *Biochemical Journal.* For the studies of Dalton's notebooks, see *A New View of the Origin of Dalton's Atomic Theory* (London, 1896), written with H. E. Roscoe; and a later work, "John Dalton's Lectures and Lecture Illustrations. Part III. The Lecture Sheets Illustrating the Atomic Theory," in *Memoirs and Proceedings of the Manchester Literary and Philosophical Society,* **59** (1915), 41-66, written with H. F. Coward.

Harden's other works include "The Composition of Some Ancient Iron and a Bronze Found at Thebes," in *Transactions of the Manchester Literary and Philosophical Society,* **41** (1897), 1-3; *Practical Organic Chemistry* (London, 1897), written with F. C. Garrett; *Inorganic Chemistry for Advanced Students* (London, 1899), written with H. E. Roscoe; "The Alcoholic Ferment of Yeast Juice," in *Proceedings of the Royal Society,* **77B** (1907),

405–420, written with W. J. Young; and *Alcoholic Fermentations* (London, 1911; rev. eds., 1914, 1923, 1932). His Nobel Prize address, "The Function of Phosphate in Alcoholic Fermentation," is in *Nobel Lectures, Including Presentation Speeches and Laureates' Biographies, Chemistry, 1922–1941* (Amsterdam, 1966), pp. 131–141.

II. SECONDARY SOURCES. The best biography is the obituary sketch by Ida Smedley-Maclean, "Arthur Harden (1865–1940)," in *Biochemical Journal,* **35** (1941), 1071–1081. See also F. G. Hopkins and C. J. Martin, "Arthur Harden, 1865–1940," in *Obituary Notices of Fellows of the Royal Society of London,* **4** (1942–1944), 3–14; C. J. Martin, in *Dictionary of National Biography, 1931–1940,* pp. 395–397; and *Nobel Lectures,* cited above, pp. 142–143.

AARON J. IHDE

HARDING, CARL LUDWIG (*b.* Lauenburg, Germany, 29 July 1765; *d.* Göttingen, Germany, 31 August 1834), *astronomy.*

Harding studied theology at Göttingen from 1786 to 1789, while also attending the mathematics and physics lectures given by A. G. Kästner. After completing his studies he served as a probationary minister in Lauenburg. Like many young men holding such a position Harding became a private tutor in 1796, when Kästner and others recommended him to Chief Magistrate A. H. Schröter, who had a private observatory at Lilienthal, near Bremen. It was well equipped for the time, with astronomical instruments constructed by Herschel, Peter Dollond, and Schröter himself.

In its short period of activity Schröter's observatory had a high reputation. The best observations of the great planets during that time were made at Lilienthal, mostly by Harding. Olbers often visited it, in 1800 with Zach. The Vereinigte Astronomische Gesellschaft which was established there included foreign scientists. This new society intended primarily to make star charts. This aim was realized only by Harding, who drew up a celestial atlas containing about 60,000 stars; this stellar chart was one of the first prepared according to scientific principles. While working on this star chart Harding discovered (1804) the third asteroid and named it Juno Georgia, to honor George III. Perhaps partly as a result of this he was transferred to the new Göttingen observatory and from 1805 was professor of practical astronomy there. While at Göttingen he observed planets, comets, and variable stars. He also discovered three comets: 1813 II, 1824 II, and 1832 II.

Harding participated in Encke's *Akademische Sternkarten* and was among the first to finish his part, hour 15–16. The first twelve volumes of *Astronomische Nachrichten* contain many short notes on his observations.

BIBLIOGRAPHY

Harding's works include *Atlas novus coelestis,* 7 vols. (Göttingen, 1808–1823); Zach's *Monatliche Korrespondenz zur Beförderung der Erd- und Himmelskunde,* **21** (1810); "Hora XV," in *Akademische Sternkarten* (Berlin, 1830); and *Kleine astronomische Ephemeriden* for 1831–1835 (Göttingen, 1830–1834), written with G. Wiesen. There are many short notices in *Monatliche Korrespondenz* and *Astronomische Nachrichten.* See Poggendorff, I, cols. 1016–1017.

A secondary source is H. A. Schumacher, *Die Lilienthaler Sternwarte* (Bremen, 1889).

H. C. FREIESLEBEN

HARDY, CLAUDE (*b.* Le Mans, France, *ca.* 1598; *d.* Paris, France, 5 April 1678), *mathematics.*

Little is known about Hardy's life. He is said to have been born in 1598 (G. Loria) or in 1605 (Claude Irson). In 1625 he was a lawyer attached to the court of Paris and in 1626 a counselor in the Châtelet. He took part in the weekly meetings of Roberval, Mersenne, and the other French geometricians in the Académie Mersenne, and was a friend of Claude Mydorge, who introduced him to Descartes. Several writers of the seventeenth century suggested methods for the duplication of the cube, including Viète, Descartes, Fermat, and Newton. Among the less well-known persons who also occupied themselves with this problem was Paul Yvon, lord of Laleu, who claimed that he had found the construction of the two mean proportionals, required in solving the problem. In addition to Mydorge and J. de Beaugrand, Hardy exposed the fallacy of Yvon's construction in his *Examen* of 1630 and again in his *Refutation* of 1638. In turn Hardy was attacked by other scholars. Owing to a lack of explicitness in statement, Fermat's method of maxima and minima and of tangents was severely attacked by Descartes. In the ensuing dispute Fermat found two zealous defenders in Roberval and Pascal, while Mydorge, Desargues, and Hardy supported Descartes.

Hardy owed his greatest fame, however, to his knowledge of Arabic and other exotic languages, and in particular, to his edition of Euclid's *Data* (1625), the *editio princeps* of the Greek text, together with a Latin translation. He is said to have translated the *Isagoge* (Tours, 1591) and the *Zetetica* (Tours, 1593) of Viète and to have occupied himself with a project for a universal language.

BIBLIOGRAPHY

I. ORIGINAL WORKS. Hardy's ed. of the *Data* was published as *Euclidis Data. Opus ad veterum geometriae au-*

torum Archimedis, Apollonii, Pappi, Eutocii ceterorumque . . . (Paris, 1625). He was author of *Examen de la duplication du cube et quadrature du cercle, cy-devant publiée à diverses fois par le Sieur de Laleu* . . . (Paris, 1630); and *Refutation de la manière de trouver un quarré égal au cercle rapportée ès pages 130 et 131 du livre nouvellement imprimé sous le titre de Propositions mathématiques de Monsieur de Laleu demonstrées par I. Pujos, et au prétendu triangle équilatéral mentionné au placard dudit sieur* . . . (Paris, 1638).

II. SECONDARY LITERATURE. On Hardy and his work, see (listed in chronological order) P. Colomiès, *Gallia orientalis* (The Hague, 1665), pp. 165–166, 259–260; C. Irson, *Nouvelle méthode pour apprendre facilement les principes et la pureté de la langue françoise* (Paris, 1667), p. 317; G. Loria, *Storia delle matematiche,* II (Milan, 1931), 309; and C. de Waard, ed., *Correspondance du M. Mersenne,* I (Paris, 1932), 187, 619, 666; II (Paris, 1937), 116, 550, 551; III (Paris, 1946), 230; IV (Paris, 1955), 322, 323; V (Paris, 1959), 136; VII (Paris, 1962), 63, 288–292; VIII (Paris, 1963), 417, 418.

H. L. L. BUSARD

HARDY, GODFREY HAROLD (*b.* Cranleigh, England, 7 February 1877; *d.* Cambridge, England, 1 December 1947), *mathematics.*

Hardy was the elder of two children of Isaac Hardy, a master at Cranleigh School, and Sophia Hall. The parents were intelligent and mathematically minded, but lack of money had precluded them from a university education. They provided an enlightened upbringing for Hardy and his sister.

The freedom to ask questions and to probe led Hardy to an early established disbelief in religious doctrine. (As a fellow of New College, Oxford, he refused to enter the chapel to take part in electing a warden.) Neither Hardy nor his sister married, and he owed much to her devoted care throughout his life, particularly in his later years.

As a boy Hardy showed all-around ability with a precocious interest in numbers. At the age of thirteen he moved from Cranleigh School with a scholarship to Winchester College, to this day a famous nursery of mathematicians. He went on to Trinity College, Cambridge, in 1896, was fourth wrangler in the mathematical tripos in 1898, was elected a fellow of Trinity in 1900, and won (with J. H. Jeans) a Smith's Prize in 1901. Success in the tripos depended on efficient drilling in solving problems quickly. Hardy, resenting the routine of the famous "coach" R. R. Webb, had the good fortune to be transferred to A. E. H. Love. No description of Hardy's development into a mathematician can be so vivid as his own:

> My eyes were first opened by Professor Love, who taught me for a few terms and gave me my first serious conception of analysis. But the great debt which I owe to him was his advice to read Jordan's famous *Cours d'analyse;* and I shall never forget the astonishment with which I read that remarkable work, the first inspiration for so many mathematicians of my generation, and learnt for the first time as I read it what mathematics really meant [*A Mathematician's Apology,* sec. 29].

Hardy flung himself eagerly into research and between 1900 and 1911 wrote many papers on the convergence of series and integrals and allied topics. Although this work established his reputation as an analyst, his greatest service to mathematics in this early period was *A Course of Pure Mathematics* (1908). This work was the first rigorous English exposition of number, function, limit, and so on, adapted to the undergraduate, and thus it transformed university teaching.

The quotation from the *Apology* continues, "The real crises of my life came ten or twelve years later, in 1911, when I began my long collaboration with Littlewood, and in 1913, when I discovered Ramanujan."

J. E. Littlewood, eight years younger than Hardy, proved in 1910 the Abel-Tauber theorem that, if na_n is bounded and $\Sigma a_n x^n \rightarrow s$ as $x \rightarrow 1$, then $\Sigma a_n = s$. The two then entered into a collaboration which was to last thirty-five years. They wrote nearly a hundred joint papers. Among the topics covered were Diophantine approximation (the distribution, modulo 1, of functions $f(n)$ of many types, such as θn^2 for irrational θ), additive and multiplicative theory of numbers and the Riemann zeta function, inequalities, series and integrals in general (for instance, summability and Tauberian theorems), and trigonometric series.

The partnership of Hardy and Littlewood has no parallel, and it is remarkable that, at its greatest intensity (1920–1931), Hardy lived in Oxford and Littlewood in Cambridge. They set up a body of axioms expressing the freedom of their collaboration, for example, "When one received a letter from the other he was under no obligation to read it, let alone to answer it." The final writing of the papers was done by Hardy.

Hardy called his discovery of Srinivasa Ramanujan the one romantic incident of his life. One morning early in 1913, he received a letter from this unknown Indian, containing a number of formulae without any proofs. Established mathematicians are exposed to manuscripts from amateurs, and Hardy could not at a glance assess it. A few hours' work convinced him that the writer was a man of genius. Ramanujan turned out to be a poor, self-taught clerk in Madras, born in 1887. Hardy brought him to England in April

1914 and set about the task of filling the gaps in his formal mathematical education. Ramanujan was ill from May 1917 onward; he returned to India in February 1919 and died in April 1920. In his three years of health and activity, he and Hardy had arrived at spectacular solutions of problems about the partition of numbers which called forth the full power of the Indian's natural insight and the Englishman's mastery of the theory of functions.

Denote by $p(n)$ the number of ways of writing n as the sum of positive integers (repetitions allowed), so that $p(5) = 7$. As n increases, $p(n)$ increases rapidly; for instance, $p(200)$ is a number of thirteen digits, a computation which in 1916 took a month. Hardy and Ramanujan established an asymptotic formula for $p(n)$, of which five terms sufficed to give the value of $p(200)$.

Hardy was a lecturer at Trinity College until 1919, when he became Savilian professor of geometry at Oxford; there he founded a flourishing school of research. For the year 1928–1929 he went to Princeton, exchanging places with Oswald Veblen. He returned to Cambridge in 1931, succeeding E. W. Hobson as Sadleirian professor of pure mathematics; he held this chair until his retirement in 1942.

Besides Littlewood and Ramanujan, Hardy collaborated with many other mathematicians, including E. C. Titchmarsh, A. E. Ingham, E. Landau, G. Pólya, E. M. Wright, W. W. Rogosinski, and M. Riesz. He had an exceptional gift for working with others, as he had for leading young men in their early days of research.

Hardy had one ruling passion—mathematics. Apart from that his main interest was in ball games, particularly cricket, of which he was a stylish player and an expert critic. Some of his interests and antipathies are revealed by this list of six New Year wishes which he sent on a postcard to a friend in the 1920's: (1) prove the Riemann hypothesis; (2) make 211 not out in the fourth innings of the last test match at the Oval; (3) find an argument for the nonexistence of God which shall convince the general public; (4) be the first man at the top of Mt. Everest; (5) be proclaimed the first president of the U.S.S.R. of Great Britain and Germany; (6) murder Mussolini.

Hardy was generally recognized as the leading English pure mathematician of his time. His writings attest both his technical power and his mastery of English prose. The photographs in *Collected Papers* show his finely cut features and something of his physical grace. His liveliness and enthusiasm are vivid in the memory of all who knew him. He received awards from many universities and academies, being elected in 1947 *associé étranger* of the Paris Academy of Sciences—of whom there are only ten from all nations in all subjects.

BIBLIOGRAPHY

I. ORIGINAL WORKS. Hardy published, alone or in collaboration, about 350 papers. A complete list is in *Journal of the London Mathematical Society,* **25** (1950), 89–101. Collected papers are being published in 7 vols. (Oxford, 1966–), edited, with valuable comments, by a committee appointed by the London Mathematical Society.

Hardy wrote four tracts published at Cambridge: *The Integration of Functions of a Single Variable* (1905); *Orders of Infinity* (1910); *The General Theory of Dirichlet's Series* (1915), written with M. Riesz; and *Fourier Series* (1944), written with W. W. Rogosinski. The last, in particular, is a model of concise lucidity.

Hardy underlined the neglect of analysis in England by writing in the preface to the 1st ed. of *A Course of Pure Mathematics* (Cambridge, 1908; 10th ed., 1952): "I have indeed in an examination asked a dozen candidates, including several future senior wranglers, to sum the series $1 + x + x^2 + \cdots$ and not received a single answer that was not practically worthless." His book changed all that. *Inequalities* (Cambridge, 1934), written with J. E. Littlewood and G. Pólya, is a systematic account and includes much material previously accessible only in journals. *The Theory of Numbers* (Oxford, 1938), written with E. M. Wright, includes chapters on a variety of topics.

Other works include *A Mathematician's Apology* (Cambridge, 1940; repr. 1967 with a foreword by C. P. Snow); *Ramanujan* (Cambridge, 1940), twelve lectures on his life and work; *Bertrand Russell and Trinity* (Cambridge, 1970), an account of a 1914–1918 controversy, showing Hardy's sympathy with Russell's opposition to the war. See especially *Divergent Series* (Cambridge, 1948), completed by Hardy shortly before his death. According to Littlewood in his foreword, "All his books gave him some degree of pleasure, but this one, his last, was his favourite."

II. SECONDARY LITERATURE. Notices on Hardy are in *Nature,* **161** (1948), 797; *Obituary Notices of Fellows of the Royal Society of London,* **6** (1949), 447–470, with portrait; *Journal of the London Mathematical Society,* **25** (1950), 81; and *Dictionary of National Biography 1941–1950* (Oxford, 1959), 358–360.

J. C. BURKILL

HARE, ROBERT (*b.* Philadelphia, Pennsylvania, 17 January 1781; *d.* Philadelphia, 15 May 1858), *chemistry.*

During his youth and until he was thirty-seven years old, Hare helped manage the family brewery in Philadelphia. He learned chemistry by independent study and by attending lectures of James Woodhouse at the University of Pennsylvania. While operating the brewery he was also professor of natural philosophy at the University of Pennsylvania medical school from 1810 to 1812.

The brewery failed around 1815 and Hare attempted, without success, to manufacture illuminating gas in New York City. Early in 1818 he became professor of natural philosophy and chemistry at the College of William and Mary for a few months, then professor of chemistry at the University of Pennsylvania medical school until 1847.

After he retired, Hare wrote a novel, *Standish the Puritan,* under the pseudonym Eldred Grayson; investigated the cause of accidental explosions of niter; and lectured and wrote on spiritualism, in which he came to believe.

Hare made his major contribution at the age of twenty, when he was still an amateur scientist. Seeking a means of producing high temperatures, he hit upon the idea of burning a mixture of hydrogen and oxygen. He devised a gasholder and oxyhydrogen blowtorch which produced a higher temperature than previously obtainable by any means. The torch made possible the melting of platinum and other substances with high melting points and formed the basis of the Drummond light and limelight.

A skillful craftsman, Hare devised ingenious apparatus for research and demonstration. His lecture hall was perhaps the best equipped in the United States. He developed the calorimotor, the deflagrator, and an electric furnace in which he produced graphite, calcium carbide, and other substances.

Few American chemists of the early nineteenth century taught more students than Hare. As a professor in the country's largest medical school for twenty-nine years, he transmitted chemistry to a proportionately large segment of the medical profession. A number of his pupils became teachers of chemistry.

BIBLIOGRAPHY

I. ORIGINAL WORKS. Royal Society, *Catalogue of Scientific Papers,* III, 177–182, lists 127 of Hare's articles. His books are *Memoir of the Supply and Application of the Blow-Pipe . . .* (Philadelphia, 1802); *Minutes of the Course of Chemical Instruction in the Medical Department of the University of Pennsylvania* (Philadelphia, 1822), which evolved into *Compendium of the Course of Chemical Instruction . . .* (Philadelphia, 1828; 4th ed., 1840); and *Engravings and Descriptions of a Great Part of the Apparatus Used in the Chemical Course of the University of Pennsylvania,* 2 vols. (Philadelphia, 1828).

II. SECONDARY LITERATURE. Edgar F. Smith, *The Life of Robert Hare an American Chemist (1781–1858)* (Philadelphia, 1917), with portrait, is the standard life of Hare but lacks references to all sources; Edgar F. Smith, *Chemistry in America. Chapters From the History of the Science in the United States* (New York, 1914), pp. 152–205, with portrait, reprints Hare's *Memoir . . . of the Blow-Pipe.* See also Wyndham D. Miles, "Robert Hare," in Eduard Farber, ed., *Famous Chemists* (New York, 1961), pp. 420–423, with portrait.

WYNDHAM DAVIES MILES

HARIDATTA I (*fl.* India, 683), *astronomy.*

Haridatta, who probably lived in south India, composed in 683 the *Grahacāranibandha,* the principal text of the *parahita* system of astronomy (see essay in Supplement), which is based on the *Āryabhaṭīya* of Āryabhaṭa I and which prevailed in Kerala until the fifteenth century. An important feature of this work is its versified table of the planetary equations employing the *kaṭapayādi* method of expressing numerals. The *Grahacāranibandha* was published by K. V. Sarma (Madras, 1954). Haridatta therein refers to his *Mahāmārganibandhana,* in which he discussed the calculation of *tithis;* this work is lost. Various opinions of Haridatta regarding astrology are cited by Govindasvāmin (*fl. ca.* 850) in the *Prakaṭārthadīpikā,* a commentary on the *Uttarakhaṇḍa* of pseudo-Parāśara's *Horāśāstra.*

BIBLIOGRAPHY

Aside from K. V. Sarma's introduction to his ed. of the *Grahacāranibandha* mentioned above, the only discussion of Haridatta is by K. Kunjunni Raja, "Astronomy and Mathematics in Kerala," in *Brahmavidyā,* **27** (1963), 118–167, esp. 123–126.

DAVID PINGREE

HARIDATTA II (*fl.* India, 1638), *astronomy.*

Haridatta, the son of Harajī, composed the *Jagadbhūṣaṇa* (see essay in Supplement) in 1638 in Mewar, Rajasthan, during the reign of Jagatsiṃha I (1628–1652); nothing else is known of him. The *Jagadbhūṣaṇa* consists of tables for computing planetary positions (utilizing the Babylonian "goal-year" periods) and solar and lunar positions. These tables are described by D. Pingree, "Sanskrit Astronomical Tables in the United States," in *Transactions of the American Philosophical Society,* n.s. **58,** pt. 3 (1968), 55*b*–59*b*; "On the Classification of Indian Planetary Tables," in *Journal of the History of Astronomy,* **1** (1970); and "Sanskrit Astronomical Tables in England," in *Journal of Oriental Research* (in press).

BIBLIOGRAPHY

Nothing has been written on Haridatta II aside from what has been mentioned above.

DAVID PINGREE

HARIOT, THOMAS. See **Harriot, Thomas.**

HARKER, ALFRED (*b.* Kingston-upon-Hull, England, 19 February 1859; *d.* Cambridge, England, 28 July 1939), *petrology.*

Harker entered St. John's College, Cambridge, in 1878 and for sixty-one years was one of its most distinguished members. Although physics was at first his principal subject, in 1884 he was appointed university demonstrator in geology at the Sedgwick Museum at Cambridge and soon became the outstanding figure among British petrologists. His earlier research was conducted chiefly in north Wales and the English Lake District. From 1895 to 1905 he combined his university work with fieldwork in Scotland for the Geological Survey. He received many honors and distinctions, most important among them being the Wollaston Medal of the Geological Society of London (its highest award) in 1922 and a royal medal of the Royal Society in 1935. Somewhat diffident and shy and not readily eloquent in speech, he left writings that are among the masterpieces of scientific literature.

Harker's original researches concerned five subjects.

1. Slaty cleavage. This was his first study, and the results of it are still authoritative. He returned to it in his later years.

2. North Wales. The igneous rocks associated with the Ordovician sedimentaries in Caernarvonshire had been mapped by the Geological Survey, but Harker described their exact petrographical nature and discussed their mutual relationships. He traced the connection between the igneous phenomena and the crustal stresses and regional cleavage of the district.

3. English Lake District. Harker made detailed surveys, largely with J. E. Marr, and petrographical examinations of two areas of plutonic and associated rocks: Shap in the south and Carrock Fell in the north. His work here threw light on the problems of igneous variation, differentiation, forms of intrusion and their association, and thermal metamorphism.

4. The islands of the Inner Hebrides. Harker's studies of the Tertiary igneous activity on the Isle of Skye and the smaller islands to the south inaugurated a new era in the investigation of igneous rock complexes. Principles of the first importance were formulated, such as those of the volcanic-plutonic-hypabyssal cycle and the nature and origin of so-called hybrid rocks. The two memoirs are enduring monuments to his great achievement. The survey work was carried on some twenty years later by a team of the most eminent geologists, who investigated all the Tertiary volcanic centers of western Scotland. In Skye, Harker studied the effects of Pleistocene ice action, emphasizing the importance of glacial erosion.

5. General works. Harker expounded the philosophical results of his research and thought in *The Natural History of Igneous Rocks.* The whole range of phenomena was examined in the light of the general principles of mathematics, physics, and chemistry, with special attention to geographical distribution and tectonic environment. *Metamorphism,* one of his latest works, is of the same caliber. Here the beautiful line drawings, a feature of all his works, are especially to be admired. In the first of his two presidential addresses to the Geological Society he reviewed the history of igneous activity in the British Isles throughout geological time. Finally, his *Petrology for Students* is a work familiar to many generations of college students, particularly treasured by those who were among his pupils.

BIBLIOGRAPHY

I. Original Works. Among Harker's more important works are "On Slaty Cleavage and Allied Rock Structures," in *Report of the British Association for the Advancement of Science* for 1885, pp. 813–852; *The Bala Volcanic Series of Caernarvonshire* (Cambridge, 1889), the Sedgwick Prize essay; "The Shap Granite and the Associated Igneous and Metamorphic Rocks," in *Quarterly Journal of the Geological Society of London,* **47** (1891), 266–328, and **49** (1893), 359–371, written with J. E. Marr; "Carrock Fell: A Study in the Variation of Igneous Rock-masses," *ibid.,* **50** (1894), 311–337, and **51** (1895), 125–148; *Petrology for Students: An Introduction to the Study of Rocks Under the Microscope* (Cambridge, 1895; 8th ed., 1954); "Ice-erosion in the Cuillin Hills, Skye," in *Transactions of the Royal Society of Edinburgh,* **40** (1901), 221–252; *The Tertiary Igneous Rocks of Skye,* Memoirs of the Geological Survey of Great Britain (London, 1904); *The Geology of the Small Isles of Inverness-shire,* Memoirs of the Geological Survey of Great Britain (London, 1908); *The Natural History of Igneous Rocks* (London, 1909); "Some Aspects of Igneous Action in Britain," in *Quarterly Journal of the Geological Society of London,* **73** (1917), lxvii–xcvi, his first presidential address to the Geological Society; *Metamorphism: A Study of the Transformations of Rock-masses* (London, 1932; 3rd ed., 1950); and *The West Highlands and the Hebrides: A Geologist's Guide for Amateurs,* J. E. Richey, ed. (Cambridge, 1941), which includes an appreciation by A. C. Seward (pp. xvii–xxiii).

II. Secondary Literature. On Harker and his work, see "Eminent Living Geologists: Alfred Harker, M.A., LL.D., F.R.S., . . .," in *Geological Magazine,* **54** (1917), 289–294; J. S. F[lett], obituary notice in *Proceedings of the Geological Society,* **96** (1940), lxix–lxxi; A. C. Seward and C. E. Tilley, in *Obituary Notices of Fellows of the Royal Society of London,* **3** (1940), 197–216; and C. E. Tilley, in *Dictionary of National Biography, 1931–1940* (London, 1949), p. 400.

John Challinor

HARKINS, WILLIAM DRAPER (*b.* Titusville, Pennsylvania, 28 December 1873; *d.* Chicago, Illinois, 7 March 1951), *physical chemistry.*

Harkins was the son of Nelson Goodrich Harkins, a pioneer in the Pennsylvania oil fields, and Sarah Eliza Draper. In 1900 he graduated from Stanford University with a B.A. in chemistry and immediately accepted a teaching position at the University of Montana, where he became a professor and chairman of the chemistry department. While associated with Montana, Harkins did graduate work at the University of Chicago (1901–1904) and at Stanford University (1905–1906). He received his Ph.D. from Stanford in 1907 and did postdoctoral study at the Technical University in Karlsruhe (1909) and at the Massachusetts Institute of Technology (1909–1910). In 1912 Harkins accepted an assistant professorship at the University of Chicago. He remained there for the rest of his life, becoming associate professor in 1914, professor in 1917, and the Andrew McLeish Distinguished Service Professor for 1935.

Among his many activities, Harkins acted as consultant to a number of private companies, the Chemical Warfare Service, and the National Defense Research Commission. From 1932 he was a member of the International Commission on Atoms. He also served as vice-president of the American Association for the Advancement of Science and was elected to the National Academy of Sciences. On 9 June 1905 Harkins married Anna Louise Hatheway, the head of the English department at Montana. They had two children, Henry Nelson, who became a surgeon, and Alice Marion, who achieved recognition as a singer.

While at Montana, Harkins published three papers on arsenic pollution in smelter smoke, in which he showed that a smelter stack spewed thirty tons per day of arsenic trioxide (and at least as much copper) over the surrounding twenty miles of pastureland. By bringing the arsenic level to 200–500 parts per million in fall grasses, this pollution killed many hundreds of sheep, horses, and cattle. Because Harkins' detailed and complete studies (supported by the Anaconda Farmer's Association) left no possible loopholes for dispute, he was recognized as an expert on smelter pollution and became a consultant to the Mountain Copper Company of California, the U. S. Department of Justice, and the Carnegie Institution.

At Chicago, Harkins began work on the structure and the reactions of atomic nuclei. The leading researchers in this newly developing science (Ernest Rutherford, Francis William Aston, Frederick Soddy, Patrick Maynard Stuart Blackett) were mostly in England and, except for T. W. Richards at Harvard, there had been little American involvement. In 1915 Harkins and E. D. Wilson published five important papers concerning the processes of building complex atomic nuclei from protons, deuterium, tritium nuclei, and α-particles. At this time the only nuclear reactions that had been studied were the decomposition reactions of radioactive nuclei, for which the Einstein equation relating mass and energy predicted the observed energies. With the Einstein equation Harkins showed the enormous energy produced in the nuclear fusion of hydrogen to produce helium, with the attendant .77 percent loss of mass; he also identified this reaction as the source of stellar energy. Harkins termed the decrease in mass in nuclear synthesis "packing effect," and showed it to be lower in complex nuclei of even atomic number (considered to be produced by condensation of α-particles) than in complex nuclei of odd atomic number (considered to be produced by condensation of a tritium or lithium nucleus with α-particles). This observation led Harkins to propose that the even-numbered elements are more stable and he demonstrated that they are the more plentiful in stars, in meteorites, and on earth. In 1919 Harkins' conclusions were confirmed by Rutherford, who bombarded various atoms with α-particles and found that of the elements so bombarded, only the odd-numbered ones lost a proton.

The Harkins and E. D. Wilson theory of atom building (1915) predicted atomic weights near units based on 16.000 for oxygen; deviations for lithium, chlorine, and many other elements were considered evidence for isotopes not yet observed. Chlorine isotope separation by diffusion was attempted in 1916, but greater success was obtained with hydrochloric acid in 1919 when 10,000 liters were processed. In February 1920 at Cambridge, Aston announced evidence from mass spectroscopy for chlorine isotopes of mass 35 and 37, while in April 1920 Harkins published a preliminary report on his evidence for chlorine isotopes of 35, 37, and 39. Aston subsequently confirmed the prediction of chlorine-39. In 1921 Harkins showed that with the diffusion process he could obtain mass differences for hydrochloric acid of one part in 645. Subsequent studies with mercury diffusion demonstrated mass differences of 180 parts per million.

Rutherford carried out his 1919 studies (the first nuclear syntheses) in a spinthariscope, which could measure only the range of nuclear particles. Harkins realized that the C. T. R. Wilson cloud chamber could allow exact determination of the energy and mass of nuclear reactions and promptly analyzed tens of thousands of α-particle tracks in nitrogen and argon by this method; he found (1923) that no collisions resulted in reactions. At Cambridge, Blackett used identical equipment and in 1925 found tracks to prove that nitrogen captured an α-particle and emitted a proton, thus synthesizing oxygen-17; Harkins confirmed this the following year.

A few months before Rutherford's prediction, Harkins in 1920 predicted the existence of the neutron. But it was not until 1932 that the neutron was actually observed, by James Chadwick at Cambridge. Immediately after Chadwick's discovery, Harkins, with David Gans and other co-workers, began investigations of nuclear reactions involving these particles. Chadwick and Rutherford contended that nuclear reactions initiated by bombardment could occur without capture of the bombarding particle, but Harkins showed evidence (measured energy losses) that in forming an excited nucleus capture always occurs; by 1936 Harkins' view was accepted.

Harkins' eighty papers on nuclear reactions and isotopes include several important contributions to theory and experiment and for some years were the only significant American contributions in this field. The great bulk of his studies, however, concern surface phenomena. On Harkins' first day at Karlsruhe in 1909 Fritz Haber greeted him with the toast, "He shall work on surface tension." Although Harkins had no interest in this subject, he soon became intrigued when he found that current measuring techniques were grossly inaccurate. Following the example of Richards—whose precision in atomic weights brought him a Nobel Prize in 1914—Harkins strove to make surface measurements a precise science. Together with F. E. Brown in 1916–1919 Harkins brought high precision to the drop weight method for the measurement of surface and interfacial tension, an easier laboratory procedure than the method of capillary height measurement perfected by Richards. Eleven years later Harkins and Hubert Fairlee Jordan achieved similar precision with the ring method. Harkins' publications remain the primary references on the drop weight and ring methods of measurement.

Precise measurements of surface and interfacial tensions allowed new interpretations and understanding. Between 1910 and 1920, when electron shifts in organic compounds had gained the attention of physical chemists, Harkins explored the relation of structure of organic molecules to their surface properties. A short time after the publication of Langmuir's landmark paper on gas adsorption, Harkins published two extensive papers (1917) on precisely measured surface tensions and interfacial tensions, versus water, for 338 different organic compounds, in which he cited evidence for oriented monomolecular films in surfaces and interfaces. Langmuir's publication, a month later, on oriented monomolecular films of insoluble polar organic molecules on water, led to competition between the two scientists. In 1920 Harkins' formalized his views on oriented monolayers at interfaces with the concepts of "work of adhesion," "work of cohesion," and the "spreading coefficient." These concepts are widely used to correlate the spreading of organic materials on water or mercury.

Harkins' series of publications on monomolecular films at liquid surfaces or interfaces stretched over a twenty-year period. Beginning in 1925 he made precise studies of the adsorption of soluble films and of film properties of insoluble films. He investigated two-component monolayers and types of organic molecules, including enzymes and polymers. In addition he applied his research on monolayers adsorbed at the oil-water interface and at the liquid-solid interface toward a better understanding of emulsions and pigment dispersions.

In about 1937 Harkins initiated a major effort in the study of gas adsorption on solid powders. These studies led to a series of papers, from 1942 to 1950, which remain basic to our present understanding of this subject. Together with George Edward Boyd, George Jura, and others, he made important and novel use of calorimetric measurements with finely divided powders. They developed the only absolute method for measuring surface areas of powders, based on the heats of immersion in a liquid of powders already equilibrated with saturated vapor of the same liquid. This method allowed calibration of relative methods such as the well-known Brunauer-Emmett-Teller (BET) method. Calorimetry was also used to measure the range of forces emanating from solid surfaces. Harkins' investigations of the total free energy change per unit area of solid surfaces during gas adsorption up to equilibrium vapor pressures (designated "equilibrium spreading pressure") form the basis of much of our knowledge of adsorption on oxides.

Although Harkins was sixty-eight when the United States entered World War II, rather than retire, he plunged into a new field of colloid chemistry, the emulsion polymerization of rubber. Together with M. L. Corrin and H. B. Klevens he developed new methods for measuring micelle formation in detergent solutions and then related the effect of structure, salts, hydrocarbons, and insoluble surfactants quantitatively to micelle formation. Harkins correlated these criteria with the conditions found to be optimum for emulsion polymerization and thus provided the fundamentals for understanding this important process.

Throughout his career, Harkins showed exceptional foresight in choosing important fields of research. His intuition in predicting phenomena, coupled with a strong drive to measure important properties with great accuracy, provided a legacy that includes basic precepts of nuclear reactions, a general outlook on surface chemistry, laboratory methods for surface studies, and a great many unequaled measurements of surface properties.

BIBLIOGRAPHY

Most of Harkins' publications are listed in his *The Physical Chemistry of Surface Films* (New York, 1952), pp. 375–390, posthumously edited by Thomas Frazer Young.

Additional information is in J. R. Partington, *A History of Chemistry,* IV (London–New York, 1964), 934, 950, 952–953, 966; Poggendorff, VIIb, 1847–1851; Gustav Egloff, *Chemical and Engineering News,* **22,** no. 10 (1944), 804–805; an anonymous article, *ibid.,* **27** (1949), 1146.

FREDERICK M. FOWKES

HARKNESS, WILLIAM (*b.* Ecclefechan, Scotland, 17 December 1837; *d.* Jersey City, New Jersey, 28 February 1903), *astronomy.*

Harkness' family immigrated to America in 1839; his father was both a Presbyterian clergyman and a physician. After graduating from the University of Rochester in 1856, Harkness worked as a journalist. He returned to Rochester, where he received the M.A. in 1861. He next turned to medicine, graduating from the New York Homeopathic Medical College in 1862 and serving briefly as a surgeon in the Civil War. Also in 1862 he joined the U.S. Naval Observatory and in 1863 was commissioned a professor in the U.S. Navy's Corps of Professors of Mathematics.

Except for service at sea (1865–1866), studying the effects of iron armor on ship compasses and terrestrial magnetism, and a brief period (1866–1867) at the Hydrographic Office, Harkness' astronomical career was spent at the Naval Observatory. In 1869 during a total solar eclipse he discovered the coronal line K 1474. Much of Harkness' work resulted from observations of the 1874 and 1882 transits of Venus. He headed the expedition to Hobart, Tasmania, to observe the 1874 transit and was in charge of reducing all the American observations. Harkness successfully devised methods and instruments for using the photographic records. Since the German and English parties had not had a similar success with photography, it was used only by the Americans and French in 1882, the latter presumably because of Harkness' defense of photographic methods. During this period he also published (1879) a theory of the focal curve of achromatic telescopes. After reducing the results of the 1882 observations, he published *The Solar Parallax and Its Related Constants* (1891), probably his principal theoretical contribution.

Harkness was much involved in the design of the present Naval Observatory building and its original equipment. From 1892 until his retirement in 1899 he was the civilian astronomical director of the observatory, an appointment made in answer to recurring criticism of the navy's administration. On the retirement of Simon Newcomb from the directorship of the Nautical Almanac in 1897, Harkness assumed that position. Fragmentary evidence suggests that Newcomb, who long sought the directorship of the observatory, did not view his successor with enthusiasm. For example, Harkness was president of the American Association for the Advancement of Science in 1893 but was never elected to the National Academy of Sciences, an honor within Newcomb's power of bestowal.

BIBLIOGRAPHY

Harkness' publications are well covered in the Royal Society *Catalogue of Scientific Papers*, VII, 909; X, 142; and XV, 643–644. The U.S. Naval Observatory records in the U.S. National Archives contain documents on his long service with that institution. An autobiographical account appears in *Science*, n.s. **17** (17 Apr. 1903), 602–604. There are no known collections of Harkness' personal papers. The Simon Newcomb Papers in the Library of Congress contain much information on the Naval Observatory during the years when Harkness was on its staff.

NATHAN REINGOLD

HARLAN, RICHARD (*b.* Philadelphia, Pennsylvania, 19 September 1796; *d.* New Orleans, Louisiana, 30 September 1843), *comparative anatomy.*

Richard Harlan was the eighth of ten children of Joshua Harlan, a wholesale grocer and merchant, and Sarah Hinchman, both Friends. He began the study of medicine with Joseph Parrish of Philadelphia, spent the year 1816–1817 as ship's surgeon on a voyage to Calcutta, and in 1818 received his M.D. degree from the University of Pennsylvania, offering a senior essay on the vital principle. Upon graduation Harlan was engaged as a demonstrator in Parrish's private anatomical school. He was elected a physician to the Philadelphia Dispensary in 1820 and from 1822 until 1838 served as a physician to the Philadelphia Almshouse.

From the beginning of his career Harlan was interested in scientific investigations. In 1821 he wrote a paper on the generation of animal heat, and in the same year, with J. B. Lawrence and Benjamin H. Coates, he presented to the Academy of Medicine a report of experiments on the process of absorption which was cited with approval when he was proposed, successfully, for membership in the American Philosophical Society in 1822. In June 1832, with Asiatic cholera threatening Philadelphia, the city's emergency sanitary board sent Harlan, Samuel Jackson, and Charles D. Meigs to Canada to study the disease and methods of treatment at Montreal and Quebec. The doctors recommended that Philadelphia erect small hospitals and emergency stations where drugs, nurses, and physicians could be found day and night and that the most infected neighborhoods be evacuated. Harlan's time at the height of the epidemic, he wrote a friend, was "usefully, at least, if not profitably employed, night and day. Cholera, cholera, cholera!!!!" A grateful city awarded him a handsome silver pitcher for his services.

Harlan was the first American to devote a major part of his time to vertebrate paleontology. In 1815 he was elected to the Philadelphia Academy of Natural Sciences, and in 1821 he was named professor of comparative anatomy in Peale's Philadelphia Museum. During the next fifteen years a steady flow of monographs came from his pen. In search of specimens he frequently explored the New Jersey marl pits; in 1829 he was on the Ohio, where, at Cincinnati, he purchased a large collection of fossils for his patron John P. Wetherill; and in 1831 he visited the mountains and caverns of Virginia. Major Stephen Long, Thomas Nuttall, Titian R. Peale, and John James Audubon sent him materials to study, the last promising on one occasion to "do my best in the Way of Tortoises and also in the way of a *Sea Cow!*" Harlan published much of the data thus collected in *Fauna Americana* (1825), the first systematic presentation of the zoology of North America. Although it described some new species, including materials collected by Long, Constantine Rafinesque, and others, it followed A. G. Desmarest's *Mammalogie* (1821–1822) so closely and extensively that reviewers rejected its claim to be an original work, charging it with numerous errors and typographical deficiencies as well. The *Fauna* was followed by *American Herpetology*, published first in the *Journal of the Academy of Natural Sciences of Philadelphia* (1827) and then separately in the same year. Chiefly for the benefit of European naturalists, Harlan prepared "Critical Notices of Various Organic Remains Hitherto Discovered in North America." In 1835 he published his collected papers as *Medical and Physical Researches.*

Harlan's career fell in a period of consolidation between Cuvier and Leidy. Working within the Cuvierian framework, he collected much new information, identified new species, and contributed significantly to taxonomic knowledge; but his achievement was limited by insufficient data, inadequate concepts, and his own haste. His most serious mistake was in classifying *Basilosaurus* among reptiles. Harlan inevitably grappled with the idea of evolution. Well acquainted with the theories of Erasmus Darwin, Lamarck, and Jules-Joseph Virey, he believed that species have existed from the beginning, are distinct and immutable, and that

> the animal kingdom is in some degree only a single animal, but varied and composed of a multitude of species, all dependent on the same origin. . . . Nature need only vary in a slight degree the numerous generations of the same plant, or of the same animal, in order to create a multitude of analogous animals, which we name *species* [*Medical and Scientific Researches,* pp. 233, 237].

He believed that the evolutionary process is continuing and entertained the notion that man might not be "the *ne plus ultra* of perfection." Why and how some species disappeared and others emerged, Harlan could not say. The process required millions of years; geology he thought was most likely to hold the answer, and he suggested that a kind of spontaneous generation might occur.

> We have every reason to conclude, that every distinction of existing species has existed from the earliest periods of the formation of the present world; and has its origin ultimately in the nature of the *soil;* every variety of which is marked by a corresponding variety in its animal and vegetable productions; and many of these are limited by geographical distribution [*ibid.,* p. 244].

Harlan visited Europe in 1833 and again from 1838 to 1840. On the second trip he read a paper to the Geological Society of London, spent much time with Richard Owen at the Royal College of Surgeons, witnessed the surgical operations of Astley Cooper, and heard Faraday lecture—"a superlatively neat manipulator, and eloquent lecturer," who "riveted the attention" of the audience. Speaking with Daguerre in Paris, he "felt as in the presence of a superior power." But after inspecting French hospitals and witnessing French surgery he came away with lessened admiration for both.

Early in 1843 Harlan commenced practice in New Orleans, where he was at once elected a vice-president of the Louisiana Medico-Chirurgical Society. He died suddenly a few months later of apoplexy. He was survived by his wife, Mrs. Margaret Hart Simmons Howell—a widow whom he had married in 1833—and by four young children, of whom the oldest, George Cuvier Harlan, became a distinguished ophthalmologist in Philadelphia.

BIBLIOGRAPHY

I. ORIGINAL WORKS. The Royal Society *Catalogue of Scientific Papers (1800–1863)*, III, 184–186, lists sixty-four papers by Harlan; many of these were reprinted in his *Medical and Scientific Researches* (Philadelphia, 1835), where, unfortunately, the place and date of original publication are not given. For Harlan's ideas on comparative anatomy and evolution, see the essays "On the Affiliation of the Natural Sciences" and "On the Successive Formations of Organized Beings," in that work.

The quarrel over *Fauna Americana* (Philadelphia, 1825) can be followed in the review in the *North American Review,* **20** (1826), 120–136—on which John Godman commented in *Journal of the Franklin Institute,* **1** (1826), 19–21, which elicited Harlan's *Refutation of Certain Misrepresentations Issued Against the Author of the 'Fauna Americana'*. . . (Philadelphia, 1826) and Godman's rejoinder in *A Letter to Dr. Thomas P. Jones* . . . (Philadelphia, 1826).

Harlan's accounts of his visit to London and Paris are contained in letters to the editors of the *Medical Examiner* (Philadelphia), **2** (1839), in which are also published several of his lectures at the Philadelphia Almshouse. A product of Harlan's Paris visit was his trans. of J. N. Gannal's *History of Embalming* . . . (Philadelphia, 1840), a useful but sometimes macabre survey and handbook on the preparation of human and animal anatomical and pathological material.

II. SECONDARY LITERATURE. George G. Simpson, "The Beginnings of Vertebrate Paleontology in North America," in *Proceedings of the American Philosophical Society,* **86** (1942), 161–164, assesses Harlan's role and position in the history of his science.

WHITFIELD J. BELL, JR.

HARPER, ROBERT ALMER (*b.* Le Claire, Iowa, 21 January 1862; *d.* Phenix, Virginia, 12 May 1946), *botany.*

Harper was the son of a Congregational minister, Almer Harper, and his wife Eunice Thompson. He grew up in a village in Illinois where he had little formal education but ample opportunity to study natural history. He worked his way through Oberlin College, received his B.A. in 1886, and then taught Latin and Greek at Gates College until 1888, when he returned to his main interest, botany. After studying at Johns Hopkins University, he took an appointment at Lake Forest College in 1889, becoming professor of botany and geology two years later. During his professorship he spent some time at Bonn under Strasburger and he also worked briefly with Brefeld; he received his Ph.D. in 1896. In 1898 he moved to the University of Wisconsin as professor of botany, and in 1911 he went to Columbia University, where he taught until his official retirement in 1930. He retained emeritus status, however, and continued research there until 1937. That year he moved to his farm at Phenix in Bedford County, Virginia.

From his arrival in New York, Harper was active in the New York Botanical Garden; he was a member of the Board of Managers (1911–1942) and chairman of the scientific directors (1918–1933). In 1899 he married Alice Jean McQueen. After her death he married Helen Sherman in 1918; they had one son.

Harper was highly regarded as a teacher, lecturer, and leader of fieldwork studies. His research ranged widely over theoretical and practical problems, but the most important was his study, almost complete by 1910, of the cytology of fungi. Influenced by Strasburger and Brefeld, his early papers were written in German. Harper investigated spore formation, illustrating his papers with excellent drawings of cells at all stages of development and clearly differentiating the free cell formation of daughter cells arising in the multinucleate mass of protoplasm of the Ascomycetes from the cleavage by constriction in the Basidiomycetes. The two processes were so different that he concluded that the Ascomycetes could not be descended from the lower fungi.

He traced the division and fusions of nuclei during the life cycle, showed that the ascocarp originates in a sexual apparatus, and found a second fusion of the included nuclei in the young ascus; but he did not clearly relate these changes to reduction division and fertilization. His views on the sexuality of fungi were at variance with Brefeld's. In 1903, with R. J. Holden, he showed that for most of the life cycle of a rust fungus the cells are uninucleate from teleutospore to sporidium and binucleate from sporidium to teleutospore. From a review of work on smut fungi he concluded that cell fusion without nuclear fusion may give benefits of larger cells with more food and better resistance. He built up a large herbarium of fungi.

Harper's later studies, published in 1920, on the inheritance of sugar and starch characteristics in corn, led him to believe that in hybrids inheritance is not through particulate pairs of characters, but that all pairs will exhibit intermediate characters. He stressed this view in his presidential address to the Botanical Society of America on the structure of protoplasm.

Harper's work at the New York Botanical Garden, although not published, was a substantial contribution to plant pathology, and he was responsible for the installation of equipment to combat insect pests and fungus diseases. He left his collection of separates and other publications to the garden.

BIBLIOGRAPHY

I. ORIGINAL WORKS. Harper's works include "Kernteilung und freie Zellbildung im Ascus," in *Jahrbuch für wissenschaftliche Botanik,* **30** (1897), 249; "Cell Division in Sporangia and Asci," in *Annals of Botany,* **13** (1899),

467–525; "Nuclear Phenomena in Certain Stages in the Development of the Smuts," in *Transactions of the Wisconsin Academy of Sciences, Arts and Letters,* **12** (1900), 475–498; "Nuclear Divisions and Nuclear Fusion in *Coleosporium sonchi-arvensis,* Lev.," *ibid.,* **14** (1903), 63–82, written with R. J. Holden; "Sexual Reproduction and the Organization of the Nucleus in Certain Mildews," *Carnegie Institution of Washington, Publication no. 37* (Washington, D.C., 1905); "Nuclear Phenomena of Sexual Reproduction in Fungi," in *American Naturalist,* **44** (1910), 533–546; "The Structure of Protoplasm," in *American Journal of Botany,* **6** (1919), 273–300; and "The Inheritance of Sugar and Starch Characters in Corn," in *Bulletin of the Torrey Botanical Club,* **47** (1920), 137–181.

II. SECONDARY LITERATURE. The most comprehensive biography of Harper is by Charles Thom in *Biographical Memoirs. National Academy of Sciences,* **25** (1949), 227–240, with portrait and bibliography. The obituary by B. O. Dodge in *Yearbook. American Philosophical Society, 1946* (1947), 304–313, contains a section on Harper's ancestry and background, with quotations of personal reminiscences by his friends; that by A. B. Stout in *Journal of the New York Botanical Garden,* **47** (1946), 267–269, deals fully with Harper's work at the garden. See also the short anonymous evaluation in *Phytopathology,* **38** (1948), 328.

DIANA M. SIMPKINS

HARPER, ROLAND McMILLAN (*b.* Farmington, Maine, 11 August 1878; *d.* Tuscaloosa, Alabama, 30 April 1966), *botany, geography, demography.*

Harper's interests were multifarious for the twentieth century: plant ecology, taxonomy, geography, and demography, as well as railroading, cemeteries, race relations, and tobacco smoking. His Ph.D. dissertation, *Phytogeographical Sketch of the Altamaha Grit Region of the Central Plain of Georgia* (1906), is an ecological classic; and his hundreds of plant records, extensively based on acute observation in the field, enriched the comprehensive writings of J. K. Small and Merritt Fernald.

Harper's paternal grandfather, William Harper, came from Kilkenny, Ireland, and settled in Ontario. His maternal grandfather, Wilhelm Tauber of Munich, was a portrait painter. Roland was the second of six children born to William Harper, who studied at the University of Munich, and Bertha Tauber. Although his Farmington boyhood friend, Clarence Knowlton, tried to interest Harper in plants, Harper later wrote that he "could not see much in botany." His boyhood interests were photography, railroading, and the physical sciences. He moved with his family to Dalton, Georgia, when he was ten, and later entered the University of Georgia as an engineering student. There he came under the influence of the zoologist John P. Campbell. Following graduate work in botany at Columbia University (Ph.D., 1905) he joined the Geological Survey of Alabama as botanist and geographer, serving chiefly but intermittently there and in the Florida Geological Survey for the next sixty-one years.

From age twenty-one Harper published over 500 titles. His writings on the natural resources of Florida (1928), the economic botany of Alabama (1913, 1928), the forests (1943) and weeds (1944) of Alabama, and plants endemic to Florida (1949) were accompanied by historical notes and annotated bibliographies. The value of these and his *Phytogeographical Sketch* were heightened by photographs of the vegetation which, with ensuing destruction of habitats, have become historical documents.

In his later years Harper contributed sociological articles, generally based on personal data gathering, to the local newspapers. "Cornbread, Appendicitis and the Birth-Rate" (1938) and "Women per Family as an Index of Culture" (1944) are typical topics. He affected a crisp editorial style. Withal, his most important contributions were in bioecology.

BIBLIOGRAPHY

I. ORIGINAL WORKS. "Autobiographical Notes [to 1900] With Special Reference to Botany, Written Mostly from Memory by Roland M. Harper, December, 1954," prepared at the request of Jack McCormick, and a partial bibliography by Jack McCormick are on file at the New York Botanical Garden. A bibliography of Harper's principal botanical publications is given by Ewan (see below). Papers selected from this are "Economic Botany of Alabama, Part 1. Geographical Report, Including Descriptions of the Natural Divisions of the State, Their Forests and Forest Industries, With Quantitative Analyses and Statistical Tables," *Monographs of the Geological Survey of Alabama,* **8** (1913); "Part 2. Catalogue of the Trees, Shrubs, and Vines of Alabama, With Their Economic Properties and Local Distribution," *ibid.,* **9** (1928); "Natural Resources of Southern Florida," in *Report of the Florida State Geological Survey,* **18** (1927), 27–206; "Forests of Alabama," *Monographs of the Geological Survey of Alabama,* **10** (1943); "Preliminary Report of the Weeds of Alabama," *Bulletin. Geological Survey of Alabama,* **53** (1944); and "A Preliminary List of the Endemic Flowering Plants of Florida," in *Quarterly Journal. Florida Academy of Sciences,* **11**, no. 1 (1949), 23–25; **11**, no. 2 (1949), 39–57; **12**, no. 1 (1950), 1–19. Harper's exsiccatae of Georgia plants are in the principal herbaria in this country and abroad. His diary, scrapbooks of newspaper clippings (classified for the earlier years), photographs, letters, and other memorabilia are at the University of Alabama.

II. SECONDARY LITERATURE. The unsigned "Scientist Making Survey of Northern Section of State," in *Arkansas*

Democrat (29 April 1923), is a contemporary portrait of Harper. A biographical sketch by Joseph Ewan, with portrait, in *Bulletin of the Torrey Botanical Club,* **95** (1968), 390–393, is based on materials furnished by Mary Susan Wigley (Mrs. Roland) Harper and Francis Harper. An "Addenda and Corrigenda" for the article, dated January 1960 and privately printed by Francis Harper, is at the New York Botanical Garden.

JOSEPH EWAN

HARPESTRAENG, HENRIK (*d.* Roskilde, Denmark, 2 April 1244), *medicine, pharmacy.*

Several thirteenth-century treatises on medical subjects are ascribed to a Henricus Dacus, or Henrik Harpestraeng, whose literary work is as well known as his life is obscure. An earlier attempt to identify him with a Maître Henry de Dannemarche who lived at Orléans in the twelfth century is impossible, in view of the obituary notice found in the *Liber daticus Roskildensis* (p. 47), which states: "Non. Apr. obiit Magister, Henricus Harpestraeng, hujus ecclesiae Canonicus MCCXLIV. qui multiplices elemosinas huic ecclesie contulit, tam in morte quam in vita sua." This proves that at the time of his death Harpestraeng was a canon of the cathedral of Roskilde, then the capital of Denmark, and that he was presumably a wealthy man. He was also commemorated in a contemporary epitaph in elegant Latin verse preserved in copies made from a now lost manuscript from the monastery at Sorö. In addition, there is no reason to disbelieve the well-founded medieval tradition that he acted as physician to King Erik Plovpenning, who reigned from 1241 to 1250. A little more can be inferred from Harpestraeng's writings, which show him to have been a remarkable medical author both in Latin and in his native tongue. This presupposes studies abroad, just as his title of *magister* points to some kind of university education. Nevertheless, in spite of his quotations from Salernitan authors, there is no definite evidence for the common belief that he studied at Salerno.

Not even a relative chronology of Harpestraeng's writings has been worked out, but it is a plausible assumption that his Latin works date from his period abroad and that his Danish manuals arose from his medical activity at home toward the end of his life.

Harpestraeng's first Latin work was *De simplicibus medicinis laxativis,* a treatise on herbs and drugs and their medical use, written in the Salernitan tradition and quoting Galen, al-Razi, Ibn Sīnā, Copho, the *Antidotarium,* Constantine the African, and others. Preserved in a single fifteenth-century manuscript (Copenhagen, G.K.S. 1654, 4°) of German provenance, it has been edited by J. W. S. Johnsson.

The *Liber herbarum,* a herbal for medical use, was written in the same vein as *De simplicibus* but also quotes the *Regimen sanitatis.* There are several fifteenth-century manuscripts (Copenhagen, A.M. 792, 4°, and G.K.S. 3457, 8°; Uppsala, D 600, 8°; Vienna, VIND. 2962, a.o.) and a number of more or less fragmentary translations into Danish, Norwegian, Swedish, Icelandic, and German. The text has been edited from the Uppsala manuscript by Poul Hauberg.

The *Remedium contra sacrum ignem,* now lost, was a therapeutical treatise on St. Anthony's fire.

The *Urte Book* was a Danish herbal, or leech book, in 150 chapters, the majority of them translated from the *De viribus herbarum* of Macer Floridus (Odo de Meung) and the *De gradibus liber* of Constantine the African. Among the numerous manuscripts one is from the thirteenth century (Stockholm, K. 48) and another from the early fourteenth (Copenhagen, N.K.S. 66, 8°). The book was extremely popular and was copied throughout Scandinavia as late as the eighteenth century. It was published by C. Molbech and later, in a critical edition, by Marius Kristensen.

Several codices contain a number of medical fragments in Danish or Swedish going back to the same source and usually considered as the scattered remains of another leech book by Harpestraeng. Until now no reconstruction of this text has been attempted.

A number of Latin fragments on phlebotomy, medical astrology, and other subjects have been ascribed to Harpestraeng, but their authenticity remains to be confirmed.

Finally, a book on gems and minerals, and a cookery book, both in Danish, were formerly ascribed to Harpestraeng but are now considered the works of an unidentified contemporary author.

As a medical author Harpestraeng showed no great originality, although he did enrich the medieval materia medica with a number of Nordic herbs unknown to the herbalists of the southern tradition, such as angelica, *Benedicta alba,* and *Benedicta ruffa.* His main importance was his establishment of European medicine in the Scandinavian countries, where his writings in the vernacular aligned popular medicine to the classical tradition. As the first scientific treatises in Danish they are of extreme linguistic interest. Through his connection with the cathedral school of Roskilde, Harpestraeng made the capital of Denmark a center of medical studies just as one generation later it became a center of astronomical research (through Peter Philomenus of Dacia) and thus an important center of learning in Scandinavia before the creation of universities in the late fifteenth century.

BIBLIOGRAPHY

I. ORIGINAL WORKS. Editions of Harpestraeng's writings are *Henrik Harpestraengs danske Laegebog,* C. Molbech, ed. (Copenhagen, 1826); "Gamalnorsk Fragment av Henrik Harpestraeng," Marius Haegstad, ed., in *Skrifter utgitt av det Norske videnskaps akademi i Oslo,* Historisk-filos. Klasse, **2,** no. 2 (1906); *Harpestraeng. Gamle danske Urtebøger, Stenbøger og Kogebøger,* Marius Kristensen, ed., 3 vols. (Copenhagen, 1908–1920); *Henricus Dacus: De simplicibus medicinis laxativis,* J. W. S. Johnsson, ed. (Copenhagen, 1914); and *Henrik Harpestraeng Liber herbarum,* Poul Hauberg, ed. (Copenhagen, 1936).

II. SECONDARY LITERATURE. See J. Brøndum-Nielsen, "Studier i Dansk Lydhistorie," in *Acta philologica scandinavica,* **4** (1929), 186–190; Poul Hauberg, "Lidt om Henrik Harpestraengs Laegebog," in *Danske Studier,* n.s. **16** (1919), 111–128; and in *Dansk Biografisk Leksikon,* IX (1936), 369–370; Marius Kristensen, *Danske Studier,* 3rd ser., **6** (1933), app., 161; L. Nielsen, *Danmarks middelalderlige Haandskrifter* (Copenhagen, 1937), pp. 148–155; A. Otto, ed., *Liber daticus Roskildensis* (Copenhagen, 1933), pp. 47, 179–186; P. Riant, "Vestigia Danorum extra Daniam," in *Danske samlinger for historie, topographi, personal- og literaturhistorie,* **2** (1866–1867), 270–271; P. Skautrup, *Det danske Sprogs Historie,* I (Copenhagen, 1944), *passim;* and E. Wickersheimer, "La véritable origine de Maître Henri de Dannemarche," in *Janus,* **37** (1933), 354–356.

OLAF PEDERSEN

HARRIOT (or **HARIOT), THOMAS** (*b.* Oxford, England, *ca.* 1560; *d.* London, England, 2 July 1621), *mathematics, astronomy, physics.*

Little is known of Harriot's early life. In 1584 he was in the service of Walter Ralegh where he had possibly been since 1580, when he finished his undergraduate studies at Oxford. Ralegh, who needed an expert in cartography and the theory of oceanic navigation, sent a colonizing expedition to Virginia in 1585, with Harriot as its scientist "in dealing with the naturall inhabitants specially imployed." He investigated their life, language, and customs and surveyed the coasts, islands, and rivers.

Harriot left Virginia in 1586, having learned, among other things, how to "drink" tobacco smoke, which he recommended in his *Briefe Report* (1588) as a cure for many complaints. When Ralegh turned his activities to Ireland and sought to colonize Munster, he leased Molana Abbey to Harriot. We do not know much about his life there, for he took care to order that all papers concerning the "Irische Accounts" be burned after his death. Although the *Briefe Report* had stressed Harriot's missionary zeal, some years later he joined a circle (Shakespeare's "School of Night") which included the atheist Christopher Marlowe and theists like Ralegh and the ninth earl of Northumberland. When, in about 1598, Harriot left Ralegh and Durham House, Northumberland gave him a yearly pension and living quarters in Sion House, Isleworth, and later (1608?) he lived in a house of his own, near the main building.

Harriot and his patron were imprisoned after the Gunpowder Plot of 5 November 1605. Although the earl was kept in the Tower of London until 1622, Harriot was released after a short time, a search of his papers having produced nothing incriminating. Subsequently he complained to Johann Kepler of impaired health; he was able nonetheless to proceed with his scientific investigations and even to undertake prolonged telescopic observations (1610–1613) of Jupiter's satellites and of sunspots. In 1613 he began to suffer from an ulcer in his left nostril. It proved to be cancerous and led to his death in 1621. He left more than 10,000 folio pages of scientific papers containing measurements, diagrams, tables, and calculations pertaining to important experimental and theoretical work in different fields.

Harriot was an accomplished mathematician who enriched algebra with a comprehensive theory of equations. By using an extremely convenient system of notation he simplified not only algebra, but also many other areas of mathematics. Among his innovations and discoveries is his proof that stereographic projection is conformal and therefore transforms rhumb lines on a sphere into equiangular helixes (logarithmic spirals) in its equatorial plane. He also made ingenious attempts to rectify and

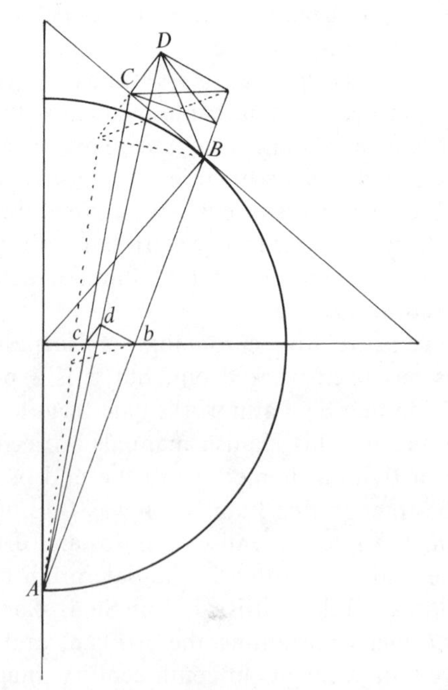

FIGURE 1. Harriot's diagram to his proof that the stereographic projection is conformal, much simplified (Add. 6789, f. 18).

square these spirals. In 1603 he computed the area of a spherical triangle: "Take the sum of all three angles and subtract 180 degrees. Set the remainder as numerator of a fraction with denominator 360 degrees. This fraction tells us how great a portion of the hemisphere is occupied by the triangle."

In about 1614 he resumed his early investigations of rhumb lines and the theory of the Mercator map, and nearly finished a table of meridional parts for this map. These computations were calculated for one-minute intervals, an enormous task that necessitated the use of sophisticated techniques of finite-difference interpolation, on which he wrote a monograph, *De numeris triangularibus.* His notational advances in this treatise are great, but even more interesting, and found only in preliminary drafts, are some symbols, including $\overset{1}{p}, \overset{2}{p}, \overset{3}{p}, \overset{4}{p}, \cdots$, for our figured numbers (or binomial coefficients)

$$\binom{n}{1}, \binom{n+1}{2}, \binom{n+2}{3}, \binom{n+3}{4}, \cdots.$$

Harriot knew that such formulas are valid even when negative integers or fractions are substituted for the number n. Not only are his apt notation and sense of structure admirable, but also the exceptional clarity of his exposition, as is evident in both finished manuscript tracts and his rough work sheets.

Harriot's *Artis analyticae praxis,* published posthumously in 1631 (in a poor edition), contains an interesting attempt at a uniform treatment of all algebraic equations, with worked-out examples of linear, quadratic, cubic, quartic, and quintic equations. Because he composed this "practice of the art of analysis" primarily for amateurs, he did not treat negative roots, but in other manuscripts he even considered "noetical" (that is, imaginary) roots, as, for example,

$$\left.\begin{array}{c} b-a \\ c-a \\ df-aa \end{array}\right| \mathrel{=\!\!\!\!\parallel} \begin{array}{l} bcdf - bdfa + dfaa + baaa \\ \quad - dcfa - bcaa + caaa - aaaa = 0000 \end{array}$$

$$\begin{aligned} a &= b \\ a &\mathrel{=\!\!\!\!\parallel} c \\ aa &= -df \\ a &\mathrel{=\!\!\!\!\parallel} \sqrt{} - df, \end{aligned}$$

a being the unknown quantity of this quartic, whose solutions appear on the right.

At times Harriot developed his mathematical deduction vertically downward, a method which may be advantageous to the mathematician, but posed a problem for the printer, as did his use of many new symbols. His symbolic shorthand and instructive examples often allowed him to dispense with explicit verbal explanation in mathematical writings. This is also apparent where he made trials of binary number systems. Unfortunately, the exact dates of Harriot's mathematical tracts and discoveries are known only in rare cases.

In his optics research it is easier to fix a chronology. No later than the early 1590's, he made a penetrating study of Ibn al-Haytham's (Alhazen's) *Optics* (Friedrich Risner's 1572 edition). To solve Alhazen's mirror problem he considered the locus of the reflection point for a spherical mirror expanding about its center. He thus anticipated Isaac Barrow, who proposed the same curve in 1669. Then Harriot investigated optical phenomena which Alhazen had neglected, or had not fully explored.

In order to establish a firm basis for the theory of burning glasses and of the rainbow, Harriot began in 1597 to measure the refraction of light rays in water and glass. He soon found that the Ptolemaic tables, then attributed to Witelo, were inaccurate. About 1601 he discovered that the *extensa* (essentially our refractive index) is the same for all angles of incidence. This enabled him to compute refraction for one-degree intervals of the angles of incidence. For water his index of refraction was cosec 48°30′ and for glass cosec 40°. In 1606 Harriot sent Kepler refraction angles and specific weights for thirteen substances, but he withheld the sine proportion from him.

Harriot also studied prismatic colors. When he looked through a prism at a white object on a dark background, it seemed to be fringed with a yellow and red border. (Blue is not mentioned.) From the breadth of the colors Harriot computed (1604) refractive indexes of the green, orange, and (extremal) red rays. By pouring liquids into hollow glass prisms, he determined analogous refractive indexes for fresh water, saturated salt water, turpentine, and spirits of alcohol. With his refraction tables Harriot calculated the *refractio caeca* (total refraction) in prisms and the path of solar rays through plano-convex lenses and glass balls. For a ray traversing water drops he found that the *arcus egressionis* (exit arc) $2r - i$ should have a maximum value

$$2 \times 40°5' - 59°17' = 20°53'.$$

Although in 1606 Harriot told Kepler that he planned a book on colors and the rainbow, his preserved manuscripts contain no statement of the exact relationship between maximum exit arc and the angular radius $R_{\prime}$ of the first rainbow, namely

$$R_{\prime} = 2 \times 20°53' = 41°46'.$$

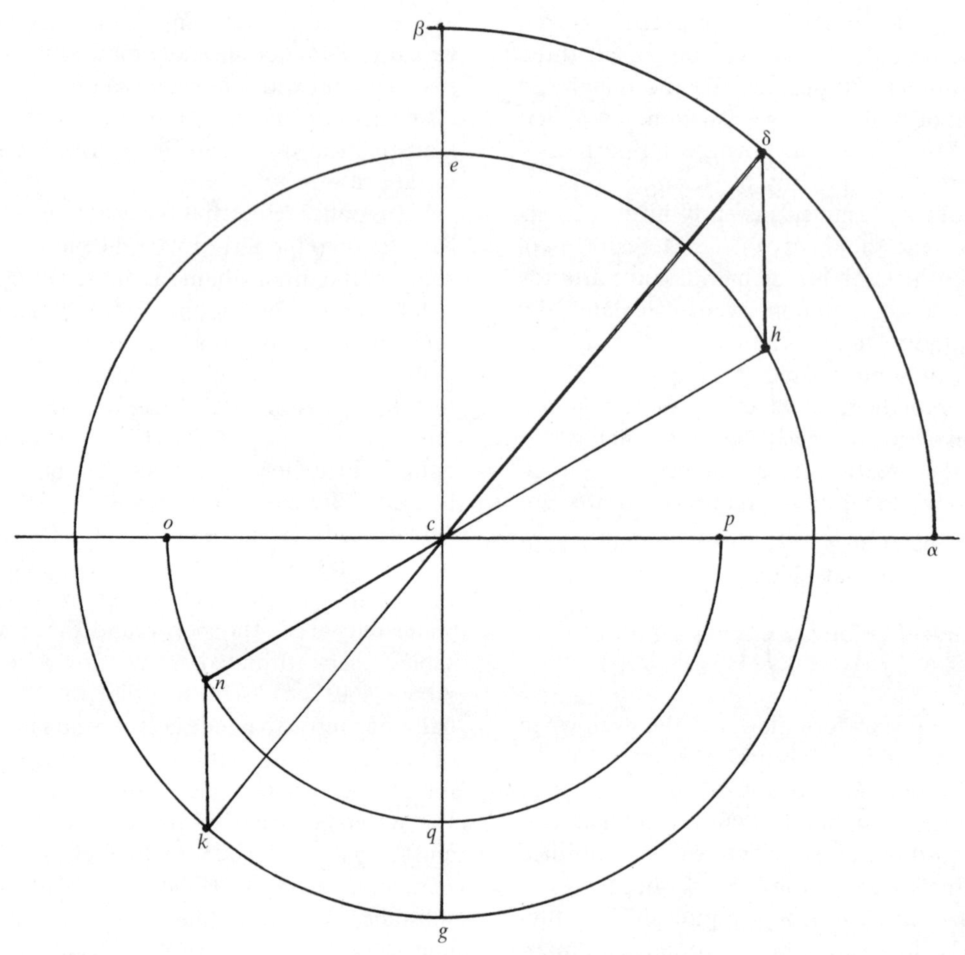

FIGURE 2. Harriot's "Regium" diagram, simplified (Add. 6789, f. 320). If k is the image point and h the eye, cn is Harriot's *linea contracta*. If h is the object point and k the eye, $c\delta$ is the *linea extensa*.

Until the early 1590's Harriot's astronomical researches centered on nautical applications. Observing from the roof of Durham House (1591–1592), he measured an angular distance of 2°56′ between the celestial north pole and the North Star. Prominent among his suggestions for improved navigational instruments was an ingenious backstaff for measuring solar altitudes. The comet of 1607 ("Halley's comet") was observed by Harriot in London and by his pupil Sir William Lower in Wales. In a letter of 6 February 1610 Lower mentions that Harriot for some years had mistrusted the "circular astronomy" of Copernicus. Accordingly, as soon as Kepler's newly published *Astronomia Nova* reached England in 1609, they both studied it eagerly. Reworking a number of Kepler's computations, they discovered many minor errors. In the summer of 1609 Harriot turned a 6X telescope on the moon. Soon after, when he heard of Galileo's findings, he began a systematic survey of the sky and to this end his assistant and amanuensis Christopher Tooke constructed better telescopes, the finest having a magnification of thirty. A few detailed moon maps, ninety-nine drawings of Jupiter with satellites, and seventy-four folios of sun disks with spots testify to Harriot's hard work and perseverance as an observing astronomer from 1610 to 1613. Despite increasing ill health he was still able to make some observations of comets in 1618.

Harriot's manuscripts reveal little about his possibly extensive chemical researches and not much about his meteorological observations. He once measured the rainfall per square foot on the roof of Durham House, and he apparently possessed a scale for wind velocity. About 1600 he may for some time have been one of Cecil's decoding experts. Of great importance are his investigations on free and resisted motion in air. From the roof of Sion House (about forty-three feet) Harriot dropped twenty bullets, releasing each when the preceding one struck the ground. He determined the total fall time by pulse beats. In order to evaluate his results theoretically, he first rejected as negligible any *gradus naturae* (initial velocity). Then, after some hesitation, he decided to take acceleration proportional to time spent and not to space traversed.

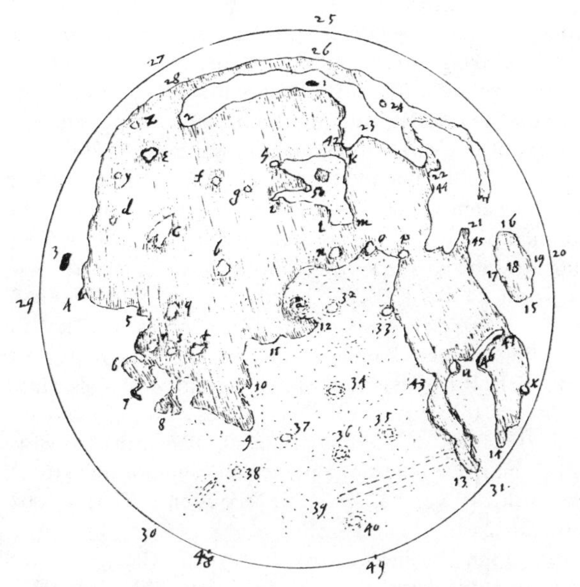

FIGURE 3. The most elaborate of Harriot's moon maps (Petworth 241).

These assumptions yielded fall spaces equivalent to thirteen feet (in other trials sixteen feet) after one second. On dropping bullets of different specific weights simultaneously, he observed no appreciable difference between bullets of iron and lead, but wax bullets lagged behind.

Harriot next turned his attention to ballistic curves, which he proved (on certain suppositions) to be parabolas with oblique axes (upright axes if air resistance could be neglected). He did not expressly say that he regarded air resistance as constant, but applied the kinematic rule of odd numbers 1, 3, 5, 7, . . ., not only for the vertical component of fall, but also "in inverse order" to the oblique component of motion subject to air resistance. Later James Gregory also proposed tilted parabolas in *Tentamina de motu projectorum* (1672), but Harriot's mathematical deduction had been more elegant. It appears from Harriot's manuscripts that he had studied printed tracts of William Heytesbury, Bernhard Tornius, and Aluarus Thomas on problems of uniform and difform change. Some of Harriot's diagrams on uniform difform motion resemble those of Nicole Oresme.

In 1603 and 1604 Harriot measured specific weights. For those metals and chemical compounds which could be obtained nearly pure, his results agree remarkably well with modern values. Near the end of his life he wrote a small treatise entitled *De reflectione corporum rotundorum* (1619) that contains two suggestive diagrams and a small web of equations, which, when unraveled, emerge as relations between the velocities of balls before and after collision. He lacked, however, any concept equivalent to the modern principle that kinetic energy is preserved in perfectly elastic collision. Apparently he considered this last treatise fundamental to understanding how atoms collide with other atoms or light globules. Two or three of his rough diagrams show light globules zigzagging between layers of atoms.

Harriot was acquainted with older mathematicians like John Dee and, according to Anthony à Wood, Thomas Allen had been his teacher at Oxford. Harriot soon surpassed these two and all other English mathematicians of his time. Apart from a few letters exchanged with Kepler, there is no documented knowledge of correspondence or personal contact with scientists of his own rank but scanty references suggest that distinguished contemporaries like William Gilbert, Bacon, and Briggs knew something of Harriot's scientific work. The mathematicians of Harriot's own circle were Nathaniel Torporley, Walter Warner, and Robert Hughes. He was also closely acquainted with such scientifically minded men as William Lower, Thomas Aylesbury, Robert Sidney, and Lord Harington.

George Chapman, the poet, praised Harriot as a universal genius and a connoisseur of poetry. Long after Harriot's death, at the time of John Aubrey, rumors still persisted of Harriot's disputes with theologians. Although his favorite maxim is said to have been *Ex nihilo nihil fit,* such heretical opinions are not prominent in his extant manuscripts, and a bookseller's bill reveals that between 1617 and 1619, Harriot bought many tracts on Christian theology. Several of his minor manuscripts deal with infinity and its paradoxes.

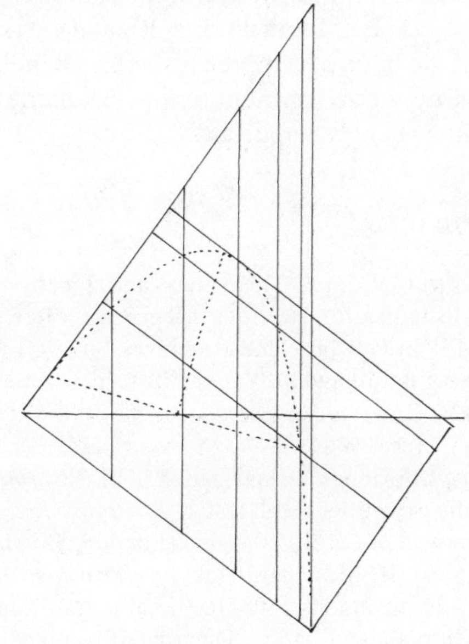

FIGURE 4. One of Harriot's ballistic parabolas with tilted axis, slightly simplified (Add. 6789, f. 64).

Harriot seldom completed his planned scientific treatises and never published any of them. This may largely be explained by adverse external circumstances, procrastination, and his reluctance to publish a tract when he thought that further work might improve it. Unlike Kepler, Harriot did not commit his inner thoughts and personal motives to paper. During the last years of his life, weakened health prevented his preparing any manuscript for the press. His will delegated this task to Torporley. Walter Warner, supervised by Aylesbury, then took over in 1627 and published the *Artis analyticae praxis,* but the planned edition of Harriot's major works came to nothing.

In the summer of 1627 Aylesbury and Warner measured refractions in several glass prisms and then calculated refraction tables modeled on Harriot's. Warner's young friend, John Pell, wrote to Mersenne on 24 January 1640 that Harriot had found the law of refraction. When, however, in 1644 Mersenne published, in *Universae geometriae . . . synopsis* (pp. 549–566), a posthumous tract by Warner on the sine law, Harriot's name was not even mentioned. Pell, who had borrowed some Harriot manuscripts from Aylesbury, later said that if Harriot had "published all he knew in algebra, he would have left little of the chief mysteries of that art unhandled" (S. P. Rigaud, *Correspondence of Scientific Men,* p. 153). After 1649, when Aylesbury was forced to leave England, Harriot's papers disappeared, but in 1784 F. X. von Zach rediscovered them under stable accounts in Petworth House, Sussex. Harriot's reputation has alternately waxed (owing to excessive praise by John Wallis and Zach) and waned (as a result of criticism by J. E. Montucla and Rigaud). More recently, on the basis of comprehensive and penetrating studies of his surviving manuscripts, his name is becoming increasingly important.

BIBLIOGRAPHY

I. Original Works. The chief—and far from fully explored—sources for Harriot's science are fifteen small Harriot MSS in Petworth House archives, Sussex, England, and eight big unordered MSS in the British Museum (Add. 6782–6789). Unless otherwise stated, the following items are British Museum MSS:

Arcticon, a textbook on navigation, and a *Chronicle* on the Virginia expedition, both lost; *A Briefe and True Report of the New Found Land of Virginia* (London, 1588), incorporated in vol. III of Richard Hakluyt's *Principall Navigations,* 3 vols. (London, 1598–1600); also translated into French, German, and Latin; "Doctrine of Nauticall Triangles Compendious," Petworth 241, vol. 6b; a tract "De rumborum ortu, natura et usu," announced in the preface to Robert Hughes, *De Globis* (1594); "Some Instructions for Ralegh's Voyage to Guiana" (1595), Add. 6708; "De numeris triangularibus," Add. 6782, fols. 107–146; "Optics and Caustics" (*ca.* 1604), Add. 6789, fols. 89–400, the most important of his works in optics; "On Ballistics," Add. 6789; numerous tables intended probably for a tract (*ca.* 1604) on specific weights, Add. 6788; drafts of chapters on algebra, not incorporated in Warner's ed. of the *Praxis,* Add. 6783; Harriot's and Lower's observations of the comet of 1607, Petworth 241, vol. 7, published in supp. I to Bradley's *Miscellaneous Works* (Oxford, 1832), pp. 511–522; "De Jovialibus planetis," ninety-nine annotated diagrams (Oct. 1610–Feb. 1612) and "Sun Spots," seventy-four fols. (Dec. 1610–Jan. 1613), Petworth 241, vol. 8; "Canon nauticus" (1614?), tables of meridional parts, Petworth 240, fols. 1–453; and "De reflectione corporum rotundorum" (1619), Petworth 241, vol. 6a, 23–31, transcription in Harley 6002, fols. 17–22r.

For John Protheroe's inventory of Harriot's MSS bundled at his death, see Add. 6789, fols. 449v–451v. Harriot's will was printed by Henry Stevens (see below), pp. 193–203, and R. C. H. Tanner (see below), pp. 244–247.

Harriot's correspondence with Kepler (1606–1609) is preserved in the Nationalbibliothek, Vienna, Codex 10703. See also *Kepler's Gesammelte Werke,* Max Caspar, ed. (Munich, 1951, 1954), vol. XV, 348–352, 365–368; vol. XVI, 31–32, 172–173, 250–251.

II. Secondary Literature. For information on Harriot, see John Aubrey, *Brief Lives,* O. L. Dicks, ed. (1962); Edward Edwards, *The Life of Sir Walter Ralegh,* 2 vols. (1868); Edward B. Fonblanque, *Annals of the House of Percy,* 2 vols. (London, 1887), privately printed (on the earl and the Gunpowder Plot, see pp. 252–329); J. E. Montucla, *Histoire des mathématiques* (Paris, 1758–1759; 2nd ed., 1799–1802); Muriel Rukeyser, *The Traces of Thomas Harriot* (New York, 1971), a nonscientific treatment; Henry Stevens, *Thomas Harriot, the Mathematician, the Philosopher, and the Scholar* (London, 1900), privately printed; John Wallis, *Treatise of Algebra, Both Historical and Practical* (Oxford, 1685), esp. chs. 53–54; and Anthony à Wood, *Athenae Oxonienses,* 3rd ed., vol. IV (1815), 459.

There are several articles by F. X. von Zach. These may be found in Bode's *Astronomisches Jahrbuch für 1788* (Berlin, 1785), 139–155, and in the first supp. to the Jahrbuch (1793); *Monatliche correspondenz zur Beförderung der Erd- und Himmels-Kunde,* **8** (1803), 36–60; and *Correspondance astronomique,* **6** (1822), 105–138. Zach's publications were severely criticized by S. P. Rigaud; see his two supplements to *Dr. Bradley's Miscellaneous Works* (Oxford, 1832–1833).

Recent books with reference to Harriot include *Correspondance du P. Marin Mersenne,* Tannery, de Waard, Rochot, eds. (Paris, 1936–1965); vol. IX, 61–65, contains an interesting letter on Harriot; vols. X and XI contain information on the law of refraction, as presented by Warner, Pell, and Hobbes; G. R. Batho, *The Household Papers of Henry Percy* (London, 1962), with scattered references to Harriot, Warner, and Torporley; R. H. Kargon, *Atomism in England From Harriot to Newton* (Oxford, 1966), 4–42, 144–150; D. B. Quinn, *Ralegh and the British*

Empire, 2nd ed. (London, 1962), 67–69, 85, 142; and D. W. Waters, *The Art of Navigation in England in Elizabethan and Early Stuart Times* (London, 1958), 546–547, 582–591.

Among the numerous recent articles on Harriot, see Jean Jacquot, "Thomas Harriot's Reputation for Impiety," in *Notes and Records. Royal Society of London,* **9** (1952), 164–187; J. A. Lohne, "Thomas Harriott," in *Centaurus,* **6** (1959), 113–121; "Zur Geschichte des Brechungsgesetzes," in *Sudhoffs Archiv für Geschichte der Medizin und der Naturwissenschaften,* **47** (1963), 152–172; "Regenbogen und Brechzahl," *ibid.,* **49** (1965), 401–415; "Thomas Harriot als Mathematiker," in *Centaurus,* **11** (1965), 19–45; and "Dokumente zur Revalidierung von Thomas Harriot als Algebraiker," in *Archive for History of Exact Sciences,* **3** (1966), 185–205; J. V. Pepper, "Harriot's Calculation of Meridional Parts as Logarithmic Tangents," *ibid.,* **4** (1968), 359–413; "Harriot's Unpublished Papers," in *History of Science,* **6** (1968), 17–40; D. H. Sadler, "The Doctrine of Nauticall Triangles Compendious. II. Calculating the Meridional Parts," in *Journal of the Institute of Navigation,* **6** (1953), 141–147; J. W. Shirley, "Binary Numeration Before Leibniz," in *American Journal of Physics,* **19** (1951), 452–454; "An Early Experimental Determination of Snell's Law," *ibid.,* 507–508; and R. C. H. Tanner, "Thomas Harriot as Mathematician," in *Physics,* **9** (1967), 235–247, 257–292.

For other sources, see George Chapman, "To My Admired and Souleloved Friend Mayster of All Essential and True Knowledge, M. Harriots," a poetic preface to *Achilles' Shield* (1598); and Nathaniel Torporley, "Corrector Analyticus: or Strictures on . . . Harriot," in J. O. Halliwell, *Collection of Letters Illustrative of the Progress of Science* (London, 1841), pp. 109–110. Letters concerning Harriot are in S. P. Rigaud, *Correspondence of Scientific Men* (Oxford, 1841); E. Edwards (see above), vol. II, 420–422; *Calendar of the Salisbury MSS at Hatfield House,* XVI (1933); XVII, 544; a letter from Torporley (1602?), Add. 6788, f. 117; letters, mostly from Lower, Add. 6789, fols. 424–449. See also Thomas Aylesbury's report to Northumberland on the publication of Harriot's *Praxis,* Birch 4396, f. 87, British Museum; and Henry Briggs's comments in *Kepler's Gesammelte Werke,* XVIII (Munich, 1959), 228–229; and in George Hakewill, *Apologie or Declaration ...,* 2nd ed. (Oxford, 1630), pp. 301–302.

Sporadic information on Harriot is in the *Calendars of State Papers, Domestic Series* and *Calendars of State Papers Relating to Ireland.* Sloane MS 2086, fols. 54–57, British Museum, contains Theodore de Mayerne's diagnosis of Harriot's cancerous ulcer with prescription for treatment. See also, in British Museum, Walter Warner's MSS bound in 3 vols, Birch 4394–4396; and Charles Cavendish's MS 6002, with transcriptions of important parts of Harriot's papers.

J. A. LOHNE

HARRIS, JOHN (*b.* Shropshire [?], England, *ca.* 1666; *d.* Norton Court, Kent, England, 7 September 1719), *natural philosophy, dissemination of knowledge.*

Harris, the son of Edward Harris, entered Trinity College, Oxford, on 13 July 1683 as a scholar and took his B.A. in 1686 and his M.A. (Hart Hall) in 1689. After leaving Oxford he took holy orders and served as vicar of Icklesham, then as rector of Winchilsea St. Thomas (1690), of St. Mildred (1708), and of Landwei Velfrey, Pembroke (1711). He held a prebend in the cathedral of Rochester (1707–1708), became curate of Strood, Kent (1711), and held other ecclesiastical posts. Harris' patron was Sir William Cowper, to whom he was chaplain. He became a fellow of the Royal Society in April 1696 and served as second secretary of that organization in 1709. He is reported to have received a B.D. from Cambridge in 1699 and did receive a D.D. at Lambeth in 1706.

Harris showed an early interest in natural philosophy and left a fragment of an autobiography which gives a picture of life at Trinity under Ralph Bathurst in the 1680's: "Lectures were here read in Experimental Philosophy and Chymistry and a very tolerable course of Mathematicks taught," especially after Harris took his first degree, for "then [Bathurst] gave him leave to teach Mathematicks" (quoted in Blakiston, *Trinity College,* p. 172). His tutor was Stephen Hunt, a fellow of Trinity from 1681 to 1689.

The published works of Harris reflect both his scientific and his theological interests. In 1698 he was Boyle lecturer and delivered eight sermons designed to confute the Hobbists and atheists and to demonstrate the consonance of science and orthodox religion. In 1697 he became a scientific controversialist, defending John Woodward against the attacks of a certain L. P.'s *Two Essays Sent in a Letter From Oxford to a Nobleman in London* (1695). Harris replied to L. P.'s alleged Hobbism in *Remarks on Some Late Papers Relating to the Universal Deluge and to the Natural History of the Earth* (1697).

During the period 1698–1704 Harris read scientific lectures at the Marine Coffee House in Birchin Lane and taught mathematics privately at his home. In conjunction with these activities he published in 1703 his *Description and Uses of the Celestial and Terrestrial Globes and of Collins' Pocket Quadrant,* which was designed to supplement his public lectures. In 1719 he published *Astronomical Dialogues Between a Gentleman and a Lady,* dedicated to Lady Cairnes, which, by his own admission, were "in Imitation of those of the excellent Mr. Fontenelle" (*Astronomical Dialogues,* p. v).

Harris' most famous work was the *Lexicon technicum,* the first edition of which appeared in 1704. This was the first general scientific encyclopedia, and for it Harris drew upon some of the greatest authorities of the day. In physics, astronomy, and mathematics he turned to Newton; in botany he consulted

John Ray and Joseph Tournefort; in other areas he drew upon Halley, Robert Boyle, Nehemiah Grew, John Woodward, John Wilkins, William Derham, and John Collins.

BIBLIOGRAPHY

I. Original Works. Harris' major works of scientific interest include *Remarks on Some Late Papers Relating to the Universal Deluge and to the Natural History of the Earth* (London, 1697); *A New Short Treatise of Algebra* (London, 1702; 3rd ed., 1714); *Description and Uses of the Celestial and Terrestrial Globes and of Collins' Pocket Quadrant* (London, 1703; 5th ed., 1720); *Lexicon technicum* (London, 1704; 5th ed., 2 vols., 1736); *Navigantium atque itinerantium bibliotheca,* 2 vols. (London, 1705); *Astronomical Dialogues Between a Gentleman and a Lady* (London, 1719; 4th ed., 1766); *History of Kent* (London, 1719); and *A Letter to the Fatal Triumvirate* [J. Freind, R. Mead, and S. Cade] (London, 1719).

Harris' Boyle lectures are in Sampson Letsome and John Nicholl, eds., *A Defence of Natural and Revealed Religion,* 3 vols. (London, 1739). In addition, Harris translated I. G. Pardies, *Short . . . Elements of Geometry* (London, 1701).

II. Secondary Literature. Thompson Cooper, "John Harris," in *Dictionary of National Biography,* repr. ed., IX, 13–14, is a good account, correct in the main. John Venn's capsule biography in *Alumni Cantabrigienses* (Cambridge, 1922) appears to contain several errors. Venn conflates "Technical Harris" with John Harris of Leicestershire, who entered St. John's College in 1684, and died in 1701. Joseph Foster's account in *Alumni Oxonienses* (London–Oxford, 1888) seems more reliable. Additional biographical material may be found in Thomas Hearne, *Remarks and Collections,* VII (Oxford, 1906), 46; and H. E. D. Blakiston, *Trinity College* (*Oxford*) (London, 1898), p. 172.

Robert H. Kargon

HARRISON, JOHN (*b.* Foulby, Yorkshire, England, 24 March [?] 1693; *d.* London, England, 24 March 1776), *horology.*

Harrison's father, Henry, a carpenter and joiner, moved to Barrow-upon-Humber, Lincolnshire, about 1697; his mother was Elizabeth Barber. He worked with his father, as well as doing surveying and clock repairing. With his brother James he made several clocks, one with a frictionless, almost continuous-impulse, grasshopper escapement and compensating bimetallic gridiron pendulum. They also used cycloidal cheeks and maintained power as Huygens had done; it is not known whether they knew of his work. The gridiron pendulum, based on experiments on the relative expansion of iron and brass, combined the two metals mechanically to cancel temperature error.

In June 1730, Harrison completed a twenty-three-page manuscript on experiments and inventions applied to clocks and outlined a timekeeper for use on ships to determine longitude. He was, of course, aware of the £20,000 prize offered by Parliament in 1714 for a reliable, accurate method of finding longitude at sea. In London to promote building this sea clock, he discussed it with George Graham, leading maker of clocks, watches, and instruments, who encouraged him with an unsecured, interest-free loan.

Returning home, Harrison produced a cumbersome marine clock using his earlier inventions and with two interconnected bar balances designed to be immune to the lurching of a ship. In 1737 the device was tested on a voyage to Lisbon. Few records survive, but on the return the landfall was identified by dead reckoning as the Start. Harrison's clock correctly indicated the ship's position to be off the Lizard, 57.5 nautical miles further west.

The Board of Longitude now granted £500 to encourage further experiment and the brothers moved to London, where a stronger and slightly smaller version of the marine clock was made. Unsatisfactory, it was not tried at sea. James returned home and John began a third machine, not ready for test until 1757. Next Harrison conceived of a radically different timekeeper in the form of a large watch—although, of course, with an entirely different escapement. As accurate as the third machine and far more convenient in size, it alone was tested on a voyage to Jamaica in 1761. On arrival, more than nine weeks later, it was only five seconds slow, about 1.25 minutes of arc, well within the 30 minutes of arc or longitude required. A second trial, to rule out a possibly fortuitous combination of circumstances, was held in 1764, and the results were also well within the limits.

Half the prize was now paid, but many previously unstipulated obstacles were raised. All four machines had to become property of the Board of Longitude; a minute description of the winner was required; it was subjected to extreme and unrealistic tests; and two more examples had to be built. Larcum Kendall, a well-known watchmaker, made one and Harrison, assisted by his son William, made another, slightly simplified. Accusations that the Board of Longitude was unfair and in some degree favored a system of lunar distances for finding longitude reached George III, who took Harrison's side. Harrison's last instrument was tested at the king's private observatory at Kew, and on the basis of its performance there, Harrison petitioned the House of Commons in 1772. The Board of Longitude, now on the defensive, dropped its opposition, and the award was made in 1773. Although soon supplanted by simpler mechanisms, the use of timekeepers to find longitude stemmed directly from Harrison's persistence and ability.

BIBLIOGRAPHY

I. ORIGINAL WORKS. The following MSS are available: untitled MS, signed and dated (Barrow-upon-Humber, 10 June 1730), with 2 pp. ink drawings, describing experiments and clock designs, Guildhall Museum, London, MS 6026 (no. 4); "Proposal for Examining Mr. Harrison's Timekeeper at Sea" (Dec. 1762), British Museum, 717K.15; "A Calculation Showing the Result of an Experiment Made by Mr. Harrison's Timekeeper . . . in a Voyage From Portsmouth to Port Royal in Jamaica . . ." (Dec. 1762), British Museum, 717K.15; "An Explanation of My Watch or Timekeeper for the Longitude . . . With Some Historical Account Coincident to My Proceedings" (7 Apr. 1763), Guildhall, MS 3972 (no. 1); and "Some Account of the Pallats etc. of My Second Made Watch for the Longitude" (3 May 1771), Guildhall, MS 3972 (no. 2).

Published works are *Remarks on a Pamphlet Lately Published by the Rev. Mr. Maskelyne . . .* (London, 1767); and *A Description Concerning Such Mechanism as Will Afford a Nice, or True Mensuration of Time; Together With Some Account of the Attempts for the Discovery of the Longitude by the Moon; and Also an Account of the Discovery of the Scale of Music* (London, 1775).

II. SECONDARY WORKS. See *Dictionary of National Biography,* XXV. See also the following, listed chronologically: Rev. Nevil Maskelyne, *An Account of the Going of Mr. Harrison's Watch . . .* (London, 1767); Thomas Reid, *A Treatise on Clock and Watchmaking* (Edinburgh, 1826); Johan Horrins (pseudonym of John Harrison, grandson of John Harrison), *Memoirs of a Trait in the Character of George III . . .* (London, 1835); R. T. Gould, *The Marine Chronometer* (London, 1923); and "John Harrison and His Timekeepers," in *Mariner's Mirror,* vol. **21,** no. 2 (Apr. 1935); and Humphrey Quill, *John Harrison, the Man Who Found Longitude* (London–New York, 1966).

EDWIN A. BATTISON

HARRISON, ROSS GRANVILLE (*b.* Germantown, Pennsylvania, 13 January 1870; *d.* New Haven, Connecticut, 30 September 1959), *biology.*

Harrison, who was to become one of the pioneers of experimental embryology, was the only son of Samuel Harrison and Catherine Barrington Diggs. Samuel Harrison was a grandson of William Harrison, who came to Philadelphia from England in 1798 under contract to the Bank of the United States to design bank notes; William's son, Ross's grandfather, was an engraver and cartographer. Samuel Harrison, Ross's father, was a mechanical engineer; he spent ten years in Russia designing rolling stock for the firm of Joseph Harrison (not related) of Philadelphia, who was under contract to the czar to build the railroad from St. Petersburg to Moscow.

Ross Harrison's education began in Germantown and was continued in Baltimore, Maryland, where his family moved during his boyhood. He entered the Johns Hopkins University in 1889; his work was sponsored by William Keith Brooks, and he received the Ph.D. degree in zoology in 1894. His doctoral dissertation was a morphological study of the development of the unpaired and paired fins of bony fishes.

During the academic year 1892–1893 Harrison studied at the university in Bonn, beginning his work on the fins of fishes under Moritz Nussbaum. He made other trips to Bonn in 1895–1896, 1898, and 1899. He received the M.D. degree in the latter year but never practiced medicine. Harrison had an excellent ear and mind for languages and spoke German as fluently as English; during the early years of his career he wrote a number of his major publications in German.

Harrison married Ida Lange at Altona, Germany, in 1896. They had five children: Richard, a cartographer; Elizabeth, a pediatrician; Dorothea, a landscape architect; Eleanor, who married Rufus Putney, Jr. (Putney later became a professor of English); and Ross, a successful businessman.

Harrison's first teaching position was at Bryn Mawr College, where during the academic year 1894–1895 he was lecturer on morphology as a substitute for Thomas Hunt Morgan, who was on a year's leave of absence. After a year's study at Bonn (1895–1896) he returned to the Johns Hopkins University in 1896 as instructor in anatomy in the medical school. He was promoted to associate in 1897 and to associate professor in 1899. In 1907 Harrison became the first Bronson professor of comparative anatomy at Yale, where he remained for the rest of his life. He was promoted to Sterling professor of biology in 1927 and became professor emeritus in 1938. From 1907 to 1938 he was head of the department of zoology.

Harrison's most important single scientific contribution was the innovation of the technique of tissue culture. It was he who first adapted the hanging drop method to the study of embryonic tissues in order to demonstrate the outgrowth of the developing nerve fiber; the first reports of these experiments, carried out from 1905 to 1907 at the Johns Hopkins Medical School, were published in 1907. At the time of these experiments there were three theories as to the mode of origin of the developing nerve fiber: (1) the cell-chain theory, which held that the nerve fiber is formed *in situ* by the cells that form the nerve sheath; (2) the plasmoderm theory, which claimed that the fiber is formed *in situ* by preformed protoplasmic bridges, under the influence of functional activity; and (3) the outgrowth theory, which maintained that the fiber is the product of the nerve cell itself. The outgrowth theory was then the most generally accepted, according to Harrison; but the main supporting evidence for it at that time was descriptive.

Harrison saw that the hypothesis could be confirmed experimentally if the nerve cell could be grown outside of the body, in the absence of sheath cells and protoplasmic bridges. Accordingly he removed portions of the nerve tube from frog embryos at stages before the fibers had formed, then studied their cellular development in hanging drops of frog lymph removed from the lymph heart and allowed to clot. In this way he could directly observe with a microscope the formation of the fiber by the nerve cell, and his observations firmly established the validity of the outgrowth theory. A contribution of vital significance not only to neurology but also to theoretical embryology, this was a final step in establishing that the cell is the primary developmental unit of the multicellular organism. "The reference of developmental processes to the cell," wrote Harrison in 1937, "was the most important step ever taken in embryology" ("Embryology and Its Relations," p. 372). His own confirmation of the nerve outgrowth theory played an important part in the analysis of developmental phenomena at the cellular level.

A number of investigators, including Julius Arnold, Gustav Born, Leo Loeb, and Gottlieb Haberlandt, had been attempting for a decade or more before the publication of Harrison's results to grow tissues or cells in isolation *in vitro* or *in vivo.* Their attempts had not been as successful as Harrison's, and it was unquestionably Harrison's experiments involving the observation of living tissues in hanging drop preparations that gave impetus to the further use of tissue and cell culture and that established it as a technique adaptable to the solution of a wide variety of problems in biology and medicine. Yet its importance in oncology, virology, genetics, and other related fields is still equaled by its importance in embryology itself. Sixty years after its first introduction into embryology laboratories, observation of the activities of cells in culture is one of the most popular pursuits of developmental biologists.

Another of Harrison's early contributions that was of great importance to the development of experimental embryology was his adoption of Born's method of embryonic grafting. In 1896 Born described the results of experiments in which he had successfully joined separated living parts of amphibian larvae. Harrison began similar experiments in 1897 in order to study the growth and regeneration of the tail of the frog larva. Born had shown that it was possible to perform fusion experiments using parts of embryos from different taxonomic families; in 1903 Harrison reported the results of experiments in which he grafted the head of the frog larva of one species to the body of a larva of a species of a different color, at the stage before the lateral line sense organ was complete. By taking advantage of the different natural pigments in the two species, he was able to observe that the sense organ developed by means of the posterior migration of the rudiment from the head into the trunk and tail. By solving a particular problem in which Harrison was interested, these experiments also served to demonstrate brilliantly the possibilities of interspecific (heteroplastic) grafting as an embryological technique. Hans Spemann, who received the 1935 Nobel Prize for physiology or medicine for his contributions to experimental embryology, acknowledged in 1936 the importance of Harrison's method of heteroplastic grafting for the experiments that led to his own theories of embryonic induction. Questions of great theoretical import to embryology had been raised by Wilhelm Roux and Hans Driesch, but the methods used by these pioneer investigators were extremely crude in comparison with Harrison's. Harrison and his students in America shared with Spemann and his students in Germany the honors for both the intellectual and the technical advances that brought the science of experimental embryology to full maturity.

During his lifetime Harrison and his students studied experimentally, principally in amphibian embryos, aspects of the development of a number of structures. Particularly noteworthy were studies on the relationships between the nervous system and the musculature. Harrison showed in 1904 that the amphibian limb could develop in the absence of the nerve supply. By means of heteroplastic grafting, he also attacked some hitherto highly elusive problems concerned with the control of growth in embryos, attaining results that could be expressed with great quantitative precision at a time when quantitative study of embryological phenomena was barely beginning. Transplanting the limb bud of a fast-growing tiger salamander larva to the flank of a slower-growing spotted salamander (1924), he showed that the limb maintained its own rate of growth; thus he could obtain a larva or adult bearing a limb far greater in size than that typical of the species. Later (1929), by performing heteroplastic transplants of the optic rudiment between the spotted and the tiger salamander, at a stage before the development of the optic nerve, he demonstrated that the size of the midbrain roof, where the optic nerve terminates in amphibians, is regulated by the size of the retina, specifically, by the number of fibers in the optic nerve, which grows into the brain from the retina. He also performed experiments (1929) on the correlative development of parts of the eye itself; by heteroplastic grafting he showed that reciprocal interactions between rudiments of the optic cup and the lens are involved in the regulation of the size of both of these components of the eye.

Another very original and significant group of experiments demonstrated the varied nature of the contributions to the embryo of the neural crest. Harrison, by a particularly ingenious set of heteroplastic transplantation experiments, proved that this structure, commonly thought to have been ectodermal in origin and significance, forms the cartilage of the gill skeleton in amphibians. One of his students, Graham DuShane, working under Harrison's guidance, demonstrated experimentally that the pigment-bearing cells of the amphibian are formed by the neural crest and not from the mesoderm, as had previously been generally believed. These results were far-reaching in their implications with respect to the old theories of germ-layer specificity, which had to be abandoned as a result of these and other data from experimental embryology.

But even more significant and original than these experiments were a series of studies on the development of the amphibian limb and its asymmetry. The vertebrate organism is bilaterally symmetrical, a number of the organs on the left side of the body, including the limbs, being mirror images of those on the right. The limb of the spotted salamander forms from a simple disk of mesoderm covered by ectoderm, and Harrison investigated the manner in which the disk becomes a right or a left limb. Harrison first demonstrated (1918) that the limb-forming potentialities of the disk are located in its mesoderm and then that the disk is, in the terminology of his day (adopted from Hans Driesch), a harmonious equipotential system. That is, any part of the rudiment, provided that it contains mesoderm, can form any part of the limb; a half-disk can form a whole limb. Next (1921) he devised an extensive series of experiments in which he grafted the disk in either normal or inverted position onto the same side of the body from which the disk had been taken, or onto the opposite side of the body; the disk can form a normal limb under any of these conditions. From the fact that a noninverted disk grafted on the opposite side of the body from which it was taken develops a limb of reversed symmetry (that is, a left limb develops on the right side, or vice versa), while an inverted disk grafted onto the opposite side develops a limb with its symmetry conforming to the side onto which it is implanted (that is, a left limb develops on the left side or a right limb on the right side), Harrison concluded that at the tail bud stage of the larvae, on which the experiments were performed, the anteroposterior axis of the limb is already determined, but not the mediolateral (transverse) or the dorsoventral. It was later shown by Harrison and his students that each of the three axes is determined in turn; for the limb, no stage has been found at which the anteroposterior axis is not determined.

Harrison later performed comparable experiments with the rudiment of the inner ear of the spotted salamander. Only preliminary reports of these experiments were published (1936, 1945), but their results demonstrated that during the development of this organ there is a stage at which none of the three axes is determined; each of the three is determined in its turn. Harrison believed that the progressive determination of the three axes must rest on some change in the orientation of the ultrastructural particles constituting the organ rudiment, and in collaboration with W. T. Astbury and K. M. Rudall he attempted in 1940 to look for evidence of such orientation by X-ray diffraction; the results were inconclusive because of the inapplicability of the method to the study of living and preserved tissues. The question as to the basis of the development of asymmetry thus remains where Harrison left it, but it is one of fundamental import and of considerable interest to molecular biology. When the answer to it is determined by methods not accessible to Harrison in his day, it will still be remembered that it was as a result of his transplantation experiments that the question could be shown to be amenable at all to experimental investigation. Harrison was interested in the intimate structure of protoplasm at least as early as 1897, and through nearly half a century of thought and experimentation he brought its investigation out of the realms of speculation and into the instrument rooms of modern molecular biology.

As an individual as well as a scientist Harrison was known for his dispassionate temperament and his calm judiciousness, and he held many important administrative offices besides the chairmanship of his department at Yale. He was an officer or member of advisory or administrative boards of many scientific and academic institutions and societies and of a number of government agencies. He was a trustee of the Marine Biological Laboratory at Woods Hole, Massachusetts, from 1908, and a member of the board of the Bermuda Biological Laboratory from 1925; he was a member and trustee of the Woods Hole Oceanographic Institution from 1930 to 1959, and his vision and foresight did much to advance oceanography to its present position among modern sciences. Harrison's most far-reaching administrative contribution was as chairman of the National Research Council during the critical years 1938–1946. He was one of the founders of the *Journal of Experimental Zoology* and was its managing editor from its beginning in 1903 until 1946. He received many honors, among them honorary degrees from Johns Hopkins, Yale, the universities of Chicago, Cincinnati, Michigan, and Dublin, Harvard, Columbia, Freiburg, Budapest, and Tübingen. He was elected to the National Academy of Sciences in 1913 and to the American Philosophical Society in the same year, and became a corresponding or honorary member of many foreign academies and societies, including the Royal Society and the French Academy of Sciences. He received a number of medals, among them the Archduke Rainer Medal of the Zoological-Botanical Association of Vienna in 1914, the John Scott Medal and Premium of the City of Philadelphia in 1925, the John J. Carty Medal of the National Academy of Sciences in 1947, and the Antonio Feltrinelli International Prize awarded by the Accademia Nazionale dei Lincei in 1956.

Harrison was exceptionally modest and objective; honors were to him less important than the establishment and maintenance of high standards of scientific endeavor. The greatest honor he would have wished would be to be remembered not only as the demonstrator of the outgrowth of the nerve fiber by a new and crucial experimental method but also as an investigator who by his intellectual acumen and technical imaginativeness contributed heavily to the origins and successful development of the important science of experimental embryology, which formed such an important bridge between the old morphology of the nineteenth century and the new molecular biology of the twentieth.

BIBLIOGRAPHY

I. Original Works. A complete bibliography of Harrison's publications is included in the memoir by Nicholas (see below). His major articles include the following: "Ueber die Entwicklung der nicht knorpelig vorgebildeten Skelettheile in den Flossen der Teleostier," in *Archiv für mikroskopische Anatomie,* **42** (1893), 248–278; "Die Entwicklung der unpaaren und paarigen Flossen der Teleostier," in *Archiv für mikroskopische Anatomie und Entwicklungsgeschichte,* **46** (1895), 500–578; "The Growth and Regeneration of the Tail of the Frog Larva. Studied With the Aid of Born's Method of Grafting," in *Archiv für Entwicklungsmechanik der Organismen,* **7** (1898), 430–485; "Ueber die Histogenese des peripheren Nervensystem bei Salmo salar," in *Archiv für mikroskopische Anatomie und Entwicklungsgeschichte,* **57** (1901), 354–444; "Experimentelle Untersuchungen über die Entwicklung der Sinnesorgane der Seitenlinie bei den Amphibien," *ibid.,* **63** (1903), 35–149; "An Experimental Study of the Relation of the Nervous System to the Developing Musculature in the Embryo of the Frog," in *American Journal of Anatomy,* **3** (1904), 197–220; "Experiments in Transplanting Limbs and Their Bearing Upon the Problems of the Development of Nerves," in *Journal of Experimental Zoology,* **4** (1907), 239–281; "Observations on the Living Developing Nerve Fiber," in *Anatomical Record,* **1** (1907), 116–118, and in *Proceedings of the Society for Experimental Biology and Medicine,* **4** (1907), 140–143; "Embryonic Transplantation and Development of the Nervous System," in *Anatomical Record,* **2** (1908), 385–410, and in *Harvey Lectures for 1907–1908* (1909), pp. 199–222; "The Development of Peripheral Nerve Fibers in Altered Surroundings," in *Archiv für Entwicklungsmechanik der Organismen,* **30,** pt. 2 (1910), 15–33; "The Outgrowth of the Nerve Fiber as a Mode of Protoplasmic Movement," in *Journal of Experimental Zoology,* **9** (1910), 787–846; "The Stereotropism of Embryonic Cells," in *Science,* **34** (1911), 279–281; "The Cultivation of Tissues in Extraneous Media as a Method of Morphogenetic Study," in *Anatomical Record,* **6** (1912), 181–193; "The Reaction of Embryonic Cells to Solid Structures," in *Journal of Experimental Zoology,* **17** (1914), 521–544; "Experiments on the Development of the Fore Limb of Amblystoma, a Self-Differentiating Equipotential System," *ibid.,* **25** (1918), 413–461; "On Relations of Symmetry in Transplanted Limbs," *ibid.,* **32** (1921), 1–136; "Experiments on the Development of Gills in the Amphibian Embryo," in *Biological Bulletin. Marine Biological Laboratory, Woods Hole, Mass.,* **41** (1921), 156–170; "Some Unexpected Results of the Heteroplastic Transplantation of Limbs," in *Proceedings of the National Academy of Sciences of the United States of America,* **10** (1924), 69–74; "Neuroblast Versus Sheath Cell in the Development of Peripheral Nerves," in *Journal of Comparative Neurology,* **37** (1924), 123–205; "The Development of the Balancer in Amblystoma, Studied by the Method of Transplantation in Relation to the Connective-Tissue Problem," in *Journal of Experimental Zoology,* **41** (1925), 349–427; "The Effect of Reversing the Medio-Lateral or Transverse Axis of the Forelimb Bud in the Salamander Embryo (Amblystoma punctatum Linn.)," in *Archiv für Entwicklungsmechanik der Organismen,* **106** (1925), 469–502; "On the Status and Significance of Tissue Culture," in *Archiv für Zellforschung,* **6** (1928), 4–27; "Correlation in the Development and Growth of the Eye Studied by Means of Heteroplastic Transplantation," in *Archiv für Enwicklungsmechanik der Organismen,* **120** (1929), 1–55; "Esperimenti d'innesto sul cestello brachiale di 'Clavelina lepadiformis' (Müller)," in *Atti dell'Accademia nazionale dei Lincei. Rendiconti,* Classe di scienze fisiche, mathematiche e naturali, 6th ser., **11** (1930), 139–146, written with Pasquale Pasquini; "Some Difficulties of the Determination Problem," in *American*

Naturalist, **67** (1933), 306–321; "Heteroplastic Grafting in Embryology," in *Harvey Lectures for 1933–1934* (1935), pp. 116–157; "On the Origin and Development of the Nervous System Studied by the Methods of Experimental Embryology (The Croonian Lecture)," in *Proceedings of the Royal Society,* **118B** (1935), 155–196; "Relations of Symmetry in the Developing Ear of Amblystoma punctatum," in *Proceedings of the National Academy of Sciences of the United States of America,* **22** (1936), 238–247; "Embryology and Its Relations," in *Science,* **85** (1937), 369–374; "Die Neuralleiste," in *Anatomischer Anzeiger,* supp. **85** (1938), 3–30; "An Attempt at an X-Ray Analysis of Embryonic Processes," in *Journal of Experimental Zoology,* **85** (1940), 339–363, written with W. T. Astbury and K. M. Rudall; "Relations of Symmetry in the Developing Embryo," in *Transactions of the Connecticut Academy of Arts and Sciences,* **36** (1945), 277–330; and "Wound Healing and Reconstitution of the Central Nervous System of the Amphibian Embryo After Removal of Parts of the Neural Plate," in *Journal of Experimental Zoology,* **106** (1947), 27–84.

Harrison's unpublished records and documents have been deposited in the archives of the Yale University Library, New Haven, Connecticut.

II. SECONDARY LITERATURE. Selected biographical notices and memoirs are M. Abercrombie, "Ross Granville Harrison 1870–1959," in *Biographical Memoirs of Fellows of the Royal Society,* **7** (1961), 111–126; A. M. Dalcq, "Notice biographique sur M. le Professeur R. G. Harrison," in *Bulletin de l'Académie r. de médecine de Belgique,* 6th ser., **24** (1959), 768–774; P.-P. Grassé, "Notice nécrologique sur Ross Granville Harrison," in *Comptes rendus hebdomadaires des séances de l'Acádemie des sciences,* **250** (1960), 2622–2623; J. S. Nicholas, "Ross Granville Harrison," in *Anatomical Record,* **137** (1960), 160–162; "Ross Granville Harrison, Experimental Embryologist," in *Science,* **131** (1960), 1319; "Ross Granville Harrison 1870–1959," in *Yale Journal of Biology and Medicine,* **32** (1960), 407–412; "Ross Granville Harrison (1870–1959)," in *Yearbook, American Philosophical Society* (1961), 114–120; and "Ross Granville Harrison 1870–1959," in *Biographical Memoirs. National Academy of Sciences,* **35** (1961), 132–162; J. M. Oppenheimer, "Ross Granville Harrison," in H. Freund and A. Berg, eds., *Geschichte der Mikroskopie. Leben und Werk grosser Forscher,* II (Frankfurt, 1965), 117–126; and "Ross Harrison's Contributions to Experimental Embryology," in *Bulletin of the History of Medicine,* **40** (1967), 525–543, repr. in J. M. Oppenheimer, *Essays in the History of Embryology and Biology* (Cambridge, Mass., 1967), pp. 92–116; and P. Pasquini, "Ross Granville Harrison," in *Acta embryologiae et morphologiae experimentalis,* **3** (1960), 119–130.

JANE M. OPPENHEIMER

HART, EDWIN BRET (*b.* Sandusky, Ohio, 25 December 1874); *d.* Madison, Wisconsin, 12 March 1953), *biochemistry, nutrition.*

The son of William Hart and Mary Hess, Hart was born on a farm. After graduation from Sandusky High School he studied chemistry at the University of Michigan, receiving the B.S. in 1897. He then became assistant chemist at the New York Agricultural Experiment Station at Geneva, where he investigated the protein in milk with L. L. Van Slyke. In 1900 he took a two-year leave of absence to study protein chemistry with Albrecht Kossel at the University of Marburg. Hart then accompanied Kossel to Heidelberg, but because of a loss of credits, he returned to the United States in 1902 without his doctorate. Once back in Geneva, he studied with Van Slyke the chemical changes which take place in the manufacture and ripening of cheese.

Hart married Annie Virginia DeMille in 1903; they had one daughter, Margaret. In 1906 he succeeded S. M. Babcock as chairman of the agricultural chemistry department at the University of Wisconsin, serving in this capacity until his retirement in 1944. Under his direction, the department came into the forefront of nutritional research during a time when the role of organic and mineral trace nutrients came to be understood. Hart's own contributions are difficult to assess, however, because of the collaborative nature of the research in which he was involved. Hart was closely familiar with the work of others and continually contributed ideas and encouragement. He was regularly involved in joint research programs with the departments of bacteriology, dairy science, poultry husbandry, and animal science. His department was unique in its success in establishing basic scientific principles while pursuing practical objectives. For example, his work on copper anemia not only had practical importance in animal feeding, but created new insights into the study of blood diseases; and the studies of single grain diets opened a broad area of nutritional deficiencies which led to knowledge of metabolic processes.

At Wisconsin, Hart continued his study of cheese curing in collaboration with bacteriologists E. G. Hastings and Alice Evans. He also developed a simple and rapid method for the determination of casein in milk. But once under the influence of Babcock, he turned his attention to the nutrition of farm animals. In 1907, with E. V. McCollum, Harry Steenbock, and George C. Humphrey, Hart undertook a four-year experiment in which sets of calves were fed presumably balanced rations derived from single plants. The failure of the animals to thrive on wheat and oat rations was a stimulus, along with animal feeding experiments elsewhere, for recognition of the vitamin concept.

At one time or another Hart worked on most of the vitamins and minerals of nutritional significance.

Of particular importance was his work on phosphorus and calcium metabolism (with McCollum and Steenbock); the role of iodine in preventing the "hairless pig" syndrome (with Steenbock); studies on rickets, particularly leg weakness in chickens (with J. G. Halpin) and irradiation of milk to enhance vitamin D content (with Steenbock); the role of iron and copper in anemia (with C. A. Elvehjem and Steenbock); the toxicity of fluorine from superphosphate fertilizer (with Paul Phillips); the essential nature of zinc in nutrition (with Elvehjem); and urea as a source of nitrogen in ruminants (with G. Bohsted).

Hart's questioning character, coupled with a genial personality, made him an effective teacher at all levels. He worked hard and expected the same of his students and faculty. Although he enjoyed sports and travel, his interests were always close to his work. After his retirement, he continued his daily rounds of the laboratories—even to the day before the heart attack which ended his life. In 1949 the University of Wisconsin honored him with the Sc.D.

BIBLIOGRAPHY

I. Original Works. Hart published almost 400 papers, nearly all of them in collaboration with students and fellow faculty members. The largest number appeared in *Journal of Biological Chemistry* and in the bulletins of the Wisconsin Agricultural Experiment Station. There is a full bibliography in C. A. Elvehjem, "Biographical Memoir of Edwin Bret Hart," in *Biographical Memoirs. National Academy of Sciences,* **28** (1954), 135–161. For the paper on the single grain experiments, see "Physiological Effects on Growth and Reproduction of Rations Balanced from Restricted Sources," in *Research Bulletin. Wisconsin Agricultural Experiment Station, College of Agriculture, University of Wisconsin,* **17** (1911), 1–131, written with E. V. McCollum, Harry Steenbock, and George C. Humphrey. Letters and other unpublished materials are held by the University of Wisconsin Archives.

II. Secondary Literature. The most complete biography of Hart is the obituary memoir by C. A. Elvehjem, *op. cit.,* pp. 117–134. See also the sketches prepared by former students for a symposium on his life sponsored by the Institute of Food Technologists in 1954: Henry T. Scott, "Edwin Bret Hart—His Life and Memories of Him," in *Food Technology,* **9** (1955), 1–4; C. A. Elvehjem, "Thirty-two Years' Association with E. B. Hart," *ibid.,* 4–7; S. Lepkovsky, "Contributions of E. B. Hart to Animal Nutrition," *ibid.,* 8; E. M. Nelson, "The Impact of E. B. Hart's Contributions on Human Nutrition," *ibid.,* 9–11; and K. G. Weckel, "E. B. Hart's Contributions to Food Technology," *ibid.,* 11–13. See also E. H. Harvey, "Edwin Bret Hart," in *Chemical and Engineering News,* **22** (1944), 435–436.

Aaron J. Ihde

HARTIG, THEODOR (*b.* Dillenburg, Germany, 21 February 1805; *d.* Brunswick, Germany, 26 March 1880), *plant physiology, forestry, entomology.*

Hartig, the son of the forester Georg Ludwig Hartig and Theodore Klipstein, spent his youth in Dillenburg and, after 1811, in Berlin. He received training in forestry in Mühlbach, Pomerania (now Poland), under the supervision of his uncle Friedrich K. T. Hartig. He continued his training in Brandenburg, and from 1824 to 1827 studied forestry under Friedrich W. L. Pfeil at the forestry institute of the University of Berlin. After graduating, Hartig was appointed *Regierungs-Referendar* in Potsdam (1831). He qualified as a lecturer in forestry at the University of Berlin and in 1835 was appointed to a nonsalaried professorship there. In 1838 he accepted a post in the forestry department of the Collegium Carolinum in Brunswick, where, simultaneously, he was promoted to *Forstrat* and made a member of the Imperial Leopoldine-Caroline Academy. After teaching at the Carolinum for many years, he retired in 1878 as *Oberforstrat.*

In 1837 Hartig established his reputation as an entomologist with the work *Adlerflügler Deutschlands.* Although he continued to concern himself with entomology in his subsequent textbooks, he more and more turned his attention to the field of plant physiology. In *Neue Theorie der Befruchtung der Pflanzen* (1842) he provided a survey of previous theories of plant fertilization and also presented his own views on the subject. In *Über das Leben der Pflanzenzelle* (1844) Hartig set forth his own nomenclature for the parts of the cell. It did not become the accepted terminology, but his attempt to describe the cell as a closed unit stimulated further investigations. In 1855 he discovered the aleurone nucleus and was the first to describe it as the basic component of cells.

Hartig's later works were concerned with the physiology and anatomy of ligneous plants. In the new edition (1851) that he prepared of his father's *Lehrbuch für Förster,* Hartig furnished the first exact description of the "descending sap flow," a decisive contribution to contemporary knowledge of the metabolic processes involved.

Besides his entomological and botanical studies, Hartig occupied himself throughout his career with practical questions of forestry. In 1834, for example, he wrote with his father *Forstliches und forstnaturwissenschaftliches Conversationslexikon.* His *Lehrbuch der Pflanzenkunde* appeared between 1841 and 1847 and his *Anatomie und Physiologie der Holzpflanzen* in 1877. Hartig also considered certain aspects of commercial forestry in several of his writings and undertook particularly profitable investigations in this area.

BIBLIOGRAPHY

I. Original Works. *Forstliches und forstnaturwissenschaftliches Conversationslexikon* (Berlin, 1834), written with G. L. Hartig; *Die Adlerflügler Deutschlands mit besonderer Berücksichtigung ihres Larvenzustandes und ihres Wirkens in Wäldern und Gärten* (Berlin, 1837); *Neue Theorie der Befruchtung der Pflanzen* (Brunswick, 1842); *Beiträge zur Entwicklungsgeschichte der Pflanzen* (Berlin, 1843); *Das Leben der Pflanzenzelle, deren Entstehung, Vermehrung und Auflösung* (Berlin, 1844).

See also *Vergleichende Untersuchungen über den Ertrag der Rothbuche im Hoch- und Pflanzenwalde, im Mittel- und Niederwald-Betriebe, nebst Anleitung zu vergleichenden Ertragsforschungen* (Berlin, 1846, 1851); *Lehrbuch der Pflanzenkunde in ihrer Anwendung auf Forstwirtschaft* (Berlin, 1841–1847); *Vollständige Naturgeschichte der forstlichen Culturpflanzen Deutschlands* (Berlin, 1852); *Kontroversen der Forstwirtschaft* (1853), with no place of publication listed; *Entwicklungsgeschichte des Pflanzenkeims, dessen Stoffbildung und Stoffwandlung während der Vorgänge des Reifens und Keimens* (Leipzig, 1858); *System und Anleitung zum Studium der Forstwirtschaftslehre* (Leipzig, 1858); *Über die Entwicklungsfolge und den Bau der Holzfaserwandung* (Vienna, 1870); *Anatomie und Physiologie der Holzpflanzen* (Berlin, 1877); *Jahresberichte über die Fortschritte der Forstwissenschaften in der forstlichen Naturkunde* (1836–1837).

His new editions of works by G. L. Hartig include *Lexikon für Jäger und Jagdfreunde,* or *Weidmännisches Conversationslexikon* (Berlin, 1859); *Kurze Belehrung über die Behandlung und Cultur des Waldes* (Berlin, 1859); *Lehrbuch für Förster* (Stuttgart, 1877); and *Lehrbuch für Jäger* (Stuttgart, 1877).

II. Secondary Literature. An obituary of Hartig appears in *Leopoldina,* **16** (1880), 70–71. See also Richard Hess, "Theodor Hartig," in *Allgemeine Forst- und Jagdzeitung,* **56** (1880), 153; *Leben hervorragender Forstmänner* (1885), 138–142; R. B. Hilf and F. Röhrig, *Wald und Weidwerk in Geschichte und Gegenwart;* R. B. Hilf, *Der Wald,* I (Potsdam, 1938), 258; Kurt Manthel, "Theodor Hartig," in *Neue Deutsche Biographie,* VII, 713; Martin Möbius, *Geschichte der Botanik* (Stuttgart, 1968).

R. Schmitz

HARTING, PIETER (*b.* Rotterdam, Netherlands, 27 February 1812; *d.* Amersfoort, Netherlands, 3 December 1885), *microscopy, zoology.*

Harting was the son of Dirk Harting, a tobacco merchant, and Jeanette Blijdenstein. The father died in 1819 and Harting, his two brothers, and one sister moved to Utrecht with their mother, to be near her family.

Harting attended school at Elburg from 1823 until 1828. He started reading medicine at Utrecht University in September 1828, obtaining the doctorate in medicine in 1835 and in obstetrics in 1837. At Utrecht he studied physics under G. Moll, chemistry under G. J. Mulder, and physiology under J. L. C. Schroeder van der Kolk.

From 1835 Harting practiced medicine in the village of Oudewater, although he wished to do research in chemistry and biology. In 1841 he was appointed professor of pharmacy at the Athenaeum in Franeker. When the Athenaeum was closed in 1843, he was transferred to the University of Utrecht but was not appointed to a chair because there was no vacancy. This proved a blessing in disguise for Dutch science, since it enabled Harting to do research in his favorite field, microscopy.

Harting's work in microscopy was both practical and theoretical. As a student he saw that the microscope was the most important instrument for the development of most sciences, including medicine, zoology, and plant physiology. While practicing medicine in Oudewater he had made his own microscope, and at Utrecht he initiated courses in practical microscopy, which were among the first in the subject at any university in the Netherlands. Harting not only cataloged the different types of microscopes belonging to Utrecht, but he also measured their optical properties of enlargement and, what is more important, their resolving power. To express his results uniformly, he introduced the one-thousandth part of the millimeter, which he called mmm (milli-millimeter), later named μ. As a result of these investigations, in 1848 he began the publication of the multivolume treatise *Het Mikroskoop,* the first full historical treatment of that subject. Some years later he taught pharmacology, plant physiology, comparative anatomy, and zoology; his colleagues included Buys Ballot and Donders.

In the Physical Institute he discovered the forgotten Leeuwenhoek microscope and the lens which Christiaan Huygens made and used in 1655 to discover the rings of Saturn and one of its satellites. (These instruments are now on exhibition at the Utrecht university museum.)

Harting also did research in geology. He studied the island of Urk (in what was then the Zuider Zee) and the valley of the Eem River. These investigations enabled him to give advice on plans for the reclamation of the Zuider Zee.

For his physiological research Harting constructed the physiometer, an instrument to facilitate the study of the swim bladders of fishes (1872). He also did some important research on pileworms in connection with the enormous damage inflicted by these teredos on the seawalls (1860). Harting was one of the first Dutch supporters of Darwin's theory of evolution. Two years before Darwin published *On the Origin of Species,* Harting gave a series of lectures which indicated that he held a theory similar to Darwin's.

Harting was also interested in anthropology and in 1861 designed the cephalograph, an instrument for measuring the dimensions of human skulls and faces. In his early years at Utrecht he cooperated with G. J. Mulder on experiments to determine the chemical nature of plant cell walls (1846). His last work on plant physiology was a spectroscopic study of chlorophyll (1855).

Harting was appointed to the chair of zoology at Utrecht, and shortly afterward wrote a textbook for his students. In order to obtain more facilities for his pupils to conduct research, he succeeded in having a Dutch subsection established in the international zoological station at Naples in 1874; he also founded a movable zoological station in the Netherlands (1876).

An excellent popularizer of scientific subjects, Harting was one of the founders of *Album der Natuur,* a periodical dedicated to the popularization of the latest results of scientific research. For several years he lectured for the Natuurkundig Gezelschap and served as its president for some time. Harting was rector of Utrecht University in 1858–1859. He crusaded for cremation and against alcoholism and spiritism. Although he retired in 1882, spending the rest of his life in Amersfoort, he remained active as president of a committee to help the Boers in their war against the British.

BIBLIOGRAPHY

A nearly complete list of Harting's publications, organized according to subject, is in *Jaarboek van de K. Akademie van wetenschappen gevestigd te Amsterdam* (1888), pp. 36–60. *Levensberichten der afgestorvene Medeleden van de Maatschappij der Nederlandsche Letterkunde . . .* (1887), pp. 176–187, contains a bibliography in chronological order.

His major publication in plant physiology is *Monographie des marattiacées* (Leiden-Dusseldorf, 1853); that on potato blight is in *Nieuwe Verhandelingen der Eerste Klasse van het K. Nederlandsche Instituut van wetenschappen,* **12** (28 Nov., 12 and 24 Dec. 1846), 203–297. On microscopy, see *Bijdrage tot de Geschiedenis der Microscopen in ons Vaderland* (Utrecht, 1846); and *Het Mikroskoop, deszelfs gebruik, geschiedenis en tegenwoordige toestand,* 3 vols. (Utrecht, 1848–1850), vol. IV, *Handleiding tot oefening in het onderzoek van plantaardige en dierlijke weefsels* (Tiel, 1854), vol. V, *De nieuwste verbeteringen van het mikroskoop en zijn gebruik* (Tiel, 1858)—vols. I–III translated into German by F. W. Thiele as *Das Mikroskop* (Brunswick, 1859; 2nd ed., 1866), facs. repro. (Brunswick, 1970).

Among his geological writings are "De bodem onder Amsterdam," in *Nieuwe verhandelingen der Eerste Klasse van het K. Nederlandsche Instituut van wetenschappen,* **5** (1852), 73–230; *Het eiland Urk, zijn bodem, voortbrengselen en bewoners* (Utrecht, 1853); "De bodem van het Eemdal," in *Verslagen en Mededeelingen der K. Akademie van wetenschappen,* **8** (1874), 282–290; "Le système eemien," in *Archives des sciences exactes et naturelles* (1875), 443–454; and "De geologische en physische gesteldheid van den Zuiderzeebodem, in verband met de voorgenomen droogmaking," in *Verslagen en mededeelingen der K. Akademie van wetenschappen,* **11** (1877), 301–325, and **12** (1878), 220–228.

On anthropology, see "Le plan médian de la tête néerlandaise masculine, déterminé d'après une méthode nouvelle," in *Verhandelingen der K. Akademie van wetenschappen,* **15** (1875), 1–22. A work on the descent of man is *De voorwereldlijke scheppingen* (Tiel, 1857), translated into German by J. E. A. Martin (Leipzig, 1859).

Zoological writings are *Handboek der vergelijkende ontleedkunde* (Tiel, 1854), trans. of Oscar Schmidt, *Vergleichende Anatomie;* "Verslag over den paalworm uitgegeven door de Natuurkundige Afdeeling . . .," in *Verslagen en Mededeelingen der K. Akademie van wetenschappen,* vol. **9** (1860); and *Leerboek van de grondbeginselen der dierkunde in haren geheelen omvang,* 3 vols. in 5 pts. (Tiel, 1862–1874).

Other works are *De macht van het kleine, zichtbaar in de vorming der korst van onzen Aardbol* (Utrecht, 1849; 2nd ed., Amsterdam, 1866), translated into German by A. Schwartz as *Die Macht des Kleinen, sichtbar in der Bildung der Rinde unseres Erdballs* (Leipzig, 1851); *Anno 2065* (Utrecht, 1865), 3rd ed., under the title *Anno 2070* (Utrecht, 1870), all eds. written under the pen name Dr. Dioscorides, translated into German as *Anno 2066* (Weimar, 1866); and *Mijne herinneringen* (Amsterdam, 1961), his autobiography, written between 1872 and 1885.

II. SECONDARY LITERATURE. For information on Harting, see the following works by A. A. W. Hubrecht: a funeral oration in *Jaarboek van de K. Akademie van Wetenschappen gevestigd te Amsterdam* (1888), pp. 1–35; and "Pieter Harting," in *De Gids,* 4th ser., **55,** pt. 1 (1886), 157–168. Other obituaries include C. H. D. Buys Ballot, in *Levensberichten der afgestorvene Medeleden van de Maatschappij der Nederlandsche Letterkunde* (1887), pp. 149–175; and H. F. Jonkman, "Pieter Harting," in *Mannen van betekenis in onze dagen* (Haarlem, 1886), pp. 319–366.

J. G. VAN CITTERT-EYMERS

HARTLEY, DAVID (*b.* Armley, Yorkshire, England, *ca.* 30 August 1705; *d.* Bath, England, 28 August 1757), *psychology.*

Hartley was born into the family of a poor Anglican country clergyman in Yorkshire. His mother, Evereld Wadsworth, died in the year David was born. His father, also called David, then married Sarah Wilkinson in 1707, by whom he had four children, but he too died while David was still a boy. After being brought up "by one Mrs. Brooksbank," Hartley attended Bradford Grammar School and in 1722 entered Jesus College, Cambridge. He studied classics, mathematics, and divinity and received his B.A. in 1726 and his M.A. in 1729. He was a fellow of Jesus from 1727 until he took leave in 1730. When he married a year later, his fellowship was terminated in accordance with the college statutes.

Although devoutly religious, Hartley had scruples against signing the articles and went into medicine instead of taking orders. He never obtained a medical degree but went to study in Newark, where he also began his practice. He then moved to Bury St. Edmunds. In 1735 he married for a second time and his wife's private fortune enabled them to settle in London. But her ill health required them to move to the health spa of Bath, where they remained until his death. His son David became a statesman and inventor.

Although Hartley was a fellow of the Royal Society, the center of his life was his medical practice, not science. He lived a simple life, devoted to the health of both rich and poor. He was an amiable and methodical man with a wide circle of friends that included Stephen Hales, bishops William Law and Joseph Butler, and Sir Hans Sloane. During his lifetime he championed a variety of causes, among them Mrs. Stephens' bogus cure for the stone (from which he had suffered as a young man), John Byrom's shorthand system, and Nicholas Saunderson's algebra textbook.

Hartley wrote one important work, *Observations on Man, His Frame, His Duty, and His Expectations,* which appeared in two volumes in 1749. It contained a systematic development of ideas which he had first set out in a pamphlet, *Conjecturae quaedam de sensu mortu et idearum generatione* (1730), and which were developed further in two small treatises on *The Progress of Happiness Deduced From Reason* (1734). The argument of Hartley's *Observations* brings together ideas from three main sources. The first is the principle of the "association of ideas" as described in the fourth edition of Locke's *Essay:* that complex ideas are formed from simple ones by repeated juxtapositions in experience. This concept had been elaborated by the Reverend John Gay in his "Preliminary Dissertation Concerning the Fundamental Principle of Virtue or Morality" (Cambridge, 1731), in which Gay attempted to deduce all man's intellectual pleasures and pains from the principle of association, in opposition to innatist theories of learning and morality.

Hartley tried to relate these speculations to a theory of the physical basis of sensation and memory derived from the "Queries" appended to Newton's *Opticks.* Newton had speculated that physical vibrations of light impinged on the retina of the eye, setting up other vibrations which traveled along the nerves to the brain. Locke's preoccupations had been epistemological, Gay's ethical, and Newton's physical. Hartley's synthesis attempted to integrate these three approachs by relating natural theology to the problem of providing a naturalistic basis for morality. He was thus led to work out a systematic psychophysiology.

The first volume of *Observations* is a tour de force which considers every significant topic in neurophysiology and human and comparative psychology, explained in terms of the development of complex ideas and habits from simple sensations and their repeated juxtapositions in experience. Mental associations were paralleled by vibrations of particles in the nervous system that persisted in the form of smaller "vibratuncles" which provided the physical basis for memory. The second volume extends the system to account for morality and the afterlife.

The significance of Hartley's work did not lie in any new empirical findings but in a set of assumptions and a framework for approaching the phenomena of life and mind. In the century following the publication of *Observations,* the work came to be seen as the fountainhead of some of the most important ideas in biological, psychological, and social thought. Viewed in a narrow perspective, it was the first published work in English to use the term "psychology" in its modern sense. Hartley's principles provided the conceptual framework for the associationist tradition in modern psychology, including learning theory and psychoanalysis. His speculations about the physiology of the nervous system laid the foundations for the dominant sensory-motor interpretation of neurophysiology and the experimental localization of functions in the cerebral cortex.

It is misleading, however, to separate the psychophysiological from the more general aspects of Hartley's influence. His book is the central document in the history of attempts to apply the categories of science, both directly and by analogy, to the study of man and society. Much of the nineteenth-century debate in Britain on man's place in nature was conducted under its influence. Considered conceptually, Hartley's was the first systematic elaboration of the explanatory principle that came to play an analogous role in the biological and human sciences to the concept of gravity or attraction in the physicochemical sciences. His unification of sensation, motion, association, and vibrations in a coherent mechanistic theory of experience and behavior provided the grounds for the secularization of the concepts of adaptation and utility. This secularization was taken up in a wide range of disciplines as a basis for accounting for cumulative ordered change through experience. It was used as a general warrant to explain changing utilities and adaptations by means of the pleasurable and painful results or consequences of actions.

Hartley's influence is perhaps best understood through the work of later theoreticians. Joseph Priestley, for example, stressed Hartley's determinism but set aside his psychophysical dualism in his publication of *Hartley's Theory of the Human Mind* (1775). This reductionist version of Hartley's theory was then placed in the service of Priestley's Unitarian philosophy of nature. Erasmus Darwin used Hartley's mechanisms as the basis for his theory of evolution and for his system of medical classification in *Zoonomia* (1794–1796). In social and political theory, William Godwin's arguments for inevitable human progress toward perfection in *Political Justice* (1793) were based on extrapolations from Hartley's ideas. The psychological, social, and political theories of the English utilitarians—especially James and John Stuart Mill—were also based on Hartleian psychology and generalizations from it.

Between 1830 and 1860 there was a convergence of various aspects of Hartley's influence. Müller drew on Hartley's motor theory of learning, which was by then gaining support from findings in experimental neurophysiology. In formulating his physiological and psychological theories Alexander Bain integrated Hartley's sensory-motor physiology with the mainstream of the English tradition of associationist psychology. Theories of evolution also drew on Hartleian mechanisms. Thus, Spencer's evolutionary theory extended associationist learning theory from the experience of the individual to that of the race. J. Hughlings Jackson applied these conceptions to the physiology and pathology of the brain, while David Ferrier applied them to the experimental localization of cerebral functions.

What had begun as the integration of corpuscular physics with empiricist epistemology and sensationalist psychology was thus reinterpreted in biological, evolutionary terms to provide the foundations for modern theories in biology, neurophysiology, human and comparative psychology, neurology, psychiatry, psychoanalysis, social and political theory, and belief in progress. It has been argued that Hartley's is the only theory of learning that has borne fruit in modern science. The list of disciplines based wholly or in part on the principles which he formulated provides some indication of the fecundity of his ideas in the history of the reification of man.

BIBLIOGRAPHY

I. Original Works. Hartley's significant works are *Conjecturae quaedam de sensu motu et idearum generatione* (London, 1730; 2nd ed., Bath, 1746), repr. in S. Parr, ed., *Tracts* (London, 1837); and *Observations on Man, His Frame, His Duty, and His Expectations,* 2 vols. (London, 1749). His clinical medical papers are listed in the *Dictionary of National Biography,* XXV (London, 1891), 68.

II. Secondary Literature. For information on Hartley or his work, see G. S. Bower, *David Hartley and James Mill* (New York, 1881); S. T. Coleridge, *Biographia Literaria* (London, 1817), chs. 5–7; E. Halévy, *The Growth of Philosophic Radicalism,* rev. ed. (London, 1952), pp. 5–34, 193, 247, 433–487; G. H. Lewes, *Biographical History of Philosophy,* 2nd ed. (London, 1857), pp. 507–511; J. Mackintosh, "Dissertation Second: Exhibiting a General View of the Progress of Ethical Philosophy, Chiefly During the Seventeenth and Eighteenth Centuries," in *The Encyclopaedia Britannica,* 8th ed. (Edinburgh, 1860), I, 378–386; J. S. Mill, "Bain's Psychology," in *Edinburgh Review,* **110** (1859), 287–321; G. Murphy, *Historical Introduction to Modern Psychology,* 2nd ed. (New York, 1949); R. C. and K. Oldfield, "Hartley's 'Observations on Man,'" in *Annals of Science,* **7** (1951), 371–381; Joseph Priestley, *Hartley's Theory of the Human Mind* (London, 1775); T. Ribot, *English Psychology* (London, 1873), pp. 35–43; H. C. Warren, *A History of the Association Psychology* (London, 1921), pp. 50–80; and R. M. Young, "Association of Ideas," in P. P. Wiener, ed., *Dictionary of the History of Ideas* (New York, in press).

Robert M. Young

HARTLIB, SAMUEL (*b.* Elbing, Prussia [now Elblag, Poland]; *d.* London, England, 10 March 1662), *science education, reform, publishing, promotion.*

Hartlib's father, a prominent merchant and dye manufacturer, was originally from Poznan, Poland; his mother was probably English. He was educated at Brieg, Silesia, which he left about 1621—probably for Cambridge, where he seems to have remained until about 1626; he did not matriculate but, presumably, pursued some course of studies. He spent the year 1627–1628 in Elbing but returned to settle in England in 1628. The following year he married Mary Burningham, who died about 1660. Hartlib's family life is little documented, but he appears to have had at least four sons and two daughters. The eldest, Samuel, born about 1631, is fairly well known; of the daughters, Mary married an alchemist and adept,

Frederick Clodius, and Nan married John Roth (or Roder) of Utrecht. (Pepys was at the wedding.) Hartlib tried to establish a school at Chichester, Sussex, in 1630, but his efforts were unsuccessful and thereafter he resided in or near London.

In official records Hartlib is described as a merchant, but there is no evidence that he acquired his income from trade. From 1645 to 1659 he received various grants from Parliament for his public services; he was given money by various private benefactors, especially for his services to education; and he may have made money from the many books he "published" (that is, edited), although he never profited from the inventions he promoted.

Initially, Hartlib devoted his efforts to the Protestant cause, especially to assisting Protestant refugees from Germany (in the midst of the Thirty Years' War), and to educational reform. In this and much else he shared the views and assisted in the work of John Dury, an ardent advocate of Protestant unity. Hartlib warmly promoted the ideas of the Czech educational reformer J. A. Comenius: he published many of Comenius' works in England, helped make possible his visit to England in 1641, and constantly advocated his views on universal education, language, government, and peace. In 1641 Hartlib published *A Description of the Famous Kingdome of Macaria,* a Utopian vision of a state in which enlightened government and true religion were supported by enlightened promotion of trade, medicine, agriculture, and the mechanic arts. Hartlib's interest in education brought him in touch with John Milton and with the mathematician (and Parliamentarian diplomat) John Pell. The success of the Parliamentary side in the Civil Wars led Hartlib and his friends to hope that something like Macaria might be brought into existence in England—what the young Robert Boyle was to call the Invisible College.

A concrete step was an attempt to establish the Office of Public Address, a scheme devised by Hartlib and Dury and intended as a public and organized version of Hartlib's many activities. Like Hartlib himself, the Office was to be in part charitable: to put the poor—especially the intellectual and religious poor—in touch with possible benefactors and to act as a labor exchange. Again like Hartlib, the Office was to act as a commercial agent, not only to purchase books and all kinds of property but also to serve as a channel of communication between English and foreign merchants. It was to maintain the sort of correspondence which Hartlib had already established with divines, educators, and scientists, and was to act as a clearinghouse for news of public affairs, new philosophical and educational ideas, inventions, experiments, and schemes. Finally and above all, it was to promote (as Hartlib longed to do) religion, education, and inventions, in a Baconian and Comenian spirit. Although Parliament ignored the scheme, so that Hartlib never had his coveted post of superintendent general, he continued to do the work of the Office of Public Address as best he could, corresponding with Pell in Switzerland, Dury in Holland and Germany, Johannes Hevelius in Danzig, John Winthrop and George Starkey in New England, and dozens more.

In addition, Hartlib tried to promote useful inventions, especially those relating to agriculture, medicine and its ancillary chemistry, and mechanics in general. His interests were so wide as to be inchoate: John Evelyn, who described him as "Master of innumerable curiosities, & very communicative" (*Diary,* 27 November 1655), learned of German stoves and how to use their heat to perfume the air, and of copying inks and devices. An even wider range of interests is revealed in the letters Hartlib exchanged with Henry Oldenburg in 1658 and 1659, the topics of which include medical and chemical receipts, perpetual motion machines, clocks, lanterns, agricultural machinery, and much else.

Hartlib had many young protégés: William Petty; the very young Robert Boyle, whom he encouraged in useful science; Petty's future rival, Benjamin Worsley (a chemist); the naturalists Arnold and Gerald Boate; and the inventors Cressy Dymock and Gabriel Plattes. These were the cornerstones of Hartlib's Invisible College—"that values no knowledge but as it hath a tendency to use," as Boyle described it. Hartlib was an endless collector of inventions, endlessly hopeful of great things that might spring from the fertile minds of the younger generation, especially hopeful that they might make life better and easier for all mankind. He only dimly comprehended Bacon's message of the possible utility of science; certainly he never understood that scientific knowledge was needed for a critical understanding of inventions or of the potentialities of science. Most of the information he collected and the discoveries he published were very minor. Yet Hartlib was a not unimportant source of communication among scientists in the decades before the formation of scientific societies, and he was a useful postal and book-buying agent as well.

BIBLIOGRAPHY

I. Original Works. Hartlib published some sixty-five works; these are listed in G. H. Turnbull, *Hartlib, Dury and Comenius. Gleanings From Hartlib's Papers* (Liverpool, 1947), pp. 88–109. Most of these were edited, collected, or translated by Hartlib; for some he wrote prefaces; others are dedicated to him.

Of the books Hartlib wrote, the most important are *A Description of the Famous Kingdome of Macaria* (London, 1641); *A Faithful and Seasonable Advice, or, the Necessity of a Correspondencie for the Advancement of the Protestant Cause* (London, 1643); *Considerations Tending to the Happy Accomplishment of Englands Reformation in Church and State* (London, 1647), the first suggestion for his Office of Public Address; and *A Further Discoverie of the Office of Publick Addresse for Accomodations* (London, 1648), which may have been written by Dury.

His best-known collection of tracts on agriculture is *Samuel Hartlib His Legacie* (London, 1651; 2nd ed., 1652; 3rd ed., 1655); it is an enlargement of a work first printed in 1650 and contains tracts written by, among others, Cressy Dymock, a minor inventor, and Robert Child, a chemical follower of J. B. van Helmont. Robert Boyle's first published paper appeared in *Chymical, Medicinal and Chyrurgical Addresses Made to Samuel Hartlib Esq* (London, 1655).

In spite of having been sorted out in the seventeenth century, when valuable material was removed, many of Hartlib's papers have survived; they are now on deposit in the library of Sheffield University through the courtesy of their owner, Lord Delamere.

A few of his letters were extracted by the recipient and are printed in A. R. and M. B. Hall, *The Correspondence of Henry Oldenburg*, I (Madison-Milwaukee, Wis., 1965).

II. SECONDARY LITERATURE. The earliest account of Hartlib's life is H. Dircks, *Biographical Memoir of Samuel Hartlib* (London, 1865); this was much corrected by F. Althaus, *Samuel Hartlib: Ein deutsch-englisches Charakterbild* (Leipzig, 1884). G. H. Turnbull, *Samuel Hartlib. A Sketch of His Life and His Relation to J. A. Comenius* (Oxford, 1920), is a useful short account; a reading of this or of Althaus is presupposed by Turnbull's *Hartlib, Dury and Comenius* (see above), which corrects details from Hartlib's papers but contains no connected biography. The best account of Hartlib's educational, promotional, and scientific activities is in R. H. Syfret, "The Origins of the Royal Society," in *Notes and Records. Royal Society of London*, **5** (1947–1948), 75–137; Miss Syfret was the first to fully identify the Invisible College with Hartlib's activities. G. H. Turnbull, "Samuel Hartlib's Influence on the Early History of the Royal Society," in *Notes and Records. Royal Society of London*, **10** (1952–1953), 101–130, contains useful references to Hartlib's MS diary.

MARIE BOAS HALL

HARTMANN, CARL FRIEDRICH ALEXANDER (*b.* Zorge, Harz, Germany, 8 January 1796; *d.* Leipzig, Germany, 3 August 1863), *mineralogy, mining, metallurgy.*

Hartmann was the son of a chief clerk at a foundry in the Oberharz, one of the most important ore mining districts in Germany. His father's activities awakened his interest in mining and metallurgy at an early age. He received his first instruction from the pastor at Zorge and at the age of ten went to Blankenburg to continue his education; there, under the supervision of his uncle, the abbot of Ziegenbein, he completed the Gymnasium course. He then attended the mining school in Clausthal, in order to learn the mining sciences, especially mineralogy.

Hartmann's studies were interrupted by his participation in the "War of Liberation" of the German states and their allies against Napoleon in 1813–1815. Discharged from the army in 1816, Hartmann became an assistant at the foundry in Zorge. Soon afterward he made several journeys in order to become familiar with other foundries, especially in Silesia. After this period of travel he went in 1818 to Berlin to continue his studies. He so excelled in mineralogy that he became an assistant to the famous professor of mineralogy Christian Weiss.

In 1821, while still a student in Berlin, Hartmann entered into a marriage which was beyond his financial means and forced him to accept a position as a bookkeeper in Rübeland, Harz, in order to provide for his family. Several years later he moved to Blankenburg, where, because of his conscientiousness and discretion, he was often entrusted with the management of important affairs. Hartmann devoted his free time to private study, concentrating on mineralogy and geology. As a result of these vigorous efforts in 1823 he was named an honorary member of the Königlich Preussische Akademie Gemeinnütziger Wissenschaften of Erfurt. Two years later the duchy of Brunswick granted him an extended leave to make scientific journeys. On one of these he went to Italy. After this educational trip Hartmann received a doctorate in jurisprudence in 1826 from the University of Heidelberg. In the same year he was named an honorary member of the Societät für die Gesamte Mineralogie of Jena. In 1827 he was made an honorary member of the Natural History Society of Edinburgh.

During this period Hartmann's first major work on mining science was published. In 1829 he was appointed commissioner of mines in Brunswick. At the same time he became an honorary member of the Erfurt Gewerbe-Verein and a member of the Prussian Verein zur Beförderung des Gewerbefleisses. A year later the Apotheker-Verein des Nördlichen Deutschland offered him an honorary membership, and in 1833 the Society for Natural Curiosities in Moscow bestowed an honorary membership upon him. In 1834 Hartmann traveled on a commission from the government of Brunswick to England and France, where he received many valuable suggestions in mining, metallurgy, mineralogy, and geology. Hartmann was also given plenipotentiary powers as Brunswick's adviser at the tariff conference in Berlin, and later he represented Brunswick at the Berlin international exhibition of 1844.

Perhaps because of such honors and his considerable technical education, furthered by an extensive knowledge of English and French language and literature, Hartmann experienced envy and ill will, and even insults from some of his superiors. Consequently, in 1841 he resigned as commissioner of mines of Brunswick and moved to Berlin, in order to work there as a technical writer on the mining sciences. In mining and metallurgy Hartmann was already a known and respected author, although he wrote and translated only to the extent that his free time allowed. Hardly settled in Berlin, in December 1841 he founded the *Berg- und hüttenmännische Zeitung, mit besonderer Berücksichtigung der Mineralogie und Geologie,* which he edited until 1858. In 1859 he began publication of the *Allgemeine berg- und hüttenmännische Zeitung,* of which he was editor-in-chief until his death.

In 1844 Hartmann became an honorary member of the Society for Mineralogy in St. Petersburg and, in 1847, member of the Gewerbeverein of Weimar. He moved to Weimar in 1845, after he had divorced his first wife and remarried. In 1854 he left Weimar and went to Leipzig. In both cities his literary production was considerable.

Hartmann did no independent research and was not noted for the originality of his writings. Nevertheless, his contribution to the literature of mining and metallurgy was great: by setting down on paper anything that could further technical knowledge he contributed to the dissemination of the latest information.

BIBLIOGRAPHY

Of Hartmann's more than 100 works, the following are his most important publications: *Handwörterbuch der Mineralogie, Berg-, Hütten- und Salzwerkskunde. Nebst der französischen Synonymie und einem französischen Register,* 2 vols. (Ilmenau, 1825), 2nd ed. under the title *Handwörterbuch der Berg-, Hütten- und Salzwerkskunde, der Mineralogie und Geognosie,* 3 vols. (Weimar, 1859–1860); *Handwörterbuch der Mineralogie und Geognosie* (Leipzig, 1828); *Lehrbuch der Eisenhüttenkunde,* 2 vols. (1833), with 2 atlases; *Lehrbuch der Mineralogie und Geologie,* 2 pts. (Nuremberg, 1835–1836); *Handbuch der praktischen Metallurgie. Nebst einem Anhange über die Anfertigung von Eisenbahnschienen,* 2 vols. (Weimar, 1837; 3rd ed., 1863); *Über den Betrieb der Hohöfen (Hochöfen), Cupolöfen . . . mit erhitzter Gebläseluft,* 6 vols. (Quedlinburg–Leipzig, 1834–1841); *Encyclopädisches Wörterbuch der Technologie, der technischen Chemie, Physik und des Maschinenwesens,* 4 vols. (Augsburg, 1838–1841); *Taschenbuch für reisende Mineralogen, Geologen, Berg- und Hüttenleute durch die Hauptgebirge Deutschlands und der Schweiz* (Weimar, 1838, 1848), with atlas and suppl.; *Grundriss der Eisenhüttenkunde* (Berlin, 1843; 2nd ed., 1852); *Handbuch der Mineralogie* (Weimar, 1843); *Handbuch der praktischen Metallurgie* (Weimar, 1847); *Geographisch-statistische Beschreibung von Californien. Nach den besten Quellen bearbeitet,* 2 vols. (Weimar, 1849); *Die neuesten Entdeckungen und Forschungs-Resultate auf dem Gebiete der gesamten Mineralogie seit dem Jahre 1843* (Weimar–Hamm, 1850); *Die neuesten Fortschritte des Steinkohlen-Bergbaues* (Quedlinburg–Leipzig, 1850); and *Die Fortschritte der Eisenhüttenkunde* (Berlin, 1851), with atlas.

See also *Vollständiges Handbuch der Eisengiesserei,* 2 vols. (Freiberg, 1847, 1853); *Vademecum für den praktischen Eisenhüttenmann* (Leipzig, 1855; 2nd ed., 1858; 3rd ed., Hamm, 1863); *Vademecum für den praktischen Bergmann* (Leipzig, 1856); *Praktisches Handbuch der Roh- und Stabeisen-Fabrikation,* 3 vols. (Leipzig, 1853–1857), with atlas; *Die Aufbereitung und Verkokung der Steinkohlen, sowie die Verkokung der Braunkohlen und des Torfes* (Weimar, 1858); *Handbuch der Bergbau- und Hüttenkunde, oder die Aufsuchung, Gewinnung und Zugutemachung der Erze, der Stein- und Braunkohlen und anderer nutzbarer Mineralien* (Weimar, 1858), with atlas; *Vollständiges Handbuch der Metallgiesserei* (Weimar, 1858); *Handwörterbuch der Berg-, Hütten- und Salzwerkskunde* (Weimar, 1860); *Berg- und hüttenmännischer Atlas, oder Abbildungen und Beschreibungen vorzüglicher Bergwerks- und Hütten-Maschinen und Apparate* (Weimar, 1860), with atlas; *Die Aufbereitung und Verkokung der Steinkohlen* (Weimar, 1861); and *Die Fortschritte des Eisenhüttengewerbes in der neueren Zeit, oder der heutige Standpunkt der Roheisen-, Stabeisen- und Stahl-Fabrikation,* 6 vols. (Leipzig, 1858–1863).

From 1842 to 1858 Hartmann edited *Berg- und hüttenmännische Zeitung, mit besonderer Berücksichtigung der Mineralogie und Geologie,* while from 1859 until his death he was editor of *Allgemeine Berg- und hüttenmännische Zeitung: Mit besonderer Berücksichtigung der Mineralogie und Geologie.* He also translated, revised, and edited about forty French and English books on mining, mineralogy, and metallurgy.

A secondary source is "Dr. Carl F. A. Hartmann," in *Der Berggeist. Zeitung für Berg-, Hüttenwesen und Industrie,* **8,** no. 103 (1863), 427–428.

M. KOCH

HARTMANN, GEORG (*b.* Eggolsheim, near Forchheim, Germany, 9 February 1489; *d.* Nuremberg, Germany, 9 April 1564), *instrument making, mathematics.*

Hartmann studied mathematics with Heinrich

Glareanus and theology at Cologne in 1510. In Italy during the summer of 1518 he became friendly with Copernicus' brother Andreas, began designing sundials, and discovered the magnetic dip. The (inaccurate) declination of six degrees which he found for Rome, probably the earliest determination on land, was revealed in a letter to Duke Albert of Prussia dated 4 March 1544, but it remained unpublished until 1831. (Robert Norman published his independent discovery in *The Newe Attractive* [1580].) Settling at Nuremberg in 1518, Hartmann designed and produced timepieces, astrolabes, globes, quadrants, armillary spheres, a star altimeter, and the caliber gauge, which he invented in 1540 to determine the weights of cannonballs from the muzzle sizes of cannons.

Hartmann was vicar of St. Sebaldus from 1518 to 1544 and in 1527 became chaplain of St. Moritz. He was friendly with Willibald Pirkheimer and Albrecht Dürer, about whose death he later reported (see E. Zinner, *Astronomische,* p. 357). From Regiomontanus' literary estate Hartmann treasured a fragment of a letter with important information; he was familiar with Regiomontanus' handwriting, his physical appearance from a portrait, and several astrolabes. In 1526, at Hartmann's request, Johann Schöner published Regiomontanus' manuscript on Ptolemy's optics, *Problemata XXIX. Saphaeae* ("Twenty-nine Problems With the Saphea"). Following Werner's death in 1528 and the dispersal of his manuscripts, Hartmann rescued two on spherical triangles and "De meteoroscopiis," which he gave to Joachim Rheticus in 1542, making a more accurate copy for himself; in 1544 Rheticus published the *De revolutionibus* chapter on triangles as *Nic. Copernici De lateribus et angulis triangulorum tum planerum rectilineorum tum sphaericorum libellus* and dedicated it to Hartmann. Hartmann published *Joh. Pisani Perspectiva communis* in 1542 and an astrological work, *Directorium,* in 1554. His unpublished "Fabrica horologium" (1527) included figures from Ptolemy's *Organum* and influenced Sebastian Münster's *Compositio horologiorum* of 1531.

BIBLIOGRAPHY

I. Original Works. The letter from Hartmann to Albert of Prussia (4 Mar. 1544), containing the report of his discovery of magnetic inclination and of the first determination of the declination on land, is repr. with a facs. in G. Hellmann, ed., *Rara magnetica 1269–1599,* Neudrucke von Schriften und Karten über Meteorologie und Erdmagnetismus, no. 10 (Berlin, 1898). The original remained unnoticed in the Royal Archives at Königsberg until published by J. Voigt, in Raumer's *Historisches Taschenbuch,* II (Leipzig, 1831), 253–366, then by H. W. Dove in *Reportorium der Physik,* II (Berlin, 1838), 129–132, and again by J. Voigt with twelve other Hartmann letters and four by Albert in *Briefwechsel der berühmsten Gelehrten des Zeitalter der Reformation mit Herzog Albrecht von Preussen 1541–1544* (Königsberg, 1841).

This correspondence provides important insight into the relationship between sovereign and scientist; Hartmann not only discusses the instruments he is making for Archduke Albert but also reports on his visits with and commissions from King Ferdinand of Bohemia and Hungary, and the apostolic envoy and Venetian *orarier,* and his correspondence with Duke Ottheinrich, who in August 1544 sent Hartmann a 1417 boxwood sundial and a commission for two ivory sundials, a brass astrolabe, and a brass armillary sphere. Details of this correspondence are described in Ernst Zinner, *Deutsche und niederländische astronomische Instrumente des 11.-18. Jahrhunderts* (Munich, 1956), p. 358. Zinner's extensive listing, pp. 362–368, of Hartmann's scientific instruments and engravings of 1523–1563 seems limited to European museums and libraries; it omits the nine astronomical charts, unaccompanied by text, in the Weaver Collection of the American Institute of Electrical Engineers, New York City.

Although Nuremberg was a printing center and Hartmann's correspondence reveals that for his instruments he engraved copperplates and printed them himself on his own presses, he put only two works into print: *Joh. Pisani Perspectiva communis* (Nuremberg, 1542), which he edited with extensive corrections and restorations (although unlisted by the Library of Congress, Columbia University and the University of Michigan each own a copy), and his astrological work *Directorium* (Nuremberg, 1554). He reproduced his writings of 1518–1528 in pre-Gutenberg style—Zinner, pp. 358–360, describes in detail "Die Wiener Handschrift Vin 12768," containing copies, completed 14 June 1526 and 19 July 1527, which Hartmann presented to Chaplain Geuder and Ulrich Stocker, and illustrations for sun-clocks for the city of Nuremberg (1526); the Weimar Landesbibliothek no. F. max. 29, a *Prachthandschrift* of 1525[?]–1527 devoted primarily to the design of sundials but including a few figures of astrolabes; Weimar Landesbibliothek no. F. 324, a copy by Hartmann with very careful figures, probably by Hartmann, of Werner's works on the spherical triangles and on the meteoroscope, the original of which Hartmann gave Rheticus for publication.

Other items of interest are an astronomical broadside, an engraving of a sundial, dated 1535, in the Houghton Library of Harvard University; a brass skaphe signed "Georgius Hartmann Noremberge Faciebat 1539," listed in the Wray sale at Sotheby's, Nov. 1959, and purchased by Dr. Weil, a dealer; and a gilt-brass dial of Ahaz, dated 1548, in a private collection in America; see Zinner, pp. 357–368. The Adler Planetarium and Astronomical Museum in Chicago has two instruments not mentioned in Zinner: (1) a gilt-brass astrolabe on which appears the inscription "GEORGIUS HARTMANN NOREMBERG FACIEBAT ANNO MDXL," the rete being a later replacement, and (2) a gilt-brass, silver, and ivory astronomical compendium of astrolabe and sundial in a box of finest goldsmith work inscribed "HARTMANN NURNBERG 1558," believed to be a gift from Emperor Charles V to Duke Emmanuel Philibert of Savoy. The catalog numbers are M-22 and A-7, respectively. Some later instruments are signed "H. G."

II. SECONDARY LITERATURE. Zinner's *Astronomische Instrumente* (cited above) is the major work. J. G. Doppelmayr, *Historische Nachricht von den Nürnbergischen Mathematicis und Kunstlern* (Nuremberg, 1730), pp. 56–58; Karl Heger, "Georg Hartmann von Eggolsheim," in *Der frankische Schatzgräber,* **2** (1924), 25–29; and K. Kupfer, "Nachtrag zu Georg Hartmann," *ibid.,* **7** (1929), 37–38, were all used by Zinner. Hellmann discusses Hartmann's discovery of the magnetic dip in the introduction to *Rara magnetica* (cited above), pp. 15–16, and attributes Hartmann's knowledge of other magnetic properties to the "Epistola Petri Peregrini de magnete" of 1269, which was probably the "alte Pergamentbuch" (parchment manuscript) Hartmann obtained in 1525, during the Peasants' War. For details of the Regiomontanus letter fragment owned by Hartmann, see Ernst Zinner, *Leben und Wirken des Johannes Müller von Königsberg* (Munich, 1938), pp. 195, 202–203.

English references to Hartmann are found scattered in Lynn Thorndike, *History of Magic and Experimental Science* (New York, 1941), V, 337, 353, 355, 364–365, 414, and VI, 60; R. J. Forbes, *Man the Maker* (New York, 1950), p. 123; and Abraham Wolf, *A History of Science, Technology and Philosophy in the 16th and 17th Centuries* (New York, 1959), I, 292. The geographer Baron N.A.E. Nordenskjold, in his *Facsimile-Atlas to the Early History of Cartography With Reproductions of the Most Important Maps Printed in the XV and XVI Centuries,* trans. from the Swedish by Johan Adolf Ekelof and Clements R. Markham (Stockholm, 1889), reasons that Hartmann, "a celebrated manufacturer of globes and cosmographical instruments," rather than Schöner probably made the unsigned terrestrial globe portrayed in Hans Holbein's *Ambassadors;* Mary F. S. Hervey, *Holbein's "Ambassadors"* (London, 1900), pp. 210–218, upholds the opposite position. The original woodcut terrestrial globe gores, twelve to a plate, reproduced by Nordenskjold, are in the New York Public Library, catalogued as "[Globe gores with Magellan's route Nuremberg? 153–?] Possibly the work of Georg Hartmann of Nuremberg."

LUCILLE B. RITVO

HARTMANN, JOHANNES (*b.* Amberg, Oberpfalz, Germany, 14 January 1568; *d.* Kassel, Germany, 7 December 1631), *iatrochemistry, medicine, mathematics.*

Hartmann, a weaver's son, worked as a bookbinder; scholarship aid enabled him to attend the university. He studied the arts, notably mathematics, at Jena, Wittenberg, and, from 1591, Marburg, from which he received a master's degree. He may also have attended the universities of Altdorf, Helmstedt, and Leipzig. A friend, the Hessian court chronicler Wilhelm Dilich, introduced Hartmann to Landgrave Wilhelm IV of Hesse-Kassel, who was interested in the natural sciences, and to the landgrave's son Moritz, who was interested particularly in alchemical metallurgical processes. In 1592 Hartmann became professor of mathematics at the University of Marburg, which was under the jurisdiction of Wilhelm IV's brother, Landgrave Ludwig. He also studied medicine and received a doctorate in this subject in 1606. In addition, in 1594 he became adviser to Landgrave Moritz in Kassel, where he taught at the court school until 1601.

Thereafter Hartmann combined his interest in mathematics, astronomy, and alchemy with medicine. Starting in 1609 he gave lectures and practical laboratory instruction on materia medica and the chemical and mineralogical preparation of medicines in the "laboratorium chymicum publicum" at Marburg. In the same year he was appointed professor of medical and pharmaceutical chemistry, in effect the first such professorship in Europe. He was several times dean and rector of the University of Marburg and was also very successful in his scientific work. By 1616 ten of his students had earned the doctorate. Following disputes with the university and the landgrave, Hartmann moved in 1621 to Kassel—nominally retaining his professorship—and became court physician, a post he lost as a result of the abdication of Landgrave Moritz in 1627. Until his death in 1631 Hartmann was professor of natural science and medicine at the new University of Kassel, which had offered courses for four years before its official opening in 1633.

Hartmann's importance is in having introduced pharmaceutical and medical chemistry into the university and in having given practical instruction in it. This new field, which had been developed in the works of Paracelsus and his disciples, was then emerging from alchemy. Yet Hartmann did not fall into alchemical speculations; instead, he sought to mediate between the Galenists and the iatrochemists. He left few writings on the practical aspects of the subject, and most of his works appeared posthumously. A glimpse of his activity is given by a laboratory journal for the year 1615.

As a physician Hartmann was not especially successful. His nickname "Theophrastus Cassellanus" derives from his Paracelsian-chemical activity. There is no doubt that his Hermetic philosophical ideas had a considerable influence during 1614–1626, which even his contemporaries called the "Rosicrucian" period; and his views, a union of animistic and vitalistic notions, reached far beyond his native land, carried by friends and students including Oswald Crollius, Johann Daniel Mylius, and Johannes Rhenanus. In addition, he corresponded with English and Polish iatrochemists and with alchemists in Prague. The many editions of his principal work, *Praxis chymiatrica,* testify to the respect that contemporaries accorded to this textbook of pharmaceutical chemistry.

BIBLIOGRAPHY

I. ORIGINAL WORKS. Hartmann's works were collected as *Opera omnia medico-chymica,* Conrad Johrenius, ed. (Frankfurt, 1684; 1690), also translated into German (1698). His individual works include *Disputationes elementorum geometricorum* (Kassel, 1600); 'Επιφυλλιδες *sive miscellae medicae cum* προϑηκη *chymico therapeutica doloris colici* (Marburg, 1606); *Philosophus sive naturae consultus medicus, oratio* (Marburg, 1609); *Disputationes chymicomedicae quatuordecim* (Marburg, 1611; 1614), also translated into English as *Choice Collection of Chymical Experiments* (London, 1682) and into German as *Philosophische Geheimnisse und chymische Experimenta* (Hamburg, 1684); *Praxis chymiatrica* (Leipzig, 1633; Frankfurt, 1634; 1671; Geneva, 1635; 1639; 1647; 1649; 1659; 1682; Leiden, 1663; Nuremberg, 1677), also translated into German as *Chymische Arzneiübung* (Nuremberg, 1678); and *Tractatus physico-medicus de opio* (Wittenberg, 1635; 1658). In addition, Hartmann prepared an edition, which was finished by his son, G. E. Hartmann, of Oswald Crollius' *Basilica chymica* (Geneva, 1635) and works of Joseph Duchesne (Quercetanus). Under the pseudonym Christopher Glückradt he commented on the *Tyrocinium chymicum* of J. Beguin (Wittenberg, 1634, 1666).

II. SECONDARY LITERATURE. The following, listed in chronological order, may be consulted: Andreas Libavius, *Examen philosophiae novae* (Frankfurt, 1615), which discusses Hartmann's ideas on vital and Hermetic philosophy; and *Appendix necessaria syntagmatis* . . . (Frankfurt, 1615), with the ch. "Censura philosophiae vitalis Joannis Hartmanni"; Friedrich W. Strieder, *Grundlage zu einer hessischen Gelehrten- und Schriftstellergeschichte,* V (Kassel, 1785), 281-289; John Ferguson, *Bibliotheca chemica,* I (Glasgow, 1906; repr. London, 1954), 365, 366; Wilhelm Ganzenmüller, "Das chemische Laboratorium der Universität Marburg im Jahre 1615," in *Angewandte Chemie,* **54** (1941), also in Ganzenmüller's *Beiträge zur Geschichte der Technologie und der Alchemie* (Weinheim, 1956), pp. 314-322; Lynn Thorndike, *A History of Magic and Experimental Science,* VIII (New York-London, 1958), 116-118; Rudolf Schmitz, "Die Universität Kassel und ihre Beziehung zu Pharmazie und Chemie," in *Pharmazeutische Zeitung,* **104** (1959), 1413-1417; J. R. Partington, *A History of Chemistry,* II (London-New York, 1961), 177-178; Rudolf Schmitz, "Naturwissenschaft an der Universität Marburg," in *Sitzungsberichte der Gesellschaft zur Beförderung der gesamten Naturwissenschaften zu Marburg,* **83-84** (1961-1962), 12-21; and Rudolf Schmitz and Adolf Winkelmann, "Johannes Hartmann (1568-1631) Doctor, Medicus et Chymiatriae Professor Publicus," in *Pharmazeutische Zeitung,* **111** (1966), 1233-1241.

R. SCHMITZ

HARTMANN, JOHANNES FRANZ (*b.* Erfurt, Germany, 11 January 1865; *d.* Göttingen, Germany, 13 September 1936), *astronomy.*

Hartmann was the son of Daniel Hartmann, a merchant, and Maria Sophia Hucke. He attended primary and secondary school in Erfurt, at which time his interest in astronomy began to develop. He continued his studies at Tübingen, Berlin, and Leipzig, where H. Bruns taught him the mathematics that were basic to his future work. Hartmann took the Ph.D. at Leipzig in 1891, then remained there to participate in making observations for the star catalog of the Astronomische Gesellschaft. He worked at the Leipzig observatory until 1896, interrupted by a period of some months that he spent at the observatory in Vienna.

In 1896 Hartmann went to the astrophysical observatory at Potsdam. He stayed there until 1909, receiving the title of observator in 1898 and of professor in 1902. At Potsdam, Hartmann was chiefly active in the fields of instrumentation and spectrography. The Potsdam observatory possessed a thirty-two-inch refractor, but the preparation of its objective for use in the new technique of photographic observation had not been successful. Hartmann was assigned to investigate the problem; he derived a new interpolative dispersion formula and with it developed a new method for testing large objectives. This formula was later important in his spectroscopic investigations.

Hartmann's other important work at Potsdam included the construction of a new spectrograph that employed a quartz prism and the investigation of the ultraviolet frequencies of previously unstudied stellar spectra. He also devised a spectrocomparator to expedite the evaluation of stellar spectra, as well as two photometric instruments, a microphotometer and a plane, or universal, photometer. His most significant work at this time, however, was the discovery of stationary calcium lines in the spectrum of δ Orionis (1904) through observing its radial velocity by the Doppler shift of its spectral lines. He thus proved the existence of interstellar matter for the first time.

Hartmann left Potsdam in 1909 to accept a position as director of the observatory and professor in ordinary at the University of Göttingen. The equipment of the observatory was obsolete and teaching was new to Hartmann; he therefore changed the emphasis of his work to concentrate on lecturing, writing, and the history of astronomy. The government of Argentina offered him the directorship of the La Plata observatory in 1911; he refused that offer, but accepted when it was repeated in 1921.

The observatory at La Plata was better equipped than the one at Göttingen, and Hartmann further improved its instrumentation. Most important, he reconstructed the thirty-two-inch reflector for the spectroscopic work which was urgently needed in southern skies. He also gave lectures in Spanish and made geophysical observations, primarily in the area of seismics. He further made a new determination of the solar parallax from the observations of the opposition of the asteroid Eros in 1931–1932.

In 1935 Hartmann returned to Göttingen, where he planned to spend his retirement evaluating scientific data. His health failed, however, and he died after a long illness. He was married to Maria Scherr; they had two sons and one daughter. He had been an active sportsman, and his interest in music continued until the end of his life.

BIBLIOGRAPHY

I. Original Works. Hartmann's works include "Apparat und Methode zur photographischen Messung von Flächenhelligkeiten," in *Zeitschrift für Instrumentenkunde,* **19** (1899), 97–103; "Interpolationsformeln für das prismatische Spektrum," in *Publikationen des Astrophysikalischen Observatoriums zu Potsdam,* **12** (1902); "Das 80-cm. Objektiv des Potsdamer Refraktors," *ibid.,* **15** (1904); "Messungen der Linienverschiebungen in Spektrogrammen," *ibid.,* **18** (1906); "Spektrokomparator," in *Zeitschrift für Instrumentenkunde,* **26** (1906), 205–217; "Tabellen für das Rowlandsche und für das internationale Wellenlängenspektrum," in *Abhandlungen der Königlichen Gesellschaft der Wissenschaften zu Göttingen,* Mathematisch-Physikalische Klasse, **10,** no. 2 (1916), 1–78; "Die astronomischen Instrumente des Kardinals Nicolaus Cusanus," *ibid.,* no. 6 (1919), 1–56, with 12 tables. Hartmann also edited and wrote some parts of *Kultur der Gegenwart,* vol. III of *Band Astronomie* (Leipzig, 1921).

II. Secondary Literature. See P. Labitzke, "Johannes Hartmann," in *Vierteljahrsschrift der Astronomischen Gesellschaft,* **72** (1937), 3–23. There is also an obituary of Hartmann in *Monthly Notices of the Royal Astronomical Society,* **97** (1937), 284–285.

Hans Christian Freiesleben

HARTREE, DOUGLAS RAYNER (*b.* Cambridge, England, 27 March 1897; *d.* Cambridge, 12 February 1958), *applied mathematics, theoretical physics.*

Hartree's chief contribution to science was his development of powerful methods of numerical mathematical analysis, which made it possible for him to apply successfully the so-called self-consistent field method to the calculation of atomic wave functions of polyelectronic atoms, that is, those which in the neutral condition have more than one electron surrounding the nucleus. These calculations involved the numerical solution of the partial differential equations of quantum mechanics for many-body systems subject to the usual boundary conditions. From the atomic wave functions it is possible to calculate the average distribution of negative electric charge as a function of distance from the nucleus. If the distribution has been correctly found for all the electrons in the atom under study, the electric field due to this distribution should lead to the original distribution, in which case the field is called self-consistent.

Hartree developed ingenious approximation methods for the rather rapid evaluation of such self-consistent fields. The corresponding wave functions and associated charge distributions are of great importance in the theoretical calculation of macroscopic properties of matter in various states of aggregation. Hartree and his collaborators evaluated wave functions for more than twenty-five different atomic species in various states of ionization. Through his stimulus and encouragement many more atoms were investigated by physicists throughout the world. Hartree's equations as generalized by V. Fock proved to be extremely valuable in theoretical calculations in solid state physics.

Hartree also applied his methods of numerical analysis to problems in ballistics, atmospheric physics, and hydrodynamics. Much of this work was of importance to Britain's war effort from 1939 to 1945. He further turned his attention to industrial control and made valuable contributions to the control of chemical engineering processes. He early became interested in machine calculation and built the first differential analyzer in Britain for the graphical solution of differential equations. He later pioneered in the introduction of digital computers and their use in the United Kingdom. He made numerous visits to the United States and shared his knowledge freely with American colleagues in the same field.

Hartree was the great-grandson on his father's side of the famous Samuel Smiles, whose book *Self Help* is an English classic. His mother, Eva Rayner, was the sister of E. H. Rayner, for many years superintendent of the Electricity Division of the National Physical Laboratory in Teddington. She was active in public service and was for a time mayor of Cambridge. Hartree's father taught engineering at the University of Cambridge and in his later years frequently collaborated with his son in his atomic calculations.

Hartree received his higher education at Cambridge University, where he was a student and later fellow of St. John's College. He did postgraduate work under R. H. Fowler and received the Ph.D. in 1926. He was elected a fellow of the Royal Society in 1932. From 1929 until 1937 he held the chair of applied mathematics at the University of Manchester. In the latter year he was appointed professor of theoretical physics there and held this post formally until 1946, although the latter years of his tenure were spent mainly on war research for the Ministry of Supply. From 1946 until his death Hartree was Plummer professor of mathematical physics at Cambridge, where he was active in promoting and operating the computing laboratory, although he continued his interest in atomic wave functions to the very end.

Hartree was universally admired for the clarity of his lectures and writings. An outstanding trait was his unselfish generosity in giving aid to others working along similar lines of research throughout the world.

Among Hartree's avocational interests was music, in which he was a proficient performer on the piano and the drums and a competent orchestra conductor. He also had a passion for railways, extending to a professional interest in signaling and traffic control. It was a distinct pleasure to take a railway journey with him since he had such an exact knowledge of the whole industry and knew how to discuss it so entertainingly.

BIBLIOGRAPHY

I. ORIGINAL WORKS. Hartree's complete bibliography includes five books and 123 articles. A complete list is given by Darwin, below. His books are *Text-book of Anti-aircraft Gunnery* (London, 1925), part only; *The Mechanics of the Atom,* his trans. and rev. of Born's *Atommechanik* (London, 1927); *Calculating Instruments and Machines* (Urbana, Ill., 1949; Cambridge, 1950); *Numerical Analysis* (Oxford, 1952; 2nd ed., 1958); and *The Calculation of Atomic Structures* (New York, 1957).

II. SECONDARY LITERATURE. On Hartree's life and work see Charles Galton Darwin in *Biographical Memoirs of Fellows of the Royal Society,* **4** (1958), 103–116; and T. S. Kuhn, J. L. Heilbron, P. Forman, and L. Allen, eds., *Sources for the History of Quantum Physics* (Philadelphia, 1960), p. 45.

R. B. LINDSAY

HARTSOEKER (or **HARTSOECKER**), **NICOLAAS** (*b.* Gouda, Netherlands, 26 March 1656; *d.* Utrecht, Netherlands, 10 December 1725), *physics, technology.*

Hartsoeker was the son of Christiaan Hartsoeker, an evangelical minister, and Anna van der Mey. Although his father wished him to study theology, Hartsoeker preferred science; he secretly learned mathematics and lens grinding. Most sources suggest that he may have studied anatomy and philosophy at the University of Leiden in 1674; a letter from Constantijn Huygens to his brother Christiaan, however, refers to him as having had no higher education, so it is possible that he was largely self-educated in his chosen fields. It is known that by 1672 he had visited Leeuwenhoek and that in 1678 he accompanied Christiaan Huygens to Paris, where he met some of the French scientists and worked for a time at the Paris observatory. In his correspondence with Christiaan Huygens from about this period, Hartsoeker claimed to have invented the technique of making small globules of glass for use as lenses for microscopes, but it is more probable that priority in this belongs to Johann Hudde.

In 1679 Hartsoeker returned to Holland, where he settled in Rotterdam and married Elisabeth Vettekeuken. He established himself as an instrument maker and wine merchant, but went bankrupt after a few years and returned to France. From 1684 until 1696 he lived in Passy, near Paris; here, with the assistance of his wife, he made lenses, microscopes, and telescopes, including some for the Paris observatory. He continued to study physics, and in 1694 published *Essai de dioptrique.*

In 1696 Hartsoeker was again in Holland, first in Rotterdam and then, the next year, in Amsterdam, where he gave instruction in physics to Peter the Great upon the visit of the Grand Embassy. He refused the czar's offer of a professorship of mathematics at St. Petersburg, however.

The town council of Amsterdam had erected a small observatory for Peter's use, and after his departure Hartsoeker was allowed to work there. It was there that Hartsoeker was visited by the count of Hesse-Kassel, to whom also he taught physics (Hartsoeker's books *Conjectures physiques* and *Suite des conjectures physiques* contain these lessons). The count of Hesse-Kassel then used his influence to secure for Hartsoeker, in 1704, a professorship of mathematics and philosophy at the University of Dusseldorf. Hartsoeker was also accorded the title Hofmathematicus des Kurfürsten von der Pfalz und Honorar-Professor von Heidelberg. He remained in Dusseldorf until 1716, and then returned to Utrecht.

Hartsoeker's career was further marked by his controversies with other scientists; as early as 1712 he had engaged in a dispute concerning the work of Leibniz, while as late as his years in Utrecht he debated the conclusions of Newton and Jakob I Bernoulli. His criticisms of Leeuwenhoek (contained in the posthumously published *Cours de physique* of 1730) are in large part ill-founded.

Of Hartsoeker's lenses, two known to be by his hand are preserved, one signed "Nicolaas Hartsoeker, pro Academia Ludg. Batav: Parisiorum 1688" in the museum of natural history in Leiden, and the other in the museum of the University of Utrecht. It is known, however, that he had made three telescopes for the Utrecht observatory at the time of Pieter van Musschenbroek's arrival in 1723.

In addition to his instrument work, Hartsoeker did research in embryology. In 1674 he recognized small "particles" in the sperm, which he at first thought to be signs of disease. Three years later he again saw these particles and showed them to Christiaan Huygens. As a result of his investigations, Hartsoeker believed that the fetus was preformed in the spermatozoon and published illustrations of the homunculus crouched there.

Hartsoeker was elected a foreign member of the Académie des Sciences in 1699 and was later also a member of the Berlin Royal Society. His work may be said to have been more honored in France than in his native Holland.

BIBLIOGRAPHY

I. ORIGINAL WORKS. Hartsoeker's books include *Essai de dioptrique* (Paris, 1694), trans. into Dutch by A. Block as *Proeve der Deursicht-Kunde* (Amsterdam, 1699); *Principes de physiques* (Paris, 1696), trans. into Dutch by A. Block as *Beginselen der Natuurkunde* (Amsterdam, 1700); *Conjectures physiques* (Amsterdam, 1706); *Suite des conjectures physiques* (Amsterdam, 1708); *Éclaircissements sur les conjectures physiques* (Amsterdam, 1710); *Nova methodus utendi maximis objectivis* (Berlin, 1710); *Description de deux niveaux d'une nouvelle invention* (Amsterdam, 1711); *Suite des conjectures physiques et des éclaircissements* . . . (Amsterdam, 1712); *Seconde partie de la suite des conjectures physiques* (Amsterdam, 1712); *Recueil de plusieurs pièces de physique où l'on fait principalement voir l'invalidité du système de Mr. Newton* . . . (Utrecht, 1722); and *Cours de physique accompagné de plusieurs pièces concernant la physique qui ont déjà paru* . . . (The Hague, 1730), which also contains "Extrait critique des lettres de feu M. Leeuwenhoek."

For his work in embryology, see *Proeve der Deursicht-Kunde,* pp. 223–229; and N. Andry, *De la génération des vers dans le corps de l'homme* (Amsterdam, 1701), with two letters by Hartsoeker.

Poggendorff gives a list of Hartsoeker's articles in various journals; many editions of the *Oeuvres complètes* of Christiaan Huygens contain Hartsoeker's correspondence with him.

II. SECONDARY LITERATURE. Christiaan Huygens' collected works, above, include many references to Hartsoeker and his work. See also M. Daumas, *Les instruments scientifiques* (Paris, 1953); M. Rooseboom, *Bijdrage tot de geschiedenis der instrumenmakerskunst in de noordelijke Nederlanden tot omstreeks 1840* (Leiden, 1950); and *Levensbeschrijving van eenige voorname meest Nederlandsche mannen en vrouwen,* II (Haarlem, 1794), 167–186.

J. G. VAN CITTERT-EYMERS

HARTWIG, (CARL) ERNST (ALBRECHT) (*b.* Frankfurt am Main, Germany, 14 January 1851; *d.* Bamberg, Germany, 3 May 1923), *astronomy.*

After graduating from the renowned Melanchthon Gymnasium in Nuremberg, Hartwig studied mathematics, physics, and astronomy at the universities of Erlangen, Leipzig, Göttingen, and Munich. In 1874 he became assistant astronomer at the observatory of the University of Strasbourg, where he obtained the Ph.D. degree in 1880. Soon afterward he was sent officially to study modern observatories in Austria, Russia, Finland, Sweden, and Denmark. In 1882–1883 Hartwig was the leader of the German astronomical expedition for observing the transit of Venus at Bahía Blanca, Argentina.

Hartwig spent the next two years as associate astronomer and lecturer at the University of Dorpat (now Tartu, Estonia). In 1886, because of his great experience in practical astronomy, he was charged with the directorship of the observatory at Bamberg, which was erected, under his supervision, with funds from the will of Carl Remeis, an enthusiastic amateur astronomer. Hartwig spent the rest of his life at the observatory in scientific contact with the astrophysical observatory in Potsdam and the nearby University of Erlangen. The latter conferred on him the title of honorary professor (1916) and the degree of D.D. *honoris causa* (1921). He married Nanette Müller in 1889.

Hartwig's work was devoted to two main branches of research: the measurement of stars and planets and the observation of variable stars. In his astrometric observations he preferred the heliometer, an instrument specially designed to measure small spherical distances with the highest precision attainable at that time. Hartwig was familiar with the heliometer from Strasbourg and Dorpat, and he had such an instrument, one of the largest ever constructed, at his disposal at Bamberg. He performed a most valuable series of measurements of the diameters of planets and of the physical libration of the moon. He also measured the positions and parallaxes of stars and, occasionally, the positions of planets and comets, two of which he discovered.

Equally valuable are Hartwig's contributions in the field of variable stars, which he observed according to Argelander's method and also, from 1913 to 1923, photographically. In 1885 he independently discovered S Andromedae, the first known extragalactic supernova. He made many series of observations of long-period variables and of U-Geminorum stars. From 1891 he published an annual catalog of variable stars with approximate ephemerides for the Astronomische Gesellschaft, and, in collaboration with Gustav Müller of Potsdam, he compiled the fundamental work *Geschichte und Literatur der veränderlichen Sterne* (Leipzig, 1918).

BIBLIOGRAPHY

I. Original Works. Hartwig's books are *Heliometrische Untersuchungen der Durchmesser von Venus und Mars,* vol. XV of Publikationen der Astronomischen Gesellschaft (Leipzig, 1879); *Beitrag zur Bestimmung der physischen Libration des Mondes aus Heliometerbeobachtungen* (Karlsruhe, 1881); *Die Physik im Dienste der Wissenschaft, der Kunst und des praktischen Lebens* (Stuttgart, 1884), written with G. Krebs *et al.;* and *Geschichte und Literatur des Lichtwechsels der bis Ende 1915 als sicher veränderlich anerkannten Sterne* (Leipzig, 1918), written with G. Müller.

Among Hartwig's papers are "Physical Libration of the Moon," in *Monthly Notices of the Royal Astronomical Society,* **41** (1881), 375; and "Katalog und Ephemeriden veränderlicher Sterne," in *Vierteljahresschrift der Astronomischen Gesellschaft,* **26-55** (1891-1920). A great number of his other papers and short notes, concerned with variable stars, the moon, and planets, were published in *Vierteljahresschrift der Astronomischen Gesellschaft,* **13-22** (1878-1887); *Astronomische Nachrichten,* **95-217** (1879-1923); *Berichte der Naturforschenden Gesellschaft in Bamberg,* **16-21** (1893-1910); and *Veröffentlichungen der Remeis-Sternwarte zu Bamberg,* 1st ser. (1910-1923); 2nd ser., **1** (1923).

Moreover, there are a great number of unpublished notes on observations of comets, lunar occultations, eclipses, variable stars, and novae, preserved at the Remeis-Sternwarte.

II. Secondary Literature. For information on Hartwig, see Cuno Hoffmeister, "Ernst Hartwig," in *Vierteljahresschrift der Astronomischen Gesellschaft,* **59** (1924), 70; E. Heise, "Ernst Hartwig," in *Deutsches biographisches Jahrbuch* (1923); and in *Astronomische Nachrichten,* **219** (1923), 185.

Konradin Ferrari d'Occhieppo

HARVEY, WILLIAM (*b.* Folkestone, Kent, England, 1 April 1578; *d.* London or Roehampton, Surrey, England, 3 June 1657), *physiology, anatomy, embryology, medicine.*

Harvey was the eldest son of Joan Halke and Thomas Harvey, a yeoman farmer and landowner who in later life engaged in commerce and rose to the gentry; five of William's six brothers enjoyed even greater success as London merchants.[1] After attending King's School, Canterbury, Harvey studied arts and medicine at Gonville and Caius College, Cambridge, from 1593 to 1599. He then completed his education at the University of Padua, the leading European medical school; among his teachers were the celebrated anatomist Girolamo Fabrici and, probably, the Aristotelian philosopher Cesare Cremonini. Upon receiving his doctorate in medicine in April 1602, Harvey returned to England and took up the practice of medicine in London; in 1609 he was appointed physician to St. Bartholomew's Hospital. In 1604 he was married to Elizabeth Browne, the daughter of Lancelot Browne, a prominent London physician; they had no children.

In 1607 Harvey was elected a fellow of the Royal College of Physicians, in whose professional and political affairs he took an active interest for the rest of his life. Among the positions he held in the college was that of Lumleian lecturer on surgery from 1615 to 1656, and in 1627 he became one of the seven elect of the College; but after 1630 his duties as royal physician increasingly curtailed his participation in the business of the college. In 1651 he donated money to the college for building and furnishing a library, which was officially dedicated in 1654; in 1656 he gave it an endowment to pay a librarian and to present an annual oration, which continues to be held in his honor.

In 1618 Harvey was appointed one of the physicians extraordinary to James I, a position which he retained after the accession of Charles I in 1625; in 1631 he was promoted to physician in ordinary and in 1639 became senior physician in ordinary. Over the years he came to be on increasingly close terms with Charles I and did not conceal his loyalty to his memory even under the Commonwealth. From 7 April to 27 December 1636 Harvey traveled in the retinue of his friend Thomas Howard, earl of Arundel, on a special royal embassy to Emperor Ferdinand II at Regensburg; this provided him with the opportunity to meet a number of prominent Continental physicians, and on a side trip to Italy he also helped to look for paintings for the royal collection. The king apparently took an interest in Harvey's scientific work and provided him with deer from the royal parks for some of his investigations. Harvey accompanied the king on his visit to Scotland in 1633 and on his Scottish campaigns of 1639, 1640, and 1641. Following the outbreak of the Civil War in 1642 Harvey seems to have been in constant attendance on the king and remained with him at Oxford from the winter of 1642 until he surrendered himself to the Scots in May 1646. In November 1646 Parliament granted Harvey's petition to attend the captive king at Newcastle; after the king was handed over to Parliament in January 1647, Harvey returned to London. Now a widower, he lived at the various residences of his brothers in and around the city and resumed his medical practice on a limited scale; he seems to have suffered only minor difficulties as a result of his royalist sympathies.

Harvey had a broad interest in literature and art as well as medicine and philosophy, and among his friends and acquaintances were Francis Bacon, Robert Fludd, George Ent, Charles Scarburgh, John Selden, Thomas Hobbes, and John Aubrey, who has left an account of him in his *Brief Lives.* Harvey seems to have been well liked by those who knew him, although he was an outspoken man and perhaps somewhat short-tempered. In his later years he suffered from gout and kidney stones and apparently was not averse to ending his sufferings by an overdose of laudanum; he is reported to have survived one such attempt in 1652, only to die of a stroke in 1657, at the age of seventy-nine.

Throughout his life, whether at London, at Oxford, or on his various travels, Harvey was an untiring observer of animal life in all its forms. The earliest evidence of his scientific activities comes from his anatomical lecture notes, written in 1616; these notes formed the basis of the anatomical demonstrations that he conducted for the College of Physicians in that year, and periodically thereafter, as part of his duties as Lumleian lecturer.[2] In preparing these notes he relied heavily on the comprehensive *Theatrum anatomicum* (1605) of Gaspard Bauhin, supplemented by an extensive knowledge of other medical and anatomical authors. As was usual at the time, his treatment of anatomy included extensive discussions of the functions of the parts; and in judging the views of earlier authorities Harvey was generally independent and critical, occasionally impatient, but not rebellious against the structure of traditional medical thought as such. Throughout his life he retained a sense of identity with his predecessors, even after he had rejected some of their most basic doctrines.

From these notes it is clear that Harvey had already begun the original investigations of the motions of the heart, respiration, the functions of the brain and spleen, animal locomotion and generation, comparative and pathological anatomy, and various other subjects that were to occupy his attention for the rest of his life. From statements in his published works it appears that he contemplated a vast research program that would lead to publications on all of these subjects, but only his works on the heart and on generation were actually seen in print. Many of his notes and manuscripts were lost when his rooms at Whitehall were sacked in 1642, and most of the rest were presumably destroyed together with his new library at the College of Physicians by the Great Fire. Aside from the lecture notes, only the rough drafts of the treatises "De musculis" and "De motu locali animalium" have survived; there are also a number of letters on scientific subjects.

Nevertheless, the publication of the relatively short *Exercitatio anatomica de motu cordis et sanguinis in animalibus* (1628), in which he announced his discovery of the circulation of the blood, was sufficient to ensure Harvey a place of first importance in the history of science and medicine. By this discovery he revolutionized physiological thought, which since antiquity had been based to an important degree on assumptions about how materials flow through the blood vessels. Beyond this, he inspired a whole new generation of anatomists who sought to emulate his methods in the study of animal functions. And, more generally still, his work was one of the major triumphs of early modern science, and thus helped to generate the enthusiasm for science that came to dominate European intellectual life during the second half of the seventeenth century.

On the other hand, Harvey's work also had important connections with the medicine and philosophy of the Renaissance. His general philosophical outlook was quite traditional, and he had little regard for the mechanical and chemical philosophies that captured the imaginations of many of his contemporaries. His reliance on observation in the study of nature was a direct outgrowth of the anatomical revival of the sixteenth century, which had broadened into an interest in comparative anatomy by the early seventeenth. He was perhaps more interested in the study of function than earlier anatomists had been; but almost from the beginning the anatomists of the Renaissance had found themselves criticizing and modifying Galen's doctrines on function as they increased their knowledge of structure, and Harvey was able to build upon their achievements. Again, Harvey made more effective use of vivisection in the study of function than had been done earlier, but such observations were not new in themselves; and on the other hand, Harvey continued to rely heavily on purely structural considerations in attempting to infer the actions of the parts. Finally, in spite of his frequent insistence on relying only on the evidence of the senses in the study of nature, a deeply theoretical, almost speculative, strain manifests itself throughout his work; indeed, in Harvey, the thinker and the observer are so intimately united that it is impossible to separate them.[3] He undoubtedly observed more, and more carefully, than had his predecessors, and without this he would not have discovered the circulation; but his originality stemmed not from the amassing of observations per se but from his remarkable gift for perceiving and pursuing the theoretical implications of his observations.

One trait which distinguished Harvey from many of his medical predecessors was his preference for the physiological doctrines of Aristotle to those of Galen. Most earlier physicians were Aristotelians in their general philosophical outlook; but since the Galenic revival in medicine of the early sixteenth century there had been a division over many physiological questions between academic physicians, who usually supported Galen, and academic philosophers, who generally defended Aristotle. Thus at Padua in the early seventeenth century the anatomist Fabrici was a fairly loyal, although not uncritical, follower of Galen with regard to physiological matters, while the philosopher Cremonini was an uncompromising exponent of Aristotle. Yet prior to Harvey there were a few physicians, most notably Andrea Cesalpino and Caspar Hofmann, who strongly supported Aristotle against Galen, and Harvey's views resembled theirs in a number of important respects, especially in the area of cardiovascular physiology. There was also an important difference, however, in that these earlier men tended to make the vindication of Aristotle the main goal of their work, whereas Harvey used Aristotle's biological writings as a reliable starting point for his own inquiries but did not shrink from disagreeing with them if it finally came to that.[4]

As Walter Pagel has emphasized, perhaps the most fundamental similarity between the biological outlooks of Aristotle and Harvey is the monistic conception of living substance.[5] In this view the soul is not a separate immaterial substance that is superadded to, and acts upon, passive matter, but is the form, or tendency to perfection, of the body; thus there is only one living thing, having both material and immaterial aspects. A corollary to this idea was Harvey's emphasis on the immanence of the vital powers in all the parts of the body, in contrast with many of his predecessors, who saw the activities of most parts of the body as manifestations of separable spirits or faculties which flow into the parts from a central source.

These trends found their most important expression in Harvey's doctrine of the primacy of the blood, which he had already begun to develop when he wrote his lecture notes in 1616 and which in later years he seems to have regarded even more highly than his discovery of the circulation.[6] For Harvey, animal life was first and foremost a property of a single homogeneous substance—blood—rather than the result of an interaction between diverse formed organs and causative agents. He took the main causative agents of traditional vitalistic physiology—spirits and innate heat—and reduced them to inseparable qualities of blood insofar as it is alive; furthermore, he thought that the soul itself inheres primarily in the blood and in a sense is identical with the blood.

In Harvey's view, the first rudiment of the embryo to be generated is a drop of blood, which already exercises the basic powers characteristic of an animal before the existence of specialized organs. It carries on nutrition and growth before the liver or other nutritive organs have been formed; it exhibits pulsatile movement before the formation of the heart or other motor organs; and if the early primordium of a chick embryo is irritated with a needle, it reacts by obscure undulations, showing that it must have a form of sensation even before the existence of the nervous system. However, the blood needs the more specialized structures as instruments for the fuller exercise of its powers, as well as to secure its own growth and preservation; and therefore it proceeds to generate the rest of the body out of itself, using an innate, unconscious Idea as its exemplar. All other parts of the body originally receive life from the blood, and thereafter the blood continues to be the main repository of heat and vitality; its constant circulation through the body serves not only to nourish all the other parts but also to sustain their heat, spirits, concoctive powers, sensibility, and contractility. Thus the blood alone is truly alive per se, and the rest of the body is an appendage that serves the blood and lives by virtue of it.

Discovery of the Circulation. Harvey's work on cardiovascular physiology began with a study of the heart and arteries, which before the discovery of the circulation were regarded as a separate functional system from the liver and veins. During the sixteenth and early seventeenth centuries the prevailing Galenic doctrines on both systems had come under frequent criticism; but whereas Galen's ideas on the veins survived more or less intact, his teachings on the heart and arteries underwent considerable modification at the hands of an influential minority. Harvey's early work was in direct continuity with this progressive tradition; when he wrote his anatomical lecture notes he had not yet discovered the circulation, but he had reached conclusions about the movements of the heart and arteries that were highly significant in their own right. It appears that for a time he considered publishing a separate treatise on this subject, and such an early work may even have formed the basis of the proem and first half of *De motu cordis.*[7] Eventually, though, Harvey became aware of certain difficulties inherent in his conception of the heartbeat, which led him to reconsider the functions of the veins and thus to discover the circulation. But even afterward both Harvey and his contemporaries continued to regard his views on the motion of the heart and arteries as quite significant in themselves, apart from the broader theory.

In the standard physiology of the sixteenth century, the chief function of the veins was to convey nutritive blood from the liver to all the parts of the body. The right ventricle of the heart was considered part of this sanguineous system, its purpose being to transmit blood from the vena cava to the pulmonary artery for the nutrition of the lungs. The idea of the centrifugal flow of venous blood from the liver was one of the most convincing elements of Galenic physiology, since it seemed to follow of necessity from the anatomy of the veins. One could trace the pathway of the nutriment from the intestines through the mesenteric veins to the liver, and it seemed obvious that from there it would flow into the vena cava and then to all the other veins of the body. Before Harvey there was considerable dissatisfaction with the failure of this view to account adequately for the important anatomical relationship between the veins and the heart; but the existence of the tricuspid valve seemed to rule out the idea that the venous blood flows outward from the heart, which was the only alternative that suggested itself. Nor did the discovery of the venous valves by Fabrici in 1574 have much effect on these ideas. Once the valves became known, it seemed clear that without them all the venous blood would collect in the lower parts of the body; the valves were thought to prevent this, but without preventing the slight downward trickle necessary to replace the blood absorbed by the parts as nutriment.

During the sixteenth century the lungs, pulmonary vein, left ventricle, and arteries were generally considered to form a separate pneumatic system, concerned with transmitting vital spirit and natural heat to the entire body. The left ventricle and arteries were also thought to ventilate the innate heat through active dilatation and contraction; thus the left ventricle inhaled and exhaled through the pulmonary vein, while the arteries inhaled and exhaled through the pores in the skin. In addition, a small amount of blood from the right ventricle was supposed to pass through the cardiac septum to the left ventricle and arteries, but this was of only secondary importance.

During the later sixteenth and early seventeenth centuries a growing minority of anatomists, such as Felix Platter and Adrian van der Spiegel, came to hold a rather different view of the heart and arteries; they looked upon them primarily as sanguineous organs and considered the transmission of blood to the arteries to be one of the most important functions of the heart. This change in thinking resulted largely from the discovery and gradual acceptance of the pulmonary circulation, which was published by Realdo Colombo in 1559. Those who accepted this concept tended to minimize or even flatly to reject the idea that the movements of the heart and arteries serve to ventilate the innate heat. It now seemed clear from the structure of the left ventricle that its principal action is to receive blood from the lungs and expel it into the aorta, just as the right ventricle transmits blood from the vena cava to the lungs. This idea did not directly conflict with the idea that some of the blood from the vena cava flows outward through the peripheral veins, since the venous and arterial blood were generally thought to serve two different purposes: nutrition and vivification, respectively.

By the time he wrote his anatomical lecture notes in 1616, Harvey had accepted this new view of the heart and had added to it a concrete understanding of the movements of the heart and arteries based on extensive vivisectional observations.[8] He originally undertook this study to settle an ancient controversy over whether the heart and arteries dilate and contract at the same time or in alternation, and whether the arteries pulsate actively or are passively distended by the impulsion of material from the heart. In antiquity techniques had been developed for exposing the heart in live animals in order to study its motions in relation to those of the arteries, and such investigations were resumed in the sixteenth century. But given the difficulty of distinguishing the movements of a rapidly beating heart, as well as other complicating factors, these observations failed to produce a clear resolution; and Harvey's contemporaries were divided in their views on the coordination between the heart and arteries. There was more general, although not unanimous, agreement that the arteries pulsate actively, largely because it was thought that all of the arteries would not dilate simultaneously if the cause of the pulse were purely mechanical.

As Harvey emphasized in the lecture notes and *De motu cordis,* he was determined to settle these disputes and therefore refused to give up in the face of initial frustration and confusion. It is also evident from both accounts that his ultimate success resulted largely from the study of the hearts of dying animals, in which the events of the heartbeat are considerably slowed down and, therefore, more easily discernible. He also studied the simpler hearts of cold-blooded animals and observed excised beating hearts, both whole and in section. In addition, he took into account a great deal of purely anatomical data in attempting to determine the action of the heart.

In choosing the questions to be answered by his observations, Harvey was much influenced by a description of the heartbeat published by Colombo in 1559.[9] In addition to maintaining that the arteries dilate when the heart is contracted, and vice versa, Colombo had focused attention on the actual nature of the heart's movement. He asserted that this consists of a more relaxed phase, during which the heart receives blood into its ventricles, and a more vigorous phase, during which it transmits what it has received. This notion went against the prevailing view that both movements of the heart are active and that, if anything, dilatation is more vigorous than constriction.

Colombo's description included a note of terminological confusion in that he referred to the more active phase of the heart's movement as "constriction" and the more passive phase as "systole," which also means constriction. Harvey was somewhat puzzled by this inappropriate use of the word "systole," but his uncertainty seems to have led him to make a very fruitful distinction between the question of the activity or passivity of the phases of the heartbeat and the question of which phase is systole and which diastole. In the lecture notes Harvey sought to establish, first of all, that the heartbeat consists of only one active movement, to which he gave the neutral designation "erection," since at this time the apex of the heart appears to be lifted up; this is followed by a completely passive relaxation. During erection the heart strikes the chest and its flesh changes from soft to hard, and at the same time the pulse of the arteries is perceived; erection is slightly preceded by the obvious contraction of the auricles, during which they expel their blood and become whiter. Indeed, the beat of the heart begins with the auricles and then proceeds to the apex of the ventricles, so that "the auricles arouse the somnolent heart."

Having established the sequence of events involved in the heartbeat, Harvey addressed himself to showing that erection, the proper motion of the heart, represents the contraction of its ventricles. A number of considerations supported this conclusion. If a beating heart is punctured, blood is forcefully expelled during erection; the heart becomes whiter during erection; the dissection of a beating heart shows that its walls become thicker during erection, which means that its cavities must become smaller. Harvey also felt that the entire structure of the heart, with its component fibers, valves, and the chordae tendineae, supported the view that its essential action is to contract and expel materials rather than to dilate and attract them.

In addition, it seemed more reasonable that the heart should contract when the arteries dilate, since there could be no passage of material from one to the other if they dilate and contract at the same time. Indeed, Harvey maintained that the pulse of the arteries is simply the result of the impulsion of blood by the heart and is not an active movement. When an artery is cut, blood is expelled from it more vigorously during the contraction of the heart; the pulsation of the pulmonary artery but not of the pulmonary vein likewise supported the view that the pulse is caused by the impulsion of blood, as did the comparative anatomy of the arteries: the more vigorous the pulse of an animal, the thicker its arteries. Harvey repeatedly compared the pulse of the arteries to the inflation of a glove; this concrete analogy, which he seems to have borrowed from Gabríele Falloppio,[10] probably helped him to overcome the standard objection that a purely mechanical impulse could not be transmitted instantly to all of the arteries.

In the lecture notes Harvey summarized his conclusions thus:

> From these things it is clear that the action of the heart insofar as it is moved is [to transfer] blood from the vena cava to the lungs through the pulmonary artery, and from the lungs to the aorta through the pulmonary vein. When the heart is relaxed, which is first, there is an entrance of blood into the right ventricle from the vena cava, and into the left ventricle from the pulmonary vein. When it is erected, or contracted, it forcefully propels [the blood] from the right [ventricle] into the lungs, and from the left [ventricle] into the aorta, whence the pulse of the arteries.[11]

Or, as he put it more succinctly, "Action: thus relaxed receives blood, contracted scups it over; the entire body of the artery responds as my breath in a glove."[12]

One of the consequences of Harvey's new view of the movement of the heart was that the amount of blood transmitted from the vena cava to the aorta at each beat had to be fairly large. Because the heart does not dilate actively, there must be an appreciable influx of blood into its ventricles to account for each cardiac diastole, and similarly the passive distention of the arteries requires that the heart expel a significant amount of blood into them at each systole; and because Harvey was convinced of the competence of the heart valves, this transfer had to be irreversible. It was to be some time before Harvey saw the full implications of so large a rate of transmission, but by 1616 it seems already to have indirectly weakened his adherence to Galen's doctrines on the veins. From a number of brief references in the lecture notes it appears that he still accepted a flow of some blood from the liver and vena cava to the peripheral veins,[13] but he attached much greater importance to that part of the blood which passes through the heart and lungs to the arteries; indeed, at one point he asserted that "the whole mass of the blood" reaches the body by this route.[14] Thus the arteries had replaced the veins as the principal blood-distributing vessels, and the vena cava was more concerned with transporting blood from the liver to the heart than from the liver to the peripheral veins.

Otherwise, Harvey noted only two major points about the veins in the lecture notes: the heart, rather than the liver, is their governing principle (*arché*); but, unlike the arteries, they do not pulsate because "they have many valves opposed to the heart" which break off the impulse caused by cardiac contraction.[15] The latter point is of some importance, for it shows that in Harvey's view the significant orientation of the venous valves was not upward in the body, as was previously thought, but inward toward the heart; this probably helped prepare him for the idea of centripetal venous flow once he saw the need for a return of blood from the arteries to the heart.[16]

In the first half of *De motu cordis* Harvey presented his conclusions about the movements of the heart and arteries in a more developed form, now bolstered by an additional wealth of vivisectional, anatomical, pathological, and embryological observations. In the proem he gives a devastating critique of Galen's doctrines on the motions of the heart and arteries, especially the idea that these motions serve a ventilating function. In chapter 1 he describes how he first took up the study of the movement of the heart and eventually decided to publish his findings. In chapters 2, 3, and 4 he presents his conclusions about the ventricles, arteries, and auricles, respectively. Chapter 5 contains a summary of his views, with emphasis on the idea that the overall action of the heart is the constant transmission of blood from the vena cava to the aorta; he aptly describes this as a "swallowing" of blood from one vessel by the other. In chapters 6 and 7 he defends the pulmonary circuit of the blood, an essential corollary to his view that the action of the four-chambered heart is to bring about such a constant transmission of blood from the vena cava to the aorta.

In chapter 8 Harvey relates how he went beyond this early work on the heart and arteries to the discovery of the circulation. From this account it seems clear that he first conceived of the centripetal flow of venous blood as a necessary consequence of his conclusions about the heartbeat, rather than as the result of a direct investigation of the veins. He realized that over a relatively short period of time the heart transmits from the veins to the arteries even more than the whole mass of the blood; the rate of transmission is in fact so large that if it took place in only one direction, the veins would soon be drained and the arteries filled to bursting. Only if blood somehow returns from the arteries to the veins at the periphery could these absurdities be avoided.

Although by 1616 Harvey had assumed that the heart transmits an appreciable portion of blood at each beat, he had not yet realized the cumulative effect of such a large rate of transmission. What brought this problem to his attention is not certain, but it is possible that his thinking was stimulated by the views of a contemporary. In a book published in 1623, Emilio Parigiano maintained that there must be a significant reflux of blood from the aorta to the left ventricle during each cardiac diastole. Among his arguments was the following: "Since the heart in systole expels the larger part of its blood into the aorta, and that in scarcely a moment of time, the aorta would always be so filled with blood that it could receive no more, while the heart . . . would be emptied in a few beats."[17] Both problems would be eliminated if there were a constant return of blood to the heart through the aortic valve. It is not certain whether Harvey actually read this work before discovering the circulation,[18] but the terms of the argument are quite similar to those that first led him to conceive of a return of blood to the heart through the veins. Parigiano himself did not attach great importance to the argument, since he accepted the need for a reflux of blood largely on other grounds, but it might have taken on new significance for Harvey against the background of his detailed study of the heartbeat and his firm belief in the effectiveness of the aortic valve.

In any case, Harvey's statement in *De motu cordis* indicates that something aroused his interest in the question of how much blood the heart transmits from the veins to the arteries and led him to undertake a searching reexamination of the action of the heart with this specific question in mind. It will be noted that he took account of anatomical as well as vivisectional factors:

> I often and seriously considered, and pondered at great length, how large would be the amount [of blood transmitted by the heart, as judged] from the dissection of live animals for the sake of experiment, from the opening of arteries, and from diverse investigations; also from the symmetry and magnitude of the ventricles of the heart, and of the vessels entering and leaving them (since Nature, who does nothing without purpose, would not have endowed these vessels with such a large proportional size without purpose); also from the elegant and careful construction of the valves and fibers, and from the rest of the structure of the heart, as well as from many other things.[19]

Harvey went on to state that it was from this inquiry into quantity that he first inferred the necessity of centripetal venous flow, which he subsequently confirmed by more direct means:

> [When I had thus considered] how large the amount of transmitted blood would be, and in how short a time the transmission would take place, I noticed that the juice of the ingested aliment could not supply [this amount] without our having the veins emptied and completely drained on the one hand, and the arteries disrupted by the excessive intrusion of blood on the other, unless the blood somehow permeates from the arteries back into the veins, and returns to the right ventricle of the heart. I began to consider whether [the blood] might have a kind of motion, as it were, in a circle [*motionem quandam quasi in circulo*], and this I afterward found to be true [*quam postea veram esse reperi*].[20]

Harvey then described the flow of blood from the left ventricle through the arteries, and back to the right ventricle through the veins, and compared this to the passage of blood through the lungs.

It is interesting that Harvey should distinguish two chief moments in the early development of his thought: his initial surmise of return venous flow as a solution to the quantitative problem and the later idea of a quasi-circular movement of the blood. To judge from his statement, it was only when he began to think of the movement of the blood precisely as circular that he was fully aware of having made an important new discovery; in other words, it appears that the metaphor of the circle played a significant role in enabling him to see through the complexity of his observations to a clear and simple conception of the movement of the blood. This is not to suggest that Harvey was looking for a circular pattern before he began thinking of venous return, but that at an early stage thereafter the possibility of a constant circular motion occurred to him and then served as the leading idea in the further clarification of his thought. That circularity per se should have caught his attention probably stemmed from the preeminence and preservative character that were attributed to circular motion in traditional natural philosophy; indeed, as Walter Pagel has shown, this tradition by itself had led a number of earlier men to associate circularity with the heart or the blood on purely speculative grounds.[21] That Harvey shared these ideas about circularity is clear from the continuation of the passage just cited, in which he related the movement of the blood to other circular and quasi-circular processes in the atmosphere and the heavens.

It is not clear how long before 1628 these developments in Harvey's thinking occurred. Some statements in *De motu cordis* seem to suggest an early date; but although Harvey made additions to his lecture notes several times during the 1620's, it was apparently not until 1627 or even later that he added a brief description of the circulation.[22] Moreover, the draft of his treatise on locomotion, written in 1627, contains numerous references to the functions of the heart but none to the circulation. Thus it may actually have been quite late that Harvey made the discovery.

In the second half of *De motu cordis* Harvey presented evidence to confirm the circular movement of the blood. First of all, in chapters 9 and 10 he strengthened the original quantitative argument by showing that the heart must expel at least some blood to the arteries at each beat and by making a rough calculation of the resulting rate of transmission. Thus, even if the amount expelled at each beat is as small as one dram and if the heart beats, say, 1,000 times in half an hour, then it still follows that in a relatively short time the heart will have transmitted more blood than the ingested aliment or the venous contents could supply. That all the blood can be rapidly evacuated from an animal by opening a large artery likewise shows how large is the rate of transmission. Furthermore, if the vena cava of a live snake is pinched, the beating heart rapidly empties itself of blood, while if the aorta is pinched, the heart soon becomes engorged with blood.

In chapters 11 and 12 Harvey sought to demonstrate by the use of ligatures that there is a passage of blood from the arteries to the veins at the periphery. If an arm is ligated so tightly that the arterial pulse is cut off, then it soon becomes pale and bloodless, while the arteries above the ligature become swollen with blood; but if the ligature is loosened sufficiently to restore the pulse, the arterial swelling above the ligature subsides and the arm becomes suffused with the blood that is allowed to flow in. Now the veins of the arm become swollen with blood—but only below the ligature, which means that this blood must flow into them from the arteries at the periphery rather than from the central veins. If such a swollen vein be opened as for bloodletting, within about half an hour most of the blood in the body can be evacuated from it; this provides an index of the amount of blood that flows from the heart into the arteries and from the arteries into the veins. Return venous flow is a necessary consequence of this rate of transfer, as well as of the rate of transfer from the veins to the arteries through the heart.

In chapter 13 Harvey went on to give a more direct demonstration of centripetal venous flow, based on the existence of the venous valves. He stressed the cardiocentric orientation of the valves and noted that a probe can be inserted inward through them but not outward. He also showed that if an arm is ligated to make the veins swell, then by drawing a finger along a vein with some pressure one can push blood inward through the valves but not outward. Furthermore, if a segment of vein is emptied by applying pressure with one finger and then squeezing the blood inward through a valve with a second, it will be apparent that it refills from the distal end of the vein when the first finger is removed. And, Harvey added, if the latter procedure be repeated a thousand times in succession, it will again be apparent on quantitative grounds that the blood must circulate.

In chapter 14 Harvey summarized his main arguments, and in chapter 15 he made some tentative suggestions about the purpose of the circulation. In chapter 16 he showed how the circulation could explain a number of previously inexplicable phenomena. Thus the ability of a localized affection, such as a snakebite, to rapidly influence the entire body results from the inward flow of a noxious substance through the veins, followed by its dispersal to the entire body through the arteries. Also, a number of problems relating to the mesenteric vessels would be eliminated by the idea of a rapid circulation of blood through them, with a gradual addition of chyle to the blood. Finally, in chapter 17 Harvey presented a wealth of anatomical evidence in support of his view, although all of this relates only to the heart and arteries rather than to the circulation as such.

The circulation was widely discussed during the twenty years following the publication of *De motu cordis,* and much of the reaction was quite favorable. The idea also found some major opponents, though, and in 1649 Harvey published *Exercitationes duae de circulatione sanguinis,* in which he replied to Jean Riolan and other critics. Among other points, Harvey sought to confirm that an appreciable portion of blood must be expelled at each beat of the heart (which in his view was the main buttress of the circulation), and that the arterial pulse is caused by this impulsion of blood. He also tried to show that the differences between venous and arterial blood are not so great as to make rapid conversions of one to the other seem implausible, and by various combinations of vascular ligation and section he gave more direct experimental demonstrations of the circulation than he had given in *De motu cordis.* Harvey also discussed aspects of the circulation in his treatise *De generatione* and in a number of letters.

A question which interested Harvey from the time when he first discovered circulation was that of its purpose, although his attempt to find an answer seems to have been complicated by his prior commitment to the idea of the primacy of the blood and by his views on respiration.[23] Most of his predecessors had regarded the heart as an inexhaustible source of life-giving heat, which it imparted to the rest of the body through the arteries, using blood or spirits as a vehicle. By 1616, though, Harvey had come to regard the blood itself as the source of heat and vitality for the rest of the body; the heart was still the most important formed organ, but only in virtue of its role in distributing blood.[24]

But when he later discovered the circulation, Harvey was faced with the question of why the blood should constantly return to its source, and an obvious answer seemed to be that it rapidly gives up its heat to the parts at the periphery and therefore returns to have it restored. For various reasons he did not think that the passage of blood through the lungs could be of major importance in restoring its vitality, and therefore he was brought back to the view that the heart is the actual source of heat in the body and the blood only its vehicle. This was what he tentatively proposed in *De motu cordis,* although the capitulation did not last for long and by the mid-1630's he seems to have been more convinced than ever of the primacy of the blood.[25] When Caspar Hofmann criticized the notion of a repeated reheating of the blood by the heart in 1636, Harvey replied that he had proposed this view only for the sake of illustration and did not wish to insist upon it.[26] He maintained then and for the rest of his life that his failure to demonstrate the purpose of the circulation was not a valid reason for denying its existence.

Nevertheless, Harvey did not lose interest in finding a purpose for the circulation that would be consistent with the primacy of the blood. At times in his treatise *De generatione* he appears to fall back on the preservative character of circular motion per se, but elsewhere in this work and the letters to Riolan he seems to have had in mind a modified version of his earlier suggestion.[27] In this view, the source of heat in the body would be a kind of internal fermentation of the blood, but one which takes place primarily in the blood concentrated in the vena cava and heart; thus, hot blood from this source would be distributed through the arteries, give up its heat, and then return to the central mass to have it restored by renewed fermentation.

The idea of an internal fermentation of the blood also provided Harvey with an additional basis for asserting the primacy of the blood over the heart. Whereas he had earlier thought of auricular systole as the event which initiates the heartbeat, in his later works he maintained that this is preceded by a self-induced swelling of the blood in the vena cava. As a result, blood is forced into the right auricle, which is irritated by the resulting distension; to rid itself of the irritating cause the auricle contracts, thereby distending and irritating the right ventricle, which also responds by contracting. The pulmonary artery is in turn distended and irritated to contraction, and a similar sequence occurs in the left heart. Thus the blood actually inaugurates its own movement, although, as Harvey conceded, one could also see the heart and blood as a functional unity which cooperates in causing the circulation and thus in preserving the life of the rest of the body.

Since antiquity ideas about the physiology and pathology of most parts of the body had been based to an important degree on assumptions about the functions of the heart and blood vessels; and therefore, by fundamentally changing the latter, Harvey pointed the way to a reform of all of physiology and medicine. At first many physicians sought to incorporate the circulation into the traditional framework with a minimum of other changes, but the middle decades of the seventeenth century saw the rise of new mechanical and chemical systems of physiology, which took the circulation as a basic assumption in the explanation of a wide range of vital phenomena; the mechanists in particular came to view the circulation of fluids through the solid parts of the body as almost the very essence of life. Subsequent developments in physiology have led to great changes in thinking about the functions of the circulation but have abundantly confirmed the importance of Harvey's discovery as the cornerstone of modern physiology and medicine.

Sensation and Locomotion. In addition to the nutritive and pneumatic systems, which he amalgamated into one circulatory system, Harvey was also interested in the third great system of classical physiology: the organs concerned with locomotion and sensation. In 1627 he began work on a treatise entitled "De motu locali animalium" ("On the Locomotion of Animals"), which was primarily concerned with applying the general principles of Aristotle's treatises on animal movement to a detailed study of muscles, nerves, and other organs involved in locomotion. Apparently he never completed the treatise, but the rough draft has survived and shows that he had begun to develop some important insights into the physiology of sensation and locomotion. These ideas were not published in Harvey's lifetime, but in his treatise *De generatione* he discussed some related themes that had a direct and significant influence on the development of the concept of tissue irritability by his younger contemporary Francis Glisson.

Harvey's predecessors generally regarded the brain as the principal organ of sensation and voluntary movement, by which they meant that the brain supplies the actual powers to carry on these activities to the sense organs and muscles through the nerves; thus, when one cuts the nerve leading to a part, one destroys its mobility or sensibility or both by cutting off its supply of the necessary faculties. The idea that sensation involves an inward movement of sense impressions to the brain was not unknown, but the function of the brain in sensation was looked upon primarily as centrifugal: the supply of the sensitive faculty. Furthermore, a fundamental distinction was made between organs of natural (involuntary) movement, such as the intestines and heart, whose power of movement is innate, and voluntary muscles, whose power of movement flows in from the brain.

For Harvey, by contrast, sensibility and contractility seem to be innate powers of the sense organs and muscles, although their preservation in these organs depends upon a constant inflow of blood, heat, and spirits through the arteries.[28] Sensation itself is primarily a function of the brain and is chiefly a centripetal process; it begins with the sensible object, passes through the sense organ and nerve, and terminates in the brain. Thus, "the use of a nerve is to communicate something sensible to the brain so that a judgement can be made," while the brain itself is above all the Aristotelian *sensorium commune,* in which we perceive that we sense, and in which the objects of the different senses are compared with each other and unified.[29]

According to Harvey, the role of the brain in voluntary movement is not to supply the motor faculty to the muscles but to act as the *maestro del coro* (choir-master), which harmonizes the movements of many individual muscles into purposeful actions.[30] If the nerve leading to a muscle is cut, what is destroyed is not its contractility but its ability to participate in useful, coordinated actions. Indeed, a cock can have its entire head cut off without its muscles ceasing to move, but the movements are disordered and without purpose; on the other hand, an individual muscle in an uninjured body can undergo spasms in no way subject to central control. Thus, for Harvey a muscle is like any other motor organ, in that it possesses contractility as long as it remains alive, a point which he expressed quite vividly by taking Aristotle's comparison of the heart to a "separate animal" and applying it to the individual voluntary muscles;[31] his conclusion that the heart may be considered a muscle implied as much a redefinition of muscle as of the heart.

In Harvey's view, the role of the brain in the coordination of voluntary movement is contingent upon its primary function as *sensorium commune.*[32] Following Aristotle, he maintained that every purposeful movement in an animal begins not from an internal impulse but from an external object that gives rise to an internal perception, then an appetite, and finally an action appropriate to the attainment of the object; thus it is because of its perception of the end that the brain can determine the action. Furthermore, Harvey seems to have thought that coordinated motion would be impossible without an inflow of sense impressions from the parts concerned; thus, he attributed the loss of coordination resulting from cutting a nerve to a destruction of the sensation of the part rather than to an interruption of motor impulses.

In the later treatise *De generatione,* Harvey proposed that in addition to the conscious sense perception that is conducted by the brain, there is an unconscious sensation that involves only the individual parts of the body.[33] He was impressed by the many instances in which involuntary actions can be evoked or altered by irritants applied directly to the parts concerned, without the irritant being consciously perceived as such. For example, an innocuous-tasting infusion of antimony can provoke the stomach to vomiting, as if the stomach itself could distinguish the harmful from the useful; similarly, the skin reacts very differently to two apparently identical pricks, one from a clean needle and one from a needle dipped in venom. Indeed, Harvey asserted, we have no other sure way of distinguishing the animate from the dead than by its movement in response to an irritant; and he argued that the ability to respond implies an ability to sense the irritant. Thus, just as we distinguish natural (involuntary) movements such as the heartbeat from animal (voluntary) movements that are made under the direction of the brain, so should we distinguish the natural sensation common to all living matter from the animal sensation that can result only from referring sense impressions to the brain. Natural sensation is "a kind of touch which is not referred to the *sensorium commune,* nor is it communicated to the brain in any way, so that in this kind of sense we do not perceive that we sense." In simpler animals having no brain, and in fetuses prior to the formation of the brain, all sensations and motions are natural, while in mature higher animals both kinds are present. Thus there is a difference between the regulated actions which muscles perform under the direction of the brain and the spastic movements which they undergo in direct response to an irritant.

Generation. Since antiquity the generation of animals had been an important subject of speculation and observation for physicians and philosophers alike. Of the two leading authorities in the field, Aristotle had made significant studies of the developing chick embryo, while Galen had described the anatomy of the reproductive organs and the fetus. Original observations in both areas began to be made again during the sixteenth century; and among Harvey's immediate predecessors the most important contributions were those of his teacher Fabrici whose studies were broadly comparative in nature. By 1616 Harvey had already begun his own lifelong study of generation, and by about 1638 he seems to have completed much of the extant *De generatione,* although it was not published until 1651; a complementary treatise on the generation of insects was lost during the Civil War.[34]

In the seventy-two exercises and eight appendixes of his long treatise, Harvey reported a wealth of observations on all aspects of reproduction in a wide variety of animal species; his attention was focused primarily on the domestic fowl and the deer as representatives of the ovipara and vivipara, respectively.[35] His description of the day-to-day development of the chick embryo was notably more accurate than earlier ones, while his direct study of viviparous generation by dissecting the uteri of hinds and does at various stages during mating and pregnancy was quite without precedent. These observations formed the basis of a critical evaluation of earlier theories of generation, especially those of Aristotle, Galen, and Fabrici; and, finding all of the latter deficient, Harvey went on to formulate the first fundamentally new theory of generation since antiquity. The originality of the theory consisted essentially in his using oviparous generation as a model for interpreting viviparous, whereas viviparous generation had previously been treated as the more fundamental type.

In accordance with the latter assumption, previous theories of generation had been largely concerned with defining the roles of semen and menstrual blood, which were considered to be the immediate precursors of the fetus in a relatively crude sense; by "semen" was meant the entire seminal mass emitted during coitus, while menstrual blood was thought to be directly incorporated into the fetus. According to Aristotle, semen emitted into the uterus by the male acts on blood supplied by the female to form the first rudiment of the fetus—the heart, which then directs the formation of the rest of the fetus from additional menstrual blood; thus there is a clear separation of efficient and material causes (paternal semen and maternal blood) at the onset of generation. In Galen's view, on the other hand, the entire fetus arises from a mixture of semen emitted into the uterus by both parents, and thereafter is nourished by menstrual blood. In both theories the egg in oviparous generation is merely the vehicle for a delayed process in which the principal roles are still played by semen and slightly altered menstrual blood. Fabrici introduced a new emphasis on the importance of the egg itself as a distinct entity which is generated by the hen and in turn generates the chick; but he did not break entirely with the view that semen and menstrual blood are the main factors underlying oviparous generation, and he accepted the Galenic theory for the vivipara.

Much of Harvey's work was concerned with eliminating the remaining vestiges of Aristotelian and Galenic doctrine from Fabrici's theory of oviparous generation, thereby shifting attention completely to the egg itself as a primary generative agent, quite distinct from parental semen and blood, on the one hand, and from the future chick, on the other. He could find no evidence that the seminal mass of the cock either enters into or even touches the eggs during their formation within the hen; furthermore, he found that for a time the hen can continue producing fertile eggs after all detectable traces of semen have vanished from her body. To Harvey this seemed to offer solid evidence that the contribution of the cock's semen to generation is indirect and incorporeal; it simply confers a certain fecundity on the hen and then plays no further role in the actual generation of the egg or the chick. Once endowed with this fecundity the hen can, entirely on her own, produce fertile eggs which will give rise to chicks resembling both herself and the cock. In trying to explain the transfer of this principle of fecundity from the semen to the hen, and from the hen to the egg, Harvey repeatedly cited the analogy of the spread of disease by contagion, in which mere exposure to a sick individual can engender within a second individual an internal principle which subsequently reproduces in him the same specific disease.

On the basis of his investigation of the reproductive tract of the hen, Harvey also rejected the view that the hen produces the egg by a slight alteration of her own blood. Instead, he maintained that the active role of the hen is confined to producing the first minute primordium of the egg; this primordium possesses its own nutritive powers, through which it actively generates the rest of the egg by the complete, substantial transformation of additional material supplied by the body of the hen.

Thus the completed egg contains no immediate parental secretion such as semen or menstrual blood, but neither, in Harvey's view, does it contain any direct rudiment of the future chick prior to the onset of generation; instead, generation involves a second substantial change in which the same principle which transformed the maternal nutriment into the egg now transforms the egg into the chick. From this it is clear to what degree Harvey regarded the egg as an individual living entity, a distinct phase in the life cycle of the species which mediates between two successive generations.[36] The entire role of the parents in generation is to produce not a chick but a fertile egg, which subsequently gives rise to a chick through its own innate powers. For Harvey an egg was "a certain corporeal substance having life in potency"; it was "of such a kind that if all obstacles are removed it will develop into the form of an animal no less naturally than all heavy things tend downward, or light things move upward."[37]

Harvey maintained that within the egg the fetus emerges gradually from a homogeneous generative fluid, beginning with one first-formed part which actively creates the remaining parts in a definite order.[38] Furthermore, in the generation of the organs three processes take place simultaneously: the qualitative differentiation of the original generative substance, the acquisition of form, and increase in size. Harvey upheld these views in opposition to many contemporary physicians, who held that the fetus is completely passive in its own formation, that generation involves merely the separation and organization of preexisting substances, and that all the parts of the fetus are sketched out simultaneously at the onset of generation and subsequently undergo only growth. In adopting the idea of the gradual emergence of the fetus, which he termed "epigenesis," Harvey followed Aristotle, although he considered a drop of blood, rather than the heart, to be the first-formed part. But the drop of blood soon gave rise to a pulsating vesicle around itself.

Harvey also disagreed with Aristotle's reliance on a clear distinction between the efficient and material causes—semen and menstrual blood—to account for the onset of generation; instead, he stressed that there is only one factor, the egg, which is that from which as well as that by which the chick is generated.[39] Generation proceeds from the cicatricula (germinal disc), which, through an inherent principle, dilates into a small portion of perfectly clear fluid; this fluid, in turn, transforms part of itself into the first actual rudiment of the chick: a drop of blood. Thereafter the vital principle inheres primarily in the blood, which proceeds to transform the rest of the egg into itself at the same time that it forms the rest of the chick out of itself. There are of course causes antecedent to the egg in generation, but these do not participate directly in the formation of the fetus, so that at no point can the model of the separate artisan and artifact be applied to the latter process. Generation is simply an unfolding of the potentialities inherent in the fertile egg, with complete continuity between the vital activities of the egg, the fetus, and the mature animal.

Harvey's study of viviparous generation began with an evaluation of the Aristotelian and Galenic theories, according to both of which the result of a fruitful coitus should be the mixing of the parental "genitures" in the uterus and the formation of the first rudiments of the fetus. However, in numerous dissections of deer from the royal parks at various intervals during and after the rutting season, as well as of other animals, Harvey could find in the uterus no prepared menstrual blood before coitus and no blood, seminal mass, or rudimentary fetus immediately afterward. Instead, he found that it is only some time after the male semen has vanished from the body of the female, and after a period in which the uterus is otherwise empty, that the first evidence of conception can be seen. Indeed, in the deer he could find no trace of the conceptus in the uterus for nearly two months after the rutting season, although it appears that in this he was misled by the unusual shape of the early embryo in the deer;[40] in other kinds of animals he found a much shorter interval.

Harvey interpreted his viviparous findings by analogy with his conclusions about oviparous generation. Instead of participating directly in the formation of the fetus, the semen of the male must confer fecundity on the female and her uterus (which Harvey regarded as the principal female generative organ in vivipara), thereby enabling the uterus to produce a fertile conceptus at some time after intercourse. This conceptus, he assumed, is the analogue of the egg rather than of the fetal chick; that is, the conceptus produced by the uterus is not the rudiment of the fetus itself but a distinct entity which subsequently gives rise to a fetus through its own powers. The viviparous conceptus differs from the egg in that it generates its fetus within the body of the mother, deriving additional nutritive material from the uterus after the generation of the fetus begins and growing together with the fetus; but in its essential characteristics it conforms entirely with the egg and can indeed be called an egg. Thus Harvey's famous dictum that "an egg is the common primordium of all animals" does not reflect his discovery of what are now considered true eggs, where they had previously been unknown; rather, it summarizes his conclusion that in all animals—indeed, in all living things—the role of the parents in generation is indirect: they produce a fertile egg, or conceptus, or seed, which subsequently produces a new animal or plant through an innate vegetative power.[41] Harvey was willing to admit the spontaneous generation of very simple animals, yet even here it is not the organism that arises spontaneously from nonliving matter but an egg or primordium that subsequently develops into the organism.

Subsequent investigation has undermined much of Harvey's theory of generation, but his views nevertheless represented a major advance over those of his predecessors. He discredited the ancient notion that the fetus arises directly from semen and menstrual blood, although the discovery of spermatozoa shortly after his death was to restore the importance of semen in a quite different sense; he sought to eliminate the model of the separate artisan and artifact from the explanation of the formation of the fetus and thus to break down the barrier between this process and the vital activities of the mature animal; and his staunch defense of epigenesis at least provided a counterbalance to the doctrine of preformation, which came to dominate embryological theory during the later seventeenth century. Finally, the principle that all animals arise from eggs has been of great importance in the history of embryology, even though its original meaning for Harvey was very different from that which it later came to have for others.

NOTES

1. Biographical information from Keynes, *The Life of William Harvey.*
2. Gweneth Whitteridge, intro. to *Prelectiones.*
3. Pagel, "Harvey Revisited," pt. 1, pp. 1–2.
4. Lesky, "Harvey und Aristoteles"; Pagel, *Harvey's Biological Ideas,* pp. 23–47, 169–209.
5. Pagel, *ibid.,* pp. 251–278.
6. *Prelectiones,* pp. 126, 142, 248–250, 256, 262, 292–294; *De generatione,* exs. 51, 52, 56, 57, 71; Curtis, *Harvey's Views on the Use of the Circulation,* pp. 64–94, 103–138.
7. *De motu cordis,* ch. 1.
8. *Prelectiones,* pp. 264–272.
9. Realdo Colombo, *De re anatomica* (Venice, 1559), p. 257.
10. Gabriele Falloppio, *De partibus similaribus,* in *Opera* (Frankfurt, 1600), II, 138.
11. *Prelectiones,* p. 270.
12. *Ibid.,* p. 272.
13. *Ibid.,* intro. by Whitteridge, p. xlvii.
14. *Ibid.,* p. 296
15. *Ibid.,* pp. 254, 258, 272.
16. According to Robert Boyle, Harvey said, in his old age, that the contemplation of the venous valves actually led him to the circulation in the first place, but this conflicts with Harvey's own direct testimony in *De motu cordis;* see Keynes, *op. cit.,* pp. 28–30.
17. Emilio Parigiano, *Nobilium exercitationum de subtilitate libri* (Venice, 1623), p. 297.
18. In his later *De generatione,* ex. 14, Harvey refers to pp. 299–303 of this work.
19. *De motu cordis,* ch. 8.

20. *Ibid.* On this passage, see Pagel, "Harvey Revisited," pt. 1, pp. 2-5.
21. Pagel, *Harvey's Biological Ideas,* pp. 89-124.
22. *Prelectiones,* intro. pp. l-li, and p. 272.
23. Curtis, *Harvey's Views, passim.*
24. *Prelectiones,* pp. 248-250.
25. Webster, "Harvey's *De generatione,*" pp. 270-274.
26. Ferrario *et al.,* "Harvey's Debate with Caspar Hofmann," p. 15.
27. *De generatione,* exs. 51, 71; *De circulatione,* in *Opera,* I, 132-138.
28. *De motu locali,* pp. 88-92, 102, 108.
29. *Ibid.,* p. 110; see also *Prelectiones,* pp. 312-314; and *De generatione,* ex. 57.
30. *De motu locali,* pp. 102-104, 108, 110, 142-150.
31. *Ibid.,* p. 110; see also pp. 40-44, 50, 94, 114.
32. *Ibid.,* pp. 34-36, 102-104, 108, 138, 148; see also Pagel, "Harvey Revisited," pt. 2, pp. 6-7.
33. *De generatione,* ex. 57; Temkin, "Glisson's Doctrine of Irritation"; Pagel, "Harvey and Glisson on Irritability."
34. Webster, *op. cit.,* pp. 262-270.
35. This account of Harvey's work on generation is based on Meyer, *An Analysis of De generatione;* Adelmann, *The Embryological Treatises of Fabricius;* and Gasking, *Investigations Into Generation,* in addition to *De generatione* itself.
36. Pagel, *Harvey's Biological Ideas,* pp. 272-276.
37. *De generatione,* exs. 26, 62.
38. Pagel, *Harvey's Biological Ideas,* pp. 233-247.
39. Pagel, "Harvey Revisited," pt. 1, p. 11.
40. Keynes, *op. cit.,* p. 346.
41. *De generatione,* ex. 62.

BIBLIOGRAPHY

I. Original Works. Harvey's main works are *Exercitatio anatomica de motu cordis et sanguinis in animalibus* (Frankfurt am Main, 1628); *Exercitatio anatomica de circulatione sanguinis* (Cambridge, 1649); and *Exercitationes de generatione animalium* (London, 1651). The early trans. of all three (London, 1653) are in some respects superior to more recent versions. The standard Latin ed. is *Opera omnia: A collegio medicorum Londinensi edita* (London, 1766); the standard English trans. is Robert Willis, *The Works of William Harvey* (London, 1847), although it is at times inaccurate. For other eds. and trans., see Sir Geoffrey Keynes, *A Bibliography of the Writings of Dr. William Harvey,* 2nd ed. (Cambridge, 1953). Translations of the treatises on the circulation by K. J. Franklin are now available in an Everyman's Library edition, *The Circulation of The Blood and Other Writings* (London-New York, 1963).

Harvey's anatomical lecture notes have been published in facsimile, with a transcription, as *Prelectiones anatomiae universalis* (London, 1886); C. D. O'Malley, F. N. L. Poynter, and K. F. Russell, *William Harvey Lectures on the Whole of Anatomy* (Berkeley-Los Angeles, 1961), is an annotated trans.; Gweneth Whitteridge has prepared a new ed. of the *Prelectiones,* as well as one of "De musculis," with intro., trans., and notes in *The Anatomical Lectures of William Harvey* (Edinburgh-London, 1964). She has also edited *De motu locali animalium* (Cambridge, 1959). The full text of Harvey's important letter to Caspar Hofmann (1636) has been published by Ercole V. Ferrario, F. N. L. Poynter, and K. J. Franklin, "William Harvey's Debate With Caspar Hofmann on the Circulation of the Blood," in *Journal of the History of Medicine and Allied Sciences,* **15** (1960), 7-21.

II. Secondary Literature. The definitive biography is Sir Geoffrey Keynes, *The Life of William Harvey* (Oxford, 1966). Kenneth D. Keele, *William Harvey, the Man, the Physician, and the Scientist* (London, 1965), provides an excellent survey of Harvey's work. For a more selective and detailed treatment of some of the main themes in Harvey's work, and for comprehensive references to the Harveian literature, see Walter Pagel, *William Harvey's Biological Ideas* (New York, 1967) and "William Harvey Revisited," in *History of Science,* **8** (1969), 1-31; **9** (1970), 1-41. Much useful information has been gathered by H. P. Bayon, "William Harvey, Physician and Biologist, His Precursors, Opponents and Successors," in *Annals of Science,* **3** (1938), 59-118, 435-456; **4** (1939), 65-106, 329-389.

On the circulation, John G. Curtis, *Harvey's Views on the Use of the Circulation of the Blood* (New York, 1915), is still of fundamental importance; related themes are developed in Walter L. von Brunn, *Kreislauffunktion in William Harvey's Schriften* (Berlin-New York, 1967); Gweneth Whitteridge, *William Harvey and the Circulation of the Blood* (London-New York, 1971), focuses on the more empirical aspects of Harvey's work.

On sensation and locomotion, see Owsei Temkin, "The Classical Roots of Glisson's Doctrines of Irritation," in *Bulletin of the History of Medicine,* **38** (1964), 297-328; and Walter Pagel, "Harvey and Glisson on Irritability: With a Note on van Helmont," *ibid.,* **41** (1967), 497-514.

On generation, see Arthur W. Meyer, *An Analysis of the De generatione animalium of William Harvey* (Stanford, 1936); Howard B. Adelmann, *The Embryological Treatises of Hieronymus Fabricius of Aquapendente* (Ithaca, N.Y., 1942), esp. pp. 113-121 on Harvey's work and its relationship to that of Fabrici; Erna Lesky, "Harvey und Aristoteles," in *Sudhoffs Archiv für Geschichte der Medizin,* **41** (1957), 289-316, 349-378; Elizabeth B. Gasking, *Investigations Into Generation, 1651-1828* (London, 1967), esp. pp. 16-36; and C. Webster, "Harvey's *De generatione:* Its Origins and Relevance to the Theory of Circulation," in *British Journal for the History of Science,* **3** (1967), 262-274.

Jerome J. Bylebyl

HARVEY, WILLIAM HENRY (*b.* Limerick, Ireland, 5 February 1811; *d.* Torquay, England, 15 May 1866), *botany.*

Harvey was the youngest of eleven children of Quaker parents. His precocious interest in natural history became concentrated on botany, particularly the study of algae; at the age of twenty-two he undertook the description of these plants for James Townsend Mackay's *Flora Hibernica* (1836). Harvey succeeded his brother as colonial treasurer at Cape Town in 1836; he published *The Genera of South African Plants* (1838) and made extensive collections of algae and angiosperms until obliged to resign in 1842 because of ill health. In 1844 Harvey was appointed keeper of the herbarium at Trinity College, Dublin, where he began work on his *Phycologia Britannica,* the first part of which appeared in 1846; the lithographs that he prepared for this and for his other publications are evidence of his ability as a botanical artist.

Harvey traveled in the eastern United States from July 1849 to May 1850, during which time he lectured in Boston and in Washington and made large collections of algae, notably from Florida; these, together with material supplied by other collectors, were described in his account of the marine algae of North America, published between 1852 and 1858. While this work was in progress, Harvey made a lengthy expedition to the southern hemisphere, the most important result of which was his *Phycologia Australica* (1858–1863).

Following his appointment to the chair of botany at Trinity College in 1856, Harvey, with the cooperation of Otto Wilhelm Sonder, began work on *Flora Capensis,* based on his South African collections; only three volumes appeared in his lifetime. He died of tuberculosis, five years after his marriage in 1861.

BIBLIOGRAPHY

I. Original Works. Harvey's writings include *A Manual of the British Marine Algae . . .* (London, 1841); *Phycologia Britannica: Or a History of British Sea-Weeds . . .,* 3 vols. (London, 1846–1851), reissued in 4 vols., each with a title page dated 1846–1851; *Nereis Australis, or Algae of the Southern Ocean . . .* (London, 1847); "Nereis Boreali-Americana: Or Contributions to a History of the Marine Algae of North America," in *Smithsonian Contributions to Knowledge,* **3** (1852), art. 4, 1–144; **5** (1853), art. 5, 1–258; **10** (1858), art. 2, 1–140; *Phycologia Australica; or, A History of Australian Seaweeds . . .,* 5 vols. (London, 1858–1863); *Thesaurus Capensis: Or Illustrations of the South African Flora . . .,* 2 vols. (Dublin, 1859–1863); and *Flora Capensis: Being a Systematic Description of the Plants of the Cape Colony, Caffraria and Port Natal,* 3 vols. (Dublin–Cape Town, 1859–1863), written with O. W. Sonder—the work was completed with an additional 4 vols. published under the editorship of Sir William Thiselton-Dyer (1896–1925); a supp. to vol. V was issued in 1933.

II. Secondary Literature. See [Lydia Jane Fisher], *Memoir of W. H. Harvey, M.D., F.R.S.* (London, 1869); Norman Moore, "William Henry Harvey," in *Dictionary of National Biography,* XXV (London, 1891), 100; R. L. Praeger, "William Henry Harvey," in F. W. Oliver, ed., *Makers of British Botany* (Cambridge, 1913), pp. 204–224; Frans A. Stafleu, *Taxonomic Literature* (Utrecht-Zug, 1967), pp. 192–193; and D. A. Webb, "William Henry Harvey, 1811–1866, and the Tradition of Systematic Botany," in *Hermathena,* no. 103 (1966), 32–45.

Michael E. Mitchell

AL-ḤASAN IBN MUḤAMMAD AL-WAZZEN. See **Leo the African.**

AL-ḤASAN IBN MŪSĀ IBN SHĀKIR. See **Banū Mūsā.**

HASENÖHRL, FRIEDRICH (*b.* Vienna, Austria, 30 November 1874; *d.* near Vielgereuth, South Tirol, Austria, 7 October 1915), *physics.*

Hasenöhrl was the son of Victor Hasenöhrl, a lawyer, and Gabriele Freiin, the Baroness von Pidall zu Quintenbach. In 1884 he entered the Theresianische Akademie, from which he graduated with high standing in 1892. Although it was a family tradition, he abandoned plans for a military career and continued his education at Vienna University, where he studied mathematics and physics under Franz Exner and Ludwig Boltzmann. After receiving his Ph.D. in 1897, Hasenöhrl spent a year at Leiden as assistant to Kamerlingh Onnes and was then appointed privatdozent at Vienna. He was awarded the Haitinger Prize of the Austrian Academy of Sciences, of which he was a corresponding member, in 1905 and that year he became associate professor at the Vienna Technical University. In 1907 he succeeded Boltzmann as professor of physics at Vienna University; Schrödinger was among his students. At the outbreak of World War I, Hasenöhrl left his university post to join the army. He was killed in the battle of Isonzo.

Hasenöhrl's first systematic research, begun for his Vienna dissertation under Exner[1] and continued in Leiden, was an experimental investigation of the temperature dependence of the dielectric constants of liquids and solids. His object was to explore the range of validity of the Mossotti-Clausius equation

$$\rho \frac{k-1}{k+2} = C,$$

with ρ the density, k the dielectric constant, and C a constant characteristic of the given material. He found, as had P. Lebedev for gases, more than reasonable agreement for a variety of substances over a significant range of temperature.

When Hasenöhrl returned to Vienna he wrote the series of papers for which he is best known, on electromagnetic radiation. Of these the most important is the prizewinning essay on the effects of radiant energy within a moving cavity.[2] Using classical theory he showed that the trapped radiation increases the kinetic energy of the motion, the effect being equivalent to an increase in the apparent mass of the cavity by the amount $\frac{8h}{3c^2}\xi_0$, a result that he soon reduced by half. In his formula $h\xi_0$ is the total radiant energy in the cavity and c is the velocity of light. Like other similar anticipations, this result was displaced by Einstein's more general theorem on the equivalence of mass and energy.

In the last years of his drastically foreshortened career Hasenöhrl turned increasingly to problems in statistical mechanics and considered their relation to the foundations of quantum theory. His most significant result was a by-product, a suggested quantum-theoretical treatment of spectral formulas like those of Balmer. The ostensible object of his paper on the foundations of the mechanical theory of heat was to consider revising classical statistical mechanics to yield Planck's and Einstein's laws for radiation and specific heats.[3] In this work he notes that Planck's simple harmonic oscillators obey the formula

$$dV = t\,dE,$$

with E an energy, t the period of the oscillator, and V the volume of phase space available to an oscillator with energy $\leqq E$. The motion within real atoms, Hasenöhrl points out, cannot be governed by a linear restoring force, and the periods of such motions are therefore energy dependent, $t = t(E)$. Nevertheless, a natural generalization of Planck's approach suggests that these motions too can occur only at discrete energies, $E_1, E_2, E_3, \cdots$, determined by the equations:

$$\int_0^{E_1} t\,dE = \int_{E_1}^{E_2} t\,dE = \int_{E_2}^{E_3} t\,dE = \cdots = h.$$

With the energy levels known, the permitted frequencies of motion are determined by

$$\nu_i = \frac{1}{t(E_i)}.$$

Hasenöhrl also saw that application of these formulas to the pendulum of finite amplitude produces a series, although not quite Balmer's. A short time later K. F. Herzfeld, then a student at Vienna, showed how the Balmer formula could be derived with the aid of special assumptions about the distribution of the positive space charge in the Thomson atom.[4]

NOTES

1. "Über den Temperaturcoefficienten der Dielektricitätsconstante in Flüssigkeiten und die Mossotti-Clausius'sche Formel," in *Sitzungsberichte der K. Akademie der Wissenschaften in Wien*, math.-naturwiss. Klasse, **105** (1896), 460–476.
2. "Zur Theorie der Strahlung in bewegten Körpern," in *Annalen der Physik*, 4th ser., **15** (1904), 344–370; corrigendum, *ibid.*, **16** (1905), 589–592.
3. "Über die Grundlagen der mechanischen Theorie der Wärme," in *Physikalische Zeitschrift*, **12** (1911), 931–935.
4. K. F. Herzfeld, "Über ein Atommodell, das die Balmer'sche Wasserstoffserie aussendet," in *Sitzungsberichte der K. Akademie der Wissenschaften in Wien*, **121** (1912), 593–601.

BIBLIOGRAPHY

In addition to the works mentioned above, see notice by Stefan Mayer, in *Physikalische Zeitschrift*, **16** (1915), 429–433, with portrait.

JOSEF MAYERHÖFER

HASSENFRATZ, JEAN-HENRI (*b.* Montmartre [now Paris], France, 27 December 1755; *d.* Paris, 24 February 1827), *chemistry*.

Hassenfratz was the eldest son of Jean Hassenfratz, called Lelièvre, and of Marie-Marguerite Dagommer; his parents ran a well-known tavern. A master carpenter at the beginning of his career, he gave courses in carpentry for five years and then was a surveyor from 1778 to 1780. In 1782 he was named a mining student, first grade, in the Service des Mines and went to central Europe to study the manufacture of steel and the exploitation of mines. In 1785 he became a deputy inspector of mines. Hassenfratz worked in Lavoisier's laboratory; published, with P. A. Adet, a new chemical notation following the nomenclature of Guyton de Morveau, Lavoisier, Fourcroy, and Berthollet; and taught physics at the École des Mines from 1786 to 1788.

A militant democrat during the Revolution, Hassenfratz was a member of the Society of 1789 and then of the Jacobin Club. By his marriage with Antoinette-Joséphine Terreux he became the brother-in-law of the deputy Baudin des Ardennes. He was for several days a member of the commune of 10 August 1792 and then was director of matériel at the Ministry of War until February 1793. He was again a member of the commune in May 1793 in order to limit the prosecution of the partisans of the Girondins. A collaborator of the Committee of Public Safety, Hassenfratz organized the manufacture of rifles; he later took refuge in Sedan to escape prosecution by the Thermidorians. Renouncing all political activity from the time of the Directory, he taught physics at the École Polytechnique until 1815 and the industrial applications of mineralogy at the École des Mines until 1822. The four volumes of the *Sidérotechnie* published in 1812 constitute his outstanding publication.

BIBLIOGRAPHY

I. Original Works. Hassenfratz's main writings are "Mémoires sur de nouveaux caractères à employer en chimie," in L. B. Guyton de Morveau, A. L. Lavoisier, C. L. Berthollet, and A. F. de Fourcroy, *Méthode de nomenclature chimique* (Paris, 1787), pp. 253–287, written with P. A. Adet; and *La sidérotechnie, ou l'art de traiter les minerais de fer pour en obtenir de la fonte, du fer ou de l'acier,* 4 vols. (Paris, 1812).

He published many letters and memoirs on chemistry and mineralogy in *Observations sur la physique* prior to 1789 and later in *Annales de chimie, Journal des mines,* and *Journal de l'École polytechnique.* The most important are "Lettre sur la matière colorante du bleu de Prusse," in *Observations sur la physique,* **28** (1786), 453–455; "Lettre sur la calcination des métaux dans l'air pur et la décomposition de l'eau," *ibid.,* **29** (1786), 305–306; "Extrait d'un mémoire sur la décomposition des pyrites dans les mines," *ibid.,* **30** (1787), 417–422; "Lettre à M. de la Métherie sur la chimie des pneumatistes," *ibid.,* 215–218, written with P. A. Adet; "Mémoire sur la combinaison de l'oxigène avec le carbone et l'hydrogène du sang, sur la dissolution de l'oxigène dans le sang, et sur la manière dont le calorique se dégage," in *Annales de chimie,* **9** (1791), 261–274; "Mémoire sur le sel marin, la manière dont il est répandu sur la surface du globe et les différents procédés employés pour l'obtenir," *ibid.,* **11** (1791), 65–89; "Explication de quelques phénomènes qui paraissent contrarier les loix des affinités chimiques," *ibid.,* **13** (1792), 3–24, 25–38; "Rapport sur la séparation de l'antimoine et de sa mine," in *Journal des mines,* **9**, no. 54 (1799), 459–471, and *Annales de chimie,* **31** (1799), 154–158; "Discours sur le cours de physique générale," in *Journal de l'École polytechnique,* 6th cahier, **2** (1799), 236–242; "Physique générale. De l'enseignement de cette science," *ibid.,* 372–408; "Premier mémoire sur les ombres colorées," *ibid.,* 11th cahier, **4** (1801), 272–283; "Lettre à A. G. Werner," in *Annales de chimie,* **49** (1804), 129–149; "Mémoire sur la propagation du son," *ibid.,* **53** (1805), 64–75, and *Journal des mines,* **17**, no. 102 (1805), 465–468; "Programme du cours des mines fait à l'École polytechnique (l'année scolaire 1806)," in *Journal de l'École polytechnique,* 13th cahier, **6** (1806), 345–371; "Mémoire sur les altérations que la lumière du soleil éprouve en traversant l'atmosphère," in *Annales de chimie,* **66** (1808), 54–62; and "Mémoire sur la colorisation des corps," *ibid.,* **67** (1808), 5–25, 113–150.

II. Secondary Literature. Unpublished sources are Archives de l'Académie des Sciences: letters from Lavoisier to Hassenfratz dated 22 July 1786, 31 Aug. 1788, 13 July 1790; Archives Nationales, $F^{14}2727^{2}$, Hassenfratz's dossier; Archives de Paris, register no. 6 of marriages 1793–1802, information on Hassenfratz's baptism; Bibliothèque de l'Académie Nationale de Médecine, minute books of the Société Royale de Médecine, no. 4 (30 Nov. 1781–14 Oct. 1783) and no. 5 (17 Oct. 1783–7 Mar. 1786); Bibliothèque de l'Institut de France, MS 2396, 2397, the papers of S. F. Lacroix; Bibliothèque Nationale, MS dept., Charavay card index, Fonds Maçonnique, dossiers on the Bon Zèle and Commanders of Mount Tabor lodges; Ministère de l'Industrie, Direction des Mines, registers of the proceedings of the Conseil Général des Mines; and Musée de la Monnaie, a letter from Balthazar Sage to Hassenfratz dated 20 July 1779.

Published sources of information are Charles Ballot, "Procès-verbaux du Bureau de consultation des arts et métiers," in *Bulletin d'histoire économique de la Révolution* (1913), 15–160; A. Birembaut, "La réintégration de Hassenfratz dans le corps des mines," in *Annales historiques de la Révolution française,* no. 173 (1963), 363–364; Gustave Laurent, "Un mémoire historique du chimiste Hassenfratz," *ibid.,* **1** (1924), 163–164; and Claude-Antoine Prieur-Duvernois, *Quelques vérités sur un proscrit qui, mieux connu, doit cesser de l'être* (Paris, 1795), published anonymously.

Arthur Birembaut

HASSLER, FERDINAND RUDOLPH (*b.* Aarau, Switzerland, 7 October 1770; *d.* Philadelphia, Pennsylvania, 20 November 1843), *geodesy.*

Hassler's career is interesting for two reasons: as an instance of the transfer of scientific skills across the Atlantic and for the study of attitudes toward science in the early United States. He was a trained European scientist who immigrated to America in 1805. Reflecting French influences, Hassler brought with him a set of metric weights and measures and an interest in the determination of the figure of the earth. A coast survey was launched in 1807 but work did not start until 1816 after Hassler had brought back books and instruments from Europe. In the interim he taught at Union College and West Point; during the later suspension of the Coast Survey, 1819–1830, he supported himself by writing textbooks and taking odd jobs. From 1832 to 1843, Hassler headed the revived Coast Survey (now Coast and Geodetic Survey) and acted as superintendent of the Office of Weights and Measures (the predecessor of the National Bureau of Standards).

No one seriously doubts Hassler's role in introducing and maintaining high professional standards in early American science, nor is there any question of his interest in expanding the Coast Survey to cover various geophysical areas, such as terrestrial magnetism and tides. What is doubtful is the assumption that Hassler was the main—if not the sole—channel by which these professional standards and areas were introduced to the United States. The sophistication of his American-trained successor, A. D. Bache, and his success in expanding the Coast Survey implies that Hassler was not the only conduit and that there was a fair degree of receptivity to such scientific work.

This question is significant because the Cajori biography, which is the standard source for Hassler, and other sources make much of his difficulties in the American environment. They rarely question Hassler's ways, which were not always tactful. In emphasizing Hassler, there is often a silence on or a downgrading of others who were active in science at the same time. For example, current research takes a much more favorable view than Cajori's of Andrew Ellicott, Hassler's antagonist in one encounter. Perhaps Hassler's difficulty was one of European style in the American environment.

BIBLIOGRAPHY

Hassler MSS are in the Coast and Geodetic Survey and National Bureau of Standards records in the U.S. National Archives. The New York Public Library has the largest collection of Hassler letters. Additional important documents are in the library of the American Philosophical Society, Philadelphia, Pennsylvania.

Still the best sources for the writings of Hassler and related contemporary works are the bibliographies in G. A. Weber, *The Coast and Geodetic Survey* (Baltimore, 1923) and *The Bureau of Standards* (Baltimore, 1925); and the extensive documentation in Florian Cajori, *The Chequered Career of Ferdinand Rudolph Hassler . . .* (Boston, 1929).

The Cajori biography is very useful because of the extensive research on which it is based. Indeed, some of the MSS cited apparently are no longer extant. Its greatest weakness is its lack of any comparable research on Hassler's contemporaries and the often uncritical treatment accorded its subject's activities.

NATHAN REINGOLD

HATCHETT, CHARLES (*b.* London, England, 2 January 1765; *d.* London, 10 March 1847), *chemistry.*

A skilled analyst, Hatchett was the effective discoverer of the element niobium (columbium). The son of a wealthy coach builder, he enjoyed luxury throughout his life. Of his education, and of the origin of his chemical knowledge, little is known. His father, although disappointed at his son's disinclination to follow his own profession, made him a generous allowance; Hatchett seems to have carried on the business after his father's death. In 1786 he married Elizabeth Collick. In about 1800 he started a small chemical manufacturing business near Chiswick and shortly afterwards took into his laboratory the young William Thomas Brande, whose family had recently moved into the neighborhood. He taught him chemistry and mineralogy and Brande eventually succeeded Davy as professor at the Royal Institution; in 1818 he married Hatchett's second daughter, Anna Frederica.

All of Hatchett's important scientific work was done in the decade 1796–1806. He was elected a fellow of the Royal Society in 1797. His analysis of the mineral now known as columbite or niobite was described in 1801. He showed that it contained a hitherto unknown metal which he called "columbium," believing that the specimen came from America; its origin is, however, doubtful. Columbium almost invariably occurs in association with tantalum, columbite and tantalite differing essentially only in the relative proportions of these metals. Both minerals were analyzed by William Wollaston (1809), who thought he had shown that the two metals were identical. Their difference was established in 1846 by Heinrich Rose, when he "rediscovered" columbium, which he called "niobium" (both names are still current, although the latter has been adopted officially). Niobium was isolated in 1864 by C. W. Blomstrand.

Hatchett acquired a reputation in Great Britain and on the Continent as a mineral analyst, but he also carried out important work on organic materials. His analysis of shell, bone, and dental enamel advanced the knowledge of the composition of these substances, and three papers describing the preparation of an artificial tanning agent contain important observations on resins. Thomas Thomson, who remarked on Hatchett's advance of vegetable chemistry, later lamented his loss to science as a result of the "baneful effects of wealth" and business cares.

BIBLIOGRAPHY

I. ORIGINAL WORKS. Most of Hatchett's papers are listed in Royal Society, *Catalogue of Scientific Papers,* III (London, 1869), 213–214. The papers mentioned in the text are "Experiments and Observations on Shell and Bone," in *Philosophical Transactions of the Royal Society,* **89** (1799), 572–581; "An Analysis of a Mineral Substance From North America Containing a Metal Hitherto Unknown," *ibid.,* **92** (1802), 49–66; and "On an Artificial Substance, Which Possesses the Principal Characteristic Properties of Tannin," *ibid.,* **95** (1805), 211–224, 285–315; **96** (1806), 109–146.

The Hatchett Diary, A. Raistrick, ed. (Truro, 1967), is an edited version of a diary in the possession of Hatchett's descendants describing a journey made in 1796 during which he visited a large number of mines, factories, and geological sites in many parts of England and Scotland; there is a short biography by the ed.

II. SECONDARY LITERATURE. The fullest biographical sketch is E. M. Weeks, "The Chemical Contributions of Charles Hatchett," in *Journal of Chemical Education,* **15** (1938), 153–158, repr. in E. M. Weeks, *Discovery of the Elements,* 7th ed., rev. (Easton, Pa., 1968), pp. 323–343, with additional material on the origin of the mineral analyzed by Hatchett. See also Thomas Thomson, *A System of Chemistry,* 3rd ed. (Edinburgh, 1807), V, 146, 237; and *History of Chemistry,* II (London, 1831), 231. "Charles Hatchett," in Sir J. Barrow, *Sketches of the Royal Society and Royal Society Club* (London, 1849), quotes letters from Hatchett's daughter, Mrs. Brande, and is mainly anecdotal; it includes a list of Hatchett's papers in *Philosophical Transactions of the Royal Society.*

E. L. SCOTT

HATSCHEK, BERTHOLD (*b.* Kirwein, Moravia, Austria, 3 April 1854; *d.* Vienna, Austria, 18 January 1941), *zoology.*

Hatschek, who came from an affluent family, studied zoology in Vienna under Claus and in Leipzig under Leuckart. He gained his doctorate in Leipzig in 1876 with the dissertation "Beiträge zur Entwicklungsgeschichte der Lepidopteren" (1877). He was greatly influenced by Haeckel, with whom he had close ties from 1876 until Haeckel's death in 1919 (Georg Uschmann, *Geschichte der Zoologie und der zoologischen Anstalten in Jena 1779–1919* [Jena, 1959], pp. 133-135).

After acquiring his Ph.D., Hatschek worked in Vienna, where he qualified as a university lecturer in 1884. Upon the recommendation of Haeckel he was appointed as ordinary professor at the Zoological Teaching Council of the German University in Prague, and from 1896 to 1925 he headed the Zoological Institute of the University of Vienna.

Hatschek married Marie Rosenthal, a portrait painter, in 1898. He repeatedly fell into periods of nervous depression which hindered the continuity of his scientific work. From time to time he occupied himself with writing short stories or with inventions, which led him to study ornithopters. He was a member of the Austrian Academy of Sciences and of the Leopoldine German Academy.

Hatschek's scientific works were influenced by the development of contemporary zoology. Questions on the development and systematic placing of animals were being investigated in the light of Darwinian theory. Within this framework he was led to investigate the growth and metamorphosis of the larva of the annelids (or trochophores) and to formulate the so-called trochophore theory (1877–1891). According to this hypothesis, the trochophore stage is typical not only for the annelids but also for the mollusks, so that a relational connection can be established between the two morphologically different animal groups ("trochophore animals").

Hatschek further came to the conclusion that the trochophores, as a transitory phase of development for the annelids and mollusks, corresponded to a permanent state in the rotifers. He saw these forms as the ancestral stock of the classification he named "Zygoneura." Haeckel's influence was evident in Hatschek's indication of the possibility of tracing back the Zygoneura to the trochophore stage and that this larva probably repeated an ancestral stage.

Hatschek later investigated the development of mesoderms in the ctenophorans, presenting a new system. He arranged the Metazoa in three main branches: Coelenterata, Ecterocoelia (or Zygoneura), and Enterocoelia (apropos of this see his *Das neue zoologische System* [Leipzig, 1911]). Hatschek's studies on the amphioxus (*Branchiostoma lanceolatum*), which he began in 1881, led to important information on the development and stratification of this animal.

He also busied himself with the larva of the petromyzons (or ammocoetes). Out of these researches arose his "Studien zur Segmenttheorie des Wirbeltierkopfes" (1906, 1909, 1929). Along with a succession of smaller works, Hatschek published a textbook, *Lehrbuch der Zoologie* (1888–1891), which, like his *Elementarkurs der Zootomie* (1896), was never completed.

BIBLIOGRAPHY

I. ORIGINAL WORKS. "Studien über Entwicklungsgeschichte der Anneliden," in *Arbeiten aus dem Zoologischen Institute der Universität Wien,* **1** (1878); "Studien über die Entwicklung des Amphioxus," *ibid.,* **4** (1882); "Über den Schichtenbau von Amphioxus," in *Anatomischer Anzeiger,* **3** (1888); *Lehrbuch der Zoologie, eine morphologische Übersicht des Tierreiches zur Einführung in das Studium der Wissenschaft,* pts. 1-3 (Jena, 1888–1891); "Die Metamerie des Amphioxus und des Ammocoetes," in *Verhandlungen der Anatomischen Gesellschaft,* **6** (1892).

See also "Studien zur Segmenttheorie des Wirbeltierkopfes": I. "Das Acromerit des Amphioxus," in *Morphologisches Jahrbuch,* **35** (1906); II. "Das primitive Vorderende des Wirbeltierembryos," *ibid.,* **39** (1909); III. "Über das Acromerit und über echte Ursegmente bei Petromyzon," *ibid.,* **40** (1909); and "Über die Mesodermsegmente der zwei Kopfregionen bei Petromyzon fluviatilis," in *Morphologisches Jahrbuch,* **61** (1929); *Das neue zoologische System* (Leipzig, 1911).

II. SECONDARY LITERATURE. Paul Krüger, "Berthold Hatschek zum 80. Geburtstage," in *Forschungen und Fortschritte,* **10** (1934), 120; "Glückwunschadresse der Akademie der Wissenschaften in Wien," in *Almanach für das Jahr 1934* (1935), pp. 257–259; Otto Storch, "Berthold Hatschek," in *Österreichische Akademie der Wissenschaften, Almanach für das Jahr 1949* (1950), pp. 284–296; Wilhelm Marinelli, "Berthold Hatschek," in *Österreichische Naturforscher und Techniker* (Vienna, 1950), pp. 90–93; Helmut Dolezal, "Hatschek, Berthold," in *Neue Deutsche Biographie,* VIII (1969), 56–57.

GEORG USCHMANN

HAUG, GUSTAVE EMILE (*b.* Drusenheim, Alsace, France, 19 July 1861; *d.* Paris, France, 28 August 1927), *stratigraphy, structural geology, paleontology.*

Haug began his study of natural history at the University of Strasbourg. He received a doctorate from that institution in 1884 and remained there for three additional years as a *préparateur.*

In 1887 Alsace experienced increasing political turmoil, a situation which prompted Haug to seek a more suitable intellectual atmosphere for the continuation of his studies. He found it in the geology laboratory of the Faculté des Sciences of Paris, which he had visited briefly in 1883 and 1884. Thus began his brilliant career at the Sorbonne under the supervision of Edmond Hébert and Ernest Munier-Chalmas. His promotion was extraordinarily rapid: he started as lecturer in 1897, was appointed adjunct professor in 1900, and became full professor in 1911. As early as 1888 Haug had been an active associate of the French Geological Survey and in 1902, after receiving several awards, he became president of the French Geological Society. His appointment in 1917 as a member of the mineralogy section of the Institut de France, succeeding Alfred Lacroix, marked the apex of his fame.

Haug's scientific activity was immense and diversified. His teaching was superb and effective, many of his students, such as the paleontologists Léon Pervinquière and Jean Boussac, becoming famous geologists in their own right. One of Haug's little-known but important activities was completing the organization of the geological collection of the Sorbonne, which thus became a first-class reference for stratigraphic and paleontological studies. This admirable working tool was undoubtedly the background for Haug's monumental *Traité de géologie,* two volumes totaling more than 2,000 pages (1907–1911). This thorough compendium of the geological knowledge of his time rapidly became one of the indispensable reference volumes of the profession. Haug's contribution combined profound erudition with sweeping synthetic views in all fields of geology. Still of fundamental importance is Haug's rule, which holds that when subsidence takes place in a geosyncline, a regression of the sea occurs over the adjacent epicontinental areas; conversely, when compression and folding begin in a geosyncline, there is a marine transgression over the epicontinental areas. Haug expressed in this law one of the fundamental relationships between tectonics and sedimentation throughout the geologic column.

Haug's treatise had been preceded by a series of large analytical works dealing with fundamental aspects of paleontology, stratigraphy, and tectonics. In the first of these fields he wrote outstanding memoirs on ammonites. His study of the morphological evolution of these cephalopods led to the very accurate stratigraphic subdivision of certain portions of the Paleozoic and Mesozoic record.

In stratigraphy, Haug investigated in great detail the horizontal variability of facies and its paleogeographic interpretation. Through this approach he unraveled the sedimentary history of the Jurassic of the Rhone basin. Then he was led to a critical examination of the concept of geosyncline. More precision was introduced into the original definition, and the model was shown to be of general application throughout the geologic column. Unquestionably Haug's memoir of 1900, *Les géosynclinaux et les aires continentales,* represents a turning point in the interpretation of the geological record through the combination of tectonics and paleogeography.

All the data collected by Haug during the detailed mapping he had undertaken since 1888 in the French Alps were published in numerous regional monographs, the two most remarkable of which deal with the stratigraphy and structure of the subalpine chains between Gap and Digne (1891) and with the structure of the high calcareous ranges of Savoy (1895). In 1903 Haug and C. W. Kilian discovered the great overthrusts of the Ubaye and Embrunais which confirmed the structure of the Western Alps as set forth by Marcel Bertrand and Pierre Termier. After an investigation of the structure of the Jura Mountains, Haug concentrated his attention on the structure of the northern calcareous ranges of Austria (near Salzburg in the Salzkammergut) and was among the first to unravel the structural pattern of the Dolomites. His revision of several quadrangle maps of southern France led to many contributions, the most outstanding of which is a memoir on the structure of Basse-Provence (1925–1930); the second part appeared posthumously.

Haug's synthetic approach to stratigraphy and structural geology was not limited to Europe. He also took a great interest in the geology of the Sahara and of North Africa.

BIBLIOGRAPHY

Haug's writings include "Les chaînes subalpines entre Gap et Digne, contribution à l'histoire géologique des Alpes françaises," *Bulletin du Service de la carte géologique et des topographies souterraines,* **3,** no. 21 (1891–1892); "Les ammonites du Permien et du Trias. Remarques sur leur classification," in *Bulletin de la Société géologique de France,* 3rd ser., **22** (1894), 385–412; "Études sur la tectonique des hautes chaînes calcaires de la Savoie," *Bul-*

letin du Service de la carte géologique et des topographies souterraines, **7,** no. 47 (1895–1896); "Études sur les goniatites," *Mémoires de la Société géologique de France,* no. 18 (1898); "Les géosynclinaux et les aires continentales. Contribution à l'étude des transgressions et régressions marines," in *Bulletin de la Société géologique de France,* 3rd ser., **28** (1900), 617–711; "Les grands charriages de l'Embrunais et de l'Ubaye," in *Comptes rendus du IX Congrès international de géologie,* I (Vienna, 1904), 493–506; "Sur les dislocations des environs de Mouthier-Haute Pierre (Doubs)," *Bulletin du Service de la carte géologique et des topographies souterraines,* **17,** no. 112 (1905–1906), written with C. W. Kilian; "Les nappes de charriage des Alpes calcaires septentrionales. Partie I. Introduction. Partie II. Alpes de Salzbourg," in *Bulletin de la Société géologique de France,* 4th ser., **6** (1906), 359–422; *Traité de géologie,* 2 vols. (Paris, 1907–1911); "Les nappes de charriage des Alpes calcaires septentrionales. Partie III. Le Salzkammergut," in *Bulletin de la Société géologique de France,* 4th ser., **12** (1912), 105–142; "La tectonique du massif de la Sainte-Baume," *ibid.,* **15** (1915), 113–190; "Contribution à une synthèse stratigraphique des Alpes occidentales," *ibid.,* **25** (1925), 97–244; and *Les nappes de charriage de la Basse-Provence.* Monographies tectoniques, 2 vols. (Paris, 1925–1930).

Information on Haug may be found in E. de Margerie, "Discours aux funérailles de Emile Haug," in *Notices et discours. Académie des sciences,* 2nd ser., **1** (1937), 157–162.

ALBERT V. CAROZZI

HAUKSBEE, FRANCIS (*b.* Colchester [?], England, *ca.* 1666; *d.* London, England, April 1713), *experimental physics, scientific instrumentation.*

Called "the elder" to distinguish him from his nephew of the same name, Francis Hauksbee is remembered for his experiments on electroluminescence, static electricity, and capillarity performed between 1703 and 1713. His discoveries were first shown at meetings of the Royal Society of London, published in a series of papers in the Society's *Philosophical Transactions,* and finally brought together in his *Physico-Mechanical Experiments on Various Subjects* (London, 1709), of which a second edition appeared posthumously (1719). Translated into Italian, Dutch, and French, Hauksbee's book was widely read in the eighteenth century. As historians of electricity from Joseph Priestley on have recognized, sustained experimentation in that subject began with Hauksbee; his demonstration that glass is a convenient and malleable material for producing frictional electricity opened the way for the work of Stephen Gray, Charles de Cisternay Dufay, and Benjamin Franklin.[1] His discoveries had a marked influence on the later speculations of Isaac Newton; and a century later Laplace turned to Hauksbee's book when he embarked on his study of capillarity.

Emerging from total obscurity, Hauksbee made his debut before the Royal Society at the meeting of 15 December 1703, the first to be presided over by the newly elected president, Isaac Newton. On this occasion Hauksbee showed a striking experiment: when mercury rushed into the evacuated receiver of his new model air pump, spilling over an inverted glass vessel, the result was a sparkling light, "a Shower of Fire descending all round the Sides of the Glasses."[2]

In succeeding months Hauksbee appeared as a paid performer before the Society, carrying out a variety of experiments with his air pump. His position was soon regularized; after 1704 he served as the Society's demonstrator or curator of experiments, although without this title which Robert Hooke and Denis Papin had enjoyed in earlier years. This arrangement was unchanged when, on 30 November 1705, Hauksbee was elected a fellow of the Society. He continued, with a diligence matched only by his ingenuity, to perform experiments at the Royal Society until the onset of his last illness early in 1713.

Of Hauksbee's origins and early life we know little. Preliminary investigations in the London records have yielded a few facts and allow some reasonable conjectures.[3] He was born about 1666, the son of one Richard Hauksbee, a draper of Colchester; in December 1678 he was apprenticed in that trade to his older brother; and by 1687, his apprenticeship completed, he had married, for in that year a daughter was born, the first of several children his wife Mary was to bear him. He was therefore in his late thirties when he first appeared before the Royal Society.

Under what circumstances Hauksbee embarked on a new career we do not know; if he served as assistant to some notable scientist, it was perhaps Papin, the curator of experiments at the Royal Society from 1684 to 1687.[4] Nor do we know who invited Hauksbee to appear before the Royal Society. It may have been Newton, for the new president clearly wished to see the Society revive its former practice of having experiments performed at its meetings. Hauksbee was already recognized as an instrument maker and experimenter of great skill; he was giving demonstrations at his house or shop in Giltspur Street by 1704, for in that year he engaged the mathematician James Hodgson to lecture for him.[5] From Giltspur Street, Hauksbee moved to Wine Office Court, Fleet Street, where he was visited in 1710 by the German traveler Zacharias Conrad von Uffenbach, and began his own public lectures, attended in that year by another German visitor, Abraham Vater.[6] Printed advertisements for these lectures show that by 1712 Hauksbee had moved to Hind Court, Fleet Street,[7] where he was living at the time of his death. His last appearance

HAUKSBEE

before the Royal Society was on 29 January 1712/13; he died late in April of that year and was buried at St. Dunstan's-in-the-West on 29 April.

Except for his demonstration in December 1703, Hauksbee's experiments during the first two years at the Royal Society were largely repetitions of earlier ones performed by Robert Boyle, Robert Hooke, and—notably—Denis Papin. Late in 1705 Hauksbee turned to those experiments that led him step by step to his investigations in electricity. It was the striking phenomenon of the "mercurial phosphorus," the subject of his first performance before the Society, that commanded his attention.

The subject was of some contemporary interest. In 1676 the French astronomer Jean Picard had noted that when he carried about a barometer in the dark, jostling the mercury, a luminosity appeared in the Torricellian vacuum at the top of the tube. The subject was further investigated by the Swiss mathematician Johann I Bernoulli, who reported that he could produce "a portable and perpetual phosphor" by shaking mercury in an exhausted glass vessel.[8]

It was this phenomenon that Hauksbee now set out to examine more closely. In the autumn of 1705 he showed the Society a number of variations of this spectacular effect, for example producing the mercurial light by forcing air upward through mercury in the evacuated receiver of his air pump. Not content with mere showmanship, he explored the precise conditions under which the light could be produced. On varying the air pressure, he found that while the glow did not appear in "so dense a medium as common air," it did not require all the air to be drawn out. He observed, too, that the greater the motion of the mercury, the stronger the light produced; clearly the cause of the light was the friction of the mercury against the glass. But was this a property peculiar to mercury, or would other substances, when strongly rubbed in a vacuum, also yield a light?

To explore this question, Hauksbee built a contrivance by which substances could be rubbed together in the receiver of his air pump. With this device he showed that when beads of amber were rubbed against woolen cloth, a light was produced that was brighter *in vacuo* than in air. By contrast, when flint and steel were struck together, no sparks were produced until air was admitted.

On 19 December 1705, Hauksbee showed the Society an experiment that was to prove a significant advance. A small glass globe was mounted on a spindle and rotated with great speed against woolen cloth attached to the tightly grasping arms of a brass spring. When this was done in the evacuated receiver, there was "quickly produced a beautiful Phaenomenon, *viz.*, a fine *purple* Light, and *vivid* to that degree, that all the included *Apparatus* was easily and distinctly discernible by the help of it."[9] But when air was let in, the light lost its color and intensity.

Although in succeeding months he showed experiments on quite different matters to the Society, Hauksbee continued to investigate the central problem in his own laboratory, devising a still more striking way of producing light by the friction of glass. Taking a glass globe about nine inches in diameter, he drew out the air and fixed the globe to a machine that gave it a swift rotary motion. When, in his darkened room, he pressed his open hand against the spinning globe, there was produced a purple light, so brilliant that "Words in Capital Letters became legible by it." Yet if he applied friction to the globe when it was full of air, the light within the globe disappeared; instead, luminous specks adhered to objects brought close to the glass, and his own neckcloth was seen to glow.[10]

In describing this experiment, Hauksbee made no reference to electrical effects. What interested him at first was that his rubbed globe, when full of air, had the mysterious "Quality of giving Light to a Body held near it." The effect was soon clarified. That a light resembling that produced by agitating mercury in a vacuum could be produced from glass led him, at a date we cannot determine, to think of briskly rubbing the upper part of a barometer tube without disturbing the mercury. When he did this, a faint light was produced. The next step was obvious: to rub larger evacuated glass tubes. The result was that a noticeable light was emitted, but there was no "giving light to a Body near it." But when hollow tubes open to the air were similarly rubbed in the dark, no luminosity appeared within them. Instead, accompanied by a faint crackling sound, a light seemed to fix on nearby objects, such as his hand or pieces of gold, brass, ivory, or wood.

In daylight equally striking effects were observed, and these Hauksbee recognized as the sort of "electric" effects described by Gilbert, Boyle, and other pioneers. No sooner were the tubes vigorously rubbed than light bits of leaf brass were drawn to them and, as suddenly, violently repelled. With his tubes, and with solid glass rods, Hauksbee observed not only electrical attraction but also the phenomenon of electrical repulsion, which, although first described by Niccolo Cabeo, had been overlooked by other early experimenters. Hauksbee also detected the electric wind, for when he held a strongly rubbed tube or rod near his face, he felt a sensation as if "fine limber hairs" were brushing against it. He found, too, that he could screen off these electrical effects by means of a piece of fine muslin.[11]

When Hauksbee presented his results to the Royal Society in November 1706, the president (Newton) remarked that "he thought those experiments evinced that Light proceeded from the subtle effluvia of the glass & not from ye gross body" used to supply the friction.[12] Although Hauksbee knew that other bodies might produce a light when rubbed in a vacuum, it was the properties of glass that henceforth became his chief concern. He built an improved version of what was, in effect, the first triboelectric generator. Taking a glass vessel "as nearly *Cylindrical* as might be," he mounted it horizontally on his machine. To study the behavior of the "luminous Effluvia" he devised a primitive electroscope, a semicircle of wire to which woolen threads were fastened so as to reach within an inch of the upper surface of his cylindrical glass globe. When this was spun, the threads were swept aside by the air; but if friction was applied to the glass, the threads straightened out, pointing toward the center of the globe. Later, Hauksbee attached a series of woolen threads to the rim of a wooden disk. This disk he placed within a glass globe filled with air; when the globe was spun and friction applied to it, the threads were seen to extend themselves outward "every way towards the circumference of the Glass."[13]

What particularly struck Hauksbee in these experiments was that when he brought his finger close to the surface of the activated globe, the loose ends of the threads were repelled. Clearly this mysterious "effluvium" could pass through glass. A dramatic experiment confirmed this inference. Hauksbee took a glass globe exhausted of its air and placed it near his machine, which swiftly spun a globe full of air. When friction was applied to the moving globe, the empty globe nearby was seen to glow.[14]

Hauksbee was not one to amuse himself—as he wrote in words that recall Newton's—"with Vain Hypotheses, which seem to differ little from Romances." Yet he was compelled to offer some conjectures about the phenomena he described with such accuracy. Like earlier observers of electric effects—like Gilbert, Boyle, and Newton—he spoke of "effluvia": of an active matter lodged in glass and other "electric bodies" and released (just how he was not certain, although the heat of friction seemed to play a part) when such bodies are strongly rubbed. This "subtile" fluid is probably particulate but certainly material; propagated with a considerable force, it shows the "powerful effects of *small* bodies, when put into brisk and vigorous motion." Although doubtless composed of minute bodies, the effluvia possess a certain continuity; they move "as it were in so many *Physical Lines,* or *Rays,*" and seem to progress "in a streight and direct track." Yet the irregular motion, now attracted and now repelled, of light bodies near rubbed glass shows that the effluvial motion "is not equable and regular, but disorderly, fluctuating and irregular." The effluvial force seems "to exert it self (as it were) by fits": sometimes as an attractive force, sometimes as a force of repulsion. In these "smaller Orbs of Matter," Hauksbee wrote, "we have some little Resemblances of the Grand Phaenomena of the Universe."[15]

Hauksbee's curiosity was wide-ranging. He published experiments on the propagation of sound in compressed and rarefied air, on the freezing of water, and on the rebounding of bodies in media of various densities. He made precise measurements of the specific gravity of different solids and, probably at Newton's urging, carefully measured the refractive index of various fluids.[16] But the most important of his later investigations were those on capillarity and surface forces.

The rise of fluids in small open tubes, which indeed had been observed *in vacuo,* was frequently recorded during the seventeenth century.[17] But Hauksbee was the first to explore the subject persistently and with care. In an early demonstration before the Royal Society he took three small tubes of different diameters, plunged them in a vessel of colored water in the receiver of his vacuum pump, and showed that the fluid rose in the tubes, whether or not the air had been withdrawn, and that it rose higher in the tubes of smaller internal diameter than in larger ones.

Hauksbee returned to the problem again in 1708, convinced that attractive forces must be at work and that the phenomenon was not peculiar to glass tubes but might be the consequence of some "Universal Establish'd Law of Nature." Later that year he demonstrated at the Royal Society that a colored liquid would rise between two flat glass plates and that the height to which it rose varied with the separation of the plates. He was able to show, too, that the phenomenon was not peculiar to glass but could be observed with plates of marble or of brass, and that fluids likewise rose through tubes filled with carefully sifted ashes. In all cases the experiments succeeded as well *in vacuo* as in the open air. And just as he varied the material of his plates, so he observed the capillary rise of liquids other than water: alcohol, turpentine, and what he called "Common Oil."[18]

One experiment was significant far beyond the

space Hauksbee devoted to describing it, for it proved that the attractive force came into play only at the inner surfaces. When he took two tubes of equal internal diameter, one with walls ten times as thick as the other, the water rose to the same height in both.

How extensively Hauksbee discussed his results with Newton and other fellows of the Royal Society we cannot know, but the Newtonian trademark is clearly evident in his book, where he speaks reverently of him. It seems clear that Hauksbee, an experimental genius but a scientific autodidact, derived his theoretical principles from Newton.

Reviewing the results obtained up to this point on the problem of capillarity, Hauksbee emphasized that all its manifestations could be reduced to the simple case of small tubes and could be explained by the same cause, which could only be attraction: "A Principle which governs far and wide in Nature, and by which most of its Phaenomena are explicable." That there is a power in nature by which the parts of matter "do tend to each other" is past dispute, for the discoveries made by Newton ("the Honour of our Nation and Royal Society") have established it beyond cavil. It operates not only in the "larger Portions or Systems of Matter" but also between minute and "insensible" corpuscles.[19] Newton has "fully determin'd and settled" the law according to which attraction acts between larger bodies. Although the law of attraction acting at minute distances between "the *smaller Portions* of Matter" has not yet been determined, capillary rise may nevertheless be "handsomely accounted for" by its action.

Such confident assertions about interparticulate attractions show that Hauksbee was familiar with the ideas Newton had published a few years before in the last new query of the Latin edition (1706) of his *Opticks.* But by the time his own book appeared, Hauksbee knew something more, since he remarks that the law of attraction at very small distances, although not yet discovered, must be different from the law operating in the case of large bodies. It is known, he says, "that the attractive Forces here [at short distances] do decrease in a greater proportion than that by which the Squares of the Distances do encrease."[20]

Of the experiments performed by Hauksbee in his last years, a number dealt with problems concerning attraction. At Newton's request he tried without notable success to discover the law of magnetic attraction,[21] and he carried out a large-scale experiment in which spherical bodies of different weight—a ball of cork, a thin glass bubble filled with mercury, a thin glass sphere—were let drop from the cupola of St. Paul's Cathedral; the time of fall was measured by counting the beats of a pendulum set in motion by an ingenious contrivance at the instant the bodies were released.[22]

In January 1711/12 Hauksbee "shewed a very Curious Experiment" in which a droplet of oil of oranges was observed to move between two closely applied glass plates. The drop was placed on one of the glass plates; the other was laid over it at a slight angle, just touching the drop and forming a wedge-shaped configuration with the plates in contact at one end. When this was done, the drop was seen to move toward the end where the plates came together; and the motion was still noted when the point of contact was raised some eight or ten degrees. It was certainly with Newton's approval, if not at his suggestion, that Hauksbee was urged to publish this paper "as relating to ye farther Discovery of the Nature of Congruity or the Agreement of the Parts of Matter."[23]

Hauksbee's next step was suggested by Newton, for we learn from the entry in the *Journal Book* for 22 May 1712 that after Newton had "directed Mr. Hauksbee" to give a progress report on his magnetic experiments,

> The President also proposed the making an Experiment of the Drop of Oyle between two Glasse-Planes in vacuo, so as to ascertain the Proportion of the Power of Gravity and [the] congruity or agreement of ye Parts, by observing at what Angle the Drop is observed to be Stationary and not to move toward the Edge of the Wedge formed by the two Glasses.[24]

This remarkable experiment Hauksbee promptly performed, although apparently not *in vacuo,* and gave the Society a written account of it on 5 June 1712. He took two glass strips, and at the midpoint of one of these he placed a drop of oil of oranges. When the other glass was laid over the first one, the oil spread out between the surfaces. But when the upper plate was raised at one end, a droplet of oil quickly formed and moved, as in his earlier experiment, toward the end where the glasses touched. When that end was raised in its turn, with the drop at various distances from the center of the plates, Hauksbee measured the elevation that arrested the motion of the drop. At the different distances from the center, as the space between the plates narrowed, the droplet was gradually compressed, becoming at first oval and then "more and more oblong," with an increasing area of surface contact between the oil and the glass. His results, summarized in a table, show that the farther the drop was from the center (and the closer to the point of contact of the glasses) the higher the two plates had to be raised to arrest the motion of the drop. As the area of contact increased, so did the attractive force opposing gravity.[25]

The experiment that occupied Hauksbee during the last months of his service to the Royal Society was suggested by another scientist of note, the mathematician Brook Taylor. In a letter to Hans Sloane, dated 25 June 1712, Taylor described an experiment which seemed to show that the curve of the surface of water between two panes of glass, inclined at a slight angle to each other, apparently resembled "the common *Hyperbola.*" But Taylor confessed that his "Apparatus was not nice enough," that is, not accurate enough, to make this certain.[26] His letter was read to the Society on 26 June, and "Mr. Hauksbee was desired to consider this Letter, and to prepare any Experiments he thinks proper."[27]

Hauksbee set to work, adopting the arrangement described by Taylor, and reported his experiments on 31 July and again after the Society's recess. He dipped two glass plates in colored water so that they formed a V when viewed edgewise. He then carefully measured the cross section of the meniscus, as we would call it, and confirmed Taylor's conjecture. Indeed, further experiments, in which the two plates were put into the water at different angles, always showed one limb of the hyperbola to be asymptotic to the surface of the water and the other to a line drawn along either of the inclined plates.[28]

As president of the Royal Society and an elder statesman, Newton could have been expected to rest on his very considerable laurels. He was, after all, in his sixties during the period of Hauksbee's activity; he had passed seventy when Hauksbee died. Yet Newton's powerful mind had not lost its edge, and his interest in science was undiminished; he rarely failed to preside over the Society's assemblies and often commented from the chair on the proceedings. The *Journal Book* leaves little doubt of his intense interest in Hauksbee's experiments: he offered criticisms and on more than one occasion suggested experiments, notably such as would elucidate the mystery of attraction, that Hauksbee should perform. What soon emerged was a unique collaboration between the venerable dean of English science and the vigorous, gifted experimenter. From Newton, Hauksbee came to understand the theoretical import of some of his discoveries; and for his part the older man relied on Hauksbee's practiced hands to test some of his conjectures. What particularly interested Newton we learn from the changes he made during this period in new editions of his two major works, the *Opticks* and the *Principia.*

When Newton brought out in 1706 the Latin version of his *Opticks,* the additions he made to the queries appended to that book reveal here and there echoes of Hauksbee's early experiments, notably those dealing with the mercurial light.[29]

In the second edition of his *Principia* (1713), in his discussion of the motion of bodies through resisting mediums, Newton made use of Hauksbee's results in dropping spheres of different weights from the cupola of St. Paul's. Moreover, Hauksbee's electrical experiments clearly inspired the cryptic concluding paragraph of the Scholium Generale, where Newton speaks of an "electric and elastic spirit" which, he says, "pervades and lies hid in all gross bodies."[30]

Further evidence of Hauksbee's influence is found in the extensive changes Newton contemplated for the second English edition of his *Opticks* (1717–1718). Sometime after 1713 Newton considered giving an account of Hauksbee's experiments in a series of new "observations" continuing the eleven of Book 3 which dealt with diffraction. Somewhat later, greatly impressed by some experiments performed at his suggestion by Hauksbee's successor as the Society's demonstrator of experiments, J.-T. Desaguliers, Newton for a moment thought of including these experiments, together with Hauksbee's, as a second part of Book 3.[31] But caution prevailed, and instead Newton incorporated this material in new queries and additions to the older ones. An account of Hauksbee's electrical experiments with the rotating glass globe and rubbed glass tubes found its place in an addition to query 8. Hauksbee's experiments on capillary rise in fine tubes and between plates of glass, as well as the experiment on the motion of a drop of oil of oranges between glass plates, are described in an extensive addition of several pages that Newton inserted into query 23/31.[32]

This second English edition of the *Opticks* discloses a remarkable revision in Newton's theory of matter. In the queries of 1706 Newton describes a world in which particles of matter move in empty space under the mysterious force of mutual attraction.[33] By 1717 he had abandoned this position and returned, with some modifications, to his earlier view that a tenuous "Aether or Aetherial elastic spirit" could perhaps best account for many of the phenomena of nature, including attraction itself. Hauksbee's experiments, as far as Newton was concerned, had made this subtle kind of matter perceptible to the senses. That bodies contain such a spirit "wch by friction they can emit to a considerable distance" and which is subtle enough to pass through glass, yet active enough to cause light to be emitted from gross bodies and produce other startling effects, Newton found "manifest" in certain phenomena "shewed to the R. Society by Mr Hawksby."[34]

NOTES

1. On this point see Abbé J. A. Nollet, *Essai sur l'électricité des corps,* 3rd ed. (Paris, 1754), p. 4.
2. *Physico-Mechanical Experiments on Various Subjects,* 2nd ed. (London, 1719), p. 9. This ed., available in a modern facs. repr. (see bibliography), will be cited here as *Experiments* (1719). For Hauksbee's first appearance, see the Royal Society's MS *Journal Book,* IV (1702–1714), p. 37. All dates in this article are given in Old Style.
3. A summary of documents relating to the Hauksbee family, most of them in the Guildhall Library, London, was compiled by Mr. D. Dawe and kindly communicated by Dr. H. Drubba of Hannover, Germany. The inferences drawn from them are the present writer's.
4. In the "Epistle Dedicatory," printed in the 1st ed. (1709) of his book, Hauksbee writes of his "Want of a Learned Education." Without supporting evidence, M. Edmond Bauer suggested that Hauksbee had been a "pupil" of Robert Boyle. See René Taton, ed., *Histoire générale des sciences,* II (Paris, 1958), 521. The chronology suggested above makes this unlikely but would fit Papin; Hauksbee's improved air pump was based on Papin's invention, and a number of his early experiments were ones performed earlier by Papin. For Hauksbee's air pump and its derivation from Papin's, see Henry Guerlac, "Sir Isaac and the Ingenious Mr. Hauksbee," in I. Bernard Cohen and René Taton, eds., *Mélanges Alexandre Koyré,* I (Paris, 1964), 240–242.
5. E. G. R. Taylor, *Mathematical Practitioners,* pp. 288, 296. John Harris, in the pref. of his *Lexicon technicum* (London, 1704), lists Hauksbee among the "Ingenious and Industrious Artificers" who make mathematical and philosophical instruments. His improved air pump was widely copied by others, including Richard Bridger, who is described as having been Hauksbee's apprentice. See W. Vream, *Description of the Air-pump* (London, 1717).
6. W. H. Quarrell and Margaret Ware, eds., *London in 1710 From the Travels of Zacharias Conrad von Uffenbach* (London, 1934), pp. 77, 168. For Vater, see J. H. S. Formey, *Éloges des académiciens de Berlin,* II (Berlin, 1757), 159.
7. One of Hauksbee's advertisements is reproduced in Lawrence Lewis, *The Advertisements of the Spectator* (London, 1909), no. 275. In the pref. to his *Course in Experimental Philosophy,* J.-T. Desaguliers contrasts Hauksbee's lectures unfavorably with those of John Keill.
8. Accounts of this episode are by Park Benjamin, *Intellectual Rise in Electricity,* pp. 453–457; and by W. E. Knowles Middleton, *The History of the Barometer* (Baltimore, 1964), ch. 13. For the influence of Bernoulli's letters on a French contemporary of Hauksbee's, see David W. Corson, "Pierre Polinière, Francis Hauksbee, and Electroluminescence: A Case of Simultaneous Discovery," in *Isis,* **59** (1968), 402–413.
9. *Experiments* (1719), pp. 30–31. Had Hauksbee succeeded in obtaining a nearly perfect vacuum, the effect would not have been produced.
10. *Ibid.,* pp. 45–49. Newton may have had these experiments in mind when he wrote Hans Sloane in September 1705, asking him "to get Mr. Hawksbee to bring his Air-pump to my house [in Jermyn Street] & then I can get some philosophical persons to see his Expts who will otherwise be difficultly got together." For this note, and a later one indicating that this private demonstration was canceled, see *Correspondence of Isaac Newton,* IV (Cambridge, 1967), 446–447, 448. Cf. John Nichols, *Illustrations of the Literary History of the Eighteenth Century,* 8 vols. (London, 1817–1858), IV, 59.
11. That Hauksbee performed some experiments with rubbed glass tubes in the spring of 1706 is suggested by a memorandum of David Gregory, dated 15 May 1706. W. D. Hiscock, ed., *David Gregory, Isaac Newton and Their Circle* (Oxford, 1937), p. 35.
12. *Journal Book,* IV, 100; entry for 6 Nov. 1706. Newton doubtless had in mind his own early experiment of 1675 showing the "effluvium" of a rubbed glass disk and its effects on light objects. See Thomas Birch, *History of the Royal Society of London,* III, 250–251.
13. *Experiments* (1719), pp. 65–75, 139–140.
14. *Ibid.,* pp. 79–82.
15. *Ibid.,* pp. 75, 81, 142.
16. Hauksbee's figures (and his table of the specific gravities of various liquids) are quoted by Joseph Priestley, *The History and Present State of Discoveries Relating to Vision, Light and Colours* (London, 1772), pp. 164–165, 481.
17. E. C. Millington, "Theories of Cohesion in the Seventeenth Century," in *Annals of Science,* **5** (1945) 253–269.
18. *Experiments* (1719), pp. 179–199.
19. *Ibid.,* pp. 200–201.
20. *Ibid.,* p. 201. Hauksbee may have been persuaded by John Keill's letter on attractions read to the Royal Society on 16 June 1708 (*Journal Book,* IV, 146) and published in the *Philosophical Transactions of the Royal Society,* no. 315 (for May–June 1708), 97–110; see Keill's "Theorema IV."
21. *Journal Book,* IV; see numerous entries for the spring of 1712. Hauksbee's first magnetic experiments were performed in collaboration with Brook Taylor, using the great lodestone belonging to the Royal Society. A paper on later experiments was published in *Philosophical Transactions,* no. 335 (for July–Sept. 1712), 506–511, and reprinted posthumously in the supp. to *Experiments* (1719).
22. *Experiments* (1719), pp. 278–281.
23. *Journal Book,* IV, 266–267. In *Experiments* (1719) he writes (p. 303): "I have since repeated the same Experiment *in Vacuo,* where, in all respects, it answer'd as in the open air."
24. *Journal Book,* IV, 294.
25. *Philosophical Transactions,* no. 334 (for April–June 1712), 473–474. Reprinted in the supp. to *Experiments* (1719), pp. 309–311.
26. The concluding part of the letter was published in *Philosophical Transactions,* no. 336 (for Oct.–Dec. 1712), 538.
27. *Journal Book,* IV, 300. The early part of Taylor's letter dealt with experiments on magnetic attraction performed with Hauksbee. See Taylor's paper in *Philosophical Transactions,* no. 344 (for June–Aug. 1715), 294–295.
28. *Journal Book,* IV, 306; later demonstrations are recorded for 30 Oct. and 6 Nov. 1712. See *Philosophical Transactions,* no. 336, pp. 539–540; and no. 337, pp. 151–154: these papers are reprinted in *Experiments* (1719), pp. 314–315, 331–333.
29. Guerlac, "Sir Isaac and the Ingenious Mr. Hauksbee," pp. 250–252.
30. *Philosophiae naturalis principia mathematica,* 2nd ed. (Cambridge, 1713), pp. 325–326. For Hauksbee's influence on the Scholium Generale see Henry Guerlac, "Francis Hauksbee: Expérimentateur au profit de Newton," in *Archives internationales d'histoire des sciences,* **16** (1963), 124–127.
31. University Library, Cambridge, MS Add. 3970 (9), described in Henry Guerlac, "Newton's Optical Aether," in *Notes and Records. Royal Society of London,* **22** (1967), 45–57.
32. See *Opticks,* 2nd ed. (1718), pp. 315, 366–369; cf. Guerlac, "Francis Hauksbee: Expérimentateur au profit de Newton," pp. 122–123.
33. The changes are carefully enumerated by Alexandre Koyré, "Les queries de l'Optique," in *Archives internationales d'histoire des sciences,* **13** (1960 [published 1961]), 15–29. For their significance see Henry Guerlac, *Newton et Epicure,* Conférence au Palais de la Découverte (Paris, 1963), and papers cited above.
34. University Library, Cambridge, MS Add. 3970 (9), fol. 626.

BIBLIOGRAPHY

I. ORIGINAL WORKS. Hauksbee's book is *Physico-Mechanical Experiments on Various Subjects. Containing An Account of Several Surprizing Phenomena Touching*

Light and Electricity (London, 1709); *Experienze fisico-mecchaniche sopra vari soggetti* . . . (Florence, 1716), the version chiefly read in France, used by the pioneer electrician, C. F. de Cisternay Dufay; *Physico-Mechanical Experiments on Various Subjects* . . ., 2nd ed. (London, 1719), which omits Hauksbee's "Epistle Dedicatory" to Lord John Somers but has a "supplement" consisting of papers Hauksbee published in the *Philosophical Transactions* after the appearance of the 1st ed.—there is a modern facs. repr., no. 90 in Sources of Science, with a useful intro. by Duane H. D. Roller (New York-London, 1970); *Natuurkundige en tuigwerkelyke ondervindingen over verscheide onderwerpen . . . . Uit het Engelich vertaalt door P. Le Clercq* (Amsterdam, 1735; repr. 1754), based on the 2nd English ed., with repr. differing only in the imprint on the title page; and *Expériences physico-méchaniques sur différens sujets* . . ., translated by M. de Brémond, 2 vols. (Paris, 1754), prefaced by a valuable "Discours historique et raisonné sur les expériences de M. Hauksbée," by Nicolas Desmarest, who points out the mutual influence of Hauksbee and Newton.

II. SECONDARY LITERATURE. Works of general value are the following (listed chronologically): *Angliae notitia* (London, 1707), p. 50, with an early printed reference to "the ingenious Mr. Francis Hauksbee, & his work on the air pump"; Robert Smith, ed., *Hydrostatical and Pneumatical Lectures by Roger Cotes,* 2nd ed. (London, 1747), lectures delivered not long after Hauksbee's death, which contain references to Hauksbee's experiments on capillarity (p. 268) and his air pump (pp. 249–250); Charles Hutton, *Mathematical and Philosophical Dictionary,* I (1796), arts. "Capillary Tubes" (p. 243), "Electricity and Electrical Force" (p. 420), and "Electrometer" (p. 423); Robert Watt, *Bibliotheca britannica,* I (London, 1824), col. 474, which gives no biographical information but merely lists Hauksbee's publications, including papers in the *Philosophical Transactions*—the list has two works actually by the younger Hauksbee; R. E. Anderson, "Hauksbee, Francis, the Elder," in *Dictionary of National Biography,* a mere summary of Hauksbee's book; and E. G. R. Taylor, *Mathematical Practitioners of Tudor and Stuart England* (Cambridge, 1967), pp. 296–297.

On Hauksbee's work in electricity, see the following (listed chronologically): G. J. 'sGravesande, *Physices elementa mathematica,* 2 vols. (Leiden, 1720–1721), with an account of electrical experiments based on Hauksbee's in II, 1–10, and an illustration of a number of Hauksbee's contrivances in pl. 1; J.-T. Desaguliers, *Course of Experimental Philosophy,* 2 vols. (1734–1744), description of electrical experiments with commentary on Hauksbee's work in I, 17–21; Charles de Cisternay Dufay, "Premier mémoire sur l'électricité—Histoire de l'électricité," in *Mémoires de l'Académie royale des sciences* (Paris, 1735), pp. 23–25, read 15 Apr. 1733, this paper included an important early account of Hauksbee's work on electricity; the unsigned "An Historical Account of the Wonderful Discoveries, Made in Germany, &c. Concerning Electricity," in *Gentleman's Magazine,* **15** (1745), 193–197, which mentions Hauksbee as the inventor of the globe generator and singles out his remark that electrostatic discharges resemble lightning, since they both produce "flame as well as light"; Daniel Gralath, "Geschichte der Electricität," in *Versuche und Abhandlungen der Naturforschenden Gesellschaft in Danzig* (Danzig, 1747), pp. 184–188; Joseph Priestley, *History and Present State of Electricity* (London, 1767), pp. 15–25, with Period II devoted to "the Experiments and discoveries of Mr. Hauksbee"; Park Benjamin, *The Intellectual Rise in Electricity* (London, 1895), pp. 457–470, the first good account; Ferdinand Rosenberger, *Entwicklung der electrischen Principien* (Leipzig, 1898), pp. 8–10; W. Cameron Walker, "The Detection and Estimation of Electric Charges in the Eighteenth Century," in *Annals of Science,* **1** (1936), 66–100, which discusses chiefly Hauksbee's thread electroscope; I. Bernard Cohen, *Benjamin Franklin's Experiments* (Cambridge, Mass., 1941), pp. 32–37; and Duane Roller and Duane H. D. Roller, "The Development of the Concept of Electric Charge," in *Harvard Case Histories in Experimental Science,* II (Cambridge, Mass., 1957), pp. 559–571.

Hauksbee's work in capillarity and surface effects is discussed in W. B. Hardy, "Historical Notes Upon Surface Energy and Forces of Short Range," in *Nature,* **109** (1922), 375–378; and E. C. Millington, "Studies in Capillarity and Cohesion in the Eighteenth Century," in *Annals of Science,* **5** (1945), 352–369.

Hauksbee and Newton are discussed in Henry Guerlac, "Francis Hauksbee: Expérimentateur au profit de Newton," in *Archives internationales d'histoire des sciences,* **16** (1963), 113–128; "Sir Isaac and the Ingenious Mr. Hauksbee," in *Mélanges Alexandre Koyré,* I. Bernard Cohen and René Taton, eds., I (Paris, 1964), 228–253; and "Newton's Optical Aether," in *Notes and Records. Royal Society of London,* **22** (1967), 45–57. On a number of points the present article supersedes these earlier papers.

HENRY GUERLAC

HAUKSBEE, FRANCIS (*b.* London, England, April 1688; *d.* London, 11 January 1763), *experimental physics, scientific instrumentation.*

Like his famous uncle, and often confused with him, the younger Francis Hauksbee in his early years made scientific instruments and gave public demonstrations on scientific subjects. Baptized on 15 April 1688, he was the son of John Hauksbee, a freeman of the Drapers' Company and the brother of the elder Francis.[1] As early as 1710 the nephew was assisting his uncle in the house in Wine Office Court, Fleet Street, where the German traveler Z. C. von Uffenbach met him.[2] By 1712 Hauksbee had moved to Crane Court, near Fetter Lane, opened his own shop, and begun a public course of experiments largely derived from those of his uncle. Explanatory lectures to accompany the experiments were given by Humphry Ditton and, after 1715, by William Whiston.[3] During his uncle's last illness in 1713, and for a short time after the elder Hauksbee's death, he was paid for performing experiments before the Royal Society.[4] But he did not succeed to his uncle's post, which went instead to the much abler J.-T. Desaguliers.

In making a reflecting telescope, in which he had a certain success,[5] Hauksbee was briefly associated with John Hadley, inventor of the optical sextant. With Peter Shaw he showed experiments with a portable chemical laboratory. He never became a fellow of the Royal Society, but in 1723 he was chosen to succeed one Alban Thomas as clerk and custodian at the Royal Society.[6] He held this post until his death at the age of seventy-five.[7] His publications, or those on which his name appears, are of little interest; for the most part they consist of announcements or outlines of courses of experiments.

NOTES

1. Summary of documents on the Hauksbee family, kindly communicated by Dr. H. Drubba of Hannover, Germany.
2. *Travels of Zacharias Conrad von Uffenbach,* p. 77.
3. Taylor, *Mathematical Practitioners,* p. 302. For an advertisement announcing and describing one of these demonstrations with Ditton, see *The Spectator,* no. 268 (7 Jan. 1712). For the collaboration with Whiston, see bibliography.
4. An entry in the Royal Society's MS *Council Book* for 24 Aug. 1713 (p. 266) reads: "Mr. Hauksbee his nephew was ordered five Guineas for his Services since his Uncle's death he giving a Receipt in full."
5. R. T. Gunther, *Early Science in Oxford,* II (Oxford, 1923), p. 332.
6. John Nichols, *Illustrations of the Literary History of the Eighteenth Century,* I, 810; IV, 506.
7. *Gentleman's Magazine,* **33** (1763), 46.

BIBLIOGRAPHY

I. Original Works. Hauksbee's writings include *A Course of Mechanical, Optical, Hydrostatical, and Pneumatical Experiments. To Be Perform'd by Francis Hauksbee; and the Explanatory Lectures Read by William Whiston* (London, n.d.), probably written by Whiston; *An Experimental Course of Astronomy Proposed by Mr. Whiston and Mr. Hauksbee* (London, n.d.); *An Essay for Introducing a Portable Laboratory by Means Whereof All the Chemical Operations Are Commodiously Performed by P. Shaw and F. Hauksbee* (London, 1731); and *Proposals for Making a Large Reflecting Telescope* (London, n.d.).

II. Secondary Literature. See R. E. Anderson, "Hauksbee, Francis, the Younger," in *Dictionary of National Biography,* which suggests, incorrectly, that he may have been the son of the elder Hauksbee; W. H. Quarrell and Margaret Ware, eds., *London in 1710 From the Travels of Zacharias Conrad von Uffenbach* (London, 1934); and E. G. R. Taylor, *Mathematical Practitioners of Tudor and Stuart England* (Cambridge, 1967), p. 302, a good short sketch.

Henry Guerlac

HAUSDORFF, FELIX (*b.* Breslau, Germany [now Wrocław, Poland], 8 November 1868; *d.* Bonn, Germany, 26 January 1942), *mathematics.*

Hausdorff's father was a wealthy merchant. After finishing his secondary education in Leipzig, Hausdorff studied mathematics and astronomy at Leipzig, Freiburg, and Berlin. He graduated from Leipzig in 1891 and five years later became a *Dozent* there. Until 1902, when he was appointed professor at Leipzig, he lived independently and devoted himself to a wide range of interests. From 1891 to 1896 he published four papers in astronomy and optics, and in the following years several papers in various branches of mathematics. His main interests, though, were philosophy and literature, and his friends were mainly artists and writers. Under the pen name Dr. Paul Mongré he published two books of poems and aphorisms; a philosophical book, *Das Chaos in kosmischer Auslese* (1898); and a number of philosophical essays and articles on literature. In 1904 he published a farce, *Der Arzt seiner Ehre,* which was produced in 1912 and had considerable success.

In 1902 Hausdorff became associate professor at Leipzig. From that time, mainly after 1904, he seems to have dealt more with set theory, at the same time gradually decreasing his nonscientific writing. In 1910 he went to Bonn as associate professor and there wrote the monograph *Grundzüge der Mengenlehre,* which appeared in 1914. In 1913 Hausdorff became full professor at Greifswald and in 1921 returned to Bonn, where he was active until his forced retirement in 1935. Even then he continued working on set theory and topology, although his work was published only outside Germany. As a Jew he was scheduled to be sent to an internment camp in 1941. It was temporarily avoided; but when internment became imminent, Hausdorff committed suicide with his wife and her sister on 26 January 1942.

Hausdorff's scientific activity contributed greatly to several fields of mathematics. In mathematical analysis he proved important theorems concerning summation methods, properties of moments, and Fourier coefficients (1921). In algebra he derived and investigated the so-called symbolic exponential formula (1906). He introduced and investigated a very important class of measures and, in connection with them, a kind of dimension which may assume arbitrary nonnegative values (1919). Both are now named for Hausdorff and are applied in particular to examination of fine properties of numerical sets.

Hausdorff's main work was in topology and set theory. Various definitions of topological spaces and related concepts had been given, mainly by Maurice Fréchet, before Hausdorff's *Grundzüge der Mengen-*

lehre appeared. The interrelations of these different approaches had not been completely recognized; and no clear way had been known to effect a gradual transition from very general spaces to those similar to spaces actually occurring in analysis and geometry.

In the *Grundzüge,* Hausdorff took a decisive step in this direction. His broad approach, his aesthetic feeling, and his sense of balance may have played a substantial part. He succeeded in creating a theory of topological and metric spaces into which the previous results fitted well, and he enriched it with many new notions and theorems. From the modern point of view, the *Grundzüge* contained, in addition to other special topics, the beginnings of the theories of topological and metric spaces, which are now included in all textbooks on the subject. In the *Grundzüge,* these theories were laid down in such a way that a strong impetus was provided for their further development. Thus, Hausdorff can rightly be considered the founder of general topology and of the general theory of metric spaces.

The *Grundzüge* is a very rare case in mathematical literature: the foundations of a new discipline are laid without the support of any previously published comprehensive work.

Hausdorff's work in topology and set theory has also brought about a number of separate results of primary importance: in topology, a detailed investigation into the basic properties of general closure spaces (1935); in general set theory, the so-called Hausdorff maximal principle (stated, although not explicitly, in the *Grundzüge*), the introduction of partially ordered sets, and several theorems on ordered sets (1906–1909); in descriptive set theory, the theorem on the cardinality of Borel sets (1916; proved independently by P. S. Alexandrov in the same year) and the introduction of the δs-operations, now often called Hausdorff operations (1927).

Hausdorff's manuscripts have not yet been fully prepared for publication, but they are not likely to provide any new scientific results.

BIBLIOGRAPHY

I. Original Works. Hausdorff's major work is *Grundzüge der Mengenlehre* (Leipzig, 1914); the 2nd ed., entitled *Mengenlehre* (Leipzig, 1927), is in fact a new book. The Russian trans. (Moscow, 1935) is a revised combination of both. Hausdorff's MSS are being published under the title *Nachgelassene Schriften,* Günter Bergmann, ed.; vols. I and II appeared in Stuttgart in 1969. Numerous papers are in *Fundamenta mathematicae, Mathematische Annalen,* and *Mathematische Zeitschrift.*

II. Secondary Literature. A short biography, an analysis of Hausdorff's work, a list of scientific papers, and a survey of his MSS are in M. Dierkesmann *et al.,* "Felix Hausdorff zum Gedächtnis," in *Jahresberichte der Deutschen Mathematikervereinigung,* **69** (1967), 51–76. An article on Hausdorff by W. Krull in "Bonner Gelehrte," in *Beiträge zur Geschichte der Wissenschaften in Bonn,* Mathematik und Naturwissenschaften (1970), pp. 54–69, contains a short biography including an account of Hausdorff's activity outside mathematics, and a detailed analysis of the *Grundzüge.* See also W. Krull, "Felix Hausdorff," in *Neue deutsche Biographie,* VIII (Berlin, 1969), 111–112. The pref. to vol. I of *Nachgelassene Schriften* includes material on Hausdorff as a university teacher and a number of short excerpts from his correspondence.

M. Katětov

HAUTEFEUILLE, PAUL GABRIEL (*b.* Étampes, Seine-et-Oise, France, 2 December 1836; *d.* Paris, France, 8 December 1902), *chemistry.*

The son of a notary, in 1855 Hautefeuille entered the École Centrale des Arts et Manufactures, where J. B. Dumas noticed him and recommended the young engineer to H. E. Sainte-Claire Deville at the École Normale Supérieure. There Hautefeuille took a doctorate in the physical sciences in 1865 and became Deville's assistant and, later, *maître de conférences.* In 1885 he was named to the chair of mineralogy at the Sorbonne and in 1895 was elected to the Académie des Sciences.

Influenced in his chemical studies by Deville's thermochemical approach, Hautefeuille was a member of that group at the École Normale Supérieure which included Henri Debray, L. J. Troost, Alfred Ditte, and F. Isambert. From 1868 to 1881 he collaborated with Troost in researches which included the conditions of transformation of cyanogen into paracyanogen and of white phosphorus into red phosphorus, and the absorption of hydrogen by sodium, potassium, and palladium. With Troost and with James Chappuis, Hautefeuille studied the allotropic relationship between oxygen and ozone, and in 1882 he obtained liquefied ozone by using Louis Cailletet's apparatus. His studies of equilibria included the dissociation of hydriodic acid (1867) and the oxidation of hydrochloric acid (1889) in air.

Hautefeuille's best-known studies were his reproductions of numerous crystallized minerals by utilizing mineral catalysts and varied temperature conditions; he carried on this research at a time when the generality of polymorphism among crystals was not fully realized. In his doctoral thesis, for example, he established that three different types of titanium dioxide—the rutile, octahedrite, and brookite crystals—could each be prepared in the laboratory from the amorphous dioxide; he also demonstrated the temperature dependence of two of the crystalline forms of silica—quartz and tridymite—and successfully produced a variety of alkaline feldspars and beryls, including the emerald. In his experiments Hautefeuille employed catalysts readily available under natural conditions, and his work confirmed the views of the French school of lithology dating from Élie de Beaumont.

BIBLIOGRAPHY

I. ORIGINAL WORKS. Hautefeuille was hesitant to publish his researches, scrupulously delaying even his notes to the Academy in order to revise them. A complete list of his publications is in Georges Lemoine's short biography, *Les travaux et la vie de Paul Hautefeuille* (Louvain, 1904), which is an extract from *Revue des questions scientifiques,* **55** (1904), 5–25. Several of his papers are reprinted in Henry Le Chatelier *et al.*, eds., *Les Classiques de la science:* vol. III, *Eau oxygénée et ozone* (Paris, 1913), and vol. VI, *La fusion du platine et dissociation* (Paris, 1914).

II. SECONDARY LITERATURE. Hautefeuille requested that no eulogies be delivered upon his death. Besides the discussion of his life and work in Lemoine, see Alfred Lacroix, "Gabriel Hautefeuille (1836–1902)," in *Figures des savants,* I (Paris, 1932), 81–89.

MARY JO NYE

HAÜY, RENÉ-JUST (*b.* St.-Just-en-Chaussée, Oise, France, 28 February 1743; *d.* Paris, France, 1 June 1822), *crystallography, mineralogy.*

The son of a poor weaver, Haüy received a classical and theological education through a scholarship to the Collège de Navarre in Paris, where, in 1764, he became a *régent.* In 1770 he was ordained a priest and was assigned a similar teaching post at the Collège Cardinal Lemoine. Encouraged by his friend Lhomond, he undertook botanical studies, but Daubenton's lectures on mineralogy at the Jardin du Roi soon turned his interest to mineralogy.

Haüy's first publications, presented to the Academy in 1781, on the crystal forms of garnet and calcspar (Iceland spar, calcite) were favorably reviewed by Daubenton and Bezout and led to his election as an associate member of the botanical class of the Academy in February 1783. In 1784 he published *Essai d'une théorie sur la structure des cristaux,* which laid the foundation of the mathematical theory of crystal structure. He left his teaching post and henceforth devoted himself entirely to the elaboration of his crystal theory and its application to mineralogical classification.

During the Revolution, Haüy showed great flexibility in response to the rapidly changing political situation; but he staunchly refused to take the oath required by the Civil Constitution of the Clergy. In 1792 he and many other members of the clergy were arrested, but he was soon released through the efforts of Étienne Geoffroy Saint-Hilaire, who had been his pupil. Having been a member of an Academy commission concerned with the metric system, Haüy became, after the dissolution of the royal academies, a secretary of the Commission on Weights and Measures. In this capacity he tried in 1793, together with Borda, to obtain the release of their fellow member Lavoisier. In 1795 Haüy began teaching courses in physics and mineralogy at the École des Mines and became a member of the newly founded Institut National des Sciences et des Arts, in the natural history and mineralogy section. In 1801 he published his main work, *Traité de minéralogie,* the first volume of which presented his crystal theory; in the three subsequent volumes he expounded his system of mineral classification. In this work he revised the nomenclature of minerals.

Haüy also did work in physics. In 1787 he published *Exposition raisonnée de la théorie de l'électricité et du magnétisme, d'après les principes d'Aepinus.* In contrast with Aepinus, he refrained from mathematical calculations and added Coulomb's recent results. Like Franklin and Aepinus, Haüy assumed one hypothetical electric fluid and one magnetic fluid, although in his later works he adhered to the two-fluid theory.

Napoleon, who in 1802, while first consul, had nominated Haüy an honorary canon of Notre Dame, in the next year ordered him to write a textbook of physics for the newly instituted lycées. This book was outstanding for its clear, methodical exposition of physics, although mathematical treatment of problems was again lacking. Like most of his contemporaries, Haüy adhered to Newton's corpuscular theory of light and to the theory that heat was caused by a "caloric matter." His own contribution to physics consisted in his researches on double refraction in crystals, on pyroelectricity in crystals (especially tourmaline and boracite), and on piezoelectricity. Haüy's *Traité de physique* brought him appointment to the Legion of Honor in 1803.

After the death of Dolomieu, Haüy became in 1802 professor of mineralogy at the Muséum d'Histoire Naturelle, where he enlarged the mineral collection (Haüy's own collection has belonged to the Muséum since 1848). In 1809 he was also appointed to the newly created chair of mineralogy at the Sorbonne. In his *Tableau comparatif* (1809) Haüy compared the results of the crystallographic and chemical determinations of mineral species. In his stubborn opposition to the notions of indefinite compounds, mixed crystals, isomorphism, and polymorphism Haüy showed that, despite his mild and pliable character, he was adamant when his deepest convictions were at stake. In 1822 he published *Traité de cristallographie,* which contained the last version of his theory and was immediately followed by the second edition of the *Traité de mineralogie,* limited to the portion on systematics. In the same year his rather uneventful life came to its end.

Haüy corresponded with many mineralogists and chemists of his time. He did no field research and avoided the problems of mineral genesis. Using the large collections at his disposal, he worked primarily in descriptive, physical, and theoretical mineralogy. He lived very frugally, supporting his brother Valentin (well-known for his activities in care of the blind), after the latter's return from Russia, and his niece and nephew. Brongniart was his successor at the Muséum and F. S. Beudant at the Sorbonne. Delafosse, who had become his assistant in 1817, gave his theory of crystal structure a more mathematical character, which was developed further by Bravais in his theory of crystal lattices.

Romé de l'Isle had deduced the various forms of the same crystal species by truncating the edges or the solid angles of the rather arbitrarily selected primitive form. Haüy established a more rigorous mathematical relationship between primary and secondary forms of the same species, and his choice of the primary form was founded on more physical grounds. The basic idea of his theory is that the primitive form of crystals of a certain species results as a nucleus from the cleavage of all their secondary forms. If mechanical division proves to be impossible, there are other phenomena—particularly striation—that reveal the nucleus. In 1793 Haüy proposed six types of primary forms: parallelepiped, rhombic dodecahedron, hexagonal dipyramid, right hexagonal prism, octahedron, and tetrahedron.

Further mathematical division of the primary forms ultimately led to the *molécules intégrantes* (which he had previously called *molécules constituantes*), the constituent molecules of the substance. These may have the shape of the primary form—as in the case of the parallelepiped—or they may differ from it—as when the octahedral nucleus of fluorite is divided into tetrahedrons with octahedral empty spaces, or when the right hexagonal prism is divided into six trigonal prisms with equilateral triangles as their bases. The values of the interfacial angles of the primitive form and the constituent molecule are characteristic and invariable for each kind of mineral, and it is assumed that the dimensions of the edges of the molecules are also constant and characteristic. Only highly symmetrical forms, such as the cube, may be common to different species: their angles and their relative dimensions are known a priori. For calcspar, too, Haüy knew the relative dimensions (1:1:1) of the cleavage rhombohedron a priori; and he calculated its angles on the assumption that its faces make a 45° angle with a horizontal plane when the axis is vertical.

Since the cleavage form of a crystal reveals only a definite value of the dihedral angles, which are fixed, according to the law of constancy of angles, but not the relative dimensions—cleavage of a crystal of sea salt, which ideally has a cube as its nucleus, will lead in most cases to an oblong rectangular prism—Haüy introduced the additional principle of symmetry. Faces that are crystallographically identical will show their equivalence when secondary faces are developed.

Whereas the primary forms were derived from the secondary ones by the physical procedure of cleavage, the reverse occurred when Haüy theoretically derived the secondary forms by stacking layers of contiguous molecules on the faces of the nucleus. Subsequent layers recede by one or two (rarely by three–six) rows of molecules in relation to the edges of the previous layer. When, for instance, the primary form is a cube and on each of its faces are superposed layers one (cubical) constituent molecule thick, each of which falls short of the edge of the preceding layer by one row of molecules, a dodecahedron with rhombic faces is formed. If each layer has two (or three or four) rows of molecules less than the previous layer on each edge, the ensuing pyramids will be less steep and a tetrahexahedron (twenty-four faces), in which each face of the cube has developed into a tetragonal pyramid, will emerge.

Similar laws of decrement may operate on the solid angles or parallel to the diagonals of the faces of the nucleus. Haüy's fundamental law of decrement states that subtractions are confined to a small number of

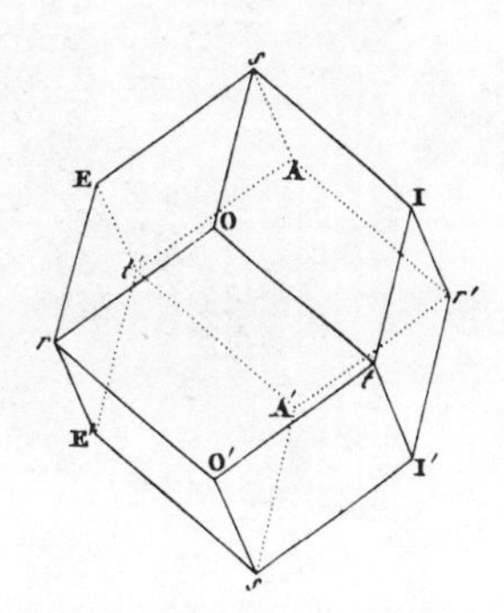

FIGURE 1. Rhombic dodecahedron.

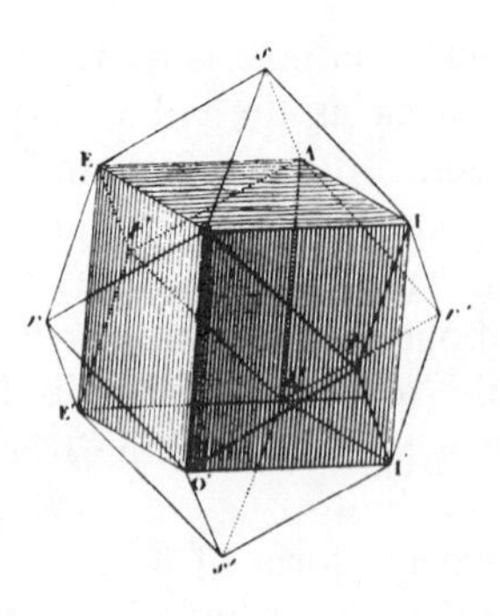

FIGURE 2. The cube as nucleus of the rhombic dodecahedron. The edges of the cube are the little diagonals of the faces of the dodecahedron.

rows of molecules, usually one or two. This statement led his successors directly to the law of rational indices (the law of rationality of intercepts), which, together with the law of constancy of angles, is fundamental to modern crystallography.

The laws of decrement are subject to the law of symmetry, which requires that the same kind of decrement be simultaneously repeated on all identical faces of the nucleus, that is, those parts of it which may be substituted for each other "without the nucleus ceasing to present the same aspect" (1815). If one face of the cube is changed, all six will undergo the same change; in a rectangular parallelepiped, however, either the two bases or the four lateral faces undergo the same change. Hemimorphic forms, such as tourmaline, caused Haüy great difficulties.

In order to discover the laws of decrement, Haüy started with regular forms: the cube of sea salt and the rhombohedron of calcspar, which have relative dimensions of 1:1:1. He pointed out that in all other cases observation gave information only on the angles

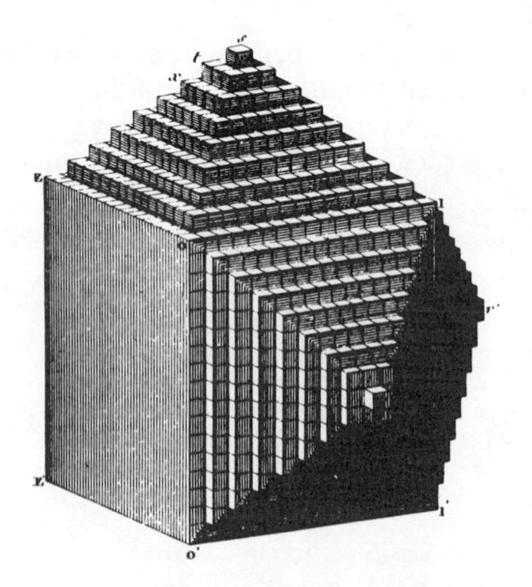

FIGURE 3. Development of a rhombic dodecahedron from a cube, according to the physical law of decrement.

and not on the dimensions, the determination of which required the theory established by the more regular forms. In order to find the relative dimensions of the edges of the molecules, a definite assumption about the decrement connected with a certain secondary form had to be made.

A very important addition to Haüy's original theory was his introduction of the notion of *molécules soustractives* (1793). Constituent molecules which do not possess the parallelepipedal form are combined to form parallelepiped units: two triangular prisms form a rhombic prism with angles of 120° and 60° on their basic faces; an octahedral space together with two adjoining tetrahedral molecules (fluorite) forms a parallelepiped. In this way all crystals may be conceived of as consisting of parallelepipedal units packed together in parallel positions so as to fill space, and all secondary forms are derived by stacking layers of molecules, according to laws of decrement, on primitive nuclei of the same form as these molecules. All crystals, then, possess a threefold periodicity along the edges of parallelepipeds. The agreement between Haüy's measurements of angles and the values calculated from the application of a certain law of decrement gave proof of the correctness of these laws as well as of the numerical value of the angles and ratios of the dimensions of the subtractive molecule.

Haüy considered the subtractive molecules to be geometrical fictions, introduced for the sake of simplification of the theory. Similarly, although the nucleus was found by physical means, he emphasized as early as 1782 that the derivation of secondary forms from the primary form did not represent the physical process of crystal growth, since even the smallest crystal may show these secondary forms (the smallest fluorite crystals are cubes, although the cleavage form is an octahedron). Moreover, he stressed that the nucleus should not be conceived of too literally, for it is found throughout the crystal. This means that his crystal theory is fundamentally static and mathematical, however much physical data and physical claims about the constituent molecules may be involved. It is at the base of the modern lattice theory developed by Delafosse, Bravais, Sohncke, Fyodorov, and Schönflies.

Haüy's belief in simplicity had some awkward consequences. He rigidly maintained, for instance, that the faces of the cleavage rhombohedron of calcspar are inclined at exactly 45° to the axis, which led to a ratio of the diagonals of the rhombic faces of $\sqrt{3} : \sqrt{2}$. He rejected the more exact measurements made by Wollaston with the reflecting goniometer in 1809, because these data would imply a less simple ratio of the diagonals ($\sqrt{111} : \sqrt{73}$) than did his theoretical values. Haüy always used the less precise contact goniometer, which made it easier to make the data conformable to the "simplicity of nature." Other crystallographers, such as H. J. Brooke (1819), inevitably criticized "the imaginary simplicity . . . supposed to exist naturally in the ratios of certain lines either upon or traversing a crystal" and the disposition to regard "the disagreement of an observed measurement" with this simplicity "rather as an error of the observation than a correction of his theoretic determination."

According to Haüy himself, he started from the observation of a hexagonal prism of calcspar which was detached from a group of crystals along a plane of the cleavage rhombohedron. Further division led him to assume rhombohedral molecules. This story masks his debt to the work of Bergman. Haüy's first publications of 1782 (on garnet and on calcspar), although correcting some errors in Bergman's publications of 1773 and 1780 on these minerals, bear the stamp of his theory, as Romé de l'Isle pointed out in 1783. Like Bergman, Haüy superposed "integrant lamellae," not "molecules," on a nucleus. On a rhombic dodecahedron of garnet he stacked rhomboidal lamellae of the same form as the faces of the nucleus, and on the rhombohedron of Iceland spar he stacked steadily decreasing lamellae, forming a scalenohedron. Bergman tried to derive both the rhombic dodecahedron of garnet and the scalenohedron of calcspar from the Iceland spar crystal, which Haüy clearly recognized in 1781–1782 as totally wrong. Nevertheless, in Haüy's theory, as in Bergman's, the lamellae decreased continuously (and not by steps), so that the laws of decrement of the superposed layers were not put forward and a rigorous deduction of secondary forms from the primary one was still lacking. Moreover, Bergman's article of 1773 had been mentioned in the 1781 manuscript of which Haüy's 1782 publication was an extract. On the other hand, although starting from the same principles, Haüy's publications of 1782 far surpassed Bergman's in the application of those principles.

In 1784, in his *Essai d'une théorie sur la structure des cristaux,* Haüy criticized Bergman's deduction of the calcite scalenohedron from the primitive form as being too vague. He proposed the notion of the crystal molecule and the laws of decrement and the constancy not only of the angles but also of the dimensions of the crystal units. He then clearly recognized the discontinuity principle: not all angles and not all inclinations of faces are possible, thus limiting the number of varieties of a crystal species. Instead of the "demi-rhombs" of his lamellar theory of 1782 he now admitted empty spaces half the size of a rhombohedral unit, since there was now at least one row less of molecules on the edges of subsequent layers.

After establishing the foundations of his crystal theory, Haüy applied it to mineralogical classification. Both Romé de l'Isle and Haüy held that the characteristic form of the constituent molecule of a compound is due to the forms, the definite proportions, and the definite arrangement of the constituent elementary particles. That is, before Proust they proposed a priori the chemical law of fixed proportions. For Haüy the mineral species was defined by a geometrical type (the form of the constituent molecule) and a chemical type (the composition of the constituent molecule); the crystallographic molecule and the chemical molecule were identical. Molecules of different species, except those of the isometric or regular system, have different forms and different composition. These ideas enabled Haüy to unite in one species minerals hitherto considered different, such as beryl and emerald, and to divide groups that had been considered varieties of the same species, such as zeolites.

Haüy's survey of the results of crystallography and chemical analysis in relation to the classification of minerals (1809) gave a detailed exposition of the successes and difficulties his method encountered. Chemical composition decided the four traditional classes in mineralogy—acidiferous (salts), earthy, nonmetallic combustible, and metallic—and the orders and genera; the form of the constituent molecule determined the species. Only with the *formes limites* (the isometric forms which may be common to different species) were the physical properties—hardness, specific weight, optical behavior—and/or the chemical composition indispensable for definition of the species.

A series of mixed crystals of calcium-iron carbonate was considered as a group of subspecies of calcium carbonate, the latter assumed to have impressed its form on the whole mixture. Haüy became involved in a controversy with Berthollet, who supposed compounds to have a variable composition. In his *Tableau comparatif* (1809), Haüy emphasized the invariability of the form and the composition of the constituent molecule of a species but was forced to admit that the definite proportions were often blurred by heterogeneous materials accidentally mixed with the compound: "Only for geometry are all crystals pure." He did not take up the problem of locating alien particles in crystals in which the polyhedral molecules normally left no interstices between them, as is the case with parallelepipedal molecules. He recognized that chemical analysis cannot decide which components are accidental and which essential, since even a small percentage of a substance may impress its form on a large percentage of "accidental" impurities; such was the case with the mixed crystals made by Beudant (1817), in which 10 percent iron sulfate "gave its form" to 90 percent copper sulfate.

Mitscherlich's discovery of isomorphism (1819) was rejected by Haüy. Like Romé de l'Isle, Haüy considered pure iron spar ($FeCO_3$) to be a pseudomorph of calcspar, and its own molecular form was considered still unknown. Consequently, he did not admit the difference of 2° between the interfacial angles of the rhombohedrons of iron carbonate and calcium carbonate found by Wollaston, although this difference would have supported in a more natural way his belief that each species has its unique characteristic form. In other cases he denied the identity of angles of substances that were believed by other scientists to be isomorphic.

Haüy admitted the polymorphism of calcium carbonate (aragonite and calcite) only reluctantly; after the discovery of the chemical identity of carbon and diamond, he concluded that geometry and physics (crystal form and properties such as specific weight and hardness) gave better criteria for distinguishing different species than did chemistry (1822). Yet in the case of two forms of titanium oxide, he said (1809, 1822) that he would regard them as different species "until new investigations . . . have unveiled the lack of agreement here existing between chemistry and crystallography."

In 1784 Haüy believed that chemistry was to play a dominant role in determining mineral species. Like his friend Dolomieu, who in 1801 considered the mineralogical species to be wholly defined by one constituent molecule, Haüy in his theory always rigidly maintained that "constant composition" defined a mineral species. Yet in practice he completely changed his view, using geometrical rather than chemical data for this purpose. The results of chemical analysis allowed divergent interpretations, all the more so since their accuracy, especially in the case of the silicates, was unsatisfactory. Consequently, in his mineralogical nomenclature Haüy could not always follow the examples of Linnaeus and Lavoisier, who had introduced a rational binomial nomenclature in their respective disciplines. With the "earths" it was impossible to establish beyond doubt the chemical type because the number and proportion of the elements essential to the constituent molecule were uncertain; and only the geometrical type (the molecular form) was known with certainty. Haüy had to resort to trivial names and omitted the division of his second class into genera. In the first and fourth classes, he classified according to Lavoisier's chemistry, although he deemed the metallic component more important than the acid part (which in Lavoisier's nomenclature decided the generic name). Instead of Lavoisier's *carbonate de chaux* there was Haüy's *chaux carbonatée;* the genus copper contained the species native copper, copper oxide, copper carbonate, and so on.

It is an irony of history that Haüy's geometrical definition of mineral species, especially in the case of the silicates, has acquired great importance in twentieth-century mineralogy through acquiring the opposite sense, that of substitution of "vicarious," isomorphic constituents.

BIBLIOGRAPHY

I. Original Works. An extensive bibliography of Haüy's writings, as well as a list of his biographies and portraits, are in A. Lacroix, "La vie et l'oeuvre de l'abbé René-Just Haüy," in *Bulletin de la Société française de minéralogie,* **67** (1944), 15–226, esp. 95–112.

Extracts of his first memoirs, presented to the Academy, are "Extrait d'un mémoire sur la structure des cristaux de grenat," in *Journal de physique,* **19** (1782), 366–370; and "Extrait d'un memoire sur la structure du spath calcaire," *ibid.,* **20** (1782), 33–39.

Besides some 130 articles, almost all on geometrical and physical crystallography and mineralogy, Haüy wrote the following longer works: *Essai d'une théorie sur la structure des cristaux appliquée à plusieurs genres de substances cristallisées* (Paris, 1784); *Exposition raisonnée de la théorie de l'électricité et du magnétisme, d'après les principes d'Aepinus* (Paris, 1787); *Traité de minéralogie,* 4 vols. and atlas (Paris, 1801; 2nd ed., rev. and enl., 1822); *Traité élémentaire de physique . . .,* 2 vols. (Paris, 1803; 2nd ed., 1806; 3rd ed., 1821); *Tableau comparatif des résultats de la cristallographie et de l'analyse chimique relativement à la classification des minéraux* (Paris, 1809); *Traité des caractères physiques des pierres précieuses, pour servir à leur détermination lorsqu'elles ont été taillées* (Paris, 1817); *Traité de cristallographie, suivi d'une application des principes de cette science à la détermination des espèces minérales,* 2 vols. and atlas (Paris, 1822).

Seventy-two letters by Haüy may be found in Lacroix (see above), pp. 113–226. More letters are in R. Hooykaas, "La correspondance de Haüy et van Marum," in *Bulletin de la Société française de minéralogie,* **72** (1949), 408–448.

II. Secondary Literature. Several articles on Haüy's life and work are in *American Mineralogist,* **3** (1918), and in *Bulletin de la Société française de minéralogie,* **67** (1944). For example, see C. Mauguin, "La structure des cristaux d'après Haüy," *ibid.,* 227–262; and J. Orcel, "Haüy et la notion d'espèce en minéralogie," *ibid.,* 265–335.

Articles by R. Hooykaas dealing with Haüy's work and the origin of his theory are "Kristalsplijting en kristalstructuur van kalkspaat I (Bergman)," in *Chemisch weekblad,* **47** (1951), 297–302; "Kristalsplijting en kristalstructuur van kalkspaat II (R. J. Haüy 1782)," *ibid.,* 537–

543; "The Species Concept in 18th Century Mineralogy," in *Archives internationales d'histoire des sciences,* **5**, no. 18-19 (1952), 45-55; "Torbern Bergman's Crystal Theory," in *Lychnos* (1952), 21-54; and "Les débuts de la théorie cristallographique de R. J. Haüy, d'après les documents originaux," in *Revue d'histoire des sciences,* **8** (1955), 319-337.

Books on the history of crystallography that treat Haüy's work are J. G. Burke, *Origins of the Science of Crystals* (Berkeley-Los Angeles, 1966), chs. 4, 5; P. Groth, *Entwicklungsgeschichte der mineralogischen Wissenschaften* (Berlin, 1926), pp. 14-57; R. Hooykaas, *La naissance de la cristallographie en France au XVIIIe siècle* (Paris, 1953), pp. 12-29; C. M. Marx, *Geschichte der Crystallkunde* (Karlsruhe, 1825), pp. 132-175; and H. Metzger, *La genèse de la science des cristaux* (Paris, 1918), pp. 80-87, 195-206.

R. HOOYKAAS

HAVERS, CLOPTON (*b.* Stambourne, Essex, England, *ca.* 1655; *d.* Willingale, Essex, England, April 1702), *osteology.*

Havers' father, Henry Havers, rector of Stambourne, was ejected under the Act of Uniformity in 1662. Clopton was admitted to St. Catharine's College, Cambridge, on 6 May 1668 but did not graduate. He studied medicine under Richard Morton, an ejected minister who is remembered for his work on phthisis (1689). He was granted an "extra license" by the Royal College of Physicians for practice outside London on 28 July 1684; graduated from Utrecht on 3 July 1685 with a thesis, *De respiratione*, from which it appears that he accepted Robert Boyle's teaching of the atomic constitution of the air; and was elected fellow of the Royal Society on 17 November 1686. He obtained the College of Physicians' full license on 22 December 1687 and practiced thereafter in London.

Between August 1689 and August 1690 Havers read to the Royal Society five discourses which formed the substance of his book *Osteologia nova, or Some New Observations of the Bones*, published in 1691. This book provided the first full description of the microscopic structure of the bone lamellae and canals, with a discussion of bone physiology. It is arranged in five sections, of which the first is the most original: (1) microscopic structure; (2) growth, physiology, and pathology; (3) the marrow; (4) the synovial glands, rheumatism, and gout; (5) cartilage. Havers wrote of his most notable observation, the "canals":

> In the bones through and between the plates are formed pores, besides those which are made for the passage of the blood-vessels, which are of two sorts; some penetrate the laminae and are transverse, looking from the cavity to the external superficies of the bone; the second sort are formed between the plates, which are longitudinal and straight, tending from one end of the bone towards the other, and observing the course of the bony strings [*Osteologia nova,* p. 43].

Havers described the small, fibrous, penetrating prolongations of the periosteum and supposed that they conveyed "nervous spirits" to the bone in order to affect its sensibility and growth; their merely connective function was demonstrated by William Sharpey in 1848. Havers also described the intra-articular synovial fringes and folds, considering them to be mucilaginous glands for the secretion of synovial fluid. An engraved plate provided a clear schema of the structure he had discovered. He also corrected Glisson's opinion that bones grow on their harder side when they develop unevenly in rickets but repeated the traditional belief that cartilage may change into bone, later corrected by Nesbitt.

A few of Havers' observations were anticipated: Anton van Leeuwenhoek had reported in 1686 that he had seen the transverse canals and Malpighi had mentioned the bone lamellae incidentally in his *Anatome plantarum* (1675); Domenico Gagliardi discussed these more fully in his *Anatomia ossium* (1689). Nevertheless, Havers' book was the first complete and systematic study. It made a marked impression, being immediately translated into Latin, reviewed at length in the *Philosophical Transactions of the Royal Society*, and praised by Giorgio Baglivi when he was elected professor of anatomy at Rome (1696). It remained the only detailed treatise on bone until Nesbitt's lectures were published in 1736, the intervening books being purely descriptive; it was not superseded until the nineteenth century.

Havers contributed two papers to the *Philosophical Transactions:* a brief case history of a patient who shed tears of blood (1693) and a study of digestion (1699) based on experiments and explaining it as a fermentation of saliva and bile. He revised John Ireton's English text for Johann Remmelin's anatomical plates in 1695 and was appointed Gale lecturer on anatomy by the Company of Surgeons in June 1698. Havers contracted to write an English text for Stephan Blankaart's anatomical plates; it was supplied after Havers' death by James Drake (1707).

Havers married Dorcas Fuller, who survived him; their children died young. He died in 1702 and was buried at Willingale, Essex, where his father-in-law was rector of the parish.

BIBLIOGRAPHY

I. ORIGINAL WORKS. Havers' writings include *De respiratione* (Utrecht, 1685); *Osteologia nova, or Some New*

Observations of the Bones and the Parts Belonging to Them, With the Manner of Their Accretion and Nutrition (London, 1691), trans. into Latin by M. F. Geuder (Frankfurt–Leipzig, 1692), posthumously repr. in English (London, 1729) and Latin (Amsterdam, 1731; Leiden, 1734); "An Account of an Extraordinary Haemorrhage at the Glandula Lachrymosa," in *Philosophical Transactions of the Royal Society,* **18,** no. 208 (1693), 51; a corrected version of Johann Remmelin, *A Survey of the Microcosme, or the Anatomie of Man and Woman* (London, 1695); *Syllabus musculorum humani corporis* (London, 1698), unsigned; and "A Short Discourse Concerning Concoction," in *Philosophical Transactions of the Royal Society,* **21,** no. 254 (1699), 233–247. An autograph letter of 25 October 1699 from Havers to Hans Sloane asks to borrow books on pleurisy: British Museum, Sloane MS 4037, fol. 348.

II. SECONDARY LITERATURE. William Sharpey describes the periosteal fibers in Jones Quain, *Elements of Anatomy,* 5th ed. (London, 1848), p. cxxxii, correcting Havers. See also M. Randelli, "Les observations microscopiques de Gagliardi et de Havers sur la structure des os," in *Comptes rendus, 85. Congrès des sociétés savantes* (Paris, 1960), pp. 601–604. K. F. Russell, *British Anatomy* (Melbourne, 1963), pp. 125–126, nos. 394–399, records the various editions of Havers' books. The *Osteologia nova* is reviewed in *Acta eruditorum* (1691), 573, and is analyzed in *Philosophical Transactions of the Royal Society,* **17,** no. 194 (1693), 544–554. Biographical articles are J. Dobson, in *Journal of Bone and Joint Surgery,* **34B,** no. 4 (Nov. 1952), 702–707, with references to the earlier literature about Havers; J. F. Payne, in *Dictionary of National Biography,* XXV (1891), 182–183; and C. B. Reed in *Bulletin of the Society of Medical History,* **2,** no. 5 (Mar. 1922), 371–388.

WILLIAM LEFANU

HAWORTH, ADRIAN HARDY (*b.* Hull, England, 19 April 1768; *d.* Chelsea, London, England, 24 August 1833), *botany, entomology.*

Haworth was born and reared in Hull, where his father, Benjamin Haworth, was a prosperous merchant and landowner. His mother, Anne Boothe Haworth, probably nurtured his interest in gardening. After attending Hull Grammar School and serving in a law office, he started to pursue a full-time career in natural history, financed by the family business interests. Haworth was married three times and had children by each wife. After living for several years in Cottingham, near Hull, he moved to Chelsea. He joined the Linnean Society in 1798. Haworth also was one of the original members of the Aurelian Society, which in 1806 dissolved and regrouped as the Entomological Society of London and subsequently merged with the Zoological Club of the Linnean Society.

Between 1803 and 1828 Haworth published *Lepidoptera Britannica,* the first comprehensive study of British butterflies and moths and the standard work for fifty years. He was author of sixty publications, primarily concerned with Lepidoptera and with succulent plants. Haworth returned to Cottingham from 1812 to 1818. During these years he helped found and arrange the Hull Botanic Garden and wrote a natural history of the parish in the form of a poem of twenty-four cantos, some of which were published in the local newspaper. In 1818 Haworth returned to Chelsea, where he remained until his death from cholera in 1833. His house in Chelsea became known as a natural history museum. In 1833 the collection contained 40,000 insects, including 1,100 species with 300 varieties of lepidopterous insects; one shell cabinet; twelve glazed cases of fish; a library of 1,600 volumes on natural history; a herbarium of 20,000 species; and over 500 species of plants in the garden.

BIBLIOGRAPHY

I. ORIGINAL WORKS. Haworth's major publications include "Rhus toxicodendron," in John Alderson, *An Essay on the Rhus toxicodendron* (Hull, 1793); *Observations on the Genus Mesembryanthemum* (London, 1794); *Prodromus Lepidopterorum Britannicorum: A Concise Catalogue . . . With Times and Places of Appearance . . .* (London, 1802); vol. VI of Henry C. Andrews, *Botanist's Repository* (London, 1803); *Lepidoptera Britannica,* 4 pts. (London, 1803–1828); *Synopsis plantarum succulentarum* (London, 1812), with supp. (1819); and *Saxifragearum enumeratio* (London, 1821). In addition Haworth published several articles in the *Transactions* of the Entomological, Linnean, and Horticultural societies. Between 1823 and 1828 he published twenty-five papers in *Philosophical Magazine.*

II. SECONDARY LITERATURE. The best and most recent biographical account of Haworth is William T. Stearn, "Biographical and Bibliographical Introduction," in *Adrian Hardy Haworth, Complete Works on Succulent Plants,* I, facs. repr. (London, 1965). Briefer and older accounts and bibliographies are G. S. Boulger in *Dictionary of National Biography,* repr. ed., IX, 246–247; and in James Britten and G. S. Boulger, *A Biographical Index of British and Irish Botanists,* rev. by A. B. Rendle, 2nd ed. (London, 1931), p. 143. Thomas Faulkner, *An Historical and Topographical Description of Chelsea and Its Environs,* II (Chelsea, 1829), 11–13, gives a contemporary view of Haworth's later years in Chelsea.

ROY A. RAUSCHENBERG

HAWORTH, WALTER NORMAN (*b.* Chorley, England, 19 March 1883; *d.* Birmingham, England, 19 March 1950), *organic chemistry.*

The second son and fourth child of Thomas and Hannah Haworth was born into a highly respected family of businessmen, lawyers, and clergymen. His father managed a linoleum factory, and Haworth obtained his first knowledge of chemistry and business from his early training there in all aspects of linoleum design and manufacture. Despite the discouragement of his family, he decided to continue his education; and in 1903, after a strenuous period of private tutoring, he entered the University of Manchester, where he studied under William Henry Perkin, Jr. He graduated in 1906 with first-class honors in chemistry.

Although he had planned to work in the chemical industry after graduation, his plans changed when he was awarded a scholarship which enabled him to study with Otto Wallach at the University of Göttingen. He received the doctorate after one year's work and returned to Manchester, where he continued his studies on terpenes. In 1911 he received a D.Sc. from Manchester.

Early that year he went to Imperial College of Science and Technology as senior demonstrator under Thomas Edward Thorpe. In 1912 he became lecturer at United College of the University of St. Andrews, where he became acquainted with the new developments in carbohydrate chemistry carried out at St. Andrews by Thomas Purdie and James C. Irvine. Terpene studies were set aside as he became fascinated with carbohydrates. During World War I he helped in a supervisory capacity with the government's production of fine chemicals and drugs.

In 1920 Haworth was appointed professor of organic chemistry at the University of Durham. He married Violet Chilton Dobbie in 1922; they had two sons. In 1925 he became Mason professor of chemistry at the University of Birmingham, succeeding Gilbert Morgan. Many of his postgraduate workers followed him from Durham, forming a nucleus of a new school of carbohydrate chemistry.

Although he suffered a breakdown of health shortly before World War II, Haworth recovered sufficiently to take an active role in the chemical part of the atomic energy project. After the war he continued his work on carbohydrates even after his retirement in 1948. His sudden death in 1950 from a heart attack followed a strenuous tour of Australia and New Zealand.

Haworth received many awards and honorary degrees during his long career. Most notable was the Nobel Prize in chemistry, which he shared with Paul Karrer in 1937, for his work on carbohydrates and for his synthesis (with E. L. Hirst in 1933) of vitamin C. He was the first British organic chemist to receive the prize.

Haworth's scientific contributions can best be divided into four main categories. His earliest studies, first published in 1908 with Perkin, involve terpenes. This work was carried out at Manchester, Göttingen, Imperial College, and for a time at St. Andrews. The last paper appeared in 1914. The investigations included derivatives of menthane and sylvestrene and condensations of aldehydes and ketones.

Haworth's first contribution on simple sugars appeared in 1915 and involved a new method of preparing the methyl ethers of sugars by use of methyl sulfate and alkali. This method proved very valuable to structural work and remained a standard procedure applicable to most sugars. Haworth next undertook structural studies of the disaccharides; the only fact known about them at the time was that two monosaccharide residues were united by loss of a molecule of water. Haworth attacked the problem by preparing the fully methylated derivatives which were then hydrolyzed by aqueous acid. Although lactose, for example, was easily characterized, sucrose was quite troublesome and required many years of patient work. Another problem solved by Haworth and his co-workers was the nature of the ring systems present in simple sugars.

In 1932 Haworth turned his attention to the problem of the structure and synthesis of vitamin C. In that year Albert Szent-Győrgyi had isolated from the adrenal cortex and from orange juice a reactive substance he named "hexuronic acid." Its identification as vitamin C did not come until later. The Birmingham group, isolating the vitamin from ample supplies of Hungarian paprika, elucidated the structure of "ascorbic acid" (the name coined by Haworth). The synthesis, the first of any vitamin, was accomplished in 1933 with the assistance of a large team of workers.

Haworth's final field of study was the polysaccharides. Very likely his earlier work on simple sugars served as a stepping-stone to the more complicated problems shown with the biologically significant polysaccharides. Two of his important contributions to this field were his early recognition of the significance of X-ray studies and his introduction in 1932 of the end-group method of studying the fine details of structure.

BIBLIOGRAPHY

Much of Haworth's research is discussed in his book *The Constitution of Sugars* (London, 1929). Other valuable historical summaries of his work are found in several of his addresses: "The Constitution of Some Carbohydrates," in *Chemische Berichte,* **65A** (1932), 43–65; "The Structure, Function, and Synthesis of Polysaccharides," in *Proceedings of the Royal Society,* **186A** (1946), 1–19, the Bakerian lecture; "Starch," in *Journal of the Chemical Society* (1946), pp. 543–549; and "Carbohydrate Components of Biologically Active Materials," *ibid.* (1947), pp. 582–589. The paper first reporting his Nobel Prize-winning synthesis is "Synthesis of Ascorbic Acid," in *Chemistry and Industry* (1933), pp. 645–646, written with E. L. Hirst.

Although as a Nobel laureate Haworth is discussed in many biographical collections, clearly the most useful detailed biography is E. L. Hirst's obituary notice, "Walter Norman Haworth," in *Journal of the Chemical Society* (1951), pp. 2790–2806.

SHELDON J. KOPPERL

IBN ḤAWQAL, ABŪ'L-QĀSIM MUḤAMMAD (*b.* Nisibis, Upper Mesopotamia [now Nusaybin, Turkey]; *fl.* second half of the tenth century), *geography.*

Information on Ibn Ḥawqal's life is far from complete. He was a merchant and possibly a Fāṭimid missionary. Beginning in 943 he traveled through much of the Muslim world: between 947 and 951 he was in the Maghrib and visited the southern limit of the Sahara and Spain. In Spain he met the Jewish physician Ḥasdāy ibn Shaprūt, vizier of ʿAbd al-Raḥmān III. The vizier gave him information on the countries of northern Europe in return for data on the Jews of the Orient and, possibly, on the Khazars. In 955 he passed through Egypt, Armenia, and Azerbaijan; in 961–969 he crossed Iraq and Persia, and from there covered Transoxiana and Khwarizm. In 973 he was in Sicily.

Ibn Ḥawqal's extant work on geography is *Kitāb al-masālik wa'l-mamālik* ("Book of Routes and Kingdoms"). Its form is that of the works called *Atlas of Islam* and its closest antecedent is the book written by al-Iṣṭakhrī (*ca.* 930), who probably led Ibn Ḥawqal to devote himself to the study of geography. Originally, Ibn Ḥawqal intended only to bring al-Iṣṭakhrī's book up to date, but the successive incorporation of new material reflected in the three revisions of the *Kitāb al-masālik* (that of 967, dedicated to Sayf al-Dawla; that of *ca.* 977, in which he criticizes the Ḥamdanids; and that of *ca.* 988) led to a new book whose descriptive portion greatly surpassed the works of earlier authors. He added details about non-Muslim towns in the Sudan, Turkey, Nubia, and southern Italy and also gave chronological precisions and much information of economic interest on raw materials, things which, as a general rule, were not mentioned in works of this nature. The stylized maps he inserted were not to be taken as exact representations of the particular lands and seas named.

BIBLIOGRAPHY

I. ORIGINAL WORKS. An inventory of MSS is in C. Brockelmann, *Geschichte der arabischen Literatur,* I (Weimar, 1898), 263, and *Supplementband,* I (Leiden, 1944), 408. The text of *Kitāb al-masālik* was published by M. J. de Goeje as *Bibliotheca Geographorum Arabicorum,* II (Leiden, 1873); and by J. H. Kramers (Leiden, 1938). The latter is the basis of the French translation by the same author, revised by G. Wiet, as *Configuration de la terre,* 2 vols. (Paris–Beirut, 1964).

II. SECONDARY LITERATURE. There are various regional studies based on the work of Ibn Ḥawqal. Lists of them can be found in the works mentioned above. See also F. Gabrieli, "Ibn Ḥawqal e gli Arabi di Sicilia," in *L'Islam nella storia* (Bari, 1966), pp. 57–67; and A. Miquel, *La géographie humaine du monde musulmane jusqu'au milieu du XIᵉ siècle* (Paris, 1967), esp. pp. 299–309; and "Ibn Ḥawqal," in *Encyclopaedia of Islam,* 2nd ed., III (1968), 810–811.

J. VERNET

HAYDEN, FERDINAND VANDIVEER (*b.* Westfield, Massachusetts, 7 September 1829; *d.* Philadelphia, Pennsylvania, 22 December 1887), *geology.*

Hayden was the son of Asa and Melinda Hawley Hayden. His father died when the boy was ten, and at the age of twelve Hayden went to live with his uncle on a farm near Rochester, Ohio. In 1847, completely without funds, he walked to Oberlin College and worked his way through, receiving his degree in 1850. Hayden then attended Albany Medical College in New York and received the M.D. in 1853. From October 1862 to June 1865 he served as a surgeon in the Union Army, assigned mostly to supervisory duties. The University of Pennsylvania appointed him professor of geology from 1865 to 1872. He married Emma C. Woodruff, daughter of a Philadelphia merchant, on 9 November 1871; they had no children. Hayden was an active member of the Academy of Natural Sciences of Philadelphia, was elected to the National Academy of Sciences in 1873, and received honorary memberships in several foreign geological societies. He remained in excellent health until 1882, when locomotor ataxia forced him gradually to abandon fieldwork and writing.

While in medical school Hayden also studied with New York state paleontologist James Hall, who sent him and Fielding Bradford Meek to collect fossils in the Badlands of the Dakotas in 1853. In 1854–1855 Hayden explored the geology of the Missouri-Yellowstone rivers area, guided occasionally by traders of the American Fur Company. He spent the 1856 and 1857 field seasons as geologist with army engineer Gouverneur Warren's expeditions to the Dakotas and the Black Hills. Hayden visited Kansas Territory with Meek in 1858 to establish the age of the lowest Cretaceous stratum in the area covered by Warren. Hayden also accompanied Captain William F. Raynolds in his exploration of the northern Rocky Mountains in 1859–1860.

Since specialization was required to handle the great numbers of fossils collected on these expeditions, Hayden sent the invertebrates to Meek, the vertebrates to Joseph Leidy of Philadelphia, and the plants to John Strong Newberry, whom Hayden had met while at Oberlin. Generally, Hayden interpreted the geology of the region himself, although he usually published in conjunction with Meek. By the Civil War these scientists had discovered a Silurian formation in the West equivalent to the Potsdam Sandstone of New York, a Permian stratum, and a group of estuary and lake deposits postdating the Cretaceous beds. Hayden presented a clear picture of the geological history of the West based on a uniformitarian premise of gradual changes analogous to modern processes: as the land rose in the vicinity of the Rockies, the Cretaceous sea drained off, leaving lakes of brackish and fresh water which received the eroded material from the new highlands during the Tertiary period. Most important, Hayden and Meek created a detailed stratigraphic column for the Cretaceous and Tertiary formations of the West; some of the names they assigned are still used by American geologists.

From 1867 to 1879 Hayden headed the U.S. Geological and Geographical Survey of the Territories and indirectly benefited American science by providing employment and experience for more than fifty scientists, including Edward Drinker Cope, Leo Lesquereux, and William Henry Holmes, as well as Meek, Leidy, and Newberry. Starting in 1872, Hayden's topographers drew contour (instead of hachured) base maps, from which quantitative, accurate cross sections and long-distance extrapolations of formations could be made by geologists. The economic importance of the Hayden survey for railroads and mining was balanced by Hayden's successful campaign (1871–1872) to set aside Yellowstone Park for the people.

Hayden spent the 1867, 1868, and 1870 field seasons refining the stratigraphic sequence for Nebraska and Wyoming, and in 1869 he made a rapid reconnaissance through the Salt Lake Basin and south along the eastern edge of the Rockies, during which he named several local formations. He studied the Yellowstone Park region in 1871 and 1872, which led him to identify two additional factors in the geological history of the West—volcanic activity in recent times and horizontal forces during the mountain-building episodes. In the reports on his fieldwork in Colorado (1873–1874) Hayden further increased the role of horizontal stresses, particularly as they related to the fault systems in the Rockies. He also expanded the role of glaciation from local influence to a major part in shaping the topography of the West. In writing about the central Rockies, Hayden presented his fullest exposition of Western stratigraphy, from Precambrian granites to Quaternary silts. As usual, he gave scant attention to metamorphic rocks in his work. Other than discovering active glaciers in the Wind River Range in 1878, Hayden produced little of direct benefit to science during the rest of his survey or during his years with the U.S. Geological Survey (1879–1886). His place in the history of American geology is, however, assured by his stratigraphic work in the 1850's and early 1870's.

BIBLIOGRAPHY

I. Original Works. No unified bibliography of Hayden's publications exists, but one can be created quickly from Charles A. White, "Memoir of Ferdinand Vandiveer Hayden. 1839 [*sic*]–1887," in *Biographical Memoirs. National Academy of Sciences,* **3** (1895), 409–413; Lawrence Schmeckebier, *Catalogue and Index of the Hayden, King, Powell, and Wheeler Surveys,* U.S. Geological Survey Bulletin no. 222 (1904), 1–37; and Max Meisel, *A Bibliography of American Natural History: The Pioneer Century, 1769–1865,* 3 vols. (Brooklyn, N.Y., 1924–1929), *passim.* Of Hayden's writings in the 1850's and early 1860's, his articles in the *Proceedings of the Academy of Natural Sciences of Philadelphia* are more useful to historians of science than are his official reports published as government documents. Hayden's annual reports for the U.S. Geological and Geographical Survey of the Territories must be read for the late 1860's and the 1870's, despite their travelogue style, and may be supplemented with his brief, occasional pieces in the *American Journal of Science.*

The largest collection of Hayden MSS is in the records of the U.S. Geological and Geographical Survey of the Territories, in Record Group 57, National Archives. In addition, relevant material can be found in the Joseph Leidy papers, Academy of Natural Sciences of Philadelphia; the George Merrill collection, Library of Congress; the Fielding Bradford Meek diaries and Spencer Fullerton Baird correspondence, Smithsonian Institution; the Edward Drinker Cope papers, American Museum of Natural History; the James Hall papers, New York State Museum, Albany; the William Raynolds diaries, Beinecke Library, Yale; and the collections of the American Philosophical Society. Other collections are listed in the U.S. Library of Congress, *National Union Catalog of Manuscript Collections* (1962–   ).

II. Secondary Literature. In addition to the biographies listed in Meisel, see the articles in the *Dictionary of American Biography* and the *National Cyclopedia of American Biography.* Three interpretations of Hayden and his work have appeared recently. Richard Bartlett, *Great Surveys of the American West* (Norman, Okla., 1962), and Thomas Manning, *Government in Science: The United States Geological Survey, 1867–1894* (Lexington, Ky., 1967), concentrate on the second half of his career. William Goetzmann, *Exploration and Empire: The Explorer and the Scientist in the Winning of the American West* (New York,

1966), analyzes Hayden's entire career, including his scientific work with Army expeditions during the 1850's.

MICHELE L. ALDRICH

HAYFORD, JOHN FILLMORE (*b.* Rouses Point, New York, 19 May 1868; *d.* Evanston, Illinois, 10 March 1925), *geodesy.*

After graduating as a civil engineer from Cornell University in 1889, Hayford joined the U.S. Coast and Geodetic Survey. Except for a brief period (1895-1898) of teaching at Cornell, he remained with the Survey until 1909, when he became director of the College of Engineering at Northwestern University in Evanston, Illinois. At Northwestern, Hayford was an active participant in various public and private commissions and was notable for his stress of the need for mathematics and broad cultural studies in engineering education. His principal scientific achievements, however, occurred during his service with the Survey, in which he succeeded C. A. Schott as head of geodetic work in 1900.

Hayford is an important member of a little-studied scientific tradition. Determining the figure of the earth requires masses of extremely precise observations, a thorough grasp of an extensive body of theory, and great skill and ingenuity in computing—a particularly crucial feature in the period before the development of electronic computers. In his great work, *The Figure of the Earth and Isostasy From Measurements in the United States* (Washington, D.C., 1909), Hayford stresses the importance of economic factors and efficiency in determining the techniques chosen (p. 46 and *passim*). His achievements therefore stem not only from scientific knowledge and mathematical skills but also from considerable managerial ability.

By introducing the use of the area method, rather than the arc method, Hayford ended an era in geodesy that dated from the seventeenth century and inaugurated the modern procedure in this field. Hayford was also the first who systematically used observations and calculations of topographical irregularities (up to 4,126 kilometers from each astronomic station) and the first to take isostasy into consideration in arriving at the figure of the earth. In 1924 the Hayford spheroid was adopted as the international spheroid of reference by the International Geodetic and Geophysical Union.

Hayford's work constituted the first demonstration of the validity of the concept of isostasy. Hayford was, of course, aware of the prior work in isostasy done by J. H. Pratt, G. B. Airy, and C. E. Dutton. Like Pratt—but unlike Airy—he and his successor in the Coast and Geodetic Survey, William Bowie, believed that the isostatic compensation is complete and local; that is, the density of blocks of the earth's crust varies laterally according to their elevation, so that the elevated land masses are less dense than other land masses and float on the subcrustal matter. Hayford postulated a uniform distribution of isostatic compensations with respect to depth and calculated the most probable value of the limiting depth as 113 kilometers.

These views were not immediately accepted by many geologists. T. C. Chamberlin, for example, had postulated a different distribution of isostatic compensation which yielded a limiting depth of 287 kilometers that, according to his planetesimal theory, was fixed at the time of the creation of the earth. Hayford contended that isostatic compensation had been increasing and was at its highest level at present. Relying on geologic evidence of changes in the surface of the earth, Hayford concluded that the earth was a failing, not a stable, structure.

Chamberlin and others could not conceive of how isostatic compensation, a vertical movement, was reconcilable with the lateral movements postulated for mountain formation and other land forms. In response, Hayford hypothesized horizontal flows of rock far beneath the surface, involving frictional heat and chemical reactions. Chamberlin, who described his approach to cosmology as naturalistic rather than mathematical, was clearly not at ease with Hayford's reasoning. On the other hand, subsequent research on isostasy has developed a "higher synthesis," incorporating Hayford's views while accounting for Chamberlin's criticisms.

BIBLIOGRAPHY

A full listing of Hayford's writings is in a useful and uncritical biography by William H. Burger, "John Fillmore Hayford, 1868-1925," in *Biographical Memoirs. National Academy of Sciences,* **16** (1933), 279-292. Although Burger frequently refers to Hayford's personal papers and Northwestern University files of great interest, inquiries to Northwestern, Hayford's children, and Burger's widow failed to uncover these documents. The United States National Archives has some Hayford materials in the U.S. Coast and Geodetic Survey records. For his relations with geologists, the T. C. Chamberlin papers at the University of Chicago Library are useful.

See also Chamberlin's review of the Tittmann and Hayford 1906 report to the International Geodetic Association in "Review of Geodetic Operations in the United States," in *Journal of Geology,* **15** (1907), 73-79, and Hayford's reply, "Comment on the Above Review by Mr. John F. Hayford," *ibid.,* pp. 79-81. The most elaborate attack on Hayford's work in isostasy is Harmon Lewis, "The Theory of Isostasy," *ibid.*, **19** (1911), 603-626.

NATHAN REINGOLD

IBN AL-HAYTHAM, ABŪ ʿALĪ AL-ḤASAN IBN AL-ḤASAN, called **al-Baṣrī** (of Baṣra, Iraq), **al-Miṣrī** (of Egypt); also known as **Alhazen,** the Latinized form of his first name, **al-Ḥasan** (*b.* 965; *d.* Cairo, *ca.* 1040), *optics, astronomy, mathematics.*

About Ibn al-Haytham's life we have several, not always consistent, reports, most of which come from the thirteenth century. Ibn al-Qifṭī (*d.* 1248) gives a detailed account of how he went from Iraq to Fāṭimid Egypt during the reign of al-Ḥākim (996-1021), the caliph who patronized the great astronomer Ibn Yūnus (*d.* 1009) and who founded in Cairo a library, the Dār al-ʿIlm, whose fame almost equalled that of its precursor at Baghdad (the Bayt al-Ḥikma, which flourished under al-Maʾmūn [813-833]). Impressed by a claim of Ibn al-Haytham that he would be able to build a construction on the Nile which would regulate the flow of its waters, the caliph persuaded the already famous mathematician to come to Egypt and, to show his esteem, went out to meet him on his arrival at a village outside Cairo called al-Khandaq.

Ibn al-Haytham, according to Ibn al-Qifṭī, soon went at the head of an engineering mission to the southern border of Egypt where, he had assumed, the Nile entered the country from a high ground. But even before reaching his destination he began to lose heart about his project. The excellently designed and perfectly constructed ancient buildings which he saw on the banks of the river convinced him that if his plan had been at all possible it would have been already put into effect by the creators of those impressive structures. His misgivings were proved right when he found that the place called *al-Janādil* (the cataracts), south of Aswan, did not accord with what he had expected. Ashamed and dejected he admitted his failure to al-Ḥākim, who then put him in charge of some government office. Ibn al-Haytham at first accepted this post out of fear, but realizing his insecure position under the capricious and murderous al-Ḥākim he pretended to be mentally deranged and, as a result, was confined to his house until the caliph's death. Whereupon Ibn al-Haytham revealed his sanity, took up residence near the Azhar Mosque, and, having been given back his previously sequestered property, spent the rest of his life writing, copying scientific texts, and teaching.

To this account Ibn al-Qifṭī appends a report which he obtained from his friend Yūsuf al-Fāsī (*d.* 1227), a Jewish physician from North Africa who settled in Aleppo after a short stay in Cairo where he worked with Maimonides.[1] Yūsuf al-Fāsī had "heard" that in the latter part of his life Ibn al-Haytham earned his living from the proceeds (amounting to 150 Egyptian dinars) of copying annually the *Elements* of Euclid, the *Almagest,* and the *Mutawassiṭāt,*[2] and that he continued to do so until he died "in [*fī ḥudūd*][3] the year 430 [A.D. 1038-1039] or shortly thereafter [*aw baʿdahā bi-qalīl*]." These words are immediately followed by a statement, of which the author must be presumed to be Ibn al-Qifṭī, to the effect that he possessed a volume on geometry in Ibn al-Haytham's hand, written in 432 (A.D. 1040-1041).

An earlier account of Ibn al-Haytham's visit to Egypt is given by ʿAlī ibn Zayd al-Bayhaqī (*d.* 1169-1170).[4] According to him the mathematician had only a brief and unsuccessful meeting with al-Ḥākim outside an inn in Cairo. The caliph, sitting on a donkey with silver-plated harness, examined a treatise composed by Ibn al-Haytham on his Nile project, while the author, being short of stature, stood on a bench (*dukkān*) in front of him. The caliph condemned the project as impractical and expensive, ordered the bench to be demolished, and rode away. Afraid for his life, Ibn al-Haytham immediately fled the country under cover of darkness, going to Syria, where he later secured the patronage of a well-to-do governor. But this account, vivid though it is, must be discarded as being unsupported by other evidence. For example, we are told by Ṣāʿid al-Andalusī (*d.* 1070) that a contemporary of his, a judge named ʿAbd al-Raḥman ibn ʿĪsā, met Ibn al-Haytham in Egypt in 430 A.H., that is, a short time before the latter died.

Ibn Abī Uṣaybiʿa (*d.* 1270) gives the name of Ibn al-Haytham as Muḥammad (rather than al-Ḥasan) ibn al-Ḥasan; and he joins Ibn al-Qifṭī's story (which he quotes in full with the omission of the last statement about Ibn al-Haytham's autograph of 432) to a report which he heard from ʿAlam al-Dīn Qayṣar ibn Abi 'l-Qāsim ibn Musāfir, an Egyptian mathematician who resided in Syria and died at Damascus in 649 A.H./A.D. 1251.[5] According to this report, Ibn al-Haytham at first occupied the office of minister at Baṣra and its environs, but to satisfy his strong desire to devote himself entirely to science and learning he feigned madness until he was relieved from his duties. Only then did he go to Egypt, where he spent the rest of his life at the Azhar Mosque, living on what he earned from copying Euclid and the *Almagest* once every year. We may add that the title of one of his writings (no. II 13, see below) appears to imply that he was at Baghdad in 1027, six years after al-Ḥākim died.[6]

IBN AL-HAYTHAM

It is unfortunate that the autobiography of Ibn al-Haytham, which Ibn Abī Uṣaybiʿa quotes from an autograph, throws no light on these different reports. Written at the end of 417 A.H./A.D. 1027, when the author was sixty-three lunar years old, and clearly modeled after Galen's *De libris propriis,*[7] it lists the works written by Ibn al-Haytham up to that date but speaks in only general terms about his intellectual development.

As cited by Ibn Abī Uṣaybiʿa, Ibn al-Haytham, reflecting in his youth on the conflicting but firmly held beliefs of the various religious sects, was led to put them all in doubt and became convinced that truth was one. When in later years he was ready to grasp intellectual matters, he decided to turn his back on the common people and devote himself to seeking knowledge of the truth as the worthiest possession that could be obtained in this world and the surest way to gain favor with God—a decision which, using Galen's expressions in *De methodo medendi,*[8] he attributed to his "good fortune, or a divine inspiration, or a kind of madness." Frustrated in his intensive inquiries into the religious sciences, he finally emerged with the conviction that truth was to be had only in "doctrines whose matter was sensible and whose form was rational." Such doctrines Ibn al-Haytham found exemplified in the writings of Aristotle (of which he here gave a conspectus) and in the philosophical sciences of mathematics, physics, and metaphysics. As evidence of his having stood by his decision, he provided a list of his writings to 10 February 1027, containing twenty-five titles on mathematical subjects (list I*a*) and forty-five titles on questions of physics and metaphysics (list I*b*).

Ibn Abī Uṣaybiʿa gives two more lists of Ibn al-Haytham's work, which we shall designate as II and III. List II, which he found attached to list I and in the author's hand, contains twenty-one titles of works composed between 10 February 1027 and 25 July 1028. Ibn Abī Uṣaybiʿa does not say whether he also copied list III from an autograph but simply describes it as a catalogue (*fihrist*) which he found of Ibn al-Haytham's works to the end of 429 A.H./2 October 1038. Nor does he specify the *terminus a quo* of this catalogue. However that may be, two things are remarkable about this last list: consisting of ninety-two titles, it includes all sixty-nine titles ascribed to Ibn al-Haytham by Ibn al-Qifṭī, with two exceptions; and in it are to be found all of Ibn al-Haytham's extant works (not fewer than fifty-five), again with only a very few exceptions. It may also be noted that the order of works in list III almost always agrees with the chronological order of their composition, whenever the latter can be independently determined from internal cross references. Thus, III 2 was written before III 53; III 3 before III 36, III 49, III 60, III 77, and III 80; III 20 before III 21; III 25 before III 31; III 26 before III 38 and III 68; III 42 before III 74; III 53 before III 54; III 61 before III 63; III 63 before III 64; and III 66 before III 77. Item III 17, however, was written before III 16 (see bibliography).

Among the subjects on which Ibn al-Haytham wrote are logic, ethics, politics, poetry, music, and theology (*kalām*); but neither his writings on these subjects nor the summaries he made of Aristotle and Galen have survived. His extant works belong to the fields in which he was reputed to have made his most important contributions: optics, astronomy, and mathematics.

Optics: Doctrine of Light. Ibn al-Haytham's theory of light and vision is neither identical with nor directly descendant from any one of the theories known to have previously existed in antiquity or in Islam. It is obvious that it combines elements of earlier theories—owing perhaps more to Ptolemy than to any other writer—but in it these elements are reexamined and rearranged in such a way as to produce something new. Ibn al-Haytham's writings on optics included a treatise written "in accordance with the method of Ptolemy" (III 27), whose *Optics* was available to him in an Arabic translation lacking the first book and the end of the fifth and last book, and a summary of Euclid and Ptolemy in which he "supplemented the matters of the first Book, missing from Ptolemy's work" (I*a* 5). These two works are now lost.

But in his major work, the *Optics* or *Kitāb al-Manāzir* (III 3),[9] in seven books, Ibn al-Haytham deliberately set out to dispel what appeared to him to be a prevailing confusion in the subject by "recommencing the inquiry into its principles and premises, starting the investigation by an induction of the things that exist and a review of the conditions of the objects of vision." Once the results of induction were established he was then to "ascend in the inquiry and reasonings, gradually and in order, criticizing premises and exercising caution in the drawing of conclusions," his aim in all this being "to employ justice, not to follow prejudice, and to take care in all that we judge and criticize that we seek the truth and not to be swayed by opinions" (Fatih MS 3212, fol. 4a r).

The book is in fact an earnest and assiduous exercise in the method outlined. Its arguments are either inductive, experimental, or mathematical, and it cites no authorities. Experiment (*iʿtibār*) in particular emerges in it as an explicit and identifiable methodological tool involving the manipulation of artificially constructed devices. (In the Latin translation of the *Optics* the word *iʿtibār* and its cognates *iʿtabara* and *muʿtabir* became *experimentum, experimentare,* and *experimentator,* respectively.) Perhaps as a result of its derivation from the astronomical procedure of testing past observations by comparing them with new ones, the method of *iʿtibār* often appears aimed at proof rather than discovery. It establishes beyond doubt that which is insecurely suggested by inadequate observations.

The *Optics* is not a philosophical dissertation on the nature of light, but an experimental and mathematical investigation of its properties, particularly insofar as these relate to vision. With regard to the question "What is light?," Ibn al-Haytham readily adopted the view ascribed by him to "the physicists" or natural philosophers (*al-ṭabīʿiyyūn*)—not, however, because that view was by itself sufficient, but because it constituted an element of the truth which had to be combined with other elements derived from "mathematicians" (*taʿlīmiyyūn*) such as Euclid and Ptolemy. In the resulting synthesis (*tarkīb*) the approach of "the mathematicians" dominated the form of inquiry, while their doctrines were altered, indeed reversed, in the light of those of "the physicists." That "the physicists" were the natural philosophers working in the Aristotelian tradition is clear enough from comparing the view attributed to them by Ibn al-Haytham with expressions and doctrines that had been current in the works of peripatetics from Alexander to Avicenna.

Light, says Ibn al-Haytham, is a form (*ṣūra,* εἶδος) essential (*dhātiyya*) in self-luminous bodies, accidental (*ʿaraḍiyya*) in bodies that derive their luminosity from outside sources. Transparency (*al-shafīf*) is an essential form in virtue of which transparent bodies, such as air or water, transmit light. An opaque body, such as a stone, has the power to "receive" or take on and make its own the light shining upon it and thereby to become itself a luminous source. This received light is called accidental because it belongs to the body only as long as the body is irradiated from outside. There are no perfectly transparent bodies. All transparent bodies possess a certain degree of opacity which causes light to be "received" or "fixed" in them as accidental light.

The light which radiates directly from a self-luminous source is called "primary" (*awwal*); that which emanates from accidental light is called "secondary" (*thānī*). Primary and secondary lights are emitted by their respective sources in exactly the same manner, that is, from every point on the source in all directions along straight lines. The only difference between these two kinds of light is one of intensity: accidental light is weaker than its primary source and the secondary light deriving from it is weaker still. All radiating lights become weaker the farther they travel. The distinction is made in transparent bodies between the accidentally fixed and the traversing light, and it is from the former that secondary light is emitted. Thus from every point of the sunlit air, or on the surface of an illuminated opaque object, a secondary light, fainter than the light coming to this point directly from the sun, radiates "in the form of a sphere," rectilinearly in all directions. (The picture is interesting since it later appears in the doctrine of the multiplication of species and it is at the basis of Huygens' principle.)

Two other modes of propagation are the reflection of light from smooth bodies and its refraction when passing from one transparent body into another. Unlike an opaque body, a smooth surface does not behave, when illuminated, like a self-luminous object; rather than "receive" the impinging light it sends it back in a determinate direction. In *Optics,* book I, chapter 3, numerous experiments involving the use of various devices—sighting tubes, strings, dark chambers—are adduced to support all of the above statements, and in particular to establish the property of rectilinear propagation for all four kinds of radiation: primary, secondary, reflected, and refracted.

Colors are asserted to be as real as light and distinct from it; they exist as forms of the colored objects. A self-luminous body either possesses the form of color or something of "the same sort as color." Like light, colors radiate their forms upon surrounding bodies and this radiation originates from every point on the colored object and extends in all directions. It is possible that colors should be capable of extending themselves into the surrounding air in the absence of light; but experiments show that they are always found in the company of light, mingled with it, and they are never visible without it. Whatever rules apply to light also apply to colors.

Some time after writing the *Optics,* Ibn al-Haytham remarked in the *Discourse* (III 60) that natural philosophers, in contrast to mathematicians, had failed to supply a definite concept of ray. In book IV of the *Optics* he had in fact tried to remedy the defect by introducing the concept of a physical ray. The underlying idea is that for a body to be able to carry the form of light it must be of certain minimal magnitude. Imagine, then, that a transparent body through which light travels is made progressively thinner by a process of division. (The operation is essentially the same as that of narrowing an aperture through which the light passes.)

Ibn al-Haytham considered that a limit would be reached after which further division would cause the light to vanish. At this limit there would pass through

IBN AL-HAYTHAM

the thin body a light of finite breadth which he calls the smallest or least light (*aṣghar al-ṣaghīr min al-ḍaw'*), a single ray whose only direction of propagation is the straight line extending through its length. A wider volume of light should not, however, be regarded as an aggregate of such minimal parts (*aḍwā' diqāq mutaḍāmma*), but a continuous and coherent whole in which propagation takes place along all the straight lines, both parallel and intersecting, that can be imagined within its width. It follows that an aperture will either be wide enough to allow only rectilinear propagation, or too small to let any light pass through; there is no room for the diffraction of light. The result of the new concept is thus an uncompromising formulation of the ray theory of light. (Compare Newton's concept of "least Light or part of Light" which accords with his interpretation of diffraction as a kind of refraction.)[10]

Theory of Vision. As employed by Ibn al-Haytham the language of forms serves merely to express the view that light and color are real properties of physical bodies. He sometimes conducted his discussion without even using the term "form" (as in the greater part of book I, chapter 3) and his experimental arguments would lose nothing of their import if that term were to be removed from them. And yet it was the term "form" that had been closely associated with the intromission theory of vision maintained in the Peripatetic tradition, whereas mathematical opticians had formulated their geometrically represented explanations in terms of "visual rays" issuing from the eye. Ibn al-Haytham adopted the intromission hypothesis as the more reasonable one and took over with it the vocabulary of forms. To this he added, as we saw, a new concept of ray that satisfied the mathematical condition of rectilinearity but was consistent with the physics of forms. His theory of vision (to be described presently) may thus be seen as one chief illustration of the program he outlined in the *Optics* (III 3), in the treatise *On the Halo and the Rainbow* (III 8), and in the *Discourse on Light* (III 60): optical inquiry must "combine" the physical and the mathematical sciences.

In chapter 5 of book I of the *Optics* Ibn al-Haytham described the construction of the eye on the basis of what had been generally accepted in the tradition of medical and anatomical writings derived from Galen's works. But he adapted the geometry of this construction to suit his own explanation of vision. In particular he assumed both surfaces of the cornea opposite the pupil to be parallel to the anterior surface of the crystalline humor, all these surfaces being spherical and having the center of the eye as common center. He placed the center of the eye

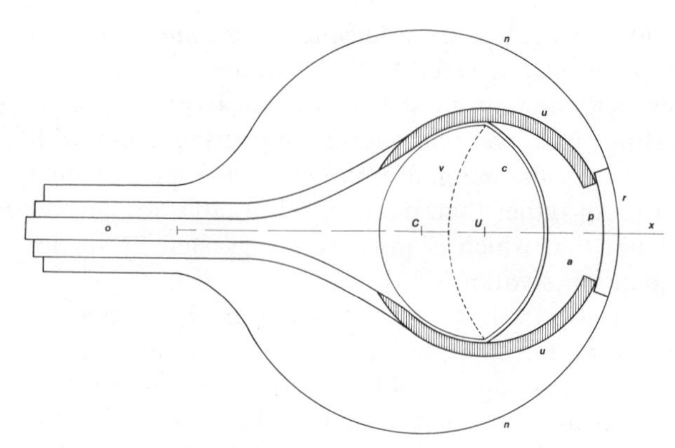

FIGURE 1. A cross section of the eye. Constructed from the text of *Kitāb al-Manāẓir*. After M. Naẓīf.

a, albugineous humor, *al-bayḍiyya; C,* center of eyeball; *c,* crystalline humor, *al-jalīdiyya; n,* exterior surface of conjunctiva, *al-multaḥima; o,* optic nerve, *al-ʿaṣab al-baṣarī; p,* uveal opening or pupil, *thaqb al-ʿinabiyya; r,* cornea, *al-qarniyya; U,* center of uvea; *u,* uvea, *al-ʿinabiyya; v,* vitreous humor, *al-zujājiyya; x,* axis of symmetry.

The axis of symmetry, passing through the middle of the pupil, the center of the uvea, and the center of the eye, goes to the middle of the optic nerve where the eyeball bends as a whole in its socket. The uvea is displaced forward toward the surface of the eye.

behind the posterior surface of the crystalline humor. The latter surface may be plane or spherical, so that the line passing through the middle of the pupil and the center of the eye would be perpendicular to it. (See Figure 1.[11]) The theory of vision is itself expounded in chapters 2, 4, 6, and 8.

Observations (such as the feeling of pain in the eye when gazing on an intense light, or the lingering impression in the eye of a strongly illuminated object) show that it is a property of light to make an effect on the eye, and a property of sight to be affected by light; visual sensation is therefore appropriately explained solely in terms of light coming to the eye from the object. As maintained by natural philosophers this effect is produced by the forms of the light and color in the visible object. But as an explanation of vision this statement in terms of forms is, by itself, "null and void" (*tantaqiḍ wa-tabṭul, destruitur*).[12]

The problem Ibn al-Haytham posed for himself was to determine what further conditions are needed in order to bring the form of an external object intact into the eye where it makes its visual effect. His solution assumed the crystalline humor to be the organ in which visual sensation first occurs—an assumption which had been current since Galen. The solution also employs the experimentally supported principle which considers the shining object as a collection of points individually radiating their light and color (or the forms of their light and color) rectilinearly in all directions.[13] In consequence of this principle any point on a visible object may be regarded as the origin of a cone of radiation with a base at the portion of the surface of the eye opposite the pupil. Since this holds for all points of the object, there will be spread over the whole of that portion the forms of the light and color of every one of these points.

Further confusion will result after the majority of these forms have been refracted upon their passage through the cornea. Ibn al-Haytham considered that for veridical perception to be possible it must be assumed that vision of any given object point can occur only through a given point on the surface of the eye, and he defined the latter point as that at which the perpendicular from the object point meets the cornea. It follows from the geometry of the eye that forms coming from all points on the object along perpendiculars to the surface of the eye will pass unrefracted through the pupil into the albugineous humor and again strike the anterior surface of the crystalline at right angles. There will then be produced on the crystalline humor a total form whose points will correspond, one-to-one, with all the points on the object, and it is this "distinct" and erect form which the crystalline humor will sense. Because the effective perpendiculars are precisely those that make up the outward extension of the cone having the center of the eye as vertex and the pupil as base (the so-called "radial cone," *makhrūṭ al-shuʿāʿ*), what we have in the end is the geometry of the Euclidean visual-ray theory.

But now the "mathematicians'" rays are strictly mathematical, that is, they are no more than abstract lines along which the light travels toward the eye—which is enough to save the geometrical optics of the ancients. As for the hypothesis that something actually goes out of the eye, it is clearly declared to be "futile and superfluous"—"Exitus ergo radiorum est superfluus et otiosus."[14] It would be absurd, says Ibn al-Haytham, to suppose that a material effluence flowing out of the eye would be capable of filling the visible heavens almost as soon as we lift our eyelids. If such effluence or visual rays are not corporeal, then they would not be capable of sensation and their function would merely be to serve as vehicles for bringing back something else from the object which itself would produce vision in the eye. But since this function is already fulfilled by the transparent medium through which light and color (or their forms) extend, visual rays are no longer of any use. (In the presence of this decisive argument it is curious that the editor of the Latin translation should misinterpret Ibn al-Haytham's remarks about preserving the geometrical property of the mathematicians' rays as an argument in support of the "Platonic" theory of *συναύγεια*, combining the intromission and extramission hypotheses.)[15]

Ibn al-Haytham managed to introduce the form of the visible object into the eye—an achievement which had apparently defeated his predecessors. But it should be noted that the "distinct form" he succeeded in realizing inside the eye is apparent only to the sensitive faculty; it is not a visibly articulate image such as that produced by a pinhole camera. In one place he ascribed the privileged role of the perpendicular rays to their superior strength. But there is another dominant idea. As a transparent body the crystalline humor allows non-perpendicular rays to be refracted into it from all points on its surface; as a sensitive body, however, it is especially concerned with those rays that go through it without suffering refraction. Veridical vision is thus due in the first place to the selective or directional sensitivity of the crystalline humor.

The vitreous humor, whose transparency differs from that of the crystalline, has still another property, namely that of preserving the integrity of the form handed down to it at its common face with the crystalline, where refraction of the effective rays takes place away from the axis of symmetry. The sensitive body (visual spirit), issuing along independent and parallel lines from the brain into the optic nerve, finally receives the form from the vitreous body and channels it back along the same lines to the front of the brain where the process of vision is completed. In the optic chiasma, where corresponding lines of the optic nerves join together, the form from the one eye coincides with that from the other, and from there the two forms proceed to the brain as one.

In book VII Ibn al-Haytham introduced what may be considered a generalization of the theory of vision already set out in book I. The form of his inquiry is the same as before: the determination of the conditions that must be assumed in order to accommodate the results of certain indubitable experiments. The experiments described here at first appear to speak against the earlier theory. A small object placed in the radial cone close to one eye, while the other is shut, does not hide an object point lying behind it on the common line drawn from the center of the eye. This means that the object point must in this case be seen by means of a ray falling obliquely, and therefore refracted, at the surface of the eye. Again, a small object placed outside the radial cone, as when a needle is held close to the corner of one eye, can be seen while the other eye is shut. Since no perpendicular can be drawn from the object in this position to any point in the area cut off from the eye-surface by the radial cone, the object must be seen by refraction.

Briefly stated (and divorced from its rather problematic, though interesting, arguments), the final doctrine intended to take all of these observations into account is that vision of objects within the radial cone is effected both by direct and refracted rays, whereas objects outside the cone are seen only by refraction. Ibn al-Haytham here maintains that sensation of refracted as well as direct forms or rays takes place in the crystalline humor, although (in accordance with the earlier theory) he states that the "sensitive faculty" apprehends them all along perpendiculars drawn from the center of the eye to the objects seen. It is this general doctrine, that whatever we see is seen by refraction,[16] whether or not it is also seen by direct rays, that, according to Ibn al-Haytham, had not been grasped or explained by any writer on optics, ancient or modern.

The main part of Ibn al-Haytham's general theory of light and vision is contained in book I of the *Optics.* In book II he expounded an elaborate theory of cognition, with visual perception as the basis, which was referred to and made use of by fourteenth-century philosophers including, for example, Ockham,[17] and which has yet to receive sufficient attention from historians of philosophy. Book III deals with binocular vision and with the errors of vision and of recognition. Reflection is the subject of book IV, and here Ibn al-Haytham gave experimental proof of the specular reflection of accidental as well as essential light, a complete formulation of the laws of reflection, and a description of the construction and use of a copper instrument for measuring reflections from plane, spherical, cylindrical, and conical mirrors, whether convex or concave. He gave much attention to the problem of finding the incident ray, given the reflected ray (from any kind of mirror) to a given position of the eye. This is characteristic of the whole of the *Optics*—an eye is always given with respect to which the problems are to be formulated. The investigation of reflection—with special reference to the location of images—is continued in book V where the well-known "problem of Alhazen" is discussed, while book VI deals with the errors of vision due to reflection.

Book VII, which concludes the *Optics,* is devoted to the theory of refraction. Ibn al-Haytham gave considerable space to a detailed description of an improved version of Ptolemy's instrument for measuring refractions, and illustrated its use for the study of air-water, air-glass, and water-glass refractions at plane and spherical surfaces. Rather than report any numerical measurements, as in Ptolemy's tables, he stated the results of his experiments in eight rules which mainly govern the relation between the angle of incidence i (made by the incident ray and the normal to the surface) and the angle of deviation d (*zāwiyat al-inʿiṭāf, angulus refractionis*) contained between the refracted ray and the prolongation of the incident ray into the refracting medium. (This concentration on d rather than the angle of refraction r—which being equal to $i - d$ he called the remaining angle, *al-bāqiya*—was also a feature of Kepler's researches.)

His rules may be expressed as follows. Let d_1, d_2 and r_1, r_2 correspond to i_1, i_2, respectively, and let $i_2 > i_1$. It is asserted that

(1) $d_2 > d_1$;
(2) $d_2 - d_1 < i_2 - i_1$;
(3) $\frac{d_2}{i_2} > \frac{d_1}{i_1}$;
(4) $r_2 > r_1$;
(5) In rare-to-dense refraction, $d < 1/2\ i$;
(6) In dense-to-rare refraction, $d < 1/2\ (i + d)$ $[d < 1/2\ r]$;
(7) A denser refractive medium deflects the light more toward the normal; and
(8) A rarer refractive medium deflects the light more away from the normal.

It is to be noted that (2) holds only for rare-to-dense refraction, and (5) and (6) are true only under certain conditions which, however, were implicit in the experiments, as Naẓīf has shown.[18] Concluding that "these are all the ways in which light is refracted into transparent bodies," Ibn al-Haytham does not give the impression that he was seeking a law which he failed to discover; but his "explanation" of refraction certainly forms part of the history of the formulation of the refraction law. The explanation is based on the idea that light is a movement which admits of variable speed (being less in denser bodies) and of analogy with the mechanical behavior of bodies. The analogy had already been suggested in antiquity, but Ibn al-Haytham's elaborate application of the parallelogram method, regarding the incident and refracted movements as consisting of two perpendicular components which can be considered separately, introduced a new element of sophistication. His approach attracted the attention of such later mathematicians as Witelo, Kepler, and Descartes, all of whom employed it, the last in his successful deduction of the sine law.

Minor Optical Works. The extant writings of Ibn al-Haytham include a number of optical works other than the *Optics,* of which some are important, showing Ibn al-Haytham's mathematical and experimental ability at its best, although in scope they fall far short of the *Optics*. The following is a brief description of these works.

The Light of the Moon (III 6). Ibn al-Haytham showed here that if the moon behaved like a mirror, the light it receives from the sun would be reflected to a given point on the earth from a smaller part of its surface than is actually observed. He accordingly argued that the moon sends out its borrowed light in the same manner as a self-luminous source, that is, from every point on its surface in all directions. This is confirmed through the use of an astronomical diopter having a slit of variable length through which various parts of the moon could be viewed from an opposite hole in a screen parallel to the slit. The treatise is a beautiful combination of mathematical deduction and experimental technique. The experiments do not, however, lead to the discovery of a new property, but only serve to prove that the mode of emission from the moon is of the same kind as the already known mode of emission from self-luminous objects. Here, as in the *Optics*, the role of experiment is in contrast to its role in the work of, say, Grimaldi or Newton.

The Halo and the Rainbow (III 8). The subject is not treated in the *Optics*. In this treatise Ibn al-Haytham's explanation of the bow fails, being conceived of solely in terms of reflection from a concave spherical surface formed by the "thick and moist air" or cloud. The treatise did, however, become one of the starting points of Kamāl al-Dīn's more successful researches.

On Spherical Burning Mirrors (III 18). In contrast to the eye-centered researches of the *Optics* the only elements of the problems posed in this treatise (and in III 19) are the luminous source, the mirror, and the point or points in which the rays are assembled. Ibn al-Haytham showed that rays parallel to the axis of the mirror are reflected to a given point on the axis from only one circle on the mirror; his remarks imply a recognition of spherical aberration along the axis.

On Paraboloidal Burning Mirrors (III 19). This refers to Archimedes and Anthemius "and others" as having adopted a combination of spherical mirrors whose reflected rays meet in one point. Drawing ably on the methods of Apollonius, Ibn al-Haytham set out to provide a proof of a fact which, he said, the ancients had recognized but not demonstrated: that rays are reflected to one point from the whole of the concave surface of a paraboloid of revolution.

The Formation of Shadows (III 36). That there were many writings on shadows available to Ibn al-Haytham is clear from his reference here to *aṣḥāb al-aẓlāl* (the authors on shadows). Indeed, a long treatise on shadows by his contemporary al-Bīrunī is extant. Ibn al-Haytham defines darkness as the total absence of light, and shadow as the absence of some light and the presence of another. He made the distinction between umbra and penumbra—calling them *ẓulma* (darkness) or *ẓill maḥḍ* (pure shadow), and *ẓill* (shadow), respectively.

The Light of the Stars (III 48). This argues that all stars and planets, with the sole exception of the moon, are self-luminous.

Discourse on Light (III 60). Composed after the *Optics*, this treatise outlines the general doctrine of light. Some of its statements have been used in the account given above.

The Burning Sphere (III 77). In this work, written after the *Optics*, Ibn al-Haytham continued his investigations of refraction, but, as in III 18 and III 19, without reference to a seeing eye. He studied the path of parallel rays through a glass sphere, tried to determine the focal length of such a sphere, and pointed out spherical aberration. The treatise was carefully studied by Kamāl al-Dīn, who utilized it in his account of the path of rays from the sun inside individual rain drops.

The Shape of the Eclipse (III 80). This treatise is of special interest because of what it reveals about Ibn al-Haytham's knowledge of the important subject of the *camera obscura*. The exact Arabic equivalent of that Latin phrase, *al-bayt al-muẓlim*, occurs in book I, chapter 3 of the *Optics*;[19] and indeed dark chambers are frequently used in this book for the study of such various properties of light as its rectilinear propagation and the fact that shining bodies radiate their light and color on neighboring objects. But such images as those produced by a pinhole camera are totally absent from the *Optics*. The nearest that Ibn al-Haytham gets to such an image is the passage in which he describes the patches of light cast on the inside wall of a "dark place" by candle flames set up at various points opposite a small aperture that leads into the dark place; the order of the images on the inside wall is the reverse of the order of the candles outside.

The experiment was designed to show that the light from one candle is not mingled with the light from another as a result of their meeting at the aperture, and in general that lights and colors are not affected by crossing one another. Although this passage occurs in book I in the context of the theory of vision,[20] the eye does not in Ibn al-Haytham's explanation act as a pinhole camera and it is expressly denied the role of a lens camera. In the present treatise, however, he approached the question, already posed in the pseudo-Aristotelian *Problemata,* of why the image of a crescent moon, cast through a small circular aperture, appears circular, whereas the same aperture will cast a crescent-shaped image of the partially eclipsed sun. Although his answer is not wholly satisfactory, and although he failed to solve the general problem of the pinhole camera, his attempted explanation of the image of a solar crescent clearly shows that he possessed the principles of the working of the camera. He formulated the condition for obtaining a distinct image of an object through a circular aperture as that when

$$\frac{m_a}{m_s} \leq \frac{d_a}{d_s},$$

where m_a, m_s are the diameters of the aperture and of the object respectively, and d_a, d_s the distances of the screen from the aperture and from the object respectively.

Ibn al-Haytham's construction of the crescent-shaped image of the partially eclipsed sun can be clearly understood by reference to Figure 2. (Because Ibn al-Haytham's own diagram shows the crescents but not the circles, the figure shown is that constructed by Naẓīf.) It represents the special case in which the two ratios just mentioned are equal. It is assumed that the line joining the centers of the two arcs forming the solar crescent is parallel to the planes of the aperture and the screen, and further that the line joining the center of the sun and the center of the circular hole is perpendicular to the plane of the latter and to the plane of the screen.

The crescents *p, q, r,* are inverted images produced by three double conical solids of light whose vertices are three different points on the aperture, and whose bases are, on the one side, the shining solar crescent, and, on the other, the inverted image. These solids are each limited by two conical surfaces of which one is convex and the other concave; and in every double solid the convex surface on one side of the aperture corresponds to the concave surface on the other. The middle crescent image *q* is produced by such a double solid having its vertex at the aperture-center; *p* and *r* have their vertices at the extremities of a diameter of the aperture. The circular images are each produced by a single cone whose vertex is a single point on the shining crescent; as many such circles are produced as there are points on the crescent sun.

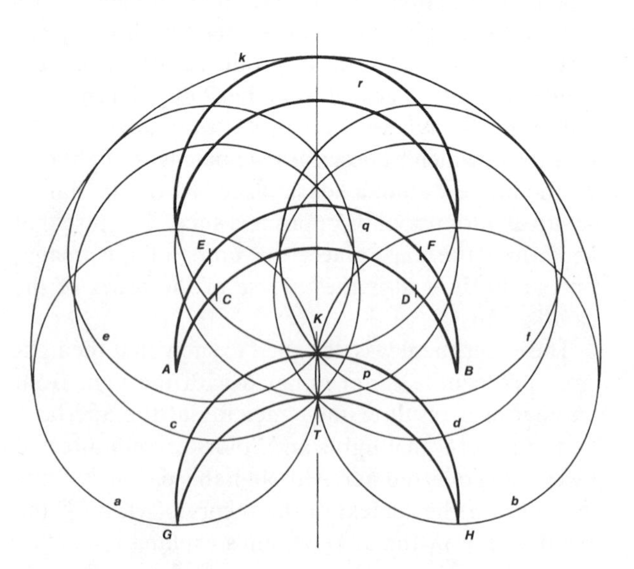

FIGURE 2. Ibn al-Haytham's construction of an inverted image of the partially eclipsed sun.

The center of each circle is therefore the point at which the axis of the cone, passing through the center of the aperture, intersects the screen. It is clear that the centers of all circles will be points on crescent *q,* and that their radii, as well as those of the arcs forming crescents *p, q, r,* will all be equal. The resultant image will therefore be bounded from above by a convex curve of which the upper part is the tangential arc of a circle whose center is the midpoint *K* of the convex arc of crescent *p,* and whose radius is twice the radius of that arc. Although circles of light will occur below arc *GTH,* they will be relatively few.

The sensible overall effect will be, according to Ibn al-Haytham, a crescent-shaped image bordered on the lower side by a sensibly dark cavity. He showed by a numerical example that the cavity will increase or decrease in size according as the ratio $m_a : m_s$ is less or greater than $d_a : d_s$. It is certain that the treatise *On the Shape of the Eclipse* was composed after the *Optics,* to which it refers. It is not impossible that, at the time of writing the *Optics,* Ibn al-Haytham was acquainted with the remarkable explanation revealed in the later work, but of this we have no evidence.

Transmission and Influence of the Optics. Of all the optical treatises of Ibn al-Haytham that have been mentioned, only the *Optics* (III 3) and the treatise *On Paraboloidal Burning Mirrors* (III 9) are known to have been translated into Latin in the Middle Ages, the latter probably by Gerard of Cremona.[21] It is remarkable that in the Islamic world the *Optics* practically disappeared from view soon after its appearance in the eleventh century until, in the beginning of the fourteenth century, the Persian scholar Kamāl al-Dīn composed his great critical commentary on it, the *Tanqīḥ al-Manāẓir,* at the suggestion of his teacher Quṭb al-Dīn al-Shīrāzī.

By this time the *Optics* had embarked on a new career in the West where it was already widely and avidly studied in a Latin translation of the late twelfth or early thirteenth century, entitled *Perspectiva* or *De aspectibus.* Of the manuscript copies that have been located (no fewer than nineteen[22]), the earliest are from the thirteenth century; but where and by whom the *Optics* was translated remains unknown. The Latin text was published by Frederick Risner at Basel in 1572 in a volume entitled *Opticae thesaurus,* which included Witelo's *Perspectiva.* In both Risner's edition and the Latin manuscripts examined by the present writer (see bibliography) the Latin text wants the first three chapters of book I of the Arabic text (133 pages, containing about 130 words per page, in MS Fatih 3212).

The Latin *Perspectiva* shows the drawbacks as well as the advantages of the literal translation which in general it is. Often, however, it only paraphrases the Arabic, sometimes inadequately or even misleadingly, and at times it omits whole passages. But an exhaustive and critical study of the extant manuscripts is needed before a full and accurate evaluation of the translation can be made. In any case there is no doubt that through this Latin medium a good deal of the substance of Ibn al-Haytham's doctrine was successfully conveyed to medieval, Renaissance, and seventeenth-century philosophers in the West. Roger Bacon's *Perspectiva* is full of references to "Alhazen," or *auctor perspectivae,* whose influence on him cannot be overemphasized. Pecham's *Perspectiva communis* was composed as a compendium of the *Optics* of Ibn al-Haytham.[23] That Witelo's *Opticae libri decem* also depends heavily on *Alhazeni libri septem* has been noted repeatedly by scholars; the cross-references provided by Risner in his edition of the two texts have served as a sufficient indication of that. But Witelo's precise debt to Ibn al-Haytham, as distinguished from his own contribution, has yet to be determined.

The influence of Ibn al-Haytham's *Optics* was not channelled exclusively through the works of these thirteenth-century writers. There is clear evidence that the book was directly studied by philosophers of the fourteenth century[24] and an Italian translation made at that time was used by Lorenzo Ghiberti.[25] Risner's Latin edition made it available to such mathematicians as Kepler, Snell, Beeckman, Fermat, Harriot, and Descartes, all of whom except the last directly referred to Alhazen. It was, in fact, in the sixteenth and seventeenth centuries that the mathematical character of the *Optics* was widely and effectively appreciated.

Astronomy. No fewer than twenty of Ibn al-Haytham's extant works are devoted to astronomical questions. The few of these that have been studied by modern scholars do not appear to justify al-Bayhaqī's description of Ibn al-Haytham as "the second Ptolemy." (The description would be apt, however, if al-Bayhaqī had optics in mind.) Many of these works are short tracts that deal with minor or limited, although by no means trivial, theoretical or practical problems (sundials, determination of the direction of prayer, parallax, and height of stars), and none of them seems to have achieved results comparable to those of, say, Ibn Yūnus, al-Ṭūsī, or Ibn al-Shāṭir. Nevertheless, some of Ibn al-Haytham's contributions in this field are both interesting and historically important, as has sometimes been recognized.

As a writer on astronomy Ibn al-Haytham has been mainly known as the author of a treatise *On the Configuration of the World* (III 1 = Ib 10). The treatise must have been an early work: it speaks of "the ray that goes out of our eyes" and describes the moon as a "polished body" which "reflects" the light of the sun—two doctrines which are refuted in the *Optics* (III 3) and in *The Light of the Moon* (III 6), respectively. The treatise was widely known in the Islamic world,[26] and it is the only astronomical work of Ibn al-Haytham to have been transmitted to the West in the Middle Ages. A Spanish translation was made by Abraham Hebraeus for Alfonso X of Castile (*d.* 1284), and this translation was turned into Latin (under the title *Liber de mundo et coelo*) by an unknown person.

Jacob ben Maḥir (Prophatius Judaeus, *d. ca.* 1304) translated the Arabic text of the *Configuration* into Hebrew, a task which was suggested to him as a corrective to the *Elements of Astronomy* of al-Farghānī, whose treatment of the subject "did not accord with the nature of existing things," as the unknown person who made the suggestion said.[27] The physician Salomo ibn Pater made another Hebrew translation in 1322. A second Latin version was later made from Jacob's Hebrew by Abraham de Balmes for Cardinal Grimani (both of whom died in 1523). In the fourteenth century Ibn al-Haytham's treatise was cited by Levy ben Gerson. Its influence on early Renaissance astronomers and in particular on Peurbach's *Theoricae novae planetarum* has recently been pointed out.[28]

The declared aim of the *Configuration* was to perform a task which, in Ibn al-Haytham's view, had not been fulfilled either by the popularly descriptive or the technically mathematical works on astronomy. The existing descriptive accounts were only superficially in agreement with the details established by demonstrations and observations. A purely mathematical work like the *Almagest,* on the other hand, explained the laws (*qawānīn*) of celestial motions in terms of imaginary points moving on imaginary cir-

IBN AL-HAYTHAM

cles. It was necessary to provide an account that was faithful to mathematical theory while at the same time showing how the motions were brought about by the physical bodies in which the abstract points and circles must be assumed to exist. Such an account would be "more truly descriptive of the existing state of affairs and more obvious to the understanding."[29]

Ibn al-Haytham's aim here was not therefore to question any part of the theory of the *Almagest* but, following a tradition which goes back to Aristotle and which had been given authority among astronomers by one of Ptolemy's own works, the *Planetary Hypotheses,* to discover the physical reality underlying the abstract theory. The description had to satisfy certain principles already accepted in that tradition: a celestial body can have only circular, uniform, and permanent movement; a natural body cannot by itself have more than one natural movement; the body of the heavens is impassable; the void does not exist. Ibn al-Haytham's procedure was then, for every simple motion assumed in the *Almagest,* to assign a single spherical body to which this motion permanently belongs, and to show how the various bodies may continue to move without in any way impeding one another or creating gaps as they moved.

The heavens were accordingly conceived of as consisting of a series of concentric spherical shells (called spheres) which touched and rotated within one another. Inside the thickness of each shell representing the sphere of a planet other concentric and eccentric shells and whole spheres corresponded to concentric and eccentric circles and epicycles respectively. All shells and spheres rotated in their own places about their own centers, and their movements combined to produce the apparent motion of the planet assumed to be embedded in the epicyclic sphere at its equator. In his careful description of all movements involved Ibn al-Haytham provided, in fact, a full, clear, and untechnical account of Ptolemaic planetary theory—which alone may explain the popularity of his treatise.

A brief look at Ibn al-Haytham's other works will give us an idea of how seriously he took the program he inherited and of its significance for the later history of Islamic astronomy. Perhaps at some time after writing the *Configuration of the World* (III 1 = *Ib* 10) Ibn al-Haytham composed a treatise (III 61) on what he called the movement of *iltifāf,* that is the movement or rather change in the obliquities (singular, *mayl*) of epicycles responsible for the latitudinal variations of the five planets (*Almagest,* XIII.2). This treatise is not known to have survived. But we have Ibn al-Haytham's reply to a criticism of it by an unnamed scholar. From this reply, the tract called *Solution of the Difficulties* [*shukūk*] *Concerning the Movement of Iltifāf* (III 63), we learn that in the earlier treatise he proposed a physical arrangement designed to produce the oscillations of epicycles required by the mathematical theory. The same subject is discussed, among other topics, in the work entitled *Al-Shukūk ʿalā Baṭlamyūs* (*Dubitationes in Ptolemaeum*) (III 64). More than any other of Ibn al-Haytham's writings, this work (almost certainly composed after the reply just mentioned) reveals the far-reaching consequences of the physical program to which he was committed.

The *Dubitationes* is a critique of three of Ptolemy's works: the *Almagest,* the *Planetary Hypotheses* and the *Optics.* As far as the first two works are concerned, the criticism is mainly aimed against the purely abstract character of the *Almagest* (this exclusiveness being in Ibn al-Haytham's view a violation of the principles accepted by Ptolemy himself) and against the fact that the *Planetary Hypotheses* had left out many of the motions demanded in the *Almagest* (a proof that Ptolemy had failed to discover the true arrangement of the heavenly bodies).

Ibn al-Haytham's objection to the "fifth motion" of the moon, described in *Almagest* V.5, is particularly instructive, being nothing short of a *reductio ad absurdum* by "showing" that such a motion would be physically impossible. Ptolemy had assumed that as the moon's epicycle moves on its eccentric deferent, the diameter through the epicycle's apogee (when the epicycle-center is at the deferent's apogee) rotates in such a way as to be always directed to a point on the apse-line (called the opposite point, *nuqṭat al-muḥādhāt*), such that the ecliptic-center lies halfway between that point and the deferent-center. The assumption implied that the epicycle's diameter alternately rotates in opposite senses as the epicycle itself completes one revolution on its deferent. But, Ibn al-Haytham argued, such a movement would have to be produced either by a single sphere which would alternately turn in opposite senses, or by two spheres of which one would be idle while the other turned in the appropriate sense. "As it is not possible to assume a body of this description, it is impossible that the diameter of the epicycle should be directed towards the given point."[30] Whatever one thinks of the argument, the problem it raised was later fruitfully explored by Naṣīr al-Dīn al-Ṭūsī in the *Tadhkira.*[31]

Perhaps most important historically was Ibn al-Haytham's objection against the theory of the five planets, and in particular against the device introduced by Ptolemy which later came to be known as the equant. Ptolemy supposed that the point from which the planet's epicycle would appear to move uniformly is neither the center of the eccentric deferent nor that of the ecliptic, but another point (the equant) on the line of apsides as far removed from the deferent-center as the latter is from the ecliptic-center. This entailed, as Ibn al-Haytham pointed out, that the motion of the epicycle-center, as measured on the circumference of its deferent, was not uniform, and consequently that the deferent sphere carrying the epicycle was not moving uniformly—in contradiction to the assumed principle of uniformity.

Although the equant had succeeded in bringing Ptolemy's planetary theory closer to observations, the validity of this criticism remained as long as the principle of uniform circular motion was adhered to. To say that the equant functioned merely as an abstract calculatory device designed for the sake of saving the phenomena was an answer which satisfied none of Ptolemy's critics, down to and including Copernicus. Nor was Ptolemy himself unaware of the objectionable character of such devices. In the *Dubitationes,* Ibn al-Haytham points to a passage in *Almagest* IX. 2 where Ptolemy asks to be excused for having employed procedures which, he admitted, were against the rules (*παρὰ τὸν λόγον, khārij ʿan al-qiyās*), as, for example, when for convenience's sake he made use merely of circles described in the planetary spheres, or when he laid down principles whose foundation was not evident. For, Ptolemy said, "when something is laid down without proof and is found to be in accord with the phenomena, then it cannot have been discovered without a method of science [*sabīl min al-ʿilm*], even though the manner in which it has been attained would be difficult to describe."[32]

Ibn al-Haytham agreed that it was indeed appropriate to argue from unproved assumptions, but not when they violated the admitted principles. His final conclusion was that there existed a true configuration of the heavens which Ptolemy had failed to discover.

It has been customary to contrast the "physical" approach of Ibn al-Haytham with the "abstract" approach of mathematical astronomers. The contrast is misleading if it is taken to imply the existence of two groups of researchers with different concerns. The "mathematical" researches of the school of Marāgha (among them al-Ṭūsī and al-Shīrāzī) were motivated by the same kind of considerations as those revealed in Ibn al-Haytham's *Dubitationes.*[33] Al-Ṭūsī, for instance, was as much worried about the moon's "fifth movement" and about the equant as was Ibn al-Haytham, and for the same reasons.[34] His *Tadhkira* states clearly that astronomical science is based on physical as well as mathematical premises. From a reference in it to Ibn al-Haytham,[35] made in the course of expounding alterations based on what is now known as the "Ṭūsī couple," it is clear that al-Ṭūsī recognized the validity of Ibn al-Haytham's physical program, although not the particular solutions offered by his predecessor.

The longest of the astronomical works of Ibn al-Haytham that have come down to us is a commentary on the *Almagest.* The incomplete text in the unique Istanbul manuscript which has recently been discovered occupies 244 pages of about 230 words each (see bibliography, additional works, no. 3). The manuscript, copied in 655 A.H./A.D. 1257, bears no title but twice states the author's name as Muḥammad ibn al-Ḥasan ibn al-Haytham, the name found by Ibn Abī Uṣaybiʿa in Ibn al-Haytham's own bibliographies, that is lists I and II. No title in list III seems to correspond to this work, but there are candidates in the other lists. The first title in Ibn al-Qifṭī's list is *Tahdhīb al-Majisṭī* or *Expurgation of the Almagest.* Number 19 in list II is described as "A book which works out the practical part of the *Almagest.*" And number 3 in list I*a* begins as follows: "A commentary and summary of the *Almagest,* with demonstrations, in which I worked out only a few of the matters requiring computation. . . ." The last title is highly appropriate to the work that has survived.

Most commentators on the *Almagest,* Ibn al-Haytham says in the introduction, were more interested in proposing alternative techniques of computation than in clarifying obscure points for the beginner. As an example he mentions al-Nayrīzī who "crammed his book with a multiplicity of computational methods, thereby seeking to aggrandize it." Ibn al-Haytham sought rather to explain basic matters relating to the construction of Ptolemy's own tables, and he meant his commentary to be read in conjunction with the *Almagest,* whose terminology and order of topics it followed. The book was therefore to comprise thirteen parts, but, for brevity's sake and also because the *Almagest* was "well known and available," Ibn al-Haytham would not follow the commentators' customary practice of reproducing Ptolemy's own text. Unfortunately the manuscript breaks off before the end of the fifth part, shortly after the discussion of Ptolemy's theories for the sun and the moon. In the course of additions designed to complete, clarify, or improve Ptolemy's arguments, Ibn al-Haytham referred to earlier Islamic writers on astronomy, including Thābit ibn Qurra (on the "secant figure"), Banū Mūsā (on the sphere), and Ibrāhīm ibn Sinān (on gnomon shadows). All diagrams have been provided and are clearly drawn in the manuscript but the copyist has not filled in the tables.

Mathematics. Ibn al-Haytham's fame as a mathematician has rested on his treatment of the problem known since the seventeenth century as "Alhazen's problem." The problem, as viewed by him, can be expressed as follows: from any two points opposite a reflecting surface—which may be plane, spherical, cylindrical, or the surface of a cone, whether convex or concave—to find the point (or points) on the surface at which the light from one of the two points will be reflected to the other. Ptolemy, in his *Optics,* had shown that for convex spherical mirrors there exists a unique point of reflection. He also considered certain cases relating to concave spherical mirrors, including those in which the two given points coincide with the center of the specular sphere; the two points lie on the diameter of the sphere and at equal or unequal distances from its center; and the two points are on a chord of the sphere and at equal distances from the center. He further cited some cases in which reflection is impossible.[36]

In book V of his *Optics,* Ibn al-Haytham set out to solve the problem for all cases of spherical, cylindrical, and conical surfaces, convex and concave. Although he was not successful in every particular, his performance, which showed him to be in full command of the higher mathematics of the Greeks, has rightly won the admiration of later mathematicians and historians. Certain difficulties have faced students of this problem in the work of Ibn al-Haytham. In the Fatih manuscript, and in the Aya Sofya manuscript which is copied from it, the text of book V of the *Optics* suffers from many scribal errors, and in neither of these manuscripts are the lengthy demonstrations supplied with illustrative diagrams.[37] Such diagrams exist in Kamāl al-Dīn's commentary and in Risner's edition of the medieval Latin translation, but neither the diagrams nor the texts of these two editions are free from mistakes. One cannot, therefore, be too grateful to M. Naẓīf for his clear and thorough analysis of this problem, to which he devotes four chapters of his masterly book on Ibn al-Haytham.

Ibn al-Haytham bases his solution of the general problem on six geometrical lemmas (*muqaddamāt*) which he proves separately: (1) from a given point A on a circle ABG, to draw a line that cuts the circumference in H and the diameter BG in a point D whose distance from H equals a given line; (2) from the given point A to draw a line that cuts the diameter BG in a point E and the circumference in a point D such that ED equals the given line; (3) from a given point D on the side BG of a right-angled triangle having the angle B right, to draw a line DTK that cuts AG in T (and the extension of BA in K), such that $KT:TG$ equals a given ratio; (4) from two points E,D outside a given circle AB, to draw two lines EA and DA, where A is a point on the circumference, such that the tangent at A equally divides the angle EAD; (5) from a point E outside a circle having AB as diameter and G as center, to draw a line that cuts the circumference at D and the diameter at Z such that DZ equals ZG; and (6) from a given point D on the side GB of a right-angled triangle having the angle B right, to draw a line that meets the hypotenuse AG at K and the extension of AB on the side of B at T, such that $TK:KG$ equals a given ratio.[38]

Obviously lemmas (1) and (2) are special cases of one and the same problem, and (3) and (6) are similarly related. In his exposition of Ibn al-Haytham's arguments Naẓīf combines each of these two pairs in one construction. It will be useful to reproduce here his construction for (1) and (2) and to follow him in explaining Ibn al-Haytham's procedure by referring to this construction. It happens that (1) and (2) contain characteristic features of the proposed solution of the geometrical problem involved. In Figure 3, A is a given point on the circumference of the small circle with diameter BG. It is required to draw a straight line from A that cuts the circle at D and the diameter, or its extension, at E, such that DE equals the given segment z.

From G draw the line GH parallel to AB; let it cut the circle at H; join BH. Let the extensions of AG, AB respectively represent the coordinate axes x, y whose origin thus coincides with A. Draw the hyperbola passing through H with x, y as asymptotes. Then, with H as center, draw the circle with radius

$$HS = \frac{BG^2}{z}$$

(HS being the side of a rectangle whose other side is z, and whose area equals BG^2). The circle will, in the general case, cut the two branches of the hyperbola at four points, such as S, T, U, V. Join H with all four points, and from A draw the lines parallel to HS, HT, HU, and HV. Each of these parallels will cut the circle circumscribing the triangle ABG at a point, such as D, and the diameter, or its extension, at another point, such as E. It is proved that each of these lines satisfies the stated condition.

As distinguished from the above demonstration, Ibn al-Haytham proceeded by considering three cases one after the other: (a) the required line is tangential to the circle, that is, A and D coincide; (b) D is on the arc AG; (c) D is on the arc AB. Despite the generality of the enunciation of lemma (1), he does not consider the case in which the line cuts the exten-

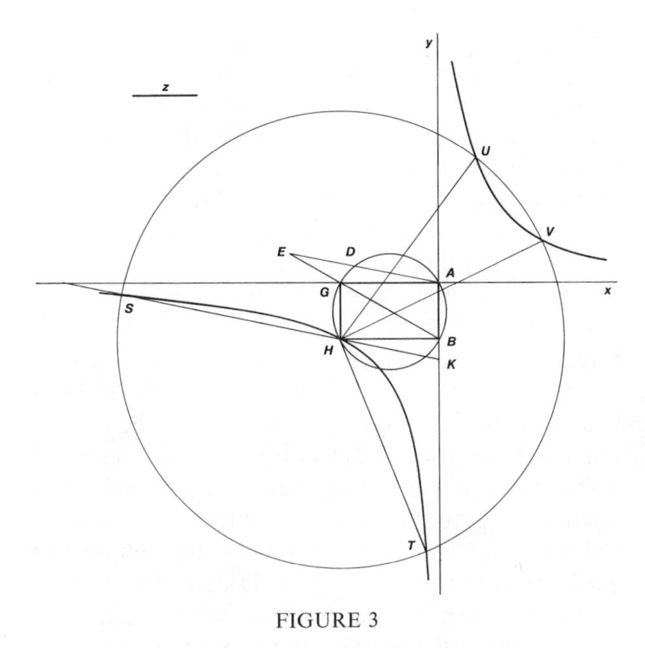

FIGURE 3

sion of *BG* on the side of *B*. Similarly in dealing with lemma (2) he separately examines three possibilities in respect of the relation of the circle *HS* to the "opposite branch" of the hyperbola: (a) the circle cuts that branch at two points; (b) the circle is tangential to it at one point; (c) the circle falls short of it. For finding the shortest line between *H* and the "opposite branch" of the hyperbola he refers to Apollonius' *Conics,* V. 34. Ibn al-Haytham did not, of course, speak of a coordinate system of perpendicular axes whose origin he took to be the same as the given point *A*. He did, however, consider a rectangle similar to *ABHG,* and described the sides of it corresponding to *AB, AG* as asymptotic to the hyperbola he drew through a point corresponding to *H* in Figure 3. For drawing this hyperbola he referred to *Conics,* II. 4.

Applying the six geometrical lemmas for finding the points of reflection for the various kinds of surface, Ibn al-Haytham again proceeded by examining particular cases in succession. Naẓīf shows that the various cases comprised by lemma 4 constitute a general solution of the problem in respect of spherical surfaces, concave as well as convex. With regard to cylindrical mirrors, Ibn al-Haytham considered the cases in which (a) the two given points are in the plane perpendicular to the axis; and (c) the general case in which the intersection of the plane containing the two points with the cylinder is neither a straight line nor a circle, but an ellipse. He described six different cases in an attempt to show that reflection from convex conical surfaces can take place from only one point, which he determined. For concave conical mirrors, he showed that reflection can be from any number of points up to but not exceeding four. And he argued for the same number of points for concave cylindrical mirrors.

Apart from the mathematical sections of the *Optics,* some twenty of the writings of Ibn al-Haytham which deal exclusively with mathematical topics have come down to us. Most of these writings are short and they vary considerably in importance. About a quarter of them have been printed in the original Arabic and about half of them are available in European translations or paraphrases. Some of the more important among these works fall into groups and will be described as such.

List III includes three works (III 39, III 55, and III 56) which are described as solutions of difficulties arising in three different parts of Euclid's *Elements.* There are no manuscripts exactly answering to these descriptions. There exist, on the other hand, several manuscripts of a large work entitled *Solution of the Difficulties in Euclid's Elements,* which does not appear in list III. It therefore seems likely that III 39, III 55, and III 56 are parts of the larger work which is listed below in the bibliography as additional work no. 1.

The object of the *Solution* was to put into effect a rather ambitious program. Unlike earlier works by other writers, it proposed to deal with all or most of the difficulties occasioned by Euclid's book, and not with just a few of them; it examined particular cases and offered alternative constructions for many problems; it revealed the "remote mathematical causes" (*al-ʿilal al-taʿlīmiyya al-baʿīda*) of the theoretical propositions (*al-ashkāl al-ʿilmiyya*)—something which "none of the ancients or the moderns had previously mentioned"; and, finally, it replaced Euclid's indirect proofs with direct ones. In this book Ibn al-Haytham referred to an earlier *Commentary on the Premises of Euclid's Elements* (III 2), and said that he meant the two works to form together a complete commentary on the whole of the *Elements.* This earlier work, restricted to the definitions, axioms, and postulates of the *Elements,* is extant both in the Arabic original and in a Hebrew translation made in 1270 by Moses ibn Tibbon. Ibn al-Haytham's interesting treatment of Euclid's theory of parallels well illustrates his approach in these two "commentaries."

In III 2 Ibn al-Haytham ascribed to Euclid the "axiom" that "two straight lines do not enclose a space [*saṭḥ*]" (his own opinion is that the statement should be counted among the "postulates"). Concerning Euclid's definition of parallel lines as nonsecant lines he remarked that the "existence" of such lines should be proved and, for this purpose, introduced the following "more evident" postulate: if a straight line so moves that the one end always touches a second straight line, and throughout this motion remains perpendicular to the second and in the same plane with it, then the other end of the moving line will describe a straight line which is parallel to the second. Ibn al-Haytham thus replaced parallelism in Euclid's sense by the property of equidistance, a procedure which had originated with the Greeks and which had characterized many Islamic attempts to prove Euclid's postulate 5.

Like Thābit ibn Qurra before him Ibn al-Haytham based his proof on the concept of motion—which procedure al-Khayyāmī and, later, al-Ṭūsī found objectionable as being foreign to geometry. The crucial step in the deduction of Euclid's postulate is the demonstration of Saccheri's "hypothesis of the right angle" by reference to a "Saccheri quadrilateral." Let *AG, BD* be drawn at right angles to *AB* (Figure 4): it is to be proved that perpendiculars to *BD* from points on *AG* are equal to *AB*, and, consequently, perpendicular to *AG*. From any point *G* draw *GD* perpendicular to *BD;* produce *GA* to *E* such that *AE* equals *AG;* draw *ET* perpendicular to *DB* produced; and join *BG, BE*. Considering, first, triangles *ABG, ABE,* then triangles *BDG, BTE,* it is seen that *GD* equals *ET*. Let *GD* now move along *DBT*, the angle *GDT* being always right. Then, when *D* coincides with *B, G* will either coincide with *A*, or fall below it on *AB*, or above it (occupying the position of *H* in the figure) on *BA* produced, according as *GD* is assumed equal to, less than, or greater than *AB*. When *D* reaches *T, GD* will exactly coincide with *ET*. During this motion, *G* will have described a straight line which, on the hypothesis that *DG* is not equal to *AB*, would enclose an area, such as *GHEA*, with another straight line, *GAE*—which is impossible. Finally, by considering in turn triangles *BDG, BDA,* and *AKB, GKD,* it is clear that *DGA* = *BAG* = a right angle. The Euclidean postulate follows as a necessary consequence.

In the larger commentary, Ibn al-Haytham reformulated postulate 5, stating that two intersecting straight lines cannot both be parallel to a third ("Playfair's axiom"), and referring to the proof set forth in the earlier, and shorter, work. It is to be noted that al-Ṭūsī's criticism (in his own work on the theory of parallels, *Al-Risāla al-Shāfiya*) of Ibn al-Haytham's attempt was based on the remarks in this larger commentary, not on the earlier proof, which al-Ṭūsī said was not available to him.[39]

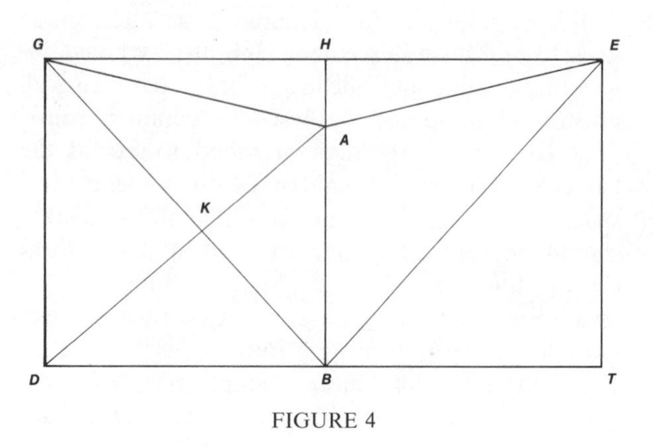

FIGURE 4

Ibn al-Haytham wrote two treatises on the quadrature of crescent-shaped figures (*al-ashkāl al-hilāliyya*) or lunes. (Their titles have sometimes been misunderstood as referring to the moon.) The second, and fuller, treatise (III 21), although extant in several manuscripts, has not been studied. From the introduction we gather that it was composed quite some time after the first (III 20, now lost), although the two works appear consecutively in list III. The treatise comprises twenty-three propositions on lunes, of which some are generalizations of particular cases already proved in the earlier treatise, as the author tells us, while others are said to be entirely new. The subject was connected with that of squaring the circle: if plane figures bounded by two unequal circular arcs could be squared, why not the simpler figure of a circle? Ibn al-Haytham put forward such an argument in a short tract on the *Quadrature of the Circle* (III 30), which has been published. The object of the tract is to prove the "possibility" of squaring the circle without showing how to "find" or construct a square equal in area to a given circle.

To illustrate his point, Ibn al-Haytham proves a generalization of a theorem ascribed to Hippocrates of Chios. The proof is reproduced from his earlier work on lunes. In Figure 5, let *B* be any point on the semicircle with diameter *AG;* describe the smaller semicircles with *AB, BG* as diameters; it is shown that the lunes *AEBH, BZGT* are together equal in area to the right-angled triangle *ABG*. On the basis of Euclid XII.2, which states that circles are to one another as the squares on the diameters, it is easily proved that the semicircles on *AB, BG* are together equal to the semicircle on the hypotenuse *AG*. The equality of the lunes to the triangle *ABG* follows from subtracting the segments *AHB, BTG* from both sides of the equation. Hippocrates had considered the particular case in which the triangle *ABG* is isosceles.[40]

Two more works which are closely related are *On Analysis and Synthesis* (*Maqāla fi 'l-taḥlīl wa 'l-tarkīb*, III 53) and *On the Known Things* (*Maqāla fi 'l-Maᶜlūmāt*, III 54). The subject matter of the latter work overlaps with that of Euclid's *Data*, which is called in Arabic *Kitāb al-Muᶜṭayāt* (Δεδομένα). Ibn al-Haytham's use of *al-maᶜlūmāt*, rather than *al-muᶜṭayāt*,

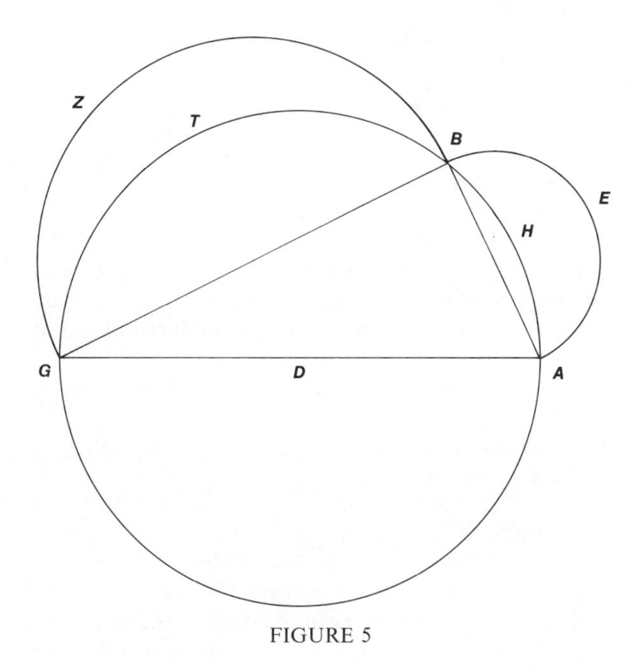

FIGURE 5

has a precedent in the Arabic translation of Euclid's book itself, where *al-maʿlūm* (the known) is regularly employed to denote the given. *On Analysis* is a substantial work of about 24,000 words whose object is to explain the methods of analysis and synthesis, necessary for the discovery and proof of theorems and constructions, by illustrating their application to each of the four mathematical disciplines: arithmetic, geometry, astronomy, and music. It lays particular emphasis on the role of "scientific intuition" (*al-ḥads al-ṣināʿī*), when properties other than those expressly stated in the proposition to be proved have to be conjectured before the process of analysis can begin.

In describing the relationship of this treatise to the one on *The Known Things* Ibn al-Haytham made certain claims which should be quoted here. The art of analysis, he says, is not complete without the things that are said to be known.

> Now the known things are of five kinds: the known in number, the known in magnitude, the known in ratio, the known in position, and the known in species [*al-maʿlūm al-ṣūra*]. The book of Euclid called *Al-Muʿṭayāt* includes many of these known things which are the instruments of the art of analysis, and on which the larger part of analysis is based. But that book does not include other known things that are indispensable to the art of analysis . . . nor have we found them in any other book. In the examples of analysis we give in the present treatise we shall prove the known things used, whether or not we have found them in other works. . . . After we have completed this treatise we shall resume the subject in a separate treatise in which we shall show the essence of the known things that are used in mathematics and give an account of all their kinds and of all that relates to them.[41]

The treatise on known things, which is extant, divides in fact into two parts, of which the first (comprising twenty-four propositions) is said to be the invention of Ibn al-Haytham himself. In 1834 L. Sédillot published a paraphrase of the introduction to this work (a discussion of the concept of knowledge) together with a translation of the enunciations of the propositions constituting both parts. There is no study of the work on *Analysis and Synthesis.* The more important of the remaining mathematical works are all available in European translations.

NOTES

1. On al-Fāsī, see Ibn al-Qifṭī, *Taʾrīkh,* pp. 392–394.
2. *Al-Mutawassiṭāt,* or intermediate books, so called because they were studied after the *Elements* of Euclid and before the *Almagest.* They included, for instance, Euclid's *Data.* Theodosius' *Spherics,* and the *Spherics* of Menelaus. See the explanation of Abu'l-Ḥasan al-Nasawī in al-Ṭūsī, *Majmūʿ al-Rasāʾil,* II (Hyderabad, 1359 A.H. [1940]), *risāla* no. 3, p. 2. The existence of a copy of Apollonius' *Conics* in Ibn al-Haytham's hand (MS Aya Sofya 2762, 307 fols., dated Ṣafar 415 A.H. [1024]) may be taken to confirm the story that he lived on selling copies of scientific texts, although the *Conics* is not one of the books mentioned in the story.
3. The expression "fī ḥudūd" could also mean "about" or "toward the end of."
4. On Bayhaqī's dates see the article devoted to him in *Encyclopaedia of Islam,* 2nd ed.
5. On Qayṣar see A. I. Sabra, "Simplicius's Proof of Euclid's Parallels Postulate," in *Journal of the Warburg and Courtauld Institutes,* **32** (1969), 8.
6. The title is "[Ibn al-Haytham's] Answer to a Geometrical Question Addressed to Him in Baghdad [*suʾila ʿanhā bi-Baghdād*] in the Months of the Year Four Hundred and Eighteen."
7. See Galen's *Opera omnia,* C. G. Kühn, ed., XIX (repr. Hildesheim, 1965), 8–61; and F. Rosenthal, "Die arabische Autobiographie," pp. 7–8. Galen's *De libris propriis* was translated into Arabic by Ḥunayn ibn Isḥāq in the ninth century.
8. See Galen's *Opera omnia, ed. cit.,* X (repr. Hildesheim, 1965), 457, ll. 11–15.
9. In the context of Arabic optics *manāẓir* is the plural, not of *manẓar* (view, appearance) but of *manẓara,* that by means of which vision is effected, an instrument of vision. One evidence for this is Ḥunayn ibn Isḥāq's Arabic translation of Galen's *De usu partium,* where *manẓara* and *manāẓir* correspond to ὄψις and ὄψεις, respectively (see Escorial MS 850, fol. 29v). *Al-Manāẓir* had been used as the Arabic title of Euclid's (and Ptolemy's) Ὀπτικά.
10. Newton, *Opticks,* bk. I, pt. 1, def. 1. See A. I. Sabra, *Theories of Light,* pp. 288–289, 310–311, n. 25.
11. The diagram of the eye in Risner's ed. of the Latin text is taken from Vesalius' *De corporis humani fabrica* (D. Lindberg, "Alhazen's Theory of Vision . . . ." p. 327, n. 30). The diagram in MS Fatih 3212 of *Kitāb al-Manāẓir,* bk. I, does not clearly and correctly represent Ibn al-Haytham's descriptions; it can be seen in S. Polyak, *The Retina* (Chicago, 1957), and in G. Nebbia, "Ibn al-Haytham nel millesimo anniversario della nascita," p. 204.
12. MS Fatih 3212, fol. 83r; *Opticae thesaurus. Alhazeni libri VII,* p. 7, sec. 14, 1. 26.

13. The importance of this principle and of its application by Ibn al-Haytham to the problem of vision has been rightly emphasized by Vasco Ronchi in his *Storia della luce,* 2nd ed. (Bologna, 1952), pp. 33–47, trans. into English as *The Nature of Light* (London, 1970), pp. 40–57.
14. *Opticae thesaurus. Alhazeni libri VII,* p. 14, sec. 23, l. 20.
15. *Ibid.,* p. 15, sec. 24: "Visio videtur fieri per *συναύγειαν*, id est receptos simul et emissos radios."
16. At least some of the Latin MSS have "reflexe." Risner's text, however, correctly reads "refracte" (*bi ʾl-inʿiṭāf*). See Vescovini, *Studi,* p. 93, n. 10.
17. Vescovini, *Studi,* p. 141.
18. See M. Naẓīf, *Al-Ḥasan ibn al-Haytham,* pp. 709–721. Unlike the Arabic, the Latin text in Risner's ed. expresses rule (5) as $d < i$ (*Alhazeni libri VII,* p. 247, ll. 8–11). Both the Arabic MSS and Risner's ed. omit the words "less than" from rule (6), and consequently express this rule as $d = 1/2\ (i + d)$[!]. The correction has been made by Naẓīf and is supported by Kamāl al-Dīn's formulation of the rule on the basis of Ibn al-Haytham's autograph (see *Tanqīḥ,* I, 7; II, 134, ll. 10–11).
19. The earliest occurrence of "al-bayt al-muẓlim" is in a ninth-century tract on burning mirrors by ʿUṭārid ibn Muḥammad al-Ḥāsib: Istanbul MS Laleli 2759, fols. 1–20. The tract is based on earlier Greek works including at least one by Anthemius of Tralles, and the term may therefore have been derived from them. See M. Schramm, "Ibn al-Haythams Stellung in der Geschichte der Wissenschaften," pp. 15–16.
20. Bk. I, ch. 5, sec. 29, p. 17 in Risner's ed. of the Latin text. The discussion continues in the Arabic for the greater part of two pages to which nothing corresponds in Risner's text.
21. M. Clagett, "A Medieval Latin Translation of a Short Arabic Tract on the Hyperbola," in *Osiris,* **11** (1954), 361.
22. D. Lindberg, *Pecham and the Science of Optics* (Madison, Wis., 1970), p. 29, n. 69. (See bibliography, "Original Works," no. III 3.)
23. *Ibid.,* p. 20.
24. Vescovini, *Studi,* pp. 137 ff.
25. See especially Vescovini, "Contributo per la storia della fortuna di Alhazen in Italia" (in the Bibliography).
26. W. Hartner, "The Mercury Horoscope . . .," esp. pp. 122–124.
27. M. Steinschneider, "Notice . . .," p. 723.
28. Hartner, *op. cit.,* pp. 124, 127 ff.
29. MS India Office, Loth 734, fol. 101r.
30. *Dubitationes* (III 64), *ed. cit.,* p. 19.
31. W. Hartner, "Naṣīr al-Dīn al-Ṭūsī's Lunar Theory," in *Physis,* **11** (1969), 287–304.
32. *Dubitationes* (III 64), *ed. cit.,* p. 39; also p. 33. The English trans. is of Isḥāq's Arabic version quoted by Ibn al-Haytham. The Greek differs only slightly from the Arabic: ". . . *οὔτε τὰ ἀναποδείκτως ὑποτιθέμενα, ἐὰν ἅπαξ σύμφωνα τοῖς φαινομένοις καταλαμβάνται, χωρὶς ὁδοῦ τινος καὶ ἐπιστάσεως εὑρῆσθαι δύναται, κἄν δυσέκθετος ᾖ ὁ τρόπος αὐτῶν τῆς καταλὴψεως* (Ptolemy, *Syntaxis mathematica,* J. L. Heiberg, ed., II [Leipzig, 1903], 212, ll. 11–14).
33. See E. S. Kennedy, "Late Medieval Planetary Theory," in *Isis,* **57** (1966), 365–378, esp. 366–368.
34. I have consulted the British Museum MSS Add. 23, 394; Add. 23, 397; and Add. 7472 Rich, the last two being al-Nīsābūrī's commentary, *Tawḍīḥ,* on the *Tadhkira.*
35. The reference is very probably to the lost tract on the movement of *iltifāf* (III 61).
36. A. Lejeune, *Recherches sur la catoptrique grecque* (Brussels, 1957), pp. 71–74.
37. Diagrams are supplied in MS Köprülü 952 (see bibliography, "Original Works," under III 3). As far as I know, this MS has not been used in studies of the *Optics.*
38. *Opticae thesaurus. Alhazeni libri VII,* pp. 142–150.
39. See al-Ṭūsī, *Rasāʾil,* II (cited in note 2 above), *risāla* no. 8, pp. 5–7. See also A. I. Sabra, "Thābit ibn Qurra on Euclid's Parallels Postulate," in *Journal of the Warburg and Courtauld Institutes,* **31** (1968), 12–32; and A. P. Juschkewitsch [Youschkevitch], *Geschichte der Mathematik im Mittelalter* (Leipzig, 1964), pp. 277–288.
40. Sir Thomas Heath, *A History of Greek Mathematics,* I (Oxford, 1921), pp. 183 ff.
41. Chester Beatty MS 3652, fol. 71r–v.

BIBLIOGRAPHY

Ibn Abī Uṣaybiʿa's lists I*a,* I*b,* II, and III of Ibn al-Haytham's works, described in the article, have been published, wholly or in part, more than once in European languages. They have recently been reproduced in a convenient form in Italian trans. by G. Nebbia in "Ibn al-Haytham nel millesimo anniversario della nascita," in *Physis,* **9** (1967), 165–214. Since practically all of Ibn al-Haytham's extant works are included in list III, they will be arranged here according to their numbers in that list. The same numbers can be used to refer to Nebbia's article, where the reader will find a useful bibliography, and to M. Schramm's book, *Ibn al-Haythams Weg zur Physik* (Wiesbaden, 1963), where many of the works listed are discussed. (The titles constituting what Nebbia calls list I*c* are in fact chapter headings of the last work in list I*b*.)

Arabic MSS of Ibn al-Haytham's works are listed in H. Suter, "Die Mathematiker und Astronomen der Araber und ihre Werke," in *Abhandlungen zur Geschichte der mathematischen Wissenschaften mit Einschluss ihrer Anwendungen,* **10** (Leipzig, 1900), no. 204, 91–95; and "Nachträge und Berichtigungen zur 'Mathematiker . . .,'" *ibid.,* **14** (1902), esp. 169–170; H. P. J. Renaud, "Additions et corrections à Suter 'Die Mathem. u. Astr. der Arab.,'" *Isis,* **18** (1932), esp. 204; C. Brockelmann, *Geschichte der arabischen Literatur,* I (Weimar, 1898), 469–470; 2nd ed. (Leiden, 1943), pp. 617–619; supp. I (Leiden, 1937), 851–854. The Istanbul MSS are more fully described in M. Krause, "Stambuler Handschriften islamischer Mathematiker," in *Quellen und Studien zur Geschichte der Mathematik, Astronomie und Physik,* Abt. B, Studien, **3** (1936), 437–532. P. Sbath, in *Al-Fihris: Catalogue de manuscrits arabes,* 3 pts. plus *Supplément* (Cairo, 1938–1940), pt. I, p. 86, cites MSS, belonging to a private collection in Aleppo, of the following works: III 6, III 8, III 48, III 60, III 65, III 67, III 68, and III 82. (I owe this reference to Robert E. Hall.)

In the list of extant original works that follows, reference will be made to Brockelmann and Krause by means of the abbreviations "Br." and "Kr.," followed by the numbers given to Ibn al-Haytham's treatises in these two authors.

Other abbreviations used are the following:

Rasāʾil: Majmūʿ al-Rasāʾil (Hyderabad, 1357 A.H. [1938]). A collection of eight treatises by Ibn al-Haytham to which a ninth, published at Hyderabad, 1366 A.H. (1947), has been added.

Tanqīḥ: Tanqīḥ al-Manāẓir . . ., 2 vols. (Hyderabad, 1347–1348 A.H. [1928–1930]). This is Kamāl al-Dīn al-Fārisī's "commentary" ("*Tanqīḥ*" means revision or correction) on Ibn al-Haytham's *Kitāb al-Manāẓir.* Vol. II has a sequel (*dhayl*) and an appendix (*mulḥaq*) which contain Kamāl al-Dīn's recensions (sing., *taḥrīr*) of a number of Ibn al-Haytham's other optical works.

I. ORIGINAL WORKS.

III 1 (Br. 28). *M(aqāla). f (ī). Hay'at al-'ālam* ("On the Configuration of the World"). A MS that has recently come to light is Kastamonu 2298, 43 fols.; unlike the India Office MS it is incomplete. For Hebrew and Latin MSS see M. Steinschneider, "Notice sur un ouvrage astronomique inédit d'Ibn Haitham," in *Bullettino di bibliografia e di storia delle scienze matematiche e fisiche,* **14** (1881), 721–736, also published as *Extrait* . . . (Rome, 1883); "Supplément," *ibid.,* **16** (1883), 505–513; and *Die hebraeischen Uebersetzungen des Mittelalters und die Juden als Dolmetscher* (Berlin, 1893), II, 559–561; F. Carmody, *Arabic Astronomical and Astrological Sciences in Latin Translation* (Berkeley, Cal., 1955), pp. 141–142; and Lynn Thorndike and Pearl Kibre, *Catalogue of Incipits of Mediaeval Scientific Writings in Latin* (Cambridge, Mass., 1963), cols. 894, 895, 1147 (the last being a Spanish trans. by Abraham Hebraeus).

The Arabic text has not been edited. A Latin version has been published from a MS of the 13th or early 14th century in Millás Vallicrosa, *Las traducciones orientales en los manuscritos de la Biblioteca Catedral de Toledo* (Madrid, 1942), app. II, 285–312; see pp. 206–208. There is a German trans. by K. Kohl, "Über den Aufbau der Welt nach Ibn al Haiṯam," in *Sitzungsberichte der Physikalisch-medizinischen Sozietät in Erlangen,* **54-55** (1922–1923), 140–179.

III 2 (Br. 8, Kr. 14). *M. f. Sharḥ muṣādarāt Kitāb Uqlīdis* ("Commentary on the Premises of Euclid's *Elements*"). Composed before III 53 and before the larger commentary on the *Elements* (see below, "Additional Works," no. 1). MSS of Ibn Tibbon's Hebrew trans. are listed in M. Steinschneider, *Hebraeischen Uebersetzungen* (cited under III 1), II, 509–510. A partial Russian trans. of this work (using a Kazan MS not recorded in Brockelmann) has been published by B. A. Rozenfeld as "Kniga kommentariev k vvedeniyam knigi Evklida 'Nachala,'" in *Istoriko-matematicheskie issledovaniya,* **11** (1958), 743–762.

III 3 (Br. 34, Kr. 15). *Kitāb al-Manāẓir* ("Optics"). All known Arabic MSS of this work are in Istanbul; see Krause. Köprülü MS 952 contains practically the whole of bks. IV, V, VI, and VII. The folios must be rearranged as follows: IV, 108r–133v; V, 2r–v, 74r–81v, 89r–107v, 134r–135v; VI, 3r–47v; VII, 1r–v, 48r–73v, 82r–88v. The reference in Brockelmann to a recension of this work in the Paris MS, ar. 2460 (Br. has 2640) is mistaken; the MS is a recension of Euclid's *Optics* which is attributed on the title page to Ḥasan ibn [Mūsā ibn] Shākir.

I have examined the following Latin MSS: Bruges 512, 113 fols., 13th c.; Cambridge University Library, Peterhouse MS 209 (= 11 · 10 · 63), 111 fols., 14th c.; Cambridge University Library, Trinity College MS 1311 (= 0 · 5 · 30), 165 fols., 13th c.; Edinburgh, Royal Observatory, Crawford Library MS 9 · 11 · 3 (20), 189 fols., dated 1269; Florence, Biblioteca Nazionale, Magliabechi CI.XX.52, 136 fols., incomplete, 15th c.; London, British Museum, Royal 12 G VII, 102 fols., 14th c.; British Museum, Sloane 306, 177 fols., 14th c.; Oxford, Corpus Christi 150, 114 fols., 13th c.; Vienna, Nationalbibliothek 2438, a fragment only from beginning of bk. I, ch. 1 (ch. 4 in Arabic text), fols. 144r–147r, 15th c. Other Latin MSS have been reported in F. Carmody, *Arabic Astronomical and Astrological Sciences,* cited under III 1, p. 140; L. Thorndike and P. Kibre, *Catalogue,* cited under III 1, cols. 774, 803, 1208; and G. F. Vescovini, *Studi sulla prospettiva medievale* (Turin, 1965), pp. 93–94, n. 10.

The only known copy of the fourteenth-century Italian trans. of the *Optics* is MS Vat. Lat. 4595, 182 fols. Like the Latin text it lacks chs. 1–3 of bk. I. It includes an Italian trans. of the *Liber de crepusculis* (see below), fols. 178r–182v.

The Latin text was published in the collective volume bearing the following title: *Opticae thesaurus. Alhazeni Arabis libri septem, nunc primum editi, eiusdem liber de crepusculis et nubium ascensionibus, item Vitellionis Thuringo-Poloni Libri X, omnes instaurati, figuris illustrati et aucti, adjectis etiam in Alhazenum commentariis a Federico Risnero* (Basel, 1572). Concerning the authorship of *De crepusculis,* see below, "Spurious Works."

Kamāl al-Dīn's commentary, the *Tanqīḥ* (cited above), does not reproduce the integral text of the *Optics,* as was at one time supposed. An ed. of the Arabic text of *Kitāb al-Manāẓir* and English trans. are being prepared by the present writer.

III 4 (Br. 51). *M. f. Kayfiyyat al-arṣād* ("On the Method of [Astronomical] Observations").

III 6 (Br. 27). *M. f. Ḍaw' al-qamar* ("On the Light of the Moon"). Composed before 7 Aug. 1031, the date on which a copy was completed by 'Alī ibn Riḍwān (Ibn al-Qifṭī, *Ta'rīkh,* p. 444). Published as no. 8 in *Rasā'il.* There is a German trans. by Karl Kohl, "Über das Licht des Mondes. Eine Untersuchung von Ibn al-Haitham," in *Sitzungsberichte der Physikalisch-medizinischen Sozietät in Erlangen,* **56-57** (1924–1925), 305–398.

III 7 (Br. 22, Kr. 18). *M.* (or *Qawl*) *f. Samt al-qibla bi 'l-ḥisāb* ("Determination of the Direction of the Qibla by Calculation"). A German trans. is C. Schoy, "Abhandlung des Ḥasan ibn al-Ḥasan ibn al-Haiṯam (Alhazen) über die Bestimmung der Richtung der Qibla," in *Zeitschrift der Deutschen morgenländischen Gesellschaft,* **75** (1921), 242–253.

III 8 (Br. 41, Kr. 19). *M. f. al-Hāla wa-qaws quzaḥ* ("On the Halo and the Rainbow"). Completed in Rajab, 419 A.H. (A.D. 1028); see *Tanqīḥ,* II, p. 279. Recension by Kamāl al-Dīn in *Tanqīḥ,* II, 258–279. A shortened German trans. of this recension is E. Wiedemann, "Theorie des Regenbogens von Ibn al Haiṯam," in *Sitzungsberichte der Physikalisch-medizinischen Sozietät in Erlangen,* **46** (1914), 39–56.

III 9 (Br. 42, Kr. 20). *M. f. Mā ya'riḍ min al-ikhtilāf fī irtifā'āt al-kawākib* ("On What Appears of the Differences in the Heights of the Stars").

III 10 = Ia10 (Br. 39, Kr. 16). *M. f. Ḥisāb al-mu'āmalāt* ("On Business Arithmetic").

III 11 (Br. 43, Kr. 21). *M. f. al-Rukhāma al-ufuqiyya* ("On the Horizontal Sundial"). This work refers to a treatise to be written later on "shadow instruments" (*ālāt al-aẓlāl*); the reference may be to III 66.

III 14. *M. f. Marākiz al-athqāl* ("On Centers of Gravity"). This is not extant but has been abstracted by al-Khāzinī in

Mīzān al-ḥikma; see the Hyderabad ed. (1359 A.H. [1940]), pp. 16–20.

III 15 (Br. 13 a, Kr. 22). *M. f. Uṣūl al-misāḥa* ("On the Principles of Measurement"). A summary of the results of an earlier work or works by Ibn al-Haytham on the subject. Published as no. 7 in *Rasā'il.* German trans. by E. Wiedemann in "Kleinere Arbeiten von Ibn al Haiṯam," in *Sitzungsberichte der Physikalisch-medizinischen Sozietät in Erlangen,* **41** (1909), 16–24.

III 16 (Br. 2, Kr. 23). *M. f. Misāḥat al-kura* ("On the Measurement of the Sphere"). Later in composition than III 17; it may be one of the works referred to in III 15.

III 17 (Br. 14). *M. f. Misāḥat al-mujassam al-mukāfī* ("On the Measurement of the Paraboloidal Solid"). See III 16. Refers to a work on the same subject by Thābit ibn Qurra and another by Wayjan ibn Rustam al-Qūhī. German trans. by H. Suter, "Die Abhandlung über die Ausmessung des Paraboloides von el-Ḥasan b. el-Ḥasan b. el-Haitham," in *Bibliotheca mathematica,* 3rd ser., **12** (1912), 289–332. See also H. Suter, "Die Abhandlungen Thâbit b. Ḳurras und Abû Sahl al-Ḳûhîs über die Ausmessung der Paraboloide," in *Sitzungsberichte der Physikalisch-medizinischen Sozietät in Erlangen,* **48–49** (1916–1917), 186–227.

III 18 (Br. 33, Kr. 10). *M. f. al-Marāya al-muḥriqa bi 'l-dawā'ir* ("On Spherical Burning Mirrors"). Published as no. 4 in *Rasā'il.* German trans. by E. Wiedemann, "Ibn al Haiṯams Schrift über die sphärischen Hohlspiegel," in *Bibliotheca mathematica,* 3rd ser., **10** (1909–1910), 293–307. See also E. Wiedemann, "Zur Geschichte der Brennspiegel," in *Annalen der Physik und Chemie,* n.s. **39** (1890), 110–130, trans. into English by H. J. J. Winter and W. ʿArafāt, "A Discourse on the Concave Spherical Mirror by Ibn al-Haitham," in *Journal of the Royal Asiatic Society of Bengal,* 3rd ser., Science, **16** (1950), 1–16.

III 19 (Br. 33). *M. f. al-Marāya al-muḥriqa bi 'l-quṭūʿ* ("On Paraboloidal Burning Mirrors"). A hitherto unrecorded MS of the Arabic text is Florence, Biblioteca Medicea-Laurenziana, Or. 152, fols. 90v–97v. It was copied in the 13th century and bears no title or author's name. Here Ibn al-Haytham mentions an earlier treatise of his on how to construct all conic sections by mechanical means (*istikhrāj jamīʿ al-quṭūʿ bi-ṭarīq al-āla*); see below, "Additional Works," no. 2. The Arabic text has been published as no. 3 in *Rasā'il.* A medieval Latin trans. as *Liber de speculis comburentibus,* probably made by Gerard of Cremona, has been published together with a German trans. from the Arabic by J. L. Heiberg and E. Wiedemann: "Ibn al Haiṯams Schrift über parabolische Hohlspiegel," in *Bibliotheca mathematica,* 3rd ser., **10** (1909–1910), 201–237. See also E. Wiedemann, "Über geometrische Instrumente bei den muslimischen Völkern," in *Zeitschrift für Vermessungswesen,* nos. 22–23 (1910), 1–8; and "Geschichte der Brennspiegel," cited under III 18. An English trans. is H. J. J. Winter and W. ʿArafāt, "Ibn al-Haitham on the Paraboloidal Focusing Mirror," in *Journal of the Royal Asiatic Society of Bengal,* 3rd ser., Science, **15** (1949), 25–40.

III 20. *Maqāla mukhtaṣara fi 'l-Ashkāl al-hilāliyya* ("A Short Treatise on Crescent-Shaped Figures"). Not extant; see III 21.

III 21 (Br. 1, Kr. 12). *Maqāla mustaqṣāt fi 'l-Ashkāl al-hilāliyya* ("A Longer Treatise on Crescent-Shaped Figures"). Composed after III 20, which it is intended to supersede, and before III 30 (*q.v.*). It may also have been written before the work listed below as "Additional," no. 1 (*q.v.*).

III 22 ([?] Br. 6). *Maqāla mukhtaṣara fī Birkār al-dawā' ir al-ʿiẓām* ("A Short Treatise on the Birkār of Great Circles"). See Wiedemann, "Geometrische Instrumente . . .," cited under III 19. See III 23. *"Birkār"* is Persian for compass. Ibn al-Haytham explains the theory and construction of an instrument suitable for accurately drawing very large circles.

III 23 ([?] Br. 6). *Maqāla mashrūḥa fī Birkār al-dawā'ir al-ʿiẓām* ("An Expanded Treatise on the Birkār of Great Circles"). See III 22.

III 25 (Br. 52). *M. f. al-Tanbīh ʿalā mawāḍiʿ al-ghalaṭ fī kayfiyyat al-raṣd* ("On Errors in the Method of [Astronomical] Observations"). Earlier in composition than III 31.

III 26 (Br. 44, Kr. 24). *M. f. anna 'l-Kura awsaʿ al-ashkāl al-mujassama allatī iḥāṭatuhā muta-sāwiya, wa-anna 'l-dā'ira awsaʿ al-ashkāl al-musaṭṭaḥa allatī iḥaṭatuhā mutasāwiya* ("That the Sphere Is the Largest of the Solid Figures Having Equal Perimeters, and That the Circle is the Largest of the Plane Figures Having Equal Perimeters"). Composed before III 38 and III 68. Refers to Archimedes' *On the Sphere and the Cylinder.*

III 28 (Br. 29). *Kitāb fī Taṣḥīḥ al-aʿmāl al-nujūmiyya, maqālatān* ("A Book on the Corrections of Astrological Operations, Two Treatises").

III 30 (Br. 9, Kr. 2). *M. f. Tarbīʿ al-dā'ira* ("On the Quadrature of the Circle"). Refers to the "book on lunes" (*kitābina fi 'l-hilāliyyāt*), that is either III 20 or III 21. There is an ed. of the Arabic text and German trans. by H. Suter, "Die Kreisquadratur des Ibn el-Haiṯam," in *Zeitschrift für Mathematik und Physik,* Hist.-lit. Abt., **44** (1899), 33–47.

III 31 (Br. 45, Kr. 25). *M. f. Istikhrāj khaṭṭ niṣf al-nahār ʿalā ghāyat al-taḥqīq* ("Determination of the Meridian with the Greatest Precision"). Composed after III 25. Points out the relevance of the subject to astrology.

III 36 (Br. 31, Kr. 7). *M. f. Kayfiyyat al-aẓlāl* ("On the Formation of Shadows"). Composed before the "Commentary on the *Almagest*" (see below, "Additional Works," no. 3) and after III 3. A recension by Kamāl al-Dīn al-Fārisī is in *Tanqīḥ,* II, 358–381. A German trans. by E. Wiedemann is "Über eine Schrift von Ibn al Haiṯam: Über die Beschaffenheit der Schatten," in *Sitzungsberichte der Physikalisch-medizinischen Sozietät in Erlangen,* **39** (1907), 226–248.

III 38 (Br. 30, Kr. 26). *M. f. Ḥall shukūk fi 'l-maqāla l-ūlā min Kitāb al-Majisṭī yushakkiku fīhā baʿḍ ahl al-ʿilm* ("Solution of Difficulties in the First Book of the *Almagest* Which a Scholar Has Raised"). This work is to be distinguished from III 64. It was composed after III 26. The name of "the scholar" appears in MS Fatih 3439, fol. 150v to be Abu 'l-Qāsim ibn [?]Maʿdān, who is otherwise unknown to me. The Title in the Fatih MS (*Ḥall Shukūk fī Kitāb al-Majisṭī yushakkiku fīhā baʿḍ ahl al-ʿilm*) does not limit the discussion to book I of the *Almagest.* The text in fact discusses, among other things, book V of Ptolemy's *Optics.*

III 39 ([?] Kr. 27). *M. f. Ḥall shakk fī mujassamāt Kitāb Uqlīdis* ("Solution of a Difficulty in the Part of Euclid's Book Dealing With Solid Figures"). This may be part of the work listed below as "Additional Works," no. 1. But see Krause, no. 27, where reference is made to a work bearing a partially similar title and of uncertain authorship. (I have not examined MS Yeni Cami T 217, 2⁰, 893 A.H., referred to by Krause.)

III 40 (Br. 3). *Qawl fī Qismat al-miqdārayn al-mukhtalifayn al-madhkūrayn fi 'l-shakl al-awwal min al-maqāla 'l-ʿāshira min Kitāb Uqlīdis* ("On the Division of the Two Unequal Magnitudes Mentioned in Proposition I of Book X of Euclid's Book"). The subject is closely connected with the so-called "axiom of Archimedes."

III 41 (Br. 23). *Mas'ala fī Ikhtilāf al-naẓar* ("A Question Relating to Parallax"). MS India Office, Loth 734, fols. 120r–120v, specifies that lunar parallax is meant.

III 42 (Br. 17). *Qawl fī Istikhrāj muqaddamat ḍilʿ al-musabbaʿ* ("On the Lemma [Used by Archimedes] for [Constructing] the Side of the Heptagon [in the Book at the End of Which He Mentioned the Heptagon]"). Composed before III 74. German trans. by C. Schoy in *Die trigonometrischen Lehren des persischen Astronomen Abu 'l-Raiḥân Muḥ. ibn Aḥmad al Bîrûnî, dargestellt nach al-Qânûn al-Masʿûdî* (Hannover, 1927), pp. 85–91.

III 43 (Br. 10, Kr. 9). *Qawl fī Qismat al-khaṭṭ alladhī istaʿmalahu Arshimīdis fī Kitāb al-Kura wa 'l-usṭuwāna* ("On the Division of the Line Used by Archimedes in His Book on the Sphere and Cylinder"). Concerned with prop. 4 of bk. II in Archimedes' work. French trans. by F. Woepcke, *L'algèbre d'Omar Alkhayyâmî* (Paris, 1851), pp. 91–93.

III 44 (Br. 46, Kr. 28). *Qawl fī Istikhrāj khaṭṭ niṣf al-nahār bi-ẓill wāḥid* ("Determination of the Meridian by Means of One Shadow").

III 46 (Br. 26). *M. f. al-Majarra* ("On the Milky Way"). German trans. by E. Wiedemann, "Über die Lage der Milchstrasse nach ibn al Haiṯam," in *Sirius,* **39** (1906), 113–115.

III 48 (Br. 24, Kr. 5). *M. f. Aḍwā' al-kawākib* ("On the Light of the Stars"). Composed before III 49. Published as no. 1 in *Rasā'il.* Abridged German trans. by E. Wiedemann, "Über das Licht der Sterne nach Ibn Al Haitham," in *Wochenschrift für Astronomie, Meteorologie und Geographie,* n.s. **33** (1890), 129–133. English trans. by W. ʿArafat and H. J. J. Winter, "The Light of the Stars—a Short Discourse by Ibn al-Haytham," in *British Journal for the History of Science,* **5** (1971), 282–288.

III 49 (BR. 37). *M. f. al-Athar alladhī* [*yurā*] *fī* [*wajh*] *al-qamar* ("On the Marks [Seen] on the [Face of the] Moon"). Composed after III 3, III 6, and III 48, to all of which it refers. German trans. by C. Schoy, as *Abhandlung des Schaichs Ibn ʿAlî al-Ḥasan ibn al-Ḥasan ibn al-Haitham: Über die Natur der Spuren* [*Flecken*], *die man auf der Oberfläche des Mondes sieht* (Hannover, 1925).

III 53 (Br. 35). *M. f. al-Taḥlīl wa 'l-tarkīb* ("On Analysis and Synthesis"). Brockelmann lists a Cairo MS. Another is Dublin, Chester Beatty 3652, fols. 69v-86r, dated 612 A.H. (1215). Composed before III 54 (to which it is closely related) and after III 2.

III 54 (Br. 11). *M. f. al-Maʿlūmāt* ("On the Known Things [Data]"). Trans. of the enunciations of its propositions by L. A. Sédillot, in "Du *Traité* des connues géométriques de Hassan ben Haithem," in *Journal asiatique,* **13** (1834), 435–458.

III 55. *Qawl fī Hall Shakk fi 'l-maqāla 'l-thāniya ʿashar min Kitāb Uqlīdis* ("Solution of a Difficulty in Book XII of Euclid's Book"). Possibly a part of the work listed below as "Additional," no. 1.

III 56. *M. f. Ḥall shukūk al-maqāla 'l-ūlā min Kitāb Uqlīdis* ("Solution of the Difficulties in Book I of Euclid's Book"). Possibly part of the work listed below as "Additional Works," no. 1.

III 60 (Br. 32, Kr. 4). *M.* (or *Qawl*) *f. al-Ḍaw'* ("A Discourse on Light"). Composed after III 3. Printed as no. 2 in *Rasā'il.* J. Baarmann published an ed. of the Arabic text together with a German trans. as "Abhandlung über das Licht von Ibn al-Haiṯam," in *Zeitschrift der Deutschen morgenländischen Gesellschaft,* **36** (1882), 195–237. (See remarks on this ed. by E. Wiedemann, *ibid.,* **38** (1884), 145–148.) A Cairo ed. by A. H. Mursī correcting Baarmann's text appeared in 1938. There is now a critical French trans. by R. Rashed: "Le 'Discours de la lumière' d'Ibn al-Haytham," in *Revue d'histoire des sciences et de leurs applications,* **21** (1968), 198–224. A recension is in *Tanqīḥ,* II, 401–407. A German trans. of this recension (*taḥrīr*) is E. Wiedemann, "Ueber 'Die Darlegung der Abhandlung über das Licht' von Ibn al Haiṯam," in *Annalen der Physik und Chemie,* n.s. **20** (1883), 337–345.

III 63 (Br. 19, Kr. 29). *M. f. Ḥall shukūk ḥarakat al-iltifāf* ("Solution of Difficulties Relating to the Movement of Iltifāf"). A reply to an unnamed scholar who raised objections against an earlier treatise by Ibn al-Haytham (III 61: "On the Movement of Iltifāf") which is now lost. In the reply Ibn al-Haytham revealed an intention he had entertained to write a critique of Ptolemy's *Almagest, Planetary Hypotheses* (*Kitāb al-Iqtiṣāṣ*), and *Optics* (MS Pet. Ros. 192, fols. 19v–20r)—almost certainly a reference to III 64.

III 64 (Br. 30). *M. f. al-Shukūk ʿalā Baṭlamyūs* ("Dubitationes in Ptolemaeum"). Composed after III 63; see preceding note. There is a critical ed. by A. I. Sabra and N. Shehaby (Cairo, 1971). English trans. of part of this work by A. I. Sabra, "Ibn al-Haytham's Criticism of Ptolemy's *Optics,*" in *Journal of the History of Philosophy,* **4** (1966), 145–149.

III 65. *M. f. al-Juz' alladhī la yatajazza'* ("On Atomic Parts"). A unique copy belonging to a private collection in Aleppo is recorded in P. Sbath, *Al-Fihris* (cited above), I, 86, no. 724.

III 66 (Br. 40, Kr. 17). *M. f. Khuṭūṭ al-sāʿāt* ("On the Lines of the Hours [i.e., on sundials]"). *"Al-sāʿāt"* has sometimes been misread as *"al-shuʿāʿāt"* (rays). The treatise refers to a work by Ibrāhīm ibn Sinān, "On Shadow Instruments." See note for III 11 above.

III 67. *M. f. al-Qarasṭūn* ("On the *Qarasṭūn*"). A unique copy belonging to a private collection in Aleppo is recorded in P. Sbath, *Al-Fihris* (cited above), I, p. 86, no. 726.

III 68 (Br. 12, Kr. 11). *M. f. al-Makān* ("On Place"). Later in composition than III 26. Published as no. 5 in *Rasā'il.* A short account is given by Wiedemann in "Kleinere Arbeiten . . .," cited under III 15 above, pp. 1-7.

III 69 (Br. 18). *Qawl fī Istikhrāj a'midat al-jibāl* ("Determination of the Altitudes of Mountains"). A longer title is *Fī Ma'rifat irtifā' al-ashkhāṣ al-qā'ima wa-a'midat al-jibāl wa irtifā' al-ghuyūm* ("Determination of the Height of Erect Objects and of the Altitudes of Mountains and of the Height of Clouds"). A German trans. is H. Suter, "Einige geometrische Aufgaben bei arabischen Mathematiker," in *Bibliotheca mathematica,* 3rd ser., **8** (1907), 27-30. A short account by Wiedemann is in "Kleinere Arbeiten . . .," cited under III 15 above, pp. 27-30.

III 71 (Br. 38). *M. f. A'midat al-muthallathāt* ("On the Altitudes of Triangles"). (An alternate title is *Khawāṣṣ al-muthallath min jihat al-'amūd* ("Properties of the Triangle in Respect of Its Altitude"). Published as no. 9 in *Rasā'il.*

III 73. (Br. 13, Kr. 3). *M. f. Shakl Banū Mūsā.* ("On the Proposition of Banū Mūsā [proposed as a lemma for the *Conics* of Apollonius]"). Published as no. 6 in *Rasā'il.* An account of it is in Wiedemann, "Kleinere Arbeiten . . .," cited under III 15 above, pp. 14-16.

III 74 (Br. 48, Kr. 30). *M. f. 'Amal al-musabba' fi 'l-dā'ira* ("On Inscribing a Heptagon in a Circle"). Composed after III 42, to which it refers. As well as referring to Archimedes it mentions al-Qūhī, whose treatise on the subject has been published and trans. by Y. Dold-Samplonius, "Die Konstruktion des regel-mässigen siebenecks nach Abû Sahl al-Qûhî," in *Janus,* **50** (1963), 227-249.

III 75 (Br. 25, Kr. 1). *M. f. Irtifā' al-quṭb 'alā ghāyat al-taḥqīq* ("Determination of the Height of the Pole With the Greatest Precision"). A German trans. is C. Schoy, "Abhandlung des Ḥasan ben al-Ḥasan ben al-Haiṯam über eine Methode, die Polhöhe mit grösster Genanigkeit zu bestimmen," in *De Zee,* **10** (1920), 586-601.

III 76 (Br. 47, Kr. 31). *M. f. 'Amal al-binkām* ("On the Construction of the Water Clock").

III 77 (Br. 33b, Kr. 32). *M. f. al-Kura 'l-muḥriqa* ("On the Burning Sphere"). Written after III 3 and III 66. A recension by Kamāl al-Dīn is in *Tanqīḥ,* II, 285-302. A German trans. of this recension is in E. Wiedemann, "Brechung des Lichtes in Kugeln nach Ibn al Haiṯam und Kamâl al Dîn al Fârisî," in *Sitzungsberichte der Physikalisch-medizinischen Sozietät in Erlangen,* **42** (1910), 15-58, esp. 16-35.

III 78 (Br. 15). *M. f. Mas'ala 'adadiyya mujassama* ("On an Arithmetical Problem in Solid Geometry").

III 79 (Br. 5). *Qawl fī Mas'ala handasiyya* ("On a Geometrical Problem"). A German trans. is in C. Schoy, "Behandlung einiger geometrischen Fragenpunkte durch muslimische Mathematiker," in *Isis,* **8** (1926), 254-263, esp. 254-259.

III 80 (Br. 20, Kr. 8). *M. f. Ṣūrat al-kusūf* ("On the Shape of the Eclipse"). Composed after III 3. A recension by Kamāl al-Dīn is *Tanqīḥ,* II, 381-401. A German trans. of the original text from the India Office MS is in E. Wiedemann, "Über die Camera obscura bei Ibn al Haiṯam," in *Sitzungsberichte der Physikalisch-medizinischen Sozietät in Erlangen,* **46** (1914), 155-169.

III 82 (Br. 21, Kr. 13). *M. f. Ḥarakat al-qamar* ("On the Motion of the Moon"). A vindication of Ptolemy's account of the mean motion of the moon in latitude.

III 83 (Br. 4). *M. f. Masā'il al-talāqī* ("On Problems of Talāqī"). These are problems involving the solution of simultaneous linear equations. There is an account by E. Wiedemann, in "Über eine besondere Art des Gesellschaftsrechnens besondere nach Ibn al Haiṯam," in *Sitzungsberichte der Physikalisch-medizinischen Sozietät in Erlangen,* **58-59** (1926-1927), 191-196.

III 92 (Br. 16). *Qawl fī Istikhrāj mas'ala 'adadiyya* ("Solution of an Arithmetical Problem"). An account is given by Wiedemann in "Kleinere Arbeiten . . .," cited under III 15 above, pp. 11-13.

ADDITIONAL WORKS. These are extant works whose titles do not appear in list III.

Add. 1 (Br. 7, Kr. 6). *Kitāb fī Ḥall shukūk Kitāb Uqlīdis fi 'l-Uṣūl wa-sharḥ ma'ānīh* ("A Book on the Solution of the Difficulties in Euclid's *Elements* and an Explanation of Its Concepts"). This seems to be a different work from Ia 1: *Sharḥ Uṣūl Uqlīdis fi 'l-handasa wa 'l-'adad wa talkhīṣuhu* ("A Commentary on and Summary of Euclid's *Elements of Geometry and Arithmetic*").

The absence of this comprehensive work from list III may perhaps be explained by supposing III 39, III 55, and III 56 to be parts of it. It refers to III 2 and also to "our treatise on crescent-shaped figures," which is either III 20 or III 21. An Istanbul MS that is not recorded in Brockelmann or in Krause is Üniversite 800, copied before 867 A.H. (1462-1463). The MS has 182 fols. but is not complete.

Add. 2. *Kalām fī tawṭi'at muqaddamāt li-'amal al-quṭū' 'alā saṭḥ mā bi-ṭarīq ṣinā'ī* ("A Passage in Which Lemmas Are Laid Down for the Construction of [Conic] Sections by Mechanical Means"). MS Florence, Biblioteca Medicea-Laurenziana, Or. 152, fols. 97v-100r. No author is named, and the "lemmas" follow immediately after a copy of Ibn al-Haytham's III 19 ("On Paraboloidal Burning Mirrors"), which also does not bear the author's name. Since Ibn al-Haytham refers in III 19 to a treatise of his on the mechanical construction of conic sections, it is very likely that the "passage" we have here is a fragment of that treatise which the copyist found joined to III 19.

Add. 3. "Commentary on the *Almagest,*" Istanbul MS, Ahmet III 3329, copied in Jumādā II 655 (1257), 123 fols. Probably written after III 36, to which it appears to refer (fol. 90r).

SPURIOUS WORKS. Ibn al-Haytham is not the author of the *Liber de crepusculis,* the work on dawn and twilight translated by Gerard of Cremona and included in Risner's *Opticae thesaurus* (see A. I. Sabra, "The Authorship of the *Liber de crepusculis,*" in *Isis,* **58** [1967], 77-85; above, III 3). An astrological work, *De imaginibus celestibus,* Vatican MS Urb. Lat. 1384, fols. 3v-26r, has also been mistakenly ascribed to him (*ibid.,* p. 80, n. 14).

Two more writings are listed in Brockelmann (nos. 49, 50) which may or may not be genuine.

I am grateful to M. Clagett for showing me a microfilm

of MS Bruges 512 (III 3) and to M. Schramm for showing me microfilms of the following MSS: Kastamonu 2298 (III 1), Üniversite 800 (Add. 1), and Ahmet III 3329 (Add. 3). For the last three MSS and for other Arabic MSS not hitherto recorded, see the appropriate volume of F. Sezgin, *Geschichte des arabischen Schrifttums* (Leiden, 1967–).

II. SECONDARY LITERATURE. Sources for the biography of Ibn al-Haytham are Ibn al-Qifṭī, *Ta'rīkh al-ḥukamā'*, J. Lippert, ed. (Leipzig, 1903), pp. 155–168 (see corrections of this ed. by H. Suter in *Bibliotheca mathematica,* 3rd ser., **4** [1903], esp. 295–296); ʿAlī ibn Zayd al-Bayhaqī, *Tatimmat ṣiwān al-ḥikma,* M. Shafīʿ, ed., fasc. I: Arabic text (Lahore, 1935), 77–80 (analysis and partial English trans. of this work by M. Meyerhof in *Osiris,* **8** [1948], 122–216, see esp. 155–156); Ibn Abī Uṣaybiʿa, *Ṭabaqāt al-aṭibbā',* A. Müller, ed. (Cairo–Königsberg, 1882–1884), II, 90–98 (German trans. in E. Wiedemann, "Ibn al-Haiṯam, ein arabischer Gelehrter," cited below); Ṣāʿid al-Andalusī, *Ṭabaqāt al-umam,* L. Cheikho, ed. (Beirut, 1912), p. 60 (French trans. by R. Blachère [Paris, 1935], p. 116). The account in Abu'l-Faraj ibn al-ʿIbrī, *Ta'rīkh mukhtaṣar al-duwal,* A. Ṣālḥānī, ed. (Beirut, 1958), pp. 182–183, derives from Ibn al-Qifṭī. In addition to the works by Suter (*Mathematiker*) and Brockelmann (*Geschichte*) already cited, see M. Steinschneider, "Vite di matematici arabi, tratte da un' opera inedita di Bernardino Baldi, con note di M.S.," in *Bullettino di bibliografia e di storia delle scienze matematiche et fisiche,* **5** (1872), esp. 461–468, also printed separately (Rome, 1874); M. J. de Goeje, "Notice biographique d'Ibn al-Haitham," in *Archives néerlandaises des sciences exactes et naturelles,* 2nd ser., **6** (1901), 668–670; E. Wiedemann, "Ueber das Leben von Ibn al Haiṯam und al Kindî," in *Jahrbuch für Photographie und Reproduktiontechnik,* **25** (1911), 6–11 (not important).

The literary relationship of Ibn al-Haytham's autobiography to Galen's *De libris propriis* is discussed by F. Rosenthal in "Die arabische Autobiographie," in *Studia arabica I,* Analecta Orientalia, no. 14 (Rome, 1937), 3–40, esp. 7–8. There is a discussion of the autobiography in G. Misch, *Geschichte der Autobiographie,* III, pt. 2 (Frankfurt, 1962), 984–991. Lists I*a* and III of the works of Ibn al-Haytham are translated from Ibn Abī Uṣaybiʿa in F. Woepcke, *L'algèbre d'Omar Alkhayyāmī* (Paris, 1851), pp. 73–76; but see H. Suter's corrections in *Mathematiker,* pp. 92–93. There is a German trans. of Ibn al-Haytham's autobiography and of Lists I–III in E. Wiedemann, "Ibn al Haiṯam, ein arabischer Gelehrter," in *Festschrift [für] J. Rosenthal* (Leipzig, 1906), pp. 169–178. M. Schramm discusses the chronological order of some of Ibn al-Haytham's works in *Ibn al-Haythams Weg zur Physik* (Wiesbaden, 1962), pp. 274–285.

The most complete study of Ibn al-Haytham's optical researches is M. Naẓīf, *Al-Ḥasan ibn al-Haytham, buḥūthuhu wa-kushūfuhu al-baṣariyya* ("Ibn al-Haytham, His Optical Researches and Discoveries"), 2 vols. (Cairo, 1942–1943)—reviewed by G. Sarton in *Isis,* **34** (1942–1943), 217–218. Based on the extant MSS of *Kitāb al-Manāẓir* and on Ibn al-Haytham's other optical works, this voluminous study (more than 850 pages) is distinguished by clarity, objectivity, and thoroughness. It is particularly valuable as a study of the mathematical sections of Ibn al-Haytham's works. M. Schramm, *Ibn al-Haythams Weg zur Physik,* is the most substantial single study of Ibn al-Haytham in a European language, and it has the merit of drawing on MS sources not previously available. In analyzing Ibn al-Haytham's attempt to combine Aristotelian natural philosophy with a mathematical and experimental approach, Schramm illuminates other important treatises of Ibn al-Haytham besides the *Optics.*

The question of mathematizing Aristotelian physics is also discussed in S. Pines, "What Was Original in Arabic Science," in A. C. Crombie, ed., *Scientific Change* (London, 1963), pp. 181–205, esp. 200–202. It is taken up afresh by R. Rashed in "Optique géometrique et doctrine optique chez Ibn al Hayṯham," in *Archive for History of Exact Sciences,* **6** (1970), 271–298. For further discussions of the concept of experiment in Arabic optics generally and in the work of Ibn al-Haytham in particular, see M. Schramm, "Aristotelianism: Basis and Obstacle to Scientific Progress in the Middle Ages," in *History of Science,* **2** (1963), 91–113, esp. 106, 112; and "Steps Towards the Idea of Function: A Comparison Between Eastern and Western Science of the Middle Ages," *ibid.,* **4** (1965), 70–103, esp. 81, 98; A. I. Sabra, "The Astronomical Origin of Ibn al-Haytham's Concept of Experiment," in *Actes du XIIe Congrès international d'histoire des sciences,* Paris, 1968, III A (Paris, 1971), 133–136.

General accounts mainly based on the *Optics* are J. B. J. Delambre, "Sur l'*Optique* de Ptolémée comparée à celle qui porte le nom d'Euclide et à celle d'Alhazen et de Vitellion," in *Historie de l'astronomie ancienne,* II (Paris, 1817), 411–432; E. Wiedemann, "Zu Ibn al Haiṯams Optik," in *Archiv für Geschichte der Naturwissenschaften und der Technik,* **3** (1910–1911), 1–53, an account of Kamāl al-Dīn's revision (*Tanqīḥ*) of Ibn al-Haytham's *Optics,* based on a Leiden MS; includes an abbreviated trans. of *Optics,* bk. I, chs. 1–3, as reported by Kamāl al-Dīn; L. Schnaasse, *Die Optik Alhazens* (Stargard, 1889); V. Ronchi, "Sul contributo di Ibn al-Haitham alle teorie della visione e della luce," in *Actes du VIIe Congrès international d'histoire des sciences* (Jerusalem, 1953), pp. 516–521; and *The Nature of Light,* a trans. of *Storia della luce* (2nd ed., Bologna, 1952) (London, 1970), pp. 40–57; H. J. J. Winter, "The Optical Researches of Ibn al-Haitham," in *Centaurus,* **3** (1953–1954), 190–210, which includes accounts of treatises other than the *Optics.*

Studies of particular aspects of Ibn al-Haytham's optical work are A. Abel, "La sélénographie d'Ibn al Haitham (965–1039) dans ses rapports avec la science grecque," in *Comptes rendus, IIe Congrès national des sciences* (Brussels, 1935), pp. 76–81 (concerned with III 49); J. Lohne, "Zur Geschichte des Brechungsgesetzes," in *Sudhoffs Archiv für Geschichte der Medizin und der Naturwissenschaften,* **47** (1963), 152–172, esp. 153–157; R. Rashed, "Le modèle de la sphère transparente et l'explication de l'arc-en-ciel: Ibn al-Haytham, al-Fārisī," in *Revue d'histoire des sciences et de leurs applications,* **23** (1970), 109–140; E. Wiedemann, "Ueber den Apparat zur Untersuchung und Brechung des

Lichtes von Ibn al Haiṯam," in *Annalen der Physik und Chemie,* n.s. **21** (1884), 541–544; "Über die Erfindung der Camera obscura," in *Verhandlung der Deutschen physikalischen Gesellschaft,* **12** (1910), 177–182; and "Über die erste Erwähnung der Dunkelkammer durch Ibn al Haiṯam," in *Jahrbuch für Photographie und Reproduktiontechnik,* **24** (1910), 12–13; J. Würschmidt, "Zur Theorie der Camera obscura bei Ibn al Haiṯam," in *Sitzungsberichte der Physikalisch-medizinischen Sozietät in Erlangen,* **46** (1914), 151–154; and "Die Theorie des Regenbogens und das Halo bei Ibn al Haiṯam und bei Dietrich von Freiberg," in *Meteorologische Zeitschrift,* **13** (1914), 484–487. Apart from Vescovini's *Studi* (see below) there is one account of Ibn al-Haytham's psychological ideas as expounded in bk. II of the *Optics:* H. Bauer, *Die Psychologie Alhazens auf Grund von Alhazens Optik dargestellt,* in the series Beiträge zur Geschichte der Philosophie des Mittelalters, 10, no. 5 (Münster in Westfalen, 1911). The physiological aspect of vision is discussed in M. Schramm, "Zur Entwicklung der physiologischen Optik in der arabischen Literatur," in *Sudhoffs Archiv für Geschichte der Medizin und der Naturwissenschaften,* **43** (1959), 289–316, esp. 291–299.

The following are concerned with the transmission of Ibn al-Haytham's optical ideas to the West; they include comparisons with Hebrew and Latin medieval, Renaissance, and seventeenth-century writers: M. Steinschneider, "Aven Natan e le teorie sulla origine della luce lunare e delle stelle, presso gli autori ebrei del medio evo," in *Bullettino di bibliografia e di storia delle scienze matematiche e fisiche,* **1** (1868), 33–40; E. Narducci, "Nota intorno ad una traduzione italiana fatta nel secolo decimoquarto, del trattato d'*Ottica* d'Alhazen, matematico del secolo undecimo, e ad altri lavori di questo scienziato," *ibid.,* **4** (1871), 1–48; and "Giunte allo scritto intitolato 'Intorno ad una traduzione italiana, fatta nel secolo decimoquarto, dell' *Ottica* di Alhazen,'" *ibid.,* pp. 137–139; A. I. Sabra, "Explanation of Optical Reflection and Refraction: Ibn al-Haytham, Descartes and Newton," in *Actes du Xe Congrès international d'histoire des sciences,* Ithaca, 1962 (Paris, 1964), I, 551–554; and *Theories of Light From Descartes to Newton* (London, 1967), pp. 72–78, 93–99 (concerned with the theories of reflection and refraction); G. F. Vescovini, *Studi sulla prospettiva medievale* (Turin, 1965), which reveals the influence of Ibn al-Haytham's *Optics* on the development of empiricist theories of cognition in the fourteenth century; and "Contributo per la storia della fortuna di Alhazen in Italia: Il volgarizzamento del MS. Vat. 4595 e il 'Commentario terzo' del Ghiberti," in *Rinascimento,* 2nd ser., **5** (1965), 17–49; D. Lindberg, "Alhazen's Theory of Vision and Its Reception in the West," in *Isis,* **58** (1968), 321–341; and "The Cause of Refraction in Medieval Optics," in *British Journal for the History of Science,* **4** (1968), 23–38. See also G. Sarton, "The Tradition of the *Optics* of Ibn al-Haitham," in *Isis,* **29** (1938), 403–406.

The following are studies relating to Ibn al-Haytham's astronomical works, particularly his treatise on *The Configuration of the World* (III 1): M. Steinschneider, "Notice sur un ouvrage astronomique inédit d'Ibn Haitham," in *Bullettino di bibliografia e di storia delle scienze matematiche e fisiche,* **14** (1881), 721–736; and "Supplément à la 'Notice sur un ouvrage inédit d'Ibn Haitham,'" *ibid.,* **16** (1883), 505–513—*Extrait du Bullettino* . . ., containing the "Notice" and the "Supplément" (Rome, 1884), includes many corrections of the earlier publications; E. Wiedemann, "Ibn al Haiṯam und seine Bedeutung für die Geschichte der Astronomie," in *Deutsche Literaturzeitung,* **44** (1923), 113–118; P. Duhem, *Système du monde,* II (Paris, 1914), 119–129; W. Hartner, "The Mercury Horoscope of Marcantonio Michiel of Venice, a Study in the History of Renaissance Astrology and Astronomy," in A. Beer, ed., *Vistas in Astronomy* (London–New York, 1955), pp. 84–138, esp. 122–127; S. Pines, "Ibn al-Haytham's Critique of Ptolemy," in *Actes du Xe Congrès international d'histoire des sciences,* Ithaca, 1962 (Paris, 1964), I, 547–550 (concerned with Ibn al-Haytham's criticism of the equant in the *Dubitationes in Ptolemaeum,* III 64); M. Schramm, *Ibn al-Haythams Weg zur Physik,* esp. pp. 63–69, 88–146.

For discussions of "Alhazen's problem," see P. Bode, "Die Alhazensche Spiegelaufgabe in ihrer historischen Entwicklung nebst einer analytischen Lösung des verallgemeinerten Problems," in *Jahresbericht des Physikalischen Vereins zu Frankfurt am Main,* for 1891–1892 (1893), pp. 63–107; M. Baker, "Alhazen's Problem. Its Bibliography and an Extension of the Problem," in *American Journal of Mathematics,* **4** (1881), 327–331; M. Naẓīf, *Al-Ḥasan ibn al-Haytham* . . ., pp. 487–589; J. A. Lohne, "Alhazens Spiegelproblem," in *Nordisk matematisk tidskrift,* **18** (1970), 5–35 (with bibliography).

A general survey of Ibn al-Haytham's work in various fields is M. Schramm, "Ibn al-Haythams Stellung in der Geschichte der Wissenschaften," in *Fikrun wa fann,* no. 6 (1965), 2–22. See also M. Naẓīf and P. Ghalioungui, "Ibn at Haitham, an 11th-Century Physicist," in *Actes du Xe Congrès international d'histoire des sciences,* Ithaca, 1962 (Paris, 1964), I, 569–571. No. 2 of the Publications of the Egyptian Society for the History of Science (Cairo, 1958) includes articles in Arabic by M. Naẓīf, M. Madwar, M. ʿAbd al-Rāziq, M. Ghālī, and M. Ḥijāb on various aspects of Ibn al-Haytham's thought. Some of these articles are reprints of previous publications. For a detailed table of contents see *Isis,* **51** (1960), 416.

Many of the European translations of Ibn al-Haytham's works, cited in the first part of the bibliography, include historical and critical notes.

A. I. SABRA

HEATH, THOMAS LITTLE (*b.* Barnetby le Wold, Lincoln, England, 5 October 1861; *d.* Ashtead, Surrey, England, 16 March 1940), *mathematics, antiquity.*

After attending the grammar school at Caistor, he went to Clifton and thence, with a foundation scholarship, to Trinity College, Cambridge, where he became a fellow in 1885 and an honorary fellow in 1920. On leaving Trinity, he entered the civil service in the department of the treasury. He retired from that service in 1926, having been awarded the C.B. (1903); K.C.B. (1909); and K.C.V.O. (1916). His academic distinctions were numerous. The University of Oxford conferred an honorary degree on him; the Royal Society elected him a fellow (1912); and he served on the council of the society. He was a fellow of the British Academy and president of the Mathematical Association from 1922 to 1923.

Heath's main interest lay in the study of Greek mathematics, for which his training in classics and mathematics at Cambridge admirably fitted him; he soon became one of the leading authorities on mathematics in antiquity. The wide range of his interest is reflected in the titles of the works he published. His *History of Greek Mathematics* is usually regarded as his most famous contribution. In *The Thirteen Books of Euclid's Elements* he made available those books of the *Elements* that had hitherto been considered unintelligible; in particular his treatment of book X is a masterpiece.

BIBLIOGRAPHY

I. Original Works. Heath's works, including his translations, are *Diophantus of Alexandria: A Study in the History of Greek Algebra* (Cambridge, 1885; rev. ed., 1910); *Apollonius of Perga: A Treatise on the Conic Sections* (Cambridge, 1896; repr., 1961); *The Works of Archimedes* (Cambridge, 1897); *The Thirteen Books of Euclid's Elements* (Cambridge, 1908; 2nd ed., 1925); *Aristarchus of Samos: The Ancient Copernicus* (Oxford, 1913); *A History of Greek Mathematics,* 2 vols. (Oxford, 1921); *A Manual of Greek Mathematics* (Oxford, 1931); *Greek Astronomy* (London-Toronto-New York, 1932); and *Mathematics in Aristotle* (Oxford, 1949), on which he was working at the time of his death.

Heath also made numerous contributions to the *Mathematical Gazette* and the *Encyclopaedia Britannica* and assisted in the preparation of the 9th ed. of the Liddell-Scott Greek lexicon.

II. Secondary Literature. On his life and work, see the obituaries in the London *Times* (18 Mar. 1940); *Proceedings of the British Academy,* **26** (1940); and *Obituary Notices of the Fellows of the Royal Society,* no. 9 (January 1941).

J. F. Scott

HEAVISIDE, OLIVER (*b.* Camden Town, London, England, 18 May 1850; *d.* Paignton, Devonshire, England, 3 February 1925), *physics, electrical engineering.*

Heaviside was the youngest of four sons of Thomas Heaviside, an artist, and Rachel Elizabeth West, whose sister Emma married Charles Wheatstone in 1847. There is no evidence that his famous uncle contributed to the education of Heaviside, who was almost entirely self-taught and who—except for a job in 1870–1874 as a telegraph operator at Newcastle-on-Tyne—lived privately, supported by his brother and later by well-wishers and by a government pension. Despite the lack of a formal education, Heaviside became expert in mathematical physics and played an important role in the development of the electromagnetic theory of James Clerk Maxwell and in its practical applications.

Having engaged in electrical experimentation since his teens, Heaviside published his first technical article when he was twenty-two. In 1873 and 1876 he proposed methods of making duplex telegraphy practical in papers in the *Philosophical Magazine.* A series of papers in *The Electrician* (1885–1887) firmly established his reputation. They were published in a commercial magazine because of the perspicacity of two successive editors, C. H. W. Biggs and A. P. Trotter, the first of whom lost his post partly as a result of his support for Heaviside, whose ideas on long-distance cable transmission were not in accord with the "official" views of W. H. Preece and H. R. Kempe at the General Post Office; their opposition sufficed to keep Heaviside's papers out of the journals of professional societies and to create difficulties for his supporters generally.

Reduced to essentials, the controversy centered on the effect of inductance in long-distance cables: Preece and Kempe held that it should be minimized, whereas Heaviside's theory unexpectedly predicted that additional inductive coils deliberately inserted at intervals would in fact improve the performance. His assertion was later shown to be correct by M. I. Pupin and others, but in a quarrel between a high government official (Preece was later knighted) and a self-educated maverick living in near penury in a Devon village it was not immediately obvious to all concerned that the latter was right. Even after the papers began appearing in *The Electrician,* understanding dawned slowly, for Heaviside's genius led him to make free and original use of mathematical tools not appreciated by even the most sophisticated contemporary—they were sometimes decades ahead of their rigorous elaboration and application to practical problems.

For example, Heaviside recognized the importance of operational calculus to the investigation of transients, anticipating later employment of Laplace and Fourier transforms in electrical engineering by use of a less rigorous system of his own devising. He developed his own form of vector notation, which resembled that proposed by his great American contemporary J. W. Gibbs. He was the first to formulate the "telegraphers' equation" for voltage V as a function of distance x, time t, and resistance R,

$$\frac{1}{C}\frac{\partial^2 V}{\partial x^2} = L\frac{\partial^2 V}{\partial t^2} + R\frac{\partial V}{\partial t},$$

where C and L are capacitance and inductance, terms that he coined (along with "impedance" and "leakance," now generally called "conductance"). This equation, whose coefficients could be optimized to produce a distortionless mode of propagation, proved to have wide application in general dynamics.

Heaviside also suggested a new system of electromagnetic units similar to the mks system now in general use and based on a proposal by Giovanni Giorgi. After the spanning of the Atlantic by radio waves in 1901 he predicted the existence of a reflecting ionized region surrounding the earth, which later became known as the Kennelly-Heaviside layer (and is now called the ionosphere), in recognition of a similar proposal made independently and almost simultaneously by A. E. Kennelly of Harvard University. Heaviside was the first to propose the theory of the steady rectilinear motion of an electric charge through the ether and is said to have predicted the increase of mass of a charge moving at great speeds.

Heaviside's fame spread, and he became something of a legend in his own lifetime. As a result of the scientific help he generously extended to all who sought it, his hermitage near Torquay became known as The Inexhaustible Cavity, even though its inhabitant sometimes lacked the money to pay his dues to professional societies. One of these societies, the Institution of Electrical Engineers, solved the problem by electing him an honorary member and, shortly before his death, awarded him its first Faraday Medal. Heaviside was elected Fellow of the Royal Society in 1891 and received an honorary doctorate from the University of Göttingen. He died alone in his seaside cottage, never having married, and is buried at Paignton.

BIBLIOGRAPHY

I. ORIGINAL WORKS. Six of Heaviside's papers in *Philosophical Magazine* were published as *Electromagnetic Waves* (London, 1889). His *Electrical Papers* were published in 2 vols. at London in 1892. His papers in *Electrician* were collected in 3 vols. entitled *Electromagnetic Theory* (London, 1893–1912; repr. 1922–1925). A new ed. of *Electromagnetic Theory* (New York, 1950) contains a critical and historical intro. by Ernst Weber.

II. SECONDARY LITERATURE. A short biography of Heaviside appears in Rollo Appleyard, *Pioneers of Electrical Communication* (London, 1930); Appleyard also contributed the entry in *Dictionary of National Biography 1922–1930,* pp. 412–414. In 1950, at London, the Institution of Electrical Engineers published *The Heaviside Centenary Volume* of articles on his work, illustrated with plates, including portraits. In 1959 the IEE published a monograph by H. J. Josephs, *The Heaviside Papers Found at Paignton in 1957.* These papers, as well as Heaviside's own library, now repose at the IEE's London headquarters. An excellent source of information on the controversy with Preece, although strongly partisan to Heaviside, is E. T. Whittaker, "Oliver Heaviside," in *Bulletin of the Calcutta Mathematical Society,* **20** (1928–1929), 216–220.

Obituaries appear in *Electrical World* (1925), and *Proceedings of the Royal Society,* **110A** (1926), xiv. See also the notices by A. Russell in *Nature,* **115** (1925), 237–238; Oliver Lodge in *Electrician,* **94** (1925), 174; and F. Gill in *Bell System Technical Journal,* **4** (1925), 349–354, with portrait. A bibliography is in Poggendorff, III, 602; IV, 601–602; and VI, 1057.

CHARLES SÜSSKIND

HECATAEUS OF MILETUS (*fl.* late sixth century and early fifth century B.C.), *geography.*

Very little is known about the life of Hecataeus (Ἑκαταῖος), son of Hegesander. He seems to have belonged to the ruling class of Miletus, since Herodotus quotes him as playing a leading role in the political deliberations of the Ionian states at the time of the Ionian Revolt, 499–494 B.C. (Herodotus V. 36, 125–126—Agathemerus calls him a "much-traveled man" [ἀνὴρ πολυπλανής], *Geographiae informatio* I.1).

Hecataeus is important as one of the earliest Greek prose writers (λογοποιοί) and especially as the author of the earliest geographical work (probably accompanied by a map, which may soon have disappeared; it was apparently not known to Eratosthenes—Strabo, *Geography,* I.1.11.). Numerous quotations from the work have been preserved by later writers; its title is given as *Periodos Gēs* (Περίοδος γῆς), simply *Periodos* (see Jacoby in Pauly-Wissowa, *Real-Encyclopädie,* col. 2671), or *Periegesis* (Περιήγησις). Most of the quotations appear in the lexicon of Stephanus of Byzantium (nearly 300 of the 335 geographical fragments listed by Jacoby come from this source—see *Fragmente,* pp. 16–47); but since this lexicon is extant in an abridged form only and Stephanus' concern was mainly with the different forms of proper names found in ancient authors, the extracts from Hecataeus are disappointingly short and give us little more than the names of various peoples, tribes, towns, rivers, mountains, harbors, islands, and so forth mentioned in the *Periegesis.*

The latter was apparently in two books—entitled "Europe" and "Asia" (including Africa)—which are commonly cited separately; thus, typical extracts are: "Massalia [Marseilles]; a Ligurian town over against Celtic territory, a colony of the Phocaeans. Hecataeus in his 'Europe'" (fragment 55 Jacoby), "Ixibatae; a tribe near the Pontus [Black Sea] bordering on the territory of Sindica. Hecataeus in his 'Asia'" (fragment 216 Jacoby), and "Hybele; a town near Carthage. Hecataeus in his 'Asia'" (fragment 340 Jacoby). The entries in the original work certainly contained more information than this, as is clear from the handful of longer fragments from Strabo (for example, fragments 102c, 119) and Herodotus (for example, fragments 127, 300, 324). The latter especially seems to have been greatly influenced by Hecataeus' work (cf. the detailed analyses made by Bunbury, Jacoby, Pearson, and Thomson), which was evidently the chief geographical work of the fifth century B.C. It is likely that when Herodotus mentions—often critically (as, for example, his scorn for the traditional circular shape of the Ionian maps in which the river Oceanus is depicted as going round the rim, IV. 36)—the beliefs of the Greeks or Ionians, he has in mind Hecataeus, who is the only authority actually cited by name and quoted verbatim by Herodotus (cf. II. 143; VI. 137). There is some evidence, however, that Hecataeus himself also criticized his predecessors, and Herodotus was no doubt influenced by this (see Jacoby, cols. 2675–2685). The *Periegesis* seems to have contained nothing on mathematical geography or geographical theory. Rather, it apparently described briefly the main features, region by region, of the (largely coastal) areas of the Mediterranean world as then known to the Greeks, more in the manner of the later *periplus,* or coastal survey, than in the connected, expository narrative form of Herodotus, Eratosthenes, and Strabo (see Jacoby, col. 2700; Thomson, p. 88).

Hecataeus also wrote a work in four books variously cited as *Genealogiae, Historiae,* or *Heroologia,* of which Jacoby prints some thirty-five fragments which indicate that it dealt with genealogical, mythographical, and ethnographical topics, with perhaps some attention to chronological questions (Pauly-Wissowa, col. 2733 ff.).

BIBLIOGRAPHY

Fragments of Hecataeus' works appear in F. Jacoby, *Die Fragmente der griechischen Historiker,* pt. 1A (Leiden, 1957), pp. 1–47; and "Hekataios 3," in Pauly-Wissowa, *Real-Encyclopädie,* VII (1912), cols. 2667–2750. He is discussed in E. H. Bunbury, *History of Ancient Geography* (London, 1879), I, 134–155; L. Pearson, *Early Ionian Historians* (Oxford, 1939), pp. 25–108; and J. O. Thomson, *History of Ancient Geography* (Cambridge, 1948), pp. 47 ff., 79 ff., 97–99.

D. R. DICKS

HECHT, DANIEL FRIEDRICH (*b.* Sosa, near Eibenstock, Germany, 8 July 1777; *d.* Freiberg, Germany, 13 March 1833), *mathematics, mechanics.*

Virtually nothing is known about Hecht's childhood, youth, or family. It may be supposed that, born in one of the most important German mining districts of the time, he became interested in mining in his early youth and obtained an education in that subject. The course of his life can be followed more exactly from 1803, when, at the age of twenty-six, he enrolled in the Bergakademie at Freiberg, Saxony. After completing his studies he took a position as overseer (mine manager) and then taught at the Freiberger Bergschule.

Hecht's predilection and talent for solving mathematical and mechanical problems resulted in his appointment in 1816 as second professor of mathematics at the Freiberg Bergakademie, where he assumed from F. G. von Busse the lectures on elementary pure mathematics and applied mathematics (mechanics). In the following year he presented a course of lectures on theoretical mining surveying, which with a few interruptions he continued until his death. Following Busse's retirement in 1826, Hecht advanced to first professor of mathematics. He soon ceased lecturing on pure mathematics and devoted his teaching activity solely to mechanics and mining machinery, especially to contemporary mechanical engineering. From this it is evident that, for Hecht, the growing union of engineering with mathematics and physics had become mechanics. It can also be clearly seen in his book *Erste Gründe der mechanischen Wissenschaften* (1819).

Hecht was not a leading figure at the Freiberg Bergakademie, but through his great industry and his strict conscientiousness in carrying out his duties he was of great help to his students. His lectures were designed less for gifted students than for those who needed extra assistance and an external stimulus in their studies. For this and for his friendly, sincere manner Hecht won many friends.

In addition to the *Erste Gründe,* Hecht's scientific activity at the Bergakademie resulted in a number of short essays in journals and widely used high school textbooks on mathematics, geometry, and underground surveying, as well as examples and tables for mathematical calculations.

BIBLIOGRAPHY

I. Original Works. Hecht's writings include *Lehrbuch der Arithmetik und Geometrie,* 2 vols. (Freiberg, 1812–1814; II, 2nd ed., 1826); *Tafeln zur Berechnung der Seigerteufen und Sohlen für die Länge der schwachen Schnur = 1* (Freiberg, 1814); *Erste Gründe der mechanischen Wissenschaften* (Freiberg, 1819; 2nd ed., 1843); *Tafel zur Berechnung der Längen und Breiten für die Sohle = 1* (Freiberg, 1819); *Von den quadratischen und kubischen Gleichungen, von den Kegelschnitten und von den ersten Gründen der Differential- und Integral-Rechnung* (Leipzig, 1824); *Beispiele und Aufgaben aus der allgemeinen Arithmetik und gemeinen Geometrie* (Freiberg, 1824); *Einfache Construction zur Bestimmung der Kreuzlinie zweier Gänge, nebst einer Anweisung, um mit Hilfe der Kreuzlinie einen verworfenen Gang wieder aufzusuchen* (Leipzig, 1825); *Nachtrag zu den ersten Gründen der Differential- und Integral-Rechnung* (Leipzig, 1827); and *Lehrbuch der Markscheidekunst* (Freiberg, 1829).

II. Secondary Literature. See "Daniel Friedrich Hecht," in *Festschrift zum hundertjährigen Jubiläum der Königl. Sächs. Bergakademie zu Freiberg* (Dresden, 1866), pp. 22–23; and "Daniel Friedrich Hecht," in Carl Schiffner, *Aus dem Leben alter Freiberger Bergstudenten,* I (Freiberg, 1935), 244–245.

M. Koch

HECKE, ERICH (*b.* Buk, Posen, Germany [now Poznan, Poland], 20 September 1887; *d.* Copenhagen, Denmark, 13 February 1947), *mathematics.*

Hecke was the son of Heinrich Hecke, an architect. He attended elementary school in Buk and high school in Posen, then studied from 1905 to 1910 at the universities of Breslau, Berlin, and Göttingen. At Berlin he worked mainly with Edmund Landau, and at Göttingen with David Hilbert. In 1910 he obtained his Ph.D. at Göttingen. Subsequently he became Hilbert's and Felix Klein's assistant and was made *Privatdozent* in 1912. In 1915 Hecke became professor at Basel. He went to Göttingen in 1918 and in 1919 to the recently established University of Hamburg, where he was a professor until his death from cancer. He was married and had a son who died young. Hecke was a member of the editorial staff of several mathematical journals and belonged to well-known learned societies.

Most of Hecke's work dealt with analytic number theory, continuing the research of Riemann, Dedekind, and Heinrich Weber. It was Hilbert who influenced him on the subject of his thesis and some additional works on an analogue of complex multiplication, namely, the construction of class fields over real quadratic number fields by adjoining certain values of Hilbert's modular functions. The findings did not meet Hecke's expectations but nevertheless yielded new results, such as an attack on the proof of the functional equation of the Dedekind zeta function. Hecke proved in 1917 that this function can be continued throughout the complex s-plane to a single pole at $s = 1$, where it is "regular" and sufficient for a functional equation of the Riemann zeta-function type.

From this Hecke deduced the decomposition laws of divisors of discriminants for the class fields of complex multiplication. He also defined the generalized Dirichlet L-series for algebraic number fields and derived a functional equation for it. The analogue of the Dirichlet prime number law for number fields followed. Further development of these methods led him to the creation and study of zeta functions $\zeta(s, \lambda)$ with characters λ; that is, Hecke's L-series, which are of fundamental importance to advanced analytic number theory. This research was continued in various directions by Emil Artin, C. L. Siegel, and J. T. Tate.

After these studies and certain related works Hecke turned in 1925 to elliptic modular functions. He systematically applied quadratic number fields to the construction of modular functions. For imaginary quadratic fields he gave an extension of a class of functions known to Klein; for real quadratic fields there arose a new type of function. Hecke was led to these through his functions $\zeta(s, \lambda)$. Hecke then dealt with the Eisenstein series of higher order, especially the partial values of the Weierstrass functions $p(z)$ and $\zeta(z)$. He determined the periods of the Abelian integrals which are received through integration of the p partial values and certain series of the imaginary quadratic number fields. The problem of the periods of the Abelian integral of the first kind concerned him again and again, especially in connection with the representation theory of finite groups.

In 1936 Hecke systematically investigated the connection, restored by the gamma integral and the Mellin integral, of the Dirichlet series with a functional equation of Riemannian type and functions belonging to a certain automorphic group. Through his "operator" T_n Hecke established a theory for the investigation of relations of modular functions to Dirichlet series with Euler's product development. He discovered new connections between prime numbers and analytic functions and new rules for the representation of natural numbers through positive integral quadratic forms of an even number of variables. In 1939 Hans Petersson proved a rule, already anticipated by Hecke, concluding a part of this theory.

Some of Hecke's works are in another field. They are related to his approach to physics, especially the kinetic theory of gases.

BIBLIOGRAPHY

I. ORIGINAL WORKS. Hecke's *Mathematische Werke,* 1 vol., Bruno Schoeneberg, ed. (Göttingen, 1959; 2nd ed., 1970), contains his journal articles. See also *Vorlesungen über die Theorie der algebraischen Zahlen* (Leipzig, 1923; 2nd ed., 1958).

II. SECONDARY LITERATURE. See W. Maak, in "Erich Hecke als Lehrer," in *Abhandlungen aus dem Mathematischen Seminar, Universität Hamburg,* **16** (1949), 1–6; O. Perron, "Erich Hecke," in *Jahrbuch der bayerischen Akademie der Wissenschaften* (1944–1948), 274–276; and H. Petersson, "Das wissenschaftliche Werk von E. Hecke," in *Abhandlungen aus dem Mathematischen Seminar, Universität Hamburg,* **16** (1949), 7–31.

BRUNO SCHOENEBERG

HEDIN, SVEN ANDERS (*b.* Stockholm, Sweden, 19 February 1865; *d.* Stockholm, 26 November 1952), *geography.*

Hedin was the son of Abraham Ludvig Hedin, town architect in Stockholm, and the former Anna Sofia Berlin. He completed his secondary education at Stockholm in 1885, received the B.S. at Uppsala in 1888, and was awarded the Ph.D. at Halle in 1892. Honorary degrees were conferred upon him by Oxford and Cambridge in 1909, Heidelberg in 1928, Uppsala in 1935, and Munich in 1943. Hedin was elected to the Royal Swedish Academy of Science in 1905 and became one of the eighteen members of the Swedish Academy in 1913; he was also an honorary member of many learned societies and holder of forty-two gold medals. He was ennobled by the king of Sweden in 1902.

In 1885 Hedin spent six months in Baku as tutor to the children of a Swedish family, acquiring a fair knowledge of Russian, Persian, and Turkish. His savings enabled him to make a four-month journey across Persia to Bushire (Bander e Būshehr), then up the Tigris to Baghdad and Kirmanshah. In 1887 he published *Genom Persien, Mesopotamien och Kaukasus,* richly illustrated with his own drawings.

In 1889 Hedin began advanced studies in geography at Berlin under the guidance of Ferdinand von Richthofen, then the foremost expert on the geology and geography of eastern Asia. Richthofen greatly influenced the future trend of Hedin's explorations, and a lifelong friendship developed between them. Hedin interrupted his studies for a year beginning in October 1890, when he served as interpreter for an embassy from the king of Sweden to the shah of Persia; he described his experiences in *Konung Oscars beskickning till Shahen af Persien år 1890.* Afterward, Hedin journeyed through the desert of Khurasan and across the Pamir to Kashgar in Chinese Turkistan, ending with a pilgrimage to the tomb of N. M. Przhevalsky on the shore of Issyk Kul. He recounted his travels in *Genom Khorasan och Turkestan* (1892–1893).

In 1892 Hedin resumed his studies in Berlin. His dissertation, entitled "Der Demavend nach eigener Beobachtung," concerned a volcano in Persia that he had ascended in 1890. His first expedition into central Asia (1894–1897) was devoted mainly to the Tarim basin and the region surrounding the source of the Tarim in the Chinese Pamir. His principal destination was the great desert of Taklamakan, a region in which (according to ancient Chinese chronicles and local folklore) there had been rich communities, now buried under masses of sand. During his crossing of the broadest part of the desert in 1896, Hedin discovered the ruins of the town of Li Hsieh on the desiccated delta of the Chira River and other ruins farther east, which had been invaded by the steadily expanding desert in the first millennium of the Christian era. The expedition ended with a journey through northernmost Tibet, Tsaidam, Ala Shan, and Inner Mongolia to Peking. The scientific results were presented in 1900 as special publication 131 of *Petermanns Mitteilungen* with a six-sheet atlas on the scale 1 : 1,000,000, drawn by B. Hassenstein. A popular account, *En färd genom Asien 1893–1897,* appeared in 1898.

During Hedin's second expedition (1899–1902) the exploration of the Tarim basin was continued with a detailed survey of the upper and middle course of the Tarim and its hydrology over a distance of some 340 miles (this figure is based on a straight-line route, while the actual distance along the river is three times that) down to Chong Köl, a region of large swamps and lakes where the greater part of the sedimentary load of the river was deposited. Hedin showed how this deposition of silt had subjected the course of the river below Chong Köl to great changes within historical time. As late as the fourth century the river flowed along the northern edge of the basin, where Hedin mapped its desiccated bed, to its ancient delta in Lob Nor, the P'u-ch'ang Hai of the Chinese chronicles. He discovered here the ruins of the town of Lou-Lan, or Kroraina, founded by the Chinese in 260 on the ancient Middle Road. A rich treasure of manuscripts collected from the ruins showed that about 330 the lower Tarim shifted into a southeasterly course, forming a new terminal lake in the southern part of the basin: Kara Koshun, discovered by Przhevalsky in 1876.

Having concluded these researches, Hedin turned to the realization of a boyhood dream, exploration of the Tibetan highland, relating the results achieved by Przhevalsky and his co-workers in eastern Tibet to those of other European explorers in western Tibet. From Charchan in the Tarim basin he ascended the Tibetan plateau and proceeded due south, with Lhasa as his goal. He did not reach Lhasa, being stopped by the Tibetans (foreigners were not allowed to enter Tibet) north of the city and forced to turn west, under military guard, toward the Ladakh frontier. But the route taken on the northern side of the great range, to which he later gave the name Transhimalaya, passed along the "Valley of the Great Lakes" and revealed the nature of this great depression, more than 600 miles in length, which in the distant past probably constituted the extension of the Indus. The personal narrative *Asien, Tusen mil på okända vägar* appeared in 1903; the monumental *Scientific Results of a Journey in Central Asia 1899–1902,* with an atlas of eighty-four map sheets, followed in 1904–1907.

The third expedition to central Asia (1906–1908) began with a four-month comparative study of the great salt desert Dasht-i-Kavir in eastern Persia, the results of which were published in 1918–1927 as *Eine Routenaufnahme durch Ostpersien.* When Hedin turned to his main task, the exploration of Transhimalaya, he met with great difficulties and the opposition of three governments (Britain, China, and Tibet). Nonetheless, he succeeded in placing a network of route surveys all along the great mountain barrier constituting the watershed between the oceanic drainage and the drainageless Tibetan plateau. Its eastern part was known already as Nien Ch'en T'ang La, and its western part as the Kailas Range. Hedin explored and mapped the unknown middle part of the range. During his crossings, eight in all, he determined the altitudes of the principal passes and revealed fundamental features of its geological structure. These explorations also took him to the source regions of the Indus, the Sutlej, and the Brahmaputra (Tsangpo), and he located the sources of their main branches. Popular narratives of this expedition are *Overland to India* (1910) and *Trans-Himalaya* (1909–1913); the scientific results are incorporated in *Southern Tibet* (1916–1922).

On all his lone journeys Hedin carried out continuous route mapping by means of carefully measured compass traverses with astronomic control. From points along the route line he sketched panoramas of the landscape with remarkable accuracy, often taking in the whole horizon. Together the route maps and panoramas give a very clear picture of the topography. This method enabled Hedin to produce, unassisted, a picture of the geomorphology of vast areas of Tibet which would otherwise have required an elaborate topographical survey.

Hedin combined his route mapping with a systematic collection of rock specimens along all routes where rocks were exposed and made notes on their appearance in the field. Therefore the geological results of his journeys in Tibet were also pioneering. He made available the first knowledge of the widespread marine transgression over the Tibetan plateau during the later part of the Cretaceous, when the plateau was reduced to low relief. Very thick marine sedimentary sequences were deposited on the plateau. After its upheaval in post-Cretaceous time, from sea level to a height of more than 23,000 feet, large parts of this sedimentary cover were removed by erosion; but there still remain fantastically sculptured ridges and imposing massifs, the multicolored layers of which are the most beautiful element in the Tibetan landscape.

In 1926, with Hedin's organization of a scientific expedition to the northwestern provinces of China, a new epoch in the history of exploration in central Asia began, not only because of the scope of the researches by a staff of experts representing most branches of geoscience but especially because of the close cooperation with Chinese scientists and scientific institutions. How this was accomplished is recounted in *History of the Expedition in Asia 1927–1935, I–III* (Sino-Swedish Expedition publications 23–25 [1943]). The expedition was originally financed by the German Lufthansa Corporation for the ground survey of a projected air route between Berlin and Peking along the central Asiatic desert belt, to be followed by flights for scientific purposes. Unfortunately, after a year and a half Lufthansa had to withdraw for political reasons. Generous grants from the Swedish government enabled the expedition to continue its work with an enlarged scientific staff for another six and a half years, two years of which were in the service and at the expense of the Chinese government—the Suiyuan-Sinkiang highway expedition (1933–1935). Although each member of the scientific staff had great freedom in choosing his field of research and usually worked independently of the others, the expedition remained an organic and effective unit in which the task allotted to each group by Hedin was defined after careful planning with the group leaders.

Now the great central Asia desert belt had been explored from Kashgar eastward to the farthest border of the Gobi. Extensive topographical surveys based on triangulation were carried out in Inner Mongolia, eastern T'ien Shan, middle K'unlun, and northwestern Tibet. This material and all other data available were brought together in *Sven Hedin Central Asia Atlas,* issued by the U.S. Army Map Service. The archaeological researches greatly increased knowledge of the extinct cultures in the eastern part of the Tarim basin and the last shifting of the Tarim back into its ancient bed at Lou Lan in 1921 and its consequences. Most important is the detailed investigation of the ruins of Kü Yen and its extensive suburbs in the delta of Edsen Gol; with its elaborate defense system, it was the northernmost outpost of the Chinese empire in the middle of the Gobi Desert during the first centuries of the Christian era. The finding of some 10,000 bamboo manuscripts there shed much light on the organization of early Chinese colonization along the ancient "silk roads." During the Suiyuan-Sinkiang highway expedition Hedin discovered the intervening link of the ancient Middle Road between Kü Yen and Lou Lan.

By 1972 fifty-one Sino-Swedish Expedition publications had appeared: nine on geography and geodesy, nineteen on geology and paleontology, nine on archaeology, two on meteorology, four on botany, and eight on ethnography.

During the two world wars Hedin's political activities drew severe criticism from many quarters. The desire of the Russian empire for access to the Atlantic was, in his opinion, a deadly threat to the Scandinavian countries; Germany, being in a geographically similar position, seemed to him Sweden's natural ally. Hedin's pamphlet *Ett varningsord* (1912) contributed to convincing the Swedish people of the necessity for a strong defense. Subsequent political developments led to implementation of many of the causes for which Hedin had become a spokesman, such as adequate armament, prolonged military service, and winter training.

Hedin combined the qualities of a great explorer, a great writer, and a skillful artist; his vigorous health, physical endurance, powerful will, endless patience, and apparently reckless courage many times brought him through seemingly hopeless situations. Using only simple means, he blazed trails through vast unknown areas, preparing the way for trained scientists. Hedin was the last of the classical explorers of the nineteenth century, but as leader of the Sino-Swedish expedition he became one of the most active representatives of the modern trend in regional geographic research.

Hedin bequethed his entire estate to the Sven Hedin Foundation, which is affiliated with the Ethnographic Museum, Stockholm, and is sponsored by the Royal Academy of Science.

BIBLIOGRAPHY

I. Original Works. A practically complete list of Hedin's publications is in W. Hess, "Die Werke Sven Hedin's," in *Sven Hedin—Life and Letters,* I (Stockholm, 1962).

Among his writings are *Genom Persien, Mesopotamien och Kaukasus* (Stockholm, 1887); the Swedish trans. of *General Prschevalskij's forskningsresor i Centralasien* ("General Przhevalsky's Explorations in Central Asia"; Stockholm, 1889–1891), with intro.; *Konung Oscars beskickning till Shahen af Persien år 1890* ("King Oscar's Embassy to the Shah of Persia, 1890"; Stockholm, 1891); *Genom Khorasan och Turkestan,* 2 vols. (Stockholm, 1892–1893); "Der Demavend nach eigener Beobachtung," in *Verhandlungen der Gesellschaft für Erdkunde zu Berlin,* **19** (1892), 304–332; *Through Asia,* translated by J. T. Bealby, 2 vols. (London–New York, 1898); "Die geographisch-wissenschaftlichen Ergebnisse meiner Reisen in Zentralasien 1894–1897," *A. Petermanns Mitteilungen aus J. Perthes Geographischer Anstalt,* spec. pub. no. 131 (Gotha, 1900); *Central Asia and Tibet,* translated by J. T. Bealby, 2 vols. (London, 1903); *Scientific Results of a Journey in Central Asia 1899–1902,* 6 vols. and 2-vol. atlas (Stockholm, 1904–1907); *Trans-Himalaya. Discoveries and Adventures in Tibet,* 3 vols. (London, 1909–1913); *Overland to India,* 2 vols. (London, 1910); *Från pol till pol,* 2 vols. (Stockholm, 1911), abbrev. English translation, *From Pole to Pole* (London, 1912); *Southern Tibet. Discoveries in Former Times Compared With My Own Researches in 1906–1908,* 9 vols. plus 2-vol. atlas of maps and an atlas of Tibetan panoramas (Stockholm, 1916–1922); *Bagdad, Babylon, Nineve* (Stockholm, 1917); *Eine Routenaufnahme durch Ostpersien,* 2 vols. plus atlas (Stockholm, 1918, 1927); *Tsangpo Lamas vallfärd,* 2 vols. (Stockholm, 1920–1922); *Mount Everest och andra asiatiska problem* (Stockholm, 1922); *My Life as an Explorer,* translated by A. Huebsch (New York, 1925); *Across the Gobi Desert,* translated by H. J. Cant (London, 1931); *Big Horse's Flight,* translated by F. H. Lyon (London, 1936); *The Silk Road,* translated by F. H. Lyon (London, 1938); *The Wandering Lake,* translated by F. H. Lyon (London, 1940); and *Sven Hedin Central Asia Atlas* (Washington, D.C., 1952–1959; Stockholm, 1969).

So far 51 Reports From the Scientific Expedition to the North-Western Provinces of China Under the Leadership of Dr. Sven Hedin (the Sino-Swedish expedition) have appeared (1937–1971).

Hedin's political writings include *Ett varningsord* ("A Word of Warning"; Stockholm, 1912); *Germany and the World Peace,* translated by G. Griffin (London, 1937); and *Sven Hedin's German Diary 1935–1942,* translated by J. Bulman (Dublin, 1951).

II. Secondary Literature. See N. Ambolt and E. Norin, "Sven Hedin's Explorations in Central Asia 1893–1908 and 1927–1935," in *Memoir on Maps, I,* Report from the Scientific Expedition to the North-Western Provinces of China no. 48 (Stockholm, 1967), 14–51; a congratulatory volume on the occasion of Hedin's seventieth birthday, *Geografiska annaler,* **17** (1935); Alma Hedin, *Mein Bruder Sven* (Leipzig, 1925); W. Hess, "Die Werke Sven Hedin's" (see above); S. Linné, "Sven Hedin and the Ethnographical Museum of Sweden, Stockholm," in *Ethnos,* **30** (1965), 25–38; G. Montell, "Sven Hedin—the Explorer," *ibid.,* 1–24; E. Norin, "Sven Hedins forskningsresor i Centralasien och Tibet" ("Sven Hedin's Explorations in Central Asia and Tibet"), in *Geografiska annaler,* **36** (1954), 9–39; and S. Selander, *Sven Hedin. Inträdestal i Svenska Akademien* (Stockholm, 1953); and *Sven Hedin. En äventyrsberättelse* ("Sven Hedin. A Tale of Adventure"; Stockholm, 1957).

ERIK NORIN

HEDWIG, JOHANN (*b.* Kronstadt [now Braşov], Transylvania [now Rumania], 8/10 December 1730; *d.* Leipzig, Germany, 7 February 1799), *botany.*

Hedwig's father, Jakob Hedwig, a town councillor, was probably a wine merchant by trade; his mother was Agnes Galles. He attended schools in his native town, in Pressburg (now Bratislava, Czechoslovakia), and Zittau, Germany. In 1752 he entered the University of Leipzig, receiving the bachelor's degree in 1756 and the M.D. in 1759. His studies included philosophy and mathematics as well as the medical sciences; and among his teachers, who were impressed with his character and abilities, were the eminent botanists J. E. Hebenstreit, C. G. Ludwig, and G. R. Böhmer. The professor of botany, E. G. Bose, invited Hedwig to lodge in his home and employed him for three years as his assistant at the hospital; without this help he might have been unable to complete his studies, for since his father's death in 1747 his family could no longer afford to pay all his expenses.

After graduation Hedwig wished to practice medicine in Kronstadt but was disappointed to find that a medical degree from the University of Vienna was required. On the advice of a friend in Chemnitz, Saxony, he decided to settle in that town, shortly afterward marrying Sophie Teller, daughter of Romanus Teller, professor of theology and minister of the Thomaskirche in Leipzig.

Although a busy and successful medical practitioner, Hedwig gave much time to the study of plants, in which he had been passionately interested since childhood. He began work at dawn and often spent several hours botanizing in the country before visiting his patients, examining his collections at the end of the day. He became increasingly occupied with the mosses and liverworts, but a shortage of books and equipment due to lack of money were at first a handicap: his only book on the lower plants was a "meagre excerpt" from J. J. Dillenius' *Historia muscorum* (1741). J. C. D. Schreber, who himself made important contributions to the knowledge of mosses, later encouraged him and provided him with books, and J. G. Köhler, inspector of mathematical instruments at Dresden, gave Hedwig an excellent compound microscope made by Rheinthaler of Leipzig. The death of his wife in 1776 and the problems of caring for his six surviving children interrupted Hedwig's botanical work for a while, but in 1778 he was persuaded by his friends to marry again, and his second wife, Clara Benedicta Sulzberger of Leipzig, worked actively to promote his scientific career as well as his personal welfare. She bore him six more children, of whom five died in early childhood and one at the age of sixteen.

In 1781, at his wife's suggestion, and in the interests of his botanical work and his children's education, Hedwig moved to Leipzig. There he continued to practice medicine and the following year published his *Fundamentum historiae naturalis muscorum frondosorum,* the first of his works to attract wide attention in Germany and abroad.

Having thus achieved international recognition after years of poverty and neglect, Hedwig's abilities now began to receive their due. In 1784 he was given charge of the military hospital at Leipzig, and in 1791 he became medical officer of the Thomasschule. In 1786 the university made him extraordinary professor of botany, and in 1789 he succeeded J. E. Pohl as ordinary professor, a post which carried with it the directorship of the botanical garden and an apartment in the Academy building. Because he did not hold the M.A. degree Hedwig was ineligible for a chair under the university regulations and was appointed only after the intervention of Friedrich August I, elector of Saxony, himself a keen botanist, who urged that a rule intended to exclude unsuitable candidates should not be used to debar a man of recognized merit.

Hedwig's fame spread widely in his own country and abroad. In 1783 he won a prize offered by the Russian Academy of Sciences for work on the reproduction of cryptogamic plants; his thesis was published the following year in St. Petersburg under the title *Theoria generationis et fructificationis plantarum cryptogamicarum Linnaei.* He was elected to membership in several German and foreign academies and scientific societies, including the Royal Society of London, of which he became a fellow "on the foreign list" in 1788. On 8 January 1797 Goethe visited Hedwig, who showed him "beautiful preparations and drawings."

During the exceptionally cold winter of 1798–1799 Hedwig continued to visit his patients as usual, without taking adequate care of his own health. Having barely recovered from a "catarrhal fever," he contracted a "nervous fever" of which he died nine days later.

Among Hedwig's several important contributions to science, the best known and probably most significant was the better understanding of the life history and reproduction of the lower plants resulting from his observations on mosses. At the time this work was being done, very little was known about the sexual reproduction of the lower plants, hence Linnaeus' name for them, Cryptogamia ("plants of the hidden marriage"). Erroneous ideas were current, mainly owing to misguided attempts to interpret the functions of the reproductive organs of ferns and mosses by reference to the relatively well-understood reproductive parts of flowers. Thus Linnaeus and most of Hedwig's contemporaries believed that because of their similarities to the pollen of the higher plants, the spores of mosses functioned as male organs and that the spore capsule (sporogonium) was therefore the equivalent of the anther.

Hedwig, in a preliminary paper, "Vorläufige Anzeige seiner Beobachtungen von den wahren Geschlechtstheilen der Moose" (1779), showed that the much smaller and less conspicuous antheridia (which he termed anthers) were the true male organs and that the minute cells emitted from them are the male gametes and fertilize the archegonia (pistilla of Hedwig). He also observed the germination of the spores and the growth from them of the filamentous protonema (cotyledons of Hedwig), which is the juvenile form of the moss "plant" or gametophyte. The views which he had already formed on the true functions of the antheridia were finally confirmed by his observation, on 17 January 1774, of "Kügelchen" (minute particles, that is, sperm cells) being discharged from the antheridia of the moss *Grimmia pulvinata.* Some years earlier a similar observation had been made on the liverwort *Fossombronia* by C. C. Schmidel, who also realized that the antheridia were the male organs but did not pursue his observations with the same thoroughness as Hedwig.

The observations of Schmidel and Hedwig did much more than merely demonstrate the true functions of the antheridia and spores in bryophytes: they were crucial in preparing the way for the fundamental work of Wilhelm Hofmeister, who eventually elucidated the life histories of the higher cryptogams and demonstrated the true homologies between their reproductive organs and those of flowering plants.

Unfortunately, Hedwig's studies of ferns were less successful than his work on mosses. He mistook the glandular hairs on the leaves of certain ferns for antheridia, not having understood the importance of the prothalli (gametophytes) which bear the true antheridia. It was left to Karl von Naegeli to discover the antheridia and sperm cells of ferns in 1844.

Hedwig's views on the sexuality of mosses did not receive immediate or unanimous acceptance: for some years after they were published, some of his contemporaries were maintaining that the mosses were "viviparous," or that they had no form of sexual reproduction; even in 1806 Ambroise Palisot de Beauvois put forward views completely at variance with the observations of Schmidel and Hedwig.

Among Hedwig's other contributions on the lower plants were his clear distinction between mosses and liverworts and his demonstration of the value, in the classification of mosses, of the peristome, the minute, elaborately constructed teeth which surround the mouth of the spore capsule. His classification of mosses, largely based on peristome characters, is embodied in his *Species muscorum frondosorum* (1801), in which all the then known species of mosses are described. This work, now used as the basis of the modern scientific nomenclature of mosses, was edited and published after Hedwig's death by C. F. Schwaegrichen, who later added several supplements.

Hedwig worked on fungi and other lower plants as well as mosses and wrote extensively about the microscopic structure of higher plants, although he frequently misinterpreted his observations and contributed little of permanent value in this field. He was the first to describe the stomata of flowering plants (although Marcello Malpighi and others had previously seen them in ferns). He observed their opening and closing and had a fair understanding of their function.

His interests were not limited to academic science. Hedwig wrote also on practical subjects, such as the liver fluke disease of sheep and the value of false acacia (*Robinia*) timber as firewood; he even wrote a reply to the inquiries of the English agricultural reformer Arthur Young on the irrigation of meadows with spring water and the cause of mildew in wheat.

Distinguished above all for exact and patiently repeated observations, Hedwig's work depended on the skillful use of dissection and the compound microscope. At first he used a simple lens magnifying 6X, but by successive improvements to the compound microscope given to him by Koehler he was eventually able to use magnifications of up to 290X. In his *Fundamentum* (ch. 2, pp. 9–11) he explains his method of dissecting mosses with the aid of needles and small knives, mounting the preparations in drops of water on glass slides—essentially the technique used today. He recorded his observations in accurate drawings. Although he did not teach himself to draw until the age of forty, the figures in his *Descriptio* (1787–1797) are among the most accurate and beautiful illustrations of mosses of his own or any other period.

Hedwig was a good teacher, and his character endeared him to his family, friends, and students. On the field excursions which he organized for his students he is said to have been indefatigable, but nevertheless the excursions were regarded as a pleasure rather than as an imposed task.

Hedwig's attitude toward his botanical work is indicated by the mottos with which he prefaced his works. The *Fundamentum* has a line from Cicero (inaccurately quoted): "Opinionis commenta delet dies, naturae iudicia confirmat" ("The passage of time obliterates the fabrications of opinion, but confirms the judgments of nature"). The motto of the *Descriptio* is a quotation from Dillenius which indicates that, like most of his predecessors, Hedwig believed that the aim of botanical research such as his own was to understand better the wisdom of the Creator.

BIBLIOGRAPHY

I. Original Works. A complete list of Hedwig's writings (32 items) by C. F. Schwaegrichen forms an app. to Hedwig's posthumous *Species muscorum frondosorum,* pp. 318–327. His major works are "Vorläufige Anzeige seiner Beobachtungen von den wahren Geschlechtstheilen der Moose und ihrer Fortpflanzung durch Saamen," in *Sammlungen zur Physik und Naturgeschichte von einigen Liebhabern dieser Wissenschaften,* **1** (1779), 259–281; *Fundamentum historiae naturalis muscorum frondosorum,* 2 vols. (Leipzig, 1782); *Theoria generationis et fructificationis plantarum cryptogamicarum Linnaei* (St. Petersburg, 1784); *Descriptio et adumbratio microscopico-analytica muscorum frondosorum nec non aliorum vegetantium e classe cryptogamica Linnaei,* 4 vols. (Leipzig, 1787–1797), sometimes called *Stirpes cryptogamicae novae; Theoria generationis et fructificationis plantarum cryptogamicarum Linnaei,* rev. and enl. ed. (Leipzig, 1798); and *Species muscorum frondosorum,* C. F. Schwaegrichen, ed. (Leipzig, 1801), repr. in the series Historiae naturalis Classica, J. Cramer and H. K. Swann, eds., with intro. by P. A. Florschütz (Weinheim-Codicote, Hertfordshire-New York, 1960).

II. Secondary Literature. See J. P. F. Deleuze, "Notice sur la vie et les ouvrages d'Hedwig," in *Annales du Muséum national d'histoire naturelle,* **2** (1803), 392–408, 451–473; H. Dolezal, "Hedwig, Johann, Botaniker," in *Neue deutsche Biographie,* VIII (1969), 191–192, which gives further references; P. A. Florschütz, intro. to repr. of Hedwig's *Species muscorum frondosorum* (see above); I. Györffy, "Zum Andenken an Joannes Hedwig am zweihundertsten Jahreswechsel seiner Geburt," in *Revue bryologique et lichénologique,* n.s. **57** (1930), 161–165, which has a photograph of the parish register with the entry for Hedwig's baptism; W. D. Margadant, "Early Bryological Literature," doctoral thesis Univ. of Utrecht (Hunt Botanical Library, Pittsburgh, Pa., 1968), see pp. 139–144 for biography and bibliographical account of Hedwig's *Species muscorum frondosorum;* J. Römer, "Aus dem Leben eines Microskopikers der Linneschen Zeit. Eine historische Studie," in *Mikrokosmos,* **2** (1908–1909), 91–97; and C. F. Schwaegrichen, "Hedwigii vita," app. to Hedwig's *Species muscorum frondosorum* (see above).

P. W. Richards

HEER, OSWALD (*b.* Niederutzwyl, St. Gallen, Switzerland, 31 August 1809; *d.* Zurich, Switzerland, 27 September 1883), *paleontology, botany.*

Heer was the son of a Protestant minister who educated his son and prepared him for university study. As a boy and young man Heer collected plants and insects in the mountains near Matt, in the canton of Glarus, where the family moved in 1811, and he exchanged samples with other collectors. Following the family tradition, he began to study theology in 1828 at Halle and took his final examinations in this course at St. Gallen. Although deeply religious, he declined to become a minister.

While at the University of Halle, Heer had been in contact with professors of the natural sciences; and from 1828 he devoted himself completely to these sciences, undertaking in 1832 the examination and cataloging of a large, private insect collection in Zurich. This work led him to decide on a scientific career. He qualified in 1834 as a *Privatdozent* in botany at the newly founded University of Zurich and at the same time undertook the direction of the botanical gardens. He was named associate professor in 1835, and in 1852 he became full professor of botany and entomology. Beginning in 1855 he also taught taxonomic botany at the Technische Hochschule in Zurich. For almost fifty years Heer engaged in rich and fruitful teaching activity. His lectures treated taxonomic botany, pharmaceutical and economic botany, and, later, paleobotany and entomology—the latter dealt particularly with beetles and fossil insects.

In his youth Heer was a tireless traveler and an excellent mountain climber. In his fortieth year he contracted pulmonary tuberculosis, which occurred frequently in his mother's family. Spending the winter of 1850–1851 on Madeira cured him for two decades, but in the winter of 1871 he became sick once again. This time he also suffered from a tubercular ailment of the leg, which put an end to all mountain excursions and confined him to his room every winter. He accepted his fate and continued to work until his death. He had remained unmarried. Heer's biographers have emphasized his integrity, truthfulness, winning goodness, and gentleness. He constantly placed himself and his scientific knowledge at the service of the community. He gave popular lectures; founded, with the botanists K. W. von Naegeli and E. A. Regel, the Verein für Landwirtschaft und Gartenbau, of which he was president for eighteen years; wrote on the extermination of the cockchafer and on the economic situation of the canton of Glarus; and for eighteen years was a member of the canton council.

The principal areas of Heer's research were paleobotany, plant geography, and entomology of living and fossil insects. His first major botanical work, on the vegetation of the canton of Glarus (1835), was the first monograph on plant geography of the Swiss Alps. Through its richness in new facts and ideas and through the precision with which the locations of the plants and their distribution in each high-altitude region were determined, this publication became a classic foundation for all later works of a similar nature. The same is true of Heer's studies on the highest limits of animal and plant life in the Alps. The first publication on this subject appeared in 1845; at the end of his life he returned to it once more, on the broadest basis, in his study *Über die nivale Flora der Schweiz* (1883).

World fame first came to Heer for his paleobotanical investigations, especially of Tertiary flora. The titles alone indicate the remarkable range of his chief works in this area: *Flora tertiaria Helvetiae* (1855–1859), *Flora fossilis Helvetiae* (1876), and *Flora fossilis arctica* (1868–1883)—the last, according to Adolf Engler, the most important paleobotanical publication which had appeared until then. In these volumes 2,632 plant species are described, including 1,627 new species. Yet Heer never considered the increase in the number of fossil plant types as the main goal of his work. Rather, he constantly sought to synthesize the countless individual observations and facts into the evolution of the plant world and its environment, especially from the Tertiary to the present.

In the *Flora fossilis arctica* Heer described rich and varied Cretaceous and Tertiary floras from the north polar regions. This rich variety led him to maintain that the Arctic was a center of new formations from which plants radiated south to America, Europe, and Asia. These migrations of new plant species and groups, which originated in the Arctic and then became closely related "vicarious" species through simultaneous differentiation, explained in a simple manner the previously mysterious similarity of tree and shrub vegetation in two such distant regions as eastern Asia and Atlantic North America. In a similar way the theory could also account for the resemblance, already recognized by Heer, of the Tertiary flora of Europe and the recent flora of eastern Asia and Atlantic North America. While this "arcto-Tertiary" flora survived in the latter areas until the present, in Europe north of the Alps it was almost completely crowded out or destroyed during the Pleistocene. Only in the Mediterranean area and on the Canary and Madeira Islands in the Atlantic did remnants of this Tertiary flora survive.

In his investigations of the Tertiary flora Heer found that even in deposits of a similar geologic age the plant remains by no means always displayed the same composition; rather, localities in, for example, Greenland, Spitsbergen, and Italy showed decided differences. Hence in the Miocene the northern limit of palm trees lay in central Germany, where evergreen forests still existed, while in Iceland at the same period deciduous trees and conifers flourished, serving to indicate a cooler climate. This poleward organization of vegetation belts led Heer to postulate corresponding climatic zones during the Tertiary. He even attempted to determine the average yearly temperatures in the various zones from the climatic requirements of the living relatives of the Tertiary trees. He likewise confirmed by means of his paleobotanical findings the fact, already known from the study of the marine mollusks, that the temperature had gradually decreased in the course of the Tertiary.

Heer's paleobotanical researches were not confined to the Tertiary. In *Flora fossilis Helvetiae* and *Flora fossilis arctica,* as well as in many articles, he described plant remains from the Carboniferous, Triassic, Jurassic, and Cretaceous periods and remarked on their paleobiogeographical and paleoclimatic significance. Heer also considered Pleistocene floras. For instance, he demonstrated by the discovery of *Betula nana* and other northern species of plants in Bovey Tracey, Devonshire, a cold glacial climate for southern England. From Pleistocene coal deposits at Dürnten and Wetzikon, Switzerland, he deduced a temperate interglacial climate for the Alps. And he found among the plant remains of the post-Ice Age Swiss lake dwellings connections with present-day cultivated plants.

As a child Heer had collected insects as well as plants and consequently was well acquainted with the close ecological relationships between the two groups of organisms. He published several works on the recent forms but increased knowledge of the world of fossil insects in a way that none of his predecessors or contemporaries had done. The chief locality he studied was Oeningen, on the Lake of Constance (southern Germany). From this Upper Miocene deposit, once a freshwater lake, he determined 826 insect species, with Coleoptera, Hymenoptera, Neuroptera (dragonfly larvae), and Hemiptera predominating. Simultaneously with Oeningen he worked at somewhat older Radoboj, in Croatia (1847–1853). He investigated the Lower Oligocene deposit at Aix-en-Provence in 1857. Of special significance was Heer's discovery of an insect fauna in the Lias at Schambelen in the canton of Aargau. In 1852 he described 143 species from this place and thus made a contribution that is still valuable to knowledge of the Mesozoic insect world.

As with fossil plants, recognition and classification of new species of fossil insects was not the ultimate goal of Heer's studies. Starting from this necessary systematic-taxonomic base he attempted to reconstruct the natural conditions of existence of the prehistoric insects, their relationships to the surrounding plant world, and their organization into life communities while making a constant comparison with the conditions of life of the related recent species. Again Heer showed himself to be a master of paleoecology. As in his paleobotanical studies, there is a consideration of biogeography. Thus, in the Oligocene insect fauna of Aix-en-Provence he recognized a Mediterranean fauna with particular suggestions of North American, Indian, and Australian characteristics.

Heer was at the same time a botanist and a zoologist, a penetrating systematist with an enormous knowledge of forms; yet he was also a biologist who sought to grasp the richness of forms and the expressions of life of both the great kingdoms of nature as a harmonious whole. Evidence for this disposition is provided by his *Urwelt der Schweiz,* an extraordinarily vivid description of the geological history and the animal and plant world of Switzerland since the Paleozoic. For its union of scientific thoroughness and polished presentation, the book belongs among the classic works of this type. It appeared in two editions (1865, 1879), the first of which was translated into French and English.

In his paleontological researches Heer took issue with the theory of evolution, especially with Darwin's ideas. He maintained that species in general were constant; nevertheless, at certain times in the history of the earth, during so-called periods of creation, species possessed the ability to bring forth resemblant species. Thus the present flora would have arisen from that of the Tertiary; the *Sequoia langsdorffi* would be the great-grandfather of the *Sequoia sempervirens,* the *Liquidambar styracifluum* would have originated from the *Liquidambar europaeum,* and so on; and, at an earlier time, the Tertiary flora would have developed from that of the Cretaceous. This whole plant world would thus have formed a great, harmonious totality in which all members stood in a genetic relationship.

Heer wrote this theory in 1855—four years before the appearance of Darwin's *Origin of Species*—in a letter to his friend C. T. Gaudin. Although with these conceptions he came very close to Darwin's theoretical ideas on descent, he decisively rejected the latter's theory of selection. On the one hand, a theory of chance contradicted his religious belief in the existence of a plan for the world made by an omnipotent Creator; on the other hand, in his investigations of fossil plants he was never able to find the "gradual and ever regularly continuing, purposeless metamorphosis of species" presupposed by Darwin. Heer was inclined to accept a relatively sudden, erratic transformation of species.

BIBLIOGRAPHY

I. ORIGINAL WORKS. Heer's writings include *Beiträge zur Pflanzengeographie* (Zurich, 1835), his diss.; "Die Insektenfauna der Tertiärgebilde von Oeningen und von Radoboj in Croatien," in *Neue Denkschriften der Allgemeinen schweizerischen Gesellschaft für die gesamten Naturwissenschaften,* **8** (1847), 1–229; **11** (1850), 1–264; **13** (1853), 1–138; "Die Lias-Insel des Aargaus," in *Zwei geologische Vorträge . . . von O. H. und A. Escher von der Linth* (Zurich, 1852), pp. 1–5; *Flora tertiaria Helvetiae,* 3 vols. (Winterthur, 1855–1859); "Ueber die fossilen Insekten von Aix in der Provence," in *Vierteljahrsschrift der Naturforschenden Gesellschaft in Zurich,* **2** (1857), 1–40; *Die Urwelt der Schweiz* (Zurich, 1865; 2nd ed., 1879), 1st ed. trans. into French as *Le monde primitif de la Suisse* (Geneva-Basel, 1872) and into English as *The Primeval World of Switzerland* (London, 1876); *Flora fossilis arctica,* 7 vols. (Zurich, 1868–1883); *Flora fossilis Helvetiae* (Zurich, 1876); and *Ueber die nivale Flora der Schweiz* (Zurich, 1883).

II. SECONDARY LITERATURE. See J. Heer and C. Schröter *et al., Oswald Heer: Lebensbild eines schweizerischen Naturforschers,* 2 vols. (Zurich, 1887), with portrait; A. Jentzsch, "Gedächtnisrede auf Oswald Heer," in *Schriften der Physikalisch-ökonomischen Gesellschaft zu Königsberg,* **25** (1885), 1–26, with complete bibliography, also in *Leopoldina,* **21** (1885), 18–20, 22–30, 42–49; K. Lambrecht and W. and A. Quenstedt, "Palaeontologi. Catalogus bio-bibliographicus," in *Fossilium catalogus,* I, pt. 72 (The Hague, 1938), 194–195; R. Lauterborn, "Der Rhein. Naturgeschichte eines deutschen Stromes," in *Berichte der Naturforschenden Gesellschaft zu Freiburg im Breisgau,* **33** (1934), 134–141; K. Mägdefrau, "Schneetälchen, Käfer und fossile Pflanzen," in *Heimat,* **67** (1959), 143–149 (with portrait); and G. Malloizel, *Oswald Heer: Bibliographie et tables iconographiques; précédé d'une notice bibliographique par R. Zeiller* (Stockholm, 1887), which contains a complete bibliography and list of the fossil animals, insects, and plants described and illustrated by Heer—references include page, plate, and figure number for each species.

HEINZ TOBIEN

HEFNER-ALTENECK, FRIEDRICH FRANZ VON (*b.* Aschaffenburg, Germany, 27 April 1845; *d.* Berlin, Germany, 7 January 1904), *engineering.*

Hefner-Alteneck was the son of art historian Jakob Heinrich von Hefner-Alteneck and Elise Pauli. After completing his education at the technical high school of Zurich, a departure from the literary-artistic traditions of his family, Hefner-Alteneck spent most of his professional life (1867–1890) as a design engineer and inventor in the employ of Siemens and Halske, electrical apparatus manufacturers in Berlin. The profitability of his inventions was rewarded by his meteoric rise in the company's technical staff and, through profit sharing, exceptional remuneration. The technical merit of his work earned him an international professional reputation. Hefner-Alteneck was a member of the Elektrotechnische Verein (president 1893–1894 and 1897–1898, honorary fellow 1900), a fellow of the Royal Swedish Academy of Sciences (1896), and a fellow of the Prussian Academy of Sciences (1901). He received an honorary Ph.D. from the University of Munich in 1897.

His first great success (1872) was the drum armature principle for dynamos, which is still in use. By placing the active conductors entirely on the peripheral surface of the rotating armature, he greatly increased the efficiency and output relative to existing designs using the Gramme-Pacinotti ring armature or the Siemens double-T armature.

Hefner-Alteneck's other inventions of major importance were a mechanical dynamometer to measure power transmitted by a drive belt, based on tension differential between the two sides of the belt (1872); a teletypewriter with alphanumeric keys to generate Morse code signals for telegraph transmission (1873); a two-current differential regulator for arc lamps which facilitated series and parallel connection of the lamps and led immediately to extensive street-lighting installations (1878); and an amyl acetate lamp as a unit of luminous intensity which, under the name "Hefner candle," was standard in Germany for more than forty years (1884). Other inventions included electric telemetering of remote water-level data; an electric servomechanism for ships' telegraphs; an automatic fire-alarm signaling system; and a method for railway block signaling. His work on the important rotating-field generator principle was done with Karl Hoffmann.

Although Hefner-Alteneck's inventions created no new fields of technology, they enormously accelerated the early growth of the electrical industry. Many of them were improvements on prior conceptions of his employer, Werner Siemens, but they were so revolutionary that Siemens himself was always first to assign credit to his design engineer.

When Siemens gave control of the company to his sons in 1890, Hefner-Alteneck rebelled at being their subordinate and requested a partnership. This was refused, and he retired on full pension at age forty-five. He continued to be active in professional societies, and from 1898 to 1904 he served on the advisory board of AEG, the German General Electric Company. His friendship with the Siemens family survived their differences, and it was while holidaying at the home of Werner's son Wilhelm that he died suddenly of a cerebral hemorrhage in 1904.

BIBLIOGRAPHY

Apart from patent records, the principal source of public information about Hefner-Alteneck's work is in twenty-five reports in the *Elektrotechnische Zeitschrift* (*ETZ*) between 1880 and 1902, concerning his presentations to the Elektrotechnische Verein. In many cases these reports were made several years after the effective dates of the inventions. Representative papers in *ETZ* include "Ueber eine neue dynamoelektrische Maschine für kontinuirlichen Strom," **2** (1881), 163–170; "Ueber elektrische Beleuchtungs-Versuche in den Strassen Berlins," **3** (1882), 443–450; "Vorschlag zur Beschaffung einer konstanten Lichteinheit," **5** (1884), 20–24; "Ueber Arbeitsmesser," **8** (1887), 514–517; and "Vorschläge zur Aenderung unseres Patentgesetzes," **21** (1900), 278–279.

The only substantial biography is Friedrich Heintzenberg, *Friedrich von Hefner-Alteneck* (Munich, 1951).

ROBERT A. CHIPMAN

HEIDENHAIN, MARTIN (*b.* Breslau, Germany [now Wrocław, Poland], 7 December 1864; *d.* Tübingen, Germany, 14 December 1949), *microscopic anatomy, microtechnique.*

Both of Heidenhain's parents were members of families which included well-known physicians and university professors. His father was the renowned physiologist Rudolf Heidenhain, who taught for thirty-eight years at the University of Breslau and was himself the son of a prominent Prussian physician; his mother, Fanny Volkmann, was the daughter of the anatomist and physiologist Alfred Volkmann, professor at the University of Halle, where Rudolf Heidenhain had studied. Four of Heidenhain's uncles were physicians, and his brother was the well-known surgeon Lothar Heidenhain, who taught and practiced in Breslau.

Heidenhain displayed interest in the natural sciences, especially geology and paleontology, while he was still a student at the Gymnasium in Breslau. Following graduation he studied biology at the University of Breslau and later in Würzburg. He then studied medicine in Freiburg im Breisgau, where he obtained the M.D. in 1890. Instead of practicing medicine, he became an assistant to the anatomist Rudolf Kölliker at Würzburg, and from there he moved to a teaching position at the University of Tübingen in 1899. He remained at Tübingen for the rest of his life and assumed the post of professor of anatomy in 1917, following the death of August von Froriep. While in Würzburg, Heidenhain married Anna Hesse, the daughter of a lawyer. The couple had three sons, all of whom they survived.

In the course of his professional life as a teacher and researcher, which spanned some fifty years, Heidenhain produced about 100 publications, including several books. He taught microscopy, embryology, and anatomy. His principal areas of research were microscopic anatomy (for example, the structure of heart and skeletal muscle, taste buds, salivary glands, and the thyroid, as well as the comparative anatomy and developmental history of the kidney) and the development of microtechnique. It is owing to his activity in the latter field that Heidenhain's name is known to all present-day histologists and cytologists. In 1891 he discovered the still widely used iron-hematoxylin staining method which bears his name. Subsequently he invented the mercuric chloride method of tissue fixation and pioneered the use of aniline dyes for the staining of tissues. Heidenhain was not satisfied with the application of then established methods of microscopic preparation to his particular research problems. Rüdiger von Volkmann, in an article honoring Heidenhain's seventieth birthday and dealing with his accomplishments in microtechnique, records the following quotation by Heidenhain: "I have always made up my own methods, just as I needed them."

Heidenhain's magnum opus is the two-volume cytology text *Plasma und Zelle* (1907–1911), which he had been invited to contribute to Heinrich von Bardeleben's *Handbuch der Anatomie des Menschen.* In this work Heidenhain attempted to develop a synthetic theory of morphogenesis based on a hierarchic arrangement of levels of organization, starting with hypothetical subcellular "protomeres" and ascending to macroscopic structures, the various levels being subject to and integrated by one set of natural laws. Although stimulated by the writings of the philosopher of biology Hans Driesch, Heidenhain rejected Driesch's metaphysical speculations and attempted to explain the general laws which govern living material on the basis of microscopic findings and researches into the developmental history of organisms and their parts. Heidenhain's efforts in the area of theoretical biology were appreciated by his contemporary Wilhelm Roux and by a subsequent generation of scientists represented by, among others, the anatomists Alfred Benninghoff, Hermann Bautzmann, and Walther Jacobj.

Heidenhain was a pleasant and congenial man, well liked by his colleagues, assistants, and students. His was not a one-track mind devoted solely to the pursuit of scientific knowledge; it was open also to other cultural concerns. He appreciated and collected old German wood carvings and potteries.

BIBLIOGRAPHY

A complete bibliography of Heidenhain's writings has been published by Walther Jacobj in *Anatomischer Anzeiger,* **99** (1952–1953), 89–94.

An appreciation of Heidenhain's accomplishments in microtechnique was published by Rüdiger von Volkmann in *Zeitschrift für Mikroskopie,* **51** (1934), 309–315. Heidenhain's student Jacobj published an obituary in *Anatomischer Anzeiger,* **99** (1952–1953), 80–89. A short biographical note on Heidenhain can be found in *Neue deutsche Biographie,* VIII (1969), 247.

MAX ALFERT

HEIDENHAIN, RUDOLF PETER HEINRICH (*b.* Marienwerder, East Prussia [now Kwidzyn, Poland], 29 January 1834; *d.* Breslau, Germany [now Wrocław, Poland], 13 October 1897), *physiology, histology.*

Among the physiologists of the second half of the nineteenth century Heidenhain has a special position as an independent worker and thinker not influenced by, and often opposed to, the modish currents of the time, especially to the tendency of oversimplification in the explanation of vital phenonomena and to the effort to reduce them to fairly simple physical and chemical processes. He was oriented more toward biological conceptions of vital phenomena than to their mathematical and physical interpretations. He did not trust preconceived opinions or theories, relying instead on the results of his experiments and on the inductive method.

Heidenhain was the eldest of twenty-two children of a physician. He very early revealed his talent and perseverance in work; his diligence in collecting plants and animals indicated his interest in the study of nature. After he completed his secondary education in his native town at the age of sixteen, he began the study of nature on an estate near his home but soon turned to medicine at the University of Königsberg; in this he was guided by his father, who had great influence on his early work and decisions. It was fairly common in Germany to attend several universities for undergraduate study, so Heidenhain went after two years to Halle, where A. W. Volkmann, one of the leading German physiologists of the time, turned his interest to physiology. After another two years he went to Berlin, where he finished his medical studies at the astonishingly early age of twenty years with a dissertation entitled *De nervis organisque centralibus cordis, cordiumque lymphaticarum ranae* (1854), which had been inspired by Emil du Bois-Reymond. In his dissertation Heidenhain refuted the opinion advanced by Moritz Schiff that the vagus nerve initiates the rhythmic contractions of the heart, demonstrating in his experiments that its function is to regulate heart activity; the automatic activity seemed to him to originate in the ganglia of the heart. Heidenhain remained with du Bois-Reymond at Berlin, working on the problem of the tonus of skeletal muscles and some other questions of nerve and muscle physiology. The results were published in his *Physiologische Studien* (Berlin, 1856).

In 1856 Heidenhain returned to Volkmann's laboratory in Halle. The following year he submitted a *Habilitationsschrift* on the determination of the blood volume in the bodies of animals and men, improving on Hermann Welcker's method. He married Volkmann's daughter Fanny in January 1859, shortly before he assumed the chair of physiology at Breslau. Students at first revolted against the "unknown and green young professor," and some older professors also showed their resentment: "One cannot have much respect for the discipline which can be represented by a twenty-five year-old teacher and master." Nevertheless, through his diligence and competence, Heidenhain soon won respect and became one of the most illustrious members of the Breslau Medical Faculty, where he broadened his research activity and maintained it almost until his death in 1897.

In Breslau, Heidenhain continued his work on muscle and nerve, still under the influence of du Bois-Reymond. His most important accomplishment was the measurement of heat production during muscle activity. Although production of heat during a longer tetanic contraction had been found since 1805 by several observers, Heidenhain was the first to detect, by direct sensitive thermoelectric measurements, a minute increase in temperature (0.001–0.005 °C.) during every simple twitch. He then studied the production of muscle heat under different conditions: for instance, the influence of the intensity of stimulation and of fatigue, and the relation of the heat produced to the work performed by a muscle lifting different loads. His most important finding was that the total energy output (heat and mechanical work) increases with increasing load (increasing active tension), an unexpected result. It showed that muscle liberated more energy when the resistance to its contraction was greater—that there is a kind of self-regulation of the energy expenditure in the working muscle—and thus that the muscle's work is very economical. When fatigue sets in, the work becomes even more economical. Thermoelectric measurement has since become an important and widely used research tool in muscle physiology, for which Heidenhain's classic work helped to form a basis.

Among Heidenhain's other important findings in muscle and nerve physiology were that of increased acid formation in the working muscle and that of special motor reactions produced by stimulation of a sensory nerve after severance and degeneration of the motor nerve (1883), the so-called pseudomotor phenomenon, which was not explained until much later.

Heidenhain used the thermoelectric method in many other investigations, but it was not sensitive enough to detect heat production in nerve or in the brain during activity induced by stimulation of a sensory nerve. The latter gave a paradoxical result found to be a consequence of vascular reactions; this led Heidenhain to further investigations of vasomotor reactions, a topic to which he and his pupils repeatedly turned. He contributed greatly to the knowledge of vascular reactions in the skin, muscles, and glands and to the problem of cardiovascular regulation in the early period of their study.

In 1867 Heidenhain began systematic studies of the physiology of glands and of the secretory and absorption processes, which remained his chief field of interest for the rest of his life. His great advantage in these studies was his knowledge and practical experience in histology. Thus he noticed morphological changes in salivary glands during their secretory activity (1866) and differentiated the serous glands from the mucous ones. He showed that secretion of saliva is largely independent of the blood flow and studied the effects of stimulation of nerves, distinguishing the "trophic" nerve fibers activating the metabolism of glandular cells from secretory fibers bringing about a passage of fluid from capillaries through the cells, washing the specific products out of the cells and into the secretory ducts. He described the difference between stimulation of parasympathetic and sympathetic nerves, the latter causing only a small secretion of a concentrated fluid. He concluded that secretion is an intracellular physiological process rather than a mechanical one.

Heidenhain noticed in the stomach two types of cells in the gastric glands and showed that one secretes the enzyme pepsin, the other hydrochloric acid. He worked out a method of a "small stomach" or gastric pouch, later improved by I. P. Pavlov, who worked for some time in Heidenhain's laboratory and always held Heidenhain in great esteem. The gastric-pouch technique has been widely used in several modified forms and has proved extremely useful in the investigation of gastric secretion and its regulation. Heidenhain also studied secretion in the pancreas, the liver, and the intestinal glands. Equally important were his studies of absorption processes in the intestine, which showed that absorption takes place from the interior of an isolated loop, even against a concentration gradient. Thus he came to the conclusion that intestinal absorption cannot be a simple physicochemical process but, rather, is a physiological activity performed by the epithelial cells.

Studies of the secretory activity of glands led Heidenhain to extensive investigations of the process of urine formation in the kidney. In 1842 William Bowman, through his study of the microscopic structure of kidney, developed a hypothesis that the renal tubules have a function in the secretion of specific constituents of urine (urea, uric acid, and so on) and that the Malpighian corpuscle might be an apparatus destined to separate the watery portion of the urine from the blood. Soon afterward Carl Ludwig, one of the four main exponents of a mechanistic physiology aimed at describing all vital phenomena in terms of simple physical and chemical processes, pointed out that the glomerulus, a tuft of capillaries projecting into the wide beginning of a long suite of renal tubules (Bowman's capsule), is a device by which a filtrate (or, rather, an ultrafiltrate without large protein molecules) may be separated from the blood plasma, and that in its further passage urine is formed by diffusion of a large proportion of water back into the capillaries surrounding the tubular system.

There had been several serious objections to Ludwig's filtration-diffusion theory; and Heidenhain, an experienced histologist, was convinced that the very complicated structure of the renal tubules, similar to that of other glands, pointed to a similarity in function. He found in his own experiments and those of others many indications that tubular cells play an active role in the formation of urine, as well as other objections to Ludwig's theory. Eventually he formulated his own theory, a modification of Bowman's. The urine was formed, he believed, by the secretory activity of tubular cells transporting (against a concentration gradient) urea, uric acid, and other specific constitutents of urine from the blood; in the tubules the constituents would be washed off by the flow of glomerular fluid. In his criticism of the filtration theory Heidenhain for a long time appeared to be right on several points, such as the role of renal tubules in performing physical work by actively transporting substances against concentration gradients. But he was clearly mistaken in rejecting two important points: the idea of glomerular filtration (ultrafiltration) and the concentration of the specific urine constituents during the passage of filtrate through the tubular system. In the latter process both physical processes (diffusion and concentration in the countercurrent system of Henle's loop) and metabolic processes (active transport) may play roles. But this transport (active resorption) is mainly in a direction opposite to that assumed by Heidenhain.

In recent reviews of research and concepts in kidney physiology the subject has been often presented as if the truth had always been on the side of Heidenhain's opponents. Urine formation in the kidney is an extremely complicated process, and it required almost eighty years after Heidenhain to trace its main features. Many hypotheses were formulated, each of them containing both correct and erroneous points. Heidenhain's chief merit was that his sober criticism, based on reliable experiments, stimulated discussion and further research at a time when renal physiology had been stagnant for about thirty years. His research contributed to the progress of knowledge more than if he had accepted a hypothesis which still required many corrections and substantial amendments.

Finally, Heidenhain's experiments on hypnotism (animal magnetism), made in an attempt to study the phenomenon scientifically, and his treatise defending the necessity and utility of vivisection in medical research should be mentioned.

Heidenhain's reliance on experiment as a sure guide in the search for truth, his technical skill, his wide-ranging work, his assiduity, and his independent thinking greatly influenced his contemporaries, especially those who had passed through his laboratory, such as Pavlov, Starling, and W. B. Cannon.

BIBLIOGRAPHY

I. Original Works. Heidenhain's publications include *Mechanik, Leistung, Wärmeentwicklung und Stoffumsatz bei der Muskeltätigkeit* (Leipzig, 1864); *Physiologie der Absonderungsvorgänge* (Leipzig, 1880), also in L. Hermann, ed., *Handbuch der Physiologie,* V, pt. 1 (Leipzig, 1883), 1–420; *Der sogenannte thierische Magnetismus. Physiologische Beobachtungen* (Leipzig, 1879; 4th ed., 1880), trans. from the 4th German ed. by L. C. Woolbridge as *Animal Magnetism. Physiological Observations* (London, 1880); and *Die Vivisektion im Dienste der Heilkunde* (Leipzig, 1879).

Heidenhain's papers were published first in the collection *Physiologische Studien* (Berlin, 1856), then in the series Studien des Breslauer Physiologischen Institutes, 4 pts.

(Leipzig, 1861–1868), and the rest mostly in *Pflüger's Archiv für die gesamte Physiologie des Menschen und der Tiere.* A partial bibliography was collected by P. Grützner, in *Pflügers Archiv,* **72** (1898), 263–265; papers published in *Pflügers Archiv* are listed in its *Registerband* for **1–30** (1885), pp. 24–26, and for **31–70** (1900), p. 20.

II. SECONDARY LITERATURE. See J. Bernstein, "Rudolph Heidenhain," in *Naturwissenschaftliche Rundschau,* **12** (1897), 606–607, and P. Grützner, "Zum Andenken an Rudolf Heidenhain," in *Pflügers Archiv,* **72** (1898), 221–265; and in *Allgemeine deutsche Biographie,* L (1905), 122–127. See also P. W. Herron, "Rudolf Heidenhain of Breslau," in *Surgery, Gynecology, and Obstetrics,* **110** (1960), 223–225.

VLADISLAV KRUTA

HEIM, ALBERT (*b.* Zurich, Switzerland, 12 April 1849; *d.* Zurich, 30 August 1937), *geology.*

From his father, a merchant, Heim gained his lifelong interest in mountain scenery, particularly as a result of a walking tour in 1865. From his mother he inherited his talent for drawing, which was a powerful vehicle for describing the complex Alpine geological structures he studied throughout his life. Concentrating on scientific subjects, he studied at the University of Zurich and then at the Zurich Institute of Technology, from which he graduated in 1869 with a dissertation on glaciers. This was his first publication, and between 1870 and 1937 he published more than 130 major articles and a dozen books and atlases, including the very influential, two-volume *Untersuchungen über der Mechanismus der Gebirgsbildung im Anschluss an die geologische Monographie der Tödi-Windgällengruppe* (1878) and the monumental, two-volume *Geologie der Schweiz* (1916–1922). The most powerful influence on his life was provided by his teacher at the Institute of Technology, Arnold Escher von der Linth, the father of Alpine geology. In 1841 Escher had interpreted the complex structure of the Glarus, south of the Wallensee, as involving colossal overfolding that caused a widespread inversion of the sedimentary sequence. Later work has shown that this is a huge nappe structure thrust from the south, but its early interpretation was hampered by the erosional removal of the thinned middle limb of the fold and by the synclinal character of its leading (northern) edge. Escher interpreted this as a double fold involving two overfolds of some 15 kilometers' displacement from the north and from the south toward an unfolded center. In reality this "center" was the spot where part of the huge single overfold had been removed by erosion, the southern "fold" being the root of the nappe and the northern "fold" the unrooted apex of the fold isolated by erosion.

After graduation Heim made geological excursions to Germany and Italy, where he made a special study of Vesuvius during its eruption of 1872. In the same year Escher died, and in 1873 Heim succeeded him as professor of geology at the Zurich Institute of Technology, also occupying the same chair at the University of Zurich from 1875. His fieldwork was concentrated in the region made famous by Escher; and between 1871 and 1885 he worked on the mapping and description of the key sheet 14 (1:100,000; Tödi-Windgällengruppe) for the Swiss Geological Commission, of which he became a member in 1888 and president in 1894. In 1878 he published *Mechanismus der Gebirgsbildung,* dedicated to Escher and containing not only the full description of the Glarus double fold which Escher had never given but also a superbly illustrated treatment of Alpine structures and mountain-building dynamics which almost immediately became the authoritative work on Alpine tectonics. Three years before the publication of the *Mechanismus* Heim married Marie Vögthin just after she became the first woman in Switzerland to be awarded a medical degree. The couple had two children.

Heim was an excellent teacher, mainly because his lectures on tectonics, mountain geology, and the geology of Switzerland were firmly based on his own field observations, mapping, and drawings. Although between 1881 and 1901, as head of the scientific section of the Zurich Institute of Technology he was much involved in administration, his field excursions and lectures made him extremely popular with his students. In 1885 Heim published his *Handbuch der Gletscherkunde;* in 1888 *Die Dislokationen der Erdrinde,* written with Emmanuel de Margerie, which formed the basis of many subsequent textbooks; in 1891 *Geologie der Alpen zwischen Reuss und Rhein* (a monograph describing the geological sheet 14 published six years earlier); and in 1894, with Karl Schmidt, the *Geological Map of Switzerland* on a scale of 1:500,000. This map was first exhibited at the International Geological Congress at Zurich and was stolen twice on the first day.

Heim was the first genuine European geological artist; his talent lay in his power to describe accurately the most complex geological structures and to illustrate them with brilliant drawings, cross sections, and models. It is ironic that, although the correct interpretation of the Glarus structure was not made by Heim, it came from a geologist who had never visited the region and knew it only from Heim's descriptions and illustrations. Even after the publication of the *Mechanismus* the difficulties facing the double-fold theory of the Glarus remained, particularly the possible erosional origin of the center of the fold and the necessity of postulating that the northern part of the double fold had been thrust from the north, whereas every other fold in the region was clearly thrust from the south. In 1884 Marcel Bertrand, after studying Heim's work and smaller-scale structures in the Belgian coalfield, proposed that the Glarus was a single overfold involving a northward displacement of at least 35 kilometers. In his 1891 monograph Heim virtually ignored Bertrand's theory and was still opposed to the idea when it was supported by Eduard Suess after a visit to the region in 1892. In the following year H. Schardt began the revolution in Alpine tectonics foreshadowed by Bertrand, by showing the reality of both superimposed, far-traveled thrust masses of northward displacement and of rootless thrust masses isolated by erosion. From this time Heim's position was constantly undermined; and when his student M. Lugeon published his synthesis of Alpine structure in 1902, it contained a letter from Heim supporting Bertrand's interpretation of the Glarus.

Heim continued with his geological mapping and, assisted by his son Arnold, produced *Geologie des Säntisgebirges* in 1905, together with a geological map of Säntis at 1:25,000. In 1910 this was followed by his geological map of the Glarner Alps at 1:50,000, also done with his son. After a deterioration of his health in 1905, Heim began to relinquish some of his teaching duties in 1908, resigning his chairs of geology three years later. He then devoted himself to the work of the Swiss Geological Commission, of which he was president until 1926, and was particularly instrumental in the completion of the 1:100,000 geological map of Switzerland. Between 1916 and 1922 Heim published his massive *Geologie der Schweiz,* which was more than a mere synthesis of the huge amount of geological work then being conducted in the Alps. With magnificently detailed diagrams he reconstructed the thicknesses and facies variations in the Alpine decken, speculated on the manner of geosynclinal deformation, showed how the rising Alpine folds furnished clastics to the northern flank of the range, identified involuted nappes, attempted to calculate the crustal shortening represented by folded structures, and produced the finest account of a national geology. Although Heim's preoccupation was with Alpine tectonics, he published on a variety of subjects including weathering, springs, landslides, the inability of glaciers to erode deep lake basins, the conditions to be expected during the drilling of the Simplon Tunnel, and Alpine erosional forms. His last major work, "Bergsturz und Menschenleben" (*Vierteljahrsschrift der Naturforschenden Gesellschaft in Zurich* [1932]), showed his interests in conservation and the relations between man and nature.

In 1931 Heim became ill with anemia. Nevertheless he continued his prolific publication until his death six years later. He was associated with more than fifty learned societies and held honorary degrees from Bern, Oxford, and Zurich.

BIBLIOGRAPHY

An extensive biography and complete bibliography of Heim's work is given by P. Arbenz in *Verhandlungen der Schweizerischen naturforschenden Gesellschaft,* **118** (1937), 330–353. An English obituary is E. B. Bailey, in *Obituary Notices of Fellows of the Royal Society of London,* **7,** no. 2 (1939), 471–474; see also M. L. Lugeon, "Memorial to Albert Heim," in *Proceedings of the Geological Society of America* for 1937 (1938), pp. 169–172, with portrait. E. B. Bailey, *Tectonic Essays: Mainly Alpine* (Oxford, 1935), provides a clear account of Heim's role in Alpine research.

R. J. CHORLEY

HEIM, ALBERT ARNOLD (*b.* Zurich, Switzerland, 20 March 1882; *d.* Zurich, 27 May 1965), *geology, geography.*

Heim was the son of Jacob Albert Heim and Maria Vögtlin. He studied geology with his father and received the Ph.D. from the University of Zurich in 1905. He lectured at the Technische Hochschule and University of Zurich from 1908 to 1911, and again from 1924 to 1928; from 1929 to 1931 he was professor of geology at Sun Yat Sen University and a member of the Geological Survey in Canton, China. He was a member of the Dirección de Minas y Geología de Argentina in Buenos Aires in 1944 and 1945; his last position was as chief geologist of the Iran Oil Company in Teheran from 1950 until 1952. Although most of his time was spent on expeditions, either commercial or more purely scientific, he maintained a residence in Zurich throughout his life.

Heim's initial fame as a geologist came from his studies of the Swiss Alps, a region in which he was active until about 1929. As early as 1905, in a lecture in Berlin, he spoke for the first time of a shoreline zone in the molasse forelands of the Glarus Alps and, through his discussion of the relationship of facies changes to the position and order of the thrust-sheets in the area, gave significant evidence for the theory of nappes. He continued these investigations in the western Säntis, where he concentrated more specifically on stratigraphy than had his father. He studied the stratigraphy of the Valangian, considered the facies changes in the Albian, and examined microscopically the clastic texture of the Cretaceous strata. Heim mapped the Churfirsten-Mattstock group (on a scale of 1:25,000) in 1906; this map served as the basis for his article "Der westliche Teil des Säntisgebirges."

Heim further recognized the erosion gulleys in the nagelfluh molasse beneath the thrust-sheets on the northern margin of the Alps, as well as the elongations (interrupted by pinchings out) and thrusting of the Säntis nappes over the Mürtschen nappe. In 1907 he first distinguished "exotic boulders" in the flysch as a stratigraphical phenomenon from the tectonic klippes. He also analyzed the nature of the facies of the Berrias-Valangian sediments of the region. In 1908 Heim realized the lithological importance of subaquatic landslides and studied, in the autochthonous Cretaceous and Eocene of the Kistenpass, the relationships of the facies to the Helvetian nappes and the course of the lines of similar facies (which he called the "isopens") and the distribution of past facies. From 1908 to 1911 he investigated the distribution of nummilitic and flysch formations, reporting on the subject in 1911.

In 1909 Heim began, in Edinburgh, the study of recent deep-sea deposits and first stated his theory that the thick ground mass of undersea limestone was chemically precipitated from lime dissolved in deep water by an increase in temperature or a decrease in pressure. In further studies of the Pacific Ocean, he pointed out that limestone rarely occurs in the deep sea undenuded by dissolution or mechanical processes. In his work on the Alpine Cretaceous he showed discontinuities representing submarine erosion and emphasized that the majority of so-called zoogenic chalks contain fossil shells as only an accessory admixture. From these data, he inferred a change in the hydroclimate. Heim published his results in 1924.

His Churfirsten monograph (which also treated the Alvier group) contained a new statement of the comparative lithology of each stratum and discussed the modes of formation of the discontinuities therein. Heim thus established the evolution and the facies relationships of nappes. He continued his Alpine investigations during World War I, working in part with his father. He made studies of the Aar massif and the lower Freiburg Alps, of asphalt occurrences in the Jura, on talc mining near Disentis, and on the Swiss phosphate deposits. He recognized petroliferous sandstones as being the first layer at the base of the varicolored molasse. In 1918 he extended his investigations to the border range of the Allgau Alps in order to treat thoroughly the stratigraphy and tectonics of the Vorarlberg region. He published his findings in 1934, in a work in which he also concerned himself with the problem of stratigraphic condensation, and compared them to data from the Appalachians. In a later publication (1938) he correlated Alpine data to his findings in the Triassic of the Himalayas and in Timor.

Heim's constant expeditionary activity retarded the publication of his geological surveys but brought him new material from all over the world. In addition to prospecting for oil, he published about 300 scientific works, including descriptions of his journeys (giving full accounts of flora and fauna and of the inhabitants of the region visited and their customs), maps, and geographical works. He also wrote on the psychology of birds and the flight of birds, bats, and insects and developed his own musical notation system to record bird and insect sounds. He developed a geological compass with a declination compensator (1913) and discussed methods and equipping of expeditions (1919 and 1930). He sought for the energy sources of the movements of the earth's crust, which he had glimpsed in variations in the earth's velocity and axes of rotation (1933). Even so, the greatest part of his work was never published.

Heim was married twice. His first wife was Anna Hartmann, whom he married in 1920; they had two sons, and were divorced in 1936. In 1949 he married Elizabeth Bertha von Brasch.

BIBLIOGRAPHY

I. ORIGINAL WORKS. A few of Heim's publications are "Der westliche Teil des Säntisgebirges," in *Beiträge zur geologischen Karte der Schweiz,* n. s. **16,** pt. 2 (1905), 313–515; "Monographie der Churfirsten-Mattstock-Gruppe," *ibid.,* n. s. **20** (1910–1918); *Sommerfahrten in Grönland* (Frauenfeld, 1910), written with M. Rikli; *Minya Gongkar. Forschungsreise im Hochgebirge von Chinesisch-Tibet* (Bern–Berlin, 1933); *Negro Sahara. Von der Guineaküste zum Mittelmeer* (Bern, 1934); *Thron der Götter. Erlebnisse der I. Schweizerischen Himalaya-Expedition* (Zurich–Leipzig, 1938), written with A. Gansser; trans. as *The Throne of the Gods* (London, 1939); "Central Himalaya. Geological Observations of the Swiss Expedition, 1936," in *Denkschriften der Schweizerischen naturforschenden Gesellschaft,* **73,** no. 1 (1939); *Weltbild eines Naturforschers. Mein Bekenntnis* (Bern, 1942, 4th ed., 1948); "Die naturwissenschaftlichen Arbeiten von A. Heim 1905–1943. Autoreferat mit Verzeichnis der Publikationen," in *Vierteljahrsschrift der Naturforschenden Gesellschaft in Zürich,* **89** (1944), supp. 3; *Wunderland Peru. Naturerlebnisse* (Bern, 1948, 2nd ed., 1957); *Südamerika. Naturerlebnisse auf Reisen in Chile, Argentinien und Bolivien* (Bern–Stuttgart, 1953); *America del Sur* (Barcelona, 1959); and "Die Fortsetzung des Verzeichnisses der Arbeiten 1944–1962," in *Journal. Schweizerische Stiftung für alpine Forschungen,* **4,** no. 11 (1962), 78–80.

II. SECONDARY LITERATURE. On Heim and his work, see A. Gansser, "Arnold Heim als Geologe," in *Journal. Schweizerische Stiftung für alpine Forschungen,* **4,** no. 11 (1962) 63–65, 76–80; "A. Heim," in *Bulletin. Vereinigung schweizer Petroleum-Geologen und -Ingenieure,* **32,** no. 82 (1965), 73–74; W. Rüegg, "Arnold Heim," in *Revista minería,* no. 70 (1965), 3–7; H. Suter, "Arnold Heim 1882–1965," in *Verhandlungen der Schweizerischen naturforschenden Gesellschaft* (1965), pp. 270–272; and R. Trümpy, in *Neue deutsche Biographie.*

WALTHER FISCHER

HEINE, HEINRICH EDUARD (*b.* Berlin, Germany, 16 March 1821; *d.* Halle, Germany, 21 October 1881), *mathematics.*

Eduard Heine, as he is usually called, was the eighth of the nine children of the banker Karl Heinrich Heine and his wife, Henriette Märtens. He was given private instruction at home and then attended the Friedrichswerdersche Gymnasium and finally the Köllnische Gymnasium in Berlin, from which he graduated in the fall of 1838. He studied one semester in Berlin and then went to Göttingen, where he attended the lectures of Gauss and Stern. After three semesters he returned to Berlin; his principal teacher there was Dirichlet, but he also attended courses of Steiner and Encke. After receiving his Ph.D. degree at Berlin University on 30 April 1842, he went to Königsberg, where for two semesters he studied with C. G. J. Jacobi and Franz Neumann. He obtained his *Habilitation* as a *Privatdozent* at Bonn University on 20 July 1844. On 13 May 1848 he was appointed an extraordinary professor at Bonn University and on 6 September 1848 a professor at Halle University, where he finally settled. In the academic year 1864–1865 he held the office of rector of the university. He was a corresponding member of the Prussian Academy of Sciences and a nonresident member of the Göttingen Gesellschaft der Wissenschaften. In 1877 he was awarded the Gauss medal. Two years earlier he turned down the offer of a chair at Göttingen University.

Heine's sister Albertine was married to the banker Paul Mendelssohn-Bartholdy, the brother of the composer. In his brother-in-law's house Heine met his future wife Sophie Wolff, the daughter of a Berlin merchant. They were married in 1850 and had four daughters and a son; one of the daughters was the writer Anselma Heine. Heine's frequent visits to his family in Berlin enabled him to discuss mathematics with Weierstrass, Kummer, Kronecker, and Borchardt. He attracted promising young mathematicians to Halle University, among them Carl Neumann, Gustav Roch, H. A. Schwarz, J. Thomas, and Georg Canton.

Heine published about fifty mathematical papers, most of them in *Zeitschrift für die reine und angewandte Mathematik.* His main fields were spherical functions (Legendre polynomials), Lamé functions, Bessel functions, and related subjects. His greatest work, *Handbuch der Kugelfunctionen,* was first published in 1861. The second edition (1878–1881) of Heine's book was still a standard compendium on spherical functions well into the 1930's if one considers the frequency with which it was quoted; that this edition has never been reprinted, however, belies this impression.

Heine's name is best known for its association with the Heine-Borel covering theorem, the validity of which name has been challenged. Indeed, the covering property had not been formulated and proved before Borel. What Heine did do was to formulate the notion of uniform continuity, which had escaped Cauchy's attention, and to prove the classical theorem on uniform continuity of continuous functions, which could rightfully be called Heine's theorem. One might, however, argue that this was the essential discovery and that Borel's reduction of uniform continuity to the covering property was a relatively minor achievement. Heine's name was connected to this theorem by A. Schoenfliess, although he later omitted Heine's name. Heine wrote a few more papers on fundamental questions. It seems not unlikely that in some way the paper on uniform continuity had its origin in the influence of Cantor, Heine's colleague at Halle.

BIBLIOGRAPHY

I. ORIGINAL WORKS. Heine's works include *Handbuch der Kugelfunctionen, Theorie und Anwendungen* (Berlin, 1861; 2nd ed., 1878–1881); and "Die Elemente der Functionenlehre," in *Journal für die reine und angewandte Mathematik,* **74** (1872), 172–188. His papers are listed in Poggendorff, I (1863), 1050, and III (1898), 606.

II. SECONDARY LITERATURE. On Heine and his work, see A. Wangerin, "Eduard Heine," in *Mitteldeutsche Lebensbilder,* III (Magdeburg, 1928), 429–436. For the theorem with Heine's name, see A. Schoenflies, "Die Entwicklung der Lehre von den Punktmannigfaltigkeiten," in 2 pts., in *Jahresbericht der Deutschen Mathematikervereinigung,* **8** (1900), 51, 109, and supp. 2 (1908), 76; both pts. were reprinted in *Entwicklung der Mengenlehre und ihrer Anwendungen,* 2nd ed., vol. I (Leipzig, 1913), 234.

HANS FREUDENTHAL

HEISTER, LORENZ (*b.* Frankfurt am Main, Germany, 19 September 1683; *d.* Bornum, near Königslutter, Germany, 18 April 1758), *anatomy, surgery, medicine.*

Heister was the son of a lumber merchant who later became an innkeeper and wine merchant. His mother, Maria Alleins, was the daughter of a merchant. He was educated at the Frankfurt Gymnasium and received additional private lessons in French and Italian. In 1702–1703 he studied at the University of Giessen and in 1703–1706 at the University of Wetzlar. When he left Wetzlar, Heister had completed the study of all subjects needed for the practice of medicine. Thereafter he went to Amsterdam, where he attended the botanical lectures of Caspar Commelin and the anatomical demonstrations of Frederik Ruysch. Amsterdam was at the time the world center for the study of exotic plants and one of the few places where anatomy could be studied by practical dissection.

In June 1707, during the War of the Spanish Succession, Heister worked in the field hospitals at Brussels and Ghent. After his return to Holland, he studied in Leiden for a short time, attending Hermann Boerhaave's lectures on chemistry and on the diseases of the eye and Goverd Bidloo's anatomical lessons. He obtained his M.D. at the University of Harderwijk in May 1708. After his return to Amsterdam, Heister gave lessons in anatomy with demonstrations on cadavers. Ruysch, the official professor of anatomy, limited himself to an hour's discussion of his anatomical preparations daily. Heister's first class consisted of ten French surgeons' apprentices, his second of German students; he lectured to each group in its own language.

In July 1709 Heister rejoined the Dutch army, this time as a field surgeon during the siege of Tournai. Later he tended those wounded in the battles of Oudenarde and Malplaquet. On 11 November 1711 he was appointed professor of anatomy and surgery at the University of Altdorf, near Nuremberg. In 1720 Heister was appointed professor of anatomy and surgery at Helmstedt. Here his teaching duties changed several times: in 1730 he was charged with the teaching of theoretical medicine and botany and in 1740 with the teaching of practical medicine and botany. He remained in Helmstedt for the rest of his life.

Heister made many minor anatomical discoveries and corrected some faulty observations of his predecessors. While a field surgeon, he discovered the true cause of cataracts: an opacity of the crystalline lens, instead of a film over the cornea, as had been believed. His main significance, though, is as a teacher and author. In Altdorf and in Helmstedt he trained a large number of surgeons and physicians. His books on anatomy, surgery, and medicine dominated the field for several generations, serving to educate thousands of surgeons and physicians throughout western Europe. Heister's main work, the *Chirurgie,* which was originally written in German, was translated into seven languages, including Latin and Japanese. Although not the first European book on surgery to be translated into Japanese, it was certainly the most successful, introducing Western methods to many Japanese surgeons.

BIBLIOGRAPHY

I. ORIGINAL WORKS. Many students earned their doctorates under Heister's guidance. Since it was then the custom for the professor, not the candidate, to write the dissertation, these dissertations (there are at least seventy-five) should be counted among Heister's work. In addition, he contributed many papers to the *Ephemerides Caesareo-Leopoldinae naturae curiosorum* and its successor, *Acta physico-medica Academiae Caesarae Leopoldino-Carolinae naturae curiosorum.* No modern bibliography of Heister's works appears to exist. The following are his most im-important books: *Dissertatio inauguralis de tunica choroidea oculi* (Harderwijk, 1708), doctoral diss.; *Oratio inauguralis de hypothesium medicarum fallacia et pernice* (Altdorf, 1710), his inaugural lecture as professor of anatomy and surgery; *De cataracta glaucomate et amaurosi tractate etc.* (Altdorf, 1711); *Apologia et uberior illustratio systematis sui de cataracta glaucomati et amaurosi contra Wolhusi ocularii parisiensis cavillationes et objectiones* (Altdorf, 1717); *Compendium anatomicum, veterum recentiorumque observationes brevissime complectens etc.* (Altdorf, 1717; other eds., some with slightly changed title, 1719, 1727, 1732; Amsterdam, 1723, 1733, 1748; Freiburg, 1726; Venice, 1730; Breslau, 1733; Nuremberg, 1736, 1737; Vienna, 1770), translated into English as *A Compendium of Anatomy, Containing a Short, but Perfect View of All Parts of the Human Body* (London, 1721, 1752), also translated into French (Paris, 1724, 1729, 1735, 1753), Dutch (Amsterdam, 1728), and Italian (Venice, 1772). *Chirurgie, in welcher alles was zur Wund Arztney gehöret, nach der neuesten und besten Art, gründlich abgehandelt wird usw.* (Nuremberg, 1718, 1719, 1724, 1731, 1739, 1743), also translated into Latin as *Institutiones chirurgicas etc.* (Amsterdam, 1739, 1750; Venice, 1740; Naples, 1759), into Dutch as *Heelkundige Onderwijzingen enz.* (Amsterdam, 1741, 1755, 1776), into English as *A General System of Surgery etc.* (London, 1750, 1768), Italian (Venice, 1765, 1770), French (Paris, 1771), and Spanish (Madrid, 1785); *Vindiciae sententiae suae de cataracta glaucomate et amaurosi adversus ultimas animadversiones atque objectiones etc.* (Altdorf, 1719); *Oratio de incrementis anatomiae in hoc seculo XVIII habita* (Helmstedt-Nuremberg, 1720), his inaugural lecture as professor of anatomy and surgery at

Helmstedt; *Infantes pro a diabolo olim suppositir habitos, revera nihil nisi Rachiticos fuisse* (Helmstedt, 1725); *Epistola de morte Silii Italici, celebris poetae et oratoris, ex clavo insanabili* (Helmstedt, 1735); *Compendium institutionum sive fundamentum medicinae* (Helmstedt, 1736; Leiden, 4 eds., incl. 1749, 1764), also translated into Dutch (Amsterdam, 1761) and German (Leipzig, 1763); *Nachricht von dem Leben und Thaten des englischen Augenarztes John Taylor* (Helmstedt, 1736); *Compendium medicae practicae, cui praemisse est dissertatio de medicinae praestantia* (Amsterdam, 1743, 1748, 1762; Venice, 1763), also translated into Spanish (Madrid, 1752) and German (Leipzig, 1763; Nuremberg, 1767); *Kleine Chirurgie oder Wund-Arztney, in welcher ein kurtzer doch deutlicher Unterricht und Begriff dieser Wissenschaft gegeben usw.* (Nuremberg, 1747, 1764 [?], 1767), also translated into Dutch (Amsterdam, 1743) and Latin (Amsterdam, 1743, 1748); *Systema plantarum generale ex fructificatione, cum regulis de nominibus plantarum a Linnei longe diversis* (Helmstedt, 1748); *Anatomisch-chirurgisches Lexicon* (Berlin, 1753); *Descriptio novi generis plantae rarissimae et speciosissimae africanae ex bulbosarum classe Brunsvigiae illustre nomen imposuit* (Brunswick, 1753), also translated into German (Brunswick, 1755); and *Medicinische, chirurgische und anatomische Wahrnehmungen* (Rostock, 1753, 1770), also translated into English (London, 1755).

The bibliography of the Japanese translations is confused. All of them are based on the Dutch trans. *Heelkundige Onderwijzingen enz.* (Amsterdam, 1755). It seems that, starting in 1792, several pts. of this book were translated under different titles and were circulated in MS only. Of these, a diss. on the dressing of wounds, *Geka Shūkō* (外科收功), translated by Ōtsuki Genkan (大槻玄幹), was published in 1814 in three maki (pts.). The work *Yōi Shinshō* (瘍醫新書), attributed to Sugita Gempaku (杉田玄白) and Ōtsuki Gentaku (大槻玄澤), published in 1822 in fifty maki, is sometimes considered to be the trans. of Heister's complete work. It is actually a collection of the MS translations mentioned above. Another trans., in 100 maki, was made in 1819 by Koshimura Tokumoto (越村徳基) under the title *Yōi seisen* (瘍醫精選) but was not printed. However, the atlas to this work, *Yōka seisen zukai* (瘍科精選図解), was published in 1820.

II. Secondary Literature. See *Apparatus librorum nec non instrumentarum chirurgicorum Laurentii Heisteri* (Helmstedt, 1760), the auction catalog of the books and instruments in Heister's estate; V. Fossel, *Studiën zur Geschichte der Medizin* (Stuttgart, 1909), pp. 111–152; E. Gurlt, "Lorenz Heister," in *Allgemeine deutsche Biographie,* XI (Leipzig, 1880), 672–676; Dr. L. [Léon Labbé?], "Laurent Heister," in *Nouvelle biographie générale,* XXIII (Paris, 1858), 806; C. P. Leporin, *Ausführliches Bericht vom Leben und Schriften des durch gantz Europam berühmten Herrn D. Laurentii Heisteri . . .* (Quedlinburg, 1725); F. Schlipp, *Laurentius Heister in seiner Bedeutung für die Augenheilkunde* (Haarlem, 1910); V. Schmieden, "Laurentius Heister, ein Beitrag zur Chirurgie," in *Zentralblatt für Chirurgie,* **51** (1924), 710–718; and J. Vossmann, *Zahnärztliches bei Laurentius Heister* (Leipzig, 1924).

P. W. van der Pas

HEKTOEN, LUDVIG (*b.* Westby, Wisconsin, 2 July 1863; *d.* Chicago, Illinois, 5 July 1951), *pathology, microbiology.*

The son of a Lutheran parochial schoolteacher, Hektoen spent his early years on a farm in a Norwegian-speaking community in Wisconsin. He attended Luther College in Decorah, Iowa, for six years and earned the B.A. degree in 1883. Having decided to study medicine, he spent the following year taking the requisite science courses at the University of Wisconsin. In 1885 Hektoen began medical studies at the College of Physicians and Surgeons in Chicago. After graduation in 1888 he was an intern at the Cook County Hospital. Here he came under the influence of Christian Fenger, a Danish-born surgeon and pathologist who brought the methods of the Vienna school to Chicago.

Hektoen's rise in the academic medical world of Chicago was rapid, and by the turn of the century he was widely known as one of the most prominent midwestern physicians. He held several posts as pathologist and professor of pathology before he was appointed to head the department at the University of Chicago in 1901, a position he held until 1933. In 1902 Hektoen was appointed director of the newly founded John McCormick Institute of Infectious Diseases. Here, and in the closely related Durand Hospital, much important research, especially on scarlet fever, was carried out in the following four decades. Owing to financial pressures, the McCormick Institute closed in 1939. It was bought by Cook County Hospital in 1943 and reopened as the Hektoen Institute for Medical Research. Through his own work at the McCormick Institute and through his teaching of a new, more scientifically based pathology that was closely integrated with biology as a whole, Hektoen greatly stimulated medical research in the Chicago area, helping to make it an important medical center. In 1903 he was offered the prestigious chair of pathology at the University of Pennsylvania. Despite the urgings of Simon Flexner—who was leaving the position in Philadelphia to go to the Rockefeller Institute—and the entreaties of William Welch and William Osler, Hektoen elected to remain in Chicago.

Hektoen turned from the study of morbid anatomy to immunology in the early years of the twentieth century and thus became one of the pioneers in this rapidly developing field. In 1905 he was the first to demonstrate that measles virus circulates in the blood during the initial thirty hours of the rash. He proved, using volunteers, that the virus can be transmitted by injection. Hektoen was also among the first to make use of blood cultures from living patients in order to aid in proper clinical diagnosis. In 1933, when Hektoen was seventy, he and his co-workers developed an important and later widely used method for prolonging the antibody-producing powers of immunizing solutions by adsorbing the antigen to aluminum hydroxide. His understanding of the immunological problems of blood transfusion led him to suggest that donors must be carefully matched to recipients in order to avoid the dangers of transfusion reactions. As a result of the scientific climate that Hektoen helped to create, the first blood bank in the United States was established at Cook County Hospital in 1937.

Hektoen's scientific career spanned more than sixty years, during which time he wrote more than 300 papers, edited numerous books, and, perhaps most significantly, served as editor of the *Journal of Infectious Diseases* from its inception in 1904 until 1941 and of the *Archives of Pathology* from 1926 to 1950. He died at eighty-eight, a much honored and revered medical leader. Throughout his long career Hektoen played important national roles through his work on American Medical Association councils, in several Chicago and national medical and scientific societies, and in his editorial positions. His many scientific honors included membership in the National Academy of Sciences and eight honorary degrees.

BIBLIOGRAPHY

I. ORIGINAL WORKS. Hektoen's many scientific papers are listed in the bibliography included in Paul R. Cannon's memoir (see below). A major book is *The Technique of Post Mortem Examination* (Chicago, 1894), intended for the many medical students who came to the postmortem demonstrations at the Cook County Hospital. With David Riesman he edited *American Textbook of Pathology* (Philadelphia, 1901). He also edited the collected works of Howard Taylor Ricketts and Christian Fenger; and with Ella M. Salmonsen he compiled *A Bibliography of Infantile Paralysis, 1789–1944* (Philadelphia, 1946).

II. SECONDARY LITERATURE. The most comprehensive biographical sketch of Hektoen was written on the occasion of his seventy-fifth birthday by his student and colleague Morris Fishbein: "Ludvig Hektoen, A Biography and an Appreciation," in *Archives of Pathology,* **26** (1938), 3–31, including a bibliography to that time. All subsequent writers have relied on this memoir. See also Thomas N. Bonner, *Medicine in Chicago 1850–1950* (Madison, Wis., 1947), esp. pp. 91–92; Paul R. Cannon, "Ludvig Hektoen, 1863–1951," in *Biographical Memoirs. National Academy of Sciences,* **28** (1952–1954), 163–197; James B. Herrick, "Ludvig Hektoen, 1863–1951," in *Proceedings of the Institute of Medicine of Chicago,* **19** (1952), 3–11; and James P. Simonds, "Ludvig Hektoen: A Study in Changing Scientific Interests," *ibid.,* **14** (1942), 284–287.

GERT H. BRIEGER

HELL (or HÖLL), MAXIMILIAN (*b.* Schemnitz [now Banská Štiavnica], Slovakia [now Czechoslovakia], 15 May 1720; *d.* Vienna, Austria, 14 April 1792), *astronomy.*

Hell was the youngest of the three sons of Matthias Cornelius Hell, the chief engineer of the royal mines at Schemnitz, then in the kingdom of Hungary. After attending the primary school at Schemnitz and the secondary school at Neusohl (now Banská Bystrica), he decided in 1738 to enter the Jesuits, which would require some twelve years of moral and scientific training. He first spent two years of probation at Trenčín, Hungary (now Czechoslovakia). Next he was sent to the University of Vienna to study philosophy, which then comprised a very wide range of disciplines, including mathematics, physics, and astronomy. J. Franz, who was in charge of teaching these sciences, soon became aware of Hell's aptitude and allowed him to participate in his astronomical observations at the Jesuit College observatory. In 1746 and 1747 Hell taught mathematics and physics in the secondary school at Leutschau, Hungary (now Levača, Czechoslovakia). He then returned to Vienna, finished his studies in theology, and in 1751 was ordained a priest. He received the Ph.D. from the University of Vienna in 1752.

Next Hell was sent by his order to Kolozsvár (now Cluj, Rumania), then part of Hungary. There he was in charge of the construction of a new college building, including an astronomical observatory. At the same time he taught mathematics, physics, and history at a secondary school. For this purpose he wrote textbooks and performed physical experiments, especially in magnetism.

Hell's scientific career proper began in 1755, when, in connection with a general reform of the University of Vienna, the Hapsburgs decided to establish a great central astronomical observatory. Its basic equipment was to be the instruments of the late imperial mathematician and geodetic surveyor, J. J. de Marinoni, who had made his house, on a relatively favorable site at the edge of Vienna, into an astronomical observatory. But since its distance of about a mile from the university was considered too far, it was decided to acquire only the instruments and to construct for them a four-story rectangular tower above the roof of the nearly completed new hall of the university. Hell, when he was appointed director of the new observatory, had to accept this improper decision. With great diligence he supervised the completion of the building, lectured in mathematics and astronomy, and was also in charge of instruction in technology, then called "popular mechanics." His main task, assumed even before it was possible to perform observations at the new observatory, was the computation and publication of an annual astronomical ephemeris similar to the *Connaissance des temps,* the only such publication then in existence. Ephemerides were of vital importance both for the imperial navy and for the merchant fleet, as well as for geodetic surveys and exact mapping of the empire. Such practical aspects led to the promotion of astronomy by the public authorities and to large attendance of Hell's lectures.

Hell was also greatly interested in pure science. Thus nearly all of the ephemeris volumes contain an appendix of scientific papers and observations by Hell himself, by his collaborators, and sometimes even by foreign astronomers. It was undoubtedly through this regular and useful series, as well as his other publications, that Hell's scientific reputation spread throughout Europe. Christian VII of Denmark and Norway invited him to undertake an expedition to the isle of Vardö, near the eastern coast of Lapland, in order to observe the transit of Venus of 3 June 1769 and the solar eclipse occurring the following day. Emperor Joseph II agreed; and Hell, accompanied by J. Sajnovics and a manservant, started for Copenhagen in the summer of 1768. The major part of the journey from Copenhagen to Vardö was by sea, and they arrived in September 1768. At the village of Vardöhus, following Hell's plans and under his supervision, the soldiers of a small royal fortress constructed a frame observatory with a separate room for living quarters. During the long winter there was time to adjust the precise clocks and the telescopes. On 3 and 4 June 1769, although the weather was somewhat cloudy, the observations could be made. The return journey was begun at the end of June, but Hell and his companions did not reach Copenhagen until 17 October. Wherever possible they performed meteorological, magnetic, and astronomical observations, the last with the aim of measuring geographic latitudes. They spent more than half a year at Copenhagen, partly to avoid traveling in winter but also so that Hell could deliver an extensive report of the results of the expedition to the Royal Danish Academy and obtain printed copies of this report. The group left Copenhagen on 22 May 1770 and reached Vienna that August.

In contrast with this successful expedition and his election as an honorary fellow of the academies of Trondheim and Copenhagen, the last two decades of Hell's life were seriously affected by the widespread but unmerited suspicion—apparently first expressed by Lalande—that he had falsified, or indeed had never made, the observations of the transit of Venus. He was also upset by the abolition of the Jesuits in 1773, although it did not formally alter his position as director of the Vienna observatory. Indeed, he continued to be assisted by his former pupils Anton Pilgram and Franz Triesnecker, ex-Jesuits like himself, in the edition of the yearly volumes of the astronomical ephemeris; he collected and analyzed all the observations of transits of Venus available to him, in order to derive a more precise value of the sun's parallax—really the best one of his time—and performed his duties at the university as before.

But the doubts regarding Hell's scientific credibility, intimately connected with the public defamation of the Jesuits then in vogue, survived him by nearly a century. About 1823 Encke, in his comprehensive evaluation of the transits of Venus of 1761 and 1769, rejected Hell's observations. Ten years later, in 1834, Karl von Littrow, by misinterpreting some rediscovered fragments of Hell's manuscripts, believed he had found direct proofs of the alleged falsification of the observations. It was not until 1883 that Simon Newcomb fully rehabilitated Hell's reputation after having scrutinized the manuscripts preserved in the Vienna observatory. The reliability of his verdict in favor of Hell is confirmed by the fact that the sun's parallax is far nearer to the true figure when Hell's observations are included with due weight, than without them.

Hell's colleagues, pupils, and friends greatly esteemed him for both his scientific and his personal qualities. He was elected to the academies of Bologna, Copenhagen, Trondheim, Göttingen, Paris, and Stockholm. Christian VII of Denmark and George III of England offered Hell honorary pensions much higher than his salary, but he refused them.

BIBLIOGRAPHY

I. ORIGINAL WORKS. Hell edited *Ephemerides astronomicae ad meridianum Vindobonensem* . . . for 1757–1768 and 1772–1792 (Vienna, 1756–1767, 1771–1791). A. Pilgram edited 3 vols. during his absence in 1768–1770 and F. Triesnecker edited 14 vols. after Hell's death.

Hell's most important papers, drawn from the appendixes to the *Ephemerides,* are concerned with the following problems: accurate theoretical prediction and subsequent observations of the transits of Venus of 1761 and 1769;

derivation of the sun's parallax from these observations; tables of the sun, moon, and planets, with corrections and additions by Hell and Pilgram; description of certain special observational techniques; series of current astronomical and meteorological observations; latitudes and longitudes of places in northern Europe; a theoretical explanation of the aurora borealis; and "Monumenta aere perenniora inter astra ponenda. Primum Serenissimo Regi Angliae, Georgio III, Altertum, Viro celeberrimo Friderico Wilhelmo Herschel."

Among Hell's numerous other works are *Adjumentum memoriae manuale chronologico-genealogico-historicum* (Vienna, 1750; 5th ed., enl., 1774); *Elementa mathematica* (Cluj, 1755); *Exercitationum mathematicarum partes tres* (Vienna-Cluj, 1755; 2nd ed., 1759; 3rd ed., 1760; 4th ed., 1773); *Elementa algebrae* (Poznan, 1760; 2nd ed., Vienna, 1762; 3rd ed., 1768; 4th ed., 1773); *Introductio ad utilem usum magnetis ex calybe* (Vienna, 1762), translated into German as *Anleitung zum nützlichen Gebrauch der künstlichen Stahl-Magneten* (Vienna, 1762); *Observatio transitus Veneris ante discum solis* (Copenhagen, 1770); and *Beyträge zur praktischen Astronomie aus den astronomischen Ephemeriden des Herrn Abbé M. Hell,* translated by L. A. Jungnitz, 4 vols. (Breslau, 1791–1794).

II. SECONDARY LITERATURE. See the following, listed chronologically: F. Triesnecker, "Monitum," in *Ephemerides astronomicae ad meridianum Vindobonensem anni 1793* (Vienna, 1792), pp. 3–5; C. L. von Littrow, *P. Hell's Reise nach Wardoe* (Vienna, 1835); C. von Wurzbach, "M. Hell," in *Biographisches Lexikon des Kaiserthums Österreich,* VIII (1862), 262–266; Poggendorff, I (1863), col. 1055; M. A. Paintner, *Historia scriptorum Societatis Jesu olim provinciae Austriacae, Hungaricae* (Vienna, 1855); C. Bruhns, "Maximilian Hell," in *Allgemeine deutsche Biographie,* X, 691–693; S. Newcomb, "On Hell's Alleged Falsification of His Observations of the Transit of Venus in 1769," in *Monthly Notices of the Royal Astronomical Society,* **43** (1883), 371–381; C. Sommervogel, "Hell, Maximilian," in *Bibliothèque de la Compagnie de Jésus,* new ed., *Bibliographie,* IV (Brussels-Paris, 1893), 238–258; F. Pinzger, *Hell Miksa Emlékezete,* 2 vols. (Budapest, 1920–1927), with a summary in German, "Erinnerung an Maximilian Hell," in vol. II; A. V. Nielsen, "Pater Hell og Venuspassagen 1769," in *Nordisk astronomisk tidsskrift* (1957), pp. 77–97; and K. Ferrari d'Occhieppo, "Maximilian Hell und Placidus Fixlmillner," in *Österreichische Naturforscher, Ärzte und Techniker* (Vienna, 1957), pp. 27–31; and "Maximilian Hell," in *Neue deutsche Biographie,* VIII (1969), 473–474.

KONRADIN FERRARI D'OCCHIEPPO

HELLINGER, ERNST (*b.* Striegau, Germany, 30 September 1883; *d.* Chicago, Illinois, 28 March 1950), *mathematics.*

Hellinger was the son of Emil Hellinger and Julie Hellinger. He grew up in Breslau, where he received the diploma of the Gymnasium in 1902. He studied at the universities of Heidelberg, Breslau, and Göttingen, where in 1907 he received the Ph.D. in mathematics. Like many outstanding mathematicians of his time, Hellinger was a student of David Hilbert. In his dissertation on the orthogonal invariants of quadratic forms of infinitely many variables, Hellinger introduced a new type of integral which is known today as the Hellinger integral. The Hilbert-Hellinger theory of forms profoundly influenced other mathematicians, in particular E. H. Moore of the University of Chicago. From 1907 to 1909, Hellinger was an assistant at the University of Göttingen. There he edited Hilbert's lecture notes and Felix Klein's influential *Elementarmathematik vom höheren Standpunkte aus* (Berlin, 1925) which was translated into English (New York, 1932).

A *Privatdozent* at the University of Marburg from 1909 to 1914, Hellinger became professor of mathematics at the newly founded University of Frankfurt am Main and taught there until 1936, when the Nazi government forced him to retire because he was a Jew. His monumental article "Integralgleichungen und Gleichungen mit unendlichvielen Unbekannten," on integral equations and equations with infinitely many unknowns, which he wrote with Otto Toeplitz over a period of many years for the *Enzyklopädie der mathematischen Wissenschaften,* has attained the status of a classic document. It first appeared in 1927, was separately published in 1928, and was reprinted in 1953.

On 13 November 1938 Hellinger was arrested and held in a concentration camp. He was released after a month and a half, with the stipulation that he leave the country immediately. In March 1939 he found refuge in the United States, where he taught mathematics at Northwestern University in Evanston, Illinois, first as lecturer and later as full professor. He acquired American citizenship in 1944. After retiring at age sixty-five, he took a position at the Illinois Institute of Technology in 1949 but fell ill that November and never recovered.

Although his main field was analysis, Hellinger also worked in the history of mathematics with Max Dehn. Hellinger's lectures were of supreme clarity. Deeply concerned with all aspects of his students' lives, he was an unpretentious, highly effective mentor.

BIBLIOGRAPHY

I. ORIGINAL WORKS. Hellinger's works include "Grundlagen für eine Theorie der unendlichen Matrizen," in *Nachrichten der Gesellschaft der Wissenschaften zu Göttingen* (1906), pp. 351–355, written with O. Toeplitz; *Die Orthogonalinvarianten quadratischer Formen von unendlichvielen Variablen* (Göttingen, 1907), his diss.; "Neue Begründung der Theorie quadratischer Formen von unendlichvielen Veränderlichen," in *Journal für die reine und angewandte Mathematik,* **136** (1909), 210–271; "Grundlagen für eine Theorie der unendlichen Matrizen," in *Mathematische Annalen,* **69** (1910), 289–330, written with O. Toeplitz; "Zur Einordnung der Kettenbruchtheorie in die Theorie der quadratischen Formen von unendlichvielen Veränderlichen," in *Journal für die reine und angewandte Mathematik,* **144** (1914), 213–238, written with O. Toeplitz; and "Die allgemeinen Ansätze der Mechanik der Kontinua," in *Enzyklopädie der mathematischen Wissenschaften,* **4,** no. 30 (1914), 601–694.

See also "Zur Stieltjesschen Kettenbruchtheorie," in *Mathematische Annalen,* **86** (1922), 18–29; "Integralgleichungen und Gleichungen mit unendlichvielen Unbekannten" in *Enzyklopädie der mathematischen Wissenschaften,* **2,** no. 13C (1927), 1335–1648, written with O. Toeplitz, repr. separately (New York, 1953); "Hilberts Arbeiten über Integralgleichungen und unendliche Gleichungssysteme," in *David Hilbert, Gesammelte Abhandlungen,* III (Berlin, 1935), 94–145; "On James Gregory's Vera Quadratura," in H. W. Turnbull, ed., *James Gregory Tercentenary Memorial Volume* (London, 1939), pp. 468–478, written with M. Dehn; *Spectra of Quadratic Forms in Infinitely Many Variables,* no. 1 in Northwestern University Studies in Mathematics and the Physical Sciences, Mathematical Monographs, vol. I (Evanston, Ill., 1941), 133–172; "Certain Mathematical Achievements of James Gregory," in *American Mathematical Monthly,* **50** (1943), 149–163, written with M. Dehn; and "Contributions to the Analytic Theory of Continued Fractions and Infinite Matrices," in *Annals of Mathematics,* **44** (1943), 103–127, written with M. Dehn.

II. SECONDARY LITERATURE. For a biography of Hellinger, see C. L. Siegel, *Zur Geschichte des Frankfurter mathematischen Seminars. Gesammelte Abhandlungen,* III, no. 81 (Berlin–Heidelberg–New York, 1966), 462–474.

WILHELM MAGNUS

HELLOT, JEAN (*b.* Paris, France, 20 November 1685; *d.* Paris, 15 February 1766), *industrial chemistry.*

Hellot came of a well-connected, middle-class family and was destined for an ecclesiastical career. He turned to chemistry after perusing the papers of his grandfather, a physician; studied under E. F. Geoffroy; and then traveled to England, where he became acquainted with various fellows of the Royal Society. Having lost his fortune in the crash following the economic manipulations of John Law, he earned his living by editing the *Gazette de France* from 1718 to 1732. Hellot was elected *adjoint chimiste* of the Académie Royale des Sciences in 1735 and was promoted to *pensionnaire chimiste supernuméraire* in 1739 and *pensionnaire chimiste* in 1743. In 1740 he was elected a fellow of the Royal Society and in the same year was appointed inspector general of dyeing, a field in which he became an authority. He married in 1750, apparently desiring domestic comforts as he grew older. In 1751 he was appointed technical adviser to the Sèvres factory, where he is said to have introduced a number of technical improvements.

Hellot's earliest researches were in pure chemistry: on the composition of ether, on metallic zinc and its compounds, on mineral acids, on phosphorus (whose method of preparation was apparently then unknown in France), and on Glauber's salt. He soon showed remarkable ability as an analytical and industrial chemist, in part through service on various commissions of the Académie Royale des Sciences. In 1740 the Academy was asked to investigate the purity of certain samples of salt from various sources; Hellot did the major share of the careful analytical work involved. In 1746 he examined standard measures, an investigation arising from a query about the exact length of the ell. In 1763 he helped investigate firedamp in mines, a problem new to France.

But Hellot's major contributions were to the chemistry and technical aspects of dyeing, and to mining and assaying. His *L'art de la teinture des laines* examines the techniques of fast dyeing materials of superior quality ("au grand teint") and fugitive dyeing of cheap textiles ("au petit teint"). He advanced a mechanical explanation for the ability of the cloth to hold the dye: the particles of dye entered the pores of the cloth, and when these pores were closed, either by the inherently astringent properties of the dye or by those of the mordant, the particles of dye were then held fast. This theory had many adherents. The major importance of this book lay in the careful discussion of techniques which made it a standard work for the remainder of the century.

Hellot's contributions to metallurgy are contained in his papers on zinc and on precious metals, but above all in his commentary on C. A. Schlütter's work, also significant as one of the first French translations of a German chemical book, a genre very popular later.

Hellot was an original and effective practical and industrial chemist, one among the first generation of French scientists to concern themselves with technology.

BIBLIOGRAPHY

I. ORIGINAL WORKS. Hellot's most important work is *L'art de la teinture des laines et étoffes de laine au grand et au petit teint, avec une instruction sur les débouillis* (Paris, 1750, 1786; Maastricht, 1772). In the English trans., *The Art of Dying Wool . . .* (London, 1789, 1901), the translator speaks of its having been "partly and poorly translated by a country Dyer, who knew but little French and no Chemistry," but I have not been able to trace this earlier work, which may never have been published. There is also a German trans., *Färbekunst* (Altenburg, 1751, 1764, 1790).

Hellot's remaining works were almost all in the form of essays published in the *Histoire et mémoires de l'Académie royale des sciences,* as follows: "Recherches sur la composition de l'éther" (1734); "Analyse chimique du zinc" (1734); "Conjectures sur la couleur rouge des vapeurs de l'esprit de nitre et de l'eau-forte" (1736); "Sur une nouvelle encre sympathetique à l'occasion de laquelle on donne quelques essais d'analyse des mines de bismuth, d'azur et d'arsenic, dont cette encre est la teinture" (1737); "Le phosphore de Kunckel et analyse de l'urine" (1737); "Sur le sel de Glauber" (1738), mainly on sulfuric acid; "Théorie chimique de la teinture des étoffes" (1740–1741); "Examen du sel de Pécais" (1740), written with Louis Lemery and C. J. Geoffroy; "Sur l'étalon de l'aune au bureau des marchands merciers de la ville de Paris" (1746) and "Sur l'exploitation des mines" (1756), both written with C. E. L. Camus; "Examen chimique de l'eau de la rivière d'Yvette" (1762), written with P. J. Macquer; "Mémoire sur les essais de matière d'or et d'argent" (1763), written with P. J. Macquer and Matthieu Tillet; and "Sur les vapeurs inflammables qui se trouvent dans les mines de charbon de terre de Briançon" (1763), written with H. L. Duhamel du Monceau and Étienne Mignot de Montigny.

Hellot also wrote the intro. to the trans. of a work on mining by C. A. Schlütter, published as *Traité des essais des mines & métaux* (Paris, 1750) and *De la fonte des mines et des fonderies* (Paris, 1753), with numerous additional comments.

II. SECONDARY LITERATURE. The chief source for Hellot's biography is the *Éloge* by Jean-Paul Grandjean de Fouchy, published in *Histoire et mémoires de l'Académie royale des sciences* for 1766. There is a short notice in F. Hoefer, *Histoire de la chimie* (Paris, 1866), II, 375–377; and a longer but confused account of his scientific contributions in J. R. Partington, *History of Chemistry,* III (London, 1962), 67–68, in large part based on the account in Thomas Thomson, *History of Chemistry,* 2nd ed., I (London, n.d.), ch. 8, 224–288. The best appraisal of his work is in Henry Guerlac, "Some French Antecedents of the Chemical Revolution," in *Chymia,* **5** (1959), 73–112.

MARIE BOAS HALL

HELLRIEGEL, HERMANN (*b.* Mausitz, near Pegau, Saxony, Germany, 21 October 1831; *d.* Bernburg, near Halle, Germany, 24 September 1895), *agricultural chemistry.*

Hellriegel was educated at the famous Saxon Fürstenschule in Grimma, then studied chemistry at the Forestry Academy in Tharandt, near Dresden. Here he established a close relationship with the agricultural chemist Adolf Stöckhardt, who soon entrusted him with the duties of an assistant. He thus received manifold stimulation for his later activity in the field of practical chemistry established by Liebig.

Stöckhardt quickly recognized Hellriegel's special capabilities and as early as 1856 arranged his appointment to the board of directors of the new agricultural research institute at Dahme in the Nieder Lausitz. There, as a young scholar of twenty-five, Hellriegel was able to devote himself to research problems in plant physiology and to concentrate on questions, first raised by Liebig, concerning the nutritive requirements of certain cultivated plants. He was especially interested in the improvement of sandy soils, often found in central and northern Germany, which presented great difficulties if used intensively. For his work, then regarded as epochal, he was named titular professor by the Saxon government in 1860.

In 1873 the government of Anhalt-Bernburg named Hellriegel director of the agricultural research institute at Bernburg. But Hellriegel, who still had no experimental equipment at his disposal, had at first to content himself with functioning as a governmental adviser on agricultural questions. He used the time to travel about the small dukedom to impress upon the peasants, through word and deed, the need for progressive agriculture.

Through the support of the German sugar industry syndicate—beet sugar had already replaced cane sugar as the preferred sweetener in Germany—in 1882 Hellriegel opened a research facility at Bernburg. There his chief undertaking was research on the conditions required by sugar beets, questions of nitrogen supply in plants playing a major role. In doing this work, which above all served to improve sandy soil and involved the experimental use of sterilized sand, Hellriegel and his colleague H. Wilfarth discovered that certain leguminous plants, cooperating symbiotically with bacteria enclosed in nodules on their roots (*Rhizobium frank*), assimilate nitrogen from the air and convert it into a utilizable bound form.

Hellriegel reported this for the first time in an address at the fifty-ninth meeting of the German Society of Scientists and Physicians at Berlin in 1886. With this pioneering discovery the success of his contemporary Albert Schultz-Lupitz in cultivating legumes, preferably the lupine, as an intermediate crop and notably in the hitherto relatively barren sandy soil, could be explained. Thereby the cultivation of intermediate crops as a scientifically investigated, systematically introduced measure for increased soil fertility acquired considerable importance.

Since Hellriegel's work was closely followed abroad, it is not surprising that he received numerous honors. He was an honorary member of the Royal Swedish

Academy of Sciences, the Royal Society of London, the Paris Academy of Sciences, and the French National Society of Agriculture. The Bavarian Academy of Sciences in Munich awarded him the Liebig gold medal.

BIBLIOGRAPHY

Hellriegel's main work is *Untersuchungen über die Stickstoffernährung der Gramineen und Leguminosen,* supp. issue of *Zeitschrift des Vereins für die Rübenzuckerindustrie im Zollverein* (1888), written with H. Wilfarth.

Secondary literature is H. Haushofer, *Die deutsche Landwirtschaft im technischen Zeitalter* (Stuttgart, 1963), pp. 163–164; O. Keune, ed., *Männer, die Nahrung schufen* (Berlin, 1952; 2nd ed., 1954); C. Leisewitz, "Hellriegel," in *Allgemeine deutsche Biographie,* L (Leipzig, 1905), 169–171; O. Lemmermann, "Die Untersuchungen Hellriegels über die Stickstoffernährung der Gramineen und Leguminosen," in *Zeitschrift für Pflanzenernährung, Dügung und Bodenkunde,* **45** (1936), 257; L. Schmitt, "Hellriegel," in *Neue deutsche Biographie,* VIII (Berlin, 1969), 488; and H. Wilfarth, "Professor Dr. Hermann Hellriegel," in *Landwirtschaftliche Presse,* no. 90 (1895).

H. SCHADEWALDT

HELMERSEN, GRIGORY PETROVICH (*b.* Duckershof, Latvia, 29 September 1803; *d.* St. Petersburg, Russia, 15 February 1885), *geology.*

After graduating from the University of Dorpat in 1825 with a Master of Sciences degree, Helmersen attended lectures in 1835–1838 at the Mining Institute in St. Petersburg and obtained a post there as professor of geognosy. He held this position for twenty-five years, serving simultaneously as class inspector and curator of the institute's museum. From 1865 to 1872 he was director of the institute. Working all his life within the Mining Department, Helmersen achieved the rank of lieutenant general in the Corps of Mining Engineers. In 1844 he was elected adjunct member, in 1847 associate member, and in 1850 full member of the Academy of Sciences in St. Petersburg. He was one of the organizers of the Geological Committee of Russia and its first director (1882).

Helmersen's scientific researches were connected mainly with regional geological investigations and were undertaken in the Urals, the Altai, central Asia, the Baltic provinces, and a number of regions of central and southern Russia. His first investigations were a study of the gold-bearing areas of the central Urals and of the geological structure of the southern part of this mountain range. The years 1830–1832 were spent abroad; Helmersen studied paleontology in several German universities and visited Austria and northern Italy. On his return he continued his work in the Urals and, somewhat later, in the Altai. In 1838–1839 Helmersen studied deposits of combustible shales in Estonia, establishing their age on the basis of their paleontological remains and investigating the possibilities of their distillation.

Helmersen devoted much attention to the study of the geological structure and mineral reserves of the central portion of European Russia, where he worked for nearly forty years. He studied coal deposits and accumulations of sedimentary iron ores of the Moscow basin. He also investigated coal deposits in the Kiev, Kherson, and Grodno guberniyas, and in the Donets and Dabrowa Gornicza coal basins. Some of his papers dealt with the geological structure of mud volcanoes and oil seeps on the Kerch and Taman peninsulas.

Helmersen played a major role in the development of geological mapping in Russia. In 1841 he compiled and published a geological map of European Russia, on a scale of thirty miles to the inch, showing the distribution of deposits belonging to different geological systems. Although very schematic, it was the first to give some idea of the location of major structures within the Russian Platform and to show Upper Paleozoic variegated deposits as an independent stratigraphic unit. In his explanatory note to the map Helmersen calls them "Permian sandstone." On his visit to Russia, Murchison designated these deposits as a new system between the Carboniferous and the New Red Sandstone: the Permian system. In 1842 this map was awarded the Demidoff Prize by the Academy of Sciences in St. Petersburg.

In 1846 Helmersen compiled a map of goldfields in eastern Siberia. On the basis of previously unknown data he substantially changed and supplemented the geological map of European Russia and of the Urals published by Murchison in 1845, publishing fundamentally new versions in 1865 and 1873. A prominent coal geologist, Helmersen was very familiar with the coal deposits of Russia and Poland. In the 1860's he directed the compilation of the first stratigraphic map of the Donets Basin. He was also interested in Quaternary glaciation and studied its traces in northern Russia and Finland. In 1857 he published an interesting paper on the origin of giant glacial kettle holes in Scandinavia.

A regular watcher of the development of theoretical concepts in natural history, in 1860 Helmersen suggested that the Academy of Sciences in St. Petersburg organize extensive paleontological research to establish which of two theories—evolution or creationism—is confirmed by geological findings. He appears to have favored evolution.

Helmersen, who published more than 130 papers, was very highly regarded in geological circles and was elected a member of numerous scientific societies in various countries. In 1879, in honor of the fiftieth anniversary of his scientific career, the Academy of Sciences in St. Petersburg instituted a prize named for Helmersen. No longer given, it was awarded for outstanding research in geology, paleontology, and geography of Russia and adjacent countries.

BIBLIOGRAPHY

Helmersen's most important papers are "Explanatory Notes to a General Map of Rock Formations in European Russia," in *Gornyi zhurnal,* no. 4, pt. 2 (1841), pp. 29–68, with map, in Russian; "Das Olonetzer Bergrevier, geologisch untersucht in den Jahren 1856, 1857, 1858 und 1859," in *Mémoires de l'Académie impériale des sciences de St.-Pétersbourg,* 7th ser., **3,** no. 6 (1860), 1–33, with map; "Chudskoe Lake and the Upstream Regions of the Narva River," in *Zapiski Imperatorskoi akademii hauk,* **7,** no. 2 (1865), 1–85, in Russian; and "Das Vorkommen und die Entstehung der Riesenkessel in Finnland," in *Mémoires de l'Académie impériale des sciences de St.-Pétersbourg,* 7th ser., **11,** no. 12 (1867), 1–13, with map.

A secondary source is F. Schmidt, "Gregor von Helmersen," in *Neues Jahrbuch für Mineralogie, Geologie und Paläontologie,* **2** (1885), 1–4.

V. V. TIKHOMIROV

HELMERT, FRIEDRICH ROBERT (*b.* Freiberg, Saxony, Germany, 31 July 1843; *d.* Potsdam, Germany, 15 June 1917), *geodesy, astronomy.*

The youngest child of Johann Friedrich Helmert, treasurer of the Johannishospitalgut of Freiberg, and of the former Christiana Friederika Linke, Helmert attended the secondary school in Freiberg and at the age of thirteen entered the Annenrealschule in Dresden, where an older brother was assistant headmaster. In 1859 he began to study engineering science at the Polytechnische Schule in Dresden. Since he was especially enthusiastic about geodesy, his teacher, August Nagel, Saxon commissioner for the Mitteleuropäische Gradmessung, hired him, while he was still a student, to work on the triangulation of the coalfield of the Erzgebirge and the drafting of the trigonometric network for Saxony. In the summer of 1863 he became Nagel's assistant on the measurement of degrees. Helmert's work on this undertaking resulted in the *Studien über rationelle Vermessungen der höheren Geodäsie* (1868), with which he received his Ph.D. from the University of Leipzig in 1867, after a year's study of mathematics and astronomy.

Helmert next worked on the establishment of a triangulation network around Leipzig under C. Bruhns and as a mathematics teacher at the *Realinstitut* run by Hölbe in Dresden; subsequently he participated in the discussion of exact standard weights conducted by the Commission on Standardization directed by Nagel. At the beginning of 1869, after declining a similar offer from the observatory at Leiden, he became an observer at the Hamburg astronomical observatory. A product of his stay in Hamburg was *Der Sternhaufen im Sternbilde des Sobieskischen Schildes* (1874).

In 1870 Helmert became geodesy instructor at the newly founded technical school in Aachen, where he was named professor in 1872. At Aachen he amassed a collection of instruments, met a busy teaching schedule, and wrote his masterpiece, *Die mathematischen und physikalischen Theorien der höheren Geodäsie,* which quickly made him known. He was an editor of *Zeitschrift für Vermessungswesen* from 1876 to 1883 and in 1877 became a member of the Scientific Advisory Council of the Prussian Geodetic Institute. He rejected offers from Córdoba, Argentina, in 1873 and from Karlsruhe in 1881.

Johann Jakob Baeyer died in 1885, and in 1886 Helmert was appointed provisional director of the Prussian Geodetic Institute in Berlin, which was connected with the central office of the Europäische Gradmessung. By the fall of 1886 he had secured the appointment of a permanent secretary for the central office, which thereby became freed for the scientific tasks of the Internationale Erdmessung. On 15 April 1887 Helmert was appointed professor of advanced geodesy at the University of Berlin and, a week later, director of the Geodetic Institute; also in 1887 he became a full member of the Prussian Commission on Standards. He suffered a stroke in 1916 and died of its effects the following year.

Helmert's first wife was Jenny Oehme, who died in 1887; his second was his niece Marie Helmert. By his second marriage he had a son, Robert.

Helmert's abilities were summed up in an obituary by O. Eggert:

> Helmert possessed to a high degree the gift of a vivid and clear delivery, which was especially easy to understand because of his straightforward style. He was able to present mathematical developments in an extraordinarily clear manner. . . . That he could hold the interest of individual students . . . is evident from a series of dissertations which resulted from his influence. Those of his students who . . . had the good fortune to come into close contact with Helmert will always gratefully recall the friendly and sympathetic support that they found during their studies under him [*Zeitschrift für Vermessungswesen,* **46** (1917), 294–295].

Helmert received many honors. In 1884 he became an honorary member of the Deutsche Geometerverein; in 1900 he became a full member of the Prussian Academy of Sciences in Berlin, and in 1903 honorary doctor of engineering of the Aachen Technische Hochschule. Besides some twenty-five German and foreign decorations, in 1912 he was awarded the Goldene Medaille für Wissenschaft.

In his dissertation Helmert developed the theory of the ellipse of error and of the middle points error; he also treated the most advantageous division of the work of measuring. In 1872 his *Die Ausgleichsrechnung nach der Methode der kleinsten Quadrate mit Anwendungen auf die Geodäsie und die Theorie der Messinstrumente* appeared. In this work Helmert introduced a new theory of equivalent observations and for the first time used the method of least squares in the examination of measuring instruments. His *Übergangskurven für Eisenbahngeleise* (1872) and *Günstige Wahl der Cardinalpunkte beim Abstecken einer Trace* (1875) resulted from his teaching at Aachen. A wealth of specialized papers served as preparation for his masterpiece, *Die mathematischen und physikalischen Theorien der höheren Geodäsie,* on which he worked from 1877.

In part 1 of that work, *Die mathematischen Theorien* (1880), Helmert demonstrated the validity of A. M. Legendre's theorem for acute triangles and treated extensively the geodesy of the sphere and the slightly oblate ellipsoid of rotation; he linked geodesy to the actual surface of the earth by means of plumb-line deflection. He discussed for the first time calculation on the ellipsoid with chords, the differential formulas for geodetic lines, and the development of series for use in the computation of distances and azimuths from geographic positions. He also considered the geodetic lines between two nearly diametrical points, the maximum values of the higher terms of Legendre's theorem, and the spherical computation of chains of triangulation. In addition he discussed the relationships between rectangular and geographical coordinates, the balancing of geodetic-astronomic measurements with regard to plumb-line deflections, and developments regarding the conclusiveness of measurements of degrees for representing the earth's shape as that of an ellipsoid of rotation.

Part 2, *Die physikalischen Theorien* (1884), discusses the shape of the earth from the standpoint of potential theory, beginning with the analytical formulation of the concept of the acceleration of gravity and the introduction of its potential. There follows a treatment of the general properties of equipotential surfaces and of their discontinuities of curvature. After a presentation of Clairaut's theorem Helmert derives the flattening of the sphere of the earth from 122 pendulum lengths. He finds it to be $1:299.26 \pm 1.26$, reduced to sea level, allowing for the condensation of the visible disturbing masses of the earth's surface to a surface parallel to the surface of the sea, at a depth of three miles. He finds gravity at sea level to be $9.7800\ (1 + 0.005310 \pm 14 \sin^2 B)$. He also investigates the perturbation effect of the five continents, considered as blunted circular cones 4,000 meters thick, on the level planes near the surface and of other disturbing masses of various shapes.

Along with the temporal changes of the level planes, Helmert discusses the disturbances of the plumb line resulting from the moon and the sun, as well as from the small movements of the earth's axis. Next he takes up the value of astronomical data for knowledge of the earth's shape. In the section entitled "Das geometrische Nivellement" he insists on taking into account the variation of gravity with geographical latitude but deems the influence of gravity anomalies to be unimportant. In treating trigonometric altimetry he also considers lateral refraction and aberration. In opposition to the trigonometric method for determining the geoid, he holds that the method of plumb-line deflections in the preparation of meridian profiles by means of closely spaced stations of latitude, for which he proposes the term *astronomisches Nivellement,* is more advantageous.

In 1886, for the Prussian portion of the Mitteleuropäische Gradmessung, Helmert provided for the establishment of the plumb-line deflections of a net of seventy points centered on Point Rauenberg, near Berlin; for the astronomical measurements of latitudes and azimuths, telegraphic measurements of lengths were, in part, also executed. In October 1887 Helmert delivered to the commissioners of the Internationale Erdmessung, in "Lotabweichungen I," the formulas and tables, with examples of their use, necessary for calculation of the plumb-line deflections. In 1890 he reported on the variations of geographical latitude in 1889, which, in 1891, he attributed to changes of position of the earth's axis. In addition, he arranged for an expedition to Honolulu and inspired the establishment of the International Bureau of Latitudes to monitor the movements of the poles.

When Sterneck measured gravity with his pendulum apparatus (constructed in 1887), Helmert made a critical evaluation of Sterneck's methods of measurement and investigated the cause of the gravitational disturbances ("Die Schwerkraft im Hochgebirge," 1890); in addition, he carried out his own measurements with this device beginning in 1892. In 1893–1894 he made test measurements with the reversible pendulum supplied by J. A. Repsold in Hamburg; in 1898 he collected the results in *Beiträge zur Theorie des Reversionspendels.* They constituted the basis for the determination of absolute gravity by Friedrich Kühnen and Philipp Furtwängler between 1900 and 1906 at the Potsdam Geodetic Institute. The value ascertained in 1906, 981.274 ± 0.003 gal, was accepted in 1909 by the Internationale Erdmessung.

Helmert was always concerned with improvements in the gravitational formula and in the reduction of determinations of gravity (1901–1904, 1915). In 1909 he calculated the value of the flattening of the earth as 1:298.3 ± 0.7 (as opposed to J. F. Hayford and William Bowie, whose figure [1912] was 1:298.4 ± 1.5). He investigated the state of equilibrium of the masses of the earth's crust (1908, 1912), the depth of the isostatic surface according to J. H. Pratt's isostasy hypothesis (1909), and the accuracy of the dimensions of Hayford's ellipsoidal earth (1911). In 1915 he made the ellipticity of the equator probable, although he was unable to complete these studies.

BIBLIOGRAPHY

I. ORIGINAL WORKS. Helmert's publications are listed in Poggendorff, III, 610–611; IV, 611–612; V, 516–517; VI, 1076. His articles appeared chiefly in *Zeitschrift für Vermessungswesen, Astronomische Nachrichten, Sitzungsberichte der Preussischen Akademie der Wissenschaften zu Berlin,* and *Verhandlungen der Internationalen Erdmessung.* Important works not mentioned in the text are *Instrumente für höhere Geodäsie* (Brunswick, 1878); *Übersicht der Arbeiten des Geodätischen Instituts unter Generallieutnant Baeyer* (Berlin, 1886); *Das Kgl. Preussische Geodätische Institut* (Berlin, 1890); *Die europäische Längengradmessung in 52° Breite von Greenwich bis Warschau. I. Hauptdreiecke und Grundlinienanschlüsse von England bis Polen* (Berlin, 1893); *Zenitdistanz und Bestimmung der Höhenlage der Nordsee-Inseln Helgoland*. . . (Berlin, 1895); "Geodäsie und Geophysik," in *Enzyklopädie der mathematischen Wissenschaften,* VI, 1, supp. 2 (Leipzig, 1910); and "Die internationale Erdmessung in den ersten 50 Jahren ihres Bestehens," in *Internationale Monatsschrift für Wissenschaft, Kunst und Technik* (1913).

II. SECONDARY LITERATURE. Obituaries include O. Eggert, in *Zeitschrift für Vermessungswesen,* **46** (1917), 281–295; O. Hecker, in *Beiträgen zur Geophysik,* **14,** no. 4 (1918); L. Krüger, in *Astronomische Nachrichten,* **204** (1917); M. Schmidt, in *Jahrbuch der Bayerischen Akademie der Wissenschaften* (1917), 53–58; and R. Schumann, in *Österreichische Zeitschrift für Vermessungswesen,* **15** (1917), 97–100. See also W. Fischer, "Helmert," in *Gedenktage des mitteldeutschen Raumes 1967* (Bonn, 1967), pp. 32–34, 63–64; Paul Gast, "Der Lehrstuhl für Vermessungskunde (Lehrstuhl Helmert)," in his *Die Technische Hochschule zu Aachen 1870–1920* (Aachen, 1920), pp. 247–250; H. Peschel, "Gendenkrede zu Helmerts 50. Todestag am 15. Juni 1967 in Freiberg," in *Vermessungstechnik,* **15** (1967), 334–340; and Rudolf Sigl, in *Neue deutsche Biographie,* VIII (1969), 497–498.

WALTHER FISCHER

HELMHOLTZ, HERMANN VON (*b.* Potsdam, Germany, 31 August 1821; *d.* Berlin, Germany, 8 September 1894), *energetics, physiological acoustics, physiological optics, epistemology, hydrodynamics, electrodynamics.*

Helmholtz was the oldest of four children. From his mother, Caroline Penn, the daughter of a Hanoverian artillery officer, he inherited the placidity and reserve which marked his character in later life. His father, August Ferdinand Julius Helmholtz, was the typical product of a romantic era. Ferdinand Helmholtz had served with distinction in Prussia's war of liberation against Napoleon and had studied philology and philosophy at the new University of Berlin before accepting a poorly paid post at the Potsdam Gymnasium. A passionate, romantic figure, he possessed an acute aesthetic sensitivity which he transmitted to his son: a profound concern with music and painting underlay much of Helmholtz' later work in sensory physiology. Ferdinand also fervently admired the philosophers Kant and J. G. Fichte; Fichte's son Immanuel Hermann Fichte was his close friend and a frequent visitor. Helmholtz' own lifelong devotion to epistemological issues was motivated by the intense philosophical discussions to which he had listened as a boy.

At the Potsdam Gymnasium Helmholtz' interests turned very early to physics, but his father did not have the money to send Helmholtz to the university, and he persuaded his son to turn to medicine, for which there existed the prospect of state financial aid. In 1837 Helmholtz obtained a government stipend for five years' study at the Königlich Medizinisch-chirurgische Friedrich-Wilhelms-Institut in Berlin. In return he committed himself to eight years' service as an army surgeon. He passed his *Abitur* with distinction and left for Berlin in September 1838.

While at the Friedrich Wilhelm Institute, Helmholtz took many courses at the University of Berlin. He studied chemistry under Eilhardt Mitscherlich, clinical medicine under Lucas Schönlein, and physiology under Johannes Müller. Although he took no courses in mathematics, he read privately the works of Laplace, Biot, and Daniel Bernoulli as well as the philosophical works of Kant. During the winter of 1841 Helmholtz began research for his dissertation under Johannes Müller and later moved into the circle of Müller's students. Chief among these were Ernst Brücke and Emil du Bois-Reymond. Confident and sophisticated, du Bois-Reymond seems to have taken the younger Helmholtz as his protégé. He and Brücke quickly won Helmholtz to their program for the advancement of physiology. With Karl Ludwig the three made up the "1847 school" of physiology. Their program reacted sharply against German physiology of previous decades. Philosophically they rejected any explanation of life processes which appealed to nonphysical vital properties or forces. Methodologically they aimed at founding physiology upon the techniques of physics and chemistry. All of Helmholtz' minor papers published between 1843 and 1847, most of which treated problems of animal heat and muscle contraction, clearly reflect the mechanistic tenets of the school.

Helmholtz received the M.D. degree in November 1842. After completing the state medical examinations he was appointed surgeon to the regiment at Potsdam. He maintained his Berlin connections, though, and in 1845 du Bois-Reymond brought the shy young doctor into the newly founded Physikalische Gesellschaft. On 23 July 1847 Helmholtz read to the society his epic memoir "Über die Erhaltung der Kraft," in which he set forth the mathematical principles of the conservation of energy.

In 1848 Brücke resigned his chair of physiology at Königsberg to accept a post at Vienna. When du Bois-Reymond refused the vacant post, Helmholtz was released from his military duty and appointed associate professor of physiology at Königsberg. Before leaving Potsdam he married Olga von Velten on 26 August 1849.

From that time on, Helmholtz led a quiet professional life of tireless labor at his research. At Königsberg he measured the velocity of the nerve impulse, published his first papers on physiological optics and acoustics, and won a European reputation with his invention of the ophthalmoscope in 1851. Both scientifically and socially the early 1850's were a period of widening horizons for Helmholtz. In 1851 he toured the German universities, inspecting physiological institutes on behalf of the Prussian government. In 1853 he made the first of many visits to England, where he formed lasting friendships with various English physicists, especially William Thomson. Despite his success at Königsberg, his situation there was not altogether happy; his wife's already delicate health was further impaired by the cold climate, and he experienced minor priority conflicts with Franz Neumann. In 1855, with the help of Alexander von Humboldt, Helmholtz obtained a transfer to the vacant chair of anatomy and physiology at Bonn.

At Bonn, Helmholtz continued his research into sensory physiology, publishing in 1856 volume I of his massive *Handbuch der physiologischen Optik.* His work took a wholly new turn with his seminal paper on the hydrodynamics of vortex motion of 1858. Helmholtz' philosophical views had begun very early to diverge from his father's idealist position, and from 1855 he began to develop these views publicly in various popular lectures. Although father and son shared epistemological interests and even held common views on the subjective nature of sensory perception, Ferdinand nevertheless remained intensely suspicious of his son's physical and empirical methods. During his years at Bonn this divergence created a strain in their frequent correspondence, although Helmholtz' letters remained dutiful and submissive.

Helmholtz was never satisfied at Bonn. Anatomy was an unfamiliar subject, and there were whispered reports to the minister of education that his anatomy lectures were incompetent. Helmholtz angrily dismissed these reports as the grumblings of medical traditionalists who opposed his mechanistic-physiological approach. At the same time he was becoming the most famous young scientist in Germany. In 1857 the Baden government offered Helmholtz a chair at Heidelberg, then at the peak of its fame as a scientific center. The promise of a new physiology institute convinced Helmholtz to accept in 1858. At the last moment the prince of Prussia intervened to persuade him to stay, but in vain.

The following thirteen years at Heidelberg were among the most productive of Helmholtz' career. He carried on his research in sensory physiology, publishing in 1862 his influential *Die Lehre von den Tonempfindungen als physiologische Grundlage für die Theorie der Musik.* His treatises on physics included "Über Luftschwingungen in Röhren mit offenen Enden" (1859) and his analysis of the motion of violin strings. The Heidelberg years also brought important changes to Helmholtz' personal life. His wife's health had declined steadily and she died on 28 December 1859, leaving Helmholtz with two small children. On 16 May 1861 he married Anna von Mohl, the daughter of Heidelberg professor Robert von Mohl. Anna, by whom Helmholtz later had three children, was an attractive, sophisticated woman considerably younger than her husband. The marriage opened a period of broader social contacts for Helmholtz. The community of physicists in England and the rest of Germany began gradually to displace the Berlin circle as the locus of his scientific interaction.

By 1860 Helmholtz had begun research for volume III of his *Handbuch der physiologischen Optik* in which visual judgments of depth and magnitude were to be treated. The study led him directly into the nativist-empiricist controversy and inaugurated a decade of intense concern with epistemological issues. The death of his father in 1858 had eliminated Helmholtz' reluctance to develop the empiricist position latent in his earlier work. He began that development in volume III of the *Handbuch* (1867), which was an extended defense of the empirical theory of visual perception. In 1868 and 1869 Helmholtz carried that position still further in his work on the foundations of geometry. He summarized his epistemology in the famous popular lecture of 1878, "Die Thatsachen in der Wahrnehmung."

By 1866 Helmholtz had completed his great treatises on sensory physiology and was contemplating abandoning physiology for physics. The scope of physiology had already become too great for any individual to encompass, he wrote in 1868, and while a flourishing school of physiology existed in Germany, German physics was stagnating for lack of well-trained young recruits. When Gustav Magnus' death in 1870 left vacant the prestigious chair of physics at Berlin, Helmholtz and G. R. Kirchhoff, his colleague at Heidelberg, became the primary candidates for the post. The Berlin philosophical faculty preferred Kirchhoff, whom they regarded as the superior teacher. When he refused the post, the nomination went to Helmholtz. Helmholtz' price was high: 4,000 taler yearly plus the construction of a new physics institute to be under his full control. Prussia readily agreed to his terms, for it was widely recognized that his call possessed great political as well as scientific significance in Prussia's bid for the leadership of southern Germany. He accepted the Berlin post early in 1871.

Helmholtz inaugurated his new position with a series of papers critically assessing the various competing theories of electrodynamic action. This work first brought Maxwell's field theory to the attention of Continental physicists and inspired the later research of Helmholtz' pupil Heinrich Hertz, who entered the Berlin institute in 1878. After 1876 Helmholtz contributed papers on the galvanic cell, the thermodynamics of chemical processes, and meteorology. He devoted the last decade before his death in 1894 to an unsuccessful attempt at founding not only mechanics but all of physics on a single universal principle, that of least action.

By 1885 Helmholtz had become the patriarch of German science and the state's foremost adviser on scientific affairs. This position was recognized in 1887, when Helmholtz assumed the presidency of the newly founded Physikalisch-technische Reichsanstalt for research in the exact sciences and precision technology. Helmholtz' friend, the industrialist Werner von Siemens, had donated 500,000 marks to the project, and he himself had been among its foremost advocates. Under his administration the Reichsanstalt stressed purely scientific research.

Although Helmholtz' productivity did not wane, his health began to fail after 1885. He had always suffered from migraine, from which he sought relief in music and mountaineering in the Alps. In old age he began to experience fits of depression which only long vacations could cure. On 12 July 1894 he suffered what appeared to be a paralytic stroke, and he died on 8 September.

Energetics. Before1847 Helmholtz' interest in force conversions had been motivated largely by physiological concerns. The mechanistic school to which he belonged demanded that the hypothesis of a unique "vital force" within the animal body be rejected as the first step to refounding physiology on chemical and physical principles. Assuming such a vital force, Helmholtz believed, was tantamount to assuming a *perpetuum mobile.* Consequently, its refutation necessitated proving that all the body heat and all the muscle force produced by the animal could be derived ultimately from the chemical force released by oxidation of its foodstuffs, with no recourse to a vital force. In this belief Helmholtz had been greatly influenced by the chemist Justus Liebig's *Die Thierchemie* (1842), in which Liebig had attempted to argue away experiments of Pierre Dulong and Cézar Despretz which seemed to refute the chemical theory of animal heat. In 1845 Helmholtz noted that these experiments were invalidated by their assumption that the heats of combustion of complex foodstuffs were equivalent to the summated heats of combustion of their constituent carbon and hydrogen. In 1845 he proved experimentally that chemical changes occur in the working muscle and, in 1848, that heat is generated by muscle contraction.

"Ueber die Erhaltung der Kraft" (1847) set forth the philosophical and physical basis of the conservation of energy. It drew heavily on the works of Sadi Carnot, Clapeyron, Holtzmann, and Joule, although it was far more comprehensive than those previous treatises. The philosophical introduction clearly illustrated the influence of Kantianism on Helmholtz' thought. Science, he began, views the world in terms of two abstractions, matter and force. The goal of science is to trace phenomena to their ultimate causes in accordance with the law of causality; such ultimate causes are unchangeable forces. We can, Helmholtz implied, know the nature of such forces virtually a priori. If we imagine matter dispersed into its ultimate elements, then the only conceivable change which can occur in the relationship of those elements is spatial. Ultimate forces, then, must be moving forces radially directed. Only the reduction of phenomena to such forces constitutes an explanation to which we may ascribe the status of "objective truth" (from the translation in Richard Taylor's *Scientific Memoirs* [London, 1853], p. 118).

That ultimate forces must be of this nature can also be inferred from the impossibility of producing work continually from nothing. That impossibility, Helmholtz demonstrated, is equivalent to the well-known principle of the conservation of *vis viva.* Assuming that principle to hold for a system of bodies in motion, Helmholtz attempted to prove that the forces under which those bodies move must be functions only of position (and hence not of velocity or acceleration) and also radially directed. If a particle m is acted on by a central force of intensity ϕ emanating from a fixed center of force and moves freely from a distance r to a distance R from that force center, then

$$\frac{1}{2}mQ^2 - \frac{1}{2}mq^2 = -\int_r^R \phi\, dr, \qquad (1)$$

where Q is the velocity of m at R and q its velocity at r. The left-hand side of equation (1) is clearly one-half the difference of the *vires vivae;* Helmholtz calls the right-hand integral the "sum of the tension forces" (*Spannkräfte*) between the distances R and r. Equation (1) remains valid when summed over the entire system of bodies and hence expresses the most general form of the principle of the conservation of energy.

Helmholtz then demonstrated how the conservation principle could be applied to various physical phenomena. The principle of the conservation of *vis viva* had already been applied to gravitation, wave motion, and inelastic collision. Previously an absolute loss of force had been assumed in inelastic collision and friction. Helmholtz argued to the contrary that the *vis viva* apparently lost in such cases is merely converted to tension forces or heat; on the latter assumption Joule had recently measured a mechanical equivalent of heat equal to 521 meter-kilograms per calorie in mks units. Helmholtz then proceeded to an extended defense of the dynamic theory of heat against the caloric theory, arguing that the free heat of a body consists in the microscopic motion of its particles, its latent heat in the tension forces between its atoms. He then introduced the equations of Clapeyron and Holtzmann for the expansion of gases. The derivation of Clapeyron's equations, he pointed out, rests upon the untenable assumption that no heat is lost when work is done by a gas in expanding. He concluded by applying the conservation principle to electrostatic, galvanic, and electrodynamic phenomena.

After 1847 Helmholtz' research interests turned for some time to sensory physiology, and he took no direct part in the subsequent development of the entropy concept or kinetic theory. Late in his career, though, he turned again to research into energy processes, this time those of the galvanic and electrolytic apparatus. In 1872 he showed that convection currents within a polarized electrolytic cell can sustain a feeble current even at voltages too low to sustain electrolytic decomposition. That phenomenon had previously seemed to violate either the conservation of energy or Faraday's laws of electrolysis. In 1877 Helmholtz attempted to predict theoretically the electromotive force of a galvanic cell for different concentrations of a salt solution. Under certain conditions the cell can be treated as a reversible cycle and the laws of Carnot and Clapeyron applied to it. The theory was in substantial agreement with experimental data by James Moser.

Helmholtz' research in physical chemistry culminated in his 1882 memoir, "Die Thermodynamik chemischer Vorgänge." Thermochemistry, especially that of Thomsen and Berthelot, assumed that the heat evolved in reactions is a direct measure of the chemical affinities at work. The occurrence of spontaneous, endothermic reactions had always presented an anomaly in this tradition, for such reactions seemed to act against the forces of chemical affinity. In 1882 Helmholtz distinguished between "bound" and "free" energy in reactions. The former is the portion of the total energy which, in accordance with the entropy principle, is obtainable only as heat; the latter is that which can be freely converted to other forms of energy. From Clausius' equations Helmholtz derived the "Gibbs-Helmholtz equation,"

$$F = U - T\frac{\partial F}{\partial T}, \qquad (2)$$

where F is the free energy, U the total energy, T the absolute temperature, and where $\partial F/\partial T$ yields the entropy. In any spontaneous reaction occurring at constant temperature and volume the free energy must decrease. Hence the free energy, not the total energy change measured by the evolution of heat, determines the direction of any reaction. Helmholtz' research had been anticipated by J. W. Gibbs, in whose formulation U in equation (2) must be replaced by the enthalpy.

This research led Helmholtz directly to his investigations into the statics of monocyclic systems and the principle of least action. In the former (1884) he demonstrated that it is possible to define certain mechanical systems the internal motions of which can be shown to obey Clausius' entropy equations. In response to an attack by Clausius he emphasized that the vibrational motion of heat does not rigorously satisfy the conditions of such a system; hence the paper constituted no mechanical derivation of the second law of thermodynamics. In the latter study (1886) he attempted to derive not only all of mechanics, but also thermodynamics and electrodynamics, from the principle of least action as formulated by Sir W. Rowan Hamilton. Although the problem dominated his attention until his death in 1894, Helmholtz achieved no satisfactory derivation. The importance of these studies lies chiefly in their influence upon Heinrich Hertz, who acknowledged his debt to Helmholtz in his *Die Principien der Mechanik* (1894).

Physiological Acoustics. Helmholtz' research in sensory physiology began in 1850, when he determined the velocity of the nerve impulse in the sciatic nerve of the frog. In 1852 he obtained more precise results through his invention of the myograph. This device, in which the muscle traces the motion of its contraction upon a rotating drum, permitted more exact measurement of the small time intervals involved than any previous method. Helmholtz' measurements yielded not only a finite velocity for nerve propagation but also the surprisingly slow one of about ninety feet per second. The result was considered a victory for the mechanistic school, for it seemed to confirm du Bois-Reymond's hypothesis that the nerve impulse consisted in the progressive rearrangement of ponderable molecules.

At the conclusion of these experiments, Helmholtz' interest turned immediately to physiological acoustics. Physicists had long known that a vibrating string produces not only a tone of its fundamental frequency f_1 but also a series of harmonics $2f_1$, $3f_1$, and so on. They also knew that two similar tones f_1 and f_2, when sounded together, would produce beats of frequency $f_1 - f_2$. After 1750 a third phenomenon had come to light: Tartini's tones, or difference tones. If tones f_1 and f_2 are sounded together, then the acute ear can sometimes hear a third tone $f_1 - f_2$ which is not a harmonic. Romieu, Lagrange, and Thomas Young all advanced the obvious "beat theory of difference tones," which held that the difference tone is a beat frequency so great that it has become a tone in itself. In 1832 G. G. Hällström noted that the harmonics of f_1 and f_2 should also beat; that is, one should hear beats or difference tones $2f_1 - f_2$, $f_1 - 2f_2$, $3f_1 - f_2$, and so on. These, however, had never been observed. Finally in 1843 G. S. Ohm advanced his law of acoustics, which asserts that the ear perceives only simple harmonic vibrations. The ear, according to Ohm, decomposes the complex sound waves which it receives into the same simple harmonic waves obtainable mathematically from Fourier analysis.

The perception of beats and difference tones seems to contradict Ohm's law, a fact which perhaps attracted Helmholtz' attention. In 1856 he demonstrated the use of resonators to isolate and reinforce upper partial tones and thus showed the existence of the higher-order difference tones predicted by Hällström. He also proved that in addition to the difference tones $mf_1 - nf_2$ there exist also very faint summation tones $mf_1 + nf_2$. Because summation tones cannot be predicted from the beat theory, Helmholtz regarded it as decisively disproved and advanced his own transformation theory. In simple harmonic motion the restoring force k on a particle m is proportional to its displacement x; but if the square of the displacement is also sensible, then $k = ax + bx^2$. If two wave trains of force $f \sin pt$ and $g \sin (qt + c)$ act on m, then the equation of motion can be written and solved by series. Helmholtz showed that the series solution of that equation contains wave functions of all frequencies mp, mq, $(mp - nq)$, $(mp + nq)$. Hence combination tones result from inharmonic distortions of the wave form, either externally in resonators (as Helmholtz showed) or in the ear at the drum-malleus junction (as Helmholtz believed); hence they do not violate Ohm's law.

Obviously all sounds of the same pitch and intensity do not sound alike. Helmholtz attributed this difference in timbre (*Klangfarbe*) mainly to the different patterns of upper partial tones, which depend on how the fundamental is produced. He advanced this theory first in connection with his fixed-pitch theory of vowel sounds. The vowel *A* sung at pitch f_1 differs from the vowel *E* sung at f_1 by the same individual, Helmholtz argued, only because the mouth serves as a variable resonator. At the vocal cords, both *A* and *E* have the same pattern of upper partials, but the different shapes of the mouth cavity reinforce different ranges of partials, giving *A* and *E* different timbres. Helmholtz also argued that timbre is independent of phase differences among the upper partial tones.

Helmholtz' greatest achievement in physiological acoustics lay in formulating the resonance theory of hearing. Like so much of his physiology, that theory rested upon Johannes Müller's law of specific nerve energies. Müller taught that the nature of the impulse carried to the sensorium by a given nerve is unique and independent of the nature of the external stimulus. Rigidly interpreted, the law seemed to require each just noticeable difference of any sensory quality to possess its own sensory receptor. Between 1850 and 1855 the microscopic anatomy of the cochlea first became known. Among the structures revealed were the rods of Corti strung out in gradually increasing size along the length of the cochlea—analogous, Helmholtz insisted, to the tuned wires of a piano. In 1857 Helmholtz boldly hypothesized that these rods function as tuned resonators. A complex sound wave transmitted to the cochlea fluid sets in sympathetic vibration those rods tuned to the frequency of its simple harmonic components. These rods in turn excite adjacent nerve endings, which transmit the impulse to the sensorium. Hence the resonance theory satisfies Müller's law and provides a physiological explanation of Ohm's law. In 1869 the experiments of Victor Hensen convinced Helmholtz that the transverse fibers of the basilar membrane, not the rods of Corti, are the cochlea resonators. With this modification the theory survived virtually unchallenged until after 1885.

Helmholtz incorporated all these results in his great work *Die Lehre von den Tonempfindungen als physiologische Grundlage für die Theorie der Musik* (1863), in which he applied his discoveries to music theory. Musicians had long known that the most perfect consonances are those whose frequencies are small whole-number ratios. Helmholtz explained this by noting that such consonances have the greatest number of coincident upper partial tones. Less perfect consonances have many slightly different upper partials, and these produce beats which are perceived as dissonance. Later editions of *Tonempfindungen* also incorporated the results of two studies of the ossicular bones carried out in 1867 and 1869, in which Helmholtz evaluated the efficiency and linearity of the ossicular chain as a transformer.

Physiological Optics. Helmholtz inaugurated his study of physiological optics with his invention of the ophthalmoscope in 1851. His friend Ernst Brücke had recently shown how the human eye could be made to glow with diffusely reflected light, like the eyes of many animals. In preparing a lecture demonstration of the phenomenon, Helmholtz realized that by means of a simple optical apparatus this reflected light could be obtained as a magnified, sharply focused image of the subject's retina. He published the mathematical theory of the ophthalmoscope with an account of the improved instrument in 1851.

Helmholtz turned to the intricate problems of color vision in 1852 with an attack on Sir David Brewster's new theory of light. Brewster had maintained the objective reality of three primary colors by supposing, in opposition to Newton, that there exist three distinct kinds of light, each of which excites in the eye one of the sensations red, yellow, or blue. Helmholtz regarded the theory as still another confusion of physical stimulus and subjective response. The experiments by which Brewster claimed to have verified his theory, Helmholtz argued, had actually led Brewster astray, for he had failed to obtain pure spectra.

In conjunction with his attack on Brewster's theory Helmholtz conducted spectrum experiments of his own. To his surprise a mixture of blue and yellow spectral lights yielded a green-tinted white, although a mixture of blue and yellow pigments yields green. From this anomaly Helmholtz elaborated the important distinction between additive and subtractive color mixtures, which he announced in 1852. Yet the same experiments led Helmholtz into serious error. When he attempted to produce white by mixing the pairs of colors which Newton's color theory predicted to be complementary, he succeeded in obtaining pure white only with yellow and indigo. He concluded that this is the only pair of complementaries and rejected Newton's color theory. This aspect of his 1852 paper was immediately attacked by H. G. Grassmann, who, like Maxwell, was attempting to develop a mathematical theory of Newton's color chart. In 1855 Helmholtz acknowledged the experimental error of his earlier paper and announced new experiments which yielded the sets of complementary colors demanded by Newton's theory.

In his paper of 1852 Helmholtz also revived Thomas Young's forgotten theory of color vision. In 1801 Young had hypothesized that each retinal nerve ending possesses three distinct color receptors, each primarily sensitive to one frequency of light. When stimulated, each receptor yields one of the subjective color sensations red, green, or violet. Hence all color sensations except the three primary ones are physiological mixtures. Ironically, Helmholtz revived Young's theory in 1852 only to refute it. He had discovered that spectral colors, when mixed, always yield a duller color of less-than-spectral saturation. Therefore the whole idea that all colors may be obtained from mixtures of three primary colors must be incorrect, he concluded, for the spectral colors, at least, can never be obtained in their full saturation by mixing any three of their number. That fact seemed to refute Young's theory; for if Young's physiological primaries are assumed to be spectral red, green, and violet, then that theory cannot explain how other spectral colors are seen in their full prismatic saturation.

Although Helmholtz dismissed Young's theory in

1852, by 1858 he had changed his mind and become its foremost advocate. In order to save Young's theory from the objections of the 1852 treatise, Helmholtz assumed that Young's physiological primaries are not spectral colors at all, but colors of far greater-than-spectral saturation. Mixtures of the three physiological primaries still undergo the loss of saturation which Helmholtz had noted in 1852, but this lower level of saturation is that of the spectral colors themselves. In this way all the spectral colors which we see can be mixed from three properly chosen physiological primaries, and Young's theory is saved.

Helmholtz' assertion that the physiological primaries possess greater-than-spectral saturation followed logically from another amendment to Young's theory. Helmholtz hypothesized that any wavelength of light, however strongly it excites one set of retinal receptors, always excites simultaneously the other two sets to a much weaker degree. It follows that any physical light, even a single wavelength corresponding to the most saturated color of the spectrum, evokes a color sensation which is not "pure" but a mixture. That mixture, even if it is a spectral color, must necessarily be less highly saturated than the physiological primaries from which it was mixed. In normal vision we never see one physiological primary color alone because there is no obvious way to stimulate one set of retinal receptors without simultaneously stimulating the other two. This fact accounts for the belief that the spectral colors are the most highly saturated which exist. In 1858, however, Helmholtz announced a method by which the pure physiological primaries could be observed approximately. In a paper on afterimages, Helmholtz pointed out that a prismatic color appears far more saturated when viewed after the retina has been fatigued by the complementary color. This fact is easily explained in the Young-Helmholtz theory by assuming that the retinal fatigue briefly inactivates two sets of color receptors. When the third set is then stimulated, we observe its corresponding color less mixed with the other two primaries than usual and hence see it as far more saturated than the spectral colors. Helmholtz regarded this experiment as striking confirmation of his amended version of Young's theory. In 1859 he further demonstrated the power of the theory by using it to explain red color blindness.

Helmholtz incorporated all these results in his *Handbuch der physiologischen Optik*, a massive work which encompassed all previous research in the field. Volume I, which appeared in 1856, contained a detailed treatment of the dioptrics of the eye which was greatly dependent on J. B. Listing's previous works. In it Helmholtz treated the various imperfections of the lens system and announced the result that the visual axis of the eye does not correspond to its optical axis. Volume I also elaborated Helmholtz' theory of accommodation and his invention of the ophthalmometer, both announced in 1855.

In volume II, Helmholtz introduced Young's theory, calling it a special application of Johannes Müller's law of specific nerve energies. He also dealt with the complex phenomena of irradiation, afterimages, and contrast, which had dominated the interest of German physiologists since Goethe's *Farbenlehre* but could be investigated only through difficult and often dangerous subjective experiments. Helmholtz defended G. T. Fechner's explanation of afterimages by the fatigue of retinal elements and advanced his own theory that contrast phenomena arise from errors of judgment and have no physiological basis. He took pains to refute all theories which tried to explain these phenomena through synesthesia or retinal induction, particularly that of Joseph Plateau. Helmholtz took no part in the development of the duplicity theory of vision, which lay outside his main area of interest.

Epistemology. Helmholtz often asserted that the task of modern philosophy is wholly epistemological, and he evinced a dislike for metaphysics. Kantianism exercised a strong influence on his thought as is obvious in his earliest papers. Later, through his physiological study, Helmholtz became convinced that sensory physiology, by revealing the processes of perception, was actually verifying and extending Kant's epistemological analysis. In Müller's law of specific nerve energies he recognized the great principle which explained the role of sense organs in transforming abstract, external stimuli into something wholly different: the immediate sensations of consciousness. Helmholtz' problem was to explain how, despite this radical transformation, we nevertheless have knowledge of the external world.

Helmholtz followed Kant in insisting that the law of causality is transcendental and a priori. In 1855 he asserted that the causal law underlies our belief in external objects, a proposition which prompted charges of plagiarism from Schopenhauer and hence cemented Helmholtz' distaste for metaphysicians. We have immediate experience, he observed, that changes occur in our sensations independent of our volitions. In order that this effect may have a cause, we postulate objects external to ourselves, which can be further analyzed into two categories: matter and force. Whether such objects actually exist must remain a metaphysical question, for both idealism and realism are wholly consistent systems. That the properties of matter and force are constant depends upon our assumption of the lawfulness of nature, which, in turn, rests upon the a priori status of the causal law. In our perception of the world, though, the conclusions we draw about the existence of external objects and forces and their interrelations do not depend upon reflection; they are instantaneous and unconscious. These highly controversial "unconscious conclusions" (*unbewusste Schlüsse*) underlie all of Helmholtz' epistemology and reveal its debt to English associationist psychology.

Because all external stimuli are mediated and radically transformed by the sense organs, our sensations are not images of external reality but tokens or signs of it. We gain knowledge about the external world by experientially discovering how our volitions can alter our sensations. Visual perception of space and localization, for example, is built up in such a way. By willing to touch an object, walk around it, or merely move our gaze we discover experientially that we can alter the accompanying retinal sensations in very regular ways. Hence, through unconscious conclusions certain retinal patterns become associated with objects localized in space, the idea of space itself having first been built up through the sense of touch. The nature of that retinal pattern matters not at all, Helmholtz insisted. All that matters is that we be capable of distinguishing one retinal point from another (in R. H. Lotze's terminology, that each point have its own "local sign") and that the proper eye movements be always able to restore the same retinal pattern.

This empiricist theory of visual perception differed radically from the alternate nativist view. Nativists held that visual perception of space and localization was not wholly learned but was in some sense innate. Johannes Müller, himself a nativist, believed that we are directly aware of the retina's extension in space and that the local signs have an intrinsic spatial meaning. In the sophisticated theory of Ewald Hering, Helmholtz' great rival, the nativist theory was extended to depth perception as well. Helmholtz devoted volume III of his *Handbuch der physiologischen Optik* (1867) to proving that the empiricist theory could explain all the phenomena of visual perception and that nativist theories, especially Hering's, were incorrect or superfluous. Helmholtz first showed that Donder's and Listing's laws, the two basic laws of eye movement, could be easily explained on an empiricist basis in accordance with his own principle of easiest orientation. But Helmholtz organized his primary refutation of nativism around the phenomena of depth perception and binocular vision.

To deal with the problem of single and double vision Johannes Müller had hypothesized that the two retinas must possess paired or "corresponding" points and that each pair of corresponding points contributes to one point of the unified visual field. This hypothesis not only explains why we do not see two visual fields (one corresponding to each eye) but also explains the existence of single and double images. If the eyes fixate on a small object in space, that object will be seen single because its retinal images fall on corresponding retinal points. Many other objects, however, will be seen slightly double, for their retinal images do not correspond. The locus of all points in space which are seen single is called the horopter curve. Its exact determination occupied Helmholtz' attention, and he and Hering published its general mathematical form independently between 1862 and 1864.

Helmholtz regarded the way in which the disparate images of two corresponding points become united into one as the crux of the nativist-empiricist dispute. Understanding how visual perception of space originates also seemed to hang upon that issue, for Charles Wheatstone's invention of the stereoscope in 1833 had revealed the dependence of visual depth perception upon binocular double vision. Many nativists maintained that the nerve fibers leading from pairs of corresponding points become anatomically united before entering the sensorium, so that the resulting single image is an organic fusion of the two different images. Helmholtz and the empiricists believed that disparate images from corresponding retinal points enter the sensorium distinct and intact, and that their union into a single image is an unconscious act of judgment dependent upon prior experience. Against the hypothesis of any organic or anatomical union, upon which he believed all nativist theories must rest, Helmholtz marshaled various observations obtained from Wheatstone's stereoscope. The eyes show surprising ability to fuse the two halves of the steroscopic image, Helmholtz noted, even though these images are different and may fall on noncorresponding points. A still greater problem for nativist theories is the phenomenon of stereoscopic luster. If an area in one half of a stereoscopic drawing is shaded white and the same area in the other half black, the fused image appears not gray but lustrous. Empiricism explains this easily as a learned response to empirical situations in which lustrous objects reflect more light into one eye than into the other. Contrary to nativist principles, the sensations excited in corresponding points clearly yield not a fusion of each but a wholly different sensation. This, Helmholtz claimed, decisively disproved the anatomical union of corresponding points upon which nativist theories depended. The union of binocular images into a single perception of depth, he concluded, must be the result of learning and experience as predicted by the empiricist theory.

Helmholtz' belief in the empirical origin of visual localization did not necessarily conflict with Kant's doctrine that space in general is a transcendental form of perception. But Helmholtz broke sharply with Kant over his claim that the axioms of geometry were also synthetic, a priori propositions. Motivated by his study of visual perception, throughout the mid-1860's Helmholtz investigated the most general analytic expressions of spatial relations. He formulated for himself the abstract mathematical concept of the extended *n*-ply manifold and became convinced that tacit assumptions of congruence and translation underlie the Euclidean axioms. In 1868, before publishing these results, he received a copy of G. F. B. Riemann's treatise of 1854, *Ueber die Hypothesen, welche der Geometrie zu Grunde liegen,* and discovered that most of his results had been anticipated. Nevertheless, he published his own work, emphasizing its one aspect that went beyond Riemann's treatment. Riemann had assumed that in any manifold the distance formula *ds* must be the square root of a homogeneous function of second degree in *dx, dy, dz,* and so on. Helmholtz, starting from the assumption of congruence, proved that Riemann's formula must follow necessarily from that assumption.

Helmholtz' interest in these problems was never that of the pure mathematician. He sought primarily to demonstrate that the Euclidean axioms presuppose the purely experiential facts of translation and congruence. Since geometries other than Euclidean can be developed from these facts, it follows that the Euclidean axioms cannot be the transcendental conditions for our perception of space, as Kant had claimed. Helmholtz' contribution to the development of non-Euclidean geometry was therefore a natural extension of his empiricist philosophical position.

Hydrodynamics. In 1858 Helmholtz published his seminal memoir "Ueber Integrale der hydrodynamischen Gleichungen, welche den Wirbelbewegungen entsprechen," important for both its physical results and its mathematical methods. His motivations for taking up this new research interest remain unclear. One motive seems, however, to have been his interest in frictional phenomena, carried over from his interest in energetics; another was his growing awareness of the power of Green's theorem.

Previously, Helmholtz began, hydrodynamics had assumed the existence of a velocity potential. Yet Euler had noted that there is fluid motion for which no velocity potential exists, including forms of rotary motion and frictional flow. If there exists a single-value velocity potential φ for a given fluid motion, then (in modern vector notation) $\nabla\varphi = \mathbf{v}$ and $\nabla \times \mathbf{v} = 0$. These conditions, Helmholtz showed, exclude the possibility of vortex motion. In cases of fluid motion where rotary motion does occur, then $\nabla \times \mathbf{v} = 2\boldsymbol{\omega}$, where $\boldsymbol{\omega}$ is the angular velocity of a given element. From this fact and from the standard Eulerian equations of motion Helmholtz obtained

$$\frac{d\boldsymbol{\omega}}{dt} = \boldsymbol{\omega}(\nabla\cdot\mathbf{v}).$$

Hence, if the fluid is ever without rotation, that is, $\boldsymbol{\omega} = 0$, then $d\boldsymbol{\omega}/dt = 0$ and the fluid can never begin to rotate. This principle became known as the conservation of vortices.

Helmholtz defined the vortex line (*Wirbellinie*) as the locus of the instantaneous axes of rotation of a rotating particle of fluid. A given vortex line, he proved, is always composed of the same particles of fluid and hence shares their motion through the fluid. He defined vortex tubes (*Wirbelfaden*) as the tubes formed by the vortex lines drawn through all points on the circumference of an infinitely small surface within the fluid. The product of the velocity of rotation and the cross section at any point of a given tube is constant; it follows that such tubes must always be closed within the fluid or terminate on its boundaries.

Helmholtz proceeded to find $\mathbf{v}$ in terms of $\boldsymbol{\omega}$, subject to the three conditions $\nabla\cdot\boldsymbol{\omega} = 0$, $\nabla\cdot\mathbf{v} = 0$, and $\nabla \times \mathbf{v} = 2\boldsymbol{\omega}$. The solution, he asserted, is

$$V_x = \frac{\partial P}{\partial x} + \frac{\partial N}{\partial y} - \frac{\partial M}{\partial z};$$

$$V_y = \frac{\partial P}{\partial y} + \frac{\partial L}{\partial z} - \frac{\partial N}{\partial x};$$

$$V_z = \frac{\partial P}{\partial z} + \frac{\partial M}{\partial x} - \frac{\partial L}{\partial y}. \qquad (3)$$

Here P functions as a scalar potential while $L, M,$ and N function as the components of the modern vector potential **A**. Hence Helmholtz had implicitly set out the Helmholtz theorem, that the velocity field is the sum of irrotational and solenoidal parts. But although Maxwell had introduced the vector potential explicitly in 1856, Helmholtz did not regard his $L,$ $M,$ and N as the components of any physical or mathematical entity. He defined each separately on a strict magnetic analogy. $L, M,$ and N are the volume integrals over the fluid space of the magnetic potential exercised on an external point x, y, z by a magnetic fluid distributed with density ω/r. In other words,

$$L = -\frac{1}{2\pi}\iiint \frac{\omega_x}{r}\, da\,db\,dc,\ M = \text{etc.} \qquad (4)$$

P is defined analogously. It follows immediately for Helmholtz that

$$\nabla \times \mathbf{v} = 2\omega - \nabla \cdot \mathbf{A},$$

and from equation (4) that $\nabla \cdot \mathbf{A} = 0$. Hence solution (3) satisfies the necessary conditions.

Equations (3) allowed Helmholtz to calculate easily the velocity induced in a particle a by a rotating particle b at distance r. He obtained the striking result that

$$d\mathbf{v}_{\text{ind}} = \frac{1}{2\pi} \frac{d\omega_b \times \mathbf{r}}{r^3}.$$

This formula, as Helmholtz pointed out, is exactly analogous to the Biot-Savart force law for the magnetic effect of currents. True to the rigorous, starkly mathematical tenor of the entire paper, Helmholtz regarded this as only a heuristic analogy; his own development was strictly kinematic and mathematical.

Although George Stokes had anticipated certain methods and results of Helmholtz, the 1858 memoir was nevertheless a tour de force; yet it seemed to attract little initial attention beyond involving Helmholtz in an insignificant controversy with Joseph Bertrand. In 1866, though, William Thomson (later Lord Kelvin) made it the basis of his theory of the vortex atom. Helmholtz had proved, Thomson noted, that his vortices share with hard atoms the properties of being conserved, undergoing collision, exerting influence on other vortices at a distance, and possessing well-defined energies. In addition, vortices have suggestive properties not possessed by hard atoms, such as the electrical and magnetic analogies demonstrated by Helmholtz. For a decade Thomson tried to develop a physics based on the assumption that atoms are tiny vortices.

Helmholtz himself never returned to vortex theory, although his previous work did influence his important paper of 1868 on discontinuous fluid motion. Finally, in collaboration with Gustav von Piotrowski, he carried out a series of complex experimental determinations (1860) of the coefficient of internal friction for various fluids.

Electrodynamics. Although Helmholtz had published earlier papers on electrodynamic phenomena, the field began to dominate his research interests only after 1870. This new direction in his research seems to have been motivated chiefly by his desire to bring electrodynamic theory into harmony with the conservation of energy. Concomitant with this purpose, Helmholtz hoped to bring order to a field which he described in 1870 as a "pathless wilderness" of competing mathematical formulas and theories. He undertook his research with three aims in mind: (1) to test the consistency of each contending theory with accepted mechanical and dynamic principles, (2) to derive differing theoretical predictions from each theory, and (3) to carry out experiments in order to decide between competing theories.

In his 1847 memoir Helmholtz had argued that ultimate forces must be conservative. He had also argued that forces cannot be conservative if the force laws expressing them contain terms involving the velocity or acceleration of the ultimate particles between which the forces work. But this argument impugned the fundamental status claimed for Wilhelm Weber's law expressing the electrodynamic force acting between two charged particles e and e'. According to Weber's law

$$F = \frac{ee'}{r^2}\left[1 - \frac{1}{c^2}\left(\frac{dr}{dt}\right) + \frac{2r}{c^2}\frac{d^2r}{dt^2}\right],$$

where r is the distance between e and e', and c is a constant. The formula involves not only the distance r but also its time derivatives; hence, according to Helmholtz, the force must violate the conservation of energy.

Helmholtz' criteria for force laws excited much opposition; his lifelong rival Clausius attacked them in 1853. The proofs upon which they rested were, in fact, incorrect; Helmholtz later acknowledged that force laws involving derivatives of distance can conserve energy, although they cannot be central and obey Newton's third law. Nevertheless, he continued to believe that the form of Weber's law implied physical inconsistencies if not explicit violation of the conservation principle. In 1870 he opened his critique of Weber's law, then the leading Continental formula for the prediction of electrodynamic effects. According to that law, Helmholtz showed, the energy of at least some systems of charges in motion is less than the energy of the same systems at rest. Hence at least some electrostatic equilibriums must be unstable. Furthermore, one can easily show from Weber's formula that two charges $+e$ and $-e$ can, under certain conditions, continue to accelerate spontaneously until their kinetic energy becomes infinite. Therefore, in both cases Weber's law predicts physical absurdities.

Helmholtz' critique provoked a running controversy, conducted with great bitterness by Weber's pupils, which lasted through the 1870's. The principals themselves found great difficulty in even understanding each other, for Helmholtz' conception was entirely of macroscopic phenomena; Weber's, of microscopic charges. The confrontation ultimately proved indecisive, yet it undermined the confidence of Continental physicists in Weber's theory and facilitated acceptance of Maxwell's theory, which replaced Weber's after 1880.

In cataloguing competing electrodynamic theories, Helmholtz also advanced a theory of his own which he believed would embrace many others as special cases. The intrinsic difficulties of electrodynamic force laws like Weber's dictated his decision to derive the force from a potential. In 1848 Franz Neumann had successfully derived all electrodynamic effects for closed currents from a potential. In 1870 Helmholtz showed that the most general form of Neumann's potential must be

$$p = -\frac{1}{2}A^2\frac{ij}{r}\left[(1+k)\,d\mathbf{s}\cdot d\mathbf{s} + (1-k)\frac{(\mathbf{r}\cdot d\mathbf{s})(\mathbf{r}\cdot d\mathbf{s}')}{r^2}\right]. \quad (5)$$

In equation (5) p represents the potential which current element $d\mathbf{s}$ exercises upon element $d\mathbf{s}'$ when $d\mathbf{s}$ carries current i, $d\mathbf{s}'$ current j; $\mathbf{r}$ represents the distance between elements $d\mathbf{s}$ and $d\mathbf{s}'$, and $A = 1/c$, where c is an undetermined, constant velocity. In equation (5) k is also an undetermined constant. For $k = -1$, equation (5) becomes simply a form of Weber's law; for $k = 1$, it becomes Neumann's potential; and $k = 0$ corresponds to Maxwell's theory. The parts of expression (5) which are multiplied by k can be written

$$-\frac{1}{2}A^2 ij\; ds\; ds' k\left(\frac{d^2 r}{ds\, ds'}\right).$$

This expression becomes zero when integrated around the full circuits S and S' if either is closed; hence for closed currents all the competing formulas are equivalent. Differences between formulas can arise only for open currents—those in which, according to Helmholtz, changes in the density of the "free electricity" occur. In 1870 there existed little experimental data on open currents.

The difficulties of open currents also arise in the propagation of electrodynamic effects in magnetic and dielectric media. Helmholtz' discussion of this topic necessitated a comparison of his theory with Maxwell's. In 1870 Maxwell's theory was little known on the Continent, for it differed radically from Continental theories. The latter assumed that a body exerted its electrodynamic action on another at a distance, independent of the intervening media. Maxwell's field theory rejected action at a distance and, as Helmholtz understood it, assumed all electrodynamic action to be propagated through contiguous, progressive polarization of a medium. On the assumption that the luminiferous ether itself is a magnetizable dielectric, Helmholtz noted, Maxwell's theory yields the striking result that electrodynamic disturbances propagate themselves in transversal waves possessing the velocity of light in free space. Like the English physicists, Helmholtz believed the existence of a dielectric ether to be strongly supported by the experiments of Faraday, especially those on diamagnetism.

Pursuing the comparison of the theories, Helmholtz first demonstrated that the derivation of a wave equation for electromagnetic propagation does not depend upon the particular assumptions of Maxwell's theory. If the polarization of the medium is taken into account and the polarization expression $\partial p/\partial t$ is introduced as one term of the current density, then Helmholtz showed how a wave equation could be derived from his own generalized potential law, even though that law rested upon the initial assumption of action at a distance. The velocity of the waves predicted by Helmholtz' wave equation depend upon the electrical and magnetic susceptibilities of free space. If these are assumed to be zero (that the ether is not a magnetizable, dielectric medium) then the velocities become infinite. If the susceptibilities are assumed to be large, then the wave velocities become finite. However, the further assumption that $k = 0$, the condition of Maxwell's theory, is required in order that the waves be wholly transverse and attain the exact velocity of light in free space. In this sense Maxwell's theory becomes a special, limiting case of Helmholtz' more general theory. Continental physicists first became acquainted with Maxwell's theory in this form, through Helmholtz' memoirs.

Like most of the Continental school, Helmholtz still distinguished in electrodynamics between inductive forces, those which tend to set in motion the electricity within a conductor, and ponderomotive forces, those which tend to set in motion the conductor itself. In 1874 he devoted a major paper to demonstrating that his generalized potential formula could serve as a potential for ponderomotive as well as for inductive forces. In this attempt he was merely generalizing and extending the earlier work of Franz Neumann. In 1845 Neumann had derived a simple induction law from Lenz's law and Ampère's expression for the ponderomotive force between current elements. Later, in 1848, he had published his more famous potential formula, equivalent to the induction law for closed currents. Neumann himself had shown that the potential formula could predict ponderomotive effects and had verified its agreement with Ampère's law for many simple cases. Helmholtz extended that verification to three-dimensional, deformable conductors and to cases of open currents.

In the course of the 1874 analysis Helmholtz

discovered a feasible method through which the various theories could be tested experimentally. Ampère's law predicted ponderomotive forces only between infinitesimal elements of conductors carrying closed currents. The ponderomotive force law derived from Helmholtz' potential annexed to Ampère's expression other terms predicting ponderomotive effects due to the free electricity accumulating at the ends of open circuits. In 1874 Helmholtz and his student N. N. Schiller carried out experiments to determine whether the end of an open current, simulated by an electrostatic discharge, would produce ponderomotive effects. They observed none, and Helmholtz reluctantly concluded that the potential law must be incorrect or that the assumptions underlying it were incomplete. In the experiment Helmholtz noted that charge was continually removed from the discharge point of the electrostatic machine through the convective motion of air particles. The potential law denied that such convection currents produced any electrodynamic effects. But if this assumption were false, he pointed out, then in addition to the ponderomotive effects produced by the open current there would be other electrodynamic effects caused by the convection current. The potential law might not then be strictly false but merely incomplete as long as it failed to take into account that effect. In 1876 Henry Rowland conducted experiments in Helmholtz' laboratory which proved that convection currents produce electrodynamic effects. Helmholtz immediately pointed out that the results of both experiments can be predicted either from Maxwell's theory or from the generalized potential law with the dielectric ether. In 1875 Helmholtz had already conducted a different experiment with similar results. He had rotated the plates of a cylindrical capacitor aligned axially in a uniform magnetic field and had observed an induced electromotive force on the plates. This effect could be predicted from the generalized potential law only by assuming that the insulating space between the capacitor plates functioned like Maxwell's dielectric ether.

Logically, the experimental evidence remained inconclusive at the end of 1876. All the major results, Helmholtz noted, could be explained by Neumann's induction law without recourse to a dielectric ether. But although Helmholtz in 1875 presented the choice between theories as still open to experimental decision, in practice he had come gradually to regard the dielectric ether as a necessity and Maxwell's theory as correct. In the Faraday lecture of 1881 he predicted the decline of action at a distance on the Continent and lent full support to Maxwell's theory. His interest in electrodynamics waned after 1876, and his work was taken up by Heinrich Hertz.

Yet Helmholtz did not accept the Maxwellian view that all current consists of the polarization of media. After 1876 his electrical research turned almost entirely to the galvanic pile, and he became firmly convinced that electricity consisted ultimately of discrete charges. In the Faraday lecture of 1881 Helmholtz set out his theory of "atoms of electricity" and his conviction that chemical forces are ultimately electrical in nature.

Conclusion. Helmholtz exerted incalculable influence on nineteenth-century science, not only through the achievements of his research but also through his brilliant popular lectures and his activity as a teacher and administrator. Helmholtz witnessed the final transition of the German universities from purely pedagogical academies to institutions devoted to organized research. The great laboratories built for him at Heidelberg and Berlin opened to him and his students possibilities for research unavailable anywhere in Europe before 1860. In many respects his career epitomized that of German science itself in his era, for during Helmholtz' lifetime German science, like the German empire, gained virtual supremacy on the Continent.

Helmholtz belonged to that brilliant and self-conscious generation of German scientists which arose in open reaction to the scientific romanticism of earlier decades. Yet—far more than they cared to admit—Helmholtz and his generation still harbored many of the preconceptions and even the program of the earlier science. Like many of his romantic predecessors, Helmholtz devoted his life to seeking the great unifying principles underlying nature. His career began with one such principle, that of energy, and concluded with another, that of least action. No less than the idealist generation before him, he longed to understand the ultimate, subjective sources of knowledge. That longing found expression in his determination to understand the role of the sense organs, as mediators of experience, in the synthesis of knowledge.

To this continuity with the past Helmholtz and his generation brought two new elements, a profound distaste for metaphysics and an undeviating reliance on mathematics and mechanism. Helmholtz owed the scope and depth characteristic of his greatest work largely to the mathematical and experimental expertise which he brought to his science. Especially in physiology that expertise, shared by few other physiologists of the day, made possible the imposing theoretical and experimental edifices that Helmholtz erected from the simplest of physiological principles. Although the biophysical program of the 1847 school did not prove wholly successful for physiology in general, in Helmholtz' field of sensory physiology it proved eminently so.

When Helmholtz abandoned physiology for physics in 1871, the former science, he complained, had already grown too complex for any individual to embrace in its entirety. At his death in 1894, that complexity had become true of virtually all fields. Helmholtz was the last scholar whose work, in the tradition of Leibniz, embraced all the sciences, as well as philosophy and the fine arts.

BIBLIOGRAPHY

I. Original Works. Helmholtz' scientific papers have been collected as *Wissenschaftliche Abhandlungen von Hermann Helmholtz,* 3 vols. (Leipzig, 1882). His lectures on popular and philosophical subjects are available as *Populare wissenschaftliche Vorträge,* 3 vols. (Brunswick, 1865–1876); and *Vorträge und Reden,* 2 vols. (Brunswick, 1884). The treatises on sensory physiology are *Handbuch der physiologischen Optik,* 3 vols. (Leipzig, 1856–1867), and *Die Lehre von den Tonempfindungen als physiologische Grundlage für die Theorie der Musik* (Brunswick, 1863); both works went through many later eds. Also available are the lectures delivered by Helmholtz during his years at Berlin: *Vorlesungen über die elektromagnetische Theorie des Lichts,* Arthur König and Carl Runge, eds. (Hamburg, 1897); and *Vorlesungen über theoretische Physik,* 6 vols. (Leipzig, 1897–1907). Most of Helmholtz' major works have been translated into English, including the treatises on physiological optics, ed. and trans. by James P. C. Southall *et al.* (Menasha, Wis., 1924–1925); physiological acoustics, trans. by Alexander J. Ellis (London, 1875); and the popular scientific lectures, trans. by E. Atkinson (London, 1881). Translations of individual memoirs appeared frequently in contemporary English journals; see the Royal Society *Catalogue of Scientific Papers,* VII, 946–947; X, 188–189; and XV, 747–748. A complete bibliography of Helmholtz' works is included in *Wissenschaftliche Abhandlungen,* III, 605–636.

II. Secondary Literature. The standard biography of Helmholtz is Leo Koenigsberger, *Hermann von Helmholtz,* 3 vols. (Brunswick, 1902–1903), which contains descriptions of Helmholtz' papers and lengthy extracts from his correspondence. An abridged trans. into English by Frances A. Welby is *Hermann von Helmholtz* (Oxford, 1906; New York, 1965). See also Emil du Bois-Reymond, *Hermann von Helmholtz, Gedächtnissrede* (Leipzig, 1897). On Helmholtz' role in the discovery of the conservation of energy see Thomas S. Kuhn, "Energy Conservation as an Example of Simultaneous Discovery," in Marshall Clagett, ed., *Critical Problems in the History of Science* (Madison, Wis., 1959), pp. 321–356; and Yehuda Elkana, "Helmholtz' 'Kraft': An Illustration of Concepts in Flux," in *Historical Studies in the Physical Sciences,* **2** (1970), 263–299. One of the few historical treatments of Helmholtz' role in sensory physiology is in Edwin G. Boring, *Sensation and Perception in the History of Experimental Psychology* (New York, 1942), *passim.* Helmholtz himself gives much excellent historical material in the *Handbuch der physiologischen Optik.* Ernst Glen Wever and Merle Lawrence, *Physiological Acoustics* (Princeton, 1954), contains an interesting modern evaluation of Helmholtz' work in physiological acoustics. Several German works treat Helmholtz' philosophical views, although they give no adequate account of their development. See Ludwig Goldschmidt, *Kant und Helmholtz* (Hamburg, 1898); and especially Friedrich Conrat, "Hermann von Helmholtz' psychologische Anschauungen," in *Abhandlungen zur Philosophie,* **18** (1904), which contains an account of his sensory physiology as well. A. E. Woodruff has published two studies of Helmholtz' electrodynamics: "Action at a Distance in Nineteenth Century Electrodynamics," in *Isis,* **53** (1962), 439–459; and "The Contributions of Hermann von Helmholtz to Electrodynamics," *ibid.,* **59** (1968), 300–311.

R. Steven Turner

HELMONT, JOHANNES (JOAN) BAPTISTA VAN (*b.* Brussels, Belgium, 12 January 1579; *d.* Brussels, 30 December 1644), *chemistry, natural philosophy, medicine, mysticism.*

Helmont was from the Flemish landed gentry. His father, Christian van Helmont, was state counselor of Brabant; his mother was Marie de Stassart, of Brussels. In 1609 he married Margerite van Ranst, of the Merode family, and through her became manorial lord of Merode, Royenborch, Oorschot, and Pellines. They had several daughters and one son, Franciscus Mercurius, who edited his father's collected works—the *Ortus medicinae* of 1648—and became known through his collaboration on the *Kabbala denudata* (edited by Knorr von Rosenroth, 1677–1684), his early attempts at teaching the deaf and dumb (1667) and orthopedic treatment of spinal deformity, his friendship with Lady Conway and Leibniz, his life as a wandering courtier and scholar, and his theosophical treatises.

Helmont's formative years were marked by growing skepticism, dissatisfaction with the traditional syllabus, and the combination of mysticism with genuine scientific research. His unorthodox career was due partly to his Flemish family background, combined with his natural enmity to the Schoolmen and Jesuits brought to Belgium following the Spanish occupation. His first course in classics and philosophy was followed from 1594 by studies in a variety of subjects from geography to law, "reaping straw and poor senseless prattle," especially in Martin del Rio's discourses against natural magic and in the study of Stoicism and medical textbooks.

After receiving the M.D. in 1599 Helmont realized the need for more than book learning in medicine. He sought this knowledge on visits to Switzerland and Italy in 1600–1602 and to France and England in 1602–1605; there may have been two London visits, one dated by himself as in 1604 and the other when he "conversed with the Queen herself," probably at the close of 1602. In spite of some medical success—for instance, during an epidemic of plague at Antwerp in 1605—and tempting offers from Ernest of Bavaria, the archbishop of Cologne, and Emperor Rudolf II, which he declined, refusing to "live on the misery of my fellow men" or to "accumulate riches and endanger my soul," he embarked on private research for seven years (1609–1616) at Vilvorde, near Brussels. On his journeys Helmont had learned as little as before and felt the need to explore the first principles of nature in order to rise above the "dung" of traditional learning. He hoped to overcome the prevalence of "useless logic" and *entia rationis* therein by "dismantling" the operations of nature and art and by promoting the seminal virtues of all things through chemistry (*pyrotechnia, per ignem*). In this and in the interest which he took in the controversy over the "weapon salve" and the magnetic cure of wounds, he was influenced by Paracelsus. This involved him in ecclesiastic prosecution for most of the rest of his life.

In 1608 Rudolf Goclenius, Protestant professor of philosophy and a believer in natural magic, published his first treatise affirming the efficacy of a pseudo-Paracelsian ointment applied not to the wound but to the weapon and acting by sympathy over long distances. Between 1615 and 1625 seven attacks and counterattacks were exchanged between Goclenius and the Jesuit Johannes Roberti, who condemned the method as "devil's deceit." In 1621 Helmont's treatise *De magnetica vulnerum . . . curatione* was published at Paris, possibly at Roberti's instigation and against Helmont's will. His argument was naturalistic: Goclenius had been wrong in omitting the presence of inspissated blood on the weapon as essential for the sympathetic effect; on the other hand, Roberti had recourse to the field most unsuitable for assessing natural phenomena—theology and activity of the devil. Helmont considered the effect to be as genuine as those of sympathy and antipathy reported in many tall stories that he related, interlarding his account with satirical invectives against the Jesuits. In 1623 Helmont's "monstrous pamphlet" was denounced by members of the Louvain Faculty of Medicine, probably at the instigation of his literary enemy Henry van Heers.

In 1625 the General Inquisition of Spain condemned twenty-seven of Helmont's "propositions" for heresy, impudent arrogance, and association with Lutheran and Calvinist doctrine. The treatise was impounded the following year, and in 1627 Helmont asserted his innocence and submission to the church before the curia of Malines, which referred the matter to the Theological Faculty of Louvain. He again acknowledged his error and revoked his "scandalous pronouncements" in 1630. Helmont was condemned by the Louvain Theological Faculty in 1633–1634 for adhering to the "monstrous superstitions" of the school of Paracelsus (that is, the devil himself), for "perverting nature by ascribing to it all magic and diabolic art, and for having spread more than Cimmerian darkness all over the world by his chemical philosophy (*pyrotechnice philosophando*)."

Helmont was placed in ecclesiastical custody for four days in March 1634, then was transferred under high security to the Minorite convent at Brussels. After several interrogations he was released but placed under house arrest. This was finally lifted in 1636, but church proceedings against him were not formally ended until 1642, two years before his death. Also in 1642 Helmont obtained the ecclesiastic imprimatur for his treatise on fever, and in 1646 his widow received his official religious rehabilitation from the archbishop of Malines. The "monstrous pamphlet," *De magnetica vulnerum . . . curatione,* was reprinted in the *Ortus medicinae,* not necessarily by Helmont's wish; it may have been inserted by his son, who was editor of the *Ortus.*

Helmont's scientific method and achievement resulted from his extensive use of the balance, quantification, and experiment. Aiming at the invisible, the semina, and forces in visible objects, Helmont applied chemical analysis to the smoke that remains after combustion of solids and fluids. He found this smoke to be different from air and water vapor in that it displays properties specific to the substance of origin. He called the "specific smoke" by the "new term gas" (from *chaos* or perhaps *gaesen,* that is, to effervesce or to ferment). It was also termed "wild" (*spiritus sylvestris*), since it could not be "constrained by vessels nor reduced into a visible body." Helmont described and identified a number of such gases, notably carbon dioxide and, in some cases, carbon monoxide, from burning charcoal, fermenting wine, mineral water, eructations, and the reaction of sulfuric acid and salt of tartar or of distilled vinegar and calcium carbonate. Others were chlorine gas from the reaction of nitric acid and sal ammoniac; a "gas pingue" from dung, the large intestine, or dry distillation of organic matter; sulfur dioxide from burning sulfur (a fatty and combustible phlogiston); the explosive gas from an ignited gunpowder mixture of charcoal, sulfur, and saltpeter; and a "vital" gas in the heart and the blood. Helmont is therefore remembered today as the discoverer of gas.

On the indestructibility of matter, Helmont stated that metals dissolved in acid are not thereby destroyed or transmuted but are recoverable in their original quantity; for instance, silver dissolved in nitric acid is comparable to a watery salt solution. One metal can precipitate another metal—for example, iron can precipitate copper from a vitriol solution—a process which before Helmont had been attributed to transmutation.

Helmont also designed advanced methods for the preparation of sulfuric acid, aquafortis (nitric acid), and in particular hydrochloric acid (*spiritus salis marini* from sea-salt and potter's clay). He studied a variety of alkali salts and was familiar with the neutralizing effect of alkali on acid (notably, following acid digestion, in the duodenum).

Chemical medicines prescribed by Paracelsus, notably mercury preparations, were improved and widely used by Helmont. He also discussed the sedative and narcotic effects of the Paracelsian "sweet spirit of vitriol" (*ether*). Helmont recognized specific gravity as an important diagnostic indicator and an aid in chemical research. He determined it for metals and notably for urine, thus replacing Leonhard Thurneisser's chemical uroscopy. Helmont devised an air thermometer-barometer, and he also used and recommended the pendulum for measuring time and for assessing the destructive powers of vacua and projectiles. In this effort he determined that the resistance of the air, the quality of the powder, the size of the bullet, and the distance of the target were significant for variations in the "swiftness, powers and proportions of motions." He realized the significance of the length—as opposed to the weight—of the pendulum and that the duration of its swings is constant.

Helmont demonstrated acid as the digestive agent in the stomach (following up hints given by Paracelsus—his *acetum esurinum* ["hungry acid"]—and by Quercetanus [Joseph Du Chesne] in 1603, and the elusive Fabius Violet [Sieur de Coquerey, possibly a pseudonym for Du Chesne] in 1635). Helmont himself came close to identifying digestive acid with hydrochloric acid. He also recognized tissue acidity as the cause of pus formation. He described the rhythmic movement of the pylorus and its directing action on digestion; the important role of bile (hitherto regarded as "excremental" and noxious) in the alkaline digestive milieu of the gut; and the combination of blood with a "ferment from the air" (*magnale*), whereby venous blood disposes of a residue that escapes through the lungs in the form of "volatile salts."

Helmont is foremost among the founders of the modern ontological concept of disease. Following Paracelsus, he denied the traditional view of the ancients who believed that diseases were due to an upset of humoral balance (*dyscrasia*) and varied according to individual mixture of humors and qualities (temperament); there were no diseases as discrete entities, but only diseased individuals.

By contrast, Helmont regarded each disease as a morbid *ens,* with a specific morbid *semen.* The latter he believed to be "fertilized" and activated by a "program of action," the morbid image or idea that it contained. This image or idea was "conceived" by the vital principle (*archeus*) of a single organ or the organism as a whole when it was irritated or perturbed by a pathogenic agent, usually from outside. Helmont visualized this agent as endowed with an *archeus* of its own, like any other object in nature, and hence able to penetrate another object, including the human *archeus.* Interaction between these *archei* produces the morbid *ens.* Although begotten by the *archeus* of the patient, the *ens* is not identical with that *archeus,* nor with the pathogenic irritant. The latter, however, "seals" the morbid *ens.* The specific disease then is the result of the conversion of the morbid idea into corporeal effects and local changes.

Through this ontological concept of each disease as a specific entity came the understanding that a variety of diseases are determined by specific pathogenic agents and by primarily local changes. Agents plus changes—the products of a complicated psychophysical interplay of vital principles—act parasitically and weaken the *archeus* so that it is no longer able to act for the common weal. Helmont's rejection of the traditional explanation of all diseases in terms of the "madness of catarrh," that is, down-flow of corrosive mucus produced by vaporized ingesta ascending and condensing in the "cold" brain, was a most conspicuous advance. He demonstrated the local nature of mucus formation and anatomical changes.

His reflections bore fruit in a number of ingenious and advanced observations, especially those concerning the various forms of asthma (the "epilepsy of the lungs"). He identified the causes of hypersensitivity in asthma, notably dust inhaled while working, food, hereditary susceptibility, climate and weather, and, above all, suppressed emotion: "A citizen being by a Peer openly disgraced and injured; unto whom he might not answer a word without the fear of his utmost ruine; in silence dissembles and bears the reproach: but straightway after, an Asthma arises" (*Ortus* [1648], p. 367). Tissue irritability, tonic and clonic muscle movement, and their independence of the brain were also carefully observed by Helmont, especially in hysteria and epilepsy, as was the association of hydrops and edema with the kidney. The changes caused by *tuberculosis* (cavities) were clearly recognized as the result of a *local metabolic change* in the air passages of the lung obstructed by inspissating ("caseous") and, finally, calcifying local secretion. Fever he declared not to be the product of humoral putrefaction, as the ancients believed, but a movement in reaction to irritation and, thus, a *natural healing process.* Consequently, Helmont rejected traditional therapy (directed against humoral imbalance as a whole), notably bloodletting and purging, and replaced it with remedies specifically considering the type of disease, the organ affected, and the causative agent, since no change in blood or humors, in heat or cold, in moisture or dryness will ever achieve the removal of the "thorn."

Helmont's discoveries and advanced scientific and medical views are embedded in his discourses on natural philosophy, cosmology, and religious metaphysics, which are not scientific and are difficult for the modern reader to comprehend—hence the ambivalence in the assessment of Helmont by historians. He is either praised as an exponent of the scientific revolution of his century or condemned as a Hermetic and an occultist. The former view is reached by selection from his works of what seems relevant today or served as a stepping-stone toward modern results and by omission of what does not. The latter view is based on a refusal even to examine his scientific and medical work, since no merit can be expected from a mind that was capable of belief in the philosophers' stone, the magnetic cure of wounds, spontaneous generation, and many other "Hermetic" tenets now recognized as unreal. Obviously neither of these views has a place in history. One must perform a synoptic analysis of the two components of Helmont's work—the scientific and the nonscientific—of how they promoted each other, and of what significance must be attached to their coexistence in terms of the original meaning of concepts that have entered science in one form or another.

A revealing example is the discovery for which Helmont is still remembered in the annals of chemistry, that of gas. For Helmont, gas was bound up with his ideas on matter, its relationship to spirit and soul, and indeed his religious cosmology as a whole. When an object was converted into gas by chemical manipulation, it had lost its shape but had lost nothing essential. On the contrary, it had retained, and now displayed, its pure essence. This essence, the gas or *archeus* of the object, was not in the object but was the object itself in a volatile—spiritualized—form. Hence gas was matter and spirit at the same time—but not simple, inert matter, which Helmont believed to be water. It was matter specifically disposed or "sealed," matter active and alive by virtue of form and function specific to it. It was spirit—but not one that was added, entering and directing matter from outside. In other words, gas represented what was specifically characteristic of each individual object; it was the material manifestation of individual specificity. Hence there were as many gases as there were individual objects. In this view, spirit and matter were regarded as two aspects of the same thing; this was a monistic and pluralistic view of a world consisting of monads (*semina*) and thus was opposed to a dualistic separation of matter and soul. Helmont believed that he had found in gas the empirical solution to the perennial problem of spirit and matter, soul and body. Seen in this light, gas was conceptually related to Aristotle's *entelecheia,* but Helmont emphasized that the latter was an *ens rationis,* a product of human reason, whereas gas was divine truth and reality that could be visualized in the test tube.

Opposing the traditional ("heathen") doctrine of the elements and regarding matter as water, Helmont seems to have been influenced by the biblical and Gnostic-alchemical tradition as well as by Nicholas Cusa. The latter—probably following an early Gnostic (pseudo-Clementine) source—had indicated that plants consist largely of water: the earth in which they grow fails to lose any weight in the process. This was demonstrated in Helmont's experiment in which a willow tree weighing five pounds was planted in 200 pounds of earth. Five years later, the weight of the tree had increased to 169 pounds while the earth had lost no weight. The influence of Cusa on Helmont's use of the balance and quantification is also shown in his examination of specific weights, a method recommended specifically by Cusa to replace the pseudo knowledge of the scholar (*orator, philosophus*) with the simple wisdom of the empiric (*idiota, mecanicus*). Helmont's general tendency to divest objects of their material cover, to "spiritualize" them, and to study the volatile nucleus reveals the influence of Neoplatonism; it is also recognizable in the vitalistic and idealistic interpretation of biological as well as chemical processes, notably of fermentation and the *ens morbi* as image or idea.

Helmont was also a follower of Paracelsus and can

be regarded as the outstanding and most successful of the second generation of Paracelsists. He implemented and advanced Paracelsian philosophy and cosmology through a series of new observations and techniques—which did not hinder him from criticizing and deviating from it on several points. For example, he rejected the interpretation of natural phenomena in terms of astrology and analogy between macrocosm and microcosm—both fundamental to Paracelsus. Moreover, Paracelsus had been familiar with acid digestion in the stomach of some animals and its improvement through the intake of acid with certain mineral waters. Helmont demonstrated that acid is the digestive factor in all animals, and he came close to identifying it with hydrochloric acid. Paracelsus used the term "chaos" (probably the etymological root of "gas") for a variety of ambient media, notably air, from which living beings derive their nourishment. He also spoke of an "essential spirit" in each individual object and of chemical manipulation whereby an inert substance could be made active (*männisch*), notably a salt that became a "violent spirit" on resolution. This may have influenced Helmont to call "certain exhalations that had been quiet before and become wild on dissolution in nitric acid or vinegar" *spiritus sylvestres.* This terminology is found in Helmont's early treatise on the waters of Spa (1624), in which he says that he calls these exhalations "wild" because they resist attempts at solidification, escaping from or breaking the glass if it is sealed before they develop. In subsequent treatises this behavior is said to be characteristic of gas, notably of carbon dioxide. Some remote influence of Paracelsus in this is therefore not unlikely. Yet it cannot be said that the latter had conceived of anything as consistent and scientific as Helmont's discovery. He had at best vague premonitions of it when he emphasized the volatility and specificity of the *arcana,* the invisible bearers of active impulses in nature.

Like Paracelsus, Helmont was not really an alchemist, although at one time he claimed to have received a specimen of the "stone" and to have accomplished transmutation. In fact he normally practiced genuine chemistry. Contrary to Paracelsus, he opposed the opinion that precipitation of one metal from a solution by addition of another metal was due to transmutation, and he gave the proper explanation of the process in scientific terms. He also dropped most of the alchemical symbolism and retained little that was "Hermetic." Nevertheless, Helmont was no scientist pure and simple. The blending of his interests and motives—scientific and nonscientific—is well shown in his ideas on biological time. Against Aristotle, he argued that time is not definable in terms of motion and succession; it is indivisible and devoid of succession, being essentially bound up with duration. This is shown in the life-span and life rhythm specific to each individual and given to the divine *semina* by the Creator. By virtue of this participation in divinity, time (*duratio*) was not different from eternity, as propounded in the Christian (Augustinian) doctrine. On the other hand Helmont showed himself influenced by St. Augustine in visualizing divine *semina* (monads) as the essential components of the universe. His skepticism toward complacent human reasoning and the application of "useless logic" to natural philosophy has also a root in Christian religion and mysticism which is equally recognizable in his fondness for dreams and visions. In these he hoped to achieve union with the object and thus with divine truth. Setting out on the search for the divine sparks in nature, Helmont found his way paved with scientific problems that provided the inescapable challenge directing him to scientific discovery.

BIBLIOGRAPHY

I. Original Works. Published in Helmont's lifetime were *De magnetica vulnerum naturali et legitima curatione contra R. P. Joannem Roberti* (Paris, 1621); for bibliographical notes see A. J. J. Vandevelde (below), pt. 2, p. 720; *Supplementum de spadanis fontibus* (Liège, 1624); see Vandevelde, pt. 2, pp. 722–723, including a bibliography of Henry van Heers, who believed that he was being criticized through Helmont's treatise; *Febrium doctrina inaudita* (Antwerp, 1642); see Vandevelde, pt. 2, p. 724; and *Opuscula medica inaudita:* I. *De lithiasi;* II. *De febribus* (2nd ed. of *Febrium doctrina inaudita*); III. *Scholarum humoristarum passiva deceptio atque ignorantia;* IIIa. *Appendix ad tractatum de febribus sive caput XVI et XVII* (not extant in 1st ed. of *Febrium doctrina inaudita*); IV. *Tumulus pestis* (Cologne, 1644); see Vandevelde, pt. 2; pp. 725–729.

Posthumously published was *Ortus medicinae. Id est initia physicae inaudita. Progressus medicinae novus, in morborum ultionem, ad vitam longam . . . edente . . . Francisco Mercurio van Helmont cum ejus praefatione* (Amsterdam, 1648), followed by the *Opuscula* (repr. from the 1644 ed.), the first collected ed. of Helmont's works. Further eds. were issued at Venice (1651), the first to have an index; Amsterdam (1652), termed the "best" ed.; Lyons (1655, 1667); Frankfurt (1682); and Copenhagen (1707).

Translations of the *Ortus* are J. Chandler, *Oriatrike or Physick Refined* (London, 1662, 1664); Jean le Conte, *Les oeuvres de Jean Baptist Van Helmont* (Lyons, 1671), selected chapters only and unsatisfactory; and Christian Knorr von Rosenroth, *Aufgang der Arztney-Kunst* (Sulzbach, 1683; repr. in 2 vols., Munich, 1971), extremely useful, since it contains commentaries and incorporates translated supplementary passages from the *Dageraed* (see below).

Translations of separate treatises from the *Ortus* are Walter Charleton, *Ternary of Paradoxes of the Magnetick Cure of Wounds. Nativity of Tartar in Wine. Image of God in Man* (London, 1650); and *Deliramenta catarrhi or The Incongruities, Impossibilities and Absurdities Couched Under the Vulgar Opinion of Defluxions* (London, 1650); J. H. Seyfried, *Tumulus pestis. Das ist Gründlicher Ursprung der Pest* (Sulzbach, 1681), largely following the text of the *Dageraed* (not mentioned in the bibliography, but a copy is in the Munich State Library and the author's possession); *Die Morgenröthe* (n.p., n.d. [mid-nineteenth century]), repr. of five treatises from Knorr von Rosenroth's *Aufgang;* Walter Pagel, "Irrwitz der Katarrhlehre. Asthma und Husten. Tobende Pleura," in his *Jo. Bapt. van Helmont* (see below), pp. 144–219; and trans. of Helmont's *On Time,* chaps. 1–46, in *Osiris* (see below), pp. 356–376.

Considered separately is *Dageraed oft nieuwe opkomst der geneeskonst in verborgen grondt-regelen der natuere* (Amsterdam, 1659; Rotterdam, 1660); see Vandevelde, pt. 1, 457; also in facs. repr. (Antwerp, 1944).

It should be noted that the *Dageraed* gives treatises in Flemish but is not a Flemish version of the *Ortus.* On the contrary, it seems to have been written earlier and compiled by Helmont himself, while the *Ortus* was posthumously arranged, edited, and prefaced by Helmont's son. It is more concise than the *Ortus,* and Helmont gives as his motive for writing in the vernacular that truth never emerges more "naked" than when offered in a simple style that makes it accessible and profitable to the common man. Why its publication should have been delayed for some fifteen years after his death is not clear (the 1615 ed., first erroneously referred to in 1826, is a ghost. Nobody has ever seen it and in it events are mentioned after 1615).

The research on which both the *Ortus* and the *Dageraed* are based goes back largely to 1609–1616. When Helmont's house was searched in 1634, no relevant MSS were found; and between 1624 and 1642 nothing was published. Thus most of the works were likely written in 1634–1640, notably during his house arrest in 1634–1636. Finally, Helmont's correspondence with Père Mersenne should be mentioned as published in Mme. Paul Tannery and Cornelis de Waard, *Correspondence du P. Marin Mersenne. Réligieux Minime,* vols. I–III (Paris, 1932–1946), with three letters in vol. II and eleven in vol. III from the years 1630–1631.

II. SECONDARY LITERATURE. Biographical and bibliographical material is found in C. Broeckx, *Commentaire de J. B. van Helmont sur le premier livre du Régime d'Hippocrate: Peri diaites* (Antwerp, 1849), one of Helmont's *juvenilia,* published from the MS for the first time; other *juvenilia* not extant elsewhere: "Commentaire de J.-B. van Helmont sur un livre d'Hippocrate intitulé: peri trophes," in *Annales de l'Académie archéol. belg.,* **8** (1851), 399–433, reprinted separately (Antwerp, 1851); "Le premier ouvrage (Eisagoge in artem medicam a Paracelso restitutam 1607) de J.-B. van Helmont," *ibid.,* **10** (1853), 327–392, and **11** (1854), 119–191, reprinted separately (Antwerp, 1854); "Notice sur le manuscrit Causa J.-B. Helmontii, déposé aux archives archiépiscopales de Malines," *ibid.,* **9** (1852), 277–327, 341–367, reprinted separately (Antwerp, 1852); "Interrogatoires du docteur J.-B. van Helmont sur le magnétisme animal," *ibid.,* **13** (1856), 306–350, reprinted separately (Antwerp, 1856); and *Apologie du magnétisme animal* (Antwerp, 1869); G. des Marez, "L'état civil de J.-B. van Helmont," in *Annales de la Société d'archéologie de Bruxelles,* **21** (1907), 107–123; Nève de Mévergnies, *Jean-Baptiste van Helmont, philosophe par le feu* (Paris, 1935), useful in its biographical section; A. J. J. Vandevelde, "Helmontiana," 5 pts., in *Verslagen en Mededeelingen. K. Vlaamsche Academie voor Taal-en Letterkunde,* pt. 1 (1929), 453–476; pt. 2 (1929), 715–737; pt. 3 (1929), 857–879; pt. 4 (1932), 109–122; pt. 5 (1936), 339–387; H. de Waele, *J.-B. van Helmont* (Brussels, 1947), reviewed by W. Pagel, in *Isis,* **38** (1948), 248–249.

Helmont's natural philosophy and chemistry are discussed in H. Hoefer, *Histoire de la chimie,* 2nd ed., II (Paris, 1869), 134–146; H. E. Hoff, "Nicolaus of Cusa, van Helmont and Boyle. The First Experiment of the Renaissance in Quantitative Biology and Medicine," in *Journal of the History of Medicine,* **19** (1964), 99–117; H. M. Howe, "A Root of van Helmont's Tree," in *Isis,* **56** (1965), 408–419, which presents the Gnostic-neo-Clementine source for the experiment with the willow tree; H. Kopp, *Geschichte der Chemie,* 4 vols. (Brunswick, 1843–1847), I, 117–127; II, 168, 241–243, 273, 344–366; III, 62–190, 227–350; IV, 380; R. P. Multhauf, *The Origins of Chemistry* (London, 1966), pp. 250–252, 285–286, 294–295, 316, 344; W. Pagel, "Helmont, Leibniz, Stahl," in *Archiv für Geschichte der Medizin,* **24** (1931), 19–59; *The Religious and Philosophical Aspects of van Helmont's Science and Medicine,* supp. to *Bulletin of the History of Medicine,* no. 2 (Baltimore, 1944), see pp. 16–26 on the wider implications of "gas"; "J. B. van Helmont (1579–1644)," in *Nature,* **153** (1944), 675; "Van Helmont; The 300th Anniversary of His Death," in *British Medical Journal* (1945), **1,** 59; "J. B. van Helmont *De tempore* and Biological Time," in *Osiris,* **8** (1949), 346–417; "The Reaction to Aristotle in Seventeenth Century Biological Thought," in *Science, Medicine and History, Essays in Honour of Charles Singer,* I (Oxford, 1953), 489–509; "The 'Wild Spirit' (Gas) of John Baptist van Helmont (1579–1644) and Paracelsus," in *Ambix,* **10** (1962), 1–13; and "Chemistry at the Cross-Roads: The Ideas of Joachim Jungius. Essay-Review of H. Kangro, J. Jungius' Experimente und Gedanken zur Begründung der Chemie als Wissenschaft," *ibid.,* **16** (1969), 100–108, includes a discussion of Helmont's interpretation of the precipitation of copper after the addition of iron to a vitriol solution; J. R. Partington, "Joan Baptist van Helmont," in *Annals of Science,* **1** (1936), 359; and *A History of Chemistry,* II (London, 1961), 209–243; C. Webster, "Water as the Ultimate Principle of Nature: The Background to Boyle's *Sceptical Chymist,*" in *Ambix,* **13** (1966), 96; and H. Weiss, "Notes on the Greek Ideas Referred to in van Helmont's *De tempore,*" in *Osiris,* **8** (1949), 418–449.

Helmont's work in medicine is treated in H. Haeser, *Lehrbuch der Geschichte der Medizin und der epidemischen Krankheiten,* 3rd ed., II (Jena, 1881), 344–363; Lester S. King, *The Road to Medical Enlightenment, 1650–1695*

(London–New York, 1970), pp. 37–62, 88–90; P. H. Niebyl, "Sennert, Van Helmont and Medical Ontology," in *Bulletin of the History of Medicine,* **45** (1971), 115–137; W. Pagel, *Jo. Bapt. van Helmont. Einführung in die philosophische Medizin des Barock* (Berlin, 1930); "The Speculative Basis of Modern Pathology. Jahn, Virchow and the Philosophy of Pathology," in *Bulletin of the History of Medicine,* **18** (1945), 1–43; "Van Helmont's Ideas on Gastric Digestion and the Gastric Acid," *ibid.,* **30** (1956), 524; "Harvey and Glisson on Irritability With a Note on Van Helmont," *ibid.,* **41** (1967), 497–514; "Harvey and the Modern Concept of Disease," *ibid.,* **42** (1968), 496–509, written with M. Winder; and "Van Helmont's Concept of Disease—To Be or Not To Be? The Influence of Paracelsus," *ibid.* (in press); W. Rommelaere, "Études sur J. B. van Helmont," in *Mémoires couronnés et autres mémoires p.p. de l'Académie royale de médecine de Belgique,* **6** (1866), 281–541, reprinted separately (Brussels, 1868); and G. A. Spiess, *J. B. van Helmonts System der Medizin verglichen mit den bedeutenderen Systemen älterer und neuerer Zeit* (Frankfurt, 1840).

Helmont's influence is the subject in Allen G. Debus, *The English Paracelsians* (London, 1965), pp. 181–183; and *The Chemical Dream of the Renaissance* (Cambridge, 1968), pp. 25 ff.; F. N. L. Poynter, "A 17th Century Medical Controversy: Robert Witty and William Simpson," in E. A. Underwood, ed., *Science, Medicine, and History. Essays in Honour of Charles Singer,* II (Oxford, 1953), 72–81; P. M. Rattansi, "The Helmont-Galenist Controversy in Restoration England," in *Ambix,* **12** (1964), 1–23; Henry Thomas, "The Society of Chymical Physitians, an Echo of the Great Plague of London," in Underwood, *op cit.,* 56–71; and C. Webster, "The English Medical Reformers of the Puritan Revolution. A Background to the Society of Chymical Physitians," in *Ambix,* **14** (1967), 16–41; "The Helmontian George Thomson and William Harvey: The Revival and Application of Splenectomy to Physiological Research," in *Medical History,* **15** (1971), 154–167.

WALTER PAGEL

HENCKEL, JOHANN FRIEDRICH (*b.* Merseburg, Germany, 1 August 1678; *d.* Freiberg, Saxony, Germany, 26 January 1744), *chemistry, mineralogy.*

The second son of Merseburg's town physician, Henckel was apparently intended for the clergy, having enrolled at the University of Jena to study theology in 1698. But he soon switched to medicine, probably to pursue his childhood interest in "the book of nature." During his medical studies he most likely attended chemistry lectures by G. W. Wedel. After a few years at Jena, Henckel proceeded to Dresden, where he first worked under the supervision of a physician engaged in chemical research and then opened his own practice. In 1711 he resumed his medical studies at the University of Halle, taking an M.D. that year under the chemist G. E. Stahl. Henckel subsequently settled in the important Saxon mining town of Freiberg, where he practiced medicine for the next eighteen years, becoming district physician in 1718 and town physician and mine physician in 1721.

In Freiberg Henckel used his leisure time to give private courses to "lovers of chemistry" and to carry out experiments. He soon became quite proficient in using heat and fire for the chemical analysis of mineral substances. In the 1720's he quickly attracted the acclaim of the German scientific world with the publication of his first major works: *Flora saturnizans* (Leipzig, 1722), an inquiry into the relations and similarities between plants and minerals; *Pyritologia* (Leipzig, 1725), an encyclopedic study of the pyrites; and *De mediorum chymicorum* (Dresden–Leipzig, 1727), an investigation of mediated reactions. Besides regaling his readers with a host of novel experiments and observations, Henckel championed limited empirical research, Stahlian chemistry, and natural religion.

Patronized by an influential noble, Henckel resigned his posts in Freiberg and went to Dresden in 1730. Two years later he used the leverage of a foreign call (possibly from the St. Petersburg Academy) to have himself appointed councilor of mines with a handsome salary and a substantial budget for investigating Saxony's mineral resources. He soon returned to Freiberg and, with state help, established a large laboratory in which he not only discharged his official duties but also carried out his published research and resumed his annual course in metallurgical chemistry. This course, which was only open to six students at a time, soon achieved renown throughout Germany and Eastern Europe for its profundity and utility. Consequently, when Henckel died in 1744, some of his disciples hoped that the course would be continued under a new teacher or, better yet, that a mining academy would be founded. Although nothing was done at the time, these hopes were eventually realized: in 1753 the Saxon government charged C. E. Gellert with teaching chemistry in Freiberg, and in 1765 it created the famous Bergakademie there.

Henckel's influence also extended to the rapidly developing science of chemistry. His exacting course did much to shape the perceptions and techniques of two significant pupils, A. S. Marggraf and M. Lomonosov. His work on pyrites and other minerals exerted a strong influence on J. H. Pott, J. G. Lehmann, and others engaged in mineral analysis. Finally, his publications, which appeared in new German editions and in English and, especially, French translations in the two decades following his death, played an important role in the spread of the Stahlian approach to chemical phenomena.

BIBLIOGRAPHY

A fairly complete bibliography of Henckel's works appears in J. R. Partington, *A History of Chemistry,* II (London, 1961), 706–707. In addition to Partington, who incorrectly gives Henckel's date of birth as 11 Aug. 1679, see Walther Herrmann, "Bergrat Henckel. Ein Wegbereiter der Bergakademie," in *Freiberger Forschungshefte: Kultur und Technik,* **37D** (1962); and *Neue deutsche Biographie,* VIII (Berlin, 1969), 515–516.

KARL HUFBAUER

HENDERSON, LAWRENCE JOSEPH (*b.* Lynn, Massachusetts, 3 June 1878; *d.* Boston, Massachusetts, 10 February 1942), *biochemistry, physiology.*

Henderson was the son of Joseph Henderson, a businessman, and Mary Reed Bosworth. He received his early education in the Salem, Massachusetts, public schools and entered Harvard University at the age of sixteen. Attracted to the study of chemistry, he was especially influenced by T. W. Richards, who taught physical chemistry. Henderson became interested in the application of physicochemical methods and principles to biochemistry, and upon graduating from college in 1898 he entered the Harvard Medical School to obtain training in the biological sciences. After receiving his M.D. in 1902, Henderson spent two years in the laboratory of the biochemist Franz Hofmeister in Strasbourg. When he returned to the United States in the fall of 1904, he went to work in Richards' laboratory at Harvard. The two men became brothers-in-law in 1910 when Henderson married the sister of Richards' wife.

In 1905 Henderson was appointed lecturer in biochemistry at Harvard, where he continued to teach until his death. His contributions to the university during his career were many and varied. He was instrumental in the founding of the department of physical chemistry in the medical school (1920), the fatigue laboratory in the Graduate School of Business Administration (1927), and the Society of Fellows (1932). The first course at Harvard dealing with the history of the sciences in general was offered by Henderson beginning in 1911. He was also largely responsible for bringing George Sarton to Cambridge in 1916. Among the honors that Henderson received were membership in the National Academy of Sciences, the American Academy of Arts and Sciences, and the French Legion of Honor.

A stout man with a red beard (which earned him the nickname of "Pink Whiskers"), Henderson loved good food and French wines. In conversation he could be quite forceful and enjoyed making dogmatic statements that stimulated his audience to respond. The code of behavior of the respectable, hard-working, thrifty Yankee guided his conduct throughout his life. In his scientific work, his strength lay in the interpretation of data and in the discovery of uniformities and generalizations. He was not a proficient experimenter, and he disliked the manipulation of complicated apparatus.

Henderson's broad outlook led him to write on philosophy and sociology as well as science. In spite of the diversity of his interests, however, his work, in retrospect, exhibits a fundamental unity. There is a marked consistency in his approach to the various fields that he studied. During the course of his research, he became impressed with the need for studying the interaction between the variables of a system and with the apparent orderliness of certain systems. His career was largely devoted to the study of the organization of the organism, the universe, and society. The emphasis in his work was always on the importance of examining whole systems.

Henderson reflected as well as contributed to an organismic, holistic trend which played an important role in the thought of the early twentieth century. The character of this trend is exemplified by such philosophies as Alfred North Whitehead's organic mechanism, Jan Smuts's holism, and the theory of emergent evolution as expounded by C. Lloyd Morgan. Organismic and holistic influences entered the social and natural sciences through the development of the functionalist school in anthropology, the gestalt theory in psychology, and organismic biology.

In his early work Henderson applied his knowledge of physical chemistry to the problem of acid-base equilibrium in the body. It was then known that the body fluids are excellent buffers, that is, they resist changes in acidity or basicity, and that this buffering ability depends upon the presence of weak acids (or bases) and their salts. Henderson derived an equation which allowed him to describe quantitatively the action of buffer solutions. The equation, published in 1908, states that

$$(\mathrm{H}^+) = k\,\frac{(\text{acid})}{(\text{salt})},$$

where (H^+) represents the hydrogen ion concentration and k represents the dissociation constant of the weak acid. This equation was converted into logarithmic form by the Danish biochemist K. A. Hasselbalch in 1916 and is now known as the Henderson-Hasselbalch equation. Although only approximately true, it still remains the most useful mathematical device for treating problems dealing with buffer solutions.

Henderson's equation made it clear that a weak acid and its salt act most effectively as a buffer at a hydrogen ion concentration equal to the acid's dissociation constant. This fact explains why carbonic acid and monosodium phosphate, along with their salts, act so efficiently in preserving the approximate neutrality of the body. These acids have dissociation constants of about 10^{-7} moles per liter, which means that they serve as excellent buffers for blood and many other physiological fluids, in which the hydrogen ion concentration is close to the same number of moles per liter.

This work greatly impressed Henderson with the "fitness" of substances like carbonic acid for various physiological processes. At about this time (1908), he became friendly with Josiah Royce and began to attend his philosophy seminars at Harvard. With Henderson's interest in philosophical problems thus stimulated, he proceeded to speculate further concerning the fitness of the inorganic environment to support life. In two books, *The Fitness of the Environment* (1913) and *The Order of Nature* (1917), he concluded that the properties of carbon dioxide, water, and carbon compounds (which he considered to be the chief constituents of the environment as far as the organism is concerned) and the properties of the elements carbon, hydrogen, and oxygen uniquely favor the evolution of complex physicochemical systems such as living beings. He could not believe that this correspondence between the properties of matter and energy and the characteristics of physicochemical systems could be due to chance. He concluded that a kind of order, or teleology, exists in nature and that the origin of this order cannot be explained in mechanistic terms. The universe has to be viewed from two complementary points of view, mechanism and "teleology" (a word which Henderson used to denote order or harmonious unity rather than design or purpose).

As an agnostic, Henderson did not draw any religious or theological conclusions from his consideration of fitness. J. D. Bernal has pointed out that facts cited by Henderson can be taken as evidence that life has to make do with what it has or it would not be here at all, rather than as an indication of some master plan in nature. Henderson's lasting contribution was to make it clear that the inorganic world has placed certain restrictions on the direction that organic evolution can take.

While he was speculating on the order of nature, Henderson was also considering the organization of the body. His studies on the complex buffer systems of the organism and on acidosis contributed greatly to the understanding of these subjects and served to focus his attention on the pattern or order of the organism. According to Henderson, the organism, like nature, had to be considered from two points of view, namely mechanism and organization. The structures and processes of the living being, which are the things that are organized, are in themselves mechanical. The concept of organization, however, is not mechanical but is a rational and teleological relationship between these parts and processes.

As far as the physiologist is concerned, Henderson felt, the investigation of biological organization basically meant the elucidation of the regulatory processes of the body, for example, the mechanisms regulating the acid-base balance which he had elaborated. In this connection, he later came to see Claude Bernard's theory of the constancy of the internal environment as an important and concrete expression of biological organization.

Henderson believed that the concept of organization taught the biologist to recognize the wholeness of the organism and the interdependence of its parts and processes. When he began his study of blood in 1919, he was convinced that every one of the variables involved in the respiratory changes of blood must be a mathematical function of all the others. As data was collected in his laboratory on the relations between the various components of blood, such as the carbon dioxide tension and oxygen, Henderson searched for a graphic device to describe the interrelations between a number of variables. Quite accidentally he stumbled upon the Cartesian nomogram, which is essentially a complex graph made by superimposing two or more simpler graphs.

He began with five experimentally determined equations involving the seven variables which he felt were necessary to explain the respiratory activity of blood. Each of these equations was expressed in terms of two independent variables, free oxygen and free carbon dioxide. A two-dimensional graph can be plotted for each of these equations, and these graphs can be combined into one figure since they all have the same Cartesian coordinates. This technique allowed him to represent all seven variables in one diagram. Each point on the nomogram has seven coordinates, so that if the value of any two variables is known, the values of the other five can be read off the chart. Henderson later learned how to transform these complex nomograms into the type of alignment chart invented by P. M. D'Ocagne, which was much easier to read. After Henderson introduced the nomogram into biology, it proved to be a useful tool for facilitating the visualization of relations between several variables, as well as for saving a great deal of computation. His description of the blood as a physicochemical system was summarized in his classic book *Blood: A Study in General Physiology* (1928).

While Henderson was writing this work, his colleague William Morton Wheeler introduced him to Vilfredo Pareto's *Trattato di sociologia generale* (1916). Henderson was very much impressed by the attempt of this Italian engineer-turned-social scientist to apply the methods of the physical sciences to sociology. Pareto's treatment of society as a system in dynamic equilibrium, similar to the organism, appealed to Henderson, who believed that society, like the body, is an organized system which possesses regulatory processes that tend to stabilize it. This doctrine, which he taught in his sociology course at Harvard, influenced the thought of such men as George Homans, Talcott Parsons, and Crane Brinton.

In his later years, particularly after reading Pareto, Henderson grew increasingly skeptical of metaphysics and came to regret the tone of certain parts of his earlier works. Although he never rejected the concept of fitness, he felt that the philosophical speculations which he had derived from this notion were meaningless. He preferred to regard the apparent existence of fitness as a basic but inexplicable fact and to speculate no further on the subject. He became fully convinced that science is only approximate, not absolute. All metaphysical statements, such as "the external world really exists," he considered nonlogical and hence meaningless for science. Conceptual schemes are used because they are convenient, but they cannot be proven true or false in the sense of facts.

BIBLIOGRAPHY

I. Original Works. A good bibliography of Henderson's published works, with about 125 entries, is Walter Cannon, "Lawrence Joseph Henderson, 1878–1942," in *Biographical Memoirs. National Academy of Sciences,* **23** (1943), 52–58. A number of items missing from the Cannon bibliography are listed in J. Parascandola, "Lawrence J. Henderson and the Concept of Organized Systems" (diss., Univ. of Wis., 1968), pp. 233–238.

Henderson's major books are *The Fitness of the Environment* (New York, 1913); *The Order of Nature* (Cambridge, Mass., 1917); *Blood: A Study in General Physiology* (New Haven, 1928); and *Pareto's General Sociology: A Physiologist's Interpretation* (Cambridge, Mass., 1935). For a selection of his sociological writings, see *On the Social System,* Bernard Barber, ed. (Chicago, 1970).

His most important articles on the regulation of neutrality include "Concerning the Relationship Between the Strength of Acids and Their Capacity to Preserve Neutrality," in *American Journal of Physiology,* **21** (1908), 173–179; and "The Theory of Neutrality Regulation in the Animal Organism," *ibid.,* 427–448. A classic review article on the whole subject of the acid-base equilibrium is "Das Gleichgewicht zwischen Basen und Säuren im tierischen Organismus," in *Ergebnisse der Physiologie,* **8** (1909), 254–325. The researches of Henderson and his co-workers on blood, summarized to a large extent in the 1928 work cited above, originally appeared in a series of ten articles bearing the general title "Blood as a Physiochemical System," in *Journal of Biological Chemistry,* **46–90** (1921–1931).

Harvard University possesses a rich collection of MS material belonging to Henderson, including correspondence, notebooks, unpublished lectures, and an unpublished autobiographical work entitled "Memories." For a description of this collection, see J. Parascandola, "Notes on Source Material: The L. J. Henderson Papers at Harvard," in *Journal for the History of Biology,* **4** (1971), 115–118.

II. Secondary Literature. The only monograph-length study on Henderson is the author's unpublished Ph.D. diss. cited above. A fairly lengthy article on Henderson's scientific work and philosophical views is J. Parascandola, "Organismic and Holistic Concepts in the Thought of L. J. Henderson," in *Journal for the History of Biology,* **4** (1971), 63–113.

The most substantial of the obituary notices is the Cannon article cited above, pp. 31–58. There are a number of more recent short biographical sketches, including Dickinson W. Richards, "Lawrence Joseph Henderson," in *Physiologist,* **1,** no. 3 (1958), 32–37; J. H. T. [John H. Talbott], "Lawrence Joseph Henderson (1878–1942), Natural Philosopher," in *Journal of the American Medical Association,* **198** (1966), 1304–1306; and Jean Mayer, "Lawrence J. Henderson—A Biographical Sketch," in *Journal of Nutrition,* **94** (1968), 1–5. See also John Edsall's excellent article in *Dictionary of American Biography* (Supplement III). A fine account of Henderson's personality is Crane Brinton, "Lawrence Joseph Henderson, 1878–1942," in E. W. Forbes and J. H. Finlay, eds., *The Saturday Club: A Century Completed, 1920–1956* (Boston, 1958), pp. 207–214.

Henderson's role in the founding of the Harvard fatigue laboratory is described by D. Bruce Dill, "The Harvard Fatigue Laboratory: Its Development, Contributions, and Demise," in *Circulation Research,* **20–21,** *Supplement I* (March 1967), I-161–I-170. His views on sociology and their influence were treated at some length in Cynthia Russett, *The Concept of Equilibrium in American Social Thought* (New Haven, 1966). For a discussion of the influence of Henderson's views on fitness, see George Wald, "Introduction," in *The Fitness of the Environment* (Boston, 1958), and Harold Blum, *Time's Arrow and Evolution* (Princeton, 1955).

On his work with buffers, see J. Parascandola, "L. J. Henderson and the Theory of Buffer Action," in *Medizinhistorisches Journal,* **6,** 297–309. Henderson's part in the founding of the Society of Fellows is described in George Homans and O. T. Bailey, "The Society of Fellows, Harvard University, 1933–1947," in Crane Brinton, ed., *The Society of Fellows* (Cambridge, Mass., 1959), pp. 1–37.

John Parascandola

HENDERSON, THOMAS (*b.* Dundee, Scotland, 28 December 1798; *d.* Edinburgh, Scotland, 23 November 1844), *astronomy.*

Thomas Henderson was the youngest of five children of a tradesman. Educated in Dundee, he was taught mathematics by the principal of the Dundee Academy, who had a high opinion of his abilities. In 1813 he began work in the local records office. In 1819 he moved to Edinburgh, where until 1831 he continued to follow a legal career and acted as secretary to the earl of Lauderdale and Lord Jeffrey.

Most of Henderson's astronomical work was done in his spare time. At Dundee he had met Sir John Leslie, William Wallace, and Basil Hall, a naval captain and well-known writer of travel books. In Edinburgh he joined the Astronomical Institution and used its Calton Hill observatory. His eyesight was very poor, but he excelled both in the practice and development of new methods of computation. His first important paper, concerning a new method of calculating occultations, was included in Thomas Young's *Nautical Almanac* from 1827 to 1831 and was also published in the *Quarterly Journal of Science,*[1] to which Henderson contributed twelve papers during the next three years.

Henderson annually visited London on business for the earl of Lauderdale. There he met many astronomers, including Sir James South, who gave him the use of his fine Camden Hill observatory. In 1827 Henderson contributed a paper to the *Philosophical Transactions of the Royal Society* on the difference in longitudes between Paris and Greenwich, and in 1830 he earned a vote of thanks from the Astronomical Society for some calculations he had done for Sir James Ross's forthcoming Arctic expedition. Through such work Henderson's name became widely known in astronomical circles. Young supported him as successor to Robert Blair in the Edinburgh chair of practical astronomy, but John Pond was elected. In 1831, however, Henderson was elected successor to Fearon Fallows as royal astronomer at the Cape of Good Hope. He resigned this post in May 1833 because of ill health and returned to Edinburgh, where in October 1834 he was more or less simultaneously made first astronomer royal for Scotland, professor of practical astronomy in the university, and director of the Calton Hill observatory.

In 1836 Henderson married the daughter of the instrument maker Alexander Adie. She died in 1842, shortly after the birth of their only child, a daughter. Henderson died of a heart disease two years later.

Henderson did not often lecture, and poor health prevented him from being a great astronomical observer, an unfortunate loss, because his computational skills were worthy of better data than he was generally able to obtain. At the Cape his instruments were not particularly impressive, comprising chiefly a ten-foot transit manufactured by the Dollond company, and a poor mural circle made by the firm of W & S Jones.[2] He had only one assistant, a Lieutenant Meadows. Working under great difficulties, they observed the transit of several thousand southern stars. Henderson did not reduce his observations until after his return to Edinburgh, and a select catalogue of declinations[3] and right ascensions[4] included only 172 stars.

Among his other observations at the Cape were those of Encke's and Biela's comets,[5] a transit of Mercury, many occultations of stars, and eclipses of Jupiter's satellites. He also observed Mars and the moon, with a view to deducing solar and lunar parallaxes, and he computed several planetary orbits.

From observations at Greenwich, Cambridge, Altona, and the Cape at the opposition of Mars (November 1832), Henderson deduced a solar parallax of 9.125″;[6] this figure was not as good as Delambre's or Thomas Hornsby's values, but of course it was appreciated that the Mars method was inferior to that using Venus. (Cf. the currently accepted solar parallax of 8.80″.) From simultaneous lunar observations at Greenwich, Cambridge, and the Cape, he deduced a lunar equatorial horizontal parallax of 57′1.8″.[7] Of the many determinations made in the early century, this figure was marginally better than the others.

Henderson's most memorable findings, however, related to the annual parallax of the bright doublet α Centauri (the third brightest star). He announced to the Royal Astronomical Society in January 1839 that declination measurements made at the Cape and reduced at Edinburgh had shown a parallax of 1.16″,[8] a figure about a quarter greater than the accepted value. It is not clear why Henderson withheld the announcement of the measurement for so long or at precisely what stage he recognized a parallactic movement. His interest in the star had been aroused because it possessed an unusually large proper motion. His delay cost him priority. F. W. Bessel announced the much smaller parallax of 61 Cygni two months earlier, while F. G. W. Struve in Dorpat announced the parallax of α Lyrae (Vega).

NOTES

1. **18** (1825), 343–347.
2. *Memoirs of the Royal Astronomical Society,* **8** (1835), 141–168 and *passim,* contains ten papers by Henderson.
3. *Loc. cit.* and **10** (1838), 49–90.
4. *Ibid.,* **15** (1846), 129–146.
5. *Ibid.,* **8** (1835), 240–243; and *Philosophical Transactions of the Royal Society,* **123** (1833), 549–558.
6. *Memoirs of the Royal Astronomical Society,* **8** (1835), 95–104.

7. *Monthly Notices of the Royal Astronomical Society,* **4** (1836–1839), 92–94.
8. *Memoirs of the Royal Astronomical Society,* **11** (1840), 61–68.

BIBLIOGRAPHY

I. ORIGINAL WORKS. Apart from works cited in the notes and a number of smaller communications in the journals mentioned, Henderson published five vols. of *Edinburgh Observations* (Edinburgh, 1838–1843), and seven more were published by his successor, Charles Piazzi Smyth (Edinburgh, 1847–1863). Reductions of his Cape observations were incomplete at his death and were never published together in one separate work. He supervised the reduction of the data for the British Association's publication of *Lacaille's Catalogue of Southern Stars* (London, 1847) but died before it was completed; John F. W. Herschel wrote the pref. For a more complete list of Henderson's many papers and shorter notes, see the Royal Society *Catalogue of Scientific Papers,* III (London, 1869), 273–275; and the obituary notice in *Memoirs of the Royal Astronomical Society* (see below, pp. 392–395).

II. SECONDARY WORKS. For sketches of Henderson's life and work, see the obituaries in *Philosophical Magazine,* 3rd ser., **27** (1844), 60–79; *Memoirs of the Royal Astronomical Society,* **15** (1844), 368–395; *Proceedings of the Royal Society,* **5** (1844), 530–532; and Philip Kelland, *Proceedings of the Royal Society of Edinburgh,* **2** (1846), 35. See also Agnes Clerke's excellent art. in *Dictionary of National Biography,* new ed., IV, 404–406. For the state of positional astronomy in Henderson's time and in relation to him, see R. Grant, *History of Physical Astronomy* (London, 1852), esp. pp. 212, 228, 551.

J. D. NORTH

HENDERSON, YANDELL (*b.* Louisville, Kentucky, 23 April 1873; *d.* La Jolla, California, 18 February 1944), *physiology.*

Henderson was the son of Isham and Sally Nielsen Yandell Henderson. His father was an engineer and the owner and founder of the Louisville *Courier-Journal.* Henderson received his B.A. from Yale in 1895 and then entered the Yale graduate school to study physiological chemistry under Russell H. Chittenden. He was a member of the naval militia from 1897 to 1899 and served as an ensign for one summer on the U.S.S. *Yale* during the Spanish-American War. In 1898 he received his Ph.D. and then undertook further studies in Germany for two years with Albrecht Kossel and Carl Voit. In 1900 Henderson returned to Yale as an instructor in physiology and spent the rest of his career there. He retired in 1938 but continued to work as professor emeritus until his death. Henderson's work brought him numerous awards and honors, including election to the National Academy of Sciences and an honorary M.D. from the Connecticut State Medical Society.

A militant man who fought strongly for his beliefs, Henderson was an active member of the Progressive party and an unsuccessful candidate for Congress in 1912 and 1914. As a scientist he did not hesitate to improvise and use crude, homemade equipment to perform his experiments. In constructing various pieces of apparatus, Henderson sometimes utilized such common objects as a child's rubber ball, the top of a tobacco tin, and a piece of garden hose. Like his friend J. S. Haldane, whom he greatly admired, he criticized attempts to explain physiological phenomena solely in terms of chemistry and physics. He was also like Haldane in his willingness to act as a subject in his own experiments.

His physiological researches were devoted almost exclusively to respiration and circulation. In 1903 Henderson began an investigation of the volume changes of the mammalian heart. This study led him to the ideas which essentially dominated his thought for the rest of his life. He became convinced that the venous return largely determines the volume of blood that the heart can pump. Failure of the circulation, he concluded, was due to failure of the mechanism controlling venous return, which he termed the "venopressor mechanism," and not to failure of the cardiac or vasomotor mechanism.

In the course of this work it was necessary to make a wide incision in the thorax of the experimental animals (dogs) in order to place the cardiometer on the heart. Under these conditions the lungs were collapsed, and air had to be blown into them to maintain respiration. On one occasion the apparatus for artificial respiration was out of order, and air had to be supplied to the lungs by means of a hand bellows attached to the trachea and operated by a janitor. Henderson noticed that the blood pressure was falling even though the heart rate was high, and he supposed that respiration was insufficient. But when the artificial respiration was increased, the animals collapsed and died even more rapidly. If artificial respiration was administered less vigorously, but still in excess, the dog passed into shock. This observation eventually led Henderson to the conclusion that the decrease in carbon dioxide which accompanies excessive pulmonary ventilation is the cause of shock. The discovery by Haldane and his co-workers in 1906 that carbon dioxide plays a role in the control of respiration reinforced Henderson's conviction of the importance of this substance in physiological processes.

Henderson's greatest contribution to science was probably in the practical application of his ideas. He introduced the technique of administering a mixture of carbon dioxide and oxygen, instead of only oxygen, as a method of resuscitation. His conviction that carbon dioxide stimulated circulation and respiration led him to use carbon dioxide-oxygen inhalation as treatment for carbon monoxide poisoning, surgical shock, asphyxia of newborn babies, and similar conditions. This technique proved very successful and saved countless lives. Henderson also became involved in the design of mine rescue apparatus and in the fixing of ventilation standards for the Holland Tunnel. During World War I he supervised the production of gas masks for the Chemical Warfare Service of the U.S. Army.

The use of carbon dioxide in resuscitation was opposed by some on theoretical grounds: it was regarded by many as a poison which had to be eliminated from the body. In addition, patients in states of carbon monoxide asphyxia, shock, and similar conditions already had a low blood-alkali content and thus appeared to be in a condition of acidosis. It was feared that administration of carbon dioxide would aggravate this condition. Henderson showed that a decrease in the blood bicarbonate level does not always involve an excess production of acids and a lowering of the blood pH. In many cases, hyperventilation causes a severe drop in the carbon dioxide tension of the blood. The blood bicarbonate is then also reduced in an effort to keep the pH from rising, but total compensation may not be achieved. Thus it is possible to have a situation in which the blood bicarbonate level is lowered but the blood pH rises instead of falls, a condition now referred to as "respiratory alkalosis." In such cases administration of carbon dioxide can be beneficial.

Henderson's theories did have their shortcomings. For example, he exaggerated the role played by the venous return in regulating the circulation. In his desire to discredit the then current theory of acidosis, he placed too much emphasis on depressed or increased breathing as cause of low or high blood pH. Factors such as the increased production of acids and the failure of the kidneys to excrete sufficient amounts of acid are causes of acidosis—in the sense of low pH—more frequently than he believed.

In the 1930's Henderson developed an elaborate theory concerning the effect of muscle tonus on physiological processes. He felt that the normal reflex muscle tonus was an extremely important factor in maintaining the respiration and circulation. Carbon dioxide acted to increase muscle tonus, and this phenomenon explained in part its effectiveness in conditions such as shock. While tonus apparently does play a role in physiological processes such as circulation, Henderson overestimated its importance.

BIBLIOGRAPHY

I. ORIGINAL WORKS. A complete bibliography of Henderson's works has not been published. A fairly extensive bibliography including all of his most important works appears in his *Adventures in Respiration* (Baltimore, 1938), pp. 288–295, a summary of his major contributions which contains a significant amount of autobiographical material. His most important scientific monograph, written in collaboration with Howard Haggard, is *Noxious Gases and the Principles of Respiration Influencing Their Action* (New York, 1927). His interest in America's liquor problem is revealed in *A New Deal in Liquor: A Plea for Dilution* (Garden City, N.Y., 1934). Two important series of papers by Henderson and his co-workers are eight papers entitled "Acapnia and Shock," in *American Journal of Physiology* (1908–1918), and twelve papers entitled "Hemato-respiratory Functions," in *Journal of Biological Chemistry* (1919–1921).

II. SECONDARY LITERATURE. The only biographies are short sketches, mostly in the form of obituary notices, including Howard Haggard, in *Year Book. American Philosophical Society* (1944), pp. 369–374; C. G. Douglas, in *Nature,* **153** (1944), 308–309; and Cecil Drinker, in *Journal of Industrial Hygiene and Toxicology,* **26** (1944), 179–180. An anonymous biographical article on Henderson also appears in *National Cyclopedia of American Biography,* XXXVI (New York, 1950), 25–26.

JOHN PARASCANDOLA

HENFREY, ARTHUR (*b.* Aberdeen, Scotland, 1 November 1819; *d.* London, England, 7 September 1859), *botany.*

Apart from some important work on the process of vegetable fertilization, Henfrey made few original contributions to botany. He was more influential as an editor, translator, and author of textbooks and manuals. In these capacities he communicated to British naturalists the dramatic developments which were taking place during his lifetime in Continental, and especially German, botany. At a time when British botanists were preoccupied with collection and taxonomy, partly as a result of the influx of exotic new species from India and other British colonial possessions, Henfrey was conspicuous as an advocate of the emerging Continental emphasis on physiological anatomy and comparative morphology.

The son of English parents, Henfrey studied medicine and surgery at St. Bartholomew's Hospital, London, where he was a favorite of Frederic Farre. In 1843, upon completing his clinical training, he was admitted to membership in the Royal College of Surgeons; but weak health dissuaded him from medical practice, and he thereafter devoted his life exclusively to botany. Elected a fellow of the Linnean Society in 1844, Henfrey was in 1847 appointed lecturer on botany at the medical school affiliated with St. George's Hospital, London. In 1854, by which time he had been elected to fellowship of the Royal Society, Henfrey was chosen to succeed Edward Forbes in the chair of botany at King's College, London. He held the chair until his death, having become in the meantime examiner in natural science to the Royal Military Academy and to the Royal Society of Arts. Henfrey's wife, Elizabeth Anne, was the eldest daughter of Jabez Henfrey. Their son, Henry William Henfrey, was a prominent numismatist.

In reporting the developments in Continental botany, Henfrey often chose sides on the leading issues of the day. He aligned himself above all with those botanists who sought to overthrow Matthias Schleiden's theories of cell development and vegetable fertilization. In his writings on cell development he accepted Hugo von Mohl's conception that vegetative cell division normally takes place by the infolding of a distinct outer layer of protoplasm, the "primordial utricle," whose role and very existence were hotly debated at the time. Like Mohl and many other botanists of the day, Henfrey effectively denied the nucleus a role in cell multiplication. On other general issues his thought often reflected that of the leading German botanists. He supported the efforts by Alexander Braun, Mohl, and others to replace Schleiden's emphasis on the cell wall with an emphasis on the protoplasmic cell contents. Although sharing his contemporaries' suspicion of *Naturphilosophie,* Henfrey did not embrace the mechanistic trend then emerging in German physiology. In *The Vegetation of Europe, Its Conditions and Causes* (1852), he joined Edward Forbes in advocating the hypothesis of special "centers of creation" for each plant species. He seemed never to doubt the doctrine of the immutability of species.

Henfrey joined the debate over vegetable fertilization at a time of great excitement about all aspects of plant reproduction. Wilhelm Hofmeister and others were developing the doctrine that sexuality extends throughout the vegetable kingdom and that the mode of reproduction is essentially the same in all plants. Henfrey confirmed in the case of ferns Hofmeister's rule that the mode of embryo production in conifers is intermediate between those of phanerogams and cryptogams. Great interest had also been generated by the attempts of Giovanni Amici and Mohl to discredit Schleiden's theory of fertilization in the flowering plants, and particularly his notion that the pollen grain was the ovule of the plant. Aligning himself quickly with Amici's school of thought, Henfrey eventually focused on the question of when germinal vesicles first appear in the embryo sac. If it could be shown that germinal vesicles existed in the embryo sac before the pollen tube reached it, then Schleiden and the other "pollinists" might at last concede that the pollen tube was not ovular but merely a fertilizing organ which conveyed spermatozoa to preexistent germinal vesicles.

In 1856, in a paper on *Santalum album,* Henfrey showed that germinal vesicles do indeed exist in the embryo sac before the pollen tube reaches it, but in the form of naked protoplasmic units rather than ordinary cells. After the pollen tube reaches the embryo sac, a cellulose coat appears on that germinal body which is to give rise to the embryo. In confirmation of this point, Henfrey showed that the germinal vesicles in ferns also lack cellulose membranes until fertilized by contact with spermatozoa. He further suggested that much of the confusion over vegetable fertilization resulted from the circumstance that in their naked protoplasmic form, the germinal vesicles are readily destroyed or altered by external agents and endosmosis. In the same year Schleiden announced that he was abandoning his theory of fertilization; and although Henfrey's work apparently had nothing to do with this change of mind, it did form part of the evidence used in confirmation and elaboration of Amici's theory of fertilization.

BIBLIOGRAPHY

I. Original Works. The Royal Society *Catalogue of Scientific Papers,* III, 275–276, lists 39 papers by Henfrey. These include a five-part report on the progress of physiological botany in *Annals and Magazine of Natural History,* 2nd ser., **1** (1848), 49–62, 124–132, 274–279, 436–443; **4** (1849), 339–348. Of the other papers, the most important are "On the Developement [sic] of Vegetable Cells," *ibid.,* **18** (1846), 364–368; "On the Reproduction of the Higher Cryptogamia and the Phanerogamia," *ibid.,* 2nd ser., **9** (1852), 441–461; "On the Developement of Ferns From Their Spores" [1852], in *Transactions of the Linnean Society of London,* **21** (1855), 117–140; "On the Developement of the Embryo of Flowering Plants," in *Report of the British Association for the Advancement of Science,* **26** (1856), Transactions of the Sections, 85–87; and "On the Developement of the Ovule of *Santalum album;* With Some Remarks on the Phenomena of Impregnation in Plants Generally," in *Transactions of the Linnean Society of London,* **22** (1856), 69–80. For citations of Henfrey's works as editor, translator, and author of textbooks, see *British Museum General Catalogue of Printed Books,* CI, cols. 692–694.

Of Henfrey's several manuals and textbooks, two enjoyed considerable success. *The Micrographic Dictionary,* which he wrote with J. W. Griffith, went through four editions between 1855 and 1881. Between 1857 and 1884 his *Elementary Course of Botany: Structural, Physiological and Systematic* went through four editions and was for a time probably the leading British textbook on botany.

Besides a number of memoirs by Karl von Naegeli, Mohl, Hofmeister, and others, Henfrey translated J. F. Schouw's *Earth, Plants and Man* (1847), Schleiden's *The Plant* (1848), Mohl's *Principles of the Anatomy and Physiology of the Vegetable Cell* (1852), Braun's *Reflections on the Phenomenon of Rejuvenescence* (1853), and J. A. Stöckhardt's *Chemical Field Lectures* (1855). For brief periods he edited *Botanical Gazette, Gardener's Magazine of Botany, Annals and Magazine of Natural History,* and *Journal of the Photographic Society of Great Britain.* With T. H. Huxley he edited and translated the natural history portion of Taylor's *Scientific Memoirs* (1853).

II. SECONDARY LITERATURE. Obituary notices are in *Proceedings of the Royal Society,* **10** (1860), xviii–xix; and *Annals and Magazine of Natural History,* 3rd ser., **4** (1859), 311–312. See also *Dictionary of National Biography,* IX, 409–410; and J. Reynolds Green, *A History of Botany in the United Kingdom From the Earliest Times to the End of the Nineteenth Century* (London, 1914), pp. 418–419.

GERALD L. GEISON

HENKING, HERMANN (*b.* Jerxheim, Germany, 16 June 1858; *d.* Berlin, Germany, 28 April 1942), *zoology.*

Henking spent most of his professional career in applied fisheries research, but he also made substantial contributions to cytology and embryology in his youth. In 1878 he began studies in zoology at the University of Göttingen and completed a dissertation on the anatomy and development of the mite *Thrombidium* in 1882 under the direction of Ernst Ehlers. Two years later he became an assistant to his former professor and, in 1886, *Privatdozent* at Göttingen. He retained this post until his transfer to the German Fisheries Association in 1892.

During these years Henking published a series of papers on the general biology and development of arachnids and did extensive studies on gametogenesis, fertilization, and embryology of insects. In the course of this work he made the first observations on what were subsequently called sex chromosomes. In the maturing germ cells of the fire wasp *Pyrrhocoris* he noted a deeply staining chromatin body which persisted throughout most of the first meiotic division. At anaphase of the second meiotic division there was a small "chromatin element" (which Henking designated "X") which, unlike the other chromosomes, did not appear to be double. This body went to one of the poles without dividing, lagging behind the other chromosomes, and led to the production of daughter cells with eleven and twelve chromosomes, respectively. Similar observations were subsequently made by other workers, but it was not until 1903 that the extra "chromatin element" was identified as a sex chromosome.

Henking's work as general secretary of the Marine and Coastal Fisheries Section of the German Fisheries Association (1892–1928) dealt with scientific fisheries research and concern for the economic and social well-being of German coastal fishermen. This agency sought to build up marine fisheries so that Germany could compete equally with other European powers in exploiting marine resources.

Henking traveled abroad extensively and studied fisheries techniques which might be adapted to the German coast. His works on the culture of oysters in the United States and on the Norwegian whaling industry were the most significant. He developed statistical procedures for estimating the size of the catch of fish and directed the first statistical survey of the German North Sea fisheries, demonstrating that there had been a great rise in the population of North Sea flounder due to curtailment of intensive fishing during World War I and that the quality and size of the catch decreased substantially when intensive fishing resumed. Henking also made pioneering studies on the migration of fish and on the effect of various ecological factors, such as the nature of the seabed on the size of the catch. His studies on the Baltic salmon, sea trout, and other salmonoids provided the basis for his subsequent successful introduction of the brook trout into the Baltic area.

In terms of fishermen's welfare, some of Henking's more significant activities were the establishment of insurance companies for coastal fishermen, the start of vocational training programs for marine fisheries, the arrangement of government-backed loans which enabled the capital-poor coastal fishermen to acquire new motorized equipment, the development of a first-aid and rescue service for fishermen, and the establishment of societies for the betterment of the social condition of coastal fishermen.

BIBLIOGRAPHY

I. ORIGINAL WORKS. An annotated bibliography of Henking's works to 1928 is in Otto Schubart, "Das literarische Werk von Hermann Henking," in *Zeitschrift für Fischerei,* **26** (1928), 311–342. Some of Henking's major papers are "Beiträge zur Anatomie, Entwicklungsgeschichte und Biologie von *Thrombidium fuliginosum* Herm.," in *Zeitschrift für wissenschaftliche Zoologie,* **37** (1882), 553–663; "Untersuchungen ueber die ersten Entwicklungsvorgänge in den Eiern der Insekten. II. Ueber Spermatogenese und deren Beziehung zur Entwickelung bei *Pyrrhocoris apterus* L.," *ibid.,* **51** (1891), 685–736;

"Norwegen's Walfang," in *Abhandlungen des Deutschen Seefischereivereins,* **6** (1901), 119–172; "Austernkultur und Austernfischerei in Nord-Amerika. Ergebnisse einer Studienreise nach den Vereinigten Staaten," *ibid.,* **10** (1907), 1–186; "Der Schollenbestand im Nordseegebiet nach Beendigung des Grossen Krieges 1914–1918. Uebersicht des Gesamtmaterials der deutschen Marktmessungen," *ibid.,* **13** (1922), 57–103; and "Die Fischwanderungen zwischen Stettiner Haaf und Ostsee," in *Zeitschrift für Fischerei,* **22** (1923), 1–92.

II. SECONDARY LITERATURE. See the following (listed chronologically): O. Schubart, "Das literarische Werk von Hermann Henking," in *Zeitschrift für Fischerei,* **26** (1928), 311–342; E. Fischer, "In Memoriam, Hermann Henking," *ibid.,* **40** (1942), 311–342; and K. Altnoder, "Prof. Dr. Geh. Reg. Rat Hermann Henking," in *Monatshefte für Fischerei* (Hamburg), n.s. **10** (1942), 93.

JAMES D. BERGER

HENLE, FRIEDRICH GUSTAV JACOB (*b.* Fürth, near Nuremberg, Germany, 19 July 1809; *d.* Göttingen, Germany, 13 May 1885), *anatomy, pathology.*

A student and the closest co-worker of Johannes Müller, Henle helped prepare the way for cytology through his studies on epithelia; created the first histology based on extensive microscopical investigations; and, through his theory of miasma and contagion, was among the precursors of modern microbiology. His father, Wilhelm Henle, was a merchant; his mother, Helena Sophia Diespeck, was the daughter of a rabbi. The social position of a Jewish family in the small town of Fürth was rather circumscribed, but increasing economic prosperity finally made possible relations with cultured circles. Henle received his first instruction at home from a private tutor; later he attended the Gymnasium at Mainz and Coblenz. His education was directed primarily toward classical and modern languages; he was also a good draftsman and was musically talented. In 1820 he suffered an attack of periostitis; it subsided but often recurred. After the family had converted to the evangelical belief in 1821, Henle for a time thought of becoming a minister. Medicine was not considered until, in Coblenz, he met Johannes Müller socially at a home musicale. Henle began his medical studies in October 1827 at the University of Bonn, where he became a member of the Burschenschaft (students' association) in the fall of 1829. Soon afterward, disappointed by the unkind behavior of other students, he severed this connection by continuing his studies at the University of Heidelberg in the spring of 1830. A year later he returned to Bonn and passed the examination for the doctorate there in August 1831.

Henle's continuing interest in anatomical investigations was rewarded by Müller's inviting him on a trip to Paris, where they met Cuvier and Dutrochet. Henle received the M.D. on 4 April 1832 at Bonn with a dissertation on the pupil membrane and the blood vessels within the eye. In March 1833 he passed the state medical examination in Berlin and immediately became an assistant to Müller, who in April 1833 was named professor of anatomy and physiology at Berlin. In the fall of 1834 Henle became Müller's prosector at the Anatomical Institute. A first attempt to qualify as lecturer failed for political reasons, since all former members of the Burschenschaft were suspected of being enemies of the state. In July 1835 Henle was arrested for this reason and detained to await trial, but through the intervention of Alexander von Humboldt and others he was released from confinement after four weeks. Meanwhile, he lost his post as prosector and, following a long investigation, was condemned to six years in prison in January 1837, yet within a few weeks he was pardoned and thus could return to his post. In the same year he qualified as lecturer in Berlin.

Beginning in the fall of 1840 Henle was professor of anatomy and physiology at Zurich, where Albert Koelliker was his prosector. From his close friendship with the clinician Karl Pfeufer there emerged the *Zeitschrift für rationelle Medicin.* In the summer of 1844 Henle became professor of anatomy and physiology at Heidelberg, along with Friedrich Tiedemann; when the latter retired, Henle also took over the direction of the Anatomical Institute. The last post of his academic career was at Göttingen, to which he was called in the late summer of 1852. He was active there for thirty-three years.

Very revealing of the romantic and sentimental young Henle is his first marriage. During his stay in Zurich he met and fell in love with Elise Egloff, who worked as a governess in the house of his friends; he set her up in her own lodgings and later arranged for his sister to educate her and give her social polish. They were married in March 1846. One son and one daughter resulted from this union, which ended barely two years later with his wife's death from tuberculosis. In August 1849 Henle married Marie Richter, the daughter of a Prussian officer; they had four daughters and one son.

Henle's health often hindered his activity, since slivers were frequently discharged at the site of the periostitis; he also suffered from neuralgia. He died of renal and spinal sarcoma.

Henle was very sociable. He loved witty conversation, encouraged home musicales and evening gatherings for reading, and was happy to open his house for concerts. His political ideas were liberal and nationalistic, but he was unable to become reconciled to Prussia's domestic politics.

Henle belonged to many scientific organizations, including the Leopoldine Academy, the Belgian Academy of Medicine (honorary member), the Bavarian Academy of Sciences of Munich (foreign member), the Petersburg Academy (corresponding member), the Swedish Academy (foreign member), the Berlin Academy (corresponding member), the Royal Society (foreign member), the Petersburg Academy of Medicine (honorary member), and the Royal Academy of Sciences (Amsterdam).

Seldom is anyone introduced to scientific work as Henle was. Johannes Müller began to edit the *Archiv für Anatomie, Physiologie und wissenschaftliche Medizin* in 1834, and it became a clearinghouse for studies oriented toward the natural sciences. Henle assumed the major share of the work of editing it and thus became familiar with current topics of biology. After undertaking comparative anatomical studies on the electric organ of the ray and on annelids, he soon turned his attention to increasingly precise microscopical research. Hints had been accumulating from all sides regarding the smallest structural elements of plant and animal organisms. The concept of the cell first became current among the botanists but was quickly extended to animals as well. Along with Gabriel Valentin, Henle was among the first authors to use the term "cell." In volume XI of the *Encyclopädisches Wörterbuch der medicinischen Wissenschaften* (1834) he had written on epidermis and epithelium but gave only a rather general account for both. His elucidations of fibrous cartilage and fatty and fibrous tissue, which appeared in 1835 in volume XII of the work, were also only general. Yet in the same year, in volume XIII (p. 125), he described the components of the gall bladder as cylindrical corpuscles, some of which appear alone and some "like basalt columns joined lengthwise, so that the chopped-off end surfaces lie in a plane. If they are turned upward under the microscope, these surfaces appear more or less angular and like cells." This description doubtless refers to the highly prismatic epithelium of the gall bladder.

In 1837 Henle presented as his *Habilitationsschrift* an investigation of the epithelium of the intestinal villi which demonstrated that he was already one of the leading experts in this field of histology. The extent of his progress is also shown by his lecture of 16 February 1838 to the Hufelandsche Medicinisch-Chirurgische Gesellschaft on mucus and pus formation. He consistently called the structural elements of the epithelium "cells"; he also described the epithelium of the urinary bladder as a form intermediate between the cylindrical and the pavement epithelium (p. 6). The dependence of the forms of the cell and of the nucleus on position and pressure in their vicinity is clearly shown (p. 7); on the other hand, Henle's observations on the origin of pus cells later proved to be false. He reported on the extension of the epithelia in the human body in 1838, distinguishing three types: pavement, cylindrical (columnar), and ciliated. Moreover, he established that they cover all the liquid-free surfaces of the body, all the inner surfaces of its canals and ducts, and all the walls of its cavities.

Henle's study of the larynx, which was highly praised by Humboldt, was a completely independent comparative anatomical work. In the last period of his activity in Berlin he studied problems in pathology that he had encountered in his editorial work on Müller's *Archiv*. By far the most important of these was his article "Von den Miasmen und Kontagien," which foresaw the bacterial nature of many diseases. At this time the term "miasma" was used for causes of disease which acted on the body from the outside, while "contagia" acted on or in the body itself. But sharp boundaries between these two should not exist, since diseases originating from miasma could become contagious. Most important was the knowledge that the carriers of disease were actually living material and that therefore the contagium, like a parasite, colonized the host body. As a result of its own powers of reproduction, even a small group of contagia suffices to cause a specific disease. Such ideas, although not absolutely new, were often considered unworthy of belief; it required more than thirty years for their acceptance.

Müller's influence on Henle expressed itself principally in the latter's comparative-anatomical and zoological investigations. With Henle's move to Zurich this source of inspiration was closed off; he continued his studies, begun in 1839 at Berlin, for his book *Allgemeine Anatomie* (1841). The first part of the work treated the chemical composition of the human and animal body; it offered nothing original. In contrast, the second part, "Lehre von den Formbestandteilen," was a major advance on Bichat's efforts of forty years before. Bichat's achievements were fully recognized by Henle, who called him the "creator of histology" (p. 122). But now the development of the microscope and the progress in the techniques of investigation offered far better possibilities, as Henle's historical sketch (pp. 134–149) shows. Yet his arrangement was incomplete: he did not place the connective and supporting substances together, as Reichert did for the first time four years later. Even the different epithelia did not yet form a unified group in Henle's system. On this he wrote (pp. 132–133): "A rational system of histology must employ the transformations of the cells as a principle of classification, so that groups of tissue can be formed according to whether, for example, the cells remain discrete or join lengthwise in rows, or expand into star shapes, or split into fibers, and so forth." Since the available information did not permit such a systematic classification, Henle satisfied himself with demonstrating only occasionally the relationship between basic elements.

It is remarkable that, despite his great experience in microscopy, he could not free himself of Schwann's error concerning cell formation. He believed that the basic material for the formation of new cells consisted of an unformed mass called cytoblastem. He did not know of the divisions within the cell; yet Remak described them in 1841, the year in which Henle's *Allgemeine Anatomie* was completed. Henle's work derives its particular importance from its constant attention to physiology—and thus to function, nourishment, development, and regeneration of the various tissues. This physiology of tissues was for Henle the foundation of general or rational pathology, "which attempts to understand the processes and symptoms of disease as the lawlike reactions of an organic substance endowed with peculiar and inalienable powers against abnormal external influences" (*Allgemeine Anatomie,* foreword, p. vii).

An important work of its time was Henle's *Handbuch der rationellen Pathologie.* It stems chiefly from his years at Heidelberg and presents pathology, one of the fundamentals of the physician's activity, again as resting on scientific knowledge; previously medicine—at least in German-speaking areas—was practiced for several decades primarily in the light of *Naturphilosophie.* Of still greater influence was his *Handbuch der systematischen Anatomie.* Composed in Göttingen over a period of sixteen years, it contains the entire contemporary knowledge of the structure of the human body and a multitude of good illustrations. If the presentation of the central nervous system is disregarded, the work may still be useful for orientation in the subject of gross human anatomy and its occasional variations. It did not become obsolete as a textbook until the functional approach to anatomy gained dominance.

Henle's name is best known today for the loop-shaped portion of the nephron named for him. His observation of it in 1862, supported by isolation preparations, was correct in itself but the interpretation was completely wrong: according to Henle, there were two loop-shaped tubules, each of which was connected at one end to a different renal corpuscle. Nevertheless, his study resulted in a new series of investigations on the kidneys through which, between 1863 to 1865, their structure was definitively determined.

BIBLIOGRAPHY

Henle's most important publications are *De membrana pupillari aliisque oculi membranis pellucentibus* (Bonn, 1832), his inaugural diss.; *Symbolae ad anatomiam villorum intestinalium, inprimis eorum epithelii et vasorum lacteorum* (Berlin, 1837); "Ueber Schleim- und Eiterbildung und ihr Verhältnis zur Oberhaut," in *Hufelands Journal der praktischen Heilkunde,* **86,** pt. 5 (1838), 3–62; "Ueber die Ausbreitung des Epitheliums im menschlichen Körper," in *Archiv für Anatomie, Physiologie und wissenschaftliche Medizin* (1838), pp. 103–128; *Vergleichend-anatomische Beschreibung des Kehlkopfes mit besonderer Berücksichtigung des Kehlkopfes der Reptilien* (Leipzig, 1839); *Pathologische Untersuchungen* (Berlin, 1840), pt. 1 also issued separately by Felix Marchand as *Von den Miasmen und Kontagien,* no. 3 in Sudhoffs Klassiker der Medizin (Leipzig, 1910) and published in English translation by George Rosen as "On Miasmata and Contagia," in *Bulletin of the Institute of the History of Medicine,* **4** (1938), 907–983; *Allgemeine Anatomie,* vol. VI of S. T. von Soemmering, *Vom Baue des menschlichen Körpers* (Leipzig, 1841); *Handbuch der rationellen Pathologie,* 2 vols. in 3 pts. (Brunswick, 1846–1853); *Handbuch der systematischen Anatomie des Menschen,* 3 vols. in 7 pts. (Brunswick, 1855–1871); and "Zur Anatomie der Niere," in *Abhandlungen der Gesellschaft der Wissenschaften zu Göttingen,* **10** (1862), 223–254.

On Henle and his work, see two writings by Friedrich Merkel: *Jacob Henle* (Brunswick, 1891), a detailed biography with an evaluation of Henle's scientific work and a complete bibliography, pp. 403–407; and *Jacob Henle. Gedächtnisvortrag* (Brunswick, 1909).

ERICH HINTZSCHE

HENRI. See **Henry.**

HENRICHSEN, SOPHUS (*b.* Kragerø, Norway, 11 November 1845; *d.* Oslo, Norway, 21 December 1928), *physics.*

Henrichsen's father, Johan Georg Henrichsen, was an office manager; his mother was Sophie Septima Moe. In 1873, the year of his marriage to Julie Adolfine Marie Forsberg, he graduated from the Institute of Physics at the University of Kristiania (now Oslo) and became amanuensis there. From 1890 to 1920 he was senior master at the technical school in Kristiania.

Henrichsen was interested in all branches of science, but his main scientific achievements were measurements of the dependence of certain physical quantities on temperature. His first scientific work was an experimental determination of the relation between temperature and electric conductivity in a sulfuric acid solution.

The dependence of the specific heat of water on temperature was measured by several physicists in the last part of the nineteenth century, but their results differed to some extent. During a stay at Gustav Wiedemann's institute at Leipzig in 1877–1878, Henrichsen carried out several such measurements by means of an improved Bunsen calorimeter. His results were considered among the most reliable of his time. With his colleague S. Wleügel he also investigated the magnetic properties of many organic liquids; their findings led to general conclusions about how the magnetic properties of such liquids depend on their chemical constitutions and temperature. For these researches he received the crown prince's gold medal from the University of Kristiania in 1887.

After 1890 Henrichsen devoted almost all his time to teaching and wrote several textbooks. He was a coeditor of a popular scientific journal, *Nyt Tidsskrift for Fysik og Kemi,* which was founded in 1896 and is still published.

BIBLIOGRAPHY

Henrichsen's most important papers are "Ueber die specifische Wärme des Wassers," in *Annalen der Physik und Chemie,* **8** (1879), 83–92; and "Ueber den Magnetismus organischer Verbindungen," *ibid.,* **45** (1892), 38–54.

A secondary source is *Norsk Biografisk Leksikon,* VIII (1932), 25–27.

KURT MØLLER PEDERSEN

HENRION, DENIS or **DIDIER** (*b. ca.* 1580; *d.* Paris [?], France, *ca.* 1632), *mathematics.*

Information on Henrion is very scarce and imprecise. The date and place of his birth are unknown. In 1613 he speaks of his youth; and since he had been an engineer in the army of the prince of Orange before settling in Paris in 1607, his birth may be placed around 1580. The date of his death is somewhat delimited by the appearance in 1632 of the French edition of Euclid's *Elements* and *Data,* under his name, sold "en l'Isle du Palais, à l'Image S. Michel, par la veusve [widow] dudit Henrion." His first name is generally indicated as Denis, although he always gave only the initial D.—except in a Latin writing of 1623, where it is given as Desiderius, the Latin form of Désiré or Didier.

Henrion's scientific activity was devoted mainly to private instruction and to the translation into French of Latin mathematical texts. From 1607 it seems to have taken place exclusively in Paris. His first work, published in 1613, is a course in elementary mathematics, in French, for the use of the nobility, that is, for the instruction of officers. Although it displays no great originality, it is a serious work, most particularly the section on geometry, which contains a group of 140 remarkable problems. Yet here, as in all of his work, Henrion drew very freely on his predecessors, especially Clavius.

Henrion's various editions of Euclid's *Elements* were really only translations of the Latin editions done by the Jesuits at Rome. He sometimes embellished them with a summary of the algebra in which this science was presented in a quite antiquated manner, without regard for the advances made by Viète, Albert Girard, and Stevin.

When the *Data* was combined with this translation of Euclid in 1632, Henrion translated the introduction by Marinus of Flavia Neapolis and the text itself from the Latin of Claude Hardy (1625). His French translation of Theodosius of Tripoli's *Spherics* (1615) was drawn from the Latin paraphrase by Clavius (1586).

Henrion's other works are in the same vein. His *Traité des logarithmes* (1626), taken from the work of Briggs, saved his name from oblivion by being the second work on the subject published in France—the first was that of Wingate (1625)—and the first written by a Frenchman.

Henrion was greatly interested in mathematical instruments, especially in the proportional divider, the invention of which he attributed to Jacques Alleaume, who had constructed several copies of it in Paris. He also described the slide rules of Edmund Gunter in the *Logocanon* (1626).

His work was not untouched by polemic. Henrion often bore a grudge against his competitors, the other translators of Euclid and writers of manuals. He was severely taken to task by Claude Mydorge regarding his notes to Jean Leurechon's *Recréations mathématiques.*

In conclusion, the body of Henrion's work, although greatly inferior to that of Hérigone, nevertheless played a not unimportant initiatory role in France.

BIBLIOGRAPHY

Henrion's own writings include *Mémoires mathématiques recueillis et dressez en faveur de la noblesse françoise,* 2 vols. (Paris, 1613–1627; 2nd ed., vol. I, 1623); *Traicté des triangles sphériques* (Paris, 1617); *L'usage du compas de proportion*

(Paris, 1618; 2nd ed., Paris, 1624; 4th ed., Paris, 1631; 5th ed., Rouen, 1637, 1664, 1680), further eds. by Deshayes (Paris, 1682, 1685); *Cosmographie ou traicté général des choses tant célestes qu'élémentaires* (Paris, 1620; 2nd ed., Paris, 1626); *Canon manuel des sinus, touchantes et coupantes* (Paris, 1623); *Sommaire de l'algèbre très nécessaire pour faciliter l'interprétation du dixiesme livre d'Euclide* (Paris, 1623), which also appears in Henrion's various eds. of Euclid; *Sinuum, tangentium et secantium canon manualis* (Paris, 1623); *Logocanon, ou Regle proportionelle sur laquelle sont appliquées plusieurs lignes et figures divisées selon diverses proportions et mesures* (Paris, 1626); *Traicté des logarithmes* (Paris, 1626); *L'usage du mecometre, qui est un instrument géométrique avec lequel on peut très facilement mesurer toutes sortes de longueurs* (Paris, 1630); and *Cours mathématique demontré d'une nouvelle méthode* (Paris, 1634).

Henrion edited or translated *Les trois livres des Éléments sphériques de Théodose Tripolitain* (Paris, 1615); *Les quinze livres des Élémens d'Euclide* (Paris, 1614, 1615; 2nd ed., Paris, 1621; 3rd ed., Paris, 1623; 4th ed., Paris, 1631; 5th ed., Rouen, 1649); *Traduction et annotations du Traicté des globes et de leur usage, de Robert Hues* (Paris, 1618); *Edition de la Géométrie et practique générale d'icelle de Jean Errard* (Paris, 1619); *Les tables des directions et profections de Jean de Montroyal, corrigées, augmentées, et leur usage . . .* (Paris, 1625; annotated new ed., Paris, 1626); *Nottes sur les Récréations mathématiques du Père Jean Leurechon* (Paris, 1627, 1630, 1639, 1659, 1660, 1669); and *Les quinze livres des Élémens géométriques . . . plus le livre des Donnez du mesme Euclide . . .* (Paris, 1632; Rouen, 1676; Paris, 1677, 1683; Rouen, 1683, 1685).

JEAN ITARD

HENRY BATE OF MALINES (*b.* Malines, Belgium, 24 March 1246; *d. ca.* 1310), *astronomy.*

In theory, one is better informed about the first thirty-five years of Henry Bate's life than about his later career because of his astrological autobiography, *Liber servi Dei de Machlinia super inquisitione et verificatione nativitatis proprie* (or *Nativitas*), written in 1280. Thus, it is known that he was born a little past midnight on the morning of Saturday, 24 March 1246, at Malines, the next-to-last child of a large family.[1] However, great importance cannot be attached to his description and characterization of himself, not so much because they are likely to be subjective as because, abiding by the laws of this genre, he could not avoid presenting a self-portrait in which he appeared to possess those traits astrologically associated with the time of his birth. This bias severely limited even the scope of his biography. He reveals only how old he was when he suffered various maladies and misfortunes and at what ages (twenty-seven and twenty-nine) he obtained his first, unspecified, ecclesiastical benefices through the intervention of "a celebrated and valorous prince."

Information gleaned from his other works, however, fills out this summary sketch: He studied in Paris, probably as a pupil of Albertus Magnus, becoming master of arts before 1274 and, perhaps, master of theology before 1301. In 1274 he attended the Council of Lyons, at which he met William of Moerbeke. He served as canon of St. Lambert's church in Liège before 1281 and then also became cantor of the chapter, no doubt with the support of his protector, Gui de Hainaut. He became involved in the disputes over the possession of the episcopal chair following its vacancy in 1291 and accompanied Gui de Hainaut, one of the claimants to the seat, to the pontifical court at Orvieto. There he remained for several months in the summer and autumn of 1292. In 1309 he apparently retired to the company of the Premonstrants of Tongerloo, where he lived out his days. He died some time after January 1310.

Bate's works include translations of astrological treatises by Abraham ibn Ezra, the twelfth-century Jewish astronomer; original astronomical and astrological writings; and a philosophical encyclopedia. The desire to make accessible to his contemporaries the astrological work of Abraham ibn Ezra dates from 1273; there exist French translations of four astrological treatises by Ibn Ezra, written at least in part during that year "at Malines in the house of Lord Henry Bate." They are *Commencement de sapience, Livre des jugements, Livre des elections,* and *Livre des interrogations.* The project was undertaken by an association comprising a Jew named Hagins, who translated the Hebrew into Latin, and a certain Obert de Montdidier, who rendered Hagins' work into French.[2] (The project also included a translation of *Livre des revolutions* of Abū Maʿshar.) It is true that reference to the translators appears only in the last lines of *Commencement de sapience,* but the other texts are too similar in style not to have been the result of the same collaboration. Bate certainly had a more responsible role in this matter than providing lodging for his authors. In any case, in 1281, at Malines, Bate himself translated into Latin another treatise of Ibn Ezra, *Liber de mundo vel seculo,* still known as *Liber de revolutionibus annorum;*[3] it is probable that Hagins assisted him.

De mundo vel seculo—which concerns the influences on earthly events that are exercised by the conjunctions of the planets, especially the "great conjunctions"—is preceded by a long introduction in which Bate defends Abū Maʿshar against the reproaches of Ibn Ezra on the question of whether astrological judgments ought to be related to the

 HENRY BATE OF MALINES

mean conjunctions of the planets (in which the planets have the same mean center) or to their true conjunctions (in which they have the same longitude).

In 1292, beginning with his stay at Orvieto and no doubt to occupy the leisure hours imposed on him by postponement of the settlement of the affair of the bishopric of Liège, Bate undertook, one after the other, the translation into Latin of five other treatises of Ibn Ezra. One of these, *Liber introductorius ad astronomiam,* had already been translated into French by Hagins as the *Commencement de sapience.* The other four texts were *Tractatus de causis seu rationibus eorum que dicuntur introductorie ad judicia, Tractatus de luminaribus* (*i.e. de Sole et Luna*) *seu de diebus criticis, Liber introductorius ad judicia astrologie,* and *Tractatus de fortitudine planetarum.*

Moreover, Bate wrote a commentary to *Liber de magnis conjunctionibus et de revolutionibus annorum mundi* of Abū Maʿshar, but this text is lost; it is known only by the long extracts from it that Pierre d'Ailly included in one of the revisions of his *Elucidarium*[4] and in his *De concordia discordantium astronomorum.*[5] These extended quotations may yet enable one to identify the text in question. It may be supposed that in it Bate did his utmost to develop the conciliation, already sketched in the introduction to the translation of *Liber de mundo vel seculo,* between the opposed theses of Ibn Ezra and of Abū Maʿshar.

In the absence of this commentary, Bate's actual astrological work is reduced to his *Nativitas,* in which he demonstrates the truth of astrology by examining the time of his birth and by confirming that deductions derived from it have been verified throughout his life.[6] Such autobiographical astrological accounts seem to have been somewhat in vogue in the thirteenth century, since Bate's *Nativitas* was written between the similar narratives by Richard de Fournival[7] (*b.* 10 October 1201) and Robert Le Febvre[8] (*b.* 18 January 1255). Bate's text begins with a long inquiry to establish, both from direct and indirect evidence, the approximate day and hour of his birth.

Bate's astronomical work is separated here from his astrological work only as an expository convenience, but in his own mind the earlier was only preparation for the later. It comprises a treatise on the astrolabe, *Magistralis compositio astrolabii;* a treatise on the equatorium (that begins "Volentes quidem vera loca planetarum coequare . . ."); and some astronomical tables, *Tabule mechlinenses.* The first two texts were printed together in 1485 by Ratdolt, along with the Latin translation of *De nativitatibus* of Ibn Ezra;[9] curiously enough, the manuscript tradition depends exclusively on this printing. Whatever the various monographs on Bate might say, his treatise on the astrolabe does not at all resemble the other medieval texts on this instrument. At the request of William of Moerbeke, Bate sent him the treatise on the instrument that he had had constructed; the astrolabe in question was quite clearly destined for astrological use and was not the pedagogical instrument employed in universities: Bate's astrolabe was not designed to elucidate the mechanism of the daily movement of the celestial vault but to facilitate the rapid acquisition of astrological data. The tympanum extends to the meridional latitude of 38°42′, which permits it to carry the complete circle of the horizon of Malines. Without dwelling on the tracings that are ordinarily found on astrolabes, Bate discusses at length the question of tracing the lines of the celestial houses. Of the three different definitions of what constitutes a celestial house,[10] he selects two in constructing the two faces of the tympanum: according to the first (to which Bate gives his preference), which makes the twelve houses from six equal divisions of the diurnal semiarc and six equal divisions of the nocturnal semiarc, the lines of the celestial houses coincide with the lines of the unequal hours. According to the other, the celestial houses are the sections of the celestial surface delimited by the twelve equal divisions of the first azimuth of the location and meeting at the two points of intersection of the horizon and the meridian; that division is fixed for the whole length of the year and is shown on the astrolabe by individual circular arcs.

Bate's equatorium is in the tradition of that of Campanus. The equatorium is an instrument used to find, in a practical manner and without any of the usual calculations, the longitude of the planets. Campanus' instrument succeeded in this task, reproducing very faithfully, with brass disks and strands of thread, the geometric resolution of the planetary movements; but it had the inconvenience of multiplying the number of brass disks. Bate devised a single disk to serve as equant for all the planets, each one possessing, moreover, its own epicycle.[11] Bate's text is unfortunately very short and often obscure.

In *Nativitas* and the tract on the equatorium Bate alludes to the astronomical tables that he constructed for the meridian of Malines. Later, in *Speculum divinorum,* he indicates incidentally that he has made three recensions of them. Only two of these have been found (MS Paris lat. 7421, and MS Paris nouv. acq. lat. 3091), but it has not been possible to determine their chronology.[12] The version in the former is the most complete, giving, with the precision of one second of arc, the hourly, daily, monthly, yearly, and pluriannual variations of the mean movements or of

HENRY BATE OF MALINES

the mean arguments of the planets and the values that they assume every twenty years. The version in the latter manuscript does not give the hourly and daily values, perhaps because the differences with those of the other version were insignificant; it presents tables of the revolutions of the years and lists of the values assumed by the auges of the planets—notably in 1285, 1290, and 1295—which are not found in MS lat. 7421. Neither of the two versions of the Malines tables contains tables of planetary equations, of which the Toledo tables gave rise to neither discussion nor revision.

Begun at the end of 1301, in honor of Gui de Hainaut, and completed well before 1305, *Speculum divinorum et quorumdam naturalium* is the most important of Bate's works. A veritable philosophic and scientific encyclopedia, *Speculum* is in twenty-three parts, each dependent on the teaching of the Faculty of Arts of Paris, both as to subject matter and style, the latter being very similar to that of *questiones disputate.* Conceived, however, away from the academic centers and consequently ignoring Étienne Tempier's censure in 1277 of Averroistic doctrines, *Speculum* abounds in exact quotations that ought (when the edition in progress is finished) to illuminate the extent and impact of sources available to a scholar at the end of the thirteenth century.[13]

On the basis of several manuscripts, some authors have credited Bate with a *Tractatus super defectibus tabularum Alfonsi;* Duhem supports this attribution, but only in part, rejecting the whole text, which begins with "Bonum quidem mihi videtur omnibus nobis astrorum . . .," as dating from 1347, and accepting only a fragment published along with large extracts of the *Tractatus* in the *Opera omnia* of Nicolas of Cusa.[14] This fragment is very short and hardly any conclusions can be drawn from it. Certainly, it cannot be asserted that Bate could have written an attack on the Alphonsine tables, since the date of their introduction to Paris remains in doubt—in spite of what Duhem thought he had proved. It may be supposed that the attribution of this fragment to Bate was the result of the attribution of *Tractatus super defectibus* which a certain tradition of the text made to him, despite the verisimilitude.

Furthermore, a few manuscripts have come to light that attribute to Bate *De diebus criticis,* which differs from the translation from Ibn Ezra noted above under the title of *Tractatus de luminaribus seu de diebus criticis.*[15]

NOTES

1. The date 18 Mar. 1244 often given comes from an error in reading by Littré, who transcribed 1244 for 1245 and did not understand, despite Bate's very explicit indications, that Bate employed the old style of dating. These indications were necessary because, for example, the diplomatic style gave the date of Easter as 3 April, while the astronomical style followed by Bate gave the date as 1 March. In order to make more precise the dates of his birth and of the death of Berthold de Malines, which latter event served to fix the birth date of his sister, Bate referred twice in his text to Sunday "in ramis palmarum": the first referred to Passion Sunday (two weeks before Easter), and the second, the first Sunday after Easter.
2. On Hagins, see the account of P. Paris, in *Histoire littéraire de la France,* XXI, 499–503.
3. P. Duhem, *Le système du monde,* II (Paris, 1914), 254–256, and VIII (Paris, 1958), 444–446. The title *Tractatus de planetarum conjonctionibus et de revolutionibus annorum,* under which *De mundo vel seculo* is sometimes cited, is provided by the first words of the text.
4. A. C. Klebs, *Incunabula scientifica et medica* (1490), 768.1, fols. e2v–e3, f3–f3v, and g3.
5. Klebs (1483), 766.1 fols. hh5v–hh8 and hh8v–ii1.
6. To the two Paris MSS (lat. 7324, fols. 24v–47, and lat. 10270, fols. 139v–177v) add Segovia 84 and Seville 5-1-38 (see G. Beaujouan, "Manuscrits scientifiques médiévaux de la cathédrale de Ségovie," in *Actes du XIe Congrès international d'histoire des sciences, Warsaw–Cracow, 1965,* III, 15–18). Bate did not write his *Nativitas* in the first person but, rather, designated himself there as "Servus Dei gloriosi."
7. Cf. A. Birkenmajer, "Pierre de Limoges commentateur de Richard de Fournival," in *Isis,* **40** (1949), 18–31.
8. Cf. E. Poulle, "Astrologie et tables astronomiques au XIIIe siècle: Robert Le Febvre et les tables de Malines," in *Bulletin philologique et historique* (1964), pp. 793–831.
9. This trans. of the *De nativitatibus* is sometimes attributed to Bate (cf. R. Levy, in the bibliography), but A. Birkenmajer ("À propos de l''Abrahismus,'" in *Archives internationales d'histoire des sciences,* **3** [1950], 378–390, esp. 386) thinks that it precedes the trans. by Hagins.
10. On the different types of division of the celestial vault into twelve houses (there were four customary ones in the Latin West), see al-Battānī, *Opus astronomicum,* C.-A. Nallino, ed., I, 246–249, and S. Garcia Franco, *Catalogo critico de astrolabios existentes en España,* pp. 77–79.
11. Cf. E. Poulle, "L'équatoire de Guillaume Gilliszoon de Wissekerke," in *Physis,* **3** (1961), 223–251, esp. 232–234.
12. Cf. M.-T. d'Alverny and E. Poulle, "Un nouveau manuscrit des Tabulae Mechlinenses d'Henri Bate de Malines," in *Actes du VIIIe Congrès international d'histoire des sciences, Florence, 1956,* pp. 355–358. See also Poulle, cited in n. 8.
13. On the *Speculum,* besides the eds. cited in the bibliography, see L. Thorndike, "Henri Bate on the Occult and Spiritualism," in *Archives internationales d'histoire des sciences,* **7** (1954), 133–140.
14. Cf. Duhem, IV (1916), 22–28, 71–72.
15. L. Thorndike, "Latin Translations . . .," p. 300.

BIBLIOGRAPHY

I. ORIGINAL WORKS. Of the trans. of the treatises of Ibn Ezra, the only ones that have been published are *De luminaribus* (Padua, 1482; Klebs 3.1), and *De mundo vel seculo* (Venice, 1507), which also contains Latin trans. by Pierre d'Abano of the treatises of Ibn Ezra already translated into French by Hagins. One of the trans. by Hagins has been published, along with the Hebrew text of Ibn Ezra and an English trans. of the latter in R. Levy and Fr. Cantera, *The Beginning of Wisdom, an Astrological Treatise by Abraham Ibn Ezra,* The Johns Hopkins Studies in Romance Literatures and Languages no. 14 (Baltimore, 1939). In the same series the list of MSS and incipits of Bate's trans. in R. Levy, *The Astrological Works of Abrahim*

Ibn Ezra, The Johns Hopkins Studies in Romance Literatures and Languages no. 8 (Baltimore, 1927), must be used with care. See also L. Thorndike, "The Latin Translations of the Astrological Tracts of Abraham Avenezra," in *Isis,* **35** (1944), 293–302; R. Levy, "A Note on the Latin Translators of Ibn Ezra," *ibid.,* **37** (1947), 153–155; and the note by P. Glorieux (see below).

The treatise on the astrolabe (Venice, 1485) appeared with *De nativitatibus* of Ibn Ezra (Klebs 4); this ed. has been reproduced in R. T. Gunther, *Astrolabes of the World,* II (Oxford, 1932), 368–376. The treatise on the equatorium was included in the 1485 ed. It has wrongly been said that there was another ed. of Bate's two texts in 1491: the latter concerns only *De nativitatibus* and not the texts of Bate.

Speculum divinorum et quorumdam naturalium was published in part, and with a table of contents, by G. Wallerand, vol. XI of Les philosophes belges: textes et études (Louvain, 1931). Another ed., by E. Van de Vyver, is being published in the collection Philosophes médiévaux; two vols., corresponding to pts. 1–3, have already appeared as vols. IV and X (Louvain, 1960, 1967).

II. SECONDARY LITERATURE. The account by E. Littré, in *Histoire littéraire de la France,* XXVI (Paris, 1873), 558–562, is completely outdated. In the absence of the publication of A. Birkenmajer's 1913 thesis, "Henri Bate de Malines, astronome et philosophe de la fin du XIIIe siècle," one must turn to the résumé he has given under the same title in *La Pologne au Ve Congrès international des sciences historiques* (Cracow, 1924).

Other accounts of Bate are P. Glorieux, *Répertoire des maîtres en théologie de Paris au XIIIe siècle* (Paris, 1933), pp. 409–411; and *La Faculté des arts et ses maîtres au XIIIe siècle* (Paris, 1971), pp. 180–182; and G. Wallerand (see above), which gives bibliographical references to older articles.

EMMANUEL POULLE

HENRY OF DENMARK. See **Harpestraeng, Henrik.**

HENRY OF HESSE (*b.* Hainbuch, Germany, 1325; *d.* Vienna, Austria, 11 February 1397), *physics, astronomy.*

Actually nothing is known of the early life or career of Henry of Hesse (variously known as Henricus Hainbuch, de Hassia, de Langenstein, Hessianus) until 1363, the date he became a licentiate for the M.A. degree at the University of Paris. In 1364, 1370, 1371, 1372, and 1373 he took part in the examinations to which the future masters of arts had to submit. He was licentiate in theology in 1375 and doctor before 4 March 1376.

Following the Great Schism of 20 September 1378, Henry remained faithful to Pope Urban VI, while the Avignon pope Clement VII was supported by the French king Charles V. In June 1379 Henry wrote his first political treatise, *Epistola pacis;* and at the end of May 1381 he wrote a second tract, *Epistola concilii pacis.* In 1381 or 1382 Henry found it expedient to leave Paris for his homeland. For a while he stayed at the Cistercian monastery of Eberbach and then went to Vienna, where he spent the rest of his life and contributed greatly to the reorganization of the University of Vienna. He died on 11 February 1397 and was buried in the church of St. Stephen.

Henry's earliest dated astronomical writing was the *Quaestio de cometa,* on the comet of 1368 and directed against the astrological treatise by John of Legnano (Bologna) on the same subject. In it he said that prognostications based on comets are worthless. Before 1373 Henry wrote two other treatises, *Tractatus physicus de reductione effectuum specialium in virtutes communes* and *De habitudine causarum et influxu naturae communis respectu inferiorum.* In the former he mentioned impetus mechanics. Like Marsilius of Inghen in his *Abbreviationes libri physicorum* and undeniably dependent on him, Henry distinguished an impetus of circular motion from an impetus of rectilinear motion. Both treatises contain some peculiar biological ideas, such as speculations about the origin of new species and the mutation of existing ones. The mathematical expression of qualitative intensity in Oresme's "art of latitudes" (there is a close relationship between the works of Oresme and Henry on natural philosophy, occult science, and astrology) led Henry to consider the possibility of the generation of a plant or animal from the corpse of another species, for example, of a fox from a dead dog.

The many astrological predictions evoked by the conjunction of Saturn and Mars in March 1373 caused Henry to reiterate his attack against the astrologers in his treatise *Tractatus contra astrologos coniunctionistas de eventibus futurorum.* Henry, likely with Oresme in mind, asserted that the foundations of astrology cannot be based on identically recurring astronomical experiences, since astronomical events are not of this type.

Henry also wrote a treatise on optics or perspective, the *Questiones super communem perspectivam,* which was largely derived from the *Perspectiva communis* of John Peckham, who is not mentioned. In it Henry referred to Euclid's *Elements,* Aristotle's *Meteorology* and *On the Heavens and the World,* and the genuine treatise of Archimedes, *On Floating Bodies,* the last of which he mentions under the title *De insidentibus in humidum.* It is unlikely that while mentioning this work he had in mind the Moerbeke translation *De insidentibus aque.* The *Questiones super communem perspectivam* consists of fifteen questions, the last of which deals with the rainbow. The maximum altitude of the iris is given as forty-two degrees, and it is stated that other colors are formed of varying proportions of white and black, light being white and opaqueness black.

BIBLIOGRAPHY

I. Original Works. The text of the *Questiones super communem perspectivam* will be in the forthcoming ed. by D. C. Lindberg and H. L. L. Busard. The work is preserved in MS Erfurt, Amplon. F. 380, 29r–40v; Biblioteca Nazionale, MS Codex S. Marci Florent. 202, convent. soppr. J.X. 19, 56r–85v; and Paris, MS Arsenal 522, 66r–88r. It is printed in a composite *Mathematicarum opus* (Valencia, 1504), 47r–65v, together with Bradwardine's arithmetic and geometry; in the latter, questions 10 ("Utrum omnis visio fiat sub angulo") and 11 ("Utrum omnes intensiones visibiles per species colorum apprehendantur sive indicentur") are missing. Herbert Prucker, *Studien zu den astrologischen Schriften des Heinrich von Langenstein* (Leipzig, 1933), in addition to discussing Henry's astrology, gives texts of the *Quaestio de cometa, Tractatus contra astrologos,* and other works.

II. Secondary Literature. A very good survey of the works of Henry of Hesse is in George Sarton, *Introduction to the History of Science,* III, pt. 2 (Baltimore, 1948), 1502–1510. Henry's *De reprobatione eccentricorum et epiciclorum* is discussed, with long extracts, in Claudia Kren, "Homocentric Astronomy in the Latin West," in *Isis,* **59** (1968), 269–281. Also of value are M. Clagett, *Nicole Oresme and the Medieval Geometry of Qualities and Motions* (Madison, Wis., 1968), pp. 114–121, which treats Henry's use of the configuration doctrine; P. Duhem, *Le système du monde,* VII (Paris, 1956), 569–575, 585–599; VIII (Paris, 1958), 483–489; X (Paris, 1959), 138–141; and L. Thorndike, *A History of Magic and Experimental Science,* III (New York, 1934), 472–510.

H. L. L. Busard

HENRY OF MONDEVILLE (*b.* Mondeville, near Caen, or Emondeville, Manche, France, *ca.* 1260 [?]; *d.* Paris, France, *ca.* 1320), *surgery, medicine.*

Henry of Mondeville (or Henricus de Mondeville, Amondavilla, Armandaville, Hermondavilla, Mondavilla, or Mandeville) studied medicine and surgery at Montpellier, Paris, and Bologna; he is regarded as a key link between Italian and French surgery and anatomy. In 1301 he served as surgeon in the armies of Philip the Fair; and for the rest of his life he served Philip, Philip's brother Charles of Valois, or Louis X and taught surgery and anatomy. In 1304 he lectured on anatomy at Montpellier, where Guy de Chauliac reported he "demonstrated" with thirteen illustrations. In 1306 he was lecturing in Paris.

Although he traveled on the king's orders or other business to various parts of France and England, Henry apparently received little income from his contacts with royalty. He complained that he had a difficult time in writing his book on surgery because of the great crowds of patients and students he had to face. Again he apparently did not benefit financially from his popularity. He began writing his book in 1306 but never completed it, possibly because from 1316 he was in ill health, probably with tuberculosis. Henry was apparently a cleric who had studied theology and philosophy; he never married, but he also never had a prebend.

His reputation is derived from *Cyrurgia,* which was not printed until the nineteenth century. His chief biographer, E. Nicaise, found some eighteen manuscript copies of Henry's *Cyrurgia* or parts of it, and at least two more have since been found. Most include only parts of his works. He originally planned to complete his *Cyrurgia* in five parts: (1) anatomy, (2) general and particular treatment of wounds, (3) special surgical pathology, (4) fractures and luxations, and (5) antidotary. The third treatise was only partially completed, the fourth was never written, and some nine of ten projected chapters were finished in the last part. His style is brief, clear, and enlightened. Although he respected authority and cited some fifty-nine different authors some 1,308 times, he did not always agree with them. Galen, who led the list with 431 citations, was regarded by Henry as neither perfect nor the final authority.

The most significant part of Henry's writing is in his second treatise on surgery, in which he followed Hugh of Lucca and Theodoric Borgognoni of Lucca in opposing deliberate efforts to make wounds suppurate. He believed that wounds should be cleaned without probing, treated without irritant dressings, and closed so that they might heal promptly. He urged surgeons to keep their instruments clean, devised improved needles and thread holders, invented an instrument to extract arrows, and removed pieces of iron from the flesh with a magnet. The most controversial aspect of his teaching is his use of anatomical illustrations which have survived in miniature copies. Loren MacKinney, the most recent investigator of the illustrations, felt that Henry had little or no influence on his successors and that although his illustrations marked a trend toward naturalization, they were not particularly accurate.

BIBLIOGRAPHY

The first printed ed. of Henry's work was *Die Chirurgie des Heinrich von Mondeville,* Julius Leopold Pagel, ed. (Berlin, 1892). It was translated into French by E. Nicaise as *Chirurgie de Henri de Mondeville* (Paris, 1893). An Old French version dating from 1314 was published by Alphonse Bos as *La chirurgie de maître Henri de Mondeville,* 2 vols. (Paris, 1897–1898). Various students of Pagel translated aspects of the *Chirurgie* into German; for a list of these see the Sarton reference below.

The most complete account of Mondeville's life and work is in Nicaise's introductory essay to the *Chirurgie*. For a summary of some of the earlier monographic literature see George Sarton, *Introduction to the History of Science*, III, pt. 1 (Baltimore, 1947), 865–873, although Sarton makes several errors in biographical details. For more recent assessments see the following, listed chronologically; Loren C. MacKinney, "The Beginnings of Western Anatomy," in *Medical History*, **6** (1962), 233–239; J. A. Bosshard, "Psychosomatik in der Chirurgie des Mittelalters, besonders bei Henri de Mondeville," in *Zürcher medizingeschichtliche Abhandlungen*, n.s. **11** (1963); Vern L. Bullough, *The Development of Medicine as a Profession* (Basel, 1966), pp. 57–64, 95; and C. Probst, "Der Weg des ärztlichen Erkennens bei Heinrich von Mondeville," in *Fachliteratur des Mittelalters: Festschrift für Gerhard Eis* (Stuttgart, 1968), pp. 333–347.

VERN L. BULLOUGH

HENRY, JOSEPH (*b.* Albany, New York, 17 December, 1797; *d.* Washington, D.C., 13 May 1878), *physics.*

Henry was born to a poor family of Scottish descent and raised as a Presbyterian, a faith he followed throughout his life. His early education was in the elementary schools of Albany and Galway, New York, where he went, prior to his father's death in 1811, to reside with relatives. Henry was apprenticed to an Albany watchmaker and silversmith a few years later. The theater was his principal interest as an adolescent, until a chance reading of George Gregory's *Popular Lectures on Experimental Philosophy, Astronomy, and Chemistry* (London, 1809) turned him to science.

In 1819 Henry enrolled in the Albany Academy and remained there until 1822, with a year off to teach in a rural school in order to support himself. The surviving Academy archives do not explain how an overage pupil gained admittance nor exactly what Henry studied. From the surviving Henry manuscripts and books we know he was schooled at the Academy in mathematics (through integral calculus), chemistry, and natural philosophy. He won the support of his principal, T. Romeyn Beck, who employed Henry as assistant in a series of chemistry lectures in 1823–1824 and later. Henry's main problem at this period was how to support himself while furthering his development as a scientist. The surviving evidence is not very clear on what he did. For an undetermined period he was a tutor in the household of the van Rensselaers and later taught the elder Henry James. Tradition has him considering the possibility of a medical career. We do know that Henry did odd surveying jobs and that in 1825 he headed a leveling party in the survey of a projected road from the Hudson River to Lake Erie. In the next year his friends attempted, unsuccessfully, to get Henry an appointment with the Topographical Engineers of the U.S. Army. Shortly afterward he was appointed professor of mathematics and natural philosophy at the Albany Academy, where he remained through October 1832, when he accepted a chair at the College of New Jersey (now Princeton University).

During these formative years at Albany, Henry was engaged in avid reading of works in science and other fields, an activity he continued throughout his life. He sometimes downgraded his Academy education by referring to himself as self-taught, an obvious reference to his efforts at self-improvement. A fair number of standard periodicals and monographs were accessible to Henry, especially since he was the librarian of both the Academy and the Albany Institute of History and Art, the local learned society. In the latter he associated with the leading citizens of the city who had enlightened, but not professional, interests in the sciences. In 1824 he read his first paper, a review of the literature on steam; by 1827 he was doing experiments on his own.

Henry's earliest known work was in chemistry, in collaboration with Lewis C. Beck, T. Romeyn Beck's brother. In 1827, when Henry started his work in electricity and magnetism, Beck was also experimenting in this area; but we have no information on the nature of these investigations. By this date Henry's reading had made him familiar with the work of Davy, Faraday, Ampère, and probably Young, whose wave theory of light influenced Henry's subsequent views. He also read and annotated Biot on electromagnetism in the Farrar translation (Cambridge, Mass., 1826) shortly after its appearance.

In his lectures at Princeton, Henry avowed that his work on electromagnetism in Albany, leading to the development of powerful electromagnets and the independent discovery of electromagnetic induction, was an application of Ampère's theories. This is not self-evident from contemporary sources, for Henry was rather reticent about stating his theoretical views; his approach was the unfamiliar one of drawing analogies with terrestrial magnetism. Throughout his career Henry was interested in terrestrial magnetism, meteorology, and other geophysical topics. In this he was part of an active research tradition of the day. When he investigated terrestrial magnetism in Albany, especially the effect of the aurora, Henry used a needle of Hansteen's which Edward Sabine had sent to James Renwick, Sr., of Columbia College. When he witnessed a demonstration of Oersted's discovery in 1826, Henry immediately saw it as a way to explain the variation of the needle. From Ampère's picture of the earth as a great voltaic pile with innumerable layers of materials producing circular currents around the magnetic axis, Henry probably conceived the idea of winding his horseshoe magnet with many strands of wire in parallel, not using a continuous strand as W. Sturgeon and G. Moll had. When Faraday met Henry in 1837, he invited his American colleague to lecture to the Royal Institution on the mathematical theory of electromagnetism, a strange request in light of the near absence of mathematics in Henry's papers but explicable if Faraday conceived of Henry as being in some sense a follower of Ampère.

Like Faraday before him (and, later, Wheatstone), Henry had by 1830 independently uncovered the sense of Ohm's law and was engaging, for example, in what we now call impedance matching. He learned of Ohm only in December 1834 and may not have had a full knowledge of the law until 1837, but clearly Henry was extremely adroit in manipulating his equipment to get desired effects of "intensity" (high voltage) and "quantity" (high amperage) at an earlier date. In the Albany experiments he wanted to design devices suitable for classroom demonstrations, that is, to get large effects from small inputs. Henry's electromagnets exemplified this on a large scale. When he applied them to demonstrate the long-predicted production of electricity from magnetism, the distinction between Henry and Faraday as experimentalists became evident. Faraday devised ingenious experimental setups to detect small effects; Henry, almost anticlimactically, devised procedures for rendering small effects grossly tangible. In connection with the experimental work on electromagnetic induction, Henry independently discovered self-induction (1832).

From the time of his transfer to Princeton late in 1832 until Henry's first European trip in 1837, there was a relative diminution of his research, undoubtedly due to the pressure of teaching duties. From 1838 until his appointment as secretary of the Smithsonian Institution in 1846, Henry was extremely active in research, not only in electricity and magnetism but also other areas of physics. His work outside electricity and magnetism is not as well-known or as consequential. Like all his research, these investigations were conducted with skill and imagination. For example, he published papers on capillarity (1839, 1845) and on phosphorescence (1841). In 1845 Henry wrote about the relative radiation of solar spots. Of particular interest in understanding his general scientific orientation are the 1846 paper on atomicity and the 1859 paper on the theory of the imponderables. He published several papers on the aurora and on heat. Henry was also greatly interested in color blindness. In his later work for the Light-House Board he did much experimental research on the propagation and detection of light and sound.

While the earliest Princeton work was a continuation of the Albany investigations, there was an enlargement of Henry's interests in electricity and magnetism. For one thing, he was most conscious of his American predecessor, Benjamin Franklin, even to the point of once using the pseudonym "F." While carefully stating that both the one- and two-fluid theories were mathematically equivalent, he always opted for Franklin, not Ampère. His work at Princeton shows particular concern for integrating the static electricity phenomena of Franklin with the most recent galvanic developments. Another strand in the Princeton experiments was Henry's efforts to explain physical phenomena in terms of wave phenomena, most likely deriving from Thomas Young.

Williams correctly notes that Henry's discovery of self-induction was important to him because it fitted his theoretical views, but that it was not particularly crucial to Faraday's concepts. The direction of Henry's thought became somewhat apparent in his 1835 paper on the action of a spiral conductor in increasing the intensity of galvanic currents. The paper started out as an affirmation of Henry's priority in the discovery of self-induction. He then combined induction proper (using Faraday's findings and his own) with self-induction to show how these produce a pattern of repulsions yielding an increased effect in spirals. He specifically linked these "magneto-electrical" results to the principles of static induction developed by Cavendish and Poisson. This explanation was then applied to Savary's report of changes of polarity when magnetic needles were placed at varying distances from a wire in which a current was being transmitted ("Mémoire sur l'aimantation," in *Annales de chimie et de physique,* **34** [1827], 5–57, 220–221). That is, currents appeared periodically in the air surrounding a current-bearing straight wire as a result of the actions of induction and self-induction. In his 1838 paper on electrodynamic self-induction Henry started out again with self-induction and also cited the Savary paper. In the 1835 paper and this later work on currents of higher orders, there is some suspicion that Henry saw these varying magnetic needles as analogous to the phenomena of terrestrial magnetism.

Henry's demonstrations in 1838 and later of the induction of successive currents of higher orders was quite in accord with these views and had considerable impact. Faraday noted in his diary for 12 November

1839 that five others in Britain besides himself had received coils from Henry like the ones used in the dynamic induction experiments. Carrying out his program of determining the relationship of static and dynamic electricity, Henry published a long paper on electrodynamic induction in 1840. In 1842 he returned to the consideration of the Leyden jar discharge discussed in his 1838 paper, noting that Savary had reported anomalies. These Henry explained as a backward and forward oscillatory discharge until equilibrium was reached as a complex resultant of inductions and self-inductions. He considered this explanation as original, but it is in Savary's paper. Henry's experiments were undertaken to confirm and explain the direction of the various currents induced by Savary from the straight wire. In this paper he reported propagating and detecting electromagnetic effects over great distance. A single spark "is sufficient to disturb perceptibly the electricity of space throughout at least a cube of 400,000 feet capacity." In this paper he also reported that lightning flashes seven or eight miles away strongly magnetized needles in his study. Similar results appear in his interesting 1848 paper on telegraph lines and lightning. As late as 1856 (diary entry of 19 January) Faraday wondered at these reports.

To explain these effects, in 1842 Henry declared himself a believer in an electric plenum. Having started with the desire to use Ampère (and Oersted) to explain terrestrial magnetism, he had first proceeded to laboratory analogues of terrestrial magnetism and of the electrical currents associated with various forms of magnetism.

In these speculations Henry was staunchly Newtonian, conceiving of astronomy as the model science and mechanics as the ultimate analytical tool. For example, although impressed by Bošković's atomism, he finally rejected it as incompatible with Newton's laws of motion on the macroscopic level. Henry could not accept Faraday's field concept because of his belief in central forces acting in a universal fluid. This view was reinforced by his differing interpretation of experiments on electromagnetic effects in vacuums. From observations of the interaction of currents and magnets, Henry expanded his earlier explanation of Savary to conclude that the currents were oscillatory wave phenomena exciting equivalent effects in an electrical plenum coincident, if not identical, with the universal ether.

Henry believed that particular disturbances originating in grosser matter produced wavelike oscillations in the plenum, whose manifestations in other, grosser bodies of matter were electricity and magnetism. He then reduced the wave phenomena to mechanical actions in the plenum. To him, static electricity was instantaneous action at a distance arising from the disturbances in the medium produced by gross matter, yielding condensations and rarefaction in the ether/plenum. Dynamic electricity was an actual transfer of part of the ether/plenum, requiring a discrete time interval to restore the equilibrium of the universal medium.

When Henry assumed the secretaryship of the Smithsonian Institution in 1846, he had fairly clear ideas of what he wanted. Certainly no one really knew what a little-known chemist, the natural son of an English duke, meant when he inserted a contingency clause in his will dedicating his estate to an institution in Washington, D.C., for "the increase and diffusion of knowledge." The debates in Congress and the press over the Smithson bequest disclose an utter confusion of aims. Basic to an understanding of Henry's ideas as a science administrator is his being a professional physicist at a time when that breed was quite rare in America. Unlike his great British contemporary, Michael Faraday, he had a good knowledge of mathematics and an appreciation of the need to generalize experimental findings into mathematical formulations. Allied with this appreciation of mathematics was a firm, scornful rejection—in print and in private writings—of crude Baconianism as a scientific method. To Henry, forming hypotheses was the essential step in research.

With views so different from the norm of his time and place, Henry arrived at a conception of the scientific community little understood by many of his contemporaries: a small group of trained, dedicated men meeting internationally recognized standards and engaging in free and harmonious intellectual intercourse among themselves. Henry, as secretary of the Smithsonian, attempted to symbolize this ideal in America. His success in forming the Smithsonian according to his ideals is a credit to his astuteness as an administrator and the broad recognition of his preeminence in the American scientific community. This success also rested on the compatibility of Henry's views with the beliefs of a significant number of educated laymen.

As secretary of the Smithsonian, Henry was not concerned with popularizing science or with education but with supporting research and disseminating findings. He consequently set great store in properly refereeing proposed publications and in furthering cooperation among scientists. One of his earliest moves was to establish an international system for exchange of scientific publications. This interest in scientific information led in 1855 to his suggestion for what later became, with modifications, the *Royal*

Society Catalogue of Scientific Papers. He had initially limited the scope to the exact sciences.

Because Henry saw scientific research neglected in America in favor of other human endeavors and was rather pessimistic about the chances of redressing the balance, he and many American scientists well into the twentieth century sounded and acted as though they were a beleaguered minority. In Henry's administration of the Smithson bequest, a contemporary reader is struck by persistent notes of alarm as the secretary fights off the attempts by Charles C. Jewett to subvert the institution into a national library and fends off well-intentioned efforts to deflect the endowment to the support of lyceum lectures and a popular museum of science, art, and curiosities of nature and human ingenuity. Not that Henry disapproved of these activities; he was, after all, one of their proponents both as secretary of the Smithsonian and as a good citizen.

Given the modest size of the bequest and the greater popular interest in nonresearch activities, Henry regarded support of research and scholarly publications as a better use of scarce funds. In reaching these conclusions in private discourses and in public justifications, he was forced to consider the relations of science to other branches of human endeavor. In Albany he had written and lectured on the relations of "pure mathematics" to "mixed mathematics" (what we now call physics) in accordance with a traditional view widely held in that day. Clearly, Henry, like many of his contemporaries, favored and looked forward to the conversion of all fields of science, and also the arts, useful and otherwise, to the status of "mixed mathematics." By this he meant an infusion of rigor, hopefully in the form of mathematics. As secretary, Henry would do what he could to promote this development across the board but would give priority to those fields at or near the desired state of intellectual development. Fiscal considerations here reinforced Henry's desire to maintain and develop a preferred intellectual model. In his writings he was impelled by his position to champion the idea of the purity of institutions, in the sense of their having specialized functions and motives. He assumed, for example, that research and popularization were incompatible in a single institution. In this belief he diverged markedly from the national practice.

Although Henry often sounded like a proponent of professional specialization, his own work and the program of the Smithsonian never had that kind of purity. Although best-known as a laboratory physicist, he was active in meteorology and other geophysical areas dominated by observation of natural phenomena; one of the institution's biggest programs was in meteorology. Henry's most original activity as secretary was to become America's leading patron of anthropology and ethnology. He read widely in these fields and, for reasons still unclear, was obviously concerned and enthusiastic. Unlike his successor, S. F. Baird, who reduced the funds for the physical sciences, Henry was careful to support research in natural history as well, despite evidences of his reservations about the value of much work in that field. After Darwin published *The Origin of Species,* Henry regarded natural selection as the best chance yet to give natural history the rigor it had lacked thus far. Rather than limiting the Smithsonian to one scientific field, he insisted on limiting its support to men of professional competence.

Henry was firmly against Smithsonian involvement in applied research, the American environment, in his view, providing more than adequate incentive for such work outside the institution. In taking this position he was not at all like the pure scientists of the next century who inhabited ivory towers; the record is replete with instances of concern with applications. What Henry was upholding was the logically anterior role of pure science, the assumption of chronological priority following naturally from that position. In the one public priority squabble of his life, with S. F. B. Morse over the telegraph, he was asserting the primacy of disinterested scientific research seeking general truths over investigations of specific practical solutions. While this assertion was quite odd to most of Henry's contemporaries, American scientists up to the present would implicitly echo him in urging greater support for pure research.

In his relations with the U.S. government Henry also struck a note persisting down to the present. Although very successful in gaining support in Congress and in the executive branch, he continually worried about political patronage forcing ill-trained men on scientific organizations and directing research into unworthy channels. Ironically, this lack of faith in both the government and the society at large inhibited Henry from buttressing the original bequest by seeking additional funds. Lacking a sound financial base dedicated to Henry's program, the Smithsonian Institution grew in directions not contemplated by him, the new growth largely obliterating his original conception. What did survive was a belief in research.

BIBLIOGRAPHY

I. Original Works. Joseph Henry's unpublished correspondence and other MSS constitute a major source for the study of his life, as well as for the various topics in which he played a significant role. Under the sponsorship of the American Philosophical Society, the National Academy of Sciences, and the Smithsonian Institution, these widely scattered documents are being gathered for publication under the editorship of Nathan Reingold of the Smithsonian Institution. They will eventually number 50,000–60,000 and include runs of scientific and personal correspondence, Henry's laboratory journals, diaries, texts of unpublished lectures and articles, and a splendid miscellany of other items. Fifteen printed vols. of selected MSS will appear; the entire body of documentation will be issued as a microfilm publication well before the printed vols. have run their course. The Henry Papers staff is describing and indexing the documents using a computer system, the first such use for a document publication.

In custody of the Henry Papers staff is Henry's personal library, containing approximately 1,200 monographic and serial titles and approximately 1,200 pamphlet titles. Many items are presentation copies and others bear annotations. Since the library includes volumes dating back to the Albany period, historians have available a splendid slice of scientific and other literature closely linked to a large body of unpublished MSS. The library is also being cataloged by computer.

Until the new ed. of Henry's scientific writings (in preparation under the editorship of Charles Weiner) appears the principal source for Henry's publications is *Scientific Writings of Joseph Henry,* 2 vols. (Washington, D.C., 1886). Still the best published bibliography is W. B. Taylor, *Memorial of Joseph Henry* (see below), pp. 365–374. Taylor's work was the basis for the *Scientific Writings.* While the Weiner ed. will not literally reproduce the text of 1886—excisions and additions are contemplated—it will largely follow the old pattern, especially in limiting the contents mainly to the scientific publications. Two restrictive boundary conditions are worth noting: (1) although some items are slated for reprinting in the Weiner ed., for practical reasons there will be no attempt to gather in all such, especially nonscientific pieces; (2) a large body of Henry writings is largely excluded from the 1886 ed. and the Weiner ed.—*Reports of the Board of Regents of the Smithsonian Institution,* 1846–1877—in which Henry wrote extensively and interestingly about his organization, the progress of science, and science's role in the American republic. In the *Scientific Writings* (and in the forthcoming ed.) is Henry's one attempt at a comprehensive treatise, often overlooked because of its misleading title. "Meteorology in Its Connection With Agriculture" (*Scientific Writings,* II, 6–402) appeared in five parts from 1855 to 1859 as appendixes to the *Report of the Commissioner of Patents.* Despite its title, it is really a general survey of the physical sciences, with only occasional attempts to relate weather and farming. Although nontechnical, the work is not really popular science as we now understand the term. The work merits careful study as a mature statement of Henry's beliefs and attitudes.

II. SECONDARY LITERATURE. The best recent discussion of Henry the scientist is in Charles Weiner's dissertation at Case Institute of Technology, "Joseph Henry's Lectures on Natural Philosophy" (Cleveland, 1965). The most recent discussion of Henry's policies at the Smithsonian Institution is Wilcomb E. Washburn, "Joseph Henry's Conception of the Purpose of the Smithsonian Institution," in Walter Muir Whitehill, ed., *A Cabinet of Curiosities* (Charlottesville, Va., 1967), pp. 106–166. Recent articles concerning Henry are L. Pearce Williams, "The Simultaneous Discovery of Electro-magnetic Induction by Michael Faraday and Joseph Henry," in *Bulletin de la Société des amis d'André-Marie Ampère,* no. 22 (Jan. 1965), 12–21; and T. K. Simpson, "Maxwell and the Direct Experimental Test of His Electromagnetic Theory," in *Isis,* **57** (1966), 411–432. Henry is discussed and some of his MSS printed in N. Reingold, ed., *Science in Nineteenth Century America, a Documentary History* (New York, 1964), pp. 59–107, 127–161, 200–225. Two recent biographies of his American contemporaries are superb for background on Henry: A. Hunter Dupree, *Asa Gray* (Cambridge, Mass., 1959); and Edward Lurie, *Louis Agassiz, a Life in Science* (Chicago, 1960). L. Pearce Williams, *Michael Faraday* (New York, 1965), has only a few references but is indispensable for an understanding of Henry, as is A. Hunter Dupree, *Science in the Federal Government* (Cambridge, Mass., 1957).

There is an extensive, older hagiographic literature whose most recent and respectable exemplar is Thomas Coulson, *Joseph Henry, His Life and Work* (Princeton, 1950). Coulson's work is based upon examination of a limited body of the extant primary sources and relies heavily upon the unpublished draft of a biography of her father by Mary Henry. It has many errors of omission and commission, its principal defect being a lack of knowledge about Henry's America. Besides Mary Henry's work, with its attempt to revive priority battles Henry never fought, the hagiographic literature has two additional sources: sentimental homages by American scientists and engineers to their distinguished predecessor and quasi-historical literature emanating from the Smithsonian. The former is best forgotten except by students of scientific mythology. Two of the works in the latter genre are still useful if used with care: W. J. Rhees, ed., *The Smithsonian Institution: Documents Relative to Its Origin and History, 1835–1899,* 2 vols. (Washington, D.C., 1901); and *A Memorial of Joseph Henry* (Washington, D.C., 1880), the former (and a companion volume of documents on the regents of the institution) for many years the principal published source on the early history of the Smithsonian. The latter is filled with undocumented bits of information on Henry's life. It and the Mary Henry draft in the Smithsonian Archives are often the only sources for the charming and possibly true stories about Henry's early years.

NATHAN REINGOLD

HENRY, PAUL PIERRE (*b.* Nancy, France, 21 August 1848; *d.* Montrouge, near Paris, France, 4 January 1905); **HENRY, PROSPER MATHIEU** (*b.* Nancy, 10 December 1849; *d.* Pralognan, Savoy, France, 25 July 1903), *astronomy, optics.*

The Henry brothers were united in their careers, as in their lives; and their work cannot be separated.

The death of the younger interrupted the work of the elder, who, consumed by grief, did not long survive him.

Following elementary studies in a Catholic school, they were accepted, each at the age of sixteen, into the Service Météorologique des Prévisions, recently created at the Paris observatory. But their astronomical vocation owed nothing to this institution; for it was with their own means, and in their own home, that they set up a small optical workshop and undertook, beginning in 1868, the construction of a thirty-centimeter mirror and its mounting. With this reflector and a secondhand clock they began to make a map of the stars in the ecliptic zone.

In 1871 the director of the observatory, Charles Delaunay, heard of their work and transferred the two brothers from the meteorological to the equatorial telescope section. The projected ecliptic map was then made more precise: to chart, on a band five degrees wide, the positions of all the stars up to the thirteenth magnitude. In the course of executing it, they had occasion to make a great number of observations of minor planets and discovered fourteen such bodies between 1872 and 1878. In 1884, when a fourth of the band had been explored and 36,000 stars recorded, an insurmountable difficulty arose: they reached the intersection of the ecliptic and the Milky Way, and the density of the stars was too great to permit a visual survey.

The aid of photography became essential. In their workshop at Montrouge the Henry brothers cut a sixteen-centimeter objective lens especially adapted to this technique and achromatized for the wavelengths to which photographic plates are sensitive. They coupled to the photographic telescope a visual guiding telescope, thus becoming able to control precisely the drive movement of the equatorial telescope during the course of exposures lasting as long as one hour. By 1884 they were making such remarkable photographs that the Paris observatory immediately commissioned them to build a large apparatus.

The first large photographic equatorial telescope was completed in 1885. It had an opening of thirty-four centimeters and a focal length of 3.40 meters; the guiding telescope, with a smaller opening, was of the same length. The performances of this apparatus revealed how much photography could assist astronomy. A negative of the Pleiades cluster, with its sharpness and wealth of weak stars, was frequently reproduced and contributed to the rise of photographic astronomy.

The Henrys' instrument was adopted in 1887 as the prototype for the international project of the Carte du Ciel. Seventeen identical instruments were built—more than half of them by the Henrys—and placed at various latitudes. The Carte du Ciel, which was completed only recently, has enabled scientists to collect a considerable number of documents whose value for the determination of stellar positions will increase every year.

The Henry brothers constructed many other devices. In particular they produced the lens of the great seventy-six-centimeter refractor of the Nice observatory. Their skill was considered incomparable.

Their careers remained strictly parallel. Prosper was named chief astronomer four years before Paul, in 1893; but in that year Paul was made head of the Service de la Carte du Ciel, a position which Prosper held seven years later. On three occasions they shared prizes awarded by the Academy of Sciences, and they were both admitted as associate members of the Royal Astronomical Society in 1889.

Unpretentious and modest, the Henry brothers sought relaxation from their duties at the observatory by working in their own workshop, where they conducted research at their own expense. They warmly received all who came to their laboratory to learn stellar photography, an art which they helped to create.

BIBLIOGRAPHY

I. Original Works. The Henry brothers published mainly observational results (minor planets, planets, comets), presented in the form of about fifty notes to the *Comptes rendus hebdomadaires des séances de l'Académie des sciences* between 1872 and 1887. The discoveries of minor planets sometimes bear just one of their names. Some results of observations were also published in *Bulletin astronomique,* vols. **1** (1884) and **3** (1886).

There are also the following works: "Sur la construction de cartes célestes, très détaillées, voisines de l'écliptique," in *Comptes rendus hebdomadaires des séances de l'Académie des sciences,* **74** (1872), 246–247; "Sur un nouveau télescope catadioptrique," *ibid.,* **88** (1879), 556–558; "Sur la suppression des halos dans les clichés photographiques," *ibid.,* **110** (1890), 751; and "Méthode de mesure de la dispersion atmosphérique," *ibid.,* **112** (1891), 377–380, published under Prosper Henry's name.

II. Secondary Literature. See Adam, Colonel Laussedat, O. Callandreau, and C. Trepied, "Discours prononcés aux obsèques de Prosper Henry," in *Bulletin astronomique,* **21** (1904), 49–58; F. W. Dyson, "Obituary, Prosper Henry," in *Monthly Notices of the Royal Astronomical Society,* **64** (1904), 296–298; P. Puiseux, Colonel Laussedat, and B. Baillaud, "Discours prononcés aux obsèques de M. Paul Henry," in *Bulletin astronomique,* **22** (1905), 97–102; and J. Rosch, "La mission des frères Henry au Pic-du-Midi pour le passage de Vénus sur le soleil (6 décembre 1882)," in *L'astronomie,* **64** (1950), 475–490.

Jacques R. Lévy

HENRY, THOMAS (*b.* Wrexham, Wales, 28 September 1734; *d.* Manchester, England, 18 June 1816), *chemistry.*

Henry was the son of a dancing master. When he left Wrexham Grammar School his family lacked the means to support a hoped-for career in the church, so he was apprenticed to a local apothecary. After completing his apprenticeship at Knutsford, Cheshire, he became assistant to an apothecary at Oxford, where he attended lectures on anatomy. In 1759 he returned to Knutsford and married Mary Kinsey; in 1764 he moved to a business in a fashionable part of Manchester.

There Henry became acquainted with several men of intellectual distinction—in particular the physician Thomas Percival, to whom he attributed his initiation into experimental science. This new interest, and his commercial ventures, claimed his attention. Percival was a Unitarian; Henry became one and was drawn into the group of Dissenters from whose enterprise arose the Manchester Literary and Philosophical Society in 1781 (he was a founder-member, one of the first joint secretaries, and president from 1805 to 1816), the short-lived College of Arts and Sciences (1783), and the Manchester Academy (1786), a successor to the famous Warrington Academy. At the last two Henry lectured on chemistry and on bleaching and dyeing.

One of the first in Britain to use chlorine for bleaching textiles, Henry pioneered the use of milk of lime to absorb the gas, thus reducing the dangers of the process. In a paper on dyeing, read in 1786, he pleaded for the application of chemical knowledge to the art and gave a correct interpretation of the role of mordants.

Henry's most profitable venture, which earned him his nickname "Magnesia," was the manufacture and sale, for medicinal purposes, of calcined magnesia following his submission for publication of "An Account of an Improved Method of Preparing Magnesia Alba" (*Medical Transactions of the Royal College of Physicians,* **2** [1772], 226–234). The manufacture of magnesia, by a process that remained almost unchanged from its beginning in 1772, provided a comfortable income for the family until 1933.

In 1773 Henry published his *Experiments and Observations,* consisting mainly of articles on the properties and uses of calcined magnesia and on putrefaction. He described experiments which appeared to uphold Macbride's observation that the putrefaction of meat and similar substances was halted by a stream of "fixed air" (carbon dioxide). Both he and Percival, Henry said, had successfully treated putrid diseases with the gas. Subsequently they questioned Priestley's reports of the toxicity of fixed air to vegetation and satisfied themselves that, on the contrary, it provided nourishment for plants. (See Thomas Percival, "On the Pursuits of Experimental Philosophy," in *Memoirs of the Manchester Literary and Philosophical Society,* **2** [1785], 326–341; and Thomas Henry, "Observations on the Influence of Fixed Air on Vegetation," *ibid.,* 341–349. See also E. L. Scott, "The 'Macbridean Doctrine' of Air: An Eighteenth-Century Explanation of Some Biochemical Processes, Including Photosynthesis," in *Ambix,* **17** [1970], 43–57.)

In 1776 Henry published the first and thus far the only complete English translation of Lavoisier's *Opuscules,* taking the opportunity to correct Lavoisier's account of some of Priestley's work. Later he translated nine of Lavoisier's essays but defended, in an appendix, his own and Priestley's belief in phlogiston. Surviving correspondence shows that, unlike Priestley, he eventually abandoned it.

Henry was elected a fellow of the Royal Society in 1775 on Priestley's recommendation. Although not a great scientist he did much to found a scientific tradition in Manchester, enhanced in his own lifetime by his son William and their mutual friend John Dalton.

BIBLIOGRAPHY

I. Original Works. Henry's main work is *Experiments and Observations on the Following Subjects: 1. On the Preparation, Calcination and Medicinal Uses of Magnesia. 2. On the Solvent Qualities of Calcined Magnesia. 3. On the Variety in the Solvent Powers of Quicklime, When Used in Different Quantities. 4. On Various Absorbents as Promoting or Retarding Putrefaction. 5. On the Comparative Antiseptic Powers of Vegetable Infusions Prepared With Lime etc. 6. On the Sweetening Properties of Fixed Air* (London, 1773); the first essay is a reprint of his 1772 paper to the Royal College of Physicians. His translations are *Essays, Physical and Chemical, From the French of Lavoisier* (London, 1776), with notes and app.; and *Essays on the Effects Produced by Various Processes on Atmospheric Air, With a Particular View to an Investigation of the Constitution of Acids* (Warrington, 1783), with a pref. on the controversy between Priestley and Lavoisier. He also published a short biography of the Swiss physiologist Albrecht von Haller, *Memoirs of Albert de Haller* (Warrington, 1783), based on his translation of the eulogy read to the French Academy of Sciences.

Henry published a number of pamphlets, the most important being *An Account of a Method of Preserving Water at Sea From Putrefaction and of Restoring to the Water Its Original Pleasantness and Purity by a Cheap and Easy Process . . .* (Warrington, 1781). His advocacy of using quicklime, to be precipitated with carbon dioxide when the water was required for consumption, was pressed upon the Admiralty but never adopted. A paper on this subject was published in the *Manchester Memoirs* (see below).

Many of Henry's papers read to the Manchester Literary

and Philosophical Society were not published; an incomplete list of those which appeared in the *Memoirs* is given in Royal Society, *Catalogue of Scientific Papers,* III (London, 1869), 292–293. Those mentioned above are "On the Preservation of Sea Water From Putrefaction By Means of Quicklime . . .," in *Memoirs of the Manchester Literary and Philosophical Society,* **1** (1785), 41–53 (read in 1781); and "Considerations Relative to the Nature of Wool, Silk and Cotton, as Objects of the Art of Dying; on the Various Preparations and Mordants, Requisite for These Different Substances; and on the Nature and Properties of Colouring Matter; Together With Some Observations on the Theory of Dying in General," *ibid.,* **3** (1790), 343–408.

II. SECONDARY LITERATURE. The main biographical source is his son William's tribute, "A Tribute to the Memory of the Late President of the Literary and Philosophical Society of Manchester," in *Memoirs of the Manchester Literary and Philosophical Society,* 2nd ser., **3** (1819), 391–429. Biographical sketches are in J. Wheeler, *Manchester: Its Political, Social and Commercial History, Ancient and Modern* (Manchester, 1836), pp. 488–493; and E. M. Brockbank, *Sketches of the Lives and Work of the Honorary Medical Staff of the Manchester Infirmary* (Manchester, 1904), pp. 72–82, with portrait. See also W. V. Farrar, K. R. Farrar, and E. L. Scott, "The Henrys of Manchester," in *Royal Institute of Chemistry Reviews,* **4** (1971), 35–47.

E. L. SCOTT

HENRY, WILLIAM (*b.* Manchester, England, 12 December 1774; *d.* Manchester, 2 September 1836), *chemistry.*

The third son of Thomas Henry—probably the most talented and certainly the most successful of the three—William Henry went first to a private school run by a Unitarian minister and then to Manchester Academy. From the age of ten an injury from which he never fully recovered, inflicted by a falling beam, prohibited him from normal boyhood activities, and he early developed a taste for study. After leaving the academy (about 1790) he became secretary-companion to Thomas Percival and began preliminary studies in medicine. He entered Edinburgh University in 1795 but left a year later to assist in his father's practice and to superintend the family manufacturing business.

Henry became a member of the Manchester Literary and Philosophical Society in 1796 and began to carry out original research in chemistry. In 1805 he returned to Edinburg and received the M.D. in 1807, submitting a dissertation on uric acid; he later specialized in urinary diseases and contributed papers to medical journals on these and allied subjects. He was elected a fellow of the Royal Society in 1808 and received the Copley Medal for papers already submitted.

Henry's first paper (1797) was a refutation of William Austin's claim to have shown that carbon was not an element (1789). In 1800 he described an attempt to determine the nature of muriatic acid (hydrochloric acid); although he obtained hydrogen and oxymuriatic acid (chlorine) on sparking muriatic acid gas over mercury, he attributed the former to water, from which he believed it was impossible to free the gas by chemical means. In 1801 Henry tried the new technique of electrolysis (he was one of the first to experiment with this) but considered it seriously limited on finding that the current could not be transmitted through gases. In 1812, after the classic researches of Gay-Lussac and Thénard in France and of Davy in England had provided the evidence for the elementary nature of oxymuriatic acid and its combination with hydrogen in muriatic acid, Henry related further experiments which appeared to favor the new views. Yet it was several years before he finally committed himself to them.

This cautious attitude toward new ideas characterized Henry's later years, although in his youth he had eagerly embraced the new chemistry of Lavoisier. At Manchester, in the winter of 1798–1799 he had given his first lecture demonstrations, firmly grounded in the new doctrines and nomenclature. His textbook, originally based on these lectures, was first published in 1801; it went to eleven editions, each larger than the previous one. Henry's *Elements* was the most popular and successful chemistry text in English for more than thirty years.

Also in 1801 Henry read his first paper to the Manchester Literary and Philosophical Society—a rebuttal of Davy's arguments against the materiality of heat; he was a lifelong calorist. In 1802 he read to the Royal Society the paper which established "Henry's law," and in 1805 there appeared the first of a series of works on the analysis of mixtures of gaseous hydrocarbons.

Stimulated by the recent trials of coal gas for lighting purposes, Henry set out to analyze various inflammable mixtures of gases obtained from coal and other materials of organic origin, with a view to determining their relative powers of illumination and to explain the differences in terms of their compositions. His investigations covered a period of more than twenty years, during which he gradually improved his analytic techniques. As well as representing a significant contribution to the progress of the gas industry, his work confirmed that of Dalton on the compositions of methane and ethylene; and their conviction that hydrogen and carbon combined only in definite proportions, to form a limited number of compounds, preceded the general acceptance of this view. In his final investigation Henry made use of the catalytic properties of platinum discovered in 1824 by Döbereiner.

The friendship of Henry with Dalton, for which he is now chiefly remembered, is best exemplified against the background of Henry's law relating to the solubility of gases: ". . . under equal circumstances of temperature, water takes up, in all cases, the same volume of condensed gas as of gas under ordinary pressure . . ." (*Philosophical Transactions of the Royal Society,* **93** [1803], 41). In 1801 Dalton had been absorbed by the problem of why an atmosphere consisting of gases of different densities did not separate into layers. His speculations gave rise to his theory of mixed gases (the embryo of the law of partial pressures), the best contemporary expression of which was given by Henry: "Every gas is a vacuum to every other gas" (Nicholson's *Journal of Natural Philosophy, Chemistry, and the Arts,* **8** [1804], 298). Henry had at first been among the many critics of the theory; but a suggestion made to him by Dalton in the light of it had enabled him to account for certain discrepancies in his solubility experiments, and he came out strongly in its defense. Dalton's own experiments on the solution of gases and the stimulus afforded by Henry's work have been seen as crucial in the development of the atomic theory (see L. K. Nash, "The Origin of Dalton's Chemical Atomic Theory," in *Isis,* **47** [1956], 101-116; and E. L. Scott, "Dalton and William Henry," in D. S. L. Cardwell, ed., *John Dalton and the Progress of Science* [Manchester-New York, 1968], pp. 220-239).

In 1809 Henry applied his analytical techniques to the composition of ammonia, confirming and refining the earlier work of Claude Berthollet, A. B. Berthollet, and Davy with respect to the proportions of hydrogen and nitrogen and showing that Davy was mistaken in thinking it contained oxygen. Later (1824) he succeeded in reconciling the results with Gay-Lussac's law and confirmed the latter's analysis of certain oxides of nitrogen.

Henry's 1824 papers were the last of any importance in experimental chemistry. Some time after this he was forced to abandon manipulative experiments because of surgical operations performed on his hands. He turned to the study of contagious diseases, which he believed were spread by chemical substances. Other adherents of this theory had tried to destroy the "contagion" by chemical reactions, usually with noxious reagents such as chlorine; Henry believed, and satisfied himself by experiment, that the "contagion" was heat-labile and could be inactivated by moderate heat. The advent of Asiatic cholera in 1831 made this work topical, and he devised a cheap and simple apparatus for disinfection by heat of clothing and other items. His hope that the method would be widely used was not realized, and his work was forgotten for many years—until long after the germ theory of disease provided a different reason for disinfection by heat.

Henry suffered from chronic ill health besides the neuralgic pains resulting from his injury, which finally became so acute as to deprive him of sleep; he committed suicide in 1836. He was highly esteemed in his day, but much of his work is a confirmation of others'—valuable but not spectacular. He tended to be cautious and unspeculative. Lacking the boldness of his friend Dalton, his commitment to the atomic theory, the initial formulation of which he had assisted, was belated and reserved. Thus, he missed the opportunity, afforded by his unique relationship with Dalton and his superiority in experimental skill, to accelerate the acceptance of a theory that was to become the very foundation of modern chemistry.

Henry is sometimes confused with his son William Charles Henry, who studied medicine at Edinburgh and chemistry under Liebig in Germany. He published a few papers, but after the shock of his father's death, he retired from medicine and science at the age of thirty-three and lived the life of a country gentleman in Herefordshire. He is generally known only for his biography of Dalton.

BIBLIOGRAPHY

I. Original Works. Henry's books are *An Epitome of Chemistry in Three Parts* (London, 1801 [2 eds.], 1803; Edinburgh, 1806; London, 1808); and *The Elements of Experimental Chemistry,* 2 vols. (London, 1810 [styled 6th ed.], 1815, 1818, 1823, 1826, 1829). The last (11th) ed. contains biographical sketches of Davy and William Wollaston. *An Estimate of the Philosophical Character of Dr. Priestley* (York, 1832), with app., a penetrating study, was read to the British Association for the Advancement of Science at York in 1831—it also appears, without the app., in *Report of the British Association for the Advancement of Science,* for 1831 and 1832 (1833), 60-71.

His scientific papers, with a few exceptions, are listed in Royal Society *Catalogue of Scientific Papers,* III (London, 1869), 293-295. Those mentioned in the text are "Experiments on Carbonated Hydrogen Gas; With a View to Determine Whether Carbon Be a Simple or a Compound Substance," in *Philosophical Transactions of the Royal Society,* **87** (1797), 401-415; "Account of a Series of Experiments Undertaken With the View of Decomposing the Muriatic Acid," *ibid.,* **90** (1800), 188-203; "A Review of Some Experiments, Which Have Been Supposed to Disprove the Materiality of Heat," in *Memoirs of the Manchester Literary and Philosophical Society,* **5** (1802), 603-

621; "Experiments on the Quantity of Gases Absorbed by Water at Different Temperatures and Under Different Pressures," in *Philosophical Transactions of the Royal Society,* **93** (1803), 29–42, 274–276; "Illustrations of Mr. Dalton's Theory of the Constitution of Mixed Gases," in Nicholson's *Journal of Natural Philosophy, Chemistry, and the Arts,* **8** (1804), 297–301; "Experiments on the Gases Obtained by the Destructive Distillation of Wood, Peat, Pit-coal, Oil, Wax etc. . . .," *ibid.,* **9** (1805), 65–74; "Experiments on Ammonia, and an Account of a New Method of Analyzing It, by Combustion With Oxygen and Other Gases," in *Philosophical Transactions of the Royal Society,* **99** (1809), 430–449; "Additional Experiments on the Muriatic and Oxymuriatic Acids," *ibid.,* **102** (1812), 238–246; "On the Action of Finely Divided Platinum on Gaseous Mixtures, and Its Application to Their Analysis," *ibid.,* **114** (1824), 266–289; "Experiments on the Analysis of Some of the Aëriform Compounds of Nitrogen," in *Memoirs of the Manchester Literary and Philosophical Society,* 2nd ser., **4** (1824), 499–517; and "Experiments on the Disinfecting Powers of Increased Temperatures, With a View to the Suggestion of a Substitute for Quarantine," in *Philosophical Magazine,* **10** (1831), 363–369; **11** (1832), 22–31, 205–207.

II. SECONDARY LITERATURE. The main biographical source is W. C. Henry, "A Memoir of the Life and Writings of the Late Dr. Henry," in *Memoirs of the Manchester Literary and Philosophical Society,* 2nd ser., **6** (1842), 99–141; biographical sketches are in J. Wheeler, *Manchester: Its Political, Social and Commercial History, Ancient and Modern* (Manchester, 1836), pp. 495–498—it was written by J. Davies and published separately as *Sketch of the Character of the Late William Henry* (Manchester, 1836)—and in E. M. Brockbank, *Sketches of the Lives and Work of the Honorary Medical Staff of the Manchester Infirmary* (Manchester, 1904), pp. 235–240, with portrait; Brockbank also includes an article on W. C. Henry, pp. 273–275, with portrait. Also see W. V. Farrar, K. R. Farrar, and E. L. Scott, "The Henrys of Manchester," in *R.I.C. Reviews,* **4** (1971), 35–47.

E. L. SCOTT

HENSEL, KURT (*b.* Königsberg, Germany [now Kaliningrad, U.S.S.R.], 29 December 1861; *d.* Marburg, Germany, 1 June 1941), *mathematics.*

Hensel was descended from a family of artists and scientists; his grandmother, the former Fanny Mendelssohn, was a sister of Felix Mendelssohn-Bartholdy. Until he was nine years old, he was educated at home by his parents; later, in Berlin, he was decisively influenced by the eminent mathematics teacher K. H. Schellbach at the Friedrich-Wilhelm Gymnasium. Hensel studied mathematics at Bonn and Berlin. Among his teachers were Rudolf Lipschitz, Karl Weierstrass, Carl Borchardt, Gustav Kirchhoff, Hermann von Helmholtz, and especially Leopold Kronecker, under whose guidance he took his Ph.D. in 1884 and qualified as *Privatdozent* at the University of Berlin in 1886. In the latter year he married Gertrud Hahn, sister of the renowned educator Kurt Hahn, known for his schools at Salem, Germany, and in Scotland. They had four daughters and a son, Albert.

After Kronecker's death, Hensel devoted many years to preparing the edition of his collected papers. In close cooperation with G. Landsberg, Hensel published his first important book, *Theorie der algebraischen Funktionen* (1902). In 1901 Hensel became full professor at the University of Marburg, where he was an extremely successful teacher and wrote the important books *Theorie der algebraischen Zahlen* (1908) and *Zahlentheorie* (1913). Also in 1901 Hensel became editor of Germany's oldest mathematical periodical, *Journal für die reine und angewandte Mathematik.* He retired in 1930 but continued to teach and advise. The following year he was awarded an honorary Ph.D. by the University of Oslo.

Hensel's scientific work was based on Kronecker's arithmetical theory of algebraic number fields. The Kronecker-Hensel method also yielded a foundation of the arithmetic in algebraic function fields. The latter foundation was developed systematically in *Theorie der algebraischen Funktionen.* The Weierstrass method of power-series development for algebraic functions led Hensel, about 1899, to the conception of an analogue in the theory of algebraic numbers: p-adic numbers. The p-adic numbers must be considered his most important discovery. In evaluating it, one must bear in mind that its conceptual base did not then exist; on the contrary, Hensel's discovery was the decisive stimulus for the development of the abstract algebraic notions required for the base: the theory of valuated fields. In his *Theorie der algebraischen Zahlen* and *Zahlentheorie,* Hensel developed p-adic numbers into a systematic theory; he also gave an application of great interest—to the classical theory of quadratic forms—and a remarkable extension of his p-adic method by the introduction of a p-adic analysis. Further developed by Hensel's pupils, especially Helmut Hasse, the p-adic method proved highly successful in the theory of quadratic forms and in the theory of algebras over number fields and is known today as the local-global principle. Hensel's method led him to many interesting results in the theory of numbers, which were published in a great many papers. At first his p-adic numbers were generally considered of no particular consequence, but he lived to see their recognition as a highly important, widely generalizable mathematical element.

BIBLIOGRAPHY

Hensel's books and contributions to books are *Theorie der algebraischen Funktionen einer Variabeln und ihre Anwendung auf algebraische Kurven und Abelsche Integrale* (Leipzig, 1902), written with G. Landsberg; *Theorie der algebraischen Zahlen* (Leipzig, 1908); *Zahlentheorie* (Berlin-Leipzig, 1913); and "Arithmetische Theorie der algebraischen Funktionen," in *Encyclopädie der mathematischen Wissenschaften,* pt. 2, sec. 5 (1921), 533-650.

Three collections of Leopold Kronecker's work, edited by Hensel, are *Werke,* 5 vols. (Leipzig, 1895-1930); *Vorlesungen über Zahlentheorie* (Leipzig, 1901); and *Vorlesungen über die Theorie der Determinanten* (Leipzig, 1903).

Hensel's most important papers are "Arithmetische Untersuchungen über die Diskriminanten und ihre ausserwesentlichen Teiler" (Berlin, 1884), his diss.; "Über eine neue Begründung der algebraischen Zahlen," in *Jahresbericht der Deutschen Mathematiker-Vereinigung,* **6** (1899), 83-88; "Die multiplikative Darstellung der algebraischen Zahlen für den Bereich eines beliebigen Primteilers," in *Journal für die reine und angewandte Mathematik,* **145** (1915), 92-113; **146** (1916), 189-215; **147** (1917), 1-15; and "Eine neue Theorie der algebraischen Zahlen," in *Mathematische Zeitschrift,* **2** (1918), 433-452.

H. Hasse's commemorative article "Kurt Hensel zum Gedaechtnis," in *Journal für die reine und angewandte Mathematik,* **187** (1950), 1-13, contains a complete bibliography.

HELMUT HASSE

HENSEN, (CHRISTIAN ANDREAS) VICTOR (*b.* Schleswig, Germany, 10 February 1835; *d.* Kiel, Germany, 5 April 1924), *physiology, marine biology.*

Hensen was the son of Hans Hensen, director of the school for the deaf and dumb at Schleswig, and his second wife, Henriette Caroline Amalie Suadicani, the daughter of the court physician Carl Ferdinand Suadicani, who had founded the lunatic asylum at Schleswig. From the two marriages of his father Hensen had eight sisters and six brothers.

From 1845 to 1850 Hensen attended the school attached to the cathedral of Schleswig, then the grammar school at Glückstadt (Holstein), where he passed the final examination in 1854. His attainments as a pupil are said to have been only mediocre. He next studied medicine for five semesters at Würzburg (under Scherer, Kölliker, Virchow) from 1854 to 1856. In Scherer's laboratory Hensen verified Claude Bernard's data on the glycogen content of the liver. He next studied for two semesters at Berlin, then in 1857-1858 at Kiel, where he passed his final examination. His sixteen-page thesis, written when he was working as a doctor in the lunatic asylum at Schleswig, deals with the possible diagnostic relationship between epilepsy and urinary secretion (1859). Soon afterward Hensen became prosector at the Institute of Anatomy at Kiel. Also in 1859 he qualified there as a lecturer in anatomy and histology. In 1864 he was appointed successor to Peter Ludwig Panum as associate professor of physiology and director of the physiology laboratory at Kiel. In 1868 he became full professor of physiology. In 1870 Hensen married Andrea Katharina Friederike Seestern-Pauly. They had two sons and two daughters. Hensen retired at the age of seventy-six. The most notable of his students in physiology were Paul Höber, Hans Winterstein, and Hans Piper.

Hensen worked mainly in physiology and marine biology. In physiology he preferred the histophysiological method and, using it, settled essential questions regarding the basic conditions for hearing and sight. In 1863 he investigated the decapod hearing organ and the morphology of the human cochlea, describing what are now called Hensen's supporting cells and Hensen's duct. He identified the fibers of the basal membrane in the cochlea as resonant corpuscles capable of vibrating. He also proved that these fibers, rather than decreasing, increase in length from base to tip of the cochlea. Hensen investigated the organ of hearing in the forelegs of grasshoppers and also in the fishes (1904). In addition he studied the structure of the cephalopod eye (1865) and the dispersion of cones in the center of the human retina. With J. C. Voelckers he histologically and experimentally investigated the nervous system for accommodation, puncturing the ciliary muscle laterally with the point of a needle. If there is accommodation, the needle moves forward. He described the light "Hensenian zone" in the Q section of the skeletal muscle fibers. In the field of embryology he observed, among other things, the formation of a torus at the beginning of the primitive furrow, or Hensen's knot (1876). He wrote lengthy summaries on the physiology of hearing (1880, 1902) and on propagation (1881). In 1878 he assumed the direction of a new institute of physiology at Kiel.

Hensen's second major area, in which he had worked as an amateur since 1863, was marine biology, that is, the investigation of the fauna of the oceans, both the microscopic plankton—a term he coined—and fishes. He developed quantitative methods for determining the amount of commercially useful fish along the coasts, thus laying the foundation for the calculation of the profitability of fisheries. About 1887 he began far-ranging plankton studies, especially quantitative investigations, using a dragnet he had invented. He calculated the quantity of plankton at various depths and in various oceans. For this purpose Hensen organized and led large plankton expeditions. He was a member, and later president, of the Prussian Commission for the Exploration of the German Waters at Kiel. He may be regarded as the originator of quantitative marine research.

Hensen was for several years dean of the Faculty of Medicine and several times was rector of the University of Kiel. He was honorary doctor of Kiel, a member of the Leopoldine Academy, and a corresponding member of the Bavarian and Prussian academies of science.

BIBLIOGRAPHY

I. Original Works. Hensen's major publications in physiology are "Über die Zuckerbildung in der Leber," in *Verhandlungen der Physikalisch-medizinischen Gesellschaft zu Würzburg,* **7** (1857), 219; "Studien über das Gehörorgan der Dekapoden," in *Zeitschrift für wissenschaftliche Zoologie,* **13** (1863), 319–412; "Zur Morphologie der Schnecke des Menschen und der Säugethiere," *ibid.,* 481–512; "Ueber das Gehörorgan von Locusta," *ibid.,* **16** (1866), 190–207; "Ueber das Sehen in der Fovea centralis," in *Virchows Archiv für pathologische Anatomie,* **39** (1867), 475–492; *Experimentaluntersuchung über den Mechanismus der Accommodation* (Kiel, 1868), written with J. C. Voelckers; "Ueber den Ursprung der Accommodationsnerven . . .," in *Albrecht v. Graefes Archiv für Ophthalmologie,* **24** (1878), 1–26, written with J. C. Voelckers; "Ueber die Accommodationsbewegung im menschlichen Ohr," in *Pflügers Archiv für die gesamte Physiologie,* **87** (1901), 355–360; and "Die Empfindungsarten des Schalls," *ibid.,* **119** (1907), 249–294.

His main writings on marine biology are "Beftreffend den Fischfang auf der Expedition," in *Jahresbericht der Kommission zur wissenschaftlichen Untersuchung der deutschen Meere in Kiel für das Jahr 1871* (1873), pp. 155–159; "Resultate der statistischen Beobachtungen über die Fischerei an den deutschen Küsten," *ibid., 1874–1876* (1878), 133–171; "Ueber die Bestimmung des Planktons oder des im Meere treibenden Materials an Pflanzen und Thieren," *ibid., 1882–1886* (1887), pp. 1–107, with four plates and a list of the specimens collected; "Einige Ergebnisse der Plankton-Expedition der Humboldt-Stiftung. Vorgelegt von E. du Bois-Reymond," in *Sitzungsberichte der Preussischen Akademie der Wissenschaften zu Berlin,* **1** (1890), 243–253; *Ergebnisse der in dem Atlantischen Ocean von Mitte Juli bis Anfang November 1889 ausgeführten Plankton-Expedition der Humboldt-Stiftung . . .*" (Kiel–Leipzig, 1892); "Das Plankton der östlichen Ostee und des Stettiner Haffs," in *Bericht der Kommission zur wissenschaftlichen Untersuchung der deutschen Meere in Kiel für die Jahre 1887–1891* (1893), pp. 103–137; "Die Nordsee-Expedition 1895 des Deutschen Seefischerei-Vereins. Über die Eimenge der im Winter laichenden Fische," in *Wissenschaftliche Meeresuntersuchungen,* n.s. **2,** no. 2 (1897), 1–97, written with Carl Apstein: "Über die quantitative Bestimmung der kleineren Planktonoganismen," *ibid.,* n.s. **5** (1901), 67–81; "Über quantitative Bestimmungen des 'Auftriebs,'" in *Mitteilungen für den Verein Schleswig-Holsteinischer Aerzte* **10,** no. 7 (1885); and "Die Methodik der Plankton-Untersuchung," in E. Abderhalden, ed., *Handbuch der biochemischen Arbeitsmethoden,* V, pt. 1 (Vienna, 1911), 637–658.

II. Secondary Literature. The only good biography is Rüdiger Porep, *Der Physiologe und Planktonforscher Victor Hensen (1835–1924). Sein Leben und sein Werk,* no. 9 in the series Kieler Beiträge zur Geschichte der Medizin und der Pharmazie (Neumünster, 1970), with numerous illustrations and good bibliography. There are obituaries by Karl Brandt in *Berichte der Deutschen wissenschaftlichen Kommission für Meeresforschung,* n.s. **1** (1925), vii–x; and J. Reibisch, in *Archiv für Hydrobiologie,* **16** (1926), i–xiv, with portrait.

K. E. Rothschuh

HENSLOW, JOHN STEVENS (*b.* Rochester, Kent, England, 6 February 1796; *d.* Hitcham, Suffolk, England, 16 May 1861), *botany.*

Henslow was the eldest of eleven children of John Prentis Henslow, a solicitor. He was educated at the Free School at Rochester and later at Camberwell in Surrey, where his inherent love of nature developed into a keen interest in natural history. In 1814 he entered St. John's College, Cambridge, and four years later graduated sixteenth wrangler; he received the M.A. in 1821. At Cambridge he studied mathematics, chemistry, and mineralogy. He was elected a fellow of the Linnean Society in 1818 and the following year a fellow of the Geological Society of London. During a geological tour of the Isle of Wight he and Adam Sedgwick engaged in discussions that later led to the formation of the Cambridge Philosophical Society, of which Henslow was a founder. His paper on the geology of Anglesea, prepared after an extensive survey of the island in 1821, was hailed as an important contribution. In 1822 he was elected to the chair of mineralogy at Cambridge.

The professorship of botany at Cambridge, to which Henslow had looked forward for many years, fell vacant in 1825. He immediately offered himself as a candidate for the position and was elected unopposed. Soon after, he resigned his chair of mineralogy and devoted himself completely to the study and teaching of botany. Systematic botany did not appeal to him; he considered it necessary only so far as it helped the study of the distribution of plants. His main interests were plant geography, morphology, and physiology. He organized botanical excursions, encouraging students to observe and study plants in their natural environment. He used his own diagrams and actual specimens at lectures to demonstrate form and structure in plants. He required students to dissect, examine, and describe the specimens they were studying. Under Henslow botany became one of the most popular subjects at Cambridge; among his students were Charles Darwin, Berkeley, R. T. Lowe, W. H. Miller, Babington, and others. Henslow recommended Darwin as naturalist for H. M. S. *Beagle* and during the five-year voyage regularly corresponded with Darwin and took care of all specimens sent by him.

Henslow's persistent efforts eventually resulted in the redevelopment of the long neglected Cambridge Botanical Garden, which he regarded as an essential adjunct to the teaching of botany. He always maintained a lively interest in museums and was directly responsible for, or contributed freely toward, their establishment in Ipswich, Cambridge, and Kew. For a number of years Henslow was a member of the senate and an examiner in botany at the University of London. He was also an active founder-member of the British Association, presiding over its natural history section on many occasions.

Henslow was ordained in 1824 and became curate of Little St. Mary's Church in Cambridge. He was appointed vicar of Cholsey in Berkshire in 1833. Four years later he received from the crown the rectory at Hitcham in Suffolk, where he moved in 1839 and resided until the end of his life. He went to Cambridge every year to deliver his lectures and he taught botany and horticulture to village children in his parish school.

In 1823 Henslow married Harriet Jenyns, daughter of the Reverend George Jenyns of Bottisham in Cambridgeshire; they had two sons and three daughters. His daughter Frances was the first wife of J. D. Hooker.

BIBLIOGRAPHY

I. ORIGINAL WORKS. Among Henslow's most important works are "Geological Description of Anglesea," in *Transactions of the Cambridge Philosophical Society,* **1** (1822), 359–452; *A Catalogue of British Plants* (Cambridge, 1829; 2nd ed., 1835); *Principles of Descriptive and Physiological Botany* (London, 1835); and *Dictionary of Botanical Terms* (London, 1857).

II. SECONDARY LITERATURE. A chronological list of Henslow's publications is given in a full length biography, L. Jenyns, *Memoir of the Rev. John Stevens Henslow* (London, 1862). A shorter list is in the Royal Society, *Catalogue of Scientific Papers,* **3** (1869). Other biographical accounts are "The Rev. Professor Henslow," in *Gardener's Chronicle and Agricultural Gazette* (1861), pp. 505–506, 527–528, 551–552; F. W. Oliver, ed., *Makers of British Botany* (London, 1913); J. R. Green, *History of Botany in the United Kingdom* (London, 1914); J. Britten and G. S. Boulger, *A Biographical Index of Deceased British and Irish Botanists,* 2nd ed. (London, 1931); and N. Barlow, ed., *Darwin and Henslow; The Growth of an Idea. Letters 1831–1860* (London, 1967).

M. V. MATHEW

HERACLIDES PONTICUS (*b.* Heraclea, *ca.* 388 B.C.; *d.* Athens, *ca.* 315 B.C.), *astronomy, geometry.*

For a complete study of his life and work see Supplement.

HERACLITUS OF EPHESUS (*fl. ca.* 500 B.C.), *moral philosophy, natural philosophy.*

Heraclitus wrote a book (see Diogenes Laërtius IX, 5), fragments of which survive in other authors of classic antiquity as quotations, paraphrases, and references. The work was apparently a collection of apothegms similar in style to the Delphic oracle, which (as he says in fr. 93) "neither states anything nor conceals it but gives a sign." The surviving fragments are full of word play and deliberate ambiguity. For ideas about the order of Heraclitus' exposition, and about the context and interpretation of particular fragments, we are dependent on later authors, who certainly quote him tendentiously. They themselves found him difficult to understand and nicknamed him "the dark one." There is very little agreement among modern scholars and philosophers on the nature of Heraclitus' thought.

Heraclitus is the first Greek philosopher to emerge as a personality. His style is unique, and he seems determined to tease his hearers with difficult challenges to their understanding, accompanied by caustic remarks about their lack of intelligence. The ancient biography (Diogenes Laërtius IX, 1) says he was an arrogant misanthrope.

In the discussion below, references are given according to the arrangement in Diels and Kranz, *Fragmente der Vorsokratiker.*

Heraclitus presents himself as the vehicle, rather than the author (see fr. 50), of a divine *logos,* which is uttered by him but is also something like a law which directs the natural world just as a city's laws, which are "nurtured by the one divine law," maintain balanced relationships among the citizens (see frs. 1, 2, 114).

The balance that is maintained in the universe is between opposites in tension with each other. "Men do not understand how, being pulled apart, it is in

HERACLITUS OF EPHESUS

accord with itself: a harmony, turning back on itself, as in the bow and the lyre" (fr. 51). The bow and the lyre have their virtue in the tension of a string pulled in opposite directions. The most striking feature of the surviving fragments is the frequent recurrence of binary oppositions. They are not the same as the "contraries" which Aristotle picked out as crucial to the theories of the early Greek *physiologoi* (*Physics A*, 4–5): the hot and the cold, the dry and the wet, and other pairs of opposed physical properties and things. It is possible to recognize the hot and the cold, the dry and the wet, in fragment 126; but the majority of the pairs are either concerned with the properties of living beings (for instance, sleeping and waking, life and death, plenty and hunger, youth and age, men and gods, health and sickness), or else they are verbal expressions (to be willing and unwilling, to be present and absent, to agree and to differ, to kindle and to quench).

Heraclitus characteristically says that these binary opposites are the same, that they are one. "The way up and the way down is one and the same" (fr. 60); "Beginning and end, on a circle's circumference, are common" (fr. 103); "Hesiod is the teacher of most men: they are convinced that he knows most—who did not know day and night; for they are one" (fr. 57); "Junctions are wholes and not wholes, agreeing and diverging, being in tune and out of tune, and out of all, one, and out of one, all" (fr. 10).

Heraclitus' *logos,* it seems, is this pattern of sameness and contrariety, manifested in the physical world and in human life. The opposites are sometimes unified by being in tension with each other, or by being at war. "War is father of all, king of all; some he reveals as gods, some as men, some he makes slaves, some free" (fr. 53); "It must be known that war is common and strife is justice and all things happen in accordance with strife and necessity" (fr. 80). Sometimes opposites are unified by being changed into each other: "In us the same is living and dead, awake and asleep, young and old, for these, transformed, are those, and those, transformed, are these" (fr. 88); "Cold things are warmed, warm cooled, wet dried, parched moistened" (fr. 126). Sometimes they are unified as correlatives: "It is sickness that makes health sweet and good, hunger satiety, tiredness rest" (fr. 111). Other modes of unification can be distinguished, but it is hard to find any systematic importance in the different modes.

It may be that for Heraclitus himself the main point was a message about the human soul, its continuity in life and death, and its connection with the divine *logos* and the "ever-living fire." Yet in the history of natural philosophy it was for the physical doctrine attributed to him that he won most fame. Plato (*Cratylus,* 402A) attributes to him the doctrine that "all things are in flux and nothing is stable," and this doctrine is taken to imply that sense perception cannot be equated with knowledge (*Theaetetus,* 181C–E). The same view of Heraclitus was taken by Aristotle (for instance, *Metaphysics A* 6, 987a29; *Physics* VIII 3, 253b9; *Topics A* 11, 104b19) and passed into common tradition.

The best direct evidence for the flux doctrine is contained in the fragments that use the images of fire and rivers. "This cosmos was made by no god or man, but always was and is and will be: ever-living fire, kindling in measures and quenching in measures" (fr. 30). Unfortunately, it is unclear what "cosmos" means here, since it is not certain that it was used in the sense of world order as early as Heraclitus; some argue that in this fragment it means any instance of order in the natural world. Ancient doxographers, taking a hint from Aristotle (*Metaphysics A* 3, 984a5), assumed that for Heraclitus fire played the same role—that of originative substance from which the whole world grew—as water for Thales, the Boundless for Anaximander, and air for Anaximenes. This assumption led to the attribution to Heraclitus of the Stoic doctrine of a periodic world conflagration (*ekpyrosis*). This attribution has had some recent defenders (especially O. Gigon), but it is more likely that Heraclitus meant to use fire as a paradigm for explaining (some or all) continuing natural processes: fire consumes things and changes them into itself, as smoke or hot vapor, and later there is condensation and the re-formation of liquids and solids. This description may well apply to such things as seasonal changes in the cosmos (see especially fr. 31); but there is some rather uncertain evidence that it also has to do with life cycles. This depends on fragments about souls (*psychai*), which seem to associate life and good functioning with a fiery state, and death with water (frs. 66, 68).

Aristotle (*Metaphysics* Γ 5, 1010a7 ff.) says that Cratylus criticized his master Heraclitus for saying that it is not possible to step twice into the same river: Cratylus thought it was impossible even once. Ancient writers took this argument to refer to a doctrine that all things are in flux and unknowable. It has recently been argued, especially by G. S. Kirk, that nothing in the relevant fragments (12, 49a, 91) requires us to think that the river analogy must apply to all things; and that the main thrust of Heraclitus' thought is not that all things change even though they seem permanent, but that the changes that do take place are measured and balanced. The tradition about the doctrine of universal flux is probably right, but there is no evidence that Heraclitus turned the doctrine into an argument to show that the natural world is unknowable.

A strange astronomy, in which the heavenly bodies are bowls of fire, is attributed to Heraclitus by the doxographers. It is very unlikely to have been intended seriously as a rival to others. When Heraclitus was placed in succession with other *physiologoi,* it was supposed that he answered the same questions as the others, and odd hints in his work were elaborated into a theory. He wrote "The sun is new every day" (fr. 6), but without context this is hard to interpret. "The sun will not overstep his measures, otherwise the Furies, ministers of Justice, will find him out" (fr. 94) appears to notice the regularity of the sun's motions but does not otherwise seem like astronomy.

Heraclitus criticized Hesiod, Pythagoras, and Xenophanes by name (fr. 40). It appears likely that he also criticized the doctrines of the Milesian school, chiefly for misunderstanding the role of opposites in the world. They believed opposites to be a secondary development from an original undifferentiated stuff; for Heraclitus, opposites and the constant tension between them were primary. Whether or not he developed a positive cosmological system of his own, the system attributed to him by Plato and Aristotle was a very important factor in Greek cosmology, as can be seen, for instance, in Plato's *Theaetetus* 181A, where thinkers are divided into "flux men" and "stationary men."

BIBLIOGRAPHY

I. Original Works. The complete Greek fragments, with German trans., are collected in H. Diels and W. Kranz, *Fragmente der Vorsokratiker,* 5th ed., I (Berlin, 1934); there are many later reprints. Other noteworthy eds. are I. Bywater, *Heracliti Ephesii reliquiae* (Oxford, 1877); R. Walzer, *Eraclito: Raccolta dei frammenti e traduzione italiana* (Florence, 1939; repr. Hildesheim, 1964); G. S. Kirk, *Heraclitus: The Cosmic Fragments* (Cambridge, 1954; repr. with corrections, 1962); R. Mondolfo, *Heráclito: Textos y problemas de su interpretación* (Mexico City, 1966); and M. Marcovich, *Heraclitus: Greek Text With a Short Commentary* (Mérida, 1967).

II. Secondary Literature. Books and articles on Heraclitus include the following, listed chronologically: O. Gigon, *Untersuchungen zu Heraklit* (Leipzig, 1935); H. Fränkel, "A Thought Pattern in Heraclitus," in *American Journal of Philology,* **59** (1938), 309–337; K. Reinhardt, "Heraklits Lehre vom Feuer," in *Hermes* (Wiesbaden), **77** (1942), 1–27; H. Fränkel, *Dichtung und Philosophie des früher Griechentums* (New York, 1951), pp. 474–505, 2nd ed. (Munich, 1963), pp. 422–453; G. Vlastos, "On Heraclitus," in *American Journal of Philology,* **76** (1955), 337–368; E. Zeller and R. Mondolfo, *La filosofia dei greci nel suo sviluppo storico* I. 4 (Florence, 1961); W. K. C. Guthrie, *A History of Greek Philosophy,* I (Cambridge, 1962), 403–492; Charles H. Kahn, "A New Look at Heraclitus," in *American Philosophical Quarterly,* **1** (1964), 189–203; and M. Marcovich, in Pauly-Wissowa, supp. X (Stuttgart, 1965), cols. 246–320.

DAVID J. FURLEY

HERAPATH, JOHN (*b.* Bristol, England, 30 May 1790; *d.* Lewisham, England, 24 February 1868), *theoretical physics, journalism.*

Although Herapath is best known as the first to work out extensive calculations and applications of the kinetic theory of gases, for most of his life he was regarded as an eccentric amateur who had unsuccessfully challenged the scientific establishment and then turned to a more profitable career as editor of a railway magazine. Occasionally given credit for his early work, Herapath did have a slight influence on later scientific developments. But because his theoretical ideas were so uncongenial to his generation, his scientific talents were mostly wasted.

Herapath was the son of a maltster, and a cousin of William Herapath, the chemist for whom the compound herapathite (used by Land in early forms of Polaroid) is named. He was largely self-educated; he learned French and was acquainted with some of the works of the great mathematical physicists of the late eighteenth and early nineteenth centuries. He seems to have absorbed their proclivity toward grand speculations in science which, on the one hand, may have led him to the kinetic theory and, on the other, been an obstacle to its acceptance by his more empirically minded countrymen.

By 1811 Herapath was engaged in researches on the theory of lunar motion. He came to the conclusion that the earth's action on the moon is greater when the earth is nearer to the sun. Following Newton's suggestion that gravitational forces might result from differences in density of the ether at various distances from massive bodies, Herapath added the notion that this variation might be connected with changes in temperature, as in the case of ordinary fluids. Thus the force of gravity would depend on temperature. But before working out a detailed application of this hypothesis to the lunar problem, Herapath was diverted by the question, what is temperature—or rather, what is heat?

He first tried to devise his theory on the accepted doctrine that heat is associated with repulsive forces between particles of a fluid, but he ran into difficulties and finally abandoned this position. Instead he concluded, in May 1814, that heat is the result or manifestation of "intestine motion." He did not claim that this idea was an original discovery; but rather that he had succeeded in giving it a better and more consequential mathematical formulation than any he had previously seen. (Apparently he was not aware of Daniel Bernoulli's brief excursion in kinetic theory in his *Hydrodynamica* [1738]. Had he known of this predecessor, his own theory would have been little different but, in advancing it, he might have benefited by Bernoulli's authority.)

Herapath derived the basic equation relating the pressure (P) and volume (V) of a gas to the mass (m) and speed (v) of its particles,

$$PV = \frac{1}{3} Nmv^2,$$

assuming that the N particles occupy altogether such a small part of the volume that each one can move freely through space most of the time, occasionally colliding with other particles or with the walls of the container. The gas pressure is thus attributed to impacts of the particles against the walls, rather than to continually acting interparticle repulsive forces, as in the Newtonian theory then generally accepted.

The main difference between Herapath's theory and that found in modern textbooks is that Herapath stressed the conservation of average momentum (mv) in collisions of particles. He assumed that the quantity of heat contained in a gas is proportional to the total momentum of all its particles, but as in Descartes's theory, this quantity is not added vectorially. He defined absolute or "true" temperature as the total momentum of a gas divided by the number of particles. Consequently he argued that when two gases or even two liquids at different temperatures are mixed, the temperature of the mixture must be calculated by averaging the "true" temperatures rather than those on the Fahrenheit or Celsius scale. He used this prediction to propose a crucial experiment to distinguish between his theory and the conventional one: If equal portions of water at 32°F. and 212°F. are mixed, the temperature of the mixture, according to Herapath's computation, should be 118.4°F., not

$$\frac{1}{2}(32^\circ + 212^\circ) = 122^\circ\text{F.}$$

Although the existing experimental data were not accurate enough to resolve this point, Herapath claimed that they confirmed his theory.

Having published a preliminary notice of his theory in the *Annals of Philosophy* in 1816, Herapath submitted a detailed account to the Royal Society in 1820. Davy, who was elected to the presidency of the Society in November of that year, was primarily responsible for the fate of the paper. Although Davy was already known as an advocate of the qualitative idea that heat is molecular motion, he found Herapath's quantitative development too speculative and complicated; he rejected the hypothesis of an absolute temperature implying an "absolute zero" of cold. Having been told that his paper would not be accepted for publication in the *Philosophical Transactions,* Herapath withdrew it and published it instead in the *Annals of Philosophy* in 1821. Five years later he launched an attack on Davy in the *Times* of London, accusing him of circulating unfounded criticisms of his experimental work, which prevented its publication. Although Davy ignored a series of letters and challenges published in the *Times,* Herapath later claimed Davy's resignation from the presidency of the Royal Society (1827) as a victory for himself.

Herapath married in 1815 and gave up his association with his father's business to start a private school of mathematics to prepare young men for the universities. Apparently this enterprise did not flourish because of his failure to establish a scientific reputation, although it is his only recorded source of income until 1832. His family responsibilities during this period were considerable since by 1837 he had eleven children, ranging in age from one to twenty-two.

In 1829 he took an interest in the promotion of Goldsworthy Gurney's steam carriages, and while this particular project failed, it encouraged him to study the rapidly expanding railways. He began to write articles on engineering and commercial aspects of the new English railway lines in 1835, and in 1836 he became editor of the *Railway Magazine and Annals of Science.* This occupation provided financial security—the magazine, later known as *Herapath's Railway and Commercial Journal,* was quite successful—and it gave him an opportunity to publish his own papers on scientific subjects. In addition to these advantages, it would appear from Herapath's numerous writings that this new career provided ample personal satisfaction. As the scientist-turned-journalist, engineer, and operations-researcher, Herapath threw himself wholeheartedly into the excitement and controversies of England's railway boom of the 1840's.

One of the first scientific papers which Herapath published in his *Railway Magazine* was a calculation of the velocity of sound in air, which he had announced at a meeting of the British Association for the Advancement of Science in 1832. This is the first known calculation of the speed of a molecule from the kinetic theory of gases. Joule, usually credited with this accomplishment, undoubtedly based his own calculation on Herapath's, who had published his computation in book III of his major work *Mathematical Physics* (London, 1847). Herapath's application of the theory of molecular speeds to the wind resistance encountered by a fast railway locomotive (1836) is also an interesting example of the explicit use of scientific principles in engineering.

Maxwell, recognizing Herapath as a precursor of his own research in kinetic theory, gave the following assessment of Herapath's work:

> His theory of the collisions of perfectly hard bodies, such as he supposed the molecules to be, is faulty. . . . This author, however, has applied his theory to the numerical results of experiment in many cases, and his speculations are always ingenious, and often throw much real light on the questions treated. In particular, the theory of the temperature and pressure and gases and the theory of diffusion are clearly pointed out ("On the Dynamical Theory of Gases," in *The Scientific Papers of James Clerk Maxwell,* II [Cambridge, 1890], 28).

While the refusal of the Royal Society to publish Herapath's paper can hardly be defended, neither can it be argued that this refusal obstructed the progress of science to any significant extent. Herapath did manage to present his theory to a scientific community that no longer accepted the Royal Society as final arbiter. His theory simply did not provide an attractive explanation of the physical phenomena of gases and heat, which, like the phenomena of radiant heat, were then considered most important.

BIBLIOGRAPHY

Herapath's major work is reprinted in *Mathematical Physics (Two Volumes in One) and Selected Papers by John Herapath* (New York, in press), with intro. by the editor, Stephen G. Brush. This reprint includes the early paper, "A Mathematical Inquiry Into the Causes, Laws, and Principal Phaenomena of Heat, Gases, Gravitation, & c," which was published in the *Annals of Philosophy,* 2nd ser., **1** (1821), 273-293, 340-351, 401-416; one of his articles on railways; and a bibliography of all known works by or about Herapath.

STEPHEN G. BRUSH

HERAPATH, WILLIAM BIRD (*b.* Bristol, England, 28 February 1820; *d.* Bristol, 12 October 1868), *medicine, chemistry.*

Herapath was the oldest son of William Herapath, a well-known analytical chemist who was professor of chemistry and toxicology at the Bristol Medical School and one of the founders of the Chemical Society. He received his higher education at London University, where he was awarded the M.B. in 1844 with honors in six different branches of medical knowledge. He became a licentiate of the Society of Apothecaries in 1843; in 1844, following his graduation, he was elected to the Royal College of Surgeons and began serving in that capacity at Queen Elizabeth's Hospital in Bristol. He received the M.D. in 1851.

Herapath published many articles in medical, chemical, and other scientific journals. These articles show that his students assisted him in some of his researches and establish a close research relationship between Herapath and both W. Haidinger and George G. Stokes, secretary of the Royal Society. Each article was often published in several periodicals, data and content unchanged, although in some cases editorial alterations were made. A number of important discoveries were reported in Herapath's articles.

The most celebrated of these discoveries occurred in 1852, when Herapath attempted to prepare polarizing capsules of large aperture. He succeeded in producing small but usable crystals of the iodosulfate of quinine (now known as herapathite), which he patented for optical use. Herapathite absorbs completely one component of polarization and transmits the other with little loss; it is usually employed in the form of small rhomboidal plates oriented in the same direction within a transparent film. Herapath also referred to this compound as artificial tourmaline and discussed its advantage in optics over the Nicol prism.

In addition to this major work, Herapath also devised new methods for detecting arsenic and other substances, designed a new combustion blowpipe for organic analyses, used the spectroscope and microspectroscope to detect bloodstains, experimented with alkaloids, and developed new techniques for pathological investigations. The broad range of his activities is indicative of his belief in the need for a close alliance between chemistry, medicine, and medical research; he attested to this belief in his lecture *On Chemistry and Its Relation to Medical Studies and Associated Sciences* (1863).

Herapath's less purely scientific works include instructions for Clifton Cleve's *Hints on Domestic Sanitation* (1848) and, in 1854, an analysis of the waters of the spa and a description of the Bristol and Clifton hot wells, which were later incorporated into the *Handbook for Visitors to the Bristol and Clifton Hotwells.*

Herapath's bibliography lists few publications of any sort after 1864, since he became ill with the disease (perhaps a form of jaundice) of which he was to die at the early age of forty-eight. Despite his illness, he continued his work on spectroscopic analysis until a few days before his death. A posthumously published memorandum reports the results of more than 250 optical analyses on the chlorophyl of various plants, including fifty-four plants in the Forth. Although short, Herapath's career was a productive one, nurtured by the interests he shared with his father, his broad medical background, his analytical skills, and most of all by his zeal for science. He was survived by his wife and six children.

BIBLIOGRAPHY

I. ORIGINAL WORKS. Among Herapath's works are "On the Optical Properties of a Newly Discovered Salt of Quinine, Which Crystalline Substance Possesses the Power of Polarizing a Ray of Light, Like Tourmaline, and at Certain Angles of Rotation of Depolarizing It Like Selenite," in *Philosophical Magazine*, 4th ser., **3** (1852), 161–173; "On the Chemical Constitution and Atomic Weight of the New Polarizing Crystals Produced From Quinine," *ibid.*, **4** (1852), 186–192; "On the Discovery of Quinine and Quinidine in the Urine of Patients Under Medical Treatment With the Salts of These Mixed Alkaloids," *ibid.*, **6** (1853), 171–175; "On the Manufacture of Available Crystals of Sulphate of Iodo-Quinine (Herapathite) for Optical Purposes as Artificial Tourmalines," *ibid.*, pp. 346–351; "Further Researches Into the Properties of the Sulphate of Iodo-Quinine (Herapathite) More Especially in Regard to Its Crystallography, With Additional Facts Concerning Its Optical Relations," *ibid.*, pp. 284–289; "Letter to Prof. Stokes—On the Compounds of Iodine and Strychnine," *ibid.*, **10** (1855), 454–455; "On the Detection of Strychnine by the Formation of Iodo-Strychnine," *ibid.*, **13** (1857), 197–198; "On the Optical Characters of Certain Alkaloids Associated With Quinine, and of the Sulphates of their Iodo-Compounds," in *Proceedings of the Royal Society*, **8** (1856–1857), 340–343; "Researches on the Cinchona Alkaloids," *ibid.*, **9** (1857–1859), 5–22; "Preliminary Notice of Additional Researches on the Cinchona Alkaloids," *ibid.*, pp. 316–321; *On Chemistry and Its Relation to Medical Studies and Associated Sciences* (Bristol, 1863); "On a New Method of Detecting Arsenic, Antimony, Sulphur, and Phosphorus, by Their Hydrogen Compounds, When in Mixed Gases," in *Report of the British Association for the Advancement of Science*, **34** (1864), Transactions Sec., 31–32; "On the Pedicellariae of the Echinodermata," *ibid.*, pp. 95–97; "On the Genus Synapta," *ibid.*, pp. 97–98; "On the Occurrence of Indigo in Purulent Discharges," in *Chemical News*, **10** (1864), 169–171; "On a New Combustion Blowpipe for Organic Analysis," in *Journal of the Chemical Society*, n.s. **2** (1864), 49–50; "On the Use of the Spectroscope and Microspectroscope in the Discovery of Blood Stains and Dissolved Blood, and in Pathological Inquiries," in *Chemical News*, **17** (1868), 113–115, 124–125; and "Memorandum of Spectroscopic Researches on the Chlorophyl of Various Plants," in *Monthly Microscopical Journal*, **2** (1869), 131–133.

II. SECONDARY LITERATURE. See Boase, *Modern English Biography Since 1850*, I (1892), 1437; *Dictionary of National Biography*, IX (1937), 615; *Illustrated London News* (24 Oct. 1868), p. 411; *Lancet* (24 Oct. 1868), **2**, 559; Poggendorff; and Royal Society *Catalogue of Scientific Papers*, III, 303; VII, 955.

CLAUDE K. DEISCHER

HERBART, JOHANN FRIEDRICH (*b.* Oldenburg, Germany, 4 May 1776; *d.* Göttingen, Germany, 14 August 1841), *philosophy, psychology, pedagogy.*

Herbart was first greatly interested in science and music, but at Jena he studied philosophy and law. He was strongly influenced by Enlightenment thought, particularly Kant's ethics and Fichte's metaphysics. Later he became a close friend of Pestalozzi. Herbart received his doctorate and qualified for lecturing at Göttingen, where he lectured on philosophy and pedagogy. In 1808 he accepted an invitation to take over Kant's chair at Königsberg, where he established the first pedagogical institute with an experimental school. He also served on various commissions responsible for the improvement of the Prussian educational system.

According to Herbart, the structure and operation of man's perception are conditioned by the changing complex of ultimate entities of reality, which he called the "reals" (*Realen*). As in the ancient theory of atoms and elements or Leibniz' monad theory, the complex structure of reality arises through a rhythmical joining (synthesis) and separation (analysis) of the reals. The behavior of these entities is determined by their tendency toward self-assertion. Hence, a dialectical struggle of opposites emerges as the "law of motion" of reality. The task of philosophy is to create a rigorous analytic-synthetic conceptual system from perceived reality.

The soul is a central totality of manifold simple reals. The ideas that appear in the soul are the result of the interplay of the "self-preservative reactions" of the reals. If in this process an idea is so thoroughly repressed that it vanishes from consciousness, it struggles to emerge from below the threshold of consciousness until it reappears as a freely moving idea (memory). Herbart held that mental processes can be described with the exactness of mathematical laws.

In Herbart's pedagogical writings each person is an individual and distinctive totality, capable of change and determination or redefinition, and therefore possessing "adaptiveness" (*Bildsamkeit*). This latter quality is especially characteristic of the moral will. Therefore, the goal of upbringing and education is the development of the personality of the whole human being. This development aims at the union of five ideas: inner freedom (harmony of moral insight and will), perfection (health of body and soul), benevolence (toward the will of others), justice (balancing of interests, respect for the rights of others), and equity (suitability of reward and punishment). Together they constitute the "virtue of self-determination." As long as insight and self-determination of the will are lacking, the desires must submit to external regulation (subordination to authority and supervision). With the growth of intellectual spontaneity the pupil's interest can be awakened through instruction and discipline.

Herbart distinguished three forms of the "interest in knowledge" (empirical, speculative, aesthetic) and three forms of the "interest in participation" (sympathetic, social, religious). The development of insight and will requires a rhythmic alternation from a probing, analytic instruction to a reflective, synthetic one. "Static" penetration leads to conceptual clarity, "progressive" penetration (association) to the increase of knowledge; static reflection yields the system of knowledge, and progressive reflection gives rise to its method. From these four fundamental concepts Herbart deduced the four formal stages of instruction. The course that the instruction takes can be demonstrative, analytic, or synthetic, according to need. A goal of discipline is to mold the interests stimulated by instruction into a totality of moving ideas (*Gedankenkreis*). In particular, instruction seeks by this means to instill within the pupil fundamental moral tenets and to form them into a conscience. With increasing age, education is first restraining, then determining, then regulating, and finally supportive, as it ends and self-education begins. With these basic concepts and requirements Herbart established pedagogy as an independent science. He was likewise a founder of educational therapy and a precursor of child psychiatry.

BIBLIOGRAPHY

I. Original Works. Collections of Herbart's writings include the following: *Sämtliche Werke,* G. Hartenstein, ed., 12 vols. (Leipzig, 1850–1852; 2nd ed., Hamburg, 1883–1892; supp. vol., 1893); *Sämtliche Werke,* K. Kehrbach, O. Willmann, and T. Fritzsch, eds., 19 vols. (Langensalza, 1887–1912; new ed., Aalen, 1964); *Pädagogische Schriften,* O. Willmann and T. Fritzsch, eds., 2 vols. (Leipzig, 1873–1875), 3rd ed., 3 vols. (Leipzig, 1913–1919); and *Pädagogische Schriften,* W. Asmus, ed., 3 vols. (Düsseldorf–Munich, 1965). Herbart's individual works include *Kleine Schriften zur Pädagogik,* T. Dietrich, ed. (Bad Heilbrunn, 1962); *Umriss pädagogischer Vorlesungen,* J. Esterhues, ed. (Paderborn, 1957; 2nd ed., 1964); *Allgemeine Pädagogik,* H. Nohl, ed. (Weinheim, 1952; 7th ed., 1965), also edited by H. Holstein (Bochum, 1966); *Aus Herbarts Jugendschriften,* H. Döpp-Vorwald, ed. (Weinheim, 1955; 3rd ed., 1965); *Hauslehrerbriefe und pädagogische Korrespondenz 1797–1807,* W. Klaffki, ed. (Weinheim, 1966); and *Kleine pädagogische Schriften,* A. Brückmann, ed. (Paderborn, 1968).

II. Secondary Literature. On Herbart's life or work, see W. Asmus, *J. F. Herbart, eine pädagogische Biographie,* 2 vols. (Heidelberg, 1968–1970); B. Bellerate, *J. F. Herbart* (Brescia, 1964); and *La pedagogia in J. F. Herbart* (Brescia, 1970); J. L. Blass, *Herbarts pädagogische Denkform* (Wuppertal, 1969); A. Brückmann, *Pädagogik und philosophisches Denken bei J. F. Herbart* (Zurich, 1961); A. Buss, *Herbarts Beiträge zur Entwicklung der Heilpädagogik* (Weinheim, 1962); H. Dunkel, *Herbart and Education* (New York, 1969); and *Herbart and Herbartianism* (Chicago–London, 1970); E. Geissler, *Herbarts Lehre vom erziehenden Unterricht* (Heidelberg, 1970); H. Holstein, *Bildungsweg und Bildungsgeschehen* (Ratingen, 1965); H. Hornstein, *Bildsamkeit und Freiheit. Ein Grundproblem des Erziehungsdenkens bei Kant und Herbart* (Düsseldorf, 1959); J. Müller, *Herbarts Lehre vom Sein* (Zurich, 1933); A. Rimsky-Korsakov, *Herbarts Ontologie* (St. Petersburg, 1903); J. N. Schmitz, *Herbart-Bibliographie 1842–1963* (Weinheim, 1964); B. Schwenk, *Das Herbartverständnis der Herbartianer* (Weinheim, 1963); K. Smirnov, *Leibniz' und Herbarts metaphysische Lehre von der Seele* (Kharkov, 1910); G. Weiss, *Herbart und seine Schule* (Munich, 1928); and H. Zimmer, *Führer durch die Herbart-Literatur* (Langensalza, 1910).

Heinrich Beck
Arnulf Rieber

HERBERT, WILLIAM (*b.* Highclere, Hampshire, England, 12 January 1778; *d.* London, England, 28 May 1847), *natural history.*

The third son of the first earl of Carnarvon and of the daughter of the earl of Egremont, Herbert was educated at Eton and at Oxford University. He received the B.A. in 1798, the M.A. in 1802, the bachelor and doctorate in civil law in 1808, and the B.D. in 1840. In 1806 he married Letitia Dorothea, daughter of the fifth viscount Allen. The couple had two daughters and two sons. Herbert served in the House of Commons in 1806–1807 and in 1811–1812. In 1814 he left politics and entered the Anglican ministry when the earl of Egremont sponsored him for a living at Spofforth, Yorkshire, a post which he held until his death. Herbert also served as head of the Collegiate Church at Manchester from 1840 to 1847.

Herbert's interests were varied and included a love of, and familiarity with, nature. He became well known for his knowledge of local birds and provided extensive notes for two editions of Gilbert White's classic *Natural History and Antiquities of Selborne.*

Plant life held the greatest attraction for Herbert.

He was a member of the Horticultural Society of London and a contributor to its publications, as well as to other journals. Herbert was a good draftsman and often did his own illustrations. Although greatly interested in plant classification, he did not propose a comprehensive system beyond the monocots. His general theory of classification had no noticeable impact during his lifetime nor afterward; but his arrangements of the Amaryllidaceae, published in 1837, established his reputation as a botanist. One hundred years later Arthur Hill of the Royal Botanic Gardens at Kew stated that before Herbert's work, the arrangements in this family were in "a state bordering on chaos" (*Herbertia,* **4** [1937], 3–4).

Herbert aimed at a "natural" classification which would reflect kinship, as contrasted with the artificial system of Linnaeus. He was more advanced in his views than was generally the case at the time, in that to him kinship meant descent from a common ancestor. He was also quite modern in his belief that development or variation had not proceeded in straight lines or at the same rate and that therefore the taxa of botanists were basically arbitrary and the result of individual judgments. Some of his recommendations for the methods of naming varieties, hybrids, and cultivated plants resemble those used in modern times.

Herbert was among the earliest in Britain to study hybridization on a large scale; and while he was particularly interested in the Amaryllidaceae, he did not limit his experiments to these or to bulbous plants. He considered hybridization a factor in evolution and provided some solid evidence in support of such a view. Charles Darwin knew Herbert and made numerous references to the latter's findings in his own works, especially in the discussion on hybridism in his "Natural Selection." Herbert specifically stated that new forms which maintain themselves in the same way as do species are produced through hybridization; and he gave backing for this assertion with his findings on some *Narcissus* and with forms of the Primulaceae. He was a pioneer in undermining the view that sterility of offspring was a valid criterion in delimiting "true" species. In addition he presented proofs that hybrids were highly variable with regard to fertility and sterility, ranging from sterility to fertility greater than parent forms. He also emphasized the role of the environment in bringing about differentiation of plant forms.

Since he worked and wrote before the thesis of natural selection was stated and before the development of Mendelian genetics, it is not surprising that some of Herbert's explanations were vague or inaccurate; but his contributions to the history of science are deserving of recognition. Indeed, Herbert provided, as C. D. Darlington so aptly phrased it, "the thin edge of the wedge which Darwin drove home" (*Herbertia,* **4** [1937], 65).

BIBLIOGRAPHY

I. Original Works. Only the more significant works are included here; a more complete list is in Guimond (see below). *An Appendix to the Botanical Register* (London, 1821) is in a sense the precursor of the *Amaryllidaceae* and lists plants that Herbert actually had in his extensive garden at Spofforth; details on this work are in Stearn (see below). "Instructions for the Treatment of the Amaryllis longifolia . . .," in *Transactions of the Horticultural Society of London,* **3** (1822), 187–196, is important in illustrating Herbert's early views on species varieties. "On the Production of Hybrid Vegetables . . .," *ibid.,* **4** (1822), 15–50, presents Herbert's early views and work. *Amaryllidaceae: Preceded by an Attempt to Arrange the Monocotyledonous Orders, and Followed by a Treatise on Cross-Bred Vegetables and Supplement* (London, 1837) is invaluable for his work in classification and important for his views on hybridization; it also furnishes examples of Herbert's ability as an artist. "Local Habitation and Wants of Plants," in *Journal of the Horticultural Society of London,* **1** (1846), 44–49, is the best reference for Herbert's views on competition between plants. "On Hybridization Amongst Vegetables," *ibid.,* **2** (1847), 81–107, is a good presentation of Herbert's views on hybridization.

Works of the Hon. and Very Rev. William Herbert, Dean of Manchester . . ., 2 vols. (London, 1842) does not contain his writings on botany and natural history but does refer to his scattered literary works and contains some of his sermons; a supplement is *The Christian* (London, 1846). Herbert's letters can be found in the correspondence of John Lindley, John Sims, and William Jackson Hooker at the Royal Botanic Gardens, Kew.

II. Secondary Literature. This list is limited to those works referred to in the text or which contain useful bibliographical data. See C. D. Darlington, "The Early Hybridizers and the Origins of Genetics," in *Herbertia,* **4** (1937), 63–69. Charles Darwin, "Natural Selection," repro. of MS of the third (long) version of *Origin of Species,* in Darwin Scientific Papers, University Library, Cambridge, is the most valuable of all of Darwin's works containing references to Herbert, in indicating possible influences or uses of Herbert's work and views in Darwin. Alice A. Guimond, "The Honorable and Very Reverend William Herbert, Amaryllis Hybridizer and Amateur Biologist" (unpublished thesis, Univ. of Wis., 1966; order no. 66-9145, University Microfilms, Ann Arbor, Mich.), deals with Herbert's career in general but emphasizes his work in biology and has an extensive bibliography of Herbert's works. Arthur Hill, "Introduction," in *Herbertia,* **4** (1937), 3–4, is an evaluation of Herbert by a botanist. Herbert F. Roberts, *Plant Hybridization Before Mendel* (Princeton,

1929), is a review of early hybridizers with a short but good account of Herbert's role, pp. 94–102. See also William T. Stearn, "William Herbert's 'Appendix' and 'Amaryllidaceae,'" in *Journal of the Society for the Bibliography of Natural History,* **2** (Nov. 1952), 375–377. Gilbert White, *The Natural History and Antiquities of Selborne,* James Rennie, ed. (London, n.d. [1832]), also edited by Edward Turner Bennett (London, 1837), is difficult to locate in the United States in the Rennie ed. but is a very good reference for Herbert's work in ornithology.

ALICE A. GUIMOND

HERBRAND, JACQUES (*b.* Paris, France, 12 February 1908; *d.* La Bérarde, Isère, France, 27 July 1931), *mathematics, logic.*

Herbrand gave early signs of his mathematical gifts, entering the École Normale Supérieure at the exceptional age of seventeen and ranking first in the entering class. He completed his doctoral dissertation in April 1929. That October he began a year of service in the French army. He then went to Germany on a Rockefeller fellowship, studying in Berlin (until May 1931) with John von Neumann, then in Hamburg (May–June) with Emil Artin, and in Göttingen (June–July) with Emmy Noether. He left Göttingen for a vacation in the Alps and a few days later was killed in a fall at the age of twenty-three.

Herbrand's contributions to mathematics fall into two domains: mathematical logic and modern algebra. He showed an early interest in mathematical logic, a subject to which French mathematicians were then paying scant attention, and published a note on a question of mathematical logic in the *Comptes rendus* of the Paris Academy of Sciences when he was hardly twenty. Herbrand's main contribution to logic was what is now called the Herbrand theorem, published in his doctoral dissertation: it is the most fundamental result in quantification theory. Consider an arbitrary formula F of quantification theory, then delete all its quantifiers and replace the variables thus made free with constants selected according to a definite procedure. A lexical instance of F is thus obtained. Let F_i be the ith lexical instance of F, the instances being generated in some definite order. The Herbrand theorem states that F is provable in any one of the (equivalent) systems of quantification theory if and only if for some number k the disjunction

$$F_1 \vee F_2 \vee \cdots \vee F_k$$

(now called the kth Herbrand disjunction) is sententially valid. (Herbrand's demonstration of the theorem contains a gap, discovered in 1963 by B. Dreben, P. Andrews, and S. Aanderaa.)

The Herbrand theorem establishes an unexpected bridge between quantification theory and sentential logic. Testing a formula for sentential validity is a purely mechanical operation. Given a formula F of quantification theory, one tests the kth Herbrand disjunction of F successively for $k = 1$, $k = 2$, and so on; if F is provable, one eventually reaches a number k for which the kth Herbrand disjunction is valid. If F is not provable, there is, of course, no such k; and one never learns that there is no such k (in accordance with the fact that there is no decision procedure for quantification theory). Besides yielding a very convenient proof procedure, the Herbrand theorem has many applications (a field explored by Herbrand himself) to decision and reduction problems and to proofs of consistency. Almost all the methods for proving theorems by machine rest upon the Herbrand theorem.

In modern algebra Herbrand's contributions are in class-field theory, the object of which is to gain knowledge about Abelian extensions of a given algebraic number field from properties of the field. Initiated by Leopold Kronecker and developed by Heinrich Weber, David Hilbert, Teiji Takagi, and Emil Artin, the theory received essential contributions from Herbrand in 1930–1931. He wrote ten papers in this field, simplifying previous proofs, generalizing theorems, and discovering important new results.

BIBLIOGRAPHY

Herbrand's logical writings have been reprinted in his *Écrits logiques,* Jean van Heijenoort, ed. (Paris, 1968); the pref. includes reference to the paper by B. Dreben, P. Andrews, and S. Aanderaa, "False Lemmas in Herbrand," in *Bulletin of the American Mathematical Society,* **69** (1963), 699–706, as well as further information about the gap in Herbrand's demonstration of his theorem. Also see *The Logical Writings of Jacques Herbrand,* Warren D. Goldfarb, ed. (Reidel, 1971).

The list of Herbrand's papers on class-field theory can be found in Ernest Vessiot's intro. to Helmut Hasse, *Über gewisse Ideale in einer einfachen Algebra* (Paris, 1934). Herbrand's *Le développement moderne de la théorie des corps algébriques* was published posthumously and edited by his friend Claude Chevalley (Paris, 1936).

JEAN VAN HEIJENOORT

HÉRELLE, FÉLIX D' (*b.* Montreal, Canada, 25 April 1873; *d.* Paris, France, 22 February 1949), *microbiology.*

D'Hérelle had an extremely cosmopolitan life. Born in Canada to a French father, who died when Félix was six, and a Dutch mother, he received his secondary education at the Lycée Louis-le-Grand in Paris.

He began his medical studies at Paris and continued them at Leiden. In 1901 d'Hérelle went to Guatemala City as director of the bacteriology laboratory of the municipal hospital and also to teach microbiology at the Faculty of Medicine. He then went to Yucatan to study the fermentation of sisal hemp. The Mexican government sent him to the Pasteur Institute in Paris to further his knowledge of microbiology; he entered in 1909 as an assistant to A. Salimbeni and remained until 1921.

While in Paris, d'Hérelle studied a bacterium which causes enteritis in acridians, *Coccobacillus acridiorum* (*Aerobacter aerogenes* var. *acridiorum*), which he had observed in Yucatan. In 1915 Roux sent him to Tunisia in an attempt to employ this microbe against locusts, but the successes achieved were not subsequently confirmed. In growing the microbe d'Hérelle had noted empty spots on the gelose culture plates and thought they resulted from a virus which accompanied the microbe and was destroying it. He then had a presentiment of "the discovery of a phenomenon of wide significance, which he linked to the battle of the organism against the diseases of the digestive tract" (P. Lépine, p. 458). Examining cultures of the dysentery bacillus in Paris in 1916, he again observed these "sterile regions" on the surface of the culture and showed that a filterable element isolated from the feces of dysentery victims completely destroyed, after several hours, a culture broth of the bacillus. On 10 September 1917 d'Hérelle presented to the Academy of Sciences, through Roux, a note entitled "Sur un microbe invisible, antagoniste du bacille dysentérique"—which he soon named "microbe bactériophage," then "bactériophage."

The phenomenon of bacteriophagy had already been observed in 1915 by an English scientist, Frederick Twort; but he did not continue his investigations, the full importance of which he did not seem to have grasped. Pierre Nicolle has rightly written: "To recognize Twort's priority is certainly not to diminish d'Hérelle's merit. Nor does one wound Twort's legitimate pride to assert that d'Hérelle, after having rediscovered the bacteriophage . . . derived the greater glory from the discovery" (*Presse médicale,* p. 350). By 1949 more than 6,000 publications had been devoted to the bacteriophage.

In 1919 d'Hérelle investigated *typhose aviaire* (fowl typhoid) in France and isolated the phages effective against its microbe. The following year he was sent to Indochina by the Pasteur Institute to study human dysentery and septic pleuropneumonia in buffaloes. In connection with the latter he perfected the techniques for isolating the bacteriophage. In 1921 he published *Le bactériophage, son rôle dans l'immunité,* which enjoyed considerable acclaim. In the same year he was appointed assistant professor at the University of Leiden, where he remained for two years. In 1923 he was associated with the Egyptian Council on Health and Quarantine as director of the Bacteriological Service, and in 1927 he was sent to the East Indies to attempt the prophylaxis of cholera by means of the bacteriophage that could cure the disease. From 1928 to 1934 d'Hérelle taught protobiology (a term created to designate the science of the bacteriophage) at Yale University. He was called upon in 1935 by the Russian government to organize institutes for the study of the bacteriophage in Tiflis, Kiev, and Kharkov. Political conditions obliged him to leave the country, and he settled in Paris, where he continued to work on the bacteriophage until his death.

D'Hérelle received many honors: the Leeuwenhoek Medal (1925), the Schaudinn Medal (1930), and the Prix Petit d'Ormoy of the Academy of Sciences (1948); he was doctor *honoris causa* of the universities of Leiden, Yale, Montreal, and Laval. He married Mary Kerr in 1893. Although a simple, affable, and apparently even-tempered man, he could, on occasion, be fiery and irascible.

With techniques that are all the more remarkable considering that quantitative methods were not yet employed in bacteriology on a large scale, d'Hérelle demonstrated the corpuscular nature of the bacteriophage that was later confirmed by electron microscopy. He also described how it attaches itself to harmful bacteria and its multiplication following their lysis. He attempted to apply phagotherapy to various human and animal infectious diseases, including dysentery, cholera, plague, and staphylococcus and streptococcus infections. This type of therapy was favored for a time, especially in the Soviet Union, but was later rejected; it has been replaced by chemotherapy and treatment with antibiotics.

Today the bacteriophage is considered to be an ultravirus and is employed in theoretical and practical studies—for example, in the diagnosis of the phagic types of the typhoid bacillus and of the paratyphoid B bacillus by means of the method developed by J. Craigie and A. Felix.

BIBLIOGRAPHY

I. Original Works. D'Hérelle's principal works are "Sur une épizootie de nature bactérienne sévissant sur les sauterelles du Mexique," in *Comptes rendus hebdomadaires des séances de l'Académie des sciences,* **152** (1911), 1413–1415; "Les coccobacilles des sauterelles," in *Annales de l'Institut Pasteur,* **28** (1914), 1–69; "Sur le procédé biologique de destruction des sauterelles," in *Comptes rendus hebdomadaires des séances de l'Académie des sciences,* **161** (1915), 503–505; "Sur un microbe invisible, antagoniste du bacille dysentérique," *ibid.,* **165** (1917), 373–375; *Le bactériophage, son rôle dans l'immunité* (Paris, 1921); *Les défenses de l'organisme* (Paris, 1923); *Le bactériophage et son comportement* (Paris, 1926); *Le phénomène de la guérison dans les maladies infectieuses* (Paris, 1938); *L'étude d'une maladie, le choléra, maladie à paradoxes* (Paris, 1946); "Le bactériophage," in *Atomes,* no. 33 (1948), 399–403; "The Bacteriophage," in *Science News,* **14** (1949), 44–59.

II. SECONDARY LITERATURE. On d'Hérelle and his work see P. Lépine, "Félix d'Hérelle (1873–1949)," in *Annales de l'Institut Pasteur,* **76** (1949), 457–460; and P. Nicolle, "Félix d'Hérelle," in *Presse médicale,* 57e année, no. 25 (1949), p. 350; "Le bactériophage," in *Biologie médicale,* **38** (1949), 233–306; and "Cinquantième anniversaire d'une grande découverte anglo-franco-canadienne en biologie: le bactériophage," in *Bulletin de l'Académie nationale de médecine,* **151** (1967), 404–409.

JEAN THÉODORIDÈS

HÉRIGONE, PIERRE (*d.* Paris [?], *ca.* 1643), *mathematics.*

Very little is known of Hérigone's life. He was apparently of Basque origin and spent most of his life in Paris as a teacher of mathematics. He also served on a number of official committees dealing with mathematical subjects, notably the one appointed by Richelieu in 1634 to judge the practicality of Morin's proposed scheme for determining longitude from the moon's motion. With the other members of this committee (Étienne Pascal, Mydorge, Beaugrand, J. C. Boulenger, L. de la Porte) he became embroiled in the ensuing controversy with Morin.

Hérigone's only published work of any consequence is the *Cursus mathematicus,* a six-volume compendium of elementary and intermediate mathematics in French and Latin. Although there is little substantive originality in the *Cursus,* it shows an extensive knowledge and understanding of contemporary mathematics. Its striking feature is the introduction of a complete system of mathematical and logical notation, very much in line with the seventeenth-century preoccupation with universal languages. Yet none of Hérigone's notational conventions seem to have become accepted, and his other works are of negligible importance.

It is as a teacher, systematizer, and disseminator of mathematics that Hérigone must be judged. As such he was no doubt a full member of the community of French mathematicians of the first half of the seventeenth century.

BIBLIOGRAPHY

I. ORIGINAL WORKS. Hérigone's only important published work is *Cursus mathematicus nova, brevi et clara methodo demonstratus,* 6 vols. (Paris, 1634–1642). There are three other "editions" of the *Cursus* (1643, 1644), but these consist of nothing but sheets from the original ed. with a few deletions and additions, and new title pages. Hérigone also published a paraphrase of the first six books of Euclid (1639), but it consists of little more than the French portion of vol. I of the *Cursus;* there is also a spurious 2nd ed. (1644).

II. SECONDARY LITERATURE. What little information there is on Hérigone has been collected by B. Boncompagni in *Bullettino di bibliografia e di storia delle scienze matematiche e fisiche,* **2** (1869), 472–476; and P. Tannery in *Mémoires scientifiques,* X (Paris, 1930), 287–289. The controversy with Morin is described by J. E. Montucla in *Histoire des mathématiques,* 2nd ed., IV (Paris, 1802), 543–545. A list of Hérigone's mathematical symbols is given by F. Cajori in *History of Mathematical Notations,* I (Chicago, 1928), 200–204, *passim.*

PER STRØMHOLM

HERING, KARL EWALD KONSTANTIN (*b.* Alt-Gersdorf, Germany, 5 August 1834; *d.* Leipzig, Germany, 26 January 1918), *physiology, psychology.*

Ewald Hering was a strong and important personality among the German physiologists of his time. In his first publications he mastered the difficult problem of visual space perception and was able to challenge the great master in that field, Hermann von Helmholtz, proposing alternative views that emphasized the physiological rather than the physical aspects of sensation. As the main representative of the phenomenological tradition, an acute observer, and a shrewd critic, Hering exerted a great influence on contemporary sense physiology and on the evolution of modern psychology. Gestalt psychology in particular owes much to him.

The son of a village parson, Hering studied medicine at Leipzig under E. H. Weber, G. T. Fechner, Otto Funke, and the zoologist J. V. Carus, with whom he went to Sicily in the winter of 1858–1859 to study the genital and excretory organs of *Alciopida,* a genus of ringed worms, which were the subject of his doctoral dissertation. From 1860 to 1865 he practiced medicine and worked as an assistant in the polyclinic directed by Ernst Wagner, and from 1862 he was also lecturer in physiology. In 1861–1864 he published a five-part study on visual space perception in which he favored the nativistic theory, arguing that each point of the retina is endowed with three local signs: one for height, one for breadth, and one for depth. Thus he found himself in opposition to Helmholtz and other empiricists, who believed that location and space forms are learned and arise with continued experience. In this treatise Hering was concerned mainly with binocular vision, the identical points of the two retinas, and the horopter.

In 1865 Hering was called to Vienna to succeed Carl Ludwig in the chair of physiology at the Military Medico-Surgical Academy, the Josephinum. There he continued his studies of binocular vision but turned also to other subjects. Most important was his discovery, with Josef Breuer, of reflex reactions originating in the lungs and mediated by the fibers of the vagus nerve: inflation of the lungs, eliciting expiration, and deflation, stimulating inspiration. These reflexes, especially the former, have an important role in the regulation of respiratory movements, which are greatly altered after the vagi are sectioned (1868). Hering and Breuer spoke of self-regulation of respiration, and the reflexes were one of the first feedback mechanisms discovered in the physiological regulations. At that time Hering also investigated the functional structure of the liver and the mechanism of the respiratory variations of blood pressure (Hering waves) due to variations of the tonus of medullary centers (1869).

In 1870 the Josephinum was abolished and Hering was appointed to succeed Jan Purkyně at the University of Prague, where he remained for twenty-five years. He devoted most of his energy to research in sensory physiology, mainly of vision, and to more general conceptions, reported in his addresses, "Über das Gedächtnis als eine allgemeine Funktion der organisierten Materie" (1870), "Über die specifischen Energieen der Nervensystem" (1884), and most important, "Zur Theorie der Vorgänge in der lebendigen Substanz" (1888). For Hering the two basic processes of life, assimilation and dissimilation, played a general role; and their mutual relation was at the base of both color vision and the function of temperature-sensing organs, as well as of the electrical phenomena in nerve and muscle, which he studied in his laboratory. In his color-vision theory Hering put forward (against the Young-Helmholtz theory of three visual substances and three colors) a three-substance, six-color theory supposing a red-green substance, a yellow-blue one, and a white-black one, each of which could be excited to respond in either dissimilation (catabolism) or assimilation (anabolism), corresponding to the sensations of white, yellow, and red or black, blue, and green, respectively. Strikingly different from the trichromatic theory—besides these three opposing pairs—was the postulate of an independent mechanism for the sensing of white and black. Thus Hering's theory avoided the criticism raised against the Young-Helmholtz theory for neglecting the uniqueness of the six principal colors (red, yellow, green, blue, white, and black) and accounted for such subjective visual phenomena as contrasts (simultaneous and successive) and mixing of colors. He drew attention to the physiological aspects of sensory functions, in contrast with physical processes, emphasizing the distinction between the stimulus (physical) and the response (physiological).

In 1895 Hering succeeded Carl Ludwig at Leipzig and remained there for the rest of his life, studying color phenomena and devising new experiments and instruments for their demonstration in support of his theory, which had been criticized for several apparent contradictions. He died before publication of the final part of his last work, begun in 1905.

Hering had the power to generalize and penetrate to basic problems, but his approach could not lead to a significant advance and his theory had little heuristic value. Moreover, many of the suppositions on which he based his theories were incorrect. The progress of knowledge required a new approach and new methods, both electrophysiological and biochemical. A new era of nerve and sensory physiology began in the early 1920's, leading to great advances in knowledge of the elementary response of the nerve fiber and nerve cell, of the organization of the neural pathways, of synaptic transmission, and of the processing of the signals at different stages of their course. Hering's idea of a double response of different receptors and, accordingly, of two kinds of signals conveyed by each nerve fiber, proved to be wrong. On the other hand, differential sensitivity of receptors and phenomena of summation and inhibition were brought to light. The three photosensitive substances postulated by the Young-Helmholtz theory were identified and their reversible photochemical reaction elucidated. Thus, at the level of peripheral receptors, the eye functions as Hering's opponents had postulated, but some of the points Hering raised against the trichromatic theory (such as yellow as the fourth basic color) can be accounted for at higher levels of the extremely complicated afferent pathway. The existence of the "on" and "off" elements in the retina recalls the two opposite effects in Hering's "substances." Different mechanisms in the retina are assumed for the sensing of luminosity (dominators) and of color (modulators), as in Hering's theory. Yet Ewald Hering, who was for almost fifty years one of the two leading men in the physiology of vision, is hardly mentioned in recent textbooks.

BIBLIOGRAPHY

I. Original Works. A bibliography of Hering's writings up to 1907 can be found in Charles Richet's *Dictionnaire de physiologie,* II (Paris, 1909), 554. Among his publications are *Beiträge zur Physiologie. Zur Lehre vom Ortsinne der Netzhaut,* 5 pts. (Leipzig, 1861–1864); *Die Lehre vom binokularen Sehen* (Leipzig, 1868); *Zur Lehre vom Lichtsinn* (Vienna, 1872); "Der Raumsinn und die Bewegung des Auges," in L. Hermann, ed., *Handbuch der Physiologie,* III, pt. 1 (Leipzig, 1879), 343–601; *Zur Theorie der Nerventätigkeit* (Leipzig), 1899); and *Grundzüge der Lehre vom Lichtsinn,* 4 vols. (Berlin, 1905–1920), translated by Leo M. Hurwich and Dorothea Jameson as *Outlines of a Theory of the Light Sense* (Cambridge, Mass., 1964). Extracts from *Beiträge . . .* and *Zur Lehre vom Lichtsinn* in English translation have been published in R. J. Herrnstein and E. G. Boring, eds., *A Source Book in the History of Psychology* (Cambridge, Mass., 1965). Hering's addresses are in *Fünf Reden von Ewald Hering,* H. E. Hering, ed. (Leipzig, 1921).

II. Secondary Literature. Obituaries include S. Garten, "Ewald Hering zum Gedächtnis," in *Pflügers Archiv für die gesamte Physiologie . . .,* **170** (1918), 501–522; C. Hess, "Ewald Hering," in *Archiv für Augenheilkunde,* **83** (1918), 89–97, and *Naturwissenschaften,* **6** (1918), 305–308; and F. B. Hofmann, "Ewald Hering," in *Münchener medizinische Wochenschrift,* **65,** pt. 1 (1918), 539–542. See also A. von Tschermak, "Ewald Hering. Zum 100. Geburtstag," *ibid.,* **81** (1934), 1230–1233. Hering's works and views in relation to modern physiology are discussed in L. M. Hurwich, "Hering and the Scientific Establishment," in *American Psychologist,* **24** (1969), 497–514. There are many references to Hering in E. G. Boring, *A History of Experimental Psychology* (New York, 1929; 2nd ed., 1950); and his *Sensation and Perception in the History of Experimental Psychology* (New York, 1942).

Vladislav Kruta

HERMANN (HERMANNUS) THE LAME (also known as **Hermannus Contractus** or **Hermann of Reichenau**) (*b.* Altshausen, Germany, 18 July 1013; *d.* Altshausen, 24 September 1054), *astronomy, mathematics.*

Hermannus was the son of Count Wolferat of Altshausen. He entered the cloister school at Reichenau on 13 September 1020 and became a monk at Reichenau in 1043. Throughout his life he suffered from an extreme physical disability which severely limited his movements and his ability to speak; hence the appellation "contractus," attached to his name since the twelfth century.

Hermannus is one of the key figures in the transmission of Arabic astronomical techniques and instruments to the Latin West before the period of translation. His familiarity with Islamic materials indicates that this knowledge had reached southern Germany by the early eleventh century. It is unlikely, though, that Hermannus knew Arabic; his devoted pupil Berthold of Reichenau, who has left a biographical sketch of his master (see Manitius, *Geschichte der lateinischen Literatur . . .,* pp. 756–777), would almost surely have mentioned this accomplishment.

Hermannus is one of the earliest Latin authors responsible for the introduction or reintroduction into the West from the Islamic world (undoubtedly Spain) of three astronomical instruments: the astrolabe, the chilinder (a portable sundial), and the quadrant with cursor. Since the thirteenth century a *De mensura astrolabii* has been ascribed to Hermannus. The first section of a second work, often called in its entirety *De utilitatibus astrolabii,* is a treatise on the astrolabe in twenty-one chapters which contains many Arabic expressions; not written by Hermannus, it was attributed to Gerbert as early as the twelfth century. N. Bubnov, the eidtor of Gerbert's mathematical works, has placed the twenty-one-chapter treatise among the doubtful works of Gerbert. The second section of the *De utilitatibus,* containing a description of the chilinder and the quadrant, is generally considered to be by Hermannus. Further evidence for his authorship lies in the subsequent paragraphs of this second section which contain an account of Eratosthenes' measurement of the circumference of the earth as reported by Macrobius, with a calculation of the earth's diameter using the Archimedean value of 22/7 for pi. These paragraphs were the subject of correspondence in 1048 between Hermannus and his former pupil Meinzo of Constance.

The *De mensura astrolabii,* which contains many latinized Arabic words, begins with a description of the fundamental circles of the base plate of the astrolabe, or *walzachora,* followed by a delineation of the rete. The astrolabe is designed for a latitude of forty-eight degrees, the latitude of Reichenau; no mention is made of the number of plates the instrument should have. Designed in the conventional manner for Western astrolabes, the dorsum contains a shadow square. This practice of expressing angles in terms of twelve points of either the inverse or the plane shadow (*umbra versa* or *umbra recta*) stemmed from Hindu sources and was transmitted through Arabic writings. The *De mensura* also contains a star table with the coordinates of twenty-seven stars expressed in right ascension and the stars' meridian altitude.

The chilinder is a portable altitude sundial designed for one latitude—forty-eight degrees in this case. Since the altitude varies symmetrically with the declination throughout the sun's yearly cycle, the surface of the dial with the hour lines is wrapped around a cylindrical column. The dial provides the time in unequal hours, that is, daylight hours derived by dividing the diurnal arc by twelve. Hermannus provides an altitude table expressed in degrees rather than inverse shadow points, as was customary later. His treatise was the first in the Latin West to describe this type of sundial, which had antecedents in Islam. Through Hermannus the chilinder became the inheritor of the *horologium viatorum* (traveler's dial) tradition first mentioned in the West in Vitruvius' *De architectura.*

The quadrant described by Hermannus is a quadrant with cursor, the "Alphonsine" type similar to that appearing in the *Libros del saber de astronomia.* It is the usual one-fourth of a circle with the margin divided into ninety degrees and has two small plates with holes on one edge for sighting and a plumb line. A cursor, inscribed with the months of the year, slides in a groove concentric to the margin. The remainder of the body of the quadrant contains the hour lines. This instrument was used to measure the sun's altitude; with the cursor it could also provide the observer's latitude and the time of day (in unequal hours).

All three instruments were widely used in the Latin West. The popularity of the astrolabe is well attested. The chilinder and quadrant with cursor also are well represented in the Latin manuscript tradition and continued to appear in printed works through the seventeenth century. It is of interest that all of these instruments were used during the Middle Ages to solve problems in mensuration as well as in pure astronomy. Hermannus' astronomical writings include a work on the length of the month (*De mense lunari*) in which he criticizes the Venerable Bede; according to Berthold, Hermannus also wrote a computus.

In mathematics Hermannus composed a treatise teaching multiplication and division with the abacus (*Qualiter multiplicationes fiant in abbaco*); the work uses Roman numerals only. He also wrote the earliest treatise on rithmomachia (*De conflictu rithmimachie*), a very complex game based on Pythagorean number theory derived from Boethius. The game was played with counters on a board; capture of the opponent's pieces was dependent on the determination of arithmetical ratios and arithmetic, geometrical, and harmonic progressions. This game, which enjoyed a considerable vogue during the Middle Ages, has been attributed to Pythagoras, Boethius, and Gerbert.

Hermannus composed an excellent world chronicle dating from the birth of Christ which was continued by Berthold and was used by later German historians, such as Manegold of Lautenbach and Otto of Freising. He was also the author of a work on music (*Opuscula musica*) containing a system of notation of musical intervals which was his own invention but had no influence, although he did make an original contribution to medieval modal theory. In addition Hermannus wrote poems and hymns.

BIBLIOGRAPHY

I. ORIGINAL WORKS. *De mensura astrolabii* and *De utilitatibus astrolabii* are available in B. Pez, ed., *Thesaurus anecdotorum novissimus,* III, pt. 2 (Augsburg, 1721), cols. 93–106, 94–139; J. P. Migne, ed., *Patrologia latina,* CXLIII (Paris, 1882), cols. 379–412; and *Gerberti, postea Silvestri II papae, opera mathematica,* N. Bubov, ed. (Berlin, 1899), pt. 2, *Gerberti opera dubia,* pp. 109–147; the *De mensura* is also reprinted in R. T. Gunther, *Astrolabes of the World,* II (Oxford, 1932), 404–408. *Regule Herimanni qualiter multiplicationes fiant in abbaco,* P. Treutlein, ed., is in *Bullettino di bibliografia e di storia delle scienze mathematiche e fisiche,* **10** (1877), 643–647. *Opuscula musica* may be found in M. Gerbert, ed., *Scriptores ecclesiastici de musica sacra potissimum,* II (St. Blasius, Belgium, 1784), 124–153; J. P. Migne, ed., *Patrologia latina,* CXLIII (Paris, 1882), cols. 413–414; and *Herimanni Contracti musica,* W. Brambach, ed. (Leipzig, 1884). For *De mense lunari,* see G. Meier, *Die sieben freien Künste im Mittelalter,* II (Einsiedeln, Switzerland, 1887), 34–46. The *De conflictu rithmimachie* is in E. Wappler, "Bemerkungen zur Rhythmomachie," in *Zeitschrift für Mathematik und Physik,* Hist. Abt., **37** (1892), 1–17. His chronicle dating from the birth of Christ is in *Monumenta germaniae historica scriptores,* V (Hannover, 1844), 67–133; see also Aemilius Ussermann, ed., *Chronicon Hermanni Contracti ex inedito hucusque codice Augiensi, una cum eius vita et continuatione a Bertholdo eius disciplo scripta,* 2 vols. (St. Blasius, Belgium, 1790), repr. in J. P. Migne, ed., *Patrologia latina,* CXLIII (Paris, 1882), cols. 55–270. The *Monumenta* text was translated into German by K. Nobbe in *Geschichtsschreiber der deutschen Vorzeit,* XI (Berlin, 1851); a 2nd ed., prepared by W. Wattenbach, appeared in vol. XLII (Leipzig, 1888).

II. SECONDARY LITERATURE. On Hermannus or his work, see M. Cantor, *Vorlesungen über Geschichte der Mathematik,* I (Leipzig, 1880), 758–761; J. Drecker, "Hermannus Contractus über das Astrolab," in *Isis,* **16** (1931), 200–219, which includes the text of *De mensura astrolabii* on pp. 203–212; P. Duhem, *Le système du monde,* III (Paris, 1958), 163–171; E. Dümmler, "Ein Schreiben Meinzos von Constanz an Hermann den Lahmen," in *Neues Archiv der Gesellschaft für ältere deutsche Geschichtskunde,* **5** (1880), 202–206; H. Hansjakob, *Hermann der Lahme* (Mainz, 1885); W. Hartner, "The Principle and Use of the Astrolabe," in A. U. Pope, ed., *A Survey of Persian Art* (London, 1939), p. 2533; C. H. Haskins, *Studies in the History of Mediaeval Science* (Cambridge, Mass., 1927), pp. 52–53; Max Manitius, *Geschichte der lateinischen Literatur des Mittelalters,* II (Munich, 1923), 756–777, 786–787; J. Millás Vallicrosa, "La introducción del cuadrante con cursor en Europa," in *Isis,* **17** (1932), 218–258; R. Peiper, "Fortolfi rythmimachia," in *Abhandlungen zur Geschichte der Mathematik,* **3** (1880), 198–227; G. Reese, *Music in the Middle Ages* (New York, 1940), pp. 137, 155; D. E. Smith, *History of Mathematics,* I (Boston, 1923), 197–200; D. E. Smith and C. C. Eaton, "Rithmomachia, the Great Medieval Number Game," in *Teachers College Record,* **13** (1912), 29–38; L. Thorndike, *A History of Magic and Experimental Science,*

I (New York, 1923), ch. 30, pp. 701, 728; W. Wattenbach, *Deutschlands Geschichtsquellen im Mittelalter,* II (Berlin, 1894), 41–47; and E. Zinner, *Geschichte der Sternkunde* (Berlin, 1931), p. 330; and *Deutsche und niederländische astronomische Instrumente des 11.–18. Jahrhunderts* (Munich, 1956), pp. 135–141, 155–156, 373–374.

CLAUDIA KREN

HERMANN, CARL HEINRICH (*b.* Lehe, near Bremerhaven, Germany, 17 June 1898; *d.* Marburg, Germany, 12 September 1961), *solid-state physics, crystallography.*

Hermann's work was done in the period during which modern crystallography and solid-state theory gathered momentum after the interruption of their beginnings by World War I. In this early period his significant contributions helped to guide the development along sound mathematical lines. Much of his work is based on the structure theory developed by Arthur Schönflies and E. S. Fedorov. This theory calls for the investigation of all possible spatially periodic arrangements of matter which differ in their internal symmetry. Any of the 230 "space groups" thus determined can serve as the repeat scheme for the arrangement of atoms in a crystal.

At the time (1925) when Hermann became interested in it, the structure theory was more than thirty years old. It had remained in a dormant, and to most physicists and crystallographers highly hypothetical, state until 1912, when Max von Laue, Walter Friedrich, and Paul Knipping gave it a realistic basis through their discovery of the diffraction of X rays traversing a crystal. Even in the early years of crystal structure analysis by W. H. and W. L. Bragg formal structure theory was disregarded. That it is now commonly used as a major tool in crystal structure determination is to no small extent the result of the simplified notations of the symmetry elements and the space groups invented simultaneously and independently by Charles Mauguin and Hermann. A combined Hermann-Mauguin nomenclature, which is well adapted to the techniques of X-ray crystal analysis, received the approval of a small international group of crystallographers meeting at Zurich in 1930 to plan the standard *Internationale Tabellen zur Bestimmung von Kristallstrukturen.* The *Tabellen,* divided into a volume on symmetry and one containing numerical tables of essential functions, appeared in 1935 with Hermann as both contributor and editor.

Another project of general benefit to crystallographers was the preparation of *Strukturberichte,* on which Hermann worked from 1925 to 1937. The first volume, written with P. P. Ewald, appeared in 1931 and covered the crystal structures known by 1928; a second volume of this series (which has been continued by others) covered structures determined between 1928 and 1932 and was published by Hermann, O. Lohrmann, and H. Philipp in 1937.

To Hermann spatial symmetry seemed to present no greater difficulty of visualization than the simpler requirements of plane geometry presented to less gifted people. In writing out the coordinates of symmetrically equivalent positions in a space group, he hardly ever took the trouble of looking them up in the tables, preferring to jot them down on the basis of internal inspection.

At the inaugural meeting of the International Union of Crystallography at Harvard in 1948, Hermann contributed an important paper on four-dimensional space groups; this was to be followed by studies of space groups in more than four dimensions. Unfortunately, these further papers were never published for lack of proof of a basic theorem which seemed reasonable and could be shown to hold in special cases but could not be established generally.

Hermann also extended Paul Niggli's notion of "Gitterkomplexe"—that is, the occurrence of the same groups of symmetrically equivalent positions in different space groups—in an attempt to establish a systematic geometrical, and chemically significant, classification of actual crystal structures (in which, of course, different atomic species cannot occupy symmetrically equivalent positions). Hermann's first attempt to list all lattice complexes in three-dimensional space groups was made in the *Internationale Tabellen;* he derived a simple nomenclature of univariant lattice complexes in 1960 from the symbols used for nonvariant complexes. This work is being continued by one of Hermann's former students, E. Hellner, and his co-workers at Marburg.

Hermann attended the Gymnasium in his hometown and then studied mathematics and physics at Göttingen. He obtained his Ph. D. under Max Born in 1923 with a thesis in which Born's newly developed theory of the optical rotatory power of crystals containing screw axes was applied to sodium chlorate. Although the numerical work was marred by wrong factors π in the transition from rational (Heaviside) to conventional charge units, this first calculation of its kind proved that the rotation of the plane of polarization of light passing through the crystal had at least one main cause in the screwlike arrangement of the atoms.

After a short period of work with Herman Mark and then with R. O. Herzog at the Kaiser Wilhelm Institut für Faserstoffe in Berlin, Hermann became assistant (and later lecturer) at the Institute of Theoretical Physics of the Technische Hochschule at Stuttgart. In this period, 1925–1937, he produced not only the *Strukturberichte* and the *Internationale Tabellen* but also papers on the effects of symmetry on higher-order tensorial properties and a study of the various kinds of statistical symmetry in noncrystalline and semicrystalline (mesomorphic) substances.

Matters of conscience were not taken lightly in Hermann's family. His father, Gerhard, and his mother, the former Auguste Leipoldt, both came from a long line of Protestant clergy. Of their six children Carl was the oldest; the next one, Grete, studied philosophy and mathematics, went into exile during the Hitler years, and later became principal of a teacher's academy in Bremen. Wilhelm became a minister after being a businessman, and one of the three daughters married a clergyman. Hermann himself married Eva Lüddecke, daughter of a clergyman. He and his wife became very active members of the German group of the Society of Friends. When the Nazis took over, his position at the Technische Hochschule of Stuttgart became untenable. R. Brill, head of the X-ray laboratory at I. G. Farbenindustrie in Oppau, offered him a job; and it was there that the well-known study of the electron distribution in diamond and other simple crystals by Brill, H. G. Grimm, Hermann, and C. Peters was done. A later paper by Hermann in conjunction with W. Schlenk, Jr., was on the determination of the structure of urea adducts with hydrocarbons (1949); Hermann's structure determination led to an entirely new concept of adduct products.

During the second half of World War II, Hermann and his wife were jailed for having listened to BBC broadcasts. After the war Hermann was appointed professor of crystallography at the University of Marburg, a post he held until his death.

BIBLIOGRAPHY

I. ORIGINAL WORKS. Hermann's writings on symmetry are "Zur systematischen Strukturtheorie," 4 pts.: I, "Eine neue Raumgruppensymbolik," in *Zeitschrift für Kristallographie,* **68** (1928), 257–287; II, "Ableitung der 230 Raumgruppen aus ihren Kennvektoren," *ibid.,* **69** (1929), 226–249; III, "Ketten und Netzgruppen," *ibid.,* 250–270; and IV, "Untergruppen," *ibid.,* 533–555; "Bemerkungen zu der vorstehenden Arbeit von Ch. Mauguin," *ibid.,* **76** (1931), 559–561; "Die Symmetriegruppen der amorphen und mesomorphen Phasen," *ibid.,* **79** (1931), 186–221, 337–343; "Kristallographie in Räumen beliebiger Dimensionszahl, I: Die Symmetrie-operationen," in *Acta crystallographica,* **2** (1949), 139–145; "Translationsgruppen in *n* Dimensionen," in *Struktur und Materie der Festkörper* (1952); and "Zur Nomenklatur der Gitterkomplexe," in *Zeitschrift für Kristallographie,* **113** (1960), 142–154.

Crystal physics and structures are discussed in "Über die natürliche optische Aktivität der regulären Kristalle $NaClO_3$ und $NaBrO_3$" (Göttingen, 1923), his diss., also in *Zeitschrift für Physik,* **16** (1923), 103–134; "Tensoren und Kristallsymmetrie," in *Zeitschrift für Kristallographie,* **80** (1934), 32–45; "Anwendung der röntgenographischen Fourieranalyse auf Fragen der chemischen Bindung," in *Annalen der Physik,* **34** (1939), 393–445, written with R. Brill, H. G. Grimm, and C. Peters; and "Die Harnstoff-Addition der aliphatischen Verbindungen," in *Justus Liebigs Annalen der Chemie,* **565** (1949), 204–240, written with W. Schlenk, Jr., with Hermann's crystal structure determination on pp. 212–216.

His books include *Strukturbericht 1913–1928,* supp. vol. 1 to *Zeitschrift für Kristallographie* (Leipzig, 1931); *Internationale Tabellen zur Bestimmung von Kristallstrukturen,* 2 vols. (Berlin, 1935); and *Strukturbericht 1928–1932,* supp. vol. 2 to *Zeitschrift für Kristallographie* (Leipzig, 1937).

II. SECONDARY LITERATURE. See P. P. Ewald, *Fifty Years of X-Ray Diffraction* (Utrecht, 1962), pp. 339, 357–360, 451, 461, 465, 689, 700; and Kathleen Lonsdale, "Obituary C. Hermann," in *Nature,* **192** (1961), 604.

P. P. EWALD

HERMANN, JAKOB (*b.* Basel, Switzerland, 16 July 1678; *d.* Basel, 11 July 1733), *mathematics.*

Hermann, the son of Germanus Hermann, a headmaster, devoted much of his time to mathematics while studying theology at Basel (bachelor's degree, 1695; master's degree, 1696; theological examination, 1701). In the last quarter of the seventeenth century mathematics, which he took up under the guidance of Jakob I Bernoulli, was characterized by the creation of the calculus and the stormy development of infinitesimal calculus. Through his exceptional ability and his zeal Hermann was able at a young age to join the small group of the most important mathematicians. In 1696 he defended Bernoulli's third dissertation on the theory of series and in 1701, through the intervention of Leibniz, became a member of the Berlin Academy with a work directed against Bernhard Nieuwentyt, a relentless critic of Leibniz's differential concept and methods. In 1707, again assisted by Leibniz, he was appointed professor of mathematics at Padua—to the same chair that Nikolaus I Bernoulli later held. The following year Hermann was accepted into the Academy at Bologna. Yet, as a Protestant in Italy, he seems not to have been completely happy; and in 1713 he gladly accepted a call—once more arranged by Leibniz—to Frankfurt-an-der-Oder.

While in Italy, Hermann composed the final version of his principal scientific work, the *Phoronomia,* which appeared at Amsterdam in 1716. This textbook—a critical analysis of which is still lacking—concerned advanced mechanics in the modern sense and was considered an important work, very favorably reviewed by Leibniz himself in the *Acta eruditorum.*

From 1724 to 1731 Hermann was connected with the flourishing Academy in St. Petersburg, where he was the predecessor of Leonhard Euler, to whom he was distantly related (he was a second cousin of Euler's mother). In addition to various papers on trajectory problems, algebraically squarable curves, and attraction, Hermann wrote volumes I and III (mathematics and fortification) of the textbook *Abrégé des mathématiques* (St. Petersburg, 1728–1730). He also gave instruction in mathematics to the grandson of Peter the Great, the future Peter II, and to Isaac Bruckner.

Homesick, Hermann repeatedly sought to obtain any reasonably suitable position in Basel (see, for instance, Johann I Bernoulli's letter of 11 November 1724 to J. J. Scheuchzer). In 1722 he received, by lottery, the professorship of ethics and natural law at Basel, but he had a substitute carry out the duties of the office until he finally returned home in 1731. No professorship of mathematics became vacant in his native city before his death—the chair was brilliantly filled by Johann I Bernoulli. Shortly before his death the Paris Academy elected him a member.

Hermann possessed a serious, calm disposition; and through his sympathetic character, objectivity, and learning he won not only the friendship of Leibniz and of Jakob I Bernoulli but also the respect of all the leading mathematicians.

Hermann's scientific importance fully justifies the decision to incorporate his works into the complete edition of Bernoulliana which is now in progress. Of the approximately 600 standard-size pages of his correspondence, about a third has been published by C. J. Gerhardt in Leibniz's *Mathematische Schriften.*

BIBLIOGRAPHY

I. ORIGINAL WORKS. Hermann's works include *Responsio ad Clar. Viri Bernh. Nieuwentijt considerationes secundas circa calculi differentialis principia editas* (Basel, 1700); *Phoronomia, sive de viribus et motibus corporum solidorum et fluidorum libri duo* (Amsterdam, 1716); and *Abrégé des mathématiques,* vols. I and III (St. Petersburg, 1728–1730). For his articles, see Poggendorff, I, cols. 1077–1078.

Some of Hermann's letters may be found in C. J. Gerhardt, ed., *G. W. Leibniz' mathematische Schriften,* IV (Halle, 1859), 253–413. Extracts of his correspondence are in *Mitteilungen der Naturforschenden Gesellschaft in Bern* (1850), pp. 118–120. A complete bibliography may be found in the Bernoulli Archives in the university library at Basel.

II. SECONDARY LITERATURE. On Hermann and his work, see (listed in chronological order) *Mercure suisse* (Oct. 1733), pp. 77–85 and (Feb. 1734) for a eulogy and list of his writings; R. Wolf, "Euler," in *Biographien zur Kulturgeschichte der Schweiz,* IV (Zurich, 1862), pp. 90 ff.; O. Spiess, ed., *Der Briefwechsel von Johann Bernoulli,* I (Basel, 1955), *passim;* J. E. Hofmann, *Ueber Jakob Bernoullis Beiträge zur Infinitesimalmathematik* (Geneva, 1956); and V. I. Lysenko, "Die geometrischen Arbeiten von Jakob Hermann," in *Istoriko-matematicheskie issledovaniya,* **17** (1966), 299–307. On Hermann and mathematics in Russia see the notice by R. Wolf in *Verhandlungen der naturforschenden Gesellschaft in Zürich,* **35** (1890), 98–99; and M. Cantor, *Vorlesungen über Geschichte der Mathematik,* III (Leipzig, 1901), *passim.*

E. A. FELLMANN

HERMANNUS CONTRACTUS. See **Hermann the Lame.**

HERMES TRISMEGISTUS, *philosophy, astrology, magic, alchemy.*

The ancient Greeks identified their god Hermes with the Egyptian Thoth and gave him the epithet Trismegistus, or "Thrice-Greatest," for he had given the Egyptians their vaunted arts and sciences. A vast literature in Greek was ascribed to Hermes Trismegistus; the cited number of works ranges from 20,000 (Seleucus) to 36,525 (Manetho).

Clement of Alexandria knew of forty-two "indispensable" books. Of these, ten dealt with the Egyptian priests and gods; ten with sacrifices, rites, and festivals; ten with paraphernalia of the sacred rites; and two were hymns to the gods and rules for the king. Four books dealt with astronomy and astrology, and six were medical in nature, concerning the body, diseases, medicines, instruments, the eyes, and women. Lactantius in the third century and Augustine in the fourth refer to the Hermetic writings and accept the legend of Hermes Trismegistus without question. Hermetic works on alchemy are cited by Zosimus, Stephanus, and Olympiodorus.

The so-called *Corpus Hermeticum,* a collection of religious and philosophical works, is best known and has received considerable attention from scholars and those interested in the occult. Most of its seventeen or eighteen works were probably written in the second century. While some Egyptian influence may be present in the pious spirit and words of the writers, the bulk of the philosophy expressed is Greek, largely Platonism modified by Neoplatonism and Stoicism. Christian thought is not evident; indeed, Augustine condemned "Hermes the Egyptian, called Trismegistus" for the idolatry and magic found in some of the writings.

The first and chief work of the *Corpus* is entitled *Poimandres.* It gives an account of the creation of the world by a luminous Word, who is the Son of God. A mystical hymn in this work was often recited by alchemists. Other works in the *Corpus* deal with the ascent of the soul to the divine when, for a chosen few, it has freed itself from the material world and become endowed with divine powers. The astrological control of man through the seven planets and the twelve signs of the zodiac is prominent.

Besides the works of the *Corpus,* a work entitled *Asclepius* exists in a Latin translation. The work, a dialogue between Asclepius and Hermes Trismegistus, is of interest for its purported description of the ancient Egyptian religion. The work was attributed, probably incorrectly, in the ninth century to Lucius Apuleius of Madauros. The original Greek title was "The Perfect Word." The *Asclepius* describes how the Egyptian idols were made animate by magic and contains a lament that the ancient religion of Egypt is to come to an end. There is also a reference to the "Son of God," a fact made much of by Lactantius.

A strong Hermetic tradition persisted in the Middle Ages. Stobaeus the anthologist (late fifth century) preserved twenty-nine excerpts of Hermetica. Michael Psellus in the eleventh century knew of the *Corpus Hermeticum,* but in the medieval mind the name of Hermes Trismegistus was usually associated with alchemy and magical talismans. Albertus Magnus condemned the diabolical magic in some Hermetic works, but Roger Bacon referred to Hermes Trismegistus as the "Father of Philosophers." Medieval chemistry was often called the "hermetic science."

The magical and philosophical literature attributed to Hermes Trismegistus received widespread currency in the Renaissance. Traditional Hermetism was erroneously considered to be of ancient Egyptian origin and thus much older than the esteemed Greek philosophers who had been influenced by Egyptian beliefs. In the fifteenth century Georgius Gemistus (Plethon) and the Platonic Academy of Florence spread the view that Hermes, a contemporary of Moses, had founded theology. The Latin *Asclepius* was printed in 1469, and Marsilio Ficino published his influential Latin translation of the first fourteen books of the *Corpus* in 1471. The Greek text of the *Corpus* was published by Adrianus Turnebus at Paris in 1554.

Both philosophical and magical Hermetism declined rapidly in the seventeenth century after Isaac Casaubon showed in 1614 that the Hermetic writings were of the post-Christian era. Hermetism continued thereafter only among the Rosicrucians and other secret societies and occult groups.

BIBLIOGRAPHY

The following are versions of the *Corpus* or parts of it: *Corpus Hermeticum,* 4 vols. (Paris, 1945–1954): I, *Corpus Hermeticum, I–XII,* text verified by A. D. Nock and translated by A.-J. Festugière; II, *Corpus Hermeticum, XIII–XVIII, Asclepius,* text verified by A. D. Nock and translated by A.-J. Festugière; III, *Fragments extrait de Stobée, I–XXII,* text verified and translated by A.-J. Festugière; IV, *Fragments extrait de Stobée, XXIII–XXIX,* text verified and translated by A.-J. Festugière, and *Fragments divers,* text verified by A. D. Nock and translated by A.-J. Festugière; A.-J. Festugière, *La révélation d'Hermès Trismégiste,* 4 vols. (Paris, 1950–1954); G. R. S. Mead, *Thrice-Greatest Hermes. Studies in Hellenistic Theosophy and Gnosis,* 3 vols. (London, 1906); Louis Ménard, *Hermès Trismégiste, traduction complète, précédée d'une étude sur l'origine des livres hermétiques* (Paris, 1925); Gustav Parthey, *Hermetis Trismegisti Poemander* (Berlin, 1854); R. Reitzenstein, *Poimandres* (Leipzig, 1904); and Walter Scott, *Hermetica,* 4 vols. (Oxford, 1924–1936).

Secondary literature includes M. Berthelot, *Les origines de l'alchimie* (Paris, 1885), pp. 39–45, 133–136, *passim;* Joannes A. Fabricius, *Bibliotheca graeca* (Leipzig, 1790), I, pt. 1, ch. 7; A.-J. Festugière, *Hermétisme et mystique païenne* (Paris, 1967); H. L. Fleischer, *Hermes Trismegistus an die menschliche Seele, arabisch und deutsch* (Leipzig, 1875); Wilhelm Kroll, "Hermes Trismegistos," in Pauly-Wissowa, *Real-Encyclopädie,* VIII, 1 (Stuttgart, 1966), 792–823; Lynn Thorndike, *History of Magic and Experimental Science,* I (New York, 1943), 288–292; and Frances A. Yates, *Giordano Bruno and the Hermetic Tradition* (Chicago, 1964). A full history of Hermetism remains to be written.

KARL H. DANNENFELDT

HERMBSTADT, SIGISMUND FRIEDRICH (*b.* Erfurt, Germany, 14 April 1760; *d.* Berlin, Germany, 22 October 1833), *chemistry.*

For a detailed study of his life and work see Supplement.

HERMITE, CHARLES (*b.* Dieuze, Lorraine, France, 24 December 1822; *d.* Paris, France, 14 January 1901), *mathematics.*

Hermite was the sixth of the seven children of Ferdinand Hermite and the former Madeleine Lallemand. His father, a man of strong artistic inclinations who had studied engineering, worked for a while in a salt mine near Dieuze but left to assume the draper's trade of his in-laws—a business he subsequently entrusted to his wife in order to give full rein to his artistic bent. Around 1829 Charles's parents transferred their business to Nancy. They were not much interested in the education of their children, but all of them attended the Collège of Nancy and lived there. Charles continued his studies in Paris, first at the Collège Henri IV, where he was greatly influenced by the physics lessons of Despretz, and then, in 1840–1841, at the Collège Louis-le-Grand; his mathematics professor there was the same Richard who fifteen years earlier had taught Evariste Galois. Instead of seriously preparing for his examination Hermite read Euler, Gauss's *Disquisitiones arithmeticae,* and Lagrange's *Traité sur la résolution des équations numériques,* thus prompting Richard to call him *un petit Lagrange.*

Hermite's first two papers, published in the *Nouvelles annales de mathématiques,* date from this period. Still unfamiliar with the work of Ruffini and Abel, he tried to prove in one of these papers the impossibility of solving the fifth-degree equation by radicals. Hermite decided to continue his studies at the École Polytechnique; during the preparation year he was taught by E. C. Catalan. In the 1842 contest of the Paris colleges Hermite failed to win first *prix de mathématiques spéciales* section but received only first "accessit." He was admitted to the École Polytechnique in the fall of 1842 with the poor rank of sixty-eighth. After a year's study at the École Polytechnique, he was refused further study, because of a congenital defect of his right foot, which obliged him to use a cane. Owing to the intervention of influential people the decision was reversed, but under conditions to which Hermite was reluctant to submit. At this time, Hermite—a cheerful youth who, according to some, resembled a Galois resurrected—was introduced into the circle of Alexandre and Joseph Bertrand. Following the example of others, he declined the paramount honor of graduating from the École Polytechnique, contenting himself with the career of *professeur.* He took his examinations for the *baccalauréat* and *licence* in 1847.

At that time Hermite must have become acquainted with the work of Cauchy and Liouville on general function theory as well as with that of C. G. J. Jacobi on elliptic and hyperelliptic functions. Hermite was better able than Liouville, who lacked sufficient familiarity with Jacobi's work, to combine both fields of thought. In 1832 and 1834 Jacobi had formulated his famous inversion problem for hyperelliptic integrals, but the essential properties of the new transcendents were still unknown and the work of A. Göpel and J. G. Rosenhain had not yet appeared. Through his first work in this field, Hermite placed himself, as Darboux says, in the ranks of the first analysts. He generalized Abel's theorem on the division of the argument of elliptic functions to the case of hyperelliptic ones. In January 1843, only twenty years old, he communicated his discovery to Jacobi, who did not conceal his delight. The correspondence continued for at least six letters; the second letter, written in August 1844, was on the transformation of elliptic functions, and four others of unknown dates (although before 1850) were on number theory. Extracts from these letters were inserted by Jacobi in *Crelle's Journal* and in his own *Opuscula,* and are also in the second volume of Dirichlet's edition of Jacobi's work. Throughout his life Hermite exerted a great scientific influence by his correspondence with other prominent mathematicians. It is doubtful that his *Oeuvres* faithfully reflects this enormous activity.

In 1848 Hermite was appointed a *répétiteur* and admissions examiner at the École Polytechnique. The next ten years were his most active period. On 14 July 1856 he was elected a member of the Académie des Sciences, receiving forty out of forty-eight votes.

In 1862, through Pasteur's influence, a position of *maître de conférence* was created for Hermite at the École Polytechnique; in 1863 he became an *examinateur de sortie et de classement* there. He occupied that position until 1869, when he took over J. M. C. Duhamel's chair as professor of analysis at the École Polytechnique and at the Faculté des Sciences, first in algebra and later in analysis as well. His textbooks in analysis became classics, famous even outside France. He resigned his chair at the École Polytechnique in 1876 and at the Faculté in 1897. He was an honorary member of a great many academies and learned societies, and he was awarded many decorations. Hermite's seventieth birthday gave scientific Europe the opportunity to pay homage in a way accorded very few mathematicians.

Hermite married a sister of Joseph Bertrand; one of his two daughters married Émile Picard and the other G. Forestier. He lived in the same building as E. Bournoff at Place de l'Odéon, and it was perhaps his acquaintance with this famous philologist that led him to study Sanskrit and ancient Persian. Hermite was seriously ill with smallpox in 1856, and under Cauchy's influence became a devout Catholic. His scientific work was collected and edited by Picard.

> From 1851–1859 Europe lost four of its foremost mathematicians, Gauss, Cauchy, Jacobi, and Dirichlet. Nobody, except Hermite himself, could guess the profoundness of the work of Weierstrass and Riemann on Abelian functions and of Kronecker and Smith on the mysterious relations between number theory and elliptic functions. Uncontested, the scepter of higher arithmetic and analysis passed from Gauss and Cauchy to Hermite who wielded it until his death, notwithstanding the

admirable discoveries of rivals and disciples whose writings have tarnished the splendor of the most brilliant performance other than his [unspecified quotation by P. Mansion].

Throughout his lifetime and for years afterward Hermite was an inspiring figure in mathematics. In today's mathematics he is remembered chiefly in connection with Hermitean forms, a complex generalization of quadratic forms, and with Hermitean polynomials (1873), both minor discoveries. Specialists in number theory may know that some reduction of quadratic forms is owed to him; his solution of the Lamé differential equation (1872, 1877) is even less well known. An interpolation procedure is named after him. His name also occurs in the solution of the fifth-degree equation by elliptic functions (1858). One of the best-known facts about Hermite is that he first proved the transcendence of e (1873). In a sense this last is paradigmatic of all of Hermite's discoveries. By a slight adaptation of Hermite's proof, Felix Lindemann, in 1882, obtained the much more exciting transcendence of π. Thus, Lindemann, a mediocre mathematician, became even more famous than Hermite for a discovery for which Hermite had laid all the groundwork and that he had come within a gnat's eye of making. If Hermite's work were scrutinized more closely, one might find more instances of Hermitean preludes to important discoveries by others, since it was his habit to disseminate his knowledge lavishly in correspondence, in his courses, and in short notes. His correspondence with T. J. Stieltjes, for instance, consisted of at least 432 letters written by both of them between 1882 and 1894. Contrary to Mansion's statement above, Hermite's most important results have been so solidly incorporated into more general structures and so intensely absorbed by more profound thought that they are never attributed to him. Hermite's principle, for example, famous in the nineteenth century, has been forgotten as a special case of the Riemann-Roch theorem. Hermite's work exerted a strong influence in his own time, but in the twentieth century a few historians, at most, will have cast a glance at it.

In Hermite's scientific activity, shifts of emphasis rather than periods can be distinguished; 1843–1847, division and transformation of Abelian and elliptic functions; 1847–1851, arithmetical theory of quadratic forms and use of continuous variables; 1854–1864, theory of invariants; 1855, a connection of number theory with theta functions in the transformation of Abelian functions; 1858–1864, fifth-degree equations, modular equations, and class number relations; 1873, approximation of functions and transcendence of e; and 1877–1881, applications of elliptic functions and Lamé's equation.

In the 1840's, and even in the early 1850's, the inversion of integrals of algebraic functions was still a confusing problem, mainly because of the paradoxical occurrence of more than two periods. Jacobi reformulated the problem by simultaneously inverting p integrals—if the irrationality is a square root of a polynomial of the $(2p - 1)$th or $2p$th degree. In the early 1840's the young Hermite was one of the very few mathematicians who viewed Abelian functions clearly, owing to his acquaintance with Cauchy's and Liouville's ideas on complex functions. To come to grips with the new transcendents, he felt that one had to start from the periodicity properties rather than from Jacobi's product decomposition. This new approach proved successful in the case of elliptic functions, when Hermite introduced the theta functions of nth order as a means of constructing doubly periodic functions. In the hyperelliptic case he was less successful, for he did not find the badly needed theta functions of two variables. This was achieved in the late 1840's and early 1850's by A. Göpel and J. G. Rosenhain for $p = 2$; the more general case was left to Riemann. In 1855 Hermite took advantage of Göpel's and Rosenhain's work when he created his transformation theory (see below).

Meanwhile, Hermite turned to number theory. For definite quadratic forms with integral coefficients, Gauss had introduced the notion of equivalence by means of unimodular integral linear transformations; by a reduction process he had proved for two and three variables that, given the determinant, the class number is finite. Hermite generalized the procedure and proved the same for an arbitrary number of variables.

He applied this result to algebraic numbers to prove that given the discriminant of a number field, the number of norm forms is finite. By the same method he obtained the finiteness of a basis of units, not knowing that Dirichlet had already determined the size of the basis. Finally, he extended the theorem of the finiteness of the class number to indefinite quadratic forms, and he proved that the subgroup of unimodular integral transformations leaving such a form invariant is finitely generated.

Hermite did not proceed to greater depths in his work on algebraic numbers. He was an algebraist rather than an arithmetician. Probably he never assimilated the much more profound ideas that developed in the German school in the nineteenth century, and perhaps he did not even realize that the notion of algebraic integer with which he had started was wrong. Some of his arithmetical ideas were carried on with more success by Hermann Minkowski in the twentieth century.

In the reduction theory of quadratic and binary forms Hermite had encountered invariants. Later he made many contributions to the theory of invariants, in which Arthur Cayley, J. J. Sylvester, and F. Brioschi were active at that time. One of his most important contributions to the progress of the theory of invariants was the "reciprocity law," a one-to-one relation between the covariants of fixed degree of order p of an mth-degree binary form and those of order m of a pth-degree binary form. One of his invariant theory subjects was the fifth-degree equation, to which he later applied elliptic functions.

Armed with the theory of invariants, Hermite returned to Abelian functions. Meanwhile, the badly needed theta functions of two arguments had been found, and Hermite could apply what he had learned about quadratic forms to understanding the transformation of the system of the four periods. Later, Hermite's 1855 results became basic for the transformation theory of Abelian functions as well as for Camille Jordan's theory of "Abelian" groups. They also led to Hermite's own theory of the fifth-degree equation and of the modular equations of elliptic functions. It was Hermite's merit to use ω rather than Jacobi's $q = e^{\pi i \omega}$ as an argument and to prepare the present form of the theory of modular functions. He again dealt with the number theory applications of this theory, particularly with class number relations for quadratic forms. His solution of the fifth-degree equation by elliptic functions (analogous to that of third-degree equations by trigonometric functions) was the basic problem of this period.

In the 1870's Hermite returned to approximation problems, with which he had started his scientific career. Gauss's interpolation problem, Legendre functions, series for elliptic and other integrals, continued fractions, Bessel functions, Laplace integrals, and special differential equations were dealt with in this period, from which the transcendence proof for e and the Lamé equation emerged as the most remarkable results.

BIBLIOGRAPHY

I. ORIGINAL WORKS. Hermite's main works are *Oeuvres de Charles Hermite,* E. Picard, ed., 4 vols. (Paris, 1905–1917); *Correspondance d'Hermite et de Stieltjes,* B. Baillaud and H. Bourget, eds., 2 vols. (Paris, 1905); and "Briefe von Ch. Hermite an P. du Bois-Reymond aus den Jahren 1875–1888," E. Lampe, ed., in *Archiv der Mathematik und Physik,* 3rd. ser., **24** (1916), 193–220, 289–310. Nearly all his printed articles are in the *Oeuvres.* It is not known how complete an account the three works give of Hermite's activity as a correspondent. The letters to du Bois-Reymond are a valuable human document.

II. SECONDARY LITERATURE. The biographical data of this article are taken from G. Darboux's biography in *La revue du mois,* **1** (1906), 37–58, the most accurate and trustworthy source. Other sources are less abundant; the exception is P. Mansion and C. Jordan, "Charles Hermite (1822–1901)," in *Revue des questions scientifiques,* 2nd ser., **19** (1901), 353–396, and **20** (1901), 348–349; unfortunately, Mansion did not sufficiently account for the sources of his quotations.

An excellent analysis of Hermite's scientific work is M. Noether, "Charles Hermite," in *Mathematische Annalen,* **55** (1902), 337–385. Others, most of them superficial *éloges,* can be retraced from *Jahrbuch über die Fortschritte der Mathematik,* **32** (1901), 22–28; **33** (1902), 36–37; and **36** (1905), 22.

HANS FREUDENTHAL

HERNÁNDEZ, FRANCISCO (*b.* Montalban, near Toledo, Spain, 1517; *d.* Toledo, 1587), *natural history.*

Hernández was physician to Philip II. He began practicing medicine at the hospital of the monastery of Guadalupe and botanized in Castile and Andalusia. By order of the king he went to Mexico, where he stayed from 1570 to 1577, studying the fauna and the flora. His series of journeys through all the territories of the viceroyalty has been reconstructed by Germán Somolinos. The development of his works is shown in the reports that he sent to Spain: in 1572 he had "drawn and painted as many as three books on rare plants . . . and almost two books on land animals and rare birds unknown to our hemisphere." In 1576 there were sixteen books on plants, minerals, and animals which he planned to take back to Spain, leaving a copy of each in Mexico. His work was deposited in the library of the Escorial.

After Hernández' death, Leonardo Recchi (Recho), a royal physician, made a summary that was published at Rome in 1628, at the expense of Prince Cesi, as *Rerum . . . Novae Hispaniae thesaurus, seu plantarum, animalium, mineralium mexicanorum historia.* Before this edition there had appeared in Mexico two abridged versions of Hernández' work. One, written by Francisco Ximénez (1615), was derived from Recho's version (known before its printing) but introduced a great number of variations; it was entitled *Quatro libros de la naturaleza y virtudes de las plantas y animales . . . .* The other (1579), based on the copy kept in Mexico, was used by Agustín Farfán for his

Tratado breve de medicina . . . . Finally, at Madrid in 1790 Casimiro Gómez Ortega published a group of Spanish manuscripts under the title *Historia plantarum Novae Hispaniae.* These publications, given the changes suffered by Hernández' manuscripts, had, until the recent edition issued by the University of Mexico, both historical and scientific interest.

The data transmitted by Hernández are a source of information on some of the species of plants and animals, such as *Canis caribeus,* that became extinct after the discovery of America.

BIBLIOGRAPHY

Hernández' work has been brought together in the *Obras completas,* 4 vols. (Mexico City, 1959–1966). In vol. I, see Germán Somolinos d'Ardois, "Vida y obra de Francisco Hernández," pp. 95–482. See also Agustín Farfán, *Tratado breve de medicina* . . . (Mexico City, 1579); and Francisco Ximénez, *Quatro libros de la naturaleza y virtudes de las plantas y animales* . . . (Mexico City, 1615).

JUAN VERNET

HERO OF ALEXANDRIA (*fl.* Alexandria, A.D. 62), *mathematics, physics, pneumatics, mechanics.*

Hero (or Heron) of Alexandria is a name under which a number of works have come down to us. They were written in Greek; but one of them, the *Mechanics,* is found only in an Arabic translation and another, the *Optics,* only in Latin. Apart from his works we know nothing at all about him.

His name is not mentioned in any literary source earlier than Pappus (A.D. 300), who quotes from his *Mechanics.*[1] Hero himself quotes Archimedes (*d.* 212 B.C.), which gives us the other time limit. Scholars have given different dates, ranging from 150 B.C. to A.D. 250, but the question has been settled by O. Neugebauer, who observed that an eclipse of the moon described by Hero in his *Dioptra* (chapter 35) as taking place on the tenth day before the vernal equinox and beginning at Alexandria in the fifth watch of the night, corresponds to an eclipse in A.D. 62 and to none other during the 500 years in question.[2] An astronomical date is the most reliable of all, being independent of tradition and opinion. The rather minute theoretical possibility that Hero might have lived long after this date I have discussed and dismissed, while I have elsewhere reviewed the whole controversy about his dates, which is now of historical interest only.[4]

The question of what sort of man he was has also been debated. H. Diels found that he was a mere artisan.[5] I. Hammer-Jensen took him to be an ignorant man who copied the chapters of his *Pneumatics* from works which he did not understand.[6] Although E. Hoppe attempted to defend Hero,[7] Hammer-Jensen maintained her opinion.[8] In 1925 J. L. Heiberg wrote: "Hero is no scientist, but a practical technician and surveyor. This view, which has been challenged in vain, was first put forth by H. Diels: [who called him] 'Ein reiner Banause.'"[9]

Such adverse judgment was based on a study of the *Pneumatics* at a time when neither the *Mechanics* nor the *Metrica* was known; and the *Pneumatics,* although by far the largest work (apart from the elementary textbooks) was neither by its contents nor form apt to inspire confidence in a serious scholar. The contents are almost exclusively apparatuses for parlor magic, and there is no discernible plan in the arrangement of the chapters. Apart from the introduction, there is no theoretical matter in the book, which consists entirely of practical descriptions.

But since then, the *Mechanics* has been published in Arabic, and a manuscript has come to light giving the *Metrica* in its original form; thus the image of Hero has changed. The *Mechanics* shows nothing of the disorder of the *Pneumatics,* consisting of an introduction, a theoretical part, and a practical part; the *Metrica* shows that Hero possessed all the mathematical knowledge of his time, while a chapter of the *Dioptra* indicates that he was familiar with astronomy. We also find that he quotes Archimedes by preference and has copied many chapters of a lost work of his on the statics of plane figures.

In the introduction to the *Pneumatics,* Diels found a quotation from Strato of Lampsacus (*fl.* 288 B.C.) and suggested that it was taken from Philo of Byzantium (*fl.* 250 B.C.), who probably took it from Ctesibius (*fl.* 270 B.C.);[10] but Philo's *Pneumatics,* which was discovered later, does not contain this passage, and a strictly accurate quotation is most likely to have been taken from the original work. The form of this theoretical introduction led I. Hammer-Jensen to assume that Hero was an ignoramus who did not understand what he copied from diverse sources; yet to me the freely flowing, rather discursive style suggests a man well-versed in his subject who is giving a quick summary to an audience that knows, or who might be expected to know, a good deal about it.

This discursive style, so very different from the concise style of the technical descriptions, is found again in the *Mechanics,* in which Hero, before giving the propositions from Archimedes' book *On Uprights,* presents the theory of the center of gravity as explained by Archimedes, not by Posidonius the Stoic, whose definition was not good enough.[11] Here again there is a strong suggestion of a teacher repeating swiftly a piece of knowledge which his students ought

HERO OF ALEXANDRIA

to know. Since we know the author as Hero of Alexandria, it seems reasonable to assume that he was appointed to the museum, that is, the University of Alexandria, where he taught mathematics, physics, pneumatics, and mechanics, and wrote textbooks on these subjects.

The *Pneumatics* can best be regarded as a collection of notes for such a textbook, of which only the introduction and the first six chapters have been given their final shape. All the chapters are uniform in style, even those taken from Philo, and eminently clear, so the idea of an ignorant compiler cannot be upheld. But there is more to be learned from the *Pneumatics.* While there is no order at all in the general arrangement of the chapters, we find here and there a short series of related chapters in which it is clear that Hero is searching for a better solution to a mechanical problem. This shows unmistakably that he was an inventor; it is therefore probable that he himself invented the dioptra, the screw-cutter, and the odometer, as well as several pneumatic apparatuses. This is all that can be learned about Hero himself.

The following works have survived under the name of Hero: *Automata, Barulkos, Belopoiica, Catoptrica, Cheirobalistra, Definitiones, Dioptra, Geometrica, Mechanica, De mensuris, Metrica, Pneumatica,* and *Stereometrica.* These can be divided into two categories, technical and mathematical. All the technical books, except the *Cheirobalistra,* seem to have been written by Hero; of the mathematical books only the *Definitiones* and the *Metrica* are direct from his hand. The others are, according to J. L. Heiberg, Byzantine schoolbooks with so many additions that it is impossible to know what is genuinely Heronian and what is not.[12]

The *Pneumatics* is by far the longest book, containing an introduction and two books of forty-three and thirty-seven chapters, respectively; but it is merely a collection of notes for a textbook on pneumatics. Only the introduction and the first six or seven chapters are finished. The introduction treats the occurrence of a vacuum in nature and the pressure of air and water; although it is written in a very prolix style with occasional digressions, the train of thought is never lost. It seems to have been written by a man very well versed in his subject, who is summarizing for students of pneumatics matters already known to them from their textbooks. Some of the theory is right, some is wrong (for instance, the *horror vacui* of nature), but it was the best theoretical explanation to be had at the time; a real understanding of the phenomenon had to wait for the experiments of Torricelli.

The first chapters, most of them taken from Philo's *Pneumatics,* describe experiments to show that air is a body, and that it will keep water out of a vessel unless it can find an outlet and will keep water in if it cannot enter. Hero goes on to siphons; but soon all order is lost, and the chapters appear haphazardly. Yet there is nothing haphazard about the chapters themselves, each of which—whether taken from Philo or a description of an apparatus seen by Hero—is written in the same concise style and according to a fixed plan, beginning with a description of the apparatus, with letters referring to a figure, then a description of how it works, then last (if necessary) an explanation. With very few exceptions it is evident that the chapters were written by Hero himself, and without exception they are very clear: each instrument can be reconstructed from the description and the figure.

The contents, on the other hand, have always been a source of puzzlement and despair for serious-minded scholars. Certainly Hero describes some useful implements—a fire pump and a water organ—but all the rest are playthings, puppet shows, or apparatuses for parlor magic. Trick jars that give out wine or water separately or in constant proportions, singing birds and sounding trumpets, puppets that move when a fire is lit on an altar, animals that drink when they are offered water—how can one respect an author who takes all these frivolities in earnest?

But Hero's treatment of these childish entertainments is quite matter-of-fact; he is interested in the way they work. In 1948 I explained this by the assumption that he was writing a handbook for the makers of pneumatic instruments, but this is not necessarily correct.[13] Hero was a teacher of physics, of which pneumatics is part. The book is a text for students, and Hero describes instruments the student needs to know, just as a modern physics textbook explains the laws governing the spinning top or the climbing monkey. Playthings take up so much of the book because such toys were very much in vogue at the time and the science of pneumatics was used for very little else. (Among the many toys of the *Pneumatics* there are even a few that use hot air or steam as a moving power, which has given rise to ill-founded speculations that the steam engine could have been invented at this time.) To this we must add that Hero was an inventor; and to a real inventor any clever apparatus is of interest, regardless of its purpose.

There is a slightly different text, found only in four manuscripts, that is generally designated Pseudo-Hero. Of seventy-eight chapters, seven have been radically changed; elsewhere the changes are only verbal corrections to clarify an already quite clear text. This text cannot have been written later than A.D. 500; therefore when the two texts agree, neither of them has been changed since then. For every chapter there is a figure, and the text in most cases begins with a reference to it, such as "Let *ABCD* be a base . . . ." Since Pseudo-Hero has the same figures as Hero, the figures cannot have been changed after A.D. 500; and there is every reason to believe that they were drawn by Hero himself. A complete set of these illustrations has been published in a reprint of Woodcroft's translation of the *Pneumatics.*[14] The *Pneumatics* was by far the most read of Hero's works during the Middle Ages and the Renaissance; more than 100 manuscripts of it have been found.

The *Mechanics,* preserved only in an Arabic version, was published in 1893 with a French translation and in 1900 with a German translation. A textbook for architects (that is, engineers, builders, and contractors), it is divided into three books. Book 1 deals with the theoretical knowledge and the practical skill necessary for the architect: the theory of the wheel, how to construct both plane and solid figures in a given proportion to a given figure, how to construct a toothed wheel to fit an endless screw, and the theory of motion. Drawing largely upon Archimedes, Hero then presents the theory of the center of gravity and equilibrium, the statics of a horizontal beam resting on vertical posts, and the theory of the balance.

Book 2 contains the theory of the five simple "powers": the winch, the lever, the pulley, the wedge, and the screw. The five "powers" are first described briefly, then the mechanical theory of each is presented and the results of a combination of the powers are calculated. Next is a chapter with answers to seventeen questions about physical problems, evidently inspired by Aristotle's *Mechanical Problems,* followed by seven chapters on the center of gravity in different plane figures and on the distribution of weight on their supports, once more from Archimedes. Book 3 describes sledges for transporting burdens on land, cranes and their accessories, other devices for transport, and wine presses; the last chapter describes a screw-cutter for cutting a female screw in a plank, which is necessary for direct screw presses.

Apart from the first chapter of book 1, which contains the *Barulkos,* the work proceeds in an orderly fashion; it shows nothing of the disorder of the *Pneumatics,* but the style is equally clear and concise, with a single exception. In book 1, chapter 24, Hero gives the theory of the center of gravity, and there he uses the same prolix and discursive style as in the introduction to the *Pneumatics.* This chapter would also seem to be a summary for students who should already know the subject. There are figures for most of the chapters; that they go back to the original Greek text can be seen from a mistake in the translation of a Greek work in one of the figures.[15] Editions of the work give only an interpretation of the figures; facsimilies have been published, with an English translation of many chapters, by A. G. Drachmann.[16] The fragments from Archimedes have been published in English with the manuscript figures.[17]

The *Dioptra* contains a description of an instrument for surveyors; it consists of a pointed rod to be planted in the ground, with two interchangeable instruments: a theodolite for staking out right angles and a leveling instrument. The description, which unfortunately is imperfect owing to a lacuna in the manuscript, covers six chapters; chapters 7–32 contain directions for the use of the two instruments in a great number of tasks. In chapter 33 Hero criticizes the *groma,* the instrument then used for staking out lines at right angles; chapter 34 describes an odometer actuated by the wheel of a car, used for measuring distances by driving slowly along a level road. Chapter 35 indicates the method for finding the distance between Alexandria and Rome by simultaneously observing a lunar eclipse in the two cities; this chapter has been thoroughly studied by O. Neugebauer.[18] There is no chapter 36, and chapter 37 is the *Barulkos,* which is also chapter 1 of book 1 of the *Mechanics;* it is out of place in both. Chapter 38 describes a ship's odometer and is certainly not by Hero.[19]

The *Belopoiika* contains the description of the *gastraphetes,* or stomach bow, a sort of crossbow in which the bowstring is drawn by the archer's leaning his weight against the end of the stock, and two catapults worked by winches; two bundles of sinews provide the elastic power to propel the arrow, bolt, or stone. The catapults are shaped like those described by Vitruvius and Philo.[20]

The *Automata,* or *Automatic Theater,* describes two sorts of puppet shows, one moving and the other stationary; both of them perform without being touched by human hands. The former moves before the audience by itself and shows a temple in which a fire is lit on an altar and the god Dionysus pours out a libation while bacchantes dance about him to the sound of trumpets and drums. After the performance the theater withdraws. The stationary theater opens and shuts its doors on the performance of the myth of Nauplius. The shipwrights work; the ships are launched and cross a sea in which dolphins leap; Nauplius lights the false beacon to lead them astray; the ship is wrecked; and Athena destroys the defiant Ajax with thunder and lightning. The driving power in both cases was a heavy lead weight resting on a heap of millet grains which escaped through a hole. The weight was attached by a rope to an axle, and the turning of this axle brought about all the movements by means of strings and drums. Strings and drums constituted practically all the machinery; no springs or cogwheels were used. It represents a marvel of ingenuity with very scant mechanical means.

The *Catoptrica,* found only in a Latin version, was formerly ascribed to Ptolemy, but is now generally accepted as by Hero. It deals with mirrors, both plane and curved, and gives the theory of reflection; it also contains instructions on how to make mirrors for different purposes and how to arrange them for illusions.

Barulkos, "the lifter of weights," is the name given by Pappus to his rendering of the *Dioptra,* chapter 37, and the *Mechanics,* book 1, chapter 1.[21] It is an essay describing how one can lift a burden of 1,000 talents by means of a power of five talents, that is, the power of a single man. The engine consists of parallel toothed wheels and is derived from the *Mechanics,* book 1, chapter 21; however, it is only a theoretical solution: parallel toothed wheels were not used for cranes during antiquity.[22] L. Nix takes *Barulkos* to be the name of the *Mechanics,* even though Pappus mentions the *Barulkos* and the *Mechanics* in the same sentence, because the Arabic name of the *Mechanics* is "Hero's Book About the Lifting of Heavy Things."[23] But since the essay is found as the first chapter of the *Mechanics* (where it does not belong), the translator would seem to have taken this title to be the title of the whole work. The *Cheirobalistra* was published in 1906 by Rudolf Schneider, who regarded it as a fragment of a dictionary dealing with catapults; it consists of six items, each describing an element that begins with the letter *K.*[24] E. W. Marsden has interpreted these chapters as a description of a sort of catapult, which he has reconstructed.[25] It is unlikely, however, that the *Cheirobalistra* is actually a work by Hero.

NOTES

1. Pappus of Alexandria, *Collectionis quae supersunt . . .,* Friedrich Hultsch, ed., III, pt. 1 (Berlin, 1878), 1060–1068.
2. O. Neugebauer, "Über eine Methode zur Distanzbestimmung Alexandria–Rom bei Heron," in *Kongelige Danske Videnskabernes Selskabs Skrifter,* **26,** no. 2 (1938), 21–24.
3. A. G. Drachmann, "Heron and Ptolemaios," in *Centaurus,* **1** (1950), 117–131.
4. A. G. Drachmann, *Ktesibios, Philon and Heron,* vol. IV of Acta historica Scientiarum naturalium et medicinalium (Copenhagen, 1948), pp. 74–77.
5. H. Diels, "Über das physikalische System des Straton," in *Sitzungsberichte der k. Preussischen Akademie der Wissenschaften zu Berlin,* no. 9 (1893), 110, n. 3.
6. I. Hammer-Jensen, "Die Druckwerke Herons von Alexandria," in *Neue Jahrbücher für das klassischen Altertum,* **25,** pt. 1 (1910), 413–427, 480–503.
7. Edmund Hoppe, "Heron von Alexandrien," in *Hermes* (Berlin), **62** (1927), 79–105.
8. I. Hammer-Jensen, "Die heronische Frage," *ibid.,* **63** (1928), 34–47.
9. J. L. Heiberg, *Geschichte der Mathematik und Naturwissenschaften im Altertum,* which is in Iwan von Müller, ed., *Handbuch der Altertumswissenschaft,* V, pt. 1, sec. 2 (Munich, 1925), 37.
10. Diels, *op. cit.,* pp. 106–110.
11. Hero, *Mechanics,* ch. 24.
12. Heiberg, *loc. cit.*
13. Drachmann, *Ktesibios . . .,* p. 161.
14. *The Pneumatics,* facs. of the 1831 Woodcroft ed., with intro. by Marie Boas Hall (London–New York, 1971).
15. A. G. Drachmann, *The Mechanical Technology of Greek and Roman Antiquity,* vol. XVII of Acta historica Scientiarum naturalium et medicinalium (Copenhagen, 1963), p. 110, text for fig. 44.
16. *Ibid.,* pp. 165 ff.
17. A. G. Drachmann, "Fragments from Archimedes in Heron's *Mechanics,*" in *Centaurus,* **8** (1963), 91–146.
18. Neugebauer, *op. cit.*
19. Drachmann, *The Mechanical Technology . . . .*
20. Vitruvius, *De architectura,* X, ch. 11; and Philo, *Belopoiika,* Greek and German versions by H. Diels and E. Schramm, in *Abhandlungen der Preussischen Akademie der Wissenschaften* for 1918, Phil.-hist. Kl., no. 16 (1919).
21. Pappus, *op. cit.,* pp. 1060 ff.
22. Drachmann, *The Mechanical Technology . . .,* p. 200.
23. Hero, *Mechanics,* introduction. pp. xxii ff.; Pappus, *op. cit.,* p. 1060.
24. Rudolf Schneider, ed. and trans., "Herons Cheirobalistra," in *Mitteilungen des kaiserlich deutschen archaeologischen Instituts,* Römische Abt., **21** (1906), 142–168.
25. E. W. Marsden, *Greek and Roman Artillery. Technical Treatises* (Oxford, 1971), pp. 206–233.

BIBLIOGRAPHY

I. ORIGINAL WORKS. *Heronis Alexandrini Opera quae supersunt omnia,* 5 vols. (Leipzig, 1899–1914), contains all Hero's works except the *Belopoiica. Automata* is published with *Pneumatica, Opera,* I. *Belopoiica* appeared as "Heron's Belopoiica Griechisch und Deutsch von H. Diels und E. Schramm," in *Abhandlungen der K. Preussischen Akademie der Wissenschaften,* Phil.-hist. Kl., no. 2 (1918); and in E. W. Marsden, *Greek and Roman Artillery. Technical Treatises* (Oxford, 1971), with English trans. and notes. *Catoptrica* is published with *Mechanica, Opera,* II, pt. 1. *Cheirobalistra,* edited and translated by Rudolf Schneider, is in *Mitteilungen des kaiserlich deutschen archaeologischen Instituts,* Römische Abt., **21** (1906), 142 ff.; and in Marsden, above. *Dioptra* is published with *Metrica, Opera,* III. *Definitiones* and *Geometrica* appear as *Heronis definitiones cum variis collectionibus Heronis quae feruntur Geometrica,* J. L. Heiberg, ed., *Opera,* IV. *Mechanica* is available as Carra de Vaux, "Les mécaniques ou l'élévateur de Héron d'Alexandrie," in *Journal asiatique,* 9th ser., **1** (1893), 386–472, and **2** (1893), 152–269, 420–514, consisting of Arabic text and French translation; and as *Herons von Alexandria Mechanik und Katoptrik,* edited and translated by L. Nix and W. Schmidt, *Opera,* II, pt. 1. *De mensuris* is published with *Stereometrica, Opera,* V. *Metrica* is available in three versions: *Herons von Alexandria Vermessungslehre und Dioptra,* Greek and German versions by Hermann Schöne, *Opera,* III: *Codex Constantinopolitanus Palatii Veteris,* no. 1, E. M. Bruins, ed., 3 pts. (Leiden, 1964)—pt. 1, reproduction of the MS; pt. 2, Greek text; pt. 3, translation and commentary; and *Heronis Alexandrini Metrica . . .,* E. M. Bruins, ed. (Leiden, 1964). *Pneumatica* can be found as *Herons von Alexandria Druckwerke und Automatentheater,* Greek and German versions edited by Wilhelm Schmidt, *Opera,* I; and *The Pneumatics of Hero of Alexandria,* translated for and edited by Bennet Woodcroft (London, 1851) and facs. ed. with intro. by Marie Boas Hall (London–New York, 1971). *Stereometrica* appears as *Heronis quae feruntur Stereometrica et De mensuris,* J. L. Heiberg, ed., *Opera,* V. *Fragmenta,* the commentary on Euclid's *Elements,* is found as *Codex Leidensis 399, 1. Euclidis Elementa ex interpretatione Al-Hadschdschadschii cum commentariis Al-Narizii,* Arabic and Latin edited by R. O. Besthorn and J. L. Heiberg, 3 vols. (Copenhagen, 1893–1911).

II. SECONDARY LITERATURE. See A. G. Drachmann, *Ktesibios, Philon and Heron,* vol. IV in Acta historica Scientiarum naturalium et medicinalium (Copenhagen, 1948), on *Pneumatics;* and *The Mechanical Technology of Greek and Roman Antiquity,* vol. XVII in Acta historica Scientiarum naturalium et medicinalium (Copenhagen, 1963), on *Mechanics;* J. L. Heiberg, *Geschichte der Mathematik und Naturwissenschaften im Altertum,* in Iwan von Müller, ed., *Handbuch der Altertumswissenschaft,* V, pt. 2, sec. 1 (Munich, 1925), on the mathematical works; and O. Neugebauer, "Über eine Methode zur Distanzbestimmung Alexandria–Rom bei Heron," in *Kongelige Danske Videnskabernes Selskab Meddelelser,* **26,** no. 2 (1938), on *Dioptra.*

A. G. DRACHMANN

HERO OF ALEXANDRIA: Mathematics.

The historical evaluation of Hero's mathematics, like that of his mechanics, reflects the recent development of the history of science itself. Compared at first with figures like Archimedes and Apollonius, Hero appeared to embody the "decline" of Greek mathematics after the third century B.C. His practically oriented mensurational treatises then seemed to be the work of a mere "technician," ignorant or neglectful of the theoretical sophistication of his predecessors. As Neugebauer and others have pointed out, however, recovery of the mathematics of the Babylonians and greater appreciation of the uses to which mathematics was put in antiquity have necessitated a reevaluation of Hero's achievement.[1] In the light of recent scholarship, he now appears as a well-educated and often ingenious applied mathematician, as well as a vital link in a continuous tradition of practical mathematics from the Babylonians, through the Arabs, to Renaissance Europe.

The breadth and depth of Hero's mathematics are revealed most clearly in his *Metrica,* a mensurational treatise in three books that first came to the attention of modern scholars when a unique manuscript copy was found in Constantinople in 1896.[2] The prologue to the work gives a definition of geometry as being, both etymologically and historically, the science of measuring land. It goes on to state that out of practical need the results for plane surfaces have been extended to solid figures and to cite recent work by Eudoxus and Archimedes as greatly extending its effectiveness. Hero meant to set out the "state of the art," and the thrust of the *Metrica* is thus always toward practical mensuration, with a resulting ambiguity toward the rigor and theoretical fine points of classical Greek geometry. For example, Hero notes in regard to circular areas:

> Archimedes shows in the *Measurement of the Circle* that eleven squares on the diameter of the circle are very closely equal [ἴσα γίγνεται ὡς ἔγγιστα] to fourteen circles. . . . The same Archimedes shows in his *On Plinthides and Cylinders* that the ratio of the circumference of any circle to its diameter is greater than 211875 to 67441 and less than 197888 to 62351, but since these numbers are not easily handled, they are reduced to least numbers as 22 to 7 [*Metrica* I, 25, Bruins ed., p. 54].[3]

That is, Hero's use of approximating values for irrational quantities arose not out of ignorance of their irrationality or of theoretically more precise values, but out of the need for values that can be handled efficiently. In the case of $\sqrt{n}$ (n non-square) he set out an iterative technique for ever closer approximation, although he himself usually stopped at the first.[4]

Book I of the *Metrica* deals with plane figures and the surfaces of common solids. It proceeds in each case by numerical example (with no specified units of measure), presuming a knowledge of elementary geometry and supplying formal geometrical demonstrations where they might be unfamiliar. Beginning with rectangles and triangles, Hero gave, in proposition I.8, the famous "Heronic formula" for determining the area of a triangle from its three sides, $A = \sqrt{s(s-a)(s-b)(s-c)}$ (the proposition is actually derived from Archimedes). He then proceeded to treat general quadrilaterals by dividing them into rectangles and triangles. *Metrica* I. 17–25 treats the regular polygons of from three to twelve sides, directly deriving the relation of side to radius in all cases except 9 and 11, where Hero appeals to the "Table of Chords" (τὰ περὶ τῶν ἐν κύκλῳ εὐθείων). For $n = 5$, 6, and 7, the relations derived are the same as those found in the Babylonian texts at Susa.[5] After discussing the circle and annulus, Hero dealt extensively with the segment (but not the sector) of a circle, offering three approximating formulas, two "ancient" (which he criticized) and his own, which treats the segment as closely approximating a segment of a parabola (area = 4/3 inscribed triangle); Archimedes' *Method* is the explicit source for the latter. For the ellipse and parabola, and for the surfaces of a cone, sphere, and spherical segment, Hero did no more than cite Archimedes' results.

Book II moves on to solid figures. Beginning with the cone and the cylinder, Hero then dealt with prisms on various rectilinear bases and with regular and irregular frustra (including the famous βωμίσκος).[6] For the sphere he turned again to Archimedes; for the torus, to Dionysodorus. He concluded with the five "Platonic" solids (regular polyhedra).

In book III the treatment of the problem of dividing plane and solid figures into segments bearing fixed ratios to one another brought Hero's work more closely in line with the pure mathematical tradition. Very similar in style and content to Euclid's *On Divisions,* the subject matter forced Hero after proposition III.9 to give up numerical calculation in favor of geometrical construction of the lines and planes sought. Nonetheless, the problem of dividing the pyramid, cone, and conical frustrum required an approximating formula for the cube root of a number.[7]

Hero's concern in the *Metrica*—to extract from the works of such mathematicians as Archimedes only the results conducive to efficient mensuration—takes full effect in the other works that have come down to us bearing his name. *Geometrica* is essentially book I of the *Metrica; Stereometrica* is essentially book II. In both cases, numerical examples are used to eliminate geometrical derivations, concrete rather than general units of measure are employed, and the Greek mode of expressing fractions yields to the then more common and familiar Egyptian mode of unit fractions. *Geodaesia* and *De mensuris* contain nothing more than excerpts from the *Geometrica.* In all these texts, it is difficult to locate precisely Hero's original contribution, for they, rather than the *Metrica,* are the texts that circulated widely, were edited frequently, and were used for instruction. That their fate conformed at least in part to Hero's intention is indicated by his *Definitiones* and *Commentary on Euclid's Elements,* both of which show clear pedagogical concerns. As Heath notes,[8] the *Definitiones,* which contains 133 definitions of geometrical terms, is a valuable source of knowledge about alternative notions of geometry in antiquity and about what was taught in the classroom; it, like the others, shows the effect of many editors.

Hero's works enjoyed a wide audience. This is clear not only from what has been said above, but also in that fragments of his works can be found in the writings of several Arab mathematicians, including al-Nayrīzī and al-Khwārizmī.

NOTES

1. Otto Neugebauer, *Exact Sciences in Antiquity,* ed. 2 (New York, 1962), p. 146.
2. See Bibliography in section above for various modern editions.
3. Tannery has suggested correcting the numerators to read 211872 and 195882, respectively; *cf.* T. L. Heath, *History of Greek Mathematics,* I (Oxford, 1921), 232–233.
4. *Metrica* I, 8 (Bruins ed., p. 41): let $N = a^2 \pm r$;

$$a_1 = \frac{1}{2}\left(\frac{N}{a} + a\right) \text{ is a first approximation for } \sqrt{N},$$

$$a_2 = \frac{1}{2}\left(\frac{N}{a_1} + a_1\right) \text{ is a second, more accurate one, and so on.}$$

On the history of this method, see Heath, II, 324, note 2.
5. Neugebauer, p. 47.
6. For a discussion, see Heath, II, 332–333.
7. Heath, II, 341–342. Hero's method must be reconstructed from a single numerical example. The best conjecture (Wertheim's) seems to be that if $a^3 < N < (a + 1)^3$, $d_1 = N - a^3$, and $d_2 = (a + 1)^3 - N$, then

$$\sqrt[3]{N} = a + \frac{(a + 1)d_1}{(a + 1)d_1 + ad_2}.$$

8. II, 314; pp. 314–316 present a summary of the contents of the *Definitiones.*

BIBLIOGRAPHY

T. L. Heath, *History of Greek Mathematics,* vol. II (Oxford, 1921), ch. 18, remains the most complete secondary account of Hero's mathematics and is the source of many of the details given above.

MICHAEL S. MAHONEY

HERODOTUS OF HALICARNASSUS (*b.* Halicarnassus, Caria, Asia Minor, fifth century B.C.; *d.* Thurii, near the site of Sybaris, southern Italy, 430–420 B.C.), *history.*

The word "history" (ἱστορίη), which is the actual title of Herodotus' book, means simply "inquiry." We have it on Plato's authority (*Phaedo* 96*a*8) that the typical work of pre-Socratic natural philosophers was called "Inquiry About Nature" (περὶ φύσεως ἱστορία), and this is confirmed by its use by Heraclitus in discussing Pythagoras (fr. 129). Although Herodotus' "inquiry" is mainly concerned with narrating and explaining the course of events (that is, with history in the modern sense), it offers much information on subjects that would now be classified as geography, ethnography, and anthropology or folklore.

Very little is known about the life of Herodotus, except that he traveled extensively in the known Mediterranean world of his day and eventually settled in Thurii. Apart from Greece and Italy, he visited Scythia, the Bosporus, Egypt, the Euphrates valley, and Babylon. Although he had much to say about Persia, it is likely that he never went there. He collected information not only about events in the recent or early history of the various peoples, but also about their religion, government, economy, and way of life in general, as well as the physical features of their lands.

His work has as its single main theme the conflict between the Greeks and Asiatics, although this theme is interrupted by numerous and sometimes long digressions. He begins with the origins of the enmity between the Greeks and Persians in the kingdom of Lydia, turning then to events in Persia. His account of Persia is broken off several times to describe the countries which came under Persian rule, especially Egypt (book 2) and Scythia (book 4). The second half of his work is a more continuous narrative of the events leading up to the defeat of the Persians by the Greeks in 480–479 B.C.

Herodotus had some merits as a researcher. He knew the value of autopsy and of direct firsthand information, and he could recognize a biased witness. He frequently gives alternative versions of an incident and ostentatiously puts it to the reader to choose between them: "My business is to record what people say, but I am by no means bound to believe it" (A. de Selincourt trans., 7.152). He is skeptical of some of the more extravagant stories that he hears, and he is astonishingly free from antibarbarian prejudice.

But Herodotus' critical powers were limited. In foreign countries he was plainly dependent on interpreters, and he did not always choose his informants well. There are many instances where he failed to find the truth, although more diligent inquiries might have revealed it. His travels were fairly cautious and comfortable—he was no great explorer. He tells us little that is useful about contemporary science or technology, and his pyschological insight into character was elementary.

It is interesting that whereas Xenophanes, Parmenides, and Empedocles retained the Homeric epic verse form for their innovating "inquiries about nature," Herodotus wrote a prose narrative to preserve "the great and marvellous things done among both Greeks and barbarians" (1.1), and, as Longinus said (13.3), remained close to Homer in spirit.

BIBLIOGRAPHY

The text of Herodotus' *History* is examined by C. Hude in *Herodoti Historiae,* 3rd ed. (Oxford, 1927). Translations of the work include A. D. Godley, ed., *Herodotus* (London-Cambridge, 1921), with original text; G. Rawlinson, *History of Herodotus* (1858); and A. de Selincourt, *Herodotus: The Histories* (Harmondsworth, 1954). For an index of Greek words with their English translations, see J. Enoch Powell, *A Lexicon to Herodotus* (Cambridge, 1938).

A bibliographic survey is provided by P. MacKendrick, in *Classical Weekly,* **47** (1954), 145–152, and in *Classical World,* **56** (1963), 269–275.

DAVID J. FURLEY

HERON. See **Hero of Alexandria.**

HEROPHILUS (*b.* Chalcedon, Bithynia, last third of the fourth century B.C.), *anatomy, physiology.*

Only scanty information concerning Herophilus' life has been preserved; the place and date of his death are unknown. We learn from Galen that he was a native of Chalcedon, which was originally a Megarian colony on the Asiatic side of the Bosphorus (III, 21K.; XIV, 683K.), that he studied under Praxagoras of Cos (VII, 585K.; X, 28K.), and that later he taught and practiced medicine at Alexandria, first under Ptolemy Soter and subsequently under Ptolemy Philadelphus. In Alexandria he lived in an environment in which the dissection of the human body did not meet with general disapproval.[1] Such an atmosphere, possibly unique in Greek cities at that time, clearly proved beneficial to the development of scientific anatomy, and Herophilus' researches significantly advanced this study. His interest in comparative anatomy is also recorded.[2]

Herophilus acquired great prestige, both as a practitioner and as a teacher of medicine, and students flocked to Alexandria to sit at his feet. He is described by Galen as a member of the "dogmatic" school of medicine, that is, he was a follower of the "logical" or "dialectical" method, as opposed to mere empiricism.[3] Although Galen at times criticizes Herophilus for being obscure, his usual assessment is that he was a great physician commendable for the soundness of his views and especially for his combination of careful observation and reasoning.

The medical writings of Herophilus were not extensive. None of his works has been preserved, but it appears that he wrote at least eleven treatises, of which three were devoted to anatomy, one to ophthalmology, one to midwifery, two each to the study of the pulse and to therapeutics, one to dietetics, and one entitled Πρὸς τὰς κοινὰς δόξας, which was evidently a polemic against commonly held medical views which he believed to be mistaken. It is recorded that he did not hesitate to question even the views of Hippocrates himself. Yet he did not depart altogether from Coan teachings but, following his master Praxagoras, based his medical theory upon the doctrine of the four humors (V, 685K.; see Celsus, *On Medicine,* introduction, ch. 15).

It was in anatomy that Herophilus made his greatest contribution to medical science, conducting important anatomical investigations of the brain, eye, nervous and vascular systems, and the genital organs. He also wrote on obstetrics and gynecology and held an elaborate quantitative theory of the pulse. Several medical terms, some still in current use, were coined by him.

The result of Herophilus' anatomical researches into the brain was that the Peripatetic confusion of the functions of the brain and the heart was corrected and that there was a reversal to the minority view, originally propounded by Alcmaeon, that the brain is the central organ of sensation and the seat of the intellect. As a result of his dissections, Herophilus was able not only to distinguish the cerebrum from the cerebellum but also to demonstrate the origin and course of the nerves from the brain and spinal cord (III, 813K.; VIII, 212K.). We are further informed that he specified the "fourth ventricle" or the "cavity" of the cerebellum as the seat of the ἡγεμονικόν (Rufus, *De corp. part. anat.* 74 [185], Daremberg and Ruelle, eds.; Aët. 4, 5, 4; Galen, *De usu. part.* IX.1, III, 667K.). This cavity in the middle of the floor of the fourth ventricle he compared to the cavity in the pens used in Alexandria (ἀναγλυφῇ καλάμου; II, 731K.), and the Latin term *calamus scriptorius* or *calamus Herophili* remains in current medical use. The three membranes of the brain were also recognized by Herophilus and designated as "chorioid" (χοριοειδῆ) because they resembled the chorionic envelope surrounding the fetus (II, 719K.). He observed and likened the meeting point of the sinuses of the dura mater to a wine press (ληνός); the Latin term *torcular Herophili* is still in use. His description of the *rete mirabile,* or retiform plexus (δικτυοειδὲς πλέγμα), as he called it (V, 155K.), at the base of the brain affords further evidence of his dissection of animals, since it does not exist in man.

Herophilus' discovery, by dissection, of the nerves and his demonstration that they originate in the brain enabled him to answer the question raised by Praxagoras: to what kind of organ the extremities of the body owe their movement. Praxagoras and Diocles had successfully distinguished between arteries and veins,[4] both believing that pneuma moved through the former and blood through the latter. Having thereby provided specific channels through which voluntary motion could be imparted to the body, Praxagoras then conjectured that certain arteries became progressively thinner, with the result that ultimately their "walls" fell together and their hollowness (κοιλότης) disappeared. To describe this final part of the artery he used the term νεῦρον—meaning, presumably, that it resembled the sinews. Galen tells us that it was by the operation of these νεῦρα that Praxagoras accounted for the movement of the fingers and other parts of the hands. Thus, although Praxagoras did not himself actually isolate and identify the nerves as such, he nevertheless played an important role in their discovery a generation later by Herophilus.

It was Herophilus, then, who transferred to the nerves the function which Praxagoras had assigned to the arteries. Rufus preserves for us the additional information that he also distinguished between the sensory and motor nerves, calling the latter not κινητικά, as Erasistratus subsequently did, but προαιρητικά (*De Corp. part. anat.,* 71–74 [184, 13 ff.], Daremberg and Ruelle, eds.). Herophilus also succeeded in tracing the sensory nerves leading from the brain to the eyes (III, 813K.) and called them "paths" (πόροι). In this instance he did not use the technical term νεῦρα but retained the more familiar terminology. Galen adds that Herophilus was of the opinion that these πόροι contained αἰσθητικὸν πνεῦμα, and elsewhere he gives the reason why Herophilus considered the optic nerve to be a particularly suitable carrier of pneuma: he had discovered by dissection that these "strings" were hollow (VII, 89K.).

Herophilus was deeply interested in the liver and, as has been seen above, he applied the methods of comparative anatomy to its study. His knowledge of its conformation was extensive and, for the most part, accurate. He observed that it differs in size and conformation in different individuals of the same species and that occasionally it occurs on the left instead of the right side (II, 570K.). The name "duodenum" (ἡ δωδεκαδάκτυλος ἔκφυσις), which he derived from its size, was applied by him to the first part of the small intestine (II, 708K.; VIII, 396K.). He was also the first to isolate the lacteals or "chyle vessels," as they are now called, thereby anticipating Gasparo Aselli, who explained their function in the seventeenth century.

Having drawn the distinction between veins and arteries, Praxagoras gave the pulse a role in diagnosis and therapeutics. This pioneer work was subsequently developed by Herophilus,[5] who wrote treatises on this subject, and it is recorded—albeit in a late testimony[6]—that he was the first to count the

HEROPHILUS

pulse by means of a clepsydra. He seems to have employed four main indications of the pulse—size, strength, rate, and rhythm (VIII, 592K.)—and to have distinguished certain cardiac rhythms as characteristic of different periods of life (IX, 463K.). In addition to normal pulses which follow natural rhythms he distinguished three divergent pulses: the pararhythmic (παραρύθμοι), the heterorhythmic (ἑτερόρρυθμοι) and the ecrhythmic (ἐκρύθμοι). Of these the first indicates only a slight divergence from normality, the second a greater, and the third the greatest (IX, 471K.). It seems likely that Herophilus was indebted to contemporary musical theory for this terminology.

Herophilus maintained that pulsation was entirely involuntary and was caused by the contraction and dilatation of the arteries in accordance with the impulse received from the heart (VIII, 702K.). He believed, erroneously, that dilatation represented the normal condition of the arteries. Actually, contraction is the return of the artery walls to their normal condition and dilatation is due to the pressure of the blood from the heart. But however that may be, he clearly recognized the connection between the heart and pulse beats. Herophilus also noticed the existence of a pulmonary systole and diastole and, upon the basis of their alternating rhythm, sought to explain the respiratory process. He believed that respiration followed a four-stage cycle comprising the intake of fresh air, the distribution of that fresh air to the body, the intake of used air from the body, and the expulsion of this used air (XIX, 318K.). His knowledge of the pulmonary blood system does not appear to have been extensive. Rufus tells us that he called the pulmonary artery the "arterial vein" (φλέψ ἀρτηριώδης) and asserted that in the lungs the veins resemble the arteries and vice versa (p. 162, Daremberg and Ruelle, eds.; see also Galen, III, 445K.). Herophilus based this assertion upon comparative thickness and estimated that the walls of an artery were six times as thick as those of a vein (see also II, 624K.).

There is no evidence that Herophilus made any significant contribution to the knowledge of the male organs of generation, but does seem to have made intensive studies in gynecology. In his treatise on midwifery (μαιωτικόν) he accurately describes the ovaries, the uterus, and the cervix, and he had observed the Fallopian tubes. Soranus tells us that Herophilus was interested in the relationship between menstruation and general health (p. 192R.; I. 29, 1 ff., Heiberg, ed.). He also preserves Herophilus' summary of the causes of difficult labor (p. 349R.; IV.1, 4 ff., Heiberg, ed.), foremost among which are held to be frequency of parturition, displacement of the embryo, and insufficient dilatation of the cervix. We also learn from Soranus that Herophilus described the causes of *uterus prolapsus* and rightly held that only the cervix, and not the entire uterus, can protrude (p. 372R.; IV.36, 1–2, Heiberg, ed.).

Although Herophilus' chief interests lay in anatomy, he also displayed a keen practical interest in other branches of medical science. He wrote a treatise on dietetics and recommended gymnastics as a means of preserving health. Although he stressed the importance of diet, regimen, and exercise, our evidence suggests that he placed too much reliance upon drugs. Celsus even goes so far as to say he would never treat any disease without medicine (*De Medicina,* V, ch. 1; see also Galen, XI, 795K.; Pliny, *Naturalis historia,* XXVI, 11).

The school founded by Herophilus had centers in both Alexandria and Laodicea. His followers neglected anatomy and to a large extent dissipated their energies in sophistry and unrewarding controversy with the rival school of the Erasistrateans. Consequently very few of them achieved preeminence in medicine.[7] Herophilus' own reputation, however, stands apart from that of his school. His true importance lies in the fact that he, together with his younger contemporary Erasistratus, laid the foundations for the scientific study of anatomy and physiology. Their careful dissections provided both a basis and a stimulus for the anatomical investigations undertaken by Galen over four centuries later.

NOTES

1. For the tradition that both Herophilus and Erasistratus vivisected humans, see James Longrigg, "Erasistratus," in *Dictionary of Scientific Biography,* IV, 382–386.
2. Galen tells us (II, 570K.) that he noted differences between the livers of men and hares.
3. For a description of these two opposed schools of medical thought, see Celsus, *On Medicine,* intro., chs. 8–35.
4. F. Solmsen, "Greek Philosophy and the Discovery of the Nerves," p. 179.
5. Some scholars (for example, Gossen, col. 1106 followed by Dobson, p. 21) have maintained that Herophilus rejected Praxagoras' belief that the pneuma moves through the arteries and maintained that these vessels were full of blood. But their standpoint is based on a misinterpretation of Galen IV, 731K.
6. Marcellinus, *De pulsibus,* xi, H. Schöne, ed. (Basel, 1907).
7. Notable exceptions among Herophilus' pupils were Demetrius of Apamea and Philinus of Cos.

BIBLIOGRAPHY

References to Galen are cited according to *Claudii Galeni opera omnia,* C. G. Kühn, ed., 20 vols. (Leipzig, 1821–1833); to Rufus, according to the ed. of C. Daremberg and E. Ruelle, *Oeuvres de Rufus d'Ephèse* (Paris,

1879; repr. Amsterdam, 1963); and to Soranus, according to the ed. of V. Rose, *Soranus* (Leipzig, 1882), and to that of J. Ilberg, *Sorani Gynaeciorum libri IV*. . . (Berlin, 1927).

See also C. Allbutt, *Greek Medicine in Rome* (London, 1921); J. F. Dobson, "Herophilus," in *Proceedings of the Royal Society of Medicine,* **18,** pts. 1-2 (1925), 19-32; H. Gossen, "Herophilus," in Pauly-Wissowa, *Real-Encyclopädie der klassischen Altertumswissenschaft,* VIII (Stuttgart, 1912), 1104-1110; E. F. Horine, "An Epitome of Ancient Pulse Lore," in *Bulletin of the History of Medicine,* **10** (1941), 209-249; W. H. S. Jones, *The Medical Writings of Anonymus Londinensis* (Cambridge, 1947); F. Kudlien, "Herophilus und der Beginn der Medizinischen Skepsis," in *Gesnerus,* **21** (1964), 1-13, and in H. Flashar, ed., *Antike Medizin* (Darmstadt, 1971), pp. 280-295; K. F. H. Marx, *Herophilus, ein Beitrag zur Geschichte der Medizin* (Karlsruhe-Baden, 1838); and *De Herophili celeberrimi medici vita scriptis atque in medicina meritis* (Göttingen, 1842); F. Solmsen, "Greek Philosophy and the Discovery of the Nerves," in *Museum Helveticum,* **18** (1961), 150 ff.; A. Souques, "Que doivent à Hérophile et à Erasistrate l'anatomie et la physiologie du système nerveux?," in *Bulletin de la Société d'Histoire de la Médecine,* **28** (1934), 357-365; F. Steckerl, *The Fragments of Praxagoras of Cos and His School* (Leiden, 1958); and G. Verbeke, *L'évolution de la doctrine du pneuma* (Paris-Louvain, 1945).

For a more comprehensive bibliography, see James Longrigg, "Erasistratus," in *Dictionary of Scientific Biography,* IV, 382-386.

JAMES LONGRIGG

HÉROULT, PAUL LOUIS TOUSSAINT (*b.* Thury-Harcourt, Normandy, France, 10 April 1863; *d.* Cannes, France, May 1914), *metallurgy.*

Héroult was born in the St. Bénin quarter of the small town of Thury-Harcourt, which is on the Orne River. His father, a tanner, later left Normandy in order to become director of a more important tannery in Gentilly, near Paris.

In 1882 Héroult was admitted to the École des Mines at Paris. He studied under Henry Le Chatelier, who communicated to him his great interest in aluminum; Le Chatelier had studied under Henri Sainte-Claire Deville, who, by acquiring the knowledge needed for the electrolytic preparation of aluminum, had established the aluminum industry.

Sketches of electrolysis tanks have been found in Héroult's course notebooks which plainly indicate the young inventor's projects. Released from military service in 1884, he went to work in the family tannery in Gentilly, where he carried out his first experiments on aluminum.

In April 1886, when he was twenty-three years old, Héroult registered his patent for the electrolysis of melted cryolite at approximately 1000° C., in a crucible lined with carbon and serving as a cathode; the melted aluminum accumulates at the bottom of the crucible. An anode of pure carbon is plunged into the bath and is burned by the oxygen liberated at its surface. This is exactly the procedure followed today.

Héroult proved to be an equally great inventor in another area of metallurgy: he is considered the creator of the method used for preparing steels in the electric furnace. In 1907 he patented a furnace in which the arc was produced between the heated scrap iron and a graphite electrode. There are many of these furnaces throughout the world, all of the Héroult type. The first direct-arc electric furnace installed in the United States was a Héroult furnace.

BIBLIOGRAPHY

See "Centenaire Paul Héroult, 1863-1963," in *Revue de l'aluminium et de ses applications* (May 1963).

GEORGES CHAUDRON

HERRERA, ALFONSO LUÍS (*b.* Mexico City, Mexico, 3 July 1868; *d.* Mexico City, 17 September 1942), *biology.*

Herrera's father, also called Alfonso Herrera, was a well-known scientist. Herrera received a degree in pharmacy in 1889 from the School of Medicine of Mexico. That year he was appointed professor of botany and zoology at the Normal School, where in 1902 he established the first chair of general biology in the country. His textbook *Nociones de biología* (Mexico City, 1904) was translated into French as *Notions de biologie et plasmogénie comparées* (Berlin, 1906).

In 1889 Herrera was also appointed assistant to the natural history department of the National Museum and from 1894 to 1897, he published catalogs of the collections of mammals, birds, reptiles and amphibia, anthropology, and fishes and invertebrates. He joined the National Medical Institute in 1890 as assistant of natural history. In 1900 he proposed the creation of a commission of parasitology at the Department of Agriculture. Appointed chief, he remained with the commission for seven years. In 1915, also at the Department of Agriculture, he organized the Direction of Biological Studies, then the largest center for biological research in the country. He was director of the center until his retirement in 1930.

Herrera was interested in problems of biological adaptation to high altitudes. He did extensive research in this field, publishing his results in *La vie sur les hauts plateaux* (1899), for which he received an award from the Smithsonian Institution.

In addition to botany, zoology, and pharmacology, Herrera studied intensively the structure and origin of living matter. This subject, the center of his research activities until his death, involved him in a series of controversies, both scientific and religious.

He was a member of many scientific societies, received the *Palmes académiques* from the French government, and was elected to the Accademia Nazionale dei Lincei.

BIBLIOGRAPHY

I. Original Works. Herrera's works include *Recueil des lois de la biologie générale* (Mexico City, 1897); *La vie sur les hauts plateaux* (Mexico City, 1899), written with Vergara Lope; *Una nueva ciencia: la plasmogenia* (Mexico City, 1911); and *Farmacopea Latino-Americana* (Mexico City, 1921).

II. Secondary Literature. For information on Herrera, see E. Beltrán, "Alfonso L. Herrera. Un hombre y una época," in *Revista de la Sociedad mexicana de historia natural,* **3** (1942), 201–210, and "Alfonso L. Herrera (1868–1968). Primera figura de la biología mexicana," *ibid.,* **29** (1968), 37–110.

Enrique Beltrán

HERRICK, CHARLES JUDSON (*b.* Minneapolis, Minnesota, 6 October 1868; *d.* Grand Rapids, Michigan, 29 January 1960); **HERRICK, CLARENCE LUTHER** (*b.* Minneapolis, 22 June 1858; *d.* Socorro, New Mexico, 15 September 1904), *comparative neurology, psychobiology.*

Clarence Luther was the oldest and Charles Judson the youngest of the four sons of Nathan Henry Herrick, a farmer and Free Baptist minister who later served as a chaplain in the Fifth Minnesota Volunteer Regiment. Their mother was Anna Strickler, a girl from Washington, D.C., who had answered Herrick's advertisement for a wife because she was attracted by the prospect of living in the West. The family income came mostly from their small farm, while a little church on the edge of Minneapolis was the center of most of their community and social life.

Both brothers attended one-room country schools. Clarence Luther Herrick then entered the Minneapolis High School in 1874 and proved to be an exceptional student, enrolling in the University of Minnesota as a freshman in 1875. He received the B.A. in 1880, having completed the six-year preparatory and college course in five years, even though he had to earn his own way as assistant in the Minnesota Geological and Natural History Survey and was absent from the university during his junior year, in which he taught in a country school.

Charles Judson Herrick had become interested in natural history early in life through the influence of his brother. He planned to enter the ministry but, soon realizing that he lacked the vocation, returned to his first love, science, and obtained the B.S. under Clarence Luther Herrick at the University of Cincinnati in 1891. (The older brother had begun his academic career in 1885 as professor of geology and natural history at Denison University, where he founded the *Bulletin of the Laboratories of Denison University;* in 1888 he resigned from Denison to accept a similar appointment at Cincinnati, where he founded the *Journal of Comparative Neurology.*)

Charles Judson Herrick's first teaching post was a one-year assignment (1892–1893) at Ottawa University, Ottawa, Kansas, where he was a professor of natural history, including physics, chemistry, biology, geology, and psychology. In the meantime, in 1891, Clarence Luther Herrick was appointed professor of neurology at the University of Chicago, but a misunderstanding led to his resignation in 1892. He was then appointed professor of biology at Denison, where, in 1893, Charles Judson Herrick became his graduate student.

Clarence Luther Herrick's career was cut short by pulmonary tuberculosis, and he resigned his post at Denison and went, in an attempt to regain his health, to New Mexico, where he worked for a while as consulting geologist and mining surveyor. He spent the rest of his life there, being second president of the University of New Mexico at Albuquerque (where he met George Ellett Coghill, whom he influenced greatly) in 1897, and manager of the Socorro Gold Mining Company from 1901 until his death in 1904.

While still at Cincinnati, Clarence Luther Herrick had launched the new science of psychobiology, which was fostered and put on an academic footing by Charles Judson Herrick, and later by Coghill. Toward this end, both brothers encouraged comparative anatomists, physiologists, and psychiatrists to coordinate their attack upon the body-mind problem, and to cooperate in advancing the interdisciplinary exchange of scientific information. Charles Judson Herrick also assumed editorship of the *Journal of Comparative Neurology* upon his brother's removal to New Mexico, as well as assuming his teaching duties at Denison. Maintaining the *Journal* strained both the health and the finances of the younger Herrick, but under his editorship it became one of the outstanding biological periodicals in America. In 1904 R. M. Yerkes joined the staff, and the name was changed to *Journal of Comparative Neurology and Psychology.*

Charles Judson Herrick remained at Denison until

1907, except for a year's study (1896) for the Ph.D. under Oliver Strong and Henry Fairfield Osborn at Columbia University; the degree was granted in 1900. His dissertation was concerned with the nerve components of bony fishes. In 1907 he became professor of neurology at the University of Chicago, where he remained until his retirement in 1934. His major work, *The Evolution of Human Nature,* was published in 1956 and summarized the advance of psychobiology.

Until Charles Judson Herrick's work on correlating nervous structure with function, very little was known in this area. Selecting species with highly adaptive modes of life, he correlated their behavior with their well-developed central nervous systems and with their highly specialized peripheral nerve peculiarities. Concomitant to these studies, he made the first analysis of the four distinct functional longitudinal columns found in all vertebrates, a contribution of inestimable value to the clinician. His analysis (covering forty years' work) of the salamander brain is the most complete account ever made of the structure of any vertebrate brain. This analysis, coupled with his studies of the developing larval brain and in strict cooperation with Coghill's studies, permitted him to formulate his concepts on "The Nature and Origin of Human Mentation," a milestone in the natural history of the body-mind problem.

That Charles Judson Herrick's thinking was both broad and deep is illustrated by a passage in the introduction to *The Evolution of Human Nature:*

> I did not devote 60 years to intensive study of the comparative anatomy of the nervous system merely to collect dead facts or add to the score of "accumulative knowledge." I wanted to find out what these animals do with the organs they have and what they do it for, with the expectation that this knowledge would help us to unravel the intricate texture of the human nervous system and show us how to use it more effectively.

Herrick's desire was realized, and he presented evidence toward the understanding of the mind: he defined psychobiology—his brother's science and his own—as "the study of the experience of living bodies, its methods of operation, the apparatus employed and its significance as vital process, all from the standpoint of the individual having the new experience."

Clarence Luther Herrick was married to Alice Keith in 1883; they had a son and two daughters. Charles Judson Herrick was married to Mary Talbot, daughter of a former president of Denison University. They had one daughter, with whom they lived in Grand Rapids following his retirement. Clarence Luther Herrick's brilliant academic and professional career ended with his early death, while Charles Judson Herrick's emeritus years were devoted primarily to propagating his specialized knowledge among those whose interest was in the philosophy and psychology of animal and human behavior. Their science of psychobiology was given its widest coverage in the writings of Adolf Meyer, one of America's pioneer biologically oriented psychiatrists. Charles Judson Herrick also wrote biographies of his brother and of Coghill.

BIBLIOGRAPHY

I. Original Works. The Herrick papers, which are deposited in the Kenneth Spencer Research Library of the University of Kansas, list 637 published works of C. Judson Herrick, which include twenty-one books, with the remainder divided equally between scientific papers and reviews of articles and books. There is also a listing of 108 unpublished articles and books. David Bodian's memoir on C. Judson Herrick, *Biographical Memoirs. National Academy of Sciences* (in press), lists his major publications. The following books by C. Judson Herrick contain, in their respective bibliographies, references to his major publications. These listings will cover adequately the fields of his endeavors: *An Introduction to Neurology* (Philadelphia, 1915; 5th ed., 1931); *A Laboratory Outline of Neurology* (Philadelphia, 1915; 2nd ed., 1920), written with Elizabeth C. Crosby; *Neurological Foundations of Animal Behavior* (New York, 1924), translated into Chinese, with extensive revision and additions, by Yu-Chuan Tsang (Peking, 1958); *Brains of Rats and Men. A Survey of the Origin and Biological Significance of the Cerebral Cortex* (Chicago, 1926); *Fatalism or Freedom. A Biologist's Answer* (New York, 1926); *The Thinking Machine* (Chicago, 1929; 2nd ed., 1960); *The Brain of the Tiger Salamander, Amblystoma tigrinum* (Chicago, 1948); "A Biological Survey of Integrative Levels," in Roy Wood Sellars *et al.*, eds., *Philosophy for the Future* (New York, 1949), pp. 222–242; *George Ellett Coghill, Naturalist and Philosopher* (Chicago, 1949); "Clarence Luther Herrick, Pioneer Naturalist, Teacher, and Psychobiologist," in *Transactions of the American Philosophical Society,* n.s. **45,** pt. 1 (1955), 1–85, which includes a list of all C. L. Herrick's published writings; and *The Evolution of Human Nature* (Austin, Tex., 1956; repr. New York, 1961), translated into Spanish by Eloy Terrón as *La evolución de la naturaleza humana* (Madrid, 1962).

His seminal article "The Nature and Origin of Human Mentation" was published posthumously with intro. and notes by Paul G. Roofe, in *World Neurology,* **2** (1961), 1027–1045.

II. Secondary Literature. The following are obituaries of C. Judson Herrick: George W. Bartelmez, "Charles Judson Herrick, Neurologist," in *Science,* **131** (1960), 1654–1655; Elizabeth Crosby, "Charles Judson Herrick," in *Journal of Comparative Neurology,* **115** (1960), 1–8; J. L. O'Leary and G. H. Bishop, "C. J. Herrick and the

Founding of Comparative Neurology," in *Archives of Neurology,* **3** (1960), 725–731; and Paul G. Roofe, "Charles Judson Herrick," in *Anatomical Record,* **137** (1960), 162–164. The biography by C. Judson Herrick, cited above, is the only available source for the life of Clarence Luther Herrick.

PAUL G. ROOFE

HERSCHEL, a family of distinguished scientists of German origin, established in England in 1757. Its most notable members, on whom separate articles follow, were **William Herschel** (1738–1822), his sister **Caroline Lucretia Herschel** (1750–1848), and his only son, **John Frederick William Herschel** (1792–1871).

The earliest known German forebear of the family was a Hans Herschel of Dresden. His son, Abraham, was the father of Isaac Herschel, the father of Caroline and William. Isaac married Anna Ilse Moritzen and they had ten children, six of whom survived. Isaac, a sometime gardener, was an oboist with the Hanoverian Foot Guards, and he gave his children a sound education at the garrison school. He educated them in music himself. He was a man of surprisingly cultivated conversation. The other children included Sophia (Griesbach), Jacob, Alexander, and Dietrich. Dietrich's daughters married into the Knipping, Richter, and Groskopff families, names which thereafter frequently occur in the Herschel family correspondence.

In 1757 William Herschel, also an oboist in a military band, took refuge in England following the defeat of the Hanoverian forces at Hastenbeck. He continued his musical career, eventually becoming an organist and the leader of an orchestra at the fashionable resort of Bath. While there he became interested in optics and astronomy and began to manufacture reflecting telescopes; for which he enlisted the aid of his brother Alexander and, more importantly, his sister Caroline, both of whom followed him to England. His sensational discovery of the planet Uranus in 1781 brought him recognition and a royal pension, later supplemented by a stipend for Caroline Herschel's services. This enabled him to devote himself to astronomy and especially to the study of star clusters, nebulae, and binary stars.

For her own part, Caroline Herschel independently discovered eight comets and three nebulae in the course of patiently and devotedly assisting her brother. She further fostered the scientific interests of her nephew, John Herschel, who enjoyed a diversified (he was a physicist and chemist as well as an astronomer) and profitable career. John Herschel traveled extensively and worked closely with both the Royal Society and the Royal Astronomical Society; he is particularly remembered for his studies and cataloging of the southern skies.

The central position that the Herschels held in British astronomy was consolidated through the work of John Herschel's sons and through the marriage of his daughters into the Maclear and Waterfield families, and finally by the work of one of his grandsons, the Reverend John Charles William Herschel. Their influence thus extended for over a century. Both William and John Herschel were Knights of the Royal Hanoverian Guelphic Order and in 1838 Queen Victoria granted John the hereditary title of baronet. The title is now extinct, and although John Herschel had twelve children, the family name today survives in England in the person of a single descendant, Miss Caroline Herschel. The badge of the Royal Astronomical Society includes a depiction of the reflecting telescope of forty-eight-inch diameter and forty-foot focal length that was built at Slough, with aid from the royal purse, under William Herschel's direction.

The Herschel family lived at Windsor, Datchet, and finally at Slough. A family home with much interesting Herscheliana is at Warfield, Bracknell, Berkshire.

HERSCHEL, CAROLINE LUCRETIA (*b.* Hannover, Germany, 16 March 1750; *d.* Hannover, 9 January 1848), *astronomy.*

Caroline Herschel was the fifth of six surviving children of Isaac and Anna Ilse Herschel, and younger sister of William. After Isaac's death in 1767 Caroline became a household drudge, and so in 1772 William fetched her to live with him in Bath, England. She then began a career as a singer; but as William became obsessed with astronomy he called for her continual assistance, oblivious to the damaging effects of this on her own career. Thus, if William was polishing a telescopic mirror, Caroline "was even obliged to feed him by putting the Vitals by bits into his mouth."

In 1782 William gave up music for astronomy and moved to the neighborhood of Windsor Castle; Caroline accompanied him, thereby ending her musical career. William now encouraged Caroline to "sweep" for comets on her own account; and in 1783, while doing this, she found three new nebulae, including the companion to the Andromeda nebula. But soon William was himself committed to sweeping for nebulae, and at night Caroline was often required to be on hand to write down his observations. In the daytime, besides managing the household and entertaining visitors, she carried out the extensive routine calculations, prepared catalogs and papers for publication, and even ground and polished mirrors; in 1787 this work was recognized with a salary of £50 from the king. Between 1786 and 1797 she discovered no fewer than eight comets, thus earning a reputation as an astronomer in her own right; and in 1798 her revision of Flamsteed's catalog of stars was published by the Royal Society.

William married in 1788; and Caroline subsequently lived in lodgings, although she continued to collaborate in his astronomical work. After William died in 1822, she returned to Hannover, where she spent the rest of her life, vigorous and alert to the end. In 1828 she received the gold medal of the Royal Astronomical Society for her manuscript reduction and arrangement of William's nebulae and star clusters; and in her old age many honors were bestowed upon her.

BIBLIOGRAPHY

I. Original Works. Caroline Herschel's *Catalogue of Stars Taken From Mr. Flamsteed's Observations Contained in the Second Volume of the Historia Coelestis, and not Inserted in the British Catalogue, With an Index to Point out Every Observation in That Volume Belonging to the Stars of the British Catalogue [and] a Collection of Errata* was published by the Royal Society (London, 1798); her "Reduction and Arrangement in the Form of a Catalogue in Zones of All the Star Clusters and Nebulae Observed by Sir William Herschel," which remained in MS, was indispensable to John Herschel's review of northern nebulae. Her more important observations of comets are reported in *Philosophical Transactions of the Royal Society,* **77** (1787), 1–3; **79** (1789), 151–153; (1792), 23–24; (1794), 1; (1796), 131–132, and repr. in *The Scientific Papers of Sir William Herschel,* J. L. E. Dreyer, ed., I (London, 1912), 309–310, 327–328, 438, 451, 528; Dreyer also includes biographical material. Other works with extensive biographical material are Mrs. John Herschel, *Memoir and Correspondence of Caroline Herschel* (London, 1876; 2nd ed., 1879); and Constance A. Lubbock, ed., *The Herschel Chronicle* (Cambridge, 1933).

II. Secondary Literature. The biographical notice by John Herschel which appeared in *Athenaeum,* no. 1056 (22 Jan. 1848), 84, is repr. with additions in *Memoirs of the Royal Astronomical Society,* **17** (1847–1848), 120–122. Excellent short biographies are Agnes M. Clerke, in *Dictionary of National Biography,* IX, 711–714; and *The Herschels and Modern Astronomy* (London, 1895), pp. 115–141. The chapters on Caroline in Marianne Kirlew, *Famous Sisters of Great Men* (London, 1906), and Helen Ashton and Katharine Davies, *I Had a Sister* (London, 1937), are derivative. Most studies of William Herschel discuss the assistance given him by Caroline.

Michael A. Hoskin

HERSCHEL, JOHN FREDERICK WILLIAM (*b.* Slough, England, 7 March 1792; *d.* Hawkhurst, Kent, England, 11 May 1871), *astronomy, physics, chemistry.*

Herschel was the only child of Sir William Herschel and the former Mary Baldwin Pitt. He was married on 3 March 1829 to Margaret Brodie Stewart. They had twelve children, including Caroline (Hamilton-Gordon), woman of the bedchamber to Queen Victoria; William James Herschel, initiator of the use of fingerprints for purposes of identification; Alexander Stewart Herschel, physicist and astronomer; and Constance (Lubbock), author of *The Herschel Chronicle.* Herschel is buried in Westminster Abbey, next to Newton.

Scions of celebrated families usually enjoy an advantage whatever career they choose. For John Herschel there was a compensating penalty in the eventual choice of his father's profession, for the son was compared with the father, to the former's detriment. William achieved outstanding fame, starting from obscurity and poverty (when he arrived in England his total wealth was one French crown piece). John lived in bourgeois affluence and had a first-class formal education and entrée to all the best-known scientists of Europe. William was a pioneer, "the father of stellar astronomy," who worked in a single field. John worked in many, sometimes within an established framework, and could be accused of dilettantism, except that he achieved distinction in each of several fields. During his life John was immensely celebrated, his name epitomizing science to the public, much as that of Einstein did in the next century. After his death there was a period of obscurity, lifting only now that northern astronomers realize that, for southern hemisphere astronomers, John occupies the same commanding innovative position that William does for those in the north.

An only child in a household devoted to astronomical observation, Herschel may have been saved from becoming a withdrawn solitary by the remarkable relationship with his Aunt Caroline, who, by the sweetness and liveliness of her nature, spanned the gap between herself and the lad forty-two years her junior. The relationship was maintained by correspondence and ended only with her death at the age of ninety-eight, soon after he had proudly sent her the massive *Results of Astronomical Observations Made During the Years 1834–38 at the Cape of Good Hope.*

The careers of many scientists have been determined by wars and other political events, and hardly at all by family circumstances. Herschel's was strongly influenced by his family. Although his early life was full of wars, although as a boy he once met Napoleon and later visited the unmarked tomb on St. Helena, although he mastered many languages, one would never guess from his writings and diaries that the world was anything but peaceful.

After a time at Eton at the age of eight, Herschel went to a private school near his home. At seventeen he entered St. John's College, Cambridge, as a Foundress scholar and read mathematics. There he made friends with George Peacock, later dean of Ely; Charles Babbage, the "irascible genius" whose unsuccessful calculating engine anticipated many modern principles; and William Whewell, the natural and moral philosopher who later became master of Trinity College, Cambridge.

All were mathematicians of great ability; and, while still undergraduates, Herschel, Peacock, and Babbage founded the Analytical Society, devoted to the introduction into the British curriculum of the advanced methods of analysis current on the Continent. A lifelong addict of puns, Herschel described part of their aims as "The replacement of the dot-age of the University by the pure d-ism of the Continent," referring to the change of notation involved. Herschel and Peacock translated Lacroix's *Traité du calcul différentiel et du calcul intégral,* which appeared in 1816; and he and Babbage published two volumes of examples, including much on finite differences, in 1820.

Herschel took the tripos in 1813, when it was still an oral examination, and was, as he wrote in his diary, "Dismissed with a flaming compliment." He was named senior wrangler and first Smith's Prizeman, with Peacock next, Babbage having withdrawn because he could not compete with Herschel. He was elected to a college fellowship in 1813 while still a B.A. and retained this distinction until his marriage. (He was to become an honorary fellow in 1867.) The fifth of a series of mathematical papers, on an application of Cotes's theorem, brought election as a fellow of the Royal Society in 1813, at the age of twenty-one. Herschel was also an expert chemist, and in 1815 he missed election to the chair of chemistry at Cambridge by only one vote. He vacillated in his choice of career. His father favored the church but Herschel first preferred the law, enrolling in 1814 at Lincoln's Inn, where he seems to have used much of his time to further his acquaintance with William Hyde Wollaston, the natural philosopher who invented the camera lucida and discovered the dark lines in the solar spectrum, and with James South, a wealthy amateur astronomer. Both men exerted strong influences on his later scientific work.

Surprisingly, Herschel accepted a minor teaching post at St. John's and, as sublector from July 1815, endured uncongenial "pupillizing" for a while. Taking his M.A. in 1816, he left Cambridge for good and embarked on a scientific career after earnest conference with his father, who, at seventy-eight, was anxious to see his astronomical work continued but felt his own powers waning. As one of his obituarists was to say, John Herschel took up astronomy out of a sense of "filial devotion."

How this was financially possible is puzzling. William was in desperate straits in 1757. His pay as royal astronomer was £200 per year. Although in his lifetime he made and sold telescopes to the value of £16,000—which must be multiplied by about forty for modern dollar values—only a fraction of it could have been profit. We do not know how wealthy his wife's first husband was, but William's youth contrasts sharply with John's affluent world of Eton and Cambridge. Although John Herschel held no permanent paid post between 1816 and 1850, he brought up a large family in comfort and disbursed large sums on his expedition to Africa. His books, for their kind, had large sales but can hardly have provided a living.

Herschel took over from his father various astronomical and instrumental techniques. The mirrors of reflecting telescopes were then made of an alloy of copper and tin called speculum metal. Herschel himself made the eighteen-inch mirror of his twenty-foot telescope. When the surface of a mirror tarnished, the mirror was replaced by another and the original was repolished and even refigured, usually by the astronomer himself. The observational techniques included the use of star sequences and "sweeping"—fixing the telescope on the meridian at a particular elevation so that, as the earth turned, all the objects at a particular declination were carried through the field of view and could be noted down. The elevation gave the declination; the sidereal time of transit, the right ascension. Used extensively by both Herschels, this technique of discovery was the one by which Uranus was discovered. A sequence of stars, a term still in use, meant a series of stars presumed to be of diverse but known magnitudes, in a given area of sky, into which other stars could be fitted by interpolation and, hence, their magnitudes estimated.

Herschel's choice of research topics was characteristically diverse. After some mathematical papers he turned to physical and geometrical optics. He studied the polarization and birefringence of crystals and worked on rudimentary spectrum analysis and on the interference of light and sound waves. He computed the forms of compound lenses and propounded "Herschel's condition," required for the production of sharp images. In 1819 he discovered that sodium thiosulfate (the photographer's "hypo") dissolved silver salts, an important fact not used practically until some decades later.

In early maturity Herschel made a number of European journeys. Charles Babbage, his companion in 1821, went because of some private affliction; Herschel may have gone because of an unhappy love affair. They went to France and there met Arago, Laplace, Biot, and Humboldt. From France they journeyed to Switzerland and Italy, where they did some very respectable mountaineering. The next year Herschel went with his old friend James Grahame, probably a fellow Johnian, and during this journey received news of his father's death. In 1824 he went again to France and Italy, returning through Germany, meeting Gay-Lussac, Poisson, Fourier, G. B. Amici, Piazzi, Encke, and K. L. Harding. He ended with a visit to his Aunt Caroline in Hannover, whither she had withdrawn to look after her relatives, quixotically renouncing all financial claims on her English family.

On his journeys Herschel made many physical and meteorological experiments and geological and other observations. In particular he used a device he called an actinometer, which consisted of a large-bulbed thermometer containing a dark liquid. He would compare the rate of rise of this liquid in the sun and in the shade and so derive a numerical measure of the solar energy.

A short journey in 1827 took him to Ireland, where he met William Rowan Hamilton, the precocious genius who shared his interests in physical optics and many other fields.

Herschel's first astronomical paper, on the computation of lunar occultations (1822), was published when he was already working in London on systematic observations of double stars with James South, the possessor of two excellent refracting telescopes. It had once been thought that a close pair of stars of differing magnitudes must result from the accidental near alignment of two similar stars at vastly different distances and that any apparent relative motion would be a parallactic effect of the motion of the earth around the sun. The pioneer work of William Herschel had demonstrated orbital motion of binary stars under mutual attraction. John continued the work, reobserving known systems and discovering new ones, with detailed study of several cases, notably Gamma Virginis, and the development of methods (1833) for the determination of orbital elements. For their catalog of 380 double stars (1824) South and Herschel received the Lalande Prize of the French Academy in 1825 and the gold medal of the Astronomical Society (1826).

Herschel's scientific life was closely bound up with two royal societies. A fellow of the Royal Society in 1813, he won its Copley Medal in 1821 and 1847, and its Royal Medal in 1833, 1836, and 1840; he served as secretary from 1824 to 1827. In 1830 a reform group within the Royal Society nominated Herschel as president. He lost by eight votes to the duke of Sussex, a son of George III. The following year the British Association for the Advancement of Science was founded; Herschel was elected its president in 1845. (A full account of the controversy is L. Pearce Williams, "The Royal Society and the Founding of the B.A.A.S.," in *Notes and Records of the Royal Society,* **16** [1961], 221-233.)

The Astronomical Society (Royal after 1831) began with a dinner attended by fourteen gentlemen, including Herschel, on 12 January 1820. Initial hostility from the senior Royal Society was placated and the new society formally established. Herschel was foreign secretary from 1820 to 1827 and in 1846–1847. He was president three times (1827–1829, 1839–1841, 1847–1849), gold medalist in 1826 and 1836, and one recipient of the series of testimonials awarded in lieu of medals in 1848 during the row over the credit for priority in discovery of the planet Neptune.

In July 1825, working with the geodesist Edward Sabine and aided by large parties of troops, Herschel collaborated with a group of French scientists to determine the longitude difference between the Greenwich and Paris observatories.

James South left England and Herschel continued astronomical observations at Slough, following his father's lead in observation of nebulae, clusters, and double stars. A monumental catalog of 2,307 nebulae and clusters, 525 being new, was issued in 1833. By 1836 he had published six catalogs of double stars, comprising 3,346 systems.

Herschel's long list of research papers includes other astronomical contributions and optical, chemical, and geological studies. Somehow he also found time to contribute long articles on a variety of topics to several encyclopedias then being published. For David Brewster's *Edinburgh Encyclopaedia* (1830) he wrote "Isoperimetrical Problems" and "Mathematics." For the *Encyclopaedia metropolitana* he produced "Light" (1827) and "Sound" (1830). In 1830 he published *A Preliminary Discourse on the Study of Natural Philosophy* as the first volume of Dionysius Lardner's *Cabinet Cyclopaedia.* In 1833, also for

Lardner, appeared *A Treatise on Astronomy,* which became his most celebrated work, *Outlines of Astronomy.*

Herschel was now nearing forty and had earned almost every possible distinction in his field. He might well have remained a solitary bachelor but for his friend James Grahame, who decided he would be better off married and even picked out the girl: Margaret Brodie Stewart, daughter of Dr. Alexander Stewart, a Presbyterian divine and Gaelic scholar, who by his two wives had had a large family. Maggie, as Herschel was to call her, was good-looking, eighteen years younger than Herschel, and possessed an extremely strong character. Grahame threw the couple together; they married in 1829, were supremely happy, and had twelve children. Maggie followed Herschel everywhere, even to the wilds of Africa, and managed all his complex affairs, even to the extent of running a household of seldom less than twenty people when she was still in her early twenties.

Herschel now conceived the idea of an astronomical expedition to the southern hemisphere, possibly delaying its execution until after his mother's death in 1832. The only possible choices of site were South America, Australia, and the Cape of Good Hope. The Cape Colony had come under British rule in 1806 as a consequence of the Napoleonic Wars. Cape Town had existed as a town since 1652 and was important as a way station for many ships en route to India. The British had established an observatory there for the "improvement of astronomy and navigation" in 1820. As the result of the work of Lacaille in 1751–1753 it had an astronomical tradition and also enjoyed the technical advantage of being in the same longitude as eastern Europe, so that cooperative observations in the same meridian were possible.

On 13 November 1833 the *Mountstuart Elphinstone* sailed from Portsmouth with the Herschel party—John, Maggie, three children, a mechanic named John Stone, and a nurse—on board. They had a twenty-foot telescope and a seven-foot equatorially mounted refractor. They landed at Cape Town on 16 January 1834, Herschel having happily beguiled the voyage with all kinds of astronomical, oceanographical, and meteorological investigations while everyone else was prostrated with seasickness. Ten days before they landed, the newly appointed director of the Cape Observatory (H.M. astronomer at the Cape), Thomas Maclear, had arrived with his family and servant; the two were to enjoy four years of happy collaboration.

Herschel leased at £225 per annum (and subsequently purchased for £3000) an eighteen-room house called "The Grove," which he named "Feldhausen" by a German approximation to its Dutch name, in the suburb of Claremont, south of Cape Town. Within six weeks he and John Stone had the reflector erected on a spot now marked by a memorial obelisk. By 1838 he had swept the whole of the southern sky, cataloged 1,707 nebulae and clusters, and listed 2,102 pairs of binary stars. He carried out star counts, on William Herschel's plan, of 68,948 stars in 3,000 sky areas. Herschel made micrometer measures for separation and position angle of many pairs. He produced detailed sketches and maps of several objects, including the Orion region, the Eta Carinae nebula, and the Magellanic Clouds, and extremely accurate drawings of many extragalactic and planetary nebulae. He observed lunar eclipses, and when Eta Carinae, an object whose nature is still not understood, underwent a dramatic brightening in December 1837, he recorded its behavior in detail. Herschel invented a device called an astrometer, which enabled him to compare the brightness of stars with an image of the full moon of which he could control the apparent brightness, and thus introduced numerical measurements into stellar photometry. Maclear provided him with accurate star positions, and he assisted Maclear in geodetic and tidal observations. He observed Encke's and Halley's comets and experimented with the actinometer and with cooking by solar heat.

Herschel and Maggie and some of the children made several trips into the nearer parts of the western Cape Colony. He helped promote exploring expeditions and galvanized the Cape Philosophical Society. His correspondence was enormous, and virtually everyone of note visited him. He drew pictures of scenery and flowers with the camera lucida, and Maggie colored some of the pictures. He did enough botany to get his name in the list of species and established systematic meteorology in the area. With several local worthies Herschel devised a new educational system for the Cape Colony, traces of which persist; and, having written memoranda from the Cape, lobbied for their acceptance when he reached home. He refused official financial aid for the expedition and was able to offer financial aid to several of his numerous brothers-in-law. On 11 March 1838 the expedition embarked on the *Windsor,* with Herschel conducting experiments throughout the voyage, and landed at London on 15 May 1838.

The newly created baronet rushed off to Hannover to see his Aunt Caroline, as well as Gauss, Olbers, and H. C. Schumacher. He produced numerous papers on topics ranging from iron meteors to variable stars to the structure of the eye of the shark. Many of these derived from his African experiences, particularly his plan for the reform of the nomenclature and boundaries of the constellations, which was ready by 1841. Herschel served on committees and commissions, including the Royal Commission on Standards (1838–1843), and as lord rector of Marischal College, Aberdeen, in 1842. He helped to organize worldwide meteorological and magnetic observations, as well as the geomagnetic expedition of James Clark Ross to the Antarctic.

From Herschel's return from Africa until the mid-1840's two special scientific preoccupations stand out: the reduction of the African results and their preparation for publication, which led to numerous relatively short papers; and the researches in photography. So expert a chemist was Herschel that he was readily able to duplicate work reported by others and to improve on it, often in a matter of days. He made the first photograph on glass (of the decrepit forty-foot telescope, destined to be dismantled and mourned in a curious ceremony on 1 January 1840) in 1839. He introduced the terms "positive" and "negative." In eleven papers on photographic topics, Herschel tested an extraordinary variety of chemicals and processes. In 1839 he reproduced a solar spectrum in its natural colors. He extended his researches into both the ultraviolet and the infrared, and discovered the "Herschel effect": the quenching effect of light of a longer wavelength on a photosensitive surface afterward exposed to a shorter wavelength. Because his interests were mainly scientific and academic, he has been deprived of credit that is now accorded to others in the practical development of photography.

In 1840 the family moved from Slough to "Collingwood," a house at Hawkhurst, Kent. Herschel was then forty-eight years old and beginning to slow down. Still to come were the remaining photographic papers, a great deal of committee work, miscellaneous astronomical papers, some investigations of the phenomena of fluorescence, and thoughts on such diverse topics as meteorology, metrology (including that of the Great Pyramid), and color blindness. The *Results* from Africa appeared in 1847. *Outlines of Astronomy* was issued in 1849, as was a *Manual of Scientific Inquiry* for the Royal Navy. Herschel found time to translate into English hexameters some of the works of Schiller, Dante's *Inferno,* and, at the end of his life, the *Iliad.*

In December 1850 Herschel took a step that is almost inexplicable unless he was motivated by either financial stringency or the example of Newton. He accepted the post of master of the mint, where he tightened up the administration and advocated decimal coinage. He sat on a royal commission investigating the curricula at Oxford and Cambridge and on a committee choosing scientific instruments for the Great Exhibition of 1851. He wrote articles on meteorology, physical geography, and the telescope for the eighth edition of the *Encyclopaedia Britannica.* It was all too much for him. He was often away from his family; he was ill with gout and depression; and he suffered a breakdown. He retired from the mint at the beginning of 1856 but still had the verve to produce a consolidated catalog of 5,079 nebulae and clusters for the *Philosophical Transactions* of 1864 and to do much of the work for the posthumous catalog of 10,300 double stars. Herschel could even write about meteors and meteorology, and on musical scales. But in Julia Margaret Cameron's photographs of him, he seems old and feeble, with all the energetic good looks of the African time quite gone. When he died in 1871, he was mourned by the whole nation, not merely as a public figure and great scientist but also as one of the last of the universalists.

BIBLIOGRAPHY

I. ORIGINAL WORKS. Books by Herschel are *A Collection of Examples of the Application of the Calculus of Finite Differences* (Cambridge, 1820); *Preliminary Discourse on the Study of Natural Philosophy* (London, 1830); *A Treatise on Astronomy* (London, 1830); *Results of Astronomical Observations Made . . . at the Cape of Good Hope, Being a Completion of a Telescopic Survey of the Whole Surface of the Visible Heavens, Commenced in 1825* (London, 1847); *Outlines of Astronomy* (London, 1849); *Essays From the Edinburgh and Quarterly Reviews, With Addresses and Other Pieces* (London, 1857); and *Familiar Lectures on Scientific Subjects* (London, 1868).

Contributions to encyclopedias include David Brewster, ed., *The Edinburgh Encyclopaedia* (Edinburgh, 1830): "Isoperimetrical Problems" (not signed), XII, pt. 1, 320–328; and "Mathematics," XIII, pt. 1, 359–383; Edward Smedley, ed., *Encyclopaedia metropolitana* (London, 1845): "Physical Astronomy," III, 647–729; "Light" (dated Slough, 12 Dec. 1827), IV, 341–586; "Sound" (dated Slough, 3 Feb. 1830), IV, 763–824; and T. S. Traills, ed., *Encyclopaedia Britannica,* 8th ed.: "Meteorology," XIV (1857), 636–690; "Physical Geography," XVII (1849), 569–647; and "Telescope," XXI (1860), 117–145.

Books translated or edited by Herschel are S. F. Lacroix, *An Elementary Treatise on the Differential and Integral Calculus,* translation (pt. 2 by G. Peacock and J. F. W. Herschel) with an appendix by Herschel and notes by Peacock and Herschel (Cambridge, 1816); *A Manual of Scientific Inquiry; Prepared for the Use of Her Majesty's Navy; and Adapted for Travellers in General,* edited by Herschel (London, 1849); and *The Iliad of Homer Trans-*

lated Into English Accentuated Hexameters by Sir J. F. W. Herschel (London, 1866).

For a list of Herschel's papers see Royal Society *Catalogue of Scientific Papers,* III (1869), 322–328; VII (1877), 965; and *Mathematical Monthly,* **3,** 220 f.

Principal repositories of Herschel documents are the Royal Astronomical Society and the Royal Society, London; St. John's College, Cambridge; the University of Texas at Austin; and the South African Public Library and the South African Archives, Cape Town.

II. SECONDARY LITERATURE. See Günther Buttmann, *The Shadow of the Telescope,* translated by Bernard Pagel, edited and with an intro. by David S. Evans (New York, 1970); Agnes M. Clerke, "Sir J. F. W. Herschel," in *Dictionary of National Biography,* IX (Oxford, 1921), 714; David S. Evans, Terence J. Deeming, Betty Hall Evans, and Stephen Goldfarb, eds., *Herschel at the Cape* (Austin, Tex.–London, 1969); and *Sir John Herschel at the Cape, 1834–1838,* special issue of *Quarterly Bulletin of the South African Library,* **12,** no. 3 (Dec. 1957), which contains an intro. by D. H. Varley; "Sir John Herschel, 1792–1871," by R. H. Stoy; "The Astronomical Work of Sir John Herschel at the Cape," by David S. Evans; and "Sir John Herschel's Contribution to Educational Developments at the Cape of Good Hope," by E. G. Pells.

DAVID S. EVANS

HERSCHEL, WILLIAM (*b.* Hannover, Germany, 15 November 1738; *d.* Observatory House, Slough, Buckinghamshire, England, 25 August 1822), *astronomy.*

Friedrich Wilhelm (William) Herschel was the third of the six surviving children of Isaac Herschel. In 1753 William Herschel joined his father's regimental band as oboist and in 1756 traveled with the band to England, where he learned the language and established musical contacts that were to prove invaluable: in 1757 Hannover was occupied by the French after their defeat of the duke of Cumberland's army, and Herschel, who because of his youth had not been formally enlisted, escaped to England with his brother Jacob.

In England, Herschel supported himself, first by copying music and later by teaching, performing, conducting, and composing. In 1766 he settled at Bath after being appointed organist to the fashionable Octagon Chapel there. By this time Herschel's inquiring mind had moved from the practice of music to its theoretical study in Robert Smith's *Harmonics,* and from there to Smith's *Opticks,* with its extensive account of the construction of telescopes and its summary of the wonders of the heavens. By 1772, when Herschel brought his sister Caroline to England, he was becoming obsessed with astronomy; and over the next decade these interests encroached increasingly on his busy sequence of musical engagements. In 1773 he was hiring telescopes and assembling others from component parts, and in September of that year he bought some secondhand equipment and began to grind his own mirrors. Throughout the rest of his astronomical career Herschel used reflecting telescopes, which avoided the problems of chromatic and spherical aberration and offered the possibility of almost indefinite increase in size.

This possibility was to become of prime importance when, after a few years of desultory observations, he directed his efforts toward understanding "the construction of the heavens," the nature and distribution of distant stars and nebulae, rather than to the study of the nearby members of the solar system, which preoccupied most astronomers of the day. He seems quickly to have realized that in order to investigate very distant (and therefore faint) objects, he would need telescopes with considerable light-gathering power, for a telescope directed to a faint object must not only magnify it but also collect enough light for the magnified image to be visible to the observer. As he put it in 1800, light-gathering power is "the power of penetrating into space." His need was therefore for reflectors with large mirrors; and as his ambitions grew, he found himself forced to undertake an increasing share of the labor of construction himself. In the grinding and polishing of large mirrors, and in the working of exquisite eyepieces, Herschel was soon without peer; and when in 1782 one of his telescopes was taken to the Royal Observatory for comparison with the instruments there, Nevil Maskelyne, the astronomer royal, conceded superiority to Herschel. For the rest of his life Herschel enjoyed the possession of telescopes which were incomparably the most powerful of the period for the study of faint objects, although he never attempted the carefully mounted and graduated instruments required for exact positional astronomy.

By 1779 Herschel had undertaken his first review of the heavens, in which he examined stars down to the fourth magnitude. In August of that year he began a second review, more systematic and extensive than the first, and concentrated (for reasons discussed below) on the discovery of double stars. On 13 March 1781, during this review, he encountered an object which his experienced eye could tell at a glance was not an ordinary star. Yet it was not one of the planets known since the dawn of history, and Herschel supposed it to be a comet. His "Account of a Comet" was read to the Bath Philosophical Society, to which he had been introduced following a chance encounter with William Watson; Watson also communicated the paper to the Royal Society. It is proof of the lead which Herschel had established between himself and other observers that he could recognize the unusual nature of the object at a glance, while they could identify it only by its slow movement relative to neighboring stars. Examination of its orbit by other astronomers showed that the object was actually a primary planet of the solar system; Herschel called it Georgium Sidus to honor George III, but it became known by the more conventional name of Uranus, proposed by J. E. Bode.

The discovery of Uranus marked a turning point in Herschel's career. His scientific isolation had already been reduced by his membership in the Bath Philosophical Society and by the visits of Maskelyne and others. But now he was world-famous as the first recorded discoverer of a planet, even if not everyone would acknowledge the magnifications he claimed for his telescopes. He was awarded the Copley Medal of the Royal Society and elected to fellowship; in May 1782 he was received by the king, and on Watson's prompting he applied for and was granted a royal pension of £200 per annum. He was to live near Windsor Castle, and his only duty was to show the heavens to the royal family from time to time.

Herschel was now able to give up music and devote his full energies to astronomy. In August 1782 he and Caroline moved to Datchet, but the land surrounding the house was flooded when the Thames overflowed. In 1785 they rented a property in Old Windsor; the landlady proved a tyrant, and the following year they moved to Slough to live in what became known as Observatory House. It was there that Herschel spent the rest of his days.

Because Herschel's pension, supplemented in 1787 by £50 per annum for Caroline's services as his assistant, was barely sufficient for his needs, he manufactured a large number of telescopes for sale; but few of these were used for serious astronomy. The work did, however, allow him to make innumerable experiments on polishing by machinery. His own favorite telescope, completed in 1783, was a reflector with a mirror of eighteen-inch diameter and twenty-foot focal length, but he made two efforts to equip himself with a larger instrument. The first attempt had taken place in 1781, when Herschel set his heart on a reflector with a mirror of three-foot diameter and thirty-foot focal length. He intended, as usual, to polish the mirror himself; but he found on inquiry that even to cast the rough disk was beyond the capacities of the local foundries. Undaunted, he converted the basement of his house into a foundry and made many experiments with metals of different compositions. A mold was prepared from horse dung, and on 11 August Herschel and his brother Alexander "cast the great mirror"; but the mold leaked and the mirror cracked on cooling. On the second attempt, with an improved composition, the molten metal ran over the flagstones, and the brothers were lucky to escape with their lives.

In 1785 Herschel successfully requested the king to finance a fresh attempt to build a large telescope. "It remained now only to fix upon the size of it, and having proposed to the King either a 30 or a 40 feet telescope, His Majesty fixed upon the largest." Four years of labor followed for Herschel and his team of workmen, during which the original grant of £2,000 was doubled and an annual allowance of £200 was also made. The mirrors of forty-eight-inch diameter were cast in London, but all other work was carried out at Slough under Herschel's direction. In mounting the mirror in the tube Herschel tilted it slightly to one side so that the observer might peer through the eyepiece directly at the mirror, without the need for additional mirrors (the "Herschelian" arrangement). The monster telescope was completed in 1789 and immediately revealed a sixth satellite of Saturn. But it was never fully satisfactory: the mirrors tarnished quickly, the structure was cumbersome to turn, and when Herschel in 1790 altered his opinion of the nature of nebulae, he thereby answered the very question the telescope's great light-gathering power may have been intended to settle. Yet it became one of the wonders of the world and a visible testimony to Herschel's mechanical ingenuity and to the scale of his cosmological ambitions.

Herschel's second review of the sky, which extended to stars of the eighth magnitude and resulted in a first catalog (1782) of 269 double and multiple stars, had been concluded late in 1781; and he at once embarked on his third and most complete review, using a higher magnification and examining all Flamsteed's stars and thousands of others besides. This review was vigorously prosecuted once Herschel was established at Datchet. Released at last from his musical duties, "I employed myself now so intirely [*sic*] in astronomical observations, as not to miss a single hour of star-light weather, for which I used either to watch myself or to keep up somebody to watch; and my leisure hours in the day time were spent in preparing and improving telescopes" (*Scientific Papers*, J. L. E. Dreyer, ed., I [London, 1912], 37). The review was completed in January 1784 and resulted in a second catalog (1785) of 434 double and multiple stars.

But now Herschel's interests were changing. There was for the present little more he could contribute to the study of double stars, and in December 1781 Watson had aroused his curiosity in the milky patches in the sky known as nebulae, by presenting him with a newly published catalog of nebulae by Charles Messier. Herschel realized that he was the privileged possessor of the most powerful instruments for the study of these mysterious objects; but Messier's catalog, even when later extended, listed little more than 100 nebulae, and Herschel decided he must again play the natural historian. And so, in October 1783, with his newly completed twenty-foot reflector, he embarked on an intensive twenty-year program of "sweeping" for nebulae and eventually raised the total of those known to 2,500. Even the move from Old Windsor to Slough was achieved without the loss of a single night's viewing.

As a professional musician, and then as an astronomer, Herschel had always thrown himself into his work with single-mindedness; there is no hint that he contemplated marriage until after the death in 1786 of a neighbor, John Pitt. Early in 1788 Herschel became engaged to Pitt's widow, Mary, and they were married in May. Herschel was nearly fifty years old. To Caroline, who had been William's constant companion in his astronomical investigations, this displacement in her brother's affections came as a bitter blow; but in time she was won over by Mary's kindliness. Herschel's daily routine continued as before, but with more frequent holidays and with reduced financial anxiety. In 1792 his only son, John Frederick William, was born. As the years passed, Herschel began to feel the physical effects of his years of unremitting toil; it was not until 30 September 1802, when Herschel was sixty-three, that the program of sweeps came to an end. In 1808 he was desperately ill; but although he never recovered his full health, he continued in his remaining years to struggle with the polishing of mirrors, to carry on his observations, and to develop his cosmogonical theories. He was knighted in 1816, and that year John left Cambridge to become his father's assistant; in this way the aged astronomer's unrivaled experience was transmitted to the next generation. Herschel died peacefully at his Slough home in 1822.

Astronomy of the Nearby Stars. The fundamental problem in sidereal astronomy in the eighteenth century was to determine the distances of stars, for this knowledge was basic to a study of their distribution in three-dimensional space. It had long been realized that the stars are self-luminous, like the sun; but all attempts to make trigonometric determinations of their distances from measurements of the apparent annual movements of stars which reflect the actual movement of earth about the sun had failed: claims to have measured these apparent movements had always proved mistaken. Yet it was clear from this failure that the apparent movements were minute and that the distances of even the nearest stars were correspondingly vast, and estimates of these distances had been made on the hypothesis that the sun and the stars are equally bright in themselves and differ in appearance only because of their different distances from the observer. Throughout his career Herschel made use of this hypothesis, which had the important merit of offering a method that might well prove practicable not only for the nearest stars but also for objects so distant that there was little hope of ever detecting their apparent annual movements.

But in his early years as an astronomer Herschel had hoped it would be possible for him to make trigonometric determinations of distances and so avoid the need for this assumption; and since he lacked the accurately positioned and graduated instruments normally thought necessary, he followed a suggestion of Galileo and other observers that a watch be kept on pairs of stars very close to each other in the sky in the expectation that the fainter member of each such "double star" would be so distant as to be, for practical purposes, a fixed point from which the apparent annual movements of the brighter (and presumably less distant) member might be measured. It was for this reason that Herschel's second sky survey was mainly in search of double stars, and his catalogs of double stars (1782, 1785, 1821) list 848 examples.

The double-star method of determining stellar distances depended upon considering the two stars' proximity in the telescope as the optical effect of a chance alignment, and it would fail if the two stars were companions in space and thus equidistant from earth. John Michell had pointed out in 1767 that since the number of double stars in the sky was too great for chance alignments to be the usual explanation, most of them must be physical companions; and in 1783 he repeated this with explicit reference to Herschel's work. And so it proved to be. In 1802 Herschel began to reexamine his doubles, and he found that in several of them the two stars had altered position relative to each other in a way that showed they were companions held together by attractive powers. After Herschel's death it was confirmed that the power was, as expected, gravitational attraction, the first proof that gravitational attraction extended beyond the solar system. It is notable that Herschel ignored the implication of his own discovery, for in some cases the two companions were of different apparent brightness although at the same distance from earth; in other words, the differences in brightness were attributable to the stars themselves, in contradiction to the hypothesis that the stars are equally bright, on which rested Herschel's chief hopes of investigating stellar distances. Herschel was not prepared to abandon the hypothesis despite this conclusive evidence to the contrary, and indeed his career in sidereal astronomy can be seen as a prolonged rearguard action in defense of the hypothesis in the face of ever increasing counter evidence.

This hypothesis did indeed offer a theoretical solution to the problem of stellar distances, but it led to practical difficulties: how was one quantitatively to compare the apparent brightnesses from which distances were to be inferred? Surprisingly, an approximate technique for comparing the sun's great brightness with that of stars had been established in 1668 by James Gregory, and the real difficulty lay in the comparison of one star with another. Newton and others had avoided the problem by making the additional assumption that the traditional magnitudes directly represented relative distances, so that a star of, for example, the sixth magnitude was six times further than a star of the first magnitude. Although this assumption could be tested against the plausibility of the resulting stellar distribution, Herschel needed it for most of his career and shrank from exposing it to the potentially destructive test. But in 1817 he developed a method of comparing the light of two stars whereby he directed similar telescopes at each star and masked the aperture of one telescope until the two stars appeared to be of equal brightness; a comparison of the apertures could then be translated into a comparison of the apparent brightnesses. Herschel could at last risk the destruction of his now redundant magnitudes hypothesis, and he easily showed that it led to an absurd stellar distribution.

One notable contribution by Herschel to the study of the stars in the vicinity of earth owed nothing to his skill as an observer and depended entirely on data freely available: his investigations of the motion of the sun and solar system through space. In 1783 Herschel published an analysis of the "proper" or individual motions of a handful of stars as listed by Maskelyne, showing that if the sun was assumed to be moving toward a point in the constellation Hercules, most of these proper motions would be explained as apparent rather than real, reflecting the movement of the observer rather as the daily movement of the stars reflects the daily rotation of the observer on earth. His result is remarkably close to modern estimates.

In 1805 and 1806 Herschel returned to the problem, this time intending to investigate not only the direction but also the velocity of solar motion. To estimate the velocity he needed to know the velocities of neighboring stars relative to the sun (these velocities to be regarded as wholly or partly a reflection of the solar velocity); and he could arrive at some knowledge of the relative velocities of these stars only by taking for each its observed proper motion (angular velocity) and multiplying it by the star's distance. Since Herschel determined the distances of stars from their apparent brightness and for this used his customary hypothesis (which in fact is not even approximately true), he obtained highly disparate results; and only with difficulty did he propose a velocity which reduced this disparity to a minimum. Even so, he was forced to explain why some bright stars appear to have no proper motion and to be at rest relative to the sun despite their presumed nearness: Herschel claimed that they were moving in company with the sun, although this resulted in his assigning these additional motions as the result of an investigation originally intended to have the opposite result, reducing the number of proper motions of stars by showing them to be optical effects only. The complexity of his two-part study defeated his contemporaries, who were unable to unravel his confused argument.

Aside from his catalogs of double stars, Herschel's main contribution to the natural history of the sun's neighbors in space took the form of catalogs of the comparative brightness of stars (1796–1799), in which each star of such a catalog was placed within a delicate sequence of stars ordered in decreasing brightness. By this means a star which was later observed to be no longer in its correct position in the sequence could be identified as variable. Evidence concerning variable stars was desirable because these variations presented a puzzle to which different physical solutions had been proposed, and several of Herschel's early papers discuss particular variable stars. In 1796 he showed from his catalogs that α Herculis varied with a period of about sixty days and so was intermediate between the very long period and very short period variable stars already known. In the same paper he joined himself to those who explained all variations as due to the rotations of stars on their axes.

The Construction of the Heavens. Herschel's most important single achievement in astronomy consisted in his development of far-reaching theories of "the construction of the heavens." Newton, Bentley, Halley, and Loys de Chéseaux had discussed whether the number of stars is finite or not; questions had been asked as to the nature of the nebulae; Thomas Wright, Kant, and Lambert had offered explanations of the Milky Way; and Kant and Lambert had speculated about higher-order systems of stars; but the necessary basis of observational evidence was almost totally lacking. It was Herschel's achievement to assemble a mass of evidence and make it the basis for bold theorizing, thus founding observational cosmology.

Galileo had confirmed earlier speculations that the Milky Way itself is composed of great numbers of stars whose light merges to give the milky appearance; and by 1781, with the publication of Messier's second catalog, the number of known nebulae or milky patches in the sky had risen to just over 100. Herschel, as we have seen, increased this number to 2,500 by twenty years (1783–1802) of systematic "sweeping"; but even before his sweeping began, the gift in 1781 of Messier's earlier catalog had aroused his interest in the problem of the nature of nebulae. According to one view, each nebula was simply a star cluster, the light of the innumerable stars merging to give the milky appearance; according to the other view, some nebulae were star clusters but others were truly nebulous and formed of a self-luminous fluid. Because of the unrivaled light-gathering power of his big telescopes, Herschel was better placed than any other astronomer to decide this question; and he quickly "saw, with the greatest pleasure, that most of the nebulae . . . yielded to the force of my light and power, and were resolved into stars." In fact Herschel had succeeded in resolving some nebulae and had convinced himself that others were "resolvable" and would be resolved with a larger telescope; from this he quickly generalized and claimed that all nebulae are star clusters.

As he considered his growing collections of nebulae, Herschel not surprisingly saw these alleged clusters as evidence of the continuing activity of clustering or attractive powers; and in three important papers (1784, 1785, 1789) he developed a cosmogony in which the universe began with stars scattered throughout infinite space: with the passage of time, and under the action of these attractive powers, the stars began to condense toward regions where their initial density had been above average; and with the further passage of time the loose, large associations of stars which had been formed gave way to fragmented and tightly packed clusters. Herschel considered that groups of nebulae which he discovered represented the fragments of larger associations of stars, and he took a similar view of the star clusters of the Milky Way.

In the 1784 and 1785 papers Herschel also inaugurated the scientific study of the Milky Way. Whether he was then aware of previous speculations concerning the structure of the Milky Way star system is uncertain. Thomas Wright of Durham in 1750 had correctly suggested, in *An Original Theory or New Hypothesis of the Universe,* that the appearance of the Milky Way as a zone of light encircling the sky is the result of our immersion in what approximates (but for Wright, only locally) to a flat layer of stars, the milkiness appearing when—and only when—we look out along the layer; but Wright had then grafted this insight onto his fundamental belief that the stars are symmetrically arranged about a supernatural center. Herschel may have encountered Wright's book during his years (1760–1766) as a performer and music teacher in the north of England; it is unlikely that by 1784, when he published his own first study of the Milky Way, he was aware of the view of Kant and Lambert that the Milky Way star system to which the sun belongs not only approximates locally to, but actually is, a flat, finite layer of stars.

Herschel adopted a similar view but went on to ask how he might chart the outline of this layer. Obviously he must first assume that his telescopes could reach the stars at the borders of our system in every direction, since otherwise the task was impossible. To go on to determine the distance to the border in any given direction, Herschel could in principle have examined the faintest star in that region and, on his customary assumption that differences in apparent brightness are entirely the result of differences in distance, calculated the distance of the star from its apparent brightness. But the measurements involved were impracticable; and instead Herschel made use of information which was actually accessible to him, the different numbers of stars visible in different fields of view. Making the assumption that within the borders of our star system the stars are spread out regularly, he included in his early sweeps a number of "gages" or star counts and with a simple mathematical formula interpreted high counts as evidence of large distances to the border in the relevant directions. Time was precious and he carried out this program only for a great circle in the sky, but the map of a cross section of our star system which resulted (1785) was proof of the power of this new technique of stellar statistics.

In later years Herschel's preoccupation with star clusters brought home to him the gulf between the observational evidence and his assumption of uniform distribution, and the completion of the forty-foot telescope revealed stars which had been inaccessible in his earlier instruments. He therefore had to abandon each of his two assumptions and the map based on them, and he admitted that in some directions at least the star system seemed "fathomless."

Herschel's identification of nebulae with star clusters provided the foundation for his theory of the construction of the heavens published in the 1780's;

for if stars could not be detected in a particular nebula, it must be because of the great distance of the nebula. In this way Herschel believed himself to have some understanding of the distances, and therefore the sizes, of nebulae; some nebulae, he believed, "may well outvie our milky-way in grandeur" (1785). He chose simply to ignore evidence of changes in certain nebulae which by their rapidity showed that the nebulae in question must be small, and therefore near. However, on 13 November 1790 he observed a nebula (NGC 1514) which he realized consisted of a central star surrounded by a luminous shell which could not be composed of stars; he admitted the existence of "true nebulosity" in a paper published the following year.

It was no longer possible for Herschel to discuss with confidence the construction of the heavens. An "unresolved" nebula might be near, small, and nebulous; or it might be a distant, vast star system. But he was able to adapt to the new evidence his earlier account of the life history of a star cluster and thus add further emphasis to the temporal element in his theorizing. Believing NGC 1514 to represent a star condensing out of the luminous matter, he published in 1811 and 1814 a theory of the development of a star cluster, beginning with "extensive diffused nebulosity" which gradually condenses into stars which in turn cluster together ever more tightly. Illustrated with many examples at every stage, these papers showed brilliantly how dynamic changes can be inferred from virtually static evidence; and Herschel concluded by characterizing the Milky Way in its present stage of dissolution as "this mysterious chronometer."

The Solar System. Although Herschel's contributions to the study of the sun, moon, planets, and comets are less significant than his investigations of the sidereal universe, he constantly interrupted his observing programs to examine these nearby objects, and nearly half of his published papers are devoted to the solar system. Herschel was no mathematician and he could not advance the mathematical analysis of planetary motions on Newtonian principles, nor did his instruments have the precision necessary for positional astronomy; but his skill as an observer and the excellence of his telescopes enabled him to contribute to the knowledge of the physical constitution of most of the principal members of the solar system.

The Sun. Herschel's interest in the sun was naturally stimulated by the realization that, of all the stars, it alone is close enough for detailed examination. He was aware of the various existing theories of the physical constitution of the sun. In a long paper published in 1795 he mentions some of them before listing his own observations and arguing that what we actually see is not the sun itself but its luminous atmosphere, which surrounds the planetlike body of the sun. Mountains on the sun, which protrude through the luminous atmosphere as dark spots, are occasionally glimpsed as sunspots. He claims that rays from the sun's atmosphere produce heat only when they act upon "a calorific medium"—which is why mountaintops on earth are cold—and so the sun itself can and does support life, and by analogy the same is true of the other stars.

In 1801, in a second long paper in which he arranged his observations according to relevant physical questions, he modified his earlier account of the sun to include in its constitution an interior layer of dark clouds not unlike our own, this layer serving to shield the solar inhabitants from the exterior, luminous layer.

By this time Herschel had extensive experience of the use of "various combinations of differently-coloured darkening glasses" in observing the sun. "What appeared remarkable," he wrote in March 1800, "was, that when I used some of them, I felt a sensation of heat, though I had but little light; while others gave me much light, with scarce any sensation of heat." This suggested experiments with a prism and thermometers which showed that radiant heat is refrangible, but in such a way that its maximum is very different from the maximum of illumination: ". . . the full red falls still short of the maximum of heat; which perhaps lies even a little beyond visible refraction." In hundreds of further experiments, Herschel confirmed the existence of invisible, infrared heat rays, and showed that heat, whether solar or terrestrial, obeys laws of reflection and refraction analogous to those of light.

In a less happy venture into the physics of light, Herschel devoted three papers (1807–1810) to investigating the cause of colored concentric rings ("Newton's rings"). Ignoring the explanation already given by Thomas Young whereby the rings result from interference between light waves, Herschel criticized Newton's theory and attempted one of his own. He brought down on his head a storm of criticism, and this may have been a cause of his poor health at this period.

The Moon. In the winter of 1779–1780 Herschel calculated the height of several lunar mountains by adapting the method of Galileo and others. This involved measuring the angular distances between the mountain and the boundary of the illuminated part of the moon, at the time when the sun's rays first reached the peak of the mountain. To make the delicate measurements Herschel used a bifilar micrometer, which he calibrated by applying it to known terrestrial objects. He concluded that the height of lunar mountains had been exaggerated and that "the generality do not exceed half a mile in their perpendicular elevation."

Herschel makes no secret of his belief that argument from analogy shows that there is "great probability, not to say almost absolute certainty," of the moon's being inhabited, and in 1787 he used analogy to interpret certain observations of three volcanoes on the moon. In the eclipse of 22 October 1790 he saw at least 150 "bright, red, luminous points," but for once refused to speculate on their cause.

Mercury. Transits of Mercury and Venus across the sun offer a means of determining the distance of the earth from the sun, a fundamental quantity in astronomy, and so the transit of Mercury on 9 November 1802 was carefully studied and timed by astronomers.

Herschel restricted himself simply to observing the appearance of the planet. He reported that "the whole disk of Mercury is as sharply defined as possible; there is not the least appearance of any atmospheric ring." He added that the planet offered a perfectly round outline, so that it was unlikely to be materially flattened at its poles.

Venus. In 1793 Herschel published a lengthy paper entitled "Observations of the Planet Venus," which was provoked by the extravagant claims made the previous year by J. H. Schröter. Schröter claimed to have observed on Venus mountains of immense height, and to have noticed that one cusp appeared blunt because of the shadow of a mountain.

In his paper Herschel demolished these claims by quoting from his own numerous observations going back to 1777 (when he had hoped to resolve a controversy over the period of rotation of the planet), and especially from the series he carried out in the spring of 1793 specifically to test Schröter's assertions. Herschel concluded by agreeing with other astronomers that Venus has a considerable atmosphere and (rightly) dismissed Schröter's mountains. He admitted that this atmosphere had defeated his attempts to determine the period of rotation of the planet (as it defeated astronomers until our own day), in contrast to the less cautious claims of Schröter and others to have established a period of about twenty-three hours.

Mars. The clearly defined markings on Mars enable the period of rotation of the planet to be determined with accuracy. Herschel's observations, published in 1781, gave by his calculations a value of 24^h 39^m 21.67^s, which closely confirmed earlier estimates; but they would have brought him within some three seconds of the modern value of 24^h 37^m 23^s if he had not neglected to apply certain corrections.

In 1784 Herschel published a lengthy paper on Mars, in which he reprinted numerous observations on the shape of the planet and on the polar regions to establish the inclination of its axis. He made an elaborate study of the white regions at the polar caps and concluded "that the bright polar spots are owing to the vivid reflection of light from frozen regions; and that the reduction of those spots is to be ascribed to their being exposed to the sun," a view which is still accepted today.

The Asteroids. Ceres, the first known celestial body in orbit between Mars and Jupiter, was discovered early in 1801, but it was soon lost in the glare of the sun and was not rediscovered until the end of the year.

Herschel first saw Ceres on 7 February 1802, and it was a week before he could detect a visible disk such as is characteristic of the appearance of a planet; evidently Ceres was very small. With his lucid-disk micrometer for comparing celestial bodies with lamps of controlled characteristics, he carried out careful observations of Ceres and then of the newly discovered Pallas, calculating their diameters to be under 200 miles and believing them to have considerable comae. As they differed so much from the known planets and comets, Herschel felt a different term would be appropriate for them, and proposed "asteroid" because even in a good telescope they resembled stars. He also forecast that more would be discovered, a prediction fulfilled with the discoveries of Juno in 1804 and Vesta in 1807.

Jupiter. Herschel's main contribution to the study of Jupiter took the form of a long paper, published in 1797, on the planet's four known satellites. As usual he began by reprinting his earlier observations. He then used his findings to argue that each satellite always returns to its original apparent brightness after every orbit around the parent planet, and that (like our moon) it will rotate on its axis in the time it takes to complete one orbit.

Saturn. Saturn exercised a special fascination for Herschel, and between 1789 and 1808 he devoted seven papers and part of an eighth to the planet, its ring, and its satellites.

On 19 August 1787 Herschel suspected he had found a sixth and previously unknown satellite, but he was not able to confirm this until 28 August 1789, when his forty-foot telescope came into commission. A few days later he found a seventh satellite. For some months he carefully tracked the satellites, establishing for Mimas and Enceladus periods within seconds of the modern values, and giving evidence to show that Iapetus rotates in its period of revolution.

He also made careful observations of the rings, which he believed to be solid. As the earth happened to be in the plane of the ring structure at the time, he compared the thickness of the ring when seen edge-on with the diameter of Jupiter's satellites; and although his estimate exceeds modern values, his method showed that the thickness did not exceed a few hundred miles.

Like other astronomers before and since, Herschel was puzzled by "luminous points" which he observed in the rings. He at first thought they were caused by irregularities in the surface of the rings, but changed his mind in 1789 when "one of these supposed luminous points was kind enough to venture off the edge of the ring, and appeared in the shape of a satellite." He tried to relate the observations of luminous spots to the movements of the known satellites, but found that numerous observations remained unaccounted-for. About twenty of these could be explained by the revolution in just over ten and a half hours of a satellite within the ring system, and in 1791 Herschel assigned this period to the rotation of the ring system (specifically to the outer ring); a result which, although substantially correct, seems to be based on illusory data.

In 1791 Herschel examined the dark region between the inner and outer rings to decide whether this region could be a genuine gap between the rings. He had already observed the southern face of the region, and now it was the northern face that was visible. Careful observations showed that the region appeared to be entirely uniform. Since (as he believed) the ring system was rotating rapidly, the evidence for uniformity was as complete as one could hope for, and he categorically asserted that a gap existed between two rings. But the conclusive test which he suggested, as to whether a star might on occasion be seen beyond and between the two rings, was not successfully made until long after his death. In the same paper he also gave micrometer measures of the breadth of each ring and of the gap.

Herschel's observations of the globe of Saturn were reported in six different papers. He discussed the belts and also the shadow of the rings (but overlooked the "Crepe Ring"), and he showed that the planet is compressed at the poles; this suggested Saturn was rotating. By a remarkably bold argument based on fluctuations in the appearance of the belts, he proposed in 1794 a rotation period of $10^h\ 16^m\ 0.4^s$, which is very close to the modern value of $10^h\ 14^m$.

Uranus. In addition to his account of the discovery of Uranus on 13 March 1781 and a letter naming the planet Georgium Sidus, Herschel published five papers on the planet and its satellites.

His search for possible satellites was not successful until 1787, when he adopted the "Herschelian" arrangement of tilting the telescopic mirror slightly in the tube and looking at it through the eyepiece directly. The resultant light-gain enabled him to discover two satellites; his determinations of the shape and size of their orbits are in close agreement with modern values. In 1788 he was able to give their synodic revolution periods, again in excellent agreement with modern values; and he also gave the first determination of the mass of the planet.

His two long papers of 1798 and 1815 were devoted almost entirely to satellites. In 1798 he made the astonishing announcement that the motion of the two known satellites of Uranus was retrograde. He also believed, as the result of numerous difficult observations, that he had discovered four additional satellites, but their existence has not been confirmed.

Comets. The discovery of comets was the prerogative of Caroline Herschel rather than her brother, but William did publish extended accounts of his observations of the great comets of 1807 and 1811, with a view to elucidating their physical nature. In discussing the 1811 comet, he suggested that such a comet had an atmosphere and, within this, nebulous matter gathered about the head of the comet. When the comet approaches the sun the nebulous matter is rarefied and suspended in the atmosphere, where it is exposed to the solar heat.

> . . . and if we suppose the attenuation and decomposition of this matter to be carried on till its particles are sufficiently minute to receive a slow motion from the impulse of the solar beams, then will they gradually recede from the hemisphere exposed to the sun, and ascend in a very moderately diverging direction towards the regions of the fixed stars.

Physical Speculations. In 1780 and 1781 Herschel read numerous papers to the Bath Philosophical Society, dealing with a variety of subjects including electricity and the nature of matter. Prompted by what he had read of the ideas of John Michell and R. J. Bošković in Joseph Priestley's *Disquisitions Relating to Matter and Spirit* (1777), Herschel supposed that each particle of matter was endowed with a system of centrally acting forces (unlike Bošković's theory of distinct zones of attractive and repulsive forces); he argued that phenomena such as the absorption and reflection of light were caused by the joint effect of the different forces. This is an extension of the Newtonian theory of atoms as surrounded by envelopes of forces.

BIBLIOGRAPHY

I. ORIGINAL WORKS. Herschel's publications consist of some seventy papers which appeared in the *Philosophical Transactions of the Royal Society* between 1780 and 1818, and one in *Memoirs of the Royal Astronomical Society,* **1** (1821). These are reprinted together with unpublished papers and biographical material in J. L. E. Dreyer, ed., *The Scientific Papers of Sir William Herschel,* 2 vols. (London, 1912). The other basic source for the historian is Constance A. Lubbock, *The Herschel Chronicle* (Cambridge, 1933), which includes many letters and private papers.

Most of Herschel's scientific manuscripts are in the library of the Royal Astronomical Society; this collection is described in J. L. E. Dreyer, "Descriptive Catalogue of a Collection of William Herschel Papers Presented to the Royal Astronomical Society by the Late Sir W. J. Herschel," in *Monthly Notices of the Royal Astronomical Society,* **78** (1917-1918), 547-554. The remaining books and papers were sold piecemeal at Sotheby's in 1958; some are now at the University of Texas at Austin and others in the Linda Hall Library, Kansas City.

II. SECONDARY LITERATURE. The most readable modern biography is J. B. Sidgwick, *William Herschel* (London, 1953); of the older biographies, that by Agnes M. Clerke (with bibliography) in *Dictionary of National Biography* is still of value.

A first introduction to Herschel's work is M. A. Hoskin, *William Herschel, Pioneer of Sidereal Astronomy* (London, 1959). Two careful surveys of all aspects of Herschel's achievement are Angus Armitage, *William Herschel* (London, 1962), and Günther Buttmann, *Wilhelm Herschel: Leben und Werk* (Stuttgart, 1961), with bibliography.

A valuable bibliography of the earlier works on Herschel is to be found in E. S. Holden, *Sir William Herschel* (London, 1881). Of these earlier works, the most important are D. J. F. Arago, "Analyse historique et critique de la vie et des travaux de Sir William Herschel," in *Annuaire publié par le Bureau des Longitudes* (1842), pp. 249-608; and F. G. W. Struve, *Études d'astronomie stellaire* (St. Petersburg, 1847), pp. 21-44.

On Herschel's debt to Galileo, see M. A. Hoskin, "Herschel and Galileo," in *Actes du XIe Congrès Internationale d'Histoire des Sciences,* **3** (Warsaw, 1968), 41-44. On Herschel's work in sidereal astronomy and observational cosmology, see M. A. Hoskin, *William Herschel and the Construction of the Heavens* (London, 1963), which reprints a selection of original papers, and A. I. Eremeeva, *Vselennaya Gershelya* (Moscow, 1966).

On Herschel's telescopes, see A. Mauer, "Die astronomischen Teleskope William Herschels," in *Orion,* **28** (1970), 5-8; V. L. Chenakal, "William Herschel's Mirror Telescopes in Russia," in *Istoriko-astronomicheskie issledovaniya,* **4** (1958), 253-340, in Russian; C. D. P. Davies, "Herschel's 18¾ in. Speculum (the '20 ft')," in *Monthly Notices of the Royal Astronomical Society,* **84** (1923-1924), 23-26; and W. H. Steavenson, "Some Eye-Pieces Made by Sir William Herschel," *ibid.,* 607-610.

On Herschel's work on infrared rays, see E. S. Cornell, "The Radiant Heat Spectrum From Herschel to Melloni. (1) The Work of Herschel and his Contemporaries," in *Annals of Science,* **3** (1938), 119-137; and D. J. Lovell, "Herschel's Dilemma in the Interpretation of Thermal Radiation," in *Isis,* **59** (1968), 46-60.

On the discovery of Uranus, see R. H. Austin, "Uranus Observed," in *British Journal for the History of Science,* **3** (1967), 275-284. On Herschel's theory of matter, see P. M. Heimann and J. E. McGuire, "Newtonian Forces and Lockean Powers: Concepts of Matter in Eighteenth-Century Thought," in *Historical Studies in the Physical Sciences,* **3** (1971).

M. A. HOSKIN

HERTWIG, KARL WILHELM THEODOR RICHARD VON (*b.* Friedberg, Germany, 23 September 1850; *d.* Schlederlohe, Germany, 3 October 1937), *biology.*

Hertwig was the son of Carl Hertwig, a merchant, and the former Elise Trapp. He was the younger and only brother of Oscar Hertwig, with whom he was educated and with whom he collaborated during his early years. The brothers began their university studies in 1868 under Ernst Haeckel at Jena, where they studied until 1871. In the autumn of 1872 Hertwig became lecturer in zoology at Jena, where his brother was made lecturer in anatomy and embryology. They became extraordinary professors in 1878 and three years later, the paths of their lives began to diverge when Hertwig went to Königsberg as professor while his brother remained at Jena. He held the same rank at Bonn in 1883 and at Munich in 1885, remaining at Munich until his retirement in 1925. He married Jula Braun in 1887; they had two sons and a daughter. Hertwig outlived his brother by many years and was active until the day before his death at the beginning of his eighty-eighth year.

Hertwig contributed to many fields of biology, both morphologically and experimentally. He was a protozoologist, an embryologist, and a cytologist. His earliest studies, begun under the influence of Haeckel, were in comparative morphology. In the late 1870's he published, together with his brother, works on the nervous system, the sense organs, and the musculature of various coelenterates. This work led them to theoretical considerations of the phylogenetic relationships of two-layered coelenterates to higher, three-layered animals. There was then much speculation as to the origins and significance of the mesoderm, both ontogenetically and phylogenetically; the two Hertwigs, in a series of studies, formulated their "coelom theory" to account for the classification and phylogeny of metazoan animals. The coelom is still used as an important taxonomic criterion.

During the next decade the two brothers collaborated on important contributions to experimental embryology. They initiated experimental studies on the chemical environment of eggs in relation to artificial hybridization, producing multipolar mitoses by the use of chemical agents. Highly important for later studies on the relative roles of cytoplasm and nucleus, their results showed that sea urchin eggs can be shaken into fragments and that both the nucleate and the nonnucleate fragments can be fertilized and can subsequently develop. During the last decade of the nineteenth century Hertwig demonstrated that sea urchin eggs, after treatment with weak solutions of strychnine, can form mitotic figures and begin to divide; this was the beginning of the studies on what Jacques Loeb soon called artificial parthenogenesis. In the 1890's Hertwig also made many studies on the cytology and life cycles of the Protozoa, particularly ciliates and heliozoans. He was especially interested in syngamy in the ciliates; and his studies on unicellular organisms, together with his wide knowledge of the zoology of higher animals, enabled him to formulate problems of broad general interest.

During the first decade of the twentieth century Hertwig emphasized the importance of maintaining constancy in the relative volumes of cytoplasm and of nucleus within the cell; when the cytoplasmic volume becomes excessive, according to his theory, the cell divides. He was led by his studies on protozoan life cycles to an interest in senescence, which he ascribed to a relative increase in nuclear volume. His studies on syngamy led also to an interest in sex and sex determination; he demonstrated that overripe frog eggs develop an excess of males, an early indication that genetic expression is subject to environmental influences. He observed in heliozoans that basophilic granules seemed to be given off by the nucleus; he called these chromidia and believed that at each mitosis they are discharged from the nucleus into the cytoplasm, to play an important role in development.

Hertwig was one of the most productive teachers in the history of zoology. His *Textbook of Zoology,* because of its broad outlook, was highly influential; but he also exerted great influence through his teaching. When he retired in 1925, 208 of his former students presented him with a testimonial; 117 of them were professors of zoology, many of them very well known.

BIBLIOGRAPHY

The principal references to obituaries of Richard Hertwig, to lists of his writings, and to evaluations of his work at the time of his sixtieth birthday are in R. Weissenberg, *Oscar Hertwig 1849–1922* (Leipzig, 1959), pp. 56–57. The articles written at the time of his seventieth birthday are in *Naturwissenschaften,* **8** (1920), 767–782. The other references given by Weissenberg are to reference works likely to be found in most sizable reference libraries.

JANE OPPENHEIMER

HERTWIG, WILHELM AUGUST OSCAR (*b.* Friedberg, Hessen, Germany, 21 April 1849; *d.* Berlin, Germany, 25 October 1922), *zoology.*

Hertwig was the elder son of Carl Hertwig and Elise Trapp. The family interest in science was keen, his father having been trained in chemistry under Liebig at Giessen. After the birth of Hertwig's brother Richard, the family moved to Mühlhausen in Thuringen. There Oscar and Richard were educated together, and on the advice of the Gymnasium head, Wilhelm Osterwald, who had taught Ernst Haeckel in Merseberg, they went to Jena and came under Haeckel's influence. In his brief but charming recollections of Haeckel, published in 1919, Oscar left no doubt as to the great debt he owed the Jena zoologist, on whose advice he forsook chemistry for medicine.

Hertwig spent from 1868 to 1888 at Jena with the exception of two short periods at Bonn. He became assistant professor of anatomy at Jena in 1878 and professor three years later. From 1888 until 1921 he occupied the first chair of cytology and embryology and directed the new Anatomical-Biological Institute. He became a member of the Leopoldina Academy in Jena and the Prussian Academy of Sciences in Berlin. Hertwig married Marie, the daughter of the teacher Wilhelm Gesenius, in 1884. They had two children, Günther (1888) and Paula (1889).

Hertwig's prize essay (Jena, 1871) and doctoral thesis (Bonn, 1872) were devoted to traditional developmental studies, but after reading Leopold Auerbach's *Organologischen Studien* (Breslau, 1874) he became deeply interested in the nature of the fertilization process. The views most widely held at that time were either that the spermatozoa make contact with the egg, thus stimulating its development by the transmission of a subtle mechanical vibration akin to the supposed action of a ferment (the contact theory advocated by G. W. Bischoff), or that they penetrate the egg and their chemical constituents become commingled with the egg yolk. Consequently, when Auerbach found two nuclei in the fertilized egg he assumed they had originated from the mixture of the chemical constituents of sperm and egg, and not from any previous nuclei.

Auerbach's work left Hertwig curious and dissatisfied. As an anatomist conscious of the role of organized structures in the cell, he expected to discover some structural continuity between the nuclear contents of the egg before and after fertilization. He was aware of the distinctive staining reactions of the nucleus and of the hereditary role that Haeckel had assigned it. He also knew that botanists had accepted Nathanael Pringsheim's observation of the penetration of the antherozoid into the oogonium in *Oedogonium.* At this juncture in Hertwig's career, Haeckel was about to take Richard Hertwig on a research trip to the Mediterranean. Hertwig rashly resigned his new assistantship at Bonn to join them, and it was on this trip that he found the sea urchin (*Toxopneustes lividus*) which so admirably suited his research purposes.

Because of its small size, finely divided yolk, and absence of any noticeable membrane, the sea urchin was remarkably transparent. Using fresh material and material fixed in acetic or osmic acid, Hertwig followed the fate of the egg and sperm nuclei with ammoniacal carmine solution. To his delight he was able to detect the presence of a remnant of the egg nucleus before and during entry of the spermatozoon and the fusion of both nuclei five to ten minutes later. Although he was mistaken in describing this nuclear structure in the egg as the nucleolus, he was able to state categorically that he had observed no breaking up of that nucleus, and hence could maintain the morphological continuity between it and the cleavage nuclei of the developing embryo. More important for the understanding of heredity, however, was his observation that only one spermatozoon is required to fertilize one egg. The entry of further spermatozoa is prevented by the formation of a vitelline membrane, which spreads around the surface of the egg starting from the cone of attraction at the point of entry of the spermatozoon.

In the winter of 1875 Hertwig wrote this work up for the *Habilitationsschrift,* which he defended the following November, when he was called upon to evaluate the statement: "The egg cell passes through no monera stage in its development." This theme was surely suggested by Haeckel, according to whom the egg passed through a nuclear-free or "monera" stage, in harmony with the principle of recapitulation. Hermann Fol and Edouard van Beneden also published papers that year on the subject of fertilization, but neither man, in interpreting his findings, freed himself completely from the chemical view in which nuclear continuity was denied. Hertwig, on the other hand, clung to this conception even after it had been superseded by the more correct theory of chromosome continuity. It was not until T. H. Morgan's work became well known in Germany that Hertwig fully accepted the modern version of this theory, which included the exchange of genes between chromosomes by crossing-over.

Hertwig returned to the subject of the egg in his famous paper, "Das Probleme der Befruchtung und der Isotropie des Eies, eine Theorie der Vererbung" (1885). Here Pflüger's evidence against the existence of a structural differentiation in the egg was rejected, and the profound influence of Naegeli's idioplasm theory on Hertwig is apparent. He called the Munich botanist's postulate of the equivalence of male and female germinal substances Naegeli's "Vererbungsaxiom," which for Hertwig became the equivalence of egg and sperm nuclei; the nuclein or chromatin (he early considered them identical) was then both the fertilizing substance and the idioplasm.

This acceptance of the genetic primacy of the nucleus contrasted with the more cautious and critical attitudes of Eduard Strasburger, Wilhelm Waldeyer, and Max Verworn. Yet Hertwig's writings were more restrained than those of August Weismann, whose uninhibited speculations Hertwig termed "weismannische Naturphilosophie." He opposed Weismann's doctrine of differentiation by selective loss of idioplasm and instead asserted the correct doctrine of the genetic equivalence of all body cells. Weismann's doctrine of ancestral germ plasms or "ids" was objectionable to Hertwig; it involved unnecessary speculation. Hertwig thought it was enough to consider the problem of increase of nuclear mass—or as Naegeli had conceived it, increase of idioplasm—which repeated fertilizations would cause. The essential function of polar bodies was therefore to remove half the nuclear substance, but not to make the resulting eggs qualitatively different. If for every character more than one determinant exists, then an equal division of the entire egg nucleus will in no way alter the hereditary constitution of the several eggs produced. These polar bodies were to be looked upon as relics of what in earlier evolutionary history were functional eggs. In 1890 Hertwig showed that spermatogenesis is equivalent to oogenesis and involves the formation from one sperm mother cell of a tetrad of spermatozoa, but unlike the egg "tetrad" all four products of the sperm tetrad are functional.

For many years Hertwig was an opponent of Weismann's doctrine that reduction division leads to qualitative differences between gametes and of Carl Rabl's and Theodor Boveri's insistence on the individuality of the chromosomes. Nevertheless, he was a lifelong exponent of the study of heredity through cytology and in 1909, at the age of sixty, with his son and daughter, he began to study the biological effects of the irradiation of eggs, spermatozoa, and embyros with X rays.

Hertwig wrote a series of papers on the germ layer theory with his brother Richard. They began by questioning the so-called specificity of the germ layers in a paper (1878) that applied this theory to the medusae. Four years earlier Haeckel had published his gastraea theory, according to which all organisms from sponges upward pass through a stage in which the embryo consists of two layers, the inner formed by invagination of the outer layer. These two layers had been recognized in medusae by T. H. Huxley in 1849. The Hertwigs rightly questioned the evidence for the existence of a third germ layer, the mesoderm, and assigned such tissues either to ectoderm or endoderm.

In their *Studien zur Blattertheorie,* parts 2 and 4 (1881, 1882), the Hertwigs tried to eliminate the confusion over the origin of mesoderm and the relation of the body cavity of vertebrates to the archenteron of the gastrula. They thought to find a universal distinction between structures arising from cells detached from existing germ layers, which they termed mesenchyme, and those arising by invagination of the endoderm, the true mesoderm. The body cavity in higher organisms, they asserted, is secondary in origin, being formed entirely by such invagination and not partly by cavity formation within solid tissue. The cells of the invaginated layer thus surround this cavity or "coelom."

In the search for phylogenetic relationships and homologies between diploblastic and triploblastic animals, the coelom theory was useful at a time when the prevailing fashion in embryology was phyletic and descriptive. It failed in that it modified one rigid scheme—the gastraea theory—by introducing another and thus did not stimulate experimental embryology.

Hertwig traveled little, rarely attended international meetings, and was known chiefly through his extensive writings. His most popular textbooks were *Lehrbuch der Entwicklungsgeschichte des Menschen und der Wirbeltiere* (1886–1880) and *Die Zelle und die Gewebe* (1893–1898), now known as the *Allgemeine Biologie,* after the title of the second edition (1906).

Hertwig did not succumb to the beguiling neatness and finality of Haeckel's biogenetic law. While Haeckel stimulated him, Naegeli profoundly influenced him. The Lamarckian, anti-Darwinian, and morphological-chemical features of Naegeli's work were all echoed by Hertwig. It was Hertwig who wrote in approval of Paul Kammerer's work, who vehemently attacked the concept of natural selection in *Das Werden der Organismen* (1916), and who upheld the chemical study of the cell while belittling the biological significance of Emil Fischer's work. Hertwig believed that proteins undergo a series of postmortem changes before they can be characterized by the techniques of the organic chemists. To learn about the proteins of the living cell the chemist must adopt more comprehensive methods and aims. He must become a biologist and above all a morphologist.

Hertwig's exposure to chemistry had been slight. His early success in routing the chemical theory of fertilization and in reestablishing the morphological conception in terms of nuclear continuity no doubt set him in the direction of histological rather than histochemical studies of the cell. He paid lip service to chemistry but never appreciated the distinction between nucleic acid and nucleoprotein; consequently, he was content to equate chromatin with nuclein and to call it a protein.

In his retrospective essays, including his magnificent "Dokumente zur Geschichte der Zeugungslehre" (1918), Hertwig emphasized the conceptual advance made in 1875 when the process of fertilization and the transfer of hereditary material became firmly associated in the nuclear theory of sexual reproduction. The demonstration of nuclear continuity and conjugation in his *Habilitationsschrift* was decisive in establishing this theory and was surely his greatest contribution to science.

BIBLIOGRAPHY

I. Original Works. A bibliography with 121 entries is included in Weissenberg's biography (see below). Hertwig wrote seven books: *Lehrbuch der Entwicklungsgeschichte des Menschen und der Wirbeltiere,* 2 vols. (Jena, 1886–1888; 10th ed., 1915); French ed. (Paris, 1891); English ed. (London, 1892); Italian ed. (Milan, 1894); *Die Zelle und die Gewebe,* 2 vols. (Jena, 1893–1898); 2nd ed. entitled *Allgemeine Biologie* (Jena, 1906; 7th ed., 1923); *Zeit- und Streitfragen der Biologie* (Jena, 1894); English ed. (London, 1896); *Die Elemente der Entwicklungslehre des Menschen und der Wirbeltiere. Anleitung und Repetitorium für Studierende und Aertze* (Jena, 1900; 6th ed., 1920); *Das Werden der Organismen. Eine Widerlegung von Darwin's Zufallstheorie* (Jena, 1916; 3rd ed., 1922); *Zur Abwehr des ethischen, des sozialen, des politischen Darwinismus* (Jena, 1918); and *Der Staat als Organismus. Gedanken zur Entwicklung der Menschheit* (Jena, 1922). Hertwig edited and contributed to *Handbuch der vergleichenden und experimentellen Entwicklungslehre der Wirbeltiere,* 3 vols. (Jena, 1901–1906).

His important papers on fertilization are "Beiträge zur Kenntniss der Bildung, Befruchtung und Theilung des thierischen Eies," in *Morphologisches Jahrbuch:* I, **1** (1876), 347–434; II, **3** (1877), 1–86; "Weitere Beiträge . . .," II, **3,** 271–279. His embryological work, with special reference to the germ layer theory, the coelom theory, and fertilization, are in *Zeitschrift für Naturwissenschaften* (Jena): "Ueber das Nervensystem und die Sinnesorgane der Medusen," **11** (1877), 355–374; "Ueber die Entwicklung des mittleren Keimblattes der Wirbelthiere," **15** (1882), 286–340; **16** (1883), 247–328; "Welchen Einfluss übt die Schwerkraft auf die Theilung der Zellen?," **18** (1885), 175–205; and "Das Problem der Befruchtung und der Isotropie des Eies, eine Theorie der Vererbung," **18** (1885), 276–318.

Among the papers he published in the *Archiv für mikroskopische Anatomie und Entwicklungsmechanik* are the following: "Vergleich der Ei- und Samenbildung bei Nematoden. Eine Grundlage für celluläre Streitfragen," **36** (1890), 1–138; "Weitere Versuche über den Einfluss der Zentrifugalkraft auf die Entwicklung thierischer Eier," **63** (1904), 643–657; "Dokumente zur Geschichte der Zeugungslehre. Eine historische Studie als Abschluss eigener Forschung," **90**, sec. 2 (1918), 1–168. His last three papers appeared in the *Deutsche medizinische Wochenschrift:* "Zur Erinnerung an Ernst Haeckel," **45** (1919), 1031; "Der jetzige Stand der Lehre von den Chromosomen," **48** (1922), 9–10; and "Die Erblichkeitslehre, ihre Geschichte und Bedeutung für die Gegenwart," **48** (1922), 1239–1240.

His researches in the use of radiation and chemical agents in the study of heredity and embryology were published mostly in the *Sitzungsberichte der Preussischen Akademie der Wissenschaften zu Berlin,* Math.-nat. Klasse. These include "Die Radiumstrahlung in ihrer Wirkung auf die Entwicklung tierische Eier" (1910), 221–233; "Neue Untersuchungen über die Wirkung der Radium-strahlung auf die Entwicklung tierischer Eier" (1910), 751–771; "Mesotherium versuche an thierischen Keimzellen, ein experimenteller Beweis für die Idioplasmanatur der Kernsubstanzen" (1911), 844–873; "Veränderung der idioplasmatischen Beschaffenheit der Samenfäden durch physikalische und chemische Eingriffe" (1912), 554–571; "Keimesschädigung durch chemische Eingriffe" (1913), 564–582; "Die Verwendung radioaktive Substanzen zur Zerstörung lebender Gewebe" (1914), 894–904. Hertwig's history of the Anatomical-Biological Institute in Berlin is in Max Lenz, *Geschichte der Königlich Friedrich-Wilhelms-Universität zu Berlin,* III (Halle, 1910), 141–154.

II. Secondary Literature. The best account of Hertwig's life and work is Richard Weissenberg, *Oskar Hertwig 1849–1922. Leben und Werk eines deutschen Biologen,* no. 7 in the series Lebensdarstellungen deutscher Naturforscher (Leipzig, 1959); the excellent bibliography by Rudolph Zaunick for this work gives a generous list of obituary notices, of which the best is by Franz Keibel in *Anatomischer Anzeiger,* **56** (1923), 372–383. In addition, Keibel's valuable discussion of Haeckel's views on the theory of recapitulation is available in "Haeckels biogenetisches Grundgesetz und das ontogenetische Kausalgesetz von Oskar Hertwig," in *Deutsche medizinische Wochenschrift,* **37** (1911), 170–172. An overly critical view of the Hertwig brothers' contribution to the germ layer theory is in J. M. Oppenheimer, *Essays in the History of Embryology and Biology* (Cambridge, Mass., 1967).

For an appreciation of the Hertwigs' fruitful suggestions in experimentation, see Fritz Balzer, *Theodor Boveri: Life and Work of a Great Biologist,* D. Rudnick, trans. (Berkeley–Los Angeles, 1967). The most authoritative treatment of Hertwig's standpoint on cytological questions is Frederick Churchill, "Hertwig, Weismann, and the Meaning of Reduction Division Circa 1890," in *Isis,* **61** (1970), 429–457.

Robert Olby

HERTZ, HEINRICH RUDOLF (*b.* Hamburg, Germany, 22 February 1857; *d.* Bonn, Germany, 1 January 1894), *physics.*

Hertz was born into a prosperous and cultured Hanseatic family. His father, Gustav F. Hertz, was a barrister and later a senator. His mother was the former Anna Elisabeth Pfefferkorn. He had three younger brothers and one younger sister. Hertz was Lutheran, although his father's family was Jewish (Philipp Lenard, Hertz's first and only assistant and afterward a fervent Nazi, conceded that one of Germany's great men of science had "Jewish blood"). At age six Hertz entered the private school of Richard Lange, a taskmaster who had no patience with error. His mother watched closely over his lessons, determined that he should be—as he was—first in his class. On Sundays he went to the *Gewerbeschule* for lessons in geometrical drawing. His skill in sketching and painting marked the limit of his artistic talent; he was totally unmusical. Very early Hertz showed a practical bent; at age twelve he had a workbench and woodworking tools. Later he acquired a lathe and with it made spectral and other physical apparatus. He had an uncommon gift for languages, both modern and ancient. He left Lange's school at fifteen to enter the Johanneum Gymnasium, where he was first in his class in Greek; at the same time he took private lessons in Arabic.

After his *Abitur* in 1875 Hertz went to Frankfurt to prepare for a career in engineering. He spent his year of practical experience there in construction bureaus, reading during his free hours for the state examination in engineering. After a short spell in 1876 at the Dresden Polytechnic, he put in his year of military service in 1876–1877 with the railway regiment in Berlin. He then moved to Munich in 1877 with the intention of studying further at the Technische Hochschule there. Since his Gymnasium days, however, he had had conflicting leanings toward natural science and engineering. While preparing for engineering he had regularly studied mathematics and natural science on the side. With his father's approval and promise of continuing financial support, he matriculated in 1877 at the University of Munich instead of at the Technische Hochschule. He was relieved at having decided on an academic and scientific career after long vacillation and was confident that he had decided rightly. To him engineering meant business, data, formulas—an ordinary life, on a par with bookbinding or woodworking—and he was uninterested. Although the Technische Hochschule had a good physics laboratory, a course of study there led to state examinations and usually to a practical career. The university by contrast promised a life of never-ending study and research, one that suited Hertz's scholarly, idealistic tastes; he knew that he wanted above all to be a great investigator.

Hertz spent his first semester at the University of Munich studying mathematics. Following the advice of P. G. von Jolly, he read Lagrange, Laplace, and Poisson, learning mathematics and mechanics in their historical development and deepening his identification with investigators of the past. Elliptic functions and the other parts of the newer mathematics he found overly abstract, believing that they would be of no use to the physicist. Although Hertz thought that, when properly grasped, everything in nature is mathematical, he was in his student days—as throughout his career—interested primarily in physical and only indirectly in mathematical problems. It was in these first months in Munich that he developed his strong, if not strongly original, mathematical talent. It was expected at this time that an intending physicist have a grounding in experimental practice as well as in mathematics, and accordingly Hertz spent his second semester at Munich in Jolly's laboratory at the university and in F. W. von Beetz's laboratory at the Technische Hochschule. He found the laboratory experience immensely satisfying, especially after his intensive mathematical studies; it was to be a lifelong pattern with him to alternate between predominantly experimental and predominantly theoretical studies. In Germany in the 1870's the ideal physicist was expected to be equally at home with mathematics and apparatus; by temperament and talent Hertz embodied the ideal.

After a year in Munich, Hertz was eager to make the customary student migration. In consultation with Beetz he decided against Leipzig and Bonn in favor of Berlin. It was a momentous decision, for it brought him together with Hermann von Helmholtz, who was to have a profound influence on him throughout his career. Immediately upon arriving in Berlin in 1878, Hertz was drawn into Helmholtz' circle of interests; he noticed an announcement of a prize offered by the Berlin Philosophical Faculty for the solution of an experimental problem concerning electrical inertia. Although he had had only one year of university study, he wanted to begin original research and try for the prize. Helmholtz, who had proposed the problem and had great interest in its solution, provided Hertz with a room in his Physical Institute, directed him to literature on the problem, and paid daily attention to his progress.

Outside the laboratory Hertz attended Kirchhoff's lectures on theoretical physics but found little new in them. He went occasionally to French plays, and he joined the crowd of officers at Heinrich von Treitschke's lectures on socialism. But he found that nothing really mattered except his research. He responded eagerly to the intensive research environment in Berlin and in German physics in general. He wrote home that his great satisfaction lay in seeking and communicating new truths about nature. Occupied any other way, he felt a useless member of society; private study as opposed to research seemed selfish and indulgent. Hertz showed himself to be an extremely persistent and self-disciplined researcher. His belief in the conformity of the laws of nature with the laws of human logic was so strong that to discover a case of nonconformity would make him highly uncomfortable: he would spend hours closed off from the world, pursuing the disagreement until he found the error. He won the Philosophical Faculty prize in 1879, earning a medal, a first publication in *Annalen der Physik* in 1880, and Helmholtz' deepening respect.

While Hertz was finishing his work on the Philosophical Faculty problem in 1879, Helmholtz asked him to try for another, much more valuable prize offered by the Berlin Academy. The prize was for an experimental decision on the critical assumptions of Maxwell's theory, a problem Helmholtz had designed expressly for his most talented student. Hertz declined, feeling that it would take him three years and that the outcome was uncertain in any case. Instead he wrote a doctoral dissertation on electromagnetic induction in rotating conductors, a purely theoretical work that took him only three months to complete. It was not a pioneering work but a thorough study of a problem that had been partially treated by many others, from Arago and Faraday to Emil Jochmann and Maxwell. He submitted his dissertation in January 1880 and took his doctoral examination the following month, earning a *magna cum laude,* a distinction rarely given at Berlin.

In 1880 Hertz began as a salaried assistant to Helmholtz in the practical work of the Berlin Physical Institute, a position he held for three years. He found the supervisory chores tedious, but they left him time to complete the research for fifteen publications and with them to begin establishing a reputation. Hertz's work in his Berlin period is difficult to summarize because of its diversity. The majority of his publications were on electricity; in addition to those on electromagnetic induction and the inertia of electricity, he published on residual charge in dielectrics and, most important, on cathode rays. In two papers in 1883 he concluded that cathode rays were not streams of electrical particles as many investigators had supposed, but invisible ether disturbances producing light when absorbed by gas. In other papers he developed a new ammeter and new hygrometer, revealing that he had retained his boyhood fascination and dexterity with instruments. His early dual attraction to engineering and physics was reflected in his research into elastic solid theory, which led to a publication in an engineering journal on a new, absolute measure of the hardness of materials. Yet another of his Berlin researches dealt with the evaporation of liquids; in this he displayed his command of thermodynamics and kinetic theory, a principal branch of nineteenth-century physics to which he did not contribute directly.

The Berlin Physical Society began meeting in the Physical Institute at the time Hertz took up his assistant's post there. He attended regularly, enjoying the sense of being at the center of German physics. He read his papers to the Society; and although he thought the discussions trival, he liked being in the company of Helmholtz, du Bois-Reymond, and other famous members.

As assistant in the Institute, Hertz came into closer relations with Helmholtz, often dining with him and his family. He sometimes found Helmholtz' halting, ponderous speech annoying, but he never doubted that Helmholtz was Germany's greatest physicist. Although his position at the Institute had great advantages—he was near Helmholtz and had at hand the finest research facilities in Germany—Hertz shared the usual ambition of wanting to advance to a regular faculty appointment. To do so, it was first necessary to be a *Privatdozent,* an unsalaried lecturer at the bottom of the university hierarchy. He did not want to be one at Berlin, for there were already too many *Privatdozenten* there. It was at this time that mathematical physics began to be recognized as a separate subdiscipline in Germany, and Hertz's opportunity came when the University of Kiel requested a *Privatdozent* for the subject. Kirchhoff recommended Hertz for the job.

In 1883 Hertz moved to Kiel, where he discovered that he was a successful lecturer; by the second semester he drew fifty students, an impressive number for a small university. The limitation of Kiel was that it had no physics laboratory. Although Hertz fitted one out in his own house, he did not get deeply into experimental work in his two years at Kiel; and it proved a source of frustration and restlessness for him. His publications from this time consisted of three purely theoretical papers: one on meteorology, one on magnetic and electric units, and one on Maxwell's electrodynamics. The last, his first deep study of Maxwell's work, was by far the most important result of his enforced isolation from laboratory work in Kiel. Ultimately important, too, for his development was his extensive reading in the philosophical writings of Dühring, Fechner, Kant, Lotze, and Mach. When Kiel offered Hertz an associate professorship in 1885, he refused it. Unlike his Kiel successor, Max Planck, he did not want a position as a purely theoretical physicist. The Karlsruhe Technische Hochschule wanted to hire him as professor of physics; once he saw the Karlsruhe Physical Institute, he knew he wanted to move.

Hertz spent four years at Karlsruhe, from 1885 to 1889. His stay began inauspiciously; for a time he was lonely and uncertain about what research to begin next. In July 1886, after a three-month courtship, he married Elisabeth Doll, the daughter of a colleague; and in November 1886 he began the experimental studies that were to make him world-famous. In the rich Karlsruhe physical cabinet he came across induction coils that enabled him to tackle the problem on Maxwell's theory that Helmholtz had set for the 1879 Berlin Academy prize. By the end of 1888 he had gone beyond the terms of Helmholtz' problem and had confirmed the existence of finitely propagated electric waves in air. All the time he was in close touch with Helmholtz, sending him his papers to communicate to the Berlin Academy for quick publication before sending them later to *Annalen der Physik.* He published a total of nine papers from his electrical researches in Karlsruhe. They drew immediate, widespread recognition, which led to another and final move for Hertz.

In September 1888 the University of Giessen tried to hire Hertz away from Karlsruhe. The Prussian *Kultusministerium* pressed him to refuse, and to consider Berlin instead, where he would go as Kirchhoff's replacement. But Hertz did not want to go back to Berlin—not yet, anyway, and definitely not as Kirchhoff's successor. At thirty-one he felt that he was too young for a major position in German physics; he felt that he would be pulled away from his researches too soon. And, as he knew from Kiel, he was not a mathematical physicist—which was what Berlin wanted. Helmholtz thought Hertz was correct in refusing, but he did not try to influence him in any way; he told Hertz that if he came to Berlin, he would find him laboratory space in the Physical-Technical Institute, the new national physical research laboratory that he headed. In December 1888 the Prussian *Kultusministerium* offered Hertz the physics professorship at the University of Bonn. He gladly accepted, more for Bonn's beautiful and quiet setting on the Rhine than for its scientific prospects. In 1889 Clark University in Worcester, Massachusetts, almost tempted him to head its new physical institute, one as splendid as Berlin's (Hertz would have gone if he had not been married); and in 1890 the University of Graz failed to entice him there as Boltzmann's successor.

Hertz moved to Bonn in the spring of 1889. He and his family took over the house where his predecessor, Rudolf Clausius, had lived for fifteen years; the continuity had precious historical significance for him. He found the Bonn Physical Institute cramped and the apparatus in a jumble, and he spent much of his time putting things in order. He had students now who worked in the Institute on his electromagnetic ideas. Hermann Minkowski, then a *Privatdozent* in mathematics, was greatly drawn to Hertz and worked in the Institute. Philipp Lenard became Hertz's assistant there in the spring of 1891. The main advantage of the Bonn position over that at Karlsruhe was that it required less teaching and left Hertz more time for research. In Bonn he continued the theoretical study of Maxwell's theory that he had begun in Karlsruhe; this research led to two classic papers on the subject, published in *Annalen der Physik* in 1890. He subsequently tried a miscellany of experiments, only one of which led to a publication: in the summer of 1891 he returned to the subject of cathode rays, studying their power of penetrating metal foils. In the spring of 1891 he began the research that would occupy him almost exclusively until his death: a purely theoretical study of the principles of mechanics inspired by Helmholtz' new work on the principle of least action. The one distraction from his mechanical study was the request at the end of 1891 by J. A. Barth, the publisher of *Annalen der Physik,* that he collect his papers on electric waves for publication in book form. Hertz dedicated the collection to Helmholtz.

Even before Hertz had finished his researches on electric waves, he began to receive international recognition. In 1888 he was awarded the Matteucci Medal of the Italian Scientific Society. In 1889 he won the Baumgartner Prize of the Vienna Academy of Sciences and the La Caze Prize of the Paris Academy of Sciences; in 1890 he won the Rumford Medal of the Royal Society, and in 1891 the Bressa Prize of the Turin Royal Academy. Between 1888 and 1892 he was elected a corresponding member of several major scientific societies, including the Berlin Academy of Sciences, the Manchester Literary and Philosophical Society, the Cambridge Philosophical Society, and the Accademia dei Lincei. He was invited to give a major address on his electric wave experiments at the 1889 Heidelberg meeting of the German Natural Scientists and Physicians. He enjoyed the sense of moving on equal terms in Heidelberg with the leading German physicists, notably Helmholtz, Kundt, Kohlrausch, Wiedemann, and Siemens. To receive the Rumford Medal he visited England, where he was feted by Crookes, Lodge, FitzGerald, Stokes, William Thomson, Strutt, and most of the other important British physicists and electrical engineers.

At the time Hertz moved to Karlsruhe he complained of toothaches; and early in 1888, in the midst of his electric wave researches, he had his teeth operated on. Early in 1889 he had all his teeth pulled out. In the summer of 1892 his nose and throat began hurting so badly that he had to stop work. At first he thought it was hay fever, and he went to the spas. But he found no cure; and from this time on, he was in almost constant pain from a malignant bone condition that his physicians did not understand well. He missed the fall semester of 1892 but taught again in the spring of 1893. He had several head operations which gave him only temporary relief; he was often depressed. He began lecturing in the fall of 1893, while working on the last stages of his book on mechanics. On 3 December 1893 he sent most of his manuscript to the press; on 7 December he gave his last lecture; on 1 January 1894 he died of blood poisoning. He was thirty-six.

Hertz left behind his wife and two daughters, Johanna and Mathilde, all of whom emigrated from Nazi Germany in 1937 to settle in Cambridge, England.

When Hertz entered physics in the 1870's, electrodynamics was in a disorganized state. Theories had multiplied in its fifty years of development, and each had its own following. In Germany the leading theories were those of Weber and F. E. Neumann. Although both theories shared the fundamental physical assumption that electrodynamic actions are instantaneous actions at a distance, they differed in their formulations and in their assumptions about the nature of electricity. Neumann's theory was one of electrodynamic potential, mathematically abstract and physically independent of atomistic assumptions. Weber's, by contrast, was above all an atomistic theory, according to which electricity consisted of fluids of particles of two signs and possessed mechanical inertia. Any pair of Weberian particles interacted through a force or potential modeled in part after Newtonian gravitational attraction; Weberian interaction differed from the Newtonian in that it depended not only on the separation of the particles but also on their relative motion.

Electrodynamic thinking in Britain was based on physical assumptions about electrodynamic actions very different from those of Weber and Neumann. Inspired by Faraday's contention that instantaneous action at a distance was illogical and that the origin of electrodynamic actions was not in particulate electric fluids but in the condition of the space or medium intervening between ponderable bodies, Maxwell constructed a new mathematical theory of the electromagnetic field.

He conceived of the field as a mechanical condition of dielectric media, the ether of free space being a special case of such media. A central contention of Maxwell's theory was that light consisted of electromagnetic waves in dielectric media. It should be remarked that in suggesting a unification of the two separate branches of physics—electricity and optics—Maxwell's theory was not unique; for as Maxwell's contemporaries Riemann and Ludwig Lorenz showed, it was possible to modify action-at-a-distance theories to yield finitely propagated electric waves analogous to light waves.

Like rational mechanics, electrodynamics had an elaborate mathematical development; but unlike rational mechanics, it had not yet found its common principles. Helmholtz characterized electrodynamics at this stage as a "pathless wilderness," and he accordingly called for experiments to test more fundamentally the assumptions of the contending theories.

Beginning in 1870, Helmholtz turned his attention to electrodynamics; his object was to bring order to electrodynamics by casting the contending theories into a form that would expose their experimentally detectable differences. For this purpose he constructed a general theory of electrodynamics; its equations included as special cases those of Weber, Neumann, and Maxwell. Helmholtz' was an action-at-a-distance theory, since it regarded dielectric polarization as the displacement of bound charges under the influence of an electric force existing independently of a medium. Helmholtz showed that the three theories agreed in their predictions of electrodynamic phenomena associated with closed currents, but that they differed in their predictions of phenomena accompanying the oscillatory surgings of electricity of unclosed currents. He emphasized that it was only by attending to the phenomena accompanying unclosed currents that a decision might be made between the competing theories and a consensus brought to this important branch of physics.

In 1871 Helmholtz was called to Berlin to take up his first professorial position in physics. His move had immense importance for the subsequent development of electrodynamics. Helmholtz now had a physical institute and physics students, and he used this institutional opportunity to pursue his program for the reorganization of electrodynamics. It was a matter of great significance to Helmholtz to bring about a consensus in electrodynamic principles; by comparison it was a matter of little significance that it was achieved through the British conception of electrodynamic action and not through the action-at-a-distance conception that Helmholtz shared with other German electricians.

To encourage experimental work in the notoriously difficult domain of unclosed currents, Helmholtz proposed for the prize of the Berlin Philosophical Faculty in 1878 a problem dealing with an implication of Weber's theory: when oscillations of electricity are set up in an unclosed circuit, Weber's hypothetical electrical inertia should reveal itself in a retardation of the oscillations. Through the experiments that Helmholtz had suggested on the self-induction of doubly wound spirals, Hertz won the Philosophical Faculty prize; he proved that the inertia of electricity is either zero or less than a very small value, thereby lending experimental support to Helmholtz' theoretical judgment of the improbability of Weber's theory.

To encourage further the experimental decision between electrodynamic theories Helmholtz proposed through the Berlin Academy of Sciences in 1879 a second prize problem, this one in connection with the behavior of unclosed circuits in Maxwell's theory. Central to Maxwell's theory was the assumption that changes in dielectric polarization yield electromagnetic effects in precisely the same manner as conduction currents do. Helmholtz wanted an experimental test of the existence of these effects or, conversely, of the electromagnetic production of dielectric polarization. Although at the time Hertz declined to try the Berlin Academy problem because the oscillations of Leyden jars and open induction coils which he was familiar with did not seem capable of producing observable effects, he kept the problem constantly in mind; and in 1886 shortly after arriving in Karlsruhe he found that the Riess or Knochenhauer induction coils he was using in lecture demonstrations were precisely the means he needed for undertaking Helmholtz' test of Maxwell's theory.

In 1884, at Kiel, Hertz had already carried out a study of Maxwell's theory. It was a theoretical response to Helmholtz' general problem of deciding between rival electrodynamic theories. Whereas Helmholtz had shown that the experimental decision lay with unclosed currents, Hertz showed that a theoretical decision could be made on the basis of predictions for closed currents. Hertz proved that Maxwell's equations were compatible with the physical assumptions shared by all electrodynamic theories and that the equations of the contending theories were not. He concluded that if the choice lay solely between Maxwell's equations and the equations of the other type of theory, then Maxwell's were clearly preferable; he did not, however, endorse Maxwell's physical interpretation of his equations, in particular Maxwell's denial of action at a distance. Indeed when Hertz returned to Maxwell's theory in Karlsruhe, he did so within the action-at-a-distance framework of Helmholtz' general theory of 1870. With it he felt more at home, less committed to unproved hypotheses than with Maxwell's theory.

Hertz's first experiments in Karlsruhe in 1886 were intended to determine the influence of dielectrics such as pitch and paraffin on the inductive communication of sparks between primary oscillatory and detector circuits. Only in 1888 did it occur to him that the center of interest in Maxwell's theory was its assertion of the finite propagation of electric waves in air. Originally Helmholtz had intended to include in the Berlin Academy problem the option of testing whether or not air and vacuum behave electromagnetically like solid dielectrics, as Maxwell's theory required them to do. But the test had seemed too difficult at the time, and it was struck from the options, only to be restored later by Hertz in his own way. It was not until after Hertz had turned to the production of electric waves in air—in fact, only after he had published his first experiments on waves—that he at last dropped Helmholtz' action-at-a-distance viewpoint; in 1889 he announced that he could describe his results better from Maxwell's contiguous action viewpoint.

Hertz knew of Helmholtz' attempt in 1871 to measure the velocity of propagation of transient electromagnetic inductive effects in air by the delay time between transmission and reception; Helmholtz' experimental arrangement was limited, and he had been able to establish only a lower limit on the velocity of about forty miles per second. Hertz did not know of G. F. FitzGerald's theoretical discussion of the possibility of producing nontransient electric waves in the ether; nor did he know of the attempts to detect electromagnetic waves in wires by O. J. Lodge, another early follower of Maxwell. It is not certain if Hertz knew of the many observations by Edison, G. P. Thompson, David Hughes, and others of the communication of electromagnetic actions over considerable distances; in any case, the observations were generally interpreted as ordinary inductions and therefore not of fundamental significance.

The influence of distance in the communication of electromagnetic actions was not significant until a theory was worked out to show its significance. Maxwell had not provided such a theory, having been mainly concerned to draw the optical rather than the invisible electromagnetic consequences of his theory. In his *Treatise on Electricity and Magnetism* (1873) he gave no theory of oscillatory circuits or of the connection between currents and electromagnetic waves. The possibility of producing electromagnetic waves in air was inherent in his theory, but it was by no means obvious and was nowhere spelled out. Hertz's proof of such waves was in part owing to his theoretical penetration into Maxwell's thought.

Hertz's proof was the result of his experimental inventiveness. He produced electric waves with an unclosed circuit connected to an induction coil, and he detected them with a simple unclosed loop of wire. He regarded his detection device as his most original stroke, since no amount of theory could have predicted that it would work. Across the darkened Karlsruhe lecture hall he could see faint sparks in the air gap of the detector. By moving it to different parts of the hall he measured the length of the electric waves; with this value and the calculated frequency of the oscillator he obtained the velocity of the waves. For Hertz his determination at the end of 1887 of the velocity—equal to the enormous velocity of light—was the most exciting moment in the entire sequence of experiments. He and others saw its significance as the first demonstration of the finite propagation of a supposed action at a distance.

Early in the course of his Karlsruhe experiments Hertz noticed that the spark of the detector circuit was stronger when it was exposed to the light of the spark of the primary circuit. After meticulous investigation in which he interposed over sixty substances between the primary and secondary sparks, he published his conclusion in 1887 that the ultraviolet light alone was responsible for the effect—the photoelectric effect. He was convinced that the effect had profound theoretical meaning for the connection of light and electricity, even though the meaning was obscure at the time. His experiments left no doubt of the reality of the effect, and soon other experimenters were studying it intensively. Hertz, however, did no more work on it, since it was a digression from his original purpose—the examination of the physical assumptions of Maxwell's theory.

Hertz followed up his determination of the finite

velocity of electric waves by performing a series of more qualitative experiments in 1888 on the analogy between electric and light waves. Passing electric waves through huge prisms of hard pitch, he showed that they refract exactly as light waves do. He polarized electric waves by directing them through a grating of parallel wires, and he diffracted them by interrupting them with a screen with a hole in it. He reflected them from the walls of the room, obtaining interference between the original and the reflected waves. He focused them with huge concave mirrors, casting electric shadows with conducting obstacles. The experiments with mirrors especially attracted attention, as they were the most direct disproof of action at a distance in electrodynamics. They and the experiments on the finite velocity of propagation brought about a rapid conversion of European physicists from the viewpoint of instantaneous action at a distance in electrodynamics to Maxwell's view that electromagnetic processes take place in dielectrics and that an electromagnetic ether subsumes the functions of the older luminiferous ether.

It was far from clear to physicists, however, precisely to what theory they were subscribing when they declared themselves followers of Maxwell. The impressive, extraordinarily rapid consensus that Hertz's experiments brought about had not fully realized the program Helmholtz had laid down twenty years before of clarifying the principles of electrodynamics. There remained the vexing question of what Maxwell's theory really meant. In two theoretical papers in 1890 Hertz set about bringing perfection of form to the theory that, in his judgment, was perfect in its physical content. The content was clear; it was that electromagnetic phenomena are caused by polarizations in a dielectric medium filling otherwise empty space. The problem was to construct a consistent form that expressed the content faithfully, that banished all suggestion of distance forces and the associated electric fluids.

The first of Hertz's theoretical papers dealt with the electrodynamics of bodies at rest. In the introduction he maintained that Maxwell's theory, as formulated in the *Treatise,* contained traces of action at a distance, the route he thought Maxwell, like himself, had taken to Maxwell's theory. To attain a consistent contiguous action theory, Hertz eliminated the vector potentials from the fundamental equations of the theory, a residue from the concept of action at a distance and a scaffolding that unnecessarily complicated the formalism. He also eliminated Maxwell's distinction between the polarization and the electric force in the free ether, a distinction intelligible only within the framework of action at a distance. In denying the existence of distance forces, Hertz asserted that the polarizations of the medium were the only things really present; and in denying the electrical fluids from which the distance forces were supposed to proceed, he treated electricity, or charge, as merely a convenient abbreviation. In Britain, Heaviside had worked on a closely parallel reformulation of Maxwell's theory since 1885; Hertz knew of Heaviside's work, but his own contained a more searching critique of the physical content of Maxwell's theory.

According to Hertz, Maxwell's equations contained everything that was secure in Maxwell's theory. This was the sense of his dictum in the introduction to *Electric Waves:* "Maxwell's theory is Maxwell's system of equations." He did not offer the dictum as a final phenomenological position; rather he meant that any search for the mechanical basis of electrodynamics should start from Maxwell's equations—or, more accurately, from Hertz's form of the equations—and that the mechanical investigations of the past were irrelevant to the present state of the science. Accordingly in 1890 Hertz postulated the equations of the theory, instead of deriving them from a mechanical model of the ether. He proposed the symmetrical relations between the electric force E and the magnetic force H in the free ether (where forces and polarizations are identical):

$$\frac{1}{c}\frac{\partial H}{\partial t} = -\text{curl } E, \qquad \frac{1}{c}\frac{\partial E}{\partial t} = \text{curl } H,$$

$$\text{div } H = 0, \qquad \text{div } E = 0,$$

where c is the speed of light. (The units are Gaussian. Hertz wrote his equations with the opposite sign because he used a left-handed coordinate system. He wrote them in components, too, rather than in vector notation.) Hertz's achievement in his first theoretical paper in 1890 was to simplify the formalism, to bring forward the logical structure of Maxwell's theory consistently interpreted as a contiguous action theory.

In his second theoretical paper, Hertz applied Maxwell's equations to moving, deformable bodies. Maxwell had not treated this problem systematically in the *Treatise* although, unknown to Hertz, he had done so elsewhere. Hertz recognized that to develop an electrodynamics of moving bodies, it was first necessary to specify whether or not the ether moves with bodies. For his part he would assume that the ether is mechanically dragged by moving bodies. The first ground for this assumption was that within the restricted domain of electromagnetic phenomena there was nothing incompatible with the idea of a dragged ether. The second ground was that its denial entailed the complication that two sets of electric and magnetic vectors had to be assigned to each point of space, one for the ether and one for the independently moving body. He recognized at the same time that a dragged ether was an unsure foundation for electrodynamics; it was incapable of explaining optical phenomena such as stellar aberration and Fizeau's experiment, phenomena which pointed to the independence of the motions of ponderable matter and the ether. He surmised that a correct theory would distinguish between the state of the ether and the state of the matter embedded in it at each point. He thought that to attempt a theory with a more probable interpretation of the ether would be premature and would require more arbitrary hypotheses than the present theory. The sole value he placed on his theory of electromagnetic forces in moving bodies was its systematic arrangement.

Hertz brought an unparalleled clarity to Maxwell's theory, organizing its concepts and its formalism so that others were able quickly to go beyond him. In underscoring the limitations of his formulation of Maxwell's theory he delineated the central problems for future research. Thus Hertz's electrodynamic theory was the last to be concerned exclusively with electrodynamic phenomena in the narrow sense. Subsequent developers of Maxwell's theory rejected Hertz's conception of the ether because of its inability to account for optical as well as electrodynamic phenomena. The most important developer was the Dutch theoretical physicist H. A. Lorentz, who constructed his electron theoretical extension of Maxwell's theory in 1892 in response to the optical insufficiency of Hertz's electrodynamics of moving bodies. In contradistinction to Hertz, Lorentz distinguished the electromagnetic field from ponderable matter by conceiving of the ether as stationary instead of dragged. This and Lorentz' other leading assumption of the molecular nature of electricity constituted the most fruitful foundation for the subsequent development of Maxwell's theory at the turn of the century.

Hertz's final years were devoted almost entirely to exploring the theoretical implications of Maxwell's electrodynamics for the rest of physics. In his 1889 Heidelberg lecture on his work on electric waves he said that from now on the ether would be the most fundamental problem in physics. Its understanding would elucidate major subsidiary problems, such as the nature of electricity, gravity, and mass. The suggestion of Hertz's work on Maxwell's electrodynamics was that a properly etherial physics would eliminate force as a fundamental concept. Hertz developed this suggestion in his last major work, his posthumously published *Principles of Mechanics*.

In a general way Hertz was guided in his mechanical studies by Mach's 1883 historico-critical analysis of mechanics, but he was once again guided specifically by problems Helmholtz had mapped out. In a series of papers in the 1880's Helmholtz had argued that a system of mechanics that included Newton's laws of motion together with the assumption of Hamilton's principle can explain all physical phenomena. Sharing Helmholtz' universalist goal for mechanics, Hertz regarded Helmholtz' work on Hamilton's principle as the furthest advance of physics. In another series of papers in the 1880's Helmholtz had constructed a mechanical analogy of the second law of thermodynamics based on monocyclic systems of hidden, moving masses. The analogy suggested to Hertz a way to reformulate mechanics without introducing forces as a fundamental concept.

Hertz accepted Kirchhoff's demonstration that mechanics can be represented in terms of three concepts alone: mass, space, and time. By contrast, the usual representations of mechanics included a fourth concept, either force or energy. Hertz explained in the introduction to the *Principles* that to construct a mechanics capable of accounting for the lawful interaction of perceptible bodies it was necessary to add a hypothesis to the three concepts. The hypothesis was that in addition to perceptible masses the universe contained hidden, moving masses bound to one another by rigid constraints. Under Hertz's hypothesis forces appeared neither in the microcosm nor in the macrocosm; the imperceptible universe was constituted of the same entities as the perceptible one.

At the head of his mechanics Hertz placed a single law of motion: the path of a system in $3n$-dimensional space is as straight as possible, subject to rigid constraints, and the system traverses the path with uniform motion. Any observable system acted upon by forces is in reality only a part of a larger force-free system that includes hidden masses. Hertz showed that the usual formulations of mechanics—Newton's, Lagrange's, and Hamilton's—can be deduced as theorems from his law of motion.

Like Helmholtz and such other contemporaries as Ludwig Boltzmann, Hertz sought to realize the historical goal of uniting the parts of physics through mechanics. Through the nineteenth century mechanics had come to pervade physics in increasingly insistent ways, and Hertz thought it was time that mechanics was given such foundations that it was exactly coterminous with physics; mechanics should no longer allow motions that do not occur in nature, nor should it exclude motions that do occur. Rejecting the view that mechanics was a branch of mathematics with unchanging principles, Hertz viewed it as the science of the actual actions and connections of nature. As such, mechanics was subject to change when the state of knowledge of physics changed—as it had with Hertz's confirmation of contiguous action in the electromagnetic ether.

Hertz opened the *Principles* with the observation that "all physicists agree that the problem of physics consists in tracing the phenomena of nature back to the simple laws of mechanics." It was one of the last times the statement could be made, and even then there were those who were disinclined to accept any longer the mechanical view of nature. The *Principles* was published on the eve of a great debate over world views, and as the most ambitious attempt to encompass all natural knowledge within mechanics it was a focus of discussion in the debate. Those, such as W. Wien and M. Abraham, who sought to derive all physics, including mechanics, from Maxwell's laws characterized their goal as diametrically opposed to that of Hertz. Of reactions to the *Principles* by others who found the mechanical world view congenial, Helmholtz' may be taken as representative. While preferring a more abstract mathematical approach in physics to Hertz's hypothesis of hidden masses, Helmholtz admired the logic, generality, and unifying objective of Hertz's mechanics. His concern was that Hertz had not troubled to provide examples of the hypothetical mechanism of hidden masses in actual mechanical problems. He thought that it would be difficult to apply Hertz's principles—as indeed it turned out to be—and that at present they constituted only an ingenious program that might have great heuristic value for future research. It seems that the heuristic value was not realized, and apart from its role in the world view debate the major importance of the *Principles* has been as a classic of nineteenth-century philosophy of science.

Hertz's chief contribution to physics was in bringing about a decision regarding the proper principles for representing electrodynamics. His experimental researches in Karlsruhe settled once and for all the long conflict in nineteenth-century physics over the merits of action at a distance versus contiguous action. After Hertz it was eccentric to continue to advocate action at a distance in electrodynamics—or for that matter in any other part of physics. By the 1870's, when Hertz began his career, thermodynamics had been secured on the basis of its two fundamental laws; but the other principal branch of physics, electrodynamics, was encumbered with a proliferating collection of competing theories, and physicists showed little will or ability to settle its fundamentals and secure an agreement. More than any other physicist, Helmholtz responded to the primary need of the discipline at this time of putting electrodynamics in order.

It was not the least of Hertz's gifts to perceive that Helmholtz had more to offer him than did Kirchhoff or any other German physicist with whom he had early contact. Hertz's relation to Helmholtz was as a disciple, but not one unduly wedded to any of Helmholtz' methods. His dependence on Helmholtz was of a different sort; it lay in his recognition of Helmholtz' sure grasp of the central, soluble problems of physics. In his brief career Hertz revealed himself not as an innovator of concepts but as one having an uncommonly critical and lucid intelligence in addressing the conceptual problems of physics that others, Helmholtz above all, had marked out.

Hertz's researches on electric waves vindicated the Helmholtzian ideal of the physicist as one whose competence embraced both experiment and mathematics. Hertz entered physics at the right time for one of his abilities to make a critical contribution; because the outstanding problem of physics was the disorderly condition of electrodynamics, what was needed was someone with the theoretical power to analyze the competing theories and with the experimental judgment to produce the evidence that would persuade the physical community that a decision between the theories had been reached.

In the last quarter of the nineteenth century many German physicists, Hertz and Helmholtz among them, were intensely concerned to bring unity to the parts of their science; and they looked to mechanics for the source of unifying concepts. Much of the interest in thermodynamics at this time centered on its mechanical foundations. Once the principles of electrodynamics, like those of thermodynamics, were secure, Hertz turned to an investigation of the mechanical foundations of an ethereal physics. Instead of inventing mechanisms for the ether, he looked at the mechanical problem from a more general point of view. Convinced that the received mechanical principles were unsuited for the task of representing contiguous action processes in the ether, he refounded the science of mechanics on alternative principles that would provide a natural mechanical basis for electrodynamics as well as for the other parts of physics.

Hertz sought a basic understanding of nature; despite his origins in engineering and despite the fact that he made his major discoveries in an engineering school while teaching technical electricity, he did not concern himself much with the practical implications of electric waves. Others soon did, however. In the early 1890's the young inventor Guglielmo Marconi read of Hertz's electric wave experiments in an Italian electrical journal and began considering the possibility of communication by wireless waves. Hertz's work initiated a technological development as momentous as its physical counterpart.

BIBLIOGRAPHY

I. Original Works. Hertz's complete scientific writings are in a 3-vol. collection under the editorship of Philipp Lenard.

I. *Schriften vermischten Inhalts* (Leipzig, 1895), translated by D. E. Jones and G. A. Schott as *Miscellaneous Papers,* P. Lenard, ed. (London, 1896). This volume contains 19 technical papers published between 1880 and 1892; Hertz's 1880 Berlin dissertation, "Ueber die Induction in rotirenden Kugeln"; and two popular lectures: his 1889 Heidelberg address, "Über die Beziehungen zwischen Licht und Elektricität," and a newspaper tribute to Helmholtz in 1891 on his seventieth birthday. Lenard's introduction is largely a series of extracts from Hertz's letters to his parents between 1877 and 1883.

II. *Untersuchungen über die Ausbreitung der elektrischen Kraft* (Leipzig, 1892), translated by D. E. Jones as *Electric Waves* (London, 1893), with pref. by William Thomson. There is a paperback reprint of the 1893 English ed. (New York, 1962). A second German ed. was published in 1894, appearing as vol. II of the collected works. The volume contains all but one of Hertz's experimental and theoretical papers on Maxwell's theory (the omitted one is Hertz's 1884 theoretical comparison of Maxwell's and the opposing electrodynamics, which is included in vol. I). The 12 papers in the volume were published between 1887 and 1890. In addition it contains an extract from a publication on rapid electric oscillations by Wilhelm von Bezold and an introduction by Hertz. The first part of the introduction is, in Helmholtz' words, a frank "inner psychological history" of Hertz's route to the experimental proof of electric waves; the second part is a theoretical analysis of the meaning of Maxwell's theory.

III. *Die Principien der Mechanik, in neuem Zusammenhange* (Leipzig, 1894), with pref. by Helmholtz, translated by D. E. Jones and J. T. Walley as *The Principles of Mechanics, Presented in a New Form* (London, 1899). A reprint of the 1899 English ed. (New York, 1956), with a new intro. by R. S. Cohen, contains a valuable bibliography of scientific, philosophical, and historical works relating to Hertz's mechanics. In his own long introduction Hertz gives a profound analysis of epistemological problems in late nineteenth-century physics.

An extraordinarily rich autobiographical source is *Heinrich Hertz: Erinnerungen, Briefe, Tagebücher,* J. Hertz, ed. (Leipzig, 1927). Aside from a short account of Hertz's childhood written by his mother in 1901, the book consists of letters from Hertz to his parents between 1875 and 1893, interleaved with copious passages from diaries that Hertz kept from childhood on. There are also several letters between Hertz and Helmholtz and two from Helmholtz to R. Lipschitz. The letters on which Lenard drew for his intro. to vol. I of his ed. of *Schriften* are published here in their entirety.

Of the considerable amount of Hertz's scientific correspondence extant, little has been published other than that in the *Erinnerungen.* But see J. Thiele, ed., "Ernst Mach und Heinrich Hertz: Zwei unveröffentliche Briefe aus dem Jahre 1890," *Schriftenreihe für Geschichte der Naturwissenschaften, Technik und Medizin,* **5** (1968), 132–134.

II. Secondary Literature. Although there is no book-length study of Hertz, there are many obituaries and historical appreciations of his life and work. In addition to the valuable introductions by Lenard, Helmholtz, and Thomson in the volumes of Hertz's collected works, the most penetrating study of Hertz by a contemporary is M. Planck, "Gedächtnisrede auf Heinrich Hertz," in *Verhandlungen der Physikalischen Gesellschaft zu Berlin,* **13** (1894), 9–29, repr. in Planck's *Physikalische Abhandlungen und Vorträge,* III (Brunswick, 1958), 268–288.

Useful later studies include P. G. Cath, "Heinrich Hertz (1857–1894)," in *Janus,* **46** (1957), 141–150; Walter Gerlach, "Heinrich Rudolf Hertz 1857–1894," in *150 Jahre Rheinische Friedrich-Wilhelms-Universität zu Bonn 1818–1968;* Armin Hermann, "Heinrich Hertz, Physiker," in *Neue Deutsche Biographie,* VIII, 713–714; J. A. W. Jenneck, "Heinrich Hertz," in *Deutsches Museum Abhandlungen und Berichte,* **1** (1929), 1–36; Max von Laue, "Heinrich Hertz, 1857–1894," in *Gesammelte Schriften und Vorträge,* III (1961), 247–256; Philipp Lenard, "Heinrich Hertz," in *Great Men of Science,* translated by H. S. Hatfield (New York, 1933), pp. 358–371; and Philip and Emily Morrison, "Heinrich Hertz," in *Scientific American,* **197** (1957), 98–106. A full summary of the facts of Hertz's career is given in O. Wenig, ed., *Verzeichnis der Professoren und Dozenten der Rheinischen Friedrich-Wilhelms-Universität zu Bonn 1818–1968* (Bonn, 1968), p. 117. There is some material on the relation of Hertz and Helmholtz in Leo Koenigsberger, *Hermann von Helmholtz* (Oxford, 1906).

The best historical analysis of Hertz's electrodynamics is given in two articles by Tetu Hirosige: "Electrodynamics Before the Theory of Relativity, 1890–1905," in *Japanese Studies in the History of Science,* no. 5 (1966), pp. 1–49; and "Origins of Lorentz' Theory of Electrons and the Concept of the Electromagnetic Field," in *Historical Studies in the Physical Sciences,* **1** (1969), 151–209; and in L. Rosenfeld, "The Velocity of Light and the Evolution of Electrodynamics," in *Nuovo cimento,* supp. **4** (1957), 1630–1669. Hertz's electrodynamics is discussed in the standard histories, especially E. T. Whittaker, *A History of the Theories of Aether and Electricity,* I, *The Classical Theories,* rev. ed. (London, 1951), pp. 319–330. Hertz's mechanics, too, is discussed in the standard histories, such as René Dugas, *A History of Mechanics,* translated by J. R. Maddox (New York, 1955), pp. 444–447.

More specialized studies are Peter Heimann, "Maxwell, Hertz, and the Nature of Electricity," in *Isis,* **62** (1970), 149–157; T. K. Simpson, "Maxwell and the Direct Experimental Test of His Electromagnetic Theory," *ibid.,* **57** (1966), 411–432; Roger H. Stuewer, "Hertz' Discovery of the Photoelectric Effect," *Actes. XIII*[e] *Congrès Interna-*

tional d'Histoire des Sciences (in press); and C. Süsskind, "Observations of Electromagnetic-Wave Radiation Before Hertz," in *Isis*, **55** (1964), 32–42; and "Hertz and the Technological Significance of Electromagnetic Waves," *ibid.*, **56** (1965), 342–345.

The Deutsches Museum in Munich has 162 letters from and 243 letters to Hertz. The collection includes substantial numbers of letters in 1880–1894 between Hertz and Bjerknes, Cohn, Drude, Elsas, FitzGerald, Heaviside, Helmholtz, König, Neesen, Poincaré, Richarz, de la Rive, Röntgen, Rubens, Sarasin, Warburg, G. Wiedemann, Wien, and Wiener. The Deutsches Museum also has the MSS of *Über die Beziehungen zwischen Licht und Elektricität* and *Die Principien der Mechanik* and of parts of a geophysical work.

RUSSELL MCCORMMACH

HERTZSPRUNG, EJNAR (*b.* Frederiksberg, Denmark, 8 October 1873; *d.* Roskilde, Denmark, 21 October 1967), *astronomy.*

Hertzsprung's father, Severin Hertzsprung, had a graduate degree in astronomy from the University of Copenhagen but, for financial reasons, decided to accept a position in the Department of Finances of the Danish government and at a very early age became director of the state life insurance company. He instilled in his son his own interest in astronomy and mathematics; but because of his awareness of the lack of financial security, he did not encourage the boy to select these fields as a career. As a result Hertzsprung decided to study chemical engineering. His interest in chemistry resulted from his study of a small book on this subject by the Danish chemist Julius Thomsen. Hertzsprung graduated from the Polytechnical Institute in Copenhagen in 1898 and spent the next several years as a chemist in St. Petersburg. In 1901 he went to Leipzig to study photochemistry in Wilhelm Ostwald's laboratory. He returned to Denmark the following year and began in earnest his study of astronomy.

During this period Hertzsprung began corresponding with the German astronomer Karl Schwarzschild, who invited Hertzsprung to visit him at Göttingen in 1909. Within a few months Hertzsprung was appointed associate professor at the university and during the same year, when Schwarzschild became the director of the astrophysical observatory at Potsdam, Hertzsprung joined him there as senior staff astronomer.

In 1919 Hertzsprung was appointed an associate director and associate professor of the observatory of the University of Leiden; he became its director in 1935. Upon retirement in 1944 he returned to Denmark, where he continued his research until 1966.

Hertzsprung received many honors for his outstanding contributions to astronomy. He was elected to eleven academies and societies in both Europe and the United States, and received honorary doctorates from Utrecht (1923), Copenhagen (1946), and Paris (1947). The Royal Astronomical Society awarded him its gold medal in 1929; in 1937 he received the Bruce Gold Medal of the Astronomical Society of the Pacific; and the city of Copenhagen honored him with its Ole Römer Medal in 1959.

Early in the twentieth century, when Hertzsprung entered the field of astronomy, study of the physical nature of stars was still in its infancy. Stellar astronomy during the nineteenth century had been directed mainly toward determining positions and motions of the stars. However, during the second half of the century pioneer work in spectroscopy had been initiated by Angelo Secchi and William Huggins, and the new technique of photography was introduced for making astronomical observations—primarily by Secchi, Warren de la Rue, and W. C. Bond. By 1875 Huggins had devised methods for photographing stellar spectra and had succeeded in determining the radial velocities of stars (that is, their motions in the line of sight) from shifts in the spectral lines.

From their visual observations of bright stars, Secchi and Huggins had already discovered that there were a few basically different types of stellar spectra and that they formed a series which distinctly connected one type to its neighbor. Secchi initially proposed four classes of spectra, and other classification schemes followed.

During the years 1890–1901 three catalogs of photographically determined stellar spectra were published by Harvard College Observatory. These form the basis for the original *Henry Draper Catalog*, in which Antonia C. Maury classified the brighter stars from the north pole to declination $-30°$ and Annie Jump Cannon classified stars (mostly brighter than fifth magnitude) south of $-30°$. Two different systems of classification were adopted in the catalog. Miss Maury used the more detailed one—twenty-two main groups, each divided into seven different indexes with the use of the letters *a*, *b*, *c*, and four double letters to indicate detailed features in the spectra. Miss Cannon used a less detailed system still used today—with the exception that subdivisions and luminosity classes have since been added.

Hertzsprung said that it was his interest in the theory of blackbody radiation and its relation to the radiation of stars that initially stimulated his interest in astronomy. The problem of the radiation of a blackbody, one that absorbs all frequencies of light and, when heated, also radiates all frequencies, had first been posed by G. R. Kirchhoff and was finally solved by Max Planck in 1900 by means of his quantum theory.

Hertzsprung, with his background as a chemical engineer and a specialist in photochemistry, was without doubt better qualified to use photography in the solution of astronomical problems than most astronomers of that period. What knowledge he needed in the basic principles of observational astronomy he obtained working with H. E. Lau, a young astronomer. Between studying the contemporary astronomical literature and observing with the telescopes at the observatory of the University of Copenhagen and at the Urania Observatory in Frederiksberg, Hertzsprung kept himself fully occupied over the next several years.

During this early period of his astronomical career Hertzsprung published two now classic papers in *Zeitschrift für wissenschaftliche Photographie,* a journal devoted to photophysics and photochemistry. Both papers, published in 1905 and 1907, were entitled "Zur Strahlung der Sterne." In examining the proper motions of stars with spectra classified by Miss Maury, he was able to show that the stars which she found to have exceptionally sharp and deep absorption lines (her index *c*) were more luminous than the rest. This discovery was the basis for measurement of luminosity by means of spectra—a method which, under the title of "spectroscopic parallaxes," has become one of the most powerful means for determining stellar distances, galactic structure, and distances to other galactic systems.

These papers also contained Hertzsprung's discovery of giant and dwarf stars. From his study of parallaxes, apparent magnitudes, proper motions, and colors, he determined that the stars could be divided into two series, one now known as the main sequence in the Hertzsprung-Russell diagram while the other constitutes the high-luminosity or giant stars. The diagram, developed from this discovery, is a plot of the intrinsic magnitude against temperature for a group of stars. It remains the cornerstone of all astronomical research related to the formation and evolution of stars. Hertzsprung's original papers did not include an illustration of the diagram, possibly because he felt his study lacked sufficient data.

Hertzsprung actually constructed the first such diagram for the Pleiades star cluster in 1906, and he took it to Göttingen in 1909. The existence of such a diagram was not generally known until the American astronomer H. N. Russell presented it at a meeting of the Royal Astronomical Society in 1913 in an address on the subject of giant and dwarf stars, based on his own independent research—unaware of Hertzsprung's earlier work.

In his 1907 paper Hertzsprung referred to the open star clusters as a method for deriving the relation between the radiation of a star and its color. Since the physical members of such a cluster would be of equal distance, or nearly so, their apparent magnitudes and colors should reveal this relation.

Before leaving Copenhagen, Hertzsprung photographed several clusters at the Urania Observatory, using coarse gratings in front of the objective of the telescope. By measuring the separation of the grating images from the central images, he obtained the effective wavelengths of the individual stars, which he used as an index for their colors. This work was continued in Potsdam, and in 1911 he published color-magnitude diagrams of the Pleiades and the Hyades—the first diagrams of this type ever to be published.

During his stay at the Mount Wilson Observatory in 1912, Hertzsprung continued cluster work on NGC 1647 and the Pleiades, using coarse gratings in front of the sixty-inch reflector, the largest telescope in the world at that time. Work on the Pleiades alone involved measurements of nearly 10,000 effective wavelengths and was only the beginning of an extensive work that was to be carried out by Hertzsprung on this cluster. Over a period of twenty years he and his associates measured positions of stellar images on 161 photographs, taken at fifteen different observatories. These measurements were made to determine the relative proper motions of 2,920 stars in the region of the Pleiades and to establish membership in the cluster. The first-epoch plates had been taken with almost identical telescopes in the early period of the *Carte du ciel* program, which started in 1887. Because of the long interval between the first- and second-epoch plates, he could not only distinguish between members and nonmembers of the cluster but also was able to determine the upper limit for the internal motions and, in this way, to estimate that the total mass of the cluster did not exceed a few hundred solar masses.

Hertzsprung also found that the magnitudes and the colors of the member stars formed a narrow sequence, a result later to be corroborated by modern photoelectric observations. As early as 1929 he noted that the brighter Pleiades members were whiter than stars of the same brightness in the solar neighborhood, and that the Pleiades differed in stellar population from the Hyades and Praesepe clusters. These differences, first noted by Hertzsprung, are now interpreted to indicate that the Pleiades are younger than the other two clusters, as well as the stars in the solar neighborhood.

Another cluster which received Hertzsprung's special attention was the Ursa Major cluster. In 1869 the English astronomer R. A. Proctor had discovered that the five bright stars in this constellation shared the same motion across the sky. That they actually shared the same motion in space was later confirmed by observation of their radial motion. In 1909 Hertzsprung noticed that two other stars, in widely separated regions of the sky, had motions directed toward the same convergent point of the sky as the five bright stars. This observation led him to make a systematic search for additional members. He succeeded in finding six among the bright stars, and two probable members. The most prominent new member was Sirius, the brightest star in the sky. This cluster of stars, sharing identical motions through space, surrounds the sun, without its being a member. The discovery by Hertzsprung resulted in a search by others for new members; to date results indicate that 135 stars are members of this remarkable cluster.

Hertzsprung's effective use of objective gratings for high-precision photographic photometry is well demonstrated by his discovery of the variability of Polaris, which had been suspected by the Dutch astronomer Antonie Pannekoek in 1891. In order to demonstrate the reality of the variability, he took nearly 1,700 exposures on 400 plates during 50 nights. He succeeded in determining the amplitude of the light variation, which was only 0.171 magnitude, with an error of only 0.012 magnitude—an accuracy in stellar photometry unheard of in 1911.

One of the principal reasons why Hertzsprung was awarded the gold medal of the Royal Astronomical Society was his determination of the distance to the Small Magellanic Cloud in 1913. The method he introduced became the basis for all measurements of very large distances in our galactic system, as well as in the expanding universe of the galaxies. The distance determination was based on a very important discovery made by Henrietta S. Leavitt at the Harvard College Observatory the previous year. She had been studying the variable stars in the Small Magellanic Cloud and had found that a relation existed between the apparent magnitude and the period of light variation of the Cepheid variables, the light variation of which can be explained by a pulsation of the star as a whole. Hertzsprung realized that the stars in the cloud could be considered to be at the same distance and that, consequently, their period of variation could actually be related to their intrinsic brightness.

The next step was to select Cepheids close enough to our sun to evaluate their distances, from which their intrinsic brightnesses could be determined. Since no Cepheid was close enough to allow a direct determination of the distance, Hertzsprung used the bright Cepheids with known proper motions. From these he deduced the mean parallactic components of their motions, and thereby their distances and their intrinsic brightnesses. It was then a simple step to compute the intrinsic brightnesses (luminosities) of the Cepheids in the Small Magellanic Cloud from their periods. His value for the distance (10,000 parsecs) was larger than any distance determined in the universe at that time (1913) but about five times smaller than the presently accepted distance. There are a number of reasons for this discrepancy, the most important being the then unknown galactic absorption.

In the same paper Hertzsprung called attention to the asymmetric distribution of the bright Cepheids with respect to the sun, an asymmetry also shared by the very hot and bright stars of spectral class Oe5. He noticed that since the least concentration was in the best-observed part of the Milky Way, the distribution could not be attributed to observational selection. He found that the center of the distribution was in the direction which was much later discovered to be the direction toward the center of our galactic system.

During World War I, Hertzsprung began a program of photographic observations of double stars to which he later devoted much of his time. The ingenious photographic method that he developed either eliminated possible systematic errors or rendered them negligible, so that the results were ten times more accurate than the conventional visual observations with a micrometer. He later made observations of this kind in Johannesburg, South Africa, assisted by two of his former students; and in 1937, when he was at the Lick Observatory of the University of California, he used the large Lick refractor for such observations. After his retirement others took plates for him, but he continued to do the measuring even past his ninetieth birthday.

Hertzsprung's other contributions to the field of double-star astronomy include his method of obtaining statistical distances (hypothetical parallaxes) for binaries of such slow orbital motion that the observed arcs are too short to permit the determination of their orbits. This method has developed into the so-called dynamical parallaxes, which has been of considerable significance in the statistical calibration of spectroscopic parallaxes.

In 1911 the English astronomer J. K. E. Halm had shown that there existed a statistical relation between the masses and the luminosities of spectroscopic binaries. Hertzsprung found the same relationship in 1915 for visual binaries and later provided the mathematical formulation in 1919—almost simultaneously with Arthur Eddington, who proved the relationship on the basis of theoretical investigations of the radiation equilibrium of the stars.

Hertzsprung returned to his early interest in the colors of the bright stars with a catalog (1922) of mean color equivalents of 734 stars brighter than the fifth magnitude and within 95° of the north celestial pole. In his reduction to a single scale of a range of color equivalents obtained by various methods, he solved the problem of finding the best linear relation between two quantities, both of which were affected by observational errors. His solution was later used by the Dutch cosmologist Willem de Sitter to discuss the velocity-distance relation for extragalactic nebulae.

In the same catalog Hertzsprung discussed the relation between color and luminosity of stars, using proper motions as distance indicators for lack of reliable parallaxes. The diagram illustrating this relationship showed, for the first time, the lack of bright stars of intermediate color, the famous "Hertzsprung gap" between the giants and main sequence stars.

Hertzsprung did not limit his study of variable stars to Polaris. His accurate light curves based upon extensive series of photographic observations of certain selected variables (S Sagittae, VV Orionis, and RR Lyrae) have only in recent years been surpassed in accuracy by photoelectric techniques.

Throughout the years 1924–1929 Hertzsprung concentrated on variable stars. In the first year and a half of this period he observed at the Union Observatory in Johannesburg with the Franklin-Adams telescope and took 1,792 plates, with a total exposure time of 638 hours. On the plates alone he made 36,000 estimates of brightness of variable stars and determined over a third of all the light curves of short-period variables published during that five-year period. He also visited Harvard for five months in 1926–1927 and made an additional 12,000 estimates of variables on the plate collection there.

Hertzsprung was deeply interested in the education of future astronomers. He said in the annual report of the Leiden observatory for 1933: "It is of importance that each student shall have the opportunity to get acquainted with as many different methods of observing as is possible with the means at his disposal, before choosing a particular branch of astronomy for his future specialty." He always emphasized that it was important to plan and execute observational programs carefully, and that great care should be exercised in drawing conclusions from empirical data.

Hertzsprung's guidance and inspiration and the example he set resulted in many of his students later occupying important positions in the astronomical world. He often said, "If one works hard, one always finds something and sometimes something important." By following this principle, Hertzsprung made contributions to astronomy which place him among the great astronomers of all time.

BIBLIOGRAPHY

I. Original Works. Among many are "Zur Strahlung der Sterne," in *Zeitschrift für wissenschaftliche Photographie*, **3** (1905), 429–442; and **5** (1907), 86–107; "On New Members of the System of the Stars β, γ, δ, ε, ζ Ursae Majoris," in *Astrophysical Journal*, **30** (1909), 135–143; "Über die Vervendung photographischer effektiver Wellenlängen zur Bestimmung von Farbenäquivalenten," in *Publikationen des Astrophysikalischen Observatoriums zu Potsdam*, **22** (1911), 1–40; "Nachweis der Veränderlichkeit von α Ursae Minoris," in *Astronomische Nachrichten*, **189** (1911), 89–104; "Über Doppelsterne mit eben merklicher Bahnbewegung," *ibid.*, **190** (1912), 113–118; "Über die räumliche Verteilung der Veränderlichen vom δ Cephei-Typus," *ibid.*, **196** (1914), 201–210; "Effective Wave-Lengths of 184 Stars in the Cluster N.G.C. 1647," in *Astrophysical Journal*, **42** (1915), 92–110; "Bemerkungen zur Statistik der Sternparallaxen," in *Astronomische Nachrichten*, **208** (1919), 89–96; "Photographische Messungen von Doppelsternen," in *Publikationen des Astrophysikalischen Observatoriums zu Potsdam*, **24**, pt. 2 (1920); "Mean Colour Equivalents and Hypothetical Angular Semi-Diameters of 734 Stars Brighter Than Fifth Magnitude and Within 95° of the North Pole," in *Annalen van de Sterrenwacht in Leiden*, **14**, pt. 1 (1922); "Effective Wavelengths of Stars in the Pleiades," in *Kongelige Danske Videnskabernes Selskabs Skrifter*, Sciences Section, 8th ser., **4**, no. 4 (1923); "On the Relation Between Mass and Absolute Brightness of Components of Double Stars," in *Bulletin of the Astronomical Institutes of the Netherlands*, **2** (1923), 15–18; "The Pleiades," in *Monthly Notices of the Royal Astronomical Society*, **89** (1929), 660–678; and "Catalogue de 3259 étoiles dans les Pléiades," in *Annalen van de Sterrenwacht in Leiden*, **19**, pt. 1 (1947).

II. Secondary Literature. See A. O. Leuschner, "The Award of the Bruce Gold Medal to Professor Ejnar Hertzsprung," in *Publications of the Astronomical Society of the Pacific*, **49** (1937), 65–81; and Rev. T. E. R. Phillips, "Address on the Award of the Gold Medal of the Royal Astronomical Society to E. Hertzsprung," in *Monthly Notices of the Royal Astronomical Society*, **89** (1929), 404–417. Obituaries include Axel V. Nielsen, "Ejnar Hertzsprung—Measurer of Stars," in *Sky and Telescope*, **35** (January 1968), 4–6, K. Aa. Strand, "Ejnar Hertzsprung, 1873–1967," in *Publications of the Astronomical Society of the Pacific*, **80** (1968), 51–56; and A. J. Wesselink, "Ejnar Hertzsprung," in *Quarterly Journal of the Royal Astronomical Society*, **9** (1968), 337–341.

K. Aa. Strand

HESS, GERMAIN HENRI (*b.* Geneva, Switzerland, 8 August 1802; *d.* St. Petersburg, Russia [now Leningrad, U.S.S.R.], 13 December 1850), *chemistry*.

Hess was noted chiefly for his thermochemical investigations which laid the groundwork for later research in chemical thermodynamics. The son of a Swiss artist, he was taken to Russia at the age of three when his father became a tutor in a rich Moscow family. In Russia, where he remained for the rest of his life, he was called German Ivanovich Gess. He took a medical degree at the University of Dorpat (now Tartu, Estonia) in 1825 and then visited the laboratory of Berzelius in Stockholm. Although he stayed only a month, he became a lifelong friend of the Swedish chemist, corresponded constantly with him, and was strongly influenced by him in his scientific career.

In 1826 Hess established a medical practice at Irkutsk, where he carried out a number of studies of Siberian mineral resources that resulted in his election as an adjunct in chemistry of the Imperial Academy of Sciences in St. Petersburg on 11 November 1828. After returning to the capital he became a full academician in 1834 and held teaching posts in most of the city's institutions of higher education.

Under the influence of Berzelius, almost all of Hess's early work was concerned with analysis of inorganic and organic substances, but he was also well aware of contemporary theoretical problems. Like most chemists of the period, he accepted Dalton's atomic theory and the law of definite proportions. He was interested in the question of the nature of affinity but did not accept Berzelius' electrochemical theory, which was then the most popular among chemists. As early as 1830 he began to think that a solution to the problem of affinity could be found by studying the quantities of heat evolved in chemical reactions.

Hess then took up the calorimetric work of Lavoisier and Laplace. He was at first uncertain whether heat was due to a vibrational motion of particles or to a material substance, caloric, but finally decided that the caloric theory was more realistic, believing that if he could find examples of the combination of caloric with chemical elements in definite proportions, he would obtain a clearer view of the nature of affinity and the inner constitution of chemical compounds. He began serious experimental studies with an ice calorimeter in 1838 and by 1840 was able to formulate his two major thermochemical laws.

His experimental studies of the heat that developed in the formation of various hydrates of sulfuric acid and in a number of neutralization reactions showed that the amount of heat was always the same, whether the reaction proceeded directly or through a number of intermediate steps. This law of the constant summation of heat was obviously a special case of the law of the conservation of energy, which had not yet been formally stated. Hess saw clearly the practical utility of his law in determining heats of reaction that could not be measured directly. His second law, that of thermoneutrality, stated that there was no heat effect when neutral salts underwent double decomposition in water solution. The explanation of this law was not given until Arrhenius published his ionic theory in 1887.

The thermochemical work of Hess was continued extensively in the second half of the nineteenth century through the studies of Thomsen and Berthelot. Both Berthelot's principle of maximum work and the thermodynamic theories of affinity which came to prevail were clearly foreshadowed in the work of Hess.

In addition to his internationally known research in thermochemistry, Hess was very influential in the development of chemistry in Russia. His text *Osnovania chistoy khimii* ("Fundamentals of Pure Chemistry") went through seven editions and did much to establish the chemical nomenclature of the Russian language. He was always interested in technological questions, and many of his students later contributed to Russia's industrial development.

BIBLIOGRAPHY

I. Original Works. Hess's papers are found chiefly in the various publications of the St. Petersburg Academy of Sciences from 1827 to 1849 and in *Annalen der Physik und Chemie* (1827–1848). The most important thermochemical papers are reprinted in *G. I. Gess. Termokhimicheskie issledovania* ("G. I. Hess. Thermochemical Investigations"; Moscow, 1958). Selections from these papers were also published as *Thermochemische Untersuchungen von G. Hess,* vol. IX of Ostwalds Klassiker der Exacten Wissenschaften (Leipzig, 1890). See also *Osnovania chistoy khimii* ("Fundamentals of Pure Chemistry"; St. Petersburg, 1831; 7th ed., 1849).

II. Secondary Literature. On Hess and his work see Y. I. Soloviev, *German Ivanovich Gess* (Moscow, 1962). An account of the students who carried on Hess's work is Z. I. Sheptunova, "Khimicheskaya shkola G. I. Gessa" ("The Chemical School of G. I. Hess"), in *Trudy Instituta istorii estestvoznaniya i tekhniki. Akademiya nauk SSSR,* **18** (1958), 75–103. An English account is H. M. Leicester, "Germain Henri Hess and the Foundations of Thermochemistry," in *Journal of Chemical Education,* **28** (1951), 581–583.

Henry M. Leicester

HESS, VICTOR FRANZ (FRANCIS) (*b.* Schloss Waldstein, Styria, Austria, 24 June 1883; *d.* Mount Vernon, New York, 17 December 1964), *physics.*

Hess was the son of Vinzenz Hess, forester to the prince of Oettingen-Wallerstein, and Serafine Grossbauer-Waldstätt. He received his early education at the Humanistisches Gymnasium in Graz, from which he graduated in 1901; from 1901 until 1905 he studied mathematics and physics with Leopold von Pfaundler at the university in that city. He took the Ph.D. at the University of Graz in 1906, remaining there to do advanced work with Franz Exner and Egon von Schweidler until 1908. In the latter year Hess became *Privatdozent* in physics at the Vienna Veterinary College, and in 1910 he was appointed assistant to Stefan Meyer at the newly founded Institute for Radium Research at the university. He was made associate professor in 1911.

When Hess joined Exner and his group in Vienna, ionization in the atmosphere was a principle area for physical research. It was generally known that free air contained electrons, and that if the electrons were removed from air sealed in a container new ones would soon be regenerated, even if the container were shielded in lead. Radioactive pollution of the walls of the container was thought to be responsible for this phenomenon at first; then the effect was attributed to gamma rays originating in the atmosphere and soil. Since the laws governing the diminution of intensity of gamma rays were known, physicists next attempted to identify the origin of those responsible for atmospheric ionization.

In 1910 Theodor Wulf, making experiments on the Eiffel Tower, observed that the ionization of the atmosphere at a height of 300 meters above a gamma-ray source is greater than that at a distance of 300 horizontal meters. He thus admitted the possibility of extraterrestrial sources for such radiation and suggested that this hypothesis might be confirmed by balloon experiments. A. W. F. E. Gockel, among others, attempted such experiments, but achieved no definite results.

Hess took up the problem stated by Wulf in 1911. He first verified the rate of absorption of gamma rays and then, with the help of the Austrian Academy of Sciences and the Austrian Aeroclub, made ten difficult and daring balloon ascensions, collecting data with improved instrumentation. He reached a height of 5,350 meters, with striking results. He was able to establish that to a height of approximately 150 meters above sea level, radiation decreased according to known laws, while at greater heights radiation increased steadily, following approximately the same laws. He found radiation at 5,000 meters to be several times greater than that at sea level, and also that radiation at all levels was the same night or day, and therefore not the result of the direct rays of the sun. He was thus able to conclude that the radiation he recorded at high altitudes entered the atmosphere from above and was, in fact, of cosmic origin. His results were verified in an extension of his experiments made by W. Kohlhörster in 1913—Kohlhörster reached a height of 9,300 meters, and recorded radiation of twelve times that at sea level—but were not acknowledged by other physicists for a number of years. ("Cosmic rays" were so named by R. A. Millikan in 1925.) In 1913 Hess himself equipped the meteorological station on Hoch Obir (2,141 meters) in Carinthia to accommodate further studies of cosmic radiation; these experiments, however, were brought to a halt by World War I.

In 1920 Hess was appointed associate professor at the University of Graz; he soon left this position to accept an offer from the U.S. Radium Corporation in Orange, New York. In the United States Hess served that organization as director of its research laboratory and also acted as consulting physicist for the Department of the Interior (Bureau of Mines). He returned to Graz in 1923 and became full professor there in 1925. After 1927 Hess was able, with the help of a number of Austrian and international organizations, to buy new equipment and to make further investigations of cosmic radiation in several parts of the Alps and the island of Helgoland. In 1931 he was further subsidized by a number of international bodies—in particular, the Rockefeller Foundation—and established a cosmic-ray observatory at an altitude of 2,300 meters on the Hafelekar Spitze, near Innsbruck. He returned to Graz in 1937, but was dismissed from his professorship in 1938, following the Nazi occupation of Austria, because of his strict Roman Catholicism.

Hess returned to the United States, where he became a professor of physics at Fordham University, in New York City, in 1938. He remained at Fordham until his retirement with emeritus status in 1956, becoming a naturalized U.S. citizen in 1944. Hess continued his experiments on the tower of the Empire State Building, at Fordham, and on voyages to South America and in the Pacific. He studied the gamma radiation of rocks, the dust pollution of the atmosphere, and also investigated the refractive indexes of mixtures of liquids. He further concerned himsef with the biomedical problems of workers who handled radium, having himself undergone a thumb amputation in 1934 as a result of an accident with radioactive substances.

Hess's discovery of cosmic radiation brought him many honors, including membership in the Austrian Academy of Sciences (1933) and the Papal Academy of Sciences; honorary doctorates from Fordham University, Loyola University, and the University of Innsbruck; the Ernst Abbe prize of the Carl Zeiss Foundation (1932); and the Austrian Medal for Science and Arts (1959). The most important honor, however, was the Nobel Prize in physics, which he shared with C. D. Anderson in 1936, on which occasion he lectured on "Unsolved Problems in Physics: Tasks for the Immediate Future in Cosmic Ray Studies." The discovery of cosmic radiation was one of the keys to the study of elementary particles in general, leading to the discovery of the positron, by Anderson in 1932, and of the μ meson by F. Neddermayer (in 1937).

Hess was married twice, to Mary Bertha Warner (*d.* 1955) in 1920, and to Elizabeth M. Hoenke in 1955.

BIBLIOGRAPHY

I. ORIGINAL WORKS. The Austrian Academy of Sciences has an unpublished list of more than 130 articles by Hess; see also his works listed in Poggendorff.

The most important reports of his discovery of cosmic rays are in *Sitzungsberichte der K. Akademie der Wissenschaften in Wien,* Mathematisch-naturwissenschaftliche Klasse, **120** (1911), 1575–1585; **121** (1912), 2001–2032; and **122** (1913), 1053–1077, 1481–1486. On his balloon experiments see "Aeronautische Radiumforschung," in *Österreichischer Aero-Club Jahrbuch,* 1911 (1912), pp. 102–108; and 1912 (1913), pp. 190–205. See also *Die elektrische Leitfähigkeit der Atmosphäre und ihre Ursachen* (Brunswick, 1926), trans. as *The Electrical Conductivity of the Atmosphere and Its Causes* (London, 1928); "Luftelektrizität," in *Müller-Pouillets Lehrbuch der Physik,* 11th ed., V (Brunswick, 1928), 519–661, written with H. Benndorf; "Das Verhalten des Bodens gegen Elektrizität und Radioaktivität des Bodens," in Edwin Blanck, ed., *Handbuch der Bodenlehre,* VI (1930), 375–396; "The Cosmic Ray Observatory in the Hafelekar (2300 Meters)," in *Terrestrial Magnetism and Atmospheric Electricity,* **37,** no. 3 (1932), 399–405; "Die Jonisierungsbilanz der Atmosphäre," in *Ergebnisse der kosmischen Physik,* **2** (1933), 95–152; "Ungelöste Probleme in der Physik," his Nobel Prize lecture, in *Les Prix Nobel en 1936* (Stockholm, 1937), pp. 1–3, trans. as "Unsolved Problems in Physics: Tasks for the Immediate Future in Cosmic Ray Studies. Nobel Lecture Dec. 12, 1936," in *Nobel Lectures. Physics (1922–1941)* (Amsterdam–London–New York, 1965), pp. 360–362; "The Discovery of Cosmic Radiation," in *Thought* (1940), pp. 1–12; *Die Weltraumstrahlung und ihre biologischen Wirkungen* (Zurich, 1940), written with Jacob Eugster, trans. as *Cosmic Radiation and Its Biological Effects* (New York, 1949); "Persönliche Erinnerungen aus dem ersten Jahrzehnt des Instituts für Radiumforschung," in *Sitzungsberichte der österreichischen Akademie der Wissenschaften,* Mathematisch-naturwissenschaftliche Klasse, **159,** sect. IIa (1950), 43–45; and "Work in the USA," in *Österreichische Hochschulzeitung* (15 Jan. 1955), p. 4.

II. SECONDARY LITERATURE. On Hess and his work, see the series of articles by Rudolf Steinmaurer, a colleague and collaborator, "Zum 70. Geburtstag," in *Acta physica austriaca,* **7** (1953), 209–215; "Zum 75. Geburtstag," *ibid.,* **12** (1959), 121 ff.; "50 Jahre kosmische Strahlung," in *Physikalische Blätter,* **18** (1962), 363–369; "Victor F. Hess, der Entdecker der kosmischen Strahlung, 80 Jahre alt," in *Acta physica austriaca,* **17** (1964), 113–120; and an obituary notice in *Almanach. Österreichische Akademie der Wissenschaften,* **116** (1966), 317–328 (with portrait). Other articles are in *Österreichs Nobelpreisträger* (Vienna, 1965), pp. 117–127; and J. G. Wilson, obituary notice in *Nature,* **207** (1965), 352.

JOSEF MAYERHÖFER

HESSE, LUDWIG OTTO (*b.* Königsberg, Germany [now Kaliningrad, U.S.S.R.], 22 April 1811; *d.* Munich, Germany, 4 August 1874), *mathematics.*

Hesse was the eldest son of Johann Gottlieb Hesse, a merchant and brewer, and his wife, Anna Karoline Reiter. He grew up in Königsberg, where he had his first contact with the sciences at the Old City Gymnasium. After obtaining his school certificate in 1832, he attended the University of Königsberg, specializing in mathematics and the natural sciences. Hesse studied mainly under C. G. J. Jacobi, who greatly stimulated his mathematical investigations. After taking the examination for headmaster in 1837 and spending a probationary year at the Kneiphof Gymnasium in Königsberg, Hesse made an educational journey through Germany and Italy. In the fall of 1838 he began to teach physics and chemistry at the trade school in Königsberg. In 1840 he graduated from the University of Königsberg and was made a lecturer there on the basis of his thesis, *De octo punctis intersectionis trium superficium secundi ordinis.* After this he lectured regularly, and in 1841 he resigned his position at the trade school. In the same year he married Maria Dulk, daughter of a chemistry professor; they had six children.

In 1845 Hesse was appointed extraordinary professor at Königsberg; he spent a total of sixteen years there as teacher and researcher. During this time nearly all his mathematical discoveries were made, and he published them in Crelle's *Journal für die reine und angewandte Mathematik.* Among those attending his lectures were Gustav Kirchhoff, Siegfried Heinrich Aronhold, Carl Neumann, Alfred Clebsch, and Sigismund Lipschitz.

Despite recognition of his scientific achievements, it was not until 1855 that Hesse received a call as ordinary professor to the University of Halle. Shortly thereafter he received an appointment to Heidelberg, which he gladly accepted, for Robert Bunsen and his former student Kirchhoff were there. From the winter of 1856 until 1868 Hesse taught in Heidelberg. During this period he wrote the widely read textbooks *Vorlesungen über analytische Geometrie des Raumes* and *Vorlesungen über analytische Geometrie*. According to Felix Klein, Hesse's methods of presenting material fortified and disseminated the feeling for elegant calculations expressed in symmetrical formulas. In 1868 Hesse accepted a call to the newly founded Polytechnicum at Munich. But only a few more years of activity were granted him, and he died in 1874 of a liver ailment. At his request, he was buried in Heidelberg, the city that had become his second home. The Bavarian Academy of Sciences, of which Hesse had become a member in 1868, arranged for the publication of his complete scientific works.

Hesse's mathematical works are important for the development of the theory of algebraic functions and of the theory of invariants. His achievements can be evaluated, however, only in close connection with those of his contemporaries. Hesse was indebted to Jacobi's investigations on the linear transformation of quadratic forms for the inspiration and starting point of his initial works on the theory of quadratic curves and planes. For proof (again influenced by Jacobi) he used the newly developed determinants, which allowed his presentation to reach an elegance not previously attained. Hesse again presented the results of these first researches when he developed his space geometry in his textbook.

In 1842 Hesse began his investigation on cubic and quadratic curves, which are closely linked to the development of basic concepts of algebra. The starting point was the paper "Über die Elimination der Variabeln aus drei algebraischen Gleichungen zweiten Grades mit zwei Variabeln." Again the problem can be traced to Jacobi. A treatise on the inflection points of cubic curves immediately followed this work. Within the framework of this treatise is the functional determinant that is named after Hesse and arises from the second partial derivative of a homogeneous function $f(x_1,x_2,x_3)$:

$$H = \begin{vmatrix} f_{11} & f_{12} & f_{13} \\ f_{21} & f_{22} & f_{23} \\ f_{31} & f_{32} & f_{33} \end{vmatrix}$$

This functional determinant has found many applications in algebraic geometry. In linear transformation of the variables x_1, x_2, x_3 into the variables y_1, y_2, y_3, $H' = A^2 \cdot H$, where A is the determinant of the matrix of the transformation and H is a covariant of f. Upon geometrically applying his first fundamental theory of homogeneous forms, Hesse obtained the result that the points of inflection of a curve C_n of the nth order are generally given as the intersection of this curve and a curve of the order $3(n - 2)$. These curves can be described by means of the Hessian determinant of C_n. Julius Plücker had previously obtained this result for C_3. With this work Hesse demonstrated how, by geometrical interpretation, the results of algebraic transformations could not only equal, but even surpass, the results of geometers.

Hesse devoted much research effort to the geometrical interpretation of algebraic transformations, admitting that he was stimulated primarily by the geometrical works of Jakob Steiner and by Plücker and Poncelet. Plücker had further discovered that the planar C_3 contains nine points of inflection, which lie on twelve straight lines in groups of three. Hesse proved that these twelve straight lines are arranged in four triple lines, each of which contains all nine points. He further demonstrated that for a complete mathematical solution of the problem an equation of the fourth degree is necessary; this was later confirmed by Aronhold.

A similar investigation of groupings was necessitated by the twenty-eight double tangents of the planar C_4. Here too Hesse's starting point was the so-called canonical representation of C_4 in the form of a symmetrical determinant of a quadruple series. By this representation of the equation of the curve, the planar problem of the double tangent can be combined with a spatial problem: eight points in space are connected by twenty-eight straight lines. If a group of planes of the second order, infinite in both directions, is drawn through these eight fixed base points, then the parameters of the conical surfaces of this group are sufficient for a condition that can be understood as the given equation of this group. This connection led to the proof that the special case of the equation of C_4 can be represented in thirty-five other ways, all markedly different from the first.

From the beginning, Hesse always sought to arrange his calculations with homogeneous symmetrical starting points, so that the algebraic course of the calculation would be the counterpart of the geometric considerations. His student Alfred Clebsch in particular has used this concept in his own work and has further expanded on it.

In England, Cayley was also working on the theory of homogeneous forms. Rivalry arose when his "Mémoire sur les hyperdéterminants" appeared simultaneously with Hesse's paper.

Hesse's teaching was also influential. In his long years as a lecturer, he continually showed his enthusiasm for mathematics, and his textbooks on analytical geometry must be seen in this context. The special forms of linear equation and of planar equation that Hesse used in these books are called Hesse's normal form of the linear equation and of the planar equation in all modern textbooks in this discipline.

BIBLIOGRAPHY

I. ORIGINAL WORKS. Hesse's collected works were posthumously published by the Math.-phys. Kl. of the Bavarian Academy of Sciences as *Gesammelte Werke* (Munich, 1897). Individual works include "Über die Elimination der Variabeln aus drei algebraischen Gleichungen zweiten Grades mit zwei Variabeln," in *Journal für die reine und angewandte Mathematik,* **28** (1844), 68–96; *Vorlesungen über analytische Geometrie des Raumes* (Leipzig, 1861; 3rd ed., 1876); *Vorlesungen über analytische Geometrie der geraden Linie* (Leipzig, 1865; 4th ed., 1909); and "Sieben Vorlesungen aus der analytischen Geometrie der Kegelschnitte," in *Zeitschrift für Mathematik und Physik,* **19** (1874), 1–67.

II. SECONDARY LITERATURE. On Hesse or his work, see Gustav Bauer, "Gedächtnisrede auf Otto Hesse," in *Abhandlungen der Bayerischen Akademie der Wissenschaften;* Alexander Brill and Max Noether, "Die Entwicklung der Theorie der algebraischen Funktionen in älterer und neuerer Zeit," in *Jahresberichte der Deutschen Mathematikervereinigung,* **3** (1892–1893), 107–565; Moritz Cantor, "Otto Hesse," in *Allgemeine deutsche Biographie,* vol. XII (Leipzig, 1880); Felix Klein, *Vorlesungen über die Entwicklung der Mathematik im 19. Jahrhundert,* vol. XXIV in Die Grundlehren der mathematischen Wissenschaften (Berlin, 1926); Franz Meyer, "Bericht über den gegenwärtigen Stand der Invariantentheorie," in *Jahresberichte der Deutschen Mathematikervereinigung,* **1** (1890–1891), 79–281; and Max Noether, "Otto Hesse," in *Zeitschrift für Mathematik und Physik,* Hist.-lit. Abt., **20** (1875), 77–88.

KARLHEINZ HAAS

HESSEL, JOHANN FRIEDRICH CHRISTIAN (*b.* Nuremberg, Germany, 27 April 1796; *d.* Marburg, Germany, 3 June 1872), *mineralogy, crystallography.*

Hessel's most important scientific contribution was his mathematical derivation, from consideration of the symmetry elements of crystals, of the fact that there can be only thirty-two crystal classes and that only two-, three-, four-, and sixfold axes of symmetry can occur. His results, published two decades before the work of Bravais, were overlooked until Leonard Sohncke drew attention to their importance in 1891.

After attending the industrial school (later the Realschule) at Nuremberg, Hessel studied science and medicine at Erlangen and Würzburg, from which he received the M.D. in 1817. He pursued further scientific studies at Munich, where he met the noted mineralogist Karl C. von Leonhard, who persuaded Hessel to accompany him as his assistant to Heidelberg. There, Hessel studied physics, chemistry, mathematics, and, in particular, mineralogy and crystallography. He received the Ph.D. in January 1821 and was called to Marburg that fall as associate professor of mineralogy and mining technology. He became full professor in 1825 and remained at Marburg until his death.

In addition to his teaching, Hessel was active in the administration of the university and served for five years as a member of the Marburg city council. He published over forty scientific books and articles, primarily in mineralogy and crystallography but also in physics, astronomy, chemistry, zoology, and botany. In 1826 Hessel demonstrated that the family of plagioclase feldspars could be considered as an isomorphous series consisting of albite and anorthite combined in all proportions, and he suggested a chemical formula for these feldspars. His results, presented in an article entitled "Ueber die Familie Feldspath" (*Taschenbuch für die gesammte Mineralogie,* **20** [1826], 289–333), did not receive contemporary attention; and this theory of the composition of the feldspars became prominent only with the work of Gustav Tschermak in 1865.

Hessel's statement of the possibility of only thirty-two crystal classes was obtained from an exhaustive analysis of the possible types of symmetry which any geometrical form might present. From a mathematical point of view, the later work of Bravais was more elegant. Hessel's results initially appeared in 1830, in an article entitled "Krystall" in *Gehler's physikalisches Wörterbuch;* and although the article was published separately in the following year, Hessel's work received no recognition among his contemporaries.

BIBLIOGRAPHY

I. ORIGINAL WORKS. Hessel's books include *Ueber positive und negative Permutationen* (Marburg, 1824); *Einfluss des organischen Körpers auf den anorganischen, nachgewiesen an Encriniten, Pentacriniten, und anderen Thierversteinerungen* (Marburg, 1826); *Krystallometrie, oder Krystallonomie und Krystallographie, besonders abgedruckt aus Gehler's physikalischem Wörterbuche* (Leipzig, 1831), new ed., edited by E. Hess (Leipzig, 1897), Ostwald's Klassiker der Exakten Wissenschaften, nos. 88 and 89; *Versuche über Magnet-Ketten und über die Eigenschaften der Glieder derselben, besonders über jene, welche ihnen angewöhnt oder auf sonstige Weise willkürlich ertheilt werden können* (Marburg, 1844); *Löthrohrtabellen für mineralogische und chemische Zwecke* (Marburg, 1847); *Die Anzahl der Parallelstellungen und jene Coincidenzstellungen eines jeden denkbaren Raumdinges mit seinem Ebenbilde und mit seinem Gegenbilde, der Regelmässigkeitsgrad der Schwerpunctes und andere bei Raumdingen in Betracht kommende Zahlen, als Merkmale für den Begriff Familie von Raumdingen nachgewiesen* (Kassel, 1853); *Die Weinveredelungsmethod des Altertums verglichen mit denen der heutigen Zeit* (Marburg, 1856); *Die merkwürdigen arithmetischen Eigenschaften der wichtigsten Näherungsreihe für die Sonnenabstände der Planeten* (Marburg, 1859); and *Uebersicht der gleicheckigen Polyeder und Hinweisung auf die Beziehungen dieser Körper den gleichflächigen Polyedern* (Marburg, 1871).

II. SECONDARY LITERATURE. See the following, listed chronologically: Leonard Sohncke, "Die Entdeckung des Eintheilungsprincips der Krystalle durch J. F. C. Hessel," in *Zeitschrift für Krystallographie,* **18** (1891), 486–498; Edmund Hess, "J. F. C. Hessel: Zur Säcularfeier seines Geburtstag," in *Neues Jahrbuch für Mineralogie,* **2** (1896), 107–122. For annotations on Hessel's work, see Hess's ed. of *Krystallometrie* mentioned above (esp. no. 88).

JOHN G. BURKE

HEURAET, HENDRIK VAN (*b.* Haarlem, Netherlands, 1633; *d.* 1660 [?]), *mathematics.*

Van Heuraet entered the University of Leiden in March 1653 as a medical student and studied mathematics under Frans van Schooten. With Christian Huygens and Jan Hudde he formed a trio of highly talented students who, under van Schooten's leadership and in touch with René François de Sluse in Liège, devised methods for tangent determinations and quadratures of algebraic curves. In a letter of December 1657 to van Schooten he reported on his results in connection with the cubic parabola $y^2 = ax^2(a - x)$ and its generalization in the "pearls" of Sluse, $y^m = kx^n(a - x)^p$.

In 1658 van Heuraet, together with Hudde, was at the Protestant academy of Saumur, where he studied the novel subject of the rectification of curves, inspired by Huygens' discovery in 1657 that the arc length of a parabola can be measured by the quadrature of an equilateral hyperbola (in modern terms, it can be expressed by means of logarithms), reported to van Heuraet by van Schooten in a letter of 28 February 1658, but only in general terms. Van Heuraet then found his own general method of rectification, which he communicated to van Schooten in a letter of 13 January 1659. Van Schooten published this letter in the Latin translation of Descartes's *Géométrie,* then being prepared for publication, under the title "De transmutatione curvarum linearum in rectas," van Heuraet's only published work and the first publication of a general method of rectification, in principle the same as the present $\int \sqrt{1 + y'^2}\, dx$. It drew attention for breaking the spell of Aristotle's dictum that curved lines could not in principle be compared with straight ones.

Van Heuraet applied his method especially to the semicubic parabola and the parabola. In a letter of 7 February 1659 to van Schooten he mentioned that he could apply his method to rotation surfaces of quadrics. Huygens and Sluse were delighted but Wallis, in a letter to Huygens (answered 9 June 1659), claimed priority for William Neile, who, he said, rectified the cubic parabola in 1657. This assertion led to the customary priority struggle. Fermat published his general rectification method in 1660—independently, it seems, of van Heuraet.

After a trip to Burgundy and Switzerland, van Heuraet reentered Leiden as a medical student in February 1659. He is mentioned in a letter from Huygens to van Schooten, dated 6 December 1659, as "subtilissimus Heuratus," but after that nothing more is heard of him.

BIBLIOGRAPHY

I. ORIGINAL WORKS. Van Heuraet's paper is in *Geometria à Renato Des Cartes . . .,* Frans van Schooten, ed. (Leiden, 1659), pp. 517–520. On pp. 259–262 van Schooten gives a construction by van Heuraet of the inflection points of a conchoid. The correspondence between van Schooten, van Heuraet, Huygens, and Sluse is in C. Huygens, *Oeuvres complètes,* II (1889); for references in other volumes, see the index.

II. SECONDARY LITERATURE. A sketch of van Heuraet's life by C. de Waard is in *Nieuw Nederlandsch biographisch woordenboek,* I (Leiden, 1911), 1098–1099. On van Heuraet's rectifications see J. E. Hofmann, "Über die ersten logarithmischen Rektifikationen," in *Deutsche Mathematik,* **6** (1941), 283–304; and M. E. Baron, *The Origins of the Infinitesimal Calculus* (Oxford, 1969), pp. 223–236. On the priority question see C. Huygens, *Horologium oscillatorium* (1673), *Oeuvres complètes,* XVII, 123, and XVIII (1934), 208–210; J. Wallis, *Tractatus duo de cycloide et de cissoide* (Oxford, 1659), *Opera,* I (Oxford, 1695), 551–553; and S. A. Christensen, "The First Determination of the Length of a Curve," in *Bibliotheca mathematica,* n.s. **1** (1887), 76–80. On Fermat's rectification see Michael Mahoney, "Fermat," in *Dictionary of Scientific Biography,* IV (1971), 572–573.

D. J. STRUIK

HEURNE, JAN VAN (or **Johannes Heurnius**) (*b.* Utrecht, Netherlands, 1543; *d.* Leiden, Netherlands, 1601), *medicine.*

Van Heurne studied medicine at Louvain and spent considerable time in Paris, where he became interested in surgery. In 1567 he traveled to Padua, the most famous center of medical education in Europe since Battista da Monte (Montanus) had introduced the teaching of medical students at the bedside. Van Heurne graduated there in 1571, returned to the Netherlands, and for twelve years practiced medicine at Utrecht.

In 1581 he was appointed professor of medicine at the University of Leiden. From the data available in the literature it appears that his lectures consisted mainly of reading from the books of Hippocrates and Galen. He made no outstanding contribution to medicine. Nevertheless his name is still mentioned, for evidently, influenced by his education in Padua, he was the first to seek to introduce bedside teaching in northern Europe.

On 4 December 1591, in the name of the medical faculty, van Heurne asked the curators of the university to make bedside teaching available to the students at Leiden. The curators delayed their response and van Heurne, a modest man, did not dare remind his superiors that he had proposed a revolutionary but necessary reorganization of the medical school curriculum. The only surviving record of his efforts is found in the archives of the University of Leiden in the form of the curators' resolution "to consider in another week the proposal of Professor van Heurne."

Forty-five years later, in 1636, Otto van Heurne, who had succeeded his father as professor of medicine at Leiden, was instructed to start teaching at the bedside.

BIBLIOGRAPHY

A complete list of van Heurne's books is in *Index Catalogue of the Surgeon General's Office,* VI (Washington, D.C., 1885), 194.

Secondary literature includes J. A. J. Barge, "Het Geneeskundig Onderwijs aan de Leidsche Universiteit in de 18e Eeuw," in *Bijdragen tot de geschiedenis der geneeskunde,* **14** (1934), 4; A. Castiglione, *Memoralia Herman Boerhaave optimi medici* (Haarlem, 1939), in Italian; J. Kroon, "Bijdragen tot de geschiedenis van het geneeskundig onderwys aan de Leidsche Universiteit (1575–1625)" (Leiden, 1911), M.D. thesis; G. A. Lindeboom, *Herman Boerhaave, the Man and His Work* (London, 1968), p. 284; and I. Snapper, *Meditations on Medicine and Medical Education, Past and Present* (New York-London, 1956).

I. SNAPPER

HEVELIUS, JOHANNES (*b.* Danzig [now Gdańsk], Poland, 28 January 1611; *d.* Danzig, 28 January 1687), *astronomy, instrument making.*

Hevelius (also known as Heweliusza, Hevel, or Hewelcke) was one of at least ten children of a prosperous brewer and property owner. Between 1618 and 1624 he was educated at a Gymnasium in Danzig; and when it was closed, he was sent to a school near Bromberg (Bydgoszcz), Poland, to acquire fluency in Polish. In 1627 he returned to the Danzig Gymnasium, where he came under the influence of Peter Krüger, a teacher of mathematics and astronomy. Krüger not only took him through the usual curriculum but also gave him private lessons in astronomy and saw to it that he learned the practical arts of instrument making and engraving.

In 1630 Hevelius went to study jurisprudence at the University of Leiden; during the voyage he made observations of a solar eclipse which he subsequently published in the *Philosophical Transactions of the Royal Society.* Besides his legal studies he acquired a further smattering of mathematics and its applications to mechanics and optics before leaving for London in 1631. From 1632 to 1634 he visited Paris, calling on Gassendi and Boulliau, and Avignon, calling on Athanasius Kircher. His letters to Krüger from this period survive. For two years he worked in his father's brewery, while studying the constitution of Danzig with a view to entering public service. In 1635 Hevelius married Katharina Rebeschke, daughter of a wealthy citizen of Danzig, and at first appears to have had little inducement to pursue his astronomical studies, despite Krüger's pleas. He observed the solar eclipse of 1 June 1639; and this year, in which he began systematic astronomical observations and which also saw the death of Krüger, was a turning point in his career.

Hevelius undertook three laborious tasks: constructing his own astronomical instruments, corresponding with many foreign astronomers, and holding civic office, first as honorary magistrate (1641) and later (1651) as city councillor (*Rathsherr*). Although his father's death in 1649 meant a further claim on his time for the day-to-day running of the brewery, it provided him with funds to build what became, for a short period, the world's leading astronomical observatory. Hevelius' first observatory was a small upper room; in 1644 he added a small roofed tower to his house and later erected a platform with two observation houses, one of which could be rotated. In 1663, the year following the death of his first wife, he married Catherina Elisabetha Koopman, his junior by thirty-six years; their three daughters lived to maturity. The daughter of a rich merchant and unusually well-educated, his second wife played a considerable part in the running of the observatory. In two plates of *Machina coelestis,* Elisabetha is represented assisting her husband in his observatory. She acted as hostess to many visiting astronomers—Halley being perhaps the best-known—and after her husband's death she edited many of his unpublished writings.

Hevelius suffered a considerable tragedy in September 1679 when, during his absence in the country, a fire destroyed his Danzig house and observatory, his instruments and the workshop for their manufacture, most of his books and papers, and his printing press. This entailed far more than a heavy financial blow, but Hevelius began to repair the damage at once, apparently having received financial help from many quarters. By August 1681 the observatory was rebuilt and reequipped, although with fewer instruments and these inferior to the ones that had been destroyed. The list of items saved is of some interest and includes most of the bound copies of his books, many of his most valuable manuscripts—including his catalogue of fixed stars, his *Globus coelestis correctus et reformatus* (which was in press), and *Prodromus astronomiae* (also approaching publication)—thirteen volumes of correspondence, and all of Kepler's manuscripts. Other works rescued were those subsequently published as *Annus climactericus* (1685) and *Firmamentum Sobiescianum sive Uranographia* (1690). There is a description of the fire in the preface to the *Annus.* Hevelius survived this catastrophe by more than seven years; but his health suffered from the shock and was not improved by a controversy with Hooke, into which he had been drawn several years earlier. He died on his seventy-sixth birthday.

Hevelius was such a punctilious publisher of his own achievements that his chief publications give a reasonably complete picture of his work. The first important work published by Hevelius was his *Selenographia: Sive lunae descriptio; atque accurata, tam macularum eius quam motuum diversorum, aliarumque omnium vicissitudinum, phasiumque, telescopii ope deprehensarum, delineatio, etc.* (1647). After a fine portrait of the author and a number of extravagantly laudatory verses by friends, Hevelius describes and illustrates an optical lathe for turning telescope lenses and gives methods for judging the parameters and qualities of lenses. His authorities are typical of the day: Witelo, Kepler, Scheiner, and Maurolico, among others. He describes Scheiner's helioscope (which he was later to modify), the microscope, and the polemoscope (the military periscope). One of his astronomical telescopes, about six feet long, is shown well-mounted mechanically—with massive ball-joints as accessories—but not equatorially. It is fitted with only a rudimentary quadrant for altitude and has no azimuth scale. Another device used with the telescope is a right-angled eyepiece for observing near the zenith. Hevelius was in the habit of using card stops with his instrument, and by their use he claimed to have perceived stars with a finite disk—a spurious appearance, of course. The largest telescope mentioned was twelve (Danzig) feet long and of approximately 50× magnification. (All lengths quoted subsequently are in units of a Danzig foot, equivalent to approximately eleven inches.)

Hevelius recounted his observations of the planets, especially Saturn, drawing it as a globe with two crescent-like handles. He recorded movements of the satellites of Jupiter, their configurations, eclipses, latitudes, and periods of revolution. He also made sunspot and eclipse observations with the helioscope, which was illustrated in both the *Selenographia* (plate L) and the *Machina coelestis* (using the same plate, now lettered "V"). A telescope pierced the center of a ball within a socket which was mounted on the wall of a darkened chamber, so that an image of the sun could be projected on blue paper pinned to a movable easel. Certain modifications of this, his first helioscope, were announced in the later book, in which they were said to have been found in 1661 by consultation with Bullialdus (Boulliau), "then one of my most valued friends." The problem of keeping the sun's disk at the same place on the easel, for protracted observation, was solved in a way of which Hevelius was inordinately proud, although it was very inelegant by comparison with the equatorial mounting known in other connections long before. An assistant controlled two screws which determined the slope of a table across which the easel moved. The method was somewhat simplified with the help of a table of the angles between the ecliptic and the vertical which Hevelius calculated for different solar longitudes and times. He might have found the more satisfactory method had he not followed Scheiner's example so closely.

The *Selenographia* proper begins with arguments disproving the ancient idea that the moon is a mirror reflecting the earth; but with the eighth chapter its contents become memorable. There Hevelius delineates and discusses the lunar markings and the movement of libration. The first lunar maps had been drawn by Thomas Harriot and Galileo almost as soon as telescopic means were available to them. Matthias Hirzgarter, in *Detectio dioptrica corporum planetarum verorum* (Frankfurt, 1643), was the first to publish a map, although an indifferent one, of the complete hemisphere. Hevelius was obliged to rely on his own observations; and the excellent engravings of *Selenographia* which resulted, done by his own hand, were judged worthy of reproduction by the fastidious Riccioli in his *Almagestum novum* (Bologna, 1651).

Hevelius gave many new names to the lunar mountains, craters, and other formations; most of them are still used. His most profitable task, though, was to draw the moon in different states of libration. He was incapable of either accounting for the multiple causes of the phenomenon or of satisfactorily formulating empirical laws to account for it; nor, *a fortiori,* did he know of the complex terrestrial and lunar motions responsible for them. Hevelius' descriptions of a librational cycle of shadow changes in the lunar details, his method of judging the libration by means of changes in the apparent (telescopic) separation of a pair of lunar details, and his introduction of rudimentary lunar coordinate systems provided a sound basis for the work of subsequent astronomers. *Selenographia* ends, aside from appendixes of various observations, with a description of a mounted lunar globe, perhaps the first of its kind, permitting the representation of librational movements. One of the conclusions of the appendixes is that the mean synodical period of solar rotation, judged from sunspot movement, is twenty-seven days. Hevelius did not, of course, appreciate the change of velocity with distance from the solar equator.

The second great work by Hevelius, not published until more than twenty years after the *Selenographia,* was *Cometographia, totam naturam cometarum, ut pote sedem, parallaxes, distantias ortum et interitum, capitum, caudarumque diversas facies . . . beneficio unius eiusque fixae et convenientis hypotheseos exhibens; etc.* (1668). An introductory engraving is doubly interesting; it depicts Hevelius sitting at a table with a cometary orbit shown as a conic section combined with a spiral, the sun at the focus of the former. By contrast, a figure of Aristotle holds an illustration of some linear and sublunary cometary paths. Below is a valuable illustration of Hevelius' house and observation platform.

Helevius devoted the first book of the *Cometographia* to the comet of 1652, showing, for example, that its parallax was not great enough for it to be sublunary. In fact he had an ingenious but inaccurate way of judging parallax and greatly underestimated the comet's distance. Later Hevelius wrote on the physical constitution of comets, but without much insight—favoring, for instance, a disklike (as opposed to a spherical) structure for the head. In books VI, VII, and XII he collected a considerable body of information, especially concerning the comets of the two preceding centuries. He supposed comets to be condensed planetary exhalations, and he believed them linked with the material responsible for sunspots, thus leading himself into obvious difficulties over velocities and orbital planes. When he questioned the physical causes of cometary motions he was barely able to pass beyond a vague and qualitative explanation in terms of impulses provided by interacting exhalations. It was by analogy with the parabolic motion of terrestrial projectiles that he decided on a fundamentally parabolic motion for comets. When in due course the idea was accepted, it was not as a result of the hypothesis of *Cometographia;* and those who have claimed priority for Hevelius are on very weak ground.

One of Hevelius' first efforts as an engraver is the frontispiece to Kircher's *Primitiae gnomonicae catoptricae* (Avignon, 1635), described by T. Przypkowski as "the richest known diagram of a reflexive sundial"; and the competence of its engraver may be explained by the great interest of Hevelius' first teacher, Peter Krüger, in gnomonics. In 1638 Hevelius designed a new type of dial, several examples of which he is thought to have made. The signed original disappeared in 1945. A fine dial by him, but of a totally different sort, is a triple mural dial on the wall of the royal palace of Wilanów, built near Warsaw for the Polish king Jan III Sobieski about 1680. In Hevelius' library were forty items on gnomonics, omitting nothing of importance published on the subject.

Hevelius undoubtedly owed the success of his observations to his skill in designing, making, and engraving instruments; and the work in which he described his techniques was of very great interest to his contemporaries. *Machina coelestis, pars prior, organographiam, sive instrumentorum omnium quibus auctor hactenus sidera rimatus ac dimensus est . . .; item de maximorum tubor constructione et commodissima directione, etc.* (1673), was followed by *Machina coelestis pars posterior, rerum Uranicarum observationes, etc.* (1679); fewer than 100 copies survived the fire.

One of Hevelius' first efforts at making large instruments was to complete a copper azimuth quadrant which Krüger had begun, the expense of which was to be met by the Danzig senate. He went on to copy several of Tycho's instruments. That he had made a wide study of earlier instruments is evident from the first book of *Machina coelestis,* in which he evaluates the accuracy of the observations of ancient and modern astronomers. Subsequent descriptions of his instruments include the following (all in the first volume of *Machina coelestis*): copper quadrant of radius three feet, wooden base with four screw feet (chap. 2); copper sextant of radius three feet, for two observers (chap. 3); copper sextant of radius four feet for a single observer (chap. 4); wooden quadrant with stand and counterpoise (chap. 5); wooden sextant of more than six feet radius, after Tycho's design (chap. 6); wooden double octant of radius eight feet, with two centers and two scales, having no alidade but movable pinnules (chap. 7); three copper quadrants of between one and two feet radius, each equipped with verniers (with thirty-one divisions, against thirty) equipped with screws for fine adjustment (chap. 8); a very fine large quadrant of five feet (for altitude) and of four feet (for azimuth), with counterpoises, pulleys, ropes, and screws for adjustment, the pinnules with two pairs of slots at right angles, all housed in an octagonal building (chap. 9); large copper vernier (with sixty-one divisions, against sixty) quadrant with mercury level, all so well counterpoised that "the slightest breath of air would cause it to turn," although it weighed 800 (Danzig) pounds (chap. 10); large brass sextant cross-membered in iron to prevent flexing, and of more than six feet radius, again very finely counterpoised with weights, ropes, and pulleys (chap. 11); copper quadrant of radius nine feet, but with scale filling only an octant, cross-membered in iron, being for use by two observers, and engraved with portraits of Hipparchus, Ptolemy, Copernicus, and Tycho (chap. 12); and portable sextant (chap. 13).

The first volume continues with a discussion in great detail of the design of pinnules, the division of instrument scales, the establishment of the meridian (and magnetic variation), and horological matters. But those chapters which were perhaps most widely read at the time (18–24) concerned his telescopes, their housings, and their mountings. Hevelius had been spurred on to build new telescopes after hearing of the discoveries (including that of the Orion nebula in 1656) made by Christian Huygens. Both men were convinced of the advantages of long-focus objectives: small chromatic and spherical aberration and high magnification with a given eyepiece, although image brightness was reduced for an extended object. Hevelius carefully described his instruments with focal lengths of 30, 40, 50, 60, 70, 140, and 150 feet (chap. 20). The problems of mounting were immense. From round tubes and tubes of box sections, he finally reduced the weight of his tube by leaving it in an open structure of narrow wooden spars, with circular rings at intervals acting as spacers (and blackened as optical stops). The larger telescopes were slung from tall masts (one of ninety feet is mentioned), and movement was effected by assistants with numerous guy ropes and pulleys. The largest seems to have had a lens about eight inches in diameter and was therefore of approximate focal ratio 1:225.

The flexing of the open frame—especially in a wind—presented the greatest mechanical problem, for the view of the objective could be almost totally obscured by the stops. Hevelius tried to solve this problem by running ropes of adjustable tension along the length of the telescope, but Halley (who sent him lenses from England) informs us that this was to no avail and that the largest telescope was useless. Problems of housing and storage were more easily solved by a prince than by a man of small means; and counting himself in the latter category, Hevelius described appropriate economies such as he had made at Sternenburg—as he called his observatory. He was often completely misled by optical imperfections, spending, for instance, much time with a micrometer measuring the diameters of spurious stellar disks. (His adaptation of Huygens' micrometer was used to better purpose for planetary diameters.)

Halley, who had first written to Hevelius as an undergraduate in 1674, visited Danzig in 1679 at the instigation of the Royal Society, in the hope of resolving amicably a violent controversy begun by Robert Hooke. Hevelius had sent copies of his *Cometographia* to several fellows of the Society, including Hooke, who had in return recommended the use of telescopic rather than plain sights on graduated instruments. The correspondence continued, with neither party yielding ground; and in reply to the *Machina coelestis,* Hooke had written his *Animadversions on the First Part of the Machina Coelestis of . . . Hevelius* (1674). There are too many imponderables for us to pronounce on the merits of the several arguments, but clearly the Danzig arguments ceased to apply as the mechanics and graduation of instruments steadily improved.

The second volume of the *Machina coelestis* contained a considerable collection of observational data and reductions of almost every sort—a mine of information, although rare, for later astronomers—but not distinguished by its organization or by any new findings of importance. As may be judged from the earlier list of instruments, he generally observed—as was then customary—the angular separations of objects; the volume contains more than 20,000 such measurements, 7,000 relating to the fixed stars. By 1685 he had prepared another large volume of observations, *Annus climactericus,* dealing principally with planets and comets. By far the most widely known of his compendia of observations, however, was published after his death by his wife: *Prodromus astronomiae exhibens fundamenta quae tam ad novum plane et correctionem stellarum fixarum catalogum construendum quam ad omnium planetarum tabulas corrigendas omnimode spectant etc.* (1690). It is a catalogue of 1,564 stars arranged alphabetically under constellation names and by stellar magnitude within constellations. Latitude, longitude, right ascension, and declination are given (the latter pair of coordinates being often miscalculated even though two assistants were employed to verify calculations). John Flamsteed, another of Hevelius' many correspondents, was later to reprint the catalogue, with a different arrangement, in volume III of his *Historia coelestis Britannica* (1725). Hevelius named eleven new constellations formed of stars not included in earlier groupings; seven of these names are still used.

An idea of Hevelius' relative accuracy may be had from a comparison of the separations of ten randomly chosen bright stars: Tycho's r.m.s. error is of the order of 1′40″, Hevelius' of 50″, and Flamsteed's (with telescopic aid) of 40″. Atmospheric refraction was an important and variable source of error in all these cases.

The *Prodromus* continued a tradition of reprinting earlier catalogues, not only of William IV, landgrave of Hesse, Riccioli, Tycho, and Ptolemy, but also of Ulugh Beg's Samarkand observatory. Illustrating the constellations of Hevelius' catalogue was a volume of fifty-six plates, possibly engraved in part by Hevelius himself: *Firmamentum Sobiescianum, sive Uranographia* (1690). Contemporary globes, such as those by G. C. Eimmart, and Gerhard and Leonhard Valk, often acknowledge Hevelius as their source. Later constellation outlines and draftsmanship also owed much to *Uranographia.*

If to Hevelius' correspondence with astronomers throughout Europe we add his published writings not mentioned above, we may form some idea of his formidable industry. He does not belong to the highest rank of theoretical astronomers, although he was the doyen of mid-seventeenth-century astronomers. His character might well be judged from the sentiments expressed on his engraved title pages, two of which stand out: "Not by words but by deeds" and "I prefer the unaided eye."

BIBLIOGRAPHY

I. ORIGINAL WORKS. Reasonably full titles of Hevelius' principal books have been given in the text. They are included with a number of lesser works, all in short-title form, in the following list; the place of publication is invariably Danzig: *Selenographia* (1647); *Excellentissimo . . . Eichstadio eclipsis solis observata* (1650); *Illustribus viris . . . Gassendo et Is. Bullialdo* (1652); *Epistolae* (1654); *Dissertatio de nativa Saturni facie* (1656); *Mercurius in sole visus* (1662); *Prodromus cometicus* (1665); *Descriptio cometae* (1666); *Cometographia* (1668); *Epistola ad Oldenburgium de cometa* (1672); *Machina coelestis,* 2 pts. (1673–1679); *Excerpta ex literis . . . ad Hevelium* (1683); *Annus climactericus* (1685); *Uranographia* (1690); and *Prodromus astronomiae* (1690). The posthumous works were often bound together. Several of the above works are available in modern facsimile editions.

For further bibliography see *Allgemeine deutsche Biographie,* XII (1880), 341–343; D. Wierzbickiego (see below); and especially L. C. Béziat, "La vie et les travaux de Jean Hévélius," in *Bullettino di bibliografia e di storia delle scienze matematiche e fisiche,* **8** (1875), 497–558, 589–669, also published separately (Rome, 1876).

Hevelius left correspondence and observations filling more than seventeen folio vols., sold nearly forty years after his death by his son-in-law to Joseph-Nicholas Delisle for 1,200 ducats. (Delisle had been called to Russia by Catherine I and was visiting Danzig en route.) See Bibliothèque de la Chambre des Députés, MS 1507, I, 36. This material passed first to the Bureau des Longitudes, Paris, and thence to the Bibliothèque Nationale and the observatory, where fifteen vols. remain. They are available on microfilm. Hevelius was responsible for preserving many of Kepler's papers, which also passed to his heirs and ultimately to Leningrad. The catalogue of Hevelius' library is in the Paris observatory, MS C, 2, 5.

II. SECONDARY LITERATURE. L. C. Béziat (see above) is a fundamental source. See also A. von Brunn, "Johannes Hevelius' wissenschaftliche Tätigkeit . . .," in *Schriften der Naturforschenden Gesellschaft in Danzig,* n.s. **13** (1911), 30–44; G. A. Seidemann, *Johannes Hevelius* (Zittau, 1864); and J. H. Westphal, *Leben, Studien und Schriften des Astronomen J. Hevelius* (Königsberg, 1820). A very good memoir in Polish is D. Wierzbickiego, "Źywot i działalność Jana Heweliusza, astronoma polskiego," in *Pamiętnik Akademii umiejętności w Krakowie, Wydzialy: Filologiczny i Historyczno-Filozoficzny,* **7** (1889), 22–78. Useful for personal detail and relations with Halley are E. F. MacPike, *Hevelius, Flamsteed and Halley* (London, 1837), pp. 1–16, 75–124; and *Correspondence and Papers of Edmond Halley* (Oxford, 1932), *passim.* For fuller details of Hevelius' instruments, see E. Zinner, *Deutsche und niederländische astronomische Instrumente des 11.–18. Jahrhunderts,* 2nd ed. (Munich, 1967), pp. 375–382. See also Tadeusz Przypkowski, "Gnomonics of John Hevelius," in *Actes du dixième congrès international d'histoire des sciences,* II (Paris, 1964), 695–697. The best comparative ed. of the star catalogues of Hevelius and others is Francis Baily, *Memoirs of the Royal Astronomical Society,* vol. **13** (1843), 296 pp. including prefaces and notes. For a facs. of *Uranographia,* with intro. especially concerning Ulugh Beg's observatory, see *Jan Hevelius, Yulduzlar osmonining atlasi,* V. P. Shcheglov, ed. (Tashkent, 1968). The place of Hevelius' telescope in the history of that instrument is discussed in H. C. King, *The History of the Telescope* (London, 1955). Still perhaps the best account of the contents of Hevelius' principal works is J. L. Delambre, *Histoire de l'astronomie moderne,* II (Paris, 1821), 435–495. For an example of the influence of *Selenographia,* see W. H. Ryan, "John Russell, R. A., and Early Lunar Mapping," in *Smithsonian Journal of History,* **1** (1966), 27–48.

J. D. NORTH

HEVESY, GYÖRGY (*b.* Budapest, Hungary, 1 August 1885; *d.* Freiburg im Breisgau, Germany, 6 July 1966), *radiochemistry, physical chemistry, analytical chemistry, biochemistry.*

Hevesy came from a family of wealthy industrialists ennobled by Franz Joseph I (he signed his name when writing in German as von Hevesy). He attended the Piarist Gymnasium in Budapest and then entered the University of Budapest, where he studied physics and chemistry. He continued his education in Berlin and Freiburg. In 1908 he received his doctorate for an investigation of the interaction between fused sodium and sodium hydroxide. Hevesy began his scientific career as an assistant to Richard Lorenz at the University of Zurich. He soon moved to the Technische Hochschule in Karlsruhe in order to study catalytic processes at Fritz Haber's institute. To his regret, Haber directed him to investigate whether molten zinc emits electrons. Because there was no one in Karlsruhe with experience in measuring radiation, Hevesy went to Rutherford at Manchester in order to acquaint himself with radioactive materials.

Rutherford's laboratory was then one of the few investigating radioactive phenomena. Many of the observations made there had the effect of unsettling the structure of classical physics and chemistry to its very foundations. There the first great generation of atomic scientists grew to maturity. Nearly all the young people then working at the laboratory became world-famous researchers. Hevesy formed an especially close friendship with Niels Bohr.

The first great surprises regarding radioactive elements had already been experienced. Rutherford had announced that radioactivity is caused by the transmutation of elements. Consequently one must conclude that atoms are composite entities, and Rutherford began to develop his conception of the planet-like structure of the atom. Many decay products of the natural radioactive series had already been identified. In his radioactive displacement rule Soddy had previously stated the relation by which radioactive daughter elements, which he named "isotopes," are listed in the periodic table. According to this rule, elements of different atomic weights must have the same location in the table. At the time it remained unclear whether isotopes are in fact fully identical in chemical terms. Rutherford himself was not certain on this point. When the laboratory received from Austria a by-product from the preparation of uranium containing so much natural lead that it absorbed the radiation of the radium D (a lead isotope) present in it, he told Hevesy: "If you are worth your salt, separate radium D from all that nuisance of lead."

Hevesy began the work necessary to fulfill this request. The separation of the radium D from the lead was not accomplished through any chemical means. He first had to realize that isotopes are not chemically separable. He continued his work at the Radium Institute in Vienna with Paneth. Although radium D cannot be separated from lead, lead can be "marked" (detected and traced) by the radiation of the admixed radium D. In 1913 Hevesy published "Über die Löslichkeit des Bleisulfids und Bleichromats" in *Zeitschrift für anorganische Chemie* (**82** [1913], 323–328), which brought him the Nobel Prize for chemistry. The introduction to this work summarizes the essential aspects of radioactive tracing:

> The fourth decay product of radium emanation, RaD, shows, as is well known, the chemical reactions of lead. If one mixes the RaD with lead or lead salts, the former cannot be separated from the lead by any chemical or physical methods; and once the complete mixing of the two materials has taken place, the concentration ratio remains the same even for arbitrarily small amounts of lead that one removes from the solution. Since RaD, as a result of its activity, can be detected in incomparably smaller amounts than lead, it can thus serve as a qualitative and quantitative proof of [the presence of] lead, to which it is attached: RaD becomes an indicator of lead.

The Nobel Prize did not come to Hevesy for thirty years (he received it in 1943). The reason for the thirty-year delay was that nearly all discoveries of Hevesy were premature. As long as scientists dealt with only the few natural radioactive isotopes, the radioactive tracing techniques possessed a very restricted range of application, as did all the other "radio tracer" and radioanalytic methods that Hevesy developed in the meantime. Their importance greatly increased when, through the invention of the production techniques for artificial radioactive isotopes, these methods found many applications.

After 1913 Hevesy contributed much to the definitive clarification of the question of isotopes ("Zur Frage der isotopen Elemente," in *Physikalische Zeitschrift,* **15** [1914], 797–804). After having unambiguously established their chemical identity, he demonstrated the identity of their electrochemical properties. He also aided H. G. Moseley in his work on the relationship of the frequency number of the Kα line and the chemical atomic number.

Hevesy—on vacation in Hungary when World War I was declared—served in the Austro-Hungarian army. After the war he was a *Privatdozent* at the University of Budapest. With Gyula Gróh he applied his "marking" method to demonstrating the autodiffusion of metal ions in the crystal lattice, tracing in particular the autodiffusion of radium D in solid lead. With Laszlo Zechmeister he showed that radioactivity was equally divided between salts crystallized from the mixture of inactive lead chloride and "labeled" lead nitrate. When a solution of a "labeled" lead salt was mixed with that of an organic lead compound, the activity was retained in the original salt. Arrhenius greeted this experiment as a significant and striking proof of his ionic theory.

The turbulent postwar political situation in Hungary impelled Hevesy to leave the country. In 1920 he went to Copenhagen and worked with his friend Bohr, who had become a professor.

In the periodic table only four spaces were not occupied. One of the missing elements possessed the atomic number 72. Element 71, lutetium, belonged to the rare earths. It had long been supposed that element 72 also belonged to that group, and thus it was sought in monazite sand, a source of rare earths. On the basis of Bohr's newly worked-out theory of electron configuration, with seventy-two electrons a new orbital should open up. Consequently, element 72 ought to be similar to zirconium rather than to the rare earths. On this supposition, Hevesy and Dirk Coster began the radiographic examination of zirconium ores and were able to demonstrate in them the line of an unknown element, which they then isolated chemically in the form of a fluoride. They named it hafnium after the Latin name of Copenhagen (*Nature,* **111** [1923], 78–79, 182).

Also in 1923 Hevesy reported that with the help of radium D he had traced the absorption of lead in plants; this was the first application of the radioactive tracer technique to biology (*Biochemical Journal,* **17** [1923], 439–445). There followed the investigation of the distribution of bismuth in the animal body (a rabbit) with the aid of an active bismuth isotope, which marked the first use of the tracer method in medical research.

In 1926 Hevesy was called to the University of Freiburg im Breisgau. There, with E. Alexander, he observed that when the elements of higher atomic number are subjected to X rays, a characteristic secondary emanation begins that can be used in the detection and determination of the element in question (*Nature,* **128** [1931], 1038–1039). In this way the method of X-ray fluorescence analysis was discovered. Because this method was not feasible with the equipment then available, it would not prevail in practice for twenty years. A further achievement during Hevesy's stay at Freiburg was the invention, with R. Hobbie, of the isotope dilution method, which enriched analytical chemistry with a completely new technique. Using it, they were able to determine the lead content of rocks (*Nature,* **129** [1932], 315).

Politics again affected Hevesy's career. After the Nazis came to power he was forced to leave Germany. He returned to Copenhagen, where he once more enjoyed the hospitality of the Bohr Institute. Until this time Hevesy had made all his discoveries with the few naturally radioactive isotopes, thus narrowing their field of possible application. In 1934 the Joliot-Curies produced the first artifically radioactive element by means of neutron irradiation, thereby beginning a period in which artificially radioactive isotopes of almost all the elements could be produced. The importance of the radioactive tracer method therefore increased rapidly. Today there is hardly a branch of science or technology in which this procedure is not used.

For analytic purposes, Hevesy immediately drew upon the Joliot-Curie method of transmuting elements through neutron irradiation. In 1935 he and Hilde Levi developed the method of neutron activation analysis, now among the most important microanalytic procedures and indispensable in testing the extremely pure materials required by modern technology. They described the first application as follows:

> We used the method of artificial radioactivity to determine dysprosium content of yttrium preparations. The procedure was the following: we mixed 0.1%, 1% etc. of dysprosium . . . and determined the intensity obtained. The yttrium sample to be investigated was then activated under exactly the same conditions and a comparison of the dysprosium activities obtained gave 1% as the dysprosium content. . . [*Kongelige Danske Videnskabernes Selskabs Skrifter,* Math. Medd., **14** (1936), 5–34].

Hevesy was the first to use an artificially produced isotope as a tracer. He produced P^{32} through neutron irradiation (1935) according to Joliot's method and immediately used the preparation to study phosphorus metabolism in rats. Thus began his extensive biochemical activity, in which he employed a great many isotopes to investigate medicochemical problems—for example, to examine the distribution of elements in the body and in carcinomas, and to study the formation of blood corpuscles, of DNA, and of other substances.

In 1942, following the German occupation of Denmark, Hevesy made a perilous escape to Sweden, where he continued his work at the University of Stockholm.

In addition to the Nobel Prize, Hevesy received other major scientific awards, including the Faraday,

Copley, and Bohr medals, the Fermi Prize, the Ford Prize, and the second Atoms-for-Peace Award.

BIBLIOGRAPHY

I. ORIGINAL WORKS. Hevesy's most important periodical publications were collected in *Adventures in Radioisotope Research,* 2 vols. (Oxford, 1962), which includes a complete bibliography and an autobiography entitled "A Scientific Career." Among his books are *Lehrbuch der Radioaktivität* (Leipzig, 1923), written with H. Paneth; and *Die seltenen Erden vom Standpunkt des Atombaues* (Berlin, 1927).

II. SECONDARY LITERATURE. See H. Levi, "George de Hevesy," in *International Journal of Applied Radiation and Isotopes,* **16** (1965), 512-524; and *Nuclear Physics,* **98** (1967), 1-24; and F. Szabadváry, "George Hevesy," in *Journal of Radioanalytical Chemistry,* **1** (1968), 97-102.

FERENC SZABADVÁRY

HEWSON, WILLIAM (*b.* Hexham, Northumberland, England, 14 November 1739; *d.* London, England, 1 May 1774), *hematology.*

Hewson, son of a country surgeon, was trained in medicine at Newcastle-on-Tyne and went in 1759 to William Hunter's anatomy school in London, where he also attended St. Thomas's and Guy's hospitals. After a winter's course at Edinburgh in 1761-1762, he became assistant and partner in Hunter's school. In 1767 he published the first practical account of paracentesis of the thorax in cases of emphysema, later admitting that this operation had been proposed by others.

During 1768-1769 Hewson read three papers to the Royal Society on his exploration of the lymphatic system in the lower vertebrates, which led to a priority dispute with Alexander Monro II; John Hunter also claimed to have preceded him. Hewson had in fact made a more complete demonstration of his subject than any of his predecessors through the previous century. He was elected a fellow of the Royal Society on 8 March 1770 and was awarded the Copley Medal in November. He continued as Hunter's resident assistant until his marriage to Mary Stevenson on 10 July 1770. Hunter proposed in 1771 to dissolve the partnership because Hewson no longer lived in the school, while Hewson claimed personal ownership of preparations that he had made while teaching there. Benjamin Franklin effected their reconciliation, but Hewson set up his own school in Craven Street in September 1772.

Hewson had reported his microscopical research on blood to the Royal Society during 1770. By well-planned experiments and precise thermometry he ascertained the role of fibrinogen and gave the first valid account of coagulation. Microscopy was little practiced because the compound microscopes of the time produced distortions and current methods of preparing tissue for examination were inadequate. Hewson relied on a single lens and devised a satisfactory means of mounting "wet" specimens. He was the first to observe the lymphocytes in the thymus and spleen and concluded that their production was the function of these glands. He republished his papers on the blood in 1771, adding a long appendix on his dispute with Monro about the lymphatics. He reported his observations on the red corpuscles in 1773, showing that they were discoid—not spherical, as was believed—but mistaking the dark center of the disk for a nucleus. He was also the first to describe clearly the three parts of the blood, components already known to contemporary anatomists.

Early in 1774 Hewson republished his papers on the lymphatics. After his death, from the effects of a dissection wound, his school and researches were continued by Magnus Falconar, who married Hewson's sister Dorothy on 7 September 1774. Falconar repeated Hewson's experiments on the spleen and thymus and in 1777 published his corroboration with a reprint of Hewson's paper on the red corpuscles. He died of phthisis on 24 March 1778, aged twenty-three; his and Hewson's joint museum was sold that October.

BIBLIOGRAPHY

I. ORIGINAL WORKS. Hewson's MSS are held in the archives of the Royal Society of London. His articles and books are listed in his *Works* (1846), pp. xlix-l; and the books alone in K. F. Russell, *British Anatomy 1525-1800 a Bibliography* (Melbourne, 1963), pp. 127-129. They include "The Operation of Paracentesis Thoracis," in *Medical Observations and Inquiries,* **3** (1767), 372-396; "The Lymphatic System in Birds, . . . in Amphibious Animals, . . . in Fish," in *Philosophical Transactions of the Royal Society,* **58** (1768), 217-226, and **59** (1769), 198-203, 204-215; "Experiments on the Blood"; "On the Degree of Heat Which Coagulates the Lymph"; "Further Remarks on the Properties of Coagulable Lymph," *ibid.,* **60** (1770), 368-383, 384-397, and 398-413, respectively; *An Experimental Inquiry Into the Properties of the Blood, and an Appendix Relating to the Discovery of the Lymphatic System* (London, 1771); "On the Figure and Composition of the Red Globules," in *Philosophical Transactions of the Royal Society,* **63** (1773), 303-323; *Experimental Inquiries; Part 2 . . . the Lymphatic System* (London, 1774); "A Letter to Dr Haygarth" in *Medical and Philosophical Commentaries* (Edinburgh), **3** (1775), 87-93, on the thymus and spleen,

1773; and *Experimental Inquiries: Part 3 . . . the Red Particles of the Blood, . . . the Structure and Offices of the Lymphatic Glands, the Thymus Gland and the Spleen,* Magnus Falconar, ed. (London, 1777).

His works were brought together as *Opera omnia . . .*, J. T. van de Wynpersse, ed. (Leiden, 1795); and *The Works of William Hewson,* edited, with intro. and notes, by George Gulliver (London, 1846).

A portrait by Vandergucht was engraved in 1780 and is reproduced in the *Works* (1846).

II. SECONDARY LITERATURE. Hewson's MS sources are in the archives of the American Philosophical Society, the British Museum (Natural History), the College of Physicians of Philadelphia, and the Royal College of Surgeons of England. For information on Hewson, see the following sources, listed chronologically: *Museum Falconarianum, a Catalogue of the Anatomical Preparations of the Late Magnus Falconar and William Hewson . . . Sold by Auction* (London, 1778); S. F. Simmons, *Account of William Hunter* (London, 1783), with a brief memoir of Hewson by his widow (pp. 38–39); J. C. Lettsom, "Memoirs of the Late William Hewson," in *Transactions of the Medical Society of London,* **1,** pt. 1 (1810), 51–63; "Correspondence," in T. J. Pettigrew, *Memoir of J. C. Lettsom,* I (London, 1817), 136–147, a memoir of Hewson by his widow; G. Gulliver, "On the Life and Writings of Hewson," in *The Works of William Hewson,* pp. xiii–xlviii; J. F. Payne, "Hewson, William," in *Dictionary of National Biography,* new ed., IX, 763–764; L. G. Stevenson, "William Hewson, the Hunters, and Benjamin Franklin," in *Journal of the History of Medicine,* **8** (1953), 324–328, documents on the quarrel of 1771–1772; M. C. Verso, "A Note on the Observations of Hewson and Falconar on the Morphology of Red Blood Cells, With an Account of Their Theory of Blood Formation," in *Medical Journal of Australia,* **2** (1957), 431–432; J. Dobson, "John Hunter's Microscope Slides," in *Annals of the Royal College of Surgeons of England,* **28** (1961), 175–188, slides acquired from Hewson's sale, with a discussion of Hewson's microscopy; and W. Dameshek, "William Hewson: Thymicologist, Father of Hematology," in *Blood,* **21** (1963), 513–516.

WILLIAM LEFANU

HEYN, EMIL (*b.* Annaberg, Germany, 5 July 1867; *d.* Berlin, Germany, 1 March 1922), *technology, metallography.*

Heyn, after graduating from the Realgymnasium in Annaberg, attended the Bergakademie at Freiberg, where Adolf Ledebur was his teacher. Several years of practical experience in the steel industry completed his education. He then taught at the Maschinenbau- und Hüttenschule at Gleiwitz, Upper Silesia (now Gliwice, Poland), until 1898, when he was called to Charlottenburg as an assistant to Adolf Martens at the Königliche Mechanisch-Technische Versuchsanstalt, from which the Königliche Materialprüfungsamt at Berlin-Dahlem developed. Here Heyn took over the task of continuing the microscopic investigations of metals and alloys begun by Martens, applying them to practical problems.

Among his publications of this period is "Die Verwendbarkeit der Metallmikroskopie für die Prüfung der Werkzeugstähle" (1901), which clearly indicates the goal of all of Heyn's later work: making scientific knowledge useful in practice. The same theme was treated in 1903 in *Die Metallographie im Dienste der Hüttenkunde,* which offered many new views on the practical application of metallography. With a lecture entitled "Labile und metastabile Gleichgewichte in Eisen-Kohlenstofflegierungen," Heyn entered into the discussion of the nature of the annealing process. He proposed to distinguish a stable iron-carbon system and a metastable iron–iron-carbide system and to include both in a double diagram, the iron-carbon equilibrium diagram generally accepted today.

Following the expansion of the Mechanisch-Technische Versuchsanstalt into the Königliche Materialprüfungsamt in 1904, Heyn became deputy director and manager of the metallography division. In an exceedingly fruitful collaboration with Oswald Bauer this division produced, until the fall of 1914, a great number of papers on interesting defects, on the constitution of steels, and on problems of nonferrous metals; these publications brought much prestige to the new institution. An article of 1911, "Über Spannungen in kaltgereckten Metallen," was fundamental in furthering the knowledge of inner stresses.

In 1901 Heyn was called to the Berlin-Charlottenburg Technische Hochschule to succeed A. von Hörmann in the chair of general mechanical technology. Despite his heavy work load at the Materialprüfungsamt, Heyn sought to fulfill his duties as a university teacher with a great sense of responsibility. At his suggestion mechanical technology, which previously had covered all the materials used in machine construction, was divided into branches and metals were treated according to their relative importance. In 1911 Heyn published his pioneering article "Der technologische Unterricht als Vorstufe für die Ausbildung der Konstrukteure," which played a major role in the organization of technological education in German colleges. In his last years Heyn worked exclusively with nonferrous metals, and with characteristic energy he founded the Deutsche Gesellschaft für Metallkunde that was to create a broader basis for the practical application of metallography.

In 1920 Heyn was appointed director of the newly

founded Kaiser-Wilhelm-Institut für Metallforschung. In December 1921 he was ceremoniously installed in this post, but a short time later he contracted a grave illness, from which he never recovered.

BIBLIOGRAPHY

I. Original Works. Encompassing the metallography of iron and of the nonferrous metals, Heyn's scientific work treats inner stresses and considers pedagogic problems. Among his many works are the following, published in the first decade of the twentieth century: "Die Theorie der Eisen-Kohlenstofflegierungen nach Osmond und Roberts-Austen," in *Stahl und Eisen,* **20** (1900), 625-636; "Einfluss des Siliziums auf die Festigkeitseigenschaften des Flussstahles," *ibid.,* **21** (1901), 460-464; "Die Verwendbarkeit der Metallmikroskopie für die Prüfung der Werkzeugstähle," *ibid.,* 977-980; "The Overheating of Mild Steel," in *Journal of the Iron and Steel Institute,* **2** (1902), 73-109, discussion on 110-145; *Die Metallographie im Dienste der Hüttenkunde* (Freiberg, 1903); "Labile und metastabile Gleichgewichte in Eisen-Kohlenstofflegierungen," in *Zeitschrift für Elektrochemie,* **10** (1904), 491-503; "Über Ätzverfahren zur makroskopischen Gefügeuntersuchung des schmiedbaren Eisens und über die damit zu erzielenden Ergebnisse," in *Metallurgie,* **4** (1907), 119-122; "Über bleibende Spannungen in Werkstücken infolge Abkühlung," in *Stahl und Eisen,* **27** (1907), 1309-1315, 1347-1358; "Zur Metallurgie des Roheisens," *ibid.,* 1565-1571, 1621-1625, written with O. Bauer; and "Über den Angriff des Eisens durch Wasser und wässrige Lösungen," in *Mitteilungen aus dem K. Materialprüfungsamt zu Berlin-Dahlem,* **26** (1908), 1-104, written with O. Bauer.

After 1910 he published "Der technologische Unterricht als Vorstufe für die Ausbildung der Konstrukteure," in *Zeitschrift des Vereins deutscher Ingenieure,* **55** (1911), 201-210, 305-308; "Über Spannungen in kaltgereckten Metallen," in *Internationale Zeitschrift für Metallographie,* **1** (1911), 16, written with O. Bauer; "Über Spannungen in Kesselblechen," in *Stahl und Eisen,* **31** (1911), 760-765, written with O. Bauer; "Untersuchungen über Lagermetalle," *ibid.,* 509-511, 1416-1422, written with O. Bauer; "Die Kerbwirkung und ihre Bedeutung für den Konstrukteur," in *Zeitschrift des Vereins deutscher Ingenieure,* **58** (1914), 383-391; "Untersuchungen über die Wärmleitfähigkeit feuerfester Baustoffe," in *Stahl und Eisen,* **34** (1914), 832-834; *Untersuchungen über Lagermetalle, ausgeführt im Kgl. Materialprüfungsamt Berlin-Lichterfelde im Auftrage des Vereins zur Beförderung des Gewerbefleisses zu Berlin* (Berlin, 1914), written with O. Bauer; "Einige weitere Mitteilungen über Eigenspannungen und damit zusammenhängende Fragen," in *Stahl und Eisen,* **37** (1917), 442-448, 474-479, 497-500; "Neuere Forschungen über Kerbwirkung, insbesondere auf optischem Wege," *ibid.,* **41** (1921), 541-546, 611-617, 700; and *Metallographie, kurze gemeinfassliche Darstellung der Lehre von den Metallen und ihren Legierungen, unter besonderer Berücksichtigung der Metallmikroskopie,* 3rd ed., revised by O. Bauer (Berlin-Leipzig, 1926).

II. Secondary Literature. See "E. Heyn†," in *Zeitschrift für Metallkunde,* **14** (1922), 97-100; and O. Bauer, "Gedächtnisrede auf E. Heyn in der Deutschen Gesellschaft fur Metallkunde, Berlin, 30 Juni 1933," in *Mitteilungen aus dem K. Materialprüfungsamt zu Berlin-Dahlem,* **40** (1922), 1-10.

Franz Wever

HEYNITZ (HEINITZ), FRIEDRICH ANTON VON (*b.* Dröschkau, near Torgau, Germany, 14 May 1725; *d.* Berlin, Germany, 15 May 1802), *mining.*

Heynitz' father was a privy councillor and royal counsel of Saxony as well as inspector of the Fürstenschule at Meissen; his mother was related to K. A. von Hardenberg, who was later chancellor of Prussia. After education at home and in Schulpforta, he began mining studies at the end of 1742 in Dresden and continued them in the following year at Freiberg, where he did practical work in mining. He also took study trips to the Erzgebirge and Bohemia.

Heynitz entered the service of Brunswick in July 1746 as associate inspector on the Mining Council at Blankenburg. In 1747 he visited mines in Sweden, and in 1749 and 1751 he became acquainted with mining in Hungary and Styria. For his achievements the Mining Administration of Brunswick named him member of the privy finance council, and on 13 August 1762, ten years after his promotion to deputy chief inspector of mines, it appointed him director of all mining in the Harz Mountains.

On 10 December 1763, Elector Frederick Christian of Saxony appointed Heynitz director of the Saxon mining industry. In this new office he exceeded all expectations, making many important contributions. His service to the Freiberg Bergakademie was especially important: for instance, as of Easter 1766, mining was taught systematically on the university level. He also became curator of the Bergakademie.

In 1768 Heynitz took over the management of the Saxon salt works. While inspecting this industry, he discovered abuses and abolished them, in recognition of which he was made chief inspector of mines. In this post too he brought about order through careful inspection and strict supervision. This involved him in intrigues at the electoral court and in disagreements with the elector, and he therefore asked to be allowed to resign. His request was granted on 24 August 1774 and he returned to Dröschkau to devote himself to extensive studies in political economy, which led to his *Essai d'économie politique.*

In October 1775 Heynitz went to Paris for a year and a half and was placed in charge of certain

Spanish mines owned by an international finance company. On 7 November 1776 the Prussian inspector general of mines, Baron Waitz von Eschen, died in Berlin. Frederick the Great considered Heynitz to be a suitable successor and offered him the post. Heynitz accepted, becoming a Prussian state minister and inspector general of mines, heading the mining and metallurgy departments. He took office on 9 September 1777.

In this position Heynitz visited all of Prussia's mining and metallurgical operations, reorganized the mining administration, began a systematic investigation of the mineral deposits, improved the management, established relief funds for the miners, and attended to the sale of the output from the mines. In addition he was responsible for the construction of turnpikes, canals (including the Klodnitz Canal), and railways, the improvement of mapping of the mines, the building of housing for miners, and the extension of the miners' health insurance system. His activity also extended to the extraction of raw materials of various kinds and their processing in factories, iron forges, steel works, and brass foundries, as well as to coinage and to manufacturing of porcelain at Berlin.

Heynitz founded the Friedrichshütte, the foundry at Gleiwitz (now Gliwice, Poland), and the Königshütte in Upper Silesia; he also restored ore mining near Tarnowitz (now Tarnowskíe Gory). He is particularly remembered for his introduction of the steam engine and the coke oven. As a result of Heynitz' efforts the Berlin Bergakademie was reorganized in 1778, the Kunstakademie received a constitution in 1790, and the Bauakademie was founded in 1798. His outstanding service in all these fields was recognized by award of the Order of the Black Eagle in 1791.

BIBLIOGRAPHY

I. ORIGINAL WORKS. Heynitz's works include *Grundriss über die Gänge und Züge am Oberharz. Copirt von Schink* (1799); *Abhandlung über die Produkte des Mineralreichs in den königl. preussischen Staaten und über die Mittel, diesen Zweig des Staats-Haushaltes immer mehr emporzubringen* (Berlin, 1786); *Tabellen über die Staatswissenschaft eines europäischen Staates der vierten Grösse nebst Betrachtungen über dieselben* (Leipzig, 1786).

II. SECONDARY LITERATURE. See E. Burisch, "Oberberghauptmann Friedrich Anton von Heinitz, der Vater der deutschen Bergleute," in *Glückauf*, **98** (1962), 28–39; "Stiftung einer Heinitz-Plakette durch die Wirtschaftsvereinigung Bergbau," *ibid.*, 44; "Friedrich der Grosse und der Freiherr von Heinitz," in E. Reimann, *Abhandlung zur Geschichte Friedrichs des Grossen* (Gotha, 1892), pp. 125–163; "Friedrich Anton von Heynitz," in C. Schiffner, *Aus dem Leben alter Freiberger Bergstudenten*, II (Freiberg, 1938), 14–16; F. Schröter, "Friedrich Anton Freiherr von Heynitz," in *Monatsschrift für Deutsche Beamte*, **16** (1892), 366–373, 411–420, 462–470; Walter Serlo, "Friedrich Anton von Heynitz," in *Männer des Bergbaus* (Berlin, 1937), pp. 67–68; and "Friedrich Anton von Heynitz (Lebensbilder zur Geschichte des Bergbaus)," in *Zeitschrift für das Berg-, Hütten- u. Salinenwesen im Preussischen Staat*, **82** (1934), 285–286; H. Spethmann, "Friedrich Anton von Heynitz zum hundertfünfzigsten Todestage," in *Glückauf*, **88** (1952), 492–495; O. Steinecke, "Friedrich Anton von Heynitz," in *Forschungen zur Brandenburgischen und Preussischen Geschichte*, XV (Leipzig, 1902), 110–158; and "Friedrich Anton von Heynitz," in *Allgemeine deutsche Biographie*, LV (Leipzig, 1910), 493–500; O. Täglichsbeck, "Heynitz oder Heinitz," in *Monatsschrift für Deutsche Beamte*, **24** (1900), 81; "Friedrich Anton Freiherr von Heinitz und seine Verdienste um den Aufschwung des schlesischen Berg- und Hüttenwesens," in Konrad Wutke, *Aus der Vergangenheit des Schlesischen Berg- und Hüttenlebens*, vol. V of *Der Bergbau im Osten des Königreichs Preussen* (Breslau, 1913), 25–90; and "Friedrich Anton von Heynitz," in *Anschnitt*, **10** (1958), 23–25.

M. KOCH

HEYROVSKÝ, JAROSLAV (*b.* Prague, Czechoslovakia, 20 December 1890; *d.* Prague, 27 March 1967), *electrochemistry.*

Heyrovský studied mathematics, physics, and chemistry in the Czech section of the Prague university (then called Charles-Ferdinand University), which he entered in 1909. There he was especially influenced by the physicists František Záviška and Bohumil (Gottlieb) Kučera and by the chemist Bohumil Brauner, who had studied in Manchester under Roscoe and was well known for his work on rare earths. Perhaps the example of Brauner and the fame of Sir William Ramsay influenced Heyrovský to continue his studies in England. He entered University College, London, in 1910, and received the B.Sc. in 1913. It was an unusual step, since at that time most Czech graduate students tended to complete their education in Germany, France, or Switzerland. Attracted to electrochemistry, a subject close to the heart of F. G. Donnan, who had succeeded Ramsay, Heyrovský began a Ph.D. thesis on the electrochemical properties of aluminum.

World War I began while Heyrovský was on holiday in Prague, and he was prevented from returning to London. After working for a short time in one of the chemical laboratories of Charles-Ferdinand University, in 1915 Heyrovský was drafted into the Austro-Hungarian army and spent the war years as a dispensing chemist and roentgenologist in a military hospital. This occupation apparently did not take up all his time and Heyrovský was still able to pursue his research interests; in the autumn of 1918 he submitted his Ph.D. thesis on the electroaffinity of aluminum.

After the war Heyrovský became an assistant to Brauner and continued to work on the chemistry of aluminum. His *habilitation* thesis, which qualified him in 1920 to become a *docent* in physical chemistry, dealt with the constitution and acidity of aluminic acid. Three papers summarizing his work on the electrochemical properties of aluminum (1920) brought him the D.Sc. from the University of London a year later.[1] Heyrovský's academic rise in Czechoslovakia was swift; in 1924 he became extraordinary professor and director of the newly established Institute of Physical Chemistry; and four years later he was appointed full professor in physical chemistry at Charles University.

Shortly after the German occupation of Prague in 1939, Czech universities were closed and their institutes and laboratories taken over by professors from German institutions. The holder of the chair in physical chemistry at the German University in Prague was J. Böhm, a former co-worker with Haber and Hevesy, "a unique character and highly qualified scientist."[2] Of mixed Czech-German parentage (his mother was Czech), he had no sympathies with Nazism and made it possible for Heyrovský to keep up with research during the occupation. Although misinterpreted by some, the actions and behavior of both scholars were honorable—in contrast with the overwhelming majority of Germans, who had to leave Czechoslovakia after World War II, Böhm remained in the country and in 1953 was elected corresponding member of the reorganized Academy of Sciences.

The reorganization of the Academy of Sciences (1952) was the culmination of a series of changes in the scientific life of Czechoslovakia which also brought about the establishment of a Central Polarographic Institute in 1950, with Heyrovský as its head. Later incorporated into the Academy of Sciences, it has been called the J. Heyrovský Institute of Polarography since 1964. Heyrovský received many honors both at home and abroad: he was the first Czech to win the Nobel Prize, awarded him in 1959 for his discovery of polarography. In 1965 he was elected a foreign member of the Royal Society.

Polarography was discovered when Heyrovský unified two somewhat disparate lines of investigation that related the principles of electrocapillarity to the measurement of electrode potentials. The historical connections of polarography with electrocapillarity date from the investigations of G. Lippmann, who set up (1873)[3] an electrochemical cell of which the polarizable electrode consisted of a mercury meniscus in a capillary; the nonpolarizable electrode was a large mercury pool at the bottom of the cell. Lippmann proceeded to examine surface tension alterations of the mercury meniscus under the influence of polarization. Its changes in elevation (proportional to the changes in surface tension), plotted against applied voltage (equal to the potential of the polarizable electrode), result in a curve, known as the electrocapillary parabola. In the light of the electric double-layer conception, the peak of the curve denotes the potential at which the mercury surface is uncharged.

This "static" approach to electrocapillary phenomena was followed by another method developed by Kučera (1903).[4] Instead of following the movements of the meniscus in the capillary, he weighed the mercury dropping from the capillary because the drop's weight is directly proportional to the surface tension. Because it involved a continuous renewal of the mercury surface, this approach was described as "dynamic." In certain cases, such as dilute electrolyte solutions, Kučera obtained parabolic curves with a secondary maximum. These anomalies did not occur with the static method, and Kučera was unable to elucidate them.

As professor of physics at the Czech university, Kučera examined Heyrovský for his doctorate. Following the examination, at Kučera's suggestion, Heyrovský undertook a systematic study of the anomalous behavior of electrocapillary curves, in the course of which his experimental and theoretical knowledge of electrochemistry became very useful. Heyrovský noticed that the addition of reducible cations to the solution caused an inflection in the electrocapillary curves at potentials close to the decomposition voltages of the cations. Instead of the rather unrewarding weighing of mercury drops at different potentials, he began to measure the current between the dropping mercury electrode and the large mercury pool which served as a reference electrode. In due course he created, in polarography, a novel method for the study of electrochemical processes.

It has been asserted that Heyrovský owed the idea of the dropping electrode to Donnan, who suggested work on the electropotential of aluminum to the young scientist. The position of aluminum in the table of electropotentials was uncertain because the metal, coated with an oxide film, did not yield reproducible measurements. In his paper on the subject (1920) Heyrovský showed that he was aware of the difficulties and, following previous attempts, decided to use amalgamated aluminum as a reversible electrode. The main problem was to devise a method to prevent the evolution of hydrogen.

Whether or not Donnan originally proposed the use of amalgam flowing out of a capillary, Heyrovský did not report on such a technique in his paper on the aluminum electrode. In fact, he adopted a type of dropping electrode employed previously by G. N. Lewis and his co-workers. Heyrovský became acquainted with the work of the American chemists on the determination of potentials of alkaline metals, published in the *Journal of the American Chemical Society* between 1910 and 1915. The significance of this work lay in the success of the American chemists in measuring the potentials of alkaline metals by means of alkali amalgam electrodes. In their experiments, the Americans had set up a special apparatus for the preparation and preservation of the dilute amalgam with the amalgam surface at the end of a capillary serving as an electrode. At first they found that the sodium amalgam surface, when placed in the sodium hydroxide, evolved considerable hydrogen. But when the amalgam surface was repeatedly renewed by allowing one or two drops of amalgam to flow out at the end of the capillary, "the surface remained clear of hydrogen for ten to twenty minutes, and showed a constant and perfectly reproducible potential within 0.1 millivolt."[5] Heyrovský was impressed by this work, done at the laboratory of physical chemistry of the Massachusetts Institute of Technology, and wrote:

> The high overvoltage of hydrogen on a mercury surface makes it possible for a dilute amalgam of a very negative metal to behave as a reversible electrode, because the evolution of hydrogen is almost entirely prevented. Lewis . . . has been able to determine the electrolytic potentials of alkali metals using dilute amalgams, and the same method has been adopted here for aluminum.[6]

Heyrovský's inclinations toward electrochemistry and his recent experience in this field transformed the study of electrocapillarity into polarography by 1921. After measuring with a galvanometer currents passing through the cell to which potentiometrically different voltages were applied, Heyrovský observed that the current-voltage curves obtained with the dropping mercury electrode represented qualitative and quantitative relationships characteristic for the solution undergoing electrolysis. As the applied voltage became greater, the current increased not continuously but in steps, reaching limiting values corresponding to the different cations or other reducible groups in the solution. On plotting the values for voltage and current, he obtained usually S-shaped curves in which the position of the polarographic curve or wave (voltage) indicated the qualitative composition of the solution and the height of the curve (current) determined its contents quantitatively.

Heyrovský reported his findings for the first time in 1922, writing in Czech; a year later he published them in English.[7] From then on, he remained in the van of experimenters with the dropping mercury electrode. He said in his Nobel lecture:

> The reason why I keep some 38 years to the electrochemical researches with the dropping mercury electrode is its exquisite property as electrode material. Its physical condition of dropping as well as the chemical changes during the passage of the electric current are well defined, and the phenomena displayed at the dropping mercury electrode proceed with strict reproducibility. Owing to the latter property the processes at the electrode can be exactly expressed mathematically.[8]

Heyrovský never tired of emphasizing that the advantages of the dropping mercury electrode depended on the considerations that the surface of mercury was renewed and that the large overvoltage on the mercury electrode prevented hydrogen deposition. Highly reproducible results are obtained with a very small amount of solution because the mean current depends only on the applied potential and is independent of time and of the direction of the polarizing voltage.

The term "polarography" was not coined until 1925. In that year Heyrovský and his co-worker Masuzo Shikata (later professor at Kyoto) published a description of an instrument which they called a "Polarograph."[9] It automatically registered the current-voltage curves or "polarograms" on a cylinder covered with photographic paper and connected to a Kohlrausch drum (originally rotated by means of a phonograph motor). It was one of the earliest automated laboratory instruments, and the first Polarograph cost only about £3 to build. The mechanic at the institute was prepared to supply a Polarograph excluding the galvanometer and source of light for £10. Before the introduction of the Polarograph the production of a polarogram often took over an hour, but the novel arrangement reduced it to fifteen to twenty minutes. It is noteworthy that with the very first instrument a high sensitivity could be achieved showing depolarizers in a concentration of 10^{-5} gram molecules per liter. According to Heyrovský, some of the later developments in the construction, although producing more complex instruments, did not make them necessarily more accurate or easier to understand.

In the early 1920's electrochemistry was not considered one of the fields offering promising new openings for research. In retrospect, it is now recognized that the polarographic investigations initiated by Heyrovský gave a new impetus to the study of electrode processes. During the first two decades or so of the twentieth century Heyrovský and a small but steadily growing band of enthusiastic pupils concentrated on the theoretical foundations of polarography, which eventually led to a more precise understanding of the polarogram.

The nature of the limiting current—that is, the current which, after reaching a maximum value, remains unaffected by an increase in voltage—was considered in some detail. Heyrovský distinguished between the migration and diffusion sides of the current resulting from the electrolysis of the solution using a dropping mercury electrode. The relationship between the migration and diffusion components of the limiting current were defined and the importance of the latter in practical polarography was explained, resulting in the working out of an equation by D. Ilkovič (1934) linking in a linear relationship the diffusion current and the concentration of the depolarizer, which is the substance reduced or oxidized at the dropping electrode.[10] Heyrovský and Ilkovič (1935) also worked out an equation for the cathodic wave which threw light on the inflection point on the wave (half-wave point) and demonstrated the importance of the corresponding potential (half-wave potential) as a constant in polarography.[11] Besides diffusion-controlled currents other types, such as adsorption currents, were observed and studied by Rudolf Brdička, Heyrovský's most distinguished pupil. It was shown that the depolarizer or some other component in the solution, when adsorbed by the dropping mercury electrode, could cause changes in polarographic currents.

An important step in theoretical polarography occurred after the recognition of the existence of kinetic currents—polarographic currents governed by the rate of chemical reactions taking place near the electrode. The theory of kinetic currents began to be worked out in the early 1940's by Brdička and K. Wiesner.[12] Heyrovský's persistent interest in the problem of hydrogen overvoltage led him to propose a mechanism for the reduction of hydrogen ions at the dropping mercury electrode, based on the classical electrochemical theroy of reversible electrode potentials and the classical kinetic theory of rate reactions.[13] He believed that the overvoltage was due to slow formation of hydrogen molecules and to an interaction of water at the electrode interface. He visualized the formation of hydrogen molecules in the three steps

$$(1) \qquad H^+ + e \longrightarrow H$$

$$(2) \qquad H + H^+ \longrightarrow H_2^+$$

$$(3) \qquad H_2^+ + e \longrightarrow H_2,$$

assuming that step (2) indicated the rate-determining reaction. Heyrovský's interpretation of the hydrogen overvoltage contributed to the understanding of catalytic hydrogen currents. These polarographic currents, observed in the presence of substances which act as catalysts, are connected with the accelerated evolution of hydrogen. Brdička's discovery (1933) that proteins containing SH groups exhibit catalytic activities accompanied by hydrogen evolution demonstrated an interesting example of a catalytic current and was developed as a polarographic test with blood sera of pathological origin (taken from tumors, for instance).[14]

By 1938 the first attempts were made in polarography to use a cathode-ray oscilloscope instead of a galvanometer. The voltage of the ordinary alternating-current supply was applied to the dropping mercury electrode, and changes of its potential were followed on the oscilloscope. But Heyrovský, who began to use this method in the early 1940's[15] and since then had studied it intensively, concluded that it was necessary to distinguish between the situation in which "the oscilloscope merely replaces the galvanometer and brings no fundamental change in the polarographic instrumentation"[16] and oscillographic polarography proper, involving methods "in which the electrode is polarized by an alternating voltage or current or by single voltage or current sweeps and for which the resulting curves are followed by means of an oscilloscope."[17] No doubt, what impressed Heyrovský about oscillographic polarography was that it reduced the time of recording the curve to fractions of seconds, a much more rapid arrangement than the ordinary polarographic method and about equally accurate. It was in the course of his studies on oscillopolarography that he found it useful to employ the streaming mercury electrode in order to obtain a steady oscillogram.

It may perhaps be useful to restate Heyrovský's definition of polarography:

> . . . polarography is the science of studying the processes occurring around the dropping-mercury electrode. It includes not only the study of current-voltage curves, but also of other relationships, such as the current-time curves for single drops, potential-time curves, electrocapillary phenomena and the streaming of electrolytes, and its tools include besides the polarograph, the mi-

croscope, the string galvanometer and even the cathode-ray oscillograph.[18]

He adhered to the view that polarography was basically "restricted to the mercury capillary electrodes."[19]

Heyrovský devoted much attention to investigations of polarographic current-voltage curves which under certain circumstances show so-called maxima of the first kind, that is, a sharp increase of current above the limiting value, followed by a sudden fall to the normal magnitude.[20] This phenomenon relates to the anomalous electrocapillary curves which Kučera asked the young Heyrovský to investigate. It is curious that the problem which catalyzed the rise of polarography is as yet not completely resolved.[21]

Not until about ten years after the first publications did the scientific community outside Czechoslovakia take notice of polarography. In 1933 Heyrovský lectured at Berkeley and other American universities as Carnegie visiting professor. A year later he had the opportunity to acquaint a Russian audience with his work when he was invited to attend the Mendeleev centenary in Leningrad. The earliest translation from Czech of Heyrovský's first book on the use of polarography (1933) appeared four years later in Russian. According to Heyrovský, the major breakthrough occurred when the German analyst Wilhelm Böttger, editor of the compendium *Physikalische Methoden der analytischen Chemie,* asked him to write on polarography for volume II, published in 1936. In 1941 Heyrovský brought out his account of the subject in German, and in the United States there appeared a series of articles by O. H. Müller in *Journal of Chemical Education* and a book by I. M. Kolthoff and J. J. Lingane which long remained the major source of systematic information on polarography for English-speaking readers.[22]

Since then interest in polarography has deepened and widened because of its extensive uses not only in electrochemical and other research but also in industrial and hospital laboratories. It has been said that polarography belongs to the "top five" analytical methods, which indicates that its international recognition derives primarily from its use in analytical practice. Certainly it was the novelty of the technique and its speed which made polarography "one of the most important methods of contemporary chemical analysis,"[23] as was pointed out in the presentation speech by A. Ölander, a member of the Nobel Committee for chemistry; clearly it was the primary reason for the award of the Nobel Prize to Heyrovský. Yet Heyrovský, who was always just as concerned with the electrochemical aspects as with the analytical ramifications of polarography, took pains to refute suggestions that it was merely a somewhat better analytical procedure. Indeed, he explicitly touched upon this point in his Nobel lecture, saying:

> We meet often with the opinion that polarography did not bring anything new into chemistry except an improvement of analytical methods. That is decidedly not so, since in the study of reductions or oxidations many otherwise inaccessible physico-chemical constants are determinable. Polarography helps the investigation of chemical structure of organic and lately even inorganic compounds. . . . Although the analytical application of polarography is highly advanced at present, the field of its utilization in basic chemical problems begins to open.[24]

Even a short biographical sketch of Heyrovský would be incomplete if it were limited to the bare outline of his contribution to electrochemistry and analytical chemistry. A consideration of his life and work reveals features interesting from the point of a more general history of science. Both the Royal Society and the *Nature* obituaries found Heyrovský unique in that during an active working life of about forty years he concentrated on the elaboration of his original discovery and remained the acknowledged leader in a continuously expanding and changing area of science. In 1959, before an audience gathered in Prague to pay homage to him as the recipient of the Nobel Prize, Heyrovský admitted the high personal cost he had paid for the award. For years he spent every free moment, including long weekends, in the laboratories and gradually gave up his many-sided interests in science, literature, music, and sports. But however great his individual devotion to polarography may have been, it could not by itself account for the widespread success of that science. It is true that from the start Heyrovský did not doubt the theoretical and practical significance of his discovery, and he decided to pursue systematically the subject of the dropping mercury electrode. At a relatively early stage he was joined by a group of investigators who recognized his undisputable, although not restraining, authority and formed the nucleus of a school of polarography whose influence eventually became worldwide.

Although not particularly keen on administration, Heyrovský early recognized that growth of scientific knowledge could not be separated from its dissemination, which had to be organized. In 1928 he investigated the publication of papers by Czech chemists and found that during the ten years of Czechoslovakia's existence as an independent state they published 163 times abroad and 235 times at home. It should be added that a considerable number of these papers appeared in both Czech and another lan-

HEYROVSKÝ

guage. The survey reinforced the position of Emil Votoček, professor of chemistry at the Czech Technical University of Prague and widely known for his researches in carbohydrate chemistry, who had proposed the founding of a Czechoslovak chemical journal for original papers written in either French or English. With the aid of the ministry of education and the patronage of the Royal Bohemian Society of Sciences (founded 1769–1771), *Collection of Czechoslovak Chemical Communications* made its first appearance in 1929 under the editorship of Heyrovský and Votoček, who also remained the publishers of the Anglo-French journal until 1947. Votoček, a gifted linguist, became responsible for the French and Heyrovský concentrated on the English section, which frequently meant that they also served as translators. Their high standards led to the *Collection* attaining international recognition. It is still flourishing, now published in German and Russian as well.

The early volumes contained many of the significant contributions to polarography by Heyrovský and his school, thus constituting an important source for the history of polarography. From the beginning it was envisioned that the journal would include a bibliography of all Czechoslovak chemical publications. In 1938 Heyrovský embarked on producing from time to time, at first in the journal and then separately, bibliographies on polarography. He persevered for years in this ambitious program, aided by J. Klumpar, O. H. Müller, J. Hrbek, J. E. S. Han, and lately above all by his wife, Marie Heyrovská, who chose to remain anonymous. Heyrovský's farsightedness insured that few other fields could compete with polarography in having from the beginning continuous and good bibliographies.

Among Czech scientists Heyrovský became second only to Purkyně. Born into a national group which spoke a language understood by practically no scientist outside the Czech community, they both became promoters of Czech science but in different ways. The activities of Purkyně in the nineteenth century and Heyrovský in the twentieth century reflected two sides of a problem which scientists belonging to small national groups or countries perennially have to face. Purkyně, who became internationally recognized on the basis of his work written in German and Latin, believed passionately that science interpreted in the national language was an indispensable part of national culture. For this reason, throughout his working life he devoted much time and energy to the creation of Czech scientific terms, to the foundation of Czech scientific periodicals—in short, to the establishment of a Czech scientific culture. A hundred years later Heyrovský was determined to demonstrate the maturity of his country's chemical science to the international scientific community. A more convincing proof than Heyrovský's own contribution could hardly have been supplied.

NOTES

1. "The Electroaffinity of Aluminium. Part I. The Ionisation and Hydrolysis of Aluminium Chloride," in *Journal of the Chemical Society (Transactions)*, **117**, no. 1 (1920), 11–26; "Part II. The Aluminium Electrode," *ibid.*, pp. 27–36; "Part III. The Acidity and Constitution of Aluminic Acid," *ibid.*, no. 2 (1920), pp. 1013–1025.
2. J. D. Cockcroft, "George de Hevesy," in *Biographical Memoirs of Fellows of the Royal Society,* **13** (1967), 141.
3. G. Lippmann, "Beziehungen zwischen den capillaren und elektrischen Erscheinungen," in Poggendorff's *Annalen der Physik und Chemie,* **149** (1873), 546–561.
4. G. Kučera, "Zur Oberflächenspannung von polarisiertem Quecksilber," in *Annalen der Physik,* **11** (1903), 529–560, 698–725, extract from his *Habilitationsschrift* (Leipzig, 1903).
5. G. N. Lewis and C. A. Kraus, "The Potential of Sodium Chloride," in *Journal of the American Chemical Society,* **32** (1910), 1462.
6. *Journal of the Chemical Society (Transactions)*, **117**, no. 1 (1920), 30.
7. "Elektrolysa se rtuťovou kapkovou kathodou," in *Chemické Listy,* **16** (1922), 256–264; "Electrolysis With a Dropping Mercury Cathode. Part I. Deposition of Alkali and Alkaline Earth Metals," in *Philosophical Magazine,* **45** (1923), 303–314.
8. "The Trends of Polarography," in *Nobel Lectures Chemistry 1942–1962* (Amsterdam–London–New York, 1964), p. 564.
9. "Researches With the Dropping Mercury Cathode, Part II. The Polarograph," in *Recueil des travaux chimiques des Pays-Bas,* **44** (1925), 496–498, written with M. Shikata.
10. D. Ilkovič, "Polarographic Studies With the Dropping Mercury Kathode. Part XLIV. The Dependence of Limiting Currents on the Diffusion Constant, on the Rate of Dropping and on the Size of Drops," in *Collection of Czechoslovak Chemical Communications,* **6** (1934), 498–513.
11. "Polarographic Studies With the Dropping Mercury Electrode. Part II. The Absolute Determination of Reduction and Depolarization Potentials," *ibid.,* **7** (1935), 198–214.
12. K. Wiesner, "Über durch Wasserstoffatome katalysierte Depolarisationsvorgänge an der tropfenden Quecksilberelektrode," in *Zeitschrift für Elektrochemie,* **49** (1943), 164–166; R. Brdička and K. Wiesner, "Polarographische Bestimmung der Geschwindigkeitskonstante für die Oxydation von Ferrohäm und anderen Ferrokomplexen durch H_2O_2," in *Naturwissenschaften,* **31** (1943), 247.
13. "Polarographic Studies With the Dropping Mercury Kathode. Part LXIX. The Hydrogen Overpotential in Light and Heavy Water," in *Collection of Czechoslovak Chemical Communications,* **9** (1937), 273–301; "The Electrodeposition of Hydrogen and Deuterium at the Dropping Mercury Cathode," in *Chemical Reviews,* **24** (1939), 125–134; *Principles of Polarography* (Prague–London, 1966), p. 235, written with J. Kůta.
14. See R. Brdička, M. Březina, and V. Kalous, "Polarography of Proteins and Its Analytical Aspects," in *Talanta,* **12** (1965), 1149–1162.
15. "Oszillographische Polarographie," in *Zeitschrift für physikalische Chemie,* Abt. A, **193** (1944), 77–96, written with J. Forejt; *Oszillographische Polarographie mit Wechselstrom* (Berlin, 1960), written with R. Kalvoda.
16. Heyrovský and Kůta, *Principles of Polarography,* p. 498.
17. *Ibid.,* p. 499.

18. "The Development of Polarographic Analysis," in *Analyst,* **81** (1956), 189.
19. *Ibid.*
20. "Betrachtungen über polarographische Maxima I. Art," in *Zeitschrift für physikalische Chemie* (Leipzig) (July 1958 [separately published]), pp. 7-27.
21. Heyrovský and Kůta, *Principles of Polarography,* pp. 429-450.
22. *Polarographie, theoretische Grundlage, praktische Ausführung mit Anwendungen der Elektrolyse mit der tropfenden Quecksilberelektrode* (Vienna, 1941); O. H. Müller, "The Polarographic Method of Analysis," in *Journal of Chemical Education,* **18** (1941), 65-72, 111-115, 172-177, 227-234, 320-329, also published as a book with the same title (Easton, Pa., 1941); I. M. Kolthoff and J. J. Lingane, *Polarography* (New York, 1941).
23. *Nobel Lectures Chemistry, 1942-1962,* p. 563.
24. *Ibid.,* pp. 582-583.

BIBLIOGRAPHY

Some of Heyrovský's publications are mentioned in the notes; a full bibliography is in *Biographical Memoirs of Fellows of the Royal Society,* **13** (1967), 182-191.

Understandably, no critical account of Heyrovský's life and work exists as yet. An obituary in Czech is by R. Brdička, in *Chemické Listy,* **61** (1967), 573-580; two in English are by J. A. V. Butler and P. Zuman, in *Biographical Memoirs of Fellows of the Royal Society,* **13** (1967), 167-182; and R. Belcher, in *Nature,* **214** (1967), 953. Much the same material has been covered in P. Zuman and P. J. E. Elving, "Jaroslav Heyrovský: Nobel Laureate," in *Journal of Chemical Education,* **37** (1960), 572, repr. in Aaron J. Ihde and William F. Kieffer, *Selected Readings in the History of Chemistry* (Easton, Pa., 1965), pp. 104-109. Marie Heyrovská, "Polarographic Literature," and other contributions by eminent experts to the Heyrovský *Festschrift, Progress in Polarography,* P. Zuman and I. M. Kolthoff, eds., 2 vols. (New York-London, 1962), contain much useful historical information.

MIKULÁŠ TEICH

HEYTESBURY, WILLIAM (*fl.* Oxford, England, *ca.* 1335), *logic, kinematics.*

Heytesbury was one of several scholars at Merton College, Oxford, during the second quarter of the fourteenth century whose writings formed the basis of the late medieval tradition of *calculationes,* the discussion of various modes of quantitative variation of qualities, motions, and powers in space and time. Other leading authors of the Merton group were Thomas Bradwardine, Richard Swineshead, and John of Dumbleton. The tradition they founded spread to the Continent in the second half of the fourteenth century and enjoyed a vogue in Italian universities during the fifteenth century and again at Paris and in the Spanish universities during at least the first third of the sixteenth century. Thereafter, however, it lost impetus with the shift of interests consequent upon the humanist movement. A question still under debate among historians of science is the precise extent of the later influence of Merton kinematics, and particularly of the Merton "mean-speed theorem," which can be used to prove that in uniformly accelerated motion starting from rest, the distances are in the duplicate ratio of the times. Other phases of the Mertonian discussions involving the mathematical concepts of limit, infinite aggregate, and the continuum as a dense set of points, as well as distinctions now treated in quantificational logic, seem to have fallen into oblivion after the sixteenth century but are anticipatory of nineteenth-century work in these areas.

Biographical information about Heytesbury, as about the other Mertonian scholars, is meager. His name, variously spelled, appears in the records of Merton College for 1330 and 1338-1339; he may have been the William Heightilbury who with other Mertonians was appointed fellow of Queen's College at its founding in 1340; in 1348, however, he was still—or once more—a fellow of Merton and by that year was a doctor of theology; finally, a William Heighterbury or Hetisbury was chancellor of the university in 1371.

Heytesbury's two best-known and most influential works—the only known ones of some length, the others being short discussions of particular questions—are his *Sophismata* and *Regule solvendi sophismata.* According to the explicit of an Erfurt manuscript (Wissenschaftliche Bibliothek, Amplon. F. 135, 17r), the *Regule* was "datus Oxonie a Wilhelmo de Hytthisbyri" in 1335; and it is probable that the *Sophismata* stems from about the same time, since the two works are closely related in content, one providing rules for the resolution of different classes of real or apparent logical fallacies and the other dealing intensively with thirty-two particular sophisms. The medieval discussion of sophisms grew out of Aristotle's *Sophistical Refutations;* but as we encounter it in Heytesbury, it has developed beyond the Aristotelian treatment in two directions. First, Heytesbury employs the *logica moderna,* a set of distinctions and word-order devices developed at the University of Paris during the thirteenth century. Second, he devotes much attention to cases and problems involving modes of purely quantitative variation in space and time.

The key innovation of the *logica moderna* was the theory of supposition, an analysis of the various ways in which a term is interpretable within a given proposition for some individual or individuals. For instance, in "That man disputes," the term "that man" is said to have discrete supposition, as referring to a single, definite individual, a *suppositum* to which one could

HEYTESBURY

point. In "Some man disputes" or "Every man disputes," on the other hand, the supposition of the subject term in either case is not discrete but common, although not in the same way. Thus, from "Some man disputes" it is permissible to descend to individual cases through an alternation: "This man disputes or this man disputes or . . .," there being no existent man who is not referred to in one of the members of the alternation. But from the statement "Every man disputes" it is permissible to descend to individual cases included under the term "man" only through a conjunction: "This man disputes, and this man disputes, and . . .," and so on. Finally, there are cases in which the descent is not possible through either an alternation or a conjunction, and in these cases the supposition is said to be confused only.

The kind of supposition of a term in any particular proposition is determined partly by the meaning of the predicate or subject term with which it is conjoined and partly by the "syncategorematic" terms included in the proposition—terms incapable of serving as subject or predicate but nevertheless influencing the supposition of the subject or predicate. Examples of syncategorematic terms are "any," "all," "some," "necessarily," "always," and "immediately."

This theory appeared, fully developed, in the works of William of Shyreswood in the middle of the thirteenth century. It seems to have derived in part from the analyses of grammarians (the first known use of the verb "to supposit" in the sense required by the theory occurs in the *Doctrinale* of the twelfth-century grammarian Alexander of Villa Dei) and in part from the Abelardian explication of universals: for Abelard a universal word gives rise only to a common and confused conception of many individuals and can come to determine a particular thing or particular things only in the context of a statement. Whatever its origins, the extensional analysis of the use of terms in discourse was much in vogue by Heytesbury's time and was used by him to reveal distinctions of structure that, in modern mathematical logic, are exhibited by means of the cross-references of quantifiers and variables.

For illustration, consider the distinction that Heytesbury makes in the *Regule* between the statement "Always some man will be" and the statement "Some man will be always." In the first statement the term "man" is preceded by the syncategorematic term "always," which according to Heytesbury has a "force of confounding" (*vim confundendi*) and thus confuses the supposition of the term that follows it. Hence the supposition of the term "some man" in the statement "Always some man will be" is confused only, and it is not permissible to descend either disjunctively or conjunctively to individual *supposita.* In the statement "Some man will be always," on the contrary, the term "some man" is not preceded by the term "always" and its supposition therefore remains determinate, so that some particular although unspecified individual is referred to. The first statement asserts the immortality of the race of mankind; the second asserts the immortality of some particular man. In the symbols of present-day mathematical logic, the first statement becomes

$$(x)\ (Ey)\ (Tx.My: \supset :Oyx).$$

(Read: "For all x there is a y such that, if x is a time and y is a man, then y occurs in x.") The second statement, on the other hand, becomes

$$(Ey)\ (x)\ (Tx.My: \supset :Oyx).$$

(Read: "There is a y such that for all x, if x is a time and y is a man, then y occurs in x.") Notationally, the distinction is one of the order of the universal and existential quantifiers, (x) and (Ey).

Because this distinction is crucial for the understanding of the modern definition of mathematical limit, it is of interest to find Heytesbury applying it to cases involving a mathematically conceived continuum. Thus, he distinguishes between the statement "Immediately after the present instant some instant will be" and the statement "Some instant will be immediately after the present instant." Once again, the distinction turns on the fact that the syncategorematic term "immediately" confounds the supposition of the term following it. Thus in the first statement the term "some instant," being preceded by the term "immediately," has confused supposition only; and it is not permissible to descend disjunctively or conjunctively to particular instants. In the second statement the term "some instant" is not preceded by "immediately" and thus has determinate supposition; the statement therefore means that, of the infinitely many instants following the present instant, there is a determinate one that will be first. Heytesbury concludes that this second statement is false, whereas the first statement is true if expounded as meaning that, whatever instant after the present instant be taken, between that instant and the present instant there is some instant. In modern symbols, with the range of the variables restricted to instants of time,

$$(i)\ (Ej)\ (Ai,i_0 \supset Bj,i_0,i).$$

(Read: "For all instants i, there is an instant j such that, if i is after the present instant i_0, then j lies between i_0 and i.") The false statement would reverse the order of the quantifiers. In effect, Heytesbury is insisting that instants in a time interval, like points on a line segment, form a dense set.

It is particularly in the two chapters of the *Regule* entitled "De incipit et desinit" and "De maximo et minimo" that Heytesbury's logical sophistication in dealing with limits and extrinsically or intrinsically bounded continua comes into play. In the first of these chapters he analyzes cases in which any thing or process or state may be said to begin or to cease to be. For instance, posing the case that Plato starts to move from rest with a constant acceleration, while at the same instant Socrates starts to move from rest with an acceleration that is initially zero but increases uniformly with time, Heytesbury concludes that "both Socrates and Plato infinitely slowly begin to be moved, and yet Socrates infinitely more slowly begins to be moved than Plato." As his explication shows, what is happening here is in effect a comparison of two infinitesimals of different order: if v_S is Socrates' velocity and v_P is Plato's, then

$$\lim_{t \to 0} \frac{v_S}{v_P} = 0.$$

The "De maximo et minimo" deals with the setting of boundaries to powers—for example, Socrates' power to lift weight or to see distant objects, or the power of a moving body to traverse a medium the resistance of which varies in some specified manner. Aristotle had flatly asserted that the boundary of a power or potency is a limiting maximum. The commentator Ibn Rushd emphasized that the incapacity of a power is bounded by a *minimum quod non;* later Schoolmen such as John of Jandun were thus faced with the question of the relation between the *maximum quod sic* and the *minimum quod non.* One thought that comes to play a role in the discussion is that no action or motion can proceed from a ratio of equality between power and resistance; this thought necessitates the assignment of the negative or extrinsic boundary, so that, for instance, Socrates' power to lift weights is to be bounded by the minimum weight that he is unable to lift. Heytesbury was not the first to consider the assignment of such extrinsic boundaries; but his formulation of rules and analysis of cases, compared with earlier discussions, shows a more exclusive concern with the mathematical and logical aspects of the problem.

An important last chapter of the *Regule* entitled "De tribus predicamentis" deals with the quantitative description of motion or change in the three Aristotelian categories of place, quantity, and quality. The principal aim of each of the three subchapters ("De motu locali," "De augmentatione," "De alteratione") is to establish the proper measure of velocity in the given category. In the case of augmentation, Heytesbury adopts a measure involving the exponential function, which had already played a role in Bradwardine's *Tractatus de proportionibus* (1328). All three subchapters exhibit the almost exclusive concern of the Mertonian *calculatores* with quantitative description of hypothetical cases.

This tendency to quantitative description had roots in earlier discussions of kinematics (as, for example, in the thirteenth-century *De motu* of Gerard of Brussels) and of what was known as "the intension and remission of forms," the variation in intensity of a quality or essence. Discussions of the latter topic prior to the fourteenth century had dealt primarily with the ontological nature of such variation; but by Heytesbury's time the Scotian assumption that intension is an additive increase had been generally accepted, and Schoolmen turned their attention to a logical or semantic question: how to denominate a subject in which the intensity of a quality varies from one point to another, or—a question treated as analogous—how to denominate or measure a motion in which the velocity varies from instant to instant of time or from point to point of the moved body. This question merges into the mathematical problems of describing different possible modes of spatial or temporal variation of intensity and of finding rules of equivalence between one distribution of intensities and another. Thus in Heytesbury's *Regule,* as in later fourteenth-century writings, any particular configuration or mode of variation of intensity in space or time is called a "latitude"; and latitudes are categorized as uniform (of constant intensity), uniformly nonuniform (the intensity varying linearly with spatial extension or time), and nonuniformly nonuniform (the intensity varying nonlinearly with spatial extension or time).

Heytesbury's *Regule* is the oldest datable writing in which we find the famous Merton rule: Every latitude uniformly nonuniform corresponds to its mean degree. Thus, if the whiteness or hotness of a body varies uniformly from an intensity of two degrees at one end of the body to an intensity of four degrees at the other end, then according to Heytesbury this latitude of whiteness or hotness is equivalent to a uniform latitude of three degrees extended over the same length. This assertion rests on the presupposition—unjustified for Heytesbury and his contemporaries by any empirical measurability—that intensities of a quality are intensities of some additive quantity. In application to local motion, since intensity of motion is measured in terms of distance traversed per unit time, and distance is an additive quantity, the Merton rule leads to testable empirical consequences. Heytesbury states the rule for local motion as follows:

> For whether it commences from zero degree or from some [finite] degree, every latitude [of velocity], provided that it is terminated at some finite degree, and is acquired or lost uniformly, will correspond to its mean degree. Thus the moving body, acquiring or losing this latitude uniformly during some given period of time, will traverse a distance exactly equal to what it would traverse in an equal period of time if it were moved continuously at its mean degree. For of every such latitude commencing from rest and terminating at some [finite] degree [of velocity], the mean degree is one-half the terminal degree of that same latitude (*Regule* [Venice, 1494], fol. 39).

The proposition implies, as Heytesbury notes, that in a uniformly accelerated motion starting from rest, the distance traversed in the second half of the time is three times that traversed in the first half—a consequence admitting of application in experimental tests. The first known assertion that the Merton theorem is applicable to free fall was made by Domingo de Soto, a Spanish Schoolman, in 1555; but it was not coupled with any attempt at empirical verification. The first experimental work on the assumption that free fall is uniformly accelerated with respect to time may have been that of Thomas Harriot, who within a few years before or after 1600 was finding the acceleration of free fall to be between 21 and 32.5 feet per second squared (for Harriot's theory of ballistics and his researches on free fall, see British Museum MS. Add. 6789, 19r–86v); in his discussion of projectile motion in the same manuscript Harriot explicitly refers to the 1494 volume that contains Heytesbury's works and commentaries thereon, so that a direct influence of the medieval treatises is here indicated.

In the case of Galileo, the evidence for direct medieval influence in his work on free fall is less clear and is still under debate. The *Juvenilia,* which may be the youthful Galileo's notes on lectures at the University of Pisa, contains references to Heytesbury and Calculator (Swineshead) and to such Mertonian distinctions as that between a *maximum quod sic* and a *minimum quod non,* and that between a uniformly nonuniform and a nonuniformly nonuniform variation in intensity (see *Le opere di Galileo Galilei,* A. Favaro, ed., I, 120, 136, 139 ff., 172). But from Galileo's letter of 1604 to Sarpi, it appears improbable that his thought on the mathematical characterization of naturally accelerated motion took its start from the Merton mean-speed theorem. According to Stillman Drake, "Galileo may have known the mean-speed rule and rejected it as inapplicable to the analysis of unbounded accelerated motion" (*British Journal for the History of Science,* **5** [1970], 42). It is at least a plausible suggestion, however, that a passing acquaintance with medieval discussions and *calculationes* involving instantaneous velocities, punctiform intensities, and different modes of variation of velocity or intensity in space or time, and also with the graphical representation of such variation that had been introduced by Oresme and was incorporated in the 1494 edition of Heytesbury's works, may have served as general preparation for the thinking that Galileo would have to do in founding his science of motion.

BIBLIOGRAPHY

I. ORIGINAL WORKS. MSS giving the *Regule solvendi sophismata* in whole or in part are Biblioteca Antoniana, Padua, Scaff. XIX, MS.407, fols. 28–32, 53–56; library of the University of Padua, MS.1123, 14c, fols. 50–65; MS.1434, 15c, fols. 1–26; and MS.1570, 15c, fols. 131–137; Bodleian, Canon. Misc. MS.221, fols. 60–82; MS.376, 15c, fols. 30–32; MS.409, A.D. 1386, fols. 1–18; and MS.456, A.D. 1467, fols. 1–43; Bruges, Stadsbibliotheek, 497, 14c, fols. 46–59; and 500, 14c, fols. 33–71; Bibliotheca Marciana, Zanetti Latin MS.310, fols. 1–3; and VIII. 38 (XI, 14), a. 1391, fols. 40–54; Erfurt, Amplonian MS.135, fols. 1–17; and Vat. Lat. MS.2136, 14c, fols. 1–32; and MS.2138, 14c, fols. 89–109. It was published at Pavia in 1481 and at Venice in 1491 (fols. 4–21) and 1494 (fols. 7–52).

The *Sophismata* exists in the following MSS: Bibliotheca S. Johannis Baptistae, Oxford, MS.198, 14c, fols. 1–175; library of the University of Padua, MS.842, fols. 1–149; and MS.1123, 14c, fols. 97–172; Bodleian, Canon. Misc. MS.409, A.D. 1386, fols. 29–98; Bibliotheca Marciana, Zanetti Latin MS.310, fols. 54–79; Paris, Bibliothèque Nationale, Latin MS.16134, 14c, fols. 81–146; and Vat. Lat. MS.2137, 14c, fol. 1 *et seq.;* and MS.2138, 14c, fols. 1–86. It was published at Pavia in 1481 and at Venice in 1491 (fols. 29–99) and 1494 (fols. 77–170).

De sensu composito et diviso is in following MSS: Biblioteca Nazionale, Florence, Cl. V, MS.43, 15c, fols. 38–44; library of the University of Padua, MS.1434, 15c, fols. 26–27; Bodleian, Canon. Misc. MS.219, A.D. 1395, fols. 4–6; Bologna University, MS.289.II.2, fols. 1–4; Bibliotheca Marciana, Zanetti Latin MS.310, fols. 49–53; and Vat. Lat. MS.2136, 14c, fols. 32–36; MS.3030, fols. 55–58; MS.3038, 14c, fols. 15–22; and MS.3065, 15c, fols. 140–143. It was published at Venice in 1491 (fols. 2–4), 1494 (fols. 2–4), and 1500 (fols. 1–23).

De veritate et falsitate propositionis was published at Venice in 1494 (fols. 183–188).

"Casus obligationis" is in MS: Bodleian, Canon. Latin MS.278, 14c, fol. 70; Bibliotheca Marciana, Zanetti Latin MS.310, fol. 96; and Vat. Lat. MS.3038, 14c, fols. 37–39.

"Tractatus de eventu futurorum" is in the MS Bibliotheca Marciana, MS.fa.300 (X,207), 14c, fols. 78–79.

"Tractatus de propositionum multiplicium significatione" is available as Bibliotheca Marciana, Latin MS.VI, 160 (X, 220), a. 1443, fols. 252–253.

The following are doubtful works: *Consequentie,* in MS as Corpus Christi College, Oxford, MS.293, fol. 337 *et seq.,* and published at Bologna; *Probationes conclusionum,* in MS as Vat. Lat. MS.2189, fols. 13–38, where the work is given the title "Anonymi conclusiones," and published at Venice in 1494 (fols. 188–203); "Regulae quaedam grammaticales," in MS as British Museum, Harleian MS.179; and "Sophismata asinina," available as: Biblioteca Nazionale, Florence, C1.V, MS.43, 15c, fols. 45–46; library of the University of Padua, MS.1123, 14c, fols. 18–22; and MS.1570, 15c, fols. 113–130; Bodleian, Canon. Latin MS.278, 14c, fols. 83–87; and Bibliotheca Marciana, Zanetti Latin MS.310, fols. 122–126.

II. SECONDARY LITERATURE. References to the relevant literature will be found in Marshall Clagett, *The Science of Mechanics in the Middle Ages* (Madison, Wis., 1959), pp. 683–698; and *Nicole Oresme and the Medieval Geometry of Qualities and Motions* (Madison, Wis., 1968), pp. 105–107; and Curtis Wilson, *William Heytesbury: Medieval Logic and the Rise of Mathematical Physics* (Madison, Wis., 1956), pp. 212–213.

A recent study of Heytesbury's work on the liar paradox is Alfonso Maierù, "Il problema della verità nelle opere di Guglielmo Heytesbury," in *Studi medievali,* 3rd ser., **7,** fasc. 1 (Spoleto, 1966), 41–74.

CURTIS A. WILSON

HIÄRNE, URBAN (*b.* Skworitz, Ingria, Sweden, 20 December 1641; *d.* Stockholm, Sweden, March 1724), *medicine, chemistry, mineralogy.*

The province where Hiärne was born was Sweden's farthest outpost against Russia, and when it was invaded by the Russians in 1656 his family was forced to flee. Urban managed to reach safety in Sweden, and in 1661 he began his medical education at the University of Uppsala. That Olof Rudbeck (the discoverer of the lymphatic vessels) and Petrus Hoffwenius were just then beginning to teach medicine at the university proved to be of great importance in Hiärne's education. Both of them saw clearly that Descartes (who was in Stockholm at that time), through his mathematical-mechanical interpretation of the world, had created a promising starting point for experimental research in nature; and that this offered an alternative to the Aristotelian-Scholastic doctrine which was being undermined by Paracelsian thought.

Hiärne was greatly impressed by the new ideas and willingly took the side of his Cartesian teachers in their first conflict, in 1663, against academic Scholasticism. He meanwhile completed his medical studies and became in 1666 personal physician to the governor-general of Livland. This situation enabled him to study abroad in Holland, England (where he was elected to the Royal Society in 1669), and France (where he graduated in medicine at Angers in 1670). The most important result of his travels was an advanced grounding in analytical and experimental chemistry, which he acquired chiefly during three years of study in Paris with the famous Christopher Glaser.

Upon his return to Sweden in 1674, Hiärne settled as a physician in Stockholm and soon acquired a considerable practice. He was elected to the Collegium Medicum in 1675. In 1684 he was appointed first personal physician to the king and in 1696, the year in which he became president of the Collegium, he was given the high honorary title of archiater. His medical practice did not prevent him from turning more and more to chemistry. It was Hiärne's expert analysis of spring water which led to the discovery of Sweden's first spa, Medevi, in 1678.

Shortly thereafter, Hiärne and several interested colleagues established a chemical research laboratory, which later became a national institution under the Board of Mines. Hiärne was appointed head of this Laboratorium Chemicum and simultaneously was named ordinary assessor at the Board of Mines (he became the board's vice-president in 1713). Hiärne set forth as the main purposes of the laboratory the examination of minerals and ores and the discovery of useful inventions. Extensive pharmaceutical research was also included in his program, and it is clear from detailed records how much Hiärne cherished the field of spagyric pharmacology. But even basic research had a place in the laboratory, he felt, and when discoveries would be made they "should be published to the greater glory of the King and the good of the fatherland."

The chemical research program was comprehensive enough to make its full realization exceedingly difficult. But Hiärne's supervision, unusual energy, and outstanding laboratory equipment brought rapid success to the venture, a success evident even in the 1680's. He had capable laboratory workers, the most able of them being Johann Georg Gmelin from Tübingen. The foundations laid by Hiärne, who envisioned the eventual creation of a viable Swedish center for advanced chemical research, proved to be enduring. Following Hiärne's death in 1724, the research program remained virtually at a standstill for several years. But as soon as a qualified successor, Georg Brandt, took over, it soon began anew to foster many of the remarkable advances in Swedish chemistry in the 1700's.

Hiärne's contributions in applied chemistry included work on improved methods for producing alum and vitriols, on impregnating agents to safeguard trees against rot, and on rust preventatives. In the field of pure chemistry he worked on problems concerning the formation of materials and the composition of bodies and ultimate particles; as his analytical method, he dissolved the substances and then tested them with different reagents and indicators which would elucidate the acid or alkaline nature of the bodies. He also studied alkalies in plants and the phenomenon in metals of increased weight through calcination. He is best known for his work on formic acid, which he produced through the distillation of ant specimens.

The lifework of Hiärne, a giant of learning, cannot readily be compressed into a short résumé. A polymath whose breadth of activity stretched over many disciplines, he did outstanding work in each of his fields and was one of the luminaries of Sweden's golden age of science. But until the great volume of written material which he left behind has been completely examined, no definitive evaluation of his work can be made.

BIBLIOGRAPHY

I. ORIGINAL WORKS. A comprehensive listing of Hiärne's works can be found in J. R. Partington, *A History of Chemistry,* III (London, 1962), 162–163; and in Sten Lindroth, in *Lychnos* (1946–1947), 51–116; within the last named essay is an itemized list of the most important archival materials concerning Hiärne.

II. SECONDARY LITERATURE. Olof Strandberg, *Urban Hiärnes ungdom och diktning* (Uppsala, 1942), a dissertation; Åke Åkerström, "Urban Hiärnes resa till Tyskland och Holland 1667," in *Lychnos* (1937), 187–211; Sten Lindroth, "Hiärne, Block och Paracelsus. En redogörelse för Paracelsusstriden, 1708–1709," *ibid.* (1941), 191–229, and "Urban Hiärne och Laboratorium Chymicum," *ibid.* (1947), 51–116; Tore Frängsmyr, *Geologi och skapelsetro. Föreställningar om jordens historia från Hiärne till Bergman* (Uppsala, 1969); and Hugo Olsson, "Kemiens Historia i Sverige intill år 1800" (Uppsala, 1971), pp. 40–71.

UNO BOKLUND

IBN HIBINTĀ (*fl.* Iraq, *ca.* 950), *astrology.*

Ibn Hibintā lived at the time of the first Buwayhid rulers of Baghdad, Aḥmad ibn Buwayh (946–949) and ʿAḍūd al-Dawla (949–982). The only work by which he is known is his vast compilation of astrological and astronomical lore entitled *Kitāb al-mughnī fi 'l-nujūm,* of which the second section only is preserved in a manuscript at Munich (MS Arab 852). The importance of this work lies entirely in the many quotations that it contains from earlier authorities, including Ptolemy (the *Planetary Hypotheses*), Dorotheus of Sidon, al-Khwārizmī, and Kanaka. One of the most interesting sections is that in which Ibn Hibintā discusses Māshā'allāh's *Fi 'l-qirānāt wa 'l-adyān wa 'l-milal,* to which he adds his own astrological interpretations of the Buwayhids' advent to power, at one point surreptitiously criticizing them, at another openly justifying their reign. Ibn Hibintā's date and location depend on these passages; there is little else in the manuscript that can be attributed to him as the original author.

BIBLIOGRAPHY

Ibn Hibintā's book was known to Ḥājjī Khalīfa—*Lexicon bibliographicum et encyclopaedicum,* G. Flügel, ed., 7 vols. (Leipzig, 1835–1858), V, 654—but is otherwise little noticed in the Arabic bibliographic and biographic tradition. The Munich MS was used by C. A. Nallino in his ed. of al-Battānī's *Opus astronomicum,* I (Milan, 1899), *passim;* and the extract from Māshā'allāh has been edited by E. S. Kennedy and D. Pingree, *The Astrological History of Māshā'allāh* (Cambridge, Mass., 1971).

DAVID PINGREE

HICETAS OF SYRACUSE (*fl.* fifth century B.C.), *astronomy.*

Hicetas (Ἱκέτας) of Syracuse was a Pythagorean, who is mentioned only twice in the doxographical tradition of late antiquity (Diogenes Laertius, VIII. 85; Aetius, III. 9. 1–2) and once by Cicero (*Academica priora* II. 39. 123). So little is known about him that even his existence as a historical person has been disputed.

Diogenes Laertius (*loc. cit.*) in his notice about Philolaus, the Pythagorean of the late fifth century B.C., states that "he was the first to say that the earth moves in a circle, but some assert that Hicetas the Syracusan was the first." Aetius (*loc. cit.*) says that "Thales and those following him said that there was one earth, Hicetas the Pythagorean that there were two, this present one and the counter-earth." Both these references, then, indicate that Hicetas was an adherent of the astronomical system connected with the name of Philolaus, according to which the earth was regarded as a planetary body circling round a central fire (the "hearth of the universe") in company with the counter-earth (supposed to be a body orbiting between the earth and the central fire with the same velocity as the earth), the moon, the sun, and the five planets. On the assumption of different orbital velocities for each of these planetary bodies, the observed motions and positions of sun, moon, and planets could be very roughly explained (for a full description, see D. R. Dicks, *Early Greek Astronomy to Aristotle* [London, 1970], p. 65 ff.).

On the other hand, Cicero (*loc. cit.*) says that according to Theophrastus (Aristotle's successor as head of the Lyceum), "Hicetas the Syracusan believes that the sky, sun, moon, stars, and in fact all the heavenly bodies stand still, and that nothing at all moves in the universe except the earth; and that because it turns and twists with great speed about its axis, all the same phenomena are produced as if the sky was in motion and the earth standing still." If Hicetas really held this view, it would prove him to be astronomically ignorant, since it would entail a complete disregard of the proper motions of the planetary bodies in the zodiac; but presumably Cicero, expressing himself with typical scientific incompetence, means no more than that Hicetas suggested that the daily phenomenon of rising and setting could as well be accounted for by assuming a stationary heaven and the axial rotation of the earth as by assuming a stationary earth and the rotation of the heavens. In that case, Hicetas was an adherent of a theory—the axial rotation of the earth—that all sources agree in attributing to Heraclides of Pontus, a pupil of Plato (see Heath, p. 251 f.). In any event, there is a clear discrepancy between the views assigned to Hicetas by Diogenes Laertius and Aetius, on the one hand, and by Cicero on the other. Tannery ("Pseudonymes antiques," in *Revue des études grecques,* **10** [1897], 127–137) suggests that Hicetas was simply a character in one of Heraclides' dialogues, based on an actual Hicetas who became tyrant of Leontini in Sicily and is mentioned in Plutarch's lives of Timoleon and Dion. This, however, seems unlikely (cf. Guthrie, I, 323–324), and, in any case, does not resolve the discrepancy in our sources.

BIBLIOGRAPHY

In addition to the works mentioned in the text, see W. K. C. Guthrie, *History of Greek Philosophy,* I (Cambridge, 1962), 323–324, 327–329; T. L. Heath, *Aristarchus of Samos* (Oxford, 1913), pp. 187–189; and E. Wellmann, "Hiketas 4," in Pauly-Wissowa, *Real-Encyclopädie,* VIII (1913), col. 1597.

D. R. DICKS

HIGGINS, BRYAN (*b.* Collooney, County Sligo, Ireland, 1737 or 1741; *d.* Walford, Staffordshire, England, 1818), *chemistry.*

Qualified as a physician, and remembered chiefly for his speculative chemical theories, it is nevertheless as an entrepreneur of fundamental research and chemical technology that Bryan Higgins most invites attention and remark. His activities in these respects nicely complement the pursuits of his better-known nephew, William. Together their lives offer important glimpses into the cornucopia of opportunities open to chemically knowledgeable residents of the British Isles in the latter part of the eighteenth century. These opportunities were often precariously established and weakly institutionalized. As population, urbanization, chemical knowledge, and manufacturing enterprise all grew rapidly in the early nineteenth century, such ill-defined forms and norms proved unequal to the demands of growth and change vociferously pursued by the propagandists for professional science. It is the more optimistic period immediately prior to such problems which is revealed in the biographies of the two Higginses. The vigor, self-confidence, and sense of chemical possibilities so immediately apparent from those biographies reflect a world of economic expansion and technical development. It was also a world not yet beset by larger social issues or the status doubts and organizational anxieties of a newly self-conscious scientific profession.

Higgins' father was a physician of considerable repute in County Sligo. Of his three sons the eldest was a merchant, and Thomas Higgins (father of William) studied arts and medicine at Edinburgh before taking an M.D. Where Bryan studied is not known. He enrolled at Leiden in October 1765, shortly before graduating there as M.D. Many other important aspects of his life remain cloaked in obscurity. He married a Miss Jane Welland of London, a lady of some means, about 1770. Beyond the fact that she bore him two daughters, no details of their family life are preserved. His settling in the metropolis may well date from the time of his marriage.

Higgins' wish and ability to engage the London beau monde with the theoretical and practical implications of natural knowledge was uncommon, especially when seen against his Irish Catholic background. In July 1774 he opened his "school of practical chemistry, wherein the pupils might have uncommon advantages, at the same time that my apparatus might be enlarged, and my experiments conducted at a common expense." This bold solution to the problem of raising the growing capital required for a career of laboratory research apparently met with success. Even so, it was necessary for the school to combine its loftier intellectual goals with more utilitarian aims.

Regular lectures were given for a number of years. An extant syllabus refers to the course as one of philosophical, pharmaceutical, and technical chemistry. As such it probably appealed to students at the London hospitals as well as to the curious, the fashionable, and the manufacturing gentlemen about town. These latter groups must have been considerable. Their recruitment was no doubt aided by the location of Higgins' school and laboratory in Greek Street, where Josiah Wedgwood had his London showrooms. The address was also conveniently close to Soho Square, the focal point for aspiring London men of science. Assured of gentlemanly support, Higgins pursued the original purposes of his school. This is apparent from *Syllabus of Chemical and Philosophical Enquiries, Composed for the Use of Noblemen and Gentlemen who have Subscribed to the Proposals Made for the Advancement of Natural Knowledge* (1776). What transpired at these and other sets of discourses and experiments for meetings of the subscribers is not known, but such activity obviously enjoyed considerable favor. Higgins came to include Samuel Johnson (a devotee of chemistry) among his acquaintances, and Edward Gibbon, Joseph Priestley, and Benjamin Franklin were among his early auditors. Finally he was emboldened to issue printed proposals for a considerably more ambitious Society for Philosophical Experiments and Conversations, in November 1793.

The Society was established in Higgins' laboratory the following January. Its chairman was Field Marshal Henry Conway, an aged but important political figure, and Thomas Young was one of the "assistants in experiments." The Society met weekly at 8 P.M. throughout the parliamentary session, the five-guinea subscription serving both to delimit the membership and to defray the cost of the apparatus and chemicals. If the tone was polite, the driving force was Higgins himself, as is immediately clear from the subsequently published *Minutes* of the apparently short-lived group.

Higgins' solution to the common problem of assembling research equipment and chemical apparatus was unusual and highly imaginative. His other activities were more routine, even including his reported (but unconfirmed) journey to Russia in the 1780's at the invitation of Catherine II. His visit to Jamaica from late 1796 to 1801, to advise on the making of sugar and rum, presumably owed something to politically well-placed friends. The actual invitation came from the Jamaica House of Assembly. They paid £1,000 a year (retroactively raised to £1,400) for his expert knowledge. His extensive suggestions culminated in *Observations and Advices for the Improvement of the Manufacture of Muscovado Sugar and Rum.* Such ad hoc technical advice was often resorted to by government, if rarely so freely rewarded, as raw-material processing and chemical manufacturing became increasingly important to an industrializing Britain and her colonies.

Like other chemists of the period, Higgins was engaged with an additional variety of practical problems that caught his own immediate interest. The behavior of mixtures of lime, sand, and water was already under theoretical discussion in his first lecture course. In 1779 he took out a patent for a cement composed of washed sand, slaked lime, limewater, and bone ash. This new combination enjoyed a modest vogue. The following year he published *Experiments and Observations Made With the View of Improving the Art of Composing and Applying Calcareous Cements.* Six years later he published *Experiments and Observations Relating to Acetous Acid, Fixable Air . . . Oil and Fuels.* In 1788 his *Synopsis of the Medical Contents of the Most Noted Mineral Waters* appeared. This leaflet served as advertisement for the waters purveyed by its publisher, one John Ellison, whose "spruce beer and mineral water machine" served fashionable London from its Whitechapel base. Higgins filed a 1781 patent suggesting chemically sophisticated and ingenious ways of manufacturing soda and potash, a 1767 patent (his first) for an oil lamp designed to look like a candle holder, and an 1802 patent for a warm air heating system.

Beside lecturing, experimenting, consulting, and advising across a broad range of chemical topics, Higgins also developed a considerable business in the manufacture and supply of reagents and chemicals. A surviving print of his Greek Street laboratory shows a room over thirty feet long, well equipped with reverberatory and melting furnaces, sand baths, and other necessary apparatus, including "several thousand flint glass and green bottles and vessels" to hold the products. Forty-foot-high chimneys apparently were necessary to disperse the fumes generated by such manufacturing operations.

Higgins' publications often interweave detailed discussion of problems in technical chemistry with aspects of his speculative theoretical views. Although he was unable to combine these conjectures with fruitful experimentation, his ideas deserve consideration. Many commentators have seen in his arguments and terminology percipient harbingers of later work in chemical atomic theory. In fact his ideas are remote from such post-Lavoisier concerns and, rather, lie within that earlier eighteenth-century tradition of theoretical and empirical inquiry which, taking its inspiration from Newton's *Opticks,* saw short-range-force explanations of the interactions of light, heat, and matter as central to any coherent natural philosophy. These subjects are dominant in Higgins' *Philosophical Essay Concerning Light* (1776). They are equally pervasive two decades later in the *Minutes* of his Society for Philosophical Experiments and Conversations. Like such speculative philosophers as Gowin Knight and Bryan Robinson before him, Higgins correctly saw that an understanding of the relations of heat, light, and matter was central to the further development of a Newtonian philosophy. His concerns for such a broad and ambitious topic, especially when cast in the phlogistic mode and based on a theory of seven elements, effectively precluded him from making tangible contributions either to chemical theory or to the allied subjects of heat and optics. Even so, his now rare publications make fascinating reading. They reveal a powerful mind actively grappling with some of the leading theoretical problems of the day.

Higgins' failure to make substantive progress in confirming his speculative ideas, coupled with the radical transformations occurring in chemical theory in the 1780's and 1790's, may have encouraged his retirement from the field. When he accepted the invitation to Jamaica in 1796, he sold his extensive accumulation of apparatus and chemicals. The "very liberal provision" settled on him by the grateful Jamaica Assembly on his return to Britain in 1801 apparently enabled him to retire to the country. He appears to have played little further part in scientific affairs, although in 1803 he did advise the Royal Institution on its chemical laboratory, at Davy's behest. Perhaps his "great affability of manner" and his unfortunate early exchange with Joseph Priestley over precedence in discovery prevented his acknowledging the priority dispute about chemical atomic theory which his nephew William later conducted with John Dalton. It was this dispute that brought Higgins to the attention of historians, but his real importance lies elsewhere.

BIBLIOGRAPHY

I. ORIGINAL WORKS. Higgins' most important writings are *A Philosophical Essay Concerning Light* (London, 1776); *Experiments and Observations Made With the View of Improving the Art of Composing and Applying Calcareous Cements and of Preparing Quicklime: Theory of These Arts and Specification of the Author's Cheap and Durable Cement, for Building, Incrustation or Stuccoing, and Artificial Stone* (London, 1780); *Experiments and Observations Relating to Acetous Acid, Fixable Air, Dense Inflammable Air, Oil and Fuels, the Matter of Fire and Light, Metallic Reduction, Combustion, Fermentation, Putrefaction, Respiration, and Other Subjects of Chemical Philosophy* (London, 1786); *Minutes of the Society for Philosophical Experiments and Conversations* (London, 1795); and *Observations and Advices for the Improvement of the Manufacture of Muscovado Sugar and Rum,* 3 pts. (St. Iago de la Vega [Spanish Town], Jamaica, 1797–1801), plus fragment of the fourth part (Jamaica, 1803).

II. SECONDARY LITERATURE. Much information on the life of Higgins, an extended discussion of his chemical ideas, and a careful (although incomplete) listing of his published works and papers are available in J. R. Partington, *History of Chemistry,* III (London–New York, 1962), 727–736. Further details of his life may be gleaned from the complex footnotes appended to J. R. Partington and T. S. Wheeler, *The Life and Work of William Higgins, Chemist* (London, 1960). Although Partington and Wheeler undertook an exhaustive search for MSS and printed information on the Higgins family, F. W. Gibbs showed that there were important sources still unexploited. His brief, only partially documented, but highly suggestive "Bryan Higgins and His Circle," in *Chemistry in Britain,* **1** (1965), 60–65, stresses the importance of patronage and personal networks in the period's chemical science and technology. All subsequent studies of Higgins have perforce been based on the extended, if less than fully reliable, account by W. K. Sullivan in *Dublin Journal of Medical Science,* **8** (1849), 465–495. The account of Higgins' laboratory is in S. F. Gray, *The Operative Chemist* (London, 1828), pp. 72–74. Higgins' theories, and his verbal claims to have discovered some gases, are fully and aggressively dealt with in Joseph Priestley, *Philosophical Empiricism: Containing Remarks on a Charge of Plagiarism Respecting Dr. H——s, Interspersed With Various Observations Relating to Different Kinds of Air* (London, 1775). The Newtonian background to Higgins' ideas is set out in A. Thackray, *Atoms and Powers: An Essay on Newtonian Matter-Theory and the Development of Chemistry* (Cambridge, Mass., 1970).

ARNOLD THACKRAY

HIGGINS, WILLIAM (*b.* Collooney, County Sligo, Ireland, 1762 or 1763; *d.* Dublin, Ireland, 30 June [?] 1825), *chemistry.*

Known chiefly for his speculative ideas on chemical combination, William Higgins is of greater interest for the insights his life offers into the emergence of chemistry as a career during the British industrial revolution. His biography thus complements that of his uncle, Bryan Higgins, the London physician and entrepreneur in technical information and pure research. William plainly lacked his kinsman's social graces. Despite an evident charm, his erratic bachelor behavior and tendency to indulge personal animosities prevented him from engaging the affections of London society. Instead he found refuge in a succession of government-supported chemical positions in

Dublin. Thanks to the combination of such scientific opportunities with family resources, he became a comparatively rich man.

The O'Higgins clan was prominent in County Sligo from medieval times. With the decline of the bardic art many of its members turned to medicine. William was apparently the second child and younger son of Thomas Higgins, a physician educated at (although not graduated from) the University of Edinburgh. Nothing is known of Higgins' early education. While still a boy he was sent to London to live with his uncle. Under the latter's guidance he developed a strong taste for, and considerable expertise in, experimental chemistry.

In the early 1780's Higgins assisted in making all the experiments detailed in Bryan Higgins' *Experiments and Observations Relating to Acetous Acid*. . . . In 1785 he undertook a mineralogical tour through England, also visiting a number of chemical manufactories. On 6 February 1786 he matriculated at Magdalen Hall, Oxford. A year later he transferred to Pembroke College, whose master, William Adams, was "considerably deep in chemistry." Among undergraduate contemporaries and active enthusiasts for natural knowledge were Davies Gilbert, future president of the Royal Society, and James Haworth, subsequently physician to St. Bartholomew's Hospital, with whom Higgins was intimately acquainted. Older friends included William Austin, professor of chemistry for a brief interlude; Martin Wall, then reader in chemistry; and Thomas Beddoes, who was Wall's successor for four stormy years. The existence of a group of such caliber indicates how tales of Oxford's scientific torpor at this time must be treated with reserve.

Higgins' initial access to and easy familiarity with these men owed much to the connections and influence of his uncle. That his own considerable chemical abilities and enthusiasm were also important may be seen in his acting as "operator" to the reader and carrying out experiments in a laboratory in the basement of the Ashmolean Museum. Despite such promising circumstances and acquaintances, Higgins abruptly left Oxford without a degree in the summer of 1788. His next four years were spent in London, where he published two editions of his most important work, the *Comparative View of Phlogistic and Anti-Phlogistic Theories* (1789, 1791). He also experimented with printing on linen and quarreled with his uncle. The two events in combination secured his interest in a fresh sphere of activity, as chemist in the new hall of the government-supported Irish Corporation of Apothecaries. The influence of Richard Kirwan (also an Irishman, Catholic, chemist, and former London resident and in addition a man of considerable means) was probably decisive on his behalf in this, as in other, appointments.

Higgins took up his position in Dublin in March 1792. He enjoyed the generous salary of "£200 a year, apartments, coals and candles." In contrast, the apothecary to the hall was paid only £80. Higgins was soon busy equipping the laboratory, attending the Royal Irish Academy (and, with greater regularity, its dining club), and acting as part-time chemist to the Irish Linen Board. In September 1794 the apothecaries authorized him an assistant at £50 per annum. Unfortunately the corporation was discovered to be in serious financial difficulties early the following year. The ambitious post of chemist was abolished, and Higgins' appointment terminated amid considerable acrimony.

Through the agency of Kirwan, Higgins was quickly reemployed as supervisor of the important Leskean cabinet of minerals recently acquired by the Royal Dublin Society. Successive acts of the Irish Parliament confirmed his position as professor of chemistry to the Society. His salary rose from the initial £100 to £300 per annum, plus fees, by 1800. As professor, Higgins conducted analyses on request, lectured to both the public and the Society, and had charge of a laboratory specially equipped to encourage his "experiments on dyeing materials and other articles, wherein chemistry may assist the arts." By 1803 the total annual expenses of his department of mineralogy and chemistry were £643 14s. 6d. These costs were defrayed mainly from the parliamentary grant to the Society. Higgins was in effect, although not in name, once more a government-maintained chemist.

The state also provided the £100 per annum Higgins regularly received as chemist to the trustees of the Linen and Hempen Manufactures of Ireland. In this capacity he did much valuable work over the years. Immediate problems included the chemistry of bleaching (especially the new use of bleaching powder) and the detection of adulterants in commercial alkalies. Higgins traveled widely in Ireland to advise local bleachers on their problems. His researches resulted in an important *Essay on Bleaching* (1799). Despite the obvious utility of, and widespread appreciation for, his services, the weakly institutionalized post of chemist to the Linen Board was abolished in due course as part of a general economy drive, like that at the Apothecaries' Hall before it. The same 1820 retrenchment threatened, but did not eliminate, the positions of Higgins and his assistant at the Dublin Society. There were, of course, particular reasons why the Irish middle class should be especially favored by successive administrations. Even so, the existence and diversity of such state-funded positions as those enjoyed by Higgins point to a far richer involvement of the Hanoverian executive with the pursuit and implications of natural knowledge than is commonly supposed.

A further facet of government involvement with science is seen in Higgins' 1803 leave of absence from the Dublin Society, which enabled him to sit on a London committee selecting a hydrometer to measure the strength of alcoholic spirits for revenue purposes. It was while in London that he met Humphry Davy, a protégé of his uncle and of his own mentor, Thomas Beddoes. Davy was one of Higgins' proposers to the Royal Society in 1806. Higgins in turn was involved in arranging Davy's highly lucrative 1810 addresses before the Dublin Society, the Farming Society, and the Linen Board. The relationship was to flourish for, from 1810 on, Davy vigorously promoted his new friend's claims to the discovery of the chemical atomic theory over those of their common rival, John Dalton.

The work on which Higgins based his claims was the *Comparative View*. Actually, the book is an interesting, if verbose and poorly structured, attempt to contrast phlogistic and antiphlogistic chemistry, to the advantage of the latter. In it Higgins hit on the idea of using arbitrary affinity numbers to reinforce his arguments. It is this which gives the work its appeal. Higgins sought to elucidate the mechanisms of possible reactions between ultimate particles of, say, sulfur and oxygen, by using diagrams of the reacting particles and the affinity forces between them. Not surprisingly, his arguments do contain among their unstated assumptions ideas on combining proportions that were later to be made explicit in chemical atomic theory. In this they are typical of much existing thought. Yet far from displaying the same concerns that drove John Dalton, the *Comparative View* is important chiefly as a brilliant and highly individualistic exploitation of dominant Newtonian assumptions about the forces of chemical affinity.

It seems to have been Davy's continuing desire to belittle Dalton's theoretical achievement that induced Higgins to assert his own priority. Once aroused, he proved a belligerent antagonist. The later *Observations on the Atomic Theory* provides an exhaustive but unconvincing account of his claims. Dalton took little notice of the controversy, finding in Thomas Thomson a more than sufficient defender of his originality against the continuing deprecations of Higgins and Davy. The priority dispute has proved unusually hardy, perhaps because Irish honor is felt to be at stake. By 1960 J. R. Partington could list more than fifty contributors to the debate. Their continuing discussions should not be allowed to obscure other, more significant aspects of Higgins' life.

BIBLIOGRAPHY

I. ORIGINAL WORKS. Higgins' most important publications are *A Comparative View of the Phlogistic and Anti-Phlogistic Theories. With Inductions. To Which Is Annexed an Analysis of the Human Calculus, With Observations on Its Origin, etc.* (London, 1789; 2nd ed., 1791); *An Essay on the Theory and Practice of Bleaching, Wherein the Sulphuret of Lime Is Recommended as a Substitute for Pot-Ash* (Dublin–London, 1799); *A Syllabus of a Course of Chemistry for the Year 1802* (Dublin, 1801); and *Experiments and Observations on the Atomic Theory, and Electrical Phenomena* (Dublin, 1814).

II. SECONDARY LITERATURE. The fundamental source for future work is J. R. Partington and T. S. Wheeler, *Life and Work of William Higgins, Chemist* (London, 1960). A mine of information on Higgins' life, work, ideas, experiments, and acquaintances, it also includes photographic reproductions of the *Comparative View* (2nd ed.) and the *Observations*. Unfortunately it does not include a bibliography of either Higgins' publications or subsequent studies. For these (in incomplete forms), consult J. R. Partington, *A History of Chemistry*, III (London–New York, 1962), 736–749. Additional information on Higgins and his context may be gleaned from such works on Dublin as H. F. Berry, *History of the Royal Dublin Society* (London, 1915); and from biographical accounts of other chemists, such as Beddoes and Kirwan. The background to Higgins' ideas may be explored in A. Thackray, *Atoms and Powers: An Essay on Newtonian Matter-Theory and the Development of Chemistry* (Cambridge, Mass., 1970). J. W. van Spronsen, "William Higgins," in *Archives internationales d'histoire des sciences*, **19** (1966), 74–77, is the most recent contribution to the literature on the priority dispute.

ARNOLD THACKRAY

HIGHMORE, NATHANIEL (*b.* Fordingbridge, England, 6 February 1613; *d.* Sherborne, Dorset, England, 21 March 1685), *anatomy, medicine.*

Son of Rev. Nathaniel Highmore, rector of Purse Caundle, Dorset, Highmore was the most distinguished member of a family that for several centuries produced clergymen, doctors, lawyers, and one well-known painter, Joseph Highmore. His most important scientific contribution is *Corporis humani disquisitio anatomica* (1651), containing the first description of the antrum of Highmore (maxillary sinus, the largest of the paranasal sinuses) and of the *corpus Highmori* (mediastinal testis). Dedicated to William Harvey, it was the first anatomical textbook to accept Harvey's theory of the circulation of the blood; its frontispiece incorporates an allegorical drawing of this new theory. Although Highmore's physiology reflects the still medieval thinking of his time, the book was accepted as a standard anatomical textbook for many years and brought the author immediate recognition in England and abroad. For instance, Johann Daniel Horst, chief court physician of Hesse-Darmstadt, in asking William Harvey (1655) to undertake a study of the lymphatic and thoracic ducts, suggested as an alternative "the most illustrious Dr. Highmore"; and Boyle spoke of Highmore as "my learned friend," quoted his experiments, and referred a knotty physiological problem to him.

Educated at Sherborne School and Trinity College, Oxford, Highmore graduated B.A. in 1635 and M.A. in 1638, then proceeded to study medicine. In 1640 he married Elizabeth, daughter of Richard Haydocke, a noted physician of Salisbury. (Highmore had probably sought practical experience with Haydocke before receiving his B.M. in 1641.) When the Civil War began in 1642, Highmore was one of a group of scientists at Trinity College, Oxford, headed by George Bathurst and William Harvey (then physician to King Charles I), who were conducting experiments on embryonic development of the chick. This study led to friendship between Highmore and Harvey and an evident agreement between them to publish the conclusions derived from their joint experiments in embryology. Highmore implied this agreement clearly in the dedication (written in 1650) of his *Corporis:* "It is now eight years since we first had it in mind to expose our careful studies . . . to the judgement of the public." Highmore and Harvey both published their results in 1651; Harvey in his *Exercitationes de generatione animalium* and Highmore in *The History of Generation* (dedicated to Robert Boyle). Highmore's *Generation* contains the first reference in English to use of the microscope, which may well have helped him to report changes in the embryonic area of the egg at a day earlier than did Harvey. The book is also notable for its careful observations and illustrations of plants, leading one modern authority to comment that Highmore's contribution to botany has not been adequately recognized.

In 1643 Highmore received his M.D. at Oxford under the "Caroline Creations" (whereby, by royal command, the university conferred degrees on those who had specially served the king's cause at the battle of Edge Hill and after). It is not known why Highmore was so honored, but one surmise is that he attended the young Prince Charles during a bout of measles at Reading in November 1642.

Fully qualified for medical practice, Highmore returned to Sherborne, where he practiced for forty years as a skillful and sought-after physician, his work marked by real concern for his patients and a commonsense approach to medicine. Despite the demands of a busy practice he found time to keep in touch with scientific thought. There was an unfulfilled suggestion to elect him a fellow of the newly formed Royal Society, and he contributed articles on medicinal springs to the Society's *Philosophical Transactions.* He also—through his essays *De passione hysterica* and *De affectionae hypochondriaca*—engaged in a controversy with the redoubtable Thomas Willis, professor of natural philosophy at Oxford.

Highmore's life was full and well rounded; internationally famous as an anatomist, loved and esteemed as a physician, he also assumed a full share of civic duties. He became a justice of the peace and county treasurer for Dorset; in Sherborne he was active in church affairs, and served for many years on the governing body of the town's historic almshouse and Sherborne School.

BIBLIOGRAPHY

I. Original Works. There are MSS (mainly unconnected medical notes) in the British Museum, Sloane and Add. MSS. Published works include *Corporis humani disquisitio anatomica* (The Hague, 1651); *The History of Generation* (London, 1651); and *Exercitationes duae . . . De passione hysterica . . . De affectionae hypochondriaca* (Oxford, 1660; 2nd ed., 1677). Possibly by Highmore is *Treatise on . . . a Plague of the Guts* (London, 1658). His articles on medicinal spas include "Some Considerations Relating to D. Witties Defence of Scarborough Spaw," in *Philosophical Transactions of the Royal Society,* **4,** no. 56 (1669), 1128–1131.

II. Secondary Literature. The only published works on Highmore are studies by J. Elise Gordon: "The Highmore Family of Dorset," in *Journal of the Sherborne Historical Society,* **3** (1966), 2 ff.; "Nathaniel Highmore, Physician and Anatomist 1613–1685," in *Practitioner,* **196** (June 1966), 851 ff.; and "Nathaniel Highmore," no. 2 of articles entitled "Two 17th Century Physicians," in *Midwife and Health Visitor,* **5** (Aug. 1969), 364 ff.

Contemporary references include Robert Boyle, *New Experiments, Physicall-Mechanicall . . .* (Oxford, 1660); J. D. Horst, *Observationem anatomicarum* (Frankfurt, 1656), for the Horst-Harvey correspondence; Robert Plot, *Natural History of Oxfordshire* (1677); Thomas Willis, *Affectionam quae dicuntur hystericae et hypochondriacae . . . contra responsionem epistolarum Nathanael Highmori M.D.* (London, 1670); and Anthony à Wood, *Athenae Oxoniensis* (London, 1692).

For references to Highmore's works and assessment of his significance, see R. T. Gunther, *Early Science at Oxford,* III (Oxford, 1937); Geoffrey Keynes, *Life of William Harvey* (Oxford, 1966); and A. T. H. Robb Smith, "Harvey at Oxford," in *Oxford Medical School Gazette,* **9,** no. 2 (1957).

Details of his life were obtained from Highmore family papers, local Dorset and Sherborne records, and communications with the University of Oxford and Trinity College, Oxford.

J. ELISE GORDON

HIKETAS. See **Hicetas.**

HILBERT, DAVID (*b.* Königsberg, Germany [now Kaliningrad, R.S.F.S.R.], 23 January 1862; *d.* Göttingen, Germany, 14 February 1943), *mathematics.*

Hilbert was descended from a Protestant middle-class family that had settled in the seventeenth century near Freiberg, Saxony. His great-grandfather, Christian David, a surgeon, moved to Königsberg, East Prussia. David's grandfather and father were judges in Königsberg. His father's Christian name was Otto; his mother's maiden name was Erdtmann. Hilbert's inclination to mathematics is said to have been inherited from his mother. From 1870 he attended the Friedrichskolleg in Königsberg; his last year of high school was spent at the Wilhelms-Gymnasium. In 1880 he took the examination for university admission. He studied at the University of Königsberg from 1880 to 1884, except for his second semester, when he went to Heidelberg. After his doctoral examination in 1884 and receipt of his Ph.D. in 1885, he traveled to Leipzig and Paris. In June 1886 he qualified as *Privatdozent* at Königsberg University. In 1892 Hilbert was appointed professor extraordinary to replace Adolf Hurwitz at Königsberg, and in the same year he married Käthe Jerosch. In 1893 he was appointed ordinary professor, succeeding F. Lindemann. He was appointed to a chair at Göttingen University in 1895, remaining there until his official retirement in 1930. In 1925 he fell ill with pernicious anemia, which at that time was considered incurable. New methods of treatment enabled him to recover, although he did not resume his full scientific activity. He died in 1943.

Königsberg, the university where Immanuel Kant had studied and taught, became a center of mathematical learning through Jacobi's activity (1827–1842). When Hilbert began his studies there, the algebrist Heinrich Weber, Dedekind's collaborator on the theory of algebraic functions, was a professor at Königsberg. In 1883 Weber left. His successor was Lindemann, a famous but muddle-headed mathematician who the year before had had the good luck to prove the transcendence of π. Lindemann displayed an astonishing seminar activity. (The notes of the Lindemann seminar are at present in the possession of Otto Volk.) Under his influence Hilbert became interested in the theory of invariants, his first area of research. At that time Königsberg boasted a brilliant student, Hermann Minkowski, two years younger than Hilbert but one semester ahead of him, who in 1883 received the Grand Prize of the Paris Academy. In 1884 Hurwitz, three years older than Hilbert and a mature mathematician at that time, was appointed professor extraordinary at Königsberg. For eight years he was Hilbert's guide in all of mathematics. In his obituaries of Minkowski and Hurwitz, Hilbert acknowledged the great influence of these two friends on his mathematical development. In 1892 Hurwitz left for Zurich and was soon followed by Minkowski. In 1902 Hilbert was reunited with Minkowski at Göttingen, where a new mathematics chair had been created for Minkowski at Hilbert's instigation.

The mathematician whose work most profoundly influenced Hilbert was the number theoretician Leopold Kronecker, although Hilbert took exception to Kronecker's seemingly whimsical dogmatism on methodological purity and hailed Georg Cantor's work in set theory, which had been criticized by Kronecker.

Hilbert's scientific activity can be roughly divided into six periods, according to the years of publication of the results: up to 1893 (at Königsberg), algebraic forms; 1894–1899, algebraic number theory; 1899–1903, foundations of geometry; 1904–1909, analysis (Dirichlet's principle, calculus of variations, integral equations, Waring's problem); 1912–1914, theoretical physics; after 1918, foundations of mathematics.

One should further mention his famous choice of mathematical problems which he propounded to the Second International Congress of Mathematicians at Paris in 1900.

At the end of a paper read at the International Mathematical Congress at Chicago in 1893, Hilbert said:

> In the history of a mathematical theory three periods can easily and clearly be distinguished: the naïve, the formal, and the critical ones. As to the theory of algebraic invariants, its founders Cayley and Sylvester are also representatives of the naïve period; when establishing the simplest invariant constructions and applying them to solving the equations of the first four degrees, they enjoyed their prime discovery. The discoverers and perfectioners of the symbolic calculus Clebsch and Gordan are the representatives of the second period, whereas the critical period has found its expression in the above mentioned theorems 6–13.

Whatever this historical tripartition means, it is obvious that Hilbert would have characterized his own numerous contributions to the theory of invariants from 1885 to 1888 as still belonging to the first two periods. Yet when he delivered his Chicago address, the theory of invariants was no longer what it had been five years before. Hilbert had perplexed his contemporaries by a revolutionary approach, nicknamed "theology" by Gordan, the "King of Invariants." What Hilbert had called Clebsch's and Gordan's formal period was the invention and the skillful handling of an apparatus, the symbolic method, which still can elicit the enjoyment of the historian who is faced with it. Hilbert's new approach was quite different: a direct, nonalgorithmic method, foreshadowing and preparing what would be called abstract algebra in the twentieth century. It has often been considered a mystery why, after his Chicago address, Hilbert left the field of invariants, never to return to it. But it should be pointed out that Hilbert was not the only mathematician to do so. It was said that Hilbert had solved all problems of the theory of invariants. This, of course, is not true. Never has a blooming mathematical theory withered away so suddenly. The theory of invariants died as a separate discipline. Hilbert had not finished the theory of invariants by solving all of its problems but, rather, by viewing invariants under a broader aspect. This often happens in mathematics. From a higher standpoint, paramount ideas can become futilities, profound facts trivialities, and sophisticated methods obsolete. Nevertheless, it is striking that the fortune of the theory of invariants changed so abruptly, that its fall was so great, and that it was caused by a single man.

In more modern terms, the theory of invariants dealt with linear groups G acting on N-space R and the polynomials on R, invariant under G. The groups actually studied at that time were mainly the linear representations of the special linear group of n-space by m-fold symmetric tensor products—in the terminology of the time, the invariants of an n-ary form of degree m. Up to that time much skill had been applied to finding and characterizing full systems of invariants. The invariants formed a ring with a finite basis, as far as one could tell from the examples available. Generally these basic invariants $I_1, \cdots, I_k$ are not algebraically independent; the polynomial relators, called syzygies, form an ideal, which again, according to the examples, has a finite ideal basis, $F_1, \cdots, F_l$. The $F_1, \cdots, F_l$ need not be ideal-independent; there can be relations $R_1F_1 + \cdots + R_lF_l = 0$ among them, so that one obtains an ideal of relators $R_1, \cdots, R_l$, or of "second-order syzygies," and so on.

When Hilbert started his work, the finiteness of a ring basis for invariants had been tackled by algorithmic methods which apply to very special cases only. Hilbert did not solve the total problem, and it still has not been solved. He also restricted himself to very special groups; explaining general methods through examples became one of the outstanding features of Hilbert's work. It is one of the reasons why he could build such a strong school.

It may be guessed that Hilbert started with the finiteness of the ideal basis of syzygies. In fact he proved the finiteness of the basis for any ideal in any polynomial ring. It was mainly this bold generalization and its straightforward proof which perplexed his contemporaries. The present formulation of Hilbert's basis theorem is as follows: The property of a ring R with one element of letting every ideal have a finite basis is shared by its polynomial ring $R[x]$. It has proved fundamental far outside the theory of invariants. Of course, it applied to the ideals of syzygies of any order as well. Moreover, Hilbert showed that the cascade of syzygies stops at last after m steps. This latter result looks like a nicety, and so it seems to have been considered for half a century, since no textbook used to mention it. Its revival in today's homological algebra is a new proof of Hilbert's prophetic vision.

Applied to the ring of invariants itself, Hilbert's basis theorem says that any invariant I can be presented in the form $A_1I_1 + \cdots + A_kI_k$ where $A_1, \cdots, A_l$ are polynomials which may be supposed of lower degree than I. If G is finite or compact, they can be changed into invariants by averaging over G. The new $A_1, \cdots, A_l$ can be expressed in the I $I_1, \cdots, I_k$ in the same way as I has been; this process is continued until the degrees of the coefficients have reached zero. This more modern averaging idea stems from Hurwitz. Hilbert himself used a differential operation, Cayley's Ω process, to reach the goal.

Further of Hilbert's results connected the invariants to fields of algebraic functions and algebraic varieties, in particular the *Nullstellensatz:* If a polynomial f vanishes in all zeros of a polynomial ideal M, then some power of f belongs to that ideal.

Other work from the same period dealt with the representation of definite polynomials or rational functions as terms of squares, a problem to which Artin made the definitive contribution thirty years later. There is also Hilbert's irreducibility theorem, which says that, in general, irreducibility is preserved if, in a polynomial of several variables with integral coefficients, some of the variables are replaced by integers. An isolated algebraic subject of later years is his investigation of the ninth-degree equation, solved by algebraic functions of four variables only and suggesting the still open problem of the most economic solving of algebraic equations.

There is no field of mathematics which by its beauty has attracted the elite of mathematicians with such an irresistible force as number theory—the "Queen of Mathematics," according to Gauss—has done. So from the theory of invariants Hilbert turned to algebraic number theory. At the 1893 meeting at Munich the Deutsche Mathematiker-Vereinigung, which Hilbert had presented with new proofs of the splitting of the prime ideal, charged Hilbert and Minkowski with preparing a report on number theory within two years. Minkowski soon withdrew, although he did read the proofs of what would be known as *Der Zahlbericht,* dated by Hilbert 10 April 1897. The *Zahlbericht* is infinitely more than a report; it is one of the classics, a masterpiece of mathematical literature. For half a century it was the bible of all who learned algebraic number theory, and perhaps it is still. In it Hilbert collected all relevant knowledge on algebraic number theory, reorganized it under striking new unifying viewpoints, reshaped formulations and proofs, and laid the groundwork for the still growing edifice of class field theory. Few mathematical treatises can rival the *Zahlbericht* in lucidity and didactic care. Starting with the quadratic field, Hilbert step by step increases the generality, with a view to a complete theory of relative Abelian fields; but from the beginning he chooses those methods which foreshadow the general principles.

At the end of the preface of the *Zahlbericht,* Hilbert said:

> The theory of number fields is an edifice of rare beauty and harmony. The most richly executed part of this building as it appears to me, is the theory of Abelian fields which Kummer by his work on the higher laws of reciprocity, and Kronecker by his investigations on the complex multiplication of elliptic functions, have opened up to us. The deep glimpses into the theory which the work of these two mathematicians affords, reveals at the same time that there still lies an abundance of priceless treasures hidden in this domain, beckoning as a rich reward to the explorer who knows the value of such treasures and with love pursues the art to win them.

It is hard, if not unfeasible, in a short account to evoke a faint idea of what Hilbert wrought in algebraic number theory. Even in a much broader context it would not be easy. Hilbert's own contributions to algebraic number theory are so overwhelming that in spite of the achievements of his predecessors, one gets the impression that algebraic number theory started with Hilbert—other than the theory of invariants, which he completed. So much has happened since Hilbert that one feels uneasy when trying to describe his work in algebraic number theory with his own terms, although it should be said that many modernizations of the theory are implicitly contained or foreshadowed in Hilbert's work.

Hilbert's work centers on the reciprocity law and culminates in the idea of the class field, where the ideals of the original field become principal ideals. The reciprocity law, as it now stands, has gradually developed from Gauss's law for quadratic residues. Hilbert interpreted quadratic residues as norms in a quadratic field and the Gauss residue symbol as a norm residue symbol. In this interpretation it can be generalized so as to be useful in the study of power residues in the most efficient way. The odd behavior of the even prime $p = 2$, which in general does not admit extending solutions of $x^2 = a \bmod p^k$ to higher values of k, is corrected by seeking solutions not in ordinary integers but in p-adic numbers, although before Hensel p-adic numbers could not occur explicitly in Hilbert's exposition. Likewise, the totality of prime spots, although not explicitly mentioned, is Hilbert's invention. In fact, to save the reciprocity law, he introduced the infinite prime spots. His formulation of the reciprocity law as $\Pi_p(\alpha/p) = 1$ foreshadowed *idèles,* and his intuition of the class field has proved an accurate guide for those who later tried to reach the goals he set.

Algebraic number theory was the climax of Hilbert's activity. He abandoned the field when almost everything had yet to be done. He left it to his students and successors to undertake the completion.

Hilbert turned to foundations of geometry. Traditional geometry was much easier than the highly sophisticated mathematics he had engaged in hitherto. The impact of his work in foundations of geometry cannot be compared with that of his work in the theory of invariants, in algebraic number theory, and in analysis. There is hardly one result of his *Grundlagen der Geometrie* which would not have been discovered in the course of time if Hilbert had not written this book. But what matters is that one man alone wrote this book, and that it is a fine book. *Grundlagen der Geometrie,* published in 1899, reached its ninth edition in 1962. This means that it is still being read, and obviously by more people than read Hilbert's other work. It has gradually been modernized, but few readers realize that foundations of geometry as a field has developed more rapidly than *Grundlagen der Geometrie* as a sequence of reeditions and that Hilbert's book is now a historical document rather than a basis of modern research or teaching.

The revival of mathematics in the seventeenth century had not included geometry. Euclid's choice of subjects and his axiomatic approach were seldom questioned before the nineteenth century. Then projective and non-Euclidean geometries were discovered, and the foundations of geometry were scrutinized anew by a differential geometry (Riemann) and the group theory approach (Helmholtz). G. K. C. von Staudt (1847) tried an axiomatic of projective geometry but, unaware of the role of continuity axioms, he failed. The first logically closed axiomatic system of projective and Euclidean geometry was Pasch's (1882), modified and elaborated by the Italian school. Hilbert is often quoted as having urged: "It must be possible to replace in all geometric statements the words *point, line, plane* by *table, chair, mug.*" But Pasch had earlier said the same thing in other words. Moreover, this was not all that had to be done to understand geometry as a part of mathematics, independent of spatial reality; one needs to understand the relations between those points, lines, and planes in the same abstract way. The insight into the implicitly defining character of an axiomatic system had been reached in the *Grundlagen der Geometrie,* but at the end of the nineteenth century it was in the air; at least G. Fano had formulated it, even more explicitly, before Hilbert. It is true that this idea has become popular thanks to Hilbert, although quite slowly, against vehement resistance.

What Hilbert meant to do in his book, and actually did, is better characterized by the following statement at the end of the *Grundlagen:*

> The present treatise is a critical inquiry into the principles of geometry; we have been guided by the maxim to discuss every problem in such a way as to examine whether it could not be solved in some prescribed manner and by some restricted aids. In my opinion this maxim contains a general and natural prescription; indeed, whenever in our mathematical considerations we meet a problem or guess a theorem, our desire for knowledge would not be satisfied as long as we have not secured the complete solution and the exact proof or clearly understood the reason for the impossibility and the necessity of our failure.
>
> Indeed, the present geometrical inquiry tries to answer the question which axioms, suppositions or aids are necessary for the proof of an elementary geometric truth; afterwards it will depend on the standpoint which method of proof one prefers.

Hilbert's goals in axiomatics were consistency and independence. Both problems had been tackled before him. Non-Euclidean geometry was invented to show the independence of the axiom of parallel lines, and models of non-Euclidean geometry within Euclidean geometry proved its relative consistency. Hilbert's approach was at least partially different; his skillfully used tool was algebraization. Algebraic models and countermodels were invoked to prove consistency and independence.

Algebraization as a tool in foundations of geometry was not new at that time. It goes as far back as Staudt's "calculus of throws," although before Hilbert it seems not to have been interpreted as a relative consistency proof. For independence proofs, algebraization had been tried, just before Hilbert, in the Italian school; but Hilbert surpassed all his predecessors. In Hilbert's work and long afterward, algebraization of geometries has proved an important force in creating new algebraic structures. Isolation and interplay of incidence axioms and continuity axioms are reflected by analogous phenomena in the algebraic models. In Hilbert's work they led to structures which foreshadow the ideas of field and skew field, on the one hand, and topological space, on the other, as well as various mixtures of both. Indeed, Hilbert taught the mathematicians how to axiomatize and what to do with an axiomatic system.

In 1904 Hilbert perplexed the mathematical world by salvaging the Dirichlet principle, which had been brought into discredit by Weierstrass' criticism. Before Weierstrass it had been taken for granted in the theory of variations that the lower bound of a functional F is assumed and hence provides a minimum. If some integral along the curves joining two points was bounded from below, a minimum curve must exist. The boundary value problem for the potential equation was solved according to the Dirichlet principle by minimizing $F(u) = \int |\text{grad } u|^2 \, d\omega$ under the given boundary conditions. After Weierstrass had shown that this argument was unjustified, the Dirichlet principle was avoided or circumvented.

Hilbert proved the Dirichlet principle by brute force, as straightforwardly as he had solved the finiteness problem of the theory of invariants. A sequence u_n is chosen such that $\lim_n F(u_n) = \inf_u F(u)$; the $|\text{grad } u_n|$ may be supposed bounded. Then a now-classic diagonal process yields a subsequence which converges first in a countable dense subset, and consequently everywhere and uniformly. Its limit solves the minimum problem. The method seems trivial today because it has become one of the most widely used tools of abstract analysis.

Hilbert also enriched the classical theory of variations, but his most important contribution to analysis is integral equations, dealt with in a series of papers from 1904 to 1910. In the course of the nineteenth century it had been learned that in integral equations the type $f - Af = g$ (where A is the integral operator and f the unknown function) is much more accessible than the type $Af = g$. Liouville (1837) once encountered such an equation and solved it by iteration. So did August Beer (1865), when trying to solve the boundary problem of potential theory by means of a double layer on the boundary; Carl Neumann mastered it (1877) by formal inversion of $1 - A$. The same method proved useful in Volterra's equations (1896). When Poincaré (1894) investigated the boundary problem $\Delta f + \lambda f = h$, turned into an integral equation $f - \lambda A f = g$ by means of Green's function, the parameter λ was analytically involved in the solution. This allowed analytic continuation through the λ plane except, of course, for certain polar singularities. To solve this kind of equation Fredholm (1900, 1902) devised a determinant method, but his greatest merit is to have more clearly understood the λ singularities as eigenvalues of the homogeneous problems.

At this point Hilbert came in. He deliberately turned from the inhomogeneous to the homogeneous equations, from the noneigenvalues to the eigenvalues—or, rather, he turned from the linear equation to the quadratic form, that is, to its transformation on principal axes. Fredholm's method told him how this transformation had to be approached from the finite-dimension case. It was a clumsy procedure and was soon superseded by Erhard Schmidt's much more elegant one (1905). With a fresh start Hilbert then coordinatized function space by means of an orthonormal basis of continuous functions and entered the space of number sequence with convergent square sums, or Hilbert space, as it has been called since. Here the transformation on principal axes was undertaken anew, first on the quadratic forms called "completely continuous" ("compact," in modern terminology) and then on bounded forms, where Hilbert discovered and skillfully handled the continuous spectrum by means of Stieltjes' integrals. The term "spectrum" was coined by Hilbert, who, indeed, must be credited with the invention of many suggestive terms. "Spectrum" was even a prophetic term; twenty years later physicists called upon spectra of operators, as studied by Hilbert, to explain optical spectra.

Hilbert's turn to the space of number sequences seems odd today, but at that time it was badly needed; Hilbert space in a modern sense was not thinkable before the Fischer-Riesz theorem (1907), and its abstract formulation dates from the late 1920's. Hilbert's approach to spectral resolution, utterly clumsy and suffering from the historical preponderance of the resolvent, was greatly simplified later, essentially by F. Riesz (1913); the theory was extended to unbounded self-adjoint operators by J. von Neumann and M. H. Stone about 1930.

Today the least studied and the most obsolete among Hilbert's papers are probably those on integral equations. Their value is now purely historical, as the most important landmark ever set out in mathematics: the linear space method in analysis, with its geometrical language and its numerous applications, quite a few of which go back to Hilbert himself.

From Hilbert's analytic period one rather isolated work, and the most beautiful of all he did, should not be overlooked: his proof of Waring's hypothesis that every positive integer can be represented as a sum of, at most, m l^{th} powers, m depending on l only.

From about 1909 Hilbert showed an ever increasing interest in physics, which, he asserted, was too difficult to be left to physicists. The results of this activity have only partially been published (kinetic gas theory, axiomatics of radiation, relativity). It is generally acknowledged that Hilbert's achievements in this field lack the profundity and the inventiveness of his mathematical work proper. The same is true of his highly praised work in the foundations of mathematics. (It is still a sacrilege to say so, but somebody has to be the first to commit this crime.) In this field even lesser merits have made people famous but, according to the standards set by Hilbert himself, his ideas in foundations of mathematics look poor and shallow. This has become clear with the passing of time. His contemporaries and disciples were much impressed, and even now it is difficult not to be impressed, by his introduction of the "transfinite" functor τ, which for every predicate A chooses an object τA such that $A(\tau A) \rightarrow A(x)$—the so-called Aristides of corruptibility, who, if shown to be corruptible, would prove the corruptibility of all Athenians. Indeed, it is a clever idea to incorporate all transfinite tools of a formal system, such as the universal and the existential quantifier, and the choice axiom into this one symbol τ and afterward to restore the finitistic point of view by systematically eliminating it. For many years the delusive profundity of that artifice led investigators the wrong way. But how of all people could Hilbert, whose intuitions used to come true like prophecies, ever believe that this tool would work? Asking this question means considering the tremendous problem of Hilbert's psychological makeup.

One desire of Hilbert's first axiomatic period was still unfulfilled: after the relative consistency of geometry he wanted to prove the consistency of mathematics itself—or, as he put it, the consistency of number theory. This desire, long suppressed, finally became an obsession. As long as mathematics is no more than counting beans, its consistency is hardly a problem. It becomes one when mathematicians start to treat infinities as though they were bags of beans. Cantor had done so in set theory, and the first to reap glory by the same kind of boldness in everyday mathematics was Hilbert. Is it to be wondered that he was haunted by the need to justify these successes?

He conceived the idea of formalism: to reduce mathematics to a finite game with an infinite but finitely defined treasure of formulas. This game must be consistent; it is the burden of metamathematics to prove that while playing this game, one can never hit on the formula $0 \neq 0$. But if a vicious circle is to be avoided, metamathematics must restrict itself to counting beans. If some chain of the game delivered $0 \neq 0$, one should try to eliminate all links involving the transfinite τ and to reduce the chain to one in which simple beans were counted—this was Hilbert's idea of a consistency proof.

From the outset there were those who did not believe this idea was feasible. Others rejected it as irrelevant. The most intransigent adversary was L. E. J. Brouwer, who from 1907 held that it is truth rather than consistency that matters in mathematics. He gradually built up a new mathematics, called intuitionism, in which many notions of classic mathematics became meaningless and many classic theorems were disproved. In the early 1920's Hermann Weyl, one of Hilbert's most famous students, took Brouwer's side. Both Hilbert and Brouwer were absolutists; for both of them mathematics was no joking matter. There must have been tension between them from their first meeting; although disguised, it can be felt in the discussions of the 1920's between a crusading Brouwer and a nervous Hilbert.

The mathematical world did not have to decide whether formalism was relevant. The catastrophe came in 1931, when Kurt Gödel proved that Hilbert's approach was not feasible. It was a profound discovery, although there had been intimations, such as the Löwenheim-Skolem paradox. Had Hilbert never doubted the soundness of his approach? All he published in this field is so naïve that one would answer "yes." But how was it possible?

Hilbert, as open-minded as a mathematician could be, had started thinking about foundations of mathematics with a preconceived idea which from the outset narrowed his attitude. He thought that something he wished to be true was true indeed. This is not so strange as it seems. It is quite a different thing to know whether mathematics is consistent, or whether some special mathematical hypothesis is true or not. There seems to be so much more at stake in the first case that it is difficult to deal with it as impartially as with the second.

At closer look, 1931 is not the turning point but the starting point of foundations of mathematics as it has developed since. But then Hilbert can hardly be counted among the predecessors, as could Löwenheim and Skolem. This is a sad statement, but it would be a sadder thing if those who know nothing more about Hilbert than his work in foundations of mathematics judged his genius on this evidence.

In 1900 Hilbert addressed the International Congress of Mathematicians on mathematical problems, saying: "This conviction of the solvability of any mathematical problem is a strong incentive in our work; it beckons us: *this is the problem, find its solutions. You can find it by pure thinking since in mathematics there is no Ignorabimus!* [*Gesammelte Abhandlungen,* III, 298]." With these words Hilbert introduced twenty-three problems which have since stimulated mathematical investigations:

1. *The cardinality of the continuum.* After a great many unsuccessful attempts the problem was solved in 1963 by Paul J. Cohen, although in another sense than Hilbert thought: it has been proved unsolvable. In the same connection Hilbert mentions well-ordering, which was accomplished by Zermelo.

2. *The consistency of the arithmetic axioms.* The history of this problem has already been dealt with.

3. *The existence of tetrahedrons with equal bases and heights that are not equal in the sense of division and completion.* The question was answered affirmatively shortly afterward by Max Dehn.

4. *The straight line as the shortest connection.* The problem is too vague.

5. *The analyticity of continuous groups.* The analyticity has been proved by small steps, with the final result in 1952.

6. *The axioms of physics.* Even today axiomatics of physics is hardly satisfactory. The best example is R. Giles's *Mathematical Foundations of Thermodynamics* (1964), but in general it is not yet clear what axiomatizing physics really means.

7. *Irrationality and transcendence of certain numbers.* From C. L. Siegel (1921) and A. O. Gelfond (1929) to A. Baker (1966–1969), problems of this kind have been tackled successfully.

8. *Prime number problems.* Riemann's hypothesis is still open, despite tremendous work. In algebraic fields it has been answered by E. Hecke (1917). Goldbach's hypothesis has successfully been tackled by L. Schnirelmann (1930), I. M. Vinogradov (1937), and others.

9. *Proof of the most general reciprocity law in arbitrary number fields.* The problem has been successfully tackled from Hilbert himself to Artin (1928) and I. R. Šafarevič (1950).

10. *Decision on the solvability of a Diophantine equation.* A rather broad problem, this has often been dealt with—for instance, by Thue (1908) and by

C. L. Siegel (1929). The general problem was answered negatively by J. V. Matijasevič in 1969.

11. *Quadratic forms with algebraic coefficients.* Important results were obtained by Helmuth Hasse (1929) and by C. L. Siegel (1936, 1951). Connections to *idèles* and algebraic groups were shown by A. Weil and T. Ono (1964–1965).

12. *Kronecker's theorem on Abelian fields for arbitrary algebraic fields.* This relates to finding the functions which for an arbitrary field play the same role as the exponential functions for the rational field and the elliptic modular functions for imaginary quadratic fields. Much has been done on this problem, but it is still far from being solved.

13. *Impossibility of solving the general seventh-degree equation by functions of two variables.* Solved by V. I. Arnold (1957), who admits continuous functions, this is still unsolved if analyticity is required.

14. *Finiteness of systems of relative integral functions.* This was answered in the negative by Masayoshi Nagata (1959).

15. *Exact founding of Schubert's enumerative calculus.* Although enumerative geometry has been founded in several ways, the justification of Schubert's calculus as such is still an open problem.

16. *Topology of real algebraic curves and surfaces.* The results are still sporadic.

17. *Representation of definite forms by squares.* This was solved by Artin (1926).

18. *Building space from congruent polyhedrons.* The finiteness of the number of groups with fundamental domain was proved by Ludwig Bieberbach (1910). A Minkowski hypothesis on the covering of space with cubes was proved by Georg Hajos (1941).

19. *The analytic character of solutions of variation problems.* A few special results have been obtained.

20. *General boundary value problems.* Hilbert's own salvage of the Dirichlet problem and many other investigations have been conducted in this area.

21. *Differential equations with a given monodromy group.* This was solved by Hilbert himself (1905).

22. *Uniformization.* For curves, this was solved by Koebe and others.

23. *Extension of the methods of variations calculus.* Hilbert himself and many others dealt with this.

> If I were a painter, I could draw Hilbert's portrait, so strongly have his features engraved themselves into my mind, forty years ago when he stood on the summit of his life. I still see the high forehead, the shining eyes looking firmly through the spectacles, the strong chin accentuated by a short beard, even the bold Panama hat, and his sharp East Prussian voice still sounds in my ears [F. W. Levi, *Forscher und Wissenschaftler im heutigen Europa,* p. 337].

This description by Levi is confirmed by many others. People who met Hilbert later were gravely disappointed.

Hilbert was a strong personality, and an independent thinker in fields other than mathematics. As an East Prussian he was inclined to political conservatism, but he abhorred all kinds of nationalist emotions. During World War I he refused to sign the famous Declaration to the Cultural World, a series of "it-is-not-true-that" statements; and when the French mathematician Darboux died during the war, he dared to publish an obituary.

Biographical sketches written during Hilbert's lifetime are more or less conventional but never Byzantine. The oral tradition is more characteristic; it has been collected by Constance Reid, who in her biography of Hilbert gives a truthful and understanding image of the man and his world. Her biography also contains a reprint of Weyl's obituary, which is the most expert analysis of his work and reflects Hilbert's personal influence on his students and collaborators: "the sweet flute of the Pied Piper that Hilbert was, seducing so many rats to follow him into the deep river of mathematics." There are more witnesses concerning Hilbert: Hilbert himself, telling about his friend Minkowski; and the list of sixty-nine theses written under his guidance, many of them by students who became famous mathematicians.

BIBLIOGRAPHY

I. ORIGINAL WORKS. Hilbert's *Gesammelte Abhandlungen,* 3 vols. (Berlin, 1932–1935; 2nd ed., 1970), includes analyses of his work and a biography by Otto Blumenthal. Not included are his *Grundlagen der Geometrie* (Leipzig, 1899; 9th ed., Stuttgart, 1962) and *Grundzüge einer allgemeinen Theorie der Integralgleichungen* (Leipzig 1912; 2nd ed. 1924).

II. SECONDARY LITERATURE. The best analysis of Hilbert's work as a whole is in Hermann Weyl, "David Hilbert and His Work," in *Bulletin of the American Mathematical Society,* **50** (1944), 612–654. See also F. W. Levi, *Forscher und Wissenschaftler im heutigen Europa, Weltall und Erde* (Oldenburg, 1955), pp. 337–347.

An analysis of his work in foundations of geometry is Hans Freudenthal, "Zur Geschichte der Grundlagen der Geometrie," in *Nieuw archief voor wiskunde,* 4th ser., **5** (1957), 105–142. The history of Hilbert's problems is discussed in P. Alexandrov, ed., *Problemy Gilberta* (Moscow, 1969); and Ludwig Bieberbach, "Über den Einfluss von Hilbert's Pariser Vortrag über 'Mathematische Probleme' auf die Entwicklung der Mathematik in den letzten dreissig Jahren," in *Naturwissenschaften,* **18** (1930), 1101–1111.

Biographical writings are Paul Bernays, "David Hilbert," in *Encyclopedia of Philosophy,* III (New York, 1967), 496–504; Otto Blumenthal, O. Toeplitz, Max Dehn, Richard

Courant, Max Born, and Paul Bernays, in *Naturwissenschaften,* **10** (1922), 67–99; Constantin Carathéodory, "Hilbert," in *Sitzungsberichte der Bayerischen Akademie der Wissenschaften zu München,* Math.-nat. Abt. (1943), 350–354; Constantin Carathéodory and Arnold Sommerfeld, "Hilbert," in *Naturwissenschaften,* **31** (1943), 213–214; G. Polya, "Some Mathematicians I Have Known," in *American Mathematical Monthly,* **76** (1969), 746–753; and Constance Reid, *Hilbert* (Berlin–Heidelberg–New York, 1970).

HANS FREUDENTHAL

HILDEBRANDT, GEORG FRIEDRICH (*b.* Hannover, Germany, 5 June 1764; *d.* Erlangen, Germany, 23 March 1816), *chemistry.*

Hildebrandt first attended the Gymnasium at Hannover and then studied pharmacy at the University of Göttingen (1780). His main interests were anatomy, physiology, and chemistry. After receiving the M.D. in 1783, he toured German factories, mines, and hospitals to gain practical experience. In 1785 he returned to Göttingen, where he became a *Privatdozent.* In the same year he was appointed professor of anatomy at the Anatomical-Surgical Institute at Brunswick. He transferred to the University of Erlangen in 1793 as professor of medicine; he later became professor of chemistry (1796) and physics (1799) there. In 1808, with J. C. F. Harletz and E. W. Martius, Hildebrandt formed the Physical-Medical Society of Erlangen.

Hildebrandt possessed a profound knowledge of anatomy, physiology, chemistry, physics, and pharmacy; and he published a great many articles on medicine, physics, and (after 1793) chemistry. At Erlangen he was the first professor of chemistry to be greatly concerned with the practical training of students.

Influenced by the writings of J. T. Mayer, Hildebrandt became an early adherent of Lavoisier's oxidation theory. He wrote a comparative and critical survey on the phlogiston and antiphlogiston theories but did not express a preference for either. In 1793 he announced that he was inclined toward the antiphlogiston theory, and his 1794 book *Anfangsgründe der Chemie* reflects that bias. Hildebrandt advanced reasonable arguments against Lavoisier's theory—for example, the impossibility of explaining the light produced by combustion—and pointed out the fallacy of Lavoisier's supposition that an acidic principle is an essential component of all acids.

That Hildebrandt was an adherent of Kant's dynamic theory of matter is shown clearly by his article "Ueber die Modification der Materie" (1805) and by his book *Anfangsgründe der dynamischen Naturlehre* (1807). The latter is one of the most complete applications in the first decade of the nineteenth century of Kant's ideas on the dynamics of chemical and physical phenomena. Hildebrandt's starting point was the dynamic system in which matter is a product of two forces: one attracting (positive) and one repelling (negative). He found this supposition far more satisfying than the atomic system but recognized that the latter could at least help in reaching a correct understanding of chemical phenomena. Hildebrandt declared emphatically that the atomistic view is only an expedient. He tried to give an explanation of all natural phenomena by means of Kant's dynamics, but he also showed that he was influenced by the more speculative concepts of Henrik Steffens and Schelling, who asserted among other things that all earthy matter is composed of polar opposites: hydrogen and oxygen, nitrogen and carbon.

Hildebrandt published much on practical chemistry, especially on the analysis of mineral waters. He wrote a book on mercury compounds (1793) and published on the nature of quicklime (1792), ammonium nitrate (1794), the composition of ammonia (1795), the preparation of pure potassium ferrocyanide and the analytical separation of iron from alum (1798), the different colors of light emitted during electrical discharges in air at low pressure (1811), the gas evolved in the deflagration of niter and charcoal (1811), and the determination of oxygen in air by nitric oxide (1815). He also wrote textbooks on pharmacology (1787) and anatomy (1789–1792).

BIBLIOGRAPHY

I. ORIGINAL WORKS. Hildebrandt's writings include *Versuch einer philosophischen Pharmakologie* (Göttingen, 1787); *Chemische und mineralogische Geschichte des Quecksilbers* (Brunswick, 1793); "Vergleichende Übersicht des phlogistischen und antiphlogistischen Systems," in Crell's *Chemische Annalen* (1793), pt. 2, 24–30, and (1794), pt. 1, 200–210; "Etwas über das antiphlogistische System der Chemie," *ibid.* (1793), pt. 2, 99–104; *Anfangsgründe der Chemie,* 3 vols. (Erlangen, 1794); *Ueber die Arzneikunde* (Erlangen, 1795); *Encyklopädie der gesammten Chemie,* 16 pts. (Erlangen, 1799–1810); *Physikalische Untersuchung des Mineralwassers im Alexanderbade bei Sichersreuth in Franken,* 2 vols. (Erlangen, 1803; 2nd ed., 1821); *Anfangsgründe der dynamischen Naturlehre* (Erlangen, 1807); and *Lehrbuch der Chemie als Wissenschaft und als Kunst* (Erlangen, 1816).

II. SECONDARY LITERATURE. An obituary note with complete bibliography is G. Bischof, "Kurzer Bericht über Hildebrandts Leben," in *Journal für Chemie und Physik,* **25** (1819), 1–16. See also J. R. Partington, *A History of Chemistry,* III (London, 1962), 638–639.

H. A. M. SNELDERS

HILDEGARD OF BINGEN (*b.* Bermersheim, Germany, 1098; *d.* Rupertsberg, near Bingen, Germany, 1179), *cosmology.*

Also called Hildegardis de Pinguia and often called St. Hildegard, Hildegard was a writer on nature and medicine (probably also a practicing "doctor"), a visionary, and transmitter and original transformer of Oriental, Judeo-Christian, and Greek cosmological and allegorical ideas. She was the tenth child of Hildebert of Vermersheim, a member of the gentry, whose estate was near Alzey on the Nahe River in the Palatinate. From 1106 to 1147 she lived at a small nunnery attached to the cloister of Disibodenberg, serving as its head from 1136. She founded her own convent on the Rupertsberg in 1147. Beginning in 1141 Hildegard followed an internal command to "write what you see and hear," that is, the visions of which she had been conscious from about 1113. After a papal inquiry she was encouraged to continue her literary and practical activities by Pope Eugene III and was enthusiastically supported by Bernard of Clairvaux. She now became the spiritual center to which popes, kings, and ecclesiastical and secular dignitaries turned for advice and augury. Her influence was felt throughout Europe, notably in France and England, and even as far as Greece and Palestine. The Holy Roman emperor, Frederick Barbarossa, submitted to her rebuke, met her at Ingelheim, granted the Rupertsberg convent an imperial letter of protection in 1163, and left it unmolested when his troops devastated the Rheingau. Although papal proceedings for canonization were instituted in 1233, it is uncertain that canonization took place.

Hildegard's mystical, visionary, and spiritual writings include *Liber Scivias* (1141–1151), a description (and illustration in a remarkable series of illuminated plates) of visions, notably of the cosmos and man's position therein; *Liber vitae meritorum* (1158–1163), a continuation of her visions, reflecting on ethics and the cosmic effects of virtue and sin; and *Liber divinorum operum* (1163–1170), on the theological significance of the cosmos. To these should be added the corpus of letters, poems ("Symphonia harmoniae coelestium revelationum"), hermeneutica, and other works.

Naturalistic and medical books include *Liber simplicis medicinae* (*Liber subtilitatum diversarum naturarum creaturarum*) (*ca.* 1150–1160), also called (although not by Hildegard) *Physica,* on plants, trees, animals, stones, metals, and elements, chiefly from the medical (curative) point of view; and *Liber compositae medicinae* (*causae et curae—de aegritudinum causis, signis et curis*), on the nature and forms of diseases and their causes, notably the forces of the cosmos—elements, winds, stars—based on an allegorical microcosmic physiology.

All the works listed above are genuine, although there have often been doubts about the naturalistic and medical books and some interpolations do exist. The cosmological motives and allegorical interpretations are identical in both "scientific" and "nonscientific" works. We possess the *testimonia* of the inventories and *necrologia* of Hildegard's convent and of Trithemius (1462–1516), who had seen the original manuscript of *Liber simplicis medicinae* there—it was listed with Hildegard's other works and he copied it for himself. Hildegard herself mentioned it as her own work in the preface to *Liber compositae medicinae* (prior to 1158).

Hildegard was a "simple" woman, typical of the unlearned mystic *idiota* who wrote down what she "saw and heard," following a command given to her by "voices." She is therefore basically original in both her spiritual and her naturalist and medical work. She is depicted as receiving her visions through the head—perhaps reflecting the Platonic idea of the seat of the soul—although she herself located the soul in the heart. This represents the biblical view rather than an Aristotelian allusion. It was too early for such an allusion in the West; and in any case such fundamental Aristotelian concepts as hyle, ether, generation, and corruption do not appear except in marginalia by copyists and in interpolated sections. Her Latin, picked up and inspired rather than properly learned, was richly interlarded with German terms and polished and scripted by her close collaborator Volmar, a monk who died in 1170.

The most important naturalist sources for Hildegard were probably folk medicine and popular tradition, notably a welter of recipes, nostrums, amulets, and magico-religious procedures, such as that for the execration of demons. In addition there was the fundamental Galenic humoralism, which formed part of the Benedictine heritage. Thus phlegm figures as the main cause of disease, since it is connected with the fall of man, who made himself more similar to earth from which he was originally formed. Just as earth brings forth good and evil herbs, good and bad humors arise in man. Flesh ulcerates and is "perforated" because Adam's blood was converted into the evil foam that serves for procreation.

Such biblical and microcosmic analogies form a kind of medicine that is indeed original and, on the practical side, partly the result of her firsthand experience in nature studies and medicine. In *Liber simplicis medicinae* the curative virtue of precious stones plays a prominent part—the devil hates them because their fire-born splendor illuminated him before he fell. In their use Hildegard followed a tradition somewhat different from that emerging later in the Paracelsian corpus. She regarded sapphire as good for the eyes and as an antiaphrodisiac, whereas it is a cure for cardiac pain in the Paracelsian corpus, in which emerald assumes the roles of Hildegard's sapphire. Carnelian (chalcedony) is a hemostatic in both traditions; but Hildegard omitted the emerald, which is also recommended as hemostatic in the Paracelsian corpus. Hildegard's use of the amethyst to treat rash is perhaps related to that stone's application in the Paracelsian corpus to plague boils.

Hildegard admitted that knowledge of nature can be derived from *magia,* including information from evil spirits, but inveighed against diabolical arts (*maleficium*), which turn knowledge to impurity and the pursuit of evil. She paid much attention to the wholesomeness of waters and the necessity to boil some of them. Arabic-Salernitan concepts are absent, as are traces of the philosophical and naturalistic trends characteristic of mid-twelfth-century Chartres, which led half a century later to those of Oxford, Paris, and Toledo.

Hildegard thus remains original in her mystical and naturalist work, the sound as well as the fantastic lore. Perhaps this judgment also applies to her ideas that all brooks and rivers derive from a large salt sea, that salt sources have more fire and virtue than ordinary water, and that soft rain is descending when the sun spends heat—analogous to men who weep for joy. Hail, on the other hand, is regarded as the "eye," that is, the eye fluid, of thunder.

Hildegard's influence was considerable in her own time and lasted far into the Renaissance, when the first printed edition of *Liber Scivias* was published by J. Faber Stapulensis in *Liber trium virorum et trium spiritualium virginum* (Paris, 1513), fol. 28r–118v, and two editions of *Liber simplicis medicinae* appeared (1533, 1544). Reference is made to Hildegard even in the Paracelsian corpus (*Fragmenta cum libro de fundamento sapientiae congruentia,* Sudhoff, ed., XIII, 334); and there are concepts common to both, although they are not necessarily derived from Hildegard or even from a common source. Trithemius praised Hildegard's naturalist and medical work as being of "wonderful and secret things of nature with fine understanding and for a mystical design." Her influence, conceptual as well as iconographical, is prominently recognizable in Agrippa von Nettesheim's *De occulta philosophia* (1531)—Agrippa was a friend and pupil of Trithemius—and particularly in the microcosmic allegorical anthropology and the pictures of Robert Fludd (1617).

BIBLIOGRAPHY

I. Original Works. Editions and translations of Hildegard's writings include two collections: *S. Hildegardis abbatissae opera omnia,* J. P. Migne, ed., *Patrologia latina,* CXCVII (Paris, 1855; 1888; 1952); and *Analecta S. Hildegardis,* J. B. Pitra, ed., *Analecta sacra,* VIII (Monte Cassino, 1882). Individual works are *Liber subtilitatum diversarum naturarum* (the *Liber simplicis medicinae,* or *Physica*), F. A. Reuss, ed., in Migne, *Patrologia latina,* CXCVII; *Die physica der heiligen Hildegard,* translated, with introduction and notes, by J. Berendes (Vienna, 1897), reprinted from *Pharmazeutische Post,* **29–30** (1896–1897); *Causae et curae,* Paul Kaiser, ed. (Leipzig, 1903); *Hildegard von Bingen, Wisse die Wege—Scivias—nach dem Originaltext des illuminierten Rupertsberger Kodex ins Deutsche,* translated and edited by Maura Böckeler (Berlin, 1928; Salzburg, 1954), with color plates, an important app., and biobibliographical notes; *Der Äbtissin Hildegard von Bingen Ursachen und Behandlung von Krankheiten* (*Causae et curae*), translated by Hugo Schulz (Munich, 1933; repr., Ulm, 1955); *Hildegard von Bingen, Heilkunde. Das Buch von dem Grund und Wesen und der Heilung der Krankheiten* (Salzburg, 1957), translated, with extensive introduction, a running commentary, and text-critical notes, by Heinrich Schipperges; *Hildegard von Bingen, Naturkunde. Das Buch von dem inneren Wesen der verschiedenen Naturen in der Schöpfung. . .* (Salzburg, 1959), trans. of *Liber simplicis medicinae,* with glossary and critical notes, by Peter Riethe; and *Hildegard von Bingen, Welt und Mensch. Das Buch "De operatione Dei" aus dem Genter Codex,* translated, with intro. and notes, by H. Schipperges (Salzburg, 1965).

II. Secondary Literature. See H. Fischer, "Die heilige Hildegard von Bingen, die erste deutsche Naturforscherin und Ärztin," in *Münchener Beiträge zur Geschichte und Literatur der Naturwissenschaften und Medizin,* **7–8** (1927), 377–538; C. Jessen, "Über Ausgaben und Handschriften der medizinisch-naturhistorischen Werke der h. Hildegard," in *Sitzungsberichte der K. Akademie der Wissenschaften in Wien,* Math.-nathist. Kl., **45**, sec. 1 (1862), 97–116; P. Kaiser, *Die Naturwissenschaftlichen Schriften der Hildegard von Bingen* (Berlin, 1901); W. Lauter, *Hildegard-Bibliographie. Wegweiser zur Hildegard-Literatur* (Alzey, 1971); H. Liebeschütz, *Das allegorische Weltbild der heiligen Hildegard von Bingen,* Studien der Bibliothek Warburg, no. 16 (Leipzig, 1930), of particular importance for establishing the authenticity of all parts of Hildegard's writings and also for tracing her sources for allegorical cosmology and *Kosmos-mensch,* notably the imagery and iconographic tradition down to Persian and Gnostic ideas; E. H. F. Meyer, *Geschichte der Botanik,* III (Königsberg, 1856; repr., Amsterdam, 1965), 517–536, with valuable app. on doubtful herbs quoted in *Liber simplicis medicinae* (*Physica*);

F. W. E. Roth, "Studien zur Lebensbeschreibung der heiligen Hildegard," in *Studien und Mitteilungen zur Geschichte des Benediktiner-Ordens,* **39** (1918), 68–118; G. Sarton, *Introduction to the History of Science,* II (Baltimore, 1931), 386–388; H. Schipperges, "Ein unveröffentlichtes Hildegard Fragment (Cod. Berol. Lat. Qu. 674)," in *Archiv für Geschichte der Medizin,* **40** (1956), 41–77; and "Zur Konstitutionslehre Hildegards von Bingen," in *Arzt und Christ* (1958), pp. 90–94; M. Schrader and A. Führkötter, *Die Echtheit des Schrifttums der hl. Hildegard von Bingen* (Cologne–Graz, 1956), a profound study of all existing MSS and their transmission; and L. Thorndike, *History of Magic and Experimental Science,* II (New York, 1923), 124–154.

WALTER PAGEL

HILDITCH, THOMAS PERCY (*b.* London, England, 22 April 1886; *d.* Birkenhead, England, 9 August 1965), *chemistry.*

Hilditch was mainly responsible for the advances in knowledge of the chemical constitution of natural fats and oils from 1925 to 1950. He received the D.Sc. from the University of London in 1911 and became a fellow of the Royal Society in 1942 and commander of the Order of the British Empire in 1952.

Both an industrial and an academic chemist, Hilditch followed the advice of his teacher Sir William Ramsay and accepted the post of research chemist for Joseph Crosfield's and Sons, soap and chemical manufacturers. He remained with Crosfield's for nearly fifteen years (1911–1925), during which time he was concerned with the catalytic hydrogenation of fats and the constitution of the less common components of commercial fats. In 1925 Hilditch was appointed the first James Campbell Brown professor of industrial chemistry at the University of Liverpool, a post he held until his retirement in 1951. His work during this quarter century constitutes Hilditch's major contribution to science. He and his students at Liverpool played a major role in transforming knowledge about the constitution of natural fats. With the help of nearly eighty students from all over the world Hilditch published more than 300 papers, dealing mainly with the component acids and glycerides of natural fats and with the experimental methods for studying these substances.

In 1925 the chemistry of fats was a neglected field. Although the chemical structure of fats had been elucidated by Chevreul in the 1820's, no other great figure appeared in this field of research until Hilditch. There was no systematic account of fats in 1925: little quantitative information was available on the component fatty acids of natural fats and none on the component glycerides. Furthermore, techniques for obtaining the fatty acids were inadequate and were nonexistent for the glycerides. By 1951 Hilditch and his students had obtained this information experimentally for a wide range of fats and oils, and their efforts stimulated others to work in this field.

Throughout this long period of work Hilditch tried to discern the underlying patterns running through animal and vegetable fats. He believed that there was a relationship between the distribution patterns of the component fatty acids and glycerides and the order of the evolutionary development of the parent organisms from which the fats were obtained. This relationship was the basis of his most important book, *The Chemical Constitution of Natural Fats* (1940), which reflected in its four editions the advances in fat chemistry made by Hilditch and his school.

Hilditch retired before both chromatographic methods and controlled enzymatic hydrolysis of fats came into general use, but he realized what might be accomplished with these methods. He had the satisfaction of knowing that his students and many other chemists were continuing his pioneering work.

BIBLIOGRAPHY

I. ORIGINAL WORKS. A bibliography of Hilditch's books and papers by R. A. Morton is in *Biographical Memoirs of Fellows of the Royal Society,* **12** (1966), 259–289. His most important book is *The Chemical Constitution of Natural Fats* (London, 1940; 4th ed. [with P. N. Williams], 1964). His other major works are *A Concise History of Chemistry* (London, 1911; 2nd ed., 1922); *The Industrial Chemistry of the Fats and Waxes* (London, 1927; 3rd ed., 1949); and *Catalytic Processes in Applied Chemistry* (London, 1929; 2nd ed., 1937).

II. SECONDARY LITERATURE. For a detailed account of Hilditch's life and career, see R. A. Morton's article mentioned above. Brief notices include F. D. Gunstone, "T. P. Hilditch, C.B.E., D.Sc., F.R.I.C.," in *Journal of the American Oil Chemists Society,* **42** (1965), 474A, 530A; P. N. Williams, "Prof. T. P. Hilditch, C.B.E., F.R.S.," in *Nature,* **208** (1965), 730–731; and W. D. Raymond, "Professor T. P. Hilditch, C.B.E., F.R.S. (1886–1965)," in *Chemistry and Industry,* **85** (1966), 251.

ALBERT B. COSTA

HILL, GEORGE WILLIAM (*b.* New York, N.Y., 3 March 1838; *d.* West Nyack, New York, 16 April 1914), *mathematical astronomy.*

In the opinion of Simon Newcomb, Hill was destined to rank "as the greatest master of mathematical astronomy during the last quarter of the nineteenth century." In 1903 Hill was ranked second after E. H. Moore by the leading mathematicians in the United States and first, tied with Newcomb, by the leading astronomers. He was honored in his lifetime by the bestowal of advanced degrees and medals and by honorary memberships in the most prestigious professional scientific societies and institutions throughout the world. Yet throughout all of this recognition he remained a simple man of the country.

Hill's father, John William Hill, was born in England while his mother, Catherine Smith, was descended from an old Huguenot family. His grandfather had been a successful engraver in London before emigrating to Philadelphia in 1816. Both Hill's father and younger brother were painters, and in 1846 his father retired to a farm in Nyack Turnpike (now West Nyack), New York. Country residence during Hill's youth was likely to carry with it grave drawbacks in the education of the young; teaching was frequently restricted to a few subjects on an elementary level. Hill was extremely fortunate, while at Rutgers College, to come under the influence of Theodore Strong, a friend of Nathaniel Bowditch, who had translated Laplace's *Mécanique céleste* into English. Strong's deep respect for tradition was reflected in the contents of his library. Hill relates that under Strong he read Sylvestre Lacroix's *Traité du calcul différential et intégral,* Poisson's *Traité de mécanique,* Philippe de Pontécoulant's *Théorie analytique du système du monde,* Laplace's *Mécanique céleste,* Lagrange's *Mécanique analytique,* and Legendre's *Fonctions elliptiques.* Hill quoted Strong as saying that "Euler is our great Master" and noted that Strong "scarcely had a book in his library published later than 1840." Poincaré said that to Strong Euler was "the god of mathematics" whose death marked the beginning of the decline of mathematics.

Hill's knowledge of the techniques of the old masters strengthened his ingenuity in the creation of new methodology. The extent of the Eulerian influence is evident in his "Researches in the Lunar Theory" (1878), which is based on an Eulerian method in its use of moving rectangular axes and the same first approximation. This device led to Hill's variational curve, the reference orbit in describing lunar motion. E. W. Brown developed the work still further for the preparation of lunar ephemerides.

After receiving the B.A. from Rutgers in 1859, Hill went to Cambridge, Massachusetts, to further his mathematical knowledge. In 1861 he joined the staff of scientists working in Cambridge on the *American Ephemeris and Nautical Almanac.* He had already begun to publish in 1859, while still at college, and his third paper, "On the Conformation of the Earth," in J. D. Runkle's *Mathematical Monthly* (1861) brought him a prize and the attention of Runkle as well. R. S. Woodward, president of the Royal Society at the time he wrote Hill's obituary notice, counted the paper as still worthy of reading and considered Hill as having become the leading contributor to the advances in dynamic astronomy during the half-century after its publication. At the *Almanac* office Hill was assigned the task of calculating the American ephemeris, work he was later authorized to continue at his home in West Nyack.

When Simon Newcomb became director of the *American Ephemeris* in 1877, he undertook the reconstruction of the theories and tables of lunar and planetary motion. Hill was induced to work on the theories of Jupiter and Saturn, known to be exceptionally difficult in the determination of their mutual perturbations. Because the *Nautical Almanac* office had meanwhile been transferred to Washington to be under the more immediate jurisdiction of the Navy Department, Hill resided there for a ten-year period beginning in 1882. His success with the Newcomb assignment represented one of the most important contributions to nineteenth-century mathematical astronomy. The calculation of the effects of the planets on the moon's motion was a particular case of the famous three-body problem, which dates back to Newton (1686).

Hill's "Researches in the Lunar Theory," published in the first issue of *American Journal of Mathematics* (1878), had, through its introduction of the periodic orbit, initiated a new approach to the study of three mutually attracting bodies. F. R. Moulton wrote in 1914 that no earlier work had approached it in practical application and no subsequent work had then surpassed it. The article became fundamental in the development of celestial mechanics.

The memoir of 1877 entitled *On the Part of the Motion of the Lunar Perigee Which Is a Function of the Mean Motions of the Sun and Moon* contains the incontrovertible evidence of Hill's mathematical genius. He was led to a differential equation, now called Hill's equation, that is equivalent to an infinite number of algebraic linear equations. Hill showed how to develop the infinite determinant corresponding to these equations.

Hill's procedures reflect his preference for the methodology of Charles Delaunay, as developed in the two-volume *Théorie du mouvement de la lune* (1860–1867), and he is said to have perfected it. Yet the methods adopted in the *Nautical Almanac* work were essentially those of P. A. Hansen, the other lunar theorist of eminence at that time.

Hill's many honors included membership in the National Academy of Sciences (1874), presidency of the American Mathematical Society (1894–1896), and the gold medal of the Royal Astronomical Society for his researches on lunar theory (1887). He was a foreign member of the Royal Society, the Paris Academy, and the Belgian Academy.

In 1898 J. K. Rees, who held the Rutherfurd chair of astronomy at Columbia University, persuaded Hill to accept the newly created lectureship in celestial mechanics. Since few students were qualified to comprehend work on that level, Hill objected to receiving pay and finally resigned in 1901. He was urged to write out his lectures, which he did very painstakingly; he gave them to Columbia but insisted on returning the money that had been paid to him.

Hill remained a recluse in West Nyack, devoted to his researches and to his large scientific library, which he bequeathed to Columbia University. Illness during the last years reduced his physical activity and a failing heart brought his career to a close.

BIBLIOGRAPHY

The Collected Mathematical Works of George William Hill, 4 vols. (Washington, D.C., 1905–1907), includes eighty-three papers and has a biographical intro. by H. Poincaré, pp. vii–xviii. A complete bibliography of Hill's papers is in Ernest W. Brown, "Biographical Memoir of George William Hill, 1838–1914," in *Biographical Memoirs. National Academy of Sciences,* **8** (1916), 275–309; and "History of the N.Y. Mathematical Society," in *American Mathematical Society Semicentennial Publications,* I (New York, 1938), 117–124, with 101 items and a complete list of his honors (p. 118).

A condensed version of Brown's memoir (see above), entitled "G. W. Hill, 1838–1914," is in *Obituary Notices of Fellows of the Royal Society,* **91A** (1915), xlii–li, repr. in *Bulletin of the American Mathematical Society,* **21** (1915), 499–511. See also E. W. Brown, "George William Hill, Mathematician and Astronomer," in *Nation,* **98,** no. 2549 (7 May 1914), 540–541; J. W. L. Glaiser, "Address Delivered by the President . . . on Presenting the Gold Medal of the Society to Mr. G. W. Hill," in *Monthly Notices of the Royal Astronomical Society,* **47** (Feb. 1887), 203–220; Harold Jacoby, "George William Hill," in *Columbia University Quarterly,* **16** (Sept. 1914), 439–442; F. R. Moulton, "George William Hill," in *Popular Astronomy,* **22,** no. 7 (Aug.–Sept. 1914), 391–400; Simon Newcomb, "The Work of George W. Hill," in *Nation,* **85,** no. 2209 (1907), 396, a letter to the editor; and R. S. Woodward, "George William Hill," in *Astronomical Journal,* **28,** no. 20 (5 June 1914), 161–162.

Columbia University Bulletin, no. 8 (July 1894), 24–25, contains a list of the materials in the course of thirty lectures on celestial mechanics given by Hill; on p. 63 of the same issue is the citation accompanying his honorary degree.

The following contain references important to Hill's work: G. D. Birkhoff, "Fifty Years of American Mathematics," in *American Mathematical Society Semicentennial Publications,* II (New York, 1938), 270–315; F. R. Moulton, *Differential Equations* (New York, 1930), pp. 224, 318, 353–354; Felix Klein, inaugural address at the general session of the Congress of Mathematics and Astronomy, Chicago, in *Bulletin of the New York Mathematical Society,* **3** (Oct. 1893), 1–3, also in *Monist,* **4** (Oct. 1893), 1–4; C. S. Peirce, "Note on Mr. G. W. Hill's Moon Theory," in *Nation,* **81** (19 Oct. 1905), 321; and review of Hill's *Collected Works, ibid.,* **85** (17 Oct. 1907), 355; E. H. Roberts, "Note on Infinite Determinants," in *Annals of Mathematics,* **10** (1896), 35–50; and D. E. Smith and J. Ginsburg, *History of Mathematics in America Before 1900* (Chicago, 1934), *passim.*

Further references are in *Dictionary of American Biography,* IX (New York, 1932), 32–33; *National Cyclopedia of American Biography* (New York, 1918), p. 388; Poggendorff, III, 631–632; IV, 639; V, 538; and *American Men of Science,* I (1906), 146.

Additional citations are found in *Encyklöpedie der mathematischen Wissenschaften,* VI (Leipzig, 1912–1926); J. J. [erwood], in *Monthly Notices of the Royal Astronomical Society,* **75** (1915); S. Newcomb, *Reminiscences of an Astronomer* (London, 1903); T. Muir, *Theory of Determinants in the Historical Order of Development,* III (London, 1920); and F. Schlesinger, "Recollections of George William Hill," in *Publications of the Astronomical Society of the Pacific,* **49** (1937).

CAROLYN EISELE

HILL, JOHN (*b.* Peterborough [?], England, 1707 [?]; *d.* London, England, 21 November 1775), *botany.*

A wide range of interests characterized Hill's activities. Among his contemporaries he was well known for his various literary entanglements and voluminous publications in science. Although these include works on medicine, zoology, and mineralogy, the majority are concerned with botany.

An apothecary, Hill developed an interest in plants as a means of supplementing his income, both by collecting for others and by concocting assorted herb remedies which he offered for sale. The latter activity earned him the epithet of "quack." His first major publication in botany appeared as a part of the three-volume *General Natural History* (1748–1752). In the second volume (1751), devoted to the plant kingdom, Hill introduced the classification system of Linnaeus to England. Several popular or semipopular works on plants followed. Many were essentially handbooks for gardeners or, like the *Useful Family Herbal,* guides to the collecting and use of herbs as medicaments. Others, like the *British Herbal* (1756) and his twenty-six-volume compendium *Vegetable System* (1759–1775), are works in taxonomic and descriptive botany intended, at least in part, for the scholarly botanist. Hill's classification, although basically Linnaean, shows the influence of Rivinus (Augustus Quirinus Bachman) in the use of the corolla as a basis for some classes.

Hill showed some interest in plant histology and physiology. For his *Construction of Timber* (1770) he prepared sections of plant stems and stained them for microscopic study. In *Sleep of Plants* (1757) he noted the effects of light on the movement of plants.

Less numerous than his botanical publications but of considerable interest are the works on mineralogy. Hill's first scientific publication was an English translation of Theophrastus' *De lapidibus,* in which he intended to clarify, expand, and correct the work of Theophrastus as well as to translate it. His method was to study both classic and contemporary works in order to clarify Theophrastus' comments. The information that he gathered and presented largely in the form of footnotes gives an interesting and far-ranging picture of eighteenth-century thought on mineralogy. His interests in mineralogy continued in the first volume of *General Natural History* (1748), which is devoted to a classification and description of the mineral kingdom. Minerals are well described, with descriptions often based on microscopic examination; and they are divided into series, classes, orders, genera, and the equivalent of species. The criteria for these categories are hazy and overlapping, but Hill does recognize the importance of crystal shape. Other works on mineralogy appeared sporadically, and in 1771 he published a manual of mineralogy.

Hill's principal achievement in zoology is the third volume of the *General Natural History,* on animals. A large section is devoted to microscopic animals, and some of the names Hill coined for these animals still stand, such as "paramecium." He also included a brief section devoted to fossil animals and demonstrated familiarity with current views on fossils. In keeping with his interest in microscopy, he revised an English edition of Swammerdam's *Book of Nature* in 1758.

Hill acquired a medical degree from St. Andrews in 1750 (probably by purchase) and published many works on medicine. Most of these reflect his apothecary and botanical interests and deal with vegetable remedies.

Hill's scientific labors were colored by his frequent satirical attacks on his contemporaries. Having failed as an actor and playwright, he engaged in penned warfare with Henry Fielding and other writers; denied membership in the Royal Society of London, he attacked that body in volumes such as his biting *Review of the Works of the Royal Society* (1751).

In addition to his other activities Hill was a contributor to the supplement of *Chambers Cyclopaedia* (1753) and editor of the *British Magazine* (1740–1750). He was married twice, first to a Miss Travers and then to the Honorable Henrietta Jones. He was a member of the Royal Academy of Sciences at Bordeaux and of the Russian Imperial Academy of Sciences in St. Petersburg. In recognition of his *Vegetable System* King Gustavus III of Sweden awarded Hill the Order of Vasa in 1774, after which Hill styled himself Sir John.

BIBLIOGRAPHY

I. Original Works. A more complete listing of some 80 works by Hill is given in Barker's article in *Dictionary of National Biography* (see below). The following works are cited in the text: *Theophrastus' History of Stones. With an English Version, and Critical and Philosophical Notes, Including the Modern History of the Gems* (London, 1746); *A General Natural History: Or New and Accurate Descriptions of the Animals, Vegetables, and Minerals of the Different Parts of the World,* 3 vols. (London, 1748–1752); *A Review of the Works of the Royal Society of London Containing Animadversions on Such of the Papers as Deserve Particular Observation* (London, 1751); *The Useful Family Herbal: Or an Account of All Those English Plants, Which Are Remarkable for Their Virtues, and of the Drugs, Which Are Produced by Vegetables of Other Countries; With Their Descriptions and Their Uses as Proved by Experience* (London, 1755); *The British Herbal: An History of Plants and Trees, Natives of Britain, Cultivated for Use, or Raised for Beauty* (London, 1756); *The Sleep of Plants and Cause of Motion in the Sensitive Plant* (London, 1757); *The Book of Nature; or the History of Insects. By John Swammerdam. Trans. by Thomas Flloyd, Revised and Improved With Notes From Reaumur and Others by John Hill* (London, 1758); *The Vegetable System, or a Series of Experiments and Observations Tending to Explain the Internal Structure, and the Life of Plants,* 26 vols. (London, 1759–1775); *The Construction of Timber, From Its Early Growth, Explained by the Microscope, and Proved by Experiments* (London, 1770); and *Fossils Arranged According to Their Obvious Characters, With Their History and Description* (London, 1771).

II. Secondary Literature. See George F. R. Barker, "John Hill," in *Dictionary of National Biography;* Lorande Loss Woodruff, "The Versatile Sir John Hill, M.D.," in *American Naturalist,* **60** (1926), 417–442; and G. S. Rousseau, "The Much-Maligned Doctor, 'Sir' John Hill (1707–1775)," in *Journal of the American Medical Association,* **212** (1970), 103–108. A new biography that will contain Hill's correspondence is being prepared by Rousseau, *A Literary Quack of London: A Life of Sir John Hill* (in press).

Patsy A. Gerstner

HILL, LESTER SANDERS (*b.* New York, N.Y., 19 January 1890; *d.* Bronxville, New York, 9 January 1961), *mathematics.*

The son of James Edward Hill and the former Ellen Sheehan, Hill attended Columbia University, receiving the B.A. *summa cum laude* in 1911 and the M.A. in 1913. He taught mathematics at the University of Montana and at Princeton until 1916, when he joined the U.S. Naval Reserve. In 1921–1922 Hill was an associate professor of mathematics at the University of Maine; in 1922 he was appointed an instructor at Yale, where he was awarded the Ph.D. in 1926 with a dissertation entitled "Properties of Certain Aggregate Functions." In 1927 Hill went to Hunter College in New York City, where he remained until his retirement except for 1945–1946, when he was a member of the faculty of the U.S. Army University at Biarritz, France.

Hill is probably best known for his mathematical approaches to cryptography and cryptanalysis, having been among the first to apply the theories and methods of matrices and linear transformations to the construction of secret codes. His work in this field, called the Hill system by A. A. Albert, was analyzed by Luigi Sacco in his *Manuale di crittografia.* Only after his death did the U.S. government reveal his associations with the code systems of the army, navy, and State Department during and after World War II. Most of his research in developing a modular algebraic cipher-code system is still unpublished and is classed as highly confidential material. It is described by H. C. Bruton, director, naval communications, as "ingenious, detailed, and complete."

BIBLIOGRAPHY

I. Original Works. Hill's published writings include "Concerning Huntington's Continuum and Other Types of Serial Order," in *American Mathematical Monthly,* **24** (1917), 345–348; "Cryptography in an Algebraic Alphabet," *ibid.,* **36** (1929), 306–312; "Concerning Certain Linear Transformation Apparatus of Cryptography," *ibid.,* **38** (1931), 135–154; "Probability Functions and Statistical Parameters," *ibid.,* **40** (1933), 505–532; "A Mathematical Checking System for Telegraphic Sequences," in *Telegraph and Telephone Age,* **24** (October 1926); **25** (April 1927); **25** (July 1927); "Properties of Certain Aggregate Functions," in *American Journal of Mathematics,* **49** (1937), 419–432, written with M. D. Darkow; and "An Algebraic Treatment of Geometry on a Spherical Surface," in *Scripta mathematica,* **3** (1935), 234–246, 327–336.

II. Secondary Literature. See Luigi Sacco, *Manuale di crittografia* (Rome, 1936); *New York Times* (10 January 1961); *New York Journal-American* (10 January 1961); *New York World-Telegram & Sun* (10 January 1961); and *Who's Who in the East* (1959), 422.

Mary E. Williams

HIND, JOHN RUSSELL (*b.* Nottingham, England, 12 May 1823; *d.* Twickenham, England, 23 December 1895), *astronomy.*

Hind was the son of John Hind, a lace manufacturer who was one of the first to introduce the Jacquard loom to Nottingham. Educated privately and at Nottingham Grammar School, he showed an interest in astronomy at an early age and when sixteen years old became a contributor to the *Nottingham Journal* and the *Atmospheric Almanac,* publishing in the latter weather predictions for 1839 and 1840. In 1840 he obtained employment with a civil engineer in London but in November of the same year, through the good offices of Sir Charles Wheatstone, was appointed to the newly formed magnetic and meteorological department of the Royal Observatory, Greenwich.

During his time at Greenwich under the astronomer royal G. B. Airy, Hind became a proficient observer with the Sheepshanks equatorial telescope and in 1844 took part in the first chronometric determination of the longitude of Valencia, Ireland. After resigning in the autumn of 1844, he was employed by George Bishop as supervisor of the latter's private observatory at Regent's Park, London. He married in 1846 and had six children.

In the course of a nine-year search for small planets, Hind discovered ten asteroids (including Iris and Flora), two comets, a variable nebula in Taurus, and several variable stars. In 1851 he accompanied Rev. W. R. Dawes to Sweden to observe the total eclipse of 28 July, when he observed "rose-coloured flames" at the sun's limb.

Hind's skill and perseverance gained him a wide reputation; tangible recognition came in the form of £100 from the Royal Bounty Fund in 1851 and an annual Civil List pension of £200 the following year. In 1853, following the death of W. S. Stratford, he was appointed superintendent of the Nautical Almanac Office (even though J. C. Adams was a candidate for the post).

His organizing and computing ability enabled Hind to carry out his official duties without relinquishing supervision of Bishop's observatory; when, on the latter's death in 1861, the instruments were moved to Twickenham by George Bishop, Jr., Hind also moved there.

Hind was in charge of the publication of the *Nautical Almanac* until 1891, when he retired. He continued his observations, despite failing health, until his death from heart disease in 1895. A regular and prolific contributor to scientific journals, Hind wrote mainly on ephemerides and comets, and was the author of several books.

Hind joined the Royal Astronomical Society in 1844, served as foreign secretary from 1847 to 1857, and was president from 1880 to 1881. He was elected a corresponding member of the Société Philomatique (1847) and of the Académie des Sciences of Paris (1851) and in 1863 fellow of the Royal Society and subsequently of the Royal Society of Edinburgh. He received the LL.D. from the University of Glasgow in 1882.

Three times recipient of the Lalande Prize, Hind numbered among his many awards for his services to astronomy the gold medals of the Royal Society, the Royal Astronomical Society, and the king of Denmark.

BIBLIOGRAPHY

Hind was a regular contributor to *Monthly Notices of the Royal Astronomical Society* from 1843 to 1890, publishing more than 150 notes or papers, principally on ephemerides and comets. "Comparison of Burckhardt's and Hansen's Lunar Tables" is an appendix to *Monthly Notices of the Royal Astronomical Society* (1890). Contributions of a more popular nature include *The Comets* (London, 1852); *An Introduction to Astronomy,* in Bohn's Standard Library (London, 1852); *The Solar System* (London, 1852); and *Illustrated London Astronomy* (London, 1853).

P. S. LAURIE

HINDENBURG, CARL FRIEDRICH (*b.* Dresden, Germany, 13 July 1741; *d.* Leipzig, Germany, 17 March 1808), *mathematics.*

The son of a merchant, Hindenburg was privately tutored at home. He later attended the Gymnasium in Freiberg, and in 1757 he entered the University of Leipzig, where he studied medicine, philosophy, classical languages, physics, mathematics, and aesthetics. Through the assistance of C. F. Gellert, one of his tutors, Hindenburg became tutor to a young man named Schoenborn, whom he accompanied to the universities of Leipzig and Göttingen. His student's distinct interest in mathematics inspired Hindenburg to become increasingly occupied with mathematical studies, and he befriended A. G. Kaestner. In 1771 he received the M.A. at Leipzig, where he became a private lecturer and, in 1781, extraordinary professor of philosophy. In 1786 he was made professor of physics at Leipzig, a post he held until his death.

Hindenburg's first scientific publications were in philology (1763, 1769); his dissertation as professor of physics was on water pumps. His earliest mathematical investigations, in which he described a method of determining by denumerable methods the terms of arithmetic series, were published in 1776.

In 1778 Hindenburg's first publication on combinatorials appeared. Through a series of papers on this subject, as well as through his teaching, he became the founder of the "combinatorial school" in Germany. Combinatorial mathematics was not new at that time: Pascal, Leibniz, Wallis, the Bernoullis, De Moivre, and Euler, among others, had contributed to it. Hindenburg and his school attempted, through systematic development of combinatorials, to give it a key position within the various mathematical disciplines. Combinatorial considerations, especially appropriate symbols, were useful in the calculations of probabilities, in the development of series, in the inversion of series, and in the development of formulas for higher differentials.

This utility led Hindenburg and his school to entertain great expectations: they wanted combinatorial operations to have the same importance as those of arithmetic, algebra, and analysis. They developed a complicated system of symbols for fundamental combinatorial concepts, such as permutations, variations, and combinations. Various authors developed this system along different lines, but its cumbersomeness soon made it outmoded. The following "central problem" of Hindenburg might be taken as characteristic of the efforts of his school: Represent a random coefficient, b_i, explicitly by means of $a_i(k = 0, 1, \cdots, m)$ in the equation

$$(a_0 + a_1x + a_2x^2 + \cdots + a_mx^m)^n = b_0 + b_1x + \cdots + b_{mn}x^{mn}.$$

The importance that Hindenburg attached to his investigations is shown by the title of the work that summarized his unified system: *Der polynomische Lehrsatz, das wichtigste Theorem der ganzen Analysis* (1796).

None of these great expectations has been realized, perhaps because Hindenburg and his followers were concerned more with the formal transformation of known results than with new discoveries. Thus the combinatorial school did not contribute to the development of the theory of determinants (Binet, Cauchy, Jacobi), although the latter made much use of the fundamental combinatorial concepts. The school's influence was limited to Germany, and no leading contemporary mathematician was a member.

Apart from founding the combinatorial school, Hindenburg was the first in Germany to publish professional journals for mathematics and allied fields. From 1781 to 1785, with C. B. Funck and N. G. Leske, he published the *Leipziger Magazin für Naturkunde, Mathematik und Ökonomie* and, from 1786 to 1789, with Johann III Bernoulli, the *Leipziger*

Magazin für angewandte und reine Mathematik. From 1795 to 1800 he edited the *Archiv der reinen und angewandten Mathematik* and the *Sammlung Kombinatorisch-analytischer Abhandlungen.*

BIBLIOGRAPHY

I. ORIGINAL WORKS. Hindenburg's writings include *Beschreibung einer neuen Art nach einem bekannten Gesetz fortgehende Zahlen durch Abzählen oder Abmessen bequem zu finden* (Leipzig, 1776); *Infinitionomii dignitatum indeterminarum leges ac formulae* (Göttingen, 1778; enl. ed., 1779); *Methodus nova et facilis serierum infinitarum ehibendi dignitates exponentis indeterminati* (Göttingen, 1778); *Novi systematis permutationum, combinationum ac variationum primae lineae* (Leipzig, 1781); and *Der polynomische Lehrsatz, das wichtigste Theorem der ganzen Analysis* (Leipzig, 1796).

II. SECONDARY LITERATURE. See M. Cantor, *Vorlesungen über Geschichte der Mathematik,* IV (Leipzig, 1908); E. Netto, in *Enzyklopädie der mathematischen Wissenschaften,* I, pt. 1 (Leipzig, 1898–1904); H. Oettinger, "Über den Begriff der Kombinationslehre und die Bezeichnungen in derselben," in (J. A. Grunert's) *Archiv der Mathematik und Physik,* **15** (1850), 271–374; and I. C. Weingärtner, *Lehrbuch der kombinatorischen Analysis, nach der Theorie des Herrn Professor Hindenburg ausgearbeitet,* 2 vols. (Leipzig, 1800–1801), which contains a list of all the important writings of the "combinatorial school" to 1800.

KARLHEINZ HAAS

HINSHELWOOD, CYRIL NORMAN (*b.* London, England, 19 June 1897; *d.* London, 9 October 1967), *chemistry.*

Hinshelwood, the only son of Norman MacMillan Hinshelwood, was educated at Westminster City School. He delayed his acceptance of a Brackenbury scholarship at Balliol College, Oxford, until 1919, in order to work for three years during World War I at Queensferry royal ordnance factory. In 1920 he was elected a fellow of Balliol College. The following year he was appointed science tutor at Trinity College, a post he held until his appointment in 1937 as Dr. Lee's professor of chemistry at Oxford University, and fellow of Exeter College. He was knighted in 1948. In 1964 he retired from his post at Oxford and became senior research fellow of the Imperial College of Science and Technology, London.

Hinshelwood is noted for his extensive and comprehensive contributions to the development of chemical kinetics, on both the experimental and theoretical levels. These studies earned him the 1956 Nobel Prize in chemistry jointly with N. N. Semenov, particularly for the elucidation of the complex system of reactions constituting the hydrogen-oxygen explosion.

In his earliest work, no doubt influenced by his work experience at Queensferry, Hinshelwood attempted to interpret the decomposition of solid mixtures containing oxidants such as potassium permanganate and ammonium dichromate. He soon turned his attention to the study of reactions occurring in the gas phase, which occupied him for the remainder of his life. One of his most important theoretical contributions to the development of chemical kinetics came from his investigation, with H. W. Thompson, of the propionaldehyde decomposition, begun in the mid-1920's. The rate was found to fall off at low pressures, whereas at higher pressures the rate of decomposition was higher than could be accounted for on the basis of Lindemann's theory of collisional activation. To Lindemann's theory Hinshelwood added the assumption that the internal energy of polyatomic molecules contributes to the activation energy. These reactions were termed "quasi-unimolecular." To account for anomalies in the slope of the reaction rate versus pressure curves for the thermal decomposition of nitrous oxide, this theory was later extended to include spontaneous and collisionally induced transitions to different internal states (for instance, triplet) of the molecule.

In the course of a series of studies on the pyrolysis of hydrocarbons, ethers, and ketones Hinshelwood uncovered the inhibiting effect of added nitric oxide and propylene. The occurrence of a limited decomposition rate in the presence of these substances was interpreted to mean that molecular and free-radical decompositions take place simultaneously, with the free-radical process being inhibited by the added gas.

During this period (middle and late 1920's) Hinshelwood turned to the investigation of the homogeneous reaction between hydrogen and oxygen in the presence and absence of various added gases. Briefly, he found that the reaction was surface-catalyzed at lower temperatures and surface-inhibited at higher temperatures. This discovery paved the way for the elucidation of the various critical explosion limits. The results, summarized in the Bakerian lecture to the Royal Society in 1946, led to the Nobel Prize ten years later.

Hinshelwood also investigated heterogeneous and homogeneous catalytic reactions, and undertook systematic kinetic studies of substituted aromatic molecules in nonaqueous solvents. The latter studies contributed to the development of L. P. Hammett's theories of energy-entropy relations among rate constants for the reactions of related series of substituted aromatic molecules.

Shortly before World War II, Hinshelwood took up the study of the kinetics of bacterial cells, selecting for his work the nonpathogenic organism *Aerobacter aerogenes.* Two lines of inquiry developed in this series of studies: the manner of growth when the bacteria are placed in a medium containing new nutrients, to which they must adapt; and the mode of inhibition of growth in the presence of antibacterial agents. These studies, which occupied more and more of Hinshelwood's attention, continued until his death and led to the development of a "network" theory of interdependent enzyme balance mechanisms in the bacterial cell. He put forth this theory to supplement currently accepted theories of mutation and selection.

BIBLIOGRAPHY

The nature and scope of Hinshelwood's contributions are well exemplified in his published books: *Kinetics of Chemical Change in Gaseous Systems* (Oxford, 1926; 4th ed., 1940); *Thermodynamics for Students of Chemistry* (London, 1926); *The Reaction Between Hydrogen and Oxygen* (Oxford, 1934), written with A. T. Williamson; *The Chemical Kinetics of the Bacterial Cell* (London, 1947); *The Structure of Physical Chemistry* (New York, 1951); and *Growth, Function and Regulation in Bacterial Cells* (Basel, 1966), written with A. C. R. Dean.

ERNEST G. SPITTLER

HIPPARCHUS OF RHODES (*fl.* Rhodes, second century B.C.), *astronomy.*

For a complete study of his life and work see Supplement.

HIPPIAS OF ELIS (*b.* Elis, Greece; *fl.* 400 B.C.), *philosophy, mathematics.*

Elis was a small state in the northwest of the Peloponnesus whose inhabitants had charge of the Olympic festival. Hippias' father was named Diopeithes,[1] but his ancestry is otherwise unknown.[2] In the Platonic dialogue *Hippias Major*[3] he is made to say that he was young when Protagoras was old, and in the *Protagoras* Plato represents him as present at a philosophic discussion with that eminent Sophist about 432 B.C.[4] The date of the birth of Protagoras is uncertain but is usually placed from 488 to 485. In Plato's *Apology,*[5] set in 399, Hippias is mentioned as a teacher of youth along with Gorgias and other famous Sophists, and may then be presumed to have been at the height of his fame. He was therefore a contemporary of Plato. His wife Platane bore him three sons; and when she was left a widow, the orator Isocrates in extreme old age took her in marriage and adopted her youngest son, Aphareus,[6] who achieved some fame as a tragic poet. Isocrates died in 338. These facts would suggest that Hippias had a long life; and the belief is made certain if, with Mario Untersteiner, the preface to the *Characters* of Theophrastus is attributed to Hippias, for he is there made to say that he has reached ninety-nine years of age.[7] The old notion that he was killed while weaving plots against his native land must be abandoned now that the correct name in the text of Tertullian has been established as Icthyas.[8]

Hippias was taught by an otherwise unknown Aegesidamus, and he emerged as a polymath who wrote and lectured over a wide range of disciplines: rhetoric, politics, poetry, music, painting, sculpture, and astronomy, as well as the philosophy and mathematics on which his fame chiefly rests.[9] The secret of his wide knowledge appears to have been an exceptional memory. According to Philostratus, he had a system of mnemonics such that if he once heard a string of fifty names, he could repeat them in correct order.[10] Most of what is known about Hippias' life and character comes from a dialogue between Socrates and Hippias recorded by Xenophon[11] and from the two Platonic dialogues that bear his name, the *Hippias Major* and *Hippias Minor.* Their authenticity has been disputed, but even if not genuine they still correctly reflect Plato's attitude; in these dialogues Hippias is represented as a naïve and humorless boaster who cannot stand up to the remorseless logic of Socrates. Xenophon's portrait is not so ruthless, but there also Hippias is reduced to silence by Socrates' arguments. Hippias was a second-generation Sophist, and Plato had no love for the Sophists as a class. Apart from more fundamental differences, Plato's aristocratic soul was offended by their professional teaching; and Hippias was especially successful in negotiating lecture fees, particularly in Sicily, although he received none in Sparta, where the law forbade a foreign education.[12]

The picture in the Platonic dialogues is no doubt a caricature; but in the light of Plato's more sympathetic treatment of other individual Sophists, there must have been enough truth in the caricature for it to be recognizable as a portrait.[13] Hippias is made to accept flattery even when laid on with a trowel, acknowledging that he had never found any man to be his superior in anything.[14] At the Olympic festival it was his custom to offer to discourse on any subject proposed to him out of those which he had prepared and to answer any questions.[15] He once appeared at the festival with everything that he wore made by himself, not merely his clothes but also a ring, an oil flask, and an oil scraper—which bears out the statement in the *Suda Lexicon* that he made self-sufficiency the end of life—and he brought with him poems, epics, tragedies, dithyrambs, and all kinds of prose works.[16]

Hippias could not have been such a figure of fun as the Platonic dialogues make him out to be, for he was frequently asked to represent his native state on missions to other states, notably Sparta.[17] He was widely traveled—two visits to Athens are recorded—and in Sicily his influence was lasting if, as Untersteiner believes, he was the mentor of Dionysius the Younger and inspired the work known as the *Dissoi logoi.*[18]

The *Suda Lexicon* tersely records that Hippias "wrote many things." None of his voluminous works has survived, but some of the titles and hints of the contents are known. His *Synagoge,* known through Athenaeus, has usually been thought, on the strength of a passage in Clement of Alexandria which seems to refer to it, to have been merely a miscellany in which he put together sayings of poets and prose writers, both Greek and foreign.[19] But Bruno Snell has advanced the theory that through this work Aristotle derived his knowledge of Thales; that the views of Thales about the All being water and about the souls of inanimate objects are thereby shown to be derived from earlier mythological speculations; and that the *Synagoge* is to be looked upon as the earliest work in both the history of Greek philosophy and the history of Greek literature.[20] If this is so, it encourages the thought that Hippias' *Nomenclature of Tribes*[21] may not have been a mere catalog but an expression of his belief in the fundamental unity of all mankind. His *Register of Olympic Victors* was no doubt a piece of Elian patriotism. It was the first such list to be drawn up; and Plutarch notes that, since it came so late after the events recorded, too much authority should not be attached to it.[22] Among his epideictic or set speeches, the one known as *The Trojan* may have been in dialogue form; in it Nestor suggests to Neoptolemus many lawful and beautiful pursuits by which he might win fame.[23] Hippias wrote an elegiac inscription for the statues made by Calon at Olympia in memory of a boys' choir from Messina drowned in crossing to Rhegium.[24] More important in its ultimate significance than any of these compositions is a work on the properties of the geometrical curve he discovered, since known as the quadratrix.

Hippias' teaching has to be reconstructed from the scattered references to him in Greek and Latin authors. Untersteiner has argued that Hippias was the author not only of the preface to Theophrastus' *Characters* but also of a spurious chapter in Thucydides (III, 84) dealing with events in Corcyra and of the epideictic speech known as the *Anonymus Iamblichi;* that the *Dissoi logoi,* a work drawing on Pythagorean and Sophistic sources, reflects the teaching of Hippias; and that the philosophical digression in Plato's seventh letter is an attack upon Hippias' doctrines.[25] If this were established, it would enable a clearer picture of Hippias' philosophy to be drawn; but Untersteiner's theories are too conjectural for any conclusions to be based on them. It is therefore to the dialogues between Socrates and Hippias as recorded by Xenophon and Plato, and to a passage in Plato's *Protagoras* which may well be an imitation of the Sophist's style, that we must look in the main for Hippias' teaching.[26]

The core of it would appear to be a distinction between *νόμος* and *φύσις*,[27] that is, between positive law and nature, with a corresponding belief in the existence of unwritten natural laws which are the same for all men in all places and at all times. Reverence for the gods and honor for parents are among such natural laws.[28] It was one of Hippias' fundamental beliefs that like is kin to like by nature, and he extended it to mean that men are neighbors and kinsmen. Positive law is a matter of human agreement and can be altered; it can be a great tyrant doing violence to human nature. It is a pity that Hippias' teaching has to be seen through the distorting mirrors of Plato and Xenophon, for he would appear to have been a progenitor of the doctrine of natural law, of the social-contract theory of the state, and of the essential unity of all mankind—in fact, no mean thinker.

It is clear from Plato's raillery that Hippias claimed proficiency in arithmetic, geometry, and astronomy,[29] and one important discovery is attributed to him: the transcendental curve known as the quadratrix.

The evidence comes from two passages in Proclus which are probably derived from Geminus. The first is "Nicomedes trisected every rectilineal angle by means of the conchoidal curves. . . . Others have done the same thing by means of the quadratrices of Hippias and Nicomedes, making use of the mixed curves which are called quadratices."[30] The second is "In the same manner other mathematicians are accustomed to treat of curves, setting forth the characteristic property of each type. Thus Apollonius shows what is the characteristic for each of the conic sections, Nicomedes for the conchoids, Hippias for the quadratrices, and Perseus for the spiric curves."[31]

Who is this Hippias? The natural assumption is that he is Hippias of Elis, who is mentioned in an earlier passage by Proclus,[32] this time in the summary of geometry derived from Eudemus, as having recorded that Mamercus (or perhaps Ameristus), brother of the poet Stesichorus, acquired a reputation for geometry. No other Hippias is mentioned by Proclus; and it is in accordance with his practice, having once referred to a person in full, to omit the patronymic on subsequent mention.[33] Hippias of Elis, as shown by the references of Plato and Xenophon, had mathematical qualifications; and among the many bearers of the name Hippias in antiquity there is no other of whom this can be said.[34] It is therefore natural to identify the Hippias who is mentioned in connection with quadratrices as Hippias of Elis; and most historians of Greek mathematics, from J. E. Montucla to B. L. van der Waerden, have done so.[35]

The objections made can easily be discounted.

1. If he made so important a discovery as the quadratrix, it has been argued, Hippias would be recorded in Proclus' "Eudemian Summary"; but the omission is accounted for by the Platonic prejudice against the Sophists, and the omission of Democritus is even more remarkable.

2. Diogenes Laertius says that Archytas was the first to use an instrument for the description of a curve,[36] and the quadratrix requires an instrument for its description. Yet, on the one hand, an indefinite number of points on the quadratrix can be obtained by the ruler and compass and, on the other hand, Diogenes is not a trustworthy guide in this matter, since (a) there is no suggestion of an instrument in Eutocius' description of the curve found by Archytas to solve the problem of doubling the cube;[37] and (b) Eratosthenes specifically states that Archytas was not able to realize his solution mechanically.[38]

3. Hippias is not mentioned by Pappus and Iamblichus in their accounts of curves used for squaring the circle;[39] but this is explained if, as seems probable, Hippias did not use the curve for that purpose but only for trisecting an angle.

It may therefore be taken that the Hippias who is mentioned by Proclus in connection with the quadratrix is Hippias of Elis; and, if so, he was its discoverer, since he preceded Nicomedes. But did he use it for squaring the circle? And did he give it the name quadratrix? This is more doubtful. Proclus implies that the curve was used by Hippias for trisecting an angle, saying nothing about squaring the circle; and those Greek authors who write about the squaring of the circle do not mention Hippias. A fundamental and obvious property of the curve is that it can be used to divide an angle in any given ratio, and therefore to trisect it; but to use it for squaring the circle is a more sophisticated matter and might not be obvious to the original discoverer. This can be seen from the way the curve is generated, as described by Pappus.[40]

Let $ABCD$ be a square and BED a quadrant of a circle with center A. If the radius of the circle moves uniformly from AB to AD and in the same time the line BC moves parallel to its original position from BC to AD, then at any given time the intersection

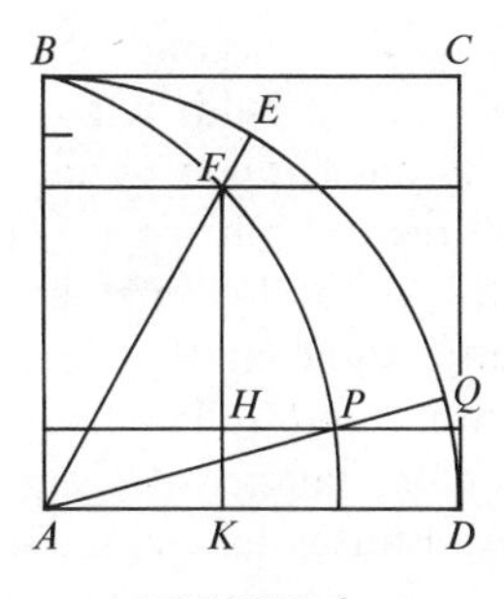

FIGURE 1

of the moving radius and the moving straight line will determine a point F. The path traced by F is the curve. If it is desired to trisect the angle EAD, let H be taken on the perpendicular FK to AD such that $FK = 3HK$. Let a straight line be drawn through H parallel to AD, and let it meet the curve at P. Let AP be produced to meet the circle at Q. Then, by the definition of the curve,

$$\angle EAD : \angle QAD = \text{arc } ED : \text{arc } QD$$
$$= FK : HK,$$

and therefore $\angle QAD$ is one-third of $\angle EAD$. It is obvious that the curve can be used not merely to trisect an angle but also to divide an angle in any given ratio; trisection is specified because this was one of the great problems of Greek mathematics when Hippias flourished.

If a is the length of a side of the square, ρ is any radius vector AF, and ϕ is the angle EAD, the equation of the curve is

$$\frac{\rho \sin \phi}{a} = \frac{\phi}{\frac{1}{2}\pi}$$

or

$$\pi\rho \sin \phi = 2a\phi.$$

The use of the quadratrix to square the circle is a more complicated matter, requiring the position of G to be known and an indirect proof *per impossibile*. (For this the article on Dinostratus may be consulted.)

The ancient witnesses can therefore be reconciled if Hippias discovered the curve and used it to trisect an angle, but its utility for squaring the circle was perceived only by such later geometers as Dinostratus and Nicomedes. In that case Hippias could not have called his curve the quadratrix, and we do not know what name he gave it. It is no objection that Proclus refers to "the quadratrices of Hippias and Nicomedes," for we have no hesitation in saying that Menaechmus discovered the parabola and hyperbola, although these terms did not come into use until Apollonius; Menaechmus would have called them "section of a right-angled cone" and "section of an obtuse-angled cone." There is, however, a more seri-

HIPPIAS OF ELIS

ous objection. From the second of the Proclus passages quoted above it could, without straining the sense, be inferred that Hippias wrote a whole treatise on the curve, setting forth its special properties; and in that case the probability increases that he was aware of its use for squaring the circle. Paul Tannery was of this opinion, and T. L. Heath thinks it "not impossible"; but on balance it seems preferable to hold, with C. A. Bretschneider and Moritz Cantor, that the circle-squaring property was discovered, and the name quadratrix given, later than Hippias.[41]

The citation of Hippias as the authority for Mamercus' mathematical proficiency has led some to suppose that Hippias wrote a history of geometry.[42] If so, it would be the first, antedating Eudemus by perhaps three-quarters of a century. But this is to read too much into the Greek word ἱστόρησεν, translated above as "related." It does not necessarily imply a full-scale treatise, but only that Hippias mentioned the fact in one of his many works.

NOTES

1. *Suda Lexicon,* "Ἱππίας," Adler ed., pt. 2 (Leipzig, 1931), Iota 543, p. 659.
2. Apuleius, *Florida* 9, Helm ed., p. 12.1.
3. Plato, *Hippias Major,* 282D–E.
4. Plato, *Protagoras,* 337C6–338B1. The scene is usually assigned to 432 B.C. but—as Athenaeus, V.218C–D, Gulick ed. (Loeb), II (London–New York, 1928), 428, points out—in antiquity Hippias could not have safely stayed in Athens until an annual truce was concluded in the archonship of Isarchus (423), and the chronology of what is presumably a fictitious gathering cannot be pressed.
5. Plato, *Apology,* 19E1–4.
6. [Plutarch], *Lives of the Ten Orators,* 838A–839C, Fowler ed. (Loeb); and *Moralia* 10, pp. 376–385 (the author makes Platane the daughter and not the widow of Hippias); Harpocration, *Lexicon,* "Ἀφαρεύς," Dindorf ed., I (Oxford, 1853), 68.18; Zosimus, *Historia nova* V, Mendelssohn ed. (Leipzig, 1887). Isocrates' marriage followed his liaison—when already an old man—with the courtesan Lagisca; hence "in extreme old age."
7. Theophrastus, *Characters,* pref. 2, Diels ed. (Oxford, 1909). See Mario Untersteiner, "Il proemio dei 'Caratteri' di Teofrasto e un probabile frammento di Ippia," in *Rivista di filologia classica,* n.s. **26** (1948), 1–25. In *I sofisti,* 2nd ed., fasc. 2, p. 115, translated by Kathleen Freeman in *The Sophists,* p. 274, he says the preface is "definitely a work of Hippias." But it is incredible that the author should have been still writing—even banalities—at the age of ninety-nine; and the figure must be treated with reserve. Perhaps there is a textual error. The preface is certainly not the work of Theophrastus; but the only reason for attributing it to Hippias is that it is such a work as the boastful Hippias of Plato's dialogues might have written, which is not a sufficiently strong ground.
8. The printed texts of Tertullian, *Apologeticum,* 46.16, until 1937 read: "et Hippias, dum civitati insidias disponit, occiditur." There was some dispute whether this referred to Hippias, son of Pisistratus; but since Tertullian is cataloging the misdeeds of pagan philosophers, there can be little doubt that the reading, if correct, would refer to Hippias of Elis. But H. Emonds, "Die Oligarchenrevolte zu Megara im Jahre 375 und der Philosoph Icthyas bei Tertullian *Apol.* 46.16," in *Rheinisches Museum für Philologie,* n.s. **86** (1937), 180–191, shows that the reading "et Hippias" has no MS authority and that "Icthyas" (Icthyas of Megara) should be substituted. Emonds has been followed by H. Hoppe (Vienna, 1939) and E. Dekkers (Tournai, 1954) in their subsequent eds.

   If the reading "Hippias" had been correct, the event could be referred, as in Untersteiner, to the war waged in 343 by the democrats of Elis, among whom Hippias might be numbered, in alliance with the surviving soldiers of the Phocian adventurer Phalaecus. With this peg gone, the case for giving Hippias an exceptionally long life is weakened, particularly if Platane is regarded as daughter and not wife of Hippias (see note 6) and the evidence for ascribing the Theophrastian preface to Hippias is regarded as unconvincing.
9. *Suda Lexicon,* "Ἱππίας." Otto Apelt, *Beiträge zur Geschichte der griechischen Philosophie,* pp. 382–384, 391–392, gives no convincing reasons for thinking that Aegesidamus is a mistake for Hippodamus of Miletus.

   Xenophon, *Memorabilia* IV.6, has Socrates apply the word "polymath" to Hippias; and Plato, *Hippias Minor,* 368B, makes Socrates call him, no doubt sarcastically, "the wisest of men in the greatest number of arts."
10. Philostratus, *Lives of the Sophists* I.11, Kayser ed., II (Leipzig, 1871), 13.27–30. See also Xenophon, *Symposium* 4.62; Plato, *Hippias Major,* 285E. According to Cicero, *De oratore* 2.86.351–354, the first to work out a mnemonic was Simonides, who is mentioned along with Hippias by Aelian, *On the Characteristics of Animals* VI.10, Scholfield ed. (Loeb), II (London–Cambridge, Mass., 1959), 22.9–13. Ammianus Marcellinus XVI.5.8, Clark ed., I (Berlin, 1910), 76.17–20, notes the belief of some writers that his feats of memory, like those of King Cyrus and Simonides, were due to the use of drugs.
11. Xenophon, *Memorabilia* IV.4.19–20.
12. *Hippias Major,* 282D–E, 283B–284C. In the former passage Hippias boasts that although Protagoras was in Sicily at the time, he made more than 150 minas—at one small place, Inycus, taking in more than 20 minas.
13. See W. K. C. Guthrie, *A History of Greek Philosophy,* III (Cambridge, 1969), 280.
14. Plato, *Hippias Minor* 364A; compare *Hippias Major* 281D.
15. Plato, *Hippias Minor* 363C.
16. *Ibid.,* 368B–C; Apuleius, *Florida* 9, Helm ed., pp. 12.3–13.6.
17. Plato, *Hippias Major* 281A–B; Xenophon, *Memorabilia* IV.4.5.
18. The visits are recorded in Plato, *Hippias Major* 281A; and Xenophon, *Memorabilia* IV.4.5. See Mario Untersteiner, "Polemica contra Ippia nella settima epistola di Platone," in *Rivista di storia della filosofia,* **3** (1948), 101–119. The text of the *Dissoi logoi* is given in Diels-Kranz, *Vorsokratiker,* II, 90, pp. 405–416, and by Untersteiner, *Sofisti,* fasc. 3, pp. 148–191.
19. Athenaeus, XIII.608F–609A, Gulick ed. (Loeb), VI (London–Cambridge, Mass., 1937), 280; Clement of Alexandria, *Stromata* VI.C.2, 15.2, Stählin ed., *Clemens Alexandrinus* (in the series *Die Griechischen Christlichen Schriftsteller*), 3rd ed., II (Berlin, 1960), 434.23–435.5. Clement is making the point that the Greeks were incorrigible plagiarists, as shown by Hippias.
20. Bruno Snell, in *Philologus,* **96** (1944), 170–182. G. B. Kerferd, in *Proceedings of the Classical Association,* **60** (1963), 35–36, has adopted and extended Snell's views, and in particular has attributed to Hippias the doctrine of "continuous bodies" mentioned in *Hippias Major* 301B–E. (This passage would seem to have anticipations of Smuts's "holism"—*τὰ ὅλα τῶν πραγμάτων.*)
21. Scholium to Apollonius of Rhodes, III.1179, *Scholia in Apollonium Rhodium vetera,* Wendel ed. (Berlin, 1935), p. 251.13–14.
22. Plutarch, *Numa* 1.6, Ziegler ed., *Vitae parallelae,* III, pt. 2 (Leipzig, 1926), 55.7–9.
23. Plato, *Hippias Major* 286A.
24. Pausanias, V.25.4, Spiro ed. (Teubner), II (Leipzig, 1903), 78.4–13. Another statue made by Calon is dated 420–410 B.C.; but this does not have much bearing on Hippias' date, since

HIPPIAS OF ELIS

his verses were added some time after the statues were made, in place of the original inscription.

25. See final paragraph of Bibliography. The *Anonymus Iamblichi* is reproduced in Diels-Kranz, *Vorsokratiker,* II, 89, 400–404.
26. Xenophon, *Memorabilia* IV.4.5–25. This passage purports to record a discussion between Socrates and Hippias in which Socrates identifies the just with the lawful—a view difficult to reconcile with Plato's Socrates—and discomfits Hippias.

    In *Protagoras* 337C–338B, Hippias mediates between Socrates and Protagoras, urging Socrates not to insist on brief questions and answers, and Protagoras not to sail off into an ocean of words. This pleases the company. In the opening sentence Plato would appear to have packed the main tenets of Hippias' thought: "Gentlemen, I look upon you all as kinsmen and neighbors and fellow citizens by nature, not by law; for by nature like is akin to like, but law, tyrant of men, often constrains us against nature."
27. Regarding these as key words, and in the fourth and fifth centuries as catch words, W. K. C. Guthrie devotes a chapter to the antithesis in *A History of Greek Philosophy,* III, 55–134.
28. Xenophon, *Memorabilia* IV.4.19–20.
29. Plato, *Protagoras* 318E; *Hippias Major* 366C–368A. The former passage deserves citation because it implies that Hippias believed in compulsory education in the quadrivium at the secondary level. Protagoras is the speaker: "The other [Sophists] mistreat the young, for when they have escaped from the arts they bring them back against their will and plunge them once more into the arts, teaching them arithmetic, astronomy, geometry and music—and here he looked at Hippias—whereas if he comes to me he will not be obliged to learn anything except what he has come for."
30. Proclus, *In primum Euclidis,* Friedlein ed. (Leipzig, 1873; repr., 1967), 272.3–10.
31. *Ibid.,* p. 356.6–12.
32. *Ibid.,* p. 65.11–15. The objection by W. K. C. Guthrie, *op. cit.,* III. 284, that it is "nearly 200 Teubner pages" earlier is not convincing.
33. He so treats Leodamas of Thasos, Oenopides of Chios, and Zeno of Sidon; and if he departs from this practice in the case of Hippocrates of Chios, it is only to avoid confusion with Hippocrates of Cos.
34. The Hippias described by the pseudo-Lucian in *Hippias seu Balneum* as a skillful mechanician and geometer is a fictional character.
35. J. E. Montucla, *Histoire des mathématiques,* I, 181; B. L. van der Waerden, *Science Awakening,* 2nd ed. (Groningen, n.d.), p. 146. Also C. A. Bretschneider, *Die Geometrie und die Geometer vor Euklides,* pp. 194–196; but H. Hankel, *Zur Geschichte der Mathematik,* p. 151, note, thought him "sicherlich nicht der Sophist Hippias aus Elis." After initial disbelief in the identification, G. J. Allman, *Greek Geometry From Thales to Euclid,* pp. 92–94, 189–193, was converted by Paul Tannery, in *Bulletin des sciences mathématiques et astronomiques,* 2nd ser., **7** (1883), 278–284; and by Moritz Cantor, *Vorlesungen über Geschichte der Mathematik,* 3rd ed., I, 193–197. After a thorough examination, A. A. Björnbo, in Pauly-Wissowa, VIII, cols. 1706–1711, accepted the identification; but Gino Loria, *Le scienze esatte nell' antica Grecia,* 2nd ed., p. 69, would say only: "Pesando dunque gli argomenti pro e contro l'identificazione, sembra a noi che i primi vincono per valore i secondi." T. L. Heath, *A History of Greek Mathematics,* I, 2, 23, 225, takes the identification for granted; but U. von Wilamowitz, *Platon,* I, 136, note, thinks that the name is so common that it is a matter of discretion; and W. K. C. Guthrie, *loc. cit.,* is undecided.
36. Diogenes Laertius VIII.iv, Cobet ed., p. 224.
37. Archimedes, Heiberg ed., 2nd ed., III, 84.12–88.2.
38. *Ibid.,* p. 90.4–11.
39. Pappus, *Collection,* Hultsch ed., pp. 250.33–252.3: "For the quadrature of the circle a certain curve was assumed by Dinostratus and Nicomedes and certain others more recent, and it takes its name from its property, for it is called by them quadratrix."

    Iamblichus as recorded by Simplicius, *In Aristotelis Categorias,* Kalbfleisch ed., p. 192.19–24: "Archimedes succeeded by means of the spiral-shaped curve, Nicomedes by means of the curve known by the special name quadratrix, Apollonius by means of a certain curve which he himself terms 'sister of the cochloid' but which is the same as the curve of Nicomedes, and lastly Carpus by means of a certain curve which he simply calls 'the curve arising from a double motion.'" When W. K. C. Guthrie, *op. cit.,* III, 284, note 2, finds significance in "the silence of Simplicius, who at *Physics* 54 ff (Diels ed.) seems to be giving as complete an account as he can of attempts to square the circle," it must be objected that Simplicius' aim in that passage was much more limited: the efforts of Alexander and Hippocrates.
40. Pappus, *op. cit.,* p. 252.5–25.
41. For references see Bibliography.
42. Kerferd, *op. cit.,* appears to hold this view.

BIBLIOGRAPHY

I. ORIGINAL WORKS. None of Hippias' many works has survived. The titles of the following are known: Ἐθνῶν ὀνομασίαι, *Nomenclature of Tribes;* Ὀλυμπιονικῶν ἀναγραφή, *Register of Olympic Victors;* Συναγωγή, *Collection;* and Τρωικός (*sc.* λόγος or διάλογος), *The Trojan.* Hippias is also known to have composed an elegiac inscription for the statues at Olympia in memory of a boys' choir from Messina drowned in crossing to Rhegium. He probably wrote a treatise on the quadratrix, of which he was the discoverer.

References to these works, and other witnesses to Hippias, are collected in H. Diels and W. Kranz, *Die Fragmente der Vorsokratiker,* 6th ed., II (Dublin–Zurich, 1970), 86, 326–334; and Mario Untersteiner, *Sofisti: Testimonianze e frammenti,* vol. VI in Biblioteca di Studi Superiori, fasc. 3 (Florence, 1954), 38–109.

It is conjectured by Untersteiner that Hippias was also the author of the preface to the *Characters* of Theophrastus; the *Anonymus Iamblichi;* and a spurious chapter in the third book of Thucydides' history, III, 84.

II. SECONDARY LITERATURE. In Greek literature the main secondary sources for Hippias are Plato, *Protagoras* 315C, 337C–338B; Plato (?), *Hippias Major* and *Hippias Minor,* Burnet ed., III (Oxford, 1903; repr., 1968); and Xenophon, *Memorabilia* IV.4.5–25, Marchant ed. (as *Commentarii*), in vol. II of Xenophon's *Works* (Oxford, 1901; 2nd ed., 1921). Other scattered references will be found in the notes.

The best recent accounts of Hippias as a philosopher are W. K. C. Guthrie, *A History of Greek Philosophy,* III (Cambridge, 1969), 280–285; and Mario Untersteiner, *I sofisti* (Milan, 1948; 2nd ed., 1967), II, 109–158, translated by Kathleen Freeman, *The Sophists* (Oxford, 1954), pp. 272–303.

Hippias' mathematical work may be studied in G. J. Allman, *Greek Geometry From Thales to Euclid* (Dublin, 1889), pp. 92–94, 189–193; A. A. Björnbo, "Hippias 13," in Pauly-Wissowa, *Real-Encyclopädie,* VIII (Stuttgart, 1913), cols. 1706–1711; C. A. Bretschneider, *Die Geometrie und die Geometer vor Euklides* (Leipzig, 1870), pp. 94–97; Moritz Cantor, *Vorlesungen über Geschichte der Mathe-*

matik, 3rd ed., I (Leipzig, 1907), 193–197; James Gow, *A Short History of Greek Mathematics* (Cambridge, 1884), pp. 162–164; T. L. Heath, *A History of Greek Mathematics,* I (Oxford, 1921), 225–230; Gino Loria, *Le scienze esatte nell' antica Grecia,* 2nd ed. (Milan, 1914), pp. 67–72; and Paul Tannery, "Pour l'histoire des lignes et surfaces courbes dans l'antiquité," in *Bulletin des sciences mathématiques et astronomiques,* 2nd ser., **7** (1883), 278–291, repr. in his *Mémoires scientifiques,* II (Toulouse–Paris, 1912), 1–18.

Among other noteworthy assessments of Hippias are the following, listed chronologically: J. Mahly, "Der Sophist Hippias von Elis," in *Rheinisches Museum für Philologie,* **15** (1860), 514–535, and **16** (1861), 38–49; O. Apelt, *Beiträge zur Geschichte der griechischen Philosophie,* VIII, "Der Sophist Hippias von Elis" (Leipzig, 1891), 367–393; W. Zilles, "Hippias aus Elis," in *Hermes,* **53** (1918), 45–56; D. Viale (Adolfo Levi), "Ippia di Elide e la corrente naturalistica della sofistica," in *Sophia* (1942), pp. 441–450; Bruno Snell, "Die Nachrichten über die Lehren des Thales und die Anfänge der griechischen Philosophie- und Literaturgeschichte," in *Philologus,* **96** (1944), 170–182; and G. B. Kerferd, in *Proceedings of the Classical Association,* **60** (1963), 35–36.

Mario Untersteiner has put forward his conjectures about Hippias in "Un nuovo frammento dell' Anonymus Iamblichi. Identificazione dell' Anonimo con Ippia," in *Rendiconti dell' Istituto lombardo di scienze e lettere,* classe di lettere, **77,** fasc. II (1943–1944), 17; "Polemica contro Ippia nella settima epistola di Platone," in *Rivista di storia della filosofia,* **3** (1948), 101–119; and "Il proemio dei 'Caratteri' di Teofrasto e un probabile frammento di Ippia," in *Rivista di filologia classica,* n.s. **26** (1948), 1–25.

IVOR BULMER-THOMAS

HIPPOCRATES OF CHIOS (*b.* Chios; *fl.* Athens, second half of the fifth century B.C.), *mathematics, astronomy.*

The name by which Hippocrates the mathematician is distinguished from the contemporary physician of Cos[1] implies that he was born in the Greek island of Chios; but he spent his most productive years in Athens and helped to make it, until the foundation of Alexandria, the leading center of Greek mathematical research. According to the Aristotelian commentator John Philoponus, he was a merchant who lost all his property through being captured by pirates.[2] Going to Athens to prosecute them, he was obliged to stay a long time. He attended lectures and became so proficient in geometry that he tried to square the circle. Aristotle's own account is less flattering.[3] It is well known, he observes, that persons stupid in one respect are by no means so in others. "Thus Hippocrates, though a competent geometer, seems in other respects to have been stupid and lacking in sense; and by his simplicity, they say, he was defrauded of a large sum of money by the customs officials at Byzantium." Plutarch confirms that Hippocrates, like Thales, engaged in commerce.[4] The "Eudemian summary" of the history of geometry reproduced by Proclus states that Oenopides of Chios was somewhat younger than Anaxagoras of Clazomenae; and "after them Hippocrates of Chios, who found out how to square the lune, and Theodore of Cyrene became distinguished in geometry. Hippocrates is the earliest of those who are recorded as having written Elements."[5] Since Anaxagoras was born about 500 B.C. and Plato went to Cyrene to hear Theodore after the death of Socrates in 399 B.C., the active life of Hippocrates may be placed in the second half of the fifth century B.C. C. A. Bretschneider has pointed out that the accounts of Philoponus and Aristotle could be reconciled by supposing that Hippocrates' ship was captured by Athenian pirates during the Samian War of 440 B.C., in which Byzantium took part.[6]

Paul Tannery, who is followed by Maria Timpanaro Cardini, ventures to doubt that Hippocrates needed to learn his mathematics at Athens.[7] He thinks it more likely that Hippocrates taught in Athens what he had already learned in Chios, where the fame of Oenopides suggests that there was already a flourishing school of mathematics. Pointing out the proximity of Chios to Samos, the birthplace of Pythagoras, Timpanaro Cardini makes a strong case for regarding Hippocrates as coming under Pythagorean influence even though he had no Pythagorean teacher in the formal sense. Although Iamblichus does not include Hippocrates' name in his catalog of Pythagoreans, he, like Eudemus, links him with Theodore, who was undoubtedly in the brotherhood.[8]

Mathematics, he notes, advanced after it had been published; and these two men were the leaders. He adds that mathematics came to be divulged by the Pythagoreans in the following way: One of their number lost his fortune, and because of this tribulation he was allowed to make money by teaching geometry. Although Hippocrates is not named, it would, as Allman points out, accord with the accounts of Aristotle and Philoponus if he were the Pythagorean in question.[9] The belief that Hippocrates stood in the Pythagorean tradition is supported by what is known of his astronomical theories, which have affinities with those of Pythagoras and his followers. He was, in Timpanaro Cardini's phrase, a para-Pythagorean, or, as we might say, a fellow traveler.[10]

When Hippocrates arrived in Athens, three special problems—the duplication of the cube, the squaring of the circle, and the trisection of an angle—were already engaging the attention of mathematicians, and he addressed himself at least to the first two. In the course of studying the duplication of the cube, he used the method of reduction or analysis. He was the first to compose an *Elements of Geometry* in the manner of Euclid's famous work. In astronomy he propounded theories to account for comets and the galaxy.

Method of Analysis. Hippocrates is said by Proclus to have been the first to effect the geometrical reduction of problems difficult of solution.[11] By reduction (ἀπαγωγή), Proclus explains that he means "a transition from one problem or theorem to another, which being known or solved, that which is propounded is also manifest."[12] It has sometimes been supposed, on the strength of a passage in the *Republic,* that Plato was the inventor of this method; and this view has been supported by passages from Proclus and Diogenes Laertius.[13] But Plato is writing of philosophical analysis, and what Proclus and Diogenes Laertius say is that Plato "communicated" or "explained" to Leodamas of Thasos the method of analysis (ἀναλύσις)—the context makes clear that this is geometrical analysis—which takes the thing sought up to an acknowledged first principle. There would not appear to be any difference in meaning between "reduction" and "analysis," and there is no claim that Plato invented the method.

Duplication of the Cube. Proclus gives as an example of the method the reduction of the problem of doubling the cube to the problem of finding two mean proportionals between two straight lines, after which the problem was pursued exclusively in that form.[14] He does not in so many words attribute this reduction to Hippocrates; but a letter purporting to be from Eratosthenes to Ptolemy Euergetes, which is preserved by Eutocius, does specifically attribute the discovery to him.[15] In modern notation, if $a:x = x:y = y:b$, then $a^3:x^3 = a:b$; and if $b = 2a$, it follows that a cube of side x is double a cube of side a. The problem of finding a cube that is double a cube with side a is therefore reduced to finding two mean proportionals, x, y, between a and $2a$. (The pseudo-Eratosthenes observes with some truth that the problem was thus turned into one no less difficult.)[16] There is no reason to doubt that Hippocrates was the first to effect this reduction; but it does not follow that he, any more than Plato, invented the method. It would be surprising if it were not in use among the Pythagoreans before him.

The suggestion was made by Bretschneider, and has been developed by Loria and Timpanaro Cardini,[17] that since the problem of doubling a square could be reduced to that of finding one mean proportional between two lines,[18] Hippocrates conceived that the doubling of a cube might require the finding of two mean proportionals. Heath has made the further suggestion that the idea may have come to him from the theory of numbers.[19] In the *Timaeus* Plato states that between two square numbers there is one mean proportional number but that two mean numbers in continued proportion are required to connect two cube numbers.[20] These propositions are proved as Euclid VII.11, 12, and may very well be Pythagorean. If so, Hippocrates had only to give a geometrical adaptation to the second.

Quadrature of Lunes. The "Eudemian summary" notes that Hippocrates squared the lune—so called from its resemblance to a crescent moon—that is, he found a rectilineal figure equal in area to the area of the figure bounded by two intersecting arcs of circles concave in the same direction.[21] This is the achievement on which his fame chiefly rests. The main source for our detailed knowledge of what he did is a long passage in Simplicius' commentary on Aristotle's *Physics.*[22] Simplicius acknowledges his debt to Eudemus' *History of Geometry* and says that he will set out word for word what Eudemus wrote, adding for the sake of clarity only a few things taken from Euclid's *Elements* because of Eudemus' summary style. The task of separating what Simplicius added has been attempted by many writers from Allman to van der Waerden. When Simplicius uses such archaic expressions as τὸ σημεῖον ἐφ' ὧ (or ἐφ' οὗ) *A* for the point *A*, with corresponding expressions for the line and triangle, it is generally safe to presume that he is quoting; but it is not a sufficient test to distinguish the words of Hippocrates from those of Eudemus, since Aristotle still uses such pre-Euclidean forms. Another stylistic test is the earlier form which Eudemus would have used, δυνάμει εἶναι ("to be equal to when square"), for the form δύνασθαι, which Simplicius would have used more naturally. Although there can be no absolute certainty about the attribution, what remains is of great interest as the earliest surviving example of Greek mathematical reasoning; only propositions are assigned to earlier mathematicians, and we have to wait for some 125 years after Hippocrates for the oldest extant Greek mathematical text (Autolycus).

Before giving the Eudemian extract, Simplicius reproduces two quadratures of lunes attributed to Hippocrates by Alexander of Aphrodisias, whose own commentary has not survived. In the first, *AB* is the diameter of a semicircle, *AC*, *CB* are sides of a square inscribed in the circle, and *AEC* is a semicircle inscribed on *AC*. Alexander shows that the lune *AEC* is equal to the triangle *ACD*.

In the second quadrature *AB* is the diameter of a semicircle; and on *CD*, equal to twice *AB*, a semi-

HIPPOCRATES OF CHIOS

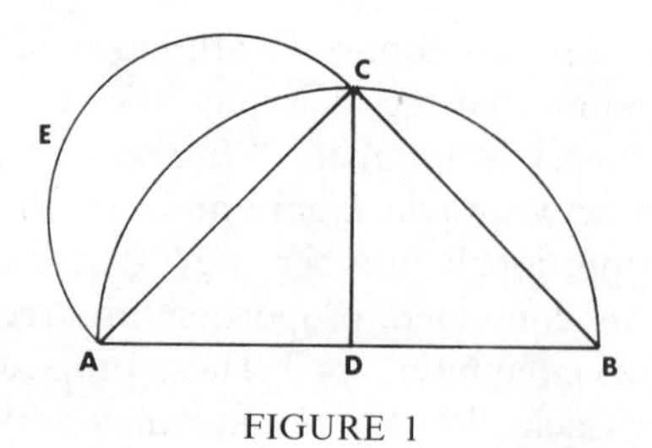

FIGURE 1

circle is described. *CE, EF, FD* are sides of a regular hexagon; and *CGE, EHF, FKD* are semicircles. Alexander proves that the sum of the lunes *CGE, EHF, FKD* and the semicircle *AB* is equal to the trapezium *CEFD*.

Alexander goes on to say that if the rectilinear figure equal to the three lunes is subtracted ("for a rectilinear figure was proved equal to a lune"), the circle will be squared. There is an obvious fallacy here, for the lune which was squared was one standing on the side of a square and it does not follow that the lune standing on the side of the hexagon can be squared. John Philoponus, as already noted, says that Hippocrates tried to square the circle while at Athens. There is confirmation in Eutocius, who in his commentary on Archimedes' *Measurement of a Circle* notes that Archimedes wished to show that a circle would be equal to a certain rectilinear area, a matter investigated of old by eminent philosophers before him.[23] "For it is clear," he continues, "that the subject of inquiry is that concerning which Hippocrates of Chios and Antiphon, who carefully investigated it, invented the paralogisms which, I think, are accurately known to those who have examined the *History of Geometry* by Eudemus and have studied the *Ceria* of Aristotle." This is probably a reference

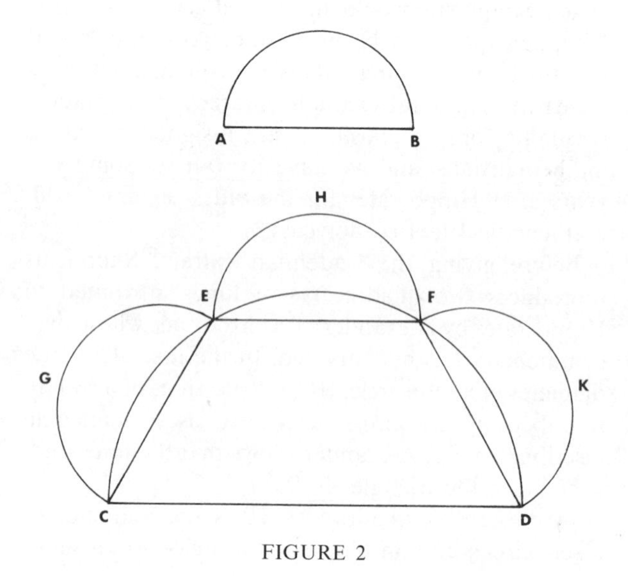

FIGURE 2

to a passage in the *Sophistici Elenchi* where Aristotle says that not all erroneous constructions are objects of controversy, either because they are formally correct or because they are concerned with something true, "such as that of Hippocrates or the quadrature by means of lunes."[24] In the passage in Aristotle's *Physics* on which both Alexander and Simplicius are commenting,[25] Aristotle rather more clearly makes the point that it is not the task of the exponent of a subject to refute a fallacy unless it arises from the accepted principles of the subject. "Thus it is the business of the geometer to refute the quadrature of a circle by means of segments but it is not his business to refute that of Antiphon."[26]

The ancient commentators are probably right in identifying the quadrature of a circle by means of segments with Hippocrates' quadrature of lunes; mathematical terms were still fluid in Aristotle's time, and Aristotle may well have thought there was some fallacy in it. We may be confident, though, that a mathematician of the competence of Hippocrates would not have thought that he had squared the circle when in fact he had not done so. It is likely that when Hippocrates took up mathematics, he addressed himself to the problem of squaring the circle, which was much in vogue; it is evident that in the course of his researches he found he could square certain lunes and, if this had not been done before him, probably effected the two easy quadratures described by Alexander as well as the more sophisticated ones attributed to him by Eudemus. He may have hoped that in due course these quadratures would lead to the squaring of the circle; but it must be a mistake on the part of the ancient commentators, probably misled by Aristotle himself, to think that he claimed to have squared the circle. This is better than to suppose, with Heiberg, that in the state of logic at that time Hippocrates may have thought he had done so; or, with Björnbo, that he deliberately used language calculated to mislead; or, with Heath, that he was trying to put what he had discovered in the most favorable light.[27] Let us turn to what Hippocrates actually did, according to Eudemus, who, as Simplicius notes, is to be preferred to Alexander as being nearer in date to the Chian geometer.

Hippocrates, says Eudemus, "made his starting point, and laid down as the first of the theorems useful for the discussion of lunes, that similar segments of circles have the same ratio as the squares on their bases; and this he showed from the demonstration that the squares on the diameters are in the same ratio as the circles." (This latter proposition is Euclid XII.2 and is the starting point also of Alexander's quadratures; the significance of what Eudemus says

HIPPOCRATES OF CHIOS

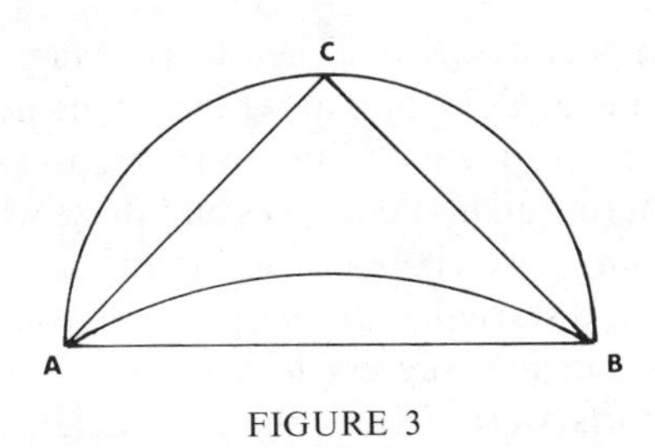

FIGURE 3

is discussed below.) In his first quadrature he takes a right-angled isosceles triangle *ABC*, describes a semicircle about it, and on the base describes a segment of a circle similar to those cut off by the sides. Since $AB^2 = AC^2 + CB^2$, it follows that the segment about the base is equal to the sum of those about the sides; and if the part of the triangle above the segment about the base is added to both, it follows that the lune *ACB* is equal to the triangle.

Hippocrates next squares a lune with an outer circumference greater than a semicircle. *BA*, *AC*, *CD* are equal sides of a trapezium; *BD* is the side parallel to *AC* and $BD^2 = 3AB^2$. About the base *BD* there is described a segment similar to those cut off by the equal sides. The segment on *BD* is equal to the sum of the segments on the other three sides; and by adding the portion of the trapezium above the segment about the base, we see that the lune is equal to the trapezium.

Hippocrates next takes a lune with a circumference less than a semicircle, but this requires a preliminary construction of some interest, it being the first known example of the Greek construction known as a νεύσις, or "verging."[28] Let *AB* be the diameter of a circle and *K* its center. Let *C* be the midpoint of *KB* and let *CD* bisect *BK* at right angles. Let the straight line *EF* be placed between the bisector *CD* and the circumference "verging toward *B*" so that the square on *EF* is 1.5 times the square on one of the radii, that is, $EF^2 = 3/2\ KA^2$. If $FB = x$ and $KA = a$, it can easily be shown that $(x + \sqrt{3/2}\,a)\,x = a^2$, so that

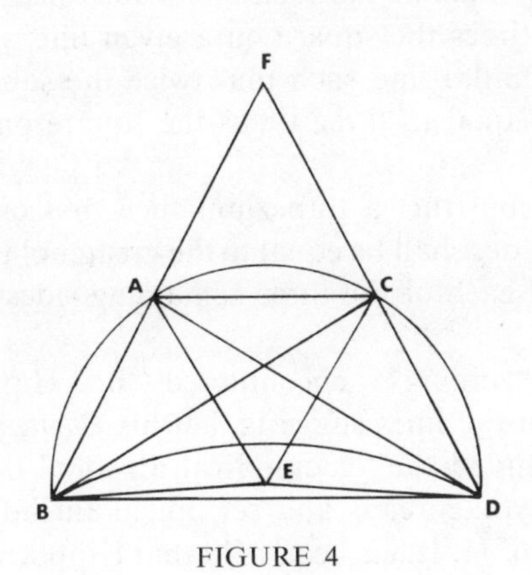

FIGURE 4

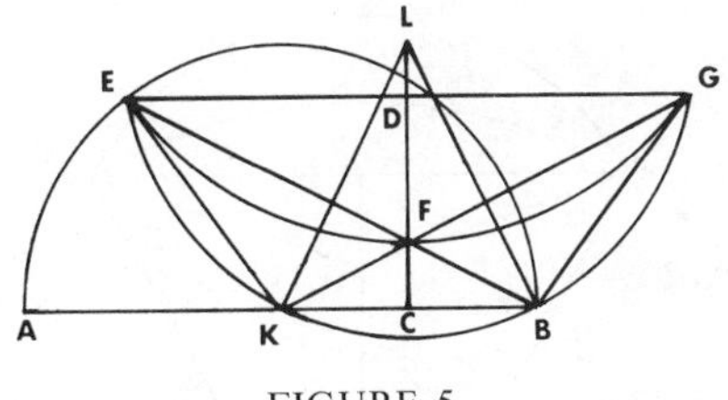

FIGURE 5

the problem is tantamount to solving a quadratic equation. (Whether Hippocrates solved this theoretically or empirically is discussed below.)

After this preliminary construction Hippocrates circumscribes a segment of a circle about the trapezium *EKBG* and describes a segment of a circle about the triangle *EFG*. In this way there is formed a lune having its outer circumference less than a semicircle, and its area is easily shown to be equal to the sum of the three triangles *BFG*, *BFK*, *EKF*.

Hippocrates finally squares a lune and a circle together. Let *K* be the center of two circles such that the square on the diameter of the outer is six times the square on the diameter of the inner. *ABCDEF* is a regular hexagon in the inner circle. *GH*, *HI* are sides of a regular hexagon in the outer circle. About *GI* let there be drawn a segment similar to that cut off by *GH*. Hippocrates shows that the lune *GHI* and the inner circle are together equal to the triangle *GHI* and the inner hexagon.

This last quadrature, rather than that recorded by Alexander, may be the source of the belief that Hippocrates had squared the circle, for the deduction is not so obviously fallacious. It would be easy for someone unskilled in mathematics to suppose that because Hippocrates had squared lunes with outer circumferences equal to, greater than, and less than a semicircle, and because he had squared a lune and a circle together, by subtraction he would be able to

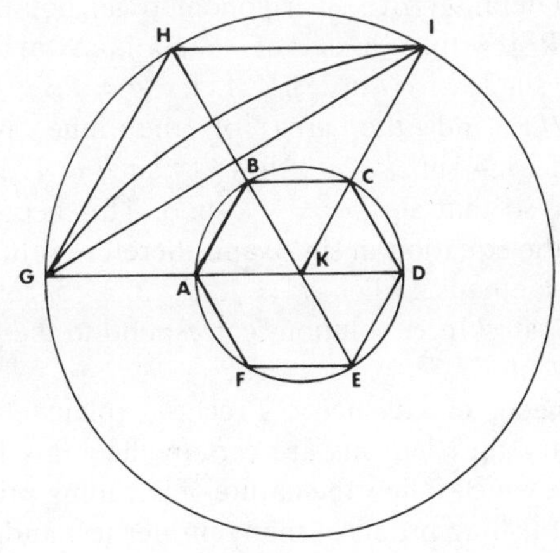

FIGURE 6

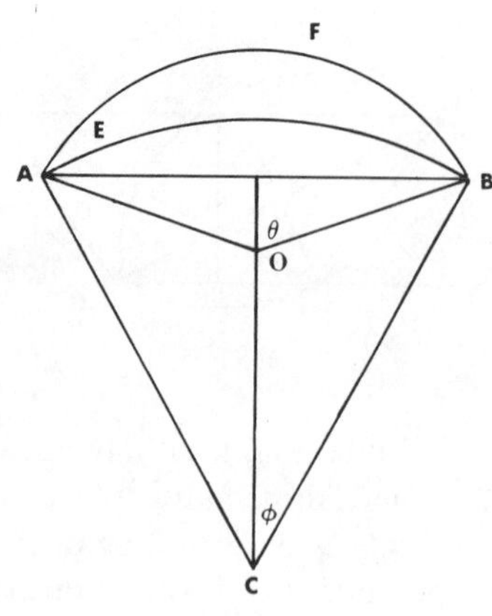

FIGURE 7

square the circle. The fallacy, of course, is that the lune which is squared along with the circle is not one of the lunes previously squared by Hippocrates; and although Hippocrates squared lunes having outer circumferences equal to, greater than, and less than a semicircle, he did not square all such lunes but only one in each class.

What Hippocrates succeeded in doing in his first three quadratures may best be shown by trigonometry. Let O, C be the centers of arcs of circles forming the lune $AEBF$, let r, R be their respective radii and θ, ϕ the halves of the angles subtended by the arcs at their centers.

$$\begin{aligned}\text{Then lune } AEBF &= \text{segment } AFB - \text{segment } AEB\\ &= (\text{sector } OAFB - \triangle\, OAB)\\ &\quad - (\text{sector } CAEB - \triangle\, CAB)\\ &= (\text{sector } OAFB - \text{sector } CAEB)\\ &\quad + (\triangle\, CAB - \triangle\, OAB)\\ &= r^2\theta - R^2\phi\\ &\quad + 1/2(R^2 \sin 2\phi - r^2 \sin 2\theta).\end{aligned}$$

It is a sufficient condition for the lune to be squarable that sector $OAFB$ = sector $CAEB$, for in that case the area will be equal to $\triangle\, CAB - \triangle\, OAB$, that is, the quadrilateral $AOBC$. In trigonometrical notation, if $r^2\theta = R^2\phi$, the area of the lune will be $1/2(R^2 \sin 2\phi - r^2 \sin 2\theta)$. Let $\theta = k\phi$. Then $R = \sqrt{k}r$ and the area of the lune is $1/2\ r^2(k \sin 2\phi - \sin 2k\phi)$. Now $r \sin\theta = 1/2AB = R \sin\phi$, so that $\sin k\phi = \sqrt{k} \sin\phi$. This becomes a quadratic equation in $\sin\phi$, and therefore soluble by plane methods, when $k = 2, 3, 3/2, 5$, or $5/3$. Hippocrates' three solutions correspond to the values 2, 3, 3/2 for k.[29]

Elements of Geometry. Proclus explains that in geometry the elements are certain theorems having to those which follow the nature of a leading principle and furnishing proofs of many properties; and in the summary which he has taken over from Eudemus he names Hippocrates, Leon, Theudius of Magnesia, and Hermotimus of Colophon as writers of elements.[30] In realizing the distinction between theorems which are merely interesting in themselves and those which lead to something else, Hippocrates made a significant discovery and started a famous tradition; but so complete was Euclid's success in this field that all the earlier efforts were driven out of circulation. What Proclus says implies that Hippocrates' book had the shortcomings of a pioneering work, for he tells us that Leon was able to make a collection of the elements in which he was more careful, in respect both of the number and of the utility of the things proved.

Although Hippocrates' work is no longer extant, it is possible to get some idea of what it contained. It would have included the substance of Books I and II of Euclid's *Elements,* since the propositions in these books were Pythagorean discoveries. Hippocrates' research into lunes shows that he was aware of the following theorems:

1. In a right-angled triangle, the square on the side opposite the right angle is equal to the sum of the squares on the other two sides (Euclid I.47).
2. In an obtuse-angled triangle, the square on the side subtending the obtuse angle is greater than the sum of the squares on the sides containing it (*cf.* II.12).
3. In any triangle, the square on the side opposite an acute angle is less than the sum of the squares on the sides containing it (*cf.* II.13).
4. In an isosceles triangle whose vertical angle is double the angle of an equilateral triangle (that is, 120°), the square on the base is equal to three times the square on one of the equal sides.
5. In equiangular triangles, the sides about the equal angles are proportional.

Hippocrates' *Elements* would have included the solution of the following problems:

6. To construct a square equal to a given rectilinear figure (II.14).
7. To find a line the square on which shall be equal to three times the square on a given line.
8. To find a line such that twice the square on it shall be equal to three times the square on a given line.
9. To construct a trapezium such that one of the parallel sides shall be equal to the greater of two given lines and each of the three remaining sides equal to the less.

The "verging" encountered in Hippocrates' quadrature of lines suggests that his *Elements* would have included the "geometrical algebra" developed by the Pythagoreans and set out in Euclid I.44, 45 and II.5, 6, 11. It has been held that Hippocrates may

HIPPOCRATES OF CHIOS

have contented himself with an empirical solution, marking on a ruler a length equal to $\sqrt{3/2}\ KA$ in Figure 5 and moving the ruler about until the points marked lay on the circumference and on CD, respectively, while the edge of the ruler also passed through B. In support, it is pointed out that Hippocrates first places EF without producing it to B and only later joins BF.[31] But it has to be admitted that the complete theoretical solution of the equation $x^2 + \sqrt{3/2}x = r^2$, having been developed by the Pythagoreans, was well within the capacity of Hippocrates or any other mathematician of his day. In Pythagorean language it is the problem "to apply to a straight line of length $\sqrt{3/2}\ a$ a rectangle exceeding by a square figure and equal to a^2 in area," and it would be solved by the use of Euclid II.6.

Hippocrates was evidently familiar with the geometry of the circle; and since the Pythagoreans made only a limited incursion into this field, he may himself have discovered many of the theorems contained in the third book of Euclid's *Elements* and solved many of the problems posed in the fourth book. He shows that he was aware of the following theorems:

1. Similar segments of a circle contain equal angles. (This implies familiarity with the substance of Euclid III.20–22.)

2. The angle of a semicircle is right, that of a segment greater than a semicircle is acute, and that of a segment less than a semicircle is obtuse. (This is Euclid III.31, although there is some evidence that the earlier proofs were different.)[32]

3. The side of a hexagon inscribed in a circle is equal to the radius (IV.15, porism). He knew how to solve the following problems: (1) about a given triangle to describe a circle (IV.5); (2) about the trapezium drawn as in problem 9, above, to describe a circle; (3) on a given straight line to describe a segment of a circle similar to a given one (*cf.* III.33).

Hippocrates would not have known the general theory of proportion contained in Euclid's fifth book, since this was the discovery of Eudoxus, nor would he have known the general theory of irrational magnitudes contained in the tenth book, which was due to Theaetetus; but his *Elements* may be presumed to have contained the substance of Euclid VI–IX, which is Pythagorean.

It is likely that Hippocrates' *Elements* contained some of the theorems in solid geometry found in Euclid's eleventh book, for his contribution to the Delian problem (the doubling of the cube) shows his interest in the subject. It would be surprising if it did not to some extent grapple with the problem of the five regular solids and their inscription in a sphere, for this is Pythagorean in origin; but it would fall short of the perfection of Euclid's thirteenth book. The most interesting question raised by Hippocrates' *Elements* is the extent to which he may have touched on the subjects handled in Euclid's twelfth book. As we have seen, his quadrature of lunes is based on the theorem that circles are to one another as the squares on their diameters, with its corollary that similar segments of circles are to each other as the squares on their bases. The former proposition is Euclid XII.2, where it is proved by inscribing a square in a circle, bisecting the arcs so formed to get an eight-sided polygon, and so on, until the difference between the inscribed polygon and the circle becomes as small as is desired. If similar polygons are inscribed in two circles, their areas can easily be proved to be in the ratio of the squares on the diameters; and when the number of the sides is increased and the polygons approximate more and more closely to the circles, this suggests that the areas of the two circles are in the ratio of the squares on their diameters.

But this is only suggestion, not proof, for the ancient Greeks never worked out a rigorous procedure for taking the limits. What Euclid does is to say that if the ratio of the squares on the diameters is not equal to the ratio of the circles, let it be equal to the ratio of the first circle to an area S which is assumed in the first place to be less than the second circle. He then lays down that by continually doubling the number of sides in the inscribed polygon, we shall eventually come to a point where the residual segments of the second circle are less than the excess of the second circle over S. For this he relies on a lemma, which is in fact the first proposition of Book X: "If two unequal magnitudes be set out, and if from the greater there be subtracted a magnitude greater than its half, and from the remainder a magnitude greater than its half, and so on continually, there will be left some magnitude which is less than the lesser magnitude set out." On this basis Euclid is able to prove rigorously by *reductio ad absurdum* that S cannot be less than the second circle. Similarly, he proves that it cannot be greater. Therefore S must be equal to the second circle, and the two circles stand in the ratio of the squares on their diameters.

Could Hippocrates have proved the proposition in this way? Here we must turn to Archimedes, who in the preface to his *Quadrature of the Parabola*[33] says that in order to find the area of a segment of a parabola, he used a lemma which has accordingly become known as "the lemma of Archimedes" but is equivalent to Euclid X.1: "Of unequal areas the excess by which the greater exceeds the less is capable, when added continually to itself, of exceeding any given finite area."[34] Archimedes goes on to say:

HIPPOCRATES OF CHIOS

> The earlier geometers have also used this lemma. For it is by using this same lemma that they have proved (1) circles are to one another in the same ratio as the squares on their diameters; (2) spheres are to one another as the cubes on their diameters; (3) and further that every pyramid is the third part of the prism having the same base as the pyramid and equal height; and (4) that every cone is a third part of the cylinder having the same base as the cone and equal height they proved by assuming a lemma similar to that above mentioned.

In his *Method* Archimedes states that Eudoxus first discovered the proof of (3) and (4) but that no small part of the credit should be given to Democritus, who first enunciated these theorems without proof.[35]

In the light of what has been known since the discovery of Archimedes' *Method,* it is reasonable to conclude that Hippocrates played the same role with regard to the area of a circle that Democritus played with regard to the volume of the pyramid and cone; that is, he enunciated the proposition, but it was left to Eudoxus to furnish the first rigorous proof. Writing before the discovery of the *Method,* Hermann Hankel thought that Hippocrates must have formulated the lemma and used it in his proof; but without derogating in any way from the genius of Hippocrates, who emerges as a crucial figure in the history of Greek geometry, this is too much to expect of his age.[36] It is not uncommon in mathematics for the probable truth of a proposition to be recognized intuitively before it is proved rigorously. Reflecting on the work of his contemporary Antiphon, who inscribed a square (or, according to another account, an equilateral triangle) in a circle and kept on doubling the number of sides, and the refinement of Bryson in circumscribing as well as inscribing a regular polygon, and realizing with them that the polygons would eventually approximate very closely to the circle, Hippocrates must have taken the further step of postulating that two circles would stand to each other in the same ratio as two similar inscribed polygons, that is, in the ratio of the squares on their diameters.

A question that has been debated is whether Hippocrates' quadrature of lunes was contained in his *Elements* or was a separate work. There is nothing about lunes in Euclid's *Elements,* but the reason is clear: an element is a proposition that leads to something else; but the quadrature of lunes, although interesting enough in itself, proved to be a mathematical dead end. Hippocrates could not have foreseen this when he began his investigations. The most powerful argument for believing the quadratures to have been contained in a separate work is that of Tannery: that Hippocrates' argument started with the theorem that similar segments of circles have the same ratio as the squares on their bases. This depends on the theorem that circles are to one another as the squares on their bases, which, argues Tannery, must have been contained in another book because it was taken for granted.[37]

Astronomy. What is known of Oenopides shows that Chios was a center of astronomical studies even before Hippocrates; and he, like his contemporaries, speculated about the nature of comets and the galaxy. According to Aristotle,[38] certain Italians called Pythagoreans said that the comet—it was apparently believed that there was only one—was a planet which appeared only at long intervals because of its low elevation above the horizon, as was the case with Mercury.[39] The circle of Hippocrates and his pupil Aeschylus[40] expressed themselves in a similar way save in thinking that the comet's tail did not have a real existence of its own; rather, the comet, in its wandering through space, occasionally assumed the appearance of a tail through the deflection of our sight toward the sun by the moisture drawn up by the comet when in the neighborhood of the sun.[41] A second reason for the rare appearance of the comet, in the view of Hippocrates, was that it retrogressed so slowly in relation to the sun, and therefore took a long time to get clear of the sun. It could get clear of the sun to the north and to the south, but it was only in the north that the conditions for the formation of a tail were favorable; there was little moisture to attract in the space between the tropics, and although there was plenty of moisture to the south, when the comet was in the south only a small part of its circuit was visible. Aristotle proceeds to give five fairly cogent objections to these theories.[42]

After recounting the views of two schools of Pythagoreans, and of Anaxagoras and Democritus on the Milky Way, Aristotle adds that there is a third theory, for "some say that the galaxy is a deflection of our sight toward the sun as is the case with the comet." He does not identify the third school with Hippocrates; but the commentators Olympiodorus and Alexander have no hesitation in so doing, the former noting that the deflection is caused by the stars and not by moisture.[43]

NOTES

1. The similarity of the names impressed itself upon at least one ancient commentator, Olympiodorus. *In Aristotelis Meteora,* Stuve ed., 45.24–25: Ἱπποκράτης, οὐχ ὁ Κῷος, ἀλλ᾽ ὁ Χῖος.
2. John Philoponus, *In Aristotelis Physica,* Vitelli ed., 31.3–9.
3. Aristotle, *Ethica Eudemia H* 14, 1247a17, Susemihl ed., 113.15–114.1.
4. Plutarch, *Vita Solonis* 2, *Plutarchi vitae parallelae,* Sintenis ed., I, 156.17–20.

HIPPOCRATES OF CHIOS

5. Proclus, *In primum Euclidis,* Friedlein ed., 65.21–66.7.
6. C. A. Bretschneider, *Die Geometrie und die Geometer vor Eukleides,* p. 98.
7. Paul Tannery, *La géometrie grecque,* p. 108; Maria Timpanaro Cardini, *Pitagorici,* fasc. 2, pp. 29–31.
8. Iamblichus, *De vita Pythagorica* 36, Deubner ed., 143.19–146.16; and, for the link with Theodore, *De communi mathematica scientia* 25, Festa ed., 77.24–78.1. The same passage, with slight variations, is in *De vita Pythagorica* 18, Deubner ed., 52.2–11, except for the sentence relating to Hippocrates.
9. G. J. Allman, *Greek Geometry From Thales to Euclid,* p. 60.
10. Timpanaro Cardini, *op. cit.,* fasc. 2, p. 31.
11. Proclus, *op. cit.,* 213.7–11. He adds that Hippocrates also squared the lune and made many other discoveries in geometry, being outstanding beyond all others in his handling of geometrical problems.
12. *Ibid.,* 212.25–213.2.
13. Plato, *Republic* VI, 510B–511C, Burnet ed.; Proclus, *op. cit.,* 211.18–23; Diogenes Laertius, *Vitae philosophorum* III.24, Long ed., 1.131.18–20.
14. Proclus, *op. cit.,* 213.2–6.
15. *Archimedis opera omnia,* Heiberg ed., 2nd ed., III, 88.4–96.27.
16. *Ibid.,* 88.17–23.
17. Bretschneider, *op. cit.,* p. 97; Gino Loria, *Le scienze esatte nell' antica Grecia,* 2nd ed., pp. 77–78; Timpanaro Cardini, *op. cit.,* fasc. 2, pp. 34–35.
18. If $a:x = x:2a$, the square with side x is double the square with side a. The problem of doubling a square of side x is thus reduced to finding a mean proportional between a and $2a$.
19. Thomas Heath, *A History of Greek Mathematics,* I, 201.
20. Plato, *Timaeus* 32 A,B, Burnet ed. With the passage should be studied *Epinomis* 990b5–991b4, Burnet ed.; and the note by A. C. Lloyd in A. E. Taylor, *Plato: Philebus and Epinomis,* p. 249.
21. Proclus, *op. cit.,* 66.4–6, in fact mentions the squaring of the lune as a means of identifying Hippocrates.
22. Simplicius, *In Aristotelis Physica,* Diels ed., 53.28–69.35.
23. *Archimedis opera omnia,* Heiberg ed., 2nd ed., III, 228.11–19.
24. Aristotle, *Sophistici Elenchi* 11, 171b12–16. Toward the end of the third century Sporus of Nicaea compiled a work known as Κηρία, or 'Αριστοτελικὰ κηρία, which was used by Pappus, Simplicius, and Eutocius; but Heiberg sees here a reference to the *Sophistici Elenchi* of Aristotle. Grammatically it is possible that "the quadrature by means of lunes" is to be distinguished from "that of Hippocrates"; but it is more likely that they are to be identified, and Diels and Timpanaro Cardini are probably right in bracketing "the quadrature by means of lunes" as a (correct) gloss which has crept into the text from 172a2–3, where the phrase is also used.
25. Aristotle, *Physics* A 2, 185a14, Ross ed.
26. Aristotle does an injustice to Antiphon, whose inscription of polygons with an increasing number of sides in a circle was the germ of a fruitful idea, leading to Euclid's method of exhaustion; Aristotle no doubt thought it contrary to the principles of geometry to suppose that the side of the polygon could ever coincide with an arc of the circle.
27. J. L. Heiberg, *Philologus,* **43,** p. 344; A. A. Björnbo, in Pauly-Wissowa, VIII, cols. 1787–1799; Heath, *op. cit.,* I, 196, note. Montucla, *Histoire des recherches sur la quadrature du cercle,* pp. 21–22, much earlier (1754) had given the correct interpretation: "Hippocrate ne vouloit point proposer un moyen qu'il jugeoit propre à conduire quelque jour à la quadrature du cercle?"
28. There is a full essay on this subject in T. L. Heath, *The Works of Archimedes,* pp. c–cxxii.
29. It was shown by M. J. Wallenius in 1766 that the lune can be squared by plane methods when $x = 5$ or $5/3$ (Max Simon, *Geschichte der Mathematik im Altertum,* p. 174). T. Clausen gave the solution of the last four cases in 1840, when it was not known that Hippocrates had solved more than the first. ("Vier neue mondförmige Flachen, deren Inhalt quadrirbar ist," in *Journal für die reine und angewandte Mathematik,* **21** 375–376). E. Landau has investigated the cases where the difference between $r^2\theta$ and $R^2\phi$ is not zero but equal to an area that can be squared, although this does not lead to new squarable lunes: "Ueber quadrirbare Kreisbogen zweiecke," in *Sitzungsberichte der Berliner mathematischen Gesellschaft,* **2** (1903).
30. Proclus, *op. cit.,* 72.3–13, 66.7–8, 66.19–67.1, 67.12–16, 20–23. Tannery (*Mémoires scientifiques,* I, 46) is not supported either in antiquity or by modern commentators in discerning a written Pythagorean collection of *Elements* preceding that of Hippocrates.
31. Heath, *op. cit.,* I, 196.
32. See Aristotle, *Posterior Analytics* Π 11, 94a28–34; *Metaphysics* Θ 9, 1051a 26–29; and the comments by W. D. Ross, *Aristotle's Metaphysics,* pp. 270–271; and Thomas Heath, *Mathematics in Aristotle,* pp. 37–39, 71–74.
33. *Archimedis opera omnia,* Heiberg ed., 2nd ed., II, 264.1–22.
34. More strictly, "the lemma of Archimedes" is equivalent to Euclid V, def. 4—"Magnitudes are said to have a ratio one to another if they are capable, when multiplied, of exceeding one another"—and this is used to prove Euclid X.1. Archimedes not infrequently uses the lemma in Euclid's form.
35. *Archimedis opera omnia,* Heiberg ed., 2nd ed., II, 430.1–9. In the preface to Book I of his treatise *On the Sphere and Cylinder* Archimedes attributes the proofs of these theorems to Eudoxus without mentioning the part played by Democritus.
36. Hermann Hankel, *Zur Geschichte der Mathematik in Alterthum und Mittelalter,* p. 122.
37. Tannery, *op. cit.,* I, 354–358. Loria, *op. cit.,* p. 91, inclines to the same view; but Timpanaro Cardini, *op. cit.,* fasc. 2, p. 37, is not persuaded.
38. *Meteorologica* A6, 342b30–343a20, Fobes ed., 2nd ed.
39. Because, like Mercury, it can be seen with the naked eye only when low on the horizon before dawn or after sunset, since it never sets long after the sun and cannot be seen when the sun is above the horizon.
40. Nothing more is known of Aeschylus. This and references by Aristotle to *οἱ περὶ Ἱπποκράτην* imply that Hippocrates had a school.
41. It is not clear how Aristotle thought the appearance to be caused, and the commentators and translators—Thomas Heath, *Aristarchus of Samos,* p. 243; E. W. Webster, *The Works of Aristotle,* III, *Meteorologica, loc. cit.;* H. D. P. Lee, *Aristotle, Meteorologica* in the Loeb Library, pp. 40–43; Timpanaro Cardini, *op. cit.,* fasc. 2, pp. 66–67—give only limited help. It is clear that Hippocrates, like Alcmaeon and Empedocles before him, believed that rays of light proceeded from the eye to the object; and it seems probable that he thought visual rays were *refracted* in the moisture around the comet toward the sun (the sun then being in a position in which this could happen), and *reflected* from the sun back to the moisture and the observer's eye (hence the choice of the neutral word "deflected"). Hippocrates believed that somehow this would create the appearance of a tail in the vapors around the comet; but since this is not the correct explanation, it is impossible to know exactly what he thought happened. It is tempting to suppose that he thought the appearance of the comet's tail to be formed in the moisture in the same way that a stick appears to be bent when seen partly immersed in water, but the Greek will not bear this simple interpretation.

    Olympiodorus, *op. cit.,* Stuve ed., 45.29–30, notes that whereas Pythagoras maintained that both the comet and the tail were made of the fifth substance, Hippocrates held that the comet was made of the fifth substance but the tail out of the sublunary space. This is anachronistic. It was Aristotle who added the "fifth substance" to the traditional four elements—earth, air, fire, water.
42. Aristotle, *Meteorologica, A*6, 343a21–343b8, Fobes ed., 2nd ed.
43. Olympiodorus, *op. cit.,* Stuve ed., 68.30–35; he reckons it a "fourth opinion," presumably counting the two Pythagorean

schools separately. Alexander, *In Aristotelis Meteorologica,* Hayduck ed., 38.28-32.

BIBLIOGRAPHY

No original work by Hippocrates has survived, but his arguments about the squaring of lunes and possibly his *ipsissima verba* are embedded in Simplicius, *In Aristotelis Physicorum libros quattuor priores commentaria,* H. Diels ed., *Commentaria in Aristotelem Graeca,* IX (Berlin, 1882). In the same volume, pp. xxiii-xxxi, is an *appendix Hippocratea* by H. Usener, "De supplendis Hippocratis quas omisit Eudemus constructionibus."

The ancient references to Hippocrates' speculations on comets and the galaxy are in Aristotle, *Meteorologicorum libri quattuor* A6, 342a30-343a20 and A8, 345b9, Fobes ed. (Cambridge, Mass., 1918; 2nd ed., Hildesheim, 1967); and in the following volumes of *Commentaria in Aristotelem Graeca:* XII, pt. 2, *Olympiodori in Aristotelis Meteora commentaria,* Stuve ed. (Berlin, 1900), 45.24-46.24, 68.30-69.26; and *Alexandri in Aristotelis Meteorologicorum libros commentaria,* III, pt. 2, Hayduck ed. (Berlin, 1899), 38.28-38.32.

The chief ancient references to Hippocrates are collected in Maria Timpanaro Cardini, *Pitagorici, testimonianze e frammenti,* fasc. 2, Bibliotheca di Studi Superiori, XLI (Florence, 1962), 16(42), pp. 38-73, along with an Italian translation and notes, and an introductory note, pp. 28-37. A less comprehensive collection is in Diels and Kranz, *Die Fragmente der Vorsokratiker,* 14th ed. (Dublin-Zurich, 1969), I, 42 (3), 395-397.

For the mathematical work of Hippocrates generally, the best secondary literature is George Johnston Allman, *Greek Geometry From Thales to Euclid* (Dublin-London, 1889), pp. 57-77, reproducing a paper which first appeared in *Hermathena,* **4,** no. 7 (Apr. 1881), 180-228; and Thomas Heath, *A History of Greek Mathematics,* I (Oxford, 1921), 182-202.

The quadrature of lunes is the subject of papers by Paul Tannery: "Hippocrate de Chio et la quadrature des lunes," in *Mémoires de la Société des sciences physiques et naturelles de Bordeaux,* 2nd ser., **2** (1878), 179-184; and "Le fragment d'Eudème sur la quadrature des lunes," *ibid.,* **5** (1883), 217-237, which may be more conveniently studied as reproduced in Tannery, *Mémoires scientifiques,* I (Paris, 1912), 46-52, 339-370. Another paper by a leading historian of early mathematics is that of J. L. Heiberg, who gave his views on the passage of Simplicius in the course of his *Jahresberichte* in *Philologus,* **43** (1884), 336-344. F. Rudio, after papers in *Bibliotheca mathematica,* 3rd ser., **3** (1902), 7-62; **4** (1903), 13-18; and **6** (1905), 101-103, edited the Greek text of Simplicius with a German translation, introduction, full notes, and appendixes as *Der Bericht des Simplicius über die Quadraturen des Antiphon und Hippokrates* (Leipzig, 1907); but Heath's criticisms, *op. cit.,* pp. 187-190, must be studied with it. There are excellent notes in W. D. Ross, *Aristotle's Physics* (Oxford, 1936), pp. 463-467. A new attempt to separate the Eudemian text from Simplicius was made by O. Becker, "Zur Textgestaltung des Eudemischen Berichts über die Quadratur der Möndchen durch Hippocrates von Chios," in *Quellen und Studien zur Geschichte der Mathematik, Astronomie und Physik,* Abt. B, **3** (1936), 411-418. The same author later dealt specifically with the passage in Simplicius, Diels ed., 66.14-67.2, in "Zum Text eines mathematischen Beweises im Eudemischen Bericht über die quadraturen der 'Möndchen' durch Hippokrates von Chios bei Simplicius," in *Philologus,* **99** (1954-1955), 313-316. A still later attempt to separate the Eudemian text from that of Simplicius is in Fritz Wehrli, *Die Schule des Aristoteles, Texte und Kommentar,* VIII, *Eudemos von Rhodos,* 2nd ed. (Basel, 1969), 59.28-66.6

Two medieval versions of Hippocrates' quadratures are given in Marshall Clagett, "The *Quadratura circuli per lunulas,*" Appendix II, *Archimedes in the Middle Ages,* I (Madison, Wis., 1964), pp. 610-626.

IVOR BULMER-THOMAS

HIPPOCRATES OF COS (*b.* Cos, 460 B.C.; *d.* Larissa, *ca.* 370 B.C.), *medicine.*

Little is known about the life of Hippocrates, although it may be stated with a fair degree of certainty that he was the son of Heraclides and Phenaretes; that he studied with his father and with Herodicus (probably of Cnidus); and that he was an Asclepiad, although there is no agreement about the term, which may either indicate a particular group of physicians or simply be synonymous with "physician." It is equally certain that he taught at Cos, traveled widely in Greece, and enjoyed exceptional fame in his lifetime. He contributed to a significant body of medical writings, but it is difficult to determine precisely which works of the corpus are actually his.

Even these few data are not indisputable. Of four surviving biographical accounts, two are *Lives,* one apparently by Soranus[1] and the other by an unknown writer of late Latin.[2] Both are brief and late; also short and even later are the biography in the *Suda Lexicon* and that by Tzetzes.[3] The four texts are marked by numerous disparities and divergent views; all are corrupted by the legend that began early to surround Hippocrates and continued to grow throughout antiquity and the Middle Ages.[4] The task of the historian is thus to distill fact from these writings, in part through comparing them with the little that remains of more ancient information; it is not surprising that philologists have reached no conclusive result. Of more recent scholarship, an extreme, and excessively skeptical, view of Hippocrates is presented by L. Edelstein in the *Real-Encyclopädie.* Following an erudite and relentless examination of Hippocratic writings Edelstein, going even further than Wilamowitz' statement that Hippocrates is a name without a book, asserts, "It is a name lacking even any accessible historical reality." A consideration of the ancient references to Hippocrates may serve to modify this statement.

Some idea of Hippocrates' renown at Athens may be had from Plato's early dialogue *Protagoras* (311b–c), which also reveals that he took paying students:

> "Tell me, Hippocrates," I said, "as you are going to Protagoras, and will be paying your money to him, what is he to whom you are going? and what will he make of you? If, for example, you had thought of going to Hippocrates of Cos, the Asclepiad, and were about to give him your money, and someone had said to you: 'You are paying your money to your namesake Hippocrates, O Hippocrates; tell me, what is he that you give him money?,' how would you have answered?"
>
> "I should say," he replied, "that I gave money to him as a physician."
>
> "And what will he make of you?"
>
> "A physician," he said.
>
> "And if you were resolved to go to Polycletus of Argos or Phidias of Athens . . ."

This would clearly indicate that the celebrity of Hippocrates as a physician was as great as that of Polycletus and Phidias as artists, although Edelstein counters this analogy with an example from Plato's *Phaedrus* (268c–d) in which Eryximacus and Acoumenes, two very obscure physicians, are named with Sophocles and Euripides. Of course, one is not expected to infer from this that these two physicians are as famous as the two great dramatists; they are mentioned in the dialogue only because of their intimacy with Phaedrus, Socrates' interlocutor, while in *Protagoras* nothing links Hippocrates to the young man seeking Socrates save that they are namesakes. Moreover the shortness of the passage cited in *Protagoras* places the names of Hippocrates and the artists close together, while no such close juxtaposition exists in the passage from *Phaedrus.*

Some fifty years later, Aristotle wrote in *Politics* (VII,1326a15–16): "Concerning Hippocrates, one could assert that the physician, and not the man, is greater than he who exceeds him in bodily size." Edelstein finds in this only a confirmation of the statement in the Brussels *Life* that Hippocrates was short of stature, but it is also clear from this sentence that the name Hippocrates, here used alone, denotes a real person, one celebrated in the medical world.

Even were it possible to deny the extraordinary fame of Hippocrates during his own lifetime, it would still be necessary to explain why he, rather than anyone else, early became legendary. It is reasonable to assume that the tenacious and varied legend, analyzed skillfully by Edelstein, became attached to a person of some prominence, rather than to a mere physician. The legendary Hippocrates appears variously as a heroic sage; as a Greek patriot who spurns the offers of Artaxerxes I; and as a friend of Democritus of Abdera, who, being thought mad by his fellow townsmen, convinces Hippocrates of his sanity. An apocryphal decree further credits him with having saved Athens from the plague. None of these stories has any basis in historical fact; indeed, it is not known whether Hippocrates ever met Democritus, for example.

Not only must it be concluded that the legend is based on a real person, and a famous one, but it must also be accepted that that person was the author of a number of famous books. It was certainly not true for his contemporaries that Hippocrates was a name with no writings attached to it, and it is true for us to only a limited degree, since we possess many medical works from the time and from the school of Cos. It is very probable that some of the most outstanding of these are by Hippocrates. Plato suggests that he wrote and circulated medical books (*Phaedrus,* 270b–c), while Diocles of Carystus (end of the fourth and beginning of the third century B.C.) would seem to allude to certain of these treatises and Ctesias of Cnidus, at the beginning of the fourth century, certainly criticized Hippocrates for an operation described in that part of the Hippocratian corpus now called *Fractures* (*Claudii Galeni Opera Omnia,* Kühn ed., XVIII, 1, 731).

It is now necessary to examine the Hippocratic corpus to determine which works might actually be attributed to Hippocrates himself. The corpus, or *Collection,* consists of about sixty medical works, the great majority of which date from the last decades of the fifth century B.C. and from the first half of the fourth. They were probably brought together in Alexandria; although the exact circumstances of their collection is unknown, we do know from Galen (Kühn ed., XVII, 1, 619) that Baccheius published the third book of the *Epidemics* there, and from Erotian that Baccheius had compiled a Hippocratic lexicon. Fragments of the latter are quoted in Erotian's own lexicon, and show that he used some twenty-three treatises from the collection.

The problem of dating the treatises is aided by internal evidence, as well as by the testimony of Diocles of Carystus and Ctesias of Cnidus. For example, the *Collection* contains a great many echoes of pre-Socratic philosophy and explicit citations of Empedocles (*Ancient Medicine* 20) and Melissus of Samos (*Nature of Man* 1), but almost nothing that shows post-Socratic thought, except in a few of the later

HIPPOCRATES OF COS

works. The people and places mentioned in *Epidemics* are also helpful. Moreover, one of the most important of the treatises, *Nature of Man,* which is very closely connected to other works of the corpus, can confidently be attributed to Polybius, the disciple and son-in-law of Hippocrates, through the testimony of Aristotle (*History of Animals* III, 3, 512b–513a). One may also note that a more general reading of Aristotle's biological works leads to the conviction that they often implicitly refer to writings in the *Collection.*

Although the date of the majority of the works in the *Collection* may thus readily be established, not all of the treatises fall within this period; some are known to be of certain later date, while a few of the texts, such as the *Oath* and *Nature of Man,* may be slightly earlier, although this is more difficult to determine. Of the later additions, *Law, Nutriment, Heart, Physician* and the first part of *Hebdomades* are Hellenistic; *Precepts* and *Decorum* are from the first or second century of the Christian era. In sum, however, it is certain that most of the *Collection* should be placed between 430 and 380 B.C., for a number of cultural, linguistic, and historical reasons. This period is that of Hippocrates' maturity and old age, which is certainly significant, but does not by itself serve to connect him with the *Collection.*

We must next examine the relationship between the *Collection* and Hippocrates' school. A number of the treatises certainly belong to the school of Cnidus, and not to that of Cos. For example, Galen (Kühn ed., XVII, 1, 888) cites a passage from the *Cnidian Maxims* that is duplicated by one in *Diseases* (II, 68); other data and parallels must lead us to catalog as Cnidian *Diseases* I, II, and III, *Internal Affections,* and *Affections.* All the gynecological treatises, too, as well as the single work *Generation-Nature of the Child-Diseases* IV, must be supposed to be at least para-Cnidian, while the philosophical presuppositions of *Regimen, Breaths,* and *On Flesh,* are incompatible with the school of Cos. Another group of treatises may with equal confidence be considered to be Coan, however. These include Polybius' *Nature of Man* and the appropriately named *Coan Prenotions,* both of which have close ties with *Epidemics* I and III; *Prognostic; Airs, Waters, Places* and *The Sacred Disease;* the originally combined *Fractures, Joints; Wounds of the Head; Regimen in Acute Diseases,* with its polemical introduction against the Cnidians and its appendix; *Aphorisms;* the two *Prorrhetics; Humors;* and several less important treatises.[5]

We thus arrive at a series of writings contemporary with Hippocrates, belonging to his school, and attributed to him—together with many others—by a tradition going back to the Hellenistic school. It would indeed be surprising if one or another of these works were not written by him, but because of the lack of adequate documentation this final attribution is the most difficult to make. It is, of course, tempting to assign to Hippocrates the finest of the writings—*Epidemics* I and III; *Prognostic; Airs, Waters, Places; The Sacred Disease; Fractures, Joints;* and *Regimen in Acute Diseases.* Although linguistic arguments have been advanced against a common authorship of these works, they are not compelling given the variety of subjects treated and the length of Hippocrates' career. In light of these considerations, vocabulary can undoubtedly vary. These six works readily fall into pairs, moreover, and it is hard to assume other than a single authorship of *Epidemics* I and III and *Prognostic;* of *Airs, Waters, Places* and *The Sacred Disease;* or of *Fractures, Joints* and *Regimen in Acute Diseases.*

There is almost unchallengeable evidence to show that of these Hippocrates wrote at least *Prognostic* and *Joints.* We know from Galen (*Corpus Medicorum Graecorum* V, 9, 2, p. 205, 6 and 270, 23) that in the third century B.C. Herophilus devoted, if not a whole book, certainly a paragraph in one of his works, "against *Prognostic* by Hippocrates." The testimony of Ctesias, moreover, supports Hippocrates' authorship of *Joints.* Galen (Kühn ed., XVIII, 1, 731), commenting on chapter 51 of *Joints,* states that Ctesias was the first to criticize Hippocrates on a specific point when he declared that the reduction of a luxation of the thigh is of little use. If we accept this evidence to show that the two works in question are by Hippocrates, it is implicit that several others must also be by him. It is further possible that he may have written other treatises that are now lost, since the *Collection* frequently mentions works that are no longer extant and the beginning of the *Regimen,* in particular, tells us that there existed a vast medical literature, of which the *Collection* is only a fraction.

A further complication in assigning specific works to Hippocrates lies in that the problem of individual authorship was undoubtedly not as acute for the Asclepiads as it is now. One may assume that literary ownership was a hazier and generally less important concept than it is today; it should be noted that chapter 1 of *Regimen in Acute Diseases* attributes the two editions of the *Cnidian Maxims* to a group of authors, which is not merely a figure of speech. There can be no doubt that the Coans adhered to the same practice. It is certain, however, that the great Coan works that Hippocrates did not write himself nevertheless reflect his thought and his teaching.

Two important documents remain to be examined;

unfortunately, neither serves to clarify Hippocrates' identity or authorship. The first of these, a passage from Plato's *Phaedrus* (270c), attributes a medical doctrine to Hippocrates, but is open to two fundamentally different interpretations:

> Socrates: And do you think you can know the nature of the soul intelligently without knowing the nature of the whole?
> Phaedrus: Hippocrates the Asclepiad says that the nature of the body can not be understood without it.
> Socrates: Yes, friend, and he was right. Still, we ought not to be content with Hippocrates, but we should examine by reason and see whether its answer agrees with his conception of nature.
> Phaedrus: I agree.
> Socrates: Then consider what reason as well as Hippocrates says about nature.

It is the "whole" that is subject to different interpretations. Philologists understand it to mean the entirety of the matter in question, while platonists, basing their arguments upon the profundity of Plato's concerns in this dialogue and such others as the *Timaeus,* consider it to mean the Whole, that is, the universe. The first interpretation is unacceptable; it is then therefore necessary to determine if the doctrine implicit in the platonic interpretation is easily deductible from Coan writings, or whether it contradicts the doctrine of Cos. It is the author's opinion that it is contradictory to them.

What one finds in the Coan writings is an emphasis on the importance to the body of such ambient effects as heat, winds, and rain, a doctrine radically different from the micro-macrocosmic philosophical system (similar to that found in the para-Cnidian *Regimen*) that Plato attributes to Hippocrates. I would suggest that this is the result of Plato's own interpretation of the Coan doctrine of ambient factors as a macrocosmic doctrine. Indeed, several passages in the extant writings could appear ambiguous in this way to a philosopher, and such distortions in no way invalidate the writings as being genuinely Hippocratic.

The second major document is a papyrus in the British Museum, which was described by Sir Frederic Kenyon in 1892 and published by H. Diels in 1893. It contains approximately 1,900 lines and dates from the second century of the Christian era. Lacunae, oversights, and errors in the text suggest that it represents a collection of notes made by a medical student for his personal use; it is certain, however, that the second section of the papyrus, which is devoted to the etiology of diseases, reproduces the work of Meno, the disciple of Aristotle whose existence and writings were known to Galen. Meno gives us the doctrines of twenty physicians, seven of whom are not known to us from any other source. Although difficult of interpretation, Meno's material concerning Hippocrates is more explicit than Plato's and may prove to be easier for specialists to agree upon.

Meno allots more space to Hippocrates (V, 35–VII, 40) than to anyone save Plato, who receives the greatest share because of the importance of the *Timaeus* for even the school of Aristotle. Moreover, like Aristotle in the *Politics,* Meno designates Hippocrates by name alone, as he does Polybius, thus confirming the eminent position that Hippocrates held in the medicine of the time. According to Meno, Hippocrates explained diseases as the result of bad air or bad diet. Air is essential for health and must circulate freely in the body; its impairment results in epidemics. Defective nourishment produces a variety of diseases; an excessive quantity or a poor quality of food leads to a tumult in the stomach and the generation of waste products from which winds arise and spread into the body to cause illness.

Resemblances between this text and *Breaths* are strong, but for our purposes they must be considered coincidental, since there is no way in which that work may be attributed to Hippocrates. *Breaths* is based on a system of cosmology such as is never present in Meno's work, but in *Nature of Man,* on the other hand, air and diet are said to be the cause of diseases, while other Coan works readily confirm the importance of food and ambient factors. It thus becomes necessary to assume that Meno is drawing upon one or more works by Hippocrates that have since been lost (as suggested by, among other things, his allusion to the plant *stratiotes,* which is not mentioned in the *Collection*). The doctrine of ambient factors is pronounced in the works that survive, and it is likewise possible to speculate that *Breaths,* too, draws upon lost treatises.

We must conclude, therefore, that we do possess a number of great medical works from the school of Hippocrates, and that Hippocrates himself was almost certainly the author of at least several of them. It would then ideally be possible to determine the nature and degree of Hippocrates' originality as a scientist, but given the limitations of our attribution, such a determination is not practicable. The essential step, therefore, is to define the contribution of the school of Cos, which is difficult in itself given the symbiosis that may be assumed to have existed among the members of a medical school and among the schools themselves.

A review of those themes which seem to be fundamental to the school of Cos should begin with its essential concept of disease processes. In an internal disease, such as is often caused by fluxes of indigestible humors, coction—a kind of slow cooking that restores equilibrium and normal properties to the disturbed humors—may occur. The disease reaches a crisis, "the decisive transformation which takes place at a given moment in the development of the disease and orients its course in a favorable direction" (Bourgey, *Observation,* p. 237). The crisis is marked by critical signs and symptoms and occurs on certain critical days in the course of the disease, although Coan speculations on these matters are more or less hazy and vary from treatise to treatise. A less frequently found notion is that of the deposit, a localized complaint that may be the forerunner of or sequel to a disease. The deposit can metastasize—travel from one part of the body to another—which may mark the transformation of one disease into another. The Coans also recognized the phenomenon of recurrence or relapse.

These basic concepts are not, however, peculiar to the school of Cos. Practically all of them may also be found in Cnidian writings, employed in a way that suggests established usage.[6] Indeed, if there is a difference on these points between the two schools, it lies in occasional differences in vocabulary; the ideas of deposit and metastasis, for example, are often differently named in Cnidian writings, although the phenomena described are obviously the same.

The doctrine of humors is found in the teachings of both schools, as it is throughout medical teachings of the times; it undoubtedly antedates Hippocrates. The Coan writings would seem to imply the four humors specified in *Nature of Man*—phlegm, blood, yellow bile, and black bile. The Cnidian treatises also embrace a theory of four humors, being in this instance those set forth in *Diseases* IV, namely water, blood, phlegm, and bile, various forms of the latter two being dominant in matters of nosology. It is thus in detail that the two schools differed.

As we have seen, ambient factors and diet are fundamental to the medical teachings of Cos; the treatise *Airs, Waters, Places,* for example, being devoted to the role of such factors as air, location, climate, and season. Cnidian writings also allude to these factors in the description and treatment of certain diseases, although unlike the Coan corpus no Cnidian treatises present a systematic exposition of them.[7] The importance of diet is stated in the Coan *Regimen in Acute Diseases, Nature of Man, Prognostic,* and even in *Fractures, Joints.* It is also mentioned in the non-Coan (or para-Cnidian) *Regimen.* The Cnidian treatises proper contain a discussion of diet as the greater part of the section devoted to therapy of each disease. The two schools are here so close in theory that despite the direct anti-Cnidian polemic of the *Regimen in Acute Diseases,* it can easily be shown that the essential elements of the dietary rules presented in this treatise are all advocated in the Cnidian writings.

Coan prognosis consisted in stating the past and present state of the disease and predicting its course after making an examination of the current symptoms but before questioning the patient (perhaps as a rationalization of shamanistic medicine).[8] This practice, too, was current at Cnidus, although we do not have a Cnidian book (such as the famous *Prognostic*) devoted to it. At any rate, the Cnidians described, often in rather different terms, a very similar procedure.[9] Both schools showed a further concern in the psychology of the patient and the effect of the psyche on the organism.[10] So, too, was the idea of the organism as an interdependent whole, which is often implicit in Coan writings (and see *Epidemics* VI, 3, 23), present in Cnidian works, as in *Diseases of Women,* where a gynecologist advises his readers "to look at the entire body" in order to decide on treatment. There was also a common concern for the study of health itself and for prophylaxis, as is evident at Cos in both *Regimen in Acute Diseases* IX and XXVIII, 1, and the last section of *Nature of Man,* but also in *Regimen* II, 4, and LXVII, where the author considers it to be one of his fundamental "discoveries."

The schools of Cos and Cnidus are also united in their repudiation of a medicine based directly on philosophical principles. The Coan writers polemicized against such tendencies in both *Nature of Man* and *Ancient Medicine;* while no such open polemic is extant among the Cnidian writings, Cnidus would seem to be as far from espousing a cosmological medicine as its rival. For this reason Celsus' remark that it was Hippocrates who separated medicine from philosophy should be taken very cautiously; this characteristic was not specific to the school of Cos. Both schools also rejected any sort of magico-religious medicine. This sort of medicine was vigorously attacked in Coan writings, especially *The Sacred Disease,* while in their practice the Cnidians would appear to have been equally hostile to it, as is made explicit in *Diseases of Girls* 1 (VIII, 468, 20–21). Both are nevertheless influenced by pre-Socratic ideas in certain of their aspects. For example, the Coan *Nature of Man* displays a philosophical orientation in many of its arguments, while the Cnidian *Generation-Nature of the Child* presents embryological conceptions similar to those of several pre-Socratic thinkers. The influence of Democritus, in particular, is certain.

A frequent generalization is that the medicine of

HIPPOCRATES OF COS

Cos was more sensitive to the patient, while that of Cnidus was more greatly concerned with the disease. In light of this it is said that the notion of the individual and the complexity of individual cases were more respected at Cos,[12] while Coan medicine placed less emphasis on therapy and Coan nosology was more apt to be general than that of Cnidus (which is said to be more concerned with the localization of the disease). Once again, this contrast is more apparent than real. It can be sustained only by taking as gospel the polemics of *Regimen in Acute Diseases* and by insisting upon the very real differences specifically between *Epidemics* I and III and *Diseases* II. A further examination of the extant works from the two schools must, however, lead to a different conclusion.

There remain to us from the school of Cos treatises devoted to one aspect or concept of medicine or to individual cases of disease; nothing comparable remains from Cnidus, which has left us only treatises on diseases. It is difficult to compare works so dissimilar in subject matter. If, for instance, there were a Coan treatise dealing explicitly with the same material as dealt with in *Diseases* II, we might arrive at legitimate conclusions. The only Coan work dealing with a specific malady is *Fractures, Joints,* and it is immediately apparent that a luxation or fracture does not have an etiology comparable with a case of phthisis or gout, and that the symptoms and treatment must be just as different. It is thus difficult to generalize about the fundamental difference in medicine as practiced at Cos and Cnidus when only the subject matter of the extant works would seem to be at issue.

Indeed, when the subject matter is similar, as in the Coan writings that present some data concerning etiology, dietetics, and therapy in specific diseases, one is struck by the resemblances between the two schools. For example, *Aphorisms* and *Epidemics* reveal the same rash positions as Cnidian gynecological writings (see Joly, *Niveau,* pp. 64–66). The appendix to *Regimen in Acute Diseases,* too, appears Cnidian in many places, while the same is true of *Epidemics* II, V, VI, and VII—which, let it be noted, differ from *Epidemics* I, III, and IV only in that they are not always restricted to noting the symptoms in the case history of one patient.

The point is often made that while fluxes of the humors were a frequent consideration in Cnidian etiology, the physicians of Cos made little use of them. Bourgey writes,

> It is very characteristic that while the most typical treatises, *Prognostic, Regimen in Acute Diseases, Fractures, Joints,* and *Aphorisms,* contain precise references to the existence of the humors, they do not speak of their circulation through the organism and do not attempt to construct on these foundations an arbitrary nosology [*op. cit.,* p. 249, n. 7].

Yet there is no reason for the author of these tracts to take up the fluxes of the humors, since they are irrelevant to his subjects (especially *Fractures, Joints*). Such fluxes are introduced most naturally into the explanation of diseases in the appendix to *Regimen in Acute Diseases* (see I; IX) and the doctrine is implied even in *Aphorisms* (see, for example, IV, 22–24) and the portions of *Epidemics* other than I, III, and IV. It is, moreover, by the flux of a humor that the author of *The Sacred Disease* explains epilepsy, while that of *Airs, Waters, Places* mentions in chapter VII a flux that comes from the head and upsets the stomach, and the nosology stated in chapter IV of *Nature of Man* is predicated on the same idea.

One ought not, however, to conclude that in these instances—and many others—the school of Cos was actually influenced by the school of Cnidus. Such influence would mean that a Coan physician would have had to consult a Cnidian whenever he considered the etiology of a disease; in therapy, too, each time a Coan describes a drug, the similarity to Cnidian texts is evident. The very number of the similarities between the schools makes such a thing unlikely.

Despite these similarities, a qualitative difference between the medicine of Cos and Cnidus has been assumed by, among others, Bourgey, who elaborated his opinion in great detail and with utmost conviction. In Bourgey's view, the medicine of Cos was an unequaled success that surpassed both the philosophical medicine embodied in *Breaths* and *Regimen* and the purely empirical and routine medicine of the Cnidians. It is my opinion that he thus overvalues Coan medicine and thereby exaggerates the differences that actually existed between it and other schools. Certainly there are analyses to be found in the philosophical medical literature that a Coan physician would be able to endorse without stricture since, for instance, the list of foods and exercises of book II of *Regimen* displays the same approach and intelligence as the best Coan medicine. The same is true of the essential elements of *Regimen* III and IV.

Bourgey further finds that Cnidian medicine was characterized by "fidelity to rather primitive practices, by the importance given to a primarily descriptive analysis—by a strict submission to facts and customs with little evidence of the intervention of critical or intelligent thought. One can find an indication of this intellectual insufficiency on the purely literary plane" (*op. cit.,* pp. 52–53). In brief, in Cnidian medicine he sees only "a crude empiricism un-

HIPPOCRATES OF COS

controlled by intelligence" (*ibid.,* p. 55). At Cos, on the other hand, observation was freed from routine to become discerning and methodical. It was, moreover, effectively combined with active and critical thought—"a superior judgment" reigned there, and "the scope of its intelligence and the penetration of its observation are manifest in a high degree" (*ibid.,* p. 63). All of this seems to me to be thrusting Cnidian medicine into the shadows so that Coan may shine brighter, since I am convinced that the two schools shared essentially the same spirit. But in order to clarify this issue it is necessary to approach the problem of evaluating the medicine of the *Collection,* and particularly that of Cos, in a somewhat different way.

Since we have seen that the two schools of Cos and Cnidus (and undoubtedly others, if we knew more about them) were similar in a number of fundamental points, we must assume that the rivalry between the schools was the result of divergences of detail. Indeed, these divergences must be grossly magnified to achieve any significance, and it is tempting to say that from the point of view of modern medicine, that great rivalry was very minor indeed. Still, it is necessary to consider these small differences in an attempt to identify the originality of the school of Cos and of Hippocrates.

The personal originality of Hippocrates is elusive. We know from Meno's testimony that he was not the founder of rational medicine, nor did he bring about a revolution in medical practice; although, as we have seen, Meno gives him more space than any other physician, he attributes to him no doctrines essentially different from those of his contemporaries. In order to account for Hippocrates' importance, one must take into consideration his prestige as a teacher, his talent as a practitioner, and his authorship of a number of treatises—which are, incidentally, a great deal more engaging than the rather monotonous writings of the Cnidians, and must have attracted a wider readership by their style and topicality. It is clear that Hippocrates did not stand alone as a genius in an intellectual desert; it is equally clear that, for reasons that we do not know precisely, he was in the front rank of the medicine of his age, as the *primus inter pares.*

We must now determine the level—and the value—of Hippocratic medicine. This pursuit is not universally acceptable, since many scholars, basically philologists and not historians of science, refuse to do so a priori, while others prefer not to apply critical methodology to a figure as highly respected as Hippocrates—even if it is not an act of actual *lèse-majesté* to judge Hippocrates' medicine in the light of our own, it at least seems to imply a lack of historical understanding. The Hippocratic *Collection* has not yet received the sort of systematic scholarly treatment that Alexandre Koyré, for example, gave to the work of Galileo. In trying to establish what was truly valuable in Hippocratic medicine, I shall confine myself to examples drawn from the school of Cos, although we have seen that there is little point in making distinctions between the medicine of Cos and that of Cnidus.

Let us first consider the excellent intentions of the followers of Hippocrates. They wished to promote a strictly scientific medicine, and they clearly thought that they had succeeded in doing so. While magic and medicine may have coexisted peacefully in Egypt, Mesopotamia, and, indeed, ancient Greece, the rational medicine of the Hippocratics (who did not necessarily invent it, although they may have thought that they did) attacked magic relentlessly. The medicine that the Hippocratics advocated, on their own example, was a rigorous rational technique and clearly a great step forward beyond its predecessors.

A number of sound Coan directives derived from this desire for rationality. Practitioners of this school were advised that "Examining the body requires sight, hearing, smell, touch, taste, and reason" (*Epidemics* VI, 8, 17) and told "To consider what can be seen, touched, and heard—what can be learned through sight, touch, hearing, smell, taste, and the understanding—what one can perceive through all the means at our disposal" (*In the Surgery* 1). In short, the Hippocratic physician was urged "To examine by reason and fact" (Loeb ed., III, 22, 2–3), a system which a late work sums up in the happy phrase "rational practice" (*Precepts* 1). (These notions of scrupulous observation were further put into a long, detailed program; see *Epidemics* I, 10.) There is little doubt that a great many of the Coan observations made on this basis, beginning with the famous *facies hippocratica* of *Prognostic,* were exact. In the interest of coordinating these observations, *Epidemics* envisages a rather elaborate method, reminiscent of Plato, "to make a synthesis of all the data concerning an illness in order to determine the similarities, then to establish between the latter new differences in order to arrive finally at a unique similarity" (Bourgey, *op. cit.,* p. 96; *Epidemics* VI, 3, 12).

This rational approach to observation and synthesis allowed the Hippocratic physicians to recognize their errors and to be aware of the methodological value of admitting and analyzing them. After having described an unsuccessful treatment, for example, the author of *Joints* adds, "I relate this on purpose, for it is also valuable to know what attempts have failed and why they have failed" (ch. XLVII; *cf. Diseases* III, 17). The same rationality led them to a firm principle of causality, one that permitted of no exception. Coan writers stated that "Each disease has a nature of its own, and none arises without its natural cause" (*Airs, Waters, Places* 22) and that "It is demonstrated that chance does not exist, for everything that occurs will be found to do so for a reason; in the face of a reason, chance visibly loses all reality" (*The Art* 6). They further accepted as self-evident the unity of the animal kingdom, man being one species among others (*Joints* 8; *The Sacred Disease* 14).

We should also briefly review the series of valuable notions previously discussed in the comparison of the Coan and Cnidian schools. These include the refusal to tie medicine to philosophical or cosmological principles, the importance of environment, and the interest taken in prophylaxis and in the psychology and mental constitution of the patient.

We must still examine Hippocratian deontology, and in this field the school of Cos set forth principles that are still valid. "To help, or at least to do no harm," writes the author of *Epidemics* I, 11; and the *Oath* known to all physicians makes this explicit:

> In whatever houses I enter, I will enter to help the sick, and I will abstain from all intentional wrongdoing and harm, especially from abusing the bodies of man or woman, bond or free. And whatsoever I shall see or hear in the course of my profession in my intercourse with men, if it be what should not be published abroad, I will never divulge, holding such things to be holy secrets.

One should also consider the medical ethics stated in chapter 6 of *Precepts,* which are worthy of consideration even though it is a late work:

> I urge you not to be too unkind, but to consider your patient's wealth and resources. Sometimes you will give your services for nothing, calling to mind a previous benefaction or your present reputation. If there should be the opportunity to serve a foreigner or a poor man, give full assistance to him; for where there is love of man there is also love of the art.

Let us now see what sort of reservations must be held concerning Hippocratic medicine. This is a subject that has occupied critics only in recent years; see, for example, F. M. Cornford's sound criticism in *Principium sapientiae* (Cambridge, 1952) and the highly skeptical remarks by L. Edelstein in *Ancient Medicine* (Baltimore, 1967), pp. 124, 405. G. Vlastos makes more concrete points, especially in a review of Cornford's book which appeared in *Gnomon* in 1955 (pp. 65–76), reprinted in *Studies in Presocratic Philosophy,* edited by Furley and Allen (London–New York, 1970) I myself discovered this review only in the later edition, after having written on the same subjects and having reached practically the same conclusions in my own *Le niveau de la science hippocratique* (Paris, 1966). Charles Lichtenthäler carries the discussion further in his *Études hippocratiques;* although he makes some pertinent analyses, his work is burdened by debatable opinions, such as that there are two divergent but complementary methods in medicine, that of Hippocrates and ours—an idea that I, for one, cannot accept.

Kudlien's critical concern is more specific. He is convinced that the medicine of the Hippocratic *Collection* was not entirely rational, and he proposes to show in it rationalized vestiges of ancient thought. He is plainly inspired in this by E. R. Dodd's *The Greeks and the Irrational* (Berkeley–Los Angeles, 1951), from which he takes in particular Gilbert Murray's notion of "inherited conglomerate." This method yields Kudlien good results, as when he discovers the mythical basis of the use of honey as a remedy for the "withering disease" (*Der Beginn,* pp. 100–106) or for the "deadly disease" mentioned in *Diseases* II, 66–67 ("Early Greek Primitive Medicine," in *Clio medica,* **3** [1968], 320–330). One can cite other instances of such survivals without referring to Dodds, as, indeed, I myself did in 1966 (*Niveau,* pp. 39, 191, 208, and 223). It is, for example, certain that the therapeutic use of excrement reveals the archaic mentality that prizes all living beings and their products. The preference for right over left is also archaic (see *Aphorisms* V, 48, and *Epidemics* VI, 2, 25).

Let us examine the preference for the right over the left as it appeared in conjunction with the idea, evident in the *Collection,* as well as in Aristotle and other Greek writers, that women are inferior. The Coan writers stated, among other things, that "When a woman is pregnant with twins, should either breast sag, she will lose one child . . . if it is the right breast, it is a male child that will be lost; if it is the left breast, a female child" (*Aphorisms* V, 38), since "the male fetus is usually on the right, the female on the left" (*ibid.,* V, 48). "At puberty, depending on which testicle develops first, the individual will father boys if it is the right one, girls if it is the left" (*Epidemics* V, 4, 21). "[The male fetus] is in the warmest place, the most solid, at the right [of the womb]; that is why males are darker and formed earlier; they move about earlier; then movement stops and they grow more slowly. They are more solid, more passionate, and more full-blooded because the location in the womb where they take form is hotter" (*Epidemics* VI, 2, 25). These errors were elaborated by Cnidian gynecologists. We cannot be sure, of course, that the belief in the inferiority of women was itself a survival of an archaic belief; indeed, it would seem more likely that it was a reflection of the society of the time.

What are certainly archaic ideas were further manifested in imitative errors. A confusion of medical and visual properties persists when the dietician of *Regimen* asserts that only white wine is diuretic (LII, 2) or that beef is less digestible than pork (XLVI, 1–2). Likewise, when he says that the eel is indigestible because it lives in the mud (XLVIII, 2), he postulates an unprovable connection between the animal and its environment. It would be arbitrary, however, to seek in these instances the survival of magical practices, particularly in light of the conscious hostility of the Hippocratic physicians to any sort of magical medicine. We may thus see that Murray's concept of "inherited conglomerate" is inadequate to deal with what is irrational in the Hippocratic *Collection.* The cataloging of the surviving evidences of an archaic mentality is not sufficient; to accept it as such leads further to an exaggeration of the truly scientific value of this school of medicine.

It is necessary to seek a broader irrationality, for which purpose we possess an excellent tool in the method of Gaston Bachelard's *Formation de l'esprit scientifique* (Paris, 1938). Although this study is devoted to biological and medical texts of the sixteenth, seventeenth, and eighteenth centuries, Bachelard's technique of psychological analysis is also fruitful in the earlier period that concerns us. This implies no criticism of Dodds's *The Greeks and the Irrational;* it is merely that in this instance Bachelard's concerns are, by analogy, more nearly our own. The methods of the two authors are by no means incompatible, moreover, and from them we may deduce that it is inevitable that survivals from the past become psychological realities. Bachelard's method is able to encompass such survivals and to couple them with psychoanalysis, in the strictest Freudian sense.

Given this persistence of a broad irrationality, the view put forth by Kudlien, among others, that a school of medical thought is necessarily truly scientific when it is clearly divorced from the magico-religious sphere, is unconvincing. I would suggest that rational medicine (an ambiguous term) is indeed one clearly separated from magic—it being understood that there is a big gap between "rational" and "scientific" medicine. This medicine is rational when compared with magico-religious medicine, but remains irrational compared to a more refinedly scientific practice. Rationality, subject to epistemology, undergoes real changes in the course of history; whether one chooses a view based on continuity or discontinuity, an acceptance of the historical nature of reason is essential.

Rather than trying to present a synthesis of the different irrational elements in Hippocratic medicine, however, I shall deal with further concrete examples. Let it be noted before we proceed that it is difficult, at this distance in history, to criticize Hippocratic statements from observed experience, since we cannot always be sure of circumstances that may have affected such observations. With this in mind, we may note that certain Hippocratic observations would seem to be questionable. For example, the *Nature of Man* states that dysentery and nosebleeds occur primarily in the spring and summer, the seasons in which bodies are hottest and reddest. Certain critics think this assertion to be the sum of long and patient observations—a technique that might almost be called statistical—but this would seem to be unlikely, since dysentery is more often a disease characteristic of late summer and autumn than of spring. The Hippocratic author has previously explained that one humor is dominant for each of the seasons. According to him, it is in spring that the blood flows most freely (a statement worthy of psychological analysis), and it is necessary for him to support this thesis. We must therefore question whether deduction has not supplanted observation.

The same Hippocratic author wrote, "If you give a man a medicine that withdraws phlegm, he will vomit up phlegm; if you give him one that withdraws bile, he will vomit up bile" (*Nature of Man* V). This argument is advanced to prove the specificity of humors, but here theory rather clearly influences observation, for it is obvious that the author believes that he sees the particular humor for which his medicine is reputedly effective. He thus combines two prejudices; his observation is short-circuited by a priori thinking, and he sees what he wishes to see.

The same tendency to make observed facts conform to preestablished convictions had a deleterious effect upon experiment. In the conscious intention of the Coan physician, experiment might have played a part comparable to the one it has today, but his results were, once again, apt to confirm his preconceptions. The author of *Airs, Waters, Places,* for instance, believes that water from melted ice or snow is bad because freezing has caused it to lose its "clear, light, sweet part." He continues, "The following experiment will prove it. In winter, pour a measured amount of water into a vessel and set it in the open, where it will freeze best; then on the next day, carry it to a sheltered place where the ice will melt best. After it has melted, measure the water again, and you will find that it has greatly diminished" (ch. VIII). Although one can accept some diminution of quantity because of evaporation, it is hard not to believe that the "greatly diminished" amount of

HIPPOCRATES OF COS

water found by the author was the result of his desire to find just such a thing. The expansion produced by the freezing, upon which one might expect any attentive observer to remark, passes unnoticed; indeed, the author's explanation of the experiment precludes this phenomenon, since he writes, "This shows proof that freezing evaporates and dries the lightest and finest part [of the water], not the heaviest and densest part, which would be impossible" (*ibid.;* and see Vlastos, *op. cit.,* p. 44).

In a similar manner the author of *Nature of the Child* presents in chapter 29 a quite fantastic account of human embryology, then proposes to verify it experimentally by observing the development of a chick up to the time of hatching through the daily examination of an egg taken from the nest of a setting hen. The design of the experiment itself is brilliant; it represents the best example of a plan for methodical observation in the *Collection.* The author, however, learned nothing from following it, since he used it only to confirm his ideas as already stated, a point that some critics have missed. He never displays any doubt of his results, whereas if he had carefully conducted the experiment he describes, he would surely have had sufficient opportunities to be perplexed. Indeed, one may even wonder if he actually followed the procedure he discusses.

One can thus see how misleading it may be to judge Hippocratic medicine solely on the basis of its declared intentions and directives. Their often rather vague formulation allows of several interpretations (for which reason it is sometimes stated that they display a truly modern mentality), so it is necessary to put them back into their context to see what they yield in actual practice.

Chapter 10 of *Nature of Man* provides a good opportunity to examine a departure in practice from the declared Hippocratic intention of observation. In its emphasis on qualities, too, it illustrates a major concern of Hippocratic medicine. Its author states that the most serious illnesses are those that affect the strongest part of the body, although what part that is is nowhere specified. If that strongest part is stricken, the entire body is affected; if the disease moves, its force will easily overcome the weaker parts. If, on the other hand, the disease occurs in a weaker part of the body, it must itself be weak, and should it move to a stronger part it will quickly be neutralized.

The author of *Airs, Waters, Places* prefers to emphasize the role of the location of a city. Here, too, what might appear to be the result of systematic observation in truth represents a deduction from a preliminary postulate. The author begins with the premise that a site exposed to hot winds has an abundance of water (ch. 3), then asserts that people living in such a site of necessity have moist heads. It is therefore to be expected that fluxes coming from the head upset their stomachs. Women, who are already moist by nature, when they live in such a locale are subject to excessive menses and sterility (since the fluxes drown semen), while frequent miscarriages are a logical corollary of sterility. Both sexes, moreover, suffer from convulsions, asthma, and epileptic fits. (It is interesting to note in this context that the author of *The Sacred Disease* states that epilepsy is the result of phlegm, which is moist, attacking the brain or throat.)

The same work further emphasizes the notion of qualities and their values in a discussion of mixed waters in chapter 9:

> People suffer especially from the stone, from gravel, from strangury, from sciatica, and from hernias when they drink waters of different kinds, or from large rivers into which other rivers empty, or coming from a lake fed by many streams of various sorts, and whenever they consume foreign waters brought from a great distance. For one water cannot be like another; some are sweet, others are salty and taste of alum, and others flow from hot springs; these waters, in a mixture, combat each other and the strongest always prevails. The strongest is not always the same; sometimes it is one, sometimes another, depending on the winds. One has its strength from the north wind, another from the south wind, and the same is true for the others.

We may here note that the concept of mixture had great importance in prescientific thinking, especially in pharmacy. Therica, for instance, is compounded of some 150 ingredients and represented, in Bachelard's felicitous phrase, the "sum of the sums" of substances. All pharmaceutical masters had to make it, moreover, and to share the product obtained among themselves; it is thus a mixture of the second power, the efficacy of which is multiplied by this very fact.

The passage from *Airs, Waters, Places* on the mixture of waters is of further particular interest since it illustrates the ambivalence of an intuitively valued idea. Here mixture is fundamentally bad; the substantial qualities must therefore war with each other instead of cooperating or blending into a beneficial whole. In language similar to that already noted in *Nature of Man,* the strongest quality wins.

It is now necessary to review the Coan theory of the four humors in a critical perspective. Although the Coan physicians again thought that their system was based on attentive observation, we must call the quality of that observation into question once again. It is more apparent that their theory of the humors represented a deduction based on a philosophical system of four elements; that of Empedocles comes

HIPPOCRATES OF COS

immediately to mind. It might also have in it elements of a rationalized ancient belief, but these are topics that require further scholarly research.

The Hippocratic physicians explained most diseases by fluxes of humors, as we have seen in passing in our discussion of *Nature of Man* and *Airs, Waters, Places. Epidemics,* more specifically, states that such humoral fluxes can start spontaneously (V, 19 and 64); the concept of flux, however, was scarcely subjected to analysis, although it is employed as primary evidence. In addition to spontaneity, the flux seems to have great freedom in direction and destination. It affects the internal organs as vessels that can absorb it, be distended by it, overflow with it, or reject it. The mechanical functions of the organs form a physics of the receptacle which, together with the qualities of the humors in motion, define the entire physiology and internal pathology of the schools of both Cos and Cnidus.

It is here important to note the role of analogy in this medical thought, a role that cannot be overemphasized. The functioning of the organism is conceived of in terms familiar from elementary mechanics, from such devices as the pump, the cupping glass, and communicating vessels. This tendency is most consciously expressed in *Generation-Nature of the Child-Diseases* IV, but it exists everywhere.[13] This use of analogy in no way parallels the use of models in contemporary science, although it is sometimes mistakenly thought to do so. The concept of model, as, for example, it is employed in nuclear physics, puts into concrete form the sum of scientific data available at given time and is an a posteriori synthesis, essentially capable of correction. The analogy of the Hippocratic authors, on the other hand, was a priori, an image that was imposed tyrannically and that rendered observation sterile. It was not subject to emendation; rather, it selected and modified available data, shaping them to itself. It was not a hypothetical resemblance; it was a fundamental identity.

Although the Hippocratic emphasis on diet and regimen was in principle sound in regard to modern medicine, here, again, contrary to what is often supposed, there is little evidence of strictly objective observation. The most striking aspect of the dietetics presented in *Regimen in Acute Diseases,* for instance, is the horror of change, perhaps the transposition of a social conservatism. The author dramatizes the least departure from the diet. He likewise exaggerates the distinctions to be made between products, giving primary importance to the differences between infusions made directly from a berry or leaf and those that have been filtered or otherwise processed. It is not difficult to guess at the psychology involved here. The physician was limited in the therapeutic possibilities available to him, in the face of common and often fatal diseases, and needed to make the most of what little he had. With only a small number of potions, he had to confer almost miraculous powers upon them. The dietetics are marked by a wide range of subtle distinctions, necessitated by their major limitations.

As for the pharmacopoeia, we know it chiefly from Cnidian works, although the traces that remain in Coan books (such as *Epidemics* II, 6, 29) indicate that here again the two schools were in close agreement. In this area what is particularly notable is a want of specificity, long lists of recipes prescribe a host of substances for a single malady and, conversely, a single substance is recommended for a host of ills. Certain of the prejudices that we have already discussed are present in the pharmacopoeia, too—such as the particular value of products made from living matter, or of human products (including "milk of a woman nursing a male baby," which was also recommended in Egypt), or rare or exotic substances.

Several Coan works are, however, less subject than others to the failings cataloged above. Of these, *Epidemics* I and III and *Fractures, Joints* are the most notable. We should not make the mistake of assuming that the author or authors of these works were more scientific in the modern sense; it is pertinent to mention that the author of *Epidemics* I and III, for example, avoided the traps into which his contemporaries fell only by drastically circumscribing his subject. His technique is to record climatic data and to follow day by day the signs and progressions of certain diseases, confining himself to external observation. One can admire the precision of this observation, but it does not go far enough. In a great many cases, the author does not even define the disease in question; in his self-imposed restriction he systematically refuses to hazard any interpretation or causal hypothesis. Since it is in the latter pursuit that the scientific mentality most readily reveals itself, one may say that if a physician limits himself to listing such factors as the place where the patient becomes ill, his vomitings, the color and consistency of his stools, his perspiration, and the fluctuations of his fever, he will perhaps attain new heights of objectivity but his work will be of little significance from the point of view of science (see Vlastos, *op. cit.,* p. 45). While we should certainly admire the author's submission to facts in this case, we must remember that this external and necessarily qualitative observation is only a very modest starting point for science per se.

The same sort of basic limitations apply to *Fractures, Joints* and *Wounds of the Head.* The works are by the nature of their subject matter confined to the questions of ascertaining fractures or luxations, and of dressing or reducing them. This is virtually the only area of medicine in which the techniques of the time—observation such as existed in this period combined with sufficient mechanical ingenuity to perfect devices for reduction or trepanning (or, simply, splints and bandages)—are adequate to produce very satisfactory results. In limiting himself to these tasks, some of which are, nonetheless, delicate, the author of these works, too, avoided the errors made by others. The author does not, for instance, consider the histological problem of the healing of broken bones; he nowhere presents an explanation of the reduction of a fracture that he describes.

In sum, then, we might say that Hippocratic medicine was rational, but that its reason was not the scientific reason of today. Indeed, I prefer to adopt Bachelard's term "prescientific." There is no question here of reproaching the physicians of the time; they brilliantly constructed an inevitable stage of medicine that we must further analyze and inventory. But to identify their medicine with ours is surely to display a lack of historical understanding. Even as prescientific medicine, however, the medicine of the Hippocratics represented a knowledge infinitely more valuable than the magic which it supplanted.

As to Hippocrates himself, if he is not all of medicine, and if we do not even know with certainty exactly what he wrote, we should nevertheless consider him to be the eminent representative of a significant stage of medicine—the stage in which war was waged on all magico-religious medical practice, and in which medicine consciously sought to become fully scientific and at least succeeded in becoming partially rational. To go further in one's praise is to fall back into hagiography, but we must note the difficulty and enormousness of the task that the Hippocratics assumed. The proof of their success may be found in that for two millennia no better work was accomplished, and often worse was done. Hippocratic medicine traversed the centuries somewhat like Aristotelian logic; and if, since the nineteenth century, the errors of the physician have been seen to be more profound than those of the logician, it is because the domain that he explored was much the more complex.

NOTES

1. Ilberg's ed. is in *Corpus Medicorum Graecorum* (1927), pp. 175–178.
2. Published by Schöne in *Rheinisches Museum,* **58** (1903), 56–66.
3. *Chiliades* VII, 944–989.
4. The same thing is true of the *Letters* and of the other apocryphal texts of vol. IX of Littré's ed.; a recent study by H. de Ley emphasizes the diversity of inspiration among these writings.
5. It is understandable that a few of the works, among them such brief texts as *Child of Eight Months* and *Use of Liquids,* cannot easily be identified as belonging to a definite school. To them I would add the major treatise *Ancient Medicine,* which is often considered purely Coan.
6. Among such references for coction, for example, are *Diseases* I, 19 (Littré VI, 174) and III, 16 (VIII, 52, 11); for the crisis, *Diseases* I, 72 (VI, 158, 22); II, 40 (VII, 56, 10), 71 (VII, 108, 24); *Internal Affections* 21 (VII, 220, 9), 27 (VII, 238, 13 and 24), 28 (VII, 240, 19); for the deposit, *Diseases* II, 27 (VII, 44), 57 (VII, 90, 7–8); *Internal Affections* 18 (VII, 212), 28 (VII, 240); for metastasis, *Diseases* I, 18 (VI, 172, 17 and 22), 29 (VI, 220, 2); *Affections* 9 (VI, 216, 16), 12 (VI, 220, 7), 19 (VI, 228, 19); *Diseases* III, 6 (VII, 124, 8): *Diseases of Women* I, 26 (VIII, 70, 9), 29 (VIII, 72, 14); and for relapse, *Diseases* II, 1 (VII, 8, 10), 12 (VII, 20, 6), 41 (VII, 58, 20); *Internal Affections* 1 (VII, 170, 15), 2 (VII, 174, 12 and 14), 44 (VII, 278, 2–4).
7. See *Diseases* II, 54 (VII, 84, 2), 55 (VII, 86, 13–14 and 22), 64 (VII, 98, 21), 66 (VII, 100, 21); *Affections* 6 (VI, 214, 8–9); *Nature of Woman* 1 (VII, 312, 10–11); *Diseases of Women* I, 11 (VIII, 42, 21), II, 111 (VIII, 238, 17); but note that the para-Cnidian *Regimen* (II, 37–38; III, 68) does so explicitly.
8. See Kudlien, "Early Greek Primitive Medicine," p. 309.
9. See *Diseases* II, 47 (VII, 72, 1), 48 (VII, 72, 16), 63 (VII, 96, 21); *Diseases* III, 15 (VII, 140, 24); *Internal Affections* 2 (VII, 172, 23), 20 (VII, 216, 18), 22 (VII, 220, 21), 27 (VII, 238, 14), 39 (VII, 262, 13); and *Diseases of Women* I, 11 (VIII, 42, 22).
10. For this aspect of Coan medicine, see *Epidemics* VI, 5, 5 and 6, 14; for Cnidian, see *Diseases* II, 72 (VII, 108–110); *Internal Affections* 48 (VII, 284–286); *Diseases of Women* I, 62 (VIII, 126); and *Diseases of Girls* 1 (VIII, 466–470).
11. See I, 11 (VIII, 42, 19 and 22), 66 (136, 2), and II, 138 (312, 3 and 7).
12. See *Joints* 8; but see also *Diseases* I, 16 (VI, 168, 25), 22 (184, 3–5), III, 17 (VII, 156, 7–8); *Nature of Woman* I (VII, 312); *Diseases of Women* I, 11 (VIII, 42, 21), II, 111 (VIII, 238–240); and *On Sterile Women* (VIII, 230, 444, 1–4).
13. O. Regenborn devotes an important essay to this ancient use of analogy, which he compares with modern hypothesis. I do not think that the comparison is valid, however; see *Niveau,* pp. 73–75.

BIBLIOGRAPHY

I. ORIGINAL WORKS. The complete critical ed., with French trans. is Émile Littré, *Les oeuvres complètes d'Hippocrate,* 10 vols. (Paris, 1839–1861), also available in photocopy (Amsterdam, 1961). This ed. has supplanted its contemporary rival: F. Z. Ermerins, ed., *Hippocratis et aliorum medicorum veterum reliquiae,* 3 vols. (Leipzig–Paris, 1859–1864).

The study of the MS tradition has since made great progress and the later eds. are better from the philological point of view, but they are far from complete. The Teubner ed. stopped after the 2 vols. edited by H. Kühlewein. Vol. I (Leipzig, 1895) contains *Ancient Medicine; Airs, Waters, Places; Prognostic; Regimen in Acute Diseases* and its *Appendix;* and *Epidemics* I and III. Vol. II (Leipzig, 1902) contains *Wounds of the Head; In the Surgery; Fractures, Joints;* and *Mochlicon.*

The Corpus Medicorum Graecorum began the ed. of the *Collection* with a volume by J. L. Heiberg (Leipzig–Berlin, 1927) which contains the *Oath; On Law; The Art; The Physician; Decorum; Precepts; Ancient Medicine; Airs, Waters, Places; Nutriment; The Use of Liquids;* and *Breaths.* Recent eds. are A. Grensemann, *On the Child of Eight Months. On the Child of Seven Months (Spurious)* (Berlin, 1968) and H. Diller, *Airs, Waters, Places* (Berlin, 1970). Other volumes are expected, notably *Nature of Man* by J. Jouanna.

The publishing unit of the universities of France has also undertaken the complete ed. The first 3 vols. to be published are by R. Joly; *Regimen* (Paris, 1967); *Generation-Nature of the Child-Diseases* IV and *The Child of Eight Months* (Paris, 1970); *Regimen in Acute Diseases* and its *Appendix, Nutriment, The Use of Liquids* (Paris, 1972).

An early partial ed. is J. E. Pétrequin, ed., *La chirurgie d'Hippocrate,* 2 vols. (Paris, 1878), with French trans. and very interesting notes. Vol. I contains the *Oath; The Physician; Wounds; Fistulas; Hemorrhoids;* and *Wounds of the Head.* Vol. II contains *In the Surgery; Fractures, Joints;* and *Mochlicon.*

The Loeb Library published 4 vols. with English trans. (Cambridge, Mass., 1923–1931). Vol. I, W. H. S. Jones, ed., contains *Ancient Medicine; Airs, Waters, Places; Epidemics* I and III; the *Oath; Precepts;* and *Nutriment.* In vol. II, W. H. S. Jones, ed., are *Prognostic; Regimen in Acute Diseases; The Sacred Disease; The Art; Breaths; Law; Decorum; The Physician;* and vol. III, E. T. Withington, ed., contains *Wounds of the Head; In the Surgery; Fractures, Joints;* and *Mochlicon.* In vol. IV, W. H. S. Jones, ed., are *Nature of Man; Humors; Aphorisms; Regimen;* and *Dreams* (and fragments of Heraclitus). The critical apparatus is greatly reduced in this edition.

There are also eds. of individual works, often with commentary: *The Art:* T. Gomperz, *Die Apologie der Heilkunst* (Vienna, 1890); *On Flesh:* K. Deichgraeber, *Hippokrates, über Entstehung und Aufbau des menschlichen Körpers* (Leipzig–Berlin, 1935); *Prognostic:* B. Alexanderson, *Die Hippokratische Schrift Prognostikon* (Göteborg, 1963); *Ancient Medicine:* A. J. Festugière, *Hippocrate. L'Ancienne médecine* (Paris, 1948); *The Heart:* J. Bidez and G. Leboucq, "Une anatomie antique du coeur humain," in *Revue des études grecques,* **57** (1944), pp. 7–40; *Breaths:* A. Nelson, *Die Hippokratische Schrift* Περὶ φυσῶν (Uppsala, 1908); the *Oath:* L. Edelstein, *The Hippocratic Oath* (Baltimore, 1943), repr. in his *Ancient Medicine,* pp. 3–63; and *The Sacred Disease:* H. Grensemann, *Die Hippokratische Schrift über die heilige Krankheit,* vol. II, pt. 1, of Ars Medica (Berlin, 1968).

Recent anthologies are H. Diller, ed., *Hippokrates-Schriften* (Hamburg, 1962); R. Joly, ed., *Hippocrate—médecine grecque* (Paris, 1964); and M. Vegetti, ed., *Opere di Ippocrate* (Turin, 1965).

II. Secondary Literature. On Hippocrates himself, see L. Edelstein, "Hippokrates. Nachträge," in Pauly-Wissowa, *Real-Encyklopädie,* supp. VI (1935), cols. 1290–1345.

The *Collection* is discussed in the following (listed chronologically): C. Fredrich, *Hippokratische Untersuchungen* (Berlin, 1899); H. Gossen, "Hippokrates," in Pauly-Wissowa, *Real-Encyklopädie,* VIII (1913), cols. 1780–1852; L. Edelstein, Περὶ ἀέρων *und die Sammlung der Hippokratischen Schriften* (Berlin, 1931); L. Bourgey, *Observation et expérience chez les médecins de la Collection hippocratique* (Paris, 1953); R. Joly, *Le niveau de la science hippocratique* (Paris, 1966); and P. Laín Entralgo, *La medicina hipocrática* (Madrid, 1970).

On the school of Cos, see K. Deichgraeber, *Die Epidemien und das Corpus Hippocraticum; Voruntersuchungen zu einer Geschichte der Koischen Arztschule* (Berlin, 1933); and H. Pohlenz, *Hippokrates und die Begründung der wissenschaftlichen Medizin* (Berlin, 1938).

Among studies on special topics are the following (listed chronologically): H. Diller, *Wanderarzt und Aitiologe. Studien zur Hippokratischen Schrift* Περὶ ἀέρων, ὑδάτων, τόπων (Leipzig, 1934); U. Fleischer, *Untersuchungen zu den pseudohippokratischen Schriften* Παραγγελίαι, Περὶ ἰητροῦ *und* Περὶ εὐσχημοσύνης (Berlin, 1939); R. Joly, *Recherches sur le traité pseudo-hippocratique Du régime* (Paris–Liège, 1961); N. Van Brock, *Recherches sur le vocabulaire médical du grec ancien* (Paris, 1961); G. H. Knutzen, *Technologie in den Hippokratischen Schriften* Περὶ διαίτηςὀξέων, Περὶ ἀγμῶν, Περὶ ἄρθρων ἐμβολῆς (Wiesbaden, 1963); J. Schumacher, *Antike Medizin* (Berlin, 1963), for its extensive bibliography; H. Flashar, *Melancholie und Melancholiker in den medizinischen Theorien der Antike* (Berlin, 1966); L. Edelstein, *Ancient Medicine* (Baltimore, 1967); F. Kudlien, *Der Beginn des medizinischen Denkens bei den Griechen* (Zurich–Stuttgart, 1967); G. Lanata, *Medicina magica e religione popolare in Grecia fino all'età di Ippocrate* (Rome, 1967), and J. Mansfeld, *The Pseudo-Hippocratic Tract* Περι῾εβδομαδων *Ch. 1–11 and Greek Philosophy* (Assen, 1971).

A very short list of articles is the following (presented chronologically): O. Regenbogen, "Eine Forschungsmethode antiker Naturwissenschaft," in *Quellen und Studien zur Geschichte der Mathematik, Astronomie und Physik,* **1,** no. 2 (1930), 130–182; W. Nestle, "Hippocratica," in *Hermes,* **73** (1938), 1–38; W. Müri, "Der Massgedanke bei griechischen Ärzten," in *Gymnasium,* **57** (1950), 183–201; K. Abel, "Die Lehre vom Blutkreislauf im Corpus Hippocraticum," in *Hermes,* **86** (1958), 192–219; R. Joly, "La question hippocratique et le témoignage du *Phèdre,*" in *Revue des études grecques,* **74** (1961), 69–92; E. Wickerscheimer, "Légendes hippocratiques du moyen âge," in *Sudhoffs Archiv für Geschichte der Medizin und der Naturwissenschaften,* **45,** no. 2 (1961), 164–175; H. Flashar, "Beiträge zur spätantiken Hippokratesdeutung," in *Hermes,* **90** (1962), 402–418; H. Herter, "Die Treffkunst des Ärztes in Hippokratischer und Platonischer Sicht," in *Sudhoffs Archiv für Geschichte der Medizin und der Naturwissenschaften,* **47** (1963), 247–290; G. E. R. Lloyd, "Who Is Attacked in On Ancient Medicine?," in *Phronesis,* **8** (1963), 108–126; H. Diller, "Ausdruckformen des methodischen Bewusstseins in den Hippokratischen Epidemien," in *Archiv für Begriffsgeschichte,* **9** (1964), 133–150; I. M. Lonie, "The Cnidian Treatises of the Corpus Hippocraticum," in *Classical Quarterly,* **15** (1965), pp. 1–30; M. Vegetti, "Il De locis in homine fra Anassagora ed Ippocrate," in *Rendiconti dell'Istituto lombardo di scienze e lettere,* **99** (1965), 193–213; F. Kudlien, "Early Greek Primitive Medicine," in *Clio medica,* **3** (1968), 305–336; J. Jouanna, "Le médecin Polybe est-il l'auteur de plusieurs ouvrages de la Collection hippocratique?," in *Revue des études grecques,* **82** (1969), 552–562; H. de Ley, "De samenstelling van de Pseudo-Hippokratische brievenversameling en haar plaats in de traditie," in *Handelingen der K. Zuidnederlandse Maatschappij voor Taal-en Letterkunde en Geschiedenis,* **23** (1969), pp. 47–80; F. Kudlien, "Medical Ethics and Popular Ethics in Greece and Rome," in *Clio medica,* **5**, no. 2 (1970), 91–121; and G. Vlastos, "Cornford's Principium Sapientiae," in D. J. Furley and R. E. Allen, eds., *Studies in Presocratic Philosophy* (London–New York, 1970), pp. 42–55, repr. from *Gnomon,* **27** (1955), 65–76.

ROBERT JOLY

HIRAYAMA, KIYOTSUGU (*b.* Miyagi prefecture, Japan, 13 October 1874; *d.* Tokyo, Japan, 8 April 1943), *celestial mechanics.*

Hirayama graduated in 1896 from the University of Tokyo, where he continued his graduate studies in astronomy. He subsequently became an assistant professor and later a full professor at the university and was simultaneously a staff member of the Tokyo Astronomical Observatory.

In 1915 Hirayama went to the United States and studied celestial mechanics under Ernest W. Brown at Yale and ephemerides at the U.S. Naval Observatory in Washington, D.C. At Brown's suggestion that a key to the problems of celestial mechanics lies in the movements of the asteroids and satellites, Hirayama worked on an explanation of the condensations and gaps of the distribution of the mean motions of asteroids. He thought that the condensations were caused by the destruction of a planet. He called a condensation (similar group) a "family" and theorized that each member of a family would have similar eccentricity, inclination, and mean motion (or orbital semimajor axis).

Among the 790 orbits of asteroids presented in the *Berliner astronomisches Jahrbuch* for 1917, Hirayama in 1918 identified three asteroid families; the number later increased to five. In 1919 he identified thirty-one asteroids of the Themis family, thirty-eight of the Eos family, twenty-three of the Koronis family, sixteen of the Maria family, and eighty-one of the Flora family.

Based on statistics as well as on the known principles of celestial mechanics, Hirayama's hypothesis was a rare theoretical accomplishment, considering the level of research in astronomy in Japan at the time. His other achievements are in latitudinal change, variable-star theory, and the history of Oriental astronomy.

BIBLIOGRAPHY

Articles by Hirayama are "Groups of Asteroids Probably of Common Origin," in *Astronomical Journal,* **31** (1918), 185–188; and "Notes on an Explanation of the Gaps of the Asteroidal Orbits," *ibid.,* **38** (1928), 147–148.

On Hirayama and his work, see Yusuke Hagihara, "Hirayama Kiyotsugu sensei o shinobite," in *Tenmon Geppō,* **36,** no. 6 (1943), 65–67; and "Hirayama Kiyotsugu sensei no omonaru kenyu ronbun," *ibid.,* 67–68.

S. NAKAYAMA

HIRN, GUSTAVE ADOLFE (*b.* Logelbach, near Colmar, France, 21 August 1815; *d.* Colmar [then Germany], 14 January 1890), *thermodynamics.*

Despite his name and inclusion in German biographical works, Hirn must be considered a Frenchman, having been born in Alsace before the annexation to Germany (1871), and having published almost exclusively in French journals. Furthermore, he must not be confused with his older brother Charles Ferdinand, inventor of the wire-rope power transmission, which Hirn perfected. The brothers, sons of a calico factory owner, were educated at home by a tutor. Hirn received some additional instruction in chemistry and physics, which he augmented through independent study. The brothers became technical directors of a mill of which the principal power source was a Boulton & Watt engine built in 1824. In the process of modernizing and meeting the necessity of driving machinery in separate buildings with the power of one engine, C. F. Hirn devised the wire-rope power transmission—the telodynamic system—an important industrial element in the latter half of the nineteenth century.

Hirn, in charge of the mechanical department, improved the economy of the steam plant by using flue gases to heat the boiler feedwater. Extensive experiments on friction, occasioned by the many lubrication problems of the mill machinery, led to the introduction of the use of mineral oil, formerly considered unfit for machinery, and the establishment of a small machine-oil business.

Hirn became one of the first to investigate the internal phenomena of the steam engine. He had five engines at his disposal, including two of about 100 horsepower—a Woolf compound and a single-cylinder machine. In 1847 he discovered the mechanical equivalent of heat (later than, but independently of, J. R. Mayer and Joule, whose priorities of 1842 and 1843 he acknowledged).

In continuing work directed at increasing the efficiency of the mill engine, Hirn established the first heat balance (1854). He showed the beneficial advantage of superheat over dry saturated steam in reducing cylinder condensation; the calculation is still known as Hirn's analysis. Furthermore, he convinced skeptics of the advantage of steam-jacketing cylinders, a practice that had been discontinued when pressures, temperatures, and speeds had risen significantly above those of Watt's era: he proved decisively that cylinder walls were active "thermal reservoirs." Hirn's *Exposition analytique et expérimentale de la théorie mécanique de la chaleur* (1862) was among the first systematic treatises on thermodynamics.

Hirn's activity was not limited to thermodynamics; the climatology and meteorology of Alsace were an interest of his later years. Before the Franco-Prussian War he had established a number of weather stations that reported to him. After his retirement in 1880, he was able to resume this work through support from the Institut de France, instrumentation for the major observatory atop his house being furnished by the Ministry of Public Works in Paris. Astronomy also engaged him, a paper on Saturn's rings causing much discussion.

BIBLIOGRAPHY

I. Original Works. Hirn's writings include *Recherches sur l'équivalent mécanique de la chaleur* (Colmar, 1858), repr. in Keller (see below), p. 29; *Exposition analytique et expérimentale de la théorie mécanique de la chaleur* (Paris, 1862; 3rd ed., 1875–1876); *Analyse élémentaire de l'univers* (Paris, 1868); "La musique et l'acoustique," in *Revue d'Alsace* (1878); *Thermodynamique* (Paris, 1881); *La vie future et la science moderne* (Paris, 1882); and *Constitution de l'espace céleste* (Paris, 1889).

He also published many articles in *Comptes rendus hebdomadaires des séances de l'Académie des sciences; Bulletin de la Société industrielle de Mulhouse; Bulletin de l'Académie r. de Belgique. Classe des sciences; Journal universel de la littérature, des sciences et des arts* (Brussels); *Gaea;* and *Bulletin de la Société d'histoire naturelle de Colmar.*

II. Secondary Literature. Biographies and obituaries of varying length are Dwelschauvers-Dery, "Reminiscences of the Life of G. A. Hirn," in *Engineering,* **49** (1890), 120–121, 174–175; "G. A. Hirn," in *Engineer,* **69** (1890), 231–234; A. Slaby, "John Ericsson und Gustav Adolf Hirn," in *Zeitschrift des Vereins deutscher Ingenieure,* **34** (1890), 1161–1168; "Gustav Adolf Hirn," *ibid.,* 232–233; Keller, "Gustav Adolf Hirn, sein Leben und seine Werke," in C. Matschoss, ed., *Beiträge zur Geschichte der Technik und Industrie,* III (Berlin, 1911), 20–60; and C. Matschoss, *Männer der Technik* (Düsseldorf, 1925), p. 117.

R. S. Hartenberg

HIRSZFELD, LUDWIG (*b.* Warsaw, Poland, 5 August 1884; *d.* Wrocław, Poland, 7 March 1954), *serology, bacteriology.*

After attending the Gymnasium in Lodz, Hirszfeld, born into a Jewish family and later a convert to Catholicism, decided to study medicine in Germany. In 1902 he entered the University of Würzburg and transferred in 1904 to Berlin, where he attended lectures in medicine and philosophy. Hirszfeld completed his doctoral dissertation, "Über Blutagglutination," in 1907, thus taking the first step in what was to become his specialty. But first he became a junior assistant in cancer research at the Heidelberg Institute for Experimental Cancer Research, where E. von Dungern was his department head. Hirszfeld soon formed a close personal friendship with Dungern which proved to be scientifically fruitful. At Heidelberg they did the first joint work on animal and human blood groups which, in 1900, had been identified as isoagglutinins by Karl Landsteiner.

Hirszfeld gradually found the working conditions at Heidelberg too confining. He wished to familiarize himself with the entire field of hygiene and microbiology, so in 1911 he accepted an assistantship at the Hygiene Institute of the University of Zurich, just after he had married. His wife, also a physician, became an assistant at the Zurich Children's Clinic under Emil Feer.

In 1914 Hirszfeld was made an academic lecturer on the basis of his work on anaphylaxis and anaphylatoxin and their relationships to coagulation; he was also named *Privatdozent.* When World War I broke out Serbia was devastated by epidemics of typhus and bacillary dysentery. In 1915 Hirszfeld applied for duty there. He remained with the Serbian army until the end of the war, serving as serological and bacteriological adviser. At this time, in the hospital for contagious diseases in Thessaloniki he discovered the bacillus *Salmonella paratyphi* C, today called *Salmonella hirszfeldi.*

After the end of the war Hirszfeld and his wife returned to Warsaw, where he established a Polish serum institute modeled after the Ehrlich Institute for Experimental Therapy in Frankfurt. He soon became deputy director and scientific head of the State Hygiene Institute in Warsaw and, in 1924, professor there. In 1931 he was named full professor at the University of Warsaw and served on many international boards. After the occupation of Poland by the

German army Hirszfeld was dismissed as a "non-Aryan" from the Hygiene Institute but, through the protection of friends, managed to do further scientific work at home until February 1941; it was, however, almost impossible for him to publish.

On 20 February 1941 Hirszfeld was forced to move into the Warsaw ghetto with his wife and daughter. There he organized anti-epidemic measures and vaccination campaigns against typhus and typhoid, as well as conducting secret medical courses. In 1943 he and his family fled the ghetto and were able to survive underground through using false names and continually changing their hiding place; his daughter died of tuberculosis in the same year.

When a part of Poland was liberated in 1944, Hirszfeld immediately collaborated in the establishment of the University of Lublin and became prorector of the university. In 1945 he became director of the Institute for Medical Microbiology at Wrocław and dean of the medical faculty. He taught at the institute, now affiliated with the Polish Academy of Sciences and named for him, until his death.

Hirszfeld received many honors, including honorary doctorates from the universities of Prague (1950) and Zurich (1951). He wrote almost 400 works in German, French, English, and Polish, many in collaboration with other well-known scholars and not a few with his wife.

Hirszfeld and von Dungern were responsible for naming the blood groups A, B, AB, and O; previously they were known as groups I, II, III, and IV. He proposed the α and β designations for isoagglutinin. In 1910–1911 Hirszfeld discovered the heritability of blood groups and with this discovery established serological paternity exclusion. During World War I he and his wife wrote works on sero-anthropology, which brought forth fundamental findings on the racial composition of recent and historical peoples. According to his so-called Pleiades theory of blood groups, the other groups probably developed from the archaic O group in the course of evolution.

Hirszfeld was the first to foresee the serological conflict between mother and child, which was confirmed by the discovery of the Rh factor. Upon this basis he developed, in the last years of his life, an "allergic" theory of miscarriage and recommended antihistamine therapy. Hirszfeld also investigated tumors and the serology of tuberculosis. His discovery of the infectious agent of paratyphoid C had far-reaching consequences for differential diagnosis.

In 1914, together with R. Klinger, Hirszfeld developed a serodiagnostic reaction test for syphilis, which did not, however, replace the Wasserman test introduced in 1906. His studies of goiter in Swiss endemic regions brought him into sharp disagreement with E. Bircher over the theory—today widely confirmed—that endemic goiters are caused by iodine deficiency in water and food, in opposition to the hydrotelluric theory.

BIBLIOGRAPHY

I. Original Works. A complete bibliography of 394 items is in Jakob Wolf Gilsohn, "Prof. Dr. Ludwig Hirszfeld" (Munich, 1965), M.D. thesis; another may be found in *Prace Wrocławskiego towarzystwa nauko-wego,* ser. B (1956), no. 6. His most important works are "Über eine Methode, das Blut verschiedener Menschen serologisch zu unterscheiden," in *Münchener medizinische Wochenschrift,* **57** (1910), 741–742, written with E. von Dungern; "Über Nachweis und Vererbung biochemischer Strukturen," in *Zeitschrift für Immunitätsforschung und experimentelle Therapie,* **4** (1910), 531–546, written with E. von Dungern; "Über unsere Modifikation der Wassermannschen Reaktion," in *Münchener medizinische Wochenschrift,* **57** (1910), 1124–1126, written with E. von Dungern; "Über Vererbung gruppenspezifischer Strukturen des Blutes," in *Zeitschrift für Immunitätsforschung und experimentelle Therapie,* **6** (1910), 284–292, written with E. von Dungern; "Über gruppenspezifische Strukturen des Blutes," *ibid.,* **8** (1911), 526–562, written with E. von Dungern; "Epidemiologische Untersuchungen über den endemischen Kropf," in *Archiv für Hygiene und Bakteriologie,* **81** (1913), 128–178, written with T. Dieterle and R. Klinger; "Studien über den endemischen Kropf," in *Münchener medizinische Wochenschrift,* **60** (1913), 1813–1814, 1814–1816, written with T. Dieterle and R. Klinger; "Über Anaphylaxie und Anaphylatoxin und ihre Beziehungen zu den Gerinnungvorgängen," in *Vierteljahrschrift der Naturforschenden Gesellschaft in Zürich,* **59** (1914), 15–34; "Über eine Gerinnungsreaktion bei Lues," in *Deutsche medizinische Wochenschrift,* **40** (1914), 1607–1610, written with R. Klinger; "Une nouvelle réaction du sérum syphilitique: la coagula-réaction," in *Semaine médicale,* **34** (1914), 360–363, written with R. Klinger; "Aus meinen Erlebnissen als Hygieniker in Serbien," in *Korrespondenzblatt für schweizer Ärzte,* **46** (1916), 513–531; "Essai d'application des méthodes sérologiques au problème des races," in *Anthropologie,* **29** (1919), 505–537, written with H. Hirszfeld; "A New Germ of Paratyphoid," in *Lancet* (1919), **1,** 296–297; "Serological Differences Between the Blood of Different Races," *ibid.,* **2,** 675–679; *Konstitutionsserologie und Blutgruppenforschung* (Berlin, 1928); "Untersuchungen über die serologischen Eigenschaften der Gewebe," in *Zeitschrift für Immunitätsforschung und experimentelle Therapie,* **64** (1929), 61–80, 81–113, written with W. Halber and J. Laskowski; *Les groupes sanguines. Leurs applications à la biologie, à la médecine et au droit,* translated by H. Hirszfeld (Paris, 1938); "Über das Wesen der Blutgruppe O," in *Klinische Wochenschrift,* **17** (1938), 1047–1051, written with Z. Kostuch; "Sur les pleiades 'isozériques' du sang," in *Annales de l'Institut Pasteur,* **65** (1940), 251–278, 386–414, written with R. Amzel; *Historia jednego zycia* ("Story of a Life"; Warsaw, 1945); and *Probleme der Blutgruppenforschung* (Jena, 1960), with an introduction by O. Prokop.

II. SECONDARY LITERATURE. The only biography in German that covers both life and work is that by Gilsohn cited above. Obituary notices include *Lancet* (1954), **1,** 987; G. Blumenthal, in *Zentralblatt für Bakteriologie, Parasitenkunde, Infektionskrankheiten und Hygiene,* **162** (1955), 1; A. Kelus, in *Schweizerische medizinische Wochenschrift,* **84** (1954), 745; H. Schlossberger, in *Zeitschrift für Immunitätsforschung und experimentelle Therapie,* **111** (1954), 269–270; and P. Speiser, in *Wiener klinische Wochenschrift,* **66** (1954), 394–395. A short biography is F. Milgrom, "Ludwik Hirszfeld, Scientist, Teacher, Humanist," in *Polish Medical History and Science Bulletin,* **3** (1960), 51–52, Polish and English summary.

H. SCHADEWALDT

HIS, WILHELM (*b.* Basel, Switzerland, 9 July 1831; *d.* Leipzig, Germany, 1 May 1904), *anatomy, histology, embryology.*

Wilhelm His was the son of Eduard His (whose name was Eduard Ochs until 1818, when he adopted his grandmother's maiden name), a merchant in Basel and member of the court of appeals. His mother was the former Katharina La Roche. His grandfather, Peter Ochs, was a well-known Swiss politician and historian. His's son, Wilhelm, Jr., who is occasionally confused with his father, was an internist in Berlin; he discovered the atrioventricular impulse-conducting system in the heart (bundle of His).

His began to study medicine at Basel in 1849 but transferred in the winter semester of 1849–1850 to the University of Bern and then, a year later, to the University of Berlin, where he studied under Johannes Müller and the embryologist Robert Remak. He attended the University of Würzburg in 1852–1853 for his clinical training, but the theoretical subjects taught there by Rudolf Virchow, Albert von Kölliker, and Franz Leydig attracted him far more than the clinical instruction. His also studied the writings of the physiologists Carl Ludwig and Hermann Lotze. He concluded his education with visits to Prague and Vienna, where he met Ernst von Brücke and the pathologist Karl Rokitansky. In the summer of 1854 he passed the physicians' examination in Basel and thus had the right to practice medicine, but not until 1855 was he able to present his dissertation, which dealt with the normal and pathological histology of the cornea. In writing it His drew upon investigations that he had begun—at Virchow's suggestion—while a student at Würzburg.

His spent the winter of 1855–1856 in Paris, where he was concerned primarily with chemistry. He attended the lectures given by Claude Bernard and met Berthelot and Brown-Séquard. Especially important for His at this time were the friendship and aid of two Swiss friends, the ophthalmologist Johann Friedrich Horner and the physicist Eduard Hagenbach. His returned to Basel, did research on his own, and in the winter semester of 1856–1857 qualified as a lecturer in anatomy and physiology. Hoping to become an assistant in the clinic of the Berlin ophthalmologist Albrecht von Graefe, His spent the summer of 1857 there as a visitor, working on the histology of the eye. During his stay he formed a close friendship with Theodor Billroth. In the fall of 1857 His was called to Basel to succeed Georg Meissner as professor of anatomy and physiology. Here, beginning in 1863, he was a member of the city parliament responsible for reorganizing the city following typhus and cholera epidemics; in 1865–1866 he was adviser on sewerage, cemeteries, and school hygiene.

In 1872 His succeeded Ernst Heinrich Weber in the chair of anatomy at the University of Leipzig. Like His in Basel, Weber had, until 1865, taught both anatomy and physiology; but in the latter year Carl Ludwig was named to the newly established chair of physiology. His's first major task was to have a new anatomy laboratory built. When it opened in April 1875, it was one of the most modern laboratories for theoretical medicine.

His was vice-chancellor of the University of Basel in 1869–1870 and of the University of Leipzig in 1882–1883. He was a cofounder and several times president of the German Anatomical Society, perpetual secretary of the mathematics and physics section of the Royal Saxon Society of Sciences, and member of the brain research commission of the International Union of Academies. In addition, he played a major role in the reorganization of the Society of German Scientists and Physicians after 1889. In 1891, as president of this society, he settled—impartially—the disputes provoked by the new statutes that Virchow had proposed for it.

His's scientific accomplishments lay in research and in the teaching of anatomy, histology, and embryology. In Basel he was at first concerned with histology and histochemistry, demonstrating, among other things, the existence of independent cornea cells. Other investigations dealt with the lymph vessels and the lymph glands and with the thymus. In 1863 he discovered the nerve plexus in the adventitia of the vessels. Toward the end of his life, when investigating embryological questions, he once again confronted histological and cytological problems. His contributed clearly formulated conceptions and doctrines to the vigorous debate over cell structure and cell division. He was especially concerned with the concept of amitosis, whose existence, in the literal sense of the word, he disputed.

In his investigations of the lymphatic system His arrived at certain ideas concerning the emergence of the body's cavities and of their boundary layers. These ideas led to his first important embryological work, *Die Häute und Höhlen des Körpers* (1865). He began his account with Remak's description of a third germ layer. Comparing it with the two others, His concluded that only those cavities formed of tissues arising from the mesoderm—vessels, serous cavities, joint cavities, synovial cavities, and connective tissue interstices—should be designated "bodily cavities." He contrasted the genuine epithelia of ectodermal and entodermal origin with the "nongenuine" epithelia, to which he gave the name they still bear, "endothelia." The latter are most distinct in the blood vessels; they are totally absent in the interstices of the connective tissues. At this period the question of the structure and origin of the connective tissue was still unresolved. Studying it from the embryological point of view, His provided new insights. Henceforth he concentrated on embryology and in the following years produced numerous descriptive works on the development of individual organs or organ systems, especially of the vertebrate central nervous system and heart. On the basis of this work he was able to show (1883) that the nerve fibers are formed through the growth of the nerve cells, which he called "neuroblasts."

His's initial reflections on the mesoblast and connective tissue led, in addition, to the theory of the "parablast." Through observations of embryonic ovaries he concluded that there is a genetic opposition between the primordial organ forms of the epithelia and the connective tissues. The latter, according to His, are formed by a distinct cell group, the "parablast," and only later grow into the embryonic disk. This theory, published in 1866, produced a sensation and led to a series of investigations the results of which at first seemed to substantiate it. Finally, though, it was realized—not least through His's own work—that the attempt to find a simple explanation of a differentiation process in early development again had to be abandoned. Following his *Über die Bildung des Lachsembryos* (1874), His furnished yet another interpretation of the processes occurring in the earliest stages of development of the vertebrate embryo: the concrescence theory. At first thought to be one of His's important discoveries, it was later seen to be untenable. Nevertheless, it proved to be fruitful, for it stimulated a great many embryological investigations. The theory asserted that at first only the rudimentary form of the head lies in the middle of the embryonic disk; the rudimentary forms of the axial portions of the body emerge on the edge of the disk and are drawn into and fused in the middle only later.

Even before the formulation of the concrescence theory His had sought to develop simple explanations of developmental processes. For example, he had published a law of growth in *Die erste Anlage des Wirbeltierleibes* (1866–1868). According to the principle of the unequal growth of the embryonic disk, which can be considered as an imperfectly elastic body, purely mechanical convolution must occur. The first primitive organ thus to emerge, he contended, is the neural furrow, which finally closes to form the neural tube. In order to reduce embryonic development to a simple law of growth, His devoted considerable attention to mathematics. His friend Eduard Hagenbach contributed to His's publication a mathematical analysis of the changes in shape that take place in an oval disk as a result of an oriented unevenness in its growth. On the basis of this conception of the general factors governing early development—according to which each organ arises at a specific location on the embryonic disk and, ultimately, on the unfertilized egg—His arrived at the principle of the embryo-forming regions.

His was thus the first scientist who sought to provide a causal explanation—in the modern sense—of embryonic development. Adolf Fick, the Swiss physiologist to whom His presented his theory of a law of growth, tried to persuade him to buttress his hypotheses with experiments; but His was reluctant to do so. Even later he clung to the method of analyzing and interpreting descriptive findings. Consequently it is not His, but Wilhelm Roux, who is considered the founder of the causal-analytic approach to embryology.

His presented his ideas on the development of the vertebrate embryo in a book dedicated to his Leipzig colleague Carl Ludwig: *Unsere Körperform und das physiologische Problem ihrer Entstehung. Briefe an einen befreundeten Naturforscher* (1874). The letters referred to in the subtitle were addressed to His's nephew Johann Friedrich Miescher, the discoverer of nucleic acid. This publication faithfully characterizes the situation then existing in ontogenetic research and theory building. It was the period of controversy over the interpretation of developmental processes, a controversy between the new mechanistic-physiological approach and the phylogenetic point of view. The debate over the so-called "biogenetic law" was especially intense. Ernst Haeckel had declared that this law, recently discovered by Fritz Müller, was the unique foundation of embryology. The fourteenth letter in *Unsere Körperform,* a not completely unpolemical rebuttal of Haeckel's theory, bore the title "Die Erklärung organischer Körperform durch das Descendenzprinzip, das 'biogenetische Grundgesetz' und seine Begründung. Unmittelbare und mittelbare Erklärung."

His was not alone in his rejection of Haeckel's view. The zoologist Alexander Goette also emphasized the difference between phylogenetic and physiological explanations of ontogeny. Unlike many of Haeckel's other opponents, Goette and His were not anti-Darwinists, yet Haeckel presented them as such in his unusually harsh polemical reply to the findings and interpretations advanced—in a generally unpolemical manner—against his position (*Ziele und Wege der heutigen Entwicklungsgeschichte* [1875]).

His's work in embryology displayed new methodological conceptions. In 1866 he constructed the first microtome, which furnished an uninterrupted series of sections. He also built an embryograph, a prismatic drawing apparatus which permitted him to make exact drawings of the microscopic sections. In addition, His encouraged the introduction of photography into anatomy and the use of lantern slides in the classroom. His series of wax models of the development of fishes, chickens, and man became especially well-known. His's use of topographical anatomy in his models also constituted an advance in anatomical instruction. And in connection with his presentation of human embryos (1880–1885) he introduced standardized charts into embryology.

His took his teaching duties very seriously; his address as vice-chancellor (1882) was entitled *Über Entwicklungsverhältnisse des akademischen Unterrichts.* His's efforts in anatomical nomenclature resulted in *Nomina anatomica,* which appeared in 1895, after many years' work. Active in anthropology as well, he and Ludwig Rütimeyer wrote *Crania Helvetica* (1864). In 1895, under His's supervision, the remains of Johann Sebastian Bach were identified and his burial place, hitherto lost and unknown, was rediscovered.

BIBLIOGRAPHY

I. Original Works. A complete list of His's publications is in the obituary notice by Fick (see below). The most important are *Die Häute und Höhlen des Körpers* (Basel, 1865); *Untersuchungen über die erste Anlage des Wirbeltierleibes* (Leipzig, 1868); *Unsere Körperform und das physiologische Problem ihrer Entstehung. Briefe an einen befreundeten Naturforscher* (Leipzig, 1874); *Anatomie menschlicher Embryonen,* 3 vols. (Leipzig, 1880–1885); *Die anatomische Nomenklatur* (Leipzig, 1895); and *Johann Sebastian Bach, Forschungen über dessen Grabstätte, Gebeine und Antlitz* (Leipzig, 1895).

II. Secondary Literature. See Rudolf Fick, "Wilhelm His," in *Anatomischer Anzeiger,* **25** (1904), 161–208; Wilhelm His, Jr., *Wilhelm His der Anatom. Ein Lebensbild* (Berlin, 1931); Eugen Ludwig, ed., *His der Ältere. Lebenserinnerungen und ausgewählte Schriften* (Basel, 1965); and *Unsere Körperform* (see above), letters 2, 13, and 14.

Hans Querner

HISINGER, WILHELM (*b.* Skinnskatteberg, Västmanland, Sweden, 23 December 1766; *d.* Skinnskatteberg, 28 June 1852), *chemistry, mineralogy, geology, paleontology.*

Hisinger's parents were Vilhelm Hising, a wealthy ironworks proprietor, and Barbara Katarina Fabrin. After his father died he was adopted by his uncle, John Hisinger. In 1786 he enrolled at the Bergskollegium; but he soon retired to estates at Baggå and Skinnskatteberg inherited from his father, where he ran the foundry and worked privately as a scientist. He was particularly familiar with the achievements of Linnaeus, Wallerius, Cronstedt, and Bergman. Hisinger became a member of the Royal Swedish Academy of Science (1804) and subsequently of a number of other learned societies. In 1788 he married Baroness Anna Märta Eleonora Taube.

Hisinger's first published paper appeared anonymously in 1789. His second work, *Samling till en minerografi öfver Sverige* (Stockholm, 1790), also published anonymously, was later enlarged (1808) and translated into German (1819, 1826). In association with Berzelius, Hisinger discovered the element cerium (1803) and investigated the effect of electric current on salt solutions (1806), thus preparing the way for the electrochemical theories of Davy and Berzelius. These studies, along with a number of investigations of the chemistry of Swedish minerals and rocks, were published in *Afhandlingar i Fysik, Kemi och Mineralogi* (1806–1818). His other contributions in this field appeared in the *Kungliga Svenska vetenskapsakademiens handlingar* (1811–1823).

Hisinger traveled widely in Sweden and Norway and he was a keen observer of the geology of the districts he visited. His observations were laid down in *Anteckningar i physik och geognosie under resor uti Sverige och Norrige* (7 fascicles [Stockholm, 1818–1840]). A further outcome of his travels was a geological map of south and middle Sweden (1832) that was enlarged to include an account of Swedish rocks with a list of their localities (Stockholm, 1834). He also wrote a *Handbok för mineraloger under resor i Sverige* (Stockholm, 1843).

In *Lethaea Svecica, seu petrificata Sveciae, iconibus et characteribus illustrata* (Stockholm, 1837, 1840, 1841), Hisinger depicted the known animal and plant fossils from Swedish deposits. Like many of his other works, this book was produced at Hisinger's own expense. His comprehensive geological, mineralogical, and paleontological collections were given to the Swedish Museum of Natural History in Stockholm.

BIBLIOGRAPHY

A short biography of Hisinger and a bibliography of his published works are in *Kungliga Svenska vetenskapsakademiens handlingar för år 1852* (1854), pp. 385–391. A bibliography also appears in Poggendorff, I, cols. 1111–1112.

GERHARD REGNÉLL

HITCHCOCK, EDWARD (*b.* Deerfield, Massachusetts, 24 May 1793; *d.* Amherst, Massachusetts, 27 February 1864), *geology.*

Hitchcock wrote extensively on the relation of science to religion, on the geomorphology of the Connecticut River Valley, on fossil tracks of extinct vertebrates, and on the metamorphosis of sediments.

The son of Justin Hitchcock and Mercy Hoyt, Hitchcock was born into a pious, respected, but poor family. Because of his father's meager income from the hatter's trade, Edward worked his way through Deerfield Academy, where his observations on the comet of 1811 marked the beginning of a lifelong career in science. While Hitchcock was preceptor of the academy (1815–1819), his eyes weakened, ending his brief career in astronomy. An encounter with Amos Eaton at Amherst, sometime during Hitchcock's Deerfield tenure, and an exchange of letters and minerals with Benjamin Silliman of Yale in 1817, turned his interest to natural history. Yale awarded him an honorary master of arts in 1818, Harvard an honorary doctor of laws in 1840, and Middlebury College an honorary doctor of divinity in 1846. Following a short-lived conversion to Unitarianism as a youth, he settled into his family's faith, Congregationalism. He studied theology at Yale in 1820(?) and acted as pastor in Conway, Massachusetts, from June 1821 to October 1825.

He was professor of chemistry and natural history at Amherst College from 1825 to 1845, having prepared himself for science-teaching by auditing Silliman's courses at Yale (October 1825–January 1826). In 1840 he co-founded, with other state geologists, the American Association of Geologists, parent of the American Association for the Advancement of Science, and in 1863 he became a charter member of the National Academy of Sciences. Hitchcock was president of Amherst College from December 1844 to November 1854 and also taught natural theology and geology there from 1845 until his death. His marriage on 13 June 1821 to Orra White resulted in six children, two of whom, Charles Henry and Edward, Jr., chose scientific careers; Charles became state geologist for Vermont, Maine, and New Hampshire. Mrs. Hitchcock illustrated her husband's works and assisted him in scientific enterprises. Despite a careful diet and zealous devotion to temperance, Hitchcock suffered from chronic intestinal and gall bladder complaints so debilitating as to hamper his geological fieldwork.

Hitchcock's scientific textbooks and popular works stemmed largely from his teaching at Amherst, while his technical geological monographs grew mostly from his appointments as state geologist of Massachusetts (1830–1833, 1837–1841) and of Vermont (1856–1861). Hitchcock was appointed in 1836 to the New York State Survey and although he resigned after only a month, his influence was evident in the survey's outcome. John Dix, who organized the survey, adopted the zoological and botanical features that characterized Hitchcock's earlier geological survey of Massachusetts, on which Hitchcock wrote a final report, *Geology of Massachusetts* (1833), the first of its kind. Dix's report, patterned after Hitchcock's, was expanded by the scientists of the New York survey into a twelve-volume work that marked a major development in geological science.

Hitchcock saw his task in Wernerian terms, using German rocks from Heidelberg in a collection at Amherst to identify the New England stratigraphic equivalents by comparing lithologies. As early as 1824 he had provided an appendix to Eaton's *A Geological and Agricultural Survey of the District Adjoining the Erie Canal,* in which he offered a different geological interpretation of the section across Massachusetts from the New York line to Boston.

Although he personally admired Charles Lyell, Hitchcock predicated his geological explanations upon the changing intensity over time of agents such as glaciation and flooding, rather than upon constant, gradual operation of such forces. He did not accept those explanations, however, which were based upon causative forces no longer in operation in the modern world. Hitchcock has therefore been proclaimed both a catastrophist and a uniformitarian. He gradually introduced a partial glacial hypothesis into his geomorphological studies, becoming a cautious admirer of Louis Agassiz after a trip to European glacial sites in 1850 convinced him of the erosional power of ice. But his important study of terraces in the Connecticut River Valley, which culminated in his *Illustrations of Surface Geology* (1857), demonstrated his reluctance to abandon entirely fluviate geology for glacial theories. In contrast to his conservatism on glacial issues, Hitchcock was an early American advocate of the thesis that heat and pressure gradually changed sediments into schist and thence possibly to granite, a theory which he felt most adequately explained his observations on metamorphosed New England conglomerates.

Hitchcock's work in paleontology focused almost exclusively on the huge footprints left by vertebrates in the Triassic sandstone of the Connecticut River Valley; he argued that these tracks, which have since been attributed to dinosaurs, were made by ancient birds. In all of his writings, particularly in *Religion of Geology* (Boston, 1851) and his sermons, Hitchcock supported a unified truth rather than a theology separate from science. In his teaching, writing, and preaching, he conceived a transcendental vision of a beneficent God more comprehensible from a fusion of theological and natural studies than from their division into separate compartments of knowledge.

BIBLIOGRAPHY

I. Original Works. Charles Henry Hitchcock lists Hitchcock's published work, including his play, newspaper articles, sermons, popular treatises, technical articles and books, and textbooks in "Edward Hitchcock," in *American Geologist,* **16** (1895), 139–149; esp. valuable are the *American Journal of Science* articles listed there. See also *Report on the Geology of Massachusetts* (Amherst, 1833); *Final Report on the Geology of Massachusetts,* 2 vols. (Amherst–Northampton, 1841); *Illustrations of Surface Geology,* Smithsonian Contributions to Knowledge, vol. IX (Washington, D.C., 1857); *Ichnology of New England,* 2 vols. (Boston, 1858); supp. on fossil footprints (Boston, 1865); and *A Report on the Geology of Vermont,* 2 vols. (Claremont, 1861).

Hitchcock's best-selling textbook, *Elementary Geology,* ran through several eds. and over thirty printings from 1840 until his death, and is thus a valuable index to his changes of thought on geological issues.

II. Secondary Literature. Hitchcock lacks an adequate biography, and therefore researchers must piece together his life from a patchwork of sources. Use of the Edward Hitchcock Papers, Special Collections Room, Robert Frost Library, Amherst College, is imperative for an accurate and complete assessment. In addition to the published biographies listed in Max Meisel, *A Bibliography of American Natural History, the Pioneer Century, 1769–1865* (Brooklyn, N.Y., 1924–1929), I, 195, and George Merrill, "Edward Hitchcock," in *Dictionary of American Biography,* see Benjamin Silliman's manuscript "Reminiscences, 1792–1862," V, Beinecke Library, Yale University; George Sheldon, *A History of Deerfield, Massachusetts* (Deerfield, 1895–1896), pp. 848–849; and Edwin H. Colbert, *Men and Dinosaurs, the Search in Field and Laboratory* (New York, 1968), pp. 37–41.

Until a similar work is written on American geologists, Charles Coulston Gillispie provides the best intellectual context for evaluating Hitchcock in *Genesis and Geology, a Study in the Relations of Scientific Thought, Natural Theology, and Social Opinion in Great Britain, 1790–1850* (Cambridge, 1951; repr. New York, 1959).

Michele L. Aldrich

HITTORF, JOHANN WILHELM (*b.* Bonn, Germany, 27 March 1824; *d.* Münster, Westphalia, Germany, 28 November 1914), *chemistry, physics.*

Hittorf is remembered primarily for his experimental work in the transport of charge by ions in electrolytic solutions and the study of electrical conduction through gases. His father was a merchant in Bonn, where Hittorf was educated and where, after a short period at the University of Berlin, he received the doctorate in 1846, having studied under Julius Plücker. A year later he became a member of the staff at the University of Bonn and at the same time he received a call, which he accepted, to become *Privatdozent* at the Royal Academy of Münster. When the school became a university, he was appointed professor of chemistry and physics (1852–1876). In 1876 a reorganization of the university permitted him to drop chemistry and he became professor of physics. Ill health made it necessary for him to retire in 1890 and he was named professor emeritus.

Hittorf had a reserved nature and professional advancement was slow. Recognition came to him only late in life. He was a corresponding member of the scientific societies of Göttingen, Berlin, and Munich and foreign member of the Royal Danish Academy of Science and Letters, as well as other foreign societies. In 1897 he received the Prussian order *pour le mérite* for science and arts. In 1898 he was elected honorary president of the Deutsche Elektrochemische Gesellschaft. Hittorf never married and lived with his younger, unmarried sister who kept house for him.

Hittorf's experimental contributions to physical science included early researches on the allotropic forms of selenium and phosphorus (1851–1865); investigations of the variations in the concentrations of electrolytes during electrolysis (1853–1859); the discovery, with Plücker, of the presence of both band and bright line spectra in the discharges of electricity through a gas at low pressures (1865); the examination and description of gaseous discharges and cathode-ray phenomena (1869–1884); and investigations of the passivity of metals (1900).

After Faraday's experimental investigations in 1834, it was accepted that the electricity passing through an electrolytic cell was carried by the movement of charged ions produced from the decomposition of the compounds making up the solution. Daniell had extended these ideas in 1839 and showed that salts were compounds not of acid anhydrides and metallic oxides as had been thought, but of metallic cations and elemental or compound acid anions. Believing that the conductivity of solutions was due to these ions, he began a study of their transference.

In 1853 Hittorf took up the problem. He extended the ideas of Daniell by reasoning in the following manner: Cations and anions exist in solutions and migrate under the influence of current through the solution. The migration of the cation toward the cathode and away from the anode, and the deposition of the anion on the positive electrode, together result in a decrease of the salt in the neighborhood of the anode. A similar analysis shows that there is also a decrease in the concentration of the salt in the neighborhood of the cathode. If the motion of the two dissimilar ions were the same, the decrease in the concentration of the salt would be the same at the two electrodes.

Hittorf developed an experimental technique which allowed him to measure the changes in concentration at the two electrodes and found that they were not the same. He concluded that the speeds of migration of the cation and the anion were different and he characterized this fact by defining "transport numbers," which specified the portion of the transport of electricity carried by each ion. He formulated the following ratios:

$$\frac{\text{Decrease of concentration around anode}}{\text{Decrease of concentration around cathode}} = \frac{\text{Speed of cation}}{\text{Speed of anion}} = \frac{u}{v}.$$

$$\text{Electricity transported by cation} = \frac{u}{u+v}.$$

$$\text{Electricity transported by anion} = \frac{v}{u+v}.$$

Hittorf called the ratios on the right transport numbers. He based his analysis on the theory of the structure of a solution proposed by Rudolph Clausius in 1857. Clausius postulated that a small number of ions are always present in a solution as the result of random thermal collisions between molecules that are occasionally energetic enough to cause a breaking of the compound into its positive and negative ions. These ions provide the transport of electric charge across the solution in the manner described by Hittorf.

Hittorf's experimental results and his interpretation of them were not immediately accepted by chemists. But Friedrich Kohlrausch used his analysis as the basis for further work in measuring the conductivity of solutions (1874), and Hittorf's research was ultimately very influential in the advances made by Arrhenius in proposing the electrolytic dissociation theory (1887).

Although the conduction of electricity through gases as well as through solutions was first studied by Faraday (1838), little other than descriptive information was available until better vacuum pumps were built. In 1855 Geissler, a glassblower and instrument maker in Bonn, built a mercury pump with which he was able to devise the first Geissler tube. In 1858 Plücker reported the results of a study at lower pressures using Geissler tubes. He found that the cathode glow would follow the "lines of force" of a magnetic field and that the glass walls of the tube fluoresced near the cathode, this fluorescence also moving in response to a magnetic field.

Following Plücker's studies, in 1869 Hittorf began a series of investigations of the discharge phenomena. He verified the effect of the magnetic field on the glow discharge and the fluorescence of the glass tube itself. He found that any solid or fluid body, whether an insulator or a conductor, when placed in front of the cathode cut off the glow. By constructing an L-shaped tube with the electrodes at the two ends, Hittorf was able to establish that the glow was confined to the arm in which the cathode was located. He concluded that the glow was generated from a point cathode and traveled in straight lines. "We will therefore speak of rectilinear paths or rays of glow, and consider any point of the cathode as the source of a cone of rays" ("Ueber die Elektricitätsleitung der Gase," in *Annalen der Physik und Chemie,* vol. **136,** pt. 1 [1869]). These results led to the brilliant researches on gaseous conduction by Crookes ten years later (1879) and the eventual identification of the cathode rays as electrons by J. J. Thomson (1897).

BIBLIOGRAPHY

I. ORIGINAL WORKS. No complete collection of Hittorf's works has been made. The series of four articles on ions, "Ueber die Wanderungen der Ionen während des Elektrolyse" (1853–1859), were reprinted in nos. 21 and 23 of Ostwald's Klassiker der Exakten Wissenschaften and the first was republished, with memoirs by Faraday and F. Kohlrausch, as "On the Migration of Ions During Electrolysis," in Harry Manly Goodwin, *The Fundamental Laws of Electrolytic Conduction* (1899).

II. SECONDARY LITERATURE. Discussions of Hittorf's work with solutions are in Wilhelm Ostwald, *Elektrochemie, ihre Geschichte und Lehre* (1896); and Harry C. Jones, *The Theory of Electrolytic Dissociation and Some of Its Applications* (1900). For a discussion of his contributions to the conduction of electricity through gases, see David L. Anderson, *The Discovery of the Electron* (Princeton, 1964), pp. 23, 54. A biography is G. C. Schmidt, *Wilhelm Hittorf* (Münster, 1924).

OLLIN J. DRENNAN

HITZIG, (JULIUS) EDUARD (*b.* Berlin, Germany, 6 February 1838; *d.* Luisenheim zu St. Blasien, Germany, 20 August 1907), *neurophysiology, psychiatry.*

Hitzig came from a distinguished Jewish family. His grandfather, also called Julius Eduard, was a well-known criminologist, publisher, and author. His father, Friedrich, was a renowned architect who designed several important buildings in Berlin, where a street bears his name. An uncle was the art historian Franz Kugler and a cousin the famous chemist and Nobel Prize winner Adolf von Baeyer.

At first Hitzig studied law but soon transferred to medicine and attended the medical school at Berlin, and for a time that at Würzburg. He had the good fortune to be taught by du Bois-Reymond, Virchow, Moritz Heinrich Romberg, and Carl Friedrich Otto Westphal. He received the M.D. in 1862 and thereafter began to practice internal medicine in Berlin; his *Habilitation* was completed in 1872.

As a result of an outstanding contribution to the knowledge of the function of the cerebral cortex, Hitzig acquired an international reputation, and in 1875 he was offered the directorship of the Burghölzli mental asylum near Zurich and a professorship at the University of Zurich. He accepted this academic advancement but retained the post for only four years. Unfortunately, his constant conflicts with the administrator of the asylum caused a national scandal. Hitzig was replaced by Forel, who described the chaotic state of the institution at the time of his arrival there—a state due mainly to Hitzig's personality. Hitzig's intentions were laudable and honorable—he tried to run the asylum well, improved the staff, dispensed with tipping, and founded a society for the relief of psychiatric patients—but his vanity and arrogance offended others and his lack of insight induced a feeling of martyrdom in himself.

From Zurich, Hitzig went to the University of Halle as professor of psychiatry and director of the Nietleben mental asylum. In 1885 he became director of a new psychiatric clinic at Halle, from which he was forced to retire on 1 October 1903, because of progressive diabetic optic atrophy, which led to blindness. He also suffered from gout. In 1866 he married Etta, daughter of the Marburg theologian Ernst Ranke, and she collaborated with Hitzig in the preparation of a book on food catering in neuropsychiatric hospitals. They had no children.

A stern, disagreeable man who displayed no emotion, Hitzig was rigid and unfriendly and seemed to enjoy being abrupt and caustic to others. He believed that controversy was essential in the scientific world and necessary for progress. Toward the end of his life he retired reluctantly from his polemics. His personality traits have been succinctly summarized as "incorrigible conceit and vanity complicated by Prussianism." His publications contain bitter polemics foreign to modern medical literature, and he was always willing to battle for a cherished notion. Opponents such as Hermann Munk and Goltz, who attacked him with equal brutality, were usually forthcoming. On the whole it was the physiologists with whom he waged constant warfare, for clinicians seemed more willing to accept his ideas.

Despite an apparently forbidding personality, it is said that Hitzig extended fatherly advice and friendship to his students and his home in Halle was a center of hospitality. He did not, however, leave a school. His contributions to medicine were to the physiology of the cerebral cortex and to clinical psychiatry.

At the beginning of the nineteenth century the phrenologists had postulated that the surface of the brain, both the cerebral and cerebellar hemispheres, was divided into specific, functional areas. But the experiments of Flourens, later shown to be erroneous, had been accepted by almost everyone as disproving this theory. In the 1860's, however, clinical evidence brought forward by Simon Alesandre Ernest Aubertin (echoing the earlier contentions of his father-in-law Jean Baptiste Bouillaud), by Broca, and in particular by J. Hughlings Jackson seemed to refute Flourens's conclusions. In 1870 Hitzig, in collaboration with Fritsch, showed conclusively that electrical stimulation of certain areas of the cerebral cortex in the dog produced movements of the contralateral limbs and that removal of these areas led to weakness of the same limbs. Their revolutionary investigations, said to have been carried out in Hitzig's bedroom, were the beginning of the electrophysiology of the cerebral cortex and of the experimental approach to the localization of function in it. Their paper "Ueber die elektrische Erregbarkeit des Grosshirns" (1870) is a classic of physiology.

This epoch-making research led Hitzig to investigate on his own the problem of localization in the cerebral cortex. In motor functions he indicated the scale of possible movements, ranging from the simplest variety to the complex psychomotor type. In 1874 he published the results of further experiments on electrical stimulation and ablation of the cortex. He was especially concerned with the visual cortex, the bilateral removal of which produced blindness or, if only one side was damaged, a contralateral visual-field defect. In this work he favored a precise and isolated localization of function and thus bitterly opposed Munk, who believed that the cortical centers acted together in aggregation, as did the special senses.

Hitzig also investigated experimentally the problem of the localization of mental processes in the brain. His findings favored a regional localization of intelligence in the frontal lobes, but as was the case with other early observers his techniques for the pre- and postoperative assessment of his animals were faulty; his behavioral methods were on the whole impressionistic. Moreover, he never reached a clear understanding of what "intelligence" signified. Nevertheless, his prestige as an investigator placed his views in opposition to the contentions of Goltz, who supported the holistic concept of brain function initiated by Flourens, in which all parts of the organ are considered functionally equivalent. In this work, too, Hitzig partially opposed Munk's aggregation theory, and he stated in 1874:

> Along with him [Munk] I believe that intelligence, or more accurately, the storage of ideas, is to be looked for in all portions of the cerebral cortex, or rather in all parts of the brain. I contend however that abstract thought must necessarily require specific organs and I locate these in the frontal brain (*Ueber die Funktionen der Grosshirnrinde,* p. 261).

Throughout his researches on cortical localization, Hitzig rejected the holistic theory and, although unsuccessful, spent much effort to disprove it. His experimental work was of the highest order, and he demanded precision and accuracy throughout; he accepted only exact data recorded under controlled conditions. He did not stray from the firm ground of fact and excluded philosophical or metaphysical speculation. In contrast to some of his contemporaries, the problem of the soul and its possible location found no place in his deliberations.

Hitzig's influence on the growing field of psychology in the nineteenth century was extensive, and he had a similar impact on psychiatry. Believing that the brain was the instrument of the mind, he attempted to place the treatment of psychiatric patients on a more scientific basis. In doing so he brought public attention to the problem of adequate care for these patients. He also influenced the development of psychiatry as a specialty and maintained that psychiatrists must also be fully qualified physicians.

BIBLIOGRAPHY

I. Original Works. Hitzig's more important books and papers are "Ueber die elektrische Erregbarkeit des Grosshirns," in *Archiv für Anatomie und Physiologie* (1870), pp. 300-332, written with G. T. Fritsch, English trans. in G. von Bonin, *The Cerebral Cortex* (Springfield, Ill., 1960), pp. 72-96, and discussed and excerpted in E. Clarke and C. D. O'Malley, *The Brain and Spinal Cord,* (Berkeley-Los Angeles, 1968), pp. 507-511; *Untersuchungen über das Gehirn* (Berlin, 1874), a collection of papers on physiological and pathological aspects of the brain, most of which had appeared elsewhere; *Ueber die Funktionen der Grosshirnrinde: Gesammelte Mittheilungen mit Anmerkungen* (Berlin, 1890), which contains Hitzig's attacks upon Hermann Munk; *Hughlings Jackson und die motorischen Rindencentren im Lichte physiologischer Forschung* (Berlin, 1901); *Physiologische und klinische Untersuchungen über das Gehirn. Gesammelte Abhandlungen,* 2 pts. (Berlin, 1904), of which the first, "Untersuchungen über das Gehirn," contains papers he had published in the 1874 collection of reprinted papers, and the second, "Alte und neue Untersuchungen ueber das Gehirn. Gesammelte Abhandlungen," contains the results of his experiments on the optic pathways; and *Welt und Gehirn* (Berlin, 1905).

II. Secondary Literature. The best biographies are those of his student R. Wollenberg, "Eduard Hitzig†," in *Archiv für Psychiatrie und Nervenkrankheiten,* **43** (1908), iii-xv; and T. Kirchoff, ed., *Deutsche Irrenärzte,* II (Berlin, 1924), 148-156. See also A. Kuntz, "Eduard Hitzig (1838-1907)," in W. Haymaker and F. Schiller, eds., *The Founders of Neurology,* 2nd ed. (Springfield, Ill., 1970), pp. 229-233, with portrait, not altogether reliable; August Forel, *Out of My Life and Work,* trans. by B. Miall (New York, 1937), pp. 112, 116, 124-125, 126-127; and Hans-Heinz Eulner, "Eduard Hitzig (1838-1907)," in *Wissenschaftliche Zeitschrift der Martin-Luther-Universität, Halle-Wittenberg,* **6** (1957), 709-712, which contains information on Hitzig's contact with others, his personality, and his stay at Halle, but does not discuss his work in detail.

Edwin Clarke

HJORT, JOHAN (*b.* Christiania [now Oslo], Norway, 18 February 1869; *d.* Oslo, 7 October 1948), *marine biology.*

Hjort was the son of Johan Storm Aubert Hjort, professor of ophthalmology at the University of Oslo, and Johanne Elisabeth Falsen. After completing preclinical medicine at Oslo, he shifted to zoology at Munich under Richard Hertwig and later went to the Zoological Station, Naples, where he worked on budding of the Ascidian genus *Botryllus.*

He pursued an academic life as lecturer in zoology at Olso until 1900, although his future as a practical fisheries biologist was presaged by his discovery in 1897 of stocks of the deep-sea prawn *Pandalus borealis,* a species with commercial potential in the Norwegian fjords.

He was founder and first director (1900–1916) of the Norwegian governmental fisheries (Bergen) and as such became highly regarded in fisheries affairs, not alone from the purely scientific and the applied aspects as related to the industry but also with respect to the fishermen's welfare. During this period Hjort made substantial contributions to marine knowledge with his "Fluctuations in the Great Fisheries of Northern Europe" (*Rapport et procès-verbaux,* **20** [1914]). He mapped the distribution and frequency of the pelagic eggs of the various cods as delimited by the temperature and salinity characteristics of North Atlantic water masses, aided in these findings by D. Damas; he was also responsible for locating spawning areas and hitherto neglected fishing banks. Small wonder that he was for forty-six years delegate from Norway to the Conseil Permanent International pour l'Exploration de la Mer, a vice-president of the council from 1920 to 1938, and its president for the last ten years of his life.

About 1900 the techniques of age determination in fishes by scale analysis were becoming established. With this tool at hand, Hjort, with Einar Lea, was able to delineate the age structure of the Norwegian herring population over several decades and to demonstrate the phenomenon of year-class dominance, that is, the successful survival of the young in nature, in any one year, in such numbers as to dominate the population as juveniles and adults over a succession of seasons. Hjort traced the spectacular 1904 year class of herring from 1907, when it entered the commercial fishery, to age fifteen in 1919; this age group was predominant in the catch over much of this span, particularly so in 1910 (77.3 percent), and even in 1919 it shared dominance with the good year class born in 1913.

Hjort's consuming interests led him in diverse directions: the sigmoid curve, yeasts, the "optimum catch," whales, philosophy, and politics and diplomacy. His role in international fishery matters included negotiations with Great Britain about the limits of territorial waters off Norwegian coasts, as well as dealings concerned with abatement of overfishing of whale stocks in the Antarctic (*Hvalrådets skrifter,* nos. 3, 7, 8, 9, 12, 14, 17, and 18 [1932–1938]).

In the field of marine ecology he is perhaps best known for the classic volume, coauthored with Sir John Murray, *The Depths of the Ocean,* the result of an expedition on the Norwegian research vessel *Michael Sars* in the North Atlantic (1910).

His admiration for the English and their culture, as well as the esteem in which he was held in scientific circles abroad, was reflected by his election to the Royal Societies of London and Edinburgh, the Geographical, Zoological, and Linnean Societies of London, as well as the Royal Irish Academy, the Paris Académie des Sciences, and the American Philosophical Society. He was professor of marine biology at Oslo from 1921 until reaching the age of retirement in 1939.

BIBLIOGRAPHY

I. ORIGINAL WORKS. Among Hjort's writings are *The Depths of the Ocean* (London, 1912), written with John Murray; "Fluctuations in the Year Classes of Important Food Fishes," in *Journal du Conseil,* **1,** no. 1 (1926), 5–38; "Essays on Population," in *Hvalrådets skrifter,* no. 7 (1933), pp. 5–152, with 6 plates; *The Restrictive Law of Population,* Huxley Memorial Lectures, Imperial College of Science and Technology (London, 1934); and *The Human Value of Biology* (Cambridge, Mass., 1938).

II. SECONDARY LITERATURE. Obituaries are "Prof. Johan Hjort, a Marine Biologist," in New York *Times* (9 Oct. 1948), p. 19; "Prof. Johan Hjort," in *The Times* (London) (12 Oct. 1948), p. 7; H. G. Maurice, "Prof. Johan Hjort, For. Mem. R.S.," in *Nature,* **162,** no. 4124 (1948), 764–766; K. A. Andersson, "Johan Hjort. 1869–1948," in *Journal du Conseil,* **16,** no. 1 (1949), 3–8; A. C. Hardy, "Johan Hjort," in *Proceedings of the Royal Society* (obit. notices), **7** (1950), 167; and Johan T. Rudd, "Minnetale over Professor Dr. Johan Hjort," in *Årbok 1949. Norske videnskapsakademi i Oslo* (Oslo, 1950), pp. 47–69.

DANIEL MERRIMAN

HOAGLAND, DENNIS ROBERT (*b.* Golden, Colorado, 2 April 1884; *d.* Oakland, California, 5 September 1949), *plant physiology.*

The son of Charles Breckinridge and Lillian May Burch, Hoagland spent his first eight years in Golden. During his later childhood he lived in Denver, where he attended East Denver High School. Majoring in chemistry, he graduated in 1907 from Stanford University. He accepted a position in 1908 as an instructor and assistant in the laboratory of animal nutrition at the University of California at Berkeley. In 1910

HOAGLAND

he was appointed assistant chemist in the Food and Drug Administration of the U.S. Department of Agriculture, working both in Berkeley and Philadelphia while engaged in a project evaluating the toxicity of aluminum, copper, and sulfur compounds as contaminants of canned foods and dried fruits.

In 1912 Hoagland entered the graduate school of the University of Wisconsin where he worked with E. V. McCollum in the Department of Agricultural Chemistry. In 1913 he received his master's degree. The next year he joined the Berkeley faculty as an assistant professor in agricultural chemistry and became associate professor in 1922. He became professor in 1927, retiring in 1949. In 1922 a division of plant nutrition within the University of California College of Agriculture was established with Hoagland as its head. He also served as president of various organizations, including the American Society of Plant Physiologists, the Pacific Division of the American Association for the Advancement of Science, the Western Society of Naturalists, and the Western Society of Soil Science, and as consulting editor of the journal *Soil Science* and of the *Annual Review of Biochemistry.* He was elected a member of the National Academy of Sciences in 1934. In 1920 Hoagland married Jessie A. Smiley of San Francisco. She died in 1933. They had three sons, Robert Charles, Albert Smiley, and Charles Rightmire.

Hoagland is best known for his research in processes of salt absorption by plants, in soil and plant interrelations, and in the utilization of various elements in soil solutions. Upon his appointment to Berkeley in 1913 as an agricultural chemist, he undertook a systematic investigation of the giant kelps that grow off the California coast. There was at the time a great need for potash fertilizers in this country as a result of the wartime cessation of imports from Germany, and Hoagland believed that the kelps might be a potential source of potassium. Although his findings were not very encouraging, the investigation raised a number of questions that were to stimulate his later research. Specifically, Hoagland was impressed by the ability of kelp to accumulate and retain large amounts of potassium, bromide, and iodide.

He found that the principal ions within the kelp cell are often contained at far higher concentrations than in the solutions in which they grow. It was therefore important to know how the solutes penetrated, not merely to equality of concentration but to accumulations of far higher concentrations.

In the work of Hoagland and his co-workers, the internodal cells of the freshwater alga *Nitella* proved particularly useful in studying active secretion of solutes. These long, multinucleate internodal cells yielded drops of liquid from their vacuoles which, when analyzed, showed that virtually all the ionic constituents present in pond water had been accumulated in the cell. This was particularly true of potassium and chloride. Having found that the strength of the solution inside the cell varied according to the season of the year and to light conditions, Hoagland proceeded to study an ion which had not previously been present in the solution and which, once absorbed by the *Nitella,* could be determined chemically with great accuracy. Using the bromide ion, Hoagland, Hibbard, and Davis (1926) showed that the accumulation of bromide ion from the external solution was a function of the intensity of light, that it occurred little if at all in darkness and was affected by the duration and intensity of light during the daily period of illumination. Hoagland then realized that the *Nitella* cell was using its light energy to bring about the absorption of the salt through its metabolic processes, and in doing so expended its energy.

Work on thin disks cut from plant storage organs such as potato and artichoke tubers by Steward, first in England, and then later in Hoagland's laboratory at Berkeley, drew attention to the role of oxygen pressure in this active accumulation process; and thus to the importance of aerobic metabolism in the process of ion intake (cf. Hoagland and Broyer, 1936; Steward and Broyer, 1934; Steward *et al.,* 1936). Absorption of ions by excised roots of barley was soon shown to present similar features. Hoagland's barley roots were grown in such a way that they contained little salt and had, therefore, a great ability to accumulate potassium bromide. This accumulation occurred when the solutions were appropriately aerated with an air stream free of carbon dioxide. An interesting discovery was that the barley roots accumulated sugar, which was supplied by the leaves during their growth; that when the roots had free access to salt, they rapidly replaced the previously accumulated sugar with the salt that would have been absorbed during growth, had it been freely available. Thus the process of accumulation of solute and salt was shown to be a normal concomitant of the active growth of cells. Metabolites elaborated in one part of the cell, or solutes from the external environment, are secreted into a vacuole as the cell grows. However, once large vacuoles have been produced which have built up within themselves a concentration of solute, there may be interchange with the external medium, entailing a more limited requirement for metabolism and active growth. These conclusions, directly derived from Hoagland's laboratory work, induced plant physiologists to appreciate that the growing cell is a working molecular machine and that Fick's diffusion law and passive permeability *sensu stricto* play a secondary role in active salt accumulation. Interestingly, biologists have yet to identify precisely how such energy is donated and how this secretory machine works.

Hoagland's *Lectures on the Inorganic Nutrition of Plants* (Prather lectures at Harvard University), published in 1948, amply summarize his views and philosophy of plant nutrition. Our current understanding of the field owes much to his significant contributions. Following the discovery that plants could be grown in water containing salts, the composition of solutions was modified in various ways. The typical inorganic culture solution widely used today is associated with Hoagland's name and was based on the proportions of macronutrients absorbed by tomatoes. This solution proved to be efficient for sand and water cultures for a wide range of plants, especially at high light intensities. Hoagland was also involved in the investigation of "little leaf," a disease of deciduous fruit trees. He traced this and other dieback symptoms in the sandy soils of California to nutritional deficiencies caused by the absence of zinc; other required trace elements were also discovered at his laboratory.

BIBLIOGRAPHY

I. ORIGINAL WORKS. Hoagland's works include "General Nature of the Process of Salt Accumulation by Roots With Description of Experimental Methods," in *Plant Physiology,* **11** (1936), 471–507, written with T. C. Broyer; and "The Influence of Light, Temperature, and Other Conditions on the Ability of *Nitella* Cells to Concentrate Halogens in the Cell Sap," in *Journal of General Physiology,* **10** (1926), 121–146, written with P. L. Hibbard and A. R. Davis.

II. SECONDARY LITERATURE. See two articles by D. I. Arnon: "Dennis Robert Hoagland 1884–1949," in *Plant Physiology,* **25** (1950), iv–xvi, abridged in *Science,* **112** (1950), 739–742; and "Dennis Robert Hoagland 1884–1949," in *Plant and Soil,* **2** (1950), 129–144, abridged in *Soil Science,* **69** (1950), 1–5. See also W. P. Kelley, "Dennis Robert Hoagland 1884–1949," in *Biographical Memoirs. National Academy of Sciences,* **29** (1956), 123–143, including a complete bibliography; *National Cyclopedia of American Biography,* XLVII (New York, 1965), 598–599; F. C. Steward and W. F. Berry, "The Absorption and Accumulation of Solutes by Living Plant Cells. VII. The Time Factor in the Respiration and Salt Absorption of Jerusalem Artichoke Tissue (*Helianthus tuberosus*) With Observations on Ionic Interchange," in *Journal of Experimental Botany,* **11** (1934), 103–119; and F. C. Steward, W. E. Berry, and T. C. Broyer, "The Absorption and Accumulation of Solutes by Living Plant Cells. VIII. The Effect of Oxygen Upon Respiration and Salt Accumulation," in *Annals of Botany,* **50** (1936), 345–366.

A. D. KRIKORIAN

HOBBES, THOMAS (*b.* Malmesbury, England, 5 April 1588; *d.* Hardwick, Derbyshire, England, 4 December 1679), *political philosophy, moral philosophy, geometry, optics.*

Thomas Hobbes, author of *Leviathan* and one of England's most penetrating philosophers, was born into an impoverished family in Wiltshire. His father, for whom he was named, was vicar of St. Mary's Church in Westport. His mother came of a yeoman family named Middleton. According to John Aubrey, the elder Thomas Hobbes was a semiliterate man: "One of the Clergie of Queen Elizabeth's time, a little learning went a great way with him and many other ignorant Sir Johns in those days." [1] We know at least that he was not a discreet individual; after a night of card playing he fell asleep in his church and was heard to utter, "Clubs is trumps." Later a more serious indiscretion caused an upheaval in the family; its effect on the child Thomas can only be guessed. Standing in front of his church, the father quarreled with a fellow parson, struck him, and was obliged in consequence to flee from Malmesbury, never to return. Thus, before he reached the age of seven, Thomas Hobbes was deprived of the society of his father; and salt was rubbed in the wound when the man his father had struck became the new vicar.

The care of the Hobbes family passed to an uncle, Francis Hobbes, a glover and an intelligent man who recognized signs of precocity in his nephew and underwrote the cost of his education. When he was seven, Hobbes was sent to school at the house of Richard Latimer, described by Aubrey as "a good Grecian." He was given a solid grounding in Latin and Greek; and at age fourteen he matriculated at Magdalen Hall (later called Hertford College), Oxford, where, however, he chafed under the restrictions of a scholastic curriculum. He preferred to "prove things after my own sense," [2] and he read deeply in areas not prescribed by his tutors. Astronomy and geography were his favorite subjects at this time.

In 1608 Hobbes, now bachelor of arts, was recommended by the principal of his college to be tutor to the son of William Cavendish, Baron Hardwicke, who later became the second earl of Devonshire. The significance of Hobbes's appointment to the Cavendish household cannot be exaggerated. The young graduate was introduced to a cultured, aristocratic world. Although his duties at first were almost menial, he was able with the passage of time to mingle with his master's guests on terms of some intimacy. In this way he came to know Ben Jonson, Lord Falkland, Sir Robert Ayton, Lord Herbert of Cherbury, and, some time later, the poet Edmund Waller, who became a particular friend. Moreover, in Chatsworth and Hardwick Hall, the great houses of the Cavendish family, Hobbes had at his disposal an excellent library in which, he said, he found the university he had missed at Oxford.

To a second branch of the Cavendish family residing at Welbeck Abbey, Hobbes owed the awakening of his interest in natural science. Sir Charles Cavendish was a skilled mathematician; and his more famous brother William, duke of Newcastle, was a scientific amateur who maintained a private laboratory and whose scientific speculations issued in such odd conclusions as that the sun is "nothing else but a very solid body of salt and sulphur, inflamed by its own motion upon its own axis."[3] Both men accepted Hobbes as a friend; and Newcastle, who had a passion for horses as well as a curiosity about optics and geometry, persuaded Hobbes to combine these interests in a curious treatise entitled "Considerations Touching the Facility or Difficulty of the Motions of a Horse on Straight Lines, or Circular," a work printed from manuscript in 1903 and described by its editor as "an irrelevant superfluity of reasoning" such as was produced by "the tailor in *Gulliver's Travels* who measures his men with the help of a sextant and other mathematical instruments."[4] It was on Newcastle's behalf that Hobbes searched the London bookshops in vain for a copy of Galileo's *Dialogues.*

In 1610 Hobbes set out on a grand tour of the Continent with his pupil. It was the year of the assassination of Henry IV of France, an event which impressed itself on Hobbes's mind as an extreme example of the chaos that follows from the abolition of sovereignty. On this first tour, through France, Germany, and Italy, Hobbes perfected his knowledge of foreign tongues and resolved, on his return, to become a scholar. In the library at Chatsworth he immersed himself in classical studies and in 1628–1629 published a brilliant translation of Thucydides' *Peloponnesian War.*

For a brief period before the Thucydides was published, Hobbes served as secretary to Francis Bacon, to whom he had been introduced by the younger Cavendish, one of Bacon's friends. Bacon had by this time been deposed as lord chancellor and was living in retirement at Gorhambury, where Hobbes accompanied him on his "delicious walkes" and where he acted as amanuensis and editorial assistant in the Latin translation of several of Bacon's *Essaies.* The connection between these two personalities is inherently interesting, but it should not be read as evidence of a Baconian influence on Hobbes's thought. Although they held some points in common, the two philosophers had worked out their ideas independently and essentially along different lines.

In June of 1628 Hobbes's master and friend, the second earl of Devonshire, died. Hobbes accepted a new appointment as tutor and cicerone to the son of Sir Gervase Clinton of Nottinghamshire, with whom he embarked, in 1629, on a second tour of Europe, to Paris, Orléans, Geneva, and Venice. It was in a library in Geneva that he first read Euclid; he was ever afterward enamored of geometry.[5] In particular, as Aubrey reports, he was attracted to the propositional character of geometry; it was a form of reasoning that fit in well with the conception of "truth" he was later to develop: that "truth" is the product of an analytical process in which definitions are placed in their proper order.

By November of 1630 Hobbes was recalled to the Cavendish family to serve as tutor in Latin and rhetoric to the next earl of Devonshire. With this young man, Hobbes, now in his forties, made his third grand tour of the Continent, the one which had the most important consequences for the development of his interest in natural science. That interest had not previously been dormant, since as Hobbes himself tells us, he had formulated a theory of light and sound as early as 1630;[6] a short manuscript tract giving a theory of sense and appetite is assigned by Dr. Frithiof Brandt to 1630. But on the third journey—to France and Italy—Hobbes made personal contact with scientific minds. In Arcetri, near Florence, he visited Galileo, whom he ever afterward held in veneration as "the first that opened to us the gate of natural philosophy universal"; and in Paris he met Marin Mersenne, the Franciscan monk in whose cell informal scientific meetings, attended by some of the best scientists of the age, took place. He also met Gassendi and Roberval; he read Descartes; and everywhere he went, he meditated on the problems of motion, which he conceived to be the principle by which a wholly material universe is to be understood.

Hobbes's deepest scientific interest was in optics. Probably this interest was awakened in him by his contact with the Cavendish circle, especially with Charles Cavendish, Walter Warner, and John Pell. A large part of the short tract of 1630 on sensation and appetite was devoted to optics; in that early work Hobbes adopted an emission or "corpuscular" theory of light, according to which there is a movement of particles of matter from the luminous source to the eye. But a letter of 1636 to William Cavendish shows that Hobbes had by this time abandoned the emission theory in favor of a mediumistic theory—light is propagated by a motion or pressure of the medium intervening between the source and the eye—and a letter of May 1640 shows that he developed the idea of the expansion and contraction of the medium as a way of accounting for the motion of the light and of the medium. He later rejected the idea of expansion and contraction because it demanded the presence of a vacuum, and a vacuum was precluded by the doctrine of plenitude in which Hobbes had come to believe.

The subtlest part of Hobbes's theory of light is his definition of a ray as "the path through which the motion from the luminous body is propagated through the medium."[7] He conceived of the propagated line of light as always normal to the sides of the ray; hence it may be thought of as a "ray front," on the analogy of a wave front.[8] What distinguishes Hobbes's conception of the ray from earlier conceptions—from that, for instance, of Descartes, who shows in his criticism of Hobbes that he entirely miscomprehended Hobbes's theory—is that for Hobbes the ray has infinitesimal elements. He accepted that light has physical dimensions but he argued that the significant feature of light, from a mathematical point of view, is its impulse or endeavor to motion; and this impulse is to be understood as the motion of infinitesimal elements. By taking this infinitesimal approach, by arguing that "we consider the width of the ray smaller than any given magnitude,"[9] Hobbes made the important transition from physical rays to mathematical rays. He himself perceived only gradually that he had introduced a new concept; but when he recognized that a shift had taken place, he abandoned the term "ray" (*radius*) and adopted the new term "radiation" (*radiatio*).[10]

These views were expressed in three manuscript treatises by Hobbes, one in English and two in Latin. The first of the Latin treatises, "Tractatus opticus," was communicated to Mersenne, who published it as book VII of the "Optics" in his *Universae geometriae* (Paris, 1644).[11] Mersenne had also published an optical treatise by Walter Warner which Hobbes had given him in Paris, and in 1641 he had published the *Objectiones ad Cartesii Meditationes,* the third "objection" of which was by Hobbes. When he returned to Chatsworth in 1637, after his third journey abroad, Hobbes continued to correspond with Mersenne on questions of physics and optics. He was now forty-nine: time, he thought, to put his ideas in order. He therefore formulated the outline of a large philosophical system, to be composed of three parts—body, man, and citizenship—and to be described in that order, since for Hobbes body or matter is the ultimate constituent of all things, including human society. Hobbes's early scientific manuscripts may be considered as preparation for *De corpore,* his formal account of the first principles of science, which he intended to put first in his system but which the pressure of events forced him to lay aside and not publish until 1655.

In the late 1630's political passions in England were boiling. Hobbes's inclinations were royalist, but he appears mainly to have been concerned by the imminent breakdown of civil order. In 1640, while Parliament and king were locked in political combat but before the outbreak of military hostilities in the Civil War, Hobbes considered it prudent for his safety to return to France. He did so and remained there for eleven years, part of that time with the duke of Newcastle in Paris. In Paris, Hobbes renewed his scientific contacts and almost immediately corresponded with Descartes about questions raised by the latter's *Méditations* and *Dioptriques.* Relations between these two proud thinkers were strained because neither was willing to concede any originality in the thought of the other.

The impulse to say something to his countrymen about politics in the hour of their travail deflected Hobbes's scientific preoccupations. In the spring of 1640, while still in England, he wrote a short treatise on politics which circulated widely in manuscript and was published in 1650 in two parts under the titles *Humane Nature* and *De corpore politico, or the Elements of Law.* In Paris he wrote *De cive,* published in 1642, a book which enjoyed international success. But *De cive* was written in Latin; and although it was separately translated into French by Samuel Sorbière and du Verdus, two of Hobbes's friends, it remained inaccessible to the general English reader. Hobbes therefore set to work on an English treatise, *Leviathan,* published in 1651. This work is justly celebrated for the brilliance, breadth, and coherence of its philosophical vision and for its concise, vigorous, and eloquent prose style.

The outlook of *Leviathan* is nominalist, materialist, and anticlerical. Hobbes believed that the universe is a great continuum of matter. It was created and set in motion by God, who is himself a material being, since the universe is utterly devoid of spirit. Of God's other attributes virtually nothing can be known. Our knowledge of the external world is derived, either directly or ultimately, from our sense impressions; and since sensory knowledge is the only knowledge we can ever have, we have no grounds for believing in the independent existence of universals or absolute ideas, or classes of things as separate entities. Human language consists of names of things and names of names, all joined by predicates. Names of names, or universals, must not be confused with names of things; universals exist in the mind and things exist in the external world; but universals are not therefore to be despised, because, being rooted in language, they play their part in the reasoning process. "Truth" for Hobbes is analytic, a product of the correct reasoning about names.

Hobbes was uncompromising in the application of his nominalist principles to ethics. He argued that ethical judgments are the products of human thought and culture. "For these words of good, evil and contemptible, are ever used with relation to the person that useth them: there being nothing simply and absolutely so; nor any common rule of good and evil, to be taken from the nature of the objects themselves." The same kind of analysis was given to the notion of justice, which Hobbes believed to have no independent or absolute existence. In Hobbes's view, justice is a function of positive law, and all law is essentially positive law. "Where there is no common power, there is no law; where no law, no injustice." Justice and injustice "are qualities that relate to men in society, not solitude," and they draw their meaning from the declared intentions and enforcements of the civil magistrate.

Such a doctrine of ethical relativism and legal positivism was profoundly offensive to orthodox opinion in the seventeenth century; in particular, it ran counter to traditional conceptions of natural law, which were conceived of as laws of eternal and immutable morality, antecedent to positive civil law, originating, as Richard Hooker had put it, "in the bosom of God." Modern scholars disagree about the meaning of Hobbes's natural law doctrine. Some commentators, such as A. E. Taylor and Howard Warrender, argue that certain obligations of the citizen and all the obligations of the sovereign to his subjects are, according to Hobbes, grounded in a natural law antecedent to civil law; on the other hand, Michael Oakeshott believes that all those prerogatives of the citizen which are immune to sovereign authority, such as the citizen's right of self-preservation, and the obligations of the sovereign himself, are rational, not moral, obligations. In this view natural law is prudential. Whichever view is correct, there can be no doubt that Hobbes cast his natural law doctrine in a secular mold.

In the same secular spirit Hobbes developed his ideas of human nature. Man is a part of material nature, so his behavior, including the behavior of his mind, can ultimately be understood by reference to physical laws. Viewed from a shorter perspective, human behavior is seen by Hobbes to be grounded in self-interest, especially in the fundamental desire to survive. Hobbes did not argue that human nature was an entity separate from human culture, but he asked his readers to imagine what life would be like in the absence of culture—in the absence, that is, of social conventions and civil restraint. This is Hobbes's famous hypothetical picture of the "state of nature." Men in this condition are rapacious and predatory; and since they are equal in the things they want and equal in their capacities to satisfy their desires, they live in a state of continuous warfare or, at the very least, in a condition of fear, their lives being then "solitary, poor, nasty, brutish, and short." The grimness of this picture is relieved for the modern reader by his discovery that Hobbes believed very strongly in the doctrine of human equality; and although Hobbes chose not to develop the democratic implications of this doctrine, his sense of human equality is wholly at variance with the precepts and practices of the modern totalitarian state.

What Hobbes feared more than the tyranny of a sovereign was anarchy, and so he constructed a model of the state in which he thought anarchy would be impossible. Moved by their fears and passions, and instructed by their reason, men would come to realize that they can be delivered from the state of nature only by the generation of a stable commonwealth. The process by which the state comes into being was not intended by Hobbes to be construed historically; in general he believed that men will come first to recognize the significance and utility of the "laws of nature"—or some twenty theorems of conduct conducive to peace which Hobbes enumerated, and which he said are summed up in the golden rule. But at this stage these theorems of peace are merely comprehended; what is required is the power to enforce them, and this power resides only in a commonwealth—and then only in its "soul," the sovereign, who must rule with absolute sway.

To achieve this condition of enforceable peace, men will make a sort of contract among themselves (but not between themselves and the sovereign) to transfer their individual powers to a central sovereign authority. Hobbes did not insist that the sovereign be a single individual; although he favored monarchy, he thought that a body of men, a parliament, or even a king and parliament working in concert could achieve the same results. The main point was that the power of the sovereign be absolute, for the slightest diminution of his power would erode the security of the citizens; and it was for their security from each other that the sovereign was brought into being. Hobbes reserves to the citizens the right of rebellion if the sovereign fails to protect their security, but he treats this question warily.

The argument of *Leviathan* does not end with these views; fully one-third of the book examines the implications of Hobbes's political philosophy in a Christian society. Hobbes recognized that a seventeenth-century audience would demand to know whether his principles conformed to the teaching of Scripture. He himself knew the Bible well, and he was able to find passages in it supporting his doctrine of absolute sovereignty; but other passages were inconvenient and there remained the question, particularly vexing in an age of religious warfare, of which of several interpretations of Scripture was the correct one. Ultimately, said Hobbes, all Scripture is subject to interpretation, there being nothing about it except its existence that is agreeable to all minds. His solution to the problem of conflicting interpretation was both political and philosophical. On the political side he adopted the ultra-Erastian position that the only interpretation of Scripture that may be publicly espoused by citizens in a commonwealth is the interpretation of the sovereign authority. The natural right which citizens, by agreement among themselves, had transferred to the sovereign included the natural right of scriptural interpretation; should they retain that right, the commonwealth would inevitably lapse into a state of nature.

Moreover, Hobbes remained philosophically skeptical about the truth of Scripture. He conceded that a core of mystery in Scripture must be accepted on faith; but the greater part of the Bible is immune to human reason. His skepticism took the form of a surprisingly modern biblical criticism in which he anticipated Richard Simon and Spinoza by calling in question the number, scope, authorship, and general authenticity of the books of the Bible.

The relationship between Hobbes's scientific ideas and outlook on the one hand and his political philosophy on the other is hard to define. The question has provoked disagreement among Hobbes's commentators. Croom Robertson thought that the whole of Hobbes's political doctrine "had its main lines fixed when he was still a mere observer of men and manners, and not yet a mechanical philosopher." Leo Strauss accepts this view, but he believes that Hobbes had cast his mature political philosophy into an alien scientific mold, which resulted in a distortion of the politics but not in any significant change of its essentially prescientific, humanistic character.

Clearly Hobbes's materialism and physics do not imply his political theory in any simple linear connection; but, as was pointed out by J. W. N. Watkins, the science implies the civil philosophy in the same way, for example, that the law of evidence has important implications for statements made by witnesses in law courts, although the law of evidence does not entail any of those statements. Watkins' treatment of this whole question is illuminating. He has shown how Hobbes came to abandon his earliest political views, set down in the introduction to his Thucydides. Those views were "inductivist"; they advocated the study of history as a guide to rational conduct. Under the shaping influence of the new scientific outlook, however, Hobbes adopted the method called resolutive-compositive, which he derived partly from Galileo, partly from Harvey, but primarily from the philosophers and scientists of the school of Padua. (Hobbes was personally acquainted with a disciple of this school, Berigardus, author of *Circulus Pisanus.*) The method is described by Hobbes in *De corpore.* It has a large Aristotelian component. Put in its simplest form, it consists of resolving whole conceptions into their constituent parts or first principles and then recomposing them. It can be seen that this method is not an instrument of discovery in any modern sense of the idea of "science"; it appears to have more usefulness in social enquiries. Hobbes assimilated it into his political theory—as in the striking example of the break-up of society into its constituent parts called the state of nature and its recomposition into a commonwealth.

Not unexpectedly, Hobbes's views in *Leviathan,* taken altogether, raised a storm of opposition. He was embroiled in controversy for the rest of his life—more, in fact, than any English thinker before or since. The first signs of opposition appeared in France before *Leviathan* was published. On the recommendation of Newcastle, Hobbes was appointed tutor in mathematics to the prince of Wales, the future Charles II. Because of fears expressed by clergymen that the prince would be contaminated with atheism, Hobbes was obliged to promise that he would teach mathematics only, and not politics or religion. And when *Leviathan* was published, no one of the English court in France liked it. Although it was absolutist, it expressed no particular bias in favor of monarchy; and it appeared to favor the Puritan regime in England when it insisted that a citizen submit to any government that can secure internal peace. Moreover, its anticlericalism and attacks on the papacy offended French Jesuits and English Catholics. For these reasons Charles ordered Hobbes to leave the English colony in France, and in 1652 the philosopher returned to England.

He stayed in London for a year and then retired to Chatsworth, where the Cavendish family treated him with affection and even a certain deference, as befitted a philosopher of international renown. But the shock inflicted by *Leviathan* on clerical and lay opinion produced a rising tide of hostile criticism, some of it intelligent and philosophical but much of it in the form of abuse.[12] Hobbes was pronounced atheist, heretic, and libertine. He was the "Monster of Malmesbury," "a pander to bestiality" whose "doctrines have had so great a share of the debauchery of his Generation, that a good Christian can hardly hear his name without saying of his prayers."[13] It is true that Hobbes had his admirers and defenders, both on the Continent and in England, including such perceptive opponents as Samuel von Pufendorf and James Harrington, who understood that *De cive* and *Leviathan* were works to be reckoned with; but the clergy of all persuasions, as well as the common lawyers and university dons, united in their opposition to Hobbes. Indeed, his doctrines were cited by the House of Commons as a probable cause of the Great Fire of 1666.

Part of Hobbes's difficulties can be traced to a controversy between himself and John Bramhall, bishop of Derry (Londonderry) and later archbishop of Armagh. The two had met in 1645 at Paris, where they debated the subject of free will. Bramhall committed his ideas to paper; Hobbes wrote a rejoinder. Both agreed not to publish what they had written, but Hobbes's side of the question was put into print without his permission in a little treatise called *Of Liberty and Necessity* (1654). Bramhall, outraged by what he considered to be Hobbes's discourtesy in ignoring his side, published in 1655 all that had passed between them. Thus was launched a controversy which continued until Hobbes had the last word with the posthumous publication of *An Answer to a Book by Dr Bramhall Called The Catching of the Leviathan* (1682). Hobbes's views were strictly determinist. A man, he said, is "free" to do anything he desires if there are no obstacles in his way; but his desire to do anything has necessary and material causes. To Bramhall this doctrine was the essence of impiety; it would deny any meaning to rewards for good actions or punishments for evil ones, thus overturning the whole apparatus of religious worship. For his part Hobbes admitted that piety might not be promoted by his doctrine, but "truth is truth" and he would not be silent.

Hobbes was not molested personally during this last period of his life because he enjoyed the protection of Charles II, although he was deeply alarmed when, sometime in the 1660's, a committee of bishops in the House of Lords moved that he be burned for heresy. He wrote, but did not publish, a short treatise in the form of a legal brief showing that the law of heresy had been repealed in the time of Elizabeth and had never been revived, so that there could be no legal grounds for executing him.[14] Nothing came of the episcopal agitation; but the king refused to license a history in English by Hobbes of the Long Parliament, published posthumously as *Behemoth,* and the crown prohibited Hobbes from publishing any other works in English on the subject of politics or religion. Not included in this ban was the Latin translation of *Leviathan,* made by Henry Stubbe and first published at Amsterdam in 1668 and at London in 1678.

A second controversy, even more absorbing of Hobbes's energy than his debate with Bramhall, was his dispute with John Wallis on questions of geometry. Wallis was a vastly superior mathematician who made important contributions to the development of the calculus. But he was an acrimonious, coarse-tempered man; in a controversy that lasted almost twenty-five years, Wallis pressed his mathematical advantage with ferocious zeal, also attacking Hobbes for what he thought were errors in Greek, for having a West Country manner of speech, for being a rustic, for disloyalty to the crown, and so on. Hobbes's replies were better mannered, but he too was capable of losing his temper. The issue between the two men was whether Hobbes had succeeded, as he claimed, both in squaring the circle and in duplicating the cube. Hobbes boldly announced success in both enterprises, although he modified his claim slightly in some of the later books written against Wallis. It should be observed that neither Hobbes nor Wallis doubted the possibility of a quadrature, a proof of its impossibility not having been discovered until the nineteenth century; moreover, the problem of the quadrature was not only venerable but had a particular vitality in the seventeenth century. Nevertheless, Wallis was able to show that Hobbes's claim of success was unfounded. Hobbes made no original contributions to geometry but, as A. De Morgan has written, though Hobbes was "very wrong in his quadrature . . . he was not the ignoramus in geometry that he is sometimes supposed. His writings, erroneous as they are in many things, contain acute remarks on points of principle."[15] Hobbes's passion for geometry derived from his analytic conception of truth. He appreciated the unity and logical structure of geometry, its freedom from verbal confusion, and its reasoning from definitions placed in their proper order.

Algebra, on the other hand, failed to attract

Hobbes. He grossly underestimated its scope and was suspicious of all attempts to "arithmetize" geometry. He thought of algebra as a minor branch of arithmetic; Wallis' "scab of symbols" simply disfigured the page, "as if a hen had been scraping there." [16] Nor did he appreciate the significance of Wallis' contributions, published in *Arithmetica infinitorum* (1655), toward the development of the differential calculus, although Hobbes's speculations in optics of an earlier stage in his life seemed to be leading him in the direction Wallis was taking.

In fact, Hobbes, in his sixties when he began his dispute with Wallis, was out of touch with the generation of rising young scientists and mathematicians. He was not opposed to experimentalism on principle, but he had no natural sympathy for it and considered that most of the experiments performed by fellows and correspondents of the Royal Society were either ill-conceived and poorly executed, or else they reached conclusions long ago arrived at by Hobbes through the use of his unaided reason. In this spirit he wrote "Dialogus physicus, sive de natura aeris" (1661), a brief but barbed attack on Robert Boyle's experiments on the vacuum pump, to which Boyle replied calmly, though forcefully, in *Examen of Mr. Hobbes, His Dialogus* (1662) and *Dissertation on Vacuum Against Mr. Hobbes* (1674). Not surprisingly, Hobbes was excluded from membership in the Royal Society, a fact which he resented, although he publicly declared that he was lucky to be out of it.

Hobbes's last years were thus clouded with controversy, but they were not without their simple pleasures and rewards. He lived comfortably on the Cavendish estates in Chatsworth and Hardwick Hall and, more frequently, in the duke of Devonshire's house on the Strand in London. He enjoyed long walks; he played tennis until he was seventy-five; and he had an abiding love of music, listening to it whenever he could and playing on his own bass viol. Capable as he was of holding his own in public controversy, and sparkling with wit in table talk, he was always gentle with people of lower rank or inferior education. He was a bachelor, but according to Aubrey he was not a "woman-hater"; and it is possible that he had a natural daughter whom he cherished.

In his eighties, mostly to amuse himself, Hobbes published translations of the *Iliad* and the *Odyssey*. And when he was ninety he published *Decameron physiologicum*, a set of dialogues on physical principles containing also a last salvo fired off against Wallis. He died of a stroke at the age of ninety-one.

NOTES

1. John Aubrey, *Brief Lives*, A. Clark, ed. (Oxford, 1898), I, 390.
2. Hobbes, *Life . . . Written by Himself* (London, 1680), p. 3.
3. Margaret Cavendish, *Philosophical and Physical Opinions* (London, 1663), p. 463; Jean Jacquot, "Sir Charles Cavendish and His Learned Friends," in *Annals of Science*, **8** (1952), 13–27, 175–191.
4. S. Arthur Strong, *A Catalogue of Letters and Documents at Welbeck* (London, 1903), p. vii.
5. G. R. De Beer, "Some Letters of Hobbes," in *Notes and Records. Royal Society of London*, **7** (1950), 205.
6. Hobbes, *Latin Works*, Molesworth, ed., V, 303.
7. *Ibid.*, pp. 221–222.
8. See, on this point, Alan E. Shapiro, "Rays and Waves," doctoral dissertation, Yale University, 1970. Dr. Shapiro has made a full study of Hobbes's optics.
9. Hobbes, *Latin Works*, V, 228.
10. Hobbes, "Tractatus opticus," British Museum, Harleian MS. 6796, ch. 2, sec. 1.
11. The two other optical MSS are "A Minute or First Draught of the Optiques," British Museum, Harleian MS 3360; and a second Latin treatise also called "Tractatus opticus," British Museum, Harleian MS 6796.
12. See Samuel I. Mintz, *The Hunting of Leviathan* (Cambridge, 1962).
13. Bishop John Vesey, "The Life of Primate Bramhall," in John Bramhall, *Works* (Dublin, 1677).
14. Samuel I. Mintz, "Hobbes on the Law of Heresy; A New Manuscript," in *Journal of the History of Ideas*, **29** (1968), 409–414; "Hobbes's Knowledge of the Law," *ibid.*, **31** (1970), 614–616.
15. A. De Morgan, *A Budget of Paradoxes* (London, 1915), p. 110.
16. Hobbes, "Six Lessons to the Professors of the Mathematics," in *Works*, VII, 316.

BIBLIOGRAPHY

I. Original Works. Hobbes's works include *De cive* (Paris, 1642); *De corpore politico, or the Elements of Law* (London, 1650); *Leviathan* (London, 1651); *De corpore* (London, 1655); *Problemata physica* (London, 1662); *Lux mathematica* (London, 1672); *Decameron physiologicum* (London, 1678); and *Behemoth* (London, 1679). The standard ed. of Hobbes's works is by William Molesworth, 16 vols. (London, 1839–1845), but it has inaccuracies and omissions. A comprehensive modern ed., to be published at Oxford, is being prepared by Howard Warrender. The standard bibliography of Hobbes's works is by Hugh Macdonald (London, 1952). Important modern eds. of *Leviathan* are by Michael Oakeshott (Oxford, 1946) and by C. B. Macpherson (Baltimore, 1968). A modern translation particularly valuable for its full annotations and attention to textual problems is François Tricaud, *Leviathan: Traité de la matière, de la forme et du pouvoir de la république ecclésiastique et civile* (Paris, 1971).

II. Secondary Literature. Contemporary biographies of Hobbes are John Aubrey, *Brief Lives*, A. Clark, ed. (Oxford, 1898); and Richard Blackbourne, in *Vitae Hobbianae auctarium* (London, 1681). The most important nineteenth- and early twentieth-century studies of Hobbes, in which biography is mingled with commentary and criticism, are G. Croom Robertson, *Hobbes* (Edinburgh, 1886); and Ferdinand Tönnies, *Hobbes, der Mann und der Denker,* 2nd ed. (Leipzig, 1912). Later twentieth-century studies of Hobbes are numerous. They include the following, listed chronologically: Frithiof Brandt, *Thomas Hobbes' Mechanical Conception of Nature* (Copenhagen, 1928); Leo Strauss, *Political Philosophy of Thomas Hobbes* (Oxford, 1936); Howard Warrender, *Political Philosophy of Hobbes* (Oxford, 1957); C. B. Macpherson, *The Political Theory of Possessive Individualism: Hobbes to Locke* (Oxford, 1962); Samuel I. Mintz, *The Hunting of Leviathan* (Cambridge, 1962); Keith Brown, ed., *Hobbes Studies* (Oxford, 1965); J. W. N. Watkins, *Hobbes's System of Ideas* (London, 1965); M. M. Goldsmith, *Hobbes's Science of Politics* (New York, 1966); and R. Kosselleck and R. Schnur, eds., *Hobbes-Forschungen* (Berlin, 1969).

SAMUEL I. MINTZ

HOBSON, ERNEST WILLIAM (*b.* Derby, England, 27 October 1856; *d.* Cambridge, England, 19 April 1933), *mathematics.*

Hobson was the eldest of six children of William Hobson, a prominent citizen of Derby, the founder and editor of the *Derbyshire Advertiser.* His mother was Josephine Atkinson. His brother, J. A. Hobson, became a well-known economist. Hobson went to Derby School and did well in languages and music as well as mathematics and science. He was brought up in a strictly religious atmosphere, from which he later broke away.

In 1874 he won a mathematical scholarship to Christ's College, Cambridge, and in January 1878 he was placed first in order of merit (the senior wrangler) in the mathematical tripos. He was elected a fellow of Christ's in the same year and spent the rest of his life in teaching and research at Cambridge. In 1883 he was elected one of the first university lecturers (as distinct from college lecturers) in Cambridge. Hobson married Selina Rosa, the daughter of a Swiss merchant, in 1882; they had four sons.

Until 1910 the mathematical life of Cambridge was unduly dominated by the famous tripos examination. The undergraduates were coached in the solving of problems, and much of the college teachers' energy went into this coaching. Many of the teachers, therefore, made no effort to break new ground in their subject. There were, indeed, famous pure mathematicians, notably Cayley and Sylvester, but they worked largely in isolation. Moreover, their interests were in formal algebra. The theory of functions, actively developed in Germany and France since the 1850's, only began to be recognized in England in the 1890's.

In 1891 Hobson published *A Treatise on Trigonometry,* which, except for Chrystal's *Algebra* (Edinburgh–London, 1886–1889), is the first English textbook of mathematical analysis. His early interests were mainly in the functions of mathematical physics, and the first of the papers on which his reputation rests, on general spherical harmonics, appeared in 1896. During the next ten years he became aware of the work of the French school (Baire, Borel, Lebesgue), realizing that this school formed the necessary foundation of the systematic theory of trigonometrical and other special functions. His *Theory of Functions of a Real Variable* (1907), together with W. H. Young's *Theory of Sets of Points* (Edinburgh–London, 1906), introduced to English readers the vital Borel-Lebesgue concepts of measure and integration. In addition, this work incorporated Hobson's own research on the general convergence theorem and convergence of series of orthogonal functions. By the age of fifty Hobson had developed into a pure mathematician. This unusually late maturity is a reflection of the existing academic conditions at Cambridge.

In 1910 Hobson succeeded A. R. Forsyth in the Sadleirian chair of pure mathematics; he was recognized as one of the leaders of English mathematics. He resigned in 1931. He had few research pupils and did not found a school. Toward the end of his life his influence was overshadowed by the rise of younger men of great analytical power, notably G. H. Hardy and J. E. Littlewood.

Hobson was a distinguished figure in the university and served on the central administrative committees. His views were progressive and, appropriately enough, he was one of the leaders in reforming the mathematical tripos and abolishing the order of merit.

BIBLIOGRAPHY

I. ORIGINAL WORKS. Hobson's works include *A Treatise on Trigonometry* (Cambridge, 1891); *Theory of Functions of a Real Variable* (Cambridge, 1907), later eds. in 2 vols; *Squaring the Circle* (Cambridge, 1913), with six excellent lectures on this classical problem; *The Domain of Natural Science* (Cambridge, 1923), the Gifford lectures in Aberdeen; and *Spherical and Ellipsoidal Harmonics* (Cambridge, 1931), with Hobson's early researches of the 1890's.

II. SECONDARY LITERATURE. See G. H. Hardy, in *Obituary Notices of Fellows of the Royal Society of London,* no. 3 (1934), with portrait; and *Dictionary of National Biography* (Oxford, 1931–1940), pp. 433–434.

J. C. BURKILL

HODGKINSON, EATON (*b.* Anderton, near Great Budworth, Cheshire, England, 26 February 1789; *d.* Manchester, England, 18 June 1861), *applied mathematics, structural mechanics.*

Hodgkinson was one of the largely self-taught British mathematicians of the eighteenth and nineteenth centuries who turned their attention to applied mathematics and applied mechanics. His research and publications were confined almost entirely to the experimental and analytical study of the theory of elasticity and the strength of materials, and he became the foremost British authority in these fields during the second quarter of the nineteenth century.

His early education was first directed toward the study of Latin, Greek, and Hebrew in preparation for the university and a clerical career. He soon displayed a strong distaste for these studies, however, and suffered severely at the hands of a stern schoolmaster. As a result, he was transferred to another elementary school, where his mathematical ability was recognized and fostered. These episodes constituted the full extent of his formal education for he was compelled to devote himself to the assistance of his widowed mother, first in working the family farm in Cheshire, and later, beginning in 1811, when the family moved to Manchester, in operating a pawnbrokerage. Hodgkinson extended his knowledge of mathematics and mechanics through private study, largely of the works of William Emerson and Thomas Simpson. In Manchester he was guided and encouraged by John Dalton; together they studied the work of Euler, Lagrange, and Laplace. Hodgkinson had no sectarian affiliations and displayed no interest in religion.

At the age of thirty-three Hodgkinson read his first paper (published two years later), a study of beam flexure. Although a correct understanding of the distribution of stresses in a flexed beam had been reached by Parent and Coulomb during the eighteenth century, the solution had remained generally unnoticed. In Great Britain it had been presented only by Robison in his *Encyclopaedia Britannica* article, "Strength of Materials," and it was from this source that Hodgkinson derived the key principle, namely that in a transversely loaded beam the summation of the tensile and compressive stresses across any section must equal zero. Only with the publication of Hodgkinson's paper in 1824 did the solution become generally known.

In a later paper (1831) Hodgkinson examined the flexural characteristics of cast iron and showed that the tensile and compressive strengths of that material are unequal; and, accordingly, for the most economical cast-iron I beam and tension and compression flanges should be unequal. His further publications covered a variety of problems in structural mechanics, including dynamical loading of beams, column theory, structural characteristics of wrought iron, and hollow girders. In general, his research was relevant to many engineering problems encountered in the rapidly developing railroad industry and, appropriately, he became in 1847 one of the commissioners appointed to study the application of iron to railway structures. The results of his investigations constituted the primary documents in the "Report of the Iron Commissioners," which appeared in 1849. During this period he also collaborated with Robert Stephenson and William Fairbairn on the design of the Britannia and Conway tubular bridges.

Hodgkinson became one of the first British autodidacts to receive a university appointment when, in 1847, he was made professor of the mechanical principles of engineering at University College, London.

BIBLIOGRAPHY

I. Original Works. Hodgkinson's most influential works include "On the Transverse Strain and Strength of Materials," in *Memoirs of the Manchester Literary and Philosophical Society,* 2nd ser., **4** (1824); "Theoretical and Experimental Researches to Ascertain the Strength and Best Forms of Iron Beams," *ibid.,* **5** (1831); the two investigations he contributed to the *Report of the Commissioners Appointed to Inquire into the Application of Iron to Railway Structures* (London, 1849); and *Experimental Researches on the Strength and Other Properties of Cast Iron* (London, 1846).

A list of many of Hodgkinson's publications is given in *Royal Society of London, Catalogue of Scientific Papers (1800–1863)* (London, 1869).

II. Secondary Literature. Extensive reviews of many of Hodgkinson's publications are contained in Isaac Todhunter, *A History of the Theory of Elasticity and the Strength of Materials,* 2 vols. (Cambridge, 1886). For a long biographical article, see Robert Rawson, "Memoir of Eaton Hodgkinson," in *Memoirs of the Manchester Literary and Philosophical Society,* 3rd ser., **2** (1865), 145–204, repr. in *Annual Report of the Smithsonian Institution* (1868), pp. 203–230.

Harold Dorn

HODIERNA, GIOVANNI. See **Odierna, Giovanni.**

HOEK, MARTINUS (*b.* The Hague, Netherlands, 13 December 1834; *d.* Utrecht, Netherlands, 3 September 1873), *astronomy.*

Hoek is known chiefly for his discovery that several comets move in the same orbit ("comet groups") and for his investigation of optical phenomena in moving bodies.

He was the son of Andries Hoek, a surgeon in The Hague, and Johanna Maria de Wit. He married a Miss G. A. Brouwer. He studied astronomy at the University of Leiden under F. Kaiser, becoming in 1856 observer at the Leiden observatory and in 1857 extraordinary professor of astronomy at the University of Utrecht. Because of bad health, which compelled him to give up his observational work, he turned to theoretical astronomy. He advised the Netherlands Shipping Company about chronometers, compasses, methods of position finding. Following the cholera epidemics of 1854–1866 he published (1867) extensive mortality tables concerning the Utrecht population.

Hoek's most important discovery (1865–1868) was that of so-called comet groups following the same orbit. Altogether he found thirty-three comets to belong to six groups. Although he erroneously believed comets to come from outside the solar system, he was right in assuming that they sometimes subdivide by fragmentation and that it is thus that the existence of groups must be explained.

Hoek also investigated optical phenomena in moving bodies (1861–1869). Fresnel had suggested, and Fizeau had found through experimentation (1851), that in an object moving with a velocity v the "ether" is carried along with a velocity

$$vk = v\left(1 - \frac{1}{n^2}\right),$$

where n is the index of refraction. By a modified setup Hoek (1868) reduced this experiment to a zero-method and confirmed the value of the convection coefficient k with a much higher precision (1.3 percent). In the theory of relativity this result is explained, without any "ether," by simple application of the velocity-addition theorem.

In Hoek's second experiment (1869), a slit is viewed through a long, horizontal column of water. As the image appears to be independent of the azimuth towards which the apparatus is oriented, with respect to the motion of the earth, Fresnel's convection factor was confirmed. Together with Oudemans, Hoek also showed that for a given substance the refractive power $(n^2 - 1)/d$ is not a constant (1864).

BIBLIOGRAPHY

I. ORIGINAL WORKS. A bibliography of Hoek's early publications (1856–1859) is found in *Annalen der Sternwarte in Leiden,* **1** (1868), 38. See also "On the Comets of 1677 and 1683; 1860 III, 1863 I, and 1863 VI," in *Monthly Notices of the Royal Astronomical Society,* **26** (1865), 1–13; "Additions to the Investigations on Cometary Systems," *ibid.* (1866), 204–208; "De l'influence des mouvements de la terre sur les phénomènes fondamentaux de l'optique dont se sert l'astronomie," in *Recherches astronomiques de l'Observatoire d'Utrecht,* **1** (1861), 1–68; "Détermination de la vitesse avec laquelle est entrainé un rayon lumineux traversant un milieu en mouvement," in *Verslagen der Koniglijke Akademie van Wetenschappen te Amsterdam,* 2nd ser., **2** (1868), 189–194, and **3** (1869), 306–313. A slightly abridged text of this is found in *Astronomische Nachrichten,* **70** (1867), 193–198, and **73** (1869), 193–200; "Recherches sur la quantité d'éther contenue dans les liquides," in *Recherches astronomiques de l'Observatoire d'Utrecht,* **2** (1864), 1–71, written with A. C. Oudemans.

II. SECONDARY LITERATURE. Some biographical details and references are found in P. C. Molhuysen and P. J. Blok, *Nieuw Nederlandsch Biografisch Woordenboek,* I (Leiden, 1911), 1118–1119. See also the *Utrechtsche Studenten Almanak* (1874).

M. G. J. MINNAERT

HOEVEN, JAN VAN DER (*b.* Rotterdam, Netherlands, 9 February 1801; *d.* Leiden, Netherlands, 10 March 1868), *comparative anatomy, natural history, anthropology.*

Van der Hoeven was the youngest son of Abraham van der Hoeven, a merchant, and the former Maria van der Wallen van Vollenhoven. After his father's death in 1803, his mother married Martinus Pruys, a physician, in 1810. In 1826 van der Hoeven married Anna van Stolk; they had seven children, of whom only a son and two daughters survived him.

Van der Hoeven's work closely reflects the cultural climate of the Netherlands in the first half of the nineteenth century. In all his works he tried to show how the infinite wisdom of the Creator is reflected in the harmony of His creatures; his anthropomorphic viewpoint starts from the assumption that man is the most perfect creature on earth—his body is the vehicle most able to carry his soul, the medium through which he obtains knowledge of his creator. Nevertheless, van der Hoeven must be considered as one of the greatest zoologists of his time: recognition of his merits is reflected by his membership in more than forty scientific academies and learned societies.

In 1819 van der Hoeven entered the University of Leiden and in 1822 obtained the Ph.D. with a dissertation on the comparative anatomy of the bony skeleton of fishes, particularly of the skull. Two years later he received the M.D. with a dissertation on diseases of the ear.

In 1822 van der Hoeven began his scientific career

as honorary custodian of the Rijksmuseum voor Natuurlijke Historie; after having finished his studies, he journeyed to Paris, where he met Cuvier and Latreille, and to Frankfurt, where he met S. T. Sömmering. When he returned to the Netherlands, he established himself as a consulting physician in Rotterdam; soon afterward he began teaching botany at a school of pharmacy, founded and directed by apothecaries, and in 1825 he was appointed lecturer in physics of the Bataafsch Genootschap. As professor of zoology at the University of Leiden, a position he held from 1826 until his death, he lectured on comparative anatomy and general zoology, anthropology (from 1830), and geology and mineralogy (from 1839). His lectures in zoology appeared in his more or less introductory *Tabula regni animalis* (1856), and in his *Handboek der Dierkunde.* Unlike most textbooks of the time, the *Handboek* starts with the lower animals and progresses to the vertebrates in order to illustrate the increasing complexity of form and function; the book's methodology is that of comparative anatomy, whereas its emphasis is on general biological principles. Thus, each taxonomic section is preceded by an introduction in which the characteristics of its members are considered. The *Handboek* shows van der Hoeven's extensive knowledge of the various animal phyla, particularly the vertebrates and the insects.

In the last phase of his life van der Hoeven summarized the best of his knowledge in his *Philosophia zoologica* (the title recalls Linnaeus' *Philosophia botanica*), an attempt to present concisely and aphoristically the contents and general principles of zoology. Separate sections deal with form, structure, and function; comparative anatomy; embryogenesis and metamorphosis; classification; and geographical distribution.

Another field of van der Hoeven's activity was the natural history of man, on which he lectured in alternate years. His special area of interest was the morphology of the skull, which he investigated by quantitative techniques. He studied skulls of a wide variety of races and possessed a great private skull collection; his catalog of the collection (1860) describes 171 skulls and 39 casts. His classic work on the morphology of the Negro (1842) contains much information on the skull. The central questions considered in this study were whether all human races belong to one species and whether all human beings stem from a single pair. To answer these questions van der Hoeven made a study of the hereditary variations and linguistic differences of the human races. These studies led him to the assumption that there must have been more than one center of human origin. In 1862 he devoted a special work to linguistic aspects of these studies.

For van der Hoeven, Cuvier was the first zoologist to have systematized all known zoological facts, and in his lectures on mineralogy and geology he showed himself to be an adherent of Cuvier's theory of catastrophism. He supposed that several creations took place in various places and at various times and that the flora and fauna of many islands were created separately (1861).

Accordingly, van der Hoeven's train of thought was at variance with any idea of evolution; most phenomena of living nature were not to be explained by theories but were immediately connected with the origin of life itself. As such, we must accept creation, although we cannot explain it. It is easier to understand that an animal was created with eyes, he stated, than to suppose that it originated from an eyeless ancestor; the first human being must have been created an adult in body and mind, for were it created a child, whose bodily and mental faculties developed gradually, it is not understandable how it could live for more than one day without the help of Providence (1860).

Van der Hoeven was also a prolific popularizer of science. He wrote a popular zoology text for young people (1868) and a natural history of the animal kingdom (1857) for a wider audience in order to demonstrate how the perfection of the Creator is reflected in his creatures, to contribute to the glory of God, and to promote useful knowledge of nature.

BIBLIOGRAPHY

I. Original Works. Before 1850 van der Hoeven published the following works: *Responsio ad quaestionem, ab ordine disciplinarum mathematicarum et physicarum anno 1819 proposita: Quaeritur, quis sit usus, qualesque dignitas anatomes comparatae in stabiliendis regni animalium divisionibus?* (Ghent, 1820), also in *Annales Academiae Gandavensis* (1821); "Responsum ad quaestionem: Quaeritur brevis et distincta expositio fabricae et functionis organi auditus in homine?," in *Annales Academiae Rheno-Trajectinae* (1822); *Dissertatio philosophica inauguralis de sceleto piscium* (Leiden, 1822), his doctoral thesis; "Redevoering over de stelling van Herder, dat de mensch een middelwezen is onder de dieren dezer aarde," in *Vaderlandsche letteroefeningen,* **2** (1822), 1–13; "Mémoire sur le genre Ornithorhynque," in *Nova acta Academia Leopoldina Carolina,* **9**, pt. 2 (1823), 351 ff., and **12**, pt. 2 (1825), 869; "Disputatio de causarum finalium doctrina, ejusque in zoologia usu," in *Provinciaal Utrechts Genootschap van kunsten en wetenschappen,* n.s. **3** (1824), 1–81; *Dissertatio pathologica de morbis aurium auditisque* (Leiden, 1824), his M.D. thesis; "Redevoering over de oorspronkelijke aarde en hare omwentelingen, zooals wij die kennen uit den tegenwoordigen toestand onzer planeet," in *Magazin voor wetenschappen, kunsten en letteren,* **3,** no. 2 (1824); *Tabula regni animalis, quam secundum alteram enchiridii sui zoölogici editionem in auditore usum scripsit* (Leiden, 1828; 2nd ed., 1856), Dutch trans. as *Tafel van het dierenrijk, met bijvoeging der kenmerken van de klassen en orden* (Leiden, 1829); *Icones ad illustrandas coloris mutationes in Chamaeleonte* (Leiden, 1831); *Beknopte handleiding tot de natuurlijke geschiedenis van het dierenrijk* (Haarlem, 1835), also in 2nd ed. with an atlas (Haarlem, 1864); 3rd ed. with new title: *Leerboek der dierkunde ten dienste van het middelbaar onderwijs* (Leiden, 1868); "Essai sur les dimensions de la tête osseuse, considérées dans leurs rapports avec l'histoire naturelle du genre humain," in *Annales des sciences naturelles,* 2nd ser., **8** (1837), zoology, 116–124; *Iets over den grooten zoogenoemden salamander van Japan* (Leiden, 1838); *Recherches sur l'histoire naturelle et l'anatomie des Limules* (Leiden, 1838); *Bijdragen tot de natuurlijke geschiedenis van den negerstam* (Leiden, 1842); *Oratio de aucta et emendata zoologia post Linnaei tempora* (Leiden, 1843); "Bijdragen tot de kennis van de Lemuridae of Prosimii," in *Tijdschrift voor Natuurlijke Geschiedenis en Physiologie,* **11** (1844), 1–48, also published separately (Leiden, 1844); *Schets van de natuurlijke geschiedenis van den mensch, ten dienste zijner lessen ontworpen* (Leiden, 1844); *Herinneringen aan eene reis naar Stockholm ter gelegenheid van de vergadering der Scandinavische natuuronderzoekers in Julij 1842* (Amsterdam, 1845), first published as a series of essays in *Vaderlandsche letteroefeningen; Redevoeringen en verhandelingen* (Amsterdam, 1846); also available in German as *Ergebnisse der Naturforschung für das Leben. Vorträge und Abhandlungen von J. van der Hoeven* (Berlin, 1848); *Beweging en verandering, dienstbaar tot instandhouding* (Leiden, 1848); *Gaan wij eene nieuwe barbaarsheid tegemoet? Eene voorspelling van Niebuhr* (Leiden–Amsterdam, 1849); *Handboek der Dierkunde,* 2 vols. (1st ed., Rotterdam, 1828–1833; 2nd ed., Amsterdam, 1849–1855); 2nd ed. trans. into German by F. Schlegel (vol. I) and R. Leuckart (vol. II) as *Handbuch der Zoologie* (Leipzig, 1850–1856), vol. II with additional material by Leuckart, trans. by Jan van der Hoeven (the son) as *Bijvoegsels en aanmerkingen behoorende tot het Handboek der dierkunde* (Amsterdam, 1856); 2nd Dutch ed. also trans. into English by W. Clark, 2 vols. (Cambridge, 1856–1858).

After 1850 van der Hoeven wrote "Bijdragen tot de ontleedkundige kennis aangaande Nautilus Pompilius, vooral met betrekking tot het mannelijk dier," *Verhandelingen der K. nederlandsche akademie van wetenshappen, Afdeeling natuurkunde,* **3,** no. 7 (1856), also trans. into German in *Archiv für Naturgeschichte,* **1** (1857), 77 ff.; into French in *Annales des sciences naturelles,* Zool. sec., 4th ser. (1856), 290 ff.; and into English in *Annals and Magazine of Natural History* (1856), 58 ff.; *Natuurlijke geschiedenis van het dierenrijk,* vol. III of J. A. Uilkens, *De volmaakheden van den Schepper in tzijne schepselen beschouwd, ter verheerlijking van God en tot bevordering van nuttige natuurkennis* (Leeuwarden, 1857); "Over de opvolging en ontwikkeling der dierlijke bewerktuiging op de oppervlakte onzer planeet in de verschillende tijdperken van haar bestaan," in *Album der Natuur* (1858), 33–48, also in *De verspreiding en bewerktuiging der dieren* (Leiden, 1858), also in English in *Annals and Magazine of Natural Science,* 3rd ser., **14** (1861), 209–221; *Berigt omtrent het mij verleende ontslag als opperdirecteur van 's Rijks Museum van Natuurlijke Historie* (Leiden–Amsterdam, 1860); *Catalogus craniorum diversarum gentium quae collegit* (Leiden, 1860); *Over natuurkundige theoriën omtrent de verschijnselen van het leven, en bepaaldelijk over Darwin's theorie aangaande het ontstaan der soorten door W. Hopkins* (Haarlem, 1860); "De geographische verspreiding der dieren; eene schets," in *Album der natuur* (1861), 368–375, also in *De verspreiding en bewerktuiging der dieren* (Leiden, 1861), 1–26; "Over de taal en de vergelijkende taalkennis, in verband met de natuurlijke geschiedenis van den mensch," in *Album der natuur* (1862), 80–94; *Philosophia zoologica* (Leiden, 1864), trans. into Italian by M. Lessona and T. Salvadori as *Filosofia zoologica* (Genoa, 1866–1867); "Considérations sur le genre Ménobranche et sur ses affinités naturelles," in *Archives néerlandaises des sciences exactes et naturelles,* **1** (1866), 1–16; *Ontleed- en dierkundige bijdragen tot de kennis van Menobranchus, den Proteus der meren van Noord-Amerika* (Leiden, 1867); and "De werken Gods. Eene reisherinnering," in *Album der natuur* (1867), 308–311.

II. SECONDARY LITERATURE. See C. P. L. Groshans, "Levensbericht van Jan van der Hoeven," in *Levensberichten der afgestorvene medeleden van de Maatschappij der Nederlandsche Letterkunde* (Leiden, 1870), pp. 52–121; P. Harting, "Levensbericht van Jan van der Hoeven," in *Jaarboek der K. akademie van wetenschappen* (1868), pp. 1–34, with a fairly complete bibliography of van der Hoeven's publications; and C. J. van der Klaauw, *Het hooger onderwijs in de zoölogie en zijne hulpmiddelen te Leiden* (Leiden, 1926), esp. pp. 8–20, 64–94.

PIETER SMIT

HOFF, KARL ERNST ADOLF VON (*b.* Gotha, Germany, 1 November 1771; *d.* Gotha, 24 May 1837), *geology, geography.*

Hoff was the son of Johann Christian von Hoff, privy councillor in Gotha, and Johanna Friederike Sophie von Avemann. After thorough private instruction at home, he attended the Gymnasium in Gotha from 1785 to 1788, when he enrolled in the University of Jena. Following his father's wishes, he studied law and diplomacy as well as history at Jena, although he was drawn to mathematics and the natural sciences. In Jena and especially at the University of Göttingen, where he transferred after two years, he undertook additional studies in physics and the natural sciences. The latter were carried out mainly under the guidance of Blumenbach, who also introduced him to geology and with whom he was to remain close friends. In 1791 he entered Gotha's civil service. He rose from *Legationssekretär* to director of the Ober-Consistorium (1829). As representative of the duchy of Gotha, he participated in diplomatic events of considerable import. He signed the Rhenish Confederation Act in 1806, took part in the Congress of Erfurt, and, in 1817 at Frankfurt am Main, signed for Gotha's entry into the Germanic Confederation. He thus proved himself in his occupation at a time when Europe and the Thuringian duchies were extremely unstable and diplomacy a task of extraordinary difficulty. It is all the more to be wondered at that Hoff simultaneously accomplished first-rate work in a totally different field—scientific research, notably in geology and geography. Indeed, he introduced a new epoch of geological study which continues still.

Hoff's development as a researcher was aided (as Reich has shown) by the lively intellectual life that prevailed in Gotha. Duke Ernst II encouraged scientific study, especially natural philosophy. Thus in 1786 he called the astronomer Zach to Gotha and built an observatory on the Seeberg, which came to be widely respected. The duke favored having at court civil servants who at the same time were scholars. Gotha's publishing firms were extraordinary for a city of 10,000 inhabitants. For example, the Ettinger firm published 800 volumes of the city's learned periodicals within twenty years. Hoff joined the circle that included his cousin Adolf Stieler and Ernst Friedrich von Schlotheim, a student of Werner who since 1793 had also been in the ducal service and who had become at the same time a leading paleontologist. Hoff devoted his free time to geology and mineralogy and utilized his official journeys to make field trips, visit major collections, and meet with experts. From Gotha he often made visits, with similarly interested friends, notably Wilhelm Jacobs, to the neighboring Thüringer Wald. There he visited mines, sometimes under the guidance of the mining director Voigt, a leading geologist and also a student of Werner; Voigt, however, had published the most solid demonstrations against the latter's Neptunian theory. Since there existed no special periodical for geology and mineralogy, Hoff founded the *Magazin für die gesamte Mineralogie, Geognosie und mineralogische Erdbeschreibung* (1801). Despite the recognition it received, it ceased publication on the death of its publisher.

In 1801 Hoff began to publish his own works, which quickly widened his reputation and earned him the friendship of Humboldt, Buch, and Goethe. He wrote numerous articles on particular problems and the first detailed descriptions of the geology of individual regions, especially of Thuringia. Of special significance was his *Das teutsche Reich vor der französischen Revolution und nach dem Frieden zu Lunéville,* of equal importance for both history and geography.

Hoff worked closely with the Gotha publishing firm of Perthes, which was at that time establishing its international reputation through the efforts of a few unselfish scholars (Hoff, for example, died poor). A map of Germany on which Stieler collaborated initiated Hoff's many-sided and significant cartographical contributions. There also appeared a collaborative travel book, *Der Thüringer Wald,* in which Hoff treated the regional geology and mineralogy and Jacobs the botany and technology. Here, as in other of his writings, Hoff presented important data on stratigraphy and sediment formation, as well as on surface formation and the history of valleys. Moreover, he wrote an important article refuting the "aqueous" origin of basalt through the description of the basalt outcrops in the vicinity of Eisenach and of their contact effects. Here Hoff, originally a Neptunist himself, pointed out that the sandstone contiguous to the basalt was "altered in just such a way as would result from heating."

All these accomplishments, which were based on careful studies and exact original observations, brought Hoff many honors and valued responsibilities. (In 1817 he shared in the reform of the University of Jena, in 1832 he was given the superintendence of all scientific and artistic collections in Gotha, and at the end of his life he was an honorary member of fifteen scientific societies.) Hoff's studies have been superseded to the extent that his name would today be known to only a few specialists had not the findings of his detailed investigations led him to formulate a revolutionary, comprehensive principle and new method of study, *Aktualismus* (actualism).

At the beginning of the nineteenth century it was widely accepted that the alterations that the earth had experienced in the course of its history were produced by sudden events of catastrophic magnitude, which in force far surpassed existing phenomena. Diluvianism, or the attribution to the Flood of geological alterations, had been such a conception. Cuvier (1769–1832) developed the catastrophist theory in a particularly consistent manner. Upheavals of the earth's crust had convulsed the sequence of rocks and destroyed the organic world, and several such catastrophes, with consequent new creations of life and intervening quiet periods, had taken place. Geological research was dependent on this theory.

The forces currently acting on the earth, then, appeared too insignificant to produce fundamental changes in the structure of the earth's surface.

At an early date Hoff presented observations which contradicted this view. For example, in 1807 he published an account of a newly formed island in the Havel River, and in 1812 he returned to Göttingen for a longer time in order to collect from the literature as many descriptions as possible of contemporary alterations in the surface of the earth. As early as 1814 he had formulated his new fundamental principle for the formation of Lower Permian sandstone conglomerates, namely, that less heed should be paid to great forces than to great periods of time. "With periods of time it is completely unnecessary to behave thriftily in the history of the earth, but one must surely do so with forces . . . this is certainly the case in geology: *gutta cavat lapidem non vi, sed saepe cadendo* (the drop hollows out the stone not by force, but by falling so often)."

The Sozietät der Wissenschaften in Göttingen had offered a prize for "the most thorough and comprehensive investigation of the alterations of the earth's crust that can be demonstrated in its history, and for the application one can make of this knowledge through research on terrestrial upheavals that lie outside the realm of history." Hoff's prize-winning work, composed in 1821, appeared in 1822 as the first volume of his *Geschichte der durch Überlieferung nachgewiesenen natürlichen Veränderungen der Erdoberfläche.* Opposing the given theme, which, as Vogelsang has emphasized, actually presupposed the catastrophist theory, Hoff insisted that one must in the first instance study the effect of those forces whose work we are able to observe today and apply this knowledge to the earliest history of the earth. Through these premises, later designated *Aktualismus,* he consciously rejected the catastrophist theory. He did not make headway at first. Not until Lyell's *Principles of Geology* attained wide dissemination as a comprehensive textbook, supported by observations throughout the world, did actualism gain ascendancy and find lasting application in countless investigations.

Lyell, to be sure, went further than Hoff, in that he denied "any progress whatsoever in the developmental history of the earth, and speaks only of perpetual transformation, where others believe they find a very gradual development from a once totally different state" (B. Cotta, 1857). Lyell's theory meant that in quantity and quality no forces and agencies other than those that we now experience have ever been active in the course of the history of the earth. Such a conclusion has more than once led to qualification or criticism of actualism in general (Andrée, Beurlen, Kaiser, *et al.*). This criticism met only Lyell's uniformitarianism, however, not actualism in Hoff's sense. Hoff himself had made allowance for this qualification in that he stated, "A limit will be found beyond which almost no known physical laws and facts will obtain. . . . But to search for these limits appears to us to be the most reasonable goal." And Johannes Walther, the most important spokesman for actualism (he named it the "ontological method"), included in his work of 1893 a chapter on the limits of this method and declared "that there have existed in all geological periods biological and physical phenomena foreign to the present. . . . Only as we become fully conscious of the limit of the ontological method will it attain its true value." From this vantage actualism has continued to justify itself. Hooykaas (1962) has also recently confirmed the distinction between Lyell's and Hoff's ideas.

BIBLIOGRAPHY

I. Original Works. Reich (see below) lists seventy-seven works by Hoff. The most important are *Das teutsche Reich vor der französischen Revolution und nach dem Frieden zu Lunéville,* 2 pts. (Gotha, 1801–1805); "Einige Bemerkungen über eine in der Havel entstandene Insel," in *Magazin naturforschenden Gesellschaft,* **1** (1807), 233–240; *Der Thüringer Wald,* 2 pts. (Gotha, 1807–1812); "Beobachtungen über die Verhältnisse des Basaltes an einigen Bergen von Hessen und Thüringen," in *Magazin naturforschenden Gesellschaft,* **5** (1810), 347–362; *Gemälde der physischen Beschaffenheit, besonders der Gebirgsformationen Thüringens* (Erfurt, 1812); "Beschreibung des Thonschiefer- und Grauwackengebirges im Thüringer- und Frankenwalde," in *Leonhards Taschenbuch für die gesammte Mineralogie,* **7** (1813), 135–137; "Beschreibung des Trümmergebirges und des älteren Flözgebirges, welche den Thüringer Wald umgeben," *ibid.,* **8** (1814), 319–438; "Merkwürdiges Vorkommen des Basaltes in der Gegend von Eisenach," *ibid.,* **15** (1821), 169–174; *Statistische geographische Beschreibung der Länder des Herzoglichen Hauses Sachsen* (Weimar, 1821); *Geschichte der durch Überlieferung nachgewiesenen natürlichen Veränderungen der Erdoberfläche,* 5 vols. (Gotha, 1822–1841); "Das Nadelöhr im Tale der Werra und Einiges über Talbildung," in *Leonhards Jahrbuch für Mineralogie* (1830), pp. 421–441; *Höhenmessungen in und um Thüringen* (Gotha, 1833); and *Teutschland nach seiner natürlichen Beschaffenheit und seinen früheren und jetzigen politischen Verhältnissen* (Gotha, 1838).

II. Secondary Literature. On Hoff and his scientific work, see K. Andrée, "Karl Ernst Adolf von Hoff als Schriftgelehrter und die Begründung der modernen Geologie," in *Schriften der Königlichen Deutschen Gesellschaft zu Königsberg,* no. 4 (1930); Karl Beurlen, "Der Zeitbegriff in der modernen Naturwissenschaft und das Kausalitätsprinzip," in *Kant-Studien,* **41** (1936), 16–37; "Die Periodizität im erd- und lebensgeschichtlichen Entwicklungsgang," in *Abhandlungen und Verhandlungen des Naturwissenschaftlichen Vereins in Hamburg,* n.s. **12** (Hamburg, 1968), 5–25; Bernhard Cotta, "Einführung," in the German trans. of Lyell's *Geology* (Berlin, 1857); Bruno v. Freyberg, *Die geologische Erforschung Thüringens in älterer Zeit* (Berlin, 1932); Helmut Hölder. "Geologie als historische Wissenschaft," in *Geol. Mitt.,* **3** (1962), 11–13; R. Hooykaas, *The Principle of Uniformity in Geology, Biology and Theology,* 2nd ed. (Leiden, 1962); Erich Kaiser, "Der Grundsatz des Aktualismus in der Geologie," in *Zeitschrift der Deutschen geologischen Gesellschaft,* **83** (1931), 389–407; Otto Reich, *Karl Ernst Adolf von Hoff, der Bahnbrecher moderner Geologie* (Leipzig, 1905), the most complete biography available; H. Vogelsang, *Philosophie der Geologie* (Bonn, 1867); Johannes Walther, *Einleitung in die Geologie als historische Wissenschaft,* 3 vols. (Jena, 1893–1894); and Karl Alfred von Zittel, *Geschichte der Geologie und Paläontologie* (Munich–Leipzig, 1899).

B. V. FREYBERG

HOFFMANN, FRIEDRICH (*b.* Halle, Germany, 19 February 1660; *d.* Halle, 12 November 1742), *medicine, chemistry.*

Hoffmann was a leading medical systematist of the first half of the eighteenth century. Although not a notably original thinker, he became a highly influential teacher and practicing physician in Germany, systematizing coherently the Galenic, iatromechanical, and iatrochemical aspects of the phenomena of health and disease. The attention he focused on the role of the nervous system in physiology and pathogenesis contributed to a gradual shift in medical approach, namely from preoccupation with so-called humors and vascular hydrodynamics to that with neuromuscular action and sensibility. This transformation was reflected in the subsequent medical systems of Cullen and John Brown.

Hoffmann (often called "the younger" to distinguish him from his father) was the son of a well-known municipal physician of Halle, who guided him during his early anatomical and chemical studies. In 1678 the younger Hoffmann went to Jena, where he studied medicine for two years under the direction of Wedel. His interest in chemistry lured him for a short period to Erfurt, where he attended the chemistry lectures of Cramer. In 1681 he received his M.D. from the University of Jena and was allowed to teach there. But the subsequent success of Hoffmann's chemistry lectures reportedly provoked the jealousy of Jena's senior faculty, and the young physician soon left for the city of Minden, where his brother-in-law provided him with an official, salaried position. Two years later, to become acquainted with the activities and methods of other European colleagues, Hoffmann embarked on a tour of Belgium, Holland, and England, where he met Boyle. He returned to the Continent in 1684 and began a successful medical career in the principality of Minden, and in 1688 was named provincial physician for Halberstadt, Saxony, famous for its mineral waters.

In 1693 Frederick III, elector of Brandenburg, chose Hoffmann to become the first professor of medicine at the new University of Halle. Hoffmann was also charged with the organization of the medical school, and his success in the new institution was immediate. His lectures on physics, chemistry, anatomy, surgery, and the practice of medicine attracted a great number of both students and physicians. Given the privilege of selecting a second professor of medicine, he brought to the university in 1694 a former fellow student at Jena, Georg Stahl, originator of the phlogiston theory.

In 1709 Hoffmann was called to Berlin by Frederick I to become the ruler's personal physician. He remained only three years at the court, which was oppressively beset with petty intriguing. After extricating himself from this situation, Hoffmann returned to Halle to resume teaching and medical practice. In 1734 Hoffmann was again summoned to the Prussian court, owing to the recommendation of Boerhaave; as physician he served Frederick William I for about eight months and then returned to Halle.

Hoffmann was a fellow of the Royal Society of London and a member of the Berlin Academy of Sciences, and in 1735 was elected to the Imperial Russian Academy of Sciences in St. Petersburg. In his last years he suffered from a pulmonary ailment, which curtailed his activities and led to his death at the age of eighty-two.

From the time of Hippocrates, physicians had sought to discover and establish the fundamental laws governing the phenomena of health and sickness. It was long believed that medicine would become a truly scientific endeavor, rather than a purely empirical craft, only through the apprehension of rational causes and an understanding of the mechanisms producing disease. This outlook led in time to formulation of an elaborate paradigm of so-called balanced and corruptible humors. For about 2,000 years this formulation sufficed to explain all of physiology and pathology.

As the Galenic concepts of soul, spirit, and faculty became obsolete in the seventeenth century, a new theoretical foundation was needed that would incorporate Harvey's theory of the circulation of the blood, new microscopic observations, and the recent discoveries in physics. Following the influential method of Cartesian mechanical philosophy, physicians began to consider the human body as a machine having constituent particles in constant motion. Hoffmann believed that the laws of mechanics could also explain the normal and pathological changes occurring in an organism. He attempted to interpret the new knowledge according to the physiological ideas of Descartes and thereby to establish a firm theoretical basis for medicine: a system of general principles capable of explaining all physiological phenomena.

For Hoffmann the organism was a machine composed of fluid and solid particles. The fluids—blood, lymph, and animal spirits—provided the continuous and appropriate movements necessary for life; hence, normal and abnormal qualities of the humors and organs were due to various kinds of chemical particles which were distributed in different proportions. The animal spirits, for example, were visualized by Hoffmann as volatile, ethereal corpuscles of matter flowing through the nervous system.

To explain the functions of the body, Hoffmann relied heavily on the Cartesian hydrodynamic schemes. According to these, the various humors flowed through the body's vessels at different rates, depending upon the diameter of the vessels. Pulsations observable on the surface of the cerebral membranes were taken as proof of the circulation of the animal spirits.

Hoffmann considered the fine material particles that constituted the "nerve spirits" to be the principal movers of the body, conferring the necessary vital motions to all other bodily fluids and solids. In these activities the ethereal spirits were directed by an *anima,* or sensitive soul, which Hoffmann conceived of as a subtle, hypothetical form of matter on which God himself had directly impressed motion. Responsible, in an Aristotelian sense, for the form of the body, this soul, or "nature," possessed mechanical powers, or virtues, responsible for the purposeful and apparently goal-directed activities of the organism; moreover, it constituted the material link with the immaterial and rational human mind created by God. Hoffmann thus seems to have provided a Neoplatonic scheme, establishing a hypothetical chain of entities between the divine mind and the coarse particles of the body and thereby bridging the strict Cartesian dualism of spirit and matter.

The *anima* was therefore for Hoffmann the first principle of motion, the directive force using the animal spirits as instruments for all vital motions, including automatic and coordinating functions. Among the powers communicated to the nervous spirits was a plastic, organizing capability that controlled nutritive processes and orderly growth.

Kurt Sprengel believed that Hoffmann had been profoundly influenced by Leibniz, who was a friend, in his conception of the animal spirits. Accordingly, Sprengel tried to interpret Hoffmann's ethereal spirits as veritable aggregates of monads which directed and coordinated their own activities according to a general plan. Hoffmann, on the other hand, envisioned animal spirits as being composed strictly of material particles, empowered by God to perform motions through mediation of the sensitive soul. In this sense, the animal spirits did carry out specific movements, in accordance with certain divinely impressed "ideas," for the preservation and normal development of the organism.

Hoffmann considered the fibers forming the vessels, heart, and muscles to have just the degree of elasticity of tension needed for optimal blood circulation. Tone in the fibers could also be influenced by the animal spirits, he held, rather than being dependent solely on immanent physical cohesiveness.

To Hoffmann, the prerequisite for good health was proper circulation of the humors, attributable to normal tension or tonus in all the fibers. Disease, on the other hand, was the result of distorted vital actions arising from impaired humoral motions and resultant changes in the solid parts of the body. Drawing upon hydrodynamic concepts, he explained a series of pathological changes as being the result of defective circulation, which in turn caused humoral obstructions and stagnations. Many of these disturbances were caused by abnormal motions of the ethereal animal spirits. These spirits increased the tone of certain fibers, producing vascular and intestinal spasms or a diminished fiber tension called atony. According to Hoffmann, local humoral stagnations led in turn to a series of chemical changes responsible for many local lesions. In addition, certain disease-producing environmental factors were brought into the body with inspired air. Hoffmann hypothesized that these miasmas, contagions, and poisonous vapors created primarily a series of blood abnormalities.

In a state of disease, the fluid and solid parts of the human machine appeared to influence each other through a *consensio,* or "sympathy," between certain organs and humors, mediated largely by the animal spirits flowing through the nerves; thus were explained certain systemic reactions to local disturbances, and vice versa.

Hoffmann's system, although extremely hypothetical, replaced in great part that of the older Aristotelian-Galenic faculties, qualities, and spirits. He followed closely the corpuscular views of contemporary physics

HOFFMANN

and chemistry and tried to interpret physiology and pathogenesis strictly in terms of matter and motion. To be sure, Hoffmann's explanations, couched in a new language, were still largely based on the older views. But the view that certain biological processes occurred in a well-coordinated and goal-directed fashion seemed incompatible with the known, contemporary mechanical models, and Hoffmann answered these difficulties by asserting that all unexplainable phenomena responded to a higher form of physics not as yet discovered.

Although its effect on the practice of medicine was then small, Hoffmann's system provided a basis on which further medical ideas and hypotheses could be formulated. His formulations of a series of general principles for understanding the human organism, as well as the formulations of other eighteenth-century systematists, led to the more precise investigations that laid the theoretical foundations of modern medicine.

Hoffmann maintained a lifelong interest in chemistry. His primary contributions here were in the investigation of mineral waters, specifically in improving contemporary analytical methods and distinguishing essential components. Hoffmann studied *spiritus mineralis* (carbon dioxide) in water, which he characterized as a weak acid intermixed with various salts. He also discerned the presence of sulfates in certain waters and clearly separated magnesia from lime. The hot springs of Carlsbad especially interested Hoffmann, who explained that the high temperatures of the waters were caused by a chemical reaction involving sulfur, iron, and oils. Another of his studies, having clear medical implications, was his description of carbon-monoxide poisoning from the fumes of burning charcoal (1716).

While taking substantial issue with Stahl's medical theory, Hoffmann accepted many of his chemical ideas, one exception being that of the existence of phlogiston in metals. Denying phlogiston, Hoffmann held that a calx was formed by the action of acids on metals.

In his work on therapeutic applications of chemical substances, Hoffmann mixed one part ether (*acidum vitrioli vinosum*) with three parts alcohol to create "liquor anodynus minerali Hoffmanni," or "Hoffmann's drops," which became a popular medical panacea.

BIBLIOGRAPHY

I. ORIGINAL WORKS. Hoffmann's voluminous writings have been collected in *Opera omnia physico-medica,* (Geneva, 1740), 6 vols. The first 3 vols. of this compilation contain his most important medical work, the *Medicinae rationalis systematicae,* originally published in 2 vols. (Halle, 1718–1720). The last 3 vols. of the *Opera omnia* include articles and monographs on medical consultations, chemical analyses, and therapeutical indications.

The first *Operum omnium physico-medicorum supplementum* (Geneva, 1754) appeared in 2 pts. This supp. vol. contains some of Hoffmann's famous works, such as *Medicus politicus* (pp. 389–422), the *Commentarius de differentia inter Friderici Hoffmanni doctrinam medico-mechanicam et Georgii Ernest Stahlii medico organicam* (pp. 423–499), and his 1695 work *Fundamenta medicinae ex principiis naturae mechanicis* (pp. 633–676).

The second and last supplement, *Operum omnium physico-medicorum supplementum secundum* (Geneva, 1760), appeared in 3 pts. It contains Hoffmann's *Opuscula physico-medica varii argumenti* and other therapeutical and chemical works.

Among the more important writings of Hoffmann available in other languages are his clinical collection *Medicina consultatoria, worinnen unterschiedliche ueber einige schwehre Casus ausgearbeitete Consilia, auch Responsa Facultatis Medicae,* 2 vols., 10 pts. (Halle, 1721–1733); and the treatise on mineral waters, *Gruendlicher Bericht von der herrlichen Wuerckung, vortrefflichen Nutzen und rechtem Gebrauch des zu Sedlitz in Boehmen neu entdeckten bittern purgierenden Brunnens* (Halle, 1725). There is also a German trans. by Auerbach, *Politischer Medicus, oder Klugheitsregeln, nach welchen ein junger Medicus seine Studia und Lebensart einrichten soll* (Leipzig, 1753).

In France, Jacques-Jean Bruhier translated into French Hoffmann's *Medicinae* as *La médecine raisonnée* (Paris, 1739) and the *Medicus politicus* as *La politique du médecin* (Paris, 1751).

In English an abridged trans. of certain parts of the *Medicinae* concerning fevers, hemorrhages, and spasmodic and atonic diseases was published as *A System of the Practice of Medicine,* 2 vols. (London, 1783), trans. by William Lewis, revised and completed by Andrew Duncan, 2 vols. (London, 1783). On mineral waters see *New Experiments and Observations Upon Mineral Waters, Directing Their Farther use for the Preservation of Health and the Cure of Diseases,* 2nd ed. (London, 1743), extracted from Hoffmann's essays on the subject and illustrated with notes by Peter Shaw. Also available is *A Dissertation on Endemial Diseases, or Those Disorders Which Arise From Particular Climates, Situations and Methods of Living,* trans. (with preface and appendix) by R. James (London, 1746). There is also a small monograph, *A Treatise on the Teeth, Their Disorders and Cure* (London, 1753), a trans. of *Historia dentium physiologicae et pathologicae pertractata* (Halle, 1698). See also Lester S. King's trans. of Hoffmann's *Fundamenta medicinae* (London–New York, 1971).

II. SECONDARY LITERATURE. The standard biography on Hoffmann is Johann H. Schulze, *Commentarius de vita Friderici Hoffmanni,* included in *Opera Omnia,* I, i–xiv. Shorter biographical sketches can be found in A. Hirsch, *Biographisches Lexikon der hervorragenden Aerzte aller Zeiten und Voelker,* 3rd (unchanged) ed., III (Munich,

1962), 256–259, and A. J. L. Jourdan, ed., *Biographie médicale, Dictionnaire des sciences médicales,* V (1822), 239–257. The latter contains a complete list of Hoffmann's writings. Additional biographical documents are W. Piechocki, "Das Testament des Halleschen Klinikers Friedrich Hoffmann des Juengeren (1660–1742)," in *Acta Historica Leopoldina* **2** (1965), 107–144; and G. Mamlock, "Koenig Friedrich Wilhelm I Briefe an den Hallenser Kliniker Friedrich Hoffmann," in *Deutsche medizinische Wochenschrift,* **37,** no. 48 (1911), 2242–2244.

Two comprehensive summaries of Hoffmann's medical system and therapeutics can be found in Kurt Sprengel, *Versuch einer pragmatischen Geschichte der Arzneikunde* (Halle, 1803), pt. 5, pp. 118–148; and Heinrich Haeser, *Lehrbuch der Geschichte der Medizin und der epidemischen Krankheiten,* 3rd ed., II (Jena, 1881), 509–519. Another perceptive sketch is Paul Diepgen, "Zum 275. Geburstage Friedrich Hoffmanns," in *Deutsche medizinische Wochenschrift,* **61,** no. 10 (1935), 389–390.

Indispensable for an understanding of Hoffmann's medical ideas are the recent publications in English by King. Among them is *The Growth of Medical Thought* (Chicago, 1963), ch. 4, pt. 4, pp. 159–174, which deals with Hoffmann's main ideas as expressed in the *Medicinae.* An analysis of Hoffmann's earlier work appeared in King, "Medicine in 1695: Friedrich Hoffmann's *Fundamenta Medicinae,*" in *Bulletin of the History of Medicine,* **43** (1969), 17–29, and in his *The Road to Medical Enlightenment 1650–1695* (London–New York, 1970), ch. 5, pp. 181–204. For a comparison of Hoffmann and Stahl see King, "Stahl and Hoffmann: A Study in Eighteenth-Century Animism," in *Journal of the History of Medicine and Allied Sciences,* **19** (1964), 118–130.

Hoffmann's chemical contributions have been summarized in Johann F. Gmelin, *Geschichte der Chemie,* II (Göttingen, 1798), 170–189, which lists 122 separate chemical publications. A more recent and valuable summary of Hoffmann as a chemist appeared in J. R. Partington, *A History of Chemistry,* II (London, 1961), ch. 19. 691–700.

For an examination of Hoffmann's theological interests see Werner Leibbrand, *Der goettliche Stab des Aeskulap,* 3rd ed. (Salzburg, 1939), ch. 13, pp. 230–236.

GUENTER B. RISSE

HOFMANN, AUGUST WILHELM VON (*b.* Giessen, Germany, 8 April 1818; *d.* Berlin, Germany, 2 May 1892), *organic chemistry.*

Hofmann's influence, as a teacher and experimentalist, on British and German chemistry was profound. He was responsible for continuing the method of science teaching by laboratory instruction that had been established and popularized by Liebig at Giessen, and for transporting it to England and to Berlin. He created his own school of chemists who were interested primarily in experimental organic chemistry and the industrial applications of chemistry, rather than in theoretical problems. Among his distinguished pupils and assistants were the Englishmen F. A. Abel, W. Crookes, H. McLeod, C. B. Mansfield, J. A. R. Newlands, E. C. Nicholson, and W. H. Perkin, and the Germans J. P. Griess, C. A. Martius, and J. Volhard.

Hofmann was the son of Johann Philipp Hofmann, the architect who enlarged Liebig's Giessen laboratories in 1839. He matriculated at Giessen in 1836, electing to study law and languages, but gradually became attracted by Liebig's chemistry classes. He obtained his doctorate in 1841 for an investigation of coal tar but was prevented from becoming Liebig's personal assistant until 1843 because of his devotion to his dying father. In the spring of 1845, through Liebig's influence, he was made *Privatdozent* at the University of Bonn. In the autumn of 1845, following guarantees of tenure at Bonn arranged by Queen Victoria's German consort, Albert, he agreed to direct the Royal College of Chemistry in London, founded by some of Liebig's English pupils as a private school for the training of agricultural, pharmaceutical, geological, and industrial chemists. Despite the financial insecurity of the college until it was absorbed by the government-sponsored School of Mines in 1853, Hofmann remained in London for twenty years. Although offered a chair of chemistry at Bonn in 1863, he preferred to accept the more commanding post at Berlin when it was left vacant by the death of Mitscherlich in the same year. After designing new laboratories for both Bonn (taken by Kekulé) and Berlin, he returned to Germany in 1865. In 1867 he was a founder of the Deutsche Chemische Gesellschaft, on the model of the Chemical Society, of which he was a prominent member. Hofmann was ennobled on his seventieth birthday.

Despite his preeminence as a practical teacher, Hofmann was personally incompetent in the laboratory; consequently he looked for, and showed extraordinary acumen in finding, unusual skills in his assistants, who devotedly performed much of his experimental work. A humorous, lovable, cosmopolitan, and intelligent man, he made a considerable fortune from his scientific work in both England and Germany. He was a good speaker in several languages, although in formal lectures or in letters he often overindulged in literary embellishment and circumlocution. His sense of style is most apparent in a large number of biographical notices and other essays on the history of chemistry (for example, his study of Liebig) which are still consulted by historians.

Hofmann had a thoroughly pragmatic attitude toward the relationship between useful experiments and theory, arguing: "As a chemical theory expands and becomes more and more consolidated, the interest attached to the individual compounds used as scaffolding in raising the structure becomes less and less, diminishing . . . in the inverse ratio of the number of compounds which the theory suggests" (*Proceedings of the Royal Society,* **11** [1860–1862], 425). On the other hand, although undoubtedly guided in his work by certain theoretical principles—particularly those of Laurent, the type theory, and the principle of homology—he was never interested in devising new theories. He remained the practical chemist.

The theme and variations of Hofmann's voluminous scientific publications were coal tar and its derivatives. In his first publication (1843) he clarified a confused situation by showing that many substances which were identified in contemporary chemical literature as obtainable from coal tar naphtha and its derivatives were all a single nitrogenous base, aniline. Both Hofmann and the iconoclastic Laurent (who visited Giessen briefly in 1843) suspected that aniline was related to phenol, which (in modern terms) Laurent correctly supposed to be a hydrate of the phenyl radical, C_6H_5. This relationship was confirmed and elucidated when they successfully converted phenol into aniline by the action of ammonia. Progressive chlorination of aniline, which could be satisfactorily explained only by Laurent's theory of direct chlorine-hydrogen substitution, gradually weakened its basicity. In this manner Hofmann found further evidence for the incredibility of Berzelius' electrochemical theory, in which such substitutions were deemed impossible without the wholesale disruption of the molecule; and he opened a way for the later reconciliation of ideas of polarity with unitary views of molecular constitution. Nevertheless, Hofmann was at first inclined to accept Berzelius' judgment that alkaloids and organic bases (like aniline) were formed by the conjugation of a hydrocarbon radical with ammonia. But by 1849 experiments on a variety of volatile nitrogenous bases led him to suspect that they were substituted ammonia compounds, as Liebig had suggested in 1837.

Wurtz's innocent preparation of the first aliphatic amines, methylamine and ethylamine, by the action of caustic soda on isocyanates in 1849 was crucial to Hofmann's change of position. In a classic paper, "The Molecular Constitution of the Volatile Organic Bases" (*Philosophical Transactions,* **140** [1850], 93–131), he showed how amines could be prepared directly from ammonia by the action of alkyl iodides; whence he concluded that organic bases were substituted ammonia in which hydrogen was replaced by hydrocarbon radicals:

Ammonia	Aniline	Ethylaniline
$\mathrm{N}\left\{\begin{array}{l}\mathrm{H}\\ \\ \mathrm{H}\end{array}\right.$	$\mathrm{N}\left\{\begin{array}{l}\mathrm{C_6H_5}\\ \mathrm{H}\\ \mathrm{H}\end{array}\right.$	$\mathrm{N}\left\{\begin{array}{l}\mathrm{C_6H_5}\\ \mathrm{C_2H_5}\\ \mathrm{H}\end{array}\right.$
Ethylamine	**Diethylamine**	**Triethylamine**
$\mathrm{N}\left\{\begin{array}{l}\mathrm{C_2H_5}\\ \mathrm{H}\\ \mathrm{H}\end{array}\right.$	$\mathrm{N}\left\{\begin{array}{l}\mathrm{C_2H_5}\\ \mathrm{C_2H_5}\\ \mathrm{H}\end{array}\right.$	$\mathrm{N}\left\{\begin{array}{l}\mathrm{C_2H_5}\\ \mathrm{C_2H_5}\\ \mathrm{C_2H_5}\end{array}\right.$

The analogy of these primary, secondary, and tertiary amines (as Gerhardt later called them) with ammonia was completed in 1851, when Hofmann prepared crystalline quaternary salts that were analogous to ammonium salts (for instance, tetraethylammonium iodide, $(C_2H_5)_4NI$). These discoveries made Hofmann's reputation and subsequently formed one of the pillars of the type theory of Gerhardt and Williamson, in which both organic and inorganic compounds were systematized and classified according to the model formula of one of four inorganic molecules—hydrogen, hydrogen chloride, water, or ammonia—by the substitution of one or more atoms of hydrogen for an equivalent atom or group. In his own researches Hofmann exploited only the ammonia type, although in his interesting and much translated textbook, *An Introduction to Modern Chemistry* (1865), he used all four types for pedagogic purposes.

Hofmann's lifelong interest in the nitrogen bases, which included the development of methods for separating mixtures of amines and the preparation of large numbers of "polyammonias" (diamines and triamines such as ethylenediamine and diethylenediamine), was extended to phosphorus bases in joint work with Cahours between 1855 and 1857. In another collaboration with Cahours in 1857, Hofmann prepared the first aliphatic unsaturated alcohol, allyl alcohol, C_3H_5OH. Subsequently, in 1868, Hofmann's investigation of its derivative, allyl isothiocyanate (mustard oil), led to the preparation of many sulfur analogs of the isocyanates and the development of an elegant and heroic method for the preparation of the disgusting isonitriles (isocyanides, or carbylamines) by the action of alkalinated chloroform on primary amines. Hofmann's love of analogy and his tenacity and thoroughness are well illustrated by his intermittent investigation of these nauseous compounds and by his twenty-year search for the lower homologue of acetaldehyde. His conviction that a methyl aldehyde (formaldehyde) must exist was rewarded in 1867 when he passed an air stream of methyl alcohol over incandescent platinum.

In 1848 Hofmann's eccentric student C. B. Mansfield devised the fundamental method of fractional distillation of coal tar for the separation of pure benzene, xylene, and toluene, thus laying the foundation for the coal tar products industries. Little was known of the detailed structures of organic compounds in the 1850's; nevertheless, there were many rational attempts to synthesize important natural products. In 1856 another of his students, W. H. Perkin, privately attempted to synthesize quinine but was led instead to the preparation, and subsequent production, of the first artificial dyestuff, aniline purple, or mauve. In view of Perkin's youth and inexperience, Hofmann tried, without success, to dissuade him from embarking on an uncertain industrial venture. Many of Hofmann's other pupils also became involved in the British dyestuffs industry, notably E. C. Nicholson, G. Maule, and G. Simpson, whose firm progressed from the manufacture of nitrobenzene and aniline to that of aniline dyes.

Hofmann was fascinated by the chemistry of dyes; and although he was not interested in the problems of large-scale industrial research, he well understood what the ideal symbiotic relationship between pure and applied research should be. It was failure to understand this, and commercial ineptitude, which allowed the initial British advantage in dyestuffs to be lost to Germany following Hofmann's return and the early retirement of his English pupils after they had made their fortunes. Hofmann's German pupils (like Martius) proved of different mettle, while German manufacturers, unlike their British counterparts, grasped the idea that the secret of commercial success lay in scientific research. Although a more complex matter, some later commentators thus attributed the decline of the British chemical industry to Hofmann's departure from London.

In 1862 Hofmann isolated from the French commercial dye fuchsine (or magenta) a triamine which was identical with the crimson solution he had obtained in 1858 when reacting carbon tetrachloride with commercially impure aniline. This dye, a derivative of triphenylmethane which Hofmann named rosaniline, could not be prepared from pure aniline, for orthotoluidine and paratoluidine had to be present, as in the commercial product. In 1863 he succeeded in displacing hydrogen in rosaniline with aniline (thus phenylating a compound for the first time) and preparing the beautiful diphenylrosaniline, or aniline blue. When alkyl groups were substituted in rosaniline, Hofmann found that an exciting range of colors from blue to violet was produced. He patented these "Hofmann's violets" (trimethylrosaniline and triethylrosaniline) in 1863 and, despite their instability, their brilliant hues enjoyed a considerable commercial success.

In the controversies between the old and modern chemical notations, Hofmann sided with innovation. From 1860 on, he adopted Gerhardt's, and subsequently Cannizzaro's, atomic weights based on the value 16 for oxygen. He also played a significant role in the dissemination of the concept of valence—the word is derived from Hofmann's term "quantivalence" (1865). As a teacher he devoted much time and skill to devising interesting lecture experiments, many of which are still used. He was the first chemist to popularize atomic models (in 1865). But such was the speed of chemical progress in the 1860's that, to some extent, Hofmann was left behind. He continued to use type formulas long after younger chemists, like Kekulé, had adopted the structural theory of carbon compounds. On the other hand, in 1865, inspired by Laurent, he suggested a systematic nomenclature for hydrocarbons and their derivatives which was adopted internationally, with modifications, by the Geneva Congress in 1892.

BIBLIOGRAPHY

I. ORIGINAL WORKS. Over 300 of Hofmann's papers are listed in the *Royal Society Catalogue of Scientific Papers,* III, VII, X, XV (London, 1867–1925), with obituaries listed in XV. Books and essays are included in *British Museum General Catalogue of Printed Books,* CV. An unpublished bibliography listing 377 items is Kathleen Mary Hammond, "August Wilhelm von Hofmann" (University of London, diploma in librarianship, 1967), copies of which are at University College, London, and Imperial College Archives, London. Hofmann wrote some fifty obituary notices, which are not listed in the aforementioned items but are indexed in Lepsius (see below).

MS collections include Imperial College Archives, London; Bayerische Staatsbibliothek, Munich (the Liebig-Hofmann correspondence, of which a German-English ed. by E. Wangermann and W. H. Brock is in progress); Chemische Gesellschaft in der D.D.R., Berlin; and Vieweg, Brunswick.

II. SECONDARY LITERATURE. The basic studies are J. Volhard and E. Fischer, *August Wilhelm von Hofmann, ein Lebensbild* (Berlin, 1902), also pub. as a special no. of *Berichte der Deutschen chemischen Gesellschaft,* **35** (1902); and the Hofmann memorial lectures by F. A. Abel, H. E. Armstrong, W. H. Perkin, and L. Playfair, in *Journal of the Chemical Society,* **69** (1896), 575–732, repr. in *Memorial Lectures Delivered Before the Chemical Society, 1893–1900,* I (London, 1901). Also useful is B. Lepsius, *Festschrift zur Feier des 50 jährigen Bestehens der Deutschen chemischen Gesellschaft* (Berlin, 1918), a special no. of *Berichte der Deutschen chemischen Gesellschaft,* **50** (1918). All the important literature on Hofmann, together with an analysis of his work, is given in J. R. Partington, *A History of Chemistry,* IV (London-New York, 1964), 432–444; and in an unpublished thesis by J. R. F. Guy, "Life and Work of A. W. Hofmann," B. S. thesis (Oxford, 1969). See also J. Bentley, "The Chemical Department of the Royal School of Mines. Its Origins and Development Under A. W. Hofmann," in *Ambix,* **17** (1970), 153–181; and, for Hofmann's industrial influences, John J. Beer, *The Emergence of the German Dye Industry* (Urbana, Ill., 1959), *passim;* and E. R. Ward, "Charles Blatchford Mansfield, Coal Tar Chemist and Social Reformer," in *Chemistry and Industry* (25 October 1969), pp. 1530–1537.

W. H. BROCK

HOFMEISTER, WILHELM FRIEDRICH BENEDIKT (*b.* Leipzig, Germany, 18 May 1824; *d.* Lindenau, near Leipzig, 12 January 1877), *botany.*

Hofmeister was the son of Friedrich Hofmeister and his second wife, the former Frederike Seidenschnur. The father was the highly successful founder of a music shop and music publishing house in Leipzig, and his home was frequented by men of the arts and sciences. After a friendship with H. G. L. Reichenbach awakened in him a serious interest in botany, he constructed a large herbarium and acquired extensive grounds in a suburb, where he established a botanical garden and built a large home. He increasingly devoted the general bookshop that he kept as an adjunct to his music business to botanical books, some of which he published himself.[1] In 1834 he was elected a corresponding member of the Bavarian Botanical Society.[2] His occupation is usually given as bookseller, but in 1837 it is listed as teacher of botany.[3] His earliest attempts to interest Wilhelm in botany were unsuccessful: the son initially preferred entomology. Also, the father's interests were largely in the systematics of plants; the son was to concern himself with their structure and function.

Wilhelm Hofmeister completed his secondary schooling in 1839. His education, although excellent, did not lead to the "classical" high school diploma, passport to a German academic career—evidence that such a career was not intended. From 1839 to 1841 he was apprenticed to a family friend, August Cranz, the owner of a music shop in Hamburg. He stayed in the Cranz home and after work was free to pursue his studies. He took language lessons and studied physical science and mathematics by himself.

Returning to Leipzig in 1841, Hofmeister became foreign correspondent in his father's business. This post initially left him ample time for study and travel. By a settlement drawn up in 1847 but not made public until 1852, the father, for a stipulated pension, turned over the major part of his business—the music enterprises—to his two sons, Adolph and Wilhelm. He retained the bookshop, "largely consisting of botanical works illustrated with copper plates . . . because of a special inclination towards natural history."[4] Later in 1852 he settled the publishing end of his natural history operation on his son-in-law, Ambrosius Abel. For the next decade Hofmeister combined, almost certainly at eventual damage to his health, a full-time business career as a music publisher with that of a research botanist. To find time for his studies he habitually rose at 5 A.M. The connection with the firm was not broken when he became professor of botany at Heidelberg in 1863.

In 1847 Hofmeister married Agnes Lurgenstein, the daughter of a Leipzig industrialist. They moved into his parents' house at Reudnitz, where eight of their nine children were born. Of these children three died in infancy and three more during Hofmeister's lifetime; he was survived by three daughters. In the period 1870–1875 his half-brother, his wife, his youngest daughter, and the two surviving sons died. In 1876, less than a year before his death, he married Johanna Schmidt, the daughter of a physician.

Hofmeister was short, dark, and extremely vivacious. His severe myopia had a considerable influence on his botanical achievements. While it was an obvious handicap in the field, it automatically turned his attention to minute detail and to the appreciation of small plants. He was exceptionally skilled in making the microscopic preparations on which his major contributions were based. He brought his face extremely close to the material and thus directly performed delicate manipulations, for which others would have to employ a dissecting microscope (which he also knew how to use to the greatest advantage). On the other hand, his poor eyesight presented major difficulties in gross experimentation and resulted in occasional explosions of rage.[5] He stubbornly refused to wear glasses, a disastrous omission in both botany and everyday life.[6]

The middle of the nineteenth century marked a period of extensive change in botany. Most botanists had been engaged in surveying and classifying the world's floras; few workers had attempted to understand the structure and function of plants. In this rapidly changing situation German botany was to play a relatively large role. Conversely, French and Anglo-Saxon botanists tended more toward classification. To be sure, some German botanists did work in systematics, classifying collections made by German travelers or "laid off" to willing, competent labor by Kew and other institutions. Yet much of the immediate stimulus for concern with structure and function had come from C. F. B. de Mirbel in France and especially Robert Brown in England; the latter's contributions to structural botany were possibly more appreciated in Germany than in England, and Hofmeister saw Brown in the light of the founding father.[7] Leadership then passed largely to Hugo von Mohl, whom Hofmeister esteemed highly.

Hofmeister began his serious botanical studies in 1841, after entering his father's business. He came under the influence of Schleiden's famous textbook, published the following year. Schleiden threw down the gauntlet to the status quo; progress was to come through the study of life history and cell structure, fields in which Hofmeister was entirely self-taught. In the area of systematics, however, he had received valuable private instruction not only from his own father but also from the two Reichenbachs, both professors of botany. H. G. Ludwig Reichenbach had made brilliant, almost visionary, contributions to the overall classification of plants (he coined the name Chlorophyta to denote an inclusive group from algae through cycads);[8] his son, H. Gustav Reichenbach, was Hofmeister's close friend.

In seeming irony, Hofmeister's first efforts in botany were directed to the demolition of Schleiden's concept of fertilization, using the methods advocated by Schleiden.[9] In flowering plants a preexisting egg cell commences development into an embryo after having been fertilized by something (now known to be a sperm nucleus) brought to it by the tip of the pollen tube. Contrary to this process, Schleiden had supported the interpretation that the embryo arises from the tip of the pollen tube (which thereby becomes the "female" component) and is induced to further development when it reaches the embryo sac. The question was then a matter of wide and tempestuous controversy. Hofmeister's first paper (1847), dealing with fertilization in the Onagraceae, is stylistically typical. Exclusive of legends, it occupies a mere three pages (six columns). There is no introduction; it starts immediately with the crucial facts that the embryo is derived from a cell preexistent in the embryo sac before fertilization and that the tip of the pollen tube can be removed without damage to the embryo sac and embryo. There is the comment that Schleiden's views obviously are not applicable to the material at hand. Hofmeister found two of Schleiden's figures inexplicable; in another the preparation would probably have been clarified with a simple touch of the dissecting needle. The implications are that the facts speak for themselves and those who need further explanation do not deserve it. The embryological observations were splendidly extended to nineteen families in Hofmeister's first book (1849). Several other embryological contributions followed during the next ten years, during which time the supporters of Schleiden's view had capitulated.

Hofmeister had already touched on the question of cell division in his first paper. His second (1848), on pollen formation, dealt largely with this phenomenon. The process of cell division was then beginning to be understood hazily; the description of nuclear multiplication was that the nucleus disappeared and two new nuclei were later formed. Hofmeister's illustrations seem to show that he was one of the earliest workers to observe chromosomes; he had no understanding of their significance.

On the basis of his reputation, the University of Rostock in January 1851 awarded the twenty-six-year-old Hofmeister an honorary doctorate of philosophy and master of liberal arts degree.[10] This unusual event preceded the publication later that year of *Vergleichende Untersuchungen . . .*, the work for which he is now remembered.

The green land plants, although of tremendous diversity, are a related whole and share a common life cycle, which in its simplest form (in ferns and mosses) is as follows. From the resistant spore there develops a plant body, the gametophyte, bearing reproductive organs, the antheridia and archegonia. The antheridia release cells which become male gametes, or sperms; a sperm then fertilizes the egg, or female gamete, contained in the archegonium. The fertilized egg develops into an embryo which grows into a second plant body, the sporophyte. Gametophyte and sporophyte constitute two dissimilar generations in determined succession in the life history, an "alternation of generations"—a misapplied phrase purloined from zoology. (Concomitant with fertilization is a doubling of the chromosome number, corrected by a reduction division during spore production. This final understanding was provided by Eduard Strasburger, after Hofmeister's death.)

At the advanced end of the scale, in the conifers and flowering plants, matters are greatly obscured by the simplification and change in proportion of some features and the superposition and intercalation of others, intelligible through a consideration of intermediate forms. Hofmeister's predecessors had futilely tried to work backward, from the complex to the simple, attempting to interpret the simpler cryptogams in terms of the complex phanerogams, which they wrongly believed they understood. Floundering and confusion persisted to Hofmeister's day. The situation was overripe for solution by someone who could arrange all the pieces in order and consider the whole. Hofmeister did.[11]

In 1849 Hofmeister published a preliminary note

in which he corrected certain gross errors of his contemporaries. He stressed that sexuality, in the sense of a fertilization process, was documented both in the cryptogams and phanerogams; and he pointed out its correct place in the life history. Further, the conifers, until then systematically misplaced, are indicated as a key connecting group;[12] their life history is briefly interpreted in terms of simpler, intermediate plants.

The message of this note probably did not reach most botanists, but there were exceptions. Mettenius promptly concurred in some of the findings.[13] Arthur Henfrey, who had published on flowering plant fertilization, also in opposition to Schleiden, published the paper in English translation.[14] Assuming the role of Hofmeister's apostle, in 1851 he gave a broad report on the issues to the British Association for the Advancement of Science.[15] Hofmeister's next publication, overlooked by his biographers, was a book review (1850) in which he gave a lucid and simple summary of his broad findings, including the first exposition of alternation of generations in the Hofmeisterian sense.

In 1851 Hofmeister's most famous work, *Vergleichende Untersuchungen,* appeared. Without a word of introduction it begins: "The mature plant of *Anthoceros* appears . . . ." The details of its structure and life history are described and copiously illustrated, entirely on the basis of original observations, followed by a brief critique of earlier work on the genus. A similar description of the next plant's life history is followed by others, in order of increasing complexity. The amount of new information is immense; the errors are minor and do not affect the overall picture. In a concluding three-page "Review" the concept of alternation of generations is explained, and the main modifications of the life history in the different groups are briefly touched upon. Clearly, a page-by-page reading is presupposed, not from the author's arrogance but from his failure to comprehend that others might be less deeply involved. The illustrations are largely from microscopic preparations; prerequisite knowledge of the gross features of the plants would enable the reader to correlate the details.

With this single publication, the core of botany passed from its Middle Ages to the modern period. The book was obviously so important that the two main German botanical journals carried very laudatory reviews by their editors.[16] Although sensing that a revolution had come, they seemed overpowered and possibly did not quite understand exactly what had happened. Not so Henfrey. He promptly wrote the sorely needed commentary, brought in the flowering plants directly, put together a plate to illustrate crucial homologies, provided the elementary textbook-type table of comparative life cycles, and brought it all from the level of the research worker down to that of the student. This work was forthwith translated into German.[17] As issued, *Vergleichende Untersuchungen* tended to collapse with heavy reference or textbook use. This explains its rarity.[18]

Two proposals, in 1852 and 1853, to prepare an English translation were unrealized. During the next decade Hofmeister published a series of supplementary papers, all of which were incorporated into a second edition,[19] which exists only in English, translated by F. Currey, secretary of the Linnean Society, and published in 1862.

Despite Henfrey's valiant efforts, English botany had remained largely unaware of the Hofmeisterian revolution. Obviously, the publication of Darwin's *Origin of Species* made a belated English translation imperative. Hofmeister's pre-Darwinian work constitutes the greatest broad evolutionary treatise in botany because it is organized on a basis of increasing complexity. The plants described do not constitute an evolutionary series any more—or less—than does the zoological sequence of amphioxus, shark, frog, lizard, pigeon, and rabbit. The interpretation of Hofmeister's work as phylogenetic has been vigorously attacked.[20] Did Hofmeister in fact see any evolutionary implications? He made some interesting pre-Darwinian statements in 1852.[21] He saw the major groups as sharply separated by unique characters. But within these groups, attempts to arrive at an ordered arrangement frequently led to an artificial separation of forms "which—to use the common phrase—are closely related. The more one's understanding progresses here, the more the truth of the old saying *natura non facit saltus* becomes apparent." Much later he was to write on selection.[22] There, relationship, "a term used by scientists of all periods," is stated to have meaning only if taken as true consanguinity. In the same paragraph the land plant groups are discussed in evolutionary terms.

In 1863 the Baden government requested faculty opinion on Hofmeister as a candidate for the vacant chair of botany at Heidelberg: "He is represented to us as one of the leading botanists in Germany, a man with the talent of a genius, highest diligence, and excellent powers of exposition, who now for the first time appears inclined to accept an academic teaching post, but for whom the offer of an appointment in Hamburg seems assured."[23] Without waiting for a response, the minister of education three weeks later made the extraordinary—and risky—appointment of this man, whose sole university connection was an honorary degree. The Hofmeisters moved to Heidelberg at the end of July 1863. Hofmeister held the professorship there until 1872, when he accepted the chair of botany at the University of Tübingen, as Hugo von Mohl's successor. Hofmeister's lectures were appreciated by more advanced students but went hopelessly over the heads of beginners. In the laboratory he was a superb teacher. Karl Goebel became his best-known student.

Hofmeister's ambitious project of a handbook of physiological (in the sense of nonsystematic) botany, of which he was the editor, dates from the early Heidelberg period. It remained incomplete, lacking a new version of the material of *Vergleichende Untersuchungen.* He did, however, complete two treatises, the first of which, *Die Lehre von der Pflanzenzelle* (1867), was original in stressing the functional aspects of cytology. It aimed at explaining phenomena such as the colloidal properties of the protoplasm and imbibition properties of the cell wall in terms of physical science. The cellular organization of the plant body is recognized as yielding "the greatest possible strength with the least possible mass." The second treatise, *Allgemeine Morphologie der Gewächse* (1868), represented a total break with the past. Not one of the 192 illustrations depicts an entire plant, or even a single mature plant organ. The stress is on organization of growing points and matters affecting the relative positions of organs. It was, in fact, the first textbook of plant morphogenesis. Hofmeister's research papers from 1859 were also mainly morphogenetic. His interest in plant movements led him to tropisms, growth movements controlled by the environment (discussed also in both of the above treatises). He decided—misled by a multiplication error to an excessive value of cell wall and tissue tension—that the response of roots to gravity was passive, a theory amply disproved before he ever became involved.

Hofmeister would have done well to withdraw from the botanical scene after having reached its pinnacle with *Vergleichende Untersuchungen.* Although monumental and full of major contributions, his subsequent work contains major errors. The phenomenal earlier successes may have led him to consider himself infallible: admitting an error apparently became impossible for him and he vilified his critics.[24] Adverse criticism, which affected his health, overwork from the dual roles of professor and commercial publisher, and the deaths in his family led to his collapse. The first of several strokes came on his fifty-second birthday, in 1876, and he had to resign his post. He died less than a year after his second marriage.

NOTES

1. See his announcements and advertisements in *Flora,* **10** (1825) and later.
2. *Flora,* **17** (1834), 233.
3. *Verhandlungen der Gesellschaft deutscher Naturforscher und Ärzte,* **15** (1838), 10.
4. His own words, from the published announcement; see *Tradition und Gegenwart.*
5. E. Pfitzer, "Wilhelm Hofmeister," p. 274.
6. See his daughter Constanze, in Karl von Goebel, *Wilhelm Hofmeister,* p. 166.
7. *Botanische Zeitung,* **17** (1859), 374. Hofmeister had visited Brown at the British Museum in 1857. His letter home—see Goebel, *op. cit.,* p. 159—tells much about both persons: "This active eighty-four-year-old 'Prince of Botanists' received me with high esteem, gave a breakfast in my honor, and dug out a good part of his curios." The occasion for the trip was a lawsuit filed by Hofmeister against the English agents of the publishing house. Hofmeister lost.
8. Hofmeister's own approach to flowering-plant systematics may reflect this influence. See Goebel, *op. cit.,* pp. 40 ff.
9. ". . . on the refutation or proof of which progress or arrest in this branch of science are first of all dependent . . ." "Ueber die Fruchtbildung und Keimung der höheren Cryptogamen" (1849), col. 793.
10. *Botanische Zeitung,* **9** (1851), col. 224. In 1867 he received an honorary M.D. from Halle. This caused him great pride, because at the same time his hero Bismarck and the field marshals Helmuth Moltke and Albrecht Roon also received honorary degrees.
11. The potential runner-up was clearly William Griffith, who covered much the same botanical material, if rather more hastily and less accurately. He died, from the effects of having crisscrossed the Indian subcontinent on foot, before he had time to arrange his notes.
12. See his earlier footnote, in *Die Entstehung des Embryo der Phanerogamen* (1849), p. 58.
13. See *Vergleichende Untersuchungen,* p. 111; and "Zur Uebersicht der Geschichte von der Lehre der Pflanzenbefruchtung" (1856).
14. *Botanical Gazette* (London), **2** (1850), 70–76.
15. On the reproduction and supposed existence of sexual organs in higher cryptogamous plants, in *Report of the British Association for the Advancement of Science,* **21** (1852), 102–123.
16. A. E. Fürnrohr, in *Flora,* **34** (1851), 765–770; D. F. L. von Schlechtendal, in *Botanische Zeitung,* **9** (1851), cols. 808–810.
17. *Annual Magazine of Natural History,* 2nd ser., **9** (1852), 441–461, pl. 17; *Tagsberichte über die Fortschritte der Natur- und Heilkunde,* Botanische Abt., no. 622 (1852), 289–296, pl. 8; no. 626 (1852), 297–304; no. 629 (1852), 205–309.
18. Of the approximately 100 extant copies, only five have been traded during the last twenty years.
19. The final changes made at this time, which were gathered for German readers in *Jahrbuch für wissenschaftliche Botanik,* **3** (1863), 259–293, indicate that he was past his prime.
20. Goebel, *op. cit.,* pp. 58–59; and W. Zimmermann, in *Repertorium novarum specierum regni vegetabilis,* **58** (1955), 286–287.
21. *Flora,* **35** (1852), 10.
22. *Allgemeine Morphologie der Gewächse* (1868), pp. 564–579, see particularly p. 569.
23. Pfitzer, *op. cit.,* p. 272; Goebel, *op. cit.,* p. 160.
24. Goebel, *op. cit.,* p. 118. See also *ibid.,* p. 62; and *Berichte der Deutschen botanischen Gesellschaft,* **30** (1912), 65, for Hofmeister's scandalous treatment of Strasburger, probably his only intellectual superior on the scene.

BIBLIOGRAPHY

I. ORIGINAL WORKS. Lists of Hofmeister's papers may be found in the footnotes of the Goebel and Pfitzer biographies (see below) and in the Royal Society *Catalogue of Scientific Papers.* Note, however, that the first paper listed in the *Catalogue* is by a W. Hoffmeister. Among his writings are "Untersuchungen des Vorgangs bei der Befruchtung der Oenothereen," in *Botanische Zeitung,* **5** (1847), cols. 785–792, pl. 8; "Ueber die Entstehund des Pollens," *ibid.,* **6** (1848), cols. 425–434, pl. 4; cols. 649–658, 670–674, pl. 6; *Die Entstehung des Embryo der Phanerogamen. Eine Reihe mikroskopischer Untersuchungen* (Leipzig, 1849); "Ueber die Fruchtbildung und Keimung der höheren Cryptogamen," in *Botanische Zeitung,* **7** (1849), cols. 793–800; review of a book by Mercklin, in *Flora,* **33** (1850), 696–701; *Vergleichende Untersuchungen der Keimung, Entfaltung und Fruchtbildung höherer Kryptogamen (Moose, Farrn, Equisetaceen, Rhizocarpeen und Lycopodiaceen) und der Samenbildung der Coniferen* (Leipzig, 1851); "Zur Uebersicht der Geschichte von der Lehre der Pflanzenbefruchtung," in *Gelehrtre Anzeigen der K. Bayerischen Akademie der Wissenschaften,* **43,** bulletin no. 7 (1856), cols. 51–56; bulletin no. 8 (1856), cols. 57–62; repr. in *Flora,* **40** (1857), 119–128, Hofmeister's own historical account of his prior discoveries; *On the Germination, Development, and Fructification of the Higher Cryptogamia, and on the Fructification of the Coniferae,* F. Currey, trans. (London, 1862), translated from the MS of the unpublished 2nd ed. of *Vergleichende Untersuchungen; Die Lehre von der Pflanzenzelle* (Leipzig, 1867); and *Allgemeine Morphologie der Gewächse* (Leipzig, 1868).

II. SECONDARY LITERATURE. See K. von Goebel, *Wilhelm Hofmeister. Arbeit und Leben eines Botanikers des 19. Jahrhunderts . . .,* vol. VIII in the series Grosse Männer. Studien zur Biologie des Genies (Leipzig, 1924), translated into English by H. M. Bower and edited by F. O. Bower (London, 1926); E. Pfitzer, "Wilhelm Hofmeister," in *Heidelberger Professoren aus dem neunzehnten Jahrhundert . . .,* II (Heidelberg, 1903), 265–358; *Tradition und Gegenwart. Festschrift zum 150 jährigen Bestehen des Musikverlages Friedrich Hofmeister* (Leipzig, 1957)—this and Virneisel differ from the other works in certain statements of fact but are obviously more accurate; and W. Virneisel, "Hofmeister, Friedrich," in F. Blume, ed., *Die Musik in Geschichte und.Gegenwart,* VI (Kassel, 1957), 574–578.

JOHANNES PROSKAUER

HOHENHEIM, BOMBASTUS VON. See **Paracelsus, Theophrastus.**

HOLBACH, PAUL HENRI THIRY, BARON D' (*b.* Edesheim, Palatinate, Germany, December 1723; *d.* Paris, France, 21 January 1789), *philosophy of science.*

Little more is known about d'Holbach's parents than that they were Germans of modest middle-class status. Under the tutelage of a wealthy parvenu uncle, Baron Franciscus Adam d'Holbach, he completed his university studies at Leiden and soon thereafter, in 1749, settled permanently in Paris, where he obtained French naturalization and, in 1750, married. On his uncle's death in 1753, he inherited a considerable fortune and his title. The famous salon that he maintained for several decades in Paris became a social and intellectual center for the *Encyclopédie,* to which d'Holbach was an important contributor, and with whose editors, Diderot and d'Alembert, he formed close ties. The *côterie holbachique,* frequented by some of the most brilliant thinkers, writers, scientists, and artists of the age, was also the foremost gathering place for the exchange of radical ideas in philosophy, politics, and science under the *ancien régime.*

D'Holbach was himself the most audacious philosophe of this circle. During the 1760's, he caused numerous antireligious and anticlerical tracts (written in large part, but not entirely, by himself) to be clandestinely printed abroad and illegally circulated in France. His philosophical masterpiece, the *Système de la nature, ou des lois du monde physique et du monde moral,* a methodical and intransigent affirmation of materialism and atheism, appeared anonymously in 1770. In the ensuing decade, he published a series of ethical and political works which criticized or attacked the absolutist monarchy, state religion, class system, administrative policies, and socioeconomic institutions of the *ancien régime.* D'Holbach's wide-ranging, militant activities on behalf of reform have been recognized as constituting a major influence toward the making of the French Revolution.

In regard to his services to science, it should be stated that d'Holbach was mainly a skillful propagator and popularizer of technical information at a time when such efforts represented a new and valuable means for promoting scientific progress. He also helped to develop and disseminate several original theories that had decisive implications not only for philosophy but for the future course of such sciences as geology, biology, and psychology.

Between 1752 and 1765, d'Holbach wrote for the *Encyclopédie* some 400 signed articles or notices and, it is estimated, at least as many that remained unsigned. The greatest part of this contribution expertly summarized the existing state of knowledge in the fields of chemistry, mineralogy, and metallurgy. D'Holbach typically gave a technological and utilitarian emphasis to his articles which, of course, dealt with subjects that invited practical applications. Concurrently, he translated a number of significant works covering the same sciences, about which the eighteenth-century French public had much to learn from German sources. Among these translations, often usefully annotated by d'Holbach, one may cite the *Minéralogie* of Wallerius (Paris, 1753); Henckel's *Introduction à la minéralogie* (Paris, 1756); C. E. Gellert's *Chimie métallurgique* (Paris, 1758); J. G. Lehmann's *Traités de physique, d'histoire naturelle, de minéralogie et de métallurgie* (Paris, 1759); and Stahl's *Traité du soufre* (Paris, 1766).

In addition to those concerning natural science, many *Encyclopédie* articles, echoing themes and arguments presented more fully in d'Holbach's subversive writings, pertained to what would now be called anthropology, ethnology, and the history of religion. In these "human sciences" d'Holbach generally interpreted whatever data were available to him in accordance with his basic aim of exposing the rampancy of superstition and fanaticism practiced by mankind in countless cults, while suggesting or drawing the appropriate parallels with Christianity. As a social scientist, he sought above all to explain how collective emotions of fear and hope in the face of the menacing, misunderstood powers of nature had inspired or encouraged every form of religious illusion from primitive magic to the abstract notion of God; moreover, he showed how governments had soon learned to exploit such human weakness and ignorance in order to tyrannize their peoples.

An entire complex of advanced scientific thought, growing out of d'Holbach's aforementioned interests, is part and parcel of the philosophy elaborated in the *Système de la nature.* In fact, d'Holbach postulated materialism and atheism squarely on the conclusion that the positive truths already attained by the science of his day, together with the triumph of the empirical method, had rendered contradictory or futile such ideas as God, the creation, soul, and immortality. He supported this revolutionary position by offering an elaborate picture of the universe as a self-sufficient, dynamic, self-creating system, made up exclusively of material elements inherently endowed with specific energies. Thereby matter, eternally in motion, produces according to regular mechanical laws the ever changing combinations that constitute nature, that is, the whole of reality.

In keeping with such a naturalistic conception of things, d'Holbach outlined an anticreationist cosmology and a nondiluvian geology. He proposed a transformistic hypothesis regarding the origins of the animal species, including man, and described the successive changes, or new emergences, of organic beings as a function of ecology, that is, of the geological transformation of the earth itself and of its life-sustaining environment. While all this remained admittedly on the level of vague conjecture, the relative originality and long-term promise of such a hypothesis—which had previously been broached only by Maillet, Maupertuis, and Diderot—were of genuine importance to the history of science. Furthermore, inasmuch as the principles of d'Holbach's mechanistic philosophy ruled out any fundamental distinction between living and nonliving aggregates of matter, his biology took basic issue with both the animism and the vitalism current among his contemporaries.

His standpoint in psychology was in accordance with the rest of his philosophical outlook. Following the example of La Mettrie's *homme machine* thesis, he claimed that all the mental, intellectual, and moral behavior of man, whom he viewed as a purely physical being, was in the first instance determined by organic structures and processes. Therefore, not only psychology, but ethics and politics as well—which d'Holbach sought to root firmly in the scientific study of man so defined—depended essentially on biology and physiology. This closely knit scheme of theories and hypotheses served not merely to liberate eighteenth-century science from various theological and metaphysical impediments, but it also anticipated several of the major directions in which more than one science was later to evolve. Notwithstanding such precursors as Hobbes, La Mettrie, and Diderot, d'Holbach was perhaps the first to argue unequivocally and uncompromisingly that the only philosophical attitude consistent with modern science must be at once naturalistic and antisupernatural.

BIBLIOGRAPHY

There is no collected edition of d'Holbach's works. The *Système de la nature* (Paris, 1770) has recently been issued in photo-reprint form, edited by Y. Belaval (Hildesheim, 1965). The best general account of Holbachian thought, including its scientific aspects, is Pierre Naville, *Paul Thiry d'Holbach et la philosophie scientifique au XVIIIe siècle* (Paris, 1943). Also see John Lough, *Essays on the Encyclopédie of Diderot and d'Alembert* (London, 1968), ch. 3, pp. 111–229; and Virgil Topazio, "D'Holbach, Man of Science," in *Rice Institute Pamphlets,* **53,** no. 4 (1967), 63–68. Two anthologies of some interest are Paulette Charbonnel, ed., *Holbach: Textes choisis; Préface, commentaire et notes* (Paris, 1957); and M. Naumann, ed., *Paul Thiry d'Holbach. Ausgewählte Texte* (Berlin, 1959).

ARAM VARTANIAN

HOLBORN, LUDWIG CHRISTIAN FRIEDRICH (*b.* Göttingen, Germany, 29 September 1860; *d.* Berlin, Germany, 19 September 1926), *physics.*

Holborn, a preeminent physicist in the field of quantitative-measurement analysis, made a particularly noteworthy contribution in this area with his precise measurements of high and low temperatures.

After attending the Realschule in Göttingen, he studied natural sciences at the university in that city from 1879 to 1884 and earned a teaching diploma in mathematics and physics. He also qualified in mineralogy and zoology but did not actually teach these subjects. From 1884 to 1889 he was an assistant to E. Schering at the geomagnetic observatory in Göttingen. In 1887 he received his doctorate for a dissertation on the daily mean values of magnetic declination and horizontal intensity, an investigation in which he carried out measurements of terrestrial magnetism. In 1903 he constructed a torsion magnetometer with F. W. G. Kohlrausch.

In 1890 Holborn changed his field of research and worked under Helmholtz, and later Kohlrausch, at the newly founded (1889) Physikalisch-Technische Reichsanstalt in Berlin; he became an actual member of the organization in 1898. Holborn became director of the thermodynamic laboratories in 1914 and representative of the president of the Reichsanstalt in 1918.

In his work on gas temperatures, Holborn determined, along with W. Wien and A. L. Day, the accuracy of measurements made with thermoelements and investigated the various fixed points, including that for oxygen. In 1901 he and F. Kurlbaum built an optical or incandescent-filament pyrometer having an adjustable brightness setting. In addition, Holborn compared the temperature scales and gave crucial support to the introduction by law, in 1924, of the thermodynamic scale in Germany. He examined the thermocaloric properties of gases and of water vapor, as well as the compressibility of gases. Holborn also plotted the isothermal lines for monoatomic gases at temperatures greater than 100°C., without, however, offering a theoretical interpretation.

BIBLIOGRAPHY

I. Original Works. Holborn wrote more than seventy scientific papers, most of which were originally published as *Mitteilungen aus der Physikalisch-Technischen Reichsanstalt.* Individual works include "Ueber die Messung höher Temperaturen," in *Annalen der Physik und Chemie,* n.s. **47** (1892), 107–134, and **56** (1895), 360–396, written with W. Wien; "Ueber die Messung tiefer Temperaturen," *ibid.,* **59** (1896), 213–228, written with Wien; "Ueber das Luftthermometer bei hohen Temperaturen," *ibid.,* **68** (1899), 817–832, and *Annalen der Physik,* **2** (1900), 505–545, written with A. Day.

See also "Über ein optisches Pyrometer," in *Annalen der Physik,* **10** (1903), 225–241, written with F. Kurlbaum (paper first read in session of the Academy of Sciences, Berlin, 13 June 1901); "Über ein störungsfreies Torsionsmagnetometer," *ibid.,* **10** (1903), 287–304, written with F. W. G. Kohlrausch; "Über ein tragbares Torsionsmagnetometer," *ibid.,* **13** (1904), 1034–1059, written with Kohlrausch; "Über die spezifische Wärme von Stickstoff, Kohlensäure und Wasserdampf bis 1400°," *ibid.,* **23** (1907), 809–845, written with F. Henning; "Über das Platinthermometer und den Sättigungsdruck des Wasserdampfes zwischen 50° und 200°," *ibid.,* **26** (1908), 833–883, written with Henning; "Über den Sättigungsdruck des Wasserdampfes oberhalb 200°," *ibid.,* **31** (1910), 945–970, written with A. Baumann; "Über die Druckwage und die Isothermen von Luft, Argon und Helium zwischen 0° bis 200°," *ibid.,* **47** (1915), 1089–1111, written with H. Schultze.

The following contributions are of special interest in the history of thermodynamics and exemplify Holborn's activity in this field: "Kalorimetrie," in *Kultur der Gegenwart,* **1,** pt. 3, sec. 3 (1915), 112–117; "Mechanische und thermische Eigenschaften der Materie in den drei Aggregatzustanden," *ibid.,* 128–153; "Umwandlungspunkte, Erscheinungen bei koexistierenden Phasen," *ibid.,* 154–178; "Die Physikalisch-Technische Reichsanstalt. Fünfundzwanzig Jahre ihrer Tätigkeit. 2. Wärme," in *Naturwissenschaften,* **1** (1913), 225–229; *Ergebnisse aus den thermischen Untersuchungen der Physikalisch-Technischen Reichsanstalt* (Brunswick, 1919), written with K. Scheel and F. Henning.

Holborn's contributions to others' works include "Mess—Methoden und Mess—Technik," in Wien and F. Harms, *Handbuch der Experimentalphysik,* I (1926), 1–329; "Elektrolytische Leitung," in L. Gräetz, *Handbuch der Elektrizität und des Magnetismus,* III (1922), and in Kohlrausch, *Das Leitvermögen der Elektrolyte* (1st ed. by Kohlrausch and Holborn, 1898; 2nd ed., Leipzig, 1916). Holborn was a standing collaborator for Kohlrausch, *Praktische Physik,* 12th–14th eds. (Leipzig, 1914–1923), and for H. Landolt and E. Börnstein, *Physikalisch-chemische Tabellen,* 5th ed. (Leipzig, 1923).

For a bibliography of Holborn's works see Poggendorff, IV (1904), 655–656; V (1926), 550; VI (1937), 1144.

II. Secondary Literature. See F. Henning, "Ludwig Holborn," in *Physikalische Zeitschrift,* **28** (1927), 157–170.

Hans-Günther Körber

HOLDEN, EDWARD SINGLETON (*b.* St. Louis, Missouri, 5 November 1846; *d.* West Point, New York, 16 March 1914), *astronomy.*

Holden designed the Lick Observatory in California and was its first director. He served as president of the University of California from 1886 to 1888 and was the principal organizer and first president of the Astronomical Society of the Pacific.

He was the only child of Edward (originally Jeremiah Fenno) Holden and Sarah Frances Singleton. Following his mother's death when he was three, Holden lived with relatives in Cambridge, Massachusetts, where he attended private school and with his cousin, the astronomer George P. Bond, made frequent visits to the Harvard College Observatory, site of the great fifteen-inch refractor.

In 1860 Holden returned to St. Louis, where he attended the Academy of Washington University, earning the B.S. from that university in 1866. During this period he lived with William Chauvenet, chancellor and professor of mathematics and astronomy at Washington University. He married Chauvenet's daughter Mary in May 1871.

After receiving the B.S., Holden entered the U.S. Military Academy at West Point as a cadet; he graduated third in his class in 1870. From October 1870 to August 1871 he served as second lieutenant with the 4th Artillery in garrison at Fort Johnson, North Carolina. He was then assigned to the Military Academy, first as assistant professor of natural and experimental philosophy (he taught mechanics, astronomy, acoustics, and optics) and later as assistant instructor in the department of practical military engineering.

Holden's first publication was "The Bastion System of Fortification, Its Defects, and Their Remedies" (1872). His next two, describing his observations of the aurora and of lightning by making use of a pocket spectroscope, appeared in the same year. During his residence at West Point, Holden became a close friend of Henry Draper, who was then pioneering in astronomical spectroscopy and photography at his private observatory in Hastings-on-Hudson. It was probably through this contact that Holden became seriously interested in astronomical research.

In March 1873, Holden resigned his commission and accepted a position at the U.S. Naval Observatory, where he assisted the eminent astronomer Simon Newcomb with the new twenty-six-inch refractor, then the largest in the world. Holden's publications while at the Naval Observatory included observations of nebulas, the surface of the sun, and the satellites of Uranus and Neptune; accounts of recent progress in astronomy; a catalogue of astronomical bibliographies in the Naval Observatory library, an "Index-Catalogue of Books and Memoires Relating to Nebulae and Clusters"; an astronomy text for high school and college students (written with Newcomb); and a biography of William Herschel.

In 1881 Holden became director of the Washburn Observatory of the University of Wisconsin, and while in that position he instituted the *Publications* of the observatory, led an eclipse expedition organized by the National Academy of Sciences, and carried out micrometer measurements of double stars and the rings of Saturn.

Holden's connection with the Lick Observatory began in 1874, when the California philanthropist James Lick established a trust for the purpose of building the world's greatest observatory. The chairman of the board of trustees consulted Henry Draper and Simon Newcomb, and he invited Newcomb and Holden to prepare plans for the observatory. Holden wrote a lengthy memorandum which became the basis for future planning. Holden was then recommended by Newcomb for the post of director. At this time "various difficulties" arose between Lick and his trustees concerning the sale of property. In the words of Holden, these difficulties were "finally settled by the resignation of the first Board." A second board was appointed, and a third, within the space of one and one-half years. Lick's death in 1876 coincided with the start of a decade of relative stability during which the observatory was constructed, with Holden acting as principal adviser. The final plan of the observatory is acknowledged to be his creation.

The observatory flourished under Holden's direction; but W. W. Campbell, who later became director and who also served as president of the university, said of Holden's directorship: "The last years of Professor Holden's administration were marred by the existence of animosities in the observatory community, and by much ill-advised criticism in the newspapers." Newcomb wrote in his *Reminiscences:*

> To me the most singular feature [of Holden's administration] was the constantly growing unpopularity of the director. I call it singular because, if we confine ourselves to the record, it would be difficult to assign any obvious reason for it. One fact is indisputable, and that is the wonderful success of the Director in selecting young men who were to make the institution famous by their ability and industry (pp. 192–193).

In February 1889, Holden organized the Astronomical Society of the Pacific. He sought to lay a foundation broad enough so that every class of member would "find a sphere of action," and this society remains nearly unique in its attempts to bring amateur and professional astronomers to a common meeting ground.

Following his resignation as director of Lick in 1897, Holden lived four years in New York, devoting much of his time to writing. His bibliography for this interval contains three elementary texts on astronomy; three books of stories for children; a book on heraldry; and articles on earthquakes, Omar Khayyám, Christianity in China, art criticism, and public schools. He occasionally wrote under the pseudonyms E. Singleton and Edward Atherton.

From 1901 until his death Holden was librarian of the U.S. Military Academy. During this interval 30,000 volumes were added, the library was catalogued, and complete bibliographies were prepared on a wide range of military subjects.

Holden was elected to the National Academy of Sciences in 1885; he was a foreign associate of the Royal Astronomical Society and a member of the American Academy of Arts and Sciences and the Astronomical Society of France. He was awarded four honorary doctorates by American universities and was buried at West Point with military honors.

BIBLIOGRAPHY

Holden's bibliography and biography, prepared by W. W. Campbell, appear in *Biographical Memoirs. National Academy of Sciences,* **8** (1919), 358–372. See also Simon Newcomb, *The Reminiscences of an Astronomer* (Boston, 1903); and Eben Putnam, *The Holden Genealogy,* 2 vols. (Wellesley Farms, Mass., 1923). Among Holden's books, the following are representative: *Sir William Herschel, His Life and Works* (New York, 1881); *Mogul Emperors of Hindustan* (New York, 1895); *Primer of Heraldry for Americans* (New York, 1898); *Elementary Astronomy* (New York, 1899); *The Family of the Sun* (New York, 1899); and *Stories From the Arabian Nights* (New York, 1900), written under the pseudonym E. Singleton.

CHARLES A. WHITNEY

HÖLDER, OTTO LUDWIG (*b.* Stuttgart, Germany, 22 December 1859; *d.* Leipzig, Germany, 29 August 1937), *mathematics.*

Hölder came from a Württemberg family of public officials and scholars. His father, Otto Hölder, was professor of French at the Polytechnikum in Stuttgart; his mother was the former Pauline Ströbel. In Stuttgart, Hölder attended one of the first Gymnasiums devoted to science and there he studied engineering for a short time. A colleague of his father's suggested that the best place to study mathematics was Berlin, where Weierstrass, Kronecker, and Kummer were teaching. When Hölder arrived at the University of Berlin in 1877, Weierstrass, lecturing on the theory of functions, had already covered the fundamentals of analysis. Hölder caught up to the class with the aid of other students' notes and was thus led to his first independent studies in mathematics.

Influenced by the rigorous foundation of analysis given by Weierstrass, Hölder developed the continuity condition for volume density that bears his name. It appeared in his dissertation (*Beiträge zur Potentialtheorie*), which he presented at Tübingen in 1882; his referee was Paul du Bois-Reymond. The Hölder continuity is sufficient for the existence of all the second derivatives of the potential and for the validity of the Poisson differential equation. These derivatives, as Arthur Korn later showed, possess exactly the same continuity properties as the density. Hölder's work on potential theory was continued on a larger scale by Leon Lichtenstein, O. D. Kellogg, P. J. Schauder, and C. B. Morrey, Jr.

Next Hölder investigated analytic functions and summation procedures by arithmetic means. He provided the first completely general proof of Weierstrass' theorem that an analytic function comes arbitrarily close to every value in the neighborhood of an essential singular point. He showed that it might be possible to compute, by repetition of arithmetic means (Hölder means), the limit of an analytic function the power series of which diverges at a point of the circle of convergence. This technique is equivalent to the one introduced by Cesàro, as Walter Schnee demonstrated.

In his *Habilitationsschrift* submitted in 1884 at Göttingen, Hölder examined the convergence of the Fourier series of a function that was not assumed to be either continuous or bounded; for such functions the Fourier coefficients had first to be defined in a new fashion as improper integrals. After qualifying as a lecturer, Hölder discovered the inequality named for him. This advance involved an extension of Schwarz's inequality to general exponents as well as to inequalities for convex functions of the type that were later treated by J. L. Jensen. After unsuccessful attempts to find an algebraic differential equation for the gamma function, Hölder inverted the method of posing the question and proved the impossibility of such a differential equation.

Hölder owed his interest in group theory and Galois theory primarily to Kronecker, but also to Felix Klein, in whose seminar at Leipzig Hölder participated soon after receiving his doctorate. To these fields he contributed "Zurückführung einer algebraischen Gleichung auf eine Kette von Gleichungen," in which he reduced an algebraic equation by using simple groups and by introducing the concept of "natural" irrationals. Here Hölder extended C. Jordan's theorem (stated in his *Commentary* on Galois) of the uniqueness of the indexes of such a "composition series," to the uniqueness of the "factor groups" that Hölder had introduced. This new concept and the Jordan-Hölder theorem are today fundamental to group theory.

With the help of these methods Hölder solved the old question of the "irreducible case." A solution of the cubic equation is given by the so-called Cardano formula, in which appear cube roots of a square root $\sqrt{D}$. For three distinct real roots $D < 0$, and therefore the quantities under the cube root sign are imaginary. The real solution is thus obtained as the sum of imaginary cube roots. Hölder showed that in this case it is impossible to solve the general cubic equation through real radicals, except where the equation decomposes over the base field.

Hölder turned his attention first to simple groups. Besides the simple groups of orders 60 and 168 already known at the time, he found no new ones with a composite order less than 200. Nevertheless, he considered his method to be "of some interest so long as we do not possess a better one suitable for handling the problem generally." Such a general method is still lacking, despite the progress and great efforts of recent years.

In further works Hölder treated the structure of composite groups having the following orders: p^3, pq^2, pqr, p^4, where p, q, r are primes, and n, where n is square-free. Finally, he studied the formation of groups constructed from previously given factor groups and normal subgroups.

While an associate professor at Tübingen, Hölder verified (in "Über die Prinzipien von Hamilton und Maupertuis") that the variational principles of Hamilton are valid for nonholonomic motions—their applicability in these cases had been questioned by Heinrich Hertz. Physicists are indebted to Hölder for this confirmation of the Hamiltonian principle, which has often been used since then in deriving differential equations of physics.

The first third of Hölder's career in research was the most fruitful. A period of depression seems to have occurred at Königsberg, where he succeeded Minkowski in 1894. He was happy to leave that city in 1899, when he accepted an offer from Leipzig to succeed Sophus Lie. In the same year he married Helene Lautenschlager, who also came from Stuttgart.

At Leipzig, Hölder turned to geometrical questions, beginning with his inaugural lecture *Anschauungen und Denken in der Geometrie* (1900). He became interested in the geometry of the projective line and undertook investigations published in his paper "Die Axiome der Quantität und die Lehre vom Mass" (1901). The topics covered in this work were, in his view, important for physics. Moreover, in 1911 he published an article on "Streckenrechnung und projektive Geometrie."

Between 1914 and 1923 this work led to the logico-philosophical studies of the foundations of mathematics which are included in *Die mathematische Methode* (1924). These philosophical inquiries attracted less attention than Hilbert's axiomatic method, but Hölder saw connections between Brouwer's intuitionism and Weyl's logical investigations and his own ideas. P. Lorenzen's recent work on logic contains ideas which are in essence similar to those of Hölder's. In his obituary on Hölder, B. L. van der Waerden wrote:

> According to Hölder one of the essential features of the mathematical method consists in constructing for given concepts, concepts of higher order in such a way that concepts and methods of proof of one stage are taken as objects of mathematical investigation of the next higher stage. This is done, for example, by first developing a method of proof and afterward counting the steps of the proof or by letting them correspond to other objects, or by combining them by means of relations [p. 161].

On the basis of this conception Hölder concluded—and recent logical investigations of Gödel fully justify his position—that "one can never grasp the whole of mathematics by means of a logical formalism, because the new concepts and syllogism that are applied to the formulas of the formalism necessarily go beyond the formalism and yet also belong to mathematics."

In his last years one of Hölder's favorite topics was elementary number theory—his third great teacher in Berlin had been Kummer. Hölder's contributions in this area appeared mainly in the *Bericht. Sächsische Gesellschaft* (later *Akademie*) *der Wissenschaften.* From 1899 he was active in the academy and for several years served as president. He was also a member of the Prince Jablonowski Society. In 1927 Hölder became a corresponding member of the Bavarian Academy of Sciences.

BIBLIOGRAPHY

I. Original Works. Hölder's books include *Anschauungen und Denken in der Geometrie* (Leipzig, 1900; Darmstadt, 1968); *Die Arithmetik in strenger Begründung. Programmabhandlung der philosophischen Fakultät* (Leipzig, 1914; 2nd ed., Berlin, 1929); and *Die mathematische Methode. Logisch-erkenntnis-theoretische Untersuchungen im Gebiet der Mathematik, Mechanik und Physik* (Berlin, 1924).

Among his papers are the following: *Beiträge zur Potentialtheorie* (Stuttgart, 1882), his diss.; "Beweis des Satzes, dass eine eindeutige analytische Funktion in unendlicher Nähe einer wesentlich singulären Stelle jedem Wert beliebig nahe kommt," in *Mathematische Annalen,* **20** (1882), 138–142; "Grenzwerte von Reihen an der Konvergenzgrenze," *ibid.,* 535–549; "Über eine neue hinreichende Bedingung für die Darstellbarkeit einer Funktion durch die Fouriersche Reihe," in *Bericht der Preussischen Akademie* (1885), 419–434; "Über die Eigenschaft der Gammafunktion, keiner algebraischen Differentialgleichung zu genügen," in *Mathematische Annalen,* **28** (1886), 1–13, "Zurückführung einer beliebigen algebraischen Gleichung auf eine Kette von Gleichungen," *ibid.,* **34** (1889), 26–56; "Über einen Mittelwertsatz," in *Nachrichten von der Gesellschaft der Wissenschaften zu Göttingen,* **2** (1889), 38–47; "Über den Casus irreducibilis bei der Gleichung dritten Grades," in *Mathematische Annalen,* **38** (1891), 307–312; "Die einfachen Gruppen im ersten und zweiten Hundert der Ordnungszahlen," *ibid.,* **40** (1892), 55–88; "Die Gruppen der Ordnungen p³, pq², pqr, p⁴," *ibid.,* **43** (1893), 301–412; "Bildung zusammengesetzter Gruppen," *ibid.,* **46** (1895), 321–422; "Die Gruppen mit quadratfreier Ordnungszahl," in *Nachrichten von der Gesellschaft der Wissenschaften zu Göttingen,* **2** (1895), 211–229; "Uber die Prinzipien von Hamilton und Maupertuis," *ibid.,* 122–157; "Galoissche Theorie mit Anwendungen," in *Encyklopädie der mathematischen Wissenschaften,* I (1898–1904), 480–520; "Die Axiome der Quantität und die Lehre vom Mass," in *Bericht. Sächsische Akademie der Wissenschaften,* Math.-nat. Klasse, **53** (1901), 1–64; "Die Zahlenskala auf der projektiven Geraden und die independente Geometrie dieser Geraden," in *Mathematische Annalen,* **65** (1908), 161–260; and "Streckenrechnung und projektive Geometrie," in *Bericht. Sächsische Akademie der Wissenschaften,* Math.-nat. Klasse, **63** (1911), 65–183.

II. Secondary Literature. The main source is the obituary by B. L. van der Waerden, "Nachruf auf Otto Hölder," in *Mathematische Annalen,* **116** (1939), 157–165, with bibliography; it also appeared in *Bericht. Sächsische Akademie der Wissenschaften,* Math.-nat. Klasse, **90** (1938), without the bibliography. See also Poggendorff, IV, 651; V, 547; VI, 1136; VIIa, 509.

Ernst Hölder

HOLMBOE, BERNT MICHAEL (*b.* Vang, Norway, 23 March 1795; *d.* Christiania [now Oslo], Norway, 28 March 1850), *mathematics.*

Holmboe was the son of a minister. After graduating from the Cathedral School in Christiania, he joined the student volunteer corps for service in the brief conflict with Sweden in 1814. In 1815 he became assistant to Christopher Hansteen, professor of astronomy at the newly created University of Christiania. He was appointed teacher at the Cathedral School in 1818; here he made his greatest contribution to mathematics by discovering and nurturing the genius of his pupil Niels Henrik Abel. Together they explored the whole mathematical literature in a quest in which the pupil soon became the leader.

In 1826 Holmboe was appointed lecturer in mathematics at the university, a move which was later criticized because it blocked the possibility of a position for Abel. Nevertheless, the friendship between the two remained undisturbed.

After Abel's death Holmboe edited his works at the request of the government; otherwise his own mathematical contributions were undistinguished. He published a number of elementary school texts which appeared in several editions. A later, more advanced calculus text was evidently influenced by Abel's research.

Holmboe was lecturer at the military academy in Christiania from 1826 until his death; during Hansteen's absence in 1828–1830 on a geomagnetic expedition to Siberia, Holmboe gave his lectures in astronomy. In 1834 he became professor of pure mathematics at the university.

BIBLIOGRAPHY

Holmboe edited Abel's writings as *Oeuvres complètes de N. H. Abel, mathématicien, avec des notes et développements* (Christiania, 1839). His advanced calculus text is *Laerebog i den höiere mathematik* (Christiania, 1849).

Oystein Ore

HOLMES, ARTHUR (*b.* Hebburn on Tyne, England, 14 January 1890; *d.* London, England, 20 September 1965), *geology, geophysics, petrology.*

Holmes came of Northumbrian farming stock and gained his early interest in earth science at Gateshead High School. Entering Imperial College, London, in 1907, he read physics under R. J. Strutt (later Lord Rayleigh) for his first degree; subsequently, under the influence of W. W. Watts, he changed to geology and graduated Associate of the Royal College of Science in 1910. Postgraduate studies with Strutt led him to investigate the application of radioactivity to geology. An expedition to Mozambique gave Holmes experience in field and petrographic work on Precambrian ultrametamorphics and Tertiary lavas and stimulated his interest in geomorphology. Thus were laid the foundations of the three main lines of research to which he was to become a major contributor: geochronology, the genesis of igneous rocks, and physical geology.

From 1912 to 1920 Holmes was demonstrator in geology at Imperial College, teaching petrology, conducting research, and writing extensively. From 1920 to 1924 he was chief geologist of an oil exploration company in Burma, returning to England to become professor of geology at the University of Durham (Durham Colleges division) in 1925. Here he refounded the department and spent some of his most productive years until he was transferred to the Regius chair of geology at Edinburgh in 1943. He retired in 1956. He was twice married: to Margaret Howe of Gateshead in 1914 and, after her death, to the distinguished petrologist Doris L. Reynolds in 1939.

Holmes made a great impact upon his times through his pioneer work on radiometric methods of rock dating, his controversial views on the origins of deep-seated rocks, and his brilliant synthesis of the contributions of geophysics and geomorphology to the understanding of the history of the earth. He was one of the most able expositors the earth sciences have ever had, writing in clear, lucid English with a gift for apt quotation and illustration. He never wished or attempted to be a public figure, but great numbers of scientists were influenced by his writings and international correspondence. He was elected a fellow of the Royal Society in 1942 and received many international honors, including the Penrose Medal of the Geological Society of America in 1956 and the Vetlesen Prize in 1964.

By 1910 the discoveries of Becquerel, the Curies, and Rutherford had revolutionized the conception of matter; and Strutt had shown that radioactive minerals are widespread in rocks. The fundamental importance of these developments for geology emerged as a result of Holmes's work. Of the various attempts to derive an "absolute" time scale for our planet, Kelvin's calculations—based on the assumption of a uniformly cooling earth and gravitational and the then known chemical sources for terrestrial and solar energies—appeared more satisfactory than those depending upon rate of denudation, rate of accumulation of sodium chloride in the oceans, or rate of sedimentation. Yet Kelvin's method allowed only twenty to forty million years for the whole of geological time—far too short, in the opinion of most geologists. Holmes showed that Kelvin's assumptions are invalidated by the availability of radioactive heat; he derived a figure of at least 1,600 million years by comparing the amounts of uranium and thorium in rocks with those of their daughter elements lead and helium, assuming a constant half-life for each of the radioactive elements.

Although some of the earliest experimental determinations were made by Holmes himself in Strutt's laboratory, the later refinements, especially the discrimination of the isotopes of uranium, thorium, and lead which became possible with the rise of mass spectrometry, were the work of other investigators. The earliest modern time scale was proposed by Holmes; and throughout his life he remained the leading figure in the field, critically discussing each new set of results as it appeared and modifying the time scale as it became necessary. A figure of 4,550 million years had become accepted as the "age of the earth" by the time Holmes's final scale was published.

The recognition that radioactive heat was available from disintegration of uranium, thorium, and potassium was shown by Holmes to have important implications for the thermal history of the earth. It was no longer safe to assume that a cooling earth was contracting; in fact, Holmes advocated cyclical expansion alternating with contraction as a means of explaining tectonic movements in the crust. He also became a strong supporter of A. L. Wegener's hypothesis of drifting continents at a time when few geologists favored the notion and was the first advocate of convection currents in the substratum (now generally called the mantle) of the earth. Paleomagnetic and paleontological evidence of large-scale movements of the continents has accumulated to an impressive degree, and the hypothesis of convection in the mantle has gained many supporters. His results in geological time, the mechanism of earth movements, and mantle convection illustrate the importance of Holmes as an innovator in fundamental geophysics.

He was also influential in petrography, through his work on techniques and systematics, and in petrology. For many years an orthodox follower of K. H. F. Rosenbusch and Alfred Harker in considering extrusive and intrusive igneous rocks to be of liquid magmatic origin, he became increasingly dissatisfied with ideas based on the limited physicochemical and thermodynamic information of the time. His collaboration with the Geological Survey of Uganda on the remarkable alkalic volcanics found in the Western Rift Valley led him to contemplate solid-state metasomatism and "transfusion" of preexisting rocks by differential introduction of fluxes of emanations. When the "granite controversy" hit its stride in the 1940's, it was natural for Holmes to be found on the side of the "soaks," as N. L. Bowen once described the advocates of metasomatism. In addition Holmes's interest in the kimberlite rocks of the diamond pipes and in eclogite as a high-pressure equivalent of basalt pointed the way to important links between petrology and geophysics.

While standing watch against German incendiary bombs at the Durham laboratories, Holmes began writing his *Principles of Physical Geology,* in which he brought together the whole range of earth processes: those of the deep interior, as revealed by geophysics; those of the crust, displayed by petrology, tectonics, and sedimentation; and those of the surface, revealed by geomorphology. The second edition of this great work, which appeared shortly before his death, is a fitting memorial to the man and his philosophy.

BIBLIOGRAPHY

I. Original Works. A full list of Holmes's writings is in *The Phanerozoic Time Scale* (London, 1964), which was published in Holmes's honor; and in the biographical memoir mentioned below. *The Age of the Earth* (London–New York, 1913; 2nd ed., London, 1937) is of historical interest and should be compared with his "A Revised

Geological Time Scale," in *Transactions of the Edinburgh Geological Society,* **17** (1959), 183–216. *The Nomenclature of Petrology,* 2nd ed. (London, 1930), and *Petrographic Methods and Calculations,* 2nd ed. (London, 1930), are standard works. The convection hypothesis is first proposed in "Radioactivity and Earth Movements," in *Transactions of the Geological Society of Glasgow,* **18** (1928–1929), 559–606. Transfusionist petrological arguments are developed in "The Volcanic Area of Bufumbira, Part II, The Petrology of the Volcanic Field . . .," in *Memoirs of the Geological Survey of Uganda,* vol. **3** (1937), written with H. F. Harwood. *Principles of Physical Geology* (London, 1944; 2nd ed., London, 1964; New York, 1965) is the most important reference.

II. SECONDARY LITERATURE. See "Award of the Penrose Medal to Arthur Holmes," in *Proceedings of the Geological Society of America* for 1956 (1958), pp. 73–74; and K. C. Dunham, "Arthur Holmes," in *Biographical Memoirs of Fellows of the Royal Society,* **12** (1966), 291–310.

KINGSLEY DUNHAM

HOLMGREN, FRITHIOF (*b.* West Ny, Sweden, 22 October 1831; *d.* Uppsala, Sweden, 14 August 1897), *physiology.*

Holmgren's father, Anders Holmgren, was rector of Motala-Vinnerstad parish; his mother was the daughter of the rector of West Ny, Anders Nordwall. Frithiof, one of twelve children, finished school at Linköping in 1849 and then went to Uppsala for medical studies in 1850–1860; his education was interrupted by periods of work as a practicing physician. Wishing to devote himself to the rising science of physiology, but knowing almost nothing about it, Holmgren went abroad in 1861. The high reputation of the Vienna School of Medicine drew him to that city, where the eminent physiologist E. W. von Brücke received him cordially and later sent him to Carl Ludwig's institute in Leipzig, where many leading physiologists of that period received their basic training in experimentation. In 1864 Holmgren returned to Uppsala. A second year abroad (1869–1870) was spent at the laboratories of Emil du Bois-Reymond in Berlin and of Hermann von Helmholtz in Heidelberg; in Paris he attended the lectures of Claude Bernard.

Inspired by du Bois-Reymond's observation (1849) of a resting current between electrodes at the front and the back of the eye, Holmgren showed (1864–1865) that this current swung in a cornea-positive direction at both onset and cessation of illumination of the (frog) eye and thus discovered the retina's electrical response to light, today's electroretinogram (ERG). It was not until 1870–1871 that he fully understood what he had recorded, believing at first that he had seen the response from the cut end of the optic nerve. But when he finally tried shifting the electrode positions on the bulb, it became obvious that the generative source of the light response was the retina itself. A little later (1873) the retinal response to light was discovered independently by James Dewar and John G. McKendrick in Edinburgh —proceeding, interestingly enough, from quite different premises.

Holmgren realized from the beginning that he had devised a new method of studying objectively the effect of light on the retina and, quite rightly, said that "a great many questions concerning physiological optics . . . can hardly be solved in any other way now known to us." Yet Holmgren himself did not embark upon any extensive study of the retina using his new method. When in 1864 he became the first professor of physiology in Sweden, much of his time was devoted to introducing and teaching the new science as an experimental discipline at Uppsala and to acquiring the necessary laboratory space. Holmgren's first institute (1867) was an apartment in the department of pathology; in 1893, four years before his death from arteriosclerosis, he created a large new institute of physiology, established through a gift of 30,000 Swedish crowns from a private donor.

A serious railway accident at Lagerlunda in April 1876 led Holmgren to suspect color blindness of the engine-driver as the cause. Although the driver himself had been killed in the accident, Holmgren began a study of the 266 employees on the Uppsala-Gävle railway line; among them he found thirteen who were color-blind, of them six green-blind. In July he presented his results at the Nordic Meeting of Physicians, which accepted his conclusions on the basis of demonstrations; by the end of the year color tests had been prescribed for railway and shipping personnel in Sweden.

Holmgren's book on color-blindness was translated into several European languages, and other countries soon followed Sweden's example in introducing tests for color-blindness. His simple method of testing was based on confusion of colors and not, as earlier methods had been, on the naming of them.

Holmgren displayed increasing interest in applied physiology and in social and cultural affairs. He campaigned for gymnastics through a society of which he was founder and president; he established a society for folk dancing; and with his wife, the former Ann Margret Tersmeden, kept open house for the students from his home county, Östergötland. The couple were devoted patriots and idealists, fighting at conservative Uppsala for the students' points of view in the cause of liberalism and freedom of thought until they were boycotted by most of their university colleagues. They also helped Artur Hazelius, the creator of Stockholm's well-known open-air museum Skansen, in his effort to preserve Sweden's rural civilization. His wife became known as a leading promoter of women's rights in Sweden.

Internationally, Holmgren was a familiar figure in physiology, much appreciated by his colleagues. His work on the retina's electrical response to light, originally published in Swedish, was translated and republished in German. Holmgren is remembered for this important discovery and for his work to prevent accidents resulting from color-blindness.

BIBLIOGRAPHY

I. ORIGINAL WORKS. For Holmgren's presentation of the discovery of the electroretinogram, see "Method att objectivera effecten af Ljusintryck på retina," in *Uppsala läkareförenings förhandlingar,* **1** (1865–1866), 177–191, German trans. in *Wilhelm Kühne, Untersuchungen des Physiologischen Instituts d. Universität Heidelberg,* II–III (1878–1882). His book on the method for detecting color blindness is *Om färgblindheten i dess förhållande till jernvägstrafiken och sjöväsendet* (Uppsala–Berlin, 1877), French trans. (1877), German trans. (1878).

II. SECONDARY LITERATURE. See two works by Ragnar Granit: *Sensory Mechanisms of the Retina* (London, 1947; repub. New York, 1963), see intro., pp. xvii–xxiii; and "Frithiof Holmgren. Minnesteckning," in *Kungliga Svenska vetenskapsakademiens årsbok, 1964* (1964), pp. 281–296.

RAGNAR GRANIT

HOMBERG, WILHELM or **GUILLAUME** (*b.* Batavia, Java [now Jakarta, Indonesia], 3 January 1652; *d.* Paris, France, 24 September 1715), *chemistry.*

Although Homberg's greatest contribution was to introduce the new scientific chemistry to the French Academy of Sciences, his life resembled that of many alchemists. His father, Johann Homberg, was originally from Saxony; upon the loss of his property in the Thirty Years' War he entered the service of the Dutch East India Company as a soldier. His mother, Barbe van Hedemard, had come to Java as the wife of a Dutch officer. Homberg, the second of four children, was made a corporal at the age of four; but his education was neglected because it was thought impossible for a European child to study in a tropical climate. When his father took the family to Amsterdam, Homberg showed great intellectual aptitude; he was soon studying law at Jena and Leipzig and was accepted as a practicing lawyer at Magdeburg in 1674. Here he discovered the fascination of botany and astronomy. Here too he met Otto von Guericke, who introduced him to experimental physics and is said to have taught him the secret of the hygrometric toy in which a figure appears in fine weather and withdraws in rainy weather. Homberg later traded this secret to Johann Kunckel in return for that of the preparation of phosphorus.

In pursuit of scientific knowledge Homberg traveled to Padua (where he studied medicine), Bologna (where he interested himself in the mysteriously phosphorescent "Bononian stone"), Rome, France, England (where he apparently worked with Boyle), and Holland. He took the M.D. degree at Wittenberg, investigated the German preparations of phosphorus and mining techniques, and worked in the chemical laboratory established by the king of Sweden. He then went to Paris, where he was supported by Colbert; and in 1682 he was converted to Catholicism. Colbert's death in 1683 left Homberg without resources, his father having disowned him both for his change of religion and for his wandering life. The gift of an ingot of "alchemical gold" by a friend is said to have permitted him to go to Rome, where he practiced as a chemist and physician while maintaining contact with French circles. The Abbé Bignon appointed him a member of the Royal Academy of Sciences in 1691. The rest of his life was spent in Paris, and his work was all done within the framework of the Academy. He was also associated with Philippe II, duke of Orléans, who in 1702 gave him a pension and laboratory, and bought for him a burning mirror made by Tschirnhausen. In 1708 Homberg married Marguerite-Angélique, daughter of the physician and botanist Denis Dodart.

In contrast with the romanticism of his external life, Homberg's intellectual existence was rational, empirical, and scientific. He had wide experience in experimental physics, publishing on the breaking of Prince Rupert's drops, the production of frictional electricity, and the expansion and contraction of substances by heat and cold. (The table of the variation of specific gravities with temperature, published in the *Mémoires* of the Academy in 1699, was used by later writers.) Like his mentor Boyle, Homberg carried the point of view of an experimental yet mechanical philosopher into the practice of chemistry and encouraged French chemists to follow. This is shown in his work at meetings of the Academy, as revealed in its *Registres de physique* (1692–1715). For example, he exploded the validity of the analyses of plants into their supposed elements or principles (salt, oil, spirit, etc.) and introduced the notion of analysis into "simple substances" (recognizable and stable chemical entities) which made possible eighteenth-century French analytical chemistry.

In his "Essais de chimie" (1702–1710) Homberg discussed the general concept of principles or elements and concluded that salt, sulfur, and mercury were not all to be found in all substances. Thus, he thought that mercury was present in metallic ores and metals but not in "fossils" (nonmetallic minerals) and salts, while organic substances had a set of principles different from those of inorganic substances. This was a step toward the modern definition of an element. Probably his most important work was on the strength of acids and the quantity of acid required to neutralize a given quantity of alkali (two papers published in 1699 and 1700). Homberg recognized that different alkalies neutralized the same acid in different proportions but believed that the relative strengths of two acids could be determined by using the same alkali in each case. He treated the question of neutralization (or dissolvability, as he called it) in quite quantitative fashion, showing that if an alkaline salt were treated with an acid, the gain in weight of the salt was an indication of the amount of acid absorbed. He came to regard specific gravity as a true indication of acid strength. Although naturally unaware of the role of gases in acid-alkali neutralizations, Homberg nevertheless understood the fundamentals of the process and thereby laid the foundation for an understanding of the nature of salts. In 1702, in his "Essai" on salt, he discussed the replaceability of metals in solution, much as Newton later did in Query 31 of *Opticks,* although without any notion of attraction.

To his contemporaries, much of Homberg's most interesting work lay in his dramatic quasi-alchemical explorations. In 1692 he published a method for making "the tree of Diana," a spectacular form of crystallization of silver salts. He published experiments on various forms of "phosphorus," including his own discovery of the luminous and explosive properties of calcium chlorate. With his burning glass he performed experiments on calcination, fusibility, and volatility. Homberg also published on pneumatics, botany, and zoology. He was one of the leading scientific spirits of the reformed Royal Academy of Sciences and highly influential in developing its course of research in the experimental sciences.

BIBLIOGRAPHY

I. ORIGINAL WORKS. All Homberg's work was published in the form of papers and essays (well over 70 in all) in the *Histoire et mémoires de l'Académie royale des sciences* (1692–1714). A list of 53 of these, mainly on chemical subjects, is at the end of his biography in the *Nouvelle biographie générale.* Among the most important are "Diverses expériences du phosphore" (1692), "Observations sur la quantité exacte des sels volatils acides contenus dans les différents esprits acides" (1699), "Observations sur la quantité d'acide absorbée par les alcalis terreux" (1700), and "Essais de chimie" (1702, 1705, 1706, 1710), on the principles of salt, sulfur, and mercury. His contributions to the daily activities of the Academy are preserved in the archives of the French Academy of Sciences.

II. SECONDARY LITERATURE. The chief source for Homberg's biography is the *éloge* by Bernard le Bovier de Fontenelle, published in *Histoire et mémoires de l'Académie royale des sciences* (1715), repr. in *Oeuvres de Fontenelle, éloges,* I (Paris, 1825), 307–319. There are good appraisals of his work in F. Hoefer, *Histoire de la chimie,* II (Paris, 1866), 298–304; and J. R. Partington, *History of Chemistry,* III (London, 1962), 42–47. The best analysis of his contribution to chemical theory is in Hélène Metzger, *Les doctrines chimiques en France* (Paris, 1923), *passim,* see esp. pp. 340 ff.

MARIE BOAS HALL

HOME, EVERARD (*b.* Hull, England, 6 May 1756; *d.* London, England, 31 August 1832), *surgery, comparative anatomy.*

Home, the son of Robert Boyne Home, an army surgeon, and Mary Hutchinson, was a king's scholar at Westminster School. John Hunter, who had married Home's sister Anne in 1771, took him as a surgical pupil at St. George's Hospital in 1773. Home qualified through the Company of Surgeons in 1778 and served at Plymouth Naval Hospital and with the army in Jamaica from 1779 to 1784. He returned to England to be Hunter's assistant in surgery, teaching, and research and was elected fellow of the Royal Society in 1785 and assistant surgeon to St. George's in 1787.

In 1792 Home married Jane Tunstall, widow of Stephen Thompson; they had two sons, the elder of whom became a naval officer, and four daughters. Home joined the army in Flanders in the spring of 1793 but returned before Hunter's sudden death on 16 October. He was appointed surgeon to St. George's, replacing Hunter, and published a short biography of him in 1794. As executor he persuaded the government to buy Hunter's museum and entrust it to the Royal College of Surgeons; he became principal curator and, from 1817, a trustee, the resident conservator being William Clift. Home was master of the Royal College of Surgeons in 1813 and its first president in 1822; he endowed the Hunterian oration, was orator in 1814 and 1822, and gave courses in comparative anatomy in 1810, 1813, and 1822. He promised a catalog of the museum but produced only a synopsis; and in 1823 he burned Hunter's manuscripts, claiming to have published all of worth in his "hundred papers in the Philosophical Transactions which form materials for a catalogue raisonné of the Hunterian collection." Enough of Hunter's writing survives to prove that Home had often published Hunter's observations as his own.

Home was appointed sergeant surgeon to the king in 1808. He attended Prince Ernest, duke of Cumberland, in 1810 after the attempt on his life and, in 1811, the prince regent (George IV), to whom he became a valued friend; he was created a baronet in 1813. Home withdrew partially from hospital practice in 1808 and retired in 1827. He was surgeon to Chelsea Hospital for Army Pensioners from 1821 and died there at the age of seventy-six.

Home was a fearless and resourceful surgeon and an excellent teacher; Benjamin Brodie was his best pupil. He conducted wide-ranging research in comparative anatomy and wrote some seventy anatomical and fifty surgical papers, largely based on Hunter's unpublished material. Brodie said of him that "his ambition to appear as a discoverer increased while his mental powers declined." His best work is in his surgical books, particularly that on treatment of ulcers on the legs, drawn from his military experience. He gave the physiological Croonian lectures at the Royal Society fifteen times between 1794 and 1826. In his last years Home lost the confidence of his colleagues through his overbearing vanity and his suspected dishonesty in the destruction of Hunter's papers.

BIBLIOGRAPHY

I. Original Works. Oppenheimer (see below) provides a full list of Home's publications. His chief writings include *A Dissertation on the Properties of Pus* (London, 1788); "A Short Account of the Author's Life," in John Hunter, *A Treatise on the Blood* (London, 1794); *Practical Observations on the Treatment of Strictures in the Urethra* (London, 1795), 2nd ed., entitled . . . *Strictures in the Urethra and Oesophagus*, 3 vols. (London, 1797–1821); *Practical Observations on the Treatment of Ulcers on the Legs* (London, 1797; 2nd ed., 1801); *Observations on Cancer* (London, 1805); John Hunter, *A Treatise on Venereal Disease*, 3rd ed. by Home (London, 1810); *Practical Observations on Treatment of the Diseases of the Prostate Gland*, 2 vols. (London, 1811–1818); *The Hunterian Oration* (London, 1814); *Lectures on Comparative Anatomy*, 6 vols. (London, 1814–1828); *Synopsis of the Hunterian Museum* (London, 1818); "Account of a New Mode of Performing the High Operation for the Stone," in *Philosophical Transactions of the Royal Society*, **110** (1820), 209–213; *The Hunterian Oration* (London, 1822), with a eulogy of Sir Joseph Banks; and *A Short Tract on the Formation of Tumours* (London, 1830).

MSS sources are Archives of the Royal College of Surgeons, Royal Society, and St. George's Hospital, London; students' notes of Home's surgical lectures are in several libraries, including those of the Royal College of Surgeons of England and the College of Physicians of Philadelphia.

There are paintings by William Beechey, engraved for Home's *Lectures*, vol. I (1814), and by Thomas Phillips, given by Home to the Royal Society; a marble bust by Francis Chantrey is at the Royal College of Surgeons.

II. Secondary Literature. Notices by contemporaries are B. C. Brodie's reminiscences of Home in his *Hunterian Oration* (London, 1837), pp. 29–31, and his *Autobiography* (1865), pp. 45–49; and W. Clift's evidence to the House of Commons Select Committee on Medical Education (1834), in *Lancet* (11 July 1835), **2**, 471–476.

Modern studies include the following (listed chronologically): J. M. Oppenheimer, *New Aspects of John and William Hunter* (New York, 1946), pt. 1, "Everard Home and the Destruction of the John Hunter Manuscripts," with appendix, "Everard Home's Publications"; D. C. L. Fitzwilliams, "The Destruction of John Hunter's Papers," in *Proceedings of the Royal Society of Medicine*, **42** (1949), 1871–1876; and J. Dobson, *William Clift* (London, 1954), which discusses the relations between Home and Clift and provides documentary evidence of Home's plagiarism.

William LeFanu

HONDA, KOTARO (*b.* Aichi prefecture, Japan, 24 March 1870; *d.* Tokyo, Japan, 12 February 1954), *physics.*

Honda was the son of Hyosaburo and Sato Honda, who were farmers. In July 1897 he was graduated from the department of physics at the College of Science, Tokyo Imperial University, and went on to study at the university's graduate school. In August 1901 he became a lecturer at the college from which he had graduated. From February 1907 to February 1911 he studied at Göttingen and Berlin, and upon his return he became a professor at the College of Science, Tohoku Imperial University.

In 1916 Honda was awarded an Imperial Academy prize for his study on iron. In May 1919 the Iron and Steel Institute (in 1922 renamed the Research Institute for Iron, Steel, and Other Metals) was made part of the Tohoku Imperial University, with Honda as its director. In 1931 he was awarded the Emperor's Prize for his invention of a method of producing K.S. magnetic steel. From June 1931 to May 1940 he was president of Tohoku Imperial University, and in 1940 he became an honorary professor there. From April 1949 to May 1953 he was president of Tokyo Science University.

Honda performed geophysical research, including a survey of seiches (surface oscillations) in lakes and swamps throughout Japan and an examination of spouting in thermal springs, but his fame is based on his study of magnetic substances as well as of the metallurgy of iron and steel. He taught many researchers in these two fields. Until 1907, under the guidance of Hantaro Nagaoka, he did research in magnetostriction, measuring the changes of magnetization and magnetostriction in iron, nickel, and cobalt at temperatures ranging from that of liquid air to 1,200° C. While in Göttingen he learned the technology of metallurgy under Gustav Tammann, particularly the method of alloying, thus laying the basis for his future contribution to the study of the physical metallurgy of steel. In 1909 Honda moved to Berlin and, under Henri du Bois, studied the effect of a change in temperature on the magnetic coefficients of elements. He measured forty-three different elements at temperatures ranging from room temperature to 1,000° C. and discovered that there is a very close relationship between the magnetic coefficient and the periodic law.

Honda used the accumulated data from his extensive measurements to arrive at very significant conclusions. After his return to Japan in 1911, with the assistance of his pupils, he made many measurements of the magnetic coefficients of gaseous bodies and, from 1914, of various chemical compounds. These studies provided much valuable material for his future study of magnetism.

Immediately before the outbreak of World War I, in order to improve shipbuilding technique, there was a great demand in Japan for basic scientific studies of iron and steel. In response to this demand Honda entered into the new field of the physical metallurgy of iron and steel. Starting from existing methods, he developed the methods of thermobalance and magnetic analysis. The focal points of his study were the transformation of steel, the tempering of steel, and the characteristic features of cementite, Fe_3C. Later he also studied nonferrous alloys.

Honda discovered the A_0 transformation of cementite and proved that what was then thought to be the A_2 transformation of iron and steel was not a true transformation (1915). He obtained these results by studying the way in which the property of one component metal affected the character of an alloy by changing the ratio of the alloy's components. Through this method he invented K.S. magnetic steel in 1917 and new K.S. magnetic steel in 1934.

Parallel with these studies Honda pursued the ferromagnetic theory based on the theory of molecular magnets of J. A. Ewing (1916–1923); and after investigating the magnetization of single crystals of iron, nickel, and cobalt, he discovered anisotropic magnetism (1926–1935).

In his extremely wide-ranging researches, Honda often noted many phenomena ignored by others at the time. For instance, in 1920 he observed the magnetic transformation point of ferric oxide, Fe_2O_3 (Morin temperature), even before Morin discovered it. He also observed the abnormality in the magnetic susceptibility curve of a few antiferromagnetic substances.

BIBLIOGRAPHY

I. ORIGINAL WORKS. Of Honda's many papers published in *Science Reports of the Tohoku Imperial University*, the major ones are "Die thermomagnetischen Eigenschaften der Elemente," **1** (1911–1912), 1–42; "On the Magnetic Transformation of Cementite," **4** (1915), 161–167, written with H. Takagi; "On the Nature of the A_2 Transformation in Iron," *ibid.*, pp. 169–214; "On K.S. Magnet Steel," **9** (1920), 417–422; and "On the Magnetisation of Single Crystals of Iron," **15** (1926), 721–753, written with S. Kaya. The results of his studies on magnetic substances are systematically presented in *Magnetism and Matter* (Tokyo, 1917); and *Magnetic Properties of Matter* (Tokyo, 1928).

II. SECONDARY LITERATURE. On Honda's life and work, see *Memories of Professor Kotaro Honda* (Tokyo, 1955), a book of recollections by his pupils; and Teijiro Ishikawa, *The Life of Kotaro Honda* (Tokyo, 1964); neither work is written in academic style, however, and both lack a bibliography.

TETU HIROSIGE

HÖNIGSCHMID, OTTO (*b.* Hořovice, Bohemia [now Czechoslovakia], 13 March 1878; *d.* Munich, Germany, 14 October 1945), *chemistry.*

Hönigschmid was the son of Johann Hönigschmid, an Austrian officer who later went into financial administration. In the course of his father's official transfers, Hönigschmid lived in various places and finished his secondary schooling in Prague. He then studied chemistry at the German University in Prague and completed graduate work in 1901 with Guido Goldschmiedt, an organic chemist. Thus his early publications are devoted to organic chemistry, although they show a tendency toward analytic chemistry as well.

Goldschmiedt encouraged Hönigschmid's natural abilities, and from 1904 to 1906, on leave of absence from Prague, he continued his studies in Paris as assistant to Moissan. There he became familiar with high temperatures and the chemistry of silicide, carbide, and boride. In 1908 he made silicide the basis of his *Habilitation* and wrote a monograph, *Carbide und Silicide.* From scientific publications he became acquainted with the work of T. W. Richards of Harvard University, who had a worldwide reputation for his exact determinations of atomic weights. Enthusiastic about this area of study, Hönigschmid in 1909 took a second leave of absence for a year in Cambridge, Massachusetts, where he made his first achievements as an atomic scientist.

As early as 1911 Hönigschmid was involved in the work of the Radium Institute that had just opened at Vienna. The precise calculation of the atomic weights of the elements of the radioactive disintegration series—radium, uranium, thorium, ionium, and lead—was then a critical problem, indispensable for the confirmation of the Rutherford disintegration theory and the displacement law of Soddy and Fajans. Hönigschmid's lead determinations showed that lead, depending on its geological origin, exhibited variable atomic weights, a discovery which was the impetus for the isotope theory.

In 1911 Hönigschmid became extraordinary professor of inorganic and analytic chemistry at the German Technical University in Prague and, later, professor. He remained connected with the Vienna Institute. In 1918 he accepted a request to head the analytic chemistry department at the University of Munich and it was here that he established his famous atomic weight laboratory, his primary interest for the rest of his life.

Hönigschmid perfected preparative and analytic methods. With a large circle of students and colleagues he successfully determined the atomic weights of some fifty elements; among these were the first weight estimates for hafnium and rhenium. Special care was given to the so-called basic elements—including silver, the halogens, potassium, sodium, nitrogen, and sulfur—which served as foundations for determining the atomic weights of other elements.

Meanwhile, mass spectrographic investigations had showed that the natural elements are mixtures of isotopes of integral atomic weights. The Prout hypothesis of the unified building material in matter was revived, and Hevesy, Brønstedt, Clusius, and others accomplished the isotopic separation of ordinary elements through physical methods. Hönigschmid's analytic atomic weight calculations were thus a welcome step toward decisiveness, confirmation, and completion.

As a result of the exclusion of Germany from the "Conseil International des Recherches," Hönigschmid, Wilhelm Ostwald, Max Bodenstein, Otto Hahn, and R. J. Meyer joined together in 1920 to form their own atomic weight commission, with Ostwald as its head. After Ostwald retired Hönigschmid took charge and was largely responsible for the eleven annual reports of the commission. In 1930 an international commission on atomic weights was formed, with Germany as a participant, and Hönigschmid worked on its reports along with G. Baxter, M. Curie, Meyer, and P. LeBeau.

At the end of World War II Hönigschmid was seriously ill. He and his wife killed themselves during the occupation when, after the destruction of the institute, they twice had to move and found the difficulties of their living conditions insurmountable.

BIBLIOGRAPHY

I. Original Works. Hönigschmid's numerous publications appear chiefly in *Sitzungsberichte der Akademie der Wissenschaften in Wien* (1901–1916); *Sitzungsberichte der Bayerischen Akademie der Wissenschaften zu München* (1920–1940); *Bericht der Deutschen chemischen Gesellschaft* (1921–1943); *Zeitschrift für anorganische und allgemeine Chemie* (1925–1945); and *Zeitschrift für Elektrochemie und angewandte physikalische Chemie* (1914–1937). The proceedings of the atomic weight commission are published in many different national journals. For a bibliography of Hönigschmid's work, see Poggendorff, V, 547–548; VI, 1137–1138; VIIa, 511.

II. Secondary Literature. See E. Zintl, "Otto Hönigschmid zum 60. Geburtstag," in *Zeitschrift für anorganische und allgemeine Chemie,* **236** (1938), 3–11, with bibliography; and L. Birckenbach, "Otto Hönigschmid," in *Chemische Berichte,* **82** (1949), xi–lxv, with portrait and bibliography.

Grete Ronge

HOOKE, ROBERT (*b.* Freshwater, Isle of Wight, England, 18 July 1635; *d.* London, England, 3 March 1702), *physics.*

The son of John Hooke, a minister, Hooke was a sickly boy; although he ultimately lived to be nearly seventy, his parents did not entertain serious hope for his very survival during the first few years of his life. His father, one of three or four brothers, all of whom found their calling in the church, intended young Robert for the ministry also; but when persistent headaches interrupted the intended program of study, his father abandoned the plan and left the boy to his own devices. What these would be was immediately manifest. When he saw a clock being dismantled, he promptly made a working replica from wood. He constructed ingenious mechanical toys, including a model of a fully rigged man-of-war which could both sail and fire a salvo. By his tenth birthday Hooke had already embraced what his biographer Richard Waller called "his first and last Mistress"—mechanics. His role in the history of science is inextricably bound to his skill in mechanics and his allied perception of nature as a great machine.

When his father died in 1648, Hooke inherited £100. Since he had displayed some artistic talent, his family packed him off to London, where his legacy was to finance an apprenticeship to Sir Peter Lely. Hooke decided to save his money; and it was his good fortune that Richard Busby, the master of Westminster School, befriended him and took him into his home. The teacher had recognized the pupil. Not only did Hooke learn Latin, the staple of the secondary curriculum, together with Greek and a smattering of Hebrew; he also discovered mathematics. By his own account he devoured the first six books of Euclid in a week, and he proceeded to apply geometry to mechanics. Nor was mathematics all. By his own account again, he learned to play twenty lessons on the organ and invented thirty ways of flying. Having exhausted the resources of Westminster, he moved on to Oxford, where he entered Christ Church as a chorister in 1653.

Apparently Hooke never took a bachelor's degree. The only Oxford degree associated with his name is the Master of Arts, to which he was nominated in 1663. Meanwhile, Oxford had given him more than a thousand degrees could match. At the time of his arrival the university was the home of the brilliant group around which the Royal Society later crystallized. John Wilkins, Thomas Willis, Seth Ward, William Petty, John Wallis, Christopher Wren, Robert Boyle—these and others, some already recognized scholars, some still students, some merely resident near the university—convened regularly for the discussion of scientific matters. Hooke soon found his place in the circle. They recognized and drew upon his talent in mechanics, and they gave him in return his introduction to the new world of thought then fomenting the scientific revolution. For a time Hooke was an assistant to Willis. Willis introduced him to Boyle, and as Boyle's assistant Hooke launched his independent career.

Typically, Hooke's initial triumphs were mechanical inventions. Although Boyle's interest focused primarily on chemistry, the report of Guericke's air pump caught his attention. He instructed Hooke to devise an improved instrument, and the modern air pump duly appeared. With the air pump and with Hooke's assistance, Boyle conducted the experiments that concluded in Boyle's law, published in 1662. As with so much, Hooke's role in the investigation is unclear. Boyle never suggested that he played any part; and Hooke, who was not reluctant to assert himself, never claimed that he did. Hooke was Boyle's paid assistant at that time, however, and the position of assistant may well have seemed to both to preclude any rights of discovery. A number of historians assign the discovery to Hooke without further ado, and almost no one wants to deny outright that he participated in it.

In 1658, at the same time that he developed the air pump, Hooke turned his attention to chronometers. It was widely recognized that an accurate portable clock could solve the critical navigational problem of determining longitude. Hooke reasoned that one might be constructed by the "use of Springs instead of Gravity for the making of a Body vibrate in any Posture." That is, by attaching a spring to the arbor of the balance wheel, he would replace the pendulum with a vibrating wheel that could be moved because it oscillated around its own center of gravity. This is, of course, the principle of the watch, and on this principle a marine chronometer with which longitude could be determined was constructed in the eighteenth century. Once again, the exact nature of Hooke's contribution to clockmaking is shrouded in mystery. About 1660 three men of means—Boyle, Robert Moray, and William Brouncker, all later prominent in the Royal Society—considered backing Hooke's invention. Should the clock have worked, the profits might well have been immense. A patent was drawn up; but before the agreement was completed, Hooke withdrew, apparently demanding of his backers assurances they were unwilling to give.

In 1674 Christiaan Huygens constructed a watch controlled by a spiral spring attached to the balance; and Hooke, suspecting that his invention had been peddled to Huygens, cried foul. Working with the clockmaker Thomas Tompion, he made a similiar watch to present to the king; and on it he defiantly engraved the assertion "Robert Hook inven. 1658. T. Tompion fecit 1675." Despite his contentions, there is no evidence that his watch of 1658, if indeed it worked, employed a spiral spring, the device of crucial importance. On the other hand, his pamphlet that pronounced Hooke's law, *De potentia restitutiva* (1678), employed a spiral spring as one example and offered a demonstration (faulty, to be sure) that the vibrations of springs obeying Hooke's Law are isochronal. It is worth adding that neither Hooke's nor Huygens' watch worked satisfactorily enough to determine longitude. Although the exact nature of Hooke's contributions cannot be determined with any assurance, knowledgeable men at the time considered him to have made important inventions in chronometry; and historians are unanimous in agreement.

In 1659 and 1660 the Oxford circle dissolved with the collapse of the Protectorate and the restoration of the Stuarts. Relieved of their academic appointments, which many of them owed to their Puritan sympathies, most of the circle moved back to London, where they continued their meetings and formalized them in November 1660. Two years later the group acknowledged the king's patronage by taking the name Royal Society. A number of the early members knew Hooke from Oxford days; and others were impressed by his first publication, a pamphlet on capillary action which appeared in 1661. As a result Sir Robert Moray proposed him for the post of curator of experiments late in 1662. With untroubled confidence the Society charged him to furnish each meeting "with three or four considerable Experiments" as well as to try such other experiments as the members might suggest.

Probably no man could have come as close to fulfilling the impossible demand as Hooke did. He provided the major portion of intellectual content at the weekly meetings. It is hard to imagine that the Royal Society would have survived the apathy that succeeded its initial burst of enthusiasm without the stimulus of Hooke's experiments, demonstrations, and discourses. Some commentators have suggested that the Society's good fortune was Hooke's calamity. Its excessive demands imposed on him a pattern of frantic activity that made it impossible for him ever to finish a piece of work. On the contrary, the tendency to flit from idea to insight without pause was Hooke's innate characteristic. He never performed so well as he did during the first fifteen years of his tenure as curator, when, with a thousand demands on his time, he poured out a continuous stream of brilliant ideas. When the demands relaxed, the temper of his mind went slack as well; and his creative period came to a close. Far from destroying him, the Royal Society provided the unique milieu in which he could function at his best.

In 1664 Sir John Cutler founded a lectureship in mechanics for Hooke; it carried an annual salary of £50. Although Hooke's initial appointment as curator had involved no remuneration, the Royal Society now appointed him to the position for life with a salary of £30, together with the privilege of lodging at Gresham College. By September 1664 he had taken up residence there in the chambers that were his home until his death. Until 1676 he was in charge of the Society's repository of rarities, and he served as librarian until 1679. In 1665 the position of Gresham professor of geometry added a further duty, and a further salary of £50. Hooke's financial position was in fact far less secure than it may appear. The Royal Society was perpetually in financial straits and unable to sustain its obligations. As for his salary as lecturer, Cutler made a career of bestowing in public benefactions that he refused in private to fulfill, and Hooke had to take him to court to obtain his due.

In 1666 another job, probably the most onerous of all in its demands on his time, came Hooke's way. The great fire of London offered a considerable opportunity to one with Hooke's technical skills. Almost on the morrow of the disaster he came forward with a plan to rearrange the city wholly by laying it out on a rectangular grid. The plan won the approval of the city fathers; although it never approached implementation, it did promote his nomination as one of three surveyors appointed by the city to reestablish property lines and to supervise the rebuilding. As surveyor, Hooke was thrown into daily commerce with Sir Christopher Wren, one of the men appointed by the royal government to the same task of rebuilding. Wren and Hooke dominated and guided the work, and cemented a friendship that lasted throughout their lives. To Hooke the position of surveyor was a financial boon, more than compensating for the uncertainty of his other income. It also provided an outlet for his artistic talents. The title "surveyor" is misleading, for if he surveyed, he also functioned as an architect. A number of prominent buildings, such as the Royal College of Physicians, Bedlam Hospital, and the Monument, were his work. Hooke's reputation as a many-sided genius has tended to focus on his manifold scientific activities. His career as an architect adds another dimension to his achievement.

The ten years following the fire constituted a period of hectic activity. The very time when the demands of his surveyorship were at their peak was also a period of productive scientific work. To be sure, Hooke's scientific career was already well launched. In 1665, the year before the fire, he had published *Micrographia,* the most important book that he produced. If not the first publication of microscopical observations, *Micrographia* was the first great work devoted to them; and its impact rivaled that of Galileo's *Sidereus nuncius* half a century before. For the first time, descriptions of microscopical observations were accompanied by profuse illustrations—another display of Hooke's artistic talent. In the public mind, Hooke's name became identified with microscopical observations; and when Thomas Shadwell wrote his wretched physico-libidinous farce, *The Virtuoso,* he modeled the leading character on Hooke. Hooke attended a performance in June 1676: "Dammd Doggs. Vindica me Deus, people almost pointed."

No amount of ignorant ridicule could dim the

book's luster. It remains one of the masterpieces of seventeenth-century science. Like Galileo's *Nuncius, Micrographia* presented not a systematic investigation of any one question but a banquet of observations with courses from the mineral, vegetable, and animal kingdoms. Above all, the book suggested what the microscope could do for the biological sciences. Hooke's examination of the structure of cork led to his coining the modern biological usage of the word "cell." (The use of the word did not entail that he had any notion of modern cytology, of course. He referred to "pores or cells"; conceived of them as passages to carry liquids for the plant's growth; and, led on by Harvey's discovery, tried to locate the valves that must obviously be present as well. Nevertheless, the later biological usage of "cell" descended directly from the *Micrographia.*) In the animal realm, he inaugurated the study of insect anatomy. His horrendous portraits of the flea and the louse, a frightening eighteen inches long, are hardly less startling today than they must have been in the seventeenth century. He examined and understood the multiple eye of the fly, and he portrayed such diverse structures as feathers and apian stings. Frequent reproduction of the *Micrographia* testifies to the unfading fascination it continues to exercise.

Hooke also used the book as a vehicle to expound his own scientific theories. A work devoted to the microscope may be excused for proposing a theory of light, however tenuously connected to microscopical observations as such. An adherent of the mechanical philosophy of nature, Hooke held light to be mechanical as well: pulses of motion transmitted through a material medium. Neither in the *Micrographia* nor in his later lectures on light, delivered before the Royal Society, did he examine the theory at any great depth; but its mere proposal suffices to enroll him among the forebears of the wave theory of light. Moreover, the specific cause that shaped the theory was a set of observations destined to play an important role in the history of optics. Initially with mica, and then with soap bubbles, layers of air between sheets of glass, and a host of analogous instances, Hooke examined phenomena of colors in thin, transparent films. He recognized that the colors are periodic, with the spectrum repeating itself as the thickness of the film increases. His theory of light intended specifically to account for such phenomena. Except in the most general terms, the theory has not survived. Yet his observations of thin films did exert an extensive influence. Both Huygens and Newton saw that the thickness of the films could be calculated from the diameters of rings formed in the layer of air between a flat sheet of glass and a lens of known curvature. Newton's experiments, stemming directly from his reading of the *Micrographia,* became the foundation of Book Two of the *Opticks,* the source of the concept of periodicity in modern optics. The demonstration of periodicity was Newton's; the original suggestion of periodicity was Hooke's.

The theory of light was also the occasion of Hooke's initial confrontation with Newton. Seven years after the publication of *Micrographia,* Newton, then an obscure young academic almost completely unknown, sent his first paper on colors to the Royal Society. As the resident expert, Hooke was called upon to comment. More than somewhat magisterially, he rejected a new conception of colors he had not taken the trouble to understand. As far as colors were concerned, Hooke's theory had offered a new version of the old idea that colors arise from the modification of light which appears white in its pristine form. He had merely proposed a mechanism to account for the modification, and he failed now to see that Newton was replacing the concept of modification with an entirely different idea. Stung to fury by Hooke's critique, Newton penned a response that was little short of savage; and Hooke was subjected to the humiliation of seeing Newton's reply published in the *Philosophical Transactions* although his critique had been private. Late in 1675, when Newton sent the Royal Society his second paper on colors, observations on thin films together with the "Hypothesis of Light," Hooke claimed—or was reported to have claimed—that all of Newton's paper was found in his *Micrographia.* On this occasion Hooke, too, wrote privately, expressing his appreciation of Newton's work rather too formally and implying that Oldenburg was intriguing against him by spreading false reports. Newton's reply accepted the explanation in similar stilted phrases. The matter dropped for the time, but the complete lack of warmth between the men is manifest from this distance.

In addition to optics, the *Micrographia* also expounded a theory of combustion. At least four men in England were actively engaged at this time in investigating combustion and exploring its analogy with respiration. It is impossible to distinguish satisfactorily the independent roles of Hooke, Boyle, Richard Lower, and John Mayow; and it is difficult to assess adequately their total work. Individuals in the group, and Hooke among them, have been hailed as precursors—virtually forestallers—of Lavoisier and the discovery of oxygen. Close analysis of the various theories does not support such a judgment. In the *Micrographia,* Hooke argued that air is "the *menstruum,* or universal dissolvent of all *Sulphu-*

reous bodies," a dissolution carried out by a salt in the air and accompanied by intense heat, which we call fire. He identified the salt with that in saltpeter, so that combustion, which usually requires air, can take place in a vacuum when saltpeter is present.

Instead of forestalling Lavoisier, who saw combustion as a chemical combination, Hooke's theory repeated the accepted view that fire is an instrument of analysis that dissolves and separates bodies. There is no occasion to scorn the insight obtained. Along with the other three men, Hooke was impressed by the analogy of combustion and respiration. He carried out experiments before the Royal Society demonstrating that a continued supply of fresh air is as essential to life as it is to fire. By opening the thorax of a dog, destroying the motion of its lungs, and then employing a bellows to maintain a stream of air which passed out of the lungs through holes that he pricked, he demonstrated conclusively that the function of respiration is to bring a constant supply of fresh air into the lungs—not to cool and not to pump, as prevailing theories held, but solely to supply fresh air. With Mayow and the others, Hooke identified the nitrous salt or spirit in the air as the ingredient essential to life. Although the conceptual expression of this insight differed radically from Lavoisier's, its significance cannot be denied; and Hooke's role in it cannot be ignored.

During the years following *Micrographia,* Hooke found time to conduct demonstrations before the Royal Society and to deliver the Cutlerian lectures despite his activities as surveyor. Part of this work extended earlier investigations—for example, both those on combustion and those on optics—but he also broke new ground. During the 1670's he published a series of six brief works which were gathered together in a single volume, the *Lectiones Cutlerianae,* in 1679. The Cutlerian lectures contain at least two important scientific discoveries. One of these was the law of elasticity to which Hooke's name is still attached—"ut tensio sic vis." That is, the stress is proportional to the strain. Hooke's law, which was implicit in much of mechanics before him, was not a major discovery. Nevertheless, no one before him had stated it explicitly. Moreover, Hooke perceived intuitively that a vibrating spring is dynamically equivalent to a pendulum; and in the lecture that announced Hooke's law, he undertook one of the early analyses of simple harmonic motion. He based it on what he referred to elsewhere as "the General Rule of Mechanicks":

> Which is, that the proportion of the strength or power of moving any Body is always in a duplicate proportion of the Velocity it receives from it . . .

That is, the "quantity of strength" employed in moving a body is proportional to the square of the velocity it receives. In many ways the passage was typical of Hooke. The demonstration foundered on its inherent confusions—although it is necessary to add that in the seventeenth century only giants such as Huygens, Leibniz, and Newton succeeded in dispelling similar confusion in dynamics. In Hooke's case, the clarity of his mechanical conceptions and the power of his analysis were not able to match his intuitive insight.

In another Cutlerian lecture, Hooke announced the three basic suppositions on which he intended to construct a system of the world corresponding to the rules of mechanics:

> First, That all Coelestial Bodies whatsoever, have an attraction or gravitating power towards their own Centers, whereby they attract not only their own parts, and keep them from flying from them, as we may observe the earth to do, but that they do also attract all the other Coelestial Bodies that are within the sphere of their activity. . . . The second supposition is this, That all bodies whatsoever that are put into a direct and simple motion, will so continue to move forward in a streight line, till they are by some other effectual powers deflected and bent into a Motion, describing a Circle, Ellipsis, or some other more compounded Curve Line. The third supposition is, That these attractive powers are so much the more powerful in operating, by how much the nearer the body wrought upon is to their own Centers.

This remarkable statement, together with others that date back to 1664, has become a major piece of evidence in the case for Hooke's claim on the law of universal gravitation. It contains two elements. On the one hand, it proposes a concept of apparently universal attraction. It is only apparently universal, however. An idea of gravitational attractions specific to each planet, forces by which they maintain the unity of their systems, was widely held in the seventeenth century. Although Hooke took a major step toward generalizing this idea, his understanding of gravitation never eliminated the notion of a force specific to certain kinds of matter and hence never reached the level of universal gravitation. Gravity, he said elsewhere, is "such a Power, as causes Bodies of a similar or homogeneous nature to be moved one towards the other, till they are united. . . ." Planets are of the same nature as the sun and hence are attracted to it. Comets are not related, and they are repelled.

Hooke himself never laid claim to the concept of universal gravitation. Rather, he asserted his propriety over the second element in the passage above, the celestial dynamics. In fact, his proposal did contain a revolutionary insight that reformulated the approach to circular motion in general and to celestial dynamics in particular. Notable in his statement is the absence of any reference to centrifugal force. Hooke was the man who first saw clearly the elements of orbital dynamics as we continue to accept them. If the principle of rectilinear inertia be granted, a body revolving in an orbit must be continually diverted from its inertial path by some force directed toward a center. When Hooke was formulating this view, Newton still thought of circular motion in terms of an equilibrium of centrifugal and centripetal forces. Moreover, it was Hooke who taught him to see it otherwise. Late in 1679 Hooke wrote to Newton, among other things asking for Newton's opinion of his proposed planetary dynamics. The correspondence is too well known to need repeating. Suffice it to say that in response to Newton's assumption of uniform gravity in a problem mechanically identical to orbital motion, Hooke stated his conviction that gravity decreases in power in proportion to the square of the distance. Hooke was always convinced thereafter that Newton had stolen the inverse square relation from him. Newton himself acknowledged in 1686 that the correspondence with Hooke stimulated him to demonstrate that an elliptical orbit around a central attracting body placed at one focus entails an inverse square force.

Nevertheless, one must beware of attributing too much to Hooke. Once again, his power of analysis could not support the brilliance of his insight. The insight cannot be taken from him. Where earlier investigations of the dynamics of circular motion had based themselves on the notion of centrifugal force, Hooke (as it were) stood the problem right side up and put it in a position to be attacked fruitfully. But his own mechanics was not adequate to that job. Although he proposed the problem of the dynamics of elliptical orbits, he acknowledged his inability to solve it; and his very derivation of the inverse square relation, on which he insisted with such vehemence, was so defective as to be ludicrous. He justified the inverse square relation, not by substituting the formula for centripetal force (which he appears not to have known) into Kepler's third law, but by a bastardized application of his own general rule of mechanics to Kepler's aborted law of velocities. Hooke did not discover or even approach the law of universal gravitation. But he did set Newton on the correct approach to orbital dynamics and, in this way, contributed immensely to Newton's later triumph.

Although one important area of Hooke's scientific activity, his study of fossils and his related contribution to geology, also figured in the *Micrographia,* its major exposition appeared only in the "Lectures and Discourses of Earthquakes," the largest section of his *Posthumous Works.* Spread over a period of thirty years, the lectures testify that geology was one of Hooke's enduring interests. Geology might almost have been created to display his talents to maximum advantage. An almost untouched field, it presented no massive volume of data to be mastered and offered few constraints to curb his facile imagination. Hooke repaid it handsomely. He provided a solution to the controversy over the origin of fossils by dividing "figured stones" into two categories—those with forms characteristic of the organism and those with forms characteristic of the substance. In regard to the latter, Hooke may be described as a protocrystallographer. He showed how the polyhedral forms of crystals (as he saw them under the microscope) could be built up from packings of bullets, the basis for the claim that he anticipated Steno in the law of constancy of interfacial angles.

In an age when the biblical account of creation made fossils with organic forms a riddle to most investigators, Hooke was remarkable for his steadfast refusal to consider them as anything but the remains of organic creatures. His refutation of the argument that they are *lusus naturae,* sports of nature produced to no purpose, is one of the classic passages of scientific argumentation in the seventeenth century. He refused to call in the Deluge to explain the presence of marine fossils far from the sea, but he concluded that the surface of the earth has been subject to vast upheavals and changes. When fossils could not be identified with existing creatures, he did not hesitate to consider the mutability of species.

One must be careful not to exaggerate the modernity of Hooke's geological ideas. Unable to destroy the preconception of a limited time span, he identified the upheavals of the surface of the earth with cataclysmic earthquakes. He has been called the first uniformitarian; quite the contrary, he was the first catastrophist. The mutations of species he conceived were limited variations under the stress of environmental change. To say as much is only to concede that Hooke could not leap from the seventeenth century into the nineteenth. With the possible exception of Steno, he was easily the most important geologist of his day. In nothing does he appear more modern than in his prescription of a program for geological study. Fossils are the "Monuments" and "Medals" of earlier ages from which the history of the earth can be reconstructed, just as the history of mankind is studied through human remains. The pursuit of Hooke's program for geology ultimately shattered the seventeenth-century preconceptions which confined his own geological theories.

Perhaps Hooke's most important contribution to science lay in the field of instrumentation. He added something to every important instrument developed in the seventeenth century. He invented the air pump in its enduring form. He advanced horology and microscopy. He developed the cross-hair sight for the telescope, the iris diaphragm, and a screw adjustment from which the setting could be read directly. He has been called the founder of scientific meteorology. He invented the wheel barometer, on which the pivoted needle registers the pressure. He suggested the freezing temperature of water as the zero point on the thermometer and devised an instrument to calibrate thermometers. His weather clock recorded barometric pressure, temperature, rainfall, humidity, and wind velocity on a rotating drum. Although it was not a scientific instrument, the universal joint was also his invention. Writing in the eighteenth century, Lalande called Hooke "the Newton of mechanics." One might add that he was the first mechanic of genius whose talent the mechanical philosophy of nature brought to bear directly on science.

The year 1677 brought significant changes to Hooke's life. The death of Henry Oldenburg led to his nomination as secretary of the Royal Society. For several years the two men had been mortal enemies. Convinced that Oldenburg had betrayed the secret of his spring-driven watch to Huygens, Hooke had publicly labeled him a "trafficker in intelligence"; but the Council of the Royal Society had come to Oldenburg's support. Now he sat in his enemy's position of power. It proved to be an empty triumph. Public success merely disguised private decline. Although he was only forty-two years old in 1677, and destined to survive another quarter of a century, Hooke had exhausted his scientific creativity. One year later the last of his Cutlerian lectures announced Hooke's law. From there on, everything was downhill.

His tenure as secretary was not successful, and he stepped down after five years. During that period he tried to continue Oldenburg's periodical—renamed *Philosophical Collections*—but he managed to bring out only seven issues in all. In 1686 Newton laid Book I of the *Principia* before the Society. Hooke was convinced that he had been robbed again, but hardly anyone listened to his protestations. And in 1687 his niece Grace, originally his ward and then his mistress through a prolonged and tempestuous romance, died. From that blow he never fully recovered. More and more he became a recluse and a cynic. A tone of bitterness pervades the small number of papers that survive from his final years. In the end he was almost bedfast. He died on 3 March 1702 in the room at Gresham College that he had inhabited for nearly forty years.

Hooke was a difficult man in an age of difficult men. His life was punctuated with bitter quarrels that refused to be settled. When he offered criticism of Hevelius' use of open sights for astronomical observations, he did it in such a way that the consequences dragged on for ten years. His conflicts with Oldenburg and Newton have already been mentioned. It is only fair to add that the other three men were at least as difficult in their own right, and that Hooke won and held the esteem and affection of such men as Boyle, Wren, and the antiquarian John Aubrey. Hooke's disposition was probably exacerbated by his physical appearance. Pepys said of him, while he was still a young man, that he "is the most and promises the least of any man in the world that ever I saw." As every description testifies, his frame was badly twisted. Add to his wretched appearance wretched health. He was a dedicated hypochondriac who never permitted himself the luxury of feeling well for the length of a full day. Hooke's spiny character was nicely proportioned to the daily torment of his existence.

As for his role in the history of science, it is impossible to avoid the commonplace assessment—that he never followed up his insights. Indeed, he was incapable of exploring them in their ultimate depths—as Newton, for example, could do. Early in his career Hooke composed a methodological essay that earnestly advocates orderly procedure and systematic coverage. It appears almost to be Hooke's judgment on himself. Typically, it remained unfinished. Waller records that in his old age Hooke intended to leave his estate to build a laboratory for the Royal Society and to found a series of lectures. He procrastinated in completing his will "till at last this great Design prov'd an airy Phantom and vanish'd into nothing." More than one of Hooke's grand designs proved an airy phantom and vanished into nothing—at least if we judge him by the standards of a Newton. Because of his claim on the law of universal gravitation, the comparison with Newton inevitably arises, but such a standard of judgment is unfair to Hooke. If he was not a Newton, his multifarious contributions to science in the seventeenth century are beyond denial; and on the crucial question of circular motion it was Hooke's insight that put Newton on the track to universal gravitation. The Royal Society honored its own wisdom when the members attended his funeral as a body.

BIBLIOGRAPHY

I. ORIGINAL WORKS. *Micrographia* (London, 1665) is readily available in reprint eds. (New York–Weinheim, 1961). Five vols. of R. T. Gunther, *Early Science in Oxford,* 14 vols. (Oxford, 1923–1945), bear the title *The Life and Work of Robert Hooke.* Vols. VI and VII contain extracts from Thomas Birch's *History of the Royal Society* that mention Hooke, extracts from Waller and Derham (see below) and from Hooke's papers in the *Philosophical Transactions,* and letters. Vol. VIII reproduces the *Lectiones Cutlerianae.* Vol. X reproduces Hooke's earliest publication, the pamphlet of 1661 on capillary phenomena, and publishes his diary for the years 1688–1693. Vol. XIII reproduces the *Micrographia. The Diary of Robert Hooke, M.A., M.D., F.R.S., 1672–1680,* Henry W. Robinson and Walter Adams, eds. (London, 1935), covers an earlier period than the Gunther diary. *The Posthumous Works of Robert Hooke, M.D., S.R.S., Geom. Prof. Gresh., &c.,* Richard Waller, ed. (London, 1705); and *Philosophical Experiments and Observations of the Late Eminent Dr. Robert Hooke, S.R.S. and Geom. Prof. Gresh. and Other Eminent Virtuoso's in His Time,* William Derham, ed. (London, 1726), are two other important sources of his work. Geoffrey Keynes has published *A Bibliography of Dr. Robert Hooke* (New York, 1960).

II. SECONDARY LITERATURE. The best contemporary sources on Hooke's life are John Aubrey's sketch in *Brief Lives,* I (Oxford, 1898), 409–416; and Richard Waller's biography, prefaced to the *Posthumous Works.* See also John Ward, *Lives of the Professors of Gresham College* (London, 1740), pp. 169–193. Hooke has recently been the subject of a more extended, perhaps excessively enamored, biography: Margaret 'Espinasse, *Robert Hooke* (London, 1956). Among the innumerable general articles on him, E. N. da C. Andrade, "Robert Hooke," in *Proceedings of the Royal Society,* **201A** (1950), 439–473, is of special importance. There is also a general discussion of his scientific career in the introduction by Richard S. Westfall to a reprint ed. of the *Posthumous Works* (New York, 1969). Mary Hesse has published two articles devoted to general aspects of his scientific thought: "Hooke's Philosophical Algebra," in *Isis,* **57** (1966), 67–83; and "Hooke's Vibration Theory and the Isochrony of Springs," *ibid.,* 433–441.

On Hooke and gravitation, see the following (listed chronologically): Philip E. B. Jourdain, "Robert Hooke as a Precursor of Newton," in *Monist,* **23** (1913), 353–385; Louise Diehl Patterson, "Hooke's Gravitation Theory and Its Influence on Newton," in *Isis,* **40** (1949), 327–341, and **41** (1950), 32–45; Alexander Koyré, "A Note on Robert Hooke," *ibid.,* **41** (1950), 195–196, a commentary on Patterson's article; and "An Unpublished Letter of Robert Hooke to Isaac Newton," *ibid.,* **43** (1952), 312–327; Johannes Lohne, "Hooke *versus* Newton," in *Centaurus,* **7** (1960), 6–52; and Richard S. Westfall, "Hooke and the Law of Universal Gravitation," in *British Journal of the History of Science,* **3** (1967), 245–261.

For Hooke's contributions to clockmaking, see A. R. Hall, "Robert Hooke and Horology," in *Notes and Records. Royal Society of London,* **8** (1950–1951), 167–177. His work on combustion is treated in D. J. Lysaght, "Hooke's Theory of Combustion," in *Ambix,* **1** (1937), 93–108; Douglas McKie, "Fire and the Flamma Vitalis: Boyle, Hooke and Mayow," in *Science, Medicine and History, Essays . . . in Honour of Charles Singer,* E. Ashworth Underwood, ed., 2 vols. (London, 1953), I, 469–488; and H. D. Turner, "Robert Hooke and Theories of Combustion," in *Centaurus,* **4** (1956), 297–310. The best discussion of Hooke's optics is in A. I. Sabra, *Theories of Light From Descartes to Newton* (London, 1967), pp. 187–195, 251–264, 276–284, 321–333; see also Richard S. Westfall, "The Development of Newton's Theory of Color," in *Isis,* **53** (1962), 339–358; and "Newton and His Critics on the Nature of Colors," in *Archives internationales d'histoire des sciences,* **15** (1962), 47–58. On Hooke as a geologist, see A. P. Rossiter, "The First English Geologist," in *Durham University Journal,* **27** (1935), 172–181; W. N. Edwards, "Robert Hooke as a Geologist and Evolutionist," in *Nature,* **137** (1936), 96–97; and a commentary on Edwards' article by Rossiter, "Hooke as Geologist," *ibid.,* 455.

RICHARD S. WESTFALL

HOOKER, JOSEPH DALTON (*b.* Halesworth, England, 30 June 1817; *d.* Sunningdale, England, 10 December 1911), *botany.*

Hooker was the second child of the botanist Sir William Jackson Hooker and Maria Sarah Turner. Glasgow High School provided him with a traditional Scottish liberal education which was broadened by his own leisure-time interest in botany and entomology. He studied medicine at Glasgow University, receiving an M.D. in 1839. His father's influence was predominant in his life; to him he owed his physical stamina, his capacity for sustained hard work, his artistic ability, and the opportunity to fulfill his youthful ambitions of becoming a great botanist and traveler.

Hooker ultimately achieved great professional eminence and many academic honors were bestowed on him. Elected a fellow of the Royal Society in 1847, he served as president of the society from 1873 to 1878 and made notable improvements in its organization and financial resources. During celebrations in 1907 to mark the bicentenary of Linnaeus' birth, the Swedish Academy of Sciences awarded him the single, specially struck Linnean Medal as "the most illustrious living exponent of botanical science." He was created C.B. in 1869 and Knight Commander of the Order of the Star of India in 1877.

Hooker was blessed with two happy marriages. In 1851 he married Frances Harriet Henslow, the eldest daughter of the Reverend John Stevens Henslow. There were six surviving children on her death in 1874. In 1876 he married Hyacinth, the only daughter of the Reverend William Samuel Symonds, and the widow of Sir William Jardine. He had two children by this second marriage.

Hooker's descriptions of three new Indian mosses, his first contributions to scientific literature, were published in his father's periodical, *Icones plantarum,* in 1837. His father's position and influence obtained for him the post of assistant surgeon and naturalist on H.M.S. *Erebus,* which, with H.M.S. *Terror,* had among its exploratory goals a determination of the position of the south magnetic pole. Under the command of Captain James Clark Ross the expedition left England in September 1839 and did not return until September 1843. The voyage encompassed the exploration of the Great Ice Barrier and of several oceanic islands, including the Falklands, Tasmania, and New Zealand.

Much of the botanical material collected on the *Erebus* voyage came from territory never before explored, and doubtless fostered in Hooker an enduring interest in taxonomy and plant geography. The results of his botanical investigations, carefully and judiciously compiled, were eventually published in *Flora Antarctica* (1844–1847), *Flora Novae-Zelandiae* (1853–1855) and *Flora Tasmaniae* (1855–1860). Known collectively under the title *The Botany of the Antarctic Voyage of H.M. Discovery Ships 'Erebus' and 'Terror,'* this great work in six quarto volumes established Hooker as a leading world botanist. Thoroughness and accuracy characterize his full Latin descriptions and detailed taxonomic notes in English. The excellent accompanying lithographs, by W. H. Fitch, were based on Hooker's field sketches and specimens. The work appeared at a critical period in biology, and the introductory essays to the *Flora Novae-Zelandiae* and *Flora Tasmaniae,* published in 1853 and 1860 respectively, are valuable indexes of the uncertainty in scientific thought at the time of the publication of Darwin's *Origin of Species* (1859). Hooker's attention was especially drawn to the taxonomic resemblances between the floras of South America, the subantarctic islands, New Zealand, and Australia. He sought to explain these similarities largely by a land bridge theory—postulating a lost circumpolar continent—and rejected Darwin's alternative hypothesis of the transport of seeds by ocean currents, winds, and birds.

In the autumn of 1843 Hooker settled with his parents at Kew, where he had access to his father's extensive herbarium and library for working out the botanical results of his Antarctic voyage. In 1845 he gave a series of botanical lectures at Edinburgh University, presumably to advance his candidature for the professorship of botany pending the death of Robert Graham, who was seriously ill. Unsuccessful in his application for the post, he accepted in February 1846 an appointment as a paleobotanist with the Geological Survey—he had earlier manifested an interest in paleobotany when he published a paper on certain Tasmanian fossil woods. His first official assignment was to prepare a catalogue of British fossil plants for an arrangement of specimens in the Geological Survey museum. He wrote many papers on fossil botany until his appointment as assistant director at the Royal Botanic Gardens, Kew, in 1855, after which the subject ceased to occupy his attention.

In 1854 Hooker wrote to Darwin:

> From my earliest childhood I nourished and cherished the desire to make a creditable journey in a new country, and write such a respectable account of its natural features as should give me a niche amongst the scientific explorers of the globe I inhabit, and hand my name down as a useful contributor of original matter.[1]

This ambition had been realized when Hooker embarked for India in November 1847. With a Treasury grant and a commission in the Royal Navy, he spent three years in northeast India, mainly in the Himalayan state of Sikkim and in eastern Nepal, engaged in botanical exploration and topographical surveying. For most of his stay in India he traveled alone, but in October 1849 he was joined by Dr. Archibald Campbell, the superintendent of Darjeeling and political agent to Sikkim. Both were arrested and imprisoned for several weeks by the Sikkim authorities.

Hooker was practically the first explorer of the eastern Himalaya since Turner's embassy to Tibet in 1789. Not only did he add much to the existing knowledge of the Indian flora but he also made detailed meteorological and geological observations, while his accurate survey work on the complex mountain terrain formed the basis of a map published by the India Trigonometrical Survey. Horticulturalists will always associate Hooker's name with the genus *Rhododendron* because he introduced a number of new species into cultivation in England. While Hooker was still in India, his father supervised the publication in 1849 of the first part of his son's *Rhododendrons of Sikkim-Himalaya,* embellished with superb color lithographs by Fitch, prepared from Hooker's sketches and dried specimens.

In 1850 he traveled in the Khasi Mountains of Assam with Thomas Thomson, a friend from his school days, with whom he collaborated on the one-volume *Flora Indica* (1855); Hooker's introductory essay to this book, containing an admirable account of the history of botany in India and the geographical distribution of the flora, was to form the basis of his masterly "Sketch of the Flora of British India" in the *Imperial Gazetteer of India* (1907). Assisted by other botanists, Hooker produced the *Flora of British India* (1872–1897) in seven volumes to replace the *Flora Indica.* This still remains a classic account in English of families, genera, and species of all Indian seed-bearing plants. It has served as a foundation for a number of regional Indian floras, but although it is inevitably in need of present-day revision, no up-to-date flora of the Indian subcontinent has yet replaced it. On the completion of this great undertaking Hooker, who had been made a knight commander of the Order of the Star of India in 1877, was made a grand commander of the order.

Hooker later wrote volumes IV and V (1898–1900) of *A Handbook to the Flora of Ceylon,* which had been left unfinished in 1896 on the death of its author, H. Trimen. During his researches on the Indian flora, Hooker became aware of the need for a taxonomic revision of the genus *Impatiens.* His last years were occupied with an intensive investigation of this genus, of which he described more than 300 species as being new.

There remains one book associated with Hooker's work in India which has an appeal for the nonbotanist, namely his *Himalayan Journals* (1854). It is a record of adventure and scientific observation, lucidly and modestly related, entirely deserving its place as a minor classic of nineteenth-century travel literature. It was dedicated to Charles Darwin.

Hooker met Darwin briefly for the first time in 1839 and within a few years they had become close friends. On 11 January 1844 Darwin confided to Hooker the direction of his thinking: "At last gleams of light have come, and I am almost convinced that species are not (it is like confessing a murder) immutable:—I think I have found (here's presumption!) the simple way by which species become exquisitely accepted to various ends."[2] For many years Hooker was kept informed of Darwin's gradual progression from tentative hypothesis to confident belief. In a letter to W. H. Harvey, written about 1860, Hooker wrote, "I was aware of Darwin's views *fourteen* years before I adopted them and I have done so *solely* and *entirely* from an independent study of the plants themselves."[3] Hooker gradually came to accept the theory of evolution on the basis of his own taxonomic work and researches on the geographical distribution of plants.

Until his introductory essay (1860) for the *Flora Tasmaniae,* however, there is in Hooker's published works no clear confirmation of Darwin's influence. In the introductory essay (1853) to the *Flora Novae-Zelandiae* Hooker tentatively advocated the permanency of species as an essential requirement for practical taxonomy. The introduction to the *Flora Indica* (1855) shows Hooker as still recognizing species as "being definite creations" but as "created with a certain degree of variability."[4] A review by Hooker in *Hooker's Kew Journal of Botany* (1856) of Alphonse de Candolle's *Géographie botanique raisonnée* (1855) revealed an ambivalence in his position. He asserted here that there was no proof for Candolle's belief "that the majority of species were created such as they now exist"; and he agreed that the "theory of transmutation [of species] accounts better for the aggregation of Species, Genera and Natural Orders in geographical areas, and for their limitation." But doubt still persisted, for he continued: ". . . unfortunately transmutation brings us no nearer the origin of species, except the doctrine of progressive development be also allowed and, as we can show, the study of plants affords much positive evidence against progressive development, and none in favour of it."[5] In 1858 Hooker and the geologist Lyell were instrumental in persuading both Darwin and Alfred Russel Wallace to agree to a joint presentation of their papers on evolutionary theory to the Linnean Society of London.

Soon after the publication of the *Origin of Species* in 1859 Hooker became a decided advocate of Darwinism, but he reached this position only after his own careful and prolonged assessment of all available scientific evidence. In his introductory essay (1860) to the *Flora Tasmaniae* he cautiously accepts the theory of evolution and natural selection. His characteristic consideration and questioning of all aspects of a problem once prompted Darwin to refer to him as "you terrible worrier of poor theorists."[6]

Hooker's eventual adherence to Darwinism came partly through the persuasions of his own phytogeographical evidence. He was one of the first botanists to offer the mutability and derivative origins of species as an explanation for the geographical distribution of plants. He was especially attracted by the similarities in the floras of widely separated regions. At first he accepted the geological theories of Lyell and Edward Forbes to explain resemblances in the flora of different Antarctic islands. Much later, however, in a lecture in 1866 to the British Association, *On Insular Floras,* he conceded that Darwin's alternative proposition of transoceanic migration offered a rational explanation; the present-day theories of continental drift were not then known. In his classic monograph, *Outlines of the Distribution of Arctic Plants* (1862), he sought to show that the vegetation of Scandinavia had migrated through Asia and America. His paper to the British Association at York in 1881 summarized his final conclusions on the geographical distribution of plants.

Hooker was appointed assistant director at the Royal Botanic Gardens in 1855, and succeeded his father as director in 1865. During his term of office Sir William Hooker had transformed the moribund gardens, formerly the private property of the British monarch. The grounds were relandscaped, new glasshouses (including the great palm house) were erected, an herbarium and library were formed, and museums of economic botany established. Hooker continued his father's program of improving the gardens: some additional avenues and walks were introduced, the rock garden was created in 1882, more glasshouses were added, and improvements were made in the water supply. The arboretum was enlarged, an improvement which included the planting of the pinetum in 1871–1872. In 1882 the Marianne North Gallery was opened for the permanent display of the botanical paintings of that indefatigable artist and traveler.

Under Hooker, Kew became an international center for botanical research. The Jodrell Laboratory, a private benefaction of his friend T. J. Phillips Jodrell, was originally built in Kew Gardens as a center for the investigation of the structure and physiology of plants. Since its foundation in 1876 many distinguished botanists have worked there. When Sir William Hooker died in 1865, he left at the Royal Botanic Gardens the nucleus of an official herbarium and library with a wish that his own vast personal collections should be added. These were purchased in 1867 and many important herbaria were added during Joseph Hooker's directorship: R. Wight's Indian plants, the mycological herbarium of the Reverend M. J. Berkeley, the mosses of W. P. Schimper, and the lichens of the Reverend W. A. Leighton. An extension to the Kew herbarium soon became necessary and the first wing was added to the building in 1877.

The links which Sir Joseph Banks had established between Kew and the British Empire were strengthened by both Sir William and Sir Joseph Hooker. Of particular help to the developing colonies were the consignments of economically useful plants first propagated at Kew. Rubber seedlings, which had originally been smuggled out of Brazil, were sent in 1876 from Kew to Ceylon where they were to become the foundation of the rubber industry of that island and later of the Malay Peninsula. The cultivation of ipecac in India was established in 1866–1867 from Kew material, and Liberian coffee plants and the West African oil palm were distributed to many plantations abroad. Hooker also added to the series of published Kew colonial floras initiated by his father.

It could be argued that Hooker's greatest service to mankind was his administration of Kew Gardens from 1865 until his retirement in 1885. The Royal Botanic Gardens are very much the creation of the two Hookers, whose energy, foresight, and organizational ability laid such firm and enduring foundations.

Joseph Hooker and the botanist George Bentham, a permanent visitor in the Kew herbarium, recognized the need for a new plant classification to replace the outdated ones of Endlicher and Meissner. Both men were well qualified to undertake such a formidable task. Their collaboration began in 1857 but unfortunately Hooker's official duties prevented him from contributing more than a third of the final work. The meticulous descriptions of the families and genera of seed-bearing plants, with full synonymy and geographical distribution, were based largely upon direct observation of the rich resources of Kew and its herbarium. The Bentham-Hooker classification, although not a phylogenetic system, is a natural one following, with modifications, the sequence of families proposed by Augustin-Pyramus de Candolle in 1819. The three-volume *Genera plantarum* (1862–1883) still remains a standard work and its Latin diagnoses are models of accuracy, clarity, and completeness.

The *Genera plantarum* was one of the many works abstracted by the *Index Kewensis* (1892 to date), an indispensable index of validly published names of flowering plants. This index owed its inception to a suggestion from Darwin for a complete list of scientific plant names, and Darwin generously helped with the cost of production of the original volumes. Work started on it in 1882 under the supervision of Joseph Hooker, for whom it was claimed that he read all the proofs containing some 380,000 specific names.

In 1870 appeared Hooker's *Student's Flora of the British Isles,* which aimed at presenting fuller information on vascular plants than that provided by existing manuals. Further editions appeared in 1878 and 1884. Hooker edited the fifth through the eighth editions (1887–1908) of G. Bentham's *Handbook of the British Flora.* He also edited two well-known botanical periodicals: *Botanical Magazine* from 1865 to 1904, writing many of the plant descriptions himself; and *Hooker's Icones plantarum,* founded by his father in 1836. For the latter periodical he edited volumes 11 to 19 (1867–1890), which were illustrated with many of his own line drawings.

His exacting duties as director at Kew Gardens did not prevent Hooker from participating in further botanical excursions abroad. In the autumn of 1860 he and Daniel Hanbury spent about two months in Palestine and Syria, where Hooker examined the history, position, and age of the famous cedar grove on Mount Lebanon. He regarded the three species of cedar found in the Himalaya, Syria, and Africa as geographical forms of one species. He contributed an account of the botany of Syria and Palestine to volume II of W. Smith's *Dictionary of the Bible* (1863).

From April to June 1871 Hooker, in the company of John Ball and George Maw, explored Morocco. The subsequent account of their journey was written mainly by Ball, but Hooker contributed the first two chapters and three valuable appendices in which he compared the flora of the Canaries and Morocco. An important discovery of the expedition was that the Arctic-Alpine flora did not reach the Atlas Mountains.

In 1871 Hooker joined his friend Asa Gray in western North America in what was to be his last major botanical expedition. Both botanists were interested in the floristic similarities of the eastern United States and eastern continental Asia and Japan. Hooker was of the opinion that the Miocene flora in western North America had been eliminated by glaciation, but that such flora had managed to survive on the eastern side of the continent and in eastern Asia.

Hooker proved himself to be highly competent in a number of botanical disciplines, but he distinguished himself most notably in taxonomy and plant geography, areas which also provided him with the evidence that made him an evolutionist. In pure morphology he will be remembered for his classic papers on *Balanophoreae* (1856), *Nepenthes* (1859), and *Welwitschia* (1863). His reasoning was instinctively inductive; he was always reluctant to commit himself to any generalization before he had examined all the available facts and tested them against his exceptionally wide experience and knowledge.

NOTES

1. F. Darwin, *More Letters of Charles Darwin,* I (1903), 70.
2. F. Darwin, *Life and Letters of Charles Darwin,* II (1887), 23.
3. L. Huxley, *Life and Letters of Sir Joseph Dalton Hooker,* I (1918), 520.
4. J. Hooker, Introduction to *Flora Indica* (1855), 20.
5. J. Hooker, "*Géographie botanique raisonnée . . .* par M. Alph. de Candolle; a Review," in *Hooker's Kew Journal of Botany,* **8** (1856), 252.
6. F. Darwin, *More Letters of Charles Darwin,* I (1903), 105.

BIBLIOGRAPHY

I. ORIGINAL WORKS. Hooker's writings include *The Botany of the Antarctic Voyage of H.M. Discovery Ships 'Erebus' and 'Terror' in the years 1839–1843 Under the Command of Captain Sir James Clark Ross:* pt. 1, *Flora Antarctica,* 2 vols. (London, 1844–1847); pt. 2, *Flora Novae-Zelandiae,* 2 vols. (London, 1853–1855); pt. 3, *Flora Tasmaniae,* 2 vols. (London, 1855–1860); *Rhododendrons of the Sikkim-Himalaya* (London, 1849–1851); *Himalayan Journals* (London, 1854); *Flora Indica* (London, 1855), written with T. Thomson; *Illustrations of Himalayan Plants* (London, 1855); "*Géographie botanique raisonnée . . .* par M. Alph. de Candolle; a Review," in *Hooker's Kew Journal of Botany,* **8** (1856), 54–64, 82–88, 112–121, 151–157, 181–191, 214–219, 248–256; *Genera plantarum,* 3 vols. (London, 1862–1883), written with G. Bentham; *Handbook of the New Zealand Flora* (London, 1864–1867); *Student's Flora of the British Islands* (London, 1870); *Flora of British India,* 7 vols. (London, 1872–1897); and "Sketch of the Flora of British India," in *Imperial Gazetteer of India,* 3rd ed., I (1907), 157–212.

II. SECONDARY LITERATURE. See L. Huxley, *Life and Letters of Sir Joseph Hooker,* 2 vols. (London, 1918); W. B. Turrill, *Joseph Dalton Hooker* (London, 1964); and Mea Allan, *The Hookers of Kew, 1785–1911* (London, 1967).

R. DESMOND

HOOKER, WILLIAM JACKSON (*b.* Norwich, England, 6 July 1785; *d.* Kew, England, 12 August 1865), *botany.*

Hooker was the son of Joseph Hooker, a merchant's clerk who collected succulents, and Lydia Vincent. To fit him for the considerable property he was to inherit from his godfather, Hooker was given a gentleman's education, first at Norwich Grammar School, where John Crome was drawing master, then at Starston Hall, where he learned estate management from Robert Paul. In 1820 he was given the LL.D. by the University of Glasgow, and in 1845 he was honored with the D.C.L. by Oxford.

His early interests were diffuse—ornithology vied with entomology, entomology with botany. Then, in 1804, when he was only nineteen, Hooker discovered a moss new to Britain that was identified as *Buxbaumia aphylla* by James Edward Smith, owner of Linnaeus' herbarium. Smith introduced Hooker to Dawson Turner of Great Yarmouth, a banker, antiquarian, and leading cryptogamist who became his patron and later his father-in-law. Turner asked Hooker to illustrate his *Historia fucorum,* a project that took thirteen years to complete. Of the finished plates, 234 of a total of 258 are by Hooker's hand.

Other honors came to Hooker early. In 1805 William Kirby, a divine and entomologist, dedicated to him and his brother Joseph a new species of *Apion* they had discovered. In 1806 he was elected a fellow of the Linnean Society, being then just twenty-one, the earliest admissible age. In 1808 Smith named for him the beautiful moss *Hookeria lucens* and its genus.

Through Turner, Hooker met Sir Joseph Banks, who in 1809 arranged for him to be included in a diplomatic mission to Iceland. He was the first to botanize there. Hooker lost all his specimens and almost lost his life, however, when the ship caught fire on the homeward journey. He was nevertheless able to publish his journal of the expedition, composed mostly from memory, in 1811. A second edition, in 1813, was dedicated to Banks, who promised to send him on another trip.

Hooker hoped to go to Ceylon, and prepared himself by copying more than 2,000 drawings of Indian plants in the India House museum; a rebellion in Ceylon made the proposed visit impossible. He then wished to go to Java in response to an offer from Lord Bathurst to pay for living plants and information about spice-bearing trees in the Dutch East Indies; his patron Turner persuaded him against that island as being notoriously malarious.

Banks was angered by this interference in his plans for Hooker, but Turner saw a better future for his protégé in England. He urged Hooker to buy a quarter share in the Turner family brewery at Halesworth, Suffolk, give up his interest in entomology, and settle down to serious botany. Hooker did all of these things and in 1815 he married Maria, his patron's eldest daughter. Of their five children, Joseph Dalton Hooker, the second-born, became a famous botanist and his father's successor.

Hooker spent twelve years at Halesworth, during which he produced his *British Jungermanniae,* with his own illustrations, widely considered to be his most beautiful work. This book established hepaticology as an independent entity and made Hooker's reputation. He also wrote four books on mosses during this period, as well as papers for the Linnean Society. London's scientific society welcomed him, and eminent foreign botanists visited him at Halesworth where his herbarium was fast becoming the largest privately held. His guests there included A. P. de Candolle, Robert Brown, Francis Boott, C. Mertens, and the eighteen-year-old John Lindley, whom Hooker started on his botanical career.

In the wake of the Napoleonic wars, times were bad, and the brewery at Halesworth began to lose money. In 1820, when the regius professorship of botany at the University of Glasgow fell vacant, Banks procured it for Hooker. Hooker left for Scotland singularly well prepared as a botanist: he knew thoroughly the plants of his native East Anglia, where three-quarters of Britain's flora was represented, and had completed his study by extensive walking tours in England, Ireland, and Scotland; he had gained a knowledge of foreign plants by studying the herbaria of Linnaeus and others, and in 1814 by a botanizing expedition on the Continent (where he met the leading European scientists, including Lamarck, Mirbel, and Humboldt, who engaged him to write the cryptogamic section of his book on South American botany); he had laid out a small but interesting botanical garden for Simon Wilkin; and he had induced new exotic plants to flower in his greenhouse at Halesworth. He had, however, never lectured, nor had he ever attended a lecture on any subject, let alone botany.

This last factor proved no drawback, and Hooker was an instant success; his charm and eloquence, his rich knowledge of plants and plant life, and, above all, his love of his subject captivated his students. He developed his own teaching materials, including a magnificent series of folio-sized colored drawings, mainly of medicinal plants, which he hung around his classroom. Since no suitable textbook existed, he wrote *Flora scotica,* published in 1821 (the year he took up permanent residence in Glasgow), in which the flowering plants were arranged according to the Linnaean system, while such orders as the cryptogams were classified according to the natural system, the first time this had been used in a book on indigenous plants. He inaugurated botanizing expeditions to the West Highlands, and to help his students understand the economic applications of plants he began the collection that later was to form the nucleus for his Museum of Economic Botany.

Hooker's lectures were so popular that they attracted private citizens and even officers from the barracks three miles away. He opened his course with a few introductory lectures on the history of botany and the general character of plant life. In succeeding lectures he devoted the first half of each hour to organography, morphology, and classification of plants, and the second half to the analysis of specimens, mainly drawn from his own herbarium.

Hooker also began a collection of lithographed illustrations of the organs of plants, which in the first edition, of 1822, were his own work. He then discovered the talents of Walter Hood Fitch, a pattern-drawer in a Glasgow calico-printing works. Under Hooker's training in botanical draftsmanship, Fitch became one of the greatest British practitioners of that art. An enlarged edition of Hooker's *Botanical Illustrations,* with plates by Fitch, was published in 1837. In the same year Hooker published *Icones Plantarum;* the thousand figures were again by Fitch.

In the light of Hooker's future career, perhaps the most significant feature of his work in Glasgow was his improvement of that city's botanic garden. He had found it a poor affair, with only 8,000 species of plants. When he left, 20,000 species grew there, and it was the equal of any garden in Europe.

In 1836 Hooker was knighted for his services to botany. In spite of his many contributions to that field, however, his most important work was yet to come. Banks, director of the royal pleasure garden at Kew, had confided to Hooker his ambition of making it "a great exchange house of the Empire, where possibilities of acclimatizing plants might be tested," as he had written to George III. Upon Banks's death in 1820 the Kew collections stood in danger of being dispersed, and Hooker began a long campaign to save Kew for the nation. He pursued the project among influential people and officialdom through several changes of government. In 1840, through the close cooperation of Lindley and John Russell, sixth duke of Bedford, Hooker's scheme to establish a national botanic garden was accepted. The gardens were given to the nation, and Hooker became their first director in 1841.

The Royal Botanic Gardens at Kew which Hooker took over comprised only eleven acres and had no library or herbarium. Hooker generously allowed botanists access to his own, which occupied thirteen rooms of his house, West Park. Since his herbarium was the largest and most valuable in private hands, West Park attracted scientists from all over the world.

In 1842 W. A. Nesfield began landscaping the great avenues, grasslands, and vistas of modern Kew. At the same time Hooker began to replan and rebuild the glasshouses, which he had found to be inadequate. His crowning achievement was the Palm House, an exquisite winged structure completed in 1848 to the design of Decimus Burton. (Other claims for the architecture of the Palm House were made for Richard Turner, in whose Dublin foundry the curvilinear ironwork was made, while Joseph Hooker, in his "Sketch of the Life and Labours of Sir William Hooker," stated that his father's ideas were contributory, as they well may have been.) By 1846, when Hooker had been at Kew for only five years, he had increased the gardens to their present size of nearly 300 acres. As a liberal innovation the gardens were opened to the public.

Hooker superintended everything and created many prospects of the gardens himself, including the lake and the walk he made to receive the Sikkim rhododendrons discovered in the Himalayas by his son Joseph. The collections were vastly increased and specimens were brought from all over the world; Hooker's former students sailed with every government expedition and were based in every quarter of the globe, while Hooker's talent for making friends secured him new exotics from private garden-owners who employed their own plant hunters. Further, Hooker was instrumental in establishing botanic gardens in the new colonies of Queen Victoria's ever-expanding empire, which provided yet more plants for Kew.

In 1847 Hooker founded the Museum of Economic Botany, the first of its kind. Here specimens of vegetable products and materials were displayed for the benefit of manufacturers, tradesmen, and craftsmen. The museum received contributions from many sources, including the 1855 French International Exhibition, the Admiralty, and the Board of Trade; it was so popular with the public, and became so crowded with exhibits, that a second building was opened in 1857, and a third in 1863.

Hooker pursued the practical applications of botany. He attacked the problem of malaria in India when, in 1859, he sent Clements Markham to Peru to acquire living plants of *Cinchona,* a cheap source of quinine. After acclimatization at Kew, the plants were established in the Nilghiri Hills. Further successful experiments at Kew resulted in the acclimatization of tussock grass, timber trees, and a number of farm crops for colonial lands, by which they were made habitable and prosperous.

In 1855 Joseph Hooker became his father's assistant at Kew, succeeding him upon his death ten years later. The nation's monument to the elder Hooker was the purchase of his herbarium and library, comprising some 4,000 volumes, a million dried plant specimens, 158 botany-class drawings, and his scientific correspondence from 1810 (bound in seventy-six quarto volumes and containing about 29,000 letters from botanists the world over).

In addition to his work as a practical botanist, teacher, and administrator, Hooker's literary output was enormous. From 1816 until 1826 he contributed most of the drawings and analyses for the new five-volume edition of *Curtis's Flora Londinensis.* From 1827 until 1845 he was wholly author and editor of *Curtis's Botanical Magazine,* and for the first ten years its illustrator. As well as editing thirty-eight volumes of this periodical Hooker was a contributor to a number of other journals. He continued to write significant books, of which his last five, on ferns, remain standard works.

Hooker was a fellow of the Royal, Linnean, Antiquarian, and Royal Geographical societies, a corresponding member of the Académie des Sciences, a companion of the Legion of Honor, and a member of almost every European and American natural science academy. William Henry Harvey, the Irish botanist, wrote of him (in a letter to Joseph Hooker), "The great secret of his success was that he deemed nothing too small for his notice, if it illustrated any fact of science or economy, and nothing too difficult to be attempted." And Asa Gray eulogized "the single-mindedness with which he gave himself to his scientific work, and the conscientiousness with which he lived for science while he lived by it" (*American Journal of Arts and Sciences,* 2nd ser., **41,** pt. 1 [1866]).

BIBLIOGRAPHY

I. Original Works. Hooker's works include *British Jungermanniae* (London, 1816); *Curtis's Flora Londinensis,* 5 vols. (London, 1817–1828); *Musci exotici,* 2 vols. (London, 1818–1820); *Muscologia britannica* (London, 1818–1827), written with Thomas Taylor; *Botanical Illustrations* (Edinburgh, 1821); *Flora scotica* (London, 1821); *The Exotic Flora,* 3 vols. (Edinburgh, 1822–1837); *Icones filicum,* 2 vols. (London, 1828–1831), written with N. K. Greville; *Flora boreali americana,* 2 vols. (London, 1829–1840); *The British Flora* (Glasgow–London, 8 eds., 1830–1860); *Genera filicum* (London, 1838–1840); *Species filicum,* 5 vols. (London, 1846–1864); *Niger flora* (London, 1849); *A Century of Ferns* (London, 1854); *Filices exoticae* (London, 1859); *The British Ferns* (London, 1861); *A Second Century of Ferns* (London, 1861); *Garden Ferns* (London, 1862); *Synopsis filicum* (London, 1865); and *Kew Gardens, a Popular Guide* (1844–1863), which went into twenty-one editions.

In addition, Hooker was wholly author of *Curtis's Botanical Magazine* from 1827 to 1845. He also founded and wrote for *Botanical Miscellany* (1830–1833) and its continuation, *Journal of Botany* (1834–1842); *London Journal of Botany* (1842–1848); and *Journal of Botany and Kew Gardens Miscellany* (1849–1857), which contained many important contributions, especially on North American botany. His *Icones plantarum,* established in 1837, is still published at Kew.

II. Secondary Literature. On Hooker and his work see Joseph Dalton Hooker, "A Sketch of the Life and Labours of Sir William Jackson Hooker," in *Annals of Botany,* **16,** no. 64 (1902), 9–221, which contains a complete chronological catalog of his works; F. O. Bower, *Makers of British Botany* (Cambridge, 1913) pp. 126–150, 227; J. Reynolds Green, *A History of Botany in the United Kingdom* (London, 1914), *passim;* and Mea Allan, *The Hookers of Kew* (London, 1967). See also the entries under his name in Britten and Boulger, *British and Irish Botanists* (London, 1931); and *Dictionary of National Biography.*

Mea Allan

HOPE, THOMAS CHARLES (*b.* Edinburgh, Scotland, 21 July 1766; *d.* Edinburgh, 13 June 1844), *chemistry.*

The successor to Joseph Black as professor of chemistry at Edinburgh University, Hope considered the teaching of science, rather than its extension by original research, to be his vocation. Nevertheless, he is remembered chiefly for his contributions to the discovery of strontium and for his conclusive demonstration that water reaches its maximum density just above its freezing point.

The third son of John Hope, regius professor of botany at Edinburgh, and the former Juliana Stevenson, Hope entered the university at the age of thirteen. He became very proficient in botany and was a strong but unsuccessful candidate for the chair when his father died in 1786. After receiving the M.D. in 1787 he became successively lecturer in chemistry, assistant professor of medicine (1789), and professor of medicine (1791) at Glasgow. In 1795 Hope was chosen by Black, whose health was failing, as assistant and potential successor at Edinburgh; Black died in 1799, and Hope gave his last series of lectures in 1843. He was elected a fellow of the Royal Society in 1810.

In 1790 Adair Crawford presented the first intimation that the mineral now called strontianite (first found near Strontian, Scotland), previously thought to be a form of barium carbonate, contained a hitherto unknown "earth." The substance was examined by a number of mineralogists over the next few years; but the fullest investigations were those, carried out quite independently, by Hope and M. H. Klaproth. Hope clearly established the intermediacy of strontia in relation to lime and baryta, foreshadowing the more explicit formulation of this particular triad by J. W. Döbereiner in 1829. The metals calcium, strontium, and barium were isolated by Davy in 1808.

The peculiar expansion of water had been noted in the seventeenth century; but Hooke, Dalton, and others were skeptical. Hope clearly showed that water is at its densest at a little above 39°F. "Hope's experiment" has become a classic and may be found in many physics textbooks.

In spite of a pompous and affected manner, Hope was a gifted and popular lecturer. His teaching was seriously weakened by a failure to provide facilities for, or to encourage, practical work.

BIBLIOGRAPHY

I. Original Works. Hope's two important papers are "Account of a Mineral From Strontian, and of a Peculiar Species of Earth Which It Contains," in *Transactions of the Royal Society of Edinburgh,* **4** (1798), 3–39; and "Experiments and Observations Upon the Contraction of Water by Heat at Low Temperatures," *ibid.,* **5** (1805),

379–405. His papers are listed in *Royal Society Catalogue of Scientific Papers,* III (London, 1869), 426–427.

II. SECONDARY LITERATURE. Biographies are T. S. Traill, "Memoir of Dr. Thomas Charles Hope, Late Professor of Chemistry in the University of Edinburgh," in *Transactions of the Royal Society of Edinburgh,* **16** (1848), 419–434; and J. Kendall, "Thomas Charles Hopé, M.D.," in *Endeavour,* **3** (1944), 119–122. The discovery of strontium is dealt with briefly in E. M. Weeks, *Discovery of the Elements,* 7th ed., rev. by H. M. Leicester (Easton, Pa., 1968), pp. 491–495; and more fully by J. R. Partington, "The Early History of Strontium," in *Annals of Science,* **5** (1947), 157–166; and **7** (1951), 95–100, which deals more specifically with Hope's contribution.

Reminiscences of Hope as a lecturer are in G. P. Fisher, *Life of Benjamin Silliman,* I (New York, 1866), 163–166; and a recent study of one aspect of Hope's career is J. B. Morrell, "Practical Chemistry in the University of Edinburgh," in *Ambix,* **16** (1969), 66–80. See also R. H. Cragg, "Thomas Charles Hope (1766–1844)," *Medical History,* **11** (1967), 186–189.

E. L. SCOTT

HOPF, HEINZ (*b.* Breslau, Germany [now Wrocław, Poland], 19 November 1894; *d.* Zollikon, Switzerland, 3 June 1971), *mathematics.*

Hopf attended school and started his university study of mathematics in his birthplace, but his studies were soon interrupted by a long period of military service during World War I. A fortnight's leave in the summer of 1917 determined his mathematical future: he ventured into Erhard Schmidt's set theory course at the University of Breslau and became fascinated by Schmidt's exposition of L. E. J. Brouwer's proof of the dimension invariance by means of the degree of continuous mappings.

In 1920 Hopf followed Schmidt to Berlin where, with topological research, he earned his Ph.D. in 1925 and his *Habilitation* in 1926. At Göttingen in 1925 he became acquainted with Emmy Noether and met the Russian mathematician P. S. Alexandroff, with whom he formed a lifelong friendship. Rockefeller fellowships enabled the two friends to spend the academic year 1927–1928 at Princeton University, where topology was fostered by O. Veblen, S. Lefschetz, and J. W. Alexander. In 1931 Hopf was appointed a full professor at the Eidgenössische Technische Hochschule in Zurich, assuming the chair of Weyl, who had gone to Göttingen.

The greater part of Hopf's work was algebraic topology, motivated by vigorous geometric intuitions. Although the number of his papers was relatively small, no topologist of that period inspired as great a variety of important ideas, not only in topology, but also in quite varied domains. He was awarded many honorary degrees and memberships in learned societies. From 1955 to 1958 he was the president of the International Mathematical Union.

Hopf was a short, vigorous man with cheerful, pleasant features. His voice was well modulated, and his speech slow and strongly articulated. His lecture style was clear and fascinating; in personal conversation he conveyed stimulating ideas. With his wife Anja, he extended hospitality and support to persecuted people and exiles.

After Brouwer created his profound "mixed" method in topology, Hopf was the first to continue Brouwer's work on a large scale. He focused on the mapping degree and the mapping class (homotopy class), which had been mere tools in Brouwer's work. Hopf set out to prove that Brouwer's mapping degree was a sufficient homotopy invariant for mappings of spheres of equal dimension (2, nos. 5, 11, 14) and in this context he studied fixed points (2, no. 8) and singularities of vector fields (2, no. 6). His initially crude and too directly geometric methods underwent gradual refinement, first by Emmy Noether's abstract algebraic influence, then through the combinatorial ideas of the American school. In 1933 his efforts culminated in the development of a complete homotopy classification by homology means of mappings of n-dimensional polytopes into the n-dimensional sphere S^n (2, no. 24).

Hopf's study of vector fields led to a generalization of and a formula about the integral curvature (2, no. 2), as a mapping degree of normal fields (1925). An extension of Lefschetz' fixed point formula (2, no. 9) was the result of work done in 1928. As a new and powerful tool to investigate mappings of manifolds, Hopf defined the inverse homomorphism (2, no. 16) using the Cartesian product of the related manifolds—a device he took from Lefschetz. In fact Hopf's 1930 paper on this subject goes back to his stay at Princeton. Not until the arrival of cohomology and the cohomology products was the inverse homomorphism better understood and more firmly integrated into algebraic topology (5).

Hopf's next great topological feat was the 1931 publication (2, no. 18) on an infinity of homotopy classes of S^3 into S^2, and the definition of the "Hopf invariant" for these mappings. As early as 1927 Hopf conjectured that the "Hopf fiber map" was homotopically essential, but the tool to prove this conjecture had still to be created: the idea of considering inverted mappings. Hopf's work on this subject was influential in W. Hurewicz' shaping the concept of homotopy groups, and in particular in his investigation (1935–1936) of homotopy groups of fiber spaces (4). H. Freudenthal, by a synthesis of Hopf's and Hurewicz' work, proved the completeness of Hopf's classification and discovered the suspension (6). From these beginnings homotopy of spheres developed after World War II into a growing field of research, to which Hopf himself had contributed (1935) the investigation of the case of mappings of S^{2n-1} into S^n (2, no. 26).

Vector fields and families of vector fields remained a concern of Hopf's. He stimulated Stiefel's work (7), which led to the discovery of what is now called Stiefel-Whitney classes, and that of B. Eckmann as well (8, 9). Hopf's 1941 paper on bilinear forms (2, no. 38) fits into the same context, as does his influential discovery (1948) of the concept of almost complex manifolds (2, no. 52), which, among other things, led to his 1958 paper with F. Hirzebruch (2, no. 66). Hopf's most important contribution to this area of mathematics is his paper, published in 1941, but begun in 1939, on the homology of group manifolds (2, no. 40), in which he proved the famous theorem that compact manifolds with a continuous multiplication with unit (now called *H*-manifolds) have a polynomial cohomology ring with all generators of odd dimension. The theorem had already been known by Lie groups methods for the four big classes. Hopf formulated the theorem in terms of homology; his tool was again the inverse homomorphism—Hopf did not like and never became fully acquainted with cohomology. He wrote a few more papers on this subject (2, nos. 41, 46) and instigated H. Samelson's 1942 investigations (10).

In 1936 Hurewicz (4) had proved that in polytopes with trivial higher homotopy groups the fundamental group uniquely determines the homology groups, raising the question of how the one determines the others. Hopf tackled the problem in his papers (2, nos. 40, 45, 49) of 1942 and 1944, which led to independent investigations of Eckmann (11), S. Eilenberg and S. MacLane (12), and Freudenthal (13). The result was the cohomology of groups, the first instance of cohomological algebra, which has since developed into a broad new field of mathematics.

Another beautiful idea of Hopf's was transferring Freudenthal's concept of ends of topological groups to spaces possessing a discontinuous group with a compact fundamental domain (2, no. 47), which Freudenthal in turn converted into a theory on ends of discrete groups with a finite number of generators (14).

Hopf was also interested in global differential geometry. With W. Rinow he contributed the concept of a complete surface (2, no. 20); with Samelson (2, no. 32) a proof of the congruence theorem for convex surfaces; with H. Schilt (2, no. 34) a paper on isometry and deformation; and he studied relations between the principal curvatures (2, no. 57).

Two beautiful papers of Hopf's that should be mentioned are one on the turning around of the tangent of a closed plane curve (2, no. 27) and one on the set of chord lengths of plane continua (2, no. 31), published in 1935 and 1937 respectively. Hopf was also interested in number theory, which he enjoyed teaching and to which he devoted a few papers.

BIBLIOGRAPHY

(1) P. Alexandroff and H. Hopf, *Topologie I* (Berlin, 1935).

(2) H. Hopf, *Selecta* (Berlin, 1964), contains an almost complete bibliography.

(3) H. Hopf, "Ein Abschnitt aus der Entwicklung der Topologie," in *Jahresbericht der Deutschen Mathematikervereinigung,* **68** (1966), 182–192.

(4) W. Hurewicz, "Beiträge zur Topologie der Deformationen. I–IV," in *Proceedings. K. Nederlandse akademie van wetenschappen,* **38** (1935), 113–119, 521–528; **39** (1936), 117–125, 215–224.

(5) H. Freudenthal, "Zum Hopfschen Umkehrhomomorphismus," in *Annals of Mathematics,* **38** (1937), 847–853.

(6) H. Freudenthal, "Über die Sphärenabbildungen, I," in *Compositio mathematica,* **5** (1937), 300–314.

(7) E. Stiefel, "Richtungsfelder und Fernparallelismus in *n*-dimensionalen Mannigfaltigkeiten," in *Commentarii mathematici helvetici,* **8** (1935–1936), 305–353.

(8) B. Eckmann, "Zur Homotopietheorie gefaserter Räume," *ibid.,* **14** (1942), 141–192.

(9) B. Eckmann, "Systeme von Richtungsfeldern in Sphären und stetige Lösungen komplexer linearer Gleichungen," *ibid.,* **15** (1943), 1–26.

(10) H. Samelson, "Beiträge zur Topologie der Gruppenmannigfaltigkeiten," in *Annals of Mathematics,* **42** (1941), 1091–1137.

(11) B. Eckmann, "Der Cohomologie-Ring einer beliebigen Gruppe," in *Commentarii mathematici helvetici,* **18** (1946), 232–282.

(12) S. Eilenberg and S. MacLane, "Relations Between Homology and Homotopy Groups of Spaces," in *Annals of Mathematics,* **46** (1945), 480–509.

(13) H. Freudenthal, "Der Einfluss der Fundamentalgruppe auf die Bettischen Gruppen," *ibid.,* **47** (1946), 274–316.

(14) H. Freudenthal, "Über die Enden diskreter Räume und Gruppen," in *Commentarii mathematici helvetici,* **17** (1944), 1–38.

(15) B. Eckmann, "Zum Gedenken an Heinz Hopf," in *Neue Züricher Zeitung* (18 June 1971).

(16) P. Alexandroff, "Die Topologie in und um Holland in den Jahren 1920–1930," in *Nieuw archief voor wiskunde,* **17** (1969), 109–127.

HANS FREUDENTHAL

HOPKINS, FREDERICK GOWLAND (*b.* Eastbourne, Sussex, England, 20 June 1861; *d.* Cambridge, England, 16 May 1947), *biochemistry.*

Hopkins was not only the father of British biochemistry but also a major contributor to biochemical thought and to experimental biochemistry throughout the world. Quiet, kindly, and mild, he had the greatest tenacity and forcefulness of character when facing challenge or opposition to the ideas in which he believed. No one was more firmly opposed than he to the vitalist thinking of many of his contemporaries and to the obscurantist attitude to which this thinking gives rise. For him the nature of protoplasm was not insolubly mysterious but something accessible to the experimental approach, something inherently comprehensible. "The use of the term protoplasm may be morphologically justified," he wrote on one occasion, "but chemically it denotes an abstraction."

His own views were perhaps most sharply crystallized in Hopkins' presidential address to the Physiology Section of the British Association for the Advancement of Science, delivered at Birmingham in 1913:

> In the study of the intermediate processes of metabolism, we have to deal, not with complex substances which elude ordinary chemical methods, but with simple substances undergoing comprehensible reactions. . . . It is not alone with the separation and identification of products from the animal that our present studies deal; but with their reactions in the body; with the dynamic side of biochemistry.

Hopkins was less concerned, except as an article of his own particular biochemical faith, with the question of whether the application of chemical methods can ultimately provide complete answers to biological problems, for that was—and still is—a problem for the future to resolve. But that biochemistry can provide significant new information on problems of this kind—had this not been clear enough from Hopkins' own work—has in the meantime become sufficiently evident to justify every article of the biochemical faith in which he so strongly believed and which he lost no opportunity to impress upon others. It was characteristic of Hopkins' department that one thought and talked in terms of dynamic events rather than of mere structure. Such an atmosphere was inevitable because, for Hopkins, "Life is a dynamic equilibrium in a polyphasic system."

Hopkins entered biochemistry at an early stage in its development, although comparatively late in his own lifetime. At school he showed no remarkable distinction except in chemistry, but he was fascinated by a microscope that had belonged to his father. "I felt in my bones," he once wrote, "that the powers of the microscope thus revealed to me were something very *important*—the most important thing I had as yet come up against; so much more significant than anything I was being taught at school." Together with an evident aptitude for chemistry, this microscope must have done much to determine his eventual scientific development.

He was brought up by his widowed mother and an unmarried uncle who, when Hopkins was seventeen, chose for him a career in the London office of a provincial insurance company. From this post he was rescued after six months by his father's cousin, Fritz Abel, who, in Hopkins' own words, "at once said 'Cambridge.'" But it was not until twenty years later that this goal was achieved.

During the intervening years he was trained as an analyst, in which capacity he worked for one of the larger railway companies and obtained his first professional qualification, the associateship of the Institute of Chemistry. He distinguished himself in the examination and was thereupon invited to become an assistant to Thomas Stevenson, expert medical jurist to the Home Office. In this capacity he became involved in several celebrated murder cases, notably those of Bartlett, Lipski, and Maybrick. In several of these his analytical skill played a large part in securing convictions.

By this time Hopkins was more conscious than ever of his need for more formal training and a university degree, which he sought and obtained as an external student at the University of London. In 1888, at the age of twenty-seven, he received a small inheritance and decided to enter the medical school at Guy's Hospital. In the course of this training he won the gold medal in chemistry and honors in materia medica—another hint of the direction he was ultimately to follow. After qualifying he worked for some years with Archibald Garrod, who became a lifelong friend and founded the then relatively new science of biochemical genetics. For a number of years Hopkins worked in the medical school by day and in a privately owned clinical research laboratory in the evenings. In September 1898, at the age of thirty-seven, he went to Cambridge at the invitation of Michael Foster, then professor of physiology.

Foster's wish was that Hopkins should undertake the teaching and development of what was then known as chemical physiology, a task which at that time meant tutoring in physiology and anatomy as well. This sort of experience has bedeviled many new entrants to the older English universities, and in Hopkins' case it led to a breakdown in 1910. Later he wrote:

> My recovery was greatly helped by an event which I count as the most outstanding among my gifts from Fortune. I heard during my illness that Trinity College had made me a Fellow and elected me to a Praelectorship in Biochemistry. . . . So far as the College itself is concerned the post carries no obligations. . . . It is my hope that in any account of my career published after my departure the generosity of Trinity College will be emphasized.

Thus it was not until the age of almost fifty that Hopkins was able to devote the greater part of his time to the development of biochemistry in the university and to his own research, although, despite difficulties and financial embarrassments in the early years at Cambridge, he had already published some thirty papers—nearly a quarter of his research output.

Hopkins made a complete recovery from his illness. Two papers appeared in 1910, and in 1912 he published what is perhaps the best-known of his works: "Feeding Experiments Illustrating the Importance of Accessory Food Factors in Normal Dietaries." Although it was known to Aristotle that raw liver can cure night blindness, and although Captain Cook was aware of the antiscorbutic properties of lime juice, it was only through Hopkins' work that the existence of vitamins became firmly and finally established. The experiments that lay behind this fundamental demonstration were, like much else of his experimental work, masterpieces of design and ingenuity and became the model for nutritional experiments for many years to come. In 1913 came his brilliant address to the British Association for the Advancement of Science at Birmingham, of which Marjorie Stephenson wrote:

> It is indeed a biochemical treatise in miniature and discloses fully and with amazing clarity Hopkins's inmost thoughts and speculations on the biochemistry of the cell. . . . It shows Hopkins at the height of his powers reviewing biochemical work from the days of Liebig onwards and interpreting it so as to build up a picture of the cell as the seat of ordered chemical events controlled in the interests of growth and function.

This address, as important a landmark in the history of biochemistry as it was in Hopkins' own intellectual development, can be read and reread today; it is in fact one that should be known by every aspiring young biochemist and, indeed, could still profitably be consulted by many of his senior colleagues.

In 1914 Hopkins became the first professor of biochemistry at Cambridge; the new department, destined to become a mecca for biochemists, was housed in makeshift accommodations until 1925. Throughout the war years he spent much time on government business, served on the Royal Society Food Committee, and became involved in many other scientific wartime activities—none of them military, for he abhorred war. Problems of food rationing and nutrition claimed much of his attention. Butter was scarce and expensive; margarine, cheap and more easily available. There was, however, considerable unease among its manufacturers regarding its nutritional value, an unease to which Hopkins' own discovery of accessory food factors contributed much. In 1917 he agreed to carry out further nutritional research on behalf of and with the support of the margarine industry, but on the understanding that he must be free to publish his results. Margarine, it soon became clear, was much inferior to butter in nutritional value because it lacked "fat-soluble A." (As Mellanby later showed, this factor has two components, now known as vitamins A and D.)

Hopkins took an active part in this work until 1920 and continued to act as a consultant to the industry for a number of years afterward. In the meantime J. C. Drummond carried out an extensive survey of natural sources of the A and D vitamins, and industrial research pushed ahead with investigations into the possibilities of introducing A and D into the commercial product. In 1926–1927 the first "vitaminized" margarines appeared in the shops, and by 1928 they had received the certificated approval of the Pharmaceutical Society. Vitamin-enriched margarine is now popular and the modern product is little, if at all, inferior to the best dairy butter from the viewpoint of calorific value and vitamin content.

After the war biochemistry became for the first time a subject for part II of the natural sciences tripos at Cambridge, and there began the great phase of expansion and development, at Cambridge in particular but in other universities as well, for which Hopkins had striven so long and so energetically. It was not until 1935 that Hopkins decided to introduce biochemistry as a subject in part I of the tripos, a decision that caused some misgivings at Cambridge and much criticism from other universities. But he was so convinced of the importance of the subject that he maintained that no student who wished to do so should be barred from studying the subject, at least on an elementary level. The innovation proved a popular and brilliant success, and elementary courses in biochemistry became widespread in English universities.

The rest of Hopkins' career can easily be summed up as a steady march from distinction to distinction. He was knighted in 1925, awarded the Copley Medal of the Royal Society in 1926, shared the Nobel Prize in physiology or medicine with Eijkmann in 1929, became president of the Royal Society in 1931, and received that most prized of all civil distinctions, the Order of Merit, in 1935.

In addition there were numerous honorary degrees from universities throughout the world. Yet near the end of his autobiography, characteristically enough, Hopkins could only say:

> My own temptation has been to try and show that it is not altogether my own fault if I have remained—what I feel myself to be, compared with many others who have received less recognition and fewer rewards—intellectually an amateur. I realise today that I know and have known no aspect of science *au fond*—I was led at a right moment to follow a path then trodden by very few and where every wayfarer was conspicuous.

Hopkins' autobiography, begun ten years before his death but never completed, shows him still active in his research, still an inspiring teacher: a quiet, calm, affectionate professor in a department most members of which owed their own distinctions largely to his early inspiration and encouragement and who revered, respected, and admired him. It often happens that a brilliant research worker is indifferent as a teacher, and in Hopkins' case elementary teaching was not his forte. Yet with the advanced classes he was superb, and his lectures were usually attended by the entire department. Often the lectures showed little sign of previous preparation—he seldom used notes in any form, instead choosing a theme that interested him at the moment and developing it as he went along. But he was best of all in discussion, formal or informal. Marjorie Stephenson wrote of him: "Never was he known to fail; by skilful suggestions and questions he turned the most unpromising material into something interesting and significant, leaving the author encouraged and sufficiently self-confident to meet the most obvious criticisms of his colleagues."

It seems likely that this success was due to Hopkins' clear mental picture of the cell as a biochemical machine; and into this scheme he was able to fit what seemed to his colleagues to be mere isolated observations, thus giving them significance. This intuitive understanding of the nature of the cell appears to have been an early development in his thinking, and it played a major part in the inspiration and encouragement he gave to his pupils.

Unlike many Continental professors Hopkins did not try to build up a school in which every student would be put to work on one or another of the professor's own problems. Any student with a worthwhile problem in mind was encouraged to follow his or her own line of thought and research, and Hopkins invariably made valuable ideas and suggestions. Frequently, having broken new ground, he would hand over even the most promising of fields to younger colleagues, many of whom later achieved much distinction through the pursuit of a line of work inherited from Hopkins.

Some idea of Hopkins' contribution to the propagation and continuation of his subject may be gained from the fact that, by the time of his death, some seventy-five of his former students occupied professorial chairs in various parts of the world.

The earliest of Hopkins' known publications (he did not himself possess a complete set of reprints or even a list of his papers) was written while he was still at school and concerned the habits of the bombardier beetle, *Brachinus crepitans.* His interest in insects led him to study the pigments of pierid butterflies. This interest remained with him and he returned to it toward the end of his life, following H. Wieland's discovery that the white pigment is a member of the pterin group and not, as Hopkins had believed, uric acid.

Hopkins' interest in uric acid was carried over into the early days of his medical research at Guy's Hospital, and his earlier training as an analyst enabled him to develop a new and superior method for its determination in urine. Although now generally superseded by colorimetric and other methods, Hopkins' procedure remained the most accurate and reliable for several decades. The effects of diet upon uric acid excretion aroused his interest in proteins and led to attempts to obtain crystalline preparations of these substances. Here again his analytical experience enabled him to improve greatly upon existing methods and to lay the foundations for new work.

Together with S. W. Cole, Hopkins went on to track down the substance responsible for the already well-known Adamkiewicz reaction of proteins and thus was led to the isolation of the amino acid tryptophan. Again the analyst's skill played a large part in devising procedures for its isolation. Determination of the structure of this new substance was carried out, and the action upon it by bacteria was investigated. This led in turn to the beginnings of bacterial biochemistry, pursued for a time by Marjorie Stephenson and

Harold Raistrick. Subsequent developments, especially in the gifted hands of Marjorie Stephenson, are well-known and form a major branch of biochemical study today.

Several miscellaneous papers on proteins followed; and Hopkins' interest then turned to nutritional studies, now that proteins could be obtained in a supposedly pure state, and he was quick to show that the newly discovered tryptophan is an indispensable dietary constituent. The nutritional roles of arginine and histidine were studied later, but in the meantime Hopkins had been much impressed by the inconsistency of the results of nutritional studies being carried out by other workers. By this time, he wrote, "I had come to the conclusion that there must be something in normal food which was not represented in a synthetic diet made up of pure protein, pure carbohydrate, fats and salts; and something the nature of which was unknown."

Young rats fed on such diets failed to grow and even lost weight unless they were given small amounts of milk daily. Hopkins concluded that milk contains "accessory food factors," which are required only in trace amounts but are indispensable for normal growth and maintenance. This led to the "vitamine hypothesis," which, although based on a series of very elegant and eloquent experiments, was hotly contested for many years. Published in 1912, the results—or at any rate Hopkins' conclusions—were still in dispute as late as 1920, although three years later most of the opposition had evaporated. For this contribution to the knowledge of nutrition he shared the 1929 Nobel Prize in physiology or medicine with Eijkmann.

Going back to the first decade of the century, we find the beginnings of yet other branches of modern biochemistry. Together with the physiologist Walter Fletcher, Hopkins undertook a series of investigations on muscle, one of the few investigations which did not directly follow the main lines of his work. It had hitherto been generally believed that the contraction of muscle is associated with the formation of lactic acid, but the evidence was more than a little unconvincing. Fletcher and Hopkins seem to have been the first to realize that all of the methods formerly used for the estimation of lactic acid involved stimulation of the muscle itself, so that as much lactic acid would be found in unstimulated controls as in stimulated muscles. It was therefore necessary to devise methods whereby lactic acid could be extracted and the amount measured without stimulation of the controls. This was achieved by using thin, small muscles, dropping them into ice-cold alcohol, and grinding the material rapidly, so that enzymatic activity was reduced to a minimum. This appears to have been the first time the necessity of stopping enzyme activity as a preliminary to chemical analysis of irritable or any other kind of tissue had been realized or even suspected.

In the hands of Fletcher and Hopkins the new technique yielded the first incontrovertible proof that muscle activity and lactic acid production are intimately associated; it led others—D. M. Needham in Hopkins' own laboratory, for example—to the early growth and development of the detailed knowledge of muscle metabolism that we possess today.

The work on muscle served not only as a starting point for the study of carbohydrate metabolism in muscle, a field which attracted such notable research workers as Parnas and Meyerhof, but also, indirectly, to the development of present knowledge of alcoholic fermentation by yeast. The latter is a process which, in the main, follows precisely the same intermediate steps as does lactic acid formation in muscle. It also paved the way for studies of fermentation and kindred processes in bacteria, so brilliantly pioneered in Hopkins' laboratory by Marjorie Stephenson.

The work on muscle emphasized the immense importance of enzymatic activity in living tissues and the extreme rapidity with which these catalysts can operate. One outcome of this was that Hopkins became interested in oxidizing enzymes, a field later developed and expanded by Malcolm Dixon, D. E. Green, and many others, again largely in Hopkins' own department. Hopkins himself became especially fascinated by the respiratory importance of -SH compounds. He was led in this direction because, he said, "I was endeavouring to discover if vitamins were to be found among sulphur-containing compounds, and was led part of the way towards the separation of the substance now described." This new substance was glutathione; and a series of papers on its isolation, structure, and biological function followed in rapid succession.

Some years later Hopkins was able to show that certain dehydrogenases are -SH-dependent enzymes. Although in the meantime similar conclusions had been reached by other investigators, Hopkins and his assistant, E. J. Morgan, made significant contributions to this field in 1938–1939. The knowledge accumulated on the importance of -SH groups in enzyme activity became of intense interest from 1939 on and played a very important part in connection with the possible use of vesicant gases by the enemy and in the development of British anti-lewisite.

Hopkins was much impressed by Lohmann's discovery that glutathione acts as a specific activator for glyoxalase, a widely distributed enzyme the function of which is still unknown. Hopkins, desirous of knowing whether glutathione is or is not widely distributed, took advantage of its activating effect upon glyoxalase to carry out a massive comparative study of the distribution of the enzyme and its cofactor, thus setting the pattern for many later comparative studies.

BIBLIOGRAPHY

A partial bibliography is in Poggendorff, VI, 1158–1159. For information on Hopkins or his work, see Ernest Baldwin and J. Needham, eds., *Hopkins & Biochemistry* (Cambridge, 1949); and Ernest Baldwin, *Gowland Hopkins* (London, 1961).

ERNEST BALDWIN

HOPKINS, WILLIAM (*b.* Kingston-on-Soar, Derbyshire, England, 2 February 1793; *d.* Cambridge, England, 13 October 1866), *geology, mathematics.*

The only son of a gentleman farmer, Hopkins had a desultory early education which included some practical farming in Norfolk. Later his father gave him a small estate near Bury St. Edmunds, but he found the task of management both uncongenial and unprofitable. After the death of his wife he sold the estate to pay off debts and to provide the means wherewith in 1822, at the age of thirty, he entered St. Peter's College (Peterhouse), Cambridge. Here he married again and his mathematical talent shone. He took the B.A. in 1827, placing as seventh wrangler, and then became a very successful private tutor of mathematics. Among his many pupils who attained high distinction were George Stokes, William Thomson (Lord Kelvin), P. G. Tait, Henry Fawcett, James Clerk Maxwell, and Isaac Todhunter. In the 1830's he was appointed a syndic for the building of the Fitzwilliam Museum.

Hopkins became intensely interested in geology about 1833, after excursions with Adam Sedgwick near Barmouth, in northern Wales. He decided that he would place the physical aspects of geology on a firmer basis, would free it from unverified ideas, and "support its theories upon clear mathematical demonstrations."[1] His mathematical models and propositions greatly impressed contemporary geologists, and in 1850 he was awarded the Wollaston Medal of the Geological Society of London for his application of mathematics to physics and geology. In 1851 and 1852 he was elected president of that society and in 1853 presided over the British Association for the Advancement of Science. He became a fellow of the Royal Society, and following his death the Cambridge University Philosophical Society founded in his honor a prize which was first awarded in 1867 and triennially thereafter.

The main written product of Hopkins' interest in pure mathematics is the two-volume *Elements of Trigonometry* (London, 1833–1847). His applications of mathematics to geology were expressed mainly in articles, the contents of which may be grouped under the following topics: crustal elevation and its effect on surface fracturing, the transport of erratic boulders, the nature of the earth's interior, and the causes of climatic change.

Hopkins attempted to explain dislocations or fractures at the earth's surface by estimating the effects of an elevatory force acting at every point beneath extensive portions of the earth's crust. From his consideration of the pressures exerted by explosive gases, vapors, and other subterranean forces upon the crust, he concluded that during crustal extension and fracturing there must originate in nearly all cases first a series of longitudinal parallel fractures and second, with continued uplift, a series of transverse dislocations at right angles to the first. This rectangular pattern of faults provided the fundamental directive lines during the elevation and formation of continents and of mountain systems. On this assumption Hopkins discussed the elevation and denudation of the English Weald and Lake District and of the Bas Boulonnais in northern France. In the Weald, a land of wide longitudinal vales at the foot of steep escarpments that are breached transversally by narrow river valleys, Hopkins concluded that the main vales and scarps were associated with longitudinal parallel fractures and that the transverse valleys were formed by dislocations at right angles to them. He admitted that he could not find true geological evidence of fracturing except perennial springs, which he assumed to be thrown out at faultlines. Today, as by the more perceptive geologists then, the Weald valleys and scarps are considered to be typical products of subaerial erosion and not of crustal fracturing.

Hopkins played an important and equally unfortunate part in the contemporary debate on the transport of erratic boulders. The aura of mathematical conclusiveness that surrounded his work caused his opinions to make a lasting impression and to be hailed as incontrovertible by his followers. At first he rejected glacial or ice transport as an explanation of the movement of erratic boulders, since it often involved "such obvious mechanical absurdities that the author considers it totally unworthy of the attention of the Society."[2] In his studies of the Lake District Hopkins postulated sudden upheavals during each of which a great mass of water, or "wave of translation," rushed down the rift valleys, rolling and sliding great boulders for long distances. The idea was welcomed by antiglacialists in Britain and by leading geologists in America, including H. D. Rogers, who in 1844 wrote:

> It has been shown by Mr. Hopkins, of Cambridge, reasoning from the experimental deductions of Mr. Scott Russell upon the properties of waves, that "there is no difficulty in accounting for a current of twenty-five or thirty miles an hour, if we allow of paroxysmal elevation of from one hundred to two hundred feet," and he further proves that a current of twenty miles an hour ought to move a block of three hundred and twenty tons, and since the force of the current increases in the ratio of the square of the velocity, a very moderate addition to this speed is compatible with the transportation of the very largest erratics anywhere to be met with, either in America or Europe.[3]

Although Hopkins' idea was wrong when applied to the transport of glacial erratics—as he himself later half admitted—in presenting it he added detail which, when applied to hydraulic work, was to prove of great value and is today known as Gilbert's sixth-power law. Assuming, as Playfair had shown, that the force of a current increases in the ratio of the square of its velocity, Hopkins calculated that "if a certain current be just able to move a block of given weight and form, another current of double the velocity of the former would move a block of a similar form, whose weight should be to that of the former in the ratio of $2^6:1$ *i.e.* of 64 to 1."[4]

Hopkins' theoretical investigations into the constitution of the interior of the earth made him "one of the most famous champions of the theory of the earth's rigidity."[5] Assuming that the earth was originally molten, he calculated from the varying effects of the sun's and moon's attraction (and especially of precession and nutation) that the solid crust of the earth had a thickness of at least one-quarter or one-fifth of its radius. This thickness, he concluded, virtually prohibited direct heat or matter transference from the molten interior to the earth's surface; and therefore volcanoes must draw their molten material from reservoirs of moderate size within the solid crust. The largely solid and rigid state of the earth was considered to be due to cooling and to great internal pressure, an opinion supported by the work of Poisson, Ampère, George H. Darwin, and Lord Kelvin. Indeed, it was on the advice of Kelvin that Hopkins in 1851 undertook at Manchester, with the help of Joule and Fairbairn, experiments that showed effectively that the fusion temperature of strata increased considerably with depth and pressure.

Hopkins' theoretical studies on the motion of glaciers and on climatic change contained nothing new except their praiseworthy quantitative precision. For example, his deductions that the most probable cause of changes of climate during geological time was the influence of alterations in the various configurations of land and sea and in ocean currents were already held by Lyell and others, but none had hitherto expressed the details in precise mathematical terms. Thus, except in the popularization of quantification and in the broader field of geophysics, Hopkins' effect on contemporary geology was frequently retrogressive rather than progressive. He was often lacking in geological insight; and it is not entirely through misfortune that his valuable sixth-power law of hydraulic traction is usually attributed to G. K. Gilbert, who applied it firmly to river flow and not to mighty waves caused by paroxysmal uplifts of mountains.

NOTES

1. W. W. Smyth, in *Quarterly Journal of the Geological Society of London,* **23** (1867), xxx.
2. "On the Elevation and Denudation of the District of the Lakes of Cumberland and Westmoreland," p. 762.
3. Address to the Association of American Geologists and Naturalists, in *American Journal of Science,* **47** (1844), 244–245; see also R. J. Chorley, A. J. Dunn, and R. P. Beckinsale, *The History of the Study of Landforms,* I, 278.
4. *Op. cit.* (1842), pp. 764–765; (1849), p. 233.
5. K. A. von Zittel, *History of Geology and Palaeontology,* p. 178.

BIBLIOGRAPHY

I. Original Works. Hopkins' writings were published, often successively in enlarged form, mainly as articles in *Transactions of the Cambridge Philosophical Society, Proceedings* and *Quarterly Journal of the Geological Society of London, Philosophical Transactions of the Royal Society,* and *Report of the British Association for the Advancement of Science.* The most important are "Researches in Physical Geology," in *Transactions of the Cambridge Philosophical Society,* **6** (1838), 1–84, mainly on crustal elevation and fracturing; "Researches in Physical Geology," in *Philosophical Transactions of the Royal Society,* **129** (1839), 381–423; **130** (1840), 193–208; **132** (1842), 43–55, on precession and nutation and their probable effect on the nature of the earth's crust and interior—see also *Report of the British Association for the Advancement of Science* for 1847 (1848), pp. 33–92; and for 1853 (1854), pp. xli–lvii; "On the Geological Structure of the Wealden District and of the Bas Boulonnais," in *Proceedings of the Geological Society of London,* **3** (1841), 363–366; "On the Elevation and Denudation of the District of the Lakes of Cumberland and Westmoreland," *ibid.* (1842), pp. 757–766, repr. in full, with map, in *Quarterly Journal of the Geological Society of London,* **4** (1848), 70–98; "On the Motion of Glaciers," in *Transactions of the Cambridge Philosophical Society,* **8**

(1849), 50–74, 159–169, which favors a rigid sliding, fracturing motion; "On the Transport of Erratic Blocks," *ibid.*, pp. 220–240; "Presidential Address," in *Quarterly Journal of the Geological Society of London,* **8** (1852), xxi–lxxx, mainly on glacial drift and temperature changes; "On the Granitic Blocks of the South Highlands of Scotland," *ibid.*, pp. 20–30, which considers that striations on rocks are due to half-floating ice; "On the Causes Which May Have Produced Changes in the Earth's Superficial Temperature," *ibid.*, pp. 56–92, a detailed paper with a map of isotherms; and "Anniversary Address," *ibid.*, **9** (1853), xxii–xcii, which attacks Élie de Beaumont's ideas on pentagonal fracturing during crustal uplift and fracturing. See also "On the External Temperature of the Earth . . .," in *Monthly Notices of the Royal Astronomical Society,* **17** (1856–1857), 190–195, which makes use of H. W. Dove's world isothermal map.

II. SECONDARY LITERATURE. See R. E. Anderson, in *Dictionary of National Biography,* XXVII (1891), 339–340; R. J. Chorley, A. J. Dunn, and R. P. Beckinsale, *The History of the Study of Landforms,* I (London, 1964), *passim,* with a portrait; J. W. Clark and T. M. Hughes, *Life and Letters of the Rev. Adam Sedgwick,* II (Cambridge, 1890), 74, 154, 323; Henry Rogers Darwin, address to the Association of American Geologists and Naturalists, in *American Journal of Science,* **47** (1844), 244–245; W. W. Smyth, in *Quarterly Journal of the Geological Society of London,* **23** (1867), xxix–xxxii; *The Times* (London) (16 Oct. 1866), p. 4; and K. A. von Zittel, *History of Geology and Palaeontology,* M. M. Ogilvie-Gordon, trans. (London, 1901), pp. 168, 178, 303.

ROBERT P. BECKINSALE

HOPKINSON, JOHN (*b.* Manchester, England, 27 July 1849; *d.* Evalona, Switzerland, 27 August 1898), *electricity, physics.*

The talents of Hopkinson, a bright student, were drawn to the engineering problems of English industry during the surge of expansion in the last quarter of the nineteenth century. The oldest of thirteen children, he began his senior studies at Owens College, Manchester, in 1865 and was awarded a D.Sc. by London University in 1870. In 1867 he was granted a scholarship in mathematics by Trinity College, Cambridge, from which he graduated in 1871 with honors. In the following year Hopkinson relinquished a fellowship there to engage in practical engineering work in optics at Birmingham. After six years he went to London to teach electrical engineering at King's College of London University and to direct the Siemens laboratory.

Hopkinson's investigations in the application of electricity and magnetism to motors and dynamos resulted in more than sixty published books and papers. As alternating current phenomena became better understood in the last decade of the century, his mathematical skills were applied to transformer and alternating current systems design, to power transmission, to hysteresis and the magnetism of steel alloys, and to compact magnetic circuits such as those in the Edison-Hopkinson dynamo which doubled the output for equal weight. These studies resulted in some forty patents in multiple-wire circuitry and rotating machines of higher efficiency.

Hopkinson's application of Maxwell's electromagnetic theories to the analysis of residual charge and displacement in electrostatic capacity led to his election as a fellow of the Royal Society in 1877. He favored coupling traction motors in series parallel, thereby providing electric railways with superior motive power. He continued as consultant to the Chance technical glassworks in Birmingham and developed improved beam designs for lightship illumination and lighthouse lenses; he also served on several commissions establishing electric light standards.

In addition to having been elected fellow of the Royal Society at twenty-nine, he was twice president of the Institution of Electrical Engineers. At the age of forty-nine Hopkinson, with three of his children, was killed in a mountain climbing accident in the Alps.

BIBLIOGRAPHY

I. ORIGINAL WORKS. Hopkinson published one book, *Original Papers on Dynamo Machinery and Allied Subjects* (New York-London, 1893). The remainder of his work consisted of papers and pamphlets published in *Proceedings of the Royal Society* and in engineering journals; these were compiled and edited in two volumes by his son, Bertram Hopkinson, who appended a fifty-eight-page biography and two portraits (Cambridge, 1901).

II. SECONDARY LITERATURE. In addition to the work by his son (see above), see Evelyn Oldenbourgh Hopkinson, *The Story of a Mid-Victorian Girl* (Cambridge, 1928). James Greig of King's College also published a critical biography in *Engineering* (13 Jan. 1950) on the occasion of the centenary of Hopkinson's birth.

BERN DIBNER

HOPPE-SEYLER, FELIX (*b.* Freiburg im Breisgau, Germany, 26 December 1825; *d.* Lake Constance, Germany, 10 August 1895), *physiological chemistry.*

Ernst Felix Immanuel Hoppe was the tenth child of Ernst Hoppe, a minister, and the former Friederike Nitzsch; there had been theologians and scholars on both sides of the family. Felix's mother died when he was six years old, and his father died three years later; the boy was raised by his brother-in-law Dr. Seyler. In 1864 he was formally adopted by his guardian and changed his name to Hoppe-Seyler. He married Agnes Franziska Maria Borstein in 1858; they had a son and a daughter.

After graduation from the Gymnasium of the orphans' home at Halle in 1846, Hoppe entered the medical school there; the following year he transferred to Leipzig, where he worked in the laboratory of the physiological chemist K. G. Lehmann and studied with the three Weber brothers, who befriended him. He completed his medical studies in 1850 at Berlin and received the M.D. in 1851, having submitted a dissertation describing a histological and chemical study of cartilage. After a year of further clinical training in Prague, he began to practice medicine in Berlin but found this uncongenial because the demands of his practice made scientific work impossible. In 1854 Hoppe became prosector in anatomy at Greifswald; because research possibilities were too limited here as well, he eagerly accepted a similar post in the new pathological institute organized in Berlin by Virchow, who made Hoppe head of the chemical laboratory and greatly encouraged his investigative efforts. There soon flowed from this laboratory a succession of papers on a variety of physiological-chemical topics. In 1860 Hoppe was appointed associate professor on the medical faculty at Berlin, and in the following year he moved to Tübingen as professor of applied chemistry at the Faculty of Medicine. After the Franco-Prussian War, Hoppe-Seyler went in 1872 to Strasbourg, then under German occupation, to become professor of physiological chemistry, the post he occupied at the time of his death.

Hoppe-Seyler's initial researches, during the 1850's, dealt largely with the improvement of analytical methods for the chemical study of biological fluids, such as blood and urine. These studies led to his significant researches on the substance he called hemoglobin, the absorption spectrum of which he described in 1862; in this work he introduced the new spectroscope of Bunsen and Kirchhoff into medical chemistry. During the succeeding years Hoppe-Seyler demonstrated that hemoglobin binds oxygen loosely to form oxyhemoglobin, which can give up its oxygen to the body tissues. He extended Claude Bernard's observations on the toxic effect of carbon monoxide by showing that this gas displaced the oxygen of oxyhemoglobin. Hoppe-Seyler's chemical and spectroscopic researches showed that treatment of hemoglobin with acid produces a material he named hemochromogen, which is readily cleaved to yield the iron-containing hematin and resembles the products formed upon the interaction of isolated hematin with various proteins. Upon treatment of hematin with strong acids, he obtained an iron-free pigment which he named hematoporphyrin. His characterization of these materials, together with the parallel work of G. G. Stokes on the reduction of oxyhemoglobin, laid the foundations of all subsequent research on the chemistry of hemoglobin and of iron-porphyrin-containing proteins, as well as on their physiological role in respiration.

Hoppe-Seyler's studies on hemoglobin were largely completed during his stay in Tübingen, where he also conducted important work on lecithin and cholesterol. He contributed to the demonstration that these two substances are widely distributed constituents of biological systems; and his student C. Diakonow added valuable chemical data to those provided earlier by Adolph Strecker on the chemical constitution of lecithin, recognized to represent a compound formed by the union of choline, fatty acids, and glycerophosphate. Hoppe-Seyler and his students also showed that lecithin is combined with proteins to form conjugated proteins (vitellins), such as those found in egg yolk. This interest in phosphorus-containing proteins led him to urge his student Friedrich Miescher to examine more closely the chemical composition of cell nuclei; Miescher's discovery of nuclein in 1869 marks the starting point of a development that led to the later recognition of the role of deoxyribonucleic acids (DNA) in heredity. Hoppe-Seyler himself established the presence of nuclein in yeast, and subsequent work during the 1880's by his assistant Albrecht Kossel provided the chemical basis for the later elucidation of the structure of the nucleic acids.

Hoppe-Seyler's personal researches at Strasbourg dealt mostly with problems relating to the nature of intracellular oxidation processes, although he continued his hemoglobin studies and also showed that the pigment of chlorophyll resembles the porphyrin of hemoglobin. During the 1870's he participated actively in the discussions concerning the nature of biological catalysis, and he advocated the theory that the hydrogen atoms of metabolites were "activated" to react with respiratory oxygen. In connection with these discussions Hoppe-Seyler and his students conducted extensive studies on the products formed in the fermentation of various substances, notably cellulose, and in the putrefaction of proteins, especially the amino acid tyrosine.

During his years in Strasbourg, Hoppe-Seyler became the leading German protagonist of the separation of physiological chemistry from medical physiology; and in 1877 he founded the *Zeitschrift für physiologische Chemie* to promote the interests of biochemistry as an active and independent area of science. Despite the opposition of some physiologists, notably Eduard Pflüger, the influence of Hoppe-

Seyler's ideas grew, in large part because of the successes achieved by his students (especially Eugen Baumann and Kossel) and by his colleague at Heidelberg, Wilhelm Kühne. These two men, through their personal researches and those of their disciples, established German biochemistry and profoundly influenced the development of this subject in other countries, especially the United States.

BIBLIOGRAPHY

I. ORIGINAL WORKS. Hoppe-Seyler's books include *Handbuch der physiologisch und pathologisch-chemischen Analyse* (Berlin, 1858; 6th ed., 1893); and *Physiologische Chemie,* 4 pts. (Berlin, 1877–1881). Some of the research reports from his Tübingen laboratory were collected in his *Medicinisch-chemische Untersuchungen* (Berlin, 1866–1871). He published about 150 articles, not including the many from his laboratory, by his students and research assistants, without his name as a coauthor. Among his most important papers are "Beiträge zur Kenntnis der Constitution des Blutes," in *Medicinisch-chemische Untersuchungen,* pp. 133–150, 363–385, 523–550; and "Ueber die Processe der Gährungen und ihre Beziehung zum Leben der Organismen," in *Pflüger's Archiv für die gesamte Physiologie,* **12** (1876), 1–17.

II. SECONDARY LITERATURE. An extensive evaluation of Hoppe-Seyler's work, as well as a list of his publications, is given by E. Baumann and A. Kossel, in *Hoppe-Seyler's Zeitschrift für physiologische Chemie,* **21** (1895), i–lxi.

JOSEPH S. FRUTON

HORBACZEWSKI, JAN (*b.* Zarubince, near Ternopol, Austria-Hungary [now R.S.F.S.R.], 15 May 1854; *d.* Prague, Czechoslovakia, 24 May 1942), *biochemistry.*

Horbaczewski studied in Vienna and in 1883, after the University of Prague had been divided into a German and a Czech section, he became extraordinary professor of medical chemistry on the Czech medical faculty, and, a year later, full professor. He was four times dean of the medical faculty and, in 1902–1903, rector of the university. Ukrainian in origin, he retained a strong interest in the fate of his nation, part of which lived in Galicia under Austrian rule. In a last effort to save the Austro-Hungarian monarchy a multinational government was formed in 1917; in it Horbaczewski headed the newly created ministry of health but resigned in July 1918. In 1925 he was elected to the All-Ukrainian Academy of Sciences in Kiev and was also offered a teaching post, which he declined because of advanced age.

Although interested in nutrition, toxicology, and even industrial chemistry, Horbaczewski contributed mainly to the chemistry and biochemistry of uric acid. In 1882 he was the first to synthesize uric acid by heating glycine and urea at 200–230°C. As a young man he succeeded where many more experienced workers had failed, and his success was only grudgingly acknowledged—although *Nature* (**27** [1882–1883], 49) hailed the synthesis as probably the most important involving urea "since Wöhler prepared it from its mineral constituents."

In a series of papers during the 1880's and early 1890's Horbaczewski investigated the origin of uric acid in mammals, including man. At first he found that uric acid was formed in the spleen pulp, which was treated with arterial blood. He connected leukocytosis with the formation of uric acid and was convinced that uric acid ultimately derived from the nuclei of the lymphatic elements of the spleen pulp. Then, following J. F. Miescher's method of separating cell nuclei, he produced the first direct experimental proof that uric acid was not a constituent of protein but was part of the cell nucleus metabolism.

BIBLIOGRAPHY

I. ORIGINAL WORKS. Most of Horbaczewski's papers appeared in *Sitzungsberichte der K. Akademie der Wissenschaften in Wien,* Math.-naturwiss. Kl. The first synthesis of uric acid is described in **86,** sec. 2 (1882), 963–964; an important paper dealing with the metabolism of purines can be found in **100,** sec. 3 (1891), 78–132. His views are summarized in *Zur Theorie der Harnsäurebildung* (Wiesbaden, 1892). Horbaczewski also published a textbook of medical chemistry, *Chemie lékařská,* 3 vols. in 4 pts. (Prague, 1904–1908).

II. SECONDARY LITERATURE. See K. Kácl, "Professor Dr. Jan Horbaczewski," in *Časopis lékařů českých,* **93** (1954), 578–580, which gives an almost complete bibliography; and M. Teich, in L. Nový, ed., *Dějiny exaktních věd v českých zemích* (Prague, 1961), pp. 344 ff., with Russian and English summary; and "K istorii sinteza mochevoy kisloty (Ot Sheele k Gorbachevskomu)" ("The History of the Synthesis of Uric Acid [From Scheele to Horbaczewski]"), in *Trudy Instituta istorii estestvoznaniya i tekhniki. Akademiya nauk SSSR,* **35** (1961), 212–244, in Russian.

M. TEICH

HORN, GEORGE HENRY (*b.* Philadelphia, Pennsylvania, 7 April 1840; *d.* Beesley's Point, New Jersey, 24 November 1897), *coleopterology.*

The son of Philip Henry Horn and the former Frances Isabella Brock, Horn earned an M.D. at the University of Pennsylvania in 1861, then served from 1862 to 1866 in the medical corps of the California Volunteers. On his return to Philadelphia he practiced medicine, especially obstetrics, for many years.

Horn's interest in zoology was aroused at the Academy of Natural Sciences of Philadelphia. His more than 200 publications extended from 1860 to 1896 and, except for two or three early items, dealt with the Coleoptera. The major influence on his life was his friendship with John Lawrence Le Conte, the leading American coleopterist of the third quarter of the nineteenth century. Horn contributed the section on Otiorhynchidae to their "Rhynchophora of America, North of Mexico" (1876) and collaborated with Le Conte on "Classification of the Coleoptera of North America" (1883).

Horn's many notable monographs on the numerous genera and families of Coleoptera dealt with the Nearctic fauna, except the Throscidae and Eucnemidae of the *Biologia Centrali-Americana* (1890). Turning to another subject, his "Synopsis of the Silphidae" (1880) and his "Genera of Carabidae" (1881) provided a brief examination of non-North American genera. The latter work was highly regarded in its day as an important contribution to the understanding of a major family. During his life he described a total of 1,583 species and varieties of Coleoptera, of which fifty-two were regarded as synonyms at the time of his death. His keys and descriptions were notable for their precision and clarity; although gradually amended, they dominated the field of determinative North American coleopterology for four or five decades after his death and still retain much of their usefulness. Horn's work made possible, in important measure, Willis Stanley Blatchley's *Coleoptera of Indiana* (1910) and furnished much of the foundation for James Chester Bradley's *Manual for the Genera of Beetles of America North of Mexico* (1930). He made three trips to Europe (1874, 1882, 1888) to meet colleagues and to study collections.

Horn packed, for transmission to the Museum of Comparative Zoology in Cambridge, Le Conte's collection, in accordance with the latter's will; and he says that there was scarcely a box that did not bring to mind memories of their long association. Horn's own collection was left to the Academy of Natural Sciences of Philadelphia.

Following Le Conte's death in 1883, Horn became the leading American coleopterist and was regarded by the British coleopterist George Charles Champion as the leading student of beetles in North America up to that time.

BIBLIOGRAPHY

I. ORIGINAL WORKS. Walter Derksen and Ursula Scheiding-Gollner (see below) provide a list of biographical notices and a bibliography of 204 of Horn's papers published after 1865. His most important single works include "Rhynchophora of America, North of Mexico," in *Proceedings of the American Philosophical Society,* **15** (1876), 1–455, written with John Lawrence Le Conte; "Synopsis of the Silphidae of the United States With Reference to the Genera of Other Countries," in *Transactions of the American Entomological Society,* **8** (1880), 219–322, plates vi–vii; "On the Genera of Carabidae With Special Reference to the Fauna of Boreal America," *ibid.,* **9** (1881), 91–96, plates iii–x; "Classification of the Coleoptera of North America," in *Smithsonian Miscellaneous Collections,* **507** (1883), 1–567, written with Le Conte; and "Fam. Throscidae and Fam. Eucnemidae," in *Biologica Centrali-Americana. Insecta. Coleoptera,* III, pt. 1 (1890), 193–257, plate x.

II. SECONDARY LITERATURE. On Horn and his work see Philip P. Calvert, "A Biographical Notice of George Henry Horn," in *Transactions of the American Entomological Society,* **26** (1898), 1–24, with portrait; Samuel Henshaw, "The Entomological Writings of George Henry Horn (1860–1896) With an Index to the Genera and Species of Coleoptera Described and Named," *ibid.,* pp. 25–72; and Walter Derksen and Ursula Scheiding-Gollner, "Index litteraturae entomologicae, Serie II," in *Die Welt-Literature über die gesamte Entomologie von 1864 bis 1900,* II (Berlin, 1965), 354–358.

MELVILLE H. HATCH

HORN D'ARTURO, GUIDO (*b.* Trieste, 13 February 1879; *d.* Bologna, Italy, 1 April 1967), *astronomy.*

Horn d'Arturo graduated from the University of Vienna in 1902 and was, successively, assistant at the observatories of Trieste, Catania, Turin, Bologna, and Rome. From 1920 he was director of the Bologna observatory, in the old university center, and supervised its complete renovation. He also had built a branch observatory near Lojano, in the Tuscan Apennines between Bologna and Florence, at an altitude of 2,600 feet, furnishing it with a Zeiss reflector of sixty-centimeter aperture. Because he was of Jewish extraction he was removed in 1938 from the chair of astronomy and the directorship of the observatory; at the end of the war he was reinstated in both these posts at the University of Bologna, from which he retired in November 1954.

A capable observer with notable technical skills, Horn d'Arturo was active in positional astronomy, statistics, cosmography, and optical astronomy. With the Lojano telescope he and his co-workers observed variable stars, gaseous and planetary nebulae, and globular clusters and investigated the apparent distribution of nebulae and of the fixed stars. In optical astronomy he demonstrated how the density of photographic stellar tracks may be measured by using the diffraction of light. He clarified the effect on vision, especially in the astigmatic eye, of the suture of the eye lens and the formation of the so-called black drop.

In instrumental techniques Horn d'Arturo conceived of a conic lens (in place of the prism lens) in which each section passing through the axis acts as an infinitely thin prism. Stellar images obtained with this instrument exhibit concentric spectral lines (circular in the case of stars on the axis); in every other instance the lines are curves of the fourth order. Horn d'Arturo devised this instrument to obtain spectra of meteors, because by properly placing the camera and, with it, the conic prism for vertical reception it is possible to cover the entire sky from the zenith to 23° above the horizon and any azimuth.

As head of the Italian expedition sent to Somaliland to observe the eclipse of 14 January 1926, Horn obtained interesting photographs of the flash spectrum and of the prominences enveloped by the corona. On that occasion he was developing one of his theories on the phenomenon of "flying shadows" and on the perpetual eastern current of the very high equatorial atmosphere.

The last years of Horn d'Arturo's scientific activity were devoted to the construction and use of his *specchio a tasselli.* With a diameter of 180 centimeters, the mirror was made up of many small mirrors arranged in a series of concentric circles. He mounted the horizontal mirror at the base of the university tower (the old Bologna observatory) so that it functioned only for the zenith. He was thus able to photograph stars to the eighteenth magnitude. Horn d'Arturo's idea, while not applied to reflectors of the standard type, has been successfully adapted for other purposes.

Horn d'Arturo founded the popular astronomical magazine *Coelum.*

BIBLIOGRAPHY

Horn d'Arturo's work was most often published in both *Pubblicazioni dell'Osservatorio astronomico della Università di Bologna* and *Memorie della Società astronomica italiana.* Among the most significant are "Il fenomeno della goccia nera e l'astigmatismo," in *Pubblicazioni dell'Osservatorio astronomico della Università di Bologna,* vol. **1,** no. 3 (1922); "Le ombre volanti," *ibid.,* no. 6 (1924); "Numeri arabici e simboli celesti," *ibid.,* no. 7 (1925); "L'eclisse solare totale del 14 gennaio 1926 osservata dalla missione astronomica italiana nell'Oltregiuba," *ibid.,* no. 8 (1926); "L'uso di una lente conica nella spettrograffia delle stelle cadenti," *ibid.,* vol. **2** (1934); "Primi esperimenti con lo specchio a tasselli," *ibid.,* vol. **3,** no. 3 (1935); "L'aggiustamento dello specchio a tasselli effettuato dal centro di curvatura," *ibid.,* vol. **5,** no. 17 (1952); and "Lo specchio a tasselli di metri 1,80 d'apertura, collocato nella Torre dell'Osservatorio Astronomico Universitario," *ibid.,* vol. **6** (1955).

An obituary is G. Mannino, L. Rosino, L. Jacchia, in *Coelum,* **35,** nos. 5–6 (1967).

GIORGIO ABETTI

HORNE (HORNIUS), JOHANNES VAN (*b.* Amsterdam, Netherlands, *ca.* 2 September 1621; *d.* Leiden, Netherlands, 5 January 1670), *anatomy.*

Van Horne was descended from a Flemish family of merchants. His father, Jacob (Jacques), was one of the first "Lords Seventeen," the directors of the Dutch East India Company; his mother was the former Margriet van der Voort. He matriculated at the University of Leiden at the age of fifteen, on 10 September 1636, for letters but later turned to medicine and is said to have assisted Johannes de Wale in his well-known studies of the circulation of the blood. He continued his medical studies at Utrecht under Willem van der Straaten, then made a study tour to Italy. At Padua, van Horne attended the anatomical lectures of Johann Vesling and took his medical degree. He also visited Naples, where he heard Marc Antonio Severino, who influenced his surgical views. On his way home, the University of Basel granted him an honorary degree. He also visited Orléans, Montpellier, and England. The period of his foreign studies covered not less than six years.

Once back in the Netherlands, van Horne asked the governors of Leiden University for permission to give anatomical demonstrations. He was appointed extraordinary professor of anatomy on 8 February 1651. After the death of Otto Heurnius in 1652 van Horne was appointed professor of anatomy and surgery on 27 January 1653.

A very learned man, with thorough knowledge of the classical and modern languages, van Horne was interested primarily in anatomy but also lectured and published on surgery. In 1652 he was the first to describe the *ductus chyliferus* (*thoracicus*) in man. As a teacher van Horne inspired Frederik Ruysch, Jan Swammerdam, and Nicolaus Steno, among others. He understood the art of making fine anatomical preparations and seems also to have prepared an anatomical atlas, which was never published.

Van Horne's friendship with the nobleman Louis de Bils, who had enriched his anatomy cabinet with fine preparations, ended in bitter polemic when the latter took advantage of van Horne's imprudent recommendation to publish a book in which he supported a fantastic theory that included the supposition that the chylus was transported directly to the liver. Van Horne was scandalized and turned in vain to his

Danish friend Thomas Bartholin for help in this struggle. His young pupil Ruysch settled the controversy with his *Dilucidatio valvularum in vasis lymphaticis et lacteis* (1665).

With the assistance of Swammerdam, van Horne investigated the ovaries. He published his observations only in a small preliminary booklet, *Prodromus.* These observations played some role in the priority dispute between Regnier de Graaf and Swammerdam.

Van Horne edited, with annotations, Leonard Botallus' *Opera omnia medica et chirurgica* (Leiden, 1660) and Galen's work in Greek and Latin on the bones (with the references of Vesalius and Eustachi to this work). His introduction to anatomy, *Microcosmus seu brevis manuductio ad historiam corporis humani* (1660), was much in demand and was translated into Dutch, German, and French. He also wrote a short introduction to surgery, in which he advised some rather crude methods of amputation of members and of the breasts.

BIBLIOGRAPHY

I. Original Works. Van Horne's writings include "De aneurysmate epistola," in Thomas Bartholin, *Anatomica aneurysmatis dissecti historia* (Panormi, 1644); *Novus ductus chyliferus, nunc primum delineatus, descriptus et eruditorum examini expositus* (Leiden, 1652); ΜΙΚΡΟΚΟΣΜΟΣ *seu brevis manuductio ad historiam corporis humani, in gratiam discipulorum* (Leiden, 1663, 1665; Leipzig, 1673; Halberstadt, 1685); ΜΙΚΡΟΤΕΧΝΗ *sive brevissima chirurgiae methodus* (Leiden, 1663, 1668; Leipzig, 1675); *Prodromus observationem suarum circa partes genitales utroque sexu* (Leiden, 1668), repr. with notes of J. Swammerdam in J. M. Hofman, *Dissertationes anatomico-physiologicae ad Jo. van Horne Microscosmum* . . . (Altdorf, 1685); *Observationes anatomico-medicae* (Amsterdam, 1674); and *Opuscula anatomico-chirurgica* (Leipzig, 1707), with annotations edited by J. G. Pauli.

II. Secondary Literature. There is no biography of van Horne. Most information is to be found in older Dutch sources, including J. Banga, *Geschiedenis van de geneeskunde en van hare beoefenaren in Nederland,* I (Leeuwarden, 1668), 436–447; and G. C. B. Suringar, "Het geneeskundig onderwijs van Albert Kyper en Johannes Antonides van der Linden. De ontleedkundige school van Johannes van Horne," in *Nederlands tijdschrift voor geneeskunde,* **7** (1863), 193–206. More recent sources are E. D. Baumann, in *Nieuw Nederlandsch Biographisch Woordenboek,* VII (Leiden, 1927), 624–625; and P. C. Molhuysen, *Bronnen tot de geschiedenis der Leidsche Hoogeschool,* III (The Hague, 1918), *passim.*

G. A. Lindeboom

HORNER, LEONARD (*b.* Edinburgh, Scotland, 17 January 1785; *d.* London, England, 5 March 1864), *geology.*

Horner was the third son of John Horner, an Edinburgh textile merchant, and younger brother of Francis Horner, the Whig politician and a founder of the *Edinburgh Review.* After attending Edinburgh University, he moved to London in 1804 as a partner in his father's business. He joined the Geological Society of London in 1808, soon after its foundation, and served as secretary from 1810 to 1814 and as president in 1845–1847 and 1860–1862. He was elected to the Royal Society of London in 1813.

In 1817 Horner returned to Edinburgh; and in 1821 he founded there a school of arts, one of the earliest examples of the Mechanics' Institute movement. In 1827 he was called to the newly founded University College, London, where he served as warden until 1831, supervising the formative years of the first English university institution to give a major place to scientific subjects. After two years' residence in Bonn for the sake of his health, Horner returned to England and was appointed to serve on the commission on the employment of children in factories; under the subsequent Factory Act (1833) he was for many years an inspector of factories. His concern as a social reformer is also reflected in published works on working-class education and on working conditions in factories. Horner married Anne Lloyd in 1806; their eldest child, Mary, married the geologist Charles Lyell in 1832.

Horner's work in promoting science-based education at all social levels was more important for the development of nineteenth-century science than was his original scientific work, although the latter was far from negligible. Horner had attended the mathematics lectures of John Playfair at Edinburgh and was greatly influenced by Playfair's geology. His two earliest papers, on the Malvern Hills (1811) and an area of Somerset (1816) in Southwestern England, show meticulous description allied to cautious Huttonian theorizing; they were written at a time when "geology" was only just beginning to become clearly distinct from "mineralogy." His presidential addresses to the Geological Society in 1846 and 1847 show strong sympathy with the Playfairian *Principles of Geology* of his son-in-law Lyell and at the same time are masterly reviews of the current progress of the science. In the 1850's Horner had the support of the Royal Society in an ambitious scheme for excavating the Nile silt around the bases of two Egyptian monuments of known historic date; he hoped to estimate the mean rate of deposition, in order to link the historical time scale to the relative time scale of geology. Although his work was criticized, he believed it gave strong evidence that even the recent geological period had lasted not less than 13,500 years and that fragments of pottery indicated almost as great an antiquity for the human race. He reiterated this conclusion in his last address to the Geological Society (1861), anticipating Lyell's *Antiquity of Man* (1863); in the same address he also gave a warm recommendation to Darwin's *Origin of Species* (1859).

BIBLIOGRAPHY

Horner's principal scientific publications are "On the Mineralogy of the Malvern Hills," in *Transactions of the Geological Society of London,* **1** (1811), 281–321; "Sketch of the Geology of the South-Western Part of Somersetshire," *ibid.,* **3** (1816), 338–384; "On the Geology of the Environs of Bonn," *ibid.,* 2nd ser. **4**, pt. 2 (1836), 433–481; "Anniversary Address[es] of the President," in *Quarterly Journal of the Geological Society of London,* **2** (1846), 145–221; **3** (1847), xxii–xc; **17** (1861), xxxi–lxxii; and "An Account of Some Recent Researches Near Cairo, Undertaken With a View of Throwing Light Upon the Geological History of the Alluvial Land of Egypt," in *Philosophical Transactions of the Royal Society,* **145** (1855), 105–138; **148** (1858), 53–92. His daughter Mary Horner Lyell edited a valuable collection of correspondence in her *Memoir of Leonard Horner, F.R.S., F.G.S., Consisting of Letters to His Family and From Some of His Friends,* 2 vols. (London, 1890).

M. J. S. RUDWICK

HORNER, WILLIAM GEORGE (*b.* Bristol, England, 1786; *d.* Bath, England, 22 September 1837), *mathematics.*

The son of William Horner, a Wesleyan minister, Horner was educated at the Kingswood School, Bristol, where he became an assistant master (stipend £40) at the age of fourteen. After four years he was promoted to headmaster, receiving an additional £10 annually. According to an account given by an "old scholar" in *The History of Kingswood School . . . By Three Old Boys* [A. H. L. Hastings, W. A. Willis, W. P. Workman] (London, 1898), p. 88, the educational regime in his day was somewhat harsh. In 1809 Horner left Bristol to found his own school at Grosvenor Place, Bath, which he kept until his death. He left a widow and several children, one of whom, also named William, carried on the school.

Horner's only significant contribution to mathematics lay in the method of solving algebraic equations which still bears his name. Contained in a paper submitted to the Royal Society (read by Davies Gilbert on 1 July 1819), "A New Method of Solving Numerical Equations of All Orders by Continuous Approximation," it was published in the *Philosophical Transactions* (1819) and was subsequently republished in *Ladies' Diary* (1838) and *Mathematician* (1843). Horner found influential sponsors in J. R. Young of Belfast and Augustus de Morgan, who gave extracts and accounts of the method in their own publications. In consequence of the wide publicity it received, Horner's method spread rapidly in England but was little used elsewhere in Europe.

Throughout the nineteenth and early twentieth centuries Horner's method occupied a prominent place in standard English and American textbooks on the theory of equations, although, because of its lack of generality, it has found little favor with modern analysts. With the development of computer methods the subject has declined in importance, but some of Horner's techniques have been incorporated in courses in numerical analysis.

Briefly, when a real root of an equation has been isolated by any method, it may be calculated by any one of several arithmetical processes. A real root r, of $f(x) = 0$, is isolated when one finds two real numbers a, b, between which r lies and between which lies no other root of $f(x) = 0$. Horner's method consists essentially of successively diminishing the root by the smaller members of successive pairs of positive real numbers.

If

$$f_1(x) \equiv a_0x^n + a_1x^{n-1} + a_2x^{n-2} + \cdots + a_n,$$

and if $x = h + y$, we have (expanding by Taylor's theorem)

$$f(h + y) \equiv f(h) + yf'(h) + \frac{y^2}{2!}f''(h) + \cdots + \frac{y^n f^{(n)}(h)}{n!},$$

$$f(h + y) \equiv f(h) + (x - h)f'(h) + \frac{(x - h)^2}{2!}f''(h) + \cdots + \frac{(x - h)^n f^{(n)}(h)}{n!}.$$

If this is written

$$f_2(x - h) \equiv c_n + c_{n-1}(x - h) + c_{n-2}(x - h)^2 + \cdots + c_0(x - h)^n,$$

the coefficients $c_n, c_{n-1}, c_{n-2}, \ldots, c_0$ in the reduced equation are given by the successive remainders when the given polynomial is divided by $(x - h)$, $(x - h)^2, (x - h)^3, \ldots, (x - h)^n$. In the original account of the method Horner used Arbogast's derivatives ($D\phi R, D^2\phi R, \ldots, D^n\phi R$). Later he dispensed altogether with the calculus and gave an account of the method in entirely algebraic terms. Successive transformations were carried out in a compact arithmetic form, and the root obtained by a continuous process was correct to any number of places. The computational schema adopted is often referred to as synthetic division. Horner suggested, correctly, that his method could be applied to the extraction of square and cube roots; but his claims that it extended to irrational and transcendental equations were unfounded.

Although Horner's method was extremely practical for certain classes of equations, the essentials were by no means new; a similar method was developed by the Chinese in the thirteenth century (see J. Needham, *Science and Civilisation in China,* I [Cambridge, 1959], p. 42). The iterative method devised by Viète (1600) and developed extensively by Newton (1669), which came to be known as the Newton-Raphson method, is applicable also to logarithmic, trigonometric, and other equations. The numerical solution of equations was a popular subject in the early nineteenth century, and in 1804 a gold medal offered by the Società Italiana delle Scienze for an improved solution was won by Paolo Ruffini (. . . *Sopra la determinazione delle radici* . . . [Modena, 1804]). Ruffini's method was virtually the same as that developed independently by Horner some years later.

BIBLIOGRAPHY

I. Original Works. Horner's writings include "A New Method of Solving Numerical Equations of All Orders by Continuous Approximation," in *Philosophical Transactions of the Royal Society,* **109** (1819), 308–335; "Horae arithmeticae," in T. Leybourn, ed., *The Mathematical Repository,* V, pt. 2 (London, 1830); and "On Algebraic Transformations," in *Mathematician* (1843).

II. Secondary Literature. Accounts of the method are given by J. R. Young in *An Elementary Treatise on Algebra* (London, 1826); and *The Theory and Solution of Algebraical Equations* (London, 1843). Augustus de Morgan described the method in sundry articles, including "On Involution and Evolution," in *The Penny Cyclopaedia,* vol. XIII (London, 1839); and "Notices of the Progress of the Problem of Evolution," in *The Companion to the Almanack* (London, 1839). See also Florian Cajori, "Horner's Method of Approximation Anticipated by Ruffini," in *Bulletin of the American Mathematical Society,* **17** (1911), 409–414.

Margaret E. Baron

HORNSBY, THOMAS (*b.* Oxford, England, 28 August 1733; *d.* Oxford, 11 April 1810), *astronomy.*

Hornsby is best remembered for his part in the foundation of the Radcliffe Observatory in Oxford. He made an accurate evaluation of the solar parallax.

The son of Thomas Hornsby of Durham, Hornsby matriculated at Corpus Christi College, Oxford, in December 1749. After taking his B.A. in 1753 and M.A. in 1757, he was elected a fellow of his college, where he built himself a small observatory. In 1763 he followed James Bradley as Savilian professor of astronomy at Oxford, where between 1766 and 1775 he gave a notable series of lectures on experimental philosophy. Their reputation is reported to have led James Watt's partner, Matthew Boulton, to arrange for his son, who was not an undergraduate, to attend them. In 1763 Hornsby was made a fellow of the Royal Society.

At Corpus Christi, Hornsby observed with a fine mural quadrant with a radius of thirty-two inches; made by John Bird, it cost £80. As Savilian professor, he used the observatory in the tower of the Schools Quadrangle, from which he observed the transit of Venus on 3 June 1769, with twelve-foot and 7.5-foot refractors (*Philosophical Transactions of the Royal Society,* **59** [1769], 172–182). He was a friend of the earl of Macclesfield and had observed the transit of Venus of 6 June 1761 from Shirburn Castle, the earl's home. From both he deduced a solar parallax of 8.78″ (*ibid.,* **55** (1765), 326–344; **61** (1771), 574–576). (The fundamental constant adopted by the Conférence Internationale des Étoiles Fondamentales at Paris in 1896, still accepted, is 8.80″.)

A printed document of 5 February 1771, signed by Hornsby as Savilian professor, recorded a petition made by him in 1768 to the earl of Litchfield and the Radcliffe trustees for the foundation of an observatory. (For a copy of S. P. Rigaud's transcript of the document, the original of which R. T. Gunther thought to be lost, see Gunther's *Early Science in Oxford,* II [Oxford, 1923], 88–89. For a copy of the original in the Bodleian Library, the shelf mark is Gough Oxf. 90.) Hornsby asked for a transit instrument, two mural quadrants, a zenith sector, and an equatorial sector—to the tune of about £1,300—to be made by the best instrument maker of the time, John Bird. He suggested that the professor of astronomy make regular observations, to be published annually, and that he give regular courses of lectures in practical astronomy.

The proposals were accepted, Hornsby was made first Radcliffe observer in 1772, and the buildings were completed by 1778. (For a plan, and list of rooms and instruments, see Gunther, *op. cit.,* pp. 90–91, 318–324.) Hornsby had persuaded Bird to use achromatic object lenses in the telescopes on his sectors and quadrants, and the lenses were made by

Peter Dollond (see Gunther, *loc. cit.,* and p. 396). The outlay on buildings and instruments was £28,000—a considerable sum. Bird's assessment of his own superbly well-divided eight-foot south mural quadrant as "by far the best instrument of the kind in the world" was undoubtedly then true. Like many other Radcliffe instruments, it is now in the Museum of the History of Science, Oxford.

Hornsby does not have any great astronomical discovery to his credit. He investigated the proper motion of Arcturus (*Philosophical Transactions of the Royal Society,* **63** [1773], 93–125); and in 1798 he pointed out that, despite the large proper motion of the double star Castor, the two components had remained at the same distance during the twenty years he had observed them. Even so, he did not suggest any physical connection between the components.

In 1783 Hornsby was made Radcliffe librarian; and in 1798—more than twenty years after having undertaken the project—he published the first volume of Bradley's *Astronomical Observations* (see his preface).

BIBLIOGRAPHY

I. Original Works. Apart from a few minor notes, such as those in connection with Oxford administration, Hornsby wrote only the five papers cited in text. He published nothing of any length, other than the Bradley ed. cited. The best collection of Hornsby manuscripts is in the Museum of the History of Science, Oxford (MSS Radcliffe 1–35, 54, 67, 71–73). For Hornsby's notes for lectures on natural philosophy, in his own hand, see Bodleian Library MS Rigaud 54. The same library has a copy of Bradley's (?) *Propositiones mechanicae,* with notes taken at Hornsby's lectures, and also a syllabus of those lectures (shelf mark Vet. A 1 c.6 [51]), from about 1770.

II. Secondary Literature. R. T. Gunther, *Early Science in Oxford,* II (Oxford, 1923), is the most useful source. See also the pref. to S. P. Rigaud, *Miscellaneous Works and Correspondence of the Rev. James Bradley* (Oxford, 1832) on the question of Hornsby's dilatory editing of Bradley. Many further but often outdated references to secondary literature are in Agnes Clerke's biography of Hornsby in *Dictionary of National Biography.*

J. D. North

HORREBOW, CHRISTIAN (*b.* Copenhagen, Denmark, 15 April 1718; *d.* Copenhagen, 19 September 1776), *astronomy.*

Horrebow was the son of Peder Nielsen Horrebow, professor of astronomy, and Anne Margrethe Rossing. He was the fourth [?] child in a family of twenty. In 1732 he was sent to the University of Copenhagen, where in 1738 he obtained his M.S. degree. He became his father's assistant at the Round Tower Observatory, working on the calendar and thus continuing a literary tradition begun by Wilhelm Lange.[1] He computed the annual almanac from 1739 to 1770. In 1743 he became a designate professor, and from 1753 he was completely in charge of his father's post at the observatory, which, in 1741, had been fully restored after the great fire of Copenhagen (1728). In 1764 Horrebow obtained the chair of astronomy. He was elected a member of the Royal Danish Academy of Sciences and Letters in 1747 and from 1769 he was a titular councillor of state. He married Anna Barbara Langhorn on 25 October 1754.

To a large extent Horrebow continued his father's work in astronomy. Eustachio Manfredi had questioned[2] the elder Horrebow's alleged determination of stellar parallaxes from Römer's observations of right ascensions of Sirius and Vega.[3] To eliminate the doubt cast on his father's arguments, Horrebow worked on parallax determination (1742–1743, 1746), confirming his father's erroneous conclusion without realizing that the effect was due to the influence of temperature variations on clocks and instruments.[4] Because Bradley's theory of aberration surpassed the said Römer-Horrebow "proof" of the motion of the earth, Horrebow, in 1751, tried to develop a new micrometer method for determining stellar parallaxes, concentrating on some of the fainter stars believed to be close to the sun because of the small range of their vortices.

Horrebow's observation of the transit of Venus (1761) has often been judged a failure, but only because of a misunderstanding concerning the correction of his clocks. On the other hand, his systematic observations of sunspots during his last fifteen years came to play a role in the later investigation of the period of sunspot activity.[5] In theoretical astronomy he maintained a constant eccentricity of the earth's orbit against Jacques Eugène Louville's theory of the decrease of eccentricity with time.

In collaboration with his brother Peder, Horrebow also dealt with meteorological subjects. He showed that a theoretical table of barometer readings corresponding to different altitudes, prepared by his father in 1751, squared better with Juan's and Ulloa's determinations of mountain altitudes in America than did other tables accessible at the time.

Horrebow also prepared textbooks in the fields of astronomy and mathematics.

NOTES

1. Wilhelm Lange, *De annis Christi libri duo* (Leiden, 1649).
2. Eustachio Manfredi, "De novissimis circa fixorum siderum

errores observationibus. Ad . . . Antonium Leprottum . . . epistola," in *De Bononiensi scientiarum et artium instituto atque academia Commentarii,* **1** (1731), pp. 612–618.
3. Peder Nielsen Horrebow, *Copernicus Triumphans, sive de parallaxi orbis annui tractatus epistolaris* (Copenhagen, 1727).
4. C. A. F. Peters, "Recherches sur la parallaxe des étoiles fixes," in *Mémoires de l'Académie impériale des sciences de St.-Pétersbourg,* 6th ser., **5** (1853), 1–180.
5. T. N. Thiele, "De macularum solis antiquioribus quibusdam observationibus Hafniae institutis," in *Astronomische Nachrichten,* **50** (1859), cols. 257–262.

BIBLIOGRAPHY

I. ORIGINAL WORKS. A full list of Horrebow's printed writings is in Niels Nielsen, *Matematiken i Danmark 1528–1800* (Copenhagen, 1912), pp. 97–99. Not included in this list is "Vindiciae aerae dionysianae, sive de annis Christi diascepsis," in vol. II of Peder Nielsen Horrebow, *Opera mathematico-physica* (Copenhagen, 1741). On the annual parallax of the fixed stars, see *De parallaxi fixarum annua ex rectascensionibus, qvam post Roemerum et parentem ex propriis observationibus demonstrat* (Copenhagen, 1747). Most of his other writings are in *Videnskabernes Selskabs Skrifter* (1751–1770), as well as in his academic dissertations. The most interesting articles are "Afhandling om fixstiernernes distance fra Jorden" in *Videnskabernes Selskabs Skrifter,* **6** (1751), 129–152; "Reflexioner anlangende veneris drabant," *ibid.,* **9** (1765), 396–403; and "Om soel-pletterne," *ibid.,* **10** (1770), 469–536. For reports on the sunspots, see *Videnskabernes Selskabs historiske almanakker* (1770–1775). His observations for the period 1767–1776 are published in Wolf's *Mittheilungen über Sonnenflecken,* **19** (1865) and **33** (1873). Unpublished papers dealing with the meteorology and geography of Iceland and Greenland and with natural philosophy are extant at the Royal Library of Copenhagen.

II. SECONDARY WORKS. For biographical information, see C. F. Bricka, *Dansk Biografisk Leksikon,* X (Copenhagen, 1936), 607–608; and Niels Nielsen's article mentioned above. Horrebow's determination of the transit of Venus is treated in Axel V. Nielsen, "Christian Horrebows observationer af venuspassagen i 1761," in *Nordisk astronomisk tidsskrift* (1957), pp. 47–50.

KR. PEDER MOESGAARD

HORREBOW, PEDER NIELSEN (*b.* Løgstør, Denmark, 14 May 1679; *d.* Copenhagen, Denmark, 15 April 1764), *astronomy.*

Born into the family of a poor fisherman, Horrebow had to work his way through grammar school and later through Copenhagen University by doing mechanical work. He was a personal assistant to Ole Römer for four years. In 1714 he was made professor at the university and director of the observatory; he held this position for fifty years, although two of his sons had to take care of his professional duties during his last years. He was a member of the academies of Copenhagen, Berlin, and Paris.

Horrebow's scientific life was shaped by two major influences. The first was his daily association during his youth with Römer, of whom Horrebow later spoke with the greatest devotion. Second, in the great fire in Copenhagen in 1728 nearly all of Römer's papers and unpublished observations were destroyed together with Horrebow's own observations; and from that time on it remained a matter of personal honor for Horrebow to describe fully Römer's scientific achievements in order to preserve them for posterity.

Horrebow's book on Römer, the classic *Basis astronomiae* (1734–1735), contained Römer's observations, which Horrebow made with his own meridian circle during three days and nights. These observations were soon being used in early determinations of proper motions.

A main problem of the times was the measurement of the annual parallax of the fixed stars. Römer had introduced a new method, well adapted for his own instruments, in which the observation of the time of transit over the meridian was central. After Römer's death Horrebow analyzed the parallax traceable in the observations. He published his results in 1727 in a book with the exultant title *Copernicus triumphans,* which was received with great interest in the astronomical world. But in 1848 C. A. F. Peters disproved Horrebow's results through a systematic run of the clocks.

In 1732, in his book *Atrium astronomiae,* Horrebow advanced a technique for determining geographical latitude, now known as the Horrebow-Talcott method since it was rediscovered a century later by the American soldier and engineer Andrew Talcott. From his collection of notes entitled *Adversaria,* it appeared possible that Römer himself had known this method; but in the light of Horrebow's known commitment to point out his teacher's contributions and of his statement that he himself found it, the naming of it for him must be regarded as justified.

Throughout his lifetime Horrebow was a fertile author. He wrote several textbooks on astronomy, mathematics, and navigation which had a considerable influence at the university and at Danish schools.

BIBLIOGRAPHY

I. ORIGINAL WORKS. Several of Horrebow's works have been collected in *Operum mathematico-physicorum,* 3 vols. (Copenhagen, 1740–1741); his *Copernicus triumphans* (reprinted in *Operum,* III) was translated into Dutch (Zutphen, 1741). See also his *Danske Skatkamer,* be-

staaende udi Grunden til Geometrien og Navigationen (Copenhagen, 1743–1746) and *Elementa philosophiae naturalis* (Copenhagen, 1748).

II. SECONDARY LITERATURE. Articles on Horrebow are *Dansk Biografisk Leksikon,* X (1936), 611–613; and J. Bernoulli, in *Recueil pour les astronomes,* supplement (Berlin, 1779), 62–71. See also C. A. F. Peters, "Recherches sur la parallaxe des étoiles fixes," in *Mémoires de l'Académie impériale des sciences de St.-Pétersbourg,* Sec. math.-phys., **5** (1848), 15–18; P. Kempf, "Ist man berechtigt, die Methode der Breitenbestimmung aus reziproken Höhen auf Römer zurückzufüren?" in *Astronomische Nachrichten,* **136** (1894), 11–14; and John E. McGrath, "A Question of Priority in Originating a Very Important Astronomical Method—Römer or Horrebow?" in *Journal of the Royal Astronomical Society of Canada,* **8** (1914), 36–40.

AXEL V. NIELSEN

HORROCKS, JEREMIAH (*b.* Lancashire, England, 1618; *d.* Toxteth Park, England, 13 January 1641), *astronomy.*

The precise date and place of Horrocks' birth are not known and the record of other biographical details is a meager one; there is good evidence, however, that his father may have been James Horrocks, a watchmaker, and his mother the former Mary Aspinwall. He grew up in Toxteth Park, then a small village about three miles from Liverpool. From 1632 to 1635 he attended Emmanuel College, Cambridge, working as a sizar for his maintenance, but he left without taking a degree. He taught himself astronomy and familiarized himself with the chief astronomical works of antiquity and of his own time.

Shortly after leaving Cambridge he befriended William Crabtree, a clothier or merchant of Broughton, near Manchester. Crabtree had studied astronomy for several years and the two young and enthusiastic friends carried on an extensive correspondence on astronomical matters that continued until Horrocks' death. Beginning in June 1639, Horrocks lived for about a year in Hoole, a village a few miles north of Liverpool, and then returned to Toxteth Park. He died suddenly, the day before an intended visit to Crabtree.

In his extraordinary and short-lived career Horrocks turned his attention to almost every aspect of astronomy. He was an assiduous and careful observer, always anxious to extend the limits of precision and to seek out and eliminate sources of possible observational error. One of his aims was to carry on the work of Tycho, but by utilizing the new opportunities available in the age of the telescope. He redetermined the astronomical constants for several planets, imaginatively investigated the problem of the scale of the solar system, improved the theory of lunar motion, began a detailed study of the tides, and theorized about the forces responsible for the motions of the planets.

As a theorist, Horrocks, although he was not in possession of the principle of inertia, represents a transition between the physical astronomy of Kepler and the fertile period 1660–1680 associated with the names of Borelli, Hooke, Halley, and Newton. His writings remained unpublished in his lifetime and the extent of his influence on his successors has yet to be explored.

In 1635 Horrocks began to compute ephemerides from Philip van Lansberge's *Tabulae motuum coelestium perpetuae* (1632). Comparing the results of his calculations with his own and Crabtree's observations, he concluded that Lansberge's tables were not only inadequate but also based on a false planetary theory. Upon Crabtree's advice he began to use Kepler's *Tabulae Rudolphinae* (1627) and soon became convinced that the tables were superior to all others and the only ones founded on valid principles. He devoted the next few years to correcting their errors and improving their accuracy.

Having some misgivings about Kepler's physical theories, Horrocks turned to the study of Kepler's works and soon became an ardent disciple. He accepted Kepler's doctrines of elliptical planetary orbits, with the sun situated in the orbital planes, and of the constant inclination of these orbits to the ecliptic. Horrocks affirmed that he had carefully and repeatedly tested Kepler's rule of the proportionality between the squares of the planetary periods and the cubes of their mean distances, and that he had found it to be absolutely true. With Kepler, he held that a planet moves more rapidly at perihelion than at aphelion and he believed planetary velocity to decrease proportionally with increasing distance from the sun. There is no mention in his surviving works of Kepler's law of areas.

Horrocks also accepted Kepler's viewpoint on the unity of celestial and terrestrial physics and his program for the creation of a celestial dynamics. He tentatively put forward a dynamical model of his own, however, which he felt eliminated some of the worst features of his master's. He started with Kepler's hypothesis that the sun moves the planets both by its rotation and by the emission of a quasi-magnetic attractive force, which becomes weaker with distance and attracts the planets as well as acting as a series of lever-arms pushing them along. The specific shape of the planetary orbit is the result of a dynamic equilibrium between a lateral (pushing) and a central force. Horrocks repudiated Kepler's idea that each planet has opposite sides "friendly" and "unfriendly" to the sun which cause it to be alternately attracted and repelled in different parts of its orbit and thus to move in an ellipse.

Possibly influenced by his reading of Galileo's *Dialogue Concerning the Two Chief World Systems,* Horrocks linked his celestial dynamics to the principles of falling bodies on earth and illustrated his conception by analogy with a pendulum. The planets may be seen as having a tendency to fall toward the sun or to oscillate about it freely, as the pendulum bob does about its mean position. But "Ye suns conversion doth turn the planet out of this line framing its motion into a circular, but the former desire of ye planet to move in a streight line hinders the full conquest of ye Sun, and forces it into an Ellipticke figure" (Manuscripts, Notebook B, fols. 16–17).

An analogy with a conical pendulum further illustrated his point. Horrocks pointed out that if a ball suspended by a string is withdrawn from its position at rest beneath the point of suspension, and given a tangential impulse, the ball will follow an elliptical path and its major axis will rotate in the direction of revolution—exactly as does the line of apsides of the lunar orbit. He further supposed a slight breeze blowing in the direction of the major axis, to support the analogy that the center of motion is in the focus of an ellipse rather than its center. According to Horrocks, therefore, and in contradistinction to Kepler, the planets tend always to be attracted to the sun and never to be repelled by it.

Horrocks' conception of gravitation and his theory of comets also differed somewhat from Kepler's. He hinted that the planets exert an attractive force on each other as well as on the sun; it is only because the sun is so massive compared to the other bodies in the solar system that it cannot be pulled from its place at the center. Originally, Horrocks proposed that comets are projected from the sun and tend to follow rectilinear paths. Like a stone thrown upward, they eventually reach a point of zero velocity and then return with accelerated motion; but since they are all the while influenced by the rotating force from the sun, they are thereby deflected into more or less circular paths. Horrocks later surmised that cometary orbits were elliptical.

In mathematical planetary astronomy, he carefully redetermined the apparent diameters of several celestial bodies, examined afresh the manner of calculating their parallaxes, and obtained improved elements for several orbits. For the horizontal solar parallax, Horrocks proposed a figure of 14″, which he arrived at by an ingenious and novel line of reasoning spiced with a dash of metaphysical speculation. It was a value not to be improved on for many years and vastly superior to Tycho's 3′ and Kepler's 59″, and even to Hevelius' 40″, a generation after Horrocks. He therefore obtained a figure for the radius of the earth's orbit of "at least . . . 15,000 semidiameters of the earth," or about 60,000,000 miles (*Transit of Venus Across the Sun,* p. 151). He reduced Kepler's estimate of the solar eccentricity, and subtracted 1′ from the roots of the sun's mean motion. Having discovered the irregularities in the motions of Jupiter and Saturn, he suggested specific corrections in the *Rudolphine Tables* for their mean longitudes and velocities, and he may have suspected that the increase in Jupiter's velocity and the decrease in Saturn's over a long span of time were periodic.

His program of correcting Kepler's tables led to Horrocks' prediction of a transit of Venus, and he became the first astronomer to observe one. Consulting the tables of Lansberge, and afterward those of Reinhold, Longomontanus, and Kepler, he learned that there would be a conjunction of Venus and the sun some time in early December 1639. The four tables differed from each other in this estimate, however, by as much as two days. Horrocks discovered a small constant error in Kepler's tables which displaced Venus about 8′ too much to the south, whereas Lansberge's erroneously elevated its latitude by a still greater amount. Correcting Kepler's error, Horrocks found that Venus would transit the lower part of the sun's disk on 4 December and wrote to Crabtree urging that they both make careful observations upon the expected date of conjunction.

Horrocks used a method of observation proposed for eclipses by Kepler and adapted to the telescope by Gassendi for the latter's observation of the transit of Mercury of 1631. The sun's light was admitted through a telescope into a darkened room so that the sun's disk was reproduced on a white screen to a diameter of almost six inches; the screen was divided along the solar circumference by degrees and along the solar diameter into 120 parts. Crabtree, observing near Manchester, saw the transit for only a few minutes and failed to record the data precisely, but his general observations proved to be in agreement with those made by his friend. Horrocks was more successful, and his analysis of his observations enabled him to correct earlier data for the planet.

Other astronomers had determined the apparent diameter of Venus as upwards of 3′, but Horrocks found it to be $1' 16'' \pm 4''$, quite close to the modern value. The transit observation also enabled him to redetermine the constants for Venus' orbit, yielding better figures for its radius, eccentricity, inclination to the ecliptic, and position of the nodes. As a result, he was also able to correct the figures for the rate of Venus' motion; he determined it to be slower by 18′ over 100 years than Kepler's tables showed.

His contributions to lunar theory, to which he turned his earnest attention in 1637, were among his most important. Following Kepler, he had as the physical cornerstone of his lunar theory the assumptions that the lunar orbit is elliptical and that many of the moon's inequalities are caused by the perturbative influence of the sun. In observation, he followed the practice initiated by Tycho of studying the moon in all its phases and not merely in the syzygies. Consequently, he was able to make improvements in the constants for several lunar inequalities, but his precepts were not reduced to tabular form until after his death. His most significant achievement in lunar theory was to account for the second inequality of longitude (evection, discovered in antiquity) by an unequal motion of the apsides and a variation in eccentricity. Depending on the moon's distance from the sun, he added to the mean position of the apogee or subtracted from it up to 12° and altered the eccentricity within a range just over 20 percent about its mean value.

Horrocks' lunar theory was first published in 1672. Tables constructed by Flamsteed were included in the edition of the following year. From observations made in 1672 and 1673, Flamsteed concluded that they were better than any then in print and Newton later proposed corrections which further improved their accuracy. Tables based on Horrocks' lunar theory continued in use up to the middle of the eighteenth century, when they were superseded by Mayer's.

Horrocks' papers remained with his family but a short time. Part of them were destroyed in the course of the English civil war, part were taken by a brother to Ireland and never seen thereafter, and still another portion was destroyed in the Great Fire of 1666. The remainder passed into the hands of an antiquary, who also managed to obtain letters by Horrocks from the Crabtree family. From the late 1650's until their eventual publication, Horrocks' manuscripts were widely circulated. The first part to be printed was his treatise on the transit, *Venus in sole visa,* which was published by Hevelius in 1662. The newly founded Royal Society assumed responsibility for publication of most of the remainder as *Opera posthuma* in 1672–1673.

BIBLIOGRAPHY

I. Original Works. Horrocks' surviving manuscripts are kept with Flamsteed's papers, vols. LXVIII and LXXVI, at the Royal Greenwich Observatory, Herstmonceux, Sussex. They are also available on film at the Public Record Office, London, and are briefly described in Francis Baily, *An Account of the Revd. John Flamsteed, the First Astronomer-Royal* (London, 1835), p. lxxiii. Horrocks' copy of Lansberge's *Tabulae perpetuae* with his corrections and marginalia is in Trinity College Library, Cambridge.

The principal published source for Horrocks' writings is his *Opera posthuma* (in some copies having the variant title *Opuscula astronomica*), John Wallis, ed. (London, 1672–1673; 1678), the text of which represents a conflation by Wallis of several treatises on the same subjects. Wallis also abridged Horrocks' letters to Crabtree and translated them into Latin. There is one important difference among the various editions. In that of 1672, the lunar theory was related in a letter of Horrocks to Crabtree dated 20 December 1638 (pp. 465 ff.) In all subsequent editions, this letter was replaced by Flamsteed's description of a letter from Crabtree to Gascoigne, 21 July 1642, explaining Horrocks' lunar theory. The *Venus in sole visa* was published with *Johannis Hevelii Mercurius in sole visus Gedani* (Danzig, 1662), pp. 111–145, from a version earlier than at least one of the texts now at Herstmonceux, together with notes by Hevelius. It has been published as *The Transit of Venus Across the Sun,* Arundell B. Whatton, trans. (London, 1859; 1868).

II. Secondary Literature. See Stephen P. and Stephen J. Rigaud, *Correspondence of Scientific Men of the Seventeenth Century* (Oxford, 1841), *passim,* especially the letters of Wallis and Flamsteed on Horrocks and Crabtree. The best recent work on Horrocks has been done by Sidney B. Gaythorpe in the following articles: "Horrocks's Observations and Contemporary Ephemerides," in *Journal of the British Astronomical Association,* **47** (1937), 156–157; "Horrocks's Observations of the Transit of Venus 1639 November 24 (O.S.)," *ibid.,* **47** (1936), 60–68, and **64** (1954), 309–315; "Jeremiah Horrocks and his 'New Theory of the Moon,'" *ibid.,* **67** (1957), 134–144; "Jeremiah Horrocks: Date of Birth, Parentage and Family Associations," in *Transactions of the Historic Society of Lancashire and Cheshire,* **106** (1954), 23–33; "On Horrocks's Treatment of the Evection and the Equation of the Centre . . .," in *Monthly Notices of the Royal Astronomical Society,* **85** (1925), 858–865. See also Betty M. Davis, *The Astronomical Work of Jeremiah Horrox,* University of London M. Sc. thesis (1967), and H. C. Plummer, "Jeremiah Horrocks and his Opera Posthuma," in *Notes and Records of the Royal Society of London,* **3** (1940), 39–52.

Among the still useful older accounts are John E. Bailey, *The Writings of Jeremiah Horrox and William Crabtree . . . Reprinted, with Additions, etc. from the* Palatine Notebook *of Dec. 1882, and Jan. 1883* (Manchester, 1883); Francis Baily, *Supplement to the Account of the Revd. John Flamsteed* (London, 1837), pp. 680–93; Jean Baptiste J. Delambre, *Histoire de l'astronomie moderne,* II (Paris, 1821), 495–514; Robert Grant, *History of Physical Astronomy from the Earliest Ages to the Middle of the Nineteenth Century* (London, 1852 [?]), pp. 420–428, 545; and Arundell B. Whatton, "Memoir of Jeremiah Horrox," the introduction to his translation *The Transit of Venus Across the Sun,* cited above, pp. 1–107.

Wilbur Applebaum

HORSFORD, EBEN NORTON (*b.* Moscow [now Livonia], New York, 27 July 1818; *d.* Cambridge, Massachusetts, 1 January 1893), *chemistry.*

Horsford had a strong Puritan background and a long New England ancestry; he was the son of Jedediah Horsford and Charity Maria Norton. His father migrated from Vermont to western New York state and combined farming with missionary activity among the Seneca Indians. In this rustic, frontier setting, Horsford early displayed an interest in nature. To cap a traditional education, he enrolled in 1837 in the Rensselaer Institute, where he studied with Amos Eaton.

For the next half dozen years following graduation Horsford attempted to apply what he had learned in a variety of occupations. He was employed on the newly established New York State Geological Survey under James Hall and was professor of mathematics and natural history at the Albany Female Academy. His interest turned to chemistry, and he experimented with the daguerreotype process in Albany. He also gave lectures in chemistry at Newark College in Delaware (later the University of Delaware), and he became a friend of John W. Webster, an early American chemist at the Harvard Medical School. Albany was then something of a scientific center and a group of friends, headed by Luther Tucker, publisher of the agricultural journal *The Cultivator,* persuaded Horsford to go to Germany to study chemistry under Liebig. In 1844 Horsford departed for Europe. He spent the next two years at Giessen under Liebig's immediate tutelage. The second American to study with him, Horsford was instrumental in the transfer of chemical skills and knowledge from Europe to America, as attested to by his many letters and journal.

On returning to America in 1847, Horsford found a ready and conspicuous outlet for his newly acquired talents when he was appointed, through the sponsorship of Webster, Rumford professor "for the application of science to the useful arts" at Harvard University. His peculiarly practical inclinations received further encouragement with the founding of the Lawrence Scientific School, to which, along with Louis Agassiz, he was promptly transferred. Horsford remained at Lawrence until his resignation in 1863, and it is here that he made his principal contributions to chemistry. On the Liebig model, Horsford developed the first laboratory in America for analytical chemistry. He became dean of the school and endeavored to establish its new scientific curricula on a sound and stable basis. He trained many men and he carried on his own practical and useful investigation in such varied fields as the use of lead pipes in Boston's water distribution, the condensation of milk, and the vulcanization of rubber.

His primary interest was in nutrition, in which he made his most promising and profitable discoveries. Probably motivated by a desire for material gain and encouraged by association with George Wilson, an industrialist who became his partner, Horsford developed a phosphatic baking powder to be used in place of yeast. On the basis of this and related products, the Rumford Chemical Company was established at what became known as Rumford, Rhode Island. The venture in industrial chemistry prospered and Horsford became rich and well known. He abandoned his academic career, but retained lifelong residence in the Harvard community at Cambridge.

The Civil War offered Horsford further opportunity to serve his country scientifically, and he approached both the Army and the Navy with proposals for the military application of chemistry. Chief among these was a compact, chemically determined ration of grain and meat, to be used by the army on the march. A trial manufacture of the ration was made under Horsford's direct supervision, but it was not successful. In 1873 he served as a United States commissioner to the Vienna Exposition and in 1876 as a juror at the Centennial Exposition in Philadelphia.

In his later years, Horsford returned to the study of Indian languages, which he had begun in his youth among the Senecas. He added to it a preoccupation with, and extensive research in, the Viking discoveries in America, on which he wrote numerous works. He was in addition an active and zealous patron of the newly founded Wellesley College.

BIBLIOGRAPHY

I. Original Works. A large part of Horsford's papers, including his letters, journals, scientific and business documents, is to be found in the Library Archives of Rensselaer Polytechnic Institute; many others are preserved in the family home, Sylvester Manor, Shelter Island, New York. For information on Horsford's life before his German training, a period on which little accurate information is available, see his letters in *Harvard College Papers,* 2nd ser., **13** (1845–1846), Harvard University Archives.

Horsford published many scientific papers in various journals, American and German, among them Silliman's *American Journal of Science* and *Proceedings of the American Association for the Advancement of Science.* Among his papers are "Untersuchungen über Glycocoll," in Liebig's *Annalen der Chemie und Pharmacie,* vol. **60** (1846), written while at Giessen; and "Value of Different Kinds of Vegetable Food, Based Upon the Amount of Nitrogen," in *Transactions of the Albany Institute* (1846). The articles of his most productive early years were privately assembled (1851) in a volume of "Original Papers," now in the Rensselaer archives. His later publications include *The Army Ration* (New York, 1864); *The Theory and Art of*

Bread-Making (Cambridge, 1861); and *A Report on Vienna Bread* (Washington, 1875).

II. SECONDARY LITERATURE. For brief biographical sketches, see *Dictionary of American Biography,* IX, 236–237, which is especially good; L. C. Newell and T. L. Davis, *Notable New England Chemists* (Boston, 1928), p. 16; H. S. van Klooster, "Liebig and His American Pupils," in *Journal of Chemical Education,* **33** (October 1956), 493 ff.; S. E. Morison, *Three Centuries of Harvard University* (Cambridge, 1930), 282, 414 ff.; and S. Rezneck, "Horsford's Marching Ration for the Civil War Army," in *Military Affairs,* **33** (1969), 249–255; and "The European Education of an American Chemist and Its Influence in Nineteenth Century America: Eben Norton Horsford," in *Technology and Culture,* **11** (1970), 366–388.

SAMUEL REZNECK

HORSLEY, VICTOR ALEXANDER HADEN (*b.* Kensington, London, England, 14 April 1857; *d.* Amara, near Baghdad, Mesopotamia [now Iraq], 16 July 1916), *neurosurgery, neurophysiology, pathology, social reform.*

Horsley was the son of John Callcott Horsley, R. A., a prominent artist, and Rosamund Haden, sister of Sir Francis Seymour Haden, the surgeon and etcher. He attended Cranbrook Grammar School and then University College Hospital Medical School (1875–1880). After qualification he spent four years in junior surgical posts and in 1884 he was made professor-superintendent of the Brown Institution, a center for human and animal physiological and pathological research. He resigned in 1890. At University College, London, he was appointed assistant professor of pathology (1882–1893) and later full professor (1893–1896). He was admitted to the Royal College of Surgeons of England in 1883 and to the Royal Society in 1886. He carried on private surgical practice and in 1885 he was appointed to the surgical staff of University College Hospital and in 1886 to that of the National Hospital for the Paralysed and Epileptic.

Horsley became professor of clinical surgery in 1899. For his contributions to medicine he was knighted in 1902. At the outbreak of war in 1914 he sought active service in the army and was eventually posted as consultant surgeon to the Mediterranean Expeditionary Force. He died of heat exhaustion, said to have been complicated by a gastrointestinal infection.

Horsley possessed outstanding intellect, creativity, inventiveness, and indefatigable and restless energy. His memory was exceptional; he was skillfully ambidextrous and he had well-developed qualities of leadership, which evoked admiration and devotion in those who worked with him. He had wide interests, both medical and social, and was an agnostic and Huxleyite. Although violent in expression and passionate in his convictions, he was also fastidious, generous, and humorous. He married Eldred, daughter of the engineer Sir Frederick Bramwell, in 1887; they had two sons and a daughter.

Horsley's many contributions to medicine fall in three general areas: experimental work, surgical innovation, and political and social reform.

His experimental work began with the study of thyroid physiology and pathology. In 1884, his experiments on the monkey led him to maintain that endemic cretinism, myxedema, and the results of surgical removal of the gland were the same condition. He thus initiated thyroid research in Britain and was the first to suggest replacement therapy in hypothyroidism using the transplant. He also carried out pioneer work on pituitary extirpation before the endocrinological function of the gland had been established, and he was one of the first to tackle a pituitary tumor surgically.

In 1886 Horsley confirmed Pasteur's discovery of a method to protect animals from rabies. In the same year he began research on localization of function in the brain. The investigations of Fritsch and Hitzig in 1870, on cerebral cortical function, had stimulated many to repeat and extend them. From 1886 to 1891 Horsley, with a series of collaborators, made important contributions to this area of neurophysiology, especially that concerning the motor cortex. From this research grew his interest in making precise experimental lesions in the deep parts of the brain, such as the cerebellum. He worked with R. H. Clarke and together they created the Horsley-Clarke stereotaxic apparatus, which only recently has become a popular and useful surgical technique in certain human disorders such as Parkinson's disease.

It was due to his demanding researches that Horsley gave up his professorship of surgery in 1906 and relinquished his charge of beds at University College Hospital. His studies were of greater importance to him than teaching or the care of general surgical cases. His post at the National Hospital and as private consultant allowed him adequate neurosurgical practice. He was one of the pioneers of brain surgery, which at the turn of the century was gradually developing into a specialty. Some have claimed that he was the most outstanding surgeon of his day, and his experimental work thoroughly prepared and fortified him for the task of advancing this new field. As in the laboratory, so in the operating theater, he was continually devising and conducting new experiments. In 1888 he published, with W. R. Gowers, an account of the first case of spinal tumor in which diagnosis led to removal and to recovery from paraplegia. His operation for trigeminal neuralgia was also an important advance.

Horsley tackled the problems of the British Medical Association with characteristic vigor and was one of the founders of its new constitution. He was equally active and outspoken in the various crusades he led; for example, in his support of temperance in alcohol, the necessity for animal experimentation, universal women's suffrage, government provision of free medical treatment for the workingman, and in his opposition to tobacco smoking. Horsley was also involved in national politics but his hatred for compromise, hypocrisy, and verbal diplomacies prevented him from gaining office. His crusade for better conditions for the wounded and sick in World War I cost him his life.

BIBLIOGRAPHY

I. Original Works. There is an unpublished bibliography of Horsley's publications (278 items) in the University of London Library, compiled by Cecilia E. Holder in 1949. The biography by Paget (see below, pp. 341–349) contains a list of 129 titles. Horsley's descendants possess most of his MSS, but some relating to his patients are in the University College Hospital Medical School Library. The University College Hospital Medical School Museum has remnants of the original Horsley-Clarke stereotaxic machine.

The following papers and books, arranged according to the order of the text, are Horsley's more important contributions: "The Brown Lectures," in *British Medical Journal* (1885), **1,** 111–115, 211–213, 419–423, on myxedema; "Note on a Possible Means of Arresting the Progress of Myxoedema, Cachexia Strumipriva, and Allied Diseases," *ibid.* (1890), **1,** 287–288, on transplantation; "Preliminary Note on Experimental Investigations on the Pituitary Body," *ibid.* (1911), **2,** 1150–1151, written with Dr. Handelsmann; and *Reports on the Outbreak of Rabies Among Deer in Richmond During the Years 1886–7* (London, 1888), written with A. C. Cope.

See also "A Record of Experiments Upon the Functions of the Cerebral Cortex," in *Philosophical Transactions of the Royal Society,* **179B** (1888), 1-45, written with E. A. Schäfer; "A Further Minute Analysis by Electrical Stimuli of the So-Called Motor-Region of the Cortex Cerebri in the Monkey (*Macacus sinicis*)," *ibid.,* pp. 205–256, written with C. E. Beevor; "On the Mammalian Nervous System, Its Functions, and Their Localisation Determined by an Electrical Method," *ibid.,* **182B** (1891), 267–526, written with F. Gotch; "On the Intrinsic Fibres of the Cerebellum, Its Nuclei and Its Efferent Tracts," in *Brain,* **28** (1905), 13–29, written with R. H. Clarke; "The Structure and Functions of the Cerebellum Examined by a New Method," *ibid.,* **31** (1908), 45–124, written with R. H. Clarke; "The Linacre Lecture on the Function of the So-Called Motor Area of the Brain," in *British Medical Journal* (1909), **2,** 125–132; *The Structure and Functions of the Brain and Spinal Cord, Being the Fullerian Lectures for 1891* (London, 1892); "A Case of Tumour of the Spinal Cord. Removal; Recovery," in *Transactions of the Medico-Chirurgical Society,* **71** (1888), 377–430, written with W. R. Gowers; "Mr. Victor Horsley and the General Medical Council," in *British Medical Journal* (1898), **1,** 225–226, and *passim;* and *Alcohol and the Human Body: an Introduction to the Study of the Subject* (London, 1907), written with Mary Sturge.

II. Secondary Literature. There are two biographies of Horsley: Stephen Paget, *Sir Victor Horsley: A Study of His Life and Work* (London, 1919), authoritative and the best source; and J. B. Lyons, *The Citizen Surgeon: a Life of Sir Victor Horsley F.R.S., F.R.C.S., 1857–1916* (London, 1966), well written but less reliable than Paget.

Other sources of information, in chronological order, are "Obituary. Sir Victor Horsley, C.B., F.R.S., M.B., F.R.C.S.," in *British Medical Journal* (1916), **2,** 162–167, with portrait; C. J. Bond, *Recollections of Student Life and Later Days. A Tribute to the Memory of the Late Sir Victor Horsley, F.R.S.* (London, 1939), the disjointed but revealing recollections of a very close friend; W. Haymaker and F. Schiller, eds., *The Founders of Neurology,* 2nd ed. (Springfield, Ill., 1970), pp. 562–566, with portrait; G. Jefferson, "Sir Victor Horsley, 1857-1916. Centenary Lecture," in *British Medical Journal* (1957), **1,** 903–910, a neurosurgeon's assessment; and A. MacNalty, "Sir Victor Horsley. His Life and Work," *ibid.,* pp. 910–916, a research colleague's opinion.

Edwin Clarke

HORSTMANN, AUGUST FRIEDRICH (*b.* Mannheim, Germany, 20 November 1842; *d.* Heidelberg, Germany, 8 October 1929), *physical chemistry.*

Horstmann stimulated the application of thermodynamics to chemical reactions when he showed that the Clausius-Clapeyron equation adequately explained the heats of dissociation of ammonium chloride upon sublimation.

He studied at the universities of Heidelberg, Zurich, and Bonn, receiving a doctorate from the university of Heidelberg in 1865. He became professor of theoretical chemistry at Heidelberg, where he remained until his death.

In 1869 Horstmann published "Dampfspannung und Verdampfungswarme des Salmiaks," which laid the basis for his contribution to theoretical chemistry. For most substances the three transitions—from solid to liquid, liquid to gas, and decomposition—occur at three different temperatures. But for sal ammoniac (ammonium chloride) the three transitions occur at the same temperature. Ammonium chloride sublimes and its molecules break up into ammonia and hydrogen chloride at the same time. Horstmann studied the vapor pressure and heat of disintegration of ammonium chloride. He attempted to determine experimentally if the vaporization is in any way dependent on pressure as is the evaporation of liquids. He found that the vapor pressure of ammonium chloride increases with temperature in the same manner as the vapor pressure of other liquids.

Horstmann then assumed that if ammonium chloride reacted to changes of pressure as did other liquids, he should be able to apply the Clausius-Clapeyron equation and calculate the heat of vaporization for ammonium chloride. He did this and found that, within experimental error, his calculated values were less than the experimental values by an amount equal to the heat of combination of ammonia and hydrogen chloride, showing that the two changes can be treated as the sum of two separate transitions.

He later extended his examination to include heats of dissociation of hydrates and carbonates.

BIBLIOGRAPHY

There is no collected works of Horstmann; his paper on the sublimation of ammonium chloride was published in *Bericht der Deutschen chemischen Gesellschaft,* **2** (1869), 137–140, repr. in no. 137 of Ostwald's Klassiker der Exakten Wissenschaften (Leipzig, 1903). For discussions of Horstmann's work, see Harry C. Jones, *The Theory of Electrolytic Dissociation and Some of Its Applications* (New York, 1900); and Wilhelm Ostwald, *Elektrochemie, ihre Geschichte und Lehre* (Leipzig, 1896).

OLLIN J. DRENNAN

HORTENSIUS, MARTINUS, also known as **Ortensius,** or **Van den Hove, Maarten** (*b.* Delft, Netherlands, 1605; *d.* Leiden, Netherlands, 7 August 1639), *astronomy.*

Hortensius' chief contributions were in the diffusion of Copernican astronomy and in his measurements of the angular size of the sun. The child of a man named Van Swaanswijk and a woman named Van den Hove, he studied mathematics with Beeckman and Snell and was a student at Leiden and Ghent from 1628 to 1630. During this period of studies, he most likely traveled to other countries, including Italy. He collaborated frequently with Philip van Lansberge in Middelburg, the Netherlands, and exchanged letters with Descartes, Mersenne, Gassendi, Huygens, and Galileo.

In 1634 Hortensius lectured on mathematics at the Amsterdam Atheneum, and in 1635 he became full professor there in the Copernican theory. He traveled often to Delft, Leiden, and The Hague, and later gave courses on nautical science, in which subject there was considerable interest. In 1638 he became a member of the commission that had to negotiate with Galileo on his method of longitude determination by observation of the satellites of Jupiter. In 1639 he was nominated professor at the Leiden university. He died shortly thereafter, leaving a natural son.

Hortensius was an autodidact in astronomy, first following Tycho, later giving serious consideration to the Copernican theory. He made observations on eclipses and on transits and endeavored to improve existing telescopes.

His findings concerning the angular diameter of the sun, mentioned in his preface to Lansberge's *Commentationes,* were vehemently criticized by Kepler in 1631, but were eventually vindicated by Hortensius in 1634. By using one of the primitive telescopes of the time and studying solar eclipses, he found the solar angular diameter to be 36′ at perigee and 33′34″ at apogee, with a ratio of 1.072. Kepler, on the other hand, believed that a telescope distorted the image and preferred to employ a small hole at the end of a long tube; he found a mean value of 30′ and a ratio 1.033 (actual values: 32′04″ and 1.034). This question was of great importance, because it was directly connected to the matter of the eccentricity of the earth's orbit, which Kepler had investigated by using observations of Mars. Kepler was vehement regarding this issue and used untenable arguments; Hortensius responded politely but suggested that Kepler might have altered the observational results in order to get agreement.

In the same preface to the *Commentationes,* Hortensius also criticized certain assertions of Tycho. Answers to his criticisms were given by Erasmus Bartholin and Longomontanus.

BIBLIOGRAPHY

I. ORIGINAL WORKS. See *Responsio ad additiunculum D. J. Kepleri praefixam Ephemeridi ejus in annum 1624* (Leiden, 1631), in which Kepler's criticisms are reproduced in full; and *Dissertatio de Mercurio in sole viso et Venere invisa* (Leiden, 1633). Hortensius' *Pleiadographia sive Pleiadum descriptio,* never published, was lost.

Hortensius also translated Philip van Lansberge, *Commentationes in motum terrae diurnum et annuum* (Middelburg, 1630), and Guil. Blaeu, *Institutio astronomica de usu globorum et sphaerarum coelestium ac terrestrium* (Amsterdam, 1634).

II. SECONDARY LITERATURE. Biographical data can be found in P. C. Molhuysen and P. J. Blok, *Nieuw Nederlandsch biografisch Woordenboek,* I (Leiden, 1911), cols.

1160–1164; and C. de Waard, *Journal tenu par Isaac Beeckman,* 4 vols. (The Hague, 1939–1953).

M. G. J. MINNAERT

HOSACK, DAVID (*b.* New York, New York, 31 August 1769; *d.* New York, 22 December 1835), *botany, medicine.*

Although Hosack's professional activities were important, his influence was more far reaching than his achievements. He was the eldest of six children of Alexander Hosack, a merchant from Elgin, Scotland, and Jane Arden Hosack, daughter of a Manhattan butcher. Educated at academies in Newark and Hackensack, he entered Columbia College in 1786 as a freshman but moved to the College of New Jersey (now Princeton University), from which he graduated with a B.A. in 1789. Following his medical studies with Nicholas Romayne in New York, Hosack studied under Benjamin Rush and Adam Kuhn at the University of Pennsylvania (M.D., 1791).

After a short medical practice in Alexandria, Virginia, he went to Edinburgh for "additional instruction." After nine months in Scotland he lived in the London area, focusing on botany. There he met William Curtis, Thomas Martyn (Regius professor of botany at Cambridge), George Pearson, and Sir Joseph Banks, and was elected a fellow of the Royal Society. He was especially favored by James Edward Smith, who presented him with duplicate specimens from the Linnaean herbarium (*cf.* Robbins, 1960) and proposed his election to the Linnean Society. When Hosack returned to the United States, he brought with him minerals later donated to Princeton University.

Hosack was the first in New York to operate for hydrocele by injection and the first American to tie the femoral artery for aneurysm. He opposed his medical colleagues on the origin and treatment of yellow fever and became a strong advocate of the contagion theory. From 1795, when he became professor of botany at Columbia College, a position he held, together with a subsequent post as professor of materia medica, until 1811, he was increasingly devoted to the development of a public botanic garden. In 1801 he founded in New York the twenty-acre Elgin Botanic Garden as a "repository of native plants, and as subservient to medicine, agriculture, and the arts." Foreign plants and seeds were received from European and West Indian correspondents. In 1811 the garden was sold to the state but was not maintained. The site, once beyond the city borders, is now marked by a plaque at Rockefeller Center. Hosack's plan to publish an "American Botany or a Flora of the United States" was also abortive but the botanical books he assembled passed to New York City's Bellevue Hospital, the founding of which (1820) he influenced.

With his protégé John W. Francis, Hosack founded the *American Medical and Philosophical Register,* which appeared in four volumes from 1810 to 1814 and in which most of Hosack's papers were reissued. Besides his classes in medicine and the writing of syllabi for them, he maintained a large practice and attended many notables, including Robert Fulton and Alexander Hamilton (he was attending surgeon at the Burr-Hamilton duel).

His fine library of four to five thousand volumes contained many presentation copies. Harriet Martineau; Joseph Sanson; David Douglas; Alexander Gordon; and Bernhard, duke of Saxe-Weimar-Eisenach, all mention Hosack's warm hospitality, particularly after he moved to Hyde Park. Gordon called him the Sir Joseph Banks of America.

Hosack married first Catherine Warner, who died in childbirth, then Mary Eddy of Philadelphia, who bore him six sons and three daughters. A third marriage, to a well-to-do widow, Magdalena Coster, in 1825, enabled him to entertain lavishly, to acquire the 700-acre estate of Samuel Bard at Hyde Park, and to establish the short-lived Rutgers Medical College.

BIBLIOGRAPHY

See Christine Chapman Robbins, "David Hosack's Herbarium and Its Linnaean Specimens," in *Proceedings of the American Philosophical Society,* **104** (1960), 293–313; and the fully documented biography "David Hosack. Citizen of New York," in *Memoirs of the American Philosophical Society,* **62** (1964), 1–246; a recent comprehensive list of publications, unpublished works, and correspondence is on pp. 212-240.

John W. Francis, Hosack's pupil, later professional colleague and lifelong friend, remarks upon him with great favor in Henry Tuckerman, ed., *Old New York, Reminiscences of the Past Sixty Years* (New York, 1865). The Francis papers are preserved in the New York Public Library, as is the MS diary of T. K. Wharton, containing numerous references to Hosack; see dates 28 July 1832, 30 March 1833, 11 Sept. 1839. Anna Murray Vail's list of 205 "Botanical Books of Dr. Hosack," in *Journal of New York Botanical Garden,* **1** (1900), 22–26, is supplemented in subsequent issues under "Library Accessions." E. J. McGuire, "The Elgin Botanic Garden and New York Literary Institution," in *United States Catholic Historical Society. Historical Records and Studies,* **4** (1906), 327–339, accounts for property sale.

JOSEPH EWAN

HOSEMANN. See **Osiander, Andreas.**

HOÜEL, GUILLAUME-JULES (*b.* Thaon, Calvados, France, 7 April 1823; *d.* Périers, near Caen, France, 14 June 1886), *mathematics, astronomy.*

Born into one of the older Protestant families of Normandy, Hoüel studied at Caen and the Collège Rollin before entering the École Normale Supérieure in 1843. He received his doctorate from the Sorbonne in 1855 for research in celestial mechanics and held the chair of pure mathematics at the Faculty of Sciences in Bordeaux from 1859 until his death.

Hoüel's reputation rests primarily on the quality and quantity of his activities in mathematical exposition. His gift for languages was used to evaluate and frequently to expound or translate important foreign mathematical writings. In the theory of complex numbers Hoüel introduced many of his countrymen to the researches of William R. Hamilton, Hermann Grassmann, Giusto Bellavitis, and Bernhard Riemann through his *Théorie élémentaire des quantités complexes* and other writings. Of greater importance were his successful efforts to overcome the long-standing failure of mathematicians to appreciate the significance of non-Euclidean geometry. Led by his own research to doubt the necessity of the parallel postulate and by Richard Baltzer to the writings of Lobachevski, Hoüel published in 1866 a translation of one of the latter's essays along with excerpts from the Gauss–Schumacher correspondence. By 1870 he had published translations of the classic writings in this area of János Bolyai, Beltrami, Helmholtz, and Riemann as well as his own proof of the impossibility of proving the parallel postulate. Hoüel also compiled logarithmic tables, worked on planetary perturbation theory, was an editor of the *Bulletin des sciences mathématiques et astronomiques,* and wrote a major text in analysis, *Cours de calcul infinitésimal.*

BIBLIOGRAPHY

I. Original Works. A bibliography of 131 items is given in Brunel (see below). His books include *Théorie élémentaire des quantités complexes* (Paris, 1874); and *Cours de calcul infinitésimal,* 4 vols. (Paris, 1878–1881).

II. Secondary Literature. Most useful is G. Brunel, "Notice sur l'influence scientifique de Guillaume-Jules Hoüel," in *Mémoires de la Société des sciences physiques et naturelles de Bordeaux,* 3rd ser., **4** (1888), 1–78. Obituary notices are *Leopoldina,* **22** (1886), 167–168; and G. Lespiault, in *Mémorial de l'Association des anciens élèves de l'Ecole normale supérieure* (Paris, 1887). See also Paul Barbarin, "La correspondance entre Hoüel et de Tilly," in *Bulletin des sciences mathématiques,* 2nd ser., **50** (1926), 50–64, 74–88.

Michael J. Crowe

HOUGH, GEORGE WASHINGTON (*b.* Tribes Hill, New York, 24 October 1836; *d.* Evanston, Illinois, 1 January 1909), *astronomy, meteorology.*

Hough's main contributions to astronomy were his discovery of 627 double stars and his floating island theory for the great red spot on the planet Jupiter, which planet he diligently observed for almost three decades. On the practical side he devised many instruments with astronomical and meteorological applications.

Born in the Mohawk Valley some thirty miles northwest of Albany, Hough was the son of Magdalene Selmser and William Hough, both of whom were descended from early German settlers. After attending schools in Waterloo and Seneca Falls (towns in the Finger Lakes region), Hough matriculated at Union College in Schenectady, where he received an M.A. degree in 1856. Two years as a school principal in Dubuque, Iowa, were followed by a year of graduate study at Harvard University. In 1859 he became assistant to O. M. Mitchel, an influential popularizer of astronomy who was then director of the Cincinnati Observatory. In 1860 Mitchel moved to Dudley Observatory in Albany, taking Hough with him; when Mitchel was recalled into the armed services (he died a major general in 1862) Hough succeeded him as director.

Hough married Emma C. Shear in 1870. He remained at Dudley until 1874, when he left to become a businessman. But in 1879 he returned to astronomy, accepting the directorship of Dearborn Observatory in Chicago. This institution was owned and run by the Chicago Astronomical Society, but located on the Douglas Park campus of the original University of Chicago (now defunct). Hough planned and supervised the removal of the observatory to the campus of Northwestern University in Evanston, Illinois, where it reopened in 1889; he continued as its director—also serving as professor of astronomy at Northwestern—until he died.

Hough began his observations of planets and double stars in Albany, but concurrently he was called upon to map star fields. To simplify the time-consuming (prephotographic) mapping techniques, he devised a machine that would print out color-coded dots on a chart directly from the telescope settings. Most of his mechanical skill at that time, however, went into meteorological instruments, such as a self-registering mercury barometer and a recording anemometer.

In 1869 Hough led Dudley's expedition to Mattoon, Illinois, to observe the total solar eclipse of 7 August. With a recording chronometer he had designed in 1865, he there obtained the first accurate timing (5.5 seconds) for the duration of "Baily's beads" preceding totality, and during the total phase of the eclipse he and other members of his party made telescopically what appears to have been the first daylight observation of meteors, presumably members of the Perseid shower.

At Dearborn, Hough began his systematic observations of Jupiter, using a micrometer to locate the various spots and bands. He also took up in earnest his search for new double stars, inspired by his association with S. W. Burnham.

In 1891 Hough received an honorary LL.D. from Union College, and in 1903 he was elected a foreign associate of the Royal Astronomical Society in London. He was also an honorary member of the Astronomische Gesellschaft in Leipzig and a corresponding member of the American Philosophical Society. His instruments won him many medals, notably at the Centennial Exposition in Philadelphia in 1876 and at the Chicago World's Fair in 1893.

BIBLIOGRAPHY

I. ORIGINAL WORKS. Hough's double-star observations were published as "Catalogue of 209 New Double Stars," in *Astronomische Nachrichten,* **116** (1887), cols. 273–304; "Catalogue of 94 New Double Stars and Measures of 107 Double Stars," *ibid.,* **125** (1890), cols. 1–32; "New Double Stars Discovered With the 18 1/2-Inch Refractor of the Dearborn Observatory, Evanston, Ill.," in *Astronomical Journal,* **9** (1890), 177–179; "Catalogue of 187 New Double Stars and Measures of 152 Double Stars," in *Astronomische Nachrichten,* **135** (1894), cols. 281–334; and "Catalogue of 132 New Double Stars and Measures of 255 Double Stars," *ibid.,* **149** (1899), cols. 65–124. For other double stars subsequently discovered by Hough, but not published by him, see Doolittle (below).

Hough's first publication on Jupiter was "On the Appearance of Jupiter, Aug. 20, 1867," in *Monthly Notices of the Royal Astronomical Society,* **27** (1867), 323. His first suggestion that the surface might be fluid appeared in *Annual Report of the Dearborn Observatory* (1881), p. 13, while his first reference to the great red spot as a floating island appeared *ibid.* (1882), p. 10. Further publications on Jupiter include "The Great Red Spot On Jupiter," in *Payne's Sidereal Messenger,* **4** (1885), 289–294; three papers with the same title, "Observations of the Spots and Markings on the Planet Jupiter, Made at the Dearborn Observatory, Northwestern University, Evanston, U.S.A.," in *Monthly Notices of the Royal Astronomical Society,* **52** (1892), 410–418; in *Astronomische Nachrichten,* **140** (1896), cols. 273–284, with plate facing col. 280; and in *Monthly Notices of the Royal Astronomical Society,* **60** (1900), 546–565; and two papers entitled "On the Determination of Longitude on the Planet Jupiter," *ibid.,* **64** (1904), 824–834, and **65** (1905), 682–687.

Many of the instruments that Hough designed (and built, usually with his own hands) are described in *Annals of the Dudley Observatory,* **1** (1866), and **2** (1871). Here also will be found tabulations and charts of his meteorological observations from 1865 through 1870. Two articles, both entitled "Description of a Printing Chronograph," in *Silliman's American Journal of Science and Arts,* 3rd ser., **2** (1871), 436–440, and in *Payne's Sidereal Messenger,* **5** (1886), 161–167, describe improvements to the original instrument of 1865. "Electrical Clock Connections for Operating the Chronograph," appeared in *Astronomy and Astro-Physics,* **13** (1894), 184–187. The electric drive Hough devised for the 18 1/2-inch Dearborn refracting telescope was described in "An Electric Control for the Equatorial," *ibid.,* 524–527, with illustration facing 521.

An account of Hough's solar eclipse expedition of 1869 appeared in *Annals of the Dudley Observatory,* **2** (1871), 296–323.

There are sixty-two entries under Hough's name in 4 vols. of the Royal Society of London, *Catalogue of Scientific Papers:* **3** (London, 1869), 446; **7** (London, 1877), 1020–1021; **10** (London, 1894), 277; and **15** (Cambridge, 1916), 952–953. This is a more complete list than appears in Poggendorff, III (Leipzig, 1898), 660, and IV (Leipzig, 1904), 667, but does not include four papers published after 1900 that are listed in Poggendorff, V (Leipzig, 1926), 560.

II. SECONDARY LITERATURE. Raymond Smith Dugan wrote the notice on Hough in *Dictionary of American Biography,* V, pt. 1 (1957), 252, which gives references to other sources of biographical information, including two obituaries by Hough's son, George Jacob Hough, in *Popular Astronomy,* **17** (1909), 197–200, with portrait facing 197, and in *Science,* n.s. **29** (1909), 690–693; and one by Thomas Lewis in *Monthly Notices of the Royal Astronomical Society,* **70** (1910), 302–304. See also Hough's entry in *American Men of Science,* I (1906), 155.

The early days at Dearborn Observatory and Hough's role in the move to Evanston are described by Philip Fox, "General Account of Dearborn Observatory," in *Annals of Dearborn Observatory, Northwestern University,* **1** (1915), 1–20.

For Hough's double-star observations, collected, annotated, and remeasured by Eric Doolittle, see *Publications of the University of Pennsylvania, Astronomical Series,* **3,** pt. 3 (1907), 1–176.

SALLY H. DIEKE

HOUGHTON, DOUGLASS (*b.* Troy, New York, 21 September 1809; *d.* Eagle River, Michigan, 13 October 1845), *medicine, geology.*

The son of Jacob Houghton, a lawyer originally from Massachusetts, Houghton grew up in Fredonia, near Lake Erie in western New York state. After attending Fredonia Academy, where he early showed a preference for science over the classics, Houghton studied medicine but in 1829 enrolled at the Rensselaer School in Troy, where he remained as an assistant after graduation. In 1830, on Amos Eaton's recommendation, Houghton was engaged to deliver courses of scientific lectures in Detroit, then a bustling frontier city.

Here Houghton settled down to a busy and varied career. Aside from his somewhat unusual scientific lecturing, he practiced medicine and engaged in profitable real estate enterprises. He served twice as mayor of Detroit and became one of the city's leading boosters. Somewhat mysteriously, Houghton returned to his earlier scientific interests in 1837, when he was named to two key positions in the recently created state of Michigan: he became the first state geologist and professor of geology, mineralogy, and chemistry at the newly established University of Michigan.

For the brief remainder of his life, Houghton carried on extensive surveys of the state, particularly in the mineral-rich Upper Peninsula. He put science to practical use on the frontier, and his reports became part of the record of American geological and geographical exploration. He won national honors and recognition for his work. At the age of thirty-six Houghton drowned in Lake Superior while on a survey. Thus ended prematurely the career of one whose combination of pragmatic and scientific qualities was peculiarly suited to early nineteenth-century America.

BIBLIOGRAPHY

I. Original Works. Best preserved are the Houghton Papers, Michigan Historical Collections, Ann Arbor, Mich.; Houghton letters and diary in the Detroit Public Library; and *Geological Reports of Douglass Houghton 1837–1845,* George N. Fuller, ed. (Lansing, Mich., 1928). In addition, Alvah Bradish, *Memoir of Douglass Houghton* (Detroit, 1889), has an appendix containing some letters and reports by Houghton as state geologist.

II. Secondary Literature. Aside from Bradish's *Memoir* (see above), there are a few brief writings about Houghton: Bela Hubbard, "Obituary: Douglass Houghton," in *American Journal of Science,* 2nd ser., **1** (1846), 150–152; and "A Memoir of Dr. Douglass Houghton," *ibid.,* **5** (1848), 217–227; Edsel K. Rintala, *Douglas Houghton, Michigan's Pioneer Geologist* (Detroit, 1954); and Helen Wallin, *Biographical Sketch of Douglass Houghton, Michigan's First State Geologist,* Michigan Geological Survey pamphlet 1 (Lansing, Mich., 1966; rev. ed., 1970). See also *Dictionary of American Biography,* IX, 254–255; and two MS articles in the archives of Rensselaer Polytechnic Institute: Franklin H. Morgan (Houghton's great-grandson), "Douglass Houghton, 1809–1845, Educator, Doctor, Geologist, Chemist, Botanist, Humanitarian, and Reluctant Politician," prepared for a commemorative meeting at Eagle River, Mich. (1 Aug. 1961); and Donald R. Hays, "Douglass Houghton, Michigan's First Chemist" (1966).

On Houghton's role in early American geology, see G. P. Merrill, *Contributions to a History of American State Geological and Natural History Surveys,* U.S. National Museum Bulletin 109 (1920), pp. 158–203; J. M. Nickles, *Geologic Literature in North America,* U.S. Geological Survey Bulletin 746 (1923), p. 529; and Alexander Winchell, "Douglass Houghton," in *American Geologist,* **4** (1889), 129–139.

Samuel Rezneck

HOWARD, LELAND OSSIAN (*b.* Rockford, Illinois, 11 June 1857; *d.* Bronxville, New York, 1 May 1950), *applied entomology.*

Howard was the son of Ossian Gregory Howard and Lucy Duham Thurber. When he was two years old, his parents moved to Ithaca, New York, where in 1873 he entered Cornell University, studying with John Henry Comstock, the eminent entomologist. In 1878 Comstock secured him a position as assistant to Charles Valentine Riley, chief of the Division of Insects in the United States Department of Agriculture and the foremost applied entomologist in the country. Upon Riley's resignation in 1894, Howard became chief of the division (later called a bureau), a position that he retained until 1927.

Howard shared in the burgeoning importance that applied entomology assumed at this time, when insects like the boll weevil (1894), the gypsy moth (1889), the San José scale (1893), and, at the turn of the century, the mosquito (in transmitting yellow fever and malaria) came to public notice. He traveled widely, visiting the bureau's field stations in the United States and keeping in touch with colleagues in Europe, as well as attending meetings and congresses. He received many honors and shared fully in the cultural and scientific life on both sides of the Atlantic.

Howard was permanent secretary of the American Association for the Advancement of Science from 1898 to 1920, was its president in 1920, and presided at the Fourth International Congress of Entomology in Ithaca, New York, in 1928.

Howard's administrative activities covered nearly every aspect of applied entomology, but he was particularly interested in biological control and medical entomology. He likewise studied the taxonomy of parasitic Hymenoptera, describing forty-seven new genera and 272 new species, as well as twenty-two new species of mosquitoes, the latter in collaboration with Harrison Gray Dyar and Frederick Knab.

Howard's contributions to scientific literature

totaled approximately 1,050. Among his more extensive and important publications were: *The Insect Book* (1901); *The House Fly* (1911); *Mosquitoes of North America* (4 vols., 1912–1917), with Dyar and Knab; *A History of Applied Entomology* (1930); *The Insect Menace* (1931); and *Fighting the Insects. The Story of an Entomologist* (1933).

Howard married Marie Theodora Clifton in 1886; they had three daughters. His wife died in 1926. A year later he retired, but continued to serve for four years as consultant on matters concerning biological control of insects. Under Howard's leadership, the entomological work of the Department of Agriculture grew from an annual budget of $30,000 to over $3,000,000. His organization of the entomological division endured for a quarter century after he retired, and today is still reflected to some degree in the entomological work of the department. During the latter years of his administration, Howard was probably the world's foremost entomologist.

BIBLIOGRAPHY

A sketch of Howard, of anonymous authorship, is "Leland Ossian Howard, 1857–1950," in *Journal of Economic Entomology,* **43,** no. 6 (1950), 958–962, with portrait. See also his autobiography, *Fighting the Insects—The Story of an Entomologist* (New York, 1933); and J. S. Wade, *et al.,* "Leland Ossian Howard, 1857–1950," in *Proceedings of the Entomological Society of Washington,* **52,** no. 5 (1950), 224–233, with portrait.

MELVILLE H. HATCH

HOWE, JAMES LEWIS (*b.* Newburyport, Massachusetts, 4 August 1859; *d.* Lexington, Virginia, 20 December 1955), *chemistry.*

The son of Francis Augustine Howe, a physician, and the former Mary Frances Lewis, Howe received the B.A. from Amherst College in 1880 and the M.A. and Ph.D. (1882) from the University of Göttingen. He was instructor of science at Brooks Military Academy, Cleveland, Ohio (1882–1883), professor of chemistry (later of physics and geology as well) at Central College, Richmond, Kentucky (1883–1894), and finally professor of chemistry and head of the department at Washington and Lee University (1894–1938). During World War II he was recalled from retirement to teach chemistry and German; he retired again in 1946. In 1883 he married Henrietta Leavenworth Marvine; they had two daughters and one son. In 1886 he received an honorary M.D. from the Hospital College of Medicine, Louisville, Kentucky, where he was professor of medical chemistry and toxicology.

Although regarded as the outstanding American authority on the platinum metals in general and an undisputed world authority on the chemistry of ruthenium in particular, Howe's magnum opus remains his *Bibliography of the Platinum Metals,* for which the American Chemical Society, Georgia Section, awarded him the Charles H. Herty Medal for the advancement of science in the southern states (1937). Aside from some miscellaneous research in organic, analytical, and inorganic chemistry, Howe's experimental work was confined to the compounds of the last-discovered and one of the least-known platinum metals—ruthenium—particularly its halide and cyanide complexes.

BIBLIOGRAPHY

I. ORIGINAL WORKS. Most of Howe's works appeared in *American Chemical Journal* and *Journal of the American Chemical Society.* His dissertation, "Über die Äthylderivate des Anhydrobenzdiamidobenzols und über ein Nitril desselben," based on research carried out under the direction of Hans Hübner, is one of his only three works in organic chemistry: the others are "A Nitrile of Anhydro-Benzdiamido-Benzene," in *American Chemical Journal,* **5** (1883), 415–418; and "The Ethyl Derivatives of Anhydro-Benzdiamido-Benzene," *ibid.,* pp. 418–424.

His major work appeared in several parts as "Bibliography of the Metals of the Platinum Group: Platinum, Palladium, Iridium, Rhodium, Osmium, Ruthenium, 1748–1896," in *Smithsonian Miscellaneous Collections,* vol. 38, no. 1084 (1897); "Bibliography of the Metals of the Platinum Group: Platinum, Palladium, Iridium, Rhodium, Osmium, Ruthenium, 1748–1917," in *Bulletin of the U.S. Geological Survey,* no. 694 (1919), compiled with H. C. Holtz; *Bibliography of the Platinum Metals 1918–1930* (Newark, N.J., 1947), compiled with the staff of Baker and Co.; *Bibliography of the Platinum Metals 1931–1940* (Newark, N.J., 1949); and *Bibliography of the Platinum Metals 1941–1950* (Newark, N.J., 1956).

II. SECONDARY LITERATURE. A discussion of Howe's life and work is G. B. Kauffman, "James Lewis Howe: Platinum Metal Pioneer," in *Journal of Chemical Education,* **45** (1968), 804–811.

GEORGE B. KAUFFMAN

HÖWELCKE, JOHANN. See **Hevelius, Johannes.**

HOWELL, WILLIAM HENRY (*b.* Baltimore, Maryland, 20 February 1860; *d.* Baltimore, 6 February 1945), *physiology.*

William Henry Howell was the son of George Henry Howell and Virginia Teresa Magruder. His family on both sides had lived in southern Maryland since early colonial times, and the Magruders owned large farms in Prince Georges County, where he and his three brothers and one sister spent their summers. He was educated in the public schools of Baltimore and in 1876 entered the Johns Hopkins University as an undergraduate, earning his A.B. in 1881 and Ph.D. in 1884. During these years Howell studied and instructed with H. Newell Martin, a noted British physiologist. His dissertation, entitled "The Origin of the Fibrin Formed in the Coagulation of Blood," was the forerunner of that research in his later years with which he made his greatest contributions to science and medicine.

In the ensuing nine years Howell taught physiology at Johns Hopkins (associate professor, 1889), the University of Michigan (professor, 1889–1892), and Harvard (associate professor, 1892). In 1893 he was recalled to Baltimore to be the first professor of physiology in the new Johns Hopkins Medical School. He served also as dean of the Medical Faculty from 1899 to 1911. In 1917, with William H. Welch, he organized the School of Hygiene and Public Health and became its director from 1926 to 1931. For three years thereafter he was chairman of the National Research Council, then retired to his laboratory at Johns Hopkins to continue his research until two days before his sudden death.

Howell was internationally known in the early years of the twentieth century as America's outstanding physiologist. At the age of twenty-seven he had been one of the founders of the American Physiological Society, and from 1905 to 1910 he served as its president. He attended many physiological congresses in Europe as the American representative on the International Committee of Physiologists and was elected to preside at the first International Physiological Congress in the United States in 1929. He received many honorary degrees both in the United States and abroad, including an M.D. from the University of Michigan and an LL.D. from the University of Edinburgh. He was elected a member of the American Philosophical Society and the National Academy of Sciences, and an honorary member of the London Physiological Society.

Howell's early contributions to physiology dealt with the circulatory system, nerve tissue, and the components of the blood. In his publications the laboratory techniques are presented in clear and meticulous detail, and the conclusions are stated with care and clarity. These attributes of his research—patience, precision, and clarity—remained characteristic of all his work, his teaching, and his writing. During his years at the Johns Hopkins Medical School Howell returned to studies of the coagulation of the blood. He was able to isolate thrombin (1910) and gave careful directions for its preparation. In 1918 he discovered the anticoagulant heparin, which he prepared from the liver and later attempted to analyze chemically. In his last years of research he proved the theory that blood platelets are formed in the lungs and was able to isolate thromboplastin in a form pure enough to be used *in vivo.* During thirty years of his work he was assisted in the laboratory by one or more members of a family of hemophiliacs, who were always loyal to his studies.

In the field of teaching Howell's best-known contribution was his *Textbook of Physiology,* which was first published in 1905 and went through fourteen editions. To two generations of medical students the textbooks presented physiology with the clarity, simplicity, and charm that characterized all his writing. Perhaps most delightful for the reader are Howell's special lectures: "The Cause of the Heart Beat" (Harvey Lecture, 1906), "The Coagulation of the Blood" (Harvey Lecture, 1916), "The Problem of Coagulation" (Pasteur Lecture, 1925), and "Hemophilia" (Carpenter Lecture, 1939).

Howell was a dedicated and able administrator, and many of his speeches contain very bold and thoughtful suggestions concerning the premedical and medical curricula of the day. He made a strong plea in 1912 for standardization of medical education throughout the country. Many of these ideas have since been adopted.

As a person, Howell was softspoken and devoted to his wife, Anne Janet Tucker, whom he married in 1887, his son and two daughters, and his eight grandchildren. He was an excellent tennis player, a good golfer, and an avid sailor. He was warmly admired by colleagues, students, and friends, and his career spanned the period in which American medicine, in a really modern sense, came of age.

BIBLIOGRAPHY

I. Original Works. Howell's numerous writings include "The Origin of the Fibrin Formed in the Coagulation of Blood," in *Studies From the Biological Laboratory, Johns Hopkins University,* **3** (1884), 63-71; "A Physiological, Histological, and Clinical Study of the Degeneration and Regeneration in Peripheral Nerve Fibres After Severance of Their Connection With the Nerve Centers," in *Journal of Physiology,* **13** (1892), 335–406, written with G. C. Huber; "An Analysis of the Influence of the Sodium, Potassium, and Calcium Salts of the Blood on the Automatic Contraction of the Heart Muscle," in *American Journal of Physiology,* **6** (1901), 181–206; *Textbook of Physiology for Medical Students and Physicians,* 14 eds. (Philadelphia, 1905–1940); "The Cause of the Heart Beat," in *Journal of the American Medical Association,* **46** (1906), 1665, 1749, the Harvey Lecture; "The Coagulation of Blood," in *Cleveland Medical Journal,* **9** (1910), 118; "The Preparation and Properties of Thrombin Together With Observations on Antithrombin and Prothrombin," in *American Journal of Physiology,* **26** (1910), 453–473; "The Condition of the Blood in Hemophilia, Thrombosis, and Purpura," in *Archives of Internal Medicine,* **13** (1914), 76–95; "Prothrombin," in *American Journal of Physiology,* **35** (1914), 474–482; "The Coagulation of Blood," in *The Harvey Lectures,* Series 12 (1916–1917), 273–324; "Two New Factors in Blood Coagulation, Heparin and Proantithrombin," in *American Journal of Physiology,* **47** (1918), 328–341, written with E. Holt; "The Problem of Coagulation," in *Proceedings of the Institute of Medicine of Chicago* (1925), Pasteur Lecture (reprint); "The Purification of Heparin and Its Presence in Blood," in *American Journal of Physiology,* **71** (1926), 553–562; "The Purification of Heparin and Its Chemical and Physiological Reactions," in *Bulletin of the Johns Hopkins Hospital,* **42** (1928), 199–206; "The Production of Blood Platelets in the Lungs," in *Journal of Experimental Medicine,* **65** (1937), 177–203, written with D. D. Donahue; "The American Physiological Society During Its First Twenty-Five Years," in *History of the American Physiological Society Semicentennial, 1887–1937* (1938), p. 1; "Hemophilia," in *Bulletin of the New York Academy of Medicine,* 2nd ser., **15,** no. 1 (1939), 3–26, the Wesley M. Carpenter Lecture; "The Isolation of Thromboplastin From Lung Tissue," in *Bulletin of the Johns Hopkins Hospital,* **76,** no. 6 (1945), 295–301.

II. SECONDARY LITERATURE. See "The Celebration of the Sixtieth Anniversary of Dr. William H. Howell's Graduation From the Johns Hopkins University," in *Bulletin of the Johns Hopkins Hospital,* **68,** no. 4 (Apr. 1941), 291–308; and "An Anniversary Tribute to the Memory of the Late William Henry Howell," *ibid.,* **109,** no. 1 (July 1961), 1–19.

ANNE CLARK RODMAN

HRDLIČKA, ALEŠ (*b.* Humpolec, Bohemia [now Czechoslovakia], 29 March 1869; *d.* Washington, D.C., 5 September 1943), *physical anthropology.*

Hrdlička was the son of Maxmilian Hrdlička, a joiner who immigrated to New York and became a factory worker, and Karolina Wajnerová. The oldest of seven children, Hrdlička went to work with his father at an early age, since the family's financial circumstances did not permit him to attend the Gymnasium. After coming to America, he worked as a laborer, but simultaneously attended the evening courses that gained him a high-school equivalency diploma. A serious illness led him to decide to study medicine and he enrolled in the New York Eclectic College, from which he graduated in 1892. In 1894 he completed further training at the New York Homeopathic College and was certified by the Maryland Allopathic Board.

Hrdlička practiced for a short time at the state hospital for the insane in Middletown, Connecticut; he left there in 1896 to study anthropology with L. P. Manouvrier in Paris. He returned to the United States in the same year and became associate in anthropology in the New York Pathological Institute, a position that he held until 1899. In the latter year, Hrdlička took charge of physical anthropology for expeditions sponsored by the American Museum of Natural History. From 1903 he was assistant curator of the physical anthropological collections at the Smithsonian Institution in Washington and from 1910 curator; in this connection he traveled extensively and personally examined many of the sites where *Pithecanthropus* had been found, as well as the sites of contemporaneous Paleolithic man. Among his wide range of physical anthropological concerns, he became an expert on the Eskimos and Indians of North America and the Indians of Central America and on the problem of the origin of human races.

In 1918 Hrdlička founded the American Association of Physical Anthropologists and its organ, *American Journal of Physical Anthropology.* He was as active in Czechoslovakian anthropological affairs, raising money for the journal *Anthropologie* (published between 1923 and 1941) at Charles University in Prague and for anatomical and anthropological institutes, as well as for the Museum of Man that is now named in his honor.

Hrdlička published the first of his major theories in "The Neanderthal Phase of Man" (*Journal of the Royal Anthropological Institute,* **57** [1927], 249–274). In this study he sought to prove that *Homo sapiens* had developed from *Homo neanderthalensis* and to show that all human races had a common origin. He presented supporting arguments drawn from anthropology, anatomy, and paleology. This work brought him the Huxley Medal of the Royal Anthropological Institute.

Hrdlička implemented this work with "The Skeletal Remains of Early Man" (*Smithsonian Miscellaneous Collections,* **83** [1930]). On the basis of his personal investigation of almost all the world sites in which *Homo neanderthalensis* had been found and of the fossils of *Homo sapiens,* Hrdlička concluded that mankind could have developed only in the Old World, since the narrow-nosed apes from which the anthropogenic series had originated were not to be found anywhere else.

Beginning in 1927 Hrdlička organized regular expeditions to Alaska and the Bering Strait. He conducted research on the contemporary population of these regions, as well as on human skeletal remains. Drawing upon ethnography, paleology, and linguistics, he formulated the theory (elucidated in *The Question of Ancient Man in America* [1937]) that America had been peopled from Asia, via the Bering Strait. He held the hypothesis that men had migrated from Kamchatka, either in primitive boats by way of the Aleutian and Komandorski Islands, or by foot across the Bering Strait itself (since the strait averages about fifty miles across, and freezes in particularly severe winters). From Alaska, then, this early population spread along the Pacific coast and large river valleys, gradually diffusing over all of North, Central, and South America.

Hrdlička was a member of all American anthropological societies and of many foreign ones. He lectured to a variety of audiences, published many scientific papers, and trained a number of subsequent workers. He died of a heart attack while preparing a new expedition to study the Indians of Mexico.

BIBLIOGRAPHY

Hrdlička's major works before 1938 include *Anthropological Investigations of One Thousand White and Colored Children of Both Sexes, the Inmates of the N. Y. Juvenile Asylum* (New York-Albany, 1900); "Divisions of the Parietal Bone in Man and Other Mammals," in *Bulletin of the American Museum of Natural History,* **19** (1903), 231–386; "Brain Weight in Vertebrates," in *Smithsonian Miscellaneous Collections,* **48** (1905), 89–112; "Contribution to the Anthropology of Central and Smith Sound Eskimo," in *Anthropological Papers of the American Museum of Natural History,* **5** (1910), 175–280; "Early Man in South America," in *Bulletin of the Bureau of American Ethnology,* **52** (1912), 1–405; written with W. H. Holmes, B. Willis, F. E. Wright, and C. N. Fenner; "The Natives of Kharga Oasis, Egypt," in *Smithsonian Miscellaneous Collections,* **59** (1912), 1–118; "The Most Ancient Skeletal Remains of Man," in *Smithsonian Report for 1913* (Washington, D.C., 1914), 491–522; "Physical Anthropology of the Lenape or Delawares and of the Eastern Indians in General," in *Bulletin of the Bureau of American Ethnology,* **62** (1916), 1–130; "Early Man in South America," *ibid.,* **66** (1918), 1–405; *Physical Anthropology; Its Scope and Aims; Its History and Present Status in America* (Philadelphia, 1919); *Anthropometry* (Philadelphia, 1920; 2nd ed., 1938); *The Old Americans. A Scientific Detailed Study of the Fathers of America and Their Children* (Baltimore, 1925); "Catalogue of Human Crania in the U.S.," in *Proceedings of the United States National Museum,* **69** (1927), 1–127; **71** (1928), 1–140; **78** (1931), 1–95; "The Neanderthal Phase of Man," in *Journal of the Royal Anthropological Institute,* **57** (1927), 249–274; "The Skeletal Remains of Early Man," in *Smithsonian Miscellaneous Collections,* **83** (1930), 1–379; "Anthropological Survey in Alaska," in *Annual Report of the Bureau of American Ethnology,* **46** (1930), 1–374; "The Humerus: Septal Apertures," in *Anthropologie* (Prague), **10** (1932), 31–96; "Ear Exostoses," in *Smithsonian Miscellaneous Collections,* **93** (1935), 1–98; and "The Pueblos, With Comparative Data on the Bulk of the Tribes of the Southwest and Northern Mexico," in *American Journal of Physical Anthropology,* **20** (1935), 235–460.

For his work after 1938, see the index to Smithsonian Institution publications. From 1918 Hrdlička edited the *American Journal of Physical Anthropology.*

K. HAJNIŠ

HUBBLE, EDWIN POWELL (*b.* Marshfield, Missouri, 20 November 1889; *d.* San Marino, California, 28 September 1953), *observational astronomy, cosmology.*

Hubble was the founder of modern extragalactic astronomy and the first to provide observational evidence for the expansion of the universe. The son of John Powell Hubble, a lawyer, and the former Virginia Lee James, he spent his early years in Kentucky and attended high school in Chicago, where his father was in the insurance business. At school he excelled both in his studies and in athletics. He won a scholarship to the University of Chicago, where he came under the influence of the eminent physicist R. A. Millikan and of the astronomer G. E. Hale, who inspired in him a love of astronomy. Hubble received a B.S. in mathematics and astronomy and also made his mark on the campus as a heavyweight boxer (he was six feet, two inches tall). A sports promoter wanted to train him to fight Jack Johnson, the world champion, but instead Hubble went to Queen's College, Oxford, in 1910 as a Rhodes scholar from Illinois.

At Oxford, Hubble first thought of reading mathematics; but after studying some of the final examination papers, he concluded that they were too specialized for his liking and instead decided to read jurisprudence. He took his B.A. in that subject in 1912. Hubble had a great love of England and was interested in the common law of the country from which his ancestors had emigrated in the seventeenth century. While at Oxford he was awarded a blue for track events and boxed in an exhibition match with the French champion, Georges Carpentier.

In 1913 Hubble returned to the United States, was admitted to the bar, and opened a law office at Louisville, Kentucky. After a short while he abandoned this career and in 1914 went to the Yerkes Observatory of the University of Chicago, where he was an assistant and a graduate student under E. B. Frost. He was awarded the Ph.D. in 1917 for a thesis entitled "Photographic Investigations of Faint Nebulae," in which he considered the classification of nebular types and concluded that planetary nebulae are probably within our sidereal system and the great spirals outside; but these questions, he said, could be decided only by instruments more powerful than those currently available.

Hubble's powers as an observer attracted the attention of Hale during a visit to Yerkes; Hale offered him a post at the Mount Wilson Observatory, where the sixty-inch reflector was then in operation and the 100-inch under construction. Meanwhile the United States had entered World War I, and Hubble had immediately enlisted as a private in the infantry. He therefore telegraphed Hale that he would accept his offer as soon as he was demobilized. He served with the American Expeditionary Force in France and rose to the rank of major. After the Armistice he remained with the American Army of Occupation in Germany until the autumn of 1919. On his return to the United States in October, he joined Hale on Mount Wilson, as he had promised. At last, at the age of thirty, he settled down to the work that was to bring him fame.

Hubble's earliest investigations at Mount Wilson were made with the sixty-inch telescope and concerned galactic nebulae. In one of his earliest papers, "A General Study of Diffuse Galactic Nebulae," he suggested a classification system based upon fundamental differences between galactic and nongalactic nebulae. He discovered many new planetary nebulae and variable stars, but the most important result of his early researches concerned the origin of the radiation from diffuse galactic nebulae. Hubble showed that they were made luminous by certain stars associated with them, the nebulosity consisting of clouds of atoms and dust not hot enough to be self-luminous. He discovered a relation between the luminosity of a diffuse galactic nebula and the magnitudes of the associated stars and showed that the gases were excited and made luminous by neighboring blue stars of high surface temperature.

The Hooker 100-inch telescope came into operational use at about the time Hubble arrived on Mount Wilson. This was a most fortunate circumstance, for the crucial contributions made to cosmology by Hubble required the full light-gathering power and resolution of this instrument. From about 1922 he turned his attention more and more to objects that we now regard as lying beyond our own stellar system.

Hubble's first great discovery was made when he recognized a Cepheid variable star in the outer regions of Messier 31, the great nebula in Andromeda, in a plate that he took on 5 October 1923. This proved to be the long-sought means of settling the problem of the status of the spiral nebulae that had puzzled astronomers for three-quarters of a century. The use of Cepheid variable stars as distance indicators had been suggested more than ten years earlier by Henrietta Leavitt of the Harvard College Observatory, and they had been used with great effect by Harlow Shapley to determine the distances and dimensions of the globular star clusters that surround the Milky Way. Hubble's discovery was the first sure indication that the Andromeda nebula lies far outside our own stellar system.

Controversy on this question had previously culminated in the famous Shapley-Curtis debate held before the National Academy of Sciences on 26 April 1920, neither side convincing the other. Curtis had argued that "the spirals are not intragalactic objects but island universes, like our own galaxy, and that the spirals, as external galaxies, indicate to us a greater universe into which we may penetrate to distances of ten million to a hundred million light-years." Shapley rejected this conclusion. He maintained that there was no reason "for modifying the tentative hypothesis that the spirals are not composed of typical stars at all, but are truly nebulous objects." The strongest argument for this view was evidence obtained by Adriaan van Maanen that Messier 101 rotated through 0.02 seconds of arc in a year and that Messier 33 and 81 rotated at comparable rates. These large angular velocities implied relativly small distances, of the order of a few thousand light-years. (The spurious nature of van Maanen's measurements was finally established in 1935, when it was conclusively shown by Hubble that they arose from obscure systematic errors and did not indicate motion in the nebulae concerned.)

By the end of 1924 Hubble had found thirty-six variable stars in M 31, twelve of which were Cepheids. From the latter he derived a distance of the order of 285,000 parsecs, or about 900,000 light-years, whereas the maximum diameter of the Milky Way stellar system was known to be in the order of 100,000 light-years. The public announcement of Hubble's discovery was made at a meeting of the American Astronomical Society in Washington, D.C., at the end of December 1924. Hubble was not present; but Joel Stebbins recalled many years later that when Hubble's paper had been read, the entire Society knew that the debate had come to an end, that the island-universe concept of the distribution of matter in space had been established, and that an era of enlightenment in cosmology had begun. Both Shapley and Curtis were present.

The way was now open for a new attack on the cosmological problem which had hitherto been the concern of theoretical investigators. Two lines of research were possible for the observer to pursue, and Hubble was a pioneer in both. On the one hand, he studied the contents and general structure of galaxies. On the other, he investigated their distribution in space and their motion. Both approaches were strongly motivated by his belief that galaxies are the structural units of matter that together constitute the astronomical universe as a whole.

Hubble was the first to introduce a significant classification system for galaxies. He presented this at the meeting of the International Astronomical Union at Cambridge, England, in 1925 and it was published the next year in the *Astrophysical Journal.* This system is the basis of the classification still used. Hubble found that most galaxies showed evidence of rotational symmetry about a dominating central nucleus, although a minority, amounting to not more than 3 percent of those he studied, lacked both these features. He called the two types "regular" and "irregular," respectively. He found that the regular galaxies fell into two main classes—"spirals" and "ellipticals"—and that each class contained a regular sequence of forms. One end of the elliptical sequence was found to be similar to one end of the spiral sequence. The spirals were subdivided into two parallel subsequences, normal and barred. The classification was essentially empirical and independent of any assumptions concerning the evolution of galaxies.

In addition to studying the shapes of galaxies, Hubble explored their contents and brightness patterns. In the nearer galaxies he discovered and studied almost every kind of intrinsically bright object known in our own system: novae, globular clusters, gaseous nebulae, super-giant blue stars, red long-period variables, Cepheids, and so on.

Despite the advance in knowledge in the last forty years, Hubble's claim to have introduced order into the apparent confusion of nebular forms and to have shown that galaxies are closely related members of a single family stands. It must be regarded as one of his most significant achievements.

During the late 1920's Hubble's main preoccupation was to determine a reliable extragalactic distance scale to the limits of observation. This was the essential preliminary to any serious investigation of the distribution of galaxies in space and its bearing on the cosmological problem. The philosophy underlying his approach to this problem had previously been summarized by him in his first detailed paper on an extragalactic system (NGC 6822), the distance of which was obtained by the Cepheid criterion. Hubble's use of the Cepheid period-luminosity law (which enabled him to regard these stars as distance indicators) was based on an appeal to the principle of the uniformity of nature. "This principle," he wrote, "is the fundamental assumption in all extrapolations beyond the limits of known and observable data, and speculations which follow its guide are legitimate until they become self-contradictory."

On this basis, Hubble proceeded to estimate the distances of galaxies beyond the "local group" in which Cepheids could be detected with the 100-inch telescope. He argued that with increasing distance one could expect the Cepheids to fade out first, then the irregular variables, then the blue giants, until only the very brightest of stars would be seen. He found that the data, although somewhat meager, indicated that the very brightest stars in late-type spirals are of about the same absolute luminosity. This upper limit of stellar luminosity appeared to be about 50,000 times that of the sun. The "brightest star" criterion of distance enabled Hubble to extend the extragalactic distance scale to about 6,000,000 light-years. In view of the criticism to which this criterion has been subjected since Hubble's day, it should be noted that he was fully aware that a risk was involved in regarding the images in question as individual stars; but he pointed out that, regardless of their real nature, the objects selected as brightest stars appeared to represent strictly comparable bodies. (In 1958 Allan Sandage showed that they are bright clouds of ionized hydrogen.)

To extend the distance scale farther, Hubble used information gained from the fact that stars could be detected in some of the spirals in the great Virgo cluster. Analysis of this large sample collection provided average characteristics of galaxies which could be used as statistical criteria of distance for more remote galaxies. For measurements of the depths of space, Hubble concentrated on the brightest members of clusters of galaxies. He regarded the clusters as so similar that the mean luminosity of the ten brightest members or even the individual luminosity of, say, the fifth-brightest member formed a convenient measure of distance. In this way he built up his distance scale to 250 million light-years.

By 1929 Hubble had obtained distances for eighteen isolated galaxies and for four members of the Virgo cluster. In that year he used this somewhat restricted body of data to make the most remarkable of all his discoveries and the one that made his name famous far beyond the ranks of professional astronomers. This was what is now known as Hubble's law of proportionality of distance and radial velocity of galaxies. Since 1912, when V. M. Slipher at the Lowell Observatory had measured the radial velocity of a galaxy (M 31) for the first time by observing the

Doppler displacement of its spectral lines, velocities had been obtained of some forty-six galaxies, forty-one by Slipher himself. Attempts to correlate these velocities with other properties of the galaxies concerned, in particular their apparent diameters, had been made by Carl Wirtz, Lundmark, and others; but no definite, generally acceptable result had been obtained. In 1917 W. de Sitter had constructed, on the basis of Einstein's cosmological equations, an ideal world-model (of vanishingly small average density) which predicted red shifts, indicative of recessional motion, in distant light sources; but no such systematic effect seemed to emerge from the empirical data. Hubble's new approach to the problem, based on his determinations of distance, clarified an obscure situation. For distances out to about 6,000,000 light-years he obtained a good approximation to a straight line in the graphical plot of velocity against distance. Owing to the tendency of individual proper motions to mask the systematic effect in the case of the nearer galaxies, Hubble's straight-line graph depended essentially on the data obtained from galaxies in the Virgo cluster. These indicated that over the observed range of distance, velocities increased at the rate of roughly 100 miles a second for every million light-years of distance (500 kilometers a second for every million parsecs).

Further progress depended on the extension of the observations to greater distances and fainter galaxies. The spectroscopic part of the work was undertaken by Milton L. Humason, Hubble's colleague at Mount Wilson. Within two years, with the aid of a new type of fast lens suitable for the difficult task of photographing the exceedingly faint spectra of remote galaxies, Hubble's law was extended to a distance of over 100 million light-years, the straight-line relationship between velocity and distance being maintained. This result has come to be generally regarded as the outstanding discovery in twentieth-century astronomy. It made as great a change in man's conception of the universe as the Copernican revolution 400 years before. For, instead of an overall static picture of the cosmos, it seemed that the universe must be regarded as expanding, the rate of the mutual recession of its parts increasing with their relative distance.

Hubble's discovery stimulated much theoretical work in relativistic cosmology and aroused great interest in fundamental papers on expanding world models by A. Friedmann and G. Lemaître that had been written several years before but had attracted little attention. The interpretation of the straight line in Hubble's graph of velocity against distance and of its slope were eagerly discussed. The constant ratio of velocity to distance is now usually denoted by the letter H and is called Hubble's constant. It has the dimensions of an inverse time—its reciprocal, according to Hubble's original determination, being approximately two (since revised to about ten) billion years. If the galaxies recede uniformly from each other, as was suggested by E. A. Milne in 1932, this could be interpreted as the age of the universe; but, whatever the true law of recessional motion may be, Hubble's constant is generally regarded as a fundamental parameter in theoretical cosmology.

In the early 1930's Humason obtained red shifts indicating velocities of recession up to about one-seventh the velocity of light. This was remarkably high for astronomical objects; and Hubble tended to prefer the neutral term "red shift" to "velocity of recession," since he believed that, although no other explanation could compete with the Doppler interpretation of the spectra, it was possible that some hitherto unrecognized principle of physics may be responsible for the effects observed. This became a central problem for him in the course of the 1930's and was one of the objectives of his detailed investigations of the distribution of galaxies. These investigations were of two kinds: surveys of large areas of the sky penetrating to moderate depths, and surveys of selected small areas to the limits of observability.

Hubble's study of the large-scale distribution of galaxies over the sky produced two important results. At first sight, this distribution appeared to be far from isotropic. No galaxies were found along the central region of the Milky Way, and outside the zone of avoidance the number of galaxies observed appeared to increase with galactic latitude. Hubble showed that these observations could be explained as the effect of an absorbing layer of diffuse matter surrounding the main plane of the Milky Way, and that when this effect was taken into account there were no significant major departures from isotropy in the distribution of galaxies. These conclusions were of great significance for the structure of our own galaxy and also for cosmology because they strengthened the case for regarding the system of galaxies as constituting the general framework of the universe.

In regard to the distribution of galaxies in depth, a preliminary reconnaissance by Hubble indicated that this was uniform. Guided by this information, surveys were made by him and by N. U. Mayall to determine the total number of galaxies in a square degree of the sky brighter than certain limiting magnitudes—for instance, nineteenth or twentieth magnitude. The analysis of these surveys presented Hubble with a difficult theoretical problem, and he enlisted the support of R. C. Tolman, a distinguished theoretical physicist and relativity expert at the California Institute of Technology, Pasadena. The crux of the problem concerned the statistical relationship between apparent brightness and distance; but the apparent brightness of a remote galaxy, corrected for all "local" effects such as the dimming due to interstellar absorption of light in our own system, depends not only on the intrinsic brightness of the galaxy but also on its red shift; and the effect of this is greater for the more remote, and therefore fainter, galaxies. (Moreover, the intrinsic brightness of a remote galaxy when the light left it may not be the same statistically as at later epochs.) The red shift, whatever its cause, diminishes the energy of the light from a galaxy and makes it appear fainter than would be the case otherwise. Moreover, the true absolute magnitude (the bolometric magnitude) depends on the total radiation of all wavelengths, whereas the magnitude registered on the photographic plate is confined to certain parts of the spectrum; and the red shift complicates the problem of converting from photographically determined apparent magnitudes to bolometric magnitudes.

As a result of his investigations with Tolman, Hubble was inclined, from about 1936, to reject the Doppler-effect interpretation of the red shifts and to regard the galaxies as stationary. He claimed that uniformity of distribution in depth was compatible with this assumption. On the other hand, if the galaxies are receding, uniformity in depth can be reconciled with the observations only if there is also a positive curvature of space, the required radius being about 500 million light-years, which was actually less than the range of the 100-inch reflector for normal galaxies. Theoretical cosmologists, notably G. C. McVittie in the late 1930's and Otto Heckmann in the early 1940's, criticized Hubble's analysis and rejected his conclusions but respected his observational achievements.

One of the curiously baffling problems concerning galaxies that engaged Hubble's attention related to the sense of rotation of spiral arms. According to some theoretical astronomers, notably Bertil Lindblad, these arms opened up in the same sense as they rotated about the nucleus, whereas other astronomers believed that they trailed. The question was difficult to resolve, because if a galaxy is seen at the right orientation to observe the arms clearly, it is not easy to tell which is the near and which the far side. With his intimate knowledge of galaxies, Hubble selected as a favorable test object NGC 3190 and in 1941 obtained the necessary spectroscopic and photographic material with the 100-inch reflector. He concluded that there was no reason to doubt that this spiral trails its arms. In the last year of his life, radio and optical evidence was forthcoming that the same situation prevails in our own galaxy.

In 1942 war again caused Hubble to divert his energies from astronomy. He had long been aware of the dangers that threatened the free world and was chairman of the Southern California Joint Fight for Freedom Committee. After the United States entered the war, he sought active service in the army but was asked instead by the U.S. War Department to become chief of ballistics and director of the Supersonic Wind Tunnel Laboratory at the Aberdeen Proving Ground, Maryland. He remained there until 1946 and was awarded the Medal of Merit for his services.

After the war Hubble devoted much time to plans relating to the Hale 200-inch telescope. He became chairman of the Research Committee for the Mount Wilson and Palomar Observatories and was largely responsible for planning the details of the Palomar Observatory Sky Survey that was made with the forty-eight-inch Schmidt telescope. Toward the end of 1949 the 200-inch was at last available for full-time observation, and Hubble was the first to use it. The first major advance after its introduction was Baade's discovery that all extragalactic distances had been underdetermined by a factor of about two. One of the reasons for this conclusion went back to Hubble's discovery in 1932 that the globular clusters in M 31 appeared to be, on the average, four times fainter than those in our own galaxy.

During the last years of his life Hubble suffered from a heart ailment. He died suddenly in 1953 from a cerebral thrombosis while preparing to go to Mount Palomar for four nights of observing.

A man of wide interests, Hubble was elected a trustee of the Huntington Library and Art Gallery in 1938. He bequeathed his valuable collection of early books in the history of science to Mount Wilson Observatory. He was a skilled dry-fly fisherman and fished in the Rocky Mountains and also on the banks of the Test, near Stockbridge, Hampshire, where he and his wife (the former Grace Burke, whom he married in 1924) used to stay with English friends.

Hubble's great achievements in astronomy were widely recognized during his lifetime by the many honors conferred upon him. He gave the Halley lecture at Oxford in 1934, the Silliman lectures at Yale in 1935, and the Rhodes lectures at Oxford in 1936. In 1948 he was elected an honorary fellow of Queen's College, Oxford, in recognition of his notable contributions to astronomy.

Hubble's work was characterized not only by his acuity as an observer but also by boldness of imagination and the ability to select the essential elements in an investigation. In his careful assessment of evidence he was no doubt influenced by his early legal training. He was universally respected by astronomers, and on his death N. U. Mayall expressed their feelings when he wrote: "It is tempting to think that Hubble may have been to the observable region of the universe what the Herschels were to the Milky Way and what Galileo was to the solar system."

BIBLIOGRAPHY

I. ORIGINAL WORKS. Hubble's Halley lecture, delivered at Oxford in 1934, was published as *Red Shifts in the Spectra of Nebulae* (Oxford, 1934). His Silliman lectures, delivered at Yale University in 1935, appeared as *The Realm of the Nebulae* (Oxford, 1936). His Rhodes memorial lectures, delivered at Oxford University in 1936, were published as *The Observational Approach to Cosmology* (Oxford, 1937). His Penrose memorial lecture was published as "Explorations in Space: The Cosmological Program for the Palomar Telescopes," in *Proceedings of the American Philosophical Society,* **95** (1951), 461–470; and his George Darwin lecture as "The Law of Red-Shifts," in *Monthly Notices of the Royal Astronomical Society,* **113** (1953), 658–666.

At the time of his death Hubble was preparing an atlas of photographs to illustrate his revised classification of the galaxies based on a careful study of the magnificent set of plates that he had accumulated between 1919 and 1948 with the sixty-inch and 100-inch telescopes at Mount Wilson Observatory. The details of his revised classification were not completed when he died; and responsibility for publication was taken by Allan Sandage, who worked with Hubble in the last years of Hubble's life. Sandage has explained his role in this publication in the following statement: "I have acted mainly as an editor, not as an editor of a manuscript but rather an editor of a set of ideas and conclusions that were implied in the notes." The work was published by Sandage as *The Hubble Atlas of Galaxies* (Washington, D.C., 1961).

Most of Hubble's original papers were published in *Astrophysical Journal* and were also issued as *Contributions From the Mount Wilson Solar Observatory.*

II. SECONDARY LITERATURE. Among the numerous biographical notices the most informative are the following: Walter S. Adams, "Dr. Edwin P. Hubble," in *Observatory,* **74** (1954), 32–35; M. L. Humason, "Edwin Hubble," in *Monthly Notices of the Royal Astronomical Society,* **114** (1954), 291–295; N. U. Mayall, "Edwin Hubble—Observational Cosmologist," in *Sky and Telescope,* **13** (1954), 78–81, 85; and H. P. Robertson, "Edwin Powell Hubble: 1889–1953," in *Publications of the Astronomical Society of the Pacific,* **66** (1954), 120–125.

G. J. WHITROW

HUBER, JOHANN JACOB (*b.* Basel, Switzerland, 11 September 1707; *d.* Kassel, Germany, 6 July 1778), *anatomy, botany.*

Huber's main contributions to science were his anatomical studies. He gave the first detailed and accurate description of the spinal cord (*De medulla spinali* [Göttingen, 1741]); but he dealt mainly with the external appearances, which he illustrated accurately, and with the accessory nerve of Willis (Clarke and O'Malley, 1968). His account of the internal features of the cord did not advance beyond those already published. To the first fascicle of Albrecht von Haller's famous *Icones anatomicae* (Göttingen, 1743–1755), Huber contributed descriptions of the uterus and spinal cord. He also studied the spinal roots and nerves, particularly the intercostal nerves and the lower cranial nerves.

As a botanist, he was an expert on the flora of the central and eastern high Alps and contributed to Haller's *Historia stirpium indigenarum Helvetiae inchoata* (Bern, 1768, 2 vols.).

Huber came from an upper-class ruling family of Basel, his father, Johann, being an apothecary. He read philosophy in Basel and then studied under the great physiologist Haller in Bern and under H. A. Nicolai, the anatomist, in Strasbourg. He received his medical degree from the University of Basel in 1733 and three years later went to Göttingen where, owing to Haller's influence, he was appointed prosector and, in 1739, professor extraordinary.

Again because of Haller's support, he was called in 1742 to the chair of anatomy and surgery in the Collegium Carolinum in Kassel. In 1748 he was appointed personal physician and privy councillor to the grand duke of Hesse. He retained this appointment and his academic post until his death. In later years Huber felt considerable enmity towards his mentor Haller, believing that he had unjustly appropriated some of his own work.

Huber was a fellow of the Royal Society of London and a member of other scientific bodies in Europe.

BIBLIOGRAPHY

I. ORIGINAL WORKS. Panckoucke (1822) lists twenty-two separate works by Huber. Adelung (1836) adds to these a group of papers published in the *Acta Physico-Medica Academicae Caesareae Leopoldino-Carolinae Naturae Curiosorum* (Nuremberg, 1727–1744), mainly on fetal and muscle anatomy, and one in *Philosophical Transactions of the Royal Society* on anatomical anomalies. Huber's most outstanding publication was *De medulla spinali speciatim de nervis ab ea provenientibus commentatio cum adjunctis iconibus* (Göttingen, 1741). The rest were of secondary importance.

II. SECONDARY LITERATURE. See Edwin Clarke and C. D. O'Malley, *The Human Brain and Spinal Cord* (Berkeley-Los Angeles, 1968), pp. 266-268, for a brief biographical sketch and extracts from Huber's book of 1741 on the spinal cord.

Each of the following biographical pieces is accompanied by a bibliography: Adelung (no initials), "Huber (Jean-Jacques)," in Dezeimeris, ed., *Dictionnaire historique de la médecine ancienne et moderne,* III (Paris, 1836), 244-246; C. L. F. Panckoucke, ed., *Dictionnaire des sciences médicales. Biographie médicale,* V (Paris, 1822) 305-306; J. M. Gesner, in F. Börner, ed., *Nachrichten von den vornehmsten Lebensumständen und Schriften jetztlebender berühmter Aerzte und Naturforscher in und um Deutschland,* I (Wolfenbüttel, 1749), 593-620.

EDWIN CLARKE

HUBER, MAKSYMILIAN TYTUS (*b.* Krościenko, Poland, 4 January 1872; *d.* Cracow, Poland, 9 December 1950), *mechanics, theory of elasticity.*

Huber's aptitude for mathematics and mechanics was already apparent during his first year of studies at the Faculty of Civil Engineering of the Lvov Institute of Technology, which he entered in 1889. His first scientific publication appeared in 1890. On obtaining his diploma, he became a teaching assistant of the Lvov Institute and studied mathematics for a year at the University of Berlin. From 1899 to 1906 he was lecturer and professor of mechanics at the Industrial High School, Cracow, later returning to Lvov as a lecturer and later professor of technical mechanics at the Institute of Technology.

Huber was drafted into the Austro-Hungarian army after the outbreak of World War I and was captured by the Russians in 1915. He was then able to continue scientific work, partly as a result of help from Stepan Timoshenko, whose textbook on strength of materials he translated into Polish. After his return to Lvov in 1918, Huber was rector of the Institute of Technology in 1921-1922; in 1928 he became director of the department of mechanics at the Faculty of Mechanical Engineering of the Warsaw Institute of Technology. In 1920 he was one of the founding members of the Academy of Engineering Sciences, of which he was president from 1928 to 1930. He became a corresponding member of the Polish Academy of Learning in 1927 and was an active member in 1934. From 1931 he was an ordinary member of the Warsaw Scientific Society.

During the German occupation of 1939-1945, when all Polish institutions of higher education were closed, Huber taught in a technical school and secretly gave instruction at the institute level. As representative of the resistance movement, he distributed financial aid to the employees of the Warsaw Institute of Technology. After the Warsaw Insurrection of 1944 he settled in Zakopane, where he directed underground technical courses. After the liberation of Poland, Huber became professor at the Gdańsk Institute of Technology, and in 1949 he moved to the Academy of Mining and Metallurgy, Cracow.

Huber's main area of scientific contribution was the theory of orthotropic (orthogonally anisotropic) plates. Work on this theory was begun in 1860 by Franz Gehring; but it remained for Huber to establish the fundamental assumptions, to give methods for the solution of variously supported plates, and to bring solutions to a form directly applicable in engineering practice—for example, in computing reinforced concrete plates. His work on plates was summarized in his 1928 lectures at the Zurich Technische Hochschule, published as *Probleme der Statik technisch wichtiger orthotroper Platten* (Warsaw, 1929).

The second major area was strength theories. In 1885 Eugenio Beltrami proposed that the critical state of deformed material may be defined by the magnitude of strain energy per unit volume. Since this hypothesis did not agree with experiments, Huber proposed in his 1904 paper "Właściwa praca odkształcenia jako miara wytężenia materiału" ("Strain Energy as a Measure of Critical State of Material") that in determination of the critical state only the energy of distortion may be considered. Richard von Mises (1913) and Heinrich Hencky (1924) independently reached conclusions analogous to those of Huber; and this experimentally confirmed and generally accepted theory is therefore known as the Huber-Mises-Hencky theory.

Huber's third important achievement relates to the concept of an absolute measure of hardness, proposed by Heinrich Hertz (1881-1882) in the solution of the case of coterminous bodies. In his doctoral dissertation, "Zur Theorie der Berührung fester elastischer Körper" (1904), Huber proved that a measure of hardness depends not only on the material but also on the shape of the bodies.

Besides these works Huber wrote more than 250 scientific publications; many of the results were included in his basic textbook *Teoria sprężystości* ("Theory of Elasticity"; Cracow, 1948-1950).

BIBLIOGRAPHY

I. ORIGINAL WORKS. Huber's works were brought together as *Pisma* ("Writings"), 5 vols. (Warsaw, 1954-1964). *Teoria sprężystości* ("Theory of Elasticity") constitutes vols.

IV and V. There are also the textbooks *Mechanika ogólna i techniczna* ("General and Technical Mechanics"; Warsaw, 1956); and *Stereomechanika techniczna* ("Technical Stereomechanics"; Warsaw, 1958).

II. SECONDARY LITERATURE. *Pisma,* I, includes a comprehensive account of Huber's scientific activities and a bibliography of his publications. See also *Polski słownik biograficzny,* ("Polish Biographical Dictionary"), X (Wrocław-Warsaw-Cracow, 1962), 74-76; and S. P. Timoshenko, *History of Strength of Materials* (New York-Toronto-London, 1953), pp. 369, 410.

EUGENIUSZ OLSZEWSKI

HUBRECHT, AMBROSIUS ARNOLD WILLEM (*b.* Rotterdam, Netherlands, 2 March 1853; *d.* Utrecht, Netherlands, 21 March 1915), *zoology, comparative embryology.*

Hubrecht's father, Paul François Hubrecht, was a banker of Dutch patrician stock; his mother, Maria Pruys van der Hoeven, came from an academic family and was a niece of the zoologist Jan van der Hoeven. Hubrecht received his training in zoology at Utrecht University under Harting and Donders, as well as in Leiden under Selenka, with whom he maintained a friendship until the latter's death. He obtained his doctorate in 1874 under Harting. In Utrecht he moved in libertine circles and always professed to be an agnostic. Hubrecht was a man of the world, lively, amiable, and witty. He spoke and wrote three foreign languages flawlessly and was an esteemed guest at international meetings. Always generous in sharing ideas as well as scientific material with colleagues, he was an ardent believer in international scientific cooperation.

Hubrecht was for many years a member of the Dutch Royal Academy of Sciences. He was also a foreign member of both the Linnean Society of London and the Zoological Society of London. He took an active part in the founding of the Dutch Zoological Station and for many years was an editor of the leading Dutch monthly *De Gids.*

Hubrecht's early interest was the invertebrates, particularly the nemerteans. His doctoral thesis (1874), which was not very significant, was based on material collected at the newly established zoological station in Naples, where he was the first Dutch guest worker and where his lifelong friendship with Anton Dohrn and Ray Lankester began. A series of more than a dozen early papers on the anatomy and development of the nemerteans (1875-1889) show Hubrecht to have been a thorough worker but still lacking the originality which characterizes his later work. That he was inclined to speculate becomes apparent from his ideas on the phylogenetic relationships between the nemerteans and the vertebrates (see the *Challenger* report, 1887), views which have long since been abandoned.

From 1875 until 1882 Hubrecht was curator of fishes at the Rijksmuseum voor Natuurlijke Historie in Leiden. Early in this period he worked with Gegenbaur in Heidelberg on the cranial anatomy of the Holocephali, the deep-sea ratfishes, but a product of his abiding interest in the invertebrates is his monograph on the primitive mollusk *Proneomenia Sluiteri* (1880). In 1878 he again spent more than six months in Naples to extend his studies on the nemerteans.

In 1882 Hubrecht succeeded Harting as professor of zoology and comparative anatomy at the University of Utrecht. During his early studies his attention had gradually shifted from adult anatomy to developmental stages. A convinced Darwinian, he was always seeking evidence for phylogenetic relationships; and he soon realized that early embryonic stages often afford more important clues than do the adult forms. Influenced, like so many of his contemporaries, by the works of F. M. Balfour, he soon came to devote all his energy to the study of comparative embryology, partly because of its interest for vertebrate evolution and partly for its own sake. It is in this field that he developed the originality and acquired the mastery that gives much of his work a lasting value, even though several of his hypotheses have since been refuted.

From about 1888 on, Hubrecht studied the early embryology and placentology of mammals. In particular he cleared up many obscure points in the development of the fetal membranes and the placenta. He began with the insectivores, which, as T. H. Huxley had suggested, occupy a central position among the mammals. In 1889 he published his first paper on mammalian embryology; it concerned the hedgehog, *Erinaceus,* of which he obtained specimens representing all early stages of development. In this paper he coined the term "trophoblast" for the outer cell layer of the early mammalian embryo, a term which remains in use.

Hubrecht soon found that he would also need material of certain tropical forms, including representatives of other "primitive" mammalian orders. After careful preparation he set out in 1890 on a journey lasting almost a year to the Dutch East Indies, from which he brought back an extremely valuable collection of gravid uteri of the tree shrew, *Tupaia* (Insectivora; now often classified with the lemuroid primates); the very rare tarsier, *Tarsius,* and the slow loris, *Nycticebus* (lemurlike animals now classified with the primates); the scaly anteater, *Manis* (Pholidota); and the flying lemur, *Galeopithecus*

(Dermoptera). Through his ability to interest others in his objectives Hubrecht was able to receive material from the East Indies for many years after his return. Later he made or organized expeditions to other areas.

Two other works of this period are the memoir "Die Phylogenese des Amnions und die Bedeutung des Trophoblastes" (1895), a masterly work even though its main thesis, the phylogenetic derivation of the mammals directly from the amphibians, is no longer held, and "Early Ontogenetic Phenomena in Mammals" (1908), which contains a synthesis of his own work and that of others carried out during the two preceding decades.

Hubrecht's brilliance of style and breadth of vision were somewhat marred by his occasional inclination to speculate too boldly and to present inferences not always firmly grounded in fact. Because of this—and because he sometimes disregarded the objections of others—he often evoked controversy. On the other hand, he bravely fought obsolete ideas and stimulated much new work. During the latter part of his life Hubrecht's influence declined as embryologists became increasingly interested in the causal and physiological aspects of development.

In 1910 Hubrecht resigned as full professor and was appointed extraordinary professor of comparative embryology, a chair founded especially for him. In 1911 he was a founder of the Institut International d'Embryologie, an international professional society still in existence; he was its first secretary. Unfortunately, ill health increasingly prevented Hubrecht from using his leisure for the extension of his research. He died of arteriosclerosis at the age of sixty-two. The Hubrecht Laboratory (International Embryological Institute) in Utrecht, founded in his memory in 1916, still houses the Hubrecht collection.

BIBLIOGRAPHY

Among Hubrecht's numerous writings published before 1900 are the following: *Aanteekeningen over de Anatomie, Histologie en Ontwikkelingsgeschiedenis van eenige Nemertinen* (Utrecht, 1874), his doctoral thesis; "Beitrag zur Kenntniss des Kopfskelettes der Holocephalen," in *Niederländisches Archiv für Zoologie,* **3** (1877), 255–276; "Proneomenia Sluiteri Gen. et Sp. N. With Remarks Upon the Anatomy and Histology of the Amphineura," *ibid.,* supp. 2 (1880); "Contributions to the Embryology of the Nemertea," in *Quarterly Journal of Microscopical Science,* n.s. **26** (1885), 417–448; "Report on the Nemertea Collected by H.M.S. Challenger During the Years 1873–76," in C. W. Thomson and J. Murray, eds., *Report on the Scientific Results of the Voyage of H.M.S. Challenger During the Years 1873–76,* XIX, *Zoology* (London, 1887); "Studies in Mammalian Embryology. I. The Placentation of Erinaceus europaeus, With Remarks on the Phylogeny of the Placenta," in *Quarterly Journal of Microscopical Science,* n.s. **30** (1889), 283–404; "Studies in Mammalian Embryology. II. The Development of the Germinal Layers of Sorex vulgaris," *ibid.,* n.s. **31** (1890), 499–562; "Studies in Mammalian Embryology. III. The Placentation of the Shrew (Sorex vulgaris, L.)," *ibid.,* n.s. **35** (1894), 481–538; "Spolia Nemoris," *ibid.,* n.s. **36** (1894), 77–126; "Die Phylogenese des Amnions und die Bedeutung des Trophoblastes," *Verhandelingen der K. akademie van wetenschappen,* 2nd sec., **4**, no. 5 (1895); "Die Keimblase von Tarsius. Ein Hilfsmittel zur schärferen Definition gewisser Säugethierordnungen," in *Festschrift zum siebenzigsten Geburtstage von Carl Gegenbaur,* II (Leipzig, 1896), 147–178; *The Descent of the Primates* (New York, 1897); and "Ueber die Entwicklung der Placenta von Tarsius und Tupaja nebst Bemerkungen ueber deren Bedeutung als haematopoietische Organe," in *Proceedings of the Fourth International Congress of Zoology* (Cambridge, 1898), app. B., pp. 343–412.

After 1900 Hubrecht published "Furchung und Keimblattbildung bei Tarsius spectrum," *Verhandelingen der K. akademie van wetenschappen,* 2nd sec., **8,** no. 6 (1902); "The Gastrulation of the Vertebrates," in *Quarterly Journal of Microscopical Science,* n.s. **49** (1905), 403–419; "Normentafeln zur Entwicklungsgeschichte des Koboldmaki (Tarsius spectrum) und des Plumplori (Nycticebus tardigradus)," *Normentafeln zur Entwicklungsgeschichte der Wirbelthiere,* no. 7 (Jena, 1907), written with F. Keibel; "Early Ontogenetic Phenomena in Mammals and Their Bearing on Our Interpretation of the Phylogeny of the Vertebrates," *Quarterly Journal of Microscopical Science,* n.s. **53** (1908); and "Früheste Entwicklungsstadien und Placentation von Galeopithecus," D. de Lange, Jr., ed., *Verhandelingen der K. akademie van wetenschappen,* 2nd sec., **16,** no. 6 (1919).

A secondary source is R. Assheton, "Dr. Ambrosius Arnold Willem Hubrecht," in *Proceedings of the Linnean Society of London* (1915), sess. 127, 28–31.

J. FABER

HUDDE, JAN (*b.* Amsterdam, the Netherlands, May 1628; *d.* Amsterdam, 15 April 1704), *mathematics.*

Jan (or Johann) Hudde, the son of a merchant and patrician, Gerrit Hudde, and Maria Witsen, was christened on 23 May 1628. He studied law at the University of Leiden around 1648, at which time—perhaps even earlier—he was introduced to mathematics by Frans van Schooten. Besides acquainting his students with the classic works of the ancient mathematicians, Schooten gave them a thorough knowledge of Descartes's mathematical methods, as published in his *Géométrie* (1637).

Hudde's contributions to mathematics were probably made between 1654 and 1663, for there is no evidence of further mathematical work after the latter year. From then on, he devoted himself to the service of Amsterdam, as a member of the city council, juror, and chancellor. In 1673 he married Debora Blaw, a widow; they had no children. On 15 September 1672 Hudde was chosen by Stadtholder Wilhelm III as one of Amsterdam's four burgomasters. He held this office until 1704, serving altogether for twenty-one of those years (intermittently with one-year hiatuses required by law). Between his terms as burgomaster, Hudde was chancellor and deputy of the admiralty. In 1680 he received the Magnifikat for his services in the administration of the civic government. His anonymous biographer depicts him as "unselfish, honest, well-educated in the sciences, with his eyes open to the general welfare."

Hudde's teacher, Frans van Schooten, often incorporated the results of his students' work in his own books. Thus, in his *Exercitationes mathematicae* (1657) there are three essays by Hudde, including a treatise written in 1654 on the determination of the greatest width of the folium of Descartes. In 1657 Hudde participated in a correspondence among R. F. de Sluse, Christiaan Huygens, and Schooten on the questions of quadrature, tangents, and the centroids (centers of gravity) of certain algebraic curves.

Schooten's edition of the *Géométrie* (1659–1661) contains two other works by Hudde. The first, *De reductione aequationum,* may have been written in 1654–1655, according to a note in the foreword. Presented in the form of a letter to Schooten, it is dated 15 July 1657. The second, *De maximis et minimis,* is dated 26 February 1658. There also exists an exchange of letters between Hudde and Huygens (1663) on problems dealing with games of chance. This enumeration comprises all of Hudde's known mathematical works; but it is recorded in the notes of Leibniz, who visited Hudde in Amsterdam in November 1676, that Hudde still had many unpublished mathematical writings, which are now lost.

In Hudde's extant mathematical works two main problems can be recognized: the improvement of Descartes's algebraic methods with the intention of solving equations of higher degree by means of an algorithm; and the problem of extreme values (maxima and minima) and tangents to algebraic curves. In the latter Hudde accomplished the algorithmizing of Fermat's method, with which he had become acquainted through Schooten.

The solution, that is, the reduction, of algebraic equations was a central problem at that time. In 1545 Ludovico Ferrari had reduced the solution of a fourth-degree equation to the solution of a cubic equation. In the *Géométrie* Descartes had combined equations of the fifth and sixth degrees into one genre and had given a method for the graphic determination of the roots. The contents of *De reductione aequationum* indicate that Hudde had originally tried to solve equations of the fifth and sixth degrees algebraically. Although unsuccessful in his attempt—and totally unaware of the reason for his failure—he at least compiled the cases in which a reduction of the degree is possible by separation of a factor. Correspondingly, Hudde also dealt with equations of the third and fourth degrees because their general solution presents great analytical difficulties. He gave the solution of the reduced cubic equation $x^3 = qx + r$ by means of the substitution $x = y + z$; he also gave the determination of the greatest common divisor of two polynomials by the process of elimination.

Hudde's rule of extreme values and tangents can be traced to Fermat. Expressed in modern terms, Fermat starts with the proposition that in the proximity of the maximum or minimum position x_0 of the function $f(x)$, $f(x_0 + h)$ is approximately equal to $f(x_0 - h)$. By expansion in terms of powers of h the linear member must, therefore, be omitted. For a rational function, which disappears at x_0, this means that x_0 is a "double" zero of the function. Proceeding from this proposition, Hudde was seeking an algorithmically usable rule for rational functions. His law states that if the polynomial $f(x) = \Sigma\, a_k x^{n-k}$ has the "double" zero of the function, then the polynomial $\Sigma\,(p + kq)a_k x^{n-k}$, with p,q arbitrary natural numbers, also has x_0 as the zero of the function. Fully stated (in Latin), the rule can be translated as "If in an equation two roots are equal and the equation is multiplied by an arithmetic progression to whatever degree is desired—that is, the first term of the equation is multiplied by the first term of the progression, the second term of the equation by the second term of the progression, and so on in regular order—then I say that the product will be an equation in which one of the mentioned roots will be found." The "double" zero of the function is, then, the zero of the greatest common divisor of the two polynomials. The greatest common divisor is found by the process of elimination that represents a variation of the well-known Euclidean algorithm. Hudde extended his dealings to include fractionalized rational functions, his method amounting to the expression (in modern terms)

$$\frac{f(x)}{g(x)} = \frac{f'(x)}{g'(x)}.$$

His rule of tangents stands in direct relation to his process of extreme values, just as most of his other works represent applications of the results of his theory of equations, that is, the rule of tangents and extreme values.

Hudde was also interested in physics and astronomy. He spent much time with the astronomer Ismael Boulliau and reported his comet observations to Huygens in 1665. In 1663 he produced microscopes with spherical lenses; in 1665 he worked with Spinoza on the construction of telescope lenses. That he also had assembled a small *dioptrica* is seen from his correspondence with Spinoza. In 1671 he sent to Huygens mortality tables for the calculation of life annuities. During the next two years Hudde was charged by the city of Amsterdam with appraising DeWitt's formulas for the calculation of life annuities.

Perhaps the most gifted of Schooten's students, Hudde was also the most strongly influenced by him. At the time of Schooten's death in 1660, Hudde felt that he commanded a comprehensive view of the basic contemporary mathematical problems. Like Descartes he held as meaningful only such mathematical problems as could be handled through algebraic equations. After 1663 he pursued mathematics only as an avocation apart from—for him—more important civic activities.

His contemporaries saw him as a mathematician of great ability. Leibniz wrote, even as late as 1697, that one could expect a solution to the difficult problem of the brachistochrone only from L'Hospital, Newton, the Bernoullis, and Hudde "had he not ceased such investigations long ago."

BIBLIOGRAPHY

I. ORIGINAL WORKS. Frans van Schooten's *Exercitationum mathematicarum libri quinque* (Leiden, 1657) contains three essays by Hudde; see Schooten's ed. of Descartes's *Géométrie, Geometria Renati Cartesii,* I (Amsterdam, 1659), for Hudde's *De reductione aequationum* and *De maximis et minimis.* Hudde's correspondence with Huygens is in the latter's *Oeuvres complètes,* 22 vols. (The Hague, 1888–1950).

II. SECONDARY LITERATURE. On Hudde and his contributions, see Karlheinz Haas, "Die mathematischen Arbeiten von Johann Hudde," in *Centaurus,* **4** (1956), 235–284—the app. contains an extensive bibliography; Joseph E. Hofmann, *Geschichte der Mathematik,* pt. 2, (Berlin, 1957), pp. 45–46, 54, 74; and P. C. Molhuysen and P. J. Blok, eds., *Nieuw Neederlandsch biografisch Woordenboek* (Leiden, 1911–1937).

KARLHEINZ HAAS

HUDSON, CLAUDE SILBERT (*b.* Atlanta, Georgia, 26 January, 1881; *d.* Washington, D. C., 27 December 1952), *chemistry.*

Hudson's career was spent almost entirely in governmental laboratories in Washington, where he trained many followers in the chemistry of the sugars. He was born of early American stock, spent his youth in Mobile, Alabama, and received the B.S. (1901), Ph.D. (physics, 1907), and Hon. D.Sc. (1947) degrees from Princeton University. His early interest was in physical chemistry, which he studied with Nernst at Göttingen and van't Hoff at Berlin. From 1928 to 1951 Hudson served in the National Institutes of Health.

Hudson and his many associates developed the stereochemistry of the anomeric sugar centers, beginning with his rules of isorotation, useful for allocation of anomeric form when proper substituents are present. This development was followed by a rule establishing the point of ring closure in aldonolactones. Hudson demonstrated that enzymic reactions follow the laws of mass action and he showed that the D-fructose unit of sucrose possesses an unusual form. He established the equation expressing the acid-base dependency of the rate of D-glucose mutarotation and from this calculated an accepted value for the ionic dissociation of water. He correlated anomeric configurations through periodate oxidation, calculated rotatory powers of unisolated anomers by the principle of maximum solubility, and synthesized the $(1 \rightarrow 4)$-β-D-linked disaccharides lactose and cellobiose. Hudson prepared many sugars and their acetates in pure anomeric forms and with the D-galactose pentaacetates, established that a sugar could exist in more than one ring form.

Hudson received many awards and he was elected to membership in distinguished scientific bodies in the United States and abroad. In his relaxed moments he was a noted bon vivant and raconteur, but when at work he was an exacting person, holding himself and his associates to high standards.

BIBLIOGRAPHY

The obituary by Lyndon F. Small and Melville L. Wolfrom in *Biographical Memoirs. National Academy of Sciences,* **32** (1958), 181–220, contains a bibliography of Hudson's publications, including posthumous works, from 1902–1955.

MELVILLE L. WOLFROM

HUDSON, WILLIAM (*b.* Kendal, Westmorland, England, 1733; *d.* London, England, 23 May 1793), *botany.*

Hudson was born and raised in Kendal, where his father kept the White Lion Inn. He was educated in the Kendal Grammar School and, on completion of his studies, was apprenticed to an apothecary on Panton Street, Haymarket, London.

Hudson proved an apt student. During his year apprenticeship he won the Apothecaries' Company's prize for botany. Between 1757 and 1758 his horizons were widened when, as resident sublibrarian of the British Museum, he studied the Sloane herbarium. Hudson was subsequently encouraged by Benjamin Stillingfleet, who introduced him to the writings of Linnaeus, to restate John Ray's *Synopsis methodica stirpium Britannicarum* in terms of the Linnean system. Thus, in 1762 he published *Flora Anglica,* which incorporated the work of other naturalists with a rearrangement of the *Synopsis.* Hudson's clear and concise language, accuracy in determining plant locations, accounts of medicinal values of the plants, and addition of valuable synonyms were very useful and popular. *Flora Anglica* quickly replaced Ray's *Synopsis* as the standard English flora and won most English naturalists over to the Linnean sexual system. In 1778 Hudson published a second, enlarged edition of his work; a reprint of the second edition appeared in 1798.

Hudson became a fellow of the Royal Society in 1761 and of the Linnean Society in 1791. From 1765 to 1771 he was director and botanical demonstrator for the Apothecaries' Garden. Growing fiscal difficulties at the garden forced Philip Miller to resign in 1770, and in 1771 Hudson tendered his own resignation as well.

Hudson's interests also included insects and mollusks, and he planned to write a *Fauna Britannica.* Unfortunately a fire in 1783 destroyed his collections, papers, and Panton Street home. Although Hudson continued his interest in natural history, his slender financial resources were not adequate to replace the loss.

Hudson never married. When his master died, Hudson took over his apothecary practice and lodged with his widow; on her death, he was joined by her daughter and son-in-law. When the residence was destroyed, they moved to a house on Jermyn Street where, after suffering for several years from what James Edward Smith describes as ulcerated lungs, and a series of paralytic strokes, Hudson died. He was interred in St. James's Church, and the remains of his collections were given to the Apothecaries' Garden in Chelsea.

BIBLIOGRAPHY

I. ORIGINAL WORKS. Hudson's only published book is *Flora Anglica* (London, 1762; 2nd, enl. ed., 1778; 2nd ed. repr. 1798). From 1768 to 1770 he published an annual "Catalogue of the Fifty Plants From Chelsea Garden, Presented to the Royal Society by the . . . Company of the Apothecaries" in *Philosophical Transactions of the Royal Society,* **58, 59,** and **60.**

II. SECONDARY LITERATURE. The most useful sources of information are articles by James Edward Smith in Abraham Rees, *Cyclopaedia,* XVIII, and G. S. Boulger in *Dictionary of National Biography,* new ed., X, 155. J. Reynolds Green, *A History of Botany in the United Kingdom* (London, 1914), pp. 271–273, sheds valuable light on the importance of Hudson's work with the Chelsea Apothecaries' Garden. Richard Pulteney, *Historical and Biographical Sketches of the Progress of Botany in England* (London, 1790), pp. 351–352, provides supplementary information about Hudson's contributions to the acceptance of Linnean taxonomy in England. Accounts of Hudson's death appear in *Annual Register: Chronicle* (1793), pp. 25–26; *Gentleman's Magazine,* **63** (May 1793), 485; and John Nichols, *Literary Anecdotes of the Eighteenth Century,* IX (London, 1815), 565–566.

For a bibliography about Hudson, see James Britten and G. S. Boulger, *A Biographical Index of British and Irish Botanists,* 2nd ed., rev. by A. R. Rendle (London, 1931), pp. 157–158, the most recent and helpful source. The *Dictionary of National Biography* (see above) is also quite valuable.

ROY A. RAUSCHENBERG

HUFNAGEL, LEON (*b.* Warsaw, Poland, 1893; *d.* Berlin, Germany, 19 February 1933), *astronomy.*

Hufnagel entered the faculty of mathematics and physics of Warsaw University in 1911. After receiving the Ph.D. at Vienna in 1919, he returned to Poland, serving as an assistant at the Free University, Warsaw, in 1921–1926. In the latter year he left Poland for Sweden, where he worked at Lund Observatory in 1926–1928. For the next two years he was Rockefeller traveling fellow at the Mt. Wilson, Lick, and Harvard College observatories. From 1930 he worked at the Astronomisches Recheninstitut, Berlin-Dahlem, and the astrophysical observatory, Potsdam.

Hufnagel's first scientific papers were in celestial mechanics. In 1919 he determined the orbit of the great September comet (1882 II), and with J. Krassowski he calculated the perturbations of the asteroid (43) Ariadne in 1925. After 1925 his major scientific work concerned stellar statistics and astrophysics. His first paper on proper motions of stars was published at Warsaw in 1925, but the most important ones were written during his two years at Lund and in Germany after 1930. In seven papers published in 1926–1933 Hufnagel considered the velocity distributions of faint stars and the influence on such distributions of accidental errors in proper motions. He cooperated on studies of this problem with K. G. Malmquist in 1933 and with F. Gondolatsch in 1931. During his stay in the United States, Hufnagel published two papers on stellar temperature and one note on galactic rotation (1929). With B. P. Gerasimovich, at the Harvard College Observatory, he investigated the semiregular variable star R Sagittae (1929). His last scientific paper, written with H. Müller (1935), concerned the absorption by interstellar clouds near the North America nebula. Hufnagel's main work in the last two years of his life was the elaboration of the first two chapters ("Grundlagen der mathematischen Statistik") of *Lehrbuch der Stellarstatistik,* edited by E. von der Pahlen (Leipzig, 1937).

BIBLIOGRAPHY

I. ORIGINAL WORKS. Hufnagel's papers include "Die Bahn der grossen September Kometen 1882II unter Zugrundelegung der Einsteinschen Gravitationslehre," in *Sitzungsberichte der Akademie der Wissenschaften in Wien,* **128** (1919), 1261–1270; "Sur les mouvements propres des étoiles," in *Bibliotheca Universitatis Librae Polonae,* **13,** fasc. A (1924); "Perturbations et tables approchées du mouvement de la petite planète (43) Ariadne," *ibid.,* **14** (1925), written with J. Krassowski: "Über eine Formel der Stellarstatistik," in *Astronomische Nachrichten,* **228** (1926), 321–324; "Zur Geschwindigkeitsverteilung schwacher Sterne," *ibid.,* **231** (1927), 297–304, and **242** (1931), 385–392, written with F. Gondolatsch; "Über die Räumliche Geschwindigkeitsverteilung der Sterne zwischen 9, und 14. Grösse," in *Meddelanden från Lunds astronomiska observatorium,* 2nd ser., vol. **5** (1927); "On the Influence of the Accidental Errors in the Proper Motions on the Velocity Distribution," *ibid.,* no. 114 (1928); "Über den Einfluss zufälliger und systematischer Fehler auf das Geschwindigkeitsellipsoid," *ibid.,* no. 123 (1930); "Note on the Galactic Rotation," in *Bulletin. Astronomical Observatory, Harvard University,* vol. **863** (1929); "Temperatures of Giants and Dwarfs," in *Circular. Astronomical Observatory of Harvard College,* vol. **343** (1929); "Note on Stellar Temperatures," in *Bulletin. Astronomical Observatory, Harvard University,* vol. **874** (1930); "The Distribution in Space of the Stars of Type A as Derived From the Draper Catalogue," in *Astronomiska Iakttagelser och Undersökningar på Stockholms Observatorium,* vol. **11,** no. 9 (1933); and "Untersuchungen über absorbierende Wolken beim Nordamerika Nebel unter Benutzung von Farbenindizes schwacher Sterne," in *Zeitschrift für Astrophysik,* **9** (1935), 331–381, written with H. Müller.

II. SECONDARY LITERATURE. Obituaries are in *Astronomische Nachrichten,* **248** (1933), 143; and *Monthly Notices of the Royal Astronomical Society,* **94** (1933), 276–277.

EUGENIUSZ RYBKA

HUGGINS, WILLIAM (*b.* London, England, 7 February 1824; *d.* Tulse Hill, London, 12 May 1910), *astrophysics.*

Huggins was the second and only surviving child (the first had died in infancy) of William Thomas Huggins, a silk mercer and linen draper in Gracechurch Street in the City of London. His mother, the former Lucy Miller, was a native of Peterborough. He was precocious and, after a short period of attendance at a small nearby school and instruction at home under the curate of the parish, he entered the City of London School at its opening early in 1837. An attack of smallpox, from which he fully recovered, led to his removal from the school shortly afterward, his education being continued by private tutors at home. Although his formal instruction was broad, including classics, several modern languages, and music, his predominant interest was in science. A gift of a microscope led to early concentration on physiology, and although at about the age of eighteen he bought his first telescope—for £15—his location in the City of London was too unsuitable for celestial observations to allow astronomy to claim much of his attention.

At about this time (1842) family circumstances led to a regretful decision to abandon his intention of going to Cambridge for a university education, and he took over the responsibility for his father's business. From then until 1854 this was his chief concern, although his spare time was almost wholly given to the microscope and the telescope. Visits to the Continent, where his knowledge of languages stood him in good stead, helped to preserve the balance of his interests.

In 1852 Huggins joined the Royal Microscopical Society and in 1854 the Royal Astronomical Society, and in the latter year he was able to dispose of the mercery business and thereafter devote his whole time to science. He removed with his parents to Tulse Hill—now a part of greater London, but then situated in the country—and in the new surroundings astronomy prevailed over microscopy as his major interest. A not unimportant factor in this choice was his sensitive nature, which made experiments on animals distasteful to him. Huggins remained at Tulse Hill for the remainder of his life, setting up an observatory equipped with instruments, partly purchased by himself and partly lent by the Royal Society, and here the whole of his astronomical researches were carried out.

His father died shortly after the removal to Tulse Hill, but his mother survived until 1868; he felt her loss keenly. In 1875 he married Margaret Lindsay Murray, of Dublin, who, although twenty-six years his junior, was an ideal partner for the next thirty-five years, taking an active part in the astronomical observations; her name is associated with his in the authorship of some of his chief publications. She seems, indeed, in this respect to have stood in a relation to her husband similar to that of Caroline Herschel to her brother William. She had also considerable artistic and musical gifts.

Huggins, although other interests ranked far below astronomy in his esteem, was by no means narrow-minded. He was an able violinist—according to his wife, "always rather an intellectual than a perfervid player"—and owned a fine Stradivarius instrument. Presumably it was the intellectual element in his musical talent that led to his contributing to the Royal Society in 1883 a paper on the proportional thickness of the strings of the violin—apparently his only publication, apart from one or two early papers on microscopical work, that was not astronomical in character. He was an expert pike fisherman and an admirer of Izaak Walton. Huggins had been brought up as a Calvinist but had never responded to this form of religion, and his views on such matters are perhaps best indicated by his wife's description of him as a "Christian unattached." For a short time in 1870 he was attracted toward the scientific study of spiritualism and corresponded with Sir William Crookes on the subject, but his experience at séances led him to the conclusion that the subject was too closely associated with trickery to merit his serious attention.

Huggins' pioneer work in astrophysics brought him many honors. In 1865 he was elected a fellow of the Royal Society, and in the following year was awarded one of its Royal Medals. The Rumford and Copley Medals of the Royal Society followed in 1880 and 1898, respectively. In 1900 he became president of the Royal Society, a position which he occupied for the customary five years. His annual addresses in this capacity were collected and published in 1906 in a volume entitled *The Royal Society, or Science in the State and in the Schools;* here they were supplemented by many illustrations and material dealing with the history of the Royal Society and closely related matters. Huggins received the gold medal of the Royal Astronomical Society, jointly with W. A. Miller in 1867 and as the sole recipient in 1885; he was president of the Royal Astronomical Society during the two sessions 1876–1878. In 1891 he was president of the British Association for the Advancement of Science, and in 1897 he was created a K.C.B. and in 1902 awarded the O.M., one of the original members of the Order of Merit, which had just been instituted. Numerous universities conferred honorary degrees on him. His financial resources, although sufficient to allow him to devote the whole of his time to astronomy, were not great; and in 1890 he was awarded a Civil List pension of £150 a year in recognition of the value of his work.

In 1908, when he felt no longer able to continue his researches, Huggins returned his instruments to the Royal Society; and they were transferred to the Solar Physics Observatory at Cambridge, where they now are. He died on 12 May 1910, following an operation.

Huggins' earliest astronomical work was on conventional lines. He formed a close friendship with W. R. Dawes, a well-known amateur observer, from whom he bought an eight-inch refracting telescope; and with this, between 1858 and 1860, he made observations of the planets. In 1859, however, Kirchhoff had shown how, from observations of the dark Fraunhofer lines in the solar spectrum, the chemical composition of the sun's atmosphere could be determined; and Huggins gave the first manifestation of one of his most marked characteristics—that of immediately perceiving the possibilities opened up by a new discovery. "This news came to me," he wrote later, "like the coming upon a spring of water in a dry and thirsty land." It at once occurred to him that this method could be applied to the stars; and he confided his idea to his friend W. A. Miller, professor of chemistry at King's College, London, who, although somewhat dubious, agreed to collaborate with him. They designed a spectroscope consisting of two dense flint glass prisms which they attached to Huggins' eight-inch telescope, and observations of stellar spectra were begun. The same idea had occurred to Rutherfurd in America, but quite independently. In order to interpret the stellar spectra it was necessary to obtain better knowledge than that which then existed of the spectra of terrestrial elements; and maps of twenty-four such spectra were prepared by Huggins, with the use of a more powerful spectroscope containing six prisms. In 1863–1864 the stellar and laboratory observations were published by the Royal Society, the general conclusion reached being that the brightest stars, at least, resembled the sun in structure, in that their light proceeded from underlying hot material and passed through an atmosphere of absorbent vapors; nevertheless, there was considerable diversity of chemical composition among the stars.

Striking as this conclusion was—much more so then than now, when it has become a commonplace—a still more sensational discovery was made in 1864. The nature of the nebulae was then quite unknown: "a shining fluid of a nature unknown to us," which was William Herschel's description of a nebula, had remained all that could safely be said on the matter. The fact that an increasing number of them had, after Herschel's time, been resolved into star clusters as more powerful telescopes became available, had led to the conjecture that all were of this character and would be so observable with instruments of sufficient resolving power. It occurred to Huggins to attempt a verification of this by observation with the spectroscope. He accordingly directed his instrument toward a planetary nebula in the constellation Draco and observed not, as he expected, a mixture of stellar spectra but a few isolated bright lines. His knowledge of laboratory spectra at once suggested the interpretation of this: the nebula consisted not of a cluster of stars but simply of a luminous gas. Other nebulae were examined; some showed similar spectra and others spectra generally resembling those of stars. It became clear that these objects, up to then regarded as identical in nature, belonged to two classes: some were clusters of stars, which would be seen as such with greater telescopic power, while others were uniformly gaseous. The bright lines observed in the gaseous nebulae, however, presented a puzzle. Hydrogen was readily identifiable, but there were other lines corresponding to nothing known on the earth; and a new element, provisionally called "nebulium," was postulated. It was not until 1927 that it was discovered by Ira S. Bowen that nebulium was ionized oxygen and nitrogen.

Huggins followed up this work by spectroscopic observations of comets and of a nova, or new star, which appeared in the constellation Corona Borealis in 1866. He showed that the radiations of three comets gave spectra containing bands coincident in position with those obtainable from a candle flame in the laboratory, and concluded that they arose from carbon or its compounds. Huggins was more attracted by the fainter than by the brighter celestial objects and gave little attention to the sun. It was accordingly his younger contemporary Norman Lockyer who discovered how to make spectroscopic observations of the solar prominences in full sunlight. On hearing of this achievement, Huggins supplemented it by simply widening the slit of the spectroscope, thus revealing a prominence in its natural form, in the light of each element that it contained, instead of merely by a narrow spectrum line.

Another example of Huggins' opportunism is afforded by his early perception of the possibility of applying the Doppler effect to the determination of the motions of the stars in the line of sight. It was in 1841 that the Austrian physicist Christian Doppler deduced on theoretical grounds that the motion of a source of sound or light—both regarded as wave phenomena—toward or away from an observer should cause a change in the frequency of reception of the waves, manifesting itself as a change of tone with sound and a change of color with light. He did not reach a full understanding of the effect of this change on stellar observations, for he thought that it would make a receding star appear redder, and an approaching star bluer, than if the star were stationary. In fact, since stellar spectra extend into the invisible regions of the infrared and the ultraviolet, all that radial motion could do would be to shift the whole visible spectrum slightly to one side or the other; its whole range of colors would still appear, leaving the resultant color unchanged. Fizeau later pointed out that, nevertheless, use could be made of the effect because the absorption lines in the spectra would partake of this general displacement; and the amount of their shift—measured by the difference of wavelength of the stellar lines and the lines of the same substances produced from stationary sources in the laboratory—would indicate the velocity of the star along the line of sight, the so-called radial velocity.

Huggins at once perceived the possibility of applying the knowledge he had obtained of the laboratory spectra of elements to the determination of such velocities. He consulted Clerk Maxwell on the theory of the matter; and after various delays in securing a sufficiently powerful spectroscope he succeeded, in 1868, in obtaining a value for the radial velocity of Sirius of 29.4 miles a second away from the sun—a figure which later, with better instruments, he amended to between 18 and 22 miles a second. This is now known to be too large, although the direction is right; but it must be remembered that only visual observations were then possible and that the attainable accuracy of measurement fell short of that which we now regard as essential for this work. The principle had been established, however, of introducing into astronomy one of the most fruitful sources of knowledge we possess concerning the structure and evolution of the universe.

Although, as has been said, these observations were visual, Huggins had not overlooked the desirability of photographing stellar spectra; and as early as 1863 he attempted to photograph the spectrum of Sirius, the apparently brightest star in the sky. But the result was poor, and he realized that the time for this refinement had not come. Satisfactory results were not obtained until 1872, by Draper; and Huggins was not slow to follow them up by extensive photographic observations of the spectra of stars bright enough for this type of examination. He also sought to apply spectroscopic photography to the detection of the solar corona in full sunlight and at first thought he had succeeded, but this hope was not confirmed. Nevertheless, he devoted his Bakerian lecture to the Royal Society in 1885 to the subject "The Corona of the Sun."

Pursuing his studies of the nebulae, Huggins came into conflict with Lockyer, another pioneer in spectroscopic astronomy. Lockyer had formed an imposing hypothesis of celestial evolution, known as the meteoritic hypothesis, a vital piece of evidence for which lay in the supposed identification of the "nebulium" green line with the head of a band, or fluting, observed in the spectrum of a magnesium spark in the laboratory. Not only was it doubtful whether, even under the admittedly unfavorable conditions of observation of nebular spectra, an extended band could appear so like a single sharp line, but also there was a slight discrepancy between the wavelength measurements of the radiations from the two sources. Huggins refused to admit their identity, and later knowledge has fully justified his skepticism.

A comparison of Huggins and Lockyer, so similar in time, place, and scientific objectives and so different in character, is inevitable. Each could serve as a type of his class—Lockyer as the adventurous and Huggins as the cautious investigator. To Lockyer, observational knowledge was merely a means to an end—the understanding of the whole course of nature. To Huggins it was an end in itself—ultimately to lead to understanding, of course, but, at the present, the beginning of a new and apparently limitless means of inquiry, to be gathered by patience and strict accuracy, uninfluenced by theoretical expectation or desire. His discovery of the gaseous nature of nebulae, which to many seemed to confirm William Herschel's conjecture that these bodies might be the parents of stars, led him to point out that such a conclusion was not safely to be drawn, since the nebulae seemed to contain very few elements and the stars many. At that time the chemical elements were regarded as eternally unchangeable; and while Lockyer, by his "dissociation hypothesis," simply brushed aside this obstacle, to Huggins it appeared insurmountable. As a contrast to Lockyer's sweeping meteoritic hypothesis, which sought to comprehend the whole universe in time and space, the following summing up by Huggins of his life's work, published in 1899 in the first of his two volumes on the work at his observatory, may be cited:

> As the conclusion of the whole matter, though there may be no reason to assume that the proportions of the different kinds of chemical matter are strictly the same in all stars or that the roll of the chemical elements is equally complete in every star, the evidence appears to be strong that the principal types of star spectra should not be interpreted as produced by great original differences of chemical constitution, but rather as successive stages of evolutional progress, bringing about such altered conditions of density, temperature, and the mingling of the stellar gases, as are sufficient presumably to account for the spectral differences observed; even though with our present knowledge a complete explanation may not be forthcoming.

In retrospect a decision between these contrasting attitudes passes into insignificance beside the recognition that the contribution of each to later progress was essential and beyond the reach of the other.

BIBLIOGRAPHY

There is no full-scale biography of Huggins. His widow intended to write a personal sketch of his life but died, in 1915, before the work was completed. The material she had prepared, after some vicissitudes, was ultimately embodied in a small volume entitled *A Sketch of the Life of Sir William Huggins, K.C.B., O.M.,* by C. E. Mills and C. F. Brooke, which was published privately (London, 1936). The authors write that "they have merely taken it upon themselves to edit the material at their disposal, and, having no knowledge of the mysteries of science, they have endeavoured as far as possible to steer clear of purely technical matters." Obituary notices in the *Dictionary of National Biography; Proceedings of the Royal Society,* **86** (1911–1912); and *Monthly Notices of the Royal Astronomical Society,* **71** (1911), 261, recount the course of his scientific work.

Huggins contributed numerous original papers to learned societies, those which he considered the more important being reprinted in *Publications of Sir William Huggins's Observatory,* Sir William and Lady Huggins, eds., 2 vols. (London, 1899–1909). Vol. I, *Atlas of Representative Stellar Spectra,* contains a history of the observatory, a comprehensive list of published papers, a description of the instruments used and the methods of observation, and an account of the later work of the observatory that had not been previously published elsewhere. There are twelve large plates, mainly of stellar spectra. Vol. II, *The Scientific Papers of Sir William Huggins,* contains reprints of published papers on the work done at the observatory from its foundation in 1856, classified under various headings and supplemented by reprints of Huggins' more important lectures and addresses.

Huggins's only other published book is a collection of his annual addresses as president of the Royal Society, *The Royal Society, or Science in the State and in the Schools* (London, 1906).

A short article contributed by him to *The Nineteenth Century* (June, 1897), entitled "The New Astronomy; a Personal Retrospect," gives an interesting account of some of his work.

HERBERT DINGLE

HUGH OF ST. VICTOR (*d.* Paris, France, 11 February 1141), *scientific classification, geometry.*

Probably from Saxony or Flanders originally, Hugh came to Paris at an early age and joined the canons regular of the abbey of St. Victor. He lectured on theology in the famous school attached to this monastery, and was its greatest representative. He wrote a very large number of exegetical, philo-

HUGH OF ST. VICTOR

sophical, and theological works which exercised a profound influence on the scholasticism of the twelfth and thirteenth centuries. The most famous of them is the *De sacramentis christianae fidei.*

Preoccupied with giving a scientific basis to the teaching of theology, Hugh wrote an introductory treatise to the sacred sciences, the *Didascalicon* or *De studio legendi,* composed before 1125. Book II of this work contains a division of philosophy which is a classification of the sciences, inspired by that of Boethius. According to Hugh, philosophy encompasses four parts: *theorica, practica* (that is, moral philosophy), *mechanica,* and *logica. Theorica* in turn is divided into *theologia, mathematica,* and *physica* or *physiologia.* The *Didascalicon* says little about *physica,* limiting itself to indicating that it is the science of nature and that it examines the causes of things in their effects and their effects in their causes. Hugh lingers a great deal longer on mathematics, to which he gives a preponderant place; it is indispensable to the knowledge of physics and ought to be studied before the latter. The word *mathematica* has two senses: When the *t* is not aspirated, this term designates "the superstition of those who place the destiny of men in the constellations" of the heavens; when the *t* is aspirated, it designates, on the contrary, the science of "abstract quantity," itself identified with the *intellectibile,* as opposed to the *intelligibile,* the object of theology. *Mathematica* thus defined is divided into four sciences, in which are recognized the four disciplines of the Carolingian quadrivium: arithmetic, the science of numbers and their properties; music, divided into music of the world (the study of the harmony of the elements, the planets, and the divisions of time), human music (the study of the body and its functions and humors, of the soul and its powers, and of the relations of the body and the soul), and instrumental music; geometry, which is subdivided into *planimetria, altimetria,* and *cosmimetria;* and finally astronomy, the subject matter of which is identical in part with that of the preceding sciences, but which is a study of the stars from the point of view of movement and time. The classification of the sciences in the *Didascalicon* gives a place not only to *theorica,* but to *mechanica* as well, that is, to the mechanical arts (the arts of clothing, armament, navigation, agriculture, hunting, medicine, and the theater). Hugh was thus the first to raise technology to the dignity of science. In this regard he was the first of a great number of the authors of the twelfth and thirteenth centuries.

The division of the sciences in the *Didascalicon* was resumed a short time later in a dialogue entitled *Epitome Dindimi in philosophiam.* The interest that he had shown for mathematics reappeared in *Practica geometriae,* the authenticity of which, sometimes contested, is now well established. Composed at about the same time as the *Didascalicon,* this treatise, which shows the influence of Macrobius and especially of Gerbert (Gerbert d'Aurillac), testifies to the state of geometry in the West before the great diffusion of Arabic science. In it Hugh presented the methods of calculating and measuring used in *altimetria* (the measurement of heights and depths), in *planimetria* (the measurement of the lengths and widths of surfaces), and in *cosmimetria,* a discipline intermediate between geometry and astronomy which is concerned with the measurement of the dimensions of the terrestrial sphere and of the celestial sphere. His descriptions of these sciences involve chiefly the properties of triangles and more precisely those of the right-angled triangle, but Hugh also described the methods that can be employed for these mensurations: surveying, measurement of shadows, use of mirrors or the astrolabe, etc. At the end of his *Practica geometriae* he alluded to an astronomical treatise which was supposed to follow, but it is not known if this is a reference to a lost work or simply to a project that Hugh never carried out.

BIBLIOGRAPHY

I. Original Works. Hugh's writings, frequently published, have been reproduced by Migne, in *Patrologia latina,* CLXXV–CLXXVII (Paris, 1854), following the ed. produced in Rouen in 1658 by the canons regular of St. Victor. This ed. sins by default and by excess; it is incomplete and contains apocryphal works, but it may still be used if one takes as a guide D. van den Eynde's *Essai sur la succession et la date des écrits de Hugues de S.-V.,* in Spicilegium Pontificii Athenaei Antoniani, XIII (Rome, 1960).

Recent editions of the scientific works are *Didascalicon, de studio legendi: A Critical Text,* C. H. Buttimer, ed., in The Catholic University of America: Studies in Medieval and Renaissance Latin, X (Washington, 1939); "Epitome Dindimi in philosophiam," R. Baron, ed., in *Traditio,* **11** (1955), 105–119; "Practica geometriae," R. Baron, ed., in *Osiris,* **12** (1956), 176–224; and *Hugonis de Sancto Victore opera propedeutica: Practica geometriae, De grammatica, Epitome Dindimi in philosophiam,* R. Baron, ed., in Publications in Mediaeval Studies. The University of Notre Dame, XX (Notre Dame, Ind., 1966).

For recent translations of his work, see *On the Sacraments of the Christian Faith* (*De sacramentis*), English vers. by R. J. Deferrari (Cambridge, Mass., 1951); J. Taylor, *The Didascalicon of Hugh of St. Victor, Translated From the Latin With Introduction and Notes,* in Records of Civilization, Sources and Studies, no. 64 (New York, 1961).

II. SECONDARY LITERATURE. Information on Hugh's work and life is in R. Baron, *Science et sagesse chez Hugues de S.-V.* (Paris, 1957); and *Études sur Hugues de S.-V.* (Paris, 1963); F. E. Croydon, "Notes on the Life of Hugh of St. Victor," in *Journal of Theological Studies,* **40** (1939), 232–253; J. Taylor, *The Origins and Early Life of Hugh of St. Victor* (Notre Dame, Ind., 1957); and R. Javelet, "Les origines de Hugues de S.-V.," in *Revue des sciences religieuses,* **34** (1960), 74–83.

Hugh's scientific thought is discussed in M. Curtze, "Practica geometriae. Ein anonymer Traktat aus dem Ende des zwölften Jahrhunderts," in *Monatshefte für Mathematik und Physik,* **8** (1897), 193–220; P. Tannery, *Mémoires scientifiques,* J. L. Heiberg, ed., vol. V, *Sciences exactes au moyen âge (1887–1921)* (Toulouse–Paris, 1922), 308–313, 326–328, 357–358, 361–368; R. Baron, "Hugues de S.-V. auteur d'une *Practica geometriae,*" in *Mediaeval Studies,* **17** (1955), 107–116; "Sur l'introduction en Occident des termes 'geometria, theorica et practica,'" in *Revue d'histoire des sciences et de leurs applications,* **8** (1955), 298–302; and "Note sur les variations au XIIe siècle de la triade géométrique altimetria, planimetria, cosmimetria," in *Isis,* **48** (1957), 30–32; L. Thorndike, "Cosmimetria or Steriometria," *ibid.,* p. 458; and J. Châtillon, "Le Didascalicon de Hugues de S.-V.," in *Cahiers d'histoire mondiale,* **9** (1966), 539–552.

JEAN CHÂTILLON

HUGONIOT, PIERRE HENRI (*b.* Allenjoie, Doubs, France, 5 June 1851; *d.* Nantes, France, February 1887), *mechanics, ballistics.*

Hugoniot was the son of a mechanical engineer; his mother was Susanne Mardin. In 1868 he entered the École Polytechnique in Paris. After completing the two-year general course, he chose military engineering as his specialty and graduated in 1872 with an appointment to the naval artillery. He held a teaching post at the École d'Artillerie de la Marine in Lorient. In 1884, on the basis of his scientific work, Hugoniot—now a captain—was made *répétiteur auxiliaire* of mechanics at the École Polytechnique and, a year later, *répétiteur.*

Hugoniot's first research, done with H. Sébert, concerned the effect of powder gases on the bore of a weapon (1882) and was based on the analysis of experimental materials. In 1884 he collaborated with Félix Hélie in preparing a revised and substantially enlarged edition of Hélie's *Traité de balistique expérimentale,* first published in 1865.

In his research in ballistics, Hugoniot made use of the results of work he had done in the mechanics of gases, and it was through this work that he gained wide recognition as one of the creators of the contemporary theory of shock waves. His theory was published in an extensive two-part memoir (pt. 1, 1887; pt. 2, 1889); its basic conclusions had been announced in a series of articles published in the *Comptes rendus* (1885–1886) of the Paris Academy.

Hugoniot's earliest work on the mechanics of gases, written with Sébert in 1884, examined a one-dimensional discontinuous flow of gas under the limiting assumption that before and after the discontinuity of parameters of flow there occurs an adiabatic process (Poisson's law). In 1885 Hugoniot developed, on a sufficiently general physical basis, the theory of discontinuous flows. It was the first theory to apply the law of conservation of energy in an obvious manner. The correspondence that he found between the pressure and the density of gas before and after discontinuity (the pressure jump)—which was called "Hugoniot's adiabatic curve"—is one of the bases of modern shock-wave theory. In 1886 Hugoniot used these findings in his polemic with G. A. Hirn on the laws of the outflow of gas from a vessel. In his mathematical research on the propagation of the shock wave in gas, Hugoniot proceeded from Monge's theory of characteristics, thus anticipating the contemporary method of analysis of supersonic aerodynamics.

These investigations had been preceded by Hugoniot's studies with Sébert in 1882 on the longitudinal vibrations of elastic prismatic beams, produced by a blow, in which he described the propagation of the disturbances by means of recurrent determinate functions, each of which describes a wave process only in the course of a determinate time interval. These researches are related to his work in mathematical analysis (1882) on the expansion of functions in series according to other functions and of functions analogous to Legendre polynomials.

BIBLIOGRAPHY

I. ORIGINAL WORKS. Hugoniot's works, in collaboration with H. Sébert, include "Étude des effets de la poudre dans un canon de 10 cm.," in *Mémorial de l'artillerie de la marine,* X (Paris, 1882); "Sur les vibrations longitudinales des barres élastiques dont les extrémités sont soumises à des efforts quelconques," in *Comptes rendus hebdomadaires des séances de l'Académie des sciences,* **95** (1882), 213–215, 278–281, 338–340; "Sur le choc longitudinal d'une tige élastique fixée par l'une de ses extrémités," *ibid.,* 381–384; "Sur les vibrations longitudinales des verges élastiques et le mouvement d'une tige portant à son extrémité une masse additionnelle," *ibid.,* 775–777; and "Sur la propagation d'un ébranlement uniforme dans un gaz renfermé dans un tuyau cylindrique," *ibid.,* **98** (1884), 507–509.

For Hugoniot's other works, see "Sur le développement des fonctions en séries d'autres fonctions," *ibid.,* **95** (1882), 907–909; "Sur des fonctions d'une seule variable analogues aux polynômes de Legendre," *ibid.,* 983–985; *Traité de* *balistique expérimentale,* 2nd ed., 2 vols. (Paris, 1884), in collaboration with Hélie; "Sur la propagation du mouvement dans les corps et spécialement dans les gaz parfaits," in *Comptes rendus hebdomadaires des séances de l'Académie des sciences,* **101** (1885), 794–796; "Sur la propagation du mouvement dans un fluide indéfini," *ibid.,* **102** (1886), 1118–1120, 1229–1232; "Sur un théorème général relatif à la propagation du mouvement," *ibid.,* **102** (1886), 858–860; "Sur l'écoulement des gaz dans le cas du régime permanent," *ibid.,* 1545–1547; "Sur la pression qui existe dans la section contractée d'une veine gazeuse," *ibid.,* **103** (1886), 241–243; "Sur l'écoulement d'un gaz qui pénétre dans un récipient de capacité limitée," *ibid.,* 922–925; "Sur le mouvement varié d'un gaz comprimé dans un réservoir qui se vide librement dans l'atmosphère," *ibid.,* 1002–1004; "Sur un théorème relatif au mouvement permanent et à l'écoulement des fluides," *ibid.,* 1178–1181; and "Sur l'écoulement des fluides élastiques," *ibid.,* 1253–1255.

For the full version of his memoirs on the mechanics of gas, see "Mémoire sur la propagation du mouvement dans les corps et spécialement dans les gaz parfaits," in *Journal de l'École polytechnique,* cahier 57 (1887), 3–97; cahier 58 (1889), 1–125.

II. SECONDARY LITERATURE. Information on Hugoniot may be found in Maurice Lévy, "Rapport sur les travaux de M. Hugoniot, capitaine d'artillerie de la marine, répétiteur de mécanique á l'École polytechnique" (17 mai 1886), MS in the Hugoniot papers in the Archives de l'Académie des Sciences, Paris. See also Z. Adamar, "Printsip Gyuygensa i teoria Yugono" ("Huygens' Principle and Hugoniot's Theory"), in *Trudy pervogo Vsesoyuznogo sezda matematikov (Kharkov, 1930)* (Moscow–Leningrad, 1936), pp. 280–283.

N. M. MERKOULOVA

HULL, ALBERT WALLACE (*b.* Southington, Connecticut, 19 April 1880; *d.* Schenectady, New York, 22 January 1966), *electron physics.*

Hull was the second of nine sons of Lewis Caleb and Frances Reynolds Hinman Hull, five of whom chose technical careers after attending Yale University. His first bent was classical: he studied Greek at Yale and after graduation taught French and German at Albany Academy (where Joseph Henry had also taught) for one year. Recognizing an enthusiasm for physics, he returned to Yale for graduate work, obtained the doctorate in 1909, and taught for five years at Worcester Polytechnic Institute in Massachusetts before his work came to the attention of Irving Langmuir and others at the General Electric Company.

Hull joined the famed General Electric Research Laboratory at Schenectady, New York, in 1914; his first work was on electron tubes, X-ray crystallography, and (during World War I) piezoelectricity. The work for which he is best known was done after the war, when he published the classic paper on the effect of a uniform magnetic field on the motion of electrons between coaxial cylinders. Versed in Greek, he coined the name "magnetron" for this configuration, which underlies the design of all subsequent "crossed-field" oscillators and amplifiers in which ultrahigh-frequency operation is achieved through control of the motion of electrons by oscillating electric and static magnetic fields at right angles to one another.

Hull's other electron tube work in the 1920's concerned noise measurements in diodes and triodes, the elimination of unwanted feedback in triodes through the introduction of a screen electrode (which marked his invention of the tetrode, independently of Walter Schottky, the acknowledged inventor), and the elimination of the destructive back bombardment of cathodes by residual gas ions, which he showed did no damage if their energy was kept below a certain value. The last project led to his invention of the thyratron, a heavy-duty, gas-filled electron tube originally intended for converting alternating current to direct current in high-power transmission; it found more immediate application in the electronic control of medium-power devices and thus led to the birth of a new branch of technology, industrial electronics.

In the 1930's Hull's interests broadened to metallurgy and glass science. Here again his researches had practical consequences, leading to the development of new alloys, such as Fernico, whose thermal and elastic properties matched those of glass sufficiently well to make strain-free glass-to-metal vacuum seals possible, a development of prime importance to the electrical industry.

Hull retired in 1950, after making additional contributions in World War II, but remained scientifically active to the end. His last publication, in 1966, came fifty-seven years after his first (a paper based on his 1909 doctoral dissertation), marking the end of an unusually long and fruitful career. He received many honors, including the Institute of Radio Engineers' Liebmann Prize (1930) and Medal of Honor (1958), and election to the presidency of the American Physical Society (1942). He was a member of the National Academy of Sciences.

In 1911 Hull married Mary Shore Walker. They had two children: a daughter, Harriet, and a son, Robert Wallace Hull, also a physicist.

BIBLIOGRAPHY

Hull's classic article on the magnetron is "The Effect of a Uniform Magnetic Field on the Motion of Electrons

Between Coaxial Cylinders," in *Physical Review,* **18** (1921), 31–57. A bibliography of his other publications follows the biography in *Biographical Memoirs. National Academy of Sciences,* **41** (1970), 215–233. Information about Hull's forebears is contained in the history published by the Hull family, *The Hull Family in America* (Pittsfield, Mass., 1913), which traces his ancestry to the seventeenth century.

CHARLES SÜSSKIND

HUMBERT, MARIE-GEORGES (*b.* Paris, France, 7 January 1859; *d.* Paris, 22 January 1921), *mathematics.*

A brilliant representative of the French school of mathematics at the end of the nineteenth century, Humbert distinguished himself primarily through his work in fields pioneered by Poincaré and Hermite.

Orphaned at a very young age, Humbert was brought up by his grandparents, industrialists in Franche-Comté. First a boarder at the Oratorian *collège* in Juilly, where he studied classics, he completed his secondary studies at the Collège Stanislas in Paris and entered the École Polytechnique in 1877. For several years he worked as a mining engineer: first in Vesoul and then in Paris, where the École Polytechnique and the École des Mines were quick to add him to their teaching corps.

Humbert earned his doctorate in mathematics in 1885. In 1891–1892 he was a laureate of the Academy of Sciences, and from then on he was well known. Elected president of the Mathematical Society of France in 1893 and named professor of analysis at the École Polytechnique in 1895, he was elected in 1901 to the Academy, filling the seat left vacant by the death of Hermite. From 1904 to 1912 he was Camille Jordan's assistant in the Collège de France and on occasion lectured in his place. Humbert then succeeded to Jordan's chair and continued the teaching of higher mathematics in that institution.

Humbert married in 1890, but his wife died a short time after the birth of their son Pierre; he remarried in 1900. A man of high moral character and intellectual rigor, Humbert was remarkably gifted not only in mathematics but also in clarity of expression and intellectual cultivation. He exerted a great influence and was able, by his discretion and objectivity, to assure respect for his religious convictions during a period of some hostility toward religion in French scientific circles.

Besides his two pedagogical works, it was through numerous memoirs (approximately 150, which have been collected) that Humbert held a major place in the mathematical discovery and production of his time. His writings were inspired by his interest in the study of algebraic curves and surfaces and were marked by the lucidity with which he related the problems encountered in this area to questions of analysis and number theory.

In his doctoral dissertation Humbert completed Clebsch's work by providing the means of determining whether a curve of which the coordinates are elliptic functions of a parameter is actually of type one. He soon noted the advantage for algebraic geometry obtained from a very general technique of representation gained by using Fuchsian functions.

Humbert familiarized himself with the work of Abel, whose theorem concerning the rational sums of certain systems of algebraic differentials he made the subject of important developments and elegant geometric applications. He then derived every possible advantage from the use of Abelian functions in geometry. In his memoir on this subject, submitted for the Academy's prize in 1892, Humbert solved the difficult problem of classifying left curves traced on hyperelliptic surfaces of type two (Kummer surfaces); but his solution excluded the case in which the four periods of the function which defines the surface are joined by a relationship with integral coefficients. Next, Humbert studied Abelian functions presenting singularities of this type and showed that these singularities are characterized by an integer.

He thus enriched analysis and gave the complete solution of the two great questions of the transformation of hyperelliptic functions and of their complex multiplication. He also pointed out the resulting consequence: the existence of a group of transformations of certain surfaces into themselves constitutes an essential difference between the geometry of surfaces and that of curves. But, most important, he completed the work of Hermite by pursuing the applications to number theory throughout his life.

The progressive alliance of geometry, analysis, and arithmetic in Humbert's works is a splendid example of how a broad mathematical education can assist discovery. The results he obtained, and with which his name remains linked, have survived the revolution of modern mathematics, although they belong to a very specialized field.

BIBLIOGRAPHY

I. ORIGINAL WORKS. All of Humbert's memoirs and articles were collected in *Oeuvres de Georges Humbert,* Pierre Humbert and Gaston Julia, eds., 2 vols. (Paris, 1929–1936), with a pref. by Paul Painlevé. Among his writings are *Sur les courbes de genre un* (Paris, 1885), his doctoral thesis; *Application de la théorie des fonctions fuchsiennes à l'étude des courbes algébriques* (Paris, 1886), repr. from *Journal de mathématiques pures et appliquées,*

4th ser., **2** (1886); and three separate notices, on C. Saint-Saëns, I.-J. Paderewski, and G. Doret, in *Fêtes musicales* (Vevey, 1913).

II. SECONDARY LITERATURE. See Emile Borel, *Notice sur la vie et les travaux de Georges Humbert* (Paris, 1922); Camille Jordan and Maurice Croiset, *Discours prononcés aux funérailles de Georges Humbert le 25 janvier 1921* (Paris, 1921); and Maurice d'Ocagne, *Silhouettes de mathématiciens* (Paris, 1928), pp. 167–172.

PIERRE COSTABEL

HUMBERT, PIERRE (*b.* Paris, France, 13 June 1891; *d.* Montpellier, France, 17 November 1953), *mathematics, history of science.*

Humbert was the son of the mathematician Georges Humbert and, like his father, attended the École Polytechnique, entering in 1910. He soon directed himself to scientific research and from 1913 to 1914 he was a member of the research class of the University of Edinburgh. The scientific and philosophical conceptions of Edmund Whittaker, the director of the class, were in accord with his own inclinations and made a deep impression on him throughout his career. Humbert's health was delicate and during World War I he was removed from combat after being wounded. He earned his doctorate in mathematics in 1918 and then began his academic career, which he spent almost entirely in the Faculty of Science at Montpellier, but which consumed only a portion of his energies.

Humbert combined his father's mathematical ability with the temperament of a humanist. He demonstrated a highly refined sensitivity to culture, devoting attention to literature and music as well as to science. Moreover, he was unsatisfied with the simple juxtaposition of knowledge and religious faith. A talented lecturer, he traveled a good deal in France and abroad. He also possessed remarkable ability for organization, which he displayed mainly in the French Association for the Advancement of Sciences and in the Joseph Lotte Association (a society of Catholic public school teachers).

The multiplicity of subjects in which Humbert was interested, and about which he contributed stimulating articles in the most diverse periodicals, is characteristic of his highly personal vocation: to promote the awakening of the intellect. In pursuit of this goal he was willing to sacrifice a certain intellectual rigor in the interest of his wide-ranging curiosity. Thus Humbert's scientific work provides no definitive advances, although it remains a valuable reference source.

In the field of mathematics, Humbert, faithful to Whittaker, directed his efforts chiefly toward the development of symbolic calculus. He also began to undertake scholarly research in the history of science, specializing in the study of seventeenth-century astronomy. He was partially influenced in this choice by his father-in-law, the astronomer Henri Andoyer. His articles on the Provençal school, whose members included Peiresc and Gassendi, revealed the resources held by the archives in Aix, Carpentras, Digne, and other localities in the south of France.

Beyond these two major areas, Humbert should be remembered for his other writings, numerous and highly varied, that remain capable of inspiring new investigations.

BIBLIOGRAPHY

I. ORIGINAL WORKS. Humbert's works in mathematics include *Sur les surfaces de Poincaré* (Paris, 1918), doctoral thesis; *Introduction à l'étude des fonctions elliptiques* (Paris, 1922); "Fonctions de Lamé et fonctions de Mathieu," Mémorial des sciences mathématiques, no. 10 (1926); "Le calcul symbolique," *Actualités scientifiques,* no. 147 (1934); "Potentiels et prépotentiels," *Cahiers scientifiques,* no. 15 (1936); "Le calcul symbolique et ses applications à la physique mathématique," Mémorial des sciences mathématiques, no. 105 (1947), rev. and enl. in a sep. pub. (Paris, 1965); *Formulaire pour le calcul symbolique* (Paris, 1950), written with N. W. McLachlan; and *Supplément au formulaire pour le calcul symbolique* (Paris, 1952), also written with McLachlan and L. Poli.

For Humbert's publications in the history of astronomy and mathematics, see "Histoire des mathématiques, de la mécanique et de l'astronomie," in Gabriel Hanotaux, ed., *Histoire de la nation française,* tome XIV, vol. 1 (Paris, 1924), written with Henri Andoyer; *Pierre Duhem* (Paris, 1932); *Un amateur: Peiresc (1580–1637)* (Paris, 1933); "L'oeuvre astronomique de Gassendi," in *Actualités scientifiques,* no. 378 (1936); *De Mercure à Pluton, planètes et satellites* (Paris, 1937); "Histoire des découvertes astronomiques," in *Revue des jeunes,* no. 16 (1948); *Blaise Pascal, cet effrayant génie* (Paris, 1947); and "Les mathématiques de la Renaissance à la fin du XVIIIe siècle," in Maurice Daumas, ed., *Histoire de la science* (Paris, 1957), 537–688.

On his contribution to the philosophy of science, see *Philosophes et savants* (Paris, 1953); and Edmund Whittaker, *Le commencement et la fin du monde, suivi de hasard, libre arbitre et nécessité dans la conception scientifique de l'univers* (Paris, 1953), translated from English by Humbert.

Articles by Humbert include "Les astronomes français de 1610 à 1667. Étude d'ensemble et répertoire alphabétique," in *Mémoires de la Société d'études scientifiques de Draguignan,* **63** (1942); "Les erreurs astronomiques en littérature"; "La mesure de la méridienne de France," in *Mémoires de l'Académie des sciences et lettres de Montpellier,* **20** (1924); **25** (1930); **27** (1932); "Spongia solis," in *Annales de l'université de Montpellier,* **1** (1943); "Claude Mydorge

(1585–1647)," *ibid.,* **3** (1945); "La première carte de la lune," in *Revue des questions scientifiques,* **108** (1931); "Le baptême des satellites de Jupiter," *ibid.,* **117** (1940); and "L'observation des halos," in *Atti dell'Accademia pontificia dei Nuovi Lincei* (1931).

Humbert wrote many other articles and memoirs which may be found in *Archives internationales d'histoire des sciences,* and *Revue d'histoire des sciences et de leurs applications,* two journals on which he collaborated.

II. SECONDARY LITERATURE. For information on Humbert, see P. Sergescu, "Notice sur Pierre Humbert," in *Archives internationales d'histoire des sciences,* **7,** no. 27 (1954), 181–183; B. Rochot, "Notice sur Pierre Humbert," in *Revue d'histoire des sciences,* **7,** no. 1, 79–80; and Jacques Devisme, *Sur l'équation de M. Pierre Humbert* (Paris–Toulouse, 1933), doctoral thesis.

PIERRE COSTABEL

HUMBOLDT, FRIEDRICH WILHELM HEINRICH ALEXANDER VON (*b.* Berlin, Germany, 14 September 1769; *d.* Berlin, 6 May 1859), *natural science.*

Humboldt's father, Alexander Georg von Humboldt, was a Prussian officer who reached the rank of major and, from 1765 to 1769, served as chamberlain to the wife of the heir to the Prussian throne. In 1766 he married a widow, Marie Elisabeth Colomb Holwede, and devoted himself to administering her estates. She herself was of middle-class Huguenot extraction and had inherited the holdings from her first husband. Not until about 1738 did Alexander Georg's father, Hans Paul, gain confirmation as one of the nobility. (Interestingly, Humboldt's baronial title was only conferred officially on the family in 1875, sixteen years after his death.)

Alexander's education and that of his older brother Wilhelm, later a statesman, linguist, and founder of the University of Berlin, was one of private tutorship. At an early age the brothers joined the circle known as the Berlin Enlightenment, with which many well-off Jewish families were associated. After 1789 Humboldt openly subscribed to French libertarian views; he lamented, for example, that the Peasant's War of 1525 had not succeeded.

From 1787 to 1792 Humboldt studied at the universities of Frankfurt an der Oder and Göttingen and at the academies of commerce in Hamburg and of mining in Freiberg, Saxony. His studies familiarized him with technology, and he also acquired a background in economics, geology, and mining science. He studied botany with particular zeal, for a time under the guidance of Karl Ludwig Willdenow. His first publication in book form in 1790 came out of a student natural history excursion. Here Humboldt attacked the theories of volcanism but without unequivocally embracing those of neptunism. Humboldt's most influential teacher in his youth was the Freiberg geologist Abraham Gottlob Werner, leader of the neptunist school opposing the plutonists. Humboldt also occupied himself in Freiberg in 1791 with antiphlogistic chemistry.

In 1790 Humboldt traveled to the Netherlands and thence to England and Paris with Georg Forster, who had been with Cook on the second world voyage and was an impassioned adherent of the French Revolution. He arrived in Paris shortly before the anniversary of the storming of the Bastille. "The sight of the Parisians, with their National Assembly and yet incomplete Temple of Liberty, to which I myself carted sand, stirred me like a vision before the soul" (to F. H. Jacobi, 3 January 1791).

Promptly after completing his studies he entered, in March 1792, the Prussian mining service and soon became a mining leader in the Prussian part of Upper Franconia. He invented safety lamps and a rescue apparatus for miners threatened with asphyxiation, himself testing these devices in dangerous experiments. Upon his own initiative and funds he founded a "free mining school" to train miners, demonstrating early his lifelong social concern. He managed to do considerable work on problems of practical mining without neglecting his scientific research. In 1793 he published a work, dating back to the Freiberg period, which he had expanded and improved. In its appendix there is a treatment of 258 "subterranean cryptogamic plants," a discussion of post-Aristotelian physiological views, and theoretical reflections. There are also descriptions of experiments in plant physiology, then in its infancy.

Humboldt, like his contemporaries, sought proof of the presupposed "life force" (*vis vitalis*). He pursued this through galvanic experiments, among them painful personal tests, hoping thereby to throw light on the "chemical process of life." The results of his investigations were published in 1797; of special note was his original attempt to draw analogies between animal and plant life processes.

During this period he managed to handle both his official duties and his studies on cohesion and universality in nature; his use of the comparative method and his working out of types were characteristic. Far from being a romantic, Humboldt was a thorough empiricist in studying general relationships in nature. For him facts, measurement, and number were the cornerstone of science, and not speculation and hypothesis. He believed in universal harmony and equilibrium in nature, and was unable to perceive the importance of oppositional forces in any development.

Humboldt traveled in 1791 from Freiberg through the Bohemian Mittelgebirge, and in 1792 and 1794 he went on inspection tours of salt mines in what is now Austria, Czechoslovakia, and Poland. In 1794 he went again to the Netherlands, partly under diplomatic auspices. Two years later he negotiated a treaty with the commander of the French troops entering Württemberg, in order to effect the formal neutralization of the Franconian principalities.

In the latter half of 1795 Humboldt, a lifelong bachelor, made an extensive trip through northern Italy and the Swiss and French Alps. He was initially accompanied by Reinhard von Haeften, an officer, and later by Karl Freiesleben, a Saxonian mining official known from his Freiberg days. The trip of 1795—in the course of which Humboldt met Alpine experts, learned about altitude effects on climate and plants, and came to recognize the evidence of the relief and the need for astronomical and geomagnetic observatories—exercised a lasting influence on him. Geomagnetism also caught Humboldt's interest early—in 1796 he discovered the magnetism of the Haidberg near Gefrees, northeast of Bayreuth—and his geomagnetic work occupied him for five decades.

The first record of Humboldt's interest in describing natural interrelationships is found in a letter (24 January 1796) to the natural scientist Pictet: "Je conçus l'idée d'une physique du monde." Humboldt, although indisputably one of the founders of geography as a science, had as his major goal a comprehensive view of nature to which the earth sciences would contribute significantly. As a Prussian government official, there would be difficulties for him in pursuing such a major undertaking, but upon his mother's death in 1796 he became financially independent. Leaving the civil service, he looked ahead to a "great journey beyond Europe."

At Jena in 1797 he concluded extensive experiments on galvanism and chemical effects on animals and plants, and also acquainted himself with anatomy. Here he renewed and deepened his earlier contacts with Goethe (whom he had met personally) and Schiller. With his wide interests, he had an immediate rapport with Goethe, but Schiller saw Humboldt as a "man of much too limited intellect." This feeling notwithstanding, Schiller published in his journal *Die Horen* Humboldt's article "The Genius of Rhodes" (1795), an allegorical tale in "semi-mythical clothing" in which, agreeing with Schiller, Humboldt endorsed the theory of the life force; Humboldt later abandoned this position.

In Jena, Humboldt also learned techniques for making geodetic and geophysical measurements, and especially for taking astronomical bearings. He later regarded such bearings to be the basis for all geography, and criticized travel by routes that were needlessly uncertain for want of correct measurements.

At the end of May 1797 Humboldt went via Dresden and Prague to Vienna to prepare for a trip to the West Indies. But his desire to see active volcanoes at first hand—inspired by a previous trip to Italy—was several times thwarted because of the political situation. He heard accounts by Viennese scholars of their travels, studied West Indies plants kept at Schönbrunn, and made a trip to Hungary. While in Salzburg at the end of October, he went on excursions with the geologist Leopold von Buch. He also practiced taking geographic bearings and made eudiometric measurements.

In April 1798 Humboldt followed his brother to Paris, where he hoped to arrange his projected transoceanic travel. The following month he read a paper before the Paris Academy, "Expériences sur le gaz nitreux et ses combinaisons avec le gaz oxygène," and later gave several lectures. Humboldt's reputation was steadily increasing; since 1793 he had been a member of the Leopoldine Carolinian Academy, and in that same year he had received the elector of Saxony's gold medal for art and science. He was present at the conclusive arc-degree measurement between Dunkirk and Barcelona. He contributed to the first relatively conclusive determination of magnetic inclination in Paris, set up galvanic experiments, and investigated the chemical composition of air.

On 20 October 1798 he left Paris with the French botanist Aimé Bonpland, his companion for the next six years. He went first to Marseilles, where he busied himself with geodetic measurements and botanic field studies, hoping to sail to North Africa. But in mid-December he went to Spain on what was virtually a "measuring expedition"; with sextant, chronometer, barometer, and thermometer, en route to Madrid by way of Valencia and Barcelona, he established data for a relief map that for the first time clearly outlined a sizable region.

In March 1799 Humboldt received permission to make a research tour through the Spanish colonies, and on 5 June he and Bonpland sailed from La Coruña. After a break in the British blockade and a stop at Tenerife, they landed on 16 July 1799 in what is now Venezuela.

He and Bonpland remained in South America until the end of April 1804. Exposed to great hardships and dangers, the two journeyed by foot, pack horse, native canoe, and sailing vessel through every conceivable type of country in what is now Venezuela, Cuba, Colombia, Peru, Ecuador, and Mexico. They recorded, sketched, described, measured, and compared what they observed, and gathered some 60,000 plant specimens, 6,300 of which were hitherto unknown in Europe. Humboldt made maps and amassed exhaustive data in countless fields—magnetism, meteorology, climatology, geology, mineralogy, oceanography, zoology, ethnography. In addition to observations on plant geography and physiognomy, he made historical and linguistic investigations. Humboldt had mutually profitable meetings with South American scholars, notably José Celestino Mutis and Francisco José de Caldas. He showed as much interest in early Indian monuments as in the current population figures, social conditions, and economic developments. He found slavery to be the greatest evil of humankind, and this remained a matter of paramount concern to him.

Humboldt navigated the Orinoco and Magdalena rivers and confirmed the bifurcation of the Casiquiare River, thereby proving the connection between the Orinoco and the Amazon. He set a new mountaineering altitude record with his ascent of Chimborazo on June 1802, although he failed to reach the summit. This trip has justly been called "the scientific discovery of America."

In 1804 Humboldt traveled to the United States, visiting Philadelphia and Washington, D.C., where he met several times with President Jefferson and members of the cabinet. He reported on his travels, his information on New Spain (Mexico) being of special interest.

After a further stop in Philadelphia and meeting with the American Philosophical Society (which elected him a member on 20 July), he set sail for France at the end of June and landed on 3 August, having been away from Europe for more than five years. Humboldt hastened to an enthusiastic reception in Paris, where he read reports of his journey in the Academy. He enjoyed his social contacts with the Parisian scientists, particularly Gay-Lussac, with whom he carried out chemical analyses of air.

It was in Paris that he also became acquainted with Simón Bolívar, with whom he was to correspond until Bolívar's death in 1830. "Humboldt has done more good for America than all her conquerors," Bolívar once said in tribute. Urging Bolívar to hold to a moderate course after his victory, Humboldt not only recommended certain natural scientists but advised the South American leader in numerous other ways. He proposed a leveling of the isthmus between Panama and the mouth of the Chagres (for more than fifty years Humboldt called for the construction of a canal linking the Atlantic and Pacific) and the furthering of science in the New World.

In March 1805 Humboldt left Paris with Gay-Lussac to see his brother Wilhelm in Rome. From Rome he went to Naples, ascended Mt. Vesuvius several times, and in September traveled via Milan, Zurich, and Göttingen to Berlin, arriving on 16 November (his stay there was cut short by Napoleon's victories at Jena and Auerstadt in 1806). At the end of 1807 he was sent on a diplomatic mission to Paris, where he remained until 1827, making trips to London, Vienna, Bratislava, and Italy. Only in Paris could Humboldt have his research findings properly evaluated by first-rank scientists, and only there could he avail himself of the best artists and technical resources.

His voluminous, never finished travel journal was published in thirty-four volumes over twenty-five years; the volumes, including some 1,200 copperplates, cost about 780,000 francs. Humboldt also recorded his travels in numerous treatises, in which he developed climatology as a science in itself; established the fields of plant geography and orography; formulated the fissure theory of volcanology; specified vegetation types; set forth concepts such as plateau, mean height of a pass, mean height of a summit, and mean temperature; and introduced the isotherm in meteorology.

Humboldt gave a major impetus to the study of the Americas. He studied the discovery and history of America and its economics and politics, particularly in Cuba and in Mexico. He addressed himself to elucidating possible connections between climate and vegetation, between altitude and fertility, between human productivity and property relationships, and between the animal and plant kingdoms. He rectified the calculations of his astronomical bearings to make them a reliable basis for maps of the regions he had visited. His geographical monographs on Cuba and Mexico represent the first treatments of geography in terms of science, politics, and economics.

During his Paris years Humboldt was not concerned solely with publishing the results of his travels. He was also preparing for a journey to Asia, where he hoped to observe and measure the ranges and volcanoes for comparison with areas of the Andes. After 1809 he spoke often of this trip, but political vicissitudes again made the planning of it uncertain. In 1818 the Prussian government guaranteed the financing of a four- to five-year trip by Humboldt to India, the Himalayas and Tibet, Ceylon, and the East Indies; and up until 1825 Humboldt made references to his forthcoming departure. Why he did not make the trip has never been adequately explained.

In 1827 Humboldt returned to Berlin. Two reasons may have prompted his return to his birthplace: his dependence on the Prussian salary (his trips had bankrupted him), and the hope of utilizing the ties between the court at Berlin and the ruling house of Russia in order to make a long-planned Siberian journey. He also returned to his home with the express purpose of raising the level of mathematics and natural sciences to the point that Berlin intellectual life would compare with that of Paris. He valued his independence, and although he served as a royal chamberlain he was not burdened with any other official posts. He was, however, an adviser on science and art and (from 1842) chancellor of the peace division of the order *pour le mérite,* positions in which he exercised no political influence.

In 1829 he set out on his Siberian trip as a guest of the Russian government. Accompanied by the naturalist Christian Gottfried Ehrenberg and the mineralogist Gustav Rose, Humboldt traveled about 9,000 miles. By now famous, he was honored everywhere. They went via Riga to St. Petersburg, from there to Tobolsk via Moscow, Kazan, and the northern Urals, then through western Siberia to the Altai Mountains on the border of Chinese Tungusic territory. He returned to St. Petersburg via the southern Urals, the Caspian Sea, Voronezh, and Moscow.

In the course of the journey Humboldt suggested that geomagnetic and meteorological stations be set up in order to reinforce his own en-route observations and measurements with systematic investigations covering larger areas. He collected, measured, and thoroughly compared relative temperatures, magnetic values, and geological, mineralogical, and biological data. His comparative methods enabled him, in one instance, to predict the existence of diamonds in the Urals, a surmise that was borne out by their discovery during his very trip.

Humboldt maintained the contacts he made on this journey with Russian scholars. During the trip he also lent his influence to the cause of Poles who had been exiled to Siberia.

Humboldt returned to Berlin before the end of 1829, and in 1830 went to Paris where, intermittently until 1847–1848 (altogether about three and a half years), he used the libraries and fulfilled diplomatic assignments. He also obtained the advice of his learned friends, especially Arago, in composing the Asian travel journals and in completing a long-worked-on history of medieval geography. The latter demonstrates his historical interests; indeed, throughout his works is the manifest conviction that scientific progress is not accidental but the result of experience and "earlier development of thought."

In the last decades of his life Humboldt collected and revised his scattered *Kleinere Schriften* (1853) and prepared the third edition of his favorite work, *Ansichten der Natur* (1849), an aesthetic presentation of research in natural science and geography and of "pictures of nature." He worked primarily on *Kosmos,* the plan for which dated from 1827–1828, when Humboldt had lectured on physical geography in Berlin. The first volume appeared only in 1845, and with the second (1847) marked a genuine popular triumph for the aged author. He wrote in a letter to Bessel, dated 14 July 1833: "It is the work of my life; it should reflect what I have projected as my conception and vision of explored and unexplored relationships of phenomena, out of both my own experience and painstaking inquiry into readings in many languages."

The *Kosmos* is a popular scientific book in the best sense of that term. The entire material world from the galaxies to the geography of the various mosses, the history of physical cosmography, the needed stimulation for nature study—he sought to present all in vivid, "pleasing" language. Volumes III through V, containing his special research findings and added material, were not equally successful; Humboldt died before completing the fifth volume. The index was prepared according to his specifications and he credited each contemporary to whom he felt indebted. The work cites over 9,000 sources and is thus an important reference for the history of science.

In the area which he especially cherished, geomagnetic measurement, Humboldt suggested in a letter (23 April 1836) to the president of the Royal Society of London the worldwide establishment of geomagnetic observatories. Gauss, with whom he corresponded, had just conceived the theory of the *intensitas vis magneticae* (1833), and it was not easy for Humboldt to see a field long his own domain become the province of a more creative mind. Humboldt nevertheless recognized his own limitations. In 1789 he had almost discovered the Gaussian addition logarithms, but he later had to confess that he could "claim for himself no serious position in the higher realms of mathematics" (letter to C. G. G. Jacobi, 27 December 1846). He saw clearly the reciprocity of mathematics with both the natural sciences and industrial application:

> Man cannot have an effect on nature, cannot adopt any of her forces, if he does not know the natural laws in terms of measurement and numerical relations. Here also lies the strength of the national intelligence, which increases and decreases according to such knowledge. Knowledge and comprehension are the joy and justification of humanity; they are parts of the national wealth, often a replacement for those materials that nature has all too sparsely dispensed. Those very peoples who are behind in general industrial activity, in

application of mechanics and technical chemistry, in careful selection and processing of natural materials, such that regard for such enterprise does not permeate all classes, will inevitably decline in prosperity; all the more so where neighboring states, in which science and the industrial arts have an active interrelationship, progress with youthful vigor [*Kosmos,* I (1845), 36].

Besides his extensive literary and court activities, Humboldt remained devoted to humanitarian causes. He was responsible for antislavery legislation in Prussia and spoke out against anti-Semitism and racism: "By asserting the unity of the human race, we also oppose every distasteful assumption of higher and lower races of man. There are more adaptive, more highly educated, and more spiritually enriched peoples, but there are none nobler than others. All are equally ordained to be free" (*ibid.,* p. 385).

Humboldt gave advice to many gifted youths along with encouragement, recommendations for awards, and often financial help. Such young scholars regarded themselves as "his children" (letter of Emil du Bois-Reymond to Karl Ludwig, 26 June 1849). Among the many people in whom he took an early interest were the mathematicians Dirichlet and Eisenstein; the explorers Moritz Wagner, Heinrich Barth, Eduard Vogel, and the brothers Schomburgk and Schlagintweit; the chemists Liebig and Mitscherlich; the physicists Poggendorff and Riess; the physiologists Müller and du Bois-Reymond; the natural scientists Louis Agassiz and Boussingault; the meteorologist Dove; the geodesist Baeyer; the astronomers Argelander, Galle, and Karl Bruhns; the Egyptologists Richard Lepsius and Heinrich Brugsch; the geophysicist Georg Erman; and the zoologist Wilhelm Peters.

Through Johann Gottfried Flügel, United States consul general in Leipzig, Humboldt followed the progress of the natural sciences in North America and remained greatly interested in the development of the United States. He nonetheless regretted that there "freedom is only a mechanism in the principle of profitability," and that indifference to slavery was prevalent. He observed, "The United States is a Cartesian spiral, sweeping away everything and yet boringly level" (letter to Varnhagen von Ense, 31 July 1854). He also complained that French rule was becoming more immoral through "administrators who have been defrauding, extorting, and using violence in Algeria" (letter to Caroline von Wolzogen, 6 May 1837). Shocked by the bloody events of March 1848, he lamented much more the subsequent period of reaction.

Humboldt was awarded honorary doctorates by the universities of Frankfurt an der Oder (1805), Dorpat (1827), Bonn (1828), Tübingen (1845), Prague (1848), and St. Andrews (1853). In 1852 he received the Copley Medal. All major academies elected him to membership, and he was a member of the illustrious Société d'Arcueil after 1807. He corresponded extensively with eminent scholars, artists, writers, and politicians of his time; indeed, over 2,000 of his correspondents are known to us. His ties to the French intellectual world were especially close. The French never forgot Humboldt's earnest intercession, during the occupation of France by the allied troops, on behalf of scientific institutions such as the Muséum d'Histoire Naturelle; nor his fight to save private property, including that of Laplace. In 1827 he became honorary president of the Société de Géographie in Paris.

Humboldt was among the first to interest astronomers in shooting stars and his method for determining the light intensity of southern stars was an original contribution to astronomy (*Astronomische Nachrichten,* **16** [1839], 225–230). He was the first to note the significant decrease of magnetic intensity with the appearance of the aurora borealis. Humboldt gave a qualitative explanation for the amplification of sound at night. (In 1955 Hans Ertel introduced with the quantitative solution the "Humboldt effect" into literature; it is the only thing in the physical sciences for which Humboldt is the eponym.) He was also the first to send guano to Europe.

Despite his accomplishments, Humboldt does not rank with the great discoverers or inventors, as he himself realized. No matter where he traveled, others had been there before him and had reported on their trips. But Humboldt saw broadly and comprehensively, and, where others perceived only isolated facts, he combined observations and saw unity in diversity. He was gifted with a quick intelligence and with boundless receptivity and powers of memory.

His deficiencies notwithstanding, Humboldt towers as a servant of worldwide science and a humanitarian. His stimulating influence on his contemporaries and on science itself, his humanistic and democratic principles, and his unshakable faith in the constant progress of mankind have remained exemplary.

BIBLIOGRAPHY

I. ORIGINAL WORKS. Unfortunately there is still no complete bibliography of Humboldt's writings. Therefore, we must still use the list which Julius Löwenberg gave in *Alexander von Humboldt: Eine wissenschaftliche Biographie,* Karl Bruhns, ed., II (Leipzig, 1872), 485–552; this list was reprinted unchanged (Stuttgart, 1960). Löwenberg had handled poorly the great problems which result from the numerous preprints, abstracts, translations, and reprints; from publications appearing in several parts; and from various forms of a single work with variations in content. In addition there are (1) an inconsistency in arrangement, (2) mistakes stemming from faulty examination in the rendering of titles, (3) listing of the same title in several places, and (4) the inclusion of writings which were not even written by Humboldt. See the review of the 1960 reprint by Fritz Gustav Lange in *Petermanns Geographische Mitteilungen,* **108** (1964), 110.

Other bibliographical sources are *Alexander von Humboldt. Bibliographie seiner ab 1860 in deutscher Sprache herausgegebenen Werke und der seit 1900 erschienenen Veröffentlichungen über ihn* (Leipzig, 1959), pp. 9–14; and Hanno Beck, *Alexander von Humboldt,* II (Wiesbaden, 1961), 347–356.

The so-called *Gesammelte Werke,* 12 vols. (Stuttgart, 1889), contain a fraction of Humboldt's writings in German.

The following are Humboldt's most important works published separately during his lifetime; and, of course, without regard to the editions in different formats, to separately published, somewhat expanded extracts, or to later printings and supplements as well as to translations: *Mineralogische Beobachtungen über einige Basalte am Rhein . . .* (Brunswick, 1790); *Florae Fribergensis specimen plantas cryptogamicas praesertim subterraneas exhibens . . .* (Berlin, 1793); *Versuche über die gereizte Muskel- und Nervenfaser . . .,* 2 vols. (Poznán–Berlin, 1797); *Versuche über die chemische Zerlegung des Luftkreises . . .* (Brunswick, 1799); *Ueber die unterirdischen Gasarten und die Mittel ihren Nachtheil zu vermindern* (Brunswick, 1799); *Ansichten der Natur, mit wissenschaftlichen Erläuterungen* (Tübingen, 1808); *Essai géognostique sur le gisement des roches dans les deux hémisphères* (Paris, 1823); *Fragmens de géologie et de climatologie asiatiques,* 2 vols. (Paris, 1831); *Asie centrale. Recherches sur les chaînes de montagnes et la climatologie comparée,* 3 vols. (Paris, 1843); *Kosmos: Entwurfeiner physischen weltbeschreibung,* 5 vols. (Stuttgart–Tübingen, 1845–1862); *Kleinere Schriften,* I, *Geognostische und physikalische Erinnerungen* (Stuttgart–Tübingen, 1853), the only vol. published; and *Atlas der kleineren Schriften . . .* (Stuttgart–Tübingen, 1853). Complete comprehension of the great American travel journals presents great difficulties to the bibliographer. Following is a survey of short titles under the various subject groups designated by Humboldt but persistently ignored by bibliographers; it is based on the folio or quarto ed. The overall title is *Voyage aux régions équinoxiales du Nouveau Continent, fait en 1799, 1800, 1801, 1802, 1803, et 1804 par Al* [*exandre*] *de Humboldt et A* [*imé*] *Bonpland . . .* (Paris, 1805–1834). Subject group I (7 vols.) includes: *Relation historique; Vue des Cordillères; Examen critique; Atlas du Nouveau Continent.* Group II (2 vols.) concerns zoology and contains *Recueil d'observations de zoologie.* Group III (3 vols.) contains the work on Mexico: *Essai politique sur la Nouvelle Espagne* and *Atlas de la Nouvelle Espagne.* To subject group IV, astronomy (3 vols.), belong *Conspectus longitudinum et latitudinum* and *Recueil d'observations astronomiques.* Group V (1 vol.) contains the work on plant geography: *Essai sur la géographie des plantes accompagné d'un tableau physique des régions équinoxiales.* The sixth and last group (18 vols.) deals with botany: *Plantes équinoxiales, Mélastomacées, Nova genera, Mimoses, Synopsis plantarum* (associated by Carl Sigismund Kunth with the travel works by means of the serial titles in the *Voyage aux régions équinoxiales . . .*) and *Graminées.*

The works to which Humboldt attributed great importance are *Essai politique sur l'île de Cuba,* 2 vols. (Paris, 1826), a greatly expanded extract from the *Relation historique; Tableau statistique de l'île de Cuba* (Paris, 1831); *Essai sur la géographie des plantes* (see above); and "Des lignes isothermes et de la distribution de la chaleur sur le globe," in *Mémoires de physique et de chimie de la Société d'Arcueil,* **3** (1817), 462–602. Humboldt exerted the greatest influence on the general public through the *Ansichten der Natur* (3rd ed., 1849) and the *Kosmos,* as well as through the extract he authorized from the *Relation historique: Alexander von Humboldt's Reise in die Aequinoctial-Gegenden des neuen Continents,* rev. and trans. by Hermann Hauff, 4 vols. (Stuttgart, 1859–1860), and through translations of his works.

Besides the publication of hundreds of single letters or of small groups of letters there are more or less comprehensive collections of correspondence. Some include several of Humboldt's correspondents—e.g., the eds. by Dézos de La Roquette, 2 vols. (Paris, 1865–1869); of E. T. Hamy (Paris, 1905); of C. Müller (Leipzig, 1928); and of D. I. Shcherbakov *et al.* (Moscow, 1962). Others contain his correspondence with one or predominantly one correspondent, such as those involving K. A. Varnhagen, von Ense, L. Assing, ed. (Leipzig, 1860); H. Berghaus, H. Berghaus, ed., 3 vols. (Leipzig, 1863); M. A. Pictet, A. Rilliet, ed. (Geneva, 1869); Count G. von Cancrin, W. von Schneider and W. Russow, eds. (Leipzig, 1869); C. K. J. von Bunsen (Leipzig, 1869); J. W. von Goethe, F. T. Bratranek, ed. (Leipzig, 1876), and L. Geiger, ed. (Berlin, 1909); C. F. Gauss, K. Bruhns, ed. (Leipzig, 1877); Wilhelm von Humboldt, F. Gregorovius, ed. (Stuttgart, 1880); W. G. Wegener, A. Leitzmann, ed. (Leipzig, 1896); F. Arago, E. T. Hamy, ed. (Paris, 1907); J. von Olfers, E. W. M. von Olfers, ed. (Nuremberg–Leipzig, 1913); F. G. Eisenstein, K-R. Biermann, ed. (Berlin, 1959); and A. Valenciennes, F. Théodoridès, ed. (Paris, 1965). But all these and others encompass only a small part of Humboldt's correspondence. The collected material, in photocopies by the German Academy of Science in Berlin, D.D.R., includes more than 10,000 pieces, among them about 5,600 unedited letters.

Humboldt's letters and MSS are scattered throughout the world. The most important owners of originals are the German Central Archives, Merseburg, D.D.R.; the German State Library, Berlin, D.D.R.; State Library for the Preservation of Prussian Cultural Possessions, West Berlin, and the Schiller National Museum, Marbach. Many public and private writings by Humboldt are also in France, the United States, the Soviet Union, and Latin America, as well as other countries.

II. SECONDARY LITERATURE. There is no bibliography of the literature on Humboldt which separates the important from the nonessential and at the same time arranges things according to subject. Therefore one must use the above mentioned bibliography (Leipzig, 1959), pp. 15-36; Hanno Beck, *Alexander von Humboldt,* II (Wiesbaden, 1961), 356-380; *Literaturzusammenstellung über Alexander von Humboldt. Schrifttum der Jahre 1805-1959,* 3rd ed. (Jena, 1959); N. G. Suchowa, *Alexander von Humboldt in der russischen Literatur* (Leipzig, 1960); and Poggendorff VIIa, supp. (1971), 295-301.

A few of the important works of the literature on Humboldt are *Mémoires Alexander von Humboldt's,* 2 vols. (Leipzig, 1861), often unjustifiably attributed to Julius Löwenberg (it must be used with great caution; it contains falsifications—but Humboldt's letters to the U.S. consul general in Leipzig, Johann Gottfried Flügel, are genuine); *Alexander von Humboldt: Eine wissenschaftlich Biographie,* Karl Bruhns, ed., 3 vols. (Leipzig, 1872); Herbert Scurla, *Alexander von Humboldt. Sein Leben und Wirken* (Berlin, D.D.R., 1955); Helmut de Terra, *Humboldt. The Life and Times of Alexander von Humboldt* (New York, 1955, 6th ed. 1968; German trans. 1956; Russian trans. 1961); *Alexander von Humboldt. Gedenkschrift zur 100. Wiederkehr seines Todestages,* Alexander von Humboldt Commission of the German Academy of Sciences (D.D.R.), ed. (Berlin, D.D.R., 1959); *Gespräche Alexander von Humboldts,* Hanno Beck, ed. (Berlin, D.D.R., 1959); Hanno Beck, *Alexander von Humboldt,* 2 vols. (Wiesbaden, 1959-1961); *Alexander von Humboldt. Studien zu seiner universalen Geisteshaltung,* Joachim H. Schultze, ed. (Berlin, 1959); Richard Bitterling, *Alexander von Humboldt* (Munich-Berlin, 1959); *Alexander von Humboldt. Vortäge und Aufsätze . . .,* Johannes F. Gellert, ed. (Berlin, D.D.R., 1960); "Beiträge zum Alexander-von-Humboldt-Jahr 1959," *Zusammenstellung von Sonderdrucken aus der wissenschaftlichen Zeitschrift, Humboldt-Universität zu Berlin,* **8** (1958-1959) and **9** (1959-1960); "Alexander von Humboldt. Seine Bedeutung für den Bergbau und die Naturforschung," in *Freiberger Forschungshefte,* **D33** (1960); V. A. Esakov, *Aleksandr Gumboldt v Rossii* (Moscow, 1960); Lotte Kellner, *Alexander von Humboldt* (London-New York-Toronto, 1963); Adolf Meyer-Abich, *Alexander von Humboldt in Selbstzeugnissen und Bilddokumenten* (Hamburg, 1967); Kurt-R. Biermann, Ilse Jahn, and Fritz G. Lange, "Alexander von Humboldt. Chronologische Übersicht über wichtige Daten seines Lebens," in *Beiträge zur Alexander-von-Humboldt-Forschung,* **1** (1968); *Alexander von Humboldt. Wirkendes Vorbild für Fortschritt und Befreiung der Menschheit.* German Academy of Sciences, Berlin, ed. (Berlin, D.D.R., 1969); *Alexander von Humboldt. Werk und Weltgeltung,* Heinrich Pfeiffer, ed. (Munich, 1969); "Numero especial dedicado a la conmemoración del bicentenario de Alejandro de Humboldt," *Islas, Revista de la Universidad Central de las Villas Santa Clara, Cuba,* **11** (1969), no. 3; and "Bicentenario de Humboldt," *Academia de Ciencias de Cuba, Serie histórica,* (1969/70), nos. 7-13.

KURT-R. BIERMANN

HUME, DAVID (*b.* Edinburgh, Scotland, 26 April 1711; *d.* Edinburgh, 25 August 1776), *philosophy, economy, political theory, history.*

His father, Joseph Home—David Hume preferred the phonetic spelling—was a country gentleman with a small estate, Ninewells, near Berwick-upon-Tweed. His mother, Catherine Falconer, was a daughter of Sir David Falconer, lord president of the Court of Session. Hume retained a lifelong admiration for the gentry, ascribing to them that "moderate scepticism" which he himself sought to foster. His father died young, in 1713, leaving Hume a small legacy on which he later could barely support himself.

Hume matriculated at the University of Edinburgh in 1723, but left three years later without taking a degree. Edinburgh was a center of Newtonian physics, and Hume most probably was taught its elements either by the mathematician James Gregory or by Newton's popularizer, Colin Maclaurin. On the philosophical side, at Edinburgh there flourished a group of ardent Berkeley disciples. The religious atmosphere was a liberal Calvinism, but at an early age, Hume told Boswell, he lost all belief in religion as a result of reading Locke and Samuel Clarke.

Following a family tradition, he set out to study law. He became convinced, however, at the age of eighteen, that he had made a great discovery which "opened up a new scene of thought," and he determined to devote himself wholly to working out his new ideas.

There is considerable controversy about the nature of Hume's "new scene of thought," but there are good grounds for believing that it at least incorporated the idea of constructing a "science of man" by applying Newtonian methods of analysis to the workings of the mind. The further development of Hume's ideas was delayed by the onset of an acute depression, which he tried to shake off by undertaking a career in business. In 1734 he abandoned business to go to France, taking up residence there at La Flèche, where Descartes had been educated. He had already taught himself French and had familiarized himself with such French sceptics as Pierre Bayle; in the extensive library at La Flèche he developed that intimate acquaintance with French philosophy which exerted so profound an influence upon him, uneasily coexisting with his Newtonianism.

HUME

Hume returned to England in 1737 with his *Treatise of Human Nature* completed. The first two books, "Of the Understanding" and "Of the Passions," were anonymously published in 1739; the third book, "Of Morals," was issued in 1740 with an important appendix containing his second thoughts. Hume was confident that the *Treatise* would create a sensation, but it was unenthusiastically received. In order to draw attention to its merits, he published what purported to be an anonymous review of the first two books as *An Abstract of a Treatise of Human Nature* (1740). As an advertising device, it failed, but the *Abstract* is a useful guide to Hume's philosophical intentions, especially interesting for the stress it lays on his associationism. Concluding that the failure of the *Treatise* was a consequence of its length and complexity, Hume henceforth expressed his ideas more fashionably—in essays and dialogues.

In 1741, with a second volume in 1742, Hume published his *Essays Moral and Political.* It is often said that Hume abandoned philosophy for economics and politics in search of literary fame. But for Hume philosophy was "the science of man," and economics, politics, history—understood as "philosophy teaching by examples"—formed for him part of it. He modified his literary style to meet the tastes of his age, but not his fundamental conception of the philosopher's task. The first book of the *Treatise* had been intended as his theory of social inquiry, his "logic"; the second book as his moral psychology; and the third as his ethics. It was now time to pass on to the other social sciences.

His new prose style having proved successful, Hume made another attempt to present his logic to the public. This time it was in an abbreviated and popular form, no longer as a treatise but as *Philosophical Essays Concerning Human Understanding* (1748), renamed in 1758 *An Enquiry Concerning Human Understanding.* This was the work which, he told his critics in an advertisement first published in the posthumous edition of 1777, should "alone be regarded as containing his philosophical sentiments and principles," his *Treatise* being, he explained, but a juvenile work.

Philosophers have been unwilling to take Hume at his word, for the *Treatise* contains a great deal of interesting philosophical analysis, especially of perception, which is not to be found in the *Enquiry.* But the *Enquiry* is in many ways the best introduction to Hume, especially in relating his philosophy to the history of scientific thought. It contains, too, a number of important essays—on miracles, on liberty and necessity, and on providence—which are not to be found in the *Treatise.*

Hume followed up the *Philosophical Essays* with his *Enquiry Concerning the Principles of Morals* (1751), an abbreviated and considerably modified version of book III of the *Treatise.* Although Hume thought it to be his best work, it has only recently received the detailed attention it deserves. At about the same time, Hume wrote the first draft of his *Dialogues on Natural Religion,* a potent criticism of the traditional arguments for the existence of God and especially of the argument from design. His friends warned Hume against publishing it; it appeared posthumously in 1779.

The *Enquiry Concerning Human Understanding* had excluded the sections on space, time, and geometry which formed part of the *Treatise.* Hume intended to write, he tells us in one of his letters, a separate work on "the metaphysical principles of geometry." He prepared for inclusion in *Four Dissertations* (1757) an essay entitled "Some Considerations Previous to Geometry and Natural Science," but the comments of Lord Stanhope, an able mathematician, dissuaded him from publishing it. Hume's talents, indeed, did not lie in that direction; the sections on space and time in the *Treatise* add little to what Berkeley had already argued. For very different reasons, he was also persuaded not to publish his essays "Of Suicide" and "Of the Immortality of the Soul"; these first appeared in an unauthorized French translation in 1770 and also in an unauthorized English edition in 1777 as *Two Essays.* He did include in the *Dissertations,* however, his "Natural History of Religion," in which he sets out to show that classical mythologies are at once more reasonable and morally more enlightened than systematic Christian theology.

Knowing that he was about to die of cancer, Hume wrote in 1776 *My Own Life,* which was first published by his literary executor Adam Smith in 1777 and which is as much an apologia as an autobiography. He died after a long illness, bravely sustained. Hume was a man of exceptional personal qualities, nicknamed in France "le bon David" and in Scotland "Saint David." Adam Smith described him as "approaching as near to the ideal of a perfectly wise and virtuous man as human frailty will admit."

Methodology. The subtitle of Hume's *Treatise* describes it as "an Attempt to introduce the experimental Method of Reasoning into Moral Subjects." Under "moral subjects" Hume includes logic, to which he assigns the task of explaining "the principles and operations of our reasoning faculty"; moral philosophy; political theory, which incorporates economics and history; and literary criticism. He sometimes wrote (as in the introduction to the *Treatise*) as if he had fulfilled the common eighteenth-century ambition to be the Newton of human nature; as if, that is, he had constructed a science of man, paralleling physical science, by relating the elements of the mind in laws of association comparable to the laws of mechanics (*Treatise,* bk. I, pt. 1, sec. IV).

Hume's important contributions to such moral subjects as economics and politics—he contributed nothing to and nowhere reveals any detailed knowledge of the physical sciences—did not depend on the use of a new method; he wrote as an intelligent and critical observer of the European scene, by no means as a methodological innovator. His approach is experimental only insofar as his explanations of social phenomena appeal to everyday human experience, rather than making use of such transcendental entities as "Providence."

As for his positive methodology, that is dependent upon, and does not go far beyond, the "Rules of Reasoning in Philosophy" which Newton had laid down in the third book of his *Principia mathematica.* Hume himself wrote of his "rules by which to judge of causes and effects" (*ibid.,* bk. I, pt. 3, sec. XV) that they are so obvious as scarcely to be worth the trouble of setting them out systematically. His importance lies not in his use or description of the experimental method, but quite elsewhere—in the doubts he raised about the rationality of the method.

His analysis of reasoning begins from a presumption universally accepted by his philosophical contemporaries, namely that what we are directly acquainted with are "perceptions in our mind," as distinct from independently existing physical objects. Hume divides these perceptions into two classes, impressions and ideas. He counts as impressions not only sensations but any operations of the mind, including the passions, which are immediately apprehended. Ideas are "the faint images of impressions"; they are what men have before their mind when they think, as distinct from when they feel.

Since there are no ideas which do not derive from impressions, anybody who uses a word which purports to refer to an idea can properly be asked from what impression that idea derives. If the idea to which the word purports to refer does not derive from any impression, the word, Hume argues, must be meaningless (*Abstract,* p. 11). This is clearly the case, he tries to show, with such familiar metaphysical words as "substance" and "essence." Hume's analysis of perception thus provides him with a powerful polemical weapon to direct against all explanations that make use of concepts not derived from experience; explanations of this kind are, in his interpretation, mere word play.

Perceptions, whether impressions or ideas, occur in spatial and temporal sequences. Furthermore, very similar sequences of perceptions—"constant conjunctions"—regularly recur. Resemblance, spatiotemporal contiguity (in the *Enquiry Concerning Human Understanding* replaced by temporal priority), and constant conjunction are, according to Hume, "*to us* the cement of the universe" (*ibid.,* p. 32). Men are able to progress from their perceptions to a belief in an orderly systematic world only by virtue of the fact that similar perceptions recur in particular ordered sequences.

Both science and common sense take it for granted, so Hume believes, that there are independently existing objects which are necessarily linked one with another (*Treatise,* bk. I, pt. 4, sec. II). Perceptions, on the other hand, depend upon the human mind for their existence and have no necessary connection with one another. Berkeley had rejected this contrast; perceptions and objects, he had argued, are identical, and science does no more than correlate perceptions. This analysis of scientific knowledge Hume dismisses, in spite of Berkeley's protestations, as a form of absolute scepticism. Berkeley's arguments, he says, if "they admit of no answer [yet] produce no conviction" (*Enquiry,* sec. XII, pt. 1). Although there are places in the *Treatise* (bk. I, pt. 2, sec. VI) where Hume writes as if he were a phenomenalist, he for the most part—particularly in the *Enquiry* (sec. XII, pt. 1)—takes it for granted that there are physical objects which give rise to perceptions in us. He does not seriously question, that is, the general world view constructed by Galileo, Boyle, Newton, and Locke: he asks, rather, what grounds we have for believing in its truth.

So long as science does no more than describe and compare perceptions no problem arises. Mathematics, according to Hume, is secure knowledge because it restricts itself to relating ideas one to another (*Treatise,* bk. I, pt. 3, sec. I). This is true, at least, of algebra and arithmetic; in the *Treatise* and the *Abstract,* although not in the *Enquiry,* Hume expresses some doubts about geometry. Nor is there any problem with what Hume calls "mental geography" so long as it confines itself to the "delineation of the distinct parts and powers of the mind" (*Enquiry,* sec. I).

In his more sceptical moods, admittedly, Hume does not allow even mathematics and "mental geography" to escape unscathed. Although the rules of mathematics are "infallible," he says, the fact remains that mathematicians themselves are properly hesitant about the validity of their proofs and fully accept them only when their colleagues do so (*Treatise,* bk.

HUME

I, pt. 4, sec. I); as for "mental geography," that breaks down when it tries to give a satisfactory account of personal identity (appendix to *Treatise,* note to bk. I, pt. 3, sec. XIV). But to carry scepticism to the point of questioning the certainty of mathematics and "mental geography," Hume suggests, is to carry it beyond the point at which it is humanly possible consistently to be a sceptic (*Treatise,* bk. I, pt. 4, sec. I).

The case is very different, Hume thinks, with what he calls matters of fact, assertions which go beyond perceptions by referring to independently existing, continuous objects and ascribing to them a necessary connection with other objects. Whenever the scientist makes a "matter-of-fact" assertion, according to Hume, he is relying upon some form of causal reasoning. Only causal reasoning can carry the mind beyond what it actually perceives to beliefs about what it has not perceived, for example, from beliefs about perceived smoke to beliefs about unperceived fire (*ibid.,* bk. I, pt. 3, sec. II). Only if causal reasoning is rational, then, can science be securely grounded.

It cannot be demonstrated, Hume is confident, either that whatever happens has a cause or that a particular occurrence is the cause of a particular effect. (Hume counts as demonstrative only those arguments which prove that it is logically impossible for the conclusion to be false.) Metaphysicians who profess to demonstrate that every event has a cause always beg the question. Every perception, Hume tells us, is distinct and separate from every other perception. There can be no contradiction, then, in supposing that a perception exists apart from any other perception, that is, without a cause (*ibid.,* bk. I, pt. 3, sec. III).

For the very same reason it is impossible, according to Hume, to demonstrate that a particular effect has a particular cause. Since perceptions are distinct and separable there is nothing in any perception, taken by itself and prior to any further experience, which logically presupposes the existence of any other perception (*ibid.,* bk. I, pt. 3, sec. VI). Our everyday experience confirms this philosophical conclusion. Prior to experience we have no way of telling how anything will behave, that fire, for example, will burn rather than thicken the human skin. Neither the effect itself, as Descartes thought, nor a power to produce the effect, as was widely presumed, is implicit in the cause; if it were, the scientist should be able simply by examining an object to discover what effects it will have, and this is impossible.

Only experience, then, enables the scientist to determine that a particular cause will have a particular effect. But experience tells him only that in the past certain similar perceptions A_1, A_2, A_3, ... have been constantly conjoined with certain other similar perceptions B_1, B_2, B_3, ... When the scientist holds that *A* is the cause of *B,* however, he ordinarily thinks of himself as being committed to something much stronger than this: that *A* is necessarily connected with *B*. Yet he has had no experience of necessary connection, as distinct from mere conjunction. Nor is there any general principle which would enable him to move from "*B* has always in the past been produced by *A*" to "*B* is necessarily produced by *A.*" It is quite easy to imagine a change in the course of nature such that *A* and *B* will no longer be constantly conjoined one with another; this is by no means a logical impossibility. Hence, Hume concludes, it is impossible to demonstrate that *B* cannot occur without *A*'s having occurred. Anybody who perceives the conjunction may be led to believe that *A* and *B* are necessarily connected, but this "being led" is a psychological fact, not a logical necessity. It is not that there is a valid inference from constant conjunction to necessary connection; the belief that *A* is necessarily connected with *B* is reducible to the fact that we habitually suppose that *A* must have happened when *B* is perceived and expect *B* whenever *A* is perceived (*ibid.,* bk. I, pt. 3, sec. XIV).

To understand scientific inference, then, we must turn to mental geography and the analysis of our mental habits, not to formal logic. The belief in any matter of fact has only two sources: the existence of a particular relationship between perceptions—constant conjunction—and the tendency of the mind to react in a certain way to constant conjunctions. That is why Hume is prepared to assert that the science of man is the fundamental science on which all other science rests; only with the help of mental geography can we explain why we hold our empirical beliefs.

If we ask, however, exactly what mental geography tells us about nondemonstrative inference, Hume's answer is by no means clear or consistent. Sometimes he says that reason (that is, empirical reasoning) is "nothing but a wonderful and unintelligible instinct in our soul" which leads us to move from past experience to expectations about the future (*ibid.,* bk. I, pt. 3, sec. XVI). This has led some commentators to assert that Hume is a naturalist who, in the manner of Pope's *Essay on Man,* bids us rely on instinct rather than reason for our fundamental beliefs. At other times, however, the responsibility for causal inferences is assigned by Hume to the imagination.

Just how sceptical is Hume's analysis of empirical inference? That, too, is a point on which he vacillates. On the one hand, he is anxious to dispute the claims of transcendental metaphysicians and theologians that they possess rationally grounded beliefs. With his eye on such opponents, he argues that it is quite absurd to go in search of remote causes for the Universe when we cannot even give a satisfactory reason for believing that a stone will fall or that the sun will rise tomorrow (*Enquiry,* sec. XII, pt. 3). A belief, he says, is nothing but an unusually vivid idea; to believe that the sun will rise tomorrow is simply to have a vivid idea that it will do so. This doctrine, too, is useful against those who argue that the moral sciences are intrinsically inferior to the physical sciences because they rest upon feeling; every form of science, Hume can reply, does so (*Treatise,* bk. I, pt. 3, sec. VIII).

On the other hand, Hume is equally anxious to destroy fanaticism and superstition. He can scarcely deny, however, that the superstitious and the fanatical have vivid ideas. He sometimes suggests, therefore, that a belief is rational provided only that it can be traced back to a constant conjunction; hence the rational justification for believing that the sun will rise tomorrow, as opposed to the irrationality of superstitious beliefs. From this perspective Hume distinguishes between demonstrations, proofs, and probabilities. It is ridiculous, he says, to declare as only a probability that the sun will rise tomorrow or that all men are mortal (*ibid.,* bk. I, pt. 3, sec. XI). Inferences from constant conjunction, he suggests, are properly describable as proofs, even though they clearly do not constitute demonstrations. But when conjunctions are irregular—*A* being only sometimes conjoined in our experience with *B*, and sometimes with something else—the proper inference is only to probabilities, since the probability of a conclusion depends upon the relative frequency of the conjunctions on which it is founded. The conclusions of the superstitious have a zero or minimal probability because they are contrary to our regular experience.

This attitude is most fully developed in Hume's critical analysis of the belief in miracles (*Enquiry,* sec. X). Hume there begins by asserting that a wise man will always proportion his belief to the evidence. A miracle is by definition a violation of the laws of nature, that is, an event which is contrary to our regular experience. The evidence in its favor, as in the case of those miracles on which the historical religions rely, is that some witness or an oral tradition tells us that the miracle happened. We are entitled to accept this testimony only, Hume says, if it would involve a greater miracle, a more manifest divergence from all past experience, to suppose that the testimony is false. Since this condition is not satisfied in the case of any recorded miracle, he says, we cannot properly treat miraculous occurrences as probable, let alone as proved.

Hume sometimes expresses his theory of "proof" in a way that links it closely with the workings of the imagination. The imagination, he tells us, has certain regular, associative ways of working, most clearly manifested in the case of causal inference. These ways we must accept as reliable and rational; to reject them is to undermine the whole foundation of our thought and action. The imagination, however, does not always work in a regular way; it has irregular and erratic tendencies which lead men into superstition. Conclusions derived from these irregular workings ought, on the face of it, to be rejected by rational men (*Treatise,* bk. I, pt. 4, sec. IV). The problem is that there exist unquestionably true beliefs—the belief in the independent existence of physical objects and the belief in personal identity, for example—which cannot wholly be explained in terms of causal inference, but which depend on the operations of irregular propensities of the imagination. So it is impossible, after all, to adopt a policy of accepting only those beliefs which are founded on constant conjunction (*ibid.,* bk. I, pt. 4, sec. VII).

In the *Treatise* especially, these considerations sometimes lead Hume to a posture of absolute scepticism, rather than the "mitigated scepticism" he generally adopts. But no man can live as an absolute sceptic (*Enquiry,* sec. XII). Mitigated scepticism, as Hume sums it up in his *Dialogues* (pts. VIII and IX), asserts simply that it is impossible to demonstrate any matter of fact and that the nature of our experience, not some a priori principle of rationality, determines what we find intelligible. Such a position is substantially that of empiricism. But it is a different matter if our fundamental beliefs turn out to rest on nothing more solid than a trick of the imagination. We have only one defense against this sceptical conclusion, Hume suggests. Nature has not left our beliefs entirely to our choice; we cannot help coming to conclusions any more than we can help breathing (*Treatise,* bk. I, pt. 4, sec. I). Mitigated scepticism is therefore useful, for it prevents us from wandering into the wilds of metaphysical speculation by impelling us to reflect on the limits of our knowledge of even everyday physical experience and relationships.

BIBLIOGRAPHY

I. Original Works. The classical edition, although an imperfect one, of Hume's works is T. H. Green and T. H. Grose, eds., *The Philosophical Works of David Hume,* 4 vols. (London, 1875). This does not include J. M. Keynes and P. Sraffa, eds., *An Abstract of a Treatise of Human Nature* (Cambridge, 1938), or Ernest C. Mossner and J. V. Price, eds., *A Letter From a Gentleman to His Friend in Edinburgh* (Edinburgh, 1967). Especially for their indexes, consult also L. A. Selby-Bigge's eds. of *A Treatise of Human Nature* (Oxford, 1888) and *Enquiries Concerning the Human Understanding and Concerning the Principles of Morals,* 2nd ed. (Oxford, 1902). The best text of the *Treatise* is the Mossner ed. (London, 1969).

See also Norman Kemp Smith, ed., *Dialogues Concerning Natural Religion,* 2nd ed., with suppl. (London, 1947). For Hume's general writings on religion see Richard Wollheim, compiler, *Hume on Religion* (London, 1963).

II. Secondary Literature. John Hill Burton, *Life and Correspondence of David Hume,* 2 vols. (Edinburgh, 1846; repr. New York, 1968), is still valuable. The best modern life is E. C. Mossner, *The Life of David Hume* (Austin, Texas, 1954; London, 1955), which includes *The Life of David Hume, Esq., Written by Himself* or, as entitled in the original MS, *My Own Life.* See also J. Y. T. Greig, ed., *The Letters of David Hume,* 2 vols. (Oxford, 1932), and Raymond Klibansky and E. C. Mossner, eds., *New Letters of David Hume* (Oxford, 1954).

It is impossible to give a straightforward, systematic, noncontroversial presentation of Hume's views. That is one of the principal themes of J. A. Passmore, *Hume's Intentions,* 2nd ed., rev. (London–New York, 1968). The most thoroughgoing commentary is N. K. Smith, *The Philosophy of David Hume* (London–New York, 1941), and the most useful introduction is D. G. C. Macnabb, *David Hume: His Theory of Knowledge and Morality* (London, 1951).

See also Charles W. Hendel, *Studies in the Philosophy of David Hume,* rev. ed. (Indianapolis, 1963), with an account of recent work on Hume; Antony Flew, *Hume's Philosophy of Belief* (London, 1961; New York, 1962), which concentrates on the *Enquiries;* H. H. Price, *Hume's Theory of the External World* (Oxford, 1940); and Farhang Zabeeh, *Hume: Precursor of Modern Empiricism* (The Hague, 1960).

John Passmore

HUME-ROTHERY, WILLIAM (*b.* Worcester Park, Surrey, England, 15 May 1899; *d.* Iffley, Oxfordshire, England, 27 September 1968), *metallurgy, chemistry.*

Hume-Rothery was the son of Joseph Hume Hume-Rothery, a lawyer, and Ellen Maria Carter. Most of his childhood was spent in Cheltenham, and while he was a schoolboy attending Cheltenham College (1912–1916) he decided on a military career. Early in 1917, a few months after he had entered the Royal Military Academy, Woolwich, he suffered an attack of cerebrospinal meningitis which left him totally deaf. He was therefore discharged from the army and subsequently entered Magdalen College, Oxford, where (following the influence of his Cheltenham science master, George Ward Hedley) he read chemistry, receiving a first-class honors degree in 1922.

Graduate work at the Royal School of Mines (under Sir Harold Carpenter) at London University turned his interest to metallurgy and led to a highly original paper on intermetallic compounds, published in 1926, the year in which he received his London Ph.D. degree. Returning to Magdalen in December 1925, he stated that he proposed "to carry on research at Oxford in intermetallic compounds and problems on the borderland of metallography and chemistry." Chemical research and the city of Oxford formed the center of his activities for the remainder of his life, although the "chemistry" in time became closer to physics.

On 28 March 1931 he married Elizabeth Alice Fea, with whose understanding help he overcame many difficulties associated with his deafness. He learned to modulate his voice, which he of course could not hear, and became an excellent lecturer. Students often served as his "ears" at large conferences, but in individual conversation his skill in lip reading, aided on occasion by his visitor's use of a pad and pencil, made his handicap almost unnoticeable. His great zest for life, combined with his ready, often puckish, sense of humor, made it easy and pleasant to exchange ideas with him on both casual and complex topics.

Hume-Rothery was an ardent fly-fisherman and an accomplished watercolorist. In the closing years of his life he began to cultivate exotic cacti, and did so with the same engrossing enthusiasm with which he took rugged country walks to seek subjects for his brush.

His work at Oxford University was supported by external research grants. Although he had many students, either undergraduate or in research, he did not have an official university appointment until 1938, when he became lecturer in metallurgical chemistry. In 1957, under pressure from the metallurgical profession, the School of Metallurgy was established at Oxford with Hume-Rothery as the first professor.

Hume-Rothery's scientific contributions are related to the principles underlying the crystal structures of alloy phases. In 1925, although the existence of various types of intermetallic phases had been shown (see summaries by Desch[1] and Giua[2]), no theory accounted for their formation. Many phases extended over a wide range of compositions (and hence were unpalatable to chemists who, a century after Dalton, believed in simple molecules), while many well-defined combinations ignored the normal rules of valency; however, the new determination of atomic arrangement in crystals by X-ray diffraction, as well as the new views of the nature of the atom and electron, had prepared the ground for a new approach.

Hume-Rothery, in 1924, saw that electrically conducting compounds must have "loose" electrons and therefore could not conform to valency rules; in his 1926 paper he pointed out—almost as an aside, sandwiched between an experimental report on the constitution of certain alloys of tin and an animadversion against the misuse of the phase rule—that body-centered-cubic β solid solutions of copper with B subgroup elements occur only when the ratio of valence electrons to atoms was in the neighborhood of 3:2; for example, CuZn, Cu_3Al, Cu_5Sn. With this first glimpse of a new field, the phrase "electron compounds" became current. The concept was soon extended by A. J. Bradley to the complex-cubic γ phases at a ratio 21:13 and by Arne Westgren to the close-packed hexagonal phases at 7:4, and thereafter to many others.

In 1934, in his most influential single paper, written with two students, Hume-Rothery pointed out that the melting points and solid solution ranges of alloys of copper or silver with many different elements became nearly identical when considered as a function of the added valence electrons (that is, atomic fraction of the solute multiplied by its valence). Moreover, making use of V. M. Goldschmidt's analysis of the structures of the elements, he showed, for the first time, the significance of the atomic size factor: Solid solutions did not form between pairs of elements whose atomic radii differed by more than 15 percent. Finally, Hume-Rothery observed that the size-related group of compounds identified by F. H. Laves (1933) or the saltlike intermetallic compounds identified by E. Zintl (1931) appeared only when the constituent elements differed greatly in electronegativity. In succeeding decades, he studied electronic and size factors in many alloy systems, notably those of the noble metals and the transition metals.

Hume-Rothery's three rules of alloy formation related immediately to theoretical work on the electron theory of metals,[3] and in particular they supported the idea of interaction between Brillouin zones and expanding spherical Fermi surfaces. Jones's calculation[4] of electron momenta at various electron concentrations in body-centered-cubic and face-centered-cubic alloys seemed to give the fundamental reason behind the observed electron to atom ratios in the α and β phases. Further refinement of the theory, however, has led to continually increasing complications, so that Hume-Rothery's original rules are still (1972) more useful as a guide to alloying behavior than is any basic mathematical theory.

Hume-Rothery was a fine experimentalist, especially noted for his accurate pyrometry on reactive materials at high temperature. Although he preferred the microscope for studying the constitution of alloys, he developed refined X-ray methods and did much to improve their interpretation.

Hume-Rothery's influence on metallurgical education was worldwide. Although his own original contributions were based on extensive knowledge of facts and an intuitive insight into their meaning, he was also an excellent interpreter of advanced work in mathematical physics. All but the first of his books were directed at undergraduate and industrial metallurgists and were expertly simplified texts, making clear to nonphysicists the new science that metallurgy was about to become. His book *The Structure of Metals and Alloys* (first published in 1936) was particularly important, and he took pride in keeping each new edition completely up to date, with the aid of collaborators in the fourth and fifth editions.

A historical essay by Hume-Rothery (1965) is revealing both of the state of the field and of his personal approach to science. Objecting to the contemporary tendency to restrict the term theory to work of a mathematical nature, he remarked that "Mendeleev's Periodic Table as a theory of chemistry . . . is more accurate than, and certainly no less fundamental than a mathematical theory of alkali metals such as that of Wigner and Seitz." He concluded

> . . . the electron theory of alloys is in an unsatisfactory state. . . . Practically nothing has been predicted by à priori calculation methods in advance of the facts, whilst the simple theories which seemed so satisfactory 20 years ago are now in great doubt. On the other hand, there is a considerable theory or generalization of facts in the form of empirical rules or principles and these have permitted some predictions to be made. The theory of alloys is thus at the stage of Kepler and not of Newton ("The Development of the Theory of Alloys," p. 346).

The empirical rules were nearly all the result of Hume-Rothery's work.

NOTES

1. C. H. Desch, *Intermetallic Compounds* (London, 1914).
2. M. Giua and C. Giua-Lollini, *Chemical Combination Among Metals* (London, 1918).
3. N. F. Mott and H. Jones, *The Theory of the Properties of Metals and Alloys* (Oxford, 1936).
4. H. Jones, "The Phase Boundaries in Binary Alloys, II. The Theory of the α, β Phase Boundaries," in *Proceedings of the Physical Society,* **49** (1937), 250–257.

BIBLIOGRAPHY

A full bibliography of Hume-Rothery's 178 papers is given in G. V. Raynor, "William Hume-Rothery, 1899–1968," in *Biographical Memoirs of Fellows of the Royal*

Society, **15** (1969), 109–139. Only the most influential are listed below:

"Researches on the Nature, Properties, and Conditions of Formation of Intermetallic Compounds . . .," in *Journal of the Institute of Metals,* **35** (1926), 295–361; "The Electronic Energy Levels of the Elements, With Special Reference to Their Connexion With the Sizes and Electronic States of Atoms in Metallic Crystals," in *Philosophical Magazine,* **11** (1931), 649–678; "The Freezing Points, Melting Points, and Solid Solubility Limits of the Alloys of Silver and Copper with Elements of the B Sub-Groups," in *Philosophical Transactions of the Royal Society,* **233** (1934), 1–97, written with G. W. Mabbott and K. M. Channel-Evans; "The Lattice Spacings of Certain Primary Solid Solutions in Silver and Copper," in *Proceedings of the Royal Society,* **157A** (1936), 167–183, written with G. F. Lewin and P. W. Reynolds; "Atomic and Ionic Radii. II. Application to the Theory of Solid Solubility in Alloys," in *Philosophical Magazine,* **26** (1938), 143–165, written with G. V. Raynor; "The Application of X-Ray Methods to the Determination of Phase-Boundaries in Metallurgical Equilibrium Diagrams," in *Journal of Scientific Instruments,* **18** (1941), 74–81, written with G. V. Raynor; "Electrons, Atoms, Metals and Alloys," in *Transactions of the American Institute of Mining Engineers,* **171** (1947), 47–62; "Applications of X-ray Diffraction to Metallurgical Science," in P. Ewald, ed., *Fifty Years of X-ray Diffraction* (Utrecht, 1962), pp. 190–211; "The Development of the Theory of Alloys," in C. S. Smith, ed., *The Sorby Centennial Symposium on the History of Metallurgy* (New York, 1965), pp. 331–346.

His books include: *The Metallic State* (Oxford, 1931); *The Structure of Metals and Alloys* (London, 1936; 2nd ed. 1944; 3rd ed. 1954; 4th ed. [with G. V. Raynor], 1962; 5th ed. [with R. E. Smallman and C. W. Haworth], 1969); *Atomic Theory For Students of Metallurgy* (London, 1946; 2nd ed., 1952; 3rd ed., 1960; 4th ed., 1962); *Electrons, Atoms, Metals and Alloys* (London, 1948; 2nd ed., 1955; 3rd ed., New York, 1963); *Metallurgical Equilibrium Diagrams* (London, 1952), written with J. W. Christian and W. B. Pearson; *Elements of Structural Metallurgy* (London, 1961); *The Structures of Alloys of Iron: An Elementary Introduction* (London, 1966).

Good accounts of the present state of understanding in the field opened by Hume-Rothery are T. B. Massalski, "Structure of Solid Solutions," ch. 4 in Robert W. Cahn, ed., *Physical Metallurgy,* 2nd ed., rev. (Amsterdam, 1970); and G. V. Raynor, "Hume-Rothery and the Development of the Science of Alloy Formation," in *Journal of the Institute of Metals,* **98** (1970), 321–329.

CYRIL STANLEY SMITH

ḤUNAYN IBN ISḤĀQ (*b.* Hira, 809–810; *d.* Baghdad, October 877), *translation, medicine.*

For a detailed study of his life and work, see Supplement.

HUNDT (HUND, CANIS), MAGNUS (*b.* Magdeburg, Germany, 1449; *d.* Meissen, Germany, 1519), *anatomy, medicine.*

Magnus Hundt the Elder is known to have been associated with Leipzig University from at least 1485. He received a bachelor's degree in 1483, a baccalaureate in medicine in 1499, and a licentiate in theology in 1504, and was a professor at Leipzig for many years. The university was removed to Meissen, near the end of his life, on account of the plague.

Hundt's best-known work, *Antropologia de hominis dignitate, natura et proprietatibus de elementis,* published in 1501, is one of the three or four earliest printed books to include anatomic illustrations. At one time, Hundt's work was looked upon as the oldest printed book with original anatomic illustrations, but that is no longer believed to be the case. His *Antropologia* included five full-page woodcuts, including two identical reproductions of the human head, which appeared on the back of the title page as well as later in the book. The woodcuts are crude and schematic and not done from nature, and although one of the woodcuts pictures the entire body and lists the various external parts, there is no attempt to equate the anatomical term with the actual representation. There is also a full page woodcut of a hand with chiromantic markings, and of the internal organs of the thorax and abdomen. Smaller woodcuts, including plates of the stomach, intestines, and cranium, are inserted throughout the text. The work gives a clear idea of anatomy prior to the work of Berengario da Carpi, and can be regarded as typifying late-fifteenth-century concepts. Hundt held that the stars exert more influence on the human body than on other composites of elements, and his book includes generalizations about human physiognomy and chiromancy as well as anatomy. He subscribed to the notion of the seven-celled uterus, which he apparently derived from Galen.

BIBLIOGRAPHY

I. ORIGINAL WORKS. Hundt's *Antropologia de hominis dignitate, natura et proprietatibus de elementis* was published by Wolfgang Stöckel ("Monacensis") at Leipzig in 1501. Hundt also edited or commented on *Introductorium in universalem Aristotelis phisician Parvulus philosophiae naturalis vulgariter appellatum* (1500), today at the British Museum, and annotated works by St. Augustine and St. Thomas Aquinas. The *Nütliches Regiment, sammt dem Bericht der Ertzney, wider etliche Kranckheit der Brust,* sometimes listed under his name, should be attributed to Magnus Hundt the Younger.

II. Secondary Literature. The best account of Hundt is in Karl Sudhoff, *Die Medizinische Fakultät zu Leipzig im ersten Jahrhundert der Universität* (Leipzig, 1909), pp. 115–121. His anatomical illustrations are discussed in Ludwig Choulant, *History and Bibliography of Anatomic Illustration,* translated and annotated by Mortimer Frank (New York, 1945), pp. 125–126.

Vern L. Bullough

HUNT, JAMES (*b.* Swanage, England, 1833; *d.* Hastings, England, 29 August 1869), *anthropology.*

James's father, Thomas Hunt, made an extensive study of the causes of stammering and developed a method of treatment that was often successful; he himself contributed greatly to James's education. Conscious of his own lack of medical training, he wished his son to study medicine and Beddoe reports that James went to Cambridge, but there is no confirmation of this. His mother's name was Mary.

Hunt lived in Hastings, where he continued his father's work; he is said to have treated some 1,700 cases of stammering. In 1854 he published the short *Treatise on the Cure of Stammering,* which included a memoir of his father. It ran to seven editions and was later expanded to give a comprehensive history of theories of stammering from classical times. He also wrote a review of the contemporary literature on the localization of the functions in the brain, with special reference to the faculty of language.

Hunt believed the chief cause of stammering to be improper use of the mouth and faulty breathing, resulting in nervousness. He found the most successful treatment to be based on analysis, with each individual patient, of the technique of voice production, followed by reeducation of muscle control and, most importantly, the building up of the patient's confidence. He noted that patients do not stammer when singing, and that Charles Kingsley, whom he later treated successfully, did not stammer when absorbed in preaching a sermon. Hunt was wholly persuaded that surgery was inadvisable for alleviating speech disabilities.

Hunt's main contribution was the impetus he lent to establishing anthropology in England as a distinct discipline. He joined the Ethnological Society in 1854 at the age of twenty-one, and from 1859 to 1862 he was honorary secretary. But he felt that its scope of study was too narrow, and in 1863 it was he who was largely instrumental in founding the Anthropological Society of London, becoming its first president. He initiated publication in 1863 of the *Anthropological Review,* which was later taken over by the society; and this caused an acrimonious correspondence concerning the new journal in the *Athenaeum* when Hyde Clarke attacked Hunt's financial management.

Hunt himself contributed several articles and unsigned book reviews to the early volumes of the *Review,* mostly on racial issues. He believed that Negroes formed a separate species and that treatment of them should take this into account. Readings of his two papers on physical and mental characteristics of the Negro before the Anthropological Society and British Association in 1863 were both followed by stormy discussions. In the course of reviews, he wrote on miscegenation, and attacked J. S. Mill's positions on race and legislation in political economy and Darwin's views on natural selection. He is also known to have done some work on the destructive effects of peat upon the human body.

The Anthropological Society published a number of significant monographs. Among these was Carl Vogt's *Lectures on Man,* the translation of which was edited by Hunt, who omitted a few passages that seemed to him not in good taste.

In helping to foster the science of anthropology in England, Hunt persuaded the British Association to set up, in 1866, a separate sub-section for the subject within the association's biological section; anthropology had previously been considered under the section for geography. In 1883 anthropology became a separate section.

BIBLIOGRAPHY

I. Original Works. Hunt's work on stammering began as a short essay with a rather longer memoir, described as "a brief act of filial piety" and entitled *A Treatise on the Cure of Stammering . . . With Memoir of the Late Thomas Hunt* (London, 1854). There was a second edition which is difficult to trace, and subsequently a third (1857), fourth (1861), fifth (1863), sixth (1865), and seventh (1870). By publication of the last edition, the memoir and testimonials had been abridged and sections on the theory and techniques of cure greatly expanded. The 1861 edition of *Stammering and Stuttering: Their Nature and Treatment* has been reprinted in facsimile (New York, 1967) with an introduction in which Elliott J. Schaffer evaluates Hunt's views in the light of later theories.

A more general work was *A Manual of the Philosophy of Voice and Speech, Especially in Relation to the English Language and the Art of Public Speaking* (London, 1859). Hunt's review "On the Localisation of the Functions of the Brain, With Special Reference to the Faculty of Language" was published in parts in the *Anthropological Review,* **6** (1868), 329–345; and **7** (1869), 100–116, 201–216.

"On the Negro's Place in Nature" was read to the Anthropological Society in 1863; an abstract of the paper and verbatim report of the two sessions of discussion were printed in *Journal of the Anthropological Society of London,* **2** (1864), xv-lvi; and the full paper was published in *Memoirs Read Before the Anthropological Society of London,* **1** (1863–1864), 1–64. Several other signed papers and anonymous book reviews are to be found in the first seven volumes of the *Anthropological Review,* including his "Introductory Address on the Study of Anthropology," inaugurating the new society, **1** (1863), 1–20; and his annual anniversary addresses on progress in anthropology.

"On the Physical and Mental Characteristics of the Negro" was recorded in abstract in *Report of the 33rd Meeting of the British Association for the Advancement of Science, held at Newcastle-upon-Tyne, 1863* (1864), 140. Hunt also edited Carl Vogt, *Lectures on Man: His Place in Creation and in the History of the Earth* (London, 1864), in which there is his editorial preface.

II. SECONDARY LITERATURE. There is a concise account of James Hunt by G. T. Bettany in the *Dictionary of National Biography,* **28** (1891), 266–267, and also an entry for his father Thomas Hunt. The appreciative *Éloge* by E. Dally in *Mémoires de la Société d'anthropologie de Paris,* 2nd ser., **1** (1873), xxvi–xxxvi, includes a bibliography of 31 items, and Hunt was remembered in his own society by the presidential address of John Beddoe in *Anthropological Review,* **8** (1870), lxxix–lxxxii. The controversy with Hyde Clarke may be traced through the index to the *Athenaeum* for 1868. There is a short entry in *A Biographical Dictionary of Modern Rationalists,* compiled by J. McCabe (London, 1920).

DIANA M. SIMPKINS

HUNT, THOMAS STERRY (*b.* Norwich, Connecticut, 5 September 1826; *d.* New York, New York, 12 February 1892), *chemistry, geology.*

Hunt's parents, Peleg Hunt and Jane Elizabeth Sterry, were both descended from Puritan stock. In 1845, after desultory schooling until the age of thirteen and numerous trivial jobs, he came to the attention of the elderly Benjamin Silliman. Silliman, struck by his Faraday-like enthusiasm for a scientific career, arranged for his son, Benjamin, professor of chemistry at Yale, to employ Hunt as a scientific assistant. The younger Silliman trained Hunt by making him analyze minerals for C. B. Adams' Geological Survey of Vermont, and his proven ability led to his appointment (1846–1872) as mineralogist and chemist to the Geological Survey of Canada, which was under the directorship of Sir William Logan and, from 1869, of A. R. C. Selwyn.

While working with the Survey, Hunt acted as part-time professor of chemistry at the University of Laval, Quebec (where he lectured in fluent French from 1856 to 1862), and at McGill University, Montreal, from 1862 to 1868. On his return to the United States in 1872 he became professor of geology at the Massachusetts Institute of Technology, and chemist to the second Geological Survey of Pennsylvania under J. P. Lesley. Hunt retired from both positions in 1878 in order to pursue geological consultancy and a literary career. In 1871 he married Anna Gale, a Canadian; six years later their childless marriage ended in separation. A frequent visitor to Europe, Hunt was personally acquainted with most of the leading English and French scientists. He was a fellow of the Royal Society and a prime mover in the creation of the American Chemical Society, the Royal Society of Canada, and the first International Congress of Geologists in Paris (1878).

Hunt, who was brought up as a Congregationalist, was converted to Roman Catholicism in Canada, but he abandoned formal religion during the 1860's for a simple deistic and poetic natural theology. An egotistical if scintillating conversationalist and lecturer, Hunt damaged his chemical and geological reputation both in America and Europe by his strident, obsessive concern for professional recognition of the priority of his innovations; while in a censorious age his personal nonconformity brought him humiliating ostracism.

The two principal influences on Hunt were the revolutionary chemistry of Laurent and Gerhardt, which he introduced to America, and the early philosophical writings of the German-American *Naturphilosoph* J. B. Stallo, which led him to the enthusiastic study of Kant, Hegel, and Oken. Hunt's polemical and priority-seeking style has led to some confusion in the literature of the history of chemistry. It will be sufficient to state here that Hunt did not invent the organic chemist's "water-type," but that like Gerhardt and Williamson he saw its possibilities after Laurent had first mentioned it; and that, although he was probably the first to propose that silica is the "carbon" of mineralogy, Hunt was not the first to define organic chemistry as "the chemistry of carbon compounds"—here he was merely following Gerhardt.

Inspired with a belief in the unity of nature, Hunt wrote speculative and transcendental works which frequently ignored facts that were inconsistent with his own or other geologists' field observations. As a chemist Hunt rejected atomism for a continuum physics in which all chemical changes were explained by interpenetration or solution, and not by the arrangements of invariant atoms. He extended Gerhardt's concept of homologous series of organic compounds to mineralogy, wherein he conceived minerals to possess "molecular weights" much greater than the current atomic theory suggested. He assumed that minerals having similar crystalline forms possessed identical equivalent volumes and hence, from analogy with gases, that their equivalent weights (or "integral weights") were proportional to their densities. Establishing this relationship enabled him, "having fixed an equivalent weight for one species, to calculate, from the densities, those of the species isomorphous with it" (*Chemical and Geological Essays* [1875], p. 440). These attempts by Hunt, Oliver Wolcott Gibbs, and others to derive minerals, like silicates, from polyacids were shown to be ineffective after the advent of X-ray crystal analysis. Hunt also developed an elaborate "natural system" of mineral classification which, despite its attractive compromise between existing systems based on either chemical or external characteristics, was not influential.

As a geologist, Hunt played a major part in Logan's elucidation of the Laurentian and Huronian systems. His own primary interest was in Paleozoic rocks, the history of which, he argued, in the absence of fossils and stratigraphic evidence, could be deduced by extrapolating from the existing mineral species that they contained the supposed prehistoric chemical conditions necessary for their origin. This so-called crenitic hypothesis was most influentially expressed in 1867 in his essay "The Chemistry of the Primeval Earth" (*Chemical News,* **15** [1867], 315–317, reprinted in *Chemical and Geological Essays*). Aware of the significance of H. Sainte-Claire Deville's work on dissociation, Hunt supposed that as the earth had cooled, familiar elements and compounds had formed. After certain climatic changes had occurred, condensed water had permeated the porous surface of the earth and dissolved chemicals which, subject to the extraordinary catastrophic conditions of the earth's hot interior, had undergone metamorphosis. These transformed materials (proto-minerals) had then been brought to the earth's surface "after the manner of modern springs," and had there been deposited as crystalline layers of granite, gneiss, or even serpentine. Other geologists found an igneous origin for these rocks more credible. But despite protracted polemics with Dana, William Logan, and David Forbes, Hunt remained stubbornly resistant to other points of view, never abandoning his modified neptunism. His inorganic "evolutionary" views, however, shorn of their controversial geological context, influenced the chemical speculations of B. C. Brodie, Jr., and Lockyer, and, through them, Crookes.

BIBLIOGRAPHY

I. Original Works. Hunt published well over 350 papers, a virtually complete list of which may be found in Douglas or Adams. Hunt's style is repetitious; his papers, full of self-citations, were reprinted by him in his books of 1875, 1886, and 1887.

Hunt's works include "Introduction to Organic Chemistry," a section appended to B. Silliman, Jr., *First Principles of Chemistry,* 25th ed. (Philadelphia, 1852), with innumerable later eds.; *Esquisse géologique du Canada, pour servir à l'intelligence de la carte géologique et de la collection des minéraux économiques envoyées à l'Exposition universelle de Paris 1855* (Paris, 1855), written with W. E. Logan, translated as *Canada at the Universal Exhibition of 1855* (Toronto, 1856); *The Geology of Canada* (Montreal, 1863), written with Logan; *Petroleum, its Geological Relations, With Special Reference to its Occurrence in Gaspé* (Quebec, 1865); *Esquisse géologique du Canada. Suivie d'un catalogue descriptif de la collection de cartes et coupes géologiques, livres imprimés, roches, fossiles, et minéraux économiques, envoyée à l'Exposition universelle de 1867* (Paris, 1867); *The Coal and Iron of Southern Ohio* (Salem, Mass., 1874); *Chemical and Geological Essays* (Boston–London, 1875; 2nd ed., Salem, 1878; 3rd ed., New York, 1890; 4th ed., New York, 1891).

See also *Special Report on the Trap Dykes and Azoic Rocks of South-Eastern Pennsylvania: Part I. Historical Introduction,* Second Geological Survey of Pennsylvania, Report E (Harrisburg, 1878); there is no part II; *Coal and Iron in Southern Ohio, the Mineral Resources of the Hockey Valley* (Boston, 1881); *Mineral Physiology and Physiography. A Second Series of Chemical and Geological Essays* (Boston, 1886; 2nd ed., New York, 1890); *A New Basis for Chemistry, A Chemical Philosophy* (Boston, 1887; 2nd ed., 1888), trans. by W. Spring as *Un système chimique nouveau,* (Paris–Liège, 1889; 3rd ed., New York, 1891), with dedication to J. B. Stallo; and *Systematic Mineralogy Based on a Natural Classification* (New York, 1891).

Other works are *Geological Survey of Canada. Report of Progress for the Year 1852–3* (Quebec, 1854). Reports for years 1853 to 1856 were published from Toronto, and for 1857 to 1872 from Montreal. In addition see *Geological Survey of Canada: Report of Progress from its Commencement to 1863,* 2 vols. (Montreal, 1863–1865).

For an important letter from Hunt to Gerhardt, written in 1847, see E. Grimaux and C. Gerhardt, *Charles Gerhardt, sa vie, son oeuvre, sa correspondance 1816–1856* (Paris, 1900), pp. 166–167. One of the more significant of Hunt's literary polemics, concerning geological chemistry or chemical geology, was with the English geologist David Forbes; see *Geological Magazine,* **4** (1867), 433–444, and **5** (1868), 49–59, 105–111, which contains references to *Chemical News.*

Collections of Hunt's letters are held at Edinburgh University Library, Scotland (Lyell papers); Royal Society, London; Columbia University Library; and the Smithsonian Institution, Washington.

II. Secondary Literature. The best obituaries are those by James Douglas, *Proceedings of the American Philosophical Society, Memorial Volume,* **1** (1900), 63–121, with photograph and bibliography; and F. D. Adams, *Biographical Memoirs. National Academy of Sciences,* **15** (1934), 207–238, with photograph and bibliography.

For Hunt as a geologist and mineralogist, see G. P. Merrill, *The First One Hundred Years of American Geology* (New York, 1924; repr. New York-London, 1964), pp. 246, 367, 410–411, 445–447, 565, 608; and E. F. Smith, "Mineral Chemistry," in C. A. Browne, ed., "A Half-Century of Chemistry in America, 1876–1926," ch. 6 of supp. to *Journal of the American Chemical Society,* **48** (1926), 79–83. For Hunt's philosophy of science see E. R. Atkinson, "The Chemical Philosophy of Thomas Sterry Hunt," in *Journal of Chemical Education,* **20** (1943), 244–245; W. H. Brock, ed., *The Atomic Debates* (Leicester, 1967), pp. 13, 24–26, 127, 156, 160, 171; and D. M. Knight, "Steps Towards a Dynamical Chemistry," in *Ambix,* **14** (1967), 190–194.

W. H. BROCK

HUNTER, JOHN (*b.* Long Calderwood, near East Kilbride, Lanarkshire, Scotland, 13 February 1728; *d.* London, England, 16 October 1793), *surgery, anatomy.*

John Hunter, youngest of the ten children of John and Agnes Hunter, received his early education at the grammar school in East Kilbride. After the death of his father, a farmer, in 1741, he remained at home and during the next six years his activities, although seemingly aimless, nevertheless provided a knowledge of animal economy that formed the basis of his later studies.

In 1748 he proposed to join his brother William, who was then becoming established as a teacher of anatomy in London, and arrived in time to assist in preparations for the autumn course of lectures. William found his brother's aptitudes promising and arranged that he should attend surgical classes at St. George's and St. Bartholomew's hospitals. He was also accepted as a pupil of William Cheselden at Chelsea Hospital.

In the summer of 1752, six months after the death of his mother, John Hunter went home to bring his sister Dorothea to London, where she lived until her marriage to the Reverend James Baillie in 1757. To improve his brother's prospects, William Hunter persuaded him to enter as a student at St. Mary's Hall, Oxford, in the summer of 1755; but apparently the instruction was of little value in John Hunter's specialized pursuits and he returned to London at the beginning of the autumn term to continue his duties in the dissecting room.

John Hunter spent eleven years working with his brother in Covent Garden, during which time he made detailed studies of the structure and use of the lymphatic vessels and of the growth, structure, and exfoliation of bone. His first paper, "The State of the Testis in the Foetus and on the Hernia Congenita," was published in William Hunter's *Medical Commentaries* (1762, pp. 75–89), with illustrations by Jan van Rymsdyck. It is in this paper that he names the *gubernaculum testis* "because it connects the testis with the scrotum, and directs its course in its descent."

He made numerous preparations from material brought to the dissecting room, obtained at postmortem examination, or from chance supply, such as the grampus caught at the mouth of the Thames in 1759 and conveyed to Westminster Bridge on a barge. His interest in the organ of hearing, particularly in fish, resulted in a fine series of specimens of this intricate structure in the skate and the cod. At this time also he described, and in many cases preserved evidence of, unusual morbid conditions: adhesion of lungs and heart to surrounding tissues, cases of aneuryism, corrosion of the stomach walls by gastric juice.

As a result of such concentrated work, his health began to suffer and, being advised to give up for a time his fascinating but dangerous pursuits, he procured an appointment on the surgical staff of the army. As England was then engaged in the Seven Years' War, he was ordered to join the expeditionary force that set sail on 29 March 1761 from Portsmouth with the intention of capturing Belle Île-en-Mer (Belleisle), a small island off the French coast. Hunter and his colleagues were kept busy treating casualties for months after the island had surrendered, and it was here that he gained much of the experience that he incorporated into his great "Treatise on the Blood, Inflammation and Gun-Shot Wounds," published in 1794, the year after his death.

After a sojourn of about a year at Belle Île-en-Mer, most of the British forces were transferred to Portugal where Hunter further developed his talents in the administration of army medical services. He also availed himself of the opportunity to study the natural history and geology of the country, continued his experiments on the organ of hearing in fish, tested the effects of hibernation on the process of digestion, and collected specimens—notably of the local lizard which has the power of regenerating its tail.

The Peace of Paris was signed in February 1763, and Hunter returned to London in the early summer, not to rejoin the Covent Garden establishment but to set up in practice in Golden Square. During the next four years he continued his experiments and research and made the acquaintance of many leading scientists and naturalists of the day, including John Ellis, Daniel Solander, and Matthew Maty. His description of the anatomy of the amphibious biped *Siren lacertina* enhanced his application to be elected a fellow of the Royal Society, which honor was accorded him on 5 February 1767. His ambition to gain a senior surgical post in a hospital, however, made it essential to have credentials other than an abundance of experience. So, at the advanced age of forty, Hunter entered as a candidate for the diploma of the Company of Surgeons and was successful at his first attempt on 7 July 1768. On 9 December he was appointed to the post made vacant by the death of Thomas Gataker, surgeon to St. George's Hospital.

When William Hunter moved from Jermyn Street in 1768, John Hunter took over the property. He already owned an attractive country residence with several acres of ground at Earl's Court, where he carried out much of his experimental work and made observations on live animals, which included leopards, deer, various birds and fish, and a bull presented to him by Queen Charlotte. It was here that he spent his honeymoon after his marriage at St. James's Church, Piccadilly, on 22 July 1771, to Anne Home, daughter of Robert Boyne Home, army surgeon. She was also the sister of the artist Robert Home, of Mary Home who married Robert Mylne, the architect, and of Everard Home. They had four children, only two of whom, John Banks and Agnes Margaretta, survived infancy.

Hunter's life was now ordered to a regular pattern. He arose very early, especially in the summer, to have the best daylight for making fine dissections, and to arrange the day's work for his assistants and pupils. His private practice and hospital duties occupied much of the rest of the day; and the evenings were usually spent in discussing interesting topics with his friends, at meetings of learned societies, or in writing notes upon his cases or subjects of research. His private practice was large, lucrative, and illustrious; many of his distinguished patients, such as William Eden, Lord Auckland, became his friends.

The list of Hunter's publications is impressive by its content and variety. His "Treatise on the Natural History of the Human Teeth" was printed in two parts, in 1771 and 1778. It was here that he mentioned briefly his experiments in transplantation of tissues, the best known of which is the human tooth fixed into a cock's comb. His "Treatise on the Venereal Disease" and his "Observations on Certain Parts of the Animal Oeconomy" appeared in 1786. Some of his experimental work was described by Everard Home, as, for example, "An Account of Mr. Hunter's Method of Performing the Operation for the Cure of Popliteal Aneurism" (*Transactions of the Society for the Improvement of Medical and Chirurgical Knowledge* [1789]).

The difficulties he had encountered in gaining his own surgical training made Hunter anxious to amend conditions for others. Even in his early days in London he would "talk anatomy" with the resident students in Covent Garden long after classes were over for the day. When he had a house of his own he began to give lectures on applied anatomy and surgery, and many of the leading surgeons and anatomists both in Great Britain and in North America owed their early training and subsequent success to John Hunter's teaching; his attention to the needs of his patients and his endeavors to devise means not only to cure but to prevent disease could not fail to appeal to the keen student. In the list of his pupils are such well-known names as John Jones, John Morgan, William Shippen, Edward Jenner, William Lynn, John Abernethy, Philip Syng Physick, and Anthony Carlisle; and through them his influence passed to succeeding generations of medical students and surgeons.

Many of Hunter's plans came to fruition when, in 1783, he purchased a fine house in Leicester Square, as well as the house behind it facing what was then Castle Street. On the intervening land a lecture room, conversazione room, picture gallery, and museum were erected. Here he was able to hold meetings of the Lyceum Medicum Londinense, a student society that he founded with George Fordyce. Each member had to read a paper at one of the weekly meetings on some original piece of research; each year a gold medal was presented for what was considered the best paper.

In the preparation, arrangement, and cataloging of his museum, Hunter had the student in mind. His was not a mere collection of curious objects, though it contained such items; It was an ordered series of specimens, largely self-explanatory, demonstrating those structures in plants and animals having special, autonomous purposes, and those designed for continuation of the species; and having a further section to show the effects of accident or disease. At a time when the scope of surgery was limited, it was of the utmost value for the student to have access to specimens obtained postmortem, which often revealed the extent to which treatment had been successful and how it might be improved.

Instruction was given on how to prepare and mount museum specimens and on the technique of making corrosion casts and models. Hunter also commissioned artists to paint pictures of unusual subjects, such as North American Indians, Eskimos, dwarfs, and examples of albinism. George Stubbs painted for him a rhinoceros, two monkeys, and a yak; the subject for the latter had been brought to England from India by Warren Hastings in 1786.

Recognition of his merit came in many forms. In 1774 he was invited to join the first board of directors of the Royal Humane Society; in 1776 he was appointed surgeon-extraordinary to King George III;

in 1783 he was made a member of the Académie Royale de Chirurgie de Paris; and two years later he succeeded David Middleton as deputy surgeon general. The Copley Medal was awarded to him in 1787, an honor that his brother never received; and in the same year he was elected a member of the American Philosophical Society. In 1790, on the death of Robert Adair, he became surgeon general, and his efforts to improve the training and status of the surgeon were extended to the army medical service. In particular, he made it known that promotion could be gained only by merit and experience.

During the last fifteen years of his life, Hunter was constantly troubled with angina. At a meeting of the board of governors held at St. George's Hospital on 16 October 1793 he suffered a severe attack, collapsed, and died. The funeral was private and the coffin was placed in the vaults of St. Martin-in-the-Fields. When it was announced in 1859 that these vaults were to be cleared, several interested persons, including Frank Buckland, urged that Hunter's remains should be reinterred in Westminster Abbey. This was accomplished on 28 March 1859, and the memorial brass on the floor of the north aisle is inscribed:

> The Royal College of Surgeons of England have placed this tablet over the grave of Hunter, to record their admiration of his genius as a gifted interpreter of the Divine Power and Wisdom at work in the Laws of Organic Life, and their grateful veneration for his services to mankind as the Founder of Scientific Surgery.

John Hunter's museum, consisting of about 14,000 specimens, was purchased by the government in 1799 and handed over to the care of the Company of Surgeons (reconstituted in the following year as the Royal College of Surgeons). Despite the depletion brought about by time and wartime destruction, several thousand original Hunter specimens can still be seen in the museum specially designed for their display in the Royal College of Surgeons of England.

BIBLIOGRAPHY

See S. R. Gloyne, *John Hunter* (Edinburgh, 1950); Jessie Dobson, *John Hunter* (London–Edinburgh, 1969); W. R. LeFanu, *John Hunter: A List of His Books* (London, 1946).

JESSIE DOBSON

HUNTER, WILLIAM (*b.* Long Calderwood, near East Kilbride, Lanarkshire, Scotland, 23 May 1718; *d.* London, England, 30 March 1783), *anatomy.*

Hunter was the seventh child of John Hunter and his wife, the former Agnes Paul. He received his early education at the grammar school in East Kilbride and at the age of thirteen was sent to the University of Glasgow, with a view to being trained for the church. For four years he studied Greek, logic, natural philosophy, and related subjects. Discovering a dislike of theology, and having become friendly with William Cullen, then commencing medical practice in Hamilton, he became Cullen's assistant. Hunter always regarded his three years in the Cullen household as the happiest of his life. Cullen advised him to attend classes at the University of Edinburgh, then to spend two or three years in London before returning to Hamilton as full partner in the practice. Accordingly, in October 1739 Hunter was enrolled as a student with Alexander Monro, professor of anatomy at the University of Edinburgh. A year later, on 25 October, he boarded the packet boat at Leith and sailed for London. For several months he stayed with William Smellie, who had settled in London during the previous year and had already built up a thriving obstetrical practice. Hunter then entered the household of James Douglas, anatomist and "man-midwife," as tutor to his son William George. He was so much influenced by the opportunities of his "darling London" that he decided not to return to the partnership with Cullen but to stay with Douglas and pursue his studies, both medical and classical, under the latter's able guidance. After Douglas' death in 1742, Hunter remained in the household as tutor, took his pupil to Paris, and attended Antoine Ferrein's course of anatomy lectures. So bright were the prospects that Hunter invited his brother James to abandon his legal studies and join him in London to embark upon a medical career; the venture proved too arduous and after a few months James was obliged to return home to Long Calderwood, where he died on 11 April 1745.

In that year, when the Company of Barber-Surgeons separated into its two specialties—the barbers retaining possession of the hall in Monkwell Street—the surgeons were obliged to relax the rules they had hitherto enforced relating to human dissection, classes on which could be held in London only within the precincts of the Company or at the Royal College of Physicians. Hunter immediately advertised in the London *Evening Post* a course of lectures on anatomy to begin on 13 October 1746, for which he charged four guineas, offering "the opportunity of gentlemen learning the art of dissecting during the whole winter session, in the same manner as at Paris." The lectures were given at his house in Covent Garden from 1746 to 1760, in Litchfield Street from 1763 to 1767, and in Windmill Street from 1768 until his death in 1783. They were an immediate success and provided one of his stable sources of income.

Hunter was well aware of the difficulties of gaining a precise knowledge of anatomical structures. Reliable textbooks were few and costly; and the practice of taking notes from the lecturer's slow dictation was not only tedious but also liable to perpetuate errors, since the notes were often passed from one group of students to the next. In order to minimize these defects, he prepared a series of specimens of gross anatomy as well as such items as corrosion casts, by means of which the ramifications of even the smallest vessels could be demonstrated. Thus over the years he compiled a permanent and accurate picture not only of normal conditions but also of the diseases and accidents current in the mid-eighteenth century. With the facilities provided by such a museum the student could continue his studies throughout the year.

On 6 August 1747 Hunter was admitted to membership in the Company of Surgeons. In the summer of the following year he again visited the Continent and was privileged to discuss anatomical techniques with Bernard Siegfried Albinus. On his return to London he was appointed deputy for Daniel Layard as a surgeon-midwife to the Middlesex Hospital, established two years previously in two houses in Windmill Street. Finding calls upon his time more pressing, he appointed John Symons of Exeter, a former pupil, to help in the school; he was succeeded in 1749 by John Hunter. William Hewson was assistant from 1760 to 1772, when William Cumberland Cruikshank was appointed to this post, becoming a partner two years later. In 1750 Hunter was granted the M.D. by the University of Glasgow; in 1752 he was elected one of the masters of anatomy at Surgeons' Hall. When he was admitted a licentiate of the College of Physicians in 1756, he applied for and was granted disfranchisement from the Company of Surgeons. In 1767 he was elected a fellow of the Royal Society, and in 1768 George III appointed him professor of anatomy to the Royal Academy. He received many other honors; but perhaps his greatest triumph was to have attended Queen Charlotte during her first pregnancy. On the day of the duke of Cornwall's birth, 12 August 1762, he wrote to William Cullen: "I am very happy, and have been so for some time. I owe it to you and thank you from my heart for the great honour I now have, and have had for some time, though very few know anything of it—I mean having the sole direction of Her Majesty's health as a child-bearing lady."

Hunter now contemplated founding "a perpetual school of Anatomy" in London, for which he offered a grant of £7,000; his books, on which he had spent more than £3,000; and his museum, which was of inestimable value. But his plan failed to gain sufficient support; and after considering whether, with Cullen's help, it would be more successful in Glasgow, he finally decided to make it a private venture. Accordingly, Hunter commissioned the architect Robert Mylne to draw plans for an anatomy school to be situated in Windmill Street. In a letter to William Cullen in 1768 he remarks that he had spent more than £6,000 on his new house. "I shall go into it," he says, "in June and hope to print off my plates of the Gravid Uterus there this summer. I shall have a printing press of my own."

Hunter had already published several short papers on various subjects; and in 1762 he had brought out his *Medical Commentaries,* which contained a defense of his attitude to some of his contemporaries, notably Alexander Monro and Percivall Pott. In an age of medical polemics, this particular dissension was aggravated by the comments of Tobias Smollett in his *Critical Review* (**9** [1758], 312). In a supplement to the *Medical Commentaries* (1764), Hunter wrote,

> Anatomists have ever been engaged in contention. And indeed, if a man has not such a degree of enthusiasm, and love of the art, as will make him impatient of unreasonable opposition and of encroachments upon his discoveries and his reputation, he will hardly become considerable in Anatomy or in any branch of natural knowledge [Introduction, p. iii].

The *Gravid Uterus,* first advertised in 1751 but not published until 1774, was a magnificent series of thirty-four engravings of elephant folio size, most of them from drawings by Jan van Rymsdyck. The printing was undertaken by John Baskerville, and the work was dedicated to the king.

For the last ten years of his life, Hunter suffered from declining health, the cause vaguely described as "wandering gout." Yet so rigid was his routine that on 20 March 1783 he insisted upon giving his introductory lecture to the course on operative surgery, collapsed during the delivery, and died ten days later. He was buried on 5 April at St. James, Piccadilly, where there is a marble tablet to his memory. He and his brother had become estranged as the result of an argument concerning the priority of discovery of the true nature of the blood supply to the placenta. They were never reconciled, and William Hunter left the property of Long Calderwood to his nephew, Matthew Baillie. The administration of the anatomy school passed to Baillie and Cruikshank. According to the terms of his will, the museum, books, pictures, and his collection of coins—one of the world's finest and most valuable—were all eventually sent to the University of Glasgow, where a suitable building was erected to house them, for which purpose he left the sum of £8,000.

During his lifetime Hunter achieved outstanding success both financially and scientifically. His portraits, by Allan Ramsay, Robert Edge Pine, Mason Chamberlin, Johann Zoffany, and Joshua Reynolds, reveal a man of elegance and spirit; the size and quality of his practice and the prosperity of his school are indications of his repute and ability.

BIBLIOGRAPHY

Sir Charles Illingworth, *The Story of William Hunter* (Edinburgh-London, 1967), gives a complete list of Hunter's works and a complete list of portraits and biographies of him.

JESSIE DOBSON

HUNTINGTON, EDWARD VERMILYE (*b.* Clinton, New York, 26 April 1874; *d.* Cambridge, Massachusetts, 25 November 1952), *mathematics.*

Huntington was the son of Chester Huntington and the former Katharine Hazard Smith. He received his A. B. and A. M. from Harvard in 1895 and 1897, and his Ph.D. from the University of Strasbourg in 1901. In 1909 he married Susie Edwards Van Volkenburgh. Almost all of Huntington's professional career was spent at Harvard University, where he was an enthusiastic and innovative teacher; one of his interests is indicated by the title—unusual in a department of mathematics—of professor of mechanics, which he held from 1919 until his retirement in 1941. His interest in teaching was also reflected in his improvement of the format of the mathematical tables that he compiled or edited.

Huntington's major scientific work was in the logical foundations of mathematics. It is now commonplace to present a mathematical theory as consisting of the logical consequences of a set of axioms about unspecified objects, assumed to satisfy the axioms and nothing more. In spite of the example of Euclid, who tried to develop geometry in this way but did not completely succeed, the thorough axiomatization of a branch of mathematics was a novelty when Huntington's career began. He constructed sets of axioms for many branches of mathematics, one of which was Euclidean geometry, and developed techniques for proving their independence (that is, that no axiom is deducible from the others) and their completeness (that is, that they describe precisely the mathematical system that they are supposed to describe). His book *The Continuum* was for many years the standard introduction to the theory of sets of points and transfinite numbers.

Huntington was interested in the applications of mathematics to many different subjects. His most influential contribution was a mathematical theory of the apportionment of representatives in Congress. The Constitution states that "Representatives shall be apportioned among the several States according to their respective numbers" but does not specify how this is to be done. In the 1920's Huntington analyzed the problem and recommended the so-called method of equal proportions; in 1941 this method was adopted by Congress.

BIBLIOGRAPHY

Huntington's writings include *The Continuum, and Other Types of Serial Order, With an Introduction to Cantor's Transfinite Numbers* (Cambridge, Mass., 1917), repr. from *Annals of Mathematics,* **6** (1905), 151–184; **7** (1905), 15–43; and "The Apportionment of Representatives in Congress," in *Transactions of the American Mathematical Society,* **30** (1928), 85–110.

R. P. BOAS, JR.

HURWITZ, ADOLF (*b.* Hildesheim, Germany, 26 March 1859; *d.* Zurich, Switzerland, 18 November 1919), *mathematics.*

Hurwitz, the son of a manufacturer, attended the Gymnasium in Hildesheim. His mathematics teacher, H. C. H. Schubert, was known as the inventor of a dazzling calculus for enumerative geometry. He discovered Hurwitz, gave him private lessons on Sundays, and finally persuaded Adolf's father, who was not wealthy, to have his son study mathematics at the university, financially supported by a friend. Before leaving the Gymnasium, Hurwitz published his first paper, jointly with Schubert, on Chasles's theorem (*Werke,* paper no. 90).

In the spring term of 1877 he enrolled at the Munich Technical University, recommended to Felix Klein by Schubert. From the fall term of 1877 through the spring term of 1879 he was at Berlin University, where he attended courses given by Kummer, Weierstrass, and Kronecker. Then he returned to Munich, only to follow Klein in the fall of 1880 to Leipzig, where he took his Ph.D. with a thesis on modular functions. In 1881-1882, according to Meissner, he turned anew to Berlin to study with Weierstrass and Kronecker. (Hilbert did not know of a second stay in Berlin.) In the spring of 1882 he qualified as *Privatdozent* at Göttingen University, where he came into close contact with the mathematician M. A. Stern and the physicist Wilhelm Weber. In 1884 Hurwitz accepted Lindemann's invitation to fill an extraordinary professorship at

Königsberg University, which was then a good place for mathematics. Among its students were Hilbert and Minkowski. Hurwitz, a few years their elder, became their guide to all mathematics and their lifelong friend. Hilbert always acknowledged his indebtedness to Hurwitz. In 1892 Hurwitz was offered Frobenius' chair at the Zurich Polytechnical University and H. A. Schwarz's at Göttingen University. He had already accepted the first offer when the second arrived. He went to Zurich and remained there for the rest of his life. He married the daughter of Professor Samuel, who taught medicine at Königsberg.

Hurwitz' health was always poor. Twice he contracted typhoid fever, and he often suffered from migraine. In 1905 one kidney had to be removed; and the second did not function normally. Although seriously ill, he continued his research.

Hurwitz' papers reveal a lucid spirit and a love of good style and perspicuous composition. Hilbert depicted him as a harmonious spirit; a wise philosopher; a modest, unambitious man; a lover of music and an amateur pianist; a friendly, unassuming man whose vivid eyes revealed his spirit.

His papers were collected by his Zurich colleagues, particularly G. Polya. Although entitled *Werke,* the edition does not include his book on the arithmetic of quaternions and his posthumous function theory. The *Werke* lists his twenty-one Ph.D. students and contains an obituary written by Hilbert in 1919 and Ernst Meissner's eulogy. All present biographical data were extracted from these contributions. Hilbert's obituary is rather disappointing—even more so if it is compared with Hilbert's commemoration of Minkowski, which rings of high enthusiasm and deep regret. Certainly Hilbert had esteemed Hurwitz as a kind man, an erudite scholar, a good mathematician, and a faithful guide. But one may wonder whether he appreciated Hurwitz' mathematics as sincerely as he appreciated its creator. Of course it is easier to write a brilliant biography if the subject is as brilliant as was Minkowski. Hurwitz was anything but brilliant, although he was as good a mathematician as Minkowski. Or, if that was not the reason, was it perhaps because Hilbert himself had changed in the ten years since he wrote Minkowski's biography, and his own productivity had come to a virtual standstill. Anyhow, because Hilbert wrote his biography, Hurwitz never got the one he deserved.

In a large part of Hurwitz' work the influence of Klein is overwhelming. Among Klein's numerous Ph.D. students Hurwitz was second to none except, perhaps, Furtwängler. Much of Klein's intuitiveness is found again in Hurwitz, although the latter was superior in the rationalization of intuitive ideas. Klein was at the peak of his creativity when Hurwitz studied with him and Klein's best work was that in which Hurwitz took a share. Klein's new view on modular functions, uniting geometrical aspects such as the fundamental domain with group theory tools such as the congruence subgroups and with topological notions such as the genus of the Riemann surface, was fully exploited by Hurwitz. In his thesis he worked out Klein's ideas to reach an independent reconstruction of the theory of modular functions and, in particular, of multiplier equations by Eisenstein principles (*Werke,* paper no. 2). Modular functions were applied by Hurwitz to a classical subject of number theory—relations between the class numbers of binary quadratic forms with negative discriminant—which had been tackled long before by Kronecker and Hermite, and afterward by J. Gierster, another student of Klein's.

The problem of how to derive class number relations from modular equations and correspondences was put in general form by Hurwitz, although the actual execution was restricted to particular cases (*Werke,* papers no. 46, 47). The problem has long remained in the state in which Hurwitz had left it; but in the last few years it has been revived in C. L. Siegel's school although, strangely enough, no attention whatsoever has been paid to Hurwitz' other, unorthodox approach to class numbers (*Werke,* papers no. 56, 62, 69, 77). It is, first, a reduction of quadratic forms by means of Farey fractions and so-called Farey polygons: on the conic defined by $x:y:z = 1:-\lambda:\lambda^2$, a pair of points $\lambda = p/q, r/s$ (p, q, r, s are integers) is called an elementary chord if $ps - qr = 1$, and such chords are taken to form elementary triangles; the reduction is carried by a systematic transition from one triangle to the next. The splitting of the conic surface into such triangles led Hurwitz in 1905 to a curious nonarithmetic infinite sum for class numbers, generalized in 1918 to ternary forms. Hurwitz also refashioned the classical expressions for class numbers into fast-converging infinite series, which, together with congruence arguments, provide easy means of computation (*Werke,* paper no. 59).

More direct products of Hurwitz' collaboration with Klein were his remarkable investigations on the most general correspondences on Riemann surfaces (*Werke,* paper no. 10), in particular Chasles's correspondence principle, and his work on elliptic σ products and their behavior under the transformation of the periods (Klein's elliptic normal curves, *Werke,* paper no. 11). For Dirichlet series occurring in class number formulas, Hurwitz derived transformations like those of the ζ function (*Werke,* paper no. 3). By means of complex multiplication he studied the development coefficients of the lemniscatic function, which look much like the Bernoulli numbers (*Werke,* paper no. 67). He also investigated the automorphic groups of algebraic Riemann surfaces of genus > 1; showed that they were finite; estimated the maximal order of automorphisms as $\leqslant 10(p - 1)$, the best value, according to A. Wiman, being $2(2p + 1)$; estimated the group order as $\leqslant 84(p - 1)$; and constructed Riemann surfaces from group theory or branching data (*Werke,* papers no. 12, 21, 22, 23, 30). Hurwitz' formula $p' - 1 = w/2 + n(p - 1)$ for the genus p' of a surface w times branched over a surface of genus p is found in *Werke,* paper no. 21, p. 376. Automorphic functions of several variables were also among Hurwitz' subjects (*Werke,* paper no. 36).

In general complex-function theory Hurwitz studied arithmetic properties of transcendents which generalize those of the exponential function (*Werke,* papers no. 6, 13), the roots of Bessel functions and other transcendents (*Werke,* papers no. 14, 17), and difference equations (*Werke,* paper no. 26). Giving a solution of the isoperimetric problem, he became interested in Fourier series, to which he devoted several papers (*Werke,* papers no. 29, 31, 32, 33). Hurwitz was the author of a condition, very useful in stability theory, on a polynomial having all its roots in the left half-plane, expressed by the positivity of a sequence of determinants (see also I. Schur, "Über algebraische Gleichungen"). He gave a proof of Weierstrass' theorem that an everywhere locally rational function of n variables should be globally rational (*Werke,* paper no. 8). He was much interested in continuous fractions, to which he devoted several papers (*Werke,* papers no. 49, 50, 52, 53, 63). He also gave a remarkable proof of Minkowski's theorem on linear forms (*Werke,* paper no. 65).

In algebraic number theory Hurwitz devised new proofs for the fundamental theorem on ideals (*Werke,* papers no. 57, 58, 60, 66). He studied the binary unimodular groups of algebraic number fields of finite degree and proved that they were finitely generated. (A survey on modern extensions of this result is found in Borel's "Arithmetic Properties of Linear Algebraic Groups.") He discovered the "correct" definition of integrity in quaternions (*Werke,* paper no. 64). In the theory of invariants he wrote several papers, among them a new proof for Franz Mertens' theorems on the resultant of n forms in n variables, in which he introduced the notion of the inertia form (*Werke,* paper no. 86). To obtain orthogonal invariants he devised the invariant volume and integration in the orthogonal groups (*Werke,* paper no. 81), which, generalized to compact groups by I. Schur and H. Weyl and complemented by the invention of Haar's measure, have become extremely powerful tools in modern mathematics.

This was one of the fundamental discoveries for which Hurwitz' name will be remembered. The other is the theorem on the composition of quadratic forms (*Werke,* papers no. 82, 89), which concerns the search for algebras over the reals with a nondegenerate quadratic form Q such that $Q(xy) = Q(x)Q(y)$. The complex numbers had been known as an example of dimension 2 for centuries; in 1843 W. R. Hamilton had discovered the quaternions, of dimension 4; and in 1845 Cayley and J. T. Graves independently hit upon the octaves, of dimension 8. Attempts to go further failed. In 1898 Hurwitz proved that the classical examples exhausted the algebras over the reals with a quadratic norm. With the increasing importance of quaternions and octaves in the theory of algebras, in foundations of geometry, in topology, and in exceptional Lie groups, Hurwitz' theorem has become of fundamental importance. Many new proofs have been given; and it has been extended several times, with the final result by J. W. Milnor that algebras over the reals without zero divisors exist in dimensions 1, 2, 4, and 8 only.

BIBLIOGRAPHY

I. ORIGINAL WORKS. Hurwitz' papers were brought together as *Mathematische Werke* (Basel, 1932). His books are *Vorlesungen über die Zahlentheorie der Quaternionen* (Berlin, 1919); and *Vorlesungen über allgemeine Funktionentheorie und elliptische Funktionen,* R. Courant, ed. (Berlin, 1922; 2nd ed., 1925), with a section on geometrical function theory by Courant.

II. SECONDARY LITERATURE. For additional information see F. van der Blij, "History of the Octaves," in *Simon Stevin,* **34** (1961), 106–125; A. Borel, "Arithmetic Properties of Linear Algebraic Groups," in *Proceedings of the [9th] International Congress of Mathematicians. Stockholm 1962* (Djursholm, 1963), pp. 10–22; A. Haar, "Der Massbegriff in der Theorie der kontinuierlichen Gruppen," in *Annals of Mathematics,* 2nd ser., **34** (1933), 147–169, also in his *Gesammelte Arbeiten* (Budapest, 1959), pp. 600–622; G. Polya, "Some Mathematicians I Have Known," in *American Mathematical Monthly,* **76** (1969), 746–753; I. Schur, "Über algebraische Gleichungen, die nur Wurzeln mit negativen Realteilen besitzen," in *Zeitschrift für angewandte Mathematik und Mechanik,* **1** (1922), 307–311; and "Neue Anwendung der Integralrechnung auf Probleme der Invariantentheorie," in *Sitzungsberichte der Preussischen Akademie der Wissenschaften zu Berlin* (1924), 189–208, 297–321, 346–355; and H. Weyl, "Theorie der Darstellung kontinuierlicher halbeinfacher Gruppen durch lineare Transformationen," in *Mathematische Zeitschrift,*

23 (1925), 271–309, and **24** (1926), 328–395, 789–791, also in his *Selecta* (Basel, 1956), pp. 262–366.

HANS FREUDENTHAL

HUSCHKE, EMIL (*b.* Weimar, Germany, 14 December 1797; *d.* Jena, Germany, 19 June 1858), *anatomy, embryology, physiology.*

Huschke was the son of Wilhelm Ernst Christian Huschke, archiater of the duke of Weimar, and the former Christina Görring. He married Emma Rostosky; they had one son and four daughters. One of the daughters married Ernst Haeckel; another, the Berlin publisher Ernst Reimer.

In 1813 Huschke began his studies at the University of Jena, then the center of *Naturphilosophie.* He was greatly influenced by Lorenz Oken, one of the most enthusiastic promoters of this philosophical trend which was so influential in German science during the first four decades of the nineteenth century.[1] In fact, Huschke can be considered one of the most direct followers of Oken's ideas and one of the links between *Naturphilosophie* and the biology of the second half of the nineteenth century. On the one hand, he transmitted his philosophical ideas—mainly through his pupil and son-in-law Ernst Haeckel—to the following generation of biologists; and on the other hand, he must be considered one of the German scientists of the mid-nineteenth century who introduced an exact methodology into the life sciences. The high esteem of his contemporaries for his scientific achievements found expression in his election to membership in many scientific academies and learned societies.

Huschke's brilliant career at the University of Jena began in 1813 with his doctoral thesis—well received by his colleagues—and continued with his inaugural dissertation in 1820. In 1823 he was appointed extraordinary professor; in 1826 he obtained an honorary professorship and became director of the anatomical institute of the university; and in 1838 he was appointed full professor of anatomy and physiology. In this position he lectured on anatomy, embryology, physiology, natural history, zoology, and medical anthropology. He directed the building of a new anatomical institute but died before it was finished.

Central in Huschke's interest was the question of the origin and development of a particular organ or function. He was especially interested in the origin and transformation of the visceral skeleton during embryogenesis (1820, 1826–1828, 1838). He often discussed these topics at meetings of the Naturforscherversammlung; in 1825 he lectured on the changes occurring in the intestines and the gills of frog larvae and, a year later, on the transformation of the branchial skeleton and the accessory blood vessels in the chicken embryo. He paid special attention to the development of the sense organs, particularly of the ear and the eye, discovering that both organs originate in a furrowlike fold of the skin (1824, 1827, 1832, 1844).[2] In his study of the genesis of the avian ear (1835), Huschke described the incisorlike folds in the *ductus cochlearis* which divide the *labium vestibulare limbus spiralis* into separate sections and which are now named for him.[3]

After 1845 the central concern of Huschke's research was the way in which soul and body form a consensus—or, in his own words, "the connection between mental faculties and the body and particularly with specific parts of the brain"—a subject which also had been the central theme of his inaugural dissertation (published in 1823).[4] Huschke's major study devoted to this subject (1854) again relied upon the genetic method. What we need, he said, is a clearer understanding of the formative process through which the various physiognomic phenomena are produced, assuming that the general laws of physiognomy can be visualized by means of genetic anatomy. Consequently, the first two parts of his study are anatomical, containing a wealth of data which Huschke used in order to introduce some new techniques of measuring the superficial parts of the brain and the surface of the bones composing the skull, and of weighing the various parts of the brain. In the third part he formulated his "physiological psychology," based on the assumption that the brain is an electric apparatus,[5] in which the psychic centers are located and that there must exist an identity of structure in the emotions and in muscle dynamics.

NOTES

1. Huschke participated in the foundation of the Deutsche Burschenschaft and in the Wartburg protests against the German Confederation; he joined enthusiastically the meetings of the Deutsche Naturforscherversammlung; and he published many important contributions in *Isis.*
2. Huschke could verify this observation with the aid of a thin hair: "Dieses gleitete hierbei in eine Oeffnung, die im dunklen Mittelpunkte des Kreises befindlich war. Nun war ich auf einmal aus aller Verlegenheit, denn ich wusste jetzt, dass die Linsenkapsel ebenso wie das ganze Auge und vorzüglich das Labyrinth des Ohrs entsteht, d.h. dass sie eine Einstülpung des äusseren Hautsystems ist" (quoted after Uschmann, *Geschichte der Zoologie* . . ., p. 13).
3. Other eponyms are Huschke's foramen—a perforation near the inner extremity of the tympanal plate; Huschke's valve—the prominent lower margin of the opening of the lacrimal ducts into the lacrimal sac; Huschke's cartilage—the vomeronasal cartilage; Huschke's canal—the duct formed by the union of the tubercles of the *annulus tympanicus;* Huschke's ligament—a fold in the peritoneum at the upper side of the lesser curvature of the stomach.

4. Huschke was not primarily interested in the causal relationship between mind and body, for he starts from the assumption that all matter has a spiritual component and that all spiritual activity is accompanied by a material component (*Schädel, Hirn und Seele des Menschen . . .*, p. 161).
5. According to Huschke, both hemispheres form a pair of battery plates, one positive and the other negative. The central coils are connected to the zero point; the front and temporal parts are the poles; the commissural system forms the moist conductors; and the *corpus callosum* joins the two electric elements. The *gyrus fornicatus*, the *fasciculus unciformis*, and the *fasciculus longitudinalis* represent the connecting wires. The hemispheres are connected to each other by means of the nervous system, forming a closed circuit.

BIBLIOGRAPHY

I. Original Works. Huschke's writings include *Quaedam de organorum respiratoriorum in animalium serie metamorphosi generatim scripta et de vesica natatoria piscium quaestiones* (Jena, 1818), his diss.; *Mimices et physiognomices fragmentum physiologicum* (Jena, 1823), his inaugural diss., also in German trans. as "Mimische und physiognomische Studien," in T. Lessing and W. Rink, eds., *Der Körper als Ausdruck,* Schriftenreihe zur Gestaltenkunde, II (Dresden, 1931); *Beiträge zur Physiologie und Naturgeschichte,* I, *Ueber die Sinne* (Weimar, 1824); "Ueber die Umbildung des Darmcanals und der Kiemen der Froschquappen," in *Isis* (1826), 615–627; "Entwicklung der Glandula thyreoidea," *ibid.,* p. 613, and (1828), p. 163; *Commentatio de pectinis in oculo avium potestate anatomica et physiologica* (Jena, 1827); "Ueber die Kiemenbögen und Kiemengefässe beim bebrüteten Hühnchen," in *Isis* (1827), pp. 401–403, see also H. Rathke's comments, *ibid.* (1828), pp. 80–85; "Ueber die Kiemenbögen am Vogelembryo," *ibid.,* pp. 160–164; "Ueber die Einstülpung der Linse," in *Zeitschrift für die Ophthalmologie,* **3** (1833), 1–29, and **4** (1835), 272–295; "Ueber die erste Entwicklung des Auges und die damit zusammenhängende Cyclopie," in *Archiv für Anatomie und Physiologie,* **6** (1832), 1–47; "Ueber die Gehörzähne, einen eigenthümlichen Apparat in der Schnecke des Vogelohrs," in *Archiv für Anatomie, Physiologie und wissenschaftliche Medizin* (1835), pp. 335–346; *De bursae Fabricii origine* (Jena, 1838), his inaugural lecture; *Lehre von den Eingeweiden und Sinnesorganen,* vol. V of S. T. von Sömmering, ed., *Vom Baue des menschlichen Körpers* (Leipzig, 1844), which contains a wealth of original observations and is also available in French as *Traité de splanchnologie et des organes des sens* (Paris, 1845); *Schädel, Hirn und Seele des Menschen und der Thiere nach Alter, Geschlecht und Raçe* (Jena, 1854); and *Ueber Craniosclerosis totalis rhachitica und verdickte Schädel überhaupt, nebst neuen Beobachtungen jener Krankheit* (Jena, 1858).

II. Secondary Literature. See E. Giese and B. von Hagen, *Geschichte der medizinischen Fakultät der Universität Jena* (Jena, 1958), pp. 457 ff.; A. Gode von Aesch, *Natural Science in German Romanticism* (New York, 1941), esp. pp. 224–239; J. Günther, *Lebensskizzen der Professoren der Universität Jena seit 1558 bis 1858* (Jena, 1858); Nicolaus Rüdinger, in *Allgemeine deutsche Biographie,* XIII, 449–451; and G. Uschmann, *Geschichte der Zoologie und der zoologischen Anstalten in Jena 1779–1919* (Jena, 1959), esp. pp. 12–14.

Pieter Smit

HUSSEY, WILLIAM JOSEPH (*b.* Mendon, Ohio, 10 August 1862; *d.* London, England, 28 October 1926), *astronomy.*

Hussey was the son of John Milton Hussey and the former Mary Catherine Severns. His education at the University of Michigan was both delayed and interrupted by the need to earn his way; he graduated in civil engineering with the class of 1889. His interest turned to astronomy and following a brief period with the U.S. Nautical Almanac Office, he returned to the University of Michigan for three years as instructor in mathematics and astronomy. From 1891 to 1892 he served as acting director of the Detroit observatory. An appointment in 1892 as assistant professor of astronomy at Leland Stanford Junior University brought Hussey to California, where he had the opportunity to work as a volunteer assistant at the newly established Lick Observatory. In 1896 he was appointed astronomer on the Lick staff, a position he held until 1905. He then accepted an appointment at the University of Michigan as professor of astronomy and director of the Detroit observatory. He continued in this position until his death.

Hussey's scientific reputation in research rests largely on his extensive discovery and measurement of double stars, made for the most part at the Lick Observatory but also at the observatory of the University of La Plata, Argentina. He was director at La Plata during 1912–1915, concurrently with his service at the University of Michigan.

In 1903 Hussey investigated possible observing sites in southern California and near Flagstaff, Arizona, for the Carnegie Institution of Washington. His enthusiastic report on Mt. Wilson was largely responsible for its selection for what subsequently became the Mt. Wilson Observatory; he likewise reported favorably on Mt. Palomar.

An ardent promoter, as early as 1903, of observatories in the southern hemisphere, Hussey visited Australia in that year to survey promising sites. His work at La Plata resulted in the increased efficiency of the observatory. His final and most important accomplishment in this area was the establishment, with financial support from R. P. Lamont, of an observatory at Bloemfontein, South Africa. Equipped with a twenty-seven-inch refractor, this observatory was designed primarily for double-star observations. On his way to Bloemfontein to oversee the installation of the telescope, Hussey died suddenly in London.

During his twenty-one years at the University of Michigan, Hussey organized the Detroit observatory on a modern and efficient basis by adding to its equipment a 37.5-inch reflector with a spectrograph suitable for astrophysical work. He established a reputation as an outstanding observatory director and as a devoted and inspiring teacher.

BIBLIOGRAPHY

I. ORIGINAL WORKS. Hussey wrote numerous technical and scientific works, of which a list of references may be found in *Astronomischer Jahresbericht,* vols. **1–28** (1899–1926). Among his more important astronomical writings are *Micrometrical Observations of the Double Stars Discovered at Pulkowa,* Lick Observatory publication no. 5 (Berkeley, Cal., 1901); "Third Series of Observations of the Satellites of Saturn," in *University of California Publications. Astronomy. Lick Observatory Bulletin,* no. 68 (1905), pp. 71–76; "Observations of One Hundred and Twenty-seven New Double Stars," *ibid.,* no. 117 (1907), pp. 124–129; "A General Account of the Observatory," in *Publications of the Astronomical Observatory of the University of Michigan,* **1** (1912), 3–34; and "Observations of the Double Stars Discovered at La Plata," *ibid.,* (1914), pp. 147–160.

II. SECONDARY LITERATURE. There are obituary notices in *Monthly Notices of the Royal Astronomical Society,* **87** (1927), 260; *Popular Astronomy,* **34** (1926), 605; and *Publications of the Astronomical Society of the Pacific,* **29** (1927), 35, which deals entirely with Hussey as a teacher.

C. D. SHANE

HUTCHINSON, JOHN (*b.* Ryton, near Newcastle-upon-Tyne, England, 1811; *d.* Fiji, Sandwich Islands, July 1861), *physiology.*

John Hutchinson carried out fundamental research on respiratory function in health and disease, and invented the spirometer, which is still used today to estimate pulmonary function accurately.

The only son of James Hutchinson, he attended London University College (now University College, London) and became a member of the Royal College of Surgeons in 1836. He first worked at the Southampton Dispensary in Leigh Street, London, but his interest centered on scientific, rather than clinical, medicine. He became a fellow of the newly established Statistical Society in 1842 and in 1846 received the degree of M.D. from the University of Giessen. He was appointed assistant physician at the Hospital for Consumption at Brompton, London, and for some years was a physician to the Britannic Life Assurance Company.

Being of a restless disposition, however, Hutchinson set out for Australia in September 1852, soon after gold deposits were discovered there. Of his activities during the eight years he spent in Victoria nothing is known, except that he made a large collection of gold-bearing rock valued at £200. In March 1861 he traveled to the Fiji Islands, where he died in July of that year. It is said that he had intended to return to England and had planned to prepare a book on his experiences of colonial and primitive life.

Hutchinson was a man of notable versatility. In addition to his scientific abilities he could play the violin with masterly skill, sculpt in bas-relief, and paint in oils. Moreover, he was an accomplished conversationalist, displaying originality, humor, brilliance, and diversity in his topics of discourse. He lectured with ease and proficiency and could talk on a variety of subjects, which he did in London and in Newcastle-upon-Tyne before the Literary and Philosophical Society.

Hutchinson studied only the mechanical aspects of respiratory function and seems to have had no interest in its chemistry, which was rapidly developing contemporaneously with his work. He investigated in particular the action of the intercostal muscles, and applied physics and mathematics to the problems of thoracic movement and pulmonary ventilation. He did this by dissection, by thoracic measurement, and by making plaster casts of cadavers.

Many before him had been interested in the biophysics of breathing, but Hutchinson was the first to offer a precise subdivision of lung volume: (*a*) breathing (that is, tidal) air, (*b*) complemental (inspiratory reserve) air, (*c*) reserve (expiratory reserve) air, and (*d*) residual air. The amount of air taken into the lungs with a single deep inspiration (the sum of (*a*) and (*b*) above) had been measured by investigators in the seventeenth and eighteenth centuries, but Hutchinson was able to quantify it accurately. He called it the "vital capacity," a term still in use, and he defined it as ". . . the greatest voluntary expiration following the deepest inspiration" (Todd's *Cyclopaedia* [1849–1852], p. 1065).

Hutchinson invented the spirometer in order to measure this volume of air. He was not the first to use such an instrument, since W. Clayfield of London had already employed a crude gasometer.[1] But Hutchinson's work was much more accurate,[2] and his machine was the forerunner of all modern methods of estimating pulmonary function. Moreover, his research was extensive and his interpretation and application of it significant.

Kentish[3] and Herbst[4] had shown that respiratory formation is modified by height, age, and disease, and Hutchinson, extending this work, proved conclusively that the vital capacity is directly related to the height of the individual (*Lancet* [1844], **1**, 567–570, 594–597). He made observations on a wide variety of subjects, including fire-brigademen, wrestlers, gentlemen, and a three-foot nine-inch dwarf, and found that lung capacity increased in an arithmetical progression of eight cubic inches for every inch of actual human height. He also demonstrated that the vital capacity decreases with aging, the accretion of excess bodily weight, or the contraction of a lung disease such as pulmonary tuberculosis, to which he gave special attention. He prepared many tables based on his researches, which involved nearly 4,000 individuals both healthy and sick, ranging from watermen to princes of the blood; and he could establish standards statistically acceptable for use, especially in life insurance work. This general scope, however, prompted one of the main criticisms of Hutchinson's work, namely that he expressed his data in averages and gave less attention to individual variations. Moreover, his plea for widespread adoption of the spirometer for diagnosis of lung disease, as well as for evaluation of the healthy state, has since been adjudged to have been not entirely justified.

Hutchinson's research, nevertheless, was important and influential, and affirms his place as a pioneer in the investigation of pulmonary physiology and pathology; his research was described by a contemporary as "... one of the most valuable contributions to physiological science that we have met with for some time" (*British and Foreign Medical Review,* **24** [1847], 327–328). He recorded his findings in a series of papers between 1844 and 1850, the best expositions of them being published in 1846 with several subsequent translations, and between 1849 and 1852 in Todd's *Cyclopaedia.* In 1852 his book *The Spirometer, the Stethoscope, and Scale-Balance* explored the various ways of quantifying physiological and clinical data, as applied to life-insurance medical examinations.

Not surprisingly, owing to his upbringing in a colliery district, Hutchinson also investigated coal mining conditions, especially accidents, and he gave valuable evidence to committees of the House of Lords on this subject. In addition he studied the general problems of fire, heat, and ventilation in industry.

NOTES

1. T. Beddoes and J. Watt, *Considerations on the Medicinal Use of Factitious Airs and on the Manner of Obtaining Them,* 3rd ed. (London, 1794[?]–1796), pt. 3, p. 103. A drawing of Clayfield's machine forms the frontispiece to H. Davy, *Researches Chemical and Philosophical: Chiefly Concerning Nitrous Oxide or Dephlogisticated Nitrous Air and Its Respiration* (London, 1800).
2. J. H. Arnett, "The Vital Capacity of the Lungs. Early Observations and Instruments," in *Medical Life,* **43** (1936), 3–6.
3. E. Kentish, *An Account of Baths, and of a Madeira-House at Bristol With a Drawing and Description of a Pulsometer ...* (London, 1814).
4. E. F. G. Herbst, "Ueber die Capacität der Lungen für Luft, in gesunden und kranken Zustände," in *Archiv für Anatomie und Physiologie,* **3** (1828), 83–107.

BIBLIOGRAPHY

I. Original Works. Hutchinson's work on respiratory functions was reported in the following: "Pneumatic Apparatus for Valuing the Respiratory Powers," in *Lancet* (1844), **1**, 390–391; "Lecture on Vital Statistics, Embracing an Account of a New Instrument for Detecting the Presence of Disease in the System," in *Lancet* (1844), **1**, 567–570, 594–597; *Contributions to Vital Statistics, Obtained by Means of a Pneumatic Apparatus for Valuing the Respiratory Powers With Relation to Health* (London, 1844); "On the Capacity of the Lungs, and on the Respiratory Functions, With a View of Establishing a Precise and Easy Method of Detecting Disease by the Spirometer," in *Medico-Chirurgical Transactions,* **29** (1846), 137–252, German trans. as *Von der Capacität der Lungen und von den Athmungs-Functionen ...* (Brunswick, 1849). An abstract with contemporary assessment is in *British and Foreign Medical Review,* **24** (1847), 327–328.

See also "Thorax," in R. B. Todd, ed., *The Cyclopaedia of Anatomy and Physiology,* IV, pt. 2 (London, 1849–1852), 1016–1087; and *The Spirometer, the Stethoscope, and Scale-Balance; Their Use in Discriminating Diseases of the Chest, and Their Value in Life-Offices ...* (London, 1852).

He also published "On Ventilation in General," in *Journal of Public Health,* **1** (1848), 231–234, 263–264, 295–298, 311–315; and **2** (1849), 3–7, 29–32, 57–61, 85–88, 115–117, 141–145, 169–175, 225–229.

II. Secondary Literature. There is no adequate biographical account of Hutchinson, but the following are valuable: "The Late Dr. John Hutchinson," in *Medical Times and Gazette,* **1** (1862), 200–201; "John Hutchinson, M.D.," in *Lancet* (1862), **1**, 240; A. Hirsch, *Biographisches Lexikon der hervorragenden Aerzte* (Vienna–Leipzig, 1886), III, 327–328; and "A Pioneer in Spirometry," in *Lancet* (1920), **2**, 563.

Edwin Clarke

HUTTON, CHARLES (*b.* Newcastle-upon-Tyne, England, 14 August 1737; *d.* London, England, 27 January 1823), *mathematics.*

Hutton was the son of a colliery worker. Largely self-educated, he rapidly acquired enough knowledge of mathematics to establish himself as a schoolmaster in Newcastle. His pupils, drawn from the families of local landowners and leading citizens, included John Scott (earl of Eldon) and Hutton's future wife, Elizabeth Surtees. Hutton carried out a local land survey

(1770) and wrote a tract on the equilibrium of bridges (1772), an elementary textbook on arithmetic (1764), and a more elaborate treatise on mensuration that was illustrated by Thomas Bewick (1767).

In 1773 Hutton was appointed professor of mathematics at the Royal Military Academy at Woolwich, where he remained for thirty-four years. He was elected to the Royal Society in 1774 and served as foreign secretary from 1779 to 1783. Hutton's resignation from office, requested by Sir Joseph Banks (then president of the Society) on the grounds that he failed to carry out his duties efficiently, led to a major attack by Horsley, F. Masères, Maskelyne, P. H. Maty, and others on Banks's management of the affairs of the Society.

Hutton wrote many papers and received the Copley Medal of the Royal Society for "The Force of Fired Gunpowder and the Velocities of Cannon Balls," published in 1778. That year he also presented a report to the Society on the mean density of the earth, deduced from Maskelyne's observations at Mount Schiehallion in Perthshire. With George Shaw and Richard Pearson he edited an abridgment of the *Philosophical Transactions* for the years 1665 to 1680.

Hutton was an indefatigable worker and his mathematical contributions, if unoriginal, were useful and practical. Throughout his life, he contributed assiduously to scientific periodicals through notes, problems, criticism, and commentary. He wrote textbooks for his pupils in Newcastle and the cadets at Woolwich; edited a great many almanacs, including the *Ladies' Diary* (1773–1818); and compiled several volumes of mathematical tables, one of which contained a comprehensive historical introduction (1785). In addition he translated from the French Montucla's four-volume edition (1778) of Ozanam's 1694 work *Recreations in Mathematics and Natural Philosophy* (London, 1803).

The *Mathematical and Philosophical Dictionary* (1795) is probably the best known of Hutton's works. Although it was criticized as unbalanced in content, unduly cautious in tone, and sometimes lacking judgment, the dictionary has served as a valuable source for historians of mathematics.

BIBLIOGRAPHY

I. ORIGINAL WORKS. Many of Hutton's contributions to the *Ladies' Diary* are included in the *Diarian Miscellany* (London, 1775). His land survey, *Plan of Newcastle and Gateshead* (1770), is now in the City Library, Newcastle-upon-Tyne. Of his scientific papers the most important are "A New and General Method of Finding Simple and Quickly Converging Series," in *Philosophical Transactions of the Royal Society of London,* **66** (1776), 476–492; "The Force of Fired Gunpowder and the Velocities of Cannon Balls," *ibid.,* **68** (1778), 50–85; "An Account of the Calculations Made From the Survey and Measures Taken at Mount Schiehallion, in Perthshire, in Order to Ascertain the Mean Density of the Earth," *ibid.,* 689–778; "Calculations to Determine at What Point in the Side of a Hill Its Attraction Will be the Greatest," *ibid.,* **70** (1780), 1–14; "On Cubic Equations, and Infinite Series," *ibid.,* 387–450; and "Project for a New Division of the Quadrant," *ibid.,* **74** (1784), 21–34.

These works and other papers, including the tract *The Principles of Bridges* (Newcastle, 1772), are brought together in *Tracts, Mathematical and Philosophical* (London, 1786) and *Tracts on Mathematical and Philosophical Subjects,* 3 vols. (London, 1812). His textbooks include *The Schoolmaster's Guide* (Newcastle, 1764); *A Treatise on Mensuration* (Newcastle, 1767–1770); *The Compendious Measurer* (London, 1784); *The Elements of Conic Sections* (London, 1787); and *A Course of Mathematics for the Cadets of the Royal Military Academy* (London, 1798–1801). See also *Mathematical and Philosophical Dictionary,* 2 vols. (London, 1795); *The Philosophical Transactions to 1800 Abridged With Notes,* 18 vols. (London, 1809); and the historical introduction to *Mathematical Tables* (London, 1785).

II. SECONDARY LITERATURE. An adequate account of Hutton's life and work is that of R. E. Anderson in *Dictionary of National Biography,* XXVIII, 351–353. Background information is in J. Bruce, *A Memoir of Charles Hutton* (Newcastle, 1823). A lengthy and eulogistic account is Olinthus Gregory, "Brief Memoir of Charles Hutton, L.L.D., F.R.S.," in *Imperial Magazine,* **5** (1823), 202–227. The local history collection in the City Library, Newcastle-upon-Tyne, contains many portraits of Hutton. A bust, executed before his death, stands in the Literary and Philosophical Society of Newcastle.

Many pamphlets (mostly anonymous) relate to the Royal Society controversy in 1784; see A. Kippis, *Observations on the Late Contests in the Royal Society* (London, 1784); *'A Friend to Dr Hutton' Writes An Appeal to the Fellows. . .* (London, 1784), anonymous; and *An Authentic Narrative of the Dissensions and Debates in the Royal Society* (London, 1785), anonymous.

MARGARET E. BARON

HUTTON, JAMES (*b.* Edinburgh, Scotland, 3 June 1726; *d.* Edinburgh, 26 March 1797), *geology, agriculture, physical sciences, philosophy.*

Hutton was the only son of William Hutton, a merchant and former city treasurer in Edinburgh, and Sarah Balfour, daughter of John Balfour, another Edinburgh merchant, whose descendants provided two professors of botany at Edinburgh University. William Hutton died in 1729 when James was three years old. His will indicates that he left the family, including Hutton's three sisters, quite well-off, and apparently Hutton was never under any pressing need to earn a living. He attended Edinburgh High School and in 1740 entered Edinburgh University as a student of the humanities. He attended the lectures given by John Stevenson on logic and rhetoric and those of the mathematician Colin Maclaurin, which included physics, experimental philosophy, and geography as well as mathematics.

It is said that Hutton enjoyed Maclaurin's lectures particularly, but his biographer John Playfair[1] states that it was to Stevenson that Hutton was indebted for his interest in chemistry, as a result of an experiment introduced into a lecture. Little information about chemistry was then available to Hutton, but he retained and developed his interest in the subject throughout his lifetime.

On leaving the university it was apparent that Hutton had an inclination for academic studies, but he was persuaded to follow an occupation more likely to provide a professional career. Consequently, in 1743 he was apprenticed to an Edinburgh lawyer. The routine of a lawyer's office was not to his liking, and he was soon released from his obligations. He then decided to study medicine, the only professional course which ensured that he would learn something more about chemistry. He reentered the university in 1744, and studied medicine there until 1747, probably attending the lectures of Andrew Plummer, professor of medicine and chemistry, who had studied under Boerhaave.

Toward the end of 1747 Hutton went to Paris, where he remained nearly two years. There, according to Playfair, "he pursued with great ardour the studies of chemistry and anatomy." Because of his interest in chemistry, he probably attended G. F. Rouelle's well-known and popular chemistry course, which also included lectures on mineralogy and geology. Thus it was possibly in Paris that Hutton first became acquainted with geology. Sometime during 1749 Hutton moved to Leiden, where he graduated M.D. in September of that year with a thesis entitled *De sanguine et circulatione microcosmi.*

After leaving Leiden at the end of 1749 Hutton spent several months in London. About this time he entered into an agreement with James Davie, an Edinburgh friend, to manufacture sal ammoniac from soot, by a method they had jointly discovered before Hutton had left Edinburgh. This undertaking operated successfully for many years and no doubt added to Hutton's income.

Hutton returned to Edinburgh in the summer of 1750. He decided against practicing medicine and chose instead to take up farming as an occupation on the small farm he had inherited from his father at Slighshouses, Berwickshire, forty miles southeast of Edinburgh. Hutton recorded that he became interested in farming some years previously after reading Jethro Tull's well-known book *Horse-Hoeing Husbandry.* The standard of farming in Scotland at that time was low and because Hutton investigated thoroughly any subject in which he became interested, before settling at Slighshouses he spent about a year (1752–1753) on a farm at Belton, near Yarmouth, in East Anglia, an area in which good farming practice prevailed. While there he made many journeys on foot into other parts of England to study agriculture and he acquired the habit of examining rock outcrops. It was in 1753, according to Playfair, that Hutton first began to study geology. As a student of farming he must have observed that in England soils vary markedly from place to place, and this may have stimulated an interest in the subject.

In 1754 Hutton spent some months traveling in Holland, Belgium, and northern France to improve further his knowledge of agriculture, and again he took the opportunity to add to his knowledge of geology. At the end of that year he moved to Slighshouses, where he spent the next fourteen years farming his land in a more scientific manner than had hitherto been customary in Scotland. So far as is known this period of his life was uneventful, except that he made a journey to northern Scotland in 1764 with his close friend George Clerk of Penicuik[2] chiefly to study geology, to which, according to Playfair, Hutton was then giving much attention. Slighshouses was an isolated farmhouse and Hutton must have lacked congenial company, although one friend, Sir John Hall of Dunglass, a man interested in both farming and science, lived in the neighborhood. The future course of Hutton's life suggests that he may have spent much time reading scientific literature, for his interests were never confined solely to geology.

In 1767 Hutton, in association with Clerk and Hall, became a member of the committee of management of a projected canal to join the Forth and Clyde rivers. He continued to take an active part in the work of the committee for some twenty years.

About 1768, after bringing his farm into good condition, Hutton was able to let it. He then moved to Edinburgh, where he spent the rest of his life, living with his unmarried sisters. There was in Edinburgh at that time a Philosophical Society (later incorporated as the Royal Society of Edinburgh). Hutton became a member and read several papers to the society, only one of which was published. Playfair states that in Edinburgh much of Hutton's time was occupied with experimental chemistry; but he published nothing on the subject until late in his career. A visitor to his apartment in 1772 recorded that "his study is so full of fossils and chemical apparatus that there is hardly room to sit down."

Hutton was by temperament both sociable and hospitable and he entered fully into the intellectual and social life of the city. Joseph Black became his most intimate friend. Others of about his own age with whom he associated closely were Adam Smith, James Lind (1736–1812), Adam Ferguson, James Burnett (Lord Monboddo), John Hope, and John Walker. Through Black he became a friend of James Watt, in whose work he took much interest. About 1781 he first met Playfair,[3] and later he befriended Sir James Hall, who attained distinction as a geologist and chemist.

In 1774 Hutton made another tour into England and Wales. He visited Birmingham and with Watt examined the salt mines in Cheshire. In an unpublished letter to George Clerk he reported that he had been studying both geology and agricultural practice during this tour; and he implied that he was now familiar with the geology of England, with the exception of Cornwall. He later obtained a report on the geology of Cornwall from Watt's son, Gregory. During Hutton's tour the elder Watt probably introduced him to some members of the Birmingham discussion group later known as the Lunar Society, for he afterward corresponded with Erasmus Darwin and Matthew Boulton.

In 1777 Hutton published in Edinburgh a small pamphlet entitled *Considerations on the Nature, Quality, and Distinctions of Coal and Culm.* Its purpose, commercial rather than geological, was to establish the claim that the low-grade stony coal (culm) then exported from Edinburgh for lime burning should qualify for a lower rate of duty. This pamphlet, and Hutton's association with the Forth and Clyde canal, suggest that the practical value of his geological knowledge was already recognized.

When the Royal Society of Edinburgh was founded in 1783, Hutton became one of its most active supporters, believing that the establishment of the Society was important for the progress of science. His active interest in geology continued and from 1785 to 1788 he visited several parts of Scotland, the Isle of Man, and England to extend his knowledge. In 1788 Hutton was elected foreign member of the French Royal Society of Agriculture. It is possibly significant that the president of the society at that time was Nicholas Desmarest.

After 1788, so far as is known, Hutton made no more field excursions; and from 1791 he was subject to recurrent illness.[4] He spent these years preparing his lesser known works on chemistry, physics, and philosophy for publication. In 1795 he published the definitive two-volume edition of his *Theory of the Earth.* His friends had previously urged him to publish this work, and he was finally prompted to do so to counter Richard Kirwan's strong criticism of the theory.[5] Finally, Hutton began the preparation of another work, the "Principles of Agriculture," but his death prevented its publication.

The variety of subjects that Hutton studied intensively, and his general way of life, indicate that he was a man interested in knowledge for its own sake, without thought of personal advancement, and his works show an overriding intent to fit all the subjects he discussed into the framework of his deistic philosophy.

The illness that led to Hutton's death was stated by Black to have been caused by stones in the bladder. The first attack in 1791 was cured by a severe operation, but a recurrence set in during 1794. Thereafter he was confined to his house, although he remained cheerful, mentally alert, and able to read and write between bouts of severe pain.

Hutton never married; he was survived by one unmarried sister, Isabella, and a natural son, James, probably born about 1747, when Hutton was still a student. His son, employed for many years in the General Post Office in London, married and raised a family. Hutton kept in contact with him, providing money when he was in need. After Hutton died, Isabella Hutton presented his geological specimens to Black, who, in turn, gave them to the Royal Society of Edinburgh, on conditions which should have ensured that they would be properly cataloged and preserved. A few years later they were transferred to the university museum, then curated by Robert Jameson. They were exhibited for a time, but ultimately disappeared and no trace of them has since been found.

Geology. Hutton's most important contribution to science was his theory of the earth, first announced in 1785. Hutton had then been actively interested in geology for fully thirty years. It is known that he had completed the theory in outline some years earlier, and according to Black, writing in 1787,[6] Hutton had formed its principal parts more than twenty years before. In essence the theory was simple, yet it was of such fundamental importance that Hutton has been called the founder of modern geology. Much has been written about the scientific and intellectual background in eighteenth-century Europe at the time Hutton formed his theory, but its novelty can only be appreciated when related to the existing state of geological knowledge.

Interest in various branches of the earth sciences was then widespread, but recognition of geology as an individual science had scarcely begun. The mining of economic minerals was one of the oldest industries, but the development of scientific mineralogy was retarded by the undeveloped state of chemistry and crystallography. Nevertheless, through mining and quarrying operations, a knowledge of stratigraphy must have been acquired locally, but it remained rudimentary because of the almost universal belief that the fossiliferous sediments had been deposited by, or during the retreat of, the Noachian flood. While fossils themselves had long aroused interest (it was recognized that some forms could not be matched by known living species) their value as chronological and stratigraphic indexes had not yet been recognized except, perhaps, over very limited local stratigraphic ranges.

Crystalline rocks such as granite and gneiss, usually found in the core of mountain ranges, were regarded as primeval in age, and the sediments, often fossiliferous, on the flanks of the mountains and in low ground were assumed to be flood deposits. This classification carried no implication that any rocks were older than the five or six thousand years allowed for in biblical chronology. By about the middle of the eighteenth century, however, one or two authors had suggested that geological time might be longer than this chronology allowed. The effects of erosion, long recognized, formed a subject for debate over whether denudation would ultimately render the earth uninhabitable, or whether it would be compensated by the elevation of new lands on which life would continue.

There existed one major gap in geological knowledge. It was unsuspected that rocks of the type now classed as igneous formed a major and widely distributed rock group, wholly distinct in origin from the sediments. The extrusion of lava from active volcanoes was looked on as a local and superficial phenomenon. After about 1740, Italian and French naturalists recognized the existence, locally, of volcanic cones and lava flows in areas where there was no record of volcanic activity in historic times; but many years passed before it was realized that volcanic activity had been worldwide, not only in historic times but in past geologic ages. The igneous origin of many rocks interbedded in, or otherwise closely associated with, the sediments was still unrecognized.

Broadly speaking, the position was that many geological observations had been made and recorded in the literature; but previous attempts to synthesize these observations into a general "theory of the earth" were unscientific and had not proved acceptable. The issue had been confused and progress retarded by a literal belief in the biblical account of creation and the universal flood.

The Theory of the Earth. Hutton's theory, or "System of the Earth," as he called it originally, was first made public at two meetings of the Royal Society of Edinburgh, early in 1785. The society published it in full in 1788, but offprints of this paper were in circulation in 1787, and possibly in 1786. The theory first appeared in print in condensed form, in a thirty-page pamphlet entitled *Abstract of a Dissertation . . . Concerning the System of the Earth, Its Duration, and Stability,* which Hutton circulated privately in 1785. The interest of this pamphlet is that it states all the conclusions which were essential to the theory as a whole. It emphasizes that even at this early date Hutton's thinking was far ahead of that of his contemporaries. For this reason, and because it is more easily comprehended than the full version, it is summarized here.

Hutton's approach in the *Abstract* is logical, but his thought is not translated into clear and incisive prose. As with almost all that he wrote (other than private letters), his style is prolix and abstruse, so that the text must be read with care to appreciate its full significance.

Hutton describes briefly his purpose in carrying out the inquiry, the methods he employed in reaching his conclusions, and the conclusions themselves. His purpose was to ascertain (*a*) the length of time the earth had existed as a "habitable world"; (*b*) the changes it had undergone in the past; and (*c*) whether any end to the present state of affairs could be foreseen. He stated that the facts of the history of the earth were to be found in "natural history," not in human records, and he ignored the biblical account of creation as a source of scientific information (a view he expressed explicitly later on). The method he employed in carrying out his inquiry had been a careful examination of the rocks of the earth's crust, and a study of the natural processes that operated on the earth's surface, or might be supposed, from his examination of the rocks, to have operated in the past. In this way, "from principles of natural philosophy," he attempted to arrive at some knowledge of the order and system in the economy of the globe, and to form a rational opinion as to the course of nature and the possible course of natural events in the future.

Hutton concluded that rocks in general (clearly he referred here to the sedimentary rocks) are composed of the products of the sea (fossils) and of other materials similar to those found on the seashore (the products of erosion). Hence they could not have formed part of the original crust of the earth, but were formed by a "second cause" and had originally been deposited at the bottom of the ocean. This reasoning, he stated, implies that while the present land was forming there must have existed a former land on which organic life existed, that this former land had been subjected to processes of erosion similar to those operating today, and that the sea was then inhabited by marine animals. He then concluded that because the greater part of the present land had been produced in this way, two further processes had been necessary to convert it into a permanent body resistant to the operations of water: the consolidation of the loose incoherent matter at the sea bottom, and the elevation of the consolidated matter to the position it now occupies.

Hutton then considered two possible methods of consolidation. The first, deposition from solution, he rejected because the materials of which ordinary sediments are composed are, with few exceptions, insoluble in water. He adopted the alternative, fusion of the sediments by the great heat which he believed to exist beneath the lower regions of the earth's crust. Heat, he claimed, was capable of fusing all the substances found in different types of sediment.

He also concluded that the extreme heat that fused the sediments must be capable of "producing an expansive force, sufficient for elevating the land from the bottom of the ocean to the place it now occupies." He supported this conclusion by stating that the strata formerly deposited in regular succession at the bottom of the ocean are now often found broken, folded, and contorted, a condition to be expected as a result of the violently expansive action of subterraneous heat.

Hutton then discussed the direct evidence of the action of heat, which he had found in the rocks themselves. He mentioned mineral veins containing matter foreign to the strata they traverse, the widespread occurrence of volcanoes, and the occurrence of what he called "subterraneous lavas." (The examples quoted here, and in the fuller version of the theory, indicate clearly that he was referring to what are now known as igneous intrusions.)

Hutton next claimed that his theory could be extended to all parts of the world, a generalization that was by then justified because similar rocks occur in other countries. He also claimed that the theory, based on rational deductions from observed facts, was not "visionary."

Finally, Hutton discussed one of the principal objects of his inquiry, the length of time the earth had existed as a habitable world, that is, in effect, the question of geological time. He rejected as humanly impracticable the possibility of estimating geological time by measuring the rate at which erosion is wearing down the land. Hence he concluded

> . . . That it had required an indefinite space of time to have produced the land which now appears; . . . That an equal space had been employed upon the construction of that former land from whence the materials of the present came; . . . That there is presently laying at the bottom of the ocean the foundation of a future land, which is to appear after an indefinite space of time. . . . so that, with respect to human observation, this world has neither a beginning nor an end [pp. 27-28].

Hutton was not prepared to be more definite than the facts allowed.

It was also in the *Abstract* that Hutton disclosed for the first time his philosophic belief that there exists in nature evidence of wisdom and design. He believed that the natural processes operating on and within the earth's crust had been so contrived as to provide for the indefinite continuance of the earth as a habitable world, providing means for the continuing existence of living beings, and that his theory provided support for this conclusion. The final paragraph of the *Abstract* includes the following statement: "Thus, either in supposing Nature wise and good, an argument is formed in confirmation of the theory, or, in supposing the theory to be just, an argument may be established for wisdom and benevolence to be perceived in nature." Hutton's theory ran counter to the belief then widely held that the present world was created by a divine being, fully populated by animal and plant life, at a time that could be measured by human records.

Hutton makes few references in the *Abstract* to the evidence on which he bases his theory. This is discussed in detail in his 1788 paper. Here, in discussing geological time, the conclusion he draws from fossils is of particular interest. He states:

> Time . . . is to nature endless and as nothing. . . . The Mosaic history places this beginning of man at no great distance; and there has not been found, in natural history, any document by which a high antiquity might be attributed to the human race. But this is not the case with regard to the inferior species of animals. . . . We find in natural history monuments [that is, fossils] which prove that those animals had long existed; and we thus procure a measure for the computation of a period of time extremely remote, though far from being precisely ascertained [pp. 215, 217].

From 1785 onward Hutton continued to collect new information to support his theory, which he published later in a two-volume work, *Theory of the Earth: With Proofs and Illustrations; in Four Parts*

(1795). In this edition the 1788 theory is restated with no essential change in the first chapter of volume I. The remainder of the two volumes deals principally with the supporting proofs and illustrations. Only two of the four parts promised on the title page were published in 1795. Hutton left an unfinished manuscript containing six chapters totaling 267 pages, evidently intended for inclusion in an additional volume of the *Theory.* These chapters, published as volume III in 1899, are of considerable interest, for they contain accounts of several of his later geological journeys. A study of the three volumes reveals the remarkable extent of Hutton's geological knowledge, the thoroughness of his investigations, and the acuteness of his observations.

The methods Hutton had employed in forming his theory were essentially the same as those employed by modern field geologists. He examined many different types of rocks, paying attention to their structural relations one to another; and he considered in detail the mineralogical and chemical composition of individual rocks. He also studied intensively the physical processes now operating on the earth's surface. In addition he examined British, European, and American literature to find support for his conclusions.

The method he employed in formulating his theory was, as he claimed, based on the principles of natural philosophy. Some of his conclusions can be described as speculative, and others were based on misinterpreted evidence, but these elements in the theory do not destroy its validity as a whole. It could be argued that Hutton's theory incorporated ideas that he had gained from other authors. This question is difficult to answer, for although he had read extensively, he seldom if ever quotes the work of another author in a manner that suggests he had made use of his ideas. More often, references are made either to correct a particular author, or to confirm Hutton's own conclusions. His originality lies in the use he made of facts and ideas, not in their sources.

The most important advance in geological science embodied in Hutton's theory was his demonstration that the process of sedimentation is cyclical in operation, a principle now accepted as axiomatic. Hutton's cycle involved the gradual degradation of the land surface by erosion; the transport of eroded matter to the sea, there to be deposited as sediments; and the consolidation of the sediments on the sea bottom, followed by their elevation to form new land surfaces, which in turn were subject to erosion. Hutton showed that this cyclic process must have been repeated an indeterminate number of times in the past, and because he could find no evidence to suggest that it might cease, he assumed that it would continue indefinitely.

In constructing his theory Hutton had used as a working hypothesis the assumption, based on his own observations, that the geological evidence provided by surface rocks provided both a key to the past and an indication of the probable future course of events. His theory formulated for the first time the general principle that some fifty years later came to be known as uniformitarianism.

In the fields of physical geology and geomorphology Hutton's views were strikingly modern. His knowledge of the processes of erosion and the agents that activate these processes, particularly river action, was thorough. His imaginative reasoning led to one remarkable conclusion about the possible action of glaciers in Switzerland. He had read in H. B. de Saussure's *Voyages dans les Alpes* (Neuchâtel, 1779) a description of scattered boulders of granite, often of immense size, which rested on limestone in the Saleve area, and had obviously been transported there from a distant source. De Saussure believed that their presence could not be accounted for by river action, and he suggested that they had been brought there by a vast debacle or general flood. Although he had not visited Switzerland, Hutton proposed a solution much nearer the truth. He suggested that in the past, when the height of the Alps had been very much greater, "immense valleys of ice sliding down in all directions towards the lower country, and carrying large blocks of granite to a great distance" (*Theory,* II [1795], 218), had transported these erratic blocks; and that in the course of time the upper parts of the mountains that had carried these glaciers had been removed by erosion. The true explanation, that the distribution of erratics of this sort had been effected by great ice sheets covering much of Europe, was not put forward until some forty years later.

Hutton also made contributions, second only in importance to his main theory, in the field of igneous geology. He was much impressed by the worldwide distribution of volcanic activity, and by the new discoveries that in some areas there occurred lavas that must have been erupted in prehistoric times. He made a detailed study of the numerous outcrops of igneous rocks in or near Edinburgh (some almost on his own doorstep), and of others in various parts of Scotland. He distinguished two types, lavas and intrusions, including among the latter both flat sheets and dykes, and he established for the first time the existence of a new class of rocks, the intrusive igneous rocks. He concluded that all igneous rocks originated in what he called the "mineral region," a subcrustal zone of undefined depth in which heat of sufficient intensity to melt rocks prevailed.

Hutton also established the igneous origin of granite, a rock hitherto classed as primeval and believed by geologists of the Wernerian school to have been deposited from water. His study of granite affords an instructive example of Hutton's acute powers of observation and reasoning and the fact that, in general, he did not reach conclusions without sound evidence to support them.

In his 1788 paper he mentioned that some of the rocks of the earth's crust are not stratified, in particular granite. He reserved judgment on the question of the origin of granite but claimed that if one species of granite could be shown to have existed in a state of fusion, then this conclusion could be extended to other varieties of the same rock. He described a particular, and quite abnormal, type of granite from Portsoy, in northeast Scotland, a specimen of which had been sent to him. He had not seen the outcrop but had been informed that it graded into granite of normal type.

This specimen (see Figure 1), illustrated in his 1788 paper, is clearly an example of the variety known as graphic granite, owing to a superficial resemblance to oriental writing evident when the rock is broken in a particular direction, perpendicular to the long axis of the contained quartz crystals which are embedded in a groundmass of feldspar. The quartz crystals then appear skeletal in form, with reentrant angles. Hutton concluded "it is not possible to conceive any other way in which these two substances, quartz and feld-spar, could thus be concreted, except by congelation [cooling] from a fluid state, in which they had been mixed" ("Theory" [1788], p. 256). That is to say the rock had cooled from a fused melt. This was a sound conclusion, for there is nothing in the appearance of the rock to suggest a sedimentary origin.

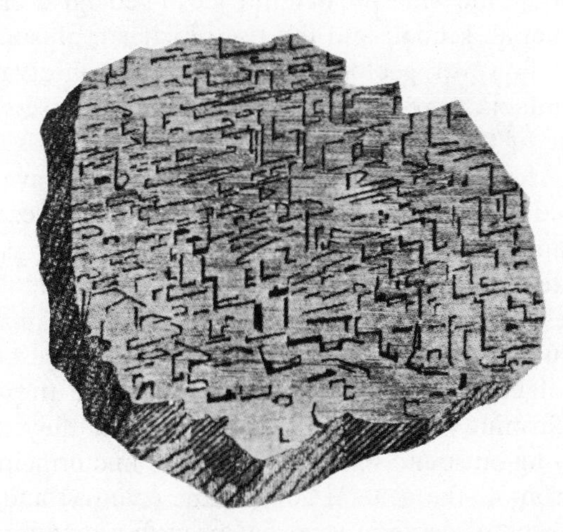

FIGURE 1. The specimen of graphic granite which led Hutton to conclude that granite is igneous in origin (*Transactions of the Royal Society of Edinburgh,* **1** [1788], plate II).

In a later paper, read to the Royal Society of Edinburgh in 1790, Hutton indicated that he had previously reserved judgment on the granite question as a whole, because he had not then decided whether granite

> . . . was to be considered as a body which had been originally stratified by the collection of different [that is, sedimentary] materials, and afterwards consolidated by the fusion of these materials; or whether it were not rather a body transfused from the subterraneous regions, and made to break and invade the strata in the manner of our whinstone or trapp ["Observations on Granite," in *Transactions of the Royal Society of Edinburgh,* **3** (1794), 77–78].

Hutton knew of the existence of foliated granite gneisses in Scotland, and he had read that such rocks were known to de Saussure in Switzerland, who had distinguished them from massive unfoliated granite. Hutton therefore suspended judgment until he had examined the margin of an outcrop of massive granite. This he did in the autumn of 1785 when visiting the duke of Atholl's estate at Glen Tilt, Perthshire. There Hutton found "the most perfect evidence, that granite had been made to break the Alpine strata, and invade that country in a fluid state. This corresponded perfectly with the conclusion which I had drawn from the singular specimen of the Portsoy granite" (*ibid.,* 79–80). Hutton made journeys into other parts of Scotland where he obtained further confirmation of his conclusion. The school of geologists who accepted Hutton's ideas about the origin of igneous rocks came to be known as "plutonists," a name first used by Kirwan.

While the thoroughness of Hutton's investigations and the ingenuity of his arguments are evident, some of his deductions and conclusions were unjustified. This was especially true in his discussions of the causes that he suggested were responsible for the consolidation and elevation of the strata. Here he was breaking new ground and attempting to solve problems that for the most part were insoluble at that time. He must have realized that if his theory was to be accepted, these problems could not be ignored. The solutions he reached were unduly influenced by the powers he attributed to the hot "mineral region" that he believed existed below both the continents and the oceans, powers for which he could produce little convincing evidence, although a source of heat was certainly present.

In discussing consolidation, although he did not consider seriously the possibility that compaction might have resulted from the pressure exerted by a thick mass of sediments, he did suggest that pressure could have driven the water out of porous rocks. Some of the rocks with which he was familiar, particularly the dynamically metamorphosed sediments in the Scottish Highlands and some unmetamorphosed limestones, were crystalline. This knowledge appears to have influenced him in reaching the conclusion that consolidation had been effected by heat. He claimed that many, although not all, the sediments had actually been fused. A difficulty inherent in this argument was that heat of the intensity he envisaged would have decomposed limestones. He dealt with this problem in the following statement:

> The essential difference, however, between the natural heat of the mineral regions, and that which we excite upon the surface of the earth, consists in this; that nature applies heat under circumstances which we are not able to imitate, that is, under such compression as shall prevent the decomposition of the constituent substances, by the separation of the more volatile from the more fixed parts [*Theory,* I (1795), 140].

Sir James Hall was later to prove experimentally that this assumption was justified.

The problem raised by Hutton's demonstration that consolidated strata had been elevated to form dry land was a formidable one. He might perhaps have evaded the issue, as others had done, by suggesting that elevation had resulted from the operation of some cataclysmic action comparable in kind to that which brought about earthquakes. Had he done so, his theory might have received less criticism, but that was not Hutton's way. He wished to get to the root of the matter. He was clearly impressed by the immense force exerted by volcanic activity, in breaking through great thicknesses of consolidated strata, followed by the eruption of lava with explosive violence; and, as he indicated in the *Abstract,* he supposed that the shattering and distortion of strata that once existed as undisturbed horizontal beds must have resulted from the action of the same force. He was also familiar with all the properties of heat known at that time, including its expansive effects on solids, liquids, and gases. He inferred, correctly, that there must exist in his "mineral region" a potential source of immense power (now it would be termed energy), and he assumed that it was heat that brought this power into action. He therefore concluded "that the land on which we now dwell has been elevated from a lower situation by the same agent employed in consolidating the strata . . . this agent is matter actuated by extreme heat, and expanded with amazing force" ("Theory" [1788], p. 266). He made no attempt to explain matters in more detail, but he qualified his conclusion by adding, "The raising up of a continent of land from the bottom of the sea is an idea that is too great to be conceived easily in all parts of its operation, many of which are perhaps unknown to us" (*ibid.,* p. 295). That Hutton failed to solve this problem, one that continues to engage the attention of geologists, is not surprising, but at least he attempted to solve it scientifically.

Reception of the Theory. It has often been stated that Hutton's theory was little understood before the publication in 1802 of Playfair's *Illustrations of the Huttonian Theory of the Earth.* This may be true, and certainly Lyell seems to have derived his knowledge of Hutton's views principally from this source.[7] Nevertheless the theory had been widely read before then, for it had already received critical notices in both British and foreign publications; translations of the *Abstract* and the 1788 "Theory" had appeared in Germany[8] and France; and the theory had received some notice at least as early as 1805 in the United States. Undoubtedly Hutton's views became quite widely known in the early years of the nineteenth century. Yet in spite of the growing interest in geology, and the rapid accumulation of factual observations, it was not until after 1830 that his theories began to gain general acceptance, largely because of Playfair's *Illustrations* and the publication of Lyell's *Principles of Geology* (London, 1830–1833). Lyell accepted most, although not all, of Hutton's views, and expounded them fully in his book; but he and his followers did not accept Hutton's conclusions on the importance of the erosional action of rivers. Some thirty years passed before geologists in both Great Britain and the United States realized that Hutton had been right.

The delay in the recognition of Hutton's work can be attributed to a variety of causes acting collectively: the natural conservatism of many geologists; reluctance to abandon belief in the biblical account of creation; the widespread influence of geologists of the Wernerian school; and the rise of catastrophism. By 1830, however, geologists, although still conservative in outlook, were much better equipped to assess the value of the Huttonian theory.

Agriculture and Evolution. Hutton must have retained an interest in agriculture long after he ceased farming, for shortly before he died he was engaged in preparing for publication a treatise entitled "Principles of Agriculture." This has survived as a manuscript of 1045 pages. Hutton stated in the preface that his objectives in writing this treatise were to assist the farming community to judge whether they were farming on sound scientific and economic principles; to promote the general good of the country; and for his own "pleasure in what has been in a manner the study of my life."

The treatise, based partly on Hutton's own experience and partly on the practice of the most successful husbandmen of his time, covers all branches of farming and animal husbandry, including implements and economics, and where appropriate, Hutton applied his scientific knowledge.

The most noteworthy part of the treatise appears in a section dealing with animal husbandry. Here Hutton outlined a theory of evolution. The question he raised was "how those varieties, which we find in every species, are procured; whether by simple propagation from original models, which had been created with the species, or whether from certain laws of variation, in the process of propagation of each species by the influence of physical causes" (p. 735). Using the dog as his example of a "species," Hutton found it "almost inconceivable" that the numerous different types of dog, "so wisely adapted to various different purposes, . . . should have arisen from the influence of external causes alone" (p. 736), unless "some intended principle in the original constitution of the animal" had operated. He then argued that without this factor, if several varieties or species of dog had existed originally, promiscuous interbreeding would have resulted ultimately in the production of a variety of dog with indefinite characteristics, a "compound species" or mongrel, and all the original varieties would probably have been lost; and we should never have seen "that beautiful illustration of design" exemplified in the different types of dog.

Hutton therefore suggested that originally the "species" had existed in only one form, and there was inherent in the constitution of the animal "a general law or rule of seminal variation" which would bring about constant changes in the animal, to a greater or lesser extent, "by the influence of external causes." Thus we should find varieties in the species "propagating for a long course of time under the influence of different circumstances, or in different situations; and we should in this see a beautiful contrivance for preserving the perfection of the animal form, in the variety of the species. . . . To see this beautiful system of animal life (which is also applicable to vegetables)" (pp. 738–739), Hutton wrote, we must consider that

> . . . in the indefinite variation of the breed the form best adapted to the exercise of those instinctive arts, by which the species is to live, will be most certainly continued in the propagation of this animal, and will be always tending more and more to perfect itself by the natural variation which is continually taking place. Thus, for example, where dogs are to live by the swiftness of their feet and the sharpness of their sight, the form best adapted to that end will be the most certain of remaining, while the forms that are least adapted to this manner of the chace will be the first to perish [p. 739].

Hutton's conclusion that there is some inherent mechanism in "species," such as seminal variation, which could lead to the establishment of animal varieties may possibly have been suggested to him by his knowledge of the animal breeding experiments carried out by the eighteenth-century agriculturist Robert Bakewell, to whom he refers elsewhere in this section of the "Principles."

Physical Sciences. Hutton's interest in the physical sciences, particularly chemistry, physics, and meteorology, extended over many years, during which he kept himself informed of their progress. Toward the end of his life he published a three-part book entitled *Dissertations on . . . Natural Philosophy,* which is of considerable interest to the historian of science. The conclusions he reached in this work were often original and sometimes supported by experiments he had carried out himself. The principal subjects discussed are meteorology, phlogiston, and the theory of matter.

Part 1 contains four dissertations on meteorology, of which three, dealing with Hutton's theory of rain and his answer to DeLuc's criticism of the theory, had been previously published by the Royal Society of Edinburgh (1788, 1790). The fourth contains a discussion on winds. Hutton attributed the origin of rain to a mixture of air currents of different temperatures, saturated or nearly saturated with moisture. His theory attracted attention for some years, including a favorable comment from John Dalton as late as 1819, although J. D. Leslie had already shown it to fail on qualitative grounds in 1813.

Part 2 is entitled a "Chymical Dissertation Concerning Phlogiston, or the Principle of Fire," a subject evidently of particular interest to Hutton. It had been the topic of a paper he read to the Royal Society of Edinburgh in 1788, following an address by Sir James Hall on Lavoisier's new chemical ideas, to which Hall had been converted after visiting Lavoisier in Paris. These papers and the accompanying discussion occupied five meetings, but they were not published.

Hutton accepted the major advances made by Lavoisier, but took the view that the concept of phlogiston had been too hastily rejected. He did not accept Lavoisier's concept of *calorique;* in fact he strongly opposed it. His view was that heat, light, and electricity were all modifications of what he called "solar substance." Hutton also considered phlogiston to be some form of the solar substance, a principle of inflammability, without gravity, which could be transferred from one substance to another. He claimed that phlogiston was actually formed by vegetative matter and decomposed during the processes of breathing and burning.

Thomas Thomson, when discussing Hutton's views on phlogiston, described him as "a man of undoubted genius," but stated also that his views were set out in a "manner so peculiar, that it is scarcely more difficult to procure the secrets of science from Nature herself, than to dig them from the writings of this philosopher."[9] Fortunately Hutton's conception of the nature and function of phlogiston has been discussed by J. A. Partington and D. McKie in sufficient detail to meet the needs of most readers.

Part 3 of the *Dissertations on . . . Natural Philosophy,* entitled "Physical Dissertations on the Powers of Matter, and Appearances of Bodies," constitutes more than half the book and contains Hutton's theory of matter. Briefly summarized, this theory suggests that to describe a body as made of small particles does not explain its nature, because if we suppose these particles to possess magnitude, we do no more than say large bodies are made of smaller bodies. Therefore the elements of a body must be something unextended. To these elements he gave the name "matter," reserving the name "body" to combinations of matter subject to powers or forces acting in various directions. He uses this conception to explain the various physical properties of bodies. Playfair emphasized the close affinity of Hutton's theory to that of Bošcović, but he states specifically there was no reason to suppose Hutton had derived his conclusions from the latter. According to Playfair, Bošcović's theory was hardly known in Scotland before 1770, whereas the earliest sketches of Hutton's theory were of much earlier date.[10]

Hutton continued his discussion of phlogiston in his last book, *Philosophy of Light, Heat, and Fire* (1794). Here he also raised the question whether there might be a species of light capable of producing heat in bodies without affecting the sense of sight. This idea, he stated, had been suggested to him by his own experience, and he hoped to test it accurately when suitable apparatus could be constructed. He proposed the use of either a prism or colored glass to produce both red and blue light, but the only experiment he actually carried out was a crude one. He adjusted the position of two sources of light, a coal fire for red light, and a flame for "compound" or white light, so that each source just permitted him to read, and he found that the amount of heat given off by the fire was much greater than that from the flame. He suggested, by analogy, that invisible light should exist, which would form a source of heat greater than that produced by the visible range of the spectrum. A few years later, William Herschel investigated the subject more thoroughly, confirming Hutton's suggestion.

Hutton's last contribution to chemistry was a paper on the "Sulphurating of Metals," read to the Royal Society of Edinburgh by a friend on 9 May 1796. The subject is discussed in terms of Hutton's ideas about light, heat, and phlogiston, and a correction is made of a conclusion he had drawn in *Light, Heat, and Fire.*

Philosophy. In 1794 Hutton published a three-volume treatise on metaphysics and moral philosophy entitled *An Investigation of the Principles of Knowledge.* This work followed on or arose out of his studies of the physical sciences. It received little notice when it first appeared, but Playfair discussed it in some detail and suggested that if the work were abridged and the obscurities removed it would deserve more attention. It has received little if any notice since Playfair's time. In the *Principles of Knowledge* Hutton acknowledged the existence of a God whom he defined as "the superintending mind . . . a Being with perfect knowledge and absolute wisdom." He considered nature as subordinate to God, and that the two terms were not synonymous, for God is infinite and unchangeable, but nature limited and changing. While he included the animal, vegetable, and mineral systems as part of nature's general design, the term "nature" properly meant the whole of that action from which, in necessarily inferring design, we learn the existence of a superintending being.

Although occasionally accused of impiety in his lifetime, Hutton was not an atheist, and may be described as a deist. In almost all that he wrote, not only on geology but on agriculture and physical subjects as well, he introduced his belief that in nature there is abundant evidence of benevolent wisdom and design. To Hutton the earth as a whole was "a machine constructed upon chemical as well as mechanical principles, by which its different parts are all adapted, in form, in quality, and in quantity, to a certain end . . . an end from which we may perceive wisdom in contemplating the means employed" ("Theory" [1788], p. 216). The earth, in Hutton's view, was evidently made for man, and once the working of the machine is understood, man will be led "to acknowledge an order, not unworthy of Divine wisdom, in a subject which, in another view, has appeared as the work of chance, or absolute disorder and confusion" (*ibid.,* p. 210).

Hutton's attitude toward the Christian religion was recorded in a brief (unpublished) manuscript entitled "Memorial Justifying the Present Theory of the Earth From the Suspicion of Impiety," which was evidently written sometime between 1788 and 1795, in answer to criticisms of his theory. In it he made no attempt to compromise with the Church, as Buffon had done. His view was that the ancient Jewish writings on which the Christian religion was founded can be accepted only insofar as they record the history of man upon earth. He denied the literal truth of the Mosaic account of creation, whose only significance, he stated, was its record that God had made all things in a certain order; and he thought it absurd to suppose that the term "day" used in that account could mean anything other than an indefinite period of time.

Hutton maintained that it was not the duty of religion to provide a history of the natural operations that had taken place on the earth in the past; but that this was the function of man, using his intellect and applying the methods of natural philosophy. He regarded the objectives of revealed religion and natural philosophy as essentially different, and saw no reason why one should interfere with the other, provided their different purposes were kept separate.

NOTES

1. This account of Hutton's life is based almost entirely on John Playfair, "Biographical Account."
2. Hutton was friendly with several members of a prominent Scottish family, the Clerks of Penicuik. His particular friend was George, second son of Sir John Clerk who had been vice-president of the Philosophical Society of Edinburgh. George Clerk was interested in mineralogy and accompanied Hutton on several of his geological excursions. On marriage to a Miss Maxwell he assumed the name of Clerk-Maxwell, and the physicist Clerk-Maxwell is his descendant. Other members of the Clerk family accompanied Hutton on later tours.
3. See Playfair, *op. cit.*, p. 74 n.
4. Playfair gives the date incorrectly as 1793.
5. See Kirwan, "Examination of the Supposed Igneous Origin of Stony Substances," in *Transactions of the Royal Irish Academy,* **5** (1793), 51–81.
6. In a letter to Princess Dashkow, then director of the Imperial Academy of Sciences, St. Petersburg, in which Black summarizes Hutton's theory; see W. Ramsay, *Life and Letters of Joseph Black, M.D.* (London, 1918), 117–125.
7. See *Life, Letters and Journals of Sir Charles Lyell,* II (London, 1881), 47–49.
8. Werner MSS at Freiberg, IX, has an abstract of the 1788 "Theory."
9. See "Chemistry, (i)" in *Supplement to the 3rd ed. Encyclopaedia Britannica* (Edinburgh, 1801), I, 287.
10. See Playfair, *op cit.*, p. 78 n.; see also R. Olson, "The Reception of Boscovich's Ideas in Scotland," in *Isis,* **60** (1969), 91–103.

BIBLIOGRAPHY

I. Original Works. Hutton's published works are the following: *Dissertatio physico-medica inauguralis de sanguine et circulatione microcosmi* (Leiden, 1749); *Considerations on the Nature, Quality, and Distinctions of Coal and Culm . . . in a Letter From Doctor James Hutton . . . to a Friend* (Edinburgh, 1777); *Abstract of a Dissertation Read in the Royal Society of Edinburgh, Upon the Seventh of March, and Fourth of April, M, DCC, LXXXV, Concerning the System of the Earth, Its Duration, and Stability* (probably Edinburgh, 1785), repr. in *Proceedings of the Royal Society of Edinburgh,* **63B** (1950), 380–382, and facs. repr. in G. W. White, ed., *Contributions to the History of Geology,* V (Darien, Conn., 1970), 1–30; *Dissertations on Different Subjects in Natural Philosophy* (Edinburgh, 1792); *An Investigation of the Principles of Knowledge, and of the Progress of Reason, From Sense to Science and Philosophy,* 3 vols. (Edinburgh, 1794); *A Dissertation Upon the Philosophy of Light, Heat, and Fire* (Edinburgh, 1794); and *Theory of the Earth: With Proofs and Illustrations,* vols. I–II (Edinburgh, 1795), facs. repr. (New York, 1959); vol. III (chs. 4–9), Sir Archibald Geikie, ed. (London, 1899), with indexes to all three vols.

Hutton's papers are "The Theory of Rain," in *Transactions of the Royal Society of Edinburgh,* **1** (1788), 42–86; "Theory of the Earth; or an Investigation of the Laws Observable in the Composition, Dissolution, and Restoration of Land Upon the Globe," *ibid.,* 209–304, facs. repr. in *Contributions to the History of Geology,* V (1970), 31–131; "Of Certain Natural Appearances of the Ground on the Hill of Arthur's Seat," in *Transactions of the Royal Society of Edinburgh,* **2** (1790), 3–11 (read to the Philosophical Society of Edinburgh, 1778); "Answers to the Objections of M. de Luc With Regard to the Theory of Rain," *ibid.,* 39–58; "Observations on Granite," *ibid.,* **3** (1794), 77–85, facs. repr. in *Contributions to the History of Geology,* V (1970), 133–139; "Of the Flexibility of the Brazilian Stone," in *Transactions of the Royal Society of Edinburgh,* **3** (1794), 86–94; and "An Examination of a New Phenomenon Which Occurs in the Sulphurating of Metals, With an Attempt to Explain That Phenomenon," *ibid.,* **4** (1798), pt. I, History of the Society, 27–[36] (misnumbered 28).

For foreign publications of Hutton's works, see "Theory of the Earth" (1788), noticed in *Magazin für das Neueste aus der Physik und Naturgeschichte,* **6,** pt. 4 (1790), 17–27, and translated in full in *Sammlungen zur Physik und Naturgeschichte,* **4** (1792), 622–725. A French trans. of the *Abstract* appeared in "Extrait d'une Dissertation sur le Système et Durée de la Terre . . . traduite de l'Anglois, par Iberti . . . suivi par les Observations du Traducteur," in *Observations sur la Physique,* **43** (1793), 3–12. N. Desmarest, *Encyclopédie méthodique. Géographie physique,* I (Paris, 1794), 732–782, contains extensive extracts from the *Abstract,* the "Theory of the Earth," and the "Theory of Rain," with commentary by Desmarest.

Hutton's extant MSS are "Principles of Agriculture" (2 quarto vols. totaling 1045 pp.), in the library of the Royal Society of Edinburgh; and the five-page "Memorial Justifying the Present Theory of the Earth From the Suspicion of Impiety," in the Fitzwilliam Museum, Cambridge, England.

Very few letters either written by or addressed to Hutton have survived. His letters to John Strange, the geologist and diplomat, in 1770 or 1771 have been published by V. A. Eyles and J. M. Eyles as "Some Geological Correspondence of James Hutton," in *Annals of Science,* **7** (1951), 316–339.

II. SECONDARY LITERATURE. The only complete account of Hutton's life is John Playfair, "Biographical Account of the Late Dr James Hutton, F.R.S. Edin.," in *Transactions of the Royal Society of Edinburgh,* **5** (1803), 39–99, facs. repr. in *Contributions to the History of Geology,* V (1970), 141–203, repr. in *The Works of John Playfair,* IV (Edinburgh, 1822), 33–118. Playfair discusses critically Hutton's published works. A few additional details about Hutton's activities as a farmer are included in "Principles of Agriculture."

W. Ramsay, *Life and Letters of Joseph Black, M.D.* (London, 1918); and *Partners in Science: Letters of James Watt and Joseph Black,* E. Robinson and D. McKie, eds. (London, 1970), contain interesting references to Hutton; especially to his illness and to his natural son. Other unpublished letters of Black in the library of Edinburgh University are also worth consulting. For brief references to Hutton in the published diaries and letters of contemporaries, see V. A. Eyles, "Introduction," in *Contributions to the History of Geology,* V (1970), xi–xxiii.

Playfair's *Illustrations of the Huttonian Theory of the Earth* (Edinburgh, 1802), repr. in *The Works of John Playfair,* I (Edinburgh, 1822), 1–514, facs. repr. of 1802 ed. (Urbana, 1956), with intro. by G. W. White, is widely used as a source of Hutton's views, but does not always present them with complete accuracy. Its publication stimulated John Murray of Edinburgh, lecturer in chemistry and supporter of Wernerian geology, to publish anonymously *A Comparative View of the Huttonian and Neptunian Systems of Geology* (Edinburgh, 1802). The *Illustrations* and *Comparative View,* translated into French and annotated by the translator, C. A. Basset, were published in one volume, *Explication de Playfair sur la Théorie par Hutton, et examen comparatif des systèmes géologiques . . . par M. Murray* (Paris, 1815).

W. H. Fitton, "A Review of Mr Lyell's 'Elements of Geology'; With Observations on the Progress of the Huttonian Theory of the Earth," in *Edinburgh Review,* **69** (1839), 406–466, includes a detailed discussion of the value of Hutton's theory in relation to contemporary geological knowledge. Fitton shows that some prominent geologists of the period, particularly in France, were either unaware of Hutton's work, or, if familiar with it, failed to acknowledge Hutton's priority when putting forward their own conclusions.

The following modern commentaries on various aspects of Hutton's work, principally his geology, may be consulted: "James Hutton 1726–1797, Commemoration of the 150th Anniversary of His Death," in *Proceedings of the Royal Society of Edinburgh,* **63B** (1950), 351–402, contains five articles on Hutton's life and work: M. Macgregor, "Life and Times of James Hutton"; E. B. Bailey, "James Hutton, Founder of Modern Geology"; G. W. Tyrrell, "Hutton on Arran"; V. A. Eyles, "Note on the Original Publication of Hutton's *Theory of the Earth* and the Subsequent Forms in Which It Was Issued"; and S. I. Tomkeieff, "James Hutton and the Philosophy of Geology," repr. from *Transactions of the Edinburgh Geological Society,* **14** (1948), 253–276. Tomkeieff notes that G. H. Toulmin, in *The Antiquity and Duration of the World* (London, 1780), expressed views in some respects similar to those in Hutton's *Theory of the Earth.* References to Toulmin and Hutton were also made by D. B. McIntyre, "James Hutton and the Philosophy of Geology," in C. C. Albritton, ed., *The Fabric of Geology* (Reading, Mass., 1963), 1–11; and G. L. Davies, "George Hoggart Toulmin and the Huttonian Theory of the Earth," in *Bulletin of the Geological Society of America,* **78** (1967), 121–124.

A useful source book and guide to Hutton's geological thought is E. B. Bailey, *James Hutton—the Founder of Modern Geology* (Amsterdam–London–New York, 1967), which contains a summary of each chapter of *Theory of the Earth,* a well-informed commentary on Hutton's ideas, and a less detailed discussion of his other works, particularly "Principles of Agriculture."

R. H. Dott, Jr., "James Hutton and the Concept of a Dynamic Earth," in C. J. Schneer, ed., *Toward a History of Geology* (Cambridge, Mass.–London, 1969), 122–141, provides a short summary and commentary on Hutton's contributions to geology.

Anthologies of Hutton's more important geological observations and conclusions, with commentaries, are J. Challinor, "The Early Progress of British Geology—III. From Hutton to Playfair, 1788–1802," in *Annals of Science,* **10** (1954), 107–148; and D. A. Bassett, "James Hutton, the Founder of Modern Geology: An Anthology," in *Geology: The Journal of the Association of Teachers of Geology,* **2** (1970), 55–76. S. I. Tomkeieff, "Unconformity—an Historical Study," in *Proceedings of the Geologists' Association,* **73** (1962), 383–401, discusses Hutton's use of unconformities as evidence for his geological theory.

Hutton's contributions to geomorphology are discussed in E. B. Bailey, "The Interpretation of Scottish Scenery," in *Scottish Geographical Magazine,* **50** (1934), 301–330; R. J. Chorley, A. J. Dunn, and R. P. Beckinsale, *The History of the Study of Land Forms* (London–New York, 1964); G. L. Davies, "The Eighteenth Century Denudation Dilemma and the Huttonian Theory of the Earth," in *Annals of Science,* **22** (1966), 129–138; and "The Huttonian Earth-Machine," in *The Earth in Decay, a History of British Geomorphology 1578 to 1878* (London, 1969), ch. 6.

R. Hooykaas, *Natural Law and Divine Miracle* (Leiden, 1959; 2nd ed., 1963), discusses Hutton's geological theory in relation to the general theory of uniformitarianism. See also his paper "James Hutton und die Ewigkeit der Welt," in *Gesnerus,* **23** (1966), 55–66.

V. A. Eyles, "A Bibliographical Note on the Earliest Printed Version of James Hutton's *Theory of the Earth,* its Form and Date of Publication," in *Journal of the Society for the Bibliography of Natural History,* **3** (1955), 105–108, gives the evidence establishing the authorship and date of publication of the *Abstract,* which was issued anonymously and undated.

The following commentaries refer to Hutton's work in

subjects other than geology: J. R. Partington and D. McKie, "Historical Studies on the Phlogiston Theory—III. Light and Heat in Combustion," in *Annals of Science,* **3** (1938), 366–370, analyzes Hutton's views on the nature of phlogiston and concludes that his theory was almost identical with that of A. Crawford; and P. A. Gerstner, "James Hutton's Theory of the Earth and his Theory of Matter," in *Isis,* **59** (1968), 26–31, and "The Reaction to James Hutton's Use of Heat as a Geological Agent," in *British Journal of the History of Science,* **5** (1971), 353–362, discusses Hutton's ideas about heat and matter in relation to his theory of the earth.

Works on the development of science in the eighteenth and early nineteenth centuries that contain numerous references to Hutton are C. C. Gillispie, *Genesis and Geology* (Cambridge, Mass., 1951), particularly valuable for its very extensive bibliography; Loren Eiseley, *Darwin's Century* (New York, 1958–London, 1959); and F. C. Haber, *The Age of the World* (Baltimore, 1959).

The most frequently reproduced portrait of Hutton is the painting by the Scottish artist Sir Henry Raeburn. Two contemporary etchings are in *A Series of Portrait and Caricature Etchings by the Late John Kay,* 2 vols. (Edinburgh, 1837). A medallion portrait by the Scottish artist James Tassie is reproduced in the English trans. of K. A. von Zittel, *History of Geology and Palaeontology* (London, 1901).

V. A. EYLES

HUXLEY, THOMAS HENRY (*b.* Ealing, Middlesex, England, 4 May 1825; *d.* Hodeslea, Eastbourne, Sussex, England, 29 June 1895), *zoology, evolution, paleontology, ethnology.*

Thomas Henry Huxley was the seventh and youngest surviving child of George and Rachel Huxley. His father taught mathematics and was assistant headmaster at a school in Ealing which Thomas Henry attended for a brief period. The regular instruction which Huxley received was minimal and lasted no more than two years. He was not considered a precocious child but did exhibit in his youth the natural ability to draw which later, in spite of no training, served him well in his zoological work. For his general education Huxley was largely self-taught; while still in his teens he read extensively, particularly in science and metaphysics, and gained a facility in reading German and French.

Huxley had early leanings toward mechanical engineering as a career, but the combination of a family of moderate means and two medical brothers-in-law led him into medicine. He attended a postmortem when he was about fourteen and may have contracted some sort of dissection poisoning which manifested itself in an apathy remedied only by a stay in the open countryside. The "hypochondriacal dyspepsia" recurrent throughout the remainder of his life he thought to have been brought on by this incident; whatever it was, he was usually cured by a spell in fresh air. In 1841 Huxley became apprenticed to one of his brothers-in-law, John Godwin Scott, who practiced in the north of London. During his apprenticeship, he continued his wide reading, attended some courses, and earned the silver medal in a botanical competition.

In September 1842 Huxley and his brother James were awarded free scholarships at Charing Cross Hospital. The lecturer on physiology, Thomas Wharton Jones, had a strong influence on Huxley's interest in physiology and anatomy and helped teach him methods of scientific investigation. Jones encouraged and aided Huxley with his first scientific paper, on the discovery of a layer of cells (Huxley's layer) directly within Henle's layer in the root sheath of hair. Huxley passed the M.B. examination at London University in 1845 and soon afterward that for membership in the Royal College of Surgeons. He applied to and was taken into the Royal Navy, being assigned to H.M.S. *Victory* for service at Haslar Hospital, where he remained until assigned to H.M.S. *Rattlesnake.*

Huxley was to join the *Rattlesnake* on a surveying voyage to the Torres Straits off Australia as ship's surgeon, not as a naturalist, a position filled by John MacGillivray. Any natural history Huxley undertook on the four-year voyage was his own affair, but it was to set the course of his career toward zoology rather than medicine. Aboard the *Rattlesnake,* Huxley's scientific equipment was minimal, consisting principally of a microscope and a makeshift collecting net. The limitation of his equipment was perhaps fortunate, as he focused his attention on the wealth of planktonic life for the study of which a steady supply of fresh specimens was necessary. Through extensive shipboard dissections and through library work in Sydney, Australia (where he also saw much of William Macleay), Huxley was able to bring some order to these minute organisms which had been simply lumped together in those two great zoological lumber rooms, Linnaeus' Vermes and Cuvier's Radiata.

The novelty of much of his material was evident to Huxley and prompted him to send several papers to the Linnean Society, about which he received no word. Somewhat disheartened, he directed in 1849 a major paper "On the Anatomy and the Affinities of the Family of the Medusae" to the Royal Society, which turned out to be the first of a series of wedges he drove into Cuvier's Radiata. By the time of the *Rattlesnake*'s return the paper had been published in the *Philosophical Transactions* and soon earned him election as a fellow of the Royal Society. Combined with several other papers it brought Huxley the Royal Medal in 1852. After his return to England in 1850 Huxley arranged for leave from active duty in order to remain in London to work on the materials he had brought back. During these several years he became very much a part of the London scientific scene, making many friendships and enlisting the support of leading scientific figures in his running battle with the Admiralty over payment for the publication of his results. The battle continued until the Admiralty became exasperated and ordered Huxley back to active duty. He refused, leaving himself without any means of support and no prospect of a scientific appointment in London, where he felt he must remain to do effective scientific work. During this period he wrote articles and miscellaneous pieces for several reviews, when the editors would pay.

Invertebrate Studies. Nearly all of Huxley's scientific effort in the period 1850–1854, during which he published about twenty scientific papers, was concentrated on the materials from the *Rattlesnake.* Working out the details and relationships of the delicate marine animals he studied set the pattern for his career and gave him a firm grasp of major zoological problems. Of his numerous publications on these invertebrates his major contributions are found in four memoirs; his 1849 paper on the Medusae, two 1851 papers on tunicates, and one in 1853 on the Cephalous Mollusca. In his paper on the Medusae (*Scientific Memoirs,* I, 9–32), Huxley made two notable contributions—recognition of this group as a coherent whole and of an embryological analogy. First, he described the structure common to the different groups of Medusae, recognizing that they all consist fundamentally of two layers, or "foundation membranes" which produce the inner and outer parts, that they all seem to lack blood and blood vessels, and that the existence of any nervous system was doubtful. He then allied with the Medusae the Hydroid and Sertularian polyps, whose structure is similarly based on the same two foundation membranes. Although it was less obvious, Huxley recognized that the complicated colonies (for example, the Portuguese man-of-war) making up the Physophoridae and Diphydae were colonies of hydralike organisms each of which had the typical Medusae double-membrane structure. The group of organisms which Huxley connected on the basis of this fundamental structure was readily accepted as one of the major groups of animals, becoming the nucleus of the Coelenterata, and as such required and received the attention of zoologists. Although its importance was perhaps not fully appreciated until after Charles Darwin's *On the Origin of Species,* the embryological analogy that Huxley drew was an even more fundamental contribution than the organization of the medusoid organisms. He concluded that the two foundation membranes are physiologically analogous to the serous and mucous layers in a typical embryo. At this time Huxley made only the comparison and did not speculate on its possible significance.

In 1851 Huxley presented to the Royal Society two major papers on tunicates: "Observations on the Anatomy and Physiology of Salpa and Pyrosoma" and "Remarks Upon Appendicularia and Doliolum, Two Genera of Tunicates" (*Scientific Memoirs,* I, 38–68, 69–79). In the *Salpa* paper Huxley confirmed earlier suggestions that this organism's life cycle passes through an alternation of solitary and chainlike colonial generations. Huxley observed a great abundance of specimens at various growth stages, from which he was able to come to the important conclusion that the solitary stage is the product of sexual generation and that the colony results from budding. This recognition contributed strongly to his theory of animal individuality in which he stated that both forms are parts or organs of a single individual, because they both develop from a single ovum. Huxley elaborated this thesis in a discourse "Upon Animal Individuality" at the Royal Institution in April 1852. He related, anatomically and systematically, all four genera of free-swimming forms covered in these two papers to the Ascidiacea, or sea squirts, and thus gathered the ascidians into a group based on their typical structure as he had done with the Medusae. The zoological position of *Appendicularia* had been most unsettled owing to its possession of a tail. Huxley demonstrated that this tail is a retained larval feature lost by most adult ascidians. The significance of this urochordal structure in relation to the vertebrate pedigree did not become evident until later.

Just as with the Medusae and the ascidians, Huxley sought and found in the Cephalous Mollusca a typical structure of which each genus and species is a modification. Briefly, Huxley started with several surface-dwelling forms with transparent shells and then dissected a wide variety of mollusks, determining their anatomical similarities on comparative grounds. Drawing heavily on the German embryologists who had studied their development, Huxley concluded that their parts are homologous and that they constitute a great group, the Cephalous Mollusca, comprising the Cephalopoda, Gasteropoda, and Lamellibranchiata all of which are modifications of a typical form, or "archetype." The paper which resulted, "On the Morphology of the Cephalous Mollusca . . ." (*Scientific Memoirs,* I, 152–193), was highly important to his contemporaries for the anatomical and systematic conclusions which Huxley reached. It is also of particular interest for its insights into Huxley's zoological methods, not only in his study of the invertebrates from the *Rattlesnake* but also in his later work on vertebrates, both fossil and recent.

Huxley opened his paper on the Cephalous Mollusca with a quotation from "the highest authority," Richard Owen, setting forth what Huxley believed to be "the true aims of anatomical investigation." In this passage Owen states that mere anatomical description is of relatively little value until the facts "have been made subservient to establishing general conclusions and laws of correlation, by which the judgment may be safely guided." Whether with invertebrates, birds, the structure of the vertebrate skull, or fossil horses, Huxley sought to establish conclusions which, no matter how general or widesweeping, were invariably firmly based on facts from his experience. Huxley's use of one of Owen's favorite words, "archetype," could lead to some confusion because of the *naturphilosophisch* and platonic connotations associated with it and because of his attitude toward the output of that mode of thinking, of which Owen's work was a part. Huxley explicitly denied any connection between his archetypes and any ideas after which organisms might be modeled. To him the word meant "the conception of a form embodying the most general propositions that can be affirmed" about the organisms under consideration.

Within any great group, such as the Cephalous Mollusca, Huxley thought that the members varied by excess or defect of the parts of the archetype. He rejected the idea of any progression from a lower to a higher type within the group and instead thought there was "merely a more or less complete evolution of one type." Here Huxley used "evolution" in its historic sense of an unfolding, or unrolling—that is, an embryological unfolding. While the manner in which he treated the Cephalous Mollusca was fundamentally the same as that used with the Medusae and ascidians, the discussion of the archetype and its modifications as a result of embryological development was a new element. There is more than a coincidental relationship between these ideas and those in the "Fragments Relating to Philosophical Zoology" which Huxley selected and translated from the works of Karl Ernst von Baer. Huxley had a broad acquaintance with the German zoological literature and held a high regard for much of it, especially in embryology and cytology; for example, see his major review "The Cell-Theory" (*Scientific Memoirs,* I, 241–278).

Vertebrate Studies. In 1854, when he succeeded Edward Forbes as lecturer in natural history at the Government School of Mines, Huxley at last had a means of support within the scientific community. Soon afterward he was appointed to the additional post of naturalist with the Geological Survey. He now had not only a scientific position but also the income needed to marry Henrietta Heathorn, whom he had met in 1847 in Sydney. They became engaged in 1849 but did not see each other again until she came to England for their marriage in 1855. Their son Leonard, the well-known teacher and writer, was the father of Julian Huxley, the biologist, Aldous Huxley, the writer, and Andrew Fielding Huxley, the physiologist.

At least in part, Huxley saw these positions as being temporary, while he awaited a post in physiology; since no such position became available, he held the same appointments for over thirty years. With considerable rapidity the focus of Huxley's attention shifted from the invertebrates to the vertebrates. This shift was induced by his duties as lecturer on natural history, which required him to prepare in unfamiliar areas of biology, combined with his duties in connection with the Geological Survey, which brought him in close contact with a range of vertebrate fossils. Although beginning with a certain distaste for fossils, he soon became deeply involved in problems in paleontology and geology. His first fossil work was in cooperation with John William Salter, identifying a variety of fossils, an experience which was to prove invaluable in the aftermath of Darwin's *Origin.* As with the invertebrates, Huxley was not concerned with species as such, but only as they led to more general zoological problems. In addition to his paleontological work, Huxley helped to organize the Museum of Practical Geology where he began, in 1855, his series of lectures to workingmen. He further developed his own regular course at the School of Mines. In addition to this he was appointed to the triennial Fullerian lectureship at the Royal Institution for 1856–1858. This sampling of his activities is indicative of the number of projects he would undertake at one time.

In the late 1850's, Huxley began a detailed study of the embryology of the vertebrates, which provided a firm base for much later work as well as strengthening his teaching. An outcome of this study was his 1858 Croonian lecture at the Royal Society, "On the Theory of the Vertebrate Skull" (*Scientific Memoirs,* I, 538–606). Huxley made an important methodological contribution to morphology by his insistence that, as suggestive as they are, comparisons of adult structures are insufficient for the demonstration of homologies. Only by studying the embryological development of the various structures from their earliest stages and determining that they follow the same path of development can we say with certainty that they are homologous. Huxley was continuing the tradition of K. E. von Baer and M. H. Rathke and, more specifically, was reviving detailed studies of the skull done by Rathke and others in 1836–1839. These neglected earlier studies had shown the inadequacies of the vertebral theory of the skull originated by Goethe, elaborated by Lorenz Oken, and developed to its fullest by Richard Owen.

Huxley's objective in his Croonian lecture was to put morphological studies on a more scientific basis, especially by the utilization of embryological criteria which could be as productive for the vertebrates as they had already been for the invertebrates. Many, and particularly Owen himself, saw this lecture as an attack on Owen, who was still considered England's preeminent anatomist. Although not intended as such, it assuredly helped to prepare the way for the disputes between Huxley and Owen after 1859. In his lecture Huxley established that the various vertebrate skulls are modifications of the same basic type and, importantly, distinguished and named the different modes by which the lower jaw is articulated to the skull, which has become an important diagnostic character. Huxley concluded that the differentiation of the skull and the vertebral column occurs at such an early stage that they could not have a common origin. He also drew an analogy between the membranous, cartilaginous, and osseous stages of the development of the skull and between the skulls of *Amphioxus,* sharks, and the higher vertebrates.

The Evolution Controversy. Those today who know Huxley know him primarily as the protagonist of evolution in the controversies immediately following the publication of *On the Origin of Species* late in 1859. Huxley was prepared for the role he was to play, since he had by then acquired a broad background in vertebrate and invertebrate zoology and in paleontology. He also had read widely in the zoological literature in English, German, and French. Inevitably, he was familiar with the various hypotheses concerning the transmutation of species, particularly those of Lamarck and Robert Chambers, both of whom he held in low opinion. Huxley's review of the 1854 edition of Chambers' *Vestiges of the Natural History of Creation* was the only one he later regretted for its "needless savagery." Before the *Origin,* Huxley was constitutionally opposed to transmutation ideas because of his critical skepticism of all theories, the same skepticism which he embodied in the term he coined, "agnosticism." Also, his belief that the natural groups of organisms were demarcated by sharp lines seemed, if valid, to negate the possibility of evolution occurring.

When the *Origin* was ready for publication, Darwin sent Huxley one of three prepublication copies, the other two going to Charles Lyell and Joseph Hooker. Darwin was not confident that Huxley would react favorably to his book but wanted him as one of its judges. Darwin and Huxley had become by then close friends, and Darwin often had drawn on Huxley's wide-ranging knowledge. On 23 November 1859 Huxley wrote to Darwin that nothing had impressed him more since his reading of Baer, praised Darwin for his new views, and warned him about the abuse he was bound to receive. Already Huxley was sharpening his claws and beak in preparation for the impending battles. Entirely by chance, Huxley had an early opportunity to praise the *Origin* publicly, although anonymously, when he was asked to review it for the London *Times* (26 December 1859). Huxley followed this with a Friday evening discourse at the Royal Institution in February 1860 and an article in the April issue of the *Westminster Review.* His February discourse "On Species and Races, and their Origin" (*Scientific Memoirs,* II, 388–394) set a model for many later defenses of the *Origin.* After discussing the varieties and species of horses and pigeons, Huxley turned to man's relation to the apes, the topic of greatest concern to his listeners and one only hinted at in the *Origin.* Without going into the full details, he argued that man differs less from the highest members of the Quadrumana than the extreme members of that group differed from one another. This was an implicit rejection of Owen's classification of the Mammalia. Moreover, and most importantly, Huxley made a strong plea, often to be repeated, for judging Darwin's work on scientific grounds, as a work in science, for "the man of science is the sworn interpreter of nature in the highest court of reason."

When the British Association for the Advancement of Science met in Oxford late in June 1860, Huxley was recognized as an able younger biologist, but his name certainly was not yet a household word. During this meeting he had two encounters important for the future of Darwin's hypothesis, as well as for his own career: one with Owen on scientific details, which was settled later, and one with Samuel Wilberforce, bishop of Oxford, which was of a more general nature. The result was Huxley's being recognized as the principal defender of the *Origin* and he thus earned the name "Darwin's bulldog." The less prolonged, although more dramatic, of these was the second, the exchange with "Soapy Sam" Wilberforce. Wilberforce had been coached in matters scientific by Owen, but he was apparently a poor learner—or Owen a poor teacher. Wilberforce made a number of scientific blunders and then directed his famous query to Huxley asking whether Huxley's ancestry from an ape was on his grandfather's or his grandmother's side. Waiting until the audience called upon him to answer, Huxley carefully corrected the Bishop's scientific errors and added that he would rather be related to an ape than to a man of ability and position who used his brains to pervert the truth. Unfortunately, there is no verbatim account of this incident, but a number of eyewitness accounts agree in substance. This episode, in addition to establishing Huxley as the principal spokesman for Darwin, gave convincing evidence that the evolutionists were not going to be cowed by the Church.

Huxley's dispute with Owen at the British Association meeting had actually begun in 1857 and was not settled finally until 1863. In 1857 Owen read before the Linnean Society a paper on the classification of the Mammalia, which he repeated substantially in his Reade lecture at Cambridge in May 1859. Cuvier had separated man, in the order Bimana, from the remainder of the primates in the order Quadrumana. Owen constructed a taxonomy based on certain characteristics of the mammalian brain, which separated man still further from the other primates and placed him in a new subclass, the Archencephala. In his system Owen argued that the human brain differed from those of all other mammals not only in degree but also in kind, and that it had certain structural characters peculiar to it. The most famous of these was a small internal ridge known as the *hippocampus minor* which became well-known to the public and gave its name to this controversy. The essential factor in Owen's taxonomy of the Mammalia was the assertion that man was zoologically distinct from all of the other mammals. After the *Origin,* Owen's system carried the additional implication that an evolutionary hypothesis valid for the other animals would not necessarily apply to man.

When, in 1857, Huxley had first become acquainted with Owen's classification of the Mammalia he doubted Owen's facts and conclusions regarding man's place in nature. Characteristically, Huxley performed a series of dissections of primate brains to satisfy himself that Owen's qualitative distinctions were not valid. Although Huxley published nothing on the brain at that time, he incorporated the information in his teaching after 1858, and it provided him with the necessary ammunition for the British Association meeting in 1860. On Thursday, 28 June, in section D, after a paper on plants by Charles Daubeny, Owen repeated his assertions of 1857 and 1859 that the brain of the gorilla, when compared with that of man, showed greater differences than existed between that of the gorilla and of the lowest of the Quadrumana, or primates. To this assertion Huxley publicly gave a "direct and unqualified contradiction" and promised to justify himself in this unusual procedure elsewhere. This he did in "On the Zoological Relations of Man With the Lower Animals" (*Natural History Review,* **1** [1861], 67–84), in which he demonstrated through references to a series of earlier studies by various authors the falsity of Owen's assertions that only in man did the cerebral hemispheres overlap the cerebellum and that only man possessed a posterior cornu of the lateral ventricle and a *hippocampus minor.* The above article was part of an extended debate with Owen (1860–1863), in which Huxley, with assistance from others and particularly from William Henry Flower's dissections, effectively showed Owen's errors regarding these cerebral characters.

By the end Huxley had demonstrated clearly that the differences between man and the apes were smaller than those between the apes and the lower primates. Therefore, man had to be considered zoologically a member of the primates. While the controversy was based on certain rather esoteric anatomical details, the results of the *hippocampus minor* debate were reported in the public press, inspired poetry and cartoons in *Punch,* and occupied a prominent place in Charles Kingsley's novel *The Water-Babies.*

Darwin had said in the *Origin* only that light would be thrown on man's relationship to his evolutionary hypothesis, and this controversy, with its public following, was instrumental in man's being considered in zoological terms and his origin as a result of the evolutionary process. Huxley also covered much of the same material on man and the other primates in lectures to various audiences and in his *Evidence as to Man's Place in Nature* (1863), which synthesized the anatomical and embryological evidence from his own work and the literature.

In addition, during the 1860's, Huxley devoted a fair share of his effort to physical anthropology, particularly on the recently discovered Neanderthal skull and various races of man and their relationships to one another. Huxley's treatment of man in zoological terms, as a topic to be considered scientifically and not emotionally, assisted strategically in the public's considering Darwin's evolutionary hypothesis on the same terms. This latter was Huxley's goal in all of his lecturing and writing on the subject of evolution. He firmly believed that if people would only look at the alternatives in a cool and reasoning manner they must recognize the fact of evolution and natural selection as its most probable mechanism.

While Huxley is best known for his defense of Darwin's hypothesis, he did not accept it uncritically and did not consider that the problem was finally settled nor that natural selection was by any means proven as the mechanism. When Huxley read the *Origin,* he was immediately convinced of the fact of evolution and thought Darwin had successfully put this ancient doctrine on a scientific basis. Huxley had long been handling the kinds of evidence that Darwin used and was well acquainted with them; he needed only for Darwin to lead the way in showing him how to arrange it. That way led through natural selection as the mechanism by which evolution had occurred. For Huxley and others natural selection provided a method for organizing their own facts. Throughout his life after 1859 Huxley maintained that natural selection was the most probable hypothesis of an evolutionary mechanism. For him it remained a hypothesis because of the lack of experimental proof. Huxley thought this proof would be the "production of mutually more or less infertile breeds from a common stock" in a selective breeding program. He thought it could be achieved "in a comparatively few years." In 1893 Huxley still thought such proof was necessary, but he was less optimistic about the time required. In his early essays on evolution, which he would stand by until the end of his life, he distinguished between morphological species, which could be demonstrated and could serve as evidence that evolution had taken place, and physiological species, which must be produced by selection to confirm natural selection as the mechanism.

After 1859 Huxley's own scientific work, as distinct from his role as Darwin's defender, had much the same character as his earlier work, with the notable difference that it was now focused principally on vertebrates, both recent and fossil. A considerable proportion of this activity was detailed, descriptive work on a single narrow problem in anatomy or paleontology—the "species work" which was always a burden to him. In several areas, however, Huxley made the same kind of broad, synthetic contributions he had made on invertebrates—"the architectural and engineering part of the business," as he called it. Huxley substantially revised the taxonomic arrangement of several groups, basing his revisions on his own observations, to which he added a broad knowledge of the relevant literature. This aspect of his work has been of value not because all of it is still considered valid, but because it posed questions and problems which stimulated further work by his followers, who, expectedly, went beyond Huxley.

Huxley did important work on all the major groups of vertebrates; but during the 1860's he was particularly interested in birds. After his Croonian lecture he began to study the development of the chick's skull. Approaching the birds as if they were all fossils, Huxley based his classification on osteological characters in what was probably the first comprehensive, comparative study (1867) of a single avian organ system (*Scientific Memoirs,* II, 238–297). His study set a model for much later avian taxonomic work which incorporated Huxley's findings. On the basis of several skeletal characters, for example, the keel and ossification of the sternum, Huxley divided the birds into three principal groups, Saururae, Ratitae, and Carinatae, the subdivisions of which were based heavily on the bony structure of the palate. In 1868, in a memoir on the anatomy of the gallinaceous birds (*Scientific Memoirs,* II, 346–373), Huxley, building particularly on P. L. Sclater and Darwin, incorporated the facts of geographical distribution into his taxonomy, making zoogeography a part of the definition of a species. He also suggested a linking of South America and Australia, a suggestion which has since received much support.

Paleontology. Some of Huxley's earliest paleontological work was on fishes from the Devonian Downton Sandstones which led him into a revision of much Devonian fish material and a memoir (1861) on the classification of the Devonian fishes (*Scientific Memoirs,* II, 421–460). Huxley was able to throw new light on many affinities, revising the work of Louis Agassiz and other early workers, by utilizing the new and rapidly growing collections of Devonian fishes and also the results of his extensive studies of piscine embryology. This memoir, and his 1866 supplement, remained a standard work on these animals for several decades. Huxley also did extensive studies of labyrinthodont Amphibia from the Mississippian and Pennsylvanian of Great Britain and similar forms from around the world (see *Scientific Memoirs,* II–III, *passim*). The most important of these were the genera *Loxomma* and *Anthracosaurus.* His elaboration of the morphology of these early tetrapods led him to place them on the borderline between the fishes, amphibians, and reptiles. Their ancestral relationship to all higher tetrapods has since been well recognized.

Closely related to his studies of birds was Huxley's interest in Mesozoic reptiles, particularly the Dinosauria. He rejected the proposed close affinity of pterodactyls to birds, the similarities being only analogous and not homologous. Huxley recognized that all the Dinosauria he had examined had strong ornithic characters in the tetraradiate arrangement of the ilium, ischium, pubis, and the femur, factors by which they differ from the majority of reptiles (see, for example, *Scientific Memoirs,* III, 465–486). At about the same time E. D. Cope recognized similar relationships. For these reptiles Huxley established (1869) the order Ornithoscelida (most of the members of which are now in the order Ornithischia), which included such forms as the *Iguanodon* (*Scientific Memoirs,* III, 487–509). On the basis of these specific similarities and more general evidence Huxley combined the reptiles and birds into the Sauropsida, one of his three great divisions of the Vertebrata; the others were the Ichthyopsida (fishes and amphibians) and the Mammalia. He later (1880) further divided the Mammalia into three groups of ascending complexity, the Prototheria (monotremes), Metatheria (marsupials), and Eutheria (placentals). These terms were meant to describe "stages of evolution" and, therefore, were more than purely taxonomic terms (*Scientific Memoirs,* IV, 457–472). These became important divisions because they were based on deep-seated anatomical characters, rather than the relatively superficial ones, such as teeth and digits, which are more closely related to the mammals' life habits.

In 1862 Huxley, as secretary of the Geological Society, was called upon to give the presidential address, in which he discussed several aspects of paleontology. He did not think that the fossil record had been able to provide evidences that modifications of any group had actually taken place through geological time or that earlier members of any long-standing group were more generalized than later ones. Eight years later, when Huxley was president of the Geological Society, he felt compelled to correct the statements of the earlier address. In the interim a substantial quantity of fossils had been discovered, among which were a number of ancestral forms of horses. From European Middle Miocene deposits came a three-toed equine, *Anchitherium,* which Huxley connected by stages to *Equus,* each member of the sequence being the result of increased specialization away from the average ungulate mammal. Huxley thought there should be Eocene predecessors of *Anchitherium* which were even less modified, and he suggested *Plagiolophus* might nearly be this form. He thought the equine pedigree would eventually be stretched back to a five-toed ancestor. One of Huxley's first visits on his American trip in 1876 was to Othniel Charles Marsh in New Haven, where he spent most of a week studying Marsh's very complete series of North American fossil horses extending back to the Upper Eocene *Orohippus.* Huxley recognized this as a more complete and more extensive series than Europe had to offer and, drawing on Marsh's insights and fossils, rejected his proposed line of equine ancestry and revised an address to be given in New York. In that lecture, primarily using sequences of teeth and limb material, Huxley presented the most complete evidence of modification having occurred through geological history. He then went on to predict that a yet more generalized form than *Orohippus* would be found; two months later he received word that Marsh had found *Eohippus,* the proposed ancestral form. This series of horses was the first extensive series which gave proof that the kinds of modifications demanded by Darwin's hypothesis had taken place and that the ancestral stages were more generalized than their more recent representatives.

Influence in Scientific Education. In addition to his extensive scientific output, of which only some high spots have been touched here, Huxley was an active teacher from 1854 until near the end of his life. Although it changed and developed under him, the lectureship on natural history at the Government School of Mines was his principal lifetime position. Until the School of Mines moved from Jermyn Street to South Kensington in 1872, to become part of the Royal College of Science, Huxley was forced to give a lecture course supplemented only by demonstrations, owing to an absence of laboratory space. After 1872 laboratory work became an integral part of his course, in which the students did the dissecting and observing to verify the facts in the text and lectures. Huxley conceived of this as an essential training in scientific method. He was fortunate in having Michael Foster and E. Ray Lankester among his first laboratory assistants. Both his lectures and laboratory classes were based on the same basic notion of types that he used in his original scientific work; Huxley thought this was the only means of bringing any logic to the myriad organic forms. In this approach to biology, Huxley was highly innovative, as he was well aware, although his method of teaching has now become a commonplace.

Huxley's teaching was by no means limited to his formal courses. He was Fullerian professor at the Royal Institution and Hunterian professor at the Royal College of Surgeons. He also gave a number of Friday evening discourses at the Royal Institution and a considerable array of special lectures on assorted topics at various locations. Of all his public speaking Huxley was most interested in the series of workingmen's lectures which he gave regularly, beginning in 1855, by which he wanted "the working classes to understand that Science and her ways are great facts for them." He was "sick of the dilettante middle class" and wanted to try his skills with the working class, who turned out in large numbers for his appearances. Huxley did not talk down to his audiences because of a firm belief that even the most complicated ideas could be understood by the majority of mankind if they were presented clearly and logically, step by step. Some of Huxley's finest addresses were to workingmen, for example, the series on man's place in nature and his 1868 "On a Piece of Chalk" (*Collected Essays,* VIII, 1–36). The latter is an excellent example of his style, which was at the root of his great success as a teacher and public speaker. This style was not dependent on the use of words or the structure of sentences but on the careful organization of ideas. Huxley's stress on clarity of thought was equally evident in the full range of his writings and was a key to their success.

In many places Huxley stressed the need for inclusion of science at all levels of education; but he did not stress science to the exclusion of history, literature, and the arts. His view of education not as an accumulation of facts but as a training and honing of all the faculties an individual might possess was the key to his conception of a liberal education. Huxley made a case for his views to various audiences: at the South London Working Men's College in "A Liberal Education; and Where to Find It"; at the opening of The Johns Hopkins University in "Address on University Education"; and on the eve of his election to the first London School Board in "The School Boards: What They Can Do, and What They May Do" (all in *Collected Essays,* III). The last outlines the program which to a great extent Huxley convinced his fellow board members to adopt under the 1870 Act of Parliament. He included the necessary disciplines of reading, writing, and arithmetic, and added physical science, drawing, singing, physical development, and domestic subjects. Surprisingly, to many of his colleagues, he also advocated studying the Bible, but without any theology, because it was great literature which embodied great morality and was the basis of three centuries of British civilization. In his service on the London School Board, Huxley proved to be one of the important shapers of primary education.

As offshoots of his teaching and often based on his lecture series, Huxley wrote several textbooks; two were designed for the general public. *Physiography* (1877) was an introduction to nature in all its aspects, what might be called general science, and *Lessons in Elementary Physiology* (1866) was a discussion of the human system written for the schools. For anatomy students Huxley wrote *The Crayfish* (1879) as an introduction to zoology and general works on both vertebrate and invertebrate animals.

Huxley's philosophic interests and writings focused on Descartes, Berkeley, and Hume, particularly on that fundamental problem of the relationship of mind and matter. Just as he conceived that the goal of education was to enable one to act, his conclusions in philosophy were the kind with which one could live. Huxley settled on a practical philosophy—an empirical idealism that recognized that all we can know of the world are affectations of the mind. He thought that such a view must be the philosophy of scientific men. He also held a practical materialism in the sense that it involved placing physical phenomena in a chain of direct causation. Huxley summarized these views in his 1868 lecture on protoplasm, "On the Physical Basis of Life" (*Collected Essays,* I, 130–165). In his discussions of education and scientific method Huxley put a strong emphasis on clear and distinct ideas, very much in the Cartesian tradition. Also, Huxley emphasized what he called a duty of doubt, an active skepticism, from which he believed freedom of thought would necessarily follow. For Huxley this freedom of thought was an essential element in the scientific process. In this context Huxley coined the term "agnosticism"—which to him embodied no belief nor implied any—when he became a member of the Metaphysical Society. For Huxley agnosticism was an attitude, a tool of the intellect, and "the fundamental axiom of modern science." It involved, positively, following one's intellect as far as it would go and, negatively, not accepting any conclusions which were not clearly demonstrable.

Huxley regarded the Bible highly, both as one of the great works of English literature and as a defense of freedom and liberty. For him it was "the *Magna Charta* of the poor and oppressed" insofar as it supported the concept of righteousness. To a great extent the Protestants had shifted the notion of infallibility from the Church to the Book, the Bible. It was in this context that Huxley first became involved in Biblical controversy, on the subject of the authority of Genesis. Huxley applied agnosticism, as a method, to this and other Biblical problems, including the divine inspiration of the New Testament Gospels and various revelations and miracles. Huxley often argued that matters of morality were independent of religion and theology, for example, in his letter to Charles Kingsley after the death of Huxley's first son, Noel.

Huxley was also a man of affairs. Between 1862 and 1884 he served on ten royal commissions investigating problems of education, fisheries, and vivisection. He held office in diverse scientific societies, particularly the Ethnological, Geological, and Royal societies, and the British Association, each of which he served as president. His role on the London School Board has been mentioned. Huxley was a member of the X-Club, which served to keep a small group of scientific friends in contact, and of that unique group the Metaphysical Society, before which he spoke on such topics as "Has a Frog a Soul?" His scientific and public works led to many honors, including the Royal, Copley, and Darwin medals from the Royal Society and ultimately appointment as a Privy Councillor. In all of his multifarious activities as scientist, educator, and public figure Huxley's success was dependent more than anything else on his clear thinking, his scrupulous weighing of all pertinent evidence, and, once he had reached a decision, on his effort to lead those around him, step by step, to see the rightness of his position.

BIBLIOGRAPHY

I. ORIGINAL WORKS. The majority of Huxley's shorter writings are available in two collections. *The Scientific Memoirs of Thomas Henry Huxley,* Michael Foster and E. Ray Lankester, eds. 4 vols. and supp. (London, 1898–1903), contains probably all of Huxley's important scientific papers as well as reports of his Royal Institution Friday Evening Discourses. Huxley selected and arranged his *Collected Essays* (London, 1893–1894), writing a preface to each of the nine volumes; the planned tenth volume was never completed. Huxley's *Diary of the Voyage of H.M.S. Rattlesnake* (London, 1935), Julian Huxley, ed., was long lost among the family's old account books. During the period he needed to write for money, Huxley, with George Busk, translated and edited *Kölliker's Manual of Human Histology* (London, 1853); with Arthur Henfrey he edited two volumes of Taylor's *Scientific Memoirs* (London, 1853–1854). The most important of Huxley's separate scientific writings was his *Evidence as to Man's Place in Nature* (London, 1863).

As introductory textbooks of zoology Huxley wrote *An Introduction to the Classification of Animals* (London, 1869) and *The Crayfish: an Introduction to the Study of Zoology* (London, 1879). His *A Manual of the Anatomy of Vertebrated Animals* (London, 1871) and *A Manual of the Anatomy of Invertebrated Animals* (London, 1877) were comprehensive treatments which served as standard textbooks.

The great bulk of Huxley's manuscripts are in the Imperial College of Science and Technology, London. These have been ably catalogued in Warren R. Dawson, *The Huxley Papers. A Descriptive Catalogue of the Correspondence, Manuscripts and Miscellaneous Papers . . .* (London, 1946). See also J. Pingree, *Thomas Henry Huxley: List of His Correspondence With Miss Henrietta Anne Heathorn, Later Mrs. Huxley, 1847–1854* (London, 1969) and *Thomas Henry Huxley: A List of His Scientific Notebooks, Drawings and Other Papers* (London, 1968).

Huxley was a prolific correspondent, and his letters may be found in many collections of papers of his correspondents.

II. SECONDARY LITERATURE. The standard source for Huxley's life is Leonard Huxley, *Life and Letters of Thomas Henry Huxley,* 2 vols. (London, 1900), which includes extensive bibliographies of his addresses, books, and scientific papers, as well as lists of honors he received and scientific societies and Royal Commissions of which he was a member. This work is weak on the period of the voyage of the *Rattlesnake,* apparently because the *Diary* of the voyage had not then been found.

Probably the best analysis of Huxley's scientific work is P. Chalmers Mitchell, *Thomas Henry Huxley. A Sketch of His Life and Work* (London, 1900); see also Michael Foster, "Obituary of T. H. Huxley," in *Proceedings of the Royal Society of London,* **59** (1896), 46–66. Most of the many biographical writings on Huxley rely heavily on the above three items. The various works devoted to the lives and letters of Charles Darwin, Joseph Hooker, and Charles Lyell are valuable for additional information and other views of events in Huxley's life.

Cyril Bibby, *T. H. Huxley: Scientist, Humanist and Educator* (London, 1959), is based on extensive original research and emphasizes Huxley's activities and writings as an educator and a public figure, topics which because of space are only touched on in the above article.

For the debate following Huxley's "On the Physical Basis of Life," see Gerald L. Geison, "The Protoplasmic Theory of Life and the Vitalist-Mechanist Debate," in *Isis,* **60** (1969), 273–292. Finally, Aldous Huxley discussed his grandfather's success as a writer in his 1932 Huxley Memorial Lecture, "T. H. Huxley as a Literary Man"; reprinted in Aldous Huxley, *The Olive Tree* (London, 1937).

WESLEY C. WILLIAMS

HUYGENS, CHRISTIAAN (also **Huyghens, Christian**) (*b.* The Hague, Netherlands, 14 April 1629; *d.* The Hague, 8 July 1695), *physics, mathematics, astronomy, optics.*

Huygens belonged to a prominent Dutch family. His grandfather, also Christiaan Huygens, served William the Silent and Prince Maurice as secretary. In 1625 his father, Constantijn, became a secretary to Prince Frederic Henry and served the Orange family for the rest of his life, as did Christiaan's brother Constantijn.

Along with this tradition of diplomatic service to the house of Orange, the Huygens family had a strong educational and cultural tradition. The grandfather took an active part in the education of his children, and thus Huygens' father acquired great erudition in both literature and the sciences. He corresponded with Mersenne and Descartes, the latter often enjoying his hospitality in The Hague. Constantijn was a man of taste in the fine arts, talented in drawing, a musician and fertile composer, and, above all, a great poet; his Dutch and Latin verse gained him a lasting place in the history of Dutch literature.

Like his father, Constantijn was actively committed to the education of his children. Christiaan and his brother Constantijn were educated at home up to the age of sixteen by both their father and private teachers. They acquired a background in music (Christiaan sang well and played the viola da gamba, the lute, and the harpsichord), Latin, Greek, French, and some Italian, and logic, mathematics, mechanics, and geography. A highly talented pupil, Christiaan showed at an early age the combination of theoretical interest and insight into practical applications and constructions (at thirteen he built himself a lathe) which characterized his later scientific work.

From May 1645 until March 1647 Christiaan studied law and mathematics at the University of Leiden, the latter with Frans van Schooten. He studied classical mathematics as well as the modern methods of Viète, Descartes, and Fermat. During this period his father called Mersenne's attention to his son's study on falling bodies, and this opened up a direct correspondence between Christiaan and Mersenne. Descartes, whose work in these years had a great influence on young Huygens, also showed an interest in and an appreciation of Christiaan's work. From March 1647 until August 1649 Christiaan studied law at the newly founded Collegium Arausiacum (College of Orange) at Breda, of which his father was a curator and where Pell taught mathematics.

Huygens did not, after his studies, choose the career in diplomacy which would have been natural for a man of his birth and education. He did not want such a career, and in any event the Huygens family lost its main opportunities for diplomatic work as a result of the death of William II in 1650. Huygens lived at home until 1666, except for three journeys to Paris and London. An allowance supplied by his father enabled him to devote himself completely to the study of nature. These years (1650–1666) were the most fertile of Huygens' career.

Huygens at first concentrated on mathematics: determinations of quadratures and cubatures, and algebraic problems inspired by Pappus' works. In 1651 the *Theoremata de quadratura hyperboles, ellipsis et circuli* [1] appeared, including a refutation of Gregory of St. Vincent's quadrature of the circle. The *De circuli magnitudine inventa* [2] followed in 1654. In the subsequent years Huygens studied the rectification of the parabola, the area of surfaces of revolution of parabolas, and tangents and quadratures of various curves such as the cissoid, the cycloid (in connection with a problem publicly posed by Pascal in 1658), and the logarithmica. In 1657 Huygens' treatise on probability problems appeared, the *Tractatus de ratiociniis in aleae ludo* [4].

A manuscript on hydrostatics [20] had already been completed in 1650, and in 1652 Huygens formulated the rules of elastic collision and began his studies of geometrical optics. In 1655 he applied himself, together with his brother, to lens grinding. They built microscopes and telescopes, and Huygens, in the winter of 1655–1656, discovered the satellite of Saturn and recognized its ring, as reported in his *De Saturni lunâ observatio nova* [3] and *Systema Saturnium* [6], respectively.

In 1656 Huygens invented the pendulum clock. This is described in 1658 in the *Horologium* [5] (not to be confused with the later *Horologium oscillatorium*) and formed the occasion for the discovery of the tautochronism of the cycloid (1659), and for the studies on the theory of evolutes and on the center of oscillation. Huygens' study of centrifugal force also dates from 1659. In these years he corresponded with increasing intensity with many scholars, among them Gregory of St. Vincent, Wallis, van Schooten, and Sluse. Studies on the application of the pendulum clock for the determination of longitudes at sea occupied much of his time from 1660 onward.

Of the journeys mentioned above, the first, from July until September 1655, brought Huygens to Paris, where he met Gassendi, Roberval, Sorbière, and Boulliau—the circle of scholars which later formed the Académie Royale des Sciences. He used the opportunity of the stay in France to buy, as did his brother, a doctorate "utriusque juris" in Angers. During his second stay in Paris, from October 1660 until March 1661, he met Pascal, Auzout, and Desargues. Afterward he was in London (until May 1661). There Huygens attended meetings in Gresham College, and met Moray, Wallis, and Oldenburg, and was impressed by Boyle's experiments with the air pump. A third stay in Paris, from April 1663 to May 1664, was interrupted by a journey to London (June to September 1663), where he became a member of the newly founded Royal Society. He then returned to Paris where he received from Louis XIV his first stipend for scientific work.

In 1664 Thévenot approached Huygens to offer him membership in an academy to be founded in Paris; Colbert proposed giving official status and financial aid to those informal meetings of scholars which had been held in Paris since Mersenne's time. In 1666 the Académie Royale des Sciences was founded. Huygens accepted membership and traveled to Paris in May of that year. Thus began a stay in Paris that lasted until 1681, interrupted only by two periods of residence in The Hague because of ill health. Huygens' health was delicate, and in early 1670 he was afflicted by a serious illness. In September, partially recovered, he left for The Hague and returned to Paris in June 1671. The illness recurred in the autumn of 1675, and from July 1676 until June 1678 Huygens again was in The Hague.

As the most prominent member of the Academy, Huygens received an ample stipend and lived in an apartment in the Bibliothèque Royale. In the Academy, Huygens encouraged a Baconian program for the study of nature. He participated actively in astronomical observations (of Saturn, for example) and in experiments with the air pump. He expounded his theory of the cause of gravity in 1669, and in 1678 he wrote the *Traité de la lumière* [12], which announced the wave, or more accurately, the pulse theory of light developed in 1676–1677. In the years 1668–1669 he investigated, theoretically and experimentally, the motion of bodies in resisting media. In 1673 he cooperated with Papin in building a *moteur à explosion,* and from that year onward he was also in regular contact with Leibniz. Huygens began his studies of harmonic oscillation in 1673 and designed clocks regulated by a spring instead of a pendulum, about which a controversy with Hooke ensued. In 1677 he did microscopical research.

In 1672 war broke out between the Dutch republic and Louis XIV and his allies. William III of Orange came to power and Huygens' father and brother assumed prominent positions in Holland. Huygens stayed in Paris, and, although he was deeply concerned with the Dutch cause, proceeded with his work in the Academy under the protection of Colbert. In 1673 he published the *Horologium oscillatorium* [10]. It was his first work to appear after he entered a position financed by Louis XIV, and he dedicated it to the French king. This gesture served to strengthen his position in Paris but occasioned some disapproval in Holland.

Huygens left Paris in 1681, again because of illness. He had recovered by 1683, but Colbert had died meanwhile, and without his support Huygens' nationality, his Protestantism, and his family's ties with the house of Orange would have engendered such strong opposition in Paris that he decided to stay in Holland. His financial position was thus not as secure but he did have an income from his family's landed property. Huygens never married. In the relative solitude of his residence in The Hague and at Hofwijck, the family's country house near Voorburg, he continued his optical studies, constructed a number of clocks, which were tested on several long sea voyages, and wrote his *Cosmotheoros* [14]. From June until September 1689 he visited England, where he met Newton. The *Principia* aroused Huygens' admiration but also evoked his strong disagreement. There is evidence of both in the *Traité de la lumière* [12] and its supplement, the *Discours de la cause de la pesanteur* [13]. Discussions with Fatio de Duillier, correspondence with Leibniz, and the interest created by the latter's differential and integral calculus drew Huygens' attention back to mathematics in these last years.

In 1694 Huygens again fell ill. This time he did not recover. He died the following summer in The Hague.

Mathematics. The importance of Huygens' mathematical work lies in his improvement of existing methods and his application of them to a great range of problems in natural sciences. He developed no completely new mathematical theories save his theory of evolutes and—if probability may be considered a mathematical concept—his theory of probability.

Huygens' mathematics may be called conservative in view of the revolutionary innovations embodied in the work of such seventeenth-century mathematicians as Viète, Descartes, Newton, and Leibniz. A marked tension is often apparent between this conservatism and the new trends in the mathematics of Huygens' contemporaries. Whereas, for example, Huygens fully accepted Viète's and Descartes's application of literal algebra to geometry, he rejected Cavalieri's methods of indivisibles. In his earlier works he applied rigorous Archimedean methods of proof to problems about quadratures and cubatures. That is, he proved equality of areas or contents by showing, through consideration of a sequence of approximating figures, that the supposition of inequality leads to a contradiction. On the other hand, he accepted Fermat's infinitesimal methods for extreme values and tangents, freely practicing division by "infinitely small"—his terminology—differences of abscissae, which subsequently are supposed equal to zero. Eventually the tediousness of the Archimedean methods of proof forced him to work directly with partition of figures into "infinitely small" or very small component figures; he considered this method to be inconclusive but sufficient to indicate the direction of a full proof. He long remained skeptical about Leibniz' new methods, largely because of Leibniz' secrecy about them.

In his first publication, *Theoremata de quadratura hyperboles,* Huygens derived a relation between the quadrature and the center of gravity of segments of circles, ellipses, and hyperbolas. He applied this result to the quadratures of the hyperbola and the circle. In the *De circuli magnitudine inventa* he approximated the center of gravity of a segment of a circle by the center of gravity of a segment of a parabola, and thus found an approximation of the quadrature; with this he was able to refine the inequalities between the area of the circle and those of the inscribed and circumscribed polygons used in the calculations of π. The same approximation with segments of the parabola, in the case of the hyperbola, yields a quick and simple method to calculate logarithms, a finding he explained before the Academy in 1666–1667.

In an appendix to the *Theoremata,* Huygens refuted the celebrated proof by Gregory of St. Vincent (*Opus geometricum* [1647]) of the possibility of the quadrature of the circle. Huygens found the crucial mistake in this very extensive and often obscure work. Gregory had applied Cavalierian indivisible methods to the summation of proportions instead of to line segments. The language of proportions was still sufficiently close to that of arithmetic for Gregory's error not to be a simple blunder, but Huygens was able to show by a numerical example that the application was faulty.

Having heard in Paris about Pascal's work in probability problems, Huygens himself took up their study in 1656. This resulted in the *Tractatus de ratiociniis in aleae ludo,* a treatise that remained the only book on the subject until the eighteenth century. In his first theorems Huygens deduced that the "value of a chance," in the case where the probabilities for a and b are to each other as $p:q$, is equal to

$$\frac{pa + qb}{p + q}.$$

He thus introduced as a fundamental concept the expectation of a stochastic variable rather than the probability of a process (to put it in modern terms). Subsequent theorems concern the fair distribution of the stakes when a game is broken off prematurely. The treatise closes with five problems, the last of which concerns expected duration of play.

In 1657 Huygens found the relation between the arc length of the parabola and the quadrature of the hyperbola. His method cannot be extended to a general rectification method, for it depends on a special property of the parabola: if a polygon is tangent to the parabola, and if the tangent points have equidistant abscissae, the polygon can be moved in the direction of the axis of the parabola to form an inscribed polygon. Huygens also employed this property to find the surface area of a paraboloid of revolution. From correspondence he learned about the general rectification method of Heuraet (1657). He found, in 1658, the relation which in modern notation is rendered by $yds = ndx$ (s: arc length; n: normal to the curve (y, x)), with which he could reduce the calculation of surface areas of solids of revolution to the quadrature of the curve $z = n(x)$; he used this relation also in a general rectification method. Some of

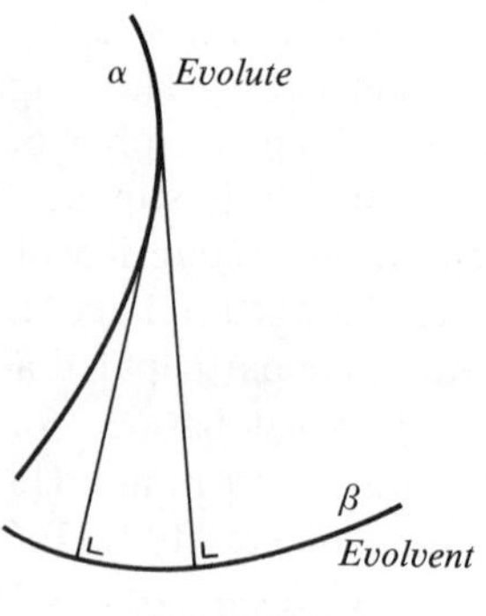

FIGURE 1

these results were published in part 3 of the *Horologium oscillatorium* [10].

In 1659 Huygens developed, in connection with the pendulum clock, the theory of evolutes (Fig. 1). The curve β described by the end of a cord which is wound off a convex curve α is called the evolvent of α, and conversely α is called the evolute of β. In part 3 of the *Horologium oscillatorium* Huygens showed, by rigorous Archimedean methods, that the tangents to the evolute are perpendicular to the evolvent, and that two curves which exhibit such a relation of tangents and perpendiculars are the evolute and evolvent of one another. Further, he gives a general method (proved much less rigorously) of determining from the algebraic equation of a curve the construction of its evolute; the method is equivalent to the determination of the radius of curvature (although Huygens only later interested himself in this as a measure of curvature) and implies, accordingly, a twice repeated determination of tangents by means of Sluse's tangent rule.

Huygens' study on the logarithmica dates from 1661; the results were published in the *Discours.* Huygens introduced this curve (modern $y = ae^x$) as the one in which every arithmetical series of abscissae corresponds to a geometrical series of ordinates. He noted its connection both with the quadrature of the hyperbola and with logarithms and pointed out that its subtangent is constant.

In the last decade of his life Huygens became convinced of the merits of the new Leibnizian differential and integral calculus through the study of articles by the Bernoullis, L'Hospital, and Leibniz, and through correspondence with the latter two. In 1691 he learned how to apply calculus in certain simple cases. Nevertheless, Huygens continued to use the old infinitesimal geometrical methods—which he applied with such virtuosity that he was able to solve most of the problems publicly posed in this period, including Leibniz' isochrone problem (1687), Johann Bernoulli's problem (1693–1694), the tractrix prob-

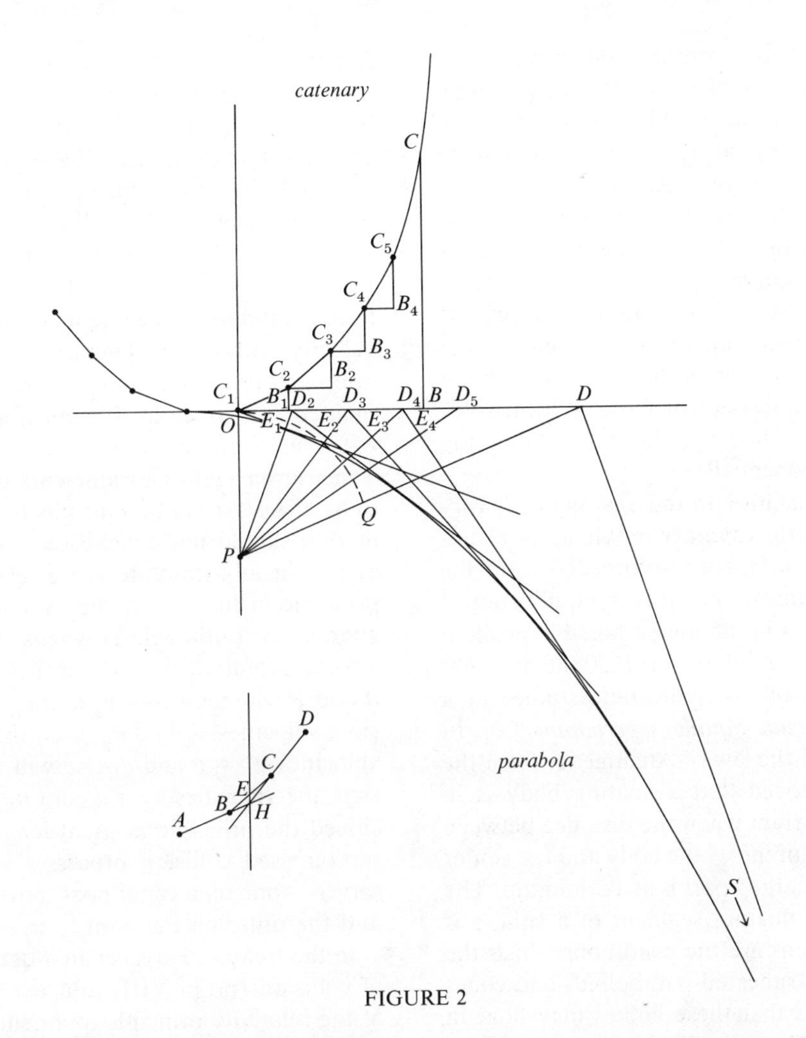

FIGURE 2

lem (1693), and the catenary problem (1691–1693). His final solution (1693) of this last problem may serve as an example of the force and style of Huygens' mathematics.

In dealing with the catenary problem, Huygens conceived the chain as a series of equal weights, connected by weightless cords of equal length. It follows from statics that every four subsequent weights A, B, C, D (Fig. 2) in the chain are disposed such that the extensions of AB and CD meet at H on the vertical that bisects BC. (Huygens had already found this result in 1646 and used it to refute Galileo's assertion that the catenary is a parabola.) By simple geometry it may now be seen that the tangents of the angles of subsequent cords to the horizontal are in arithmetical progression. Huygens further conceived (Fig. 2) the chain $C_1C_2C_3C_4 \cdots$ (the lowest link being horizontal) stretched along the horizontal axis, to become $C_1D_2D_3D_4 \cdots$. Point P on the vertical through C_1 is chosen such that $\angle C_1PD_2 = \angle C_2C_1B_1$. (As Huygens knew, it can be proved that in the limit C_1P is equal to the radius of curvature in the vertex of the chain.) As the tangents of $\measuredangle D_iC_1P$ are obviously in arithmetical progression, $\angle D_iC_1P$ must be equal to $\angle C_{i+1}C_iB_i$. Introducing normals D_iE_i on $D_{i+1}P$, it follows that the triangles $D_iD_{i+1}E_i$ are congruent with $C_iC_{i+1}B_i$, so that the chain is stretched, as it were, together with its series of characteristic triangles.

Considering the abscissa C_1B and the ordinate BC of a point C on the catenary, it is clear that

$$C_1B = \sum C_iB_i = \sum D_iE_i$$

and that

$$BC = \sum B_iC_{i+1} = \sum E_iD_{i+1}.$$

Huygens now imagines the interstices to be "infinitely small," so that C_1 coincides with the vertex O of the catenary, and he takes $OD = \widehat{OC}$. It is then clear that $\Sigma E_iD_{i+1} = QD$, if $PQ = PO$, so that the ordinate BC is equal to QD. To evaluate the abscissa OB, Huygens extends the normals D_iE_i and remarks that they are the tangents of a curve $\widehat{OS}$, which has the property that the normals PD_i on its tangents D_iE_i

meet in one point. This determines the curve $\widehat{OS}$ as a parabola; by the theory of evolutes ΣD_iE_i is equal to the arc length $\widehat{OS}$ of the parabola minus the tangent SD, so that the abscissa OB is equal to $\widehat{OS} - SD$. This result, in combination with the previously found equality $BC = QD$, makes possible the geometrical construction of corresponding ordinates and abscissae of the curve. The construction presupposes the rectification of the parabola, which, as Huygens knew, depends on the quadrature of the hyperbola. Thus his solution of the catenary problem is the geometrical equivalent of the analytical solution of the problem, namely, the equation of the curve involving exponentials.

Statics and Hydrostatics. In the treatment of problems in both statics (the catenary problem, for example) and hydrostatics, Huygens proceeded from the axiom that a mechanical system is in equilibrium if its center of gravity is in the lowest possible position with respect to its restraints. In 1650 he brought together the results of his hydrostatic studies in a manuscript, *De iis quae liquido supernatant* [20]. In this work he derived the law of Archimedes from the basic axiom and proved that a floating body is in a position of equilibrium when the distance between the center of gravity of the whole body and the center of gravity of its submerged part is at a minimum. The stable position of a floating segment of a sphere is thereby determined, as are the conditions which the dimensions of right truncated paraboloids and cones must satisfy in order that these bodies may float in a vertical position. Huygens then deduced how the floating position of a long beam depends on its specific gravity and on the proportion of its width to its depth, and he also determined the floating position of cylinders. The manuscript is of further mathematical interest for its many determinations of centers of gravity and cubatures, as, for example, those of obliquely truncated paraboloids of revolution and of cones and cylinders.

Impact. Huygens started his studies on collision of elastic bodies in 1652, and in 1656 he collected his results in a treatise *De motu corporum ex percussione* [18]. He presented the most important theorems to the Royal Society in 1668, simultaneously with studies by Wren and Wallis; they were published, without proofs, in the *Journal des sçavans* in 1699 [9]. Since Huygens' treatise is a fundamental work in the theory of impact and exhibits his style at its best, it is worth describing in some detail.

Huygens' theory amounted to a refutation of Descartes's laws of impact. Indeed, Huygens' disbelief in these laws was one of the motivations for his study. Descartes supposed an absolute measurability of velocity (that is, a reference frame absolutely at rest). This assumption is manifest in his rule for collision of equal bodies. If these have equal velocities, they rebound; if their velocities are unequal, they will move on together after collision. Huygens challenged this law and in one of his first manuscript notes on the question, remarked that the forces acting between colliding bodies depend only on their relative velocity. Although he later abandoned this dynamical approach to the question, the relativity principle remained fundamental. It appeared as hypothesis III of *De motu corporum,* which asserts that all motion is measured against a framework that is only assumed to be at rest, so that the results of speculations about motion should not depend on whether this frame is at rest in any absolute sense. Huygens' use of this principle in his impact theory may be described algebraically (although Huygens himself, of course, gave a geometrical treatment) as follows: If bodies A and B with velocities v_A and v_B acquire, after collision, velocities u_A and u_B, then the same bodies with velocities $v_A + v$ and $v_B + v$ will acquire, after collision, the velocities $u_A + v$ and $u_B + v$. Huygens discussed the principle at great length and as an illustration used collision processes viewed by two observers—one on a canal boat moving at a steady rate and the other on the bank.

In the treatise, Huygens first derived a special case of collision (prop. VIII) and extended it by means of the relativity principle to a general law of impact, from which he then derived certain laws of conservation. This procedure is quite contrary to the method of derivation of the laws of impact from the axiomatic conservation laws, which has become usual in more recent times; but it is perhaps more acceptable intuitively. In the special case of prop. VIII the magnitudes of the bodies are inversely proportional to their (oppositely directed) velocities ($m_A : m_B = |v_B| : |v_A|$), and Huygens asserts that in this case the bodies will simply rebound after collision ($u_A = -v_A$, $u_B = -v_B$). To prove this, Huygens assumed two hypotheses. The first, hypothesis IV, states that a body A colliding with a smaller body B at rest transmits to B some of its motion—that is, that B will acquire some velocity and A's velocity will be reduced. The second, hypothesis V, states that if in collision the motion of one of the bodies is not changed (that is, if the absolute value of its velocity remains the same), then the motion of the other body will also remain the same.

The role of the concept of motion (*motus*) as used here requires some comment. Descartes had based his laws of impact partly on the theorem that motion is conserved, whereby he had quantified the concept of motion as proportional to the magnitude of the body and to the absolute value of its velocity ($m|v|$). Huygens found that in this sense the *quantitas motus* is not conserved in collision. He also found that if the velocities are added algebraically, there is a law of conservation (namely, of momentum) which he formulated as conservation of the velocity of the center of gravity. But for Huygens the vectorial quantity $m\vec{v}$ was apparently so remote from the intuitive concept of motion that he did not want to assume its conservation as a hypothesis. Nor could he take over Descartes's quantification of the concept, and thus he used a nonvectorial concept of motion, without quantifying it, restricting himself to one case in which motion is partly transferred, and to another in which it remains unchanged.

Huygens now deduced from hypotheses III, IV, and V that the relative velocities before and after collision are equal and oppositely directed: $v_A - v_B = u_B - u_A$ (prop. IV). To derive proposition VIII, he drew upon three more assertions: namely, Galileo's results concerning the relation between velocity and height in free fall; the axiom that the center of gravity of a mechanical system cannot rise under the influence of gravity alone; and the theorem that elastic collision is a reversible process, which he derived from proposition IV. Huygens considered the velocities v_A and v_B in proposition VIII as acquired through free fall from heights h_A and h_B, and supposed that the bodies after collision are directed upward and rise to heights h'_A and h'_B. Because the collision is reversible, the centers of gravity of the systems (A, B, h_A, h_B) and (A, B, h'_A, h'_B) must be at the same height, from which it can be calculated that $u_A = -v_A$ and $u_B = -v_B$. Proposition VIII is now proved, and by means of the relativity principle the result of any elastic collision can be derived, as Huygens showed in proposition IX. Finally, he deduced from this general law of impact the proposition that before and after collision the sum of the products of the magnitudes and the squares of the velocities of the bodies are equal (conservation of $\Sigma\ mv^2$).

Optical Techniques. Working with his brother, Huygens acquired great technical skill in the grinding and polishing of spherical lenses. The lenses that they made from 1655 onward were of superior quality, and their telescopes were the best of their time. In 1685 Huygens summarized his technical knowledge of lens fabrication in *Memorien aengaende het slijpen van glasen tot verrekijckers* [17]. In *Astroscopia compendiaria* [11], he discussed the mounting of telescopes in which, to reduce aberration, the objective and ocular were mounted so far apart (up to twenty-five meters) that they could not be connected by a tube but had to be manipulated separately.

Geometrical Optics. As early as 1653 Huygens recorded his studies in geometrical optics in a detailed manuscript, *Tractatus de refractione et telescopiis* [16]. He treated here the law of refraction, the determination of the focuses of lenses and spheres and of refraction indices, the structure of the eye, the shape of lenses for spectacles, the theory of magnification, and the construction of telescopes. He applied his theorem that in an optical system of lenses with collinear centers the magnification is not changed if the object and eye are interchanged to his theory of telescopes. He later used the theorem in his calculations for the so-called Huygens ocular, which has two lenses. He began studying spherical aberration in 1665, determining for a lens with prescribed aperture and focal length the shape which exhibits minimal spherical aberration of parallel entering rays. He further investigated the possibility of compensating for spherical aberration of the objective in a telescope by the aberration of the ocular, and he studied the relation between magnification, brightness, and resolution of the image for telescopes of prescribed length. These results were checked experimentally in 1668, but the experiments were inconclusive, because in the overall aberration effects the chromatic aberration is more influential than the spherical.

About 1685 Huygens began to study chromatic aberration. He did not start from his own experiments, as he usually did, but rather began with the results of Newton's work; he had first heard of Newton's theory of colors in 1672. Huygens confirmed the greater influence of chromatic as compared with spherical aberration, and he thereby determined the most advantageous shapes for lenses in telescopes of prescribed length.

About 1677 Huygens studied microscopes, including aspects of their magnification, brightness, depth of focus, and lighting of the object. Under the influence of Leeuwenhoek's discoveries, with his microscope he observed infusoria, bacteria, and spermatozoa. In consequence he became very skeptical about the theory of spontaneous generation.

Astronomy. With the first telescope he and his brother had built, Huygens discovered, in March 1655, a satellite of Saturn, later named Titan. He determined its period of revolution to be about sixteen days, and noted that the satellite moved in the same plane as the "arms" of Saturn. Those extraordinary appendages of the planet had presented astronomers since Galileo with serious problems of interpretation; Huygens solved these problems with the hypothesis that Saturn is surrounded by a ring. He arrived at this solution partly through the use of better observational equipment, but also by an acute argument based on the use of the Cartesian vortex (the whirl of "celestial matter" around a heavenly body supporting its satellites).

Huygens' argument began with the premise that it is a general feature of the solar system that the period of rotation of a heavenly body is much shorter than the periods of revolution of its satellites, and that the periods of inner satellites are smaller than those of outer satellites. This is the case with the sun and the planets, with the earth and the moon, and with Jupiter and its satellites. In the same way the "celestial matter" between Saturn and its satellite must move so that the parts near the planet—including the "arms"—will have a period of revolution about equal to the period of rotation of the planet and much shorter than the sixteen days assigned to the satellite. In the period of Huygens' observations in 1655–1656, no alteration was observed in the aspects of the "arms," a phenomenon which could be explained only if the matter forming the "arms" was distributed with cylindrical symmetry around Saturn, with its axis of symmetry—the axis of the vortex—perpendicular to the plane of the satellite and of the "arms" themselves. Therefore, the "arms" must be considered as the aspect of a ring around Saturn. In his further calculations, Huygens established that this hypothesis could also be used to explain the observed long-term variations in the aspect of the "arms."

In March 1656 Huygens published his discovery of Saturn's satellite in the pamphlet *De Saturni lunâ observatio nova* [3], in which, to secure priority, he also included an anagram for the hypothesis of the ring. (After decoding, this anagram reads "Annulo cingitur, tenui, plano, nusquam cohaerente, ad eclipticam inclinato"—"It is surrounded by a thin flat ring, nowhere touching, and inclined to the ecliptic.") The full theory was published, after some delay (1659), in *Systema Saturnium* [6], together with many other observations on the planets and their satellites, all contributing to an emphatic defense of the Copernican system.

Of Huygens' further astronomical work, one should mention the determination of the period of Mars and the observation of the Orion nebula. He described the latter, in *Systema Saturnium,* as the view through an opening in the dark heavens into a brighter region farther away. He also developed micrometers for the determination of angular diameters of planets.

Pendulum Clock. In the winter of 1656–1657 Huygens developed the idea of using a pendulum as a regulator for clockworks. Galileo had strongly maintained the tautochronism of the pendulum movement and its applicability to the measurement of time. Pendulums were so used in astronomical observations, sometimes connected to counting mechanisms. In cogwheel clocks, on the other hand, the movement was regulated by balances, the periods of which were strongly dependent on the source of motive power of the clock and hence unreliable. The necessity for accurate measurement of time was felt especially in navigation, since good clocks were necessary to find longitude at sea. In a seafaring country like Holland, this problem was of paramount importance. Huygens' invention was a rather obvious combination of existing elements, and it is thus not surprising that his priority has been contested, especially in favor of Galileo's son, Vincenzio.

There is no question of Huygens' originality, however, if one acknowledges as the essential point in his clock the application of a freely suspended pendulum, whose motion is transmitted to the clockwork by a handle and fork. The first such clock dates from 1657, and was patented in the same year. In the *Horologium* Huygens described his invention, which had great success; many pendulum clocks were built and by 1658 pendulums had been applied to the tower clocks of Scheveningen and Utrecht.

Huygens made many theoretical studies of the pendulum clock in the years after 1658. The problem central to such mechanisms is that the usual simple pendulum is not exactly tautochronous. Its period depends on the amplitude, although when the amplitudes are small this dependence may be neglected. (This problem was recognized in the first applications of Galileo's proposal.) There are three possible solutions. A constant driving force would secure constant amplitude, but this is technically very difficult. The amplitude may be kept small, a remedy Huygens applied in the clock he described in the *Horologium,* but then even a small disturbance can stop the clock. The best method, therefore, is to design the pendulum so that its bob moves in such a path that the dependence of period on amplitude is entirely eliminated. Huygens tried this solution in his first clock, applying at the suspension point of the pendulum two bent metal laminae, or cheeks, along which the cord wrapped itself as the pendulum swung. Thus the bob did not move in a circle but in a path such that—it could be argued qualitatively—the swing was closer to being tautochronous than in the usual pendulum.

In 1659 Huygens discovered that complete independence of amplitude (and thus perfect tautochronism) can be achieved if the path of the pendulum bob is a cycloid. The next problem was what form to give the cheeks in order to lead the bob in a cycloidal path. This question led Huygens to the theory of evolutes of curves. His famous solution was that the cheeks must also have the form of a cycloid, on a scale determined by the length of the pendulum.

Huygens also studied the relation between period and length of the pendulum and developed the theory of the center of oscillation. By this theory the notion of "length" of a pendulum is extended to compound pendulums, so that Huygens could investigate how the period of a pendulum can be regulated by varying the position of an additional small weight on the arm. These studies form the main contents of Huygens' magnum opus, the *Horologium oscillatorium* [10] (1673). After 1673 Huygens studied harmonic oscillation in general, in connection with the tautochronism of the cycloid. He developed the application of springs instead of pendulums as regulators of clocks—a question on which he engaged in priority disputes with Hooke and others. Huygens also designed many other tautochronous balances for clocks.

Huygens considered the determination of longitudes at sea to be the most important application of the pendulum clock. Here the main difficulty was maintaining an undisturbed vertical suspension. Huygens designed various apparatus to meet this problem, some of which were tested on sea voyages after 1663. Huygens discussed these experiments in *Kort Onderwijs aengaende het gebruyck der Horologien tot het vinden der Lenghten van Oost en West,* a manual for seamen on how to determine longitudes with the help of clocks. Clocks tested on later expeditions (for example, to Crete in 1668–1669 and to the Cape of Good Hope in 1686–1687 and 1690–1692) were not really successful.

Simple Pendulum: Tautochronism of the Cycloid. In 1659, in a study done on the ordinary simple pendulum, Huygens derived a relation between the period and the time of free fall from rest along the length of the pendulum. His result, which he published in part 4 of the *Horologium oscillatorium,* is equivalent to $T = 2\pi\sqrt{l/g}$. In deriving the relation, Huygens used a certain approximation which discards the dependence of the period on the amplitude. The error thus introduced is negligible in the case of a small amplitude. In a subsequent investigation, Huygens posed the question of what form the path of the pendulum bob should have, so that the approximative assumption would cease to be an approximation and would describe the real situation. He found a condition for the form of the path related to the position of the normals to the curve with respect to the axis; and he recognized this as a property of the cycloid, which he had studied in the previous year in connection with a problem set by Pascal. He thus discovered the tautochronism of the cycloid—"the most fortunate finding which ever befell me," he said later. He published his discovery, with a scrupulously rigorous Archimedean proof, in the second part of *Horologium oscillatorium.*

Center of Oscillation. Huygens began his studies on the center of oscillation in 1659 as part of his work on the pendulum clock. By 1669 he had formulated a general computation rule applicable to all sorts of compound pendulums (*Horologium oscillatorium,* part 4). He showed that the period of a compound pendulum depends on the form of the pendulous body and on the position of the axis (Fig. 3). The theory of the center of oscillation determines this dependence by establishing the length λ of the simple pendulum that oscillates isochronously with the compound pendulum. The center of oscillation of the compound pendulum is the point O which lies at distance λ from the axis on the line through the center of gravity Z, perpendicular to the axis. If one assumes all the mass of the pendulum to be concentrated in O, the simple pendulum thus formed (with the same axis) will have the same period as the compound one.

In determining centers of oscillation Huygens proceeded from two hypotheses. The first, which he also used in deriving laws of impact, asserts that the center of gravity of a system, under the sole influence of gravity, cannot rise; the second, that in the absence of friction the center of gravity of a system will, if the component parts are directed upward after a descent, rise again to its initial height. Huygens further supposed that the latter hypothesis also applies if during the movement the links between the component parts are severed.

Huygens' determination of centers of oscillation can now be represented as follows: The compound pendulum (Fig. 3) consists of small parts with weight g_i, whose distances to the axis are α_i. The center of gravity Z has distance ζ to the axis; λ is the length of the isochronous simple pendulum, whose bob in initial position (the amplitudes of both pendulums being equal) is at height h above its lowest position; passing this lowest position it has velocity v. It is now obvious that in moving from the initial to the lowest position, the center of gravity Z descends over $\frac{\zeta}{\lambda}h$, a height to which it will therefore ascend again. Huygens now imagines that at the moment of passing the lowest position, all the linkages between the parts are severed. These parts then have velocities

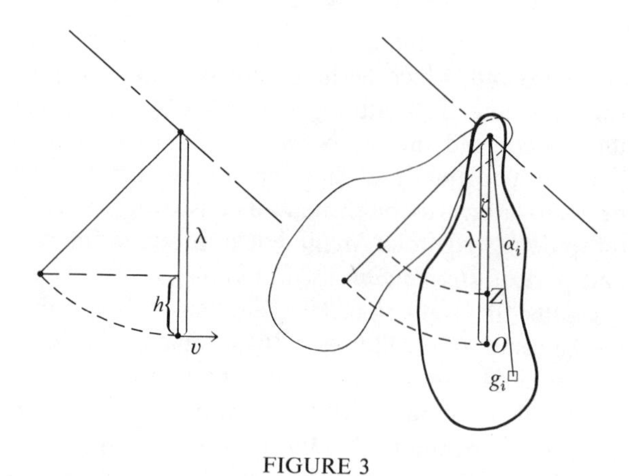

FIGURE 3

$v_i = \frac{\alpha_i}{\lambda} v$, with which they can, when directed upward, ascend to heights h_i. Now according to Galileo's law of falling bodies, v_i^2 is proportional to h_i; velocity v corresponds to height h, so that

$$h_i = \frac{v_i^2}{v^2} h - \frac{\alpha_i^2}{\lambda^2} h.$$

If all the parts are directed upward and arrested at their highest positions, the center of gravity will be at height $\frac{\Sigma g_i h_i}{\Sigma g_i}$; the second hypothesis asserts that this height is equal to $\frac{\zeta}{\lambda} h$.

Thus, $$\frac{\zeta}{\lambda} h = \frac{\Sigma\, g_i \frac{\alpha_i^2}{\lambda^2} h}{G},$$

with $$G = \Sigma g_i,$$

hence $$G\zeta\lambda = \Sigma g_i \alpha_i^2.$$

This, then, is Huygens' general computation rule for the center of oscillation. More recently, the final term $\Sigma g_i \alpha_i^2$, rendered as $\Sigma m_i \alpha_i^2$, has been called the "moment of inertia," but Huygens did not give it a separate name. Huygens determined the centers of oscillation of compound pendulums of many types; he applied complicated geometrical transformations to interpret $\Sigma g_i \alpha_i^2$ as being a quadrature, a cubature, or dependent on the center of gravity of certain curvilinear areas or bodies. He also derived the general theorem which asserts that with respect to different parallel oscillation axes of one pendulum, the product $\zeta(\lambda\text{-}\zeta)$ is constant and that, consequently, if the center of oscillation and the axis are interchanged, the period remains the same.

In the fourth part of *Horologium oscillatorium,* Huygens also discussed the possibility of defining a universal measure of length by using the length of a simple pendulum having a period of one second, an idea he had first developed in 1661. The advantage of such a method of measurement is that it is not affected in the case of bodies subject to wear or decay, while the theory of the center of oscillation makes it easy to verify the measure itself. In this connection Huygens again mentioned the relation between period and time of fall along the pendulum length, which he had determined as being equivalent to $T = 2\pi\sqrt{l/g}$. He does not, however, touch upon the possibility that the acceleration of free fall is dependent on the geographical position because of the centrifugal force of the earth's rotation. Strangely, he had in 1659 already recognized this possibility, which invalidates his definition of a universal measure of length. But he apparently did not think that the effect occurred in reality, a view which he sustained even after having heard about Richer's observations in Cayenne; indeed, it was only by reports on experiments in 1690–1692 that Huygens was convinced of the actual occurrence of this effect.

Centrifugal Force. In 1659 Huygens collected in a manuscript, *De vi centrifuga* [19] (1703), the results of his studies on centrifugal force, which he had taken up in that year in his investigations on the cause of gravity. He published the most important results, without proofs, in *Horologium oscillatorium.* The fundamental concept in Huygens' treatise is the conatus of a body, which is its tendency to motion and the cause of the tension in a cord on which the body is suspended or on which it is swung around. The conatus of a body is measured by the motion that arises if the restraints are removed; that is, in the case of bodies suspended or swung, if the cords are cut. If these motions are similar—if, for instance, both are uniformly accelerated—then the two conatus are similar and therefore comparable. If the motions that arise are the same, then the two conatus are equal.

Huygens showed that for bodies suspended on cords and situated on inclined planes, the conatus, measured in this way, are indeed proportional to the forces which the theory of statics assigns in these cases. He remarked that the motions arising when the restraints are removed must be considered for only a very short interval after this removal, since a body on a curved plane has the same conatus as a body on the corresponding tangent plane; this obtains although the motions which they would perform are approximately the same only in the first instants after release. What was probably the most important result of this study for Huygens himself was his conclusion that centrifugal force and the force of gravity are similar, as is evidenced by the property of horizontal circular motion. After the cutting of the cord, the body will proceed along the tangent with a uniform motion, so that with respect to an observer participating in the circular motion, it will recede in the direction of the cord; it will recede in such a way that, in subsequent equal short-time intervals, the distance between observer and body will increase with increments approximately proportional to the odd numbers $1, 3, 5, \cdots$.

The motion of the swung body when released is thus similar to the motion of free fall, and the conatus of suspended and swung bodies are therefore similar and comparable. Huygens compared them by calculating for a given radius (length of cord) r, the velocity v with which a body must traverse the horizontal circle to cause in its cord the same tension as if it were suspended from it. For this to be the case, the spaces traversed in subsequent equal short increments of time in free fall and in release from circular motion must be the same (that is, the conatus must be the same).

Using the law of falling bodies in the form of the relation $v(t) = 2s(t)/t$, it can be deduced that the required velocity v must be the velocity acquired by a body after free fall along distance $s = r/2$. Huygens then deduced from geometrical arguments that the centrifugal conatus is proportional to the square of the velocity and inversely proportional to the radius. These results were later summarized in the formula $F = \frac{mv^2}{r}$—which formula, however, differs significantly in its underlying conceptions from Huygens' result, since its standard derivation involves a measure of the force of gravity by the Newtonian expression mg and since it assimilates centrifugal to gravitational force by the common measure involving the second derivative of the distance–time function. In Huygens' treatment, the notion of "acceleration" as a measurable quantity is entirely absent, and the similarity of the two different forces is a demonstrandum rather than an axiom.

Fall and Projectiles. In the second part of *Horologium oscillatorium* Huygens gave a rigorous derivation of the laws of unresisted descent along inclined planes and curved paths, these being the laws which he applied in his proof of the tautochronism of the cycloid. In this derivation he made use of an earlier investigation (1646), in which he had dealt with Galileo's law of falling bodies, by considering that such a law has to be scale-free. He also made use of a study of 1659 in which he derived the law of falling bodies from the principles of relativity of motion.

In 1659 Huygens also made experiments concerning the distance which a freely falling body traverses from rest over a period of one second. This is the form in which the physical constant now indicated by the gravitational acceleration g occurs in the work of Huygens and his contemporaries. By means of the relation between period and length of the simple pendulum, derived in the same year, he found for this distance the value of fifteen Rhenish feet, seven and one-half inches, which is very close to the correct value. Huygens published this result in *Horologium oscillatorium,* part 4.

In 1668 Huygens studied fall and projectile motion in resisting media, a subject on which he had already made short notes in 1646 and 1659. He supposed the resistance, that is, the change of velocity induced by the medium in a short time interval, to be proportional to the velocity. By considering a figure in which the velocity was represented by an area between a time axis and a curve, Huygens was able to interpret vertical segments of the area perpendicular to the axis as the changes in velocity in the corresponding time interval. These changes are calculated as combinations of the acceleration of gravity and the deceleration by the medium. Thus a certain relation between the area and ordinates of the curve is known, and Huygens recognized this relation as a property of the logarithmica which he had studied extensively in 1661. In that way he found the velocity-time relation (and consequently the distance-time relation) in this type of retarded motion without having explicitly introduced acceleration as a distinct quantity.

But by 1669 Huygens had become convinced by experiments that the resistance in such media as air and water is proportional to the square of the velocity. This induced him to make a new theoretical study of motion in resisting media. Huygens derived a property of the tangents of the curve which represented the velocity-time relation in this case. The determination of the curve was now a so-called inverse tangent problem (equivalent to a first-order differential equation). Huygens reduced it to certain quadratures, but no solution as simple as that for the other case of resistance could be found. Huygens published these results in 1690 in a supplement to the *Discours.*

Concepts of Force. Huygens' study of resisted motion shows that, although he did not accept a Newtonian force concept as a fundamental mechanical principle, he was quite able to perform complicated calculations in which this concept occurs implicitly. In that study, however, he left undiscussed the question of the cause of the forces. His researches on harmonic oscillation (1673–1674) illustrate how unnatural it was for Huygens to disregard this question. Huygens' starting point was the tautochronism of the cycloid. He remarked that a force directed along the tangent, which can keep a body at a certain point P on a cycloid in equilibrium, is proportional to the arc length between P and the vertex of the cycloid. He concluded from this that, in general, if the force exerted on a body is proportional to its distance to a certain center and directed toward that center, the body will oscillate tautochronously (that is, harmonically) around that center.

Before coming to this conclusion, however, Huygens stated emphatically that in such an instance the force exerted has to be independent of the velocity of the body (otherwise the property of the force in the case of the cycloid cannot be extended to the case of bodies moving along the curve). He added that this condition of independence will be satisfied if the agent that causes the force (gravity, elasticity, or magnetism, for example) has infinite or very great velocity. This argument appears again in his studies on the cause of gravity. He also expressly formulated the hypothesis that equal forces produce equal motions regardless of their causes. Only under these presuppositions could Huygens accept the conclusion that proportionality of force and distance yields harmonic oscillation. He applied the argument to springs and torsion balances, and he designed numerous ingenious apparatus for tautochronous balances for clocks. He further studied in this connection the vibration of strings.

Huygens also took a critical position toward Leibniz' concept of force. Although in his collision theory he had found that the sum of the products of the quantity of matter and the square of the velocity is conserved, he did not consider mv^2 to be the quantification of a fundamental dynamical entity (what Leibniz called *vis viva*). In Huygens' opinion, Leibniz failed to prove both the existence of a constant *vis viva* and the proportionality of this entity to mv^2. On the other hand, Huygens liked the idea that a force, or "power to lift," is conserved in mechanical systems, as is indicated by a note in his manuscripts of 1693. This is not surprising since the principle on which most of his mechanical theories are founded—namely, that the center of gravity of a mechanical system cannot rise of its own force—can be shown to be equivalent to the principle of conservation of energy. In support of his principle, Huygens sometimes argued that a mechanical *perpetuum mobile* would otherwise be possible, a conclusion he considered absurd. This view is understandable in its turn because (as we have seen) so many of Huygens' basic ideas in mechanics derived from the pendulum and from the Galilean notion of constrained fall.

Mechanistic Philosophy. Huygens' studies on light and gravity (as well as his few researches on sound, magnetism, and electricity) were strongly influenced by his mechanistic philosophy of nature. In the preface of his *Traité de la lumière,* Huygens described a "true philosophy" as one "in which one conceives the causes of all natural effects by reasons of mechanics." In his view, the motions of various particles of matter and their interactions by direct contact are the only valid starting points for philosophizing about natural phenomena. In this he was following Descartes, and if one wants to view this as the essence of Descartes's thought, then Huygens may be called a Cartesian.

There are marked differences between Huygens and Descartes in the actual working out of this philosophy, however. Of these, the most important is that Huygens rejected Descartes's complete trust in the power of reason to attain truth. Complete certainty, according to Huygens, cannot be achieved in the study of nature, although there are degrees of probability; the determination of these degrees requires that the philosopher use good sense. Huygens assigned a most important role to experience and experiment in the discovery and verification of theoretical explanations. He also accepted the intercorpuscular vacuum—in regard to which his philosophy is nearer to Gassendi's than to Descartes's.

According to Huygens, the particles of matter move in the vacuum. These particles are homogeneous, being one kind of matter and differing from each other only in shape and size. The *quantitas materiae* is therefore proportional to the content of the particles or, equivalently, to the space occupied by them. The weight of ordinary bodies is proportional to their *quantitas materiae* because the collisions of ethereal particles that cause gravity have effects proportional to the magnitudes of the colliding particles. This may be considered to mark one of the first insights into the difference between mass and weight.

Huygens explained differences in specific gravity of ordinary bodies as differences in the density of matter. The great variety of specific gravities in nature led him to suppose large interspaces, or "pores," between the component particles of bodies and to attribute an important role to the forms of these interspaces. In Huygens' view the particles are completely hard and, in collision, completely elastic. They are indivisible and keep the form in which they were created. They move in right lines or, in the case of vortices, in circles; they can influence each other's motions only by direct contact.

Huygens' mechanistic explanations of natural phenomena thus consisted in showing that, given a certain combination of shapes, magnitudes, number, and velocities of particles, processes occur which manifest themselves macroscopically as the phenomena under consideration.

In the course of working out a pattern of size relations between particles, Huygens came to the conclusion that four or five discrete classes of particles exist. Particles of the same class are approximately equal in form and magnitude. The classes are differentiated by the magnitudes of the particles, those of one class being much smaller than those of the preceding class and much larger than those of the class following.

The particles of the first class are the components of the ordinary bodies and of the air. They move slowly, and Huygens used suppositions about their forms in his explanations of cohesion and coagulation. He considered sound to be vibrations in ordinary bodies and in the air. The particles of the second class form the "ether," and the phenomena of light may be explained by shock waves in this medium. In some ordinary bodies, the spaces between the particles of the first class are so formed that the ether particles can traverse them freely: these bodies are transparent. The particles of the third class are the carriers of magnetic phenomena, and those of the fourth class form the "subtle matter" which causes gravity. (It is not clear whether Huygens supposed a fifth class between the third and the fourth classes to account for electrical phenomena.) Particles of the fourth class move very rapidly in circular paths around the earth; they are so small that they can pass through the "pores" of all ordinary bodies and are scarcely hindered by the particles of the other classes. In Huygens' explanation of gravity as caused by the motion of these particles, as well as in his explanation of magnetism, the concept of vortex plays a fundamental role.

Huygens' adherence to a strongly geometrical approach to problems in infinitesimal mathematics prevented him from making the definitive innovations in the infinitesimal calculus that Newton and Leibniz did. Similarly, his strict adherence to mechanistic principles prevented his achieving results in mechanics comparable to Newton's revolutionary work. Huygens immediately realized the importance of Newton's *Principia,* but he also strongly opposed Newton's use of attractive force as a fundamental explanatory principle. Force, in the Newtonian sense, could never count as a fundamental mechanical principle for Huygens. The occurrence of such forces always required a further, mechanistic explanation for him.

It is important to emphasize the role of Huygens' mechanistic vision in his studies and the reasons which led him to defend this vision so strongly against Newton. First of all, it is remarkable that in Huygens' early work the mechanistic point of view is of importance only as a source of inspiration rather than as a principle of explanation. The special hypotheses on which Huygens based his studies on collision, centrifugal force, motion of pendulums, and statics were not substantiated by mechanistic arguments, nor did Huygens seem to think this should be done. There is no mechanistic philosophy in the *Horologium oscillatorium.*

It would seem that only after his removal to Paris (1666) did Huygens come to emphasize strongly the necessity for strict mechanistic explanations and to combat the supposition of occult qualities—among which he counted attraction—that some of the members of the Academy applied rather freely. His most important reason for taking this position was, no doubt, that he simply could not accept a phenomenon as properly explained if he could not imagine a mechanistic process causing it. As further reasons we must consider the impressive results that he gained precisely by applying this mechanistic point of view. Huygens' discovery of Saturn's ring was directly connected with the vortex theories; and his study of centrifugal force, which showed that the centrifugal tendency (conatus) of particles moving in circles is indeed similar to the centripetal tendency of heavy bodies, supported the explanation of gravity as the effect of a vortex. Finally, Huygens formulated the wave theory of light, which constituted a mechanistic explanation of refraction and reflection, and which he applied in a masterly fashion to the refractive properties of Iceland spar.

The publication, in 1690, of the *Traité de la lumière* and its supplement, the *Discours,* must be seen as Huygens' answer to Newton's *Principia.* In these works Huygens opposed his mechanistic philosophy to Newton's *Philosophia naturalis.* The wave theory of light and its application to the refraction in Iceland spar are an effective mechanistic explanation of natural phenomena, equal in mathematical sophistication and elegance to Newton's explanation of the motion of the planets. Huygens' explanation of gravity dealt with fundamental problems that Newton avoided and left unsolved. Finally, Huygens' treatment of motion in resisting media proved that he could achieve the same results as Newton in this difficult subject although with different methods.

Wave Theory of Light. Light, according to Huygens, is an irregular series of shock waves which proceeds with very great, but finite, velocity through the ether. This ether consists of uniformly minute, elastic particles compressed very close together. Light, therefore, is not an actual transference of matter but rather of a "tendency to move," a serial displacement similar to a collision which proceeds through a row of balls. Because the particles of the ether lie not in rows but irregularly, a colliding particle will transfer its tendency to move to all those particles which it touches in the direction of its motion. Huygens therefore concluded that new wave fronts originate around each particle that is touched by light and extend outward from the particle in the form of hemispheres. Single wave fronts originating at single points are infinitely feeble; but where infinitely many of these fronts overlap, there is light—that is, on the envelope of the fronts of the individual particles. This is "Huygens' principle."

About 1676 Huygens found the explanation of reflection and refraction by means of this principle; his theory connected the index of refraction with the velocities of light in different media. He became completely convinced of the value of his principle on 6 August 1677, when he found the explanation of the double refraction in Iceland spar by means of his wave theory. His explanation was based on three hypotheses: (1) There are inside the crystal two media in which light waves proceed. (2) One medium behaves as ordinary ether and carries the normally refracted ray. (3) In the other, the velocity of the waves is dependent on direction, so that the waves do not expand in spherical form, but rather as ellipsoids of revolution; this second medium carries the abnormally refracted ray. By studying the symmetry of the crystal Huygens was able to determine the direction of the axis of the ellipsoids, and from the refraction properties of the abnormal ray he established the proportion between the axes. He also calculated the refraction of rays on plane sections of the crystal other than the natural crystal sides, and verified all his results experimentally.

Although the completeness of Huygens' analysis is impressive, he was unable to comprehend the effect that we now recognize as polarization, which occurs if the refracted ray is directed through a second crystal of which the orientation is varied. Huygens described this effect in his first studies on the crystal, but he could never explain it. These results are included in the *Traité de la lumière,* which was completed in 1678; Huygens read parts of it to the Academy in 1679.

Gravity. Huygens' explanation of gravity developed the ideas of Descartes. He presupposed a vortex of particles of subtle matter to be circling the earth with great velocity. Because of their circular movement these particles have a tendency (conatus) to move away from the earth's center. They can follow this tendency if ordinary bodies in the vortex move toward the center. The centrifugal tendency of the vortex particles thus causes a centripetal tendency in ordinary bodies, and this latter tendency is gravity. The space which a body of matter vacates, under the influence of gravity, can be taken by an equal quantity of subtle matter. Hence the gravity of a body is equal to the centrifugal conatus of an equal quantity of subtle matter moving very rapidly around the earth.

This argument led Huygens to study centrifugal force in 1659. In his investigations he proved the similarity of the centrifugal and the gravitational conatus, a result that strengthened his conviction of the validity of the vortex theory of gravity. The study also enabled him to work out this theory quantitatively, since given the radius of the earth and the acceleration of gravity he could calculate the velocity of the particles; he found that they circle the earth about seventeen times in twenty-four hours.

Huygens developed this theory further in a treatise presented to the Academy in 1669. Since the cylindrically symmetrical vortices posited by Descartes could explain only a gravity toward the axis, Huygens imagined a multilaterally moving vortex—in which the particles circle the earth in all directions—by which a truly centrally directed gravity could be explained. The particles are forced into circular paths because the vortex is held within a sphere enveloping the earth, and bounded by "other bodies," such that the particles cannot leave this space. The boundary of the gravitational vortex was supposed to be somewhere between the earth and the moon, because Huygens thought the moon to be carried around the earth by a uniaxial vortex (the so-called *vortex deferens*). Later, convinced by Newton of the impossibility of such vortices, he supposed the gravity vortex to extend beyond the moon.

Galileo's law of falling bodies requires that the acceleration which a falling body acquires in a unit of time be independent of the velocity of the body. This independence is the greatest obstacle for any mechanistic explanation of gravity, for the accelerations must be acquired during collisions, but the change of velocity of colliding bodies is dependent on their relative velocities. On this problem Huygens argued that, because the velocity of the vortex particles is very great with respect to the velocity of the falling body, their relative velocity can be considered constant. Thus, in effect he argued that Galileo's law of falling bodies holds only approximately for small velocities of the falling body.

Huygens never discussed the fundamental question

raised by this explanation of gravity—namely, how, by means of collisions, a centrifugal tendency of the particles of the subtle matter can transfer a centripetal tendency to heavy bodies.

In the *Discours,* the treatise of 1669 is reiterated almost verbatim, but Huygens added a review of Newton's theory of gravitation, which caused him to revise his own theories somewhat. He resolutely rejected Newton's notion of universal attraction, because, as he said, he believed it to be obvious that the cause of such an attraction cannot be explained by any mechanical principle or law of motion. But he was convinced by Newton of the impossibility of the *vortices deferentes,* and he accepted Newton's explanation of the motion of satellites and planets by a force varying inversely with the square of the distance from the central body. According to Huygens, however, this gravity is also caused by a vortex, although he did not dwell on the explanation of its dependence on the distance.

Cosmotheoros. Huygens did not believe that complete certainty could be achieved in the study of nature, but thought that the philosopher must pursue the highest degree of probability of his theories. Clearly Huygens considered this degree to be adequate in the case of his explanations of light and gravity. It is difficult for the historian to assert how plausible, in comparison with those explanations, Huygens considered his theories about life on other planets and about the existence of beings comparable to man. These theories were expounded in his Κοσμοθέωρος, *sive de terris coelestibus, earumque ornatu, conjecturae* [14].

The argument of the book is very methodically set forth, and its earnestness suggests that Huygens did indeed assign a very high degree of probability to these conjectures. Huygens' reasoning is that it is in the creation of life and living beings that the wisdom and providence of God are most manifest. In the Copernican world system—which is sufficiently proved as agreeing with reality—the earth holds no privileged position among the other planets. It would therefore be unreasonable to suppose that life should be restricted to the earth alone. There must be life on the other planets and living beings endowed with reason who can contemplate the richness of the creation, since in their absence this creation would be senseless and the earth, again, would have an unreasonably privileged position. In further discussion of the different functions of living organisms and rational beings, Huygens came to the conclusion that, in all probability, the plant and animal worlds of other planets are very like those of the earth. He also surmised that the inhabitants of other planets would have a culture similar to man's and would cultivate the sciences.

In the second part of *Cosmotheoros,* Huygens discussed the different movements of the heavenly bodies and how they must appear to the inhabitants of the planets. He took the occasion to mention new advances in astronomy. In contrast to most other Huygensian writings, *Cosmotheoros* has had wide appeal and a broad readership, and has been translated into several languages.

Conclusion. In the period bounded on one side by Viète and Descartes and on the other by Newton and Leibniz, Huygens was Europe's greatest mathematician. In mechanics, in the period after Galileo and before Newton, he stood for many years on a solitary height. His contributions to astronomy, time measurement, and the theory of light are fundamental, and his studies in the many other fields to which his universal interest directed him are of a very high order.

But Huygens' work fell into relative oblivion in the eighteenth century, and his studies exerted little influence. There is thus a marked discrepancy between Huygens' actual stature as a natural philosopher and the influence he exerted. This is due in part to his extreme reluctance to publish theories which he considered insufficiently developed or which did not meet his high standards of adequacy and significance. For this reason his work on hydrostatics, collision, optics, and centrifugal force were published too late to be fully influential. It is also clear that Huygens did not attract disciples: he was essentially a solitary scholar.

Other reasons for Huygens' limited influence must be sought in the character of his work. His infinitesimal-geometrical mathematics and his studies in mechanics and the theory of light, inspired by his mechanistic philosophy, were culminations that defined limits rather than opening new frontiers. Even his early studies in mechanics, based on hypotheses that we can recognize as equivalent to conservation of energy, served as a basis for later work to only a limited extent—although it is true that one may consider the eighteenth-century researches in mechanics, so far as they were centered around the Leibnizian concept of *vis viva,* to be continuations of Huygens' approach. The Newtonian notion of force became the fundamental concept in mechanics after publication of the *Principia;* Huygens' work could not easily be incorporated into this new mechanics, and it was only much later that the two different concepts could be synthesized.

Huygens' work nonetheless forms a continuously impressive demonstration of the explanatory power of the mathematical approach to the study of natural phenomena, and of the fertility of its application to the technical arts. His magnum opus, *Horologium oscillatorium,* stands as a solid symbol of the force of the mathematical approach and was recognized as such by Huygens' contemporaries. Compared to the relatively simple mathematical tools which Galileo used in his works, the wealth of mathematical theories and methods that Huygens was able to apply is significant, and herein lies the direct and lasting influence of his work.

BIBLIOGRAPHY

I. Original Works. For a complete list of the works of Huygens which appeared before 1704, see *Oeuvres* XXII, 375–381 (see below). Here we recapitulate the writings discussed above:

1. *Theoremata de quadratura hyperboles, ellipsis et circuli ex dato portionum gravitatis centro, quibus subjuncta est 'Εξέτασις Cyclometriae Cl. Viri Gregorii à St. Vincentio,* Leiden, 1651 (*Oeuvres* XI).

2. *De circuli magnitudine inventa. Accedunt ejusdem problematum quorundam illustrium constructiones,* Leiden, 1654 (*Oeuvres* XII).

3. *De Saturni lunâ observatio nova,* The Hague, 1656 (*Oeuvres* XV).

4. *Tractatus de ratiociniis in aleae ludo,* in F. van Schooten, *Exercitationum mathematicarum libri quinque,* Leiden, 1657 (Latin trans. of [7] by van Schooten).

5. *Horologium,* The Hague, 1658 (*Oeuvres* XVII).

6. *Systema Saturnium, sive de causis mirandorum Saturni phaenomenôn, et comite ejus planeta novo,* The Hague, 1659 (*Oeuvres* XV).

7. *Tractaet handelende van Reeckening in Speelen van Geluck,* in F. van Schooten, *Mathematische Oeffeningen begrepen in vijf boecken,* Amsterdam, 1660 (also published separately in the same year; *Oeuvres* XIV).

8. *Kort onderwijs aengaende het gebruyck der Horologien tot het vinden der Lenghten van Oost en West,* 1665 (*Oeuvres* XVII).

9. *Règles du mouvement dans la rencontre des corps,* in *Journal des sçavans,* 1669 (*Oeuvres* XVI).

10. *Horologium oscillatorium, sive de motu pendulorum ad horologia aptato demonstrationes geometricae,* Paris, 1673 (*Oeuvres* XVIII); a German trans. in the series *Ostwald's Klassiker der Exakten Wissenschaften,* no. 192 (Leipzig, 1913).

11. *Astroscopia compendiaria, tubi optici molimine liberata,* The Hague, 1684 (*Oeuvres* XXI);

12. *Traité de la lumière, où sont expliquées les causes de ce qui lui arrive dans la Reflexion & dans la Refraction, et particulièrement dans l'étrange Refraction du Cristal d'Islande.* (*Avec un Discours de la Cause de la Pesanteur*), Leiden, 1690 (*Oeuvres* XIX); there is a German trans. in *Ostwald's Klassiker,* no. 20 (Leipzig, 1903).

13. *Discours de la cause de la Pesanteur* appears in [12] (*Oeuvres* XXI).

14. *Κοσμοθεωρος, sive de terris coelestibus, earumque ornatu, conjecturae,* The Hague, 1698 (*Oeuvres* XXI).

15. B. de Volder and B. Fullenius, ed., *Christiani Hugenii Opuscula Posthuma* (Leiden, 1703).

16. *Tractatus de refractione et telescopiis,* MS originating from 1653, was later changed and amplified many times. One version is published under the title *Dioptrica* in the Volder and Fullenius edition and another version in *Oeuvres* XIII.

17. *Memorien aengaende het slijpen van glasen tot verrekijckers,* MS originating from 1685, published in *Oeuvres* XXI. A Latin trans. was published in Volder and Fullenius.

18. *De motu corporum ex percussione,* MS originating from 1656, published in *Oeuvres* XVI. A German trans. appeared in *Ostwald's Klassiker,* no. 138 (Leipzig, 1903).

19. *De vi centrifuga,* MS originating from 1659, published in *Oeuvres* XVI, a German trans. existing in *Ostwald,* no. 138, Leipzig, 1903. Like [18] this is also found in Volder and Fullenius.

20. *De iis quae liquido supernatant,* MS originating from 1650, appears in *Oeuvres* XI.

In his will, Huygens asked Volder and Fullenius to edit some not yet published MSS, which resulted in their posthumous edition [15].

Two further publications of Huygens' writings, edited by G. J. 'sGravesande, are [21] *Christiani Hugenii Opera Varia* (Leiden, 1724) and [22] *Christiani Hugenii Opera Reliqua* (Leiden, 1728). Little more than a century later, P. J. Uylenbroek edited Huygens' correspondence with L'Hospital and Leibniz in [23] *Christiani Hugenii aliorumque seculi XVII virorum celebrium exercitationes mathematicae et philosophicae* (The Hague, 1833).

In 1882, the Netherlands Academy of Sciences at Amsterdam organized a preparatory committee for a comprehensive ed. of Huygens' works. In 1885 it was agreed that the Society of Sciences of Holland at Haarlem would take responsibility for the publication. The undertaking resulted, after more than sixty years of editorial commitment, in what may be considered the best edition of the works of any scientist, the *Oeuvres complètes de Christiaan Huygens, publiées par la Société Hollandaise des Sciences,* 22 vols. (The Hague, 1888–1950).

The first ten vols. comprise Huygens' correspondence, the subsequent ones his published and unpublished scholarly writings, of which the most important are accompanied by a French trans. Vol. XXII contains a detailed biography of Huygens by J. A. Vollgraff.

The editors in chief were, successively, D. Bierens de Haan, J. Bosscha, D. J. Korteweg, and J. A. Vollgraff. Among the many collaborators, C. A. Crommelin, H. A. Lorentz, A. A. Nijland, and E. J. Dijksterhuis may be mentioned. The editors adopted a strict code of anonymity, which was broken only in the last volume.

II. Secondary Literature. While Huygens' work is easily accessible in the *Oeuvres,* there exists relatively little secondary literature about him. We may mention [24]

P. Harting, *Christiaan Huygens in zijn leven en werken geschetst* (Groningen, 1868); [25] H. L. Brugmans, *Le séjour de Christiaan Huygens à Paris et ses relations avec les milieux scientifiques français, suivi de son journal de voyage à Paris et à Londres* (Paris, 1935); and [26] A. Romein-Verschoor, "Christiaen Huygens, de ontdekker der waarschijnlijkheid," in *Erflaters van onze beschaving* (Amsterdam, 1938–1940), written with J. Romein.

The only recent separately published scientific biography of Huygens is [27] A. E. Bell, *Christian Huygens and the Development of Science in the Seventeenth Century* (London, 1947). On the occasion of the completion of the *Oeuvres* edition, there appeared [28] E. J. Dijksterhuis, *Christiaan Huygens* (Haarlem, 1951).

J. A. Vollgraff, who by editing the last seven vols. of the *Oeuvres* acquired a thorough knowledge of Huygens' life and works, has written a book about Huygens which has not been published. The private typescript will be transferred to the Leiden University Library.

H. J. M. Bos.

HYATT, ALPHEUS (*b.* Washington, D.C., 5 April 1838; *d.* Cambridge, Massachusetts, 15 January 1902), *invertebrate paleontology, zoology.*

Alpheus Hyatt, an influential evolutionist and co-founder of the neo-Lamarckian theory, was the descendant of an old Maryland family. After a year at Yale, he went to Harvard in 1858 to study with Louis Agassiz. He graduated from the Lawrence Scientific School of Harvard in 1862. He married Ardella Beebe (1867) and after serving in the Union army during the Civil War became professor of zoology and paleontology at the Massachusetts Institute of Technology, a position he held until 1888. In 1877 he was appointed professor of biology at Boston University and remained there until his death. He became custodian in 1870 and curator in 1881 of the Boston Society of Natural History. In 1875 Hyatt was elected to the National Academy of Sciences. He founded the marine laboratory at Annisquam, Massachusetts (later moved to Woods Hole), and was a co-founder of the American Society of Naturalists.

Although he was primarily a prolific specialist on the systematics and evolution of ammonoids ("Genesis of the Arietidae" [1889]) Hyatt also published extensive works on gastropods and bryozoans and wrote an important treatise on North American sponges (1877). He was working on the evolution and zoogeography of Hawaiian tree snails at the time of his death. It is as an evolutionary theorist, however, that he is best known. He was among the gifted group of students who broke with their mentor Agassiz and embraced evolutionary theory soon after 1859.

Hyatt was not a Darwinian. He granted natural selection an executioner's role in removing the unfit, but he did not see how it could create the fit. Moreover, he thought he could detect repeated patterns of directed change in the fossil record that could not be the result of adaptation to changing environments. He believed that evolution could lead to increasing complexity of organization only if variation were intrinsically directed toward advantageous states (rather than being random in direction, as the Darwinians thought).

To produce this variation, he accepted the Lamarckian postulate that organisms could pass on to their offspring the advantageous characters that they had acquired during their lifetimes. Hyatt and the vertebrate paleontologist E. D. Cope were the leading exponents of this so-called neo-Lamarckian school. They believed that most important new characters arose from the mechanical activity of animals themselves (for example, that the astragalus of even-toed ungulate animals developed from pressures of contact in sustained running) and that this is why structure is so well adapted (in an engineer's sense) to function. Hyatt's extended argument (1894) for the origin of the ammonites' "impressed zone"—that it arose from pressures of contact with its own outer whorls—was surely the most influential case ever made for this belief.

It is often said that the neo-Lamarckians accepted only this side of Lamarckism, rejecting or ignoring Lamarck's perfecting principle and his distinction between vertical progress up the ladder of life and horizontal side-branches as adaptations to specific environments (eyeless moles and long-necked giraffes). This interpretation is not correct. Both Hyatt and Cope distinguished progressive evolution, which they regarded as the addition of stages to an ancestral ontogeny, from specific alterations of existing ontogenies. The mechanism of addition, and therefore of evolutionary progress, is the principle of recapitulation. Cope and Hyatt both formulated this principle independently in 1866, the same year that Haeckel announced it in his *Generelle Morphologie der Organismen.* (While the evolutionary interpretation was new, the principle dates back to ancient Greek science.) The addition of new stages depends upon a "law of acceleration" that "makes room" for them by shortening ancestral ontogenies. The law of acceleration operates continuously to transfer the adult stages of ancestors to earlier and earlier steps of a descendant's ontogeny (with new steps being added at the end of growth). Thus, the sequence of embryonic stages parallels the sequence of ancestral adults, and phylogeny can be read from ontogeny. Hyatt used this principle of recapitulation (often incautiously as an absolute a priori) to reconstruct the history of ammonoids.

But since it is not natural selection, what determines the sequence of new stages in phylogeny? In attempting to answer this question, Hyatt made his most imaginative and original contribution to evolutionary thought—his "old age" theory (see especially his 1880 work). Species, as individuals, have a determined cycle of youth, maturity, and old age leading to extinction. Early in its history, a species adds the vigorous features of its phyletic youth and prospers. Later it adds the degenerate features of its phyletic senescence (the incorporation of inadaptive states, an anti-Darwinian tenet) and eventually succumbs. This theory of "racial senescence" was fairly popular, especially among paleontologists, until the formulation in the 1930's of the "modern synthesis" of evolutionary theory.

BIBLIOGRAPHY

See "On the Parallelism Between the Different Stages of Life in the Individual and Those in the Entire Group of the Molluscous Order Tetrabranchiata," in *Memoirs of the Boston Society of Natural History,* **1** (1866), 193-209; "Revision of the North American Poriferae," *ibid.,* **2** (1875-1877), 399-408, 481-554; "The Genesis of the Tertiary Species of *Planorbis* at Steinheim," in *Anniversary Memoirs of the Boston Society of Natural History* (1880); "Genesis of the Arietidae," in *Memoirs of the Museum of Comparative Zoology at Harvard College,* **16** (1889); "Phylogeny of an Acquired Characteristic," in *Proceedings of the American Philosophical Society,* **32** (1894), 349-647.

An obituary notice by W. K. Brooks is in *Biographical Memoirs. National Academy of Sciences,* **6** (1909), 311-325.

STEPHEN JAY GOULD

HYLACOMYLUS. See **Waldseemüller, Martin.**

HYLLERAAS, EGIL ANDERSEN (*b.* Engerdal, Norway, 15 May 1898; *d.* Oslo, Norway, 28 October 1965), *physics.*

Hylleraas (the name is taken from the farm where he was born) was the son of Ole Andersen, a schoolteacher, and the former Inger Rømoen. The youngest of eleven children, he grew up in the rural community of Engerdal. Following elementary school he worked for a few years as a logger. In 1918 he entered the University of Christiania (now Oslo), where he studied mathematics and physics. After his graduation in 1924 he worked for two years as a high school teacher in Oslo. Articles on double refraction in monoaxial crystals earned him a fellowship that enabled him to spend 1926-1928 in Göttingen, working under Max Born. These were the decisive years in the formation of quantum mechanics, and the ideas and challenges that faced him in this period determined the course of Hylleraas' entire scientific career. The next two years were spent partly in Oslo and partly in Göttingen, and in 1931 Hylleraas was made a member of the Christian Michelsen Institute in Bergen. In 1937 he followed Vilhelm Bjerknes as professor of theoretical physics at the University of Oslo, a chair he still occupied at his death. After World War II he was one of the Norwegian representatives at the Nordisk Institut for Teoretisk Atomfysikk (NORDITA) and in the Centre Européen de la Recherche Nucléaire (CERN). He spent 1947-1948 and 1962-1963 in the United States, at Princeton and the University of Wisconsin. Vigorous and hardworking until the day of his death, Hylleraas died of a heart attack in 1965.

Aside from Sommerfeld's *Atombau und Spektrallinien,* which Hylleraas called "our student bible," it was Born's *Dynamik der Kristallgitter* that had the strongest influence on his early development. When he went to Göttingen in 1926, his intention was to continue his work in crystal lattice theory, which he in fact did for some time. Yet by 1926 Born had already moved into the new field of quantum mechanics, and it was only after some hesitation that Hylleraas followed his master. He had already earned a reputation as a very gifted mathematical physicist, and at Born's suggestion he attacked the problem of the ionization energy of the ground state of the helium atom. The Bohr-Sommerfeld theory had predicted the impossible value of about 28 electron volts, as against the experimental 24.46 electron volts; and it was thought that the helium problem would be the first real test of the Schrödinger equation. Hylleraas' method of attack was significant for two reasons: first, the variational methods he introduced were largely his own and have since become standard techniques; second, to manage the very extensive calculations he used an electric Mercedes-Euklid calculating machine. This was probably the first time that machine calculation played an important part in physics; it has since become a standard mode of scientific activity.

Hylleraas arrived at a value of 24.35 electron volts for the ionization energy; and this result was, as he put it:

> . . . greatly admired and thought of as almost a proof of the validity of wave mechanics also in the strict numerical sense. The truth about it, however, was in fact that its deviation from the experimental value by an amount of one tenth of an electron volt was on the spectroscopic scale quite a substantial quantity and might as well have been taken to be a disproof ["Reminiscences"].

In 1929 he refined his own method through the introduction of a new set of generalized coordinates and managed to achieve full agreement between theory and experiment.

Another spectacular early coup was the demonstration in 1930 of the theoretical stability of the negative hydrogen ion, although with characteristic modesty Hylleraas attributed this demonstration to Hans Bethe: "He—not I—is the father of that curious little child, the strange particle H^-, which for a while appeared to be recognized nowhere, neither in heaven nor on earth." A decade later the existence of H^- in the solar atmosphere was definitely established.

Although Hylleraas always considered the helium atom and the negative hydrogen ion his special domains, over which he never really relinquished his hegemony, he contributed heavily and fundamentally to other areas of the quantum theory of atoms, molecules, and crystals. A very fine article on the wave mechanical treatment of lithium hydride (1930) is an amazing demonstration of the power of the Schrödinger equation and has remained a tour de force of twentieth-century physics. In the following years Hylleraas extended the application of wave mechanics to beryllium, boron, and carbon, most of this in connection with the experimental work of the Swedish spectroscopist Bengt Edlén. From the years 1935–1937 there is a set of fundamental articles on the energies, potential distributions, and spectra of diatomic molecules. Starting in the late 1930's Hylleraas also contributed to nuclear physics, although most of his work in this field was never published. There are three long articles (1939–1943) on problems of tidal theory that fall outside his main area of interest. During the period 1945–1965 much of Hylleraas' work was directed to secondary activities: reorganization of the University of Oslo and of the physics program, teaching, and editing. He still managed to turn out a number of significant articles, notably on scattering, on relativistic electron theory, and on spinors. During this time either he or one of his students kept the theory of the helium atom and the hydrogen ion up-to-date with the continuing experimental refinements.

As a physicist Hylleraas never possessed a transcending genius like, for instance, that of Bohr. His ability was mathematical, and indeed, he very nearly became a mathematician. His ingenuity and tenacity in forcing the mathematical solution of problems was amazing and recalls Sommerfeld, who was one of his heroes. The elegance and usefulness of many of Hylleraas' methods—and indeed the extent to which they are now part of physics—are insufficiently appreciated. In his belief in the efficacy of a numerical and computational approach to physics, often combined with the use of calculating machines, he directly anticipated what is perhaps the main structure of modern science.

A modest, kind, and soft-spoken man, Hylleraas revealed a simplicity and humanity no doubt derived from a happy childhood, of which he spoke glowingly; and he always retained close ties with his native community. He trained two generations of theoretical physicists in Norway, and perhaps in retrospect his best efforts were directed toward teaching. He never shirked that part of his responsibility, and he worked long and fruitfully as a popularizer. In Norwegian physics he ranks second only to Bjerknes.

Hylleraas received a number of honorary degrees, memberships, and prizes. In 1963 the University of Florida arranged a symposium on atomic and molecular physics in his honor, and it was this event that probably brought him the greatest pleasure.

BIBLIOGRAPHY

I. ORIGINAL WORKS. Hylleraas' "Reminiscences From Early Quantum Mechanics of Two-Electron Atoms" is in "Proceedings of the International Symposium on Atomic and Molecular Quantum Mechanics in Honor of Egil A. Hylleraas," in *Review of Modern Physics,* **35,** no. 3 (1963), 421; his *Matematisk og Teoretisk Fysikk,* 4 vols. (Oslo, 1950–1952) is now available in an American ed., *Mathematical and Theoretical Physics* (New York, 1970).

II. SECONDARY LITERATURE. A short biography of Hylleraas, including a comprehensive bibliography, by O. K. Gjøtterud is in *Nuclear Physics,* **89** (1966), 1–10. The best biography and evaluation is H. Wergeland, "Egil A. Hylleraas 15.5. 1898–28.10.1965," in *Fra Fysikkens Verden,* **28** (1966), 1–10.

PER STRØMHOLM

HYPATIA (*b.* Alexandria, Egypt, *ca.* 370; *d.* Alexandria, 415), *mathematics, philosophy.*

Hypatia, the first woman in history to have lectured and written critical works on the most advanced mathematics of her day, was the daughter and pupil of the mathematician Theon of Alexandria. It is believed that she assisted him in writing his eleven-part treatise on Ptolemy's *Almagest* and possibly in formulating the revised and improved version of Euclid's *Elements* that is the basis of all modern editions of the work. According to Suidas she composed commentaries not only on the *Almagest* but also on Diophantus' *Arithmetica* and Apollonius' *Conic Sections.* None of them survives.

Although accurate documentation of Hypatia's

activities is lacking, it is known that she lectured in her native city on mathematics and on the Neoplatonic doctrines of Plotinus and Iamblichus and that about A.D. 400 she became head of the Neoplatonic school in Alexandria. Her classes attracted many distinguished men, among them Synesius of Cyrene, later bishop of Ptolemais. Several of his letters to Hypatia are extant. They are full of chivalrous admiration and reverence. In one he asks her how to construct an astrolabe and a hydroscope.

In spite of her association with Synesius and other Christians, Hypatia's Neoplatonic philosophy and the freedom of her ways seemed a pagan influence to the Christian community of Alexandria. Prejudice was strengthened by her friendship with Orestes, Roman prefect of the city and political enemy of Cyril, bishop of Alexandria. The mounting hostility culminated in her murder by a fanatic mob. None of her writings was preserved; but the general loss of Hellenic sources must be blamed on repeated book-burning episodes rather than on lynching. The great Alexandrian library had been burned by Roman soldiers long before Hypatia's day, and during her lifetime the valuable library in the temple of Serapis was sacked by an Alexandrian mob.

Hypatia has been the subject of much romantic drama and fiction, including the 1853 novel *Hypatia, or New Foes With an Old Face,* by Charles Kingsley. Such works have perpetuated the legend that she was not only intellectual but also beautiful, eloquent, and modest.

BIBLIOGRAPHY

See T. L. Heath, *History of Greek Mathematics,* II (Oxford, 1921), 528–529; A. W. Richeson, "Hypatia of Alexandria," in *National Mathematics Magazine,* **15,** no. 2 (Nov. 1940), 74–82; Socrates Scholasticus, *Ecclesiastical History,* VII (London, 1853), 15; *Suidae Lexicon,* Ada Adler, ed., I (Leipzig, 1928), 618; B. L. van der Waerden, *Science Awakening* (New York, 1961), 290.

EDNA E. KRAMER

HYPSICLES OF ALEXANDRIA (*fl.* Alexandria, first half of second century B.C.), *mathematics, astronomy.*

Hypsicles is attested, by the more definitive manuscripts, to be the author of what has come to be printed as book XIV of Euclid's *Elements.* In the preface to that book he states that Basilides of Tyre came to Alexandria, where he engaged in mathematical discussions with Hypsicles' father. Together they studied a tract by Apollonius of Perga on the dodecahedron and the icosahedron inscribed in the same sphere, and found the treatment unsatisfactory.

Later, presumably after his father's death, Hypsicles himself found what would appear to have been a revised, more accurate version in wide circulation. Taken together, these facts suggest that Hypsicles' father was an older contemporary of Apollonius, living at Alexandria. As Apollonius died early in the second century B.C., the middle point of Hypsicles' activities may be placed at about 175 B.C.

The so-called book XIV, like book XIII, is concerned with the inscription of regular solids in a sphere. Hypsicles proves a proposition, which he attributes to Aristaeus (who was probably not the author of *Five Books Concerning Solid Loci*), that the same circle can be described about the pentagonal face of a regular dodecahedron and the triangular face of a regular icosahedron inscribed in the same sphere. He proves, as had Apollonius before him, that the volume of the dodecahedron bears the same relation to the volume of the icosahedron as the surface of the former bears to the surface of the latter, because the perpendiculars to the respective faces are equal; and that both the ratios are equal to the ratio of the side of the inscribed cube to the side of the dodecahedron.

Arabic traditions suggest that Hypsicles also had something to do with the so-called book XV of the *Elements,* whether he wrote it, edited it, or merely discovered it. But this is clearly a much later and much inferior book, in three separate parts, and this speculation appears to derive from a misunderstanding of the preface to book XIV.

One other work by Hypsicles survives, the *Anaphorikos* (Ἀναφορικός), or *On the Ascension of Stars.* Although quite brief, probably truncated, and based on a false assumption, it is noteworthy in being the first work in which the ecliptic is divided into 360 parts or degrees. He writes,

> The circle of the zodiac having been divided into 360 equal arcs, let each of the arcs be called a spatial degree, and likewise, if the time taken by the zodiac circle to return from a point to the same point is divided into 360 equal times, let each of the times be called a temporal degree [*Die Aufgangszeiten der Gestirne* 55–59, De Falco, ed., p. 36].

This division into 360 parts was almost certainly borrowed from Babylonia, and the *Anaphorikos* is therefore testimony to the existence of links between Greek and Babylonian astronomy in the second century B.C.

Hypsicles posits for himself two problems. Given the ratio of the longest day to the shortest day at any place, how long does it take any given sign of the zodiac to rise there? Second, how long does it take any given degree in a sign to rise? The practical object of this investigation may have been, as T. L. Heath conjectures, to tell the time at night. But the problem came really within the province of spherical trigonometry, which was not developed until Hipparchus. Ptolemy later solved it with the help of his table of sines (*Syntaxis mathematica,* bk. 2, J. L. Heiberg, ed. [Leipzig, 1898], ch. 8, pp. 134–141); and Hipparchus had no doubt solved it before Ptolemy, for Pappus of Alexandria (*Collection* VI 109, Hultsch ed., 600.9–13) refers to calculations "by means of numbers" appearing in Hipparchus' book *On the Rising of the Twelve Signs of the Zodiac.* This method of solution was not open to Hypsicles, which is further confirmation of his date.

The longest day, Hypsicles says, is the time during which Cancer, Leo, Virgo, Libra, Scorpio, and Sagittarius rise (14 hours at Alexandria), and the shortest is the time in which Capricornus, Aquarius, Pisces, Aries, Taurus, and Gemini rise (10 hours); and as their ratio is 7:5, the former signs take 210 temporal degrees and the latter 150. He assumes that the quadrants Cancer-Virgo and Libra-Sagittarius take equal times to rise, 105 temporal degrees, and that the quadrants Capricornus-Pisces and Aries-Gemini each require 75 degrees. He further assumes that the times taken by Virgo, Leo, Cancer, Gemini, Taurus, and Aries form a descending arithmetical series, and that the times for Libra, Scorpio, Sagittarius, Capricornus, Aquarius, and Pisces are in the same series.

With the help of three lemmas concerning arithmetical progressions which he has proved at the outset of his book, Hypsicles shows that Virgo and Libra take $38^\circ\ 20'$ to rise, Leo and Scorpio 35°, and so on, the common difference being $3^\circ\ 20'$. He goes on to prove that each spatial degree takes $0^\circ\ 0'\ 13''\ 20'''$ less (or more) than its predecessor to rise.

As Hypsicles' assumption that the times of rising form an arithmetical progression was erroneous, his results were correspondingly in error. But his tract was a gallant attempt to solve the problem before trigonometry provided the right way. The *Anaphorikos* has probably survived by reason of having been included in the collection of ancient Greek texts known as *The Little Astronomy.* It was translated into Arabic toward the end of the ninth century; the translation is variously ascribed to Qusṭā ibn Lūqā and Isḥāq ibn Ḥunayn, but it was in any case considerably altered by later writers. From Arabic it was translated into Latin by Gerard of Cremona (*ca.* 1150) as *Liber Esculei De ascensionibus.* The first printed edition, in Greek and Latin, by Jacobus Mentelius (Paris, 1657) remained the only one until that of K. Manitius in 1888, and this has in turn been superseded by the critical edition of De Falco and Krause (1966).

Hypsicles is cited by Diophantus of Alexandria in *De polygonis numeris* (*Diophanti Alexandrini opera omnia,* I, P. Tannery, ed. [Leipzig, 1893–1895], 470.27–472.4) as the author of the following definition:

> If there be as many numbers as we please beginning from 1 and increasing by the same common difference, then, when the common difference is 1 the sum of all the numbers is a triangular number; when 2, a square number; when 3, a pentagonal number, and so on, the number of angles being called after the number which exceeds the common difference by 2 and the sides after the number of terms, including 1.

In modern notation, the nth a-gonal number (1 being the first) is

$$\frac{1}{2}n\{2 + (n-1)(a-2)\}.$$

From this reference by Diophantus it is presumed that Hypsicles must have written a book, since lost, on numbers. According to Achilles Tatius (*Introductio in Aratum,* E. Maass, ed., *Commentariorum in Aratum reliquiae,* Berlin, 1898, p. 43.9), Hypsicles also wrote a book on the harmony of the spheres; it has not survived.

BIBLIOGRAPHY

I. ORIGINAL WORKS. *Hypsiclis liber, sive Elementorum liber XIV qui fertur, Euclidis opera omnia,* J. L. Heiberg and H. Menge, eds., V (Leipzig, 1888), 1–67; *Des Hypsikles Schrift Anaphorikos nach Überlieferung und Inhalt kritisch behandelt* (Programm des Gymnasiums zum heiligen Kreuz in Dresden), Karl Manitius, ed. (Dresden, 1888), including an introduction, Greek text, and Gerard of Cremona's Latin translation; V. De Falco and M. Krause, eds., *Hypsikles: Die Aufgangszeiten der Gestirne,* in *Abhandlungen der Akademie der Wissenshaften zu Göttingen,* Phil.-hist. Klasse, 3rd ser., no. 62 (1966), with an introduction and valuable interpretation by O. Neugebauer; this has Greek text, scholia, and translation by De Falco, and Arabic text and German translation by Krause.

II. SECONDARY LITERATURE. A. A. Bjørnbo, "Hypsikles 2," in Pauly-Wissowa, IX (1914), cols. 427–433. See also T. L. Heath, *The Thirteen Books of Euclid's Elements,* 2nd ed. (Cambridge, 1925; repr. New York, 1956), I, 5–6; III, 512–519; and *A History of Greek Mathematics* (Oxford, 1921), I, 419–420; II, 213–218; and Jürgen Mau, "Hypsikles," in *Der kleine Pauly,* II (Stuttgart, 1967), cols. 1289–1290.

IVOR BULMER-THOMAS

HYRTL, JOSEPH (*b.* Kismarton, Hungary [now Eisenstadt, Austria], 7 December 1810; *d.* Perchtoldsdorf, Austria, 17 July 1894).

In the middle and latter parts of the nineteenth century anatomy became the most important of the basic sciences on which medicine drew, and Hyrtl was one of the anatomists responsible for this development. His scientific reputation stemmed especially from his *Lehrbuch der Anatomie des Menschen mit Rücksicht auf physiologische Begründung und praktische Anwendung,* first published at Prague in 1846. In this well-written and clearly organized work he emphasized the material that was most important for the practitioner, rendering it entertaining through historical and etymological digressions. The book went through twenty editions and was translated into virtually every major language. A year later (1847) Hyrtl published his *Handbuch der topographischen Anatomie,* which likewise was widely read. He boasted that through this book he had introduced topographical anatomy into the German-speaking world and had made it an independent discipline.

Hyrtl also won worldwide recognition through his achievements as a technical anatomist. Delighting in precise work and trained in the tradition of an anatomy based on preparations, he was able through the skillful exploitation of his position as a university teacher to establish a virtual monopoly in the production and sale of special anatomical preparations. His microscopic injection preparations were considered unexcelled and, through exchange or purchase, reached all major anatomical museums. His corrosion preparations, made by injecting vessels and bone cavities with stiffening material and then destroying the surrounding soft tissue or bones by maceration, brought high prices. Hyrtl thus revived and further developed the source of morphological instruction that Frederik Ruysch had originally discovered. With this method he made investigations in comparative anatomy, for example, of the mammalian inner ear—from the mouse to the elephant. Because of the macroscopic and morphological orientation of his research Hyrtl satisfied himself with corrosion specimens of the labyrinth and left the histological elucidation of the terminal auditory apparatus in the cochlea to his student Alfonso Corti.

By 1813 Hyrtl had moved to Vienna from Eisenstadt, where his father had been oboist in the orchestra of Nicholas, Prince Esterházy. He became a choirboy in the palace chapel and thus a student at the state boarding school. After completing his secondary education he studied medicine in Vienna. He was encouraged by the anatomist Joseph Berres, becoming the latter's prosector in the summer of 1833. In 1835 he earned his doctorate with a dissertation in the history of medicine entitled *Antiquitates anatomicae rariores,* in which, already, he stressed the necessity of giving anatomical instruction a clinical orientation and considered physiological experiments on animals to be unproductive.

In May 1837 Hyrtl was summoned to Prague as professor of anatomy. He returned to Vienna in 1845 to occupy the chair left vacant by the death of Berres. With great diligence he enlarged the demonstration collections, published numerous special investigations, and, in addition to his regular lectures, gave courses in applied anatomy for physicians.

Hyrtl became increasingly isolated in Vienna. His ambition and irascibility made him an extremely difficult colleague and finally cut him off from any close professional ties. Most of all, the physiologist Ernst Brücke became conscious of Hyrtl's haughty manner and contempt for physiological experimentation. It is indicative of Hyrtl's position in the Viennese medical faculty that in the thirty years of his membership, he was never elected dean. He was, however, elected rector for the academic year 1864–1865, during which the university's five-hundredth anniversary celebration took place. Although ambitious enough to accept, Hyrtl was soon confronted with the many difficulties resulting from the domestic political situation. Consequent embitterment may have contributed to Hyrtl's decision to retire in 1874, while still in full possession of his intellectual and physical powers. He settled in Perchtoldsdorf, where he spent the next twenty years working on publications on the history of anatomical nomenclature that are still of value, such as *Onomatologia anatomica* (1880). He founded an orphanage in Mödling and, since he was childless, made this institution the sole heir to the fortune he had earned from his textbooks and anatomical preparations.

Hyrtl was the most scintillating anatomy teacher of the nineteenth century. His expressive voice could impart surging pathos, solemn dignity, lucid objectivity, nonchalant malice, or cutting asperity as the occasion demanded. Along with Brücke's lectures, his were the best attended at the faculty of medicine.

BIBLIOGRAPHY

I. ORIGINAL WORKS. A complete bibliography can be found in M. Holl, "Joseph Hyrtl," in *Wiener klinische Wochenschrift,* 7 (1894), 557–559, repr. in Franz Wolf and Gottfried Roth, *Professor Josef Hyrtl* (Vienna, 1962), pp. 138–147. His works include *Antiquitates anatomicae rariores* (Vienna, 1835); *Lehrbuch der Anatomie des Menschen mit Rücksicht auf physiologische Begründung und prak-

tische Anwendung (Prague, 1846; 20th ed., Vienna, 1889); *Handbuch der topographischen Anatomie und ihrer praktisch medizinisch-chirurgischen Anwendungen,* 2 vols. (Vienna, 1847; 7th ed., Vienna, 1882); *Das Arabische und Hebräische in der Anatomie* (Vienna, 1879); and *Onomatologia anatomica. Geschichte und Kritik der anatomischen Sprache der Gegenwart* (Vienna, 1880; repr. Hildesheim, 1970).

II. SECONDARY LITERATURE. See Erna Lesky, *Die Wiener medizinische Schule im 19. Jahrhundert* (Graz–Cologne, 1965), pp. 240–251, with bibliography; W. S. Miller, "Joseph Hyrtl, Anatomist," in *Bulletin of the Society of Medical History of Chicago,* **3** (1923), 96–108; Viktor Patzelt, "Joseph Hyrtl. Sein Werk nach 100 Jahren," in *Anatomischer Anzeiger,* **103** (1956), 160–175, with bibliography; and Johannes Steudel, "Joseph Hyrtl," in *Medizinische Welt,* **18** (1944), 462–465. The *Index-Catalogue of the Library of the Surgeon General's Office,* 2nd ser. **7,** 772, refers to numerous obituaries.

J. STEUDEL